AA002417

2006 IEEE International Conference on Semiconductor Electronics

Kota Kinabalu, Malaysia
29 November- 1 December 2006

Volume 1 of 2

IEEE Catalog Number: 06EX1443
ISBN: 0-7803-9730-4

Copyright © 2006 by The Institute of Electrical and Electronics Engineers, Inc.
All Rights Reserved

Copyright and Reprint Permissions: Abstracting is permitted with credit to the source. Libraries are permitted to photocopy beyond the limit of U.S. copyright law for private use of patrons those articles in this volume that carry a code at the bottom of the first page, provided the per-copy fee indicated in the code is paid through Copyright Clearance Center, 222 Rosewood Drive, Danvers, MA 01923.

For other copying, reprint or republications permission, write to IEEE Copyrights Manager, IEEE Operations Center, 445 Hoes Lane, Piscataway, New Jersey USA 08854. All rights reserved.

IEEE Catalog Number: 06EX1443

ISBN: 0-7803-9730-4

Additional Copies of This Publication Are Available from:

IEEE Service Center
445 Hoes Lane
Piscataway, NJ 08854
IEEE Service Center
445 Hoes Lane
Piscataway, NJ 08854
Phone: (800) 678-IEEE
                (732) 981-1393
Fax: (732) 981-9667
E-mail: customer-service@ieee.org

International Advisory Committee

Prof. Dr. Vijay Arora	:	Wilkes University, USA
Prof. Dr. Hiroshi Iwai	:	Tokyo Institute of Technology, Japan
Prof. Dr. Nico de Rooij	:	University Neuchatel, Switzerland
Prof. Dr. Arokia Nathan	:	University of Waterloo, Canada
Prof. Dr. Cary Yang	:	Santa Clara University, USA
Prof. Dr. Peter Houston	:	Sheffield University, UK
Prof. Dr. Jong Duk Lee	:	Seoul National University, Korea
Prof. Dr. Edward Chang	:	National Chiao Tung University, Taiwan
Prof. Dr. Jang Kyoo-Shin	:	Kyungpook University, Korea
Assoc. Prof. Dr. Jung-Chih Chiao	:	University of Texas at Arlington, USA
Assoc. Prof. Dr. Francis Tay Eng Hock	:	National University of Singapore, Singapore
Dr. Lerwen Liu	:	Zyvex Corporation, USA
Assoc. Prof. Dr. Joydeep Dutta	:	Asian Institute of Technology, Bangkok, Thailand
Dato' Ahmad Kabeer b. Mohd. Nagoor	:	AKN Technology Berhad, Malaysia
En. Abdul Wahab Abdullah	:	MIMOS Berhad, Malaysia
Prof. Dr. Muhammad Yahaya	:	Universiti Kebangsaan Malaysia, Malaysia
Prof. Dr. David Nagel	:	George Washington University, USA
Assoc. Prof. Dr. S.F. Yoon	:	Nanyang Technological University, Singapore
Prof. Dr. S.M. Sze	:	National Chiao Tung University, Taiwan
Dr. Masbah R.T. Siregar	:	LIPI, Indonesia
Prof. Dr. C.I.M. Beenakker	:	DIMES, Netherlands
Prof. Dr. Muhamad Rasat Muhamad	:	Universiti Malaya, Malaysia
Dato' Dr. Mohd. Ariffin Hj. Aton	:	SIRIM Berhad, Malaysia
Prof. Dr. Peter Ashburn	:	University of Southampton, UK

Organizing Committee

Chairman
Prof. Dr. Burhanuddin Yeop Majlis

Technical Chairman
Mr. Richard Keating

Secretary
Mrs. Badariah Bais

Treasurer
Mr. Rahman Wagiran

Members
Prof. Dr. Muhamad Mat Salleh *(Universiti Kebangsaan Malaysia)*
Prof. Dr. Sahbudin Shaari *(Universiti Kebangsaan Malaysia)*
Prof. Dr. Tou Teck Yong *(Universiti Multimedia Malaysia)*
Assoc. Prof. Ibrahim Ahmad *(Universiti Kebangsaan Malaysia)*
Assoc. Prof. Dr. Razali Ismail *(Universiti Teknologi Malaysia)*
Assoc. Prof. Dr. A. H. M. Zahirul Alam *(International Islamic University Malaysia)*
Dr. Azlan Aziz *(Universiti Sains Malaysia)*
Dr. Ghazali Omar *(Infineon Technologies)*
Dr. Roslina Mohd. Sidek *(Universiti Putra Malaysia)*
Dr. Mohd. Nizar Hamidon *(Universiti Putra Malaysia)*
Dr. Abdullah Chik *(Universiti Malaysia Sabah)*
Hj. Abdullah Lin *(Silterra Malaysia Sdn. Bhd.)*
Mr. Ronnie C. H. Ng *(Intel Microelectronics (M) Sdn. Bhd.)*
Mr. Ismahadi Syono *(MIMOS Berhad)*
Dr. Hua Younan *(Chartered Semiconductor, Singapore)*
Dr. Awangku Abdul Rahman Pgn. Hj. Yusof *(Universiti Malaysia Sarawak)*

Secretariat
Chairman ICSE2006
Electron Devices Chapter, IEEE Malaysia Section
c/o Institute of Microengineering and Nanoelectronics (IMEN)
Universiti Kebangsaan Malaysia
43600 Bangi, Selangor
MALAYSIA
Phone: +603 8925 9080
Fax: +603 8925 9080 / 8925 0439
Email: eds@vlsi.eng.ukm.my

International Conference on Semiconductor Electronics
November 29-December 1, 2006

This ICSE2006 conference is aimed at bringing together scientists and engineers to discuss various issues and trends in the field of semiconductor technology. Malaysia is centrally located in the world's fastest growing economy region of Asia-Pacific and the growth of semiconductor industry in this region has been tremendous. It is timely that this periodically held IEEE international conference becomes an important avenue for addressing worldwide concerns on the technology.

Scope of Conference

Semiconductor Devices and Integrated Circuits

- Device physics
- Device characterization and testing
- Device modeling, simulation and design
- Reliability and failure analysis
- VLSI and MMIC circuits design
- Process / manufacturing technology
- IC packaging
- Materials and new fabrication technologies for VLSI structures
- Training and human resources development in microelectronics industry

MEMS Sensors, Actuators and Nanoelectronics

- MEMS design and development
- MEMS modeling
- MEMS materials and manufacturing
- MEMS fabrication process
- MEMS packaging
- MEMS actuators
- Bio-MEMS, RF MEMS and MOEMS
- Nano-Optics

Message From ICSE2006 Conference Chairman

 *Selamat Datang* to Kuala Lumpur and ICSE2006.

On behalf of the Organising Committee, I have great pleasure in welcoming all of you to ICSE2006. This is the seventh ICSE organized by the Electron Devices Chapter of IEEE Malaysia Section and technically co-sponsored by Electron Devices Society. Over the last 14 years, the ICSE has become the preeminent international forum on semiconductor electronics. Our coverage embraces all aspects of the semiconductor technology, from materials issues and device fabrication, MEMS and microsensors, photonics technology, integrated circuit design (VLSI and RFIC), IC testing, semiconductor manufacturing, and system applications.

The ICSE provides the ideal forum to present your latest results in semiconductor related issues. The proceedings of ICSE2006 consists of two keynote papers as well as 228 contributed papers from all over the world. We are delighted to have four prominent speakers in microelectronics. Prof. Dr. David J. Nagel, from The George Washington University who will speak on The Micro-Scale Structures and Nano-Scale Materials for Chemical and Biological Sensors and Dr. Lerwen Liu from Zyvex Corporation, USA on Nanotechnology. The other two invited speakers are Prof. Dr. Kukjin Chun and Dr. Natarajan Mahadeva Iyer. This will further enhance the program and I am sure participants will find the experience worthwhile. This time we divide papers into two main category, Semiconductor Devices and Integrated Circuits, and MEMS Sensors, Actuators and Nanoelectronics. The conference proceedings will serve as a permanent record and a valuable reference for many microelectronics and MEMS experts.

I would like to express my gratitude to Telekom Research and Development Sdn. Bhd., MIMOS Bhd. and Universiti Kebangsaan Malaysia (UKM) for the sponsorship and IEEE Electron Devices Society for the technical co-sponsorship.

To all participants, I hope we can mutually gain benefits and new knowledge through fruitful discussion while making new contact with other participants. To overseas participants, I wish you a pleasant stay in this country and we will endeavor to make your stay here gratifying and enjoyable as possible.

Finally, my gratitude to all those who have helped to make this conference a reality, specifically to my committee members for their efforts to ensure the success of this conference. *Terima kasih.*

PROF. DR. BURHANUDDIN YEOP MAJLIS
ICSE2006 CHAIRMAN AND ELECTRON DEVICES CHAPTER CHAIR

Table of Contents

Micro-Scale Structures and Nano-Scale Materials for Chemical and Biological Sensors..1
David J. Nagel

Electrical Characterization of Sub-100nm Features in Semiconductor Devices..6
Lerwen Lui

Implementation of 6kV ESD Protection for a 17GHz LNA in 130nm SiGeC BiCMOS..7
D. Linten, M.I. Natarajan, S. Thijs, S. Van Huylenbroeck, S. Xiao, G. Carchon, S. Decoutere, M. Sawada, T. Hasebe, G. Groeseneken

Optimization of the Temperature Sensor Position for MEMS Gas Flow Meters 13
E. AbbaspourSani, D. Javan

Non-Crossing Differential Capacitive MEMS Accelerometer with Electrostatic Spring Tuning 16
Ahmad Alabqari Ma' Radzi, Burhanuddin Yeop Majlis

Fabrication of Platinum Membrane on Silicon for MEMS Microphone..21
Azrul Azlan Hamzah, Burhanuddin Yeop Majlis, Ibrahim Ahmad

Effect of Annealing on the Sol-Gel Derived SrBi4Ti4O15 Thin Films for Piezoelectric Pressure Sensors..26
Nor Azlian Abdul Manaf, Muhamad Mat Salleh, Muhammad Yahaya

Fabrication Study of Solid Microneedles Array Using HNA..32
Norazreen Abd Aziz, Burhanuddin Yeop Majlis

Effects of Mechanical Geometrics on Resonance Sensitivity of MEMS Out-of-Plane Accelerometer..37
A. Manut, M.I. Syono

Structure Design and Fabrication of an Area-changed Bulk Micromachined Capacitive Accelerometer..41
Badariah Bais, Burhanuddin Yeop Majlis

The Effect of Design Parameters on Static and Dynamic Behaviors of the MEMS Microphone..47
Bahram Azizollah Ganji, Burhanuddin Yeop Majlis

Deep Trenches in Silicon Structure using DRIE Method with Aluminum as an Etching Mask 53
Bahram Azizollah Ganji, Burhanuddin Yeop Majlis

Performance Improvement of OTFTs using Double Layer Insulator 60
Dong-Wok Park, Cheon An Lee, Keum-Dong Jung, Byeong-Ju Kim, Byung-Gook Park, Hyungcheol Shin, Jong Duk Lee

Simulation of a Novel Lateral H Structure PIN InGaAs Photodiode with Consistent Electron Drift Length 64
Esther Loo Chee Hong, Sahbudin Shaari, Burhanuddin Yeop Majlis

Distributed CATV Inputs in FTTH-PON System..70
Tan Fent Fent, Sahbudin Shaari, Burhanuddin Yeop Majlis

On the Use of a Mixed-Mode Approach For MEMS Testing 74
Md. Fokhrul Islam, M.A. Mohd. Ali

A Novel Model for Membranes of Micropumps Partially Loaded with Electromagnetic Forces 78
Fu-Shin Lee, Chih-Hsiung Chen

Investigation of the Torsion and Bending Effects on Static Stability of Electrostatic Torsional Micromirrors..84
Ghader Razazadeh, Faraz Khatami, Admadali Tahmasebi

Electromechanical Behavior of Microbeams with Piezoelectric and Electrostatic Actuation 90
Ghader Rezazadeh, Ahmadali Tahmesebi, Vahid Molodpour

The Gas Sensing Potential of Nanocrystalline SnO2 and In2O3 Powders Prepared by Mechanical Milling..95
Goib Wiranto, Adiseno, I Dewa P Hermida, Roberth V. Manurung, Slamet Widodo, Masbah R.T. Siregar

vii

Table of Contents

Reduction of Turn-On Voltage in a Singe Layer Structured Organic Light-Emitting Diode Using Nanocomposites SiO2:PHF 101
Tengku Hasnan Tengku Aziz, Muhamad Mat Salleh, Mursyidah Umar, Muhammad Yahaya

Development of an Integrated Miniaturized Mulit-Ion Flow Cell System for Water Quality Measurement 105
Hiskia, Masbah R.T. Siregar, Robeth V.M.

Design and Fabrication of Micromachined Gas Sensors 110
Hiskia, Masbah R.T. Siregar, Moch. Muljono

An Active Integrated Spiral Antenna System for UWB Applications 114
M. Jalali, A. Abdipour, A. Tavakoli, G. Moradi

Analysis of a Bilaminar Circular Piezoelectric Actuator for Micropumps 118
Juliana Johari, Burhanuddin Yeop Majlis

Degradation of Single Layer MEH-PPV Organic Light Emitting Diode (OLED) 124
Nurjuliana Juhari, Wan Haliza Abd. Majid, Zainol Abidin Ibrahim

Micro-fabricated Square Interwinding Transformer Using Surface Micromatching 128
Jumril Yunas and Burhanuddin Yeop Majlis

Fabrication of a Single Carbon Nano Tube for Use in Nanolithography of MOSFET Gate 134
J. Karamdel, N. Talebi, M. Sattari, J. Derakhshandeh, F.L. Ayatollahi, A. Farrokhi, A.R. Hadi

Development of Integrated Detection System for Shock Vibration by Using MEMS Accelerometer 139
Khairul Anuar Bin Abd Wahid, Ishak Abdul Azid, Aftanasar Md Shahar and Othman Sidek

Organic Light Emitting Diode (OLED) Using Different Hole Transport and Injecting Layers 146
Mohd Khairy Othman, Muhamad Mat Salleh, Abdule Fatah Awang Mat

How the Optical Properties of Au Nanoparticles are Affected by Surface Plasmon Resonance 150
Yen Hsun Su, Wei Hao Lai, Lay Gaik Teoh, Hao En Hong, Min Hsiung Hon

Gas Sensing Properties of ZnO:Al Thin Films Prepared by RF Magnetron Sputtering 154
Lay Gaik Teoh, Hong Ming Chen, Yen Hsun Su, Wei Hao Lai, Shih Min Chou, Min Hsiung Hon

Factors Affecting the Growth of Carbon Nanotubes 158
Lee Cheng Choo, Roslan Md. Nor

Effect of Self-Weight and Non-Rigidity on the Bending Characteristics of Surface Micromachined MEMS Test Structures 163
Hing Wah Lee, Mohd. Ismahadi Syono

Low Temperature Carbon Nanotube Fabrication Using Very High Frequency-Plasma Enhanced Chemical Vapour Deposition Method 167
Sukirno, Satria Zulkarnaen Bisri, Lilik Hasanah, Mursal, Ida Usman, Adi Bagus Suryamas, Thomas Alfa Edison

Design and Modeling of Micromachined Condenser MEMS Loudspeaker Using Permanent Magnet Neodymium-Iron-Boron (Nd-Fe-B) 172
Fatima Lina Ayatollahi, Burhanuddin Yeop Majlis

A Hydrogel-based Microvalve for Insulin Delivery Application 179
Masoomeh Tehranirokh, Badariah Bais, Burhanuddin Yeop Majlis

Application of Full Factorial Design Method in MEMS Capacitive Thermal Sensor Sensitivity 184
Mehdi Mehrban, Shahriar Kouravand, Ghader Rezazadeh, Ali Donyavi

Design of a High Sensitivity Structure for MEMS Fingerprint Sensor 189
Mitra Damghanian, Burhanuddin Yeop Majlis

High-Precision Thickness Control of Silicon Membranes Using Etching Techniques 197
Mohsen Nabipoor, Burhanuddin Yeop Majlis

Table of Contents

Aluminum Based Two-Port-Clamped-Clamped Resonators..200
Mustafa Al_Khusheiny and Burhanuddin Majlis

The Use of Photoluminescence Spectra of TiO2 Nanoparticles Coated with Porphyrin Dye Thin Film for Grading Agarwood Oil..205
Nurul Huda Yusoff, Muhamad Mat Salleh, Muhammad Yahaya, Mat Rasol Awang

Study on the Effect of Q-Switched Nd:Yag Laser Interaction with Al in Variable Magnetic Field................210
Rabia Qindeel, Noriah Bidin, Yaacob Mat Daud

OXADM Multiplex Protection Scheme for Bidirectional Path Switched Ring................................215
Muhammad Syuhaimi Bin Ab-Rahman, Hayati Hussin, Abang Annuar Ehsan, Sahbudin Shaari

Inductively-Tuned K-Band Distributed MEMS Phase Shifter..220
Saeid Afrang, Burhanuddin Yeop Majlis

Crosstalk Enhancement in Multiplexer/Demultiplexer Based Arrayed Wavelength Grating in Dense Wavelength Division Multiplexing..225
Salah Elfaki, Abdeen Abdel Kareem, A.B. Mohammed, Sahbudin Shaari

Analytical Model Studying of a Novel Tunable Capacitor Based on Bimetallic Thermal Actuator................230
Shahriar Kouravand, Mehdi Mehrban, Ghader Rezazadeh, Mehdi Sabet

Uniformity Study of GaAs-based Vertical-Cavity Surface-Emitting Laser Epiwafter Grown by MOCVD Technique..235
Mohd Sharizal Alias, Paul O. Leisher, Kent D. Choquette, Khairul Anuar, Dominic Siriani, Sufian Mitani, Mohd Razman Y., Abdul Fatah A.M.

Electro-Opto Characteristics of 850 nm Oxide-Confined Vertical-Cavity Surface-Emitting Lasers................239
Mohd Sharizal Alias, Paul O. Leisher, Kent D. Choquette, Khairul Anuar, Dominic Siriani, Sufian Mitani, Mohd Razman Y. and Abdul Fatah A.M.

Efficiency and Spectral Characteristics of 850 nm Oxide-Confined Vertical-Cavity Surface-Emitting Lasers..243
Mohd Sharizal Alias, Paul O. Leisher, Kent D. Choquette, Khairul Anuar, Dominic Siriani, Sufian Mitani, Mohd Razman Y., Abdul Fatah A.M.

2D Silicon-based Photonic Crystals..248
Y.K. Sin, K. Ibrahim

Calculation of Quantum Efficiency for Resonant Cavity Photodiodes using the FDTD Method................252
Mohammad Soroosh, Mohsen Jalali, Mohammad Kazem Moravvej-Farshi

Static Quasi 3D Thermal Simulation of Ion Implanted Vertical Cavity Surface Emitting Lasers................256
E. Sooudi, V. Ahmadi, M. Ebnali Heidari, M. Soroosh

The Effect of Annealing on the Performances of the White Organic Light Emitting Diode (OLED)................261
Suhaila Sepeai, Muhamad Mat Salleh, Muhammed Yahaya

Considering RFID Inmate Tagging Application to Enhance Prison Management................................265
Mohd. Suhaimi Selamat, Burhanuddin Yeop Majlis

Challenges in Implementing RFID Tag in a Conventional Library..270
Mohd. Suhaimi Selamat, Burhanuddin Yeop Majlis

Investigating the Performance of RF MEMS Switches..275
Hieng Tiong Su, Iganacio Llamas-Garro, Michael J. Lancaster, Martin Prest, Jae-Hyoung Park, Jung-Mu Kim, Chang-Wook Baek, Yong-Kweon Kim

Comparison of Electronic Transport Parameter of CNT(10,10)/CNT(17,0) and CNT(5,5)/CNT(8,0) Carbon Nanotube Metal-Semiconductor On-Tube Heterojunction..279
Sukirno, Satria Zulkarnaen Bisri, Irmelia, Lilik Hasanah, Adi Bagus Suryamas, Ida Usman, Mursal

Table of Contents

See-saw Type RF MEMS Switch with Fine Gap Vertical Comb..284
Sungchan Kang, Hyeon Cheol Kim, Kukjin Chun

Numerical Modeling of a Diffusion-Based In0.53Ga0.47As Lateral PIN Photodiode for 10 Gbits/s Optical Communication Systems 288
P Susthitha Menon, Kumarajah Kandiah and Sahbudin Shaari

The Effect of Surface Microstructure on The Response of Titanium Dioxide Coated with Cobalt-Porphyrin Thin Films Towards Gases in Quartz Crystal Microbalance Sensor 293
Syariena Arshad, Muhammad Mat Salleh, Muhammad Yahaya

Synthesis and Characterization of CuO Nanowires 298
Chang Fu Dee, Muhammad Yahaya, Muhamad Mat Salleh, Burhanuddin Yeop Majlis

Analytical Modeling of Optical Cross Add and Drop Multiplexing Switch 302
Muhammad Syuhaimi Bin Ab-Rahman, Abang Anuar Ehsan, Hayati Husin, Sahbudin Shaari

Modelling and Analysis of a Transformer based MEMS Piezoelectric Vibration Type Microgenerator 306
Luay Yassin Taha, Burhanuddin Yeop Majlis, Ahmad Al Ali

A Novel Successive Interference Cancellation Scheme in OCDMA System 311
Tawfig Eltaif, Hossam M. H. Shalaby, Sahbudin Shaari

Simulation of Piezo-Resistive Metal Gauge on Rectangular Membrane for Low Pressure Application 316
Mohd Yunus Hamid, Thangamani U, P R Vaya

Design and Modeling of an Electromagnetic Levitated and Actuated Micromotor 321
W.S.H. Wong, W.K.S. Pao, D. Holliday, P.H. Mellor

Fabrication of Polymer Light Emitting Diodes with ITO/PVK:PBD:DPVBi:DCJTB/Al Structure 326
Yap Chi Chin, Muhammad Yahaya, Muhamad Mat Sellah

Homodyne Linear Crosstalk Impact in an Array Waveguide Router as an OADM for WDM Networks 330
Yasin M. Karfaa, M. Ismail, Abbou F.M., S. Shaari

Avalanche Multiplication and Excess Noise Factor of Heterojunction Avalanche Photodiodes 335
A.H. You, L.C. Low, P.L. Cheang

Faults Detection Approach for Self-Testable RF MEMS...340
Syed Zahidul Islam, Wallace Wong, Su Hieng Tiong, and Mohd. Alauddin Mohd Ali

MEMS Switch Material Dependency on Designing a Reconfigurable Antenna 345
A.H.M. Zahirul Alam, Md. Rafiqul Islam, Sheroz Khan, Azlani Bt. Mohd. Nat, Manis Mulyany Bt. Abdul Razak

Surface and Composition Reactivity of Pt/GaN Catalytic Contact as Schottky Barriers Gas Sensor 350
Abdo Yahya Hudeish, Azlan Abdul Aziz

Metals/GaN Catalytic Contact Properties for Hydrogen Gas Sensor Applications.................................353
Abdo Yahya Hudeish, Azlan Abdul Aziz

A Low-Cost CMOS Reconfigurable Receiver for WiMAX Applications..357
Adiseno, G. Wiranto, T.M.S. Soegandi

Characterization of Contact Etching Profile for 0,35 um Analog Mixed Signal Product Development.....363
Ahmad Sabirin Zoolfakar, Azlina Mohd Zain

Comparison of the Growth Si-based Crystalline Silicon Carbide (SiC) by Chemical Vapor Deposition (CVD) using Carbon Monoxide (CO) and Treated Carbon Dioxide 368
A.Y.K. Lim, K. Ibrahim

Self Heating Characterization of 32V MOSFETs Using Pulsed Gate Measurement...............................372
Alhan Farhanah Abd Rahim, Albert Victor Kordesch, Yusman Mohd Yusof

Design of 100nm Single-Electron Transistor (SET) by 2D TCAD Simulation...377
Amiza Rasmi, Uda Hashim, Abdul Fatah Awang Mat

Table of Contents

Effect of Rapid Thermal Annealing (RTA) on n-Contact of 980 nm Oxide VCSEL..........383
Khairul Anuar M.S., Mohd Sharizal A., S.M. Mitani, Mohamed Razman Y., Awang Mat A.F., P.K. Choudhury

Development of SiC/MgO Distributed Bragg Reflector using RF Magnetron Sputtering Technique..........388
Khairul Anuar M.S., Hariyadi Soetedjo, Mohd Sharizal A., Sufrian S.M., Goh Boon T., Richard R., Saadah A.R., Mohamed Razman Y., Abdul Fatah A.M.

Temporal Partitioning of Tasks on a Heterogeneous Reconfigurable Architecture..........392
Arjumand Yaqoob, M. Ashraf Chughtai

Packet Synchronization Structure with Peak Detection Algorithm for MB-OFDM UWB..........398
Aymen M. Karim, Masuri Othman, Edmond Zahedi

Source-Coupled Logic (SCL): Operation and Delay Analysis..........402
Mohamed Azaga, Masuri Othman

Photopthermal Study of Ceramic ZnO Doped with Y2O3..........407
Azmi Zakaria, Zahid Rizwan, Mansor Hashim, Abdul Halim Shaari

Grain Size Effect on LTCC Tape Performance as Substrate for Microelectronic Devices..........412
Azmi Ibrahim, Rosidah Alias, Che Seman Mahmood, Sabrina Mohd Shapee, Mohamed Razman Yahya, Abdul Fatah Awang Mat

Particle Size Analysis of Barium Titanate Powder by Slow-Rate Sol-Gel Process Route..........416
Balachandran R., Yow HK, Jayachandran M., Wan Yusmawati Wan Yusof, Saaminathan V.

Rapid Crystallization by Microwave Heating for Barium Strontium Titanate Powders Prepared Using Slow-Rate Sol-Gel Technique..........420
Balachandran R., Yow H.K., Zaidina Bt Mohd Daud, Saaminathan V.,

Soft IP Integration and Reuse Challenges in Intel Entry Level Network Processor..........424
Wai Mun Ng, Kok Sing Yap, Kean Hong Boey

Delayering of Gate Poly in Stack & Split Gate Memory Structure..........428
Bridger K.S. Wong and Chan Sieng Fong

High-Performance In0.52Al0.48As?in0.6Ga0.4As Power Metamorphic HEMT for Ka-Band Applications..........432
Chia-Yuan Chang, Edward Yi Chang, Yi-Chung Lien, Yasuyuki Miyamoto, Szu-Hung Chen, Li-Hsin Chu

Development of the Vacuum Spark as an EUV Source for Next Generation Lithography..........435
Chew Soo Hoon, Wong Chiow San

TDR Single Ended and Differential Measurement Methodology..........439
Chan Kim Lee, Jiun Kai Beh, Jimmy Huat Since Huang

Contact Hole Printing in Binary Mask by FLEX Technique..........444
Cheong Yew Shun, Ko Bong Sang, Mohd Jeffery Bin Manaf, Kader Ibrahim, Dr. Zul Azhar Zahid Jamal

Porous Silicon Dioxide Synthesized Using Photoelectrochemical (PEC) Wet Etching..........448
L.S. Chuah, C.W. Chin, Z. Hassan, H. Abu Hassan

Characteristics of Thermally Treated Contacts on Porous Silicon Based Metal-Semiconductor-Metal (MSM) Photodector Structures..........452
L.S. Chuah, C.W. Chin, Z. Hassan, H. Abu Hassan

Softbake and Post-exposure Bake Optimization for Process Window Improvement and Optical Proximity Effect Tuning..........456
C.Y. Liau, E.K. Yet, C.H. Lee, Ivy Tan, Christopher Loo, B.C. Lee, Y.K. Ng, W.B. Sheu

FPGA Implementation of an Optimized Coefficients Pulse Shaping FIR Filters..........463
Mohamed Almahdi Eshtawie, Masuri Othman

A 10-Bit 50-MSPS Pipelined CMOS ADC..........468
Mohamad-Faizal Hashim, Yuzman Yusoff and Mohd. Rais Ahmad

Table of Contents

Performance of OCDMA Systems Using AND Subtraction Technique 473
S.A. Aljunid, Feras N. Hasson, M.D.A. Samad, M.K. Abdullah, M. Othman, S. Shaari

Pyroelectric Properties of Polyvinylidene Fluoride (PVDF) by Quasi Static Method 477
Gan Wee Chen, Wan Haliza Abd. Majid

Effects of Substrate Temperature on the Properties of Hydrogenated Nanocrystalline Silicon Thin Film Grown by Layer-by-Layer Technique 481
Goh Boon Tong, Saadah Abdul Rahman

SIMS Analysis of Gate Oxide Breakdown Due to Tungsten Contamination 486
D. Gui, Z.X. Xing, Z.Q. MO, Y.N. Hua, S.P. Zhao

Evaluation of Stopping Power of Photo-resist to Ion Implantation by Using SIMS 490
D. Gui, Z.X. Xing, Z.Q. Mo, Y.N. Hua, S.P. Zhao

Design and Analysis of an Ultra Wideband Matrix Mixer 494
R. Hallaji, A. Abdipour, G. Moradi

Design and Analysis of Insulate Gate Bipolar Transistor (IGBT) with P+SiO2 Collector Structure Applicable to High Voltage to 1700 V 498
Han-Sin Lee, Yo-Han Kim, Ey-Goo Kang, Man-Young Sung

Process Optimization of p+LDD in 130nm Process Technology using TCAD Simulation 502
Hani Noorashiqin Abd Majid, Muhamad Rasat Muhamad, Albert Victor Kordesch, Chew Soon Aik

Effect of N2 and O2 Anneal Gas Ratio For Low Resistance p-Type ZnO Formation 507
Haslinda Abdul Hamid, Mat Johar Abdullah, Azlan Abdul Aziz, Naif H. AL-Hardan, Siti Azlina Rosli

Electrical Properties of p-Type Al-N Codoped ZnO Thin Films 511
Haslinda Abdul Hamid, Mat Johar Abdullah, Azlan Abdul Aziz, Siti Azlina Rosli

A Highly Compatible Architecture Design for Optimum FPGA to Structured-ASIC Migration 515
Hee Kong Phoon, Matthew Yap, Chuan Khye Chai

Effect of Mesa Spacing on the Electrical Properties of Mesa Isolation in High Electron Mobility Transistor Structures 520
Hesly Afida Hashim, Mohd. Khairy Othman, Mohd. Nizam Osman, Asban Dolah, Mohamed Razman Yahya

Characterization of Si3N4 Metal-Insulator-Metal (MIM) Capacitors for Monolithic Microwave Integrated Circuits (MMIC) Applications 524
Lee Hock Guan, Mohd Nizam Osman, Asban Dolah, Ahmad Ismat Abdul Rahim, Mohamed Razman Yahya, Abdul Fatah Awang Mat

A New High Resolution Frequency and Phase Synthesis Method based on 'Flying-Adder' Architecture 529
Hossein Gharaee, Elham Tathesari

A Digital Implementation for UWB Impulse Radio Transciever 533
Hossein Gharaee, Abdolreza Nabavi, Babak Bornoosh, S. Mehdi Fakhraei

Studies on Failure Mechanism of Al Fluoride Oxide-AlxoyFz on Microchip Al Bondpads 537
Hua Younan, Zhao Siping, Rao Ramesh, Li Kun

Studies on A Sample Preparation Method for HR-SEM and Application in Failure Analysis of Trench TEOS Gauging Measurement in Wafer Fabrication 542
Zhao Siping, Hua Younan, Mo Zhiqiang, Cho Jie Ying

An Integrated 2.4GHz CMOS Class F Power Amplifier 546
Huang Min Zhe, Abu Khari Bin A'ain, Albert Victor Kordesch

Solder Joint Strenght of Lead Free Solders under Multiple Reflow and High Temperature Storage Condition 550
Ibrahim Ahmad, Azman Jalar Burhanuddin Yeop Majlis, Eu Poh Leng, Yong Soo San

Table of Contents

A Study on Inter-Metallic Compound Formation and Structure of Lead Free SnAgCu Solder System......................554
I. Ahmad, Ho Huey Jiun, Eu Poh Leng, B.Y. Majlis, A. Jalar, R. Wagiran

Design and Simulation of 50 nm Vertical Double-Gate MOSFET (VDGM)......................558
Ismail Saad, Razali Ismail

Design and Simulation of a High Performance Lateral BJTs on TFSOI......................563
Ismail Saad and Razali Ismail

Weak point and improvement of CMOS Schmitt Trigger Circuit used in Microcontroller about ND-mode ESD......................567
Jae-Seong Jeong, Jung-Min Lee, and Sang-Deuk Park

Integration of 8051 With DSP in Xilinx FPGA......................571
A.J. Salim, M. Othman, M.A. Mohd Ali

Increased Capacitance Density with Metal-Insulator-Metal - Metal Figner Capacitor (MIM-MFC)......................576
Kalavathi Subramaniam, Albert Victor Kordesch, Mazlina Esa

Considerations on the C-V Characteristics of Pentacene Metal-Insulator-Semiconductor Capacitors......................581
Keum-Dong Jung, Byung-ju Kim, Byeong-Ju Kim, Cheon An Lee, Dong-Wook Park, Byung-Gook Park, Hyungcheol Shin, Jong Duck Lee

Characteristics of Serial Peripheral Interfaces (SPI) Timing Parameters for Optical Mouse Sensor......................585
M.K. Md Arshad, U. Hashim, Chew Ming Choo

Design of Experiment (DOE) For Thickness Reduction Of GaAs Wafer Using Lapping Process......................592
Mohd Khairy Othman, Asban Dolah, Nurul Afzan Omar, Mohamed Razman Yahya

Study of Flow Visualizationg in Stacked-Chip Scale Packages (S-CSP)......................595
M. Khalil Abdullah, M.Z. Abdullah, S. Kamarudin, Z.M. Ariff, P. Hussin, J.J. Antony, H. Haroon, M.R. Saad, M. Manikam

An FIB Method Using Progressive Multi-Cut Technique & Application in Failure Analysis of Wafer Fabrication......................600
Khoo Ley Hong, Hua Younan, Zhao Siping, Mo Zhiqiang

I/O Process Optimization to Cover Wide Range Operation Voltage......................603
Deb Kumar Pal, Kenny Sabri, Margaret Ting Leh Kee, Son Jin Yeong, Park Hyun Suck

High-Speed Hybrid Parallel-Prefix Carry-Select Adder Using Ling's Algorithm......................607
Lakshmanan, Ali Meaamar, Masuri Othman

Study on Alignment Capability and Overlay Performance in 130nm BEOL Lithography Process......................612
Lau Siau Yen, Suhana Mohd Said, Norhayati Soin, Kader Ibrahim, Ko Bong Sang

Studies on Electron Penetration Versus Beam Acceleration Voltage in Energy-Dispersive X-Ray Microanalysis......................619
Sharon Lee, Hua Younan, Zhao Siping, Mo Zhiqiang

Effect of Post Annealing Treatments on the Characteristics of Ohmic Contacts to n-Type InN......................623
L.S. Chuah, Z. Hassan, H. Abu Hassan

Nanoporous InN Films Synthesized using Photoelectrochemical (PEC) Wet Etching......................627
L.S. Chuah, Z. Hassan, F.K. Yam, H. Abu Hassan

TEM Characterization of Nickel Silicide Process......................631
K. Li, E. Eddie, S.P. Zhao

Investigation and Failure Analysis of "Flower-like" Defects on Microchip Aluminum Bondpads in Wafer Fabrication......................635
Hua Younan, Lim Yeow Kheng, Zhao Siping, Rao Ramesh, Tan Jaun Boon

Plane-view Transmission Electron Microscopy for Advanced Integrated Circuit......................639
Pan Liu, K. Li, Eddie E.R., Siping Zhao

xiii

Table of Contents

Application of Focus Ion Beam Circuit Edit in Failure Analysis 643
S.K. Loh, Teo HT, S.P. Neo, Z.G. Song, C.K. Oh

Circuit Debug using Time Resolved Emission (TRE) Prober-A Case Study 646
Houn Wai Wong, Pit Fuh Low, Vee Kin Wong

FPGA Implementation of a Canonical Signed Digit Multiplier-less based FFT Processor for Wireless Communication Applications 650
Mahmud Benhamid, Masuri Othman

Low Temperature Heteroepitaxial Growth of 3C-SiC on Si Substrates by Rapid Thermal Triode PLasma CVD using Dimethylsilane 655
Abdul Manaf Hashim, Kanji Yasui

The Influence of Doping Concentration, Temperature, and Electric Field on Mobility of Silicone Carbide Materials 660
Mazhar Tayel, Ayman El-Shawarby

A 0.18um, 1.8-V CMOS High Gain Fully Differential Opamp Utilized in Pipelined ADC 665
Ali Meaamar, Masuri Bin Othman, Omid Shoaei

Low-Voltage, High-Performance Current Mirror Circuit Techniques 670
Ali Meaamar, Masuri Othman

VLSI Implementation of 1/2 Viterbi Decoder for IEEE P802.15-3a UWB Communication 675
Meilana Siswanto, Masuri Othman, Edmond Zahedi

Integrated LC VCO Compatible with Memory Process for Gigahertz Clock Generation 680
Minsoek Choi, Youngho Jung, Yeo Jo Yoon, Young-Wug Kim, Hyungcheol Shin

A Novel DC-Coupled, Single-Ended to Differential, Transimpedance Amplifier Architecture Based on gm-boosting Technique 682
M. Jalali, M.K. Moravvej-Farshi, A.R. Nabavi

Analysis of Airborne Boron and Phosphorus Contaminations on Wafer Surface by TOF-SIMS 686
Mo Zhiqiang, Gui Dong, Hua Younan, Zhao Siping, Xing Zhenxiang

Study of Porous Silicon Fabricated by Pulsed Anodic Etching of n-Si (100) 689
N.K. Ali, M.R. Hashim, A. Abdul Aziz

The Effects of High Temperature Storage on Lead Free Solder Joint Material Strength Using Pull Test Method 694
Muhammad Najib Harif, Ibrahim Ahmad, Azami Zaharim

Failure Analysis Approach in Memory Failure of SOI Devices 699
S.P. Neo, S.K. Loh, Z.G. Song, S.P. Zhao

Front End Defects on Deep Submicron Devices 702
S.P. Neo, S.K. Loh, Z.G. Song, S.P. Zhao

Failure Analysis of a Unique Poly Defect 705
S.P. Neo, Z.G. Song, C.K. Oh, S.P. Zhao

Trenched MOSFET Vgs Uniformity Improvement through Furnace Loading Procedure 708
H.S. Ng, A.F. Yee, Christopher Loo, Y.K. Ng, W.B. Sheu

Simulation of Flash Memory Characteristics based on Discrete Nanoscale Silicon 713
C.Y. Ng, J.I. Wong, M. Yang, C.L. New, T.S. Khor, T.P. Chen

Realisation of a Differential Multiplier-Divider based on Current Feedback Amplifiers 717
Nihit Bajaj, Jivesh Govil

A SEM Based Technique To Detect Pin-holes In As-Deposited/As-Grown Dielectrics 722
Nitin R. Kamat, Oh Chong Khiam, Zhao Si Ping

xiv

Table of Contents

A Study of Yield Loss In Copper Back-End Process Due To Stress and Poor Adhesion of the Thin Films 726
Nitin R Kamat, Manni Lal, P.B. Sahoo, Zhao Si Ping

Electrical Analysis of High Temperature SAW Resonator Packages 734
Mohammad Hadi Shahrokh Abadi, Mohd. Nizar Hamidon, Roslina Sidek, Mina Malekzadeh

Alignment Mark Architecture Effect on Alignment Signal Behavior in Advanced Lithography 741
Normah Ahmad, Uda Hashim, Mohd Jeffery Manaf, Kader Ibrahim Abdul Wahab

Device Characteristics of HEMT Structures based on Backgate Contact Method 749
Norman Fadhil Idham M., Nurul Afzan O., Hariyadi Soetedjo, Ahmad Ismat A.R., Idris Sabtu, Mohamed Razman Y., Abdul Fatah A.M.

Effect of Indium Content in the Channel on the Electrical Performance of Metamorphic High Electron Mobility Transistors 752
Norman Fadhil Idham M., Ahmad Ismat A.I., Rasidah S., Asban D., Mohamed Razman Y., Abdul Fatah A.M.

Design of an RF BJT-Low Noies Amplifier at 1GHz 756
Sharifah Fatmadiana Wan Muhamad Hatta, Norhayati Soin

A 10GHz Reconfigurable UWB LNA in 130nm CMOS 760
Parviz Amiri, Hossein Gharaee, Abdolreza Nabavi

Physical-Based SPICE Model of CMOS STI y-Stress Effect 764
Philip Beow Yew Tan, Albert Victor Kordesch, Othman Sidek

Optimized Clamp Deployment with Simulation and Characterization in Full-Chip ESD (Electro-Static-Discharge) Design 768
Chuah Cheow Theng, Othman Sidek

Dependence of Texture in Al Bondpads on Ta/TaN Bilayer Barrier and its Correlation to Optical Reflectivity in 0.13um IC Technology 773
Lee Yuan Ping, Ramesh Rao Nistala, Hua Younan, Mousumi Bhat

Meeting the Challenges of Elemental Analysis in 90nm & Beyond Technologies - Case Studies of Scanning Auger Nanoprobe 777
Nistala Ramesh Rao, Tan Chin Wang, Song Zhigang, Hua Younan, Zhao Siping

Improved Booth Encoding for Reduced Area Multiplier 782
Razaidi Hussin, Ali Yeon Md. Shakaff, Norina Idris, Rizalafande Che Ismail, Afzan Kamarudin

A Novel Method to Design Interstage Matching Network in the Smith Chart 785
M. Rezvani Abkenari, M. Tayarani, A. Abdipour, H. Kiumarsi

Dependence of Radio Frequency Power on Optical, Chemical Bonding and Photoluminescence Properties of Hydrogenated Amorphous Carbon Nitride Films 790
Richard Ritikos, Goh Boon Tong, Rozidawati Awang, Siti Meriam Abdul Gani, Saadah Abdul Rahman

High Performance Complex Number Multiplier Using Booth-Wallace Algorithm 795
Rizalafande Che Ismail and Razaidi Hussin

Effects of High Dose BF2+ Implant on the Improvement of P+ Contact Resistance 800
Mohd Rofei Mat Hussin, Norazah Abd. Rashid, Richard Keating

A 0.8V Operational Amplifier Using Floating Gate MOS Technology 804
Rohan Sehgal

Design of Single-Stage Folded-Cascode Gain Boost Amplifier for 100m V 10-bit 50MS/s Pipelined Analog-to-Digital Converter 809
Rohana Musa, Yuzman Yusoff, Tan Kong Yew, Mohd Rais Ahmad

Modification and Modeling of Ni/Si Interface for Photodetector Applications 814
M.R. Hashim, M.Z.M. Yusoff

Table of Contents

An Enhancement of Decimation Process using Fast Cascaded Integrator Comb (CIC) Filter 820
Rozita Teymourzadeh, Masuri Bin Othman

Effects of Metal Work Function and Operating Temperatures on the Electrical Properties of Contacts to n-type GaN 825
S.M. Thahab, H. Abu Hassan, Z. Hassan

Power Deduction in Digital Signal Processing Circuit Using Inventive CPL Subtractor Circuit 829
C. Senthilpari, K. Diwakar, Ajay Kumar Singh

Design and Synthesis of Mobil Robot Controller Using Fuzzy 834
Md. Shabiul Islam, Md. Anwarul Azim, Md. Saukat Jahan, Masuri Othman

Development of a Fuzzy Logic Controller Algorithm for Air-conditioning System 839
Md. Shabiul Islam, Md. Shakowat Zaman Sarker, Kazi Ashique Ahmed Rafi, Masuri Othman

Design and Nonlinear Analysis of High-Grain and Broad-band Distributed Power Amplifier with Traveling-Wave Gain Stages by Harmonic Balance Method 844
S. Asadi, A. Abdipour, A. Tavakoli, G. Moradi

Silicon Chip Removal Technique Using Wet Etching Process for Failure Analysis on Multi-Chip Packages (MCP) 847
Shamsul Mohamed, Rodzaki Saad

Pulse Generation with Reduced Ringing for Ultra Wide Band Applications in Indoor Wireless Communication 852
Sheroz Khan, A.H.M. Zahirul Alam, Mohammad Rafiqul Islam, Othman O Khalifa, Ismail Adam

Application of Autosched AP Simulation Model in Wafer Fab 855
Victor Siow Yuen Tien, Yeo Eng Teck, G. Devandran A/L Govindasamy, Shahrul Kamaruddin

Characteristics of RIE SF6/O2/Ar Plasmas on n-Silicon Etching 860
Siti Azlina Rosli, Azlan Abdul Aziz, Haslinda Abdul Hamid

Highly Chemical Reactive Ion Etching of Silicon in CF4 Containing Plasmas 865
Siti Azlina Rosli, Azlan Abdul Aziz, Haslinda Abdul Hamid

Photolithography Process Improvement for Thick Implant Resist Using 120 C Post-Apply Bake 870
S.I. Yet, E.C. Goh, Faith Lim, A.E. Ling, B.C. Lee, Y.K. Ng, W.B. Sheu

A Versatile HSPICE Electro-Opto-Thermal Circuit Model for Vertical-Cavity Surface-Emitting Lasers 875
E. Sooudi, V. Ahmadi, M. Soroosh

Adopting Electroabsorption Modulator for the WLAN 802.11a Radio over Fibre System 880
Syamsuri Yaakob, Mohd Azmi Ismail, Romli Mohamad, Mohamed Razman Yahya, Abd. Fatah Awang mat, Mohd. Ridzuan Mokhtar, Hairul Azhar Abdul Rashid

Failure Analysis of Pitting Problem on Microchip Al Bondpads in Wafter Fabrication 885
Tai Eve, Hua Younan, Rao Ramesh, Zhao Siping

Auger PID Characterization of Threshold Voltage Shift and Application in Bond pad Monitoring of Wafer Fabrication 890
Marcus Tan Y.H., Hua Younan, Ling Timothy, Zhao Siping

Studies on Failure Mechanism of ET High Via Resistance in Wafer Fabrication 893
Hua Younan, Tan Sock Khim, Li Kun, Zhao Siping

The Effect of Al and Pt/Ti Simultaneously Annealing on Electrical Characteristics of n-GaN Schottky Diode 896
Tariq Munir, Azlan Abdul Aziz, Mat Johar Abdullah

Epilayer Thickness and Doping Density Variation Effects on Current-Voltage (I-V) Characteristics of n-GaN Schottky Diode 901
Tariq Munir, Azlan Abdul Aziz, Mat Johar Abdullah

Table of Contents

A VLSI Design Framework with Freeware CAD Tools 905
Y.K. Teh, F. Mohd-Yasin, M.B.I. Reaz, A. Kordesch

Implementation of Internal Mixed Signal ESD Protection onto RFID Transponder IC 910
M.K. Khaw, F. Mohd-Yasin, Y.K. Teh, M.B.I. Reaz

Device Design Consideration for Nanoscale MOSFET Using Semiconductor TCAD Tools 915
Teoh Chin Hong, Razali Ismail

Nonlinear Stability Analysis of Microwave Oscillators Using Lyapunov Function 920
H. Vahdati, A. Abdipour, A. Mohammadi

Conductivity of Cubic GaMnN Grown on Undoped GaN Layers 924
V K Sundaramoorthy, C.T. Foxon, S.V. Novikov, K.W. Edwards, I. Harrison

A Realistic March-12N Test and Diagnosis Algorithm for SRAM Memories 928
Wan Zuha Wan Hasan, Masuri Othman, Bambang Sunaryo Suparjo

Characterization of Strained Silicon MOSFET Using Semiconductor TCAD Tools 933
Wong Yah Jin, Ismail Saad, Razali Ismail

The Growth of III-V Nitrides Heterostucture on Si Substrate by Plasma-Assisted Molecular Beam Epitaxy 937
F.K. Yam, Z. Hassan, L.S. Chuah, N. Zainal, C.W. Chin, S.M. Thahab, M. Hussein

The Energy Band Gap of AlxGa1-xN Thin Films as a Function of Al-Mole Fraction 942
S.S. Ng, F.K. Yam, Z. Hassan, H. Abu Hassan

Queue Time Impact on Defectivity at Post Copper Barrier Seed, Electrochemical Plating, Anneals and Chemical Mechanical Polishing 947
Yasmin Abdul Wahab, Anuar Fadzil Ahmad, Zaiki Awang

Analysis of the Output Noise Voltage in CMOS Image Sensor Readout Circuit 953
Youngchang Yoon, Hochul Lee, Byung-Gook Park, Jong Duk Lee, Hyungcheol Shin

Electrical Characteristics of 100 MeV 28Si Implantation in GaAs 956
Yousuf Pyar Ali, A.M. Narsale, Othman Sidek, A.R. Damle, B.M. Arora

The Study of Pt/porous GaN Schottky Contact for Hydrogen Sensing 960
F.K. Yam, Y.P. Ali, Z. Hassan, N.H. Mohd. Noor, C.W. Chin

The Characterization of KrF Photoresists and the Effect of Different Chromophore Bulkiness on Line Edge Roughness (LER) for Submicron Technology 964
Ahmad Yusri Mohamed Bakri, Mohd. Jeffery Manaf, Kader Ibrahim Abdul Wahab, Ibrahim Bin Ahmad

Simulation Study on the Performance of SiC-GTO 974
Muhamad Zahim Sujod, Hiroshi Sakata

Failure Analysis of NSOP Problem Due to Al Fluoride Oxide on Microchip Al Bondpads 979
Zhao Siping, Hua Younan, Rao Ramesh, Li Kun

Studies on A New Sela-FIB Sample Preparation Method and Its Application in Failure Analysis of Wafer Fabrication for 110nm Technology Node and Beyond 986
Zhao Siping, Hua Younan, Er Eddie, Khoo Ley Hong

Synchrotron Radiation X-ray Diffraction and X-ray Photoelectron Spectroscopy Investigation on Si-based Structures for Sub-Micron Si-IC Applications 990
Zhe Chuan Feng, Li-Chi Cheng, Chu-Wan Huang, Ying-Lang Wang, T.R. Yang

A Highly Linear CMOS Down Conversion Double Balanced Mixer 994
Kumar Munusamy, Zubaida Yusoff

Solder Bump Strength and Failure Mode of Low-k Flip Chip Device 1000
Zulkarnain Endut, Ibrahim Ahmad, Gary Lee How Swee, Horazham Mohd Sukemi

Table of Contents

Wavelength Shifting in the Fiber Bragg Grating (FBG) based Encoder and Decoder Modules for SAC-OCDMA System..1005
Z. Zan, M.K. Abdullah, S.A. Aljunid, R.K.Z. Sahbudin, M.H. Yaacob, M. Mokhtar, S. Shaari

Effects of the Power Differences in the AND-Subtraction Detection Technique in SAC-OCDMA System Performance..1010
Z. Zan, S.A. Aljunid, M.K. Abdullah, R.K.Z. Sahbudin, M.H. Yaacob, M. Mokhtar, S. Shaari

Pulse Power Failure Model Of Power MOSFET Due To Electrical Overstress Using Tasca Method........1015
Nur Syakimah Ismail, Ibrahim Ahmad, Hafizah Husain, Shirly Chuah

New I-V Model For AlGaN/GaN HEMT At Large Gate Bias..1019
Ali El-Abd, M. Abdel Aziz, Abdel Aziz Shalby, Salah Khamis

Simulated Dielectric Characteristics of Pt/BST/Ni-Fe/Cu Multilayer Capacitor Stack for Storage Application..1024
Balachandran R., Yow HK, Manickam RM, Saaminathan V

Numerical Analysis of Filamentation in Conventional Double Heterostructure and Quantum Well High-Power Broad-Area Laser Diodes..1028
Amireh Seyedfaraji, Vahid Ahmadi, Mahyar Noshiravani, Farzan Gity

Monte Carlo Simulation of Surface Annealing Before Epitaxial Growth..1033
Chang Fu Dee, Burhanuddin Yeop Majlis

Analysis of Poly Resistor Mismatch..1037
Philip Beow Yew Tan, Albert Victor Kordesch, Othman Sidek

Modeling of Polyimide MIM Capacitors for Applications in Planar Monolithic Microwave Integrated Circuits..1039
R. Sanusi, A.I. Abdul Rahim, M.N. Osman, N. Kushairi, A. Rasmi, N.F.I. Muhammad, M.R. Yahya, A.F. Awang Mat

Simulation of InGaN Multiple Quantum Wells (MQWs) Light Emitting Diodes (LEDs)..1043
S.M. Thahab, H. Abu Hassan, Z. Hassan

TCAD Simulation of STI Stress Effect on Active Length for 130nm Technology..1047
Wan Rosmaria Wan Ahmad, Albert Victor Kordesch, Ibrahim Ahmad, Philip Tan Beow Yew

Simulation of Electromigration Test Structures With and Without Extrusion Monitors..1051
Verena Hein, Gisbert Hoelzer, Torsten Schroeter, Yvonne Yeo, Tan Hong Mui

Micro-Scale Structures and Nano-Scale Materials for Chemical and Biological Sensors

David J. Nagel, *Member IEEE*
Department of Electrical and Computer Engineering
The George Washington University
2033 K Street NW Suite 340J
Washington DC USA
Email: nagel@gwu.edu

Abstract. **Semiconductor and other materials with dimensions on the scale of micrometers or nanometers are being made into chemical and biological sensors. Besides being small, such devices are highly capable, low power and relatively inexpensive. They have many applications in personal health and safety, and for control of industrial processes.**

I. INTRODUCTION

Initially, information was generated only by people. Then, in the past half-century, computers became very important sources of information. Now, sensors also originate information, which is increasingly available and significant. Sensors have a long history, with temperature measuring devices being invented almost 400 years ago. Over the last century, an increasing number of sensors have been developed to measure physical, chemical and biological conditions in the world. Now, there are over 100 types of sensors sold by over 3000 companies [1].

In the past two decades, two large and related trends have had major impacts on sensors. The first is *miniaturization* to the micrometer and nanometer scales. The abilities to make very fine-scale structures, which were developed in the semiconductor industry, lead to the development of methods for micro-machining. Remarkable structures made of silicon and other materials can be produced now with critical dimensions of micrometers. Small cantilevers, other resonators and thin but tough membranes are primary examples. Devices made by micro-machining methods are generally termed Micro Electro Mechanical Systems (MEMS).

It is even possible to make such structures with nanometer-scale structures, which are called NEMS. Both MEMS and NEMS perform many functions, with sensing being the dominant use. Such devices do some earlier functions better, for example, deploying airbags in cars. In other cases, they enable new functions, such as rotation angle rate determination to stabilize the Segway Human Transporter.

The second trend for sensors in the recent past is *integration*, which is one of the fundamental characteristics of the modern world. The integration of technologies into systems enables fast terrestrial and space communications, and travel by land, sea and air. Cell phones are familiar examples of the integration of technologies. Today, the integration of technologies is fundamental to life and commerce. It has inestimable value.

Chip-scale micro-computers, which became available in the 1980s, were integrated then with old style sensors to produce "smart sensors". In the 1990s, micro-machining technologies were applied to make commercial MEMS sensors, especially accelerometers. In this decade, these very small and low-power sensors were put with both micro-controllers and chip-scale radios to make wireless sensor networks. The resulting short-range "mesh" networks have the remarkable capability to self-organize either for their initial configuration or in response to events in the network. Such networks have many applications. They are another example of the possibility and benefits of the integration of different device technologies into systems to produce useful new information.

In the past two decades, nano-materials have been developed and commercialized.

0-7803-9730-4/06/$25.00 ©2006 IEEE

Many have unprecedented properties because of their small dimensions. Semiconductor quantum dots and carbon nanotubes are primary among nano-materials. So, now it is not only possible to integrate diverse devices to produce remarkable systems. We can also integrate new materials with small dimensions onto small cantilevers and other microstructures to produce new sensors.

II. CHEMICAL AND BIOLOGICAL SENSORS

Because nano-scale materials are sensitive to the molecular level, the new devices from the integration of those materials with MEMS are especially useful for chemical and biological sensors. Sensitive, fast, compact, inexpensive and disposable sensors for diverse chemicals and bio-molecules have many applications in research, clinical medicine and security. They are especially attractive for prompt point-of-care medical diagnostics, as an alternative to sending samples to a laboratory and waiting a day or more for the results.

In general, chemical detectors for molecules in the gas phase, and biological detectors for molecules in an aqueous medium, are most important. Sensors that detect molecules, whatever their type and phase, must provide two functions, *recognition* and *transduction*. The first step insures that the molecule of interest is sensed specifically. The second makes it possible for the recognition event to be turned into something that can be recorded.

Many nano-scale materials and processes have been demonstrated for both recognition and transduction in bio-chemical sensors. They are described in numerous papers and books. We sought to reduce this plurality of options to a few graphics that would show relationships between them. They are presented and described in the rest of this section. Then, the next section gives some examples of miniature chemical and biological sensors based on microstructures and nano-materials. The concluding section deals with some opportunities and realities regarding such sensors.

We first consider the materials and mechanisms for sensing chemicals in the gas

phase, usually in air. There is already a large market for such chemical sensors. They are in systems with applications ranging from quantification of alcohol in the breath to the control of industrial processes.

There are two fundamental types of chemical sensors, those that work on all molecules simultaneously and others that perform sensing sequentially. These are indicated in Figure 1. The sensors that interact with all molecules at once either depend on the characteristics of materials or employ spectral details to separate signals from different molecules. The materials listed have some degree of chemo-selectivity, that is, they respond more or less differently to different vapor molecules. Sensors based on them generally employ neural networks, which have to be trained, to separate the effects due to various materials in the sample. Two types of semiconductor devices fall into this category. Chemo-selective polymers can be put on SAW or MEMS devices, so that some aspect, usually their resonant frequency, changes when target molecules are absorbed into the polymers. Optical and mass spectroscopic methods are now done in parallel with multiple detectors all responding together. Once, they were scanned and would be classed as sequential sensors. Ion mobility spectroscopy (IMS) still detects molecules sequentially after their field-induced drift in the air within the instrument prior to

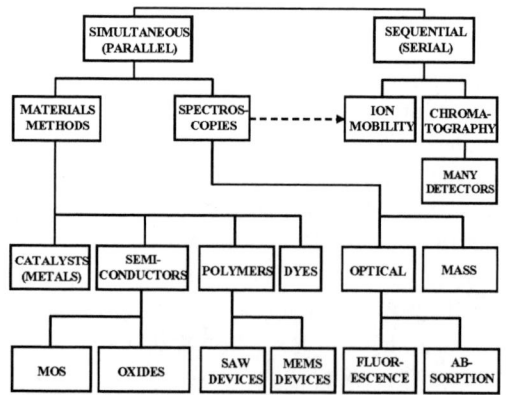

Figure 1. Methods and materials employed for recognition and transduction in *chemical* sensors. MOS = metal-oxide-semiconductor transistors and SAW = surface acoustic wave.

their detection. In chromatography, repeated adsorption and desorption occurs faster for light molecules than for heavy ones, and spreads the molecules out in space and time.

Biosensors seek to quantify particular molecules in complex wet samples. Glucose sensors are over 40 years old. However, many biosensors are recent developments because of MEMS and nano-materials.

In biological sensors, the recognition and transduction steps are generally more distinct than in chemical sensors. Figure 2 gives the means used for molecular recognition in biosensors. As for chemical micro-sensors there are several alternative means for recognizing particular molecules. The top row in Figure 2 gives the processes for making or increasing the recognition materials in the second row. DNA is amplified using the polymerase chain reaction (PCR). Many modern biosensors use DNA. SELEX stands for **S**ystematic **E**volution of **L**igands by **EX**ponential enrichment. It is used to isolate and amplify artificial molecules of particular use from large pools of candidate molecules. Aptamers are sequences of DNA or RNA, typically 15 to 60 base pairs long, that have desirable binding sites for recognizing target molecules in a sample. Polymers can play two recognition roles, either as substrates for molecular imprinting to produce binding sites (MIPS) or gels through which analyte

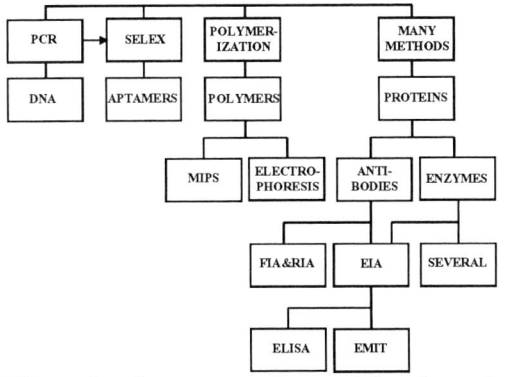

Figure 2. Processes and materials of use in *biological* sensors for *recognition* of particular molecules. Acronyms are explained in the text.

molecules are separated by electric fields (Electro-Phoresis). Proteins, which are crucial to life processes, are also central to biosensors. Antibodies are the foundation of animal immune systems and enzymes are organic catalysts. They both have very specific recognition capabilities, which are the basis of several recognition techniques: flow injection analysis (FIA), radio immunoassays (RIA), enzyme immuno assays (EIA), enzyme labeled immuno sorbent assay (ELISA) and the enzyme multiplied immunoassay technique (EMIT).

Many mechanisms are available for transduction in biosensors, as shown in Figure 3. Electrical and optical methods are most widely used. Many of these techniques are widely familiar. Others include FIA, fluorescence resonance energy transfer (FRET), upconverting phosphor technology (UPT) and surface enhanced Raman scattering (SERS). Some transduction mechanisms for biosensors exploit mass differences before and after recognition. Frequency shifts are used for biosensors as they are for chemical sensors. Magnetic techniques exploit giant magnetoresistant (GMR) and other magnetic materials.

It is clear that there are numerous combinations of the recognition and transduction elements that can be used in biosensors. A few examples are given in the next section.

III. SOME DEMONSTRATED AND COMMERCIAL CHEMICAL AND BIOLOGICAL SENSORS

Many chemical and biosensors that include either or both micro-structures and nano-materials have been developed, and some are commercialized. A few examples are provided here to illustrate what is possible.

Micro-machined hot plates have many uses, including chemical sensing. Figure 4 gives one example [2]. Typically, a layer of tin oxide with particle sizes in the sub-micrometer range is deposited on the thin membrane with an embedded heater. The resistance of the surface material depends on both the type and amount of organic molecules adsorbed from the vapor that the sensor experiences. Tin oxide chemical sensors are on the market [3].

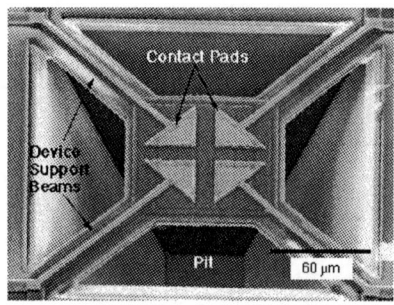

Figure 4. Photograph of a hotplate made by surface and bulk micro-machining. The sensing layer of nano-materials is applied over the contacts, and its vapor-dependent resistance is measured.

Another nano-sensor for small organic vapor molecules has been developed [4]. A cross section of it is given in Figure 5. The region between micro-electrodes is filled with coated gold nano-particles. When vapor molecules are absorbed for recognition, the distance between particles and the resistivity both increase for transduction.

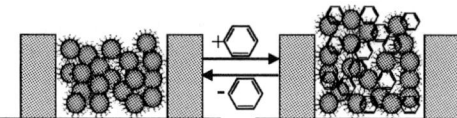

Figure 5. Schematic of a chemical sensor with nano-particles between electrodes on a surface.

Some bio-sensor devices and arrays are available commercially. The widely-advertised and available blood sugar sensors for diabetics are bio-sensors. In them, an enzyme recognizes glucose molecules and catalyzes reactions that release charges to be measured amperometrically (electrically).

The company Affymetrix sells arrays of bio-sensors where the recognition is done with DNA molecules and the transduction is optical fluorescence [5]. Nanogen offers bio-sensor arrays, also with DNA recognition, but using electrochemical detection [6]. Figure 6 shows a photograph of the heart of one of the Nanogen arrays.

IV. OPPORTUNITY AND LIMITATIONS

There are many combinations of the diverse micro-structures (especially cantilevers and resonators) and the many nano-materials (most notably chemo-selective semi-conductors and polymers for chemical sensors, and DNA, antibodies and enzymes with electrical and fluorescent readouts for biosensors). So, a question arises: is there a *strategy* for designing such sensors? There is one major factor common to both chemical and biological sensors. In both cases, the essential condition is how the number of analyte molecules of interest available to a sensor compares with the minimum detection limit (MDL) of the sensor. Are there enough molecules to detect their presence and quantify them?

Most commercial chemical sensors are based on particular properties of materials, as noted in Figure 1. It is usually difficult to lower the MDL of such devices. Then, the strategy is to present more molecules to the sensor, that is, to get above the MDL, by use of a pre-concentrator ahead of the sensor. It is more difficult to pre-concentrate bio-molecules, so techniques that amplify the target molecule, especially PCR for DNA, or which have very low MDL are of great interest. The MDL can be lowered by means of recognition molecules that can be reused. This is the case with enzymes, which can catalyze numerous reactions, so that more molecules are produced and available for transduction. An alternative is to use an optical transduction mechanism that gives many measurable photons from the same molecule. Fluorescence is the primary technique, where a dye molecule or quantum dot attached to the result of the recognition step can produce numerous photons. The detection of individual photons by using low noise and efficient photo multipliers also lowers the MDL, but such equipment is more bulky and generally best suited to a laboratory rather than factory or field use.

Because of the large increases in funding for nanotechnology in the past decade, and the wide variety of materials now available for use in sensors, there are many opportunities. This applies to individual nano-materials, many of which have been used for sensors already.

But, it is especially true for composites of nano-materials that have much less explored. Figure 7 shows the possibilities for pair-wise composites of nano-materials with zero, one, two and three dimensions (from top to bottom). The application of new composite nano-materials to chemical and biosensors is still in its early stages.

Some cautions regarding very small chemical and biological sensors are appropriate. It is fundamentally necessary to control the flows of material down to the nano-scale and the flow of signals up from that scale. Small materials in small devices might work with small samples. But, this can lead to small signals, so statistical and other noise sources provide a limitation. The calibration of small sensors can be challenging, and they can be very sensitive to interferants and have short shelf lives due to diffusion over even short distance. Even though sensors can be small, they usually require larger ancillary components, such as controllers, memories, radios and, particularly, batteries. Hence, there must be an overall balance in the size and capabilities of the components in a sensor system. Similarly, the use of miniature sensors in arrays can require sophisticated valving and control systems in order to expose specific sensors to analytes at particular times. These realities temper enthusiasm for chemical and bio-sensors based on MEMS and nano-materials. However, their small size, high performance and low cost are often compelling, so that their utility can be expected to increase steadily.

REFERENCES

[1] C. A. Grimes, E. C. Dickey, and M. V. Pishko (Editors), "Encyclopedia of Sensors", Am. Sci. Publ., 10 Volumes (2005)

[2] J. Suehle, R. E.Cavicchi., M. Gaitan and S. Semancik, "Tin Oxide Gas Sensor Using Micro-Hotplate by CMOS Technology and in-situ Processing," IEEE-Electron Device Letters, Vol. 14, 118-120 (1993).

[3] Figaro Engineering, Inc., http://www.figaro.co.jp/en/top.html

[4] A.W.Snow, H. Wohltjen and N.L. Jarvis, http://www.nrl.navy.mil/content.php?

P=02REVIEW45

[5] Affymetrix, Inc., http://www.affymetrix.com/

[6] Nanogen, Inc., http://www.nanogen.com/

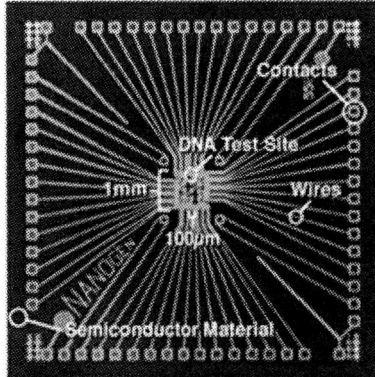

Figure 6. Micrograph of the test sites and electrodes in an array of bio-sensors made by Nanogen.

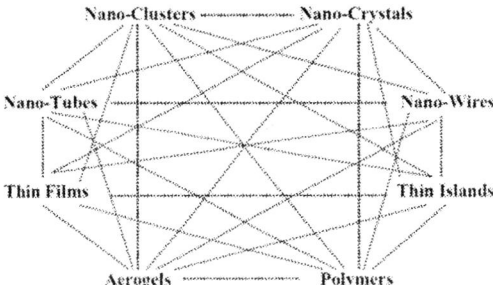

Figure 7. Possible binary combinations of nano-materials for use in sensors and other applications.

Electrical Characterization of Sub-100nm Features in Semiconductor Devices

Lerwen Liu
Asia-based Business Development Consultant of Zyvex Corporation (USA)
E-mail : lliu@zyvex.com

Abstract - **Zyvex (www.zyvex.com) is the first molecular nanotechnology company with a vision of developing adaptable, affordable, molecularly precise manufacturing. New tools are required to manipulate nanoscale objects and Zyvex has built a family of nanomanipulators capable of flexible 3D nanomanipulation, The sProber™ and nProber™, which are used by major semiconductor companies worldwide, are capable of probing 90, 65, and 45nm node technology with 5nm resolution of movement and cartesian xy,z motion. These systems are installed in scanning electron microscopes (SEMs) or dual beam focused ion beam (FIBs) providing a clean, stable environment for probing and electrical characterization, Zyvex's nanoprobers (consisting of 4, 5, 6 and 8 probes with probe tip size smaller than 50nm) are capable of characterizing transistors in die and at the contact level, finding and characterizing non visual fails, which are difficult to diagnose with other conventional techniques. Zyvex's nanomanipulator system can also probe copper, gold, and aluminum metal 1 layers. Zyvex's four probe system has been employed to characterize suspected failing transistors that makeup SRAM, DRAM, flash RAM and logic devices. Very recently, Zyvex has developed a novel 8 probe system combined with a FEI SEM) to probe an entire SRAM bit cell at the metal 1 layer. The nProbe is also completely encoded and includes many upgrades for higher throughput and easier to use. In this talk, Dr Liu will provide an overview on Zyvex Nanotechnology R & D efforts and its nanoprobing technologies for semiconductor failure analysis as well as next generation device characterization.**

0-7803-9730-4/06/$25.00 ©2006 IEEE

Implementation of 6kV ESD Protection for a 17GHz LNA in 130nm SiGeC BiCMOS

D. Linten*, M.I. Natarajan[1], S. Thijs, S. Van Huylenbroeck, S. Xiao, G. Carchon, S. Decoutere,
M. Sawada[2], T. Hasebe[2] and G. Groeseneken[+]

IMEC vzw, 75 Kapeldreef, Leuven, B-3001, Belgium

[1] SilTerra Malaysia Sdn. Bhd., Lot 8, Phase 2, Kulim Hitech Park, 09000, Kulim, Kedah, Malaysia

[2]Hanwa Electronics Ind. Co. Ltd., Wakayama, Japan.

*also at: Electrical Engineering Dept., Vrije Universiteit Brussel, Belgium

[+]also at: Electrical Engineering Dept., Katholieke Universiteit, Leuven, Belgium

Abstract **A systematic implementation of ESD protection for a 17GHz LNA in 130nm SiGeC BiCMOS technology is presented. The ability to achieve pre-silicon ESD reliability confidence is demonstrated through the comparison of HBM ESD simulations and measurements on the circuit. The inductor-based ESD protection methodology achieves more than 6kV HBM ESD robustness for the LNA, the highest ever reported in a similar technology.**

I. INTRODUCTION

Due to the increasing demand of data rate and bandwidth, operation frequencies at Ku-band are proposed for wireless local area network (WLAN) and 4G communications. These applications bring in new challenges for the circuit designers. In addition to the design implementation challenges in a given technology, assuring the circuit reliability is more critical since the systems using these circuits are close to human contact and in everyday use such as a mobile phone or ultra mobile computers, etc. One of the major reliability concerns in deep sub-micron technologies is the well known Electrostatic discharge (ESD) robustness.

A number of previous reports highlighted the challenges involved in implementing suitable and sufficient ESD protection for CMOS and BiCMOS technologies [1]-[3]. Further, it has also been reported that a number of RF circuits are even marketed without any ESD protection which essentially confirms the conflicting performance requirements and specifications for simultaneous RF and ESD functionalities. Various ESD protection methods such as circuit-ESD co-design, full/partial plug-and-play ESD protection, etc., have been proposed but these are mostly for sub-10GHz applications [1]-[5].

This paper presents, for the first time, the implementation details of an ESD protection strategy for a 17GHz Low Noise Amplifier (LNA) in state-of-the-art 130nm SiGeC BiCMOS technology. The ESD protection of the RF input is based on a plug-and-play inductor approach [3], box A in Fig. 1. The measured ESD robustness of the LNA is 6kV HBM, which is the highest ESD robustness ever reported from a similar circuit and technology node.

In the next section, the 130nm SiGeC BiCMOS process used for the fabrication of the LNA is briefly presented. Section III describes the LNA functional and ESD

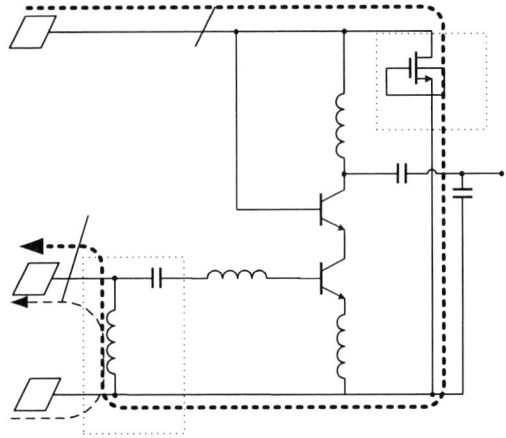

Fig. 1 Schematic of the ESD-protected 17GHz LNA

protection design. The results and analysis of the measurements are presented in Section IV.

II. 130NM SIGEC BICMOS TECHNOLOGY

The LNA is processed in a 130nm SiGeC BiCMOS technology. The HBT has f_T/f_{max} values of 205/275GHz [6]. The HBT module is embedded into a 130nm RF-CMOS baseline process with 65nm physical gate length devices and a heavily nitrided DPN gate oxide with EOT equal to 1.5nm. High quality ONO MIM capacitors are available between the top two metal interconnect layers in the three-metal level copper BEOL. High-quality inductors are realized using thin-film wafer level packaging (WLP) techniques, and are termed as Above-IC inductors [6].

III. ESD-PROTECTED LNA DESIGN

A. Unprotected LNA design

The unprotected LNA is a cascoded common emitter amplifier with inductive emitter degeneration, and inductive load, shown in Fig. 1 without boxes A and B. Both input and output are matched to 50 $\tilde{}$

An unprotected LNA is very sensitive to ESD stress events on its RF input. The BJT transistor Q1 in the 17GHz LNA is a two-emitter finger device with an emitter length of 14□m and width of 0.13□m. The forward BE diode of Q1

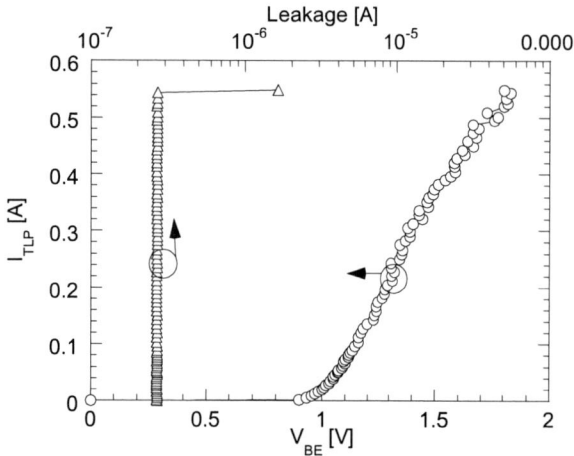

Fig. 2 TLP measurement results of an unprotected 17GHz LNA, when stressed between RF input and ground.

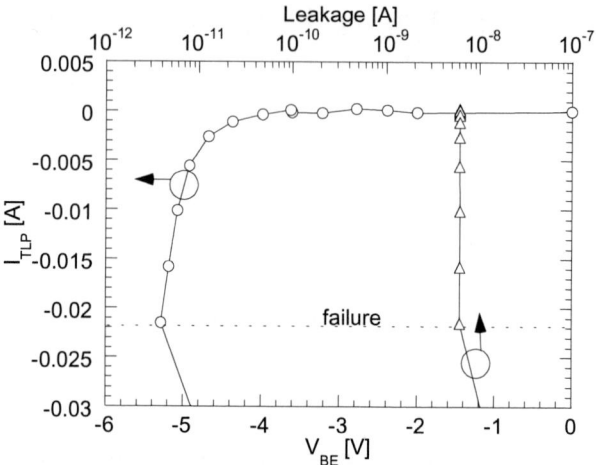

Fig. 3 TLP measurement results of an unprotected 17GHz LNA, when stressed between ground and RF input.

fails at 0.5A 1.8V Transmission line pulse (TLP), as shown in Fig. 2. The reverse BE diode fails at -20mA -5V TLP, as shown in Fig. 3.

When compared to a 90nm CMOS technology, with a MOS device EOT of 1.5nm, an unprotected LNA fails due to gate-oxide failure when the gate voltage exceeds +4.1V/-5.1V during 100ns [3]. Clearly the BJT device is less sensitive (-30V HBM) to ESD stress than its CMOS counter part (4.1V TLP). However, this is still insufficient for a production environment, where typically 2kV HBM robustness is required for RF pins.

B. ESD protection for a 17GHz LNA

The ESD design target is 6kV HBM robustness of the 17GHz LNA. The preferred ESD protection solution in the IC industry is always a plug-and-play approach. This means that the ESD protection elements are added after the core LNA design. The advantage in such an approach is that ESD IP blocks can be delivered together with the design kit for the technology.

A classical approach for RF ESD protection is the dual-diode ESD protection. In order to provide a 6kV HBM robustness the diodes need to be sized 60x2□m^2 in this technology. At RF, each diode represents a capacitive load of 121fF. Such an extra capacitive load can still be tolerated in a 5GHz LNA design without degrading its RF performance [1]. However, for designs at 17GHz, the shunt impedance generated by the ESD protection is divided by a factor 3.4 (132□ at 5 GHz, 32□ at 17GHz). This will shunt the incoming RF signal at 17GHz to V_{SS}, and destroy the RF performance of the LNA.

An alternative solution at these high frequencies is the use of a plug-and-play inductive ESD protection approach [4], where an inductor L_{ESD} and the ac-coupling capacitor C_c are added after the core LNA design, shown in box A in Fig. 1. The inductor is selected in order to provide a high impedance for the RF signal, and a short circuit for the ESD stress events, without affecting the LNA performance.

As a worst case stress condition, an 8kV HBM simulation has been performed, see Fig. 4, with the RF input stressed negative towards ground (V_{SS}^+-IN). The current that flows through the inductor L_{ESD} (1nH) results in a voltage overshoot at the RF input of -12V, see Fig. 4. This overshoot is coupled to the base of Q1. The diode of the BE junction clamp this voltage overshoot below 2V, by draining 60□A which is far below its failure threshold of 20mA. This shows that under V_{SS}^+-IN stress, pre-silicon ESD simulation predict >8kV ESD robustness. Similarly, pre-silicon ESD simulations when stressing the RF input pin positive and negative with respect to V_{SS} predict an ESD robustness of at least 8kV.

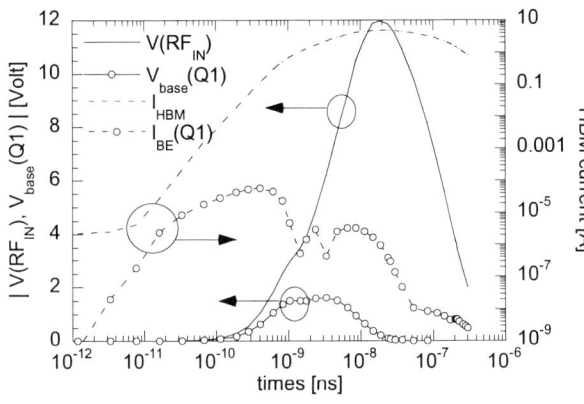

Fig. 4 8kV HBM simulation stressed IN to V_{SS} negative (V_{SS}^+-IN).

In the case of a CMOS LNA [4] the voltage overshoot can easily destroy the gate oxide of the amplification transistor. Adding small clamping diodes as 'plug-and-play', clamp the voltage across the gate oxide to a safe level below the gate oxide breakdown. However, in this BiCMOS design, the function of the clamping diodes is achieved by the intrinsic base-emitter junction in the bipolar Q1.

A grounded-gate NMOS device, M_{ESD} in Fig. 1, is used as a power clamp. Based on an accurate model of the power clamp, valid in the

ESD current and voltage domain, [8], the clamp has been sized to handle a 6kV HBM stress. Further, during a 6kV HBM stress between V_{DD} and IN, the voltage across the base-emitter and the collector-emitter should not reach breakdown voltage.

The simulated voltages at various nodes of the circuit during a 6kV HBM ESD event between V_{DD} and IN are shown in Fig. 5. A voltage peak of 17V is seen between V_{DD} and IN, and is the sum of the voltage drops across various

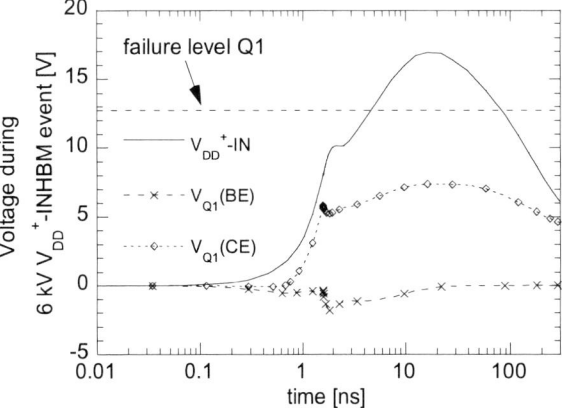

Fig. 5 Simulated voltages during a 6kV HBM ESD event between VDD and IN (V+DD-IN).

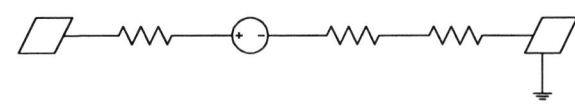

Fig. 6 V_{DD}^+-IN ESD current path.

elements: the power clamp (with an on resistance $R_{on, PC}$ and a holding voltage V_H), the V_{SS}-bus resistance (R_{VSS}), and the series resistance of L_{ESD}, as shown in Fig. 6. A maximum voltage of 7V is observed between the collector and emitter of Q1, which is far below the 13V breakdown voltage. These simulation suggest again a pre-silicon ESD reliability confidence of achieving 6 kV HBM.

IV. MEASUREMENTS RESULTS

The micrograph of the LNA is shown in Fig. 7. Above-IC inductors L_{ESD}, L_G, L_S, and L_{Load} have an inductance value of 1, 0.75, 0.2 and 0.5nH, with a quality factor (Q) of 17, 22, 20, and 12 at 17GHz, respectively. The total chip area including the bond pads is 1460x970□m².

The inductor L_{ESD} consumes an area of $300\text{x}320\square\text{m}^2$. Applying the inductive ESD solution on a 5GHz LNA [5], results in an area penalty of $620\text{x}620\square\text{m}^2$ for a 3nH Above-IC inductor. This clearly demonstrates the area reduction of the ESD solution with increased operation frequency of the LNA.

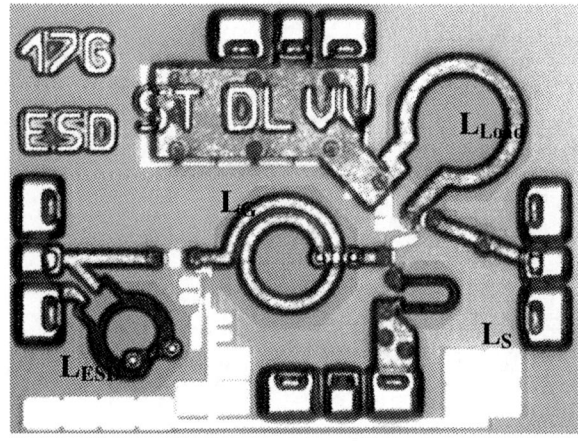

Fig. 7 Micrograph of the LNA.

A. RF measurement results

The LNA draws 4mA from a 1.8V supply. Measured S-parameters are shown in

Fig. **8**. A good input and output matching is achieved at the targeted operating frequency. The measured noise figure is plotted in Fig. 9. The ESD protection (L_{ESD}) contributes around 1dB of noise figure. The RF measurements show that, even with the ESD protection added, a good RF functionality can be achieved.

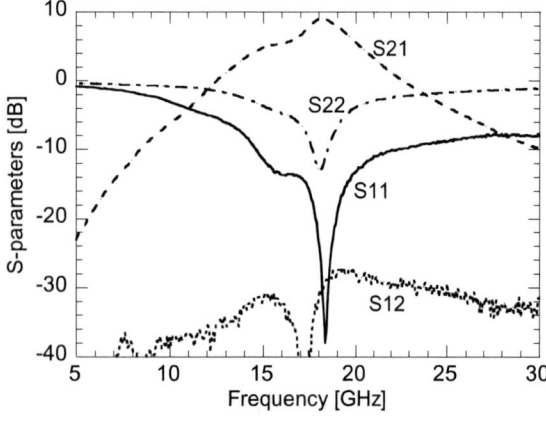

Fig. 8 Measured S-parameters.

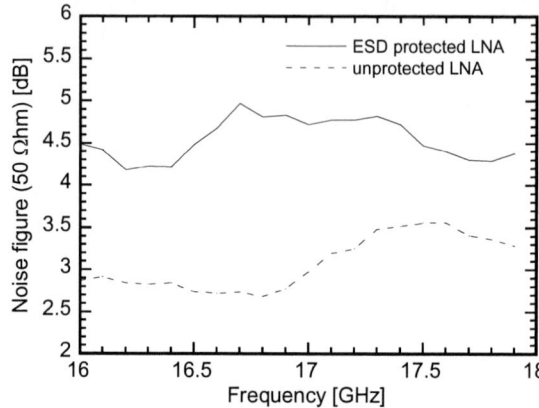

Fig. 9 Measured 50$\square$ noise figure.

B. ESD measurement results

On-wafer TLP and HBM measurements have been performed (Hanwa HED-T5000, HED-W5000) and are summarized in Table 1.

The failure current of the inductor L_{ESD} is above 10A TLP. As expected from the ESD simulations in Fig. 4, the RF input can handle more than 8kV HBM (> 10A TLP) (equipment measurement limit) when stressed positive or negative with respect to V_{SS}. A worst case ESD stress event is an ESD stress between V_{DD} and V_{SS} ($V_{DD}{}^+\text{-}V_{SS}$), and V_{DD} and the RF input ($V^+{}_{DD}\text{-}$IN), where both only a 6kV HBM and 3.2A TLP is measured.

TABLE 1 SUMMARY TLP AND HBM MEASUREMENT RESULTS.

TLP [A] | HBM [kV]	zapped pin (positive stress)		
ground pin	IN	VSS	VDD
IN		> 10 | > 8	3.2 | 6
VSS	> 10 | > 8		3.2 | 6
VDD	> 10 | > 8	> 10 | > 8	

In the ESD current path between V_{DD} and V_{SS}, only the power clamp is present. TLP IV measurements on a stand-alone power clamp, shown in

Fig. 10, reveal that it fails at ~3.2A. This indicates that the failure in the LNA during $V_{DD}{}^+\text{-}V_{SS}$ ESD stress combination is due to the failure of the power clamp.

When stressing positive between V_{DD} and the RF input IN ($V^+{}_{DD}\text{-}$IN), the same failure level of 6kV or 3.2A is observed, as shown in

Fig. 10. The ESD current path can be modeled as shown in Fig. 6: the ESD current flows through the power clamp, which has a holding voltage V_H of 4.36V and an on-resistance $R_{ON,PC}$

of 0.56Ω through the ground bus with a resistance R_{VSS} of around 0.2Ω and finally through the inductor L_{ESD} which has a series resistance $R_{ON, LESD}$ of 0.4Ω This once again confirms that weak element in the ESD protection is the power clamp which fails at a current level of 6kV HBM or 3.2A TLP.

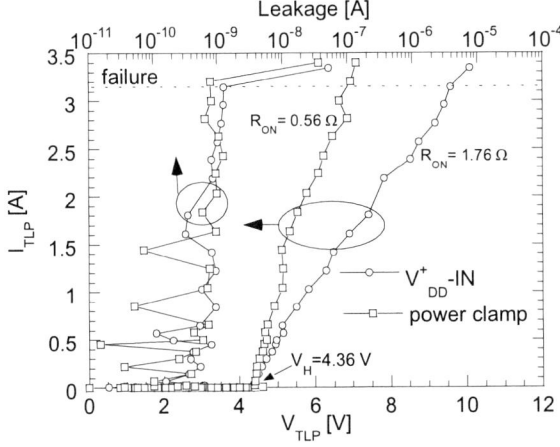

Fig. 10 TLP measurements of the grounded-gate NMOS power clamp and the ESD path between V_{DD} and the RF input pin.

The ESD robustness can easily be increased to 8kV by increasing the size of the power clamp. An 8kV V^+_{DD}-IN stress simulations, see Fig. 11, shows that when the size of the power clamp is doubled, a lower voltage stress of 6.5V appears across the collector-emitter of Q1. This voltage level will not damage the device.

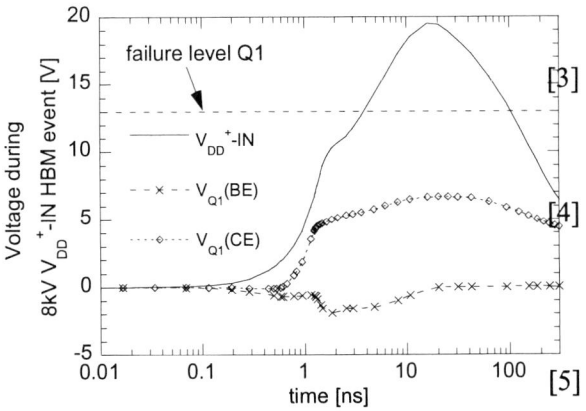

Fig. 11 Simulated voltages during a 8kV HBM ESD event between V_{DD} and IN (V^+_{DD}-IN).

VII. CONCLUSION

BiCMOS LNAs are less sensitive towards ESD stress events than CMOS LNAs. Nevertheless, an appropriate ESD protection is required to meet industry standards.

In this paper, the systematic design of an ESD protection for a 17GHz BiCMOS LNA is presented. Since a classical plug-and-play dual-diode ESD protection is not feasible at this high operating frequency, an inductive plug-and play ESD protection is used. Using accurate models for the ESD protection devices, valid at the ESD voltage and current domain, pre-silicon ESD reliability confidence is demonstrated.

The LNA has an ESD robustness of 6kV HBM, while preserving good RF performance. To the authors' knowledge, this design is the first ESD-protected 17GHz LNA with the highest ESD robustness ever reported in a similar technology node.

ACKNOWLEDGEMENT

The authors would like to thank the Institute for the Promotion of Innovation through Science and Technology in Flanders (IWT-Vlaanderen) for their support, the IMEC PLINE for processing of the wafers, the IMEC thin-film line for the above-IC passives.

REFERENCES

[1] Ph. Jansen, et al. ``RF ESD Protection strategies - The design and performance trade-off challenges'', CICC 2005, pp. 489-496.

[2] P. Leroux, et al, "High-performance 5.2GHz LNA with on-chip inductor to provide ESD protection", Electronics Letters, Volume 37, Issue 7, March 2001, pp. 467 – 469.

[3] S..Hyvonen, et al. "An ESD-protected 2.45/5.25GHz dual-band CMOS LNA with series LC loads and a 0.5V supply", RFIC 2005, pp. 43-46.

[4] S. Thijs, et al., "Implementation of Inductor Based ESD Protection for 5.5GHz LNA in 90nm RF CMOS - Concepts, Constraints and Solutions", EOS/ESD Symposium 2004, pp.40-49.

[5] D. Linten, et al., "Low-power low-noise highly ESD robust LNA, and VCO design using Above-IC inductors", CICC 2005, pp. 497-500.

[6]   S. Van Huylenbroeck, et al., "Lateral and Vertical Scaling of a QSA HBT for a 0.13□m 200GHz SiGe:C BiCMOS Technology", BCTM 2004, pp 229-232, 2004.

[7]   G. Carchon, et al., "Wafer-level packaging technology for High-Q On-chip Inductors and Transmission Lines", in IEEE Trans. MTT, vol. 52, no.4, pp. 1244–1251, April 2004.

[8]   V. Vassilev, et al., "Advanced modeling and parameter extraction of the MOSFET ESD breakdown triggering in the 90nm CMOS node technologies", EOS/ESD symposium 2004, pp. 98-106.

Optimization of the Temperature Sensor Position for MEMS Gas Flow Meters

E. AbbaspourSani and D. Javan
Urmia University, Urmia, Iran
e.abbaspour@mail.urmia.ac.ir

Abstract— **A bulk micromachined structure for a thermal type gas flow meter is proposed. It is considered as the thermal type device and consists of a micro heater and two temperature sensors situated at both sides of the heating element. The sensor works on the bases of displacement of temperature profile around the heating element with the gas flow. The heater and the temperature sensors are assumed to be situated on a stacked (SiO_2 / Si_3N_4) thermally isolated membrane. The effect of the distance of the temperature sensor from the heating element on the output signal is investigated using Ansys/Flotran software. The simulation results for a specified device provide the optimum distance between the central heating element and two sensing elements in the range of 200um.**

I. INTRODUCTION

Flow meters are used to measure the rate of displacement of a gas or liquid with respect to time. There is a growing demand for gas flow measurement in the industrial, environmental, automotive and medical applications. In consideration of possible flow sensing principles in macroscopic flow meters (electromagnetic field, ultrasonic, coriolis force, differential pressure, thermal…), the thermal flow measurement and differential pressure detection have revealed to be most promising for micromachined sensors technology [1].

Although differential pressure detection is very appropriate for liquid flow measurement, it is, however, less applicable for gas flow. In this case, thermal flow sensors are preferred because they are more sensitive and also they create low pressure drop.

Thermal flow sensors are basically thermally isolated structures which carry heater and temperature sensors [2]. In this paper the design and simulation of a calorimetric thermal flow sensor has been presented. The working principle of the designed sensor is based on the displacement of a temperature profile created by a heating element due to the presence of incoming gas flow.

II. DEVICE STRUCTURE AND OPERATION

A schematic diagram of the micromachined type die of the proposed gas flow meter is shown in Fig.1. The die can be easily fabricated by the well established bulk micromachining technology. It consists of three main elements, one central heater and two temperature sensing resistors. These resistors are considered on a micromachined oxide /nitride membrane which provides thermal isolation between the resistors and the silicon substrate. There are additional resistors other than the heater and temperature sensors on the membrane. These resistors monitor the incoming gas temperature and can be used for the electrical out put signal conditioning. The die that performs the gas flow measurement, will be simply called the sensor through out this paper.

The operation of the sensor is based on the theory of convective heat transfer. The resistive heating element is situated in mid point between the temperature sensors and heated up to about $120 \ c^\circ$ by the applied voltage. When gas flows over the device, it distributes the temperature around the sensing elements created by the heater. This temperature asymmetry caused by the forced convection effect is measured by the two resistive sensing elements. Due to gas flow over the sensor, the temperature of one of the sensing elements increases while the other cools down. The temperature difference between two sensing resistors changes the conductivity of the sensing resistors, which can be converted to an electrical output signal. This conversion can be easily accomplished by an external Wheatstone bridge, taking the sensing elements as the two resistors of the bridge and the other two external resistors to be added as the bridge balancing elements.

0-7803-9730-4/06/$25.00 ©2006 IEEE

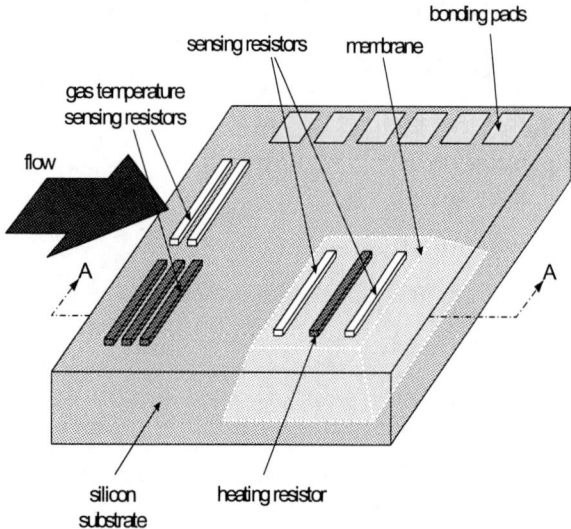

Figure 1. Schematic view of the designed gas flow sensor

III. FINITE ELEMENT METHOD

Finite element method is a very powerful technique for modeling complex structures and enables us to solve system equations for our designs. This method has been widely used for the characterization of mechanical, thermal and electrical behavior of the micro machined sensors [6].

The simulation of the proposed structure was performed employing the Ansys/Flotran package using "fluid" elements, which are appropriate for flow analysis and provide the values of velocity, pressure and the temperature of the fluid to all directions. The following boundary conditions were applied to the model:
- No-slip (zero velocity) conditions are applied all along the walls including the walls of sensor chip.
- A zero relative pressure is applied at outlet of the flow channel.
- The y component of the flow velocity is set to zero every where in the flow channel.

The loads applied to the model were:
- Gas flow velocity (v) is to vary from 0 to 1 m/s at the inlet of the flow channel.
- The temperature of the silicon substrate is the same as the inlet gas flow temperature.
- The temperature difference between the heating element and the gas in the channel is constant during the simulations.

The simulations were performed under static conditions (without flow) and then under the gas flow. The temperature profile on the membrane under the static conditions is shown in Fig.2. It can be seen that the temperature profile around the heating element is symmetrical.

IV. SIMULATION RESULTS

Extensive FEM simulations with different

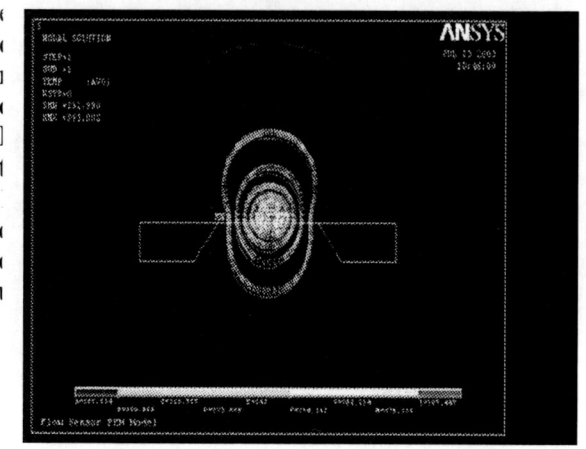

Fig.2. Symmetrical temperature profile on the membrane, v=0m/s

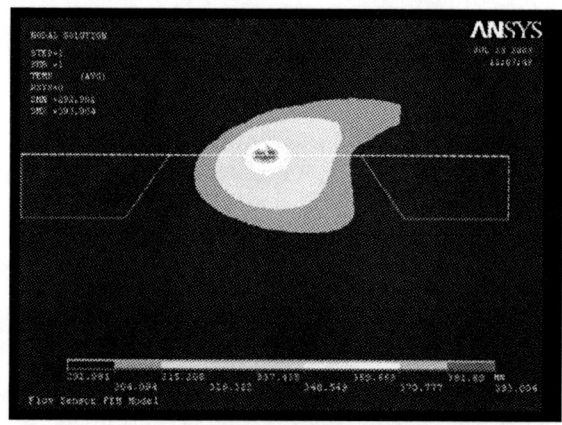

Fig.3. Displacement of the temperature profile due to gas flow, v=1m/s

The temperature variation at each point of the membrane depends on the point distance from the central heater. When the flow velocity increases, the upstream side of the heating element is cooled but the downstream side is heated. In addition to this it is also clear that cooling effect in the upstream side is stronger than the heating effect in the downstream side. Therefore, we expect that the response of the sensor will be mainly determined by the upstream sensing element. Fig.4 shows the result of temperature difference (ΔT) versus gas flow velocity for six different values of separation between the heater and the sensing resistors (X). As it is evident from the plots, for the specified dimension and conditions of our proposed gas flow meter, the temperature difference (ΔT) increases as X increases from 115 up to 200. It reaches to a maximum at 200 and then decrease as X is increased. In other words the optimum distance between the heater and the sensing elements in our device was 200. It is obvious that this distance will be changed for the different device dimensions.

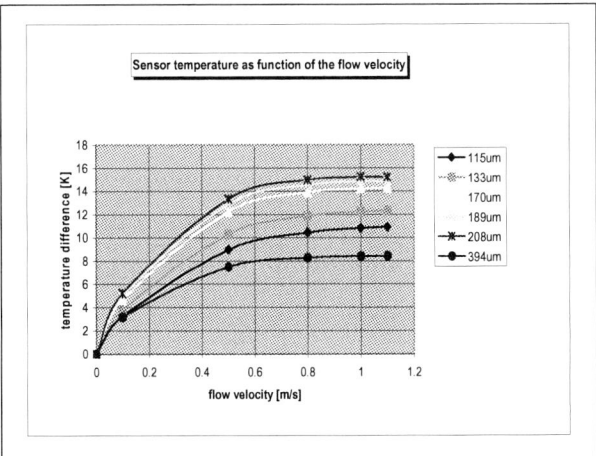

Fig.4. Temperature difference vs. flow velocity for various heater/sensor separation distances

V. CONCLUSION

A micromachined thermal flow sensor for measuring gas flows with a velocity range of 0-1 m/s has been proposed. The device was simulated using thermal, fluid FEM method and Ansys/Flotran software. It is determined that for a specified dimension and conditions of the gas flow meter, the temperature difference (ΔT) increases as the distance between the heater and sensing elements (X) increases from 115 up to 200. It reaches to a maximum at 200 and then decrease as X is increased. In other words the optimum distance between the heater and the sensing elements in our case was 200. This method can be applied to the other device dimension and conditions for determining the optimum separation between the central heater and the heat sensors.

REFERENCES

[1] M. Ashauer, H. Glosch, F. Hedrich, N. Hey, H. Sandmaier and W. Lang, "Thermal flow sensor for liquids and gases based on combinations of two principles," *Sensors and Actuators A 73, Issues 1-2, pp. 7-13, 1999*

[2] M. Elwenspoek, "Thermal flow micro sensors," *CAS 99, Volume 2, pp. 423-425, 1999*

[3] A. G. Nassiopoulou and G. Kaltsas, "Porous silicon as an effective material for thermal isolation on bulk crystalline silicon," *Phys. Stat. sol. pp 307-310, 2000*

[4] BW. Van oudheusden, "Silicon thermal flow sensors," *Sensors and Actuators A 30, pp 5-26, 1992*

[5] C. Rossi, E. Scheid and D. Esteve, "Theoretical and experimental study of silicon micromachined microheater with dielectric stacked membranes," *Sensors and Actuators A 63, pp. 183-189, 1997*

[6] C. Rossi, P. T. Boyer and D. Esteve, "Realization and Performance of thin SiO2/Si3N4 membrane for microheater applications," *Sensors and Actuators A 64, pp. 241-245, 1998*

Non-Crossing Differential Capacitive MEMS Accelerometer with Electrostatic Spring Tuning

Ahmad Alabqari Ma' Radzi* and Burhanuddin Yeop Majlis, *SMIEEE*

Institute of Microenginering and Nanoelectronics (IMEN)

Universiti Kebangsaan Malaysia, 43600 UKM Bangi, Selangor, MALAYSIA

e-mail: abqari@kuittho.edu.my, burhan@vlsi.eng.ukm.my

*Current address: Faculty of Electrical and Electronic, Kolej Universiti Teknologi Tun Hussein Onn (KUiTTHO), 86400 Parit Raja,Batu Pahat, Johor, MALAYSIA

Abstract- **This paper presents the design and simulation of non-crossing differential capacitive MEMS accelerometer with electrostatic spring tuning. The accelerometers were designed with critical damping for 10 g range, mechanical sensitivity of 0.012 μm/g and thickness of 40 μm which is compatible with bulk micromachining process requirements. Simulations were carried out using *IntelliSuite 7.2*. Results show that the open loop accelerometer with mechanical spring of 52.3 N/m could be tuned by negative electrostatic spring using 10 V, 7.5 V and 5V to produce -22 N/m, -12.5 N/m and -5.6 N/m respectively with 12 pairs of electrostatic finger with 1.8 μm gap, and capacitance sensitivity is increased from 20 fF/g to 40 fF/g while the dynamic range decreased to 5.8 g. The designed force-balanced accelerometer has capacitance sensitivity of 10.67 fF/g, mechanical spring of 62.2 N/m and uses 8 pairs of electrostatic finger to produce -17.7 N/m, -9.9 N/m and -4.4 N/m respectively using 10 V, 7.5 V and 5 V. Hence, the dynamic range could be tuned down from 17 g to 14 g and capacitance sensitivity increased to 12 fF/g. The fabricated devices are also shown in this paper. The designed accelerometer with electrostatic spring tuning therefore enable us to tune to the suitable spesifications with the needed g measurement.**

I. INTRODUCTION

Most MEMS accelerometers use the capacitive type configuration to sense acceleration. This is due to their high sensitivity, good dc response and noise performance, low drift, low temperature sensitivity, low power dissipation and simple structure [1-5]. In the capacitive type

of acceleration sensing, differential capacitance is widely used to get a linear output. The main advantage of the non-crossing type compare to the crossing type is that it can use a three-mask fabrication process [6]. However, the non-crossing type introduces an inherent nonlinearity relationship with the displacement of the proof mass. In order to improve the linearity, the movement of the mass is limited to a small range.

MEMS accelerometer with stiffness tuning consists of parallel plate with branched comb-fingers that acted as the negative electrostatic spring has been fabricated [5]. The mechanical stiffness is reduced only along the sensing direction, thus an accelerometer that have been tuned with a low frequency has an improved resolution. Meanwhile, force-balanced approach could reduce the proof mass displacement, increased of the dynamic range and the output linearity is improved [7-9]. By having two electrodes with different polarity for feedback will ensure that the electrostatic force be linear to the applied feedback voltage and feedback voltage will be linear to acceleration [5,7].

II. DESIGN THEORIES

Considering two parallel plate with gap (*d*), given an applied voltage (*V*) and the capacitance (*C*) between the plates, the electrostatic force generated between the two parallel plates can be calculated by taking the derivative of the energy in the direction of motion. Hence for *x* direction, the electrostatic force is given by:

$$F_{e,x} = \frac{\partial W_e}{\partial x} = \varepsilon_0 \frac{AV^2}{2d^2} \qquad (1)$$

We adopted straight pairs configuration of finger and electrode for electrostatic spring

0-7803-9730-4/06/$25.00 ©2006 IEEE

tuning. The effective negative electrostatic spring constant, k_e, is obtained by differentiating proofmass displacement, x, through the electrostatic force equation for straight parallel plate with tuning gap, d_t, and is given by [4]:

$$k_e = -\frac{\partial F}{\partial x} = -N_t \varepsilon_0 l t V_t^2 \left(\frac{1}{(d_t + x)^3} + \frac{1}{(d_t - x)^3} \right)$$
.....(2)

where N_t is the number of tuning pairs, l is the length, t is the thickness and V_t is the tuning voltage applied. Thus, at equilibirium, $x=0$, the effective electrostatic spring constant is given by:

$$k_e = -\frac{\partial F}{\partial x} = -2N_t \varepsilon_0 l t V_t^2 \left(\frac{1}{d_t^3} \right) \qquad (3)$$

which is used for the force balanced accelerometer since the proofmass position is nulled by the close-loop operation.

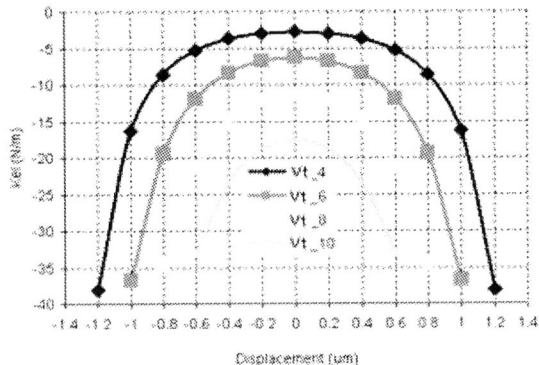

Figure 1: Effective electrostatic spring constant of varying V_t as a function of proofmass displacement

One major problem with electrostatic forces is that they are always attractive and nonlinear because they are proportional to the voltage squared and inversely proportional to the gap squared. Figure 1 shown the effective electrostatic spring constant of varying V_t as a function of proofmass displacement. When displacement is larger, the negative electrostatic spring constant increased quadratically. The displacement must not exceed our critical value of mechanical spring constant to prevent the value of the total spring constant becomes negative and the system will be unstable. Thus, we have to limit our stiffness tuning operation for a very small proofmass displacement to

maintain the stabil operation with better linearity responses.

Force-balanced configuration applied to non-crossing accelerometer with stiffness tuning is shown in Figure 2 by adding two pairs of finger and electrode for electrostatic force balanced. To achieve a linear, negative feedback, it is necessary to superimpose a feedback voltage, V_f, on a bias voltage, V_b, on both electrodes which results in a net electrostatic force on the proofmass as follows [10]:

$$F_e = F_1 - F_2 = \frac{N_f \varepsilon_0 A}{2} \left[\begin{array}{c} (V_b + V_f)^2 \left(\frac{1}{(d_0 - x)^2} - \frac{1}{(D_0 + x)^2} \right) \\ -(V_b - V_f)^2 \left(\frac{1}{(d_0 + x)^2} - \frac{1}{(D_0 - x)^2} \right) \end{array} \right]$$
.....(4)

where d_0 is the small gap and D_0 is the big gap between the electrodes. Under closed-loop operation, the displacement of proof mass will be small, therefore it can be assumed that $x^2 \ll d_0^2$ [5,7,10]. It is desirable to have the electrostatic force to be linear with the feedback voltage [5,6]. The feedback voltage, V_f can be calculated as a function of applied acceleration and is given as follows [10]:

$$V_f = \frac{d_0^2 D_0^2}{2 A N_f \varepsilon_0 V_b [D_0^2 - d_0^2]} ma \qquad (5)$$

where m is the mass, a is the acceleration and N_f is the number of feedback pairs . The bias voltage, V_b, is chosen according to desired dynamic range of the accelerometer and the minimum required bias voltage, V_b, to superimpose the feedback voltage can be calculated as[10]:

$$V_b = \sqrt{\frac{2}{N_f \varepsilon_0 A} mad_0^2 D_0^2 \left[\frac{1}{D_0^2 - d_0^2} \right]} \qquad (6)$$

When there is an acceleration along the x-direction, the sense capacitance C_1 and C_2 will change. The differential capacitance, ΔC can be written as $C_1(x) - C_2(x)$, which is given by:

$$C_1(x) = N_s \varepsilon A \left(\frac{1}{d + x} + \frac{1}{D - x} \right) \qquad (7)$$

$$C_2(x) = N_s\, \varepsilon A \left(\frac{1}{d-x} + \frac{1}{D+x} \right) \qquad (8)$$

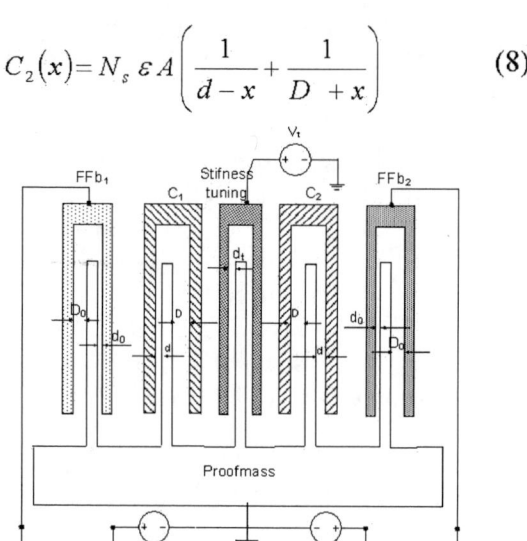

Figure 2: Schematic of non-crossing force balanced MEMS accelerometer with electrostatic spring tuning.

where N_t is the number of sense pairs. Hence, the capacitance sensitivity as a function of g acceleration can be calculated as:

$$S_{\Delta C} = \frac{C_1(x) - C_2(x)}{g} \qquad (9)$$

III. RESULTS AND DISCUSSIONS

A. Open-loop accelerometer with stiffness tuning

An open-loop non-crossing accelerometer with stiffness tuning was designed with a critical damping for ±10g range with mechanical sensitivity of 0.012 µm/g and a capacitive sensitivity of 20 fF/g. The weight of the proof mass is 6.38 x 10^{-08} kg with a mechanical spring constant of 52.30 N/m. The electrostatic spring tuning consists of 12 pairs of electrode fingers with gap dimension of 1.8 µm.

Based on simulations that were done, from Figure 3 we could observed the net proofmass displacement of varying V_t as a function of g acceleration for 100 g range. The nonlinearity of electrostatic force will lead to snap-down or pull-in effect which spontaneously force electrode touch the proofmass fingers. Figure 4 shown the plot of proofmass displacement as a function of g acceleration which we limited our operation at 0.12 µm. The mechanical stiffness could be tuned by negative electrostatic spring using 10 V, 7.5 V and 5V to respectively produce -22 N/m, -12.5 N/m and -5.6 N/m, and calculations show that the dynamic range that can be achieved with stiffness tuning is lowered

respectively to 5.8 g, 7.7 g and 9 g. As a results, the capacitance sensitivity increased from 20 fF/g up to 40 fF/g with maximum 10 V of V_t applied. Simulation results deviates from the theoretical values by 6%. Simulation plots has nonlinearity value of 0.80, 0.65 and 0.43 respectively for electrostatic spring using 10 V, 7.5 V and 5V, while their calculated value are 0.06, 0.02 and 0.016.

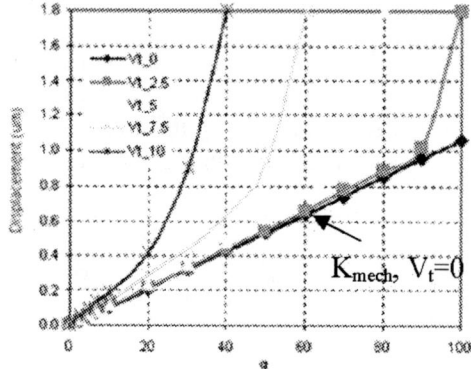

Figure 3: Net proofmass displacement of varying V_t as a function of g.

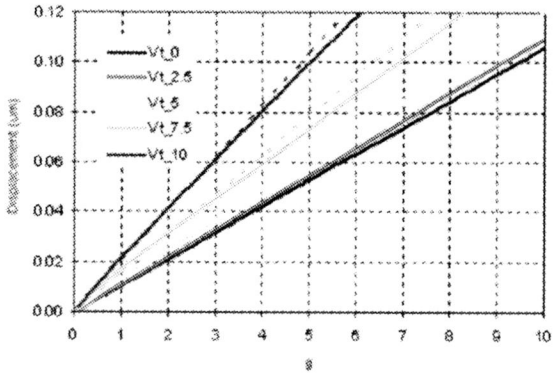

Figure 4: Simulations plot of proofmass displacement as a function of g with the dashed line shown is the calculated value.

B. Force balanced accelerometer with stiffness tuning

The designed force-balanced non-crossing accelerometer with stiffness tuning has dynamic range of ±17g, mechanical sensitivity of 0.012 µm/g and capacitive sensitivity of 10.67 fF/g. The weight of the proof mass is 7.56 x 10^{-08} kg with a mechanical spring constant of 62.2 N/m. The electrostatic spring tuning consists of 8 pairs of electrode fingers with gap dimension of 1.8 µm. Force balanced operation uses of 10 pairs of electrode fingers each for positive and negative polarity with $d_0 = 1.8$ µm and $D_0 = 3.6$ µm.

Force balanced operation increased our dynamic range from ±10g to ±17g with displacement constraint of 0.12 µm. From Figure 5, it is seen that the nonlinearity improved with higher bias voltage and increased with higher dynamic range. Figure 6 shown the net proof mass displacement of varying V_b as a function of g acceleration. We set our nonlinearity unit to be below 1%, thus our force balanced will be operated with V_b 14 g and V_b 12 g in the dynamic range of ±17g and ±15g respectively. Therefore, by calculations the nonlinearity were 0.60 and 0.84 respectively for V_b 14 g and V_b 12 g.

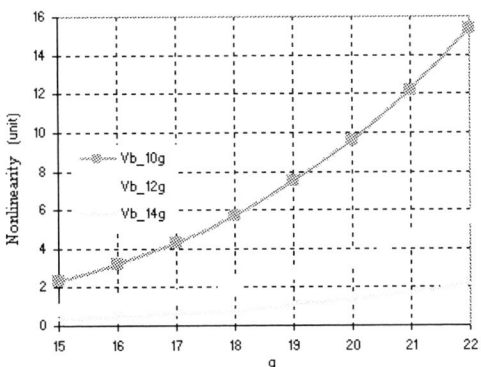

Figure 5: Nonlinearity value of varying V_b as a function of dynamic range.

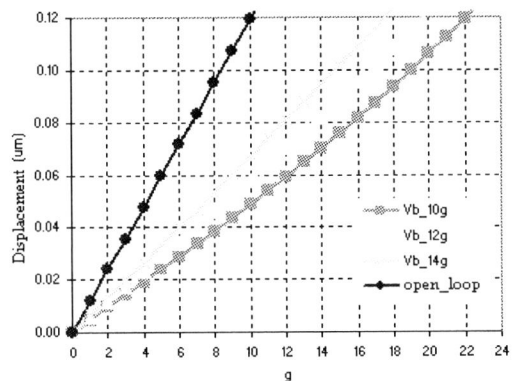

Figure 6: Net proofmass displacement of varying V_b as a function of g with $V_t = 0$ V

Figure 6 shown the plot of proofmass displacement as a function of g acceleration with stiffness tuning of 10 V applied. The mechanical stiffness was tuned by negative electrostatic spring using 10 V to produce -17.7 N/m. Simulations show that the dynamic range that can be achieved with stiffness tuning is decreased from 17 g to 14 g, and from 15 g to 14 g respectively for operation with V_b 14 g and V_b 12 g. As a results, the capacitance sensitivity increased from 10.67 fF/g up to 12 fF/g with maximum 10 V of V_t applied and V_b 14 g. Simulation results deviates from the theoretical values by 8%. Simulation plots of electrostatic

spring using 10 V has nonlinearity value of 1.23, and 0.95 respectively for operation with V_b 14 g and V_b 12 g, while their calculated value are 0.19, 0.76. This shown the improved nonlinearity value with stiffness tuning operation.

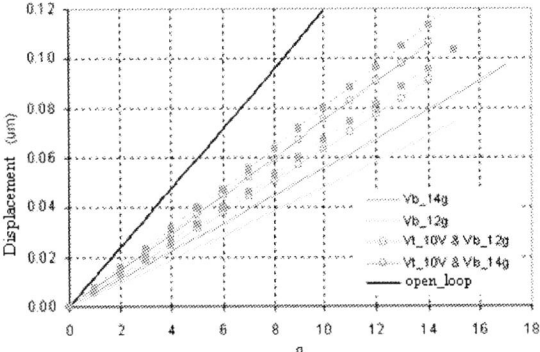

Figure 7: Simulations plot of proofmass displacement as a function of g with the dashed line shown is the calculated value.

C. Devices fabrication

The designed accelerometer is fabricated utilizing a three-mask process. The main fabrication concept is using two types of substrate i.e., silica glass and silicon single crystal. The substrates then intimate using wafer bonding process. The fabrication process is done in collaboration with MEMSTECH and SENZFAB. The three-masks process could lower the fabrication cost effectively. Figure 8 shows the fabricated of the designed accelerometers shown by Leica microscope with amplification of 94 times from its original dimension.

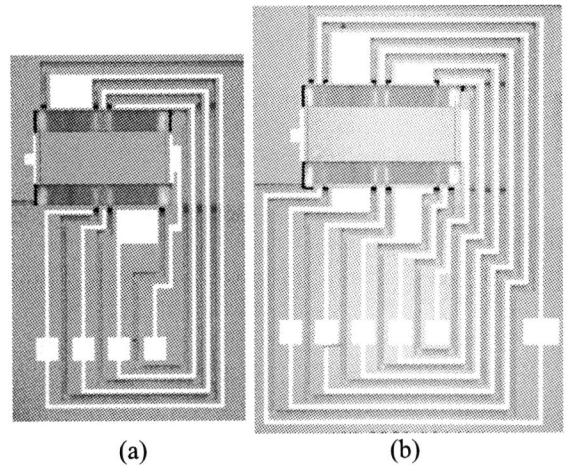

(a)   (b)

Figure 8: Fabricated accelerometers with electrostatic stifness tuning which (a) open loop, and (b) force balanced.

IV. CONCLUSION

An open loop and force-balanced non-crossing differential capacitive MEMS accelerometer with electrostatic spring tuning design and analysis were presented. Electrostatic spring acted as the negative spring constant that could tuned the mechanical stiffness of an accelerometer. Thus, the capacitance sensitivity is increased but the dynamic range is decreased to maintain its stabil operation. By using electrostatic force-balanced, the dynamic range can be improved considerably. The designed accelerometer with electrostatic spring tuning enable us to arrange the performance to the spesifications needed. Fabrication cost is lowered by using three-mask process as an advantage of using non-crossing capacitance electrodes.

ACKNOWLEDGEMENT

This project is financially supported by Ministry of Science, Technology and Inovation of Malaysia under IRPA program: Development of MEMS Technology for Automotive Application and Kolej Universiti Teknologi Tun Hussein Onn.

REFERENCES

[1] Y. Nemirovsky and O. Bochobza-Degani, "A methodology and model for the pull-in parameters of electrostatic actuators", *J. Microelectromech. Syst.*, vol. 10, No.4, pp. 601-615, Dec. 2001.

[2] R. Puers and D. Lapadatu, "Electrostatic forces and their effects on capacitive mechanical sensors", *Sensors and Actuators A*, vol. 56, pp. 203-210, 1996.

[3] Y. Zhou, "Layout synthesis of accelerometer", Carnegie Mellon University, August 1998.

[4] G. Zhang, "Design and Simulation of a CMOS-MEMS Accelerometer", Carnegie Mellon University, August 1994.

[5] K.Y. Park *et. al*, "Capacitive type surface-micromachined silicon accelerometer with stiffness tuning capability", *Sensors and Actuators A*, vol. 73, pp. 109-116, 1999.

[6] A.S. Tamsir, F. Saharil and B.Y, Majlis, "Spring Constant and von-Misses Stress of a Non-Crossing Differential Capacitive Accelerometer", *Proc. 2003 IEEE National Symposium on Microelectronics (NSM2003)*, Malaysia, pp. 247-251, 2003.

[7] S. Mutlu, "Surface Micromachined Capacitive Accelerometer with Closed-loop Feedback", University of Michigan.

[8] G. Maysam and R. Rajagopalan, "A Capacitive Accelerometer Design for Eliminating Picture Vibrations in Video Cameras", University of Michigan.

[9] R.P. van Kampen *et. al.*, "Application of electrostatic feedback to critical damping of an integrated silicon capacitive accelerometer", *Sensors and Actuators A*, vol. 43, pp. 100-106, 1994.

[10] Ahmad Alabqari M. R., Badariah B., and Burhanuddin Y. M., "Design Consideration of Force-Balanced Non-Crossing Differential Capacitive MEMS Accelerometer," Proc. 2005 IEEE National Symposium on Microelectronics(NSM05), Kuching, pp. 326-330.

Fabrication of Platinum Membrane on Silicon for MEMS Microphone

Azrul Azlan Hamzah, Burhanuddin Yeop Majlis, *SMIEEE,* and Ibrahim Ahmad, *MIEEE*
Institute of Microengineering and Nanoelectronics (IMEN)
Universiti Kebangsaan Malaysia
43600 Bangi, Selangor, MALAYSIA
E-mail: burhan@vlsi.eng.ukm.my,

Abstract — **Platinum membrane with silicon nitride layer is fabricated and analyzed. The membrane, which is designed for MEMS microphone application, is fabricated using sputter platinum and CVD silicon nitride. Membranes with sandwich layer of platinum-nitride-platinum with thickness of 6.35 μm are successfully fabricated. Deflection of the fabricated membrane corresponding to given pressure is measured using Tencor surface profiler. It is observed that deflection at its center is proportional to applied pressure for pressure between 20 Pa to 200 Pa. Average center deflection for applied pressure of 200 Pa is measured to be 0.41 μm. The fabricated platinum membrane is deemed suitable for MEMS microphone application due to its linear deflection response in acoustic pressure range.**

I. INTRODUCTION

STRUCTURAL membrane is a crucial part for MEMS devices like pressure sensor and microphone. The membrane is designed to deflect corresponding to the applied pressure on its surface. For these devices, capacitance value between a fixed plate and a deflectable membrane changes corresponding to the magnitude and frequency of membrane's deflection. These capacitance values are then converted to applied pressures or acoustic signals by a signal processing IC. Metal membrane is preferred for these applications due to its better capacitance response compared to that of silicon [1].

Metal membrane is commonly fabricated by physical deposition techniques such as sputtering and evaporation. Usually, metal membrane is formed by removing the pre-deposited sacrificial layer underneath the deposited metal layer, thus releasing the membrane structure [2]. The membrane could also be made free standing by etching into the silicon substrate underneath the deposited metal layer [3].

Alternatively, metal membrane could also be formed by sputter deposition of metal on a very thin silicon nitride membrane, as will be introduced in this paper. The silicon nitride membrane is initially formed by KOH thru etching of double side nitride coated silicon wafer. The metal layer is then deposited either on the KOH etch side or the back side of the membrane. Nitride membrane is then optionally removed, leaving a free standing metal membrane structure. Alternatively, metal could be deposited on both sides of the nitride layer forming metal-nitride-metal sandwich membrane structure, as depicted in figure 1 below.

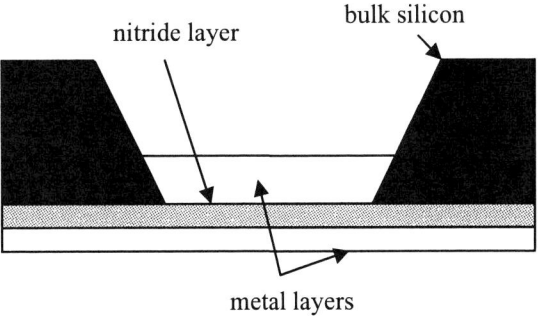

Figure 1: Cross section of nitride membrane with metal sandwich.

Membranes fabricated using this technique could be ideally used as vibrating diaphragm for MEMS microphone. For this application, membrane's wafer is bonded to a silicon wafer with cavities forming the capacitive layers needed for MEMS microphone [4]. Platinum is chosen as the membrane material in this study due to its excellent electrical conductivity, thus providing a good capacitance response [1].

0-7803-9730-4/06/$25.00 ©2006 IEEE

II. THEORY

A square membrane can be modeled as a rectangular plate clamped on all sides. Using Ritz's and energy method, surface deflection (W) can be formulated as follows:

$$W = \sum_{m=1}^{\infty} \sum_{n=1}^{\infty} c_{mn}\left(1-\cos\frac{2m\pi x}{a}\right)\left(1-\cos\frac{2n\pi y}{b}\right) \quad (1)$$

where a and b are plate's length and width respectively, and c_{mn} is Navier's constant [5]. For rectangular plate, the first term c_{11} can be formulated as:

$$c_{11} = \frac{p_0 a^4}{4\pi^4 D}\left(\frac{1}{3 + 3\left(\frac{a}{b}\right)^4 + 2\left(\frac{a}{b}\right)^2}\right) \quad (2)$$

The second and subsequent terms can be dropped from calculation resulting calculation error of less than 1.5%, thus considered negligible. The term D can be formulated as follows:

$$D = \frac{E_{poly} t^3}{12(1-v^2)} \quad (3)$$

p_0 is external pressure applied perpendicularly on the plate, E_{poly} is plate's Young's modulus, t is plate's thickness, and v is plate's Poisson ratio. It should also be noted that W_{max} calculation using these methods assume small deflection ($W<t$) on plate's surface.

The acoustical pressure range for MEMS microphone is between 20 Pa and 50 Pa [1]. A capacitive MEMS microphone usually has a 1 µm to 2 µm gap between the movable and the stationary plate. Thus, deflection in the range of 0.1 µm to 0.5 µm is sufficient to produce usable capacitance signals. The movable membrane's material is preferably a highly conductive metal for better capacitance response. Square membrane with width 1560 µm is chosen as an exemplary ideal size for a MEMS microphone. Using equation (1), (2), and (3), it can be calculated that for a platinum membrane with $a = b = 1560$ µm, a thickness of about 6 µm would result in deflection of 0.11 µm for an applied pressure of 50 Pa. Thus, platinum membrane with thickness 6 µm is chosen for fabrication.

III. FABRICATION

Using the technique introduced in this paper, one could virtually fabricate membrane of any kind of material as long as the material could be deposited by either sputtering or evaporation technique. However, this study would focus on fabrication of metal membrane deposited using DC sputtering technique. Platinum is chosen due to its high electrical conductivity, thus producing better capacitance output for our chosen application.

Fabrication of the membrane is initialized by pattering and opening square holes on nitride layer on either side of a double side CVD nitride coated (100) silicon wafer. The wafer used for this experiment is 375 µm thick (100) type silicon, with both side coated with 0.35 µm CVD silicon nitride. Photoresist is used as mask layer to pattern nitride during BOE etching. Taking wafer thickness into account, the mask window is made bigger by 272 µm on each side to achieve desired width at the bottom of the wafer after KOH anisotropic etching. Detailed KOH mask compensation calculation could be obtained from the work of Mimiwaty [6]. The wafer is then thru etched in 35% (wt) KOH solution at 80°C. The etching process takes approximately 8 hours and 20 minutes to complete. After KOH etching, the backside nitride layer will be left as membrane on the square silicon holes as depicted on figure 2 below.

Figure 2: Optical microscope picture showing a nitride membrane on a thru etched silicon substrate.

In the next step, several types of membrane could be fabricated according to one's application requirements. A single layer metal membrane could be formed by sputter depositing metal layer on the back side of the nitride layer to the required thickness. The nitride layer is then removed by BOE etching from the front side (KOH etch side). One could also fabricate multi metal layer membrane by further depositing insulator material followed by metal on the back side. Alternatively, one could readily fabricate metal sandwich membrane by depositing metal layer on both sides of the nitride membrane as exemplified below. Note that for this type of membrane, different metal could be used on each side of the nitride. The silicon nitride layer, which is not removed, acts as insulating layer between the metal layers. The process flow for fabricating different types of metal membrane is summarized in table 1 below.

Process	Nitride with one side metal	Nitride with sandwich metal	Single layer (or more) metal
KOH etch thru silicon	yes	yes	yes
Metal sputter on nitride backside	yes	yes	yes (could be multiple layers)
Metal sputter on KOH etch side	no	yes	no
Nitride removal	no	no	yes

Table 1: Process flow for fabricating various types of membrane.

IV. RESULTS AND DISCUSSIONS

Platinum-nitride-platinum sandwich membranes of thickness 6.35 μm was successfully fabricated, as depicted in figure 3 through 5 below. The membrane is fabricated by depositing platinum on both sides of nitride layer using DC sputtering with forward current of 40 mA for a total deposition time of 6 hours. The membrane's width is 1562 μm, measured using Tencor surface profiler. A break across the membrane reveals that the membrane has no crystal orientation (Figure 4).

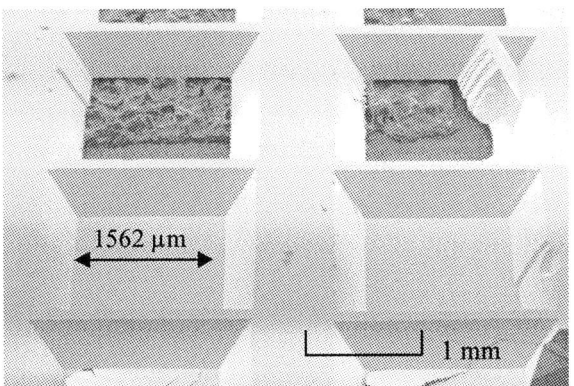

Figure 3: SEM micrograph showing array of KOH etched square holes. Bottom row of etched squares have platinum-nitride-platinum membranes.

Figure 4: SEM micrograph showing a platinum-nitride-platinum membrane broken for thickness measurement.

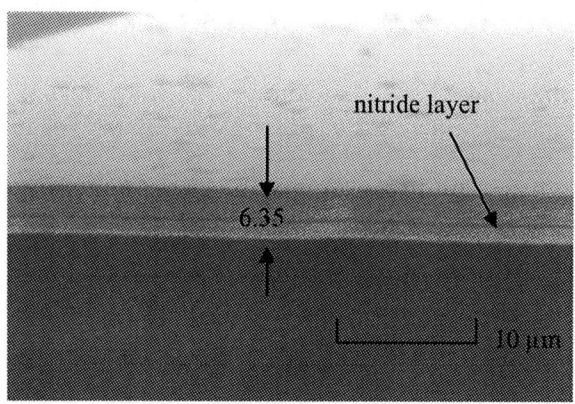

Figure 5: SEM micrograph showing a close-up view of the platinum-nitride-platinum membrane. The nitride layer is visible in between the platinum layers.

A surface profiler scan across the membranes on KOH etched side revealed that the membranes are thicker on its center compared to its sides by about 0.25 μm. This is due to uneven sputtering of metal on silicon nitride's surface. Since KOH etch walls around the membrane is about 400 μm high, they shadow the membrane and reduce metal deposition on the edges of the membrane. Therefore, deposited metal thickness is maximum at the center of the membrane where it is furthest from any KOH wall, and reduces steadily towards the edge (Figure 6). However, this thickness variation is less than 4% of its average thickness, thus the membrane is still considered flat and usable for our application.

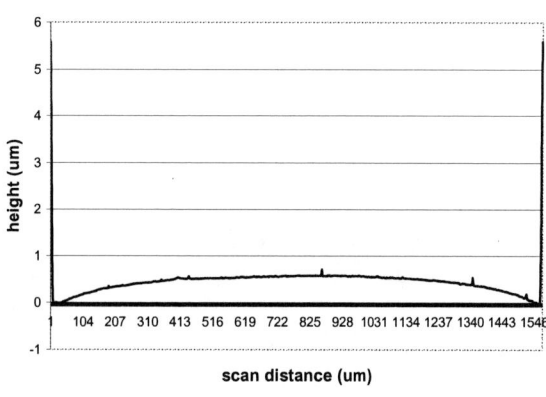

Figure 6: Cross sectional profile of the top side (KOH etched side) of a platinum-nitride-platinum membrane.

Tencor surface profiler was used to apply variable force on the surface of the membrane during measurement scan. The force values are readily converted into pressure since membrane area is known. Deflection values are plotted against scan distance for various pressure values to determine deflection shape of the membrane under applied pressure. The deflected shapes under various applied pressure are shown in figure 7 below.

Seven samples are then taken to characterize deflection behavior of the membranes under various applied pressure. Deflection at the center of the membrane is measured for each of the seven samples for pressure ranging from 20 Pa to 200 Pa at 20 Pa increment. Deflection values obtained are plotted against applied pressure as depicted in figure 8 below. It is observed that deflection of the membrane is linearly related to applied pressure for pressure between 20 Pa and 200 Pa. All the samples showed very consistent results, with maximum deflection at 200 Pa of about 0.41 μm (Figure 8).

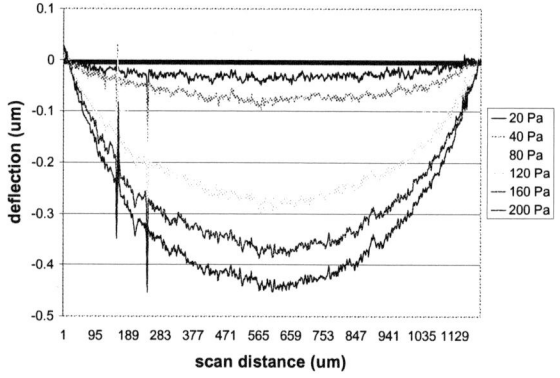

Figure 7: Measured deflection across membrane's length for various applied pressures.

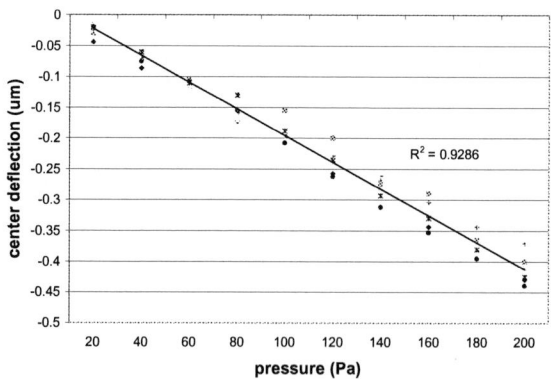

Figure 8: Deflection at membrane's center vs. applied pressure.

IV. CONCLUSION

Platinum-nitride-platinum sandwich membrane of thickness 6.35 μm is successfully fabricated. From deflection test conducted, it is observed that the membrane deflects linearly with applied pressure for pressure range of 20 Pa to 200 Pa. Since acoustical pressure range needed for MEMS microphone is 20 Pa to 50 Pa, the fabricated membrane is suitable for this application. The platinum membrane's working range could be further adjusted by controlling its thickness by varying deposition time. It is also observed that the fabricated membrane is sufficiently flat for MEMS microphone application.

REFERENCES

[1] B. Ganji and B. Y. Majlis, *"A Highly Sensitive MEMS Capacitive Microphone Using Polysilicon Diaphragm"*, International Conference on MEMS and Nanotechnology, Kuala Lumpur, Malaysia, 2006.

[2] C. L. Goldsmith, et. al, *"Performance of Low-Loss RF MEMS Capacitive Switches"*, IEEE Microwave and Guided Wave Letters, Vol. 8, No. 8, August 1998.

[3] J. Miao, et. al, *"Design Considerations in Micromachined Silicon Microphones"*, Microelectronics Journal, Volume 33, Issues 1-2, pp. 21-28, January 2002.

[4] Mohamed-Gad-el-Hak, *"The MEMS Handbook, 2nd edition"*, CRC Press, 2005.

[5] A. C. Ugural, *"Stresses in Plates and Shells"*, McGraw-Hill, 1981.

[6] M. Mohd Noor, *"An Experimental Studying Anisotropic Wet Etching for Silicon Membrane Fabrication"*, MSc. Thesis, Universiti Kebangsaan Malaysia, 2003.

Effect of Annealing on the Sol-Gel Derived $SrBi_4Ti_4O_{15}$ Thin Films for Piezoelectric Pressure Sensors

Nor Azlian Abdul Manaf[1], Muhamad Mat Salleh[1], *Member, IEEE* and Muhammad Yahaya[2]
[1]Institute of Microengineering and Nanoelectronic (IMEN),
[2]School of Applied Physics, Faculty of Science & Technology,
Universiti Kebangsaan Malaysia
43600 Bangi, Selangor
Email: *mms@pkrisc.cc.ukm.my*
myahya@pkrisc.cc.ukm.my

Abstract This paper reports the effect of annealing on fabrication of Strontium Bismuth Titanate $SrBi_4Ti_4O_{15}$ (SBT) thin films for piezoelectric pressure sensors. The SBT films and capacitance devices with structure of $Al/TiO_2/SBT/TiO_2/SiO_2/Si$ were fabricated using sol-gel technique. The microstructure of SBT thin films have been systematically studied in as-prepared (un-annealed) condition as well as after annealing at 400, 500, 600 and 700 ^0C. The general trend seems to indicate that annealed temperature show better piezoelectric properties. X-ray diffraction patterns reveal changes of crystalline structure after annealing at 600 ^0C temperature. Another important parameter is dielectric constant, which is found toward higher value within annealing temperature. For the sensor device measurement, the SBT thin film pressure sensors were tested by pneumatic loading method at pressure range between 0 to 450 kPa. It found that the sensors sensitive to the applied pressure and the response recovered back when the pressure were removed. The 600 ^0C annealed pressure sensor demonstrates better repeatability compared to the others. The results indicated that the sensor performance was affected by the structure of the film. A crystalline structure gives an optimum response towards pneumatic pressure. However, the too high of annealing temperature may cause a catastrophic failure of the pressure sensor response, hence the existing of crack and inter diffusion between top and bottom electrode. The correlation between annealing process with structure of SBT pressure sensors and the piezoelectric properties will be discussed.

I. INTRODUCTION

Pressure sensors are one of the most common micro-sensors devices used in various industrial such as automotive, medical, military, aeronautical, hydraulic, instrumentation, process and industrial control [1]. Among the pressure or stress sensing materials, piezoelectric material is famous to be used because of the special capabilities such as fast response, ruggedness, high stiffness, ability to measure the quasi-static pressure and good thermal stability. All these features are essential in the dynamic pressure measurement, which is very important in various industrial areas. At present, lead zirconate titanate (PZT) based ceramic is most widely applied in piezoelectric thin film sensor because of its large piezoelectric coefficient [2,3]. However the evaporation of toxic lead during the fabrication of their ceramic causes an environmental problem. Therefore, there is an increasing interest of investigating lead free of piezoelectric materials to replace PZT based piezoelectric ceramics.

Strontium Bismuth Titanate $SrBi_4Ti_4O_{15}$ (SBT) is is a kind of lead free perovskite-type piezoelectric with a relatively high Curie temperature ($T_c= 620^0C$), good piezoelectricity and thermal stability [4]. SBT is one of the family called Aurivillius phase generally formulated as $Bi_4A_{m-1}B_mO_{3m-1}$. Their crystal structure can be regarded as a regular intergrowth of $(Bi_2 O_2)^{2+}$ layers and $(A_{m-1}B_mO_{3m-1})^{-2}$ perovskite where A can be a mono, di or trivalent element allowing dodecahedral coordination, B is a transition element suited to octahedral coordination and m is an integer which represents the number of perovskite [5].

The deposition methods to obtain SBT thin films are sol gel technique, metal organic, chemical vapor deposition, RF magnetron sputtering [6], and laser ablation [7]. Sol gel offers an economical technique, able to obtain homogeneous films with large area, excellent composition control, low processing cost and high purity. Therefore, sol gel technique is widely acknowledged to be advantageous over physical deposition techniques. When using sol-gel technique to prepare the ferroelectric thin film, an annealing step is require to promote crystallization of the SBT materials. A crystalline structure will gives an optimum response towards pneumatic pressure. Therefore, it may effectively improve the response of the sensing element. This paper reports the fabrication of SBT thin films without annealed and annealed at 400, 500, 600 and 700 ^{0}C as pressure sensor by using sol gel technique.

II. EXPERIMENT

The solution of SBT was prepared by sol gel technique. The source materials are bismuth acetate $Bi(CH_3COO)_2$, strontium acetate $Sr(CH_3COO)_2$ and titanium butoxide $Ti(OC_4H_9)_4$. Initially, the powder of bismuth, barium and strontium were dissolved in acetic acid containing 20% volume of deionize water at room temperature. A clear and transparent solution is obtained. In order to improve the stability of the solution, 0.1 ml of ethanol amine was added to the solution with constant stirring. Finally the solution titanium (IV) was dropped wisely to the solution in a proper molar ratio with constant stirring at room temperature to form a clear and yellowish transparent solution. The flowchart of the whole synthesis process of precursor solution is shown in Fig 1.

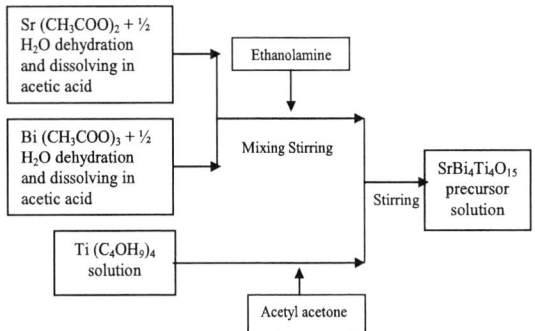

Fig. 1 Flow chart of the synthesis process of SBT solution

The prepared SBT solution was ready to be used for spin coating. The SBT films were spin coated on silicon wafer at 400 rpm for 30 second followed by heating process at 300 ^{0}C in air for 15 minutes. The purpose of heating process is to evaporate the organic solvents. For annealing, the sample was heat treat in air at 400, 500, 600 and 700 ^{0}C for 2 minutes.

The microstructures of the films have been studied to investigate the effect of annealing on the properties of the films. The crystal structure of the SBT thin film is examined using an X-ray diffractometer (XRD) (Cu Kα radiation). Dielectric measurements carried out using impedance spectroscopy Solarton-Schlumberger. Atomic force microscope (AFM) is used to measure the roughness of the surface film. For the homogeneities and grains size, its have been characterized by Scanning Electron Microscope (SEM).

The SBT pressure sensors were fabricated with the structure of Al/TiO$_2$/SBT/TiO$_2$/SiO$_2$/Si. The raw material is silicon wafer and the silicon oxides were deposited by electron beam evaporation with thickness of 1500 Å respectively. Ruthenium oxide as bottom electrode was deposited by RF magnetron sputtering for 2 hours. In order to reduce the crystallization temperature for the formation of SBT ferroelectric phase, the buffer layer, TiO$_2$ was then deposited on these thin films. Another purpose for the insertion of the buffer layer is to reduce hole injection layer and avoid the inter diffusion between film and electrode. The SBT thin films were spin coated on TiO_2/ RuO_2/ SiO_2/Si substrate followed by heat treatment at 300^0C. These spin coating and heating process was repeated five times to produce multilayer thin films. For annealing, the sample was heat treated in air at 400, 500, 600 or 700 ^0C for 2 minutes. To complete the fabrication of the pressure sensor, Al was deposited as a top electrode. The structure of the fabricated SBT thin film pressure sensor is shown in Fig. 2

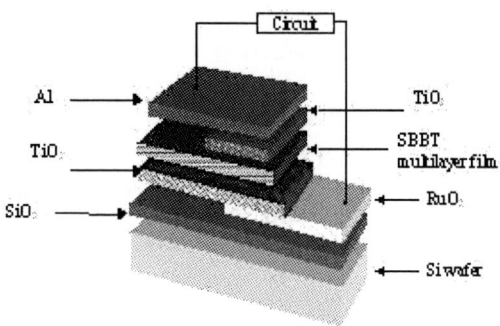

Fig. 2 The fabricate SBBT pressure sensor

The piezoelectric response of the sensor was measured using pneumatic loading method as shown in Fig 3. This measurement system has been divided into three parts: an air compressor system, a pressure vessel and an electrometer as data storage system. The pressure measurement was made from 0 to 450 kPa at room temperature. The air was compressed by the air compressor and imparted to the sensor, which was mounted inside the pressure vessel. The voltage that was generated by the sensor was collected by the electrometer.

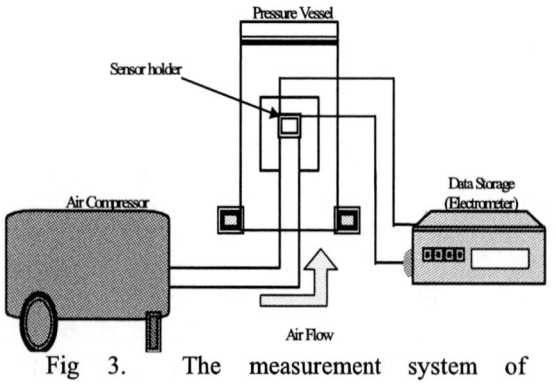

Fig 3. The measurement system of piezoelectric response.

III. RESULTS AND DISCUSSION

A. Thin Film Characterization

Fig 4. shows the SEM images of the surface morphology of SBT thin films. The image indicated that an annealed film was in orderly forms, where the nano grains were formed with the clear grain boundaries. This feature is very important, since the electromechanical responds can be arose from the orderly oriented grains. It also showed that the grains size of SBT thin films increased with the annealing temperature. It means that the films exhibit a pronounce tendency to grain growth. This is reasonable since the coalescence of particles that caused by thermal effect. The average grain size and film thickness are presented in Table 1.

AFM images are illustrated in Fig 5. and the surface roughness, R_a was calculated from the AFM images that shown in Table 1. The value of Ra indicated the surface morphologies of SBT thin film are influence by annealing process. The annealed film tends to be smooth, thin and flat although the grains size as shown by SEM image is increased.

Fig 6 showed the XRD pattern of the SBT thin film before and after annealing. It was found that the un-annealed and low annealed film is in amorphous phase and it began to converse to crystalline phase after annealed at 600 ^0C. The SBT film indicated a perovskite peaks of $SrBi_4Ti_4O_{15}$ at (112) with random crystalline orientation.

The dielectric properties of SBT thin film measured in terms of the dielectric constant. The value of the dielectric constant was calculated from the capacitance of the film using Debye distribution formula, measured at 1 kHz at an oscillation level of 20mV. The dielectric constant of un-anneal SBBT film is 72.81 and it found to be increase within temperature to 187.4 for 700 ^0C annealing SBT thin film. The higher dielectric constant is due to the increasing number of domain population and its wall motion [9] and the better crystallization of the thin film after annealing process.

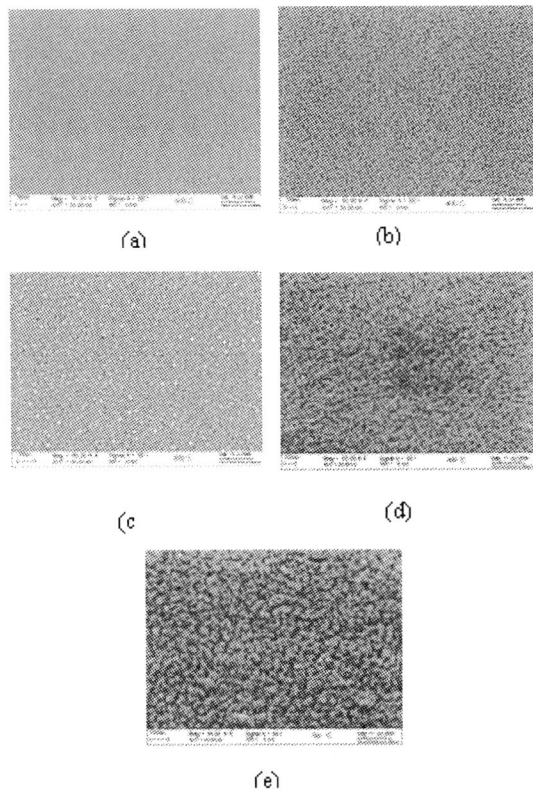

(a)　　　　(b)

(c)　　　　(d)

(e)

Fig. 4 SEM photo of surface morphology of SBT thin film (a) un-anneal and anneal at (b) 400^0C, (c) 500^0C, (d) 600^0C and (e)700^0C.

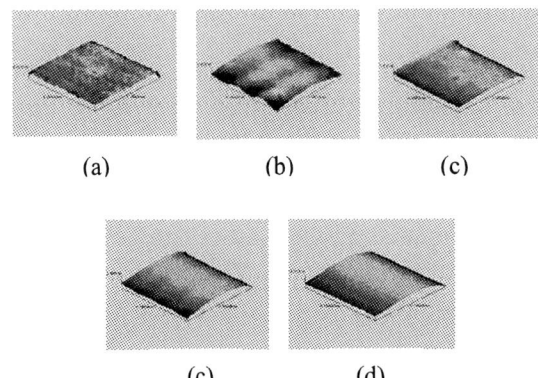

(a)　　　　(b)　　　　(c)

(c)　　　　(d)

Fig 5. AFM images of the SBT thin films (a) un-annealed and annealed at (b) 400 ^{0}C, (c) 500 ^{0}C, (d) 600 ^{0}C and (e) 700 ^{0}C

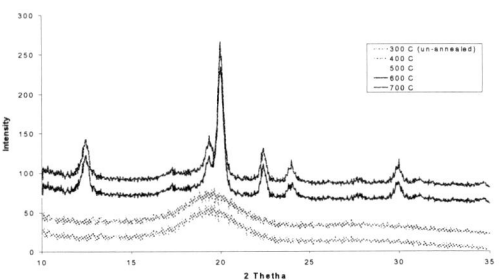

Fig 6. XRD patterns of $SrBi_4Ti_4O_{15}$ thin films for un-annealed and annealed from

Annealing Temp. ^{0}C	Roughness Ra (nm)	Dielectric constant, ε	Average grains size (nm)
Un-annealed	10.48	72.8	18.96
400	4.70	116.3	16.15
500	4.15	178.5	17.43
600	3.39	181.4	25.10
700	3.32	187.4	36.72

Table 1. The physical properties of SBT thin films with various temperatures

B. Measurement of piezoelectric response

The piezoelectric measurement system was designed using the concept of direct piezoelectric effect. When pressure is applied to piezoelectric materials, they create a strain or deformation in the materials, due to the deflection of the lattice in a naturally piezoelectric quartz crystal [7].

Fig 7. shows the piezoelectric response of the SBT thin film sensors measured at pressure range from 0 to 450 kPa at room temperature. It is showed that the sensors were sensitive to the applied pressure and the responses were recovered back when the pressure is removed. The measurements were made for seven cycles at pressure 450 kPa. The sensors give responds to the applied pressure. The 600 ^{0}C annealed pressure sensor showed the best repeatability compared to others since the similar respond maintained until the seventh cycles. The SBT pressure sensor achieved a linear characteristic response between 50 kPa to 450 kPa pressure load. It was shown in Fig 8. The sensitivity of the sensors is represented by the slope of the graph and the R^2 is the correlation coefficient. The graph demonstrates that the sensitivity of the pressure sensor is increased with annealing temperature.

It was found that the performance of the pressure sensors could be influenced by the internal structure of SBT thin films. The origin of the generated voltage is produced from the non-symmetry arrangement of positive and negative charged of atoms in their crystal structure. The as prepared SBT film is amorphous and was crystallized in perovskite phase after annealing at 600^{0}C. The crystallized structured enhanced the electromechanical response in the ferroelectric materials.

Although the film annealed at 700 ^{0}C has higher sensitivity and linearity properties, this film has poor repeatability. The film contains larger grains, and the grains maybe broken down due to friction among them under applied pressure.

Based on the SEM result, it is suggests that nanostructure of the films would play important role to generate the high voltage and stable electromechanical response. When the grain size increase, the domain that carrying the dipoles will be increase as well [7,8]. In this kind of situation, the domains might have chances to be aligned in the preferential direction. The preferential oriented grains may become the active source for the piezoelectricity and generate great differential in electromechanical response, hence perform better sensitivity.

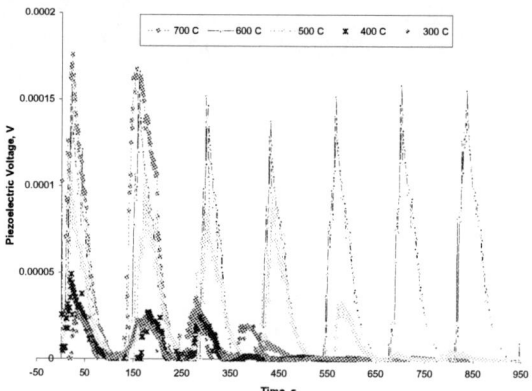

Fig 6. Piezoelectric response of SBT thin film pressure sensor measured at 450 kPa pressure load for un-annealed and annealed at 400, 500, 600 and 700.

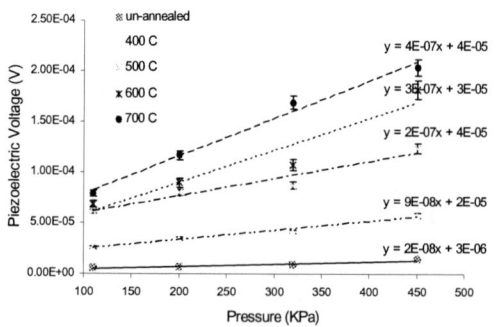

Fig 7. Graph of linearity for pressure sensor performance.

Temperature ^{0}C	Sensitivity mV/kPa
Un annealed	0.02
400	0.09
500	0.20
600	0.30
700	0.40

Table 2. The sensitivity value of the pressure sensors.

IV. CONCLUSION

The microstructure and piezoelectric responds of $SrBi_4Ti_4O_{15}$ thin films pressure sensors have been systematically studied in as-prepared (un-annealed) condition as well as after annealing at 400, 500, 600 and 700 ^{0}C. The XRD results indicated that the SBT film starts to converse from amorphous phase to crystalline after annealing at 600 ^{0}C. The dielectric constant and grain size were found toward higher value with annealing temperature. The sensitivity of the sensors toward applied pressure was tested by pneumatic loading method. All sensors gave responds to the applied pressure. The 600 ^{0}C annealed sensor is shows better repeatability compare to others. It can conclude that the performance of SBBT thin films pressure sensors were affected by the microstructure properties of the film.

ACKNOWLEDGEMENT

The authors would like to thank the Malaysia Ministry of Science, Technology and Innovation for the grant IRPA 03-02-02-0020-SR0003/07-06 project.

REFERENCES

[1] D. Tandaske, *Pressure sensors: selection and application,* Marcel Deaker, pp. 73-75, 271-277 (1994).

[2] J.P. Kim, Y.S. Yang, S.H. Lee, H.J. Joo & M.S. Jang. "Microstructure and dielectric properties of lanthanum bismuth titanate thin films", *Journal of the Korean Physical Society* 35 , pp. S1202-S1205 (1999).

[3] C. J Lu, S. B. Ren, J.S Liu, H.M Shen and Y. N Wang, "The effect of grain size on domain structure in unsupported $PbTiO_3$ thin thin films", *Journal Physics: Matter*, pp. 8011-8016. (1996).

[4] D. Xie, W.Pan and H. Shi "Synthesis and characterization of $Sr_{1-x}Ba_xBi_4Ti_4O_{15}$ ferroelectric materials", *Material science and Engineering B*, (2003).

[5] Y. Xu. "Ferroelectric materials and their application.", *North-Holland* (1991).

[6] M. Yamaguci and T. Nagotomo, "Preparation and properties of $Bi_4Ti_3O_{12}$

thin film grown at low substrate temperature", *Thin Solid Film,* p. 294 (1999).

[7] W. Wu, K. Fumoto, Y. Oishi,, M. Okuyama and Y. Hamakawa, "Preparation of bismuth titanate thin film", *Japanese Journal of Applied Physics,* pp. 5141-5145 (1995).

[8] T.L. Jordan d Z. Ounaies, "Piezoelectric ceramics characterization", ICASE, NASA Langley Research Center, Hampton, Virgina. (2001).

[9] S.B Ren, C.J Lu, J.S Liu, H.M Shen and Y. N Wang, "Size-related ferroelectric domain structure transition in a polycrystalline $PbTiO_3$ thin film", *Physical Review B*, p 14337, (1996).

[10] X.H Liu., Z.G. Yin & J.M. Liu, "Microstructure and electrical properties of ferroelectric $Pb(Zr_{0.53}Ti_{0.47})O_3$ films on Si with TiO_2 Buffer layers", *J. Phys.: Condens. Matter* 12, pp. 9198-9194 (2000).

[11] N. Azlian Manaf, M . M. Salleh and M. Yahaya. 2006. " Fabrication of Sr_x $Ba_{(1-x)}Bi_4Ti_4O_{15}$ thin films for piezoelectric pressure sensor", *Journal of Solid State Science and Technology Letters*, pp. 93-94 (2006).

Fabrication Study of Solid Microneedles Array Using HNA

Norazreen Abd Aziz and Burhanuddin Yeop Majlis *SMIEEE*
Institute of Microengineering and Nanoelectronics (IMEN)
Universiti Kebangsaan Malaysia
43600 Bangi, Selangor, Malaysia
Email: burhan@vlsi.eng.ukm.my

Abstract This paper presents an approach for the process development of the out of plane silicon microneedles array fabrication. This study utilizes the wet etching technology using HNA to build a structure of sharp-tipped microneedles; biconvex-conical shaped. The height of the fabricated microneedle is 41.8μm and the tip radii is 1.4 μm. Investigation on the effect of varying the mask opening window had been carried out. The design approach and the fabrication process are well explained here.

I. INTRODUCTION

As an alternative for oral and bolus injection, transdermal drug delivery (TDD) has been developed rapidly to transport drug across the blood stream. Microneedles is one of a new approach of TDD that act as an interface between microsystem and human body-skin . It promises 'painless' as minimally invasive and increases the effectiveness of drugs. In addition, the micron size of the needles causes less compression of the tissue and pain receptors. Arranging needles in array allows injection to a wider distributed area and gives better liquid absorption by the tissue.

Microneedle can be divided into two categories based on their general design. In plane microneedles have shaft that are parallel to the substrate surface and generally result in one row of needle which does not allow drug distribution over large area. The study of in plane microneedle is beyond the scope of this research work. While out of plane microneedles have perpendicular shaft which can obtain large arrays of microneedles. Works will only concentrate on the out of plane microneedle type. Different shapes and material have been fabricated based on either dry etch or wet etch technologies. Following discussion will concentrate on technologies used to fabricate out of plane microneedle.

Henry et al [1] conducted a study on the use of microfabricated microneedles to enhance drug delivery. An array of solid microneedles are designed to increase the skin permeability as a means of enhancing drug diffusion. The fabrication process employed combination of chromium mask and Reactive Ion Etching technique using SF_6/O_2 plasma (known as the Black Silicon Method). These microneedles array are quite similar to the 'drug patch' style available in the market.

Fig 1 Solid Silicon Microneedle using Black Silicon Method [1]

McAllister et al [2] developed microfabrication techniques to make hollow microneedles out of different materials such as silicon, polymer, metal and glass. They used Henry's solid microneedles as a master to form a mold for symmetrically tapered needles. The first hollow out of plane microneedle having openings in the shaft rather than at the tip, were introduced by Griss et al [3]. Their design structures are less prone to clogging and the large size of the side openings allow a large area for fluid to flow. Based on two DRIE process followed by RIE isotropic etching, the resulting needles are extremely sharp and robust.

Fig 2 Side opened microneedle fabricated by Gris et al. [3]

Stoeber et al [4] produced hollow microneedle by superimposing a DRIE anisotropic etch (for channel) with dry/wet isotropic etch which defines the shape of the needle. Various shape and well-defined needle tip were fabricated by Gardernier et al [5]. Combination of DRIE and anisotropic KOH wet etching allows channel opening positioned freely by means off-centering the needle tip. Davis et al [6] used a modified-LIGA process to micromachined molds out of polyethylene terephtalate using an ultraviolet laser. They coat the molds with nickel and released the resulting metal by etching the polymer mold. A novel approach to microneedle fabrication, Park et al [7] produced microneedles out of biodegradable polymer using PDMS molds as a master structure. This fabrication method is expected to be suitable for rapid scale-up for inexpensive, mass production. Recently Ji et al [8] designed needle with macroporous tip which has potential for biological application such as carrying drugs. Dry etching technology by STS's inductively coupled plasma was used to obtain pyramidal structure. Electrochemical etching in MeCN/HF was carried out in order to generate the porous tip of the needle.

II. DESIGN OF SOLID MICRONEEDLE ARRAY

A. Skin Anatomy

Figure 3 shows the cross section of the skin-transdermal drug delivery. Stratum Corneum (SC) which is at approximately 20μm thick contains dead tissues, acts as a barrier to penetration or transport. The Viable Epidermis (VE) is typically located between 20-100μm underneath the skin surface, composed of living cells contains few nerves[9]. Drug will be injected through microneedle into the VE which later diffuses into the blood vessel.

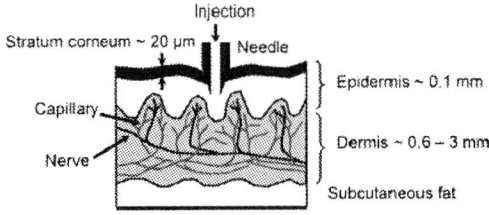

Fig 3 Cross section of skin anatomy-transdermal needle injection [4]

B. Design Requirement

In order to deliver drugs, the length of the needles must be designed in the range between 20-100μm. Sharp-tipped needle is another requirement issue, as it can lead to easy penetration and less force for insertion. Due to the small dimensions, the fluid volume of drugs to be delivered is very small . To get a large volume of drugs, the needle must be arranged in array.

The shape of the needle tip is developed using isotropic etching. Isotropic behavior; etch in all crystallographic directions at the same rate [10]. Thus, resulting in a smooth underetch semi-rounded profile. In this behavior, it seems that the meeting point of two underetching semi-rounded profile will produce the outer needle shape with sharp tip. Ideally, the lateral etching occurs at about the same rate as vertical etching, therefore the mask diameter must be twice than the needle height. Six mask samples of 8x8 arrays of 80μm diameter with various opening window size (70, 220, 370, 520, 670, 820) is used in this work. Investigation is carried out only for circular mask.

C. Wet Etch Technology

Wet chemical etchings offers higher selectivity and higher etch rates compared to plasma etch. For silicon isotropic etching, the most commonly used etchant are the mixtures of hydroflouric acid (HF), nitric acid (HNO_3) and acetic acid (CH_3COOH).

Referring to the Iso-Etch Curve developed by Robin and Schwartz [11], we choose a composition 20%HF, 50%HNO$_3$ and 30%CH$_3$COOH which gives etch rates at about 25µm/min. The combination of the above chemical is called HNA. Table 4.8 [10], summarized the masking material for acidic etchant. It is shown that LPCVD Si$_3$N$_4$ is a good mask for deeper HNA isotropic etching.

III. EXPERIMENTAL PROCEDURE

Figure 4 shows the schematic of fabrication process using the circular mask shape. A standard (100) silicon wafer, 600um thick with 200nm deposited silicon nitride (LPCVD) over the substrate is used as a starting material. A photoresist layer (AZ1500 2000 rpm) is coated onto the nitride layer. Then, samples are exposed by using mask aligner and hard bake at 120°C for 5 minutes. The exposed photoresist is removed by soaking the wafer in liquid developer, AZ300MIF leaving the desired circular array of photoresist on the nitride layer. The wafer is immersed into 6:1 BHF (Buffered Hydroflouric) that etched the uncovered nitride pattern. The present photoresist is then removed by soaking the wafer into acetone which left the circular array of nitride as a mask for the next step. The silicon wafer is etched isotropically by dipped them into the HNA at room temperature without any agitation. Etching was performed until the nitride mask falls off due to under etching.

IV. RESULT AND DISCUSSION

All of the six samples were immersed in the HNA for 6 minutes. The fabricated samples was characterized for the lateral etching and vertical etching profile. From the characterization result it is found that the mask with 70µm opening window size produced the optimum sharp tip. Figure 5 shows the SEM images of the resulting needles using mask with opening window size, 70µm. As a comparison, it is seen that the mask with opening window size, 220µm does not produce a sharp tip as in Figure 6 .

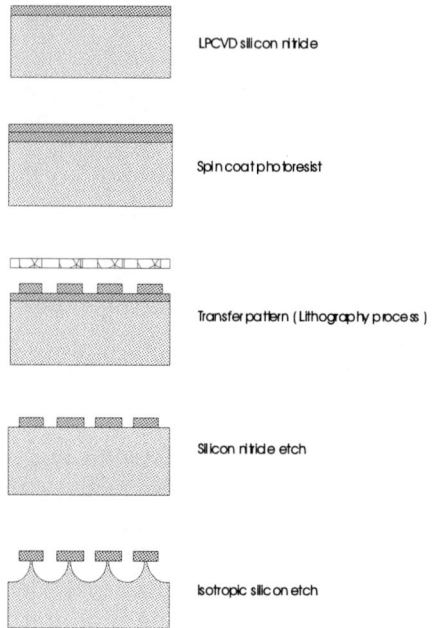

Fig 4 Process flow for the fabrication of solid silicon microneedle

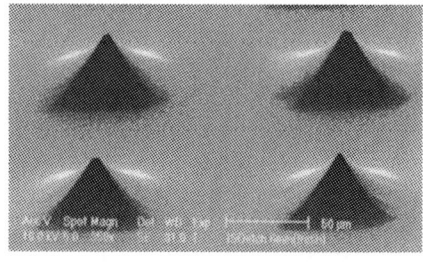

Fig 5 SEM images of fabricated needles using 70µm opening window size

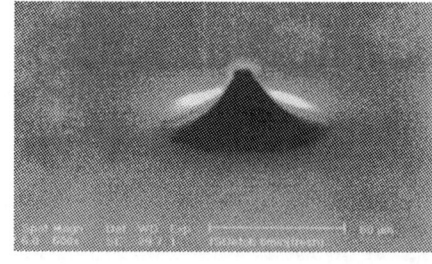

Fig 6 SEM images of fabricated needles using 220µm opening window size

The effect of the etch rates on the opening window size is presented in Figure 7. From Figure 7, it is seen that by widening the opening window, the etch rates is slower. This is due to the exposed surface area is large and therefore it will take longer time to get the sharp tip.

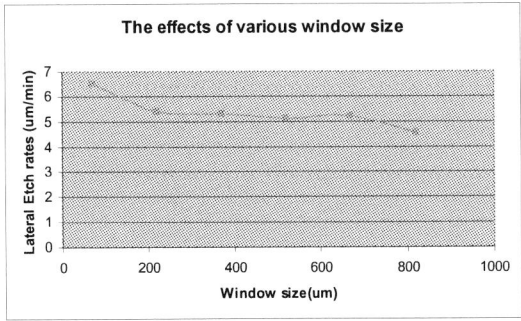

Fig 7 The effect of the etch rates on the opening window size

Figure 8 shows the SEM images of fabricated microneedles array using mask, 80μm diameter with 70μm opening window size. The microneedles was immersed in the HNA at about 6 minute, gives the etch rate at 6.5μm/min. With the developed process described above, a biconvex-conical shape needles with 1.4μm tip radii and length approximately 41.8μm were achieved. The needle tip obtained is less sharp compared to Henry's [1] microneedles. Sharp tips could be achieved by using wet etch process with agitation. Longer needles could be obtained by combining the wet etch process with DRIE which will be carried out later. The characterization of the mask size will be further investigate together with the process parameter including temperature, agitation rate and etchant composition.

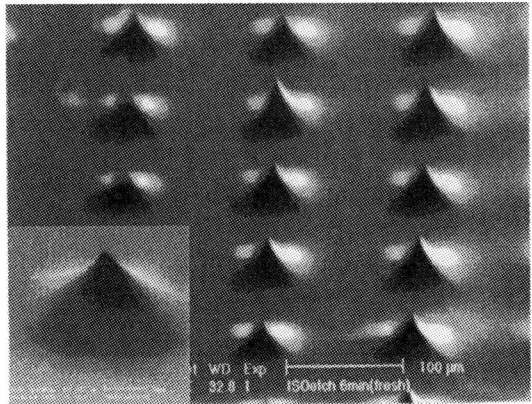

(a)

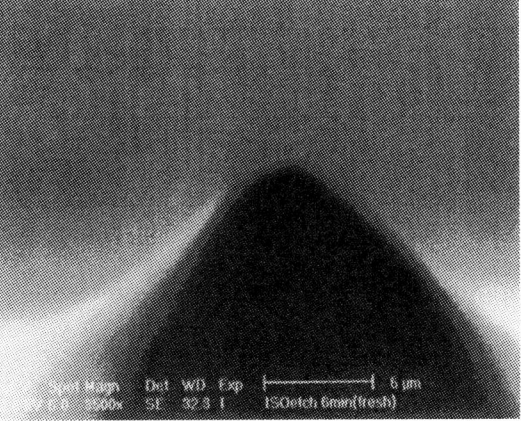

(b)

Figure 8 SEM images of fabricated solid silicon microneedles array. (a) A section of 8x8 array of micorneedle. (b) Close-up view of single tip microneedle

V. CONCLUSION

In this paper, a MEMS-based solid microneedle arrays are designed and fabricated. A 8x8 array biconvex-conical shape microneedles were isotropically fabricated using HNA. The average height of the needles is 41.8μm and the diameter of the tip radii is 1.4μm. The etch rates is dependant on the opening window size. In the future, the process parameters such as temperature, agitation rate and etchant composition will be studied and optimized.

REFERENCES

[1] S. Henry, D. V. McAllister, M.G. Allen and M. R. Prausnitz, "Microfabricated Microneedles: A Novel Approach to Transdermal Drug Delivery," *Journal of Pharmaceutical Sciences*. Vol. 87, no 8, pp. 922-925 (1998).

[2] D. V. McAllister, P. M. Wang, S. P. Davis, J-H Park, P. J. Canatella, M.G. Allen and M. R. Prausnitz, "Microfabricated needles for transdermal delivery of macromolecules and nanoparticles: Fabrication methods and transport studies," *Proc. Of the Natioanl Academy of Sciences of the USA*. Vol. 100, no 24, pp. 13755-13760 (2003).

[3] P. Griss and G. Stemme, "Side-Opened Out-of-Plane Microneedles for Microfluidic Transdermal Liquid Transfer," *J. of Microelectromechanical Sys*. Vol. 12, no. 3, pp. 296-301 (2003).

[4] B. Stoeber and D. Liepmann, "Arrays of Hollow Out-of-Plane Microneedles for Drug Delivery," *Journal of Microelectromechanical Systems*. Vol. 14, no. 3, pp. 472-479 (2005).

[5] H. J. G. E. Gardeniers, R. Luttge, E. J. W. Berenschot, M. J. Boer, S. Y. Yeshurun, M. H. R. Oever and A. Berg, Silicon Micromachined Hollow Microneedles for transdermal Liquid Transport," *Journal of Microelectromechanical Systems* Vol. 12, no. 6, pp. 855-862 (2003).

[6] S. P. Davis, J-H Park, W. Martanto, M.G. Allen and M. R. Prausnitz, "Hollow Metal Microneedles for Insulin Delivery to Diabetic Rats," *IEE Trans. On Biomedical Engineering* Vol. 52, no 5, pp. 909-915 (2005).

[7] J-H Park, M.G. Allen and M. R. Prausnitz, "Biodegradable polymer microneedles: Fabrication, mechanics and transdermal drug delivery," *Journal of Controlled Release* Vol. 104, pp. 51-66 (2005).

[8] J. Ji, F. E. H. Tay, J. Miao and C. Iliescu, " Microfabricated microneedles with porous tip for drug delivery," *Journal of Micromechanics and Microengineering* Vol. 16, pp. 958-964 (2006).

[9] O. Johansson, L. Wang, M. Hilliges and Y. Liang, "Intraepidermal nerves in human skin: PGP 9.5 immunohistochemistry with special reference to the nerve density in skin from different body regions," *J. of Periph. Nervous System* Vol. 4, no. 1, pp. 43-52 (1999).

[10] M. J. Madou, *Fundamentals of Microfabrication-The Science of Miniaturization*, CRC Press (2002).

[11] H. Robbins and B. Schwartz , "Chemical Etching of Silicon-II: The System HF, HNO_3, H_2O and $HC2C3O2$," *Journal of Electrochemical Soc*. Vol. 107, pp. 108-11 (1960).

Effects of Mechanical Geometries on Resonance Sensitivity of MEMS *Out-of-Plane* Accelerometer

A. Manut, *Member, IEEE* and M.I. Syono

MEMS Group, MIMOS Berhad, Technology Park of Malaysia, 57000 Kuala Lumpur, Malaysia

E-mail: azrif.manut@mimos.my , matek@mimos.my

Abstract **Mechanical geometry such as flexure's beam width, beam length, shin length and movable plate's size variations have effects on the resonance frequency because it's change the values of spring's constant and mass of the structure. This paper presents the effects of these mechanical geometries on resonance of MEMS out-of-plane accelerometer using Architect module in CoventorWare. Resonance frequency of the accelerometer is important because it determines the working frequency bandwidth of the accelerometers. Sensitivity analysis is carried out to determine which geometry most affects the device performance and needs to be controlled. An out-of-plane accelerometer with a fixed and a movable plate suspended by four flexures is used. Sensitivity by analysis shows resonant frequency decreases by 1.422% with 1% increase in beam length while the lowest changes is -0.061% of resonant frequency with 1% increase in shin length.**

I. INTRODUCTION

MEMS accelerometers play an important role in the field of sensors. Accelerometers already penetrated the world market since early 1980s [1]. In the automotive industry, accelerometer already have a well established market where they are used to activate safety systems including air bags, in vehicle stability systems and in electronic suspensions. The accelerometer market is expanding from its base in the automotive industry to industrial and consumer applications which require high sensitivity, low-*g* accelerometers [2]. It is estimated that the revenue for MEMS application devices is growing from USD$6.5 billion in 2004 to USD$10 billion in 2010[8].

Capacitive-type sensing is one-way to sense acceleration. The majority of MEMS-based accelerometers in the market today utilize this sense mechanism due to their high sensitivity, low noise, low temperature sensitivity and low power dissipation characteristics [3-5]. Capacitive-based accelerometer are frequently designed based on either the gap-changed or area-changed principle. Most of the presented capacitive accelerometers used a gap-changed capacitor structure [6]. In this study, distance-changed capacitive sensing is chosen to design the accelerometer.

Resonance frequency is an important factor for performance of the device. Resonant frequency is most dependant on the geometry of the flexure and mass of the device. Geometries on the flexure such as beam width, beam length, and shin length affected the spring's constant of the flexure thus changing the resonance frequency. Size of the plate also affects the resonant frequency by altering its mass. Changing one of these geometries will give different impacts on changes of resonance frequency.

In this paper, the study of effects of the mechanical geometry on sensitivity of frequency resonants of MEMS out-of-plane accelerometer is presented. Sensitivity analysis is performed to asses the impact of changes geometries characteristic and to determine which components have the greatest effects on overall design performance.

II. THEORY

The accelerometer used in this study is based on the principle of gap-changed. The accelerometer has a 2 μm thick polysilicon proof mass and suspended 1.5 μm above the stationary electrode by means of crab-leg flexures that are anchored onto the substrate. The thickness is fixed to meet our process requirements. The proof mass consists of perforated holes for smooth etching of sacrificial layer. Under acceleration, the proof mass will move along the z-direction towards the substrate. Crab-leg type is used as flexures.

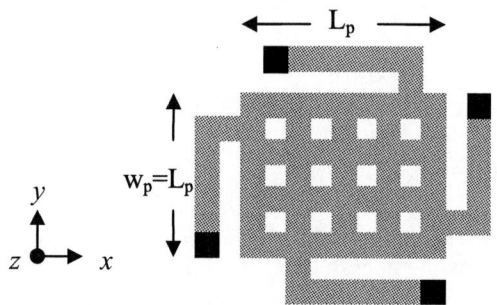

Fig. 1 Schematic illustration of the design of out-of-plane accelerometer

The accelerometer can be modelled as system composed of proof mass, a spring and a damper. This is a 2nd order dynamic mechanical system. The differential for the displacement z_m as a function generated by the acceleration, F is given by the following:

$$F = ma = m\frac{d^2 z_m}{dt^2} + b\frac{dz_m}{dt} + kz_m$$

where m is the mass of the proof mass, a is the applied acceleration, b is the damping coefficient and k is the spring constant of the flexures.

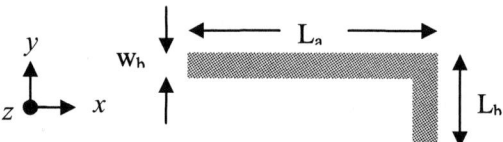

Fig. 2 Crab-leg beam flexure

The spring constant of the flexure depends on the material and geometrical aspects of the structure. Figure 2 shows the crab-leg beam flexure stucture used in this study. The vertical displacement along the z-axis is obtained from small deflection theory and the spring constant is derived as follows [7]:

$$k_z = \frac{48\alpha^2\left(\beta L_a + \alpha L_b\right)\left(\alpha L_a + \beta L_b\right)}{\left(\begin{array}{l}\alpha^2\beta L_a^5 + 4\alpha\beta^2 L_a^4 L_b + \alpha\beta^2 L_a^4 L_b + 4\alpha^2\beta L_a^3 L_b^2 + \\ 4\alpha^2\beta L_a^2 L_b^3 + 4\alpha^3 L_a L_b^4 + \alpha\beta^2 L_a L_b^4 + \alpha^2\beta L_b^5\end{array}\right)}$$

where
$$\alpha = EI_x$$
$$\beta = GJ_{a,b}$$

where E is the Young's modulus, I_x is the moment of inertia of the beam and truss respectively, G is the torsion modulus of elasticity and $J_{a,b}$ is the constant torsion.

Resonant frequency is a critical factor for the device sensitivity. Poor resonant frequency value will result in small operation bandwidth thus will limits the operation of the device. Resonant frequency is calculated using

$$f_r = \frac{1}{2\pi}\sqrt{\frac{k}{m}}$$

The equation above, shows that resonant frequency, f_r is proportional to k and inversely proportional to m.

Sensitivity analysis is an alternate and very effective method to study the impact of design parameters on a given performance parameter such as resonance frequency. Sensitivity can be written as [9]:

$$Sensitivity = \frac{\delta F}{\delta p} = \frac{\Delta F / F}{\Delta p / p}$$

where p is the nominal value of the perturbed parameter, Δp is the amount by which the parameter is perturbed, F is the nominal value of the performance measured and ΔF is the amount by which the performance measure changes in response to the parameter perturbation. The results are usually normalized to provide meaningful comparisons.

III. SIMULATION RESULTS AND DISCUSSION

Simulations are carried out using Architect module in CoventorWare 2005. Fig. 3 shows the schematic design of the structure.

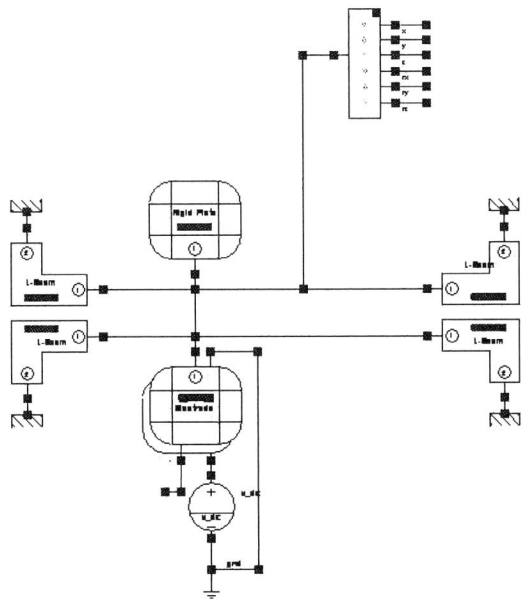

Fig. 3 Schematic design of the structure in Architect module of CoventorWare2005.

Fig. 4 shows the resonance frequency for x, y and z axis. Because the z axis is the dominant axis, it has lower resonance frequency than x and y axis.

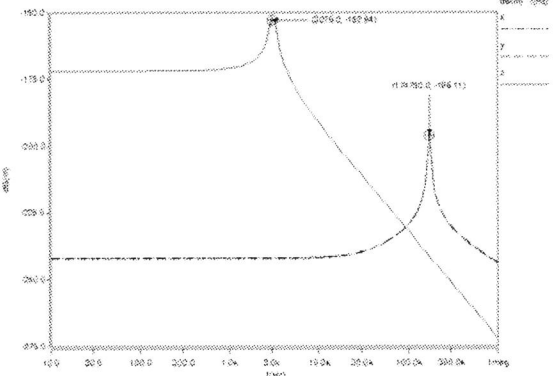

Fig. 4 Resonance frequency for x, y and z axis

By varying beam length L_b from 100 μm to 400 μm, we can see that resonance frequency in z axis decrease, in this case from 3.8 kHz to 32.7 kHz when the length is decrease as shown in Fig. 5.

Fig. 6 shows resonance frequency for various beam width w_b. From the graph, one can see that with the increase in beam width the resonance frequency also increases.

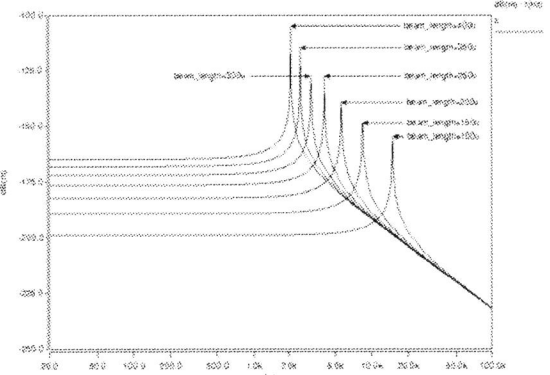

Fig. 5 Resonance frequency for varying beam length L_b.

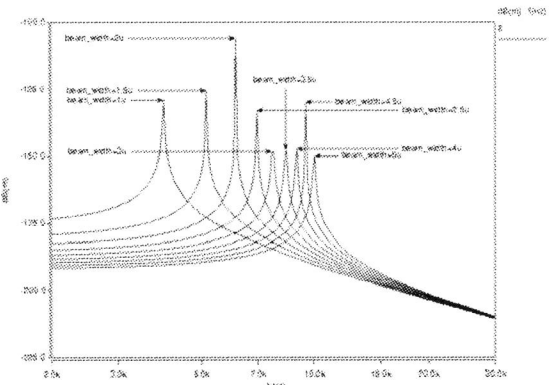

Fig. 6 Resonance frequency for varying beam width, w_b.

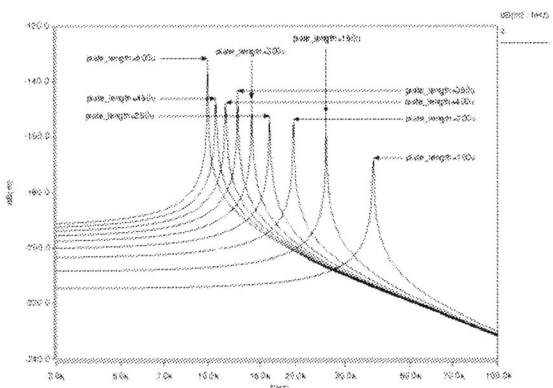

Fig. 7 Resonance frequency for varying plate length, L_p.

By increasing plate length as shown in Fig. 7, resonance frequency decreases. This is also true when we vary shin length as shown in Fig. 8. Shin length was varied from 1 μm to 5 μm.

Geometry Name	Perturbation (%)	ΔF in Hz
beam_length	0.04	-3.0
shin_length	2	-3.6
beam_width	10	170.3
plate_length	0.067	-0.8

IV. CONCLUSION

The effect of mechanical geometry on sensitivity of resonance frequency of MEMS out-of-plane accelerometer is presented. From the sensitivity analysis, beam length is most sensitive geometry with sensitivity of 1.427 and the least sensitive is shin length but beam width will effect most the resonant frequency when all mechanical geometries changed 0.2 μm.

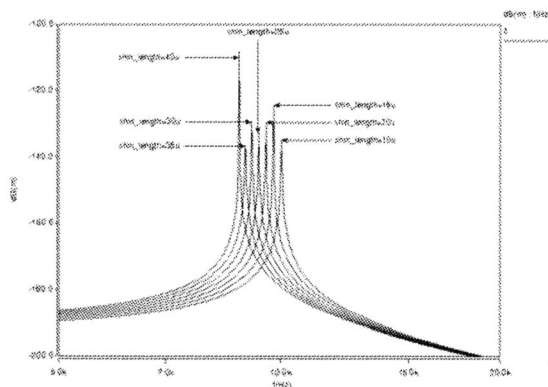

Fig. 8 Resonance frequency for varying shin_length, L_b.

Sensitivity analyses were done by simulation. Table 1 shows the sensitivity of the geometries to resonance frequency in z axis.

Table 1 Sensitivity report

Geometry Name	Sensitivity
beam_length	-1.427
shin_length	-0.057
beam_width	0.554
plate_length	-0.661

From Table 1, we can concluded that design geometry beam length is most sensitive. A sensitivity of 1.427 means that a 1% pertubation of the beam length results in a 1.427% decrease in investigated resonance frequency. But one must keep in perspective of absolute changes in geometry parameters. A 1% change in beam length of 300μm is 3μm, its not same value with 1% change of 2μm beam width which is 0.02μm.

We scale the sensitivities above to a common absolute perturbation to identify those dimension whose changes affects one or more parameters outside desirable performance regime. Table 2 shows frequency change due to geometry variations.It is obvious that width of the beam causes a problem. A change of 0.2μm in beam width results in a change of 170.3 Hz in the resonance frequency. While the change in 0.2 μm in beam length only decrease 3.0 Hz of resonance frequency. Thus, this analysis has immediately given us an indication that small changes in beam width will effect most of resonance frequency.

Table 2   Resonance frequency change due to geometry variation.

REFERENCES

[1]. Jiri Marek and Hans-Peter Trah, "Sensors for Automotive Technology," *Wiley-VCH GmbH & Co*, 2003.

[2]. N. Yazdi, F. Ayazi and K. Najafi, "Micromachined Inertial Sensors", *Proceedings of the IEEE*, vol. 86, no. 8, pp. 1640-1659 (1998).

[3]. N.Yazdi, A. Salian and K. Najafi, "A High Sensitivity Silicon Accelerometer with a Folded Electrode Structure", *Journal of Microelectromechanical Systems*, vol. 12, no. 4, pp. 479-486 (2003).

[4]. F.E.H. Tay, J. Xu, V.J. Logeewaran, "A Differential Capacitive Low-g Microaccelerometer with mg resolution", *Sensors and Actuators A*, vol. 86, pp 45-51 (2000).

[5]. B. Li, D. Lu and W. Wang, "Micromachined Accelerometer with area-changed capacitance", *Mechatronics*, vol. 11, pp. 811-819 (2001).

[6]. B. Bais and B. Y. Majlis, "Mechanical Sensitivity Enhancement of an Area-changed Capacitive Accelerometer by Optimization of the Device Geometry", *Proceedings of DTIP of MEMS and MOEMS*, Montreux (2004).

[7]. G.K. Fedder, "Simulation of Microelectromechanical Systems", *Ph.D thesis,* University of California, Berkeley, 1994.

[8]. Bourne, M.(analyst), "MEMS Market and Technologies", *In-Stat*, July 2004

[9]. Coventor ver.2005, 'Gyroscope Design Tutorial', March 2004.

Structure Design and Fabrication of an Area-changed Bulk Micromachined Capacitive Accelerometer

Badariah Bais, *Member, IEEE* and Burhanuddin Yeop Majlis, *Senior Member, IEEE*
Institute of Microengineering and Nanoelectronics (IMEN)
Universiti Kebangsaan Malaysia, 43600 Bangi, Selangor, MALAYSIA
Email: badariah@vlsi.eng.ukm.my, burhan@lsi.eng.ukm.my

Abstract — In this paper, a lateral capacitive MEMS accelerometer for low-g applications was designed and fabricated. The accelerometer is chosen to be based on an area-changed detection scheme. Area-changed accelerometers can provide alternatives in designing high sensitivity and low mechanical noise floor sensors as it is capable of providing a large proof mass. Based on the calculation of sensitivity and basic resonance frequency, three kinds of accelerometers were designed and optimized. Bulk silicon is chosen as the proof mass material and a three-mask micromachining technology was adopted to fabricate the active structures of the accelerometer. Deep RIE and wafer bonding techniques are utilized as it offers solutions in fabricating the thick proof mass and high aspect ratio sensing element structures with small sensing gaps to achieve high sensitivity and low noise performance.

I. INTRODUCTION

The development of low-g MEMS accelerometer has gained much interest lately. Low-g accelerometer has potential applications in the automotive field such as vehicle stability enhancement, roll-over detection, inclination/theft detection, and vehicle dynamics such as potential wheel skid while turning. Other applications include tilt, vibration and shock measurements. However, the major drawback in the design of low-g accelerometer is the difficulty in achieving high sensitivity and low mechanical noise.

Majority of low-g accelerometers use the capacitive-sensing scheme of detection and most of the accelerometers presented in current research use a varying air-gap principle to sense the acceleration in order to achieve high sensitivity [1-5]. The high sensitivity detection is made possible by having narrow air gaps and

sensing elements with a large overlap area. However, there are a few disadvantages with the structures having narrow air gaps and a large sensing area. One, since capacitance change is proportional to the square of the air gap, there is an inherent nonlinearity from the varying air-gap structure. Second, since the movement of the proof mass must also be limited in a small range in order to improve the linearity, the differential capacitance tends to be very small. This in turn makes the design of the detection circuit difficult.

An alternative approach is to use multiple pairs of interdigitated electrode fingers and a varying overlap area method of capacitance change. For an area-changed capacitive accelerometer, the differential capacitance is shown to vary linearly with the overlapped area of the proof mass and fixed electrodes underneath and independent of the air gap between the proof mass and the electrodes [6-7]. The coplanar movement of the proof mass and the fixed electrodes with respect to each other offers several advantages such as the operational range of the device is not limited to the thickness of the air gap, the pull-in voltage is increased considerably since the sensing direction is parallel to the air gap of the sensing element fingers and damping can be made low since the damping is dominated by the viscous drag and not from the squeeze-film damping between the sense fingers.

Surface micromachined accelerometers have been widely used for low-g applications since it offers the advantages of monolithic circuit integration and the means to provide venting holes without complicating the fabrication process [8-9]. However, the thickness of surface-micromachined accelerometers is limited by fabrication constraints. This makes the sensing mass very small which results in low sensitivity and high thermal mechanical noise. Bulk micromachining technology is a very promising approach to fabricate these accelerometers as its thickness can be made as thick as necessary in order to meet its performance requirements [10-11].

In this paper, a lateral capacitive MEMS accelerometer for low-g applications was designed and fabricated. The accelerometer is chosen to be based on an area-changed detection scheme. Area-changed accelerometers can provide alternatives in designing high sensitivity and low mechanical noise floor sensors as it is capable of providing a large proof mass which is critical in low-g applications. Bulk silicon is chosen as the proof mass material and a three-mask fabrication process flow is developed. Deep RIE and wafer bonding techniques are utilized to build the thick proof mass and high aspect ratio fingers to achieve high sensitivity and low noise performance.

II. SENSOR STRUCTURE AND DESIGN

The accelerometer used in this study is shown in Fig. 1. The accelerometer is silicon-based and uses the area-changed principle of capacitive detection scheme. It consists of a 40 μm thick silicon proof mass suspended 3 μm above the glass substrate by means of folded suspension beams that are anchored onto the substrate. The proof mass consists of many fingers connected in parallel. Stationary electrodes composed of differential comb fingers are placed directly under the mass. Under acceleration, the mass will move along the y-direction parallel to the substrate.

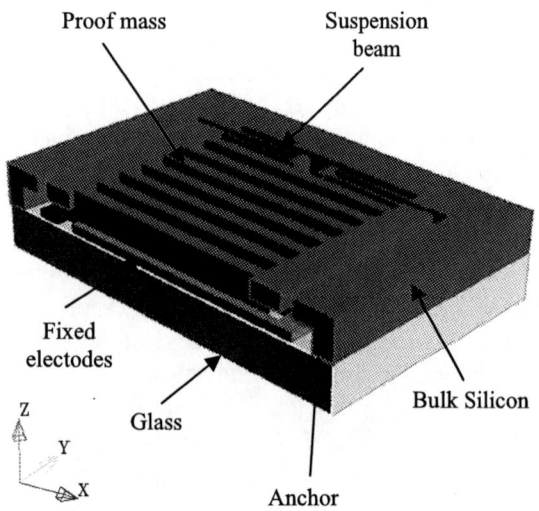

Fig. 1 3D view of the area-changed accelerometer.

The accelerometer can be modeled as a mass-spring-damper system whereby the dynamic behavior of this lumped element system can be described by the second-order dynamic differential equation of motion. The displacement, Δy_m as a function of force generated by the acceleration, F is given by the following:

$$F = ma = m\frac{d^2\Delta y_m}{dt^2} + b\frac{d\Delta y_m}{dt} + k\Delta y_m \qquad (1)$$

where m is the mass of the proof mass, a is the applied acceleration, b is the damping coefficient and k is the spring constant of the suspension beam. When an acceleration, a is applied to the accelerometer, the proof mass is displaced and exerts a force, F on the suspension beam. This force causes the suspension to deflect a distance, Δy_m. For frequencies below the mechanical resonance of the mass-spring system, a linear relationship in (1) can be assumed as:

$$\frac{\Delta y_m}{a} = \frac{m}{k} \qquad (2)$$

The undamped resonant frequency is given by:

$$\omega_o = \sqrt{\frac{k}{m}} \qquad (3)$$

where ω_o is the natural frequency of the accelerometer in the direction of applied acceleration. From eq. (2), it is seen that for a given acceleration, the rate of change of displacement is proportional to the mass of the proof mass and inversely proportional to the spring constant. Therefore, using a heavier proof mass and a reduced spring constant of the suspension beam, a higher rate of displacement of the proof mass per g is obtained. Hence, a smaller spring constant is needed to improve the device mechanical sensitivity. However, from eq. (3), lowering the spring constant will result in a lower resonance frequency which limits the bandwidth. This is the trade-off in performance of the accelerometer.

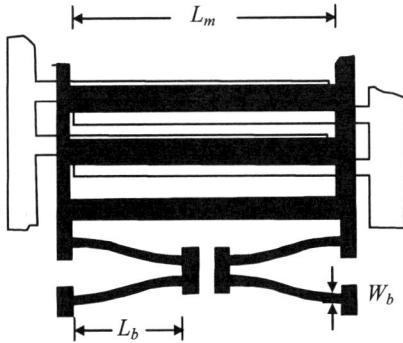

Fig. 2 Top view of the accelerometer.

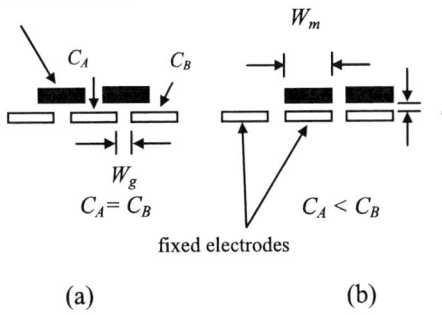

Fig. 3 Working principle of the accelerometer (a) no acceleration (b) with acceleration.

The overall sensitivity of an accelerometer is determined by its mechanical sensitivity (μm/g) and capacitive sensitivity (F/μm). In this study, the width of the proof mass bars, W_m, is set to be the same as the width of the fixed electrode fingers and the length of the mass bars, L_m, be the same as the length of the fixed electrode fingers under the mass bars and the vertical distance between the mass and the fixed electrodes is set as d as shown in Fig. 2 and Fig. 3. If there is no acceleration, the proof mass is at the balance position. At this situation, the left capacitor C_A and the right capacitor, C_B in this device are the same as shown in Fig. 3. Therefore, the static capacitance, C_S is written as:

$$C_A = C_B = C_S = N \frac{\varepsilon L_m (W_m - W_g)}{2d} \qquad (4)$$

where N is the number of the fingers in the proof mass, ε is the permittivity and W_g is the gap between the fixed electrodes. When acceleration

is applied to the structure, the beam deforms and so the mass deviates from the balance position. Under applied acceleration, the changed capacitance C_A and C_B are as follows:

$$C_A = N \frac{\varepsilon (W_m - W_g - 2\Delta y_m) L_m}{2d} \qquad (5)$$

$$C_B = N \frac{\varepsilon (W_m - W_g + 2\Delta y_m) L_m}{2d} \qquad (6)$$

Therefore, the differential capacitance can be written as below:

$$\Delta C = C_A - C_B = \frac{N \varepsilon L_m}{2d} (4\Delta y_m) \qquad (7)$$

Hence, the sensitivity of the accelerometer can be expressed as:

$$Sens = \frac{\Delta C}{\Delta y_m} \frac{\Delta y_m}{g} = 9.8 \frac{4C_s}{(W_m - W_g)} \frac{m}{k} \qquad (8)$$

Folded suspensions with straight truss are employed in the design and are attached to each end of the movable proof mass and can be made very compliant. This design also provides low cross-axis sensitivity. Previous analysis shows the cross-axis sensitivity to be less than 0.7% [12]. Mechanical noise can be reduced by increasing the size of the proof mass.

The minimum detectable signal is mainly determined by the mechanical noise which can be reduced by increasing the size of the proof mass. The theoretical minimum detectable signal, a_{min}, due to the Brownian motion [13] can be expressed as follows:

$$a_{min} = \sqrt{\frac{4k_b T \omega_o}{mQ}} \qquad (9)$$

where k_b is the Boltzmann constant, T the absolute temperature, ω_o the resonant frequency, m is the mass of the proof mass and Q is the quality factor of the accelerometer.

The proof mass structure should be as large as possible in order to obtain a high sensitivity and low mechanical noise. Large proof mass can be achieved by increasing the thickness and/or increasing the surface area which will increase the capacitance value. However, the thickness of the proof mass can only be increased up to the limitation of the

DRIE capability of the fabrication center in order to obtain vertical etching profiles of the proof mass fingers. Therefore, the design of the accelerometer needs to be optimized to achieve high sensitivity and minimum mechanical noise based on the allowable fabrication process parameters. In this study, the thickness of the proof mass is limited to a maximum of 40 μm and the minimum dimension of the proof mass and suspension beam widths and gap are fixed to 2 μm. The accelerometer is to be operating in the range of ±5g.

III. FABRICATION PROCESS

The main fabrication process basically consists of different modules such as glass processing, silicon processing, anodic bonding of the two substrates, chemical mechanical polishing (CMP), and deep reactive ion etching (DRIE). The sequence of the proposed fabrication process is shown in Fig. 4. First, the glass wafer is initially clean and prepared for deposition of the Cr/Au. Then, the metal is patterned to define the fixed electrodes structure. Cavity of 3 μm is then formed in the silicon substrate. Then the glass wafer and silicon are anodically bonded. After bonding, the silicon wafer is thinned to the desired thickness of 40 μm by using Chemical Mechanical Polishing (CMP). Finally, the movable proof mass structure is patterned using Deep Reactive Ion Etching (DRIE).

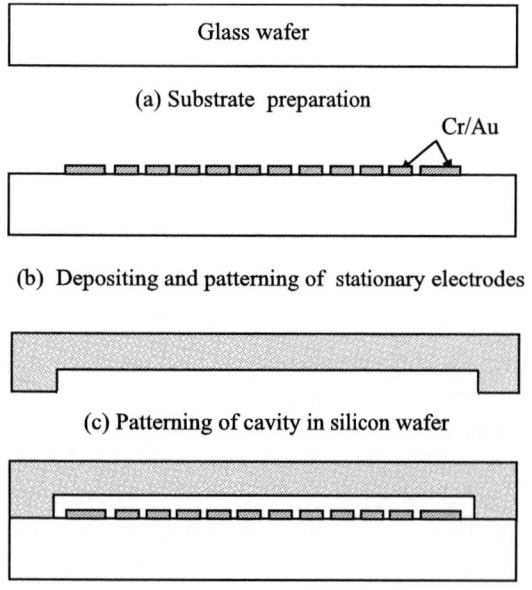

(a) Substrate preparation

(b) Depositing and patterning of stationary electrodes

(c) Patterning of cavity in silicon wafer

(d) Silicon anodic bonding

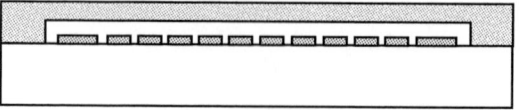

(e) Thinning of silicon

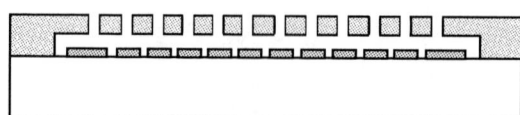

(f) Patterning of the movable proof mass structure

Fig. 4 Schematic of the fabrication steps of the accelerometer.

IV. RESULTS AND DISCUSSION

Various area-changed accelerometers were designed with a wide performances as given in Table 1.

Table 1 List of different area-changed accelerometers and their calculated performances.

Specification	Type1	Type2	Type3
Proof mass length (μm)	1600	1600	1600
Proof mass width (μm)	22	6	22
Movable mass (μg)	128	41	128
Suspension beam length (μm)	560	566	560
Suspension beam width (μm)	2	2.3	3
Thickness of proof mass (μm)	40	40	40
Spring constant (N/m)	0.63	0.92	2.12
Resonant frequency (Hz)	350	760	650
Rest capacitance (fF)	1180	236	1180
Sensitivity (fF/g)	474	103	141
Range	±5g	±5g	±5g
Mechanical noise floor (μg/√ Hz)	0.3	0.53	0.3

The accelerometer was fabricated with three masks bulk micromachining fabrication process in order to define the main design structures namely one on the glass to form the fixed electrodes and metal interconnects and two on the silicon to create the cavity and to define the sensor structure and electrodes. Fig. 5 shows an SEM micrograph of a typical device. Fig 6 is a close-up view of the supports of the proof mass fingers. A folded suspension beam can be seen along with movable proof mass sensing fingers.

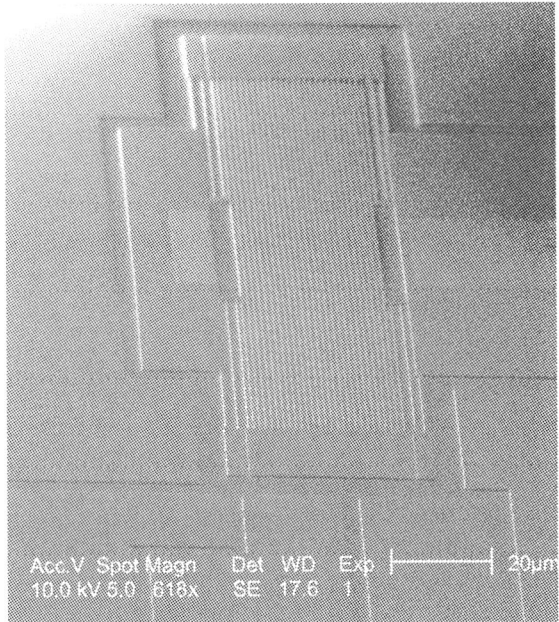

Fig. 5 SEM micrograph of a type2 area-changed accelerometer.

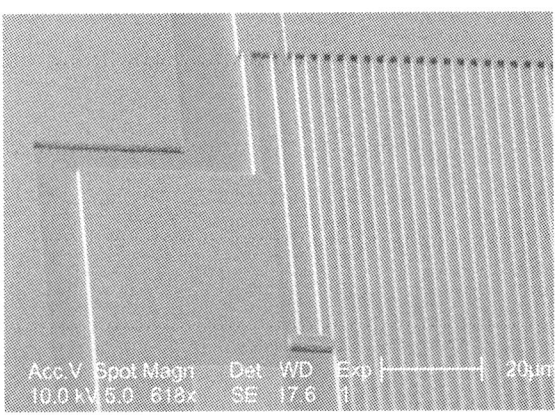

Fig. 6 Close-up of the folded suspension beams and the proof mass fingers.

Fig. 7 shows the optical micrograph of the fixed electrode structure from the glass side along with the contact connection to the bond pad while Fig. 8 shows the detail of the interdigitated comb finger structures of the fixed electrode.

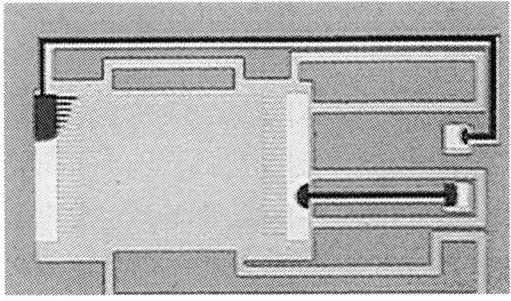

Fig. 7 Optical micrograph of the fixed electrode structure from the glass side.

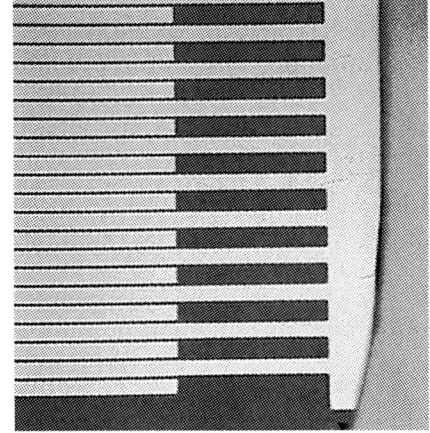

Fig. 7 Close-up view of the fixed electrode structure showing the comb structures.

Measurement made on the fabricated geometrical structures of the proof mass and the fixed electrodes shows good agreement with the designed dimensions with deviations of up to 5%. The thickness of the proof mass was measured to be 27 µm which is 68% of the original height. The reduction in the thickness will affect the dynamic behaviour of the accelerometer. However, its static capacitance which is dependent on the area will remain the same as designed.

V. CONCLUSION

The structure design and fabrication of a bulk-micromachined area-changed capacitive accelerometer for low-g applications was presented. Theoretical modeling was presented and verified with the simulation results. A folded suspension design with rigid truss that offers low spring constant and low cross-axis sensitivity was chosen. Various parameters were studied and optimized in order to achieve high sensitivity and low mechanical noise. Varying the design parameters of the device geometry highlighted in the analysis, a wide performance range can be achieved, thus being suitable for broad range of applications. A 3-mask bulk micromachining wafer bonding fabrication process was utilized to produce this accelerometer. Silicon-on-glass was used to achieve low mechanical noise with high sensitivity while maintaining a simple structure. The fabricated accelerometer shows fairly good agreement with the designed parameters.

ACKNOWLEDGEMENT

This project is financially supported by Ministry of Science, Technology and Innovation of Malaysia under IRPA program: Development of MEMS Technology for Automotive Application.

REFERENCES

[1] N. Yazdi,, K. Najafi and A.S. Salian, "A High-Sensitivity Silicon Accelerometer with a Folded-Electrode Structure," *J. of Microelectromechanical Systems*, vol. 12 no. 4, pp. 479-486 (2003).

[2] W. Henrion, L. DiSanza, M. Ip. S. Terry and H. Jerman, "Wide Dynamic Range Direct Digital Accelerometer," in *4th Tech. Dig. of the IEEE Solid-State Sensors and Actuator Workshop*, pp. 153-157 (1990).

[3] K.H.-L. Chau, S.R. Lewis, Y. Zhao, R.T. Howe, S.F. Bart, and R.G. Marcheselli, "Integrated Force-balanced Capacitive Accelerometer for Low-g Applications," *Sensors and Actuators A*, vol.54, pp. 472-476 (1996).

[4] F.E.H. Tay, J. Xu, V.J. Logeeswaran, "A Differential Capacitive Low-g Microaccelerometer with mg Resolution," *Sensors and Actuators A,* vol. 86, pp. 45-51 (2000).

[5] A. Selvakumar. A multifunctional silicon micromachining technology for high performance microsensor and microactuators. *PhD Thesis.* University of Michigan (1997)

[6] B. Li, D. Lu and W. Wang, "Micromachined Accelerometer with Area-changed Capacitance," *Mechatronics,* vol. 11, pp. 811-819 (2001).

[7] B. Ha, Y. Oh and C. Song, "A Capacitive Silicon Microaccelerometer with Force-Balancing Electrodes." *Jpn. J. Applied Physics*, vol. 37, pp. 7052-7057 (1998).

[8] B.P. van Drieënhuizen, N.I. Maluf, I.E. Opris and G.T.A. Kovacs, "Force-Balanced Accelerometer with mG Resolution, Fabricated using Silicon Fusion Bonding and Deep Reactive Ion Etching," *1997 Int'l. Conf. Solid-State Sensors and Actuators,* Chicago, June 16-19, pp. 1229-1230 (1997).

[9] C. Lu, M. Lemkin and B. Boser, "A Monolithic Surface Micromachined Accelerometer with Digital Output," *IEEE J. Solid State Circuits,* vol. 30, no. 12, pp. 1367-1373 (1995).

[10] J. Chae, H. Kulah & K. Najafi, K. An in-plane high-sensitivity, low-noise micro-g silicon accelerometer. *Proc. 16th IEEE Int. Conf. on MicroElectroMechanical Systems (MEMS2003),* pp. 466-469 (2003)

[11] Z. Xiao *et. al.*, "Laterally Capacity Sensed Accelerometer Fabricated with Anodic Bonding and High Aspect Ratio Etching," *Proc. 10th Int'l Conf. Solid-State Sensors and Actuators (TRANSDUCERS '99),* Sendai, Japan, pp. 1518-1521 (1999).

[12] Badariah Bais and Burhanuddin Yeop Majlis, "Suspension Design Analysis on the Performance of MEMS Area-changed Lateral Capacitive Accelerometer," *Proc. 2004 IEEE Int'l Conf. on Semiconductor Electronics (ICSE2004),* Kuala Lumpur, Malaysia, pp. 335-339 (2004).

[13] T.B. Gabrielson, "Mechanical-thermal Noise in Micromachined Acoustic and Vibration Sensors," *IEEE Transactions on Electron Devices*, vol. 40, no. 5, pp. 903-909 (1993).

The Effect of Design Parameters on Static and Dynamic Behaviors of the MEMS Microphone

Bahram Azizollah Ganji and Burhanuddin Yeop Majlis, Senior Member, IEEE
MEMS Laboratory, Institute of Microengineering and Nanoelectronics (IMEN)
Universiti Kebangsaan Malaysia, 43600 Bangi, Selangor, MALAYSIA
Email: baganji@vlsi.eng.ukm.my, burhan@vlsi.eng.ukm.my

Abstract **This paper describes the effect of design parameters of MEMS capacitive microphone on the static and dynamic behaviors. The aim is to develop the microphones with high sensitivity and flat frequency response. The high sensitivity can be obtained by changing the initial stress of diaphragm, σ_r, diaphragm size, a, diaphragm thickness, t, back plate thickness, h, air gap thickness, d, back plate hole radius, r, surface area fraction occupied by the holes, a., and bias voltage. The equivalent electrical circuit method has been used to predict the structure behaviors and the microphone performance. The optimized structure has a diaphragm thickness of 0.8 μm, a diaphragm area of 2.43 mm², an air gap of 4.0 μm and a 1.0 μm thick back plate with acoustical ports. The device shows maximum sensitivity of 47.9 mV/Pa, with a high frequency response extending to 18 kHz.**

I. INTRODUCTION

MOST of the silicon microphones are based on capacitive principle because of the high sensitivity, flat frequency response, small size, low noise level and low power consumptions [1]. The capacitive microphone consists of a thin, flexible diaphragm and a rigid back plate. The most successful devices use poly silicon as the diaphragm material because of its low intrinsic stress. This stress is very important because it determines the diaphragm sensitivity and its resistance to warpage.

In this paper we use low-stress poly silicon as the diaphragm, a silicon nitride perforated back plate, an n-type silicon substrate and the metal contacts. The equivalent electrical circuit method has been used to predict the structure behaviors and the microphone performance. Optimal diaphragm edge width, thickness, and air gap are determined for maximum sensitivity subject to pull-in voltage and processing constraints.

Results show that in compared with the previously reported by P. C. Hsu [2] and S. Chowdhury [3] this microphone has advantages of high sensitivity and high frequency response in hearing range.

II. MICROPHONE DESIGN

The high-sensitivity microphones for the proposed acoustical sensor are of the capacitive type and can be fabricated as a single structure using MEMS technology. The microphone is to be fabricated using a diaphragm, an air gap and a back plate with acoustical ports, as shown in Fig. 1. When biased oppositely, the diaphragm and the back plate constitute an air core parallel plate capacitor. When the acoustical wave strikes on the diaphragm, it causes the diaphragm to vibrate; accordingly that changes the capacitance due to changing air gap. The capacitance change causes a time varying current.

A well-designed subminiature microphone should be easy to manufacture, should deliver a large output signal despite of its small size, should have a flat frequency response and a low noise level. A large microphone output signal is obtained, if the open loop microphone sensitivity ($S_{open} = S_m.S_e$) is high. The mechanical sensitivity is defined as the increase in the deflection of the diaphragm's center, dw, resulting from an increase in the pressure, dP, acting on the diaphragm, $S_m=dw/dP$.

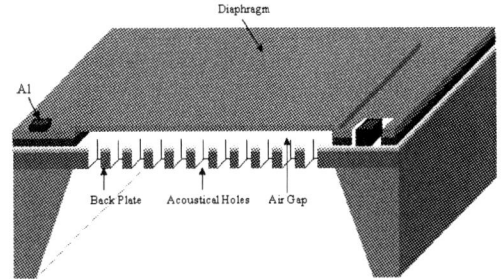

Fig. 1 Schematic description of a MEMS capacitive microphone with flat diaphragm

The electrical sensitivity $S_e=dV/dS_a$ with dS_a being the change in thickness of the air gap and dV the resulting change in the voltage across the air gap can be expressed by $S_e= E_a= V_b/ dS_a$ if the electrical field E_a in the air gap is assumed homogenious, where V_b is the bias voltage. S_e becomes large if V_b is large and dS_a is small [4].

III. ANALYSES OF THE MICROPHONE USING EQUIVALENT CIRCUIT METHOD

The dynamic behavior of the microphone can be calculated using equivalent analog electrical model with lumped parameters [3]. The equivalent circuit of the proposed microphone is given in Figure 2. The acoustic force, F_{sound}, and flow velocity of air, V_m, are modeled as equivalent voltage and current sources, respectively. The air radiative resistance is defined as R_r, and the air mass is defined as M_r, The diaphragm mechanical mass is M_m, and its compliance is C_m. The air gap and back vent losses are represented by viscous resistances R_g and R_h, respectively. The air gap compliance is given by C_a [2].

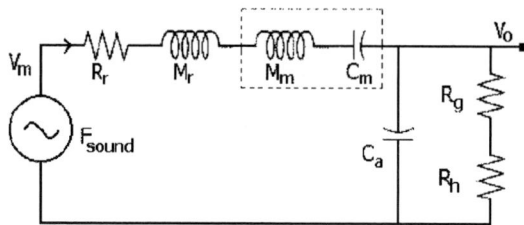

Fig. 2 Equivalent electrical circuit of the condenser microphone

The first resonant frequency of the diaphragm is obtained by:

$$ f_{res} = \sqrt{\frac{1}{\rho}\left(\frac{D\pi^2}{a^4}+\frac{T}{2a^2}\right)} \qquad (1) $$

where D is flexural rigidity of the diaphragm, ρ is mass per unit area (area density) of the diaphragm for a diaphragm thickness, a is the diaphragm edge width, and T is tension of diaphragm, that it is proportional with residual stress of the diaphragm , σ_r, and diaphragm thickness, t, ($T= \sigma_r t$).

The total equivalent mechanical impedance, Z_t, following from the equivalent electrical circuit that is shown in figure 2 can be obtained as:

$$ Z_t = R_r + j\omega(M_r + M_m)+\frac{1}{j\omega C_m}+\frac{R_g+R_h}{1+j\omega(R_g+R_h)C_a} \qquad (2) $$

The open-circuit output voltage V_0 of the microphone under the assumption of constant charge on the condenser plates is given by:

$$ V_o = \overline{E}x = \frac{\overline{E}V_m}{J\omega} = \frac{\overline{E}F}{J\omega Z_t} = \frac{\overline{E}PA_{eff}}{J\omega Z_t} \qquad (3) $$

where $\bar{E}$ is the electrical field in the air gap, x is the average diaphragm deflection, $V_m = j\omega x = F/Z_t$ is the velocity of the diaphragm, $F = PA_{eff}$ is the force on the diaphragm and P is the sound pressure. Consequently, the sensitivity of the microphone is given by dividing the output signal by the applied acoustic sound pressure, yielding the following expression:

$$ S = \frac{|V_o|}{P} = \frac{EA_{eff}}{|J\omega Z_t|} = \frac{Ea^2}{|J\omega Z_t|} = \frac{V_b a^2}{|J\omega d Z_t|} \qquad (4) $$

where V_b is the bias voltage between the diaphragm and the back plate and d is the average air gap distance.

A goal in our design is the maximization of sensitivity subject to fabrication and principal design variables, such as, initial stress of diaphragm, diaphragm size, a, diaphragm thickness, t, back plate thickness, h, air gap thickness, d, back plate hole radius, r, surface area fraction occupied by the holes, α, and bias voltage. At low frequencies, the sensitivity of the microphone can be approximated as:

$$ S_0 = \frac{32V_b a^2}{\pi^6 T d} \qquad (5) $$

The pull-in voltage for a fully clamped square diaphragm under tension is approximately [5] given by:

$$ V_P = \sqrt{\frac{6d^2}{5\varepsilon_0}\left[C_1\frac{t\sigma_r}{\hat{a}^2}\left(\frac{d}{3}\right)+C_2(v)\frac{t\hat{E}}{\hat{a}^4}\left(\frac{d}{3}\right)^3\right]} \qquad (6) $$

where $\hat{a} = a/2$, the quantities, v is poison's ratio, C_1 and C_2 are numerical parameters and their values are given by:

$$C_1 = 3.45, \text{ and } C_2 = 1.994(1 - 0.271v)/(1 - v) \quad (7)$$

In [6], it was shown that for clamped diaphragms, a fringe field correction is required for the Young's modulus as given by:

$$\hat{E} = E/(1 - v^2) \quad (8)$$

where $\hat{E}$ is the effective Young's modulus and E is young's modulus of elasticity. If t<0.01a, V_p reduces to:

$$V_p = \sqrt{\frac{8C_1}{5}} \sqrt{\frac{Td^3}{\varepsilon_0 a^2}} \quad (9)$$

With substitution of Eq. (9) in Eq. (5), the sensitivity can be expressed in terms of the pull-in voltage as:

$$S_0 = \frac{0.18373}{\varepsilon_0} \left(\frac{V_b}{V_p^2}\right) d^2 \quad (10)$$

From the above equation, we can see that sensitivity is a function of d^2 and is proportional to the ratio of the bias voltage to the square of the pull-in voltage. Thus, by minimizing V_p and maximizing d the sensitivity can be increased. The static capacitance between the diaphragm and the back plate can be expressed as:

$$C = \frac{\varepsilon_0 a^2}{d} \quad (11)$$

In order to maximize C, either a can be increased or d can be decreased. If one parameter (a or d) is maximized based on physical constraints like fabrication, or size requirements, etc, the other one (a or b) can be fixed. Due to fabrication constraints associated with micromachining, efficient etching requirements for the sacrificial layer sets an upper limit on the size of d. Thus, a becomes a variable and the specific capacitive readout circuit used determines the lower limit of a. Now V_p can be expressed as in terms of C as:

$$V_p = \frac{64}{7} \sqrt{\frac{2}{45}} \sqrt{\frac{Td^2}{C}} \quad (12)$$

For a desired C, the pull-in voltage can thus be minimized by decreasing the tensile force per unit length of the diaphragm.

The design to maximize the sensitivity of a MEMS capacitive microphone consists of:

- Finding the optimum air gap thickness permitted by fabrication process.
- Minimizing the tensile force per unit length of the diaphragm by using a low stress material and applying special techniques like phosphorous doping, high temperature annealing, etc.
- Defining a minimum diaphragm size to satisfy the constraints associated with the hearing instrument dimensions.
- Maximizing the ratio of holes to back plate areas by increasing the acoustic hole density to increase the cutoff frequency.

IV. RESULTS AND DISCUSSION

Initial stress, σ_r, is an important factor in the modeling and simulation of the microphone. It can be seen from figure 3 that, in cases where the initial stress is reduced (so sensitivity is increased), the resonance frequency will also reduce. From the above analysis, it can be concluded that there is a dilemma between the high sensitivity and high resonance frequency. For all the diaphragms, In order to satisfy most of the microphone, the first resonance frequency of the diaphragm should be well above 18 kHz.

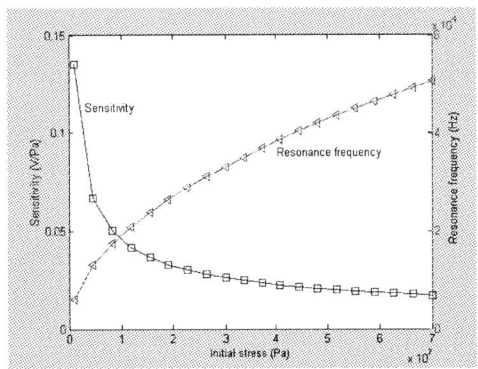

Fig. 3 Microphone Sensitivity and Resonance Frequency of the diaphragm at different initial stress levels

In fig. 4, the pull-in voltage as a function of the initial stress, T, is shown. It can be seen that, the pull-in voltage can be minimized by decreasing the initial stress of the diaphragm. The calculated sensitivity of a typical microphone at 1 kHz as a function of bias voltages is shown in Figure 5. In fact, the electrical sensitivity increases with the bias voltage. However, the bias voltage cannot be increased without limit. At a certain pull in voltage, the diaphragm collapses to the back plate.

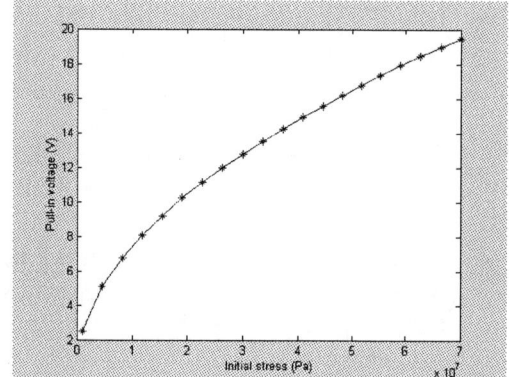

Fig. 4: Pull-in voltage (V_p) versus initial stress (T) at f=1 KHz

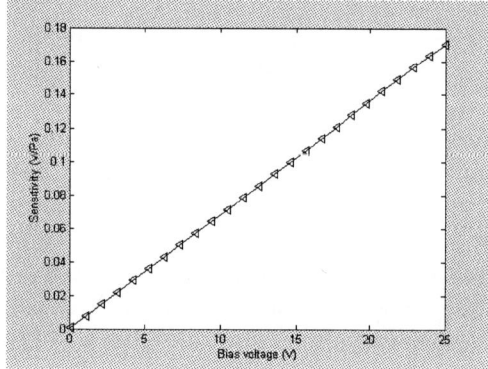

Fig. 5 Microphone sensitivity at different bias voltages

According to Eq. (5), a high sensitivity requires the use of a low stress material to construct a thin membrane with a large surface area (see Fig. 6) and a small air gap between the diaphragm and the back plate. The desired miniaturization is in contrast to a large membrane area and also, technology sometimes limits the upper size. Therefore, it might be advantageous to use an array of smaller microphones instead of one large one.

Fig.7 shows the pull-in voltage versus diaphragm length. It can be seen that, the pull-in voltage decreases, when the diaphragm width is decreased. The sensitivity and the resonance frequency of the diaphragm can be optimized by choosing appropriate structure parameters. The thicker the diaphragm, the higher the resonance frequency, but the lower the mechanical sensitivity (see Fig. 8).

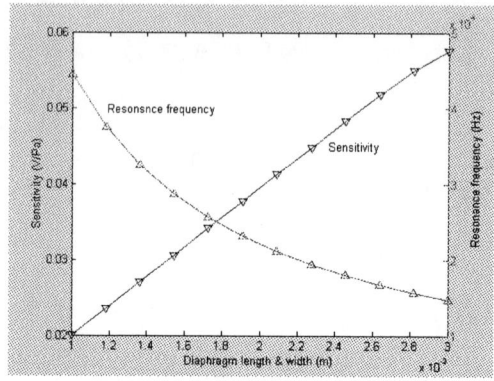

Fig. 6 Sensitivity and resonance frequency versus diaphragm width.

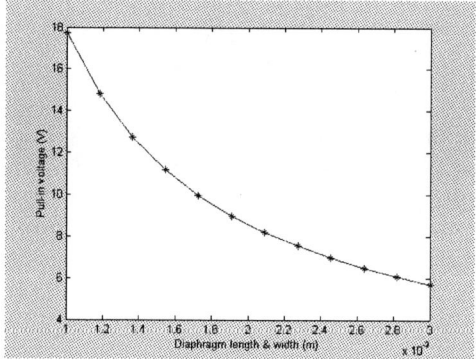

Fig. 7 Pull-in voltage versus diaphragm width.

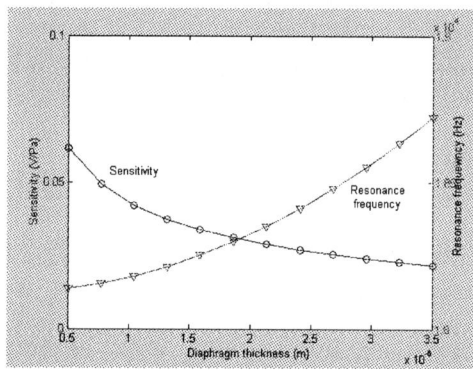

Fig.8 Sensitivity and resonance frequency versus diaphragm thickness

Fig. 9 shows the relationship of pull-in voltage and diaphragm thickness. As can be seen, the pull-in voltage, V_p, decreases when the diaphragm thickness decreases. The air between the electrodes offers a mechanical resistance to the membrane deflection. If the back electrode is realized as a perforated plate an oscillating membrane causes air streaming laterally through the air gap towards the holes. The narrow air gap offers a high resistance to the streaming air caused by friction of neighboring layers of air due to the viscosity of air according to the following equation:

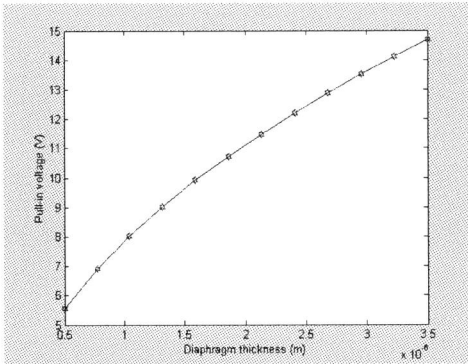

Fig. 9 Pull-in voltage (V_p) versus diaphragm thickness(t) at f=1 KHz

$$R_{gap} = \frac{1}{a^2 . n} . \frac{12\mu}{d^3 . \pi} . [\frac{A}{2} - \frac{A^2}{8} - \frac{LnA}{4} - \frac{3}{8}]$$

(13)

where n the ventilation hole density, μ the viscosity of air and A the surface fraction occupied by the ventilation holes. In Eq. (13), the air resistance of the air gap, R_{gap}, has a close relation to the density of the air holes in the back plate and the size of the air gap. Although increasing the air gap can effectively reduce the air resistance and thus increase sensitivity (see Fig. 10), the microphone capacitance may become too small to be measured. Hence increasing the density of air holes is an effective method. From figure 10, it can also be seen that the pull-in voltage is decreased when the air gap thickness is decreased.

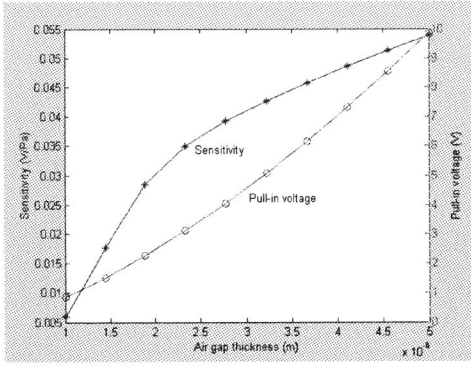

Fig. 10 Sensitivity and pull-in voltage versus air gap thickness

The shape of the frequency response of the microphone is determined by the damping and resonance behavior of the microphone structure, which depends mainly on the size and stress of the diaphragm. Damping is caused by losses in

the diaphragm and the viscous losses associated with the air streaming in and out of the air gap. The perforation of the back plate to create acoustical ports provides a means to control streaming losses and therefore the damping characteristics of the microphone structure. A low damping property can be obtained using a highly perforated back plate.

Upper limit of the operating frequency can be adjusted by the density and size of the holes in the back plate. The lower limit of the frequency can be adjusted by the size of the static pressure hole in the diaphragm.

Fig.11 shows the frequency response at different number of back plate holes, n. It can be seen that the cut of frequency increases when the acoustic holes density is increased. Because the viscous losses at high frequencies, due to the compression of air in the air gap, is decreased.

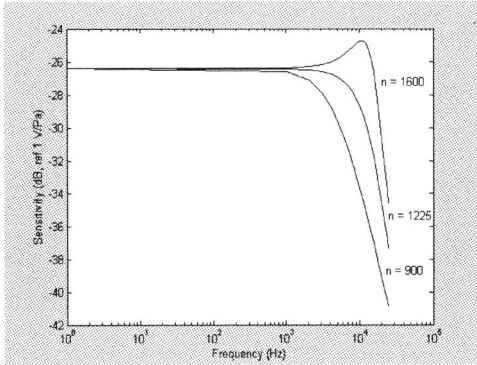

Fig. 11 Sensitivity frequency response at different number of back plate holes (a) n=900, (b) n=1225, (c) n=1600.

Fig. 12 shows the sensitivity versus back plate hole size, r. It can be seen that, sensitivity increases, by increasing the back plate hole size until 30 μm. The above it, acoustical hole size doesn't affect on the microphone sensitivity.

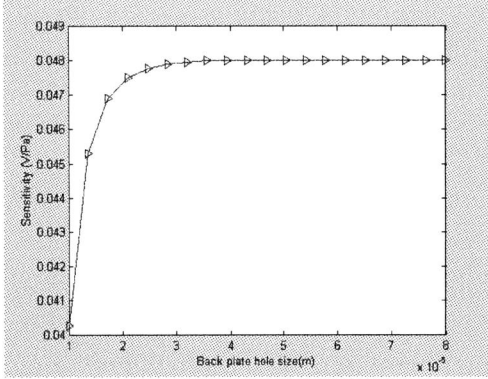

Fig. 12 Sensitivity versus back plate hole size.

At higher frequencies, the air flowing in and out of the air gap causes the thin back plate to vibrate in phase with the diaphragm, and results in decreasing the microphone output signal. A rigid back plate is a prerequisite of good performance. A deposited stiff back plate can only be achieved if the thickness is large, i.e. in excess of at least 5 µm. Such thickness is difficult to achieve by deposition because internal stresses cause cracks and deposition times are often long. The use of a stress-free monocrystalline back plate of approximately 20 µm is stiff enough to prevent a reduction in mechanical sensitivity at higher frequencies.

Fig. 13 shows the sensitivity versus back plate thickness at different acoustical hole sizes. It can be seen that, for acoustical hole size under 30 µm, the sensitivity decreases, when the back plate thickness is increased. For acoustical hole size of 30 µm and above it, the back plate thickness doesn't affect on the sensitivity.

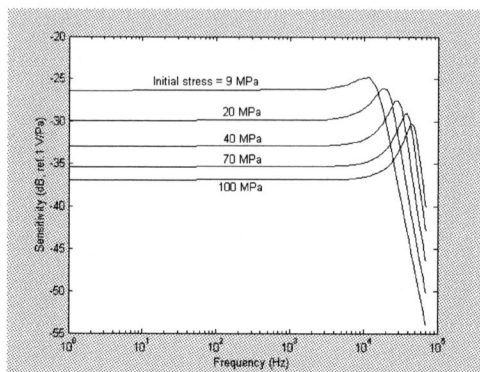

Fig. 14 Frequency response of the microphone at different initial stress levels

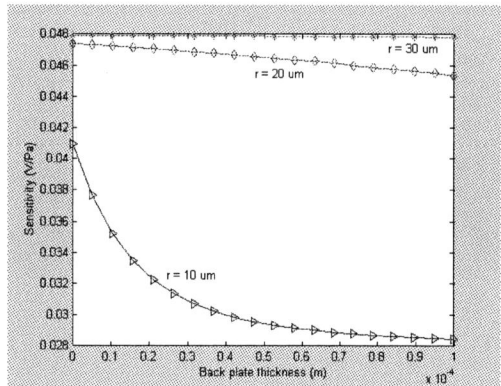

Fig. 13 Sensitivity versus back plate thickness at different acoustical hole size.

Based on these criteria the optimum design parameters for the microphone are as following: diaphragm thickness of 0.8 µm, initial stress of 9 MPa, diaphragm area of 2.43 mm^2, an air gap of 4.0 µm and a 1.0 µm thick back plate with the back plate hole size of 30 um^2, and area ratio of holes to back plate is %24.4. The device shows maximum sensitivity of 47.9 mV/Pa (-26.38 dB), with a high frequency response extending to 18 kHz (see Figure 14).

V. CONCLUSION

The analysis of a high sensitivity capacitive microphone has been presented in this paper using a Poly-Si diaphragm suspended above a perforated silicon nitride back plate. The equivalent electrical circuit method has been used to predict the dynamic behaviors and the microphone performance. The optimized structure has a diaphragm thickness of 0.8 µm, a diaphragm area of 2.43 mm^2, an air gap of 4.0 µm and a 1.0 µm thick back plate with acoustical ports. The device shows maximum sensitivity of 47.9 mV/Pa with a flat frequency response extending to 18 kHz.

REFERENCES

[1] Jing Chen, Litian liu, Zhijian Tan, Yang Xu, and Jun Ma , "On the single- chip condenser miniature microphone using DRIE and back side etching techniques" *Sensors and Actuators A 103*, 42-47 (2003).

[2] P. C. Hsu, C. H. Mastrangelo, and K. D. Wise, "A high density polysilicon Diaphragm condenser microphone", *The 1998 MEMS conference, IEEE,* pp. 580-585 *(1998).*

[3] S. Chowdhury, G. A. Jullien, M. A. Ahmadi, and W. C. Miller, "MEMS acousto- magnetic components for use in a hearing instrument" *SPIE's Symposium on design, Test, Integration, and Packaging of MEMS/MOEMS,* (2000).

[4] W. Kronast, "Single chip condenser microphone using porous silicon as sacrificial layer for the air gap", EEE, (1998).

[5] S.Chowdhury, M. Ahmadi, and W. C. Miller, "Nonlinear Effects in MEMS Capacitive Microphone Design"*Proceedings of the International Conference on MEMS,NANO and Smart Systems (ICMENS'03), IEEE,* (2003).

[6] P.M. Osterberg, and Stephan D. Senturia, "M-TEST: A Test Chip for MEMS Material Property Measurement Using Electrostatically Actuated Test Structures" *Journal of Microelectromechanical System, IEEE,* VOL. 6, NO. 2, pp. 107-118, (1997).

Deep Trenches in Silicon Structure using DRIE Method with Aluminum as an Etching Mask

Bahram Azizollah Ganji and Burhanuddin Yeop Majlis, Senior Member, IEEE
MEMS Laboratory, Institute of Microengineering and Nanoelectronics (IMEN)
Universiti Kebangsaan Malaysia, 43600 Bangi, Selangor, MALAYSIA
Email: baganji@vlsi.eng.ukm.my, burhan@vlsi.eng.ukm.my

Abstract In this paper, we describe our experience with deep reactive ion etching (DRIE) of silicon to depths ranging from 10 μm to more than 250 μm for MEMS applications, including MEMS microphone. The DRIE was produced in oxygen-added sulfur hexafluoride (SF_6) plasma, with sample cooling to cryogenic temperature. We used an inductively coupled plasma reactive ion etcher and Al mask. A 90 min etching experiments using etching gas SF_6 of 60 standard cubic centimeters per minutes (sccm) with oxygen (13 sccm) were performed by supplying RF power of 5 W to an ICP of 600 watts. For the Al mask, an etch rate of 5.44×10^{-3} nm/min was achieved. By controlling the major parameters for plasma etch, anisotropic etch profiles and smooth etched surfaces, an etch rate of 2.85 microns per minute and a high selectivity of 5.24×10^5 to the Al etch mask have been obtained. An etch depth of 257 μm was demonstrated. Our experiments show that Al is a best mask material for very deep trenches etch.

Keywords: DRIE, Deep trench, silicon structure, Al mask, etch rate, selectivity

1. INTRODUCTION

DEEP-REACTIVE ion etching (DRIE) is an important process for the fabrication of many silicon-based microelectromechanical systems (MEMS) [1]–[4]. This method has several attractive features. First, this process creates high-aspect ratio (the depth to width) structures with vertical side walls and good depth control. Second, the etching rate is independent of crystal orientation and has been reported to be as high as 4μm/min [5], which is superior to chemical wet etching techniques. Third, micron scale structures can be etched using patterned photoresist as the mask [6], [7]. Fourth, unlike thick-photolithography and its associated edge-

bead problem [8], DRIE can produce uniform structures over the entire wafer surface. Fifth, DRIE is a more easily accessible technique than LIGA (LIthographie-Galvanoformung-Abformung, i.e., lithography-electroforming-molding) [9]. Sixth, the problems associated with wet etching process using KOH etching such as sticking can be avoided. This is especially important for processing the microphones since the diaphragms are thin in the range of few μm. One can see the etching profiles are vertical thanks to the high aspect ratio of silicon using deep RIE instead of the slope etching profiles using KOH etching. In this way, the chip size of each device is reduced. We estimate a half chip area to be used compared to the device by KOH etching. That means we can double the number of devices fabricated on one wafer. Seventh, it is believed that using deep RIE can make the process more compatible with CMOS and enables the integration of MEMS devices with IC on the single chip [10].

MEMS structures may have features as small as 1 micron in width and trench up to several hundred microns in depth. This kind of structure can ideally be obtained with the technique of deep reactive ion etching (DRIE). DRIE is an extended development of the conventional RIE but is operating in high density plasma. This advancement shows several advantages over the RIE method and therewith overcame some limitations that holding back the usefulness of the RIE in the bulk micromachining of silicon [11]. In this paper, we report the deep RIE of silicon structure with depth of more than 250 μm for MEMS used in capacitive microphone.

2. CARRIER FOR DEEP RIE OF MEMS STRUCTURE

Deep RIE decreases mechanical strength of the wafer. The functional structure itself is often fragile. Therefore, fragility of the wafer and/or functional structure is a typical problem that has to be addressed at deep RIE. Often it was

0-7803-9730-4/06/$25.00 ©2006 IEEE

necessary to use a special carrier to complete a deep RIE. Our experience shows that deep RIE using a carrier is an essential need in MEMS fabrication. We have used silicon carrier wafers. Typically, the silicon carrier wafer etches at a fairly rapid rate (about 2 μm/min) because Si readily reacts to form a volatile product with atomic fluorine. Therefore, it is necessary to cover the silicon carrier with nonreactive materials such as, photoresist, polymer or metal (Al) before deep RIE Photoresist or polymer often has to be deposited using special technique in order to get good coverage of the carrier. Still cracks and/or defects in the photoresist/polymer layer allow He leak into the chamber. The leak may significantly change parameters of the process or even stop it. Metal (Al) layer can be used in cases when metal can be removed after deep RIE without affecting device metallization. Different methods of wafer mounting on carrier have been used. The simple way is to attach samples to silicon carrier wafers using grease or a drop of photoresist [12, 13].

3. MASKING MATERIALS FOR DEEP RIE APPLICATIONS

Masking material for deep RIE can also be a challenge. Photoresist can be used as a mask for deep RIE. Selectivity to photoresist as low as 40:1 has been observed for wafers etched with the carrier wafer. Experiments show that different photoresists have significantly different etch rates. Requirements for an ideal masking material for deep RIE can be listed as follows [12].

- High selectivity that would allow using a thin layer of the material as a mask for deep RIE.
- Capability to form a thin uniform film on the silicon wafer without open area or bumps at the periphery.
- Either photosensitive material or the capability to pattern it using thin photoresist as a mask.
- Thermal stability of the mask.
- No cracking after baking/process steps.
- Capability to strip the material after processing.
- No side effect (like grass in the trenches during deep RIE).

We have found positive photoresists as a mask for deep RIE: AZ 1500, AZ 5214E, AZ 7212, AZP4620 exhibit stronger resistance against DRIE than negative resists such as SU-8

2007. Metal (typically Al) is also used as a mask for deep RIE. Al is easily attacked by %10 HF.

4. ETCHING EQUIPMEN AND CONDITIONS

Deep trench etching of silicon microstructure is achieved by using the RIE system from Oxford Instrument (Plasmalab System 100). CF_6 and O_2 gases were introduced into an etching chamber for generating plasma, and He gas was used for cooling a substrate. In addition, the equipment had a variable valve controlling the pressure inside the chamber during etching. The system produces high density of reactive ions with inductive couple plasma (ICP) source. Together with liquid nitrogen (LN_2) cooling of the sample, highly anisotropic profile with high etching rate can be created.

DRIE conditions are defined by many parameters such as gas-flow rates, RF power, inductive couple plasma (ICP) source power, chamber pressure and stage (bottom electrode) temperature that affect the etch performance, in particularly the etch rate, structure profile, selectivity of silicon to mask material and uniformity of etch pattern throughout the sample. Typical values set for these parameters are listed in Table 1 [11, 14, 15].

Table 1: Conditions for DRIE of silicon structure

Parameters	Value
SF_6	60 sccm
O_2	13 sccm
ICP	600 Watt
Camber Pressure	10 mTorr
RF Power	5 Watt
Stage Temperature	-115 C°
He	10 Torr

5. FABRICATION

Figure 1 illustrates an eight-step fabrication-process flow for the DRIE of silicon structure. A 400-μm-thick (100)-oriented silicon wafer is used for the substrate.

In step (1), a 300 nm-thick Al, which is a masking material for DRIE, is evaporated on surface of the silicon wafer. In step (2), approximately 1.5 μm of thin positive photoresist (AZ-1500) was spin-coated on the Al surface. The resist was pre-baked for 90 sec in a 100 °C on the hotplate. In step (3), it was exposed for 10 sec using a contact mask aligner with a UV light source. In step (4), it was developed for 60 sec in

AZ300MIF, the photolithography was completed with a postbake for 2 minutes in a 120 C° ambient. In step (5), Al was wet etched for 17 sec in 10% hydrofluoric acid (HF) solution. In step (6), the Photoresist layer was removed in acetone. In step (7), a 300 μm deep cavity was fabricated in the silicon wafer by the DRIE system. In step (8), the Al thin film is removed by 10%HF solution.

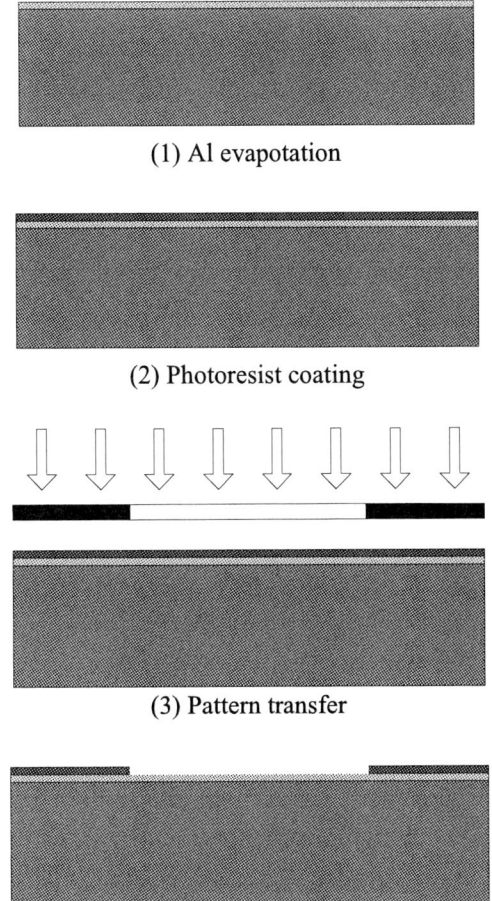

(1) Al evapotation

(2) Photoresist coating

(3) Pattern transfer

(4) Photoresist development

(5) Wet etching Al

(6) Photoresist removal

(7) DRIE of silicon wafer

(8) Al removal

Fig. 1: Fabrication process flows for DRIE of silicon wafer

6. EXPERIMENTAL RESULTS

6.1. Etch Rate and Selectivity Measurements

For this study deep etching was performed on the silicon wafer using a Plasmalab System 100 ICP 180 at different RF powers. Other base etch recipe used are as listed in Table 1. A series of experiments were performed to determine the etch rate and selectivity of the some masking materials such as positive resists (AZ 1500, AZP4620), negative resist (SU-8 2007), and metal (Al). The etch profile as well as the depth of the silicon structure are inspected under the Scanning Electron Microscope (SEM, Philip XL30). With this, the etch rate of silicon can be calculated Table 2 shows the etch rate of silicon at different RF powers. It can be seen that by increasing the RF power, the etch rate increases, and by increasing the etch time, the etch rate decreases.

Table 2. Etch rate of Si at different RF power and etch time

Etch deep	RF power	Etch time	Etch rate
17.5 μm	5 W	5 min	3.5 μm/min
257 μm	5 W	90 min	2.85 μm/min
39.5 μm	50 W	10 min	3.95 μm/min

Thicknesses of masking materials, before etching, were measured with Tencor P-12 (Disk Profiler), and after etching, were measured with a thin film mapping system (Filmetrics F50), in order to determine the amount of mask layer being removed. The etch rate of mask materials and the selectivity of the etching can be determined by comparing the etching ratio of silicon to mask layer. The results are shown in Table 3. Experiments show that different materials have significantly different etch rates. We achieved the best results with Al as a mask layer for very deep trenches etch. We also found that the positive photoresist exhibit stronger resistance against RIE than negative resist.

Table 3. Etch rate and selectivity of the masking materials

Material	Before Etch	After Etch	RF Power	Etch Time	Etch Rate	Selectivity
AZ1500	1.45 µm	0.898 µm	5 W	5 min	0.11 µm/min	31.82
AZP4620	6.68 µm	44.53 nm	5 W	60 min	0.11 µm/min	25.9
SU-8 2007	11.5 µm	6.4 µm	50 W	10 min	0.51 µm/min	7.745
Al	250.49 nm	250 nm	5 W	90 min	5.44×10^{-3} nm/min	5.24×10^{5}

6.2. Deep Etching of Silicon using the DRIE Method

A piece of silicon structure patterned using DRIE was examined with SEM (Fig. 2(a)). The height of the trenches in this image is about 17.5 µm. The mask was 1.48 µm positive photoresist (AZ 1500), and the etching time was 5 minutes.

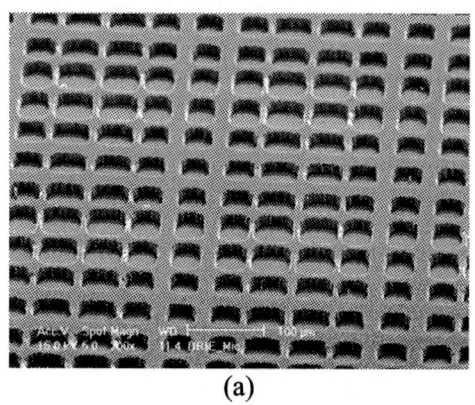

(a)

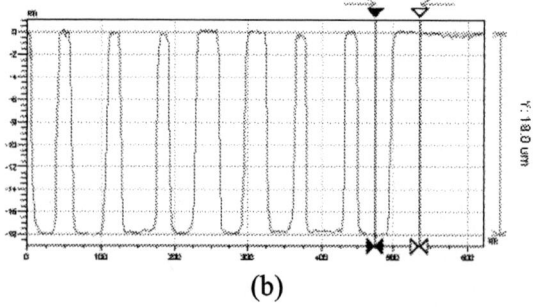

(b)

Fig. 2. (a) SEM image of silicon trenches, (b) surface profile of a silicon structure measured with OSP method

The other parameters are as listed in Table 1. To measure the trench depth, the same silicon structure was examined using optical surface profilometry (OSP); the result is shown in Fig. 2(b). The trench depth measured by OSP is about 17.7 µm, it is similar to the SEM measurement. The structure shown in Fig. 3 was then viewed by scanning electron microscopy (SEM) to obtain the angle of the sidewall profiles. Both the inside at the top (I_{Top}) and bottom (I_{Bottom}) were measured along with the total height of the etched structure (D_{Si}). The angle of the inside sidewall is calculated by [16]:

$$\text{Inside angle} = \tan^{-1}\left[\frac{(I_{Top} - I_{Bottom})/2}{D_{Si}}\right] \quad (1)$$

The angles obtained will be negative for a re-entrant profile and positive for a positive profile. The height of the trenches in this image is about 17.7 µm while the width of trenches is about 31.7 µm at the top and 30 µm at the bottom. The aspect ratio is 0.57 and the angle of the sidewall of trenches is about 2.75 °.

Figure 4 shows an SEM of 250 x 250 µm² pit etched to a depth of 39.5 µm using the baseline etch process (60 sccm SF_6, 13 sccm O_2, 600 watts ICP, 50 watts RF). The mask was 11.5 µm negative photoresist (SU8-2007), and the duration of the etch was 10 min. The etch rate of SU-8 2007 is about 0.51 µm/min. Figure 5 shows the surface profile of a typical well in the sample shown in figure 4. A highly uniform etch is obtained using the baseline deep RIE process.

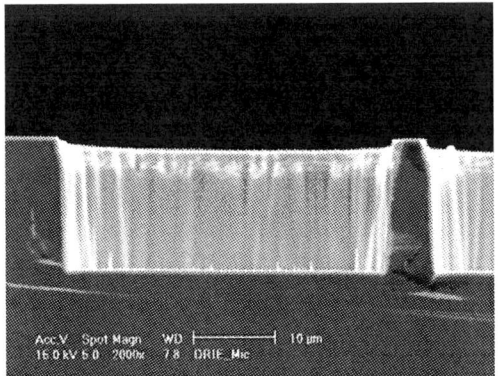

Fig. 3. SEM showing the structure used to measure the side wall angle

Fig. 4. SEM image of 250 x 250 μm^2 pit in silicon

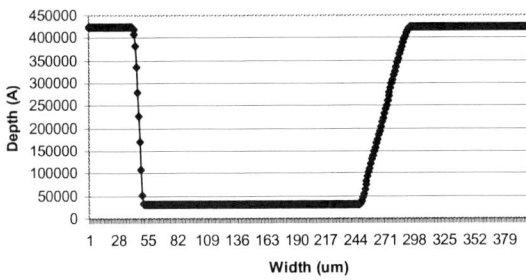

Fig. 5. Surface profile of the bottom of a well etched in the sample

Fig. 6 (a) shows the SEM picture of the silicon structure and trenches etched using the optimized DRIE process. The parameters of DRIE system used to control the etch characteristics are according to Table 1. After 90 min, 257 um deep was obtained. The etch rate of silicon is almost 2.85μm/min, and the etch selectivity to the mask is also ∞. Deeper trenches are possible by increasing the etch time. The etch mask was 345 nm thick Al which was evaporated and patterned using 10% HF. After the etching process was finished, the Al on the silicon was removed with 10% HF. In Fig. 6 (b), the width of trenches is about 3.32 mm the top and 3.42 mm

at the bottom. The gap profile was re-entrant with an angle of -0.01 °. It can be also seen that the depth of the well is uniform.

With flow rate of SF_6 fix, the etch profiles of silicon structures changed from negative to positive tapered as the flow rate of O_2 increased. The oxygen has played a role to form the passivation layer at the sidewall. When too low the flow rate, the layer form is too thin to protect the etching wall and the profile appeared negative (Fig. 3). Whiles, too much of oxygen present, positive taper (Fig. 6(b)) would be produced. Either of these cases is not preferable in the fabrication of MEMS devices where vertical features are required. In the worse case, the thin structure (as small as 1 micron) may also be destroyed in a negatively profile recipe. A vertical profile is therefore highly desired.

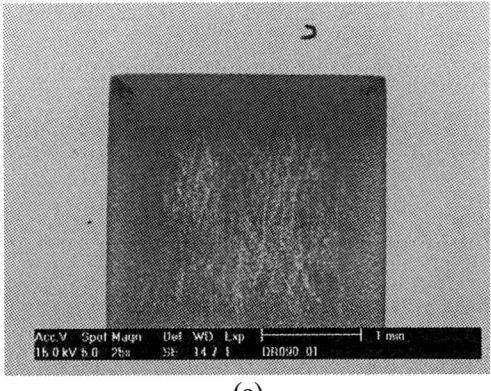

(a)

(b)

Fig. 6 (a) SEM image of 3 x 3 mm^2 trench in silicon structure, (b) Cross-sectional SEM image of silicon trench

Fig. 7 (a) shows the sidewall surface and Fig. 7 (b) shows the bottom surface of the silicon trench. We found that the Al mask often causes grass formation.

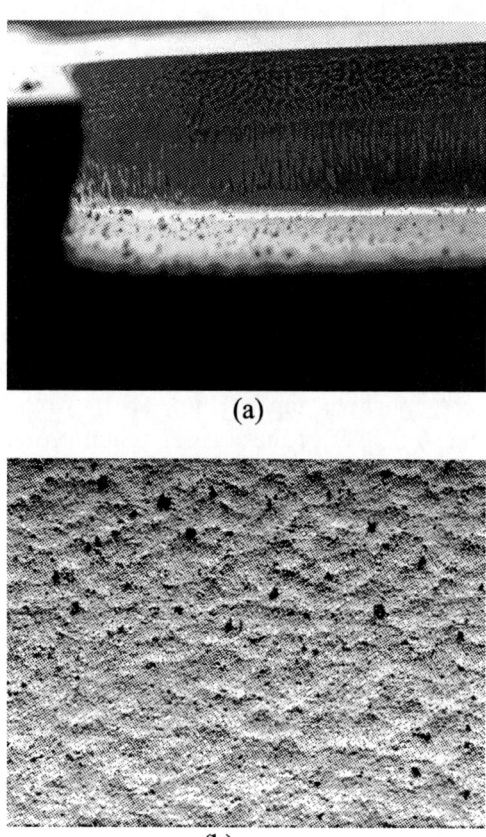

(a)

(b)

Fig. 7 (a) SEM image of sidewall surface, (b) SEM image of bottom surface of the silicon trench.

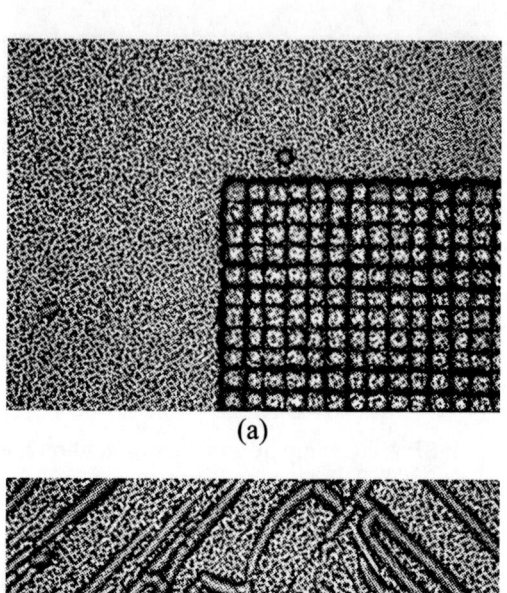

(a)

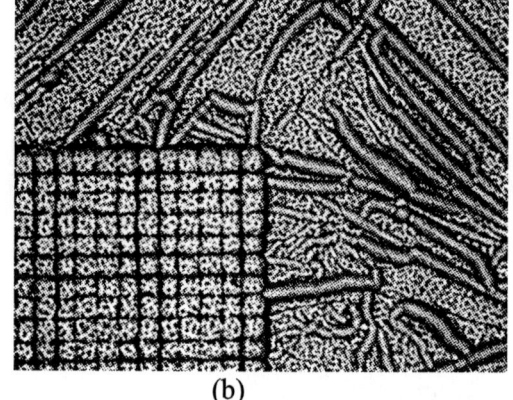

(b)

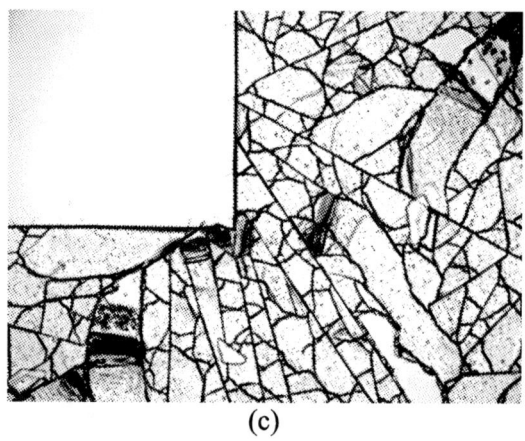

(c)

fig.8 SEM image of the silicon structure using 50 W RF, Masking layer: (a) 345 nm Al on sample for 20 min, (b) 1.72 μm positive photoresist (AZ 7212) on sample, (c) 7 μm positive photoresist (AZP4620) on carrier for 30 min.

Study has also been carried out to investigate the effect of RF power on etch rate and selectivity. Fig. 8 shows SEM image of silicon wafer using 50 watts RF power, with all other parameters identical to the Table 1. The etch mask was 345 nm thick Al, and the duration of the etch was 30 min. It can be seen that Al mask was removed and the surface of silicon wafer was damaged. It is indicating that the RF power has the major affect on etch rate of silicon and Al. Stronger RF power will not only be able to produce more reactive ions for etching but also provide more energy for the ions. This is because of the stronger RF field. Under the higher RF field, ions strike more heavily onto the sample. The physically etching by high energy ions bombardment has less selectivity to the material being etched, i.e. both silicon and Al used to protect silicon were also being removed away. Therefore, higher selectivity can be achieved by lower the RF power. In our study, decreased the RF power to 5 watts has produced selectivity infinite. This is a good value for deep trench silicon fabrication, because a thin Al mask is sufficient for producing very deep trenches in silicon wafer.

7. CONCLUSION

We performed more than 250-μm-deep reactive ion etching (DRIE) of silicon for microelectromechanical systems used in microphone. We used an inductively coupled plasma reactive ion etcher, which enables high density plasma generation and high speed pumping. Sulfur hexafluoride mixed with oxygen was used as an etching gases and Al was used as a mask. A 90 min etching experiments using etching gases SF_6 (60 sccm) and O_2 (13 sccm) were performed by supplying RF power of 5 W to an ICP of 600 W and a temperature of -115 °C. This yielded an etched depth of 257 μm, a maximum etch rate of 2.85 μm/min, vertical sidewalls, smooth etched surfaces, and high selectivity to the etch mask. Due to the high aspect ratio of etching profile using DRIE, the density of the MEMS devices fabricated on single wafer can be doubled, which contributes to reduction of the fabrication coast per device chip.

REFERENCES

[1] L. Fu, J. M. Miao, X. X. Li, and R. M. Lin, "Study of deep silicon etching for micro-gyroscope fabrication," *Appl. Surface Sci.*, vol. 177, no. 1-2, pp. 78–84, 2001.

[2] H. Toshiyoshi, K. Isamoto, A. Morosawa, M. Tei, and H. Fujita, "A 5-volt operated MEMS variable optical attenuator," in *Proc. Transducers' 2003*, vol. 8–12, Boston, MA, Jun. 2003, pp. 1768–1771.

[3] C. Iliescu and J. Miao, "One-mask process for silicon accelerometers on Pyrex glass utilizing notching effect in ICP DRIE," *Electron. Lett.*, vol. 39, no. 8, pp. 658–659, Apr. 2003.

[4] M. Chabloz, Y. Sakai, T. Matsuura, and K. Tsutsumi, "Improvement of sidewall roughness in deep silicon etching," *Microsyst. Technol.*, vol. 6, no. 3, pp. 86–89, Feb. 2000.

[5] A. A. Ayon, B. Braff, C. C. Lin, H. H. Sawin, and M. A. Schmidt, "Characterization of a time multiplexed inductively coupled plasma etcher," *J.Electrochem. Soc.*, vol. 146, no. 1, pp. 339–339, Jan. 1999.

[6] Z. L. Zhang and N. C. MacDonald, "A RIE process for submicron, silicon electromechanical structures," *J. Micromech. Microeng.*, vol. 2, no. 1, pp. 31–31, Mar. 1992.

[7] J. W. Coburn and H. F. Winters, "Plasma etching—A discussion of mechanisms," *J. Vac. Sci. Technol.*, vol. 16, no. 2, pp. 391–403, Mar.1979.

[8] Y.-J. Chuang, F.-G. Tseng, and W.-K. Lin, "Reduction of diffraction effect of UV exposure on SU-8 negative thick photoresist by air gap elimination," *Microsyst. Technol.*, vol. 8, no. 4–5, pp. 308–313, Aug. 2002.

[9] J. X. Gao, L. P. Yeo, M. B. Chan-Park, J. M. Miao, Y. H. Yan, J. B. Sun, Y. C. Lam, and C. Y. Yue, "Antistick Postpassivation of High-Aspect Ratio Silicon Molds Fabricated by Deep-Reactive Ion Etching", Journal of Microelectromechanical systems, Vol.15, No.1, February 2006.

[10] J. Miao, R. Lin, L. Chen, Q. Zou, S. Y. Lim, S. H. Seah, "Design considerations in micromachined silicon microphones", Microelectronics Journal 33, pp. 21-28, Published by Elsevier Science Ltd.2002.

[11] Burhanuddin Yeop Majlis, SrMIEEE, Yuen-Yee Wong, and K. Sooriakumar, "Optimization of Deep Trench Etching Process for Silicon MEMS Structure Using Deep Reactive Ion Etching", *NSM 2005 Proc. 2005*, Kuching, Malaysia.

[12] N. Belov, N. Khe, "Using Deep RIE for Micromachining SOI Wafers", *2002 Electronic Components and Technology Conference, IEEE*, 2002.

[13] Glenn Beheim and Carl S. Salupo1, "Deep RIE Process for Silicon Carbide Power Electronics and MEMS", *Published in Materials Research Society Symposium Proceedings* Vol. 622 © 2000 Materials Research Society, 2000.

[14] T Akashil, Y Yoshimura, and S. Higashiyama, "Deep reactive ion etching of pyrex glass using a bonded silicon wafer as an etching mask", *IEEE*, 2005.

[15] T. Akushi, I M Kanamaru, A. Kazumu, Y Itou, I M Horino, K. Fukudu, T. Ishikawa, T. Huradu, and R. Okada' "Fabrication of a 35-channel optical scanner integrated by passive- self alignment using through-holes precisely formed by DRIE", *IEEE*, 2004.

[16] C.M. Waits, B. Morgan, M. Kastantin, R. Ghodssi, "Microfabrication of 3D silicon MEMS structures using gray-scale lithography and deep reactive ion etching", *Sensors and Actuators A* 119, 245–253, 2005.

Performance Improvement of OTFTs using Double Layer Insulator

Dong-Wook Park, Cheon An Lee, Keum-Dong Jung, Byeong-Ju Kim, Byung-Gook Park, *Member, IEEE,* Hyungcheol Shin, *Senior Member, IEEE,* and Jong Duk Lee, *Member, IEEE*

Inter-University Semiconductor Research Center (ISRC)
and School of Electrical Engineering, Seoul National University
San 56-1, Shinlim-dong, Gwanak-gu, Seoul 151-742, Republic of Korea
Phone: +82-2-880-7279, Fax: +82-2-882-4658, E-mail: principia31@daum.net

Abstract **Organic thin-film transistors (OTFTs) with improved performance are fabricated using SiO₂/cross-linked PVA double layer insulator. The improved field-effect mobility, on-current, off-current, on/off ratio, and subthreshold slope are 0.12 cm²/V·sec, 9.2×10^{-6} A, 2×10^{-12} A, 4.6×10^{6}, and 0.4 V/dec, respectively. In addition, a negligible threshold voltage shift is observed in the device. The performance improvements are attributed to good leakage characteristic of SiO₂ and high-k characteristic of cross-linked PVA. Based on the proposed OTFTs, an organic inverter working as high as 1 kHz is implemented.**

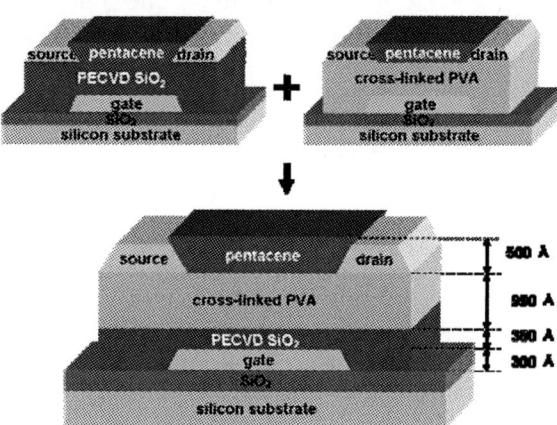

Fig. 1. Structure of fabricated bottom-contact pentacene OTFTs. The OTFT with double layer insulator is proposed to take the advantages of etch layer.

I. INTRODUCTION

Organic Thin-Film Transistors (OTFTs) are expected to have several advantages over inorganic TFTs. The greatest expectation is that the properties of organic materials permit the fabrication of high-performance OTFTs on flexible substrates using low-cost processing method. Thanks to the intensive investigation, a wide range of applications such as flexible displays, radio-frequency identification tags, and organic circuits are demonstrated [1-3].

In OTFTs, the insulator material is as important as the organic semiconductor. For the inorganic insulator, silicon dioxide (SiO₂) is broadly used because of its reliable performance such as low leakage characteristic. For the organic insulator, poly-4-vinylphenol (PVP), polyimide, and poly(vinyl alcohol) (PVA) are widely studied [4-6]. Among the various organic insulator, cross-linked PVA is one of the promising materials because it has a high dielectric constant and photosensitivity [6,7].

However, relatively large leakage current and hysteresis have been serious problems of cross-linked PVA [6,8]. In this work, to improve the leakage and hysteresis characteristic, the SiO₂/cross-linked PVA double layer insulator is proposed and the effect is analyzed.

II. DEVICE FABRICATION

The structure of fabricated bottom-contact OTFT is depicted in Fig. 1. The devices are started with an oxidized silicon substrate. The titanium (Ti) gate electrode is evaporated and patterned by photolithography and wet etch. For the double layer insulator, the SiO₂ is deposited by plasma-enhanced chemical vapor deposition (PECVD), and as a second insulator layer, 2 % PVA solution mixed with ammonium dichromate photo-sensitizer is spin-coated and cross-linked by UV exposure [7]. The gate dielectric is patterned by photolithography and

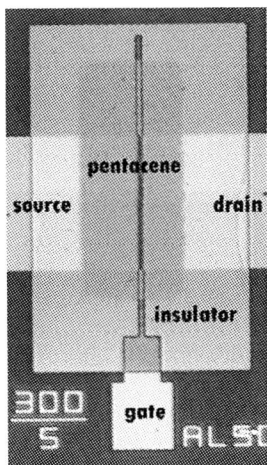

Fig. 2. Top view of fabricated OTFT. Each layer is patterned by using photolithography or shadow mask. The channel length and width are 5 μm and 300 μm, respectively.

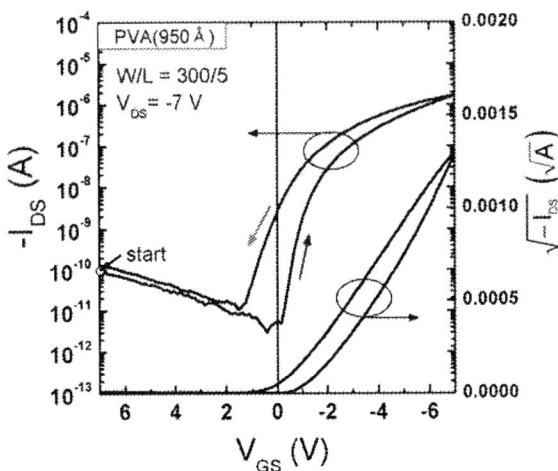

Fig. 3. Transfer characteristics of the OTFT with cross-linked PVA single layer insulator. Relatively large off-current and hysteresis are observed.

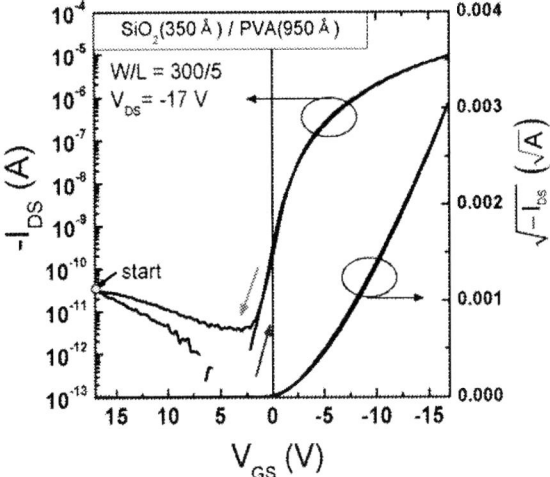

Fig. 4. Transfer characteristics of OTFT with SiO$_2$/cross-linked PVA double layer insulator. The reduced off-current and hysteresis are observed.

dry etch. The gold (Au) source/drain electrodes are patterned by photolithography and lift-off process. The device is completed using thermal evaporation of pentacene at high vacuum ambient of around 10^{-8} Torr. The substrate temperature is 80 ℃, and the active region is patterned by a shadow mask.

Fig. 2 shows the top view of completed OTFT. Each layer is patterned by using photolithography or shadow mask. The channel length and width are 5 μm and 300 μm, respectively. The typical I-V characteristics of the devices are measured with a probe station and semiconductor parameter analyzer 4156C (Agilent Technologies) in a light-shielded environment.

III. RESULTS AND DISCUSSION

The electrical parameters such as on-current, off-current, on/off ratio, subthreshold slope (SS) and field-effect mobility are extracted from the transfer characteristics. The field-effect mobility is calculated at the saturation regions from the slope of the plot of $|I_{DS}|^{1/2}$ versus V_{GS} curve based on the equation (1):

$$I_{DS}(sat) = \mu C_i \frac{W}{2L}(V_{GS} - V_T)^2 \quad \cdots \quad eq.\ (1)$$

where C_i is the capacitance per unit area of the gate dielectric, and V_T is the threshold voltage.

Fig. 3 shows the transfer characteristics (I$_{DS}$-V$_{GS}$ curve) of OTFT with cross-linked PVA single layer insulator. As a reference data, the on-current, off-current and the on/off ratio are 1.7×10^{-6} A, 1×10^{-11} A, and 1.7×10^{5}, respectively. In addition, the threshold voltage (V$_T$) shift

dependent on gate voltage sweep direction is observed. During the sweep V$_T$ shift of 1.2 V is measured. Because of the V$_T$ shift which might be originated from hysteresis of the device, the value of SS and field-effect mobility is hard to be extracted in this case.

On the other hand, Fig. 4 shows the reduced off-current and hysteresis using SiO$_2$/cross-linked PVA double layer insulator. The improved on-current, off-current and the on/off ratio are 9.2×10^{-6} A, 2×10^{-12} A, and 4.6×10^{6}, respectively. The V$_T$ shift is observed as 0.2 V which is nearly negligible. In this case, the extracted value of SS and field-effect mobility are 0.4 V/dec and 0.12 cm^2/V·sec.

The increment of on-current is attributed to role of cross-linked PVA which is known as

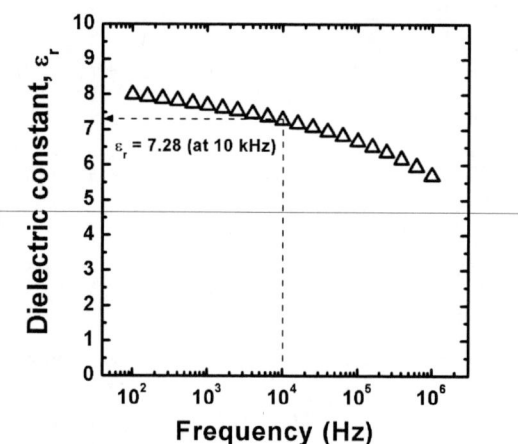

Fig. 5. Dielectric constant of SiO_2/cross-linked PVA double layer insulator. Relatively large dielectric constant of 7.28 is extracted at 10 kHz.

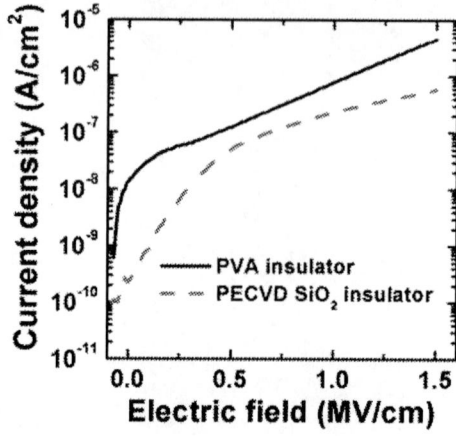

Fig. 6. Leakage current density measured in metal/insulator/metal (MIM) capacitor. The SiO_2 shows better leakage characteristics than that of cross-linked PVA.

high-k material. To verify the high-k characteristic, the dielectric constant of SiO_2/ cross-linked PVA double layer insulator is extracted from the measured capacitance. Fig. 5 shows the dielectric constant as a function of frequency. Relatively high dielectric constant of 7.28 is extracted at 10 kHz. Considering that the dielectric constant of SiO_2 is 3.9, the high-k characteristic of cross-linked PVA is confirmed from the result.

The reduced off-current is attributed to good leakage characteristic of SiO_2 layer. Fig. 6 shows the leakage current density of each insulator. The metal/insulator/metal (MIM) capacitors are fabricated for the measurement. The values are compared at the same electric field. The leakage current density of SiO_2 under 1 MV/cm is 2.3×10^{-7} A/cm^2 and that of cross-linked PVA is 7.8×10^{-7} A/cm^2. According to the results, SiO_2 makes a favorable comparison with cross-linked PVA in terms of leakage characteristics. To reduce the leakage current, it can be a possible method to adopt SiO_2 insulator as a supplementary.

In addition to the decreased off-current, the proposed OTFT shows considerably reduced hysteresis. Without SiO_2 layer, the OTFT shows large threshold voltage shift in the transfer curve as depicted in Fig. 3. However, the OTFT with SiO_2 layer shows the negligible threshold voltage shift as illustrated in Fig. 4. According to our research, the main mechanism of hysteresis is charge injection from the gate electrode to cross-linked PVA insulator [8]. In the proposed OTFT, the SiO_2 layer can block the injection charges, so that the hysteresis is considerably reduced.

An organic inverter is demonstrated using the proposed double layer insulator. Shown in Fig. 5(a) and 5(b) are the voltage transfer curve (VTC) and transient characteristics of the inverter. The inset of Fig. 7(a) is the schematic of inverter which consists of an enhancement-mode driver and an enhancement-mode load. The inverter is measured with a 1 kHz input clock frequency when the supply voltage is 15 V. The output voltage swing is around 10 V. The

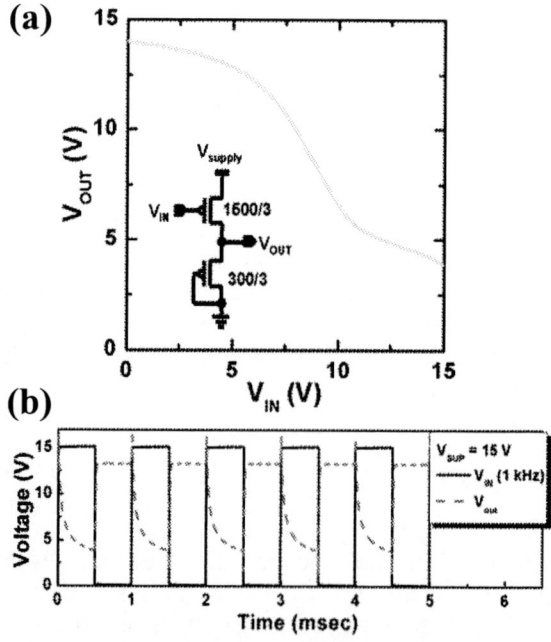

Fig. 7. Voltage transfer curve(a) and transient characteristics(b) of an organic inverter. The schematic diagram is depicted in the inset of fig. (a). The inverter is operated with a 1 kHz clock frequency.

operation frequency of 1 kHz is relatively fast compared to the results in literatures [3,9]. The double layer insulator described in this work can be used to improve the performance. However, the improvement of transient characteristic and output voltage swing is still required. The performance can be limited by low mobility of OTFT. Threshold voltage control is also important in organic circuits. Eventually, various studies on organic devices and circuits are expected to lead to organic electronics with sufficient performance.

IV. SUMMARY

Organic TFTs showing improved performances are fabricated. The performance improvement is attributed to the high-k characteristic of cross-linked PVA insulator and good leakage characteristic of SiO_2 insulator. Finally, an organic inverter working at a clock frequency of 1 kHz is demonstrated.

ACKNOWLEDGEMENT

This work was supported by "Samsung SDI - Seoul National University Display Innovation Program".

REFERENCES

[1] M. Mizukami, N. Hirohata, T. Iseki, K. Ohtawara, T. Tada, S. Yagyu, T. Abe, T. Suzuki, Y. Fujisaki, Y. Inoue, S. Tokito, and T. Kurita, "Flexible AM OLED Panel Driven by Bottom-Contact OTFTs," *IEEE Electron Device Lett.* Vol. 27, no. 4, pp. 249-251 (2006).

[2] P. F. Baude, D. A. Ender, M. A. Haase, T. W. Kelley, D. V. Muyres, S. D. Theiss, "Pentacene-based radio-frequency identification circuitry," *Appl. Phys. Lett.* Vol. 82, pp. 3964-3966 (2003).

[3] B. Yoo, A. Madgavkar, B. A. Jones, S. Nadkarni, A. Facchetti, K. Dimmler, M. R. Wasielewski, T. J. Marks, and A. Dodabalapur, "Organic Complementary D Flip-Flops Enabled by Perylene Diimides and Pentacene," *IEEE Electron Device Lett.* Vol. 27, no. 9, pp. 737-739 (2006).

[4] H. Klauk, M. Halik, U. Zschieschang, G. Schmid, W. Radlik, W. Weber, "High-mobility polymer gate dielectric pentacene thin film transistors,*" J. Appl. Phys.* Vol. 92, pp. 5259-5263 (2002).

[5] S. Pyo, H. Son, K. Choi, M. Yi, S. Hong, "Low-temperature processable inherently photosensitive polyimide as a gate insulator for organic thin-film transistors," *Appl. Phys. Lett.* Vol. 86, pp. 133508 (2005).

[6] Th. B. Singh, N. Marjanovic, G. J. Matt., N. S. Sariciftci, R. Schwodiauer and S. Bauer, "Nonvolatile organic field-effect transistor memory element with a polymeric gate electret," *Appl. Phys. Lett.* Vol. 85, pp. 5409-5411 (2004).

[7] C. D. Sheraw, J. A. Nichols, D. J. Gundlach, J. R. Huang, C. C. Kuo, H. Klauk, T. N. Jackson, M. G. Kane, J. Campi, F. P. Cuomo, B. K. Greening, "An Organic Thin Film Transistor Backplane for Flexible Liquid Crystal Displays," *58th Device Research Conference(DRC) Conf. Dig.*, pp. 107 (2000).

[8] C. A. Lee, D. W. Park, S. H. Jin, I. H. Park, J. D. Lee, and B.-G. Park, "Hysteresis mechanism and reduction method in the bottom-contact pentacene thin-film transistors with cross-linked poly(vinyl alcohol) gate insulator," *Appl. Phys. Lett.* Vol. 88, pp.252102 (2006).

[9] H. Klauk, M. Halik, U. Zschieschang, F. Eder, D. Rohde, G. Schmid, and C. Dehm, "Flexible Organic Complementary Circuits," *IEEE Tran. Elec. Dev.* Vol. 52, pp.618-622 (2005)

Simulation of a Novel Lateral H Structure PIN InGaAs Photodiode with Consistent Electron Drift Length

Esther Loo Chee Hong, Sahbudin Shaari and Burhanuddin Yeop Majlis
Institute of Micro Engineering and Nanoelectronics (IMEN)
Universiti Kebangsaan Malaysia
43600 UKM Bangi, Selangor, MALAYSIA
Email: esther@vlsi.eng.ukm.my, sahbudin@vlsi.eng.ukm.my

Abstract: We investigated the microscopic view of electron captured in a lateral PIN InGaAs photodiodes which was implemented on an H electrode structure. The results illustrated a near ideal generated current with minimal delay time. We assumed the structure was implemented in a 10Gbps, 30Gbs and 50Gbps network with a PIN sensitivity of 30dBm with incident light power at 0.1mW. The sizes of the electrodes range from 7μm to 8μm at 0.8 and 0.5 intrinsic-junction ratio respectively. The fastest response 1,210,900,945 A/W of the device was obtained through a simulator study.

I. INTRODUCTION

Simulation and modeling are well established methods in producing near ideal parameters for new devices without compounding on cost and time. However, available simulators were too dependant on the devices material properties to expound on the possibilities of utilizing new conceptual electrode structures. Despite that, many breakthrough have been done to increase the efficiencies of the PIN photodiodes without resorting to extreme changes in electrodes structures, chief among them are the creations of lateral trench photodetectors [1], side illuminated PIN photodetectors [2] and so forth.

Nonetheless, a remodified version of a space saving fractal mathematical structure were tested out in a simulated environment of an in-house program, H-Gen. H-Gen was created by using MATLAB with a strong emphasis on the generation of the H-electrodes. Analytical calculation was performed on microscopic level in order to fully justify the usage of such a radical design.

The need for high speed optical communications is funneling the research and development of novel photodiodes, light sources and interconnects [2, 6]. However PIN structures are among the most common structures for photodiodes. It comprises of a structure with an n intrinsic region sandwiched between an electron rich and holes rich junctions. PIN diodes have small collecting areas, low dark current and extremely high bandwidths ($\approx$10GHz), due to shorter gap transit time [3]. By applying a reverse bias voltage into the structure, the resulting depletion region enables incident light to generate current. Among the popular PIN electrodes design were the interdigitated design which incorporates intervening $p+$ fingers with $n+$ fingers as was illustrated in Fig 1. Advantages of implementing such a design were its efficient usage of the active area. Nevertheless, the bandwidths of interdigitated metal-semiconductor-metal (MSM) detectors and lateral PIN photodiodes are still limited by the absorption layer thickness as the electric field intensity decays rapidly below the shallow or surface electrodes [2,4].

Therefore, the H structure comes into the picture, Fig. 2; with its efficient methods of maximizing active space and a sponge-like ability to capture incident light manages to create a condition of near consistent drift length throughout the device. This in turn produces a unique set of data which contradict conventional structures with an impressive response factor [7]. However, unlike Djuric and Krstajic [6], we did not take diffusion into account in order to simplify the analytical model.

0-7803-9730-4/06/$25.00 ©2006 IEEE

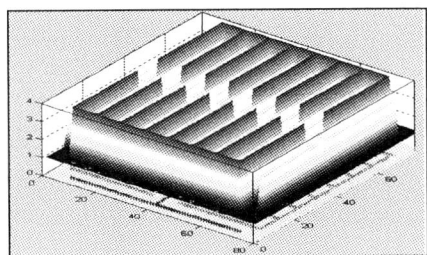

Fig. 1 : A 3D model of an interdigitated PIN photodiode. The red electrodes are the positive electrode while the yellow is the negative electrode.

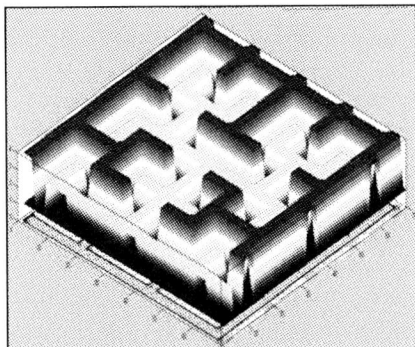

Fig. 2 : A 3D model of an H space PIN photodiode with the complexities factor of 1.

II. SIMULATION MODELS

An in house program, H-Gen was created to test out the idea of H-space electrodes whereupon H-space electrodes were created based on a mathematical model (1),

$$f(x), f(y)|_{P,N} = \pm \Delta\varphi\left(\sum_{N=0}^{n}\left(C_N x^N, B_N y^N\right) - \left(f_m(x), f_m(y)\right)\right) \quad (1)$$

with C_N and B_N as sign states based on the H boundry of (2) and (3) while the functions $f_{in}(x)$ and $f_{in}(y)$ are remodified Hilbert functions which could diverges into infinity. Therefore m is the boundary constraint which controls the complexities of the divergence.

$$C_N = \begin{cases} -1 & 0 < x < \dfrac{N}{4} \\ +1 & \dfrac{N}{4} < x < \dfrac{N}{2} \\ +1 & \dfrac{N}{2} < x < \dfrac{3N}{4} \\ -1 & \dfrac{3N}{4} < x < N \end{cases} \quad (2)$$

$$B_N = \begin{cases} -1 & 0 < x < \dfrac{N}{4} \\ -1 & \dfrac{N}{4} < x < \dfrac{N}{2} \\ +1 & \dfrac{N}{2} < x < \dfrac{3N}{4} \\ +1 & \dfrac{3N}{4} < x < N \end{cases} \quad (3)$$

$$f_m(x) = \sum_{n=1}^{N}\left(\sum_{n=1}^{N-1} 2(x_n+1)(x_n^{13}+x_n^{8}+2x_n^{6}+3x_n^{4}+3x_n^{2}-x_n+3)\right)^4 \quad (4)$$

$$f_m(y) = \sum_{n=1}^{N}\left(\sum_{n=1}^{N-1} 2(y+1)(3y_n^{14}+2y_n^{12}-y_n^{10}-y_n^{9}+y_n^{8}+y_n^{6}-y_n^{5}+y_n^{4}+2y_n^{2}+3)\right)^4 \quad (5)$$

N=1,2,..m.

In this case, we set the complexities of the H electrodes as m=4. This gives us an ideal active area of 1mm^2. The formula of obtaining the length of the device is as shown in (6) with $\Delta\Phi$ as the intrinsic-juction electrodes ratio while N_{in} and N_{ele} represents the number of intrinsic region and electrodes per single side. Since m, the complexities factor is 4, N_{ele} and N_{in} are 65 and 64 respectively.

$$N_{ele}(d_{size})+N_{in}(d_{size}/\Delta\Phi) = D \quad (6)$$

The incident light used here are coherent laser light operating at a network speed of 10Gbps, 30Gps and 50Gps respectively at the distance of 50km. We assumed the net power of the light were 0.1mw by the time it appears on the device. To obtain the proper configuration for the incident light, we utilized formula (7) to obtain the time for a single pulse, T with BPS as network speed used in the simulations. Since we are only using ½ of a single pulse set, Δt = ½ T. Therefore to obtain the number of frames the Δt generates, we divide Δt with the frame period of 3.00×10^{-14}s since the frequency of the frames interval were assumed to be operating at 300THz. This is as illustrated in Fig 3.

$$T = 1/(BPS) \quad (7)$$

Table 1. Types of network and the amount of photons generated by a single pulse of 0.1 mW.

Network speed	N (Nframe × n)	Δt (ps)	Nframe
10Gps	31,207	50	1667
30Gps	10,423	16.7	507
50Gps	6,241	10	300

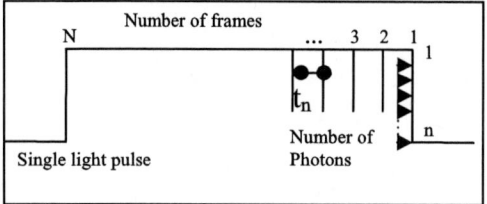

Fig. 3. Model of a light pulse. With n number of photons per frames for N frames.

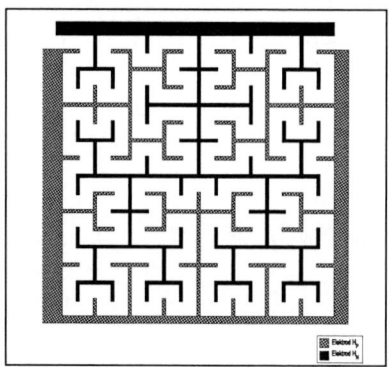

Fig. 4. A model of the H electrode with a 0.8 intrinsic-junction ratio. The red structure here represents the positive electrodes while the black represents the negative electrodes.

Fig. 5. A model of the H electrode with a 0.5 intrinsic-junction ratio. The red structure here represents the positive electrodes while the black represents the negative electrodes.

Since we assumed the power of the incident light is 0.1mW, the number of photons, N was easily obtained by using formula (8) with hv equivalent to 1.60218×10^{-19}. The parameters are shown in Table 1.

$$P = \frac{Nhv}{\Delta t} \qquad (8)$$

The light pulse is then directed into the H electrode device. The H electrode structure was placed on an $InGa_{0.47}As_{0.53}$ substrate with a donor concentration of 10^{19} and an acceptor concentration of 10^{16}. The reverse bias voltage used in here is 3.3V. The sizes of the electrodes used were 7μm to 8μm at 0.8 and 0.5 intrinsic-junction ratio respectively as illustrated in Fig 4 and Fig. 5.

In order to increase the accuracy of the readings, the simulator was created to be time and space dependant. Finally, a set of current-time graphs were analyze accordingly.

III. RESULTS

The H electrode structure with the complexities factor of 4 is as illustrated below in Fig. 6 while Fig. 7 depicts the flux lines that was generated between the electrodes and acted as a drift guide to the electrons and holes. The incident light applied here was as shown in Fig. 8. The set of incident light of the device is shown in detail in Fig. 9.

Each model is recorded accordingly in Table 2 with I-J ratio stands for intrinsic-junction ratio. The current generated by the self same light were duly analyzed in subsequent current-time graphs, Fig. 10, 11, and 12 for the two sets of electrodes configurations in the same network.

From the graphs, we can determine the responsitivities and time delay of each of the devices through the slopes and the width of each of the models. The results are recorded accordingly in Table 3 and Table 4

Table 2 Sets for each simulated InGaAs PIN photodiode.

I-J Ratio	Network		
	10Gbps	30Gbps	50Gbps
0.5	Model 1	Model 3	Model 5
0.8	Model 2	Model 4	Model 6

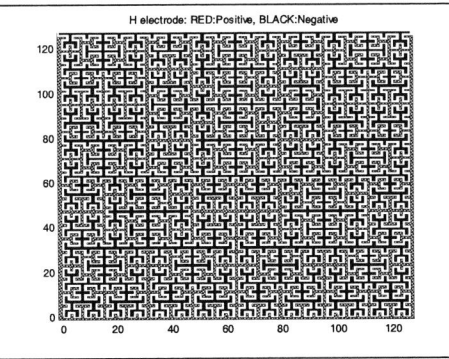

Fig. 6 : A 2D structure of the H electrodes with the complexities of 4.

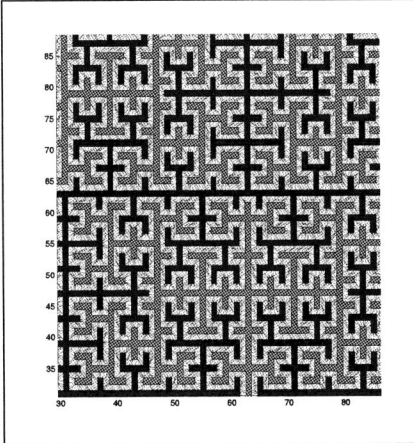

Fig. 7: Flux plot of the electric field generated between the electrodes acted as a guide to the incident light.

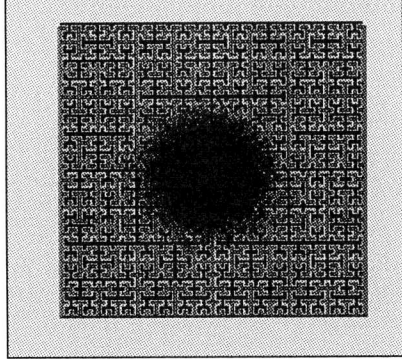

Fig. 8 : An example of incident light on the device with 0.5 I-J ratio.

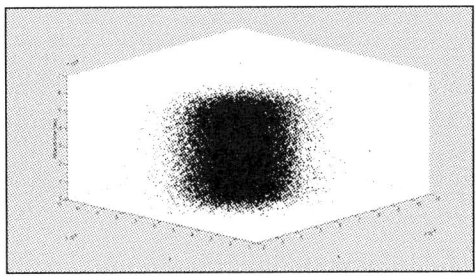

Fig. 9 :. A 3D view of the incident light pulse.

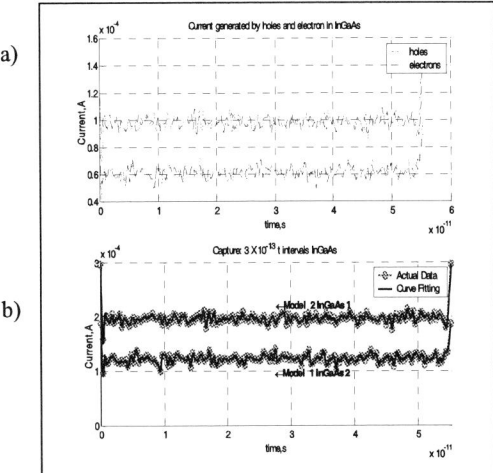

Fig. 10 : The current time plot for Model 1 and Model 2 based on Table 2. a) The current generated in the electrodes by the incident holes and electrons. b) The combination of the current generated by holes and electrons, with the comparisons between two models for the 10Gbps network.

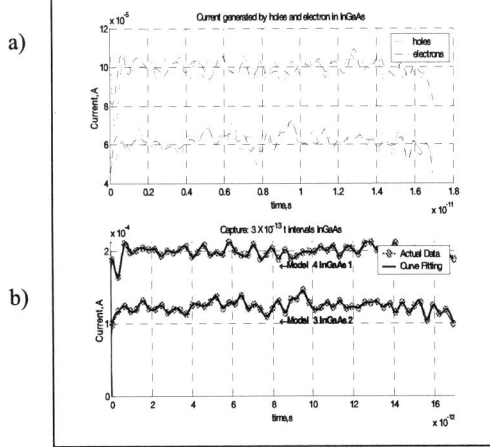

Fig. 11 : The current time plot for Model 3 and Model 4 based on Table 2. a) The current generated in the electrodes by the incident holes and electrons. b) The combination of the current generated by holes and electrons, with the comparisons between two models for the 30Gbps network.

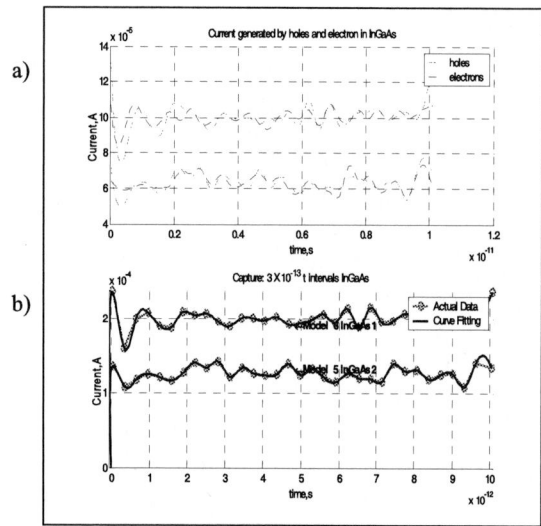

Fig. 12 : The current time plot for Model 5 and Model 6 based on Table 2. a) The current generated in the electrodes by the incident holes and electrons. b) The combination of the current generated by holes and electrons, with the comparisons between two models for the 50Gbps network.

Table 3 The response and pulse width for each models' generated current.

Models	Pulse width (s)	Response (A/s)
1	5.585×10^{-11}s	501,376,306
2	5.585×10^{-11}s	1,896,776,608
3	1.706×10^{-11}	372,292,682
4	1.730×10^{-11}	811,804,878
5	10.170×10^{-12}	924,400,199
6	10.380×10^{-12}	1,210,900,945

Table 4 Time difference between the current pulse width and the incident light pulse.

Models	Light pulse (s)	Time difference
1	5.548×10^{-11}	0.037×10^{-11}
2	5.548×10^{-11}	0.037×10^{-11}
3	1.685×10^{-11}	0.021×10^{-11}
4	1.685×10^{-11}	0.045×10^{-11}
5	9.950×10^{-12}	0.022×10^{-11}
6	9.950×10^{-12}	0.043×10^{-11}

From Tables 3 and 4, we noticed that the broadening of pulses is extremely short which is in the range of 0.002 ps to 0.004ps. Apart from that, the amazing response shown on Table 3 clearly demonstrates the effectiveness of the H electrodes.

IV.   CONCLUSIONS

Since, the H electrodes were a subset product of the fractal model; we discovered a fair consistency in the drift distance of the electrodes as the electrodes fingers were placed generously on the active surface of the device. The order of placement which adhere strictly to the H boundary rules enable the generation of finite size H electrodes. Therefore, in a certain sense, the H electrodes were merely a combined multitude of interdigitated electrodes of various sizes.

Nevertheless, from the current-time graph generated, there was a consistency in the generation of current from the incident light. H electrodes clearly exhibit a rare sponge like effect of capturing incident light and despite such novelty, the production of the electrodes are not impossible. In fact, since the electrodes were created for lateral PIN photodiodes, simple planar fabrication techniques are suffced to produce a workable model.

However, the drawback of using this design is the apparent complexities in the electrodes whereupon a single tear or break on any of the fingers will affect the overall output immensely. Therefore, the H-Gen simulator plays an important role in producing simulated results in real-time environment.

V.   REFERENCES

[1] Yang, Min, Kern Rim, Rogers, D.L. et al., "A High Speed, High Sensitivity Silicon Lateral Trench photodetector", IEEE Electron Device Letters, Vol. 23, N0.7, July 2002. P. 395-397.

[2] Magnin,V., Giraudet,L., Harari, J. et al, "Design, optimization, and fabrication of side-illuminated p-i-n photodetectors with high responsivity and high alignment tolerance for 1.3 and 1.55µm wavelength use", Journal of Lightwave Technology, Vol. 20, No. 3, March 2002. P. 477-488.

[3] Sze, S.M., "Physics of Semiconductor Devices", 2nd ed. New York: Wiley,1981.

[4] Karp, Sherman and Gagliardi, R.M., "Optical Communications", 2nd ed. New York: Wiley, 1995.

[5] Kyoung-Hwan Oh, Seong-June Jo, Soo-Kang In et al, "Improved speed performance of tensile-strained $In_xGa_{1-x}As$ (x<0.53) photodetectors utilizing a patterned substrate growth.", Indium phosphide and Related material Conference, IPRM. 2002 P. 135-138.

[6] Djuric, Z., Krstajic, P., Smiljanic, M. et al., "Influence of Carrier Diffusion on a Response of a Resonant- Cavity Enhanced Detector", IEEE Journal of Quantum Electronics, Vol. 30. No.2 Feb. 2002, P.197-201.

[7] Loo,E; Shaari, S. Majlis, B.Y., "Novel modification of an interdigitated lateral pin photodetector electrode design based on photon absorption characteristics.", Proceeding SPIE - The International Society for Optical Engineering, August 2005. Vol. 5881. P 189-195.

Distributed CATV Inputs in FTTH-PON System

Tan Fent Fent, Sahbudin Shaari, *Member, IEEE* and Burhanuddin Yeop Majlis, *Member, IEEE*
UKM-TM Microelectronics Research Centre
Faculty of Engineering
Universiti Kebangsaan Malaysia
43600 Bangi, Selangor, MALAYSIA
Email: fent@vlsi.eng.ukm.my , sahbudin@vlsi.eng.ukm.my

Abstract **This is about CATV system on FTTH-PON network with a new idea dealing with multiple input points of the video broadcasting station by using the same wavelength of 1550 nm. The externally modulated intensity of different sources does not produce any interference that became an advantage of this system. The main purpose of this study is to determine the possibility of broadcasting of multiple inputs CATV with the lengths of 20 km despite the only video input.**

I. INTRODUCTION

Fiber-to-the-home or FTTH - A system growing so rapidly that fulfills the high demand of users despite of it's generally too expensive deployment. The expensive optical transceiver at each subscriber and the high costs of the headend, which includes the outside plant, are the main reason for the high cost of FTTH including CATV. The low-power budget system enables the network to be shared by approximately 16 to 32 subscribers [1]. FTTH network also show that fiber optic is becoming reality with the ability to deliver triple play services (voice, video and data) including CATV over a passive optical network (PON) distribution via a single optical fiber to the home.

CATV or cable TV that formerly known as community antenna television is one of the promising services in FTTH due to its low cost and compatibility with current TV sets. Besides, multi-channel CATV distribution system can be easily applied by using sub-carrier modulation (SCM) techniques in which television signal can be directly launched into TV set input after

optoelectrical detection [2]. In other words, the subscriber TV can directly connect to the video service through single 75ohm coaxial connector available in ONT.

So far, CATV distribution is collected from various sources such as broadcast cable television and direct broadcast satellite (DBS) television and delivered to the FTTH network from the central office (CO) of the system. In this paper, there is a bold attempt where CATV distribution is not only be transmitted from the CO of the system but at any point of the feeding line by creating more than one entry point. This entry point can be of any point between the CO and the receiver such as ONU, ONT or passive splitter of the system. This paper is reporting the performance study at the video channels from multiple CATV input at various feeding point. Multiple CATV inputs of several channels with different radio frequencies are fed into various points of the feeding line simultaneously onto the same wavelength of 1550 nm.

II. THEORY

There are three types of television standards throughout the world that includes NTSC, PAL, and SECAM. These standards can vary from one country to another and the three are not compatible with each other. For instance, Malaysia is using PAL B standard while the United States is using NTSC standard. However, a multichannel distribution system in fiber optic uses the conventional amplitude modulated vestigial sideband (AM-VSB) with subcarrier multiplexing (SCM) because this is a standard format for

0-7803-9730-4/06/$25.00 ©2006 IEEE

television signal in which the channels can be directly launched at TV set input after optoelectrical detection [3].

CATV system that may follow any of the standards has driven an enormous market despite the strict noise and distortion requirement for multichannel AM-VSB or frequency-modulated (FM) video transmission system [2]. For instance, the unwanted spurious frequency components (distortions) generated in multichannel analog CATV transmission system due to the nonlinearities of transmitters, receivers, amplifiers and other operating devices [4].

Ever since multichannel is easy to achieve by using SCM techniques, each and every analog video channel of 6 MHz is upconverted to a narrow channel at high frequency [2] of radio frequency (RF) range. The upconversion can be any type of the analog or digital modulation applied. The resulting subcarriers signals will be allocated separately in the RF and combined using a microwave power combiner. The electrical signals are added to the electro-optic device where the laser diode is biased to intensity modulate. This process is done in the OLT together with other services by WDM techniques. Hence, CATV service is easily broadcasted to the user via single fiber cable and up to the length of 20 km. At the same time, the end user can receive the video channels with the compatible TV sets where it comes with appropriate demodulation and filtering.

Figure 1 illustrates a typical FTTH-PON system. The network access from Central Offices is connected to the remote end by optical splitter with downstream of 1490/1550nm and upstream of 1310nm. CATV system is a unidirectional service and uses only downstream of 1550 nm to broadcast TV channels to users. At the receiver or ONU, the optical signal is converted back into electrical signals and distributed in PON topology where user will receive the video channel with proper filtering (tuning).

III. PROPOSED MULTI-CATV INPUT

Multi-CATV input at any point of the feeding line is using exactly the conventional method of FTTH-PON as described above. First of all, groups of CATV broadcasting stations are formed. In this case, four groups will be formed with four channels each. The formation of a group is easy using the simplest techniques of SCM to combine four channels of different RF (radio frequency) range. Each video channel of 6 MHz must modulate a different RF sub-carrier in order to travel simultaneously. A continuous-wave laser diode is then used to drive the electrical signal to optical signal travelling in 1550nm is combined into a Mach-Zehnder modulator so as to exploit the interference and the fringe pattern that resulted from optical path differences. This combination of devices is considered as per entry point to the feeding line as illustrated in Figure 2.

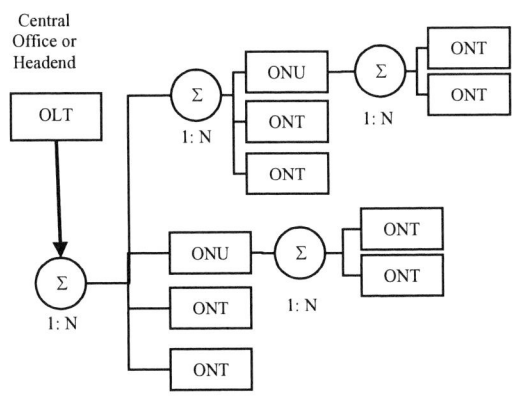

Fig 1 FTTH-PON System

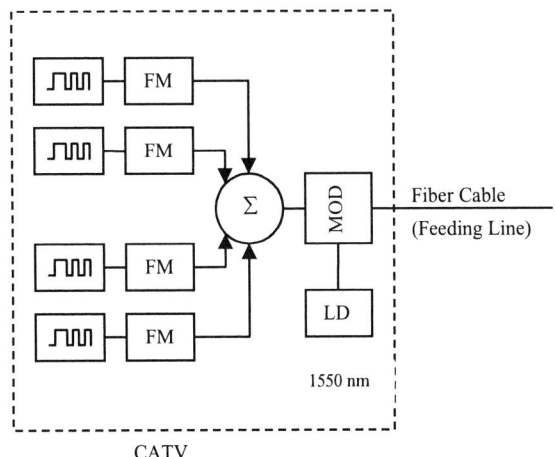

Fig 2 CATV created from four single channel

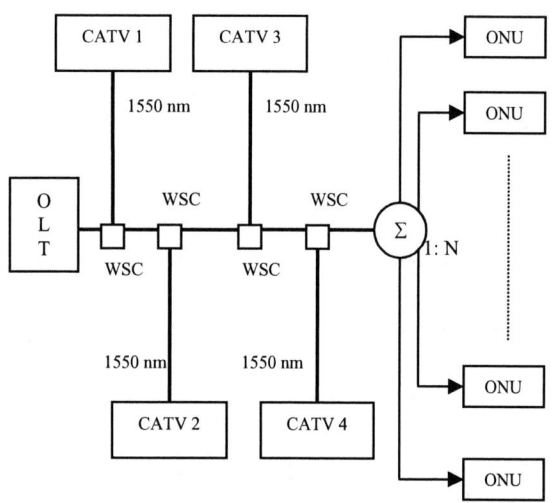

Fig 3: Multi-CATV Inputs in FTTH-PON System

Multiple entry points of CATV broadcasting are achieved by feeding this combination at a desired point along the feeding line using Wavelength Selection Coupler (WSC) within the desired length of 20 km. This is as illustrated in Figure 3. Although there is more than one CATV broadcasting station occurs in this system, it is identical to previously described CATV system because it uses only one wavelength of 1550nm. It is also a plus point to this system whereby the signals in each CATV has an independent external modulation. The CATV acts as an independent broadcasting station that feeds into the feeding line between the CO and OLTs or ONUs. This is certainly very convenient and flexible for broadcasting station that is nearer to the end user but further away from the CO.

Additionally, the nearer the broadcasting station placed to the OLT the better signal quality can be obtained.

IV. RESULTS AND DISCUSSION

Figure 3 shows an FTTH-PON network with multi-CATV inputs that is coupled into the fiber optic line by using WSC. The network is then simulated. The specific devices used for the

simulation are such as distributed feedback laser (DFB) and Mach-Zehnder modulator to bias the signal into 1550 nm, single mode fiber (CorningSMF28_1550) as the medium of transmission, and PIN Photodiode as the receiver. Besides, other components such as modulator, demodulator, filters, combiner, and splitter are also being used in the simulation. Each of these components parameters are designed in order to permit the input signals to be transmitted and received over a given distance with an acceptable signal quality. It is necessary that each and every channel of the CATV system is assigned with different carrier frequency as allocated in radio frequency range.

First of all, we obtain the transmitted electrical power spectrum of four channels from a group. The result is as illustrated as Figure 4.

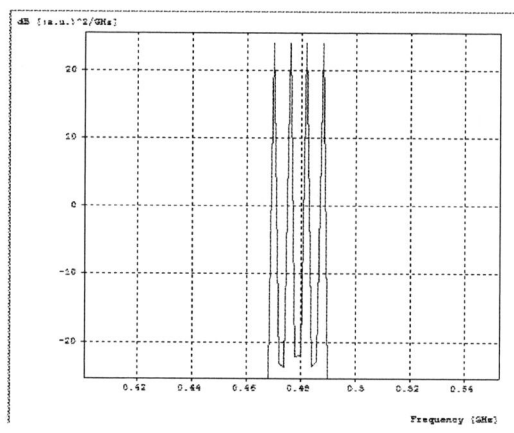

Fig 4 Electrical power spectrum of transmitted signal from a group of four channels

From the figure, we observed the electrical power spectrum generated from CATV by linearly combining four AM-VSB. It shows clearly all four electrically modulated channels of different radio frequencies of 470 – 494 MHz. These signals will be optically modulated by using a DFB and a Mach-Zehnder modulator at 1550 nm and transmitted through a SMF transmission medium. This process is repeated in every CATV as in Figure 3 before it is fed into the fiber optic cable.

The results of the optical spectrum of the system are observed to be varying with the distance and after combining the feeding line due to the fiber loss and coupling are shown in Figure 5. It illustrates the optical spectrum of the system at the end of the transmission before the pin diode. The optical spectrum increases as it is coupled into the fiber line due to overlapping with other CATV signals at the same wavelength 1550 nm.

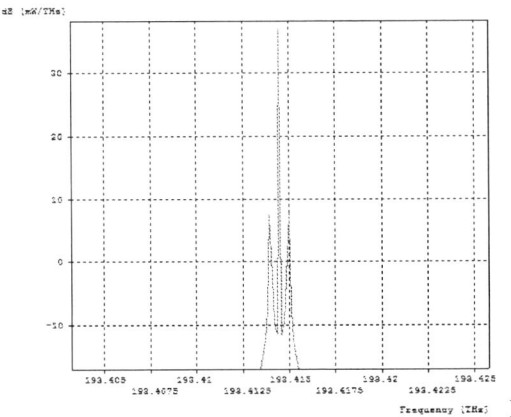

5 Optical Spectrum of all the transmitted channels

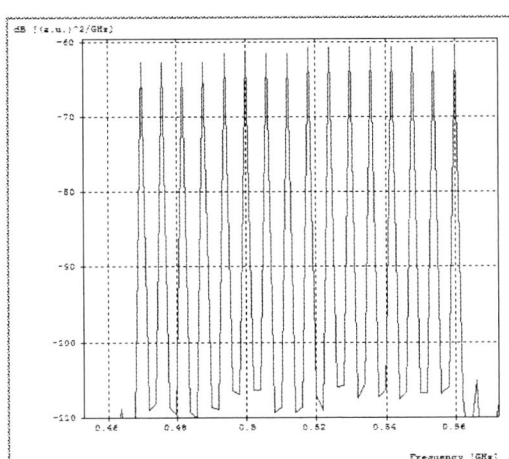

Fig 6 Power Spectrum of received signals after the photodetection

Figure 6, shows all 16 channels received after the optoelectrical detection. All the results obtained are purely the non-amplified system. In order to get a better power for the system, erbium doped fiber amplifier (EDFA) is suggested to be used in future.

V. CONCLUSION

We have presented the result of a multi-CATV input by using FTTH-PON system and SCM techniques. It is shown that more than one broadcasting station can be fed into the feeding line at different point by using one optical carrier i.e.at 1550 nm. This is an idea which can be applied toward setting up a number of CATV stations along the FTTH-PON access feeding lined rather than a single CATV station or input near thecentral office. However, it is also applicable in a network with long fiber lines. CATV with similar operating wavelength will be so cost effective in terms of services and maintenance and this will reduce the overall cost for video signals to reach homes or end user in FTTH access network.

REFERENCE

[1] Anton H. H. Tan, "SUPER PON – A Fiber to the Home Cable Network for CATV and POTS/ISDN/VOD as Economical as a Coaxial Cable Network," *IEEE J. Lightwave Technol,* vol. 15, pp. 213-218, Feb. 1997.

[2] Thomas E. Darcie, "Subcarrier Multiplexing for Lightwave Networks and Video Distribution Systems," *IEEE J. on Selected Areas in Communication,* vol. 8, pp. 1240-1248, Sept. 1990.

[3] M. Jaworski, M. Marciniak, "Counteracting of Stimulated Brilliouin Scattering in Externally Modulated Lightwave AM-CATV Systems," *IEEE LFNM,* pp. 71-73, May 2000.

[4] C. K. Chan, L. K. Chen, "A Correction Scheme for Measurement Accuracy Improvement in Multichannel CATV systems," *IEEE Transaction on Broadcasting,* vol. 42, pp. 122-129, June 1996.

[5] R. Gross, R. Olshansky, "Multichannel Coherent FSK Experiments Using Subcarrier Multiplexing Tecniques," *IEEE J. Lightwave Technol,* vol. 8, pp. 406-415, Mar. 1990.

[6] R. Hui, B. Zhu, R. Huang, C. Allen, K. Demarest, and D. Richards, "Subcarrier Multiplexing for High-Speed Optical Transmission," *IEEE J. Lightwave Technol,* vol. 20, pp.417-426, Mar. 2002.

On the use of a Mixed-Mode Approach For MEMS Testing

Md. Fokhrul Islam, *Student Member, IEEE* and M. A. Mohd. Ali, *Member, IEEE*
Department of Electrical, Electronic and System Engineering
Faculty of Engineering
Universiti Kebangsaan Malaysia
43600 Bangi, Selangor, MALAYSIA
Email: engr_tutul96@yahoo.com , mama@vlsi.eng.ukm.my

Abstract — **In the testing environment, test patterns are generated using techniques such as exhaustive, pseudo-random, deterministic and weighted random testing. Using deterministic testing technique, huge amount of memory space and lengthy testing time are required to generate and store large number of test patterns. On the other hand, pseudo-random technique reduces the number of test patterns but cannot achieve complete fault coverage. Hence primitive polynomial linear feedback shift register (LFSR) based pseudo-random and deterministic techniques have recently been proposed to be used simultaneously. This has been referred to as the mixed-mode approach. This paper introduces the adaptation of the mixed-mode test technique for MEMS testing.**

I. INTRODUCTION

The development of testing methodologies for MEMS is extensively making use of the results obtained in the standard IC's test field [1]. The increasing need for multi-functional sensor and actuator systems that are capable of real-time interaction with both electrical and non-electrical environments has led to the development of a very broad class of "Microsystems". Microsystems are heterogeneous since they are based on the interactions of multiple energy domains that can include electrical, mechanical, optical, thermal, chemical and fluidic. Reliable manufacture of affordable microsystems naturally requires the use of cost-effective test methods that distinguish malfunctioning systems from good ones. The multi-domain nature of microsystems makes them inherently complex for both design and test. The growing use of microsystems in life-critical applications such as air-bags, bio-

sensor and communications creates a significant need for high reliability that cannot be achieved without the use of robust test methods [2].

Most MEMS sense physical signals (acceleration, force, pressure, radiation …) and convert them into electrical signals processed by the associated electronics. In most cases, this operation is described by means of a first or second-order transfer function, that is linear to a large extent, and that can be verified using a mixed-mode test scheme. Since MEMS are essentially analog devices, we are approaching their test as an extension of the field of analog and mixed-signal electronic testing [3]. This paper introduces the application to MEMS testing of the mixed-mode test methodology which is adaptation of the previously introduced technique for mixed-signal devices [4].

II. THE TEST METHOD

The technique is based on impulse response (IR) evaluation using mixed-mode test sequences. The IR of a system provides enough information about the functional evaluation of the system. From theory we know that the output of a system is equal to the input signal convoluted with the system IR. These facts were exploited in [5] in order to present a pseudo-random testing approach for mixed-signal circuits.

In Fig. 1 and in the associated equations we may describe mathematically how to get the IR of the digital MEMS under test (MUT). The input to the digital MUT is mixed mode patterns (MMP) signal $x(n)$ in Equ. 1. The auto correlation value of $x(n)$ is given in Equ. 2 and we may well state that it is almost equal to a perfect impulse as shown in Equ. 3. Equ. 4 shows that the output of a system is equal to the input signal convoluted with the system IR. The expression of the cross correlation of the input signal $x(n)$ with the output signal $y(n)$ is presented in Equ. 5. We obtain the desired result

0-7803-9730-4/06/$25.00 ©2006 IEEE

h(n) by substituting the auto correlation value of x(n) in Equ. 5.

$$x(n) = MMP \qquad (1)$$

$$\Phi_{ss}(n) = 1 \qquad (n = 0) \qquad (2a)$$

$$\Phi_{ss}(n) = -\frac{1}{L} \qquad (0 < n < L) \qquad (2b)$$

$$\Phi_{ss}(n) \approx \delta(n) \qquad (3)$$

$$y(n) = x(n) \otimes h(n) \qquad (4)$$

$$\Phi_{sy}(n) = x(n)\Phi y(n) =$$
$$= x(n)\Phi[x(n) \otimes h(n)] =$$
$$= [x(n)\Phi x(n)] \otimes h(n) =$$
$$= \Phi_{ss}(n) \otimes h(n) \qquad (5)$$

$$\Phi_{sy}(n) \approx h(n) \qquad (6)$$

Fig. 1 Mathematical derivation of IR.

The architecture of our mixed-mode testing technique consists of three steps: 1) modeling the MUT as a digital system by embedding the MUT between a DAC and an ADC, 2) applying digital mixed-mode test patterns to the modeled digital system, and 3) constructing the signature set (the cross correlation between the input and the output responses) for classification of the faulty and fault-free circuit. Fig. 2 shows the overall block diagram of the mixed-mode test approach.

Pseudo-random testing of MEMS has been introduced earlier in [6]. Unfortunately, random-pattern-resistant (r.p.r.) faults which limit the fault coverage cannot be detected with this approach. In the proposed mixed-mode testing, primitive polynomial linear feedback shift register (LFSR) based pseudo-random and deterministic test patterns will be used simultaneously. The deterministic patterns will be used to detect the r.p.r. faults that the pseudo-random patterns miss. Fig. 3 shows an equivalent circuit of an RF MEMS filter that will be used as a test bench to assess the effectiveness of the proposed technique. In order to evaluate the percentage of misclassification, an evaluation set of faulty and fault-free instances will be generated for the equivalent circuit.

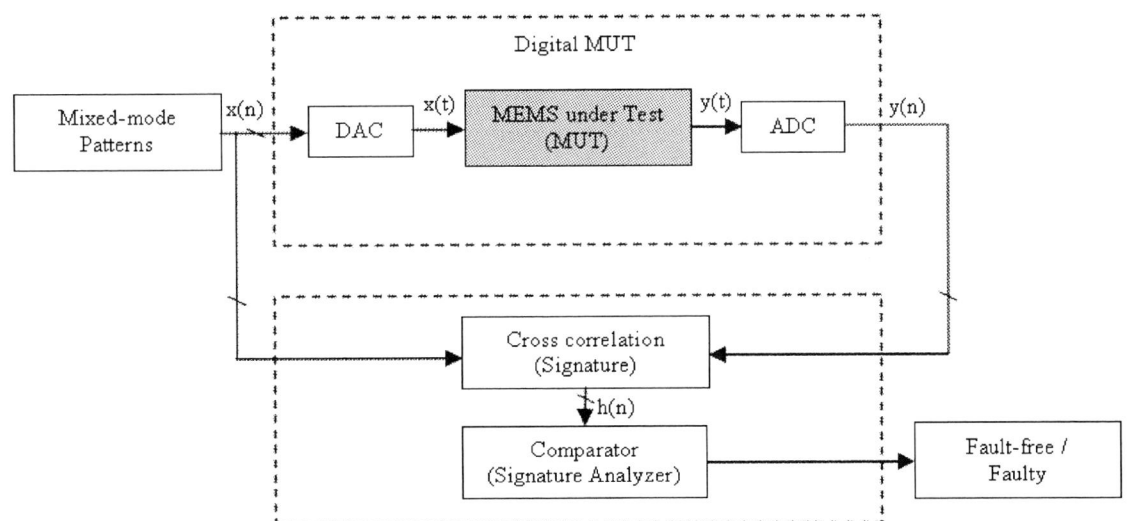

Fig. 2 Block diagram of the test approach.

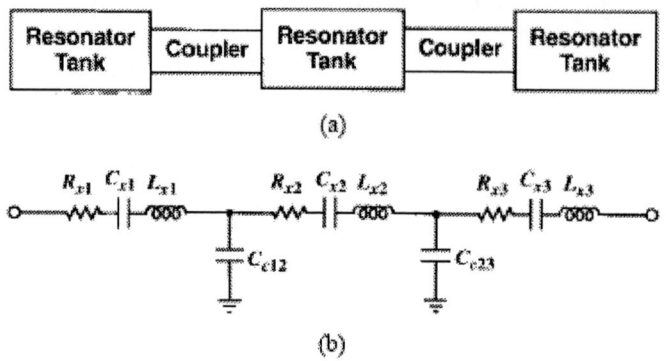

(a)

(b)

Fig. 3 (a) A general topology for coupled resonator filter. (b) One electrical circuit implementation for the network topology of (a) using series *RLC* tanks and shunt capacitor couplers.
Source: K. Wang et. al.

III. HARDWARE FOR MIXED-MODE TEST PATTERN GENERATOR

In the simplest case, mixed-mode testing can be performed by using an LFSR to generate pseudo-random patterns to detect the random pattern testable faults and then loading deterministic test patterns for the r.p.r. faults from a ROM. The problem with this approach is that the size of the required ROM is often prohibitive [7].

Instead of storing the test patterns themselves in a ROM, techniques have been developed for storing LFSR seeds that can be used to generate the test patterns [8]. The LFSR that is used for generating the pseudo-random patterns is also used for generating the deterministic patterns by reseeding it with computed seeds. Since the seeds are smaller than the test patterns themselves, they require less ROM storage. One problem is that for a normal LFSR with a fixed feedback polynomial, it may not always be possible to find a seed that will generate the required deterministic test pattern.

A new mixed-mode BIST scheme is described in [9]. Deterministic test patterns that detect the r.p.r faults are embedded in a pseudo-random sequence of bits generated by an LFSR. This is accomplished by altering the pseudo-random sequence of bits by adding logic at the LFSR's serial output to "fix" certain bits. As illustrated in Fig. 4, logic is added to generate a bit-fixing sequence that alters the pseudo-random sequence by causing certain bits to be fixed to either a '1' or a '0'.

There are two types of test pattern generators: 1) test-per-scan and 2) test-per-clock. The new approach can be used for either a 'test-per-scan' or 'test-per-clock' scheme and is capable of providing complete (100%) fault coverage with a reasonable test length. In the proposed mixed-mode testing approach, in order to optimize the mixed-mode patterns, complete test pattern generator (TPG) application will be considered based on test-per-clock scheme.

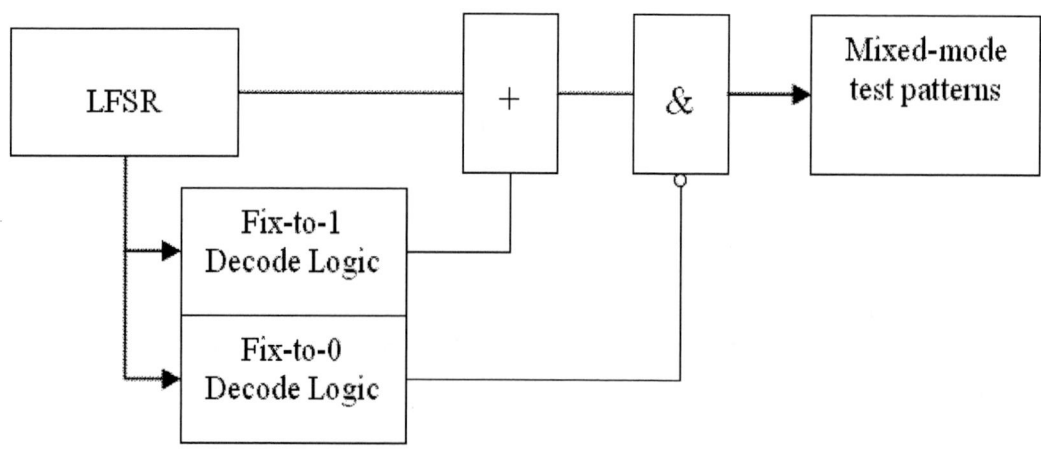

Fig. 4 Logic for altering the pseudo-random bit sequence.
Source: N. A. Touba et. al.

IV. CONCLUSIONS

We have introduced a testing technique based on IR evaluation using mixed-mode test sequences for MEMS testing. In the proposed mixed-mode testing, primitive polynomial LFSR based pseudo-random and deterministic test patterns will be used simultaneously, which is capable of complete (100%) fault coverage. An equivalent circuit of RF MEMS filter will be used as a test bench to assess the effectiveness of the proposed technique.

REFERENCES

[1] V. Beroulle, Y. Bertrand, L. Latorre and P. Nouet, "On the use of an oscillation-based test methodology for CMOS Micro-Electro-Mechanical Systems", Proceedings of the 2002 Design, Automation and Test in Europe Conference and Exhibition (2002).

[2] N. Deb and R. D. Blanton, "Built-In-Self-Test of MEMS Accelerometers", Journal of Micro-Electro-Mechanical Systems, vol. 15, no. 1, Feb (2006).

[3] B. Courtois, S. Mir, B. Charlot and B. Lubaszewski, "From microelectronics to MEMS testing", ISRN TIMA-RR-00/1001-FR, TIMA Laboratory, France (2000).

[4] C. Y. Pan and K. T. Cheng, "Pseudorandom testing for mixed-signals circuits", IEEE Trans. On Computer-Aided Design of Integrated Circuits and Systems, Vol. 16, No. 10, pp. 1173-1185 (1997).

[5] F. Corsi, C. Marzocca and G. Matarrese, "Defining a BIST-oriented signature for mixed-signal devices", In Proc. IEEE Southwest Symposium on Mixed-Signal Design, Las Vegas 22-26 February, pp. 202-207 (2003).

[6] A. Dhayni, S. Mir, L. Rufer and A. Bounceur, "Pseudorandom functional BIST for linear and nonlinear MEMS", Proceedings of the 2006 Design, Automation and Test in Europe (DATE06), pp. 664-669 (2006).

[7] V. K. Agarwal and E. Cerny, "Store and generate Built-In testing approach", Proc. of FTCS-11, pp. 35-40 (1981).

[8] B. Koenemann, "LFSR-coded test patterns for scan designs", Proc. of Europian Test Conference, pp. 237-242 (1991).

[9] N. A. Touba and E. J. McCluskey, "Bit-fixing in pseudorandom sequences for scan BIST", IEEE Transactions on Computer-Aided Design of Integrated Circuits and Systems 20(4): 545-555 (2001).

[10] K. Wang and C. T.-C. Nguyen, "High-order medium frequency micromechanical electronic filters", Journal of Micro-Electro-Mechanical Systems, vol. 8, no. 4, Dec (1999).

A Novel Model for Membranes of Micropumps Partially Loaded with Electromagnetic Forces

Fu-Shin Lee and Chih-Hsiung Chen
Mechatronic Engineering Department
Huafan University
Taipei, Taiwan, R.O.C.
E-mail : fslee@huafan.hfu.edu.tw

Abstract **The object of this paper is to establish a theoretic model for four-edge-clamped micro-pump membranes locally excited by electromagnetic forces. The studied membrane is to be fabricated as a 25 μm thick polyimide (PI) film, which includes a central portion of Ni-Fe metal layer. Its deflection distributions are formulated through ingenious modification and manipulations through available models, and theoretical solutions for the PI membrane with the prescribed particular boundary and load conditions are finally reached. Furthermore, coupled-field finite element (FEM) simulations for the membrane deflection distributions caused by the external electromagnetic field are also practiced. It is inspirational that errors between simulation results obtained from the FEM simulations and calculation outcomes from the novel approach are within 2%, which applies the same to the well-known ANSYS and COMSOL models.**

I. INTRODUCTION

MICROELECTROMECHANICAL (MEMS) have played an important role in fabricating critical sensors and actuators since it was developed. Originally, MEMS technique was primarily employed on fabricating micro-sensors. However, the technique has been directed to realize vital micro-actuators recently. On the other hand, micro-actuators characterize in their high responses, great driving powers, handy manipulations, and excellent accuracies. Micro-pump is a kind of commonly studied micro-actuators [1] and is usually classified into electromagnetic [2][3], piezoelectric [4], thermo- actuating [5], electrostatic [6], or shape memory alloy [7] categories according to the basic actuation principles involved. Each type of micro-pump features its design as well as fabrication characteristics in various

applications, and may be adequately adopted in its particular micro-system. Among all essential keystones to lay out for studying micro-pump behaviors, the way a membrane of a micro-pump can deflect is a crucial issue to investigate since it concerns the pumping efficiency, pumping frequency, and pumping resolution for the fluid transported.

For this reason, it is aimed to develop a model to theoretically assess deflection distributions of a partially loaded polyimide (PI) membrane, which has its four edges clamped in a micro-pump chamber and a confined area actuated by an external electromagnetic field. The PI membrane studied is actuated by electromagnetic forces generated between a Ni-Fe film on the membrane and a excited coil underneath. The correctness of the developed model is justified through executing Finite Element Method (FEM) simulations on the well known ANSYS and COMSOL workbench. Comparisons of the theoretical approach and simulations results conclude that the analytical approach could outperform the FEM simulation means. Furthermore, the novel formulation would be verified by means of experimental measurements on that particular kind of membranes, which would be obtained from MEMS fabrications in the upcoming research schedule.

II. THEORETICAL APPROACH

In the first place, geometry and load conditions of the micropump to be fabricated are described for the membrane to perform necessary pumping functions. A polyimide (PI) membrane shown in Fig. 1 for a designated micropump is partially loaded by an uniform electromagnetic force within an confined area, $-u/2 \leq x \leq u/2$ and $-v/2 \leq y \leq v/2$, and the area of the whole

0-7803-9730-4/06/$25.00 ©2006 IEEE

membrane is limited in $-a/2 \le x \le a/2$ and $-b/2 \le y \le b/2$. In addition, edges of the membrane are all fixed according to a practical design to fabricate the micropump using MEMS technology.

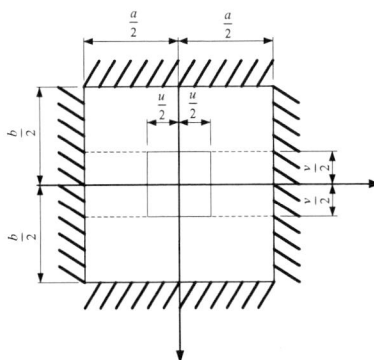

Fig. 1 Polyimide (PI) membrane with a central Ni-Fe film exerted by electromagnetic forces

Boundary conditions of the partially PI membrane are listed as following.

1. $w = 0$ for $x = \pm a/2$ and $y = \pm b/2$

2. $\partial w/\partial x = 0$ for $x = \pm a/2$ 　　　　p

3. $\partial w/\partial y = 0$ for $y = \pm b/2$

However, formulas in the literature to estimate membrane deflection distributions are available using model I and model II in Table I, in which common boundary and load conditions for the analyzed models are prescribed. On the contrary, it demands to derive an analytical formula in this study now to estimate the PI membrane deflections under the stated particular boundary and load conditions, and it needs ingenious modifications on the available theoretical models.

For model I, which is four-edge simply supported with a uniform distributed force exerted over a confined area $u \times v$ in Table I, an external load exerted on the membrane is defined as P, and the distributed force as $q = P/(u \times v)$ on the confined area. The deflection of the membrane in Model I [8] is

$$w = \frac{1}{\pi^4 D} \sum_{m=1}^{\infty} \sum_{n=1}^{\infty} \frac{a_{mn}}{\left(\dfrac{m^2}{a^2} + \dfrac{n^2}{b^2}\right)^2} \cos\frac{m\pi x}{a} \cos\frac{n\pi y}{b}, \quad (1)$$

with the indices $m, n = 1, 2, 3, L, \infty$, and $D = Eh^3/12(1-\upsilon^2)$ as stiffness of the membrane. Note that E, h, and υ are its Young's modulus, thickness, and Poisson's ratio, respectively, and a_{mn} is defined as

$$a_{mn} = \frac{16q}{mn\pi^2} \sin\frac{m\pi u}{2a} \sin\frac{n\pi v}{2b}. \quad (2)$$

Table I   Boundary and load conditions for models

Model Types	Model I	Model II	Model III
Boundary condition	Four-edge simply supported $w = 0$, $\partial w^2/\partial x^2 = 0$ and $\partial w^2/\partial y^2 = 0$ for 4 edges.	Four-edge simply supported $w = 0$ for 4 edges, and $\partial w^2/\partial x^2 \big|_{x=\pm a/2} = 0$	Four-edge clamped PI membrane $w = 0$, $\partial w/\partial x = 0$ and $\partial w/\partial y = 0$ for 4 edges
Load condition	Uniform pressure over a confined area $u \times v$	Symmetric loading moments $-D\dfrac{\partial w^2}{\partial y^2}\Big|_{y=\pm\frac{b}{2}} = f_{1,2}(x)$	Uniform pressure over a confined area $u \times v$

As a result, deflection of the membrane in model I is concluded as

$$w = \frac{16q}{\pi^6 D} \sum_{m=1}^{\infty} \sum_{n=1}^{\infty} \frac{1}{mn\left(\dfrac{m^2}{a^2} + \dfrac{n^2}{b^2}\right)^2} \left(\begin{array}{c} \sin\dfrac{m\pi u}{2a}\sin\dfrac{n\pi v}{2b} \\ \times \cos\dfrac{m\pi x}{a}\cos\dfrac{n\pi y}{b} \end{array} \right), (3)$$

and its edge slopes at $y = \pm b/2$ are expressed as

$$\left(\frac{\partial w}{\partial y}\right)\Bigg|_{y=\pm b/2} = \mp \frac{16q}{\pi^5 Db} \sum_{m=1,3,5}^{\infty} \sum_{n=1,3,5}^{\infty} \frac{(-1)^{(n-1)/2}}{m\left(\dfrac{m^2}{a^2} + \dfrac{n^2}{b^2}\right)^2},$$
$$\times \left(\sin\frac{m\pi u}{2a}\sin\frac{n\pi v}{2b}\cos\frac{m\pi x}{a} \right) \quad (4)$$

where only $m, n = 1, 3, L, \infty$ are used due to the nature of its symmetric boundary conditions.

For model II in Table I, a four-edge simply supported membrane features in its symmetrically loaded moments on the respective edges. The governing equation therein for the membrane deflection $w_1(x, y)$

79

is

$$\frac{\partial^4 w_1(x,y)}{\partial x^4} + 2\frac{\partial^4 w_1(x,y)}{\partial x^2 \partial y^2} + \frac{\partial^4 w_1(x,y)}{\partial y^4} = 0. \quad (5)$$

Its general solution is assumed as

$$w_1 = \sum_{m=1}^{\infty} C_m \left(\begin{array}{c} \dfrac{m\pi y}{a}\sinh\dfrac{m\pi y}{a} \\[2mm] -\alpha_m \tanh\alpha_m \cosh\dfrac{m\pi y}{a} \end{array} \right) \cos\dfrac{m\pi x}{a}, \quad (6)$$

where $\alpha_m = \dfrac{m\pi b}{2a}$. Considering a specific set of x-axis symmetric moments

$$f_1(x) = f_2(x) = \sum_{m=1}^{\infty} E_m \cos\frac{m\pi x}{a}\sin\frac{m\pi}{2} \quad (7)$$

exerted on the edges $y = \pm\dfrac{b}{2}$ as shown in Fig. 2, where $m, = 1, 2, \mathrm{L}, \infty$. Consequently, the deflection $w_1(x,y)$ in (6) is expressed [8] as

$$w_1 = -\frac{a^2}{2\pi^2 D}\sum_{m=1,3,5,\ldots}^{\infty} E_m \frac{(-1)^{(m-1)/2}}{m^2 \cosh\alpha_m}\cos\frac{m\pi x}{a} \quad (8)$$
$$\mathbf{g}\left(\frac{m\pi y}{a}\sinh\frac{m\pi y}{a} - \alpha_m \tanh\alpha_m \cosh\frac{m\pi y}{a}\right).$$

Its slopes at the membrane edges are derived as

$$\left(\frac{\partial w_1}{\partial y}\right)\Bigg|_{y=\pm b/2} =$$

$$\mathrm{m}\frac{a}{2\pi D}\sum_{m=1,3,5}^{\infty} E_m \frac{(-1)^{(m-1)/2}}{m}\cos\frac{m\pi x}{a}. \quad (9)$$
$$\mathbf{g}\left(\tanh\alpha_m + \frac{\alpha_m}{\cosh^2\alpha_m}\right)$$

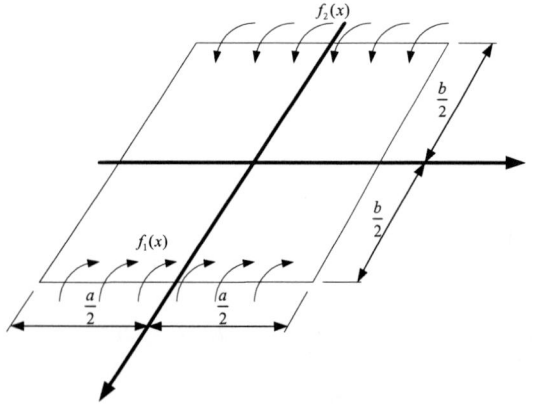

Fig. 2 Four-edge simply supported membrane

with symmetric moment loadings

By analogy, it is assumed a simply supported membrane is only loaded at $x = \pm a/2$ by another specific set of y-axis symmetric moments

$$f_3(y) = f_4(y) = \sum_{m=1}^{\infty} F_m \cos\frac{m\pi y}{b}\sin\frac{m\pi}{2}. \quad (10)$$

Hence, its membrane slopes are

$$\left(\frac{\partial w_2}{\partial y}\right)\Bigg|_{y=\pm b/2}$$
$$= \mathrm{m}\frac{4a^2}{\pi^2 Db}\sum_{\hat{m}=1,3,5}^{\infty}\frac{F_{\hat{m}}}{\hat{m}^3}\sum_{m=1,3,5}^{\infty}\frac{m(-1)^{(m-1)/2}}{\left(\dfrac{a^2}{b^2}+\dfrac{m^2}{\hat{m}^2}\right)^2}\cos\frac{m\pi x}{a}, \quad (11)$$

with $\beta_m = {m\pi a}\big/{2b}$ and $\hat{m} = 1, 3, \mathrm{L}, \infty$.

Considering the loading and boundary conditions for the PI membrane to be fabricated, as listed in Model III of Table I, it is evident that the deflection distributions of the PI membrane should fulfill the four-edge clamped boundary conditions and a confined area loading situation. Hence, it is ingenious to derive the demanded deflection of the PI membrane from superposing the deflections in Model I and Model II, once the following superposed boundary conditions are satisfied at $y = b/2$, or

$$\left(\frac{\partial w}{\partial y}\right)\Bigg|_{y=b/2} + \left(\frac{\partial w_1}{\partial y} + \frac{\partial w_2}{\partial y}\right)\Bigg|_{y=b/2} = 0. \quad (12)$$

Substituting (4), (9) and (11) into (12), it is managed to have the following expression for each m-index as considering the same factor $\cos(m\pi x/a)$ used in the expressions, and the designed PI membrane is rectangular with symmetric moments $E_m = F_m$ loaded for the x-axis as well as for the y-axis.

$$\frac{E_m}{m}\left(\tanh\alpha_m + \frac{\alpha_m}{\cosh^2\alpha_m}\right) + \frac{8m}{\pi}\sum_{\hat{m}=1,3,5}^{\infty}\frac{E_m}{\hat{m}^3}\frac{1}{\left(1+\dfrac{m^2}{\hat{m}^2}\right)^2} \quad (13)$$

$$= -\frac{32qa^2}{\pi^4 m}\sin\frac{m\pi u}{2a}\sum_{n=1,3,5}^{\infty}\frac{(-1)^{(n-m)/2}}{(m^2+n^2)^2}\sin\frac{n\pi u}{2a}$$

III. IMPLEMENTATION OF DERIVED FORMULA

Material as well as geometry properties for the PI membrane designed are shown in Table II, together with an approximate electromagnetic force [9] proposed. Only four coefficients of E_m are intended to be solved in (13) for simplicity. As a result, it leads to

$$
\begin{bmatrix}
1.80326 & 0.0763944 & 0.0188349 & 0.00713014 \\
0.0763944 & 0.404522 & 0.0330425 & 0.0158966 \\
0.0188349 & 0.0330425 & 0.225466 & 0.0162759 \\
0.00713014 & 0.0158966 & 0.0162759 & 0.155849
\end{bmatrix}
\begin{bmatrix}
E_1 \\ E_3 \\ E_5 \\ E_7
\end{bmatrix}, \quad (14)
$$

$$
=
\begin{bmatrix}
0.304292 \\
-0.002716 \\
-0.000109709 \\
0.00000783125
\end{bmatrix} K
$$

where $K = -\dfrac{4qa^2}{\pi^3}$. Finally, the equation in (14) is solved to obtain

$$
\begin{bmatrix}
E_1 \\ E_3 \\ E_5 \\ E_7
\end{bmatrix}
=
\begin{bmatrix}
0.1705 \\
-0.0381 \\
-0.0089 \\
-0.0029
\end{bmatrix} K
\qquad (15)
$$

for expressions in (7) and (10), which stand for the extra external moments needed to be exerted on the membrane in Model II to balance the distributed loading on a confined area in Model I, as to satisfy the overall loading and boundary conditions in Model III.

Table II Membrane material and geometry properties

Areas of load	$u = 3000\,\mu m$, $v = 3000\,\mu m$
Young's modules of membrane	$E = 2.5 GPa$
Density of membrane	$\rho = 1420\,kg/m^3$
Poisson's ratio of membrane	$\nu = 0.34$
Thickness of membrane	$25\,\mu m$
Electromagnetic load	24.98 Pa
Boundary conditions	four edges clamped

With the numerical data for E_m obtained, the total deflection distributions of the four-edge clamped PI membrane in Model III, which is exerted by a distributed electromagnetic force over a confined area $u \times v$, is successfully formulated. Hence, the central deflection of the loaded PI membrane in Model III is

$$
\left. (w_{\max}) \right|_{x=y=0} = \left. (w) \right|_{x=y=0} + \left. (w_1) \right|_{x=y=0} + \left. (w_2) \right|_{x=y=0}
$$
$$
= 0.00084795 \frac{qa^4}{D}. \qquad (16)
$$

Assuming the confined loaded area of the PI membrane, $u \times v = 3000\,\mu m \times 3000\,\mu m$, is fixed in Model III, (16) is exercised for various membrane dimensions. The area of the PI membrane varies from $a \times b = 3000\,\mu m \times 3000\,\mu m$ to $a \times b = 10000\,\mu m \times 10000\,\mu m$, and the calculated center deflections using the derived formula in this research are listed in Table III.

Table III Theoretical PI membrane deflections (μm)

$a = b$ (μm)	3000	5000	7000	9000
Center deflection	0.695	4.356	11.349	21.396

IV. FINITE ELEMENT SIMULATIONS

Using an ANSYS FEM model, deformations due to the prescribed electromagnetic force exerted on the confined area of the PI membrane are illustrated in Fig. 3 for $a = b = 5000\,\mu m$ and $u \times v = 3000\,\mu m \times 3000\,\mu m$. Center deflections of the simulated PI membranes are tabulated in Table IV.

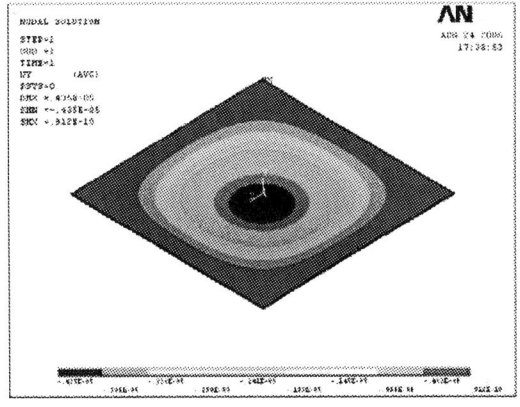

Fig. 3 ANSYS simulation result for PI membrane

Table IV ANSYS deflection simulation results (μm)

$a = b$ (μm)	3000	5000	7000	9000

Center deflection	0.695	4.35	11.3	21.4

Similarly based upon a COMSOL FEM model, deformations due to the prescribed electromagnetic force exerted on the confined area of the PI membrane are illustrated in Fig. 4 COMSOL simulation result for PI membrane for $a = b = 5000\ \mu m$ and $u \times v = 3000\mu m \times 3000\mu m$. Center deflections of the simulated PI membranes are tabulated in Table V.

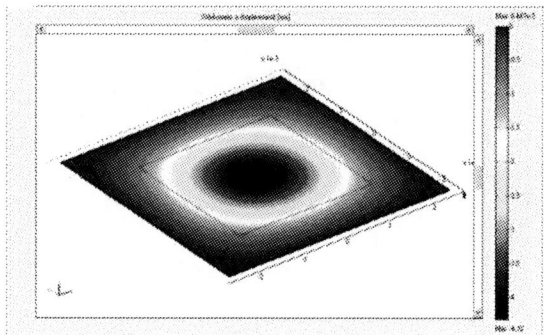

Fig. 4 COMSOL simulation result for PI membrane

Table V COMSOL deflection simulation results(μm)

$a = b$ (μm)	3000	5000	7000	9000
Center deflection	0.690	4.32	11.135	21.129

V. COMPARISIONS AND DISCUSSIONS

The deflections calculated from employing the theoretic formulations and deflections obtained through conducting the ANSYS as well as COMSOL FEM simulations are tabulated in Table VI for comparisons.

It is observed that the error percentages between the theoretical calculation results and ANSYS simulation results are within 1%, and the differences between the theoretical solutions and COMSOL simulation outcomes are limited in 2%. Consequently, all the calculations based upon the theoretical approach and simulation results based upon the FEM models are plotted on Figure 5.

Table VI   Center deflections of the PI membrane

$a = b$ (μm)	Theory Cent. deflec. (μm)	ANSYS (FEM) Cent. deflec. (μm)	Error (%)	COMSOL (FEM) Cent. deflec. (μm)	Error (%)
3000	0.695	0.692	0.505	0.69	0.792
5000	4.356	4.35	0.160	4.32	0.849
7000	11.349	11.3	0.438	11.135	1.892
9000	21.396	21.4	-0.019	21.129	1.248

VI. CONCLUSIONS

A novel approach to determine deflection distributions of a partially loaded four-edge clamped PI membrane theoretically is proposed in this research. The analytical solution for deflections of the electromagnetically loaded membrane is derived through ingeniously manipulation and superposition of available models with adequate boundary and load conditions. The comparisons between theoretical approach and FEM simulations confirm the validity of the proposed formulation for a partially loaded membrane with its four edges clamped.

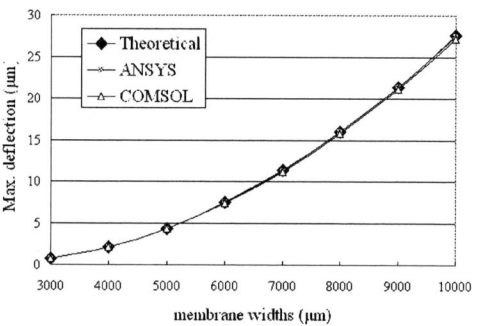

Figure 5   PI membrane center deflections

ACKNOWLEDGEMENT

The authors are grateful to the National Science Council for partial financial support under contract No. NSC 92-2212-E-211-005 and contract No. NSC 94-3113-E-211-001.

REFERENCES

[1] N.-T. Nguyen, X. Huang, and T. K. Chuan, "MEMS-Micropumps: A review", *Journal of Fluids Engineering*, Vol. 124, pp. 384-392, 2002.

[2] Q. Gong, Z. Zhou, Y. Yang, and X. Wang, "Design, optimization and simulation on microelectromagnetic pump", Sensors and Actuators A, Vol. 83, pp. 200-207, 2000.

[3] D. de Bhailis, C. Murray, M. Duffy, J. Alderman, G. Kelly, and S.C. O. Mathuna, "Modeling and analysis of a magnetic micro actuator", *Sensors and Actuators* A, Vol. 81, pp. 285-289, 2000.

[4] M. Koch, N. Harris, R. Maas, A. G. R. Evans, N. M. White, and A. Brunnschweiler, "A novel micropump design with thick-film piezoelectric actuation", Meas. Sci. Technol., Vol. 8, pp. 49-57, 1997.

[5] C. Grosjean, and Y. C. Tai, " A thermopneumatic Peristaltic Micro-pump", *Inter. Conf. on Solid-State Sensor and Actuators*, pp. 1776-1779, 1999.

[6] M. T. A. Saif, B. E. Alaca, and H. Sehitoglu,

"Analytical modeling of electrostatic membrane actuator for micro pumps", IEEE *Journal of Microelectromechanical System*, Vol. 8, pp. 335-345, 1999.

[7] W. L. Benard, H. Kahn, A. H. Heuer, and M. A. Huff, "Thin film shape-memory alloy actuated micropump", *Journal of* MEMS, Vol. 2, pp. 245-251, 1998.

[8] S. Timoshenko and S. Woinowsky-Krieger, *Theory of Plates and Shells*, McGraw-Hall, 1959.

[9] Fu-Shin Lee, Chih-Hsiung Chen, Jyh-Ling Lin, and Chiang-Chao Liao, "Design and Prototyping of an Electromagnetically Excited Micro-pump", *Chinese Society of Sound and Vibration*, 2004.

Investigation of the Torsion and Bending effects on Static Stability of Electrostatic Torsional Micromirrors

Ghader Rezazadeh, Faraz Khatami, Ahmadali Tahmasebi
Mechanical Engineering Department.
Urmia University,
Urmia, Iran
Email: g.rezazadeh @mail.urmia.ac.ir

Abstract— In this paper the electromechanical behavior of a torsional micromirror was investigated using of a static model with considering torsion and bending characteristics of micro-beams. A set of nonlinear equations based on the parallel plate capacitor model was derived to represent the relationships between the applied voltage, torsion angle, and vertical displacement of the torsional micromirror. Step by Step Linearization Method (Newton's Method) was used to calculate the rotation angle and vertical displacement of the micromirror due to the applied voltage. This method is fast and gave acceptable and accurate results which were in good agreement with the experimental data.

I. INTRODUCTION

Today MEMS have evoked great interest in the scientific and engineering communities. Things behave substantially differently in the micro domain. Due to many advantages of MEMs, microoptoelectromechanical systems (MOEMS) including optical switches, microscanners, digital micromirror device (DMD) etc, are investigated for emerging related fields like all-optical telecommunication networks. In the MOEMS systems electrostatic micromirror is used widely as an actuator. Various micromirror devices have been reported for various applications in open literatures. Based on their motion types, micromirrors can be simply classified into four categories: deformable micromirror, movable micromirror, piston micromirror, and torsional micromirror. The torsional micromirror has been widely used for applications because of its good dynamic response and small possibility of ad-

hesion, for instance in digital projection displays, spatial light modulators and optical crossbar switches. When the vertical displacement of the torsional micromirror reaches a gap of 10% between the micromirror and electrode plate, the phase of reflected light beam is changed by half and even one wavelength of the lights. Thus, it seriously affects the design and usage of relevant optical applications, and may even result in the wrong design, i.e., in spatial light modulators. Therefore, accurately analyzing of the torsional micromirror characteristics is very important to optical devices and applications. Hence in this paper, a torsional micromirror is presented. First, the parallel-plate model along with an effective torsional spring coefficient of the structure to estimate the pull-in voltage is considered. When the vertical displacement and torsion angle of the torsional micromirror are within the values of the same order, the bending and torsion effect concurrently affect the static characteristics of the torsional micromirror, so as the second consideration, a parallel-plate model by assuming the effect of torsional and bending characteristics is considered. In order to solve the nonlinear governing equations the step by step linearization method (Newton's method) is applied. This method is easy, fast and reliable to predict relationship between voltage, angle, and displacement and is used to calculate pull-in voltage of the torsional micromirror. The obtained results due to the effects of torsion and bending are compared well with experimental data.

II. MODEL DESCRIPTIONS AND ASSUMPTIONS

Figure 1 is a schematic 3D view of the micromirror and Fig. 2 represent the cross-sectional view of the micromirror considering, a) torsion effect, b) torsion and bending effect. As it is shown in the foregoing figures, assume a micromirror plate with length b, width a_3, thickness t, beneath the micromirror, there are two electrodes on the substrate, a_1 is the distance between the axes of rotation to the edge of the electrode and a_2 is width of the electrode. The moving plate is suspended by two similar torsion beams with length l, width w, thickness t, Young's modulus E, shear modulus G. In order to model the electrostatic torque, it is assumed that the plates are infinitely wide, so fringing fields (fields at the edges of the plates) are neglected, the deformation of the micromirror plate is very small, and the vertical displacement of the micromirror is mainly attributed to the deflection of the microbeams, so the micromirror can be regarded as a rigid body [1]. It is assumed that the torsion beams have negligible residual stress and stress stiffening effects. Any non-uniformity in the electric field due to the curvature is neglected [2]. The micromirror can be driven to rotate by adding potential between the micromirror and one electrode, and can rotate in the reverse direction if potential is introduced between the micromirror and another electrode instead. When the rotation angle is small it's assumed that $\sin\theta \approx \theta$ and $\cos\theta \approx 1$. This will give rise to less than 1% error even at $10°$.

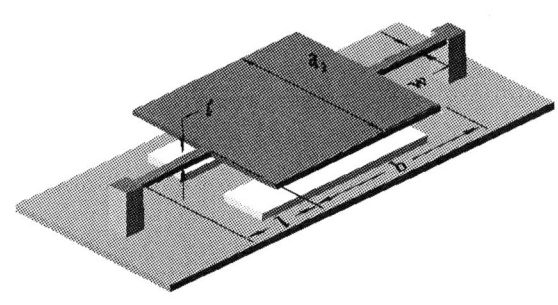

Fig. 1. Schematic 3D view of the torsional micromirror

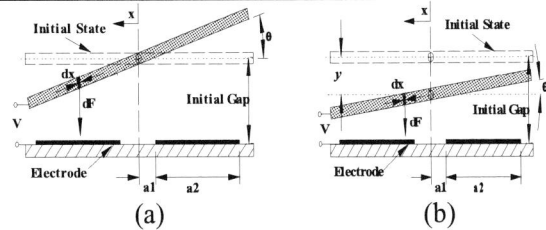

(a)   (b)

Fig.2. Cross sectional view of the torsional micromirror, a) Model considering torsion effect, b) Model considering torsion and bending effects.

III. MATHEMATICAL MODELING

For modeling by considering the torsion effect of the microbeams, It is assumed that the effective area of the plate is A and for an element $dA = bdx$, the distance between the rotational plate element and electrode is considered as $u = g - x\theta$, when θ is the rotation about z axis and g is the initial gap between plates. Hence the total electrostatic torque can be written as

$$T_E(\Theta, V) = \beta_1 V^2 \Omega_1(\Theta) \qquad (1)$$

where

$$\Omega_1(\Theta) = \frac{1}{\Theta^2}\left(\begin{array}{c} \dfrac{\alpha\Theta}{(1-\alpha\Theta)} - \dfrac{a\alpha\Theta}{(1-a\alpha\Theta)} \\ + \ln(\dfrac{1-\alpha\Theta}{1-a\alpha\Theta}) \end{array} \right) \qquad (2)$$

$$\beta_1 = \frac{b\varepsilon}{2\theta_0^2}, \ a = \frac{a_1}{a_2}, \ \alpha = \frac{a_2}{g}\theta_0, \ \Theta = \frac{\theta}{\theta_0}$$

Parameter θ_0 is the critical rotation angle that the micromirror plate takes to touch the electrodes plate and is defined as $\theta_0 = \sin^{-1}\left(\dfrac{2g}{a_3}\right) \approx \dfrac{2g}{a_3}$. The micromirror plate is suspended by beams and therefore is subjected to a mechanical torque opposing the electrostatic torque (considering no bending effect). The beam mechanical elastic torque can be written as [3]:

$$T_M = K_T \theta, \ K_T = \frac{c_2 w t^3 G}{l} \qquad (3)$$

Where w is the width, t is the thickness and G is the elastic shear modulus of the microbeam. The coefficient c_2 depends on the ratio (w/t). In order to study the equilibrium position the following function as the difference of electrical and mechanical torque is introduced:

$$\Phi(\Theta, V) = T_E - T_M \qquad (4)$$

At the equilibrium position, the electrostatic torque and the mechanical elastic torque are equal, meanwhile:

$$\Phi(\Theta, V) = 0 \qquad (5)$$

The rotation angle of the micromirror can be obtained by solving the nonlinear Eq. (5) at a specific applied voltage. For sufficiently low voltages, there are two physical exhibits of the rotation angle, where only one of them is stable. For a certain voltage, the two solutions of Eq. (5) coincide and pull-in phenomenon occurs. For voltages above the pull-in voltage, the electrostatic torque is greater than the mechanical torque for any angle [4]. The pull-in state is found at maximum of $V(\Theta)$. Using of the implicit function theorem and Eq. (5), to reach the maximum value of the $V(\Theta)$, the equation $\frac{\partial \Phi(\Theta, V)}{\partial \Theta} = 0$ should be satisfied. By Solving them, simultaneously, the pull-in parameters pull-in voltage ($V_{Pull-in}$) and angle of rotation (Θ) of micromirror can be calculated.

For modeling by considering the effects of torsion and bending of the microbeams, It can be seen from Fig. 2.b that $u = g - (y + x\theta)$, where y is the vertical displacement of the micromirror. Hence the total electrostatic torque as a function of non-dimensional vertical displacement, normalized rotation angle and applied voltage can be written as follow

$$T_E(\Theta, Y, V) = \beta_2 V^2 \Omega_2(\Theta, Y) \qquad (6)$$

where

$$\Omega_2(\Theta, Y) = \frac{1}{\Theta^2}\left(\begin{array}{c} \dfrac{\alpha\Theta}{(1 - Y - \alpha\Theta)} \\ - \dfrac{a\alpha\Theta}{(1 - Y - a\alpha\Theta)} \\ + \ln(\dfrac{1 - Y - \alpha\Theta}{1 - Y - a\alpha\Theta}) \end{array} \right) \qquad (7)$$

$$\beta_2 = \frac{b\varepsilon}{2\theta_0^2}, \quad Y = \frac{y}{g}.$$

The torsion and bending rigidity of the microbeams have the same order of magnitude, therefore the electrostatic force causes the microbeams to deflect and affects on the micromirror parameters and can not be neglected. Thus the total applied electrostatic force on the micromirror plate can be noted as follows:

$$F_E(\Theta, Y, V) = \beta_3 V^2 \Omega_3(\Theta, Y) \qquad (8)$$

whereas

$$\Omega_3(\Theta, Y) = \frac{1}{\Theta}\left(\begin{array}{c} \dfrac{1}{(1 - Y - \alpha\Theta)} \\ - \dfrac{a}{(1 - Y - a\alpha\Theta)} \end{array} \right) \qquad (9)$$

$$\beta_3 = \frac{b\varepsilon}{2g\theta_0}$$

The mechanical elastic beam torque can be defined as Eq. (3) and the mechanical elastic beam force can be defined as follow

$$F_M = K_b y, \ K_b = \frac{24EI}{l^3} \qquad (10)$$

where K_b is the effective bending stiffness of the microbeams. The parameter I is the moment of inertia of the rectangular cross-section area of the microbeam. To study the stability of the micromirror, the following functions are introduced

$$\Psi_1(\Theta, Y, V) = T_E - T_M \qquad (11)$$
$$\Psi_2(\Theta, Y, V) = F_E - F_M \qquad (12)$$

Hence at the equilibrium position the following relationships must be satisfied

$$\Psi_1(\Theta, Y, V) = 0 \qquad (13)$$
$$\Psi_2(\Theta, Y, V) = 0 \qquad (14)$$

The angle of rotation and vertical deflection of the micromirror can be calculated using the nonlinear Eqs. (13) and (14) at a given applied voltage. The pull-in state is found at maximum of $V(\Theta, Y)$. Using of the implicit function theorem and Eqs. (13) and (14), to reach the maximum value of the $V(\Theta, Y)$, the following equation should be satisfied:

$$\begin{vmatrix} \dfrac{\partial \Psi_1(\Theta, Y, V)}{\partial \Theta} & \dfrac{\partial \Psi_1(\Theta, Y, V)}{\partial Y} \\ \dfrac{\partial \Psi_2(\Theta, Y, V)}{\partial \Theta} & \dfrac{\partial \Psi_2(\Theta, Y, V)}{\partial Y} \end{vmatrix} = 0 \qquad (15)$$

Solving Eqs. (13), (14) and (15) simultaneously, the pull-in voltage ($V_{Pull-in}$), angle of rotation (Θ) and vertical deflection (Y) of micromirror can be calculated.

IV. NUMERICAL SOLUTION

Because of nonlinearity of foregoing equations, the solutions are complicated and elaborating. In order to solve them, it is tried to linearize them. Because of considerable value of Θ and Y respecting to initial gap and especially because of considerable value of applied voltage relative to its pull-in value, the linearizing respect to initial state, may cause to appear some considerable errors [5]. Therefore, to minimize the value of errors, step by step by increasing the applied voltage is proposed. It is proposed that the Y_i and Θ_i are the non-dimensional vertical displacement and normalized rotation angle

86

of micro mirror due to applied voltage V_i. To model of electrostatic micromirror considering the torsion effect using of Taylor's series and its expansion for function $\Phi(\Theta, Y)$ about Θ_i, V_i, we have

$$\Phi(\Theta_{i+1}, V_{i+1}) = \Phi(\Theta_i, V_i) + \begin{bmatrix} \dfrac{\partial \Phi}{\partial \Theta} & \dfrac{\partial \Phi}{\partial V} \end{bmatrix}_{\Theta_i, V_i}$$

$$\begin{bmatrix} \delta\Theta_i \\ \delta V \end{bmatrix} + \frac{1}{2}\begin{bmatrix} \delta\Theta_i & \delta V \end{bmatrix} \begin{bmatrix} \dfrac{\partial^2 \Phi}{\partial \Theta^2} & \dfrac{\partial^2 \Phi}{\partial \Theta \partial V} \\ \dfrac{\partial^2 \Phi}{\partial V \partial \Theta} & \dfrac{\partial^2 \Phi}{\partial V^2} \end{bmatrix}_{\Theta_i, V_i}$$

$$\begin{bmatrix} \delta\Theta_i \\ \delta V \end{bmatrix} + \dots \qquad (16)$$

It should be noted that both of (Θ_i, V_i) and (Θ_{i+1}, V_{i+1}) represent the equilibrium state of the micromirror, so they satisfy $\dfrac{\partial \Phi(\Theta, V)}{\partial \Theta} = 0$. Using of the first order approximation of Eq. (16) we have

$$\Theta_{i+1} = \Theta_i - \left(\frac{\partial \Phi}{\partial \Theta}\Big|_{\Theta_i, V_i}\right)^{-1}\left(\frac{\partial \Phi}{\partial V}\Big|_{\Theta_i, V_i}\right)\delta V \qquad (17)$$

Now using of Eq. (17), the angle of rotation at each step of a given applied voltage can be obtained.

To model of electrostatic micromirror considering the effect of torsion and bending using Taylor's series expansion for function $\Psi_1(\Theta, Y, V)$ and $\Psi_2(\Theta, Y, V)$ about (Θ_i, Y_i, V_i), the following relationship can be gained:

$$\Psi_1(\Theta_{i+1}, Y_{i+1}, V_{i+1}) = \Psi_1(\Theta_i, Y_i, V)$$

$$+ \begin{bmatrix} \dfrac{\partial \Psi_1}{\partial \Theta} & \dfrac{\partial \Psi_1}{\partial Y} & \dfrac{\partial \Psi_1}{\partial V} \end{bmatrix}_{\Theta_i, Y_i, V} \begin{bmatrix} \delta\Theta_i \\ \delta Y_i \\ \delta V \end{bmatrix} +$$

$$+ \frac{1}{2}\begin{bmatrix} \delta\Theta_i & \delta Y_i & \delta V \end{bmatrix} \begin{bmatrix} \dfrac{\partial^2 \Psi_1}{\partial \Theta^2} & \dfrac{\partial^2 \Psi_1}{\partial \Theta \partial Y} & \dfrac{\partial^2 \Psi_1}{\partial \Theta \partial V} \\ \dfrac{\partial^2 \Psi_1}{\partial Y \partial \Theta} & \dfrac{\partial^2 \Psi_1}{\partial Y^2} & \dfrac{\partial^2 \Psi_1}{\partial Y \partial V} \\ \dfrac{\partial^2 \Psi_1}{\partial V \partial \Theta} & \dfrac{\partial^2 \Psi_1}{\partial V \partial Y} & \dfrac{\partial^2 \Psi_1}{\partial V^2} \end{bmatrix}_{\Theta_i, Y_i, V}$$

$$\begin{bmatrix} \delta\Theta_i \\ \delta Y_i \\ \delta V \end{bmatrix} + \dots$$

$$(18)$$

and

$$\Psi_2(\Theta_{i+1}, Y_{i+1}, V_{i+1}) = \Psi_2(\Theta_i, Y_i, V)$$

$$+ \begin{bmatrix} \dfrac{\partial \Psi_2}{\partial \Theta} & \dfrac{\partial \Psi_2}{\partial Y} & \dfrac{\partial \Psi_2}{\partial V} \end{bmatrix}_{\Theta_i, Y_i, V} \begin{bmatrix} \delta\Theta_i \\ \delta Y_i \\ \delta V \end{bmatrix} +$$

$$+ \frac{1}{2}\begin{bmatrix} \delta\Theta_i & \delta Y_i & \delta V \end{bmatrix} \qquad (19)$$

$$\begin{bmatrix} \dfrac{\partial^2 \Psi_2}{\partial \Theta^2} & \dfrac{\partial^2 \Psi_2}{\partial \Theta \partial Y} & \dfrac{\partial^2 \Psi_2}{\partial \Theta \partial V} \\ \dfrac{\partial^2 \Psi_2}{\partial Y \partial \Theta} & \dfrac{\partial^2 \Psi_2}{\partial Y^2} & \dfrac{\partial^2 \Psi_2}{\partial Y \partial V} \\ \dfrac{\partial^2 \Psi_2}{\partial V \partial \Theta} & \dfrac{\partial^2 \Psi_2}{\partial V \partial Y} & \dfrac{\partial^2 \Psi_2}{\partial V^2} \end{bmatrix}_{\Theta_i, Y_i, V} \begin{bmatrix} \delta\Theta_i \\ \delta Y_i \\ \delta V \end{bmatrix} + \dots$$

It must be noted that (Θ_i, Y_i, V_i) and $(\Theta_{i+1}, Y_{i+1}, V_{i+1})$ both represent the equilibrium state of the micromirror so they satisfy the Eqs. (13) and (14). Using the first order approximation for Eqs. (18) and (19), we have:

$$\begin{bmatrix} \Theta_{i+1} \\ Y_{i+1} \end{bmatrix} = \begin{bmatrix} \Theta_i \\ Y_i \end{bmatrix} - \begin{bmatrix} \dfrac{\partial \Psi_1}{\partial \Theta} & \dfrac{\partial \Psi_1}{\partial Y} \\ \dfrac{\partial \Psi_2}{\partial \Theta} & \dfrac{\partial \Psi_2}{\partial Y} \end{bmatrix}^{-1} \begin{bmatrix} \dfrac{\partial \Psi_1}{\partial V} \\ \dfrac{\partial \Psi_2}{\partial V} \end{bmatrix}\delta V \qquad (20)$$

Hence using of the Eq. (20) at each step of the applied voltage, the angle of rotation and vertical displacement of the micromirror can be obtained.

V. NUMERICAL RESULTS AND DISCUSSION

In this section, first it is tried to find the best step size for applying voltage for the Step by Step Linearization Method. Then related figures and tables for comparison between the torsion effect with torsion and bending effect results are proposed. The geometrical and material properties of the model are listed in Table 1 as:

Table 1: Parameters of the electrostatic micromirror

Items	Parameters	Values
Material properties	Shear modulus (Gpa)	66
	Young's modulus (Gpa)	170.28
Micromirror	Width, a_3 (μm)	100
	Length, b (μm)	100
Torsion beam	Length, l (μm)	65
	Width, w (μm)	1.55
	Thickness, t (μm)	1.5

87

Electrode	Coefficient, c_2	0.1406
	Width, a_1 (μm)	6
	Width, a_2 (μm)	84
	Gap, h (μm)	2.75

Table 2 shows the obtained results for torsion effect, and torsion and bending effect considerations, containing calculated pull-in voltage for different step sizes of applied voltage. It can be seen that decreasing of the step size of the applied voltage causes pull-in voltage change, up to a point where favorite accuracy is satisfied and after that, by changing the step size of applied voltage, pull-in voltage remains constant.

Table 2: The obtained pull-in voltages for different steps of the applied voltage

Step size of the applied voltage (V)		1	0.5	0.1	0.05
Pull-in voltage (V)	Torsion effect	20	20	19.8	19.8
	Torsion and bending effect	18	17.5	17.4	17.4

Based on the results of Table 2, the optimum step size for applied voltage may be taken account as $0.05(V)$. Figure 3. shows the rotation angle versus the applied voltage. As it is seen, the calculated results due to the considering of the torsion and bending effects are in good agreement with the experimental results of Ref.[6], where the difference between results of consideration of the torsion effect respect to the experimental ones may not be neglected.

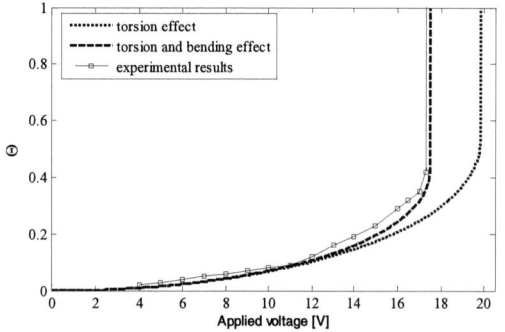

Fig. 3. Comparison between calculated and experimental results [6] for the rotation angle.

The calculated results and experimental data are listed in Table 3 which shows that the calculated pull-in voltage due to the consideration of torsion and bending effects is in good agreement with the experimental data.

Table 3: Comparison between calculated results with experimental data at the pull-in point

Pull-in characteristics		Θ	E_1 (%)	V (V)	E_2 (%)	Y	E_3 (%)
Experimental results [6]		0.4198	.	17.4	.	0.0778	-
Calculated results for	Torsion effect	0.5187	23.5	19.8	13.8	-	-
	Torsion and bending effect	0.4224	0.6	17.5	0.6	0.0763	1.2

Notes: $E_1 = [(\Theta - \Theta_{exp})/(\Theta_{exp})] *100\%$, $E_2 = [(V - V_{exp})/V_{exp}] *100\%$, and $E_3 = [(Y - Y_{exp})/Y_{exp}] *100\%$.

V. CONCLUSION

The primary of this work was to derive the relationships between the applied voltage, torsion angle, and vertical displacement, using the parallel-plate capacitor model for the torsional micromirror, then the normalized governing nonlinear equations were solved by Step-by-Step Linearization Method which was fast, reliable and easy to calculate the pull-in voltage with approximately small difference respecting to the experimental data for the foregoing model. The results showed that assuming the effects of the torsion and bending gives better results than the only torsion consideration.

REFERENCES

[1]. J. P. Zhao, H.L. Cheu, J.M. Huang, A.Q. Liu, A study of dynamic characteristics and simulation of MEMS torsional micromirrors, Sensors and Actuators A, Vol. 120, pp. 199-210, 2005.

[2]. J.M. Huang, A.Q. Liu, Z.L. Deng, Q.X. Zhang, J. Ahn, A. Asundi, An approach to the coupling effect between torsion and bending for electrostatic torsional micromirrors, Sensors and Actuators A, vol. 115, pp. 159-167, 2004.

[3]. S.P. Timoshenko, J.N. Goodier, Theory of Elasticity, 3d ed, McGraw-hill New York, sec 109, 1970.

[4]. O. Degani, E. Socher, A. Lipson, T. Leitner, D. J.Setter, S. Kaldor, Y. Nemirovskey, Pull-in study of an Electrostatic torsion Micromirror, Journal Of Microelectromechanical Systems, VOL. 7, NO. 4, DECEMBER, 1998.

[5]. Gh. Rezazadeh, A. Tahmasebi and Mikhail Zubtsov, Application of Piezoelectric Layers in Electrostatic MEM Actuators: Controlling of Pull-in Voltage, Journal of Microsystem Technologies, Vol.12, No. 12. , pp 1163-1170, 2006.

[6]. J.M. Huang, A.Q. Liu, Z.L. Deng, Q.X. Zhang, J. Ahn, A. Asundi, An approach to the coupling effect between torsion and bending for electrostatic torsional micromirrors ,Sensors and Actuators A, vol. 115, pp. 159-167, 2004.

Electromechanical Behavior of Microbeams with Piezoelectric and Electrostatic Actuation

Ghader Rezazadeh, Ahmadali Tahmasebi, Vahid Molodpour
Mechnical Engineering Department.
Khoy Azad University,
Khoy, Iran
Email: g.rezazadeh @mail.urmia.ac.ir

Abstract— **In this paper, a novel micro cantilever MEM switch is considered with piezoelectric layers that bounded to the microbeam. This developed model is actuated by applied voltage to the electrostatic electrodes and piezoelectric layers. Respect to the applying of the actuation voltage to the piezoelectric layers and electrostatic electrodes, three methods can be used to actuate this switch. The governing equation is derived and in order to linearize the nonlinear governing equations; step by step linearization method is used and generalized differential quadrature is applied to solve the linear systems of equations. The obtained results for a simple micro cantilever MEM switch are in good agreement with other simulation results to demonstrate the feasibility of proposed numerical methods. Respect to any of three actuation methods, the numerical results show that the proposed model can be actuated electrostatically and piezoelectrically, which of course the switching voltage is less than the original model.**

I. INTRODUCTION

MEMS actuators have their share of problems, such as high driving voltage, relatively low speed and low power handling. In recent years, many efforts have been done to solve these problems. Researchers have used some methods to decrease the actuation voltage. One of them is decreasing the air gap between the fixed plate and the beam. Another method is increasing the electrostatic area, and the third one is decreasing the spring constant of the beams. Meanwhile up to now, the researches aren't very focused on the dual

actuation of MEM switches to decrease pull-in voltage and actuate the MEM switches. In the dual actuation systems of MEMS, the second actuation may be done by means of piezoelectric material or magnetic actuation. The corresponding afford of this work is based on piezoelectricly dual actuation system. In this paper, a novel model of micro-cantilever switch with piezoelectric layers is considered. Because of its structures, this model is actuated by electrostatic force and so the piezoelectric force. Meanwhile, the piezoelectric layers can make external moment and so the distribute electrostatic force. Due to the proposed structure, three methods are available to actuate the switch. The first method is based on the applying a bias voltage to the piezoelectric layers and an actuation voltage to the electrostatic electrodes. The second one is vice versa than the previous one. In the last method, the actuation voltage applies to the piezoelectric layers and electrostatic electrodes, simultaneously. Because of nonlinear nature of the problem, step by step method (Newton's method) is used to linearize the governing equations. The linearization forms of the equations are solved by generalized differential quadrature (GDQ) method. The numerical results due to a simple micro cantilever beam are compared well with other simulation results. Respect to any of three actuation methods, the simulation results reveal that the pull-in voltages of the proposed model when are actuated both electrostatically and piezoelectrically is considerable less than a simple model.

II. ELECTROMECHANICAL BEHAVIOR OF CANTILEVER MICROBEAM WITH PIEZOELECTRIC LAYERS

As it is shown in the Fig. 1, assume a beam with thickness h, width b, length L, density ρ and

0-7803-9730-4/06/$25.00 ©2006 IEEE

isotropic with Young's modulus E that piezoelectric layers have bonded on its top and bottom surfaces. The piezoelectric layers are located throughout the beam length. And have thickness h_1, density ρ_0, Young's modulus E_P and equivalent piezoelectric coefficient for one-dimensional problem $\bar{e}_{31}$. Suppose that x is the coordinate along the length of the beam with its origin at the left end, and $u(x)$ is the deflection of the beam, defined to be positive downward.

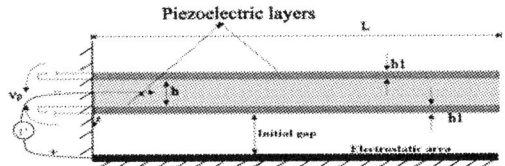

Fig. 1. Schematic of Cantilever MEM actuator with piezoelectric layers

The stress in the piezoelectric layers accounting for electro-mechanical effects can be expressed as [1]:

$$\bar{\sigma}_x = -E_P z \frac{d^2 u}{dx^2} - \bar{e}_{31} E_e \qquad (1)$$

where the over-bar represents the parameters in the piezoelectric layer and E_e is the electric field in the piezoelectric layer. In this model, two piezoelectric layers are bounded top and bottom through out the length of the beam. The forces in the piezoelectric layers are opposed to each other direction due to the difference direction of applied voltage to the each piezoelectric layers V_p. Hence, the axial force P moment M_y in the beam section are given as:

$$P = \int_{-h/2}^{h/2} \sigma_x b dz + \int_{h/2}^{h/2+h_p} \bar{\sigma}_x b dz +$$
$$\int_{-h/2-h_p}^{-h/2} \bar{\sigma}_x b dz = 0 \qquad (2)$$

$$M_y = 2\int_{0}^{h/2} z\sigma_x b dz + 2\int_{h/2}^{h/2+h_p} z\bar{\sigma}_x b dz \qquad (3)$$

Thus, the total moment can be obtained as:

$$M_y = 2\int_{0}^{h/2} -\widetilde{E} z^2 \frac{d^2 u(x)}{dx^2} b dz +$$
$$2\int_{h/2}^{h/2+h_p} (-E_P z \frac{d^2 u(x)}{dx^2} - e_{31} E_e) z b dz \qquad (4)$$

By integrating of the Eq. (4), we have:

$$M_y = -\widetilde{E} I \frac{d^2 u(x)}{dx^2} - E_P I_P \frac{d^2 u(x)}{dx^2} - e_{31} E_e b(h h_P + h_P^2) \qquad (5)$$

whereas $I_P = 2(\frac{b h_P^3}{12} + b h_P (\frac{h+h_P}{2})^2)$.

Substituting $E_e = V_P / h_P$ into Eq. (5) the bending moment can be written as follow:

$$M_y = +\eta \frac{d^2 u(x)}{dx^2} - M_P \qquad (6)$$

whereas $\eta = \widetilde{E} I + E_P I_P$ and $M_P = e_{31} V_P b(h + h_p)$.

On the other hand, the moment can be written based on the distributed electrostatic force and considering the fringing field effect, thus $M_y = M_e$ and:

$$M_e(x,u,V_e) = \frac{\varepsilon_0 b V_e^2}{2}$$
$$\int_x^L \left(1 + 0.65 \frac{g_0 - u(\xi)}{b}\right) \frac{(\xi - x) d\xi}{(g_0 - u(\xi))^2} \qquad (7)$$

where V_e is the applied voltage between the movable/ground plates on the fixed substrate, ξ is a dummy variable of x, g_0 is the initial gap between the movable/fixed plates and ε_0 is the dielectric (permittivity) of air. The governing equation of proposed switch can be written as follow:

$$\eta \frac{d^2 u}{dx^2} = M_t \qquad (8)$$

Whereas the total moment can be expressed as: $M_t = M_e + M_P$.

III. NUMERICAL SOLUTION

Because of nonlinearity of derived equation, the solution is complicated and time consuming, so in order to solve it, it is tried to linearized it. Because of considerable value of u respect to initial gap especially when the applied voltage increasing, the linearizing respect to u, may causes to appear some considerable errors. Therefore, to minimize the value of errors, the method of step by step increasing the applied voltage is proposed [2].

The Taylor's series can be written for M_t^{i+1} as follow:

$$M_t^{i+1} = M_t^i + \left(\frac{\partial M_t}{\partial V_e}\right)^i \delta V_e$$
$$+ \left(\frac{\partial M_t}{\partial V_p}\right)^i \delta V_p + \left(\frac{\partial M_t}{\partial u}\right)^i \psi + O(2) \qquad (9)$$

The higher order of the Taylor's series can be neglected to linearize the total moment due to the next step of the applied voltages. Thus, the linear coupled total moment can be written as:

$$M_t^{i+1}\left(x, u^{i+1}, V_e^{i+1}, V_p^{i+1}\right) = \frac{\varepsilon_0 b\left(V_e^i\right)^2}{2}$$
$$\int_x^L \left(1 + 0.65 \frac{g_0 - u^i(\xi)}{b}\right) \frac{(\xi - x)d\xi}{\left(g_0 - u^i(\xi)\right)^2}$$
$$+ \frac{\varepsilon_0 b\left(V_e^i\right)^2}{2}$$
$$\int_x^L \left(2 + 0.65 \frac{g_0 - u^i(\xi)}{b}\right) \frac{(\xi - x)d\xi}{\left(g_0 - u^i(\xi)\right)^3} \psi(\xi) \qquad (10)$$
$$+ \frac{\varepsilon_0 b}{2}$$
$$\int_x^L \left(1 + 0.65 \frac{g_0 - u^i(\xi)}{b}\right) \frac{(\xi - x)d\xi}{\left(g_0 - u^i(\xi)\right)^2} \left(2\delta V_e \, V_e^i\right)$$
$$+ e_{31} b (h + h_p) \delta V_P$$

Using of Eq. (8) at the i-th step, the following equation to calculate the ψ can be expressed as:

$$\eta \frac{d^2\psi}{dx^2} - \frac{\varepsilon_0 b\left(V_e^i\right)^2}{2}$$
$$\int_x^L \left(2 + 0.65 \frac{g_0 - u^i(\xi)}{b}\right) \frac{(\xi - x)d\xi}{\left(g_0 - u^i(\xi)\right)^3} \psi(\xi)$$
$$= \frac{\varepsilon_0 b\left(2\delta V_e \, V_e^i\right)}{2} \qquad (11)$$
$$\int_x^L \left(1 + 0.65 \frac{g_0 - u^i(\xi)}{b}\right) \frac{(\xi - x)d\xi}{\left(g_0 - u^i(\xi)\right)^2}$$
$$+ e_{31} b (h + h_p) \delta V_P$$

The nonlinear Eq. (8) is converted to the linear Eq. (11), therefore implying any numerical method and imposing the boundary conditions, the Eq. (11) may be discretized into N nodes and then by solving linear system of algebraic equations, the ψ can be obtained at the each step of applied voltages.

Using of GDQ method [3], the corresponding Eq. (11) can be expressed as:

$$\eta \sum_{k=1}^N w_{jk}^{(2)} \psi_k - \frac{\varepsilon_0 b\left(V_e^i\right)^2}{2}$$
$$\sum_{m=j}^N \left(2 + 0.65 \frac{g_0 - u_m^i}{b}\right) \frac{(\xi_m - x_j)\Delta}{\left(g_0 - u_m^i\right)^3} \psi_m$$
$$= \frac{\varepsilon_0 b\left(2\delta V_e \, V_e^i + \left(\delta V_e^i\right)^2\right)}{2} \qquad (12)$$
$$\sum_{m=j}^N \left(1 + 0.65 \frac{g_0 - u_m^i}{b}\right) \frac{(\xi_m - x_j)\Delta}{\left(g_0 - u_m^i\right)^2}$$
$$+ e_{31} b (h + h_p) \delta V_P$$

Whereas [4]:

$$w_{jk}^{(1)} = \begin{bmatrix} 0 & 0 & \cdots & 0 & 0 \\ 0 & w_{22}^{(1)} & \cdots & w_{2,N-1}^{(1)} & w_{2,N}^{(1)} \\ \vdots & \vdots & \ddots & \vdots & \vdots \\ 0 & w_{N-1,2}^{(1)} & \cdots & w_{N-1,N-1}^{(1)} & w_{N-1,N}^{(1)} \\ 0 & w_{N,2}^{(1)} & \cdots & w_{N,N-1}^{(1)} & w_{N,N}^{(1)} \end{bmatrix},$$

$$w_{jk}^{(2)} = \begin{bmatrix} w_{11}^{(1)} & w_{12}^{(1)} & \cdots & w_{1,N-1}^{(1)} & w_{1,N}^{(1)} \\ w_{21}^{(1)} & w_{22}^{(1)} & \cdots & w_{2,N-1}^{(1)} & w_{2,N}^{(1)} \\ \vdots & \vdots & \ddots & \vdots & \vdots \\ w_{N-1,1}^{(1)} & w_{N-1,2}^{(1)} & \cdots & w_{N-1,N-1}^{(1)} & w_{N-1,N}^{(1)} \\ w_{N,1}^{(1)} & w_{N,2}^{(1)} & \cdots & w_{N,N-1}^{(1)} & w_{N,N}^{(1)} \end{bmatrix}$$

$$\cdot \left[w_{jk}^{(1)}\right]$$

and $\Delta = L/(N-1)$

The locations of the sampling points on the beam herein are determined using the so-called Chebyshev–Gauss–Lobatto distribution, which is given in [5].

$$x_j = \frac{1}{2}\left[1 - \cos\frac{j-1}{N-1}\pi\right] \text{ for } j = 1, 2, \ldots, N \qquad (13)$$

IV. RESULTS AND DISCUSSION

In this section, first it is tried to compare the proposed algorithms with the results predicted by CoSolve simulation software [6] and closed form 2D model [6] to demonstrate the feasibility of it and finding the best step size for applying voltage and number of grid points for the GDQ method. The geometrical and material properties of used model in the Ref. [6] are listed in Table 1 as:

Table 1-Geometrical and material properties of the MEM actuator and piezoelectric layers

	MEM Actuator	piezoelectric layer (PZT-4)
Length	150 μm	150 μm
Width	50 μm	50 μm
Height	3 μm	0.01 μm
Young's modulus	169 (GPa)	78.6 (GPa)
Poisson's ratio	0.06	0.3
Mass density	2331 (Kg/m^3)	7500 (Kg/m^3)
$\bar{e}_{31}$	-	-9.29
g_0 (initial gap)	1 μm	
ε (Permittivity of air)	8.854187×10^{-12} (F/m)	

The Table 2 shows the results of proposed method and the predicted results of CoSolve simulation software [6] and closed form 2D model [6]. The result were gained after finding the best step size for applied voltage and number of grid points for GDQ method.

Table 2- The calculated pull-in voltages for simple cantilever microbeams

Calculated result	CoSolve simulation [6]	Closed form 2D model [6]	Δ_1	Δ_2
17	16.9	16.8	0.59 %	1.19 %

with:

$$\Delta_1(\%) = \frac{ABS(Obtained\ Re\,sults - CoSolve\ Simulation)}{CoSolve\ Simulation} * 100$$

$$\Delta_2(\%) = \frac{ABS(Obtained\ Re\,sults - Closed\ form\ 2D\ model)}{Closed\ form\ 2D\ model} * 100$$

To find the best step size for applying voltage and number of grid points for the GDQ method, some sample grid points with step size for the applying voltage are used. The results are listed in the Table 3 and Table 4.

Table 3- The obtained pull-in voltages with 30 grid points for different step size for applying voltages

Step size for applying voltage (V)				
2	1	0.5	0.1	0.05
20	18	17.5	17	17
42	39	38.5	37.9	37.9

Table 4- The obtained pull-in voltages with 0.1 volt step size for applying voltage for different number of grid points for GDQ method

Number of grid points for the GDQ method				
5	7	9	11	14
17.6	17.1	17	17	17
39.2	38.3	38.1	37.9	37.9

As they are presented in Table 3 and 4, the best results can be obtained for 0.1 (V) and 14 grid points for step size of applying voltage and GDQ method, respectively.

In the first consideration, a bias voltage is applied to the piezoelectric layers. Figure 2 show the influence of the bias voltage to the pull-in voltage.

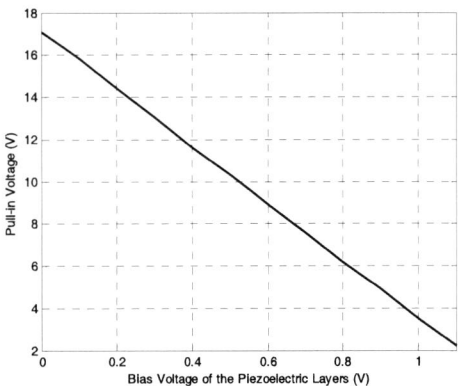

Fig. 2. Pull-in voltage versus the applied bias voltage to the piezoelectric layers

The calculate pull-in voltage when there is not any bias voltage applied to the piezoelectric layers is about 17.1(V). In this case, when bias voltage is about 1.1 (V), the pull-in voltage become 2.2 (V), which means 82.6 % decreasing of switching voltage.

In the second consideration, a bias voltage is applied to the electrostatic areas. Figure 3 show the influence of the bias voltage to the pull-in voltage. The calculate pull-in voltage when there is not any bias voltage applied to the electrostatic areas is about 0.61

(V). In this case, when bias voltage is about 10 (V), the pull-in voltage become 0.45 (V), which means 38.1 % decreasing of switching voltage.

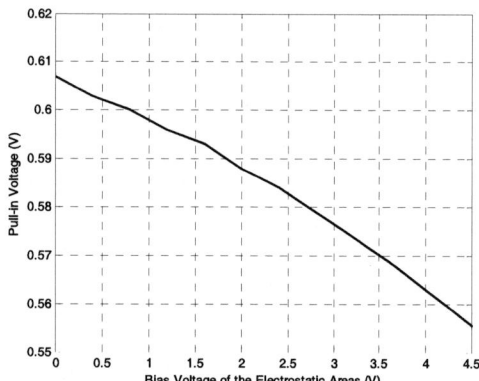

Fig. 3. Pull-in voltage versus the applied bias voltage to the electrostatic area

For simultaneous actuation, when the applied voltage to the piezoelectric layers and electrostatic areas are the same, the pull-in voltage is about 0.6 (V), which means 92.8 % decreasing in switching voltage respect to the original microbeam with no piezoelectric layers. Figure 4 show the deflection of the proposed model versus the length of the microbeam due to the different applied voltage to both piezoelectric layers and electrostatic areas.

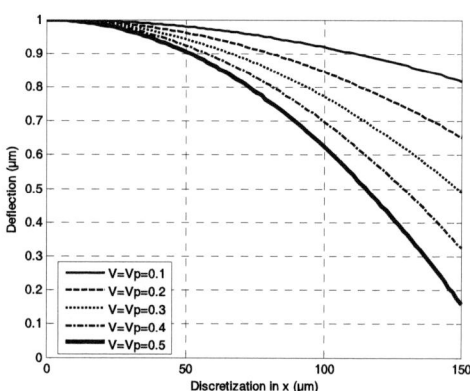

Fig. 4. Deflection versus the length of the microbeam

V. CONCLUSION

The primary of this work was about to develop a model with piezoelectric layers which were applied a moment in the cross section ob the microbeam, to enhance the switching voltage of MEM actuators. In the mentioned model, a dual actuation system was used. The nonlinear governing equations were linearized using of step by step linearization method, then the generalized differential method as a great means to discrete the linear forms of equations, was applied. The mentioned model was treated with three considerations. The first one was about an assumed bias voltage to the piezoelectric layers, which the actuation voltage was applied to the electrostatic areas. The next one was like pervious one where the bias voltage and actuation voltage were applied to the electrostatic areas and piezoelectric layers, respectively. The last consideration was about simultaneous actuation due to the applied voltage to the piezoelectric layers and electrostatic areas. The numerical results for the simple micro cantilever beam showed that the used numerical solutions are so reliable, fast and easy to program. The calculated results of the mention model and three considerations showed that in the dual actuation systems where piezoelectric layers are used, the switching voltages were tremendously decreased.

REFERENCES

[1]. E.F. Crawley and J. de Luis, 1987, "Use of Piezoelectric Actuators as Elements of Intelligent Structures," *AIAA Journal*, 25(10): 1373-1385.

[2]. Gh. Rezazadeh, A. Tahmasebi and Mikhail Zubtsov, Application of Piezoelectric Layers in Electrostatic MEM Actuators: Controlling of Pull-in Voltage, Journal of Microsystem Technologies, Vol.12, No. 12. , pp 1163-1170, 2006.

[3]. Chen W, Shu C, He W, Zhong T. "The application of special matrix product to differential quadrature solution of geometrically nonlinear bending of orthotropic rectangular plates." Journal of Computers and Structures 2000;74:65–76

[4]. Wang X, Bert CW." A new approach in applying differential quadrature to static and free vibration analysis of beams and plates." Journal of Sound and Vibration 1993;162(3):566–72.

[5]. Bert CW, Malik M. Free vibration analysis of tapered rectangular plates by differential quadrature method: a semi-analytical approach. Journal of Sound and Vibration 1996;190(1):41–63.

[6]. P. Osterberg. Electrostatically Actuated Microelectromechanical Test Structures for Material Property Measurement. PhD thesis, MIT, September 1995.

The Gas Sensing Potential of Nanocrystalline SnO_2 and In_2O_3 Powders Prepared by Mechanical Milling

Goib Wiranto, *Member IEEE*, Adiseno, I Dewa P Hermida, Roberth V. Manurung, Slamet Widodo, and Masbah R. T. Siregar

Research Centre for Electronics and Telecommunications
Indonesian Institute of Sciences (LIPI)
Jl. Sangkuriang, Bandung 40135
Ph. 62 22 250 4661 Fax. 62 22 250 4659
Email. goib@ppet.lipi.go.id

Abstract - The use of nanocrystalline metal oxide powders to increase the sensitivity of gas sensors has been the subject of this research. Of particular interest to our application in environmental monitoring is In_2O_3 and SnO_2, which are known, respectively, to have a high sensitivity to oxidising polutan gases such as NO_2 and O_3, and reducing pollutant gases such as CO and NH_3. Preparation of undoped nanocrystalline SnO_2 and In_2O_3 powders by mechanical milling via centrifugal action have been conducted. The technique used has allowed the reduction of grain sizes from 3 – 5 μm to below 100 nm with no contaminating carbon content, as confirmed by SEM and EDS spectra analysis. Furthermore, the FTIR spectra indicated that the SnO_2 nanopowders had a strong band at 671 cm^{-1} and In_2O_3 at 601 cm^{-1}. The resulting nanowpowders were then mixed with α-based terpinol to produce a thick film paste as an active material of gas sensors, and applied to an alumina platform consisting of AgPt heater and electrode tracks.

I. INTRODUCTION

ONLINE monitoring of gaseous pollutants is becoming more and more important as the need for high quality outdoor air is increasing. Such monitoring requires gas sensors and telemetric system capable of delivering information about pollutant parameters at real time from the designated measurement stations [1]. Among other types, perhaps gas sensors based on metal oxide semiconductors have been the most widely studied and applied in such environmental monitoring [2]. This is usually due to their simplicity and reliability. In addition, the availability of thick film technology has made their fabrication process very cost-effective, and thus many commercially available gas sensors have been produced using this technology.

Examples are Figaro TGS series gas sensors for the detection of CO, NH_3, H_2S, CO_2, and air contaminants.

A lot of work, however, still needs to be done to improve the sensitivity of gas sensors based on metal oxide semiconductors. In previous works, many have shown that the use of nanosized particles in the fabrication of thick-film gas sensors have significantly increased their sensitivity [3-5]. Although the detection limit can be decreased to below the allowable threshold set by international standards, the reproducibility of nanoparticles regarding their chemical composition and surface contamination remains problematic. On the other hand, the perfect control of nanoparticles surface as well as a good knowledge on the chemistry of the gas-solid interface are important in optimising the gas sensor performances.

The use of nanocrystalline metal oxide powders consisting of In_2O_3 and SnO_2 for gas sensors has been the subject of this research. Thick film technology will be used in the fabrication of the sensor structure, including the metal oxide paste and the subsequent screen printing of the heater, electrode, and sensitive layer on an alumina substrate. In this paper, the preparation of the nanocrystalline In_2O_3 and SnO_2 powders using mechanical milling will be presented, taking into account the effect of rotation speed, time, and sintering temperature.

II. THEORETICAL BACKGROUND

It has been known that the mechanism of gas detection by sensors based on semiconducting materials is due to variation in electrical conductivity caused by adsorption of gases on the semiconductor surface [6]. Thus, the surface of the semiconducting materials plays a major role compared to the bulk. Its reactivity to gases can be increased by increasing the spesific surface area of the semiconducting materials. By reducing the particle size of the semiconducting materials down to nanometer scale, a large increase in the spesific surface area can be obtained [7].

In thick film technology, the sensing layer is very porous and consists of numerous interconnected metal oxide grains in the form of single crystals or polycrystalline agglomerates. However, the high porosity enables the ambient gases to access the intergranular connections. As a results, the gas interaction can take place at the surface of individual grains, at the boundaries between grains, and at the interface between grains and electrode and substrates.

There are two kinds of reactivity that may take place at the semiconductor surface with the presence of gases. The first reactivity is due to the presence of oxidising gases such NO_2 and O_3, in which case oxygen adsorption at the semiconductor surface forms negatively charged oxygen species. As a result, the electrical conductivity of n-type semiconductor decreases.

The second reactivity is due to the presence of reducing gases such as H_2, NH_3, CO, and hydrocarbon in which case oxygen adsorption at the semiconductor surface causes electron injection to the conduction band. As a result, electrical conductivity of n-type semiconductor increases. This oxygen adsorbtion in ionic form can be described in the following reactions

$$O_2 + 2e^- \rightarrow 2O^-_{ads} \qquad (1)$$

$$O_2 + e^- \rightarrow O_2^-{}_{ads} \qquad (2)$$

III. Experimental

2.1 Powder preparation

Nanosized SnO_2 and In_2O_3 powders were prepared by mechanicall ball milling process of centrifugal-type. The process was done at room temperature. Initially, each semiconducting metal oxide powder, with average grain size of 3 – 5 μm, was mixed with acetone. The amount of SnO_2 powder was 35 g, and that of In_2O_3 was 15 g. The milling process was then performed with stainless steel balls with diameter of 12 mm. The number of balls used was 12, and the speed of rotation was maintained at 250 rpm. The milling process was done for 15 hours with 15 minutes paused after every hour.

Upon completion of the milling process, the nanopowder was let to dry in air and then half of each powder was sintered at 700 °C for 8 hours. Analysis of the nanopowders was then performed using SEM, EDS, and FTIR. To make a printable metal oxide paste, the nanopowders were mixed with α-based terpinol. To obtain a suitable viscosity, the α-based terpinol was added dropwise to the nanopowders followed by constant stirring. Figure 1 shows the steps involved in the preparation of the thick film paste based on the metal oxide nanopowders.

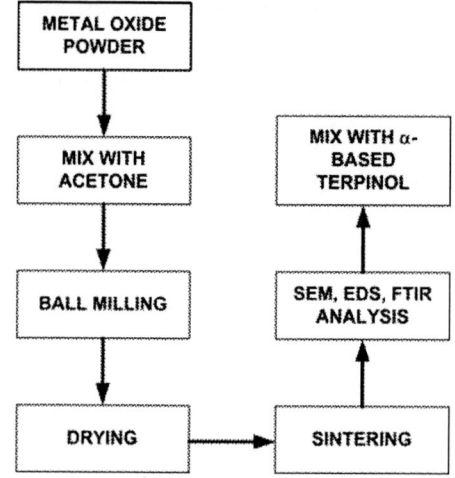

Fig. 1 The flowchart summarising the steps in the preparation of the thick film paste based on SnO_2 and In_2O_3 nanopowders

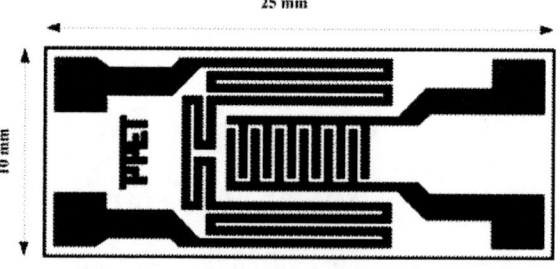

Fig 2. Layout of the gas sensor platform on Al_2O_3 substrate

2.2 Fabrication of the gas sensor platform

The gas sensor platform has been fabricated on an alumina (96% Al_2O_3) substrate. The size of the alumina substrate was 2 x 2 inch with a thickness of 0.7 mm on which up to 10 sensor platform can be fitted with dimension as shown on Figure 2. Each platform consists of an interdigitated AgPt electrode for measuring the sensor resistance and heater track for keeping the sensor at the operating temperature.

Initially, the alumina substrate was mechanically drilled on each pad as a means for the sensor interconnection with the electronic circuitry. The high operating temperature of the sensor has restricted the use of solder-based interconnection. An AgPt paste was then screen printed, dried, and fired.

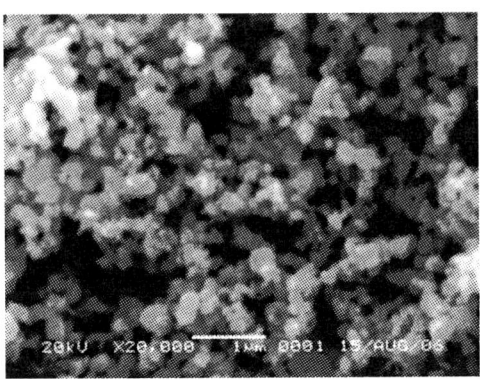

Fig 3. SEM picture of SnO_2 nanopowders, with an average nanograin size of less than 100 nm

IV. RESULTS AND DISCUSSION

4.1 Tin oxide nanoparticles

Figure 3 and 4 show the SEM pictures of the SnO_2 powders before and after sintering at 700 °C. As can be seen, an average grain size of less than 100 nm has been achieved. Although a complete particle reduction to nanoscale can not be obtained, but still a significant reduction of more than 97% from its original grain size of 3-5 µm is achievable with this technique. The actual quality of the nanopowder depends on many factors, such as the size of the original grain, the weight of the powder, ball-to-powder mass ratio, rotation speed, angle of collision, milling atmosphere, milling time, and milling temperature [8].

The effect of sintering at 700 °C on the SnO_2 powders is apparent from the formation of agglomarates on the nanopowders. This is in accordance with the previous work [9], which showed that increasing the temperature and time of sintering process will increase the average grain size, although the effect of sintering time is not significant at high temperatures above 600 °C.

The fact that the original SnO_2 powders were mixed with acetone before the milling process has made the resulting nanopowders have a high carbon content even after drying. The EDS spectra in Figure 5 shows that the SnO_2 nanopowders contain 16.82% of C. After sintering at 700 °C for 8 hours, all the C content can be completely removed. This is shown in the EDS spectra on Figure 6.

Fig 4. SEM picture of SnO_2 nanopowders with the same average grainsizes but showing the formation of agglomerates

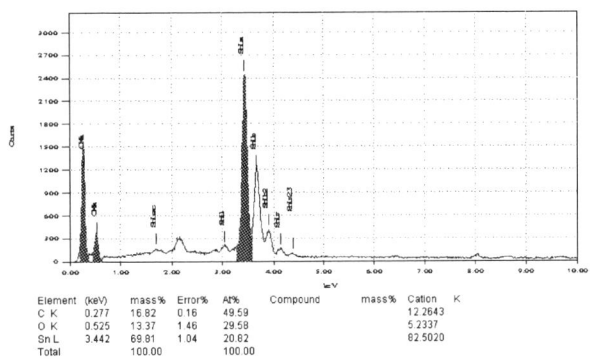

Fig 5. EDS spectra of the SnO_2 nanopowders before sintering at 700 °C

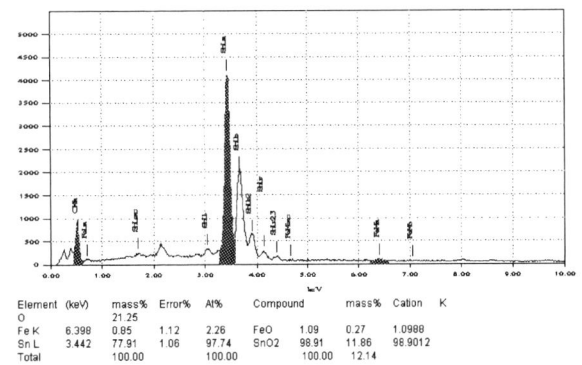

Fig 6. EDS spectra of the SnO_2 nanopowders with the C content completely removed

The main problem of using the centrifugal ball milling process is related with powder contamination. The EDS spectra on Figure 6 shows that a small amount of Fe (0.85%) present in the SnO_2 nanopowder after sintering. The present of Fe in the form of FeO (1.09%) could be due to particles left in the oven chamber. The addition of FeO in SnO_2 and its effect to the sensor response to gases will be interesting as indicated in the previous work [10].

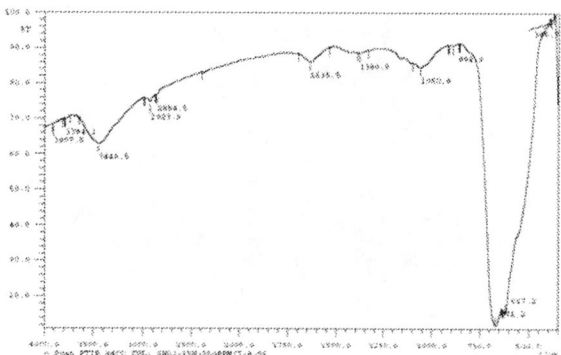

Fig 7. FTIR transmission spectrum of the SnO_2 nanopowders without sintering at 700 °C

Further analysis of the SnO_2 nanopowders has been performed using Fourier Transform Infrared (FTIR) transmission spectroscopy. FTIR analysis brings information on the interatomic bonds constituting the bulk, the chemical nature of the surface bonds, the possible presence of contaminating species on the surface, the nature of the adsorption sites, the free carrier density, etc [11]. As such, it is possible to correlate the surface reactions with the changes in electrical conductivity of the nanopowders upon reaction with gases, and this can be an important analysis to obtain the gas sensing potentiality of the nanopowders.

Figure 7 shows the FTIR spectrum of SnO_2 nanopowders before sintering process at 700 °C. The main chemical species present on the surface of the SnO_2 nanopowders are hydroxyl (OH) groups. They adsorb in the highest wave number region (peak at 3448.5 cm^{-1}) of the spectrum. The lower bands with the peaks at 1060.8 cm^{-1}, 1380.9 cm^{-1}, and 1635.5 cm^{-1} are problably attributed to hydroxide, molecular water, and Sn-O-Sn bonds. The SnO_2 bands have the peaks around 671.2 cm^{-1}. The FTIR spectrum of SnO_2 powders after sintering at 700 °C did not show a significant difference with that before sintering.

Fig 8. SEM picture of the In_2O_3 nanopowders milled at 250 rpm for 15 h

Fig 9. SEM picture of the In_2O_3 nanopowders sintered at 700 °C for 8 h

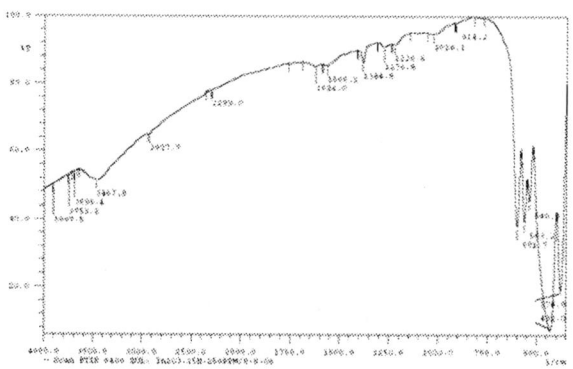

Fig 10. Infrared transmission spectrum of the In_2O_3 nanoparticles

4.2 Indium oxide nanoparticles

Figure 8 shows the SEM picture of In_2O_3 powders after ball milling at 250 rpm for 15 h. Unlike the case of SnO_2, the resulting particle sizes vary from less than 100 nm to 1 µm. It is suspected that the difference is caused by the higher grain size of the starting In_2O_3 materials. The effect of sintering the at 700 °C for 8 hours is similar to SnO_2. As shown in Figure 9, the nanopowders begin to agglomerate.

The FTIR spectrum of the In_2O_3 particles is presented in Figure 10. Like the case of SnO_2, the band in high wavenumber (peak at 3467.8 cm^{-1}) area correspond to surface hydroxyl (OH) groups. Several peaks at the lower wave numbers (from 1026.1 to 1624.0 cm^{-1}) are due to the surface reaction of the nanoparticles with the atmosphere.

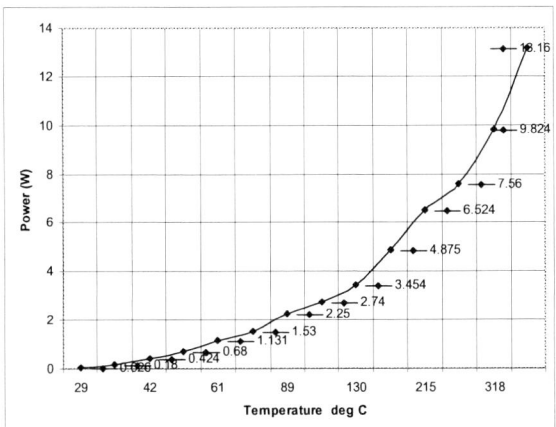

Fig 11. Temperature versus power characteristics of the AgPt thick film heater

4.3 Characteristics of the thick film heater

One of the requirements for metal oxide based gas sensors is their high operating temperature, which is normally between 100 – 300 °C. The temperature of gas sensors influences their sensitivity, selectivity, and response time. In this research, the choice of using AgPt as the heater and electrode materials have been based on the trade-off between cost and performance. An ideal heater and electrode materials would be platinum and gold, respectively. However, the cost of platinum and gold pastes for thick film technology is almost ten times that of silver. Unfortunately, silver is known to have an electromigration at high temperatures. Therefore for many applications, AgPt paste seems to be a good option.

The heater measurement was done by applying constant currents (Kenwood Reg. DC Power Supply PD18-30AD) and simultaneously measure the voltage drop and temperature (Lutron Dual Termometer TM 915A) across the heater tracks. Figure 11 shows resulting temperature versus power characteristics of the

AgPt thick film heater. Notice that in order to maintain the heater temperature between 100 – 300 °C, the amount of power required is around 3 – 9 Watts, which is typical for thick film technology.

V. CONCLUSION

In this paper, the preparation of SnO_2 and In_2O_3 nanopowders for gas sensing application has been described. The method of preparation has been based on the use of ball milling process of centrifugal type. With the rotaion speed of 250 rpm for 15 h, the grain sizes of both metal oxide powders were able to be reduced to less than 100 nm. Despite the presence of very small contaminating species, the EDS and Infrared spectra showed that the chemical nature of both nanopowders were stable, even after sintering at 700 °C for 8 h. Thus, the method of producing such nanopowders is potentially useful for gas sensor fabrication.

VI. FUTURE WORK

The work presented in this paper is the first step towards the design and fabrication of gas sensor system for environmental monitoring, in which the main parts consist of gas sensors and on-line telemetric system. Therefore many aspects still need to be done including characterisation of these nanomaterials for their responses to polutant gases, and the use of MEMs technology in the fabrication of the sensor platform. The work on this particular aspect is still in progress, and the results will be published elsewhere.

ACKNOWLEDGEMENT

This work is financially supported by LIPI Competitif Project No. 27.04/SK/KPPI/II/2006 on Development of New Methods in the Design and Fabrication of Gas Sensor System for Environmental Monitoring.

REFERENCES

[1] W. Tsujita, A. Yoshino, H. Ishida, and T. Moriizumi, "Gas sensor network for air-pollution monitoring", *Sensors. Actuators B*, vol. 110, pp. 304-311 (2005).

[2] N. Yamazoe, "Towards innovations of gas sensor technology", *Sensors. Actuators B*, vol. 108, pp. 2-14 (2005).

[3]   T. Mochida, K. Kikuchi, T. Kondo, H. Uono, and Y. Matsuura, "Highly sensitive and selective H_2S gas sensor from r.f. sputtered SnO_2 thin film", *Sensors. Actuators B*, vol. 24-25, pp. 433-437 (1995).

[4]   U. Kersen, "The gas-sensing potential of nanocrystalline SnO_2 produced by a mechanochemical milling via centrifugal action", *Appl. Phys. A: Material Sci. Process.*, vol. 75, pp. 559-563 (2002).

[5]   Korotcenkov, "Gas response control through structural and chemical modification of metal oxide films: state of the art and approaches", *Sensors. Actuators B*, vol. 107, pp. 209–232 (2005).

[6].   D. Kohl, "Function and applications of gas sensors", *J. Phys. D: Appl. Phys.*, vol. 34, pp. R125–R149 (2001).

[7]   S. A. Hooker, "Nanotechnology advantages applied to gas sensor development", *Proc. The Nanoparticles 2002 Conf.*, pp. 1-7 (2002).

[8]   U. Kersen, "The gas-sensing potential of nanocrystalline SnO_2 produced by a mechanochemical milling via centrifugal action", *Appl. Phys. A.*, vol. 75, pp. 559-563 (2002).

[9]   M. Sriyudthsak and S. Suphotina, "Humidity-insensitive and low oxygen dependence tungsten oxide gas sensors", *Sensors. Actuators B*, vol. 113, pp. 265-271 (2006).

[10]   K. Arshak and I. Gaidan, "Gas sensing properties of ZnFe2O4/ZnO screen-printed thick films", *Sensors. Actuators B*, vol. 111-112, pp. 58-62 (2005).

[11]   M. –I. Baraton and L. Merhari, "Electrical behavior of semiconducting nanopowders versus environment", *Rev. Adv. Mater. Sci*, vol. 4, pp. 15-24 (2003).

Reduction of Turn-On Voltage in a Single Layer Structured Organic Light-Emitting Diode using Nanocomposites SiO$_2$:PHF

Tengku Hasnan Tengku Aziz, Muhamad Mat Salleh, Mursyidah Umar and Muhammad Yahaya
Institute of Microengineering and Nanoelectronics (IMEN)
School of Applied Physics, Faculty of Science & Technology
Universiti Kebangsaan Malaysia
43600 Bangi, Selangor, MALAYSIA
Email: mms@pkrisc.cc.ukm.my

Abstract **Polymer light-emitting diode with ITO/PHF/Al structure has been fabricated, where PHF is poly (4,4'-diphenylene diphenylvinylene). The original device has turn-on voltage at 18.0 V. A reduction of turn-on voltage of this device is achieved by using the nanocomposites layer consisting of PHF and SiO$_2$ nanoparticles as an emitting layer in a single structured ITO/nanocomposite/Al polymer light emitting diode. The SiO$_2$: PHF was prepared by mixing 1.0 ml of PHF with 0.05 ml of SiO$_2$ colloidal solution. It was found that the spin-coated nanocomposites has reduced the OLED turn-on voltage to 9.0 V.**

I. INTRODUCTION

Organic light emitting diodes, OLEDs have recently gained a lot of interest in both academic and commercial fields owing to its potential applications for next generation flat panel color display [1-2]. Great efforts have been made to improve the performance of organic electronic such as reducing the operating voltage, besides improving the lifetime and stability.

There are several techniques to reduce the turn-on voltage of OLEDs; such using the hole injecting [3-4] or hole transport [5] layers or ITO anode treatment [6-7] due to the inefficiency of hole injection from bare ITO. As an alternative the usage of nanocomposite-emitting layer may reduce the turn-on voltage, as it will enhance the ITO-organic film interface. In recent years, there has been increasing interest in combining the nanotechnology advances with organic devices. Uses of different organic-inorganic nanocomposites in OLEDs have been reported [8-9].

This paper reports the effect of using PHF-SiO$_2$ nanoparticles nanocomposites on the performance of the devices with a structure of Al/PHF/ITO where ITO Indium Tin Oxide coated on the glass substrate as an anode, PHF is poly (4,4'-diphenylene diphenylvinylene) as a blue emitting layer and Al is aluminum as cathode.. SiO$_2$ is one of the metal oxides that exhibit unique dielectric and chemical properties that can be utilized as a charge accelerator.

It was found that the PHF-SiO$_2$ nanoparticles nanocomposites has reduce the turn-on voltage of the original OLED devices from 18.0 V to 9.0 V.

II. EXPERIMENTAL

The configuration of the fabricated structure for polymer light-emitting diode for the original device was ITO/PHF/Al while for the polymer-metal oxide device was ITO/nanocomposites SiO$_2$: PHF/Al (Fig.1)

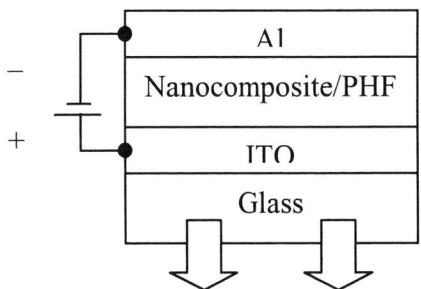

Fig. 1 Structure of the OLEDs device

0-7803-9730-4/06/$25.00 ©2006 IEEE   101

The nanocomposite emitting layer of PHF and SiO_2 nanoparticles were prepared through a direct mixing of PHF solution and SiO_2 nanoparticle solution. The nanoparticles SiO_2 were prepared by dissolve 0.014 gram of 50 nm nanoparticle SiO_2 powder in 4.0 ml of ethanol, EtOH. The mixture then was stirred for 16 hours. 0.05 ml of the mixture then dropped into 10 mg/ml of PHF in toluene. The SiO_2: PHF solution was then deposited on the ITO coated glass substrate with a sheet resistance of 5 Ω/m^2 using spin coating technique. The typical spinning speed and spinning time used were 3000 rpm and 40 s respectively. An aluminum layer as cathode was deposited onto the emitting layer through a mask by electron gun evaporation technique from a Molybdenum crucible at a chamber pressure of 2.5×10^{-5} mbar, yielding active areas of 0.71 cm^2. The current-voltage of the device was measured using Keithley 238 source measure unit. The electroluminescent spectra were measured by HR2000 Ocean Optic spectrometer. The PHF and nanocomposite films were also deposited by the same method on Si wafer. The surface morphology of the film was recorded at the area 7000 nm × 7000 nm using atomic force microscopy, AFM. All measurement was done at the room temperature. The photoluminescence properties of the nanocomposites were measured using the Perkin Elmer LS 55 Luminescence Spectrometer.

III. RESULTS AND DISCUSSION

Fig. 2 shows the photoluminescence spectra of PHF and PHF-SiO_2 nanocomposites thin films. The form and position of peaks in the spectra curves for both films are almost similar, except that the intensity for nanocomposites film is higher. This indicates that the presence of transparent SiO_2 nanoparticles do not affected the optical properties of organic PHF polymer. However, the SiO_2 nanoparticles have increased the number of PHF molecules deposited in the film, since they provide more deposition surface area.

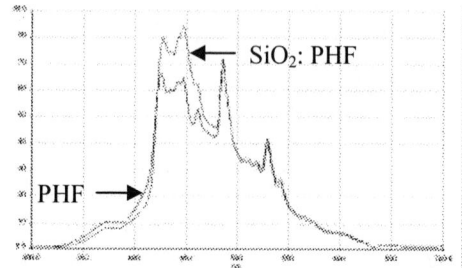

Fig.2 Photoluminescence graph of the devices

This was further confirmed with the electroluminescence curves of the devices as shown on Fig. 3, where the nanocomposites device has higher intensity.

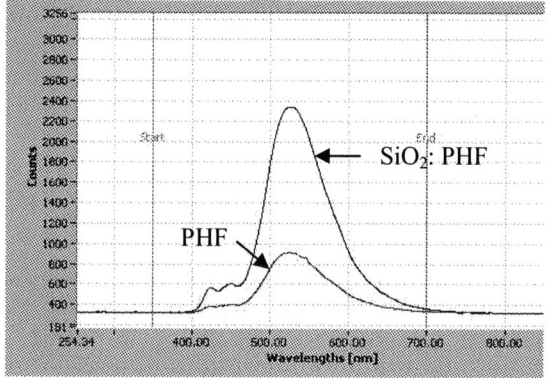

Fig. 3 Electroluminescence graph of the devices

Fig. 4 shows the current-voltage (*I-V*) curves of the PHF and PHF:nanocomposites devices. The *I-V* curve is the first indicator to determine whether the device is able to emit light or not. The device that shows rectifier diode behavior is able to produce light. All the OLED devices showed rectifier diodes. The turn-on voltage could not be determined directly from the *I-V* curve. Most researchers determine the turn-on voltage using a tunneling model of the Fowler-Nordheim theory [5]. In this model the tunneling current, *I* can be estimated by the equation:

$$I \propto F^2 \exp\left(\frac{-b}{F}\right)$$

where *F* is the external electric field and *b* is a parameter that depends on the barrier shape. The data from *I-V* curves were used to plot $\ln(I/F)$ versus $1/F$. The turn-on voltages were then estimated from these plots. It was found that the turn-on voltage nanocomposites device has reduced the OLED turn-on voltage from 18.0 V to 9.0 V.

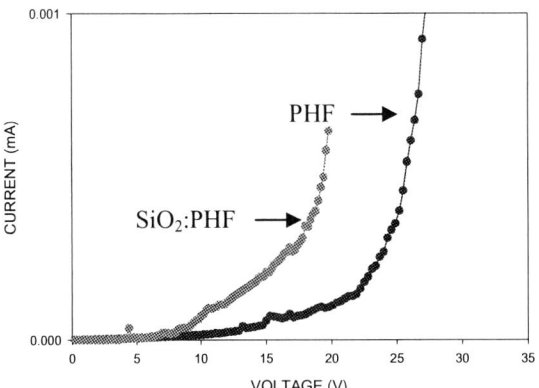

Fig. 4 Current-Voltage (I-V) characteristic of the OLED devices

In order to understand the reason for the voltage reduction in of the OLED device, the morphology of the polymer and nanocomposites thin film was study using AFM. The PHF and nanocomposites PHF: SiO_2 were spin coated on silicon substrates. The morphology of the films was recorded at area 700×700 nm^2 (Fig.5). The average roughness average for PHF-SiO_2 nanoparticles film was 3.64 nm while for original PHF film was 5.67 nm. It is believed that the SiO_2 nanoparticles has smoothened the PHF:SiO_2 nanocomposites film [8], as the polymer will bond into the homogenous SiO_2 nanoparticles.

We believed that low turn-on voltage and high

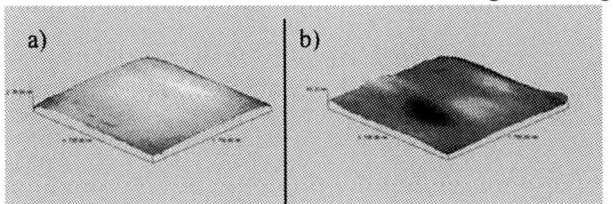

Fig.5 AFM image for a) nanocomposites b) PHF

efficiency of the nanocomposite SiO_2: PHF device is due to the adhesion of SiO_2 nanoparticles to the anode and the cathode as the smoother film will

creates a optimum heterojunction between the anode and cathode that attribute to apparently contradictory effects; improved of holes or electrons injection that leads to low turn on voltage and improved the performance [10-11].

IV. CONCLUSION

The ITO/SiO_2: PHF/Al and ITO/PHF/Al in the device structured of OLED devices have been successfully fabricated and characterized. It was found that the presence of nanoparticles will lower the turn-on voltage and improved the device performance.

ACKNOWLEDGEMENT

This work has been carried out with the support of the Malaysian of Science, Technology and Innovation, under the IRPA grant 03-02-006-SR007/04-04.

REFERENCE

[1] Berntsen, A., Croonen, Y., Liedenbaum, C., Schoo, H., Visser, R., Vleggaar, J & Weijer, P. 1998. Stability of polymer LEDs. *Optical Materials* 9: 125-133.
[2] Forrest, S.R., Burrows, P.E., Shen, Z., Gu, G., Bulovic, V & Thompson, M.E. 1997. The stacked OLED (SOLED): a new type of organic device for achieving high-resolution full-color displays. *Synthetic Metals* **91**: 9-13
[3] Zhang Zhi Feng, Deng Zhen-Bo, Liang Chun-Jun, "Organic light-emitting diodes with a nanostructured TiO_2 layer at the interface between ITO and NPB layers", *Displays,* Vol. 24, No. 4-5, pp. 231-234(2003).
[4] S. M. Tadayyon, H. M. Grandin, K. Griffiths, P. R. Norton, H. Aziz and Z. D. Popovic, "CuPc buffer layer role in OLED performance: a study of the interfacial band energies", *Organic Electronics*, Vol. 5, No. 4, pp. 157-166 (2004).
[5] Parker, I.D. 1994. Carrier tunneling and device characteristics in polymer light emitting diodes. *Journal of Applied Physics* **75**(3): 1656-1666.[6] C. N. Li, C. Y. Kwong, A. B. Djurisic, P. T. Lai, P. C. Chui, W. K. Chan, S. Y. Liu, "Improved performances of OLEDs with ITO surface treatments", *Thin Solids Film*, Vol. 477, pp. 57-62 (2005)
[7] X. Buwen, S. Yafeng, M. Meng and L. Chuannan, "Enhancement of hole injection with an ultra-thin Ag_2O

modified anode in organic light-emitting diodes", *Microelectronics Journal*, Vol. 36, No. 2, pp.105-108 (2005).

[8] C.C Oey, A.B. Djurisic, C.Y. Kwong, C.H. Cheung and W.K. Chan, "Nanocomposite hole injection layer for organic device applications", *Thin Solid Films* Vol. XX, pp. 1-6 (2005)

[9] S.A. Carter, J.C. Scott & P.J. Brook, "Enhanced luminance in polymer composite light emitting devices", *Applied Physics Letter* 71, pp. 11451147

[10]] Z.B. Deng and X.M. Ding, "Enhanced brightness and efficiency in organic electroluminescent devices using SiO_2 buffer layer.

[11] Shengwei Shi and Dongge Ma, "A pentacene-doped hole injection layer for organic light-emitting diodes", *Semiconductor Science and Technology*, Vol. 20, pp. 1213-1216 (2005)

Development of an Integrated Miniaturized Multi-Ion Flow Cell System for Water Quality Measurement

*Hiskia, #Masbah R.T. Siregar, *Robeth V. M

*Research Centre for Electronics and Telecommunication, Indonesian Institutes of Sciences
#Research Centre for Physics, Indonesian Institutes of Sciences
Indonesian Institutes of Sciences
Phone: +62 22 2504660; Fax: +62 22 250 4659
E-mail: hiskia@ppet.lipi.go.id

Abstract **An integrated miniaturized multi-ion flow cell system suitable for portable chemical analysis equipment and made using thick-film manufacturing technology have been developed. The system was to be applied initially to the determination of conductivity, temperature and pH concentrations in aqueous solutions, but is expected to have application to a much larger number of species. It is shown that these sensors can be used for the development of a low-cost system for the determination of the ion activity in a simple way by flow injection analysis measurement. The system is compact and consists of an integrated thick-film multi-ion flow cell having heaters, sensors, miniaturised pumps, hydrodynamic injector and an electronic controller.**

I. INTRODUCTION

Over the past few years, methods of detecting water pollutants have been based on conventional means, raging from the use of 'lacmus' paper for pH measurement to glass electrodes for ion detection. Such means has been known for lack of accuracy, robustness, and compatibility with other sensing elements.

With the advancement in microfabrication technology, however, it is now possible to construct an analytical system, which is not only capable of simultaneously detecting multiple parameters, but also small size and accurate. This will make it favourable in performing in-situ and real time environmental pollutant monitoring.

There has been a rapid growth in the development of field-portable analytical instrumentation capable of in-situ and real-time measurement. Recently, researchers have begun to explore the use of FIA in portable field analyzers [1,2]. Micro-machining techniques have been used recently in combination with processes such as piezoelectric pumping and microvalves to produce a new generation of miniaturised flow injection analyzers. Further, developments in flow injection analysis are concerned with multiple sensors, microarrays of sensors, micromachined flow systems, portable FIA, submersibles, remote monitoring FIA, and μFIA [2,3].

The significant advantage of the flow injection system is the possibility of miniaturisation and integration of practically all components of such systems, leading to micro total analysis systems (μTAS) [1-4]. A system that can improve the reliability and speed with which analyses are made, reduce the necessary sample size and analysis time, dramatically reduce the reagent consumption and allow access to novel types of analysis techniques [5-10]. The first step in integration and miniaturisation of μTAS has been taken in some laboratories to integrate the sensors, actuators and microelectronic components into a complete miniaturized chemical analysis system [7,11,12].

The use of thick-film fabrication has the potential to allow, in a variety of applications, the integration on the one substrate of electronic circuits together with the sensing element, heaters, and microchannels. Thick-film technology is suitable also for applications where extreme miniaturisation is not necessary. It is especially suited to disposable sensor applications. Furthermore, the application of mass production techniques in the field of miniaturized

0-7803-9730-4/06/$25.00 ©2006 IEEE

electrochemical sensors results in low cost devices that is highly reproducible. For these reasons, this work involves investigation into the use of an integrated miniaturised multi-ion flow cell system to improve the overall system performance. The aim of this project is to develop a sensitive, selective and robust thick film multi-ion microsensor incorporated into a flow injection analysis system for field monitoring of water quality parameters. It will be applied initially to the determination of conductivity, temperature and pH, with later application to other multiple parameters.

In this present research, a modular miniaturised chemical analysis system (μTAS) was developed. It consisted of a thick-film multi-ion flow cell and sensors combined with a planar miniaturised pump (Figure 1). This concept is based on a flow injection potentiometric sensing scheme. Initial work concentrated on the design of a flow cell, heater, sensors, miniature pump, and electronic controller. Interfacing to a personal computer (PC) was also envisaged.

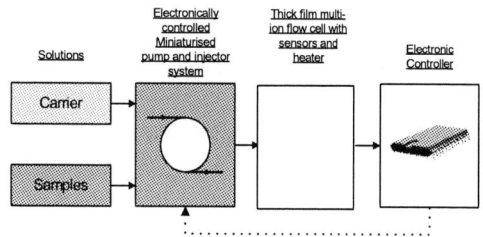

Fig. 1. Schematic diagram of the integrated miniaturised multi-ion flow cell system

II. THE SYSTEM MODULE

2.1 The miniaturized pump

In developing a μTAS, design and fabrication of suitable microfluid components (micropumps, microchannel, microvalves, reactions chambers, filters and mixers) is an essential step toward the integrated system. For a fully functional system careful selection of suitable pumping components is essential. In this research a specially designed (constant and rippleless) miniature planar pumps with electromagnetic driven actuators were developed for the delivery of the sample, carrier, and reference solutions through the flow cell.

The pump was of the reciprocating type and consisted of three main building blocks: an electromagnetic actuator, a pump chamber with a flexible pump membrane which acted as a capacitor (C), and passive circular silicon based check valves. The pump functioned by means of a rare earth magnet driven piston pressing or releasing a diaphragm, which was integrated with dual check valves. Due to the deflection of the pump membrane, the volume of the pump chamber changed. By means of two check valves, the liquid was periodically sucked in (through one valve) and forced out (through the other valve), thus creating unidirectional flow.

A novel approach in this project was the incorporation of electronic control of the pumps to achieve an accurate four phase pumping process. Four pumps were interconnected as a single pumping module according to the principles of rippleless pumps (Figure 2). Electromagnetic actuation was realized with a rare earth magnet (Neodymium iron boron, grade N35) plunger. The force developed by these actuators depends on the applied voltage/current and on the number of turns in the coils.

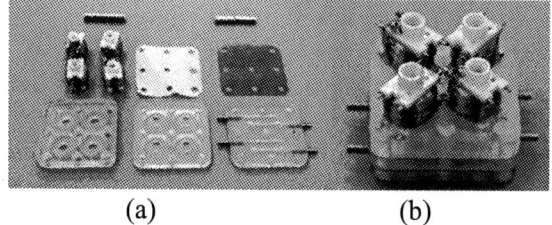

(a)   (b)

Fig. 2. Photograph of a planar miniaturised pump
(a) showing the pump component
(b) assembled system

The pump body consisted of a five layer sandwich: polycarbonate body for the first layer (29.5x 29.5x 5 mm), thin silicone rubber diaphragm membrane for the second layer (29.5x 29.5x 0.1 mm), polycarbonate as pump main volume for the third layer (29.5x 29.5x 3 mm), polymer (silicone) valves for the fourth layer (29.5x 29.5x 0.3 mm), and polycarbonate as tubing connection for the fifth layer (29.5x 29.5x 4 mm).

The miniaturized 4-phase pump has been evaluated and has the following specifications:
Electronically controlled flow rate to maintain constant, rippleless flow
The developed pump was self-priming under the applied conditions.
Low power consumption (about 10 watt)

DC power supply (12 to 22 volts, square wave signal compared to piezoelectrically driven pumps which need about 100 V).

Flow rate 1.5 to 3.5 ml min^{-1} which is higher than silicon type micropumps

Simple to operate

Chemical inertness : polycarbonate body and silicon based valve membrane

Pumps analyte in the liquid state

dimensions : 29.5 mm long x 29.5 mm wide x 25 mm high

Low maintenance, and

Low cost

2.2 Hydrodynamic Injector

Hydrodynamic injection is based on the principle of intermittent pumping; a fixed volume of sample is permitted to flow into a conduit and is intercalated into the carrier stream by what might be termed a stopped-flow technique. The pumps are activated in sequence, alternately and intermittently pumping carrier and sample solution. When the sample pump is on, the carrier pump is off and the sample zone is formed in the tubing. When the sample pump is off, the carrier pump on, the sample is pushed by carrier solution towards the detector. The waste line is fed back through the pump in order to block the flow path. Some kind of timing device (an automatic system) is needed to control the stopping and starting of the two pumps.

In this present research, the hydrodynamic injector technique was selected because it is simple, and makes it possible to inject precisely defined volumes of sample into the FIA system. Two 3-way valves (South Bend Controls Inc. USA) were used in order to block the flow path.

2.3 Totally integrated thick film flow cell

The integrated thick film multi-ion flow cell was specially designed to achieved the following objectives: (1) minimum dispersion, (2) minimum dead-volume, (3) good contact of the sample plug with each electrode integrated with the gradient chamber, (4) well defined and stable flow across each of the electrodes, and (5) efficient use of the sample plug by making the electrodes large relative to the total cell area.

The basic concept of the flow-cell developed to meet the above specification is illustrated in Figure 3. The main components were the microchannel, chamber, heater, and the sensors.

The flow cell consisted of two ceramic substrates; the top substrate contained microchannel, chamber, sensors and tubing connector, while the bottom substrate consisted of microchannel, chamber, heater and tubing connector.

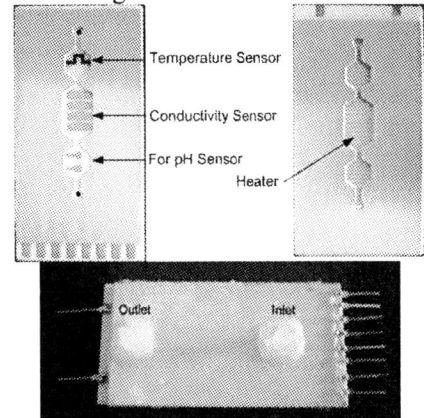

Fig. 3. Thick film multi-ion flow cell

The conductivity sensor used was a gold paste (Dupont 8352), the temperature sensor used was a thermistor paste(ESL NTC-2114), the pH sensor used was a pressed pellet Antimony electrode (Sb_2O_3)and the reference electrode used was a silver/silver chloride (Ag/AgCl). The total internal volume of the flow cell was around 85 mm^3.

The flow-cell had one inlet for periodically introducing the sample or carrier solution, via flow injection and one outlet. The integrated thick film multi-ion flow cell has been evaluated and has the following specifications:

Fabricated using thick film technology process

Multi-ion detection capability.

Laminar flow because of the design of the straight channel shape and choice of flow rates

Chemically resistant to the sample components through the use of high temperature thick film dielectric paste (ESL 4608).

Internal volume : about 85 μl

Flow rate capability : up to 32.5 ml/min

Type of solution handled : aqueous based

Tested operating temperature range : up to 36 ^{0}C

Temperature sensor : 20 ^{0}C to 40 ^{0}C

Conductivity sensor : 1 μS s/d 2000 μS

pH sensor : 2 to 11

Although flow cells of the type developed in this research can be fabricated to include a wide range of electrode materials to measure different ionic species, the author chose to fabricate electrodes only for conductivity, temperature and pH measurement. Following evaluation of individual components a complete miniaturised chemical analysis system unit was assembled and tested.

2.4 Electronic controller

A novel approach in this research was incorporation of electronic control of the pump to achieve an accurate four phase pumping process. The electronic controller consisted of a timing control for hydrodynamic injection, automated sample injection (3 injections for each sample), a 4-phase clock generator for the pumps, and an instrumentation amplifier for the sensors.

The design and objectives of the electronic controller were based around the operation of a 4-phase pump to maintain a constant and rippleless flow, including control or setting of the pump speed (flow rate) and the operation of the on/off valves. The prototype of a timing controller for hydrodynamic injector operation is shown in Figure 4.

Fig. 4. Prototype of timing controller for hydrodynamic injector

III. RESULTS AND DISCUSSION

In this present research, a modular miniaturised chemical analysis system was designed and evaluated for application in the determination of conductivity, temperature and pH of solution. The first objective was to evaluate the response characteristics of the gold electrode as conductivity sensor, the thermistor electrode as temperature sensor and the antimony (Sb) as pH sensor, in a flow analysis system.

The gold electrode showed good sensitivity to conductivity of solution in the range of 1 μS s/d 2000 μS. The thermistor electrode showed good linearity in the range of 20 ^{0}C to 40 ^{0}C. The antimony (Sb) electrode showed good sensitivity to pH in the range of pH 2 to pH 11. The antimony electrode exhibited responses between 48 - 50 mV / pH. Baseline to baseline time (Δt) was about 2 minutes. Antimony (Sb) exhibited adequate sensitivity for use as a pH sensor, but has several disadvantages: the electrode potential is influenced by (1) the level of oxygen dissolved in solution, and (2) any oxidants present [13]. In this research, the antimony electrode was only evaluated with standard pH buffers and needs further testing with real samples to evaluate its response to interfering ions. Bartlett et al. (1985) reported two further disadvantages, a variable temperature coefficient over the sensor's 0 to 11 pH span of useful measurement, and the need for frequent standardisation, due possibly to changes in electrode surface characteristics with time [13]. However, these pose no real problems in the system developed in this research.

IV. CONCLUSION

The design, development and evaluation of integrated miniaturised multi-ion flow cell system have been described. The responses tested were within acceptable ranges, demonstrating the successful design and fabrication of a miniaturised chemical analysis system product suitable for the measurement of conductivity, temperature and pH of aqueous solution. Practical benefits to be derived from a miniaturised system include the ability to analyze small volume samples with increased speed compared to traditional analysis systems. The system developed has the capability for application to the determination of a wide variety of analytes, with potential uses in the industrial process control, water quality monitoring and other environmental monitoring areas. The system developed can be used as a portable analyzer system.

ACKNOWLEDGMENT

The authors wish to acknowledge the contribution of Mr. Aam Muharam from Institute Teknologi Nasional (ITENAS), for helping in the development of the timing control circuit, Mrs. Lilis Retnaningsih and Mrs. Lia Maulani of the PPET LIPI for their assistance and technical support in laboratory work, always cheerfully given and also, thank for the financial support of the DIPA LIPI 2005.

REFERENCES

[1] Alexander, P. W., Benedetto, L. T. D., Dimitrakopoulos, T., Hibbert, D. B., Ngila, J. C., sequeira, M., and Shiels, D., *"Field-portable flow-injection analysers for monitoring of air and water pollution"*, Talanta, 43, pp. 915-925 (1996).

[2] Alexander, P. W., Dimitrakopoulos, T., and Hibbert, D. B. *"A Metallic Tungsten Sensor for On-line Monitoring of pH in a Portable Flow Injection Analyser"*, RACI 14th Australian Sysmposium on Analytical Chemistry, Adelaide, pp. 329.

[3] Haswell, S. J. *"Development and Operating Characteristics of Micro Flow Injection Analysis System Based on Electroosmotic Flow."* Analyst, 122(January), pp. 1R-10R (1997).

[4] Trojanowicz, M., and Alexander, P. W. *"Portable Flow-Injection Systems For Field Testing."* Biosensors for direct Monitoring of Environmental Pollutants in Field, D. P. Nikolelis, U. J. Krull, J. Wang, and M. Mascini, eds., Kluwer Academic Publishers, pp. 173-184 (1998).

[5] Berg, A. V. d., and Bergfeld. *"Development of mTAS concepts at the MESA research intitute"*, Analytical Methods and Instrumentation, TAS'96 conference, Basel, pp. 9-15 (1998).

[6] Dempsey, E., Diamond, D., Smyth, M. R., Urban, G., Jobst, G., IsabellaMoser, Verpoorte, E. M. J., Manz, A., Widmer, H. M., Rabenstein, k., and Freaney, R. *"Design and development of a miniaturised total chemical analysis system for on-line lactate and glucose monitoring in biological samples"*, Analytica Chimica Acta, pp. 346, 341-349 (1997).

[7] Branebjerg, J., Fabius, B., and Gravessen, P. *"Application of Miniature Analysers: From Microfluidic Components to mTAS."* Micro Total Analysis Systems, MESA Reasearch Institute, University of Twente, The Netherlands, pp. 141-151.

[8] Drost, S., Wormann, W., Ross, B., Koster, O., Konz, W., Edler, B., Schuhmann, W., Meixner, L., and Ferrenti, R. *"Microanalytical System for Environmental Control."* TRANSDUCERS '97, International Conference on Solid-State Sensors and Actuators, Chicago, pp. 931-934 (1997).

[9] Manz, A., Graber, N., and Widmer, H. M. *"Miniaturized total chemical analysis systems: a novel concept for chemical sensing."* Sensors and Actuators, B, 1, pp. 244-248 (1990).

[10] Manz, A., Verpoorte, E., Raymond, D. E., Effenhauser, C. S., Burggraf, N., and Widmer, H. M. *"m-TAS: Miniaturized Total Chemical Analysis Systems."* Micro Total Analysis Systems, MESA Reasearch Institute, University of Twente, The Netherlands, pp. 5-23.

[11] Schoot, B. H. v. d., Jeanneret, S., Berg, A. v. d., and Rooij, N. F. d. *"A Modular miniaturized chemical analysis system."* Sensors and Actuators B, 13-14, pp. 333-335 (1993)..

[12] Spiering, V., Bergveld, P., Elwenspoek, M., and Berg, A. v. d. *"A msensor array in fluidic systems for space applications."* National Sensor Conference, Delft, The Netherlands, pp. 143-146.

[13] Bartlett, J. D., Cherry, R. H., Connolly, S., Corrao, P. M., Durham, R. M., staff., E., Erk., G. F., Hammerstein, C., John N. Harman, I., and Houser, E. A. *"Analytical and Testing Instrumentation."* Process Instrument and Controls Handbook, D. M. Considine and G. D. Considine, eds., McGraw Hill Book Company, 6.1-6.201 (1985).

Design and Fabrication of Micromachined Gas Sensors

*Hiskia, #Masbah R.T. Siregar, *Moch.Muljono
*Research Centre for Electronics and Telecommunication, Indonesian Institutes of Sciences
#Research Centre for Physics, Indonesian Institutes of Sciences
Phone: +62 22 2504660; Fax: +62 22 250 4659
E-mail: hiskia@ppet.lipi.go.id

Abstract **With the advent of microfabrication technology, efforts have been made to make miniaturized gas sensor. This paper describes the use of micromachining technology in the design and fabrication of closed-membrane-type gas sensor. Micromachined gas sensors consist of an active area that comprises a platinum (Pt) heater, interdigitated gold (Au) electrodes and a gas sensitive layer (tin oxide) situated at the centre of a thin membrane which itself is supported by an outer frame of silicon (micro-heater). The micromachined gas sensors comprise the advantages of microsize, structures in identical, highly uniform, and low power consumption. The micromachined gas sensor developed initially to determine carbon monoxide (CO), with later application to others gases.**

I. INTRODUCTION

Recently, air monitoring and control is of increasing important for protection from environmental pollution. The air is a mixture of gases. Depending on the environment, it may contain various chemical pollutants, e.g. SO_x, NO_x, NH_3, H_2S, CO or volatile organic compounds. Some of these pollutants are either explosive or highly toxic like carbon monoxide (CO). CO is one of the main gases of interest in the field of gas sensor applications. Living in the contemporary industrialized world requires the protection against the influence the need of these hazardous substances, what induces the need for continuous monitoring of the specified compounds. However the precise instrumental analyses are very expensive and time consuming. The achievements in microelectronics provided the development and massive production of low-cost semiconductive sensors, based on gas sensitive materials changing its resistance when hazardous substances appear in the air [1].

The Si-based micromachined gas sensors are very promising because their fabrication procedure can be compatible with the Integrated circuit (IC) process, and this also makes it possible to integrated sensors and signal conditioning together. Micromachining technology was applied towards the development of specialized micro-hotplate devices in the early 1990s. The idea to markedly reduce the power consumption of gas sensors by means of micro-hotplate technology was published by the group at NIST in 1993. In the past decade, activities directed towards development of low power consuming micromachined gas sensors have increased substantially, including entry of the first commercial devices into the market. However, a continuing drawback of such implementations has been the fragility of the thin dielectric membrane used to achieve thermal isolation of the microhotplate from the rest of the chip [2-4].

Semiconductor oxides like SnO_2 and In_2O_3 are now to be excellent chemoresistive materials for gas detection. Eventually the tin dioxide became the dominating gas sensitive material and the best-understood type among the oxide-based gas sensors. Thick film SnO_2 sensors have been proved to be cheap, highly sensitive sensors for a variety of gases (depending on operation temperature and doping). Eventough, highly selective and sensitive SnO_2 sensors are not available yet. It is well known however that sensors selectivity can be fine-tuned over a wide range by applying the appropriate SnO_2 crystal structure and morphology, dopants, contact geometries, operation temperature or mode of operation, etc [5-8].

This paper was focused on the use of micromachining technology in the design and fabrication of closed-membrane-type gas sensor. Micromachined gas sensors consist of an active area that comprises a heater, interdigitated electrodes and a gas sensitive layer (tin dioxide) positioned at the centre of a dielectric thin membrane.

II. EXPERIMENT

A. Interdigitated electrodes

Micromachined gas sensors are processed by technique also used for chip technology. In design process, the use of computer-aided design in developing mask sets for foundry runs greatly simplifies the fabrication of multiple device variations. The size and geometry of contacts, for example, can be easily adjusted in the mask layout. In this research, there are seven mask has been used to develop micromachined gas sensor. There are as follows; top cavity, micro-heater, isolator, electrode, sensing element and bottom cavity which are draw by Corel Draw version 11. The structure of a micromachined gas sensor element illustrated in figure 1.

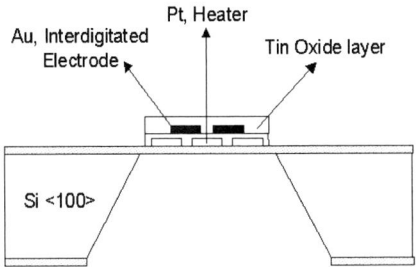

Fig. 1. Structure of Micromachined gas sensor element

Metal interdigitated electrodes, deposited on top of the formed membranes in the so-called active area, make the measurement of the gas sensitive layer resistance possible. The interdigitated electrode thereby located underneath the sensing film (tin dioxide layer), and the material for those electrodes are gold.

B. Micro-heater element

The property of change of conductivity after gas exposure that exhibits these materials is very weak at room temperatures. A maximum response temperature ranges from 200^0C to 400^0C and depends on gas, sensor material and doping. Therefore it is needed to design a heater that able to working in the range 25^0C to 400^0C.

The heater design is crucial for control of the temperature distribution along the active area.

Consequently, Pt-micro-heater structure was designed and located on Si_3N_4 membrane in order to improve the heat transfer, low power consumption and the Pt-heater enables the operation of the sensor at a well-controlled temperature distribution. In figure 2, shows micro-heater and interdigitated electrode layout.

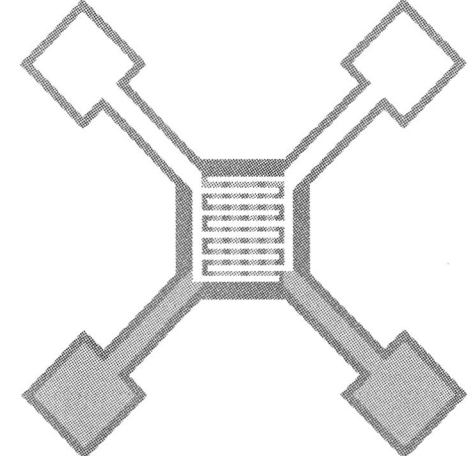

Fig. 2. The layout of Pt micro-heater (grey) and Au interdigitated electrode (yellow) was placed on the membrane (blue)

C. Closed-membrane

Basically there are two different structures for micromachined gas sensors using metal oxide as the material. One is the closed-membrane type gas sensor (back-side etched) and the other the suspended-membrane type (front-side etched) gas sensor. It is useful to greatly improve the thermal response and decrease power consumption of the sensors [2, 9]. In this research, the close-membrane-type is formed by anisotropic CsOH-etch from the backside.

D. Sensing layer

There are several metal oxide deposition techniques can be applied such as sputtering, thermal evaporation, spray pyrolysis, and screen printing. Spray pyrolysis refers to the thermal decomposition of gaseous species at a hot susceptor surface. The spray deposition scheme is particularly attractive because of its relatively fast rate (> 1000 Å/sec) and because it does not require a vacuum. In this research, the sensitive layer tin dioxide (SnO_2) is deposited onto a thin dielectric membrane by spray coating technique.

The process flow of micromachined gas sensor sketched in figure 3.

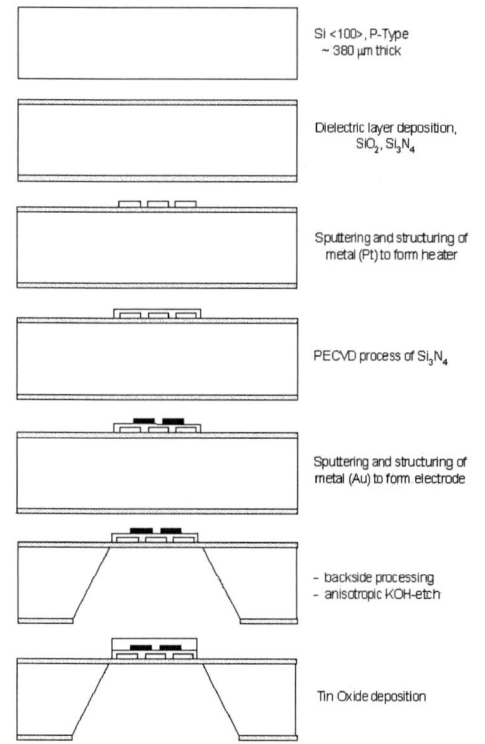

Fig. 3. Schematic process flow for the micro-machined gas sensor

III. RESULTS AND DISCUSSION

The micromachined gas sensor was developed based on closed-membrane-type. It is formed by anisotropic etching (CsOH) of silicon from the backside. The process flow for formation of the membrane is in either case fully IC-compatible, however a disadvantages lies in the need of double-side alignment for the bulk silicon etch from the backside, and the sloped side-walls obtained by anisotropic etching. The photograph of the SiO_2, Si_3N_4 membrane layer is shown in figure 4.

Fig. 4. The photograph of the SiO_2, Si_3N_4 membrane layer

The fabrication process described as follows. The starting material was a 3 inch, 380 μm-thick, p-type, <100> oriented, double polished silicon wafer. Subsequently, oxidized by dry thermal oxidation for 72 hours at 1100^0C to obtain 1.5 μm of SiO_2. Followed by a 1.5 μm thick Si_3N_4 film obtained by PECVD from ammonia and dischlorosilane. Then a 100 nm-thick heater of platinum is deposited by sputtering. The micro-heater is patterned by means of photolithography process. The heater is covered by a 1 μm-thick Si_3N_4 dielectric layer in order to avoid electrical shortcuts between the micro-heater and the interdigitated electrodes. Gold interdigitated electrodes with a thickness of 100 nm were deposited on the Si_3N_4. The space between electrodes is 30 μm. The photograph of micro-heater and interdigitated electrode are shown in figure 5 and 6.

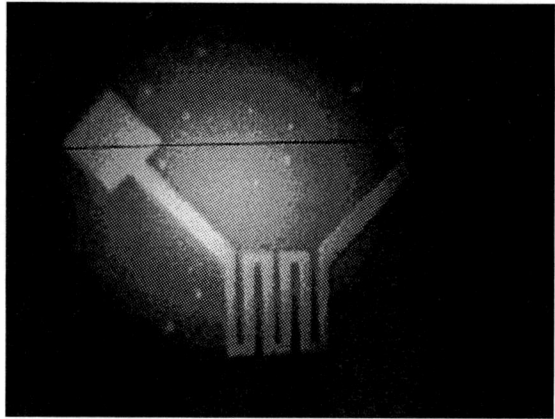

Fig. 5. The photograph of micro-heater

The deposition of the sensing layer is the most crucial part in the preparation of gas sensors. Normally the deposition is carried out as the last process step in the fabrication of a micromachined gas sensor. This way poisoning of standard equipment with tin oxide can be avoided and the gas sensing film can be protected from uncontrollable modifications during later process steps.

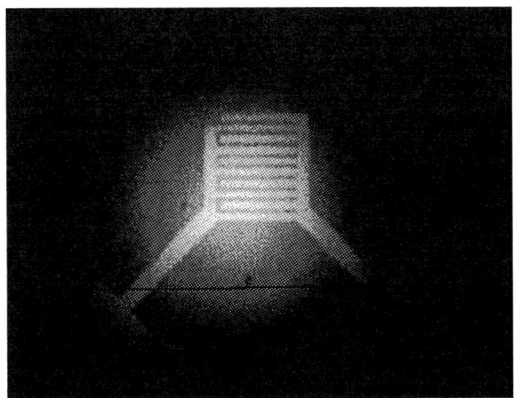

Fig. 6. The photograph of interdigitated electrode

IV. CONCLUSION

Micromachined gas sensor has been fabricated using microfabrications technology. It developed initially to determine carbon monoxide (CO), with later application to others gases. The micromachined gas sensors consist of the advantages of microsize, structures in identical, highly uniform, and low power consumption. The characterization of sensor responses against CO concentrations and heater performance will be conducted further. The size of sensor was 2 mm X 2 mm, and the active area was 680 μm x 680 μm.

ACKNOWLEDGEMENT

We would like to thank all member of Micro-device project for their encouragement and valuable discussions. This work is supported by KNMRT -RISTEK, Indonesia.

REFERENCES

[1] Licznerski B.,"*Thick-film gas microsensors based on tin dioxide*", Faculty of Microsystem Electronics and Photonics, Wrocław University Of Technology, 11/17 Janiszewskiego St., 50-372 Wrocław, Poland (2004),

[2] Isolde Simon, Nicolae Barsan, Michael Bauer, Udo Weimar,"*Micromachined metal oxide gas sensors: opportnities to improve sensor performance", Sensors and Actuators B, 73, pp. 1-26* (2001).

[3] S. Semancik, R.E Cavicchi, M.C Wheeler, J.E.Tiffany, G.E. Poirier, R.M. Walton, J.S. Suehle, B. Panchapakesan, D.L. DeVoe, "*Microhotplate platforms for chemical sensor*

research", Sensors and Actuators B, 77, pp 579-591 (2001).

[4] D.Vincenzi, M.A. Butturi, V.Guidi, M.C. Carotta, G. Martinelli, V.Guarnieri, S.Brida, B. Margesin, F.Giacomozzi, M. Zen, G.U. Pignatel, A.A. Vasiliev, A.V. Pisliakov, "Development of a low-power thick film gas sensor deposited by screen-printing technique onto a micromachined hotplate", Sensors and Actuators B, 77, pp 95-99 (2001).

[5] Hann S."*SnO2 thick film sensors at ultimate limits: Performance at low O2 and H2O concentrations; Size reduction by CMOS technology- DISSERTATION*", der Fakultät für Chemie und Pharmazie der Eberhard-Karls-Universität Tübingen, Germany (2002).

[6] Hyee-Jung Lee, Jae-Hoon Song, Young-Soo Yoon, Tae-Song Kim, Kwang-Joo Kim, Won_Kook Choi "*Enhancement of CO sensitivity of Indium Oxide-based Semiconductor Gas Sensor Through Ultra-thin Cobalt Adsorption*", Sensors and Actuators B, Vol. 79, pp. 200-205 (2001).

[7] Steven R. Davis, Alan V. Chadwick, John D. Wright, "*The effect of crystallite and dopant migration on the carbon monoxide sensing characteristics of nanocrystalline tin oxide based sensor materials*", Journal of Materials Chemistry, 8 (9), pp. 2065-2071 (1998).

[8] Ulo Kerson, "The Gas-Sensing Potential Of Nanocrystalline SnO2 Produced By Different Chemical Reactions and Milling Conditions", Report 70, Helsinki University of Technology Department of Electrical and Communication Engineering Laboratory of Electromechanics (2003).

[9] Yaowu Mo, Yuso Okawa, Motoshi Tajima, Takehito Nakai, Nobuyuki Yoshiike, Kazauki Natukawa "Micro-machined gas sensor array based on metal film micro-heater", Sensors and Actuators B, 79, pp. 175-181 (2001).

An Active Integrated Spiral Antenna System for UWB Applications

M. Jalali , A. Abdipour , A. Tavakoli and G. Moradi
Microwave and Wireless Communications Research Lab., Electrical
Engineering Department, Amirkabir University of Technology
(Tehran Polytechnic), Tehran, Iran
Email: jalali@cic.aut.ac.ir

Abstract— **This paper describes the implementation of an active integrated spiral antenna for UWB (Ultra-Wideband) applications. By integrating a low–noise amplifier with antenna, the voltage standing wave ratio (VSWR) can be kept low for a large bandwidth, resulting in an improved spiral antenna performance for UWB. For the designed circuit in the UWB frequency range (3.1-10.6 GHz), the gain of the LNA is better than 13dB, its noise figure is less than 3dB, S_{11} and S_{22} are less than -13dB, respectively.**

I. INTRODUCTION

As wireless communication applications require more and more bandwidth, the demand for wideband antennas increases as well. The spiral antenna has wide bandwidth, *i.e.,* good spectral efficiency compared to other planar antennas [1]. These antennas are based on Archimedes principle for a spiral, which can have many shapes depending on the design goals. Theoretically a spiral antenna with an infinite number of turns with optimal spacing between arms has infinite spectral efficiency and bandwidth. Practically we need to deal with the fact that the unlimited size is not possible, and the turns cannot be too close to each other without suffering the gain. Therefore the antenna should be integrated with a low-noise amplifier. As a very common application of UWB systems, RFID (radio frequency identification) [2, 3] is a technology that facilitates the tracking of objects, primarily for inventory tracking. RFID could improve business processes such as inventory management and efficiency in supply chain management.

UWB [4], is a wireless technology to provide a specification for a low complexity, low cost, low power consumption and high data-rate short-range wireless connectivity among devices. UWB has numerous eminent

advantages and is a good candidate of the wireless technologies that could be implemented in RFID systems:

- Low cost:
UWB is a pulse-based communication technology. Its transceiver could be greatly simplified, compared with carrier-based technologies. Hence, cost of UWB could be much lower than the traditional wireless communication technologies.
- High penetration ability:
The frequency spectrum of UWB is very wide. The low frequency components of the spectrum enable the signal penetrate deeply. Therefore, RFID tags could be placed inside the cargo.

Therefore the antenna should be integrated with a low-noise amplifier because the VSWR can be kept low for a large bandwidth, resulting in an improved spiral antenna performance for UWB.

Design and simulation were done using ADS (Advance Design System from Agilent). The main reason for choosing ADS is its ability to simulate systems and active components where the antenna is integrated.

II. OVERVIEW OF ANTENNA SYSTEMS

A. Passive Antenna Solutions

The spiral antenna is a popular frequency independent antenna. Fig. 1 shows the geometry of our proposed structure to utilize one monofilar spiral antenna for UWB.

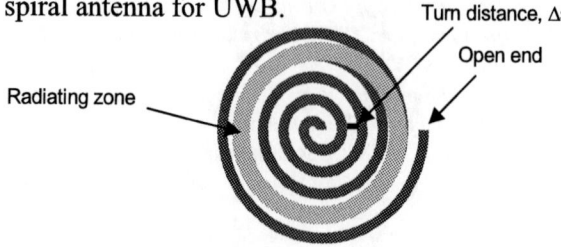

Fig. 1 Geometry of a monofilar spiral

As shown in Fig.1, the circumference of the radiation zone determines the radiation frequency. The circumference should be $>2\lambda$ [5]. If the current flowing backwards is smaller than that flowing forward, there is a net current flow through the spiral so radiation occurs.

Feeding to the spiral can be done either from the center or from the outside. In our design the feeding is done by a via to the center of the spiral (Fig.2).

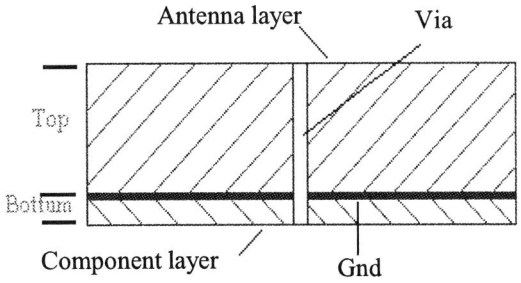

Fig. 2 Cross section of the three layer PCB.

As shown in Fig. 2, a ground plane is used to shield components from the antenna radiation. The input impedance depends on the line width together with the distance to the ground plane, because the characteristic impedance of the spiral arm is dependant to the line width as in the case of a microstrip line. The real part of the input impedance can be controlled by the line width, while the imaginary part is more difficult to be controlled.

B. Substrate

Table 1. Three layer PCB parameters

Dielectric top	RO4350B	h=1.524 mm
Dielectric bottom	RCC	h=0.05 mm
Dielectric	all layers	3.48 ± 0.05
Dissipation factor	all layers	0.004
Metal thickness	RO4350B(Top)	$18\ \mu$ m
Metal thickness	bottom	$25\ \mu$ m
Metal	all layers	$5.8 \times 10^{'}$ s/m
Surface roughness	all layers	0.001 mm

Table 1. shows material parameters used for simulations. Rogers RO43XX series are used for high frequency PCBs suitable for radio frequency (RF) modules[6].

C. VSWR Considerations

Fig. 3 shows the layout and VSWR simulation of a monofilar spiral antenna.Two radii with three different turn distances for each size of the antenna were designed and simulated. It is shown in all two of the simulation in Fig. 3b-c that a more dense turn distance (Δ s) provides a higher spectral efficiency. As mentioned in the introduction, in theory the infinite number of turns gives the infinite spectral efficiency and increase of radius improves VSWR. Futhermore, using of bigger radius increases the calculations/simulations time and antenna size. However, owing to the chosen line width and the need of spacing between turns and limited number of turns, the bandwidth is limited for this antenna. By integrating a low–noise amplifier with antenna, the VSWR can be kept low for a large bandwidth.

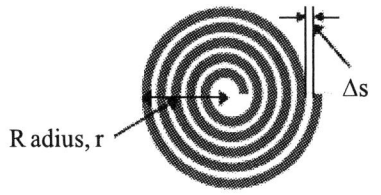

(a) Layout

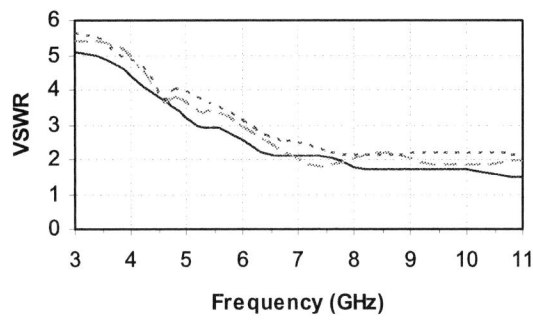

(b) r=40 mm

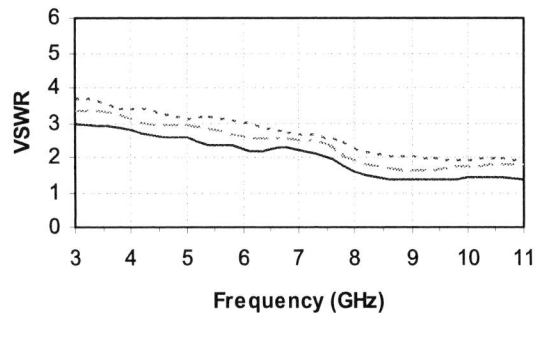

(c) r=50 mm

Fig. 3 Monofilar spiral antenna of 50Ω input impedance: (a) Layout, (b) VSWR simulation, r=40 mm, (c) VSWR simulation, r=50 mm.

III. DESIGN OF WIDEBAND AMPLIFIER

Wideband amplifiers are used in many radio frequency (RF) and high-data rate communication systems including satellite transceivers, pulsed radar systems, optical receivers, *etc*. For such applications, a distributed amplifier topology is often used because it can overcome the classical amplifier gain-bandwidth tradeoff by combining the outputs from several active gain elements in an additive fashion.
Traveling wave amplifier (TWA) was chosen among different topologies. Its advantages are high bandwidth and relatively high gain, though noise figure is a little higher. Schematic of a 4-stage TWA is shown in Fig. 4.

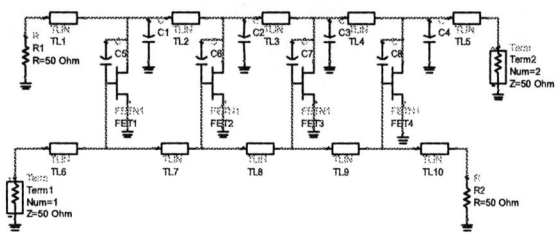

Fig. 4 A schematic of a 4-stage distributed amplifier

In conventional wideband amplifier, the gain-bandwidth product usually limits the high frequency gain. TWA overcomes this limitation. Gate capacitance of transistor is absorbed as a part of the transmission lines. Because of losses of transmission lines, there is an optimum number of stages for the gain and bandwidth, which can be approximated with the following equation:

$$N_{opt} = \frac{\ln(\frac{A_g}{A_d})}{A_g - A_d} \qquad (1)$$

where A_g and A_d correspond to the gate and drain lines attenuation respectively.

Based on parameters of the transistors, the optimum number of stages was calculated to verify (1). In these simulations, the stages number amplifiers were varied from 3 to 5. Distance between stages remained constant. The goal was to achieve as wide bandwidth as possible with minimum gain of 13 dB.
The LNA with four stages was observed to give the best bandwidth. Results are shown in Fig. 5
The length of the transmission lines (distance) between stages is optimized to get maximum

gain and bandwidth. Then, number of stages was remained constant (N=4) and distance between stages was constant.

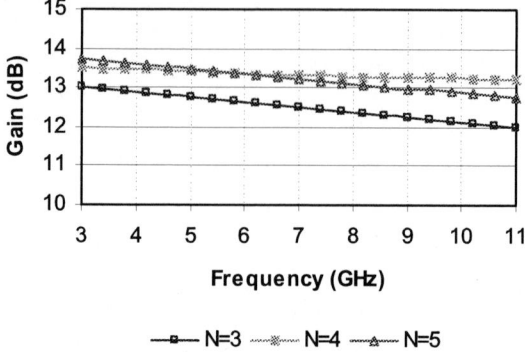

Fig. 5 The effect of number of stages in TWLNA on gain and bandwidth

The minimum noise figure and noise figure of the LNA is shown in Fig. 6. Noise figure is less than 2 in the whole frequency range. Fig. 7 shows the input matching and output matching performance of the LNA, which are denoted by S_{11} and S_{22}. They are less than -13dB in all of the frequency range. Thus input and output VSWR are less than 2.

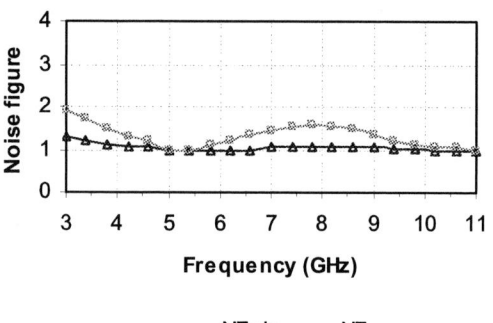

Fig. 6 Minimum noise figure and noise figure of the TWLNA circuit

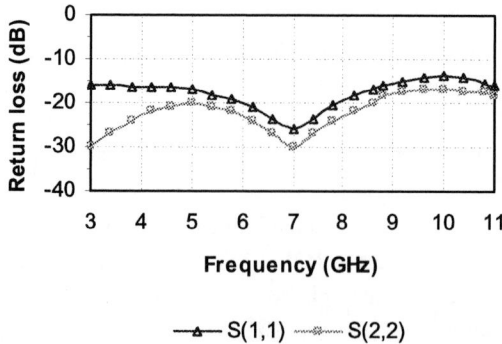

Fig. 7 Input (S_{11}) and output (S_{22}) reflection coefficients simulation.

The performance of the multi-band LNA can be further improved with several ways. One significant improvement would be use of specific-designed transistors. In the current design, the transistors are commercial (ATF-137XX) so that their performance or physical size cannot be changed. With specific-designed transistors, we could tune the width of the gate.

IV. RESULTS

After the predefined development, we integrated the different receiver components with the radiating element. The schematic of an active integrated spiral antenna is shown in Fig. 8.

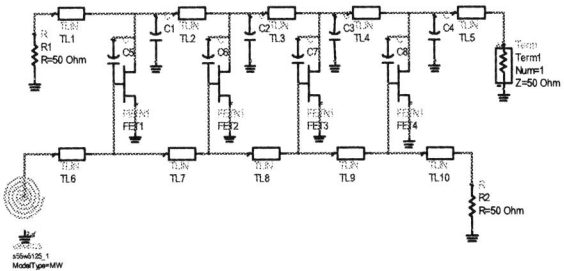

Fig. 8   Schematic of an active integrated spiral antenna

Fig. 9 shows the VSWR of the active receiving antenna to simulation values of the passive element. The curves indicate a VSWR less than 1.5 for all frequency band.

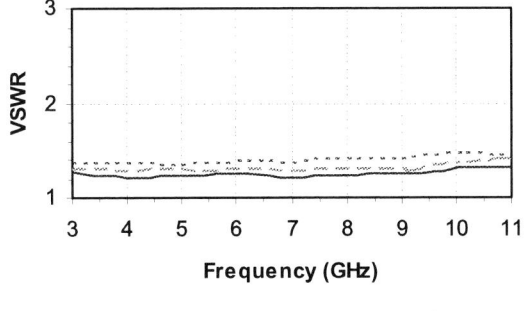

(a) r=40 mm

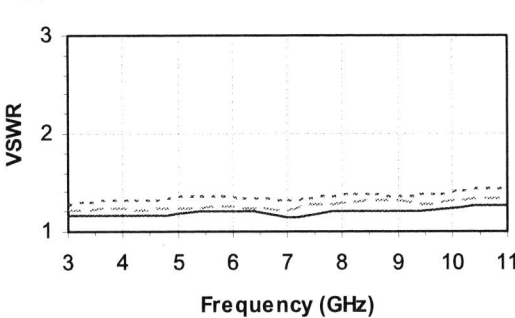

(b) r= 50mm
Fig. 9 VSWR simulation of the active antenna: (a) r=40 mm, (b) r=50 mm.

V. CONCLUSION

A monofilar spiral antenna for UWB by an input impdance of 50 Ω and different radiuses was designed and simulated.

- It is shown that one monofilar spiral antenna with a radius of 40mm covers partially the UWB frequency band, *i.e.*, above 5.1 GHz at VSWR<3.

- It is shown that one monofilar spiral antenna with a radius of 50mm covers the entire UWB frequency band at VSWR<3.

A traveling wave LNA is implemented for UWB RFID. In the frequency range from UWB, the gain of the LNA is better than 13 dB and its fluctuation is less than 0.3 dB. Noise figure is less than 3 dB. S_{11} and S_{22} of the LNA are less than -13 dB in this frequency range.

The studied active antenna was an integration of a distributed amplifier and a monofilar antenna. It is shown that this antenna is sufficient to cover the hole UWB frequency band, *i.e.*, 3.1-10.6 GHz at VSWR<1.5. This structure is a good condition for UWB systems.

ACKNOWLEDGEMENT

This work is supported in part by Iran Telecommunication Research Center (ITRC).

REFERENCES

[1] E . Gschwendtner, D. Loffler, W. Wiesbeck, "Spiral antenna with external feeding for planner applications" *Africon, 1999 IEEE*, vol. 2, 28 Sept.-1Oct. pp. 1011-1014, 1999.

[2] R. Glidden, C. Bockorick, S. Cooper, C. Diorio, D. Dressier, V. Gutnik, C. Hagen, D. Hara, T. Hass, T. Humes, J. Hyde, R. Oliver, O. Onen, A. Pesavento, K. Sundstrom, M. Thomas, "Design of ultra-low-cost UHF RFID tags for supply chain applications" *IEEE Communications Magazine*, vol. 42 ,pp. 140-151 Aug. 2004.

[3] X. Duo, T. Torikka, Li. Rong Zheng, M. Ismail, H. Tenhunen, E. Tjukanoff, "A DC-13GHz LNA for UWB RFID applications" *Norway, 2004 IEEE Norchip Conference*, Nov. 2004.

[4] G. Roberto Aiello and Gerld D. Rogerson, "Ultra-Wideband wireless Systems" *IEEE Microwave Magazine*, vol. 4, pp. 36-47, Jun. 2003.

[5] H. Nakano, Y. Okabe, H. Mimaki, J. Yamauchi, "A Monofilar Spiral Antenna Excited Through a Helical Wire" *IEEE Transactions on Antennas and Propagation*, vol. 51, no. 3, March 2003.

[6] M. Karlsson and S. Gong, "An integrated spiral antenna system for UWB",*Proc. IEEE 35th European Microwave Conf*, Paris, France, Oct 2005, pp 2007-2010.

Analysis of a Bilaminar Circular Piezoelectric Actuator for Micropumps

Juliana Johari, *Member IEEE* and Burhanuddin Yeop Majlis, *Senior Member IEEE*
Institute of Microengineering and Nanoelectronics
Universiti Kebangsaan Malaysia
43600 Bangi, Selangor, MALAYSIA
E-mail : burhan@vlsi.eng.ukm.my

Abstract — This paper presents a general approach to the problem of modeling diaphragm micropumps. This study utilized the finite element method to optimize the deflection of a bilaminar circular plate which consist of a single piezoelectric disc as an actuator and bonded to an elastic diaphragm of certain dimensions. Optimum actuator dimensions were fixed for a determined diaphragm dimensions.

I. INTRODUCTION

Micropumps are the basic element for many microfluidic devices, whose primary aim is the controlled handling of small doses of fluid. Typical flow rates are in the range of $\mu l s^{-1}$. Micropumps, which sizes are two orders of magnitude smaller than macroscopic pumps, can be actuated by piezoelectric, thermal or electromagnetic forces.

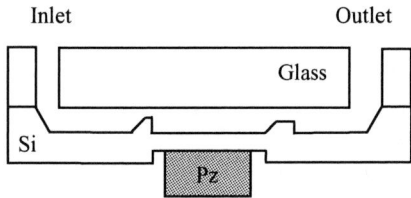

Figure 1. A schematic diagram of a piezoelectric micropump

In the microelectromechanical systems (MEMS), a piezoelectric actuator has been widely used in micropumps and other actuation devices. A piezoelectric actuator is chosen because it can produce extremely fine position changes down to the sub nanometer range. The smallest changes in operating voltage are converted into smooth movements. Motion is not influenced by friction or threshold voltages [1] [2].

Piezoelectric bilaminar structure is applied in the design. The simplest bilaminar structure is a layer of piezoelectric element bonded to one side of a passive elastic diaphragm. When an electric field is exposed to the piezoelectric, its element strains transversely and radially. The radial strain causes the surface of the passive diaphragm to expand or contract, which causes the entire bilaminar structure to bend.

Much work has been carried out to predict and optimize the behaviour of multiple-layered bilaminar piezoelectric in the Cartesian domain. Such analysis is vital to the understanding of bilaminar behaviour, but analysis based on Cartesian geometry cannot necessarily be applied to the circular case.

Dobrucki and Pruchnicki studied an asymmetrical bilaminar circular using a finite element method. An analytical model was presented which focused on the displacements of the bilaminar plates with the piezoelectric plate covering the entire diaphragm plate. They assumed free boundary conditions in their work [4]. Yanagisawa and Nakagawa presented an analytical equation to optimize the radius of a resonating piezoelectric actuator with a limited number of thickness ratio and only one choice of material constants [5]. Chee presented a very comprehensive review of analytical and numerical approaches to model and optimize the piezoelectric actuator behaviour [3].

Morris view that the simple case of a single circular piezoelectric element bonded to a circular diaphragm has not been analyzed in closed form for static actuation. In order to obtain a closed form analytical solution, major assumptions such as the entire passive diaphragm to be covered by a

0-7803-9730-4/06/$25.00 ©2006 IEEE

piezoelectric layer and no transverse shear have to be made. Such assumptions will result in a considerable disagreement between the analytical results and experimental results [6].

For this reason therefore a finite element method has to be the main methodology to the analysis and optimization of the piezoelectric microactuator.

II.FINITE ELEMENT ANALYSIS

1) Model Formulation

A two dimensional finite element model of a bilaminar circular plate is developed as in Figure 2. Idealized edge condition is considered fixed-edge. The idealized model used is to calculate the effects of the diaphragm radius-to-thickness ratio, α and also the actuator-to-diaphragm, Young's modulus ratio, β on the optimum actuator dimensions.

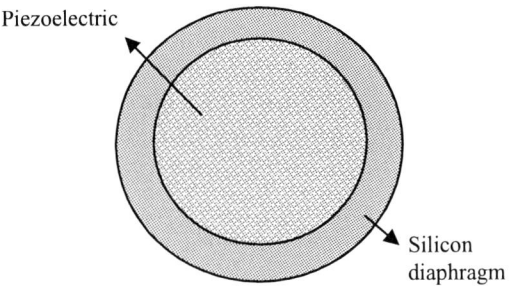

Figure 2. A single piezoelectric disc bonded on a silicon layer in a micropump actuator.

The thickness and radius of the piezoelectric has been chosen to a fixed value, as it is easily available and purchased. The thickness is fixed at 100 μm with a radius of 10 mm. In [8], the simulation carried out shows that for a fixed 0.1 mm thick piezodisc, a silicon diaphragm thickness of 32 μm is predicted to be optimal.

Figure 3 shows the relationship between the membrane thickness and the maximum deformation obtained from the analytical studies. The calculated results give impression that the thinner the diaphragm, the larger is the

deformation magnitude. Therefore, the thickness of the coupling parts should be as thin as possible.

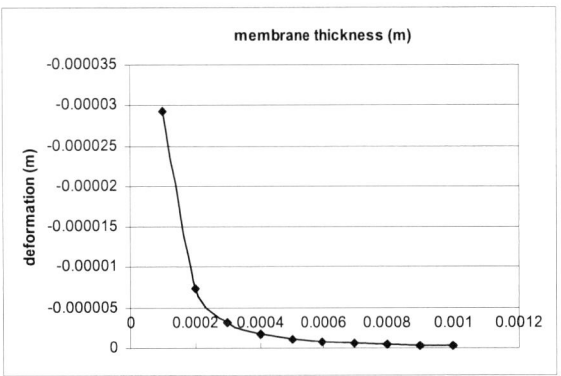

Figure 3. The relationship between the diaphragm thickness and the maximum deformation

The disc type piezoelectric actuator is chosen because they have relatively large deformation and fast mechanical response, but the generated pressure is small [9]. The chosen piezodisc is Philips PXE5 manufactured by Morgan ElectroCeramics. It is the hard, strong, higher charge coefficients and higher permittivity type; i.e. Produce higher displacement.

The piezoelectric disc is axially glued to the metallic plate (in this case the silicon diaphragm) and is polarized in the thickness direction.

Since many applications require displacements far greater than are possible with simple PXE ceramics operating in the d_{31} and d_{33} (piezoelectric charge constants) can be made possible.

PXE ceramics operating as actuators will operate in the d_{31} and d_{33} mode below the resonant frequency can produce into relatively low displacements of maximum 0.2 mm.

When a certain driving voltage is applied to the piezoelectric disc, the piezoelectric disc causes the diaphragm to oscillate. The deformation magnitude is proportional to the input voltage. A high input voltage can drive the deformation magnitude larger therefore; can produce a high flow output. To select a high driving voltage as an input draw some disadvantages in practical applications for it can destroy the system.

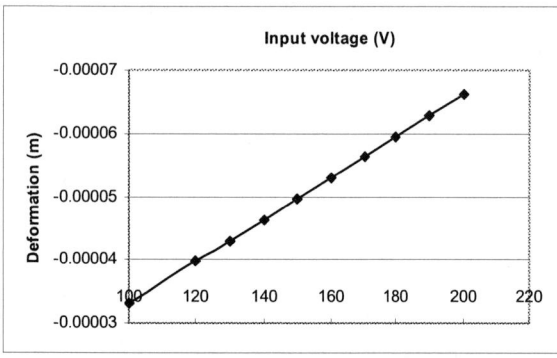

Figure 4. The relationship between the driving voltage and the membrane deformation magnitude

The analysis is justified when the operating frequencies are much less than the resonant frequency of the piezodisc actuator which is normally over 1 MHz. As in the driving element of a micropump for liquid transport, the typical operating frequency is in the lower kiloHertz range, while the resonance is 10 – 30 times higher.

The finite element analysis software CoventorWare Version 2005 is used to realize the model. Since the material constants are in C/N (μm/V) we have to select the piezo-electric strain and the coupling matrix would be d.

The piezoelectric equations written in the Strain-Charge form are:

$$[\varepsilon] = [D]^{-1}_{[E]=o}[\sigma] + [d]^{+}[E] \tag{1}$$

$$[\sigma_Q] = [d][\sigma] + [\xi]_{[\sigma]=0}[E] \tag{2}$$

where

$[\varepsilon]$ is the strain matrix μm /μm

$[\sigma]$ is the stress matrix μN/μm^2

$[E]$ is the electric field vector μN/pC, V/μm

$[\sigma_Q]$ is the electric charge density displacement vector, pC/μm^2

$[d]$ is the PZE-strain coupling matrix, pC/μN, μm/V

$[D]^{-1}_{[E]=0}$ is the material compliance matrix when there is no electrical field; i.e $[E] = 0$, μm^2 /μN

$[\xi]_{[\sigma]=0}$ is the electrical permittivity (dielectric constant) matrix when there is no stress; i.e. $[\sigma] = 0$, pC/Vμm, pF/μm

Piezoelectric charge constants, d in the order of x, y ,z, xy, zx and yz.

$$\begin{bmatrix} 0 & 0 & -215 \times 10^{-6} \\ 0 & 0 & 0 \\ 0 & 0 & 500 \times 10^{-6} \\ 0 & 0 & 0 \\ 515 \times 10^{-6} & 0 & 0 \\ 0 & 0 & 0 \end{bmatrix}$$

Piezoelectric permittivity, ε

$$\begin{bmatrix} 15.9 \times 10^{-9} & 0 & 0 \\ 0 & 15.9 \times 10^{-9} & 0 \\ 0 & 0 & 18.6 \times 10^{-9} \end{bmatrix}$$

For the passive diaphragm, Silicon is used with a Young's Modulus, E of 169 GPa and Poisson's ratio, v of 0.23.

2) Optimization

A routine in the CoventorWare with specific solver MemMech for piezoelectric analysis is used to optimize the actuator-to-diaphragm radius, r ratio and thickness ratio, t.

Constraints were set on the applied voltage, as the piezoelectric electric field strength limit is 300 V/mm. The driven voltage for the model is set at 150 V as not to destroy it due to heat practically but this will be tested later in the experimental test. The chosen voltage also relevant for it gives a reasonable maximum deflection.

3) Dimensional analysis

The choice of proper geometrical and elastic properties of a membrane is the first step in designing a micropump.

The optimization of the piezoelectric actuator and the silicon membrane for the fixed-edge conditions is carried out for several different values of α. This is to minimize the number of parameters needed to describe the results and also to maximize the usefulness.

The restrictions in this analysis are; first it is only concentrate on the piezoelectric PXE5 as the actuator and silicon as the diaphragm material (the investigation of every possible material is beyond the scope of this study); and secondly the piezoelectric diameter and thickness are fixed while the silicon diaphragm thickness is also fixed. Only the diaphragm radius is varying.

III. FEA RESULTS AND DISCUSSION

1) *Optimization of the bilaminar circular plates with fixed-edge support*

The displacement of the circular plate is at first compared to the analytical prediction to verify the accuracy of the FEA program shows agreement. The value of β is 0.32 and applied to all models.

For an equal diameter of piezodisc and silicon diaphragm, the maximum deflection is at 2.5 µm in the central region. The small deflection rather contribute to the value of α. In Figure 5, the centre maximal deflection shows an oval shape with the slope more towards the z-axis.

As the dimension for the diaphragm is equal to that actuator, it seems that the stress patterns are maximized at the fixed edge of the circular plate's shown in Figure 6. The stress is at 100 MPa.

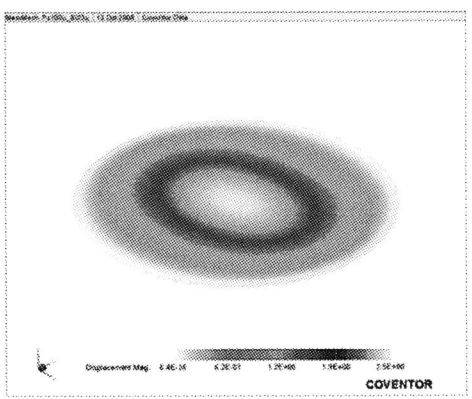

Figure 5. The Silicon diaphragm with an equal diameter of 10 mm with the piezoelectric disc.

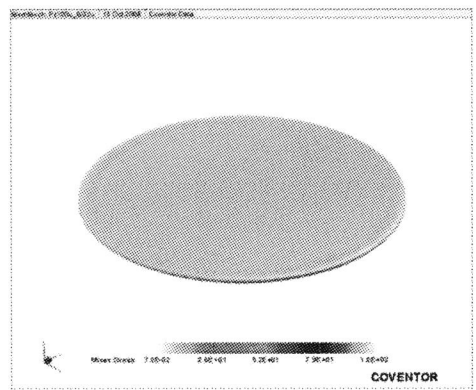

Figure 6. The stress patterns in an equal diameter circular plate.

By increasing the value of α, the maximum deflection will also increase.

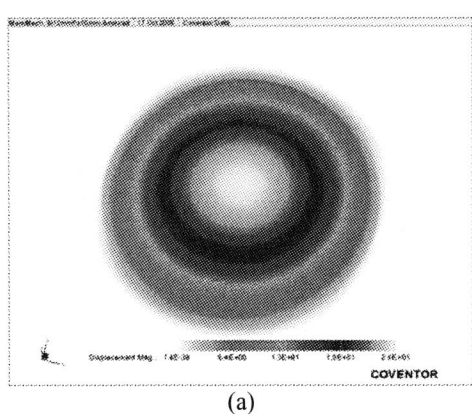

(a)

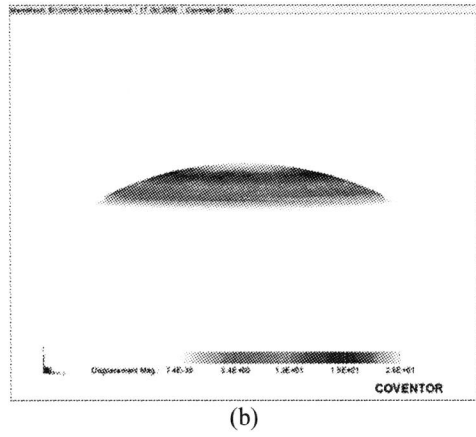

(b)

Figure 7. The Silicon diaphragm with a diameter of 12 mm and thickness 32 µm. (a) frontal view and (b) the side view with exaggeration.

Figure 7, shows the analysis of Si diaphragm with a diameter of 12mm, the maximum deflection at the centre shows a good circle shape which indicates that the maximum deflection is centrally symmetrical without sloping to other axis.

The maximum deflection occurs at 26 μm. The deflection value is important in the later design of the full system of the reciprocating displacement micropump.

Figure 8 shows the pattern of the stress which the maximum stress, 81 MPa is at the edge of the piezodisc, while the minimum stress is at the fixed edge of the silicon diaphragm.

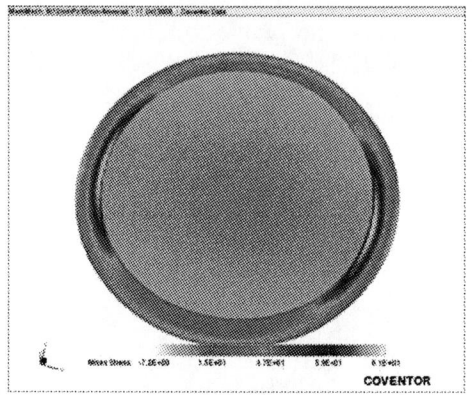

Figure 8. The maximum stress is at edge of the piezodisc.

Further investigation is done in the increment of silicon diaphragm diameter as to look at the maximum deflection at the central region. From the analysis (does not attached here as space consuming), we can observed that with the increase of silicon diameter and fixed piezoelectric disc, the area which unbonded to the piezodisc tend to oscillate. This is not needed in the structure as we do not want other than the concerned area to deflect.

The FEA results show good observation on the maximum deflection of the bilaminar circular plate in obtaining the optimize value on the α.

IV. ANALYTICAL MODEL OF SIMPLY SUPPORTED BILAMINAR CIRCULAR PLATE

An analytical model is developed to include the effect of realistic, passive-diaphragm edge support. A silicon-edge support, corresponding to the piezoelectric actuator, is modeled using the equations given in Equation (3).

There is a difference in diameter between the piezodisc and the pump diaphragm in the micropump design. For this purposes, a simplified model is illustrated in Figure 9. The model consists of a stack made up of a diaphragm and a piezodisc with an equal diameter.

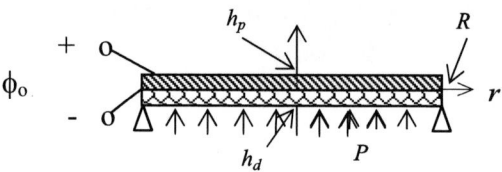

Figure 9. A simply supported bilaminar plate under the influence of an electric field and a single-sided pressure [Stemme].

The analytical formula for the axial deflection of a bilaminar plate composed of piezoelectric and silicon can be derived as:

$$w(r) = \frac{R^2 - r^2}{h_p^2 \left[3\left(1 - \alpha^2 \eta\right)^2 - 4\left(1 + \alpha\eta\right)\left(1 + \alpha^2 \eta\right)\right]}$$
$$\cdot \left[3d_{33}\left(1 + \alpha\right)\alpha\eta\varphi_o\right] \qquad (3)$$

where

$$\eta = s_{11}^E \cdot E \cdot \frac{\left(1 - v_p^2\right)}{\left(1 - v_m^2\right)}$$

R = overall radius (m)
h_p = the piezodisc thickness (m)
α = the ratio of the diaphragm thickness to the piezodisc thickness
ϕ_o = the applied electric voltage (V)
v = Poisson's ratio
d_{33} = the piezoelectric charge constant
s_{11} = the elastic compliance (m^2/N)
E = the Young's modulus of the diaphragm material (N/m^2)

The displacements of the actuators are associated with the piezoelectric charge coeffiecient, d_{33}.

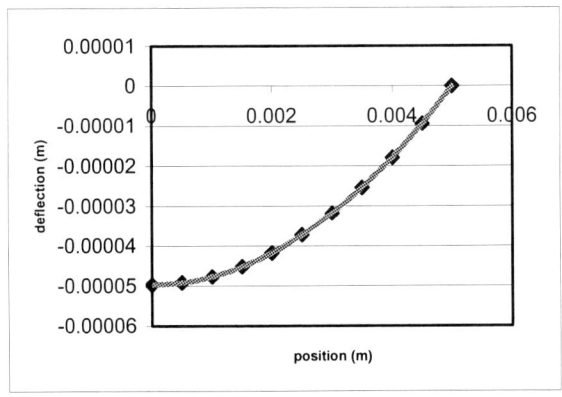

Figure 10.shows the deformation curves of the structure in the concave form of the membrane under experimental conditions (silicon membrane thickness: 32 μm; piezoelectric disc thickness: 100μm; Voltage: 150V).

From (3), we get the maximum deflection of 50μm, at $r = 0$.

V. CONCLUSIONS

By comparing the results for a fixed-edge from the finite element analysis and also the results from the analytical studies for the simply supported edge plates (based on the same physical parameters), we see that the maximum deflection for a simply supported plate is greater than that for the plate with the fixed-edge. Actually, many support mmbers tolerate some degree of flexibility, and a condition of true edge fixity is especially difficult to obtain. Finite Element Analysis certainly indicates that great improvement in the optimization of the bilaminar circular priezoelectric actuator in a micropump design. In either case, the analysis of a generic pump diaphragm is informative.

REFERENCES

[1] D. J. Laser and J. G. Santiago, "A review of micropumps," Journal Micromech. Microeng. 14. R35-R64 (2004)

[2] N. T. Nguyen, "MEMS-Micropumps:A Review," Journals of Fluid Engineering Vol.24 Ed.2 384-392 (2002)

[3] C.Y.K. Chee, L.Tong, G. P. Steven, "A review on the modeling of piezoelectric sensors and actuators incorporzted in intelligent structures," J. Intell. Mater. Syst. Struct. 9 3-19 (1998)

[4] A. B. Dobrucki, P. Pruchnicki, "Theory of piezoelectric axisymmetric bimorph," Sens. Actuators A 58 203-212 (1997)

[5] T. Yanagisawa, Y. Nakagawa, "Determination of optimum dimension for unimorph type piezoelectric loudspeaker," Trans. Ins. Electron. Informat. Commun. Eng. A J. 76 1261-1269 (1993)

[6] C. J. Morris, F. K. Forster, "Optimization of a circular piezoelectric bimorph for a micropump driver," J. Micromech. Miroeng. 10 459-465 (2000)

[7] Shifeng Li, Shaochen Chen, "Analytical analysis of a circular PZT actuator for valveless micropumps

[8] W. Wijngaart, H. Andersson, P. Enoksson, K. Noren and G. Stemme, "The first self-priming and bi-directional valve-less diffuser micropump for both liquid and gas," IEEE 0-7803-5273-4/00 (2000)

[9] M. Anjanappa, R. Angara and L. Si, "Micro/MEMS Actuators for microfluidic applications," Proceedings of ISSS 2005 SE-64-71 (2005)

[10] A. C. Ugural, "Stresses in Plates and Shells," Mc Graw Hill, New York 1981

Degradation of Single Layer MEH-PPV Organic Light Emitting Diode (OLED)

Nurjuliana Juhari[a], Wan Haliza Abd. Majid[b] and Zainol Abidin Ibrahim[b]

[a]Kolej Universiti Kejuruteraan Utara Malaysia (KUKUM), Kompleks Pengajian KUKUM, Jejawi,
02600 Arau, Perlis
email: juliana_juhari@yahoo.com
[b]Solid State Laboratory, Department of Physics, Faculty of Science,University of Malaya,
50603 Kuala Lumpur
[b]email: q3haliza@um.edu.my,drzai@um.edu.my

Abstract The degradation process of a single layer electroluminescence (EL) polymer MEH-PPV Organic Light Emitting Diode (OLED) with the MEH-PPV thickness of 57±3 nm is discussed. Typical structure of OLED fabrication is Al/MEH-PPV/ITO (Indium Tin Oxide). Electroluminescence (EL) spectrum indicates that the emission of MEH-PPV device is the yellow orange color. The device degrades by days as demonstrated by the increased in the turn on voltage obtained from *I/V* curves. The Scanning Electron Microscope (SEM) images show some bubbles emerge on the surface of the device after an electric field was applied to it.

I.   INTRODUCTION

Electroluminescence in conjugated polymer was initially discovered by Burroughes et al in 1990 [1] by using poly (p-phenylene vinylene) or PPV which produces green emission. In this work, Poly[2-methoxy-5-(2'-ethylhexyloxy)-1, phenylene vinylene] (MEH-PPV) with 200,000 molecular weight was used in the fabrication of organic light emitting diode (OLED). Organic Light Emitting Diode (OLED) consists of a multilayer structure of an emitting layer sandwiched between a transparent anode, and a metallic cathode [1]. MEH-PPV as the emitting layer used in this work is a very attractive conjugated polymer that can be used in the application of display technology [2]. This material produces orange color when an electric field is applied across the active layer. One of the advantages of this conjugated polymer is its solubility in common organic solvent and its ability to produce a good optical quality film through a simple processing technique such as spin coating. One important factor related to device fabrication is lifetime of the cell due to oxidation effect which is investigated in this report. Other factors suggested which contribute to degradation mechanisms for OLEDs include, crystallization of the amorphous organics films [3], contact delamination, dark sport formation, quality of the polymer materials, and changes in morphology at the metal/organic interfaces. These have been identified as causes of device instability [2,4]. According to Scott et al [5], Indium Tin Oxide (ITO) provides source of oxygen which causes oxidation. The oxidation process quenches the luminescence, and as a result the carrier mobility of the device is depleted. Encapsulating the device can eliminate the degradation process. Another way to reduce the degradation is by coating polyanaline (PANi) between ITO and the hole-transport layer in order to avoid penetration of oxygen into the emissive layer of the device.

II.   EXPERIMENT

The red color, MEH-PPV used in this work was purchased from H.W. Sands with purity of > 99.9%.Typical structure of OLED device is Al (Aluminium)/MEH-PPV/ITO (Indium Tin Oxide) as shown in Figure 1. The MEH-PPV was then coated onto ITO substrate with spin speed of 4k rpm for 10 seconds. The thickness of MEH-PPV film as measured by an ellipsometer was 57±3 nm. The sample was dried in a vacuum chamber at 40°C for 2 hour. After 2 hours, the aluminium as the top electrode of the device was deposited onto the polymer layer in a high vacuum chamber (1x10^{-6} Torr) to form a complete single layer device of an active area of 23.64 ± 0.04 mm^2. The device was then applied with an electric field of 10 V to measure its *I-V* characteristic. Electrical data were obtained from Keithly 236 Source Measurement Unit (SMU). The electrical measurements were taken for 9 days. The SEM images were obtained after the device has been applied with an electric field.

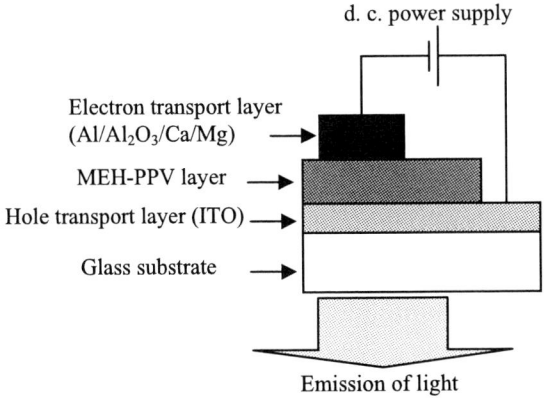

Figure 1: Configuration configuration of the ITO/MEH-PPV/Al device

III. RESULTS AND DISCUSSION

EL spectrum of MEH-PPV polymer layer prepared from chloroform solution for solution concentration of 1 mgml^{-1} and spun at constant speed of 4k rpm is presented in Figure 2. The electroluminescence spectrum in Figure 2 indicates that emission spectrum of MEH-PPV is in the yellowish region. The EL peak wavelength is 570 nm which can be assigned to transition of (0-0) vibration bands [6] and the corresponding emission energy is 2.1 eV.

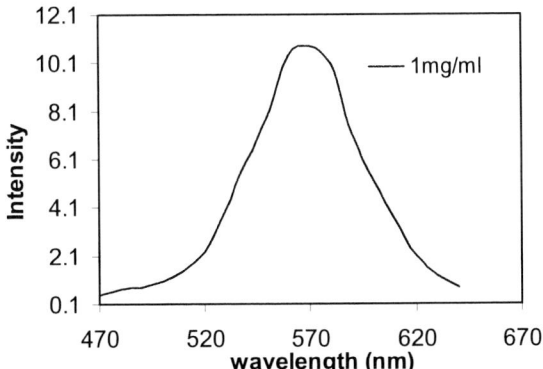

Figure 2: EL spectrum of MEH-PPV film

Figure 3 shows the *I-V* curve of the device taken on different days. The measurement was repeated until 9 days to observe the degradation experienced by the device. Day 1 shows the turn on voltage of 5.5 ± 0.1V and the device produces higher current output compared to the other day. After day 2, the turn on voltage suddenly increases significantly to 5.8 ± 0.1V, and finally to 6.5 ± 0.1V on day 9. The current output is drastically reduced from day 1 to day 9. This is probably due to the formation of non-emissive species during OLED operation [7]. Another cause of degradation is probably due to the formation of oxide layer when the aluminum is heated up by the thermal evaporator. According to Shen at al [8], the excited state reaction with water and oxygen can leads to quench-site formation which degrades the efficiency of the device. Scott in 1995 [5] explained that oxidation of MEH-PPV lead to luminescence quenching and reduction in conjugation. As a result the carrier mobility of the device would be depleted. OLED is very sensitive to oxygen [9,10] and oxidation process has been identified as the key contributor to the degradation of the active materials during device operation.

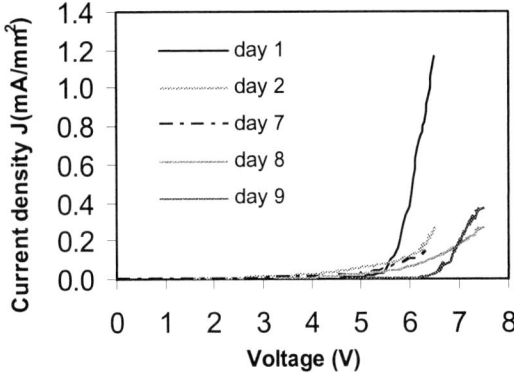

Figure 3: : Comparison of turn on voltage for spin coated MEH-PPV from solution of 1 mgml^{-1} taken during day1 to day 9 of fabrication

Figure 4 (i) and (ii) represents the SEM images taken with two different scales. Figure 4 (i) represents the full image of the active device and Figure 4 (ii) represents the enlarged image of the square pattern in Figure 4 (i). The SEM images of MEH-PPV device were taken after applying an electric field for one day. The white circle in Figure 4 (ii) shows an irregular object of the device.

The irregular object is most probably a dust particle as reported in reference [11]. When an electric field of 10 V was applied to the device, the bubble patterns were formed as indicated by the black circle in Figure 4 (ii). Considering the high current density needed to operate the device, it is probably that the bubble pattern arose due to the gas formation in the organic layer caused by local Joule heating [12]. Through the local joule heating, the molecule in the emissive layer is inter-diffuse which leads to decrease in local conductivity [13]. From $J=\sigma E$, where J is current density, E is electric field and σ is conductivity, if conductivity decreases, current density is expected to decrease as indicated by results shown in Figure 3.

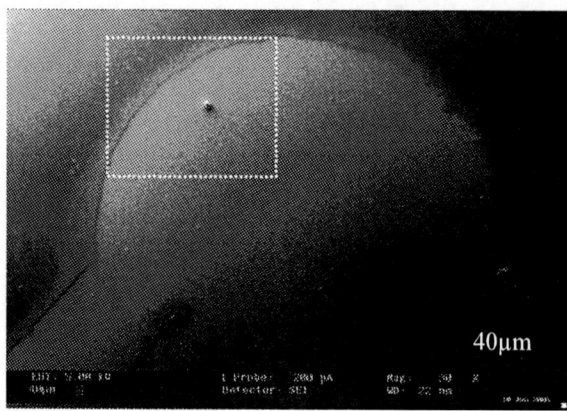

Figure 4 (i): The SEM images of the device structure of ITO/MEH-PPV/Al with scale 40μm after applying an electric field in day 1.

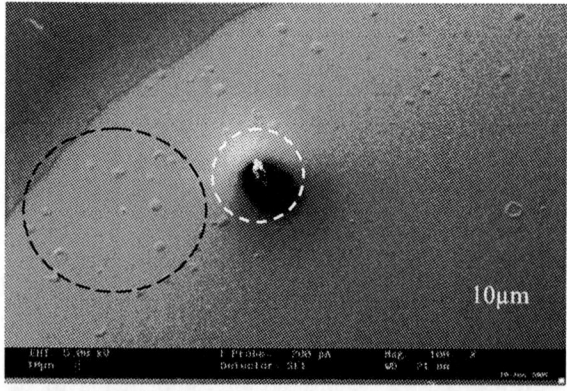

Figure 4 (ii): The enlarged SEM images of the square pattern in Figure 3 (i) of the device with resolution 10 μm

IV.   CONCLUSIONS

Electroluminescence (EL) spectrum is in the yellowish region with emission energy of 2.1 eV which corresponds to peak wavelength at 570 nm. Device degradation is related to oxidation process which occurs in the polymer layer. The turn on voltage of the device increased as the device degraded by days and drastically reduced the current density after day 1. SEM indicates that the deposited polymer electrode surface is affected by the application of the electric field.

V.   ACKNOWLEDGEMENT

The authors would like to thank the Government of Malaysia for providing research Grant under IRPA RMK8 : 09-02-03-1028, Vote F and KUKUM Short Term Grant: 9003-00026.

REFERENCES

[1]   Burroughes,H., Bradley, D.D.C., Brown, A.R., Marks, R.N., Mackay, K., Friend, R.H., Burns, P.L and Holmes, A.B., 1990. Light Emitting Diode based on conjugated polymers. *Nature*, 347, 539-541.

[2]   Liu,Jie, Shi,Y, Ma, L., and Yamg, Yang, 2000. Device performance and polymer morphology in polymer light emitting diodes: The control of device electrical properties and metal/polymer contact. *Journal of Applied Physics,* 88(2), 605-609.

[3]   Becker, H., Burns, S.E., and Friend, R.H., 1997. Effect of metal films on the photoluminescence and electroluminescence of conjugated polymers. *Physical Review B,* 56(4), 1893-1905.

[4]   Aguilhon, L., Bourgoin, J-P., Barraud, A., and Hesto, P., 1995. Thin film organic channel field effect transistor. *Synthetic Metals*, 71, 1971-1974.

[5]   Scott, J.C., Kaufman, J.H., Brock, P.J., DiPietro, R., Salem, J., and Goita, J.A., 1995. Degradation and failure of MEH-PPV light emitting diodes. *Journal Applied Physics*, 79(5), 2745-2750.

[6]   Huser, Thomas, Yan, Ming, and Rothberg, Lewis J., 2000. Single chain spectroscopy of conformational dependence of conjugated polymer photophysics. *Proceedings of the National Academy of Sciences*, 97(21), 11187-1119.

[7] Kafafi, Zakya H., 2000. Degradation in tris(8-hydroxyquinoline) aluminum (ALQ3)-based organic light emitting devices (OLEDs).*In:* Aziz, Hany, Popovic, Z.D., Hu, Nan-Xing, DosAnjos, P., and Ioannidis, A., eds. Proceedings of Organic Light Emitting Materials and Device IV, San Diego 31 July-2 August 2000. The International Society for Optical Engineering, 251-255.

[8] Kafafi, Zakya H., 2000. Degradation mechanisms in Organic Light Emitting Diode. *In:* Shen, J., Wang, D., Langlois, E., and Yang, J., eds. Proceedings of Organic Light Emitting Materials and Device IV, San Diego 31 July-2 August 2000. The International Society for Optical Engineering, 238-242

[9] Atreya, M., Li, S., Kang, E.T., Neoh, K.G., Ma, Z.H., Tan, K.L., and Huang, W., 1999. Stability studies of poly (2-methoxy-5-(2'-ethyl-hexyloxyl)-p-phenylene vinylene) [MEH-PPV]. *Polymer Degradation and Stability*, 65, 287-296.

[10] Czerw, R., Carroll, D.L., Woo, H.S., Kim, Y.B., and Park, J.W., 2004. Nanoscale observation of failures in organic light-emitting diodes, *Journal of Applied Physics*, 96(1), 641-644.

[11] Nguyen, Thuc-Quyen, Martini, I.B., Liu, J., and Schwartz, B.J., 2000. Controlling interchain interactions in conjugated polymers: The effects of chain morphology on exciton-exciton annihilation and aggregation in MEH-PPV films. *Journal Physics Chemistry*, 104, 237-255.

[12] Czerw. R., Carroll, D.L., Woo, H.S., Kim, Y.B., and Park, J.W., 2004. Nanoscale observation of failures in organic light-emitting diodes *Journal of Applied Physics*, 96(1), 641-644.

[13] Kafafi, Zakya H., 2000. Electroluminescence devices incorporating a new oxadiazole derivative. *In:* Jung, G.Y., Wang, C., Pearson, C., Bryce, M.R., Samuel, I.D.W., Petty, M.C., eds. Proceedings of Organic Light Emitting Materials and Device IV, San Diego 31 July-2 August 2000. The International Society for Optical Engineering, 307-315.

Micro-fabricated Square Interwinding Transformer using Surface Micromachining

Jumril Yunas and Burhanuddin Yeop Majlis, *SMIEEE*
Institute of Microengineering and Nanoelectronic
Universiti Kebangsaan Malaysia
43600 Bangi, Selangor, MALAYSIA
Email : burhan@vlsi.eng.ukm.my

Abstract A new micro-transformer with multi layer interwinding coil structure was fabricated by surface micromachining technology, which represents an innovated fabrication concept in developing a monolithically micro-transformer. The preliminary fabrication process resulted in a controlled and reproducible process. Voltage gain by open loaded secondary coil of higher than -6 dB achieved at the frequency range from 3 to 8 MHz, shows a good prospect for the fully integrated MEMS passive devices. Design, fabrication and experimental results are described and discussed in this paper.

I. INTRODUCTION

The need of fully integrated on chip passive devices equipment has been increased rapidly due to the increasing demand of micro circuitry for the portable and handheld electronics. Some passive devices such as transformers and inductors usually require large foot print areas and are located discrete far from its active circuitry, which have been the main factor of high losses and low efficiency in the circuit. But by increasing the switching frequency of the ICs (Integrated Circuits), it is now possible to integrate transformer with its active circuitry on a substrate and at the same time enables minimizing the size of the devices [1].

The implementation of two coupled spiral coils operating at high frequency still remains problematic in silicon based IC. The influences of the silicon substrate, insulating and metal conductor layer to the transformer characteristic become dominant and cause the increase of the skin and eddy current effects. However, by using high ratio metal trace and air core instead of magnetic core can solve some of the problems and the size of the transformer can be reduced significantly [2,3].

Several fabrication techniques of the planar spiral structures have been reported. Several planar spiral transformers, such as integrated planar coils, planar coreless coil and stack structured planar spiral coils were developed using thick and thin film technology [4,5,6,7]. Meanwhile, Donnell et al have presented some types of micro-fabricated coupled coil using MEMS approach designed for application of miniature low-voltage, high-current power converters, which shows much promise for fully integrated high frequency power devices [8,9,10].

In spite of achieving high quality and coupling factor, the requirement of complex and expensive fabrication techniques are still the fundamental problem addressed in those reports.

In this paper we present the investigation in to the fabrication of planar micro-transformers using MEMS surface micromachining approach which is directed to the integration of the transformers in Si based IC's.

First part of the paper includes the design to determine the geometrical and technological construction of the coils. In the second part we introduce the possibility of using a simple CMOS IC process fabrication required to develop the micro-transformer structure. In the final part of the paper we report the measurement results of the fabricated devices.

II. DESIGN

The transformer design used in this work is a square spiral planar coil, because of its simple geometry and higher inductances relative to other types such as circular, meander and hoop type [11]. Figure 1a (top) shows the geometry of square

0-7803-9730-4/06/$25.00 ©2006 IEEE

spiral coils on the top metal layer. Two separated coils are wound side by side representing a planar interwinding structure. The geometry of the structure is characterized by a maximum outer dimension d_o 2.2 mm and an inner d_i 0.5 mm. The coil has a width of 100 μm and separated by a space of 50 μm.

Fig 1b (bottom) shows the stack structure of the transformer. The bottom metal layer is function as to connect the inner side of the coil with the pads outside of the coil enabling connections with the measurement instruments. Metal in this layer has the same coil structure as the top layer. Both layers are separated by a sufficient thick oxide layer to ensure isolation between the metal. The inner side of each coil initially is connected with an output pad through *via* connection.

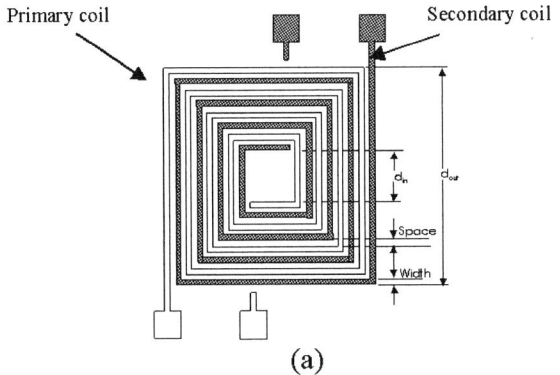

(a)

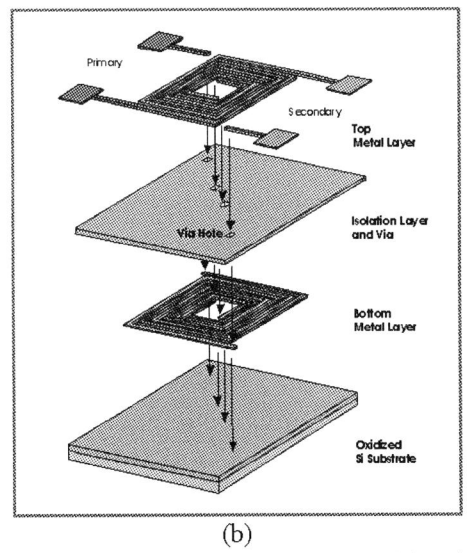

(b)

Fig. 1 Geometry of the transformer (a) and the structure of stack transformer model (b)

The separated planar coils are physically coupled through air. The primary side is connected

with the signal generator (source), while the secondary side with load. The AC source from the primary coil produces the changing of the magnetic fields which is then coupled with the secondary coil. Induction will then be created in the material of the secondary coil via the electromagnetic coupling. If the coupling is high enough, the transfer of electrical energy from the primary side to the secondary side will be established.

The geometrical configuration and technological structure of the transformer are summarized in the table 1.

Tabel 1. Geometrical and technological parameter configuration

	Parameter		Size [μm]	Spec
Geometry	Outer dimension (d_O)		2200	
	Inner dimension (d_i)		500	
	Width (w)		100	
	Space (s)		50	
	Via dimension		80x80	
Layer Structure	Substrate	Si	$t_{Si}= 360$	μr=11.9 ρ=1-10 Ωcm
	Oxide layer	SiO$_2$	$t_{ox1} = 500$	μr=1.3
	Bottom Metal	Al	$t_{M1} = 500$	ρ=2.7 mΩcm
	Oxide isolation	SiO$_2$	$t_{ox2} = 800$	μr=1.3
	Top metal	Al	$t_{M2} = 500$	ρ=2.7 mΩcm

III. FABRICATION PROCESS

For the fabrication work presented in this paper, the main concerns are to minimize the structure of the transformer, simplify the process steps and also cost. For these reasons, a stack structure of inter winding planar coil are developed and fabricated using surface micromachining process. The major advantages with this process are that it is simple, needs only 3 masks and compatible with CMOS IC's post process sequence.

As a starting material, we used an oxidized Si substrate in order to minimize the parasitic effect of the substrate. The oxide layer has a thickness of 500 nm with approximately 1% uniformity.

Next, the bottom metal layer is deposited on a planar oxide surface. The material used as the conductor material is Aluminum (Al) which is deposited using vacuum evaporation method. By 4 times evaporation of the Al, we obtained a thickness of approximately 0.5 μm.

In our experiment, we have also tried to evaporate the Al sources by maximum 8 times deposition steps, and therefore a 1.4 μm Al was achieved. The figure 2 shows the metal profile after several times evaporation.

The next step was resist coating for patterning of bottom metal structure using standard photo resist AZ1500 (shippley). The bottom coil pattern structure was then produced after 1st photolithography step and followed with Al etching using $H_3PO_4:HNO_3:H_2O=80:5:20$ for about 4 minutes.

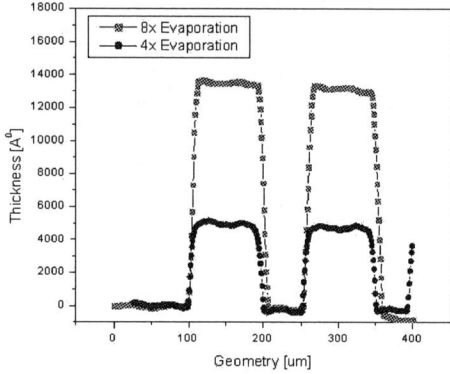

Fig. 2 The profile of the multi metal layer after evaporation

Figure 3 shows an optical microscope image of the bottom metal pattern obtained after metal etching. It is observed that pattern is produced properly. The over etch decreased the effective surface of the trace about 10%.

Fig. 3 Microphotograph of bottom metal coil after completing liftoff

Next step is the deposition of the isolation layer which was deposited using spin on glass method (SOG). The SOG material (from Honeywell)) is coated on the sample at 4000 RPM and 30 seconds, achieving an oxide thickness of 0.8 μm. Care must be taken by backing process during this step to avoid crack due to the high thermal stress.

A via window is formed by patterning the resist AZ1500 in the 2nd lithography step. The dimension of the via is approximately 80 μm x 80 μm. Figure 4 shows the image of via holes before (a) and after completion of SiO_2 etching (b) in 10% HF for 10 seconds and 1% HF for 20 seconds. The resist patterning of the via and via etching are the most critical steps in this process. So, care was taken during this step to avoid misalignment and attacking of HF to bottom Al.

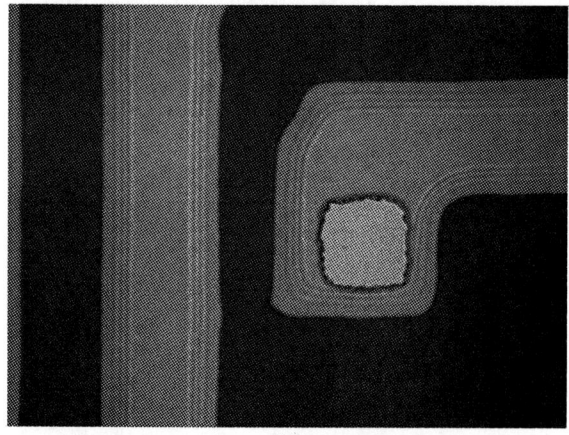

(a)

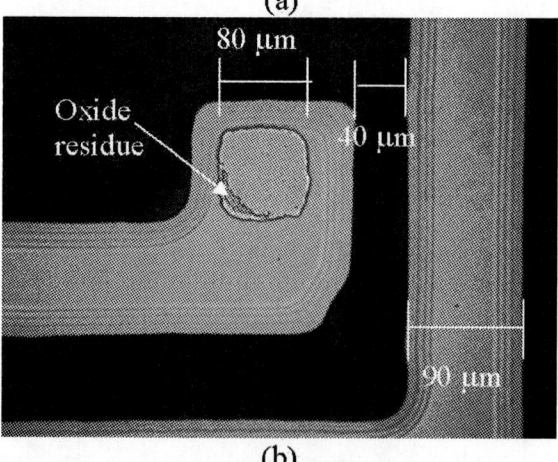

(b)

Fig, 4 Microphotograph of via window (a) Via Pattern after lithography, (b) via window after oxide etching and resist removal

Next, the top metal is produced using liftoff technique. The resist thickness is 3 micron enabling lift-off a 0.5 microns thick metal layer. The produced metal trace has 110 μm width, since the space 40 μm. Therefore, the produced metal trace deviate about ±10% compared to our design due to low printing resolution of the mask and also because of the long ramping of the resist side wall. By 3 times coating resist made from AZ1500 (Shipley) we achieved a 1.8 um resist layer, which is suitable for liftoff a 0.6 um metal layer (see figure 5).

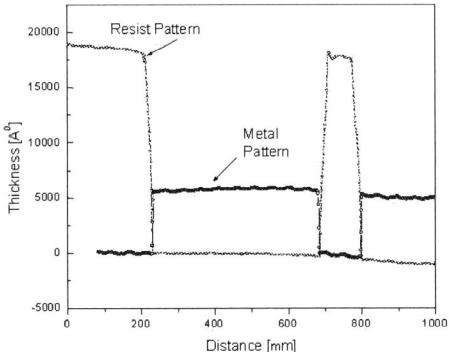

Fig. 5 The profile of the thick resist pattern enabling lift-off

In other test, it was also observed that the use of the image reverse technique using resist AZ5214E results a lower deviation and sharper resist side-wall. Another alternative for conductor patterning using thick resist AZP4620 is still in progress, which should allows a deposition of metal with thickness > 3 μm.

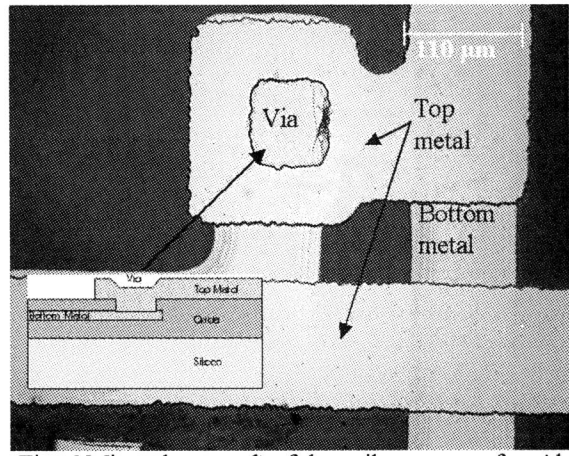

Fig. 6 Microphotograph of the coil structure after Al etching

After Al evaporation, the *via* holes are filled, enabling connection between bottom and top metal. Easy lift-off was obtained after leaving the sample in aceton for 10 minutes.

Beside the use of liftoff technique, metal etching can still be considered to use. However care must be taken during aligning the *via* and the top metal pattern to avoid the damage of the bottom metal trace during the etching process. Figure 6 shows an optical microscope image of the bottom metal pattern obtained after completing metal lift-off and resist removal.

IV. EXPERIMENTAL RESULT AND DISCUSSION

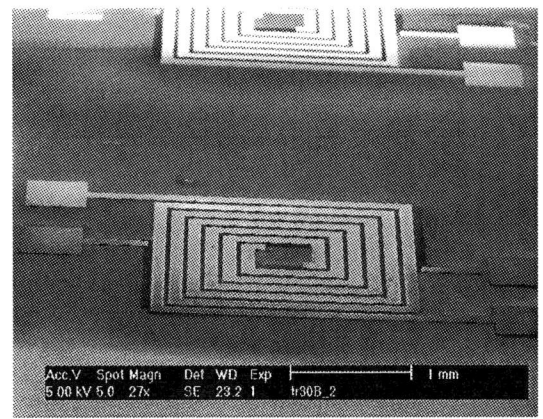

Fig. 7 SEM Image of fabricated structure

Several types of interwinding planar transformers are fabricated by surface micromachining. A sample of fabricated device can be seen in Figure 7.

Table 2 shows the characteristics of fabricated transformers from three various transformer types, which are measured using a LCR meter (Agillent 4284A)

Type A : single layer planar interwinding coil.
Type B,C : double layer planar interwinding coil with different top metal thickness.

Table 2. Series Inductance and resistance characteristics of fabricated device

Type	Turn	d_{Al} top [μm]	Lp* [nH]	Ls* [nH]	Rp* [Ω]	Rs* [Ω]
A	3	0.27	233	207	44	45
B	2 x 3	0.5	380	220	36.2	31.7
C	2 x 3	1.4	670	510	17.6	15.1

* Measured at 5 kHz.

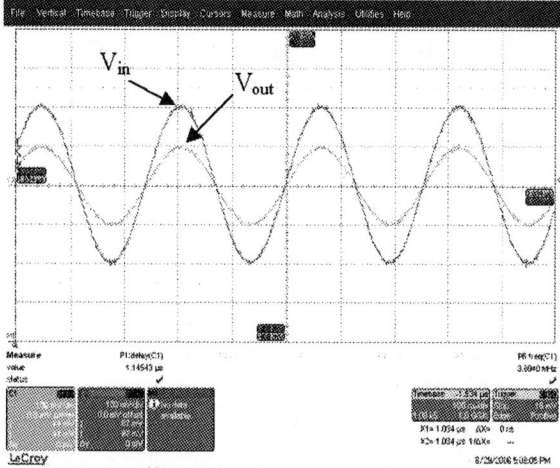

Fig. 8 Oscillograph of input V_i and output V_o for transformer of type A, operating at frequency 4 Mhz

The voltage gain of the transformer is measured in a probe station. The Input signal is a sinusoidal signal with $V_i = 400mV_{p-p}$. The input signal is generated by a signal generator (Fernell) operating at frequency up to 11 MHz. The output voltage signal is measured using Le Croy Oscilloscope. The output voltage signal $V_o = 200mV_{p-p}$ is obtained at 4 MHz, where the phase difference remains 0^0 (for the case of the transformer type A), as shown in Figure 8.

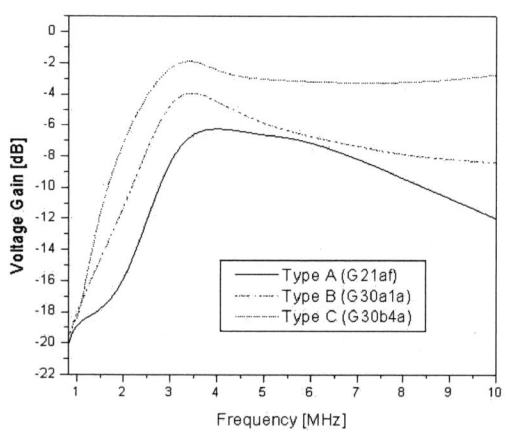

Fig. 9 Frequency dependence gain Characteristic of the transformer type B

The gain characteristic (input and output voltage ratio for no load case) versus frequency is shown in Figure 9. For an operating frequency up to 10 MHz, the voltage output of the transformers shows frequency dependence. By operating it in the low frequency below 1 MHz, the gain is very low. It increases rapidly by increasing the frequency until it reaches the optimum gain and for the frequency above 10 MHz the curve decreases due to the increase of the skin effect and parasitic capacitances.

In all samples, the optimal gains are found at the frequency range from 3 to 8 MHz, as shown in Figure 9. It is also clearly shown, that monolayer transformer achieved relatively lower gain. The lower gain resulted by monolayer transformer (Type A) due to high series resistance and low self inductance indicates to a high voltage drop at the primary side and a high flux leakage in the air gap. Hence, the multi metal structure or thick metal can be considered to achieve a high transformer performance. By applying a stack coil structure, the gain can be increased up to 5.3 dB. Furthermore, transformer with 1.4µm thick metal shows a maximum gain of −0.72 dB, which corresponds to a voltage ratio of 0.92 by a winding ratio of 1. This result exhibits a much better gain characteristic than the previous report [6].

V. CONCLUSION

Surface micromachining process of interwinding transformer by various multi layer structure is introduced. This structure required multi metal layer deposition and provides a successive via patterning and interconnection which are compatible to CMOS IC process. By using this method, some preliminary interwinding coils with stack and monolayer structure have been successfully fabricated at IMEN laboratory. Although no magnetic core is used, a maximum voltage gain of −0.72 dB is obtained at 3 MHz. To conclude, the transformer works efficiently at a frequency range between 3 and 8 MHz. This transformer is useful for application in contact less power conversion and signal isolation.

ACKNOWLEDGMENT

The authors would like to thank the Malaysian Ministry of Science, Technology, and Innovation for sponsoring this work under IRPA project "Development of MEMS Technology for Automotive Applications".

REFERENCES

[1] S. Hayano, "Development of Film Transformer," *IEEE Transaction On Magnetic*, Vol.30, No.6 (1994).

[2] J. M. Bustillo, "Surface micromachining for microelectromechanical systems," *Proceedings of IEEE*, Vol. 86, No. 8 (1998).

[3] S.C.Tang, "Coreless planar PCB transformers-A fundamental concept for signal and energy transfer," *IEEE transaction on Power Elecronics*, Vol. 15, No. 5 (2000).

[4] M. Mino, "Switching converter using thin film microtransformer with monolithically integrated rectifier diodes," *IEEE Trans. Magn.*, Vol. MAG-32, pp.291–296 (1996).

[5] K.I. Arshak, "Development of high frequency coreless transformer using thick film polymer technology," *Microelectronics Journal*, Vol. 30 (1999).

[6] J.Y.Park, "Packaging compatible microtransformers on a silicon substrate," *IEEE Transaction on Advanced Packaging*, Vol. 26, No. 2 (2003).

[7] A. Zolfaghari, B. Razavi, "Stacked inductors and transformers in CMOS technology," *IEEE Journal of Solid State Circuits*, Vol. 36, No.4 (2001).

[8] T.O. Donnell, "Thin Film Transformers for Future Power Conversion", *IEEE* (2004).

[9] S Prabhakarana, "Microfabricated coupled inductors for integrated power converters", *Journal of Magnetism and Magnetic Materials*, 290–291 (2005).

[10] M. Brunnet, "Design study and fabrication techniques for high power density microtransformers", *IEEE* (2001).

[11] K. Kawabe, " Planar Inductor", *IEEE Transaction On Magnetics*, Vol.MAG-20, No.5 (1984).

Fabrication of a Single Carbon Nano Tube for use in Nanolithography of MOSFET Gate

J.Karamdel, N.Talebi, M.Sattari[1], J.Derakhshandeh[2], F.L.Ayatollahi[3], A.Farrokhi, A.R.Hadi[1]

(1) Islamic Azad University - South Tehran Branch, E.E. Department, Faculty of Engineering
No. 209 North Iranshahr Ave. Tehran, Iran

(2)TUDelft University,Faculty of EEMCS, Feldmannweg 17,2600 GB Delft,The Netherlands

(3)F. O. E, Multimedia University, Jalan Multimedia 63100 cyberjaya , Selangor Malaysia

j-karamdel@azad.ac.ir – jkaramdel@yahoo.com – linaa@mmu.edu.my

Abstract: **Vertically aligned carbon nanotubes were grown by DC-PECVD on (100) cleaned Si using a mixture of Methane and hydrogen as the feed gases. For having single and small CNTs, we have used a refinement process, by means a seed layer is patterned by the standard lithography method. It forms islands of Cobalt for subsequent growth of nanotubes. By consecutive repetition of exposure/growth steps single and isolated islands of cobalt can be formed so that single standing vertical tubes are fabricated. The grown nanotubes are encapsulated by an insulating Tio_2 layer and in result; Straight lines with width between 40 to 120 nm have been successfully drawn by these CNTs. The applied voltage between anode and cathode for electron emission from CNTs is 100 volts in 100μm distance. This technique has been applied on P-type (100) silicon substrates for the formation of the gate region of N-MOSFET devices. The resulted transistor has a drive current of 286 $\mu A/ \mu m$ in V_{GS}=250mV and a maximum g_m equal to 2100 mS/mm with sub-threshold slope of 100mv/decade.**

I. INTRODUCTION

The most important thing in the fabrication of nano-scale MOSFET transistors is the formation of ultra fine gates, and it requires high resolution lithography and patterning. Among the different methods of nano lithography such as, ion beam lithography, extreme deep ultra violet lithography [1-2-4] and electron beam writing [5], the last one is most suitable technique for generating patterns in the scales about 10 nm. The scanning of the beam over an electro-resist-coated sample leads to the development of the desired pattern on the

samples [5] for fabricating nano-devices. However, electron beam writers are expensive and because of high electron energy they have some negative effects on PMMA.

On the other hand Carbon Nano Tubes can be used as field emission devices. Field emission is the extraction of electrons from the tip of a CNT by tunneling through the surface potential barrier under a strongly applied electric field. This property can be used as a source of electron for nano-writing. Several techniques are being exploited for the growth of single-wall and multi-wall nanotubes. Among them seems Plasma enhance chemical vapor deposition is a favorite method to grow nanotubes in a vertical arrangement [6-9].

Recently several researches have been done on vertical growth of carbon nanotubes using a PECVD technique where a mixture of acetylene or ammonia and hydrogen was used to carry the main constituents for the growth of CNT's [8, 9]. For having a fine and high accuracy lithography one needs a single and narrow emission of electrons, therefore we used a new technique for generating it. By means we used the first nanotubes for lithography of next stage cobalt's seeds' patterning. Following the growth of the single and desired tubes, they are coated with an insulating layer Tio_2 and the encapsulated tubes are then polished to open up the top side of the tubes. The electrical characteristics of the tubes, as well as their application for sub-micron and nano-lithography are addressed. This novel approach has been used to form lines with less than 80 nm width suitable for the gate of nano-metric MOSFET devices. The electrical characteristics of the realized transistors will be presented. The lithography technique proves to be a suitable

0-7803-9730-4/06/$25.00 ©2006 IEEE

approach for making transistors at sub-80 nm scale. In the following sections, we present the CNT growth, followed by the preliminary results for lithography. Finally, the process fabrication of MOS transistors and their electrical characteristics will be described.

II. EXPERIMENT

Basically the process of carbon nano tubes growth starts with evaporation of Cobalt layer on (100) silicon substrates. Silicon substrates are cleaned using RCA#1 solution and coated with 10–20Å thick cobalt layer using e-beam evaporation and at ambient temperature. Co-coated samples are then patterned using standard photo-lithography. The specimens were then placed in the DC-PECVD chamber (Fig.1). The vacuum in the quartz reaction chamber is achieved and maintained at a level of 10^{-4} Torr by a turbo-mechanical and rotary dual-stage pump, and mass flow controllers are used to control the level of the supplied gasses.

was critical in the final growth properties of the nanotubes.

Immediately after the two steps pretreatment, methane gas was introduced into the chamber with a flow rate of 5 sccm for 20 min in order to initiate the growth of CNTs. The flow hydrogen gas remained constant at 30 sccm. Growth of the CNTs was performed over substrate temperatures ranging from 600 to 650 °C and at a pressure of 1.8 Torr.

Fig. 2. SEM images of CNTs grown at a temperature of 650 °C and plasma power density of 4.5 W/cm2

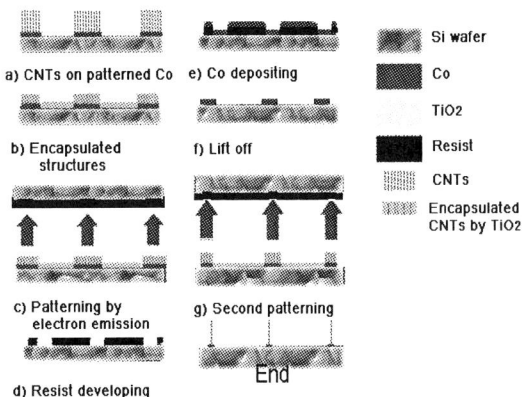

Fig.3. The schematic picture showing the refinement process for the growth of single and isolated nanotubes. The cobalt layer is first patterned and CNTs are grown. In each patterning/growth step, the size of the islands becomes smaller and eventually (after four runs), one can achieve single standing vertical tubes.

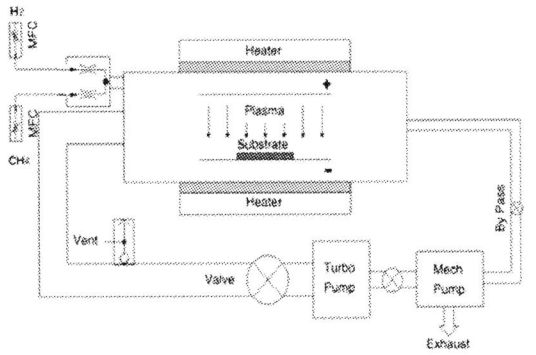

Fig. 1- Schematic diagram of the DC-PECVD growth reactor of the carbon nanotubes

After that a dual-step pre-growth treatment of the surface was performed. First, the surface of the sample was treated with a hydrogen blow at a flow rate of 30 sccm for 15 min with the substrate temperature maintained at 550 °C. This step was followed by a 15-min treatment of hydrogen plasma (plasma current of 30 mA) at the same temperature of the growth, which ranged between 550 °C and 650 °C, in order to create nano island of Co with diameters ranging from 10 nm to 50 nm. It was observed that the initial Co-island formation

Scanning electron microscopy was prepared on a CAMSCAN2300 at an electron accelerating voltage of 25 kV. All images were generated using the secondary electrons.Fig.2 shows the SEM images of CNTs.

As explained before for having the desired CNTs we used the first nanotubes for lithography of next stage cobalt's seeds' patterning. Fig. 3 shows the refinement process in which circularly patterned islands are initially used as the Co-seed for the growth of CNTs, and by consecutive exposure/growth steps, smaller and smaller islands are formed. The resulted CNTs are now in isolated single-standing form and they can be used to draw single lines in a desired shape and place. Fig. 4 shows the evolution of a single-standing isolated tube from clusters of CNTs. After the CNT's are grown on the silicon substrate, they are coated with an insulated layer of titanium oxide as shown in Fig. 5. The deposition of TiO_2 is achieved by means of atmospheric pressure CVD technique with a mixture of O_2 and $TiCl_4$ as the main precursor.

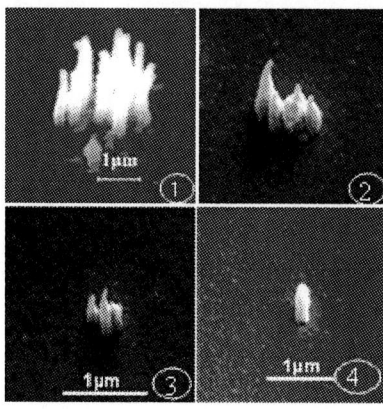

Fig. 4- SEM images of nanotubes obtained in refinement steps.

The electrical and Physical characteristics of this layer have been investigated and reported before [10]. This CVD technique allows a conformal over-coating of the surface as well as the nanotubes. Since the nanotubes are coated by TiO2, a step of polishing is necessary to expose the buried nanotubes tip. The exposed CNT is encapsulated and by one step of plasma ashing in oxygen it is cleaned and becomes ready for proper electron emission.

In an attempt to realize MOSFET devices, we have used this nano-writing approach to form the critical gate of the transistors. Fig. 6 shows the schematic of the fabrication process, indicating five main steps in forming the gate. Once the gate is defined

using nano-writing approach, the metal gate is used to define the source and drain areas.

Doping is achieved by a Ge/Sb alloy deposition with a thickness of 2nm leading to a layer with a sheet-resistance of $2K\Omega/\square$. Gate oxide is 5nm thick, grown thermally with an inversion Cox of $0.7\mu F/cm2$ and the gate length is 80 nm.

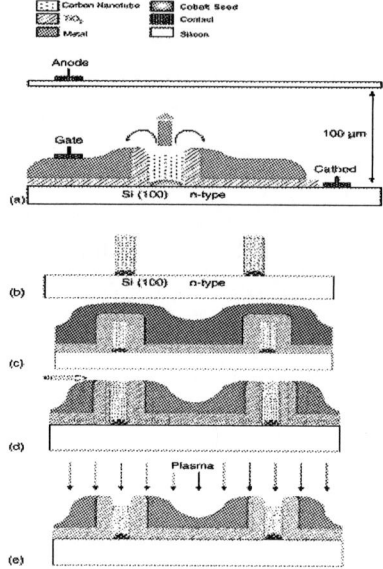

Fig.5- The schematic diagram shows the overall operation of the self-defined electron emission sources and their fabrication process flow. The emission of electrons from the negatively biased cathode will end up with tracks of electrons on the oppositely placed, resist-coated substrate (a). The fabrication steps are the growth of nanotubes from cobalt islands (b), followed by a conformal deposition of TiO2 (c), polishing of the top insulating layer (d) and finally oxygen plasma ashing (e) to expose the buried tip.

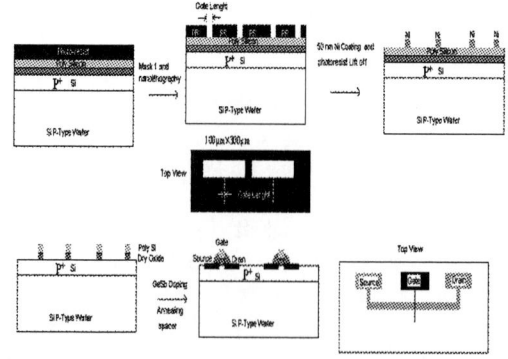

Fig. 6 The schematic of the transistor fabrication process

III. RESULTS AND DISCUSSION

Fabricated CNTs are used for drawing nano-scale lines. Fig. 7 collects the lines are drawn using this approach with a width varying between 40 and 120nm. The width of the lines is controlled by the outer diameter of the CNTs.

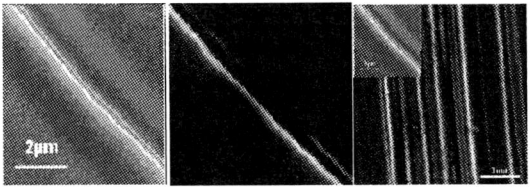

Fig. 7 The SEM images corresponding to the lines are drawn using the encapsulated carbon nanotubes on the resist-coated substrates.

After drawing lines the photo-resist is developed in its standard developing solution and subsequent steps for the formation of MOSFET transistor are followed. The current-voltage characteristics of the device are plotted in parts (a) to (b) of Fig.8, evidencing a proper behavior. The device shows a sub-threshold slope of 100mV/decade and a threshold voltage of 80mV. Also an effective mobility of 350cm^2/V-s and a drive current of 286μA/μm in V_{GS}=250mV are extracted for this small geometry transistor. Also DIBL was measured to be 150mV/V (not shown here). These results suggest that the device has a controllable short channel effect. Fig.8 (b) shows the saturation transconductance (gm) curve of the 80 nm nMOS transistor with a peak "gm" equal to 2100mS/mm.

IV. CONCLUSION

The fabrication of nano-scale transistors requires formation of a gate followed by doping incorporation. The gate length which is the main obstacle in achieving nano-scale transistors has been realized by a combination of photo-lithography and our recently developed CNT-based nano-lithography to draw 40-120 nm-wide lines to be used for subsequent doping steps. By improvements of the lithography process this method can be used to fabrication of PMOS transistors and possibly small integrated circuits.

Also trying to use the inner hollow tube of the CNT's can be used to form features in the sub-ten-nanometer scale by means of hydrogen ion writing.

(a)

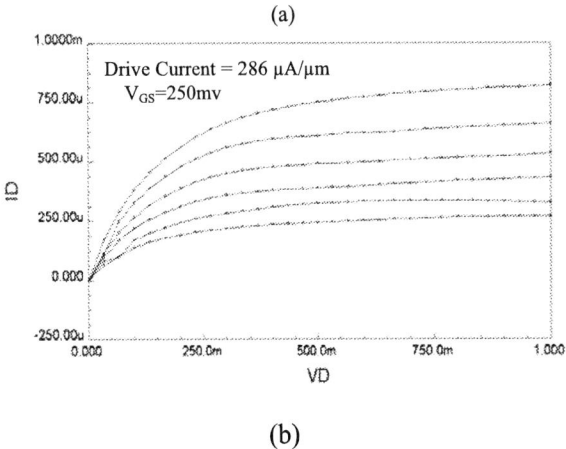

(b)

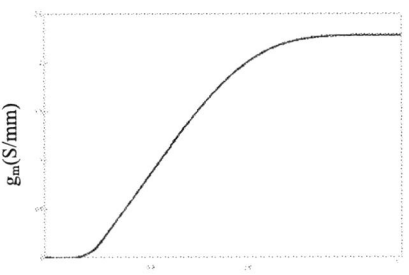

Fig.8- Id-Vd family, Subthreshold Id-Vg characteristics and Saturation transconductance (gm) curves of the 80nm NMOS transistor

ACKNOWLEDGEMENT

The authors would like to thank the Malaysian Ministry of Science, Technology and Environment for sponsoring this work under project IRPA 03-02-02-0015 and Telekom Malaysia Berhad for providing the facilities at the UKM-TM Microelectronics Research Centre.

REFERENCES

[1] C.C. Wu, Y.K. Leung, et al., A 90 nm CMOS Device Technology With 228 High Speed, General Purpose, and Low Leakage Transistors For System On Chip Applications, IEDM, 2002, p. 65. 230

[2] R. Chau, J. Kavalieros, et al., 30 nm Physical Gate Length CMOS Transistors with 1.0 ps n-MOS and 1.7 ps p-MOS Gate Delays, IEDM, 2000, p. 45.

[3] A. Chatterjee, R.A. Chapman, et al., Sub-100 nm Gate Length Metal Gate NMOS Transistors Fabricated by a Replacement Gate Process, IEDM, 1997, p. 821.

[4] T. Ghani, et al., 100 nm Gate Length High Performance/Low Power CMOS Transistor Structure, IEDM, 1999, p. 415.

[5] B. Yu, et al., 15 nm Gate Length Planar CMOS Transistor, IEDM, 2001, p. 937

[6] H.W.Ch. Postman, T. Teepen, Z. Yao, M. Grifoniu, C. Dekker, Science 293 (2001).

[7] A. Javey, Q. Wang, A. Ural, Y. Li, H. Dai, Nano Letters 2 (9) (2002) 929–932.

[8] J. Koohsorkhi, H. Hoseinzadegan, S. Mohajerzadeh and M.D. Robertson, 2004 Device Research Conference, Notre Dame University, Indiana, Conference Digest, p:55-56 (2004).

[9] S H. Hesamzadeh, B. Ganjipour, S. Mohajerzadeh, a. Khodadadi, Y. Mortazavi and S. Kiani, *Carbon*, vol. **42**, p:1043 (2004).

[10] B. Arvan, A. Khakifirooz, R. Tarighat, S. Mohajerzadeh, A. GoodarziE. Asl. Soleimani, E. Arzi, Mater. Sci. Eng. B 109 (2003)17–23.

[11] Martel R, Derycke V, Appenzeller J, Wind S, Avouris Ph. Proceedings of the 39th Conference on Design Automation, p:94-98 (2002).

[12] J. Derakhshandeh, Y. Abdi, S. Mohajerzadeh, H. Hosseinzadegan,E. Asl. Soleimani, H. Radamson,' Fabrication of 100 nm gate length MOSFET's using a novel carbon CNT-based nano-lithography' Materials Science and Engineering MSB 10430 1–5 ,2005

Development of Integrated Detection System for Shock Vibration by using MEMS Accelerometer

Khairul Anuar Bin Abd Wahid[1], Ishak Abdul Azid[1]*, Aftanasar Md Shahar[2] and Othman Sidek[3]

[1] School of Mechanical Engineering

[2] School of Aerospace Engineering

[3] School of Electrical and Electronic Engineering, Universiti Sains Malaysia, Penang

Email *- ishak@eng.usm.my

Abstract **In this paper an integrated system is developed to detect shock vibration by using MEMS accelerometer. The interface circuit which converts the signal from an accelerometer capacitive element to a pulse signal for digital signal processing is built. Changes in the capacitance between fixed and movable plates are induced by acceleration from shock vibration. Interface circuit incorporating this capacitor then extracts the acceleration signal and converts it to digital or analog form using microcontroller microprocessor. GUI is developed to make the integration data system more interesting and user friendly. The sensing of MEMS and monitoring software are integrated together to provide real time acquire data and graphical visualization for monitoring, recording and analysis purposes.**

I. INTRODUCTION

Micro-electro-mechanical-system (MEMS) technology is one of the most exciting areas of study. The principles of MEMS technology are based on mechanical properties of the silicon to create moving structures. Combined with microelectronics, signals generated by the moving structures give the output signal to create a new generation of sensors [1]. MEMS accelerometer is an electromechanical device that will measure acceleration forces. These forces may be static, like the constant force of gravity when in standstill or not moving, or they could be dynamic - caused by moving or vibrating the accelerometer. By measuring the amount of static acceleration due to gravity, the acceleration force with respect to the earth can be found.

There are three different types of MEMS accelerometers; piezo-resistive [2], tunneling [3], and capacitive [4-7]. The capacitive accelerometer is chosen for this system because it provides high sensitivity, good DC response, low drift, low temperature sensitivity, low power dissipation, and simple structure [8]. As it is known, MEMS accelerometer is a sensor which converts acceleration from motion or gravity to an electrical signal [9]. MEMS accelerometer consists of fixed and movable finger. Acceleration is sensed by measuring the capacitance of the structure which varies in proportion to changes in acceleration [10].

The interface circuit incorporating with the capacitive element then extracts the acceleration signal and converts it to either digital [11-12] or analog form [13-16]. Using microcontroller processing, the available signal will be extracted and calibrated through the program using Code Designer Lite and compiled using Pro Basic Compiler.

Using MEMS accelerometer as shock vibration detector, this can be used for various conditions and applications such as in nature disaster [17]. In this case, MEMS accelerometer provides the data that is useful to monitor and upgrade the alert system. Furthermore, shock vibration detection system can also be used to detect fall [18]. This situation is applicable in medical field area and security system. For medical field area, MEMS accelerometer has been used to monitor a range of human movement especially for the critical convict. Besides that's its also applicable for security system such as installing the MEMS accelerometer inside the sensitive electronic subsystem like hard

0-7803-9730-4/06/$25.00 ©2006 IEEE

disk. If the hard disk falls, then MEMS accelerometer will detect the falling force whether it is critical or not. If it is critical, then the hard disk will turn off to protect the hard disk.

In short, there are many researches being carried out to fully utilize the potential of using MEMS in medical areas [19-20]. Therefore, hopefully the developed shock vibration detection system will fit in either of the system'application based on shock and vibration detection.

II. SYSTEMS OVERVIEW

The integrated detection system for shock vibration consists of MEMS micro accelerometer to sense for shock vibration and microcontroller to compute the micro accelerometer signal. The system overview of the shock vibration detector is shown in Figure 1.

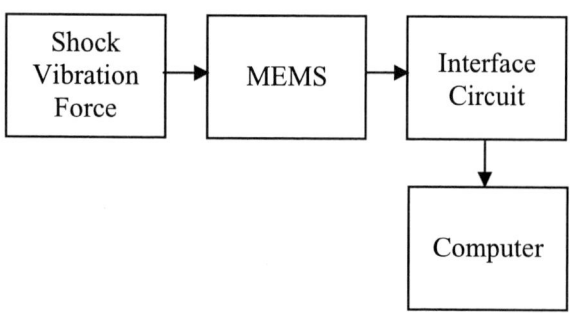

Fig. 1   System overview of the shock vibration detection system

The force occurs during the shock vibration will cause capacitive element inside the MEMS micro accelerometer to be in motion. The motion capacitive element will then drive the electrical signal output. The microcontroller as microprocessor is used to compute the output electrical signal. The system developed can plot the signal in real-time.

III. SYSTEM CONFIGURATION

This system uses the integration interface circuit between MEMS accelerometer, computation signal circuit and programming. All this subsystems are joined together to achieve a successful shock vibration detection system. Figure 2 shows the MEMS accelerometer ADXL202 used in this detection system.   It is a dual axis capable of

measuring both positive and negative accelerations to a maximum level of $\pm 2g$. It can measure both dynamic and static accelerations.

Fig. 2   MEMS accelerometer produced by Analog Devices

Figure 3 shows an internal circuit of ADXL202 accelerometer inside IC packaging. This MEMS accelerometer is chosen to measure the shock vibration because of its small size, low costs and reliable performance. The sampling rate depends on the dynamic force from shock vibration and the bandwidth of this MEMS accelerometer is from 0.01 Hz to 5 kHz. The outputs are digital signal which are proportional to the acceleration in each two sensitive axes. These outputs are measured directly with a microcontroller and require no A/D converter. If voltage outputs are desired, a voltage output proportional to acceleration is available from X_{FILT} and Y_{FILT}.

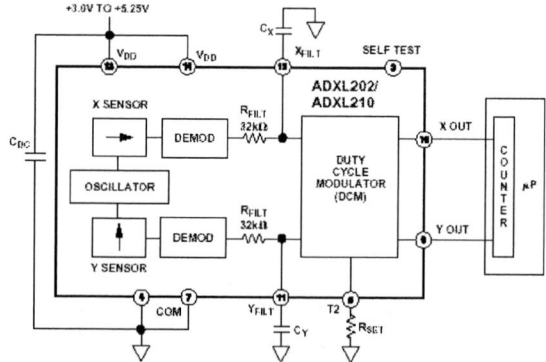

Fig. 3 Internal circuit of MEMS accelerometer

When MEMS is forced by shock vibration, electrical signal output will give voltage form through x-out and y-out axis. This electrical signal will be computed by microprocessor microcontroller. Figure 4 shows the core of microcontroller circuit for the integration with the MEMS circuits. Since this microcontroller uses the EPIC software, the oscillator mode can be selected.

Therefore, PIC microcontroller can be operated in four different oscillator modes which are *low-power crystal, crystal, high speed-crystal and resistor.*

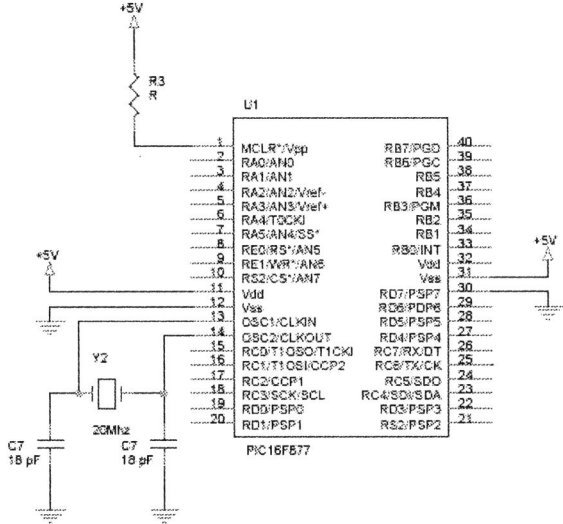

Fig. 4 Microcontroller Core Circuit

PIC16F877 is used because it is easy to program since it uses 35 single word instruction based on CMOS FLASH –based 8 bit microcontroller. These microcontroller features 256 bytes of EEPROM data memory, self programming, and 8 channels of 10-bit analog to digital converter. I/O pin is used to detect the signal from the accelerometer. This microcontroller is integrated with the MEMS accelerometer ADXL202. Figure 5 shows the basic integrating connection between microcontroller core circuits with MEMS accelerometer circuit.

To display the electrical signal into visualization, the sub system is interfaced with the integration computer system monitoring. The signal from output microcontroller is transmitted to the PC via serial communication cable. For this system, both digital and analog outputs are considered to be outputs for shock vibration detection system. The microcontroller is used to compute the output signal from MEMS accelerometer. To extract the signal from MEMS accelerometer, PICBasic Pro Compilers is used as the high level language program that is packaged with CodeDesigner Lite software. CodeDesigner Lite is an *integrated development environment* (IDE) which is used for writing and programming PIC microncontroller. CodeDesigner Lite is an advanced text editor capable of calling the PICBasic Pro Compilers and the EPIC software. Figure 6 shows the circuit diagram used for the serial communication.

The developed system combines these shock data from MEMS accelerometer and integrates with the computer system monitoring. Using the DB-9 as a cable between PC communication and microcontroller, MAX 232 is used as a buffer driver. It protects the microprocessor microcontroller from the possible damage from static. LM7805 is used for voltage regulator to ensure proper voltage level and to prevent damage to the circuit. LM7805 is a cheap passive regulator component and easy voltage regulation. This regulator makes the shock vibration system easy to be carried out by just using the +9VDC battery.

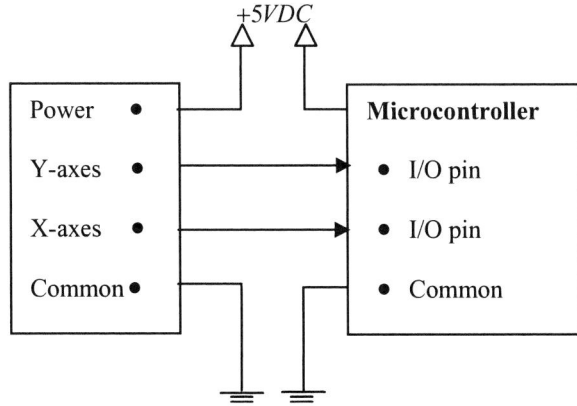

Fig. 5 Integration between Microcontroller Circuit and MEMS Circuit

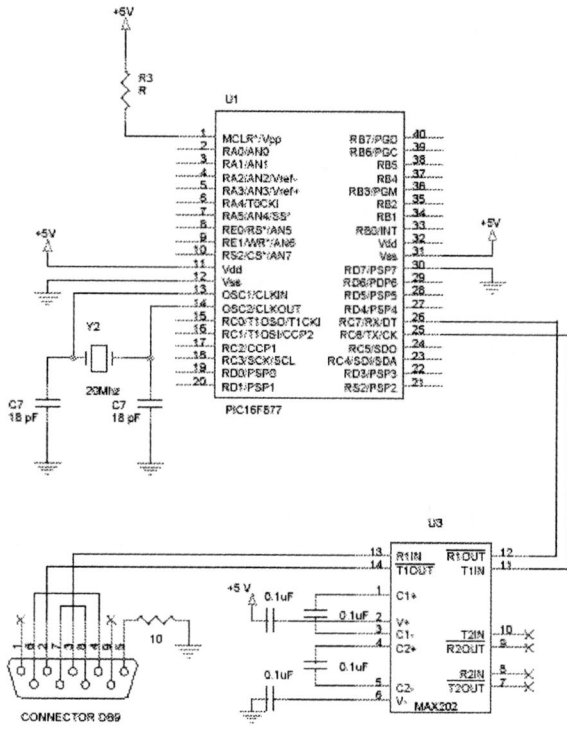

Fig. 6 Circuit diagram for serial communication

IV. RESULTS AND DISCUSSIONS

Figure 7 and 8 show the first experimental result to check the linearity for x-axis and y-axis of shock vibration system against the changing of tilt angle from 0^0 to 90^0. The output voltage from x-axis and y-axis were recorded at every 10^0.

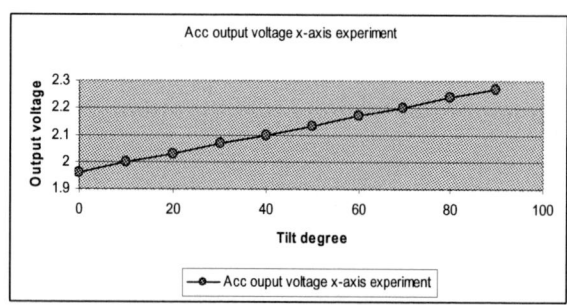

Fig. 7 Accelerometer output voltage x-axis against tilt degree

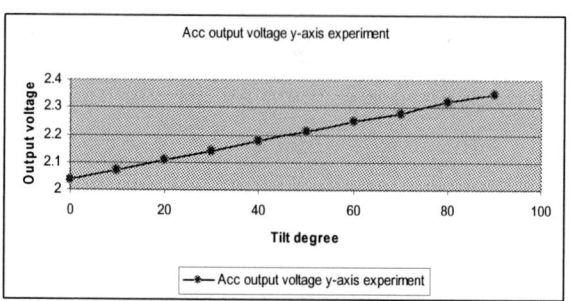

Fig. 8 Accelerometer output voltages y-axis against tilt degree

This system is ideally tested under different tilt angle and the output voltage is proportional with the change of tilt degree. The GUI is developed to show the visual of the shock vibration signal. Table 1 and 2 showed the data for MEMS accelerometer outputs voltage via x-axis and y-axis.

Table 1: Accelerometer outputs voltage x-axis

Degree of tilt	Acc. outputs voltage x-axis experiment
0	1.96
10	2
20	2.03
30	2.07
40	2.1
50	2.13
60	2.17
70	2.2
80	2.24
90	2.27

According to application note ADXL202 produced by Analog Devices, the equation to determine the angle by using voltage output via each axis is given by:

$$\theta = \arcsin\left[\left(V_{(out)} - V_{(zero\ g)}\right)/(1g \times Scale\ factor\left(\frac{V}{g}\right))\right]$$

142

Table 2: Accelerometer outputs voltage via y-axis

Degree of tilt	Acc. outputs voltage y-axis experiment
0	2.04
10	2.07
20	2.11
30	2.14
40	2.18
50	2.21
60	2.25
70	2.28
80	2.32
90	2.35

$V_{(out)}$ is the voltage value that measured each 10^0 and $V_{(zero\ g)}$ is the voltage value when MEMS accelerometer is at $0g$ position. Referring to the ADXL202 datasheet, scale factor available is $312mV/g$. This parameter was used to calculate the change of degree when the output voltage was applied for every 10^0 for each axis. Figures 9 and 10 show the comparison change of tilt degree via experiment and calculation. Both results show the linearity of the changed voltage during tilting angle. However, experiment result shows a better result in terms of the linearity.

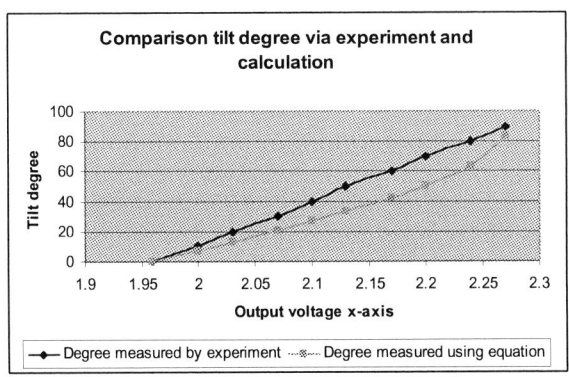

Fig. 9 Comparison tilts degree via experiment and calculation for x-axis.

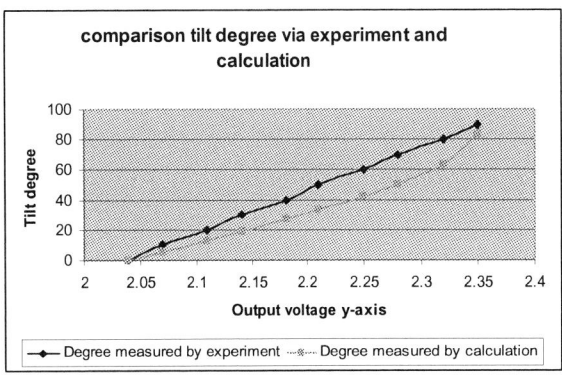

Fig. 10 Comparison tilts degree via experiment and calculation for y-axis

The second experiment is set to test the MEMS shock vibration sensing system. The shock vibration system is communicated with personal computer linked with a DB-9 cable. The MEMS is placed at the flat surface with the static position. The x-axis and y-axis is set at the 0g conditionally. MEMS accelerometer is rotated within 0^0 to 180^0 angle, which produces a sinusoidal signal like graph. Then MEMS accelerometer is set static again and at the 0g condition. In the rest position, MEMS accelerometer is then applied with the shock rapidly. The signal visualized on the screen shows the shock signal. Figure 11 shows the shock vibration interface before using high pass filter. It shows two signals produced by x-axis and y-axis. According to the signal output, those signals are proportional with the acceleration and the real time visualization.

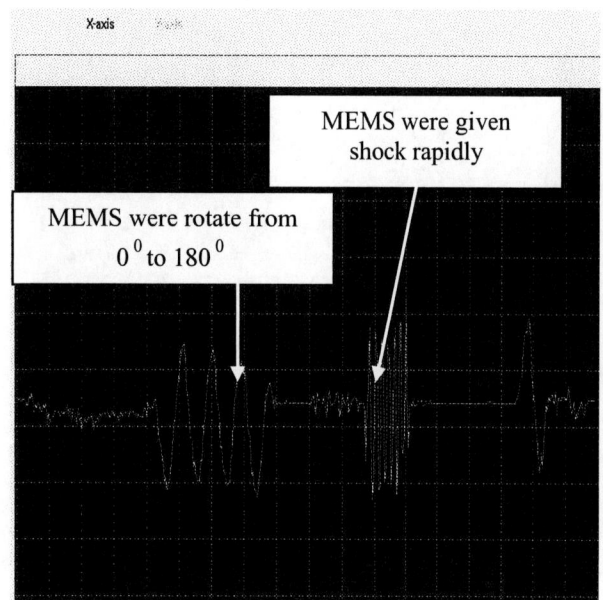

Fig. 11 Shock vibration interface before add-up high pass filter

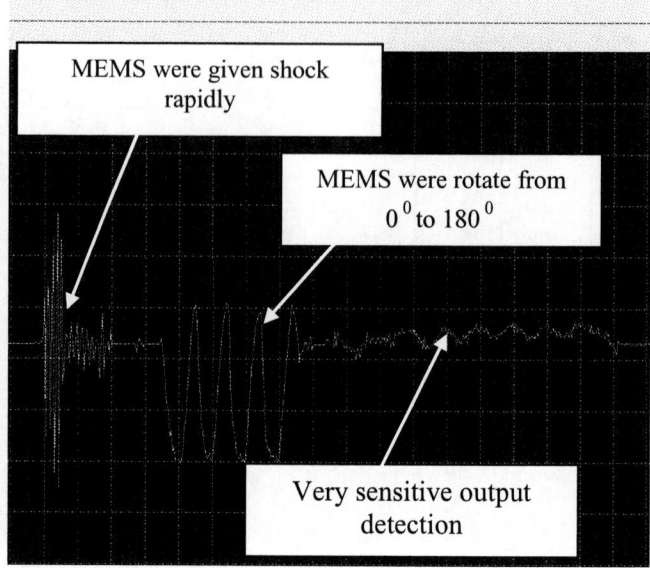

Fig. 12 Shock Vibration Interface after add-up high pass filter

Generally for shock vibration sensing, signals used is between 10 and 200 Hz. Since the response of the ADXL202 is extends to 5 kHz, a band pass filter is added to remove out the band signals. Since the high pass filter is added up on ADXL202, the shock vibration sensing output is tested again to compare the smoothness of the reading output so that the signal output is accurate and sensitive. Figure 12 shows the result for shock vibration sensing system after the addition of the high pass filter. Using the same method to test the shock vibration detection system, the output has improve its accuracy and sensitivity.

V.   CONCLUSIONS

In this paper, a shock detection system is developed by using ADXL202 MEMS accelerometer as a sensor. Microcontroller is successfully used to compute the signal from output accelerometer. The developed system used analog high pass filter. The design system is small, lightweight and easy to carry. This developed shock detection system can accurately detect and correlate the force applied.

ACKNOWLEDGMENT

This work was conducted under IRPA research grant. The support from Universiti Sains Malaysia is gratefully acknowledged.

REFERENCES

[1] Tai-Ran Hsu "MEMS and Microsystems Design and Manufacturer" Mc-Graw Hill, International edition, 2005.

[2] L.M. Roylance and J.B. Angell, " A batch-fabricated silicon accelerometer," *IEEE trans. Electrons Devices,* vol. ED-26, pp.1911-1917, Dec.1979.

[3] C. Yeh and K. Najafi, "A low-voltage bulk-silicon tunneling based microaccelerometer," *Tech Dig. IEEE IEDM,* Washington DC. Dec.1995 pp593-596.

[4] J.S Chae, H. Kulah, K. Najafi, "A High Sensitivity Silicon On-Glass Lateral μG Microaccelerometer", *Nanospace 2000,* Houston, TX.

[5] N.Yazdi, K.Najafi, "An All-Silicon Single-Wafer Fabrication Technology for Precision Microaccelerometers", *Proc. Transducer 97,* pp. 1181-1184.

[6] Hung Chi Hung, Tomoyoki Enomoto, Masanobu Shinozuka "Real time visualization of structural response with wireless MEMS sensor" 13th World Conference on Earthquake Engineering, paper no. 121, 2004.

[7] H.G Rodney, K.H. Tan, and T. Lachman, "Tilt Sensing and Measurement System Using MEMS Sensor" The 2006 International Conference on MEMS and Nanotechnology, 2006.

[8] N. Yazdi, F. Ayazi, and K. Najafi, "Micromachined inertial sensors", *Proc of the IEEE*, vol.86, pp.1640-1659, Aug 1998.

[9] James Doscher "Accelerometer Design and Applications" Technical Note, Micromachining Avengelist, Analog Devices.

[10] Ki-Ho han and Young-Ho Cho, "Self balanced navigation grade capacitive microaccelerometer and their performance for varying sense voltage and peressure", *Member IEEE*, Journal of Microelectromechanical System vol.12, No.1 February 2003, pp11-19.

[11] W. Yun R. Howe and P. Gray, "Surface micromachined digitally force balanced accelerometer with integrated CMOS detection circuitry, "*IEEE Solid State Sensor and Actuator Workshop,* Hilton Head Island, SC June 21-25,1992.pp126-131.

[12] W.Henrion, L. Disanza, M. I S. Terry and H. Jerman, "Wide dynamic range direct digital accelerometer," *IEEE Solid-State Sensor and Actuator Workshop,* Hilton Head Island, SC June 4-7,1990. pp 153-157.

[13] S.Suzuki,*et al,* "Semiconductor capacitive type accelerometer with PWM electrostatic servo technique" *Sensors and Actuators,* A21-A23, 1990, pp. 316-319.

[14] Y. Matsumoto and M.Esahi, "Low drift integrated capacitive accelerometer with PLL servo technique," *Transducer'93,* pp. 826-829.

[15] Lj. Ristic , R. Gutteridge, J.Kung, D. Koury, B. Dunn and H. Zunino "A capacitive type accelerometer with self test feature based on a double pinned polysilicon structure," *Transducer'93*, pp810-813.

[16] Keith Hoffman, Richard Varuso and Date Fratta, "The use of Low costs MEMS accelerometer for the Near-Surface Monitoring of Geotechnical Engineering Systems" *GeoCongress 2006 Conference,* Atlanta, GA.

[17] Jerome P. Lynch, Kincho H. Law, Anne S. Kiremidjian, Ed Carryer, Thomas W. Kenny, Aaron Partridge and Arvind Sundararajan "Validation of a wireless modular monitoring system for structures" *SPIE's* 9th *Annual*

International Symposium on Smart Structures and Materials, San Diego, CA, USA March 17-21, 2002.

[18] Kara E. Bliley, Daniel J. Schwab, David R. Holmes, Paul H. kane, James A. Levine, Erik S. Daniel and Barry K. Gilbert, "Design of a compact system using a MEMS Accelerometer to measure body posture and ambulation", *Proceedings of the 19th IEEE Symposium on Computer-Based Medical System (CBMS'06).*

[19] Yuriko Tsuruoka, Tamura Yoshiyasu, Shibasaki Ryosuke and Tsuruoka Masako, "Analysis of walking improvement with dynamic shoe insoles, using two accelerometers", *Science Direct Journal,* 2005.

[20] Davey T.W. Fong, Joe C. Y. Wong, Alan H. F. Lam, Raymond H.W Lam and Wen J. Li, "A wireless motion sensing system using ADXL MEMS accelerometer for sports science applications" *Proceeding of the 5th World Congress on Intelligent Control and Automation,* June 15-19, 2004, Hangzhou P.R. China.

Organic Light Emitting Diode (OLED) Using Different Hole Transport and Injecting Layers

Mohd Khairy Othman [a,b], Muhamad Mat Salleh [a] and Abdul Fatah Awang Mat [b]

[a]Institute of Microengineering and Nanoelectronics (IMEN), Universiti Kebangsaan Malaysia, 43600 Bangi Selangor, Malaysia
[b]TM Research & Development Sdn Bhd, UPM-MTDC, Lebuh Silikon, 43400 Serdang, Selangor.

E-mail : mohd_khairy@yahoo.com

Abstracts **This paper reports the various structures and performances improvements using different hole transporting layer in OLED based on DPVBi as emitter. Here the indium tin oxide (ITO) used as an anode, copper pthalocyanine (CuPc) as the hole injecting layer, PEDOT:PSS and poly-9-vinylcarbozole (PVK) as hole transporting layer, 4,4'-bis(2,2'diphenilvinyl)-1,1'-biphenyl (DPVBi) as the blue emitting layer and aluminum (Al) as the cathode. The CuPc and DPVBi were prepared by thermal evaporation while the PEDOT:PSS and PVK films were prepared using spin coating technique. The effect of inserting additional layer of CuPc, PVK and PEDOT:PSS between anode and the emitting layers was analyzed through the current-voltage (IV) curves and the electroluminescence spectra. The additional layer structure was found to increase the maximum luminance compared to that one of single layer device. The used of PVK as hole transporting layer has improved the diode properties of the device and able to prevent the device from short circuits. The optimized DPVBi layer thickness was observed at 56 nm and the insertion of 10 nm CuPc hole injecting layer show the device reduce it turn on voltage from 7.0 V to 6.5 V**

I. INTRODUCTION

Since 1960s there have been many studies in organic electroluminescence. The studies lead to more knowledge of identifies and understand organic basic processes. In 1987, Tang and VanSkyle [1] are the first to obtain a green light of luminance properties from Alq3 as organic film. The polymer electroluminescence device (PLED) was then discovered by the Cambridge University [2].

In recent years a continuous improvement of organic light emitting diodes (OLEDs) performance (polymer or small molecules) has been progressively developed. The preparation and purification of molecules and it deposition processes have been achieved better and better [3].

In this study, we realized some multilayer structures which have improved transport of holes and electron blocking properties. First we studied a monolayer structure with DPVBi as emissive layer and then some multilayer structures with a hole transport and hole injecting properties of copper pthalocyanine (CuPc), poly-9-vinylcarbozole (PVK) and polyethylenedioxy-thiophene: polystyrene (PEDOT:PSS). Fig. 1 show the molecular structures used in this study.

II. EXPERIMENTAL

The emissive material of DPVBi was purchased from American Dye Source Inc and was used directly without further treatment. The ITO coated glass used as an anode was purchased from Merck Balzers with sheet resistance of 40 Ω/m^2. Prior the deposition of organic thin films, the ITO substrate was cleaned with acetone, 2-propanol and rinsed with DI water for about 15 minutes. The substrate was then dried with N_2 and baked for about 20 minutes at 70°C using vacuum oven.

The PVK polymer was diluted with dicloroethane solvent while the PEDOT:PSS and it solvent was filtered using 0.45 µm filtration method. The PVK and PEDOT:PSS solution were deposited onto the substrates by spin coating technique while the CuPc and DPVBi materials was deposited by thermal evaporation. The deposition of DPVBi organic films was

0-7803-9730-4/06/$25.00 ©2006 IEEE

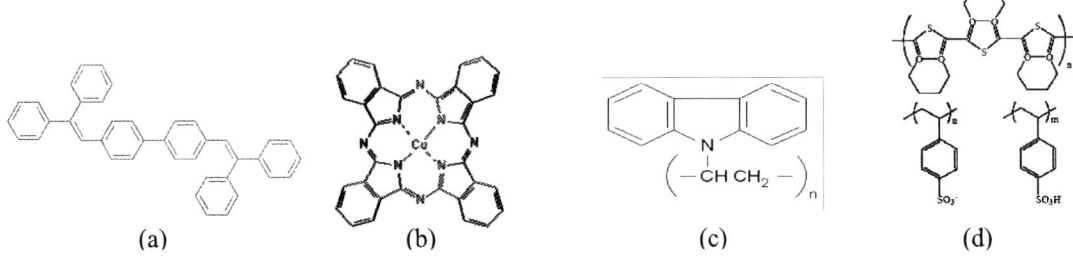

Fig. 1 Molecular structure of a) DPVBi, b)CuPc, c) PVK and d) PEDOT:PSS

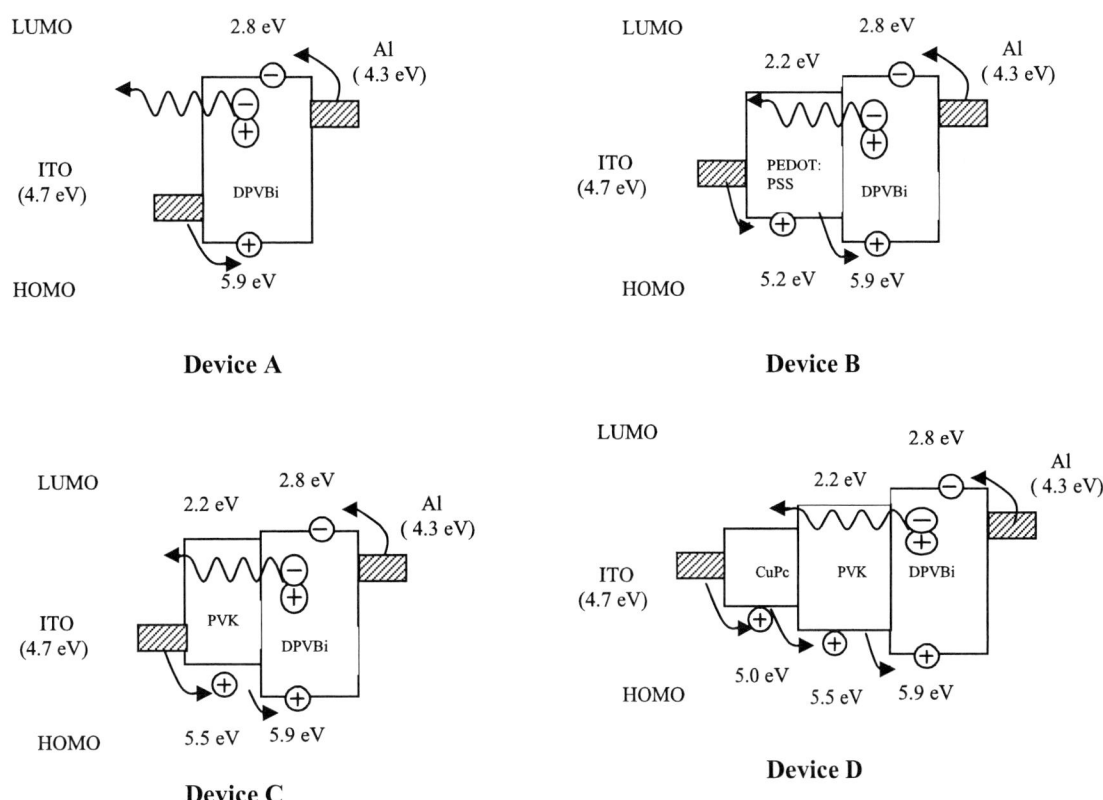

Fig. 2 Schematic energy diagrams of the various structures of Device A, B, C and D.

performed by thermal evaporation at 2.5 x10⁻⁵ mbar using resistively heated tungsten boat and the deposition rate was controlled at 0.5-1 A°/sec. The DPVBi thickness was fixed at 56 nm while the PVK, PEDOT:PSS and CuPc layer was kept at thickness of 100 nm, 98 nm and 10 nm respectively. Lastly, the top electrodes of aluminum were then deposited immediately after the DPVBi deposition without breaking the vacuum.

The performance of the devices were analyzed through the current-voltage (IV) curve using Keithley 238 source measurement unit and

the electroluminescence (EL) spectra was measured by HR2000 Oceans-Optic spectrophotometer. All measurements were carried out inside a glove box with the capability of controlling less than 0.01 ppm of moisture and oxygen content.

The four different device structures are as below:

- Device A: ITO/DPVBi/Al
- Device B: ITO/PEDOT:PSS/DPVBi/Al
- Device C: ITO/PVK/DPVBi/Al
- Device D: ITO/CuPc/PVK/DPVBi/Al

147

Fig.2 shows the energy diagram for device A which is the basic OLED used as reference to show the limit of the monolayer structures. To improve the monolayer OLED performance, we inserted PVK (device B) and PEDOT:PSS (device C) as comparison to achieve the best hole transport layer in OLED structures. Then, in order to obtain the best results we optimized the DPVBi thickness (20, 42, 56, 65 nm) of device B. For device D, we tried to improve the performance of the device by introducing a hole injecting layer (HIL).

III. RESULTS & DISCUSSION

Fig. 3 shows the IV curve of the A, B and C devices. It was found that the device C with PVK as the hole transporting layer showed better diode rectifier property compared to those of devices A and B.

This was explained from the energy diagram in fig 2. For the devices of A and B, holes were injected directly to DPVBi without any barrier (device A) and with minimum energy barrier (device B). While in device C, holes were injected into DPVBi after passing through an optimum energy barrier that leads to better control of charge accumulation and electron-hole recombination.

This indicated that without sufficient energy barrier of HTL material, the charge crossing the DPVBi layer is unable to control the recombination and causing fast degradation of device. This also will lead to poor diode properties that may short circuit the devices.

After identified the best hole transport layer (device C), the devices were further optimized based on the variation of the DPVBi layer thicknesses. The results are shown in Fig. 4.

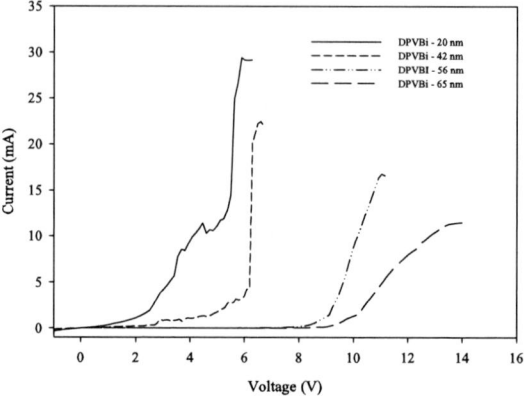

Fig. 4 Current-voltage curves of devices C with different thicknesses of DPVBi layer.

The devices with 56nm-thickness and 65nm-thickness showed better diode properties. Meanwhile, as shown in fig. 5, electroluminescence intensity produced by the devices is increasing with the thickness of the DPVBi layer until it reached the optimum intensity at 56nm-thickness of DPVBi. The thinner DPVBi layer showed less recombination of electron–hole in DPVBi layer zone. The result indicated that 56 nm DPVBi was optimized thickness for ITO/PVK/DPVBi/Al device structures.

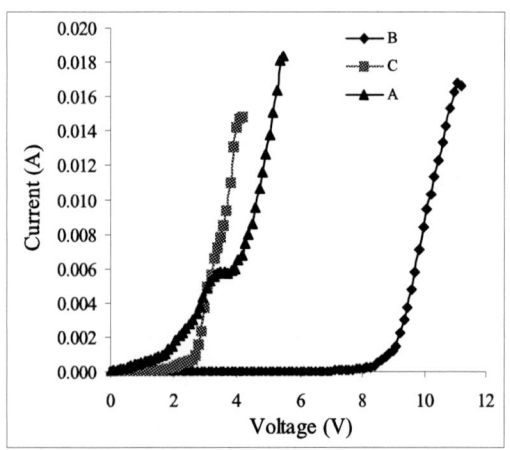

Fig. 3 Current-voltage (IV) curves for single layer devices (device A).

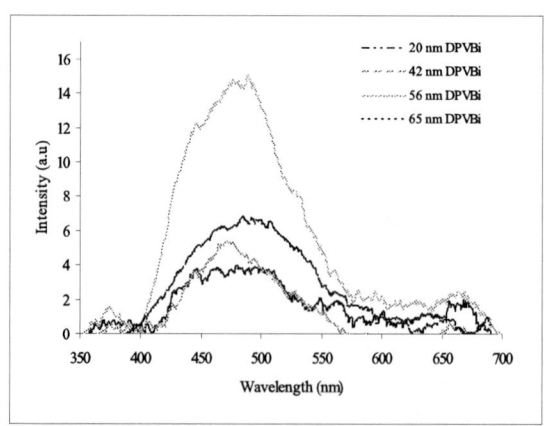

Fig. 5 Electroluminescence intensities of ITO/PVK/DPVBi/Al devices with variation of DPVBi thicknesses.

148

The performance of the 56 nm DPVBi device is improved further by inserting CuPc layer between ITO and PVK. This ITO/CuPc/PVK/DPVBi/Al device has reduced the turn on voltage from 7.0 V to 6.5 V. The comparison of IV curves between devices C and D is shown in fig. 6. It means that the insertion of CuPc layer further enhanced the hole injecting capability from ITO to PVK.

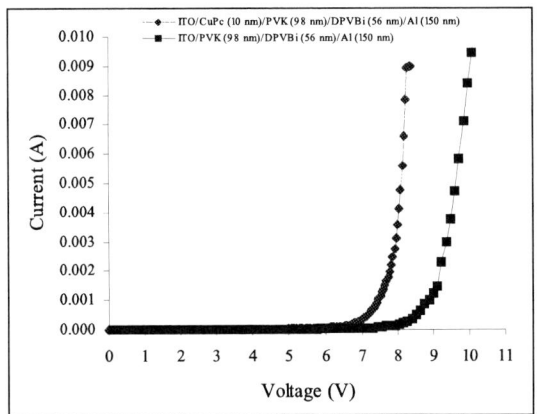

Figure 6. The current voltage comparison of an optimum structure of device C and device D

IV. CONCLUSION

OLED devices with multilayer structures have been realized. The used of PVK as hole transporting layer has improved the device diode properties and able to prevent device from short circuits. The optimized DPVBi layer thickness was found at 56 nm. The insertion of 10 nm CuPc hole injecting layer resulted a lower device turn on voltage at 6.5 V.

ACKNOWLEDGEMENT

This works has been carried out with the support of Malaysian Ministry of Science, Technology and Innovation, under the IRPA grant 03-02-02-0067-SR0007/04-04

REFFERENCES

[1] Tang, C. W.,VanSlyke, S. A. 1987. Organic electroluminescent diodes. *J. Appl. Phys*. 51(12): 913-915.

[2] Burroughs, J.H., Bradley, D.D.C., Brown, A.R., Marks, R.N., Mackay, K., Friend, R.H., Burns, P.L., and Holmes, A.B. (1990). Light-emitting diodes based on conjugated polymers, Natures,.*539-541*

[3] Shaheen, S.E., Jabbour, G.E., Morrell, M.M, Kawabe, Y., Kippelen, B., Peyghambarian, N., Nbor, N.F.,Schlaf, R., Mash, E.A., Armstrong, N.R. 1998. Bright blue organic light-emitting diode with improved color purity using a LiF/Al cathode. *J. Appl. Phys*. 84(4): 2324-2327

How the Optical Properties of Au Nanoparticles are Affected by Surface Plasmon Resonance

Yen Hsun Su[1], Wei Hao Lai[1], Lay Gaik Teoh[2*], Hao En Hong[1], Min Hsiung Hon[1]

[1] Department of Materials Science and Engineering, National Cheng Kung University,
No. 1, Ta-Hsueh Road, Tainan 70101, Taiwan

[2] Department of Mechanical Engineering, National Pingtung University of Science
and Technology, No. 1, Shuehfu Road, Neipu, Pingtung 91201, Taiwan
*Email: n5888107@mail.npust.edu.tw

Abstract – **Gold nanoparticles (Au NPs) have been attracting more attention because they have many color varieties in the visible region based on plasmon resonance, which is due to the collective oscillation of the electrons at the surface of the nanoparticles. We prepared 6 nm Au NPs to modify the surface of the glass substrate. Surface plasmons resonance of Au NPs in toluene is between 500 nm and 600 nm. When Au NPs are modified on the glass substrate, the peak of surface plasmons resonance of Au NPs is shifted. We employed spectral ellipsometry to detect optical properties. Then the characteristics of surface plasmons resonance of Au NPs is determined by reflective index. The performance of surface plasmons resonance of Au NPs on the glass substrate is simulated and shown.**

I. INTRODUCTION

METAL nanoparticles are of great interests because of their unique electronic, optical, and magnetic properties [1-5]. For example, nanoparticles of noble metals such as gold, silver, and transition metals, have been attracted more attention [6-8] because they have many color varieties in the visible region based on plasmon resonance, which is due to the collective oscillation of the electrons at the surface of the nanoparticles [9, 10]. Gold nanoparticles (Au NPs) is a major research area in the field of unusual chemical and physic properties, such as catalytic activity, novel electronics, optics, magnetic properties, biosensors, and their potential application in solar cell.

II. EXPERIMENT

Au NPs were synthesized by chemical reduction method. 3 mM of $HAuCl_{4(aq)}$ 30 ml was added into 5 mM of tetraoctylammonium (TOAB) bromide 80 mL in toluene. Besides, 0.05 M $NaBH_{4(aq)}$ 25 ml was added in the former solution. The color of the solution was turned from yellow to black. Au NPs were modified by tetraoctylammonium bromide. Mercapto-propyl-tri-methoxy-silane (MPTMS) was used as a molecular linker connecting the substrate with Au NPs by SAM process. The substrate is immersed in solution ($NH_4OH:H_2O_2:H_2O=5:1:1$) to make modified OH⁻ on the surface for 3 hrs. Then the substrate is immersed in 1% MPTMS toluene for 1 day. The substrate is immersed in Au NPs colloid for 1 day. Au NPs were modified on substrate.

III. RESULTS AND DISCUSSION

The shape of Au NPs modified by TOAB is close to spherical shape in Fig. 1a. The size is about 6 nm. The nano-beam diffraction pattern of Au NPs is standard indexed diffraction pattern for face center cubic (FCC) crystal in the [011] beam direction in Fig. 1b. From TEM observation, Au NPs belong to poly-crystal. Some of diffraction point, not following the [011] beam direction, are contributed by other grains.

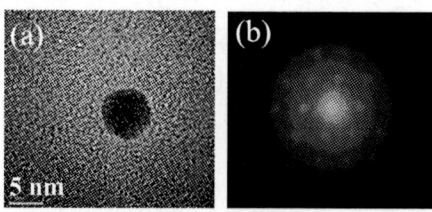

Fig. 1 TEM images of morphology and nano-diffractive pattern of Au NPs in toluene.

The electron charges on the material boundary would perform coherent fluctuations called surface plasma oscillations or surface plasmons of Au. The peaks appear between 500 nm and

600 nm are caused by radiative surface plasmons of Au [11, 12].

According to Mie theory [13], the resonances denoted as surface plasmon were relative with the onset of quantum size and shape effects of Au NPs. This decrease of intensity of the surface plasmon as particle size decrease is observed. Moreover the blue-shift of surface plasmon is as particle size decrease. The frequency of surface plasmon of not spherical Au NPs exist two or three bands, depending on their shape. The resonances peak is observed from 500 nm to 600 nm in Fig. 2. Because of the spherical shape, there is one resonances peak.

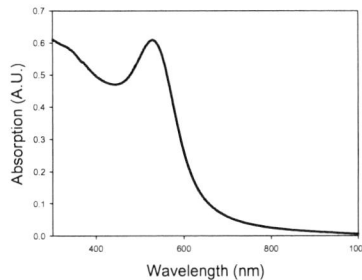

Fig. 2 Absorption of Au NPs in toluene.

Atomic force microscopic image, we knew Au NPs are on the glass substrate. The high of Au NPs layer is about 5 nm, which responses to the TEM resolution.

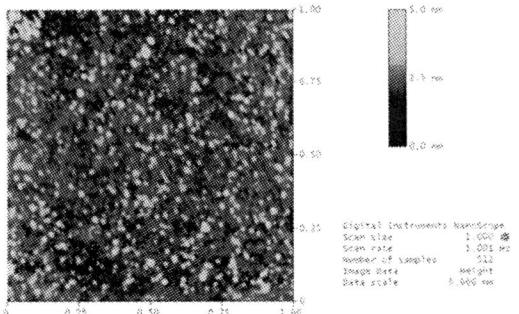

Fig. 3 AFM image of Au NPs on glass substrate.

Fresnel equation is used to calculate the optical properties of Au NPs layer by ellipsometry. Ellipsometry is used as a powerful and sensitive tool to characterize the optical properties of the layer. The optical property, reflective index, is calculated from the Frensnel equation. The theory of Fresnel equation for ellipsometry is described as follows:

$$\tan \Psi \exp(i\Delta) = \frac{R_p}{R_s} = \rho(\lambda, d, N_0, N_1, N_2) \tag{1}$$

where the ellipsometry angles, Δ (phase delay) and Ψ (phase shift), are related to the reflection coefficients. Rp and Rs are the reflection coefficients of p- and s- polarized light which can measure characteristics of SPR. λ is the wavelength of the incident light. d is the thickness of the film. N_0, N_1, and N_2 are the complex reflective indices of ambient, layer, and substrate, respectively. The function ρ [14] refers to a multi-layer reflection model based on Fresnel equations. Simultaneously determining optical properties, the ellipsometric experimental data were fitted by mathematical model data. The best fit to the experimental data was determined by minimizing the mean-square-error (MSE) [14].

$$MSE = \frac{1}{2K - M} \sum_{j=1}^{K} \left[(\Psi_j^{mod} - \Psi_j^{exp})^2 + (\Delta_j^{mod} - \Delta_j^{exp})^2 \right] \tag{2}$$

where K is the number of (Ψ and Δ) pairs, M is the number of the model parameters, and the superscripts mod and exp refer to model and experimental values. The less the value of MSE are, the more accurate the data are. In generally, the MSE is best to be less than 15. The data will have best physical means. Then, the dielectric complex can be driven by reflective index which get from ellipsometric data.

$$\varepsilon_1 = \sqrt{n^2 - \kappa^2} \tag{3}$$

$$\varepsilon_2 = 2n\kappa \tag{4}$$

where n is the reflective coefficient and k is the extinction coefficient.

Because Au NPs layer is a kind of composite layer, Au NPs and void, the effective media approximate (EMA) model is employed to calculate the optical properties of Au NPs. The size of Au NPs is infinitesimally small relative to the wavelength of light, the optical properties of Au NPs is characterized by quasi-static regime.[15] The Maxwell-Garnett effective medium theory can be used to qualitatively describe the quasi-static regime as follows.

$$\frac{\varepsilon_{eff} - \varepsilon_m}{\varepsilon_{eff} + 2\varepsilon_m} = f \frac{\varepsilon - \varepsilon_m}{\varepsilon + 2\varepsilon_m} \tag{5}$$

where ε_{eff} is the dielectric complex of effective media, ε_m is the dielectric complex of metal material, ε is the dielectric complex of void (=1+0i).

Therefore, the optical properties of Au NPs are calculated in Fig. 4. The MSE value is less 0.001.

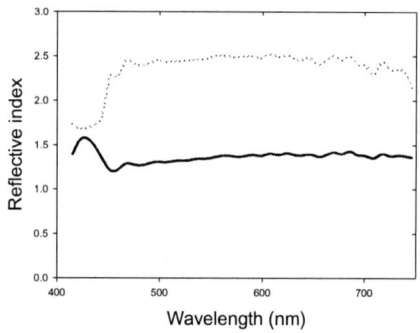

Fig. 4 Calculations of optical properties of Au NPs on glass substrate.

The absorption of light by an optical medium is quantified by its absorption coefficient (α). The relation between absorption coefficient and extinction coefficient is as follows.

$$\alpha = \frac{4\pi k}{\lambda} \qquad (6)$$

where λ is the free space wavelength of the light. The reflectivity (R) depends on both n and k and is given as follows.

$$R = \frac{(n-1)^2 + k^2}{(n+1)^2 + k^2} \qquad (7)$$

The absorptive and reflective condition is simulated in Fig. 5.

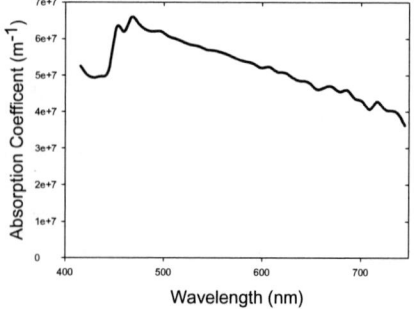

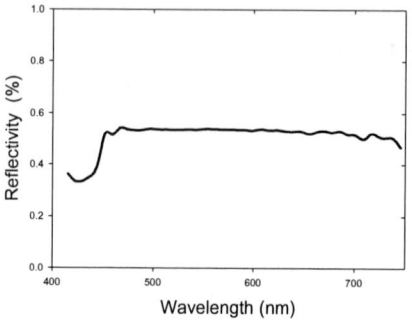

Fig. 5 Simulations absorptive spectrum and reflective spectrum of Au NPs layer.

IV. CONCLUSION

We prepare 6 nm Au NPs in toluene. The structure of Au NPs belongs to FCC structure. Surface plasmons resonance of Au NPs in toluene is about 560 nm. Further the Au NPs is modified on the glass substrate by self-assemble monolayer, which is observed by AFM. Spectral ellipsometry is employed to determine the shift of surface plasmons resonance of Au NPs on the glass substrate. We achieve to calculate the properties of Au NPs on glass substrate. The performance of surface plasmons resonance of Au NPs either on the glass substrate or in toluene is achieved to simulate and display.

ACKNOWLEDGEMENT

This work financially supported by the National Science Council of Taiwan, the Republic of China, grant No. NSC 94-2120-M-006-006, is gratefully acknowledged. The authors would like to thank the Center for Micro/Nano Science and Technology, National Cheng Kung University, Taiwan, for support through equipments and cooperation.

REFERENCES

[1] L. J. Sherry, S. H. Chang, G. C. Schatz, R. P. V. Duyne, B. J. Wiley and Y. Xia, 'Localized Surface Plasmon Resonance Spectroscopy of Single Silver Nanocubes,' *Nano Lett.* vol. 5, pp. 2034-2038 (2005).

[2] E. S. Kwak, J. Henzie, S. H. Chang, S. K. Gray, G. C. Schatz and T. W. Odom, 'Surface Plasmon Standing Waves in Large-Area Subwavelength Hole Arrays,' *Nano Lett.* vol. 5, pp. 1963-1967 (2005).

[3] S. H. Chang, S. K. Gray and G. C. Schatz, 'Surface plasmon generation and light transmission by isolated nanoholes and arrays of nanoholes in thin

metal films,' *Opt. Express* vol. 13, pp. 3150-3165 (2005).

[4] C. Hubert, A. Rumyantseva, G. Lerondel, J. Grand, S. Kostcheev, L. Billot, A. Vial, R. Bachelot, P. Royer, S. H. Chang, S. K. Gray, G. P. Wiederrecht and G. C. Schatz, 'Near-Field Photochemical Imaging of Noble Metal Nanostructures,' Nano Letters vol. 5, pp. 615-619 (2005).

[5] L. Yin, V. K. Vlasko-Vlasov, A. Rydh, J. Pearson, U. Welp, S. H. Chang, S. K. Gray, G. C. Schatz, D. B. Brown and C. W. Kimball, 'Surface plasmons at single nanoholes in Au films,' Appl. Phys Lett. vol. 85, pp. 467-469 (2004).

[6] S. Link and M. A. El-Sayed, 'Spectral properties and relaxation dynamics of surface plasmon electronic oscillations in gold and silver nanodots and nanorods,' *J. Phys. Chem. B* vol. 103, pp. 8410-8426 (1999).

[7] R. C. Jin, Y. Cao, C. A. Mirkin, K. L. Kelly G. C. Schatz, and J. G. Zheng, 'Photoinduced Conversion of Silver Nanospheres to Nanoprisms,' *Science* vol. 294, pp. 1901-1903 (2001).

[8] J. J. Mock, M. Barbic, D. R. Smith, D. A. Schultz and S. Schultz, 'Shape effects in plasmon resonance of individual colloidal silver nanoparticles,' *J. Chem. Phys.* Vol. 116, pp. 6755-6759 (2002).

[9] R. A. Ferrell, 'Predicted Radiation of Plasma Oscillations in Metal Films,' *Phys. Rev.* vol. 111, pp. 1214-1222 (1958).

[10] R. H. Ritchie, H. B. Eldridge, 'Optical Emission from Irradiated Foils. I,' *Phys. Rev.* vol. 126, pp. 1935-1947 (1962).

[11] G. L. Hornyak, C. J. Patrissi and C. R. Martin, 'Fabrication, Characterization, and Optical Properties of Gold Nanoparticle/Porous Alumina Composites: The Nonscattering Maxwell-Garnett Limit,' *J. Phys. Chem. B* vol. 101, pp. 1548-1555 (1997).

[12] A. C. Templeton, J. J. Pietron, R. W. Murray and P. Mulvaney, 'Solvent Refractive Index and Core Charge Influences on the Surface Plasmon Absorbance of Alkanethiolate Monolayer-Protected Gold Clusters,' *J. Phys. Chem. B* vol. 104, pp. 564-570 (2000).

[13] W. P. Wuelfing, F. P. Zamborini, A. C. Templeton, X. Wen, H. Yoon and R. W. Murray, 'Monolayer-Protected Clusters: Molecular Precursors to Metal Films,' *Chem. Mater.* vol. 13, pp. 87-95 (2001).

[14] Y. S. Shon, S. M. Gross, B. Dawson, M. Porter and R. W. Murray, 'Alkanethiolate-Protected Gold Clusters Generated from Sodium *S*-Dodecylthiosulfate (Bunte Salts),' *Langmuir* vol. 16, pp. 6555-6561 (2000).

[15] R. M. A. Azzam and N. M. Bashara, Ellipsometry and Polarized Light, North-Holland Personal Library, Oxford (1989).

Gas Sensing Properties of ZnO:Al Thin Films Prepared by RF Magnetron Sputtering

Lay Gaik Teoh*, Hong Ming Chen, Yen Hsun Su[1], Wei Hao Lai[1], Shih Min Chou[1], Min Hsiung Hon[1]

Department of Mechanical Engineering, National Pingtung University of Science
and Technology, No. 1, Shuehfu Road, Neipu, Pingtung 91201, Taiwan
[1] Department of Materials Science and Engineering, National Cheng Kung University,
No. 1, Ta-Hsueh Road, Tainan 70101, Taiwan
*Email: n5888107@mail.npust.edu.tw

Abstract – **The ZnO:Al thin films were prepared by RF magnetron sputtering on Si substrate using Pt as interdigitated electrodes. The structure was characterized by XRD and SEM analyses, and the ethanol vapor gas sensing as well as electrical properties have been investigated and discussed. The gas sensing results show that the sensitivity for detecting 400 ppm ethanol vapor was ~20 at an operating temperature of 250□. The high sensitivity, fast recovery, and reliability suggest that ZnO:Al thin film prepared by RF magnetron sputtering can be used for ethanol vapor gas sensing.**

I. INTRODUCTION

GAS monitoring devices are in demand for a rapidly growing range of applications. Metal oxide-based chemical sensors have been used extensively for the detection of toxic pollutant gases, combustible gases and organic vapors. The main advantages of chemical sensors are their low price, small size, high sensitivity, and low power consumption. Ethanol is one of the most commonly used and widespread alcohols, and thus there is a need to develop sensors for its detection. The most common application of ethanol sensors is as a breath analyzer, since the ethanol vapor in human breath is said to be correlated with the concentration in the blood.

Recently, gas sensors based on the semiconducting metal-oxides such as SnO_2 and ZnO [1-3] have been found to be very useful for detecting ethanol vapor. Pure and doped ZnO films have been investigated as sensors for O_2 [4], H_2 [5], NO_x [6], and ethanol [7]. One of the requirements of the gas sensors is low power consumption, because the sensors need to reliably and continuously work. A low resistance material has lower driving power when it is used as a sensor. Appropriate donor doping can produce the electronic defects that increase the influence of O_2 partial pressure on the conductivity. Nanto et al. showed that a lower operating temperature may be achieved by the doping effect, and a significant resistance change can be obtained in the doped ZnO rather than the undoped ZnO sensor, which resulted in a higher sensitivity [8].

Several techniques have been used to prepare doped ZnO films, such as RF magnetron sputtering [9], chemical vapor deposition [10], sol-gel [11], and spray pyrolysis [12]. In this study, the Al doped ZnO (defined as ZnO:Al) thin film sensors deposited by RF magnetron sputtering using Pt as the electrode are reported. X-ray diffraction (XRD), scanning electron microscopy (SEM), and conductivity measurements were used to characterize the microstructure and electrical properties of ZnO:Al gas-sensing films that were deposited on Si substrate.

II. EXPERIMENT

Pt electrodes with interdigital structure were deposited on Si substrate by a RF magnetron sputtering process. The ZnO:Al sensing film was sputtered over the interdigitated electrodes. The target is a 3 in. 98 wt % Zn-2 wt% Al alloy, and the distance between the target and substrate was held at 45 mm. A mixed Ar + O_2 gas was introduced into the chamber and metered by mass flow controllers for a total flow rate fixed at 15 cm^3/s. Sputtering deposition was carried out at various RF powers and a pressure of 3 x 10^{-3} torr. Before deposition, the chamber was evacuated to an ultimate background pressure of 10^{-6} Torr, and then a presputtering process was employed to clean the target surface.

The surface and cross-section morphologies were characterized by a scanning electron microscope (FE-SEM, Philips, XL-40FEG). X-

0-7803-9730-4/06/$25.00 ©2006 IEEE

ray diffractometer (RIGAKU-D-max) with Cu Kα radiation was used to determine the crystallographic structures. The as-deposited films were applied for the measurement of gas sensing properties. The gas sensing properties were evaluated at various operating temperatures, from 120 to 250□, by measuring the changes of resistance of the sensor in air and in ethanol gas. The sensitivity in the experiment was defined as

$$S=R_a/R_g \qquad (1)$$

where R_a is the sample resistance measured at ambient environment while R_g is that under the test gas.

III. RESULTS AND DISCUSSION

The ZnO:Al thin films sputtered onto Si substrates were used as an ethanol gas sensor. Fig. 1 represents the XRD diffraction patterns of ZnO:Al film of thickness 100 nm deposited at various RF powers and a pressure of 3×10^{-3} Torr on Si substrates. XRD analysis reveals that all films exhibit only the (0002) peak, indicating that they have ZnO (0002) preferred orientation, implying a c-axis growth perpendicular to the substrate surface. This was due to the lowest surface free energy [13] of the most densely packed (0002) planes in the wurtzite ZnO structure. With increasing RF power the locations of the measured diffraction angle do not change significantly and the dominant (0002) peak becomes sharper, indicating the well-established c-axis orientation of ZnO:Al films. This suggests that the crystallinity of the resulting film increases and the grain size becomes larger with increasing RF power. The grain size of ZnO:Al film was estimated using the Debye-Scherrer equation [14] and the full width at half maximum (FWHM) of (0002) peaks is shown in Fig. 1. The calculated grain sizes of the films were more or less uniform for all films, ranging from 51 to 69 nm, for the ZnO:Al films sputtered at the RF powers of 100~140 W, respectively.

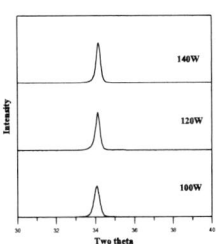

Fig. 1 XRD diffraction patterns for ZnO:Al films deposited at various RF powers.

The SEM surface and cross-section morphologies for the film deposited at 120 W are shown in Fig. 2. It is seen that the deposits are flat and smooth in morphologies, while the cross-section micrograph reveals a columnar structure as it grew on the substrate. The columnar morphology of films deposited on silicon substrate helps enhance the chemical interaction of ethanol gas with ZnO:Al film, which strengthens the output signal and sensitivity of the sensor.

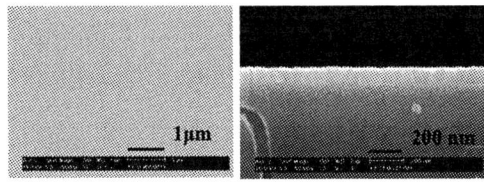

Fig. 2 SEM surface and cross-section morphologies for ZnO:Al film deposited at RF power of 120 W.Etch rate against RF power.

Sensitivity of a ZnO:Al thin film to 400 ppm ethanol vapors from 120 to 250°C is shown in Fig. 3(a), and it can be seen that the sensitivity increases with increasing operating temperature. The characteristic behavior of the resistance decrease of the sensor upon detecting ethanol vapors is typical for n-type semiconductor oxide gas sensors. Fig. 3(b) presents the response and recovery transients of the sensor to 400 ppm ethanol vapors at 250°C. The response and recovery time (time for 90% of resistance change) for ethanol vapors /air mixture for 400 ppm concentration at 250°C is about 2-4 min, with good reproducibility. The calibration curve shown in Fig. 4 indicates that the relation between sensitivity and concentration is linear, which is a benefit for an actuator to respond to detect different concentrations of combustible gases and organic vapors.

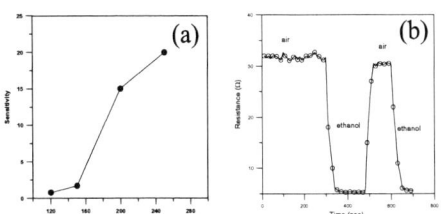

Fig. 3(a) Relationship between operating temperature and sensitivity for ZnO:Al thin film at 400 ppm,
(b) Electrical response of the sensor at operating temperature of 250°C.

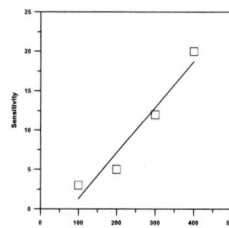

Fig. 4 Sensitivity variation as a function of ethanol concentration for ZnO:Al thin film as operated at 250□.

The gas sensing mechanisms normally accepted for semiconductor sensors assume that the oxygen adsorded on the surface of the oxide removes some of the electronic density and thus decreases the material's conductivity. When reduction gas molecules come into contact with this surface, they may interact with this oxygen, leading to an inverse charge transference [15]. Upon the return of the electrons to the conduction band, conductivity increases. This utilizes the gas-induced resistance variations in potential barrier height at grain boundaries (i.e. changes in thickness of the space charge layer) to detect ethanol vapors in air. For polycrystalline substances, grain boundaries contribute most of the resistance. It may also be concluded that the surface conductivity of a semiconducting oxide crystal depends on the electron concentration near the surface, which in turn is affected by the nature of the chemisorbed species. The sensitivity characteristics of the sensor are associated with a modification of the ZnO surface by the electron transfer mechanism. In this study, the reason for a decrease in the resistance may be due to the oxidation of the ethanol vapors upon coming in contact with the ZnO:Al film surface, which liberates free electrons and H_2O. The atmospheric oxygen chemisorbs on the surface of the ZnO:Al film as O_2^- or O^-, removing an electron from the conduction band of the ZnO:Al semiconductor, developing a depletion region on the surface. Ethanol vapors react with the chemisorbed oxygen and reinject the carrier, thereby reducing the resistance of the ZnO:Al material. The possibility of a reaction of ethanol with the ZnO:Al sensing layers can be explained as two oxidation states [16]:

$$C_2H_5OH_{(g)} + [O] \rightarrow CH_3CHO + H_2O \quad (2)$$

$$C_2H_5OH_{(g)} + [O] \rightarrow C_2H_4 + H_2O \quad (3)$$

in which [O] represents the surface oxygen ions.

The first reaction is a process initiating the oxidation by the dehydrogenation to CH_3CHO intermediate, and the second reaction is initiated by the dehydration to C_2H_4. But the selectivity for the two reactions is initiated by the acid–base properties of the oxide surface. The dehydrogenation process is more probable on the oxide surface with basicity property, while the dehydration is favoured on the acid surface [16]. The intermediate products, acetaldehyde and ethylene, are subsequently reduced to CO_2 and H_2O. At higher temperature, the depletion region created by the chemisorbtion of oxygen on the surface extends more deeply, providing larger scope for more gaseous elements to be adsorbed, thereby giving a better response. Also, the hydroxyl group desorbs at higher temperatures [17]. Thus for lower temperature operation (< 150□), the surface of the sensor does not get completely desorbed, which causes a smaller change in resistance. ZnO:Al has thus been found to markedly promote the sensitivity to ethanol vapor.

IV. CONCLUSION

In this study, the structures and sensing properties of ZnO:Al films as an ethanol vapor gas sensor obtained by RF magnetron sputtering system were investigated. The structural characteristics reveal that flat and well-defined columnar films with c-axis textured were formed. The film exhibited good sensitivity to the ethanol vapors with quick response-recovery characteristics, it was found that the sensitivity for detecting 400 ppm ethanol vapor was ~20 at an operating temperature of 250□. ZnO:Al seems to be a promising semiconducting material for the detection of ethanol vapor.

REFERENCES

[1] B. P. J. D. Costello, R. J. Ewen, N. Guernion and N. M. Ratcliffe, 'Highly sensitive mixed oxide sensors for the detection of ethanol,' *Sens Actuators B* vol. 87, pp. 207-210 (2002).

[2] D. F. Paraguay, M. Miki-Yoshida, J. Morales, J. Solis and L. W. Estrada, 'Influence of Al, In, Cu, Fe and Sn dopants on the response of thin film ZnO gas sensor to ethanol vapour,' *Thin Solid Films* vol. 373, pp. 137-140 (2000).

[3] P. Ivanov, E. Llobet, X. Vilanova, J. Brezmes, J. Hubalek and X. Correig, 'Development of high

sensitivity ethanol gas sensors based on Pt-doped SnO_2 surfaces,' *Sens Actuators B* vol. 99, pp. 201-206 (2004).

[4] G. Sberveglieri, S. Groppelli, P. Nelli, F. Quaranta, A. Valentini and L. Vasanelli, 'Oxygen gas-sensing characteristics for ZnO(Li) sputtered thin films,' *Sens. Actuators B* vol. 7, pp. 747-751 (1992).

[5] T. Yamazaki, S. Wada, T. Noma and T. Suzuki, 'Gas-sensing properties of ultrathin zinc oxide films,' *Sens. Actuators B* vol. 14, pp. 594-595 (1993).

[6] J. Muller and S. Weissenrieder, 'ZnO-Thin film chemical sensors,' *Fres. J. Anal. Chem.* vol. 349, pp. 380-384 (1994).

[7] I. Stambolova, K. Konstantinov, S. Vassilev, P. Peshev and Ts. Tsacheva, 'Lanthanum doped SnO_2 and ZnO thin films sensitive to ethanol and humidity,' *Materials Chemistry and Physics* vol. 63, pp. 104-108 (2000).

[8] H. Nanto, T. Minami and S. Takata, 'Zinc-oxide thin film ammonia gas sensor with high sensitivity and excellent selectivity,' *J. Appl. Phys.* vol. 60, pp. 482-484 (1986).

[9] J. Hu and R. G. Gordon, 'Textured aluminum-doped zinc oxide thin films from atmospheric pressure chemical-vapor deposition,' *J. Appl. Phys.* vol. 71, pp. 880-890 (1992).

[10] O. Takai, M. Futsuhara, G. Shimizu, C. P. Lungu and J. Nozue, 'Nanostructure of ZnO thin films prepared by reactive rf magnetron sputtering,' *Thin Solid Films* vol. 318, pp. 117-119 (1998).

[11] M. Ohyama, H. Kouzuka and T. Yoko, 'Sol-gel preparation of ZnO films with extremely preferred orientation along (002) plane from zinc acetate solution,' *Thin Solid Films* vol. 306, pp. 78-85 (1997).

[12] J. L. Van Heerden and R. Swanepoel, 'XRD analysis of ZnO thin films prepared by spray pyrolysis,' *Thin Solid Films* vol. 299, pp. 72-77 (1997).

[13] Y. E. Lee, Y. J. Kim and H. J. Kim, 'Thickness dependence of microstructural evolution of ZnO films deposited by rf magnetron sputtering,' *J. Mater. Res.* vol. 13, pp. 1260-1265 (1998).

[14] B. D. Cullity, *Elements of X-ray Diffraction*, Second ed., Addison-Wesley, Reading MA, pp 162 (1978).

[15] M. R. Islam, N. Kumazawa and M. Takeuchi, 'Chemical sensor based on titanium dioxide thick film:Enhancement of selectivity by surface coating,' *Appl. Surf. Sci.* vol. 142, pp. 262-266 (1999).

[16] S. Matsushima, T. Maekawa, J. Tamaki, N. Miura and N. Yamazoe, 'Role of Additives on Alcohol Sensing by Semiconductor Gas Sensor,' *Chem. Lett.* vol. 18, pp. 845-848 (1989).

[17] D. Kohl, 'Surface processes in the detection of reducing gases with SnO_2-based devices,' *Sens. Actuators B* vol. 18, pp. 71-113 (1989).

Factors Affecting the Growth of Carbon Nanotubes

Lee Cheng Choo and Roslan Md. Nor
Department of Physics, Faculty of Science, University of Malaya,
50603 Kuala Lumpur, MALAYSIA
Email: cc_lee81@yahoo.com and rmdnor@um.edu.my

Abstract **Carbon nanotubes were prepared by using a hot filament chemical vapour deposition (HFCVD) system. Methane diluted in hydrogen was used as the reaction gas and nanosized nickel particles were used as the catalyst. Studies on the effect of catalyst preparation and growth conditions were conducted. The carbon nanotube produced under different conditions were characterized using field emission scanning electron microscopy (FESEM), energy dispersive x-ray (EDX), X-ray diffraction (XRD), Raman spectroscopy and transmission electron microscopy (TEM).**

I. INTRODUCTION

Carbon nanotubes (CNTs) have attracted extensively research interest since their discovery in 1991 by Sumio Iijima [1] due to their unique and interesting mechanical, chemical and electrical properties and wide range of potential applications [2-5]. Various methods have been used to produce carbon nanotubes such as arc discharge, laser ablation and chemical vapour deposition (CVD) [6]. Hot filament CVD is one of the most commonly used catalytic methods for synthesizing CNTs. The advantages of HFCVD are its flexibility, simplicity, low cost and the possibility to produce large area of CNTs [7]. In this technique, a filament is utilized for decomposing the hydrocarbon-hydrogen gases to produce active radicals which are crucial for CNTs formation. The growth conditions such as the substrate temperature, gas composition and pressure, and growth time are essential in producing high yield and high purity of CNTs. Besides that, catalyst preparation and hydrogen thermal treatment will also influence the final structure and morphology of the CNTs. In this work, we report the results of parametric studies of CNTs growth using the HFCVD method.

II. EXPERIMENT

Nickel catalysts were deposited onto silicon substrates prior to the growth of carbon nanotubes. The substrates were treated by soaking the silicon into a 10% hydrofluoric acid solution for 45 minutes to etch away the native oxide layer. The substrates were then cleaned with acetone in an ultrasonic bath to remove any contaminants on the surface. Salt solution i.e. nickel (II) sulphate 6-hydrate ($NiSO_4.6H_2O$) was used as nickel source. The colloidal solution of 20mM $NiSO_4$ was prepared by dissolving the appropriate amount of the nickel salt into 100ml distilled water. The mixture was ultrasonically stirred for 30 minutes to obtain a clear solution. The silicon substrates were then dipped into the colloidal solution and calcined in an electrical tube furnace at 500°C for 120 minutes at a heating rate of 20°C per minute.

After being deposited with the salt solution, the silicon substrate was transferred into a hot filament reactor chamber. The chamber was then pumped down to 10^{-4}mbar by using a turbo-molecular pump backed by a rotary pump. The sample was heated in H_2 ambient (100sccm) for 15 minutes at 700°C. Deposition of CNTs was conducted by flowing methane gas into the chamber. The gas mixture was dissociated by using a tungsten filament which is located 1cm above the sample and is powered by a voltage source.

In this work, two growth conditions were investigated. The variation of substrate temperature and gas pressure was in the range of 700 to 900°C and 5 to 10mbar, respectively. Other growth parameters such as the gas composition and growth time were kept constant at 100:20 (volumetric flow rate of H_2:CH_4) and 30 minutes, respectively.

III. RESULTS AND DISCUSSION

The results obtained from the analysis of carbon nanotubes prepared at different substrate temperatures and gas pressures are presented in the light of measurements using FESEM (Philips Quanta 200F), EDX (LEICA S440), XRD (Siemens D5000, CuKα, 0.154nm), Raman

0-7803-9730-4/06/$25.00 ©2006 IEEE

spectroscopy (Renishaw Ramascope 2000, Ar^+, 514.5nm) and TEM (LEO, 120kV).

1. Catalyst Preparation

Figure 1 shows the effect of hydrogen thermal treatment on the formation of nanosized nickel particles from thin layer of $NiSO_4$.

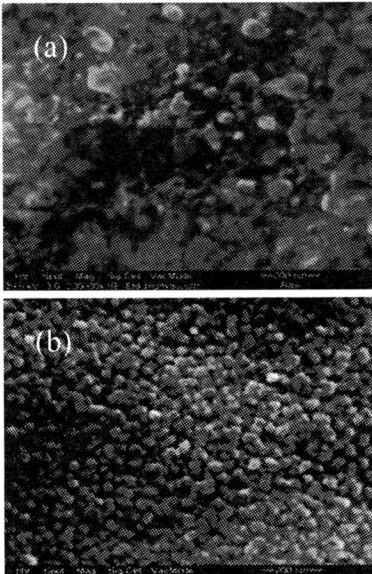

Figure 1: FESEM images of the Ni-Si samples: (a) before and (b) after H_2 thermal treatment.

There is noticeable distinction between the un-treated and H_2-treated Ni-Si samples. Oxide nanoparticles were expected in the un-treated sample. The bright spot shows that the sample was charging due to the non-conducting nickel oxide. It is also noticeable that the distribution of the nanoparticles was not uniform and the size of the particles is ranged from 20 to 150nm.

After the H_2 thermal treatment process, the nickel films were seen to break into smaller islands. The breaking of nickel films into island may be due to the difference of surface energy and thermal coefficient between nickel and silicon during the elevated heating temperature [8]. Nickel oxides were reduced to nanosized particles in the hydrogen environment at elevated temperature.

When the heating temperature was increased to 700°C, an increase of nanosized nickel particles density was observed. The nanoparticles were distributed homogeneously and the size of the particles was around 30nm. Due to the elevated temperature, the nickel particles were seen to agglomerate into relatively larger particles and the spacing between them was found to be very small. The agglomeration may

possibly due to the migration of nickel particles on the silicon surface [9].

Based on our present work, we discovered that nanosized nickel particles suitable for CNTs growth can be fabricated by thermal reduction of thin nickel salt films on silicon substrate.

2. Growth Conditions

In this work, the effect of substrate temperature and gas pressure on the quality of the deposited CNTs was investigated. These parameters are essential in determining the structure, diameter and quality of the deposited products. Figure 2 and Figure 3 show the Raman spectra of the deposited CNTs obtained at different substrate temperature and gas pressure, respectively.

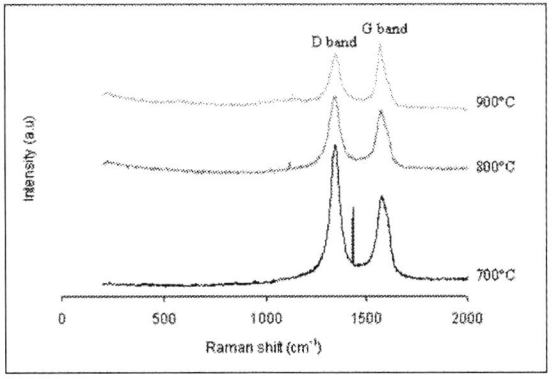

Figure 2: Raman spectra of the deposited CNTs synthesized at different substrate temperature.

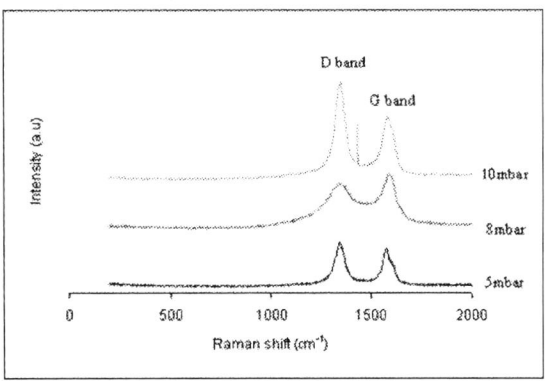

Figure 3: Raman spectra of the deposited CNTs synthesized at different gas pressure.

There are two prominent peaks exhibited in the Raman spectra i.e. D band ($\sim 1340 cm^{-1}$) and G band ($\sim 1580 cm^{-1}$). The D band corresponds to the sp^3 bonding as well as tube curvature and defects such as amorphous carbon whereas G band corresponds to sp^2 bonding of crystalline graphene sheets. The intensity ratio of D and G band can be used to express the purity index or tube crystallinity [10-11]. The position, intensity and width of the peaks were analyzed using Lorentzian curve fitting.

From the Raman spectra, it can be seen that the intensity of D band and the I_D/I_G ratio decrease with the increasing of substrate temperature. These results show that better crystallinity or quality of CNTs was produced at higher temperature with lesser amount of amorphous carbon or other carbonaceous impurities. At elevated temperature, more H radicals were produced which would then etch the amorphous carbon and other disordered carbon structure efficiently [12-13]. Higher temperature also results in higher amount of deposited products due to the increase in reaction rate. In addition, both the dissociation and hydrogenation rate of methane gas were high at elevated temperature [14].

The study on the effect of gas pressure showed that the quality of CNTs produced decreases with increasing of pressure except for the case at 8mbar. The deposited products produced at 8mbar results in higher content of amorphous carbon which can be observed by investigating the width of D band. D bandwidth of greater than 100cm^{-1} show that most of the products produced is amorphous carbon [10]. When the gas pressure and the diffusion rate are inappropriate, CNTs are not likely to form and most probably will appear in the forms of amorphous carbon [15]. For the case of 5 and 10mbar, the intensities of D and G band decrease with increasing of pressure. This is because lower pressure provides slower reaction rate which would then result in better quality of CNTs [16]. In addition, higher pressure tends to produce higher density due to the large quantity of carbon species formation which then leads to immense production of carbon products [14].

Figure 4 and Figure 5 illustrate the XRD spectra of the deposited CNTs produced at different substrate temperature and gas pressure, respectively. The diffraction peaks show the presence of carbon materials along with other catalyst impurities. From the XRD spectra for both substrate temperature and gas pressure, the graphite-like (002) peak was observed at approximately 26.5°. Two prominent peaks at about 44.5° and 51.8° and a less intense peak at about 76.4° was ascribed to the (111), (200) and (220) plane of nickel metal, respectively. Several other peaks i.e. approximately at 28°, 48° and 56° also appeared in the spectra which attributed to the (111), (220) and (311) plane of silicon, respectively. The hkl plane for each element and

its corresponding 2θ angle were referred from the JCPDS file.

CNTs produced at 700°C and 900°C displayed stronger (002) peaks as compared to that of 800°C. However, the peak at 700°C has a broader width compared to the one at 900°C. Intense nickel peaks were also found at the temperature of 900°C. The existence of these graphite and nickel peaks indicates that the deposited products contained more crystalline carbon and residual metal, respectively [10].

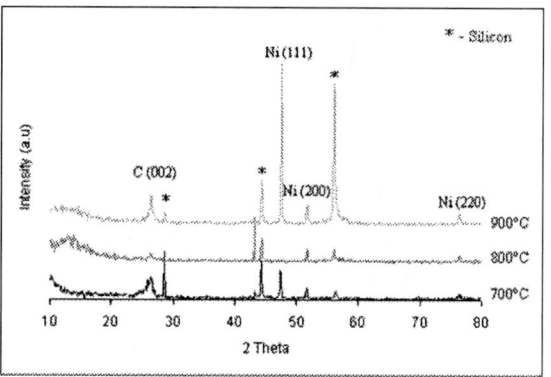

Figure 4: XRD spectra of the deposited CNTs synthesized at different substrate temperature.

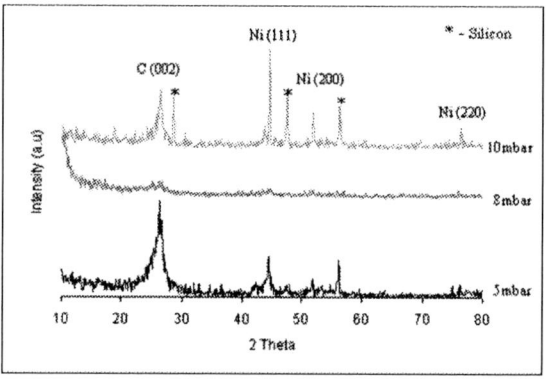

Figure 5: XRD spectra of the deposited CNTs synthesized at different gas pressure.

The same observation also applies to the different gas pressure. Deposited products obtained at 8mbar exhibit weaker graphite-like and nickel peaks. Therefore, the inequitable value of substrate temperature and gas pressure was found to influence the chemical reactions between the hydrocarbon gas and the metal catalyst. From the results, higher substrate temperature and lower gas pressure were found to produce better quality of CNTs and these outcomes are in agreement with the Raman analysis.

According to some researchers, the mean diameter of the CNTs can be estimated through the use of Scherrer expression [10]. The half width of the graphite peak is used to calculate the average diameter of the deposited CNTs. The tubes diameters were found to be in the range of 10 to 30nm, which is in agreement with the size of the catalyst particles. The results show a large diameter variation which probably due to the interpretation of the graphite peak. The calculated tube diameter may not be precise because the observable graphite peak may comprise both carbon nanotubes and residual graphite. In order to obtain a more define tube diameter, analytical techniques such as transmission electron microscopy (TEM) is preferred.

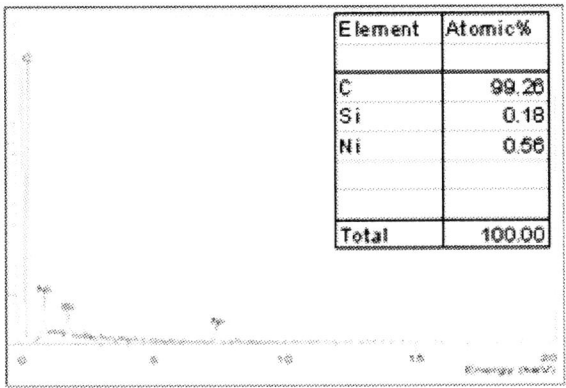

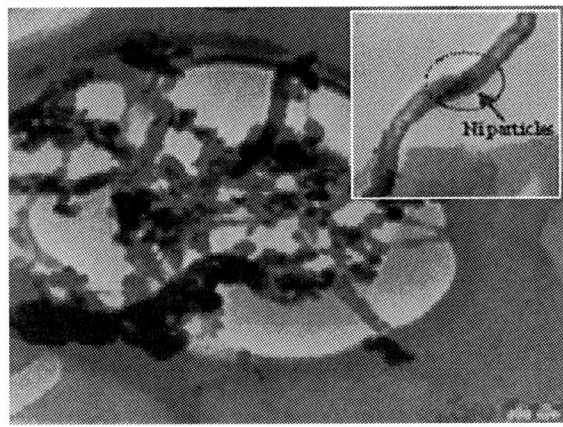

Figure 6: EDX pattern and TEM images of the deposited tube products synthesized preset growth conditions.

Figure 6 illustrates the EDX pattern and TEM images of the deposited CNTs synthesized at preset growth conditions. EDX characterization shows the amount of carbon, nickel and silicon. Carbon is the major element in the deposited products with 99.2%, followed by nickel with 0.6% and silicon with 0.2%. There was no oxygen signal appeared in the EDX spectrum and this results complement the XRD analysis that only crystalline nickel occurs instead of nickel oxide. The EDX spectrum also illustrated that no other chemical impurities occurs in the sample.

TEM images displayed mixtures of carbon materials i.e. tubes, fibers, particles, etc. in the deposited products. Carbon nanotubes can be easily differentiated from other carbon materials by observing the hollow inner wall that are several nanometers in thickness. The diameter of the carbon nanotubes produced is ranged from 10 to 30nm, which is in agreement with the size of the catalyst particles. The nickel particles were also clearly observed in the close-up image (inset), as marked with arrow. The particles can be seen at the tip as well as inside the tube structures.

CONCLUSIONS

In summary, carbon nanotubes were successfully produced by using HFCVD method. A variation of carbon structures were synthesized on nickel catalyst using methane diluted in hydrogen. Hydrogen thermal treatment plays a vital role in the formation of nanosized nickel particles. Higher substrate temperature and lower gas pressure were found to produce better quality or crystallinity of CNTs.

ACKNOWLEDGEMENT

The authors would like to thank the Ministry of Science, Technology and Innovation of Malaysia for the research funding (09-02-03-0229EA229) and University of Malaya for the financial support.

REFERENCES

[1] S. Iijima, Helical Microtubules of Graphitic Carbon, Nature 354 (1991), 56-58.
[2] L. Forro and C. Schonenberger, Physical Properties of MWNT, Topics Appl. Phys. 80 (2001), 329-390.
[3] B. I. Yakobson and P. Avouris, Mechanical Properties of CNTs, Topics Appl. Phys. 80 (2001), 287-329.
[4] S. G. Louie, Electronic Properties, Topics Appl. Phys. 80 (2001), 287-329.
[5] P. M. Ajayan and O. Z. Zhon, Applications of Carbon Nanotubes, Topics Appl. Phys. 80 (2001), 391-425.
[6] Y. Ando, X. Zhao, T. Sugai and M. Kumar, Growing CNTs, Materialstoday (2004).
[7] J. Bonard, Thin Solid Films (2005), 1-7.
[8] X. Chen, R. Wang, J. Xu, and D. Yu, TEM Investigation on the Growth Mechanism of CNTs Synthesized by HFCVD, Micron 35 (2004), 455-460.

[9] Y. Huh, J. Y. Lee, J. Cheon, Y. K. Hong, J. Y. Koo, T. J. Lee and C. J. Lee, Controlled Growth of CNTs over Cobalt Nanoparticles by Thermal CVD, J. Mater. Chem. 13 (2003), 2297-2300.

[10] T. Belin,and F. Epron, Characterization Methods of CNTs: A Review, Materials Science and Engineering (2005).

[11] P. K. Chu and L. Li, Characterization of Amorphous and Nanocrystalline Carbon Films, Mater. Chem. and Phys. 96 (2006), 253-277.

[12] M. R. Maschmann, P. B. Amama, A. Goyal, Z. Iqbal, R. Gat and T. S. Fisher, Parametric Study of Synthesis Conditions in Plasma Enhanced CVD of High Quality SWNTs, Carbon 44 (2006), 10-18.

[13] S. A. Moshkalyov, A. L. D. Moreau, H. R. Guttierrez, M. A. Cotta and J. W. Swart, CNTs Growth by CVD using Thin Film Nickel Catalyst, Mater. Sci. and Eng. B 112 (2004), 147-153.

[14] G.Y. Xiong, Y. Suda, D. Z. Wang, J.Y. Huang and Z. F. Ren, Effect of Temperature, Pressure, and Gas Ratio of Methane to Hydrogen on the Synthesis of Double-Walled CNTs by CVD, Nanotechnology 16 (2005), 532-535.

[15] A. Hussain, Ferric-Sulfate-Catalyzed HFCVD Carbon Nanotube Synthesis, Nanotechnology 14 (2003), 925-930.

[16] T. D. Makris, R. Giorgi, N. Lisi, L. Pilloni, E. Salernitano, F. Sarto and M. Alvisi, CNTs: Growth by HFCVD: Effect of the Process Parameters and Catalyst Preparation, Diamond and Related Materials 13 (2) (2004), 305-310.

Effect of Self-Weight and Non-Rigidity on the Bending Characteristics of Surface Micromachined MEMS Test Structures

Hing Wah Lee, Member, *IEEE* and Mohd. Ismahadi Syono, Member, *IEEE*
MEMS Department, MIMOS Berhad
Technology Park Malaysia,
57000 Kuala Lumpur, Malaysia
Email: hingwah.lee@mimos.my

Abstract - **This work studies the bending characteristics due to the effects of self-weight and non-rigidity induced on the surface micromachined (SMM) micro-electro-mechanical systems (MEMS) test structures. By using the available information of the linear bending stiffness for MEMS, the bending characteristics of the SMM MEMS cantilever test structure can be evaluated analytically. Results obtained from the analytical modeling have been compared with results from finite element analysis (FEA) using CoventorWare where the present analytical modeling showed good agreement with results obtained through FEA, establishing the methodology superiority. It is apparent that the finite anchor stiffness presence due to the non-rigid effect yields significant difference on the deflections of the SMM MEMS test structures while the effect of self-weight is negligible. It is thus deemed a necessity that the anchor's flexibility should be incorporated in order to achieve good accuracy of the analytical modeling. This study also showed that the present analytical modeling is restricted to be employed only in the linear range. This is due to the fact that as the displacement increases, the geometrically nonlinear effect becomes conspicuous corresponding to the small deflection theory used to generate the analytical modeling.**

I. INTRODUCTION

Surface micromachined (SMM) micro-electro-mechanical systems (MEMS) test structures are commonly used to measure thin film properties and accurate knowledge of these properties is critical for assessing the long-term reliability of

MEMS devices. Micro-sized cantilever structures have been popularly employed to obtain mechanical properties such as the modulus of elasticity and linear bending stiffness. At the same time, the bending characteristics of the structures can be evaluated through computer simulation [1,2], analytically [3-5] or experimentally [6-8].

It has been a common practice to neglect the effect of self-weight and non-rigidity while analyzing the bending characteristics of the SMM cantilevers. Researchers made assumptions that those effects would not have much effect on the overall characteristics and performance of the sensing process. However, the emergence of SMM cantilevers as mechanical biosensors for detection of pH [9], protein [10], DNA [11] and viruses [12] necessitate the needs for greater accuracy and sensitivity in the measurement scheme. For instance, the weight of a single baculovirus is only 1.5×10^{-15}g [12]; which will yield displacements much smaller than displacements due to the weight of the cantilever itself. Thus it is only sensible that all possible contributing means to the deflections of the sensor are taken into considerations so that the difference in the bending during detection is generated purely by the virus.

The present work will address the issues of self-weight and non-rigidity on the bending characteristics of the SMM MEMS test structures. The test structures of interest are the aluminum cantilever. Initially, the linear bending stiffness of the cantilevers will be obtained through CoventorWare simulation. Subsequently, the anchor stiffness can be evaluated by feeding the bending rigidity and the linear bending stiffness into the general analytical model proposed in this work. With all the parameters known, the

0-7803-9730-4/06/$25.00 ©2006 IEEE

deflections caused by the effects of self-weight and other applied load can then be found accurately.

The proposed analytical model is useful because it also offers a way to accurately predict the anchor stiffness and the bending rigidity of the test structures after each fabrication process. This can be achieved by evaluating the linear bending stiffness of the structures experimentally rather than using simulation results. It is important considering the fact that the bending rigidity of a material often differs from the tabulated value after it undergoes the fabrication process.

II. ANALYTICAL MODEL OF SMM CANTILEVER

Initially, anchor stiffness is found from the linear bending stiffness and bending rigidity. Once anchor stiffness is obtained, displacements caused by the applied load can be evaluated.

Fig. 1 shows a SMM cantilever with length l and the applied uniform load of P. The equivalent free body diagram (FBD) for the analytical model of the SMM cantilever with finite anchor stiffness, K_θ is described in Fig. 2 while a rigid cantilever is shown in Fig. 3. The anchor stiffness should be considered since the anchor itself will start deforming as transverse displacement of the cantilever commences.

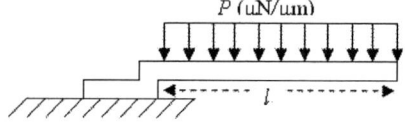

Fig. 1 SMM cantilever.

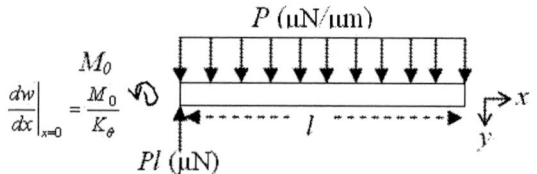

Fig. 2 FBD of SMM cantilever with anchor stiffness, K_θ.

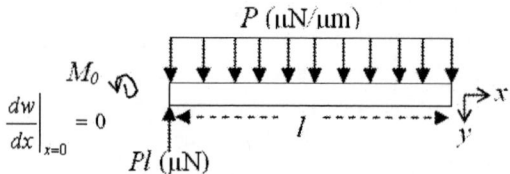

Fig. 3 FBD of rigid cantilever.

Considering the cantilever beam and using the Bernoulli-Euler beam bending theory [5], the beam curvature, d^2w/dx^2 relates to the bending moment, M as follows:

$$\frac{d^2w}{dx^2} = \frac{M}{EI} \tag{1}$$

where EI is the bending rigidity.

The bending moment at location x along the beam axis is given by:

$$M = P(l-x)\frac{(l-x)}{2} \tag{2}$$

Since the anchor is not perfectly rigid, rotation exists and the slope can be expressed as:

$$\left.\frac{dw}{dx}\right|_{x=0} = \frac{M_0}{K_\theta} \tag{3}$$

where M_0 is the moment at the anchor

Using Eqs. (1) – (3), the displacement of the beam, w in the y-direction can be evaluated as:

$$\frac{w}{P} = \frac{l^4}{8EI} + \frac{l^3}{2K_\theta} \tag{4}$$

The linear relationship between the load and displacement for an elastically deformed cantilever beam is determined by the linear bending stiffness, K which can be given as:

$$P = Kw \tag{5}$$

Substituting Eqs. (5) into (4), the anchor stiffness can then be given as:

$$K_\theta = \frac{l^3}{2}\left(\frac{1}{K} - \frac{l^4}{8EI}\right)^{-1} \tag{6}$$

The value of the anchor stiffness will be specific for a given beam length, width and thickness. Table 1 summarizes the analytical models used to obtain deflections for the SMM cantilever for both the rigid and non-rigid scheme.

Table 1 Analytical model to obtain deflections for SMM cantilever for both rigid and non-rigid scheme.

Scheme	Displacement (y-direction)
Analytical (Non-Rigid)	$\dfrac{w}{P} = \dfrac{l^4}{8EI} + \dfrac{l^3}{2K_\theta}$
Analytical (Rigid)	$\dfrac{w}{P} = \dfrac{l^4}{8EI}$

III. SIMULATIONS

The present work includes numerical modeling using CoventorWare simulation tool to validate the cantilever analytic model as described in Table 1. The SMM cantilever is meshed using the Manhattan bricks method with optimized elements length of 2μm×2μm×1μm in the x–y-z direction respectively. Table 2 shows the displacement obtained for the SMM cantilever with width of 10μm, length of 750μm and thickness of 1μm for variations in element sizes (length) while Fig. 4 shows the meshed model of the SMM cantilever.

Table 2 Displacement of cantilever for variations in the element sizes for meshing in CoventorWare.

Element size - length (μm)	Displacements (μm)
5	0.144467720
4	0.143211716
3	0.140022513
2	0.139948734
1	0.139948734

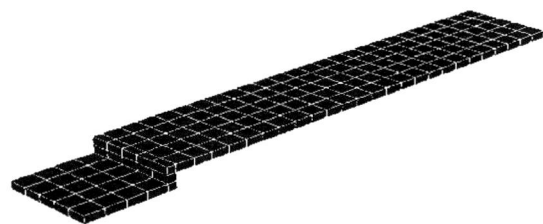

Fig. 4 Meshed model of SMM cantilever

IV. RESULTS AND DISCUSSIONS

The displacements of the SMM cantilever due to the uniform applied load will be investigated for the three different schemes mentioned earlier;

which are the analytic model (rigid, non-rigid) and numerical simulation (CoventorWare). Both the effects of self-weight and non-rigidity on the displacements will be discussed accordingly.

Table 3 shows the effect of self-weight on the aluminum cantilever beam of width 10μm and thickness 1μm with length that varies in the range of 50–250 μm; subjected to an applied uniform load of 0.01μN/μm. From Table 3, it is clear that the deviation increases as the length of the cantilever increases. This is comprehensible since the weight of the beam itself increases for longer beam. Due to the effect of self-weight, cantilever beam with the length of 250μm creates additional displacements of almost 1.5nm which are significant especially when the SMM cantilever structure is used for biosensing application involving detection of nano-sized particles. Thus, it will be more desirable to design aluminum SMM cantilever structure with length of less than 50μm for biosensing application.

Table 3 Comparison in the displacement obtained for different beam length with applied load of 0.01μN/μm due to the effect of self-weight.

Length (μm)	Displacement (μm)		Absolute deviation (μm)
	Self-weight		
	With	Without	
50	0.13722	0.13722	0
100	1.96788	1.96783	0.00005
150	9.85728	9.85705	0.00023
200	30.7328	30.7321	0.00070
250	73.3442	73.3428	0.00140

Meanwhile, Fig. 5 and 6 illustrate the load-displacement relationship of SMM cantilever for different length. The figures show that the present analytical model agrees well with the numerical simulation solutions, establishing the proposed analytical model superiority. This demonstrates that the difference in the displacement comes directly from the effect of the anchor stiffness, asserting claim that the flexible anchor significantly affects the kinematics of the SMM structures. Therefore, the presence of the finite anchor stiffness should be treated as an important factor as long as SMM structures indispensably involve the anchor geometries.

The present analytical model will not be as accurate for large displacements as shown in Fig. 6. This is due to the fact that as displacement increases, the geometrically nonlinear effect becomes conspicuous corresponding to the small deflection theory used to generate the analytical modeling. The difference between the analytical model (non-rigid) and numerical simulation is apparent for large deflections outside the linear range. Hence the proposed analytical model will only be applicable within the linear range.

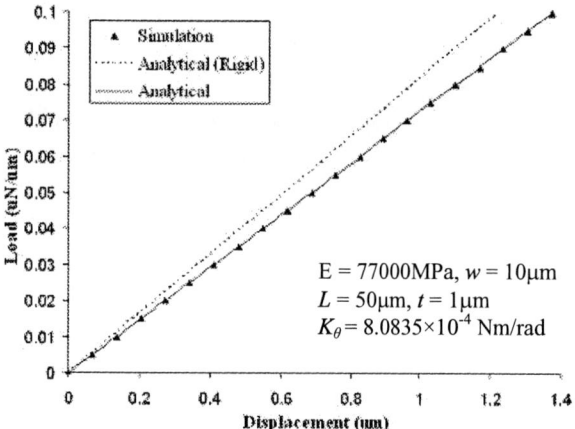

Fig. 5 Load-displacement relationships of SMM cantilever with the length of 50μm.

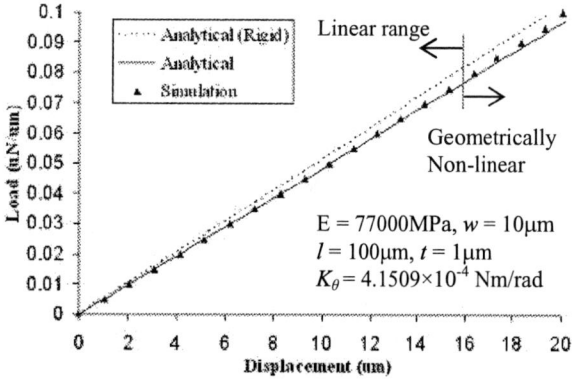

Fig. 6 Load-displacement relationships of SMM cantilever with the length of 100μm.

V. CONCLUSION

The proposed analytical model incorporating the effects of the flexible anchor predicts the displacements of the SMM cantilever with higher accuracy as compared to conventional solution. However, the analytical model will only be applicable within the linear range. The effects of

self-weight will be negligible for cantilever with the length of less than 50μm.

REFERENCES

[1] S. Greek and N. Chitica, "Deflection of surface-micromachined devices due to internal, homogeneous or gradient stress," *Sensors and Actuators*, vol. 78, pp. 1-7 (1999).

[2] E. Peiner and L. Doering, "Force calibration of stylus instruments using silicon microcantilevers," *Sensors and Actuators*, vol. 123-124, pp. 137-145 (2005).

[3] J. A. Palesko and D. H. Bernstein, *Modeling MEMS and NEMS*, CRC Press (2002).

[4] S. P. Timoshenko and J. N. Goodier, *Theory of Elasticity*, McGraw-Hill (1934).

[5] D. L. Logan, *A first course in the finite element method using Algor*, Brooks/Cole (2001).

[6] S. Carlsson and P. L. Larsson, "On the determination of residual stress and strain fields by sharp indentation testing. Part II. Experimental investigation," *Acta Mater*, vol. 49, pp. 2193-2203 (2001).

[7] S. Suresh and A. E. Giannakopoulos, "A new method for estimating residual stresses by instrumented sharp indentation," *Acta Mater*, vol. 46, no. 16, pp. 5755-5767 (1998).

[8] J. H. Lim, S. C. Yeon, Y. K. Jeon, J. G. Kim and Y. H. Kim, "Nano-indentation method (micro twisting test) for the measurement of the Poisson's ratio of MEMS thin films," *Sensors and Actuators*, vol. A108, pp. 20-27 (2003).

[9] N. I. Abu-Lail, M. Kaholek, B. LaMattina, R. L. Clark and S. Zauscher, "Micro-cantilevers with end grafted stimulus-responsive polymer brushes for actuation and sensing," *Sensors and Actuators*, vol. B114, no. 1, pp. 371-378 (2006).

[10] J. Fritz, M. L. Baller, H. P. Lang, H. Rothuizen, P. Vettiger, E. Meyer, H-J Guntherodt, Ch. Gerber and J. K. Gimzewski, "Translating biomolecular recognition into nanomechanics," *Science*, vol. 288. no. 5464, pp. 316-318 (2000).

[11] G. Wu, H. Ji, K. Hansen, T. Thundat, R. Datar, R. Dote, M. F. Hagan, A. K. Chakraborty and A. Majumdar, "Origin of nanomechanical cantilever motion generated from biomolecular interactions," *Proceedings of the National Academy of Sciences of the USA*, vol. 98, no. 4, pp. 1560-1564 (2001).

[12] B. Ilic, Y. Yang and H. G. Craighead, "Virus detection using nanoelectromechanical devices," *Applied Physics Letters*, vol. 85, no. 13, pp. 2604-2606 (2004).

Low Temperature Carbon Nanotube Fabrication using Very High Frequency-Plasma Enhanced Chemical Vapour Deposition Method

Sukirno[1], Satria Zulkarnaen Bisri[1], Lilik Hasanah[1,2], Mursal[1,3], Ida Usman[1,4], Adi Bagus Suryamas[1], Thomas Alfa Edison[1]

[1] Laboratory for Electronic Material Physics, Faculty of Mathematics and Natural Sciences – Institut Teknologi Bandung (ITB). Jl. Ganesha 10 Bandung 40132 Indonesia

[2] Department of Physics, Universitas Pendidikan Indonesia (UPI)., Universitas Pendidikan Indonesia, Jl. dr. Setiabudhi 229 Bandung 40154 Indonesia

[3]Department of Physics, Universitas Syiah Kuala, Banda Aceh, Indonesia

[4]Department of Physics, Universitas Haluoleo, Palu, Indonesia

Contact e-mail: sukirno@fi.itb.ac.id

Abstract **It is explained one of the carbon nanotube research progress in Indonesia. It is being attempted a carbon nanotube growth using the modifications of existing home made PECVD system, which are Very-High Frequency PECVD (VHF-PECVD) and Hot-Wire VHF-PECVD (HW-VHF-PECVD). They are catalytic growth processes, which various metal catalyst thin films were growth on the Silicon substrate, such as Fe catalyst grown by using *dc*-Unbalanced Magnetron Sputtering Method and Al as well as Ni catalyst grown by high vacuum thermal evaporation method. SiO$_2$ buffer layer was also grown beneath the Al catalyst layer. Some post-treatments for the catalyst thin films were also conducted, which were post-annealing process under various atmosphere conditions, N$_2$, NH$_3$ and free air. By using a single gas source, which was methane as the source of carbon, a carbon nanotube fabrication has been attempted at relatively low temperature, 400°C. It was expected that the high value of plasma frequency could lead to higher dissociation rate of methane to produce much carbon radicals. Meanwhile, the usage of hot-wire filament was expected to pre-decompose some the methane gas before reach the reactor in order to produce some of hydrogen radicals, which might be needed for CNT growth process. From morphological characterization results, it has been achieved some carbon sub-microstructure, in 100 nm of grain size and vertically aligned. From the composition characterization results, it has been confirmed that carbon atoms fraction were quite significant and the catalytic growth** **mechanism was happened. Furthermore, from the SEM characterization results of the samples, which was grown on the buffered catalyst, showed that a long aligned horizontal carbon nanotube has been grown.**

I. INTRODUCTION

Carbon Nanotubes (CNT) studies have become great research interest since its first discoveries and now, it becomes one of five the hottest research topics in Physics [1]. The unique nature of CNT properties led to wide range applications. One of the biggest remaining issues in CNT research is of the fabrication process. Its wide range application potentials demand various fabrication techniques, which are more efficient as well as better, in term of yielded CNT quality.

So far many techniques have been performed in synthesizing CNT, such as by using laser vaporation techniques [2], CVD-based process, and even electro-deposition process [3]. Chemical Vapour Deposition (CVD) process has been widely used to synthesis various kinds of CNT, either single-walled or multi-walled. Most of them were performed in a relatively high deposition temperature condition. Some attempt have been made to growth CNT in low temperature condition by using Thermal-CVD technique [4]. Besides that, most of the processes were using many types of precursor gaseous mixture. It has been attempted a utilization of single precursor gas to synthesis CNT, in large scale production, at relatively high temperature thermal-CVD process [5].

0-7803-9730-4/06/$25.00 ©2006 IEEE

In term of the CNT fabrication process by using PECVD process, which is one type of CVD-based process, Meyyappan et al. [6] has reported that the yielded CNT is a high quality aligned nanotube, for the deposition temperature of 600°C-800°C.

In this paper, it is reported that CNT fabrication attempt by using one of PECVD-based technique, which is Hot Wire-Very High Frequency-Plasma Enhanced Chemical Vapor Deposition (HW-VHF-PECVD) at relatively low deposition temperature. By using PECVD, which is operated at higher frequency, it is expected that there might be differences on the yielded CNT properties. Besides that, it is only used a single gas as precursor of the deposition process.

This paper is organized by explaining first how the experiment was performed by using home-made PECVD system. In the term of the experiments, it will also be explained how the catalyst layer were deposited by using various available technique. Afterwards, it will be shown how the morphological characterizations results as well as the composition characterizations results. Discussions and analyzes of the results will be explained.

II. EXPERIMENT

The CNT growth, which has been done, is a catalytic growth process. It means that there are some substances which were used as catalyst layer in order to make the CNT grown on the substrate, especially for the Single-Walled CNT. Most of the catalysts used are transitional metal catalyst [7] such as Fe, Ni, Mo, Mn, Co as well as Pt. The catalyst will act as probing material which led the CNT growth since the carbon atoms will be diffused on active catalyst particle [7-8]. Those catalyst substances must be deposited on top of the substrate before the actual growth process of CNT by using PECVD process. In this experiment, it is used several different metal catalysts, which are Fe, Al, Ni and Ni/Al, in order to observe the differences of the growth results.

Two different processes were done in order to grow different metal catalyst. The *dc*-Unbalanced Magnetron Sputtering (*dc*-UBMS) was used in order to growth Fe catalyst [9]. Meanwhile, Ultra-High Vacuum Thermal Evaporation technique is used in order to growth Ni, Al and Ni/Al catalyst.

The n-type silicon (100) wafer was cleaned by immersing it in several liquid in order to remove dirt as well as the undesired oxide. The liquids, which are used as cleanser, are acetone p.a. and methanol p.a. Besides that, as the etcher of the undesired oxide, it is used 20% fluoric acid. Pure water (Q2) is used to rinse the substrate in between two immersions in different liquids. The cleaned substrates are dried by using nitrogen stream.

There are two different treatments on the substrate before the deposition process of metal catalyst. For the substrate, which Fe catalyst was grown on it, it was used as prepared silicon (100) wafer. On the other hand, the other substrates was treated by oxidizing the silicon (100) wafer in order to make 100 nm thick SiO_2 layer. The oxidizing process was done in electric furnace at 1000°C at room pressure with 100 sccm of oxygen gas flow. The purpose of the existence of SiO_2 layer is in order to prevent any forming silicides which might be happened during the metal catalyst deposition [4]. The formed silicides might affect the diffusion process of carbon atoms on the catalyst particle, which led to inhibition of CNT growth.

The deposition of Fe metal catalyst was performed at 400°C of deposition temperature, with 1.6×10^{-2} torr base pressure, which is increased during deposition at 5.4×10^{-2} torr. As feed gas for target bombardment, it is used argon at 100 sccm flow rate. It was applied 600 volt *dc,* which result current as big as 0.024 A, to glow discharge the argon gas to become plasma, with plasma power of 14.4 watt. This sputtering process was performed in 3 hours.

There are two different Al catalysts, which were prepared on the substrate that varied in term of thickness, as function of deposition time. By using this high vacuum thermal evaporator, all of the depositions are performed at 10^{-5} torr. The first Al catalysts are grown in 7 seconds of deposition time; meanwhile the second types are grown in 15 seconds. The Ni catalysts, both for Ni catalyzed substrate and Ni/Al tandem catalysts are grown in 15 seconds of deposition time. Those different catalysts are post treated by thermal annealing method at 400°C and 600°C under nitrogen or NH_3 atmosphere.

It was used the home made PECVD system, which was modified in term to provide the hot wire component inside the reactor, either the hot wire filament or the hot wire cell. Besides that the system also operated at 70 MHz of plasma frequency, instead of 13.56 MHz. By higher frequency used, it is expected that the decomposition rate of the precursor gas become higher, which led to the increasing number of radicals that able to be deposited as the carbon nanostructure material. The existence of hot wire cell is in purpose to pre-decompose the precursor gas before it flow into the cavity between the electrodes and the substrate holder. With this pre-decomposition, it is expected that there will be some hydrogen radicals present, which is needed in most of CNT growth process. Meanwhile, by using hot wire filament in the cavity, it is expected that the plasma temperature would increased as well as some differences in term of decomposition type. By this, the deposition rate would be increased as well as some existence of hydrogen radicals.

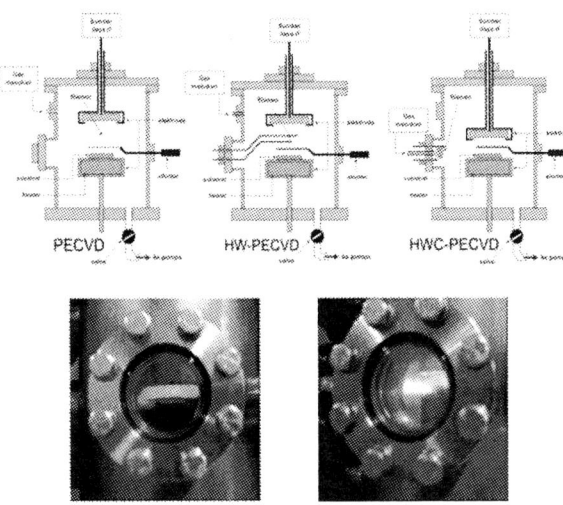

Fig. 1 The comparison of VHF-PECVD, HW-VHF-PECVD and HWC-VHF-PECVD (above) and the differences between the plasma formed between the VF-PECVD (bottom left) and HW-VHF-PECVD (bottom right)

By using VHF-PECVD and HWC-VHF-PECVD, the depositions were performed at 400°C deposition temperature, with 400 mtorr deposition pressure. It is used a single gas precursor, which is pure methane (CH_4) that flow on 100 sccm rate. The plasma power used is 80 watt for VHF-PECVD and 50 watt for HWC-VHF-PECVD. The growth processes were performed in 150 minutes time for VHF-PECVD and 45 minutes for HWC-VHF-PECVD. Meanwhile, by using HW-VHF-PECVD, the plasma power was in 50 watt, with growth process time was 120 minutes. Both HWC and HW filament are operated at 800°C.

III. RESULTS AND DISCUSSION

The resulted samples were characterized by using JEOL low vacuum SEM operated at 20 kV in order to characterize the morphological structure of the samples. Besides that, one of the samples, which were grown by using HWC-VHF-PECVD, has been characterized by using Ntegra AFM in semi-contact mode with 20 nm AFM tip size. In order to analyze the composition of the structure, it has been performed an Energy Dispersion Spectroscopy (EDS), which were done by the similar equipment for SEM characterization.

Fig. 2 AFM scan images of samples which grown using HW-VHF-PECVD on Fe catalyzed substrate (Fe/Si) at 400°C growth temperature and single gas source, CH_4. (a). 5 μm x 5 μm scan size, (b). 2.5 μm x 2.5 μm scan size, (c). 3-dimensional extrapolation of (b) result using Gwyddion image processor.

From the AFM characterization of the sample that grown by using HWC-VHF-PECVD method, as seen in Fig. 2, it could be inferred that some submicrostructure or nanostructure has been synthesized as pillars. In some certain points, there are sign of nanostructure which bundled and forming big granule that the AFM could perceived. From the data processing, it is obtained that the distance between the deepest trench and the highest peak is more than 300 nm, which the actual distance might be bigger than that value. From the EDS result, it could be inferred that the synthesized sample has a very big proportion of carbon atom. The presence of the silicon signature comes from the substrate. However, it also stated that some of the Fe catalyst are oxidized which led to FeO, so that in some area, the CNT growth are inhibited, since this proportion of Fe is not disappeared or decreasing after the CNT growth.

From the SEM characterization of the samples which was grown by VHF-PECVD on an Al/SiO$_2$/Si substrate, as seen in Fig.3, it could be perceived that a signature of a horizontal aligned CNT is present. The structure is about 2 micron long with no more than 50 nm width. From the EDS characterization, it could be obtained that this structure is a carbon nanostructure. However, the density of the synthesized structure is very low. It is suspected that the horizontal structure synthesized since the Al catalyst density are very low, grown in only 7 seconds by using high vacuum thermal evaporator. As consequences, it is possible that nucleation of carbon molecules from two different catalyst particle are connected each other. However, this appearance still needed to be further investigated by another technique of characterization in order to confirm whether the structure is CNT or not in term of its molecular symmetry.

substrate before and after the PECVD process, it could be inferred that the catalytic process happens since only in the area where there are catalyst particles, the carbon structure formed. The EDS result shows that the carbon atom proportion is large meanwhile the Ni atom proportion decrease due to the catalytic growth process.

The sample which was grown on Al catalyzed substrate shows that some carbon sub-microstructure formed. The similar thing was happened on the sample that was grown on the tandem Ni/Al catalyst. However, the morphological of the grain was not as homogeneous as the sample which was grown on Ni catalyst. It is suspected that on the Al catalyst, the formed carbon nanostructure tends to be horizontal, instead of making vertical structure. However, from the EDS characterization result, it could be elucidate that the carbon atom proportion in this samples is bigger than on the Ni catalyzed sample.

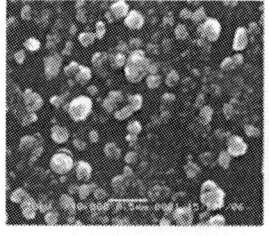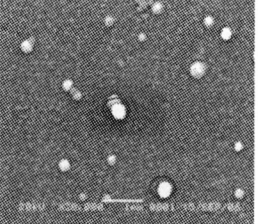

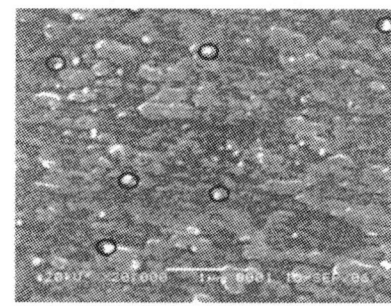

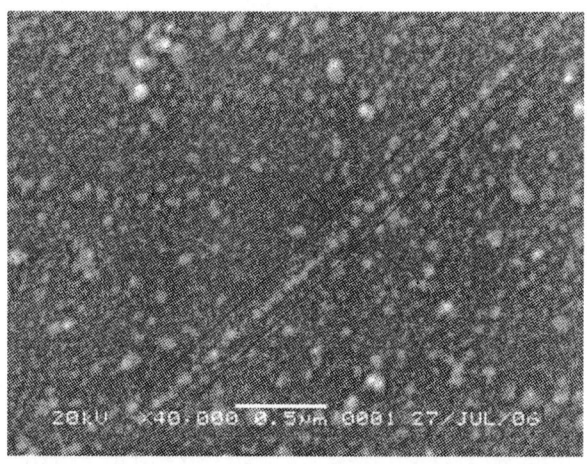

Fig. 3 SEM image of possible horizontally aligned carbon nanotube grown by VHF-PECVD on Al/SiO$_2$/Si catalyzed substrate at 400°C from a single gas source of methane.

Fig. 4 SEM results of samples that were grown by using HW-VHF-PECVD on Ni/Al/SiO$_2$/Si substrate (left), Ni/SiO$_2$/Si substrate (right) and Al/SiO$_2$/Si (bottom) at 400°C from a single gas source of methane. The signatures are marked with red circles.

Meanwhile, the results of the samples that were grown by using HW-VHF-PECVD methods show some other different results. On the substrate which was catalyzed by Ni, it could be seen that there might some high pillar structure form, which is suspected as bundled CNT. Since the sample is characterized in term of its surface morphology, the contrast between dots and their background might indicate a high vertically structure. However, it could be inferred that on that samples some selective growth is happened. In comparison with the catalyzed

From the explanations, it could be concluded that there are some possibilities to grow CNT at relatively low deposition temperature by using various modification of PECVD system that operated in higher frequency as well as by adding hot wire or hot wire cell component. By that method a catalytic growth process has been observed, which is the one of identities of CNT

170

growth process. By varying the used modified process, it could inferred that the addition of hot-wire component, both cell and filament, results pillar shape structure. The role of the catalyst is very important in this process. By varying the type of the catalyst, it could be concluded that Fe catalyst and Ni catalyst are most likely to be used as catalyst particle in order to make pillar structure shape. Meanwhile, by using Al catalyst, the grown structure tends to be horizontally formed.

However, the homogeneity as well as the reproducibility of this method still remaining as question while further investigation is still being done. Further characterization still needed to be done in order to identify the type of the formed nanostructure as CNT in term of its molecule symmetry.

ACKNOWLEDGEMENT

The author would like to thank Dr. Alexander Yaluvenko from NT-MDT, Russia for the AFM Characterization and Mr. Wikanda from PPGL-Bandung for the SEM and EDS characterizations. This research is funded by ITB Frontier Research Grant 2006.

REFERENCES

[1] J. Giles, "Top Five in Physics", *Nature* 441, 265 (2006).

[2] F. Kokai, K. Takahashi, D. Kasuya, T. Ichihashi, M. Yudasaka, S. Iijima. "Synthesis of single-wall carbon nanotubes by millisecond-pulsed CO_2 laser vaporization at room temperature", *Chem. Phys. Lett.* 332 (2000) 449-454.

[3] D. Zhou, E.V. Anoshkina, L. Chow, G. Chai, "Synthesis of carbon nanotubes by electrochemical deposition at room temperature", *Carbon* 44 (2006) 1013-1024.

[4] M. Cantoro, et.al. "Catalytic Chemical Vapor Deposition of Single-Walled Carbon Nanotubes at Low Temperatures". *Nano Lett.* 6 (2006) 6 1107-1112.

[5] A.M. Cassell, J.A. Raymakers, J. Kong, H. Dai. "Large Scale CVD Synthesis of Single-Walled Carbon Nanotubes". *J. Phys. Chem. B* 103 (1999) 6484-6494.

[6] M. Meyyappan, L. Delzet, A. Cassell, D. Hash. "Carbon Nanotube Growth by PECVD: a review". *Plasma Sources Sci. Technol.* 12 (2003) 205-216.

[7] M. Lin, J.P.Y. Tan, C. Boothroyd, K.P. Loh, E.S. Tok, Y.-L.Foo, "Direct Observation of Single-Walled Carbon Nanotube Growth at the Atomistic Scale". *Nano Lett.* 6 (2006) 3 449-452.

[8] Y.H. Lee, S.G. Kim, D. Tomanek. "Catalytic Growth of Single-Wall Carbon Nanotubes: An *Ab Initio* Study". *Phys. Rev. Lett.* 78 (1997) 12 2393.

[9] Sukirno, S.Z. Bisri, R.Y. Sari, L. Hasanah, Mursal, I. Usman, A.B. Suryamas, Darsikin, "Catalytic carbon nanotubes fabrication using Very-High Frequency Plasma Enhanced Chemical Vapour Deposition", *Proc. ITB of Science and Technology* (2006) in press.

Design and Modeling of Micromachined Condenser MEMS Loudspeaker using Permanent Magnet Neodymium–Iron–Boron (Nd–Fe–B)

Fatima Lina Ayatollahi, Burhanuddin Yeop Majlis, SM IEEE
Institute of Microengineering and Nanoelectronics
Universiti Kebangsaan Malaysia
43600 Bangi, Selangor, MALAYSIA
Email: linaayat@vlsi.eng.ukm.my, burhan@vlsi.eng.ukm.my

Abstract   There has been some work in recent years to produce MEMS speakers. Geometries have included diaphragm structures, cantilever beams, and thermally actuated domes. The diaphragm skeleton is built out of the metal and glass layers of the CMOS chip, greatly reducing the layering ad patterning required, compared to other MEMS designs. "Brute multiplicity" is useful to circumvent the poor linearity of the individual elements; the number of elements determines the distortion of the total sound produced, not the quality of the transducer. The objective of this research focused on design a new way to produce the sound in rang of human hearing using micro speaker (on MEMS scale) and study concerned about possibility of MEMS speaker fabrication.

I. INTRODUCTION

The field of MEMS is based on the use of microfabrication techniques, such as surface and bulk micromachining, to produce microsensors, microactuators, and microsystems. This technology, driving from the microelectronics industry, allows the realization of small, smart device (the loudspeaker and the control/readout electronics can be integrated on the same chip) at a very low price.

Ten percent of the general population suffers from hearing loss and that number rises to 35% for people over 65 years old. People who are 25–45-year-old make up 42% of the hearing-impaired population while people who are older than 65 make up 45%. Nowadays, new hearing devices such as digital hearing aids and cochlear implants have significantly improved the quality of life for hearing-impaired people [1][2][3].

It is clear that the miniaturization of electrical devices has profoundly changed human society. This microelectronic revolution has been driven by the ability to manufacture electrical components, of ever decreasing size and increasing complexity, in a parallel or batch-fabrication process [4] [5]. The associated reduction in cost, size, and power consumption, of microelectrical systems, coupled with their increased functional capabilities, has allowed microelectrical systems to become ubiquitous and indispensable. The enormous demand for cheaper, faster, and better microelectrical systems has prompted the integrated-circuit (IC) industry to develop an impressive collection of fabrication processes, equipment, and materials.

Traditional sound reconstruction techniques use a single to few analog speaker diaphragms with motions that are proportional to the sound being created. Louder sound is generated by greater motion of the diaphragm and different frequencies are produced by time-varying diaphragm motion. The practical considerations of an analog speaker, however ultimately limit its performance, particularly the frequency response and linearity. [6]

Several requirements are essential to digitally reconstruct sound using an array of microspeakers:

- The acoustic response of a single microspeaker should be fast and on the order of 10's of microseconds. This response speed will limit the sampling rates that can be used in converting digital information to an analog acoustic waveform.
- The acoustic response must be repeatable over time and uniform across all speaklets in the array.

0-7803-9730-4/06/$25.00 ©2006 IEEE

- Designing the array on the same chip will minimize process variations during fabrication, controlling the overall speaklet uniformity. The acoustic energy from multiple speaklets must add linearly.

II. MAGNETIC MATERIALS

As Ferromagnetic materials, soft or hard, have the ability to present a very high magnetization. Soft magnetic materials, or permeable materials, exhibit small coercivity and low saturation field, thus acquiring a magnetization only when a field is applied in the surrounding of the material. Hard magnetic materials, or permanent magnets, present large coercivity and high saturation field, thus keeping a high magnetization without any applied field [7]. Torques and forces are produced on a magnetic material when surrounded by a magnetic field.

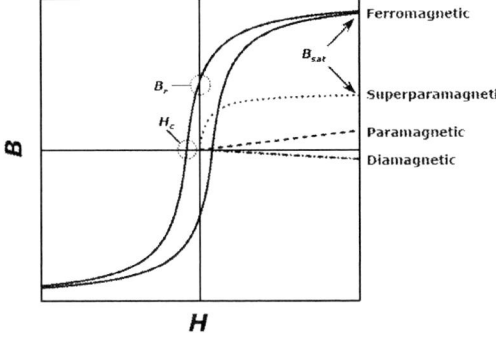

Figure 1: Magnetization curves for dia-, para-, superpara- and ferromagnetic materials showing the shape and approximate relative magnitude of the magnetization.

In a uniform magnetic field, the interaction between the magnetization and the external field only generates a torque that tends to align the direction of magnetization with the direction of the external magnetic field. Both theoretical and finite-element analyses are applied to the modeling, design and optimization of the force-displacement characteristic of the curved permanent magnet mechanism.

A magnetization curve is generic hysteretic plot of magnetization as a function of magnetic field for a ferromagnetic material. Fig 1

Permanent Magnet actuators are more flexible than electromagnetic actuators because of several reasons which are high force, large gaps and zero-voltage operation. But it is more difficult to fabricate.

Magnetic materials can be classified according to their magnetic susceptibility $\chi = M / H$ and relative permeability $\mu_r = (\chi / \mu_0 + 1)$ into several categories: ferromagnetic, ferromagnetic, antiferromagnetic, paramagnetic, diamagnetic, and superconducting materials.

Listed in Table I are the typical ranges of χ / μ_0 for each category of magnetic material and examples of each are identified. of these, ferromagnetic materials have found the most use in magnetic microsensors, microactuators, and microsystems. Their high relative permeability amplifies small magnetic fields into large flux densities for microsensors and their high saturation magnetization can generate strong fields for microactuation. [8]

Table 1: Magnetic Materials Classification.

Category	χ/μ_0	Examples
Ferromagnetic	10^7 to 10^2	Ni, Fe, Co, NiFe, NdFeB
Ferrimagnetic	10^2 to 10^1	Fe_3O_4, ferrites, garnets
Antiferromagnetic	small	MnO, NiO, $FeCO_3$
Paramagnetic	10^{-3} to 10^{-6}	Al, Cr, Mn, Pt, Ta, Ti, W
Diamagnetic	-10^{-6} to -10^{-3}	Ag, Au, C, H, Cu, Si, Zn
Superconducting	-1	$YbBa_2Cu_3O_x$

Hard magnetic or permanent magnetic materials would be more appropriate to be used to realize high-force actuators and sensitive magnetometers. For example, hard magnetic materials with a high remanent magnetization M_r can be conveniently used in bi-directional (push-pull) microactuators [8].

In addition, microactuators driven by off-chip coils could be activated with lower fields and hence lower power levels if a hard magnetic material is used instead of a soft magnetic material. However, except for a few instances hard magnetic materials have not been used extensively in MEMS [9].

The primary reason for this has been the lack of readily available and reliable deposition and micromachining processes. Various hard magnetic materials can be prepared by metallurgical processes (e.g., sintering, pressure bonding, injection molding, casting, extruding, and calendering), vacuum processes (e.g., evaporation, sputtering, MBE, CVD), and electrochemical processes (e.g., electroless deposition and electrodeposition).

III. PROBLEM DESCRIPTION

The requirements set for a loudspeaker for a hearing instrument are summarized in design section. The loudspeaker system consists mainly of two volumes, the ear canal and the loudspeaker itself. The dimensions of the ear canal and the loudspeaker are small compared to the wavelength in the considered frequency range; hence the acoustic pressure due to the sound pressure in the ear canal is approximated as quasistatic. Thus, this loudspeaker is comparable to a pump. Many publications are available on this type of microsystem actuator [10] [11], but low supply voltages and power consumption are usually not limiting factors.

IV. DESIGN

There have been several approaches in designing the MEMS loudspeaker in recent years [16], [17]. This paper is representing a new design of microloudspeaker in terms of fabrication process and coefficient performance.

The device consists of a micromachined membrane in a silicon wafer with an electroplated coil bonded to a back plate with two small cylindrical and ring magnets among each other with opposite magnetization direction vectors. Electromagnetic actuation is chosen in this design because of its large driving force over a large air gap and a low driving voltage as well as potential for high acoustic fidelity and high efficiency. The coil carries the audio frequency signal and the magnetic fields from the stationary permanent magnet and coil interact to produce the required vertical force on the coil. The subsequent oscillation of the coil causes the membrane to compress the air and produce the mechanical sound wave.

The Lorentz force determines the driving force in the current-carrying coil, and is given by:

$$\vec{F} = I\vec{l} \times \vec{B} \qquad (1)$$

In micro-machined microphones [16],

diaphragms are generally made from silicon or silicon nitride. However, the demands on a loudspeaker diaphragm are much more severe than those on a microphone diaphragm. For this reason a new, less brittle material needed to be used for the membrane.

The chosen material was Polyimide [13]. This may be spun on to the substrate and freed later with etching of an underlying cavity. The vibration of diaphragm of the loudspeaker behaves like a second-order system, and is described by [14]:

$$M\frac{d^2\omega}{dt^2} + R\frac{d\omega}{dt} + \frac{\omega}{C} = F \qquad (2)$$

where M is the total mass including the coil, the diaphragm and air load, R is mechanical resistance due to dissipation in the air load and C is compliance of the suspension system. When the coefficients (M, R, C) are constant and the system is subjected to a harmonic force $F = F_0 \cos\omega t$, a particular solution to equation (4) is obtained [14]:

$$\omega = \omega_0 \cos(\omega t - \phi) \qquad (3)$$

where the displacement of the vibration ω_0 has phase lag φ, and is expressed by:

$$\omega_0 = \frac{F_0}{\sqrt{(1/C - M\omega^2)^2 + \sqrt{(\omega/R)^2}}} \qquad (4)$$

$$\phi = \tan^{-1}\frac{C\omega}{R(1 - MC\omega^2)}$$

However, in real cases, C and F change with displacement ω, caused by the nonlinear behavior of the mechanical suspension, inhomogeneous distribution of the magnetic flux, or nonlinear vibration modes at high frequency. The nonlinear vibration will produce undesired harmonic components when a sinusoidal signal is applied, and make the acoustic output distorted. Harmonic distortion should be avoided in the design. Since the dimension of the microspeaker and ear cavity is small compared to the wavelength of sound, the sound pressure is distributed uniformly in the volume. The pressure change is proportional to the volume displacement of the diaphragm, and is expressed by [14]:

$$dP = \frac{-1.4P_0}{V_0}dV \qquad (5)$$

where P_0 is the pressure of the atmosphere and V_0 is the volume of the ear cavity (which is about 2 cc). The generated sound pressure level (SPL) is defined as

$$SPL = 20 \log_{10} \frac{dP}{P_{ref}} \ (dB) \qquad (6)$$

Where $Pref$ is 20 μPa.

For a cylindrical magnet of height z_0 and radius a, magnetized with a uniform dipole moment M (per unit volume) in the z direction, the external field distribution B at the point (r, θ, z) could be derived by the equivalent surface current $J = \nabla \times M$ [12],

$$B_n = \frac{\mu_0 J}{2\pi} \int_{-z_0/2}^{z_0/2} \frac{1}{[(a+r)^2 + (z-z')^2]^{1/2}} \times$$
$$\left[\frac{a^2 - r^2 - (z-z')^2}{(a-r)^2 + (z-z')^2} E(k) + K(k) \right] dz' \qquad (7)$$

$$B_t = \frac{\mu_0 J}{2\pi} \int_{-z_0/2}^{z_0/2} \frac{z-z'}{[(a+r)^2 + (z-z')^2]^{1/2}} \times$$
$$\left[\frac{a^2 + r^2 + (z-z')^2}{(a-r)^2 + (z-z')^2} E(k) - K(k) \right] dz' \qquad (8)$$

where $K(k)$ and $E(k)$ are the complete elliptic integrals of the first and second kinds with

$$k = \sqrt{\frac{4ar}{(a+r)^2 + (z-z')^2}}$$

Since the product $\vec{I} \times \vec{B}$ varies with the position on the diaphragm, the force will not be proportional to the current, and distortion will be produced. In our design, only a single loop of voice coil is used to drive the diaphragm. The wider the coil width the greater the diaphragm excursion and the better the heat dissipation; nevertheless, the tradeoff is the generation of harmonic distortion. To maximize the driving force on the coil, both permanent magnets were integrated into our device to induce magnetic flux to be perpendicular to the current-carrying coil as shown in Fig 2.

The loudspeaker diaphragm is made from polyimide because of its excellent properties, including high breakdown voltages, high thermal stability and low thermal expansion coefficient. Compared to the conventional membrane materials such as nitride, polysilicon and silicon, polyimide has smaller Young's modulus (i.e. 60 to 100 times smaller), and the simple and low temperature fabrication processing steps.

For better frequency response and efficiency, the mass of the moving part of the loudspeaker needs to be minimized. In our design, only a thin flexible membrane with a single coil formed the sound generation plate, and the relatively heavy magnets were located on the static part of the device.

V. SIMULATIONS AND RESULTS

Finite Element Method Magnetics (FEMM) is a finite element package for solving 2D planar and 3D axisymmetric problems in low frequency magnetics and electrostatics. In this research FEMM has been used to simulate both density flux and force to optimize the gap size. The gap size has been chosen in order to maximize tangential flux density which will induce the electromagnetic force to the coil.

Magnetization between cylindrical and ring magnets has been simulated by FEMM. Polar axisymmetric problems are enough to be solved in a cross sectional plane including the z axis. The result is shown in Fig. 2 including the flux density and contours.

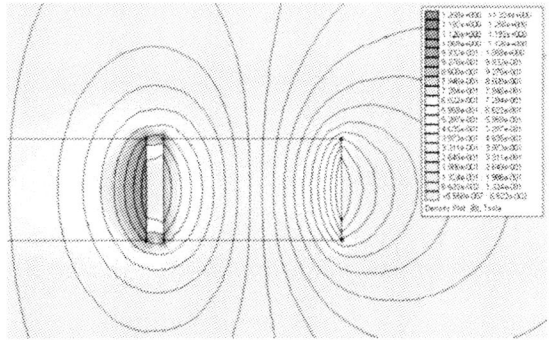

Figure 2: Flux density and contours for magnetization of permanent magnet Neodymium–Iron–Boron (Nd–Fe–B).

It is possible to observe that the maximum flux is higher near the air gap edges which its exact place will show the real position of the coil to get maximum vibration of membrane. The greater amount of tangential flux is desired to get electromagnetic force which can be optimized by varying the gap size considering the 50 um distance above the permanent magnets as a safe and applicable distance for the membrane.

Fig. 3 shows the behavior of the tangential component of the magnetization across the horizontal line 50 um above the permanent magnets. The peak point of B_t gives a certain amount for calculating electromagnetic force between permanent magnets and coil. Induced magnetization of B_t which is 0.59 T is used in next steps.

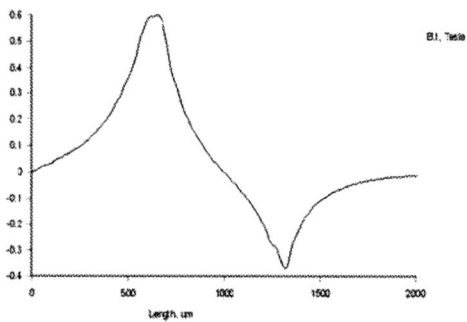

Figure 3: B_t curves for magnetization of permanent magnet NdFeB, result taken in 50 um above permanent magnets, using FEMM.

B_n curve has been shown in Fig 4 describes the amount of normal component of magnetic flux density which does not assist vertical desired force. Since B_n has a small value in the radius which the peak point of B_t occurs, it does not affect the membrane vibration with harmonics or forces in not the desirable directions.

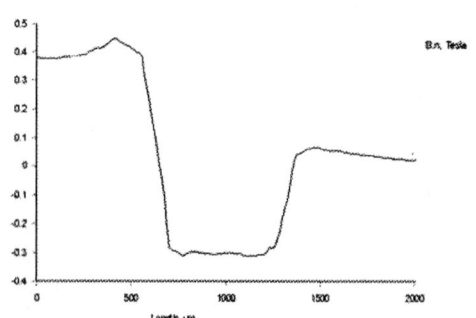

Figure 4: B.n curves for magnetization of permanent magnet NdFeB

In order to maximize B_t, structures with different gap sizes have been simulated using FEMM and the output B_t amounts have been measured. As shown in Fig 5, the smaller gap sizes give the bigger desirable flux density in the appropriate direction.

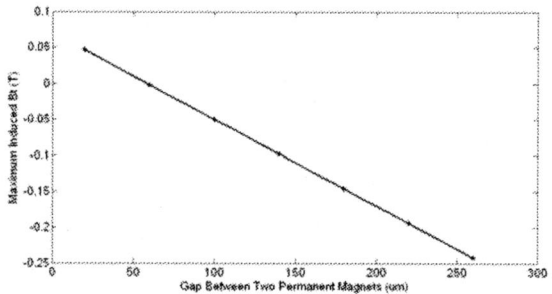

Figure 5: Maximum measured B_t due to increasing of gap size between cylindrical and ring permanent magnets.

The exact place of the coil is determined by the results of simulation using Coventor. The position of the gap should be determined using the gap size obtained in the previous step. Due to the fact that there is no soft magnet mounted around the coil as shown in [17], limiting the position of the coil on the membrane, it is possible to change the coil radius to get the maximum displacement of air volume. The effect of gap position on membrane deflection and upcoming displaced air volume is shown in Fig 6.

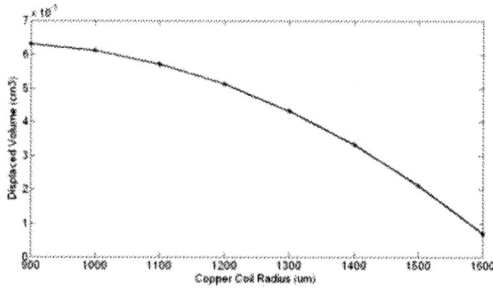

Figure 6: Membrane displacement according to different radius of cupper coil subjected to NdFeB.

This selected gap size will determine the exact coil position on the membrane to achieve the best performance of the hearing aid device. Fig 6 specifies the fact that the smaller coil radius causes more concentrate of force on the central parts of the membrane leading to a greater displaced air volume. The lowest limit of the coil radius is forced by the fabrication limitations of the cylindrical and ring permanent magnets.

The schematic deflection of membrane and coil subjected to magnetic flux of NdFeB is shown in Fig 7. This deflection is maximized in the central area of the membrane following by a flat pattern in this area which is predictable. As the electromagnetic force causing the deflection applied to the coil having electric current, the deflection has such a pattern. Coil current is used to make the vibration of membrane in hearing human range.

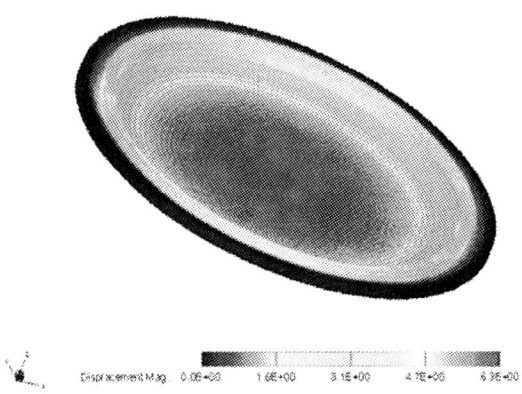

Figure 7: Schematic deflection of membrane with coil and subjected to magnetic flux of NdFeB, using Coventor.

As it was mentioned in the design discussion, the membrane boundary is bonded to the silicon substrate, which should not have any deflection.

Having a membrane radius size of 1750 um and thickness of 5 um and coil radius of 900 um with 50 um width and 5 um thickness, applying a 200 mA current to the coil will generate maximum deflection of the membrane as 16.53 um and the displaced air volume as 6.23×10^{-5} cm^3; therefore, simulating results have proved by theory and calculations. The calculated SPL in this case would be 107 dB showing a better performance compared to other works [17].

VI. CONCLUSION

This work contributes a new design in MEMS

loudspeaker using two permanent magnets as a cylindrical and ring among each other. Removing the soft magnet has simplified the fabrication process, reduced cost and improved performance. Development of this particular structure in terms of suitable fabrication process will be the next step to be achieved.

ACKNOWLEDGEMENT

This work is financially supported by Malaysian Ministry of Science, Technology and Innovation under the project title "MEMS Devices and Sensing Microstructures"
.

REFERENCES

[1] K.J. Dormer, M.A. Phillips, Auditory prostheses: implantable and vibrotactile devices, IEEE Eng. Med. Biol. Mag. 6 (1987) 36–41.

[2] S.U. Ay, F.G. Zeng, B.J. Sheu, Hearing with bionic ears, IEEE Circuits Devices Mag. 13 (1997) 18–23.

[3] K.D.Wise, K. Najafi, Fully-implantable auditory prostheses: restoring hearing to the profoundly deaf, Tech. Digest, International Electron Devices Meeting, 2002, pp. 499–502.

[4] Brett M. Diamond, John J. Neumann, Jr., and Kaigham J. Gabriel, Digital Sound Reconstruction Using Arrays of CMOS-MEMS Microspeakers, 2002 IEEE pp. 292-295.

[5] Richard C. Jaeger, Introduction to Microelectronic Fabrication, Addison-Wesley, Reading, Massachusetts, 1993.

[6] Stanley Wolf and Richard N. Tauber, Silicon Processing for the VLSI Era, Lattice Press, Sunset Beach, California, 1986.

[7] J. W. Judy, and N. Myung, "Magnetic materials for MEMS", MRS Workshop on MEMS Materials, Miami, Florida, (August 19-21, 2001).

[8] H. J. Cho, S. Bhansali, and C. H. Ahn, "Electroplated thick permanent magnet arrays with controlled direction of magnetization for MEMS application", Journal of Applied Physics, vol. 87, no. 9, pt. 1-3, pp. 6340-6342 (2000).

[9] T.-S. Chin, "Permanent magnet films for applications in microelectromechanical systems", Journal of Magnetism and Magnetic Materials, vol. 209, no. 1-3, pp. 75-79 (2000).

[10] Zhang W et al 1996 A bi-directional magnetic micro-pump on a silicon wafer Solid-State Sensor and Actuator Workshop, 1996

[11] Ji J et al 1991 Microactuation based on thermally-driven phase-change Transducers '91 Cat No 91CH2817-5, pp 1037–40

[12] Scheeper, P. R; Olthuis, W; Bergveld, P. "A Review Of Silicon Microphones". Sensors and Actuators A. Vol. 44, p1-11, 1944.

[13] C. Shearwood, M. Harradine, T. S. Birch and J. C. Stevens, "Applications Of Polyimide Membranes To MEMS Technology". Microelectronic Engineering. Vol. 30, p547-550, 1996.

[14]   Fawwaz T. Ulaby, Electromagnetics for Engineers, Pearson international Edition 2005 by Pearson Edfucation,Inc Upper Saddle River, NJ 07458.

[15]   Thomson W T and Dahleh M D 1998 Theory of Vibration with Applications (Englewood Cliffs, NJ: Prentice-Hall)

[16]   Harradine M A, Birch T S, Stevens J C and Shearwood C 1997A micromachined loudspeaker for the hearing impaired IEEE Int. Proc. Transducers'97 (Chicago, USA) pp 429–32

[17]   Ming-Cheng Cheng,Wen-Sheh Huang and Star Ruey-Shing Huang, A silicon microspeaker for hearing instruments, 2004 IOP Publishing Ltd Printed in the UK.

A Hydrogel-based Microvalve for Insulin Delivery Application

Masoomeh Tehranirokh, Badariah Bais, *Member, IEEE* and Burhanuddin Yeop Majlis, *Senior Member, IEEE*
Institute of Microengineering and Nanoelectronics (IMEN)
Universiti Kebangsaan Malaysia
43600 Bangi, Selangor, MALAYSIA
Email: burhan@vlsi.eng.ukm.my

Abstract A hydrogel-actuated microvalve that responds to changes in the concentration of glucose in an external liquid environment is described. The microvalve consists of a thin hydrogel, sandwiched between a stiff porous membrane and a flexible silicon rubber diaphragm. Swelling of the hydrogel forces the membrane to deflect therefore boss causes the valve to be opened. Simulations have been done by FEM analysis and the results show enough deflection to open the valve and let the fluid flow through the microchannel to obtain 130 microliter/min for output flow.

I. INTRODUCTION

The process for delivering a drug is as important as the actual activity of the drug in determining the therapeutic effect. For optimum therapeutic effect, the right amount of a drug needs to get to the right place at the right time. Consequently, advanced drug delivery formulations have been developed over the past 20 years. In drug delivery systems microvalves are a key component that have been developed for active control of drugs.

The future of microvalves will depend on demands for small size, low weight, low power consumption, high reliability, low cost, and most importantly the right application.

The stimuli-responsive hydrogel microvalve is chosen to deliver insulin when the blood glucose level is higher than the normal concentration. The use of responsive hydrogel materials to regulate flow eliminates the need for external power, external control, and complex fabrication schemes.

Hydrogels consist of a broad range of polymers with high water content and stimuli-responsive hydrogels are able to undergo volumetric changes in response to chemical changes in their local environment. Volume transitions in hydrogels have been realized. Also hydrogels have been found to control fluid transport in the xylem of plants. The reversible ionization initiates an osmotic pressure gradient causing the volume expansion or contraction of the hydrogel via the movement of water into and out of the gel. Different hydrogels have been developed to respond to a wide variety of signals including pH, temperature, light, glucose, antigens, electric field and magnetic field. However, they have not found widespread use in macro scale systems due to their relatively long response times (e.g., hours to days at the millimeter scale). Since the responsive nature of the hydrogel is limited by diffusion of chemical signals into the gel matrix, decreasing the size of the hydrogel will decrease the response time. The improved time response facilitates the use of responsive hydrogels in many practical applications including flow control elements in drug delivery systems [1].

In recent years, a valved responsive hydrogel microdispensing device with integrated pressure source by D.T. Eddington et al. has been fabricated that hydrogel array generates a storable pressure source [2]. A. Baldi et al. fabricated a hydrogel-actuated environmentally sensitive microvalve for active flow control [3] with an opening response time as low as 7 min obtained using a 30-micrometer hydrogel in response to a sudden pH change and also a microstructured silicon membrane with entrapped hydrogels for environmentally sensitive fluid gating [4] that the response time for the temperature-sensitive microvalve is 10 s, while the pH- and glucose-sensitive microvalves response times are 4 min and 10 min, respectively.

0-7803-9730-4/06/$25.00 ©2006 IEEE

In this paper, optimization of the A. Baldi et al. microvalve for insulin delivery was done. To achieve this design, the structure of the microvalve was changed from a normal opened valve to a normal closed valve (Fig. 1). A phenylboronicacid-based hydrogel was used to construct a smart microvalve that responds to the changes in the glucose level [5].

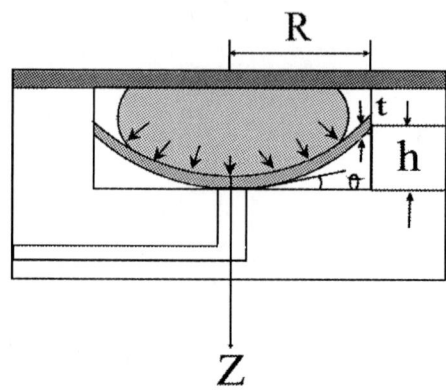

Fig. 2. Thin circular membrane under uniform pressure of swelling hydrogel

Force equilibrium requires:

$$\pi r^2 p + 2\pi rT \sin\theta = 0 \qquad (1)$$
$$\sin\theta \approx dz/dr$$

Thus:

$$z = -\frac{p}{4T} r^2 \qquad (2)$$

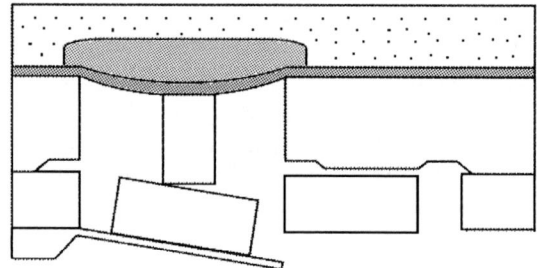

Fig. 1. Schematic drawing of the microvalve cross section

In the following section, hydrogel swelling behavior and microfluidic mechanics in microchannels is studied and the FEM simulation of the diaphragm deflection and microchannel part is performed and presented.

II. HYDROGEL SWELLING BEHAVIOR

A thin circular membrane that is clamped along the perimeter and subjected to a net uniform pressure exerted by the hydrogel actuator is shown in (Fig.2). Bending moments are assumed to be negligibly small and the membrane force is assumed to be uniform across the thickness of the membrane and the membrane material is isotropic and linearly elastic at small strain [6].

Under uniform pressure, the net force on the membrane is perpendicular to a section of radius that is given by $\pi r^2 p$ and the restoring force due to tension in the membrane is $2\pi rT \sin\theta$, where T denotes tension per unit length of the membrane.

The strain along the radial direction is given by:

$$\varepsilon_r = \frac{\left(\int_0^R \sqrt{1+z'^2}\, dr - R\right)}{R} \approx \frac{p^2 R^2}{24T^2} \qquad (3)$$

$$T = \sigma_r t = \frac{\varepsilon_r E}{1-v^2} t = \left(\frac{1}{24}\frac{E}{1-v^2} tp^2 R^2\right)^{1/3} \qquad (4)$$

$$z_{max|r=R} = -\frac{1}{4}\left[\frac{24(1-v^2)p}{E}\right]^{1/3}\frac{R^{4/3}}{t^{1/3}} \qquad (5)$$

where z is the deflection, E is the Young's modulus, v is the Poisson's ratio, t is the membrane thickness and σ_r is the radial stress.

In order to obtain large deflection, using either a thinner membrane or a membrane with larger diameter is possible:

$$z_{max} \propto (R^4/t)^{\frac{1}{3}} \qquad (6)$$

The pressure, p generated by the hydrogel actuator is unknown. But if Z_{max} can be measured experimentally, then the pressure can be determined as:

$$p = -\frac{8z_{max}^3 Et}{3(1-v^2)R^4} \qquad (7)$$

III. Microfluidic Mechanics In Microchannels

The classical solution of the problem of steady laminar flow in straight ducts (Fig. 3) is based on a number of assumptions on flow conditions [7]:

- The flow is generated by a force due to a static pressure in the fluid.
- The flow is stationary and fully developed, i.e. it is strictly axial.
- The flow is laminar.
- The Knudsen number is small enough so that the fluid is a continuous medium.
- There is no slip at the wall.
- The fluids are incompressible Newtonian fluids with constant viscosity.
- There is no heat transfer to/from the ambient medium.
- The energy dissipation is negligible.
- There is no fluid/wall interaction (except purely viscous).
- The walls are straight.
- The micro-channel walls are smooth.

In this case, the problem of developed laminar flow in a straight duct reduces to integrating the equations.

Navier-Stokes equations simplify to:

$$\mu(\partial^2 u / \partial y^2) = \frac{dp}{dx} \qquad (8)$$

where $u = u(y)$ is the longitudinal component of velocity (Fig. 4), p is the pressure, μ is the dynamic viscosity and x is directed along the micro-channel axis.

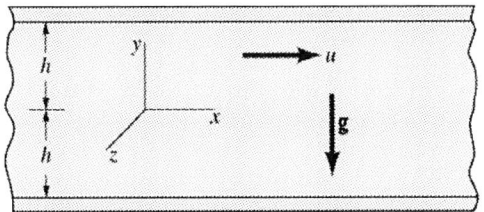

Fig. 3. Steady and laminar fluid flow in x direction through parallel plates

Applying boundary conditions (no-slip conditions at $y = \pm h$), eq. (8) can be solved as:

$$u = \frac{1}{2\mu}(\frac{\partial p}{\partial x})(y^2 - h^2) \qquad (9)$$

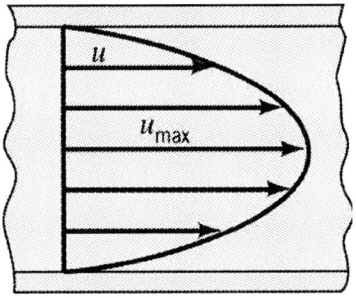

Fig. 4. Velocity profile of fully development flow

According to standard theory, the stationary laminar and fully developed flow through a channel with a constant cross section is described by [8]:

$$\Phi = \frac{1}{C_R}\frac{A^2}{\eta l}\Delta P \qquad (10)$$

where ΔP denotes the pressure drop at the channel, A the channel cross section, l the channel length and η is the viscosity of the fluid. C_R is introduced as a geometry coefficient, which summarizes the influence of the shape of the cross section.

A trapezoidal shaped microchannel with length of 2 mm and width 100 um is considered. For KOH-etched trapezoidal microchannel, C_R is given by:

$$C_R = \frac{12-1.38a+4a^2}{a-0.85a^2+0.28a^4} \qquad (11)$$

where a is the ratio of the height to the average width.

IV. SIMULATION RESULTS

The mechanical behavior of the circular bossed diaphragm with a thickness of 20 um and diameter of 1800 um for membrane and a central boss with a diameter of 500 um was evaluated. Fig 5 shows the deflection behavior of the bossed diaphragm under 1 kPa applied pressure condition.

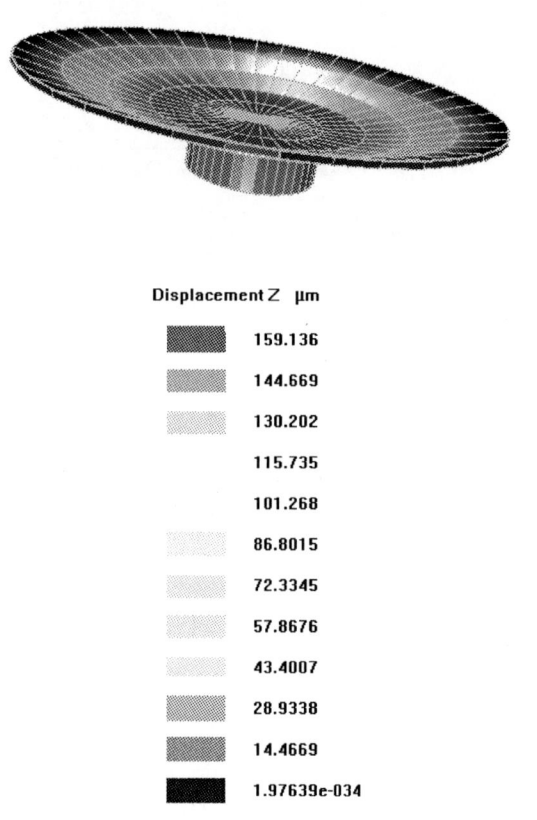

Fig. 5. Deflection of the bossed membrane under 1 kPa pressure.

Simulation shows enough deflection for opening the microvalve. Fig. 6 demonstrates deflections versus various applied pressure. It is seen that for an applied pressure of 1 kPa, a deflection of about 160 um is achieved.

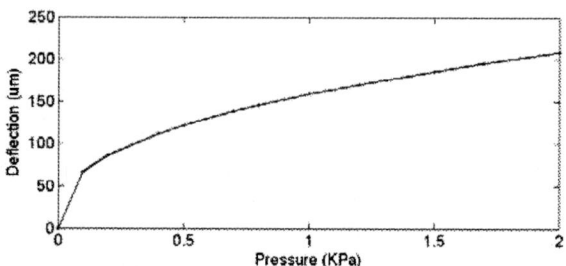

Fig. 6 Diaphragm deflection versus various applied pressure

The fluid pressure along the length of the microchannel was simulated and shown in Fig. 7.

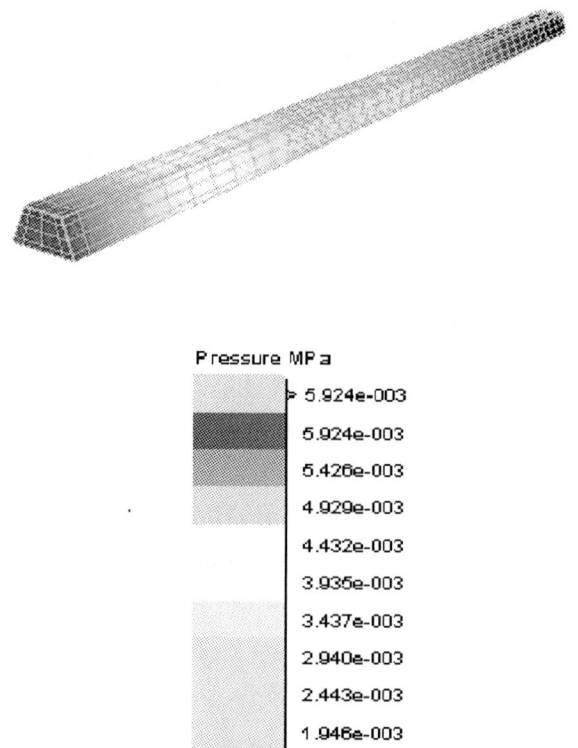

Fig. 7. Pressure of fluid along the microchannel

Microfluidic simulations for velocity was also done and it is shown that a good flow rate of up to 130 microliter/min output for an applied 5 kPa inlet pressure can be obtained.

V. CONCLUSION

Due to the low concentration of blood glucose, insulin should be delivered as a drug and a normally stimuli-responsive hydrogel microvalve can be used for this application without any external power. Simulation results show that silicon rubber as a material for membrane gives good deflection for opening the microvalve. The results also indicate that the fluids flow through a microchannel and a sufficient flow rate at the output of the microvalve can be achieved. Process development, flow simulations with insulin and fabrication will be considered in the future study.

ACKNOWLEDGEMENT

This work is financially supported by Malaysian Ministry of Science, Technology and Innovation under the project titled "MEMS Devices and Sensing Microstructures".

REFERENCES

[1] D. T. Eddington and D. J. Beebe, *"Flow control with hydrogels,"* Advanced Drug Delivery Reviews 56, pp. 199–210, 2004.

[2] D. T. Eddington and D. J. Beebe, *"A Valved Responsive Hydrogel Microdispensing Device with Integrated Pressure Source,"* IEEE Journal of Microelectromechanical Systems, VOL. 13, NO. 4 , pp. 586-593, Aug. 2004.

[3] A. Baldi, Y. Gu, P. E. Loftness, R. A. Siegel and B. Ziaie, *"A Hydrogel-Actuated Environmentally Sensitive Microvalve for Active Flow Control,"* IEEE Journal of Microelectromechanical Systems, VOL. 12, NO. 5, pp. 613-621, Oct. 2003.

[4] A. Baldi , M. Lei, Y. Gu, R. A. Siegel and B. Ziaie, *"A microstructured silicon membrane with entrapped hydrogels for environmentally sensitive fluid gating,"* Sensors and Actuators B 114, pp. 9–18, 2006.

[5] R. A. Siegel, Y. Gu, A. Baldi and B. Ziaie, *"Novel Swelling/Shrinking Behaviors of Glucose binding Hydrogels and their potential use in a Microfluidic Insulin Deliveru System,"* Macromol. Symp. 207, pp. 249-256, 2004.

[6] R. H. Liu, Q. Yu and D. J. Beebe, *"Fabrication and Characterization of Hydrogel-Based Microvalve ,"* IEEE Journal of Microelectromechanical Systems, VOL. 11, NO. 1, pp. 45-53, 2002.

[7] G. Hetsroni , A. Mosyak, E. Pogrebnyak and L.P. Yarin, *"Fluid flow in micro-channels,"* International Journal of Heat and Mass Transfer 48, pp. 1982–1998, 2005.

[8] M. Richter, P. Woias, and D. Weil, *"Microchannels for applications in liquid dosing and flow-rate measurement,"* Sensors and Actuators A, pp. 480-483, 1997.

Application of Full Factorial Design Method in MEMS Capacitive Thermal Sensor Sensitivity

Mehdi Mehrban, Shahriar Kouravand, Ghader Rezazadeh and Ali Donyavi
Mechanical Engineering Department
Urmia University, Urmia, Iran
Email: m.mehrban@gmail.com

Abstract **In this paper sensitivity of MEMS capacitive thermal sensors based on deflection of a bimetallic cantilever beam was investigated. Using design of experiment method, and applying a 2^5 factorial design the effect of factors in this sensor was calculated and main factors that affect on sensor's sensitivity were identified. Analysis of variance for the main factors of design and their interactions were studied for their significance.**

I. INTRODUCTION

MEMS technologies for fabricating tiny sensors and actuators are being developed successfully throughout the world. The field of MEMS is large and growing, with numerous means reported for both sensing and actuation on-chip. Compared to conventional systems, MEMS are more lightweight, smaller and less power consuming. All these qualities make MEMS very promising devices for space applications. However, to be space compatible, they need a robust behavior towards harsh environmental conditions, such as vacuum, temperature cycles, and irradiation. Thermal sensors are one of the most applications of the MEMS sensors and MEMS capacitive thermal sensor based deflection of bimetallic cantilever beam is the one of the last type of it. These microsensors were used to measure the temperature rising due to the change of a comb drive capacitance. Bimetallic microbeams can provide sensitive structure for fabrication of temperature sensors that cover the wide range of temperature based on these structures [1]. Sensitivity of this sensor depends on several factors. Effect of each factor on sensitivity may be not equal. With investigation of these factors effect the sensitivity of thermal sensors can be optimized. Several methods can be used for calculate the effect of each factors on sensitivity of this thermal sensor. One of the most applied methods

for these purposes is Full Factorial Design method (FFD) that is used to seek the active factors which affect on the system performance, and aid decision making in such a way that the optimum solution can be found. The traditional approach to experimental work is to vary one factor at a time and to hold all other factors fixed. This method does not produce satisfactory results in a wide range of experimental settings. If interaction exists between the factors, there is no guarantee that the final set of operating conditions will be the optimum [2]. Box and Meyer [3] stated that highly factorial designs are powerful tools for identifying important factors to improve the system performance. Madu [4] had already applied 2^{4-1} fractional factorial experiment design to analysis the management of maintenance floats. Murugabaskar and Huang [5] had worked on the application of full factorial design or fractional factorial design. They concluded that it is an efficient tool for analyzing the system performance.

In this article sensitivity of MEMS capacitive thermal sensors based on tip deflection of bimetallic cantilever beam was investigated. Using 2^5 full factorial designs with five main factors and two levels for them, the effect of factors and their interaction that affect on sensitivity of thermal sensor was studied.

II. MEMS CAPACITIVE THERMAL SENSOR STRUCTURE

Fig. 1 shows a thermal sensor whose sensitivity was studied in this work. Bimetallic cantilever beam was deflected because of different temperature expansion coefficient of selected materials due to temperature rising. Due to the tip deflection of bimetallic cantilever beam, the effective surfaces of comb drive plates and its equivalent capacitance was changed. Which means the influence of temperature on the capacity of the system can be easily measured [1]. This model is a Micro Electro Mechanical

System that consists of a bimetallic cantilever beam and one comb drive that is jointed at the tip of the cantilever beam. This system acts as a MEMS capacitive thermal sensor.

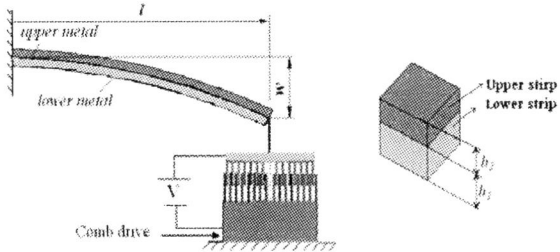

Fig. 1 A MEMS capacitive thermal sensor [1].

By assumption that the width of bimetallic cantilever beams are same, the equation which indicates the relationship between tip deflection of bimetallic cantilever beam with respect to temperature rising is derived as follow [1]:

$$w = \frac{3nh_1h_2(h_1+h_2)(\alpha_1-\alpha_2)\Delta T}{h_1^4 + n^2h_2^4 + nh_1h_2(6h_1h_2+4h_1^2+4h_2^2)}l^2 \quad (1)$$

where n, h_1, h_2, α_1, α_2, ΔT and l are ratio of young's modulus of upper layer on young's modulus of lower layer, height of lower layer, height of upper layer, thermal expansion coefficient of lower material, thermal expansion coefficient of upper material, temperature rising and length of bimetallic cantilever beam. To calculate the capacity of the comb drive due to change of its surfaces, the following equation can be used [1]:

$$C = 2N\frac{\varepsilon_0 B}{g}(d_0 - \frac{3nh_1h_2(h_1+h_2)(\alpha_1-\alpha_2)\Delta T}{h_1^4 + n^2h_2^4 + nh_1h_2(6h_1h_2+4h_1^2+4h_2^2)}l^2) \quad (2)$$

Where N is the number of combs, g is the gap, B is the width of plates of comb drive capacitor and d_0 is the initial effective height of comb drive plates (Fig. 2). Here sensitivity can be defined as a function of capacitance and temperature as below:

$$S = \frac{C}{\Delta T} \quad (3)$$

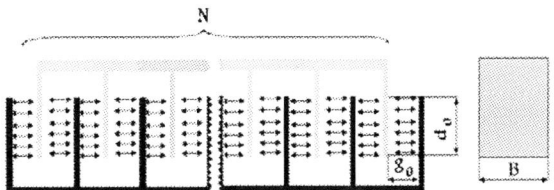

Fig. 2 Schematic of the sensor's comb drive

Dimensional parameters of sensor's comb drive are $d_0 = 10\,\mu m$, $g = 25\,\mu m$, $B = 15\,\mu m$, $N = 14$.

These parameters are the same as those for the sensor that designed in [1].

III. FULL FACTORIAL DESIGN

Factorial design is an optimization process which is used for preliminary evaluation of the experimental variables of a system. It allows the determination of their effects and significances [6]. In this article experiments were carried out using 2^5 factorial design. Maximum and minimum selected levels of each factor were shown in Table1.

Table 1 Factors and their levels used in the factorial design.

Variable	Low level (-)	High level (+)
(A): α_1 - Thermal expansion coefficient of lower material (k^{-1})	2.6	4.6
(B): α_2 - Thermal expansion coefficient of upper material (k^{-1})	14.3	23.6
(C): h_1 - height of lower strip (μm)	1.8	4.0
(D): h_2 - height of upper strip (μm)	1.8	4.0
(E): l- Length of bimetallic cantilever beam (μm)	300	500

In this work all possible combination of factor levels were investigated. For this purpose, 32 run using Eq. 3 by applying given values of factors showed in Table 1, were done in order to calculate the sensor sensitivity. The results of experiments and their corresponding sensitivity are shown in Table 2.

185

Table 2 Design matrix and the results of sensitivity.

Experiment	Factors					Sensitivity (S) ($fF/^\circ C$)	Experiment	Factors					Sensitivity (S) ($fF/^\circ C$)
	A	B	C	D	E			A	B	C	D	E	
1	-	-	-	-	-	0.620	17	+	-	-	-	-	0.531
2	-	-	-	-	+	1.201	18	+	-	-	-	+	0.946
3	-	-	-	+	-	0.485	19	+	-	-	+	-	0.465
4	-	-	-	+	+	0.817	20	+	-	-	+	+	0.763
5	-	-	+	-	-	0.453	21	+	-	+	-	-	0.380
6	-	-	+	-	+	0.729	22	+	-	+	-	+	0.527
7	-	-	+	+	-	0.443	23	+	-	+	+	-	0.403
8	-	-	+	+	+	0.701	24	+	-	+	+	+	0.589
9	-	+	-	-	-	0.866	25	+	+	-	-	-	0.727
10	-	+	-	-	+	1.903	26	+	+	-	-	+	1.502
11	-	+	-	+	-	0.644	27	+	+	-	+	-	0.626
12	-	+	-	+	+	1.302	28	+	+	-	+	+	1.201
13	-	+	+	-	-	0.554	29	+	+	+	-	-	0.437
14	-	+	+	-	+	1.000	30	+	+	+	-	+	0.684
15	-	+	+	+	-	0.553	31	+	+	+	+	-	0.491
16	-	+	+	+	+	1.000	32	+	+	+	+	+	0.834

Where *A, B, C, D* and *E* respectively are the experimental factors that represent α_1, α_2, h_1, h_2 and *l*. The main effect for each factor and their interaction effects can be calculated using the Table 2 and using this concept that the effect of each factor is the change in response by a change in the level of the factor [7]. Based on this concept, the effect of a factor can be calculated as (4).

Considering signs and results in Table 2, main effect and interaction effects of factors can explain in figure 3. In this figure the effects of all of the factors and their interactions were shown.

As an example figure 4 illustrates that there is an interaction effect between the thermal expansion coefficient of lower strip (α_1) and the height of upper strip (h_2) on sensor sensitivity.

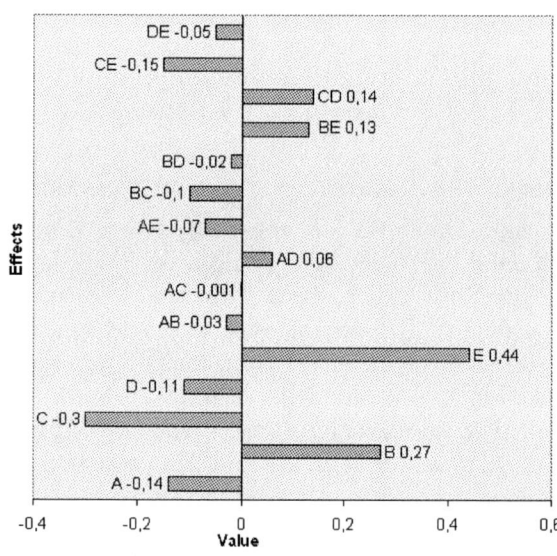

Fig. 3 The effect of factors and their interactions on sensitivity.

$$\text{Effect of a factor} = \frac{\sum \text{responses at high levels} - \sum \text{responses at low levels}}{\text{half the number of runs in the experiment}} \qquad (4)$$

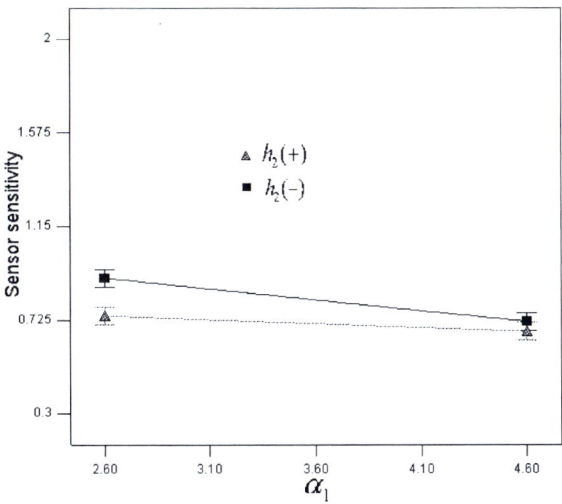

Fig. 4 Interaction effect of Thermal expansion coefficient of lower strip (α_1) and height of upper strip (h_2).

IV. ANALYSIS OF VARIANCE (ANOVA)

In the 2-level factorial designs, normal plot can be used in order to choose significant effects of factors. If the effects represent a sample from a normal population, we would expect to see them form an approximate straight line on a normal probability plot of the effects. Usually only a few effects turn out to be important. They show up as outliers on the normal probability plot. The ordered effects plotted on half normal probability plot are shown in Figure 5. It is seen that more effects and interactions lie approximately along a straight line, while all main factors and their interactions that involve AD, AE, BC, BD, CD

and CE are not in the direction of the crossed straight line and can be significant factors.

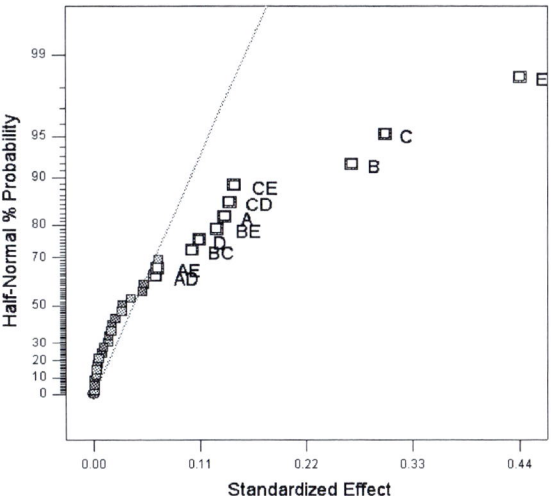

Fig. 5 Half normal plot of the effects.

These ANOVA are based on reduced models was carried out by first fitting the full model according to the constraints of the design and eliminating non-significant terms. Analysis of variance for this model is shown in Table3.
The Model F-value of 57.04 implies that the model is significant and there is only a 0.01 percent chance that a critical F-value could occur due to noise. Also P-values less than 0.05 indicate model terms are significant so in this case all selected main factors and AD, AE, BC, BE, CD, CE interactions is significant that effect on the sensitivity.

Table 3 ANOVA result for sensitivity effected factors.

Source	Sum of Squares	Degree of freedom	Mean Square	F-value	P-value
Model	3.680	11	0.330	57.04	< 0.0001
A-α_1	0.150	1	0.150	24.96	< 0.0001
B-α_2	0.570	1	0.570	97.15	< 0.0001
C-h_1	0.730	1	0.730	123.78	< 0.0001
D-h_2	0.095	1	0.095	16.18	0.0007
E-l	1.540	1	1.540	262.52	< 0.0001
AD	0.032	1	0.032	5.53	0.0291
AE	0.034	1	0.034	5.86	0.0251
BC	0.082	1	0.082	13.89	0.0013
BE	0.130	1	0.130	22.05	0.0001
CD	0.160	1	0.160	26.79	< 0.0001
CE	0.170	1	0.170	28.69	< 0.0001
Residual	0.120	20	0.059	-	-
Total	3.800	31	-	-	-

V. CONCLUSION

In the present work sensitivity of MEMS capacitive thermal sensors was studied, and by applying the FFD method effective parameters that affect on sensor sensitivity were distinguished. It was seen that E factor which represents length of bimetallic cantilever beam has maximum effect and height of upper strip has minimum effect on sensor sensitivity. In addition CE interaction that represents the interaction of length of bimetallic cantilever beam and height of lower strip has maximum effect on sensor sensitivity between other interactions. Using ANOVA for obtain the significant factors it was distinguished that the all main factors and some interactions have significant effect on sensor sensitivity.

REFERENCES

[1] Sh. Kouravand, Gh. Rezazadeh, M. Sabet and A. Tahmasebi, "MEMS capacitive micro thermometer based on tip deflection of bimetallic cantilever beam," *journal of Sensors & Transducers,* vol. 70, pp. 637-644 (2006).

[2] F. T. S. Chan and H. K. Chan, "Simulation Analysis of a PCB Factory using Factorial Design – A Case Study," *Int. Journal of Adv Manu. Techno.*, vol. 21, pp. 523–533 (2003).

[3] G. E. P. Box and R. D. Meyer, "Finding the active factors in fractionated screening experiments," *Journal of Quality Technology*, vol. 25, no. 2, pp. 94–104 (1993).

[4] C. N. Madu, "Simulation analyses of a maintenance float shop," *International Journal of Production Economics*, vol. 29, pp. 149–157 (1993).

[5] S. Murugabaskar and W. V. Huang, "Simulation analysis with group screening," *International Journal of Computers and Industrial Engineering*, vol. 25, pp. 25–28 (1993).

[6] Adriana C. Ferreira, Maria das G. A. Korn and Sergio L. C. Ferreira, "Multivariate Optimization in Preconcentration Procedure for Manganese Determination in Seawater Samples by FAAS," *Microchimica Acta*, vol. 146, pp. 271–278 (2004).

[7] D. C. Montgomery, Design and analysis of experiments, 6th Edition, *John Wiley*, New York (2004).

Design of a High Sensitivity Structure for MEMS Fingerprint Sensor

Mitra Damghanian, Burhanuddin Yeop Majlis, SM IEEE
Institute of Microengineering and Nanoelectronics, Universiti Kebangsaan Malaysia
43600 Bangi, Selangor, MALAYSIA
Email: burhan@vlsi.eng.ukm.my

Abstract — **A novel capacitive pressure sensor structure for fingerprint sensor/imager application is described which has higher sensitivity and linearity compared to all reported prototypes. The protrusion also has got a new shape to achieve the maximum sensitivity. Simulations have been done by FEM analysis and the results show the great improvements in output parameters. The technique can be applied to other capacitive structures as well.**

I. INTRODUCTION

Among MEMS pressure sensors, a pressure range from zero to 2MPa is related to tactile sensors which usually work in the frequency range of 0~20Hz. Capacitive fingerprint imagers, a narrower branch of tactile sensors, offer many advantages over existing technologies such as optical imagers, thus they are a promising alternative due to their low power consumption, small size and no turn on temperature drift. High sensitive fingerprint imagers are required in various applications related to biometric extraction and identification. Today, the need for identifying users is becoming more and more necessary for several typical operations such as access control, workstation login, electronic banking or even emergency medical situations.

Some capacitive structures for fingerprint application have been prototyped successfully as in [1], [2] and [3] but not commercially manufactured because of not sufficient qualifications such as resolution, sensitivity, linearity, or considerably high cost of the product, explaining the necessity of improving these parameters.

Although fingerprint imagers are placed in the tactile sensors category, the density of sensitive cells is much higher than the density required in

tactile sensors used in robotic applications. This high density will bring several challenges such as wiring problems and heat dissipation as well as the more complicated fabrication processes to have a reasonable product yield.

Nowadays it has become a proven matter to build the sensing structure of a fingerprint sensor on top of the circuit needed for measurement, conversion and transmission of the signal. Otherwise, no wiring opportunity is assumable because of the high density of the sensor array. Thus, as an obligation, the fabrication process of the sensor part should be compatible with the regular CMOS fabrication process or at least not destructive for the underlying circuit layer. It also minimizes the parasitic capacitances which might be very disturbing according to the small cell capacitance. To have a general view about how small cell capacitances are, it can be mentioned that the typical capacitance for each cell in existing designs with a cell size about $50\,\mu m \times 50\,\mu m$ is in the order of a few femto-Farad and the amount of ΔC in the active range of the cell usually doesn't exceed 10 percent of the primitive capacitance that means around some tenth to one femto-Farad. This cell size is considered according to the typical size of ridges and valleys on a finger tip as shown in Fig. 1.

Increasing $\Delta C / C$ as a critical parameter, will improve SNR and also can help to reduce the amount of the supply voltage which is becoming an advantage nowadays according to the consequent less power consumption and heat generation that is a big challenge in such a high density application. In this paper improvements have been made on the structure of the pressure sensor and as a result a better sensitivity, linearity and overall performance of the device has been

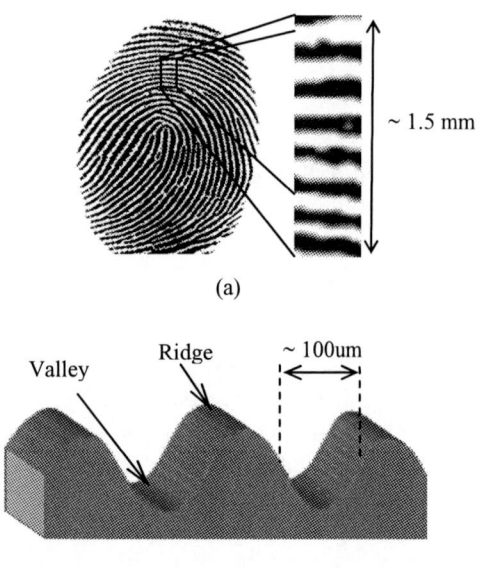

(a)

(b)

Fig. 1 (a) A typical fingerprint magnified for better understanding of the size of the lines (b) schema of the fingertip surface

achieved. The draw back of this improvement is requirement of a more accurate fabrication process.

In the following part, bending mechanics of thin films will be briefly studied and the effect of adding protrusion on the film will be noted, which lights up the working principals of the designed structure. FEM simulation of the sensitive designed structure and a comparison between existing structures and the designed structure are the subsequent parts which will be summarized in the final part.

II. BENDING OF THIN FILMS

As it is desired to make a high density array of the pressure sensitive cells with the most efficient usage of the consumed area, it is more desirable to use rectangular geometry instead of the circular one. Rectangular shape is also easier to optimize and fabricate than other polygons. Furthermore, by using a square diaphragm rather than just a rectangular shape, the device will not be sensitive to the random direction of the ridges and valleys of the fingerprint (Fig. 1).

Analysis of the static deflection of plates is well documented in the classical plate bending literature. Many researchers have developed small and large deflection analysis for constant thickness plates.[4]

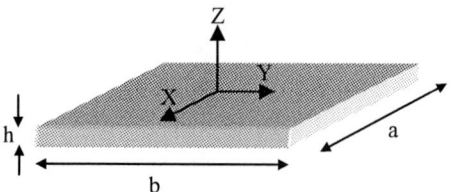

Fig. 2 Geometry of the diaphragm / wide beam

A) FOUR SIDES CLAMPED PLATE

As shown in Fig. 2, a rectangular plate is considered in parallel with Cartesians axis and with all four sides clamped. The deflection caused by applying pressure to the uniform elastic plate surface can be determined through the solution of this equation [5]:

$$\frac{Eh^3}{12(1-v^2)}\Delta^2 \omega(x,y) - P(x,y) = 0 \quad (1)$$

where E denotes the Young's modulus, h is the thickness of the plate, v is the Poisson ratio, $\omega(x,y)$ is the deflection of the plate, $P(x,y)$ is the pressure applied to the surface of the plate and Δ is the two-dimensional Laplace operator:

$$\Delta = \frac{\partial^2}{\partial^2 x} + \frac{\partial^2}{\partial^2 y} \ (Cartesian \ co-ordinates) \ (2)$$

Boundary conditions in the case of a four sides clamped rectangular diaphragm as shown in Fig. 2 would be:

$$\begin{cases} \omega(0,y) = \omega(x,0) = \omega(a,y) = \omega(x,b) = 0 \\ \omega'(0,y) = \omega'(x,0) = \omega'(a,y) = \omega'(x,b) = 0 \end{cases} \quad (3)$$

As equation (1) does not have a closed form solution even for a uniform pressure applied to a square diaphragm, it should be tried to solve using numerical solutions such as FEM methods or by means of some approximations.

The following analytical form can describe approximately the deflection of a square diaphragm [5]:

$$\omega(x,y) = \omega_0(1-\cos 2\pi x/a)(1-\cos 2\pi y/a) \quad (4)$$

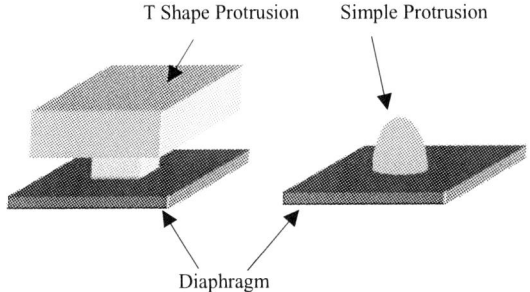

Fig. 3 Different shapes of central protrusion

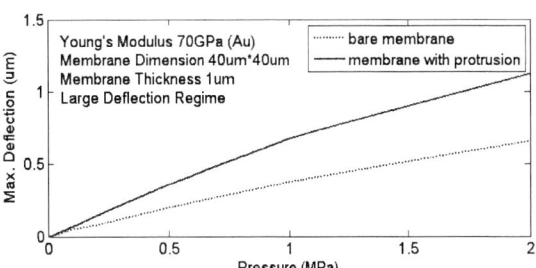

Fig. 4   Deflection increases using protrusion on top of the membrane

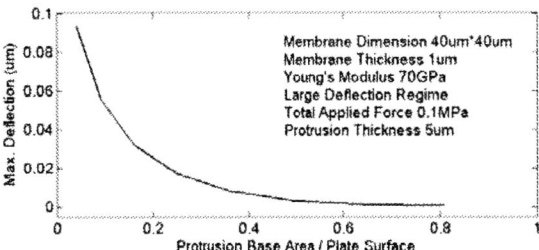

Fig. 5   Membrane maximum deflection as a function of the square shaped protrusion base area

where a is the side length of the plate and ω_0 follows a relation which is:

$$\frac{Pa^4}{Eh^4} = \frac{4.13\omega_0}{(1-v^2)h} + \frac{3.41a^2\sigma_i\omega_0}{h^3E} + \frac{1.98(1-0.585v)\omega_0^3}{(1-v)h^3} \quad (5)$$

where σ_i is the initial tension. Thus the maximum deflection would happen at the centroid and the amount of that deflection using approximate solution would be:

$$\begin{cases} \omega_{max} = -\alpha \dfrac{Pa^4}{Eh^2} \\ \alpha = 0.0138 \ (Square\ membrane) \end{cases} \quad (6)$$

The amount of maximum stress also can be calculated as follows:

$$\begin{cases} \sigma_{xx(max)} = \sigma_{yy(max)} = -\beta \dfrac{Pa^2}{h^2} \\ \beta = 0.3078 \ (Square\ membrane) \end{cases} \quad (7)$$

Consider that the maximum stress arises not from the tension in the membrane but from the bending close to the clamped edges and will occur at the centroid of them.

As it is desirable to maximize sensitivity which means somehow more diaphragm deflection under the same applied force, it would be useful to solve the deflection equations related to a square elastic diaphragm with a central protrusion (Fig. 3).This protrusion will concentrate the applied force on the center of the diaphragm hence enlarges the maximum deflection. As mentioned before, analytical solutions are not available for such a

case unless assuming the force concentrated on just the central point which is not truly the real case. Using FEM simulations is the most common method to give the sense of how deflection happens and the amount of it. Fig. 4 compares the FEM results of a simple elastic square diaphragm deflection under uniform applied pressure and the one having simple protrusion faced to the equal force. Simulations have been done by Intellisuite software. For better understanding, all physical parameters such as size, thickness and material are the same. The protrusion base area is selected as one tenth of the diaphragm surface which is a practical situation [6], [7]. It also can be observed that the maximum deflection will increase while the protrusion base area gets smaller (Fig. 5). Obviously there are fabrication constraints as well as force concentrating ability of protrusion that limit the size of the protrusion base area and its thickness as it should be in practical range. Also the height of protrusion should be optimized to get the maximum sensitivity. It has been shown that a T shaped protrusion (Fig. 3) acts more efficiently according to the overall sensitivity of the device and also contamination protection compared to the case using a simple protrusion [7].

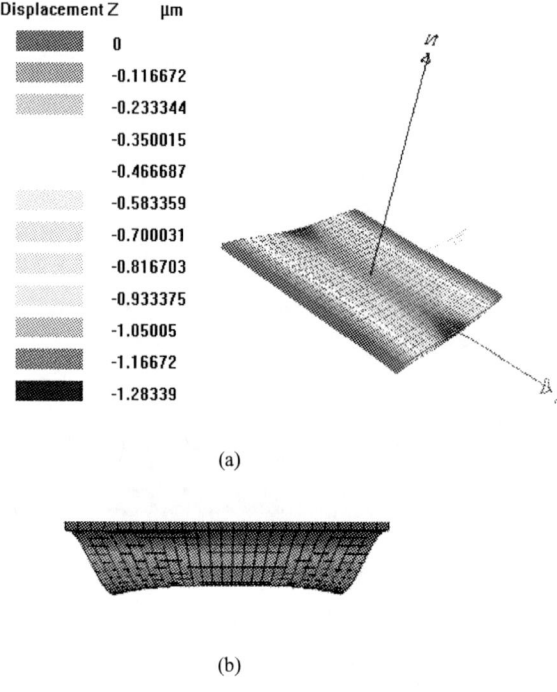

(a)

(b)

Fig. 6 a) Deflection of a wide beam under uniform 1MPa applied pressure   b) Side view of the exaggerated deformation of the wide beam free

B) WIDE BEAM

Now the deflection of a wide beam faced to the same pressure is considered. A beam is supposed wide when $a \geq 5h$ [8], where a is the width and h is the height of the beam. The later assumed beam dimensions are not within the Bernoulli-Euler limit thus it does not allow neglecting the shear stresses near the fixed supports and approximating the stress in the beam as purely uniaxial along the beam length [8]. Deflection of such a fixed-fixed wide beam faced to a uniform pressure can be determined by solving the same equation as equation (1). The only difference is in boundary conditions which would be as follows:

$$\begin{cases} \omega(x,0) = \omega(x,b) = 0 \\ \omega'(x,0) = \omega'(x,b) = 0 \\ \omega''(0,y) = \omega''(a,y) = 0 \\ \omega'''(0,y) = \omega'''(a,y) = 0 \end{cases} \quad (8)$$

Width and length of the beam are assumed as a and b respectively. The edges in parallel with the x axis are also considered as fixed edges (Fig. 2).

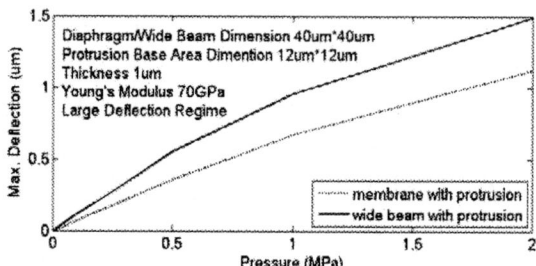

Fig. 7   Maximum deflection comparison of a membrane and a wide beam

It should be point out that the maximum deflection of the wide beam under uniform applied pressure will not occur at the centroid point, but in the middle of the free edges. The case is shown in Fig. 6 for a typical $50\,\mu m \times 50\,\mu m \times 1\,\mu m$ wide beam under a 1MPa uniform applied pressure. Deformation of the free edges is not desirable. It is neither because of the stress limitations, nor reaching the touch mode region, but because of the nonlinearity. It even helps the sensitivity but in a nonlinear way.

A protrusion on top of the wide beam can prevent deformation of the free edges very efficiently. In this case it is possible to compare maximum deflections of the diaphragm and the wide beam. Fig. 7 shows that a wide beam with a simple protrusion can have 25% larger maximum deflection than a membrane under the same conditions.

III. SENSITIVITY AND LINEARITY

Usually there is a trade off between sensitivity and linearity in a capacitive cell structure. When using diaphragm as the second electrode, for better linearity it is possible to increase the thickness of the diaphragm. The thicker the diaphragm, a better linearity is achieved. On the other hand, sensitivity will decrease as the thicker diaphragm has the smaller deflection faced to the same pressure. It will be shown that a wide beam can improve both the sensitivity and linearity at the same time under the same applied pressure.

Sensitivity, S, of a capacitive pressure sensor in its operating pressure range can be analytically defined as the slope of a straight line obtained by the least square linear curve fitting over its $C - P$ characteristics plot. Based on this straight line, linearity, L, in percentage could be defined as:

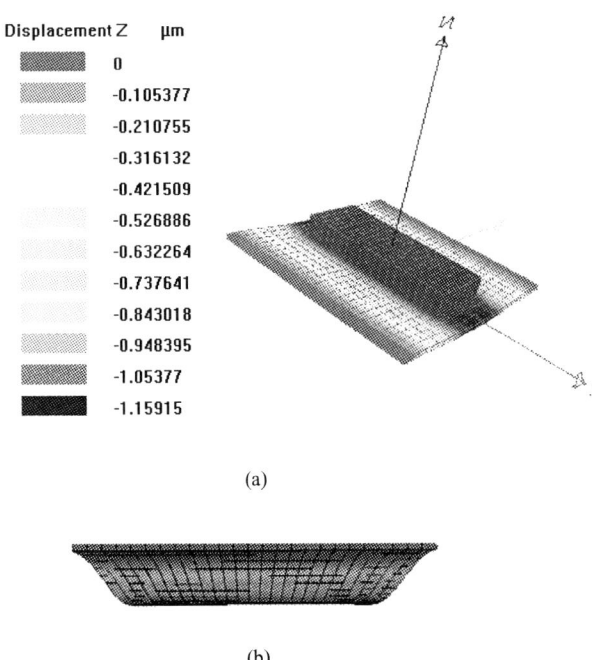

(a)

(b)

Fig. 8 a) Deflection of a wide beam with protrusion faced to the same force as Fig. 6. b) Side view of the exaggerated deformation of the wide beam. Free edges deformation is considerably less than Fig. 6.

$$L = 100 - \frac{|\Delta C_+| + |\Delta C_-|}{2(C_{\max} - C_{\min})} \times 100 \quad (9)$$

Where ΔC_+ and ΔC_- are the maximum and minimum capacitance deviation from the straight line respectively and $C_{\max}$ and $C_{\min}$ are the maximum and minimum capacitance in pressure operation range. Capacitance in this case neglecting fringe effects can be obtained from integration of a series of parallel differential capacitive elements each having relation of a simple parallel plate capacitor:

$$C = \iint\limits_{\substack{upper \\ electrode \\ surface}} \frac{\varepsilon_0 \varepsilon_r dxdx}{g - w(x,y)} \quad (10)$$

Where g denotes the cavity depth and ε_0 and ε_r are electrical permittivity of air and dielectric respectively. Yet, as $w(x,y)$ can not be analytically defined, capacitance also should be determined by means of proper FEM tools or some

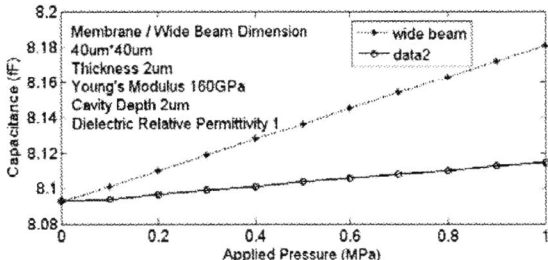

Fig. 9 Comparison of C-P diagrams of a bare diaphragm and a wide beam

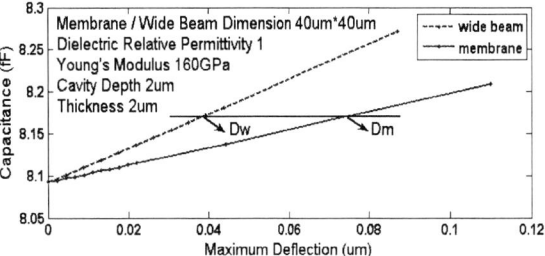

Fig. 10 Capacitance versus deflection in diaphragm and wide beam.

approximations.

Larger deflection in wide beams compared to that of diaphragms will straightly result larger capacitance changes caused by the same force. Fig. 9 shows a comparison between the bare diaphragm and wide beam capacitance changes when all other design parameters including primitive capacitances are the same. An analytical comparison between sensitivities is possible with the comparison of the corresponding S values. Calculating S_W and S_D with *MATLAB* using *polyfit* function, related to the sensitivity of wide beam and diaphragm respectively shows that a wide beam can make a more sensitive capacitive cell according to its larger deflection behavior. According to Fig. 9, these amounts are $S_W = 0.0882 > S_D = 0.0225$ which gives a near 4 times better sensitivity response in this case. The improvement is much greater than the difference between the amount of maximum deflection measurements in previous section because this max deflection in the diaphragm happens only in the central region but in the wide beam it happens in nearly all the axial points (Fig. 6) indicating a larger capacitance variation. It can be observed that using a protrusion on top of the movable electrode will improve the sensitivity as it will increase the deflection of the movable electrode. Consider two points, D_W and

D_m in Fig. 10, where wide beam has the same changes in capacitance compared to the usual diaphragm with the same physical dimensions. Although maximum deflection is smaller in wide beam, it is distributed along its symmetry axis (x axis in Fig. 8) resulting a larger sensitivity.

Applying the data used in Fig. 9 to equation (9) will generate linearity percentages of wide beam, L_W and diaphragm, L_D. In this case $L_W = 98.75 > L_D = 95.45$ are calculated which shows a considerable improved linearity for wide beam. The amount of $C_{\min}$ is the same for wide beam and diaphragm since both are considered to have the same physical size and parameters which directly defines the resting capacitance which is also the minimum one. Parameter $C_{\max}$ is larger in wide beam because of its bigger and at the same time axial deflection behavior. Deviations are also in the same order in both cases. Thus all variations will increase the amount of nonlinearity ratio denominator, $C_{\max} - C_{\min}$ in the wide beam, resulting linearity improvement.

IV. STRESS

As a general assumption in the design of capacitive sensor with the aid of deforming thin films, maximum stress of the thin plate in the working range should not exceed 60% of the material elastic limit according to the safety margin. The farther is the distance from this elastic limit, the design is more reliable. In this design the thin plate is also considered not to have any initial stress although it is not truly the case.

It is mentioned that maximum stress in a diaphragm will appear in the center of each clamped edge while in a wide beam it happens near the end points of them, Fig. 11. Although a wide beam has a larger induced stress under the same pressure because of its larger deflection compared to that of a diaphragm, it would be useful to compare the maximum stress in a diaphragm and its analogous wide beam while having the same center deflection, as shown in Fig. 12. The case is interesting because under the same maximum deflection, wide beam has smaller induced stress and simultaneously larger capacitance change.

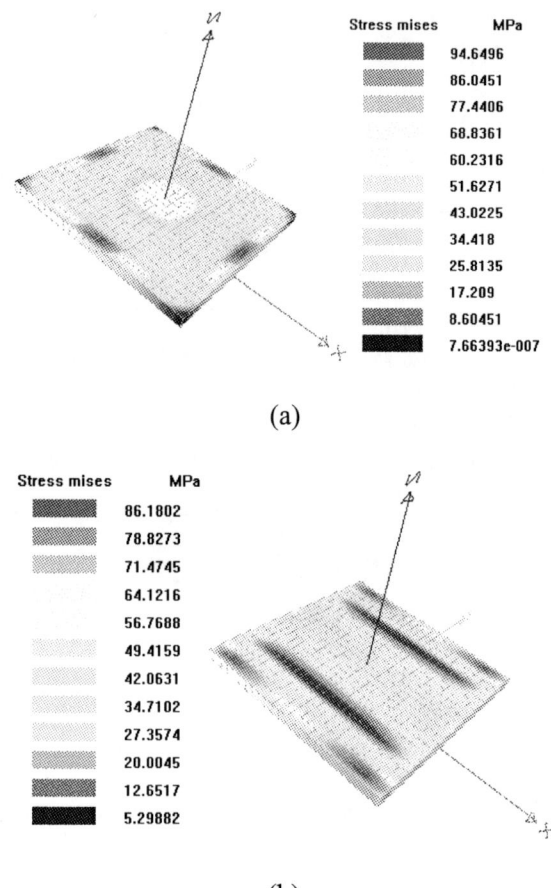

(a)

(b)

Fig. 11 Stress patterns in a)diaphragm and b)wide beam faced to the uniform pressure. Both (a) and (b) have the same maximum deflection.

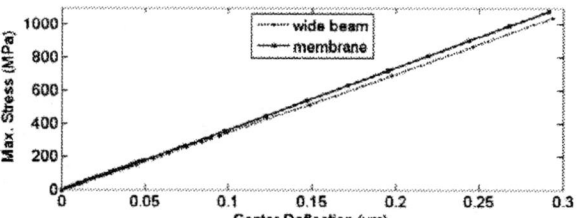

Fig. 12 Maximum stress comparison between a diaphragm and a wide beam having the same center deflection.

V. DESIGNED STRUCTURE

According to all better performance of the wide beam compared to the usual diaphragm form, a new designed structure for the pressure sensing cells is suggested. The schematic view of the structure is presented in Fig. 13 which shows an individual absolute capacitive cell. The final

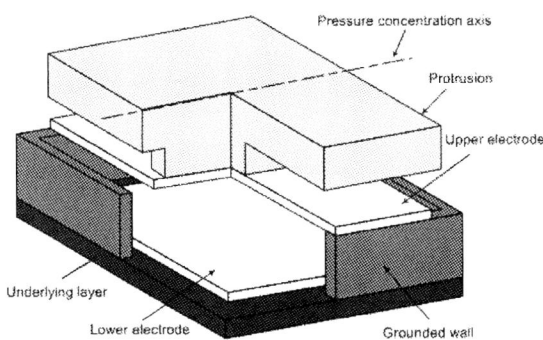

Fig. 13 Schematic structure of an individual capacitive cell. The coating polyimide is not shown for more clarity.

fingerprint imager would be a high density array of such cells. A $50\mu m \times 50\mu m$ size, which means a 500 dpi resolution, has been selected for each cell because a ridge on a finger surface is about $100\mu m$ to $400\mu m$ wide. Maximum applied pressure by a fingertip is normally less than 1MPa and the structure is designed to remain in its linear region applying this amount of pressure. In small deflection regime which the device is intended to work in according to the perfect linearity, maximum deflection of the plate should not exceed 30% of its thickness [9].

Upper and lower electrodes with $2\mu m$ thickness the same as $4\mu m$ high grounded walls are made from Au on top of the CMOS circuit layer. There would be a $2\mu m$ cavity depth as the space between capacitor electrodes. The space between the wide beam and the grounded walls can be used as etch holes to easily etch the sacrificial layer under the beam. A thin polyimide layer is also coating the upper electrode as the sealing layer to avoid contamination. This polyimide coating does not limit the bending of the wide beam as its Young's modulus is around $6MPa$ and that of the Au electrode is around $60MPa$ thus the overall bending is limited by the Au wide beam.

As it has been decided to change the membrane to a wide beam, the shape of the protrusion has also been changed to reach the higher sensitivity. The base area of the protrusion has got a rectangular shape instead of the previously square one. Modified T shape protrusion is also made from the polyimide material in the next step.

Thickness of the protrusion is selected to be 10um with a $3\mu m$ primitive spacing according to [7]. This protrusion concentrates the applied force on the axial line of the wide beam. Protrusion base area is selected $40\mu m \times 10\mu m$ with respect to the maximum applied pressure. With this size, device will remain in its small deflection regime and certainly in its elastic range. The upper part of protrusion has a $40\mu m \times 40\mu m$ area size with a $10\mu m$ thickness.

Using closed form model for wide beams pull-in voltage published in [8] and [10], a more than some hundreds pull-in voltage is calculated and the result is confirmed using FEM simulations. This large amount arisen from the small cell size and thick membrane will eliminate any concern about bias voltage of the capacitive cells needed for readout circuit.

FEM simulations show the device linearity equal to 97.956% and sensitivity equal to $1.2482\,fF/MPa$ in the $0 \sim 1MPa$ operating range and device will stand up to 5 times greater shock pressure without exceeding elastic limit. Maximum stress induced in $1MPa$ applied pressure is $227MPa$ which has a good safety margin compared to the about $1GPa$ gold electrode elastic limit.

VI. SUMMARY

FEM analysis indicates that great improvement in sensitivity and linearity is accessible without exceeding elastic limit if the usual diaphragm structure could be replaced with a wide beam in capacitive fingerprint imager structure. The new shape of the protrusion also helps to concentrate the applied pressure on the wide beam maximum deflection axis as well as preventing edge deformation.

Process development, device fabrication and real performance measurements are the next steps towards authentication of the design.

ACKNOWLEDGEMENT

This work is financially supported by Malaysian Ministry of Science, Technology and Innovation under the project title "MEMS Devices and Sensing Microstructures".

REFERENCES

[1] R. J. De Souza, K. D. Wise, "A Very High Density Bulk Micromachined Capacitive Tactile Imager," Int. Conf. on Solid-state Sensors and Actuators, Chicago, June 16-19, 1997.

[2] K. Machida, S. Shigematsu, H. Morimura, Y. Tanabe, N. Sato, N. Shimoyama, T. Kumazaki, K. Kudou, M. Yano, H. Kyuragi, "A Novel Semiconductor Capacitive Sensor For a Single Chip FingerPrint Sensor/Identifier LSI," IEEE Trans. On Electron Device s, Vol. 48, No. 10, Oct. 2001.

[3] T. Fojimori, Y. Hanaoka, K. Fujisaki, N. Yokoyama, H. Fukuda, "Fully CMOS Compatible On-LSI Capacitive Pressure Sensor Fabricated Using Standard Back-End-Of- Line Processes," 13th Int. Conf. On Solid-State Sensors, Ctuators and Microsystems, Seoul, Korea, Jun. 5-9, 2005.

[4] DiGiovani M., Flat and Corrogated Diaphragm Design Handbook, Marcel Dekker, Inc., New Yourk, 1989

[5] Elwenspoek, M., *Mechanical Microsensors*, Spinger-Verlag Berlin Heidelberg 2001

[6] N. Sato, S. Shigematsu, H. Morimura, M. Yano, K. Kudou, T. Kamei, K. Machida, "Novel Surface Structure and Its Fabrication Process for MEMS Fingerprint Sensor," IEEE Trans. On Electron Devices, Vol. 52, No. 5, May 2005.

[7] N. Sato, K. Machida, H. Morimura, S. Shigematsu, K. Kudou, M. Yano, H. Kyuragi, "MEMS Fingerprint Sensor Immune to Various Finger Surface Conditions," IEEE Trans. On Electron Devices, Vol. 50, No. 4, Apr. 2003.

[8] P. M. Osterberg and S. D. Senturia, "M-TEST: A test chip for MEMS material property measurement using electrostatically actuated test structures," J. Microelectromech. Syst., Vol. 6, No. 2, pp. 107–118, Jun. 1997.

[9] S. Beeby, G. Ensell, M. Craft, N. White, "Mems Mechanical Sensors," ARTECH HOUSE 2004, ISBN: 1-58053-536-4

[10] S. Chowdhury, M. Ahmadi, W. C. Miller, "Pull-In Voltage Study of Electrostatically Actuated Fixed-Fixed Beams Using a VLSI On-Chip Interconnect Capacitance Model," Journal of Microelectromech. Syst. Vol. 15, No. 3, Jun. 2006.

High-Precision Thickness Control of Silicon Membranes Using Etching Techniques

Mohsen Nabipoor and Burhanuddin Yeop Majlis, *Senior Member IEEE*
Institute of Microengineering and Nanoelectronics (IMEN)
University Kebangsaan Malaysia (UKM)
43600 Bangi, Selangor, Malaysia
Email: mohsen@vlsi.eng.ukm.my burhan@vlsi.eng.ukm.my

Abstract **A visual method is demonstrated for fabrication of silicon membranes by deep reactive ion etching (DRIE) and wet etching techniques. A DRIE cavity is created on silicon substrate closed to the membrane recess, and the backside of the wafer is etched by a wet etching process until it reaches the bottom of the DRIE cavity. Both isotropic and anisotropic wet etching with a loose control of temperature and concentration could be used. Because of the high accuracy etch rate of the silicon by DRIE, the depth of the cavity could be defined accurately and the fabricated membrane thickness would be precise.**

I. INTRODUCTION

Use of silicon membranes as sensing or actuating elements is very common in many silicon micro structures. It is very important to have silicon membranes with high precision dimensions to produce reliable devices with minimum variation in sensitivity, performance and functionality. Thin silicon membranes have been fabricated by bulk micromachining of silicon wafer with a very precise thicknesses [1-2]. A P+ layer is normally used to act as an etch stop against anisotropic wet etching and helps to control the membrane thickness accurately. The P+ layer can be created by diffusion process or ion implantation which in both processes the depth of the P+ layer from the surface of the silicon substrate doesn't exceed a few microns [3]. To create slightly thicker silicon diaphragms, i.e. 10 μm thick and above, the above mentioned technique can not be used anymore and the straightforward method for fabricating such thick diaphragms is time control wet etching of the silicon substrate. However, the etch rate of silicon in wet isotropic and anisotropic chemicals is very sensitive to process parameters such as the temperature and the solution concentration which leads to a big variation in diaphragm thickness over different devices and different wafers.

One way to place the etch stop layer in depths more than a few microns is to growth an epitaxy layer of silicon after the diffusion process [4]. The method that we used to fabricate a thick silicon membrane however is simpler. It occupies a DRIE process to create a cavity closed to the diaphragm recess and watch the wet etching of the backside of the wafer and stop the etching process whenever the process reaches the bottom of the DRIE cavity so the other side of the wafer is visible. This method is inexpensive and time-saving compared to other methods and yields in good process repeatability.

II. METHODOLOGY

For a capacitive pressure sensor suitable for measuring tire pressure of light vehicles (up to 100 PSI), a 1 mm × 1 mm square capacitor with 2-6 μm gap and a thick silicon membrane of 12-30 μm is required [5]. In this experience we tried to create a capacitor with 3.5 micron gap and a square silicon membrane of 1 mm length and 24 micron thick. The 3.5 micron gap was created by anisotropic wet etching of silicon in potassium hydroxide (KOH) as illustrated in Figure (1-a). A 270 nm thick wet oxide was grown, patterned and used as the mask for KOH etching. Then a DRIE process was performed to create the 28 μm depth cavity (Figure 1-b). After fabricating the capacitor electrodes and bonding the silicon and the glass wafers together (Figure 1-c), the loose control anisotropic etching of silicon from backside can be performed until the process reaches the bottom of the DRIE cavity (Figure 1-d). Because of the high accuracy of the etch rate of silicon in DRIE process, the result membrane has a good uniformity and precise thickness. To decrease the cost of the process, one can replace the DRIE process with an anisotropic or isotropic wet etching. Because of the small depth of the cavity compared to wafer thickness, the depth variation of the cavity created by wet etching could also kept small if the wet process parameters, i.e. temperature and time are controlled very accurately.

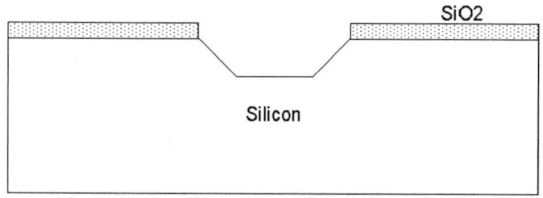

(a) Anisotropic etching of Silicon in KOH

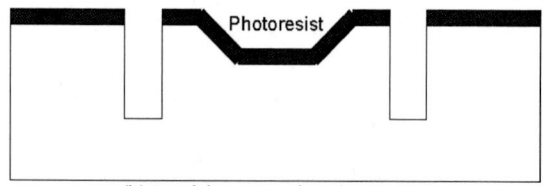

(b) Making trenches by DRIE

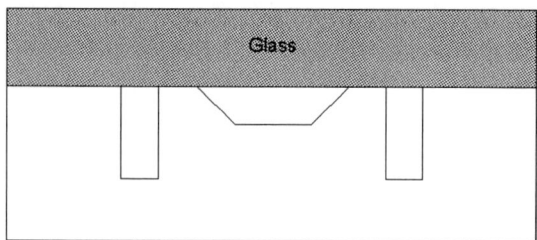

(c) Anodic bonding of silicon to glass

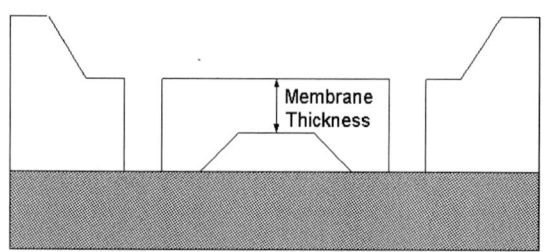

(d) Backside anisotropic etching of Silicon

Figure 1: Simplified process flow for fabricating thick silicon membranes.

III. RESULTS AND DISCUSSION

The Scanning Electron Microscopy (SEM) image of a device after creation of 3.5 μm KOH recess and 28 μm depth DRIE cavity is shown in Figure (2).

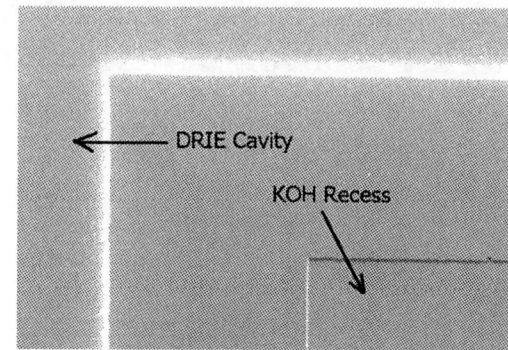

Figure 2: SEM image of the device after KOH and DRIE etching.

The measured depth of 8 recesses created by KOH etching on 2 silicon chips is illustrated in graph of Figure (3). The average depth is 3.4 μm with a standard deviation of 0.29 μm which is around 8.5 % of the mean value.

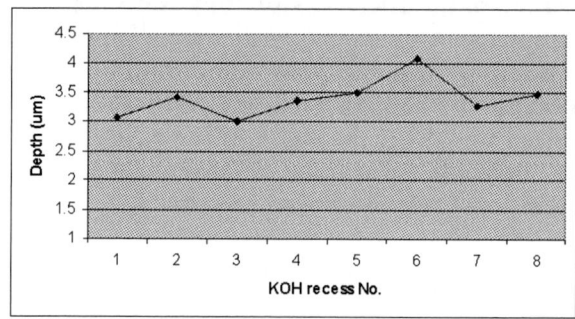

Figure 3: Depth of 8 KOH etched recesses on silicon

On the other hand, average depth of 12 DRIE trenches was measured as 28 μm with a standard deviation of 0.63 μm which is around 2.25 % of the mean value. These results show that the accuracy of DRIE process is higher than the accuracy of KOH etching. The depth of DRIE trenches on silicon is shown in graph of Figure-(4).

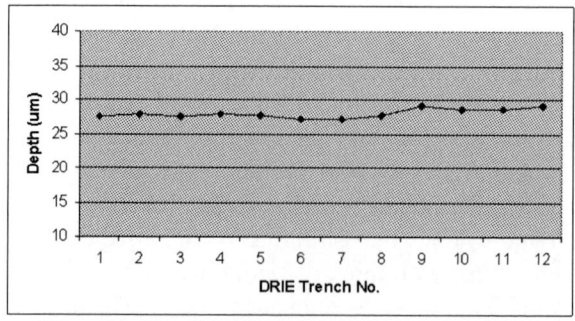

Figure 4: Depth of 12 DRIE trenches on silicon

After backside etching of the silicon in KOH and visual observation of the sample, the process was stopped when the bottom of DRIE cavity was etched and the other side of the silicon wafer was visible. The size of the cavity was chosen big so that it could be observed with naked eyes. Figure 5 shows the photograph of the sample from backside of silicon taken by an optical microscope.

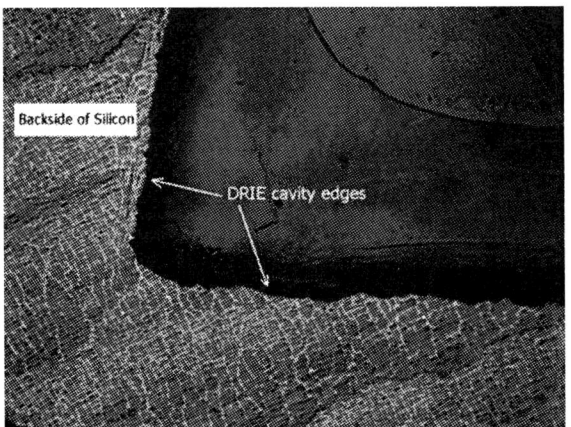

Figure 5: Backside etching of silicon until reaching the bottom of DRIE cavity.

IV. CONCLUSION

A visual method to precisely control the thickness of silicon membranes was illustrated in which DRIE trenches are used to define the membrane thickness. By this method, the process variations in wet etching of silicon don't cause much variation in membrane thickness. Although in this experiment we used DRIE process, one can perform isotropic or anisotropic wet etching of silicon to create the cavity which is a much cheaper process compared to DRIE. Because of the short length of wet process in creating the cavity, it could be controlled very easily and the depth variation of the cavity can be maintained within acceptable tolerances.

ACKNOWLEDGMENT

This work has been supported by the Malaysian Ministry of Science, Technology and Innovation under the project title "Development of MEMS Technology for Automotive Application".

REFERENCES

[1]   A.D. Dehennis and K.D. Wise, "A Wireless Microsystem for the Remote Sensing of Pressure, Temperature, and Relative Humidity," IEEE J. of MEMS Vol. 14, No. 1, Feb. 2005

[2]   S. Chatzandroulis. D. Tsoukalas, and P. A. Neukomm, "A Miniature Pressure System with a Capacitive Sensor and a Passive Telemetry Link for Use in Implantable Applications,"IEEE Journal of MEMS, Vol. 9, No. 1, March 2000.

[3]   S. Franssila, Introduction to Micro Fabrication, Wiley Publication, 2004.

[4]   C.H. Mastrangelo, X. Zhang, and W.C. Tang, "Surface-Micromachined Capacitive Differential Pressure Sensor with Lithographically Defined Silicon Diaphragm," IEEE Journal of MEMS, Vol. 5, No. 2, June 1996, pp. 98-105.

[5]   M. Nabipoor and B.Y. Majlis, "A New Passive Telemetry LC Pressure and Temperature Sensor Optimized for TPMS," J. of Physics, 2006.

Aluminum based Two-Port-Clamped-Clamped Resonators

Mustafa Al_Khusheiny and Burhanuddin Majlis, SMIEEE
Institute of Microengineering and Nanoelectronic
Universiti Kebangsaan Malaysia
43600 Bangi, Selangor, Malaysia
Email: mustafa1@vlsi.eng.ukm.my, burhan@vlsi.eng.ukm.my

Abstract - **In this paper a new structure of clamped-clamped µresonator (CCMR) of resonance frequency (f_o) of 180 kHz, is designed, modeled, fabricated and tested, using aluminum as structural material. *IntelliSuite* simulator is used to model the mechanical properties of the new MEMS resonator using a static displacement analysis and to get the optimum values of the beams parameters. The effective spring constant and mass of the resonator were calculated using a special proposed simulator, based on Mapple, besides using it to model the mechanical parameters into equivalent electrical circuit for the resonator, and determine the electrical properties just by giving the physical dimensions of the µresonator. Surface micromachining technology was used to fabricate the proposed MEMS resonator in IMEN's Clean Room.**

I. INTRODUCTION

Clamped-Clamped Resonators (CCR's) play an important role in realization the low power MEMS Filters, Oscillators and Mixers [1,2,3,4]. CCR's can be used in communication because of their relatively high spring constant which enables large dynamic range and power handling, besides their ease of manufacturing. The µmechanical resonators offer the potential for the very high Q's in the context of conventional IC processes; as their Q determines the insertion loss and phase noise respectively. The resonance frequency, motional resistance and the Q-factor is the most characteristics for the CCR, and these three characteristics is largely depend on the physical dimensions of the CCR.

In this paper a two-port-clamped-clamped MEMS resonator of resonance frequency of 180 kHz is modeled, and implemented. *IntelliSuite* simulator is used to model the mechanical properties of the CCMR using a static displacement analysis; the effective spring constant and mass of the resonator were

calculated besides calculating the effects due to mechanical non-linearity two-port-clamped-clamped resonator. While a new software using Mapple was built up, by which, the electrical parameters of the MEMS resonator was obtained; just by giving the mechanical geometries of the two-port-clamped-clamped microresonator, and the value of the applied dc-bias

II. TWO-PORT CLAMPED-CLAMPED RESONATORS STRUCTURE & OPERATION

Fig.1 shows the schematic diagram of the two-port clamped-clamped beam resonator. The µresonators used in this work have a configuration with separate drive and sense electrodes. The drive electrode is composed of two symmetrical parts around the centered one ,which is represents the sense electrode; the area of the two parts of the drive electrode is equal to that of the sense one, and this is simplify the working of the microstructure as we will see later.

The operation of the microstructure is described as follow; when an AC voltage is applied to the drive electrode, and a DC voltage, known as the polarization voltage, is applied to the µresonator beam, the microstructures is driven into resonance, resulting in an electrostatic force between the electrode and the beam that consequently causes the vibration of the µresonator. The capacitance change between the sense electrode and the beam due to the movement of the beam generates a current in the sense electrode. The microstructure vibrating (motion) is detected by sensing, producing an ac current at the sense electrode of the device. When the frequency of the applied drive voltage is equal to the mechanical natural frequency of the resonator, motion of the microstructure and consequently the output ac current of the device reach their maximum values. The amplitude of vibration at the resonance frequency is Q times (Q=quality factor) larger than the DC displacement amplitudes with the same drive voltage value [5,6].

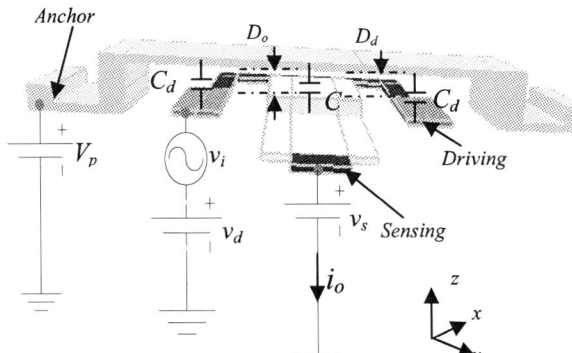

Fig. 1 Operating configuration of a two-port clamped-clamped beam μresonator, with driving and sensing electrodes

By changing the dimensions of the resonator, a different resonance frequency for the resonator can be achieved. When a dc bias voltage V_p is applied to the conductive layer of the beam, while an ac excitation signal $v_i = V_i \cos(\omega_i t)$ is applied to the underlying drive electrode. In this configuration, a dominant force component is generated at ω_i, which drives the beam into mechanical resonance when $\omega_i = \omega_o$, creating a dc-biased (via V_p) time-varying capacitance between the sense electrode and resonator beam, and sourcing an output current $i_o = V_p(\partial C_s / \partial z)(\partial z / \partial t)$

III. TWO-PORT-CLAMPED-CLAMPED-BEAM DESIGN

A. Resonance Frequency Design

In general, the resonator beam width W_r is governed by transducer and length-to-width ratio design considerations. Thus, the length L_r becomes the main variable with which to set the overall resonance frequency.

For LF designs, for which beam length is large with respect to its thickness dimension, the shear displacements and rotary inertias are not efficient, and the effect of the anchor areas of the resonator on the resonance frequency is not appear. So to obtain the accurate beam resonance frequency for LF μmechanical resonators, *Euler–Bernoulli* method is more suitable to be used here [7]:

$$f_n = \frac{(k_n L_r)^2}{2\pi L_r^2}\sqrt{\frac{EI}{\rho S}},$$

$$\cos(k_n L_r) = \frac{1}{\cosh(k_n L_r)} \tag{1}$$

For fundamental mode $k_1 L = 1.875 \rightarrow$

$$f_{nom} = \frac{1}{2\pi}\sqrt{\frac{k_r}{m_r}} = 1.03\sqrt{\frac{E}{\rho}}\left(\frac{h}{L_r^2}\right) \tag{2}$$

Where the f_{nom} is the resonance frequency without applying any voltage between the resonator beam and its electrode, and $S = W_r h$ is the cross-sectional area of the beam, E is the Young's modulus of the resonator material, ρ is its density, and the geometric dimensions L_r and h are shown in Fig. 1.

When the support beams are designed as described above, the expression for resonance frequency f_o of the beam in Fig. 1 will actually be as:

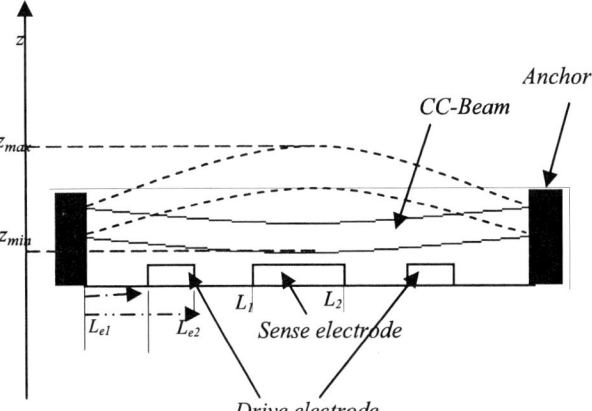

Fig. 2 Two-port clamped-clamped μresonator cross μsectional schematic

$$f_o = f_{nom}\sqrt{1 - \frac{k_e(y)}{k_m(y)}} \tag{3}$$

Where $k_e(y)$, $k_m(y)$ are the electrical and mechanical stiffness of the CC-mixed beam at the location y, and $k_e(y)/k_m(y)$ is a parameter representing the combined electrical-to-mechanical stiffness ratio integrated over the electrode width W_e, and can be expressed as [3]:

$$\left\langle \frac{k_e(y)}{k_m(y)} \right\rangle = \int_{L_{e1}}^{L_{e2}} \frac{dk_e(y')}{k_m(y')} = 2*\int_{L_{e1}}^{L_{e2}} \frac{(V_{pd}^2 \varepsilon_o W_r}{[d(y)]^3 k_m(y)}dy \tag{4}$$

where L_{e1} and L_{e2} are shown in Fig. 2. L_r is the beam length .and $d(y)$ represent the deviation of the beam.

B. Equivalent Circuit Model for Two-Port CC-Beam Resonators

Fig. 3 presents the equivalent circuit used in this work, in which transformers model both electrical and mechanical couplings to and from the resonator, which itself is modeled by a core of LCR circuit with element values corresponding to actual values of mass, stiffness, and damping which is given by:

$$c_r(y) = \frac{k_m(y)}{2\pi f_{nom} Q_{nom}} \qquad (5)$$

where $c_r(y)$ is the damping factor at location y; and Q_{nom} is the quality factor of the two-port-CC beam without applying any voltage between beam and electrode.

The dynamic stiffness of the resonator is: $k_r(y) = k_m(y) - k_e(y)$, where $k_e(y)$ is the electrical spring constant, and $k_m(y)$ is the mechanical spring constant of the resonator without applying any external force on it.

Practically the transformed LCR circuit of the resonator elements values are given in the next equations:

$$L_x = \frac{m_{re}}{\eta^2}, \qquad C_x = \frac{\eta^2}{k_{re}}, \qquad R_x = \frac{c_{re}}{\eta^2} \qquad (6)$$

The subscript e denotes the electrode location at $y = Lr/2$, and R_x is the motional resistance seen a cross the electrical-to-resonator gap at resonance.

In this model, the turn ratios of the sense and drive port depend on the area of each of the sense and drive electrodes, and can be expressed as follow:

$$\eta_s = V_{ps}\left(\frac{\partial C}{\partial z}\right)_{sensor-electorde} \qquad \eta_d = V_{pd}\left(\frac{\partial C}{\partial z}\right)_{drive-electorde} \quad (7)$$

where V_{ps} and V_{pd} are the voltage applied on the sense and drive port respectively. While $\partial C_d/\partial z$ and $\partial C_s/\partial z$ are the change of the capacitance with respect the electrode-to-beam-gap for the driving and sensing respectively, and A_d, A_s are the effective area of both the drive and sense electrodes, assuming $A_d = A_s$ for simplifying the calculations; thus

$$\frac{\partial C_d}{\partial z} \cong \frac{A_d}{D_d^2}, \frac{\partial C_s}{\partial z} \cong \frac{A_s}{D_o^2}, \quad so \quad \eta_s \approx \eta_{d=}\eta \qquad (8)$$

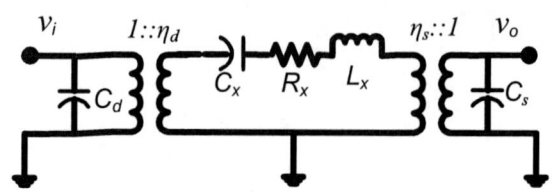

Fig. 3 Alternative demonstration of the equivalent small signal model for a two port micromechanical beam resonator.

C. Pull-In Voltage V_{PI}

When the applied dc-bias voltage V_p is sufficiently large, the device will be failed due to either the breaking of the beam under the large force, which is produced by the high dc voltage, or due to pulling down the resonator beam onto the electrode, and this leads to destruction of the device due to excessive current passing through the now shorted electrode-to-resonator path.

To calculate the pull-in voltage V_{pI}, the resonance frequency is set to be equal to zero in equation (3) and solve for finding V_p; but here unlike previous paper [3], the deviation of the beam is taken in our calculations; so while V_p is changed, the deviation of the beam $d(y)$ will be changed and as a result, the output of equation (4) will be changed, so in order to get the accurate value of the Pull-in-Voltage, a certain closed loop in calculation should be used for each value of V_p to find the deviation $d(y)$, then solve for equation(3)=0.

IV. RESULTS AND ANALYSIS

Micromechanical two-port-CC resonator is modeled, designed and fabricated as detailed in section II and III, using the surface micromachining technology, the sacrificial oxide thickness in this process is 0.9 μm and the Silicon-nitride thickness, beneath the electrodes, is 0.28 μm. The two-port-clamped-clamped beam resonators used in this work is shown in Fig. 4, while the preferred bias, excitation, and sensing circuitry are shown in Fig. 5.

TABLE I Fabricated two-port-clamped-clamped MEMS resonator design parameters, and Performance Summery, using Aluminum as structural material.

Input Parameter	Value			Unit
Beam length, L_r	300*			μm
Beam width, W_r	40*			μm
Electrode width, W_e	60*			μm
Beam thickness, h	3.23*			μm
Electrode-to-resonator Gap, D_o	0.9*			μm
Poisson Ratio ,v	0.36*			......
Young Modulus, E	64*			GPa
Density of Al ,ρ	2700*			Kg/m^3
Output Parameter	^	^^	**	
Resonator Mass m_{re} (y=Lr/2)	33.87			ng
The resonance frequency f_o	~180	179.97	181.4	kHz
The applied dc voltageV_p	2	2	2	volts
The Pull-In Voltage V_{PI}	3	4.5	4	Volts
CCR Quality Factor, Q_{res}	27***	18277#		
R_x		4.913		$M\Omega$
C_x		0.0098		fF
L_x		79413		H
Resonator Stiffness k_{re} (v=Lr/2)		43.3		N/m

*Input for both simulating and experimental,
^^ Calculated by my own software. ^ Measured.
** Simulated by *IntelliSuite*
***Operating under normal pressure. # operating under 20mTorr

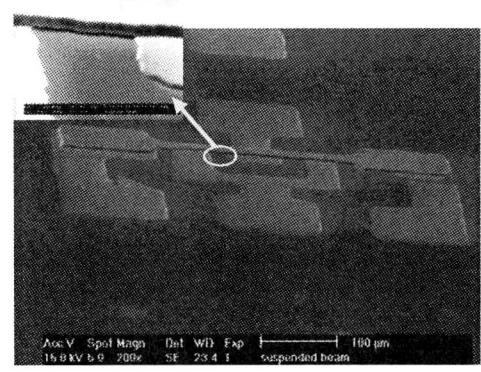

Fig. 4 SEM of two-port clamped-clamped resonator showing the small gap (0.9 μm).

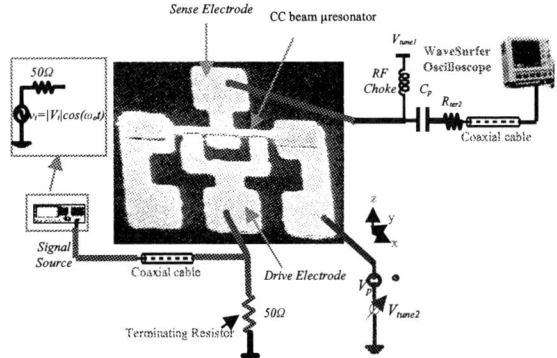

Fig. 5 Perspective view schematic of a two-port-μmechanical CC resonator, along with the preferred bias,excitation, and sensing circuitry.

Table I shows the dimensions of the reported device and its designed, simulated and measured parameters, it is with 180 kHz resonance frequency. There were some variations from the measured and the simulated values which were obtained by *IntelliSuite* and the proposed software. We believe that much of the difference was caused by the residual stresses in the two-port-cc beam structure which was not considered in our theoretical calculation of the designed frequency. Table I, also shows the experimental and theoretically calculated resonator parameters used in Spice simulation. For calculation of the equivalent motional resistance of the resonators, it has been assumed that the polarization voltage is the difference between the dc bias which is applied on the beam and *rms* value of the ac-input at the drive electrode, mathematically; $V_{dc-bias}=(V_p-rms$ value of the ac input).

The measured loaded mechanical quality factor Q_{load} of the two-port-CC beam, operating in atmospheric pressure was around 27. The proposed software was used to calculate the Q_{load} under low pressure of 1 Torr, which was 6072 and 18277 under 20 mTorr; this difference is due to the increased damping caused by the air flow around the beam. Thus, as the device is brought under vacuum, the Q and magnitude of the resonant peak increases, so the resonators work well in air but a vacuum of less than 1 Torr is required for optimum performance [8].

Fig. 6 shows the output of the 180 kHz two-port-CC-beam of (Fig. 4) at the a point near to its resonance frequency, while Fig. 7 shows the output voltage of the MEMS resonator at frequency of 163 and 200 kHz respectively. From the figures we can see that the resonator target (pass) the signal of the frequency near to its resonance frequency only.

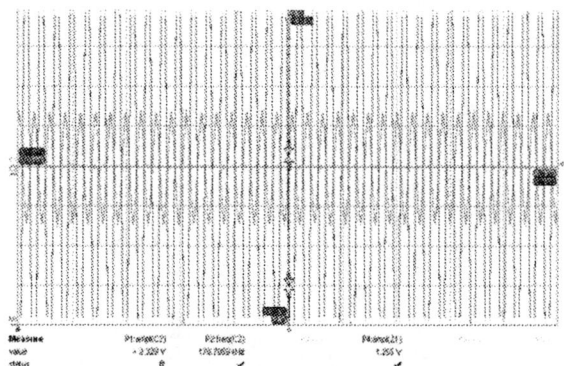

Fig. 6 The output of the ~180 kHz- two-port-clamped-clamped resonator under 2 volts dc-bias

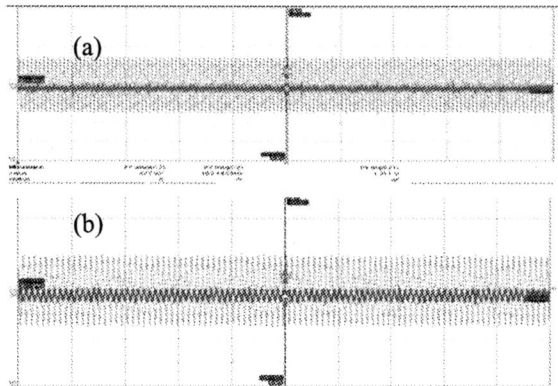

Fig. 7 The output of the ~180 kHz- two-port-clamped-clamped resonator.

While, Fig. 8 shows the Spice simulation results using the equivalent circuit model presented using the proposed software for the fabricated MEMS resonator with resonance frequency of 180 kHz.

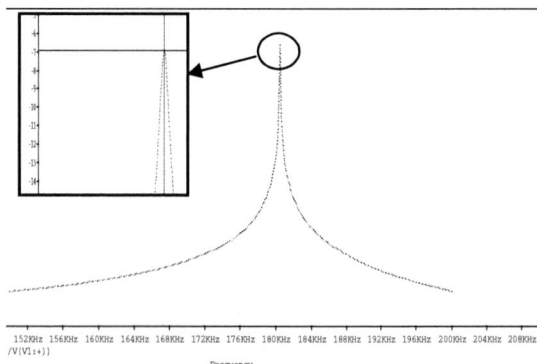

152KHz 156KHz 160KHz 164KHz 168KHz 172KHz 176KHz 180KHz 184KHz 188KHz 192KHz 196KHz 200KHz 204KHz 208KHz
/V(V1:+))
Frequency

Fig. 8 Simulated frequency spectrum for a ~180 kHz two-port-clamped-clamped resonator, using Spice software depending on the electrical model which is obtained by the proposed software.

V. CONCLUSIONS

In this paper, a novel proposed design of two-port-CC MEMS µresonator of resonance

frequency of 180 kHz was developed and achieved in the IMEN Clean Room and the results were very near to the simulated one.

In order to get high Q for the resonator it should be operated under vacuum (less than 1 Torr) in order to decrease the damping factor through the vibration of the beam.

In general, the resonance frequency of this clamped-clamped beam depends upon many factors, including geometry, structural material properties of its materials, stress, surface topography, and the magnitude of the applied dc-bias voltage V_P. The effect of the anchor areas on the resonance frequency was not appear because the dimension of the resonator is large with respect to its thickness.

ACKNOWLEDGEMENT

This work is supported by the Ministry of Science, Technology and Innovation of Malaysia under IRPA program; Development of MEMS Technology for Automotive and RF Applications.

REFERENCES

[1] K. Wang, "VHF Free-Free Beam High-Q Micromechanical Resonators," *J. Microelectromechanical Systems*, pp.347-360 (2000).

[2] A.-C. Wong, H. Ding, and C. T.-C. Nguyen, "Micromechanical mixer+filters," in *IEEE Int. Electron Devices Meeting Tech. Dig.* San Francisco, CA, pp. 471–474 (1998).

[3] J. R. Clark "High-Q HF microelectromechanical filters," *IEEE J. Solid-State Circuits*, pp. 512–526 (2000).

[4] M. Al_Khusheiny, B. Y. Majlis, "Designing and Modeling A Mems If Bandpass Filter," *proceeding of (ICMN'06)*, Malaysia, pp 271-279,2006.

[5] B. Y. Majlis, & M. Al_Khusheiny, "Capacitivly Coupled Band–Pass MEMS Filter Based on Clamped-Clamped Resonators," *Asia-Pacific Conference of Transducers and Micro-Nano Technology—APCOT 2006*, pp. 1-4 (2006).

[6] H. A. C. Tilmans and R. Legtenberg, "Electrostatically driven vacuum encapsulated polysilicon resonators, Part II: Theory and performance," *Sens. Actuators A*, vol. 45, pp. 67–84 (1994).

[7] S. P. Timoshenko, *Vibration Problems in Engineering*, (1990).

[8] B. Piekarski, D. DeVoe, M. Dubey, R. Kaul & J. Conrad, "Surface Micromachined Piezoelectric Resonant Beam Filters," *Sensors and Actuators A: Physical*, vol. 91, no. 3, pp. 313-320(8) (2001).

The Use of Photoluminescence Spectra of TiO$_2$ Nanoparticles Coated With Porphyrin Dye Thin Film for Grading Agarwood Oil

Nurul Huda Yusoff[a], Muhamad Mat Salleh[a] and Muhammad Yahaya[b] and Mat Rasol Awang[c]

[a]Institute of Microengineering and Nanoelectronics,
Universiti Kebangsaan Malaysia, 43600 Bangi, Selangor
[b]School of Applied Physics, Faculty of Science & Technology,
Universiti Kebangsaan Malaysia, 43600 Bangi, Selangor
[c]Malaysia Institute of Nuclear Technology Research (MINT),
43000 Kajang, Selangor, MALAYSIA.
Email: mms@pkrisc.cc.ukm.my

Abstract This paper explores the possibility using nanostructure thin film of TiO$_2$ nanoparticles coated with porphyrin dye based on fluorescence technique to grade agarwood oil. The sensing material was prepared using synthesized of TiO$_2$ nanoparticles colloid is in a sol-gel form. Then the nanoparticles were coated with dye, Iron (III) meso tetraphenyl porphine chloride. The coated nanoparticles were deposited on quartz substrate using self-assembly through dip coating technique. The sensing properties of the thin film toward five grades of agarwood oil were studied using luminescence spectrometer. It was found that the thin film produced different emission spectra peaks for different grades of agarwood oil. Hence the thin film potentially be use as sensing material for grading agarwood oil and others nature product for the future.

I. INTRODUCTION

Agarwood oil is one of the most valuable non-timber products harvest from tropical forest. The oil is extract from the wood of Agar. It can be used as a perfume, an aroma therapy and an essential oil or as an aid for the deepest meditation. There are many grades of agarwood oil, and the highest quality oil is extremely expensive. In fact, agarwood oil is one of the most expensive natural products in the world, with prices ranging from 100 USD to 2000 USD per kilogram depending on the extractive content and the stage of maturity.

Hence, it is very important to have a sensor for grading these oils.

Recently, sensor base on optical sensing technique has received growing interest as an alternative to sensors based on electrical resistance which operates at high temperature. Various techniques have been employed, which include fluorescence [1], UV-VIS absorption [2], refractive index [3] and surface plasmon resonance (SPR) [4]. In this paper, we explore the possibility of using fluorescence technique in developing gas sensor to detect the presence of five grades of agarwood oil. This technique offers several advantages over other methods such as sensitivity, high efficiency and specificity. Fluorescence technique has been sought for huge potential applications in food quality control [5], medical [6] and environmental [7]. The technique also has been used for measuring various gaseous levels, including sulfur dioxide [8], chlorine [9-10] and nitrogen dioxide [11].

The sensing material that we used was TiO$_2$ nanoparticles coated with porphyrin dye. Porphyrin is considerable interest for development of gas sensor system. Porphyrin has shown a high sensitivity and selectivity toward a wide spectrum of organic gases [12]. Their interesting features in the interactions with the gaseous molecules have been widely described elsewhere [12-13]. The used of porphyrin will modify TiO$_2$ nanoparticles structure and enrich its properties as gas sensor. In this research we were interested in developing sensing material that can produce different emission spectra and ease for chemical identification process.

II. EXPERIMENTAL

Materials used for synthesis TiO_2 nanoparticles in the experiment were titanium (IV) ethoxide (TEOT), kalium chloride (KCl) and ethanol. Porphyrin dye used for sensing material was Iron (III) meso tetraphenylporphine chloride purchase from Strem chemical. Poly-L-Lysine (PLL) with $m_w = 70,000$ which obtained from Aldrich was used to charge substrate surface into positive charge.

TiO_2 nanoparticles colloid was synthesized from titanium (IV) ethoxide in ethanol with addition of kalium chloride (KCl) as stabilizer [14] with some modifications. Firstly, ethanol was mixed with KCl and stirred for one hour to get a complete mixture. Then, titanium ethoxide was dropped wisely to the admixture solution in glove box under nitrogen atmosphere and humidity at 15%. Acetylaceton was dropped into the solution at room tempreture and stirred for two hours to form a clear, transparent yellowish solution. TiO_2 nanoparticles coated with porphyrin dye were prepared as follow: The Iron (III) meso tetraphenyl porphine chloride was dissolved in toluene at a concentration of 0.2 mg/ml. Then, porphyrin dye was coated on TiO_2 nanoparticles at ratio 2:1 and stir for one hour at room temperature to obtain a complete mixture of porphyrin dye colloid and TiO_2 nanoparticles colloid.

TiO_2 nanoparticles coated with porphyrin dye were self-assembled prepared on quartz substrate. The clean quartz substrate was immersed in polycation solution for 30 minutes to charge the surface into positive charge. Using KSV Dip coating system, the substrate was dipped into the TiO_2 nanoparticles coated with dye colloid. The substrate was kept in the colloid for 15 minutes and lifted up at a constant speed of 15 mm/min. At the first stage, one bilayer of TiO_2 nanoparticles coated with dye thin film was formed on the substrate surface. The prepared thin film was annealed at 200°C for one hour. The annealing process is to reduce recovery time of the thin film after being exposed to agarwood oil. The formation of nanostructure thin film was examined using scanning electron microscopy (SEM) and atomic force microscopy (AFM).

The sensing properties of TiO_2 nanoparticles coated with porphyrin dye thin film toward five grades of agarwood oil were studied using Perkin Elmer luminescence spectrometer. The thin film is place inside sample holder of analysis chamber. Then, a small piece of filter paper is dipped to the agarwood oil liquid and place near the thin film. The agarwood oil molecules will interacted with the thin film surface and produce different photoluminescence spectrum. The peak position of this spectrum was used to identify the grade of agarwood oil. This procedure was repeated until all grades of agarwood oil were analyzed.

III. RESULTS & DISCUSSION

Thin film of TiO_2 nanoparticles coated with porphyrin dye was deposited onto quartz substrate using self assembly technique. Fig.1 shows the scanning electron microscopy (SEM) image of self assembled thin film. The thin film surface was found to be constituted by homogeneous nanoparticles with an average diameter of 19.88 nm.

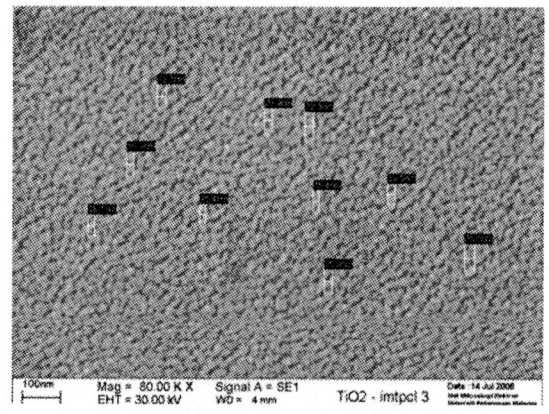

Fig. 1 SEM image of TiO_2 nanoparticles coated with porphyrin dye thin film

Fig. 2 shows AFM image of TiO_2 nanoparticles coated with porphyrin dye thin film. The film surface arranged of a large number of hills. The hilly structure on the surface may suitable for some gaseous molecules to be absorbed.

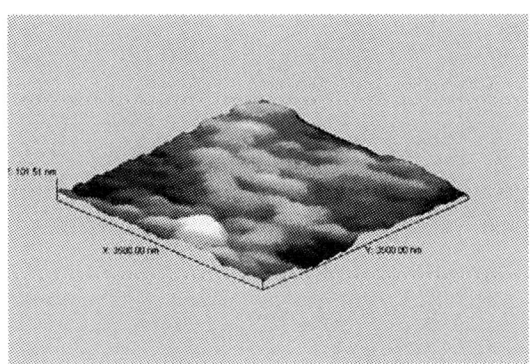

Fig. 2 A three-dimensional image of TiO$_2$ nanoparticles coated with porphyrin dye thin film.

The agarwood oil was graded with label from C1 to C5. Table 1 summarizes the chemical compounds contained in each agarwood oil grades. The chemicals were analyzed using chromatography.

When exposed to five grades of agarwood oil, the sensing film undergoes change in their fluorescence emission spectra. Fig. 3 shows emission spectra of TiO$_2$ nanoparticles coated with porphyrin dye thin film in presence of air and five grades of agarwood oil that can be determined from their distinct emission peaks. The changes emission peak of the thin film from the original either to red shift and blue shift depended on changes in electronic states between the ground state and excited state of the porphyrin molecules. This change could be due to the interaction of porphyrin molecules with agarwood oil molecules that causes rearrangement of the electrical dipole in the porphyrin compounds of the thin film [12]. A certain circumstance may be created by this activity is the changes on the energy gap between highest occupied molecular orbital (HUMO) and lowest unoccupied molecular orbital (LUMO). Thin film before interacting with agarwood oil molecules possesses a gap between these two orbital that determines its initial emission upon the excitation light due to the excited transition of the electrons to the ground state. When an excitation light source absorbed by the thin film molecules which interact with gas molecules, it may emit light at different wavelength peaks.

The sensing experiments using the same thin film were repeated for many times. It was found that no significant different of the results. This means that the sensor has repeatability

property and suitable to be use as sensing material for grading of agarwood oil.

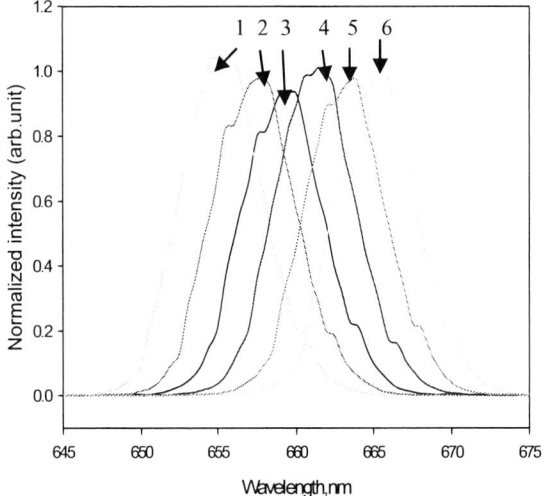

Fig. 3 The emission spectra of TiO$_2$ coated with porphyrin dye thin film in present of air for (3) and agarwood oil label as C1 for (1), C2 for (2), C4 for (4), C5 for (5) and C3 for (6).

Table 1 Agarwood oil grades from chromatography analysis

Gaharu wood grades	CAS No	Chemical Name
C1		
	99.62.7	1,3-bis (1-methylethyl)-Benzene
	17066.67.0	Beta-Salinenes or decahydro-4a-methyl-1-methylene-7-Naphthalene
	17066.67.0	Alpha-Humulene or 2,6,6,9-tetramethyl-1, 4,8-Cycloundecatriene
	6753.98.6	Alpha-Humulene or 2,6,6,9-tetramethyl-1, 4,8-Cycloundecatriene
	50407.04.0	1-methoxy-2-methyl-1-phenyl-1-propene
	489.86.1	Guaiol
	1209.71.8	2-Naphthalenemethanol, 1,2,3,4,4a, 5,6,7-octahydro-.alpha.,alpha.,4
	. .0	Murolan-3, 9(11)-diene-10-peroxy
	127.47.9	Retinol, acetate or Retinol, acetate, all trans or All-trans-retin
	74744.51.7	4,Hexadecen-6-yne, (E)

ID	Compound
1139.30.6	Caryophylene oxide or 5-Oxytricyclo [8.2.0.0(4,6)-] dodecane, 4,12,1
77171.55.2	(.)-Caryophyllene oxide or (.)-5-Oxatricyclo [8.2.0.0(4,6)] dodecane
77171.55.2	(.)-Spathulenol or 1H-Cycloprop e azulen-7-ol, decahydro-1, 1,7-trim
6753.98.6	Alpha-Caryophyllene
C2	
3691.12.1	Azulene, 1,2,3,4,5,6,7,8a-octahydro-1,4-dimethyl-7-(1-methylethenyl)
3691.11.0	Azulene, 1,2,3,4,5,6,7,8a-octahydro-1,4-dimethyl-7-(1-methylethenyl)
639.99.6	Cyclohexanemethanol, 4-ethenyl-.alpha.,.alpha.,4-trimethyl-3-(1-met
489.86.1	Guaiol
1209.71.8	2-Naphthalenemethanol, 1,2,3,4,4a,5,6,7-octahydro-.alpha.,alpha.,4
473.16.5	2-Naphthalenemethanol, 1,2,3,4,4a,5,6,8a-octahydro-.alpha.,alpha.,
3691.11.0	Delta-Guaiene or Azulene, 1,2,3,4,5,6,7,8,8a-octahydro-1,4-dimethyl
55700.81.7	3-(acetyloxy)-(3.beta) Lanost-24-ene-11-one
. .0	17-(1,5-Dimethylhexyl)-10,13-dimethyl-3-styrylhexadecahydrocyclopen
54630.68.1	17-hydroxy-2-methyl-,(2.beta., 5.beta., 17.beta)-Androstan-3-one
1139.30.6	(.)-Caryophyllene oxide or (.)-5-Oxatricyclo [8.2.0.0(4,6)] dodecane
1139.30.6	Azulene, 1,2,3,4,5,6,7,8-octahydro-1,4-dimethyl-7-(1-methylethenyl)
22567.17.5	Azulene, 1,2,3,4,5,6,7,8a-octahydro-1,4-dimethyl-7-(1-methylethenyl)
54630.68.1	1H-Cycloprop[e] azule, 1a, 2,3,5,6,7,7a, 7b-octahydro-1, 1,4,7-tetram
C3	
639.99.6	Cyclohexanemethanol, 4-ethenyl-.alpha.,.alpha.,4-trimethyl-3-(1-met
35845.67.1	2,2,8,8-tetramethyl-3,6-Nonadien-5-one orNonadiyn-5-one,2,2
127.47.9	Retinol, acetate or Retinol, acetate, all trans or All-trans-retin
1460.73.7	Agaruspirol or Agarospirol
6753.98.6	Alpha-Caryophyllene
502.99.8	3,7-dimethyl-1, 3,7-Octatriene or Ocimene or 2,6-Dimethyl-1, 5,7.oc
1209.71.8	2-Naphthalenemethanol, 1,2,3,4,4a, 5,6,7-octahydro-.alpha.,alpha.,4
489.86.1	Guaiol or 5-Azulenemethanol, 1,2,3,4,5,6,7,8-octahydro-.alpha.,alpha
C4	
. .0	2-isopropenyl-(+)-2-Carene
. .0	2-isoprppenyl-(.)-2-Carene
639.99.6	Elemol
639.99.6	Cyclohexanemethanol, 4-ethenyl-.alpha.,.alpha.,4-trimethyl-3-(1-met
489.86.1	Guaiol or 5-Azulenemethanol, 1,2,3,4,5,6,7,8-octahydro-.alpha.,alpha
1139.30.6	Caryophyllene oxide
51371.47.2	Globulol
127.47.9	Retinol, acetate or Retinol, acetate, all trans or All-trans-retin
35845.67.1	2,2,8,8-tetramethyl-3, 6-Nonadien-5-one orNonadiyn-5-one, 2,2
22417.84.1	3-(acetyloxy)-(3.beta) Lanost-24-ene-11-one
53126.63.9	Satratoxin
1209.71.8	2-Naphthalenemethanol, 1,2,3,4,4a, 5,6,7-octahydro-.alpha.,alpha.,4
C5	
1139.30.6	(.)-Caryophyllene oxide or (.)-5-Oxatricyclo [8.2.0.0(4,6)] dodecane
489.86.1	Guaiol
1209.71.8	2-Naphthalenemethanol, 1,2,3,4,4a, 5,6,7-octahydro-.alpha.,alpha.,4
21657.90.9	Hedycaryol
639.99.6	Elemol
88588.48.1	8S,13-Cedran-diol
51371.47.2	Globulol
127.47.9	Retinol, acetate or Retinol, acetate, all trans or All-trans-retin
. .0	2,methyl-9-(propo-1-en-3-ol-2-yl)-Bicyclo[4.4.0]dec-2-ene-4-ol
6753.98.6	Alpha-Humulene or 2,6,6,9-tetramethyl-1, 4,8-Cycloundecatriene
. .0	Spiro(3,17-diacetoxyandrostane)[2,2'](1',3'-

	dithiane)
77171.55.2	(.)-Spathulenol or 1H-Cycloprop e azulen-7-ol, decahydro-1, 1,7-trim
17066.67.0	Beta-Salinenes or decahydro-4a-methyl-1-methylene-7-Naphthalene
56588.33.1	Cholestane-3, 6,7-triol,(3.beta.,5.alpha.,6.beta.,7.beta.)

IV. CONCLUSION

Thin film of TiO_2 nanoparticles coated with porphyrin dye has ability to differentiate five grades of agarwood oil base on fluorescence spectra. Hence the thin film potentially be use as sensing material for grading agarwood oil and others nature product for the future.

ACKNOWLEDGEMENT

This project has been carried out with support of Ministry of Science, Technology and Innovation of Malaysia under the IRPA 03-02-02-0020-SR0003/07-7-06 grant.

REFERENCES

[1] M.G. Boron, R. Narayanaswamy and S.C. Thorpe, "Luminescence porphyrin thin film for NOX sensing," Sens Actuators B 11, p.195-199 (1993).

[2] T. Tanaka, A. Guilleux, T. Ohyama, Y.Y. Maruo and T.Hayashi, "A ppb-level NO2 gas sensor using coloration reaction in porous glass," Sens. Actuator B 56, p. 247-253 (1999).

[3] A. Brandenburg, R. Edelhauser and F. Hutter, "Integrated optical gas sensor using organically modified silicates as sensitive films," Sens. Actuator B 11, p. 361-374 (1993).

[4] P.S. Vukusic and J.R. Samble, "Cobalt phthalocyanine as basis for the optical sensing of nitrogen dioxide using surface plasmon resonance," Thin Solid Film 221, p. 311-317 (1992).

[5] M.S. Kim, A.M. Lefcourt, and Y. R. Chen, "Multispectral fluorescence imaging technique for nondestructive food safety inspection," Proceedings of SPIE - The International Society for Optical Engineering, vol. 5271, p. 62-72 (2004).

[6] M.Z.Krecicka, T. Krecicki, M. Fraczek, E.B. Pawlik and T.Zatonski, "Autofluorescence laryngoscopy in the diagnosis of laryngeal cancer-early results," Otolaryngologia polska. The Polish otolaryngology, vol. 59, issue 2, p. 195-199 (2005).

[7] S.W. Hong, K.H. Kim, J. Huh, C. H. Ahn, and W.H. Jo, "Design and synthesis of a new pH sensitive polymeric sensor using fluorescence resonance energy transfer," Chemistry of Materials, vol 17, issue 25, p. 6213-6215 (2005).

[8] T.M.A. Razek, M. J. Miller, S.S.M. Hassan, and M. A. Arnold, "Optical sensor for sulfur dioxide based on fluorescence quenching," Talanta 50, p. 491–498 (1999).

[9] T.E. Brook, and R. Narayanaswamy, "Immobilization of ruthenium tris-biphyridyl complex for chlorine gas detection," Sensors and Actuators B 38-39, p.195-201 (1997).

[10] M.G. Baron, R. Narayanaswamy, and S. C. Thorpe, "A kineto-optical method for the determination of chlorine gas," Sensors and Actuators B 29, p.358-362 (1995).

[11] D. Y. Sasaki, S. Singh, J. D. Cox, and P.I. Pohl, "Fluorescence detection of nitrogen dioxide with perylene/PMMA thin films," Sensors and Actuators B 72, p.51-55 (2001).

[12] Akrajas, M. M. Salleh and M. Yahaya, "Enriching the selectivity of metalloporphyrins chemicals sensors by means of optical technique," Sensor and Actuators B, vol. 85, p.191-196 (2002).

[13] M.M. Salleh, Akrajas and M. Yahaya, "Optical sensing of volatile organic compounds using metalloporphyrin complexes Langmuir-Boldget films," Intl. J. of Nonlin. Sci & Num. Simul, vol.3, p.461-464 (2002).

[14] S. E. Assmann, J. Widoniak and G. Maret, "Synthesis and characterization of porous and nonporous monodisperse colloid TiO_2 particles," Chem. Mater 16, p. 6-11 (2004).

Study on the Effect of Q-Switched Nd:Yag Laser Interaction with Al in Variable Magnetic Field

Rabia Qindeel, Noriah Bidin and Yaacob Mat Daud

Physics Department, Faculty of Science
Universiti Teknologi Malaysia, 81310 Skudai, Johor, Malaysia.
Tel: +607-5534096   Fax: +607-5566162
E-mail: plasma_qindeel@yahoo.com, noriah@dfiz2.fs.utm.my, ymd@dfiz2.fs.utm.my

Abstract **Dynamic and confinement of laser generated Aluminum plasma plumes expanding across a variable magnetic field have been investigated. A Q-switched neodymium-doped yttrium aluminum garnet laser with 1064 nm and 8 ns pulse width was used to generate plasma that was allowed to expand across 0.1 – 0.8 T magnetic field. The expansion of laser-produced plasma is characterized using constant laser power and in variable magnetic field. A Pulnix CCD video camera was used to visualize and record the activities in the focal region. The horizontal and vertical expansions of plasma plume were measured by using Matrox Inspector 2.1 software. The results showed that the plasma is expanding vertically and horizontally with respect to the magnetic field. The Al plasma interaction produced spectrum in the range of UV to visible light.**

I. INTRODUCTION

In the field of inertial fusion, confinement of expanding target plasma using a magnetic field offers a potential mean to slow high-energy particles before they implant in surrounding structures. The presence of magnetic field during the expansion of laser-produced plasma may initiate several interesting physical phenomena [1-10], including conversion of plasma thermal energy into kinetic energy, plume confinement, ion acceleration, emission enhancement and plasma instabilities. Pulsed laser induced plasma has very short temporal existence and transient in its nature, with a fast evolution of the characteristic parameters that heavily dependent on irradiation conditions such as incident laser intensity, irradiation spot size, ambient gas composition and pressure. These parameters vary drastically with axial or radial distance from the target surface under the same irradiation conditions [11].

High-speed photographic recording of expanding laser plasmas using framing image-converter cameras is a well-established technique [12-13]. More recently, the technique has been refined by the introduction of CCD (Charged Coupled Device) cameras that allow direct digital recording, thus greatly facilitating computer processing of the images as well as providing improved sensitivity. Such imaging devices have been widely used in the area of pulsed laser plasma plume formation. Despite the technique's ease of implementation, the retrieval of quantitative information characterizing the plume expansion from the imaged is usually complex. By narrowing down the spectral range of the imaged light to that of a known emission line, it becomes possible to track the evolution of the corresponding excited state. This can be carried out by placing a tuned interference filter in front of the camera, in which the measured intensities are integrated over the line profile [14].

In the present work high-speed imaging technique is used to capture the image of plasma plume expansion. The production of plasma in magnetic field obtained by irradiating Al target with a pulsed Nd:YAG laser is investigated and discussed.

II. THEORY

Magnetism is the fundamental force that determines the character, motion and shape of ionized matter (plasma). From the equation of motion,

$$F = m\frac{dv}{dt} = q \ (v \times B) \qquad (1)$$

one can calculate that, in the presence of magnetic field B, a charge particle q is accelerated in the direction perpendicular to both velocity, v and B. This forms a circular motion about an imaginative guiding centre of the particle. The sense of rotation is opposite for positive and negative particles when viewed in the direction of B. The electron producing auroras are thus rotating counter clockwise when viewed from the ground in the northern hemisphere. They are also in helicoidal trajectories because of additional velocity component parallel to B. A single fluid magnetohydrodynamic equation for generalized form of Ohm"s law [15] in fully ionized plasma can be written as

$$E + (V \times B) = \eta J + \frac{(J \times B)}{en_e} \qquad (2)$$

where E and B are the electric and magnetic field in the plasma, J the charge current density, η the resistivity, n_e the electron density, and V the mass flow velocity. During one-dimensional flow in the absence of current, one can write :

$$E_y = V_x . B_z. \qquad (3)$$

It indicates that during the expansion of plasma across the magnetic field, there will be a charge separation due to Lorentz force in the direction perpendicular to V and B. In the presence of this electric field in the plasma, there will be a drift velocity in the direction perpendicular to E as well as B, and as a result there will be a current J in the plasma plume. Direction of this current will be opposite on the front and inside of the plasma. This current will generate a $J \ x \ B$ force in the direction of plasma expansion near the front where as it will be in opposite direction inside the plasma. This $J \ x \ B$ force will contribute in generating fast particles from the plasma whereas the bulk of the plasma will be decelerated [16].

III. METHODOLOGY

A Q-switched Nd:YAG laser model HY200 Lumonics was used as a source of energy. The fundamental wavelength of the laser is 1064 nm and 8 ns pulse duration. The IR laser was focused by using convex lens with focal length of 200 mm. The focused beam was targeted on Aluminum

material. The dimension of the rectangular Al block was 1 cm x 1.4 cm. The target material is placed ~1 cm at the pole edges which have diameter 75 mm creating variable magnetic field along the plume expansion direction. The separation between the two poles edges of magnetic fields was kept at a distance of 1.5 cm. The magnetic field was aligned vertically with respect to the in coming beam and could be verified in the range of 0.1 T – 0.8 T. It means that the IR laser was focused perpendicular to the magnetic field. The laser was operated with frequency of 10 Hz, during alignment. He-Ne laser was coaxial with IR laser for an ease of alignment. The IR laser can be triggered externally using home made trigger control unit. The external trigger unit (ETU) was designed to synchronize the Nd:YAG laser and the camera. The activities at the focal region were visualized by using a Pulnix TM – 6EX CCD video camera. The camera was interfaced to the personal computer. The event of the plasma formation was captured and analyzed via the software of Matrox version 2.1 software. The energy of the laser beam was monitored by using photodiode digitized on the power meter. The whole experimental setup is depicted in Figure 1.

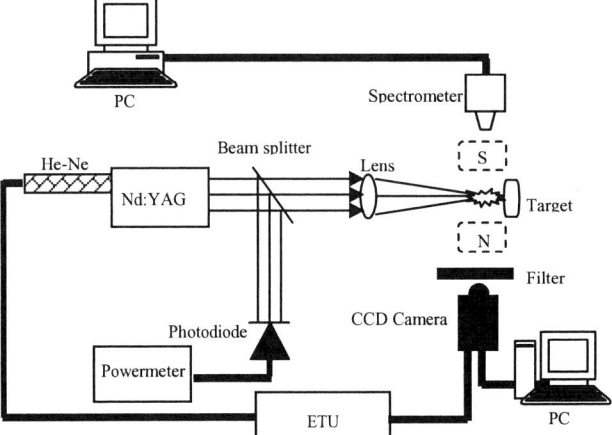

Figure 1: Schematic diagram of plasma plume in magnetic field

IV. RESULTS AND DICSUSSION

When IR laser was focused on an Al target by 200 mm focal length lens, plasma was generated. The plasma was formed on the target surface at atmospheric pressure and room temperature. The expansion of plasma formation was studied via changing the magnetic field. CCD camera was used to grab the images. Typical images of the plasma plume on target sequenced in the presence of the magnetic field recorded after the breakdown are shown in Figure 2.

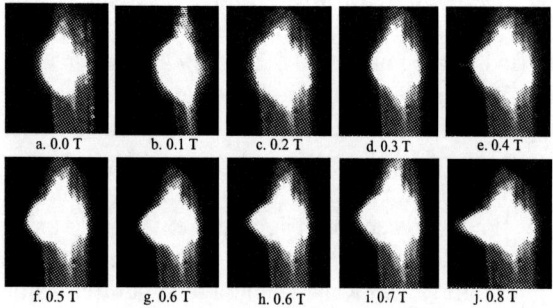

Figure 2: Aluminum plasma plume expansion in variable magnetic field 0.1 T – 0.8 T and without magnetic field (a). Magnification factor of 10x.

Two types of plasma shape are observed from the images of Figure 2. First shape, the plasma is expanding horizontally to the laser beam axis and the other, plasma is expanded vertical to the laser beam. Considering horizontal plasma expansion; with laser energy kept constant at 125 mJ any plasma expansion is not due to energy delivered to focal region. Since the plasma is generated in the magnetic region, the force of motion or any related physical phenomenon subjected to be due to the field. Initially the hemispherical plasma was formed with respect to the target surface. As the magnetic field increases, the plasma is expanded elongated towards the incoming beam. When laser was focused on the target, the laser beam evaporates and ionizes the material, and then plasma plume is formed. Initially the laser intensity exceeds the ablation threshold of the target; the laser beam evaporates and ionizes material, creating a plasma plume above the material surface. Initially the atoms, molecules and ions undergo collisions in the high-density region near the target forming the so-called Knudsen layer, creating a highly directional expansion perpendicular to the target. Ions in the expanding plasma may be accelerated as high as few hundred eV. The existing magnetic field enhances the acceleration of particle removal hence the plasma can move faster and further away. As a result the plasma expands protruding or elongated along the beam axis, subsequently expanding in horizontal part.

Now let us consider the vertically expanding plasma. We know that magnetic field inhibits the motion of charged particles, which are forced, to spiral around field lines. Consequently, some of the charged particles move along the field lines. This means that magnetic field can impose structure on plasma. They can also insulate a region of plasma, as conduction across field lines is very poor. As a result, we observed that some parts of the plasma expand in vertical position. Both of these plasma expansions were quantified in terms of area. The collected data are plotted as area of plasma versus magnetic field as illustrated in Figure 3.

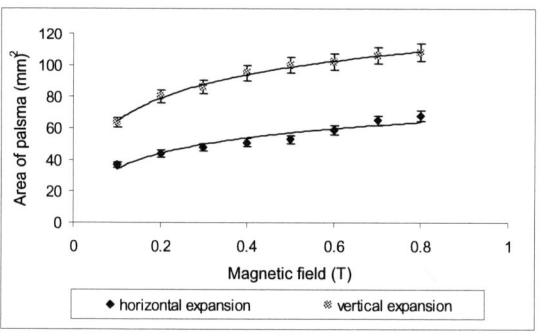

Figure 3: Area of plasma upon magnetic filed

The area of plasma was measured in terms of horizontal and vertical expansion of plasma in magnetic field. In Figure 3 the top curve is represents the vertical expansion of plasma along the magnetic field while the bottom curve represents the horizontal expansion of plasma. The plasma expands rapidly thus creating relatively big as well a large area of plasma. Strong laser plasma interaction creates an additional high pressure, kinetic energy region fuelling further expansion of the plume. Both curves indicate that the area of the plasma gradually increase with the magnetic field. The experimental results show the magnetic field lines have significant influence on the formation of plasma. Vertical plasma is found greater than horizontal plasma, which is in contrast with the plasma formation without magnetic field. However, the magnetic field used in this expansion is not strong enough to totally change the structure of plasma. The plasma plume without magnetic field is shown in Figure 2a.

Beside area of plasma plume, the plasma interaction with Al was also observed by spectrum analyzer. Ablations of Al create an intensely luminous plume that expands normal to the target surface. The typical spectrum produced due to the plasma interaction with Al is shown in Figure 4. In Figure 4, the emission bands of Al plasma are produced in the range of 360 to 600 nm with majority at 390 nm and 400 nm. Maximum intensity of Al plasma is in the range of 70-80 and 40-50.

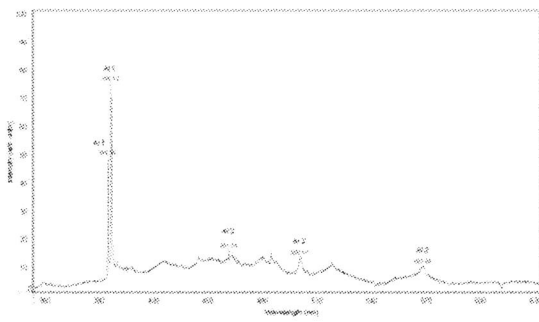

.

Figure 4: Spectrum of Al plasma beam surface The

The intensity of Al-1 lines are more pronounced between 394.35 nm and 396.12 nm, while Al-2 461.14 nm, 500.17 nm and 567.84 nm have peaked at relatively low intensity. This intensity corresponds to the high ionization level for Al-1 relative to Al-2. The energy of photons at high intensity are 8.38 eV and 8.35 eV while at low intensity they are 7.17 eV, 6.61 eV and 5.82 eV respectively. The plasma beams pose the spectrum of UV to visible light. Surrounding the plasma fume in air cooled and confinement enhances electron collision and excitation hence the invisible UV emission from the excited states is increased in intensity. On the other hand, at longer wavelength energy of photon is decreased, the intensity is low and the radiations are in visible region (green light).

Let us observe the effect on the target after interaction. Aluminum target used have free electrons behaving like gas of electrons inside fixed metal ions. When laser radiation falls on metal, most of the energy is absorbed by free electrons. Electrons convert their energy by collision that cause of production of photons (lattice vibrations quantized particles) or thermal energy into the material. The absorbed thermal

energy melted and ablated out the target material. As a result deep dent was observed. The example of exposed materials after examined under photomicroscope is depicted in Figure 5.

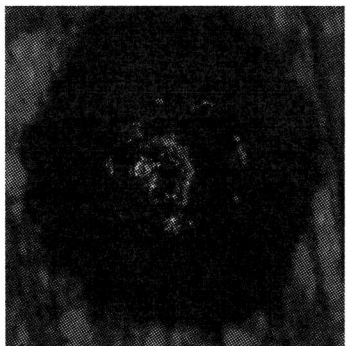

Figure 5: Damaged on Al target after exposure; Magnification factor of 5x.

Almost circular damage with 1.211 mm diameter is observed after plasma interaction with Al. The damage contains dark spot, which indicates the depth. The higher amount of removal particles leaves a dent with a deepest part at the center. The shape and roughness of damage surface is due to material properties when irradiate by high power laser.

V. CONCLUSION

When a Q-switched Nd:YAG laser was focused on Al target, plasma was generated. The dynamic expansion of plasma was studies based on variable magnetic field. The plasma was expanded in two directions that is vertically and horizontally. Each expansion is due to the motion of charge in-lined with the field and Lorenz force that accelerated the charge to move faster and further away from the target. Both expansions are found gradually increased with the magnetic field. The luminosity of the plasma plume created spectrum in the range of UV to visible light with majority of beams is in bluish UV region of 390 nm. The result of particle removal from surface of the target after interaction with laser plasma is circular dent of diameter 1.211 mm, which is deepest at the center.

ACKNOWLEGEMENT

The authors would like to express their special thanks to Prof. M. Asghar Hashmi from Islamia

University Bahawalpur, Pakistan and Dr Zuhairi Bin Ibrahim from UTM for their suggestions and discussions, the government of Malaysia for supporting this project through IRPA grant and to UTM for supporting the progress of the project.

REFERENCE

[1]   S. S. Harilal, B. O'Shay, M. S. Tillack, C. V. Bindhu and F. Najmabadi (2004). "Confinement and dynamics of laser-produced plasma expanding across a transverse magnetic field" Physical Review E **69,** 026413.

[2]   B.H. Ripin, E. A. McLean, C. K. Manka, C. Pawley, J. A. Stamper, T. A. Peyser, A. N. Mostovych, J. Grun, A. B. Hassam, and J. Huba, (1987). Phys. Rev. Lett. **59,** 2299

[3]   S. Okada, K. Sato, and T. Sekiguchi, (1981). Jpn. J. Appl. Phys. **20,** 157.

[4]   L. Dirnberger, P. E. Dyer, S. R. Farrar, and P. H. Key, (1994). Appl. Phys. A: Solids Surf. **59,** 311.

[5]   D. W. Koopman, (1976). Phys. Fluids **19,** 670.

[6]   M. Van Zeeland, W. Gekelman, , S. Vincena and G. Dimonte, (2001). Phys. Rev. Lett. **87,** 105001

[7]   S. Sudo, K. N. Sato and T. Sekiguchi (1978). J. Phys. D **11,** 389.

[8]   R. G. Tuckfield and F. Schwirzke, (1969). Plasma Phys. **11,** 11.

[9]   G. Jellison and C. R. Parsons, (1981). Phys. Fluids **24,** 1787.

[10]   F. S. Tsung, G. J. Morales, and J. N. Leboeuf, (2003). Phys. Rev. Lett. **90,** 055004

[11]   S. S. Harilal, C. V. Bindhu, C. Riju, Issac, V. P. N. Nampoori, and C. P. G. Vallabhan (1997). "Electron density and temperature measurement in laser produced carbon plasma" J. Appl. Phys. **82** (5) 2140 – 2146.

[12]   J. F. Ready, Effects of High Power Laser Radiation, Academic Press, Orlando, (1971).

[13]   T. P. Hughes. Plasma and laser light, Adam Hilger, London, (1975).

[14]   Wlliam Whitty, and Jean-Paul Mosnier. Appl. Surf. Science **127-129** 1035-1040. (1998).

[15]   F. F. Chen. (1974). "Introduction to plasma physics and controlled fusion" New York: Plenum Press.

[16]   Neogi, A. & Thareja R. K. (1999). Physics of plasma **6,** 365.

OXADM Multiplex Protection Scheme for Bidirectional Path Switched Ring

Muhammad Syuhaimi Bin Ab-Rahman, Hayati Hussin, Abang Annuar Ehsan and Sahbudin Shaari,
Member, IEEE, *Member*, SPIE

Photonic Technology Laboratory,
Institute of Microengineering and Nanoelectronics,
Universiti Kebangsaan Malaysia,
43600 UKM Bangi,
Selangor, MALAYSIA
Email: syuhaimi@vlsi.eng.ukm.my

Abstract—This paper describes our recent approach toward advanced optical layer networks through the development of optical cross add and drop multiplexing (OXADM). The OXADM node focuses on providing functionally such as transport, multiplexing, routing, supervision, terminating and survivability in optical layer with ring and mesh topologies. OXADM provide survivability through restoration against failures by means of multiplex protection for BPSR configuration. Restoration schemes in OXADM node are more reliable and more efficient compare to conventional OADM and OXC device due to internally routing mechanisms and no 'drop and re-add function' is needed to route the signal to alternative path or processed the signals in both electronic and optical domain as practice in conventional nodes today (1). No difference in output power and BER as compare to pass through signal. The network has undergone performance test at 2.5 Gbps (OC-48) with BER less than 10^{-9}.

Index Term – OXADM, survivability, multiplex protection, BPSR

I. INTRODUCTION

Providing recovery scheme at the optical layer becomes inherently attractive as the network throughput increases. Reconfigurable nodes support recovery actions as protection switching in the optical domain. The first step towards a survivable optical layer has been the use of WDM rings. These rings can be divided into unidirectional and bidirectional signal flows, the former using dedicated protection and the latter using shared protection (3). Linear and multiplex protection schemes are designated for dedicated protection while ring protection is designated for shared protection.

In path switching, restoration of traffic is handled by the source node and the destination node. A Unidirectional Path Switched Ring (UPSR) that uses dedicated protection is referred to as a Dedicated Protection Ring (DPRing). A two-fiber DPRing uses one fiber as the working fiber and the other fiber as the protection fiber (5)(6). To increase the capacity of transmission, the protection fiber can also be used as working line that is employed in OXADM ring network (3). The ring is referred as Bidirectional Path Switched Ring (BPSR).

II. LINEAR AND MULTIPLEX PROTECTION

OXADM provides a linear protection scheme to implement the dedicated protection in unidirectional path switched ring network (UPSR). In path switching, restoration of traffic is handled by the source node and the destination node. Dedicated protection normally activates when one of the transmission line breakdown. When a link failure occurs within the ring, the signal will be switched to the alternative route as illustrated in Figure 2 below. The restoration is significant for ensuring signal flows continuously [3][4]. The device is compatible with types of link usage 1:1. The accumulation feature will support the shared protection to be performed in case of two different set of wavelengths going to east and west links. The restoration mechanism in BLSR is called as multiplex protection. Figure 1 shows the OXADM restoration schemes that are implemented in UPSR and BPSR.

Figure 2 depicted the activation of dedication protection when failure occurs between Node 2 and Node 3. The affected node will switch the signals to protection route. The switching performed within the optical layer will be able to achieve high speed restoration against the failure/degradation of cables, fibers, and optical amplifiers. It is important to avoid huge losses of data and great influence upon a large number of users over a wide area.

Figure 3 and 4 depicted the output power and BER performance at 10 nodes with 70 km span for

bypass and dedicated protection condition. No changes in output power and BER performance were observed which shows no degradation in the BER, as confirmed by comparison of these two simulation results of the bypass and multiplex protection states. The degradation of output power after node 5 is observed due to the not compensating between the losses and amplifiers gain. The gain of first pre amplifier and post amplifier are 3 dB and 22 dB respectively, then 7 dB and 21 dB values are applied after node 5. This will lead to the degradation of the received power at every node.

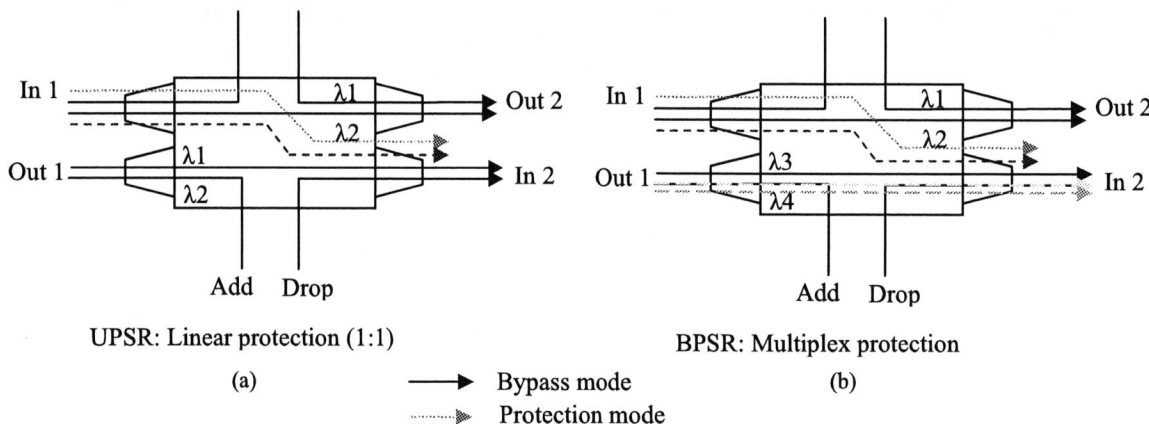

Fig. 1 OXADM restoration strategies: (a) linear protection mechanism (b) multiplex protection mechanism.

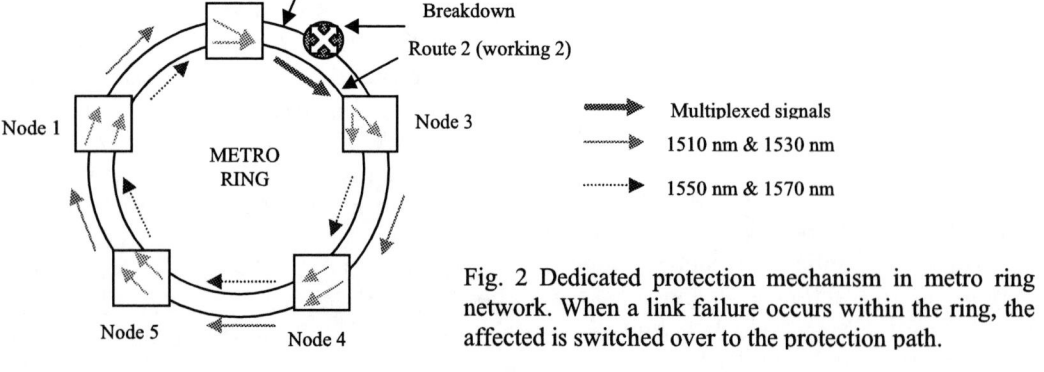

Fig. 2 Dedicated protection mechanism in metro ring network. When a link failure occurs within the ring, the affected is switched over to the protection path.

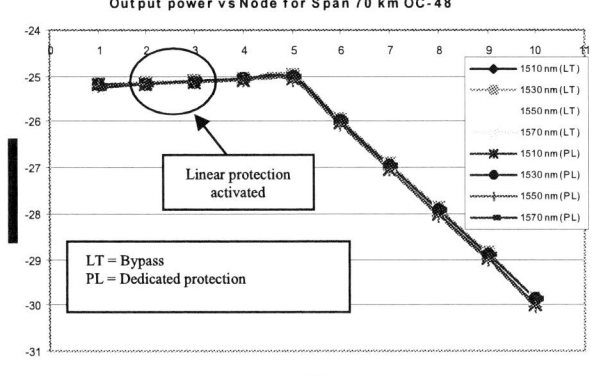

Fig. 3 The output power at 10 nodes with 70 km span for bypass and dedicated protection mechanisms which occurs in between node 2 and

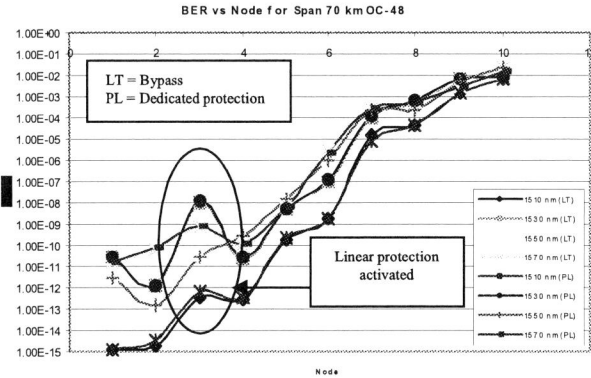

Fig. 4 The BER performance at 10 nodes with 70 km span for bypass and dedicated protection mechanisms which occurs in between node 2 and node 3.

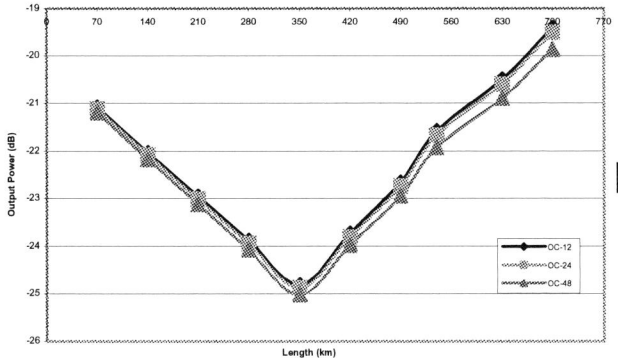

Fig.. 5 The output power at 10 nodes with multiplex protection mechanism which occurs in between node 2 and node 3 at three different bit rates.

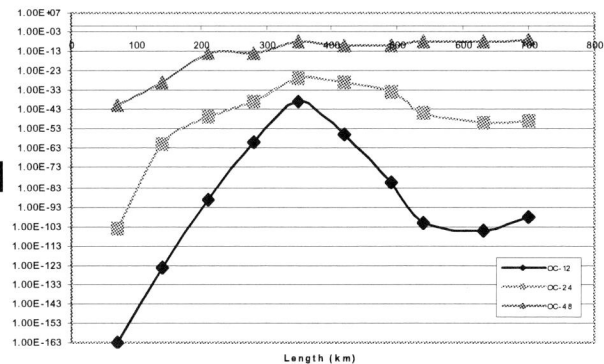

Fig. 6 The BER performance at 10 nodes with multiplex protection mechanism which occurs in between node 2 and node 3 at three different bit rates.

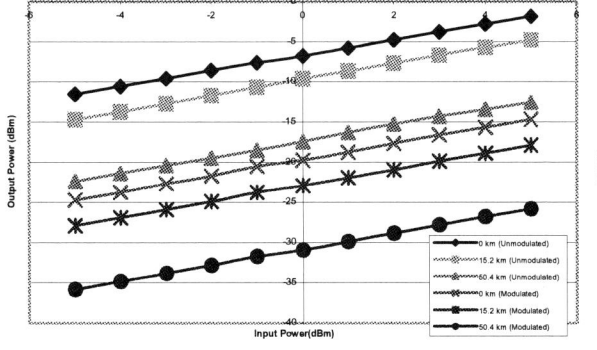

Fig. 7 The modulated and unmodulated output power at different input power in multiplex protection mechanism at three different distances in point to point configuration.

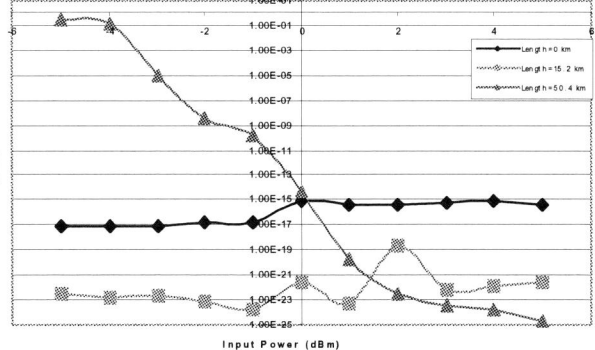

Fig. 8 The BER performances at different input power in multiplex protection mechanism at three different distances in point to point configuration.

217

The BLSR network had been tested at three different bit rates; OC-12, OC-24 and OC-48. Few changes in output power and BER performance were observed which shows degradation in the BER as shown in Figure 5 and Figure 6. This can be compensated by increasing the receiver sensitivity.

Hardware tested by using tunable light source in point to point configuration will generate two types of signals; modulated and unmodulated signal. The magnitude differences at three difference distances are depicted in Figure 7. The BER performance at difference input power under multiplex protection condition is shown in figure 8 at sensitivity of -30 dBm. At the insertion loss of 6 dB, the maximum length can be achieved in point to point configuration is 71 km without regeneration. Figure 9 shows the experimental set up for point to point configuration using two OXADM nodes.

III. CONCLUSION

This paper described our recent approach toward advanced ring metropolitan network through the development of an optical cross add and drop multiplexer (OXADM) switch. The OXADM focuses on providing survivability through restoration against failure by means of multiplex protection that are activated for BPSR configuration. The BER characteristics were measured at 2.5 Gbps (OC-48) and no degradation was observed, as confirmed by a comparison of these simulation results with those obtained from systems without restoration activation (pass through). Restoration scheme in OXADM node is more reliable and efficient compare to conventional device due to internally routing mechanisms and no 'drop and re-add function' is needed to route the signal to an alternative path when failure occurs in working path.

Though successful proposal have been obtained, it is apparent that there are still many issues which have to be solved in order to establish complete optical layer routing and supervision are also expected to be integrated in the optical layer. For the global interworking of the optical layer network, a development which satisfies the requirements of international standardization such as ITU-T of important. UKM photonics will continue to make efforts toward the development of advanced customer access network through realizing the restoration system.

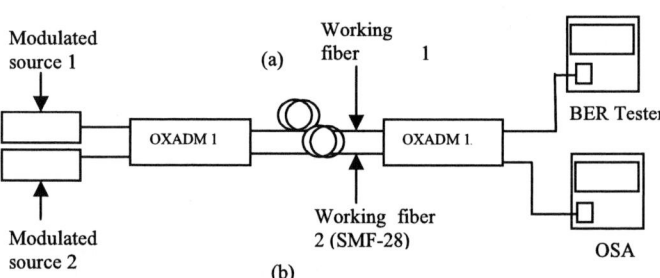

Fig. 9 (a) Experimental set up (b) Block diagram for point to point testing using OXADM nodes connected by two fibers.

REFERENCE

[1] A. Tzanakaki, I. Zacharopoulus & I. Tomkos. Optical Add/Drop Multiplexers and Optical Cross-Connects for Wavelength Routed Network. *ICTON 2003*, pp 41-46 (2003).

[[2] H. Tsushima, S. Hanatani, T. Kanetake, J.A. Fee, and S.a. Liu, Optical cross-connect system for survivable optical layer networks, *Hitachi Review*, Vol. 47, No. 2. pp 85-90 (1998).

[3] L. Eldada & J.v. Nunen. Architecture and Performance Requirements of Optical Metro Ring Nodes in Implementing Optical Add/Drop and Protection Functions. *2000 Telephotonics Review*, (2000).

[4] P. Arijs, R. Meersman, W. V. Parys, A. Tanzi, M. Pierpaoli, F. Bentivoglio & P. Demeester, Architecture and design of optical Channel Protected Ring Networks. *IEEE Journal Lightwave Technology*. 2001, Vol. 19. No. 1. pp 11-22 (2001).

[5]   T. Shiragaki, N. Henmi, T. Kato. Optical Cross-Connect System Incorporated with Newly Developed Operation and Management System. Journal on Selected Areas in communications., Vol. 16, No. 7. pp 1179-1189 (1998).

[6]   http://www.acterna.com/global/products

Inductively-Tuned K- Band Distributed MEMS Phase Shifter

Saeid Afrang and Burhanuddin Yeop Majlis, *Senior* Member IEEE

MEMS Laboratory

Institute of Microengineering and Nanoelectronics (IMEN)

University Kebangsaan Malaysia (UKM)

43600 Bangi, Selangor, Malaysia

s.afrang@vlsi.eng.ukm.my, burhan@vlsi.eng.ukm.my

Abstract This paper presents a new method in the MEMS phase shifters using inductors and inline MEMS bridges. There is LC serial and parallel configuration in the un-actuated and actuated position respectively. The resonance frequencies due to these configurations causes an increasing in impedance change and then increasing in the phase shift in the K- band frequency range. For a unit cell at 20 GHz the Un-actuated simulation results in a reflection coefficient -26.5dB, an insertion loss of -0.185dB and the phase constant of -23.5 degree. The actuated simulation results in a reflection coefficient -13.5dB, an insertion loss of -0.18dB and the phase constant of -6 degree. The spacing in the proposed structures is S=250e-6. This spacing is smaller than similar phase shifters using only capacitors at 20GHz. Then the governing equations for the impedance, Bragg frequency and the periodic spacing between the bridges are derived.

I. INTRODUCTION

The distributed phase shifter consists of a high impedance line capacitively loaded by the periodic placement of discrete varactors [1]. By applying a single bias voltage on the line, the distributed capacitance can be changed, which in turn changes the phase velocity of the line and creates a true-time delay phase shift.

Analog phase shifters using coplanar waveguide (CPW) lines with distributed microelectromechanical system (MEMS) bridges demonstrated broad-band characteristics with low loss [1]. However, since there was a limit on the control range of the bridge height before the bridge snaps, the analog phase shifter showed relatively small phase shift. This problem was solved by operating the MEMS bridges in the digital mode [2]–[8], where two distinct capacitance states (ON and OFF) were defined with a high ratio. Digital phase shifters with this approach allow large phase shift and low sensitivity to electrical noise. In the published CPW digital phase shifters [2], [3], a small metal–insulator–metal (MIM) capacitor or metal-air-metal(MAM)in series with the MEMS bridge capacitor was used to reduce the total shunt capacitance seen by the line, resulting in an acceptable return loss for both switching states over a wide band.

This work, presents new concept of distributed MEMS true time delay phase shifters using both inductors and capacitors. As it is known, the inductor changes the phase of a signal same as the capacitor. In this method transmission line is loaded by the periodic placement of series combination of inductors and capacitors in the un-actuated position and parallel combination of inductors and capacitors in the actuated position. In the un-actuated position, due to serial configuration between inductor and the capacitance of movable MEMS bridge, there is resonance frequency in the structure. The impedance near this resonance frequency is less than the similar structures with only capacitively loaded phase shifters. In the actuated position, due to parallel configuration between inductor and inherent capacitance of transmission line, there is resonance frequency in the structure. The impedance near this resonance frequency is more than the similar structures with only capacitively loaded phase shifters. As a result, due to a large impedance change between un-actuated and actuated position the phase shifting in the proposed structure is more than the similar structures with

0-7803-9730-4/06/$25.00 ©2006 IEEE

only capacitively loaded phase shifters. But in the proposed structure inherent capacitance of the inductor increases return loss in the actuated position. This problem is solved by decreasing the spacing of a unit sell.

II. Ka- BAND MEMS PHASE SHIFTERS DESIGN AND MODELING

Coplanar wave-guide (CPW) transmission line has been used for the design. Fig. 1 shows one unit cell MEMS phase shifter on a glass substrate.

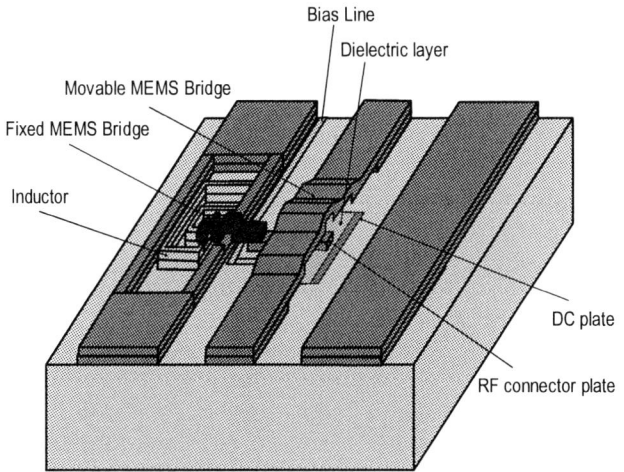

Fig. 1 The schematic of proposed structure(unit cell)

There are two MEMS bridges, two capacitors and one inductor in the structure. One of the bridges is fixed and another bridge is movable. The movable bridge is inline and defined in the CPW center conductor. The fixed bridge connects the inductor to AC capacitor. The inductor is placed in the gap, inside the ground plane. One end of the inductor is connected to fixed MEMS bridge and another end is grounded. There are two capacitors under the movable MEMS bridge. The plate under the dielectric layer with the movable MEMS bridge make DC capacitor. The DC voltage is applied to this capacitor. Another capacitor is AC capacitor. The plate on the dielectric layer with the movable MEMS bridge make AC capacitor.

The general model for a periodically loaded t-line is shown in Figure 2.

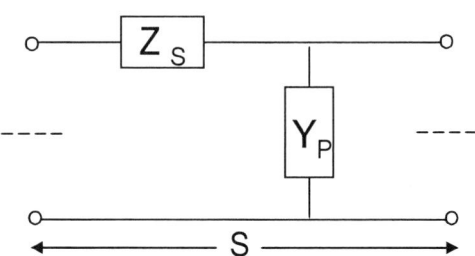

Fig.2. Layout of a DMTL constructed using a CPW line

The characteristic impedance for a DMTL constructed using a CPW line is

$$Z_o = \sqrt{\frac{Z_s}{Y_p}} \cdot \sqrt{1 + \frac{Z_s \cdot Y_p}{4}} \qquad (1)$$

The Bragg frequency is the frequency point at which the periodic filter structure of the distributed loaded line causes the line impedance to become zero and thus there will be no power from one port to other. Therefore from the impedance equation and using the equation

$$Z_s \cdot Y_p = -4 \qquad (2)$$

The Bragg frequency can be defined.

The phase shift per unit length is found from the change in the phase constant and is given by:

$$\Delta\varphi = \frac{360S.f.Z_o \cdot \sqrt{\varepsilon_{r,eff}}}{c}(\frac{1}{Z_u} - \frac{1}{Z_d})\frac{degree}{section} \qquad (3)$$

Where ε_{eff} is the effective dielectric constant of the unloaded line, and Z_u and Z_d are the up-state and down-state loaded line impedance values. The phase shift is determined by the impedance change of the DMTL.

There are two different equivalent circuits for the un-actuated and actuated position in the proposed structure. Therefore, there are different equations for the impedance, Bragg frequency and periodic spacing between the bridges in both cases.

A. Un-actuated Position

Figure 3 shows the equivalent circuit of the structure in the un-actuated position. In this circuit, sL_t and sC_t are the inductance and capacitance of the transmission line respectively, and R_I, L_I and C_I are the resistance, inductance and the capacitance of the inductor respectively. The AC capacitance of movable MEMS bridge is shown by C_b.

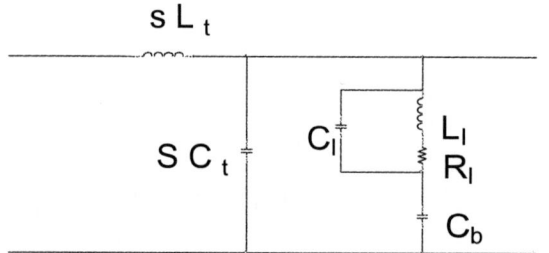

Fig.3.The equivalent circuit in the un-actuated position

Using general model for the proposed structure in the un-actuated position the loaded t-line impedance is given by:

$$Z_u = \sqrt{\frac{-\omega s L_t R_I.C_u + j s L_t(1-\omega^2 L_I.C_u)}{-\omega(s C_t.R_I.C_u + C_I.C_b.R_I) + j\left[s C_t(1-\omega^2 L_I C_u) + C_b(1-\omega^2 C_I L_I)\right]}}$$
$$\times\sqrt{1+\frac{-\omega^2 s L_t\left[s C_t(1-\omega^2 L_I C_s) + C_b(1-\omega^2 C_I L_I) + j\omega R_I(s C_t.C_s + C_b.C_I)\right]}{4(1-\omega^2.L_I.C_s + j\omega R_I.C_s)}}$$

(4)

Where

$$C_s = C_I + C_b \qquad (5)$$

Solving equation 2 the Bragg frequencies in the un-actuated position is defined as:

$$\omega = 2\sqrt{\frac{C_s}{sL_t(sC_t.C_s + C_b.C_I)}} \qquad (6)$$

The lowest impedance in the proposed structure occurs in the un-actuated position. Solving the impedance and Bragg frequency equations for the

separation, s, assuming the Bragg frequency is defined at the lowest impedance of interest (Z_u) results in:

$$S = \frac{Z_u.c}{\pi.f_B.Z_o.\sqrt{\varepsilon_{r,eff}}} \; meters \qquad (7)$$

B. Actuated Position

When the voltage is applied to the movable MEMS bridge, the bridge moves down and makes required contact with the connector plate. Therefore, the ungrounded end of the inductor is connected to the center line. In this condition, the inductor is connected in parallel to the CPW center line.

Figure 4 shows the equivalent circuit of the structure in the actuated position

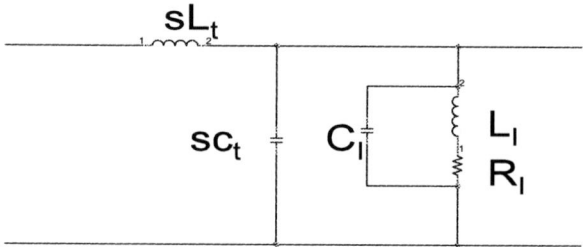

Fig.4.The equivalent circuit in the actuated position

Using general model for the proposed structure in the actuated position the loaded t-line impedance is given by:

$$Z_d = \sqrt{\frac{-\omega^2 L_I.sL_t + j\omega R_I.sL_t}{1 - \omega^2 L_I.C_d + j\omega R_I.C_d}} \times$$
$$\sqrt{1 + \frac{-\omega^2.sL_t.R_I.C_d + j\omega.sL_t(1-\omega^2.L_I C_d)}{4(R_I + j\omega L_I)}}$$

(8)

Where

$$C_d = sC_t + C_I \qquad (9)$$

Using equation 2 the Bragg frequency is defined as

$$\omega = \sqrt{\frac{4L_I + sL_t}{sL_t.L_I.C_d}}$$

222

(10)

For $sL_t \ll 4L_I$, the Bragg frequency is as below:

$$\omega = \frac{2}{\sqrt{sL_t \cdot C_d}} \qquad (11)$$

III. Ka- BAND UNIT CELL MEMS PHASE SHIFTER SIMULATIONS

The unit cell of the proposed K-band phase shifter is calculated and simulated at 30 GHz using MATLAB and EM3D softwares respectively. The coplanar waveguide line with the dimensions of G/W/G = 150/150/150 μm is used. The inductor simulation and calculation results in a inductivity $L_I = 1.4$ nH , a resistivity $R_I = 1.5\Omega$ and the inherent capacitive $C_I = 20$ fF . The substrate is a 400 μm thick glass with the dielectric permittivity of $\varepsilon_r = 3.8$ and the bridge capacitance $C_b = 15$ fF. Figures 5 and 6 show simulation results of the impedance and the phase in the un-actuated position.

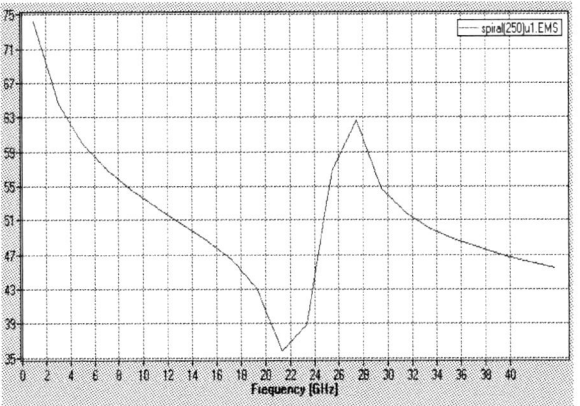

Fig.5. Impedance characteristic in the un-actuated position

Figure 5 shows impedance characteristic of the structure in the un-actuated position. As it is seen, at the frequencies lower than the first resonance frequency the impedance is decreased.

Figure 6 shows phase diagram in the un-actuated position. This diagram shows the phase constant of -23.5 degree at 20 GHz. In the un-actuated position simulation results

show the return loss of -26.5dB at 20GHz. This result shows the structure has very low reflection coefficient in the working area. The insertion loss in this case is -0.18dB.

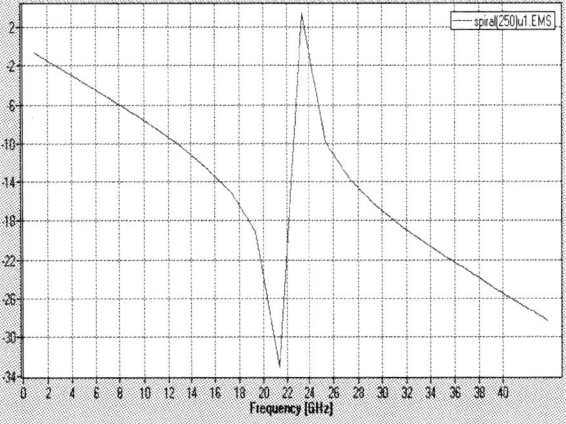

Fig.6. Phase diagram in the un-actuated position

Figures 7 and 8 show simulation results of the impedance and the phase in the actuated position.

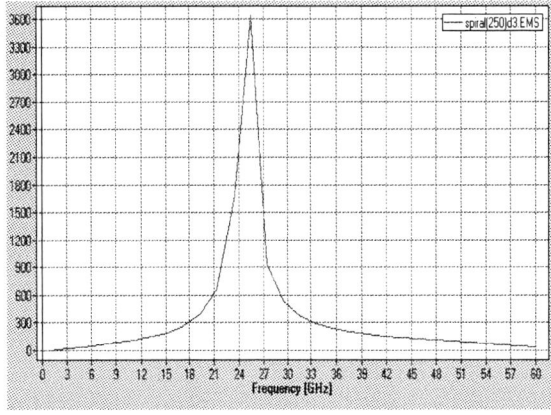

Fig.7. Impedance characteristic in the actuated position

As figure 7 shows, the impedance at the frequencies near the resonance frequency is increased. At this region due to increasing in the impedance the phase is decreased.

Figure 8 shows phase diagram in the actuated position. This diagram shows the phase constant of -6 degree at 20 GHz. In the actuated position simulation results show the return loss of -13.5dB at 20GHz and the insertion loss of -0.188dB

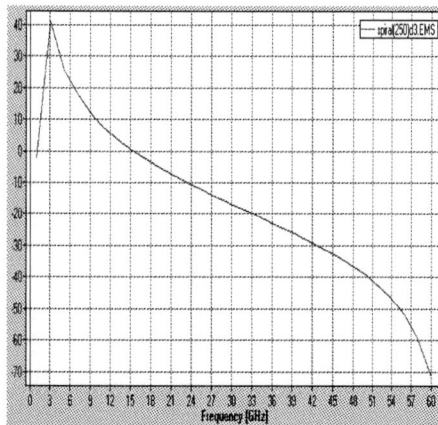

Fig.8. Phase diagram in the actuated position

CONCLUSION

The unit cell of the proposed K-band phase shifter was simulated at 20 GHz using EM3DS software. The spacing in the proposed structure is 250μm this spacing is very smaller than similar phase shifters using only capacitors. The Un-actuated simulation results in a reflection coefficient -26.5dB, an insertion loss of -0.18dB and the phase constant of -23.5 degree. The actuated simulation results in a reflection coefficient -13.5dB, an insertion loss of -0.18dB and the phase constant of -6 degree

ACKNOWLEDGEMENT

This work has been supported by the Malaysian Ministry of Science, Technology and Innovation under the project title "MEMS Devices and Sensing Microstructures"

REFERENCES

[1] N. S. Barker and G. M. Rebeiz, "Optimization of distributed MEMS transmission-line phase shifters— U- band andW-band designs," *IEEE Trans. Microwave Theory Tech.*, vol. 48, pp. 1957–1966, Nov. 2000.

[2] A. Borgioli, Y. Liu, A. S. Nagra, and R. A. York, "Low- loss distributed MEMS phase shifter," *IEEE Microwave Guided Wave Lett.*, vol. 10, pp. 7–9, Jan. 2000.

[3] J. S. Hayden and G. M. Rebeiz, "2-bit MEMS distributedX-band phase shifters," *IEEE Microwave Guided Wave Lett.*, vol. 10, pp. 540–542, Dec. 2000.

[4] Y. Liu, A. Borgioli, A. S. Nagra, and R. A. York, "K-band 3-bit low-loss distributed MEMS phase shifter," *IEEE Microwave Guided Wave Lett.*, vol. 10, pp. 415–417, Oct. 2000.

[5] J. S. Hayden, A. Makczewski, J. Kleber, C. L. Goldsmith, and G. M. Rebeiz, "2 and 4-bit DC-18 GHz microstrip MEMS distributed phase shifters," in *IEEE MTT-S Int. Microwave Symp. Dig.*, May 2001, pp. 219–222.

[6] B. Pillans, S. Eshelman, A. Malczewski, J. Ehmke, and C. L. Goldsmith, "Ka-band RF MEMS phase shifters," *IEEE Microwave Guided Wave Lett.*, vol. 9, pp. 520–522, Dec. 1999.

[7] M. Kim, J. B. Hacker, R. E. Mihailovich, and J. F. DeNatale, "A DC-to-40 GHz four-bit RF MEMS true-time delay network," *IEEE Microwave Wireless Comp. Lett.*, vol. 11, pp. 56–58, Feb. 2001.

[8] A. Malczewski, S. Eshelman, B. Pillans, J. Ehmke, and C. L. Goldsmith, "X-band RF MEMS phase shifters for phased array applications," *IEEE Microwave Guided Wave Lett.*, vol. 9, pp. 517–519, Dec. 1999.

Crosstalk Enhancement in Multiplexer/Demultiplexer Based Arrayed Wavelength Grating in Dense Wavelength Division Multiplexing

Salah Elfaki[1], Abdeen Abdel Kareem[1], A. B. Mohammed[2] and Sahbudin Shaari[3]

[1]Electronics Dept, Faculty of Electrical Engineering, Sudan University of Science & Technology, Sudan

[2] Faculty of Electrical Engineering, Universiti Teknologi Malaysia, Johor, Malaysia.

[3]Institute of Micro Engineering and Nanoelectronics,Universiti Kebangsaan Malaysia, Selangor, Malaysia

Email. salahelrofai@yahoo.com., sahbudin@vlsi.eng.ukm.my

Abstract **In this paper we describe the performance of the Multiplexer/Demultiplexer (MUX/DEMUX) in the Dense Wavelength Division Multiplexer (DWDM) technology based on arrayed waveguide grating (AWG). The accumulated crosstalk in large-scale AWG solved by the cascade-connection of small AWGs. We propose two branches of 64 channels each have two stages of AWG cascaded as second and third stages. We introduce an interleaver filter based on AWG to produce odd and even optical channels (64) as an input to each branch, which relax the system by doubling the space between channels.**

Key words: Crosstalk, demultiplexing, wavelength division-multiplexing.

I. INTRODUCTION

Recently, with rapid increase in demand for large optical transmission capacity, dense wavelength division multiplexing (DWDM) systems became attractive systems for huge and high-speed data transmission methods [1]. AWG based MUX/DEMUX has many advantages including small size, high reliability, low cost, and high ports count [2]. AWG is a key component for various optical devices as multiplexers/demultiplexers, Add-drop multiplexer, and multiwavelength receiver [3-4]. Large-scale AWG performance decreases according to increases of channels due to the crosstalk accumulation. The crosstalk accumulated can be reduced by cascaded connection of small AWG.

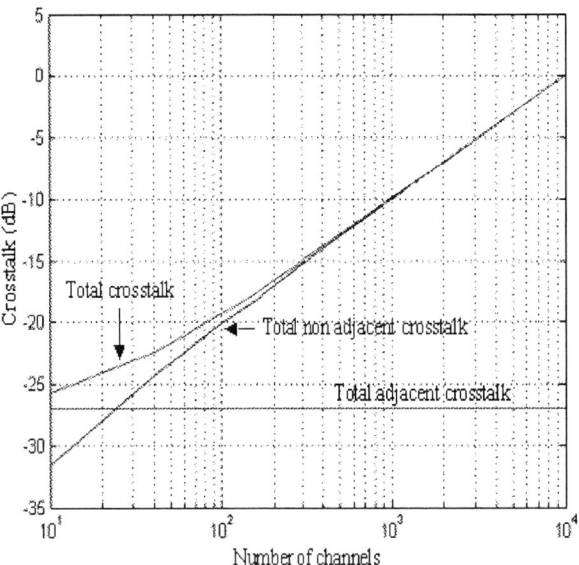

Fig.1 Shows the total crosstalk of conventional AWG compared with the total adjacent crosstalk and total nonadjacent crosstalk.

II. CROSSTALK AND PROPOSED SOLUTION

A crosstalk in conventional AWG has a large-scale limitation. The crosstalk of the AWG based on multiplexer/demultiplexer can be classified in two types' adjacent crosstalk, and nonadjacent crosstalk The DWDM signal affected by the total crosstalk, which is the sum of the two types.

$$X_{TOT} = 2X_{adj} + (N-3)X_{nonadj} \qquad (1)$$

Where N is the number of channels of AWG, X_{adj} and X_{nonadj} are the crosstalk from the adjacent and nonadjacent channels respectively. For using silica conventional AWG, the level of adjacent crosstalk is about -30dB, and typically -40dB for nonadjacent crosstalk [5-6]. On the other hand the crosstalk can be found from the ratio of the transmittance of two optical channels $\dfrac{T(\lambda_i)}{T(\lambda_K)}$. In the fiber optics systems the transmittance is the fraction of light transmitted through the system [7].

$$2X_{adj} + (N-3)X_{nonadj} = \sum_{i=K-1,K+1} \frac{T(\lambda_i)}{T(\lambda_K)} + \sum_{i=1,2,\dots K-2,K+2,\dots N-1,N} \frac{T(\lambda_i)}{T(\lambda_K)} \qquad (2)$$

From Fig.1. We can conclude that the worst case of the total crosstalk is -15dB therefore we expect that AWG channels value limited to several hundred. Also the main component that affects the total crosstalk is the non-adjacent crosstalk.

Crosstalk level required in large-scale AWG can be adjusted by reducing the total nonadjacent crosstalk to be less than -40dB. That means for two stages the non-adjacent crosstalk should be around -80dB for ideal double filtration these value reasonable for 1000 channels [6]. For the future we need more than two stages of cascaded AWG, to cover the large band ranged from 1300nm up to 1600nm due to the contribution of zero-water-peak fiber. The non-adjacent crosstalk should be around -120dB for the three stages.

III. PROPOSED AWG CASCADE CONNECTION DESIGN

To meet the crosstalk requirement for large –scale AWG, we propose a MUX/DEMUX with small AWG cascade in three stages, configure in figure (3).

The first stage is an AWG based on an interleaver filter. That means its outputs are even channels and odd ones in other words the out put should be double the input spacing, these technique for making the input to the second stage more relax. The second stage composes of two branches cascading to the first stage using PLC - PLC technique, the same cascading done for stage three to stage two.

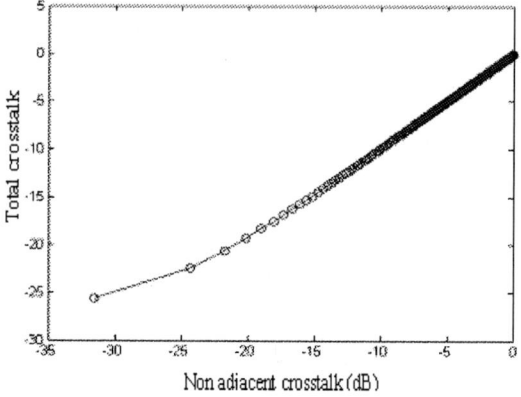

Fig.2 Estimated total crosstalk as a function of its total nonadjacent crosstalk.

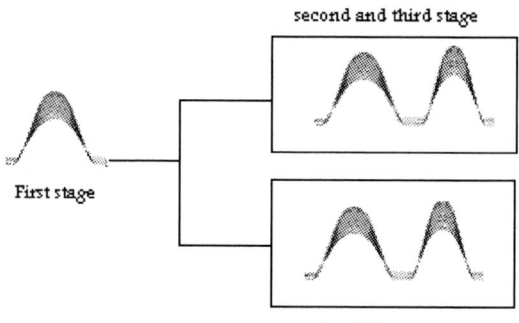

Fig.3 Shows the proposed AWG cascaded connection

The requirement for designing the cascaded AWG, is to minimize the circuit and optimized the bandwidth of the second and the third AWG.

A. Minimize the circuit

Using the planner lightwave circuit (PLC) techniques as PLC-PLC attached between the first stage and the second one also the same attachment between the second and the third stage of the first branch, and exact done for the second branch. Optimizing the waveguides to provide a low insertion loss [6]. Using PLC-PLC can reduce the effects of AWG center wavelength mismatching between the PLCs. Usually PLC- PLC configuration enables us to measure the characteristics of each AWG.

B. Bandwidth optimization of cascaded AWGs

For design the bandwidth of AWG of the second stage and the third stage, we have to consider that the bandwidth of each stage must be wide but not too broadband; otherwise it will not reduce the non-adjacent crosstalk. We should consider the differences between the center wavelengths of the first AWG and the AWGs of the two cascaded stages, and optimize the bandwidth to absorb the center wavelength differences.

The optimization of the bandwidth of the two cascaded stages can be done by estimated of the spectral characteristics of the cascaded AWG as a function of the bandwidth, and the center wavelength differences due to the following calculations. We assumed the transmittance in linear units $T_1(\lambda)$ for first stage AWG, which has a Gaussian spectral (4) where λ, λ_1, and σ_1 are the wavelength, the ith channel wavelength, and the bandwidth respectively, the second stage center at channel i = k. the lower limit of the spectrum is Xnonadj,1 this means the first stage nonadjacent crosstalk is Xnonadj,1. We set σ_1 so that the first stage AWG adjacent crosstalk is Xadj,1, in which

$$\sigma_1 = \Delta\lambda\sqrt{-In(Xadj,1)} \qquad (3)$$

Where $\Delta\lambda$ is channel spacing.

$$T_1(\lambda) = \exp\left[-\left(\frac{\lambda - \lambda_K}{\sigma_1}\right)^2\right] for$$
$$|\lambda - \lambda_K| \le \sigma_1\sqrt{-In\left(Xnonadj,1\right)}$$
$$T_1(\lambda) = X_{nonadj,1} for$$
$$|\lambda - \lambda_K| \rangle \sigma_1\sqrt{-In\left(X_{nonadj,1}\right)} \qquad (4)$$

We assumed the transmittance $T_2(\lambda)$ and $T_3(\lambda)$ for the second, and the third stages. Where σ_2 and σ_3 are the bandwidths of stage number two and three of proposed design respectively and $\partial_2\lambda, \partial_3\lambda$ are the center wavelength differences from the channel wavelength (λ_K).

$$T_2(\lambda) = \exp\left[-\left(\frac{\lambda - \lambda_K - \partial_2\lambda}{\sigma_2}\right)^2\right] for$$
$$|\lambda - \lambda_K| \le \sigma_2\sqrt{-In\left(X_{nonadj,2}\right)} \qquad (5)$$
$$T_2(\lambda) = X_{nonadj,2} for |\lambda - \lambda_K| \rangle \sigma_2\sqrt{-In\left(X_{nonadj,2}\right)}$$

$$T_3(\lambda) = \exp\left[-\left(\frac{\lambda - \lambda_K - \partial_2\lambda - \partial_3\lambda}{\sigma 3}\right)^2\right] for$$
$$|\lambda - \lambda_K| \le \sigma_3\sqrt{-In\left(X_{nonadj,3}\right)} \qquad (6)$$
$$T3(\lambda) = X_{nonadj,3} for |\lambda - \lambda_K| \rangle \sigma_3\sqrt{-In\left(X_{nonadj,3}\right)}$$

Calculated the product of the transmittance of the three AWGs to estimate the cascaded transmittance T (λ_C) and the spectral characteristics AWG, also calculated σ_C which is the bandwidth of the cascaded AWG.

$$T_c(\lambda) = T_1 \times T_2 \times T_3$$

$$T_c(\lambda) = \exp\left[\begin{array}{c} -\left(\frac{\lambda - \lambda_K}{\sigma_1}\right)^2 \left(\frac{\lambda - \lambda_K - \partial_2\lambda}{\sigma_2}\right)^2 \\ \left(\frac{\lambda - \lambda_K - \partial_2\lambda - \partial_3\lambda}{\sigma_3}\right)^2 \end{array}\right]$$
$$for |\lambda - \lambda_K| \le \sigma_1\sqrt{-In\left(X_{nonadj,1}\right)}$$

$$T_c(\lambda) = X_{nonadj,1}\exp\left[-\left(\frac{\lambda - \lambda_K - \partial_2\lambda}{\sigma_2}\right)^2\right]$$
$$for \sigma_1\sqrt{-In\left(X_{nonadj,1}\right)} \langle |\lambda - \lambda_K| \le \sigma_2\sqrt{-In\left(X_{nonadj,2}\right)}$$

$$T_c(\lambda) = X_{nonadj,2}\exp\left[-\left(\frac{\lambda - \lambda_K - \partial_2\lambda - \partial_3\lambda}{\sigma_3}\right)^2\right]$$
$$for \sigma_2\sqrt{-In\left(X_{nonadj,2}\right)} \langle |\lambda - \lambda_K| \le \sigma_3\sqrt{-In\left(X_{nonadj,3}\right)}$$
$$T_c(\lambda) = X_{nonadj,1} \times X_{nonadj,2} \times X_{nonadj,3}$$
$$for |\lambda - \lambda_K| \rangle \sigma_3\sqrt{-In\left(X_{nonadj,3}\right)} \qquad (7)$$

$$\sigma_c = \frac{\sigma_1\sigma_2\sigma_3}{\sigma_1\sigma_2 + \sigma_1\sigma_3 + \sigma_2\sigma_3} \qquad (8)$$

IV. SIMULATION WORK

Using the WDM_PHASER, which is an advanced design and modeling tool for design and simulate our proposed configuration of cascaded AWGs. Defining the parameters for scanning and then assign the range of parameter values and the number of simulation steps. To evaluate the device as a multiplexer/demultiplexer, the wavelength is assigned as the scan variable, and the power in each output port is plotted as a function of wavelength at the input port. Fig.4. Shows the spectral of the output power vs. the wavelength for 8 channels, 16 channels, 32 channels and 64 channels each of 100GHZ spacing. Table.1 shows the most important scan results (crosstalk) of our unit design (AWG) before cascading and after cascading which considered as a real data.

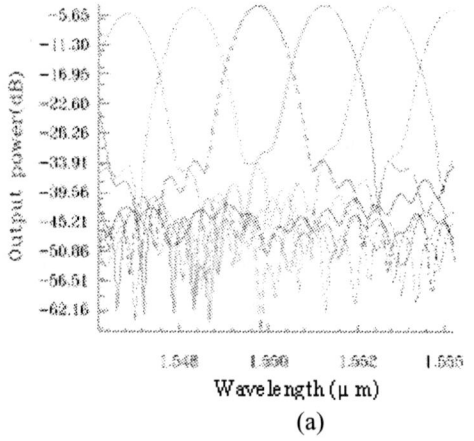

(a)

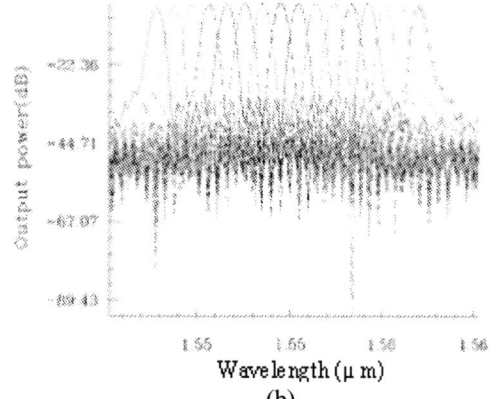

(b)

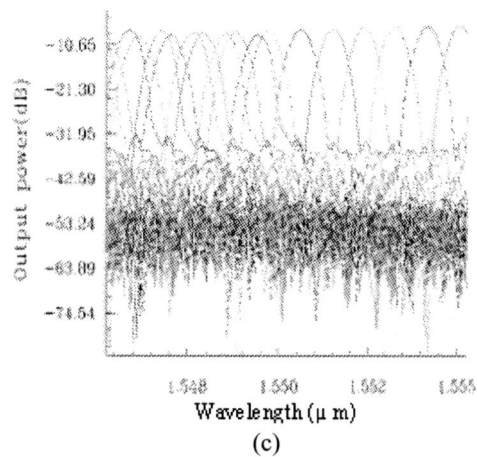

(c)

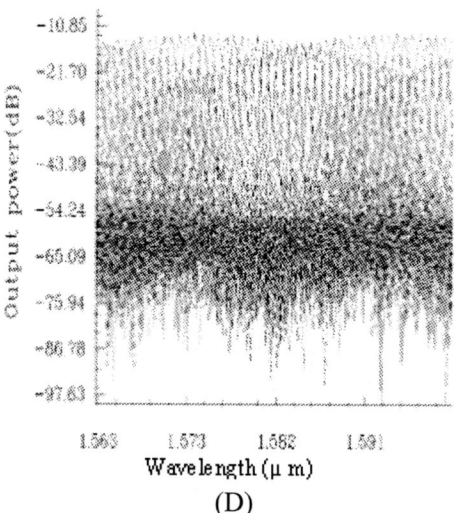

(D)

Fig.4 shows the spectral response of AWG for both output power vs. Scan parameters of (A-8ch, B-16ch,C-32ch, D-64ch) for 100GHZ spacing.

From the table 1 we conclude that the cascade reduce the crosstalk, in table 2 we get the real data results from simulation design, used for solving our new proposed equations (6,7,8), compares with theoretical calculations give an excellent agreement between calculations and simulation results.

Table 1 shows the AWG results of different simulation channel number of (100GHZ).

Simulation channel No.	No. Of channel	Crosstalk before cascading	Crosstalk after cascading
8	2	-33.7	-44.1
8	3	-33.7	-48.9
8	4	-33.1	-44.4
16	2	-30.3	-45.6
16	3	-29.6	-47.5
16	11	-35.9	-47.9
16	14	-30.9	-43.3
32	1	-34.8	-49.4
32	2	-34.3	-44.2
32	20	-34.9	-54.1
32	26	-33.8	-53.5
32	27	-32.9	-47.2
64	1	-39.6	-67.9
64	2	37.8	-64.2
64	27	-32.5	-62.9
64	51	-32.4	-65.7
64	58	-33.5	-65.7
64	61	-34.6	-59.2

Table 2 shows real simulation data for proposed three stages AWGs cascaded design.

No. of channel	No. of stage	Central frequency in GHZ	Central frequency in nm	Channel spacing in GHZ	Channel Spacing in nm	Output channel Bandwidth in GHZ
16	1	193412.9	1550	49.729428	0.39852881	67.220468
8	2	193913.32	1546	100.13001	0.79829992	134.6576
8	3	193412.9	1550	200.97866	1.6106315	271.66771

V. CONCLUSION

In summary we have investigated the cascading connection for AWGs. We found that cascaded connection reduces the crosstalk, also we design three stages of AWG cascaded, for enhance the crosstalk by controlling its nonadjacent crosstalk.

ACKNOWLEDGEMENT

The authors would like to thank Photonics Research Group at Universiti Technologi Malaysia (UTM), University Kebangsaan Malaysia (UKM), and University Putra Malaysia (UPM) for supporting this work.

REFERENCES

[1] J.I. Hashimoto, T. Takagi, T. Kato, G. Sasaki, M. Shigchara, K. Murashima, M.Shiozaki, and T. Iwashima, "Fiber Brag-Grating External Cavity Semiconductor Laser (FGL) Module for DWDM Transmission," Journal of lightwaves technology. Vol 21. No.9, September 2003

[2] T. Lang, J-Jun He, and S. He, "Cross-order Arrayed Waveguide Grating Design for Triplexers in Fiber Access Networks,"IEEE photonics Technology Letters, Vol.18, No.1,January 2006

[3] Y. Hibino, "Recent Advances in High-Density and Large-Scale AWG Multi/Demultiplexers With Higher Index-Contrast Silica-Based PLCs," IEEE Journal of selected topics in quantum electronics, Vol. 8, No. 6, November/December 2002.

[4] J. Cho, D. Han, J. H. Song, and S. Jung, "Crosstalk Enhancement of AWG Fabricated by Flame Hydrolysis Deposition Method," IEEE Photonic Technology letters,Vol.17,No.11, November2005.

[5] S. Kamei, A. Kaneko, M. Ishii,T. shibata, Y. Inoue, and Y. Hibino, "Crosstalk Reduction in Arrayed-Waveguide Grating Multiplexer/Demultiplexer Using Cascade Connection,"Journal of lightwave technology,VOL.23, No. 5, May2005

[6] P. J. Winzer, M.Pfennigbauer, and R. J. Essiambre, "Cherent Crosstalk in Ultradense WDM Systems," Journal of lightwave technology, vol. 23,No.4,April 2005

[7] J.N. Dowing, " Fiber-optic communications," Thomson, copyright © 2005.

Analytical Model Studying of a Novel Tunable Capacitor Based on Bimetallic Thermal Actuator

Shahriar Kouravand[1], Mehdi Mehrban[2], Ghader Rezazadeh[3], Mehdi Sabet[4]

[1,2,3,4]Mech. Eng. Dept. Urmia University, Urmia, Iran

Email: shahriar.kouravand@gmail.com

Abstract **This paper presents a novel tunable capacitor with bimetallic thermal actuator in micro scale which its design was based on the tip deflection of bimetallic cantilever beam. Thermo electro mechanical governing equations were derived. Design steps are described and calculated results show that using of the proposed tunable capacitor can be achieved large tuning ratio.**

I. INTRODUCTION

Daily, domain of micro-electro-mechanical system (MEMS) applications is extending approximately in all of sciences. Accuracy, low price and small dimension of this products cause to researchers consider to this area. The tunable capacitor is one of the most important and active branches in the area of applying MEMS technology to passive and active components, such as inductors, switches and filters. In most researchers work, wide range of actuation mechanism has been used to tuning the capacitors such as electrostatic and thermal actuators [1-6].Young, et al were presented a tunable capacitor consists of a thin sheet of aluminum suspended in above the substrate for monolithic low-noise voltage controlled oscillators (VCOS) [1]. Theoretically, a tuning range of up to 50% can be achieved. Dec et al presented two new tunable capacitors with two and three plates respectively. For the two plate tunable capacitor, the fixed plate was anchored under the suspended plate, while with the three plates tunable capacitor; the suspended plate is placed in the middle of the three plates [2]. The measurements of the tuning range of the two and three plates tunable capacitors are 1.5:1 and 1.87:1, respectively. Another tunable capacitor with one suspended top plate and two fixed bottom plates has been proposed by Jun Zou

et al [3]. One of the two fixed plates and the top plate form a variable capacitor, where as the other fixed plate and the top plate are used to provide electrostatic actuation for capacitance tuning. The controllable tuning range of 68% has been achieved experimentally. Instead of the parallel plate style tunable capacitor, J. Jason Yao et al have presented an interdigitated "comb" structure tunable capacitor [4,5]. With the special fabrication process, the continuous and controllable tuning range is up to 300%. A tunable capacitor can also be built using an electro-thermal actuating mechanism. The advantages of tunable capacitors actuated by thermal actuators Compared to electrostatic are avoiding the static charges collecting on the plates, Improvement in the reliability of tuning, approximately linear capacitance tuning, Lower driving voltages [6]. The tunable capacitor with an electro-thermal actuator has some disadvantages such as lower tuning and more space requirements. A series mounted MEMS tunable capacitor with an electro-thermal actuator was reported by Zhiping Feng et al [6]. The electro-thermal actuator has been used to drive the top plate of the tunable capacitor to move vertically. The tuning range of this type capacitor was reported to be 2:1. Here, a new system to tuning the capacitor is introduced that is compact. Differential expansion of a laminate made of two materials of unequal thermal expansion coefficients, the bimetallic or bimorph can be used to amplify deflection. A tunable capacitor based on deflection of bimetallic beam was designed. Using bimetallic materials as a cantilever beam that coupled with a capacitor can create a thermal actuators (Fig. 1). Thereby by controlling the operating temperature in system, the micro bimetallic beam deflection can control, so the charge of capacitor can be tuned.

II. MODEL DESCRIPTION

The proposed model is shown in Figure1. This model is consists of a bimetallic cantilever beam and one capacitor that is jointed at the tip of bimetallic cantilever beam. Temperature of bimetallic beam is controlled with a heater subjected to a voltage v. Heater and bimetallic cantilever beam surrounded by a heat shield. This shield isolates the bimetallic beam from fluctuations of ambient temperature.

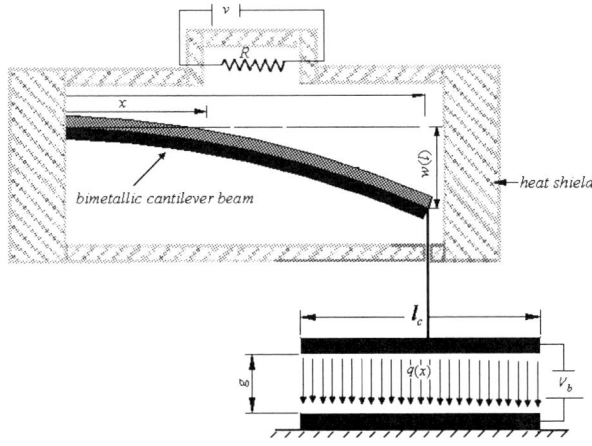

Fig. 1 Schematic of the tunable capacitor based on bimetallic thermal actuator

When the internal temperature of the shield is changed the bimetallic cantilever beam deflects because of different temperature expansion coefficient of bimetallic materials. Due to its tip deflection, gap of the capacitor changes, it means that by temperature rising in shield, capacitance of the system can be tuned.

III. MATHEMATICAL MODELING

Figure 2 shows an element of bimetallic cantilever beam. Assume a beam with length l, thickness h, width b, expansion coefficient α, cross sectional area A and isotropic with Young's modulus E, whereas for upper micro beam sub indicate 1 is used and for the other one 2 is used. Suppose that x is the coordinate along the length of the beam with its origin at the left end, and z is the coordinate along the cross section with its origin at the neutral axis of cross section. $w(x)$ is the deflection of the

beam, and u is the displacement along x axis. In this figure dash lines is whereabouts neutral axis.

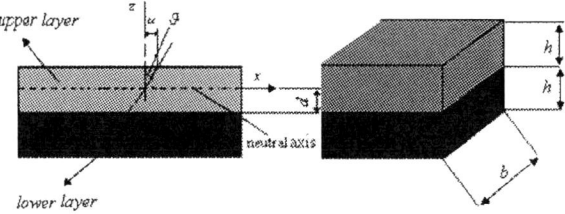

Fig. 2 An element of bimetallic cantilever beam

For selected micro beams, l/h is usually large enough to neglect the shear deformation and using of strain definition [7], the total strain at the given cross section can be written as:

$$\varepsilon_{tot} = z \frac{d^2 w}{dx^2} \qquad (1)$$

Total strain at x direction at a given cross section of the beam, is the sum of thermal and mechanical strains, thus:

$$\varepsilon_{tot} = \varepsilon_m + \varepsilon_T \qquad (2)$$

where:

$$\varepsilon_T = \alpha \Delta T \qquad (3)$$

With

$$\Delta T = T - T_0 \qquad (4)$$

Where ε_T and ε_m are the thermal and the mechanical strains, respectively, ΔT is the rising temperature which is to be measured respect to the initial temperature T_0. Substituting Eqs. (1) and (3) into Eq. (2), the following equation can be obtained:

$$\varepsilon_m = z \frac{d^2 w}{dx^2} - \alpha \Delta T \qquad (5)$$

Using of Hook's low and Eq. (5), the relationship between the stress and the strain based on Euler-Bernoulli beam theory can be expressed as below:

$$\sigma = Ez\frac{d^2w}{dx^2} - E\alpha\Delta T \qquad (6)$$

The axial force respect to the equilibrium condition along the x axis is given as:

$$\int_{A_1}\sigma\, dA + \int_{A_2}\sigma\, dA = 0 \qquad (7)$$

Substituting the Eq. (6) into Eq. (7):

$$\int_{-d}^{h-d} b(E_1 z\frac{d^2w}{dx^2} - E_1\alpha_1\Delta T)dz +$$
$$\int_{-h-d}^{-d} b(E_2 z\frac{d^2w}{dx^2} - E_2\alpha_2\Delta T)dz = 0 \qquad (8)$$

where d is distance of neutral axis from contact surface of two materials. The Eq. (8) can be reduced to:

$$\frac{d^2w}{dx^2}\left(\frac{(1-n)h - 2d(1+n)}{2}\right) -$$
$$\Delta T(\alpha_1 + n\alpha_2) = 0 \qquad (9)$$

where $n = \dfrac{E_2}{E_1}$. The bending moment $M(x)$ at a given section is:

$$\int \sigma\, zdA = M(x) \qquad (10)$$

Substituting Eq. (6) into Eq. (10):

$$\frac{d^2w}{dx^2}\left(\int_{-d}^{h-d} z^2 dz + \int_{-h-d}^{-d} nz^2 dz\right) -$$
$$\int_{-d}^{h-d}\alpha_1 z\Delta Tdz - \int_{-d}^{h-d} n\alpha_2 z\Delta Tdz = \frac{M(x)}{E_1 b} \qquad (11)$$

By integrating of the Eq. (11) and Substituting value of d from Eq. (9) into this equation and simplify it, the final expression which indicates the relationship between deflections of bimetallic cantilever at a given rising temperature can be written as follow:

$$\frac{d^2w}{dx^2} = \frac{12\Delta Tn(\alpha_1 - \alpha_2)}{h(n^2 + 14n + 1)} +$$
$$\frac{12(1+n)M(x)}{E_1 bh^3(n^2 + 14n + 1)} \qquad (12)$$

Moment for this model, in a given section is:

$$M(x) = \frac{\varepsilon_0 A V_b^2}{2(g_0 - w(l))^2}(l - x) \qquad (13)$$

where ε_0, A, V_b, g_0 and $w(l)$ are respectively permittivity of air, area of the effective capacitor plate, initial gap and tip deflection of cantilever. Substituting Eq. (13) into Eq. (12) and by integration of both side of Eq. (12) in term of x, expression that indicated relation between beam deflection and temperature rising is:

$$w(x) = \frac{6\Delta Tn(\alpha_1 - \alpha_2)}{h(n^2 + 14n + 1)}x^2 +$$
$$\frac{6(1+n)\varepsilon_0 A V_b^2}{E_1 bh^3(n^2 + 14n + 1)(g_0 - w(l))^2}(\frac{lx^2}{2} - \frac{x^3}{6}) \qquad (14)$$

Finally, using of Eq. (14), the expression for calculating the capacitance with respect to temperature rising is derived as follow:

$$C = \frac{\varepsilon_0 A}{g_0 - w(l)} \qquad (15)$$

IV. RESULTS

To have a large deflection and large tuning ratio, two materials with high difference in their thermal expansion coefficients are considered (Gold and Silicon). Physical and geometrical properties of beam and capacitor are shown at Table 1. Thickness and width of two materials are assumed same together. Calculated results were compared well with results in references [8] at the Table2.

232

Table 1. Design variable for compare with Ref. [8]

Design variable	value
α_1	$14.3 \times 10^{-6} \ k^{-1}$
α_2	$2.6 \times 10^{-6} \ k^{-1}$
E_1	$80 \times 10^9 \ Nm^{-2}$
E_2	$122 \times 10^9 \ Nm^{-2}$
b	$90 \, \mu m$
V_b	0
h	$3 \, \mu m$
g_0	$80 \, \mu m$
L	$500 \, \mu m$
ε_0	$8.854 \times 10^{-12} \ fm^{-1}$

Table 2. Comparison of calculated tip deflection of proposed model with results of Ref. [8]

Temperature rising (c^0)	Tip deflection in this work (μm)	Tip deflection value in [8] (μm)	$\Delta\%$
10	3.615400	3.615410	0.00028
30	10.846222	10.846230	0.00007
50	18.077043	18.077050	0.00004
80	28.923273	28.923280	0.00002
100	36.154094	36.154099	0.00001

Where:

$$\Delta(\%) = \frac{ABS\left(Obtained \ \text{Re} sults - \text{Re} sults \ of \ \text{Re} f[8]\right)}{\text{Re} sults \ of \ \text{Re} f[8]} * 100$$

Now for simulation of model design, materials with geometrical properties according Table 3 are chosen. A bias voltage was applied for capacitor and it was kept constant. Figure 3 shows the deflection of the bimetallic cantilever at the different temperature rising. Thus, the gap changes, which causes the capacitance of the system changes. The change in capacitance of the proposed model due to the temperature rising can be easily found.

Table 3. Design variable for typicality simulation

Design variable	value
b	$20 \, \mu m$
h	$2.5 \, \mu m$
g_0	$3.5 \, \mu m$
L	$100 \, \mu m$
B	$20 \, \mu m$
L_C	$50 \, \mu m$

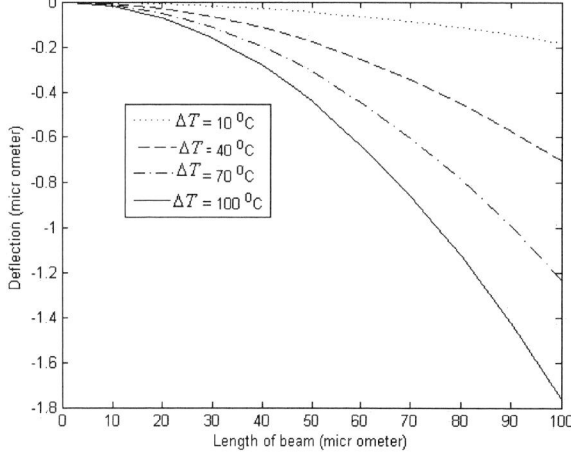

Fig. 3 Deflection of the cantilever versus the temperature rising

As it is shown in Fig.3, capacitance changes nonlinearly with respect to the temperature rising. The changed capacitance value can be used to create a novel tunable capacitor with bimetallic thermal actuator.

Fig.4 shows that with rising temperature to 100°C tuning range changes from 0.25 fF to 0.53 fF. It's nature that as the temperature increase, turning range increased too but while temperature increases more than 100°C, changing of capacitance will be very fast that can be due to Pull-in phenomena. Here maximum range of temperature changing 100°C was chosen to prevent Pull-in phenomena occurring.

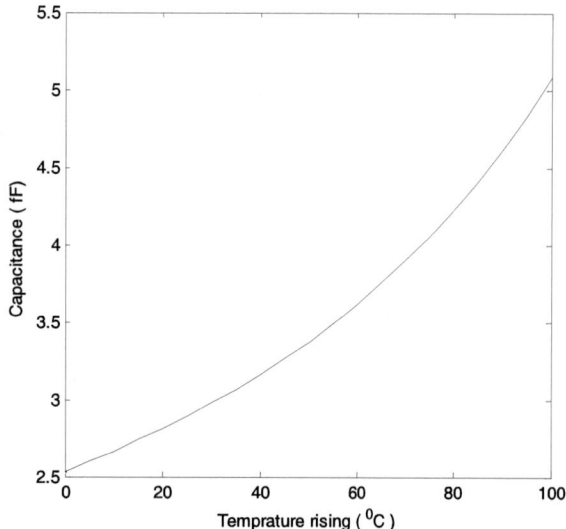

Fig. 4 Capacitance diagram due temperature rising

V. CONCLUSION

In this work was presented the design of a novel tunable capacitor using a bimetallic thermal actuator. The governing equation of the model was derived and solved analytically. Calculated results show that with increasing the temperature, capacitance increases nonlinearly. The proposed tunable capacitor can be used in many fields of MEMS applications.

REFERENCE

[1] D. J. Young and B. E. Boser, "A micromachined variable capacitor for monolithic low-noise vcos," in Tech. Digest of Solid-state sensor and actuator workshop Hilton Head Island, SC, pp. 86—89. 1996.

[2] Dec and K. Suyama, "Micromachined eletro-mechanically tunable capacitors and their applications to rf ic's," IEEE transactions on microwave theory and techniques, vol. 46, no. 12, pp. 2587—2596, 1998.

[3] Ch. L. J. Zou and J. S. Aine, "Development of a wide tuning range mems tunable capacitor for wireless communication systems," in Device research conference 2000, Conference digest 58th DRC, pp. 111—113, 2000.

[4] J. B. Yoon and C. T.-C. Nguyen, "A high-q tunable micromechanical capacitor with movable dielectric for rf application," in Technical Digest, IEEE Int. Electron Devices Meeting, San Francisco, California, pp. 489—492, 2000.

[5] R. Anderson, J. J. Yao, S.T. Park and J. DeNatale, "A low power/low voltage electrostatic actuator for rf mems application," in Solid-state sensor and actuator workshop Hilton Head Island, SC, pp. 246—249. 2000. 79

[6] B. Su, K. F. Harsh, K. C. Gupta, V. Bright, Zh. Feng, W. Zhang and Y.C. Lee, "Design and modeling of rf mems tunable capacitors using electro-thermal actuators," in Microwave Symposium Digest, IEEE MTT-S International, pp. 1507—1510. 1999.

[7] S. Timoshenko, Strength of material, Part1, New Yourk: Van Nostrand, 1930.

[8] Sh. Kouravand, Gh. Rezazadeh, M. Sabet and A. Tahmasebi, MEMS capacitive micro thermometer based on tip deflection of bimetallic cantilever beam, journal of Sensors & Transducers, Vol.70, pp.637-644, 2006

Uniformity Study of GaAs-based Vertical-Cavity Surface-Emitting Laser Epiwafer Grown by MOCVD Technique

Mohd Sharizal Alias, *Member, IEEE*, Paul O. Leisher [*], *Student Member, IEEE*, Kent D. Choquette [*], *Fellow, IEEE*, Khairul Anuar, Dominic Siriani [*], Sufian Mitani, Mohd Razman Y. and Abdul Fatah A. M.

Telekom Research & Development (TMR&D)
Idea Tower, UPM-MTDC, Lebuh Silikon
43400 Serdang, Selangor, MALAYSIA
Email: sharizal@tmrnd.com.my

[*] Electrical & Computer Engineering Department
University of Illinois at Urbana-Champaign (UIUC)
Urbana, 61801, Illinois, USA

Abstract In this paper, a highly uniform 850 nm VCSEL epiwafer was grown by using MOCVD technique. The VCSEL quantum wells exhibit a peak emission wavelength of 839.5 nm. The grown epiwafer exhibits a Fabry-Perot cavity resonance at 846.5 nm. Various VCSEL devices fabricated from the center to the edge of the VCSEL epiwafer show similar trend for the L-I and output spectral characterizations. Devices fabricated at the edge of the epiwafer exhibits different characterization trends due to the growth limitations.

I. INTRODUCTION

Vertical-cavity surface-emitting laser (VCSEL) structures typically consist of more than 200 epitaxially grown layers of III-V compound semiconductor material. Thus a VCSEL wafer is commonly called an "epiwafer" due to this high number of monothically grown epitaxial layers. Like other III-V materials based devices, VCSEL epiwafers can be grown by Molecular Beam Epitaxy (MBE) techniques [1] or Metal-Organic Chemical Vapor Deposition (MOCVD) techniques [2].

In general, the VCSEL growth process is very challenging since precise thickness and composition control of the numerous epitaxial layers is required. Additionally, in order to reduce the device series resistance generated in the Distributed Bragg Reflector (DBR) mirrors, the composition and doping concentrations across these DBR interfaces require a special grading scheme [3-6]. MBE offers only abrupt grading schemes, such as step-graded and chirped-graded, which are inferior compared to the continuous grading scheme such as linear-graded and parabolic-graded that can be employed by MOCVD [7]. Continuous grading schemes yield lower series resistance in the DBRs of a VCSEL.

In this paper, an 850 nm VCSEL epiwafer with parabolic grading scheme is grown by MOCVD. Numerous VCSEL devices are then fabricated on various epiwafer positions to investigate the epiwafer uniformity.

II. DEVICE STRUCTURE & EXPERIMENT

The 850 nm VCSEL epiwafer was grown by MOCVD with the composition and doping being parabolically graded. The active region consists of three GaAs quantum wells. Alternating high- and low-refractive index layers of $Al_{0.9}Ga_{0.1}As$/ $Al_{0.15}Ga_{0.85}As$ form the bottom 34.5-period n-type and top 24.5-period p- type DBR mirror. A layer with high aluminum alloy concentration ($Al_{0.98}Ga_{0.02}As$) layer is introduced for the purpose of selective oxidation. A cross-sectional schematic of the 850 nm VCSEL epiwafer is shown in Figure 1.

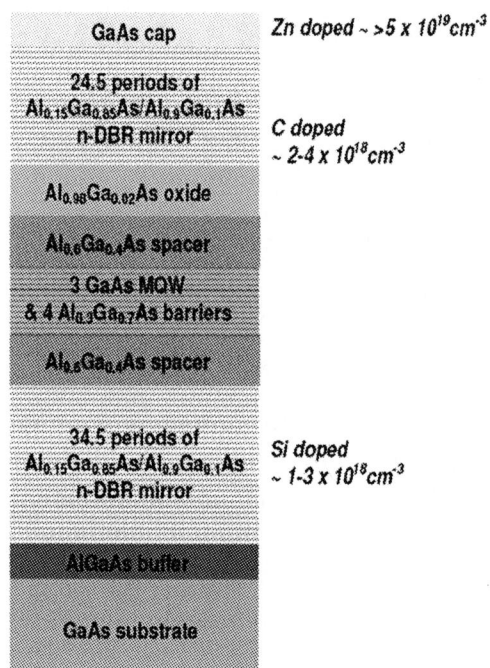

Fig. 1. Schematic of the 850 nm VCSEL epiwafer.

The VCSEL fabrication was completed as follows. After metallization of backside and top ring ohmic contacts on the 850 nm VCSEL epiwafer, square-shaped mesas are formed by optical lithography. Device mesas are etched by using Inductively Coupled Plasma-Reactive Ion Etching (ICP-RIE). The oxide layer is then selectively oxidized by wet oxidation. Figure 2 illustrates a cross-sectional schematic of the completed 850 nm VCSEL device after device fabrication.

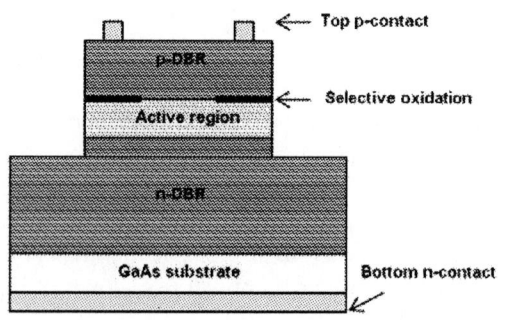

Fig. 2. Schematic of 850 nm VCSEL device.

In order to characterize the VCSEL epiwafer uniformity, a number of VCSEL devices were fabricated from the center to the edge of the

epiwafer and assigned label P1 (center) through P5 (edge), with P2, P3 and P4 devices in between, as shown in Figure 3. The 850 nm VCSEL epiwafer was characterized using photoluminescence (PL) technique. As for the VCSEL device measurement, the light output versus current (L-I) characteristics were measured by on-wafer probing connected with a semiconductor parameter analyzer and Si photodetector. The output spectra were obtained using an optical spectrum analyzer.

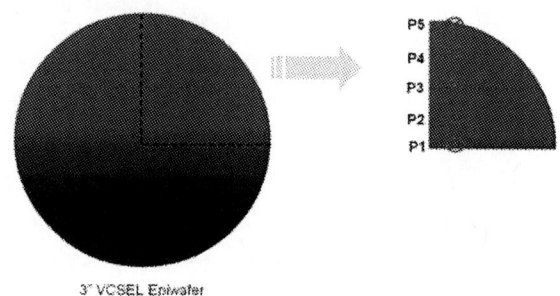

Fig. 3. 850 nm VCSEL devices fabricated at various positions on the VCSEL epiwafer.

III. RESULTS & DISCUSSION

Figure 4 shows the GaAs quantum wells emission from the VCSEL active region as measured using PL. The emission peak is at 839.5 nm and is intentionally designed lower than the nominal 850 nm VCSEL operating wavelength. This is done because to compensate for the red-shift of the gain spectrum as the device is heated under continuous-wave electrical injection.

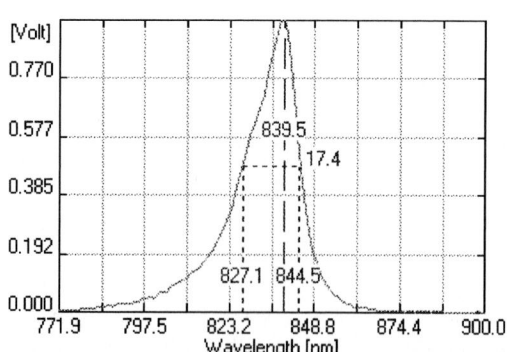

Fig. 4. QW emission wavelength of the 850 nm VCSEL epiwafer as measured using PL.

Figure 5 shows the DBR mirror reflectivity spectrum of the 850 nm VCSEL epiwafer. It exhibits a symmetric and flat-topped stopband, which indicates a high quality growth for the $Al_{0.9}Ga_{0.1}As/$ $Al_{0.15}Ga_{0.85}As$ DBR mirrors. The DBR demonstrates single longitudinal cavity resonance or Fabry-Perot (F-P) dip at 846.5 nm wavelength and a stopband width of 79.6 nm. The F-P dip determines the VCSEL emission wavelength.

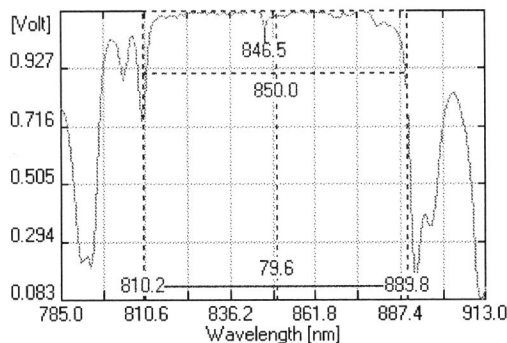

Fig. 5. DBR reflectivity spectrum of the 850 nm VCSEL epiwafer as measured using PL.

The quantum wells emission wavelength and reflectivity measurements were performed across the entire wafer and mapped using the PL, as shown in Figure 6. The mapping plots show only minor variations (± 3 nm and ± 16 nm) for the peak emission wavelength and F-P dip, respectively. The results from these mapping measurement show that the 850 nm VCSEL epiwafer is highly uniform.

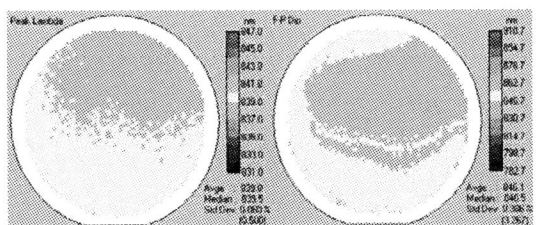

Fig. 6. PL mapping of the peak QW lambda and the DBR F-P dip for the 850 nm VCSEL epiwafer.

Further investigation of the epiwafer uniformity was carried out by characterizing the devices which were fabricated on various positions of the epiwafer (P1 to P5). Figure 7 shows the L-I

characteristics of the fabricated 36 μm mesa VCSEL devices. Devices at P1, P2, P3 and P4 exhibit nearly identical L-I characteristics with an average threshold current of 0.75 mA and average maximum output power of 3.1 mW. The VCSEL device at location P5 (corresponding to the epiwafer edge) exhibits different characteristic.

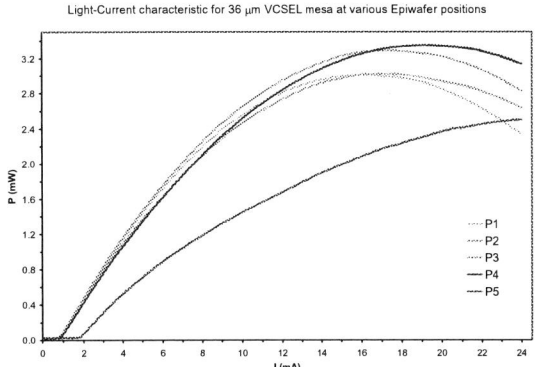

Fig. 7. L-I characteristics of 850 nm VCSEL at various epiwafer positions.

A similar trend was observed in the output spectral characteristics measured for the VCSEL devices, as shown in Figure 8. Devices at position P1 through P4 lase with a fundamental wavelength of approximately 848 nm. The VCSEL device exhibited a fundamental wavelength of 843 nm.

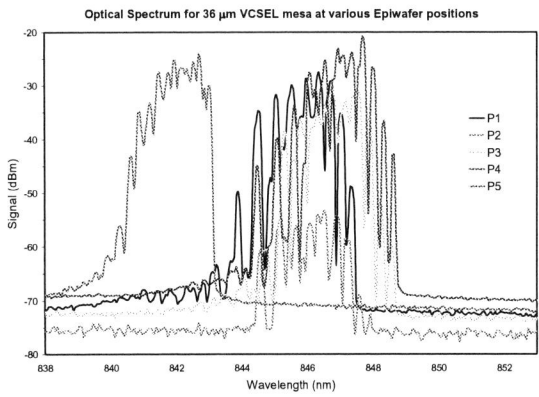

Fig. 8. Spectra characteristic of 850 nm VCSEL at various epiwafer positions.

The results of the L-I characteristics and output spectra characteristics measurements are reploted in Figure 9.

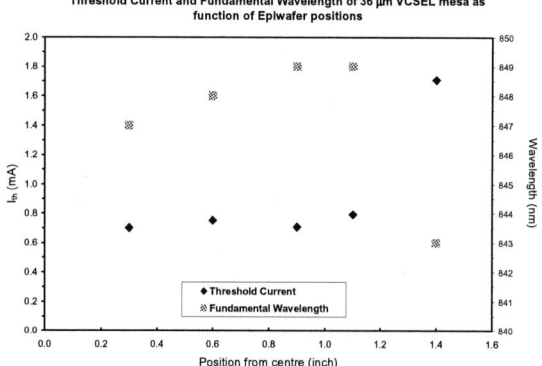

Fig. 9. I_{th} and spectra characteristic at various epiwafer positions for the 850 nm VCSELs.

Figure 9 clearly shows that the VCSEL devices at positions P1 through P4 have nearly identical characteristics, showing a high-degree of MOCVD growth uniformity for the 850 nm VCSEL epiwafer. Only the device at P5 exhibits significantly different characteristics since it is located at the very edge of the epiwafer. The P5 VCSEL has a slightly different cavity length, which affects the electro-opto properties and lasing wavelength of the VCSEL device. The inferior P5 device characteristics are due entirely to growth limitations, which are common for all kinds of epitaxial growth, including MOCVD and MBE.

IV. CONCLUSION

In this paper, we present uniformity analysis results for a high-quality 850 nm VCSEL epiwafer grown by MOCVD. Photoluminescence and reflectivity mappings for this wafer illustrate high uniformity across the entire wafer. Further investigation was carried out by fabricating and characterizing VCSEL devices from the center to the edge of the VCSEL epiwafer, in order to provide further uniformity details. Only devices fabricated at the very edge of the epiwafer exhibited significantly different characteristics. This effect was attributed to growth limitations.

ACKNOWLEDGEMENT

This work is sponsored by Telekom Malaysia Berhad through project number R05-0604-0. We would like to thank the researchers and management of Telekom Research & Development for their technical aid and constant support. Also, we would like to acknowledge the Photonics Device Research Group of University of Illinois at Urbana-Champaign for their scientific collaboration during this project.

REFERENCES

[1] K. –Y. Cheng, "Molecular Beam Epitaxy Technology of III-V Compound Semiconductors for Optoelectronic Applications," *Proc. IEEE,* vol. 85, no. 11, pp. 1694-1714 (1997).

[2] J. J. Coleman, "Metal Organic Chemical Vapor Deposition for Optoelectronic Devices," *Proc. IEEE,* vol. 85, no. 11, pp. 1715-1729 (1997).

[3] R. S. Geels, S. W. Corzine, J. W. Scott, D. B. Young and L. A. Coldren, "Low threshold planarized vertical-cavity surface-emitting lasers," *IEEE Photonics Technol. Lett.*, vol. 2, pp. 234-236 (1990).

[4] K. Tai, L. Yang, Y. H. Wang, J. D. Wynn and A. Y. Cho, "Drastic reduction of series resistance in doped semiconductor distributed Bragg reflectors for surface-emitting lasers," *Appl. Phys. Lett.,* vol. 56, pp. 2496-2498 (1990).

[5] E. F. Schubert, L. W. Tu, G. J. Zydzik, R. F. Kopf, A. Benvenuti and M. R. Pinto, "Elimination of heterojunction discontinuities by modulation doping," *Appl. Phys. Lett.,* vol. 60, pp. 466-468 (1992).

[6] P. Zhou, *et. al,* "Low series resistance high efficiency GaAs/AlGaAs vertical-cavity surface-emitting lasers with continuously graded mirrors grown by MOCVD," *IEEE Photonics Technol. Lett.*, vol. 3, pp. 591-593 (1991).

[7] C. Wilmsen, H. Temkin, and L. A. Coldren, Vertical-Cavity Surface-Emitting Lasers: Design, Fabrication, Characterization and Applications, Cambridge University Press (1999).

Electro-Opto Characteristics of 850 nm Oxide-Confined Vertical-Cavity Surface-Emitting Lasers

Mohd Sharizal Alias, *Member, IEEE*, Paul O. Leisher [*], *Student Member, IEEE*, Kent D. Choquette [*], *Fellow, IEEE*, Khairul Anuar, Dominic Siriani [*], Sufian Mitani, Mohd Razman Y. and Abdul Fatah A. M.

Telekom Research & Development (TMR&D)
Idea Tower, UPM-MTDC, Lebuh Silikon
43400 Serdang, Selangor, MALAYSIA
Email: sharizal@tmrnd.com.my

[*] Electrical & Computer Engineering Department
University of Illinois at Urbana-Champaign (UIUC)
Urbana, 61801, Illinois, USA

Abstract In this paper, an MOCVD grown of VCSEL with an operating wavelength of 850 nm is fabricated and characterized. The sample includes numerous oxide aperture sizes, allowing a thorough investigation of the electrical and optical characteristics and overall device performance. Low threshold current operation <1mA was achieved for oxide apertures smaller than 10 μm. Analysis of the VCSEL performance as a function of the oxidize aperture sizes is also reported.

I. INTRODUCTION

Since their breakthrough in last decade, Vertical-Cavity Surface-Emitting Lasers (VCSEL) have dominated the semiconductor laser applications in short-haul optical communication, especially at an operating wavelength of 850 nm. Although much of the current research deals with long wavelength VCSEL (1330 to 1550 nm), there is still significant interest in 850 nm VCSEL. Much of this work has focused on improving the device performance by utilizing such methods as oxide confinement [1], surface-relief etching [2], extended optical cavity [3], increased fundamental mode gain [4], hybrid implant/oxide structures [5], composite resonators [6] and photonic crystal structures [7].

In this paper, 850 nm VCSEL devices with various etched mesa sizes and corresponding oxide aperture sizes are fabricated and characterized. The light-current (L-I) and current-voltage (I-V) characteristics for each of these 850 nm VCSEL devices is measured. The electro-opto properties of the devices with different oxide apertures are also investigated.

II. DEVICE FABRICATION & EXPERIMENT

The VCSEL structure consists of a bottom 34.5-period n-type Distributed Bragg Reflector (DBR) mirror, an active region of three GaAs quantum wells, and a top 24.5-period of p-type DBR mirror; all grown by Metal-Organic Chemical Vapor Deposition (MOCVD). Both DBR mirrors are composed of alternating high- and low-refractive index layers of $Al_{0.9}Ga_{0.1}As/$ $Al_{0.15}Ga_{0.85}As$ with one $Al_{0.98}Ga_{0.02}As$ layer was introduced underneath the top p-DBR mirror for the purpose of selective oxidation.

The 850 nm VCSEL fabrication began with the definition of a top ring contact pattern using standard optical lithography. Ti/Au (15 nm/150 nm) was then evaporated, and lifted off, to form ohmic contacts to the p-type top DBR. The backside (n-type) ohmic contact was formed by evaporation of AuGe/Ni/Au (40 nm/20 nm/150 nm). These evaporations were performed using both thermal and electron beam evaporation. In order to isolate the various VCSEL devices and expose the high Al layer for oxidation, mesa structures were defined as follows. First, SiO_2 was deposited by Plasma-Enhanced Chemical Vapor Deposition (PECVD) and subsequently patterned using standard optical lithography and CF_4 Reactive Ion Etching (RIE). This patterned dielectric was then used as the mask for the mesas which were etched by Inductively Coupled Plasma RIE (ICP-RIE). The high Al layer was then selectively oxidized by wet oxidation to define the oxide and the SiO_2 mesa mask was removed by CF_4 RIE. Figure 1 shows the cross-section schematic of a single 850 nm VCSEL device after fabrication. VCSEL devices with different mesas sizes were fabricated on the single 850 nm VCSEL epiwafer sample as shown in Figure 2.

grounded through the probe station chuck. Current is injected by the Keithley 236 DC current source, and precisely varied by the Agilent 4156C semiconductor parameter analyzer. To determine the optical power output, the emission beam from the lasing VCSEL is measured using a Si photodetector which is also connected to the semiconductor parameter analyzer. The differential series resistance for the VCSEL device was obtained from the slope of I-V curve.

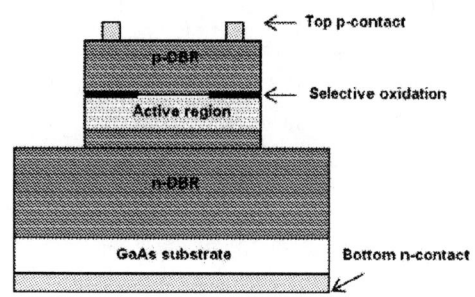

Fig. 1. Schematic of 850 nm VCSEL device.

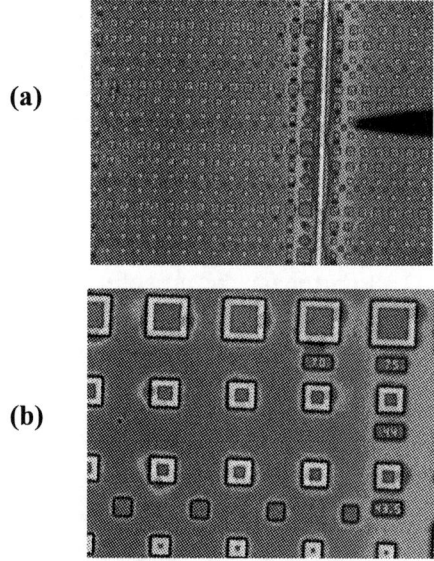

(a)

(b)

Fig. 2. (a) Various 850 nm VCSEL devices fabricated and (b) the VCSEL mesas sizes in detail.

Device measurements for the L-I and I-V characterization were measured using on-wafer probing. The probe station set-up schematic is illustrated in Figure 3. Electrical contact to the VCSEL device is established by probing the top DBR contact, with the backside of the sample

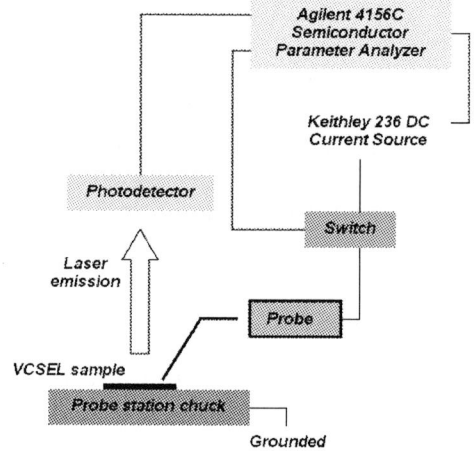

Fig. 3. Schematic of VCSEL measurement set-up.

III. RESULTS & DISCUSSION

Figure 4 exhibits typical L-I characteristic for the 850 nm oxide VCSEL devices with various etched mesa sizes.

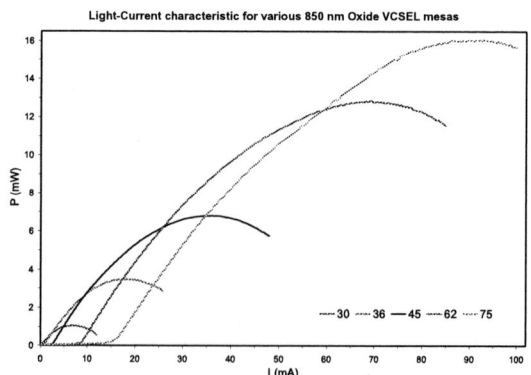

Fig. 4. L-I characteristics of 850 nm VCSEL devices with various mesas sizes.

Threshold currents ranging from 0.5 mA to 15 mA and maximum output powers from 0.9 mW to 16 mW were achieved with increasing mesa size

240

by these devices. This is expected since larger mesas will have larger oxide apertures which define the lasing modes. Higher output power is due to increased active region area, but this comes at a cost of increased threshold current. In contrast, smaller VCSEL mesas have smaller oxide apertures which require lower total current to achieve lasing threshold.

The I-V characteristics for the 850 nm VCSEL devices are shown in Figure 5 for various mesa sizes. As the VCSEL mesas sizes increases, the slope of the I-V curve decreases, indicating a reduction in differential series resistance. This reduction is due to the increasing cross-sectional area of the oxide aperture.

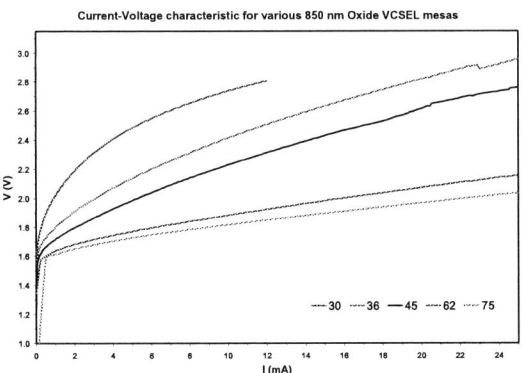

Fig. 5. I-V characteristics of 850 nm VCSEL devices with various mesas sizes.

In order to determine the oxide aperture dimensions, electrical probing is used to test which mesa sizes are not conducting (pinched-off). For this sample, it was found that VCSEL with mesa sizes 27 μm and below were pinched-off. Thus, each of the VCSEL mesas characterized for the L-I and I-V characteristics (30, 36, 45, 62 and 75 μm) has a 3x3 μm^2, 9x9 μm^2, 18x18 μm^2, 35x35 μm^2, 48x48 μm^2 oxide aperture, respectively.

VCSEL electro-opto characteristics as a function of oxide apertures were analyzed in detail. Figure 6 illustrates the threshold current versus the VCSEL oxide aperture size. Since laser threshold gain depends on threshold current density, as opposed to total threshold current, higher current injection is required for VCSEL with larger aperture sizes to achieve lasing.

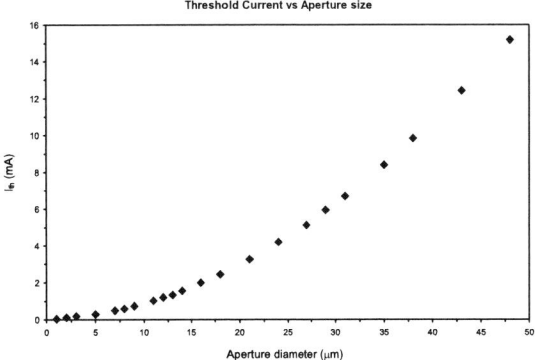

Fig. 6. Threshold current as function of aperture size for the 850 nm VCSEL devices.

Figure 7 shows the threshold current density as a function of aperture size for the 850 nm VCSEL devices. Because these devices were fabricated on the same wafer and processed in parallel, threshold gain, and therefore threshold current density, should in theory remain constant regardless of aperture size. The departure from this constant trend for small oxide apertures (from 3x3 μm^2 to 10x10 μm^2) is primarily due to increased optical scattering loss off the small oxide apertures.

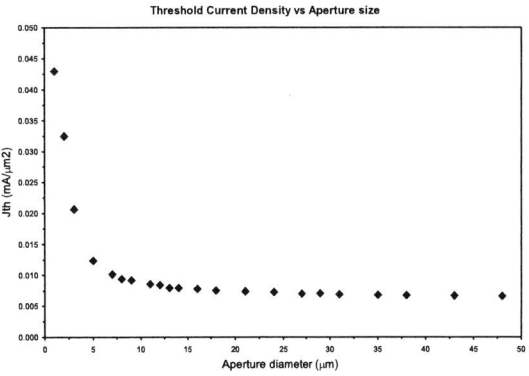

Fig. 7. Threshold current density as function of aperture size for the 850 nm VCSEL devices.

The maximum output power and differential series resistance as a function of oxide aperture size is shown in Figure 8 and Figure 9, respectively. In Figure 8, the maximum output power increases for increasing VCSEL apertures. This is because more lasing modes are confined in larger oxide apertures. The inverse dependence of series resistance on VCSEL aperture size, as shown in Figure 9, is due increasing cross-

241

sectional conductive area. This is in agreement with Ohm's law.

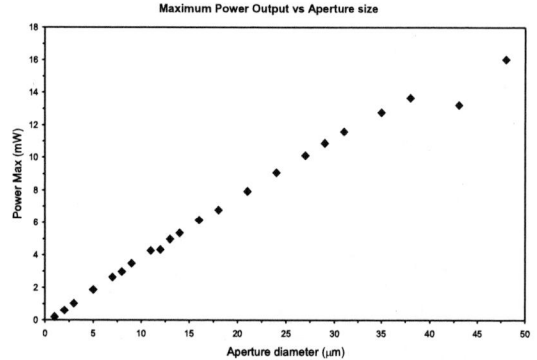

Fig. 8. Maximum power output as function of aperture size for the 850 nm VCSEL devices.

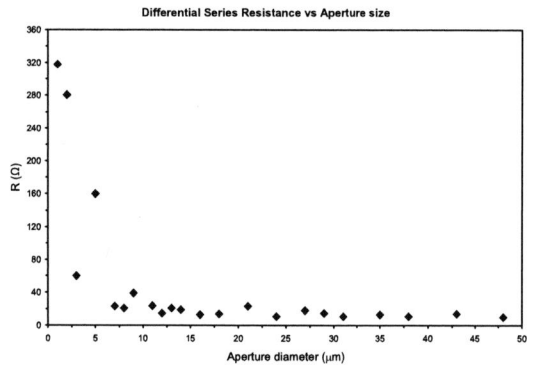

Fig. 9. Differential series resistance as function of aperture size for the 850 nm VCSEL devices.

IV. CONCLUSION

The electro-opto characteristics of VCSEL devices fabricated on the same 850 nm epiwafer with various oxide aperture sizes was analyzed in this work. The results show that the fabricated VCSEL devices perform well, indicating a high-quality epiwafer was growth. Low threshold current and low differential series resistance was demonstrated by the fabricated 850 nm VCSEL devices.

ACKNOWLEDGEMENT

The work is sponsored by Telekom Malaysia Berhad through project number R05-0604-0. We

are grateful to the researchers and the management of Telekom Research & Development for the technical aid and constant support. Also, we would like to acknowledge the Photonics Device Research Group of University of Illinois at Urbana-Champaign for tremendous scientific collaboration during the project.

REFERENCES

[1] K. D. Choquette, R. P. Schneider, Jr., K. L. Lear, and K. M. Geib, "Low threshold vertical-cavity lasers fabricated by selective oxidation," *Electron Lett.*, vol. 30, pp. 2043-2044 (1994).

[2] H. Martinsson, J. A. Vukusic, M. Grapbherr, R. Michalzik, R. Jager, K. J. Ebeling and A. Larsson, "Transverse mode selection in large-area oxide-confined vertical-cavity surface-emitting lasers using a shallow surface-relief," *IEEE Photonics Technol. Lett.*, vol. 11, pp. 1536-1538 (1999).

[3] H. Unold, S. W. Z. Mahmoud, R. Jager, M. Kicherer, M. C. Riedl, and K. J. Ebeling, "Improving single-mode VCSEL performance by introducing a long monolithic cavity," *IEEE Photonics Technol. Lett.*, vol. 12, pp. 939-941 (2000).

[4] K. D. Choquette, K. M. Geib, R. D. Briggs, A. A. Allerman, and J. J. Hindi, "Single transverse mode selectively oxidized vertical cavity lasers;" in *Proc. SPIE*, K. D. Choquette and S. Lei, Eds., 2000, 3946, pp. 230-233.

[5] E. W. Young, K. D. Choquette, S. L. Chuang, K. M. Geib, A. J. Fischer, and A. A. Allerman, "Single-Transverse-Mode Vertical-Cavity Lasers under Continuous and Pulsed Operation," *IEEE Photonics Technol. Lett.*, vol. 3, no. 9, pp. 927-929 (2001).

[6] A. J. Fischer, K. D. Choquette, W. W. Chow, A. A. Allerman, D. K. Serkland, and K. M. Geib, "High single-mode power observed from a couple-resonator vertical-cavity laser diode," *Apply. Phys. Lett.*, vol. 79, no. 25, pp. 4079-4081 (2001).

[7] N. Yokouchi, A. J. Danner, and K. D. Choquette, "Two-dimensional photonic crystal confined vertical-cavity surface-emitting lasers," *IEEE J. Sel. Top. Quantum Electron.*, vol. 9, no. 5 pp. 1439-1445 (2003).

Efficiency and Spectral Characteristics of 850 nm Oxide-Confined Vertical-Cavity Surface-Emitting Lasers

Mohd Sharizal Alias, *Member, IEEE*, Paul O. Leisher [*], *Student Member, IEEE*, Kent D. Choquette [*], *Fellow, IEEE*, Khairul Anuar, Dominic Siriani [*], Sufian Mitani, Mohd Razman Y. and Abdul Fatah A. M.

Telekom Research & Development (TMR&D)
Idea Tower, UPM-MTDC, Lebuh Silikon
43400 Serdang, Selangor, MALAYSIA
Email: sharizal@tmrnd.com.my

[*] Electrical & Computer Engineering Department
University of Illinois at Urbana-Champaign (UIUC)
Urbana, 61801, Illinois, USA

Abstract — In this work, a number of 850 nm vertical-cavity surface-emitting lasers (VCSELs) with varying oxide aperture sizes are fabricated and characterized to study the device efficiency and spectral characteristics. Differential quantum efficiencies up to 28% corresponding to wall-plug efficiencies of 15%, and multi-mode output spectrum rang of 845 nm to 850 nm were measured for a number of these devices. Additionally, the efficiency characteristics and spectral as a function of oxide aperture size for these 850 nm VCSELs are analyzed and explained.

I. INTRODUCTION

Vertical-cavity surface-emitting lasers (VCSELs) with operating wavelengths at 850 nm are widely commercialized for short-haul optical data communication. There is currently much research interest in VCSEL technology at longer wavelengths (1330 to 1550 nm) for long-haul optical communication application. However, there still exists a need for improving device performance at shorter wavelengths (850 nm) by methods such as oxide confinement [1] surface-relief etching [2], extended optical cavity [3], increased fundamental mode gain [4], hybrid implant/oxide structure [5], composite resonators [6], and photonic crystals [7].

In this paper, oxide-confined 850 nm VCSEL devices with various oxide aperture sizes have been fabricated. We report the VCSEL device efficiency and spectral characteristics with the intention of studying the oxide aperture effects on the device performance

II. DEVICE FABRICATION & EXPERIMENT

The oxide 850 nm VCSEL epiwafer structure was grown by Metal-Organic Chemical Vapor Deposition (MOCVD). The epiwafer consists of three GaAs quantum wells which form the active region and is sandwiched between the bottom n-type and top p-type DBR mirrors. Alternating high- and low-refractive index layers ($Al_{0.9}Ga_{0.1}As/ Al_{0.15}Ga_{0.85}As$) form the bottom 34.5-periods n-DBR mirror and the top 24.5-periods p-DBR mirror. A high aluminum concentration alloy layer ($Al_{0.98}Ga_{0.02}As$) was introduced in the p-DBR mirror for selective oxidation. From this epiwafer, VCSEL devices with numerous mesas (and corresponding oxide aperture sizes) were fabricated to investigate the 850 nm VCSEL device efficiency and spectral characteristics as a function oxide aperture size.

The VCSEL fabrication proceeded as follows. First, Ti/Au (15 nm/150 nm) and AuGe/Ni/Au (40 nm/20 nm/150 nm) were evaporated to form the top p- and backside n-metal contacts, respectively. Using standard optical lithography, mesas were defined in an SiO_2 mask which was previously deposited by Plasma-Enhanced Chemical Vapor Deposition (PECVD) and then etched by Inductively Coupled Plasma Reactive Ion Etching (ICP-RIE). These mesas serve the dual purpose of

0-7803-9730-4/06/$25.00 ©2006 IEEE   243

device isolation and exposing the high Al-containing layer for selective oxidation. The oxide apertures were formed by wet oxidation. Because oxidation rate is independent of mesa size, each different mesa size uniquely corresponds to one oxide aperture size. Figure 1 shows the schematic of a single fabricated 850 nm VCSEL device on the epiwafer. Many different mesa sizes were fabricated so that the device characteristics as a function of oxide aperture size could be studied.

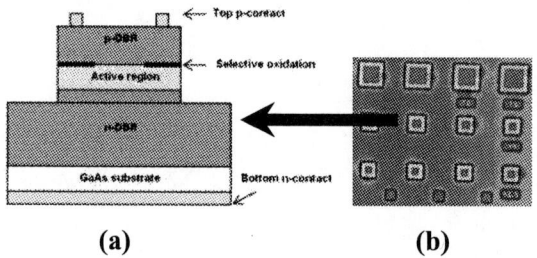

(a)   (b)

Fig. 1. The fabricated oxide 850 nm VCSEL in (a) schematic single device and (b) various mesa and aperture sizes on the VCSEL epiwafer.

The measurement set-up used to characterize the VCSEL device efficiencies and output spectra is illustrated in Figure 2. Current is supplied by the Keithley 236 DC current source or Agilent 4156C semiconductor parameter analyzer through the probe and backside contact. To measure the output power, the lasing VCSEL beam is detected by a Si photodetector which is also connected to the semiconductor parameter analyzer.

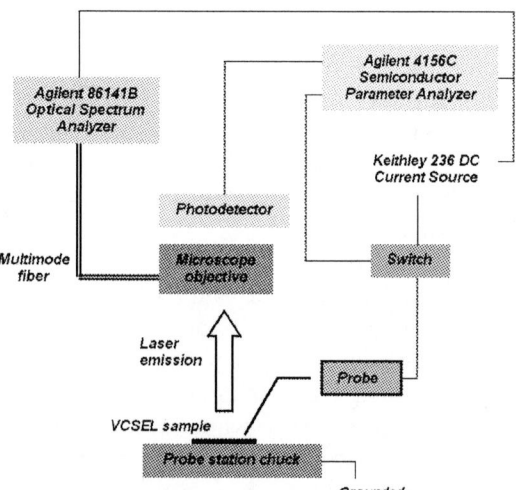

Fig. 2. Schematic of VCSEL measurement set-up.

The VCSEL device efficiency is determined from the light-current (L-I) characteristic. The differential quantum efficiency is calculated as follows [8]

$$\eta_D = \frac{dP/\hbar\omega}{dI/e} \cong \frac{dP}{dI}\frac{1}{E_g} \qquad (1)$$

where e is the electron charge, E_g is the bandgap energy, and $\hbar\omega$ is the photon energy. The wall-plug efficiency, or power conversion efficiency, is calculated using two methods. The theoretical power conversion efficiency is specified as [9]

$$\eta_{wp} = \eta_D \frac{E_g}{V_b}\left(1-\frac{I_{th}}{I}\right) \qquad (2)$$

where η_D is the differential quantum efficiency and V_b is the voltage bias. The power conversion efficiency can also be calculated using Equation 3, where P_{out} is optical power output, I is current and V is voltage, respectively.

$$\eta_{wp} = \frac{P_{out}}{P_{in}} = \frac{P_{out}}{IV} \qquad (3)$$

To analyze the device spectral properties, the VCSEL laser beam is coupled into a multimode fiber through the microscope objective of the probe station. The multimode fiber is then connected to the Agilent 86141B optical spectrum analyzer.

III. RESULTS & DISCUSSION

Figure 3 illustrates an example of a typical light-current-voltage (L-I-V) characteristic for 850 nm oxide VCSEL device. This VCSEL device has mesa size of 38x38 μm^2 with corresponding oxide aperture size of 11x11 μm^2. From the L-I curve, the VCSEL threshold current, I_{th} is approximately at 1 mA with maximum output power, P_{max} around 4.25 mW. The I-V curve shows that the turn-on voltage, V_{on} is about 1.5 V. This value approximately corresponds to the bandgap energy, E_g of the active region which is at 1.46 eV. The differential series resistance, R_s can be obtained from the slope of the I-V curve, and is found to be approximately 24 Ω. VCSELs typically exhibit high series resistances, relative to other semiconductor laser diodes, due to the steep heterojunction barriers of the multiple DBR interfaces used.

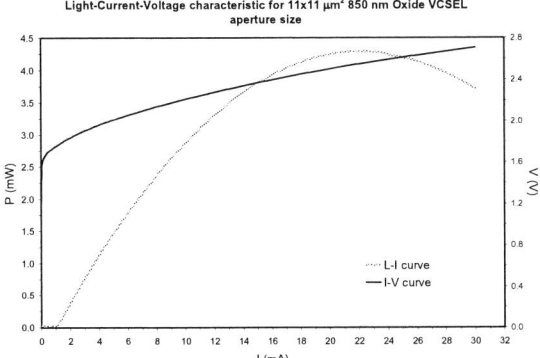

Fig. 3. L-I characteristic for 850 nm oxide VCSEL with 11x11 μm² oxide aperture.

As described in the previous section, the differential quantum efficiency, η_D can be calculated from equation (1). Using the L-I curve example of the 850 nm oxide VCSEL with 11x11 μm² oxide aperture, the differential quantum efficiency of this device is found to be 27%. This parameter indicates the efficiency of a VCSEL in converting the injected electron-hole pairs to photons emitted from the device. Figure 4 illustrates the differential quantum efficiency of various 850 nm oxide VCSELs fabricated with different oxide aperture sizes, ranging from 1x1 μm² to 48x48 μm². VCSELs with smaller oxide apertures show decreased differential quantum efficiency due to increased scattering loss which results from the small oxide aperture.

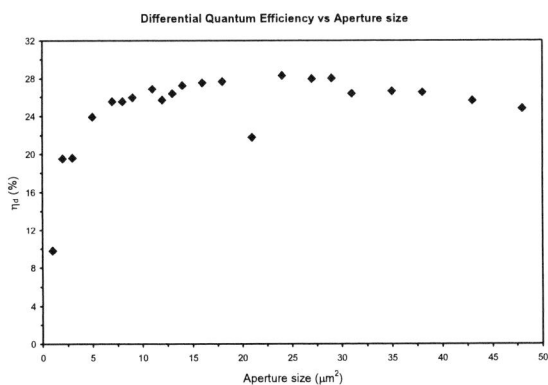

Fig. 4. Differential quantum efficiency for 850 nm VCSEL devices with various oxide apertures.

Figure 5 illustrates the wall-plug efficiency as a function of input current for the 850 nm oxide VCSEL with 11x11 μm² oxide aperture as

calculated using equations (2) and (3). The wall-plug efficiency denotes the total input electrical power to output optical power efficiency for the VCSEL. Both calculations reveal a maximum wall-plug efficiency of 14.5% for the 11x11 μm² oxide aperture VCSEL.

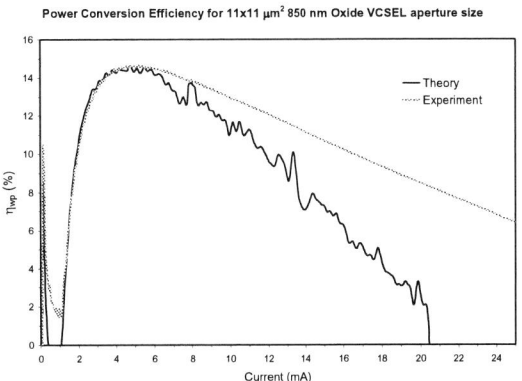

Fig. 5. Wall-plug efficiency for 850 nm oxide VCSEL with 11x11 μm² oxide aperture.

The wall-plug efficiency is calculated for the fabricated devices with oxide apertures from 1x1 μm² to 48x48 μm² and plotted in Figure 6. The smaller oxide aperture VCSELs exhibit higher wall-plug efficiencies compared to the larger oxide apertures VCSELs. This is because the smaller oxide aperture devices require less electrical current to lase and less heat is generated from the common self-heating phenomena in VCSEL.

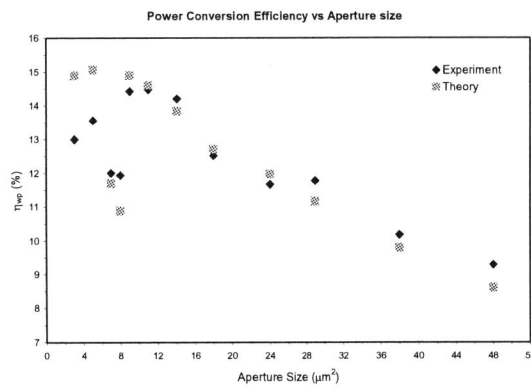

Fig. 6. Wall-plug efficiency for 850 nm VCSEL devices with various oxide apertures.

Figure 7 shows an example of output spectral characteristic for 850 nm VCSEL with 3x3 μm² oxide aperture at different current levels (from $0.5 \times I_{th}$ to $20 \times I_{th}$). The VCSEL operates with multiple transverse modes from threshold through rollover, which is a common characteristic in oxide VCSELs.

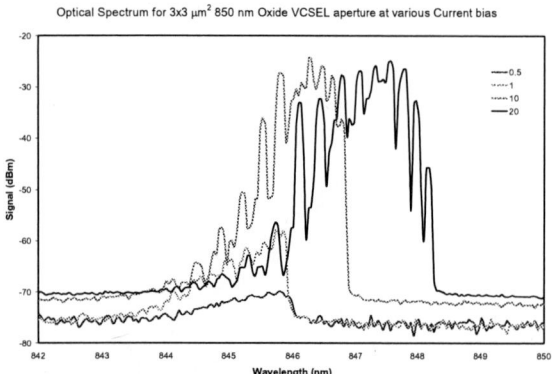

Fig. 7. Spectral characteristics for 850 nm oxide VCSEL with 3x3 μm² oxide aperture at different threshold current bias.

Figure 8 illustrates optical near-field images of the output laser emission from the same VCSEL device at different current injection levels, from threshold to rollover.

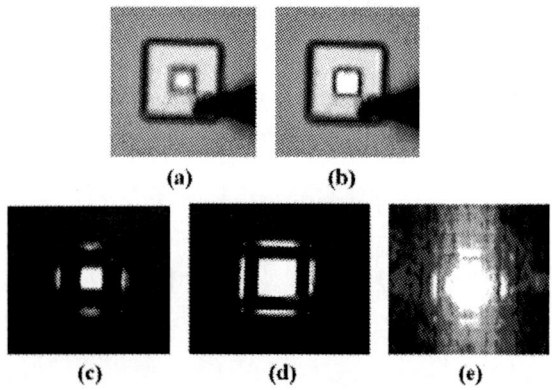

Fig. 8. Laser emission from the surface of the 850 nm VCSEL device with 3x3 μm² oxide aperture at (a) threshold, (b) and (c) above threshold, (d) higher current, and (e) before rollover.

In Figure 9, the fundamental lasing emission wavelength as a function of injected current is shown. The fundamental lasing wavelength increases as the injected current increased. This red-shift phenomenon is due to the increasing heat generated from the VCSEL as higher current is injected.

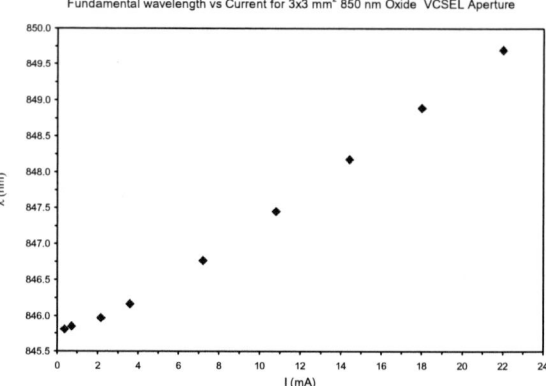

Fig. 9. Fundamental wavelength analysis for oxide 850 nm VCSEL devices.

IV. CONCLUSION

In this paper, 850 nm VCSELs with differing oxide aperture sizes were fabricated in order to study the device efficiency parameters. Details of device efficiency such as differential quantum efficiency and wall-plug efficiency are studied, showing efficient VCSEL devices were fabricated. Output spectral characteristics were also presented, indicating typical multimode operation for these devices.

ACKNOWLEDGEMENT

The work is sponsored by Telekom Malaysia Berhad through project number R05-0604-0. We are grateful to the researchers and management of Telekom Research & Development for technical aid and constant support. Also, we would like to acknowledge the Photonics Device Research Group of University of Illinois at Urbana-Champaign for scientific collaboration during the project.

REFERENCES

[1] K. D. Choquette, R. P. Schneider, Jr., K. L. Lear, and K. M. Geib, "Low threshold vertical-cavity lasers fabricated by selective oxidation," *Electron Lett.,* vol. 30, pp. 2043-2044 (1994).

[2] H. Martinsson, J. A. Vukusic, M. Grapbherr, R. Michalzik, R. Jager, K. J. Ebeling and A. Larsson, "Transverse mode selection in large-area oxide-confined vertical-cavity surface-emitting lasers using a shallow surface-relief,"

IEEE Photonics Technol. Lett., vol. 11, pp. 1536-1538 (1999).

[3] H. Unold, S. W. Z. Mahmoud, R. Jager, M. Kicherer, M. C. Riedl, and K. J. Ebeling, "Improving single-mode VCSEL performance by introducing a long monolithic cavity," *IEEE Photonics Technol. Lett.*, vol. 12, pp. 939-941 (2000).

[4] K. D. Choquette, K. M. Geib, R. D. Briggs, A. A. Allerman, and J. J. Hindi, "Single transverse mode selectively oxidized vertical cavity lasers;" in *Proc. SPIE*, K. D. Choquette and S. Lei, Eds., 2000, 3946, pp. 230-233.

[5] E. W. Young, K. D. Choquette, S. L. Chuang, K. M. Geib, A. J. Fischer, and A. A. Allerman, "Single-Transverse-Mode Vertical-Cavity Lasers under Continuous and Pulsed Operation," *IEEE Photonics Technol. Lett.*, vol. 3, no. 9, pp. 927-929 (2001).

[6] A. J. Fischer, K. D. Choquette, W. W. Chow, A. A. Allerman, D. K. Serkland, and K. M. Geib, "High single-mode power observed from a couple-resonator vertical-cavity laser diode," *Apply. Phys. Lett.*, vol. 79, no. 25, pp. 4079-4081 (2001).

[7] N. Yokouchi, A. J. Danner, and K. D. Choquette, "Two-dimensional photonic crystal confined vertical-cavity surface-emitting lasers," *IEEE J. Sel. Top. Quantum Electron.*, vol. 9, no. 5 pp. 1439-1445 (2003).

[8] Y. Suematsu, and A. R.Adams, *Handbook of Semiconductor Lasers and Photonic Integrated Circuits*, Chapman & Hall (1994).

[9] K. Iga, "Surface-emitting lasers-Its birth and generation of new optoelectronics field," *IEEE J. Select. Topics Quantum Electron.*, vol. 6 no. 6, pp. 1201-1215 (2000).

2D Silicon-based Photonic Crystals

Y. K. Sin and K. Ibrahim

Nano-Optoelectronics Research and Technology Laboratory
School of Physics
Universiti Sains Malaysia
11800 Minden, Pulau Pinang, MALAYSIA
Email: willsyk@yahoo.com and kamarul@usm.my
Tel: 604-6533599, Fax: 604-6579150

Abstract In the present paper, fabrication of a silicon-based two-dimensional photonic crystal based on electron beam lithography (EBL) and reactive ion etching (RIE) was demonstrated. In addition the structural color of fabricated crystal was obtained with a simple experimental setup consisted of a light source and a digital camera. From the result, the structural color of PhCs was underlying on the concept of reflection grating.

I. INTRODUCTION

RECENTLY Photonic crystals (PhCs) become intense interest among photonic industry because their applications promised a better performance in optical components like light-emitting diodes, waveguides and optical fibers [1]-[3]. PhCs are materials with periodically patterned in refractive index, which prohibited the propagation of certain range of frequencies. The prohibited frequencies are known as photonic bandgap; and this phenomenon is analogous to electronic bandgap where electrons interact with crystalline solids [4].

Since Yablonovitch [5] and John [6] had introduced this novel class of engineered dielectric materials in 1987, numbers of study are carried out on the fabrications and characterizations of PhCs. A variety of materials such as semiconductors, oxides and polymers have been utilized in the fabrication process for different structures. In our study, silicon was used to fabricate a two-dimensional crystal.

Silicon is the dominant material in microelectronics but only a passive material in the field of photonics due to the indirect bandgap characteristic. With the fabricated PhCs structures, silicon can be turned into an active photonic material in photon confinement and light localization [7]. From these points of view, we can visualize a combination of photonic and electronic integrated circuits which fabricated on a same silicon wafer in our future.

Structural color is defined as the resulting color when light interacts with a system's physical structure. This phenomenon is ruled by four mechanisms such as interference, scattering, refraction, and diffraction [8]. Periodicity of the designed PhCs structures basically can be said to be a diffraction grating pattern. Hence the structural color of PhCs is based on the mechanism of diffraction in reflection mode.

Characterizations of PhCs normally are the study of photonic bandgap where transmittance and reflectance spectrum from the crystal structures are indicated. But lately some scientists attempt to relate the structural color effect with the photonic bandgap in characterization method [9]-[11]. In this method, spectroscopic ellipsometry is the tool for measuring the reflectance spectrum from PhCs at different angle of incidence and reflection.

In this paper we demonstrate the fabrication of a 2D silicon-based PhC obtained by electron beam lithography (EBL) and reactive ion etching (RIE). Furthermore we show the structural color of the fabricated PhC with the assistance of a light source and a digital camera.

II. EXPERIMENT

In our work, p-type (111) oriented silicon wafer was used to fabricate 2D PhCs structures. A 200nm-thick layer of silicon oxide was grown with thermal oxidation to act as masking layer in RIE process. After that, the e-beam patterning of polymethyl methacrylate (PMMA) was made by spin-coating a 200nm-thick resist layer onto the silicon oxide. The resulting structure formed in PMMA after the exposure of EBL was a periodic pattern of submicron square windows. Consequently, isotropic wet etching was performed through these windows into the oxide layer by buffered HF solution (7 NH_4F: 1HF).

0-7803-9730-4/06/$25.00 ©2006 IEEE

Following this silicon was etched via the PMMA and silicon oxide mask by RIE. In the final stage, residual from the masking layer was removed and a 2D silicon-based PhC was fabricated.

Refer to Fig.1, a simple experimental setup consisted of a light source and a digital camera was used to study the structural color effect of the fabricated crystal structure. In this setup, the light source was fixed at the position normal to the plane of PhC. Hence the angle of incidence, θ_m, was equal to 0° and this would simplify the general grating equation.

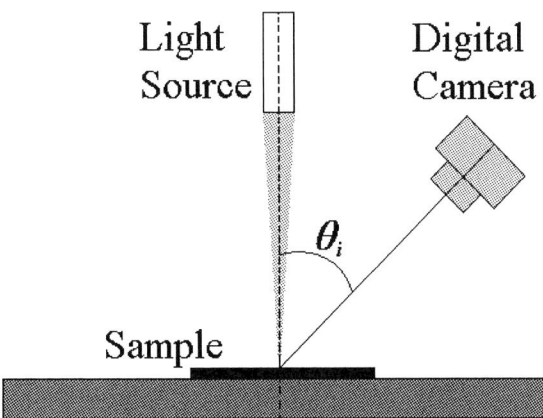

Fig.1 Experimental setup consisted of a light source and a digital camera which used to observed the structural color of fabricated crystal structure at different angles of reflection, θ_i.

III. RESULTS AND DISCUSSION

From scanning electron microscopic (SEM) images shown in Fig. 2 and 3, we noticed that fabricated structures were air holes in approximately round shape with diameter at ~550nm. This is because the square windows which written in PMMA were rounded during

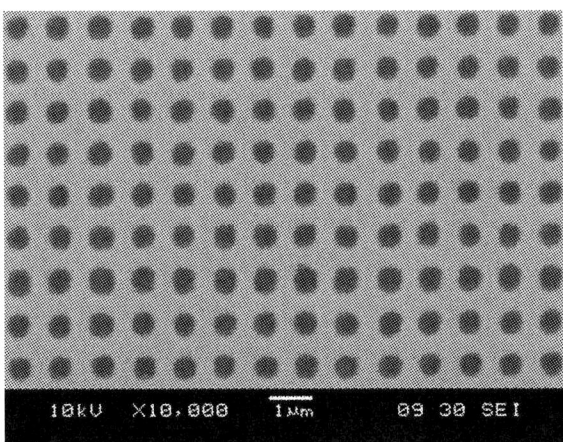

Fig.2 SEM images of fabricated PhCs structures at magnification ×10,000.

Fig.3 SEM images of fabricated PhCs structures at magnification ×10,000 with 30° tilt.

the isotropic wet etching process in silicon oxide layer. In addition distance between two air holes were found in the range of 900-950nm.

The general grating equation is as given as follow [12]:

$$a(\sin\theta_m - \sin\theta_i) = m\lambda \qquad (1)$$

In this equation, θ_m and θ_i represent angle of incidence and reflection respectively; as well as a and λ represent the lattice constant and wavelength. In addition the values of m specify the order of the various principal maxima. According to the experimental setup and the fabricated crystal structure, reflectance spectrum was calculated with the values of a and θ_m equal to 900-950nm and 0° respectively. The resulting spectrums were shown in Table 1.

Table 1 Reflectance spectrums from fabricated PhC which calculated with grating equation.

Angle of reflection, θ_i	Wavelength (nm)	Structural Color
30°	450 ~ 475	Indigo + Blue
35°	516 ~ 545	Blue + Green
40°	579 ~ 611	Yellow + Orange
45°	636 ~ 672	Red
50°	689 ~ 728	Red
55°	737 ~ 778	Red
60°	779 ~ 823	Red (1st order)
	390 ~ 411	Violet (2nd order)

The spectrums reflected from fabricated crystal structures and bulk silicon observed by a digital camera at different angles of reflection was shown in Table 2. No difference was observed between the spectrums from bulk silicon at any angle of reflection. But the reflectance spectrum from silicon-based PhC was varied from short wavelength to long wavelength while the angle of reflection was increased. Comparing the result in Table 1 and 2, the reflectance spectrum from fabricated silicon-based PhC matched the calculated spectrum with the grating equation.

Table 2 Reflectance spectrums from fabricated silicon-based PhC (left) and bulk silicon (right) taken with digital camera at different angle of reflection, θ_i.

Angle of reflection, θ_i	Silicon-based PhC	Bulk Silicon
30°		
35°		
40°		
45°		
50°		
55°		
60°		

IV. CONCLUSION

As a conclusion a 2D silicon-based PhC has been fabricated successfully. From the observation, an optical characteristic of silicon is changed with the appearance of PhCs structures. Additionally we showed that the structural color of the crystals is relied on the concept of reflection grating. In the future, PhCs structure will enrich the applications of silicon in optical components.

ACKNOWLEDMENT

The authors would like to thank the Malaysian Ministry of Science, Technology and Environment for sponsoring this work under project IRPA RMK-8 Strategic Research and National Scientific Fellowship; as well as Universiti Sains Malaysia for providing the facilities at the Nano-Optoelectronics Research and Technology Laboratory.

REFERENCES

[1]   S. Fan, P. R. Villeneuve, and J. D. Joannopoulus, "High extraction effiency of spontaneous emission from slabs of photonic crystals", *Physical Review Letters*, vol. 78, p. 3294 (1997).

[2]   P. Rigby, "A photonic crystal fiber", *Nature*, vol. 396, pp. 415-416 (1998).

[3]   K. Aoki, H. Hirayama, and Y. Aoyagi, "A novel three-dimensional photonic laser and fabrication of three-dimensional photonic crystals", *RIKEN Review*, vol. 33, pp. 24-27 (2001).

[4]   J. D. Joannopoulus, P. R. Villeneuve and S. Fan, "Photonic crystals: putting a new twist on light", *Nature*, vol. 386, pp. 143-149 (1997).

[5]   E. Yablonovitch, "Inhibited spontaneous emission in solid-state physics", *Physical Review Letters*, vol. 58, pp. 2059-2062 (1987).

[6]   S. John, "Strong localization of photons in certain disordered dielectric superlattices", *Physical Review Letters*, vol. 58, p. 2486 (1987).

[7]   L. Pavesi, Z. Gaburro, L. D. Negro, P. Bettotti, G. V. Prakash, M. Cazzanelli and C. J. Oton, "Nanostructured silicon as a photonic material", *Optics and Lasers in Engineering*, vol. 39, pp. 345-368 (2003).

[8]   P. Vukusic, "Structural Color", in J. A. Schwarz, C. I. Contescu and K. Putyera (ed), *Dekker Encyclopedia of Nanoscience and Nanotechnology*, Marcel Dekker Inc., New York, pp. 3713-3722 (2004).

[9] C. I. Hsieh, H. L. Chen, W. C. Chao and F. H. Ko, "Optical properties of two-dimensional photonic-bandgap crystals characterized by spectral ellipsometry", *Microelectronic Engineering*, vol. 73-74, pp. 920-926 (2004).

[10] M. Ahles, T. Ruhl, G. P. Hellmann, H. Winkler, R. Schmechel and H. von Seggern, "Spectroscopic ellipsometry on opaline photonic crystals", *Optics Communications*, vol. 246, pp. 1-7 (2005).

[11] C. H. Lin, H. L. Chen, W. C. Chao, C. I. Hsieh and W. H. Chang, "Optical characterization of two-dimensional photonic crystals based on spectroscopic ellipsometry with rigorous coupled-wave analysis", *Microelectronic Engineering*, vol. 83, pp. 1798-1804 (2006).

[12] E. Hecht and A. Zajac, *Optics*, Addison-Wesley Publishing Company, Inc. (1976).

Calculation of Quantum Efficiency for Resonant Cavity Photodiodes using the FDTD Method

Mohammad Soroosh[1], Mohsen Jalali[2], Mohammad Kazem Moravvej-Farshi[2]

1. Engineering Department, Azad University, Gonabad Branch, Khorasan Razavi, Iran
2. Electrical Engineering Department, Tarbiat Modarres University, Tehran, Iran
Email: msoroosh@modares.ac.ir , Fax: 98-535-7255005, Tel: +98-535-7255001

Abstract **Using the finite difference time domain method (FDTD), we solve the Maxwell equations in resonant cavity photodetector. Then we obtain the optical field density and quantum efficiency. Also, the effect of bragg space and reflector are calculated.**

I. INTRODUCTION

Resonant cavity enhanced photodetectors can have quantum efficiencies close to unity. This is achieved by utilizing reflectors around the active region. The photons make multiple passes across the active region, improving the probability of absorption [1].

Several models are presented for avalanche photodiodes [2-6]. These can't describe the resonant cavity structure. The FDTD method is a powerful computational tool. It gives time-domain data, useful for transient analysis, and can yield frequency-domain data via fourier transforms. The FDTD simulation automatically and self-consistently takes into account the full-wave effects of distributed electromagnetic wave coupling, radiation, ground loops and ground bounce.

Given the variety of materials that FDTD can model, very complex and detailed structures may be analyzed with FDTD. These capabilities have made FDTD an attractive and robust method for solving a number of electromagnetic interaction problems [7].

We simulate the optical field distribution in resonant cavity photodetector, using the FDTD. In our simulation, we solve the Maxwell equation in the cavity and brag regions. The absorption condition and reflectors effect are set for simulation. Also, one dimensional simulation is applied to the device.

II. RESONANT CAVITY PHOTODETECTOR

A typical resonant cavity photodetector consists of a thin intrinsic absorbing layer sandwiched between bragg-mirrors with n-type and p-type spacer layers. The cavity length is frequently of the order of a wavelength or less.

The optical field that builds up inside the cavity due to the multiple reflections from the top and bottom mirrors gives rise to a standing wave within the cavity whose interaction with the absorption layer results in high absorption at the resonant frequency. In high-Q cavities, the resonance dies off very rapidly with only a slight detuning from the central wavelength, leading to a narrow-line width response [1].

III. THE FDTD METHOD

Consider a region of space which is source-free and has constitutive electrical parameters that are independent of time. Then, using the MKS system of units, Maxwell's curl equations are given by [7]:

$$\nabla \times E = -\mu(\partial H / \partial t) \qquad (1)$$

$$\nabla \times H = \sigma E + \varepsilon(\partial E / \partial t) \qquad (2)$$

where E is the electric field in volts/meter, H is the magnetic field in amperes/meter, ε is the electrical permittivity in mhos/meter, μ is the magnetic permeability in henrys/meter, and σ is the conductance in ohms/meter.

We denote a space point in a rectangular lattice as [2]:

$$(i, j, k) = (i\Delta x, j\Delta y, k\Delta z) \qquad (3\text{-}a)$$

$$F^n(i, j, k) \equiv F(i\Delta x, j\Delta y, k\Delta z, n\Delta t) \qquad (3\text{-}b)$$

where Δx, Δy, and Δz are the lattice space increments in the x, y, and z coordinate directions, Δt is the time increment, and i, j, k, and n are integers. Rather than positioning the field components as in Yee's cell shown in Fig. 1.

We use the centered finite difference expressions for the space and time derivatives that are both simply programmed and second order accurate in the space and time increments respectively [2].

0-7803-9730-4/06/$25.00 ©2006 IEEE   252

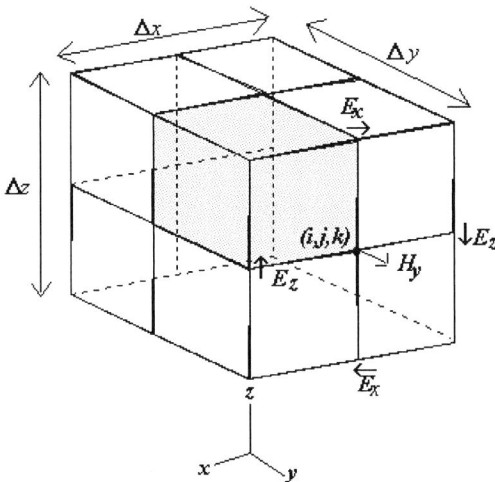

Fig. 1 Yee lattice.

$$\frac{\partial F^n(i,j,k)}{\partial x} = \frac{F^n(i+1/2,j,k)}{\delta} - \frac{F^n(i-1/2,j,k)}{\delta}$$
(4)

$$\frac{\partial F^n(i,j,k)}{\partial t} = \frac{F^{n+1/2}(i,j,k)}{\Delta t} - \frac{F^{n-1/2}(i,j,k)}{\Delta t}$$
(5)

To achieve the accuracy of (4), and to realize all of the required space derivatives of the system of (1) and (2), we positioned the component of E and H about a unit cell of the lattice.

To achieve the accuracy of (5), the evaluated E and H at alternate half time steps. The following are sample finite difference time stepping expressions for a magnetic and an electric field component resulting from these assumptions:

$$E_x^{n+1/2}(k) = \frac{\left(1 - \frac{\Delta t.\sigma}{2\varepsilon_0\varepsilon_r}\right)}{\left(1 + \frac{\Delta t.\sigma}{2\varepsilon_0\varepsilon_r}\right)} E_x^{n-1/2}(k) -$$

$$\frac{0.5}{\varepsilon_r\left(1 + \frac{\Delta t.\sigma}{2\varepsilon_0\varepsilon_r}\right)} [H_y^n(k+1/2) - H_y^n(k-1/2)]$$
(6)

$$H_y^{n+1}(k+1/2) = H_y^n(k+1/2) - \frac{\Delta t}{\mu_0 \Delta z}$$
(7)

$$\left[E_x^{n+1/2}(k+1) - E_x^{n+1/2}(k)\right]$$

The cell size Δx and Δt are as following:

$$\Delta x \approx \frac{3 \times 10^8}{10 f_{max}}, \quad \frac{1}{\sqrt{\varepsilon_0 \mu_0}} \frac{\Delta t}{\Delta x} = \frac{1}{2}$$
(8)

where f_{max} is the maximum frequency of wave.

We simplify the resonant cavity photodetector as shown in Fig. 2.

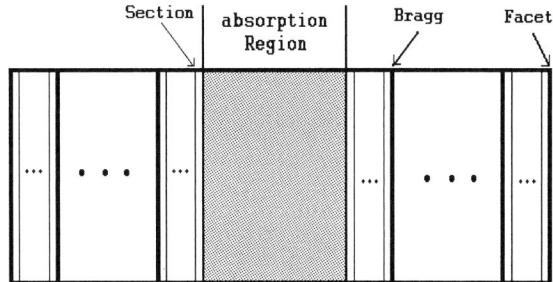

Fig. 2 Schematic of resonant cavity photodetector.

A common sense approach tells us that acceptable boundary conditions are as following:

$$E_x^n(1) = E_x^{n-2}(2), E_x^n(k_f) = E_x^{n-2}(k_f - 1)$$ (9)

where k_f is the final section.

When a plane wave traveling in medium 1 strike a medium 2, the fraction that is reflected is given by the reflection coefficient Γ. This is determined by the intrinsic impedances η_1 and η_2 of the respective media [7]:

$$\Gamma = (\eta_2 - \eta_1)/(\eta_2 + \eta_1)$$ (10)

To examine the accuracy of the model, we compare the results obtained from the model with the other work [8,9]. As an example, a AlGaAs/GaAs/InGaAs resonant cavity photodetector has been simulated [10].

Fig. 3 compares the optical field density of simulation with the other work [8,9]. From Fig. 3, it is seen that our simulation is good agree with the other simulation [10].

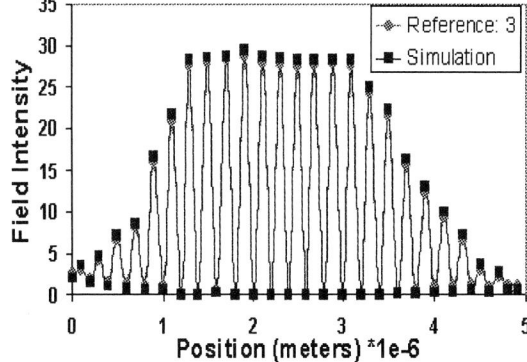

Fig. 3 Optical field distribution in a resonant cavity photodetector [10].

Fig. 4 shows the optical field with absorption region width [10]. As the width is increased, the optical field is reduced because the optical power is constant and loss is increased.

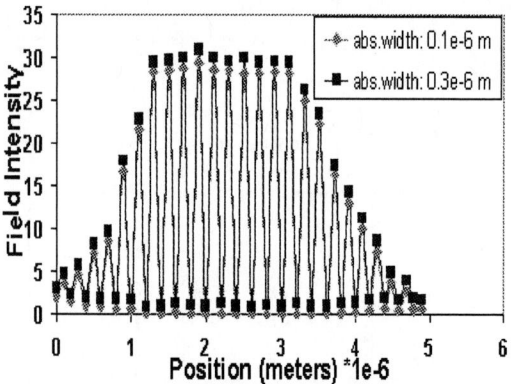

Fig. 4 Optical field distribution versus the absorption region width [10].

As shown in Fig. 5, the more reflection from the braggs causes the intensity of optical field is increased. The quantum efficiency is an important factor of photodetectors. The quantum efficiency is defined the number of electron-hole pairs contributing to photo-induced current divide to the number of incident photons.

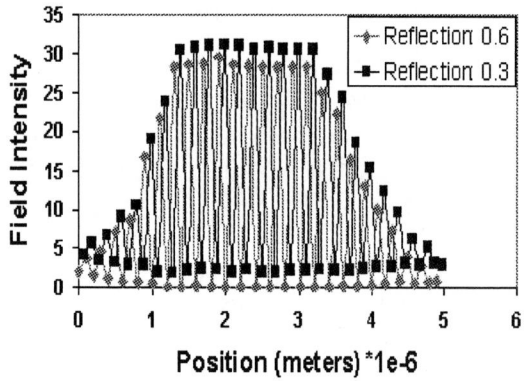

Fig. 5 Optical field distribution versus the reflection coefficient variation [10].

Fig. 6 shows the quantum efficiency with different normalized absorption coefficients.

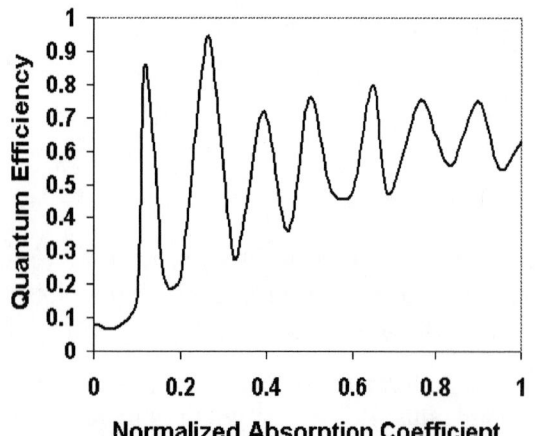

Fig. 6 Quantum efficiency as a function of the normalized absorption coefficient [10].

The reflection of surface causes the quantum efficiency is decreased (Fig. 7).

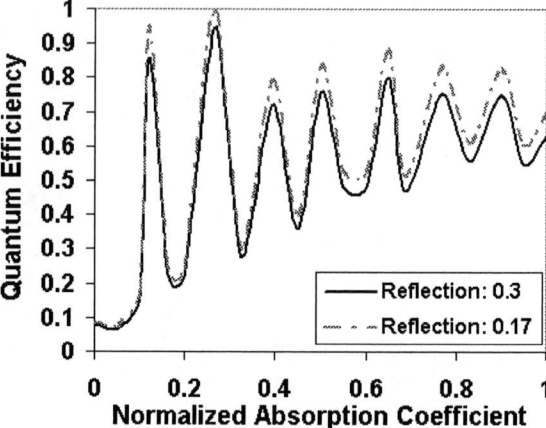

Fig. 7 Quantum efficiency as a function of the normalized absorption coefficient with different reflection coefficient [10].

Also, the effect of the bragg space on the quantum efficiency is calculated (Fig. 8). The space between the braggs causes the resonant frequency is changed.

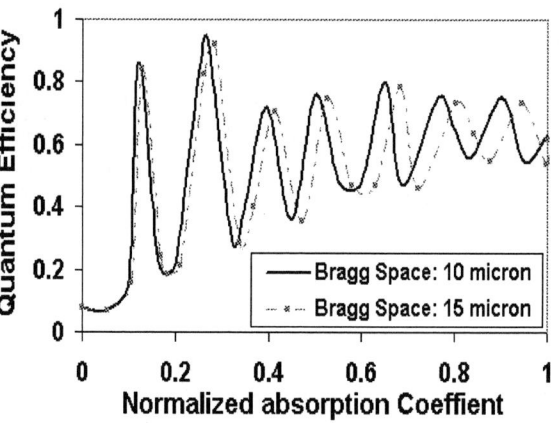

Fig. 8 Quantum efficiency as a function of the normalized absorption coefficient with different bragg space [10].

IV. CONCLUSION

The FDTD method is developed in terms of time and space in which we can visualize how the beam propagates and builds up in the RCE-type structures. The finite difference time domain (FDTD) method is robust, flexible, and efficient to solve the propagation problems.

ACKNOWLEDGMENT

The authors would like to acknowledge the useful discussions from Dr. S. Dehestani.

REFERENCES

[1] H. S. Nalwa, *Photodetectors and Fiber Optics*, Academic Press (2001).

[2] M. Soroosh, A. Zarifkar, M. Razaghi and M. K. Moravvej-Farshi, "Separate absorption and multiplication avalanche photodiode (SAM-APD) model for circuit simulation," *IEEE Gulf International Conference (GCC)*, pp. 606-611 (2004).

[3] C. Groves, J. P. R. David, G. J. Rees and D. S. Ong, "Modeling of avalanche multiplication and noise in heterojunction avalanche photodiode," *J. Appl. Phys.*, vol. 95, pp. 6245-6251 (2004).

[4] A. Banoushi, M. R. Kardan and M. Ataee Naeini, "A circuit model simulation for separate absorption, grading, charge, and multiplication avalanche photodiodes," *J. Solid State Electron.*, vol. 49, pp. 871-877 (2005).

[5] M. M. Hayat and G. Dong, "A new approach for computing the bandwidth statistics of avalanche photodiodes," *IEEE Trans. Electron Dev.*, vol. 47, pp. 1273-1279 (2000).

[6] J. Jou, C. K. Liu, C. M. Hsiao, H. H. Lin and H. C. Lee, "Time delay circuit model of high speed p-i-n photodiodes," *IEEE Photon. Technol. Lett.*, vol. 14, pp. 525-527 (2002).

[7] D. Sullivan, *Electromagnetic Simulation Using FDTD Method*, IEEE Press (2000).

[8] J. P. Kim and A. M. Sarangan, "Simulation of resonant cavity enhanced (RCE) photodetectors using the finite difference time domain (FDTD) method," *J. Optic Express*, vol. 12, pp. 4829-4834 (2004).

[9] M. Soroosh, and A. Zarifkar, "Simulation of optical field for resonant cavity photodetectors using the finite difference time domain," *IEEE Conf. Laser and Fiber Optical Networks Modeling (LFNM)*, vol. 2, pp. 291-293 (2005).

[10] M. Gokkavas, B. M. Onat and E. Ozbay, "Design and optimization of high speed resonant cavity enhanced schottky photodiodes," *IEEE J. Quantum Electronics*, vol. 35, pp. 208-213 (1999).

Static Quasi 3D Thermal Simulation of Ion Implanted Vertical Cavity Surface Emitting Lasers

E. Sooudi, V. Ahmadi, M. Ebnali Heidari, and M. Soroosh

Dept. of Elect. Engineering, Tarbiat Modares University, Tehran, Iran

P.O.Box: 14155-143

E-mail: sooudi_e@alum.sharif.edu; v_ahmadi@modares.ac.ir;

Abstract - **In this paper, we simulate static thermal behavior of a gain guided VCSEL by solving quasi 3D heat equation. Several heat sources such as Joule heating, reabsorption of spontaneous emission and nonradiative recombination in different vertical positions of the device are considered. We use inhomogeneous thermal conductivities and see reduction of temperature in whole device than the case of homogenous. Moreover, by increasing current spreading, thermal lensing increases. However, increasing the aperture radius causes broadening of profile and lowering of temperature peak. We also found that Joule heating of p-DBR has critical role in temperature distribution. We also analyze the effect of top DBR period numbers on threshold current and active layer temperature peak.**

I. INTRODUCTION

Vertical-cavity Surface-Emitting Lasers (VCSELs) have been attracting much attention in recent years because of their numerous advantages. They are very attractive laser sources for a wide range of applications in optical communications (specially short haul), optical recording, and optical interconnection. Gain guided (ion implantation) VCSELs have been used as the first commercial 850 nm structures since 1997 [1].

However, the behavior of a VCSEL is affected by the temperature rise within it which influences the laser emission wavelength, dynamic characteristics, optical wave guiding, the threshold current and efficiency.

Thermal modeling of gain guided VCSELs with homogeneous thermal conductivity using Green function technique was presented before [2], [3]. In this paper, using a quasi 3D reasonably self-consistent thermal model with finite difference solution technique; we simulate effects of inhomogeneous thermal conductivities

and various structure parameters in temperature distribution in the device.

II. MODEL THEORY

A. Structure

The gain guided structure schematic is shown in Fig. 1. The contact is ring type and device is top emitting. The device consists of multi QW (3 layers) active layer (GaAs/Al$_{0.3}$Ga$_{0.7}$ As), two half-wavelength spacer layers, 19.5 periods AsAs-Al$_{0.16}$Ga$_{0.84}$As top p-DBR, and 27.5 periods bottom n-DBR with the same material structure [2].

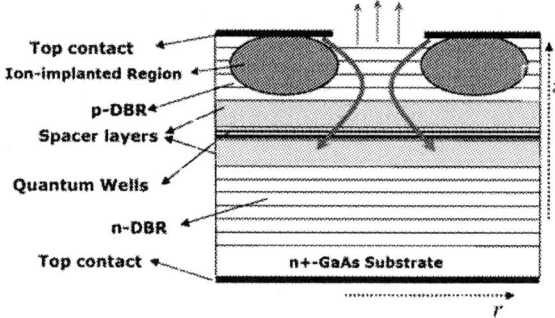

Fig. 1. Schematic of gain guided structure [2]

B. Thermal Model

Thermal heat equation in cylindrical coordinate can be written as (1):

$$\sigma_{ver.} \frac{\partial^2 T(z)}{\partial z^2} + \frac{1}{r}\frac{\partial}{\partial r}\left(r\sigma_{hor.}\frac{\partial T(r)}{\partial r}\right) + Q(r,z) = 0 \quad (1)$$

Where $T(r,z)$ is temperature distribution in vertical (z) and lateral (r) directions, $\sigma_{ver.}$ and $\sigma_{hor.}$ are thermal conductivity in vertical and lateral position, and $Q(r,z)$ is thermal power density. Joule heating in two DBRs and spacers, nonradiative recombination and reabsorption of spontaneous emission in QWs are heat sources considered. Heat sources in QWs can be written as (2), [2]:

0-7803-9730-4/06/$25.00 ©2006 IEEE   256

$$Q_{QW}(r) = \frac{1}{d_{act.}}\Big[V(r)I(r)(1-\eta_{sp}) +$$

$$V(r)j_{th}\eta_{sp}(1-f_{sp})\Big] + \frac{1}{d_{act.}}\Big[V(r)(I(r)-I_{th}(r))\times$$

$$(1-\eta_i)(1-\eta_{sp})\Big] + V(r)(\frac{I_{in}}{\pi r_s^2} - j_{th})(1-\eta_i)\eta_{sp}(1-f_{sp})\Big] \quad (2)$$

where $d_{act.}$ is total QW thickness, r_s is radius of the device, $I_{th}(r)$ is threshold current density distribution, j_{th} is threshold current density, I_{in} is total input current, η_{sp} is spontaneous quantum efficiency, η_i is stimulated emission quantum efficiency, f_{sp} is coefficient assumed for the escape of spontaneous emission from the active layer, and $V(r)$ is voltage drop across the active layer. We assume that reabsorption happens spatially uniform and nonradiative recombination distribution is proportional to current profile.

For two DBRs and spacer layers thermal power density related to Joule heating can be written as (3):

$$Q_{DBR}(r) = \rho_{x-DBR}I(r)^2 \quad (3)$$

$$Q_{spacer}(r) = \rho_{spacer}I(r)^2 \quad (4)$$

Where, ρ_{x-DBR} and ρ_{spacer} are resistivity of top and bottom DBRs (which are different) and spacer layers.

For current distribution, both Poisson and continuity equations should be solved self consistently in whole of device. However, this method substantially increases the memory and is out of the scope of our model. Alternatively, we use an analytical approach proposed in [4], which has been used extensively by the others [2], [3] and [5]:

$$I(r) = \begin{cases} I_0 \exp[(r-s)/r_1] & r \le s \\ I_0 \exp[-(r-s)/r_2] & r > s \end{cases} \quad (5)$$

where s is radius of implanted aperture, and r_is are parameters employed for modeling the effect of current funneling, r_1, and spreading, r_2. I_0 is maximum current density and is calculated from (6):

$$I_0 = \frac{I_{in}}{\int_0^s 2\pi r \exp[(r-s)/r_1]dr + \int_s^{r_s} 2\pi r \exp[-(r-s)/r_2]dr} \quad (6)$$

where, I_{in} is total injected current.

Voltage drop across active layer can be determined using simple diode approximation through (7) [2]:

$$V_{qw} = \begin{cases} \dfrac{k_B T}{q}\left(\ln \dfrac{I}{I_s} + \dfrac{r-s}{r_1} \right) & r \le s \\ \dfrac{k_B T}{q}\left(\ln \dfrac{I}{I_s} - \dfrac{r-s}{r_2} \right) & r > s \end{cases} \quad (7)$$

where k_B is Boltzmann constant, T is temperature, q is electron charge, and I_s reverse current density. Note that, for multilayer sections such as DBR or spacers, thermal conductivity is not equal in lateral and vertical positions. So, we use (8) and (9) for calculating vertical and horizontal thermal conductivities [1]:

$$\sigma_{hor.} = \frac{\sum\limits_{i=1}^{n} \sigma_i z_i}{\sum\limits_{i=1}^{n} z_i} \quad (8)$$

$$\sigma_{ver.} = \frac{\sum\limits_{i=1}^{n} z_i}{\sum\limits_{i=1}^{n} \dfrac{z_i}{\sigma_i}} \quad (9)$$

Where σ_i and z_i are thermal conductivity and thickness of each layer. It is worth noting that, for graded structures (spacer layer) sigma is converted to integral in (8) and (9). Moreover, $\sigma_{ver.}$ is less than $\sigma_{hor.}$, therefore, we expect unlike behavior of temperature distribution than simple equal $\sigma_{ver.}$ and $\sigma_{hor.}$.

In all of the simulations, η_{sp}, η_i, and f_{sp} values are 0.9[8], 0.9[2], and 0.95[8] respectively.

C. Numerical Solution

The method of numerical solution used is finite difference. By substituting difference equations to differential ones, we will have simultaneous linear equations solved by successive over relaxation (SOR) technique. Boundary conditions for thermal equation are:

$$\frac{\partial T(r)}{\partial r}\Big|_{r=0} = \frac{\partial T(r)}{\partial r}\Big|_{r=r_{max}} = 0 \quad (10)$$

$$\frac{\partial T(r)}{\partial z}\Big|_{z=z_{max}} = 0 \quad (11)$$

These boundary conditions assure that no heat run away from top and lateral borders of device and temperature over $r=0$ is symmetric. Moreover, we assume that bottom of device's temperature is equal to the sink:

$$T(r,0) = T_0 \quad (12)$$

It should be noted that, j_{th} and I_{th} used in (2) are calculated separately through a comprehensive electro-opto-thermal model in [6], which is based on [2]. T_0 is room temperature.

III. SIMULATION RESULTS AND DISUSSIONS

In this section, we present some simulation results and specially concentrate on the effects of current spreading parameters on temperature distribution.

3D distribution of temperature is shown in Fig. 2. It can be seen that by increasing the distance from center of the device, temperature decreases in all vertical positions. Here, we see clearly that all of the boundary conditions are satisfied, while this is difficult to be obtained by Green function method used in [2].

For better illustrating the effect of inhomogeneous layers on simulation results, we compare temperature profile in active layer for homogenous and inhomogeneous thermal conductivities in Fig. 3. As a result, we perceive reduction of temperature for inhomogeneous case. The reason for this difference is that, because of higher lateral thermal conductivity, heat diffuses more in lateral position of whole device and temperature peak reduces.

In Fig. 4 temperature rise in active layer for different values of current spreading parameter, r_2, is shown. It has to be mentioned that increasing r_2 is equivalent to more current spreading in the lateral position of active layer. In other words, less thickness of implanted region causes more broadening of current profile. Because of the fact that, more spreading of current rises threshold current; therefore, there is more temperature rise in active layer because of more dissipated thermal power (specially in p-DBR), as a result, thermal lensing strengthens in active layer.

Temperature profile in the center of device across vertical position is shown in Fig. 5. Since resistivity of GaAs based p-DBR is greater about two order of magnitude than n-DBR, and current density is higher because of small aperture size, Joule heating in p-DBR is high and vital in determining temperature rise and we see that maximum temperature is at the top of device.

By increasing aperture size, both parameters r_1 and r_2 will change. r_1 models funneling phenomenon in current density in region below the current aperture due to ring type contact [2],

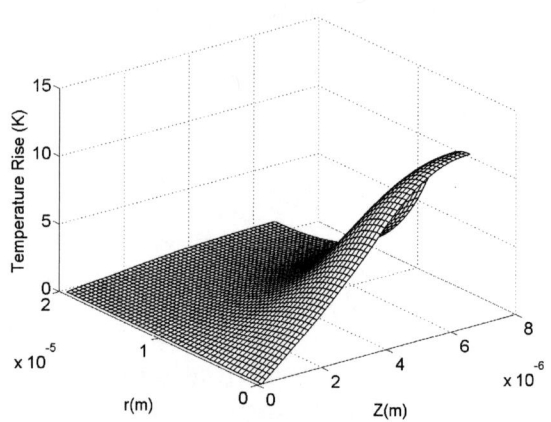

Fig. 2. 3D distribution of temperature rise device for $1.5I_{th.}$, r_1=80μm, r_2=5μm, and s=5μm

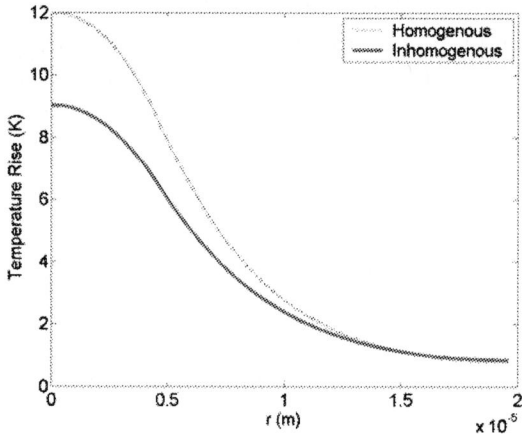

Fig. 3. Temperature rise in active layer with and without inhomogeneous thermal conductivities for $1.5I_{th.}$, r_1=80μm, r_2=5μm, and s=5μm.

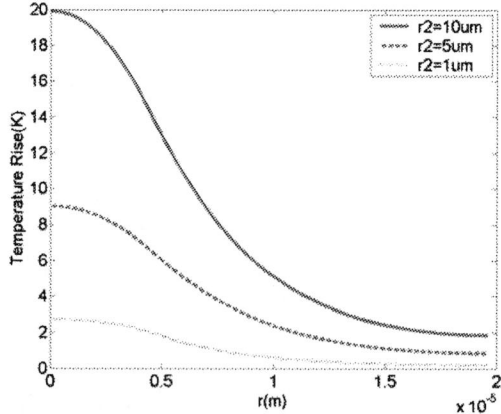

Fig. 4. Temperature rise in active layer for different current spreading values $1.5I_{th.}$, $r1$=80μm, s=5μm, and r_2=1, 5 and 10μm.

[7]. Therefore, current spreading and funneling strengthens by increasing aperture size. In Fig. 6, broadening of temperature rise in active layer is illustrated. Moreover, we see that by increasing

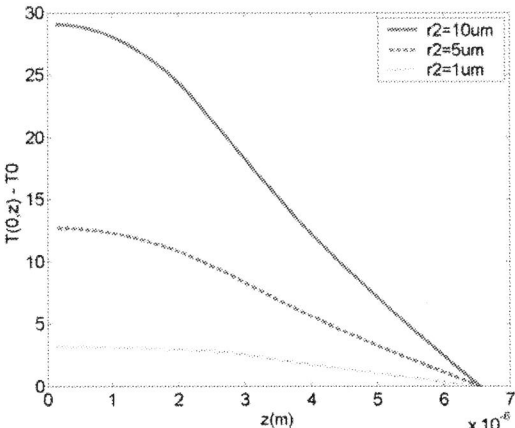

Fig. 5. Temperature rise in center of device in vertical position for $1.5I_{th.}$, $r_1=80\mu m$, $s=5\mu m$, and $r_2=1, 5, 10\mu m$..

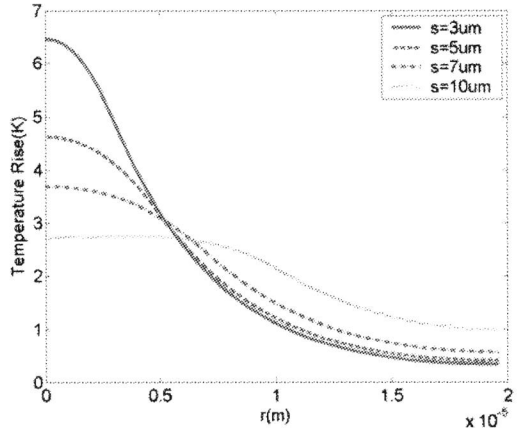

Fig. 6. Temperature rise in active layer for different aperture radius $1.5I_{th.}$, $r_1=90\mu m$, $s=3\mu m$, $r_2=1.7$ μm, $r_1=80\mu m$, $s=5\mu m$, $r_2=2.9$ μm, $r_1=40\mu m$, $s=7\mu m$, $r_2=4$ μm, and $r_1=20\mu m$, $s=10\mu m$ and $r_2=5.7\mu m$.

aperture size, despite the higher threshold current temperature peak lowers. This result agrees well with the work [3] as well.

In Fig. 7, 3D distribution of $Q(r,z)$ is shown. It ca be understood that because of higher current cross-section (increasing aperture radius), current density in p-DBR reduces, thus, Joule heating power in p-DBR decreases. Furthermore, higher funneling and spreading current in active layer leads to quite more uniformity in lateral distribution of current and thermal power density. Consequently, as the reverse of previous case, enlarging the aperture size, lowers temperature rise in whole device. Also, thermal

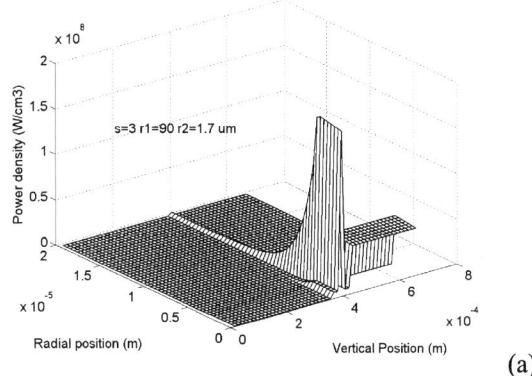

(a)

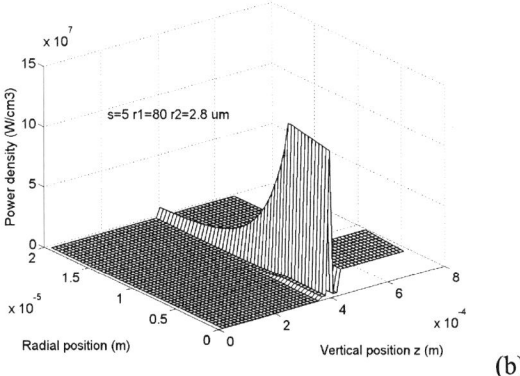

(b)

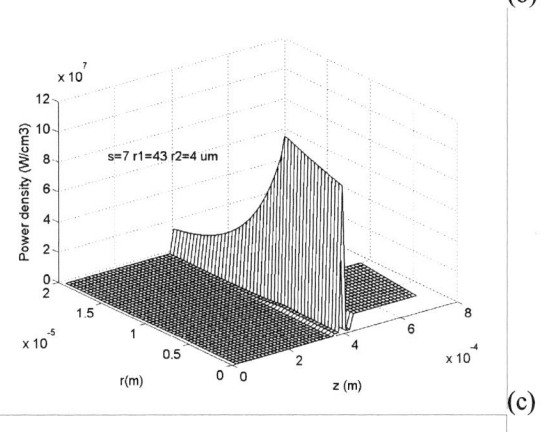

(c)

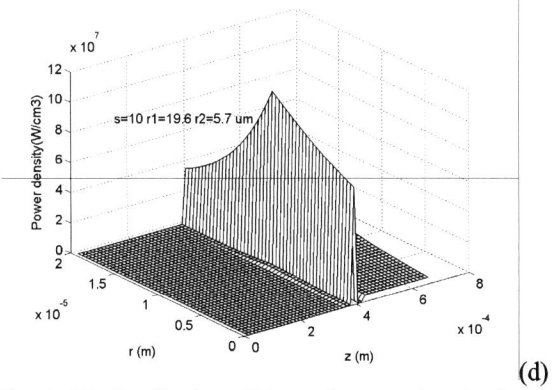

(d)

Fig. 7. 3D distribution of thermal power density for different aperture radius and $1.5I_{th.}$ (a) $r_1=90\mu m$, $s=3\mu m$, $r_2=1.7$ μm, (b) $r_1=80\mu m$, $s=5\mu m$, $r_2=1.7$ μm, (c) $r_1=90\mu m$, $s=3\mu m$, $r_2=1.7$ μm, and (d) $r_1=90\mu m$, $s=3\mu m$, $r_2=1.7\mu m$.

lensing is not noticeable in high current apertures.

Finally, in Fig. 8 we study the effect of period number of top DBR on the active layer temperature peak and threshold current. For each period number, effective mirror reflectivity (using TMM method) and cavity length is calculated. Then, substituting these parameters in a comprehensive numerical model for this device, we determine the threshold current. By calculating threshold current for each case temperature rise is determined. It is found that, by increasing number of periods, because of less threshold current, temperature peak decreases. In contrast, beyond a definite number of periods, by maintaining a constant threshold current, Joule heating in top DBR rises because of more electrical resistance and device temperature is increased. Consequently, for a fixed structure, there are always an optimum number of DBR layers for less threshold current and a reduced amount of active layer temperature peak.

IV. CONCLUSION

In this paper, using quasi-3D heat equation solved by finite difference, we simulated static thermal behavior of an ion implanted VCSEL. Three main heat sources considered here were: Joule heating, re-absorption of spontaneous emission and nonradiative recombination. Considering inhomogeneous thermal conductivities, we observed reduction of temperature in whole device compared with the homogenous case. In addition, by increasing current spreading, thermal lensing phenomenon arises. In contrast, increasing the aperture radius led to broadening of profile and lowering of temperature rise. We also found that Joule heating of p-DBR has important role in temperature distribution and thermal lensing. Moreover, we observed that by proper choosing of top DBR period number, a reduced amount of threshold current and active layer temperature peak is obtained.

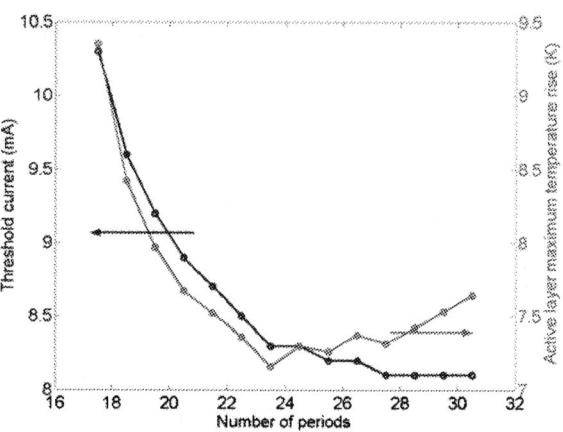

Fig. 8 Threshold current and temperature rise versus number of periods of top DBR, for $1.5I_{th.}$, $r_1=80\mu m$, $s=5\mu m$, and $r_2=10\mu m$.

REFERENCES

[1] S. F. Yu, Analysis and design of vertical cavity surface emitting lasers, John Wiley & Sons, Hoboken, New Jersey, 2003.

[2] Y.-G. Zhao and J. G. Mcinerney, "Transverse-mode control of vertical-cavity surface-emitting lasers," *IEEE J. Quantum Electron.*, vol. 32, no. 11, pp. 1950-1958, Nov. 1996.

[3] K. Tastavridis, I.H. White, J.M. Rorison, and R.V. Penty, "Theoretical investigation of the evolution of the transverse modes in vertical cavity surface emitting lasers," *Proc. SPIE*, vol. 3946, pp. 117-128, 2000.

[4] N. K. Dutta, "Analysis of current spreading, carrier diffusion, and transverse mode guiding in surface emitting lasers," *J. Appl. Phys.*, 68, 1961-1963, 1990.

[5] S.F.Yu, C.W. Lo, "Influence of transverse modes on the dynamic response of vertical cavity surface emitting lasers," *IEE Proc. Optoelectron.*, vol. 143, no. 3, pp. 189-194, Jun. 1996.

[6] E. Sooudi, "Modeling and Analysis of Operation and Characteristics of a Quantum Well VCSEL," M.Sc. Thesis, Tarbiat Modares University, Tehran, Iran, March 2006.

[7] W. Nakwaski, "Current spreading and series resistance of proton-implanted vertical-cavity top-surface-emitting lasers," *Appl. Phys. A: Mat. Sci. Proc.*, vol. 61, pp. 123-127, 1995.

[8] G. Chen, M. A. Hadley and J. S. Smith, "Pulsed and continuous-wave thermal characteristics of external-cavity surface-emitting laser diodes," *J. Appl. Phys.*, 76(6), pp. 3261-3271, Sep. 1994.

The Effect of Annealing on the Performances of the White Organic Light Emitting Diode (OLED)

Suhaila Sepeai[a], Muhamad Mat Salleh[a], *Member, IEEE,* Muhammad Yahaya[b]

[a]Institute of Microengineering and Nanoelectronics (IMEN)
[b]School of Applied Physics, Faculty of Science & Technology
Universiti Kebangsaan Malaysia
43600 Bangi, Selangor, MALAYSIA
Email : mms@pkrisc.cc.ukm.my

Abstract **A white organic light emitting diode, OLED is one of the approaches to obtained full colour flat screen display. OLED devices with the structure of ITO/PHF:rubrene/Al have been fabricated where poly (9,9-di-n-hexylfluorenyl-2,7-dyl), PHF used as blue light emitting host and 5,6,11,12-tetraphenyl-napthacene, rubrene as an orange dye dopant. Indium tin oxide, ITO used as anode and aluminium, Al as cathode. The white OLED obtained by varying the concentration of rubrene. The problem occurred by using a dopant is, the turn–on voltage is increased with the dopant concentration. This paper reports the effect of annealing process to the devices at 50 °C to 150 °C. It was found that the annealing process has reduced the turn-on voltages and increased the brightness. Meanwhile, the colour of light emission was deviated from white to yellowish white.**

I. INTRODUCTION

Organic light emitting diodes (OLEDs) are thin film devices in which organic materials are sandwiched between two electrodes. These devices emit light when electricity is passed through them. Organic materials have been considered for the fabrication of practical electroluminescent devices indeed of having extremely high fluorescent quantum efficiencies in the visible spectrum [1,2]. OLEDs have gained much attention because their attractive characteristics and potential applications to flat panel displays, such as mobile phones, PDA, small area handheld telecom devices to large area displays due to their high luminance, low fabrication costs and ease to fabricate large area devices. Current research on OLEDs is focusing on

the integration of OLEDs into full colour, flat panel display and one of the approaches is to use white OLEDs with colour filter array.

We have fabricated single OLED devices that utilized an emitting layer consisting poly (9,9-di-n-hexylfluorenyl-2,7-dyl), PHF as blue light emitting host and 5,6,11,12-tetraphenyl-napthacene, rubrene as red-orange dopant. This blue colour doped with an orange dye with a variation of rubrene concentration to produced a white light emission.

Although we have successful obtained a white light emitting devices, we found that the devices have high turn-on voltage, in the range of 14.0 V - 18.0 V. This paper reports an attempts to reduce turn-on voltage of the devices through annealing process. Many research groups are considered to insert hole transport layer [3-5] or electron transport layer [6-8] to reduce the turn-on voltage. Some were utilized a low work function metal as electrode such as Li, Ca, or Mg [8-11]. However, these devices have complicated fabrication process, phase separation, crystallization, and oxidation for low work function metal.

II. EXPERIMENTAL

OLED devices with a structure of ITO/PHF:rubrene/Al has been fabricated (Fig.1) where poly (9,9-di-n-hexylfluorenyl-2,7-dyl), PHF as blue light emitting host and 5,6,11,12-tetraphenyl-napthacene, rubrene as red-orange dye dopant. ITO is indium tin oxide that coated on glass substrate as an anode. An aluminium, Al used as cathode. The ITO with a sheet resistance of 50 Ω/m^2 was obtained from Merck Balzers. The PHF (Mw=300000 g/mol) was purchased from H.W Sands Corp., while rubrene (Mw=532.69 g/mol) was purchased from Aldrich. The chemical structure of PHF and rubrene are shown in Fig. 2 and Fig. 3, respectively.

The ITO was etched and patterned to serve as anode. PHF and rubrene were dissolved in toluene with a typical weight ratio is 10:1. The solution was heated at 75 °C and spin-coated onto the ITO glass with spinning speed at 3000 rpm for 40 s.

An aluminium layer as a cathode was deposited onto PHF doped with rubrene layer through a mask by the electron-gun evaporation technique from a Molybdenum crucible at a chamber pressure of 2.5 x 10^{-5} mbar. The annealing process of the devices was done after Al deposition in a vacuum oven. The annealing temperatures were at 50 °C, 100 °C and 150 °C for 1 hour. Then the devices were purged with hydrogen flow for 30 minutes to avoid oxidation on Al layer.

The performances of the devices were analyzed through the current-voltage (I-V) curve by a Keithley 238 source measurement unit and the electroluminescent spectra were measured by HR2000 Ocean Optic spectrometer. The morphology of the annealed PHF doped with rubrene thin films were analyzed using Atomic Force Microscope, AFM.

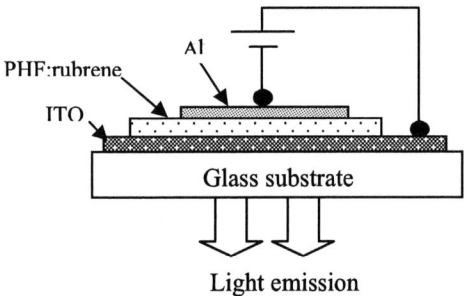

Fig. 1 Structure of the OLED device.

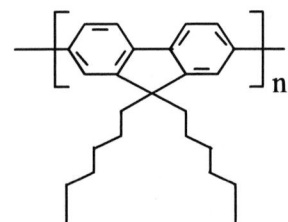

Fig. 2 Molecular structure of PHF.

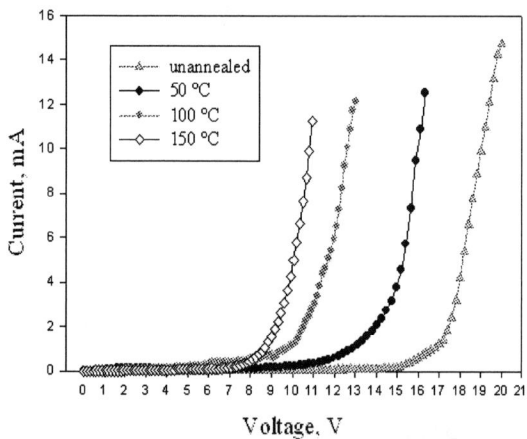

Fig. 3 Molecular structure of rubrene.

III. RESULTS AND DISCUSSION

Fig. 4 shows the current-voltage characteristic of the annealed device with a variation of annealing temperatures from 50 °C to 150 °C. The I-V curve is the first indicator to determine whether the device is able to emit light or not. The device that shows a rectifier diodes behaviour is able to produce light. All the OLED devices showed rectifier diodes behaviour. Using Fowler-Nordheim tunneling theory, the turn-on voltage can be calculated from the I-V curve [12]. The turn-on voltages for device that annealed at 50 °C, 100 °C and 150 °C are 11.0 V, 9.0 V and 8.0 V respectively.

Fig. 4 Current-voltage (I-V) characteristic of the annealed devices.

The electroluminescent, EL spectra of ITO/PHF:rubrene/Al annealed device is shown in Fig. 5. The intensity is increased accordingly with the annealing temperature.

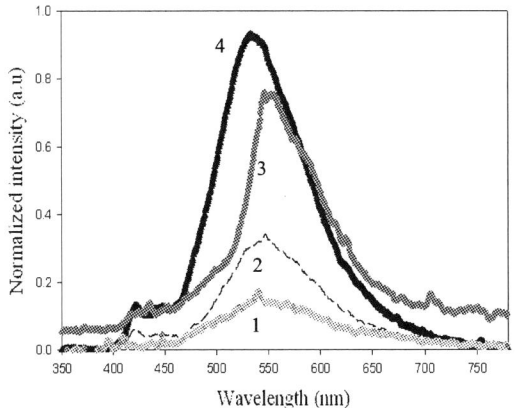

Fig. 5 The emission spectra of (1) un-annealed, annealed at (2) 50 °C, (3) 100 °C and (4) 150 °C devices.

Table 1 summarizes the turn-on voltages and brightness of un-annealed and annealed devices. All brightness data were taken at operation voltage of 17.0 V. It is observed that the brightness gradually increased while the turn-on voltages, V_{on} decreased with the increasing of annealing temperature.

Table 1 The performance of the annealed devices.

Annealing temperature (°C)	V_{on} (V)	Brightness (cd/m^2)
un-annealed	15	766
50	12	1570
100	9	3252
150	8	9042

Table 2 CIE coordinates of the un-annealed and annealed devices at various temperatures.

Annealing temperature (°C)	CIE coordinates	
	x	y
un-annealed	0.3	0.33
50	0.47	0.41
100	0.45	0.43
150	0.41	0.45

Table 2 shows the CIE coordinates of white light before and after annealing. The white point in CIE coordinate defined as (0.33,0.33). It was

observed that the light emission from the device was deviated from white colour to yellowish white after annealing process. Further research to obtained a low turn-on voltage and white light emission device by adjusting the rubrene concentration is in progress.

To explain how annealing process affected the performance of the devices, we studied the surface morphology of PHF doped rubrene thin films on quartz substrate. These thin films have been annealed at the same temperature as the annealed devices. The results are shown in Fig. 6 and summarized in Table 3. It was found that the annealed films are smoother than the un-annealed film. The annealing process has enhanced the interface adhesion between thin films and both electrodes. Hence, more carriers were injected into devices with a minimum energy, means more recombination occurred in thin films. This will increased the brightness and reduced the turn-on voltage.

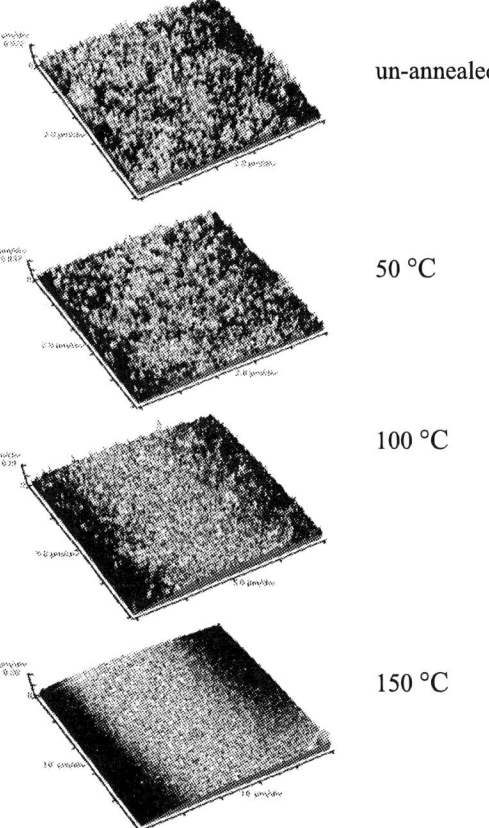

un-annealed

50 °C

100 °C

150 °C

Fig. 6 The AFM images of un-annealed, annealed at 50 °C , 100 °C and 150 °C PHF doped rubrene thin film.

Table 3 Surface roughness of the un-annealed and annealed devices at various temperature.

Annealing temperature (°C)	Surface roughness (nm)
un-annealed	6.78
50	5.96
100	5.72
150	4.28

IV. CONCLUSION

White OLED device with single layer structure of ITO/PHF:rubrene/Al has been fabricated. It was found that the annealing process has reduced the turn-on voltages and increased the brightness of the OLED devices.

ACKNOWLEDGEMENT

This work has been carried out with the support of the Malaysian Ministry of Science, Technology and Innovation, under the IRPA grant 03-02-006-SR0007/04-04.

REFERENCES

[1] K.H. Drexhage, and F.P. Schafer, *Topics in Applied Physics:Dye Lasers*. Vol.1, p.144 (1977).

[2] H.Gold, and K. Ventkataraman, The Chemistry of Synthetic Dyes. Vol. 5, p. 535 (1971).

[3] J. Thompson, R.I.R. Blyth, M.Mazzeo, M. Anni, G. Gigli, R. Cingiolani, "White light emission fromblends of blue-emitting organic molecules: A general route to the white organic light emitting diode?", *Appl. Phys. Lett*. 79, p. 560-562. (2001).

[4] Dipti Gupta, M.Katiyar, Deepak, "Energy transfer and morphology study of a new iridium based cyclometalated phosphorescent complex",*Optical Materials*. 28, pp. 295-301. (2006).

[5] Liu Zugang and Helena Nazare, "White organic light emitting diodes emitting from both hole and electron transport layers", *Synthetic Metals*. 111-112. pp. 47-51 (2000).

[6] Fawen Guo and Dongge Ma, "High efficiency white organic light emitting diodes based on double recombination zones", *Optical Materials*. 28, pp. 966-969 (2006).

[7] Hongjin Jiang, Yan Zhao, Boon Siew Ooi, Yuwen Chen, Terence Wee, Yee Loy Lam, Jingsong Huang, Shiyong Liu, "Improvement of organic light-emitting diodes performance by the insertion of a Si_3N_4 layer", *Thin Solid Films*.363.pp.25-28.(2000).

[8] X.Y. Zheng, W.Q. Zhu, Y.Z. Wu, X.Y.Jiang, R.G.Sun, Z.L.Zhang, S.H.Xu, "A white OLED based on DPVBi blue light emitting host and DCJTB red dopant", *Displays* 24, p. 121-124 (2003).

[9] Tsun-Ren Chen, Rong-Hong Chien, Anchi Yeh, Jhy-Der Chen, "Synthesis, characterization and electroluminescence of B(III) compounds: BPh_2(2-(2-quinolyl)naptho[b]imidazolato) and BPh_2(2-(2-quinolyl)benzimidazolato)", *Journal of Organometallic Chemistry*. 691. pp.1998-2004.(2006)

[10] Xiaoming Wu, Yulin Hua, Zhaoqi Wang, Jiajin Zheng, S.Y, K.Wu, Songliu, Feijian Zhu, Xia Niu, "Multi-color display and its model for a white OLED combined with optical color filters", *Optics*.(2005).

[11] Shizuo Tokito, Toshimitsu Tsuzuki, Fumio Sato, Toshiki Ijima, "High efficiency blue and white phosphorescent organic light-emitting devices", *Current Applied Physics*. 5. pp. 331-336. (2005).

[12] Parker, I.D. *Journal of Applied Physics*. **75** (30). pp. 1656-1666. (1994)

Considering RFID Inmate Tagging Application to Enhance Prison Management

Mohd. Suhaimi Selamat, Burhanuddin Yeop Majlis, SMIEEE
Institute of Microengineering and Nanoelectronics (IMEN)
Universiti Kebangsaan Malaysia
43600 Bangi, Selangor, MALAYSIA
Email: mssuop888@vlsi.eng.ukm.my, burhan@eng.ukm.my

Abstract   **Security is one of the most important aspects of prison management, which includes security of inmate, staff as well as the public. This paper studies two inmate tagging solutions, which work effectively to enhance numbers of correctional institution in the U.S.. They are Clincher® wristband identification from Precision Dynamics Corporation (PDC) and the TSI Prism™ real-time tracking from Alanco / TSI Prism, Inc. (TSI). Features and advantages of the inmates tagging systems are highlighted.**

I. INTRODUCTION

Malaysian prison department mission states: "*to protect the society by providing safe detention and effective correction program*" and among its objectives is "*to ensure security and safety of inmates, staff and the public*" [1]. Mass escape, strike, abduction, and assaults [3] are continuous challenges face by the department, hence, inmate management and monitoring is utmost important task. At the ratio of one officer to two hundred inmates (1:200) [2], the task would be more difficult.

While current overall growth drivers for RFID include security access control, asset management, supply chain, transportation and distribution [7], inmate management and monitoring are another niche application of RFID technology. The omnipresent barcode labels that triggered a revolution in identification systems some considerable time ago are being found to be inadequate [8] to fulfill unique requirements of prison operation.

The air interface con-contact data transfer between a tag and its interrogator (reader/programmer) is seen as a fundamental feature of RFID, the transfer invariably being achieved without the need for obstacle-free, line-of-sight alignment [2]. Distinctive features of RFID technology includes:

- Non-line-of-sight
- Robust and rugged
- Fast reading speed
- Simultaneous multiple reading
- Secured and tamper proved
- Programmable [6]

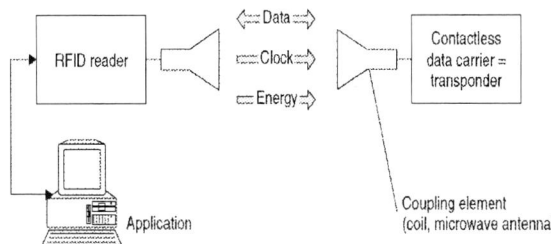

Figure 1 Main Component of RFID System: Reader and Transponder (Tag) [7].

In addition to sharing RFID features, Clincher® photo ID wristband and TSI Prism™ real-time tracking RFID systems have two main system components: RFID reader with antenna and transponders or tags. Specific features of the system outlines in this paper distinctively show their unique applications and advantages.

II. CLINCHER PHOTO ID WRISTBAND

RFID Clincher wristband is built on passive Taxes Instruments Tag-it™ RFID technology operates on 13.56 MHz frequency [10]. Clincher is an improved version of the first inmate ID, which was developed by PDC in 1976 based on Tag-it™ technology. Tag-it™, which is designed for various forms of people access and identification, can be worn as an ID badge, wrist-Band, ankle-band to promote   automatic data capture and

0-7803-9730-4/06/$25.00 ©2006 IEEE

identification [10]. Tag-it inlay is laminated with Clincher wristband to form a smartband or "*gegelang pintar*".

Figure 2 Clincher Photo ID Wristband [5].

Clincher available in laminated styles that accommodate black-and-white or color photos as well as inmate data. As video imaging is quickly becoming a standard features in identification, Clincher is designed to be fully compatible with video imaging systems [5], Figure 2. Unlike IDs which attached to clothes such as a badge, Clincher remains on the inmate at all times (during transportation, exercise, showers, sleep, visiting, etc.), ensuring maximum security [5], Figure 3.

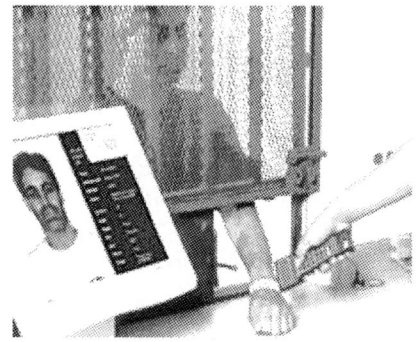

Figure 3 Clincher Wristband and Portable Reader [5].

The wristband, which features passive RFID for data storage, photo as visual identification, and come in variety of colors, is suitable for the following applications in prison:
- Inmates Identification
- Security and access control
- Tracking Inmates
- Classification of inmates

On the other hand, Tag-it™ inherits passive RFID property, which has limited reading range up to 1 meter [10]. Smaller tags such as Clincher wristband, generally has lower read distance [10]. As a result, readers and antenna shall be installed at strategic location or handheld devices [10] shall be used for certain applications.

III. TSI Prism™ REAL-TIME TRACKING

TSI Prism™ real-time tracking is develop based on Motorola technology [6], which was then improvise by TSI to develop RFID tag for prison inmates as well as officers. The tag operates at 902-928 MHz transmit signals every 2 to 10 seconds which includes its ID, reader(s) information, officer alarms, tamper, and battery status [6]. From these signal TSI Prism™ systems track and record the location and movement of inmates and officers in real-time. Even though the technology was initially developed using proprietary tracking [6], new generation of TSI Prism is compatible with 802.11b WiFi technology, which operates at 2.4 GHz.

Figure 4 TSI Prism Inmate Tag [6].

TSI Prism™ unique feature of real-time tracking offers continuous and comprehensive prison monitoring by providing:
- Continuous tracking and evidence retention
- Movement controls without physical barriers
- Spatial Separation of subjects
- Automatic Identification of bystanders
- Location and duress alarms with location
- Secure identification tags for various purposes
- Assists investigation [6]

TSI Prism™ offers real-time tracking at the expense of one main disadvantage: cost. While the real-time tracking technology is an excellent solution and widely accepted in the U.S., implementing the system in a small prison with modest number of inmates requires some justifications.

Figure 5 TSI Prism Officer Tag [6].

Characteristics	Clincher ®	TSI Prism ™
Operating Frequency	13.56 MHz	902-928 MHz, 2.4 GHz
RFID Type	Passive	Active
Power	None	Rechargeable
Tag Housing Material	Polyethylene & Polyester	Hypo-allergenic ABS Plastic
Tamper	Tamper-Evident	Tamper-Proof
Wearability	Wrist mounting	Belt and wrist mounting
Reusability	Use once	Reusable
Memory and ID	2 k bits, 64 bits ID	48 bits ID
Reading	50 tags/s	2-10 second
Color	8 Different Colors	Black
Compatibility	IS0 15693	802.11b
Reading Range	Max. 1 m	100 – 2500 m square
Tracking Method	Zone	Real-time Location System (RTLS)

Table 1 Clincher® vs. TSI Prism™ Characteristic Comparisons [5][6].

Features and advantages of the inmates tagging systems were highlighted; Table 1 summarizes Clincher and TSI Prism™ characteristics. The characteristics of RFID technology were chosen to suite its specific application.

Passive, low frequency RFID is chosen for Clincher® tags, to meet its short reading range for identification purposes. On the other hand, TSI Prism™ requires active and operated in higher frequency in order to implement real-time tracking in wide area.

Short lifetime of Clincher® tags is limited by its constructed material (polyethylene compound), therefore it is designed for one-time-use only. TSI

Prism™ is constructed with higher durability is reusable and its limitation on battery lifespan is compensated by its rechargeable feature.

Both tagging system could be used for tracking, which demonstrates in Figure 6. Clincher® uses zone tracking method, while TSI Prism™ utilizes Real Time Location System (RTLS) [12] tracking.

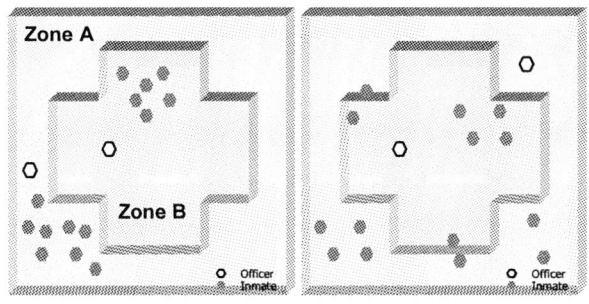

(a) Zone Tracking   (b) Real-Time Tracking

Figure 6 Zone Tracking versus Real-Time Tracking. (a) Zone Division Tracking. Illustrates Clincher® Tracking. (b) Real-time Tracking, Illustrates TSI Prism™ Tracking.

Zone tracking requires strategic location of RFID readers to separate between defined zones, which track movement entering and leaving between zones. Figure 6 (a) illustrates how the tracking method separates two groups in zone A and zone B. Enhance algorithm could to used to distinguish inmates and officer based on their ID number. In more complex implementation zone tracking is combined with anti-pass-back features and space-time restriction. The limited reading range often contributes to less accuracy on zone tracking performance.

Real-time tracking on the other hand, provides exact location of everyone in the compound as shown in Figure 6 (b). Long range RFID antennas are placed to provide full coverage the area. The tracking mechanism is using active RFID tags, which transmit signal to the antennas as frequent as two seconds. Therefore the exact location of inmates could be determined and refreshed every two second as well.

Clincher® is design for identification and related applications; while TSI Prism™ is specifically design for real-time tracking and monitoring every movement in the prison. Hence, Clincher® come in different colors and designed to incorporate printed information such as photo and names for visual identification and classification. However, TSI Prism™ identification features are incorporated in its real-time software and used for tracking purposes. Further research works on the system effectiveness and suitability are suggested to be studied before considering RFID inmate tagging implementation.

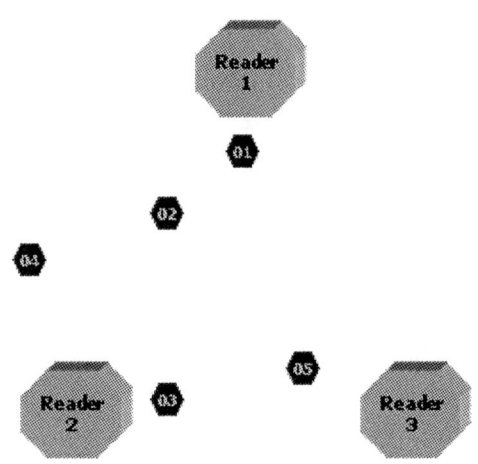

Figure 7 RFID Tracking System Design. Showing three long range RFID reader and 5 RFID active tags.

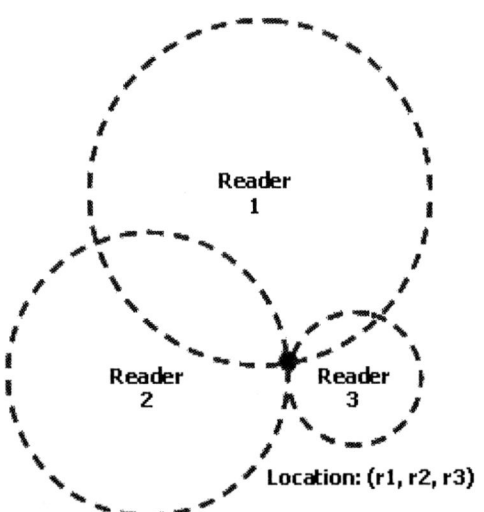

Figure 8 Triangulation Using Three RFID Long Range Reader.

In addition, new tracking methods could be studied to develop more effective and economical tracking application in addition to several popular tracking methods, which includes Global Positioning System (GPS) Based Location Tracking, Location by Wireless Ethernet, the Bat Ultrasonic Location System, the Cricket Location

Support System, the bat system (BAT), the Active Badge system, Radar and Cellular Location [11].

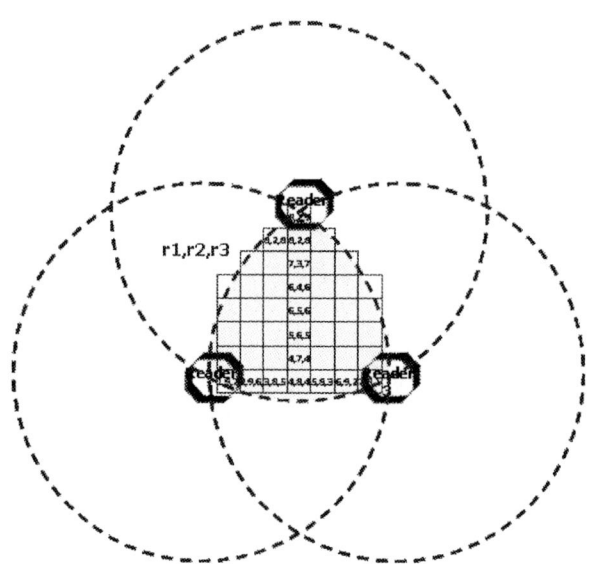

Figure 9 Suggested Triangulation Mapping.

RFID tracking system design shown in Figure 7 is proposed for the study. Three RFID long range readers are required for this experiment. The location tracking could be mapped by calculating distance between tag and the three readers as illustrated in Figure 8. Figure 9, suggest radial triangulation mapping (r1,r2,r3) of the system.

IV. CONCLUSION

In conclusion, if prison department manages to overcome the inmate tagging application implementation in prison should be considered to assist and enhance security and safety of inmates, staff and public in prison. In addition, new tracking methods could be studied to develop more effective and economical tracking application

ACKNOWLEDGEMENT

This work is financially supported by Malaysian Ministry of Science, Technology and Innovation under the project title "MEMS Devices and Sensing Microstructures"

REFERENCES

[1] "Misi", Laporan Tahunan 2003, Jabatan Penjara Malaysia. 2003.

[2] "Jabatan Penjara Memerlukan Sekurang-Kurangnya 20,000 Warden Tambahan", Berita Harian, Khamis 17 Mac 2005.

[3] "Keselamatan, Jabatan Penjara Malaysia", Ibu Pejabat Penjara Malaysia, 2004

[4] "Bahagian Keselamatan", Laporan Tahunan 2002, Jabatan Penjara Malaysia.

[5] http://www.pdcorp.com/law-enforcement/photo_id_bands.html, 16 October, 2006.

[6] http://www.alanco.com, 16 October, 2006.

[7] Anthony Furness, "Understanding RFID – A Guide to Radio Frequency Identification Technologies and Applications", Vicarage Publication Ltd, 2000.

[8] Klaus Finkenzeller, "RFID Handbook: Fundamentals and Applications in Contactless Smart Cards and Identification", Second Edition, 2003

[9] Nadeem Raza, Viv Bradshaw, Matthew Hague, Microlise Systems Integration Limited, "Applications Of RFID Technology", IEE, Q 1999.

[10] "Radio Frequency Identification System", Texas Instruments, 2000

[11] "A Comparative Study of Radio Frequency-Based Indoor Location Sensing Systems", Manapure, S.S., Darabi, H., Patel, V., Banerjee, P., Networking, Sensing and Control, 2004 IEEE International Conference on Volume 2, 2004.

[12] "MICLOG RFID Tag Program Enables Total Asset Visibility", Buckner, M., Crutcher, R., Moore, M.R., Whitus, B., MILCOM 2002. Proceedings Volume 2.

Challenges in Implementing RFID Tag in a Conventional Library

Mohd. Suhaimi Selamat, Burhanuddin Yeop Majlis, SMIEEE
Institute of Microengineering and Nanoelectronics (IMEN)
Universiti Kebangsaan Malaysia
43600 Bangi, Selangor, MALAYSIA
Email: mssuop888@vlsi.eng.ukm.my, burhan@eng.ukm.my

Abstract **— Operating a library involves in keeping track large number of resources such as books and magazines. Radio Frequency Identification (RFID) technology has been promoted in recent years as an alternative technology in improving asset management in a library. The RFID tags were applied to replace bar code and magnetic stripe functions as identification and anti-theft detection. This paper is written based on an actual implementation of RFID tagging system in a library one of local university in Malaysia. It focuses on three main challenges during the implementation which relates to system integration, parallel operation with existing system, and procedure changes. In conclusion, implementing RFID tag in a library faced many challenges; however the library could harness the technology advantages in improving its operation.**

I. INTRODUCTION

The omnipresent barcode labels that triggered a revolution in identification systems some considerable time ago are being found to be inadequate in an increasing number of cases [1]. One example of such cases is asset management in a library. Barcode labels, which have been used to tag library assets such as book and CDs were considered less effective comparing with new tagging system using Radio Frequency Identification (RFID). The air interface con-contact data transfer between a tag and its interrogator (reader/programmer) is seen as a fundamental feature of RFID, the transfer invariably being achieved without the need for obstacle-free, line-of-sight alignment [2]. Other distinctive features of RFID technology includes:

- No need line of sight

- Robust and rugged
- Fast reading speed
- Simultaneous multiple reading
- Secured and tamper proved
- Programmable [6]

It is an added advantage of the technology which overcomes shortcomings of barcode based systems. Therefore, RFID technology has been promoted in recent years as an alternative technology in improving asset management in libraries. The RFID tags were applied to replace bar code and magnetic stripe functions as identification and anti-theft detection. Similar implementation was reported by Prof. Anthony Furness in 2000 as case study titled: "New RFID smart label system installed for 'digital' library at University of Nevada, Las Vages" [1].

II. RFID SYSTEM

This paper is written based on an actual implementation of passive mid range RFID tagging system in a conventional library of a local university in Malaysia. The overall project was completed June, 2006.

OBID ID ISC.MR101 mid-rage RFID reader is used in the implementation. The reader coupled with pad antenna OBID ID ISC.ANT240 is capable to read multiple tags within 30 cm range. It operates at 13.56 MHz and reads all ISO 15693 and ISO 18000-3 standard tags [3]. Figure 1 shows physical look of the reader and antenna.

The reader and antenna formulates RFID main subsystems as shown in Figure 2: self-check station, staff station, conversion station and anti-theft gate. Self-check station, staff station, and conversion station were connected through Ethernet network, which linked with Library Management System (LMS) server. Anti-theft gate

0-7803-9730-4/06/$25.00 ©2006 IEEE

is installed at the entry point of the library in parallel with existing anti-theft panel.

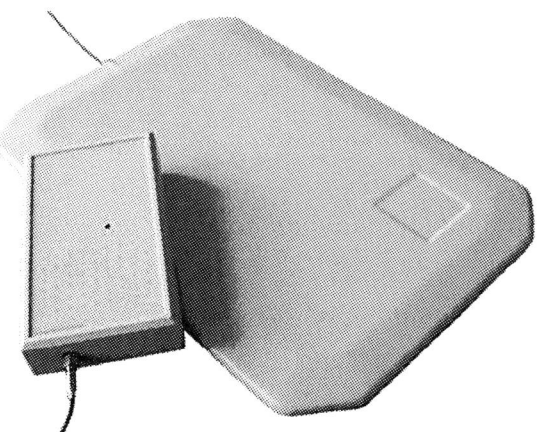

Figure 1 OBID ID ISC.MR101 RFID Reader and Antenna.

In addition to RFID reader and antenna, another important component is RFID tag or label. The label used is of identification/data-key carrier (Type ID) [1], which has combination of data and electronic article surveillance (EAS). Both data and EAS are programmed using the same device, the RFID reader. The label comes in reel form of 1800 units per reel. Internally the label equipped with Phillips I-Code SL chip with 1024 bits programmable memory [5].

The overall implementation process involved installation of subsystems, converting the books, applying RFID tags, system integration, and testing. Figure 5, 6, and 7 show some of the implementation processes.

This paper focuses on three main challenges during the implementation which relates to system integration, parallel operation with existing system, and procedure changes.

III. IMPLEMENTATION CHALLENGES

The first challenge was to integrate RFID tagging system with the Library Management System (LMS), which used barcode system. Before the RFID subsystems could work with the LMS, it is necessary to establish the same communicate protocol between the LMS and the subsystems. In order to establish this, system engineers from both parties have to work together to synchronize the

communication protocol in LMS and RFID system.

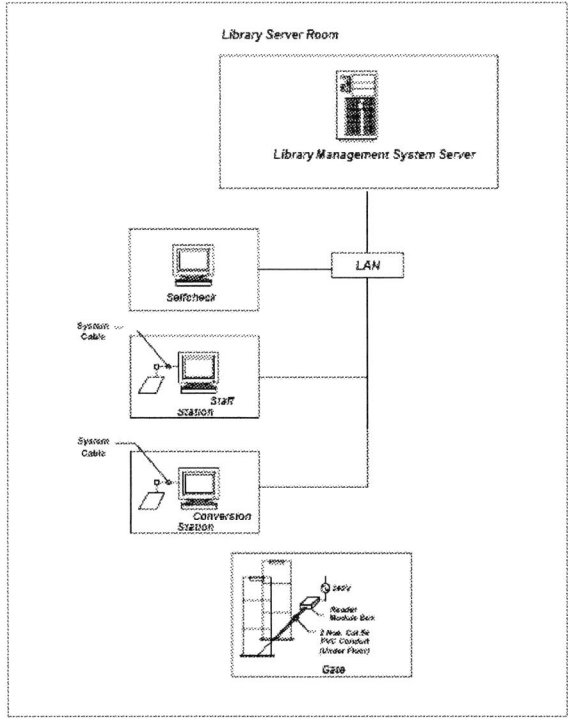

Figure 2 Integrated Library System. Showing self-check station, staff station, conversion station, anti-theft gate, and connection with Library Management System.

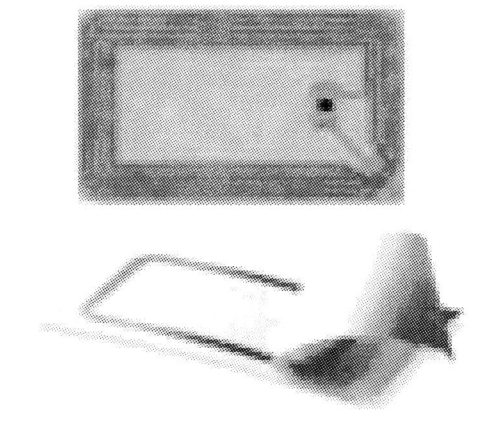

Figure 3 RFID label.

When communication was establish, the RFID sub-systems were then tested to work in parallel with barcode system. It is necessary to have both systems running in parallel so that library could operate normally while conversion process is in progress, Figure 7. Allowing the RFID tagging system co-exists and operates with existing system creates new problems.

Figure 4 RFID Tagged Book.

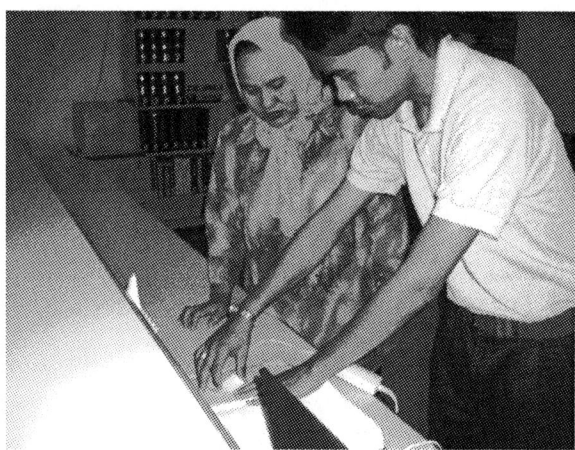

Figure 5 Book Tagging. Showing library staff applying RFID tag.

During the conversion process, in order to check RFID tagged books as well as books with barcode, both anti-theft gates were activated and installed in parallel, Figure 8. Borrowing and returning process has to go through both security activation and deactivation process. To minimize the effect of the problem, the conversion process was expedited and running double-shift. As the result, all books in the library were tagged in less than one week and conventional anti-theft detection was deactivated and the library operated only on RFID tagging system.

Figure 6 Barcode Scanning. Barcode is scanned during conversion process.

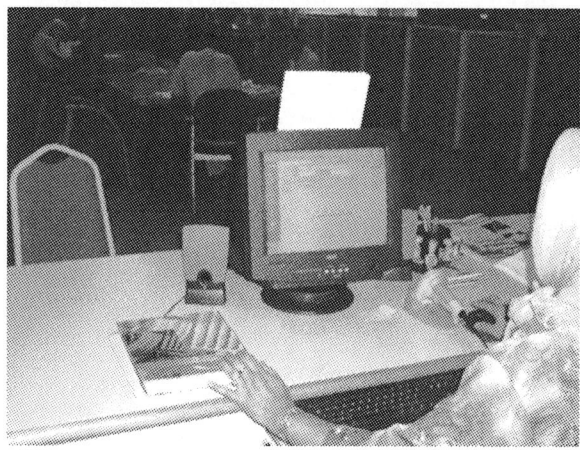

Figure 7 Conversion Station. Showing staff performing conversion.

Figure 8 RFID Anti-Theft Gate. The gate is installed in parallel with existing gate.

Before the system is smoothly operational, some procedures relate to borrowing and returning books need to be altered to suite the new tagging system.

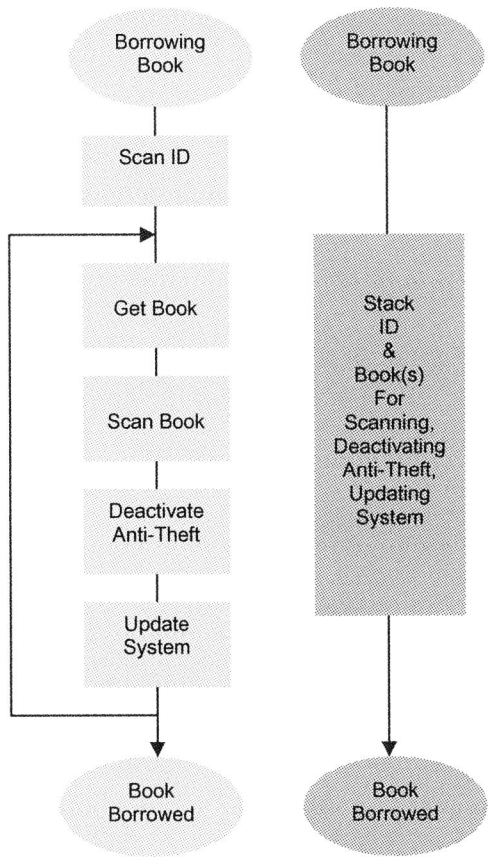

(a) Conventional System   (b) RFID Tagging System

Figure 9 Conventional versus RFID Tagging Borrowing Book Flowchart.

Figure 9 and 10 shows flow comparison between conventional and RFID tagging book borrowing and retuning. Usually in conventional system book has to be scan and its anti-theft features need to be deactivated one by one. After RFID tagging system implementation books and ID could be stack for scanning, anti-theft deactivation at one go. In addition this process could be perform with or without library staff assistant when self check counter is used, Figure 11. The self-check counter makes borrowing returning book process more automated with less involvement of librarians. Therefore they could focus on more effective work to better serves for the library.

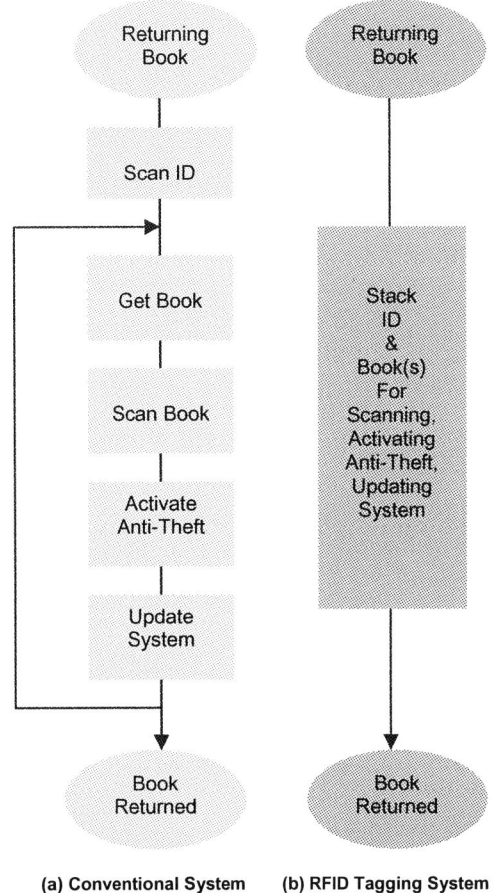

(a) Conventional System   (b) RFID Tagging System

Figure 10 Conventional versus RFID Tagging Retuning Book Flowchart.

Figure 11 Self-Check Station. Allowing student to self-check for borrowing and returning books.

In addition to suggesting new operational procedures, the library staffs were trained on total operation of RFID tagging system and the new work flow. Series of trainings were also conducted to the students and lecturers especially on using the self-check station, Figure 11. This successful project is one step closer towards creating more user friendly environment as a flexible and dynamic workspace or Sm@rtLibrary [4]. It managed to overcome the main historically barrier to such implementation which has been cost justification for a particular application [6] and proved that RFID technology is trusted by local librarians in improving their library operation. Further study on RFID implementation can be studied to improve filing management and important document storage.

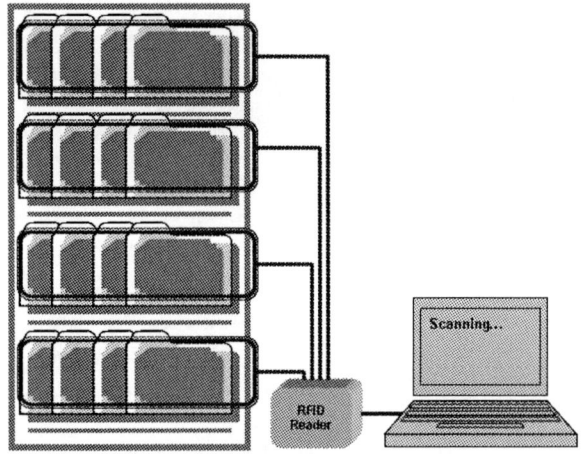

Figure 12 RFID Filing System Proposed Design.

Figure 12 illustrate the proposed design, which consists of four antennas used to scan document tagged with RFID. The antennas periodically scan documents and update their status.

%%001,029,UKM3.1/32/207/Jld,Pengesahan Pendaftaran Kemasukan Pelajar P32781,IN**

%%<Cabinet No>, <Folder No>, <Document No>, <Content Summary>,<Status>**

Figure 13 Suggested Data Format of RFID Filing System.

In addition to periodical scanning, user may perform document searching from the system. Content of the document could be stored in the RFID tag to help the searching process. Security control on the status could be implemented for example only document with status "OUT" is allowed to be removed from the cabinet. If any document with status "IN" is removed the system will produce alarm signal. Additional features such as access authorization could also be implemented. Only restricted people should get retrieve high classified document. He/she should present his/her RFID card on the shelf antenna for authorization. Hence, document movement in and out of the cabinet could be traced and secured.

IV. CONCLUSION

In conclusion, implementing RFID tag in a library would face many challenges, which relates to system integration, parallel operation with existing system, and procedure changes; however the library could harness the advantages of RFID technology and dynamically improve its operation.

ACKNOWLEDGEMENT

This work is financially supported by Malaysian Ministry of Science, Technology and Innovation under the project title "MEMS Devices and Sensing Microstructures".

REFERENCES

[1] Klaus Finkenzeller, "RFID Handbook: Fundamentals and Applications in Contactless Smart Cards and Identification", Second Edition, 2003

[2] Anthony Furness, "Understanding RFID – A Guide to Radio Frequency Identification Technologies and Applications", Vicarage Publication Ltd, 2000.

[3] "OBID Advanced Reader Technologies Specification", FEIG Electronic

[4] Marie-Luise Moschgath, Jorg Hahner, Rolf Reinnema, "Sm@rtLibrary - An Infrastructure for Ubiquitous Technologies and Applications", 2001 IEEE.

[5] "ISO Book Label I-Code", http://www.telepen-barcode.co.uk/images/productdownloads/RFID%20 Labels.pdf, Codeco™ Specifications RFID Labels, October, 2006

[6] Nadeem Raza, Viv Bradshaw, Matthew Hague, Microlise Systems Integration Limited, "Applications Of RFID Technology", IEE, Q 1999.

Investigating the Performance of RF MEMS Switches

Hieng Tiong Su[1], Ignacio Llamas-Garro[2], Michael J. Lancaster[3], Martin Prest[3], Jae-Hyoung Park[4],
Jung-Mu Kim[5], Chang-Wook Baek[6], and Yong-Kweon Kim[5]

[1] School of Engineering, Swinburne University of Technology (Sarawak Campus),
Sarawak, Malaysia. *Email: suht@ieee.org.*
[2] Large Millimeter Telescope, National Institute for Astrophysics, Optics and Electronics, Mexico.
[3] School of Electronic, Electrical and Computer Engineering, The University of Birmingham, UK.
[4] LG Electronics Institute of Technology, Seocho-Gu, Seoul, Korea.
[5] School of Electrical Engineering and Computer Science, Seoul National University, Seoul, Korea.
[6] School of Electrical and Electronics Engineering, Chung-Ang University, Korea.

Abstract—**The performance of Micro-Electro-Mechanical System (MEMS) metal switches were investigated at wide temperature range. Measurements were carried out using cryogenic probe station and S-parameters were taken using a network analyser for frequencies up to 20 GHz. A total of 28 switches were evaluated. The investigation shows a 50% increase in the actuation voltage and a decrease in the percentage of operational switches as the temperature was reduced to 10 K. At room temperature the best isolation (when open) was 30 dB at 10 GHz with an insertion loss of 0.14 dB (when closed). Measurement accuracy was reduced at low temperature, however, isolations and insertion losses were similar to room temperature values.**

Index Terms— **RF MEMS, metal switch, reliability, low temperatures, actuation voltage**

I. INTRODUCTION

Radio Frequency Micro-Electro-Mechanical Systems (RF MEMS) has attracted world-wide research interest for the past few years. This is mainly due to the compatibility in terms of the fabrication technologies already used in semiconductor industries. RF MEMS technologies have shown potential applications in many industries, for instance, wireless communication applications. The advantages of RF MEMS devices are their small size, integration capability and superior performance. RF MEMS switches are prime candidates to replace conventional GaAs FET and p-i-n diode switches in RF and microwave communication systems. Using RF MEMS switches, RF circuits such as variable capacitors, tuneable filters, phase shifters and signal routers have been

demonstrated [1]. Depending on the type of MEMS switches, low insertion loss when closed and high isolation when open have been reported [2, 3]. Because of low current is needed to operate the MEMS switches, they consume very low power; in addition, they show very linear transmission characteristics with extremely low signal distortion, making them suitable for modern radar and communication application [4].

In this paper, we report our recent wide temperature range measurements of RF MEMS switches, as part of our assessment to determine feasibility of integrating similar switches with superconducting microwave devices [5] in order to achieve dynamic microwave device tuning.

II. STRUCTURE OF THE METAL SWITCH

The layout and cross-section of the metal switches are shown in Fig. 1(a) and 1(b) respectively. The metal switches were fabricated on a 0.52 mm thick quartz substrate with a dielectric constant of approximately 3.78. The switches consist of two electrostatically actuated nickel pads and a small central gold plated nickel RF contact pad. The RF contact pad is electrically isolated and physically connected to the actuation pads by a thin dielectric membrane of silicon nitride. The switch actuates on a gold 50-Ω coplanar waveguide (CPW) of thickness 3 μm, and signal line width 50μm. Below the RF contact pad of the switch is a 5 μm gap in the signal line. The RF contact pad bridges this gap by direct contact when the actuation voltage is applied. A dielectric layer prevents shorting between the actuation pad and the ground electrodes. The height of the RF contact pad above the transmission line, before actuation, is 1.2 μm. The two coplanar ground planes are connected via air-bridges. The

0-7803-9730-4/06/$25.00 ©2006 IEEE

electrostatic actuation bias was applied between the actuation pads and the CPW ground planes using dc probes. More details of the fabrication process (of a similar metal switch) can be found elsewhere [6]. Without the biasing lines, the total area of the switch is about 0.5 mm × 0.32 mm.

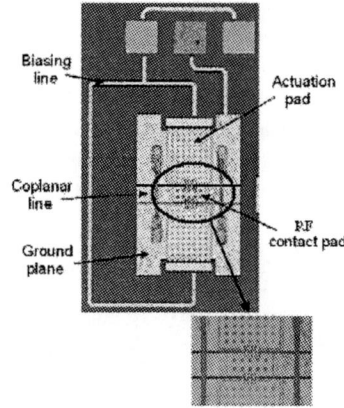

(a) A photograph of the metal switch

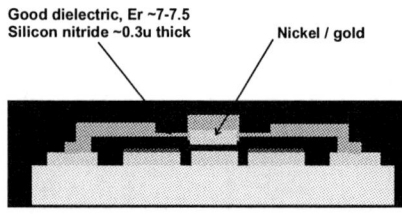

(b) Side-view of the metal switch

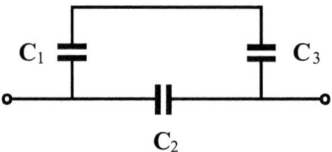

(c) Capacitive network representation of the switch
Fig.1. Layout of the metal switch.

When in the OFF state, the switch can be represented by a capacitive network as shown in Fig. 1(c). C_1 and C_3 represent the gap between the RF contact pad and the CPW signal line which is equal on both sides, therefore $C_1 = C_3$. C_2 results from the 5 µm gap in the signal line. The OFF state capacitance $C_{off} = (C_1/2 + C_2)$, effectively representing the OFF state isolation. For high OFF state isolation, C_{off} is required to be small.

III. MEASUREMENTS

Measurements of the metal switches were carried out using a cryogenic probe station. Two special 90-degree microwave bent probes [7] with a pitch size of 150 µm were used as they allowed clearance (due to their long-legs) for the dc biasing probes to make contact with the biasing pads.

Measurements were carried out for metal switches on a die. The die contains about 80 metal switches, arranged in rows and columns. These switches were designed with two different RF contact pad sizes: 100 µm x 50 µm and 50 µm x 50 µm. There are also different geometries in terms of the width and length of the actuation pads, as shown in Fig. 2.

The die was glued to the sample holder of the probe station using silver paste. This prevents movement when probing the metal switches and provides a good thermal contact. However, it also creates a ground plane for the coplanar structure but the effect is assumed negligible considering the thickness of the quartz substrate (0.52mm), which is much larger than the slot width (between the signal line and ground plane) of 5 µm.

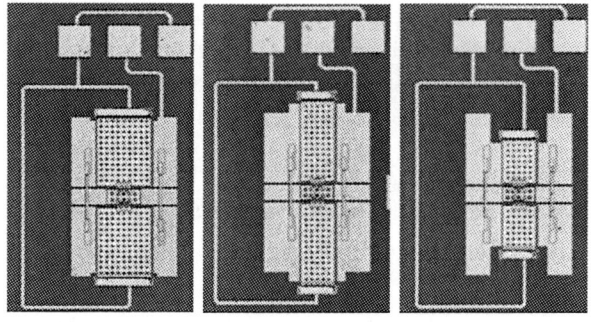

Fig. 2. Metal switches with different actuation pad sizes, all with RF contact pad size of 100 µm x 50 µm.

The S-parameters were measured using an Agilent 8722ES network analyzer. Calibration were carried out using Cascade Microtech software (WinCal); the calibration technique used was Thru-Reflect-Reflect-Match (LRRM).

Fig. 3 shows measured performance of some metal switches at 290 K for frequencies up to 30 GHz. These results were compared with the results obtained using Sonnet simulations [8], and they showed good agreement. In the OFF state, the metal switch with a smaller RF contact pad size shows better isolation, about 30 dB at 10 GHz. Switches labeled C_2R_3 and C_6R_4 actuated at 38 V and 43 V, respectively. The measured insertion losses at

10 GHz were 0.16 dB and 0.14 dB, respectively, which were higher than the simulated insertion loss of 0.04 dB.

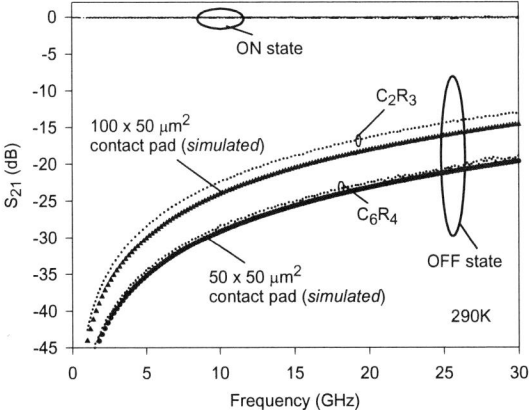

Fig. 3. Comparison between simulated results and the measured performance of the metal switches at 290 K.

To avoid damaging the metal switches, initial measurements of the metal switches started at 290 K. This allowed investigation of the performance of the metal switches as the temperature was decreased. At several temperature intervals, actuation voltages and S-parameters of the metal switches were recorded. Fig 4 shows the recorded actuation voltages for a number of metal switches over the temperature range from 290 K to 10 K.

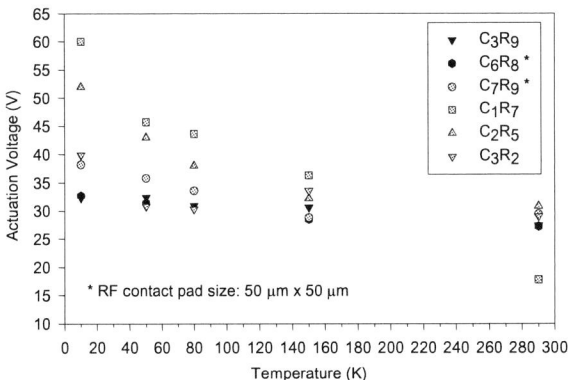

Fig. 4. Measured actuation voltages for a number of switches as a function of temperature.

Note that the actuation voltages of the metal switches have increased at low temperatures. For the working switches, most of them actuated at a bias voltage between 30 V and 60 V. It was observed that the switches become more difficult to actuate at low temperatures. At temperatures 10 K and 50 K, a few switches showed high insertion losses (up to 5 dB) and showed no further change when the bias voltage

was increased; which implied that the RF contact pads were not fully closed. Goldsmith [9] attributes the increase in actuation voltage to an increasing stress in the actuators as a result of the difference in thermal expansion coefficients between the metal and the quartz substrate.

We observed that metal switches with larger actuation pad tended to actuate at a lower bias voltage as compared with metal switches with smaller actuation pad. This is attributed to the higher pull-in force encountered by the larger pad for the same bias voltage applied.

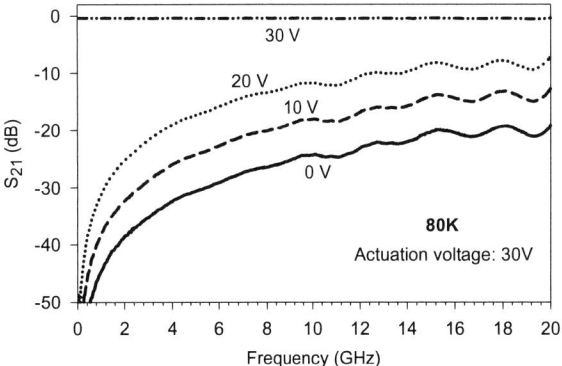

Fig. 5. Measured S_{21} response for biases up to 30 V (switch C_3R_2).

Fig. 5 shows performance of one of the switch at 80 K for applied biases up to 30 V. As can be seen before pull-in voltage, S_{21} changes when bias voltage increases, which is due mainly to the increase in the capacitance value exists between the RF contact pad and CPW signal line. Reducing the applied bias voltages will bring the S_{21} curve back to the original position except at certain voltage level close to the actuation voltage. Fig. 6 shows the values of S_{21} at 10GHz and 80K for applied biases both at increasing and decreasing modes. Note that the switch remained closed at about 19 V at the decreasing mode. This phenomenon was also observed for other switches and could be due to: (1) surface deformation of the insulating dielectric layer between the actuation pad and the ground electrode resulted after pull-in and (2) residue charge (image charge) remained on both of the actuation pads and the ground electrodes.

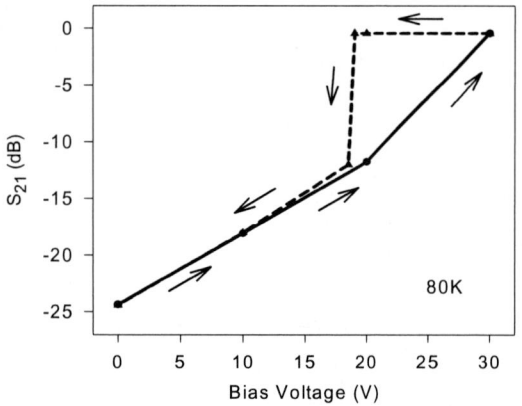

Fig. 6. Measured S_{21} at 10 GHz as a function of applied bias in increasing and decreasing modes.

It is interesting to compare the percentages of working switches at low temperatures. In this experiment, a total number of 28 switches were measured. Fig. 7 shows the percentages of the switches that actuated at temperatures 290 K, 150 K, 80 K, 50 K and 10 K. At 290 K, about 82% of the switches actuated, which decreased to 39% at 80 K and further at lower temperature.

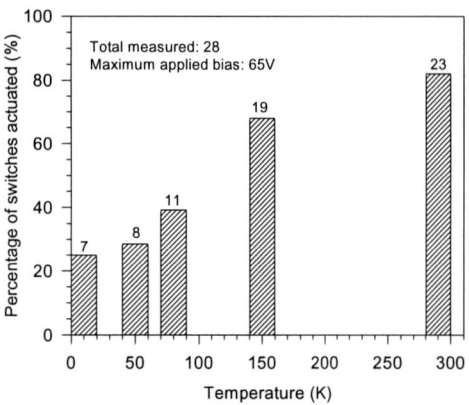

Fig. 7. Different percentages of the working switches recorded at different temperatures.

IV. CONCLUSION

Low temperature performance of MEMS metal switches has been investigated. Measurements showed an increase in the actuation voltage as the temperature went down to 10 K. It was found that the operational percentage of these switches decreases at low temperature. It should be noted that at present no special design considerations were observed in order to optimise this figure and it is expected that this could be improved.

ACKNOWLEDGMENT

The authors would like to thank Dr. Phe Suherman for providing technical support of using the cryogenic probe station during measurements. This work was supported by the Birmingham University Portfolio Award.

REFERENCES

[1] S. Lucyszyn, "Review of radio frequency microelectromechanical systems technology", *IEE Proc.-Sci. meas. Technol.* vol. 151, No. 2, March 2004.

[2] Z. J. Yao, S. Chen, S. Eshelman, D. Denniston, and C. Goldsmith, "Micromachined low-loss microwave switches", *IEEE J. Microelectromech. Syst.*, vol. 8, pp. 129-134, June 1999.

[3] J. B. Muldavin and G. M. Rebeiz, "High-isolation CPW MEMS shunt switches: Modeling and design", *IEEE Trans. Microwave Theory Tech.*, vol. 48, pp. 1045-1056, June 2000.

[4] F. De Flaviis, R. Coccioli, and T. Itoh, "Non-linear analysis and evaluation of distortion introduced by MEM switches in reconfigurable antenna systems", in *Proc. IEEE AP-S Int. Symp.*, Davos, Switzerland, 2000.

[5] Harold Weinstock and Martin Nisenoff, "Microwave Superconductivity", *NATO Science Series, Series E:Applied Sciences*, vol. 375, 2001.

[6] Jae-Hyoung Park, Sanghyo Lee, Jung-Mu Kim, Hong-Teuk Kim, Youngwoo Kwon, and Yong-Kweon Kim, "Reconfigurable Millimeter-Wave Filters Using CPW-Based Periodic Structures With Novel Multiple-Contact MEMS Switches", *Journal of Microelectromechanical Systems*, vol. 14, no. 3, pp. 456-463, June 2005.

[7] GGB Industries Inc., Florida (USA), www.picoprobe.com.

[8] Sonnet Software, Inc. version 9, 2005.

[9] Charles L. Goldsmith, and David I. Forehand, "Temperature Variation of Actuation Voltage in Capacitive MEMS Switches", *IEEE Microwave And Wireless Components Letters*, vol. 15, no. 10, pp. 718-720, Oct 2005.

Comparison of Electronic Transport Parameter of CNT(10,10)/CNT(17,0) and CNT(5,5)/CNT(8,0) Carbon Nanotube Metal-Semiconductor On-Tube Heterojunction

Sukirno[1], Satria Zulkarnaen Bisri[1], Irmelia[1], Lilik Hasanah[1,2], Adi Bagus Suryamas[1], Ida Usman[1,3], Mursal[1,4]

[1] Laboratory for Physics of Electronic Material, Faculty of Mathematics and Natural Sciences – Institut Teknologi Bandung (ITB). Jl. Ganesha 10 Bandung 40132 Indonesia
[2] Department of Physics, Universitas Pendidikan Indonesia (UPI)., Universitas Pendidikan Indonesia, Jl. dr. Setiabudhi 229 Bandung 40154 Indonesia
[3] Department of Physics, Universitas Haluoleo, Palu, Indonesia
[4] Department of Physics, Universitas Syiah Kuala, Banda Aceh, Indonesia

Contact e-mail: sukirno@fi.itb.ac.id

Abstract – Carbon Nanotubes research is one of the top five hot research topics in physics. It is because of its unique properties and functionalities, which leads to wide-range applications. One of the most interesting potential applications is in term of nanoelectronic device. There is a possibility to found some unique structure, where different carbon nanotubes are connected coaxially. It has been modeled carbon nanotubes heterojunction, which was built from two different carbon nanotubes, that one is metallic and the other one is semiconducting. There are two different carbon nanotubes metal-semiconductor heterojunction. The first one is built from CNT (10,10) as metallic carbon nanotube and CNT (17,0) as semiconductor carbon nanotube. The other one is built from CNT (5,5) as metallic carbon nanotube and CNT (8,0). All of the semiconducting carbon nanotubes are assumed to be a pyridine-like N-doped. Those two heterojunctions are different in term of their structural shape and diameter. It has been calculated their charge distribution and potential profile, which would be useful for the simulation of their electronic transport properties. The calculations are performed by using self-consistent method to solve Non-Homogeneous Poisson's Equation with aid of Universal Density of States calculation method for Carbon Nanotubes. The calculations are done by varying the doping fraction of the semiconductor carbon nanotubes. It is obtained that the charge are distributed almost evenly along the semiconducting carbon nanotubes and the potential profile peaks in the vincinity of semiconducting carbon nanotubes center position, with some valley-shapes that show some sign of charge confinements nearby. However, from the comparison of two different heterojunctions, it could be inferred that the geometrical aspects of the heterojunction building blocks has effect on their electronic transport parameter. It is also obtained the calculation results of the electron tunneling transmission coefficient that transported through the heterojunction, which has energy lower than the potential barrier value.

I.   INTRODUCTION

Since the first time it was discovered in 1991 by Sumio Iijima [1], carbon nanotube (CNT) has been object for intense scientific research and recently, engineering research; moreover, it also become the most influencing materials in nanotechnology development [2]. It is because of its unique and remarkable properties including its mechanical, electronic and chemical properties; so that this material has wide-range applications including in nanoelectronics, lightweight-superstrong materials, renewable-energy-developments and also in medicine[2-6].

One of the most interesting potential applications is in term of nanoelectronic devices.

0-7803-9730-4/06/$25.00 ©2006 IEEE

There is a possibility to found some unique structure, where different carbon nanotubes are connected coaxially [3].

It is possible to connect carbon nanotubes to become on-tube structures by introducing the presence of topological defects, in form of pentagonal or heptagonal ring, in addition of hexagonal ring in honeycomb graphene structure. On-tube structures which could be formed are independent of the CNTs electronic properties; therefore, there could be CNT metal-semiconductor junction, CNT metal-metal or CNT semiconductor-semiconductor junction and metal-semiconductor-metal CNT junction [as well as T-shaped metal-semiconductor-metal CNT junction [7], which is different kind of junction that is not on-tube junction. On-tube connected structures of CNT may show behavior as new nanoscale devices. Therefore, study about properties of connected CNT is very important for future nanoscale device developments.

In this paper, it will be explained the comparison of electronic transport parameter of carbon nanotube metal-semiconductor on-tube heterojunction. It has been modeled carbon nanotubes heterojunction which was built from two different carbon nanotubes, that one is metallic and the other one is semiconductor. There are two different carbon nanotubes on-tube metal-semiconductor heterojunction which was studied in this research. The first one is built from CNT (10,10) as metallic carbon nanotube and CNT (17,0) as semiconductor carbon nanotube. The other one is built from CNT (5,5) as metallic carbon nanotube and CNT (8,0). All of the semiconducting carbon nanotubes are assumed to be a pyridine-like N-doped.

This paper devoted into 3 chapters which includes introduction, geometrical modeling and simulation of electron transport as well as results and discussion.

II. GEOMETRICAL MODELING AND SIMULATION OF ELECTRON TRANSPORT OF HETEROJUNCTIONS

The carbon nanotube heterojunction geometrical structure is simply generated by connecting the two individual carbon nanotubes, which become the building block [8]. Since the aim is to make an on-tube heterojunction, the axis center of each nanotube must be positioned parallel and aligned.

In the program, the metal carbon nanotube is used as the first nanotube which has the exact coordinates. And then, the atomic coordinates of the semiconductor carbon nanotubes was translated as long as the length of the metallic carbon nanotube with an addition of 1 angstrom. Afterward, the semi conductor carbon nanotube atomic coordinates is moved approaching the metallic carbon nanotube with a step of 0.01 angstrom. The aim is to find out the right distance between both nanotubes, so the length of atomic bonds among the edge atoms are match with the carbon-carbon atomic bond, which is 1.40 angstrom until 1.44 angstrom.

From the method above [9], the carbon nanotube heterojunction atomic coordinate could be obtain as in figure 1. In the junction tapering area, it was formed pentagonal, heptagonal, and trigonal-like hexagonal rings instead of normal hexagonal ring whereas in the part of individual nanotube. These rings shapes are possible to be formed in the nanotube. In this study, realizing that it is no more than 1 angstrom in length, the joint structure length is neglected for the calculation of the heterojunction electronic properties.

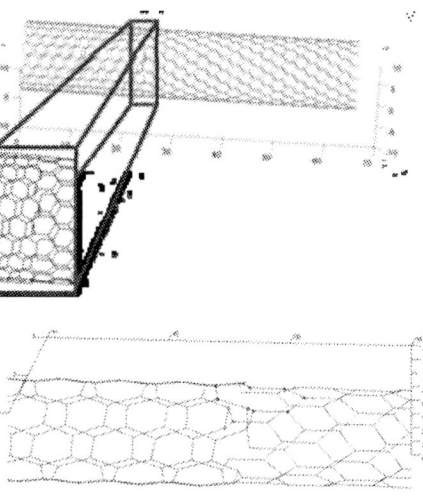

Fig. 1 The geometry construction of the carbon nanotube on-tube metal-semiconductor heterojunction, which constructed from CNT (10,10) and CNT (17,0) (above) and from CNT (5,5) and CNT (8,0) (under)

The charge distribution in the semiconductor carbon nanotube is calculated by using one-sided of the way how to calculate the charge distribution in a *p-n* junction carbon nanotube [13]. It is stated that in the *n-type* part of carbon nanotube, the charge distribution per carbon atom as a function of position could be stated as

$$\sigma(z) = \frac{e}{\varepsilon} f - \frac{e}{\varepsilon} \int_{\varepsilon_\eta(z)}^{\infty} n(E,z) F(E) dE \quad (1)$$

To calculate the charge distribution, the density of states used is the Universal Density of States [14], instead of the method used to find the individual carbon nanotube density of states. The reason why Universal Density is used is because of the density of states needed is a density of states which could be dependent on position in the nanotube, the bias voltage applied as well as the potential profile of the heterojunction itself.

By using the Universal Density of States [14], the Van Hove Singularity values are presumably could be shifted by the appearance of potential. Consequently, the Van Hove Singularity factor becomes [15]

$$g(E, \varepsilon_\eta) \to g(E, \varepsilon_\eta - e\psi(z)) \quad (2)$$

where the $\psi(z)$ is the appeared electric potential, which might be varied as function of position, and e is the charge of an electron. Therefore,

$$g(E, \varepsilon_\eta - \psi(z)) = \begin{cases} |E| / \sqrt{E^2 - (\varepsilon_\eta - e\psi(z))^2} & , |E| > |\varepsilon_\eta - e\psi(z)| \\ 0 & , |E| < |\varepsilon_\eta - e\psi(z)| \end{cases}$$

$$(3)$$

Meanwhile, potential profile in semiconductor CNT is calculated from charge distribution which is obtained by using Green Function for electrical potential by using approximation that CNT is a charged cylinder.

$$\psi(z) = \frac{1}{4\pi} \int_{-\infty}^{\infty} 2\pi R \frac{\rho(z')}{\varepsilon \sqrt{R^2 + (z - z')^2}} dz' \quad (4)$$

The self-consistent method is used to calculate eq. (1) and eq. (4) simultaneously. The result of the calculation itself is the charge distribution per square area as well as the potential profile, while the input is the guessed potential profile.

The integration method used in this calculation is the Gaussian Quadrature Method where the number of the mesh is 1000. The weighing factor is calculated by Mathematica 5 package. The rate of the convergence is calculated from averaging of the potential profile result which its previous result over the space. The result will become more convergent when the ratio is become closer to one.

The calculation is performed using standard parameters [16] which are in the room temperature (300 K), 3.9 in electrical permittivity

of CNT, 1.44 Å of a_{C-C}, 2.50 eV of $V_{pp\pi}$ as well as no voltage bias. The other parameters are obtained from identifying properties of each CNT, such as the length of semiconductor CNT, its diameter as well as the number of atoms in the nanotube rings. The calculation is executed by using MATLAB and using several kinds guess potential. The calculation is also performed with various doping fraction.

III. CALCULATION RESULTS AND DISCUSSION

From the self-consistent calculation, it is obtained the result of charge distribution which could be seen in figure 2. The figure is plotted charge distribution that has been normalized with the maximum value. Generally, from the plot, it could be seen that the charge almost distributed evenly in the semiconductor CNT. These things happen even though the doping fraction was varied. One thing that makes different is the increasing number of doping fraction, the more value of the number of the electrons for each carbon atoms. It could be seen from the value indicated by the vertical axis.

However, the result of the charge distribution calculation has a good agreement with the gold-semiconductor CNT-gold junction Park, et.al [16]. obtained that in the gold-semiconductor CNT-gold heterojunction, the charge distribution has valleys nearby the joint between the metal and the CNT where the charge distribution is almost similar along the nanotube.

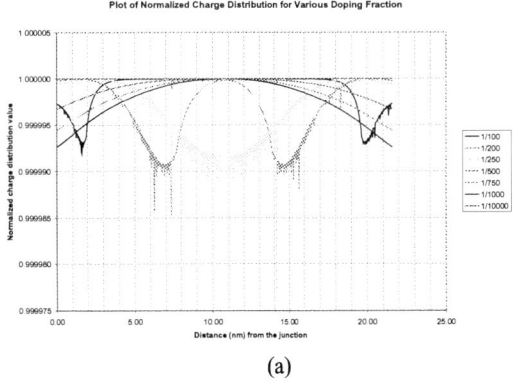

(a)

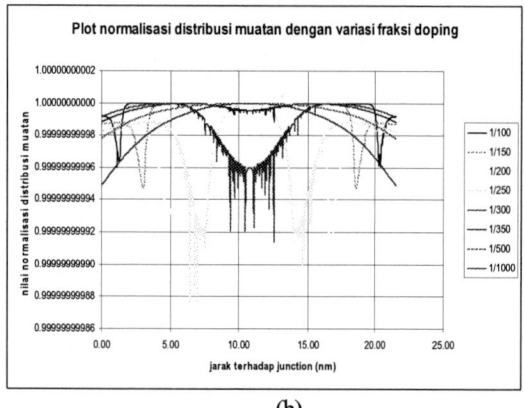

(b)

Fig. 2. The plot of normalized charge distribution of the CNT(10,10)/CNT(17,0) heterojunction (a) and CNT(5,5)/CNT(8,0) heterojunction (b)

From the potential profile result (figure 3) [10], it can be inferred that the potential formed has it peak in the vicinity of semiconductor-CNT center point, with non-zero potential at both ends. This may make a barrier structure for the electron which flow from metallic CNT through the semiconductor CNT. Even though the potential profile looks to be symmetric, the exact values of the potential at both ends are different. At the surface of the junction, the potential value is smaller than the other.

From the comparison of charge distribution and potential profile of both carbon nanotube metal-semiconductor heterojunctions, which are CNT(10,10) /CNT(17,0) and CNT(5,5)/CNT(8,0), it is obtained some similar anomalies as well as some differences. The similarity is the anomaly that is formed valley-shapes in charge distribution for bigger doping fraction, which means electron are accumulated at the edge and the middle of nanotube but there are areas formed valley-shapes that bordered them. The smaller doping fraction given, electrons are more accumulated in the middle of nanotube. In potential profile, tops of potential happen in the middle of semiconductor CNT and the bigger doping fraction, the highest potential barrier formed.

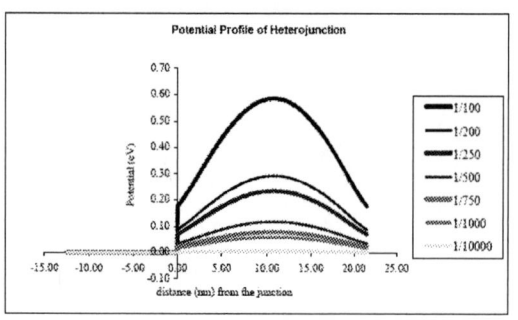

(a)

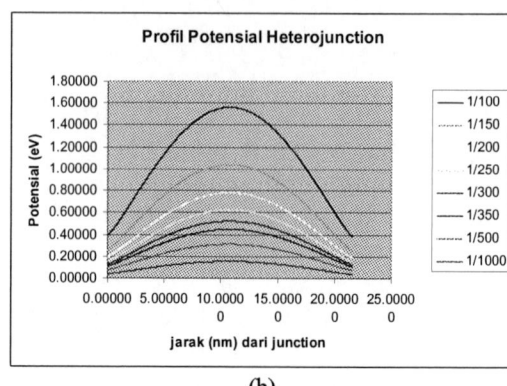

(b)

Fig. 3. . The plot of profile potential of the CNT(10,10)/CNT(17,0) heterojunction (a) and CNT(5,5)/CNT(8,0) heterojunction (b)

The effect of doping fraction gives difference of potential height. The bigger doping fraction is, the higher potential profile will be. This could happen since the number of electron which formed the barrier becomes bigger because the bigger number of doping fraction.

Otherwise, differences could be seen in the quantity of the result. The highest value of potential barrier for CNT (5,5)/(8,0) is bigger than CNT (10,10)/(17,0) heterojunction. It is because of bigger semiconductor CNT bandgap. In the other hand, charge distribution occurred in CNT (5,5)/(8,0) heterojunction is not abrupt as in CNT (10,10)/(17,0). This thing happened because there are less atomic carbon involved in the device which means less atomic carbon contributed. Beside that, if it is seen from influence of doping fraction in charge distribution for both difference devices, it will be obtained that there is different boundary where charge distribution shape is not similar. In CNT (10,10)/(17,0) heterojunction, 1/250 doping fraction gives charge distribution shape that consist of only one valley which positioned in the middle of nanotube. Whereas, the bigger value of doping fraction gives two valley-shapes. On the other hand, in CNT (5,5)/(8,0) the only one valley-shape is found for 1/350 doping fraction. These things are estimated happened because of the difference of Van Hove Singularities. Beside that, there is possibility of different property caused of different geometrical factors.

ACKNOWLEDGEMENT

This paper is a part of "Simulation and Fabrication of Carbon Nanotube part II" funded ITB Frontier Research Grant 2006.

REFERENCES

[1] S. Iijima, *Nature* (London) 354, 56 (1991).

[2] J. Giles, "Top Five in Physics", *Nature* 441, 265 (2006).

[3] R. Saito, G. Dresselhaus dan M.S. Dresselhaus. *Physical Properties of Carbon Nanotubes.* Imperial College Press (1998).

[4] T. Nakanishi, A. Bachtold dan C. Dekker. "Transport through the Interface between a Semiconducting Carbon Nanotube and a Metal Electrode", *Phys. Rev. B* **66** 073307 (2002).

[5] Y. Xue and M.A. Ratner. "Schottky Barrier at Metal-Finite Semiconducting Carbon Nanotube Interfaces", *arXiv: cond-mat/0312546v1* (2003).

[6] N. Tzolov, B. Chang, A. Yin, D. Straus dan J.M. Xu. "Electronic Structure of the Contact between Carbon Nanotube and Metal Electrodes". *Phys. Rev. Lett.* **92** 075505 (2004).

[7] M. Menon and D. Srivastava. "Carbon Nanotube 'T Junctions': Nanoscale Metal-Semiconductor-Metal Contact Devices". *Phys. Rev. Lett.* **79** 22 4453 (1997).

[8] S. Z. Bisri, L. Hasanah, dan Sukirno. "Charge Distribution an Potential Profile in Nanometer Metallic Carbon Nanotube-Semiconducting Carbon Nanotube Heterojunction". *Proc. Asian Phys. Symp. 2005* 315.

[9] S. Z. Bisri, L. Hasanah, A. B. Suryamas, Sukirno. "Simulation of carbon nanotube on-tube metal-semiconductor heterojunction electronic properties". *submitted to Indonesian Journal of Physics 2006.*

[10] R. Czerw, et.al. "Identification of Electron Donor States in N-Doped Carbon Nanotubes", *Nano Lett.* **1** (9) 457-460 (2001).

[11] T. Durkop, B.M. Kim and M.S. Fuhrer, "Properties and Application of High Mobility Semiconducting Nanotubes", *J. Phys. : Condens. Matter* **16** (2004) R553-R580.

[12] S. Zulkarnaen. "Preliminary Studies of Electronic Transport in nanometer Metallic Carbon Nanotube and Semiconductor Carbon Nanotube Heterojunction". *Final Project, Physics Department of ITB* (2006).

[13] F. Lĕonard and J. Tersoff. "Novel Length Scales in Nanotube Devices," *Phys. Rev. Lett.* **83** 24 5174 (1999).

[14] J.W. Mintmire dan C.T. White. "Universal Density of States for Carbon Nanotubes", *Phys. Rev. Lett.* **81** 12 2506 (1998).

[15] M. Shiraishi dan M. Ata, "Work Function of Carbon Nanotube", *Carbon* **39** (2001) 1913-1917.

[16] N. Park, D. Kang, S. Hong and S. Han. "Pressure-dependent Schottky Barrier at the Metal-Nanotube Contact," *Appl. Phys. Lett.* **87**, 013112 (2005).

See-saw Type RF MEMS Switch with Fine Gap Vertical Comb

Sungchan Kang, Hyeon Cheol Kim and Kukjin Chun, *Member, IEEE*

School of Electrical Engineering and Computer Sciences
Seoul National University
301-1053(#038), Kwanak P.O. box 34, Seoul, 151-742, Korea
Tel: +82-2-888-6475, Fax: +82-2-874-5979
Email: kangsc@mintlab.snu.ac.kr

Abstract - This paper reports RF MEMS switch that has low actuation voltage and high isolation characteristics. We propose see-saw type RF MEMS switch based on a single crystalline silicon structure with fine gap vertical comb. Low voltage actuation and high isolation characteristics are key features to be solved in electrostatic RF switch design. Since these parameters in the conventional parallel plate MEMS switch design are in trade-off relation, both requirements cannot be met simultaneously. In vertical comb design, however, the actuation voltage is independent to the vertical separation distance between the contact electrodes. Then, we can design the large separation distance between contact electrodes to get high isolation. We have designed and fabricated a RF MEMS switch which has -46dB isolation loss at 5GHz, -0.9dB insertion loss at 5GHz and 40V operation voltage.

I. INTRODUCTION

Several universities and companies have developed RF MEMS switches in the last decade. The major research effort is the outstanding demonstrated RF performance of the MEMS switches from dc to 100 GHz compared to p-i-n diodes or FET transistor. [1] But these studies have only limited their focus on the RF performance of MEMS switches and have provided little information on several important phenomena directly related to the inherent electromechanical characteristics of RF MEMS switches. [2] So we

have developed the switch that has good RF performance and low actuation voltage simultaneously. Reducing the actuation voltage of MEMS switches may not only broaden the range of their possible applications, but also significantly enhance their performance.

We design the switch that have low actuation voltage and high isolation characteristics using vertical comb.

The air gap between contact metals determines isolation property and control voltage in conventional parallel plate switch. In vertical comb design, the control voltage is only dependant on the gap between the combs and independent to air gap between the contact electrodes. If the fine gap (<1.5 µm) between the combs can be patterned, we can get the low voltage operation and high isolation property simultaneously.

In conventional vertical comb structure design, the actuation voltage is in the order of 10~100 volts. In telecommunication application, however, the voltage should be below 10 V. To reduce the actuation voltage and get the sub-micron gap, we have utilized reduced-gap Si deep etching process and developed a new process to overcome align limitation of the vertical comb.

II. STRUCTURES

Figure 1 shows the schematic view of the proposed metal-to-metal (Au) contact switch. It consists of three parts, which are the bottom substrate, the movable silicon plate, and the capping substrate for packaging. The structure of the switch is made of single crystalline silicon for reliability and high productivity [3]. The thickness

0-7803-9730-4/06/$25.00 ©2006 IEEE

of the silicon layer is 50um and that of the upper and the lower comb is 40um. Then, the movable structure can actuate 10um in the vertical direction. The actuation voltage of the proposed switch is designed to be 6V for wireless application. The gap between contact metal and signal line is designed to be 5um in distance and the isolation loss of proposed switch at 5GHz is 42dB.

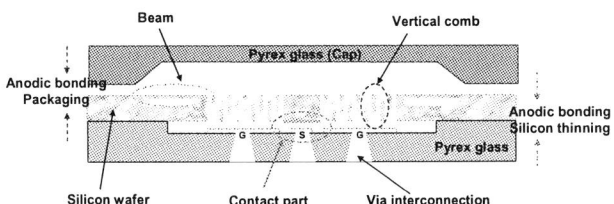

Fig. 1 Schematic view of the proposed see-saw type RF MEMS switch

III. OPERATION

The proposed switch utilizes electrostatic forces that act between combs. [4] The electrostatic forces are created when a voltage is applied between the combs causing them to attract. The force developed by combs increases with voltage difference (V), the number of comb teeth (N), and the width of the teeth (w), and decrease as the combs are further apart (g). So, we can design the switch that has constant force regardless of air gap between the contact electrodes. And, if we decrease gap between combs, low voltage operation of switch is possible.

The operation force of the vertical combs is

$$F = \frac{1}{2}\frac{\varepsilon w}{g}(2N)V^2 \qquad (1)$$

IV. SIMULATION

We used ANSYS for mechanical simulation and HFSS for electromagnetic simulation. [5] The resonant frequency of the structure is 2.7KHz. The isolation loss is below -37dB and the insertion loss is less than -0.12dB at the 10GHz. And pull-in voltage of proposed switch is 6V. Figure 2 show electromagnetic simulation data.

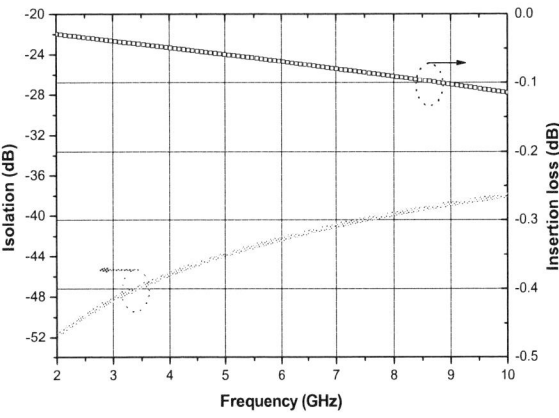

Fig. 2 HFSS simulation results: Isolation and insertion loss

V. FABRICATION

The bottom substrate and the movable silicon plate are fabricated separately. After fabrication of each part, the assembly process is performed using anodic bonding process.

The process procedure is illustrated in fig. 3. After the bonding step (e), we can align the upper and lower comb pattern using the front side alignment only and this technique can allow a sub-micron vertical comb formation.

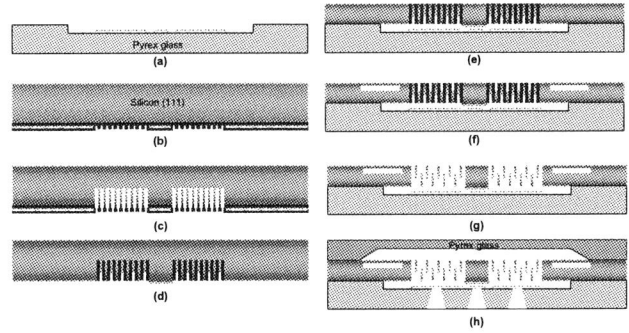

Fig. 3 Fabrication Process: (a) Air-gap formation and CPW patterning on bottom glass wafer, (b) Deep Silicon etching mask patterning using reduced-gap formation, (c) Si deep etching, (d) Contact metal formation and vertical comb patterning, (e) Anodic bonding and silicon wafer thinning, (f) Beam pattering and vertical comb patterning, (g) Release, (h) Packaging

We can get reduced gap using TEOS dep. & etch. Fig. 4 shows that cross section of the sub-micron gap by reduced gap formation. And Fig. 5

285

shows that cross section of 50um Si deep etch, which have 1 :40 of aspect ratio.

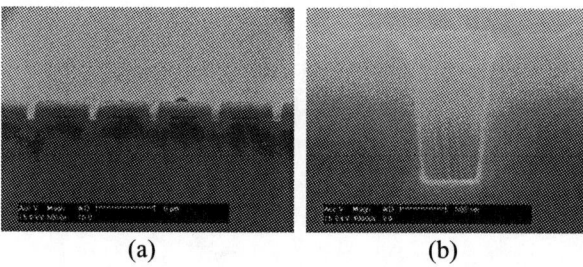

(a)   (b)

Fig. 4 Sub-micron gap

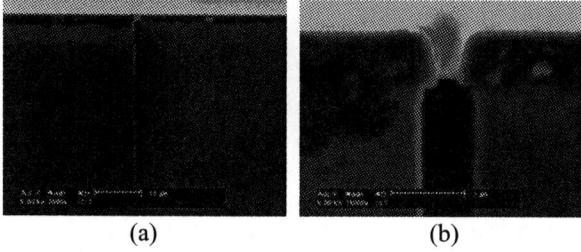

(a)   (b)

Fig. 5 After deep Si etching (1 :40 aspect ratio)

Figure 6 shows the fabricated see-saw type RF MEMS switch. And Fig. 7 shows completed see-saw type RF MEMS switch, contact part, torsion bar and vertical combs.

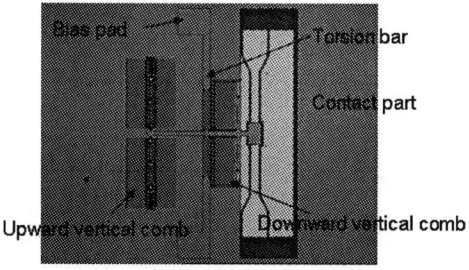

Fig. 6 Fabricated See-saw type RF MEMS switch with vertical comb

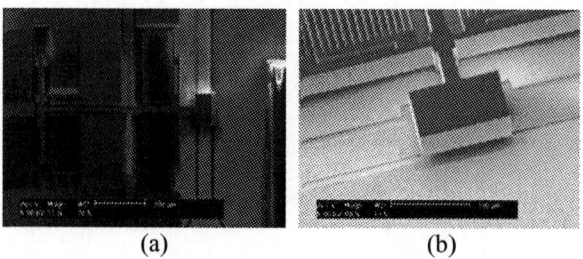

(a)   (b)

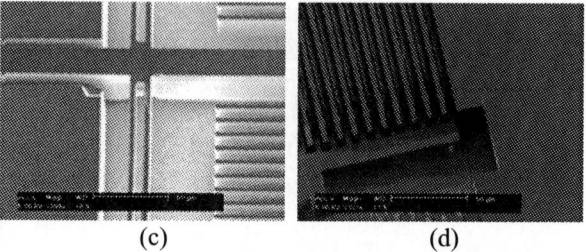

(c)   (d)

Fig. 7 Fabrication results: (a) See-saw type RF MEMS switch, (b) Contact part, (c) Torsion bar, (d) Fine gap vertical comb.

VI. MEASUREMENT RESULTS & DISCUSSION

Figure 8 shows RF characteristics of fabricated RF MEMS switch. The fabricated see-saw type RF MEMS switch with vertical comb has -46dB isolation loss at 5GHz and -0.9dB insertion loss at 5GHz. The operation voltage of fabricated RF MEMS switch is 40V. The fabricated switch has 40V operation voltage which is higher than design. The reason is that the thickness of fabricated torsion bar is 10um while the design was 3um. We can achieve sub-10V operation voltage by the process revision.

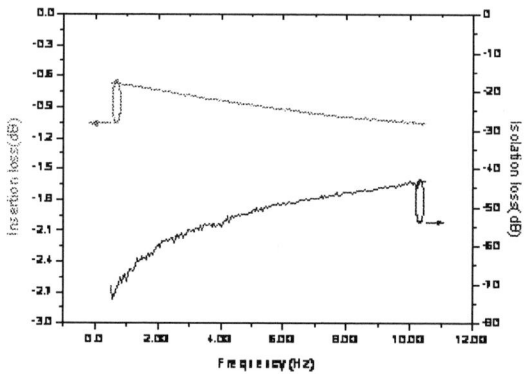

Fig. 8 Isolation and insertion loss characteristics of fabricated see-saw type RF MEMS swtich

VII. CONCLUSION

We have designed and fabricated see-saw type RF MEMS switch which has -46dB isolation loss at 5GHz, -0.9dB insertion loss at 5GHz and 40V operation voltage. To reduce the actuation voltage and get the fine gap, we have utilized the reduced-gap silicon deep etching process and developed a new process to overcome align limitation of the vertical comb. Additional package and process for the interconnection will be performed.

ACKNOWLEDGEMENT

The authors wish to acknowledge the assistance and support of the Center for Advanced Transceiver Systems and the Ministry of Commerce, Industry and Energy of Korea.

REFERENCES

[1] Gabriel M. Rebeiz, Jeremy B. Muldavin, "RF MEMS Switches and Switch Circuits", IEEE Mircrowave Magazine, pp.59-71, December 2001

[2] Dimitrios Peroulis, Sergio P. Pacheco, Kamal Sarabandi, Linda P. B. Katehi, "Electromecahnical Considerations in Developing Low-Voltage RF MEMS Switches", IEEE Transactions on Microwave Theory and Techniques, Vol. 51, No. 1, pp. 259-270, January 2003.

[3] M. Sakata, Y.Komura, T.Seki, K.Kobayashi, K. Sano, S. Horiike, "Micromachined relay which utilized single crystal silicon electrostatic actuator," 12th IEEE International Conference on. (MEMS), pp.21-24, 1999

[4] Hector J. De Los Santos, Georg Fischer, Harrie A.C. Tilmans, Joost T.M. van Beek, "RF MEMS for Ubiquitous Wireless Connectivity", IEEE Microwave Magzine, pp.36-49, December 2004

[5] Il-Joo Cho, Taeksang Song, Sang-Hyun Baek, Euisik Yoon, "A Low-Voltage and Low-Power RF MEMS Series and Shunt Switches Actuated by Combination of Electromagnetic and Electrostatic Forces", IEEE Transactions on Microwave Theory and Techniques, Vol. 53, No. 7, July 2005

Numerical Modeling of a Diffusion-Based In$_{0.53}$Ga$_{0.47}$As Lateral PIN Photodiode for 10 Gbit/s Optical Communication Systems

P Susthitha Menon, Kumarajah Kandiah and Sahbudin Shaari, *Member, IEEE*
Photonics Technology Laboratory (PTL)
Institute of Micro Engineering and Nanoelectronics (IMEN)
Universiti Kebangsaan Malaysia
43600 UKM, Bangi, Selangor, MALAYSIA
Email: susi@vlsi.eng.ukm.my, sahbudin@vlsi.eng.ukm.my

Abstract A novel purely diffusion-based In$_{0.53}$Ga$_{0.47}$As lateral PIN photodiode was successfully modeled. Device dimensions are 12 x 1.8 µm^2 with electrode spacing of 1.5 µm and width of 1 µm. The effective intrinsic region width is ~0.2 µm. The 2D modeled device achieved responsivity of 0.765 A/W and -3dB frequency of 10.9 GHz and is able to cater for 10 Gbit/s optical communication system networks. The device was biased at -2V and illuminated with a 5 W/cm^2 optical spot power at a wavelength of 1.55 µm. SNR ratio was recorded at 31.2 dB.

I. INTRODUCTION

Photodetectors are key components in every lightwave communication system either in the well-established 2.5 Gb/s (STM-16/OC-48) to 10 Gb/s (STM-64/OC-196) systems or the emerging 40 Gb/s or 160 Gb/s systems. PIN photodiodes (PD) based on In$_{0.53}$Ga$_{0.47}$As/InP heterostructures are the most suitable devices for these applications due to their high efficiency and their capability for high-speed operation [1-2].

The lateral PIN photodiode (LPP) is proposed as an alternative design to the conventional vertically stacked PIN photodiode (VPP) in terms of ease of fabrication and cost-savings catering for 2.5 to 10 Gb/s communication systems [3]. Development of a LPP is also advantageous for the fabrication of monolithic integration of photodiodes with well-known preamplifier transistor circuits such as FET, HEMT or HBTs to form optical integrated receiver modules [4].

A LPP is created in such a way that the *p, i* and *n* regions are all in one plane. It has the same properties as the VPP where detection is made

through the intrinsic layer but the fabrication method is much simpler. LPPs only require standard CMOS processes such as diffusion or ion implantation to create the *p* and *n* regions whereas VPPs generally require costly epitaxial techniques such as MOCVD (metalorganic chemical vapor deposition) and MBE (molecular beam epitaxy) to build up the appropriate layers [5-8].

The development stage of InGaAs technology can incur high costs due to its complexity. Therefore optimized device simulations prior to fabrication provides the advantages of reduced development cost and time in comparison to an exclusive experimental methodology [1].

This paper presents the key parameters and analysis of the modeling of a diffusion-based In$_{0.53}$Ga$_{0.47}$As lateral PIN photodiode using a commercial numerical simulator which utilizes a two and three-dimensional and drift-diffusion approach [9].

II. THEORY AND DEVICE MODELING

The material parameters needed for the simulations were taken from periodical literature. Carrier mobilities and minority carrier lifetime are known to be strong functions of the doping level. Electron and hole mobility data were fit to the empirical expression for doping dependence suggested by Caughey-Thomas [10-11]:

$$\mu = \mu_{min} + \frac{\mu_{max} - \mu_{min}}{1 + (\frac{N}{N_{ref}})^{\alpha}}$$

(1)

where μ_{min} and μ_{max} are the minimum and maximum mobilities of electron and holes, N_{ref} is the critical doping density above which ionized impurity scattering becomes dominant and α is the exponent that controls the slope at $N \cong N_{ref}$. Similarly, the minority carrier lifetime was fit to the empirical expression as proposed by Conklin et al. [12] based on the form :

$$\tau(ns) = 10^{\beta - \gamma \log N} \qquad (2)$$

where β and γ are fitting parameters and N is the carrier doping density. Model parameters for the SRH lifetime model used in this simulation was obtained via curve-fitting methodology of equation 2. The band gap and band discontinuities as well as the optical properties of InGaAs were taken from [13-14].

Several sources of noise are inherent in the process of photon detection in a lateral PIN photodiode. Among them are quantum noise, dark current noise and thermal noise. Fig. 1 shows the noise sources and disturbances at an optical receiver [15].

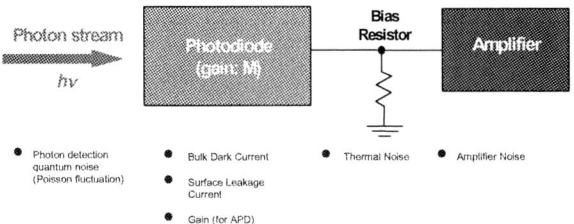

Fig.1 Noise sources and disturbances at an optical receiver.

III. SIMULATION RESULTS

The diffusion-based InGaAs LPP consists of a p-type InGaAs well and an n-type InGaAs well diffused into an unintentionally low-doped n-type InGaAs substrate. The InGaAs substrate is 1.7 μm thick and is covered with a 0.1 μm-thick oxide as a passivation layer. Gold is used as the electrode material and electrode finger width is 1 μm. The device dimensions are 12 x 1.8 μm². The junction depth for both the p+/n+ wells are 0.8 μm respectively and lateral diffusion per well is 0.7 μm at each side of the well. The distance between electrodes is 1.5 μm and the effective intrinsic region width is ~0.2 μm. The InGaAs region was unintentionally doped with silicon to obtain background doping of 1e11 atoms/cm³. The

compensation ratio θ (N_A/N_D) is set at 0.4 where N_D = 1e19 atoms/cm³.

Fig. 2 shows the device upon illumination of a gaussian beam with spectral width of 5 μm, optical spot power of 5 W/cm² and wavelength, λ=1.55 μm. Reverse bias voltage is at -2 V. The power was increased gradually from 1 W/cm² until 10 W/cm² and the resultant photo-IV curve is shown in Fig. 3.

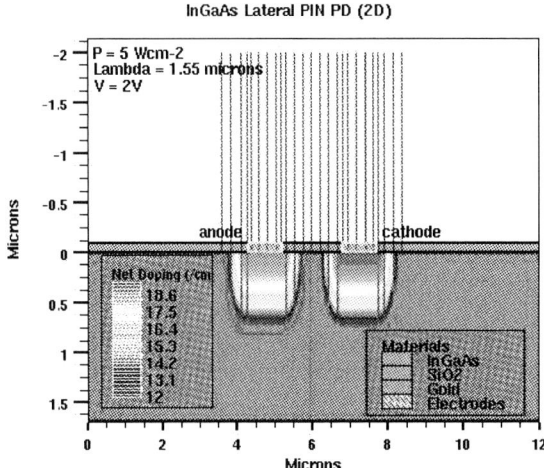

Fig.2 Two-dimensional InGaAs Lateral PIN Photodiode structure under illumination.

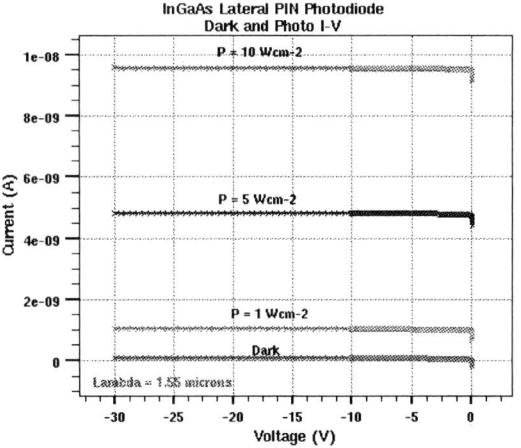

Fig.3 Dark and Photo-IV curve of the device.

The gold electrodes were set to be opaque so that no optical power is absorbed beneath this region.

The device achieved responsivity of 0.765 A/W and external quantum efficiency of 61.1 %. The -3dB frequency of the device is 10.9 GHz. Therefore, the device qualifies to be used in an

OC-192 SONET/STM-64 SDH network system where the bit rate is at 9.95 Gbit/s.

The device structure was extended to three-dimensional design with additional z axis of 1 µm as in Fig. 4. Again, upon illumination of a gaussian beam on the surface of the device, the 3D structure exhibited responsivity and -3dB frequency values of 0.41 A/W and 6.75 GHz respectively. The reduction in values is due to the 3D software inability to simulate gaussian beams with σ=0.05 as was used in the 2D structure. Therefore the Gaussian beam used in the 3D structure was larger at σ=1 µm hence a wider beam width.

A lenslet was added to the top portion of the beam to focus the beam to the device intrinsic region as in Fig. 4. The lenslet index was fixed at 1.57 and radius of 3.03 µm. A higher responsivity compared to the 3D device without lenslet was recorded at 0.65 A/W and the -3dB frequency is 4.86 GHz. Again, instead of a gaussian beam, a uniform plane wave was used to illuminate the device due to 3D software shortcomings. Fig. 6 and Fig. 7 show the comparison in responsivity and -3dB frequency curves of the 2D and 3D devices.

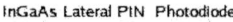

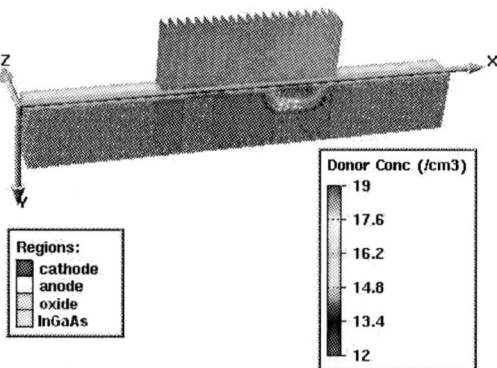

Fig. 4 Three-dimensional device structure of the InGaAs lateral PIN PD

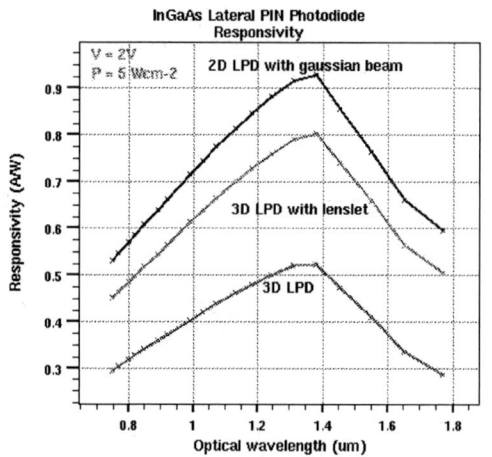

Fig. 5 Three-dimensional device structure of the InGaAs lateral PIN PD with a lenslet above the device.

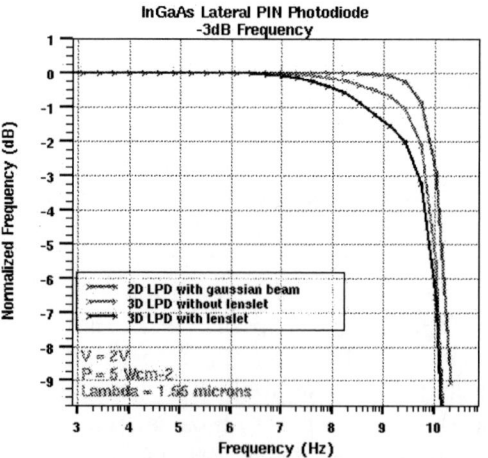

Fig. 6 Comparison in responsivity between the devices

Fig. 7 Comparison in -3dB frequency between the devices

The 2D LPP was connected to a load resistor of 50 Ω in order to calculate the inherent noises as shown in Fig. 8. Assuming no leakage current and using a reverse bias voltage of 2V and optical spot power of 5 W/cm^2, the different noises were calculated as shown in Table 1. Other parameters of the device are λ=1.55 μm, DC photocurrent I_P= 4.79 x 10^{-9} A, AC photocurrent i_P=4.78 x 10^{-9} A, bandwidth B=10.9 GHz, dark current I_D=3.3 x 10^{-11} A and load resistor, R_L= 50 Ω. In a PIN PD, the gain (M) and noise figure F(M) is equivalent to 1 and room temperature (T=300K) was used. The thermal noise has the largest impact on the total noise current compared to the quantum or dark current noise. The calculated signal-to-noise ratio (SNR) based on the total noise current is 1325.47 (31.2 dB).

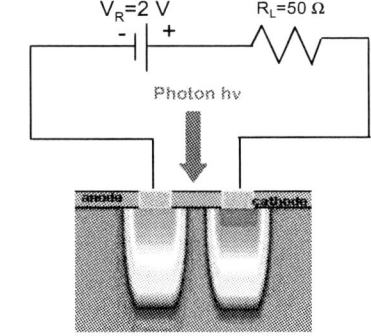

Fig. 8 In$_{0.53}$Ga$_{0.47}$As Lateral PIN Photodiode (LPD) connected to load resistor and reverse bias voltage

Parameter	Equation	Noise (A^2/Hz)
Quantum Noise	$\left\langle i_Q^2 \right\rangle = 2qI_p BM^2 F(M)$	1.67 x 10^{-17}
Dark Current Noise	$\left\langle i_{DB}^2 \right\rangle = 2qI_D BM^2 F(M)$	1.15 x 10^{-19}
Thermal Noise	$\left\langle i_T^2 \right\rangle = 4K_B TB / R_L$	3.61 x 10^{-12}

Table 1 Noise sources in a In$_{0.53}$Ga$_{0.47}$As Lateral PIN Photodiode

Fig 9 shows the simulated noise sources from the device. The total noise current comprises of generation-recombination (quantum) current, electron diffusion current and hole diffusion current at a bandwidth of 10.9 GHz is 5.79 x 10^{-24} A^2/Hz. This value is much smaller than the calculated total noise current. The noise current

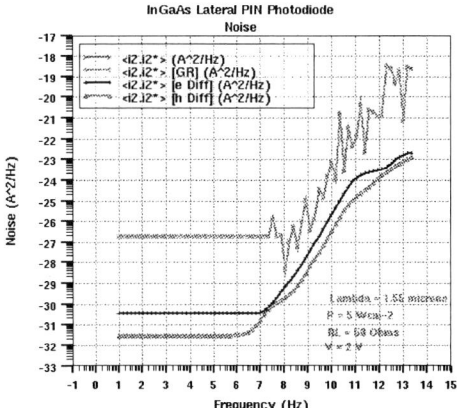

Fig. 9 Simulated total noise current in a InGaAs LPP.

shows increment when the bandwidth is increased accordingly.

IV. CONCLUSION

A novel purely diffusion-based In$_{0.53}$Ga$_{0.47}$As lateral PIN photodiode was successfully modeled. The 2D modeled device achieved responsivity of 0.765 A/W and -3dB frequency of 10.9 GHz and is able to cater for 10 Gbit/s optical communication system networks. The device was biased at -2V and illuminated with a 5 W/cm2 optical spot power at a wavelength of 1.55 μm. SNR ratio was recorded at 31.2 dB.

V. ACKNOWLEDGMENT

The authors would like to acknowledge the support of the Malaysian Ministry of Science, Technology and Innovation for funding this work under IRPA grant 03-02-02-0069-EA231.

REFERENCES

[1] B. Jacob, B. Witzigmann, M. Klemenc and C. Petit, "A TCAD methodology for high-speed photodetectors," *Solid -State Electronics* vol 49, pp. 1002-1008 (2005).

[2] E. Budianu, M. Purica, E. Rusu and S. Nan, "Heterostructures on InP substrate for high-speed detection devices over large spectral range (0.8-1.6 um)," *Microelectronic Engineering* vol 51-52, pp. 393-400 (2000).

[3] Y.-W. Chen, W.-C. Hsu, Y.-J. Chen, "Low Dark Current InGaAs(P)/InP p-i-n Photodiodes," Proceedings of the Sixth Chinese Symposium in Optoelectronics, pp. 95-98 (2003).

[4] M. Bitter, "Indium phosphide/indium gallium arsenide pin-photodiode arrays for parallel optical interconnects and monolithic InP/InGaAs pin/HBT optical receivers for 10-Gb/s and 40-Gb/s," *ProQuest Dissertations and Theses* (2001).

[5] P. S. Menon and S. Shaari, "Device characteristics of an $In_{0.53}Ga_{0.47}As/InP$ lateral PIN photodiode," *IEEE National Symposium on Microelectronics,* pp 381-385 (2005).

[6] A. A. Ehsan, S. Shaari and B. Y. Majlis, "Silicon lateral p-i-n PD for OEIC," *Proc. 2001 IEEE National Symposium on Microelectronics*, pp. 316 (2001).

[7] B. Jalali, P.D. Trinh, S. Yegnanarayanan, F. Coppinger, " Guided Wave optics in silicon-on-insulator technology," *IEE Proc-Optoelectronics*, vol 143, No. 5 (1996).

[8] B. Lee, I. Yun, "Effect of different etching processes on edge breakdown suppression for planar InP/InGaAs avalanche photodiodes," *Microelectronics Journal*, vol 33, pg. 646 (2002).

[9] SILVACO International, *ATLAS User's Manual*, Version 5.8.0.R. USA, SILVACO International Incorporated (2004).

[10] D. M. Caughey, and R.E. Thomas, "Carrier mobilities in silicon empirically related to doping and field," *Proceedings of the IEEE,* vol 55, pp. 2192-2193 (1967).

[11] S. Datta, S. Shi, K.P. Roenker, M.M. Cahay, W.E. Stanchina, "Simulation and Design of InAlAs/InGaAs pnp heterojunction bipolar transistors," *IEEE Trans. on Electron Devices*, vol 45, no. 8, pp 1634-43 (1998).

[12] T. Conklin, S. Naugle, S. Shi, K.P. Roenker, S.M. Frimel, T. Kumar, M.M. Cahay, "Inclusion of tunneling and ballistic transport effects in an analytical approach to modeling of NPN InP based heterojunction bipolar transistors," *Superlattice Microstuctures.* vol 18, pp 1-12 (1996).

[13] S. Adachi, *Physical properties of III-V semiconductor compounds: InP, InAs, GaAs, and InGaAsP*, Wiley Interscience, New York (1992).

[14] S. Srivastava, "Simulation Study of InP-Based Uni-Travelling Carrier Photodiode," *University of Cincinnati*, (2003).

[15] Lecture Notes "Optical Receivers Theory and Operation", Electrical and Computer Engineerng, Ryerson Univeristy (2004).

The Effect of Surface Microstructure on The Response of Titanium Dioxide Coated with Cobalt-Porphyrin Thin Films Towards Gases in Quartz Crystal Microbalance Sensor

[1]Syariena Arshad, [1]Muhammad Mat Salleh, Member, IEEE and [2]Muhammad Yahaya
[1]Institute of Microengineering and Nanoelectronics (IMEN)
[2]School of Applied Physics
Faculty of Science and Technology
University Kebangsaan Malaysia
43600 Bangi, Selangor, MALAYSIA
Email: mms@pkrisc.cc.ukm.my

Abstract A single-layer and multi-layer titanium dioxide (TiO$_2$) coated with dye porphyrin thin films were prepared on Quartz Crystal Microbalance (QCM) using sol-gel dip coating method and were tested for sensing of volatile organic compounds (VOCs). The porphyrin was 2,3,7,8,12,13,17,18-octaethyl-21H,23H-porphine cobalt (II). The sensing sensitivity was based on the change in the fundamental frequency of the QCM upon exposure towards three vapor samples, namely ethanol, acetone and 2-propanol. It was found that the thin films were sensitive towards all vapors and the sensing sensitivity was affected by the number of film layers. Although the three-layer film exhibited higher frequency response compared with the other films that have small number of layers, the sensing properties of this film are not repeatable and less selective. The performance of the QCM sensor was depended on microstructure of the thin film which was varied through the number layers.

I. INTRODUCTION

Volatile organic compounds (VOCs) have been known as major pollutants in environment, release from various industrial process, vehicles exhaust fumes and environmental activities. Some of them are carcinogens and cause bad health impact to human and nature. There is an increasing public concern about the presence of environmental pollutants and safety of those living and working close to their sources. Thus, the

requirement for cheap and reliable sensors is urgent.

Quartz crystal microbalance (QCM) sensors are widely studied for determining VOCs [1-2]. QCM is a piezoelectric quartz crystal which oscillates at a resonance frequency determined by several factors, including the mass of the crystal. When the electrodes are connected to an oscillator and AC voltage is applied over the electrodes the quartz crystal starts to oscillate due to piezoelectric effect. The principle of the sensor is that the QCM frequency will decrease when exposed to the VOC. The decrease in frequency is proportional to the increase in mass due to the presence of gas adsorbed on its surface, according to the Sauerbrey law:

$$\Delta f = - k\, S_m \Delta m \qquad (1)$$

Where k is geometric factor for the fraction of the active device area being perturbed, S_m is the device specific constant and Δm is changed in mass/area on the device surface.

Quartz crystal sensors to detect gas are generally fabricated by coating the QCM surface with sensitive thin film. Organic chemical, for example metalloporphyrins are used by many researchers as sensing element because of their high sensitivity, selectivity and good recoverability [1-4]. The main features of porphyrins are they can be chemically modified by simple attachment of atoms or ligands in the centre and the peripherals sites of the molecular skeleton.

In recent years, there have been many efforts to improve the response properties of a sensing element to the presence of VOCs. In order to increase the sensor sensitivity, the sensing element is constituted with nanoparticles system. The characteristic of gas adsorption depends on the surface area of the sensing film. Titanium dioxide (TiO$_2$) nanoparticles have been studied intensively as gas sensor material because of their high surface area that is able to improve the film sensitivity [5]. Nanostructure thin film gives larger surface area and provides high sensitivity at low gas concentrations [6]. Thus nanoparticles titanium dioxide coated with dye-porphyrins thin films are a promising candidate for the sensing film.

In this study, we fabricated different layers of TiO$_2$ coated dye-porphyrin thin films. The metalloporphyrin used was 2,3,7,8,12,13,17,18-octaethyl-21H,23H-porphine cobalt (II), CoOEP. The different number of layer in thin film may produce different surface microstructure of the films. Hence, we may study the effect of microstructure on sensing performance of QCM system. The thin films were prepared using sol-gel technique and the VOCs used for sensing study were ethanol, acetone and 2-propanol.

II. METHODOLOGY

The TiO$_2$ sol was prepared by using titanium (IV) ethoxide, TEOT (obtained from Aldrich), kalium chloride, KCl and ethanol (received from MERCK). Firstly, KCl powder was dissolve in deionized water and added into ethanol with stirring. Then TEOT was dropped wisely into the solution under nitrogen in a glove box. Acetylacetone (supplied by MERCK) as stabilizer was added into the solution and was stirred for 2 hours at room temperature. Nanoparticles TiO$_2$ sol were successfully prepared. To the TiO$_2$ sol, Poly-L-Lysine, PLL (molecular weight =70,000 and supplied by Aldrich) was added drop wise and keeps stirring to charge the TiO$_2$ nanoparticles into positive charge. Separately, 0.2 mg/ml metalloporphyrin solution was prepared and was added drop wise into the TiO$_2$ solution. Both solutions were stirred for several hours to obtain a complete mixture of porphyrin and TiO$_2$ nanoparticles. The volume ratio of porphyrin to TiO$_2$ was 2.

The QCM substrate used was 10 MHz AT-cut quartz crystal (13.5mm in diameter) with gold electrodes (5 mm in diameter) on each side obtained from ICM manufacturing. The thin film was deposited onto QCM substrate using sol-gel

dip coating method. The substrate were dipped in TiO$_2$ coated dye porphyrin solution for 30 minutes and pulled out with slow and uniform pulling rate of 40 mm/min. A thin and uniform film of TiO$_2$ coated with dye porphyrin was formed on the substrate and dried with nitrogen at room temperature. The dipping steps were repeated to obtain two-layer and three-layer thin films. Surface analysis was conducted with atomic force microscopy (AFM).

Fig. 1 shows an experimental set-up for the bulk acoustic sensor system. The QCM coated with thin film was positioned in the sensor chamber. A power voltage of 4.5 V was supplied to the oscillator circuit. Nitrogen gas was purged into the sensor chamber for 30 minutes to remove the air out of the chamber. The detection of VOCs; 2-propanol, acetone and ethanol were started by exposing the thin film to the vapor with the flow rate of 10 ml/min. The gas flow rate was controlled by mass flow controller (MFC). The sensitivity of the thin films to the VOCs were studied by measuring the QCM frequency changes using FC-7015U universal frequency counter and the data were transferred to the computer. The recovery time was recorded starting from when the VOC vapor is cut off and pure nitrogen is purge into the system.

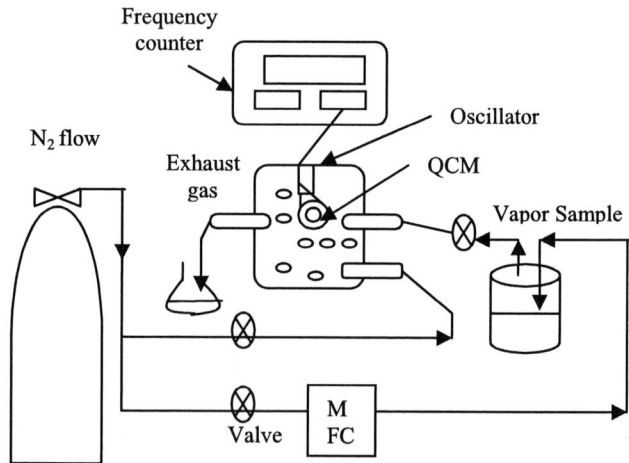

Fig. 1 The bulk acoustic system sensor set-up.

III. RESULTS AND DISCUSSION

The single-layer, two-layer and three-layer thin films had been successfully deposited on QCM substrate. Fig. 2 illustrates three dimensional images of the thin films taken in the area of 7,000 x 7,000 nm^2. The surface analysis of the image revealed that the roughness of single-layer, two-layer and three- layer thin films were 20.29 nm, 16.40 nm and 57.21 nm respectively. It was observed that the additional of film layers affected the surface microstructure. The single layer thin film shows the present of large number of small hills that were distributed to all over the film surface. For the two-layer film the distance between small hills and valleys become smaller because the valley were filled by the nanoparticles. Thus the surface structure of this film becomes smoother. The addition of the third layer made the surface profile become rougher with the presence of large hills, deep valleys and flat areas.

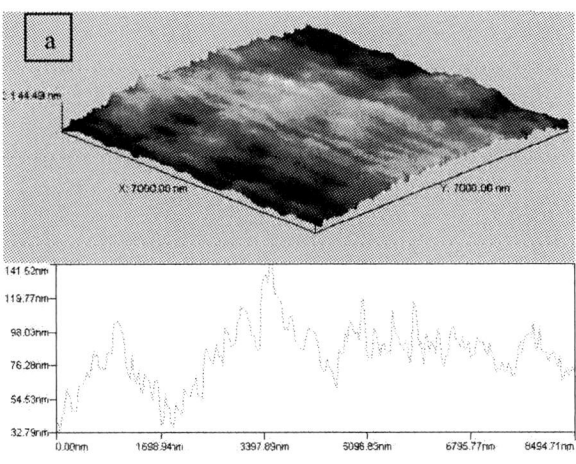

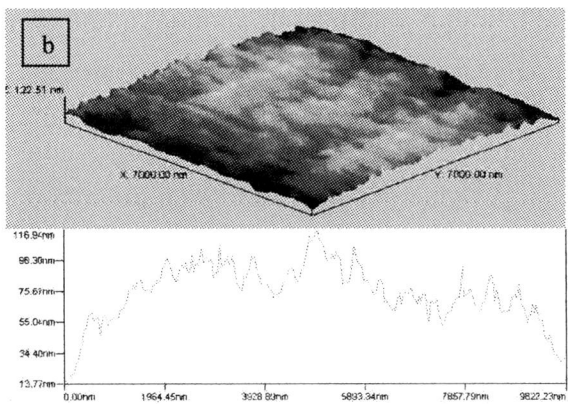

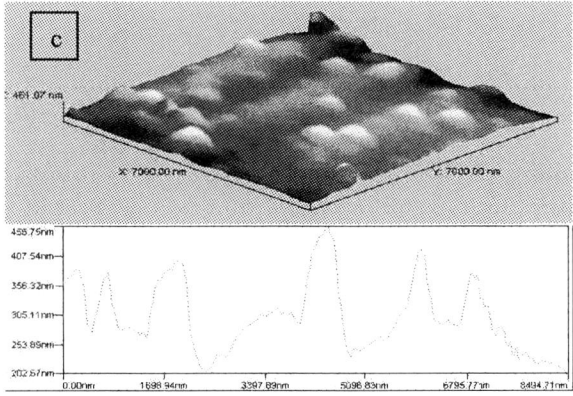

Fig. 2 The 3-dimensional image of (a) single-layer, (b) two-layer and (c) three-layer CoOEP thin films.

Fig. 3 shows the response of single-layer, two-layer and three-layer thin films towards acetone vapor. When the thin films were exposed to the vapor, the frequency of QCM was decreased to a certain values depend on the thin film sample. It was observed that the change of the frequency is increasing with the number of layer. After the vapor was removed from the system, the frequency response was recovered back to the original frequency. However, for the three-layer thin film the recoverability was good only up to two cycles. Poor recoveries for the three-layer thin film may be due to some vapor molecules were trapped in the deep valley of the thin film surface. The vapor molecules were not easily sweep away from the surface even after the pure nitrogen was blow into the system. For the film with rough surface structure, the availability of the adsorption sites on the surface may be high but the vapor molecules were not easily sweep away from the surface. Hence its recovery time may increases and the QCM was difficult to recover back to its original frequency.

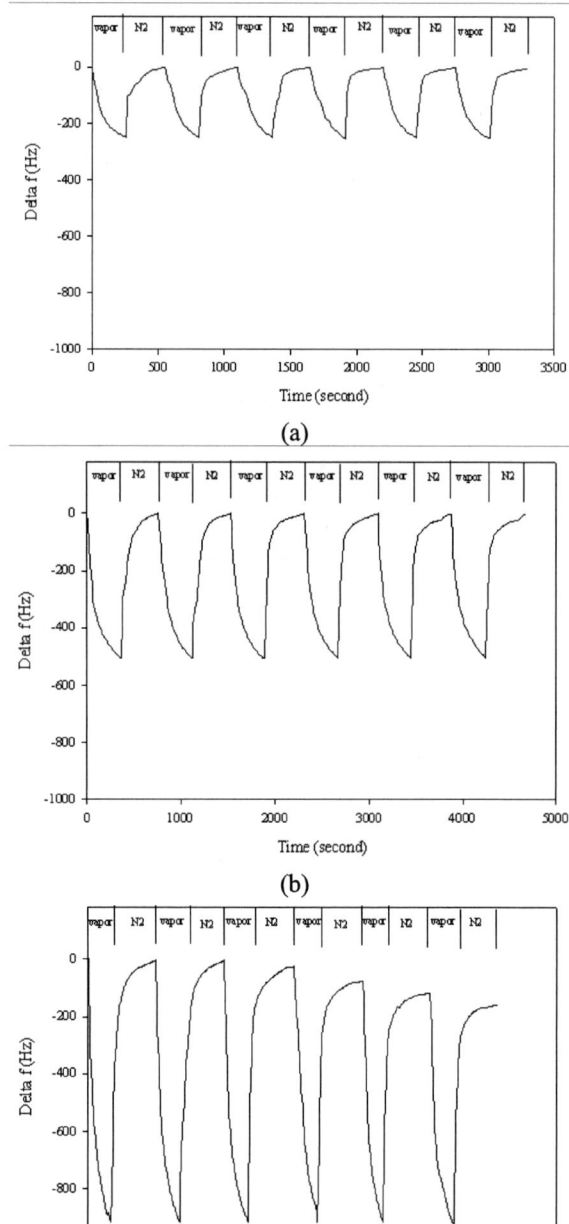

Fig. 3 Response of (a) single-layer, (b) two-layer and (c) three-layer CoOEP thin film towards acetone vapor.

The sensing experiment was repeated for other two organic vapors, ethanol and 2-propanol. The shapes of frequency response for both vapors were similar to that observed for the thin films that were exposed to acetone vapor,

except that the sensitivity was different for different vapor.

Fig. 4 shows the variation of response for ethanol, acetone and 2-propanol vapors towards different layer of CoOEP thin films. It was observed that the CoOEP thin films have selectivity properties towards these gases. According to equation (1), the frequency change depends on increasing of mass due to the presence of gas adsorbed on the film surface. For the single-layer and two-layer thin films, it shows that 2-propanol (molecular weight of 60) had highest frequency response compare with acetone (molecular weight of 58.1) and ethanol vapor (molecular weight of 46.07). The smooth thin film surface was suitable for the interaction of high molecular weight vapor. But the frequency responses of ethanol vapor for both films were higher than acetone vapor. This could be due to strong interaction between ethanol molecules with the thin films surface. Meanwhile, the rough surface structure was found to be sensitive for the low molecular weight vapor. The vapor molecules with low molecular weight tend to easily sweep away from the film surface and the presence of high hills had prevented the vapor molecules from easily sweep away. For the three-layer thin film, the frequency response of ethanol vapor was the highest, but the frequency response of acetone and 2-propanol were almost the same. This thin film may demonstrate the highest frequency response, but it is less selective and could not discriminate vapor with small different molecular weight.

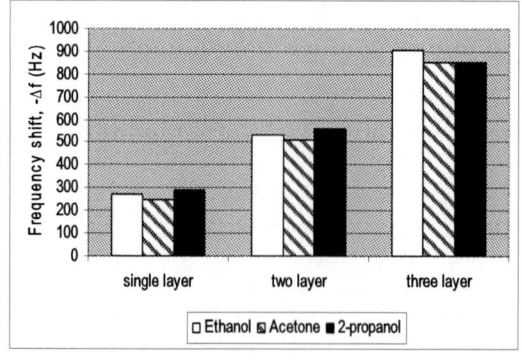

Fig. 4 The variation response for ethanol, acetone and 2-propanol vapors towards different layer of CoOEP thin films.

IV. CONCLUSION

The different layer of 2,3,7,8,12,13,17,18-octaethyl-21H,23H-porphine cobalt (II) thin films were found to be sensitive towards the present of three vapor samples; ethanol, acetone and 2-propanol. It was observed that the additional of film layer affect the surface microstructure. The additional of film layer can increased the frequency response but the recovery time was longer. The three layers thin film demonstrated highest frequency response but it could not recovered after second cycles and could not discriminated small different molecular weight vapors. From the experimental result, it can be concluded that the two layers thin film was the most suitable thin film to detect VOCs vapor as it exhibit higher frequency response than single layer thin film and has good recoverability compare with three layers thin film.

V. ACKNOWLEDGEMENTS

The authors would like to thanks to the MOSTI for IRPA 03-02-02-0020-SR003/07-06 grant.

REFERENCES

[1] S. Arshad, M. M. Salleh, M. Yahaya, "Metalloporphyrins thin film for detection of volatile organic compounds in bulk acoustic system" *NSM Proceeding 2005, Kuching,* pp. 318- 321 (2005).

[2] R. Mohamed, M. M. Salleh and M. Yahaya. "The sensing selectivity of metalloporphyrin thin films towards organic vapours in bulk acoustic system", *NSM2003 Proceeding*: 52-55 (2003).

[3] P. Montmeat, S. Madonia, E. Pasquinet, L. Hairault, C. P. Gros, J. M. Barbe, R. Guilard, "Metalloporphyrins as sensing material for quartz-crystal microbalance nitroaromatics sensors" *IEEE Sensor Journal,* Vol 5, No. 4, August 2005 pp. 610-615 (2005).

[4] Akrajas Ali Umar , Muhamad Mat Salleh and Muhammad Yahaya, "Enriching the selectivity of metalloporphyrins chemical sensors by means of optical technique" *Sensors and Actuators B* 4277 pp. 1-6 (2002.)

[5] J.H. Kim, S.H. Kim, S. Shiratori, "Fabrication of nanoporous and hetero structure thin film via a layer-bi-layer self assembly method for a gas sensor" *Sensors and Actuators B,* pp. 241-247 (2004).

[6] M. C. Carrota, M. ferroni, D. Gnani, V. Guidi, M. Merli, G. Martinelli, M.C. Casale, M. Notaro, "Nb-doped $TiO2$ as thick film gas sensors for environmental monitoring" *Sensors and Actuator B, Chem. 58* pp. 310-317 (2005).

[7] M. Kikuchi, S. Shiratori, " Quartz crystal microbalance (QCM) sensor for CH_3SH gas by using polyelectrolyte-coated sol-gel film" *Sensors and Actuators B 10,* pp. 564-571 (2005).

[8] A. Z. Simoes, A. H. M. Gonzales, A. Ries, M.A. Zaghete, B.D. Stojanovic, J.A. Varela, "Influence of thickness on crystallization and properties of $LiNbO_3$ thin films" *Materi al Characterization 50,* pp. 239-244 (2003).

[9] M. Mascini, A. Macagnano, M. D. Carlo, R. Paolesse, B. Chen, P. Warner, A. D'Amico, C.D. Natale, D. Compagnone "Piezoelectric sensors for dioxins: a biomimetic approach" *Biosensors and Bioelectronics 20,* pp. 1203-1210 (2004).

Synthesis and Characterization of CuO Nanowires

Chang Fu Dee, Muhammad Yahaya, Muhamad Mat Salleh and
Burhanuddin Yeop Majlis (*SMIEEE*)

Institute of Microengineering and Nanoelectronics (IMEN),
Universiti Kebangsaan Malaysia
Email: cfdee@vlsi.eng.ukm.my

Abstract – **Synthesis of copper oxide nanowires was done heating up copper wires in wet ambient air at 400 and 500°C. The existence of nanowires was confirmed by SEM images and EDX spectroscopy. Nanowires were not formed in nitrogen ambient. The diameters of synthesized nanowires are between 30 to 160 nm and lengths up to 39 μm. SEM image shows that CuO nanowires were formed on top of the oxide grains. Vapor–solid growth mechanism is also suggested for the growth of this nanowire.**

I. INTRODUCTION

Nanostrutures materials such as nanotube [1], nanorods, nanobelts, nanodots, nanowalls, nanowires, nanoflowers, nanorings, nanocombs etc. have been synthesized in the laboratories. Those materials will served as building block for future high performance devices [2]. CuO nanowires are among one of the nanostructures used to improve sensitivity of gas sensor [3]. The gas sensing ability could be improved by increase the active surface area. In this paper, we will discuss an easy method for mass production of the copper oxide nanowires from copper wires obtained from computer motherboard used for intercomponent connection.

II. EXPERIMENTAL DETAILS

Copper wires with diameter 125 μm in average after the protected plastics were mechanically peeled off were heated up in wet air and nitrogen ambient. The wet air was prepared by passing through compressed air to an aquarium-stone air bubble generator to capture water vapour before passing through the sample in a quartz tube of a funnel. Copper wires with average length of 1 cm were heated at temperature 400 and 500°C separately in wet air ambient. The same experiment was carriered out in nitrogen ambient at 500°C for 9 hrs. Characterization was done by SEM (Philips XL30) for the samples heated for 1 hour and every subsequence 2 hrs to 9 hrs.

III. RESULT AND DISCUSSION

Figure 2(a)–(e) and (i)–(v) show the nanowires grown at temperature 500 and 400°C respectively in wet air ambient for difference period of times which are 1, 3, 5, 7 and 9 hrs. The formation of these CuO nanowires were confirmed by EDX spetroscopy as shown in Fig. 1. There is not much difference on the density of the nanowires grown. Nevertheless, for the first 1 hr heating at 400°C, there are only nanowhiskers formed. The strength for most of the nanowhiskers is not strong enough to hold the nanowires in straight. The average diameter is around 50 nm. The same phenomena didn't observed for samples synthesized at 500°C. All nanowires have almost the same diameter from top to the root and no sharp tips on top of them. The average diameter is 120 nm.

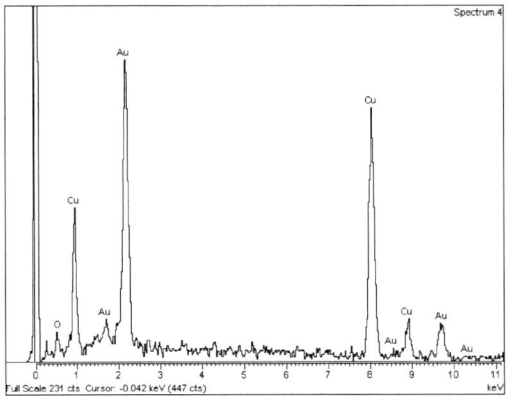

Fig. 1 EDX spectroscopy confirmed the formation of CuO nanowires

The process started by peeling of the surface and formation of oxide grains as shown in schematic diagram in Fig. 3. It is because Cu is weaker than its oxide [4], formation of oxide causes the stress on the surface and it peeled off from the bulk. Nanowires grown from the oxide grains (Fig. 4). It was also noticed that the grains formed on both external and internal side of the peeled-off surface, but only the external side has nanowires. The diameters of synthesized nanowires were between 30 to 160 nm and lengths up to 39 μm.

There are two common models used to explain the formation of CuO nanowires, which are vapor-liquid-solid (VLS) and vapor-solid (VS) model. The typical characteristic of the VLS model is the presence of metal particles on the end of grown nanostructures. SEM images and EDX spectroscopy show that there was no capping catalyst at the end of the CuO nanostructures. Moreover, there was no catalyst used for the synthesis process of CuO, so VS model was more suitable for this growth.

During the heating process, the Cu atoms were evaporated from the surface. The Cu atoms combine with oxygen and form the CuO

nuclei. The increase in concentration of the CuO nuclei will force it to growth in upward direction to form nanorod.

Figure 5 (a) to (c) show the SEM images of copper wire after heated in the nitrogen ambient for 5 hrs. The surface of the wire didn't peel off as we can see in all the samples annealed in wet air ambient. The copper melted and coalesced to granule. The granule gathers to form in certain patterns. Formation of these patterns needs to be further studied. No nanowire formed for all the samples heated in nitrogen ambient.. It provides the information that water is needed to the growth of CuO nanowires.

IV. CONCLUSION

In conclusion, copper oxide nanowires have been successfully synthesized by heating of copper wire in the wet air. The diameters of synthesized nanowires were between 30 to 160 nm and lengths up to 39 μm. This was confirmed by the EDX spectroscopy. This method of provides the mass production of nanowires. Further research need to be carried out in controlling the growth direction. This is important when the nanowires need to be arranged in the array form to increase the gas sensing capability.

(a) 500°C, 1hr

(i) 400°C, 1 hrs

(b) 500°C, 3 hrs

(ii) 400°C, 3hrs

(c) 500°C, 5 hrs

(iii) 400°C, 5hrs

(d) 500°C, 7 hrs

(iv) 400°C, 7hrs

(e) 500°C, 9hrs

(v) 400°C, 9hrs

Fig. 2   Copper Oxide nanowires grown at 400 and 500C for 1 to 9 hrs.

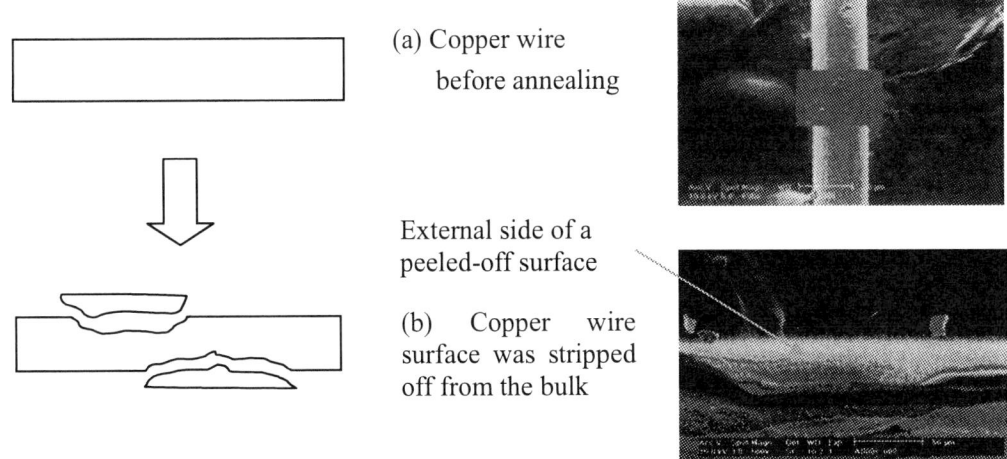

(a) Copper wire before annealing

External side of a peeled-off surface

(b) Copper wire surface was stripped off from the bulk

Fig. 3   Copper wires surface were peeled off and the nanowires grow on the external surface

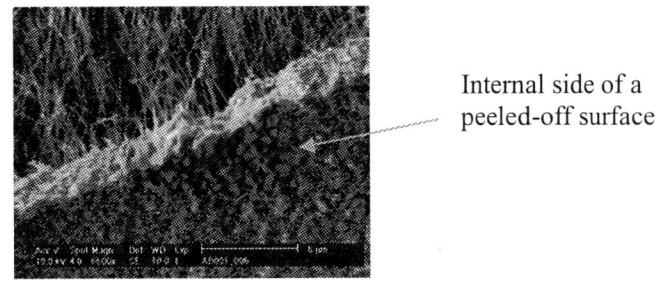

Internal side of a peeled-off surface

Fig. 4   Formation of nanowires from the oxide grains

(a)   (b)   (c)

Fig. 5   SEM images of copper wire after annealed at 500°C for 5 hrs in nitrogen ambient. The surface of the copper wire didn't peel off from the bulk and there was no nanowires formed.

REFERENCE

[1] S. Ijima, "Helical microtubules of graphitic carbon," *Nature*, vol. 354, pp. 56, 1991.

[2] A. S. Edelstein and R. C. Cammarata, *Nanomaterials: synthesis, properties and application*. Philadelphia: Insititute of Physics, 1996.

[3] K. Sawicka, M. Karadge, A. Prasad, and P. I.

Gouma, "Oxidation synthesized CuO nanowires for gas sensing applications," *Microscopy and Microanalysis*, vol. 10, pp. 360-361, 2004.

[4] C. H. Xu, C. H. Woo, and S. Q. Shi, "Formation of CuO nanowires on Cu foil," *Chemical Physics Letters*, vol. 399, pp. 62-66, 2004.

Analytical Modeling of Optical Cross Add and Drop Multiplexing Switch

Muhammad Syuhaimi Bin Ab-Rahman, Abang Anuar Ehsan, Hayati Husin and Sahbudin Shaari,
Member, IEEE, *Member*, SPIE

Photonic Technology Laboratory
Institute of Microengineering and Nanoelectronics
Universiti Kebangsaan Malaysia
43600 UKM Bangi
Selangor, MALAYSIA
Email: rahul_devraj@hotmail.com

Abstract: **The optical cross add and drop multiplexing (OXADM) device described in this paper increases the capacity, flexibility and reliability of the trunk-line optical network used for data communication. It features two different bandwidths which operate in working and protection line in a ring configuration. The OXADM has the same features of the conventional OADM node but with more efficiency and reliability. The OXADM prototype system focuses on providing survivability through restoration against failures, such as cable/fiber cut in optical layer with ring topology. Two types of restoration scheme have been proposed to ensure data flow continuously by means of linear/multiplex protection and ring protection. This paper describes the analytical modeling of OXADM and studies their limitation as compared with conventional OXC device.**

Keywords: OXADM, analytical, power penalty, OXC.

I. INTRODUCTION

OXADMs are elements that provide capability to add and drop function and cross connecting traffic in the network similar to OADM and OXC [1]. OXADM consists of three main subsystems; a wavelength selective demultiplexer, a switching subsystem and a wavelength multiplexer. Each OXADM is expected to handle at least two distinct wavelength channels each with a coarse granularity of 2.5 Gbps of higher (signals with finer granularities are handled by logical switch node such as SDH/SONET digital cross connects or ATM switches. The device has minimum two input and two output ports which be connected to the optical trunks.

There are eight ports for add and drop functions which are controlled by four lines of thermal-optical switches. The other four lines thermal optical switches will be used to control the wavelength routing function between two different paths (Fig. 1).

The function of OXADM includes terminating node, drop and add, routing, multiplexing and also provides mechanism of restoration for point to point, ring and mesh metropolitan and also customer access network in FTTH [1][2][3][4].

With the setting of the thermal-optic switch configuration, the device can also be programmed to function as other optical devices such as multiplexer, demultiplexer, coupler, wavelength converter (with fiber grating filter configuration), OADM, wavelength round about and etc. for the single application.

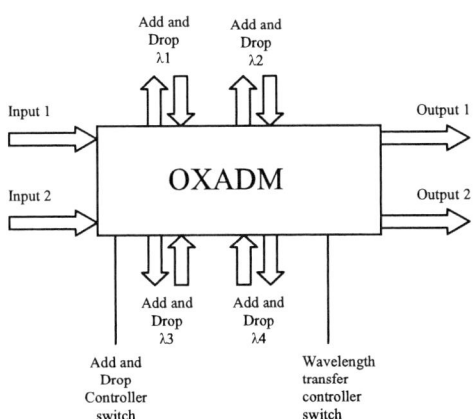

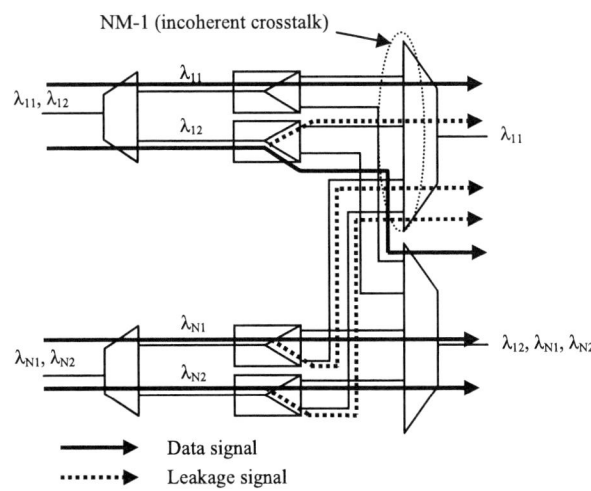

Fig. 1 The block diagram of optical cross add and drop multiplexing (OXADM)

II. ANALYTICAL MODEL

In ring topology, OXADMs are located in the nodes, which have more than two switching directions in ring networks. The function of OXADM is to flexibly switch the wavelengths among the different input and output ports. Because of the OXADM's imperfect performance, the insertion loss and crosstalk are induced (figure 2). The contribution of crosstalk are coming from the number of input/output ports connected to the fiber lines and the number of wavelengths per lines. Due to the presence of crosstalk that are additional signal power is required to maintain a specified bit error rate (BER) in the presence of a particular impairment. Expressed in decibels, the power penalty is defined as [5]

$$\text{Power-Penalty} = 10 \log_{10}[P_{imp}/P_{no\ imp}] \quad (1)$$

P_{imp} = Power required with impairment
$P_{no\ imp}$ = Power required without impairment

Assuming aligned polarization, the probability density function (PDF) for the resultant aggregate interference is approximately Gaussian, which leads to a theoretical power penalty [5]:

III. RESULTS AND DISCUSSION

Figure 3 depicted the power penalties at every cascaded 2x2 OXADM node with the number of operating wavelength M=2, 4, 6, 8. The

$$P_{Penalty} = -5\log_{10}[1-4x\sigma^2_{RIN}xQ^2] \quad (2)$$

Where Q is the Q factor corresponding to the reference BER, σ^2_{RIN} is the autocovariance of the beat noise resulting from interferometric intensity noise.

Note there is an error floor, corresponding to the BER, at $4Q^2\sigma^2_{RIN}=1$, where the power penalty tends to infinity. It is impossible to achieve BER smaller than the error floor because of the nature of the crosstalk [5]. Here we apply the worst case used as in Fig. 2 for the power penalty caused by incoherent crosstalk contributions, this crosstalk is induced when optical propagation delay differences between signal and crosstalk contributions in an OXADM exceed the coherent time of the laser [5]. Assuming the OXADM is fully loaded, each signal passing through the OXADM will be interfered by MN-1 crosstalk contribution, which is leaked by the demultiplexer/multiplexer pairs. Based on the above assumptions, the power penalty from crosstalk contributions in one OXC is given in Equation 3.
The maximum power penalty (pp) caused by MN-1 incoherent crosstalk contributions from one OXADM is:

$$\text{Max(pp)}_{OXADM}=-5\log_{10}[1-4xMax(\sigma^2_{RIN})x\ Q^2]$$
$$= -5\log_{10}[1-4\ x\ \varepsilon\ x\ (MN-1)\ x\ Q^2] \quad (3)$$

increment of number of operating wavelength will increase the power penalties. The same profile can also be seen with the incremental of port numbers (N=2, 4, 6, 8) as shown in Figure 4, due to the autocovariance of the beat noise resulting from interferometric intensity noise in OXADM is $\text{Max}(\sigma^2_{\text{RIN}}) = \varepsilon(MN-1)$.

Fig. 5 shows the maximum of nodes that can be achieved with the increment of number of ports in OXADM. In point to point configuration, the maximum ports at two operating wavelengths is 43 ports but at three operating wavelengths the figure decreases to 29, as compare to OXC device with 86 ports and 85 ports respectively.

The increments of number of channel also give much effect to the number of OXADM nodes that can be cascaded. With 2 ports, the maximum channels that can be operated are 43 channels but in OXC the figure is double. The maximum channels that can be handled by OXADM at 4 ports are 26 channels compare to that of OXC which is 84 channels. Although the accumulation features decrease the scalability of OXADM but its application is widely.

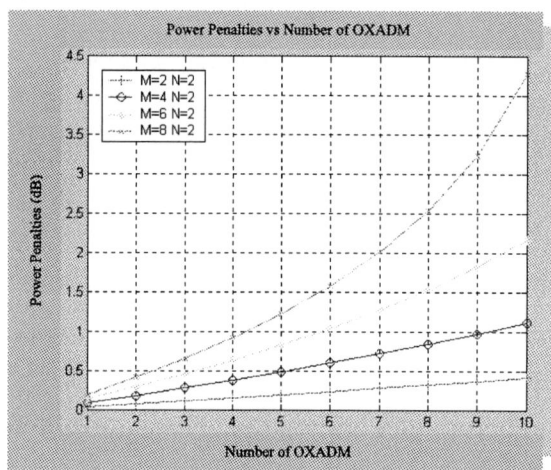

Fig. 3 Power penalties at cascaded 2x2 OXADM with the increment of number of wavelength.

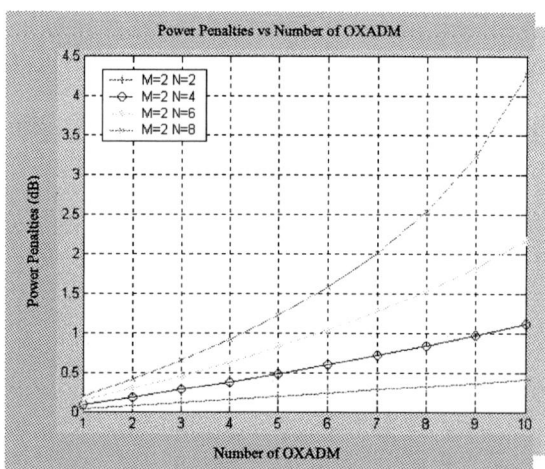

Fig. 4 Power penalties at cascaded OXADM with the increment of port numbers.

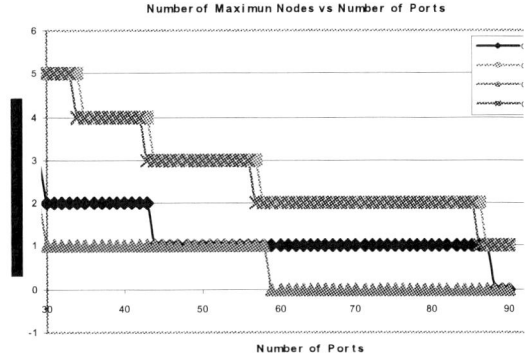

Fig. 5. Number of maximum nodes associated with the number of ports in OXADM nodes as compare to the existing OXC nodes.

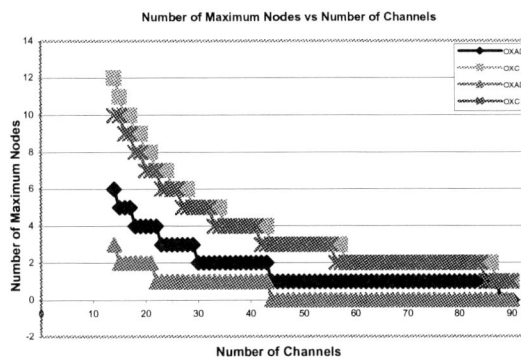

Fig. 6. Number of maximum nodes associated with the number of channels in OXADM nodes as compare to the existing OXC nodes.

IV. CONCLUSION

The paper describes the analytical modeling of OXADM and studies their limitation as compared to conventional OXC device. With the MN-1 crosstalk contribution, the power penalty is twice higher than OXC that has M+N-2 crosstalk contribution but the OXADM nodes provides more flexibility, reliability and widely application [4][5]. OXADM nodes focus in providing survivability through restoration in optical ring network [1]. Accumulation feature increases the efficiency and survivability of OXADM device as availability and flexibility nodes in ring and mesh network, security switch in FTTH, wavelength management in FTTH CO and decoder in CDMA technologies [1][2][3][4].

REFERENCES

[1] Mohammad Syuhaimi Ab-Rahman, Abang Anuar Ehsan & Sahbudin Shaari. 2006. Mesh upgraded ring in metropolitan network using OXADM. ICOCN/ATFO 2006 (China).

[2] Mohammad Syuhaimi Ab-Rahman Abang Anuar Ehsan & Sahbudin Shaari. 2006. Self-protection passive optical network using access control system (ACS). ICOCN/ATFO 2006 (China).

[3] Mohammad Syuhaimi Ab-Rahman, Hayati Husin, Abang Anuar Ehsan & Sahbudin Shaari. 2006. OXADM Multiplex Protection Scheme for Bidirectional Path Switched Ring. ICCE 2006 (Vietnam).

[4] Mohammad Syuhaimi Ab-Rahman & Sahbudin Shaari. 2006. OXADM restoration scheme: Approach to optical ring network protection. 4th IEEE ICON 2006 (Singapore).

[5] Stevens. 2005. Impact of routing optimization strategy on the transmission performance of wavelength routed transparent optical network.

http://cui.phy.stevens-tech.edu/lab2005/pdf/thesis/YH-thesis.pdf

Modelling and Analysis of a Transformer based MEMS Piezoelectric Vibration Type Microgenerator

Luay Yassin Taha, *Student Member, IEEE* , Burhanuddin Yeop Majlis, *SMIEEE* and Ahmad Al Ali[1]
Institute of Microengineering and Nanoelectronics (IMEN), Universiti Kebangsaan Malaysia,
43600 Bangi, Selangor, Malaysia
[1]Centre of Excellence & Research, Emirates Aviation College, Dubai, United Arab Emirates
Email: burhan@vlsi.eng.ukm.my , dr.alali@emirates.com

ABSTRACT   **This work is focused on the feasibility of using the transformer type linear model in describing both the static and the dynamic behaviour of the piezoelectric microgenerator subjected to an external vibration force in the z-axis, i.e, 33 mode. Using this model, the microgenerator output parameters (Voltage & Power) are studied according to the input parameters (Force and frequency) for a certain piezoelectric type and dimensions. Many 2D plots are shown to explain how the input and the piezoelectric material choice can be varied to obtain the optimum output parameters. MatLab programs are written to simulate the model for No load (open circuit) and the load cases. A transient analysis is also carried out using the Multisim software to test the linearity of the model.**

I. INTRODUCTION

The generalized constitutive equations for a linear piezoelectric material in reduced-matrix form are given by:

$$\{S\} = [s^E]\{T\} + [d]\{E\} \tag{1}$$

$$\{D\} = [d]\{T\} + [\varepsilon^T]\{E\} \tag{2}$$

where {S} is the six-dimensional strain vector, {T} is the vector of stresses induced by the combined mechanical and electrical effects, {D} is the three-dimensional electric displacement vector, {E} is the electric field vector, $[s^E]$ is the six by six compliance matrix evaluated at constant electric field, [d] is the three by six matrix of piezoelectric strain coefficients, and $[\varepsilon^T]$ is the three by three dielectric constant matrix evaluated at constant stress [1]. Since our work is focused only on the analysis of the 33

mode, the above equations can be simplified further into the reduced one dimensional form along the z-axis:

$$S_3 = s^E_{33} T_3 + d_{33} . E_3 \tag{3}$$

$$D_3 = d_{33} T_3 + \varepsilon^E_3 E_3 \tag{4}$$

Where T_3 is the stress applied along the z-axis, i.e., 3; and the measured S_3, D_3, and E_3 are all along the same axis, d_{33} and s^T_{33} are the z-axis piezoelectric strain coefficients and the compliance evaluated at constant electric field, respectively, and ε_3^T is dielectric constant matrix evaluated at constant stress long the z-axis. For more simplicity, the symbols S, s, T, d, T, D, and ε are being used during this work.

The reduced form coupled equations are used in the realization of piezoelectric microgenerators. Some realizations were reported in the literature using the cantilever beam [1 - 3]. Other realization is also possible based on applying an incident Force in the z-axis to the piezoelectric material via a compressing mass [4 - 5].

Fig.1 shows the piezoelectric microgenerator realization model used in our work. Fig.2 shows one cylindrical shape piezoelectric element. The analysis is carried out by applying a force with certain frequencies along the z-axis, then recording the output parameters for different input & system variations. MatLab programs are written to help in calculations and plotting in 2D & 3D. A linearity checkout is needed for the model. This is carried out using the Multisim transient analysis feature.

0-7803-9730-4/06/$25.00 ©2006 IEEE

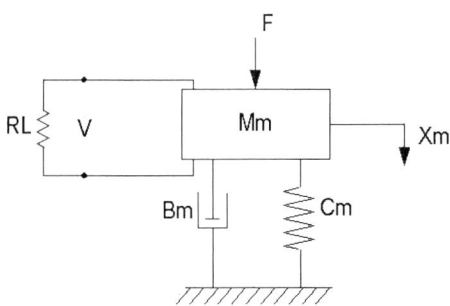

Fig.1 Piezoelectric microgenerator realization model

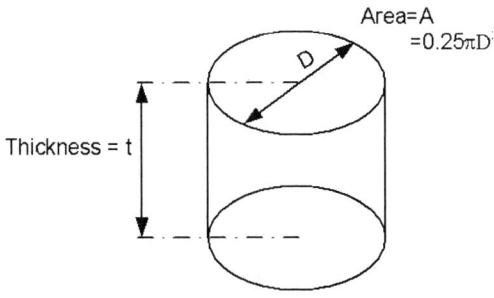

Fig.2 One cylindrical shape piezoelectric element

II. THE TRANSFORMER MODEL

It has been proven that the transformer model can be used – based on the electromechanical analogy - to describe the mechanical to electrical conversion (or vice versa) [1]. Stephen R. Platt, Shane Farritor, and Hani Haider proposed a complete electrical circuit model based on the transformer model. The input force is assumed to be sinusoidal type [6]. Fig.3 (a) shows the transformer model. Cm is the piezoelectric compliance, Cp' is the blocked capacitance of the piezoelectric element and is equal to $C_p - d^2/C_m$, where C_p is the free capacitance ($C_p = \varepsilon^T A / t$). Fig.3 (b) shows the electrical circuit equivalent to fig.3 (a) after referring to the secondary side and adding the effect of the piezoelectric mass and damping (R_{em} & L_{em}). The coupling (phi) is equal to the transformer ration N2/N1 and is equal to $- C_m/d$. R_L is the external load resistance. This circuit will be used in our analysis next section. The model requires a deep analysis to study the behaviour of the piezoelectric microgenerator when varying the input parameters, dimensions, and the material type.

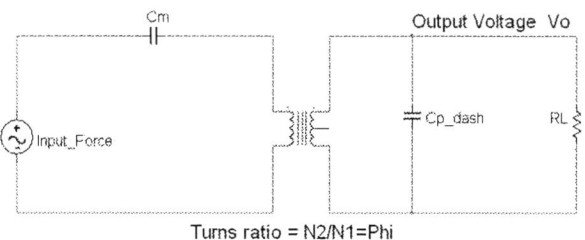

Fig.3 (a)

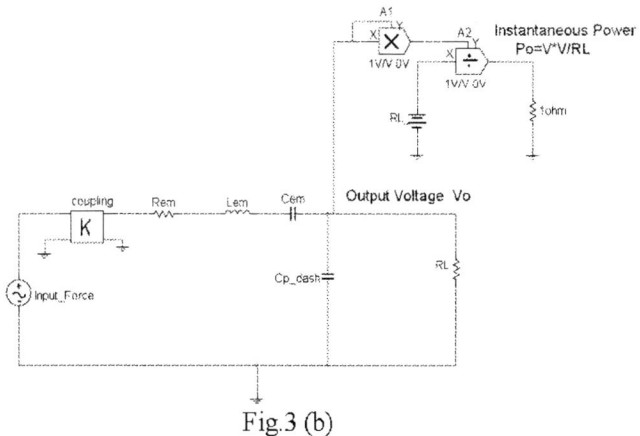

Fig.3 (b)

Fig.3 (a) the transformer model,
(b) Referred to secondary simplification circuit

III. OPEN CIRCUIT (NO LOAD) ANALYSIS

When the piezoelectric output terminals are opened, D=0. Substitute this in eq.4 yields to:

$$0 = dT + \varepsilon E \qquad (5)$$

Now E= V_{op}/t (V_{op} =open circuit voltage, and t = thickness, see fig.2) , T = F /A (F = Input force, A = Area). $\varepsilon = \varepsilon_o . \varepsilon_r$ (ε_o = dielectric constant of the free space = 8.85 x 10 $^{-12}$ F /m) and ε_r = piezoelectric dielectric constant).
Rearranging Eq. 5 according to E, T, and ε to obtain the final open circuit voltage formula:

$$V_{op} = \frac{d . F . t}{\varepsilon_o . \varepsilon_r . A} \qquad (6)$$

Thus the open circuit voltage is directly proportional to the piezoelectric coupling constant d, force F and thickness t , and inversely proportional to the dielectric constant ε_r and cross sectional area A. Fig. 4 shows a MatLab Plots of V_{op} versus the input force and the input stress. The graph is linear and shows reasonable output voltage for all bands of the selected thickness (0.1mm-1mm). Care should be taken not to exceed the maximum allowable stress for the piezoelectric material as this causes depolarization and hence material will loose its piezoelectric property. In our work the choice is for Morgan Electro Ceramic product type PXE 5, having a maximum mechanical stress of 2.5 MPa [7]. Going to fig.4, it is clear that the boundary limits of 2.5MP produces voltage bands from about 8 V (for t =0.1mm) up to about 78V. Thus the thickness to area ratio (TAR) is also required to be studied. Fig.5 illustrates the effect of changing the thickness and hence the thickness to area ratios TARS on the open circuit voltage for different values of the input force (up to the maximum of 50N). Again the choice of dimension must not alter the maximum limit of the input stress as declared.

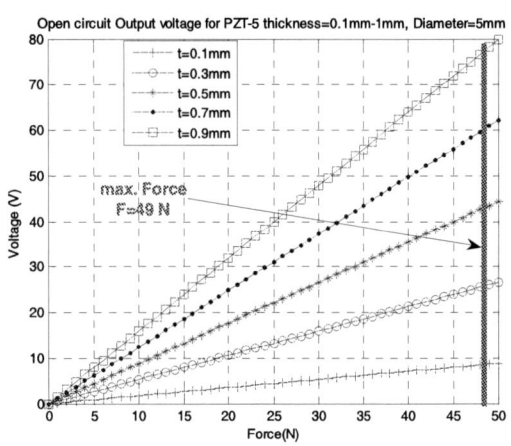

Fig. 4 (a)

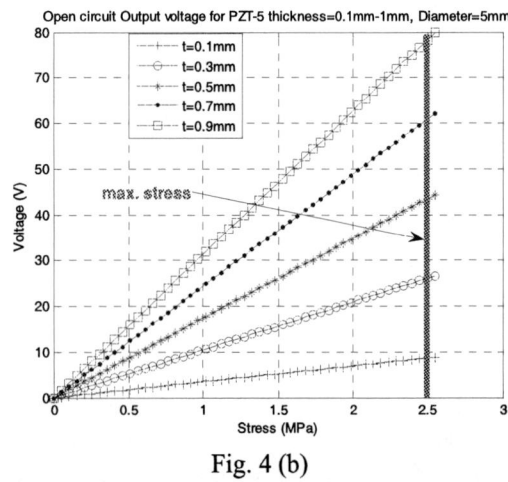

Fig. 4 (b)

Fig.4 Open circuit output voltage, (a) versus the input force, (b) versus the input stress

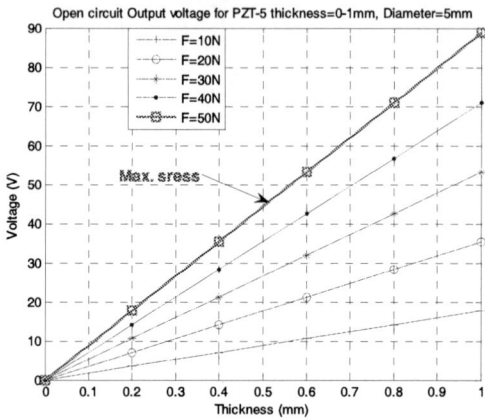

Fig.5 (a)

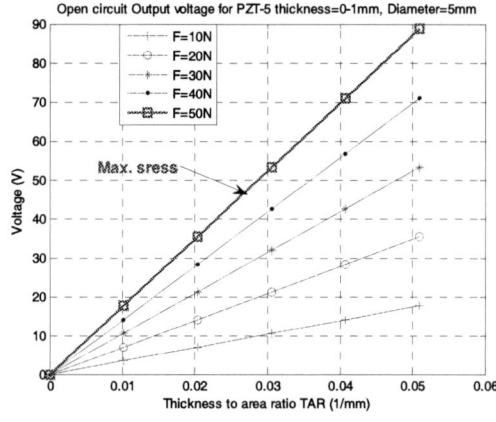

Fig.5 (b)

Fig.5 Open circuit output voltage, (a) versus the thickness, (b) versus TAR, for input forces from 10N-50N

IV. LINEARITY TEST

Assuming that '**P**' denotes an operator that relates the piezoelectric generator output Vo with the input force F consisting of three different forces F1, F2, and F3, such that F = F1+F2+F3, and **P**[F] = Vo, **P**[F1] =Vo1, **P**[F1] =Vo2, and **P**[F3] =Vo3; then the piezoelectric generator is linear if and only if it satisfies the following superposition relationships :

$$\mathbf{P} \, [\, F1 + F2 + F3] = \mathbf{P}[F1] + \mathbf{P}[f2] + \mathbf{P}[F3] \qquad (7)$$
$$Vo = Vo1 + Vo2 + Vo3 \qquad (8)$$

Multisim transient analysis is used for this purpose. The three forces are modelled as three voltages after the conversion using the transformer turns ratio (phi). First the three inputs are passed to the model simultaneously (F1+F2+F3). The output is recorded and named as Vo, see fig. 6(a). Then the individual forces F1, F2, and F3 are passed to the model. The outputs are recorded as Vo1, Vo2, and Vo3 and added together using a summing circuit. The output is named as Vo'. Fig. 7 shows the transient plot for the two outputs. It is clear that the two outputs takes the same shape but with a certain difference in the peak values. The positive portions of the output Vo is seen above Vo' while the negative portion is seen down Vo', this will compensate the difference between them if the average is calculated. In general there is a good similarity between the two outputs.

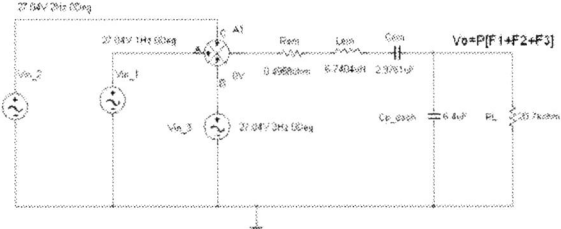

Fig. 6 (a) Measurement of the output voltage Vo from three input forces applied simultaneously

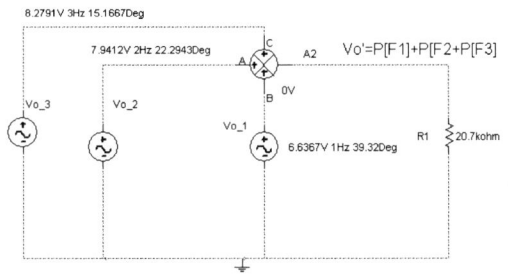

Fig. 6 (b) measurement of the output voltage Vo' by adding the individual outputs from the three forces

Fig.6 Linearity test

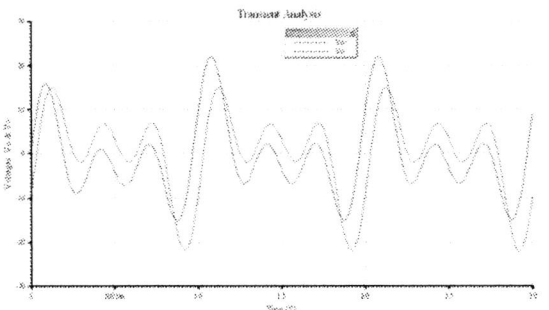

Fig.7 Linearity test results, the two outputs Vo & Vo' are similar

V. LOAD ANALYSIS

The piezoelectric generator output terminals are now connected to a load resistance RL. The analysis includes: the input force variation for different frequencies but with constant load resistance, the load resistance variation for different input forces but with constant frequency. The piezoelectric generator parameters are taken from [6] for comparison issues.

Fig.8 shows a MatLab plot for the result. It is clear that at low frequencies there is a significant change at the output voltage & power but this is not when increasing the frequency above 100Hz.Thus increasing the frequency is not a major factor for the improvement of the microgenerator output voltage & power.

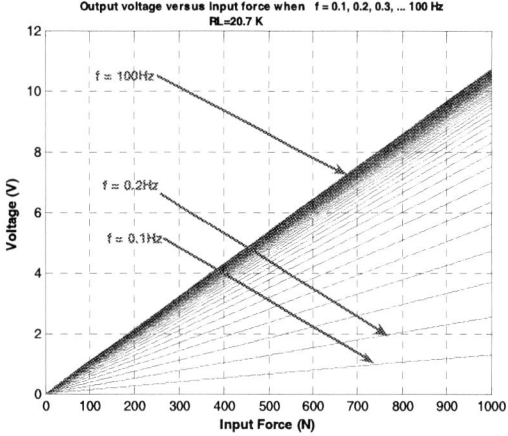

Fig. 8 (a)

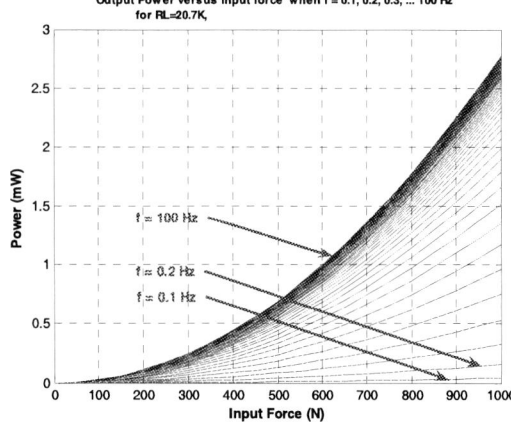

Fig. 8 (b)

Fig.8 Output voltage versus Input force for different frequencies and constant load

Fig.9 shows a MatLab plot for the result. The output voltage goes to the maximum when increasing the load resistance. This is correct as the output approaches the open circuit value when RL goes to infinity. The power curve takes the shape of the standard curve and the results for F=800N is identical as with [6]. Both power & voltage branches are shifted up thus getting higher values when increasing the input force. This is correct but with care because of the depolarization problem discussed before.

must be taken into consideration to overcome depolarization. The thickness to area ratio is an important factor that controls the design of the output voltage. The frequency factor becomes less significant and effecting the output voltage & power (above 100Hz).The model accepts the principle of superposition but with some differences in the shape & peak voltage.

REFERENCES

[1] S Roundy and P Kwright, "A piezoelectric vibration based generator for wireless electronics"; Smart materials and structures, 13 (2004) 1131–1142

[2] N M White, P Glynne-Jones and S P Beeby, "A novel thick-film piezoelectric microgenerator"; Smart materials and structures , 10 (2001) 850–852

[3] Y.B. Jeon , R. Sood, J.-h. Jeong , S.-G. Kim, "MEMS power generator with transverse mode thin film PZT"; Sensors and Actuators A 122 (2005) 16–22

[4] Chao-Nan Xu, Morito Akiyama, Kazuhiro Nonaka, and Tadahiko Watanabe," Electrical Power Generation Characteristics of PZT piezoelectric ceramics; IEEE transactions on Ultrasonic, Ferroelectrics, and Frequency control, vol. 45, no. 4, July 1998

[5] Chok Keawboonclinay and Thomas G. Engel, " Electrical power generation characteristics of Piezoelectric generator under Quasi-Static and dynamic stress conditions"; IEEE transactions on ultrasonic, ferroelectrics, and frequency control, Vol. 50, No. 10. October 2003

[6] Stephen R. Platt, Shane Farritor, and Hani Haider, "On Low-Frequency Electric Power Generation With PZT Ceramics" IEEE/ASME transactions on mechatronics, vol. 10, no. 2, April 2005

[7] Morgan Electro Ceramics , Introduction Piezoelectric Ceramics, Technical information, 2001 May 16.

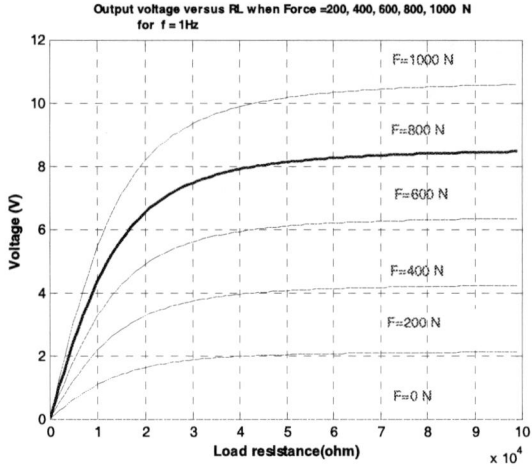

Fig. 9 (a)

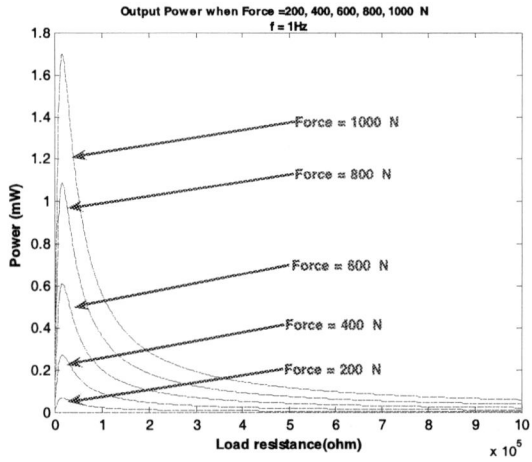

Fig. 9 (b)

Fig.9 Output voltage versus the load resistance for different Input force and constant frequency

VI. CONCLUSION

The transformer linear model is tested for both no load and load conditions, and seen to be appropriate for linear system analysis of piezoelectric microgenerator. The stress factor

A Novel Successive Interference Cancellation Scheme in OCDMA System

[1]Tawfig Eltaif, S*tudent Member, IEEE*, [2]Hossam M. H. Shalaby, *Member, IEEE,* [1]Sahbudin Shaari, *Member, IEEE*
[1]Photonics Technology Laboratory (PTL), Institute of Micro Engineering and Nanoelectronics,
Universiti Kebangsaan Malaysia, 43600 UKM Bangi, Selangor, MALAYSIA
[2]Elect. Eng. Dept, Fac. of Engineering, Alexandria University, Alexandria 21544, EGYPT.
E-mail: tefosat@ieee.org

Abstract – **In DC/CDMA systems, overcoming near/far effects is imperative for satisfactory performance. One way to combat the near/far effect is to use stringent power control. Another approach is multiuser detection (MUD). In addition to mitigating the near/far effect, MUD has more fundamental potential of raising capacity by canceling MAI. In this paper we going to present MUD scheme's known as successive interference cancellation (SIC). Where DS-OCDMA receiver was studied based on the SIC technique without hard limiter. Also we mention to use this technique with hardlimiter placed before the nondesired users.**

Index Terms - **optical code-division multiple access (OCDMA), multiple-access interference (MAI), hard limiter (HL), optical orthogonal codes (OOCs), Successive interference cancellation (SIC).**

I. INTRODUCTION

Code Division Multiple Access (CDMA) has emerged as the technology of choice for wireless and optical transmission, because it provides a number of attractive features over other multiple access schemes – time division multiple access (TDMA) and wavelength division multiple access (WDMA) to meet the high capacity, which requires each user to use different wavelength to transmit. When the number of users is large or the channel access is bursty, OCDMA offers an advantage over WDMA, due to the OCDMA characteristics of graceful degradation as the number of users increase. Other advantages of CDMA are flexibility to user allocation and security against unauthorized users. As the communication channel is not subdivided into time or frequency slots, CDMA allows random access to an indefinite number of users. As

additional users subscribe to the system, they can be given unique codes and then access the channel without the need to synchronize with any other user. The capacity of CDMA systems employing the conventional matched filter detector at the receiver end is often limited by the interference due to the other users in the system, known as multiple-access interference (MAI). MAI is introduced in CDMA systems due to the inability to maintain complete orthogonally of users' signature sequences over the hostile communication channels. [1]- [5]

Due to these advantages, (CDMA) has been the focus of much research in the last twenty years, primarily in the radio frequency domain, and also in the optical domain, where the traditional method to recover the data at the receiving end of an optical CDMA system is to use an optical correlator followed by a photodetector and a decision device [1].

To enhance the performance of the correlation receiver, the simplest techniques are used, where Salehi and Brackett used an optical hard limiter before the correlator at the receiver side [3]. Although the performance of their receiver (which involved an ideal photodetector) was improved, Kwon [6] has shown that such improvement becomes insignificant or more realistic systems, e.g., with avalanche photodiodes (APDs).With the implementation of double optical hardlimiters before and after the correlator at the receiving end, Ohtsuki [7], [8] was able to significantly improve the error-rate performance. Recently, Shalaby has proposed a new receiver model, called a chip-level receiver [9]. This receiver does not require the optical hardlimiters or the correlator in its implementation; hence, it is much more practical than the correlation receiver with hardlimiters.

0-7803-9730-4/06/$25.00 ©2006 IEEE

The other studies on interference-cancellation techniques are inspired by radio frequency (RF) communications, such as multiuser detection [1], parallel interference cancellation (PIC) [4], [10]–[14], serial interference cancellation [15], or turbo codes [16]. These techniques are more complex than the HL techniques, but they are more efficient. In [4] and [14] Shalaby *et al.* have presented several interference-cancellation techniques which significantly improve the performance for a modified prime sequence. Here in this paper we are going to design new method of interference cancellation schemes to mitigate MAI for OCDMA, based on the use of HL device and SIC.

The paper unfolds as follows; section II describes the general direct detection OCDMA system. Also the popular detection options are described in other subsections: correlation detector, optimal detector. The SIC system performance analysis is presented in section III. Finally we give our conclusion in section IV.

II. DS-OCDMA SYSTEM DESCRIPTION

A. System Structure

In our system we consider an incoherent, DS-OCDMA system, where the system consist of N users, labeled $n = 1, 2 \ldots$ N, using on-off keying (OOK) modulation to transmit binary data via optical channel for each user. And specific sequence code for each user. In particular, OOC will be used as the signature codes [1] in this paper. Where it's a family of (0,1) sequences of length F and weight W which satisfy that λ_a, λ_c are equal to 1, with good auto- and cross correlation enables the effective detection of the desired signal. The nth user has the spreading code

$$c_n(t) = \sum_{k=-\infty}^{\infty} c_{k,n} PT_c(t - kT_c) \qquad (1)$$

Where $c_{k,n} \in \{0,1\}$, T_c refers to the chip duration, and $P_\tau(t)$ is the rectangular pulse in $[0, \tau]$ with unit amplitude. We consider an ideal synchronous case, i.e., $\tau_k = 0$. It has been shown in [3] that the synchronous case is the worst case. Let:

$$b_n(t) = \sum_{i=-\infty}^{\infty} b_{i,n} PT(t - iT) \qquad (2)$$

Refer to the binary data of the nth user where $b_{i,n} \in \{0,1\}$ and T is the bit duration. Then we can

say that the intensity signal of the nth user is $S_n(t) = A_n b_n(t) c_n(t)$ where A_n is the signal strength of the nth user. In this case in the receiver side we can get the signal $r(t)$ as sum of the user's signals. As follows:

$$r(t) = \sum_{n=1}^{N} A_n b_n(t - \tau_n) c_n(t - \tau_n) + \rho \quad (3)$$

Where τ_n are the relative delay, and we consider an ideal synchronous case, $\tau_n = 0$. The term ρ represents the noise signal due to the dark current.

B. Conventional Correlation Receiver

It is the simplest receiver structure. Assuming the desired user is j, and as described in [3], that the receiver signal $r(t)$ in case of this receiver is multiplied by the code sequence corresponding to the desired user $c_j(t)$. Then after the T period we can get the decision value $Y_i^{(j)}$.

$$Z_i^{(j)} = \int_o^T r(t).c_j(t).dt$$

$$Z_i^{(j)} = W.b_i^{(j)} +$$

$$\sum_{\substack{k=1 \\ k \neq j}}^{N} b_i^{(j)} . \int_0^T c_k(t).c_j(t).dt \qquad (4)$$

In the above equation the second part refer to the big problem faced the researchers in OCDMA called MAI. In the next stage the value $Z_i^{(j)}$ is compared with the decision threshold level θ, to get the estimation of the transmitted bit $b_i^{(j)}$. Where in this receiver, the photon counts over all mark positions of the underlying code are collected to form one decision Variable Y:

$$Y = \sum_{i=1}^{w} Y_i \qquad (5)$$

It is clear that a threshold θ is set. And if the collected photon counts in one bit time is less than θ, "0" is declared, otherwise "1" is declared to be sent.

Also it has been shown in [3] that the error probability $p_b(\theta)$ for the ideal synchronous case is:

$$p_b(\theta) = \frac{1}{2} \sum_{i=\theta}^{N-1} \binom{N-1}{i} \left(\frac{W^2}{2F} \right)^i \left(1 - \frac{W^2}{2F} \right)^{N-1-i}$$

$$(6)$$

C. Conventional Correlation Receiver with hard limiter (CCR with HL)

It's an optical device [3] placing before the tapped delay line, which used to reduce the interference effect, by reducing the receiver optical power. An ideal OHL function is defined as:

$$g(x) = \begin{cases} 1, \ldots\ldots x \geq 1 \\ 0, \ldots\ldots 0 \leq x \leq 1 \end{cases} \qquad (7)$$

Also in [2] has been shown that for ideal chip synchronous case the error probability is:

$$P_b(\theta) \leq \frac{1}{2}\binom{W}{\theta}\prod_{i=0}^{\theta-1}\left(1 - \left(1 - \frac{W^2}{2F}\right)^{N-1-i}\right) \qquad (8)$$

III. SUCCESSIVE INTERFERENCE CANCELLATION

In this section we will explain a new design interference cancellation scheme which called successive IC (SIC), where the interference due to all nondesired users is estimated in order to be removed from the received signal. SIC is a multiuser detection (MUD) technique typically employed in code division multiple access (CDMA) communication systems to improve the capacity and overall throughput of the system.

A. SIC Algorithm

Operation of the system is simple, where its works by attempting to detect and demodulate the strongest user signal currently present in the overall received signal. Here the strongest user is not known beforehand, but is detected from the strength of the correlations of each of the user's chip sequence with the received signal. The correlation values can be getting from the bank of the correlator. As shown in Fig 1, the block diagram of SIC receiver without hard limiter. In the next stage after this user has been detected and demodulated; its contribution to the original signal is regenerated and subtracted from the overall received signal to get a new received signal. Then we can conclude that the algorithm repeats except the strongest user from the new received signal (that one has one less user signal) is detected, demodulated, regenerated, and subtracted.[17]-[18]. Then at the end we can say one by one the strongest received signals are subtracted

from the original signal till all users have been detected, and demodulated. Fig.2 shows the flowchart of this process, and in general algorithm the successive cancellations are carried as follows:

i)   Recognize the strongest signal (one with maximum correlation value).
ii) Decode the strongest user.
iii) Regenerate the strongest users' signal using its chip sequence.
iv) Cancel the strongest user.
v)   Repeat (until all users are decoded or a permissible number of cancellations are achieved).

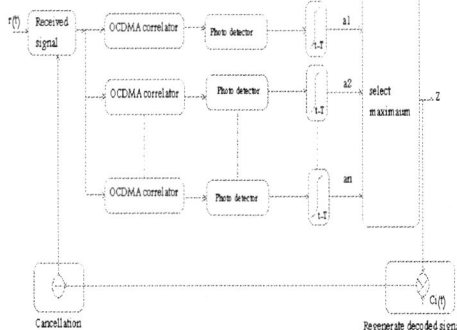

Fig. 1 SIC Receiver block diagram.

SIC with hardlimiter block diagram is shown in Fig3. In details analysis the general equations that describe SIC are presented in next section.

B. General equations

Assuming the signal coming to the receiver is $r(t)$ in eq.(3). And for OOCs (F, W, 1, 1) and for synchronous case, and for N simultaneous users, we consider the threshold level θ_T ($0 < \theta_T \leq W$). In a general case we look for the error probability, which can be written as follows:

$$P_b = \frac{1}{2}P([E/0] + [E/1]) \qquad (9)$$

As we mentioned previously the main function of this system based in maximum cross correlation between the users, then the effect n^{th} user's signal on the first receiver is denoted by $I_n^{(1)}$.

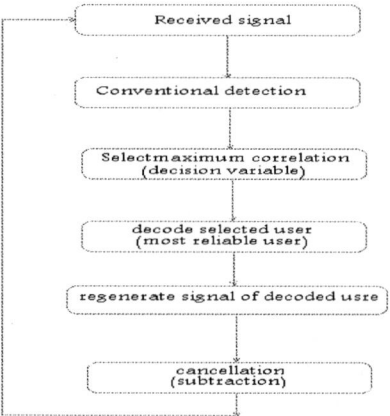

Fig. 1 Flow chart of interference cancellation schemes.

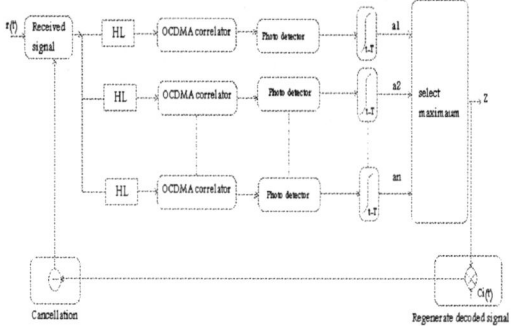

Fig. 3 SIC receiver block diagram with hardlimiter

We define the cross correlation between user i^{th} and the nth user as:

$$I_{n,i}(\tau_{n,i}) = \frac{1}{T}[\int_0^T c_n(t-\tau_{n,i}).c_i(t)dt] \quad (10)$$

Where $\tau_{n,i}$ is the time delay between the nth user relative to the ith user.

Hence, Z_1, the output of the first user's correlator at time T, can be written as:

$$Z_1 = \frac{1}{T_c}\int_0^T r(t).c_1(t-\tau_1).dt$$

$$= \frac{1}{T}\int_0^T [\sum_{n=1}^N A_n b_n(t-\tau_n)c_n(t-\tau_n) + \rho_1].c_1(t).dt \quad (11)$$

Then the decision on the bit is then made by using the decision variable.

$$Z_1 = b^{(1)}W + l_1 \quad (12)$$

Where $b^{(1)}W$ refer to the desired signal term of the first user, and the second term can be defines as:

$$l_1 = \sum_{n=2}^N b_n I_{n,1}(\tau_{n,1}) + \rho_1 \quad (13)$$

It is assumed that users are detected in order of decreasing signal strength such that user 1 will always correspond to the strongest user. Once this user has been detected and modulated, the result is used to regenerate the user signal. Then the regenerated signal is subtracted from the original signal. The correlation value is used for cancellation:

$$r_1(t) = r(t) - Z_1.c_1(t-\tau_1)$$

$$= \sum_{n=2}^N A_n b_n(t-\tau_n).c_n(t-\tau_n)$$

$$+ \rho_1 - l_1.c_1(t-\tau_1) \quad (14)$$

Now for the second strongest user, we have *(N-2)* interfering signals moreover some noise due to imperfect cancellation. In following decision statistic for user 2 after canceling user 1:

$$Z_2 = b^{(2)}W + l_2 \quad (15)$$

And l_2 is defined as:

$$l_2 = \sum_{n=3}^N b_n I_{n,2}(\tau_{n,2}) + \rho_2 - c_1 I_{1,2}(\tau_{1,2}) \quad (16)$$

In general for the jth cancellations, we get:

$$r_j(t) = r_{j-1}(t) - Z_j.c_j(t-\tau_j) \quad (17)$$

Where Z refers to the correlation after (j-1)[st] cancellation, then the decision variable for the (j+1)[st] user is given by:

$$Z_{j+1} = b_{j+1}W + l_{j+1} \quad (18)$$

And l_{j+1} are given by:

$$l_{j+1} = \sum_{n=j+2}^N b_n I_{n,j+1}(\tau_{n,j+1}) + \rho_{j+1} - \sum_{i=1}^j l_i I_{i,i+1}(\tau_{i,i+1}) \quad (19)$$

In the above expression, the first term is MAI of the uncancelled users, second term is noise signal due to the dark current, and the third term is due to the cumulative noise from imperfect cancellation.

IV. CONCLUSION

Through this work we have presented a new proposal in interference cancellation based in OCDMA, called successive interference cancellation (SIC). The interference cancellation is never perfect, and the residual cancellation errors propagate because of the successive nature of the decoding. In fact, these residual errors are the principle capacity-limiting issue in SIC systems. In future work we need obviously to analyze the performance of SIC; also we need to analyze the practical implementation of the interference canceller.

ACKNOWLEDGEMENT

I would like to thank Dr. Salah Elfaki, University of Science & Technology, Sudan for helpful discussions.

REFERENCE

[1] M. B. Pearce and B. Aazhang, "Multiuser Detection for Optical Code Division Multiple Access Systems," *IEEE Trans. Commun*, vol. 42, no. 21314. (1994).

[2] J. A. Salehi, "Code division multiple-access techniques in optical fiber networks—Part I," *IEEE Trans. Commun.*, vol. 37, no. 8, pp. 824–842. (1989).

[3] J. A. Salehi and C. A. Brackett, "Code division multiple-access techniques in optical fiber networks—Part II: Systems performance analysis," *IEEE Trans. Commun.*, vol. 37, no. 8, pp. 834–842. (1989).

[4] H. M. H. Shalaby, "Synchronous fiber-optic CDMA systems with interference estimators," *J. Lightw. Technol.*, vol. 17, no. 11, pp. 2268–2275. (1999).

[5] C. Goursaud, A. Julien-Vergonjanne, C. Aupetit-Berthelemot, J. P. Cances, and J. M. Dumas, "DS-OCDMA Receivers Based on Parallel Interference Cancellation and Hard limiters," *IEEE Transactions Commun*, vol. 54, no. 9. (2006).

[6] H. M. Kwon, "Optical orthogonal code-division multiple-access system—Part I: APD noise and thermal noise," *IEEE Trans. Commun.*, vol. 42, pp. 2470–2479. (1994).

[7] T. Ohtsuki, "Performance analysis of direct-dectection optical asynchronous CDMA systems with double optical hard-limiters," *IEEE J. Lightwave Technol.*, vol. 15, pp. 452–457.(1997).

[8] T. Ohtsuki, "Direct-detection optical asynchronousCDMAsystems with double optical hard-limiters: APD noise and thermal noise," in *Proc. IEEE Global Telecommunications Conf. (GLOBECOM '98)*, Sydney, Australia, pp. 1616–1621. (1998).

[9] H. M. H. Shalaby, "Chip-level detection in optical code-division multiple access," *IEEE J. Lightwave Technol.*, vol. 16, pp. 1077–1087. (1998).

[10] D. Divsalar and M. K. Simon, "Improved CDMA performance using parallel interference cancellation," in *Proc. Mil. Commun. Conf.*, pp. 911–917. (1994).

[11] R. Fantacci, A. Tani, and G. Vannuccini, "Performance evaluation of an interference cancellation receiver for noncoherent optical CDMA systems," in *Proc. GLOBECOM*, vol. 3, pp. 2823–2827. (2002)

[12] C. Goursaud, N. M. Saad, Y. Zouine, A. Julien-Vergonjanne, C. Aupetit-Berthelemot, J. P. Cances, and J. M. Dumas, "Parallel multiple access interference cancellation in optical DS-CDMA systems," *Ann. Telecommun.*, vol. 9–10, pp. 1053–1068. (2004).

[13] C. Goursaud, A. Julien-Vergonjanne, Y. Zouine, C. Aupetit-Berthelemot, J. P. Cances, and J. M. Dumas, "Improvement of parallel interference cancellation technique with hard limiter," in *Proc. IEEE Globecom, Photon. Technol. Commun.*, CD-ROM. (2005)

[14] H. M. H. Shalaby and E. A. Sourour, "Co-channel interference cancellation in optical synchronous CDMA communication systems," in *Proc. IEEE 3rd Int. Symp. Spread Spectrum Tech. Appl.*, Oulu, Finland, pp. 579–583. (1994).

[15] C. Goursaud, A. Julien-Vergonjanne, Y. Zouine, M. Morelle, C. Aupetit-Berthelemot, J. P. Cances, and J. M. Dumas, "Serial interference cancellation receiver for optical CDMA systems," in *Proc. IEEE Conf. ITST*, Brest, France, pp. 403–406. (2005).

[16] C. Argon and S. W. McLaughin, "Optical OOK-CDMA and PPMCDMA systems with turbo product codes," *J. Lightw. Technol.*, vol. 20, no. 9, pp. 1653–1663. (2002).

[17] P. Patel and J. Holtzman, "Analysis of a Simple Successive Interference Cancellation Scheme in a DS/CDMA System," *IEEE journal on selected areas in commun*, vol. 12, no. 5. (1994).

[18] P. Patel and I. Holtzman, "Analysis of a simple successive interference cancellation scheme in DS/CDMA system using correlations," in *Proc. GLOBECOM* Houston, TX. (1993).

Simulation of Piezo-Resistive Metal Gauge on Rectangular Membrane for Low Pressure Application

Mohd Yunus Hamid, Thangamani U, P R Vaya

Centre for Artificial Intelligence, School of Engineering and Information TechnologyUniversiti Malaysia Sabah, Locked Bag No. 2073, 88999 Kota Kinabalu, Sabah, Malaysia Tel: +60-8-832-0000 x 3536, Fax: +60-88-320348, E-mail: thanga_m@yahoo.com

Abstract **Piezo-resistive metal gauge on rectangular membrane design and its simulation for low pressure application is presented in this paper. Small deflection analytical equations are derived for both simply supported and clamped edge boundary conditions. The design and orientation of grid pattern on rectangular membrane is based on ANSYS simulation results. It is used to find out maximum strain locations to achieve high sensitivity. Maximum of 0.3509 micro strain and maximum resistance change in grid = 90.6271 micro-ohm are achieved for an applied load of 1mPa. The sensitivity of the gauge is 0.35με/mPa.**

I. INTRODUCTION

A strain gauge converts force, pressure, tension, weight in to a change in electrical resistance which can then be measured. Metal gauges are robust, reliable, and cost effective. The operational principle of metal gauge is based on the fact that any electrical conductor changes its resistance with mechanical stress. It is due to changes in conductor cross section and resistivity caused by micro structural changes. The size of the strain gauge is usually decided by the space available or by the topography of the strain field. There are no physical constraints in choosing the gauge dimension [1]. The equations are derived based on small deflection theory [2]. Finite element tool, ANSYS, is used to find maximum strain deformation location on the membrane. It allows choosing the desired gauge orientation on membrane to achieve maximum sensitivity. Materials like Polyimide, Polystyrene, and Silicon are studied as a substrate for gauge design. However, Polyimide is found to develop maximum strain (See Fig. 1) and hence it is chosen as a substrate for the gauge. The alloy 'Constantan' is a commonly used Piezo-resistive metal gauge material and its high peel strength makes the gauge less sensitive to mechanical damage during installation [3]. With its ease of handling and its suitability for use over the temperature range from -195°C to +175°C polyimide is an ideal substrate for general purpose static and dynamic stress analysis. In the present study, analytical equations are derived for substrate for supported at middle edges and at the fixed edges and grid on the substrate.

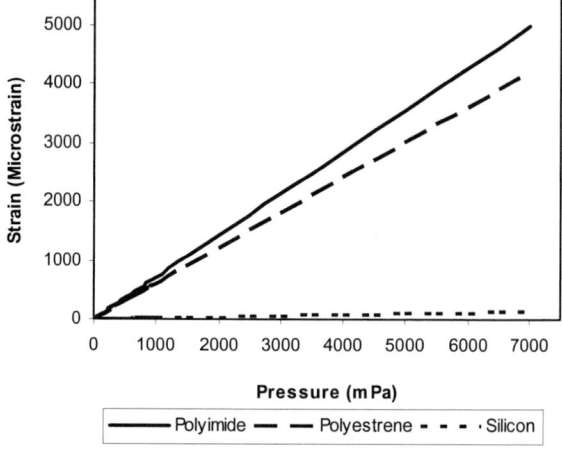

Fig.1 Different Substrate Materials

II. ANALYTICAL EQUATIONS FOR GRID ON SUBSTRATE

The change in resistance of the grid on a rectangular membrane due to applied pressure is calculated by strain developed in membrane and strain passed on to the grid as discussed in the following subsections [4]. The schematic of grid on rectangular membrane is shown in Fig.2.

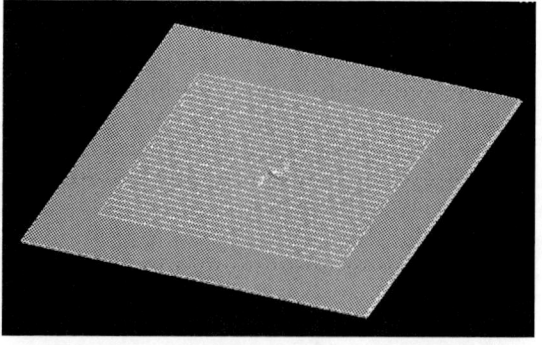

Fig. 2 Schematic of Grid on Rectangular Membrane

0-7803-9730-4/06/$25.00 ©2006 IEEE

A. STRAIN ON RECTANGULAR MEMBRANE

Strain distribution on rectangular membrane depends on its geometry, boundary conditions and applied load. There are two different cases discussed in this paper. They are i) supported edges and ii) clamped edges. The applied pressure P is constant over entire membrane.

Supported Edge: The maximum strain developed at center of the membrane supported at the middle of the edges due to applied pressure is given by [5]

$$\varepsilon_{max} = \frac{\beta P w^2}{Y t^2} \quad (1)$$

where β=0.1110 when the aspect ratio is 2. w is the width of the membrane. P is the applied pressure, t is the thickness of the membrane, and Y is the Young's modulus of the membrane material.

Clamped Edge: The maximum strain developed at center of the membrane fixed at the edges due to applied pressure is given by [5]

$$\varepsilon_{max} = \frac{\beta_1 P w^2}{Y t^2} \quad (2)$$

where β_1=0.12472 when the aspect ratio is 2.

B. STRAIN ON GRID

The resistance of a grid is given by

$$R = \rho \frac{l}{A} \quad (3)$$

$$A = WT \quad (4)$$

where, L is the length of the grid, W is the width of the grid, T is the thickness of the grid, and ρ is the resistivity of the grid. Under strain the rate of changes in R

$$\frac{dR}{R} = \frac{dl}{l} - \frac{dA}{A} + \frac{d\rho}{\rho} \quad (5)$$

$$\frac{dA}{A} = \frac{dW}{W} + \frac{dT}{T} \quad (6)$$

where dW and dT written in terms of axial strain is given by $dW = -\gamma \varepsilon_a W$ & $dT = -\gamma \varepsilon_a T$

$$\frac{dR}{R} = (1 + 2\gamma)\varepsilon_a + \frac{d\rho}{\rho} \quad (7)$$

where γ is the Poisson's ratio $\left(= -\dfrac{Lateral\ strain, \varepsilon_l}{Axial\ strain, \varepsilon_a} \right)$

An axial stress produces both the axial and lateral strain. Axial strain, $\varepsilon_a = \dfrac{\Delta l}{l}$ and Lateral strain, $\varepsilon_l = \dfrac{dA}{A}$. A fractional change in resistivity $\dfrac{d\rho}{\rho}$ is due to piezoresistivity. In metals, $\dfrac{d\rho}{\rho}$ is related to fractional change in volume, $\left(\dfrac{dV}{V}\right)$ as [6].

$$\frac{d\rho}{\rho} = K \frac{dV}{V} = K\left(\frac{dl}{l} + \frac{dA}{A}\right) \text{ as } V = lA$$

$$\frac{d\rho}{\rho} = K\varepsilon_a(1 - 2\gamma) \quad (8)$$

where K is the Bridgman constant, 1.13≤K≤1.15 substitute equation (8) in to (7)

$$\frac{dR}{R} = \varepsilon_a \{1 + K + 2\gamma(1 - K)\} \quad (9)$$

Equation (9) is used to calculate the change in the grid resistance due to applied strain.

C. CHANGE IN THE GRID RESISTANCE IN TERMS OF STRAIN DEVELOPED FROM MEMBRANE

Supported Edge: Change in the grid resistance on membrane supported at middle edge is obtained by equation (9) and (1).

$$dR = R\left\{\frac{\beta P w^2}{Y t^2}\right\}[1 + K + 2\gamma(1 - K)] \quad (10)$$

It is a function of strain developed from rectangular membrane, Poisson's ratio of the material, and Bridgman constant.

Clamped Edge: Changes in the grid resistance for fixed edge membrane is given by equation (9) and (2).

$$dR = R\left\{\frac{\beta_1 P w^2}{Y t^2}\right\}[1 + K + 2\gamma(1 - K)] \quad (11)$$

where, R is the resistance of the gauge, S is the sensitivity of the gauge, and ε is the strain of the grid obtained by ANSYS simulation.

III. SIMULATION AND RESULTS

Finite element analysis technique is used to find out the strain on the strain gauge [7]. Design tool Pro ENGINEER (wildfire) is used to design the substrate and the grid pattern. A substrate having length = 8mm, width = 4mm, and thickness = 0.05 mm is chosen for this analysis. ANSYS Solid brick element having 8 node 185 structure is used to design substrate as well as the grid. The brick element is generally used for bending strain and stress analysis. The substrate is simulated for supported edge and clamped edge boundary conditions. The maximum and minimum strained locations for clamped edge membrane are marked as shown in Fig.3. In Fig.4 the maximum strain locations for supported edges are at long side edges. Since the minimum strained locations are at center, gauge design on center area is insensitive. The substrate mesh element matrix size is 80×40 divisions. Because of limitation at convergence, mesh element matrix size should not exceed 80, 40. Successive iteration result shown in Fig.5 should be less than 5% to get maximum strain location at center is shown in Fig.6. The area of maximum strained region i.e., active region for grid design is found to be 2.6mmx1.2mm is shown in Fig.6. A grid having length = 2.6mm, width = 0.04444mm, and thickness = 0.003mm is designed by Pro ENGINEER design tool and assembled on the maximum strained area found (Fig.6) is shown in Fig.7. The substrate and the grid are meshed independently while maintain the continuity.

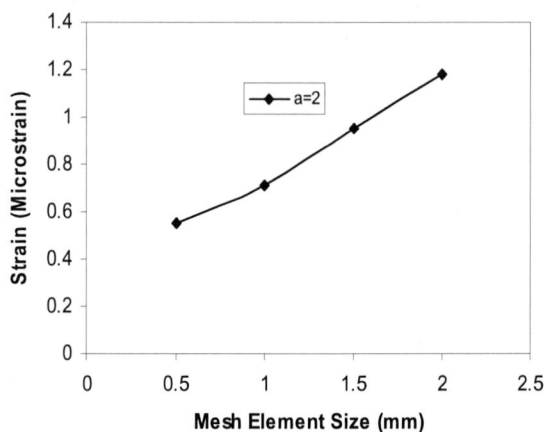

Fig. 5 Strain Variation Vs Mesh Element Size

The variation of mesh element size with strain is shown in Fig.5. Though the strain value increases linearly with increase of element size, its convergence limits the mesh element size, i.e., the successive iteration of element size result must be less than 5%.

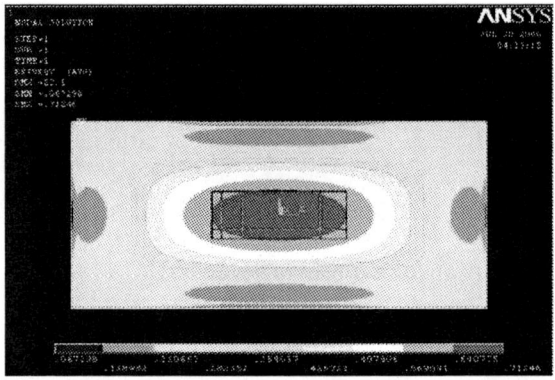

Fig. 6 Maximum Strain developed Area represented by rectangle box

Fig.6 shows the area chosen for grid design. This strain deformation region allows us to get maximum sensitivity.

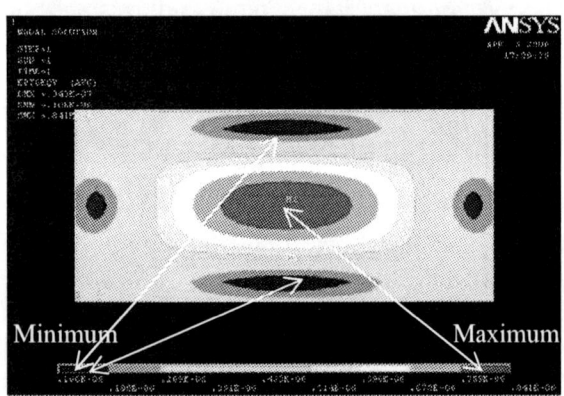

Fig. 3 Strain Distribution: Clamped edges

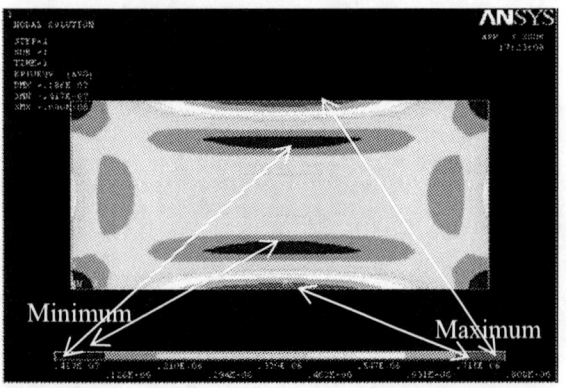

Fig. 4 Strain distribution: Supported edges

Table. 1 Area of Different contours

Colour	Length, mm	Width, mm
Red	2.6	1.2
Brown	3.6	1.6
Yellow	4.4	2
Yellow 2	5	2.2
Green	5.6	2.6

Table.1 shows area of different strain deformation contours. The different regions of area are represented by colour contour plot as shown in Fig.6.

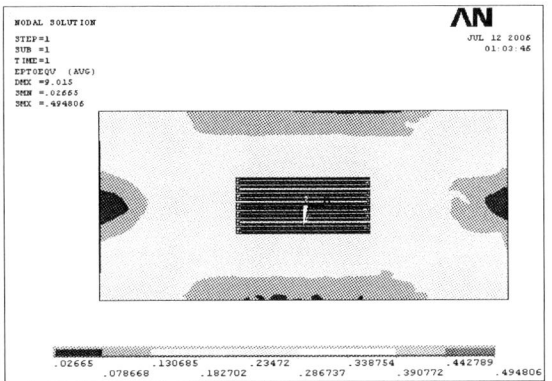

Fig. 7 Grid Design and its Strain Distribution

Fig.7 shows design of grid on maximum strain deformation region and its contour plot. Its strain values on grid, substrate, and both of them are given in table.1

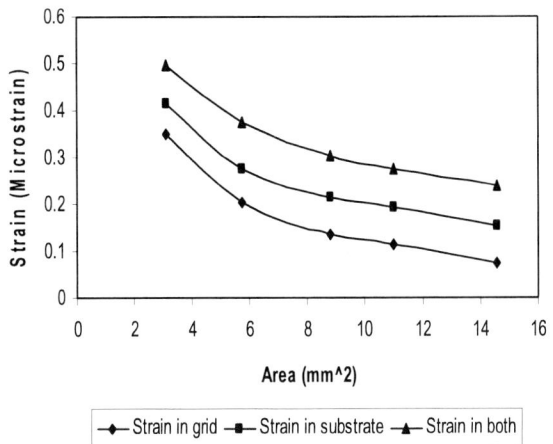

Fig. 8 Strain Variation Vs Contour Area

The grid, substrate, and the combination of grid & substrate strain decreases when the area of grid pattern and its corresponding contour increases as in table 1.

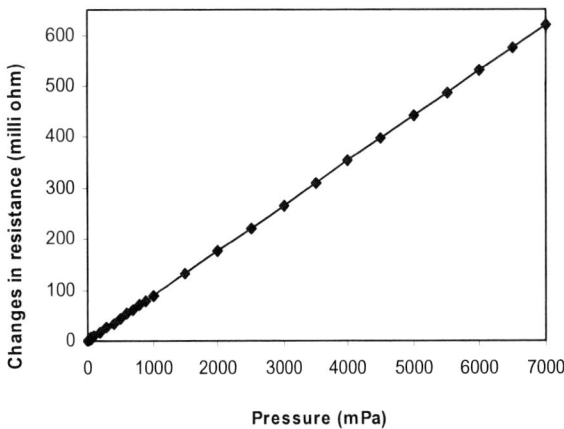

Fig. 9 Resistance Change in Grid Variation Vs Applied Load

The linear variation of resistance-changes with applied load is shown in Fig.9. The change in the gauge resistance, ΔR, is calculated by substituting the strain in grid, in equation (11). Table.2 gives strain values of gird, substrate, and combination of grid and substrate.

Table.2 Strain in grid, substrate, and the combination of grid & substrate

Pressure (mPa)	Strain in grid (με)	Strain in substrate (με)	Strain in grid and substrate together (με)
1	0.350981	0.412684	0.494806
10	3.51	4.127	4.948
20	7.02	8.254	9.896
40	14.039	16.507	19.792
60	21.059	24.761	29.688
80	28.079	33.015	39.585
100	35.098	41.268	49.481
200	70.196	82.537	98.961
300	105.294	123.805	148.442
400	140.393	165.074	197.923
500	175.491	206.342	247.403
600	210.589	247.61	296.884
700	245.687	288.879	346.364
800	280.785	330.147	395.845
900	315.883	371.415	445.326
1000	350.981	412.684	494.806
1500	526.472	619.026	742.209
2000	701.963	825.368	983.613
2500	877.453	1032	1237
3000	1053	1238	1484
3500	1228	1444	1732
4000	1404	1651	1979

4500	1579	1857	2227
5000	1755	2063	2474
5500	1930	2270	2721
6000	2106	2476	2969
6500	2281	2682	3216
7000	2457	2889	3464

IV. CONCLUSION

Piezo-resistive metal gauge on rectangular membrane for low pressure application has been designed and simulated. Maximum strained locations are identified on substrate and grid pattern is designed over it. The simulation results shows resistance-change of gauge is linear to applied pressure. Analytical equations are derived to find out the resistance-change in the gauge due to applied strain. The design of gauge on rectangular membrane with clamped edge boundary conditions is suitable for low pressure applications.

ACKNOWLEDGMENT

The authors wish to thank Prof. Dr. N. Siva Prasad and Mr. Prajwal Nimmagadda, IIT MADRAS, INDIA, for their helpful discussion in Pro ENGINEER and ANSYS.

REFERENCES

[1] Force and mass determination by strain gauge. David Joyce and Michael Scott. Published by Hottinger Baldwin Messtechnik

[2] William P. Eaton, Fernando Bitsie, James H.Smith, David W. Plummer. A New Analytical Solution for Diaphragm Deflection and its Application to a Surface-Micromachined Pressure Sensor. Internationa Coference on Modeling and Simulation of Microsystems, MSM 99 (1999).

[3] A.L. Window. Strain Gauge Technology, 2nd ed. Elsevier Applied Science (1982).

[4] W K Schomburg, Z Rummler, P Shao, K Wulff, and L Xie. The Design of Metal Strain Gauges on Diaphragms. IOP Publishing J.Micromech Microeng. Vol.14, pp.1101-110 (2004)

[5] Warren C.Young., Richard G.Budynas. Roark's Formulas for Stress and Strain, 7th ed. McGraw-Hill.

[6] Chris Field and David M.Beams. Project Guise: Introduction to Strain Gauges, Dept. of Electrical Engineering, University of Texas, Tyler (2005)

[7] Li Cao, Tae Song Kim, Susan C.Mantell, and Dennis L.Polla. Simulation and Fabrication of Piezoresistive Membrane type MEMS Strain Sensors.

Design and Modeling of an Electromagnetic Levitated and Actuated Micromotor

W. S. H. Wong[†], W.K.S. Pao[†], D. Holliday[‡] and P.H. Mellor[‡]

[†]School of Engineering, Swinburne University of Technology (Sarawak Campus)
93576, Kuching, Sarawak, MALAYSIA
[‡] Department of Electrical & Electronic Engineering, University of Bristol
Bristol BS8 1UB, UNITED KINGDOM
Email: wwong@swinburne.edu.my

Abstract **The design of an electromagnetic levitated and actuated micromotor is presented. The micromotor is consists of a counter-circular tracks that provide stable levitation force and radial parallel tracks that produce motoring rotational torque. Using simple filamentary representation of different features of the micromotor design, analytical expressions of the electromagnetic force produced by the features are derived and are used as foundation for the micromotor design. Finite element modeling is then used to analyze the actual micromotor design layout. From both the analytical and modeling results, the effects of different design features are presented. Finally, a macro-scale demonstrator is constructed to corroborate the micromotor design.**

I. INTRODUCTION

SINCE the 80's several actuation technologies has been developed for MEMS micromotor. Fan *et. al.* first demonstrated the electrostatic-driven micromotor in 1989 [1]. When voltage is applied to the stator poles, the potential difference between the stator and rotor poles generates an electrostatic force that tries to increase the capacitance between them, thus aligning the poles. By careful application of voltages to the respective stator poles in sequence, a continuous rotating torque can be produced on the rotor [2, 3]. The main drawback of the design however is the excessive friction between the center pin-bearing used to hold the rotor and the rotor itself. Due to the large surface area to volume ratio at micro-scale dimension, large torque is required to overcome the friction. The friction also causes the rotor to wear out rapidly [4]. Mehregany *et. al.* address this issues in the development of the harmonic side-drive or 'wobble' micromotor where the rotor wobbles around a center bearing,

reducing frictional contact between them [5]. Frechette *et. al.* in 2001 has proposed a contactless gas lubricated bearing system for the micromotor for micro-turbine application [6]. However, complicated bonding processes are required to put the system together.

In piezoelectric actuated micromotor, a rotating wave (bending and extensional) is generated on the stator made up of piezoelectric material by applying voltage to the stator in sequence. The friction between the rotor which is in physical contact with the stator translates the rotating wave into rotary movement of the rotor. This physical contact between the rotor and stator bring about large rotating torque on the rotor and large rotor holding torque without any energy consumption [7-9]. However, processing and depositing suitable thin piezoelectric film can be complicated [8].

Electromagnetic actuated micromotor has also received much attention and early designs were demonstrated in [10] and [11]. Similar to most macro size electromagnetic induction motor, an electromagnetic field is produced by applying alternating current to the stator coil. Voltage will then be induced in the rotor, causing current to flow and produce its own electromagnetic field. Subsequently, the interaction between the stator and rotor magnetic fields generates rotating torque on the rotor. However, due to the complication in stator coil microfabrication, the stator coil of the micromotor is limited to planar construction instead of cylindrical structures that surrounds the rotor typical in macro size induction motor. In the early design, a center pin bearing is used to hold the rotor. In their design, Williams *et. al.* removed the requirement for a physical bearing, replacing it with repulsive electromagnetic force to levitated and hold the rotor at the centre on top of the planar stator [12,

0-7803-9730-4/06/$25.00 ©2006 IEEE

13]. This 'contactless' bearing eliminates surface friction thus enabling high speed operation, and allows free rotor rotation about multiple axis. It was envisaged that the contactless micromotor will be suitable for microgyroscope application [13].

This paper further investigates the design of the contactless bearing electromagnetic micromotor and proposed some design enhancements to improved performance, particularly the rotor driving torque. In the following sections, the design of the micromotor is presented. Using simple filamentary representation of different features of the micromotor design, analytical expressions of the levitation and rotating force are presented. Finite element modeling is then used to analyze the actual micromotor design layout. From both the analytical and modeling results, the effects of different design features are presented. Finally, a macro-scale demonstrator will be presented to corroborate the micromotor design.

II. MICROMOTOR DESIGN

Figure 1a shows the top view of a 4-phase electromagnetic micromotor stator whilst figure 1b shows the side view of both the rotor and stator. The rotor is a conducting disc. On the stator, the inner circular track provides the levitation force whilst the outer circular track provides the centering force. Together, they act as contactless bearing for the rotor, eliminating friction and thus enabling high-speed operation. The radial parallel tracks at the 4 perpendicular voltage feeds provide the rotational force. The stator rests on top of a ferromagnetic backing plane which increases the magnetic coupling between the stator and rotor, enhancing force.

Extensive study has been carried out by *William et. al* on the levitation and centering force produced by the circular tracks. From a simple parallel filamentary circular coils representing the stator (bottom coil) and rotor (upper coil) shown in figure 2, the vertical repulsive force between them was derived as:

$$F = \frac{\mu^2 r^2 \left[\ln(8r/z) - 2\right]}{z} \frac{\omega^2 L_2}{R_2^2 + \omega^2 L_2^2} I_1^2 \quad (1)$$

when an alternating current of magnitude I_1 and frequency ω is applied to the bottom coil. R_2 and

L_2 are the resistance and self-inductance of the upper coil. From the equation, the force acting on the upper coil is proportional to the square of the current magnitude but inversely proportional to the coils separation. Also, when ω is much lesser than R_2/L_2, the force is proportional to the square of the frequency. At higher frequency, the force levels off. Further study using finite element modeling and experiment on the micromotor have shown that the rotor levitation force varies in similar manner.

William *et al.* has also conclude that if the rotor edge lies between the inside edge of the outer circular track and about two-thirds of the width of the track, the rotor will be stable and centered when levitated. Maximum centering force is achieved when the edge of the rotor lies about one-third of the width of the track. It was also established that the centering force reduces the levitation force as the outer track is working in opposition to the inner track.

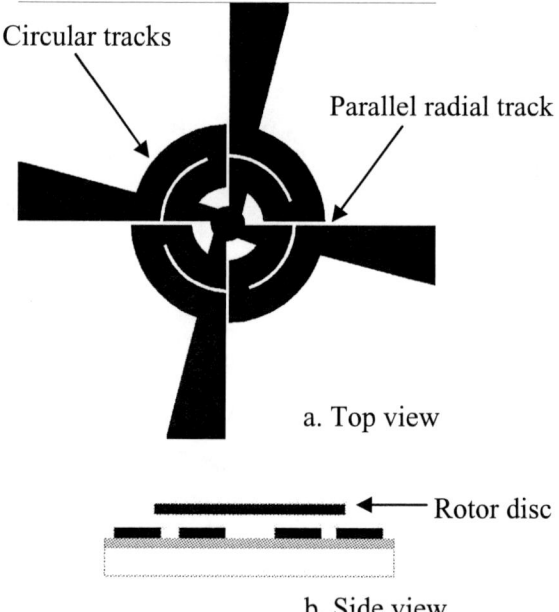

a. Top view

b. Side view

Fig. 1 Micromotor top and side view

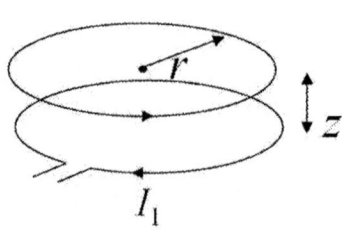

Fig. 2 Filamentary representation of the micromotor

In the study of rotational force, *William et. al* employed two filamentary conductors separated by a distance of $2x_o$ to represent the parallel radial tracks. Figure 3 shows the rotor and filaments from the side view of the micromotor. Using this simple model, it was found that the lateral force on the rotor; torque; is proportional to the square of the excitation current I_0 and sine of the phase angle φ between the current flowing in the radial tracks. At low frequencies, the torque is proportional to the excitation frequency but the torque levels off at higher frequencies. Finally, the torque increases as levitation height h reduces.

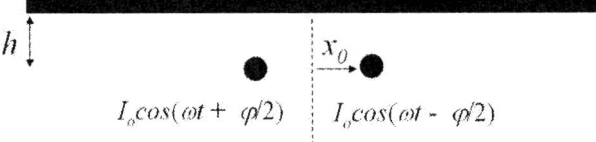

Fig. 3 Filamentary representation of a radial track.

To further analyse the rotational force on the rotor due to radial tracks, a 2D planar Finite Element Analysis is employed. Looking at the levitated micromotor system from the side view, over a small perimeter and a small radial depth, the curvature of the rotor and stator are ignored. The levitated rotor, the stator tracks and ferrite backing plane are thus modelled as slabs in 2D plane, as shown in figure 4. The rotor is made of aluminium, and the stator tracks are made of copper. In between the stator tracks and the backing plane, a layer of insulation made of silicon oxide is placed but it is modelled here as air for its insulating properties.

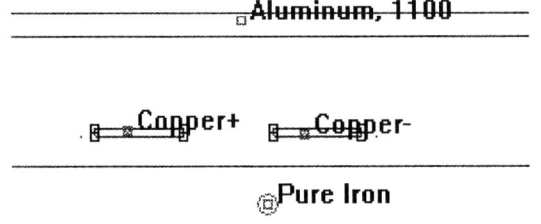

Fig. 4 Model used for FEA of rotational force.

The result of both rotational force and levitation force produced by the parallel tracks are shown in figure 5. When the current flowing between the tracks are 180° apart, the levitation and rotational forces are both near zero. However, unlike the rotational force, the levitation force does not change direction after 180°. Maximum levitation force but zero rotational force is produced when the currents are in phase. When the currents are 90° out of phase, maximum rotational force is generated. Since the primary function of the parallel radial tracks is to provide rotational force, the phase between the current is set to 90° at all times. At low frequency, the rotational force is proportional to frequency. However, due to skin effect, the resistance of the rotor increases at higher frequency, causing both levitation and rotational forces to fall off slowly as frequency increases.

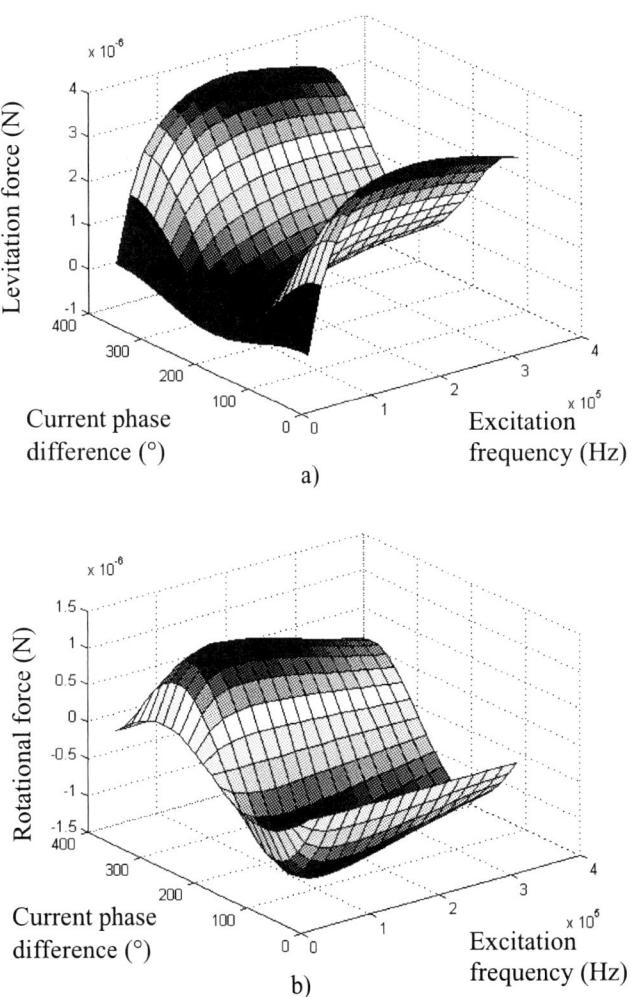

Fig. 5 Levitation and rotational forces produced by parallel radial tracks.

Analysis was also carried out to study the effect of track separation on the forces. From figure 6, as the track separation increases, the magnitude of the rotational force increases until it peaks at about 6 µm of separation. Any separation more than 6 µm would decrease the rotational force. Therefore, careful selection of the track selection is necessary to optimize the rotational force.

In the earlier stator designed by *William et. al* there was effectively a third radial parallel tracks, as shown in figure 7. Finite element analysis was carried out to study the effect of this third track. Shown in figure 8, it is concluded that the third radial track disrupts the rotational force and should thus be removed.

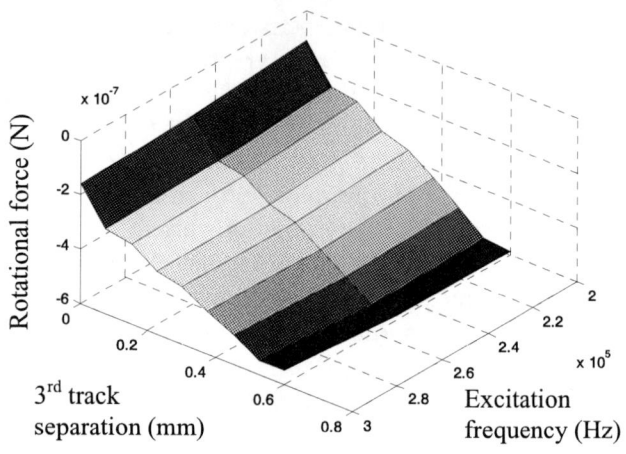

Fig. 8 Effect of the third radial track

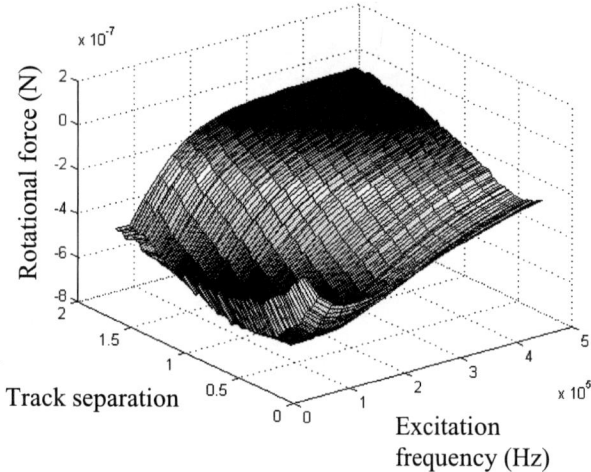

Fig. 6 Effect of the parallel track separation

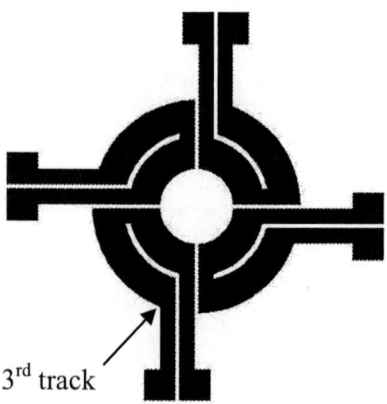

Fig. 7 Earlier proposed stator layout with a third parallel radial track.

III. EXPERIMENT SETUP

A macro-scale demo with aluminum rotor size of 4mm has been build. The stator shown in figure 1a is made using copper etched from a PCB. A 200 kHz $2A_{pk-pk}$ current is driven into the stator and the results was levitated rotating rotor.

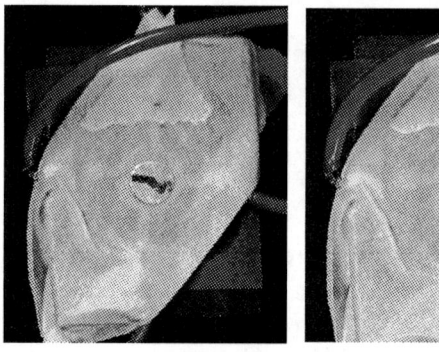 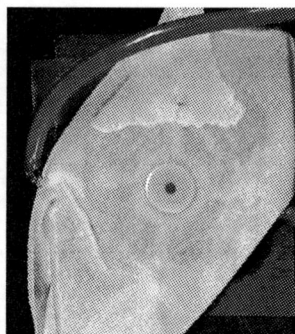

a. Non rotating   b. levitated & rotating

Fig. 9 Macro-scale micromotor demo

V. CONCLUSION

The different design features of a micromotor for optimum levitation and rotational forces have been shown. Particularly for the rotational force, the stator radial tracks separation and removal of the third destructive track are crucial for optimum force. In future, a micro-scale demo will be fabricated to test the different design features. Mechanical considerations for high-speed operation such as aerodynamics drags and mode of vibration will be looked into.

REFERENCES

[1] Fan, L. S., Tai, Y. C. and Muller, R. S., "IC – Processed Electrostatic Micromotors," *Sensors and Actuators*, vol. 20, pp41-47 (1989).

[2] Alladi, D. J., Nagy, M. L. and Gaverick, S. L., "An IC for closed-loop control of a micromotor with an electrostatically levitated rotor," *ISCAS '99*, vol 6, pp.489-492 (1999).

[3] Neugebauer, T.C., Perreault, D.J., Lang, J.H. and Livermore, C., "A six-phase multilevel inverter for MEMS electrostatic induction micromotors," *IEEE Transactions on Circuits and Systems II*, vol. 51, no. 2, pp. 49 – 56 (2004).

[4] L.S. Tavrow, S.F. Bart and J.H. Lang, "Operational characteristic of microfabricated electric motors," *Sensors and Actuators*, vol. A 35, pp33-44, 1992.

[5] Mehregany, M., Nagarkar, P., Senturia, S. D. and Lang J. H., "Operation of Microfabricated Harmonics and Ordinary Side-Drive Motors," *Proc. IEEE MicroElectroMechanical System Workshop*, pp.1-8 (1990).

[6] Frechette, L.G., Nagle, S.F., Ghodssi, R., Umans, S.D., Schmidt, M.A. and Lang, J.H., "An electrostatic induction micromotor supported on gas-lubricated bearings," *Proc. MEMS 2001*, pp290 – 293 (2001).

[7] Kaajakari V., Rodgers S. and Lal A., "Ultrasonically driven surface micromachined motor," *Proc. MEMS 2000*, pp40 – 45 (2000).

[8] Friend J., Nakamura K. and Ueha S., "A piezoelectric micromotor using in-plane shearing of PZT elements," *IEEE Trans. on Mechatronics*, vol. 9, no. 3, pp467-473, (2004).

[9] Flynn A.M., Tavrow L.S., Bart S.F., Brooks R.A., Ehrlich D.J., Udayakumar K.R.and Cross E., "Piezoelectric micromotors for microbots", *Journal of Microelectro-mechanical System*, vol. 1, no. 1, pp14-18 (1992).

[10] Guckel H., Christenson T.R., Skrobis K.J. and Klein J., "Planar rotational magnetic micromotors," *Int. Journal of Applied Electromagnetics in Materials,* vol. 4, no. 4, pp337-382 (1994).

[11] Chong A. H., Yong J.K. and Allen M.G., "A planar variable reluctance magnetic micromotor with fully integrated stator and coils," *Journal of Microelectromechanical Systems,* vol.2, no. 4, pp165-173 (1993).

[12] C. Sherwood, C. B. Williams, P.H. Mellor, R. B. Yates, M.R.J. Gibbs and A.D. Mattingley, "Levitation of a micromachined rotor for application in rotating gyroscope," *Electronics Letters*, vol. 31, pp.1845-1846 (1995).

[13] C. B. Williams, C. Sherwood, P.H. Mellor and R. B. Yates, "Modeling and testing of frictionless levitated micromotor," *Sensors and Actuators*, vol. A 61, pp.469-73 (1997).

[14] Meeker, D., "FEMM: Finite Element Method Magnetics User's Manual," (2003).

[15] Bart, S. F., Mehregany, M., Tavrow, L.S. and Lang J.H., "Electric Micromotor Dynamics", *IEEE Trans on Electron Devices*, 39 (3): 566-575 (1992).

Fabrication of White Polymer Light Emitting Diodes with ITO/PVK:PBD:DPVBi:DCJTB/Al Structure

Yap Chi Chin[1], Muhammad Yahaya[1], *Member, IEEE* and Muhamad Mat Salleh[2], *Member, IEEE*
[1]School of Applied Physics, Faculty of Science and Technology
[2]Institute of Microengineering and Nanoelectronics (IMEN)
Universiti Kebangsaan Malaysia
43600 UKM Bangi, Selangor, MALAYSIA
Email: mms@pkrisc.cc.ukm.my

Abstract White polymer light emitting diodes (PLEDs) with the configuration of ITO/PVK:PBD:DPVBi:DCJTB/Al were fabricated where indium tin oxide (ITO) was used as anode, poly(9-vinylcarbazole) (PVK) as polymeric host, 2-(4-biphenylyl)-5-phenyl-1,3,4-oxadiazole (PBD) as electron-transporting molecule, 4,4'-bis(2,2-diphenylvinyl)-1,1,-biphenyl (DPVBi) as blue dopant, 4-(dicyanomethylene)-2-t-butyl-6(1,1,7,7-tetramethyljulolidyl-9-enyl)-4H-pyran (DCJTB) as red dopant, and aluminium (Al) as cathode. The emitting layer was prepared using spin-coating technique. The white light emission was obtained through the combination of orange emission from DCJTB and the blue emission from DPVBi. The device using the 0.06 wt% DCJTB exhibited the CIE coordinates (0.38, 0.33) closest to the standard CIE coordinates for white light emission (0.33, 0.33) and its current turn-on voltage was approximately 13V.

I. INTRODUCTION

Organic light emitting diode (OLED) is a thin film device in which the emitting organic material is sandwiched between two electrodes. It will emit light when electricity is passed through it. OLEDs using either small molecules or polymers, have attracted considerable interest due to their easy processability, flexibility, low cost, low operating voltages, wide viewing angles, tunability of the color emission, fast response time, and ease of forming large area [1].

Recently, white OLEDs have attracted intense research interest due to their potential applications in full color displays combined with a color filter, in backlights for liquid crystal displays as well as general illumination purpose. There have been many approaches to develop the

white light emitting devices [2-7]. In this paper, single-layer dye-doped devices were reported because of their relatively simple fabrication and relatively stable color coordinates with respect to the bias voltage [6].

Zheng et al. reported a white OLED fabricated by co-evaporating DCJTB and DPVBi in a multilayers structure [7]. However, this process required the evaporation of two organics simultaneously, which makes the deposition process difficult to control. In contrast, it is quite easy for polymeric systems since a polymer layer doped with multiple kinds of fluorescent dyes can be fabricated by solution coating methods. For that reason, the blends of PVK, which served as polymeric host, DPVBi as blue fluorescent dye and DCJTB as red fluorescent dye were used to produce white light.

PVK is one of the most frequently used polymeric hosts, due to its excellent film-forming and hole-transporting properties. Since PVK is mainly hole-transport, owning very limited electron-transport capabilities, it is expected that in the single-layer structure, due to the highly unbalanced transport properties of holes and electrons, the recombination of carriers will be inefficient and the carrier recombination zone will also be closer to the cathode where the luminescence quenching occurs [8]. Therefore, an electron-transporting agent, PBD was doped into polymer layer to improve the charge balance in the emitting layer. This approach has also been reported in literature [9].

White PLEDs with the configuration of ITO/PVK:PBD:DPVBi:DCJTB/Al have been successfully fabricated. Combining the orange emission from DCJTB and the blue emission from DPVBi gave the white light. The device using the 0.06 wt% DCJTB exhibited the CIE coordinates (0.38, 0.33) closest to the standard CIE coordinates for white light emission (0.33, 0.33).

0-7803-9730-4/06/$25.00 ©2006 IEEE

II. EXPERIMENT

The host polymer PVK, having a high weight-average molecular weight of 1,100,000 g/mole and the electron-transporting molecule, PBD were purchased from Aldrich Chemical Company. The blue dopant, DPVBi and the red dopant, DCJTB were purchased from American Dye Source Inc. and Elight Corporation, respectively. All materials were used as received without further purification.

The single-layer PLEDs with ITO/PVK:PBD:DPVBi:DCJTB/Al structure were fabricated as shown in Fig. 1. The ITO-coated glass substrates were etched and patterned to serve as anode. The substrates were cleaned with 2-propanol and acetone in an ultrasonic bath each for 15 minutes. 1 mL of 1,2-dichloroethane solutions contained 10 mg of PVK, 10 mg of PBD, 0.6 mg of DPVBi and variable amount (0.001-0.006 mg) of DCJTB were prepared. The solutions were then spin-coated onto the ITO with a typical spinning speed and time at 2000 rpm for 40 s. Lastly, 150 nm aluminium was deposited as cathode by using electron gun evaporation technique.

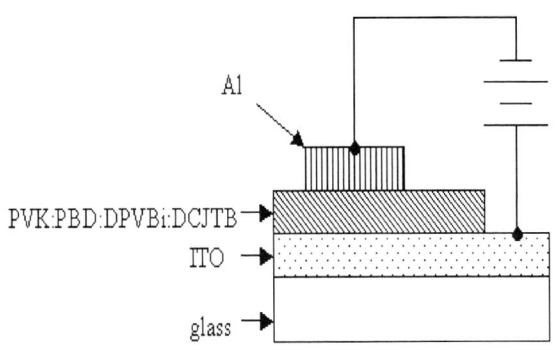

Fig. 1 The PLED with
ITO/PVK:PBD:DPVBi:DCJTB/Al structure.

The absorption and photoluminescence (PL) properties were investigated by depositing the corresponding thin films on pre-cleaned quartz substrates with Perkin Elmer LAMBDA 900 UV-VIS spectrophotometer and Perkin Elmer LS55 luminescence spectrometer, respectively. In addition, Keithley 238 source measurement unit was used to measure the electrical characteristics of the devices, while the EL spectra were obtained with Ocean Optic HR2000 spectrometer. All measurements were carried out at ambient atmosphere.

III. RESULTS AND DISCUSSION

Fig. 2 shows the absorption and PL emission spectra of PVK/PBD host, DPVBi (5 wt% in PMMA) and DCJTB (5 wt% in PMMA) thin films. The maximum absorption wavelength for DPVBi and DCJTB is 349 nm and 505 nm, respectively, while the emission of PVK/PBD, DPVBi and DCJTB peak at 428 nm, 437 nm and 606 nm, respectively.

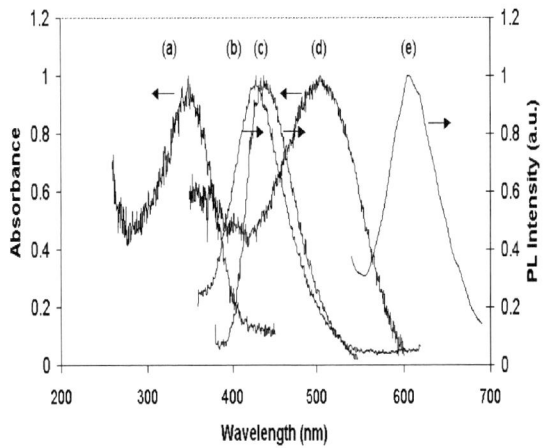

Fig. 2 Absorption of DPVBi (a) and PL of DPVBi (c) in PMMA host matrix (5 wt%), absorption of DCJTB (d) and PL of DCJTB (e) in PMMA host matrix (5 wt%), and PL of PVK:PBD (b) in thin films.

Fig. 3 shows the EL spectra of the devices with various doping concentration of DCJTB at applied voltage of 27 V. The weight percentage of DCJTB (wt%) was determined as the ratio of DCJTB to PVK. With increasing DCJTB concentration, the relative intensity of the blue emission decreased and that of the red emission increased. Fig. 4 gives the CIE coordinates of (0.22, 0.20), (0.28, 0.25), (0.30, 0.26) and (0.38, 0.33) for 0, 0.01, 0.03, and 0.06 wt.% DCJTB-doped devices, respectively. Among all the EL devices tested in this experiment, the device using the 0.06 wt% DCJTB exhibited the CIE coordinates (0.38, 0.33) closest to the standard CIE coordinates for white light emission (0.33, 0.33).

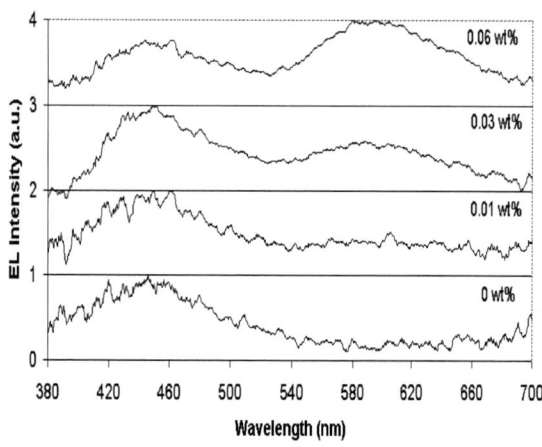

Fig. 3 EL spectra of devices with different DCJTB doping concentration at 27 V.

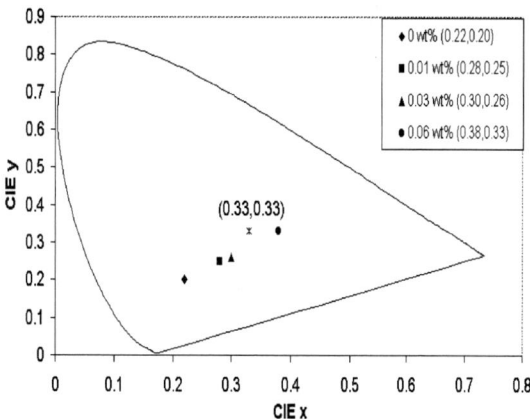

Fig. 4 CIE coordinates of devices at 27 V.

It was found that the shapes of EL spectrum are different from those of PL spectrum at the same DCJTB doping ratio (Fig. 5). At 0.06 wt% DCJTB loading level, one can hardly see the PL from DCJTB, but for EL the emission from DCJTB is very strong. This phenomenon is common in the doped systems. The main reason for the observation of this phenomenon lies on different excitation mechanisms of PL and EL. The PL spectra are excited by photons. There is no carrier injected in this process, the emission spectra of a doped system are only determined by the energy transfer. However, in EL spectra, besides energy transfer, carrier trapping also plays an important role. It is not a good overlap for absorption spectrum of DCJTB and emission spectra of DPVBi or PVK/PBD as shown in Fig. 2. This suggests that the energy transfer from PVK/PBD host or DPVBi to DCJTB was inefficient. As a result, carrier trapping became the dominant excitation mechanism. This also

can be deduced from the energy level scheme as shown in Fig. 6 [7, 10]. Since the LUMO (Lowest Unoccupied Molecular Orbital) and HOMO (Highest Occupied Molecular Orbital) energy levels of DCJTB are located inside the energy gap of other materials, some of the electrons and holes will be trapped and recombine directly in DCJTB molecules when electric field is applied. It was also found that for 0 wt% DPVBi-doped device, the blue emission disappeared in its EL spectrum (not shown here). This implies that the blue emission was originated from DPVBi and not from PVK/PBD host. As a result, the white light was generated by combining the orange emission from DCJTB and the blue emission from DPVBi.

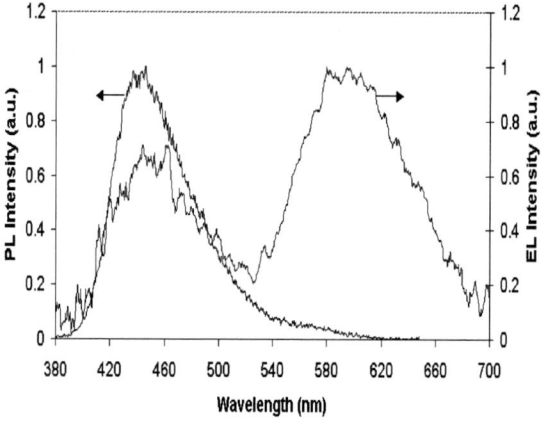

Fig. 5 PL and EL spectra of PVK:PBD:6 wt% DPVBi:0.06 wt% DCJTB blend film

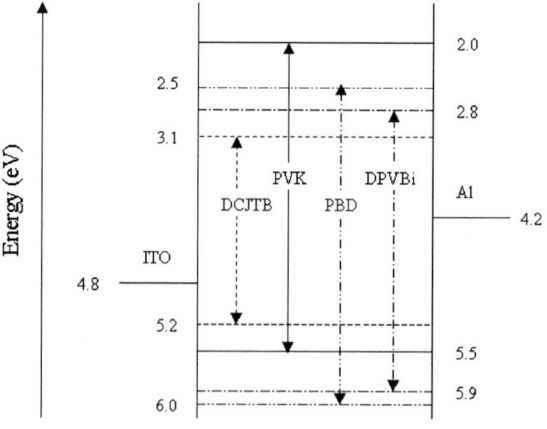

Fig. 6 Energy level diagram of ITO/PVK:PBD:DPVBi:DCJTB/Al device.

Fig. 7 shows the current-voltage characteristics of the white light emitting devices at room temperature. The forward current was found to increase with increasing forward bias voltage, and the curve has the same characteristics as those of a typical diode. The current turn-on voltage for 0, 0.01 and 0.03 wt% DCJTB-doped devices was around 18 V. The current turn-on voltage for 0.06 wt% DCJTB-doped device was the lowest (about 13 V). For 0, 0.01 and 0.03 wt% DCJTB-doped devices, the driving voltage increased slightly with increasing DCJTB concentration. This result was consistent with the assumption that DCJTB act as traps for both holes and electrons. However, this assumption was not longer correct for 0.06 wt% DCJTB-doped device. The exact mechanism is under investigation. After optimizing the devices structure, such as adjusting the thickness of emitting layers and introducing hole and electron injection layers etc., further reduce in turn-on voltage is expected.

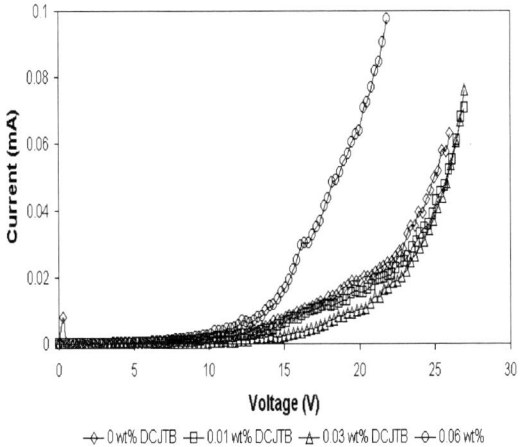

Fig. 7 Current-voltage characteristics of devices at different DCJTB doping concentrations.

IV. CONCLUSION

White PLEDs with the structure of ITO/PVK:PBD:DPVBi:DCJTB/Al have been successfully fabricated. Mixing the orange emission from DCJTB and the blue emission from DPVBi produced the white light emission. The device using the 0.06 wt% DCJTB exhibited the CIE coordinates (0.38, 0.33) closest to the standard CIE coordinates for white light emission (0.33, 0.33) and its current turn-on voltage was approximately 13V.

ACKNOWLEDGEMENT

This work has been carried out with the support of Malaysian Ministry of Science,

Technology and Innovation, under the IRPA grant 03-02-02-0067-SR0007/04-04.

REFERENCES

[1] D. Gupta, M. Katiyar and Deepak, "Various approaches to white organic light emitting diodes and their recent advancements," *Optical Material* vol. 28, no. 4, pp. 295-301 (2006).

[2] G. K. Ho, H. F. Meng, S. C. Lin, S. F. Horng, C. S. Hsu, L. C. Chen and S. M. Chang, "Efficient white light emission in conjugated polymer homojunctions," *Applied Physics Letters* vol. 85, no. 20, pp. 4576-4578 (2004).

[3] Q. F. Xu, H. M. Duong, F. Wudl and Y. Yang, "Efficient single-layer "twistacene"-doped polymer white light-emitting diodes," *Applied Physics Letters* vol. 85, no. 16, pp. 3357-3359 (2004).

[4] J. I. Lee, H. Y. Chu, S. H. Kim, L. M. Do, T. Y. Zyung and D. H. Hwang, "White light emitting diodes using polymer blends," *Optical Material* vol. 21, no.1-3, pp. 205-210 (2002).

[5] C. L. Chao and S. A. Chen, "White light emission from exciplex in a bilayer device with two blue light-emitting polymers," *Applied Physics Letters* vol. 73, no.16, pp. 426-428 (1998).

[6] J. T. Lim, M. J. Lee, N. H. Lee, Y. J. Ahn, C. H. Lee and D. H. Hwang, "Efficient white light emitting devices with a blue emitting layer doped with a red dye," *Current Applied Physics* vol. 4, no. 2-4, pp. 327-330 (2004).

[7] X. Y. Zheng, W. Q. Zhu, Y. Z. Wu, X. Y. Jiang, R. G. Sun, Z. L. Zhang and S. H. Xu, "A white OLED based on DPVBi blue light emitting host and DCJTB red dopant," *Display* vol. 24, no. 3, pp. 121-124 (2003).

[8] J. Kido, K. Hongawa, K. Okuyama and K. Nagai, "Bright blue electroluminescence from poly(N-vinylcarbazole)," *Applied Physics Letters* vol. 63, no. 19, pp. 2627-2629 (1993).

[9] G. E. Johnson, K. M. McGrane and M. Stolka, "Electroluminescence from single layer molecularly doped polymer films," *Pure and Applied Chemistry* vol. 67, no. 1, pp. 175-182 (1995).

[10] C. C. Wu, J. C. Sturm, R. A. Register, J. Tian, E. P. Dana and M. E. Thompson, "Efficient organic electroluminescent devices using single-layer doped polymer thin films with bipolar carrier transport abilities," *IEEE Transactions on Electron Devices* vol. 44, no. 8, pp. 1269-1281 (1997).

Homodyne Linear Crosstalk Impact in an Array Waveguide Router as an OADM for WDM Networks

Yasin M. Karfaa, *Student Member, IEEE*, M. Ismail, *Member, IEEE*, Abbou F. M*, *Member, IEEE*,
and S. Shaari, *Member, IEEE*,
Dept. of Electrical, Electronics and Systems Engineering
University Kebangsaan Malaysia, Selangor, Bangi - 43600
*Alcatel Network System, 50450, Kuala Lumpur, Malaysia
Email: yasin_m_k@yahoo.com, mahamod@eng.ukm.my

*Abstract--***Performance analysis is carried out to evaluate the bit error rate (BER) performance limitation of a WDM transmission system imposed by crosstalk due to array waveguide grating routing multiplexer/demultiplexer. Power penalty evaluated at a BER of 10^{-9} shows a significant impact of crosstalk on the number of wavelengths that can be connected to the AWG. Penalty is higher at higher input power, higher number of channels and higher bit rates used for transmission.**

Index terms-- **wavelength division multiplexing, waveguide grating router, bit error rate.**

I. INTRODUCTION

MODELING Optical Network provides the capability of easy adaptation to changes in the network traffic requirements. The network dimensions are limited by a number of effects such as optical crosstalk in the switch matrices and by the polarization dependent loss in the optical components, amplifier spontaneous emission, laser saturation, fiber nonlinearities, reflections, jitter accumulation, and signal bandwidth narrowing caused by filter concatenation [1]. Hence it is necessary to estimate for advance the limits to the number of building blocks that can be cascaded. With the help of theory, a statistical model for the bit error rate (BER) at the receiver can be done.

An array waveguide (AWG, shown in fig.1) has several uses as a multiplexer, demultiplexer, though it can be done using MZIs interconnected in suitable fashion, but, despite its polarization dependence [2] that means to mitigate were reported recently; it is preferable to use an AWG,

that is relative to MZI chain has lower loss, flatter passband spectral response (good for wavelength control), operates on wide temperature ranges (0 to 85 degrees Celsius) and easier to realize on an integrated-optic substrate together with photodetectors. However, the arrayed-waveguide grating (AWG) used to simplify the add/drop process in OADMs generates crosstalk due to leakage from other channels into the desired channel. The dominant source of system degradation is identified as being heavily dependent upon the number of wavelength channels employed. Increasing possible number of channels improves power penalty levels significantly whilst allowing a relaxation in the crosstalk requirements. Investigating the crosstalk due to an array waveguide is part of the process of improving the performance of WDM optical networks [3].

II. CROSSTALK ANALYSIS

Consider the N wavelength Optical Add-Drop Multiplexer (OADM) shown in fig.1 and fig.2, where m channels are added-dropped and n channels are passed (m + n = N). An advantage of this design is the use of a single AWG to perform both demultiplexing and multiplexing functions in the OADM [4]. At the output port there are three types of channels leakage: those that are entering the OADM from the network (N-1), those that are added-dropped (either of them is m), and those that are passing to the next node (n). Since the later is happening twice for non-ring networks, so, the related crosstalk term is doubled (2n). The analytical equations for WDM AWG router are as depicted in this section. In the expressions the input signal power to the AWG is defined as P_{in}, I_1

is the average current of the bit '1' state, I_o is the average current of bit '0' state, I_{XT1} is the bit '1' state for crosstalk current, I_{XT0} is the bit '0' state for crosstalk current.

A worst-case scenario is assumed with the interferers being in the same state of polarization as the signal. Given a high extinction ratio beating only occurs between '1's and not '0's. With the fixed decision-threshold setting, the decision-threshold is set at the ideal threshold $I_1/2$ (photocurrent when the signal bit is 'mark', and cannot be changed after the crosstalk is added). If the system performance can be guaranteed in the worst case, it will be better in other cases.

III. MATHEMATICAL MODEL

Considering the detector shot noise and receiver noise while ignoring the second amplifier after the AWG, the bit-error-rate (BER) of an AWG based OXC switch can be written as [5]:

$$BER_{WORST\_CASE} = \frac{1}{8}\left[\begin{array}{c} erfc\left(\frac{1}{\sqrt{2}}\frac{I_1 + I_{XT0} - I_D}{\sigma_{1\_0}}\right) + erfc\left(\frac{1}{\sqrt{2}}\frac{I_D - I_{XT0} - I_0}{\sigma_{0\_0}}\right) + \\ erfc\left(\frac{1}{\sqrt{2}}\frac{I_1 + I_{XT1} - I_D}{\sigma_{1\_1}}\right) + erfc\left(\frac{1}{\sqrt{2}}\frac{I_D - I_{XT1} - I_0}{\sigma_{0\_1}}\right) \end{array}\right] \quad (1)$$

The total noise current at the output of the receiver low-pass filter consists of detector shot noise the amplified spontaneous emission (ASE) noise, the thermal noise due to receiver pre-amplifier, crosstalk noise terms or the beat noise terms. Noise variances are as following [3]:

$$\sigma_{1\_0}^2 = \sigma_{th}^2 + \sigma_{shot}^2 + 2qR_d(P_{rec} + P_{ASE} + P_{XT0})B_e \quad (2)$$

$$\sigma_{0\_0}^2 = \sigma_{th}^2 + \sigma_{shot}^2 + 2qR_d(P_{ASE} + P_{XT0})B_e \quad (3)$$

$$\sigma_{1\_1}^2 = \sigma_{th}^2 + \sigma_{shot}^2 + 2qR_d(P_{rec} + P_{ASE} + P_{XT1})B_e \quad (4)$$

$$\sigma_{0\_1}^2 = \sigma_{th}^2 + \sigma_{shot}^2 + 2qR_d(P_{ASE} + P_{XT1})B_e \quad (5)$$

Where σ_{1_0}, σ_{0_0}, σ_{1_1}, σ_{0_1} represent the square roots of variances of noise when the signal bit '1' is interfered by crosstalk bit '0', the signal bit '0' is interfered by crosstalk bit '0', the signal bit '1' is interfered by crosstalk bit '1', the signal

bit '0' is interfered by crosstalk bit '1' respectively. The shot noise is to be added to the

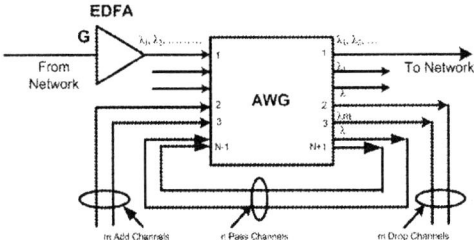

Fig. 1 An array wave-guide for routing in the WDM Network.

crosstalk for bit '1' only and not to bit '0', and is defined in (8). σ_{th}^2 is power of thermal noise defined in (6), q is electronic charge = 1.602×10^{-19}C, R_d is the receiver responsivity (at 1A/W

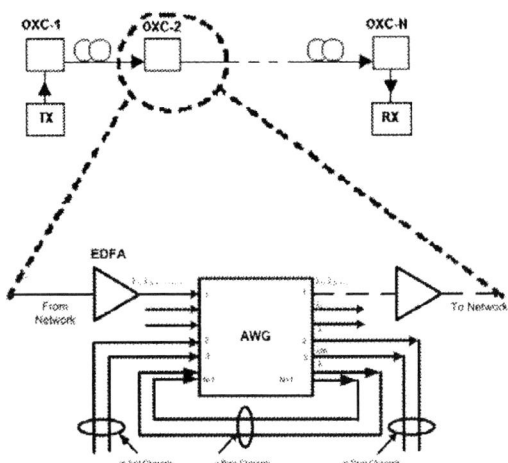

Fig. 2 A node of an optical cross-connect of type array waveguide.

for this paper), B_e is the electrical bandwidth of the receiver's filter (B_e = 10GHz in this paper. This equals to the bit rate used). P_{in} is assumed to be 0 dBm, which is 1 mW. P_{rec} is the power appears at the output of the receiver low pass filter, and it varies according to the losses, but, it is less than P_{in}. Decision threshold current is given by (11). The thermal noise is given by:

$$\sigma_{th}^2 = \frac{4KT}{R_L}B_e \quad (6)$$

K: Boltzman Constant = 1.380658 x 10^{-23}, T: Temperature in Kelvin (T=300 Kelvin in this paper), R_L: receiver front end load (R_L=50Ω in this paper). The amplified spontaneous emission power is defined by:

$$P_{ASE} = h\nu F_{ASE}(G-1)B_e \qquad (7)$$

h: Plank's constant = 6.6261 x 10^{-34}, v: center frequency at wavelength λ=1.55μm, F_{ASE}: spontaneous emission factor (inversion factor). (F_{ASE}=3 for this paper), G: Amplifier gain. (G=15 for this paper). The total shot noise of the receiver follows as:

$$\sigma_{shot}^2 = 2qR_d P_{rec} B_e \qquad (8)$$

The photocurrent for transmitted bit '1' is given by:

$$I_1 = 2R_d P_{rec} \qquad (9)$$

The photocurrent for transmitted bit '0' is given by:

$$I_0 = 0 \,(\text{an off state}) \qquad (10)$$

The threshold current I_D is given by:

$$I_D = \frac{\sigma_{0\_1} I_1 + \sigma_{1\_1} I_0}{\sigma_{0\_1} + \sigma_{1\_1}} \qquad (11)$$

If P_{XT1} represents the crosstalk power due to bit '1', and P_{XT0} is the crosstalk power due to bit '0', then, the corresponding crosstalk currents are given by:

$$I_{XT1} = R_d P_{XT1} \qquad (12)$$

$$I_{XT0} = R_d P_{XT0} \qquad (13)$$

L is the insertion loss for the AWG and a typical value is 4 dB and it is fixed and will not change with node number; S is the AWG crosstalk suppression with a used value of 22 dB. The amplified spontaneous emission ASE noise will change and accumulate from node to another node, but, the effects starts only at the end of first node.
Let P_{in1} be the input to the AWG when bit '1' is transmitted, P_{in0} be the input to the AWG when bit

'0' is transmitted, P_{o1} be the signal power at the photo-diode when all the channels are transmitting bit '1' at the same time (worst case), P_{o0} be the signal power at the photo-diode when transmitting bit '0' (off state), P_A be the power of the added channels.

The detected power at the photodiode with the availability of crosstalk is found by further extending the formula of [3] and [6]:

$$P_{out} = P_1 + P_2 + P_3 - 2\sqrt{P_1 P_2} - 2\sqrt{P_1 P_3} - 2\sqrt{P_2 P_3} \qquad (14)$$

Where P_{out} is the output power at the photo-detector, P_1 is the output power for the channels under study that consists of desired signal and the crosstalk signal due to interaction between the desired signal with the same wavelength in the added channel; P_2 is the crosstalk signal of added channels that have different wavelength from the desired signal; P_3 is the crosstalk signal of the passed channels to the network through AWG; $\sqrt{P_1 P_2}, \sqrt{P_1 P_3}, \sqrt{P_2 P_3}$ are the beat terms to count for the interaction between every two of the above three signals. The crosstalk and signal terms at the output are:

$$P_1 + P_2 + P_3 = P_{out1\_XT} \qquad (15)$$

Assuming all wavelength channels carry bit '1', the output power terms are expressed as given by:

$$P_{out1\_XT} = R_d^2 \left(LP_{in} + 2L\sqrt{SP_{in}P_A} + LSP_A \right)^2 +$$
$$R_d^2 \left(LSP_{in} + 2LS\sqrt{P_{in}P_A} + LSP_A \right)^2 + \qquad (16)$$
$$+ R_d^2 \left(LSP_{in} + 2LSP_{in}\sqrt{LS} + L^2 S^2 P_{in} \right)^2$$

Equation (16) is not counting for the numbers of signals for added/dropped and passed channels. Then when considering (m-1) added channels, and (n) passed channels with double for non-ring network, the output power summed with crosstalk signals is expanded to:

$$P_{out1\_All\_XT} = R_d^2 \left(LP_{in} + 2L\sqrt{SP_{in}P_A} + LSP_A \right)^2 +$$
$$R_d^2 \left(LSP_{in} + 2LS\sqrt{P_{in}P_A} + LSP_A \right)^2 (m-1) \qquad (17)$$
$$+ R_d^2 \left(LSP_{in} + 2LSP_{in}\sqrt{LS} + L^2 S^2 P_{in} \right)^2 2n$$

The equation (17) is still not counting for beat terms between every two out of the three terms of (17), therefore, with the beat terms of crosstalk, the new output power is obtained:

$$
\begin{aligned}
P_{out1} = & R_d^2 \left(LP_{in} + 2L\sqrt{SP_{in}P_A} + LSP_A \right)^2 + \\
& R_d^2 \left(LSP_{in} + 2LS\sqrt{P_{in}P_A} + LSP_A \right)^2 (m-1) \\
& + R_d^2 \left(LSP_{in} + 2LSP_{in}\sqrt{LS} + L^2S^2P_{in} \right)^2 2n. \\
& - 2R_d^2 (LP_{in} + 2L\sqrt{SP_{in}P_A} + LSP_A)(LSP_{in} + \\
& 2LS\sqrt{P_{in}P_A} + LSP_A)\sqrt{m-1} \\
& - 2R_d^2 (LP_{in} + 2L\sqrt{SP_{in}P_A} + LSP_A) \times \\
& (LSP_{in} + 2LSP_{in}\sqrt{LS} + L^2S^2P_{in})\sqrt{2n} \\
& - 2R_d^2 (LSP_{in} + 2LS\sqrt{P_{in}P_A} + LSP_A) \times \\
& (LSP_{in} + 2LSP_{in}\sqrt{LS} + L^2S^2P_{in})\sqrt{2n(m-1)}
\end{aligned}
\tag{18}
$$

Where P_{in} is the input signal power, P_A is the added channel signal power and both are assumed to be 1 mW for simplicity; L and S are transmission factors (<1, and they are defined early). Since this AWG is connected to a network that is not a ring network, so, the pass channels powers are multiplied by 2 because the pass channels signals are flowing in the AWG for two times with each time there is a crosstalk penalty. Let P_{out}^{XTF} is the output of wavelength channel under study λ_1 when the AWG is carrying only wavelength channel λ_1 (such as when there is no crosstalk or crosstalk free), and P_{out1}^{XTA} is the real output of the same channel, i.e. when there is crosstalk (so, $P_{out1}^{XTA} = P_{out1}$).

Equation (18) is valid under assumption that all wavelength channels including wavelength channel λ_1 carries bit '1'. Since wavelength channel λ_1 may carry bit '1' or bit '0' at any instant of time, (18) has to be modified. If wavelength channel λ_1 carries bit '0', (18) reduces to:

$$
P_{out0} = R_d^2 (LSP_A)^2 (1 - 2\sqrt{m-1})
\tag{19}
$$

$$
P_{out0}^{XTF} = 0
\tag{20}
$$

Where P_{out0} is the real output for bit '0' (with crosstalk is available), and P_{out0}^{XTF} is the ideal output

for bit '0' if there is no crosstalk. The crosstalk model for the AWG is used to derive bit error rate (BER) that is expressed in (1) for the transmission link considering the detector shot noise and receiver noise. The crosstalk powers for bit '1' and bit '0' are expressed as [3]:

$$
P_{XT1} = P_{out}^{XTF} - P_{out1}
\tag{21}
$$

$$
P_{XT0} = -P_{out0}
\tag{22}
$$

IV. RESULTS

The matltab program calculations produce the graphic relation that appears in the Fig 3 below, which represents the bit error rate (BER) versus received power P_{rec} for various numbers of nodes. The diagram shows that for any number of nodes between 10 to 2000, the BER is too high for lower values of received power (all the values below -9 dBm are giving high BER) for any number of nodes, then, with increasing the received power, the curve improves but faster for less number of nodes.

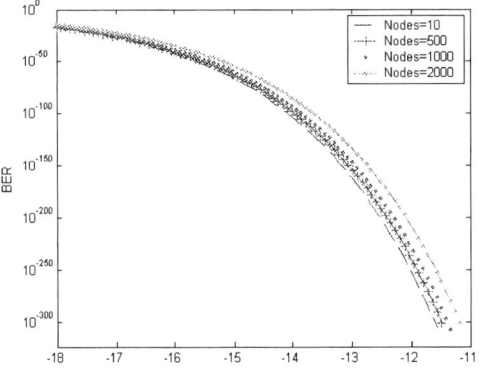

Fig. 3 BER versus received power in presence of crosstalk for varying nodes.

V. CONCLUSION

The crosstalk due to the optical cross-connect (AWG) in the WDM optical networks increases imposing power penalty as the number of nodes increases which means more users and more load in the network will reason larger power penalties. For lower received power the BER is high, and so, the network gets high values of BER in both of the cases: large number of nodes and low received power.

ACKNOWLEDGMENT

The authors would like to thank Dr. S. P. Majumder from Department of Electrical and Electronic Engineering, Bangladesh University of Engineering and Technology (BUET), Dhaka-1000, Bangladesh, for the support in getting good references, and Mr. Hairul Azahar from Faculty of Engineering, Multimedia University, Cyberjaya, Selangor, Malaysia, for cooperating in terms of Matlab works.

REFERENCES

[1]   J. L. Miller., E. Friedman., "Optical Communications Rules of Thumb", *McGraw-Hill,* (2003).

[2]   Rajiv Ramaswami., Kumar N. Sivarajan., "Optical Networks: A Practical Perspective", *2ⁿᵈ edition, Academic Press,* (2002).

[3]   S. P. Majumder, and S. Dey, "BER Performance Degradation Due to Component Crosstalk of an Arrayed Waveguide Grating and FBG-OC Based WDM Cross-Connect", *Proc. Annual IEEE IINDICON Conference (INDICON 2005)*, pp. 97-100, (2005).

[4]   Gemelos, S.M.; Wonglumsom, D.; Kazovsky, L.G., "Impact of Crosstalk in an Arrayed-Waveguide Router on an Optical Add-Drop Multiplexer", *IEEE Photon. Technol. Lett.*, Vol. 11, No 3, , pp. 349-351, (1999).

[5]   G. P. Agrawal, Fiber Optic Communication Systems, *3ʳᵈ edition, John Wiley & Sons,* (2002).

[6]   T. Gyselings, G. Morthier, and R. Baets, "Crosstalk Analysis of Multiwavelength Optical Cross Connects", *J. Lightwave Technol.,* Vol. 17, No. 8, pp.1273-1283, (1999).

Avalanche Multiplication and Excess Noise Factor of Heterojunction Avalanche Photodiodes

A. H. You[1,*], L. C. Low[1], P. L. Cheang[2]

[1]Faculty of Engineering and Technology, Multimedia University, Jln Ayer Keroh Lama, 75450 Melaka, MALAYSIA

[2]Centre for Foundation Studies and Extension Education, Multimedia University, Jln Ayer Keroh Lama, 75450 Melaka, MALAYSIA

*Email: ahyou@mmu.edu.my

Abstract A Monte Carlo (MC) model to compute the statistics of avalanche multiplication and excess noise factor in heterojunction avalanche photodiode (HAPD) is presented. The proposed model is able to simulate the multiplication gain and excess noise factor incorporating the dead-space effect, band-edge discontinuity and hole to electron ionization ratio in HAPDs. The dead-space effect is included in our model, which has been shown to play an important role in reducing noise in homojunction APDs. It is shown that the dead-space effect also reduces the avalanche noise in heterojunction devices. We demonstrate that the dead-space effect and feedback impact ionization are the dominant effects to improve the excess noise factor in HAPDs.

KEYWORDS: heterojunction avalanche photodiode, impact ionization, multiplication gain, excess noise factor, dead-space effect

I. INTRODUCTION

AVALANCHE photodiodes (APDs) are famously known as detectors in long-haul fiber optic systems due to their advantage of high internal gain generated by avalanche multiplication. However, multiplication gain in APDs depends on the noise produced by the nature random process of impact ionization. The conventional McIntyre theory [1] predicts that a low excess noise can only be achieved in semiconductor materials with a large difference in hole and electron ionization coefficients and that carrier types with a large ionization coefficient should be used to initiate multiplication. Unfortunately, most of the APDs that operate at 1.3 and 1.55 μm for fiber optic communication systems utilize those materials

with nearly equal α and β that cause large excess noise. In order to overcome this problem, the band-edge discontinuity at the heterojunctions of multiple quantum well structures (MQWs) [2] has been proposed to solve the excess noise by alter the ionization coefficients in order to obtain low noise in APDs. The enhancement of ionization coefficient occurs in MQW structure when the carriers fall into the narrow gap material to gain the energy from band-edge discontinuity. Consequently, lower excess noise is achieved with decreasing of hole-to-electron ionization coefficient ratio.

The heterostructure APD was first proposed by Chin *et al* [2] to minimize the excess noise factor. An electron or hole may gain extra energy when cross over a band-edge discontinuity which would enhance the impact ionization. Proper selection of the heterojunction material layer will determine the degree of enhancement in ionization coefficient. The enhancement is supported by the later measurements and numerical work on multilayer structure devices by researchers [3,4]. However, Chia *et al* [5] found that no enhancement of ionization coefficients was observed in $Al_xGa_{1-x}As/GaAs$ MQW APDs. Their measurements suggested that the excess energy gained by the carriers crossing the heterojunction interface is negligible. Their statement has been supported by Wang *et al* [6,7] with the measurements on avalanche noise in submicron III-V heterojunction. Later, it has been proved by Kwon *et al* [8] that low noise could be achieved in heterojunction APDs (HAPDs) when the finite initial energy of carriers entering the multiplication region.

Recently, Groves *et al* [9] developed a simple Monte Carlo (MC) model to simulate the avalanche process in heterojunction by using the idea of localization of ionization process [10] to reduce the excess noise. The results show that there is considerable disagreement in the

0-7803-9730-4/06/$25.00 ©2006 IEEE

literature of ionization coefficient enhancement in heterojunction device. Hence, further investigations are required to resolve the discrepancies found in the HAPDs. It becomes complex with nonlocal effects which seems to become very important in determining multiplication in thin structures recently. Okuto *et al* [11] pointed out that ionization coefficients should take into account the dead space, defined as the minimum distance that a carrier must travel after each ionizing event to gain sufficient energy for the subsequent impact ionization.

We develop the MC model to study the avalanche multiplication and noise of HAPDs, particularly in the thin multiplication devices. We use our model to study the band-edge discontinuity and dead-space effect in HAPD for all device lengths. A significant reduction in noise is expected as the device shrinks. We demonstrate the dead-space effect and the effects of feedback impact ionization in HAPDs.

II. MONTE CARLO MODEL

A sophisticated MC model is proposed to study the avalanche multiplication and excess noise factor in HAPD. Double-carrier multiplication in a multiplication region extending from $x = 0$ to w with the present of electric field is divided into two layers as shown in Fig. 1. The multiplication region, w is considered as a normalized multiplication region in this general model. The first and second layers denoted the multiplication region extending from $x = 0$ to 1 and $x = 1$ to 2, respectively. Since the heterojunction consists of two different semiconductor materials, α_1, β_1, k_1 and α_2, β_2, k_2 are denoted as the electron- and hole-impact ionization, and hole to electron ionization ratio in the first- and second-layer. These six parameters depend strongly on the semiconductor material properties that are useful in designing the HAPDs. It is flexible to engineering the heterojunction to achieve the low noise and high gain APD by varying the ionization ratio in this model. The band-edge discontinuity is placed between the first and second region that is located at junction, $x = 1$. Hence, p_e and p_h are the probabilities for electron and hole to cross over the heterojunction. These probabilities depend on the electron thermionic emission and hole trapping at the interface. It is possible also to include the tunneling and reflection of the carriers at the interface.

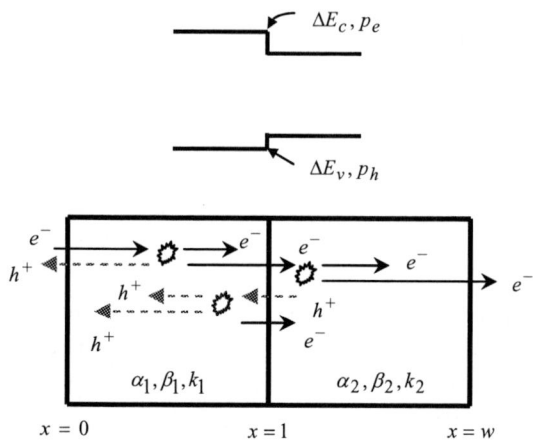

Fig. 1 The impact ionization across the HAPD for both electron and hole multiplication. Band diagram shows the band-edge discontinuity of heterojunction.

An electron injected at $x = 0$ travels in the x-direction and cross the junction along the multiplication region. Impact ionization occurs after the electron travels a random ionization distance, l_e, which includes the dead space. An electron hole pair is generated upon ionization, so that two electrons and a hole replace the original electron. The two electrons behave in a statistically identical but independent manner. On the other hand, the hole travels in the $-x$-direction and ionizes after traveling a random ionization distance, l_h, results in two holes and an electron. The electrons and holes repeat the process when they travel through the multiplication region in the first layer. The multiplication continues until all possible carriers have left the multiplication region. Those electrons travel in the first layer might have a probability to enter the second layer, which depends on the p_e at the junction.

An electron enters the second layer undergo the multiplication with α_2, β_2 and k_2. The multiplication manner is different from that of the first layer since the material properties are different in these two layers. The electrons and holes repeat the ionization process when they travel through the multiplication region in the second layer. There is a possibility for the generated holes to feedback into the first layer at the probability of crossing the heterojunction,

p_h at the junction. In the subsequent process, the electrons and holes bounce in between the first and second layers until all the carriers have left the device.

The probability for an electron or hole to undergo impact ionization after traveling a distance, x, in uniform electric field, E, includes the dead space, d. The dead space in heterojunction includes the band-edge discontinuity is written as $d = d^* - \Delta E / eE$, where $d^* = E_{th} / eE$ is the dead space in homojunction device, ΔE is the energy difference at interface and E_{th} is the threshold energy. In our statistics of impact ionization, a random electron or hole ionization path length, l, can be generated by substituting a uniformly distributed random number, r, between 0 and 1. Each trial in the simulation terminates when all carriers have left the multiplication region. Each injected carrier gives rise to a multiplication, M, in each layer which is a random variable, owing to the stochastic nature of impact ionization. M for each trial is recorded to build reliable statistics by performing an ensemble average over many trials. Spatially determined statistics such as $< M >$ and F can be calculated using

$$< M > = \frac{1}{N} \sum_{i=1}^{N} M_i \qquad (1)$$

and

$$F = \frac{\sum_{i=1}^{N} M_i^2}{N < M >^2}, \qquad (2)$$

where M_i is the multiplication results from trial i and N is the total number of trials in the simulation. The multiplication is iterated to and fro between the first and second layers to include all the secondary ionization events. This usually requires 10^6 trials to obtain the successive values of $<M>$ and F in our simulation. The total multiplication and excess noise factor for the heterojunction device is the sum of these values contributed from the first and second layers.

III. RESULTS AND DISCUSSION

There are ten parameters used to adjust the multiplication and noise in our Monte Carlo model. They produce a more realistic phenomenon in HAPD. Figure 2 shows the mean multiplication and excess noise factor for different hole probabilities crossing the heterojunction. The mean multiplication and excess noise factor increases as the hole probability increases. Hole probability, p_h is determined by the valence band discontinuity, ΔE_v and controls the degree of feedback ionization in the heterojunction.

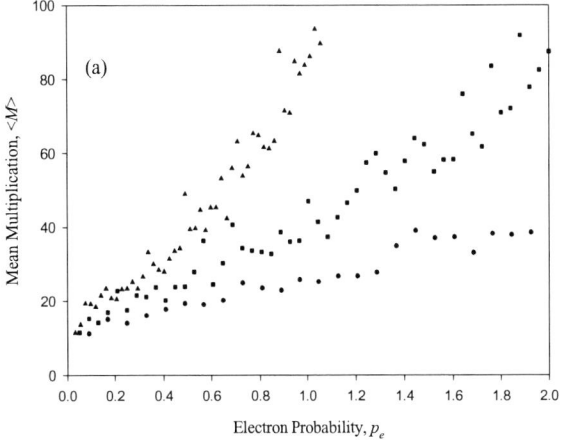

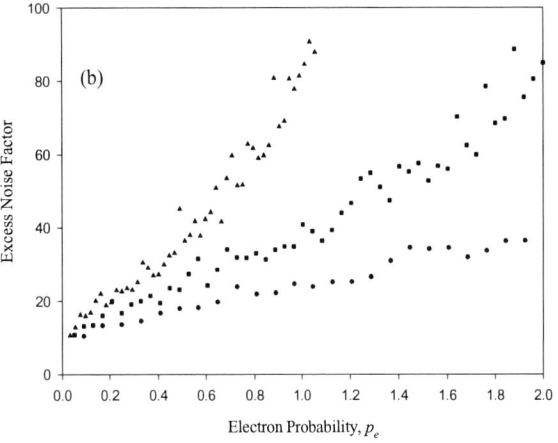

Fig. 2 (a) Mean multiplication and (b) excess noise factor versus electron probability, p_e with $k_1 = k_2 = 0.5$, $\alpha_1 = \alpha_2 = 0.5$, $d = 0$ and, for the hole probabilities, $p_h = 0.5$ (circle), 1.0 (square) and 2.0 (triangle).

The probabilities of electron and hole at the interface are used to control the number of carriers to cross the heterojunction, which depends strongly on the band-edge discontinuity. However, the excess noise factor seems to increase simultaneously with the increase of multiplication in the device. For a fixed electron probability, p_e, similar trends for mean multiplication and excess noise factor are shown in the figures. There is no obvious variation of noise observed in Fig. 2. It shows that the probabilities of electron and hole do not play a major role in reducing the noise at interface in HAPDs. Instead, it only controls the number of ionization events happen in the device.

In fact, a lot of works [12,13] has been conducted on thin homojunction APDs to produce a low excess noise factor by considering the dead-space effect. Our MC model incorporated the dead-space effect for both electron and hole is used to study the excess noise factor in HAPD. Since the dead space in heterojunction is included with the band-edge discontinuity, the excess noise factor is expected slightly higher than that of homojunction device. Significant reduction of excess noise factor occurs when the dead-space increased up to $0.3w$ in the thin heterojunction device as shown in Fig. 3. When the device length is thinner, a large electric field results and increases the dead space to multiplication length ratio, d/w in the thin device. The large dead space in thin multiplication gives more deterministic impact ionization that causes the reduction of noise in the thin heterojunction. The impact ionization seems to be localized also in the thin heterojunction device.

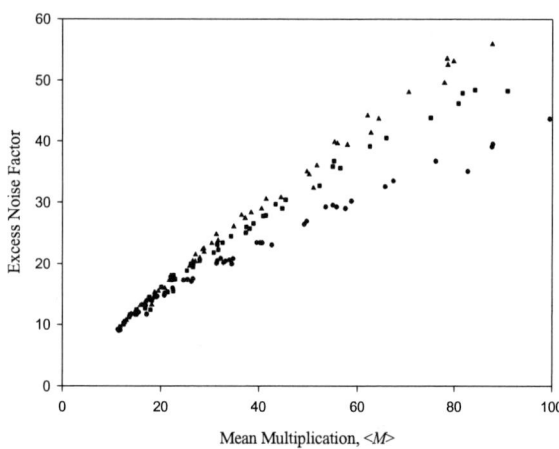

Fig. 3 Excess noise factor versus mean multiplication with $k_1 = 5.0$, $k_2 = 0.1$, $\alpha_2 = 0.8$, $p_e = 1.5$, $p_h = 0.5$ and for $d = 0.3w$ (circle), $0.1w$ (square) and 0 (triangle).

Figure 4 shows the excess noise factor depends on the hole to electron ionization ratio, k. Less noise is observed for the case $k_1 < k_2$ as compare to the case $k_2 < k_1$. It is due to the small hole impact ionization in the first layer that may give a less excess noise factor for the case $k_1 < k_2$. Therefore, it is suggested that the first layer or the initiated multiplication edge layer should have a lower k ratio relative to the next

adjacent layers. The structure is purposely designed to limit the feedback ionization, which is dominated in the heterojunction device. It is also shown that the k_1 ratio in the first layer controls the overall performance of the HAPDs. As shown from our simulation results in Fig. 3, when $k_1 = 5$ is chosen in the first layer, the noise in the heterojunction device is reduced significantly although it is for the case $k_1 > k_2$.

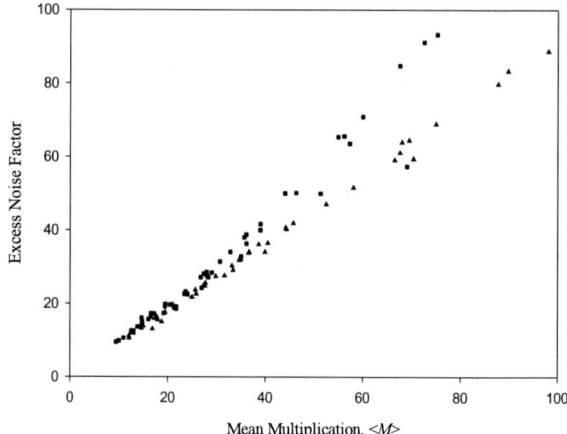

Fig. 4 Excess noise factor versus mean multiplication for $\alpha_2 = 0.5$, $p_e = p_h = 1.0$, $d = 0$, (k_1=1.0 and k_2=0.1) (square) and (k_1=0.1 and k_2=1.0) (triangle).

IV. CONCLUSION

We developed an MC model to simulate multiplication gain and excess noise factor in HAPD. The proposed model involves ten adjustable parameters to design the engineering structure of HAPDs. The multiplication gain and excess noise factor of electron-initiated multiplication depends on various parameters are clearly demonstrated in our results. Our model is able to predict the multiplication gain and excess noise factor in thinner HAPDs. It is shown that the dead-space effect and the effects of feedback impact ionization are predominant in heterojunction devices. The HAPDs performance is found to be depended on the material properties in the first layer.

ACKNOWLEDGEMENT

This work is funded by ASM, SAGA Fund 3rd Round (Physics), P91c and partly supported by MMU internal funding, PR/2006/0625.

REFERENCES

[1] R. J. McIntyre, 'Multiplication Noise in Uniform Avalanche Photodiodes,' IEEE Trans. Electron. Device., 1966, Vol. ED-13, No. 1, pp. 164-168.

[2] R. Chin, N. Holonyak, G. E. Stillman, J. Y. Tang and K. Hess, 'Impact Ionization in Multilayered Heterojunction Structures,' Electron. Lett., 1980, Vol. 16, pp. 467-469.

[3] F. Capasson, W. T. Tsang, A. L. Hutchinson and G. F. Williams, 'Enhancement of Electron Ionization in a Superlattice: A New Avalanche Photodiode with a Large Ionization Rate Ratio,' Appl. Phys. Lett., 1982, Vol. 40, pp. 38-40.

[4] T. Kagawa, H. Iwamura and O. Mikami, 'Dependence of the GaAs/AlGaAs Superlattice Ionization Rate on Al Content,' Appl. Phys. Lett., 1989, Vol. 54, pp. 33-35.

[5] C. K. Chia, B. K. Ng, J. P. R. David, G. J. Rees, R. C. Tozer, M. Hopkinson, R. J. Airey and P. N. Robson, 'Multiplication and Excess Noise in Multilayer Avalanche Photodiodes,' J. Appl. Phys., 2003, Vol. 94, pp. 2631-2637.

[6] S. Wang, J. B. Hurst, F. Ma, R. Sidhu, X. Sun, X. G. Zheng, A. L. Holmes, A. Huntington, L. A. Coldren and J. C. Campbell, 'Low-noise Impact Ionization Engineered Avalanche Photodiodes Grown on InP Substrates,' IEEE Photon. Technol. Lett., 2002, Vol. 14, pp. 1722-1724.

[7] S. Wang, F. Ma, X. Li, R. Sidhu, X. G. Zheng, X. Sun, A. L. Holmes and J. C. Campbell, 'Ultra-low Noise Avalanche Photodiodes with a 'centered-well' multiplication region,' IEEE J. Quantum Electron., 2003, Vol. 39, No. 2, pp. 375-378.

[8] O. H. Kwon, M. M Hayat, S. Wang, J. C. Campbell, A. Holmes, Y. Pan, B. E. A. Saleh and M. C. Teich, 'Optimal Excess Noise Reduction in Thin Heterojunction Al$_{0.6}$Ga$_{0.4}$As-GaAs Avalanche Photodiodes,' IEEE J. Quantum Electron., 2003, Vol. 39, No. 10, pp. 1287-1294.

[9] C. Groves, J. P. R. David, G. J. Rees and D. S. Ong, 'Modeling of Avalanche Multiplication and Noise in Heterojunction Avalanche Photodiodes,' J. Appl. Phys., 2004, Vol. 95, pp. 6245-6251.

[10] D. C. Herbert, C. J. Williams and M. Jaros, 'Impact Ionization and Noise in SiGe Multiquantum Well Structures,' Electron. Lett., 1996, Vol. 32, pp. 1616-1618.

[11] Y. Okuto and C. R. Crowell, 'Ionization Coefficients in Semiconductors: A Nonlocalized Property,' Phys. Rev. B, 1974, Vol. 10, No. 10, pp. 4284-4296.

[12] P. Yuan, K. A. Anselm, C. Hu, H. Nie, C. Lenox, A. L. Holmes, B. G. Streetman, J. C. Campbell, R. J. McIntyre, 'A New Look at Impact Ionization – Part II: Gain and Noise in Short Avalanche Photodiodes,' IEEE Trans. Electron Devices, 1999, Vol. 46, pp. 1632-1639.

[13] A. H. You and D. S. Ong, 'Avalanche Multiplication and Noise Characteristics of Thin InP p$^+$-i-n$^+$ Diodes', Japan. J. Appl. Phys., 2004, Vol. 43 (11A), pp. 7399-7404.

Faults Detection Approach for Self-Testable RF MEMS

Syed Zahidul Islam, Wallace Wong, Su Hieng Tiong, and *Mohd Alauddin Mohd Ali
Swinburne University of Technology (Sarawak Campus)
Jalan Simpang Tiga, Kuching Sarawak, Malaysia
*Universiti Kebangsaan Malaysia, UKM, Bangi, Malaysia
Email: zsyed@swinburne.edu.my, wwong@swinburne.edu.my, hsu@swinburne.edu.my
*E-mail: mama@vlsi.eng.ukm.my

Abstract **Efficient built-in or external test strategies are becoming essential in Micro-Electromechanical Systems (MEMS), especially for high reliability and safety critical applications. This paper describes self-testable and self-reparable RF MEMS fault testing approach. Simulation results show the effects of switches faults in MEMS.**

I. INTRODUCTION

MEMS are small electromechanical devices such as switches, motors and transducers with features from a few microns to a few millimeters in size [1]. The low cost, small device form factor and integrability promised by MEMS have driven its rapid development and widening applications in RF communication, optic and biotechnology [2,3]. But many difficulties preventing large volume of MEMS production are strongly related to test, including manufacturing costs, quality and reliability verification. Manufacturable MEMS product must be testable, in order to ensure and prove the required quality and reliability levels. MEMS device must be tested as an entity after manufacturing, with all modules in interaction and after packaging. The great diversity of MEMS structures and their working principles, various defect sources, multiple field coupling, as well as the essential analog features, make MEMS testing very challenging.

Many RF MEMS capacitive switches exhibit low loss, low power consumption, and measurable intermodulation distortion compared to conventional semiconductor based switching devices [4,5]. Figure 1 shows the layout view of a single ohmic shunt RF-MEMS switch and the bottom is cross-section of the electrodes configuration below the suspended gold plate. MEMS switches reported so far are based on silicon, GaAs or similar semiconductor materials fabricated by micro-machining techniques. To achieve fault simulation and testability analysis in these devices, it is necessary to model both the mechatronic and electrical elements within the same simulation environment to ensure the efficient injection and analysis of faults.

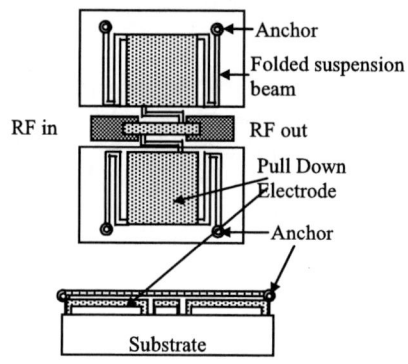

Fig 1 Single ohmic shunt RF-MEMS switch

II. RELATED WORKS

The issue of MEMS reliability and testing is addressed initially in 1997 where attempts were made to map inertial capacitive MEMS structural defects and the resulting behaviors with contaminations during the fabrication process [6]. In their work, a combination of fabrication process simulator and finite element analysis tool was used to study the impact of structural defects on the behavior of the MEMS devices. It was concluded from the study that some defects could cause catastrophic failures whilst other defects have negligible effects at first, but may have some serious implications in the long run. Therefore, whilst the current industrial practice of characterizing and checking the functionality of MEMS devices may be sufficient to screen defective devices, there are certain failures that are transparent to those tests. This is unacceptable especially in safety-critical applications where MEMS devices must be reliable all the time, or show some indication before they are completely unusable. *Blanton* et al. [6] also suggested that the MEMS defects could be categorized according to the faulty behavior classes e.g. surface micromachined

0-7803-9730-4/06/$25.00 ©2006 IEEE

sensors. Then, with the available fault models, test methodologies could be developed to ensure the reliability of the MEMS devices. Subsequently, their work has lead to the development of Contamination and Reliability Analysis of Micro ElectroMechanical Layout (CARAMEL) in 1999, a CAD tool for analyzing the impact of contamination particles on MEMS systems behavior [7]. Recently, the team is developing more comprehensive MEMS fault models based on analysis of actual physical failure mechanisms with the eventual aim to enable accurate high-level abstraction for CAD modeling and simulation [8]. Their study is limited to MEMS capacitive accelerometer and resonators which exhibit most of the basic elements of capacitive MEMS.

Courtois et al. attempted to model faulty MEMS behavior using analog Hard Description Language (A-HDL) [9, 10]. Instead of modeling the actual physical defects, "mutants" and "saboteurs" were used to represent defects and therefore the approach lacks realism. "Mutant" represents parametric deviations while "saboteurs" represents missing or extra components. The approach is easily integrated into existing VLSI simulation tools to perform overall electronics -electromechanical interacting system.

With more understanding of the possible MEMS defects and their effect on the system response, work on a Built-In-Self-Test (BIST) MEMS has started recently. *Blanton* et. al. has developed an algorithm to detect faulty MEMS microaccelerometer on a modularized design, where the sensing elements of the microaccelerometer are divided into standalone segments [11]. By applying certain test pattern (a combination of ON and OFF modules), the modules that are faulty could be detected. In [12], *Xiong* et. al. proposed a new layout for the microaccelerometer to cover wider range of symmetry and asymmetry defects. However, no experiment results are available at present. *Xiong* et. al. also mooted the idea of Built-In-Self-Repair (BISR) technique that enables the microaccelerometer to switch to redundant modules once defects are detected.

III. DEFECTS AND FAILURE MECHANISMS

MEMS are adversely influence by the three sources of errors which are shown in Figure 2 using Venn diagram [13]. The design area represents mistakes in translating specifications into a working design. The modeling area represents modeling deficiencies which is incorrect operation due to imprecise or inadequate models of MEMS and their environment of operation. The third and final area represents the negative impact that manufacturing errors have on sensor operation. Errors in this area include statistical variations in device geometries, integrity of material properties and point defects. The shaded error area is due to the lack of MEMS-specific CAD tools, insufficient MEMS manufacturing experience and lack of precise models.

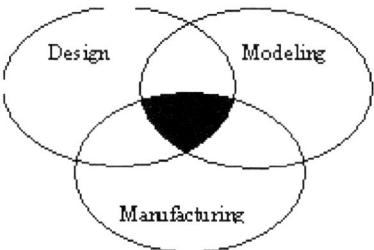

Fig. 2 Venn diagram showing MEMS errors [13]

Study on chip field rejects and the actual technological process steps can be used to build a list of potential failure mechanisms [14] and defects which can occur in these devices. Generally, it is possible to consider:

- Failures due to an inadequate CMOS process: hillocks, voids, alkaliion migration, dielectric breakdown, electro-migration, etching steps etc.
- Failures due to an inadequate silicon bulk post-process etching: incomplete or excessive etching, unwanted etching of metal parts, deposition of undesired residuals.
- Failures due to incorrect operation: acceleration, shocks, vibration, temperature, harsh environment, dust, particles, over-range voltages.

MEMS devices fabricated today are tested functionally. Being essentially analog devices, their test is being approached by means of techniques from the analog test rather than digital test domain. In general, a MEMS device must be tested as an entity after manufacturing with all modules in interaction and after packaging. For instance, additional stresses due to packaging can significantly impact device behavior. Since encapsulation of MEMS is often a very critical issue, sometimes accounting for more than 80% of the overall cost, testing a device should also be considered before packaging, screening out defective devices as early as possible.

Functional testing of large-volume chips embedding MEMS risks being extremely expensive. As in microelectronics, the search for generating cost-effective structured tests for large-volume chips targeting realistic faults and defects is important. Methodological studies for the derivation of the most common defects in surface micromachining fabrication processes are shown in Kolpekwar and Blanton [15]. Technology process simulation with injected realistic contaminations is performed, showing that this can result in a wide variety of defective structures.

Both stiction and etch variation defects are present on the surface of the central mass for bulk-micromachined accelerometer and microresonator which is very vulnerable to stiction defect due to the small capacitance gap there [11]. The fixed capacitance plates of a capacitive MEMS device may also contain some defects. Depending on their locations and extents, such defects may be fatal and lead to device failure. Thus, an effective test must be able to detect all defects in the fixed capacitance plates.

IV FAULT MODELING APPROACH

A. Fault Model of Configurable RF MEMS Filter

The proposed work intends to address the reliability issue of RF MEMS using reconfigurable RF MEMS filter as research subject. Previously, many authors such as *Lee* et al. [16] have only investigated the operational reliability of the capacitive switch used in RF MEMS in different environment through simulation and experiments, but no proper mapping has been made between possible physical defects with the device's response. The faulty behavior of RF MEMS as a result of defects may be very different from those of capacitive MEMS discussed before as RF signal is going through the device and not mechanical displacement typical of capacitive MEMS. Figure 3 shows an example of a proposed RF MEMS filter structure to be studied. The filter consists of half wavelength resonators and is to be constructed using microstrip technology. Different conductor length is added to each end of the resonator using RF MEMS switch in order to change the resonant frequency. Figure 4 shows simulated resonant frequency of a single resonator, with and without turning ON the switches. The resonator is simulated on a quartz

substrate and the line width is 200 μm. Note that the amount of frequency tuning depends on the order of the switches are turned ON.

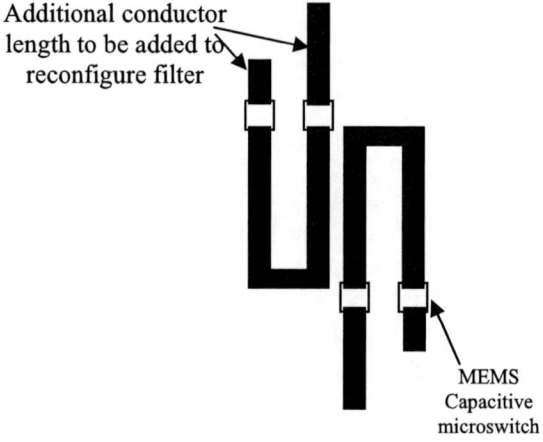

Fig 3 Top view of RF MEMS Configurable Filter

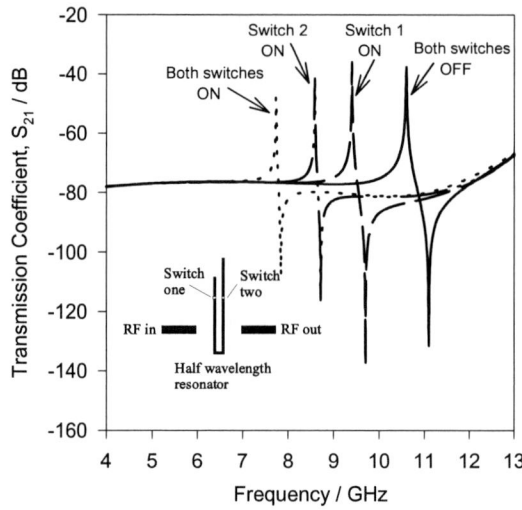

Fig 4. Simulated resonant frequency to investigate the amount of frequency shift due to adding extra conductor length to the end of the resonator.

Figure 5 shows simulated results for possible passband frequency tuning for a 3-pole filter (shown in inset) achieved using RF MEMS switches. The original filter bandwidth is 2 % centered at about 10.6 GHz. To tune the passband frequency, at least three switches are required in this design. The amount of tuning depends on the consecutive switches turned ON. If any of the switches is faulty, the frequency tuning of the resonator will be affected and hence results in bad filter response. Figure 6 shows as an example of the frequency response as a result from a faulty switch.

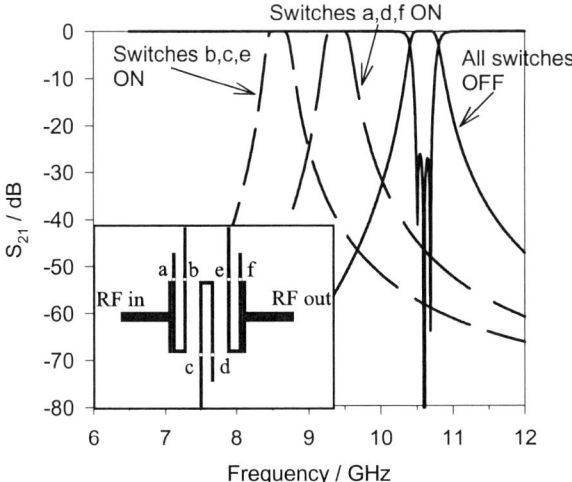

Fig 5. Simulated results for possible passband frequency tuning for a 3-pole filter.

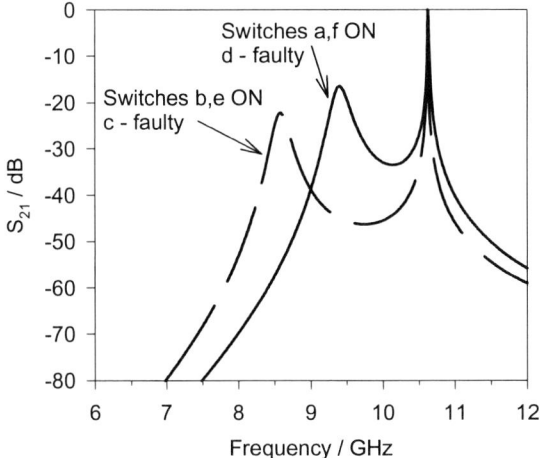

Fig 6 The frequency response as a result from a faulty switch.

B. Designed for testability – Non invasive Testing

ONE of the main causes of failure of RF MEMS device such as that shown in Fig 3 is due to microswitch malfunction; often due to stiction [17]. Usually, the switch is tested by checking the continuity between the two terminals of the switch. However, for RF MEMS, this means that there must be a physical connection between the conductor and testing terminals. Unlike other capacitive MEMS, this is not acceptable as any foreign structures connected to the conductor will change the conductor's frequency response. Thus it is proposed that a non-contacting capacitive sensor as that shown in Figure 6 be used to inspect the switch. Using this method, the capacitance between a non-contacting and sufficiently far pair of conductors will be measured to check the operation of the switch. If the switch is turned ON, the equivalent

capacitance of the structure is that of Figure 7a and when it is OFF, Figure 7b results.

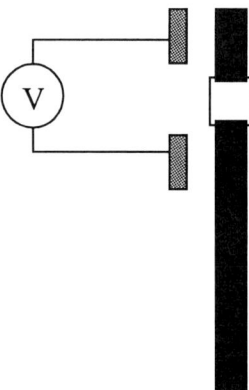

Fig 6 Inspecting switch

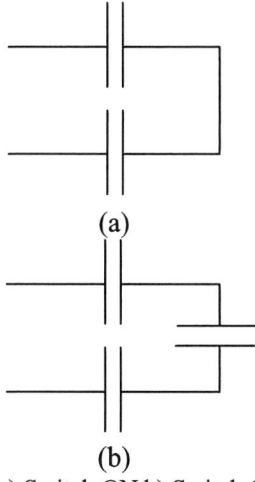

(a)

(b)

Fig 7. a) Switch ON b) Switch OFF

With the availability of the faulty behavior – defect mapping and microswitch inspector, the design of the RF MEMS filter can be further improved to include Design-for-Testability (DFT) features. This is the stepping stone for development of Built-In-Self-Test (BIST) and Built-In-Self-Repair (BISR) for MEMS which enhances its reliability and therefore the appeal of MEMS for critical applications.

C. Test Pattern for High Number of MicroSwitch

The structure such as that shown in Figure 2 could be extended to include more conductor elements and thus increases the number of microswitches involved. Instead of analyzing every possible combinations of individual switch, a test pattern can be generated to provide quick testing with 100% coverage. The techniques used are derived from VLSI testing methodologies as the microswitch failure resembles the stuck-at error common in VLSI.

343

IV. CONCLUSION

Several fault modeling of self-testable and self-repairable capacitive MEMS are investigated. Based on reliability, proper mapping of physical defects is figured out in RF MEMS testing research environment. Simulation results on original and faulty MEMS switching proved the significant impact on variation frequency response. To overcome this effect, non-contacting capacitive sensor can be one of the solutions to inspect faulty switches.

REFERENCES

[1] Marc J. Madou, "Fundamentals of Microfabrication: The Science of Miniaturization", CRC Press, 2002.

[2] "MEMS poised to move into the mainstream", EE Times, CMP Media LLC, Pp.129-143, September 2002.

[3] Frost & Sullivan, "Malaysian E&E Industry Upgrade Study", commissioned by MIMOS Berhad, August 2003.

[4] Yao JJ. RF MEMS from a device perspective. J Micromec Microeng 2000; 10(4): R9-R38.

[5] Rebeiz GM, Muldavin JB. RF MEMS switches and switch circuits. IEEE Microwave Mag 2001:59-71.

[6] A. Kolpekwar and R.D. Blanton, "Development of a MEMS Testing Methodology," in *Proc. International Test Conference*, vol. 3, pp.923-931, 1997.

[7] A. Kolpekwar, Tao Jiang and R.D. Blanton, "CARAMEL: Contamination and Reliability Analysis of MicroElectromechanical Layout," Journal of Micro ElectroMechanical Systems, vol. 8, no. 3, Sep 1999.

[8] Tao Jiang and R.D. Blanton, "Inductive Fault Analysis of Surface Micromachined MEMS," IEEE Trans. on Computer-Aided Design of Integrated Circuits and Systems, vol. 25, no. 6, pp52-68, June 2006.

[9] S. Mir, L. Rufer and B. Courtios, "On-Chip testing of embedded transducer," in *Proc. 17th International Conference on VLSI Design*, 2004.

[10] B. Charlot, S. Mir and B. Courtois, "Fault Simulation of MEMS using HDLS," Journal of Modeling and Simulation of Microsystems, vol. 2, no. 1, pp35-42, 2001.

[11] N. Deb and R.D. Blanton, "Built-In Self-Test of MEMS Accelerometers," Journal of MicroElectroMechanical Systems, vol. 15, no. 1, Feb 2006.

[12] Xingguo Xiong Yu-Liang Wu Jone, W.-B., "A dual-mode built-in self-test technique for capacitive MEMS devices," IEEE Trans. on Instrumentation and Measurement, vol. 54, no. 5, pp1739-50, Oct 2005.

[13] T. Jiang and R. D. Blanton, "Inductive Fault Analysis of Surface –Micromachined MEMS" 21 May 2005.

[14] K. Hofmann, J.M. Karam, B. Courtios, and M. Glesner. Generation of HDL-A code for non-linear behavioral models. In IEEE International Workshop on Behavioral Modeling and simulation BMA'97, Washington D.C. USA, October 1997.

[15] A. Kolpekwar and R.D. Blanton, "Development of a MEMS Testing Methodology," IEEE Test Conference, Washington, D.C., pp.923-931, Nov. 1997.

[16] S. Lee, R. Ramadoss, M. Buck, V.M. Bright, K.C. Gupta and Y.C. Lee, "Reliability Testing of Flexible Printed Circuit-Based RF MEMS capacitive switches," Journal of Microelectronics Reliability, vol. 44, pp. 245-250, 2004.

[17] J.H. Park, S.H. Lee, J.M. Kim, H.T.Kim, Y.W. Kwon and Y.K. Kim, "Reconfigurable Millimeter-Wave Filters Using CPW-Based Periodic Structures with Novel Multiple-Contact MEMS Switches," Journal of Microelectromechanical Systems, vol. 14, no. 3, June 2005.

MEMS Switch Material Dependency on Designing a Reconfigurable Antenna

A.H.M. Zahirul Alam, *Member, IEEE* Md. Rafiqul Islam, *Member, IEEE,*
Sheroz Khan, *Member, IEEE,* Azlani Bt. Mohd. Nat and Manis Mulyany Bt. Abdul Razak
Faculty of Engineering, International Islamic University Malaysia
P.O. Box 10, 50728 Kuala Lumpur, MALAYSIA
Email: zahirulalam@iiu.edu.my

Abstract **We report on reconfigurable antenna performance using RF MEMS switches dependency on material used for MEMS structure. Telfon as a MEMS structural material gives triple-band antenna operation at 17.21, 23.54 and 29.27 GHz with return loss -12.04, -19.37 and -17.34 dB, respectively. The design is performed using 3D electromagnetic simulators considering ideal MEMS switches. There is no shift of resonance frequency that occurred at 23.54 GHz when both the MEMS switches are either ON or OFF.**

I. INTRODUCTION

RECONFIGURABLE multi-band antennas are attractive for many military and commercial applications where it is required to have a single antenna that can be dynamically reconfigured to transmit or receive on multiple frequency bands. Such common-aperture antennas result in considerable savings in size, weight and coast. They find application in space-based radar, communication satellites, electronic intelligence, and aircraft and many other communications and sensing applications. A number of different reconfigurable antennas, planar and 3-D have been developed. Some of them were developed for radar applications [1]-[2] and other planar antennas were designed for wireless devices [3]-[4]. Reconfigurable slot antennas were designed for UHF [5]. Reconfigurable patch antennas were also designed to operate in both L and X bands [6].

Radio frequency microelectrical mechanical systems (RF MEMS) is an emerging technology that promises the potential of revolutionizing RF and microwave system implementation for the next generation of telecommunication applications. Its low-power, excellent RF performance, large tuning range, and integration capability are the key characteristics enabling system implementation with potential improvements in size, cost, and increased functionality. They are normally built on high-resistivity silicon wafers, gallium arsenide (GaAs) wafers, and quartz substrates using semiconductor microfabrication technology with a typical four-to six-mask level processing. [7]–[11]. Typical example of MEMS based antenna are reported in [12]-[13]. In this paper, we proposed a triple band reconfigurable antenna using RF MEMS switches that can be fabricate in easy process steps. The effect of material used for MEMS switches is described when both the switches are either OFF or ON.

II. RECONFIGURABLE MEMS ANTENNA DESIGN

The schematic diagram of the reconfigurable antenna is shown in Fig. 1. It consists of three patches where the feeding configuration exists in the center patch. The patch antenna design was supported with a model built using a high frequency structure simulator based on finite elements modeling (FEM). A tool with 3D modeling capabilities was necessary due the fact that, for small ground planes, the antenna behavior depends on the ground size.

The two critical steps in designing the patch antenna were the definition of the patch dimensions and the feeding configuration. The patch dimensions have direct influence on the operating frequency and on the antenna gain. The difficulty to predict accurately the patch dimensions is related to the fringing fields together with the small size of the ground plane used.

The antenna feeding should be designed carefully since it must provide a correct impedance matching. At high-signal frequencies it is necessary to design a feeding line with specific characteristic impedance. Also, that line must be connected in a point of the antenna where the input impedance is the same than the feed-line characteristic impedance. The patch

0-7803-9730-4/06/$25.00 ©2006 IEEE

antenna was fed with a microstrip line connected to a point inside the patch where the input impedance is 50 Ω. The MEMS contacts are formed on alumina or Teflon or GaAs or SiN substrates which forms cantilever beam on the antenna surface.

A patch antenna is consists of a segmented patch and MEMS switches are constructed on a substrate of Rogers as shown in Fig. 1. The patch segments of the segmented patch can be electrically connected to every supplementary by the MEMS switches. MEMS switches are formed on either alumina or Teflon or GaAs or SiN material.

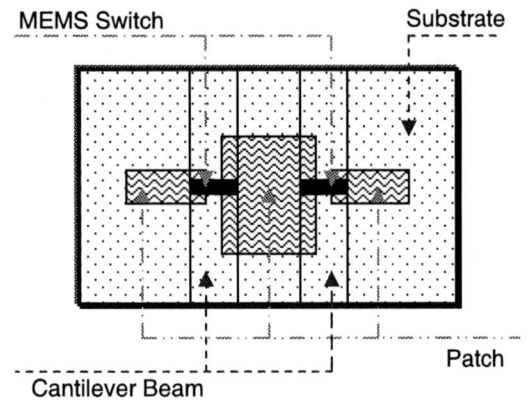

Fig. 1 Schematic diagram of MEMS reconfigurable antenna.

III. RESULTS AND DISCUSSIONS

In order to study the effect of MEMS switches materials which form cantilever beam, various dielectric constant materials were chosen by considering substrate material as Rogers. The dimension of the wing patches were maintained as 5×2 mm after optimization. The optimization procedure is described in the following section. Then, the comparison of the results is analyzed for various MEMS bridge materials when both the switches are either ON or OFF.

Wing Patch Dimension 5×1.22 mm²

It is observed that three are three resonant frequencies depending on the MEMS switches states, both switches either ON or OFF as shown in Fig. 2. The figure shows that when both switches are OFF, there are three frequency bands. The lower frequency at 17.21 GHz with return loss -19.24 dB, the centre frequency response at 23.54 GHz with -24.49 dB return loss

and the upper frequency at 30.78 GHz with return loss of -12.23 dB. On the other hand, when the both switches are ON, t there are two frequencies band reacted. The frequency is 20.53 GHz with -16.03 dB return loss and frequency of 23.39 GHz with -10.27 dB return loss.

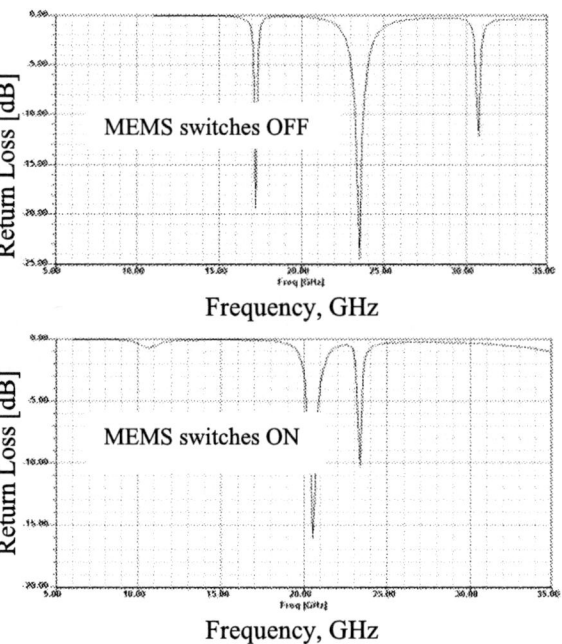

Fig. 2 Return loss of the antenna when sitches are either OFF or ON for wing patch dimension 5×1.22 mm.

Wing Patch Dimension 5×2 mm²

It is observed that there are three resonant frequencies depending on the MEMS switches states, both switches either ON or OFF as shown in Fig. 3. The figure shows that when both switches are OFF, there are three frequency bands. The lower frequency at 16.61 GHz with return loss -11.47 dB, the centre frequency response at 23.39 GHz with -25.14 dB return loss and the upper frequency at 28.37 GHz with return loss of -10.19 dB. On the other hand, when the both switches are ON, the antenna behaves like a single patch antenna with resonant frequency at 23.39 GHz and return loss -19.38. It is found that there is no shift of resonant frequency which occurs at 23.39 GHz irrespective of switches states.

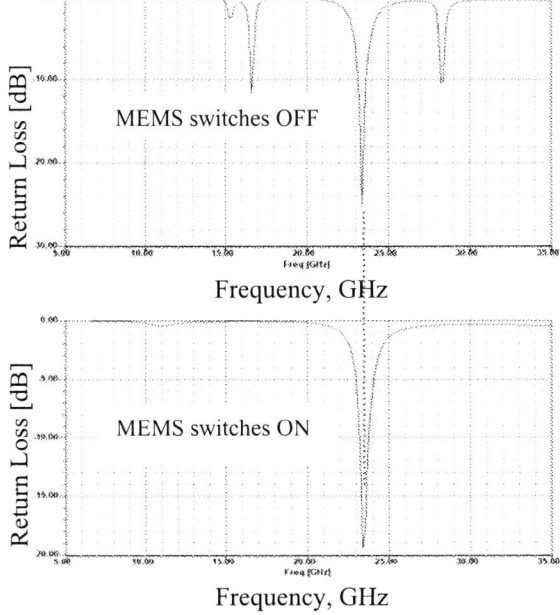

Fig. 3. Return loss of the antenna when sitches are either OFF or ON for wing patch dimension 5×2 mm.

Wing Patch Dimension 5×3 mm²

There is also three frequencies response emerged when the both switches are off when the width of the wings are increased as shown in Fig. 4. The return losses are -5.44, -19.04 and -

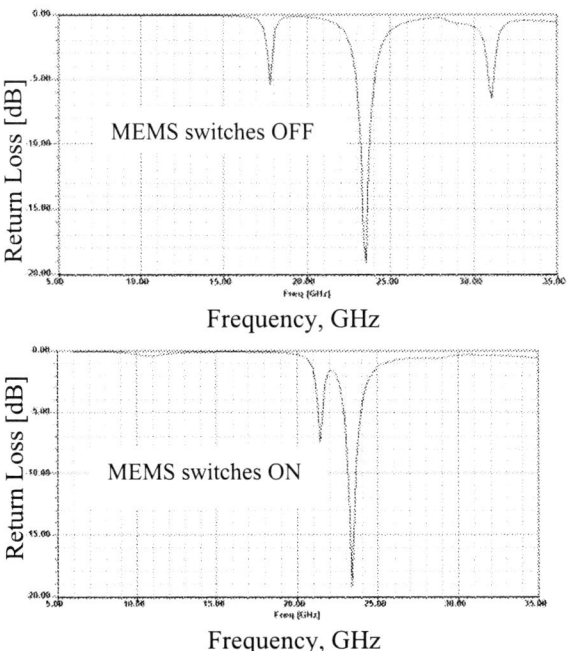

Fig. 4. Return loss of the antenna when sitches are either OFF or ON for wing patch dimension 5×3 mm.

6.42 dB at 17.81, 23.54 and 31.08 GHz respectively when both the switches are OFF. However, when the both switches are ON, there are also two frequencies band will response which are 21.43 GHz with return loss is -7.4 dB and 23.39 GHz with -19.21 dB of return loss.

Wing Patch Dimension 2×1 mm²

When the wing patch dimension is reduced to 2×1 mm, there is only one frequency response when both switches are either ON or OFF and the resonance occurred at 23.39 GHz. However, the insertion loss is reduced to -23.35 dB from -20.83 dB when both switches are ON.

Alumina as a MEMS Bridge Material

MEMS bridges were designed by choosing alumina material with permittivity 9.4. It is observed that there are three resonant frequencies depending on the MEMS switches states whether both the switches are either ON or OFF. The Fig. 5 shows that there are three frequency bands when both the switches are OFF. The lower frequency at 17.21 GHz with return loss -12.04 dB, the centre frequency response at 23.54 GHz with -19.37 dB return loss and the upper frequency at 29.27 GHz with return loss of -17.34 dB. On the other hand, when both the switches are ON, the antenna behaves like a single patch antenna with resonant frequency at 23.54 GHz and return loss -20.85 dB. It is found that there is no shift of resonant frequency which occurs at 23.54 GHz irrespective of the switches states.

Teflon as a MEMS Bridge Material

MEMS bridges were also designed by choosing Teflon material with permittivity 2.1. In this case, there are no changes of performance of antenna in terms of resonant frequencies compared to Alumina. However, lower return losses are observed for lower and centre resonant frequencies as shown in Fig 6. The return losses are -14.67, -23.07 and -11.48 dB at 17.21, 23.54 and 29.27 GHz respectively when both the switches are OFF. For both switches ON, the return loss is -23.61 dB.

the switches are OFF. For both switches ON, the return loss is -23.61 dB.

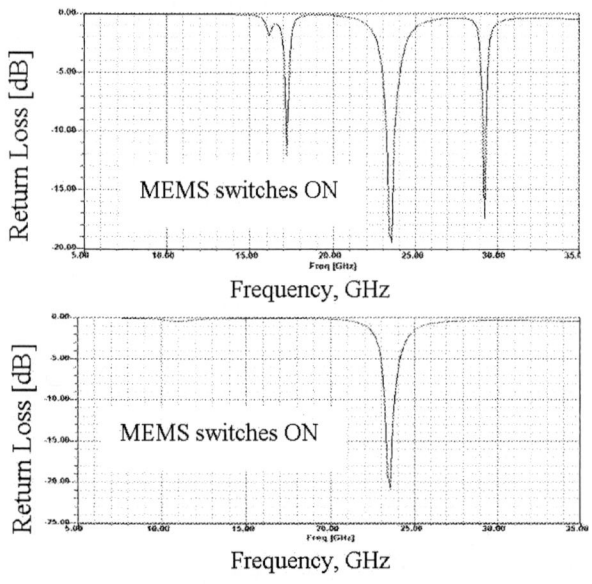

Fig. 5 Return loss of the antenna for alumina as MEMS bridge material.

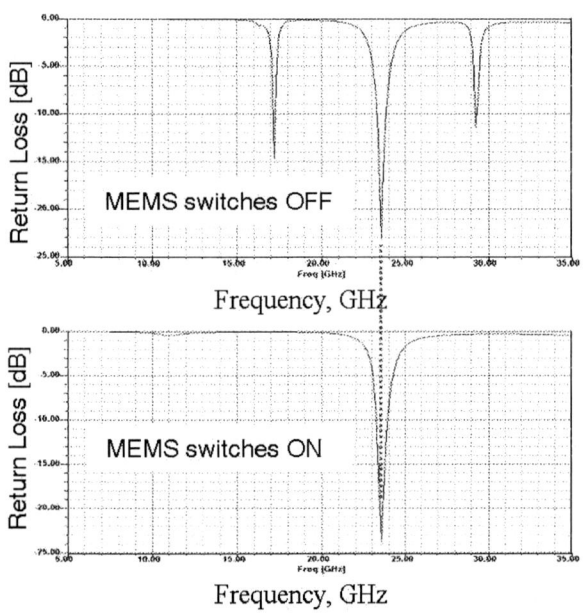

Fig. 6. Return loss of the antenna for Teflon as MEMS bridge material.

Comparison of Different MEMS Bridge Material

Fig. 7 shows the variation of return loss of the antenna at resonant frequencies for different types of materials as a MEMS bridge when both the switches are OFF. It is observed that there are three resonant frequencies for all the materials. The lower, centre and higher resonant

occurs almost of the same points irrespective of the materials used which proves that the MEMS bridge material types does not influence the frequency of operation. However, the insertion losses are different. The alumina gives better performance compared to other materials for all the frequency range.

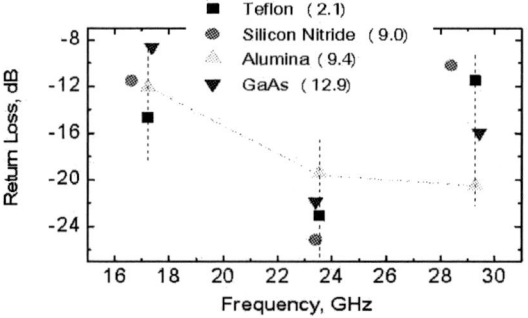

Fig. 7 Return loss of the antenna for different types of MEMS bridge materials when the switches are OFF.

Return loss and resonant frequency of the antenna for different types of MEMS bridge materials are shown in Fig. 8 when both the switches are ON. It is observed that Teflon as a MEMS bridge material gives lower insertion loss compared to other materials.

However, Alumina has been chosen as the best materials for MEMS bridge compared to Teflon, Si and GaAS considering return losses when both the switches are either ON or OFF.

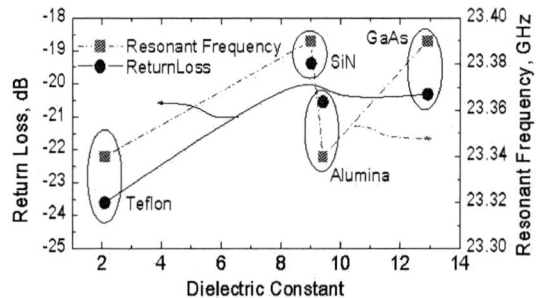

Fig. 8 Return loss and resonant of the antenna for different types of MEMS bridge materials when the switches are ON.

IV. CONCLUSIONS

Antenna performance has been optimized by determining the suitable material of MEMS structure. A reconfigurable triple-band antenna operating at 17.21, 23.54 and 29.27 GHz using MEMS switches with low insertion loss was achieved by selecting alumina for MEMS

348

structure. Single resonant frequency occurred when both the MEMS switches are ON confirmed that the antenna behaves like a single patch antenna. There is not much shift in resonance frequency by changing the MEMS structure material. However, MEMS material has greatly effect the performance of the antenna in terms of insertion loss. The proposed reconfigurable antenna can be implemented in easy fabrication process steps by sandwich method.

REFERENCES

[1]. K.Tomiyasu, "Conceptual reconfigurable antenna for 35 GHz high-resolution spaceborne synthetic aperture radar", *IEEE Transactions on Aerospace and Electronic Systems*, Vol. 39, Issue: 3 pp. 1069 – 1074 (2003).

[2]. J.T. Aberle, Sung-Hoon Oh, D.T. Auckland, and S.D. Rogers, "Reconfigurable antennas for wireless devices" *IEEE Antennas and Propagation Magazine*, Vol.45, Issue: 6, pp. 148 – 154 (2003).

[3]. G.H. Huff, J. Feng, Shenghui Zhang, G. Cung, and J.T. Bernhard "Directional reconfigurable antennas on laptop computers: Simulation, measurement and evaluation of candidate integration positions", *IEEE Transactions on Antennas and Propagation*, Vol. 52, Issue:12, pp. 3220 – 3227 (2004).

[4]. W.H. Weedon, W.J. Payne, and G.M. Rebeiz, "MEMS-switched reconfigurable antennas" *IEEE Antennas and Propagation Society International Symposium*, Vol. 3, pp. 654 – 657 (2001).

[5]. D. Peroulis, K. Sarabandi and Linda P.B. Katehi, "Design of reconfigurable slot antennass" *IEEE Transactions on Antennas and Propagation*, Vol. 53, No. 2, pp. 645 – 654 (2005)

[6]. C. E. Tong, and R. Blundel, "An annular slot antenna on a dielectric half-space", *IEEE Transactions on Antennas and Propagation*, Vol. 2, no.7, July 1994, pp. 967 – 974.

[7]. J. J. Yao and M. F. Chang, "A surface micromachined miniature switch for telecommunications applications with signal frequencies from DC up to 4 GHz," *Proc. Transducers*, pp. 384–387 (1995).

[8]. D. Hyman, J. Lam, B. Warneke, A. Schmitz, T. Y. Hsu, J. Brown, J.Schaffner, A. Walston, R. Y. Loo, M. Mehregany, and J. Lee, "Surface-micromachined RF MEMS switches on GaAs substrates," *Int. J. RF Microwave Computer-Aided Eng.*, vol. 9, pp. 348–61 (1999)

[9]. D. Hah, E. Yoon, and S. Hong, "A low-voltage actuated micromachined microwave switch using torsion springs and leverage," *IEEE Trans. Microwave Theory Tech.*, vol. 48, pp. 2340–2345 (2005).

[10]. J. Y. Park, G. H. Kim, K.W. Chung, and J. U. Bu, "Monolithically integrated micromachined RF MEMS capacitive switches," *Sens. ActuatorsA, Phys.*, vol. A89, no. 1-2, pp. 88–94 (2001).

[11]. S. P. Pacheco, L. P. B. Katehi, and C. T. C. Nguyen, "Design of low actuation voltage RF MEMS switch," *MTT-S Int. Microwave Symp.*, vol. 1, pp. 165–168 (2000).

[12]. B.A. Cetiner, H. Jafarkhani, J.Y. Qian, H.J. Yoo, A. Grau, and F. DeFlaviis, "Multifunctional reconfigurable MEMS integrated antennas for adaptive MIMOsystems," *IEEE Commun. Mag.*, vol.42, no.12, pp.62-70 (2004).

[13]. G. H. Huff and J. T. Bernhard, "Integration of Packaged RF MEMS Switches With Radiation Pattern Reconfigurable Square Spiral Microstrip Antennas", *IEEE Transactions on Antennas and Propagation*, Vol. 54, Issue: 2, pp. 464 – 469 (2006).

Surface and Composition Reactivity of Pt/GaN Catalytic Contact as Schottky Barriers Gas Sensor

Abdo Yahya Hudeish[1] and Azlan Abdul Aziz[2]

[1]Physics Department, Hodeidah University Hodeidah, Yemen
[2]School of Physics, Universiti Sains Malaysia, 11800 Penang, Malaysia
Email: ahodeish@yahoo.com; lan@usm.my

Abstract **Platinum (Pt) Schottky barrier formed on n-GaN is investigated for nitrogen (N_2), hydrogen (H_2) and methane (CH_4) gas sensing applications. The samples are studied for its sensing performance using standard current-voltage forward biased characterisation and these data are correlate to contact grain size and porosity using X-ray diffraction and scanning electron microscope. It is shown that sensor sensitivity not only depends on barrier lowering compound formed at the metal semiconductor junction, but also on the size of the grain and the grain porosity. As such, Pt/n-GaN gas sensor sensitivity is dependent on type of gases and the temperature of the sensor. Maximum sensor sensitivity is achieved at $T_{device} = 250°C$ with broad sensitivity plateau value from 100°C until 350°C showing the usefulness of Pt/nGaN sensor device at lower operating temperature (<400°C).**

I. INTRODUCTION

It is known that the electrical resistivity of a semiconducting oxide can be modified by adsorption of gases from the ambient [1]. This property has been used in semiconductor sensors for the detection of inflammable and toxic gases [2-4]. The semiconductor gas sensors offer good advantages with respect to other gas sensor devices (such as spectroscopic or optic systems), due to their simple implementation, low cost and good reliability for real-time control systems [5].

In recent years, the III-nitrides group semiconductor has been extensively researched and proved as good candidate to replace Si and SiC in high temperature gas sensing applications [6]. However, in high temperature application, problems such as material thermal stability, contact formation and general safety remains as main obstacle in progress of producing gas sensor with equivalent sensitivity at lower

sensing temperature. Furthermore, contact grain size and its pore structure have a major effect on the properties in polycrystalline materials and their full characterisation should be the first step in the study of materials.

In this paper, we will demonstrate the feasibility of sensor operation on GaN at much lower temperature range, from 100°C up to 500°C, by using an optimized annealing process. In addition, we will investigate the sensitivity and thermal stability of GaN with Pt catalytic metals within mentioned temperature range.

We will also examine the microstructures of the contact end-products by scanning electron microscope, SEM to obtain quasi-three dimensional information on the grain shape, size and pore sizes. The investigation on the variation of the electrical resistance will be in the presence of three different gases: nitrogen (N_2), hydrogen (H_2) and methane (CH_4).

II. EXPERIMENTAL DETAILS

The n-GaN grown on sapphire (Al_2O_3) substrate was used as sample for gas sensor. The GaN was first treated for 3 min in 1:1 HCl: H_2O, process step needed to remove any organic contaminants and oxides layer on GaN surface. Catalytic contacts of Pt were formed by using dc magnetron sputtering via shadow mask. Circular contact dimension sputtered was at 100 nm in thickness and approximately 300 nm in diameter. Distance between contact is approximately 1 mm. For better device performance, post-annealing process was performed at 600°C for 10 min [7].

Device were then mounted on sample holder made from a piece of ceramics plate, over layered with Au bond-pads for wire bonding. Connections from device to sample holder was achieved using Au wire by wedge-type wire bonder as shown in Fig. 1. The microstructure and phase composition were performed by scanning electron microscopy (SEM) and X-ray diffraction (XRD). Average grain size was determined by the linear intercept technique from SEM micrograph on fracture surface. For the gas sensing measurements, the sensor element was mounted on a heater and placed in a homemade chamber capable of controlling the different gas concentrations. The schematic of the sensor assembly is as shown in Fig. 1. Gas sensing properties were investigated at various operating temperatures from 25 to 500°C. The experiments were performed with three test gases: nitrogen (N_2), hydrogen (H_2) and methane (CH_4).

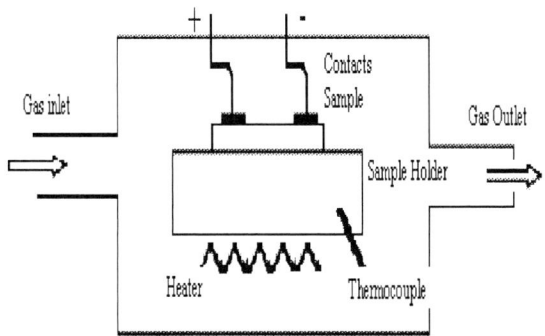

Fig. 1: Schematic diagram of the testing chamber used for sensor characteristics measurement.

The sensitivity, S, is defined as the ratio

$$S = \frac{R_a - R_g}{R_a} \qquad (1)$$

where R_a and R_g are the sensor resistances in air and in presence of the test gas, respectively.

III. RESULTS AND DISCUSSION

Figure 2 shows the X-ray diffraction peaks for the sample annealed at 600°C in air for 10 minutes. All the peaks belong to the spinel nitride. No other separate phase oxides could be identified by X-ray diffraction. The lattice constant was found to be 8.326±3 Å, a slight increase from the lattice constant of Ga_3Pt (8.320 Å).

Scanning electron micrographs for the sample is presented in Fig. 3. This figure shows that sample consists primarily of irregularly shaped of 1 to 6 µm aggregates of fine (0.1 µm) particles. It can be seen that the crystallite size of the sample is extremely fine, on the order 100 to 500 nm. The material is characterised by high intergranular porosity (about 40%). The bulk

density was calculated to be 3.11 g/cm^3. Many large and small pores are present in all material. The gas sensitivity depends largely on the microstructure, such as grain size, surface area and pore size.

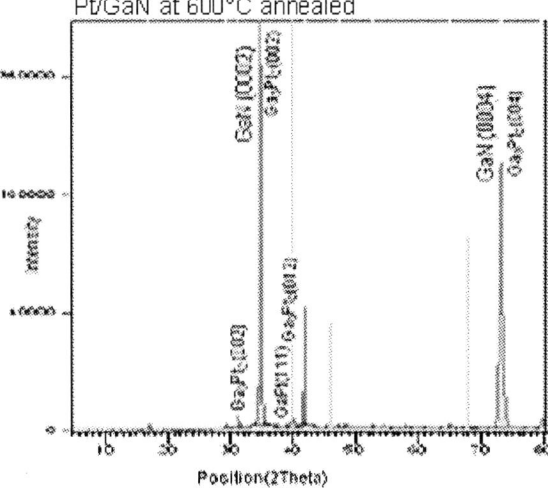

Fig. 2: Typical XRD spectra for the studied sample of Pt/GaN after each gas sensing experiment.

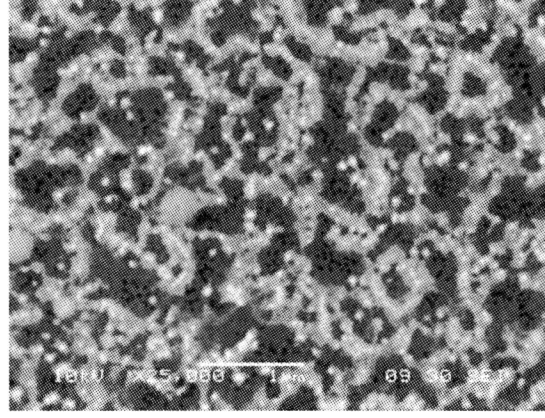

Fig. 3: Typical SEM micrograph for the studied sample of Pt/GaN after each gas sensing experiment.

Figure 4 shows the gas sensing measurements for nitrogen, hydrogen and methane at various temperatures, between 25 and 500°C. It is clear from the graph that the sensitivity of the Pt/n-GaN to hydrogen is the best. As expected the sensitivity increases with increasing the operating temperature and reaches a maximum value around 250°C and decreases further with temperature. The maximum sensitivity to nitrogen and methane is much lower. Usually, the gas sensing mechanism depends on the work temperature, because this mechanism is thermally activated [1].

Data from Fig. 4 on H_2 sensitivity curve also suggest that due to excess H_2 atoms available to form barrier-lowering effect and readily interact at the boundary area of Pt/GaN surface grain structure, may be responsible for very high gas sensitivity, S achieved. For methane, H_2 availability is at significantly less percentage.

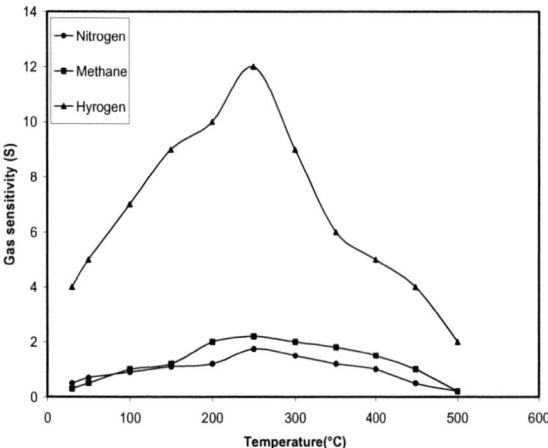

Fig. 4: Gas sensing profile of the Pt/GaN for nitrogen, methane and hydrogen.

As a result, Pt/GaN device sensitivity is most likely only due to effect of barrier lowering. Results from N_2 sensing experiment revealed a typical low response over the whole range of sensing temperature. As such, temperature variations have very little impact on our device N_2 and methane gas sensing and its sensitivity, whereas the opposite is true for H_2 gas sensing performance.

IV. CONCLUSION

Sensitivity studies on Pt/n-GaN for nitrogen, hydrogen and methane have shown that the electrical resistivity of this sample with catalytic metal is more sensitive to hydrogen and its sensitivity depends on the temperature. It is shown that sensor sensitivity not only depends on barrier lowering compound formed at the metal semiconductor junction, but also on the size of the grain and the grain porosity. As such, Pt/n-GaN gas sensor sensitivity is dependent on type of gases and the temperature of the sensor.

ACKNOWLEDGEMENT

This work was conducted under IRPA RMK-8 Strategic Research grant. The support from Universiti Sains Malaysia is gratefully acknowledged.

REFERENCES

[1] L. Satyanarayana, K. M. Reddey, and S. V. Manorama, Sensors and Actuators B 89, 62 (2003).

[2] A. Ratna Phani, S. Manorama, and V. J. Rao, Appl. Phys. Lett. 66, 2619 (1995).

[3] Duk-Dong Lee, Dae-Sik Lee, IEEE Sensors Journal 1, 214 (2001).

[4] O. V. Safonova, G. Delabonglise, B. Chenevier, A. M. Gaskov, and M. Labeau, Materials Science and Engineering C 21, 105 (2002).

[5] C. Bittencourt, E. Lobet, M. A. P. Silva, R. Lauders, L. Nieto, K. O. Vicaro, J. E. Sueiras, J. Calderer, and V. Correig, Sensors and Actuators B 6992, 1 (2003).

[6] N. S. Chen, X. J. Yang, E. S. Liu, and J. L. Huang, Sensors and Actuators B 66, 178 (2000).

[7] Z. Lin, H. Kim, J. Lee, and W. Lu, App. Phys Lett; 84,1585 (2004).

Metals/GaN Catalytic Contact Properties for Hydrogen Gas Sensor Applications

Abdo Yahya Hudeish[1] and Azlan Abdul Aziz[2]
[1]Physics Department, Hodeidah University, Hodeidah, Yemen.
[2]School of Physics, Universiti Sains Malaysia, 11800 Penang, Malaysia.
Email: ahodeish@yahoo.com, lan @usm.my

Abstract **Platinum (Pt), Palladium (Pd) and Nickel (Ni) as catalytic contact are successfully deposited on GaN substrates using dc sputtering method to be utilized as device for sensing hydrogen gas. The catalytic contact formation are studied in terms of its surface morphology and compound formation using Atomic force microscopy (AFM) and High resolution X-ray diffraction (HRXRD) respectively, revealing the crystalline properties and uniform deposition of fine grains of the catalytic films. Sample roughness varied from 13.1 (Ni), 18.9.4 (Pd) to 25.4 nm (Pt), increases in parallel to its hydrogen sensing response capability. The formation of hydride compound on Pd and Ni catalytic layer and Ga_3Pt_2 on Pt layer are found to be significant in controlling the sensor response. Device fabricated using Pt as catalytic metal give smallest surface roughness value and capable of forming compound Ga_3Pt_2 with GaN substrate, identified as the main reason in reducing Schottky barrier height of the catalytic metal/GaN. As such, sensor produced using Pt is shown to have the best sensor response performance. Furthermore, surface roughness and compound formation on catalytic metal can be linked directly to sensor performance.**

I. INTRODUCTION

The interest in sensors, especially microsensors, dramatically increased during the last decade of the 20th century. This is due to decreasing prices of materials for microsensors, more reliable fabrication technology, and numerous choices in the design of microsensors. As such, device, either MOSFET, MOS capacitor or Schottky diode based on material such as Si and SiC has been extensively developed as H_2 sensor. However, the mechanisms of sensor are still unclear, so the possibilities for new gas sensor concepts are hardly expanding scientifically [1].

Group III-Nitrides semiconductors are one of the leading candidate for blue light emitting diode and electronic device for high frequency and high power application. Devices based on this material are also ideal for gas sensing application since the Fermi level is not pinned in III-Nitride film. Moreover, III-Nitride materials, specifically GaN have excellent thermal conductivity and chemical resistance, which makes device fabricated on GaN have great potentials for sensors in harsh environments [2-4].

Based on previous work by Duxstad *et al.* [5] using Rutherford backscattering spectroscopy (RBS) on changes in metal/GaN film morphology, we wish to extend the scope of their study to include application of metal/GaN as catalytic contact, capable of sensing H_2 gas using atomic force microscopy (AFM) and high-resolution x-ray diffraction (HXRD).

II. EXPERIMENTAL DETAILS

GaN n-type samples grown on sapphire (Al_2O_3) substrates were used. The carrier concentration of the samples is approximately $3\times10^{17} cm^{-3}$. Prior to metallization process, the native oxide was removed with NH_4OH: $H_2O=1:20$ solution, followed by dipping in a HF: $H_2O=1:50$ solution. Boiling aqua regia (HCl: $HNO_3 = 3:1$) was used to chemically etch and clean the samples. Pt, Pd and Ni film with thickness 100nm were deposited by Edward 306 onto the n-GaN at backing pressure of 5×10^{-5} Torr. The device cross-section through GaN on sapphire with Pt as catalytic contact is as shown in Fig. 1.

Based on previous work on process optimisation for metal/GaN contact, all the metal/GaN sensors were annealed at 600°C for 15 minutes under Ar (99.99 % purity) [6]. H_2 sensing measurements were conducted in a

0-7803-9730-4/06/$25.00 ©2006 IEEE

homemade chamber under 2 percent H_2 in N_2 at T=400°C.

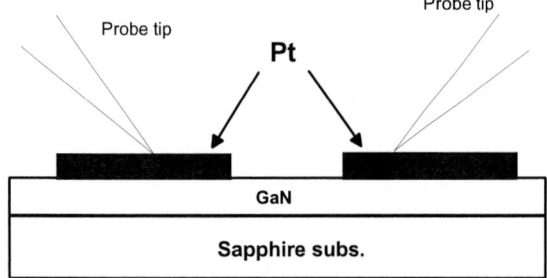

Fig. 1: Proposed gas-sensing mechanisms underlying the function of catalytic metal gas sensor.

During device sensing experiment, electrical characterizations were conducted for sensor resistivity and sensitivity. Results from this work have already been published elsewhere [7-9]. After each sensing setup, surface characterisation was performed on the GaN surface using atomic force microscopy (AFM) from Surface Imaging System, GmbH to see the change in surface roughness for different thickness of sensors. Phase formations at the surface were then characterized using PANalytical High resolution X-ray (HXRD), performed by Cu K_α radiation.

III. RESULTS AND DISCUSSION

AFM morphology as in Fig. 2 shows that, there is a noticeable change in surface morphology for different catalytic metal (Pt, Pd and Ni) upon hydrogen exposure. The comparison of the AFM figures shows a visible change in the roughness for catalytic metal. The surface roughness generated from a scan of a sensor sample Pt/n-GaN, Pd/n-GaN and Ni/ n-GaN are observed for the thickness 100 nm with hydrogen exposure at 200°C. The sample of Pt surfaces show the highest whereas sample of Ni surface show lowest roughness respectively. The surface roughness is expected to increase when the species deposition takes place on the surface.

The rms roughness for sample of Pt/ n-GaN is 25.4 nm and the difference between the lowest and highest points on the surface is 75.5 nm. For sample of Pd/n-GaN, shown in Fig. 2, the surface roughness was found to be 18.9 nm and the difference between the lowest and highest points on the surface was 68.2 nm. Sample of Ni/n-GaN is also shown in Fig. 2(c). In this case, roughness was found to be 13.1 nm and the difference

between the lowest and highest points on the surface was found to be 64.5 nm. When we correlated these findings to previous work electrical data, we can conclude that samples with least surface roughness have the least H_2 sensing capability.

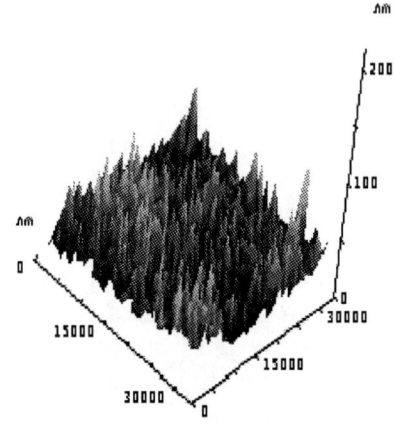

(a) Pt/GaN

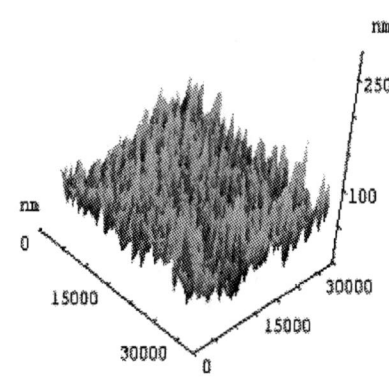

(b) Pd/GaN

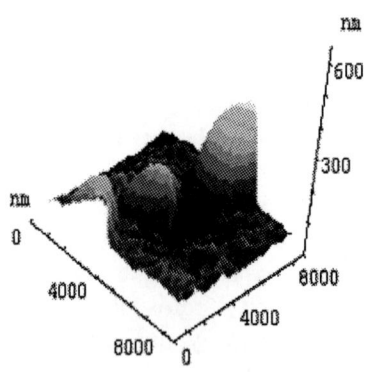

(c) Ni/GaN

Fig. 2: AFM of the Pt/GaN, Pd/GaN and Ni/GaN surface (annealed at T=600°C) after H_2 gas sensing exposure.

Data from XRD identifies the compound responsible for H_2 sensing performances. Figure 3 shows the XRD spectra of Pt/n-GaN, Pd/n-GaN and Ni/n-GaN after hydrogen exposure at T=400°C with main peaks ranging from $10°$ $<2\theta<90°$. The typical peaks corresponds to GaN (0002) at $2\theta= 34.74°$ and GaN (0004) at $2\theta= 73.24°$ were observed for all samples.

For Pt/n-GaN, compound such as (Gallium platinum) Ga_3Pt_2 (002) at $2\theta= 34.74°$, Ga_3Pt_2 (012) were detected at $2\theta= 41.92°$ and Ga_3Pt_2 (004) at $2\theta= 73.24°$, respectively. When the samples are test upon hydrogen exposure with high operating temperature and annealed at 600°C, formation of compound will give low Schottky barrier heights. This conclusion is consistent with the result of the Ga_3Pt_2 contact.

For the case of Pd/n-GaN the peaks of compounds of (Palladium hydride) PdH such as PdH (101) at $2\theta= 41.77°$, PdH (120) at $2\theta= 72.87°$ and PdH (112) at $2\theta= 73.06°$ were detected after annealed samples, lead to the composition of hydride metal layers at the metal film surface or to desorption of water and hydroxyl group at defects.

For the case of Ni/n-GaN the peaks intensity of Ni (002) at $2\theta= 41.62°$ and GaNi (100) appears at $2\theta= 31.13°$ and at $2\theta= 30.92°$. The hydride NiH formation was detected (111) at $2\theta= 41.62°$. Again, the hydride metal layer formation is most likely due to the effect of water and hydroxyl desorption at the surface.

All samples, after repeated exposure to H_2 gas sensing environment at various device sensing temperature, exhibited similar and uniform surface properties, indicating the thermal and chemical stability of the metal/GaN catalytic contact device.

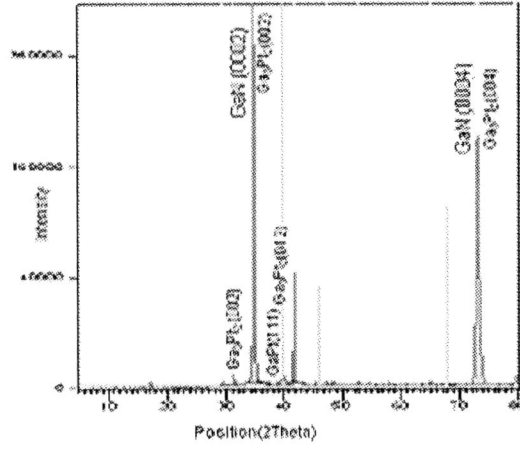

(a) Pt/GaN

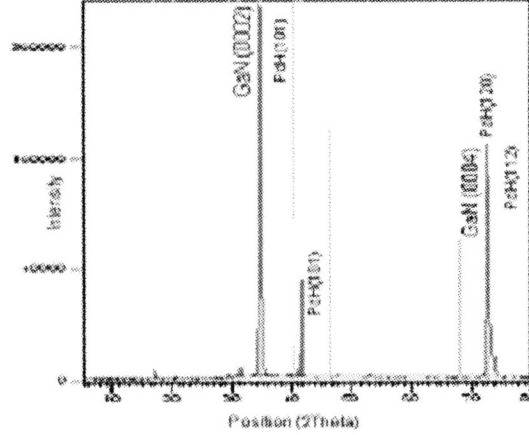

(b) Pd/GaN

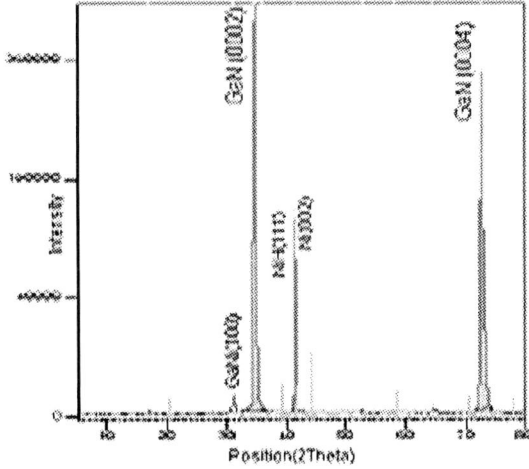

(c) Ni/GaN

Fig. 3: XRD spectra of Pt/GaN, Pd/GaN and Ni/GaN samples (annealed at 600°C) after H_2 gas sensing exposure.

IV. CONCLUSION

We have successfully fabricated and characterized contact properties of metal/GaN catalytic sensor. The metal/GaN catalytic contact surface property was shown to be a strong function of sensing sensitivity. XRD data on contact composition of metal/GaN shows that the formation of hydride compound on Pd and Ni catalytic layer and Ga_3Pt_2 on Pt layer are significant in controlling the sensor response. Device fabricated using Pt as catalytic metal produced smallest surface roughness value and capable of forming compound Ga_3Pt_2 with GaN substrate, identified as the main reason in reducing Schottky barrier height of the catalytic metal/GaN.

ACKNOWLEDGEMENT

This work was conducted under IRPA RMK-8 Strategic Research grant. The support from Universiti Sains Malaysia is gratefully acknowledged.

REFERENCES

[1] E. Miko, Proceedings of the Sensor Technology Conference 2001. 14, 212 (2001)

[2] H. Morkoç, and S. N. Mohammad, Science Magazine, 267, 51 (1995)

[3] B. P. Luther, S. D. Wolter, and S. E. Mohney, *Sens. Actuators B*. 56, 164. (1999)

[4] J. Schalwig, G. Muller, U. Karrer, M. Eickhoff, O. Ambacher, M. Stutzmann, L. Gorgens, and G. Dollinger, *Appl. Phys. Lett.* 80, 1222 (2002)

[5] K. J. Duxstad, E. E. Haller, and K. M. Yu, *Journal of Applied Physics*. 81, 3134 (1997)

[6] A. Y. Hudiesh, A. Abdul Aziz and Z. Hassan, Proceedings of the 2004 IEEE International Conference on Electronics Materials and Packaging 2004 – EMAPS2004, 574. (2004)

[7] A. Y. Hudiesh, A. Abdul Aziz, Z. Hassan, and K. Ibrahim, Proceedings of the 2004 IEEE International Conference on Semiconductor Electronics-ICSE 2004, 52 (2004)

[8] A. Y. Hudiesh, A. Abdul Aziz and Z. Hassan, Proceedings of the National Seminar of Science Technology and Social Sciences; 177 (2004)

[9] A. Y. Hudiesh, Z. Hassan, and A. Abdul Aziz, *Technical Journal* – School of Elec. Eng. & Electronics, USM, vol. 10, 41 (2004)

A Low-Cost CMOS Reconfigurable Receiver for WiMAX Applications

Adiseno, G. Wiranto, *Member, IEEE* and T.M.S. Soegandi, *Member, IEEE*
Research Centre for Electronics and Telecommunication
Indonesian Institute of Sciences (PPET-LIPI)
Jl. Sangkuriang Komp LIPI Gdg 20
Bandung-40135, Indonesia
E-mail : adiseno@ppet.lipi.go.id

Abstract This paper presents a low cost solution for lower frequency WiMAX CPE receivers. A receiver which consists of wideband LNA, switching mixer, VCO and reconfigurable low-pass filter have been designed and simulated in different WiMAX receive bands such as in the 2.3-GHz band, in the ISM band (2.4-GHz), and in the 2.5-GHz band, as well as in different channel bandwidths, ranging from 1.75 MHz up to 10 MHz. The zero-IF architecture is chosen as it suits for receivers using OFDM signals with null DC subcarrier. The LNA and switching mixer have been fabricated in 0.18μm RF-CMOS technology and the measurement results show that these building blocks have a combined RF-to-IF gain of 20-dB, a DSB noise figure of 3.5-dB, an RF-to-IF IIP_2 and IIP_3 higher than +20-dBm and 0-dBm, respectively. The circuit consumes 16-mA from 1.8-V supply and occupies a die area of 0.42×0.50 mm^2. Simulation results of the negative-Gm VCO show that it can oscillate at the desired frequencies with phase-noise performance of better than − 125-dBc/Hz at a frequency offset of 3.5 MHz. A reconfigurable 4th order of Gm-C low-pass filter (LPF) has been designed using two 2nd order biquad g_m-C LPFs and its simulation results show that − 3-dB frequency corners can be adjusted to 1.75-MHz, 3.5-MHz, 7-MHz and 14-MHz. Both VCO and LPF are simulated using 0.18μm RF-CMOS technology.

I. INTRODUCTION

WORLDWIDE Interoperability for Microwave Access (WiMAX) is a wireless metropolitan access network (W-MAN) technology whose standard for fixed and nomadic users, well-known as IEEE 802.16-2004, is introduced in 2004 [1]. It addresses the first-mile/last-mile connection in W-MAN, focuses on the efficient use of bandwidth between 2 and 66 GHz, primarily between 2 and 11 GHz, and defines a medium access control (MAC) layer that supports multiple physical layer specifications customized for the frequency band of use. WiMAX systems offer true differentiated broadband services at minimal cost. They let thousands of users share capacity for data, voice and video [2].

There are several RF-bands targeted as working frequencies, including the 2.3-GHz, the 2.5-GHz and the 3.5-GHz bands for licensed frequency spectrum, as well as in the 2.5-GHz and the 5.8-GHz for license-exempt frequency spectrum. It provides different quality of services (QoS) to adjust different radio-link qualities maintaining the connection. The system uses therefore different modulation techniques and different channel bandwidth. It employes, however, the OFDM technique to efficiently use limited available frequency spectrum [1].

Radio solution, in the other side, offers fixed performance, i.e. working in a specified RF-band, a channel bandwidth and a given set of figures of merite [3]-[5]. Reconfigurable RF systems that support different requirements are thus highly desired. Existing solutions such as found in dual or tri-band mobile phones [6] are not suitable as those simply place each of the systems beside each other on a single chip.

These considerations put limits on the design of a receiver front-end. In this paper, building blocks which consists of wideband LNA, switching mixer, VCO and reconfigurable low-pass filter have been designed, simulated and for LNA and mixers, measured. These building blocks can be used to form reconfigurable RF systems.

In section II, basic considerations are given which are derived from system level study. The circuit theory, upon which the presented circuits

0-7803-9730-4/06/$25.00 ©2006 IEEE

are designed, is described in section III. Section IV presents measurement and simulation results of circuits described in section III. Finally, the results are summarized in section V.

II. BASIC CONSIDERATIONS

The main focus in this research is designing a reconfigurable receiver which supports different requirements. A study of WiMAX standard is therefore a good starting point to derive the required specifications to meet different requirements. Table I lists some specifications that are required in designing WiMAX receivers for lower RF bands, assuming a zero-IF architecture is used.

Rx band [Hz]	B(f) [Hz]	Modulation
2.15-2.162G 2.3-2.36G 2.4-2.48G 2.5-2.69G 3.4-3.6G 5.725-5.85G	1.75M 3.5M 7M 14M	BPSK-1/2 QPSK-1/2 QPSK-3/4 16QAM-1/2 16QAM-3/4 64QAM-2/3 64QAM-3/4

Table I: WiMAX receiver specifications summary.

In addition to Table I, noise figure of the receiver should be less than 7 dB, while the IIP_3 and IIP_2 should be higher than -10 dBm and +20 dBm, respectively. The phase-noise of the frequency synthesizer should be lower than -70 dBc/Hz, -90 dBc/Hz, -110 dBc/Hz and -125 dBc/Hz at the offset frequency of 10-kHz, 100-kHz, 1-MHz and 3.5-MHz, respectively.

WiMAX systems use OFDM signals with 200 subcarriers, 8 pilot subcarriers, null subcarriers for guardbands (28 for lower frequency and 27 for higher frequency) and the DC subcarrier (see Fig. 1).

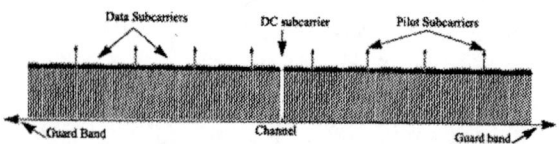

Fig. 1 OFDM frequency description in WiMAX.

Therefore, DC-offsets problem that usually occurs in zero-IF receivers can easily be minimized by removing signals at DC. The zero-IF architecture suits thus very well in WiMAX

receivers.

The receiver front-end consists of a frequency-agile block and a wide-band down-converter. The frequency-agile block may be implemented using switchplexers [7] or tunable active RF-band filters [8]. For our purposes we assume a suitable solution is available. The wide-band down-converter consists of a wide-band LNA with wide-band input matching and switching mixers.

Fractional-N synthesizer was chosen as the frequency synthesizer. This architecture eliminates loop-bandwidth problem as found in classical integer-N synthesizer. It has a fast lock transient and low close-in phase noise. Its fractional spurs increases, however, adjacent channel distortion. The VCO in this frequency synthesizer is based on a negative-g_m oscillator.

The filter specifications have been derived to meet the WiMAX specification. The corner frequency of the filter has to be tunable from about 875 kHz up to 7 MHz which are half of channel bandwidths as listed in Table I. In order to meet new and upcoming requirements between these limits the filter is tunable by means of an analogue tuning voltage, so that tuning over a wide frequency range is possible. The system specifications indicate the need of a 5th order Butterworth filter to fulfil the demands, where the adjacent channel rejection requirement and intersymbol interference have been considered. The used Butterworth characteristic has less than < 120 ns of group delay for a corner frequency of 7 MHz. The first pole of the filter is already part of the mixer output stage. Hence, for the I and Q path of the receiver a 4th order Butterworth filter has to be implemented, respectively, consisting of two 2nd order filters. The zero-IF receiver is shown in Fig. 2.

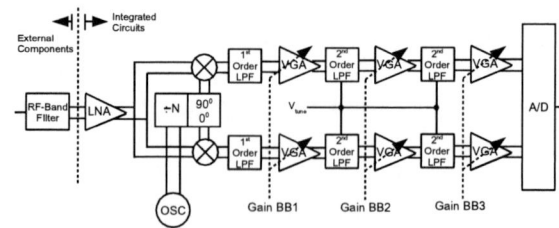

Fig. 2 A zero-IF WiMAX receiver.

III. CIRCUIT DESIGN

Dual-loop resistive feedback is used to achieve wide-band input matching at the LNA. This dual-

358

loop feedback senses the output current, feeds it back to the input, where it is compared with the input current to determine the current gain in the first loop and with the input voltage to determine the voltage-to-current gain in the second loop. The input impedance is, thus, the ratio between the second-loop and the first-loop gains [9]. This can be made almost independent to the input bias current, provided an overall high loop gain can be achieved. The output current of the LNA is connected to switching mixers which are stacked below the LNA. The combined conversion RF-to_IF gain is thus the gain of the LNA times $2/\pi$ which is the conversion gain of the switching mixer. The circuit is shown in Fig. 3.

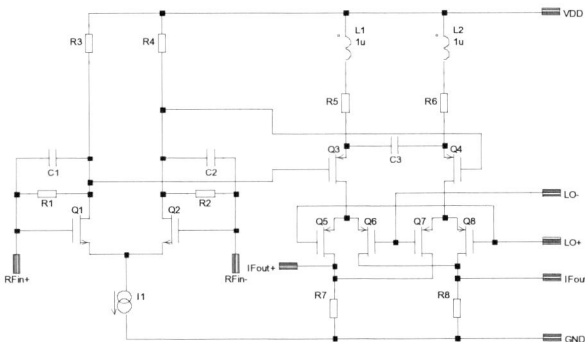

Fig. 3 A wideband LNA with switching mixers.

A negative-g_m oscillator is designed for the VCO. This oscillator does not need any extra circuitry to maintain its amplitude as the amplitude is proportionally related to $1/g_m$. The capacitance which is usually connected in parallel with the inductance to form the LC-tank is divided into 2 similar varactors. Connecting this way will lower the parasitic capacitance to the ground. To increase the tuning range in order to use it in multiband VCO, several switched capacitor arrays are connected in parallel with varactors. The higher RF bands are desired to tune, the more capacitor arrays are switched on. The circuit is shown in Fig. 4.

For the implementation of the filter the g_m-C approach [10] has been chosen because of the easy tuning capability by varying the gvalue of the transconductors. The g_m-C filter also has a low noise floor but the ability to handle large signals is limited. A biquad circuit realization for the 2nd order filter is used because of its advantages in design and layout. As will be shown later similar circuit blocks can be used to implement both transfer functions. All filter stages operate with one common bias generating

circuit which improves the matching between the filter stages over the tuning range.

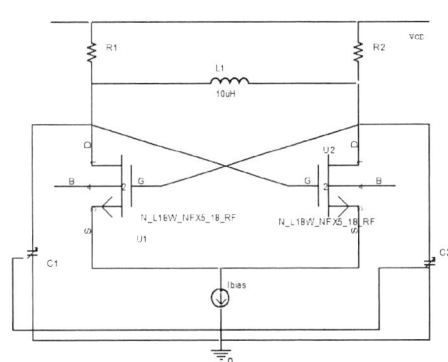

Fig. 4 A negative-g_m oscillator.

The topology of the second-order biquad g_m-C low-pass filter (LPF) is shown in Fig. 4 The transfer function of this biquad is

$$H\ s\ =\frac{V_{out}}{V_i}=\frac{\dfrac{g_{m1}}{g_{m4}}\dfrac{g_{m3}g_{m4}}{C_1 C_2}}{s^2\ \dfrac{g_{m2}}{C_1}s\ \dfrac{g_{m3}g_{m4}}{C_1 C_2}} \quad (1)$$

Recognizing the common transfer function of second-order LPF, one obtains the corner frequency and the quality factor Q of the biquad as,

$$_0=\frac{g_m}{C} \qquad Q=\frac{g_m}{g_{m2}} \quad (2)$$

assuming $g_{m1}=g_{m3}=g_{m4}=g_m$ dan $C_1=C_2=C$. Six of the eight OTAs can be designed to be identical, while remaining OTAs can easily be adapted only with changing the OTA current.

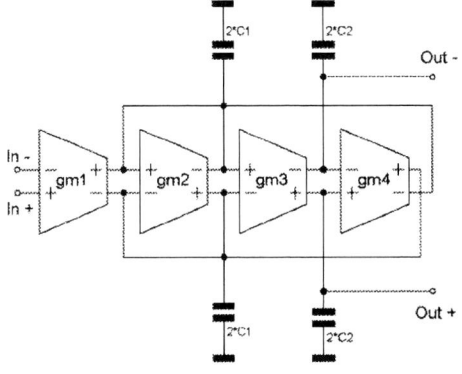

Fig. 5 A 2nd order biquad LPF.

A CMOS fully differential folded-cascoded OTA has been chosen for the OTAs (Fig. 6). Transistors T2, T6, T7, T9, T8, T10 and T11 acts as current sources with current I. Higher current flows through T4 and T5. The transconductance g_m is proportionally related to squared root function of the current I. This property is used to change the corner frequency of the filter.

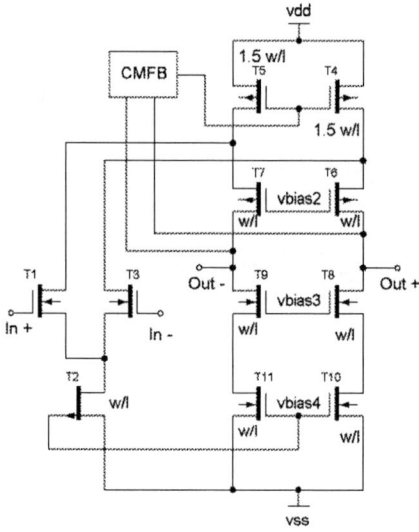

Fig. 6 A fully differential folded-cascoded OTA

A tuning range of corner frequencies of about 1:3 can be achieved by varying the current through OTA with ratio 1:10. The requested tuning range of filter is 1:8 (875 kHz to 7 MHz). Additionally switched capacitor arrays at the output are used to cover the required channels.

IV. MEASUREMENT AND SIMULATION RESULTS

The LNA and mixers were implemented in 0.18 μm RF-CMOS technology (Fig. 7). Several QFPP packaged chips were measured in the 2.3-GHz, 2.5-GHz and 3.5-GHz bands. Different baluns are used at the input for these RF-bands. The conversion gain, DSB NF and S11 of the LNA and switching mixers are shown in Fig. 8.

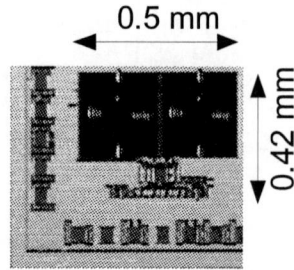

Fig. 7 A die-photo of the LNA and switching mixers

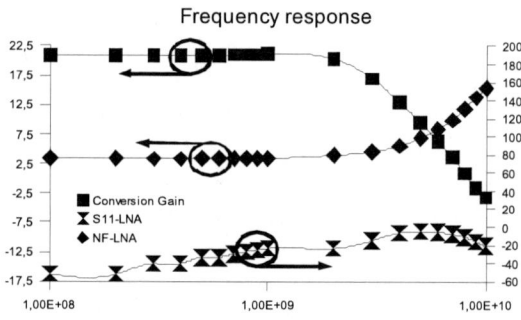

Fig. 8 Frequency response of the LNA and mixers.

The −3-dB bandwidth of the LNA is at 3.07 GHz. However, the conversion RF-to-IF gain in the 3.5-GHz band is 16.3 dB which is still acceptable for WiMAX applications. The DSB NF and S11 of the LNA and the mixers in that RF-band, which are 6 dB and −15 dB, respectively, are still to below WiMAX system specification requirements. Fig. 9 shows the second and third intermodulation in the 2.5-GHz band. A +22 dBm of IIP_2 and +1.5 dBm of IIP_3 is achieved.

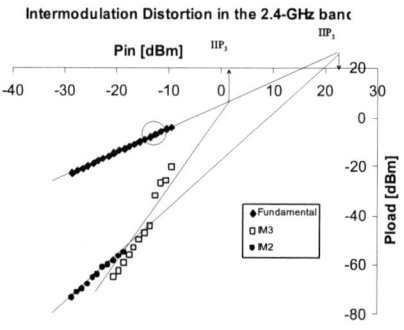

Fig. 9 Intermodulation distortion in the 2.5-GHz band.

The negative-g_m oscillator was simulated using RF transistors available in 0.18 μm RF-CMOS technology kit. The VCO can be tuned to oscillate up to 3.6 GHz. Simulation results in phase-noise performance at that frequency (Fig. 10) show that for frequency offsets of 100 kHz and 1 MHz the achieved phase-noise is slightly higher than required in the specification, while for other frequency offsets, it fulfills the

360

WiMAX requirements. With a proper PLL design, especially placing poles of the rest of the PLL system in the right places, the phase-noise of the VCO determines the phase-noise of the frequency synthesizer. The output voltage swing of the VCO is 1.7V which would be enough to drive the switching mixers to obtain highest conversion gain and lowest phase-noise.

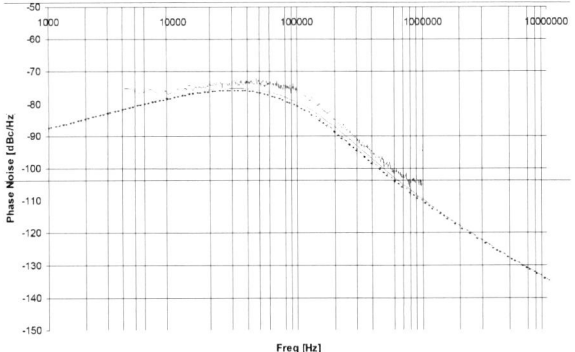

Fig. 10 Phase-noise simulation of the VCO

The reconfigurable LPF was simulated using 0.18 μm RF-CMOS technology and its simulation results are shown in Fig. 11. The complete range can be covered by switching on additional capacitor arrays at the output. Four different tuning voltages are simulated for minimum and maximum capacitance. The filter can handle input voltages of about 700 mVpp using 3.3V transistors with a thicker polysilicon gate available in 0.18 μm RF-CMOS technology kit.

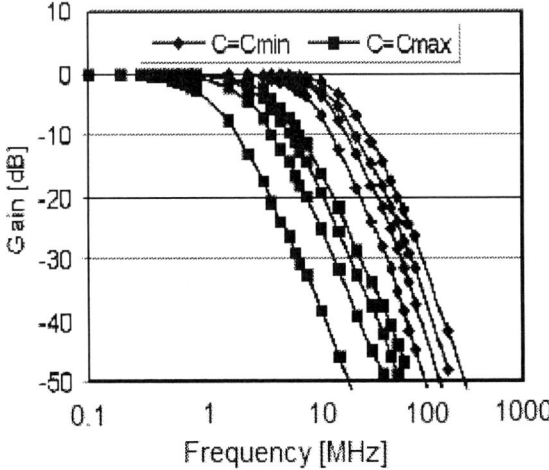

Fig. 11 Frequency response of the reconfigurable LPF

V. SUMARRY

The design, measurement and simulation results of wideband LNA, switching mixers, VCO and

reconfigurable LPF for WiMAX systems have been presented. The receiver uses a zero-IF architecture as OFDM signals with a null DC subcarrier are employed for WiMAX application. The circuit was implemented in 0.18 μm RF-CMOS technology.

The conversion gain, DSB NF and intermodulation distortion requirements of LNA and switching mixers fulfill the WiMAX specifications. The –3-dB bandwidth of the LNA is indeed less than 3.5GHz, but the gain, NF and intermodulation distortion at that frequency is still acceptable for WiMAX systems. As wideband matching at the input was employed, only single circuitry is needed to cover different RF-bands, leading to a low-cost solution.

The simulated phase-noise performance is better than WiMAX specifications require except for two frequency offsets, i.e. at 100 kHz and 1 MHz. The phase-noise at these frequency offsets are equal to –77 and –80 dBc/Hz, respectively.

Simulation results of reconfigurable LPF shows that the circuit can provide the required filtering action with a corner frequency of 875 kHz up to 7 MHz. The filter can handle input voltages up to 700 mVpp.

REFFERENCES

1) IEEE standards, IEEE Std 802.16-2004, Part 16: Air Interface for fixed BWA systems.
2) IEEE 802.16 working group, http://www.ieee802. org/16/pub/backgrounder.html.
3) E. Duvivier et. al., "A fully integrated zero-IF transceiver for GSM-GPRS quad-band application", IEEE JSSC vol. 38, pp. 2249-2257, Dec 2003.
4) Y. Le Guillou et. al.,"Highly integrated direct conversion receiver for GSM/GPRS/EDGE wit on-chip 84-dB dynamic range continous-time ΣΔ ADC", IEEE JSSC vol. 40, pp 403-411, Feb 2005.
5) P. Rossi et. el.,"A variable gain RF front-end, based on a voltage-voltage feedback LNA, for multi-standard applications", IEEE JSSC vol. 40 , pp. 690-697, March 2005.
6) R. Magoon et. al., "A single chip quad band (850/ 900/1800/ 1900 MHz) direct conversion GSM/ GPRS RF transceiver with integrated VCO's and fractional-N synthesizer", in ISSCC 2002 Dig., vol. 2, pp. 232-23, 2002
7) Hikita et al., "RF-circuit configurations and new SAW duplexers for single-end dual-band cellular radios", IMS Dig., vol. 4, pp.

1445-1448, 1999.

8) Y. Wu *et al.*, "RF bandpass filter design based on CMOS active inductors", IEEE Trans. CAS-II, vol. 50, no. 12, pp. 942 - 949,2003

9) E.H. Nordholt, "Design of high-performance negative-feedback amplifiers", Elsevier Science Publishing Company Inc., 1983.

10) U. Stehr *et. al.*,"A fully differential CMOS integrated 4th order reconfigurable gm-C lowpass filter for mobile communication", Proc. ICECS2003, 2003.

Characterization of Contact Etching Profile for 0.35um Analog Mixed Signal Product Development

Ahmad Sabirin Zoolfakar*, Azlina Mohd Zain[†]
*Technology Development, Microelectronics and Semiconductor,
Email: ahmads@mimos.my
[†]Process Engineering, Wafer Fabrication Operation
Email: amz@mimos.my
MIMOS Berhad, Malaysia
Telephone: (603) 8996 5000, Fax: (603) 8657 9441

Abstract— An Analog Mixed Signal (AMS) device incorporating polysilicon insulator polysilicon (PIP) capacitor and polyresistor sub-modules has been fabricated using 0.35um CMOS technology. Addition of the analog sub-modules has introduced topography height difference between the PIP capacitor region and the MOS transistor region. The resultant topography had resulted in variation of the pre-metal dielectric (PMD) layer thickness where contact holes will be formed. Topography height difference had also resulted in PMD layer thickness variation between one location to another. The thickest PMD layer was located on the MOS transistor active region, while the thinnest PMD layer was on the polyresistor region. Etching contact holes on such topography along with different etch stop materials is challenging. Hence, contact etching process needs to be optimized to ensure contact holes at all locations are cleared without too much overetching of the underlying layer. Furthermore, the etched profile needs to be slightly tapered to achieve good barrier metal step coverage. A new contact etching process that fulfilled the device requirements with good contact resistance parameters has been developed. The process optimization experiments and the electrical test results of the contact resistance are presented and discussed in this paper.

I. INTRODUCTION

Today's application needs of more complex and high performance systems have lead to innovations in the integrated circuits (IC's) architecture. There are few applications where analog functions cannot be replaced with their digital counterparts regardless of the advances in technology [1]. Thus, designers have merged the analog and digital circuitry on the same IC known as Analog Mixed Signal (AMS) devices. AMS technology allows capacitors, as well as resistors, transistors, diodes and other circuit elements to be formed together in semiconductor integrated circuits. Capacitors formed within digital integrated circuits typically provide storage locations for individual bits of digital data. One type of capacitor which is widely used is the polysilicon-insulator-polysilicon (PIP) capacitor.

A PIP capacitor fabricated within an integrated circuit is typically formed horizontally on a semiconductor wafer, with two electrodes sandwiching a dielectric layer parallel to the wafer surface. Polysilicon is used as the top and bottom electrode material, separated by a thin dielectric layer.

Contact openings in PMD layer are used for connecting the first metal to the devices, in this case, the MOS transistor, capacitor and polyresistor analog devices. Contact is acting as a bridge between the front-end and the back-end modules in VLSI fabrication. It is critical to achieve good contact resistance at the MOS transistor regions to ensure no delay in these devices [2].

II. ISSUES IN CONTACT HOLE ETCHING

Incorporation of the PIP capacitors and resistors sub-modules in the standard CMOS technology had caused topography height difference. This is because the PIP capacitors and polyresistors reside on the field oxide (FOX) area which is at higher ground compared to the substrate level. Furthermore, the double polysilicon structure of the PIP capacitor had made the resultant topography of the PMD

0-7803-9730-4/06/$25.00 ©2006 IEEE   363

insulation layer even worse. Figure 1 illustrates the topography of the PMD layer on the capacitor region, the polyresistor region and the MOS transistor region. As a result, the thickness of the PMD layer also varies from the logic device region to the analog device region, with the PMD on the MOS transistor area being the thickest.

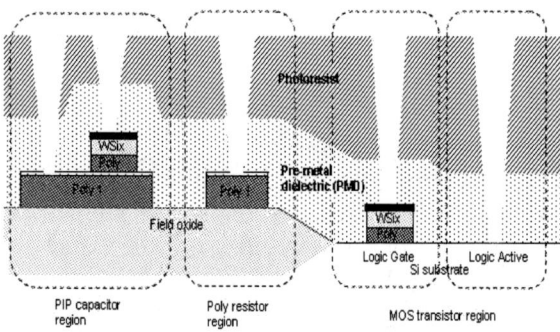

Fig.1 Cross-sectional representation of the PIP capacitor, polyresistor and MOS transistor region

The thickness variation had caused difficulty in etching contact holes into the PMD layer. Contact holes are formed into the PMD film by performing a dry etching process, exposing the films up to top plate capacitor, bottom plate capacitor, polyresistor, gate electrode and the source/drain region or better known as active region. It is important to ensure that the contact hole etching process has cleared the PMD insulation layer at all regions simultaneously. The contact resistance will increase significantly even if a very thin layer of the PMD material remained in the contact hole [2]. Another difficulty in etching the contact holes is that the different types of materials the etching process has to etch through and stop on (etch stop materials). The etching process has to etch through ONO layer on top of PIP poly 1 and polyresistor, stopping on polysilicon. On PIP poly 2 and logic polygate, the etching has to stop on tungsten silicide (WSix) while on logic active area the etching has to stop on the silicon substrate. These etch process requirements calls for a good etch selectivity i.e. etch rate ratio of the etched material to the underlying films or the masking material.

Another condition that contact etch process needs to fulfill is the etched profile i.e. shape of the contact hole. A good contact etch profile is essential to ensure good barrier metal step coverage over the contact hole [3]. This means an etched profile with a slight inclination (taper)

of the sidewall with top corner rounding is preferred to achieve good metal step coverage.

Another important parameter that needs to be control is the critical dimension (CD) of the contact holes. The CD size will also affect the contact resistance value [4]. The contact CD variation is caused by the topography height difference of the PMD layer. This is due to the lithography problem when patterning on high and low topography. Etching very small contact holes of size less than 0.5um along with PMD layer thickness variation will face etch microloading effect or RIE lag phenomena. High aspect ratio (height/width) contact hole etches slower than low aspect ratio contact hole [3]. This means contact etching process cannot clear the oxide at all locations simultaneously at a given etch time. A certain percentage of overetch time could be inserted in the etching process step to solve this issue but this requires a high etch selectivity to minimize underlying film loss. Meeting all of the contact etch process requirements is a challenge to the etch process engineer. In this paper, the optimization process involved in developing the contact etching recipe for 0.35 AMS technology is discussed and the electrical test results are presented.

III. EXPERIMENTAL

Contact hole etching process is performed using a Tokyo Electron Limited (TEL) plasma etcher. Plasma etching process consists of ionized gas particles which chemically react with the film on the wafer surface. The by-products of the reaction is volatile under plasma condition which is enables them to be pumped out from the system [4]. The TEL plasma etcher is designed in such a way that the etching is also assisted by high energy ions. This allows the etching process characteristic to be anisotropic i.e. etching rate in vertical direction is faster than the horizontal direction. This type of etching is also known as reactive ion etching (RIE) [2].

In this experiment, a standard 0.5um CMOS contact etching process was chosen as the baseline process recipe condition. This was because the 0.35um CMOS PMD layer planarization scheme follows exactly as the 0.5um CMOS process flow. The baseline recipe consisted of CF_4, CHF_3, Argon and O_2 gases in the ratio of 6:4:60:1, at 300mTorr pressure and 900Watt RF power. The CF_4 gas is the primary gas etchant, while CHF_3 gas promotes a protective film layer that increases the etch selectivity to the underlying layer. The Argon gas provides ion bombardment and promotes chemical reaction. Addition of a small amount of oxygen gas (O_2) is to minimize CD loss i.e. reduction in critical dimension size between pattern and after etching process [5]. A fifty percent overetching is performed after the main etch step to ensure complete removal of the oxide film at the thickest PMD layer. The baseline etching process recipe consists of two etch steps:

Step 1: Main etch (ME); etch time 1 minute 37 seconds

Step 2: Over etch (OE); 50%

The main etch step is to remove the bulk of the PMD oxide layer while the overetch step is to remove the remaining oxide on areas where the PMD thickness is thicker.

A short loop test wafers that comprised the PMD layer and litho contact pattern were prepared for the experiment. Typically, the PMD layer consists of an undoped silicate glass (USG) and boron and phosphorous doped silicate glass (BPSG). The PMD layer was then reflowed (annealed) at 850°C. Contact holes pattern was performed by lithography and the litho contact CD was then measured by CDSEM. The etch process time was calculated based on the etch rates of the annealed USG and BPSG films. A preliminary test run test using the standard 0.5um baseline recipe was performed to gauge the etch depth and its profile. The individual thickness of the USG and BPSG films that make up the PMD layer were measured by FIB and FESEM to better estimate the etch process time. In order to control the etch depth and CD of the contact holes, a new three-step-etching process condition was introduced with two main etch step namely ME1, ME2 and overetch, OE.

The ME1 etching rate is higher for better CD bias control. This step is used to etch a certain depth of the contact hole with CD control. The ME2 step has a slower etching rate for better selectivity to silicon purpose. The ME2 process condition will also introduce a slight taper of the sidewall. Another new etching process step named as corner rounding step (CRS) was introduced to minimize cusping effect of the top corner of the contact hole. Two Argon gas flow set points in the CRS steps i.e. 100sccm and 300sccm flow were experimented.

A total of eight experimental runs were performed with some parameter changed in each run. The summary of the optimization experimental runs are summarized in Table 1. Upon completion of each experimental runs, the photoresist mask layer was removed by plasma ashing process followed by a solvent cleaning step to remove any polymer residues that reside in the bottom of the contact holes. The test wafers were then sent to Failure Analysis lab for FIB and FESEM cross-sections of the contact holes at the MOS transistor and analog regions.

TABLE 1. Summary Of Experimental Runs

Run	Process Step	Contact etching process condition
1	1 – Main etch 1 2 – Main etch 2 3 – Overetch	CF4/CHF3/Ar/O2 30/20/300/5 sccm, 300mTorr, 900Watt, time 1:26s CF4/CHF3/Ar 30/20/300 sccm, 300mTorr, 900Watt, time 46s 30%
2	1 – Main etch 1 2 – Main etch 2 3 – Overetch	CF4/CHF3/Ar/O2 30/20/300/5 sccm, 300mTorr, 900Watt, time 1:26s CF4/CHF3/Ar 30/20/300 sccm, 300mTorr, 900Watt, time 46s 50%
3	1 – Main etch 1 2 – Main etch 2 3 – Overetch	CF4/CHF3/Ar/O2 30/20/300/5 sccm, 300mTorr, 900Watt, time 1:26s CF4/CHF3/Ar 30/20/300 sccm, 300mTorr, 900Watt, time 46s 80%
4	1 – Main etch 2 – CRS 3 – Main etch 2 4 - Overetch	CF4/CHF3/Ar/O2 30/20/300/5 sccm, 300mTorr, 900Watt, time 1:26s O2/Ar 40/100 sccm, 1600mTorr, 900Watt, time 15s CF4/CHF3/Ar 30/20/300 sccm, 300mTorr, 900Watt, time 46s 30%
5	1 – Main etch 1 2 – Main etch 2 3 – Overetch 4 - CRS	CF4/CHF3/Ar/O2 30/20/300/5 sccm, 300mTorr, 900Watt, time 1:26s CF4/CHF3/Ar 30/20/300 sccm, 300mTorr, 900Watt, time 46s Overetch 30% O2/Ar 40/100 sccm, 1600mTorr, 900Watt, time 15s
6	1 – Main etch 1 2 – CRS 3 – Main etch 2 4 - Overetch	CF4/CHF3/Ar/O2 30/20/300/5 sccm, 300mTorr, 900Watt, time 1:26s O2/Ar 40/300 sccm, 1600mTorr, 900Watt, time 15s CF4/CHF3/Ar 30/20/300 sccm, 300mTorr, 900Watt, time 46s Overetch 30%
7	1 – Main etch 1 2 – Main etch 2 3 – Overetch 4 - CRS	CF4/CHF3/Ar/O2 30/20/300/5 sccm, 300mTorr, 900Watt, time 1:26s CF4/CHF3/Ar 30/20/300 sccm, 300mTorr, 900Watt, time 46s Overetch 30% O2/Ar 40/300 sccm, 1600mTorr,

		900Watt, time 15s
8	1 – Main etch 1 2 – Main etch 2 3 – Overetch 4 - CRS	CF4/CHF3/Ar/O2 30/20/300/5 sccm, 300mTorr, 900Watt, time 1:26s CF4/CHF3/Ar 30/20/300 sccm, 300mTorr, 900Watt, time 46s Overetch 30% O2/Ar 40/300 sccm, 1600mTorr, 900Watt, time 15s

IV. RESULTS AND DISCUSSION

A. Physical Cross Section

A preliminary test run using the standard baseline contact etch recipe resulted in significant overetching of the underlying film especially at the MOS transistor region (Figure 2). The WSix film on top of poly 2 layer had almost been etched through while about 120nm of silicon had been etched through at the active region. The same effect could also be seen at the capacitor and polyresistor regions where the poly 1 and poly 2 films were etch almost halfway through.

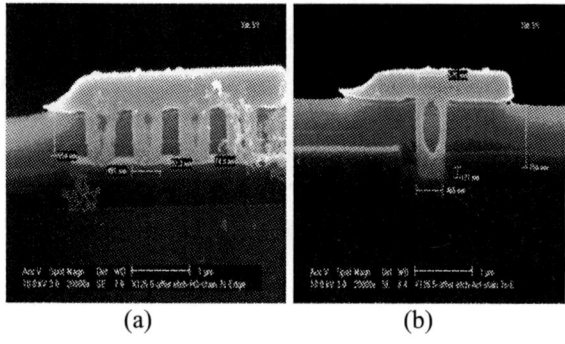

(a)   (b)

Fig. 2 Cross-sections of contact holes using 0.5um CMOS baseline contact etching process condition

The overetching depth in the MOS transistor regions needs to be strictly controlled as this would affect the contact resistance value of contact on polygate as well as on the active source and drain area. Failure to meet the contact resistance specification would result in failure of the transistor operation. Judging from past experience in 0.5um CMOS process development, the silicon overetching at the active region should not be more than 500A in order to achieve a good contact resistance value [7].

The conclusion made from the preliminary run was that the 0.5um CMOS baseline contact etching process time was not optimum for the smaller size of contact holes even though the same PMD layer scheme was adopted for the 0.35um process. This was perhaps due to less

pattern density that resulted in higher etching rate of the film. Another factor that could contribute to the overetching condition is the over estimation of the etching process time. The actual PMD thickness at the transistor region after the reflow process was probably not as initially deposited. The ME1 etching process time was later recalculated based on the actual PMD thickness measured through FESEM cross-sections.

The new estimated etching time was reduced to 1 minute 26 seconds for the ME1 step. A test run was performed using ME1 step only to check the resulting etching depth. FESEM cross-section showed that there was about 180nm of PMD film remaining at the logic active region. It was decided that the next etching step to clear the remaining PMD layer should have a high selectivity to silicon to prevent too much overetching.

A new etching step ME2 with a better selectivity to silicon was introduced in the next test run with 30% overetch time to clear the PMD in the bottom of the contact hole (Run 1). Two other runs were performed using a 50% (Run 2) and 80% (Run 3) overetch time for evaluation purpose. Results showed that a 30% overetch time is the optimum etch time for the contact etching without removing too much of the underlying film at all regions. However, the etch profile showed some degree of cusping at the top corner of the contact holes.

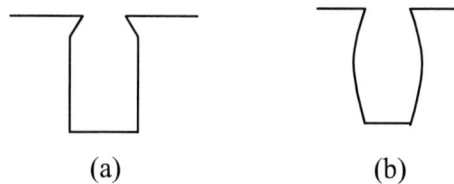

(a)   (b)

Fig. 3 Illustration on the cusp (a) and bow profile (b)

Cusp shape profile is not preferred since it will cause barrier metal overhang during metallization process [4]. To improve this condition, a new etch step, Corner Rounding Step, CRS with 100 sccm Argon flow was introduced to remove the cusping effect. Initially, it was thought that by inserting the CRS step after the ME1 step could remove the cusping effect. However, results from Run 4 did not show any desired improvement. Instead, a slight bowing profile (curved sidewall) was observed. Another trial run employing the CRS step was performed by inserting it in the final step of the etching process (Run 5). Again, the result was not successful as there was hardly any improvement on the profile. Figure 3 illustrate the difference of cusp and bow profile.

The current CRS step process condition was not strong enough for the top corner rounding effect. It was thought that more physical bombardment effect could be achieved by increasing the Argon gas flow. Therefore another attempt to employ the CRS step was performed with 300sccm Argon flow. Run 6 and Run 7 were performed by inserting the new CRS step process condition in the second and final step respectively. FESEM cross-section results showed that a good contact etch profile was obtained by employing the new CRS step process condition in the final etch step. The etch profile showed only 499A of silicon overetch in the contact active region with no cusping at the top corner of the contact holes. A slight taper of the contact profile gave a good barrier metal lining in the contact hole and tungsten plug filled-up the contact hole perfectly (Figure 4).

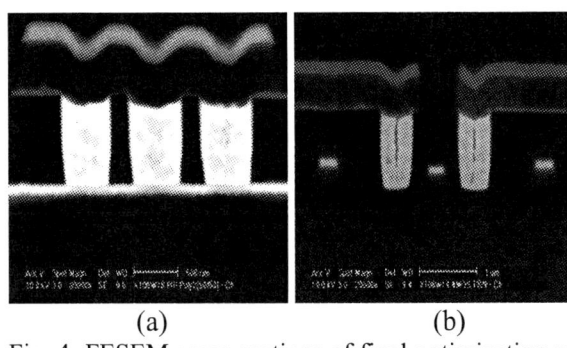

(a)   (b)

Fig. 4 FESEM cross-sections of final optimization of the contact holes at MOS transistor on Poly 2 (gate electrode) (a) and on active (b)

B. Electrical Contact Resistance Results

In order to determine the effectiveness of the optimized contact etch recipe, the recipe was tested on full AMS device wafers for electrical measurement. Contact resistance measurement was carried out on the wafers to test the opening of the contact holes [6]. Generally, good contact holes will introduce low contact resistance. Table 2 shows the electrical contact resistance specification and result for both active and capacitor region. From the table, one can observed that the contact recipe was well optimized since all the contact resistance results are within the specification range.

ITEM	SPECIFICATION	ACHIEVED
0.4um N+ Contact	< 100ohm/hole	68ohm/hole
0.4um P+ Contact	< 150ohm/hole	140ohm/hole
0.4um Poly2 Contact	< 10ohm/hole	7ohm/hole

V. CONCLUSION

A new contact etching process condition that fulfilled the requirements of the 0.35um analog mixed signal product has been developed. A three-step etching process that consisted of main etch 1 (for etch depth and CD control), main etch 2 (with high selectivity to silicon) and a 30% overetch were found to give the best contact etch profile. A slightly tapered etch profile that allows a good metallization process with minimum overetching of the underlying layer has been achieved. Electrical test results also showed that the contact resistance values achieved were within the specified range.

ACKNOWLEDGEMENT

The authors would like to thank the Failure Analysis Department team for their efforts in producing the FESEM and FIB cross-section images.

REFERENCES

[1] Behzad Ravazi, Design of Analog CMOS Integrated Circuits, McGRAW-HILL, 2001.
[2] Stanley Wolf and Richard N. Tauber, Silicon Processing for the VLSI Era, Lattice Press, 1986.
[3] Stephen A. Campbell, The Science and Engineering of Microelectronic Fabrication, Oxford University Press, 1996.
[4] S. R. Wilson, Clarence J. Tracy and John L. Freeman, Handbook of Multilevel Metallization for Integrated Circuits, Noyes Publications 1993.
[5] Tokyo Electron Limited, TEL-Etch Multi Chamber System Unity 85PE Rev.3.2.4 Process Manual, April 2000.
[6] Dieter K. Schroder, Device Measurements, John Wiley & Sons, 1998.
[7] MIMOS Berhad, 0.5um CMOS C5 Process Parameters, Mysem, 2005.

TABLE 2. Electrical Contact Resistance Specification

Comparison of the Growth Si-based Crystalline Silicon Carbide (SiC) by Chemical Vapor Deposition (CVD) using Carbon Monoxide (CO) and Treated Carbon Dioxide (CO₂)

A.Y.K. LIM[1] and K. IBRAHIM[2]
[1,2]NOR Laboratory, Department of Physics,
University of Science Malaysia,
11800 Penang, Malaysia.
E-mail: [1]alexlimyk@hotmail.com; [2]kamarul@usm.my

Abstract **Silicon carbide (SiC) has received special attention in recent years because of its remarkable properties. This work presents the investigation on the growth of Si-based SiC using carbon monoxide (CO) compared to the treated carbon dioxide (CO₂) as reported earlier. Experiments results has revealed the existence of Si-C bond and the bond formed on silicon (Si) surface through the characterization using X-ray diffraction (XRD) and raman spectroscopy (RS). Thickness study is carried out show that growth using carbon monoxide has a thicker layer of SiC at the same growth condition compared to treated carbon dioxide. The reflective index (RI) of the growth SiC was measured. This growth technique is promising and shows great potential of producing relatively desirable quality SiC films for electronic devices fabrication.**

I. INTRODUCTION

SILICON CARBIDE, one of the known promising fabrication material for electronic devices, has received special attention in recent years because of its remarkable properties such as wide band gap, high thermal conductivity, high breakdown field and high saturation drift velocity. There have been numerous efforts, some of which are still on going; to grow crystalline SiC layers on top of large-area wafer substrate materials such as silicon or sapphire [1].

One of the factors that have inhibited the development of SiC device technology is the absence of low-cost SiC substrates. Reproducible wafers of reasonable consistency, size, quality, and availability at acceptable cost are a prerequisite for commercial mass production of semiconductor electronics [2]. The motivation for the heteroepitaxial growth of SiC on Si

surface has been to provide relatively inexpensive and large-area substrate of SiC for electronic devices. However, the growth techniques of SiC crystals are relatively immature and device fabrication technologies are not yet sufficiently developed to the degree required for widespread use.

There are several examples of work that have been done by chemical vapor deposition (CVD) using separate precursors for Si and C. For example, Nishino et al, reported a single crystal film of 3C-SiC were achieved by CVD on silicon substrate [001] but has hide defect density [3]. Powell et al. used SiH_4 and propane to grow single-crystalline SiC on Si surface at 1360C [4]. Nagasawa and Yamaguchi have also reported the growth of 3C-SiC on 111 oriented Si using SiH_2Cl_2 and C_2H_6 by low pressure chemical vapor deposition (LPCVD) at a temperature of 1000C [5].

This paper present the continuous work on investigation of the growth of Si-based silicon carbide using carbon monoxide (CO) and compared to previous results using the treated carbon dioxide (CO_2)[2]. In the experiment at setup, either CO_2 or CO is used as the precursor for the atmosphere pressure chemical vapor deposition (APCVD) process to grow the SiC. CO_2 gas is shocked under a specially made UV light treatment box but not the CO gas. The objective of the treatment using the UV is to create unstable bond of C-O.

It is expected that the growth using CO gas helps improvement in growth thickness SiC growth. The polytypes of crystalline structure of SiC is influenced by the effects of temperature and precursor flow rate to the substrate. At the temperature setting proposed in the experiment, cubic type of SiC is envisioned to be grown as 3C-SiC is known as a low-temperature polytypes owing to its higher probability of appearance at low temperatures [6].

0-7803-9730-4/06/$25.00 ©2006 IEEE

II. EXPERIMENT

A p-type silicon wafer with orientation of [100], cut into squares of 5mm by 5mm is used. Standard RCA cleaning technique is utilized for cleaning the samples then removal of oxide by HF. After rinsing the samples in deionized water and drying, they were sent for atmosphere pressure chemical vapor deposition (APCVD) process for the growth. The APCVD process utilizes the Nabertherm R70/9 furnace and it has an Edwards digital mass flow meter attached to it. The precursor is pumped into the system through a gas control system. The digital flow meter is used to monitor the pressure and gas flow inside the furnace. The samples will be exposed and a desired layer will be growth on the samples.

An ultraviolet-rays (UV-rays) box is added and connected to the furnace in the setup when CO_2 gas is used. It is made by special stainless steel chamber with two tubes of UV lamp (up to 16W) are fixed on the inner surface of the cover is built. The special lamps emit UV-rays with 184.9nm, which has a very short wavelength in the VUV range. The ideal here is to provide treatment to the carbon dioxide when the gas flows through it. UV-rays will decompose the carbon dioxide bond and will exist in unstable C-O bond [7].

For the experiment of using CO gas, UV ray setup box is disconnected from the system where the flow control system is directly connected to the furnace chamber. Fig. 1 below shows the setup of the experiment.

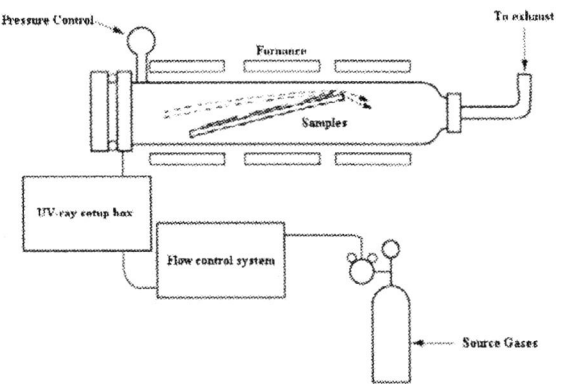

Fig. 1 APCVD with UV-ray box setup

To make a meaningful comparison study, the growth was fixed at the temperature of 900C and flow rate of 0.5L/min for 1.5 hours. Several characterization techniques were employed to investigate the growth SiC films which include thickness and RI measurement, X-ray diffraction (XRD) spectra study and raman spectra (RS) plotting

III. RESULTS AND DISCUSSION

Filmmetric F20 is used for the measurement of the thickness and the reflective index (RI) value for the growth SiC. This optical based equipment allows a non-contacting technique for the measurement. To increase the accuracy, several spot where taken for measurement in order to obtain the average value. From the results, it is found that the thickness of SiC grown using CO gas is significant thicker then using CO_2 gas. The key difference here likely is due to the total amount of existed unstable C-O bond that is going to react with Si to form SiC. The higher the concentration of existence of the CO will better promote the growth of SiC.

From the measurement, the average thickness grown is 70nm-100nm by using treated CO_2 gas and ~600-650nm by using CO gas (Fig. 2). It is 6x greater in magnitude of the growth thickness or the growth rate by using CO gas versus treated CO_2 gas.

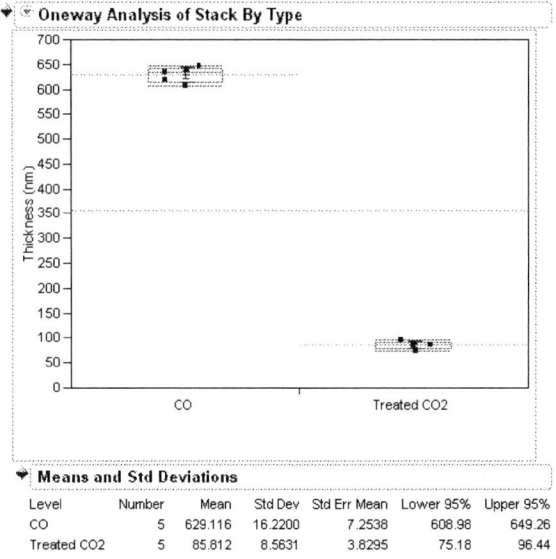

Level	Number	Mean	Std Dev	Std Err Mean	Lower 95%	Upper 95%
CO	5	629.116	16.2200	7.2538	608.98	649.26
Treated CO2	5	85.812	8.5631	3.8295	75.18	96.44

Fig. 2 Thickness measurement (side by side comparison for treated CO_2 versus CO)

The RI is measured at the wavelength of 638nm and the results are in the range of 2.5-2.7 which is closed to the optical properties of the SiC [8].

X-ray diffraction (XRD) experiments were performed by means of a PANanalytical X'Pert

PRO MRD system equipped with an incident beam optics, diffracted beam optics and detectors for high speed data collection. The phase analysis of the sample was collected with continuous scan step of 0.2 degree for 3s from 35 to 90 degree.

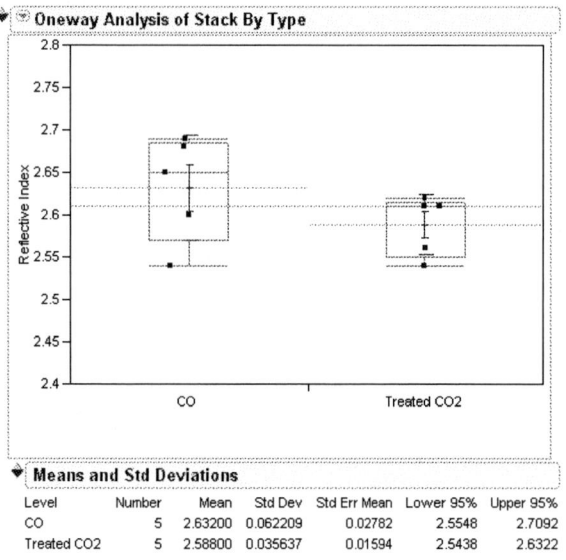

Means and Std Deviations

Level	Number	Mean	Std Dev	Std Err Mean	Lower 95%	Upper 95%
CO	5	2.63200	0.062209	0.02782	2.5548	2.7092
Treated CO2	5	2.58800	0.035637	0.01594	2.5438	2.6322

Fig. 3 Reflective Index measurement (side by side comparison for treated CO_2 versus CO)

XRD spectrums of a sample growth using treated CO_2 and CO are shown in Fig.4. Both spectra reveal a significant peak at $2\theta=\sim41°$ (primary order reflection) while for CO, there is additional small peak at $2\theta=\sim90°$ (secondary order reflection). All these peaks are referred to the cubic type (3C) of crystalline SiC [9]. It is noted also that the orientation of the overgrowth and the substrate is parallel (cubic-SiC//Si) and the evident is strong here that the overgrowth has a strong preferred orientation along the [100] direction. In general, the height of a peak is strongly dependent on the thickness of the sample. Layer of SiC growth grown using treated CO_2 can be too thin that the peak of $\sim90°$ cannot be detected.

For sample grown using CO, there is others peaks that correspond to others polytypes but not shown here as it is be too weak and from the results it is confirmed that 3C SiC is the dominant structure grown. Si has a peak at $2\theta=\sim69°$

Raman spectra (RS) were recorded using a model of Jobin Yvon LabRAM HR800-UV Raman Microscope (wavelength of Cu Kα ~ 1.5405 Å) equipped with a double monochromator and a charge-coupled detector

(CCD) cooled with liquid nitrogen. The samples analyzed at room temperature and the raman spectrum is taken at several different spots on samples to gauge the homogeneity of the growth samples.

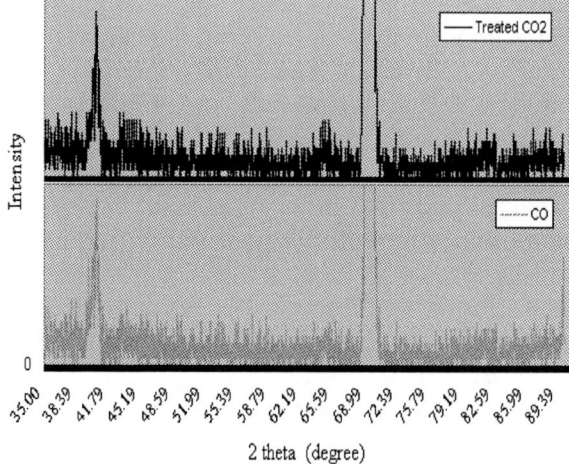

Fig. 4 X-ray diffraction spectra 2θ (degree) scans of SiC growth on Si (CO2 versus CO)

Both either the growth using treated CO_2 or CO exhibit the same profile. Fig. 5 shows the raman spectra of the SiC growth on Si.

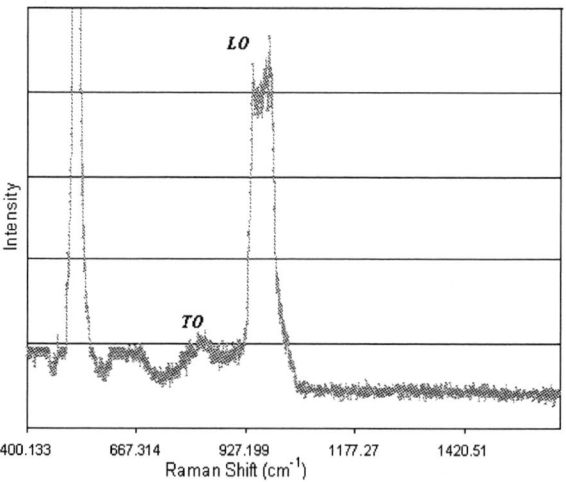

Fig. 5 Raman spectra of the SiC growth on Si (similar profile for growth using treated CO2 and CO)

The crystalline Si raman mode at $521cm^{-1}$ is revealed and amorphous Si-Si raman mode at $480cm^{-1}$. The most important raman mode indicated here is the Si-C. The band observed in the region of $900-1000cm^{-1}$ is associated with Si-C bonds for the LO mode [10]. Also, in this region, it is noted that the existence of the second-order raman band of silicon co-exists at $\sim970cm^{-1}$. A weak TO mode appears with intensity of only a few percent of the LO band intensity showed in the range of 780cm^{-1} to 800cm^{-1} [11]. In additional, there is no raman mode of 1300-1600cm^{-1} which is performed by C-C bond detected.

In consequences, based on the XRD and RS results, microcrystalline structure of Si-C thin film is formed on the oriented Si surface.

IV. CONCLUSION

Silicon carbide (SiC) was grown on the p-type silicon [100] by atmosphere pressure chemical vapor deposition process using treated carbon dioxide or carbon monoxide. The growth thickness improve when existence of higher concentration of C-O bond. SiC growth is confirmed by the X-ray diffraction and raman spectra analysis. The technique is promising and shows great potential of producing relatively desirable quality SiC films. More study is being carried out to further characterize SiC growth and improvement in the purity of crystalline structure grown.

ACKNOWLEDGEMENT

The authors would like to thank the Malaysian Government for sponsoring this work under the IRPA grant. Also, to En. Muthalib and Cik Ee BC for equipment setup, En. Azhar for supporting the X-ray diffraction measurement, Mr. Kong PQ for raman spectra analysis and all the USM NOR Laboratory assistants for the support.

REFERENCES

[1] P. G. Neudeck, "Silicon Carbide Electronic Devices," Encyclopedia of Materials: Science and Technolgy, 9, K. H. J. Bushchow, R. W. Cahn, M. C. Flemings, B. Ilschner, E. J. Kramer, and S. Mahajan, Eds. Elsevier Science, 8508-8519 (2002).

[2] S.madapura, A.J.Steckl and M.Loboda, "Heteroepitaxial Growth of SiC on SI(100) and (111) by Chemical vapor deposition Using Trimethylsilane", J. Electrochem. Soc., 146, 3, 1197-1202(1999)

[3] Nishino,S. J.A. Powell and H.A. Will. "Production of large-area single-crystal wafers of cubic SiC for semicondutor devices", Applied Physics Letters 42(5), 460-462 (1983)

[4] J.A. Powell, L.G. Matus, and M.A Kuzmarki, "Growth and characterization of cubic SiC single-crystal films on Si", J. Electrochem. Soc.,134, 1558(1987).

[5] H. Nagasawa and Y.Yamaguchi, "Growth of 3C-SiC on Si(111) using SiH$_2$Cl$_2$ and C$_2$H$_6$ by LPCVD ", Springer Proceeding in Physics, Springer-Verlag, Berlin,71, 40 (1992)

[6] H. Matsunami and T.Kimoto, "Step-controlled epitaxial growth of SiC: high quality homoepitaxy", Material Science and Engineering, R20, 125-126(1997)

[7] C. Schulz, J.D. Koch, D.F. Davidson, J.B. Jeffries and R.K. Hanson, "Ultraviolet absorption spectra of shock-heated carbon dioxide and water between 900 and 3050K", Chem. Phys. Lett., 355, 82-88 (2002).

[8] G.L. Harris, "Properties of SiC", London UK: Inst of Electrical Engineers, 15-20(1995).

[9] K.W. Lee, K.S. Yu and Y. Kim, "Heteroepitaxial growth of 3C-SiC on Si (001) without carbonization, J. Crystal Growth, 179,153-160(1997).

[10] J. C. Burton, L. Sun, M. Pophristic, S. J. Lukas, F. H. Longa, Z. C. Feng and I. T. Ferguson, "Spatial characterization of doped SiC wafers by Raman spectroscopy", J. Appl. Phys, 84,11,6268-6273(1998).

[11] S. Nakashima and H. Harima, "Characterization of defects in SiC crystals by Raman Scattering", W.J. Choyke, H. Matsunami and G. Pensl, Eds. Springer: 586-605(2004).

Self Heating Characterization of 32V MOSFETs using Pulsed Gate Measurement

Alhan Farhanah Abd Rahim[1,2], Albert Victor Kordesch[1], *Senior Member,IEEE* and Yusman Mohd Yusof[1], *Member,IEEE*

[1]Silterra Malaysia Sdn Bhd, Kulim Hi-Tech Park, 09000 Kulim, Kedah.
[2]Faculty of Electrical Engineering, Universiti Teknologi MARA, 13500 Pulau Pinang.
Email: alhan_farhanah@silterra.com, al_kordesch@silterra.com

Abstract - **High voltage (HV) MOSFETs are extremely interesting for automotive, medical ultra-sound, flat panel LCD displays, RF and many other applications. One of the key phenomena related to the operation of HV devices is the self heating effect (SHE). Intuitively SHE represents the heating of the device due to its internal power dissipation. SHE results in a reduction of the drain current and the negative output conductance effect. In this work we measure the self heating effect of 32V MOSFETs fabricated in Silterra's 180nm HV CMOS process technology. We extract the equivalent thermal impedance of the device (thermal resistance, R_{TH} and capacitance, C_{TH}). These parameters are needed for circuit simulation for circuit design. A pulsed gate measurement is used in order to be able to measure the device characteristics when it is still "cold." The pulse width is typically 2μs. The influence of pulse duration is analyzed.**

I. INTRODUCTION

The aggressive reduction of feature size has lead to increased power dissipation per unit area in almost all modern silicon technologies, including integrated power devices and high voltage electronics, SOI and VLSI/ULSI. The characterization of both electrical and thermal effects and their interaction is necessary for understanding the device operation but also to precisely predict the safe operating area of advanced HV devices. Despite recent efforts, [1-5] to characterize SHE, using the reduction of the drain current and negative output conductance effect [1,2] in silicon, SOI and DMOS devices, there is still a need for a simple and efficient extraction procedure for the measurement of thermal resistance and capacitance for use in advanced HV

IC simulation. The aim of this work is to use the pulsed gate measurement technique[6] to characterize various sizes of HV devices fabricated in a 180nm HV CMOS technology. In this method, an extraction of the thermal parasitic network of the HV devices is done from the measurement of 'cold' and 'hot' devices. An analysis of the gradual impact of SHE pulsed measurement parameter pulse width on device electrical characteristics is presented. Finally we intend to use HSPICE simulations to validate the method and demonstrate that a thermal resistance model is useful for accurate simulation of HV MOS circuits.

II. MEASUREMENT SETUP

Various pulsed measurements aimed to access the HV and SOI MOSFET characteristics without SHE have been reported in previous work[5].

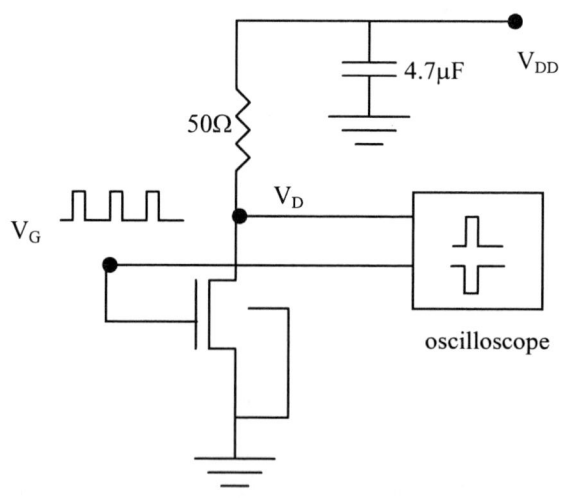

Fig. 1 Schematic diagram of our measurement setup used for SHE investigation in HV MOSFET.

In general, such measurement setup requires a dedicated and expensive configuration since they

0-7803-9730-4/06/$25.00 ©2006 IEEE

cannot be achieved with just a semiconductor parameter analyzer because microsecond pulses are required to avoid SHE. Herewith, we used a simple test setup able to measure the characteristics of the HV devices without SHE (fig 1) and investigate SHE as a function of the measurement parameters (pulse duration and duty factor) and device width. A pulse generator, Agilent 81110A is used to turn on and off the transistor by applying a square signal on the gate. The drain is biased through a fixed resistance by an Agilent HP4156B semiconductor parameter analyzer. A digital memory oscilloscope that triggers on the signal applied to the gate (V_G) is used for monitoring the drain voltage of the transistor (V_D). The supply line capacitance (C_{DD}) is used to reduce the peaks and stabilize the V_{DD} at constant level voltage when transition occurs. A value of 4.7uF has been used successfully in our particular setup.

III. EXPERIMENTAL RESULTS

The measurement setup relies on *on-off* transitions of the HV transistor. The oscilloscope triggers on the rising edge of the gate signal while biasing V_{DD} voltage and the drain are synchronously monitored on the other channel. The gate voltage is measured using DC coupling and AC coupling was used for V_D. A typical response of the transistor is shown in fig. 2.

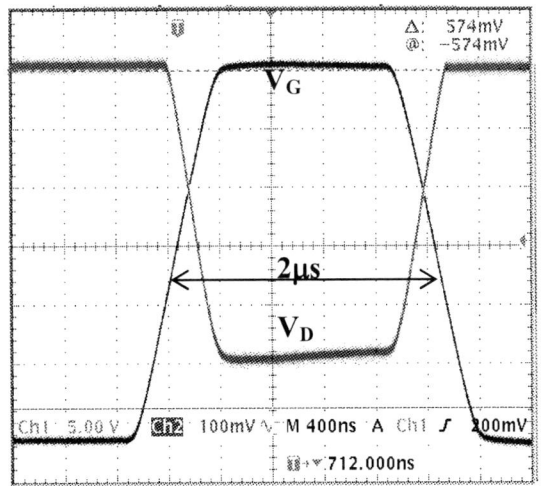

Fig 2 Dynamic response of HV NMOS to typical gate pulse (pulse width and period are: 2µs and 200µs respectively) recorded with experimental setup shown in fig 1. The V_G pulse is 32V and ΔV_D = 480mV.

The drain current is calculated from $I_D= \Delta V_D/50\Omega$. Fig 3 compares the output characteristics of the HV MOSFET with DC and pulse measurement, measured at VG = 32V and T = 25°C. It can be seen from the figure that for DC a negative conductance occurs because I_D drops at higher V_D caused by the self-heating of the device. As the device heats up, its channel mobility decreases, which in turn lowers its drain current.The pulsed measurement shows that SHE is practically eliminated at $V_D = V_G = 32V$. The I_D is increased by a factor of 1.07 with the pulsed measurement.

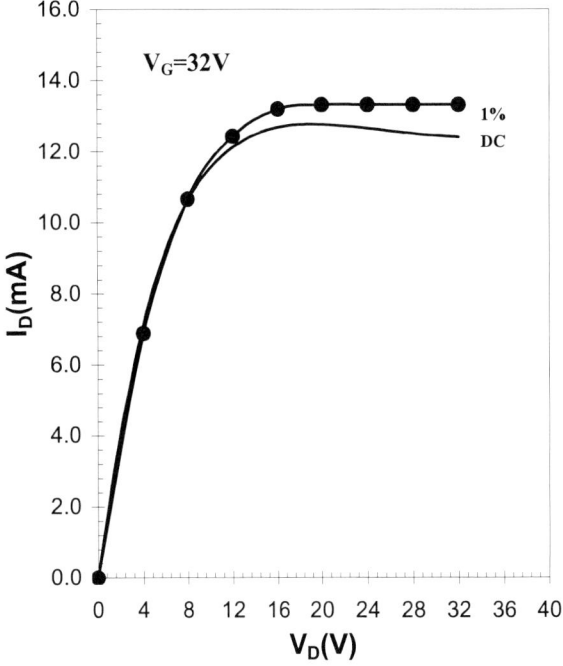

Fig. 3 Output characteristics, I_D-V_D, of HV NMOS (W/L=25µm/5.5µm) for DC(solid line) and pulsed(symbol) measurement with duty cycle 1% (pulse width = 2µs and period = 200µs) measured at T=25°C.

Different pulse widths were used to study the dependence of the output characteristics on SHE effect (please refer to fig 4). We set the rise and fall times to 500ns to reduce ringing. For long channel HV MOSFETs, the pulse width applied on. For pulses much shorter than 2µs, we were not able to measure I_D. It can be seen from fig 4 that SHE is practically eliminated at t=2µs.

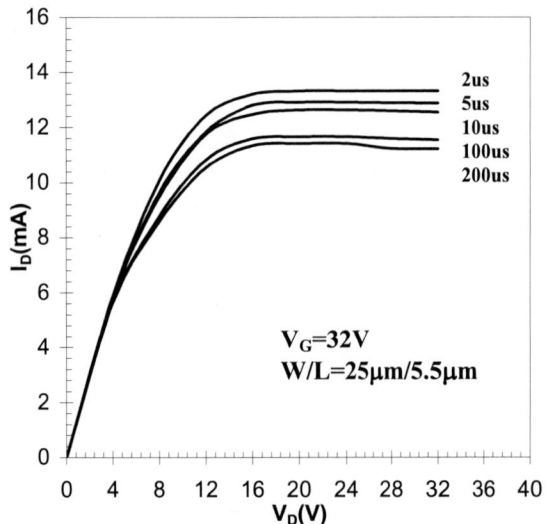

Fig. 4 Comparison of Pulse IV characteristics of HV NMOS for different pulse width (2μs, 5μs, 10μs, 100μs and 200μs).

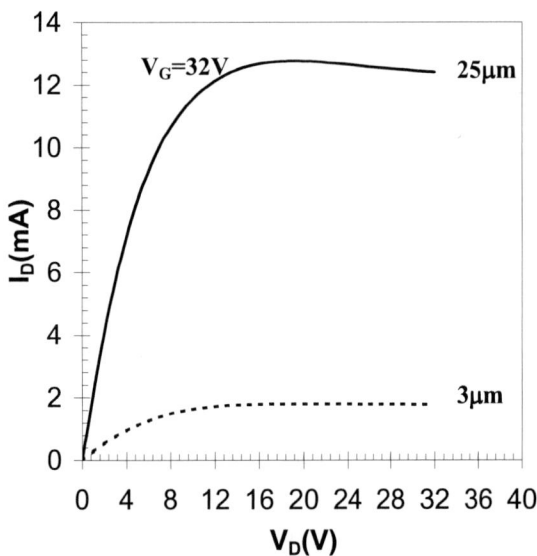

Fig. 5 Comparison of DC IV characteristics of HV NMOS for different device widths, 25μm(solid line) and 3μm(dotted line), with L=5.5μm and at T=25°C.

Fig 5 shows the behavior of I_D-V_D for different widths at V_G=32V and T=25°C. The figure shows that SHE is slightly higher for W=25μm as the negative slope of I_D-V_D is more pronounce than for device with W=3μm. Hence it can be said that for these bulk HV devices, the dominant heat evacuation is down through the single crystal silicon substrate. Fig 6 shows temperature rise as a function

of different widths measured at V_D=V_G=32V. This figure shows that the temperature rise is almost the same for W=25μm and 3μm transistors. Hence it can be said that the narrower device has almost the same SHE as the wider device. This shows that for our HV devices, the ratio of Power/Area is constant because I_D scales with device width,W.

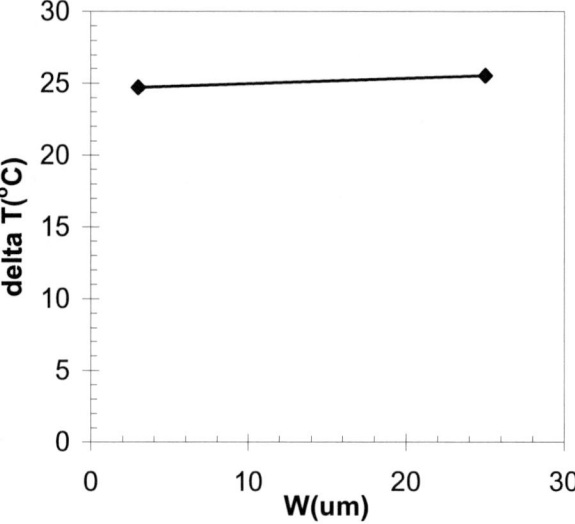

Fig 6 Temperature rise, delta T as a function of different widths at V_G=V_D=32V and L=5.5μm.

The conventional way to model SHE is based on the use of a thermal parasitic network that includes a temperature-dependent thermal resistance, R_{TH} and a thermal capacitance, C_{TH}[7-10]. The main objective is then to extract R_{TH} and C_{Th} values from the experimental impact of SHE on output, I_D-V_D, characteristics. In this approach, the device and wafer temperatures are treated as two node-voltages, while the power in the device act like a current generator. In steady state condition the temperature difference(due to SHE) is given by:

$$\Delta T = R_{TH}P_D \qquad (1)$$

Where R_{TH} is the thermal resistance and P_D is the injected power. We assume that R_{TH} is approximately constant over the small range of temperature.

When the drain voltage and consequently the drain current start increasing, the temperature will start rising with time constant related to the overall heat conductivity of the device. The mobility is to first order, the most affected parameter, and when the temperature rises it decreases and hence also the drain current.

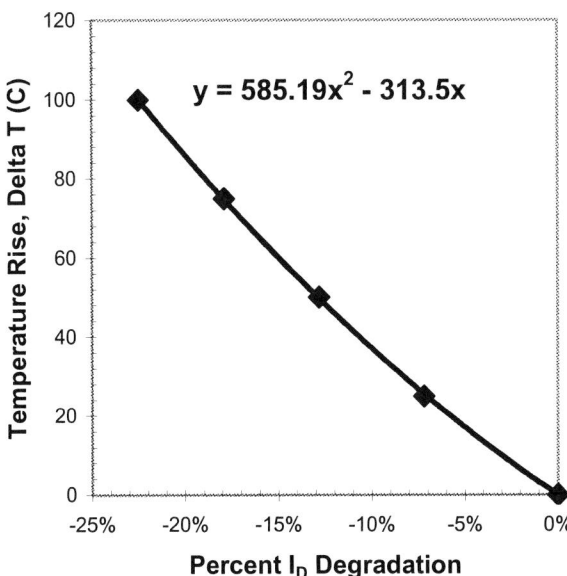

Fig. 7 Temperature rise of the device as function of I_D degradation for HV NMOS(W/L=25μm/5.5μm) at V_G=32V, V_D=32V.

Referring to fig 3, at V_D = 32V, I_D from pulse and DC measurement are 13.32 mA and 12.4mA respectively which gives the value of percent I_D degradation about –6.9%.

Fig 7 shows the temperature rise of the device as a function of percent I_D degradation at $V_D=V_G=32V$. From this figure, the delta T for percent I_D degradation of –6.9% is 24.4°C. Using equation (1), the thermal resistance for delta T = 24.4°C is 61.59^{0}C/W.

Fig 8 shows percent I_D degradation as a function of V_D for different temperatures. It can be seen that the decrease of I_D is not depend much on the external temperature changes at higher V_D. Hence we believe that R_{TH} for our devices can be approximated by a constant over this temperature range.

Fig 9 shows a transient drain current characteristics of the transistor at T=25°C. It shows that I_D drops as time increases. We believe that the transient decrease in drain current can be attributed to the self-heating of the device.

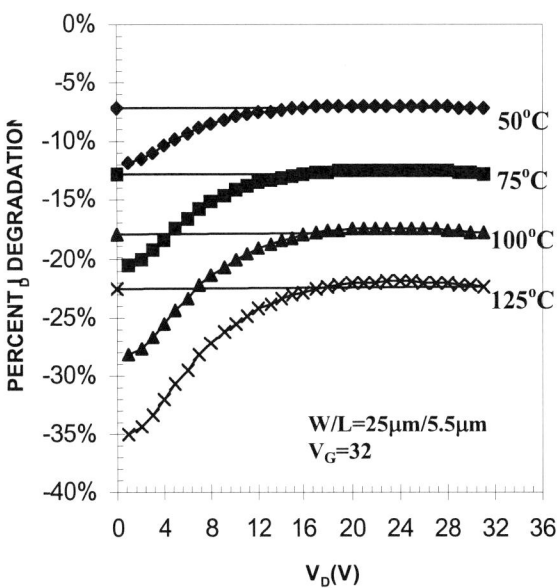

Fig. 8 Percent I_D degradation as function of V_D for different temperatures.

Since all surrounding electrical RC time constants are in the ns range, we can only attribute our measured time constant to thermal heating of the transistor volume.

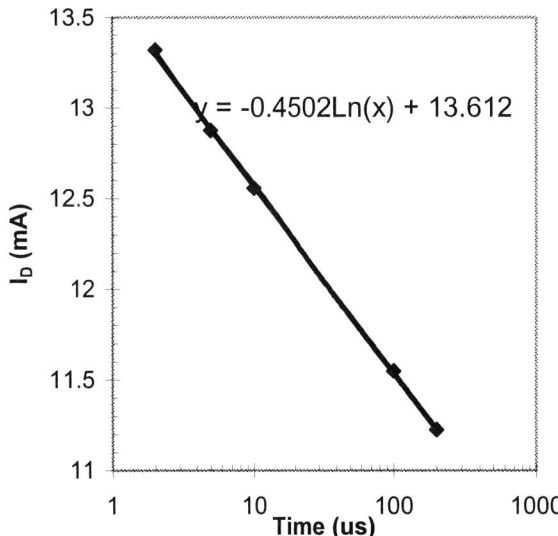

Fig. 9 Transient drain-current characteristics of HV NMOS (W/L=25μm/5.5μm).

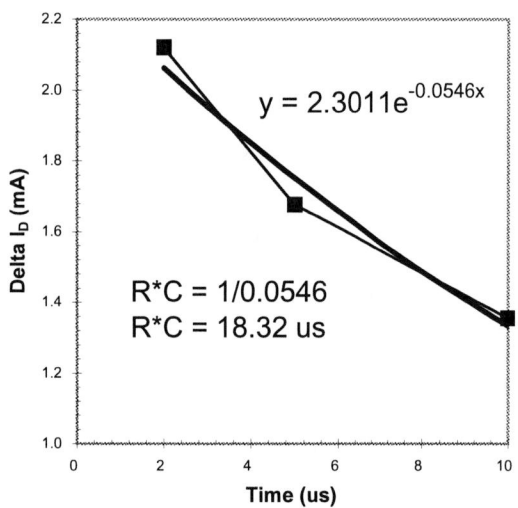

$$y = 2.3011e^{-0.0546x}$$

$$R*C = 1/0.0546$$
$$R*C = 18.32 \text{ us}$$

Fig. 10 Time constant for self heating of HV NMOS(W/L=25μm/5.5μm).

Fig 10 shows delta I_D as a function of time. The time constant of the device is extracted from the exponential curve where the equation is :

$$y=2.3011e^{0.0546x} \qquad (2)$$

The inverse of the exponential (0.0546) is the thermal time constant, τ_D of the device, 18.32μs. Hence the extracted thermal capacitance of the device is 297.4nF calculated from below equation:

$$C_{TH} = \tau_D/R_{TH} \qquad (3)$$

IV. Conclusions

An investigation of SHE effects in HV MOSFET based on pulsed gate measurement has been presented. It has been shown that the pulse width is important to define SHE free measurement conditions. We have shown that by using pulsed gate technique the negative conductance effect can be avoided. It is also shown that the temperature rise is almost the same for different widths. This shows that for our HV devices, the ratio of Power/Area is almost constant because I_D scales with device width. An extraction of device thermal time constant has been done by measuring the 'hot' and 'cold' devices. The extracted thermal resistance and capacitance from this device are 61.59°C/W and 297.4nF respectively. It is shown that thermal resistance of this device does not depend strongly on temperature within this range.

Acknowledgment

The author would like to thank members of Silterra Malaysia Sdn Bhd for supporting and contributing to the research work in this paper.

References

[1] D. Sharma, J. Gautier, G. Merckel, " Negative Dynamic Resistance in MOS Devices ", in IEEE J. Solid State Circuits, Vol. SC-13, No. 3, pp.378-380, June 1978.

[2] W. Redman-White, M. S. L. Lee, B. M. Tenbroek, M. J. Uren, R. J. T. Bunyan, " Direct Extraction of MOSFET Dynamic Thermal Characteristics from Standard Transistor Structures Using Small Signal Measurements", in Electronics Letters, vol. 29, No 13, pp. 1180-1181, June 1993.

[3] D. J. Walkey, T. J. Smy, T. Macelwee and M. Maliepaard, "Compact Representation of Temperature and Power Dependence of Thermal Resistance in Si, Inp and GaAs Substrate Devices using Linear Models", Solid State Electronics, Vol 46, pp. 819-826, June 2002.

[4] Wei Jin, S. K. H. Fung, W. Liu, P. C. H. Chan, C. Hu, "Self Heating Characterization for SOI MOSFET based on AC Output Conductance", Technical Digest IEDM 1999, pp. 175-178, 5-8 Dec 1999.

[5] K. A Jenkins, J. Y.-C. Sun and J. Gautier, "Characteristics of SOI FET's Under Pulsed Conditions", in IEEE Transactions on Electron Devices, Vol 44, No 11, November 1997.

[6] C. Anghel, R. Gillion, and A.M. Ionescu, " Self Heating Characterization and Extraction Method for Thermal Resistance and Capacitance in HV MOSFETs", in IEEE Electron Device Letters, Vol 25, No 3, March 2004.

[7] Y. Tsividis, *Operation and Modeling of the MOS Transistor*, Oxford University Press, New York, NY, 1999.

[8] W. Jin, W Liu, S.K.H. Fung, P. C. H. Chan, and C. Hu,"SOI Thermal Impedance Extraction Methodology and Its Significance For Circuit Simulation," IEEE Trans Electron Devices, vol. 48, pp. 730-736, April 2001.

[9] L.T. Su, D. Antoniadis, N. D. Arora, S. doyle, and D.B. Krakauer, "SPICE model and parameters for fully depleted SOI MOSFETs including self-heating", IEEE Electron Device Lett., Vol 15, pp. 374-376, Oct 1994.

[10] BSIMSOI[online],Available:http://www-device.eecs.berkeley.edu.

Design of 100nm Single-Electron Transistor (SET) by 2D TCAD Simulation

[a]Amiza Rasmi, [b]Uda Hashim, [a]Abdul Fatah Awang Mat

[a]Microelectronics & Nano Technology Unit, Telekom Research and Development Sdn. Bhd.,
Idea Tower I, UPM-MTDC, Technology Incubation Centre One, Serdang, Selangor, Malaysia
[b]Kolej Universiti Kejuruteraan Utara Malaysia, Perlis, Malaysia
Email: amiza@tmrnd.com.my

Abstract **One of the great problems in current large-scale integrated circuits (LSIs) is increasing power dissipation in a small silicon chip. Single-electron transistor (SET) which operate by means of one-by-one electron transfer, small size and consume very low power are suitable for achieving higher levels of integration. In this paper, SET is designed with 100nm gate length and 10nm gate width is successfully simulated by Synopsys TCAD. The power of SET device that obtained from simulation is 3.771 x 10^{-9} Watt for fixed current and 3.3565 x 10^{-9} Watt if fixed the gate voltage, V_G, and the capacitance of this device is 0.4297 aF. These results were achieved at room temperature operation.**

I. INTRODUCTION

Single-electron transistor (SET) is a key element in single electronics where device operation is based on one-by-one electron manipulation utilizing the Coulomb blockade effect and can be made very small (nanometer scale). However, SET has low voltage gain, high input impedances, and is sensitive to random background charges [1]. This makes it unlikely that SET would ever replace field-effect transistors (FETs) in applications where large voltage gain or low output impedance is necessary.

SETs can potentially take the industry all the way to the theoretical limit of electrons for computing applications by allowing the use of a single electron to represent a logic state [2]. The first application for SET could be for the memory and special applications in metrology, such as primary thermometers and supersensitive electrometers [2].

In this paper, some introduction of single-electron transistor (SET) is discussed in the next section. The process and device simulation is then discussed in the section 3, before concluding.

II. SINGLE-ELECTRON TRANSISTOR (SET)

The most fundamental three-terminal single-electron devices (SEDs) are called single-electron transistor (SET) [3-5]. SET is always three-terminal devices with gate, source, and drain as shown in Figure 1, unlike quantum dots (QDs) and resonant tunneling devices (RTDs) which may be two terminal devices without gates [6]. The SET is expected to be a key device for future extremely large-scale integrated circuits because of its ultra-low power consumption and small size.

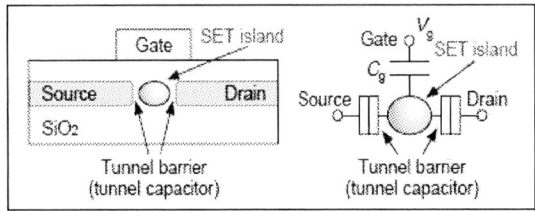

Figure 1: Schematic structure and equivalent circuit of a single-electron transistor (SET) [7]

As shown in the Figure 1, the device must have a small island together with a gate electrode coupled to the island with gate capacitance, C_g. Source and drain electrodes are attached to the island via a tunnel barrier. In addition, SET is a device whose operation relies on single electron tunneling through a nanoscale junction. The operation principle of SET is shown in Figure 2.

0-7803-9730-4/06/$25.00 ©2006 IEEE   377

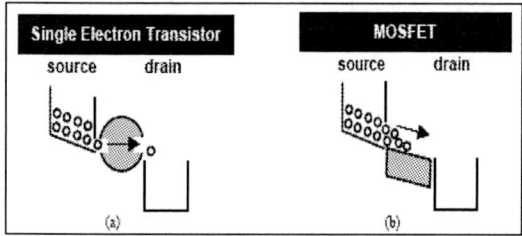

Figure 2: Transfer of electrons is (a) one-by-one in Single Electron Transistor (SET), which is in contrast with (b) conventional MOSFET where many electrons simultaneously participate to the drain current [8-9].

From Figure 2, many electrons (1000-10, 000 electrons) [10] simultaneously participate from the source to the drain current in the conventional MOSFETs. On the contrary, the electrons in SET devices are transferred one-by-one through the channel [11]. These features are suitable for achieving higher levels of integration. In addition, SET has unique features that conventional MOS transistors do not have. The first one is SET can have many gates (gate and back gate) [11]. The other one is oscillatory conductance as a function of a gate voltage [12]. Therefore, it is generally assumed that single-electron devices have potential to be much faster than conventional MOSFETs.

III. RESULTS AND DISCUSSION

Process and device simulation is commonly use for the design of new very large scale integration (VLSI) devices and processes. Simulation programs serve as exploratory tools in order to gain better understanding of process and device physics [13]. In this research, the Synopsys TCAD tools Taurus TSUPREM-4 [14] and MEDICI [15] is utilized for process simulation and device simulation.

The process simulation is to define the SET device structure and process parameter. In the process simulation, the mask layout is become as an input data. The mask layout consist of four mask layers namely source layer, polysilicon gate layer, contact layer, and metal layer as shown in Figure 3.

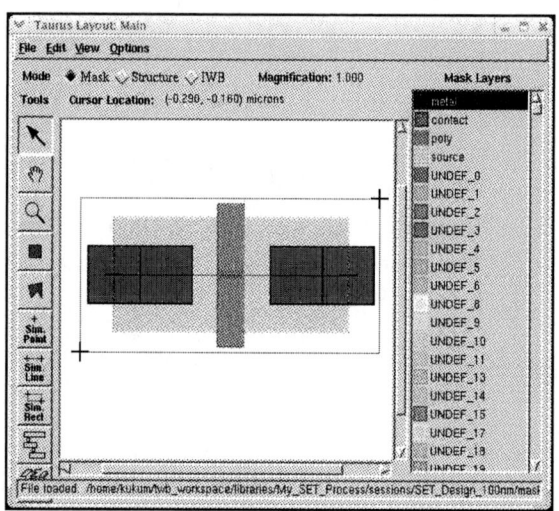

Figure 3: Mask layout

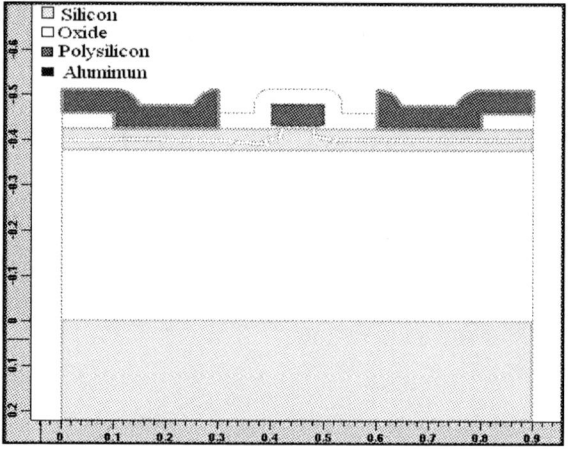

Figure 4: Cross-section view of SET device

Figure 4 shows the cross-section of SET device with the gate length is 100nm and the gate width is 10nm. The four electrodes such as source, drain, gate and substrate is defined.

The device simulation is to define the device characteristics such gate characteristics and drain characteristics. The output data from the process simulation is used as an input data for device simulation. This device simulation is done by using the Taurus Medici.

For gate characteristic, the drain current, I_D as a function of the gate voltage, V_G is shown in Figure 5.

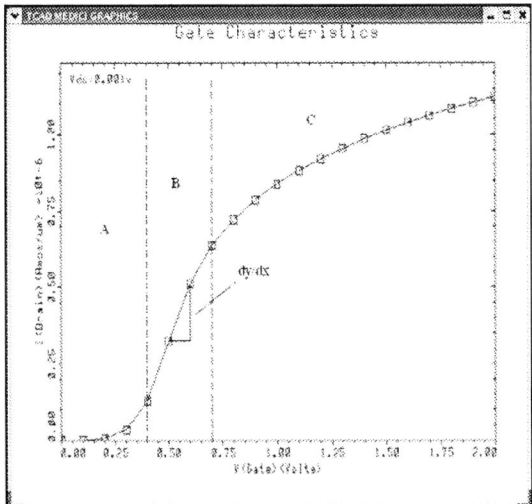

Figure 5: Drain current, I_D as a function of the gate voltage, V_G.

Here, the drain voltage was 1 mV, the source and substrate (back gate) voltages were fixed at 0 V and the device temperature was 300K.

As shown in Figure 11, the graph is divided into three sections. In section A (from V_G at 0 V to V_G = 0.4 V), the drain current is increased immediately. At V_G = 0. 4 to V_G = 0.7 V (section B), the drain current increased linearly. While, in the section C when V_G is 0.7 V to 2.0 V, the drain current increased slowly.

Additionally, this graph is also given some information about this device. Firstly, the linear slope of this exponential graph is calculated. The linear slope is given by

$$\frac{dy}{dx} = \frac{y_2 - y_1}{x_2 - x_1} \qquad (1.1)$$

From the I_D-V_G graph (Figure 5),

$$\frac{dy}{dx} = \frac{(5.2166 \times 10^{-7} - 2.8303 \times 10^{-7}) A/\mu m}{(0.6 - 0.5)V} = 2.3864 \times 10^{-6} A/\mu m V$$

From this calculation, the linear slope is 2.3864 x 10^{-6} A/μmV. From the Ohm law,

$$V = IR, \qquad (1.2)$$

the resistance, R of SET device can be calculated. Based on the Ohm law, the resistance, R is

$$R = \frac{V_G}{I_D} = \frac{1}{\left(2.3864 \times 10^{-6}\right) \times 0.1 \mu m} \mu m V/A = 4.1904 \times 10^6 \Omega$$

From the calculation, the SET device resistance, R is 4.1904 x 10^6 Ω. The power, P of SET device is given below,

$$P = VI \qquad (1.3)$$

From Equation (1.2), $V = IR$, so the power is

$$P = I^2 R \qquad (1.4)$$

From I_D-V_G graph as shown in Figure 5, let say I = 3.0 x 10^{-7} A/μm hence

$$P = [(\frac{3.0 \times 10^{-7} A}{\mu m} \times 0.1 \mu m)^2] \times [4.1904 \times 10^6 \Omega]$$

$$P = 3.771 \times 10^{-9} Watt$$

The power, P for SET device is 3.771 x 10^{-9} Watt for a fixed current. If voltage is fixed, let say V = 0.5 V, I = 2.8302 x 10^{-7} A/μm (get from Figure 5) and R = 4.1904 x 10^6 Ω,

$$P = [(\frac{2.8302 \times 10^{-7} A}{\mu m} \times 0.1 \mu m)^2] \times [0.4190 \times 10^6 \Omega]$$

$$P = 3.3565 \times 10^{-9} Watt$$

So, power, P =3.3565 x 10^{-9} Watt. From the two calculations of power, the power, P of SET device is very low (10^{-9} Watt).

The threshold voltage, V_{TH} is the minimum value required for tunneling as shown below [16,17]

$$V_{th} = \frac{e}{C} \qquad (1.5)$$

From Equation (1.5), the capacitance, C of SET device can be calculated. From the I_D-V_G graph, V_{TH} = 0.3728 V at V_G = 0.5 V and e = 1.602 x 10^{-19} C. Hence, the capacitance, C is

$$C = \frac{e}{V_{TH}} = \frac{1.602 \times 10^{-19} C}{0.3728 V} = 0.4297 \times 10^{-18} F$$

The calculation showed capacitance, C = 0.4297x10^{-18} F or 0.4297 aF. As a result, the capacitance, C from the simulation result is 0.4297 aF at 300 K. Based on the previous experimental result done by Wasshuber, the capacitance, C is 3 aF at 300 K [18]. Another result done by Takahashi *et.al.* [19] is $C = 2$ aF at room temperature. Compare to two of the three results, the result from the simulation is smaller than the previous experiment done by others.

The charging energy, E_C for SET is given below,

$$E_C = \frac{e^2}{2C} \qquad (1.6)$$

From Equation (1.5), C = 0.4297 x 10^{-18} F. Hence, the charging energy for this system is

$$E_C = \frac{(1.602 \times 10^{-19})^2}{2(0.4297 \times 10^{-18})}$$

$$E_C = \frac{2.9863 \times 10^{-20}}{1.602 \times 10^{-19}} eV$$

$$E_C = 0.1864 eV = 186.4 meV$$

The calculation showed the charging energy, E_C for SET system is 186.4 meV at 300 K. Based on the previous reported about SET [20]; the charging energy is increased to 173 meV, which means that the transistor is able to operate at 300K. As a result, the value of charging energy from this calculation is nearly to the previous reported. Hence, the SET in this project is operated at 300 K.

In addition, the threshold voltage at 0.600 V gate bias is 0.3728 V and the channel length of the SET device is 61.27 nm. The simulation result shows that the sub threshold slopes for SET at 0.10 V gate bias is 112.3 mV/decade. The leakage current (I_{OOF}) at 0.0 V gate bias is 4.1715×10^{-12} A/µm and the drive current, I_{ON} is 0.034 µA/µm at $V_G = V_D = 1.5$ V.

For drain characteristic, the drain current, I_D as a function of the drain voltage, V_D is shown in Figure 6. Here, the gate voltage, V_G was 1.5 V, the source and substrate (back gate) voltages were fixed at 0 V and the device temperature was 300 K.

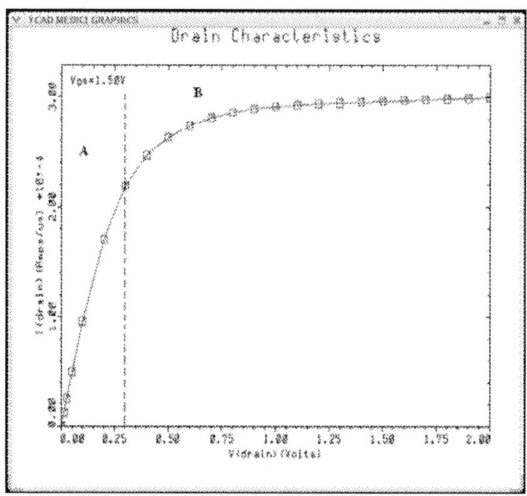

Figure 6: Drain current, I_D as a function of the drain voltage, V_D.

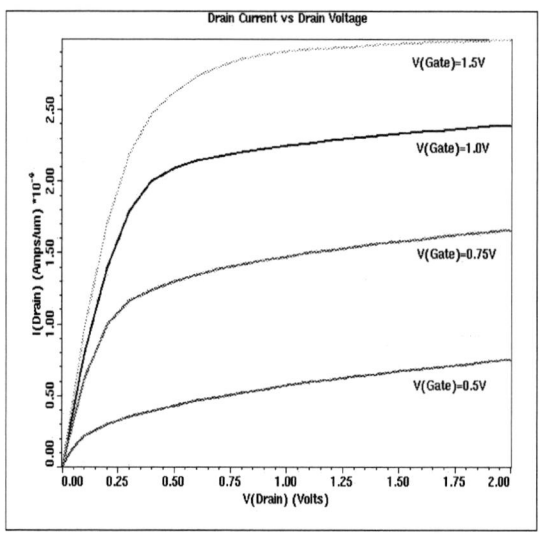

Figure 7: Drain current, I_D as a function of the drain voltage, V_D at various gate voltages, V_G.

Figure 7 shows the drain current as a function of a drain voltage at various gate voltages, V_G. From the graph, increasing the gate voltage, V_G the drain current, I_D will also increased. The saturation occurred at $V_D = 0.30$ V to $V_D = 2.0$ V.

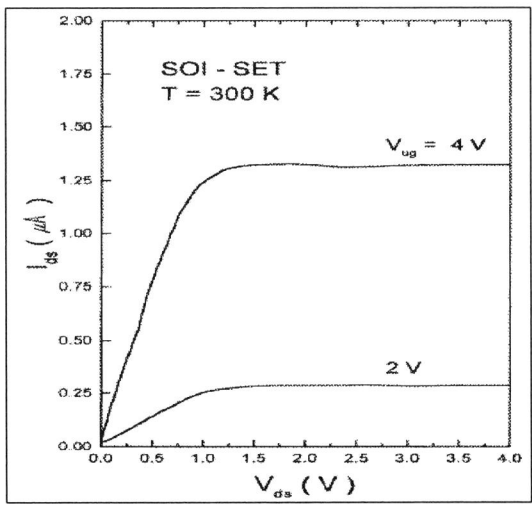

Figure 8: The I_D-V_D characteristics for different values of the gate voltage at 300 K [21].

Figure 7 and 8 shows the I_D-V_D characteristics at various gate voltage, V_G and temperature at 300K are obtained from simulation and previous experiment. The curve from the simulation is quite similar to the previous experiment.

IV. CONCLUSIONS

Among various single-electron devices (SEDs), the single-electron transistor (SET) is the most fundamental. The SET simulation is successfully simulated using Synopsys TCAD simulation tools.

From the simulation, the threshold voltage for SET device is 0.3728 V and the channel length is 61.27 nm. The SET device is operated at 300K (room temperature operation). From the calculation, the charging energy, E_C of SET system is 186.4 meV. The capacitance, C of SET device is 0.4297 aF and the power, P of SET device is 3.771 x 10^{-9} Watt for fixed current and 3.3565 x 10^{-9} Watt if fixed the gate voltage. The power, P of this SET device is obtained from the I_D-V_G graph with fixed the resistance, R. The resistance, R is 4.1904 x 10^6 Ω.

The value of capacitance that obtained in this simulation is smaller than the previous experiment and the charging energy is higher than the previous reported. Ultimately, this research has utilized the process and device simulation tools as an alternative for an actual SET fabrication process.

ACKNOWLEDGEMENT

The authors would like to thank to Northern Malaysia University College of Engineering and Government of Malaysia for granted this project through Intensification of Research in Priority Areas (IRPA).

REFERENCES

1. Peter Hadley,Günther Lientschnig,and Ming-Jiunn Lai, *Single-Electron Transistors*, Department of Nanoscience, Delft University of Technology.
2. Christopher Wasshuber, *Single Electron Transistor Advances Revealed*, Electronicstalk, (2003) www.electronicstalk.com/news/tex/tex544.html
3. P. Hadley, Günther Lientschnig, and Ming-Jiunn Lai. 2002. Single-Electron Transistors. *Proceedings of the International Symposium on Compound Semiconductors*, 1-8.
4. Stephen John Robinson. 2003. *Conventional Microelectronic Processing for Single-Electron Transistors. Unpublished Degree Dissertation*, College of the University of Illinois at Urbana-Champaign.
5. Konstantin K. Likharev. 1987. Single-Electron Transistors: Electrostatic Analogs of the DC Squids. *IEEE Transaction on Magnetic,* Vol. 23, No. 2, 1142-1145.
6. David Goldhaber-Gordon, Michael S. Montemerlo, J. Christopher Love, Gregory J. Opiteck, and James C. Ellenbogen. 1997. Overview of Nanoelectronic Devices. *Proceedings of the IEEE*, Vol. 85, No. 4, 521-540.
7. Yasuo Takahashi, Yukinori Ono, Akira Fujiwara, and Hiroshi Inokawa. 2004. Silicon Single-Electron Devices. *NTT Technical Review*, Vol. 2, No. 2, 21-27.
8. A.M. Ionescu, M.J. Declercq, S. Mahapatra and K. Banerjee. 2002. Teaching Microelectronics in the Silicon ICs Showstopper Zone: a Course on Ultimate Devices and Circuits: Towards Quantum Electronics. *4th European Workshop on Microelectronics Education (EWME 2002)*, 1-4.
9. Ken Uchida, Junji Koga, Ryuji Ohba, and Akira Toriumi. 2003. Programmable Single-Electron Transistor Logic for Future Low-Power Intelligent LSI: Proposal and Room-Temperature Operation. *IEEE Transaction on Electron Devices*, Vol. 50, No. 7, 1623-1630.
10. A.T. Tilke, F.C. Simmel, R.H. Blick, H. Lorenz, and J.P. Kotthaus. 2001. Review: Coulomb Blockade in Silicon Nanostructures. *Progress in Quantum Electronics*, Vol. 25, 97-138.

11. Yasuo Takahashi, Yukinori Ono, Akira Fujiwara, and Hiroshi Inokawa. 2003. Development of Silicon Single-Electron Devices. *Physica E*, Vol. 19, 95-101.

12. Yasuo Takahashi, Yukinori Ono, Akira Fujiwara, and Hiroshi Inokawa. 2002. Silicon Single-Electron Devices for Logic Applications. Plenary Invited Papers, *32ⁿᵈ European Solid-State Device Research Conference (ESSDERC 2002)*, 61-68.

13. Siegfried Selberherr. 1993. Technology Computer-Aided Design. *Southern African Journal of Physics*, Vol. 16, No. 1/2, 1-5.

14. Synopsys. 2004. *Taurus TSUPREM-4 User Guide.* Synopsys Inc., USA.

15. Synopsys. 2004. *Taurus Medici: Medici User Guide.* Synopsys Inc., USA.

16. J. Hoekstra and R. H. Klunder. 1999. Single Electron Transistor Circuit Simulation. Microelectronics Advanced Research Initiative (MEL-ARI) Answers, *Technical Report July 1998-July 1999*, 1-20.

17. I. G. Neizvestnyi, O. V. Sokolova, and D. G. Shamiryan. 1999. Single Electronics Part 1. *Russian Microelectronics*, Vol. 28, No. 2, 67-88.

18. Christoph. Wasshuber. 2001. *Computational Single-Electronics.* Springer-Verlag Wien New York, 1-146

19. Y. Takahashi, M. Nagase, H. Namatsu, K. Kurihara, K. Iwdate, Y. Nakajima, S. Horiguchi, K. Murase, and M. Tabe 1995. Fabrication Technique for Si Single-Electron Transistor Operating at Room Temperature. *Electronics Letters*, Vol. 31, No. 2, 136-137.

20. Belle Dumé. June 2003. Room Temperature Single-Electron Devices Made Easier. *Science*, PhysicsWeb.
http://physicsweb.org/articles/news/7/6/16/1

21. B T Lee, J W Park, K S Park, C H Lee, S W Paik, S D Lee, Jung B Choi, K S Min, J S Park, S Y Hahn, T J Park, H Shin, S C Hong, Kwyro Lee, H C Kwon, S I Park, K T Kim, and K-H Yoo. 1998. Fabrication of a dual-gate-controlled Coulomb Blockade Transistor based on a Silicon-On-Insulator Structure. *Semiconductor Science and Technology*, Vol. 13, 1463-1467.

Effect of Rapid Thermal Annealing (RTA) on *n*-Contact of 980 nm Oxide VCSEL

Khairul Anuar M.S., Mohd Sharizal A. *Member IEEE*, S.M. Mitani, Mohamed Razman Y., Awang Mat
A.F., and P. K. Choudhury* *Senior Member IEEE*

Microelectronic and Nano Technology Programme
UPM-MTDC, UPM, 43400 Serdang, Selangor, Malaysia
*Faculty of Engineering, Multimedia University
63100 Cyberjaya, Selangor, Malaysia
Email: anuark@tmrnd.com.my , sharizal@tmrnd.com.my

Abstract – **The paper deals with the development of Ni/Au/Ge/Au ohmic contacts for the fabrication of VCSELs to be operated in the 980 nm of the electromagnetic (EM) spectrum. The VCSEL structures are grown by the process of molecular beam epitaxy (MBE) whereas the contacts are deposited by electron beam evaporator. The *n*-contact metallization has been performed along with RTA before as well as after the fabrication of the VCSEL structure, and the effect of RTA treatment on the grown VCSEL has been studied in the different cases.**

I. INTRODUCTION

The development of ohmic contact in VCSELs and many other semiconductor devices has been of prime importance as it greatly affects the device performance. In order to drive the device, low-resistance ohmic contacts are most desirable as it not only provides a medium for current injection, but also the generation of heat is minimized in this case. Such low-resistance ohmic contacts are of paramount importance particularly in the case of tiny electronic devices as it drastically affects the device performance, particularly for high-speed operation.

The development of different types of low-resistance ohmic contacts has been reported by the investigators [1–2]. It is found that nickel (Ni) is generally used for the formation of *n*-type ohmic contact in the case of GaAs with AuGeNi alloy [3]. Further, Ni atom can function as a *p*-type dopant in GaAs material with the activation energy of 0.2 eV [4]. The use of palladium (Pd) is also reported as a good candidate for the formation of low-resistance ohmic contact on *p*-type GaAs. However, for *p*-GaAs material, the use of NiPdAu alloy is also appeared in the literature [5]. In all such cases related to the metallic contact developments, the process of annealing plays a vital role as it essentially governs the smoothness of the surface morphology.

In the present communication, the performance of a multilayered metallic structure of Ni/Au/Ge/Au alloy has been reported for the operation of VCSEL in the 980 nm region of the EM spectrum. The RTA has been used in different forms – (i) the development of the *n*-contact through RTA treatment before the fabrication of VCSEL structure, (ii) the development of the *n*-contact through RTA treatment after the fabrication of VCSEL structure, and (iii) the development of the *n*-contact without undergoing the process of RTA; in this case, the VCSEL is fabricated before the deposition of the contact metallization. Thus, the paper essentially presents a qualitative study of such ohmic contacts with their implementation in the case of VCSELs. As it is well-known that the *L-I-V* characteristics primarily determine the performance of VCSELs, in the present paper, such characteristics are studied for the samples fabricated by following the three different processes as discussed above. In particular, *I-V* and *L-I* characteristics are emphasized. The results show that the process of annealing greatly affects the performance of the device.

II. EXPERIMENT

MBE V80 is used to grow the 980 nm oxide VCSEL epilayers. The VCSEL structure consists of 15 periods of *p*-DBR mirrors, 3 quantum wells (InGaAs) and 27 periods of *n*-DBR mirrors. Fig. 1 shows the SEM cross section of the 980 nm VCSEL structure, where we can observe the *p*-

DBR, oxide layer, InGaAs quantum well and apart of n-DBR. The epilayers (VCSEL wafer) are processed using standard microelectronic fabrication technique [6] that involves various steps such as lithography, pattern transfer, etching, thin film deposition, and wet oxidation process. For metallization contact (n- and p-types), electron beam evaporator is used to deposit this stacked structure. Ti/Pt/Au 35nm/35nm/320nm is used as the p-contact.

Three different cases are studied viz. (i) the n-metallic contact is deposited at first during the fabrication process, and the sample is then treated with RTA at 400°C for 180 seconds (named as RTA First), (ii) the n-contact is formed in the last step of fabrication followed by the process of RTA as that in (i) (named as RTA Last), and (iii) the sample is not allowed to undergo the process of RTA (named as No RTA). Mesas with aperture 27×27 μm² are used, and studies have been made of the electrical and the optical properties of such VCSELs that include threshold current (I_{th}), threshold current density (J_{th}), slope efficiency (S.E.), and external quantum efficiency η_d.

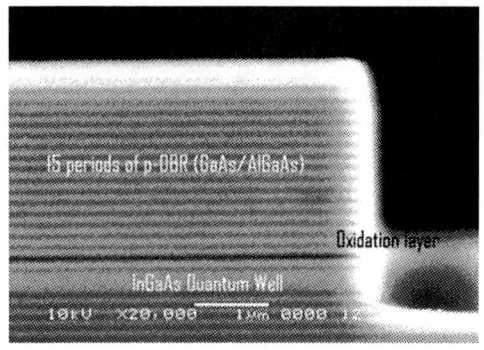

Fig. 1 SEM cross-section of 908 nm VCSEL

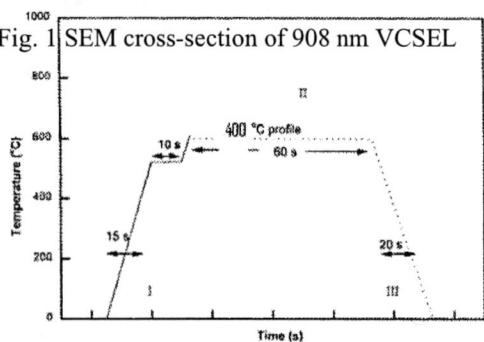

Fig. 2 Temperature profile for RTA process

RTA was carried out in nitrogen ambient using a commercial JetFirst100 rapid thermal annealing system. The system consists of an aluminium process chamber, IR lamps in two zones with maximum power of 38 W and a low

temperature pyrometer to monitor temperatures from 100°C to 1000°C. Fig. 2 shows the temperature profiled used in this work. Region I, II and III are the ramping-up, annealing and the ramping-down stages, respectively. A special characteristic of stage I include a pre-stab sub-stage, where the temperature profile during the ramping up stage is given some time to stabilize. This ensures that the control of the temperature profile is maintained at the optimum condition. In this work, the samples are annealed at 400°C for 180 seconds, ramping up time is 15 seconds and that of ramping down is 20 seconds. Electrical and optical properties of VCSEL are obtained using standard power-electrical-voltage measurement (L-I-V) setup. External quantum efficiency (η_d) and wall plug efficiency or power conversion efficiency for the VCSELs (which are deduced from the L-I-V characteristics) is calculated as given [7, 8] below:

$$\eta_d = \frac{\Delta P}{\Delta I}\frac{q\lambda}{hc} \qquad (1)$$

$$\eta_{conv} = \eta d \frac{Eg}{Vop}\left(1 - \frac{I_{th}}{I}\right) \qquad (2)$$

where q is the electronic charge, λ is the wavelength, h is Planck's constant, c is the speed of light, and E_g is the bandgap energy (eV). For this 980 nm VCSEL with $In_{0.18}Ga_{0.82}As$ quantum wells, the bandgap energy is 1.147 eV [9].

III. RESULTS AND DISCUSSION

Figs. 3(a) and 3(b), respectively, show the I-V and the dV/dI-I characteristics for the VCSEL device with Ni (5 nm)/Au (5 nm)/Ge (30 nm)/Au (90 nm) n-type ohmic contact when the samples undergo different RTA processes. The annealing temperature is kept as 400°C. From fig. 3 it is observed that the differential resistance of the device is significantly decreased when the n-contact metallization is performed at first during the fabrication process of the VCSEL structure (RTA First). The differential resistance slightly increases when the sample does not undergo the RTA treatment (No RTA). However, the sample with RTA Last shows high differential resistance.

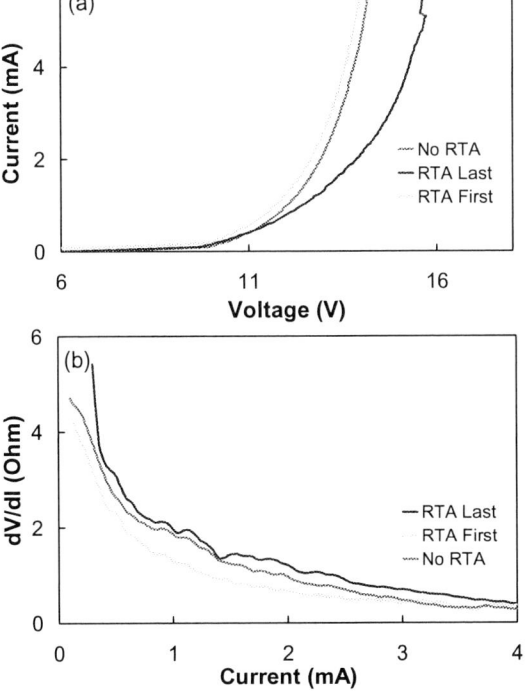

Fig. 3 *I-V* Curves (a) and *dV/dI* (b) characteristics

The series resistance characterization represents the growth quality of each multilayer interface in VCSEL devices. VCSELs with poor multilayer interface (such as too abrupt or step-index) will exhibit high series resistance. Also, a high series resistance value for VCSELs/laser diodes can result from low quality ohmic contact depositions on both sides of the device. But, in our case, we believe the high series resistance, as observed corresponding to the case of RTA Last, is due to poor multilayer interface that occurred for this sample.

Similar improvements due to the reduction of the differential resistance can also be observed in the *P-I* characteristics (fig. 4) for the same VCSEL samples. The annealing temperature is kept as the same, i.e. 400°C. Table 1 presents the observed electrical and electro-opto properties for the VCSEL samples.

Low threshold current requirement has been one of the features of a good VCSEL device. We observe that the required threshold current becomes minimum (1.20 mA) for sample with RTA First compared to the case of RTA Last, which is 2.65 mA. While for No RTA sample, the threshold is 1.38 mA. We also notice that the threshold current density J_{th} (for mesa aperture 27×27 μm²) is high for RTA Last compared to RTA First and No RTA cases. Threshold current density directly indicates the quality of the

semiconductor material which is used in the fabricated device. It is desirable for the lasing device to exhibit low threshold current density.

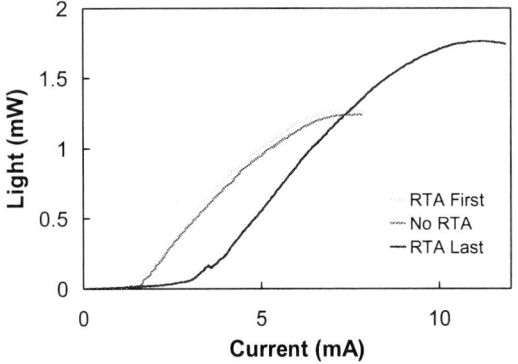

Fig. 4 *L-I* curves for RTA First, RTA Last and No RTA

Table 1: Electrical and electro-opto properties

	RTA First	RTA Last	No RTA
I_{th} (mA)	1.20	2.65	1.38
J_{th}	1646	3635	1893
S.E.	0.25	0.23	0.22
η_d (%)	1.04	0.9	0.43

For an efficient VCSEL device, the output power should be high with very low injection current. We observe from fig. 4 that the required threshold current is more for the cases of metallization without RTA and the metallization with RTA Last, and the maximum output power is obtained when the *n*-metallization is performed at the last during the fabrication process. Therefore, one has to consider a kind of tread off between these two factors, i.e. the output power and the differential resistance. On one hand, the differential resistance should be as much less as possible, which we get when the contact metallization is done at the first stage of fabricating the VCSEL structure, and the metallization without RTA, whereas on the other hand, the high output power is achieved when the metallization is performed at the last stage of fabrication process. However, the variation in annealing temperature would finally affect the performance of the device, and this study (with the use of different annealing temperatures) is expected to take up in a future communication.

To study the buried interfaces degradation, as we suspect, a graph of external quantum efficiency (*dP/dI*) versus *I* is plotted, as shown in fig. 5. An increase of leakage is observed when the threshold current increases (for RTA Last), and the external quantum efficiencies remain unaffected for all the samples. This shows that the sample with RTA Last has experienced buried interfaces degradation [10]. This is attributed to the effect of temperature and pressure during the treatment of RTA on the mesa of the VCSEL structure.

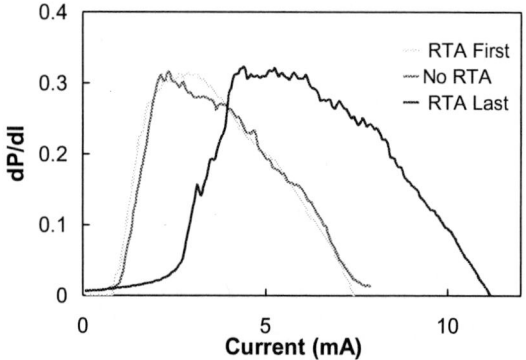

Fig. 5 Plots of *dP/dI* (η_d) vs. current

Wall plug efficiency or power conversion efficiency (η_{conv}) is also one of the important parameters to determine the performance of VCSELs. Fig. 6 shows the plot of the wall plug efficiency of the device against the injection current. Generally, because of their small active volume in VCSEL, the output power tends to be relatively low as compared to high-power edge emitting lasers [11]. We observe that the wall plug efficiency is the maximum for the cases of RTA First and No RTA, and for RTA Last, η_{conv} is 0.8%. Again, this is attributed to the buried interfaces degradation of the sample with RTA Last which exhibits high threshold current, and consequently, decreasing the η_{conv}.

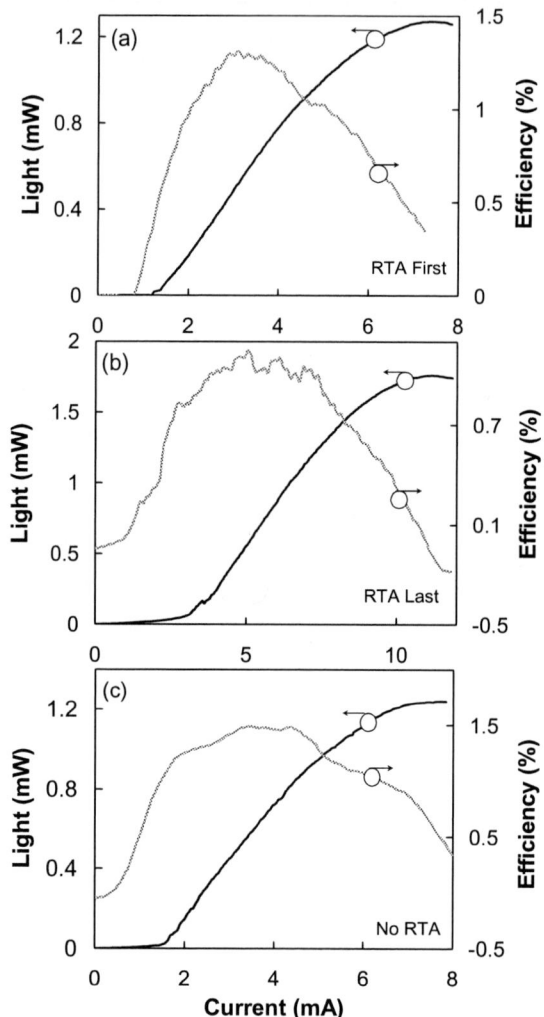

Fig. 6 Plots of η_{conv} vs. *I* for (a) RTA First, (b) RTA Last and (c) No RTA

IV. CONCLUSION

The foregoing discussion primarily deals with the formation of the *n*-contact for the 980 nm VCSEL using Ni/Au/Ge/Au alloy. Results are compared for the cases of contact formation before as well as after fabricating the VCSEL structure, and also, the treatments with and without RTA. The inference can be drawn that the process of RTA helps to minimize the contact resistance (I_{th}). We also observed buried interfaces degradation for samples treated with RTA at the last step of fabrication. It is concluded that, so far as the operational features of the VCSEL is concerned, some sort of tread off among the VCSEL parameters are to be considered, viz. threshold current, output power, and wall plug efficiency. The maximum efficiency is obtained when the contact metallization, following the RTA treatment, is done at the first step of fabrication of VCSEL structure.

ACKNOWLEDGEMENT

The authors would like to express their sincere thanks to J. Viheriala, S. Suomalainen, L. Rajala and M. Pessa of the Optoelectronic Research Center (ORC), Tampere for their help and support.

REFERENCES

[1] H. J. Gopen, A.Y.C. Yu, "Ohmic contacts to epitaxial pGaAs*," *Solid State Electron.,.* 14, pp. 515-517 (1971).

[2] T. Sanada, O. Wada, "Contact resistances of Au-Ge-Ni, Au-Zn and Al to III-V Compounds," *Jpn. J. Appl. Phys.* 52, pp. L491-L494 (1981).

[3] M. Heiblum, M.I. Nathan, C.A. Chang, "Characteristics of AuGeNi Ohmic Contacts to GaAs," *Solid State Electron.,* 25, pp. 185-195 (1982).

[4] Chih-Hung Wu, Sen-Mao Liao, Kai-Cheng Chang, "Ni/Pd/Au ohmic contact for p-GaAs and its application in red RCLED," *Materials Science and Engineering.,* B 117, pp. 205-209 (2005).

[5] H.R. Grinolds, G.Y. Robinson, "Au/Be Ohmic Contacts to p-type InP," *Solid State Electron.,* 23, pp. 973-985 (1982).

[6] K.D. Choquette, K.L. Lear, R.P. Schneider, Jr. K.M. Geib, J.J. Figeil and R. Hull, "Fabrication and performance of selectively oxidized vertical-cavity lasers," *IEEE Photon. Technol. Lett.,* vol. 7, pp. 1237-1239 (1995).

[7] K. S. Mobarhan, "Test and Characterization of Laser Diodes: Determination of Principal Parameters," *Newport Application Note: Fiber Optics & Photonics* (1).

[8] Y. Suematsu, and A. R.Adams, "*Handbook of Semiconductor Lasers and Photonic Integrated Circuits,* " Chapman & Hall, 1994

[9] Pallab, Bhattacharya "*Properties of Lattice-Matched and Strained Indium Gallium Arsenide* London: INSPEC, the Institution of Electrical Engineers, 1993

[10] Paolo Montangero Avon Technologies, Italy, "Edge Emitting Laser Reliability," Short Course 2 IEEE ISLC 2006.

[11] K.D. Choquette and H. Q, Hou, "Vertical Cavtiy Surface Emitting Lasers: Moving from Research to Manufacturing,"Proc. IEEE Vol. 85 No. 11, pp.1730-1739 (1997)

Development of SiC/MgO Distributed Bragg Reflector using RF Magnetron Sputtering Technique

Khairul Anuar M. S., Hariyadi Soetedjo, Mohd Sharizal A. *Member, IEEE,* Sufian S. M, Goh Boon T.*, Richard R.,* Saadah A. R.*, Mohamed Razman Y., and Abdul Fatah A.M.

Microelectronic and Nano Technology Programme,
Telekom Research and Development Sdn. Bhd., Idea Tower, UPM-MTDC,
43400 Serdang, Selangor, Malaysia.
*Solid State Research Lab, Physics Department,
University Malaya, Lembah Pantai, Kuala Lumpur, Malaysia
E-mail : anuark@tmrnd.com.my , sharizal@tmrnd.com.my , saadah@um.edu.my

Abstract – **A Bragg mirror structure is an essential part for vertical cavity surface emission laser (VCSEL) applications. High optical reflectance at required stopband width is one of major concern by means of application requirements. For this purpose, Bragg mirrors consisting of SiC/MgO multilayers have been developed using an RF magnetron sputtering technique at room temperature. These structures have been characterized using various measurement techniques like ellipsometry, reflectance spectroscopy, Fourier transform infrared (FTIR) spectroscopy, scanning electron microscopy (SEM) and X-ray diffraction (XRD) technique. From these measurements, it was confirmed that mirror materials were deposited on the substrates. For application requirement, the DBR mirror structure fabricated using only a seven period SiC/MgO multilayered structure produced the expected stop-band at 850 nm wavelength with high reflectivity of 95%.**

I. INTRODUCTION

A Bragg mirror also known as Distributed Bragg Reflector, DBR (or quarter wave mirror) is a structure which consists of an alternating sequence of layers of two different optical materials, with each optical layer thickness corresponding to one quarter of the wavelength for which the mirror is designed. DBR mirrors are particularly important because they can be integrated in optoelectronic semiconductor devices. For such purpose, the electrically active part of the device can be grown on top of a buried DBR mirror, or even in between two mirrors forming a resonator structure. Examples of such devices are vertical cavity surface emitting laser (VCSEL) [1-3], resonance cavity light emitting diode, and organic light emitting diode (RCLED, RCOLED) [4-5], solar cells with enhanced efficiency [6], wavelength selectivity photodetectors [7], surface Plasmon technology [8], Fabry Perot filters [9] and reflection modulators [10].

In VCSELs technology, three types of DBR are widely used. The first one is developed using semiconductor compound such as GaAs/AlAs and $Al_{0.16}Ga_{0.84}As/Al_{0.92}Ga_{0.08}As$. The second type, a hybrid DBR uses a combination of a semiconductor compound and a dielectric material. An example of this type of DBR is $GaAs/Al_2O_3$. The third type is made using a combination of dielectric structures such as SiO_2/TiO_2, SiO_2/Si_3N_4, and $ZnSe/CaF_2$. A hybrid DBR offers some advantages over the semiconductor compounds DBR such as the ability to have high index contrast. This enables a high reflectivity mirror with large stop-bands to be developed using only a few period of the multilayered structure. For example Si/SiO_2 DBR needs only five periods (ten layers) to achieve a 99% reflection as compared to $Al_{0.16}Ga_{0.84}As/Al_{0.92}Ga_{0.08}As$ which needs twenty four periods (forty eight layers) to produce the same reflectivity at a given wavelength.

Dielectric DBR usually are grown using plasma enhanced chemical vapor deposition (PECVD), electron beam evaporation or sputtering technique. For crystalline dielectric DBR, metal organic chemical vapor deposition (MOCVD) and molecular beam epitaxy (MBE) are commonly used. In this paper, we present results leading to the development of a hybrid SiC/MgO DBR grown using the RF magnetron sputtering technique. The DBR developed is a SiC/MgO multilayer structure of seven periods with SiC and MgO thicknesses of 123.5 nm and

0-7803-9730-4/06/$25.00 ©2006 IEEE

81.3 nm respectively. The objective is to produce a DBR structure with a high reflectivity stop-band.

II. EXPERIMENT

The SiC/MgO multilayer structure was deposited using a conventional Edwards RF Magnetron Auto 306 co-sputtering system. The sputtering targets used were α-SiC and MgO targets with purity of 99.9% purchased from Super Conductor Material Inc.New Jersey. The target was 1.5 cm thick with a diameter of 3.5 cm. Si (111) and glass slides with dimensions of 2.5cm by 2.0 cm were used as substrates. Standard RCA (Radio Corporation of America) cleaning technique was utilized for cleaning the Si substrates and the glass substrates were cleaned with *DECON* detergent. Pure argon (99.9%) was introduced into the chamber after the chamber was evacuated at a pressure below x10^{-6} mbar. The distance between the target and substrate was fixed at 7.0 cm. The targets were sputtered for 10 minutes prior to the deposition process to remove any impurities. A shutter covering the substrates was used to protect the substrates during the cleaning process. To enhance the uniformity of the deposition, the substrates were rotated at 3.5 rotations per minutes. The sputtering voltages used for SiC and MgO were 640 (RF power of 250 W) and 510 V (RF power of 222 W) respectively. The films were deposited at room temperature. The thickness of deposited layer was monitored using crystal quartz balance which was calibrated using an ellipsometer. The average deposition rate for SiC and MgO is 7.17 Å/min and 4.58 Å/min respectively as shown in Fig. 1.

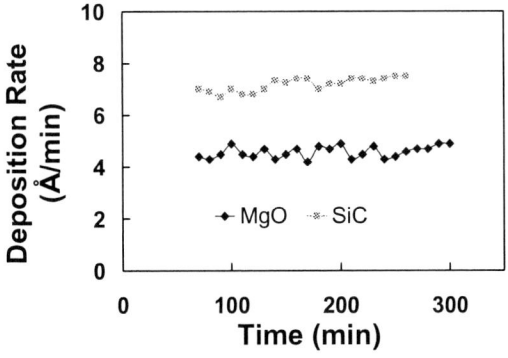

Fig. 1 Deposition rate against time for SiC and MgO

Single individual layer of SiC and MgO was characterized using a Perkin-Elmer System 2000 Fourier Transform Infrared (FTIR) spectrometer and SIEMENS D5000 X-ray diffractometer (XRD). The DBR structure fabricated was characterized using a Field Effect Scanning Electron Microscope (FESEM) and a Jasco 570 UV-Vis-Nir spectrophotometer in the reflectance mode scanned from 2400 nm to 190 nm with resolution of 2 nm.

III. RESULTS AND DISCUSSION

Fig 2 shows the infrared absorption spectrum for a single SiC layer measured at a resolution of 1 cm^{-1}. The spectrum shows three main absorption bands with peaks observed at 665, 808 and 1056 cm^{-1} which correspond to Si-H wagging, C-H transverse optical (TO) [11] and Si-O-Si stretching vibration modes respectively. The presence of the Si-O-Si stretching vibration mode is attributed to the interstitial oxygen impurities in the Si substrate [13, 14]. The SiC TO mode at 808 cm^{-1} shows that α-SiC (4H-SiC, 6H-SiC) is formed using this deposition technique instead of β-SiC (3C-SiC) which has a characteristic phonon mode at 780 cm-1 [15, 16]. The broad peak also suggests that high quality SiC film has been obtained.

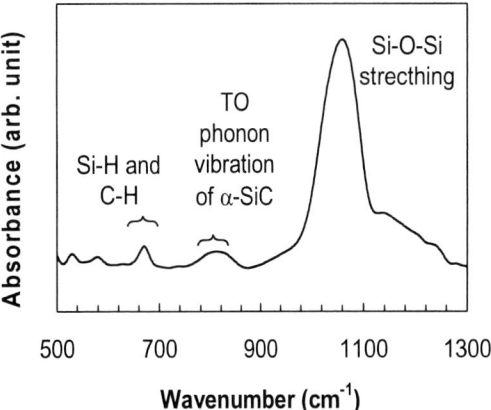

Fig. 2 FTIR absorption spectra of SiC on silicon substrate

Fig 3 shows the X-ray diffraction patterns for the single layer SiC and MgO films deposited on c-Si. Two peaks are observed at 2θ angles of 41.05 and 57.95 degrees which correspond to SiC (200) [17] and Si (222) [18] reflections respectively. The Si (222) peak is deduced to be from substrate material. The XRD pattern for the MgO film shows the protrusion of a sharp peak at 2θ angle of 36.40 degrees which correspond to MgO (111) reflection. The Si (222) reflection is also observed here. The XRD patterns observed show that the films deposited are nanocrystalline as the XRD peaks are very small. Tang et al, [19] in their work could not observe any peak related to SiC in their films deposited at room temperature. Fig 4 shows the surface morphology for the SiC (a) and MgO films (b) observed by FESEM. The images confirm the nanocrystalline structure of both the SiC and MgO films.

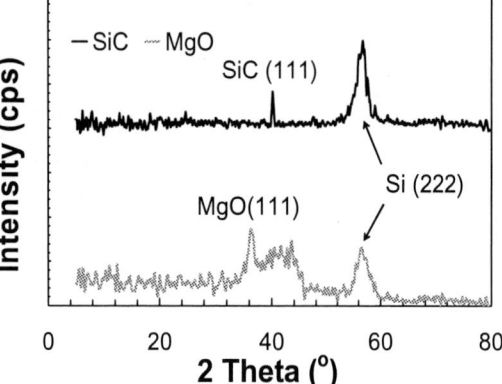

Fig. 3 SiC and MgO films on Si substrate as observed under X-RD

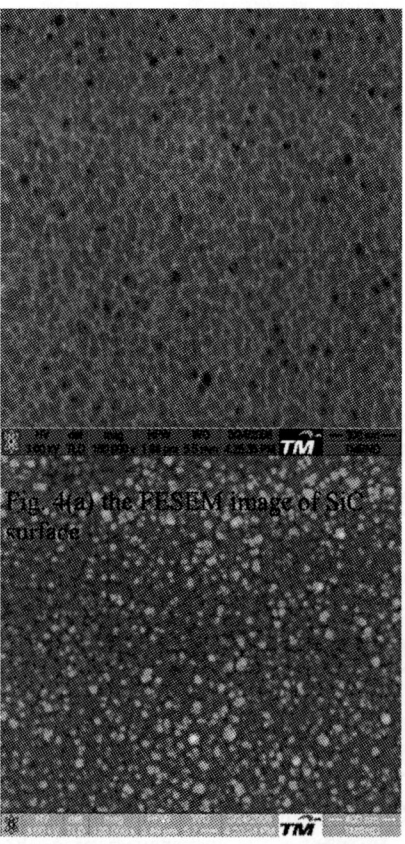

Fig. 4(a) the FESEM image of SiC surface

Fig. 4(b) the FESEM image of MgO surface

Simulation on the reflection for first period SiC/MgO layered structure has been done using the Transfer Matrix Method (TMM) and applied in the HS_Design Simulator version 1.0. The basic equation for TMM approach is:

$$R = \left(\frac{\left(n_H / l_H\right)^{2N} - 1}{\left(n_L / l_L\right)^{2N} + 1} \right)^2 \qquad (1)$$

where R is total reflection, N is number of layers, n_H and n_L is high refractive index layer and low refractive index layer respectively. In our work, the high refractive index layer is SiC with $n = 3.17$ with thickness of 81.3 nm and the low refractive index layer is MgO with n = 2.19 with thickness of 123.5 nm.

The reflection spectrum for the first period of SiC/MgO layers (experimental and simulation) is presented in Fig 5. From the reflection spectra, the intensity of reflection for first period is 85% while for the intensity obtained from the simulated spectrum is 60.5%. The base width of the interference curves are 582 nm and 1449 nm for the experimental and simulated spectrum respectively. The significant variation in the experimental and simulated spectra could be due to variation of the physical and chemical properties of sputtered SiC and MgO from the ideal properties used in the simulation. Fig 6 shows SEM cross section SiC/MgO structure. This image confirms the formation of almost abrupt layers of SiC and MgO.

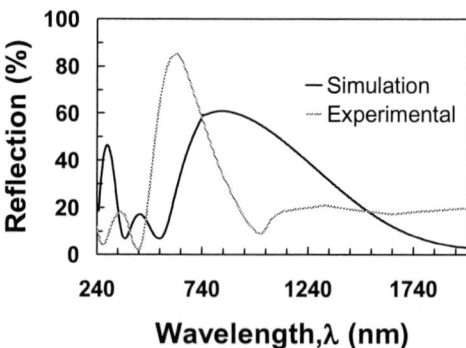

Fig. 5 Reflection spectra for first period SiC/MgO (experimental and simulation)

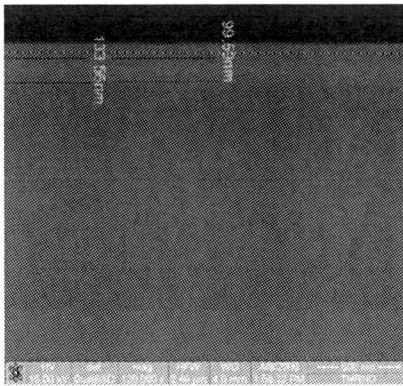

Fig. 6 SEM cross section of SiC/MgO

Fig 7 shows the reflection spectrum for seven periods of SiC/MgO structure. As observed, a stop-band appears at about 850 nm wavelength in the reflection spectrum with a base width of 260 nm. High reflection of 96.5% is achieved in the middle of the stopband. Because of the high contrast in the refractive index of SiC and MgO ($\Delta n \approx 0.79$), only a seven period DBR structure is needed to obtain a high reflectivity.

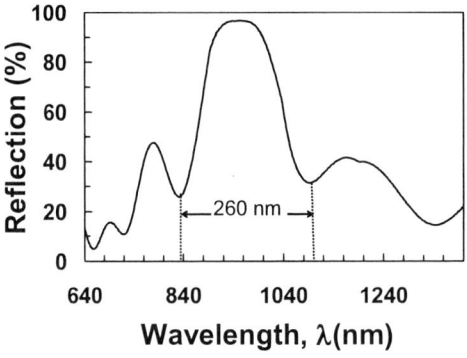

Fig. 7 Reflection spectrum for seven periods of SiC/MgO

IV. CONCLUSION

The SiC and MgO films grown using RF Magnetron Sputtering Technique at room temperature have low average growth rates of 7.17 Å/min and 4.58 Å/min respectively. The SiC film produced by this technique at room temperature is confirmed to be good quality α-SiC. Both the SiC and MgO films have nanocrystalline structures. The DBR structure produced from these two component materials is able to produce a high reflectivity stop-band at wavelength of 850 nm with only a seven period structure.

REFERENCES

[1] K. Iga, F. Koyama, S. Kinoshita, IEEE Journal of Quantum Electronics, 24 (1998) 1845

[2] C.F. Schaus, H.E. Schaus, M.Y.A. Raja, S.R.J. Brueck, Electronics Letters 25 (1998) 538

[3] J.L. Jewell, K.F. Huang, K. Tai, Y.H. Lee, R.J. Fischer, S.L. McCall, A.Y. Cho, Applied Physics Letters 55 (1989) 424

[4] N.E.J. Hunt, E.F. Schubert, R.A. Logan, G.J. Zydzik, Applied Physics Letters 61 (1992) 2287

[5] Organic example

[6] V.M. Andreev, Semiconductors 33 (1999) 942

[7] S.S. Murtaza, J.C. Campbell, J.C. Bean, L.J. peticolas, Applied Physics Letters 65 (1994) 795

[8] Example of surface palsmone.

[9] C.R. Pidgeon, S.D. Smith, Journal of Optical Society of America 54 (1964) 1459

[10] K.K. Law, R.H. Yan, J.L. Merz, L.A. Coldren, Applied Physics Letters 56 (1990) 1886

[11] Z.D. Sha, X.M. Wu, and L.J. Zhuge, Vacuum 79 (2005)

[12] Z.D. Sha, X.M. Wu and L.J.Zhuge,"The structure and optical properties of SiC film on Si (111) substrate with a ZnO buffer layer by RF-magnetron sputtering technique" In Press Physics Letters A

[13] L.J. Zhuge, X.M. Wu, Q. Li, W.B. Wang, S.L. Xiang, Physica E 23 (2004) 86

[14] Z.D. Sha, X.M. Wu, L.J. Zhuge, Physics Letter A 346 (2005) 186

[15] C. Tan, X.L. Wu, S.S. Deng. G.S. Huang, X.M. Bao, Physics Letter A310 (2003) 236

[16] L.S. Liao, X.M Boa, Z.F. Yang, N.B. Min, Appl. Physics Letter 66 (1995) 2382

Temporal Partitioning of Tasks on a Heterogeneous Reconfigurable Architecture

Arjumand Yaqoob and M. Ashraf Chughtai Member IEEE
Embedded System Design Group
Faculty of Electrical Engineering
University of Engineering & Technology, Lahore 54890 PAKISTAN
Email: ayaqoob@uet.edu.pk

Abstract -Partitioning of tasks on a heterogeneous Reconfigurable Architecture (RA) is a constrained placement problem. FPGAs being reconfigurable devices can implement spatial tasks that fit in the device's finite state and operational resources. The partitioning problems for heterogeneous system need to be scheduled at early stages of the design process. We present a technique to partition a task into temporally interconnected sub-tasks. These temporal partitioned tasks are then implemented on the target RA to yield efficient and cost effective results for partitioning applications.

I. INTRODUCTION

Reconfigurable Computing (RC) combines programmable hardware with processors to exploit the strengths of hardware and software. In the computing paradigm a hardwired system design doesn't facilitate flexibility, as the complete system requires redesigning to incorporate minor design changes. On the other hand, the general-purpose software based systems lack design optimization and inefficient use of resources. [1] The RC is the most suitable solution for compute intensive tasks in terms of efficient and fast solutions having advantage of flexibility and zero non-recurring engineering (NRE) cost over Application Specific Integrated Circuits (ASICs) as well.

From system point of view a heterogeneous system consists of a processor and a configurable device like FPGA. This kind of system level model is most suitable for solving the computing problems that need hardware as well as software executions [2]. Thus, there comes to be a need of partitioning an application according to the hardware and software constraints of the problem.

This is well evident that the partitioning of tasks on this system model is coupled with timing constraints i.e temporal execution of tasks at the partitioning stage is an important aspect. The aspect of temporal partitioning problems is the exploitation of the heterogeneous RA for solving task assignments. A general form of heterogeneous RA shown in Fig.1.

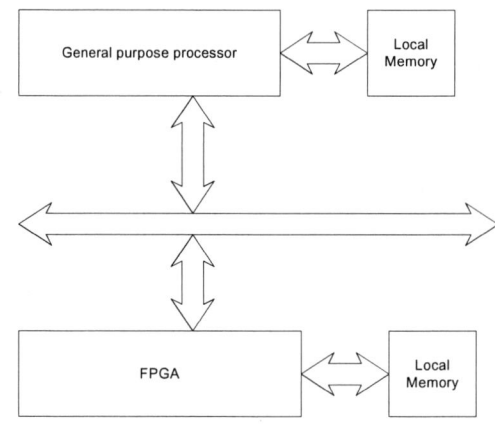

Fig.1 illustrates a general-purpose microprocessor attached with a configurable device.

General-purpose microprocessor provides a silicon medium that can be configured to solve any computational task. A set of operations that can be executed in time is called temporal or serial computation. The processor is a sequential processing element that can be configured to solve a computation in time, whereas a configurable device is a parallel device that can run operations concurrently [3]. The placement of tasks on a processor is a serial or temporal problem while the placement of tasks on a configurable device is a parallel or concurrent problem [4]. In a heterogeneous system, the programmable device having the capability of spatially partitioning the

0-7803-9730-4/06/$25.00 ©2006 IEEE   392

tasks is to exploit most compute intensive part of the application. The processor is required to partition the tasks for implementation on the system. The temporal partitioning of tasks at the design input stage is a serial execution problem that is to be addressed at the partitioning stage. The spatial partitioning of tasks is handled at the design stage when the partitioning of tasks has completed and concurrent or parallel tasks are identified for execution on configurable device.

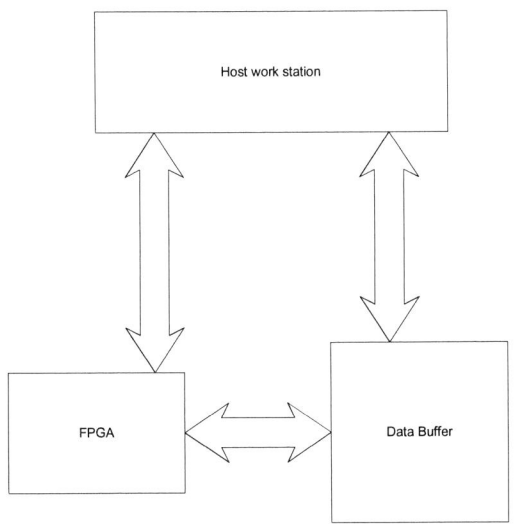

Fig.2. Reconfigurable Architecture

In this computing paradigm the number of instructions/operations required by a task determines the granularity of a task. Thus depending upon the application a most suitable partitioning of tasks can be determined. An application consists of some potential for spatial computation and some restrictions that require temporal computation i.e in the case of some computational part that require serial implementation.

In this paper we have proposed a temporal partitioning method for a RA. The architectural model used to apply the partitioning technique shown in Fig.2, consists of single FPGA and a data buffer interfaced to FPGA and host processor. This reconfigurable model is part of the Reconfigurable and Adaptive Computing Environment (RACE) [5]. RACE is the standard architectural model in order to design solutions for different applications under these constraints. The host workstation is responsible for the effective communication

between the data buffer and FPGA, to implement the temporal partitioning.

The task graph representation is a method to model tasks. We have used the representation of task graph in the temporal partitioning, which is similar to a *data flow graph* (DFG), in which sub tasks can be represented as nodes and their dependencies as edges. In this regard all the nodes with same level in the graph constitute potential candidates for spatial computation. The nodes connected by edges should be executed sequentially i.e they require temporal computation. In this way every application is examined for spatial flexibility and for temporal constraints, and independent tasks can be executed spatially whereas every temporal partitioning represents a configuration on FPGA. The orders of these partitions represent temporal constraints. In this process of determining the temporal partitioning constraints all intermediate data is stored in the associated data buffer.

Several research efforts have been made to partition and map a computational task onto spatially interconnected reconfigurable processing elements [5] [6]. The objective is to exploit the parallelism in the application by mapping it to spatially interconnected elements. These processing machines [5] and [6] and some other computing environments have practically demonstrated the performance gains equivalent to super computers. These above systems are typically characterized by their logic gate capacities and their interconnection pattern. The interconnection bandwidth directly affects the logic utilization of FPGA [7].

The objective of this paper is to establish a temporal mapping methodology by formulating an application so that an application of arbitrary size can be implemented on a small RA. In section 2 a possible execution model for partitioning of tasks on a heterogeneous RA is explained. In section 3 the problem formulation is described, which is followed by the discussion on implementation details and results on proposed heterogeneous RA, using Xilinx © standard tools and products. [8]

II. EXECUTION MODEL

The execution model is shown in Fig.3. The data buffer is used for storing all the intermediate data during the execution of the temporal partitions.

Hence every temporal partition reads in data from the buffer, performs computation and stores the result back in the buffer. In order to handle this data transfer a finite state machine has been used as a controller [4].

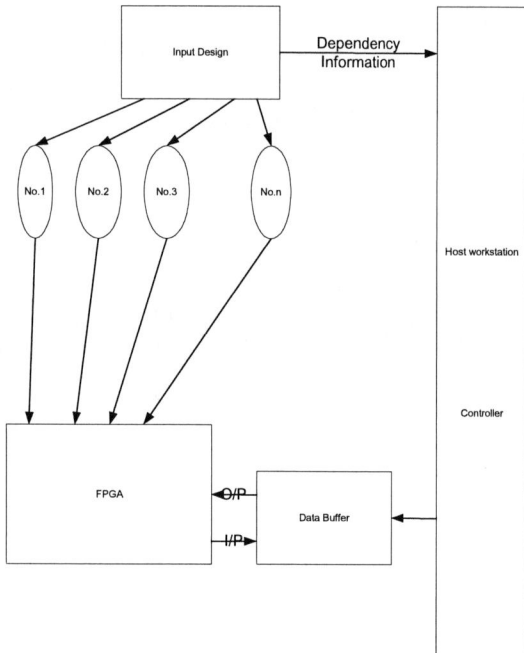

Fig.3. Execution Model

There are n temporal partitions and each partition reads its inputs from the data buffer and writes its outputs to data buffer. The finite state machine controller residing in the host machine handles the data transfer. The finite state machine model of controller is shown in Fig.4. This model is suitable for effective run time execution of tasks.

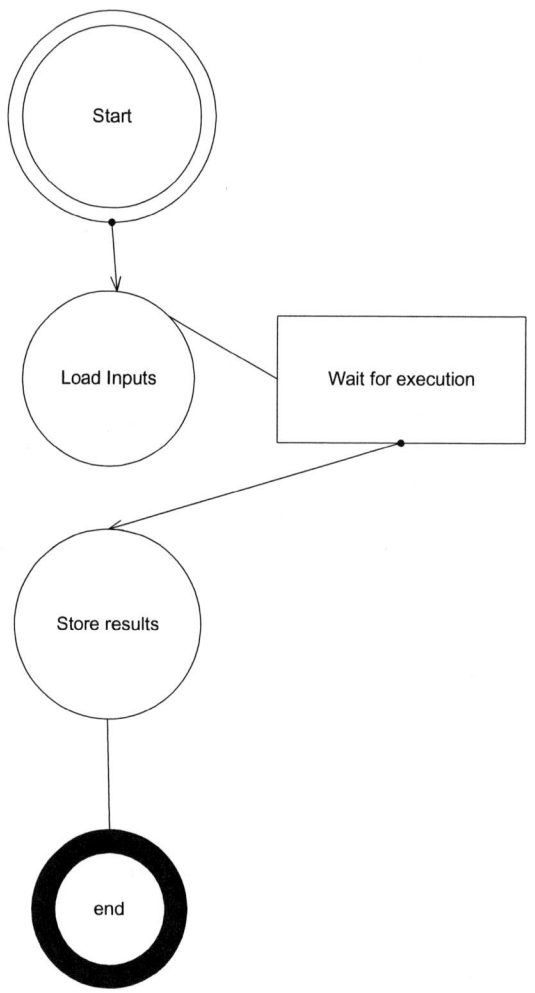

Fig.4. FSM for Controller

III. PROBLEM FORMULATION AND DESIGN METHODOLOGY

The input design specification is preprocessed and transformed into a DFG model. The DFG is represented by G= (V, E, W, D) where for each $v_i \in V$, there exists a $w_i \in W$ and $d_i \in D$, where $1 \leq i \leq n$. The w_i and d_i represent size and delay associated with the node v_i, and for $1 \leq i \leq n$ the n represents number of temporal partitions. A node v_i in the graph symbolizes a functional block in the input application. The set of edges E, constitute the data flow between the functional blocks. The associated size and delay of functional blocks, as its characteristics are technology dependent, and they are modeled to Xilinx XC3S200 FPGA [8]. The size, wi_i of functional block, represented by the node v_i is expressed in terms of number of configurable logic blocks (CLBs).[9]

The temporal partitioning is performed under the constraint of available area on the FPGA. A node v_i can be executed iff all of its predecessor nodes have been executed. The nodes with same level and the independent nodes can be executed spatially i.e in parallel. The dependent nodes are executed in As soon as Possible (ASAP) manner in the algorithm for temporal partitions, as shown below, but in the case of a node vi having In-degree zero then all the predecessor of v_i Pred(v_i) will have fixed ASAP levels. The complexity of this process can be $O(|V| + |E|)$, if $|V|$ and $|E|$ denote the size of node set and edge set in the graph.

A. TEMPORAL PARTITIONING

The temporal partitioning is employed to exploit this parallelism in order to partition the tasks. In this methodology level of each node denoted by Lev(v_i) is associated with the Total_Cost of the process. The Size(v_i) of each node v_i is associated with the Area_Filled on the FPGA. Area_Filled keeps track of the currently occupied area on the FPGA. The control operation in this process is also associated with the FSM_Cost of the process, which is also the part of Total_Cost.

A DFG model of the partitioning methodology can be explained with the help of directed graph model. This kind of model is here represented in Fig.5.

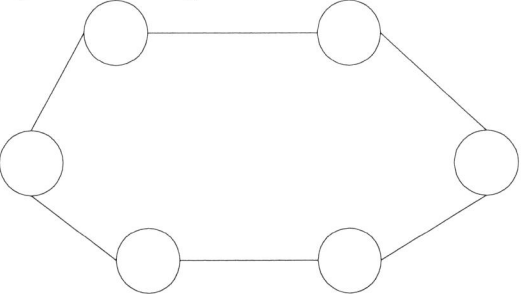

Fig.5. Data flow model for partitioning methodology

The directed graph model explains the partitioning process according to the constraints. Since ASAP level is that level represented by a node which has no predecessors. The level of a node v_i is assigned according to the Max level, that actually transforms with FSM cost for its implementation. The Total cost is assigned to a task process that is according to all the constraints previously assigned.

Any Lev(v_i) is examined for Max_Level, that denotes the maximum ASAP level in the input design. There is a partition number when, which has all ASAP levels. The algorithm for this process is described as below as Fig.6.

For each node vi with Level(v_i)=1 do
Partition(v_i) ⟸ 1
End For each
i ⟸ 2
Lev ⟸ 2
Area_Filled ⟸ 0
While (Lev ≤ Max_Level)
For each node v_i with Level(v_i) =Lev do
If (Area_Filled+Total_Cost)≤ Size of RU and
Total_Cost ⟸ FSM_Cost(v_i) then
    Partition(v_i) ⟸ i

If Area_Filled ⟸ Area_Filled + Total_Cost
  End If
  Else
i⟸i+1

Partition(v_i)⟸i
Area_Filled⟸Total_Cost
End Else
End For each
Lev⟸Lev+1
End While

Fig.6: Temporal Partition Algorithm

The algorithm calculates and fills the available area on the FPGA in increasing order of ASAP levels. All the nodes with ASAP level of 1are assigned to partition 1.Hence all the successive temporal partitions are dependent on partition number 1. The algorithm fills available area on the FPGA in increasing order of ASAP levels.

IV.   IMPLEMENTATION AND RESULTS

The temporal partitioning of tasks has been applied for implementation using the target architecture in order to exploit the design methodology adopted. The total number of temporal partitions implemented are according to our target application, which runs on the heterogeneous RA, we have implemented this architecture using Digilent© Spartan-3 platform having Xilinx

XC3S200 FPGA on it. The platform provides a capacity of 200,000 gates and 4320 logic cells. We have used the autonomous scheme by implementing the task allocation approach to apply the temporal partitioning of tasks on heterogeneous RA. That scheme has been adopted from our previous work of dynamic scheduling and allocation of tasks on a reconfigurable system [10]. The EDK 7.1v platform tool of xilinx has been used to set up this configuration on the Spartan-3, by implementing a 32-bit RISC processor core of MicroBlaze.[11]. Then by the application of temporal configurations through the system soft processor core, which contains the FSM controller and is responsible for organizing the configurable memory architecture of data buffer; the desired temporal partitioning configurations have been achieved. The ISSI© IS61LV25616AL-10T configurable array having 256Kx32 bytes of memory array has been used [12]. The application run time depends on the communication bandwidth between host processor and FPGA and on the configuration bandwidth of the configurable device. The implemented task allocation approach for temporal partitioning of tasks has achieved a configuration time of 96MBPS using the standard Ethernet MAC © core, with the configuration access time on configurable device using the IMPACT© which has a value 9.87nsec[13].

The results of application of the temporal partitioning methodology on the implementation platform are reported in the form of time required to partition an application from the input design. The application selected was the task allocation approach, which can partition the tasks according to hardware/software co-designing of the application. In our work the application of temporal partitioning for an application has resulted in an improved application synthesize time as compared with task allocation approach methodology. This has also resulted in the cost effectiveness for the design, since we have achieved this improvement on a low cost platform. The reported results have been obtained by comparison of synthesize and configuration time from UCF (user constraint file) generated by the software tool and compared with the previous results. This comparison has been performed by using the quantify tools for the system [14]. The increased bandwidth in time we got with a factor of 2.23 in this technique as a result of temporally partitioning an application at the input stage of the design.

V. CONCLUSIONS

A temporal partitioning methodology for implementation of tasks on a heterogeneous RA has been described. The results have been obtained by applying a task allocation methodology on the target RA. The target architecture provides facilitation for temporal partitioning at the design input stage by using RA. The architecture adopted for this temporal partitioning methodology has been implemented using the standard platform of Xilinx for obtaining the implementation results. The implementation methodology uses the temporal partitioning method to achieve the desired results. The implementation results have shown the effective improvement in design synthesizes and implementation due to exploitation of temporal constraints. The results are also cost effective as the implemented soft cores for processing and data storage have been taken from Xilinx ©standard core generation process, that use Xilinx core connect technology. The complete system comprise of a low cost board, which is capable to handle these constraints for implementation The future work in this regard is aimed for implementing some more dense and compute intensive algorithmic applications for tasks level exploitation using temporal constraints. This will help in modifying the design methodology for temporal constraints.

ACKNOWLEDGEMENT

We are thankful to Department of Electrical Engineering, University of Engineering & Technology, Lahore for providing us the resources during the whole period of this work. Our thanks are also extended to the faculty members of EED, UET for their valuable suggestions and discussions.

REFERENCES

[1] Hauser J, Wawrzynek J; "Garp: A MIPS Processor with a Reconfigurable Co-processor." Procedings of the IEEE Symposium on FPGAs for Custom Computing Machines, Napa Valley,CA,1997, pp 12-21.

[2] Katherine Compton and ScottHauck;"Reconfigurable Computing: A survey of systems and Software." ACM Computing Surveys, Vol.34, No.2 June 2002 pp 171-210.

[3] Andre Dehon; "The Density Advantage of Configurable Computing." IEEE Transactions on Computer Vol.5, No.2, April 2000 pp41-49.

[4] Karthikya M, Gajjalal P, DineshB;"Partitioning in Time: A paradigm for Reconfigurable Computing." Kluwer Academic Publishers, 1998.

[5] Doug Smith and Dinesh Bhatia. RACE: Reconfigurable and Adaptive Computing Environment. In Field Programmable Logic: Smart Applications, New paradigms and compilers, pp 87-95. Springer verlag, Berlin, Sept.1996.

[6] Duncan A. Buell, Jeffrey M.Arnold and Walter J. Kleeinfield; "Splash 2, FPGAs in Custom Computing Machines.IEEE Computer Society Press,1996.

[7] Andre Dehon, Joshua Adam, Michael Dehorimier e.t; "Design Patterns for Reconfigurable Computing." In Proceedings of IEEE Symposium on Field-Programmable Custom Computing Machine (FCCM 2004), April 20-23, 2004.

[8] Xilinx Inc. www.xilinx.com/productsupport.

[9] Xilinx Inc. Spartan3 Field Programmable gate Arrays, June 2002.

[10] M.Ashraf Chughtai, Arjumand Yaqoob; "An Approach to Task Allocation for Dynamic Scheduling in Reconfigurable Computing Systems." In Proc. of 9th IEEE Int'l Conference on Multi Topic INMIC '05 Dec. 23-24.

[11] Xilinx Inc. www.xilinx.com/ise/embedded/mb_ref_guide.pdf

[12] ISSI Inc. www.issi.com/pdf/61LV25616AL.pdf

[13] Xilinx Inc. Tri Mode Ethernet Media Access Controller. www.xilinx.com/bvdocs/ipcenter/data_sheets/tri_mode_eth_mac_gs_ug139.pdf

[14] Rational Software Cooperation. Quantify user's Guide, Version 3.1,1997.

Packet Synchronization Structure with Peak Detection Algorithm for MB-OFDM UWB

Aymen M. Karim, Masuri Othman, Edmond Zahedi
Institute of Microengineering and Nanoelectronics
National University of Malaysia
43600 BANGI, Selangor Darul Ehsan, Malaysia
aymen77eng@gmail.com, masuri@vlsi.eng.ukm.my, ezahedi@vlsi.eng.ukm.my

Abstrac In Multi-band Orthogonal Frequency Division Multiplexing (MB-OFDM) wireless personal area networks (WPAN), a robust signal detection algorithm is required for practical hardware implementation. In the literature, the signal detection algorithms for OFDM are based on correlating the received signal with a time delayed replica of itself. The result of correlation is then compared with a threshold value. The selection of the threshold value is very important for correct packet detection. Because of the characteristics of the channel environment, the determination of a threshold value is not a trivial task. In this paper, a structure for packet synchronization in MB-OFDM is proposed which utilizes a peak detector. In this case, the determination of the threshold value is eliminated by finding the peak values of the correlation value using our proposed algorithm.

I. INTRODUCTION

Orthogonal Frequency Division Multiplexing (OFDM) has received a great interest by being adopted in many standards such as DAB, DVB, IEEE 802.11a, and ETSI BRAN HiperLAN2 due to its robustness against the multi-path channel environment and its suitability to be implemented by FFT [1]. Multi-Band OFDM Ultra-Wideband (MB-OFDM UWB) achieves rates of 110 Mbits/second at a distance of 10 meters by dividing the UWB spectrum (3.1 to 10.6 GHz) into 528 MHz wide sub-bands and using OFDM modulation to transmit the information in each sub-band [2].

Due to the sensitivity of OFDM signals to the timing errors [3], a reliable synchronization algorithm is required to perform packet detection and to determine symbol timing. In general, synchronization algorithms are based on sending a periodic training sequence. The receiver will correlate the incoming signal with a delayed

replica of it (the delay is selected according to the period time). The start time of the packet is determined by comparing the peak values with a threshold value. Because of the characteristics of the channel environment, the determination of the threshold value is not a trivial task and should be chosen adaptively.

In this paper, we propose a structure for packet detection by the training sequence proposed in a proposal for IEEE 802.15 [2], our structure utilizes an algorithm for peak detection, the output of the correlator will not be compared with a threshold value, instead it will go into the peak detection block. In this case, the determination of the threshold value is eliminated.

This paper is organized as follows. In section II, the training sequence proposed in [2] is reviewed. In section III, the packet detection scheme which will be used in our structure is presented. The proposed algorithm is explained in section IV. Finally, the conclusion is given in section V.

II. TRAINING SEQUENCE IN THE IEEE 802.15-03

The Physical Layer Convergence Protocol (PLCP) preamble in the IEEE 802.15-03 proposal consists of three parts [2]: packet synchronization sequence, frame synchronization sequence, and the channel estimation sequence (Figure 1). The packet synchronization sequence consists of 21 periods denoted as {PS0, PS1, …, PS20}.

Each PS sequence is constituted by repeating the sequence in Table 1 sixteen times according to the pattern in Table 2. Each piconet uses a distinct time domain sequence (4 piconets). Then, after repetition, each sequence is pre-appended by 32 zero samples and by appending a guard interval of 5 zero samples to constitute one PS sequence.

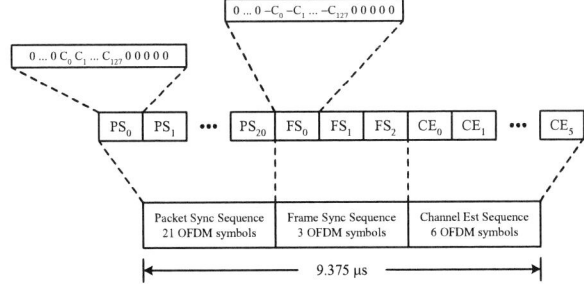

Fig. 1 Standard PLCP preamble format.

This part of the preamble is used for packet detection and acquisition, coarse carrier frequency estimation, and coarse symbol timing. The second part consists of three periods denoted as {FS0, FS1, FS2} and is used to synchronize the receiver algorithm within the preamble. Finally, the last part consists of six periods denoted as {CE0, CE1, …, CE5} and is used to estimate the channel frequency response, for fine carrier frequency estimation, and fine symbol timing.

TABLE I. TIME DOMAIN SEQUENCES

1	1	-1	-1	-1	1	1	-1	1
2	1	-1	1	1	-1	-1	-1	1
3	1	1	-1	1	1	-1	-1	-1
4	1	1	1	-1	-1	1	-1	-1

TABLE II. REPETITION PATTERNS

1	+	+	+	+	-	+	+	+	-	-	+	-	+	-	+	+
2	+	-	-	-	-	-	+	-	+	-	-	+	+	-	-	+
3	+	+	-	-	-	+	-	-	-	+	-	-	+	-	+	+
4	+	-	-	+	-	+	-	-	+	+	-	-	-	-	-	+

III. PACKET DETECTION SCHEME

In [4], the following timing metric is proposed,

$$\Lambda_\varepsilon(d) = \left(\frac{L}{L-1} \frac{|P(d)|}{E(d)} \right)^2 \quad (1)$$

where

$$P(d) = \sum_{k=0}^{L-2} b(k) . \sum_{m=0}^{M-1} r^*(d+kM+m).r(d+(k+1)M+m) \quad (2)$$

$$E(d) = \sum_{i=0}^{M-1} \sum_{k=0}^{L-1} |r(d+i+kM)|^2 \quad (3)$$

and $b(k) = p(k) \times p(k+1), k = 0,1,...,L-2$

In the above equations, $\{p(k) : k = 0,1,...,L-1\}$, is the repetition pattern, r are the received samples, M is the length of the time domain sequence, and L is the length of the repetition pattern.

Synchronization is achieved for,

$$\hat{d} = \arg\max_d (\Lambda_\varepsilon) \quad (4)$$

where $\hat{d}$ is the estimated time index.

If L=2, and the repetition pattern is $p = [++]$, the timing metric will resemble the above timing metric proposed by Schmidl and Cox [5].

Figure 2 shows the output of the correlator for the timing metric in [2] applied to a sequence consisting of three PS's.

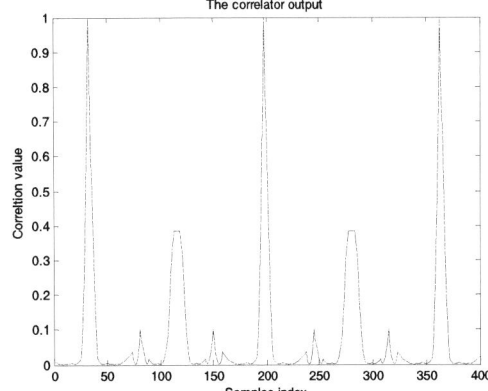

Fig. 2 The correlator output according to the specification in the IEEE 802.15-03 proposal.

IV. PROPOSED DETECTION SCHEME

The proposed structure in this paper is divided into two structures, the autocorrelator structure, and the peak detection structure.

A. The autocorrelator

The solution given by equation (2) is the optimal solution to find the correlation for the sequences PS's. The calculation of $P(d)$ can be divided into two steps. First we find $R(d)$, where,

$$R(d) = \sum_{m=0}^{M-1} r^*(d+m).r(d+m+M) \quad (5)$$

Then, we find $P(d)$ where,

$$P(d) = \sum_{k=0}^{L-2} b(k)R(d+kM) \quad (6)$$

$R(d)$ can be calculated iteratively as below,

$$R(d+1) = R(d) + (r^*_{d+M}r_{d+2M}) - (r^*_d r_{d+M}) \quad (7)$$

reducing the computations to one complex multiplication, one addition, and one subtraction.

Figure 3 illustrates the proposed structure to implement equation (2). The upper part is used to find $R(d)$, the value of $R(d)$ obtained from the upper part is then passed to the lower correlator to find $P(d)$. The lower part of the correlator is used to delay the value of $R(d)$ by 8 samples for several times. The result of the delay is passed to a 2's complement block. The 2's complement block operation depends on the sequence $b(k)$. If $b(k)$ is positive, the signal will be passed without taking its 2's complement, otherwise, the 2's complement of the signal will be found.

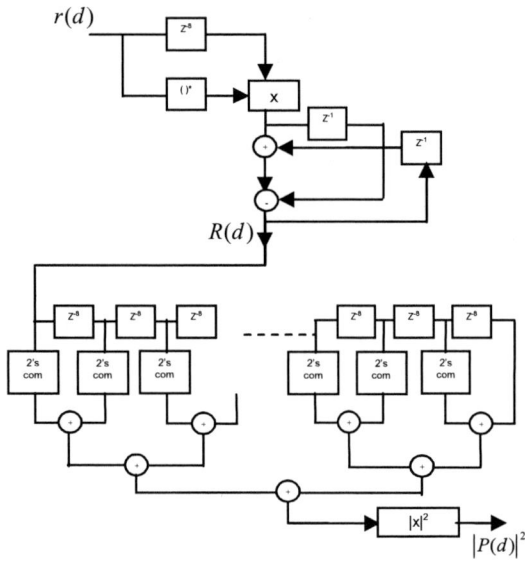

Fig. 3 The proposed structure.

Finally, the squared absolute value of $P(d)$ ($|P(d)|^2$) is passed to the structure for finding the peak.

B. Peak detection

The proposed algorithm is stimulated by the algorithm proposed in [6]. There, the instantaneous peak detector is used to find an instantaneous peak of the input coming from a differentiator, and the group peak detector is used to detect the falling slopes of the incoming signal. If the group peak detector finds a falling edge at the same time as the instantaneous peak detector finds a relative peak, this means that the detected peak is actually an absolute peak. Our algorithm (Figure 4) has two blocks similar to the instantaneous peak detector in [6], but the first block in our structure is used to find the rising slopes of $|P(d)|^2$ during a specified number of samples, and the second is used to find the falling slopes of $|P(d)|^2$.

The peak detector consists of two registers, two comparators, and tow counters. The maximum value of the upper counter depends on how many samples the rising slope takes to reach the peak (12 samples), and the maximum value of the down counter depends on how many samples the falling slope takes to go down (6 samples). An interesting feature in $|P(d)|^2$ (Figure 5) is that it does not show any flat area like the metric proposed in [5]. Therefore, it could be fed directly to our algorithm without the need for a differentiator being compared with the previous sample recorded in the upper register. If $|P(d)|^2$ is smaller than or equal to the previous recorded sample, the counter will be reset and the register will be updated to contain the new sample. If it is bigger than the previous recorded sample, the counter will be triggered increasing its count by one and the register will be updated to contain the new sample. If this situation persists, the counter will reach its maximum value and send an overflow signal to enable the lower structure. In Figure 4 the lower structure operation is the same as the upper part, the only differences being the counter maximum value and the comparator direction. If the situation persists ($|P(d)|^2$ in the lower part is less than the last recorded sample), the counter will reach its maximum value and set an overflow flag which will signal that a peak is detected.

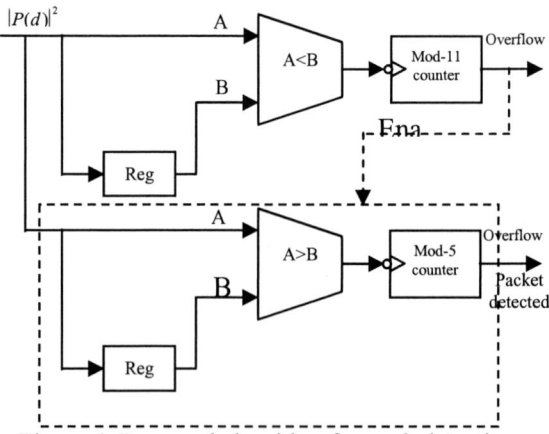

Figure 4. Proposed algorithm for peak detection.

The results of the peak detection algorithm implemented on a sequence of three PS's is illustrated in Figure 5 (noise free condition). It is clear that the packet will be detected about 6 samples after the starting of the peak (down counter maximum value).

400

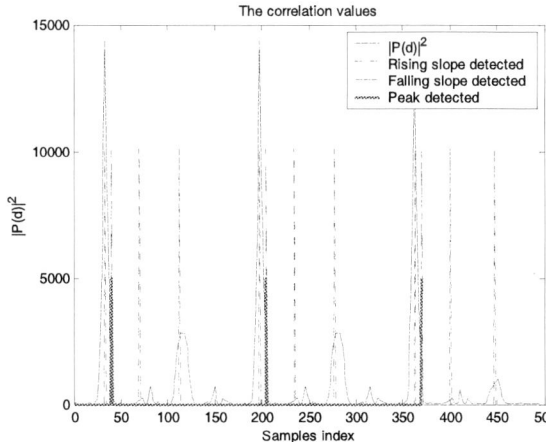

Figure 5. Illustration of the packet detection procedure.

V. CONCLUSION

In this paper, an algorithm for packet detection in MB-OFDM UWB is proposed, the correlation of the incoming signal is divided into two parts, upper part and lower part, where in the upper part the correlation is calculated iteratively.

Moreover, the determination of the threshold value for comparison is eliminated by a peak detection algorithm.

REFERENCES

[1]   S. B. Weinstein, P. M. Ebert, "Data transmission by frequency division multiplexing using the discrete Fourier transform," *IEEE Trans. Commun.,* Oct. 1971, Vol. COM-19, pp. 628-634.

[2]   A. Batra et al., "Multi-band OFDM Physical Layer Proposal for IEEE 802.15 Task Group 3a," IEEE P802.15-03/268r1-TG3a, September 2003.

[3]   C. R. N. Athaudage, "BER sensitivity of OFDM systems to time synchronization error", *Proceedings of the 8th IEEE International Conference on Communication Systems (ICCS-2002)*, November 25-28, 2002, Singapore, pp. 42-46.

[4]   Minn, H., V.K. Bhargava, and K.B. Lataief, "A Robust Timing and Frequency Synchronization for OFDM Systems", *IEEE Transactions on Wireless Communications,* Jul 2003, pp. 822 - 839.

[5]   Schmidl, T.M., and C. Cox, "Robust Frequency and Timing Synchronization for OFDM", *IEEE Transaction on Communications,* Dec 1997, pp. 1613 - 1621.

[6]   M. Krstic, A. Troya, K. Maharatna, "Optimized Low-power Synchronizer Design for the IEEE 802.11a Standard", *Proceedings of the IEEE ICASSP'03*, vol. II, pp. 333-336, Hong Kong (P.R. China), April 2003.

Source-Coupled Logic (SCL): Operation and Delay Analysis

Mohamed Azaga ,Student Member, IEEE and Masuri Othman.

VLSI Design Center, Fakulti Kejuruteraan, Blok Inovasi 2, UKM, 43600 Bangi, Selangor, Malaysia

mohazaga@vlsi.eng.ukm.my, masuri@vlsi.eng.ukm.my

Abstract - **This work describe and present the analysis for the Source-Coupled Logic (SCL) inverter. The SCL inverter circuit model and its operation is defined. The analysis for the SCL is carried from the point of view of input/output voltage characteristics, and the effect of noise margin. Finally, the inverter gate delay model is described, and the effect of biasing current on the delay is shown. All simulation is done based on the 0.18μ Silterra process, and using Cadence Spectre simulation platform. The result shows that, the delay of the SCL inverter is decreased as biasing current increase.**

I. INTRODUCTION

The recent development in VLSI technology has allowed rapid growth in the design and production of portable electronic systems. One of the limiting factors in the deployment of these devices is the battery life time and the power consumption of the circuitry. As the chip density increases, it is also desirable to integrate analog and digital circuitry onto the same die. This integration however faces difficulty in the designed high of precision analog circuitry in the presence of digital noise [1].

SCL logic style seems to be promising in both reducing power consumption and providing an analog friendly environment. The Source-Coupled Logic (SCL) has shown reduction of about two order of magnitude. The low switching noise feature of SCL is obtained at the cost of static power dissipation; a design strategy of SCL gates is required to meet specifications, while keeping power consumption as low as possible. Moreover, design criteria to consciously manage the power- delay tradeoff are required [2].

This paper begins with the discussion of theory, the architecture, and the operation of the SCL logic. In section III, the SCL is characterized from point of input/output voltage, noise margin, and delay model. The simulation results are shown in each corresponding section. Finally, we conclude with general remarks.

II. SCL : PRINCIPLE OF OPERATION

SCL is a dual rail logic circuit that use both the variable and its complement (A, Ā) as an input pair. The output of a dual rail circuit is also a pair (E, Ē) that drives the next gate(s) in the logic cascade. However, dual rail logic interprets the difference (E-Ē) as the logic variable instead of just one or the other. When viewed at the level of Boolean algebra, the use of both the variable and it's complement is superfluous; the result is the same as that found using a single-rail circuit. Moreover, dual rail networks are more complicated to wire [3].

The circuit schematic of SCL inverter gate, shown in Fig. 1, is made up of an NMOS source-coupled pair having transistors working in the saturation or cut off region, that approximate well the behavior of a voltage-controlled current switch. The biasing current (I_{ss}) is steered to one of the two output branches and converted into a differential output voltage by two PMOS transistor working in the linear region (Active load pull-up resistance) [2]. The logic function of the SCL is implemented by the logic block connected between the active load (PMOS) and the current source (I_{ss}) . For an inverter/buffer, the logic block is the differential pair constructed by NMOS transistors M1 & M2 [4].

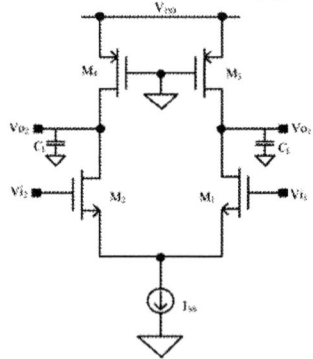

Fig. 1 SCL Inverter

The SCL gate uses PMOS active load, but other types of load, such as physical resistor or a diode-connected NMOS/PMOS could be used. However, resistor load is not normally chosen since large silicon area needed and its parasitic capacitance can be high. For the second type of load, the output levels will loss the threshold

0-7803-9730-4/06/$25.00 ©2006 IEEE   402

voltage furthermore; the MOS diode load is slower than the PMOS active load for practical bias currents [5].

The operation of the SCL logic is based on the input differential pair circuit. The two inputs control the flow of current through the two branches of the differential pair. For example, if V_{GS} (M2) is higher than V_{GS} (M1), the current I_{D2} exceeds the current I_{D1}. Therefore, the output voltage V_{o2} begins to drop until it reach steady sate , where the current going through PMOS active load (M4) matches the I_{D2}. In mean time V_{o1} is charged to V_{DD} through M3.

The output voltage swing V_{swing} is defined as voltage difference between V_{o1} and V_{o2} at steady state. The amount of current passing through the ON branch (M2) controls the delay of the logic gate transition (1→0), while the PMOS active load (M3) controls the charging of the output nodes (0→1 transition) [4].

Defining ΔV as the voltage drop of M3(M4) due to the drain current equal to I_{ss}, the logic swing of the gate , V_{swing}, equal $2\Delta V$.

To achieve best performance, all current must pass through the ON branch and the load resistance (PMOS) should be small in order to reduce the RC delay. This guarantees that the voltage is $V_{DD} - I_{ss}.R_D$, where I_{ss} is the current flowing through current source, and R_D is the PMOS equivalent linear resistance [4].

III.   SCL: CHARACTERISTICS

A. Input-Output transfer voltage

As conversion from current-to-voltage in the SCL inverter is performed by the two PMOS transistors M3-M4, both of which have a source-gate voltage equal to V_{DD} and a much smaller source-drain voltage (in order of hundred milivolts). Therefore, transistors M3-M4 work in the triode region, and can be modeled as an equivalent linear resistor R_D [5].

Using the standard BSIM3v3 MOSFET model [6], under the static condition, PMOS transistor can be suitably approximated by an equivalent linear resistance R_D given by :

$$R_D = \frac{R_{int}}{1 - \dfrac{R_{DS}}{R\,int}} \quad (1)$$

where R_{DS} = (R_{DSW} * 1e-6)/W_p models the source/drain parasitic resistance which depends on the empiric model parameter R_{DSW} as well as the PMOS transistor effective channel width W_p. R_{int} is given by:

$$R_{int} = \left[\mu_{eff,p} C_{OX} \frac{W_p}{L_p} \left(V_{dd} - |V_{th,p}| \right) \right]^{-1} \quad (2)$$

that represent the intrinsic resistance of PMOS transistor in the linear region (i.e., it does not account for the parasitic drain/source resistance). In equation (2), the $\mu_{eff,p}$ represents the effective hole mobility, parameter L_p is the PMOS effective channel length, C_{ox} is the oxide capacitance per area, and $V_{th,p}$ is the threshold voltage [2].

The output voltage $V_o(V_i)$ SCL inverter can be evaluated by substituting the equivalent resistance R_D in (1).

Thus, the differential output voltage V_o is equal

$$V_o = V_{o1} - V_{o2} = -R_D (i_{D1} - i_{D2}). \quad (3).$$

The minimum differential input (V_i) required to fully switch the entire tail current I_{ss} to one side is give by [7]:

$$I_{ss} = \frac{1}{2} \mu_n C_{ox} \frac{W}{L} V_i^2 \rightarrow V_i = \sqrt{\frac{2 I_{ss}}{\mu_n C_{ox} \dfrac{W}{L}}} \quad (4),$$

which gives output transfer characteristics:

$$v_o(v_i) = \begin{cases} R_D I_{ss} & if \quad v_i < -\sqrt{\dfrac{2 I_{ss}}{\mu_n C_{OX} W_n / L_n}} \\[3em] -v R_D I_{ss} \sqrt{\dfrac{\mu_{eff,n} C_{OX} W_n}{I_{ss} L_n} \left(\dfrac{\mu_{eff,n} C_{OX} W_n}{2 I_{ss} L_n} v_i \right)^2} & if \quad |v_i| \le \sqrt{\dfrac{2 I_{ss}}{\mu_n C_{OX} W_n / L_n}} \\[3em] -R_D I_{ss} & if \quad v_i > \sqrt{\dfrac{2 I_{ss}}{\mu_n C_{OX} W_n / L_n}} \end{cases} \quad (5),$$

where $V_i = V_{i1} - V_{i2}$ (Differential input). From (5), $V_{OL} = -R_D I_{ss}$ & $V_{OH} = R_D I_{ss}$, then the logic swing is equal to [5]:

$$V_{swing} = V_{OH} - V_{OL} = 2 R_D I_{ss} = 2 \Delta V \quad (6).$$

From (5) , that output voltage swing (V_{swing}) is a function of R_D for fixed I_{ss} . R_D is controlled by the aspect ratio of PMOS , so for desired V_{swing}, the aspect ratio can be varied accordingly .

To verify the theoretical analysis, the simulation was carried out on the SCL inverter gate. For evaluate the V_{swing} , DC simulation is done by varying the aspect ratio of the PMOS with fixed I_{ss} =10μA. The results are taken from V_{o1} and V_{o2} (Fig. 2) and differential output voltage is shown in Fig. 3.

Fig. 2 & 3 shows the output pf SCL inverter as function of differential input (V_i) , when (V_i) is greater certain value the SCL inverter fully switch to either side and passing all I_{ss} to that side. The value of output voltage is function of R_D and it is controlled by the aspect ratio of PMOS.

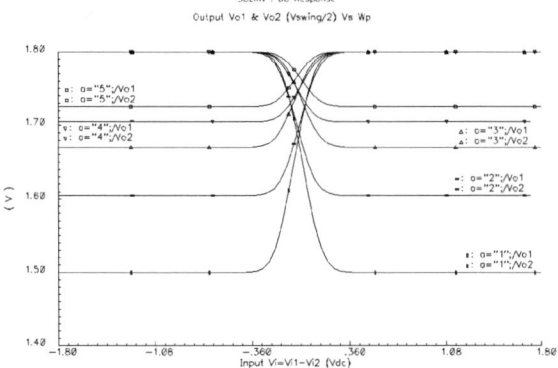

Fig. 2 Single end output voltage varying with the aspect ration of PMOS.

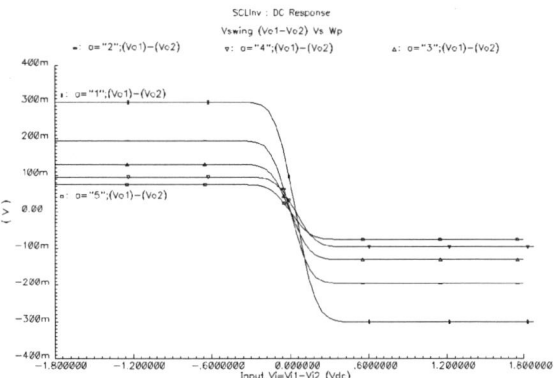

Fig. 3 Differential output voltage varying with the aspect ration of PMOS.

The V_{swing} /2 ($I_{ss}.R_D$) must be kept low enough to ensure the NMOS transistors M1-M2 are not in the triode region . In particular, when the gate voltage of an NMOS transistor is high (i.e. equal to V_{DD}), the drain voltage is equal to V_{DD}- $I_{ss}.R_D$, the triode region can be avoided if the gate–drain voltage V_{GD} is lower than threshold voltage by:

$$V_{GD} = V_{DD} - \left[V_{DD} - R_D \cdot I_{SS}\right] = R_D \cdot I_{SS} \leq V_{th,n} \,, (7)$$

which imposes an upper bound to $I_{ss}.R_D$, and hence the logic swing as given by (6) [5].

B. Noise Margin (NM)

Due to the symmetrical property, the logic threshold is equal to zero (V_{LT}=0), and the associated small-signal voltage gain is $g_{m,n}.R_D$, (where $g_{m,n}$ is the small-signal transconductance of transistor M1-M2 with $I_{D1,2} = I_{ss}$ /2) [2].

Since $v_{i1} = v_{i2} = v_{o1} = v_{o2} = V_{DD} - \Delta V/2$, and $I_{D1,2} = I_{ss}$ /2, when the gate is biased around logic threshold, voltage V_{DS} of transistor M1-M2 is equal to their V_{GS}. Hence, the resulting expression of the voltage gain A_V is [2]:

$$A_V = g_{m,n} R_D = \Delta V \sqrt{\mu_{eff,n} C_{ox} \frac{W_n}{L_n} \frac{1}{I_{ss}}} \qquad .(8)$$

The NM is equal to NM_L (for Low-Logic) and similar to NM_H (for High-Logic) due symmetrical property, which is defined as NM_H = V_{OHmin} - V_{IHmin} ($NM_L = V_{ILmax} - V_{OLmax}$) where V_{ILmax} and V_{IHmin} are the input voltage values such that $\partial v_o / \partial v_i = -1$.$V_{OLmax}$ and V_{OHmin} are the corresponding output voltages (i.e. V_{OLmax} =Vo(V_{IHmin}) and V_{OHmin} =Vo(V_{ILmax})) [5]. By differentiating (5) for v_i and setting it to -1, V_{IHmin} results is

$$V_{IH\,min} = \sqrt{\frac{2Iss}{\mu_{eff,n} C_{OX} \frac{W_n}{L_n}} - \frac{Iss}{2\mu_{eff,n} C_{OX} \frac{W_n}{L_n}} \frac{1}{A_V^2}\left(\sqrt{1+8A_V^2}+1\right)}$$

$$V_{IH} \cong \sqrt{\frac{2I_{SS}}{\mu_{eff,n} C_{OX} \frac{W_n}{L_n}}\left(1 - \frac{1}{\sqrt{2}A_V}\right)} \qquad (9),$$

where $A_V \rangle\rangle 1/\sqrt{8}$ has been assumed. Approximating V_{OHmin} to -ΔV leads to the following expression of NM :

$$NM = \Delta V\left(1 - \frac{\sqrt{2}}{A_V}\sqrt{1 - \frac{1}{\sqrt{2}A_V}}\right) \cong \Delta V\left(1 - \frac{\sqrt{2}}{A_V}\right)(10),$$

where $A_V \rangle\rangle 1/\sqrt{2}$ was assumed. The value of NM is proportional to half the logic swing [5]. Simulation were performed by setting V_{swing} to 600 mV, I_{ss} to 10 uA and varying the NMOS aspect ratio (dimensions in nm). The effect of A_V is shown in Fig. 4, which shows how $\partial V_o / \partial V_i$ changes with W_n. A_V increases for low values of NMOS aspect ratio then asymptotically tends to a constant value as illustrated in Fig. 5. The SCL inverter gate output delay increases linearly as

W_n increases thus limiting us to small W_n, as shown in Fig. 6.

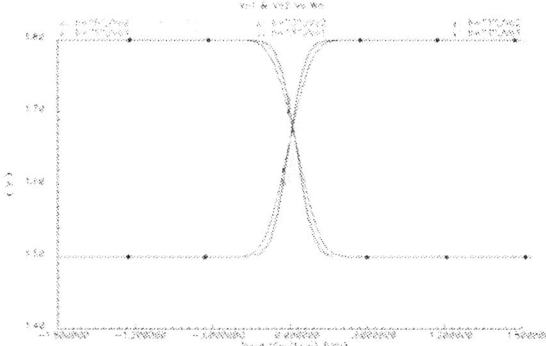

Fig. 4 Change of $\partial V_o / \partial V_i$ with different W_n.

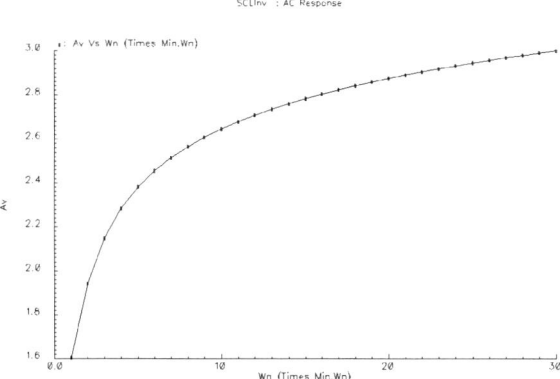

Fig. 5 A_V versus NMOS aspect ratio.

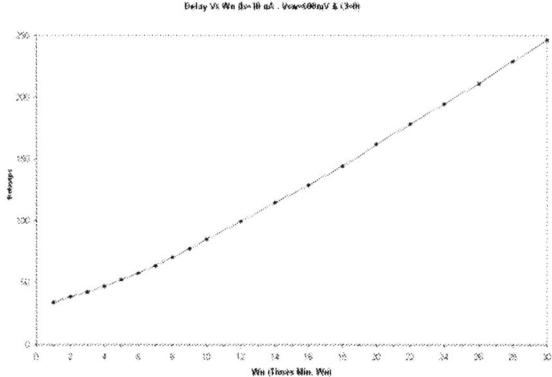

Fig. 6 SCL inverter delay versus W_n.

C. Delay Model

To model the propagation delay, $\tau_{PD,SCL}$, of the SCL inverter , it is useful to observe that NMOS transistors work in the saturation region most of the time , and their source voltage is the same for both input logic values (it's fixed by the NMOS transistor in the ON state). Thus, after linearization the circuit around the logic threshold $v_i = 0$, the half-circuit concept applies, since the circuit is symmetrical and it's input is differential [8].

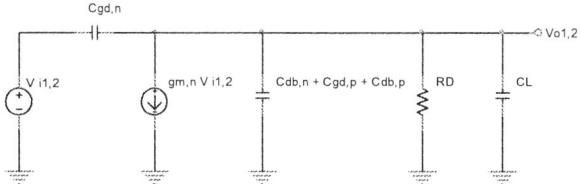

Fig. 7 Equivalent linear half-circuit of the SCL inverter.

Fig. 7 shows the small-signal model of the half-circuit concept. The circuit has time constant (τ) that can be evaluated by applying the open-circuit time constant method [9]. It gives the resulting delay as 0.69τ, assuming a step input waveform and neglecting the high-frequency zero. Hence, the propagation delay $\tau_{PD,SCL}$ of the SCL gate is given by [5]:

$$\tau_{PD,SCL} = 0.69 * R_D(C_{gd,n} + C_{db,n} + C_{gd,p} + C_{db,n} + C_L) \quad (11)$$

The NMOS capacitance $C_{gd,n}$, in (11) is evaluated in the saturation region. Thus its value equal to the overlap capacitance $C_{gd0}.W_{eff}$ between the gate and the drain. Junction capacitances $C_{db,n}$ and $C_{db,p}$ can be linearized by modifying their value in a zero-bias condition via coefficients K_j according to [10]. Capacitance $C_{gd,p}$ is equal to the sum of the overlap contribution $C_{gd0}W_{eff,p}$ and the intrinsic contribution associated with the channel charge of the PMOS transistors working in the linear region, $C_{gd,p,int}$. In particular, we can adopt the BSIM3v3 capacitance model [6], which express capacitance $C_{gd,p,int}$ as the derivative of charge flowing into drain Q_D with respect to the voltage V_D [8]:

$$C_{gd,p,int} = \frac{\partial Q_D}{\partial V_D} \cong \frac{3}{4} A_{bulk,max} WLC_{OX} . \quad (12)$$

This derivation assumes the gate, source, and bulk voltage are constant, and using $A_{bulk,max}$ (maximum bulk charge effect) slightly greater than unity and assuming

$$V_{SD} \ll V_{SG} - |V_{th}| / A_{bulk}.$$

To validate the delay model, the bias current was varied from 10 to 100 μA, the transistor aspect ratios were sized to obtain the typical value V_{swing} = 600 mV, A_V=1.94, and the load capacitance C_L was set to 0 F, 100 fF and 1 pF, respectively. Fig. 8, Fig. 9, and Fig. 10 show the results. As expected, the delay is decreased by increasing the bias current I_{ss} and asymptotically tends to a constant value.

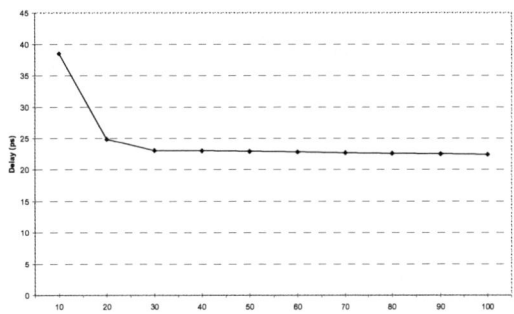

Fig. 8 Delay versus bias current I_{SS} with $C_L = 0$.

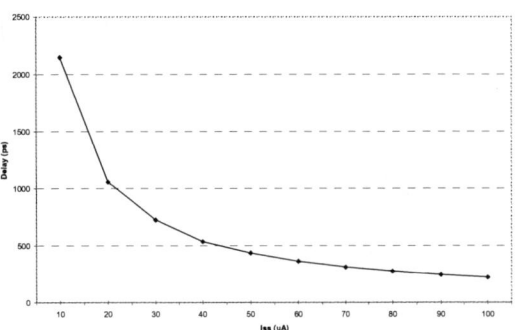

Fig. 9 Delay versus bias current I_{SS} with $C_L = 100f$ F.

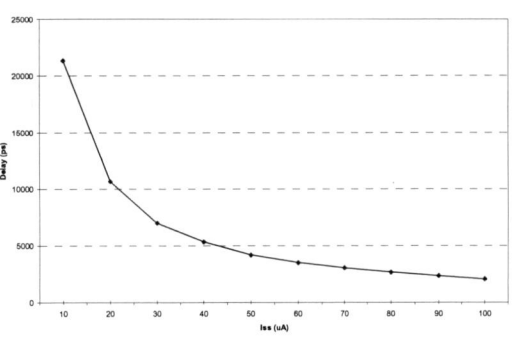

Fig. 10 Delay versus bias current I_{SS} with $C_L = 1p$ F

IV. CONCLUSION

In this work the basic theory and operation of SCL logic are presented. The delay of inverter SCL gate is validated by Spectre simulation using a 0.18μ silterra CMOS process.

Different parameters which affect the SCL logic are characterized. The SCL delay is decreased by increase of biasing current. The output voltage swing is small which reduce the dynamic power consumption. The SCL logic draws a fixed static current from the supply.

The other advantages of SCL which inherited from it's differential design include common mode noise rejection, insensitivity to process change, and friendliness to neighboring analog circuit components.

REFERENCES

[1] Jason Musicer, and Jan Rabaey, " MOS current mode logic for low power, low noise CORDIC computation in mixed-signal environments", Proc. ISLPED 2000, pp.102-107, 2000.

[2] Massimo Alioto, and Gaetano Palumbo, "Design strategies for source coupled logic gates", IEEE Transactions on circuit and systems-I: Fundamental theory and applications, Vol. 50, No. 5, May 2003.

[3] John P. Uyemura," CMOS logic circuit design", Dordrecht : Kluwer Academic Puplisher ,2001.

[4] Mohamed W. Allam, and Mohamed I. Elmasry, "Dynamic Current Mode Logic (DyCML): A new low-power high-performance logic style", IEEE Journal of solid-state circuits, Vol.36, No. 3, March 2001.

[5] Massimo Alioto, and Gaetano Palumbo, " Model and design of bipolar and MOS current-mode logic: CML, ECL and SCL design circuit", Dordrecht : Springer, 2005.

[6] Y. Cheng, and C. Hu., "MOSFET Modeling & BSIM3 User's guide", Boston, MA : Kluwer, 1999.

[7] Michael M. Green, and Ullas Singh, " Design of CMOS CML circuits for high-speed broadband communication", Proc. of the Int. Circuits and Systems symposium, ISCAC '03, Vol. 2. May 2003, pp. 204-207.

[8] Massimo Alioto, Gaetano Palumbo, and S. Pennisi, " Modeling of source coupled logic gates", Int. Journal circuit theory application, Vol. 30, No. 4, pp. 459-477, 2002.

[9] B. Cochrun and A. Grabel, " A method for the determination of the transfer function of electronic circuit", IEEE Transactions circuit theory, Vol.CT-20, pp. 16-20., Jan. 1973.

[10] J. Rabaey, " Digital integrated circuit (A design prospective)", Englewood Cliffs, NJ: Prentice-Hall, 1996.

Photopthermal Study of Ceramic ZnO Doped with Y_2O_3

*Azmi Zakaria, Zahid Rizwan, Mansor Hashim, Abdul Halim Shaari,
W. Mohmood Mat Yunus.
Department of Physics,
Faculty of Science, Universiti Putra Malaysia,
43400 Serdang, Selangor, MALAYSIA.
E-mail*: azmizak@fsas.upm.edu.my

Abstract **A powerful tool, photopyroelectric spectroscopy, for examining the optical properties as non-radiative recombination for the semiconductor materials is used for the Y_2O_3 doped ZnO ceramics. The ceramic (ZnO-x, x=0.4-1.6 mol% of Y_2O_3) was sintered at isothermal temperature, 1175 and 1275 °C for 1 hour to investigate its optical properties. The PPE spectroscopy is used to study the energy band-gap of the ceramic. The wavelength of incident light, modulated at 9 Hz, is kept in the range of 300 to 800 nm and the photopyroelectric spectrum with reference to the doping level is discussed. The band-gap energy is estimated from the plot $(\rho h v)^2$ vs $h v$ and is about constant at 2.82 eV for the samples sintered at 1275 °C at all doping levels of Y_2O_3. The energy band-gap decreases from 2.88 to 2.82 eV with the decrease of Y_2O_3 mol% at the sintering temperature of 1175 °C. The steepness factor σ_A (in A region) and σ_B (in B region) which characterizes the slop of exponential optical absorption is discussed with reference to the doping level of Y_2O_3. The X-ray diffractometry shows that the crystal structure of ZnO doped with different mol% of Y_2O_3 remains to be of hexagonal type and the secondary phase is Yttrium-rich. Microstructure and compositional analysis of the selected areas are analyzed using SEM and EDAX which shows the Yttrium-rich phase coexist in the grain boundaries and nodal points. The relative density is decreased with the increase of Y_2O_3 mol% indicating the increase in porosity. The grain size decreases with the increase of Y_2O_3 mol% which shows that Y_2O_3 acts as a grain inhibitor.**

I. INTRODUCTION

Zinc Oxide (ZnO) a white polycrystalline solid material is a n-type semiconductor material with a large energy band-gap, 3.2 eV [1]. It crystallizes into a wurtzite structure, and is a complete hexagonal closed-packed (hcp) lattice with oxygen atoms inserted into the zinc hcp-lattice. It is widely used in the manufacturing of paints, rubber products, cosmetics, pharmaceuticals, floor covering, plastics, textiles, ointments, inks, soap, batteries, and also in electrical components such as piezoelectric transducers, phosphors, gas sensors and varistors [2, 3].

Varistors are extensively used as protective devices to regulate transient voltage surges of unwanted magnitudes [4]. The exact role of many additives in the electronic structure of ZnO varistors is uncertain. ZnO based varistor is formed with other metal oxides of small amounts such as Bi_2O_3, Co_3O_4, Cr_2O_3, MnO, Sb_2O_3 etc. These additives are the main tools that are used to improve the non-linear response and the stability of ZnO varistor [5]. Varistor effect (a highly nonohmic behavior in the *I-V* characteristics), which can be explained by mechanism involving the grain boundaries and the associated defect concentration gradients [6]. The distribution of vacancies and impurities as well as their behavior during annealing treatments appear as one of the factors that determine the electrical properties of ceramic ZnO. Much work has been done in *I-V* studies on ZnO based varistor by previous workers [5, 7]. It is necessary to get information on optical absorption of ceramic ZnO doped with different metal oxides for the investigation on electronic states of ceramic ZnO and doped impurities during sintering process and in this paper we discuss the photopyroelectric spectroscopy of the Y_2O_3 doped ceramics ZnO to study the behavior of Y_2O_3.

0-7803-9730-4/06/$25.00 ©2006 IEEE   407

II. MATERIALS AND METHODS

ZnO (99.9% purity, Alfa Aesar) was doped with Y_2O_3 (99.9 % purity, Alfa Aesar) where Y_2O_3 varies as 0.4, 0.8, 1.2, 1.6 mol %. The 24 hours ball milled powder of each mole percent was pre-sintered at temperatures 750 °C for 2 hours. Then each sample was ground and polyvinyl alcohol (1.4 wt %) was mixed as a binder. The dried powder was pressed under a force of 800 kg cm^{-1} to form a disk of 10 mm diameter with 1 mm thickness. Finally the pellets were sintered at 1175, 1275 °C for 1 hour in air at heating and cooling rate of 3 °C min^{-1}. Density of all the sintered samples was measured by geometrical method [8]. The mirror like polished samples was thermally etched for the microstructure analysis and the average grain size was determined by the grain boundary-crossing method. The disks of each sample were ground to make a fine powder for the photopyroelectric (PPE) spectroscopy and XRD analysis. Cu K$_\alpha$ radiation with PANAalytical (Philips) X'Pert Pro PW1830 was used for X-ray diffraction, and the XRD data were analyzed by using X'Pert High Score software for the identification of the crystalline phases.

PPE spectroscopy is a non-radiative tool [9] to study optical properties of the materials. Pyroelectric (PE) film transducer is used to detect the PE voltage signal caused by the periodic heating in the sample. The measurement of PPE signal amplitude by the PPE spectrometer system has been described elsewhere [10]. In the system, a light beam from 1 kW Xenon arc lamp (Oriel 6921) was mechanically chopped at 9 Hz for scanning wavelengths range of 300 to 800 nm. Prior the PPE measurement, the fine powder sample was ground in deionised water and then a few drops of each mixture were dropped on the 1.5 cm^2 aluminium foil and dried in air to form a thin sample layer on the foil. The foil was placed in contact to polyvinylidene difluoride PE film sensor [11] using a very thin silver conductive grease. The true sample spectrum was normalized with carbon black PPE spectrum. In determining the energy band-gap (E_g), it was assumed that the fundamental absorption edge of doped ZnO is due to the direct allowed transition. The optical absorption coefficient β varies with the excitation light energy hv [12] and is given by the expression, $(\beta hv)^2 = C (hv - E_g)$ near the band gap, where hv is the photon energy, C is the constant independent of photon energy, and E_g is the direct allowed energy band-gap. The PPE signal intensity ρ is directly proportional to β, hence $(\rho hv)^2$ is related to hv linearly. From the plot of $(\rho hv)^2$ versus hv, the value of E_g is obtained by extrapolating the linear fitted region that crosses photon energy axis.

III. RESULTS AND DISCUSSION

XRD analysis, Fig. 1, of the ceramic shows that there are only two phases ZnO and Y_2O_3 at both temperatures 1175 and 1275 °C. It is observed that the peaks of the secondary phase Y_2O_3 is more prominent at the higher doping level of the Y_2O_3 mol %. The Y_2O_3 peaks are at he angles 22.069°, 29.308°, 33.306°, 43.715°, 51.168° (ref. code 00-020-1412 and 00-039-1064).

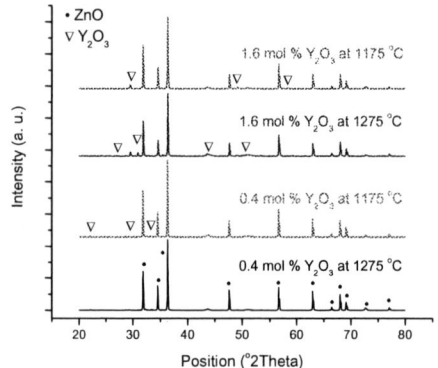

Fig. 1: XRD pattern at 0.4 and 1.6 mol % of Y_2O_3

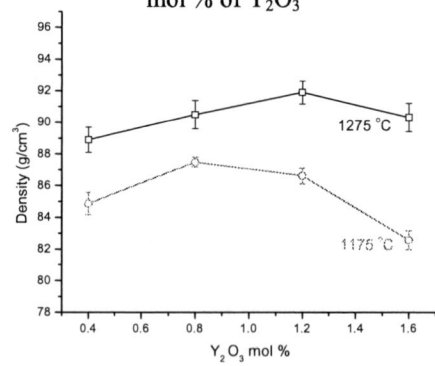

Fig. 2 Variation of density with doping level of Y_2O_3

Density of the ceramic doped with Y_2O_3, Fig. 2, increases with the increase of Y_2O_3 mol % up to 0.8 mol % of Y_2O_3 at both temperatures. The value of density is 84.86 % of the theoretical density and increases to 87.5 % at 0.8 mol % of Y_2O_3 indicating a decrement in pores. Density is decreased to 82.57 % with the further increase in the Y_2O_3 for the temperature of 1175 °C, indicating that pores are increasing with the increase of Y_2O_3.

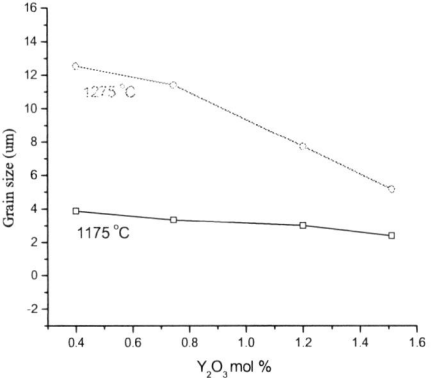

Fig. 3: Grain size dependence on Y_2O_3

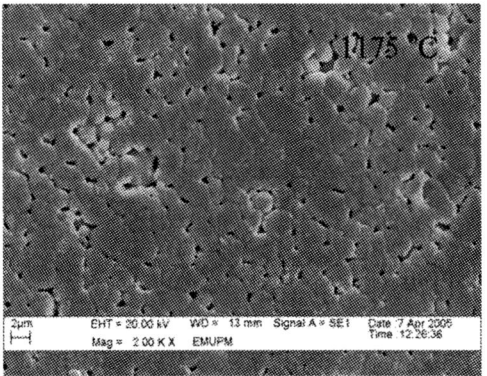

Fig. 4: SEM micrograph at 1.6 mol %Y_2O_3

Fig. 5: SEM micrograph at 1.6 mol %

For 1275 °C, the value of the density increases slowly from 88.9 to 91.9 % with the increase of mol % of Y_2O_3 and then decreases to 90.32 % at the 1.2 mol % of Y_2O_3 which indicates the pores are minimum at the 1.6 mol % of Y_2O_3. The grain size is 12.52 µm, Fig. 3, and is decreased to 5.17 µm with the increase of Y_2O_3 but the density, Fig. 2, is increased with the increase of Y_2O_3 indicating the decrease in the pores. The grain size is reduced from 3.87 to 2.37µm with the increase of Y_2O_3 but the density is maximum at 0.8 mol % of Y_2O_3 indicating the decrease in the pores at 1175 °C sintering temperature. Further increase of the Y_2O_3 reduces the grain size as well as the density indicating the increase of the pores at the lower sintering temperature. The reduction in the grain size with the increase of the Y_2O_3 indicates the Y_2O_3 is a grain suppressor at the both sintering temperatures.

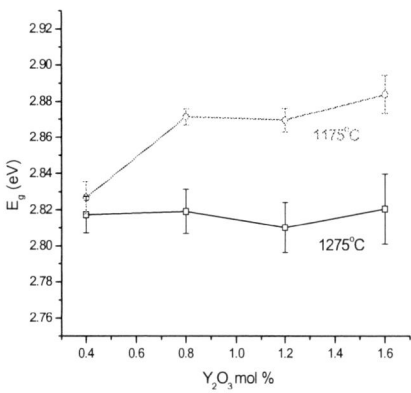

Fig. 6: Energy band gap dependence on Y_2O_3

The SEM micrograph is shown in Fig. 4 and Fig. 5, the clusters of Y_2O_3 are very clear and are verified by the EDAX analysis. Y_2O_3 is segregated mostly at the nodal points and at the grain boundaries. A very small peak in the EDAX analysis was found at the surface of grain, this may be due to the interstitials of Y_2O_3.

The Energy band gap (E_g), Fig. 6, of the Y_2O_3 doped ZnO ceramic is reduced to 2.82 eV and is about constant at all doping levels of Y_2O_3 for the sintering temperature of 1275 °C. It is expected that this decrease in E_g, may be due to the growth of interface states by Y^{+3} ions in the grain boundaries as well as in the interior of the grains. Although Y^{+3} has large ionic radius (0.93 Å) than Zn^{+2}(0.74

Å) so a limited substitution in the lattice is possible as Y^{+3} behaves as acceptor. The value of energy band gap is reduced to 2.83 eV at the 0.4 mol % Y_2O_3 for the at 1175 °C sintering temperature. The energy band gap increases to 2.88 eV with the further increase of the Y_2O_3. It is expected that the growth of interface states decreases. This decrease in the interface states may be due to the creation of the lattice defects as an effect of the limited substitution of Y^{+3} in the lattice [13].

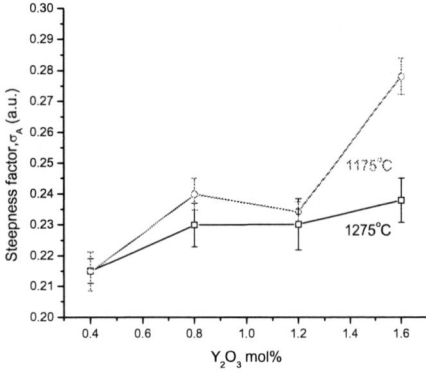

Fig. 7: Dependence of the steepness factor (σ_A) on Y_2O_3

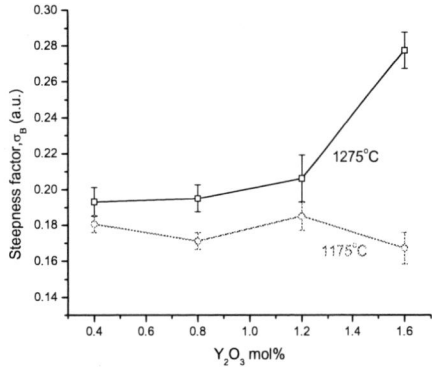

Fig. 8: Dependence of the steepness factor (σ_B) on Y_2O_3

The steepness factor σ_A (in A-region), Fig 7, which characterizes the slop of exponential optical absorption increases with the increase of Y_2O_3. σ_A increases at both temperatures and is more prominent at the lower sintering temperature, 1175 °C. This indicates the growth of interface states produced by the Y^{+3} ions as it is expected from the energy band gap Fig. 6. The value of σ_A is decreased with the increase of the sintering temperature from 1175 to 1275 °C. This indicates

the increase of growth of interface states in the grain boundaries as well as in the interior of the grains.

The steepness factor σ_B (in B-region), Fig. 8, decreases slightly with the doping level of Y_2O_3 at the sintering temperatures of 1175 °C. The decrease in σ_B indicates the increase in the defects due to the substitution of the Y^{+3} in the Zn lattice. The value of σ_B is about constant up to the 1.2 mol % doping level Y_2O_3 at the sintering temperature of 1275 °C indicates the saturation of the defect states. The value of the σ_B seems to be abnormal at the 1.6 mol % of Y_2O_3.

IV. Conclusion

In XRD analysis shows that no secondary phase was found in the ceramic. EDAX analysis shows the Y_2O_3 is segregated at the grain boundaries and as well as at the nodal points. The maximum density, 92% of theoretical density was found at 1.2 mol % of Y_2O_3. Grain size decreases with the increase of Y_2O_3 indicates the Y_2O_3 serves as a grain inhibitor. Energy band gap is reduced to a minimum value of 2.82 eV and is about constant at all doping level of Y_2O_3 for the 1275 °C sintering temperature as well as Y_2O_3 produces the defect states in the ceramic ZnO.

Acknowledgement

The authors would like to thank the Ministry of Science, Technology and Innovation of Malaysia (MOSTI) for the financial support of this work under IRPA Grant No. 02-02-04-0132-EA001

References

[1] Gupta T.K. (1990). *Application of zinc oxide varistors.* J. Am. Ceram. Soc. **73**(7): 1817-40

[2] Lin H.M., S.J. Tzeng, P.J. Hsiau, W.L. Tsai (1998). *Electrode effects on gas sensing properties of nanocrystalline zinc oxide.* Nanostruct. Mater. **12**, 465-77

[3] Look D.C. (2001). *Recent advances in ZnO materials and devices.* Materials Science and Engineering B **80**, 383-387

[4] David R. Clarke. (1999). *Varistor ceramics.* J. Am. Ceram. Soc. **8**, 485-501

[5] Eda K. (1989). *Zinc Oxide Varistors.* IEEE Elect.

Insul. Mag. **5**, 28-41

[6] Einzinger R. (1979). Grain junction properties of ZnO varistors. Appl. Surf. Sci. **3**, 390-408

[7] Choon-Woo Nahm (2003). *Electrical properties and stability of praseodymium oxide based ZnO varistor ceramics doped with Er_2O_3.* J. Eu. Ceram. Soc. **23**, 1345-53

[8] Wang J.F., Wen-Bin Su, Hong-Cun Chen, Wen-Xin Wang, and Guo-Zhong Zang (2005). *(Pr, Co, Nb)-doped SnO_2 Varistor Ceramics.* J. Am. Ceram. Soc. **88**(2), 331-34

[9] Minamide A., M. Shimaguchi, Y. Tokunaga (1998). *Study on Photopyroelectric Signal of Optically Opaque Material Measured by PVD Film Sensor.* Jpn. J. Appl Phys. **37**, 3144-47

[10] Mandelis A. (1984). *Frequency-domain photo-pyroelectric spectroscopy of condensed phases (PPES): A new, simple and powerful spectroscopic*

technique. Chem. Phys. Lett. **108**, 388-92

[11] Tam A.C. and H. Coufal (1983). *Photoacoustic generation and detection of 10-ns acoustic pulses in solids.* Appl. Phys. Lett. **42**, 33-35

[12] Toyoda T., H. Nakanishi, S. Endo, T. Irie (1985). *Fundamental absorption edge in semiconductor $CdInGaS_4$ at high temperatures.* J. Phys. D Appl. Phys. **18**, 747-51

[13] Choon-Woo Nahm (2003), Byoung-Chul Shin, Byong-Hyeon Min, *Microstructure and electrical properties of Y_2O_3-doped $ZnO-Pr_6O_{11}$-based varistor ceramics.* Materials Chemistry and Physics. **82** 157-64

Grain Size Effect on LTCC Tape Performance as Substrate for Microelectronic Devices

Azmi Ibrahim[1], Rosidah Alias[1], Che Seman Mahmood[2], Sabrina Mohd Shapee[1], Mohamed Razman Yahya[1] and Abdul Fatah Awang Mat[1]

[1]Microelectronics and Nano Technology Program, TM Research & Development Sdn. Bhd., Idea Tower I & II, UPM-MTDC Lebuh Silikon, 43400 Serdang, Selangor, Malaysia.

[2]Materials Technology Group (MTEC), Industrial Technology Division, Malaysia Institute For Nuclear Technology Research (MINT) Bangi, 43000 Kajang, Selangor, Malaysia

Email: iazmi@tmrnd.com.my

Abstract: As substrates for microelectronic devices, LTCC tape must have excellent combination of dielectric and thermal properties. This paper reports the grain size effect on the dielectric and thermal performance of alumina based LTCC tape. LTCC tape was prepared using tape-casting technique, then fired up to $850^{\circ}C$. The dielectric and thermal performance of the fired LTCC tape was carried out using impedance analyzer and thermal analyzer respectively. It was found that the reduction in grain size effectively reduces the thermal diffusivity value and increases the dielectric constant value of the LTCC tape. The grain size effect on the dielectric and thermal behavior can be explained by the changing of the grain microstructure.

Keywords: LTCC tape, grain size, thermal, dielectric, microstructure

I. INTRODUCTION

LTCC technology is well known in the field of microelectronic for offering advanced approaches to miniaturization of electronics packaging. This technology is widely applied in telecommunication industry to fulfill the present and future demand for smaller, cheaper and more reliable telecommunication devices by producing compact multilayer structures with buried passive components [1].

One of the most important parts in LTCC technology is the development of the LTCC tape materials. To fulfill the requirements of substrates and packaging materials for microelectronic applications, high thermal diffusivity and low dielectric constant LTCC tape are required [2].

High thermal diffusivity is important for reducing heat generated by high operating speed in microelectronic circuits [3]. For example most of wireless MMIC systems operate at 40 GHz and above for many applications. Therefore the development of LTCC tape that posses high thermal diffusivity has attracted huge attention.

Low dielectric constant material is required in order to reduce the signal delay in device as clearly shown by the equation 1 [4]:

$$T_d = \frac{l \times \sqrt{\varepsilon_r}}{c} \qquad (1)$$

where

T_d: Time delay
l: signal transmission length
ε_r: dielectric constant of substrate
c: speed of light

In order to meet the requirement for microelectronic application the research on the development of new LTCC materials with superior thermal and dielectric properties attracts many researchers [2,5]. The aim of this paper is to report the thermal and dielectric properties of LTCC tape produced from different grain size of alumina.

0-7803-9730-4/06/$25.00 ©2006 IEEE

II. EXPERIMENT

LTCC tapes consist mainly of alumina and glass powders were prepared using tape casting technique [6]. Two types of alumina powder with the initial grain size of 0.4μm and 0.15μm were chosen for this work. The prepared LTCC tape was sintered using a heating profile up to 850°C. The samples density was calculated conventionally based on mass and volume measurements. The microstructure of the samples was studied using Scanning Electron Microscopy (SEM). Meanwhile, thermal and dielectric properties measurements were carried out using Nanoflash LFA 447 Thermal Analyser and Agilent 4291 B Impedance materials analyzer respectively.

III. RESULTS & DISCUSSION

LTCC tape samples of 0.1 to 0.2mm green thickness were succesfully prepared and the sintered tape density, as shown in Table 1, increases as the grain size of the alumina decreases. This result indicates that, small grain size cause better arrangement of the grains and to become closer to each other. As a result it can reduce pores between grains to make the sample become dense.

Table 1. Density of LTCC tape samples

Figure 1 shows the microstructure of

Alumina grain size, μm	Density, g/cm³
0.40	1.40
0.15	2.15

LTCC tape samples that was studied using SEM. The micrographs clearly show the obvious different in grains sizes between the two tapes. However, due to agglomeration, the grains were present in agregate of various sizes that distributed inhomogeneously through out the tapes.

Thermal diffusion property of the samples was measured by using the Nanoflash LFA 447 thermal analyzer at temperature in the range of 25 to 100°C. It was found that the thermal diffusivity values were 1.0 and 0.4 respectively for the tape with bigger and smaller sized alumina grains at

room temperature. However, this value was decreased with the increasing of operation temperature up to 100°C as shown in Figure 2. This phenomenon was explained by *Lawrence* [7] that due to decreasing of the phonon pathway with the increment of operation temperature.

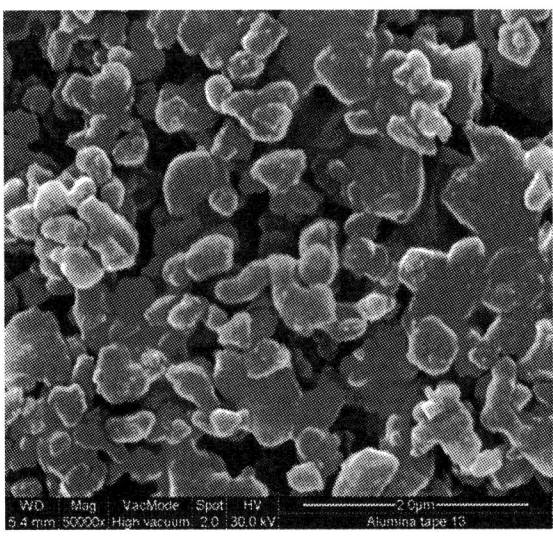

(a)

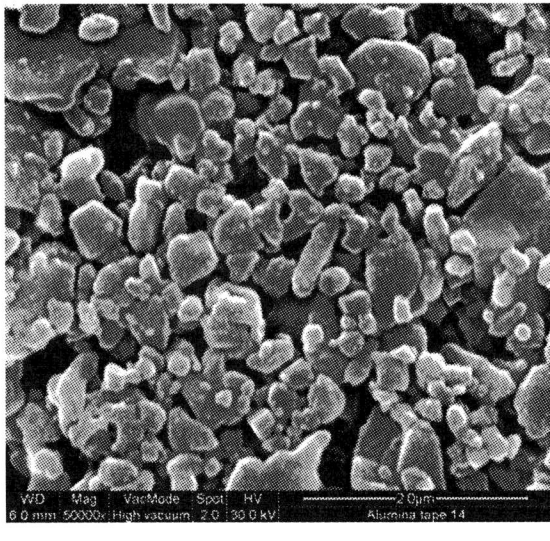

(b)

Figure 1. SEM micrographs of LTCC tape sample made from alumina (a) 0.4μm and (b) 0.15μm

From Figure 2 it was also found that decreasing the grain size of alumina has significantly reduced the thermal diffusion properties of the LTCC tape samples. Similar trend was reported by *R. Vaben & D. Stover* [8], when the decrease of the grain size of SiC ceramic, reduces its thermal transport properties due to the increasing amount of phonon-phonon scattering at the grain boundaries compared to phonon-phonon scattering in fined grained materials.

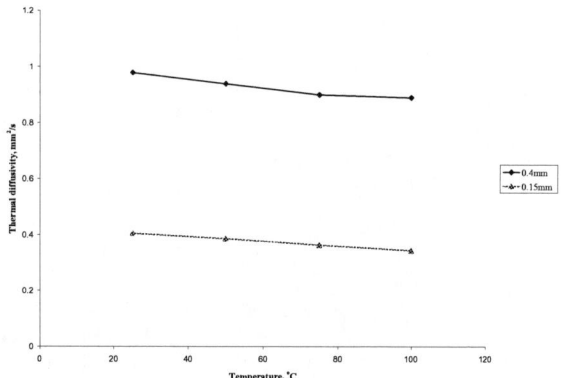

Figure 2. Thermal diffusivity of LTCC tape samples made from alumina grain size of 0.4μm and 0.15μm

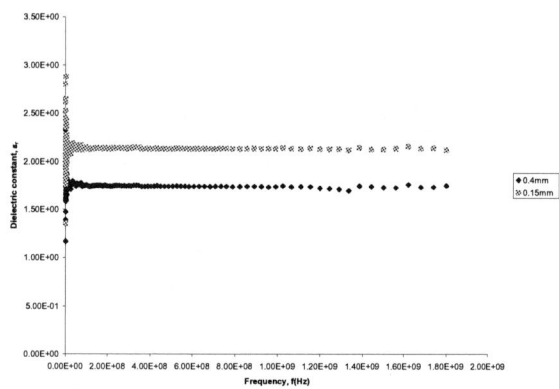

Figure 3. Dielectric constant of LTCC tape samples made from alumina grain size of 0.4μm and 0.15μm

Dielectric properties of the alumina substrates samples were measured using Agilent 4291 B Impedance materials analyzer at frequency in the range of 1.0 to 1.8GHz. Figure 3 show the dielectric constant of the sintered LTCC tape samples. It was found that both samples show good stability of the dielectric constant under the increasing of frequency up to 1.8GHz. The observed dielectric constant values are about 1.8 and 2.2 respectively for the tape with bigger and smaller sized alumina grains. This value is relatively lower compared to commercial LTCC

material that exhibit dielectric constant from 3.9-10 or more [9].

IV. CONCLUSION

Alumina based LTCC tape samples, successfully prepared made from alumina of different grain sizes, after sintering at 850°C were found to possess good electrical properties for microelectronic applications. By decreasing the alumina grain size the thermal diffusivity of the final tape products decreases and the dielectric constant value increases. The change in dielectric and thermal behavior of the products is due to the changing in grain microstructure of the LTCC tape material.

V. ACKNOWLEDGEMENT

The authors wish to thank Telekom Malaysia Bhd. for their funding support under project R05-0606-0 and Material Technology Group of Malaysia Institute for Nuclear Technology Research (MINT) for their technical support.

VI. REFERENCES

[1] M. Lahti & V. Lantto. Passive RF band-pass filters in an LTCC module made by fine-line thick film paste. *Journal of the European Ceramic Society*, 21, pp. 1997-2000 (2001).

[2] J. Ma & Z. He, "Thermal, electrical and mechanical properties of layered substrates for microelectronic applications," *Thin Solid Films*, Vol. 462-463, pp. 477-480 (2004).

[3] Sherwani. N, Yu. Q & Badida. S. *Introduction to Multichip Modules.* John Wiley Inc. USA, (1995).

[4] A. Takashi, A. Nagai, & A. Mukoh. *IEEE/ISHM '90 Syposium-Italy*, pp. 196-205 (1990).

[5] A. Stiegelschmitt, A. Roosen, C. Ziegler, S. Martius & L.P. Schmidt. Dielectric data of ceramic at high frequencies. Journal of European Ceramic Society, 24, pp. 1463-1466, (2004).

[6] Z. Yuping, J. Dongliang & P. Greil. Tape casting of aqueous Al₂O₃ slurry. *Journal of European Ceramic Society*, 20, pp. 1691-1697, (2000).

[7] Lawrence. H.V.V. *Seramik Fizik Untuk Jurutera* (Translated by Zainal Arifin Ahmad). DBP. Kuala Lumpur (1991).

[8] R. Vaben & D. Stover. Processing and properties of nanophase ceramics. *Journal of Materials processing Technology.* 92-93, pp. 77-84, (1990).

[9] J. Mazierska, M.V. Jacob, A. Harring, J. Kupra, P. Barnwell & T. Sims. Measurements of loss tangent and relative permittivity of LTCC ceramics at varying temperatures and frequencies. *Journal of European ceramic Society*, 23, pp. 2611-2615, (2003).

Particle Size Analysis of Barium Titanate Powder by Slow-Rate Sol-Gel Process Route

Balachandran R[1], *Member, IEEE,* Yow HK[1], *Senior Member, IEEE,* Jayachandran M [2], Wan Yusmawati Wan Yusof [3], Saaminathan V[1]

[1]Faculty of Engineering, Multimedia University, Cyberjaya, 63100, Selangor, Malaysia
[2]Science officer, Department of Chemistry, University Putra Malaysia, Serdang, Malaysia
Email: balachandran.ruthramurthy@mmu.edu.my
Tel: ++60-3-83125427, Fax: ++60-3-83183029

Abstract **In this research work, different compositional ratios of Barium Titanate [BT] powders have been prepared using slow-rate gellification by sol-gel route. The as-prepared materials have been calcinated at 300° C for 8 hours to remove the presence of carbon, and subsequently annealed at 700°C for one hour for the phase formation. After annealing, the finely grinded powders were characterized by X-Ray diffractometer [XRD], Atomic Force Microscope [AFM] and Nano particle size analyzer [NPSA]. The XRD pattern shows that the material was BT with polycrystalline phase. It exhibited tetragonal phase with crystalline size in the range of 97 nm. The lattice constants 'a' and 'c' were perfectly matched with the standard reports. The surface morphology study by AFM shows that the materials were homogenously fine grains. The average particle sizes reduce from 750 nm to 78.49 nm when the concentration of barium salt decreases. This observation clearly indicates that the tuning of the particle sizes is possible by controlling the compositional ratios of barium and titanium. The particle distributions of the samples have also been analyzed by nano particle size analyzer. In this work, samples with equal amount of barium and titanium exhibited particle size distribution down to 100 nm range, with good structural and morphological properties.**

I. INTRODUCTION

The reduction in the size of the existing oxide material by the use of Silicon Dioxide [SiO_2] has progressed to the extent that any further thickness reduction will result in a significant increase of leakage current that could eventually lead to circuit breakdown.[1] In recent years, BT have been considered as dielectric materials for the capacitor in DRAM.[2] BT is a compound oxide material made of Barium and Titanium. It has a high dielectric constant with a theoretical value of 300 when compared to low-k dielectric SiO_2 which has dielectric constant value of only about 4 [3]. Among the Ferroelectric, Photorefractive, Piezo electric, Electro luminescent and temperature dependency, BT could be considered as an oxide material for the capacitor dielectric with high dielectric constant (up to 300), low leakage current, fast dielectric response, low dielectric loss and long life time. [4] For the miniaturization and high capacitance of multilayer ceramic capacitors, it is necessary that the particle diameter of $BaTiO_3$ powders be less than 100 nm and have a narrow size distribution with high purity.[2]

There are reports on the preparation of BT powder using various techniques such as solid-state grain growth [5], co-precipitation [3], spray-pyrolysis method [6], Electrochemical method [7], and sol-gel method [2]. Sol-gel method has been considered for the preparation of nano BT powder because it allows homogeneous distribution of elements on a molecular level with low temperature sintering capability, can easily adjust the material composition which would tune the electrical and dielectrical properties of the material and requires low capital cost. [8]

In this research work, to control the reaction of precursor materials during sol and gel formation, small diameter bio-micro syringe has been used to inject one of the precursor solutions, Titanium Iso-propoxide mixed with acetic acid for controlling the particle size. The as-prepared gel has been heated through conventional electrical furnace for calcinations and annealing to get BT powder.

II. EXPERIMENTAL PROCEDURE

In this work, Barium Acetate (Aldrich) and Titanium (IV) Iso-propoxide (Aldrich 205273) were taken as starting materials for the synthesis. Stochiometric ratios of Barium Acetate were dissolved in deionized water and were heated slowly until the starting materials were completely dissolved. Subsequently, the Titanium Iso-propoxide mixed with acetic acid is injected slowly by bio-micro syringe. The diameter of the bio-micro syringe is 1 μm. This slow addition enables better control of the reaction to get the transparent gel. Ethylene Glycol ethyl (CP-R & M-UK) was added to improve the stability of the solution. Four different molar ratios of Barium Acetate and Titanium Iso-propoxide for this work have been taken as 0.5:0.5, 0.60:0.40, 0.65:0.35 and 0.70:0.30. Once the precursor solutions are completely dissolved, small amount of polymer agents and sol initiator are added to get the transparent BST gel.

The flow chart for the preparation of the BT powders using slow-rate gellification by sol-gel route is as shown in Fig. 1.

After ageing for one day, in order to dry the sample, the transparent gel was subjected to conventional furnace heating. The gels were calcinated at 300° C for 8 hours to remove the presence of carbon and subsequently annealed at 700°C for one hour for the phase formation and they were finely grinded by agate-mortar to obtain BT powder.

The finely grinded powders were characterized by XRD, AFM and NPSA.

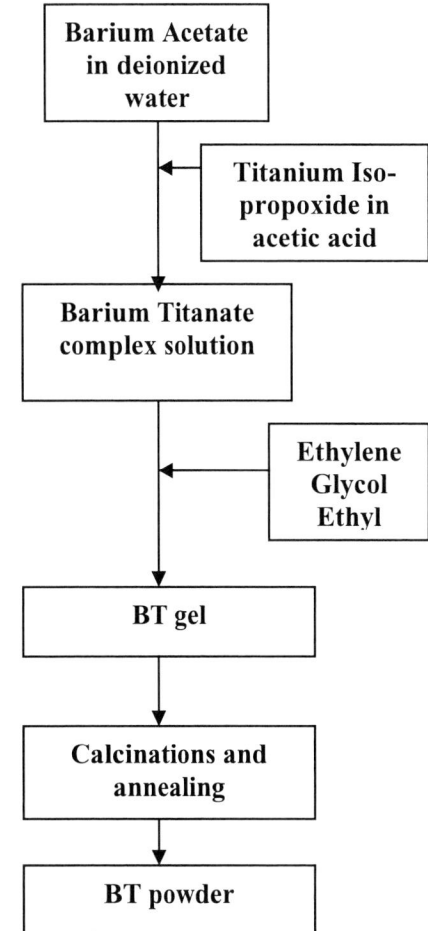

Fig. 1 Flow chart showing the preparation of BT powders using sol-gel method

III. RESULTS AND DISCUSSION

The finely grinded BT powders were subjected to XRD analysis. The pattern of finely grinded BT powders with equal molar ratio of Barium and Titanium is as shown in Fig. 2.

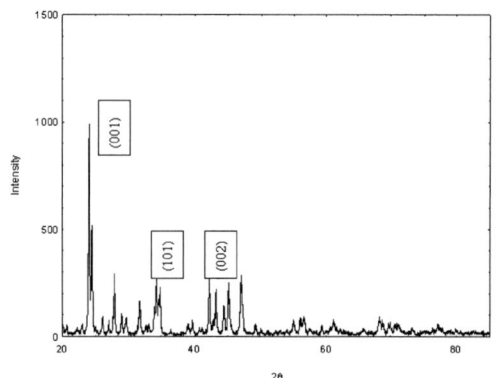

Fig.2 XRD pattern of Barium titanate with molar ratio of 0.5 and 0.5

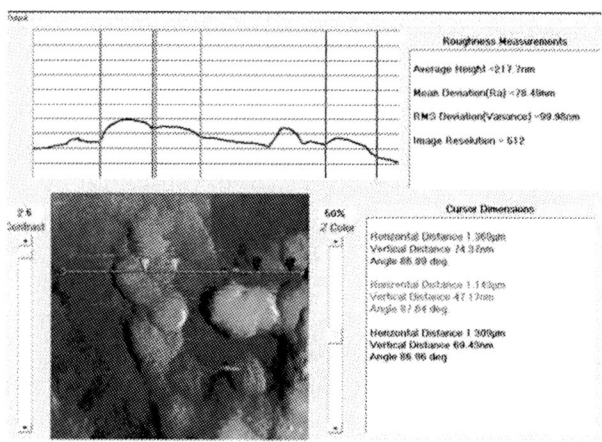

Fig. 3a

The XRD pattern shows that the material was BT with polycrystalline phase. It exhibited tetragonal phase with crystalline size in the range of 97 nm. The lattice constants 'a' and 'c' were perfectly matched with the standard reports. [9]

Subsequently, the BT powders were analyzed for AFM. The mean deviation of the particles for the various compositions is as shown in Table 1.

No.	Compositional ratio of Barium to Titanium	Mean deviation of the particle (nm)
1.	0.50/0.50	78.49
2.	0.60/0.40	116.1
3.	0.65/0.35	214.6
4.	0.70/0.30	750.0

Table 1 Mean deviation of size of the particles for various compositions of Barium and Titanium

From Table 1, it could be understood that the particle size could be varied by the molar ratio of the Barium and Titanium. The AFM graphs of equal molar ratio of barium and Titanium are as shown in Fig. 3a.and 3b

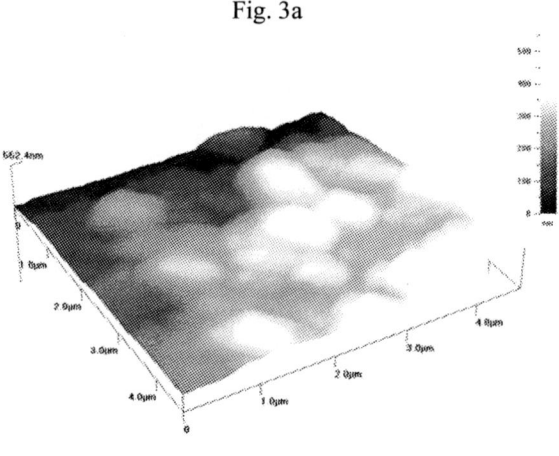

Fig. 3b

AFM graphs of Barium titanate with molar ratio of 0.5 and 0.5

The surface morphology study by AFM shows that the materials were homogenously fine grains. The average particle sizes reduce from 750 nm to 78.49 nm when the concentration of barium salt decreases. This observation clearly indicates that the tuning of the particle sizes is possible by controlling the compositional ratios of barium and titanium.

The particle distributions of the samples have also been analyzed by NPSA. The nano particle size analyzer result for equal molar ratio of Barium and Titanium is as shown in Fig. 4.

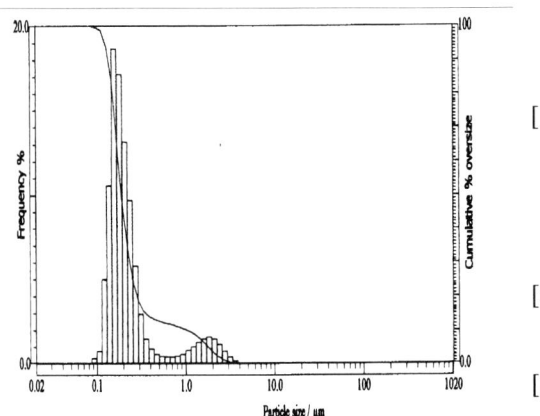

Fig. 4 Particle size distribution of Barium titanate for equal amount of Barium and Titanium

From the NPSA, it is understood that BT powder with equal amount of barium and strontium exhibited particle size distribution down to 100 nm range.

IV. CONCLUSION

In this research work, the tuning of the BT particle sizes could be controlled by the molar ratios of the Barium and Titanium using slow-rate sol-gel route. The mean deviation of the particle sizes were analyzed using AFM and found that the particle sizes were tuned from 754 nm to 78.49 nm and it was cross-checked from the results of the Nano particle size analyzer. The XRD pattern showed that prepared BT powders exhibited polycrystalline with tetragonal phase and as the particle size decreases, the corresponding variations were observed in the FWHM peaks.

ACKNOWLEDGMENT

The authors would like to thank Multimedia University, (Cyberjaya Campus), Malaysia for sponsoring this work under internal research fund PR/2005/0512 and Dr T Saravanan, Scientist, IGCAR, India for his help on some technical matters.

REFERENCES

[1] C. Tung-Sheng, V. Balu, S. Katakam, L. Jian-Hung, and J. C. Lee, "Effects of Ir electrodes on barium strontium titanate thin-film capacitors for high-density memory application," *Electron Devices, IEEE Transactions on*, vol. 46, pp. 2304-2310, 1999.

[2] Tomoya Ohno, Daisuke Suzuki, and H. S. Ida, "Size Effect for Barium Titanate Nano-particles†," *KONA*, vol. 22, pp. 195-200, 2004.

[3] T.-T. Fang, H.-B. Lin, and J.-B. Hwang, "Thermal Analysis of Precursors of Barium Titanate Prepared by Coprecipitation," *Journal of the American Ceramic Society*, vol. 73, pp. 3363-3367, 1990.

[4] T. Bang-Hung, S. Heidger, and J. A. Weimer, "Sputtered barium titanate and barium strontium titanate films for capacitor applications," 2000.

[5] T. Yamamoto and T. Sakuma, "Fabrication of Barium Titanate Single Crystals by Solid-State Grain Growth," *Journal of the American Ceramic Society*, vol. 77, pp. 1107-1109, 1994.

[6] W.-N. Wang, I. Wuled Lenggoro, K. Okuyama, Y. Terashi, and Y.-C. Wang, "Effects of Ethanol Addition and Ba/Ti Ratios on Preparation of Barium Titanate Nanocrystals Via a Spray Pyrolysis Method," *Journal of the American Ceramic Society*, vol. 89, pp. 888-893, 2006.

[7] P. Bendale, S. Venigalla, J. R. Ambrose, E. D. Verink, and J. H. Adair, "Preparation of Barium Titanate Films at 55 deg;C by an Electrochemical Method," *Journal of the American Ceramic Society*, vol. 76, pp. 2619-2627, 1993.

[8] G.Pfaff, "Synthesis and characterization of $MTiO_3$(M = Mg, Ca, Sr, Ba) sol-gel," *J.Mater.Chem.,*, vol. 2, pp. 591, 1992.

[9] C. Muralidhar and P. K. C. Pillai, "XRD studies on barium titanate (BaTiO3)/polyvinylidene fluoride (PVDF) composites " *Journal of Materials Science*, vol. 23, pp. 410-414, 2005.

Rapid Crystallization by Microwave Heating for Barium Strontium Titanate Powders Prepared using Slow-Rate Sol-Gel Technique

Balachandran R[1], *Member, IEEE,* Yow HK[1], *Senior Member, IEEE,* Zaidina Bt Mohd Daud[2], Saaminathan V[1]

[1]Faculty of Engineering, Multimedia University, Cyberjaya, 63100, Selangor, Malaysia
[2]Science officer, Department of Chemistry, University Putra Malaysia, Serdang, Malaysia
Email: balachandran.ruthramurthy@mmu.edu.my
Tel: ++60-3-83125427, Fax: ++60-3-83183029

Abstract In this research work, different compositional ratios of Barium Strontium Titanate [BST] powders have been prepared using slow-rate gellification by sol-gel route and ex-situ microwave heating method. The finely grinded powders prepared were analyzed using Thermo Gravity Analysis [TGA], Differential Thermal Analysis [DTA] and characterized by X-ray Diffractometer [XRD]. TGA result shows that the organic solvents and unwanted impurities were removed at 117.89°C and decomposition of carbonate took place at 205.19°C. In addition, the material for 0.5 M ratio of Barium acetate and Strontium acetate and 1 M of Titanium Iso-propoxide started crystallization at 438.93° C to yield BST. From the DTA results, it was found that when the microwave heating time was increased from 10 minutes to 25 minutes, the temperature at which endothermic reaction occurs has been shifted to lower values from 110°C to 83.413 °C. These values matched those of the TGA results. The crystalline size was estimated from FWHM result to be in the range of 70 nanometer. The major peaks in the XRD analysis of the sample correspond to (100), (110), (111) and (200) were obtained for BST and agreed with American Standard for Test Measurement [ASTM] data. The materials exhibited polycrystalline pattern with "d" values that matched with standard reports.

I. INTRODUCTION

BST is a compound oxide material made of Barium Titanate and Strontium Titanate. It covers wide range of Curie temperature and can be used to tune the high dielectric constant ranging from 300 to 1250 by changing the composition 'x' from 0 to 1 [$Ba_xSr_{1-x}TiO_3$]. [1] From the characteristic properties include the following: Polarizability, Piezoelectricity, Pyroelectricity, Electro-optic activity. In particular, it can be used as a gate oxide material in capacitor [2] with high dielectric constant and low leakage current density which requires less frequent refreshing and high dielectric break strength. [1]

There are reports on the preparation of BST powder using various techniques such as solid-state reaction,[3] co-precipitation [4]and sol-gel method.[5] Sol-gel method has been considered for the preparation of nano BST powder because it allows homogenous distribution of elements on a molecular level with low temperature sintering capability, requires low capital cost and can easily adjust the material composition which [6] would tune the electrical and dielectrical properties of the material.

In this research work, to control the reaction of precursor materials during sol and gel formation, small diameter bio-micro syringe has been used to inject one of the precursor solutions, Titanium Iso-propoxide mixed with acetic acid for controlling the particle size. The as prepared gel has been heated using microwave oven for heating transparent gel with different time duration in order to reduce the total preparation time when compared to conventional electrical furnace.

II. EXPERIMENTAL PROCEDURE

The flow chart for the preparation of Barium Strontium Titanate powder using slow rate gellification by sol-gel route is as shown in Fig. 1. In this research work, Barium Acetate (Aldrich), Strontium Acetate (Aldrich) and Titanium (IV) Iso-propoxide (Aldrich 205273) were taken as starting materials for the synthesis. Stochiometric ratios of Barium Acetate and Strontium Acetate were dissolved in deionized water and were heated slowly until the starting materials were completely dissolved. Subsequently, the Titanium Iso-propoxide mixed with acetic acid is injected slowly by bio-micro syringe. The diameter of the bio-micro syringe is 1 μm. This slow addition enables better control of the reaction to get the transparent gel. Ethylene Glycol ethyl (CP-R & M-UK) was added to improve the stability of the solution. Two samples have been prepared with the molar ratios of Barium Acetate, Strontium Acetate and Titanium Iso-propoxide as 0.35, 0.65,1 and 0.5, 0.5,1 respectively. Once the precursor solutions are completely dissolved then small amount of polymer agents and sol initiator are added to get the transparent BST gel.

After ageing for one day, in order to dry the sample at room temperature, the gel was subjected to microwave heating [National model no. NN-C2000P/NN-C780P with microwave power output of 1000W] at 2.45 GHz. Microwave heating of the gel was considered, compared to conventional electrical furnace, in order to save the time and power consumption. The gel was heated at different time duration from 10 minutes to 25 minutes and they were finely grinded by agate-mortar to get BST powder.

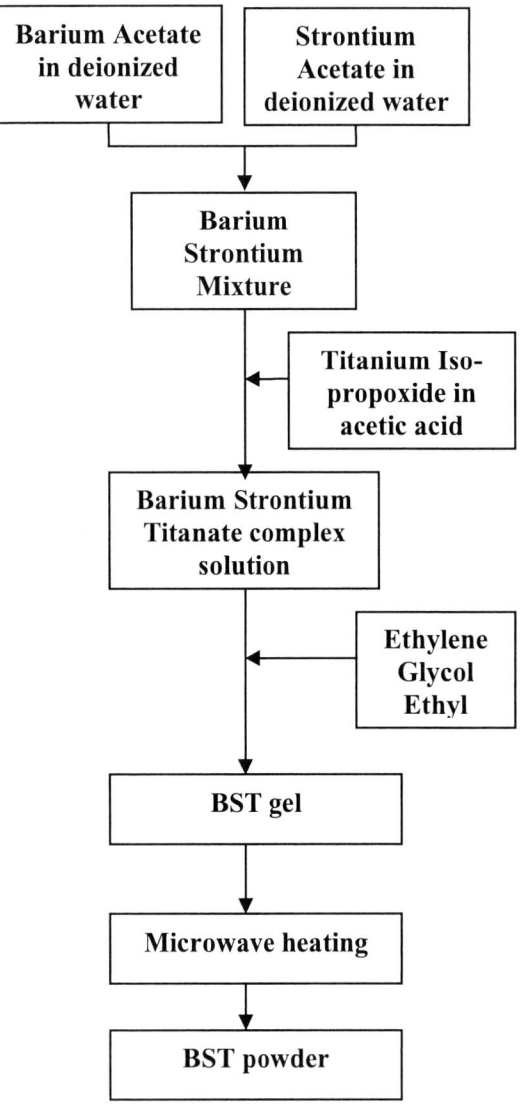

Fig. 1 Flow chart showing the preparation of BST powder using sol-gel method

III. RESULTS AND DISCUSSION

The finely grinded powders were subjected for TGA (Mettler Toledo Model No. TGA/SDTA851e with Software: SW -8.10) from 25° C to 1000° C at the rate of 10° C/min. The TGA results for $Ba_{0.35}Sr_{0.65}TiO_3$ and $Ba_{0.5}Sr_{0.5}TiO_3$ powder microwave heated from 10 to 25 minutes are as shown in Fig. 2a and Fig. 2b. The BST powders heated for 25 minutes show a considerable weight loss. The TGA result for $Ba_{0.5}Sr_{0.5}TiO_3$ heated at 25 minutes is as shown in Fig.3. It shows that the organic solvents and unwanted impurities were removed at 117.89°C and decomposition of carbonate took place at 205.19°C. In addition, the material started crystallization at 438.93°C to yield BST.

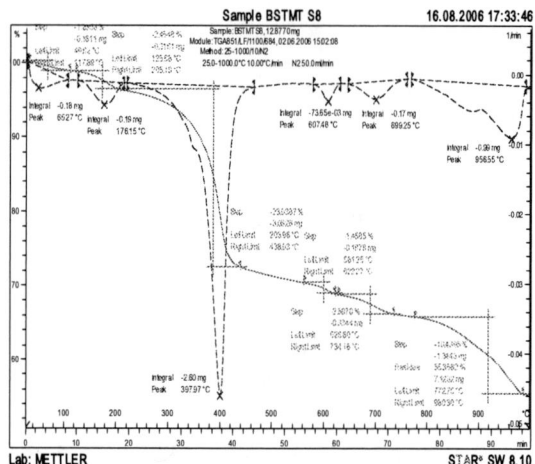

Fig.3 TGA analysis of $Ba_{0.5}Sr_{0.5}TiO_3$ powder microwave heated for 25 minutes

Subsequently the BST powders have been subjected to DTA. (Perkin Elmer Model: DTA 7). The DTA result of $Ba_{0.5}Sr_{0.5}TiO_3$ powder microwave heated for 25 minutes is as shown in Fig.4. From the DTA results obtained for various microwave heating duration, it was found that when the microwave heating time was increased from 10 minutes to 25 minutes, the temperature at which endothermic reaction occurs has been shifted to lower values from 110°C to 83.413 °C. These values matched those of the TGA results.

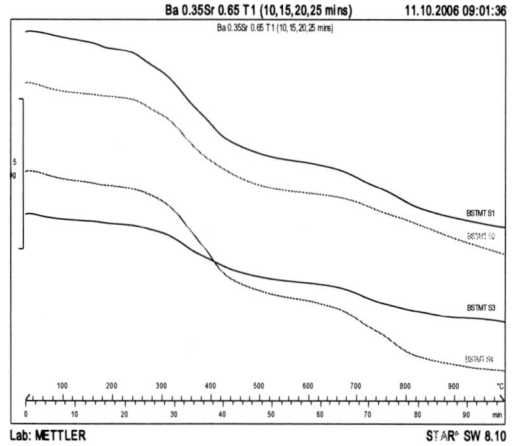

Fig.2a TGA analysis of $Ba_{0.35}Sr_{0.65}TiO_3$ powder microwave heated from 10 to 25 minutes

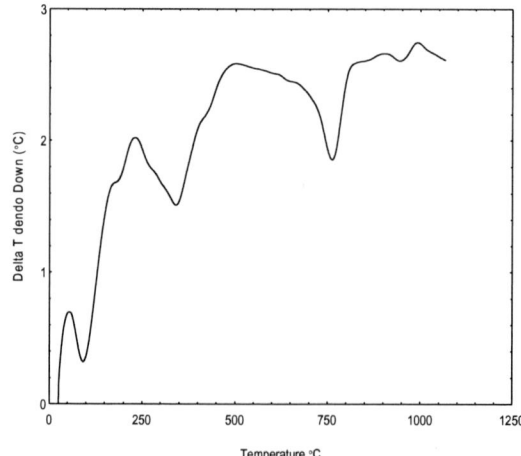

Fig.4 DTA analysis of $Ba_{0.5}Sr_{0.5}TiO_3$ powder microwave heated for 25 minutes

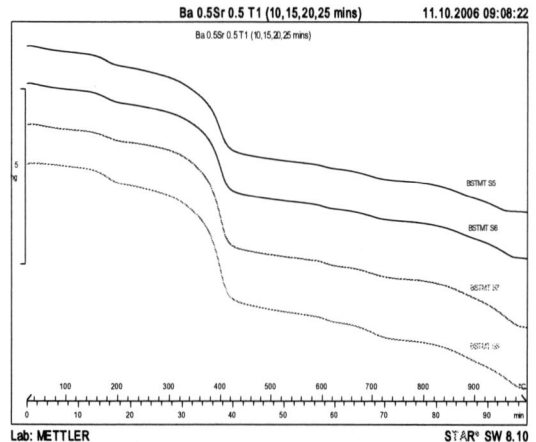

Fig.2b TGA analysis of $Ba_{0.5}Sr_{0.5}TiO_3$ powder microwave heated from 10 to 25 minutes

The microwave heated BST powders have then been analyzed using XRD (SHIMADZU Model: XRD-6000). The crystallite size (D) calculated using Scherer's formula using FWHM result which was to be in the range of 70 nanometer for $Ba_{0.5}Sr_{0.5}TiO_3$ heated at 25 minutes.

The major peaks in the XRD analysis of the sample [$Ba_{0.5}Sr_{0.5}TiO_3$ heated at 25 minutes] correspond to (100), (110), (111), (200) with few other minor peaks were obtained from the result that corresponds to unknown elements with low intense.

The materials exhibited polycrystalline pattern with "d" values that matched with standard reports. The XRD pattern of $Ba_{0.5}Sr_{0.5}TiO_3$ powder microwave heated for 25 minutes is as shown in Fig.5.

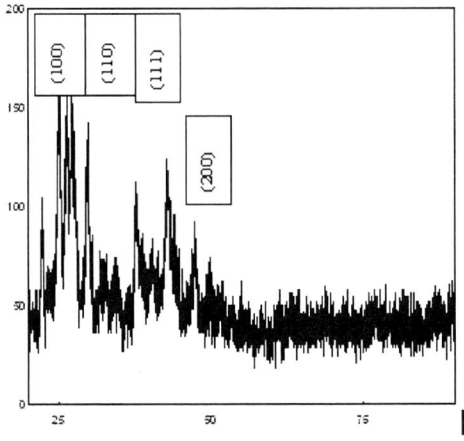

Fig. 5 XRD of $Ba_{0.5}Sr_{0.5}TiO_3$ powder microwave heated for 25 minutes

V. CONCLUSIONS

In this present work, two different compositional ratios of BST powders have been successfully synthesized using slow-rate gellification by sol-gel route and ex-situ microwave heating method. The microwave heating has been identified as an effective method to heat the samples for calcinations and sintering. From the two compositional ratios of BST prepared, the transparent gels heated for 25 minutes showed higher percentage of weight loss when compared to that heated for 10,15 and 20 minutes respectively. Between 0.35:0.65:1 and 0.50:0.50:1 ratio, it was found that crystallization occurred at earlier temperature for 0.5:0.5:1 composition than 0.35:0.650:1. From the DTA results, it was found that when the microwave heating time was increased from 10 minutes to 25 minutes, for both

compositions, the temperature at which endothermic reaction occurs has been shifted to lower temperature values. These values matched those of the TGA results. The crystalline size was estimated from FWHM result which to be in the range of 70 nanometer. The materials exhibited polycrystalline pattern with "d" values that matched with standard reports. Thus, slow-rate sol-gel technique with rapid crystallization is shown to have improved the time and crystallization temperature. This crystallization process could be adopted for the industrial bulk preparation of the oxide material.

ACKNOWLEDGMENT

The authors would like to thank the Multimedia University, (Cyberjaya campus), Malaysia for sponsoring this work under internal research fund PR/2005/0512 and Dr T Saravanan, Scientist, IGCAR, India for his help on some technical matters.

REFERENCES

[1] D. Tahan, A. Safari, and L. C. Klein, "Sol-gel preparation of barium strontium titanate thin films," 1994.

[2] S. Lahiry, V. Gupta, K. Sreenivas, and A. Mansingh, "Dielectric properties of sol-gel derived barium-strontium-titanate ($Ba0.4Sr0.6TiO_3$) thin films," *Ultrasonics, Ferroelectrics and Frequency Control, IEEE Transactions on*, vol. 47, pp. 854-860, 2000.

[3] Teoh Wah Tzu, A. F. M. Noor, and Z. A. Ahmad, "Formation of Barium Strontium Titanate powder by solid state reaction using different calcination temperatures," *USM*, 2002.

[4] F. Schrey, "Effect of pH on the Chemical Preparation of Barium-Strontium Titanate," *Journal of the American Ceramic Society*, vol. 48, pp. 401-405, 1965.

[5] I. R. Abothu, P. M. Raj, D. Balaraman, M. D. Sacks, S. Bhattacharya, and R. R. Tummala, "Low-cost embedded capacitor technology with hydrothermal and sol-gel processes," 2004.

[6] G.Pfaff, "Synthesis and characterization of MTiO3(M = Mg, Ca, Sr, Ba) sol-gel," *J.Mater.Chem.,*, vol. 2, pp. 591, 1992.

Soft IP Integration and Reuse Challenges in Intel Entry Level Network Processor

Ng, Wai Mun; Yap, Kok Sing; Boey, Kean Hong, *MIEEE*

Communications Infrastructure Group
Intel Microelectronics (M) Sdn. Bhd., Bayan Lepas FIZ Phase III, Penang, MALAYSIA
Email: wai.mun.ng@intel.com; kok.sing.yap@intel.com; kean.hong.boey@intel.com

Abstract **Often a new product evolves from its predecessors with enhanced performances and added features. To reduce re-development time and effort, IP-based reuse methodology is used. This paper presents the challenges encountered in integrating IPs of multiple sources from two design cases in an Intel-based Network Processor**

I. INTRODUCTION

IP Reusability for fast and flawless integration plays a crucial role in today's fast changing technology. The complexity of today's IC has become very challenging for chip designers to design with a high quality chip in a short turnaround cycle to meet the very competitive market. [1] Conventional approach of "designing from scratch" requires more design efforts and investment which renders IP reusability an attractive solution and has been the trend of today's design approach. Reusable IP is now a common methodology used to improve design cycles but how effective and efficient is an IP for reuse remains a challenging question. A high quality IP is not only a matter of having superior speed and area performances across a wide range of application platforms but of equal importance is the interface and integration employability.

In this paper, we present two design case studies, a DDR controller and a USB controller, to discuss IP integration methodology and integration challenges. These two designs represent the units that reused IPs from multiple predecessor designs and external IP in a mix and match manner. We address the lack of key ingredients in the success of IP reuse.

II. DDR Controller

A. Overview of the DDR Controller

The DDR Cluster in question consists of memory controller unit (MCU), PAD DDR Controller logic (PAD_DDR_LOG) and pad ddr io as shown in Figure 1. The DDR Memory controller unit is reused from a previous project whilst the PAD DDR Controller logic and the PAD DDR IO buffers are reused from another different project. This becomes an integration of IP from a mixture of different sources.

The only IP that can not be reused from any of the previous project is the analog buffer used in the PAD DDR IO. The analog buffer is a "design from scratch" effort and is not discussed further in this paper. The complexities of the DDR cluster reuse increases when IPs from different sources are used.

B. Memory Controller Unit

Due to the difference in project requirements between the predecessor projects and the present project, 2 issues were discovered in the course of validation:

   i) 16 bit mode endianess
   ii) different type of lockup issues

In the case of (i), the projects that the component was inherited from operated at either 32-bit or 64-bit mode. As a result, the 16-bit mode endianess issue was not discovered in previous validation efforts.

In (ii), the issue was discovered in the present project and not in previous project because of the difference in validation environment and parameters.

These 2 scenarios highlight the relationship between the maturity of an IP and the past projects in which it was used. In addition to it's unit-level validation, IPs often undergo extensive validation in the projects it was used before, therefore an understanding of the features and validation environment of those projects can usually provide useful indication on the maturity of the IP in question.

By fixing all the above issues of MCU, the quality of the block increases and this contributes to a higher quality of IP for future reuse.

C. DDR Pad Logic

The DDR Pad Logic is reused from a project that supported only buffered memory mode and 64 bit data transfer so minor RTL modifications were required to support unbuffered memory and 16/32 bit modes. For DDR DFT, minor modification to reduce some calibration features and added more debug registers.

The product family line of the present project did not need all the complicated features of the original PAD DDR Logic so with all this RTL changes will increase the reusability for this product family.

D. Test Bench Changes

In DDR MCU unit-level validation environment, various test functions had to be modified to support 16-bit mode operation. In the present project, we noticed that significant efforts would be needed to validate DDR pad controller logic in multi blocks environment (cluster level), rendering the decision for full chip environment to validate the DDR pad controller logic. The other advantage of shifting validation efforts to full chip environment, is that this manner enhances the portability of ADMS simulation on the DDR PAD IO buffers with integrated DDR II.

E. DDR Cluster Integration Complexity

During the DDR cluster integration, insufficient knowledge base and information on some of the calibration modes introduced some challenges to the starting phase of the project. This was exacerbated by people movements in the team. As a result, significant time was spent on understanding the design by going through RTL files, followed by proper documentation – either adding further details to existing documentation or creating new documentation. While the importance of documentation is hardly ever disputed, the type information inside can vary greatly.

III. USB BLOCK

A. Background and Components Build of USB2.0

The implementation of USB2.0 comprises of three major components, i.e. the USB2.0 EHCI Controller, the Digital PHY Transceiver and the Analog PHY[2]. In our design, each of these blocks is independently sourced as soft IP from different vendors, a mixture of internal and external sources. IP sourcing requires complete understanding of features sought, interfacing and specifications requirements and performance needs.

USB2.0 in our design is the one single design unit that is potentially at risk for failure. This is due to the differences in interfacing capability, incomplete design features and unstable IP in design environment, leading to problems associated with integration flows.

B. USB2.0 Design and Implementation

Our USB2.0 design mimics very much to the SoC environment. With the IP reuse technology, one would expect a plug n' play concept but from our implementation, oftentimes, the changes has to be made to the IP as enhancement, at least in the following area (i) testability, (ii) fixed reset schemes and clocking methodology (iii) design-for-test capability and generic parameterized interfacing capability.

C. USB2.0 Validation Environment

Four categories of pre-silicon validation methodology had been applied to verify the functionalities of the USB2.0. At unit level, a full regression test suites was used to stress the USB2.0 stacks and protocols. Besides stressing the USB2.0 under normal operating functionalities, we have built-in features to validate the behavior of the USB2.0 under error conditions and recovery. Fig. 2 shows the basic block diagram of the unit level USB2.0 validation environment.

Full Chip Validation is built with basic directed tests for connectivity checks and basic error conditions verifications and randomized regressions to stress USB2.0 AHB interfaces in a multi-masters environment. This is particularly crucial if the design implements two identical and independent high performance USB2.0s accessing to the same bus whereby overall system performance can be degraded.

Silicon validation of USB2.0 on emulation platform is also used to ensure the quality of the USB2.0 and the digital PHY. SV is targeted to drive basic USB protocol transactions, to test configurations of USB2.0 Control and Status Register, to determine the functional correctness of USB2.0 data structures and to introduce USB2.0 error conditions that are not do-able neither at unit nor FC level validations.

Last but not the least, ADMS mixed-mode simulation is used to run simple test cases to verify the USB2.0 data flow with respect to the Analog PHY, which is different from the previous three validation's categories in which digital PHY has been used. Due to the inherent long runtimes associated with USB2.0 protocol on ADMS environment, a full simulation with USB2.0 protocol is not possible at the time of this writing.

IV. RESULTS

DDRII and USB2.0 designs have been discussed. Both designs incorporated IPs of different origins. In each of these design cases, it has been observed that the integration flow is not as simple as plug n' play of IPs, but much efforts are needed to tailor the IPs into new design environment. IPs are seldom or never made generic or parameterized enough to cater across many application platforms. Missing functional DFT features, uncontrollability over IP test suites and lacking of IP being validated as a whole but mainly as standalone with BFMs to meet standard governing protocol specifications are oftentimes the problems the designers face during integrations and validations of IP of multiple sources.

Our pre-silicon validation strategy has proven to be a big win having successfully verified all design conditions and constraints without strains to the design schedules despite the integration hurdles. The criticality (gating tape-out) of these designs is brought down from High to Low and the successful A0-steppings on these designs proved the accomplishment of the team in integration and validation methodologies.

V. CONCLUSIONS

In this paper, we have demonstrated that IP reuse is not simply plug and play as much as we would like it to be. A great amount of effort is still required to enhance reusability. For a successful IP integration and reuse, it is essential that the reuse ingredients are implemented. Some of the reuse and integration must-have that we found in the course of carrying DDR II and USB2.0 design in this project are the availability of comprehensive IP technical support, full design documentations, qualified IP with interfacing compatibilities, built-in DFT capabilities, full-compliancy to standards and a comprehensive test plan and diversified tests coverage. In short, issues with reusing IP will be greatly reduced if the IP used has been prepared for reusability. Making a block to reuse would require more manpower and time. However, this would be made up when the IP is reused in another project. IP reuse is becoming a popular design approach and we could only receive the optimum benefits from IP reuse if the reused IP has high reusability grade and high quality.

ACKNOWLEDGEMENTS

The authors would like to thank both the DDR and USB team members for their concerted efforts, contributions and drives to meet the schedules and resolve issues in a timely manner. Special thanks to SH Wong and JG Khor for their times in reviewing this paper and valuable feedbacks.

REFERENCES

[1] System Level SOC Design Unit4: IP selection, IP libraries, Physical Issues, Pittsburgh Digital Greenhouse

[2] MindShare, Don Anderson, Universal Serial Bus System Architecture (2nd Edition)

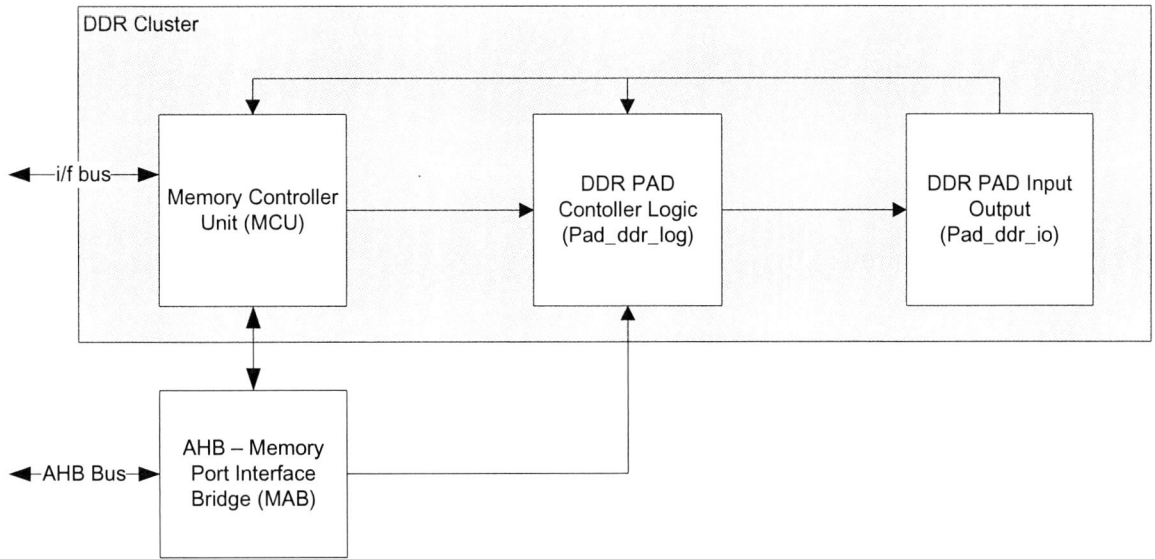

Figure 1: DDR Cluster

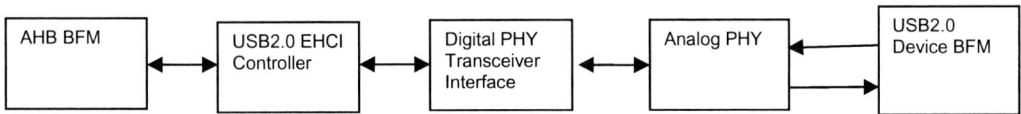

Figure 2: USB 2.0 Unit Level Test bench Configuration

Delayering of Gate Poly in Stack & Split Gate Memory Structure

Bridger KS Wong & Chan Sieng Fong
X-FAB Sarawak Sdn. Bhd.
1 Silicon Drive, Sama Jaya Free Industrial Zone,
93350 Kuching, Sarawak, Malaysia.
Email: bridger.wong@xfab.com
Tel: +60.82.354887

Abstract **The work presented here shows a new technique for removing the gate ploy on top and keeping floating gate (FLGT) below intact in both stack and split gate memory structure. The experiment shows that TMAH solution plus other supporting chemical will be able to provide us a good final sample.**

I. INTRODUCTION

The non-volatile memory like flash and embedded flash has stack gate structure and split gate structure consisting of dual poly [1]. To perform reverse engineering to find the fault or defect in 1st poly of stack gate structure or to remove the gate poly (GP) on top and keeping floating gate (FLGT) below intact in split gate structure has always been a challenge.

Various techniques involving chemical or chemical-mechanical have been used but face various limitations [2]. Now we have successfully developed a technique that can remove gate poly without affecting FLGT. We will be able to use this technique locate any fault or defect on FLGT in the future from top planar rather from side cross-section that has limited area to investigate.

II. CURRENT TECHNIQUES & LIMITATIONS

Techniques using chemical-mechanical like lapping is used to remove the top GP layer while the bottom FLGT layer is sometimes required for failure analysis. Other techniques using wet or dry chemical like HF extensively damages the FLGT.

The current techniques involving plasma etching, acid etch or lapping have some limitations below:

1. Difficult to control the exact removal of GP
2. Difficult to control uneven poly in Split Gate Memory Structure
3. Unevenness and uniformity issue and affects both type of memory structures

III. EXPERIMENT

We used both Stack Gate Memory Structure (Fig. 1) and Split Gate Memory Structure (Fig. 2) for our evaluation. The GP structure is obtained after delayering the metal, inter-metal dielectric (IMD), inter-layer dielectric (ILD) and passivation layers on top of the full layer devices.

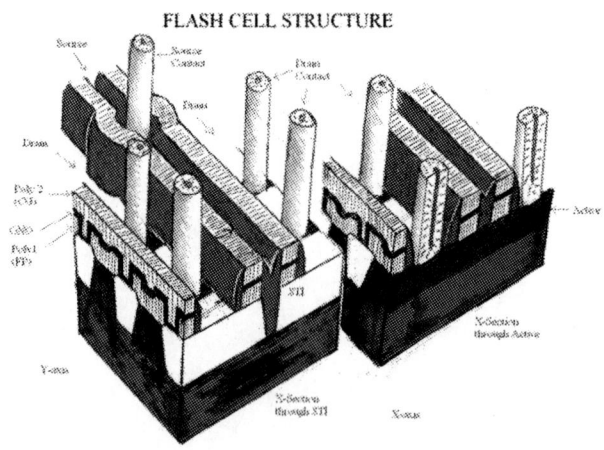

Fig. 1 Stack gate memory structure (3-D View)

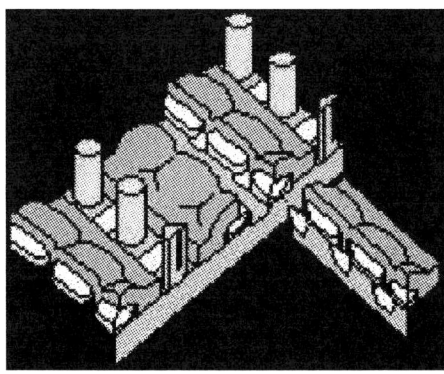

Fig. 2 Split gate memory structure (3-D view)

We evaluate with
1. 2.38% and 20% concentration tetra-methyl ammonium hydroxide (TMAH)
2. 80% concentration phosphoric acid (H3PO4)
3. 2.5% concentration hydrofluoric acid (HF)

We run many different evaluations starting with 2.38% TMAH and moving to higher concentration. We also evaluated at room temperature before moving up to 100C and 200C. We also try 2.5% HF and 80% H3PO4 to assist at certain step.

TMAH has high selectivity for poly and low selectivity for oxide or nitride, similar like what potassium hydroxide (KOH) will remove polysilicon and remains silicon covered with an oxide layer [3]. US rinse will shake free the poly 2 remains. Hot H3PO4 is good to remove nitride remains while maintaining poly.

Our structures have either oxide or nitride around the poly 1 to protect it and thus have to select an alkali material like TMAH for good uniformity [3]. Otherwise the poly 1 will be etched off by the TMAH during the poly 2 removal.

IV. RESULT FOR STACK GATE MEMORY STRUCTURE

We have summarized our result based on the series of pictures and labels. A → B indicated A process follows by B process. The GP is completely removed. However there is also reaction with the active as can be seen of deeper trench.

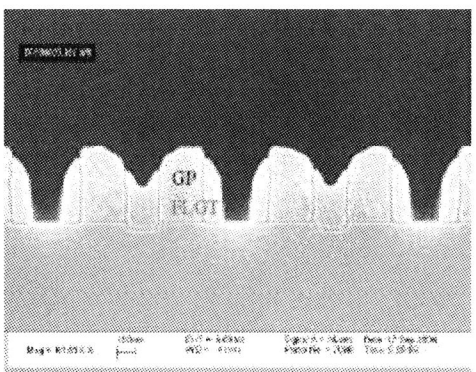

Fig. 3 Memory structure (x-SEM view)

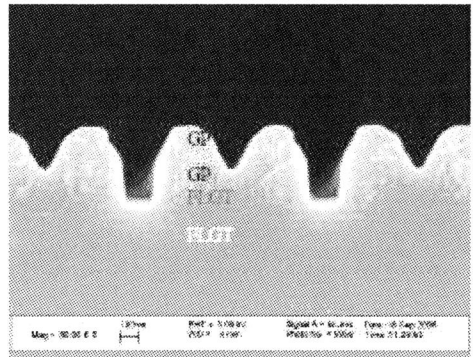

Fig. 4 Post TMAH (x-SEM View)

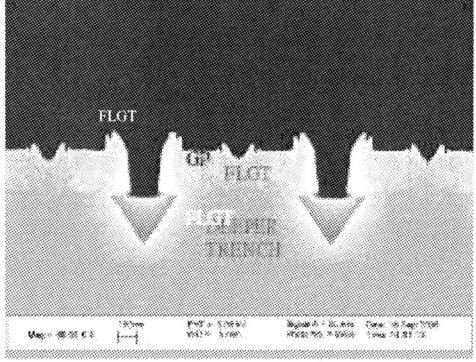

Fig. 5 Post HF →TMAH (x_SEM view)

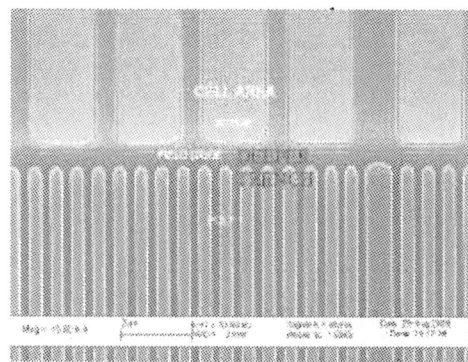

Fig. 6 Post TMAH (top view)

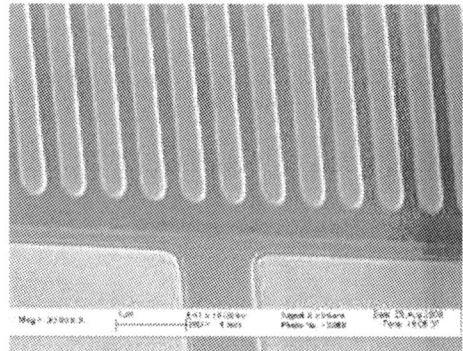

Fig. 7 Post TMAH (tilt view)

V. RESULT FOR SPLIT GATE MEMORY STRUCTURE

We have summarized our results base on the series of pictures and labels. A → B indicated A process follow by B process. Our evaluation is based on the GP and FLGT layers after delayering metals, IMD, ILD and passivation layers above.

The finding is showed that there is GP starting flaking using 6 min 100C 20% TMAH. Alternatively we can also use 1 min 200C 20% TMAH to accomplish the removal. This is followed by 5 min US and we find GP is completed removed. Although US is good to peel off the GP after TMAH etch, too long rinse will potentially damage the FLGT. Therefore we only use maximum of 5 min.

We found that there is still a *"white shell"* surrounding FLGT. This is suspected due to the undoped GP or oxide. We evaluate the following split using HF and phosphoric acid to clear the *"white shell"*. 7 min HF would be ideal to accomplish full removal and left with only floating poly.

Using 5 min HF we found there is still 50% of "white shell" surrounding FLGT. Going to 10 min, we found complete removal of the "white shell" but you can also observe excessive deep trench at active region.

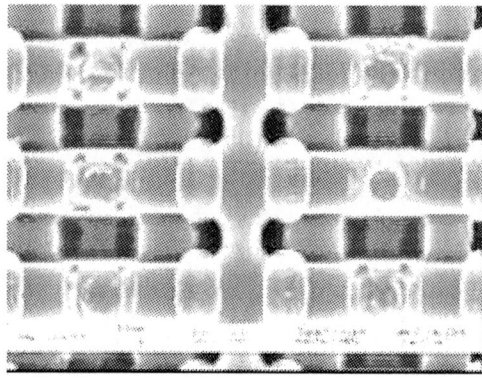

Fig. 8 Post 1 min 200°C 20% TMAH → 5 min US → 7 min HF (top view)

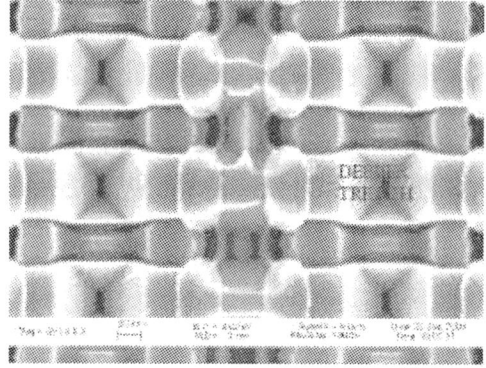

Fig. 9 Post 1 min 200°C 20% TMAH → 5 min US → 10 min HF (top view)

VI. DISCUSSION

The hot TMAH solution plus ultrasonic rinse with pre-treated HF acid is very suitable to remove the GP for Stack Gate Memory Structure. We have achieved reasonably good GP removal for Split Gate Memory Structure using hot TMAH plus ultrasonic rinse plus HF dip. We are now able to examine the FLGT for any defect or particle clearly.

However there is some etch impact on the active as can be observed deeper trench formed for both Stack Gate Memory Structure and Split Gate Memory Structure. This will not be a concern if the area of interest is the FLGT rather than the active.

430

We found 1.5 min 200C 20% TMAH is suitable for Stacked Gate Structure and 5 min ultrasonic if required. If there is SiN on top, dip in 2.5 % HF for 2 min.

Based on our evaluation for Split Gate Structure, 6 min 100C 20% TMAH or 1.5 min 200C 20% TMAH is suitable for Split Gate Structure. 5 min ultrasonic (US) and 7 min 2.5% HF have to be applied to complete remove the poly 2.

We also noted that there is some lateral etch effect on the FLGT that is unfortunately unable to resolve at this time and we control this by keeping HR time to minimum.

ACKNOWLEDGEMENT

The authors would like to record their sincere appreciations to their management for their approvals and support of this paper and to Stephen Wee Gim Heng, Angela Deborah and Lawrence Chin Pao Liang for assisting in this evaluation. The 3-D view of memory structures are taken with courtesy of Alvin Chan Tze Ping.

REFERENCES

[1] S.K. Ghandhi, VLSI Fabrication Principles, John Wiley and Sons, p. 482-487, 490 (1983)
[2] T.W. Lee, "A Review of Wet Etch Formulas for Silicon Semiconductor Failure Analysis", Proceedings of the International Symposium for Testing and Failure Analysis, p. 319-331 (1996)
[3] Christopher L. Henderson, Daniel L. Barton, Edward I. Cole Jr., Michael P. Strizich, "Chemical and Mechanical Deprocessing", 21st Century Product Analysis, Semitracks Inc p. 485, 489-490, 507-510 (2001)

High-Performance $In_{0.52}Al_{0.48}As/In_{0.6}Ga_{0.4}As$ Power Metamorphic HEMT for Ka-Band Applications

[1]Chia-Yuan Chang, [1]Edward Yi Chang, *Member, IEEE*, [1]Yi-Chung Lien, [2]Yasuyuki Miyamoto, [1]Szu-Hung Chen, [1]Li-Hsin Chu

[1]Dept. of Materials Science and Engineering, National Chiao Tung University, Hsinchu 300, Taiwan, R.O.C.

[2]Dept. of Physical Electronics, Tokyo Institute of Technology, Tokyo 152-8552, Japan

Corresponding Author: [1]Prof. Edward Yi Chang, E-mail: edc@mail.nctu.edu.tw ,

Tel: (886)-3-5712121ext.52971, Fax: (886)-3-5745497

Abstract · A 70-nm $In_{0.52}Al_{0.48}As/In_{0.6}Ga_{0.4}As$ power MHEMT with double δ-doping was fabricated and evaluated. The device has a high transconductance of 827 mS/mm. The saturated drain-source current of the device is 890 mA/mm. A current gain cutoff frequency (f_T) of 200 GHz and a maximum oscillation frequency (f_{max}) of 300 GHz were achieved due to the nanometer gate length and the high Indium content in the channel. When measured at 32 GHz, the 4×40 μm device demonstrates a maximum output power of 14.5 dBm with P1dB of 11.1 dBm and the power gain is 9.5 dB. The excellent DC and RF performance of the 70-nm MHEMT shows a great potential for Ka-band power applications.

I. INTRODUCTION

FOR High frequency communication system applications such as communication satellites, radar, mobile millimeter-wave communication, and smart munitions, high-performance power amplifiers are required in the emission part. Due to superior low noise and power performances in the millimeter-wave range, InAlAs/InGaAs metamorphic HEMT (MHEMT) is a good alternative to pseudomorphic HEMT (PHEMT) on GaAs or lattice-matched HEMT on InP [1]. Although, PHEMT grown on GaAs substrate has demonstrated excellent output power density at 60 and 94 GHz in the previous work [2]. The power gain and power added efficiency (PAE) were limited by the low Indium content of the pseudomorphic InGaAs channel. On the contrary, InP-based HEMTs have shown excellent high frequency characteristics by reducing the gate length (L_g) to sub-100nm range [3], [4]. However, the advantages of InP-based HEMTs, such as higher electron saturation velocity,

higher conduction band discontinuity and lower access resistance, also can be achieved with MHEMT that can be grown on less expensive and larger size GaAs substrate [5]–[7]. In this work, a 70-nm $In_{0.52}Al_{0.48}As/In_{0.6}Ga_{0.4}As$ power MHEMT with double δ-doping structure was processed and evaluated. The device demonstrates excellent DC and RF performances at Ka-band and shows great potential for the millimeter-wave power applications.

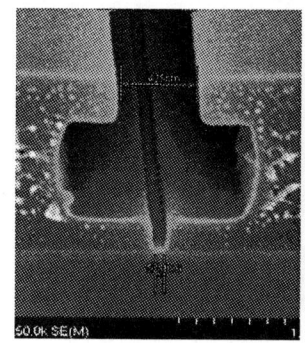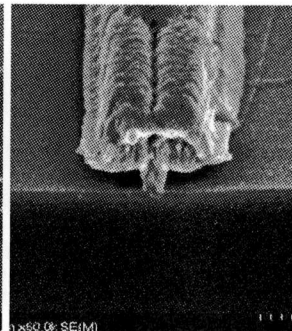

Fig. 1 Cross-sectional SEM images of the (a) resist profile and the (b) 70-nm T-gate of the MHEMT.

II. EXPERIMENT

The epitaxial structure of the MHEMT was grown by molecular beam epitaxy (MBE) on 3-inch semi-insulating GaAs substrate. The structure from bottom to top consists of an InAlAs buffer layer, a Si δ-doping layer, an $In_{0.52}Al_{0.48}As$ spacer, an $In_{0.6}Ga_{0.4}As$ channel layer, an $In_{0.52}Al_{0.48}As$ spacer, a Si δ-doping layer, an $In_{0.52}Al_{0.48}As$ barrier layer, and a Si-doped $In_{0.53}Ga_{0.47}As$ cap. The double δ-doping structure and the $In_{0.6}Ga_{0.4}As$ channel layer of the MHEMT are designed to provide higher carrier concentration and superior electron transport properties.

The mesa isolation was done by wet

0-7803-9730-4/06/$25.00 ©2006 IEEE   432

chemical etch. Source and drain Ohmic metals were formed with Au/Ge/Ni/Au. The T-shaped gate was carried out in the 50-KeV JEOL electron beam lithography system (E-beam) using conventional tri-layer E-beam resist with two steps exposure. The tri-layer resist system of

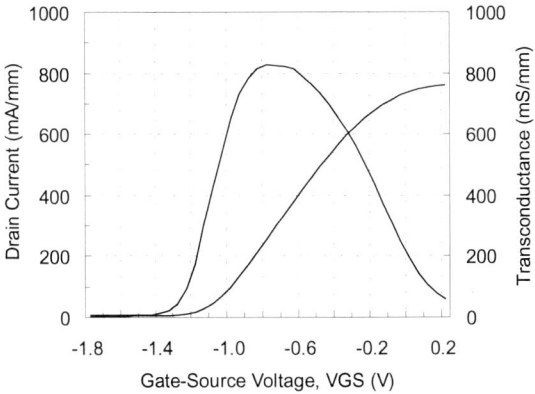

Fig. 2 Current-voltage characteristics of the 2×40 μm MHEMT.

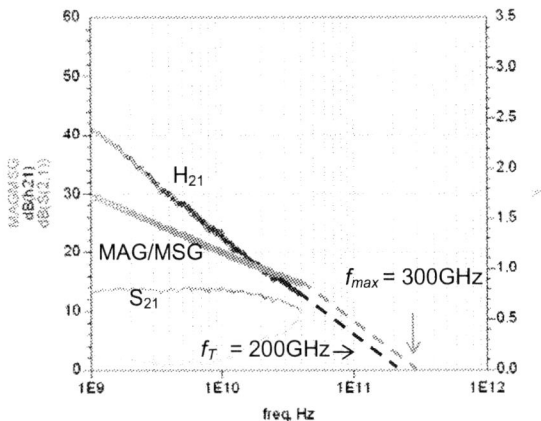

Fig. 3 Frequency dependence of the current gain (H_{21}) and *MAG/MSG* of the power MHEMT.

ZEP-520/PMGI/ZEP520 was used for the E-Beam lithography and shown in Fig. 1(a). The Ti/Pt/Au was evaporated as gate metal. The gate length of the T-shaped gate was 70nm as shown in Fig. 1(b). Finally, a 100-nm-thick silicon nitride was deposited as passivation layer using PECVD method.

III. RESULTS AND DISCUSSION

Fig.2 shows the current-voltage characteristics of the 2×40 μm MHEMT. The fabricated $In_{0.52}Al_{0.48}As/In_{0.6}Ga_{0.4}As$ MHEMT shows a maximum drain-source

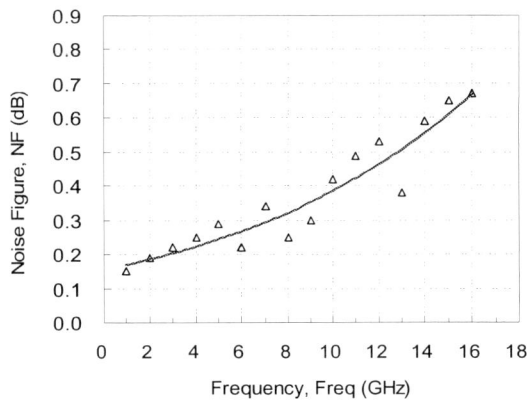

Fig. 4 Noise Figure (NF) of the 2×40 μm MHEMT measured from 1 to 16 GHz.

current of 890 mA/mm and transconductance of 827 mS/mm. The high current density was due to the double δ-doping structure which provided higher carrier concentration and superior electron transport properties in the $In_{0.6}Ga_{0.4}As$ channel.

S-parameter measurement was done from $1 - 40$ GHz by using a vector network analyzer with an on-wafer configuration. Fig. 3 shows the frequency dependence of the current gain (H_{21}) and power gain (MAG/MSG) for the 2×40 μm MHEMT with gate and drain bias of $- 0.6$ V and 1.5 V. The f_T and f_{max} of the MHEMT are 200 GHz and 300 GHz, respectively, by extrapolating H_{21} and MAG/MSG using least-squares fitting with a -20 dB/decade slop. The H_{21} is 13 dB at 40 GHz and the MAG/MSG is 15 dB at 40 GHz. Fig. 4 shows the noise figure (NF) of the MHEMT from 1 to 16GHz. The minimum NF was below 0.67 dB up to 16 GHz. This superior behavior is attributed to the low access resistance, larger drain current and high transconductance across a wide range of gate Bias.

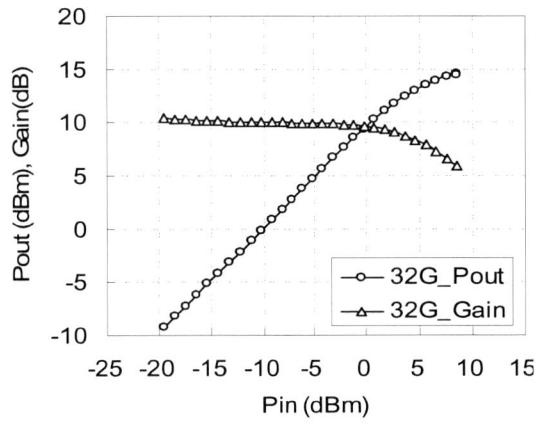

Fig. 5 Measured 32-GHz power performance of the 4×40 μm power MHEMT at drain bias of 2.5 V.

Further, the power performance of a 4 × 40 µm gate width device was measured at 32 GHz by load-pull systems for Ka-band application. The measured result is shown in Fig. 5, at drain bias of 2.5 V. With the tuner impedance matched for maximum power, the device showed the maximum output power of 14.5 dBm and P1dB of 11.1 dBm with 9.5 dB power gain at 32 GHz. This high power gain is attributed to the high Indium content at the channel and the short gate length. Overall, the MHEMT exhibits comparable RF performances to the InP-based HEMT due to the appropriate epi-structure design and the short gate length.

IV. CONCLUSION

The $In_{0.52}Al_{0.48}As/In_{0.6}Ga_{0.4}As$ power MHEMT with double δ-doping structure and 70nm T-gate has been designed and fabricated. The MHEMT developed showed excellent DC and RF performances and demonstrated great potential for power applications at Ka-band and millimeter-wave range.

ACKNOWLEDGEMENT

The authors would like to acknowledge the assistance and support from the National Science Council, and the Ministry of Economic Affairs, Taiwan, R.O.C., under the contracts: NSC 95-2752-E-009-001-PAE and 94-EC-17-A-05-S1-020. A part of this work was supported by "Nanotechnology Support Project" of the Ministry of Education, Culture, Sports, Science and Technology (MEXT), Japan.

REFERENCE

[1] M. Zaknoune, M. Ardouin, Y. Cordier, S. Bollaert, B. Bonte, and D. Théron, "60-GHz power performance $In_{0.35}Al_{0.65}As$–$In_{0.35}Ga_{0.65}As$ metamorphic HEMTs onGaAs," IEEE Electron Device Lett., vol. 24, no. 12, pp. 724–726, 2003.

[2] M.-Y. Kao, P. M. Smith, P. Ho, P.-C. Chao, K. H.G. Duh, A. A. Jabra, and J. M. Ballingall, "Very high power-added efficiency and low-noise 0.15-µm gate-length Pseudomorphic HEMT's," IEEE Electron Device Lett., vol. 10, no. 12, pp. 580–582, 1989.

[3] Y. Yamasjita, A. Endoh, K. Shinohara, K. Hikosaka, and T. Matsui, "Pseudomorphic $In_{0.52}Al_{0.48}As/In_{0.7}Ga_{0.3}As$ HEMTs with an ultrahigh f T of 562 GHz," IEEE Electron Device Lett., vol. 23, no. 10, pp. 573–575, 2002.

[4] L. D. Nguyen, A. S. Brown, M. A. Thompson, and L. M. Jelloian, "50-nm self-aligned-gate pseudomorphic AlInAs/GaInAs high electron mobility transistors," IEEE Trans. Electron Devices, vol. 39, pp. 2007–2014, 1992.

[5] D.–W. Tu, S. Wang, J. S. M. Liu, K. C. Hwang, W. Kong, P. C. Chao, and K. Nichols, "High-performance double-recess InAlAs–InGaAs power metamorphic HEMT on GaAs substrate," IIEEE Microwave Guided Wave Lett., Vol. 9, pp. 458–460, 1999.

[6] C. S. Whelan, W. E. Hoke, R. A. McTaggart, S. M. Lardizabal, P. S. Lyman, P. F. Marsh, and T. E. Kazior, "Low noise $In_{0.32}(AlGa)_{0.68}As$–$In_{0.43}Ga_{0.57}As$ metamorphic HEMT on GaAs substrate with 850 mW/mm output power density," IEEE Electron Device Lett., vol. 21, pp. 5–8, 2000.

[7] D. C. Dumka, H. Q. Tserng, M. Y. Kao, E. A. Beam, III, and P. Saunier, "High-performance double-recessed enhancement-mode metamorphic HEMTs on 4-In GaAs substrates," IEEE Electron Device Lett., vol. 24, no. 3, pp. 135–137, 2003.

Development of the Vacuum Spark as an EUV Source for Next Generation Lithography

Chew Soo Hoon and Wong Chiow San
Plasma Research Laboratory, Physics Department, Science Faculty,
University of Malaya, 50603 Kuala Lumpur, MALAYSIA
Tel: 603-79674395, Fax: 603-79674146
Email : caseycsh@perdana.um.edu.my

Abstract **Extreme Ultraviolet Lithography (EUVL) which requires a radiation in a 2% wavelength band around 13.5 nm is expected to be the Next Generation Lithography (NGL) system. A 13.5 nm EUV source is needed to satisfy the demand for the production of semiconductor chips with critical dimensions of 50 nm and below. Nowadays, plasma based EUV sources such as laser produced plasmas and gas discharges are considered internationally by many as the practical light sources. Recently, much progress has been made in vacuum spark discharges as they seem to offer an alternative with much higher conversion efficiency into EUV photons. The vacuum spark (UMVS-III) being a compact pulsed plasma discharge has been investigated in this laboratory as a possible EUV source. An extension of the earlier research work on X-ray production by the vacuum spark to the EUV region is carried out.**

I. INTRODUCTION

THE application of Extreme Ultraviolet Lithography (EUVL) in large scale semiconductor chip manufacturing requires a light source of wavelengths between 10 and 20 nm [1]. Narrow spectral bands particularly around the central wavelength of 13.5 nm are of main interest because of the available highly reflective multilayers used for imaging a mask onto a semiconductor wafer [1]-[3].

Several EUV sources are being considered such as synchrotrons, laser produced plasmas, gas discharge produced plasmas and vacuum sparks [4]-[7]. However, EUV sources based on plasma are commercially attractive since they are simple in operation, compact and cost effective. Various working elements including tin, lithium, xenon, oxygen, beryllium, silicon and titanium have been investigated for the generation of EUV radiation in the 13.5 nm wavelength region

[6],[8]-[10]. Recently, the development of EUV sources have been centered on the tin plasmas from vacuum sparks [4],[8]. It is reported that tin has a much higher conversion efficiency owing to many of its transitions in the region of interest [9],[11]. Vacuum sparks have long been known as compact pulsed plasma sources capable of producing intense X-ray from solid targets [12]-[14]. The spectrum of the radiation is dependent on the type of the anode materials. By using suitable anode materials, the vacuum spark will allow generation of characteristic emission in the EUV spectral range. Thus, the generation of EUV radiation from vacuum spark can be a very attractive alternative. In this work, the possibility of using two anode materials, copper (Z=29) and tin (Z=50), have been investigated in a 370 J vacuum spark, which exhibit strong emission in the spectral range of interest.

II. EXPERIMENTS

The experimental setup of UMVS-III used for generating EUV radiations is schematically illustrated in Fig. 1. The vacuum spark is powered by a single 1.85 µF Maxwell capacitor. The system consists of a cathode and anode mounted on holders, which are placed in a vacuum chamber. The chamber is continuously pumped to a pressure of 5.0×10^{-4} mbar by using a diffusion pump backed by a rotary pump. The cathode is a stainless steel disk with an aperture (5.0 ± 0.3 mm) along the axis and tapered towards the replaceable anode. The spacing between the electrodes is set at 1.5 mm. The electrons released from the triggering pin are accelerated in a high electric field favored by the transient hollow cathode effect towards the anode. The bombardment of electrons at the anode will vaporize some of the anode materials which are heated by the discharge to form plasma that generates light emission in the UV, EUV, soft X-ray and hard X-ray region.

0-7803-9730-4/06/$25.00 ©2006 IEEE   435

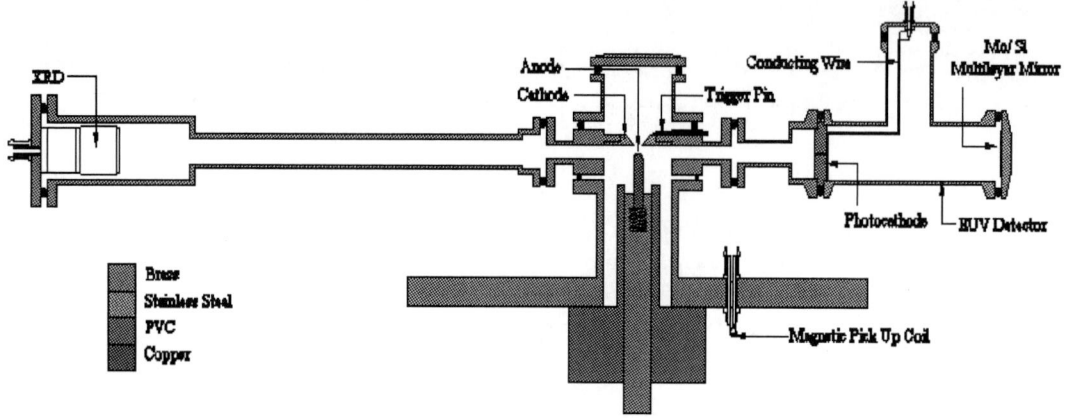

Fig. 1 The experimental setup of the vacuum spark system.

For preliminary studies, the device is operated in single shot mode where the EUV radiation is emitted in a small dose in each pulse. A magnetic probe is implemented to monitor the rate of change of current. To monitor the EUV emission at the wavelength of 13.5 nm, a detector based on the principle of photoelectric effect is employed. The EUV detector [15] makes use of a Mo/Si multilayer coating mirror to selectively reflect the EUV radiation at the wavelength of 13.5 nm. The Mo/Si multilayer reflective mirror has a peak reflectivity near 65 % at the wavelength of 13.5 nm when the angle of incidence is 5 degrees off normal, as shown in Fig. 2.

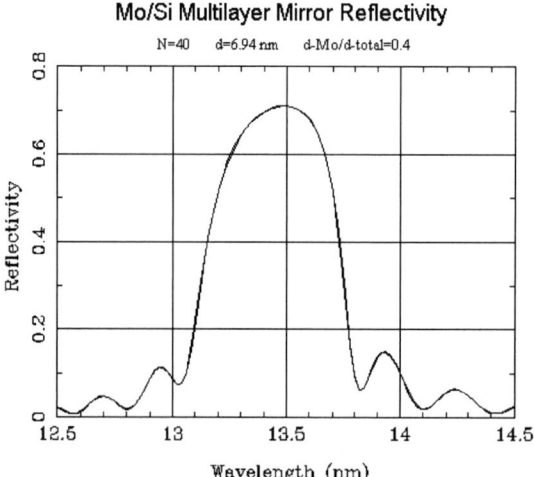

Fig. 2 Near normal incidence reflectivity of a Mo/Si multilayer mirror as a function of wavelength.

A photocathode is used to detect the EUV photons reflected from the mirror and the corresponding photocurrent generated will be recorded by an oscilloscope. Hence, by utilizing a Mo/Si multilayer coating mirror in the design of the EUV detector, the time evolution of EUV emission from the vacuum spark plasma can be observed. In addition, an X-ray diode (XRD) that utilizes a thick aluminium cathode and a PIN diode (Quantrad model 100-PIN-250) are used as X-ray detectors to measure X-ray emission.

III. RESULTS AND DISCUSSION

The working pressure throughout the experiments using copper and tin anodes was maintained at 5.0×10^{-4} mbar. Since the EUV radiation is strongly absorbed by the air, the pressure has to be below 10^{-3} mbar. The discharge voltages were varied from 5 kV to 20 kV for both experiments.

Fig. 3 (a) shows the dI/dt, EUV, XRD and PIN diode signals obtained for a copper discharge at 10 kV. By considering the discharge current to be a source of under-damped LCR circuit, the circuit inductance is calculated to be 56.3 nH and circuit resistance is 26.6 mΩ. The initial dI/dt is about 2×10^{11} As^{-1}. Both the soft X-ray pulse recorded by the XRD and the EUV detector are observed to follow the discharge current. Due to the low inductance of the circuit, the current rises quickly and causes strong heating of the plasma. The near 13.5 nm EUV emission resulted from copper plasma in the vacuum spark can be contributed by the process called as line radiation. It is reasonable to believe that the characteristic line radiations at around 13.5 nm are attributed to the multiple ionization of Cu^{9+}, Cu^{10+} and Cu^{11+} species in the copper emission lines (see Fig. 3 (b)). Cu^{11+} specie is dominant at an electron temperature of around 46 eV. At this electron temperature, the continuum emission also peaks at around 13.5 nm. This can be predicted from the population density distribution computed assuming the Coronal Equilibrium (CE) model. Upon careful examination of the EUV and XRD signals, both

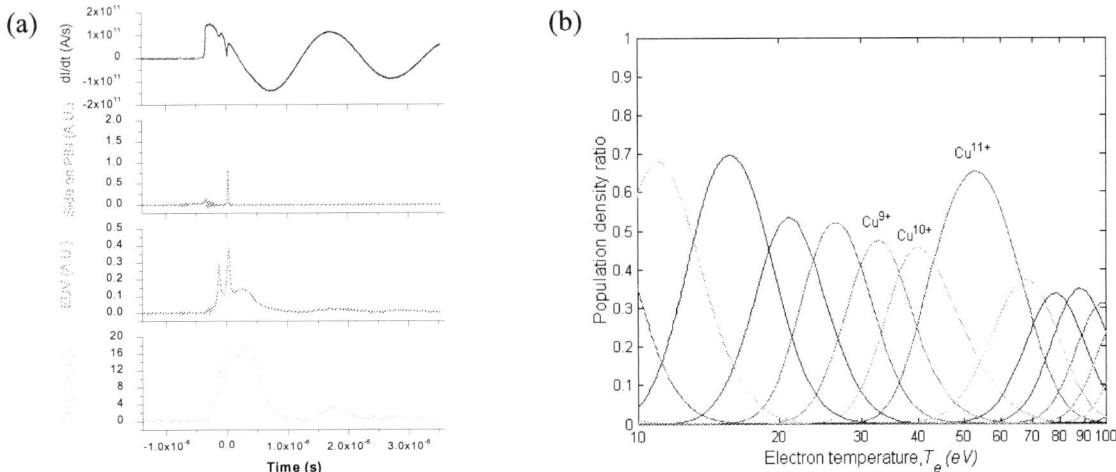

Fig. 3   (a) dI/dt, EUV signal and X-ray intensity at 10 kV for copper anode, (b) The population density ratio of copper ionic species versus electron temperature calculated from the Coronal Equilibrium Model.

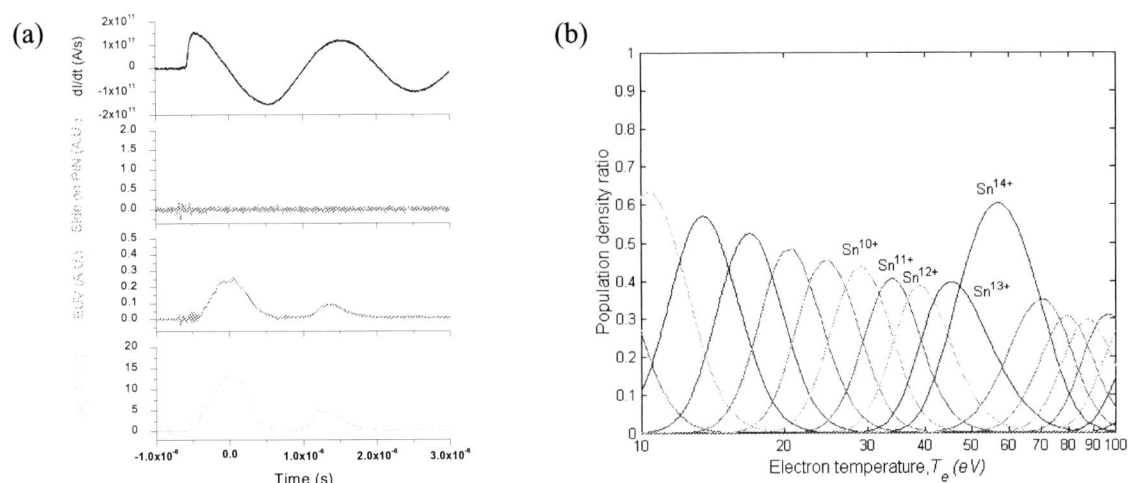

Fig. 4   (a) dI/dt, EUV signal and X-ray intensity at 10 kV for tin anode,   (b) The population density ratio of tin ionic species versus electron temperature calculated from the Coronal Equilibrium Model.

light intensities reach their peaks in about 650 ns after the breakdown of the vacuum spark and decay gradually. The peak of the EUV emission is obtained after the peak of the dI/dt signal indicating that the most intense EUV radiation is emitted at this time. At this instant following the dips of dI/dt, the current drops and the plasma starts to cool down and decay. The observation of the dips shows that the magnetic compression or pinching by the Lorentz force is likely to happen after the breakdown. It is noted that the EUV signal has a duration of about 1.5 μs. In addition, it is clear that the strong peaks from the EUV detector and XRD signals coincide with the dI/dt dips. It is also observed that there is a hard X-ray pulse detected from the side on PIN in this case.

A tin vacuum spark discharge at 10 kV is shown in Fig. 4 (a). In this discharge, no dip is being observed in the dI/dt signal. Also, no sharp peaks are observed in the XRD and EUV signals. Furthermore, it can be seen that there is no hard X-ray pulse detected from the PIN diode. This indicates that the plasma has not been heated enough to give emission in the hard X-ray region. However, the EUV detector registers a broad pulse of the duration 1.5 μs. Tin (Z=50) is more advantageous compared to copper because of its many transitions arising from Sn^{10+} - Sn^{14+} ions at the wavelengths near 13.5 nm (see Fig. 4 (b)). These prominent species are predicted for electron temperatures of approximately from 30 to 60 eV [16]. The pulse shapes of different discharge voltages are found to be identical in the EUV signals. Furthermore, the EUV emission signal is observed to correspond well with the XRD signal.

In general, significant EUV emissions are observed for discharges with voltages of 5 kV to 20 kV in this series of experiments for copper and tin discharges. It is noted that the EUV output is enhanced as the charging voltage is increased. The measurement of the signals from the magnetic probe and EUV detector show that the EUV radiation occurs in the first cycle of the discharge current.

IV. CONCLUSION

EUV emissions from both copper and tin plasmas of the vacuum spark have been investigated. Time evolution measurements of the EUV emission using a EUV detector suggest that besides being an intense X-ray source, the vacuum spark may also be utilized as a pulsed EUV source for Next Generation Lithography. Faster discharge must be produced to achieve a good generation of EUV emissions. In order to obtain more information on the characteristic EUV emission lines, time resolved spectroscopic measurements will be carried out.

ACKNOWLEDGEMENT

The authors wish to express their gratitude to the Ministry of Science, Technology and the Environment, Malaysia for funding this research project under IRPA Grant 090203-0224 (EA 0224) and Vote-F. They are also grateful to Mr. Jasbir Singh for his technical assistance.

REFERENCES

[1] V. Banine and R. Moors, "Plasma sources for EUV lithography exposure tools," *J. Phys. D: Appl. Phys.* vol. 37, pp. 3207-3212 (2004).

[2] D. T. Attwood, "Applications of short wavelength radiation: soft X-ray microscopy and EUV lithography," *J. Phys. IV France, vol.* 11, Pr2-443-449 (2001).

[3] U. Stamm, "Extreme ultraviolet light-sources-state of the art, future developments, and potential applications," *RIKEN Review No. 5: Focused on Laser Precision Microfabrication,* pp. 63-70 (2003).

[4] J. Jonkers, "High power Extreme Ultra-Violet (EUV) light sources for future lithography," *Plasma Sources Sci. Technol.* vol. 15, S8-S16 (2006).

[5] S. Düsterer, H. Schwoerer, W. Ziegler, C. Ziener and R. Sauerbrey, "Optimization of EUV radiation yield from laser-produced plasma," *Appl. Phys. B,* vol. 73, pp. 693-698 (2001).

[6] E. S. Wyndham, M. Favre, H. Chuaqui, P. Choi, A. M. Lenero and J. S. Diaz, "Reproducibility of a titanium plasma vacuum

spark discharge," *IEEE Trans. Plasma Sci.,* vol. 33, no. 5, pp. 1662-1667 (2005).

[7] K. Deguchi and T. Haga, "Proximity X-ray and Extreme Ultraviolet Lithography," *C. R. Acad. Sci. Paris: Challenges in Microelectronics, Série IV,* pp. 829-842 (2000).

[8] E. R. Kieft, J. J. A. M. van der Mullen, G. M. W. Kroesen, V. Banine and K. N. Koshelev, "Stark broadening experiments on a vacuum arc discharge in tin vapor," *Phys. Rev. E,* vol. 70, 066402 (2004).

[9] J. White, P. Hayden, P. Dunne, A. Cummings, N. Murphy, P. Sheridan and G. O' Sullivan, "Simplified modelling of 13.5 nm unresolved transition array emission of a Sn plasma and comparison with experiment," *J. Appl. Phys.,* vol. 98, 113301 (2005).

[10] A. Egbert, B. Mader, B. Tkachenko, A. Ostendorf, C. Fallnich, B. N. Chichkov, T. Mißalla, C. Schürmann, K. Gäbel, G. Schriever and U. Stamm, "Compact electron-based extreme ultraviolet source at 13.5 nm," *J. Microlith., Microfab., Microsyst.,* vol. 2, no. 2 , pp. 136-139 (2003).

[11] S. A. George, C. S. Koay, K. Takenoshita, R. Bernath, M. Al-Rabban, C. Keyser, V. Bakshi, H. Scott and M. Richardson, "EUV spectroscopy of mass-limited Sn-doped laser micro-plasmas," *Proc. SPIE- Emerging Lithographic Technologies IX,* vol. 5751, pp. 779-788 (2005).

[12] F. Wu, W. C. Tang, K. W. Witpszo and E. Panarella, "The vacuum spark and spherical pinch x-ray/ EUV point sources," *Proc. SPIE- Emerging Lithographic Technologies III,* vol. 3676, pp. 410-420 (1999).

[13] C. S. Wong, C. X. Ong, S. P. Moo, and P. Choi, "Characteristics of a vacuum spark triggered by the transient hollow cathode discharge electron beam," *IEEE Trans. Plasma Sci.,* vol. 23, no. 3, pp. 265-269 (1995).

[14] C. S. Wong, "Pulsed plasma x-ray and EUV sources," *Mal. J. Sci.,* vol. 21A, pp. 65-70 (2002).

[15] L. M. Yap, "Characteristics of a pulsed capillary discharge EUV source," *M.Sc. Appl. Phys. (course work report),* University of Malaya, Kuala Lumpur (2005).

[16] H. S. Poh, private communication.

TDR Single Ended and Differential Measurement Methodology

Lee, Chan Kim, Beh, Jiun Kai and Huang, Jimmy Huat Since
Intel Microelectronic (M) Sdn. Bhd.
FIZ, 11900 Bayan Lepas, Penang, Malaysia
E-mail : chan.kim.lee@intel.com

Abstract — **As the frequency of the high speed interfaces in computer platforms and systems continue to ramp due to architectural advancement and strong market demands, electronic packages have been stretched to its limit to support these interfaces. Furthermore, due to increase in the complexities in the package, signal integrity, manufacturing concern and package size reduced, designers are facing a lot of signal quality issue caused by impedance mismatch. To ensure signal integrity, it is necessary to understand and control impedance in the transmission environment through which the signals travel. Mismatches and variations can cause reflections that decrease signal quality as a whole. TDR (Time Domain Reflectometry) is the most common tools for verification and analysis of the transmission properties of high-speed systems and components. It is able to locate signal path discontinuities that cause reflections, verify traces characteristic impedance and estimate traces length. Hence, this paper is to discuss about TDR set up and type of TDR measurement procedure and methodology for single ended and differential pair interface in the package. There are several key elements that determine the precision of TDR measurement such as the edge speed of the stimulus pulse produced by the TDR, bandwidth of the channel used to received the pulse from the DUT, calibration and deskew process. Correlation between measurement data vs simulation data is performed in order to ensure the accuracy of the data collected. The simulation method is done by using Ansoft 3D and 2D simulation tools to characterized package interconnect such as traces, wirebond, VIA/PTH and ball structures. The output file (Spice File, .sp) will be important to Ansoft Designer software for TDR simulations.**

I.   INTRODUCTION

TIME Domain Reflectometry (TDR) measurements allows us to understand the behavior of a device (Package or PCB board) or a system, to debug, to verify our simulated model, etc. Measurement data give the most accurate behavior of a device, but is not straightforward to achieve accurate measurement data especially at higher frequencies (>1GHz). During any measurements, it is necessary to de-embed the effects of any instrumentation or fixtures such as cables, probes, PCB traces or more separate from the device. Calibration standards, procedure are usually required to de-embed the effect of these fixtures properly from TDR measurements. Furthermore, the accuracy of the measurements also depends on the quality of calibration standards and the accurate characterization of these standards. TDR measurements are commonly used for Package or Board characterization to determine characteristic impedances (Z_o or Z_{diff}) or traces and therefore the interconnect discontinuities can be observed and windowed effectively. TDR measurement required good time resolution to observed electrically small devices.

II. THEREOTICAL BACKGROUND

TDR measures the reflections that results from a signal traveling through a medium, it can be package, circuit board or cable. TDR instrument send a pulse to the DUT, compares the reflections from the unknown transmission environment to those produce by a standard impedance. Figure 1 is showing the simplified block diagram.

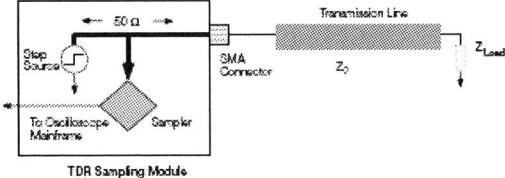

Fig. 1 Block diagram of TDR circuit

TDR measurement is based on impedance ratio. It can be describe in terms of Reflection Coefficient, ρ (rho). The coefficient ρ is the ratio of the reflected pulse amplitude to the incident pulse amplitude.

$$\rho = V_{reflected} / V_{incident}$$

For a fixed termination Z_L, ρ also can be expressed in terms of the transmission line characteristic Impedance, Z_o and the load impedance Z_L.

$$\rho = (Z_L - Z_o) / (Z_L + Z_o)$$

With the formula above, we can derive the matched load, a short circuit and an open load. ρ has a range of values from +1 to -1, with 0 representing matched load, +1 representing open load and -1 representing short.

When Z_L is equal to Z_o, the load is matched. Hence, $V_{reflected}$ wave is equal to 0 and ρ is 0, there is no reflection.

$$\rho = V_{reflected} / V_{incident} = 0/V = 0$$

If is short circuit, Z_L will be read 0, the reflected wave is equal to the incident wave, but opposite in polarity.

$$\rho = V_{reflected} / V_{incident} = -V/V = -1$$

As for open circuit, Z_L is infinite. The reflected waveform is equal to the incident waveform and of the same polarity. The ρ value is +1.

$$\rho = V_{reflected} / V_{incident} = V/V = 1$$

III. EQUIPMENT SETTINGS/PROCEDURES

TDR Setting

The combination of the Agilent 86100 series Infiniium DCA mainframe, the left module 54754A Differential TDR module, right module 86118A 70GHz dual electrical module with remote sampling and Picosecond Pulse Labs (PSPL) Differential & Common mode TDR/TDT source 4022 provides a good solution for TDR/TDT measurements. PSPL 4022 module increases the signal risetime from 35ps to ~9ps. The effective rise time of the input step at the end of the cables was measured to be 30ps. TDA Systems' IConnect was used to extract Characteristic Impedance profile (Z-line) of the DUT in time domain. Figure 2 and 3 shows the TDR module connection and the whole system with video systems.

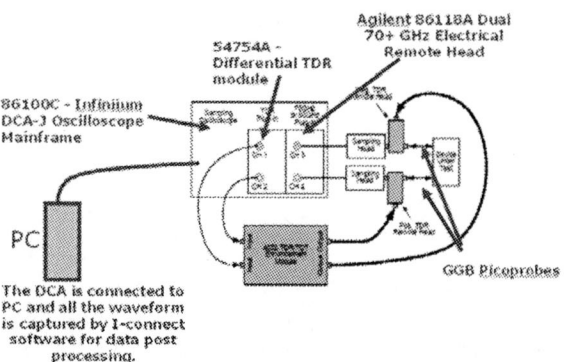

Fig. 2 TDR module connection

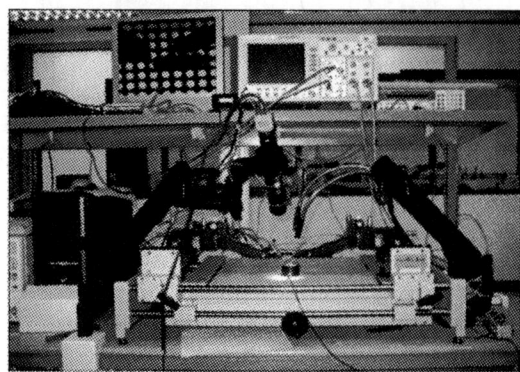

Fig.3 TDR station with video systems.

TDR Procedures

To obtain accurate test results, errors introduced from module, cables or probes must be correctly removed. The procedures needed before performing any measurement are:
1) Module calibration
2) De-skew process
3) Calibration

Single ended measurement does not require de-skew process.

Module Calibration

It is important to perform both module calibrations (left and right module) to ensure the module is at nominal setting.

De-skew Process

True TDR measurements required both stimulus at the acquisition window well matched in timing and step response. Both receiver and transmitter are usually adjustable to obtain matched TDR step timing to yield a valid TDR measurement. The best way to reduce this problem is to minimize the interconnect length and use matched cables. Below are the steps required to obtained valid differential or common mode stimulus. Figure 4 is showing TDR set-up for deskew process.

1) Compensate skew of receiver different module heads

This purpose of this step is to compensate for time delay differences between two remote module heads including cables to TDR heads.

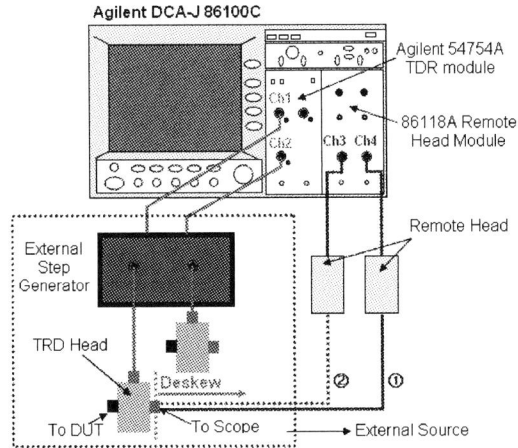

Fig. 4 TDR set-up

2) Compensate skew between Incident Channels to DUT

The purpose of this step is to compensate for the time delay difference between the TDR module's output and the TO DUT ports on the TDR heads.

3) Compensate skew between reflected channels in TDR head

These steps compensate for time delay differences between the To Scope and To DUT ports in the TDR head.

4) Compensate skew of differential probes

These steps compensate for the time delay difference resulting from probes with pico probes connected to TDR head at DUT section.

Calibration

Calibration on all the connection is required after completed module calibration and de-skew process. Since this is the picoprobing measurement, a calibration substrate is required in this process. A short, load reference process will removes systematic errors to provide accurate results. This process is important to eliminate cable losses and noise introduced in the systems.

Single ended or differential measurement can be performed once completed all the procedures mentioned above.

IV.   CHARACTERISTIC   IMPEDANCE   DATA
COLLECTION

Wirebond-PBGA package (DUT)

The DUT used is a four layer package which consist of gold wirebond connection, microstrip traces, plane, via/ball and plating bar. The package is built using FR4 and designed to have effective dielectric constant of approximately 4.25. The DUT are accessed using Picoprobes at the package pin side.

TDR measurement

A 2 port differential TDR measurement is performed on the package. The TDR of a reflect "open" waveform was stored as an open reference. As the picoprobes landed to the DUT, reflection measurements of DUT were taken. Both waveforms were captured in TDA Systems' IConnect software. Figure 5 is showing an Open and DUT waveform.

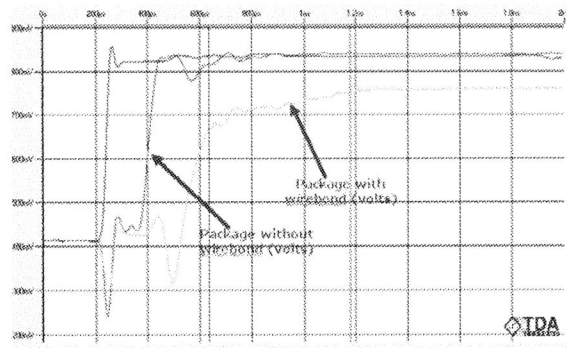

Fig. 5 is showing the Open reference waveform (RED) and the DUT waveform with 2 different package.

The Z-line feature in IConnect software was used to extract package/board characteristic impedance. Figure 6 shows the Z_{diff} results for the package (DUT). The TDR plot is probing at the same differential pair signal on 3 different packages. The packages used in this measurement are:

1) Full interconnect of the package (Wirebond, traces, via/ball and plating tail).
2) Package without plating bar connection.
3) Package without wirebond connection.

The TDR plot of the first package allows us to overview the characteristic impedance of the whole package interconnect. Each package interconnects characteristic impedance attributes can be identified clearly after completed second and third package measurements. From the TDR plot, capacitive vias, characteristic impedance of the traces and plating bar, inductive wirebond and Cdie effect can be captured clearly.

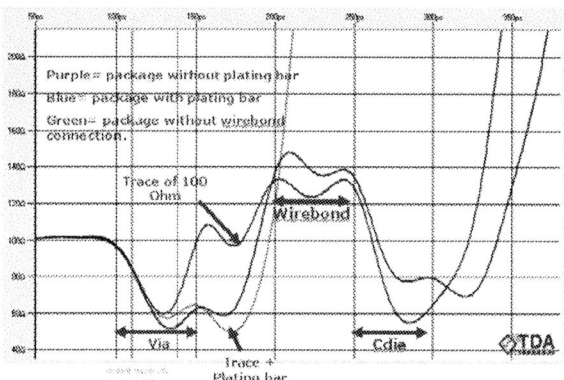

Fig. 6 Differential TDR plot

V. CORRELATION PROCESS

This measurement data can be used to correlate, debug or verify simulated package models. Each package interconnect is simulated using appropriate simulation tools. The package wirebond, via & ball is modeled with actual dimension using AnsoftLink™, Ansoft Q3D Extractor®. The package traces and plating bar is modeled using Ansoft Q2D to obtain RLGC model of the transmission line. AnsoftLink™ is a tool used to extract 3D model with actual dimension from package layout. The same 3D model is then imported, modeled and analyzed in Ansoft Q3D to generate parasitic RLC value. Figure 7 & 8 is showing the actual package model from Q3D and Q2D simulation tools.

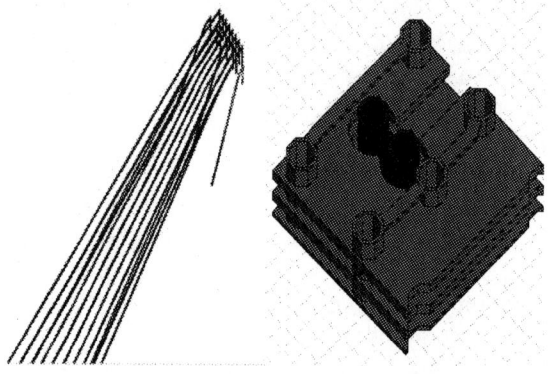

Fig. 7 Actual dimension of Q3D package model (wirebond, via model)

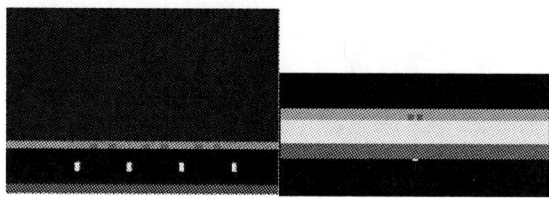

Fig. 8 Q2D trace and plating bar model. Cross sectional view

In order to study the package characteristic impedance, Ansoft Q3D and Q2D model is exported into .sp file format. All these .sp files are exported into a TDR simulation tools namely Ansoft Designer/Nexxim. The appropriate stimulus (incident signals) and trace length set up is important during the simulation in order to obtain well correlated results. The TDR setup in Nexxim software is shown in figure 9. All Q3D and Q2D package interconnect model are shown in schematic block manner. This simulation software can produce different risetime of stimulus to correlate with measurement stimulus.

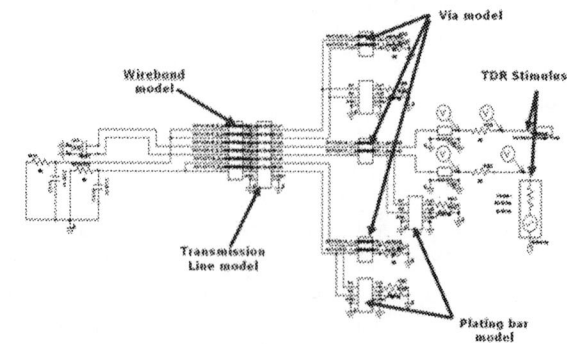

Fig. 9 TDR simulation set-up

442

Referring to figure 10, it is the measurement taken on the package without plating bar and comparing with simulated results. This open stub plating tail will cause some reflected energy and changing the trace impedance. By removing this plating tail, we are able to capture the trace impedance independently and allow us to compare with trace simulated result. All other package interconnects characteristic impedance are well matched with measurement results.

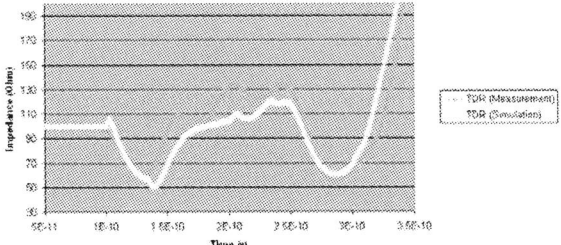

Fig. 10 TDR simulation plot vs measurement plot

Another alternative way to verify the package model simulation is to determine the trace propagation delay. The derivation of propagation delay of the trace is [1]:

$$T_d = \sqrt{L_o\,C_o}$$

Figure 11 is showing two set of measurement results on two different signals with different trace length (10.4mm and 7.09mm).The measured time difference between these 2 signals is ~40ps. To correlate this measurement results, parasitic L and C from Q2D simulation is used for propagation delay calculation shown below:

$\sqrt{LC}$=5.074e-7*9.896e-11= 7.08ns/m or 7.08ps/mm

Length difference= 3.31mm, 2*3.31=6.62mm

Time travel= 6.62mm*7.08ps/mm

=46ps

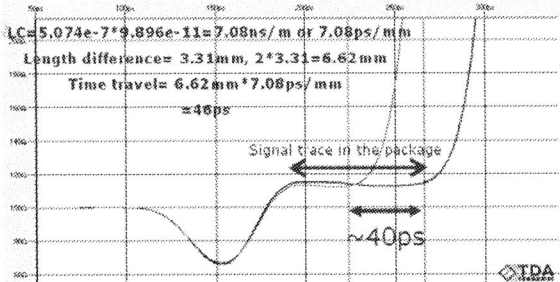

Fig. 11 Trace measurement results

VI. CONCLUSION

In this paper, we described the TDR measurement methodology for obtaining Z_o or Z_{diff} of a package and compared results with simulation results. The simulated package models matched closely with measurement results. The advantage of TDR measurement is clearly described in this paper. TDR/TDT measurements are much more straightforward to acquire. TDR gives immediate feedback if a good contact is establish between the probe and the trace during reference and DUT measurements, while VNA it is not transparent to the user during calibration. Lastly, Time Domain waveforms are much more intuitive to understand, to determine the effects of different aspects of a particular DUT.

ACKNOWLEDGEMENT

Thanks to Ahmad Jalaluddin bin Yusof from Intel Microelectronics, ETMS department for providing valuable information and feedback during TDR measurement set up.

REFERENCES

[1] Michael Leddige, "Transmission Line Fundamentals" A reference to Intel Corporation internal training material.

Contact Hole Printing in Binary Mask by FLEX Technique

Cheong Yew Shun[1], Ko Bong Sang[1], Mohd Jeffery Bin Manaf[1], Kader Ibrahim [1],
Dr. Zul Azhar Zahid Jamal[2],
[1]SilTerra Malaysia Sdn Bhd,
[2]Northern Malaysia University College of Engineering University (KUKUM)
Tel: +60126768955, Email: yewshun_cheong@silterra.com

Abstract **The shrinking of contact opening is inevitable, as the technology progresses. The most obvious shrink strategies, wavelength reduction and phase-shifting masks (PSM) offer the most significant improvements to meet these demands, but at a hefty price. By using focus latitude exposure technique (FLEX) it is possible to print down contact hole with reasonable processing latitude using 248nm wavelength lithography and binary mask. A few parameters were varied to obtain suitable condition. Several feature sizes were printed and measured to get the proper combination of the print bias and focal distances with acceptable process margin. Verification on production wafer is needed for different pitch of feature sizes and fine-tuning on the focal distance for optimum results.**

I. INTRODUCTION

PATTERN formation in optical lithography had become more critical as the technology node advances. It became challenging to print smaller feature size with better or acceptable process latitude. Various techniques had been tried for better process latitude. This is including hardware modification of the scanner or stepper tool itself, which involves huge, cost and need higher technical knowledge. Applying different photoresist, tuning the track parameters, illumination condition, reticle and so on are easier to implement [1]. Combination of two or more techniques is the trend of today's lithography to get better performances.

As the year past by, the price of phase shift mask (PSM) reticle increases drastically for higher technology node. Difficulty in maintenance of PSM reticle is also one of the major problems. Using binary mask is the better way for pattern printing in term of cost and maintenance. But the concern is the aerial image formation of the binary mask is not as good as PSM for higher technology node [2].

This paper is to study the capability of this technique, focus latitude exposure technique (FLEX), with the use of binary mask for higher technology node which has been applied in I-line scanner in lower technology node [5][9]. From the literatures [7][9][10], this technique is to increase the focus latitude that provide larger process window [4]. There are three major factors that effect the success of this technique. First is the percentage dose for each focal plane that to be used compare with traditional technique. Second is the number of focal planes or exposure needed for printing the patterns. Last will be the focal distance between the focal planes.

II. EXPERIMENT

The whole experiment had been carried out at SilTerra wafer fabrication facility. Bare silicon wafers were used for this purpose instead of stacked oxide wafer. TEL Track was used in this experiment for coating and developing processes. ASML 248nm scanner was used to perform the focus exposure matrix (FEM) on the wafers by varying dose and focus. Printed wafers were measured at a KLA-Tencor CD SEM (scanning electron microscope).

Isolated hole patterns with different feature size are the main objective of this experiment. The numerical aperture (NA) used in this study was fixed at 0.70 and partial coherence factor (σ) is 0.60. Number of exposure was also fixed to two and varying the two focal distances in micrometer. This is to simplify the experiment plan and put more focus on the focal distance and dose to be used.

Feature sizes that were used in this experiment are 150nm, 160nm, 170nm, 180nm, 190nm and 200nm. With this, different mask bias will be applied on it.

III. RESULTS AND DISCUSSION

Even though it is suggested in literature [9] that there is no limit on number of focal planes for contact hole and more focal plane will give larger DOF margin. In this experiment, the

0-7803-9730-4/06/$25.00 ©2006 IEEE

number of focal planes was fixed at two. This not only simplifies the work but it reduces the resources and eliminates uncertain factors. Besides fixing the number of focal planes, the focal distances were also fixed to 0.14um, 0.3um and 0.6um at the beginning. Further split were carried on for the feature size that mostly meet the FLEX concept. Table 1 shows the condition for the first split and table 2 is the split condition for the successful feature size. The dosage values are the same for each focal plane.

Table 1, Split table for the first condition.

Mask CD	reference dosage	focal distance
150	40.9	0.14
		0.3
160	40.9	0.14
		0.3
170	20.5	0.14
		0.3
		0.6
	40.9	0.14
		0.3
180	20.5	0.14
		0.6
	40.9	0.14
		0.3
		0.6
190	20.5	0.14
		0.6
200	20.5	0.14
		0.3
		0.6

Table 2, Split table for further split condition.

Mask CD	reference dosage	focal distance
180	40.9	0.9
		0.75
	31.5	0.6
		0.9

The split tables were changed from time to time according the result condition. The split for certain feature size were stopped to minimize the experiment step. Figure 1, 2 and 3 show the judgment criteria in FLEX technique. The condition in figure 1 is under focal distance where it shows convergent shape on the chart, lower DOF margin. The focal distance needed to be extended or change of feature size. In the figure 2, the focal distance maintains the same but the feature size had changed to bigger size, from 170nm to 180nm. Hence, it shows flatter than the previous chart. It has bigger DOF margin.

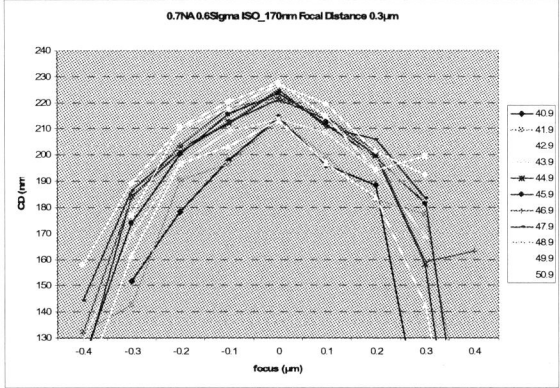

Figure 1, Bossung curve for focal distance 0.3μm for feature size 170nm. (Under focal distance)

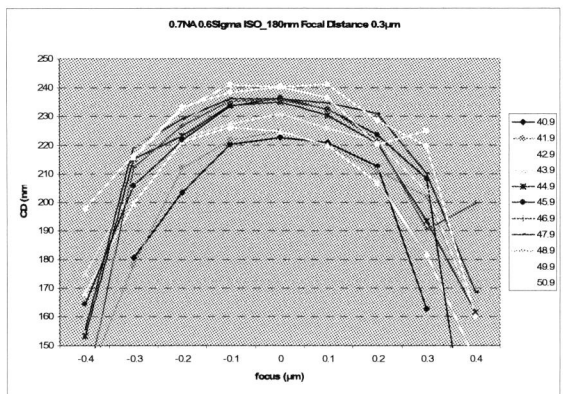

Figure 2, Bossung curve for focal distance 0.3μm for feature size 180nm. (Near to optimum focal distance)

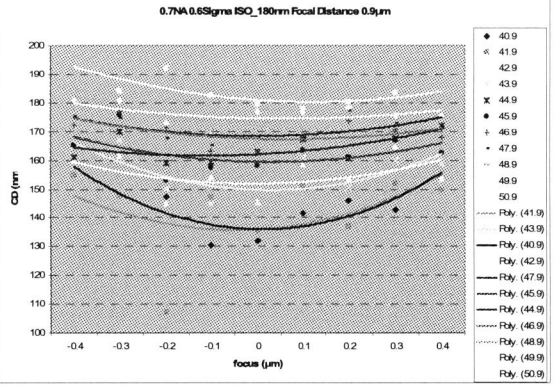

Figure 3, Bossung curve for focal distance 0.9μm for feature size 180nm. (Over focal distance)

Even though the figure 3 shows the best result compared to previous two, but few lines near the 150nm CD show a "smile" shape. This means the target CD near 150nm and below the DOF margin is not optimum. To print down smaller CD size the focal distance has to be decreased.

All the split conditions in table 1 did not give good result except for 180nm feature, which shows bigger DOF margin. The results were shown in figure 4. The focal distance for first figure is 0.14μm and follows by 0.3μm and 0.6μm.

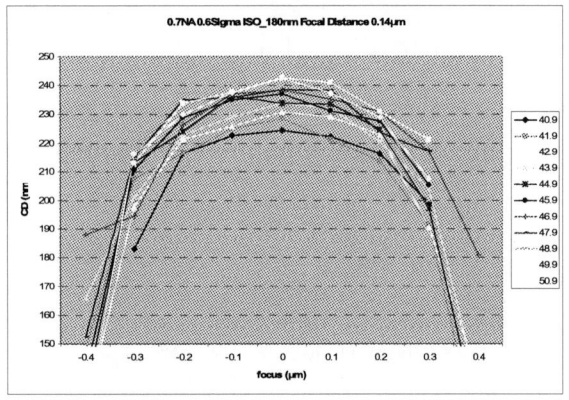

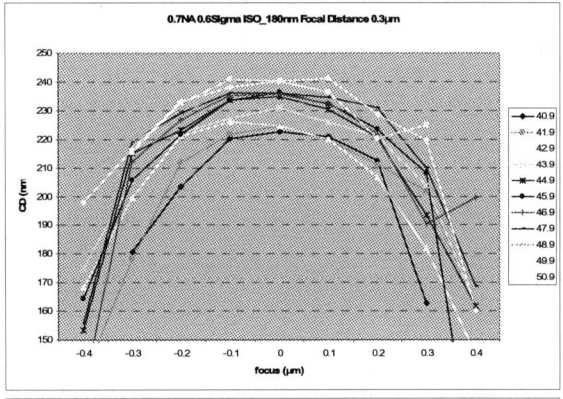

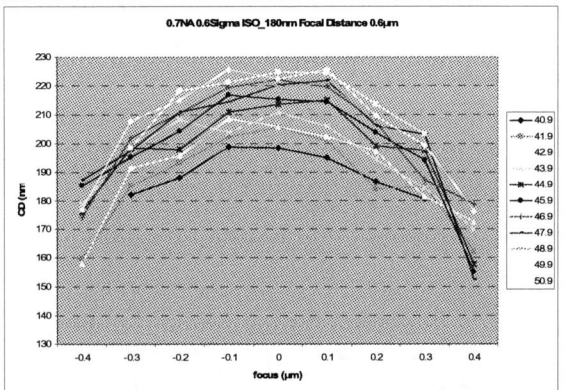

Figure 4, Best result from the split table 1. Focal distance follows by sequence, which is 0.14μm, 0.3μm and 0.6μm.

In table 2, the experiments were continued on feature size 180nm with different focal distance and dosage. As shown in figure 3, the focal distance will be less than 0.9μm for feature 180nm if the targeted CD is around 150nm. From here, the optimum focal distance will be in between 0.6μm and 0.9μm. By applying the

medium value 0.75μm which shown in figure 5, it gives better result.

After that, the reference dosage was reduced to explore the possibility to print down smaller CD. As shown in figure 6, it can be print down to 150nm with larger DOF margin.

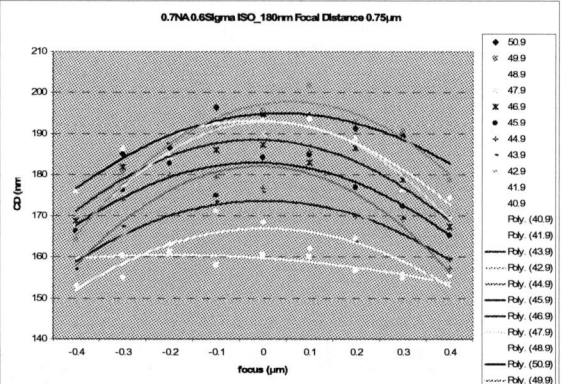

Figure 5, Bossung curve with focal distance 0.75μm on feature 180nm.

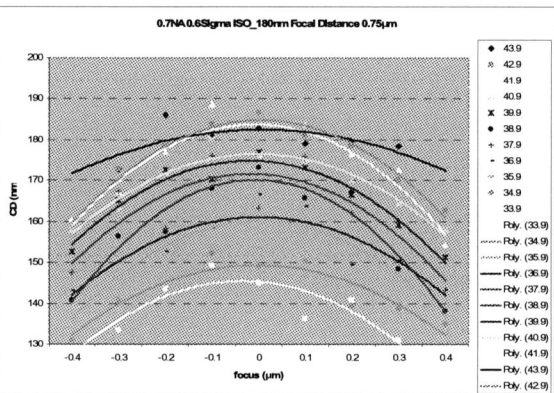

Figures 6, Bossung curve with focal distance 0.75μm on feature 180nm by dosage reduce.

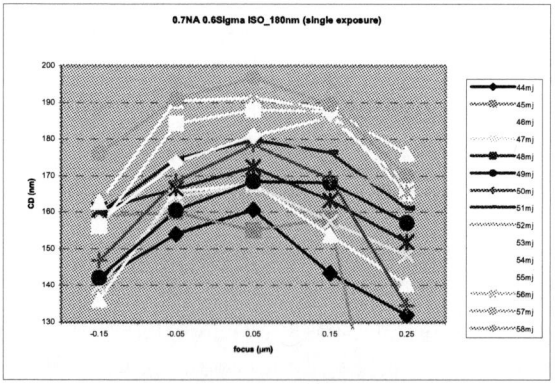

Figure 7, Bossung curve for single exposure.

From the figure 7, it shows that for target CD of 180nm in single exposure, the dosage is 51mj and has DOF margin of 0.2μm. When comparing to FLEX technique with the same target CD the dosage is 42mj each with DOF margin over

0.4μm. It is clear that the FLEX technique give bigger DOF margin [7] and percentage dose for each focal plane is about 80% of the single exposure dose.

Even though this technique can print down smaller feature size with larger DOF margin prior switch to PSM or smaller wavelength, process thru put is a big concern.

IV. CONCLUSION

The purpose of this experiment is to get the suitable focal distance by trying various feature sizes. The result shows the best focal distance is 0.75μm for lower CD size. The result also had shown the possibility to print 150nm by using mask CD 180nm with DOF margin ~0.6μm. Further experiments are needed for different feature sizes and optimization of the focal distance for certain target CD.

ACKNOWLEDGMENT

The authors of this paper would like to thank SilTerra Malaysia Sdn Bhd for sponsoring this work and for providing the facilities at the SilTerra wafer fab in Kulim, Kedah.

REFERENCE

[1] H. J. Levinson, *Principles of Lithography*, SPIE Press 2001

[2] J. R. Sheats, *Microlithography: Science and Technology*, Marcel Dekker, Inc., 1998

[3] H. Xiao, *Introduction to Semiconductor Manufacturing Technology*, Prentice-Hall, 2001.

[4] PROLITH workbook chapter 2, Depth of Focus.

[5] Chris A. Mack, "Understanding Focus Effects in Submicron Optical Lithography, Part 3: Methods for Depth-of-Focus Improvement", *Optical/Laser Microlithography V*, SPIE Vol. 1674, pp. 272-284.

[6] Chris A. Mack and Patrick M. Kaufman, "Mask Bias in Submicron Optical Lithography", *Journal of Vacuum Science & Technology Society*, Vol. B6, No. 6, Nov/Dec1988

[7] S. Manakli, Y. Trouiller, P. Schiavone, "Combination multiple focal planes and PSM for sub 120nm node with KrF lithography: study of the proximity effects", Microelectronic Engineering, 61-62 (2002) 123-132.

[8] H. Fukada, R. Yamanaka, T. Terasawa, "Characterization of Super-Resolution Photolithography", IEDM, IEEE, 1992.

[9] Hiroshi Fukuda, Norio Hasegawa and Shinji Okazaki, "Improvement of Defocus Tolerance in A Half-micron Optical Lithography by the Focus Latitude Enhancement Exposure Method: Simulation and Experiment", J. Vac. Sci. Technol. B7 (4), Jul/Aug 1989.

[10] Hiroshi Fukuda, Akira Imai, Tsuneo Terasawa, "New Approach to Resolution Limit and Advanced Image Formation Technique in Optical Lithography", IEEE Electron Devices, Vol 38. No. 1, Jan 1991.

Porous Silicon Dioxide Synthesized using Photoelectrochemical (PEC) Wet Etching

L. S. Chuah[1], C. W. Chin[2], Z. Hassan, H. Abu Hassan
School of Physics, Universiti Sains Malaysia, 11800 Minden, Penang, Malaysia
Email: [1]chuahleesiang@yahoo.com, [2]takeshi_ccw@yahoo.com, zai@usm.my
Tel: 604-6533673, Fax: 604-6579150

Abstract Porous SiO_2 can be used as a template to reduce substrate-induced stress, similar to porous GaN. Such a regrowth method may reduce the defect density in the epitaxial layer leading to high quality stress free layer on porous template. The samples were prepared on silicon (Si) wafers, (111)-oriented, with n-doping. After standard cleaning steps, SiO_2 of 1200 Å thickness was prepared by thermal oxidation of the Si at 1000° C for 1.50 hours. The wafer was then cleaved into few pieces. To prepare porous structures by photoelectrochemical (PEC) method, the samples were dipped into a mixture of hydrofluoric acid (HF): water: ethanol under different etching durations. Structural properties of porous SiO_2 have been investigated by scanning electron microscope (SEM). Elemental composition of the sample was identified using energy dispersive x-ray (EDX) analysis. Fourier transform infrared reflectance (FTIR) spectroscopy was used to characterize the chemical species and chemical bonding state.

I. INTRODUCTION

Amorphous silicon dioxide (SiO_2) is widely used for surface passivation in solar cells. In microelectronics, porous silicon dioxide are used for segregation for the diffusion of a shallow junction [1], which control the diffusion depth of the junction by adjusting the porosity and the pore size of segregation. They are also utilized as hosts for nanometer composite materials or sensors [2].

Although Si and SiO_2 are dissimilar materials, there is a surprisingly low defect density (less than one per 10^4 interface atoms) over large areas [3], resulting in a nearly ideal interfacial behaviour.

The reasons for the low defect densities are not yet completely understand. One possible explanation is a transition layer between silicon and thermal SiO_2. Atomically flat interfaces from silicon to amorphous SiO_2 are observed by transmission electron microscopy (TEM) in most cases, but there are some reports of crystalline SiO_2 phases existing in the dioxide near the interface.

Ordered SiO_2 on Si (100) substrate has been modelled as quartz [4-5], cristobalite [4], or tridymite [6]. Even when a structurally flat interface without any apparent interface layer has been observed by high-resolution transmission electron microscopy (HRTEM), most X-ray photoelectron spectroscopy (XPS) measurements have shown that a suboxide region exists, confined within a few monolayers, i.e. less than 3Å thick [7]. On the other hand, crystalline SiO_2 up to 20 Å thick has been detected by X-ray diffraction (XRD) [8] at interfaces formed by thermal oxidation but without any structural observation of a thick crystalline SiO_2 by HRTEM.

Porous semiconductors have been widely studied in the last decade, primarily due to the potential for intentional engineering of properties not readily obtained in the corresponding crystalline precursors as well as the potential applications in optoelectronics, chemical and biochemical sensing. Among porous semiconductors, porous silicon receives enormous attention and has been investigated most intensively; however the instability of physical properties has prevented it from large scale application.

Porous SiO_2 can be used as a template to reduce substrate-induced stress for heteroepitaxy growth, similar to porous GaN. Such a regrowth method may reduce the defect density in the epitaxial layer leading to high quality stress free layer on porous template.

0-7803-9730-4/06/$25.00 ©2006 IEEE

In this work, investigation of the structural and optical properties of porous SiO_2 prepared by photoelectrochemical (PEC) method was carried out using various post-deposition analysis.

II. EXPERIMENT

The samples used were silicon (Si) wafers, (111)-oriented, with n-doping. After standard cleaning steps, SiO_2 of 1200 Å thickness was prepared by thermal oxidation of the Si at 1000°C for 1.50 hours. The wafer was then cleaved into few pieces. To prepare porous structures by photoelectrochemical (PEC) method, the samples were dipped into a mixture of HF: H_2O: Ethanol (1:1:10) for 10, 30 and 45 minutes. Typical photoelectrochemical (PEC) wet etching apparatus are schematically shown in Fig. 1.

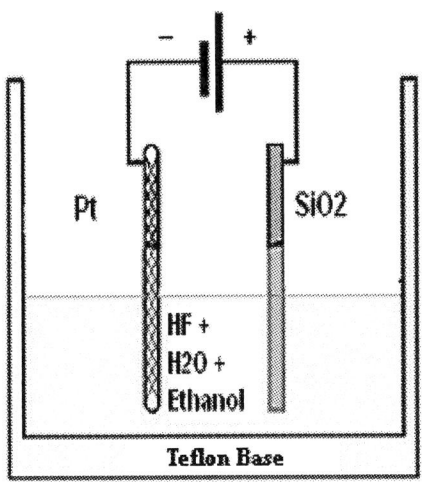

Fig.1. Schematic of photoelectrochemical (PEC) wet etching apparatus.

In order to achieve significant hole current in n-type Si, external illumination under an UV lamp with 500 W power is required. The anodization is performed in potentiostatic (voltage-controlled) mode. Structural properties of porous SiO_2 have been investigated by scanning electron microscope (SEM) (JOEL JSM-6460LV). Energy dispersive X-ray analysis (EDX) was used to identify the elemental composition of the sample. Fourier transform infrared reflectance (FTIR) spectroscopy

was used to characterize the chemical species and chemical bonding state.

III. RESULTS AND DISCUSSION

Porous SiO_2 was fabricated by electrochemical anodization in ethanoic hydrofluoric acid (HF). Ethanol is often added to facilitate evacuation of H bubbles, which develop during the process. To form porous SiO_2, the current at the SiO_2 side of the SiO_2/electrolyte interface must be carried by holes, towards the interface. A hypothesized chemical reaction which could describe the anodization is

$$SiO_2 + 6HF + H_2O \rightarrow H_2SiF_6 + 4H_2O + 2H^+ + 2e^-$$
(1)

Fig. 2 shows SEM images of the porous SiO_2 samples generated under different durations. SEM images show that average pore size for sample was around 2.5 to 15 µm. For the 10 min sample, the etching was in the initial stage, pores started to form and only circular shape was observed. For the 30 min sample, the surface became relatively rough, the SiO_2 layer exhibits triangular-like uniform microarray which appears to be denser on the outer surface. Triangular shaped pores illustrate an orientation dependence of the photo-anodization process of the (111) Si material, probably due to the known anisotropy of Si dissolution. For samples etched for 30 minutes or longer, the surface morphology was found to be similar, suggesting that an etching saturation was reached, at this stage; high densities of triangular pores were observed.

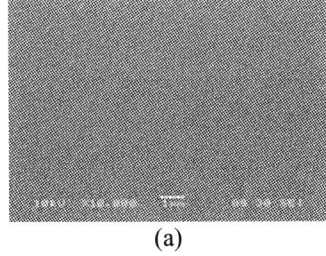

(a)

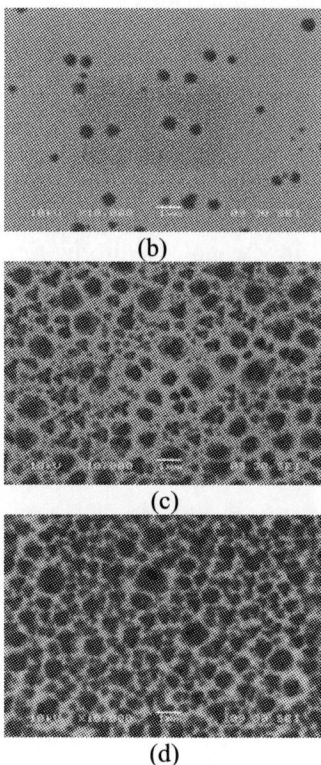

Fig. 2   SEM images of the samples etched under
different duration. (a) As grown, (b) 10 min,
(c) 30 min, (d) 45 min

Fig 3 show the energy dispersive x-ray analysis
(EDX) spectra of the porous SiO_2 samples
generated under different durations. One of the
most useful features of SEM analysis is the EDX
tool. An accessory to an SEM, this analytical tool
allows simultaneous non-destructive elemental
analysis of a sample. It is a technique used for
identifying the elemental composition of a sample.
The peak intensity corresponds to the
concentration of the element in the sample (the
higher the concentration, the higher the peak). Si
and O peaks were observed in all of the spectra.

The evolution of the FTIR spectra of porous SiO_2
film with various etched durations is presented in
Fig. 4. The Si–O stretching mode (1080 cm^{-1}) and
Si–O rocking vibration (460 cm^{-1}) are observed in
as grown sample. As can be seen, after etched for
10 min, the intensity of the Si–O stretching mode

and Si–O rocking vibration significantly decrease.
At etched durations of 30 and 45 min, the intensity
of the Si–O stretching mode significantly decrease
and the Si–O rocking vibration disappear.

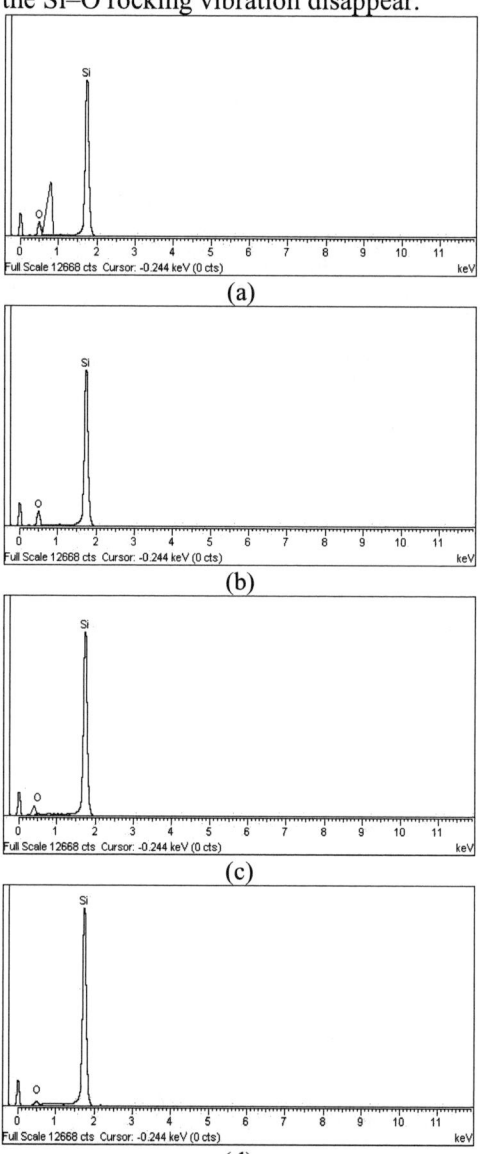

Fig. 3. EDX images of the samples etched under
different duration. (a) As grown, (b) 10 min, (c) 30 min,
(d) 45 min

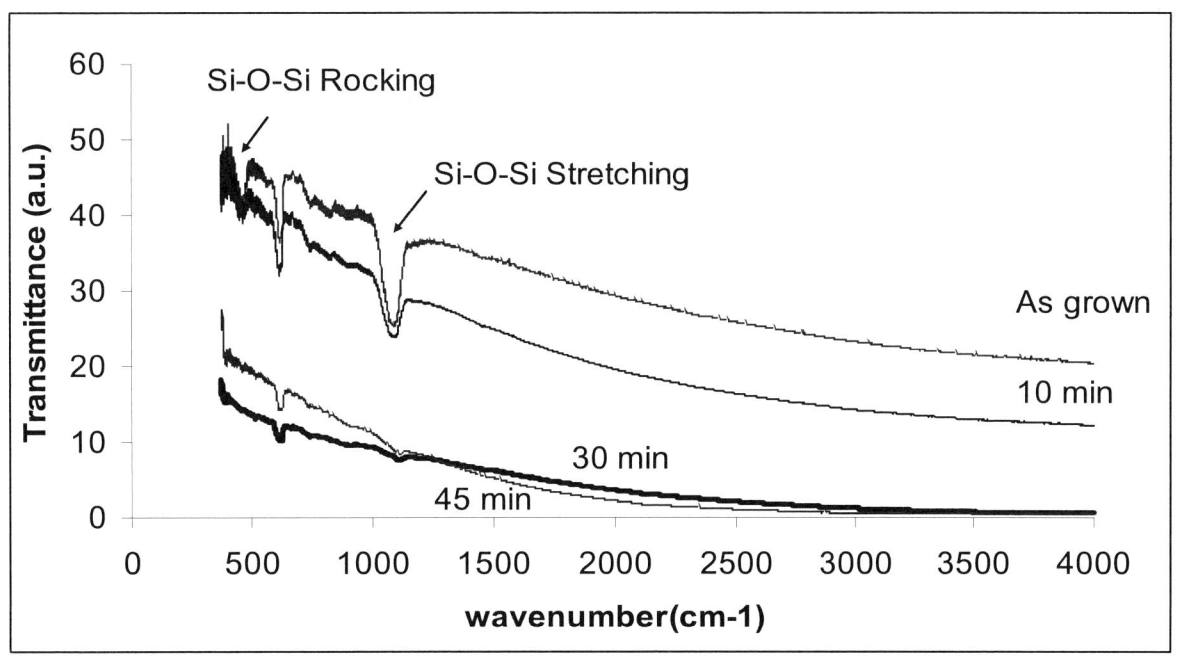

Fig. 4. FTIR spectra of porous SiO$_2$ film with various durations.

IV. CONCLUSION

Porous silicon dioxide (SiO$_2$) on silicon were prepared by photoelectrochemical wet etching. SEM images show that average pore size for sample was around 2.5 to 15 μm. Energy dispersive x-ray analysis (EDX) spectra has confirmed the presence of silicon and oxygen for all of the etched samples. As shown in FTIR spectra, the Si-O peak were observed.

ACKNOWLEDGEMENT

The authors would like to acknowledge financial support from IRPA RMK-8 Strategic Research and Universiti Sains Malaysia

REFERENCES

[1] M.H. Jo, H.H. Park, D.J. Kim, S.H. Hyun, S.Y. Choi, J.T Paik, J.Appl. Phys. 82, pp. 1299 (1997).

[2] S.S. Wang, H.L. Liu, L.Y. Zhang, X. Yao, J. Xi'an, Jiaotong University 29(9), pp57 (1995).

[3] S. C. Witczak, J. *S.* Suehle, and M. Gaitan. *Solid-State Electron.* vol 35, pp. 345 (1992).

[4] C.Kaneta and T. Yamasaki. *Micro electronic Engineering.* vol 48, pp. 117 (1999).

[5] T. Yamasaki, C. Kaneta, T. Uchiyama, T. U&, and K. Terakura. *Phys. Rev. B* vol 63, pp. 115314 (2001).

[6] A. Ourmazd, D. W. Taylor, J. A. Rentschler, and J. Bevk. *Phys. Rev. Lett.* vol 59, pp. 213 (1987) .

[7] M. T. Sieger, D. A. Luh, T. Miller, and T.-C. Chiang. *Phys. Rev. Lett.* vol 77, pp. 2758 (1996).

[8] P. Ii. Fuoss, L. J. Norton, S. Brennan, and A. Fiscber-Colbrie. *Phys. Rev. Lett.* vol 60, pp. 600 (1988).

Characteristics of Thermally Treated Contacts on Porous Silicon Based Metal–Semiconductor–Metal (MSM) Photodetector Structures

L. S. Chuah[1], C. W. Chin[2], Z. Hassan, H. Abu Hassan

School of Physics, Universiti Sains Malaysia, 11800 Minden, Penang, Malaysia
Email: [1]chuahleesiang@yahoo.com, [2]takeshi_ccw@yahoo.com, zai@usm.my
Tel: 604-6533673, Fax: 604-6579150

Abstract To date, little work has been done on porous silicon-based MSM photodetectors. Porous silicon (PS) was obtained on n-type silicon (111) using photoelectrochemical etching in HF. Microstructural investigation has been done by scanning electron microscope (SEM) and X-ray diffraction (XRD) measurements. We have found that the PS consists of a regular silicon microarray with triangular geometry. From the X-ray diffraction scan, PS shows a broadening of the full width at half maximum with respect to the as-grown epilayer. In this work, PS-based MSM photodetectors (photodiodes) with nickel (Ni) Schottky contacts were fabricated and characterized. The application of thermal treatment to the contacts at various annealing temperatures (500–700 °C) was investigated. Electrical characterization was performed by current–voltage (*I–V*) measurements. Morphological characterization was performed by atomic force microscopy (AFM) measurements.

I. INTRODUCTION

FOR light detection, various types of solid-state photodetectors have been reported, such as p–n junction, metal–semiconductor–metal (MSM) structure and Schottky diodes. Among these structures, MSM photodetectors have attracted much interest due to their fabrication simplicity. Photodetectors operating in the short wavelength ultraviolet (UV) region are important devices that can be used in various commercial and military applications. For example, these photodetectors can be used in space communications, ozone layer monitoring and flame detection. Until very recently, the primary means of UV light detection was the use of silicon photodiodes.

The possibility of producing optoelectronic devices has attracted a great deal of attention towards porous silicon (PS) since the discovery of its room temperature photoluminescence [1] and electroluminescence [2]. Although the research has been mainly focused on the photo- and electroluminescent properties, it was found a few years ago that PS can be efficiently employed in the development of photodetectors and solar cells [3].

Other important advantages of using porous silicon in optical detection devices are that PS-based interference filters can be developed to match the desired optical properties which avoid the use of extra anti-reflection coatings [4-5], and that there exists the possibility of employing the photoluminescent properties of PS to convert ultraviolet and blue light into longer wavelength light with better quantum efficiency in silicon solar cells and photodiodes.

In the case of the development of solar sensors and cells, other important advantage of using PS is that the bandgap of PS may be adjusted for optimum light absorption [6]. In the present work, porous Si-based MSM photodetectors (photodiodes) with nickel (Ni) Schottky contacts were fabricated and characterized. The application of thermal treatment to the contacts at various annealing temperatures (400–700 °C) was investigated

II. EXPERIMENT

The samples were prepared using silicon wafers, (111)-oriented, with n-doping. After standard cleaning steps, to prepare porous structures by photoelectrochemical (PEC) method, the samples were dipped into a mixture of HF:Ethanol (1:1) for 2 minutes with current densities of 50mA/cm^2 and subsequently well rinsed in de-ionized water. In order to achieve significant hole current in n-type Si, external illumination under an UV lamp with 500 W power is required. Typical photoelectrochemical (PEC) wet etching apparatus are schematically shown in Fig. 1.

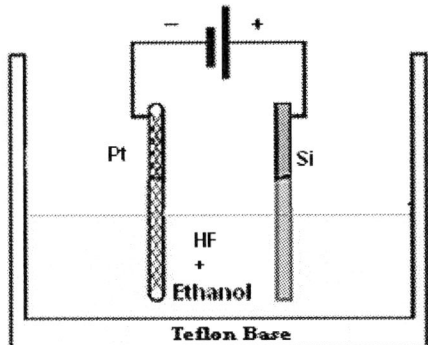

Fig.1. Schematic of photoelectrochemical (PEC) wet etching apparatus.

The anodization is performed in potentiostatic (voltage-controlled) mode. Structural properties of PS have been investigated by scanning electron microscope (SEM) and X-ray diffraction (XRD). PS-based MSM photodetectors (photodiodes) with nickel (Ni) Schottky contacts were fabricated and characterized. Our photodetectors are the metal–semiconductor–metal (MSM) photodiodes with both interdigitated contacts (electrodes) forming Schottky barriers.

The fingers width is 230 μm and the finger spacing is 400 μm. The length of each electrode is about 3.3 mm, and it consists of 4 fingers at each electrode. Each electrode has four fingers as shown in Fig. 2. The metal that was used for forming both interdigitated Schottky contact electrode was thermally evaporated.

The fabricated photodiodes were then annealed at temperatures from 400-700 °C in a conventional tube furnace in flowing nitrogen environment. For the samples (photodiodes) annealed at temperatures from 500 °C, the annealing duration was 15 min, while the 600 °C samples were annealed for 5 and 2 min for the 700 °C samples. The electrical properties of the photodiodes were analyzed by means of I–V characteristics of the devices.

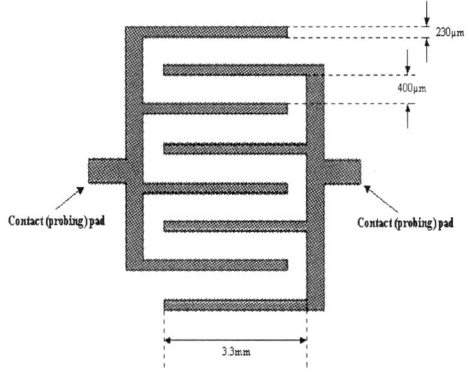

Fig. 2.

III. RESULT AND DISCUSSION

Fig. 3 shows the plan view and cross-sectional view of PS prepared on n-Si. As shown in these figures, the silicon layer exhibits triangular-like uniform microarray which appears to be denser on the outer surface. The thickness of the PS cross section is about 13μm. Triangular microvoids formed by the rigid triangular shaped microarray network of 300-400μm sizes is clearly seen in the plan view SEM image. The characteristic faceted, triangular shaped pores illustrate an orientation dependence of the photo-anodization process of the (111) Si material, probably due to the known anisotropy of Si dissolution.

(a)

(b)

Fig 3 SEM micrographs of porous silicon layers on n-Si (111) by PEC-etching (a) plan-view specimen (b) cross-sectional specimen.

Fig. 4 shows the XRD phase analysis scan of porous and as-deposited silicon. The intensity data was collected by performing ω (sample angle) - 2θ (detector angle) scan at a range of different values. The peaks at about 28.48° and 58.92° correspond to (111) and (222) diffraction peaks of Si. Porous samples exhibited broader FWHM than the as grown sample for (111) diffraction planes.

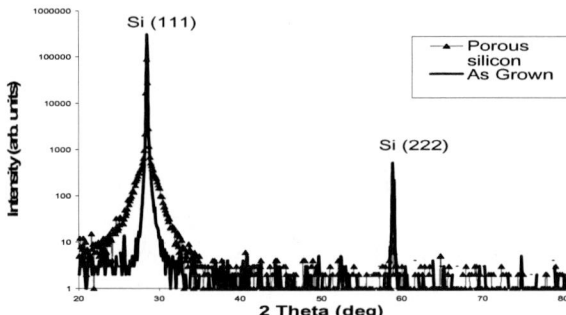

Fig. 4 XRD phase analysis scan of porous silicon and as-deposited silicon.

The Schottky contact properties of the MSM photodiodes can be closely described by the equation below [7-8]

$$I = I_0 \exp\left(\frac{eV}{nkT}\right)\left[1 - \exp\left(\frac{-eV}{kT}\right)\right], \qquad (1)$$

where I is the current, I_0 is the saturation current, V is the bias voltage, and n is the ideality factor. The expression for the saturation current, I_0 is

$$I_0 = SA^*T^2 \exp\left(\frac{-e\Phi_b}{kT}\right), \qquad (2)$$

where Φ_b is the barrier height, S is area of the Schottky contact and A^* is the effective Richardson coefficient. Equation (1) can be rewritten as

$$\frac{I\exp(eV/kT)}{\exp(eV/kT) - 1} = I_0 \exp(eV/nkT). \qquad (3)$$

At T ≤ 370K and when $V \le -0.5$V, equation (3) can be simplified to

$$I\exp\left(\frac{eV}{kT}\right) = I_0 \exp\left(\frac{eV}{nkT}\right), \qquad (4)$$

Here, the plot of ln [$I \exp (eV/kT)$] vs V will give a straight line with the slope = e/nkT and y-intercept at ln I_0.

By referring to Table 1, under light illumination, we found that for high temperature annealing (600 and 700°C), the ideality factor of the samples determined from equation (4) increased. We suspect the diffusion of Ni metal layer into the samples away from the porous silicon surface has resulted in a degraded metal-semiconductor contact. Since the samples were still hot at the time they were taken out of the furnace, a great deal of diffusion of the metal layer will still take place.

Fig. 5 shows the I-V characteristics of the sample annealed at 500°C under illumination with white light. From equation (4), the value of n is calculated to be 1.002. This value is quite near to unity, thus indicating the high quality of the Schottky contact under investigation and the absence of a thick interfacial layer. The current does not saturate when the reverse bias increases.

Very similar I-V curves of the MSM Schottky contacts can be found in the literature [9]. This behavior of the reverse Schottky contact is explained by the barrier height dependence on electric field in the depletion region, and hence, on the applied bias, and this is not taken into account in Eq. (1). It is known that metal-semiconductor barrier height decreases due to image force lowering. The barrier lowering depends on the applied voltage and modifies the shape of the I-V curve both in forward and in reverse directions. However, for reverse bias voltages, the barrier lowering is more substantial due to stronger electric field in the depleted region of the reverse-biased contact.

Table 1: Ideality factor of MSM structures based on porous samples annealed at different temperatures

Temperature (°C)	Duration, min	Ideality factor, n
500	15	1.002
600	5	1.101
700	2	1.115

There are also other mechanisms, which modify the shape of the I-V characteristics of the Schottky contact under high bias voltage, like in particular, tunneling effects and carrier recombination.

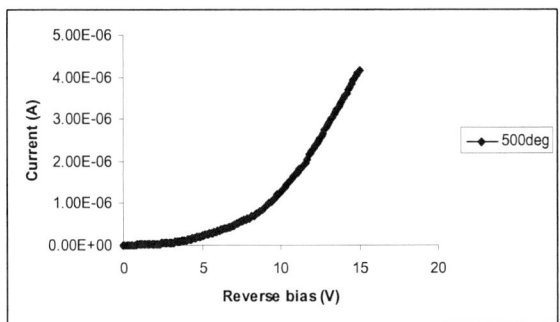

Fig. 5 Current-voltage (*I-V*) characteristics of the porous silicon annealed at 500°C.

Fig. 6 shows the atomic force microscopy image of the nickel contact on the MSM structures based on porous samples annealed at different temperatures. For the annealed samples, the smallest root mean square (rms) roughness value came from the 500°C (Fig. 6a) samples with a value of 8.52 nm, followed by the 600°C (Fig. 6b) and 700°C (Fig. 6c) samples with a value of 15.52 and 16.72 nm, respectively.

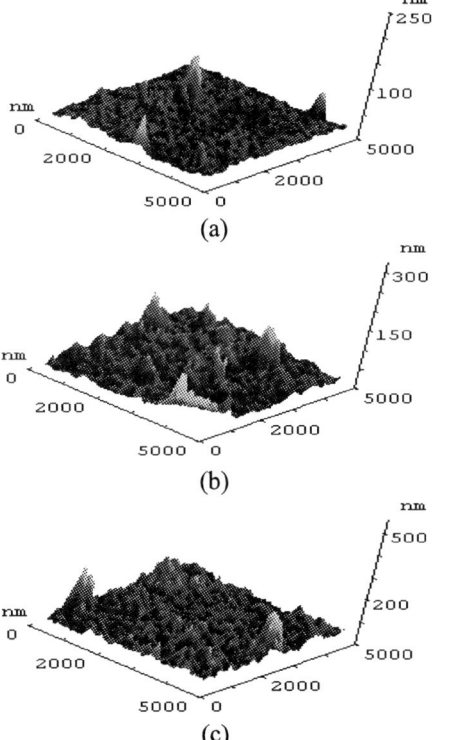

Fig. 6. Atomic force microscopy image of the nickel contact on the MSM structures based on porous samples annealed at different temperatures.(a) 500°C, (b) 600°C and (c) 700°C.

IV. CONCLUSION

A photoelectrochemical (PEC) wet etching method has been developed to produce porous silicon. SEM images show that the porous silicon consists of a regular Si microarray with triangular geometry. From the X-ray diffraction scan, porous silicon shows a broadening of the full width at half maximum with respect to the as-grown epilayer. The application of thermal annealing treatment to our Ni/porous silicon MSM photodiodes at various annealing temperatures (500–700°C) was investigated. High temperature annealing treatment leads to the degradation of the metal–semiconductor contacts of the photodiodes. Electrical and morphological characterization was performed by current–voltage (I–V) and atomic force microscopy (AFM) measurements, respectively.

ACKNOWLEDGEMENT

The authors would like to acknowledge financial support from IRPA RMK-8 Strategic Research and Universiti Sains Malaysia.

REFERENCES

[1] L. T. Canham, *Appl. Phys. Lett.* 57, pp. 1046 (1990).

[2] N. Koshida, H. Koyama, *Appl. Phys. Lett.* 60, pp. 347 (1992).

[3] J.P. Zheng, K.L. Jiao, W.P. Shen, W.A. Anderson, H.S. Kwok, *Appl. Phys. Lett.* 61, pp. 459 (1992).

[4] M.G. Berger, C. Dieker, M. Thonissen, L. Vescan, H. Luth, H. Munder, W. Theiß, M.Wernke, P. Grosse, J. *Phys. D Appl. Phys.* 27, pp. 1333 (1994).

[5] R.J. Martı́n-Palma, P. Herrero, R. Guerrero-Lemus, J.D. Moreno, J.M. Martı́nez-Duart, *J. Mat. Sci. Lett.* 17, pp. 845 (1998).

[6] Y.S. Tsuo, Y. Xiao, M.J. Heben, X. Wu, F.J. Pern, S.K. Deb, Proc. *23rd IEEE Photovoltaic Specialist Conference, IEEE* pp. 287, (1993).

[7] E.H. Rhoderick, R.H. Williams, *Metal–semiconductor Contacts*, Second ed., Oxford University Press, New York, 1998, pp. 39.

[8] V.L. Rideout, *Solid-State Electron.* 18, pp. 541 (1975).

[9] M. Ito and O. Wada, IEEE *J. Quantum Electron.* 22, pp. 1073 (1986).

Softbake and Post-exposure Bake Optimization for Process Window Improvement and Optical Proximity Effect Tuning

C.Y Liau, E.K Yet, C.H Lee, Ivy Tan, Christopher Loo, B.C Lee, Y.K Ng, W.B Sheu
X-FAB Sarawak Sdn. Bhd.
1 Silicon Drive
Sama Jaya Free Industrial zone
93350 Kuching
Sarawak, Malaysia
Email: chuyaw.liau@xfab.com (Phone: +60.82.354645)

Abstract We have shown that process effects induced by extending the softbake (SB) and post-exposure bake (PEB) temperature in the process flow of chemically amplified photoresists can lead to significant improvements in Depth-of-Focus (DOF) and Exposure latitude (EL) and small geometry printing capability (resolution). Through careful optimization of SB and PEB temperature, dense line and space structures of 160nm and below can be printed with substantially big process margin, using binary masks and 248nm lithography under the half annular illumination mode. Besides, we have also shown that the optical proximity effect, namely the non-linearity, proximity bias and line-end shortening in specific is tunable by changing the SB and PEB temperatures. The main objective of this study is to demonstrate how, using 248nm lithography with binary masks and with a moderate Resolution Enhancement Technique (RET); the process latitude can be improved besides minimizing the impact from optical proximity effect.

I. INTRODUCTION

Photolithography remains the driving and enabling technology in the semiconductor industry to pattern integrated circuits with ever decreasing feature sizes. In optical projection lithography, the resolution, R of a lens at the diffraction limit is given by the expression

$$R = k_1(\lambda/NA), \quad \text{---------------} \quad (1)$$

Where λ and NA are wavelength and numerical aperture of the exposure tool respectively, and k_1 is

an empirical constant. The NA is the sine of half of the angle of the image-forming cone of light at the image. In practice, k_1 depends on lens aberrations, illumination conditions (degree of coherence and intensity distribution in the aperture plane), mask (e.g., whether phase shifted masks are used), geometrical shapes (or spatial frequencies), exposure tool conditions, resists and process. Resolution can be improved by 3 ways: by shortening the exposure wavelength, by increasing the numerical aperture, and by decreasing the value of k_1. Today, tool vendors and process developers are pushing k_1 to less than 0.5. The smaller the k_1, the narrower the process windows are.

On the other hand, the depth of focus (DOF) is given by the expression

$$DOF = k_2(\lambda/NA^2), \quad \text{---------------} \quad (2)$$

Where k_2 is also and empirically determined constant. Due to the inverse square dependency on NA, the depth of focus is extremely shallow. For this reason, high-NA lenses are always associated with stringent requirements on planarization techniques for resists (top surface imaging or multiplayer resists) and processes. Eliminating the NA between Equations (1) and (2), we obtain

$$DOF = (k_2/k_1^2)(R^2/\lambda), \quad \text{---------------} \quad (3)$$

This equation shows explicitly that at the same NA and same lens resolution, a lower k_1 gives a larger DOF. This is the motivation for exploring lower k_1 under the existing exposure wavelength. In low k_1, high-spatial frequency components of the mask spectrum carry a sizeable fraction of the transmitted light energy. These components are not captured by low-pass pupil. As a result, images are distorted from original patterns because of the loss of these high-frequency components. There are primarily 4 phenomena of image shape distortion, namely, non-linearity, proximity effect, line shortening and corner rounding [1].

Non-linearity refers to changes of the nominal CD not reflected linearly in the printed image. For large dimensions, a unit change in nominal CD results in a unit change in the printed CD, but non-linearity becomes increasingly an issue when the CD decreases. In addition to increased line width variation, non-linearity can also lead to nonprinting of critical features. Proximity effect refers to features with the same nominal CD printing differently because of environmental variations. As the pitch size increases, separation between adjacent lines widens; the imaged line width varies as spacing changes. This type of distortion results in increased across-chip line width variation (ACLV).

Line shortening is the phenomenon whereby the printed length of the rectangle is less than the nominal length when the width prints on target. Line shortening results primarily from diffraction, mask pattern rounding, and the diffusion of chemical species in photoresist. Optical diffraction is a major contributor at low k_1 imaging. As the CD decreases, line shortening increases dramatically. This behavior is of critical concern because of its impact on overlay budget and circuit density. The fourth type of image distortion is corner rounding. The high-frequency components of a sharp corner are filtered out by the pupil, resulting in a rounded image.

Today, most fabrication facilities use chemically amplified photoresists, complex and highly tuned formulations of a polymer film loaded with photoacid generators (PAGs), and other additives. With exposure of the photoresist film through a mask, the PAG releases acidic protons. A post-exposure bake is then applied and the acid protons diffuse and catalyze a deprotection reaction on the polymer that alters its solubility in an aqueous base developer solution. [2,3] While the semiconductor manufacturing industry is focusing on good control in cost and manufacturability in the midst of downscaling of designed line width and pitches, the cost of reticle and exposure tool type has become increasingly significant. Hence, it is inevitable that intensive evaluations are being done to develop reliable and

robust process that does not need to go for lower exposure wavelength and phase shifted mask while shrinking the optical feature size.

In this paper, we report on the significant process improvements that were achieved by tuning the softbake (SB) and post-exposure bake (PEB) temperature in the process flow of a low Ea chemically amplified photoresists. It was noticed that the lithographic process latitudes, for both the DOF and exposure latitude (EL) would have significant toggle with SB and PEB temperature. Meanwhile, it was also noticed that by changing these temperatures, the optical proximity effect (OPE) was found to be tunable.

II. EXPERIMENTAL SETUP AND METROLOGY METHODOLOGIES

The imaging experiments discussed in this paper were performed using a Nikon NSR-S205C 248nm scanner. The 248nm photoresist used in all experiments was PEK112A5 (550nm) from Shin Etsu. ARC applied was BARC DUV42P-11 from Brewer Science of 130nm thickness and the wafer substrate used was on 120nm/75nm Oxide/Nitride stack. Exposures were performed using a binary mask (BIM), which has various test patterns. Wafers were processed based on the conditions shown in **Table 1**.

Table 1: Process experiments carried out.

Scanner Illumination	Soft Bake (SB)	Post Exposure Bake (PEB)
NA 0.75 ½ Annular σ 0.68	90°C 90s	120°C 90s
NA 0.75 ½ Annular σ 0.68	110°C 90s	100°C 90s
NA 0.75 ½ Annular σ 0.68	110°C 90s	120°C 90s
NA 0.75 ½ Annular σ 0.68	110°C 90s	130°C 90s

Both the BARC DUV42P_11 (130nm after softbake of 240°C 60s) and photoresist, PEK112A5 (550nm after softbake of 90°C 90s and

110°C 90s respectively) were coated on a TEL Clean Track Act 8. The hard bake condition used was 110°C for 90s. CD measurements were performed on Hitachi CDSEM S-9260A (Settings: I_p 5.6pA; Vac = 600V; Mag = 200k; measurement algorithm: linear, threshold 50%, baseline start point 2).

For process window calculation, focus-exposure matrix CD data were introduced into in-house process window calculation software. Process windows were calculated and evaluated using a +/- 10% Spec on the CD target. Analysis of the process windows was done by fitting ellipses into the windows (assuming only random errors) and by making use of generic files. Top SEM image profile was also inspected; hence bridging and poor resist profiles were filtered out although the CD was in spec.

Patterns with acceptable profile and no bridging will be assigned as "1" in the generic file while those with bridging or poor resist profiles will be assigned as "0" in the generic files. Putting the generic window and the FEM process window on top of each other would then result in an overall process window that tells one which process latitudes can be tolerated if one wishes to print an acceptable pattern.

The optical proximity effect (OPE) can be analyzed by measuring the test patterns as follows:-

a. Non-linearity, CD measurements were done on iso patterns (line or space) with varying nominal CD.
b. Proximity bias, CD measurements were carried out on patterns with a fixed nominal CD but varying in pitch sizes
c. Line end shortening, CD measurements were done on the head to head spacing (iso or dense structures) with different nominal space CD and nominal line CD.

III.   RESULTS AND DISCUSSION

Direct comparison was made between the various test conditions. This section, summarizing the most important test results of this comparison, is divided into two subsections. In the first part (subsection III.I), the focus and exposure latitude would be discussed. The second part of this section

(subsection III.II) deals with the line width and line length control aspects.

III.I. Focus and Exposure Latitude

Fig. 1 and 2 summarize the process performance for both soft bake (SB) and post-exposure bake (PEB) temperature on PEK112A5 resist.

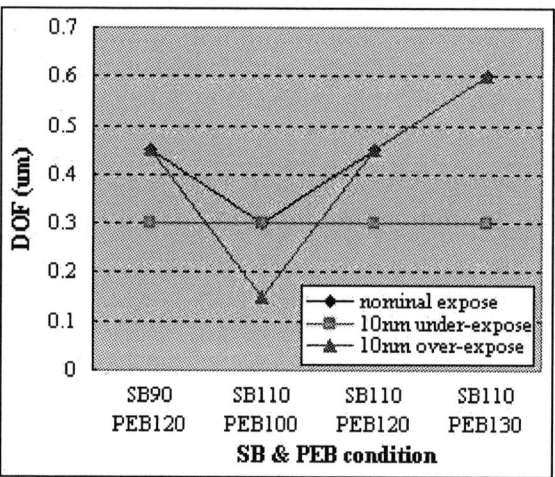

Fig. 1: DOF Vs SB and PEB temperature conditions

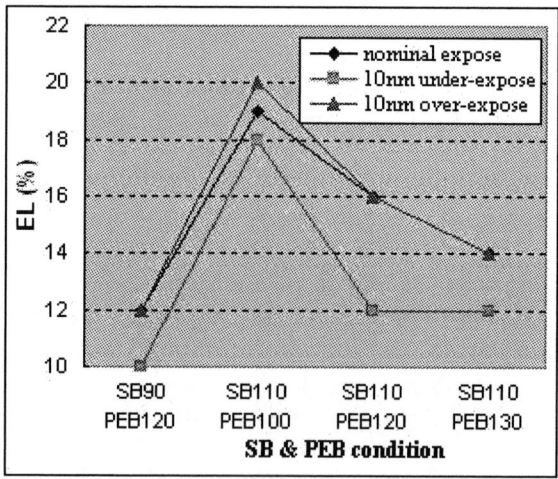

Fig. 2: EL Vs SB and PEB temperature conditions

In general, improvement observed in Depth of Focus (DOF) with PEB temperature is in line with what has been reported earlier by C.Y Liau et. al for patterning iso and semi-dense space structures [4]. SB temperature change however has no or negligible impact on DOF as evident in Fig 1.

Fig 2 shows that the exposure latitude (EL) increases with SB temperature. This can be explained by densification of resist film. When the soft bake temperature increases, more solvent content is evaporated from the film; hence the proton diffusion length is cut short in the densely packed film under the same PEB temperature condition. The smaller proton diffusion length results in lower sensitivity and hence increased EL. It was however found out that the EL degraded with PEB temperature. As PEB temperature increases, the proton diffusion length and diffusion rate increase thus larger area will be acidified that results in increased resist sensitivity. The improved sensitivity of the photo resist will translate into poorer EL.

The summary of the process latitude with mask bias is shown in Fig 3. It is evident from the graph that both EL and DOF increase dramatically when going from 10nm under-exposure to 10nm over-exposure. No significant improvement can be observed between nominal exposure and +10nm over-exposure strategy. The reduced DOF can be attributed to the fast-diminished contrast at the far defocus region as a result of lower light intensity applied under the under-exposure condition passing through the mask opening. The CD vs. Dose slope increases under the under-expose condition results in increased resist sensitivity and hence degrading the EL. Besides, it was also observed that the resolution improves as it goes from under-expose technique to over-expose technique judging from the fact that bigger line CD (and hence smaller space CD) can be defined without causing pattern bridging.

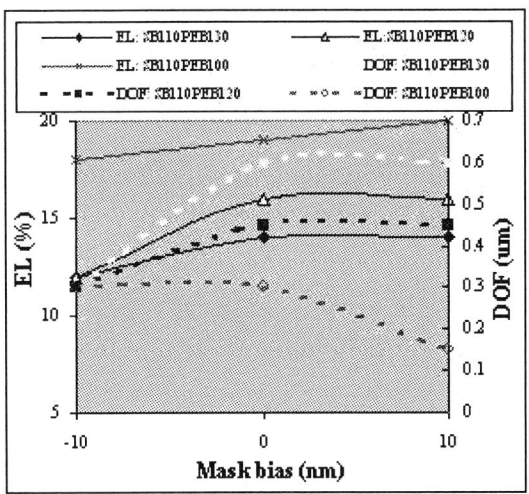

Fig. 3 : Process Latitude Vs Mask Biasing

III.II. Line width and Line Length Control

III.II.a. Non-Linearity

Fig. 4 shows measurement of isolated line with various nominal line widths. Generally SB and PEB temperature change has no significant impact to the non-linearity. The temperature split conditions fit well to the linearity curve for nominal line widths in 150 to 300nm range.

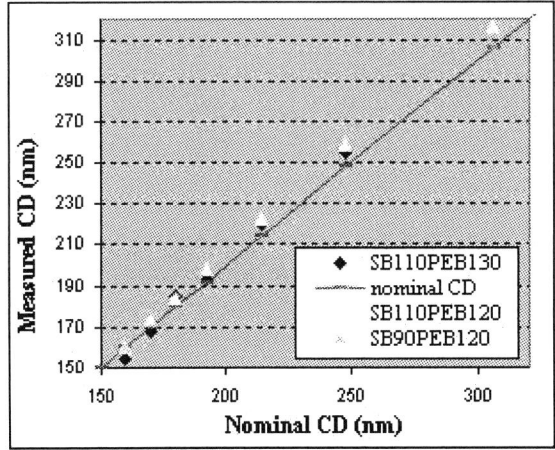

Fig. 4 : Linearity plot of isolated line under different process conditions

The iso space measurement results however show a much bigger toggle as shown in Fig 5.

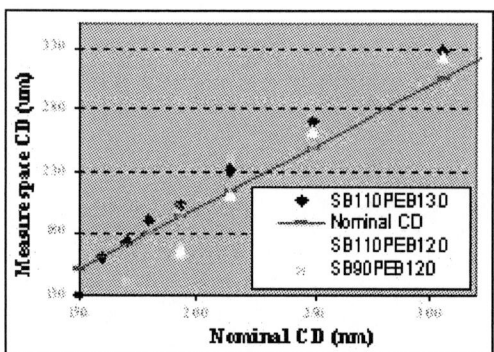

Fig.5: Linearity plot of isolated space under different process conditions

Pattern bridging is found for nominal space width < 170nm for SB90PEB120 condition. Surprisingly, for SB110PEB120 process condition, only nominal space of > 190nm can be defined properly. This also implies k_1 factor increases as soft bake temperature increase due to deteriorating resolution. One of the probable explanations for this phenomenon is again attributed to higher film densification when soft bake temperature increases which retards deprotection rate of the blocking units by photo generated acid.

The resolution limit can be improved by increasing the PEB temperature as evident from the graph that shows that even with the nominal space of 150nm can now be patterned though with a smaller CD value. The improvement has been explained which is due longer diffusion length and higher diffusion rate which enhance the chemical amplification to convert the resin in the resist to acid at this low exposure area of the aerial image [4]. Besides, it can also be deduced that the resolution improvement obtained by increasing 10°C of PEB temperature far surpasses the effort of decreasing 20°C of soft bake temperature. However, more detailed studies need to be carried out to identify for other drawbacks under the raised PEB temperature condition.

III.II.b. Proximity Bias

It is well known that the printed dimension of line will depend on the proximity of the line to other structures. Fig 6 shows measurement of isolated lines under the various process conditions, at 160nm, with adjacent spaces ranging from 1:1 to 1:7 times of the nominal line width. It can be seen

that the deviation from nominal line width is most serious using a lower soft bake temperature. The SB110PEB130 again shows that it is more superior in resisting the OPE though a substantial optical proximity correction (OPC) is still needed.

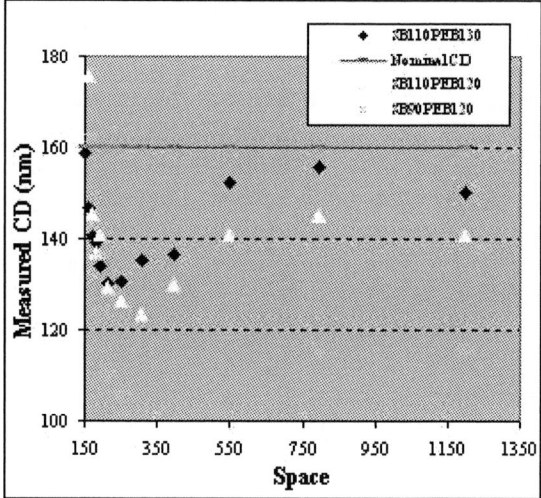

Fig.6: Proximity bias of 160nm isolated line under different process conditions

Same observation can be obtained from impact of bake condition splits to the proximity effect on dense pattern. Fig. 7 shows the measurement results of dense lines at 160nm with pitch size ranges from 320 to 400nm. Data is consistent with that of iso line.

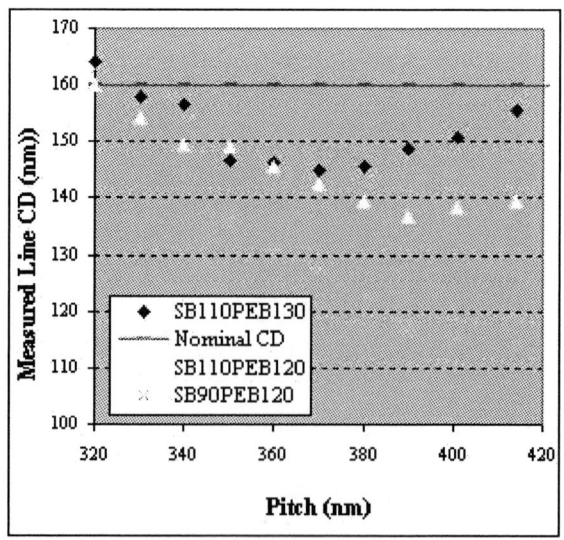

Fig.7: Proximity bias of 160nm dense lines under different process conditions

III.II.c. line end shortening

A continuing concern is line-end shortening (LES) since the required overlap of gate over active area of a design rule does not scale from previous process generations, as do other rules. Fig. 8 shows that the line-end shortening is serious for PEK112A5 resist under various process conditions. It is worthwhile noting that the line-end pullback changes linearly with head to head spacing. This response is essential, as it makes possible for line end shortening correction to be done accurately. However, under the soft bake temperature of 90°C, the line end bias is found to be double that of elevated soft bake temperature of 110°C. This bias reduces by half when it comes to dense head to head structures (Fig. 9). It is interesting to note that the pullback does not change significantly for 110°C SB condition when the environment changes from iso to dense structures. This again implies that the increased soft bake temperature has helped to resist line end shortening more effectively under the iso environment. Though hammer head or serifs can be placed at the line end structures to correct for the pullback, an overcompensated serifs or hammerhead (especially for dense line and space structures) may result in line bridging and thus lowering down the process window of certain critical photo layers. Hence, it is important to minimize the pullback as much as possible.

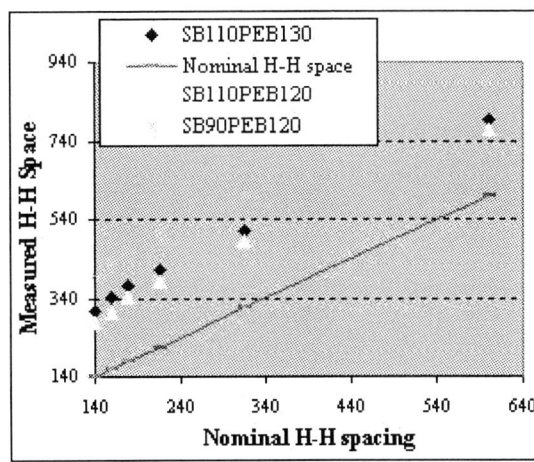

Fig. 9: Head-to-head spacing of 160nm printed dense line with drawn space ranges from 140nm to 640nm

Fig. 10 and 11 show the head to head line-end pullback at a range of isolated line dimensions for H-H spacing of 160 and 600nm respectively. As can be noted the bias escalates as when the nominal line width gets smaller and starts to narrow down as the lines gets wider. The line-end pullback from SB 90°C is so serious that it may render it hard to be compensated. Smaller pullback seen from soft bake temperature of 110°C is again attributed to film densification, which shortens the diffusion length of the photo-generated acid. Meanwhile, increasing PEB temperature though increases the pullback, is not as severe as seen in the lowered soft bake temperature condition.

The line-end pullback bias increases slightly as the H-H spacing increases from 180 to 600nm for SB90PEB120 and SB110PEB120 conditions. Surprisingly, the line-end pullback bias appears to be independent of the H-H spacing under fixed nominal line width when increasing PEB temperature from 120°C to 130°C as evident from Fig 10 and 11.

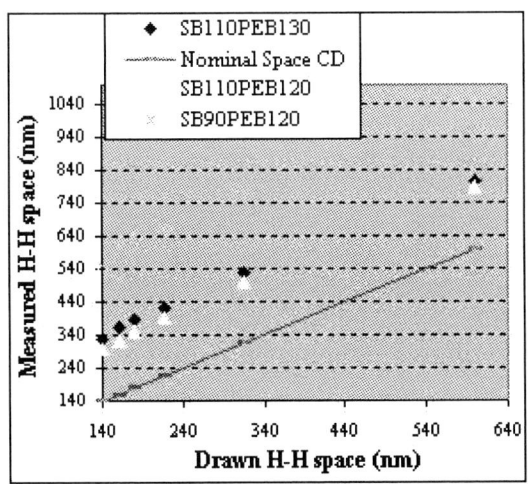

Fig. 8: Head-to-head spacing of 160nm printed isolated line with drawn space ranges from 140nm to 640nm

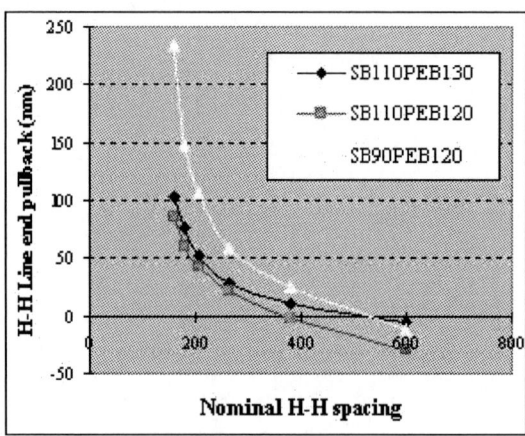

Fig.10: Line end pullback at a range of printed isolated line CD when head-to-head space is 180nm.

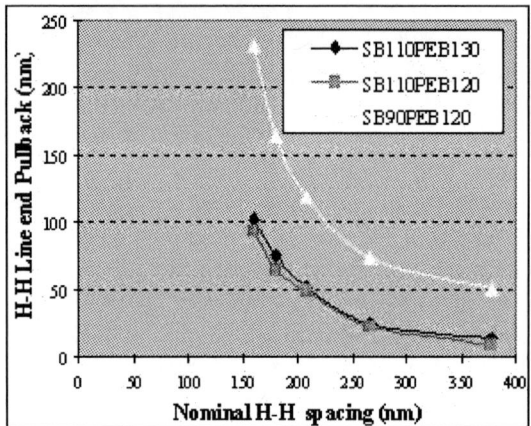

Fig.11: Line end pullback at a range of printed isolated line CD when head-to-head space is 600nm.

IV. CONCLUSION

DOF can be improved by increasing PEB temperature however with slight deterioration in EL for line structures. Increasing the SB temperature through film densification to shorten the diffusion length of the acid protons can offset the deterioration in EL. Process window is found to be inferior when choosing an under-expose process on dense structures patterning for a fixed printed CD. No significant gain is found when going from nominal exposure method to over-exposure method. Increasing PEB temperature is also found to have improved the resolution limit on the other hand, a slight deterioration is found with an increased in SB temperature. It is however found out that the improvement in resolution obtained through a unit PEB temperature increase far surpasses that of a unit decrease in SB temperature.

The change in SB and PEB temperature has no significant toggle in non-linearity. Proximity bias can be improved by increasing the softbake temperature. Use of low SB temperature is most effective in controlling the line-end shortening and head to head line end pullback. This contribution is believed to be mainly attributable to densification of film density, which limits the diffusion length of proton acids.

V. REFERENCES

1. M. Born and E. Wolf, Principle of Optics, 6th Edition, Pergamon Press, Oxford, 400~502 (1983).
2. C.P. Ausschnitt, A.C. Thomas, T.J. Wiltshire, IBM J. RES & DEV., 41(1/2), 21 ~ 37 (1997).
3. W.D. Hinsberg, F.A. Houle, M.I. Sanchez, G.M. Wallraff, IBM J. RES & DEV., 45(5), 667 ~ 682 (2001).
4. C.Y Liau, C.H Lee, J. T Kang, S.W Yoon, C Loo, B Seow, W.B Sheu, Semicond. Sci. Technol. 20, 693 ~ 698 (2005).

ACKNOWLEDGEMENT

The authors wish to thank 1st Silicon (Malaysia) Sdn. Bhd. for the support and funding of this study.

FPGA Implementation of an Optimized Coefficients Pulse Shaping FIR Filters

Mohamed Almahdi Eshtawie, *IEEE student Member*   Masuri Othman

Institute of Microengineering and Nanoelectronics
Universiti Kebangsaan Malaysia
43600 Bangi, Selangor, MALAYSIA
E-mail: eshtawie@vlsi.ukm.eng.my

Abstract - **This paper presents the design and FPGA implementation for different order pulse shaping finite impulse response (FIR) filters. In this paper, the coefficients of the implemented filters have been modified with an optimization algorithm proposed in an earlier work. The use of this algorithm results in reducing the number of non-zero coefficients used to represent the filter's frequency response. Reducing the number of non-zero coefficients optimizes the implementation process especially when dealing with high order filters and when using lookup table (LUT) based techniques such as distributed arithmetic (DA). The designs have been downloaded to Xilinx Virtex-II FPGA and encouraging results were obtained. Hence, high-speed multiplierless design with a minimized number of arithmetic operations for different order pulse shaping FIR filters is achieved.**

I. INTRODUCTION

DIGITAL signal processing chips are the most common approach to implement digital filtering algorithms. On the other hand, application specific integrated circuit (ASIC) are the approach for higher rates. However, recently due to advances in their technology, field programmable gate arrays (FPGAs) have been applied to a variety of applications traditionally reserved for ASICs. They are well suited to datapath designs such as those encountered in digital filtering. FPGA approach in digital filter implementation has the advantage of higher sampling rate than are

available from traditional DSP chips, and lower cost than an ASIC for moderate volume applications. The rest of the paper is organized as follows: Section II gives a brief background on FIR filter design, DA and some related previous work. Section III discusses the design and methodology. The result is presented in section IV and the conclusion is given in section V.

II. BACKGROUND

Digital filters are typically used to modify attributes of signal in the time and frequency domain through the process called linear convolution [1]. This process is formally described by the following formula [2].

$$y[n] = x[n] * f[n] = \sum_k x[k] \cdot f[n-k]$$
$$= \sum_k x[k] \cdot c[k] \qquad (1)$$

Where $c[k]$ are filter's coefficients. When the filter coefficients do not change over time, then digital filters are generally classified as being FIR or IIR filters. The output of an FIR filter of order or length N to an input time sample $X[n]$ is given by a finite version of convolution sum:

$$y[n] = \sum_{k=0}^{N} x[k] \cdot c[k] \qquad (2)$$

Equation (2) implies that the FIR filter is realized by a large number of adders, multipliers and delay elements. This hardware

0-7803-9730-4/06/$25.00 ©2006 IEEE

requirement however imposes restrictions over their very large scale integration implementation (VLSI) especially when the filter order is high and when using conventional arithmetic algorithms to perform the linear convolution process. Therefore, an efficient hardware implementation of filter's structure is possible by optimizing the implemented adders and multipliers.

Due to the enormous occupied area of FIR filters with a large number of taps, hardware-reusing architectures such as time-multiplexing architectures as in [3] and [4] and a distributed arithmetic (DA) approach based on bit-serial access [5], where arithmetic operation is not lumped in familiar fashion such as a multiplier but is distributed, have been widely adopted for implementation [6]. Using multiplierless algorithms such as distributed arithmetic (DA) technique [5] results in a completely different filter architecture. In [7], the DA technique with its basic architecture has been used in designing and implementing high speed raised cosine FIR filter. In fact, the architecture presented in [7] suffers from the drawback of the exponential growth of the DA lookup table (LUT) size with the filter order. In [8], this defect is eliminated by proposing a new architecture for the LUT so that its size is not dependent on the filter order anymore. With the new LUT architecture, we have been able to design different order FIR filters without pileup with its size. Additional improvement is added to the architecture proposed in [8] when we proposed an algorithm for optimizing (reducing) the number of non-zero coefficients used to represent the phase and frequency response of an FIR pulse shaping filter. When reducing the number of non-zero coefficients, the VLSI implementation of different high order pulse shaping filter became more practical and easy task. In contrast, with the original filter coefficients representation, the VLSI implementation is not applicable when using the basic DA architecture [5]. In multiplierless algorithm such as DA the only arithmetic operation implemented to get the filter output is the addition. Therefore, the reduction process is pointed to the number of adders and addition operation used in the design. As a consequence, the overall speed,

area, and power consumption of the designed system will also be optimized.

III. DESIGN AND METHODOLOGY

A. Distributed Arithmetic

The distributed arithmetic architecture used in this work is based on the online structure proposed in [8]. Fig. 1 shows this architecture.

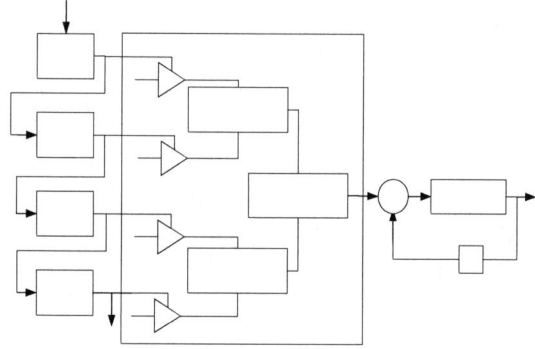

Fig. 1. The online LUT architecture.

B. Pulse Shaping Filters

In digital communication, pulse-shaping filters allow the transmission of pulses with negligible intersymbol interference (ISI). Therefore, these filters must have a frequency response with sufficient selectivity and attenuation to suppress noise and interference in adjacent channels. In particular, the 3GPP standard [9] proposes $1/T_c = 3.84e - 6s^{-1}$ and α=0.22. In this case, the bandwidth occupied by the modulated raised cosine with these parameters is $1.22 \times 3.84 = 4.68MHz$ which is less than $5MHz$ separation between adjacent channels of W-CDMA. Therefore the value of α used in designing the raised cosine pulse shaping filter in this paper is 0.22.

Designing an ASIC or FPGA based digital filter has main steps starting with developing the filter performance specifications and choosing the architecture for implementation, passing through some other steps, and ending up with synthesis and prototyping of the designed filter. Fig. 2 summarizes the design flow of digital filters.

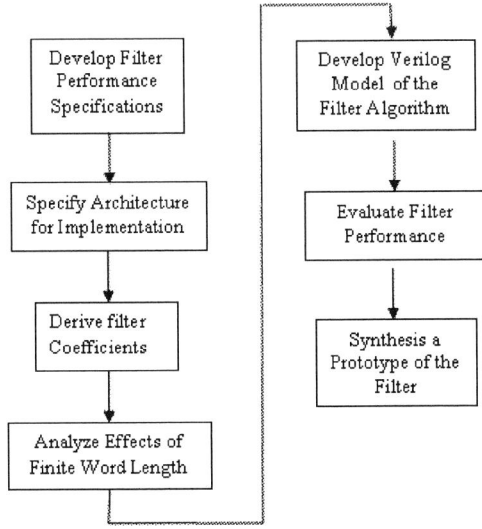

Fig. 2 Design flow for digital filter

C. Optimization Algorithm

The algorithm used in designing the different order pulse shaping filter is based on minimizing the number of non-zero coefficients used to represent the filter's phase and frequency response, so that the need for arithmetic operations i.e. addition operation is minimized. The key point when applying this algorithm is to have a steady state performance in the passband region while maintaining almost the same transition band in the filter frequency response.

D. System Verification

As a real time application, each filter is examined with a stream of input data. The expected output of each of the designed filter is achieved with this verification step. Fig. 3 shows the general block diagram proposed for performing 8-tap FIR filter operation verification step. In this structure, the ROM memory is used to store the input data. This memory is accessed with the address value outputted from its previous module (module mid). Based on the address obtained a specific value (location content) is readout of memory. The ROM output value is sent to the shift registers unit and is loaded in parallel to the first register. At each clock pulse, the output of the shift registers unit is the active address or location of the online LUT by which the value of the filter output is calculated.

IV. RESULTS

This section describes the results obtained with the methodology mentioned in the previous section. The online architecture enabled us to design high order raised cosine pulse shaping filter without facing the problem of constructing a very large size memory. For example when designing 100-tap FIR filter with the DA basic architecture, a LUT of the size 2^{100} is to be precalculated and constructed. The problem will be worst if we go for higher orders such as 256, and 512-tap FIR filters.

The optimization algorithm is applied to different order FIR filters. The simulation result of the algorithm when applied to 512-tap raised cosine FIR filter is shown in Fig. 4.

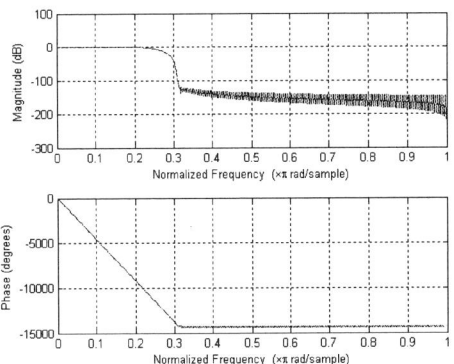

Fig. 4a Phase and frequency response of the filter with its original coefficients

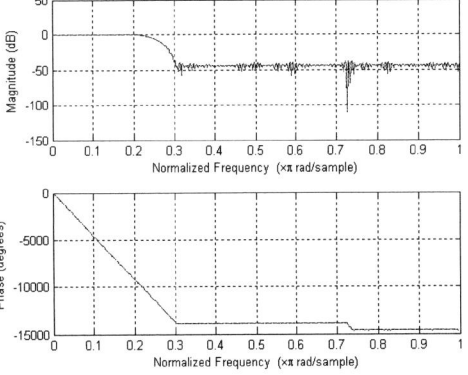

Fig. 4b Phase and frequency response when applying the optimization algorithm

Table I (column 2) shows the total number of coefficients used to represent the filter frequency and phase response after applying the optimization algorithm, whereas column 3 shows the number of non-zero coefficients.

Here the maximum number of the carry lookahead adders needed for each design is equal to: (non zero coefficients – 1)

Table I

Filter order	Total number of coefficients	Number of non zero coefficients
16	12	8
32	20	12
64	20	12
128	24	14
256	36	16
512	40	22
1024	40	28

The Verilog hardware description language (HDL) is used to design the different order pulse shaping filters i.e. 512, 256, 128, 64, 32, 16 and 8-tap filters. The Verilog codes have been synthesized with the Xilinx Synthesis Technology (XST). The results show that, when using the online LUT architecture, the time taken to synthesize the Verilog code for the highest designed order filter i.e. 512-tap is few minutes whereas, the time taken to synthesize the code for 14-tap FIR filter with a precalculated LUT (basic DA architecture) is several hours. Table II shows some of the results given in the Xilinx synthesis technology report.

Table II Parameters Obtained synthesis report

Filter order	Maximum frequency	Minimum period
8-tap	155.473 MHz	6.436 ns
16-tap	146.800 MHz	6.812 ns
32-tap	113.636 MHz	8.800 ns
64-tap	110.400 MHz	9.058 ns
256-tap	99.389 MHz	10.062 ns
512-tap	99.305 MHz	10.055 ns

Table III presents some more results obtained when implementing the design. The Verilog code is simulated using ModelSim XE II/starter 5.7g. Fig. 5 illustrates the output waveform obtained for 128-tap FIR filter. Finally, all designed filters have been successfully downloaded to Xilinx Virtex-II FPGA FG456 and the results obtained for each of them is the expected output.

V. CONCLUSION

This paper presented a successful FPGA implementation of different order pulse shaping FIR filters. The coefficients of the designed filters have been manipulated with an optimization algorithm so that their frequency and phase response is represented with a minimum number of non zero coefficients. In addition, the online DA LUT proposed in a previous work is also implemented and show to give encouraging results.

REFERENCES

[1] U. Meyer-Baese, "Digital Signal Processing With Field Programmable Gate Arrays", Second Edition, Springer Verlag, Berlin,2004,

[2] Mariusz Rawski, Pawel Tomaszewicz, Henry Selvaraj, Tadeusz Luba, "Efficient Implementation of Digital Filters with Use of Advanced Synthesis Methods Targeted FPGA Architecture", Proc. IEEE 8th Euromicro conference on digital system design (DSD'05),

[3] J. R. Choi, L.H. Jang, S. W. Jung, and J. H Choi, "Structured design of a 288-tap FIR filter by optimized partial product tree compression," IEEE J. Solid-State Circuits, vol. 32, pp. 468-476, Mar. 1997.

[4] C. J. Nicol, P. Larsson, K. Azadet, and J. H. O'Neill, "A low-power 128-tap digital adaptive equalizer for broadband modems," IEEE J. Solid-State Circuits, vol. 32 pp. 1777-1789, Nov. 1997.

[5] Stanley A. White, "Applications of distributed arithmetic to digital signal processing: A tutorial review," IEEE ASSP magazine July, 1989.

[6] Kyung-Saeng Kim and Kwyro Lee, "Low-power and area efficient FIR filter implementation suitable for multiple tape," IEEE Trans. On VLSI systems, vol. 11, No. 1, Feb. 2003.

[7] Mohamed A. Eshtawie and Masuri Othman," Designing of a high speed raised cosine FIR filter Using a distributed Arithmetic technique," proceeding of Int. Conf. on Computer and Communication Engineering 2006, vol. 2 pp 1025-1029, 9-11 May 2006 Kuala Lumpur, Malaysia.

[8] Mohamed A. Eshtawie and Masuri Othman," On-Line DA-LUT Architecture for High-Speed High-Order Digital FIR Filters," paper status is published in the tenth IEEE international conference on communication systems (IEEE ICCS 2006), 30-1 Nov. 2006 Singapore.

[9] 3rd Generation Partnership Project, Technical Specification Group Radio Access Networks. TS 25.101 V3.6.0(2001-03).

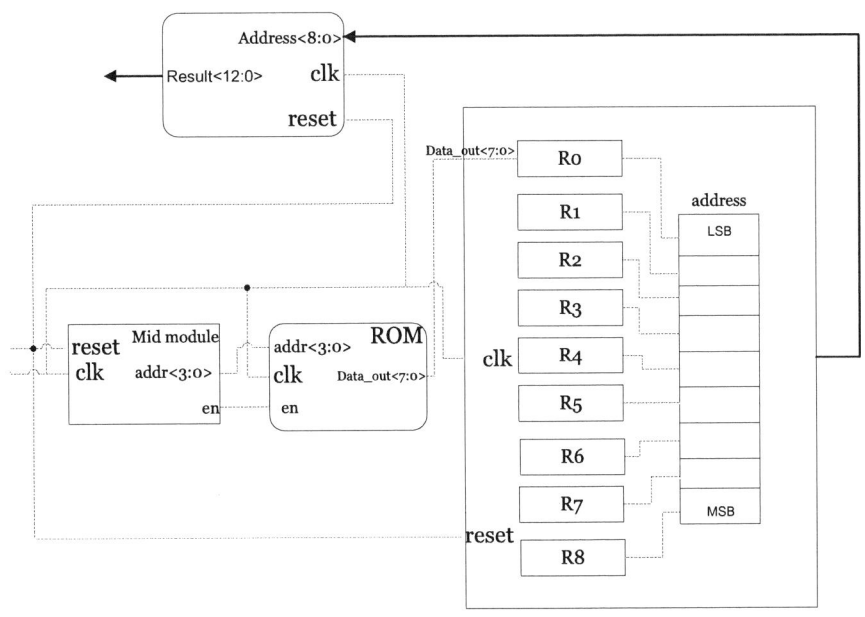

Fig. 3 General Block diagram of 8-tap raised cosine pulse shaping filter

Table III Parameters Obtained when Implementing the Design

Filter order	No of Slice Flip Flop	Total memory usage	Equivalent Gate Count for Design	Number of occupied Slices
8-tap	157	74412 KB	67812	127 (8%)
16-tap	225	75436 KB	68527	176 (11%)
32-tap	357	78508 KB	69634	250 (16%)
64-tap	425	83180 KB	70685	323 (21%)
128-tap	662	92908 KB	72875	470 (30%)
256-tap	807	107180 KB	73996	548 (35%)
512-tap	1678	143404 KB	80604	952 (54%)

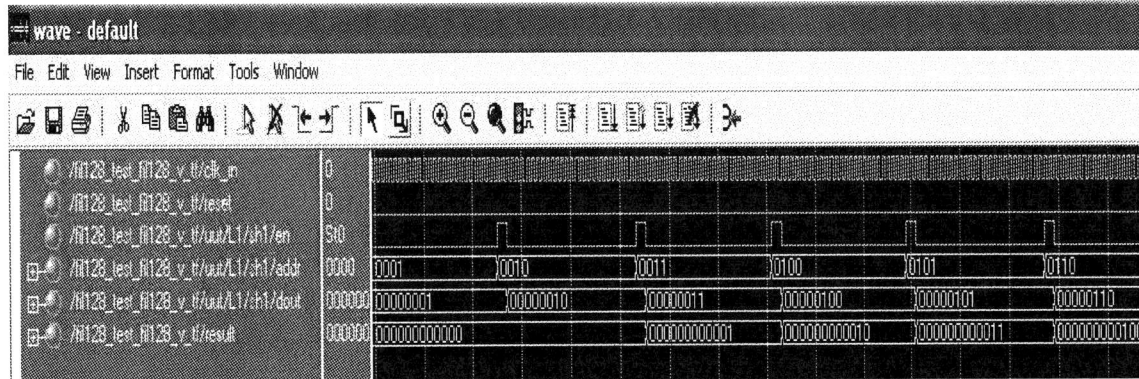

Fig. 5 The ModelSim output waveform obtained from 128-tap filter

A 10-Bit 50-MSPS Pipelined CMOS ADC

Mohamad-Faizal Hashim, Yuzman Yusoff and Mohd. Rais Ahmad, *Member, IEEE*
AMS IC Design Group, Microsystems, MIMOS Berhad, Kuala Lumpur, Malaysia
Email: m_faizal@mimos.my

Abstract - **This paper presents a 10-bit 50-MSPS pipelined ADC targeted to 0.35um CMOS technology. The main characteristics of pipelined ADC such as signal to noise and distortion ratio (SNDR), spurious free dynamic range (SFDR), differential non-linearity (DNL), integral non-linearity (INL) and power consumption are simulated in Hspice®. In this simulation, a full-scale of Nyquist-frequency sine-wave input is used. The results show the designed pipelined ADC achieves a SNDR of 58dB, SFDR of 70dB, maximum differential nonlinearity (DNL) and integral nonlinearity (INL) are less than 0.5 least significant bit (LSB) and a power consumption of 350-mW.**

I. INTRODUCTION

Analog-to-digital converter (ADC) consists of transforming a continuous-time and continuous amplitude, or analog signal, into a discrete-time and discrete amplitude, or digital signal. The resulting limited rate number representation enables storage, processing and transmission of the signal with much more flexibility than when in its analog form.

The selection of ADC architecture is primarily driven by speed, power consumption and resolution. For high speed and low power applications such as wireless communication, pipelined ADC is the best choice, whereas, for high resolution and low speed requirement, over-sampling architecture is preferred. Flash architecture gives the best speed but can only be used when high power consumption is acceptable.

The design of a pipelined ADC involves a number of considerations such as optimum values of capacitors, resolution of different stages, circuit technique to implement the stages and optimum values for the bias currents of operational transconductance amplifiers (OTAs). Each of these choices affects the capacitive load on each stage, which determines the power dissipation of the ADC.

This paper describes design of a 10-bit 50-MSPS pipelined CMOS ADC at Nyquist input frequency. The paper is organized as follows. In the section II the pipelined ADC architecture is discussed. Section III describes the circuit design and operation of the pipelined ADC. Section IV shows some simulation results, and section V and gives the conclusions.

II. PIPELINED ADC ARCHITECTURE

Fig. 1 shows a block diagram of a conventional multistage pipelined ADC. The whole system consists of cascaded stages. Each stage contains a sub-ADC, sub-DAC and a sample-and-hold amplifier (SHA). The use of SHA allows all stages to process residues from different samples simultaneously; therefore, the output residue only depends on the speed of each stage.

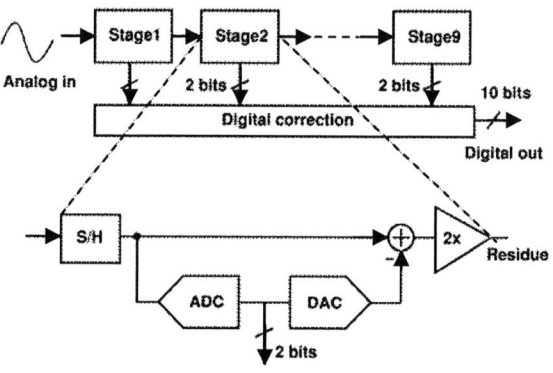

Fig. 1 Pipelined ADC architecture

The sum of the resolutions of all the stages is usually made greater than the output resolution to introduce redundancy. The use of redundancy and digital error correction to overcome the effects of comparator and SHA offsets on the ADC linearity is well known [1]–[7]. The essence of these techniques is that the output of each stage consists of two parts: a low-resolution digital output and the amplified analog residue. As long as the analog residue is accurate, errors in the digital outputs can be sensed and corrected. As a result, the main accuracy limitations with redundancy and digital error correction stem from the linearity of the sub-DACs and the accuracy of the gain in the SHAs.

In this work a pipelined 1.5-bit/stage architecture is used. A low per-stage resolution of 1.5 bits is chosen for several reasons. With low resolution in each stage, the inter-stage gain is small (2 V/V), and hence the inter-stage amplifier will have larger bandwidth and better performance efficiency. Consequently the circuit can be operated at higher frequencies. Using 1.5-bit/stage with digital error correction offsets smaller than voltage reference (Vref) divided by four (Vref/4) can be corrected at each stage and no missing code happens. This large correction range relaxes the requirements of the comparator. Moreover, power is minimized with lower per-stage resolution [2], [3]. However, lower resolution per stage results in increased latency and increased noise from later stages due to the low inter-stage gain.

The architecture implemented consists of nine stages. The 1.5 bit/stage used in the first eight stages outputs 1.5 bits and is used to generate two decision levels which are represented by two bits. Using digital error correction, the eight stages effectively contribute one bit each. The signal range both in the input and output is from negative voltage reference (-Vref) to positive voltage reference (+Vref). Nominally, the comparator decision levels are set to –Vref/4 and +Vref/4 and the sub-ADC output codes for the three regions are "00", "01", and "10". Since no digital error correction can be done after the last stage, the last stage is built with three comparators and contributes two effective bits.

III. CIRCUIT DESIGN

A. Gain Stage

The gain stage structure operates on a two phase clock, ϕ_1 and ϕ_2 which corresponds to a 10ns clock period. Fig. 2 shows the circuit implementation of the gain-stage. During the first phase, the input signal Vin whose range is from within -Vref to +Vref is applied to the input of the sub-ADC, which has thresholds at +Vref/4 and -Vref/4.

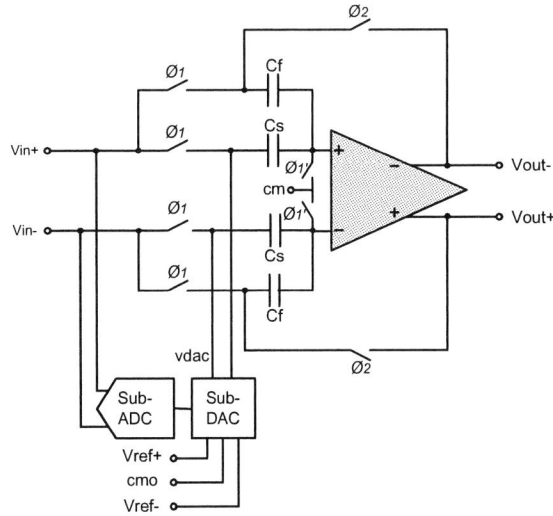

Fig. 2   Gain-stage circuit

Simultaneously, Vin is applied to sampling capacitors C_s and C_f. At the end of the first clock phase, ϕ_2 Vin is sampled across C_s and C_f, and the output of the sub-ADC is latched. During the second clock phase, closes a negative feedback loop around the op-amp, while the top plate of is switched to the Sub-DAC output. This configuration generates the stage residue at Vout. The output of the sub-ADC is used to select the Sub-DAC output voltage Vdac through an analog multiplexer. Vdac is capacitively subtracted from the residue, such that

$$V_o = \left(1 + \frac{C_s}{C_f}\right)V_{in} - V_{ref} \ ; \text{ if } V_{in} > \frac{V_{ref}}{4} \qquad (1)$$

$$V_o = \left(1 + \frac{C_s}{C_f}\right)V_{in} \ ; \text{ if } +\frac{V_{ref}}{4} \le V_{in} \le -\frac{V_{ref}}{4} \ (2)$$

$$V_o = \left(1 + \frac{C_s}{C_f}\right)V_{in} + V_{ref} \ \text{ if } V_{in} > -\frac{V_{ref}}{4} \qquad (3)$$

In this design, equal values of sampling capacitor (C_s) and feedback capacitor (C_f) are used in order to achieve a gain of 2 in the transfer function. Lastly, the OTA must settle to better than 0.01% accuracy in one clock phase (one half-cycle). It is this settling time that limits the overall pipeline throughput.

B. Operational Transconductance Amplifier

In each stage, an operational transconductance amplifier (OTA) is needed to provide gain and bandwidth. As sampling rates increase, the amplifier has less time to provide an accurate output. The amplifier settling time is proportional to the unity gain frequency. Settling time of the OTA will affect the speed of the ADC. Settling time itself is the duration it takes for the output of op-amp to reach a final value to within predetermined tolerance when excited by a small signal [8]. A longer settling time implies that the rate of processing analog signals must be reduced. For precision, high-speed data converter design we should use an inverting op-amp technology where the inputs of the op-amp remain a fixed voltage [9]. However, this approach may not be able to reduce noise in the circuit as it only has a single-ended differential output. Hence, a fully-differential topology becomes a necessity in a high-speed ADC design [9].

In this project, we design a two-stage-fully-differential OTA (Fig. 3) to be used in each stage of a 10-bit pipelined A/D converter. The first stage is a folded-cascode amplifier with PMOS input transistor. This folded-cascode OTA offers self-compensation and good input-common range. Cascoding is a well known means to enhance the DC gain of an amplifier without degrading its high-frequency performance [11]. The second stage is a class A amplifier with current mirror.

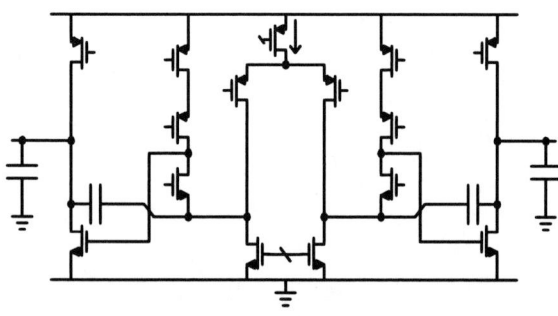

Fig. 3   Schematics of the OTA

Adding more stages introduces instability in amplifier feedback configuration. To overcome this problem, frequency compensation methods such as Miller or Cascode compensation must be utilized [10].

C. Dynamic Comparator

Fig. 4 shows the structure of a resistive dividing latch comparator used in this pipelined ADC. Transistors M9-M12 form an input comparing circuit and M1-M8 form a cross couple latch. As Ø is low, the circuit works in the reset mode where transistor M1 and M4 are conducting and transistor M5 and M6 are cut off. Both differential outputs are forced to VDD and no current path exists between the supply voltages. Therefore the power consumption is only due to VDD charging the two output capacitors.

When Ø is high, the circuit works in the regeneration mode. M1 and M4 are cut off, and M5 and M6 are turned on. In this mode, the circuit can compare the input voltages by using input transistors operated in the triode region. These transistors act like voltage control resistors and give the conductance of the left and right input branches g_L and g_R as

$$g_L = k_n \left(\frac{W_{10}}{L}(V_i^+ - V_T - V_{ds9,10}) \right.$$
$$\left. + \frac{W_9}{L}(V_R^- - V_T - V_{ds9,10}) \right), \tag{4}$$

and

$$g_L = k_n \left(\frac{W_{12}}{L}(V_i^+ - V_T - V_{ds11,12}) \right.$$
$$\left. + \frac{W_{11}}{L}(V_R^- - V_T - V_{ds11,12}) \right), \tag{5}$$

where V_T is the threshold voltage and V_{ds9} - V_{ds12} the drain-source voltage of corresponding transistor. By setting $g_L = g_R$, the input threshold voltage that makes the output change its state can be solved by

$$V_{i+} - V_{i-} = \frac{W_B}{W_A}(V_{R+} - V_{R-}) \tag{6}$$

This equation assumes $W_A = W_9 = W_{12}$ and $W_B = W_{10} = W_{11}$. The threshold of the comparator can be adjusted to the desire level by dimensioning the transistor width W_A and W_B.

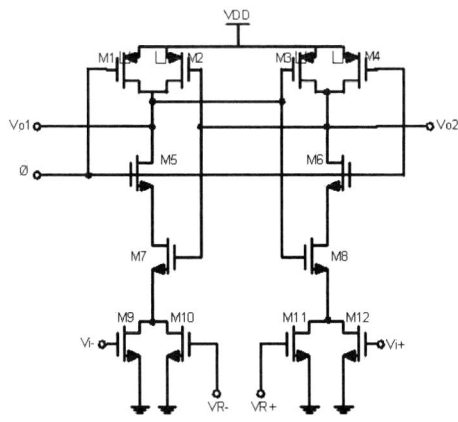

Fig. 4   Schematic of the dynamic comparator

IV. SIMULATION RESULTS

To measure the AC characteristics of OTA, an ideal CMFB is used, and the load from feedback, next stage and parasitic capacitance is added. The OTA is measured in closed loop and the result is shown in Fig. 5.

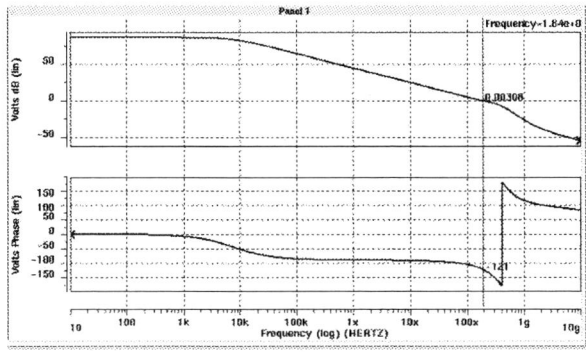

Fig. 5 Gain and Phase margin of OTA

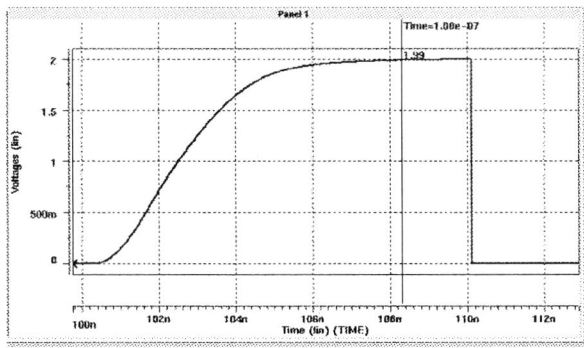

Fig. 6 Settling time of OTA

The OTA settling time is obtained when it is applied in switched capacitor circuitry and real capacitive CMFB is added. The transient simulation is used to obtain the following settling plotting (Fig. 6).

The dynamic linearity of the 10-bit 50-MS/s pipelined ADC was simulated by analyzing a fast Fourier transform (FFT) of the output codes for a single input sine wave. The simulations were performed in HSPICE. The peak signal to noise and distortion ratio (SNDR) is 59dB at 50 MHz sampling rate for a full-scale 2V p-p Nyquist sine wave input. The calculated ENOB from SNDR result is equivalent to 9.5 bit. The simulated spurious-free dynamic range (SFDR) is 70dB. Fig. 7 shows the FFT plot of the converter where a full-scale input Nyquist-frequency sine wave is sampled at 50 MHz.

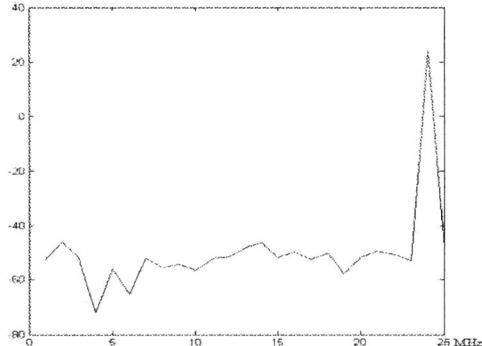

Fig. 7   An FFT plot of the output for sampling rate fs=50MHz and a Nyquist-frequency input

The most important measures of static or DC-linearity of ADC are integral nonlinearity (INL) and differential nonlinearity (DNL). These properties indicate the accuracy of a converter and include the errors of quantization, nonlinearities, offset and noise. The ramp input signal was used to simulate the DNL and INL. Maximum DNL of 1 least significant bit (LSB), and INL of 1LSB are achieved. The power consumption at a clock frequency 50-MHz was 350-mW from 3.3-V supply. Table 1 summarizes the simulated results.

Table 1 : Summary of ADC performance

Technology	0.35-μm CMOS
Voltage Supply	3.3-V
Resolution	10 bits
Conversion Rate	50 MS/s
Input Range	2V differential
ENOB	9.34
DNLmax	1 LSB
INLmax	1 LSB
SNDR	58dB@Nyquist frequency
SFDR	69dB@Nyquist frequency
Power dissipation	350-mW

V. CONCLUSION

In this paper, a 10-bit 50-MSPS pipelined analog-to-digital converter has been presented. The designed pipelined ADC architecture has eight 1.5-bit residue gain stages and 2-bit backend flash stage. HSpice simulation employing BSIM3v3.1 model parameters of a 0.35um CMOS confirm the 10-bit resolution of the 3.3-V 50-MSPS ADC. Simulated values of the signal-to-noise-and-distortion and spurious-free-dynamic range values with full scale Nyquist-frequency input are 58dB and 69dB respectively. The total power consumption of this ADC is 350-mW.

VI. REFERENCES

[1] B. Nejati and O. Shoaei,"A 10-bit, 2.5-v, 40msample/s, pipelined analog-to-digital converter in 0.6-um cmos", *Circuits and Systems, 2004.* Volume 1, 23-26 May 2004 Page(s):I-73 - I-76 Vol.1

[2] D.W. Cline, P.R. Gray, "A power optimized 13-b 5 Msamples/s pipelined analog-to-digital converter in 1.2 μm CMOS", *IEEE JSSC*, vol. 31, no. 3, pp. 294-303, March, 1996.

[3] S.H. Lewis, "Optimizing the stage resolution in pipelined, multistage, analog-to-digital converters for video-rate applications*", IEEE Circuits and Systems II*, vol. 39, no. 8, pp. 516-523, Aug., 1992.

[4] A. Abo, "*Design for reliability of low-voltage, switched-capacitor circuit*", Ph.D. Thesis, University of California, Berkeley, May, 1999.

[5] M. Taherzadeh-Sani, R. Lotfi, O. Shoaei, "An analytical approach to estimate the dynamic non-linearity parameters in a pipelined ADC", *Proc. of European Solid-State Circuits Conference, ESSCIRC*, Estoril, Portugal, 2003.

[6] P.T.F. Kwok, H.C.Leung, "Power optimization for pipeline analog-to-digital converters", in *IEEE Trans. On Circuits & Systems-II*, vol.46, pp.549-53, May 1999.

[7] H. Lewis, "Optimizing the stage resolution in pipelined, multistage, analog-to-digital converters for video-rate applications", *in IEEE Trans. Circuits &Systems-II*, vol.39, No.8, pp.516-523, Aug. 1992.

[11] Allen and Holberg, "*CMOS Analog Circuit Design*", 2nd Edition, Oxford University Press, 2002.

[12] R. Jacob Baker, "*CMOS Mixed-Signal Circuit Design*", IEEE Press Series, Wiley, 2002, pp. 316.

[13] Behzad Razavi, "*Design of Analog CMOS Integrated Circuits*", McGraw-Hill, 2001.

[14] Johns, D., and Martin, K., "*Analog integrated circuit design*",(John Wiley & Sons,1997).

Performance of OCDMA Systems Using AND Subtraction Technique

S.A. Aljunid[1], Feras N. Hasson[2], M.D.A.Samad[3], M.K.Abdullah[3], M. Othman[3] and S. Shaari[2]

[1]School of Computer and Communication Engineering, Northern Malaysia University College of Engineering Block A, Kompleks Pusat Pengajian KUKUM, Jalan Kangar-Arau, 02600 Jejawi, Perlis.
[2]Photonics Technology Laboratory (PTL), Institute of Micro Engineering and Nanoelectronics (IMEN), Universiti Kebangsaan Malaysia, 43600 UKM, Bangi, Selangor, Malaysia
[3]Photonic Laboratory, Department of Computer System and Communication, Faculty of Engineering, University Putra Malaysia, 43400 UPM, Serdang, Selangor.
Email: syedalwee@kukum.edu.my

Abstract - **A new detection scheme, namely AND subtraction technique is proposed and presented in this paper. The theory is being elaborated and experimental results have been done by comparing Double-Weight (DW) code against the existing code, Hadamard. In this paper we have proved that AND subtraction technique gives better Bit Error Rates (BER) performance than Complementary subtraction technique against the received power level.**

I. INTRODUCTION

In optical CDMA systems, the detection process affects the design of transmitters and receivers. In general, there are two basic detection techniques namely coherent and incoherent. While coherent detection refers to the detection signals with knowledge of the phase information of the carriers, incoherent detection refers to the case without such knowledge. Alternatively a system consisting of unipolar sequences in the signature code, is called incoherent system. A system that uses bipolar codewords is called a coherent system. Because incoherent detection does not need phase synchronization, hardware complexity of the system is reduced. This is the main reason why we have chosen incoherent detection in this research

In an incoherent CDMA system, each user is assigned a distinct codeword as its address signature based on the spectral amplitude only. When a user wants to transmit data bit one, it sends out a codeword corresponding to the address signature of the intended receiver. At the receiver,

all the codewords from different users are correlated. If a correct codewords arrives, an autocorrelation function with a high peak results. For incorrect codewords, cross-correlation functions are generated and they create Multiple Access Interference (MAI). Multiple Access Interference (MAI) can be reduced by using subtraction technique. The most common subtraction technique is the Complementary subtraction technique, which is also known as balanced detection technique [1-2].

In most researches [2-5], complementary method has been used at the receiver side to recover the original signal. In this paper, we introduce a new approach called AND subtraction technique. The purpose of this new subtraction is to reduce the receiver complexity and at the same time to improve the system performance.

II. THEORY FOR AND SUBTRACTION TECHNIQUE

In AND subtraction technique, the cross-correlation $\theta_{\overline{XY}}(k)$ is substituted by $\theta_{(X\delta Y)Y}$, where $\theta_{(X\delta Y)}$ represents the AND operation between sequences X and Y. For example, let $X = 0110$ and $Y = 0011$ and therefore $(X$ AND $Y) = 0010$. Example of an AND receiver is shown in Table 1.

0-7803-9730-4/06/$25.00 ©2006 IEEE

TABLE 1 : Example of AND Subtraction Techniques for DW and Hadamard Codes

	Double-Weight Code				Hadamard Code			
	C1	C2	C3	C4	C1	C2	C3	C4
User 1 (X)	0	1	1	0	1	0	1	0
User 2 (Y)	0	0	1	1	1	1	0	0
$(X \times Y)$	0	0	1	0	1	0	0	0
$\sum (X \times Y)$	1				1			
User 1 $(\overline{X})$	1	0	0	1	0	1	0	1
User 2 (Y)	0	0	1	1	1	1	0	0
$(\overline{X} * Y)$	0	0	0	1	0	1	0	0
$\sum (\overline{X} * Y)$	1				1			
Z_{AND}	0				0			

At the receiver,

$$Z_{AND} = \theta_{XY}(k) - \theta_{(X\&Y)Y}(k) = 0 \qquad (1)$$

Equation (1) shows that, with AND subtraction technique, the multiple access interference or the interference from other channels can also be cancelled out. This subtraction technique can be implemented with any OCDMA codes, but for comparison purposes, the DW code and Hadamard code are used as an example in Table I. Also shown in Table 1 is the operation involved in both codes.

III. EXPERIMENTAL SETUP

The experimental setup for two channels OCDM system using DW code [6] is shown Figure 1. In the transmitter section, each site consisted of six components: two Pseudo Random Bit Sequence (PRBS) generators, two non-return-zero (NRZ) pulse generators, two laser diodes, couplers, splitters and external modulators. The laser diode sources launched an optical power of 5 dBm with a linewidth of 10MHz. The two laser diodes had wavelengths of $\lambda_1 = 1552.511$ nm and $\lambda_2 = 1552.007$ nm which represented DW code sequence of 011 as Channel 1. However, channel two consisted of only one wavelength (i.e $\lambda_2 = 1552.007$), which overlapped, with one of the weights of Channel 1. Only one light source used for Channel 2, instead of two light sources. This was considered sufficient because the results would only be taken for Channel 1. Thus, even if the other weight of Channel 2 was used, it served no real purpose because it will be filtered out anyway at the receiver of Channel 1. This was the reason that only the overlapping part was used.

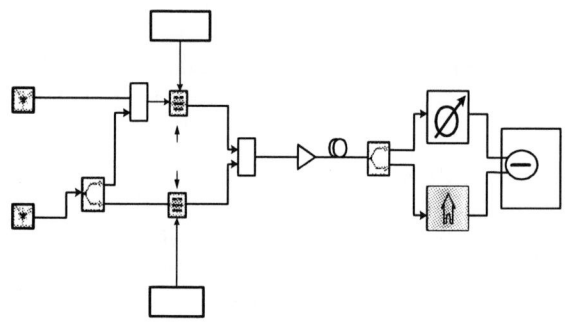

Fig.1 Experimental Setup for OCDM System Using DW Code.

IV. EXPERIMENTAL RESULTS

Figures 3 and 4 show the encoder spectrum of coding sequence 011 (Code 1) and 010 (Code 2) respectively, while Figure 5 shows the combined spectrum inside the fiber core. An increase of power can be observed at the last chip due to the overlapping of chips from the two codes. Figure 6 and 7 show the eye patterns before and after the AND subtraction for Channel 1.

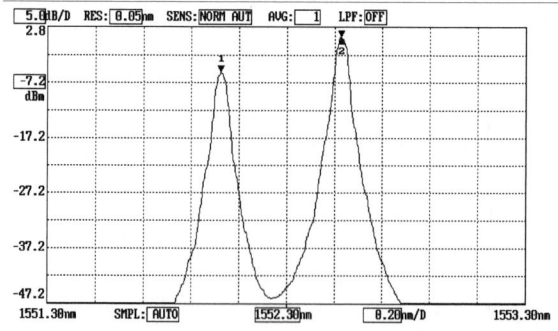

Fig. 2 Spectrum for Output Encoder Code 1 (011).

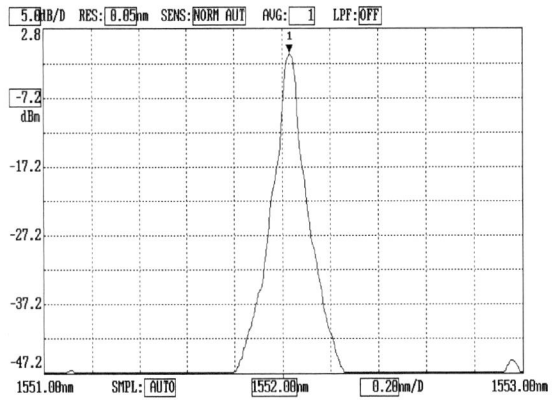

Fig. 3 Spectrum for Output Encoder Code 2 (010).

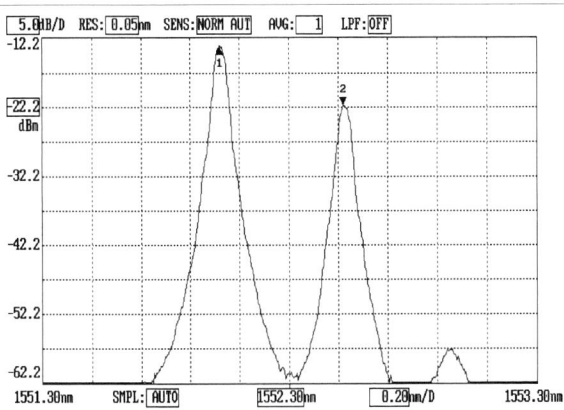

Fig. 4 The Combined Spectrum inside Fiber.

Figure 5 shows the result from experiments using the two different OCDMA codes; DW and Hadamard taken against the received power level. The experiment results for Hadamard code was adopted from [7] for performance comparison purposes. It is shown that, the OCDM system using DW codes performed better than the system using Hadamard code, although the Hadamard code had a lower bit rate which is 155 Mbps. While for DW codes, 622 Mbps has been used for the transmission rate. However, note that the results for Hadamard was based on Complementary subtraction technique, while our results for DW used AND subtraction. Thus, the inferior performance of Hadamard code could have been attributed to the code structure itself and the detection scheme.

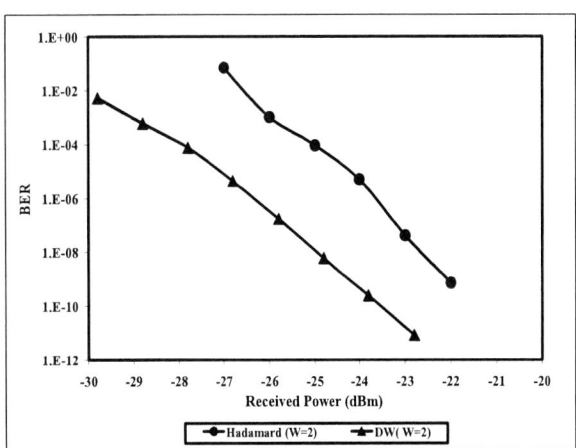

Fig.5 BER versus Received Power for Hadamard Code at Bit Rate 155 Mbps and DW Codes at Bit Rate 622 Mbps.

V. CONCLUSIONS

The spectral codes were intentionally chosen to have an overlapped spectral bit, to show the successful signal recovery via newly proposed subtraction technique. This technique is shown here to be able to minimize the effect of the chip interference. The DW code has shown superior performance compared to Hadamard code. Based on the result from the experiment, although the DW and Hadamard codes has a same weight of two but it shows that DW code performed better than the Hadamard at higher bit rate. This is one of the major advantages of DW code, with minimal number of weights, the cost and the complexity of the system can be reduced for metropolitan network application.

REFERENCES

[1] L. Nguyen, B. Aazhang, and J. F. Young, "All-Optical CDMA With Bipolar Codes," Electronic Letters, Vol. 31, No. 6, 469–470, March 1995.

[2] E. D. J. Smith, R. J. Blaikie, and D. P. Taylor, "Performance Enhancement of Spectral-Amplitude-Coding Optical CDMA Using Pulse-Position Modulation," IEEE Transaction on Communications, Vol. 46, 1176–1185, September 1998.

[3] Raymond M. H. Yim, Jan Bajcsy, and Lawrence R. Chen, "A New Family of 2-D Wavelength–Time Codes for Optical CDMA With Differential Detection," IEEE Photonics Technology Letters, Vol. 15, No. 1, 165-167, January 2003.

[4] Djordjevic, I.B., Vasic, B., "Novel Combinatorial Constructions Of Optical Orthogonal Codes For Incoherent Optical CDMA Systems," Journal of Lightwave Technology, Vol. 21, Issue: 9, 1869 – 1875, September 2003.

[5] Zaccarin, D., Kavehrad, M., " Performance Evaluation of Optical CDMA Systems Using Non-Coherent Detection and Bipolar Codes," Journal of Lightwave Technology,Vol. 12 , Issue: 1, 96–105, January 1994.

[6] S. A. AlJunid, M.. Ismail, B. M. Ali, A. R. Ramli, M. K. Abdullah, "A New Family of Optical Code Sequences For Spectral-Amplitude-Coding Optical CDMA Systems", IEEE Photonics Technology Letters, Vol. 16, No.10, 2383-2385, October 2004.

[7] C.F..Lam, T.K. Tong Dennis, M.C.Wu, "Experimental Demonstration of Bipolar Optical CDMA System Using a Balanced Transmitter and Complementary Spectral Encoding", IEEE Photonics Technology Letters, Vol. 10, No. 10, 1504-1506, October 1998.

Pyroelectric Properties of Polyvinylidene Fluoride (PVDF) by Quasi Static Method

Gan Wee Chen and Wan Haliza Abd. Majid
Solid State Research Laboratory
Physics Department
University of Malaya
50603 Kuala Lumpur, MALAYSIA
Email: wcgan1981@yahoo.com.sg , q3haliza@um.edu.my

Abstract — Our main objective is to investigate the behavior of polyvinylidene fluoride (PVDF) polymers when they are subjected to electric field in different temperature range. Initially, the PVDF powders were dissolved in acetone with different concentration and the PVDF films were then prepared by spin coating technique. The pyroelectric coefficients, *p* was calculated by using the Quasi Static method with different heating rates. The PVDF thin films prepared from different solution concentrations exhibit different pyroelectric coefficients. The results show that the difference solution concentrations of the sample and temperature heating rates will influence the value of pyroelectric coefficient.

I. INTRODUCTION

Recently, pyroelectric materials are attracting much attention because they are widely used as temperature sensors and IR-light detectors [1]. The discovery of strong piezoelectricity in polyvinylidene fluoride (PVDF) by Kawai in 1969 has attracted much attention to these polymers [2]. PVDF is a high performance engineering thermoplastic featuring unique properties. PVDF shows an excellent behavior, both mechanically and physically. It is also chemically inert toward most acids, organics aliphatic and aromatic compounds, solvents, oxidants, halogens and alcohol [3].

PVDF is one of the famous semicrystalline polymers for its pyroelectric, piezoelectric and ferroelectric properties. PVDF can form a different crystal depending on the condition of the crystallization [4]. It is the polymer material which has been classified as insulator or dielectric. The investigation of dielectric and pyroelectric properties through PVDF will provide an important knowledge to an understanding of the structure of polymer material.

II. EXPERIMENT

The PVDF powders were supplied by Sigma-Aldrich Pte Ltd. Initially, the PVDF powders were dissolved in acetone with different concentrations which are 40 mg/ml, 60 mg/ml and 80 mg/ml. The solutions were then agitated in the ultrasonic bath at 70°C for 22 minutes to ensure the powders were completely dissolved. PVDF films were prepared by spin coating (2000rpm for 20 seconds) the polymer solution onto the glass substrate which was previously coated with aluminum electrodes (150nm thick). The film was kept overnight at room temperature and then annealed at 120°C for 2 hours to remove the solvent. An aluminum top electrode (20nm thick) was then thermally evaporated on the top surface of the film to produce the desired metal-insulator-metal (MIM) structure [5].

The PVDF thin film was poled for 30 minutes at room temperature by applying a high electric field in order to attain the pyroelectric activity. The poling electric field was 20kV/mm. After that, the thin films were immediately placed into the vacuum chamber to measure the pyroelectric current, I_p with the triangle waveform temperature range (25°C to 50°C). The pyroelectric current, I_p was determined using the experimental set-up shown in Fig. 1. The triangle waveform temperature range (25°C to 50°C) was generated by using LakeShore temperature controller. The pyroelectric current, I_p which was generated when the PVDF films are heated and cooled repeatedly was measured by Keithley 617 electrometer. The IEE-488 GPIB interface was used to allow the computer to control and receive

0-7803-9730-4/06/$25.00 ©2006 IEEE

data from both instruments which are mentioned above.

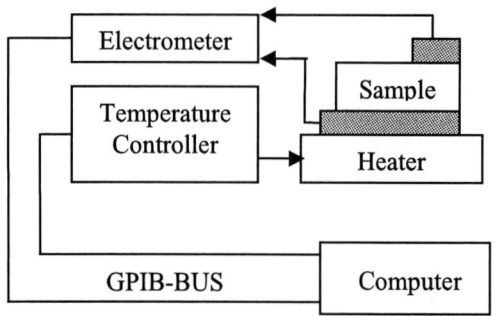

Fig. 1 Schematic drawing of the experiment set-up for Quasi Static pyroelectric characterization.

Basically, the most common way that has been used to measure pyroelectric coefficient is quasi-static method and dynamic temperature method. The quasi-static method utilize temperature variations under nearly thermal equilibrium conditions whereas the dynamic temperature method use fast temperature changes caused by (i) a pulsed or (ii) a periodically modulated heat flux [6]. In our study, we emphasized on the quasi-static method to measure the pyroelectric coefficient. In this method, we set the temperature of the sample to increase at a constant rate in order to generate a triangle temperature waveform while recording the short-circuited pyroelectric current. The heating rates which have been used in this experiment are 0.01 $^{\circ}$C/s, 0.03 $^{\circ}$C/s, 0.07 $^{\circ}$C/s, 0.10 $^{\circ}$C/s and 0.12 $^{\circ}$C/s.

III. RESULTS AND DISCUSSION

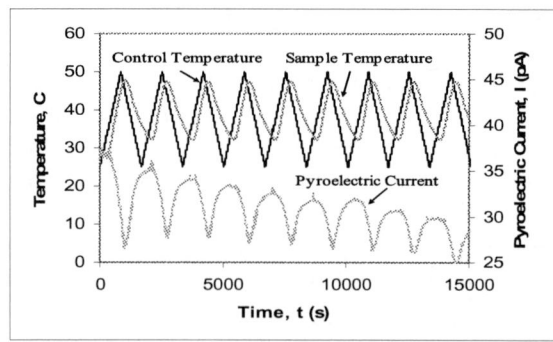

Fig 2 Pyroelectric current is decayed with time at the heating rate of 0.03 $^{\circ}$C/s

Fig 2 shows that the pyroelectric current is decayed with time when the PVDF thin film was

poled at 20kV/mm. The PVDF thin film was immediately placed into the chamber to measure the pyroelectric current at the heating rate of 0.03°C. We have noticed that the amplitude of the pyroelectric current at the beginning is larger compare to the amplitude of the pyroelectric current at the end of the measurement after 4 hours. The damping oscillation figure of this square waveform pyroelectric current is most probably due to high quantity of the surface charges on the PVDF thin film. Poling process will realign the polar axes of the thin film crystallites and as a result there will be a polarization field exists inside the sample. When such polarization occurs, one side of the molecule acquires a positive charge and the opposite side has a negative charge of the same magnitude [7]. However, the charges on the surface of the sample are not returned to their stable energy state after the poling process. An injection of charge carrier from the electrodes into the layer of dielectric becomes dominant when the sample is poled. These injected carriers may remain at the grain boundaries, dislocation sites and other interfaces (called crystalline-amorphous boundaries in semicrystalline polymers) [8]. Therefore, the magnitude of the pyroelectric current is decayed with time. This problem can be solved by short circuiting the sample inside the chamber for 1 day after poling process. Short circuiting and annealing the sample for 17 hours at 50°C can also eliminate the contribution of thermally stimulated current in subsequent pyroelectric measurement. However, at sufficiently high annealing temperature, the charges may be detrapped and causes the polarization inside the sample to decay [5, 8].

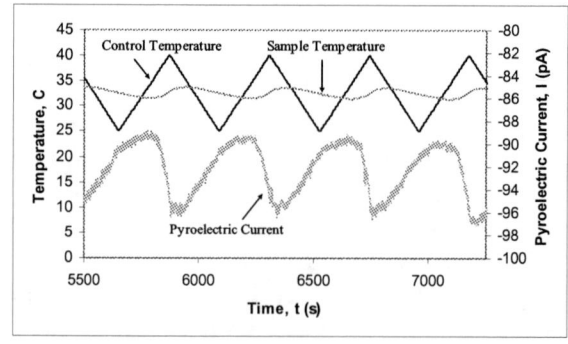

Fig 3 Measurement of the pyroelectric current for the sample concentration of 80 mg/ml at the heating rate of 0.07 $^{\circ}$C/s

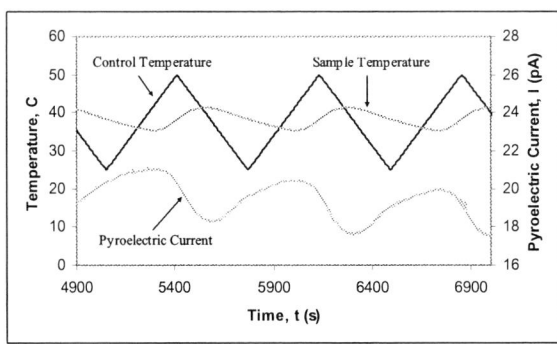

Fig 4 Measurement of the pyroelectric current for the sample concentration of 60 mg/ml at the heating rate of 0.07 °C/s

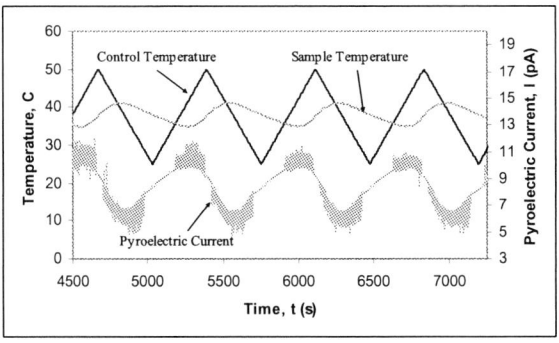

Fig 5 Measurement of the pyroelectric current for the sample concentration of 40 mg/ml at the heating rate of 0.07 °C/s

Fig 3, 4 and 5 show the measurement of the pyroelectric currents for the sample concentration of 80 mg/ml, 60 mg/ml and 40 mg/ml at the heating rate of 0.07 °C/s respectively. From the figures above, we notice that the amplitude of the pyroelectric current which is also defined as peak-to-peak current, I_{p-p} increase with respect to the increase in the solution concentration.

An expression for pyroelectric coefficient is obtained by measuring the short-circuit current, I_p. The rate of change of polarization can be represented by pyroelectric current, I_p if the surface charge is allowed to flow. The pyroelectric current, I_p is expressed as

$$I_p = pA\frac{dT}{dt} \qquad (1)$$

where p is the pyroelectric coefficient, A is the effective area of the sample and dT/dt is the rate of temperature change. From equation (1), p can be evaluated if I_p, A and dT/dt are known. Therefore if a triangle temperature waveform is

applied to the sample, the square waveform would be expected to be obtained. The rate of the temperature change can be calculated from the gradient of the triangle wave of the sample temperature, while the pyroelectric current can be obtained from the amplitude of the squarewave. Practically, an ideal square waveform current is not observed due to the thermal mass of the pyroelectric sample substrate [9].

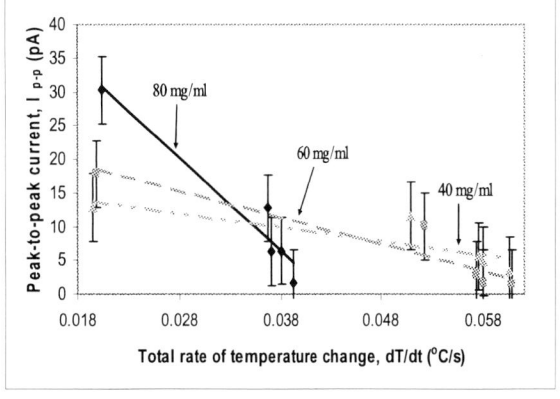

Fig 6 The peak-to-peak current, I_{p-p} is plotted as a function of total rate of temperature change, dT/dt (°C/s)

Fig 6 shows the peak-to-peak current, I_{p-p} plotted as a function of total rate of temperature change, dT/dt (°C/s). From Fig 6, we found that the pyroelectric coefficients obtained from the sample with concentration of 80 mg/ml, 60 mg/ml and 40 mg/ml are $p \approx -24\mu Cm^{-2}K^{-1}$, $-10\mu Cm^{-2}K^{-1}$ and $-9\mu Cm^{-2}K^{-1}$ respectively. A negative sign of pyroelectric coefficients are presented due to the trend of the graph I_{p-p} against dT/dt. The value of I_{p-p} decreases with the heating rate of the sample. This result is similar to the technical data reported by Solvay & Cie which also shows negative sign of pyroelectric coefficient [10]. According to the result in Table 1, the sample concentration of 80 mg/ml exhibits the highest pyroelectric coefficient among the samples follow by 60 mg/ml and 40 mg/ml sample. Hence, we can assume that the higher the sample concentration, the higher the pyroelectric coefficient, p. In addition to the discussion above, it is also reported that the thickness of the sample also influences the value of the pyroelectric coefficient, p of PVDF thin film [11].

Sample Concentration	Pyroelectric Coefficient, p (μCm^{-2}K^{-1})	Temperature Range, $^{\circ}$C
80mg/ml	-24.37	25-40
60mg/ml	-9.98	25-50
40mg/ml	-9.20	25-50

Table 1 Value of pyroelectric coefficient due to the sample concentration

IV. CONCLUSION

The pyroelectric coefficient, p is found to be higher in the higher sample concentration. The results also indicated that the higher heating rate which is supplied to the sample decreased the value of the pyroelectric current, I_p. In future work, the pyroelectric coefficient of PVDF as a function of electric field will be carried out to form integrated pyroelectric sensors and arrays. The ferroelectric properties of PVDF will also be investigated.

ACKNOWLEDGEMENT

The authors would like to thank the Malaysian Ministry of Science, Technology and Environment for sponsoring this work under project IRPA 09-02-03-1028 and Department of Physics, University of Malaya providing the facilities at the Solid State Research Laborator

REFERENCES

[1] H. Amorin, J. Portelles, F. Guerrero, J. Perez, J.M. Siqueiros, "Dielectric Hysterisis and Pyroelectricity in LSBNT Ferroelectric Ceramic System," *Solid State Communications*, 113 p.581-585 (2000).

[2] T. R. Jow and P. J. Cygan, "Investigation of Dielectric Breakdown of Polyvinylidene Fluoride using AC and DC methods," *IEEE International Symposium on Electrical Insulation*, Baltimore, MD USA, June 7-10 (1992).

[3] J. Kulek, Cz. Pawlaczyk, E. Markiewicz, "Influence of poling and ageing on high frequency dielectric and piezoelectric response of PVDF-type polymer foils," *Journal of Electrostatics*, Volume 56, Issue 2, p.135-141 (2002).

[4] J. P. Youn, S. K. Yong, C. Park, "Micropatterning of semicrystalline poly(vinylidene fluoride) (PVDF) solutions," *European Polymer Journal* 41 p.1002-1012 (2005).

[5] Q. Q. Zhang, H. L.W. Chan, C. L. Choy, "Dielectric and pyroelectric properties of P(VDF-TrFE) and PCLT-P(VDF-TrFE) 0-3 nanocomposite films," *Composites*: Part A 30 p.163-167 (1999).

[6] Q. Andrea and H. Jurg, "Analysis of the Polarization Distribution in a Polar Perhydrotriphenylene Inclusion Compound by Scanning Pyroelectric Microscopy," *J. Phys. Chem. B*, 102, p.4277-4283 (1998).

[7] J. W. Smith, *Electric Dipole Moments*, Butterworths Scientific Publications (1955).

[8] C. Dias, M. Simon, R. Quad and D. K. Das-Gupta, "Measurement of the pyroelectric coefficient in composite using a temperature-modulated excitation," *J. Phys. D: Appl. Phys.* 26, p.106-110 (1993).

[9] W. H. Abd. Majid, *Pyroelectric Activity In Cyclic And Linear Polysiloxane Langmuir-Blodgett Film*, University of Sheffield (1994).

[10] Technical Data, Piezoelectric SOLEF PVDF polyvinylidene fluoride films, Solvay & Cie., Rue du Prince Albert 44, 1050 Bruxelles, Belgium.

[11] F. M. Luiz, A. M. Jose, K. S. Walter, "Study of Pyroelectric Activity of PZT/PVDF-HFP Composite," *Materials Research*, Vol. 6, No. 4, p.469-476 (2003).

Effects of Substrate Temperature on the Properties of Hydrogenated Nanocrystalline Silicon Thin Film Grown by Layer-by-Layer Technique

Goh Boon Tong and Saadah Abdul Rahman

Solid State Research Laboratory, Physics Department,
Faculty of Science, University of Malaya,
50603 Kuala Lumpur, MALAYSIA
Email: gohbt@um.edu.my , saadah@um.edu.my

Abstract **Nanocrystalline silicon (nc-Si) thin films prepared by layer-by-layer (LBL) technique were studied. The LBL technique involves periodic interruption of the deposition process whereby the silane diluted in hydrogen plasma discharge is stopped for a fixed period of time during which the growth surface is treated with hydrogen plasma discharge. This technique controls the crystallite size in the film structure. The films are grown at different substrate temperatures ranging from room temperature to 400 °C. The dependence of the optical and structural properties of the films on the substrate temperature is investigated using optical transmission, X-ray diffraction (XRD) and Fourier transform infrared (FTIR) spectroscopy technique. The effects of substrate temperature were studied and these properties showed strong dependence on the substrate temperature. Initial increase in the substrate temperature to 200 °C resulted in a blue shift in the absorption edge but higher substrate temperature shifted it back towards lower energy. Higher substrate temperatures also lowered the hydrogen content in the film and increased the concentration of monohydride bonds in the film structure. Increase in crystallinity was observed with increase in substrate temperature.**

I.   INTRODUCTION

Hydrogenated nanocrystalline silicon (nc-Si:H) thin films are attractive materials especially in photoluminescence devices in the visible region due to the quantum size effects of Si nanocrystallites embedded in amorphous silicon matrix [1]. Presence of Si

nanocrystallites widens the band gap and enhances photoluminescent intensities of these materials. Luminescent properties of these films are reported to be strongly dependent on crystallite size and volume fraction of nanocrystallites Si presented in the films [2, 3].

Usually, most researchers fabricate nc-Si:H thin films by continuous plasma enhanced chemical vapour deposition (PECVD) using silane highly diluted in hydrogen [4]. However, the crystallite size and concentration are not controllable by the preparation conditions. In this work, we study the effects of substrate temperature on the growth rate, optical and structural properties of nc-Si:H films grown by LBL technique.

II.   EXPERIMENT

The films were deposited on glass and crystal silicon (c-Si) substrates by layer-by-layer (LBL) technique using a home-built RF-PECVD (13.56 MHz) system. The schematic diagram of reaction chamber of this system is shown in fig. 1. LBL process was performed by periodically alternating the deposition process of nc-Si:H layer with the surface hydrogen plasma treatment process for a period of 5 minutes and 3 minutes respectively. This process was done by periodically stopping the silane flow into the reaction chamber during the hydrogen plasma treatment process. Five sets of nc-Si:H films were prepared at 5 different substrate temperatures i.e. room temperature, 100 °C, 200 °C, 300 °C and 400 °C.

The optical absorption spectra of the nc-Si:H films were measured using a JASCO V570 UV-VIS-NIR spectrophotometer. The film thickness and refractive index were determined from interference fringes of the spectra. The optical energy gap, E_g was deduced from the Tauc's plot, $(\alpha h\nu)^{1/2}$ versus $h\nu$ where α and $h\nu$ are the absorption coefficient and the photon energy respectively [5]. The X-ray diffraction (XRD) spectra were obtained using the SIEMENS D5000 X-ray diffractometer. The average size was estimated from the full-width at half-maximum (FWHM) of the diffraction peak using Sherrer's formula [6]. The structural properties were also investigated from the Fourier transform infrared (FTIR) spectra of the films measured using a Perkin-Elmer System 2000 FTIR spectrometer. The hydrogen content, C_H and the microstructure parameter, R of the film were determined from the Si-H wagging and Si-H/Si-H$_2$ stretching bands at around 630 cm^{-1} and 2000 cm^{-1} respectively using techniques described in our earlier works frequently used by researchers working on hydrogenated amorphous silicon [7].

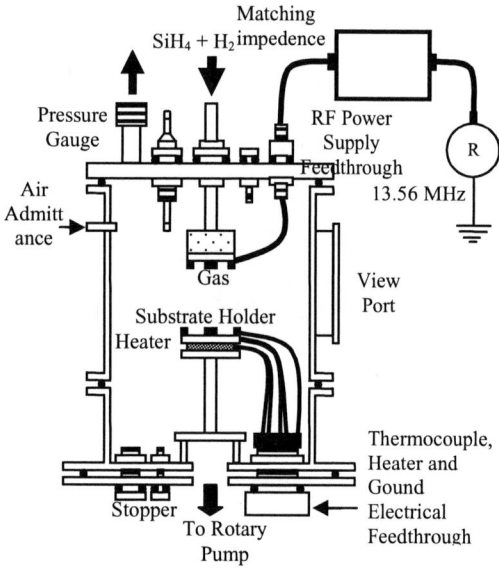

Fig. 1 Schematic diagram of the RF-PECVD reactor for LBL plasma deposition.

III. RESULTS AND DISCUSSION

In the layer-by layer (LBL) deposition technique, during the deposition process both SiH$_4$ and H$_2$ gas are dissociated followed by secondary reactions of reactive species with the SiH$_4$ and H$_2$ molecules. During the H plasma treatment process when the SiH$_4$ flow is stopped, only H$_2$ molecules are dissociated and H etching process proceeds on the growing surface [8].

Fig. 2 shows the variation of the growth rate and refractive index with substrate temperature while fig. 3 shows the plots of the optical energy gap, E_g and Tauc's slope, $B^{1/2}$ versus substrate

temperature. The growth rate and the refractive index increase with increase in substrate temperature to 100 °C. These parameters show no variation with increase in substrate temperature to 200 °C. With further increase in substrate temperature to 300 °C, both the growth rate and refractive index show a further increase. When the substrate is increased to 400 °C, the growth rate continues to increase while the refractive index drops to a lower value. The E_g and $B^{1/2}$ values of the nc-Si:H films prepared at substrate temperature of 100 °C increase significantly to high values as compared to the film prepared at room temperature. A significant drop in the E_g value is observed for the film prepared at substrate temperatures of 300 °C and 400°C. The $B^{1/2}$ value decreases significantly only for the film prepared at substrate temperature of 400 °C.

For the film prepared at low substrate temperatures (100 – 200 °C), the deposition rate is stable due to the increase in hydrogen etching effect during the hydrogen plasma treatment process. The high E_g and $B^{1/2}$ values for the films prepared at these substrate temperatures are evidence of quantum confinement effects as a result of formation of nanosized Si grains in the film structure. When the substrate temperature is increased to 300 and 400 °C, the hydrogen etching rate is reduced and deposition rate continues to increase.

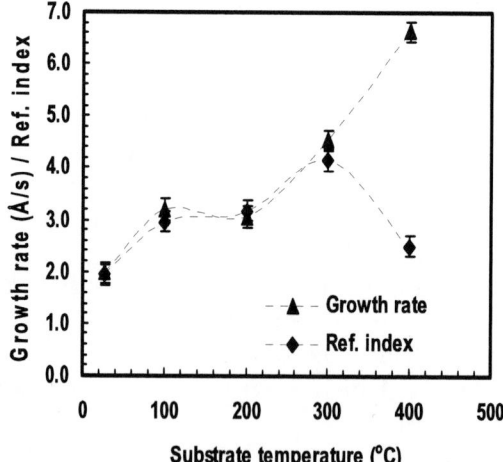

Fig. 2 Variation of growth rate and refractive index of nc-Si:H films prepared by LBL technique with substrate temperature.

The decrease in E_g value for the films prepared at substrate temperature of 300 °C with high $B^{1/2}$ value indicates the disappearance of quantum confinement effect as the grain size has increased. At 400 °C, a highly disordered film structure is indicated by the low E_g and $B^{1/2}$ value. The decrease in the refractive index for the film prepared at 400 °C is due to a less compact film structure which can be due to formation of voids as dangling bond concentration are increased since hydrogen incorporation is more difficult at this substrate temperature.

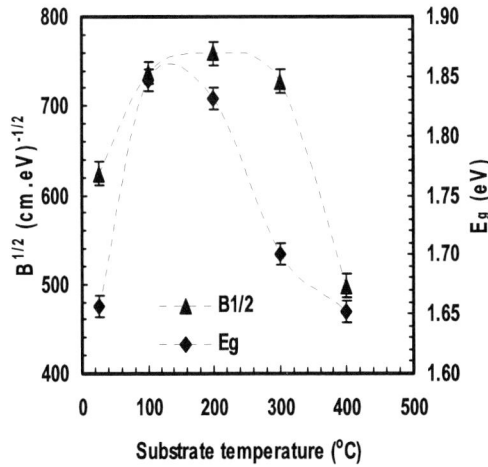

Fig. 3 Variation of $B^{1/2}$ and E_g of nc-Si:H films prepared by LBL technique with substrate temperature.

Fig. 4 shows the XRD patterns of nc-Si:H films deposited on c-Si substrates prepared by LBL technique at different substrate temperatures. Typical XRD spectra of nc-Si:H films usually show three peaks at scattering angles 28.5 °, 47.8 ° and 56.2 ° corresponding to the Si (111), (220) and (311) orientations respectively [9-11]. From fig. 4, only one sharp peak is present at 2θ angle of 56 ° representing the Si (311) orientation for the films prepared at substrate temperatures below 400 °C. This peak shows a small protrusion from a broad peak produced by the merging of the Si (220) and Si (311) orientation peaks. A broad Si (311) orientation is observed for the films prepared at substrate temperature of 400 °C. These results show that increase in substrate temperature to 300 °C increase crystallinity in the film structure and crystallinity drops when the substrate temperature is increased to 400 °C.

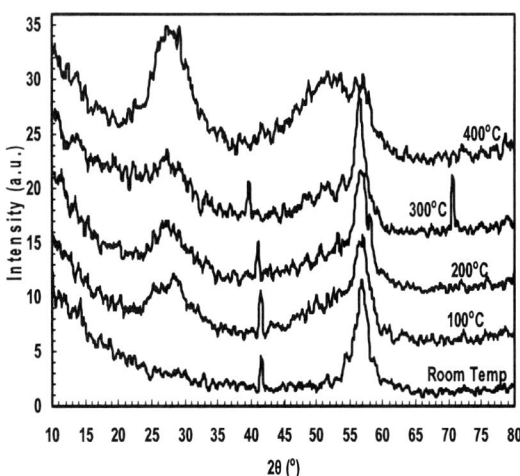

Fig. 4 XRD patterns of nc-Si:H films deposited on c-Si substrates prepared by LBL technique at different substrate temperatures.

Fig. 5 shows the average grain size of nc-Si:H films prepared by LBL at different substrate temperatures calculated from the Si (311) orientation peak. The average grain size is largest for the film prepared at 300 °C (6.2 ± 0.2 nm) as compared to the average grain size of the other films (~3.5 – 4.5 nm). This explains the decrease in the optical energy gap for this film (fig. 3) with the $B^{1/2}$ value remaining high. No evidence of quantum confinement effects is present in this film as the crystallites form a continuous structure and are not embedded in amorphous silicon matrix structure.

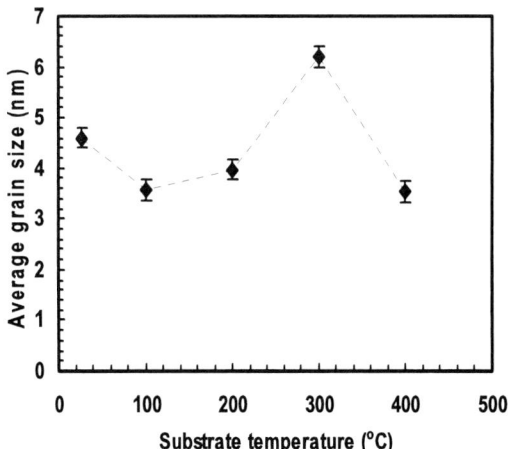

Fig. 5 Average size of nc-Si:H films deposited on crystal silicon substrates prepared by LBL technique at different substrate temperature.

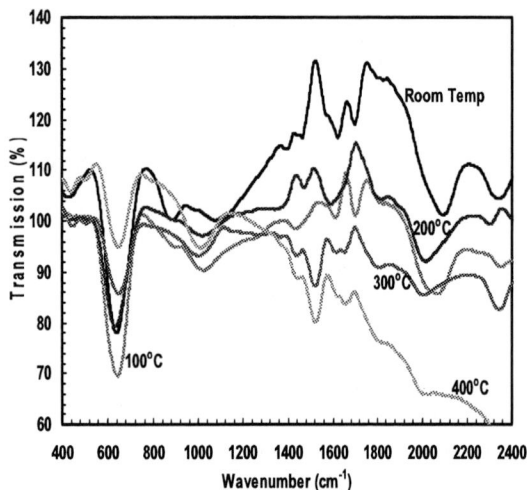

Fig. 6 FTIR transmission of nc-Si:H films prepared by LBL technique at different substrate temperatures.

Fig. 6 shows the FTIR spectra of the LBL nc-Si films prepared at different substrate temperatures. A typical FTIR spectrum of nc-Si:H films usually consist of peaks around 630, 880 and 2000 – 2100 cm^{-1} which corresponds to Si-H wagging, $(Si-H_2)_n$ bending and Si-H/Si-H$_2$ stretching bands respectively [12, 13]. Table 2 tabulates the various component of FTIR spectra obtained from absorption peak analysis of the spectra in the range 500 – 2500 cm^{-1}. The presence of Si-H$_2$ bending scissor absorption band at 880 cm^{-1} for the film prepared at room temperature to 300 °C indicates the presence of $(Si-H)_2$ bonds for these films. The presence of Si-H$_2$ bonds in these films suggests evidence of nanocrystallinity in the structure of these films. The shift in the Si-H/Si-H$_2$ stretching band towards lower wavenumber is consistent with decrease in the hydrogen content in the films prepared at higher substrate temperature as shown in fig. 7.

Table 1: Summary of various component of IR spectrum obtained from absorption peak analysis of the spectrum in the range 500 – 2500 cm^{-1}.

Variable parameter	Si-H wagging	Si-H$_2$ bending-scissoring	Si-H/Si-H$_2$ or Si-H polyhydride stretching
T(°C)	640 ± 1 cm^{-1}	880 ± 1 cm^{-1}	2000-2100 ± 1 cm^{-1}
room temp	639	883	2090
100	643	896	2064
200	634	shoulder	2013
300	648	shoulder	2006
400	648	-	2013

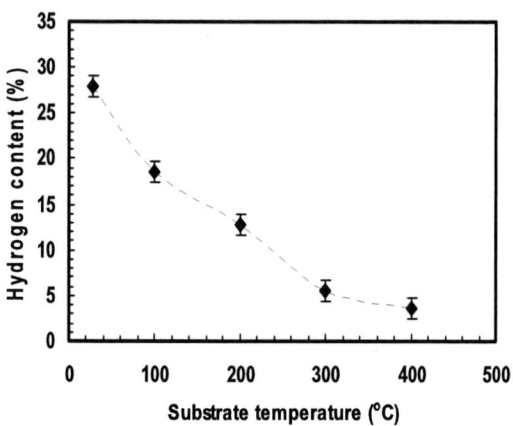

Fig. 7 Variation of hydrogen content of nc-Si:H films prepared by LBL technique with substrate temperature.

IV. CONCLUSION

The effects of substrate temperature on the film properties of nc-Si:H films produced by LBL technique using RF PECVD system have been studied. The growth rate of nc-Si:H films deposited by LBL technique increases with increase in substrate temperature. Hydrogen etching effects during the hydrogen plasma treatment is most dominant at substrate temperatures of 100 and 200 °C. Films produced at these substrate temperatures have high concentration of Si-Si bonds and shows evidence of quantum confinement effects. High substrate temperature of 400 °C forms a highly disordered film structure.

ACKNOWLEDGEMENT

This work was supported by the by the Ministry of Science, Technology and Innovation, Malaysia and University Malaya under IRPA research grant 09-02-03-0222-EA222 and IPPP/ UPDit/Geran(PPP) Research Fund P0166/2006A respectively. Also we would like acknowledge Mr. Muhamad Aruf from Physics Department, University of Malaya, Kuala Lumpur for operating the SIEMENS D5000 X-ray diffractometer.

REFERENCES

1. A.S. Gudovskikh, J.P. Kleider, V.P. Afanasjev, A.Z. Kazak-Kazakevich and A.P. Sazanov, "Investigation of nc-Si inclusions in multilayer a-Si:H films obtained using the layer by layer

technique", J. Non-Cryst. Solids, vol. 338-340, pp. 135 (2004).

2. X.-N. Liu, S. Njng, L.-C. Wang, G.-X. Chen and X.-M. Bao, "Photoluminescence of nanocrystallites embedded in hydrogenated amorphous silicon films", J. Appl. Phys., vol. 78, pp. 6193 (1995).

3. Y.H. Wang, J. Lin and C.H.A. Huan, "Structural and optical properties of a-Si:H/nc-Si:H thin films grown from Ar-H_2-SiH4 mixture by plasma-enhanced chemical vapour deposition", Mater. Sci. and Eng. B, vol. 104, pp. 80 (2003).

4. Nihed Chaabane, Pere Roca I Cabarrocas and Holger Vach, "Trapping of plasma produced Nanocrystalline Si particles on a low temperature substrate", J. Non-Cryst. Solids, vol. 338-340, pp. 51 (2004).

5. J. Tauc, *Amorphous and Liquid Semiconductors*, Plenum Press, (1974).

6. H.P. Klung and L.E. Alexander, *X-ray Diffraction Procedures*, Wiley, (1974).

7. H. Rozairi and S.A. Rahman, "High Deposition Rate Thin Film Hydrogenated Amorphous Silicon Prepared by d.c. Plasma Enhanced Chemical Vapour Deposition of Helium Diluted Silane", *ICSE2002 Proc.*, pp. 300 (2002).

8. Akihisa Matsuda, "Microcrystalline silicon: Growth and device application", J. Non-Cryst. Solids, vol. 338-340, pp. 1 (2004).

9. Chun-Yu Lin, Yean-Kuen Fang, Shih-Fang Chen, Ping-Chang Lin, Chun-Sheng Lin, Tse-Heng Chou, Jenn Shyong Hwang and Kuang I. Lin, "Growth of nanocrystalline silicon thin film with layer-by-layer technique for fast photo-detecting applications", Mater. Sci. & Eng. B, vol. 127, pp. 251 (2006).

10. S. Ray, S. Mukhopadhyay, T. Jana and R. Carius, "Transition from amorphous to microcrystalline Si:H: effects of substrate temperature and hydrogen dilution", J. Non-Cryst. Solids, vol. 299-302, pp. 761 (2002).

11. Y. He, C. Yin, G. Cheng, L. Wang and X. Liu, "The structure and properties of nanosize crystalline silicon films", J. Appl. Phys. vol. 75, pp. 797 (1994).

12. Sumita Mukhopadhyay, Amartya Chowdhury and Swati Ray, "Substrate temperature dependence of microcrystalline silicon growth by PECVD technique", J. Non-Cryst. Solids, vol. 352, pp. 1045 (2006).

13. Atif Mossad Ali, Takao Inokuma and Seiichi Hasegawa, "Structural and photo-luminescence properties of Nanocrystalline silicon films deposited at low temperature by plasma-enhanced chemical vapor deposition", Appl. Surf. Sci., Article in Press, (2006), Available online at www.sciendirect.com.

SIMS Analysis of Gate Oxide Breakdown
Due to Tungsten Contamination

D. Gui, Z.X. Xing, Z.Q. Mo, Y.N. Hua and S.P. Zhao, *Snr Member, IEEE*

Chartered Semiconductor Mfg Ltd
60 Woodlands Industrial Park D Street 2
Singapore 738406
Tel: 65-63604108, Fax: 65-63624978
Email: guid@charteredsemi.com

Abstract The gate oxide is the most fragile element of metal-oxide-semiconductor (MOS) transistor. Metallic contamination in the gate oxide leads to high leak current and even gate oxide breakdown. In this paper, we have investigated a failure case of NMOS gate oxide breakdown using secondary ion mass spectrometry (SIMS) because of its excellent sensitivity. The SIMS depth profiles at the test pad in the scribe line showed that the gate oxide breakdown was caused by tungsten (W) contamination. Further study indicated that W contaminated wafers during n-poly implantation by the re-deposition from the supporting disk of implanter. Based on the SIMS results, measures have been suggested to reduce the W contamination.

I. INTRODUCTION

THE mainstream of ultra-large scale integration (ULSI) technology is complementary metal-oxide-semiconductor (CMOS)-based silicon process due to its low power consumption [1]. As its name states, MOS has a unique element, 'oxide' namely 'gate oxide' here, distinguishing itself from other types of transistor structures, e.g. bipolar transistor. The gate oxide is the most fragile element of a transistor because it is an extremely thin layer of material biased during normal operation. Scaling of the gate oxide thickness causes the oxide to be more susceptible to gate current leakage, defect creation, and ultimately dielectric breakdown. Among the root causes of the gate oxide failure, metal contamination is the most common one. The presence of metallic contamination is known to have a negative impact on the performance of MOS device, e.g. increasing the leak current and reducing the minority carrier lifetime as well as gate oxide quality [2]. The transition metals are one major contamination in the wafer fabrication because of their high diffusion coefficient [3]. To improve the line yield and device reliability, it is necessary to monitor and determine the metallic contamination including the transition metals in the starting materials and the processes, which requires a diverse array of process characterization tools. Among them, secondary ion mass spectrometry (SIMS) is a critical tool of characterizing the dopant profile and contamination due to its excellent sensitivity and high depth resolution [4]. However, SIMS faces a big challenge in small area analysis for IC failure analysis. The test area is usually the test pad located in the scribe line or other structure with smaller size. To scan the primary beam over the small test pad, the spot size of the primary beam should be much smaller than the test area, which will limit the primary beam current. Furthermore, the small volume of material limits the total available atoms of trace elements. These two factors result in poor sensitivity in small area SIMS analysis, compared to large area analysis in blanket wafers [5].

In this paper, we will present the investigation on a failure case of gate oxide breakdown. The signatures imply that the failure is related to selective doping process. SIMS depth profiling was performed at electrical test (E-test) pad located in the scribe line using Cameca Wf SIMS machine. The root cause has been identified as tungsten (W) contamination in the gate oxide. Further study showed that the W was introduced by n-poly implantation. The process has been improved on the basis of the SIMS results.

0-7803-9730-4/06/$25.00 ©2006 IEEE

II. EXPERIMENTAL

The product wafers were sorted as 'good' or 'bad' wafers by the end of line (EOL) E-test. SIMS depth profiling was carried out on CAMECA IMS Wf system. Before SIMS experiment, the wafers were delayered to expose poly Si layer using diluted hydrofluoric acid (HF). A primary O_2^+ beam with 5 keV net impact energy was used to obtain W profiles from poly Si till Si substrate. To avoid any mass interference to W, a technique of energy window shift was employed to reduce the W background. The SIMS depth profiling was carried out in the in-line measurement pad structures located in the scribe line. Therefore, the primary beam was confined to scan over an area within 50μm x 50μm. The secondary ions were detected at the center part of the scan area with a diameter of 12 μm to avoid the crater effect.

Short loop blanket wafers were depth profiled to find out the source of W. The scan area and detection area are 200μm x 200μm and 62μm in diameter, respectively. The depth scale was calibrated using an *in situ* laser interferometer. And the relative sensitive factor for W was calibrated using a W implanted standard sample with a known dose.

III. RESULTS AND DISCUSSION

The failure was sorted out by EOL E-test. The failure has following signatures: 1). All failed devices are NMOS. 2). The gate oxide breakdown voltage decreases from –4.8V, the normal value, to –1.6V. The signatures imply that the failure is related to selective doping process. The machine communality study showed that the n-ploy implantation of the all failure wafers was made just after the wafers with WSi_x layer. It is reasonable to suspect that the failure is caused by the W contamination. Therefore two product samples, sorted as 'good' and 'bad' at the EOL E-test, were depth profiled using SIMS to confirm this suspicion. The results were shown in the Fig. 1 and Fig. 2. It is interesting to note that the surface W peak in Fig. 1 and Fig. 2 is the WSi_x residual after delayering. Thanks to the enhancement effect of oxygen on the positive ion, the gate oxide position can be easily marked at the Si peak at the depth of around 150nm. Compared to the W

profile of good sample, the bad one obviously has W contamination in the gate oxide region. The transition metal contamination can reduce the height of the potential barrier at electrode-SiO2 interface, which results in the gate oxide breakdown [3]. Hence one can conclude that the root cause of this failure is W contamination in the gate oxide.

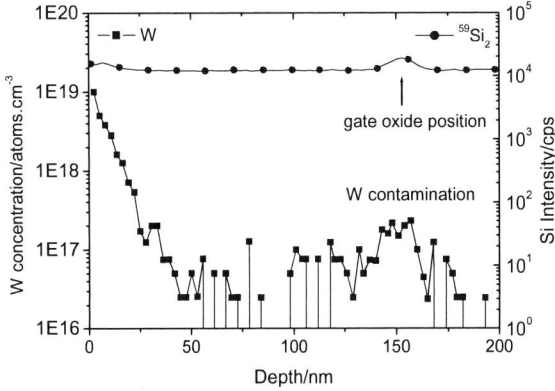

Fig.1 SIMS depth profiles of bad sample. W contamination was detected in the gate oxide.

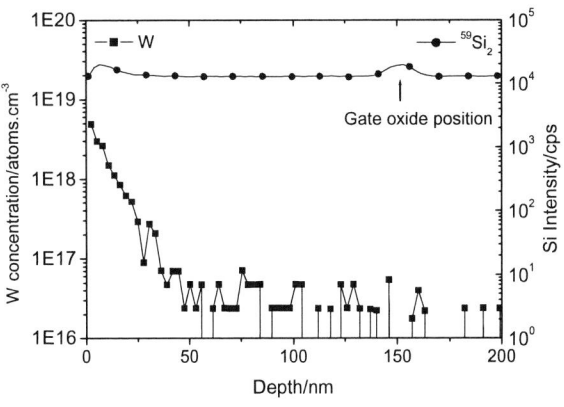

Fig.2 SIMS depth profiles of good sample. No obvious W contamination was detected in the gate oxide.

It is well known that the implantation is not an extremely clean process. The contamination may be introduced onto the wafer in the form of either energetic 'co-implantation' or not-energetic re-deposition from the implanter parts, e.g. disc supporting the wafers [6]. To confirm that W is indeed introduced during n-poly implantation, an experiment has been designed to determine the W contamination on the wafer surface after n-poly implantation. The W level was checked at different seasoning conditions, listed in the Table 1. The typical results are shown in Fig. 3 and Fig. 4. It is evident that n-poly implantation indeed introduced W contamination.

Table 1 Seasoning conditions for n-poly implant

Condition	Description
A	run 3 lots wafers without WSi_x layer, followed by n-poly implant 0.5hr
B	run 4 lots wafers with WSi_x layer, followed by n-poly implant 0.5hr
C	condition B, followed by n-poly implant 0.5hr
D	condition C, followed by n-poly implant 0.5hr
E	condition D, followed by n-poly implant 0.5hr
F	clean disk, followed by n-poly implant 0.5hr

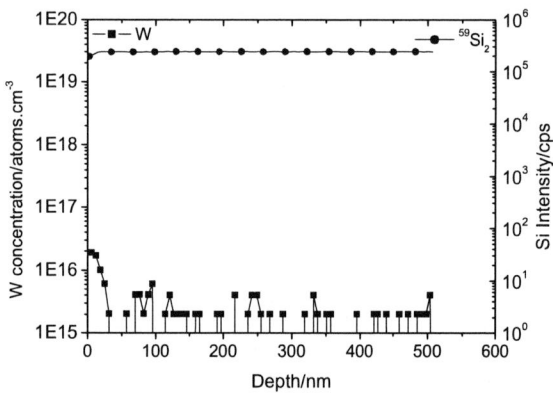

Fig.3 SIMS depth profiles of wafer implanted with implanter c under seasoning condition A.

We also compared the W level of three implanters under the same conditions listed in Table 1. From the results shown in Table 2, implanter c has highest W contamination level. The reason is that the implanters have different disk shape. Implanter c has a round shape disk, whereas implanter a and b have bar shape disk. The round disk provides bigger area to accommodate W. Furthermore, W dose decreases

with the seasoning time of wafers without WSi_x layer, which provides a method of cleansing the disk to reduce the W contamination.

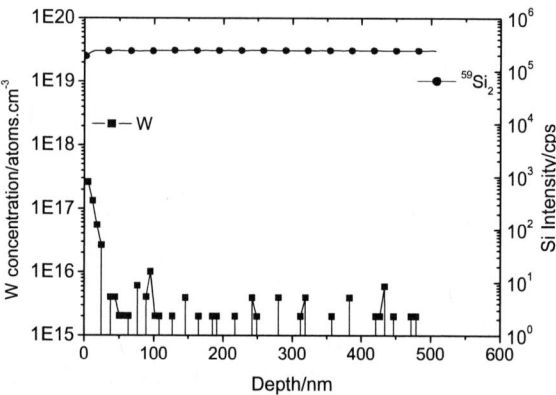

Fig.4 SIMS depth profiles of the wafer implanted with implanter c under seasoning condition B. W contamination was observed.

Table 2 W dose on the wafer surface under different seasoning condition.

Condition	W dose/atoms.cm^{-2}		
	implanter a	implanter b	implanter c
A	2.890E10	3.547E10	1.258E11
B	7.225E10	7.833E10	2.457E11
C	5.188E10	5.775E10	1.959E11
D	4.116E10	5.362E10	1.673E11
E	4.068E10	4.524E10	1.377E11
F	3.511E10	3.485E10	1.009E11

. To reduce W contamination, the n-poly implantation on the wafer with WSi_x layer should be performed on the implanter with bar-shape disk. The cleansing disk with wafers without WSi_x layer would be helpful for further reducing W contamination.

IV. CONLUSION

In this work, we have investigated a case of NMOS gate oxide breakdown using CAMECA Wf SIMS. The failure was caused by W contamination in the gate oxide, which was introduced via re-deposition from the implanter disk during n-poly implantation. Implanter with bar-shape disk was suggested to be used for the implantation on the wafers with WSi_x layer to reduce the possibility of W contamination. The cleansing disk with wafers without WSi_x layer would be helpful for further reducing W contamination.

ACKNOWLEDGEMENT

The authors would like to thank Ms. Soh Chye Lin with Chartered Semiconductor Manufacturing Ltd for the useful technical discussion.

REFERENCE

[1] P. Chatterjee, "ULSI market opportunities and manufacturing challenges", *IEDM Tech. Dig. 11* (1991)

[2] S.Q. Hong, T. Wetterroth, H. Shin, S.R. Wilson, D. Werho, T.C. Lee and D.K. Schroder, " Improvement in gate oxide integrity on thin-film silicon-on-insulater substrates by lateral gettering", *App. Phys. Lett.* Vol. 71(23), p. 3397 (1997)

[3] D.L. Kwong, "Si device processing", in R. W. Cahn (ed.) Materials science and technology: a comprehensive treatment, Vol. 16 Processing of semiconductors / Vol. ed.: K.A. Jackson, VCH Verlagsgesellschaft, Weinheim, 1996

[4] F.A. Stevie, R.G. Wilson, D.S. Simons, M.I. Current and P. C. Zalm, "A Review of SIMS Characterization of Contamination Associated with Ion Implantation," *J. Vac. Sci. Technol. B,* vol. 12, 2263-79 (1994)

[5] H. Ogata, "Gigabit devices and technologies: SIMS challenges of the near future, in *Proceedings of SIMS X,* 1997, p. 3

[6] D. Gui, Y.N. Hua, Z.X. Xing and C.W. Tan, "SIMS Analysis of Boron Cross Contamination in Al Implantation Process of Wafer Fabrication", Proceedings of IEEE NSM 2005, Kuching, Malaysia, November 21st- 24th, 2005, p.116

Evaluation of Stopping Power of Photo-resist to Ion Implantation by Using SIMS

D. Gui, Z.X. Xing, Z.Q. Mo, Y.N. Hua and S.P. Zhao, *Snr Member, IEEE*

Chartered Semiconductor Mfg Ltd
60 Woodlands Industrial Park D Street 2
Singapore 738406
Tel: 65-63604108, Fax: 65-63624978
Email: guid@charteredsemi.com

Abstract: **The stopping power of photo-resist is an important parameter to define the photo-resist thickness. In this paper, we have developed a novel method to evaluate the stopping power of the photo-resist. This method is to directly determine the implantation profile in the photo-resist using secondary ion mass spectrometry (SIMS) due to its excellent sensitivity and high depth resolution. We have obtained the depth distribution of dopants, such as boron, arsenic, phosphorous and indium, in the photo-resist using CAMECA Wf SIMS machine. The results obtained with the novel method are straightforward and unambiguous, compared to the conventional method. The safe photo-resist thickness has been optimized based on the SIMS results.**

I. INTRODUCTION

IN wafer fabrication, photo-resist plays a crucial role in forming integrated circuit (IC) patterns. The IC patterns on the mask are reproduced on the photo-resist and finally transferred onto the silicon (Si) wafers in a series of lithography and etch process. Not only the circuit or device pattern itself, the patterns on the photo-resist define also the area of selective doping. Nowadays, ion implantation is the mainstream technology of selective doping. The fact that implantation share the same photo-resist with other processes [1], e.g. lithography, put stringent requirements on the photo-resist parameters. One of them is photo-resist thickness. On one hand, lithography requires the photo-resist to be as thin as possible, since the lithography resolution decreases with the photo-resist thickness as well as the radiation wavelength

[2]. This requirement becomes more challenge as the IC dimension continues to shrink. On the other hand, implantation requires the photo-resist to be as thick as possible to avoid any penetration of the implantation through the photo-resist itself to counter dope the area under the photo-resist. The undesirable penetration will lead to the drift of the device parameters or even failure. Therefore the photo-resist thickness should be optimized on the basis of its stopping power.

Conventionally the stopping power is evaluated by determining the implantation penetration in the Si substrate after striping the photo-resist. The 'safe' photo-resist thickness can be defined as the minimum thickness, with which the penetration dose is less than the control limit. To obtain the safe thickness, the penetration dose is determined with a series of photo-resist thickness. Although this method provides a simple measurement of the penetration dose in the Si wafer using SIMS, the results can be ambiguous because dopant profile near the surface region may be not only the penetration but also the surface contamination introduced during the sample preparation. In this paper, we will present a novel method to evaluate the stopping power of the photo-resist. This method is to directly determine the implantation profile in the photo-resist using SIMS. Compared to the conventional method, this novel method is straightforward and unambiguous. However, depth profiling the photo-resist is a challenging job due to the fact that the photo-resist is dielectrics, which results in charging effect during SIMS depth profiling [3]. Without proper charge neutralization, it is either impossible to get the dopant profile in the photo-resist or, if we do get, the results will be unreliable. With optimized charge neutralization,

0-7803-9730-4/06/$25.00 ©2006 IEEE

we have obtained the depth distribution of dopants, such as boron (B), arsenic (As), phosphorous (P) and indium (In), in the photo-resist using CAMECA Wf SIMS machine. The SIMS results provide useful information for setting the suitable photo-resist thickness.

II. EXPERIMENTAL

The experiments were carried out on Cameca IMS Wf machine. A primary O_2^+ beam with 3keV net impact energy was used to depth profile electropositive elements, B and In, in the positive mode for better sensitivity. To avoid crater effect, the primary beam scanned over an area of 200μm x 200μm and the secondary ions were detected from an area with a diameter of 62μm. For the dopants of As and P, a primary Cs^+ beam with 9keV net impact energy was used. The scan area is 150μm x 150μm, whereas the analysis area is 33μm in diameter.

To neutralize the charges generated by the SIMS sputtering, the samples were coated with a thin layer of platinum (Pt) for good contact with sample stage. To get reliable SIMS results, the sample surface should have no net charge, which does not change the energy distribution of the secondary ions. Therefore, the electron beam current should be finely controlled to realize the optimized charge neutralization condition.

III. RESULTS AND DISCUSSION

Conventionally, the stopping power of the photo-resist is evaluated by determining the dopant's penetration dose or profile in the substrate

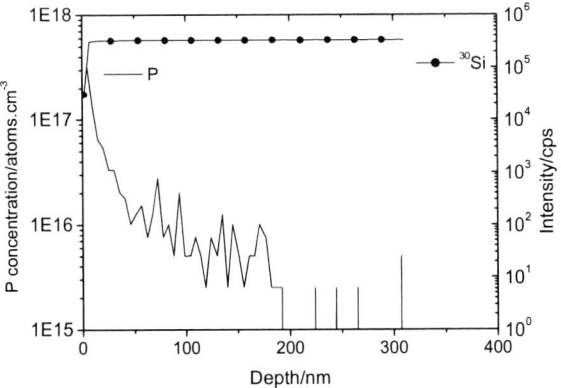

Fig.1 P penetration profile in the Si substrate

after stripping the photo-resist. If the photo-resist is not thick enough, the implantation will penetrate through the photo-resist. Fig.1 shows an example of P penetration profile.

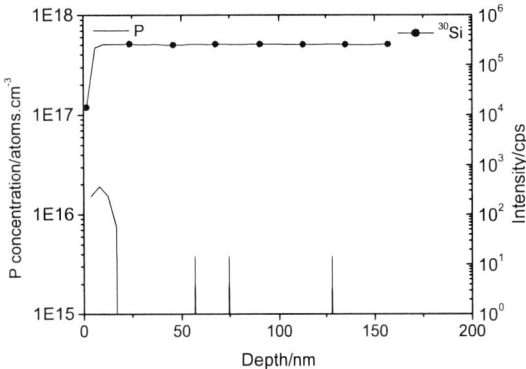

Fig.2 P profile in the Si substrate after stripping the photo-resist. P was implanted at 1.8MeV through 3.5μm photo-resist.

However, the results sometimes can be ambiguous because dopant profile near the surface region may be not only the penetration but also the surface contamination introduced during the sample preparation. As shown in the Fig. 2, a low concentration and shallow P profile was detected on the Si wafer surface, which was implanted with 1.8MeV P through 3.5 μm photo-resist. The evidence is not so strong to make a conclusion that the P profile is really caused by penetration or surface contamination. To ensure the profile is not a penetration, a novel method was proposed to evaluate the stopping power of the photo-resist. A sample with a 1.8MeV P implantation in 4μm photo-resist was depth profiled by using 9keV Cs^+ beam. The $^{12}C^-$ and $^{30}Si^-$ were monitored as reference ions to identify the interface between photo-resist and Si substrate. From the profiles shown in Fig.3, the P implantation is well confined in the photo-resist layer. Therefore, the P profile in Fig. 2 is indeed surface contamination. In other words, the photo-resist thickness is safe. The safe thickness of the photo-resist can be defined as the minimum thickness, with which the penetration dose or concentration is smaller than the control limit. If the control limit is 1E15 at/cm³, the safe thickness is 2μm, assuming that the P profile follows Gaussian distribution.

491

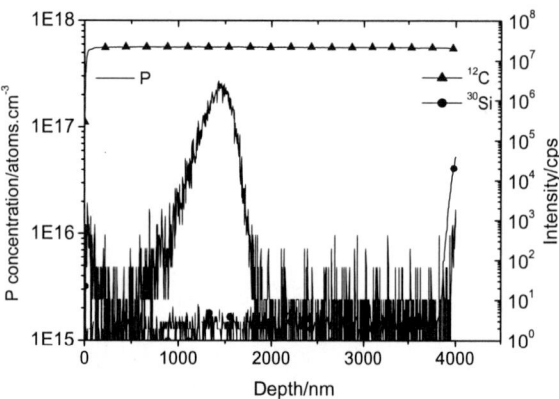

Fig.3 P profile in 4μm photo-resist. P was implanted at 1.8MeV with a dose of 1.3E13 at/cm^2.

Similarly, this method can also apply to the implantation of other dopants, such as As, B and In, in photo-resist, as shown in Fig.4-6. It is notable that the for better sensitivity As was detected as

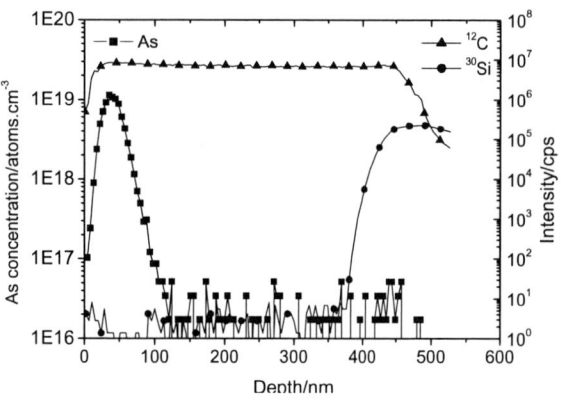

Fig.4 As profile in 450nm photo-resist. As was implanted at 45keV with a dose of 3.8E13 at/cm^2.

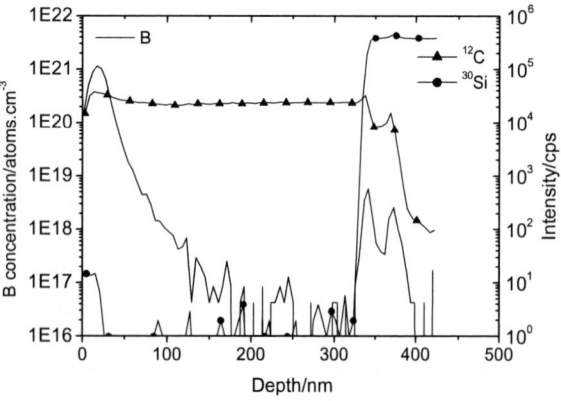

Fig.5 B profile in 350nm photo-resist. B was implanted at 8keV with a dose of 2.5E15 at/cm^2.

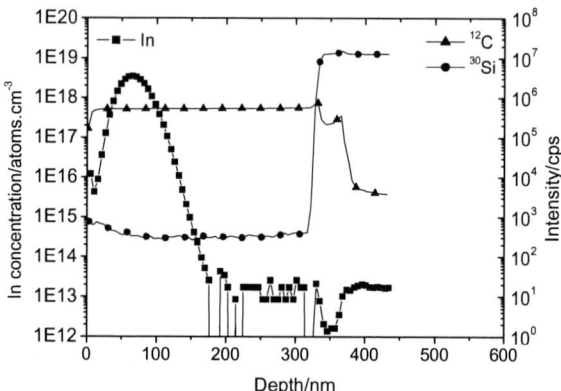

Fig.6 In profile in 350nm photo-resist. In was implanted at 80keV with a dose of 1.5E13 at/cm^2.

'As$^-$' rather than 'AsSi', since the Si concentration in the Photo-resist is very low.

To obtain the safe thickness of the photo-resist, the dopant profile should be extrapolated using Guassian function when the control limit is set below the SIMS detect limit   For the given implant dopant and energy in the work, the safe thickness for As, B and In are 150nm, 270nm and 148nm, respectively, if the control limit for all elements is 1E15 at/cm^3.   It is notable that the stopping power is dependent on the composition of the photo-resist. Each type of photo-resist should be independently evaluated on its stopping power of certain implantation.

The actual photo-resist thickness in the process should be set as a suitable value taking account of safe thickness for blocking implantation and lithography resolution. Another factor should be taken into account is the topography of the patterned wafer, which requires a minimum thickness to ensure that the photo-resist is uniformly coated on the wafer [4].

IV. CONLUSION

The stopping power to implantation is one of the key parameters of the photo-resist. In this paper, we have presented a novel method to evaluate the stopping power by directly determining the dopant profiles in the photo-resist using SIMS. For the given implant energy and dose, the safe thickness has been obtained with Cameca Wf for the dopant of B, P, As and In. Compared to the conventional method, this novel method is more straightforward and unambiguous. The photo-resist thickness can be then optimized by balancing the requirements of lithography resolution and blocking the undesired implantation.

ACKNOWLEDGEMENT

The authors would like to thank Mr. Zhang Xiaolin with Chartered Semiconductor Manufacturing Ltd for the useful technical discussion.

REFERENCE

[1] T.C. Smith, "Photoresist and particulate problem", in: J.F. Ziegler (ed.), Handbook of ion implantation technology, Elsevier Science Publishers B.V., 1992

[2] L.F. Thompson, C.G Willson and M.J. Bowden, "Inroduction to Microlithography", *ACS Symp. Ser. 219*. Washington: ACS.

[3] D. Gui, Y.N. Hua, Z.X Xing and S.P. Zhao, "Study on Potassium Contamination in SOI Wafer Fabrication Using Dynamic SIMS", Proceedings of 13th IPFA 2006, Singapore, July 5th - 7th, 133-136

[4] R Leuschner and G. Pawlowski, "Photo-lithography", in R. W. Cahn (ed.) Materials science and technology: a comprehensive treatment, Vol. 16 Processing of semiconductors / Vol. ed.: K.A. Jackson, VCH Verlagsgesellschaft, Weinheim, 1996

Design and Analysis of an Ultra Wideband Matrix Mixer

R. Hallaji , A. Abdipour and G. Moradi

Microwave, mm Wave and Wireless Communications Research Lab., Electrical
Engineering Department, Amirkabir University of Technology
(Tehran Polytechnic), Tehran, Iran
Email: Reza_hallaji@ aut.ac.ir, abdipour@aut.ac.ir and GhMoradi@aut.ac.ir

Abstract- **Ultra wideband matrix mixer using well known harmonic balance (HB) technique is analyzed. Matrix structure gives a suitable conversion gain. The proposed structure contains FET elements in each row and column, and causes an improved performance including conversion gain and output power. Mixer performance is analyzed in Ultra wide band (UWB). The structure is used in UWB circuits and systems such as RFID cards.**

I. INTRODUCTION

Active mixers are used in modern microwave/ mmwave wave receivers such as mobile and satellite communication circuit. There are many works on signal analysis of such FET mixers [1,2]. Most of them are based on HB method for signal analysis (*e.g.* the gain and stability) of nonlinear circuits. For the circuits containing one large input signal and one (or more) small signal(s), the large-signal-small-signal (LSSS) method is used. In this approach after large signal analysis, a perturbational analysis is applied and a small signal linear multifrequency (SSLMF) model is obtained [2].

Consider a lumped component nonlinear circuit that is described by nonlinear ordinary differential equations (ODEs) in the time domain. Most often these equations are written in the modified nodal admittance (MNA) formulation [3].

$$C\dot{\chi}+G\chi+f(\chi)+u=0 \qquad (1)$$

where C and G are $N_x \times N_x$ matrices, x is a column vector of unknown circuit variables, and u is a vector of independent sources.

For steady-state analysis, we must either assume that the circuit is under periodic excitation or that the circuit is autonomous and generates periodic output. In both cases, solution vector x is periodic with fundamental frequency corresponding to period τ

$$x(t + \tau) = x(t). \qquad (2)$$

Equation (1) with boundary conditions (2) can be solved by expanding nonlinear ODEs (1) into a nonlinear algebraic equation for the expansion coefficients of x using an approach that is well known as the method of moments [4]. Suppose that we have an expansion basis $\{u_k\}|_{k=1}^m$ so that we can write a best approximation for x as follows:

$$x = \lim_{m\to\infty} \sum_{k=1}^{m} X_k v_k. \qquad (3)$$

In order for the solution to satisfy boundary conditions (2), expansion basis must satisfy these boundary conditions as well. In other words, expansion basis must be periodic. Let us assume that $[x_l]$ is a discrete vector containing values of x sampled in the time domain at time points $[t_l]$, $l=\overline{1...N_t}$ and that basis $\{v_k\}$ is periodic and has a pair of forward T and inverse $\tilde{T}$ discrete transforms associated with it as follows:

$$X = T[x_l] \qquad [x_l] = \tilde{T}X. \qquad (4)$$

The nonlinear term in (1) can be represented in the following form:

$$F(X) = Tf(\tilde{T}X) \qquad (5)$$

Equation (1) can then be written in the transform domain as a nonlinear matrix equation

$$\hat{C} DX + \hat{G} X + F(x) + U = 0 \qquad (6)$$

where $\hat{C}, D$ and $\hat{G}$ are $N_t N_x \times N_t N_x$ matrices. $\hat{C}$ and $\hat{G}$ are obtained from C and G, respectively, by taking their tensor product with an $N_t \times N_t$ identity matrix. We denote the left-hand side of (6) as $\phi(X)$ and write it as

$$\phi(X) = \left(C \hat{D} + \hat{G} \right) X + F(X) + U = 0. \qquad (7)$$

Matrix in (6) and (7) is a projection of the derivative operator d/dt onto space spanned by $\{v_k\}$ as follows:

$$\left[D_{i,j} \right] = \left\langle \frac{d}{dt} v_i, v_j \right\rangle. \qquad (8)$$

The solution of (7) is usually performed using Newton iterations. Assuming $X^{(0)}$ as the initial guess for X, the linear matrix equation to be solved at each step becomes

$$J\left(X^{(i)}\right)\left(X^{(i+1)} - X^{(i)}\right) = -\phi\left(X^{(i)}\right) \qquad (9)$$

where $X^{(i)}$ is the solution of the i th iteration, $\phi(X)$ is defined by (7), and $J(X)$ is the Jacobian of $\phi(X)$

$$J(X) = [J_{kl}(X)] = \left[\frac{\partial \phi_k}{\partial X_l} \right], \qquad k,l = \overline{1,\ldots.(N_t N_x)}. \qquad (10)$$

Substituting (7) into (10) and applying a chain rule, we obtain the following expression for computing the Jacobian [5]:

$$J(X) = \hat{C} D + \hat{G} + T \left[\frac{\partial f_k}{\partial x_l} \right] \tilde{T}. \qquad (11)$$

The Jacobian is computed as a sum of the following three matrix components:

1) $\hat{C} D$;

2) $\hat{G}$;

3) $T \left[\dfrac{\partial f_k}{\partial x_l} \right] \tilde{T}.$

Sparsity of the Jacobian becomes equal to the sparsity of the densest of these three components. Matrices C and G result from the MNA formulation and typically have a rather sparse structure. Matrix of derivatives D will have a sparse structure only if the chosen basis allows sparse representation of the derivative operator, *i.e.* most of the elements in (8) vanish. This naturally happens if $\{v_i\}$ has local support (local support for basis functions means local support for their derivatives and, therefore, D becomes a band-limited matrix).

Sparsity of the third component in (11) depends primarily on the sparsity of the forward and inverse transform matrices T and $\tilde{T}$, as $[(\partial f_k)/\partial x_l]$ for time-invariant systems is just a block matrix consisting of diagonal blocks.

For simplicity, we will first consider a scalar case of (1) where both $\hat{C}$ and $\hat{G}$ matrices in (7) can safely be assumed as being diagonal.

II. FOURIER BASIS: HB FORMULATION

For many years now, Fourier basis has been the natural choice for solving the steady-state analysis problem. Fourier basis for the solution of (1) is usually constructed on an interval that ensures periodicity of the solution and includes $2N_f + 1$ basis functions with base frequencies that are multiples of the fundamental frequency in the circuit [6]

$$\{v_i\} = \{1, \cos\omega t, \sin\omega t, \cos 2\omega t, \sin 2\omega t, \ldots, \cos N_f \omega t, \sin N_f \omega t\}. \qquad (12)$$

Since complex exponents are natural eigenfunctions of the derivative operator, the derivative matrix D in this basis becomes a diagonal matrix in real Schur form with base frequencies on the main diagonal

$$D = \omega \begin{bmatrix} 0 & & & & & & & & & \cdots & \\ & 0 & 1 & & & & & & & \cdots & \\ & -1 & 0 & & & & & & & \cdots & \\ & & & 0 & 2 & & & & & \cdots & \\ & & & -2 & 0 & & & & & \cdots & \\ & & & & & 0 & 3 & & & \cdots & \\ & & & & & -3 & 0 & & & \cdots & \\ \cdots & \cdots & \cdots & \cdots & \cdots & \cdots & \cdots & \cdots & \cdots & \cdots & \cdots \\ & & & & & & & \cdots & 0 & k \\ & & & & & & & \cdots & -k & 0 \end{bmatrix}. \qquad (13)$$

The transform matrix T has dimensions of $N_t \times (2N_f + 1)$ with N_t being the number of time points and N_f being the number of frequencies. This matrix has the structure shown in (14) at the next page. If

$$T = \begin{bmatrix} 1 & \cos(\omega t_0) & \sin(\omega t_0) & \dots & \cos(N_f \omega t_0) & \sin(N_f \omega t_1) \\ 1 & \cos(\omega t_1) & \sin(\omega t_1) & \dots & \cos(N_f \omega t_1) & \sin(N_f \omega t_1) \\ \dots & \dots & \dots & \dots & \dots & \dots \\ 1 & \cos(\omega t_{N_t-1}) & \sin(\omega t_{N_t-1}) & \dots & \cos(N_f \omega t_{N_t-1}) & \sin(N_f \omega t_{N_t-1}) \end{bmatrix}$$

(14)

$$N_t = 2N_f + 1 \qquad (15)$$

Then T is a square matrix, which is nonsingular with a proper choice of time-sampling points. If more restrictions are imposed on the time-sampling points, T can also be made orthogonal as follows:

$$\tilde{T} = T^{-1} = T^T. \qquad (16)$$

This matrix clearly is dense, which would suggest $O(N^2)$ operations for computing Fourier coefficients in (4). This cost can be reduced to $O(N \log N)$ by applying the fast Fourier transform (FFT) algorithm for computing the T and T^{-1} operators.

III. NUMERICAL RESULTS

Matrix structures have very useful applications in different microwave/ mmwave circuits such as amplifiers and mixers [7, 8]. A simple version of our proposed matrix mixer is shown in Fig.1.

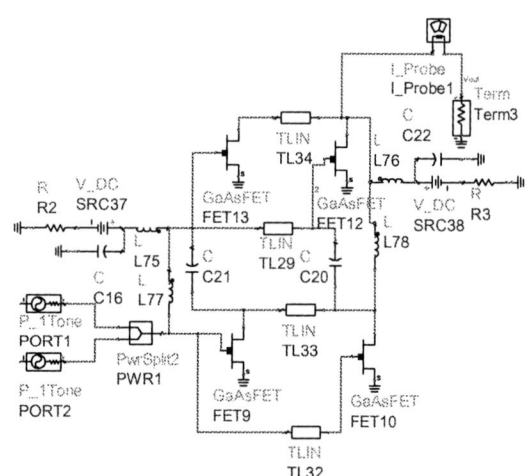

Fig. 1 The simple model of the matrix mixer

The mixer was configured for down-conversion with local oscillator (LO) input at 6 GHz, RF input 7 GHz and IF output at 1 GHz. inputs and outputs were matched to 50-Ω active impedance at their respective frequencies.

LO power was kept constant at +10 dBm and RF power was kept constant at -10dBm. For evaluating the circuit performance, we have simulated the conversion gain versus the LO Power. The simulation results show that lower LO Power leads increasing the basic harmonic power and decreasing the higher harmonic ones. This result is shown in Table.1.

Table.1. The output power at different frequencies

LO power (dBm) / Freq(GHz)	+10	0	-10
12	-2.45	-15.32	-34.46
14	-13.97	-41.06	-40.4
15	-33.16	-76.88	-83.59

ACKNOWLEDGEMENT

This work is supported in part by Iran Telecommunication Research Center (ITRC).

REFERENCES

[1] S. A. Maas., *"Microwave mixers,"* Artech House Inc., 1990.

[2] S. A. Maas., *"Nonlinear microwave circuits,"* Artech House Inc., 1994.

[3] J. Vlach and k. singhal, *"Computer Methods for Circuit Analysis and Design,"* New York: Van Nostrand, 1983.

[4] D. G. Dudley, *"Mathematical Foundations for Electromagnetic Theory,"* Piscataway, NJ: IEEE Press, 1994.

[5] K. S. Kundert, G. B. Sorkin, and A. Sangiovanni-Vincentelli, "Applying harmonic balance to almost-periodic circuits," *IEEE Trans. Microwave Theory Tech.*, vol. MTT-36, pp. 366–378, Feb. 1988.

[6] P. J. C. Rodrigues, *"Computer-Aided Analysis of Nonlinear Microwave Circuits,"* Norwood, MA: Artech House, 1998.

[7] A. Abdipour, and G. Moradi, "Improvement of signal and noise performances of a matrix amplifier," *International Symp. Electron Device for Modeling and Optoelectronic applications,* Austria, 339-343, Nov. 2001.

[8] G. Moradi, A. Abdipour, F. Farzaneh and A. Ghorbani, "Signal and noise improvement of travelling wave FET mixers," *International Conference on Microwave and mm-Wave Technology*, Beijing, China, Sept. 2000, pp 76-79.

Design and Analysis of Insulate Gate Bipolar Transistor (IGBT) with P+/SiO₂ Collector Structure Applicable to High Voltage to 1700 V

Han-Sin Lee, Yo-Han Kim, Ey-Goo Kang and Man-Young Sung
Department of Electrical Engineering, Korea University
1, 5-ka, Anam-dong, Sungbuk-ku
Seoul 136-713, KOREA
E-mail : hanlee@korea.ac.kr

Abstract   **In this paper, we propose a new structure that improves the on-state voltage drop along with the switching speed in Insulated Gate Bipolar Transistors(IGBTs), which is widely applied in high voltage semiconductors. The proposed structure is unique that the collector area is divided by SiO₂ regions, whereas in existing IGBTs, the collector has a planar P+ structure. The process and device simulation results show remarkably improved on-state and switching characteristics. The current and electric field distributions indicate that the segmented collector structure increases the electric field near the SiO₂ edge which leads to an increase in electron current and finally a decrease in on-state voltage drop to 30% ~ 40%. Also, since the area of the P+ region decreases compared to existing structures, the hole injection decreases which leads to an improved switching speed to 30%.**

I. INTRODUCTION

Insulated Gate Bipolar Transistors(IGBTs), widely used as high voltage semiconductors to inverters and motor drivers, has better characteristics for high voltage blocking and on-state voltage drop, however has weak turn-off switching characteristics compared to Power MOSFETs[1]. Also, it is difficult to improve the on-state voltage drop and turn-off switching characteristics simultaneously due to the trade-off relationship of the two characteristics. New designs of IGBTs have been proposed to improve the on-state voltage drop and turn-off switching characteristics by using Field Stop[2], Super Junction[3] and striped Anode[4] etc. In the case of striped anode, the P+ collector region is composed of a sandwich structure with high doping and low doping regions for reducing the hole injection efficiency. As a result, the

switching speed is improved, however the on-state resistance is increased by the amount of reduction of total hole injection. In this paper, we propose a new collector structure named High Density Electron IGBT (HDE-IGBT) which improves the on-state resistance and switching characteristics at the same time. HDE-IGBT has segmented P+ regions between the trenched SiO₂ to reduce the amount of hole injection amount. In this paper, the characteristics of HDE-IGBT are analyzed, and a new model of HDE-IGBT is suggested by process and device simulation.

II. THE CONCEPT OF P+ /SiO₂ COLLECTOR

Fig. 1 represents the proposed High Density Electron IGBT (HDE-IGBT). (a) is the conventional IGBT and (b) is the proposed structure. In the case of HDE-IGBT, compared to the conventional IGBT, the P+ region of collector area is segmented by locally trenched SiO₂ regions.

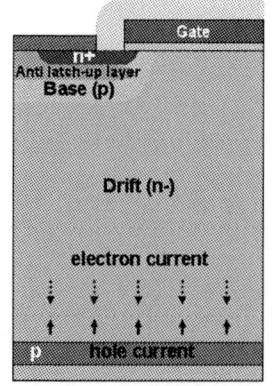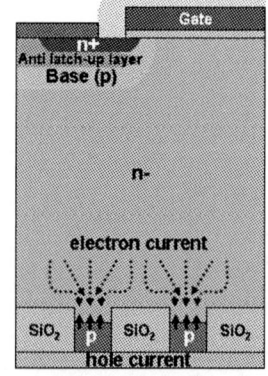

Fig. 1 The proposed High Density Electron-IGBT (HDE-IGBT) structure. (a) Conventional IGBT structure (b) Proposed IGBT structure

In the case of HDE-IGBT, it is possible to control the hole injection efficiency by adjusting the P+ area, which induce improvement of switching speed and decrease on-state resistance by enlarging the electron concentration near the oxide edge. The general operation method of HDE-IGBT is the same to that of the conventional IGBT, which is, the injection of high-density holes from the P+ collector into the N- drift region and electron current from N+ emitter into the N- drift. However, the injected hole concentration of HDE-IGBT is much less than that of conventional IGBT because of the decreased area of the P+ collector. The decreased amount of holes causes faster switching speed since the holes act as minority carriers in IGBT. In addition, the electron current can be increased around the SiO_2 corner by concentration of electric field, which induces a reduction of on state resistance. As a result, not only the on-state characteristics but also the turn-off characteristics can be improved simultaneously.

In this paper, we carried out some simulations with the process (TSUPREM4) and device (MEDICI) simulator for analysis of conventional IGBTs and the HDE-IGBT. The device structure used in this research is for 1700 V operation and its doping concentration is shown in Table 1. The analyzed parameters are breakdown voltage, on state voltage drop and turn off speed.

Table 1 Doping concentration of simulated IGBT

	Emitter (N+)	Anti latch-up layer(P+)	Base (P)	Drift (N-)	Collector (P+)
Concen-tration (cm^{-3})	2×10^{19}	1×10^{19}	1×10^{17}	5.5×10^{13}	5×10^{17}

III. ELECTRICAL CHARACTERISTICS

TSUPREM4 and MEDICI simulators were used to compare the electrical characteristics of IGBTs. Fig. 2 represents the on-state I-V curve of conventional IGBT and HDE-IGBT. In this figure, the P+ area size of HDE-IGBT(W×L) is set to 0.5×0.5 μm^2. The on-state voltage drop (Vcesat @Ice=100 A/cm²) of HDE-IGBT is 2.5 V, which is smaller about 0.6 V than that of the conventional IGBT. The HDE-IGBT has superior on-state characteristics compared to the conventional IGBT.

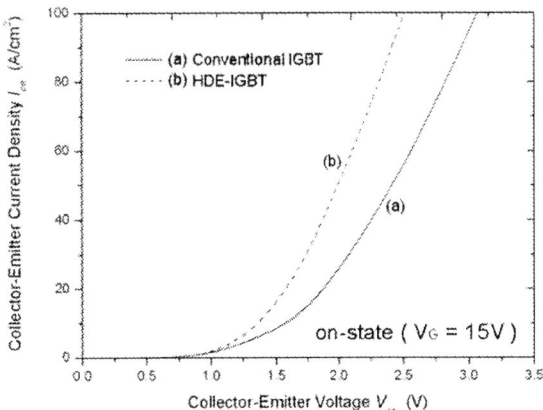

Fig.2 The on-state current-voltage characteristics

High voltage blocking capability is an important factor because IGBT is usually used in high voltage applications. Fig. 3 shows the I-V characteristics of IGBTs in turn off state. As can be seen from the figure, the breakdown voltage of HDE-IGBT is about 20 V lower than that of conventional IGBT, which seems to be caused by increase of carriers. However, the HDE-IGBT has sufficient voltage for 1700 V devices. In fig. 2 and fig. 3, a reduction in Vcesat can be achieved with a little drawback of the decrease in breakdown voltage.

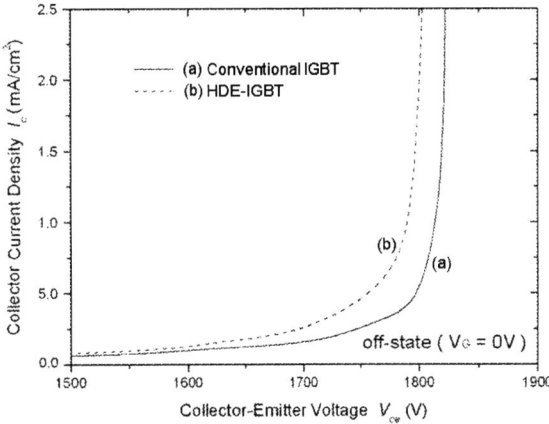

Fig.3. The off-state current-voltage characteristics

The greatest advantage of HDE-IGBT lies in the fact that both the on-state resistance and turn-off switching characteristics can be improved at the same time. Fig. 4 shows the turn-off transient time (T_f) as a function of on-state voltage drop between the conventional IGBT and HDE-IGBT.

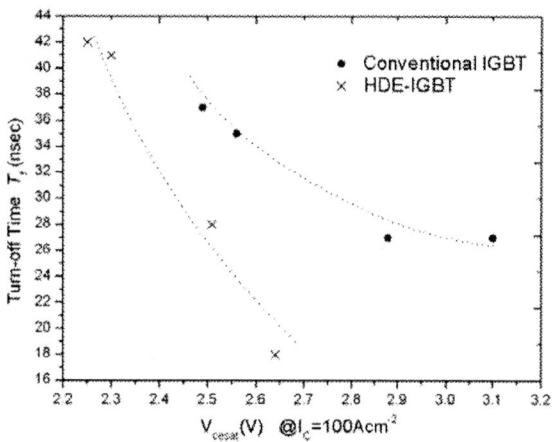

Fig. 4 Turn off transient time as a function of on-state voltage drop.

In the figure, the curve, which is known as the trade-off relation between the on-state resistance and switching speed, represents that both Vcesat and T_f performance improve at the same time in the HDE-IGBT.

IV. DISCUSSION

For analyzing the reason why the on-state voltage drop and switching characteristics are simultaneously improved, simulations on the carrier distribution of the collector region were accomplished.

Fig. 5 shows the 3-dimensional plot of the electron current density of conventional IGBT and HDE-IGBT.

X-axis and Y-axis represent the width of collector and the distance from the gate respectively, and Z-axis represents the current density. As can be seen from the figure, the electron current density near the separated collector area is increased significantly.

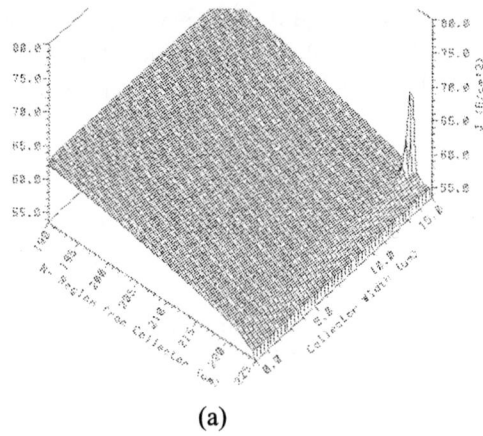

(a)

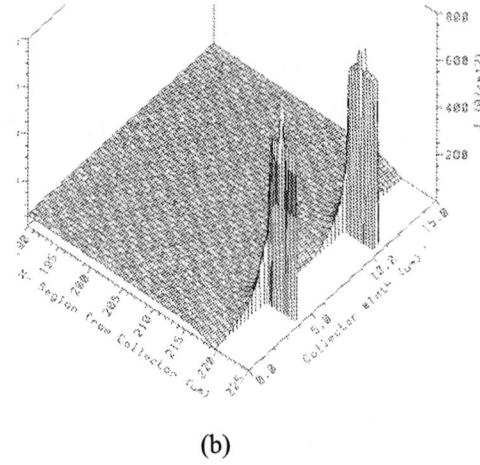

(b)

Fig. 5 The 3D-electron density near the collector area. (a) Conventional IGBT structure (b) Proposed HDE-IGBT structure

Fig. 6 shows the electron current density with respect to the collector width. The current density of the conventional IGBT is about 57 A/cm^2 at whole area of collector. However the current density of HDE-IGBT has a maximum peak of 220 A/cm^2 near the oxide edge. In other words, the separated collector structure induces the increased electron current in spite of decreasing hole current. The increased electron current causes the reduction of on-state voltage drop, and the decreased hole current results in an improvement of switching characteristics.

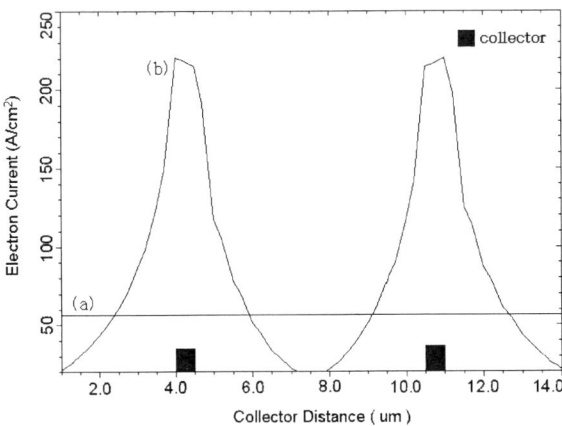

Fig. 6 Electron current density near collector area.
(a) Conventional IGBT (b) HDE-IGBT

Evaluation of the electric field was done to find out the reason to increased current, as shown in Fig. 7. It can be seen that the electric field is concentrated at the corners of SiO_2 region, near the P+ region. Thus, it can be concluded that the increased electron current, which is the reason for on-state voltage drop, is caused by the high electric field near the corners of SiO_2 region. In addition, the hole injection efficiency is decreased due to the reduced P+ collector regions, which effectively improve the switching characteristics.

In conclusion, the simultaneous improvement of the two characteristics in HDE-IGBT can be explained by increase of electron near the oxide corner and decrease of hole caused by reduced P+ area.

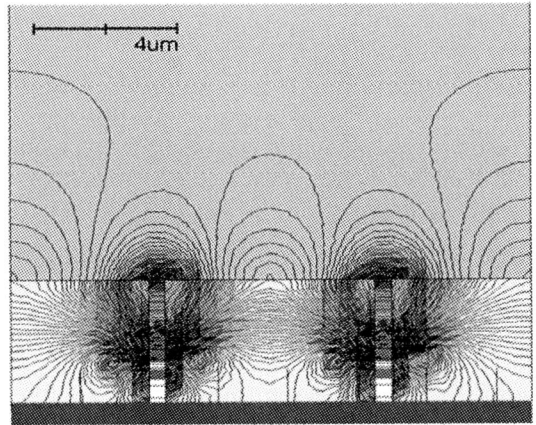

Fig. 7 Electric field contour near the HDE -IGBT collector area.

V. CONCLUSION

The proposed structure named HDE-IGBT has been designed and analyzed by simulation. With the new structure, it is possible to improve the on state voltage drop (Vcesat) and turn off time (T_f) at the same time. The simultaneous improvement of Vcesat and T_f used to be a difficult problem because of their trade-off relationship. However, the separated SiO_2 structure and the reduced P+ area makes it possible to improve Vcesat and T_f by an increment of electron caused by concentrated field around the SiO_2 corners with reducing the hole injection efficiency. In the case of HDE-IGBT, the on-state voltage drop is reduced to 30~40 % and turn-off switching speed to 30 % compared to conventional IGBTs. Applying this proposed structure to semiconductor products may causes the increase of process complexity. However, the outstanding characteristics are far more appealing and it is expected to compensate the drawback of difficult process steps.

ACKNOWLEDGEMENT

This research was accomplished by project named "Development on Power Semiconductor for Electric Power Generation and Industrial Inverter Applications", supported by The Korea Ministry of Commerce, Industry and Energy.

REFERENCE

[1] B. J. Baliga, *Power Semiconductor Devices*, PWS, 1996

[2] T. Laska, M. Münzer, F. Pfirsch, C. Schaeffer, and J. Schmidt, "The field stop IGBT (FS IGBT)–a new power device concept with a great improvement potential," *Proc. 12th ISPSD*, pp. 335-358, 2000.

[3] F. D. Bauer, "The super junction bipolar transistor: a new silicon power device concept for ultra low loss switching applications at medium to high voltages," *Solid-State Electronics*, Vol. 48, no.5, pp. 705-714, 2004.

[4] N. Luther-King, et al., "Striped anode engineering: a concept for fast switching power device," *Solid-State Electronics*, Volume 46, Issue 6, Pages 903-909, 2002

Process Optimization of p+LDD in 130nm Process Technology using TCAD Simulation

Hani Noorashiqin Abd Majid[a,b] *Student Member, IEEE*, Muhamad Rasat Muhamad[b], Albert Victor Kordesch[a], *Senior Member IEEE*, Chew Soon Aik[a], *Member IEEE*

[a]Device Modeling Department, Silterra (M) Sdn Bhd, Kulim Hi-Tech Park, Kulim, Kedah.
[b]Department of Physics, Faculty of Science, Universiti Malaya, 50603 Kuala Lumpur.
Email: hani_noorashiqin@silterra.com , rasat@um.edu.my , al_kordesch@silterra.com

Abstract – **The objective of this work is to study the effect of pMOSFETs Lightly Doped Drain (p+LDD) using two different species of dopants; Boron and Boron Di-Flouride using TCAD Simulation. The result from this simulation will be compared with electrical measurements on silicon wafers. In our experiments, we used the same dose for both of the dopants, at Dose A. We only systematically split the energy for BF2 into three splits. 2D simulation results using TSUPREM-4 and MEDICI help us to understand the mechanism involved. Our results show improvement for the junction profile by comparing BF2 and Boron. It has been reported before, that incorporating Fluorine from BF2 dopant is very useful for the shallow junction devices [1]. This beneficial effect of a BF2 p+LDD will make improvement on the device performance. Since the experiment is comparing the simulation data with the experimental data, it will be very useful for the process engineer to tune their process implant parameters for a future technology.**

I. INTRODUCTION

Microelectronics device manufacture encompasses the fabrication, testing and simulation, of structures utilized in a variety of application specific integrated circuits. The fabrication of transistors, traditionally from elemental silicon wafers is conducted in a clean room environment utilizing various types of equipment. Process simulation is a critical element in facilitating the optimization of fabrication stages, confirming test results and theoretical models and providing physical insight into structural operation [2]. TSUPREM-4 and MEDICI are the process simulators we use. Each

phase of device manufacture is replicated to a reasonable degree of accuracy.

Process simulation output includes two dimensional structure profiles, and the extraction of structural and electrical parameters. Structural information includes doping profiles, impurity concentrations, and layer thickness whereas electrical extraction includes current vectors, electron and hole concentrations and potential contours. Metal oxide semiconductor field effect transistor (MOSFET) is one of the typical device structures that can be simulated.

Device simulation can be used to replicate processing steps as well as analyzing the behavior of the finished structure. Results are categorized as simulated or measured. MEDICI is a device simulator. Device models are imported into circuit simulations and characteristics extracted under various conditions.

The objective of this paper is to introduce the comparison of the simulation data with the experimental data. We used the optimized process factor affecting p+LDD by using BF2 instead of Boron and to tune this process by splitting this new implant into different energy experiments. We show that with the correct energy of BF2 used in PLDD we can achieve almost the same effect as with the standard Boron implant and this plays a major role in order to optimize a 130nm process.

II. EXPERIMENT

The 130nm CMOS process that was implemented in our work consists of several process steps [4]. First the shallow trench isolation (STI) is formed, then the wells are implanted and a 2nm gate oxide is grown. The polysilicon is then deposited. Phosphorus implant is doped in the NMOS area. The gate was then etched and at this point the splits for this work were performed. The LDD is implanted using Boron at Dose A with Energy A for a standard reference PMOS and it is compared with various types of BF2 energy splits with the same dose as reference (Dose A). Three different energies are used (High, Medium, Low), see Table 1, in order to tune up the correct process to match the standard process. A pocket implant is then used in order to compensate the short channel effect, using Arsenic. The shallow S/D is implanted next using Boron at 3KeV (PMOS). As mentioned earlier, the above process is fabricated on the silicon wafer and also in the 2D simulator.

Splits	A	B	C
Energy (KeV)	Low	Medium	High

Table 1 Experiment splits for different BF2 implant energies.

Investigation of the BF2 incorporation in p+LDD implant was done using experiment data and simulated data. We studied the saturated drain current and threshold voltage in order to optimize the 130 nm process by using BF2 instead of Boron in the p+LDD process.

III. RESULTS AND DISCUSSION

In this work, we want to see the significant changes, differences and to compare the experiment data with the simulation data. We start from the simulation result, from Figure 1; we plot the profile overview using TV2D in TSUPREM-4. It is observed clearly in this plot that the junction depth using Boron is much deeper (0.28 nm) than BF2 (0.25 nm). This shows that Boron which is implanted using Energy A, diffuses more into the well compared to BF2 dopant which used different energy (Low, Medium, High) but it is shallower.

From this TV2D plot, we can also see a significant change in lateral diffusion of Boron [7]. When the PLDD is doped with BF2, the lateral diffusion is increased with the increase of energy although there is no change in junction depth. This proved that BF2 dopant is useful for the shallow junction device [8].

(a)

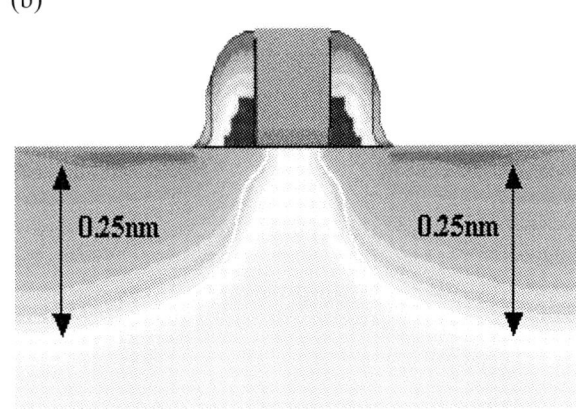

(b)

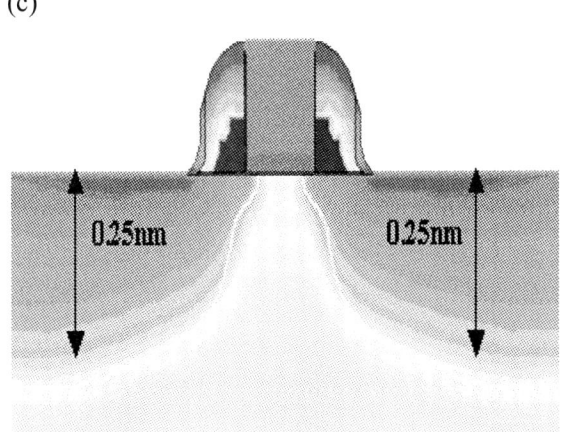

(c)

(d)

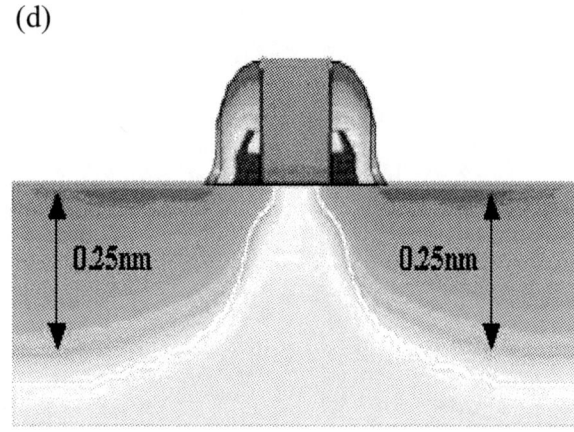

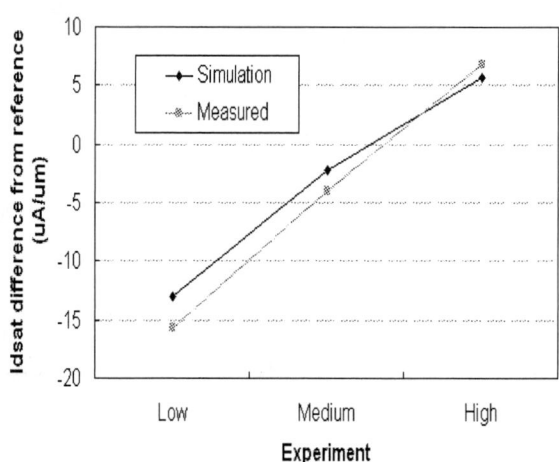

Fig. 1 Boron profiles and junction depth for different PLDD dopant and energy. (a) B11, Energy A (b) BF2, Low Energy (c) BF2, Medium Energy (d) BF2, High Energy.

Figure 2 Figure shows the Idsat different between simulation data and experimental data from the reference value.

Comparing our simulation results with the experimental data, there is a reasonable match, from Table 2 and Table 3. What we can say here is that we can use simulation experiments using TCAD in order to predict and tune our process parameters. In theory, TCAD could be used to design a new process that also works on the first try.

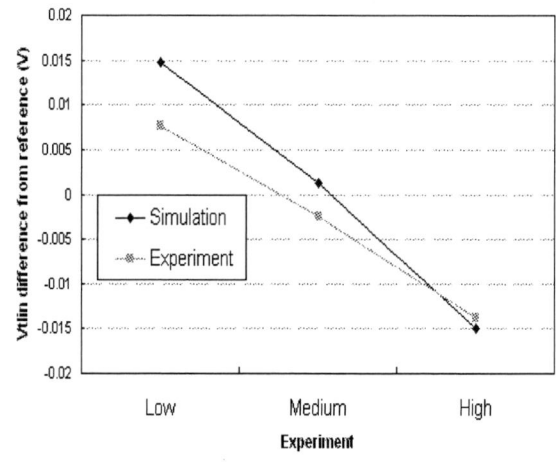

Idsat (µA/µm)	Simulation Data	Experimental Data
Boron (Energy A)	-222	-268
BF2 (Low)	-209	-252
BF2 (Medium)	-220	-264
BF2 (High)	-228	-275

Table 2 Saturation current simulation data compared to experimental data with drawn gate length is 0.13µm.

Figure 3 Figure shows the Vtlin difference from the reference value for three different BF2 implant energies.

Vtlin (V)	Simulation Data	Experimental Data
Boron (Energy A)	-0.244	-0.296
BF2 (Low)	-0.259	-0.304
BF2 (Medium)	-0.246	-0.294
BF2 (High)	-0.250	-0.282

Table 3 Threshold voltage simulation data compared to experimental data with drawn gate 0.13µm.

All the simulation results are quite close to the optimum value obtain from the experiment. The simulator accuracy is sufficient to get better results and prediction for our process.

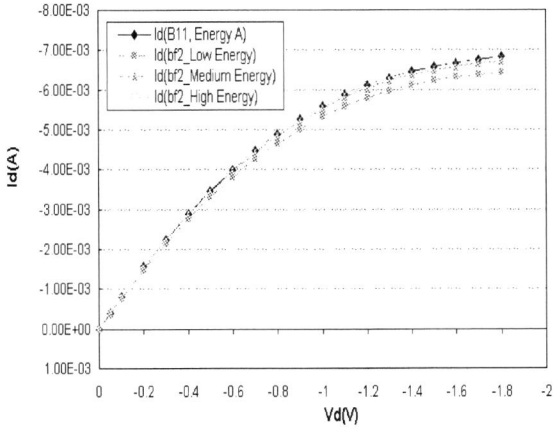

Fig. 4 Drain Current (Id) versus Drain Voltage (Vd) for three different BF2 energy splits compared with Boron.

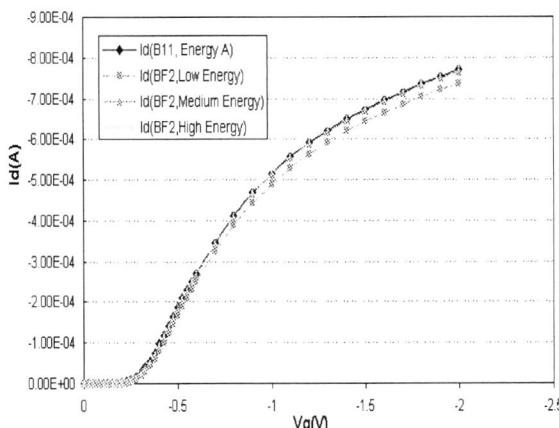

Fig 5 Drain Current (Id) versus Gate Voltage (Vg) for three different BF2 energy splits compared with Boron.

From Figure 2, we can see the simulation results for 0.13µm pMOSFET which we can predict that the BF2 implant energy that has the same effect as Boron in standard module is at medium energy. 4 and Figure 5 shows the Id-Vd and Id-Vg relation from the simulation results by MEDICI [5] for 0.13µm pMOS indicating the linear and saturation regions.

The comparison of the experimental data with the simulation data using this p+LDD splits experiments can be regarded as satisfactory. When designing a short channel CMOS, there are some scaling limiting factors that need to be taken care of such as short channel effect (SCE) and hot electron effect. Therefore, shallow junction of drain/source and LDD structure are implemented to suppress the short channel effect. The TSUPREM-4 is very useful software to simulate the fabrication process. While MEDICI simulator is used to simulate the electrical characteristics. The simulation results that are compared with experimental, and the results are still in the allowed range.

IV. CONCLUSION

In this experiment, we tried 3 different energy splits to see the matching data from the simulator and the experimental data. Each split has different implant energy for BF2 dopant. From the results, we noticed that if the PLDD is implanted with BF2, the effect is almost the same as implanted with Boron. The best and suggested BF2 implant energy that has the same effect as the Boron implant is at medium energy. Further investigation was done to see the same effect of this energy for BF2 by plotting the Id versus Vg curve and Id versus Vd curve. Within this plot, we predict which implant energy has the same effect as the Boron implant energy at energy A by looking at the Idsat and Vt obtained. The closest value that matches the Boron is the medium energy split. From the MEDICI electric field profile also, we believe that BF2 implant using medium energy is best as it has the same effect as the Boron dopant. We also proved that by using the simulator, TSUPREM-4 and MEDICI, we could tune the process parameter easily without having to fabricate transistors, which utilizes a lot of equipment and money. Process simulation also is a critical element in facilitating the optimization of fabrication stages, confirming test results and theoretical models and providing insight into structural operation. Each phase of device in this simulator is replicated to a high degree of accuracy.

ACKNOWLEDGEMENT

The authors acknowledge the members of Device Modelling Group, Silterra (M) Sdn. Bhd. (Yusman Mohd Yusof, Norhafizah Che May, Philip Tan, Mohd Fahmi Muhsain and Izahan Syemylona Ishak) for the unlimited contribution and support.

REFERENCES

[1] James R. Pfiester, Frank K. Baker, Thomas C. Mele, Hsing-Huang Tseng, Philip J. Tobin, James D. Hayden, James W. Miller, Craig D. Gunderson, Louis C. Parrillo, *IEEE Trans. Electron Devices*, 37, 1842 (1990).

[2] D.Y. Lee, H.C. Lin, M.F. Wang, M.Y. Tsai, T.Y. Huang, and T. Wang, in *Proceedings of Solid State Devices and Materials*, 204 (2001).

[3] G.S. Samudra, B.P. Seah and C.H. Ling, "Determination of LDD MOSFET Drain Resistance from Device Simulation," *Solid State Electronics*, Vol.39, No.5, pp.753-758, 1996.

[4] Synopsys, "TAURUS TSUPREM-4 User's Guide 2006,"

[5] Synopsys, "TAURUS MEDICI User's Guide 2006,"

[6] Da-Yuan Lee, Tiao-Yuan Huang, Horng-Chih Lin, Wan-Ju Chiang, Guo-Wei Huang, Tahui Wang, "Effects of Process and Gate Doping Species on Negative Bias Temperature Instability of p-Channel MOSFETs", *Journal of The Electrochemical Society*, 151 (2), 2004.

[7] Fu-Cheng Wang, Constantin Bulucea, "BF2 and Boron Double-Implanted Source/Drain Junctions for Sub-0.25µm CMOS Technology", *IEEE Electron Device Letters*, Vol.21, No.10, 2000.

[8] C. –Y. Kang et al., "Effects of nitrogen implantation in silicon for shallow p+-n, junction formation," *Appl. Phys. Lett.*, vol.74, p. 1833, Mar.1999.

Effect of N_2 and O_2 Anneal Gas Ratio For Low Resistance p – Type ZnO Formation.

Haslinda Abdul Hamid, Mat Johar Abdullah, Azlan Abdul Aziz, Naif H. AL – Hardan, Siti Azlina Rosli
Nano- Optoelectronics Research and Technology Laboratory (N.O.R),
School of Physics,
Universiti Sains Malaysia,
11800 Minden, Pulau Pinang, MALAYSIA
E-mail: haslinda.zm05@student.usm.my, matjohar@usm.my

Abstract **P – type conduction in ZnO thin films was realized by codoping method. ZnO thin films were prepared on silicon (111) substrates by DC magnetron sputtering using pure zinc disk as target and underwent heat treatment at 300 °C for 1 hr. Results indicated that the co doped p – type ZnO had the lowest resistivity of 3.412×10^{-3} Ω.cm with a carrier concentration of 1.54×10^{22} cm^{-3}.**

I. INTRODUCTION

ZINC oxide (ZnO) is a II – VI compound semiconductor with a wide direct band gap of 3.3 eV and a hexagonal wurtzite structure [1]. It is widely used material in various applications such as gas sensors, UV resistive coatings, piezoelectric devices, varistors, surface acoustic wave (SAW) devices and transparent conductive oxide electrodes [2]. ZnO has also attracted attention for its possible application in short wavelength light emitting diodes (LEDs) and laser diodes (LDs) because the optical properties of ZnO are similar to those of GaN [3]. ZnO, however, has proven to be difficult to be doped as p – type ZnO because ZnO exhibits an asymmetry in its ability to be doped n – type or p – type, which is called the 'unipolarity'. The theoretical prediction for the realization of p – type ZnO by codoping donor and acceptor was proposed by Yamamoto *et. al.* [4]. Doping with impurities such as Ga, In and Al can increase the conductivity of ZnO. It was reported that the Al dopant concentration plays an important role in the conductivity of aluminium doped zinc oxide (AZO) and most researchers agree that 2 at. % of Al in AZO ensures the highest conductivity [5]. Many techniques such as reactive magnetron sputtering [6], pulsed laser deposition (PLD) [7],

chemical vapour deposition (CVD) [8] and dip – coating technique [9] have been used for deposition. Among these techniques, the magnetron sputtering is most often used due to its high growth rate and large area film uniformity. In this study, p – type ZnO thin films were prepared using Al – N – codoping method in the ambient of O_2 – N_2 mixture while underwent heat treatment and their properties are measured.

II. EXPERIMENT

Al – N codoped ZnO thin films were prepared on silicon (111) substrates by Edwards 306A DC magnetron sputtering using pure zinc metal disk (99.99 % purity) supplied by plasmaterial mixed with two rods of aluminium (99.99 % purity) as shown in fig. 2. The silicon substrates were cleaned by standard wet cleaning procedures and dipped in 2 % HF for 10 seconds after cleaning, then rinsed in DI water and dried out by blowing with nitrogen before loading into the sputtering system. The sputtering gas was pure argon with the base pressure of 2×10^{-5} Torr and increased to 3×10^{-3} Torr during the process. Before the deposition, the target was pre – sputtered for 5 minutes to remove contaminants on its surface as shown in fig. 1. The Zn films were prepared with power of 100 W. After deposition, the films were annealed at 300 °C for 1 hr in different ratios of nitrogen and oxygen ranging from 30 % to 70 % that controlled by Nabertherm tube furnace. The surface morphology and microstructure were investigated by scanning electron microscopy (SEM) from JEOL model JSM – 6460 LV attached with EDS unit model INCAx – sight from Oxford instruments. X – ray diffractometer from X'Pert PRO was used to determine the crystallographic

0-7803-9730-4/06/$25.00 ©2006 IEEE

structure using Cu as anode material with Kα radiation (λ = 1.54 Å). Electrical characteristics were measured by using Accent HL 5500 Hall System.

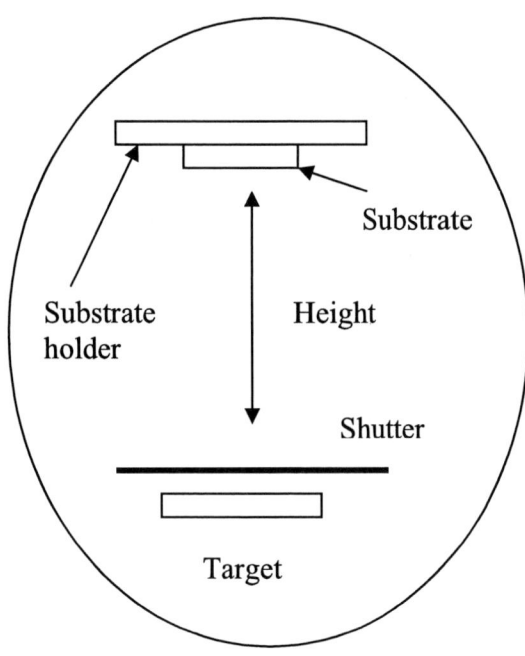

Fig. 1   The sputtering configuration used for deposition of ZnO

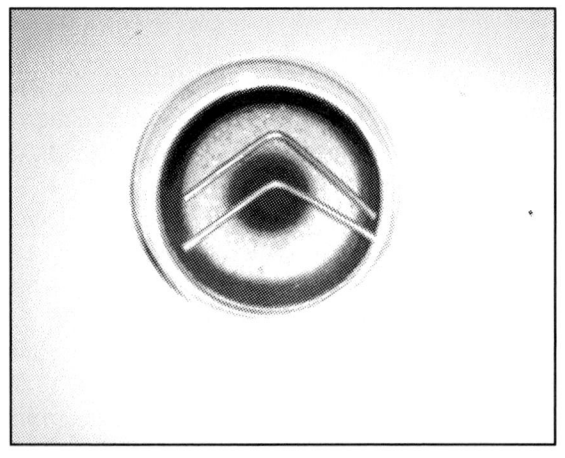

Fig. 2   Top view of Zn target with two rods of aluminium.

III. RESULTS AND DISCUSSION

Fig. 3 illustrates the typical XRD profiles of the Al – N codoped ZnO films deposited on silicon substrates with pure oxygen and with different nitrogen partial ratios of 30 %, 50 % and 70 %. It can be seen that after the heat treatment, Zn films underwent a partial transformation and have been converted to ZnO. The ZnO films are polycrystalline with hexagonal close packed crystal lattice (a= 3.2427, c= 5.1948 Å) and exhibit three main peaks at 2θ= 31.84°, 34.50° and 36.34°, which correspond to (100), (002) and (101) direction of the hexagonal ZnO structure, respectively. It should be noted that no other phases corresponding to AlN, Al_2O_3 or Zn_3N_2 are detected in these films although the peak of aluminium is detected and there have been reports on improvement in crystallinity of ZnO on doping with group III elements. Fig. 4 shows the variation of the full width at half maximum (FWHM) of the (101) peak at different nitrogen ratio. The FWHM for pure oxygen was 0.43° and when the nitrogen ratio was kept at 30 % the FWHM increased to the highest peak of 0.94° and decreased to 0.63° at 50 % of nitrogen ratio. The FWHM was then broadened with a small value changed to 0.79°. The FWHM of XRD depends on the crystalline quality of each grain size and distribution of grain orientation [10]. According to that statement, the initial decreased of FWHM can be explain that the crystalline quality of grains become poor due to the higher density of native defects and the increase of FWHM from 0.43° to 0.94° and 0.63° to 0.79° in the films indicate that the concentrations of the defects are relatively low. The mean grain size could be estimated by the Scherer formula [11]:

$$D = \frac{0.94\,\lambda}{B \cos \theta}$$

where D, λ, B and θ are the mean grain size, the x – ray wavelength (0.154nm), the FWHM of the (101) peak and the diffraction angle, respectively.

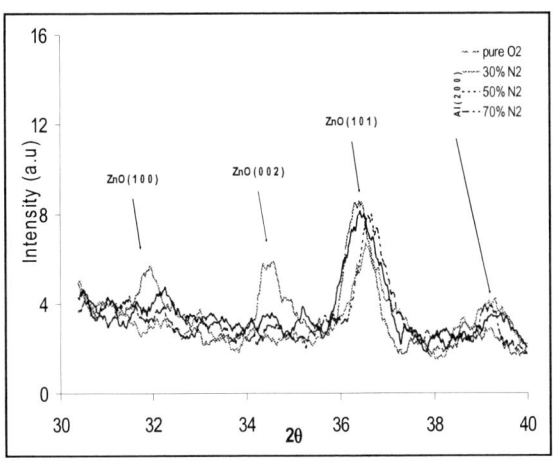

Fig. 3 XRD spectra of samples pure oxygen and annealed at different ratio of nitrogen.

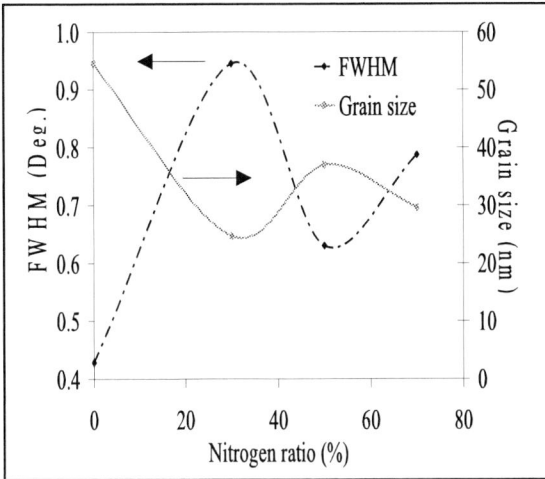

Fig. 4 The variation of the FWHM and grain size at pure oxygen and different ratios of nitrogen.

Table 1. The Hall Effect data of the samples annealed at 300 °C for 1 hr.

N_2 (%)	Resistivity (Ω.cm)	Mobility ($cm^2/V\ s$)	C.concentration (cm^{-3})	Type
Pure O_2	3.876×10^{-2}	0.407	-3.95×10^{20}	n
70	3.412×10^{-3}	0.119	$+1.54 \times 10^{22}$	p
50	4.962×10^{-3}	0.154	$+8.16 \times 10^{21}$	p
30	2.280×10^{-2}	0.225	-1.22×10^{21}	n

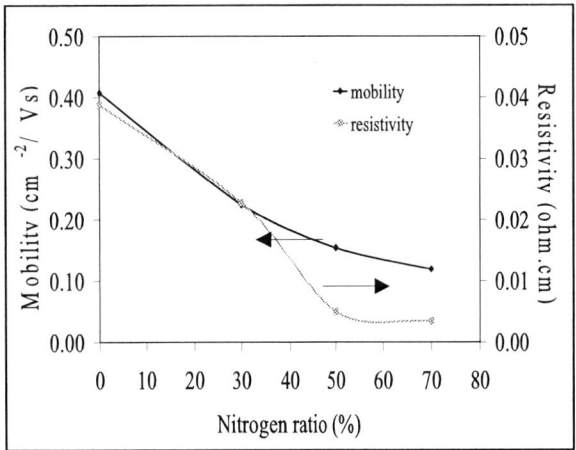

(a)

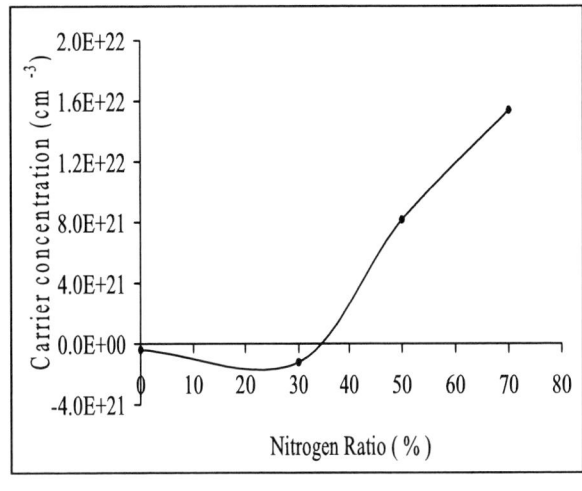

(b)

Fig. 5 Variation of electrical resistivity, mobility (a) and carrier concentration (b) of pure oxygen and codoped ZnO on nitrogen.

By increasing the nitrogen ratio from 30 % to 70 %, the mean grain size varied between 25 - 37 nm. The results of Hall measurements on ZnO films are summarized in Table 1. It shows the dependence of electrical resistivity, Hall mobility, carrier concentration and carrier type of Al – N codoped ZnO thin films on nitrogen partial ratio. Films deposited at 70 % and 50 % of nitrogen ratio showed p – type conduction whereas for pure oxygen and 30 % of nitrogen ratio samples showed n – type conduction. As the nitrogen ratio decreased, most of the residual nitrogen atoms will almost be replaced by the oxygen atoms almost be replaced by the oxygen atoms, leading to the n – type conversion with higher resistivity. Fig. 5 shows the dependence of electrical resistivity, Hall mobility and carrier concentration of Al – N codoped ZnO thin films on nitrogen partial ratio, respectively. The highest carrier concentration and the lowest resistivity are both obtained at nitrogen partial ratio of 70 % while the highest value of Hall mobility appears at pure oxygen ratio. From fig. 5 (a), increasing the nitrogen ratio, resulted in the decrease of the resistivity due to the activated nitrogen atoms which serve as acceptor. In the mechanism of co doping, the N acceptor dopant substitutes for O and the incorporation of Al donor dopant changes the chemical states of nitrogen in the ZnO films. Moreover, the strong acceptor – donor attractive interaction overcomes the repulsive interactions between the acceptors to reduce the Madelung energies and to enhance the incorporation of acceptors that lead to the formation of an acceptor – donor complex in the band gap [9]. Furthermore, the excess Al atoms as interstitial atoms exist in the films which act as scattering centres can increase the resistivity. As the nitrogen partial ratio increases, the decreasing mobility of co doped films probably caused by the large amounts of defects due to high concentration of dopants which acts as scattering centres. The solubility of N in the obtained films can be greatly increased due to the existence of Al, so p – type ZnO thin films with higher concentration can be achieved with the Al – N codoping method.

IV. CONCLUSION

It is concluded that p – type ZnO films can be achieved by using the codoping technique. Codoped films showed a decrease resistivity with the highest ratio of N_2. Hall measurements proved that the carrier concentration were increased from 3.95×10^{20} cm^{-3} to 1.54×10^{22} cm^{-3} whereas Hall mobility is decreased from 0.407 to 0.119 cm^2/V s for the increase of N_2 partial ratio.

ACKNOWLEDGEMENT

The support from Universiti Sains Malaysia through short term grant is gratefully acknowledged.

REFERENCES

1) D. R. Lide, Handbook of Chemistry and Physics, 71st ed, CRC, Boca Raton, FL, 1991.
2) R. P. Ried, E. Kim, D. M. Hong and R. S. Muller, Jrnl of Microelectrochemical Systems, 2 p. 111 – 120 (1993).
3) Y. Ito, K. Kushida, K. Suguwara and H. Takeuchi, IEEE Transaction on Ultrasonics, Ferroelectrics and Frequency Control, 42 p. 316 – 323 (1995).
4) M. Komatsu, N. Ohashi, I. Sakaguchi, S. Hishita and H. Haneda, Appl. Surface Science, 189 p. 349 – 352 (2002).
5) T. Yamamoto and H. K. Yoshida, Jpn. J. Appl. Phys. 38 p. 166 (1999).
6) H. L. Hartnagel, A. L. Dawar, A. K. Jain, C. Jangadish, Semiconducting Transparent Thin Films, IOP (1995).
7) J. M. Bian, X. M. Li, X. D. Gao, W. D. Yu, L. D. Chen, Appl. Phys. Lett. 84 p.541 (2004).
8) T. Yamamoto and H. K. Yoshida, Physica B, 155 p. 302 – 303 (2001).
9) Yoshino K, Hata T, Kakeno T, Komaki H, Yoneta M, Akaki Y and Ikari T, Phys. Status Solidi C, p. 0626 (2003).
10) N. Matsumi, O. Fukuoka, M. Tazawa, M. Sataka, Surf. Coat. Technol. 196 p. 50 – 55 (2005).
11) Lim W. T and Lee C H, Thin Solid Films, 353 p.12 (1999).

Electrical Properties of p–Type Al – N Codoped ZnO Thin Films

Haslinda Abdul Hamid, Mat Johar Abdullah, Azlan Abdul Aziz, Siti Azlina Rosli
Nano-Optoelectronics Research and Technology Laboratory (N.O.R),
School of Physics,
Universiti Sains Malaysia,
11800 Minden, Pulau Pinang, MALAYSIA
E-mail: haslinda.zm05@student.usm.my, matjohar@usm.my

Abstract **Aluminium doped zinc thin films were deposited on silicon substrates using direct current (DC) magnetron sputtering in argon atmosphere. These films were then annealed in nitrogen – oxygen mixed gases for 1 hour at temperature range of 200 °C to 500 °C. P – type conduction of Al – N codoped ZnO thin films were obtained for all temperatures with a high hole concentration of 1.85×10^{22} cm^{-3}.**

I. INTRODUCTION

ZINC oxide (ZnO) is a II – VI compound semiconductor with a wide direct band gap of 3.3 eV and a hexagonal wurtzite structure [1]. It is widely used material in various applications such as gas sensors, UV resistive coatings, piezoelectric devices, varistors, surface acoustic wave (SAW) devices and transparent conductive oxide electrodes [2]. ZnO has also attracted attention for its possible application in short wavelength light emitting diodes (LEDs) and laser diodes (LDs) because the optical properties of ZnO are similar to those of GaN [3]. ZnO, however, has proven to be difficult to be doped as p – type ZnO because it exhibits an asymmetry in its ability to be doped n – type or p – type, which is called the 'unipolarity'. The theoretical prediction for the realization of p – type ZnO by co doping donor and acceptor was proposed by Yamamoto *et. al.* [4]. Doping with impurities such as Ga, In and Al can increase the conductivity of ZnO. It was reported that the Al dopant concentration plays an important role in the conductivity of aluminium doped zinc oxide (AZO) and most researchers agree that 2 at. % of Al in AZO ensures the highest conductivity [5]. Many techniques such as reactive magnetron sputtering [6], pulsed laser deposition (PLD) [7], chemical vapour deposition (CVD) [8] and dip – coating technique [9] have been used for deposition. Among these techniques, the magnetron sputtering is most often used due to its high growth rate and large area uniformity. In this study, p – type ZnO thin films were prepared using Al – N codoping method in the ambient of $O_2 - N_2$ mixture while underwent heat treatment.

II. EXPERIMENT

Al – N codoped ZnO thin films were prepared on silicon (111) substrates using Edwards 306A DC magnetron sputtering. Pure zinc metal disk (99.99 % purity) supplied by plasmaterial was used as the target together with two rods of aluminium (99.99 % purity) as shown in fig 1. The silicon substrates were cleaned by standard wet cleaning procedures and dipped in 2 % HF for 10 seconds after cleaning, then rinsed in DI water and dried out by blowing with nitrogen before loading into the sputtering system. The sputtering gas was pure argon with the base pressure of 2×10^{-5} Torr and increased to 3×10^{-3} Torr during the process. Before the deposition, the target was pre – sputtered for 5 minutes to remove contaminants on its surface as shown in fig 2. The Zn films were prepared with power of 100 W. After deposition, the films were annealed at different temperatures from 200 °C to 500 °C for 1 hour in nitrogen - oxygen gas mixture in Nabertherm tube furnace. Electrical characteristics were measured by Accent HL 5500 Hall System. The existence of ZnO is verified by XRD and presented in another paper.

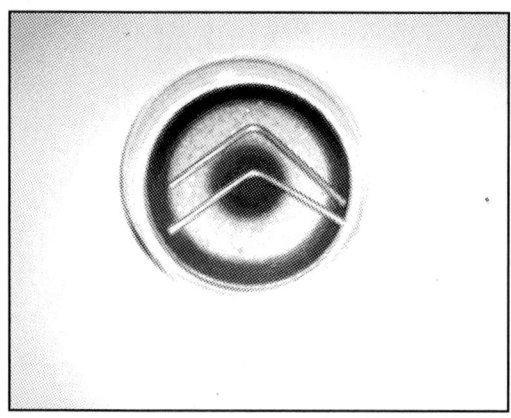

Fig. 1 Zinc target with two rods of aluminium.

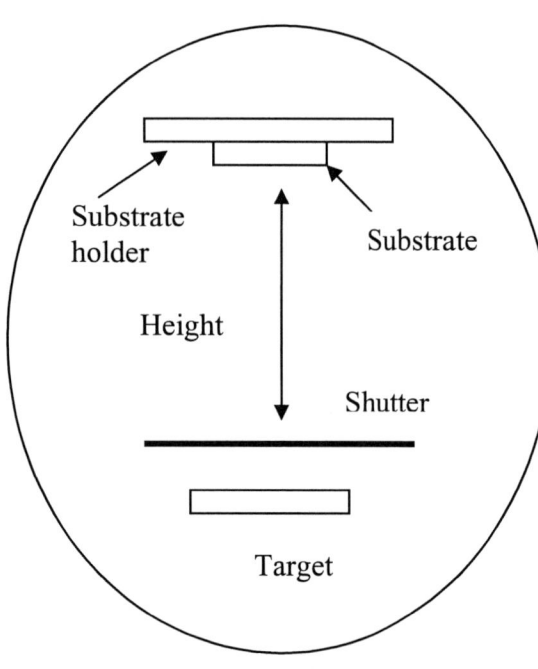

Fig. 2 The sputtering configuration used for

deposition of ZnO.

III. RESULTS AND DISCUSSION

A Hall measurement was performed by Accent HL 5500 Hall system using a van der pauw four point configuration. Fig 3 shows that ohmic contacts were made by aluminium metal at four corners of square shaped samples. Excellent linearity was found from I – V curve indicating good ohmic contacts between aluminium and Al – N codoped ZnO as shown in fig 4. The magnetic field was 0.320 Tesla and the optimal current was determined automatically by the Hall system. The results of the Hall Effect measurements obtained are listed in Table 1.

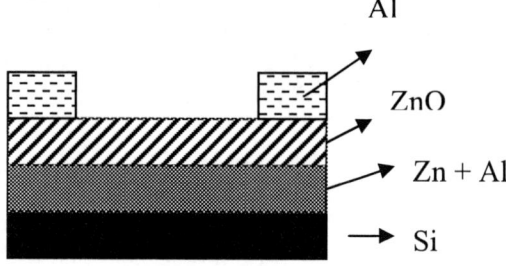

Fig. 3   Schematic illustration of ZnO for I – V measurements.

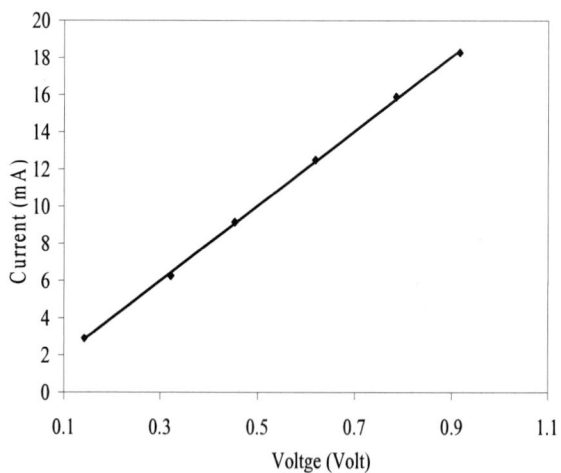

Fig.4   I – V curves of ZnO thin film on the silicon substrate.

Table. 1 The electrical properties of different temperatures that annealed for 1 hr.

Annealing Temperatue (°C)	Resistivity (ohm.cm)	Mobility (cm²/Vs)	Carrier Concentration (cm⁻³)	Carrier type
200	0.00153	0.220	1.85×10^{22}	p
300	0.00496	0.154	8.16×10^{21}	p
400	0.09907	1.100	5.72×10^{19}	p
500	0.17230	4.190	8.64×10^{18}	p

It appears that p – type conductivity can be achieved at all growth temperature as shown in Table 1. For the change of the carrier concentration as shown in fig 6, both the total N atoms in the films and the activated nitrogen atoms which serve as acceptor should be considered. As the annealing temperature increased, it reduces the N acceptor incorporation in the ZnO thin films thus decreasing the carrier concentration. So, the amount of the acceptor doping into ZnO films is small, thus resulting in a low hole concentration from about 10^{22} cm⁻³ to 10^{18} cm⁻³. As such it is very important to have low growth temperature for enhancement of the N incorporation in the formation of Al – N bonds.

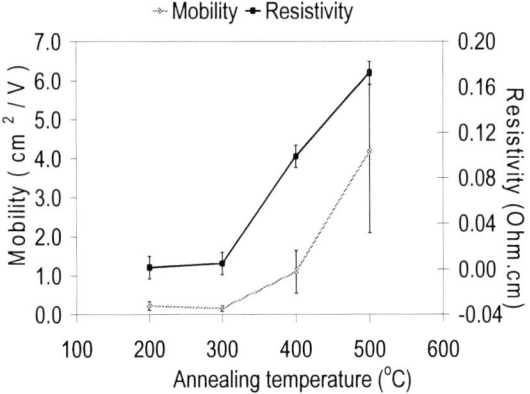

Figure 3. Hall mobility and resistivity for ZnO films.

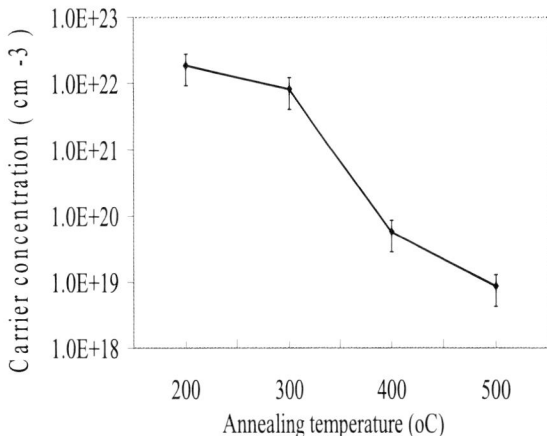

Fig. 4. Carrier concentrations for ZnO films.

The Hall mobility measurements showed that the films were p – type. From fig 5, the Hall mobility of codoped films initially decreased at 300 °C and then increased until 4.190 at 500 °C due to the decreased impurity scattering.

Furthermore, the excess Al atoms as interstitials atoms that exist in the films which act as scattering centers can increase the resistivity (fig 5). The solubility of N in the obtained films can be greatly increased due to the existent of Al, so p – type ZnO thin films with higher concentration can be achieved with the Al – N codoping method.

IV. CONCLUSION

In summary, we have presented Al – N codoped method to realize p – type ZnO films at all growth temperatures. From the above analysis, it is believed that the codoping with nitrogen and aluminium induces the formation of Al – N bonds, as proposed by theoretical analysis, which may be responsible for the p – type conduction of Al – N co doped films obtained.

ACKNOWLEDGEMENT

The support from Universiti Sains Malaysia through short term grant is gratefully acknowledged.

REFERENCES

1) D. R. Lide, Handbook of Chemistry and Physics, 71st ed, CRC, Boca Raton, FL, 1991.

2) R. P. Ried, E. Kim, D. M. Hong and R. S. Muller, Jrnl of Microelectrochemical Systems, 2 p. 111 – 120 (1993).

3) M. Komatsu, N. Ohashi, I. Sakaguchi, S. Hishita and H. Haneda, Appl. Surface Science, 189 p. 349 – 352 (2002).

4) T. Yamamoto and H. K. Yoshida, Jpn. J. Appl. Phys. 38, p.166 (1999).

5) H. L. Hartnagel, A. L. Dawar, A. K. Jain, C. Jangadish, Semiconducting Transparent Thin Films, IOP (1995).

6) J. M. Bian, X. M. Li, X. D. Gao, W. D. Yu, L. D. Chen, Appl. Phys. Lett. 84 p. 541 (2004).

7) T. Yamamoto and H. K. Yoshida, Physica B, 155 p. 302 – 303 (2001).

8) Yoshino K, Hata T, Kakeno T, Komaki H, Yoneta M, Akaki Y and Ikari T, Phys. Status Solidi C, p. 0626 (2003).

9) N. Matsumi, O. Fukuoka, M. Tazawa, M. Sataka, Surf. Coat. Technol. 196 p. 50 – 55 (2005).

A Highly Compatible Architecture Design for Optimum FPGA to Structured-ASIC Migration

Hee Kong Phoon, Matthew Yap, Chuan Khye Chai
Altera Corporation (M) S/B
11900 Bayan Lepas, Penang, MALAYSIA
Email: hkphoon@altera.com , myap@altera.com , ckchai@altera.com

Abstract - Structured-ASIC design provides a mid-way point between FPGA and cell-based ASIC design for performance, area and power, but suffers from the same increasing verification burden associated with cell-based design. In this paper we address a structured ASIC architecture fabric directly tie to FPGA prototype and functional in-system verification with a clean migration path to structured ASIC. Our goal is to leverage the power/delay/area benefits of structured ASIC technology vs. FPGA with a simple flow which maintains the benefits of FPGAs for ease of test, prototyping, characterization and pre-verification. We will go over the introduction of FPGA to structured-ASIC migration, the architecture of the logic fabric follow by the Lcell to Hcell mapping methodology which can eliminate the need of complicated verification effort and overview of the CAD flow.

I. INTRODUCTION

Today's hardware designers are faced with difficult decisions arising from conflicting design efficiency and time-to-market pressures. Cell-based ASICs offer the best density, performance and power characteristics, but have long design times, high NRE costs, and difficult verification cycles that limit their use to high-volume applications. FPGAs offer zero NRE-cost but significantly higher unit cost and generally poorer density, performance and power compared to ASIC.

A less-often cited benefit of FPGAs is design methodology. Because FPGAs are pre-fabricated in high-volume by the FPGA vendor, DSM issues, variation, manufacturability and test, timing and power characterization, signal-integrity analysis etc. are performed by the vendor and

highly amortized across thousands of designs. This means that FPGAs are generally "verified by construction" and the designer can limit his verification efforts to functional verification and timing analysis, avoiding not just risk but many expensive tools. Most FPGA designers, in fact, use re-programmability as an integral part of their test methodology and perform significant amounts of their test in-system---a mask re-spin to an ASIC designer is analogous to a software iteration for an FPGA designer.

Recently, structured ASICs or "neo-gate-arrays" have been proposed as a hardware model that bridges the gap between ASIC efficiency and FPGA flexibility. A structured ASIC consists of a base array of relatively simple logic structures in a regular fabric that is hard-wired for most processing layers, but can be targeted at a specific application by customizing only the top several processing steps (metal or via layers).

The advantages of a structured ASIC steals from both sides: The efficiency can be close to that of an ASIC while the NRE costs are significantly minimized – at 90nm the NRE cost for a structured ASIC partial mask-set runs about 1/10 that of a full mask set for a cell-based ASIC. Recent keynotes [1] and panel sessions [2] have dealt with arguments over the efficacy of Structured ASIC.

In this paper we propose a structured ASIC fabric and design methodology which more closely ties the ease of design and verification to that of FPGAs, without sacrificing efficiency. Our thesis is based on the idea that verification pressures and DSM design issues will force the majority of structured ASIC designs to be conversions from FPGA prototypes, and thus the key methodological issues are in facilitating this flow.

The remainder of the text is organized as follows. In Section II and III, we introduce the

0-7803-9730-4/06/$25.00 ©2006 IEEE

FPGA logic cell and structured ASIC HCell and HCell Macro. The methodology for mapping between Lcell and Hcell is shown in Section IV. Then, Section V and VI describe the overview of CAD flow that migrates from FPGA to structured ASIC. Finally, we conclude in Section VII.

II. FPGA LOGIC CELL VERSUS HCELL

SRAM-based FPGA logic cells normally consist of k-input LUTs and flip-flops. The basic functionality of a logic cell is to implement k-input combinational functions and optionally registered the function's output. For example, Figure 1 shows the basic logic unit in Altera's Stratix II device [3], [4], called ALM ("Adaptive Logic Module"). An ALM consists of a 6-LUT, 2 arithmetic units and two flip-flops. The ALM can be configured to be a 6-LUT or two smaller logic cells through appropriate configuration SRAM programming and optionally registered the outputs. The LUT and the flip-flop in the same logic cell can also be used separately. In this paper, we will use this ALM as an example, but the same methodology is applicable to other 4-LUT architectures such as Xilinx Virtex or Altera Stratix devices.

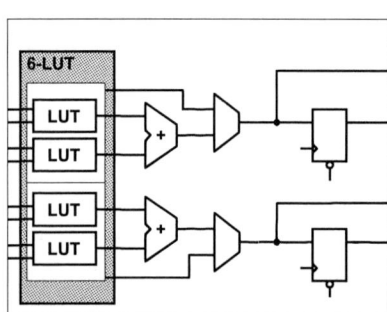

Fig. 1 Simplified FPGA Logic Cell

HCell is the term we used to represent the basic logic unit which forms the logic fabric in the HardCopy II structured ASIC base array. It is similar to the FPGA logic element in the sense that the logic fabric consists of a regular pattern formed by tiling the basic cells in a two-dimensional array. The difference is that the HCell has no overhead for configuration and routing interface. The example used in this paper is assumed to be finer granularity.

A wide range of HCell candidates can be used, from fine-grained NAND/NOR gates, inverters and multiplexers to coarse-grained LUTs

and flip-flops. Some of the HCell examples are shown in Figure 2. In a structured ASIC logic fabric, an array of basic cells and general purpose routing tracks are pre-laid down on the lower layers of the chip. Specific via or metal layers then form via programming options or metal programming options to customize the generic array into specific functionality.

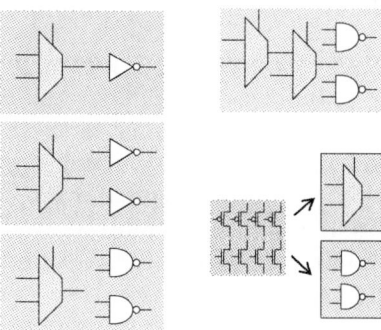

Fig. 2(a) Fine-grained HCell Examples

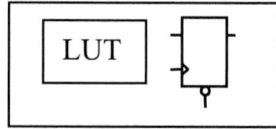

Fig. 2(b) Coarse-grained HCell Examples

III. HCELL MACROS

A library of HCell macros is created to represent pre-optimized, pre-characterized ASIC library cells. Each HCell macro consists of one or more HCells arranged in one or two dimensional arrays. A group of HCells together emulate a given FPGA combinational logic cell, DFF, or arithmetic block. The macro defines the number of HCells needed, the configuration of each HCell used, and the connectivity among these HCells. The configuration and the connectivity is done using lower level layers to minimize the impact on routing resources for P&R and achieve better density. Figure 3 illustrates the sample HCell macros in the library.

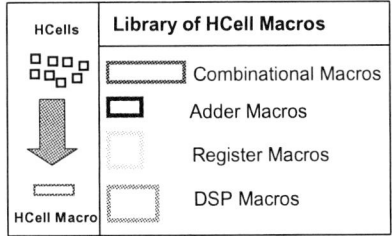

Fig. 3 Sample HCell Macros

The HCell and HCell macros is the 1:1 correspondence between FPGA prototype logic cells and structured cells. Though it could be slightly more efficient to re-synthesize entire cones of logic and blocks into HCells, but maintaining all names and functional boundaries would simplify the verification efforts and outweigh the area savings. Since the conversion of FPGA LEs and routing to Hcells results in a 10:1 or more decrease in logic area the contribution to overall die area is dramatically smaller than I/O, RAM and other hard blocks in the base die.

IV. FPGA LOGIC CELL TO HCELL MAPPING

Combinational HCell macros are created by re-synthesizing FPGA logic cells into HCell base logic structures. For a given FPGA logic cell, optimization is done to compact the logic by removing some of the FPGA logic cells that are not used for a given function, thus reducing the area cost, the logic delay and power dissipation. The optimization method used is similar to Binary Decision Diagram reduction [5]. More sophisticated optimization can also be done on each FPGA logic cell so that the resulting HCell macro meets the desired density, performance, and power cost metrics. Since the Hcell macro library is pre-designed and the optimization problem is relatively small, we can improve results with longer runtime when necessary. A complete example of these optimizations in the re-synthesis flow is shown in Figure 4 if the Hcell architecture consists of a single MUX, then the 6-input LUT FPGA cell implementing the function below is reduced to a 3 HCells macro.

!D*C+D*E*!F*A+D*E*F*!B

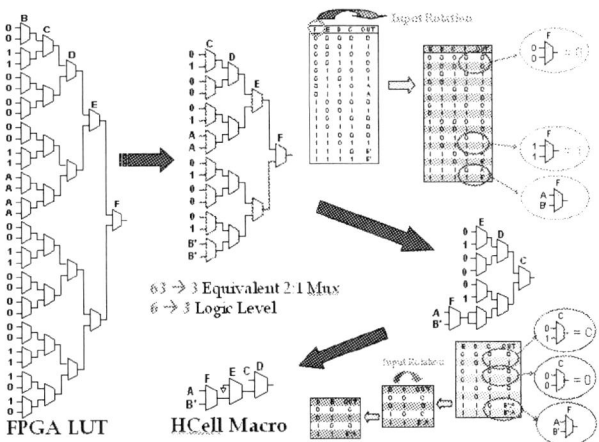

Fig. 4 Optimization of a 6-input LUT

Relative to the combinational FPGA logic cells, we use only a small number of HCell macro registers corresponding to all possible types of register used in FPGA. These registers are built out of HCells to further reduce the area cost with respect to the FPGA logic fabric since we will only build the right type and right number of registers based on the FPGA design.

Figure 5 illustrates the logic cell to macro concept in more detail. Each of the logic portion, the register and the arithmetic portion of the FPGA cell are transformed into an appropriate HCell macro in the library. Figure 6 shows specific examples: a 2-LUT and 3-LUT use one HCell each, a 4-LUT and 6-LUT become HCell macros with 2 and 3 HCells respectively and a DFF with CLR is built from an HCell macro with 2 HCells.

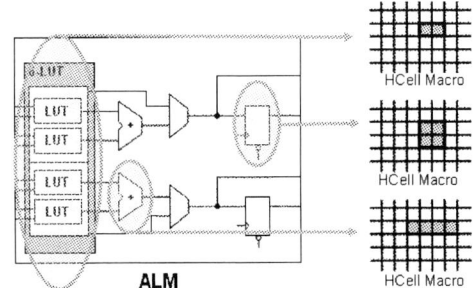

Fig. 5 FPGA Logic Cell to HCell Macro Mapping

517

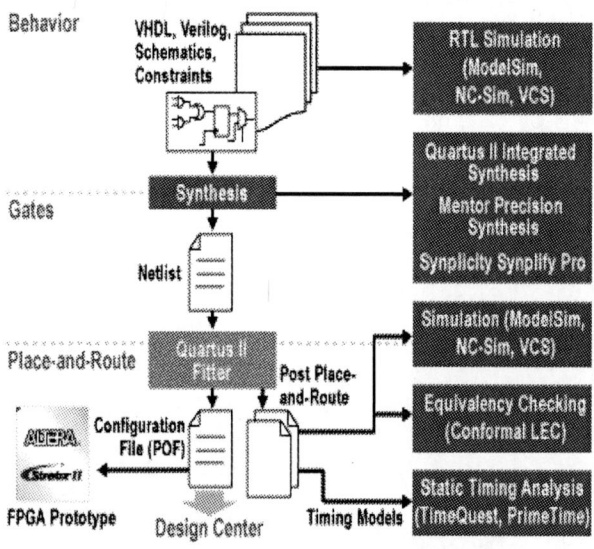

Fig. 6 LUT & Flip-Flop generation with 2:1 MUX HCell

V. UNIFIED FRONT-END DESIGN FLOW

The overall flow can be categorized into 2 major portions, consisting of front-end and back-end flow. The later will be discussed in next section. The front-end design flow unifies the FPGA and structured ASIC designs into one environment. The centerpiece of this design environment is Altera's Quartus II software, which includes features to make designing for structured ASICs as easy as designing for an FPGA.

Fig. 7 Structured ASIC Front-end Design Flow

The following highlights the major steps of the said-mentioned structured ASIC front-end design flow (see Figure 7):

1. Start with RTL and/or schematic design source and a complete set of timing constraints which constrain every path of the design
2. Select a pair of companion FPGA and the structured ASIC devices
3. Synthesize the design with synthesis tools
4. Run fitting (place and route) for both devices, in which the structured ASIC device is an early estimation on fit-ability correlating to backend results
5. Run static timing analysis on both devices and verify timing constraints are met. The timing for the structured ASIC device is an estimate based on the cell placement and global route
6. Submit design for structured ASIC conversion only when the timing constraints are met for both devices and functionality proven in FPGA prototype

VI. INDUSTRY STANDARD BACK-END DESIGN FLOW

Once the design is fully verified in-system with a FPGA, generate the necessary design files in Quartus II software, and hand-off to Altera's Design Center. The design center's experienced ASIC design engineers use the industry-standard design methodology and EDA tools for the structured ASIC device back-end physical implementation (see Table 1).

Design Flow Steps	EDA Tools
Design for Testability (DFT) Insertion	Synopsys DFT Compiler
Test Vector Generation	Synopsys TetraMax ATPG
Clock Tree Synthesis (CTS) and Global Signal Insertion	Synopsys Astro
Timing and Signal-Integrity Driven Place and Route	Synopsys Astro
Post-Layout Parasitic Extraction	Synopsys Star-RCXT
Static Timing/Crosstalk/Noise Analysis	Synopsys PrimeTime SI
Physical Verification	Synopsys Hercules and

	Mentor Graphics Calibre
Formal Verification	Cadence Conformal

Table 1 Structured ASIC Back-end Design Flow Steps

Above mentioned back-end design flow leads the ASIC industry in terms of turn-around time. From netlist handoff to design tapeout, the turn-around time is only one to two months, depending on design complexity. Quartus II software forward-annotates cell placements for the back-end final routing. Since timing closure must be achieved for both the FPGA and the structured ASIC device in Quartus II software prior to netlist hand-off, the final timing closure in the back-end is relatively easy to achieve when compared to a standard-cell ASIC flow. In a typical standard cell ASIC flow, design verification continues even after netlist handoff, causing multiple iterations of verification, functional engineering change orders (ECOs), and place & route merge, further delaying the schedule. Since the structured ASIC design is fully verified in-system with an FPGA, no functional ECOs are needed in the back-end flow. This ensures that the quoted back-end design turn-around time is predictable, and avoids the schedule risk often associated with the standard cell ASIC flow.

VII. CONCLUSION

In this paper, the proposed logic fabric consists of a single HCell structure that forms the basis for all logic cells from the FPGA design. Multiple Hcell could be grouped into a HCell macro which implement the functionality directly corresponding to a single FPGA combinational logic cell or register. The 1:1 correspondence between FPGA prototype logic cells and structured macro cells which maintaining all names and functional boundaries would simplify the verification efforts. The methodology used is independent of the FPGA logic fabric and exact HCell implementation.

This unique structured ASIC design flow is low risk, low cost, easy to use, and easy to adapt. It facilitates the seamless migration from FPGA to structured ASIC devices and enables fast time-to-market. Along with the predictable back-end design turn-around time, the structured ASIC

solution is becoming a true alternative to the long design time, high cost, and high risk associated with the standard-cell ASICs.

REFERENCES

[1] R. Camposano, "Will the ASIC Survive", Keynote Address at SBCCI 2004, p5.

[2] El-Gamal, J. Cohn, A. Kahng, I. Bolsens, A. Broom, C. Hamlin, P. Magarshack, Z. Or-Bach and L. Pileggi, "Fast, Cheap and Under Control: The Next Implementation Fabric", in Proc. DAC 2003, pp. 354-355.

[3] M. Hutton et.al., "Improving FPGA Performance and Area Using an Adaptive Logic Module", in Proc. FPL 2004, pp.135-144.

[4] D. Lewis et.al, "The Stratix II Logic and Routing Architecture", in Proc. FPGA 2005, pp. 14-20.

[5] Fabio SOMENZI, "Binary Decision Diagrams", Department of Electrical and Computer Engineering, University of Colorado at Boulder.

Effect of Mesa Spacing on the Electrical Properties of Mesa Isolation in High Electron Mobility Transistor Structures

Hesly Afida Hashim, Mohd.Khairy Othman, Mohd.Nizam Osman, Asban Dolah and Mohamed Razman Yahya

TM Research & Development Sdn Bhd, UPM-MTDC, Lebuh Silikon, 43400 Serdang, Selangor.
E-mail : hesly@tmrnd.com.my

Abstract **This paper report effects of variation spacing of mesa isolation on pHEMT substrate in order get an optimum isolated area. The multilayer pHEMT substrate was used as a substrate and wet etching techniques have been applied to form the islands. The citric mixture used is $C_2H_8O_7:H_2O_2:H_2O$ with ratio of 4:1:1. The mesa spacing used in this study was varies at 570, 792 and 835 µm. The etch time for each spacing was fixed at 3 minutes. The electrical effect of mesa isolation spacing was characterized through current-voltage curve. It was found that the 570 µm mesa spacing shows optimum current value for mesa isolation. This indicated that 570 µm mesa spacing etch with citric acid mixture of 4:1:1 ratio for 3 minutes etch time can produce optimum current for application of pHEMT device.**

I. INTRODUCTION

The AlGaAs/InGaAs/GaAs psuedomorfic HEMT (pHEMT) is recognize as a workhorse for millimeter wave circuits today. The pHEMT's combination of low noise, high efficiency and good linearity along with its ability to deliver outstanding performance over a wide range of bias conditions have made the device of choice for a host of RF front end applications. [1]A primary process for semiconductor fabrication is wet etching which uses chemical to etch away desired materials.

The mesa etching processes are used to isolate masked areas from one another to avoid short circuiting the fabricated devices via conducting layers.[2] In order to form the mesa structures, highly anisotropic etching is wanted. Good etching profile is needed to produce a well isolated mesa islands. [3]

In this study, we try to study the effect of different spacing parameter towards the device isolations characteristics.

II. EXPERIMENTAL

The multilayer pHEMT sample was grown using molecular beam epitaxy on GaAs substrates. Lithography method was used to transfer pattern on the samples using a mesa structure mask with varying mesa spacing of 570 µm, 792 µm and 835 µm. The etch time for each spacing was fixed at 3 minutes using Citric Acid with the ratio of 4:1 towards the Hydrogen Peroxide. To confirm the depth of mesa etching, Veeco NT1100 optical profiler was used.

The electrical characteristic of the devices were analyzed through the current-voltage (IV) curve using Keithley 238 source measurement unit. The sample is placed on the stage in the probe station where the two adjacent mesa islands are probed and tested for their isolation properties.

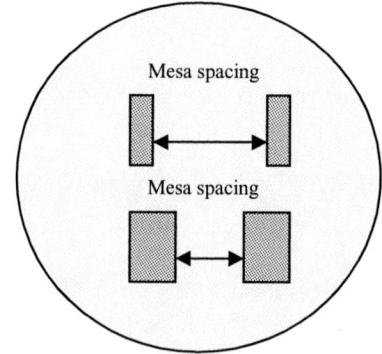

Figure 1: Example of Mesa Spacing Pattern

0-7803-9730-4/06/$25.00 ©2006 IEEE

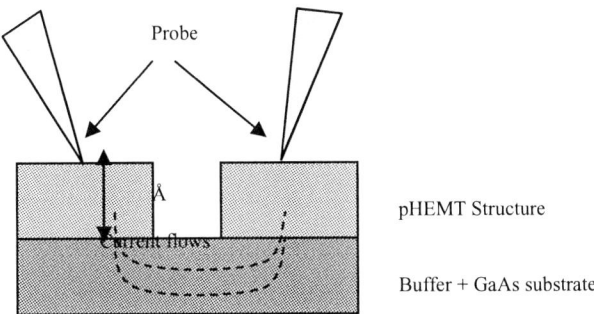

Figure 2: Probing for Current-Voltage Characteristics

Figure 1 shows the example of mesa etching spacing pattern formed using lithography technique on the wafer. Figure 2 shows the probing technique used to measure the current-voltage using Keithley 238 source measurement unit.

III. RESULTS & DISCUSSION

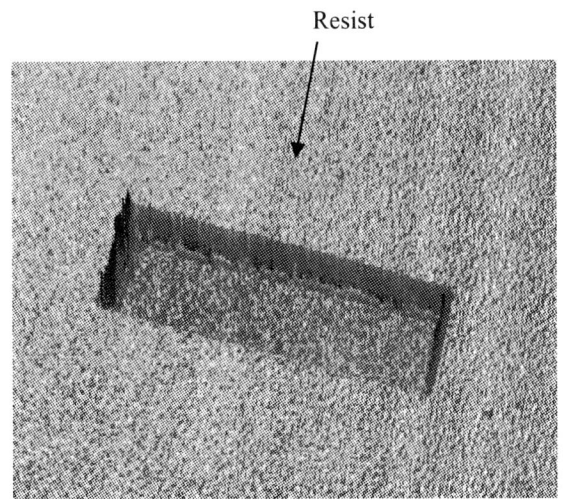

Figure 3(a): Mesa Resist Before Etch

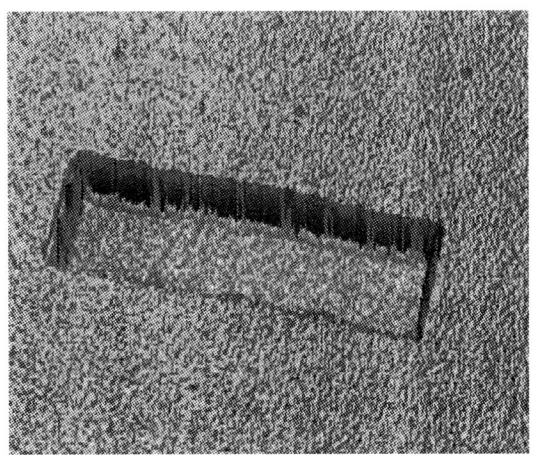

Figure 3(b): Mesa Trench Formed After Etch

Figure 3(a) shows a 3D picture taken using the optical profilometer of the mesa resist pattern prior to mesa etching for 835µm mesa spacing. Where as Figure 3(b) shows the mesa trench formed as a results of the 3 minutes citric acid etched (prior to resist strip).

The measured mesa trench depth obtained is 907Å for all the three different mesa spacing.

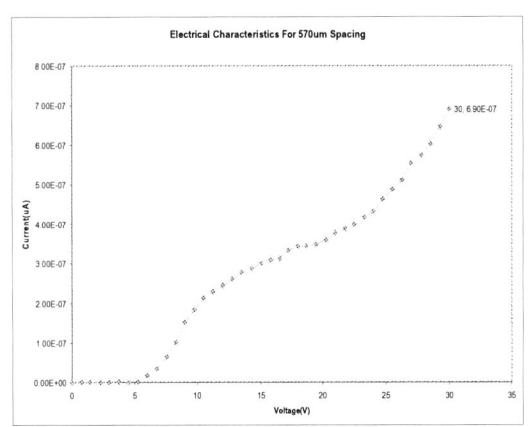

Figure 4: Electrical Characteristics for 570µm Spacing

Figure 4 shows the current-voltage characteristics for the mesa trench with mesa spacing of 570µm. It can be seen that current starts to flow at a voltage of approximately 5.25V. As the voltage is increased from 5.25V to 9.75V, the current increased to a value of 1.8×10^{-7}A. A further increased of the voltage bias to 30V produces a current of 6.9×10^{-7}A.

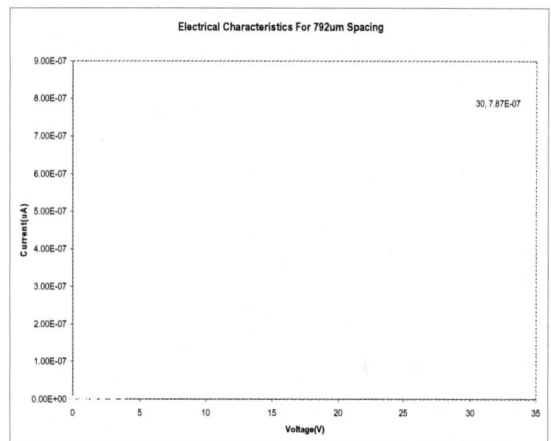

Figure 5: Current-Voltage Characteristics for 792μm Spacing

Figure 5 shows the current-voltage characteristics for the mesa trench with mesa spacing of 792μm. It can be seen that current starts to flow at a voltage of approximately 4.55V. As the voltage is increased from 4.55V to 12.75V, the current increases to a value of 3.94×10^{-7} A. A further increased of the voltage bias to 30V produces a current of 7.87×10^{-7}A.

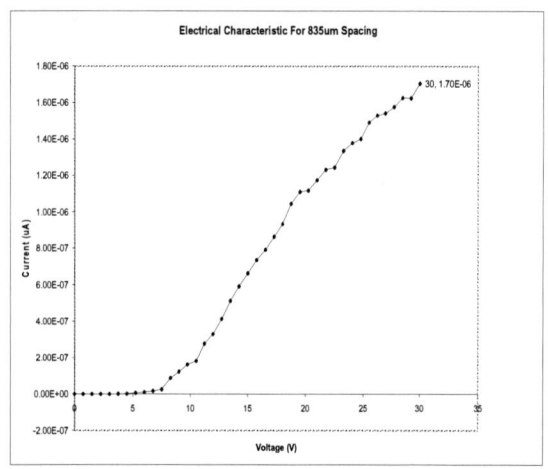

Figure 6: Electrical Characteristics for 835μm Spacing

Figure 6 shows the current-voltage characteristics for the mesa trench with mesa spacing of 835μm. It can be seen that current starts to flow at a voltage of approximately 6V. As the voltage is increased from 6V to 19.5V, the current increases to a value of 1.1×10^{-6} A. A further increased of the voltage bias to 30V produces a current of 1.7×10^{-6}A.

Comparison of Figures 4, 5 and 6 shows that the 3 different mesa spacings have very similar current-voltage characteristics. It is shown that there is a markedly increase in output current when the voltage bias is increased from 4.55V to 6V. A further increase in voltage bias causes a larger increase in output current for all 3 mesa spacings.

Mesa spacing	Current at 30V, A
570μm	6.90×10^{-7}
792μm	7.87×10^{-7}
835μm	1.70×10^{-6}

Table 1: Current Value at 30V for Different Mesa Spacing

Table 1 shows the summary of current values obtained at 30V for the 3 different mesa spacings. These results shows that the lowest current obtained is for the mesa structure with a mesa spacing of 570μm, i.e. with a current of 6.90×10^{-7} A. This implies that a smaller mesa spacing gives better isolation in terms of output current. However these results require further investigations in order to find a physical explanation to this new phenomenon.

IV. CONCLUSION

The variations of mesa spacing on pHEMT substrate give different isolations characteristics. The spacing affects the isolations results. In this study it is shown that the smallest mesa spacing of 570um generates the smallest output current of 6.90×10^{-7}. Whereas larger mesa spacings of 792 and 835um gives larger output currents of 7.87×10^{-7} and 1.7×10^{-6} A respectively. These results require further investigation to determine the physical explanation for this new phenomenon.

ACKNOWLEDGEMENT

The author would like to thank Telekom Malaysia for sponsoring the work through research project R05-0607-0 and Institute of Micro Engineering and Nano electronics (IMEN) for the support in laboratory facilities.

REFFERENCES

[1] Xin Cao, Yiping Zeng, Lijie Cui, Meying Kong, Liang Pan, Baogiang Wang, Zhanping Zhu, Journal of Crystal Growth 227-228(2001) 127-131

[2] I.Hallakoun, T.Boterashvili, G.Bunin, Y.Shapira, 2000 GaAs Mantech (2000) 25-27

[3] Lee Hock Guan, Ashaari Yusof, Asban Dolah, Mohd. Sazli Jusoh, Mohamed Razman Yahya, Research Journal, Issue No.8, 2005 35-39

Characterization of Si_3N_4 Metal-Insulator-Metal (MIM) Capacitors for Monolithic Microwave Integrated Circuits (MMIC) Applications

Lee Hock Guan, Mohd Nizam Osman, Asban Dolah, Ahmad Ismat Abdul Rahim, Mohamed Razman Yahya and Abdul Fatah Awang Mat

Telekom Research and Development Sdn. Bhd.,
Idea Tower I & II, UPM-MTDC, Technology Incubation Centre One,
Lebuh Silikon, 43400 Serdang, Selangor Darul Ehsan.

E-mail: lee@tmrnd.com.my

Abstract **The fabrication of the MIM capacitors is illustrated in the paper. The effective capacitor areas are designed at 60 x 60, 240 x 240 and 380 x 380 μm^2. From the s-parameter measurements, the fabricated capacitors showed capacitive capability up to frequency of 40GHz. Strong parasitic effect was observed at low frequency while attenuation effect was observed at high frequency.**

I. INTRODUCTION

Capacitor is widely used in MMIC's technology design where the capacitance is included in MMIC circuits in any of four basic configurations i.e. an open-circuit transmission line, coupled lines or interdigitated capacitors, Schottky diodes, and metal-insulator-metal (MIM) capacitors. Coupled lines and open-circuit transmission lines provide fairly low capacitance values. For these two capacitor types, the capacitance is dependent on the electrical length of the transmission lines. Therefore, the capacitance is highly frequency dependent. The advantage of these capacitors is that they are easily fabricated since they require only a single metal layer. The most popular type of capacitor for MMICs design is MIM capacitor because of the high capacitance per unit area can be obtained. Therefore, smaller and less costly circuits can be designed. A schematic of an MIM capacitor is shown in Fig. 1. The structure is composed of two metal plates separated by a thin layer of dielectric material. Typically, the dielectric material overlaps the first metal layer and the upper metal layer has a smaller area than the lower metal layer. This configuration helps to minimize fringing fields to ground and shorts between the upper and lower capacitor plates.

Usually the air bridge shown on Fig. 1 is not required in the design. It is often designed in to further minimize parasitic capacitance. The typically dielectric used is silicon nitride Si_3N_4, of thickness ranges from 0.1 to 0.4 mm. in which it also has been used in the encapsulation of MMIC fabrication process. SiO_2 and Ta_2O_5 are other type dielectric films used too. Since the dielectric layer is substantially thinner than the substrate thickness, MIM capacitors exhibit significant fringing effects, which are a function of the perimeter. Careful experimentation to determine the magnitude of this effect for specific process parameters, such as dielectric type and thickness, is essential for any stable process. Test capacitors should also be included on the wafer for in-process verification [1, 2].

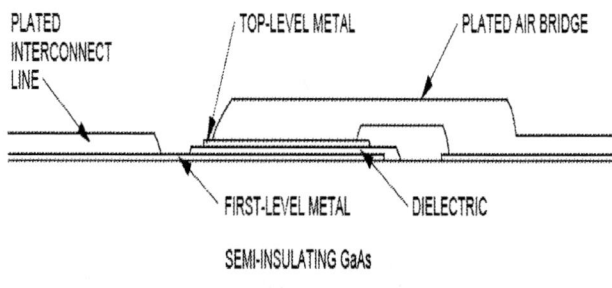

Figure 1: MIM capacitor using air bridge [3].

The model circuit of the MIM capacitor is shown in Fig. 2. The capacitor is represented by C-prime and Res components, whereas metal 2 and metal 3 is represented by inductances and capacitances of L_M2, C_M2, L_M3 and C_M3 [3, 4, 5].

0-7803-9730-4/06/$25.00 ©2006 IEEE

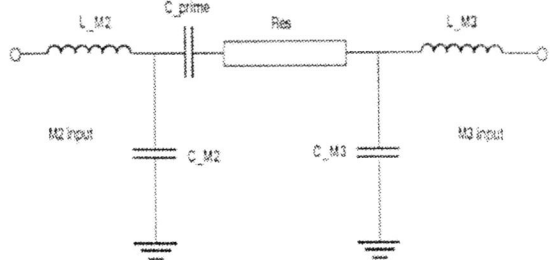

Figure 2: Equivalent circuit of MIM capacitor

II. EXPERIMENTAL PROCEDURE

The fabrication of the capacitor was done using silicon as the substrate material. Three masks were used, i.e. (1) bottom contacts, (2) Si_3N_4 etching and (3) top contacts. Both masks for the contacts consist of Radio Frequency (RF) pads for the RF characteristics measurement. Aluminum was used as the metal contact for both top and bottom contacts, which was deposited using thermal evaporator. The thickness of the Aluminum is estimated to be around 0.6μm. Silicon Nitride is sandwiched between these two contacts as the dielectric material. The deposition of the Silicon Nitride was done in sputtering system. The thickness is about 960Å. The fabrication steps are summarized in Fig. 3. Upon completion of the capacitors, the S-parameter measurement was carried out using HP 8722 network analyzer sweeping from 100MHz to 40 GHz to determine the RF characteristics of the capacitors [6, 7].

III. RESULTS AND DISCUSSION

Fig. 4 shows the measured S-parameters of the sputtered Si_3N_4 MIM capacitors with and three capacitor sizes which are 60 x 60, 140 x 140 and 380 x 380μm². From the s-parameter measurement, the fabricated capacitors are showing capacitive effect throughout the frequency range from 100MHz to 40 GHz. Although the capacitors are showing capacitive effect throughout the swept frequency, further analysis need to be carried out to determine its suitability to operate at the designated frequency. From the analysis, the capacitors showed reduction in the capacitance values as the frequency increases as illustrated in Fig. 5.

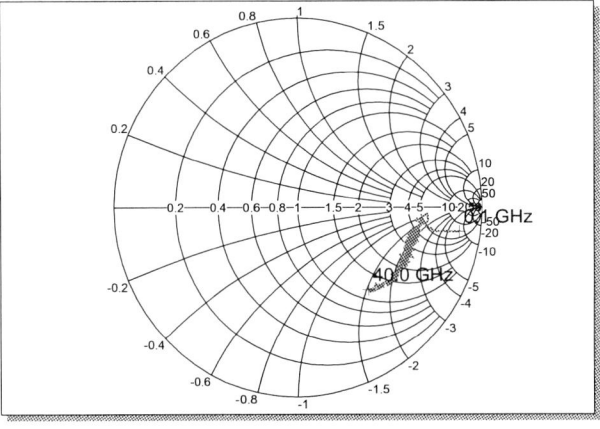

(a)

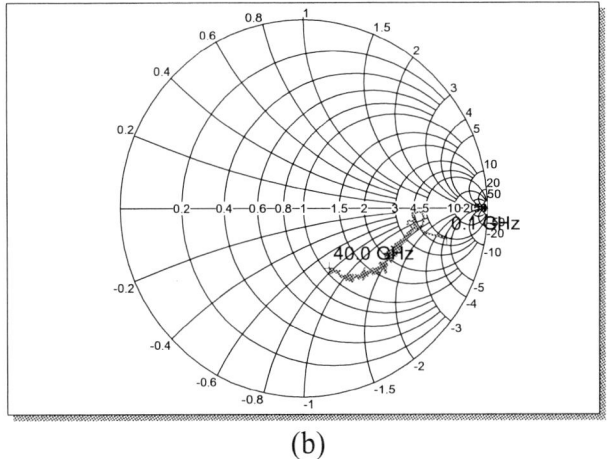

(b)

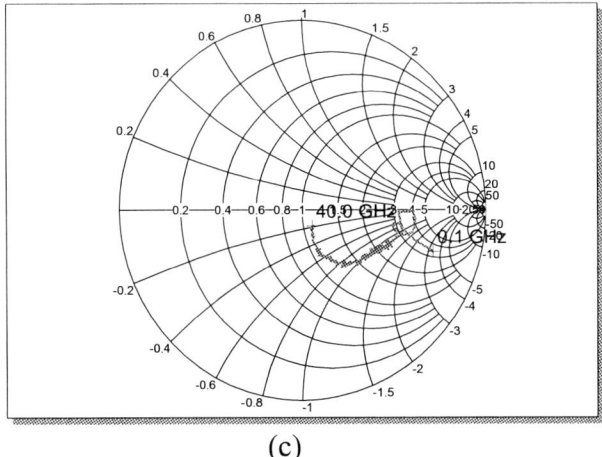

(c)

Figure 4 S-parameter (S_{11}) of MIM capacitors with various sizes. (a) 60 x 60, (b) 140 x 140 and (c) 380 x 380μm².

One attribute of the capacitance reduces as the frequency increases is due to slow carriers respond time at high frequency resulting no capacitance obtained [8, 9] Thus, capacitance dropped as the frequency increases. In addition, the trend of the graphs also signifies that high attenuation of the capacitance at high frequency as the capacitance is practically near to zero, which is the natural phenomenon. Hence, the capacitors might not be suitable to be operating at high frequency. In Figure 6, the capacitance versus dimensions is illustrated for both low and high frequency. For low frequency (Figure 6 (a)), the capacitance showed reduction as the dimensions increases where else the capacitance suppose to increase according to the capacitance formula stated in Equation (1) [10].

$$C = \frac{\varepsilon_o \varepsilon_r A}{d} \qquad (1)$$

where C is the capacitance; ε_o is the permittivity in vacuum; ε_r is dielectric constant of the insulator; A is the area of the capacitor and d is the thickness of the insulator.

Because of the usage of Aluminum as the metal contacts, the electrodes are easily get oxidized, consequently high parasitic capacitance was obtained at low frequency. As a results, the dominancy parasitic effect at low frequency on both the contacts area was observed, which was represented by L_M2, C_M2, L_M3 and C_M3 shown in Fig. 2. It is likely that these parasitic effect response better at low frequency as at high frequency (Fig. 6 (b)), the capacitance showed increment as the dimension increases. This further strengthens the fact that the parasitic effect is dominant at low frequency because at high frequency, slow parasitic carriers cannot response to the high frequency changes [8-10], therefore the capacitance obtained increases as been stated in Equation (1). On the other hand, the grounding of the devices during measurement would also cause the parasitic effect to be dominant. Since the capacitors were fabricated without via-hole grounding, therefore the parasitic effect is significant and quantifies [8-10].

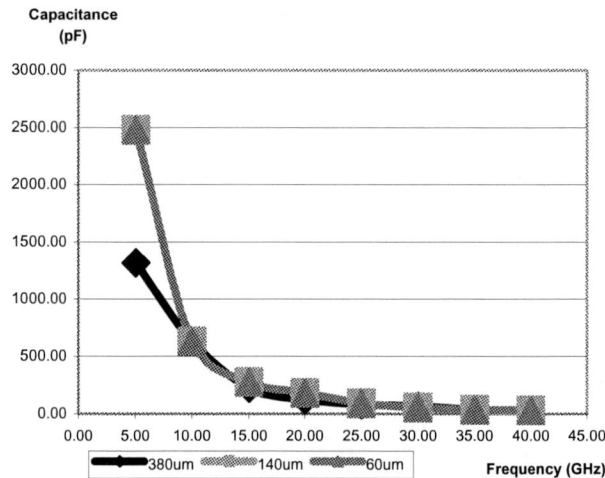

Figure 5: Capacitance vs. frequency corresponds to dimensions

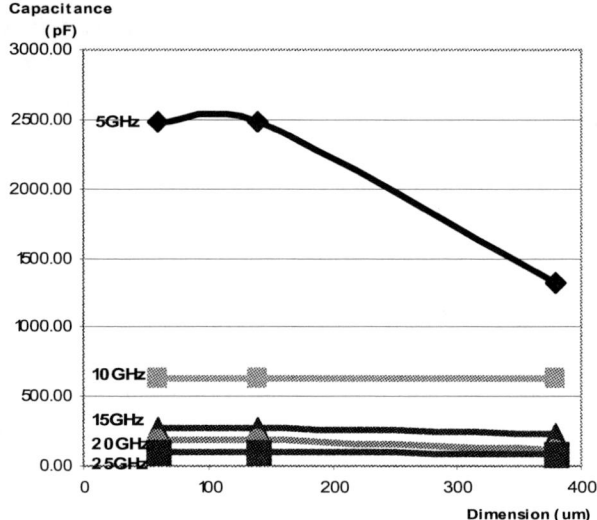

Figure 6 (a): Capacitance vs. dimensions corresponds to low frequency

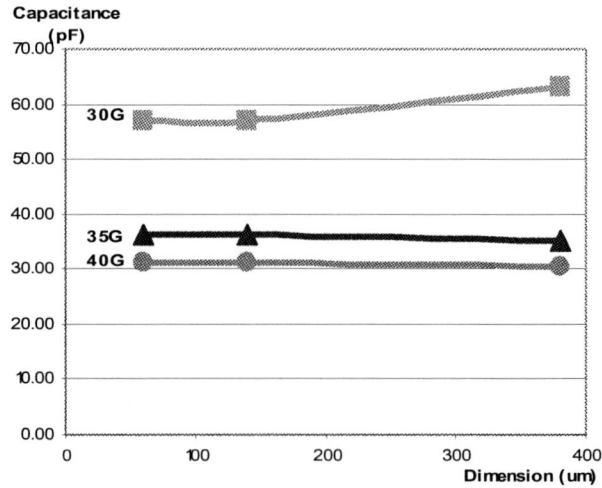

Figure 6 (b): Capacitance vs. dimensions corresponds to high frequency

IV. CONCLUSION

Results of preliminary experimental fabrication and characterization of Si_3N_4 MIM capacitor was presented. Aluminum was used as the metal contacts and sputtered silicon nitride Si_3N_4 as the dielectric materials. S-parameter measurement shows that the capacitor is capacitive throughout the swept frequency i.e. from 100MHz to 40 GHz. However, due to strong parasitic effect at low frequency and high attenuation at high frequency, the fabricated capacitors are yet appropriate to be used in MMIC circuits. Further improvement need to be carried out in the fabrication processes and materials used as the metal contacts in order to reduce the oxidation of the metal contacts which cause the parasitic component in the measurement. Via-hole grounding should be incorporate in the next fabrication to reduce parasitic and improve the grounding during the measurement.

ACKNOWLEDGMENT

The author would like to thank Telekom Malaysia for sponsoring the work through research project R05-0607-0 and Institute of Micro Engineering and Nanoelectronics (IMEN) for the support in laboratory facilities. Special thanks to Dr Lim Soo King from Universiti Tunku Abdul Rahman for his consultancy and advice on this work.

REFERENCE

[1] Giancarlo Bartolucci, Franco Giannini, Ernesto Limiti, and Steven P. Marsh, "MIM Capacitor Modeling: A Planar Approach", *IEEE Transactions on Microwave Theory and Techniques,* Vol. 43, No.4, pg 901-903 (1995).

[2] Jae-Hak Lee, Dae-Hyun Kim, Yong-Soon Park, Myoung-Kyu Sohn, and Kwang-Seok Seo, "DC and RF Characteristics of Advanced MIM Capacitors for MMIC's Using Ultra-Thin Remote-PECVD Si_3N_4 Dielectric Layers", *IEEE Microwave and Guided Wave Letters* Vol.9, No.9 pg 345-347 (1999).

[3] Sammy Kayali, George Ponchak, Roland Shaw, "GaAs MMIC Reliability Assurance Guideline for Space Applications", *JPL Publication 96-25*, pg.58 (1996).

[4] I.D. Robertson, S. Lucyszyn,, "RFIC and MMIC Design and Technology", *The Institution of Electrical Engineers, London*, pg 90-93 (2001).

[5] Jeffrey A. Babcock, Scott G. Balster, Angelo Pinto, Christoph Dirnecker, Philipp Steinmann, Reiner Jumpertz, and Badih El-Kareh, "Analog Characteristics of Metal–Insulator–Metal Capacitors Using PECVD Nitride Dielectrics", *IEEE Electron Device Letters,* Vol.22, No.5, pg230-232, (2001).

[6] C. H. Ng, K. W. Chew, and S. F. Chu, "Characterization and Comparison of PECVD Silicon Nitride and Silicon Oxynitride Dielectric for MIM Capacitors", *IEEE Electron Device Letters,* Vol.24, No.8, pg 506-508, (2003).

[7] Zhichun Wang,, Jan Ackaert, Cora Salm,, Fred G. Kuper, Marnix Tack, Eddy De Backer, Peter Coppens, Luc De Schepper, and Basil Vlachakis, "Plasma-Charging Damage of Floating MIM Capacitors", *IEEE Transaction on Electron Devices,* Vol.51, No.6, pg1017-1023, (2004).

[8] Ph Lombard, J.D. Arnould, O. Exshaw, H. Eusebe, Ph. Benech, A. Farcy and J. Torres, "MIM Capacitors Model Determination and Analysis of Parameters Influence, *IEEE ISIE, June 20-23, 2005, Dubrovnik, Croatia* (2005).

[9] Dieter K. Schoder, "Semiconductor Material and Device Characterization", John Wiley & Sons, Inc, pg 420-427, 455-484 (1998).

[10] "Applications and Design of Thin Film Capacitors", *MIC Corporation Report*, (1995).

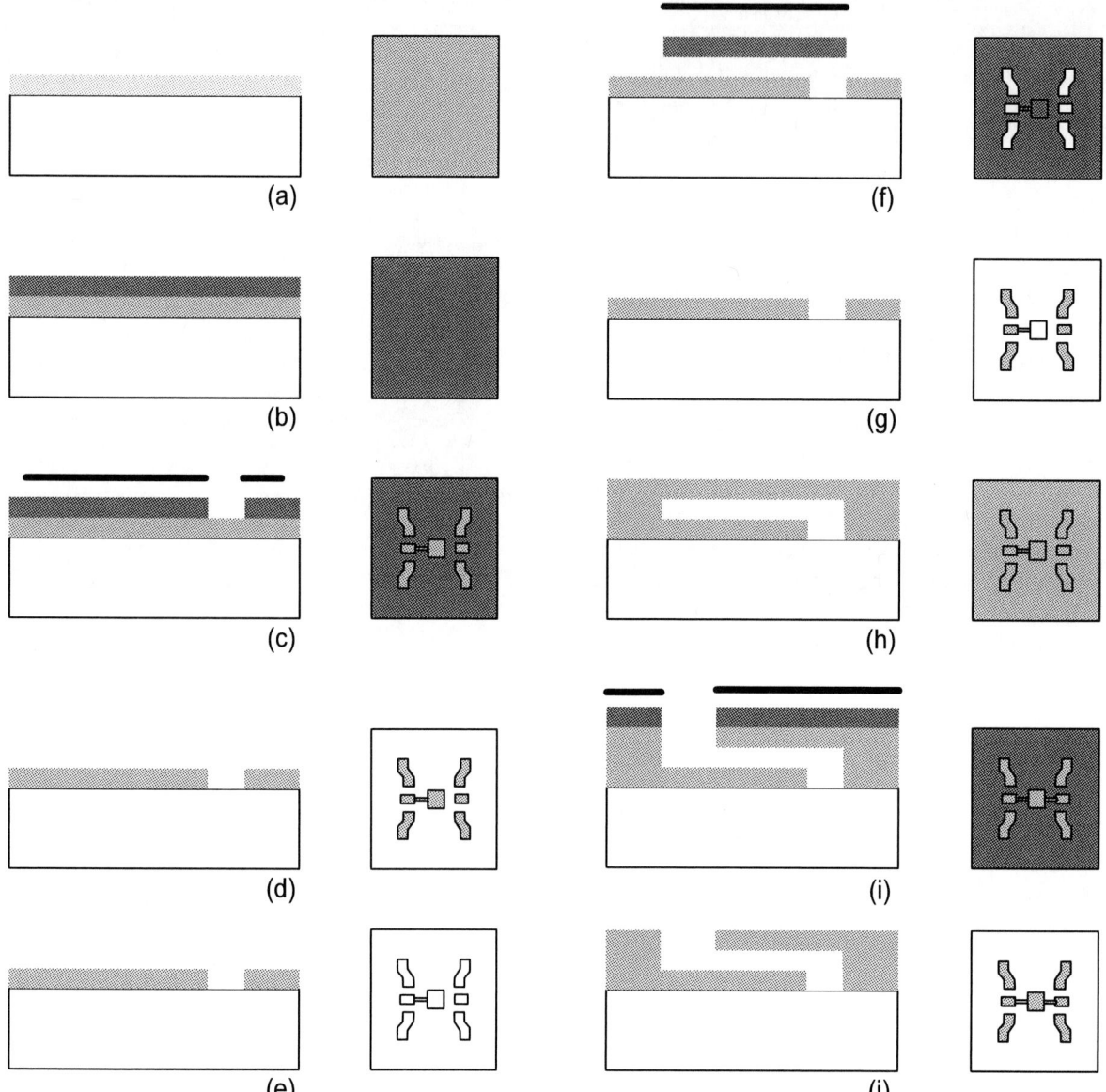

Fig. 3 Fabrication processes of MIM capacitor with cross sectional and top view.
(a) deposition of metal; (b) & (c) photolithography to etch metal for bottom contact; (d) formation of the bottom contact; (e) deposition of Si₃N₄ layers; (f) & (g) photolithography to etch Si₃N₄ layers on RF pads; (h) deposition of metal-top contact; (i) photolithography to etch metal for op contacts; (j) completed MIM capacitors.

Author Index

A

A., Mohd Sharizal 383, 388
A'ain, Abu Khari Bin 546
Abadi, Mohammad Hadi Shahrokh 734
AbbaspourSani, E. 13
Abd, Ali El- 1019
Abdipour, A. 114, 494, 785, 844, 920
Abdullah, M. Khalil 595
Abdullah, M.K. 473, 1005, 1010
Abdullah, M.Z. 595
Abdullah, Mat Johar 507, 511, 896, 901
Abkenari, M. Rezvani 785
Ab-Rahman, Muhammad Syuhaimi Bin 215, 302
Adam, Ismail 852
Adiseno, ... 95, 357
Afrang, Saeid 220
Ahmad, Mohd. Rais 468
Ahmad, Anuar Fadzil 947
Ahmad, I. 554
Ahmad, Ibrahim 21, 550, 694, 1000, 1015, 1047
Ahmad, Ibrahim Bin 964
Ahmad, Mohd Rais 809
Ahmad, Normah 741
Ahmad, Wan Rosmaria Wan 1047
Ahmadi, V. 256, 875
Ahmadi, Vahid 1028
Aik, Chew Soon 502
Al_Khusheiny, Mustafa 200
Alam, A.H.M. Zahirul 345, 852
AL-Hardan, Naif H. 507
Ali, Ahmad Al 306
Ali, and Mohd. Alauddin Mohd 340
Ali, M.A. Mohd 74, 571
Ali, N.K. 689
Ali, Y.P. 960
Ali, Yousuf Pyar 956
Alias, Mohd Sharizal 235, 239, 243
Alias, Rosidah 412
Aljunid, S.A. 473, 1005, 1010
Amiri, Parviz 760
Antony, J.J. 595
Anuar, Khairul 235, 239, 243
Ariff, Z.M. 595
Arora, B.M. 956
Arshad, M.K. Md 585
Arshad, Syariena 293
Asadi, S. 844
Awang, Mat Rasol 205
Awang, Rozidawati 790
Awang, Zaiki 947
Ayatollahi, F.L. 134
Ayatollahi, Fatima Lina 172
Azaga, Mohamed 402
Azid, Ishak Abdul 139
Azim, Md. Anwarul 834

Aziz, A. Abdul 689
Aziz, Azlan Abdul 350, 353, 507
Aziz, Azlan Abdul 511, 860, 865, 896, 901
Aziz, M. Abdel 1019
Aziz, Norazreen Abd 32
Aziz, Tengku Hasnan Tengku 101

B

Baek, Chang-Wook 275
Bais, Badariah 41, 179
Bajaj, Nihit 717
Bakri, Ahmad Yusri Mohamed 964
Beh, Jiun Kai 439
Benhamid, Mahmud 650
Bhat, Mousumi 773
Bidin, Noriah 210
Bisri, Satria Zulkarnaen 167, 279
Boey, Kean Hong 424
Boon, Tan Jaun 635
Bornoosh, Babak 533

C

Carchon, G. 7
Chai, Chuan Khye 515
Chang, Chia-Yuan 432
Chang, Edward Yi 432
Cheang, P.L. 335
Chen, Chih-Hsiung 78
Chen, Gan Wee 477
Chen, Hong Ming 154
Chen, Szu-Hung 432
Chen, T.P. 713
Cheng, Li-Chi 990
Chin, C.W. 448, 452, 937, 960
Chin, Yap Chi 326
Choi, Minsoek 680
Choo, Chew Ming 585
Choo, Lee Cheng 158
Choquette, Kent D. 235, 239, 243
Chou, Shih Min 154
Choudhury, P.K. 383
Chu, Li-Hsin 432
Chuah, L.S. 448, 452, 623, 627, 937
Chuah, Shirly 1015
Chughtai, M. Ashraf 392
Chun, Kukjin 284

D

D., Asban 752
Damghanian, Mitra 189
Damle, A.R. 956
Daud, Yaacob Mat 210
Daud, Zaidina Bt Mohd 420

Author Index

Decoutere, S. 7
Dee, Chang Fu 298, 1033
Derakhshandeh, J. 134
Diwakar, K. 829
Dolah, Asban 520, 524, 592
Dong, Gui 686
Donyavi, Ali 184

E

Eddie, E. 631
Eddie, Er 986
Edison, Thomas Alfa 167
Edwards, K.W. 924
Ehsan, Abang Annuar 215
Ehsan, Abang Anuar 302
Elfaki, Salah 225
El-Shawarby, Ayman 660
Eltaif, Tawfig 311
Endut, Zulkarnain 1000
Esa, Mazlina 576
Eshtawie, Mohamed Almahdi 463
Eve, Tai 885

F

F., Awang Mat A. 383
Fakhraei, S. Mehdi 533
Farrokhi, A. 134
Feng, Zhe Chuan 990
Fent, Tan Fent 70
Fong, Chan Sieng 428
Foxon, C.T. 924

G

Gani, Siti Meriam Abdul 790
Ganji, Bahram Azizollah 47, 53
Gharaee, Hossein 529, 533, 760
Gity, Farzan 1028
Goh, E.C. 870
Govil, Jivesh 717
Govindasamy, G. Devandran A/L 855
Groeseneken, G. 7
Guan, Lee Hock 524
Gui, D. 486, 490

H

H., Marcus Tan Y. 890
Hadi, A.R. 134
Hallaji, R. 494
Hamid, Haslinda Abdul 507, 511, 860, 865
Hamid, Mohd Yunus 316
Hamidon, Mohd. Nizar 734
Hamzah, Azrul Azlan 21
Harif, Muhammad Najib 694

Haroon, H. 595
Harrison, I. 924
Hasan, Wan Zuha Wan 928
Hasanah, Lilik 167, 279
Hasebe, T. 7
Hashim, Abdul Manaf 655
Hashim, Hesly Afida 520
Hashim, M.R. 689, 814
Hashim, Mansor 407
Hashim, Mohamad-Faizal 468
Hashim, U. 585
Hashim, Uda 377, 741
Hassan, H. Abu 448, 452, 623, 627, 825, 942, 1043
Hassan, Z. ..448, 452, 623, 627, 825, 937, 942, 960, 1043
Hasson, Feras N. 473
Hatta, Sharifah Fatmadiana Wan Muhamad 756
Heidari, M. Ebnali 256
Hein, Verena 1051
Hermida, I Dewa P 95
Hiskia, ... 105, 110
HK, Yow 416, 1024
Hoelzer, Gisbert 1051
Holliday, D. 321
Hon, Min Hsiung 150, 154
Hong, Esther Loo Chee 64
Hong, Hao En 150
Hong, Khoo Ley 600, 986
Hong, Teoh Chin 915
Hoon, Chew Soo 435
HT, Teo 643
Hua, Y.N. 486, 490
Huang, Chu-Wan 990
Huang, Jimmy Huat Since 439
Hudeish, Abdo Yahya 350, 353
Husain, Hafizah 1015
Husin, Hayati 302
Hussein, M. 937
Hussin, Razaidi 795
Hussin, Hayati 215
Hussin, Mohd Rofei Mat 800
Hussin, P. 595
Hussin, Razaidi 782
Huylenbroeck, S. Van 7

I

I., Ahmad Ismat A. 752
Ibrahim, Azmi 412
Ibrahim, K. 248, 368
Ibrahim, Kader 444, 612
Ibrahim, Zainol Abidin 124
Idris, Norina 782
Irmelia, ... 279
Islam, Md. Fokhrul 74
Islam, Md. Rafiqul 345
Islam, Md. Shabiul 834

Author Index

Islam, Md. Shabiul 839
Islam, Mohammad Rafiqul 852
Islam, Syed Zahidul 340
Ismail, Razali 563
Ismail, M. 330
Ismail, Mohd Azmi 880
Ismail, Nur Syakimah 1015
Ismail, Razali 558, 915, 933
Ismail, Rizalafande Che 782, 795

J

Jahan, Md. Saukat 834
Jalali, M. 114, 682
Jalali, Mohsen 252
Jalar, A. 554
Jamal, Dr. Zul Azhar Zahid 444
Javan, D. 13
Jeong, Jae-Seong 567
Jin, Wong Yah 933
Jiun, Ho Huey 554
Johari, Juliana 118
Juhari, Nurjuliana 124
Jung, Keum-Dong 60, 581
Jung, Youngho 680

K

K., Yow H. 420
Kamaruddin, Shahrul 855
Kamarudin, Afzan 782
Kamarudin, S. 595
Kamat, Nitin R 726, 722
Kandiah, Kumarajah 288
Kang, Ey-Goo 498
Kang, Sungchan 284
Karamdel, J. 134
Kareem, Abdeen Abdel 225
Karfaa, Yasin M. 330
Karim, Aymen M. 398
Keating, Richard 800
Kee, Margaret Ting Leh 603
Khalifa, Othman O 852
Khamis, Salah 1019
Khan, Sheroz 345, 852
Khatami, Faraz 84
Khaw, M.K. 910
Kheng, Lim Yeow 635
Khiam, Oh Chong 722
Khim, Tan Sock 893
Khor, T.S. 713
Kim, Byeong-Ju 60, 581
Kim, Byung-ju 581
Kim, Hyeon Cheol 284
Kim, Jung-Mu 275
Kim, Yo-Han 498
Kim, Yong-Kweon 275

Kim, Young-Wug 680
Kiumarsi, H. 785
Kordesch, A. 905
Kordesch, Albert Victor 1037, 1047
Kordesch, Albert Victor 372, 502, 546, 576, 764
Kouravand, Shahriar 184, 230
Kun, Li 537, 893, 979
Kushairi, N. 1039

L

Lai, Wei Hao 150, 154
Lakshmanan, ... 607
Lal, Manni 726
Lancaster, Michael J. 275
Lee, B.C. 456
Lee, B.C. 870
Lee, C.H. 456
Lee, Chan Kim 439
Lee, Cheon An 60, 581
Lee, Fu-Shin 78
Lee, Han-Sin 498
Lee, Hing Wah 163
Lee, Hochul 953
Lee, Jong Duck 581
Lee, Jong Duk 60, 953
Lee, Jung-Min 567
Lee, Sharon 619
Leisher, Paul O. 235, 239, 243
Leng, Eu Poh 550, 554
Li, K. 631, 639
Liau, C.Y. 456
Lien, Yi-Chung 432
Lim, A.Y.K. 368
Lim, Faith 870
Ling, A.E. 870
Linten, D. 7
Liu, Pan 639
Llamas-Garro, Iganacio 275
Loh, S.K. 643, 699, 702
Loo, Christopher 456, 708
Low, L.C. 335
Low, Pit Fuh 646
Lui, Lerwen 6

M

M., Abdul Fatah A. 239
M., Abbou F. 330
M., Abdul Fatah A. 235, 243, 388, 749, 752
M., Jayachandran 416
M., Norman Fadhil Idham 749, 752
M., Robeth V. 105
M., Sufrian S. 388
Mahmood, Che Seman 412
Majid, Hani Noorashiqin Abd 502
Majid, Wan Haliza Abd. 124

Author Index

Majid, Wan Haliza Abd. 477
Majlis, Burhanuddin 200
Majlis, Burhanuddin Yeop 128
Majlis, Azman Jalar Burhanuddin Yeop 550
Majlis, B.Y. 554
Majlis, Burhanuddin Yeop ...16, 21, 32, 41, 47, 53, 64, 70
Majlis, Burhanuddin Yeop 118, 172, 179, 189, 197
Majlis, Burhanuddin Yeop 220, 265, 270, 298, 306
Majlis, Burhanuddin Yeop 1033
Malekzadeh, Mina 734
Manaf, Mohd Jeffery 741, 964
Manaf, Mohd Jeffery Bin 444
Manaf, Nor Azlian Abdul 26
Manikam, M. 595
Manurung, Roberth V. 95
Manut, A. 37
Mat, A.F. Awang 1039
Mat, Abd. Fatah Awang 880
Mat, Abdul Fatah Awang 377, 412, 524
Mat, Abdule Fatah Awang 146
Meaamar, Ali 607, 665, 670
Mehrban, Mehdi 184, 230
Mellor, P.H. 321
Menon, P Susthitha 288
Mitani, S.M. 383
Mitani, Sufian 235, 239, 243
Miyamoto, Yasuyuki 432
MO, Z.Q. 486, 490
Mohamad, Romli 880
Mohamed, Shamsul 847
Mohammadi, A. 920
Mohammed, A.B. 225
Mohd-Yasin, F. 910
Mokhtar, M. 1005, 1010
Mokhtar, Mohd. Ridzuan 880
Molodpour, Vahid 90
Moradi, G. 114, 494, 844
Moravvej-Farshi, M.K. 682
Moravvej-Farshi, Mohammad Kazem 252
Muhamad, Muhamad Rasat 502
Muhammad, N.F.I. 1039
Mui, Tan Hong 1051
Muljono, Moch. 110
Munir, Tariq 896, 901
Munusamy, Kumar 994
Mursal, ... 167, 279
Musa, Rohana 809

N

Nabavi, A.R. 682
Nabavi, Abdolreza 533, 760
Nabipoor, Mohsen 197
Nagel, David J. 1
Narsale, A.M. 956
Nat, Azlani Bt. Mohd. 345

Natarajan, M.I. 7
Neo, S.P. 643, 699, 702, 705
New, C.L. 713
Ng, C.Y. 713
Ng, H.S. 708
Ng, S.S. 942
Ng, Wai Mun 424
Ng, Y.K. 456, 708, 870
Nistala, Ramesh Rao 773
Noor, N.H. Mohd. 960
Nor, Roslan Md. 158
Noshiravani, Mahyar 1028
Novikov, S.V. 924

O

O., Nurul Afzan 749
Oh, C.K. 643, 705
Omar, Nurul Afzan 592
Osman, M.N. 1039
Osman, Mohd Nizam 524, 520
Othman, M. 473, 571
Othman, Masuri 398, 402, 463, 607, 650
Othman, Masuri 670, 675, 834, 839, 928
Othman, Masuri Bin 665, 820
Othman, Mohd Khairy 146, 592, 520

P

Pal, Deb Kumar 603
Pao, W.K.S. 321
Park, and Sang-Deuk 567
Park, Byung-Gook 60, 581, 953
Park, Dong-Wok 60
Park, Dong-Wook 581
Park, Jae-Hyoung 275
Phoon, Hee Kong 515
Ping, Lee Yuan 773
Ping, Zhao Si 722, 726
Prest, Martin 275

Q

Qindeel, Rabia 210

R

R., Ahmad Ismat A. 749
R., Balachandran 416, 420, 1024
R., Eddie E. 639
R., Richard 388
R., Saadah A. 388
Radzi, Ahmad Alabqari Ma' 16
Rafi, Kazi Ashique Ahmed 839
Rahim, A.I. Abdul 1039
Rahim, Ahmad Ismat Abdul 524
Rahim, Alhan Farhanah Abd 372

Author Index

Rahman, Saadah Abdul 481, 790
Ramesh, Rao 537, 635, 885, 979
Rao, Nistala Ramesh 777
Rashid, Hairul Azhar Abdul 880
Rashid, Norazah Abd. 800
Rasmi, A. 1039
Rasmi, Amiza 377
Razak, Manis Mulyany Bt. Abdul 345
Razazadeh, Ghader 84
Reaz, M.B.I. 905, 910
Rezazadeh, Ghader 90, 184, 230
Ritikos, Richard 790
Rizwan, Zahid 407
RM, Manickam 1024
Rosli, Siti Azlina 507, 511, 860, 865

S

S., Khairul Anuar M. 383, 388
S., Rasidah 752
Saad, Ismail 558, 563, 933
Saad, M.R. 595
Saad, Rodzaki 847
Sabet, Mehdi 230
Sabri, Kenny 603
Sabtu, Idris 749
Sahbudin, R.K.Z. 1005, 1010
Sahoo, P.B. 726
Said, Suhana Mohd 612
Sakata, Hiroshi 974
Salim, A.J. 571
Salleh, Muhamad Mat 26, 101, 146, 205, 261, 298
Salleh, Muhammad Mat 293
Samad, M.D.A. 473
San, Wong Chiow 435
San, Yong Soo 550
Sang, Ko Bong 444, 612
Sanusi, R. 1039
Sarker, Md. Shakowat Zaman 839
Sattari, M. 134
Sawada, M. 7
Schroeter, Torsten 1051
Sehgal, Rohan 804
Selamat, Mohd. Suhaimi 265, 270
Sellah, Muhamad Mat 326
Senthilpari, C. 829
Sepeai, Suhaila 261
Seyedfaraji, Amireh 1028
Shaari, Sahbudin 288
Shaari, Abdul Halim 407
Shaari, S. 330, 473, 1005, 1010
Shaari, Sahbudin 64, 70, 215, 225, 302, 311
Shahar, Aftanasar Md 139
Shakaff, Ali Yeon Md. 782
Shalaby, Hossam M. H. 311
Shalby, Abdel Aziz 1019

Shapee, Sabrina Mohd 412
Sheu, W.B. 456, 708, 870
Shin, Hyungcheol 60, 581, 680, 953
Shoaei, Omid 665
Shun, Cheong Yew 444
Sidek, Othman 139, 764, 768, 956, 1037
Sidek, Roslina 734
Sin, Y.K. 248
Singh, Ajay Kumar 829
Siping, Zhao 537, 542, 600, 619, 635, 686
Siping, Zhao 777, 885, 890, 893, 979, 986
Siregar, Masbah R.T. 105, 110
Siregar, Masbah R.T. 95
Siriani, Dominic 235, 239, 243
Siswanto, Meilana 675
Soegandi, T.M.S. 357
Soetedjo, Hariyadi 388, 749
Soin, Norhayati 612, 756
Song, Z.G. 643, 699, 702, 705
Sooudi, E. 256, 875
Soroosh, M. 256, 875
Soroosh, Mohammad 252
Su, Hieng Tiong 275
Su, Yen Hsun 150, 154
Subramaniam, Kalavathi 576
Suck, Park Hyun 603
Sujod, Muhamad Zahim 974
Sukemi, Horazham Mohd 1000
Sukirno, ... 167, 279
Sundaramoorthy, V K 924
Sung, Man-Young 498
Suparjo, Bambang Sunaryo 928
Suryamas, Adi Bagus 167, 279
Swee, Gary Lee How 1000
Syono, M.I. 37
Syono, Mohd. Ismahadi 163

T

T., Goh Boon 388
Taha, Luay Yassin 306
Tahmasebi, Admadali 84, 90
Talebi, N. 134
Tan, Ivy 456
Tan, Philip Beow Yew 764, 1037
Tathesari, Elham 529
Tavakoli, A. 114, 844
Tayarani, M. 785
Tayel, Mazhar 660
Teck, Yeo Eng 855
Teh, Y.K. 905, 910
Tehranirokh, Masoomeh 179
Teoh, Lay Gaik 150, 154
Teymourzadeh, Rozita 820
Thahab, S.M. 825, 937, 1043
Theng, Chuah Cheow 768

Author Index

Thijs, S. 7
Tien, Victor Siow Yuen 855
Timothy, Ling 890
Tiong, Su Hieng 340
Tong, Goh Boon 481, 790

U

U, Thangamani 316
Umar, Mursyidah 101
Usman, Ida 167, 279

V

V., Saaminathan 416, 420, 1024
Vahdati, H. 920
Vaya, P R 316

W

Wagiran, R. 554
Wahab, Kader Ibrahim Abdul 741, 964
Wahab, Yasmin Abdul 947
Wahid, Khairul Anuar Bin Abd 139
Wang, Tan Chin 777
Wang, Ying-Lang 990
Widodo, Slamet 95
Wiranto, G. 357
Wiranto, Goib 95
Wong, Bridger K.S. 428
Wong, Houn Wai 646
Wong, J.I. 713
Wong, Vee Kin 646
Wong, W.S.H. 321
Wong, Wallace 340

X

Xiao, S. 7
Xing, Z.X. 486, 490

Y

Y., Mohamed Razman 383, 388, 749, 752
Y., Mohd Razman 235, 239, 243
Yaacob, M.H. 1005, 1010
Yaakob, Syamsuri 880
Yahaya, Muhammad 26, 101, 205, 261, 293, 298, 326
Yahya, M.R. 1039
Yahya, Mohamed Razman 412, 520, 524, 592, 880
Yam, F.K. 627, 937, 942, 960
Yang, M. 713
Yang, T.R. 990
Yap, Kok Sing 424
Yap, Matthew 515
Yaqoob, Arjumand 392
Yasin, F. Mohd- 905

Yasui, Kanji 655
Yee, A.F. 708
Yen, Lau Siau 612
Yeo, Yvonne 1051
Yeong, Son Jin 603
Yet, E.K. 456
Yet, S.I. 870
Yew, Philip Tan Beow 1047
Yew, Tan Kong 809
Ying, Cho Jie 542
Yoon, Yeo Jo 680
Yoon, Youngchang 953
You, A.H. 335
Younan, Hua 537, 542, 600, 619, 635, 686, 773
Younan, Hua 777, 885, 890, 893, 979, 986
Yunas, Jumril 128
Yusof, Wan Yusmawati Wan 416
Yusof, Yusman Mohd 372
Yusoff, M.Z.M. 814
Yusoff, Nurul Huda 205
Yusoff, Yuzman 468, 809
Yusoff, Zubaida 994

Z

Zaharim, Azami 694
Zahedi, Edmond 398, 675
Zain, Azlina Mohd 363
Zainal, N. 937
Zakaria, Azmi 407
Zan, Z. 1005, 1010
Zhao, S.P. 486, 490, 631, 699, 702, 705
Zhao, Siping 639
Zhe, Huang Min 546
Zhenxiang, Xing 686
Zhigang, Song 777
Zhiqiang, Mo 542, 600, 619, 686
Zoolfakar, Ahmad Sabirin 363

2006 IEEE International Conference on Semiconductor Electronics

Kota Kinabalu, Malaysia
29 November- 1 December 2006

Volume 2 of 2

IEEE Catalog Number: 06EX1443
ISBN: 0-7803-9730-4

Copyright © 2006 by The Institute of Electrical and Electronics Engineers, Inc.
All Rights Reserved

Copyright and Reprint Permissions: Abstracting is permitted with credit to the source. Libraries are permitted to photocopy beyond the limit of U.S. copyright law for private use of patrons those articles in this volume that carry a code at the bottom of the first page, provided the per-copy fee indicated in the code is paid through Copyright Clearance Center, 222 Rosewood Drive, Danvers, MA 01923.

For other copying, reprint or republications permission, write to IEEE Copyrights Manager, IEEE Operations Center, 445 Hoes Lane, Piscataway, New Jersey USA 08854. All rights reserved.

IEEE Catalog Number: 06EX1443

ISBN: 0-7803-9730-4

Additional Copies of This Publication Are Available from:

IEEE Service Center
445 Hoes Lane
Piscataway, NJ 08854
IEEE Service Center
445 Hoes Lane
Piscataway, NJ 08854
Phone: (800) 678-IEEE
            (732) 981-1393
Fax: (732) 981-9667
E-mail: customer-service@ieee.org

International Advisory Committee

Prof. Dr. Vijay Arora	:	Wilkes University, USA
Prof. Dr. Hiroshi Iwai	:	Tokyo Institute of Technology, Japan
Prof. Dr. Nico de Rooij	:	University Neuchatel, Switzerland
Prof. Dr. Arokia Nathan	:	University of Waterloo, Canada
Prof. Dr. Cary Yang	:	Santa Clara University, USA
Prof. Dr. Peter Houston	:	Sheffield University, UK
Prof. Dr. Jong Duk Lee	:	Seoul National University, Korea
Prof. Dr. Edward Chang	:	National Chiao Tung University, Taiwan
Prof. Dr. Jang Kyoo-Shin	:	Kyungpook University, Korea
Assoc. Prof. Dr. Jung-Chih Chiao	:	University of Texas at Arlington, USA
Assoc. Prof. Dr. Francis Tay Eng Hock	:	National University of Singapore, Singapore
Dr. Lerwen Liu	:	Zyvex Corporation, USA
Assoc. Prof. Dr. Joydeep Dutta	:	Asian Institute of Technology, Bangkok, Thailand
Dato' Ahmad Kabeer b. Mohd. Nagoor	:	AKN Technology Berhad, Malaysia
En. Abdul Wahab Abdullah	:	MIMOS Berhad, Malaysia
Prof. Dr. Muhammad Yahaya	:	Universiti Kebangsaan Malaysia, Malaysia
Prof. Dr. David Nagel	:	George Washington University, USA
Assoc. Prof. Dr. S.F. Yoon	:	Nanyang Technological University, Singapore
Prof. Dr. S.M. Sze	:	National Chiao Tung University, Taiwan
Dr. Masbah R.T. Siregar	:	LIPI, Indonesia
Prof. Dr. C.I.M. Beenakker	:	DIMES, Netherlands
Prof. Dr. Muhamad Rasat Muhamad	:	Universiti Malaya, Malaysia
Dato' Dr. Mohd. Ariffin Hj. Aton	:	SIRIM Berhad, Malaysia
Prof. Dr. Peter Ashburn	:	University of Southampton, UK

Organizing Committee

Chairman
Prof. Dr. Burhanuddin Yeop Majlis

Technical Chairman
Mr. Richard Keating

Secretary
Mrs. Badariah Bais

Treasurer
Mr. Rahman Wagiran

Members
Prof. Dr. Muhamad Mat Salleh *(Universiti Kebangsaan Malaysia)*
Prof. Dr. Sahbudin Shaari *(Universiti Kebangsaan Malaysia)*
Prof. Dr. Tou Teck Yong *(Universiti Multimedia Malaysia)*
Assoc. Prof. Ibrahim Ahmad *(Universiti Kebangsaan Malaysia)*
Assoc. Prof. Dr. Razali Ismail *(Universiti Teknologi Malaysia)*
Assoc. Prof. Dr. A. H. M. Zahirul Alam *(International Islamic University Malaysia)*
Dr. Azlan Aziz *(Universiti Sains Malaysia)*
Dr. Ghazali Omar *(Infineon Technologies)*
Dr. Roslina Mohd. Sidek *(Universiti Putra Malaysia)*
Dr. Mohd. Nizar Hamidon *(Universiti Putra Malaysia)*
Dr. Abdullah Chik *(Universiti Malaysia Sabah)*
Hj. Abdullah Lin *(Silterra Malaysia Sdn. Bhd.)*
Mr. Ronnie C. H. Ng *(Intel Microelectronics (M) Sdn. Bhd.)*
Mr. Ismahadi Syono *(MIMOS Berhad)*
Dr. Hua Younan *(Chartered Semiconductor, Singapore)*
Dr. Awangku Abdul Rahman Pgn. Hj. Yusof *(Universiti Malaysia Sarawak)*

Secretariat
Chairman ICSE2006
Electron Devices Chapter, IEEE Malaysia Section
c/o Institute of Microengineering and Nanoelectronics (IMEN)
Universiti Kebangsaan Malaysia
43600 Bangi, Selangor
MALAYSIA
Phone: +603 8925 9080
Fax: +603 8925 9080 / 8925 0439
Email: eds@vlsi.eng.ukm.my

International Conference on Semiconductor Electronics
November 29-December 1, 2006

This ICSE2006 conference is aimed at bringing together scientists and engineers to discuss various issues and trends in the field of semiconductor technology. Malaysia is centrally located in the world's fastest growing economy region of Asia-Pacific and the growth of semiconductor industry in this region has been tremendous. It is timely that this periodically held IEEE international conference becomes an important avenue for addressing worldwide concerns on the technology.

Scope of Conference

Semiconductor Devices and Integrated Circuits

- Device physics
- Device characterization and testing
- Device modeling, simulation and design
- Reliability and failure analysis
- VLSI and MMIC circuits design
- Process / manufacturing technology
- IC packaging
- Materials and new fabrication technologies for VLSI structures
- Training and human resources development in microelectronics industry

MEMS Sensors, Actuators and Nanoelectronics

- MEMS design and development
- MEMS modeling
- MEMS materials and manufacturing
- MEMS fabrication process
- MEMS packaging
- MEMS actuators
- Bio-MEMS, RF MEMS and MOEMS
- Nano-Optics

Message From ICSE2006 Conference Chairman

 *Selamat Datang* to Kuala Lumpur and ICSE2006.

On behalf of the Organising Committee, I have great pleasure in welcoming all of you to ICSE2006. This is the seventh ICSE organized by the Electron Devices Chapter of IEEE Malaysia Section and technically co-sponsored by Electron Devices Society. Over the last 14 years, the ICSE has become the preeminent international forum on semiconductor electronics. Our coverage embraces all aspects of the semiconductor technology, from materials issues and device fabrication, MEMS and microsensors, photonics technology, integrated circuit design (VLSI and RFIC), IC testing, semiconductor manufacturing, and system applications.

The ICSE provides the ideal forum to present your latest results in semiconductor related issues. The proceedings of ICSE2006 consists of two keynote papers as well as 228 contributed papers from all over the world. We are delighted to have four prominent speakers in microelectronics. Prof. Dr. David J. Nagel, from The George Washington University who will speak on The Micro-Scale Structures and Nano-Scale Materials for Chemical and Biological Sensors and Dr. Lerwen Liu from Zyvex Corporation, USA on Nanotechnology. The other two invited speakers are Prof. Dr. Kukjin Chun and Dr. Natarajan Mahadeva Iyer. This will further enhance the program and I am sure participants will find the experience worthwhile. This time we divide papers into two main category, Semiconductor Devices and Integrated Circuits, and MEMS Sensors, Actuators and Nanoelectronics. The conference proceedings will serve as a permanent record and a valuable reference for many microelectronics and MEMS experts.

I would like to express my gratitude to Telekom Research and Development Sdn. Bhd., MIMOS Bhd. and Universiti Kebangsaan Malaysia (UKM) for the sponsorship and IEEE Electron Devices Society for the technical co-sponsorship.

To all participants, I hope we can mutually gain benefits and new knowledge through fruitful discussion while making new contact with other participants. To overseas participants, I wish you a pleasant stay in this country and we will endeavor to make your stay here gratifying and enjoyable as possible.

Finally, my gratitude to all those who have helped to make this conference a reality, specifically to my committee members for their efforts to ensure the success of this conference. *Terima kasih.*

PROF. DR. BURHANUDDIN YEOP MAJLIS
ICSE2006 CHAIRMAN AND ELECTRON DEVICES CHAPTER CHAIR

Table of Contents

Micro-Scale Structures and Nano-Scale Materials for Chemical and Biological Sensors...1
David J. Nagel

Electrical Characterization of Sub-100nm Features in Semiconductor Devices..6
Lerwen Lui

Implementation of 6kV ESD Protection for a 17GHz LNA in 130nm SiGeC BiCMOS..7
D. Linten, M.I. Natarajan, S. Thijs, S. Van Huylenbroeck, S. Xiao, G. Carchon, S. Decoutere, M. Sawada, T. Hasebe, G. Groeseneken

Optimization of the Temperature Sensor Position for MEMS Gas Flow Meters 13
E. AbbaspourSani, D. Javan

Non-Crossing Differential Capacitive MEMS Accelerometer with Electrostatic Spring Tuning 16
Ahmad Alabqari Ma' Radzi, Burhanuddin Yeop Majlis

Fabrication of Platinum Membrane on Silicon for MEMS Microphone..21
Azrul Azlan Hamzah, Burhanuddin Yeop Majlis, Ibrahim Ahmad

Effect of Annealing on the Sol-Gel Derived $SrBi_4Ti_4O_{15}$ Thin Films for Piezoelectric Pressure Sensors....................26
Nor Azlian Abdul Manaf, Muhamad Mat Salleh, Muhammad Yahaya

Fabrication Study of Solid Microneedles Array Using HNA 32
Norazreen Abd Aziz, Burhanuddin Yeop Majlis

Effects of Mechanical Geometrics on Resonance Sensitivity of MEMS Out-of-Plane Accelerometer.............................37
A. Manut, M.I. Syono

Structure Design and Fabrication of an Area-changed Bulk Micromachined Capacitive Accelerometer 41
Badariah Bais, Burhanuddin Yeop Majlis

The Effect of Design Parameters on Static and Dynamic Behaviors of the MEMS Microphone...................................47
Bahram Azizollah Ganji, Burhanuddin Yeop Majlis

Deep Trenches in Silicon Structure using DRIE Method with Aluminum as an Etching Mask 53
Bahram Azizollah Ganji, Burhanuddin Yeop Majlis

Performance Improvement of OTFTs using Double Layer Insulator 60
Dong-Wok Park, Cheon An Lee, Keum-Dong Jung, Byeong-Ju Kim, Byung-Gook Park, Hyungcheol Shin, Jong Duk Lee

Simulation of a Novel Lateral H Structure PIN InGaAs Photodiode with Consistent Electron Drift Length 64
Esther Loo Chee Hong, Sahbudin Shaari, Burhanuddin Yeop Majlis

Distributed CATV Inputs in FTTH-PON System...70
Tan Fent Fent, Sahbudin Shaari, Burhanuddin Yeop Majlis

On the Use of a Mixed-Mode Approach For MEMS Testing 74
Md. Fokhrul Islam, M.A. Mohd. Ali

A Novel Model for Membranes of Micropumps Partially Loaded with Electromagnetic Forces 78
Fu-Shin Lee, Chih-Hsiung Chen

Investigation of the Torsion and Bending Effects on Static Stability of Electrostatic Torsional Micromirrors...84
Ghader Razazadeh, Faraz Khatami, Admadali Tahmasebi

Electromechanical Behavior of Microbeams with Piezoelectric and Electrostatic Actuation 90
Ghader Rezazadeh, Ahmadali Tahmesebi, Vahid Molodpour

The Gas Sensing Potential of Nanocrystalline SnO_2 and In_2O_3 Powders Prepared by Mechanical Milling................95
Goib Wiranto, Adiseno, I Dewa P Hermida, Roberth V. Manurung, Slamet Widodo, Masbah R.T. Siregar

vii

Table of Contents

Reduction of Turn-On Voltage in a Singe Layer Structured Organic Light-Emitting Diode Using Nanocomposites SiO2:PHF......101
Tengku Hasnan Tengku Aziz, Muhamad Mat Salleh, Mursyidah Umar, Muhammad Yahaya

Development of an Integrated Miniaturized Mulit-Ion Flow Cell System for Water Quality Measurement......105
Hiskia, Masbah R.T. Siregar, Robeth V.M.

Design and Fabrication of Micromachined Gas Sensors......110
Hiskia, Masbah R.T. Siregar, Moch. Muljono

An Active Integrated Spiral Antenna System for UWB Applications......114
M. Jalali, A. Abdipour, A. Tavakoli, G. Moradi

Analysis of a Bilaminar Circular Piezoelectric Actuator for Micropumps......118
Juliana Johari, Burhanuddin Yeop Majlis

Degradation of Single Layer MEH-PPV Organic Light Emitting Diode (OLED)......124
Nurjuliana Juhari, Wan Haliza Abd. Majid, Zainol Abidin Ibrahim

Micro-fabricated Square Interwinding Transformer Using Surface Micromatching......128
Jumril Yunas and Burhanuddin Yeop Majlis

Fabrication of a Single Carbon Nano Tube for Use in Nanolithography of MOSFET Gate......134
J. Karamdel, N. Talebi, M. Sattari, J. Derakhshandeh, F.L. Ayatollahi, A. Farrokhi, A.R. Hadi

Development of Integrated Detection System for Shock Vibration by Using MEMS Accelerometer......139
Khairul Anuar Bin Abd Wahid, Ishak Abdul Azid, Aftanasar Md Shahar and Othman Sidek

Organic Light Emitting Diode (OLED) Using Different Hole Transport and Injecting Layers......146
Mohd Khairy Othman, Muhamad Mat Salleh, Abdule Fatah Awang Mat

How the Optical Properties of Au Nanoparticles are Affected by Surface Plasmon Resonance......150
Yen Hsun Su, Wei Hao Lai, Lay Gaik Teoh, Hao En Hong, Min Hsiung Hon

Gas Sensing Properties of ZnO:Al Thin Films Prepared by RF Magnetron Sputtering......154
Lay Gaik Teoh, Hong Ming Chen, Yen Hsun Su, Wei Hao Lai, Shih Min Chou, Min Hsiung Hon

Factors Affecting the Growth of Carbon Nanotubes......158
Lee Cheng Choo, Roslan Md. Nor

Effect of Self-Weight and Non-Rigidity on the Bending Characteristics of Surface Micromachined MEMS Test Structures......163
Hing Wah Lee, Mohd. Ismahadi Syono

Low Temperature Carbon Nanotube Fabrication Using Very High Frequency-Plasma Enhanced Chemical Vapour Deposition Method......167
Sukirno, Satria Zulkarnaen Bisri, Lilik Hasanah, Mursal, Ida Usman, Adi Bagus Suryamas, Thomas Alfa Edison

Design and Modeling of Micromachined Condenser MEMS Loudspeaker Using Permanent Magnet Neodymium-Iron-Boron (Nd-Fe-B)......172
Fatima Lina Ayatollahi, Burhanuddin Yeop Majlis

A Hydrogel-based Microvalve for Insulin Delivery Application......179
Masoomeh Tehranirokh, Badariah Bais, Burhanuddin Yeop Majlis

Application of Full Factorial Design Method in MEMS Capacitive Thermal Sensor Sensitivity......184
Mehdi Mehrban, Shahriar Kouravand, Ghader Rezazadeh, Ali Donyavi

Design of a High Sensitivity Structure for MEMS Fingerprint Sensor......189
Mitra Damghanian, Burhanuddin Yeop Majlis

High-Precision Thickness Control of Silicon Membranes Using Etching Techniques......197
Mohsen Nabipoor, Burhanuddin Yeop Majlis

Table of Contents

Aluminum Based Two-Port-Clamped-Clamped Resonators..200
Mustafa Al_Khusheiny and Burhanuddin Majlis

The Use of Photoluminescence Spectra of TiO2 Nanoparticles Coated with Porphyrin Dye Thin Film for Grading Agarwood Oil..205
Nurul Huda Yusoff, Muhamad Mat Salleh, Muhammad Yahaya, Mat Rasol Awang

Study on the Effect of Q-Switched Nd:Yag Laser Interaction with Al in Variable Magnetic Field..210
Rabia Qindeel, Noriah Bidin, Yaacob Mat Daud

OXADM Multiplex Protection Scheme for Bidirectional Path Switched Ring..215
Muhammad Syuhaimi Bin Ab-Rahman, Hayati Hussin, Abang Annuar Ehsan, Sahbudin Shaari

Inductively-Tuned K-Band Distributed MEMS Phase Shifter..220
Saeid Afrang, Burhanuddin Yeop Majlis

Crosstalk Enhancement in Multiplexer/Demultiplexer Based Arrayed Wavelength Grating in Dense Wavelength Division Multiplexing..225
Salah Elfaki, Abdeen Abdel Kareem, A.B. Mohammed, Sahbudin Shaari

Analytical Model Studying of a Novel Tunable Capacitor Based on Bimetallic Thermal Actuator..230
Shahriar Kouravand, Mehdi Mehrban, Ghader Rezazadeh, Mehdi Sabet

Uniformity Study of GaAs-based Vertical-Cavity Surface-Emitting Laser Epiwafter Grown by MOCVD Technique..235
Mohd Sharizal Alias, Paul O. Leisher, Kent D. Choquette, Khairul Anuar, Dominic Siriani, Sufian Mitani, Mohd Razman Y., Abdul Fatah A.M.

Electro-Opto Characteristics of 850 nm Oxide-Confined Vertical-Cavity Surface-Emitting Lasers..239
Mohd Sharizal Alias, Paul O. Leisher, Kent D. Choquette, Khairul Anuar, Dominic Siriani, Sufian Mitani, Mohd Razman Y. and Abdul Fatah A.M.

Efficiency and Spectral Characteristics of 850 nm Oxide-Confined Vertical-Cavity Surface-Emitting Lasers..243
Mohd Sharizal Alias, Paul O. Leisher, Kent D. Choquette, Khairul Anuar, Dominic Siriani, Sufian Mitani, Mohd Razman Y., Abdul Fatah A.M.

2D Silicon-based Photonic Crystals..248
Y.K. Sin, K. Ibrahim

Calculation of Quantum Efficiency for Resonant Cavity Photodiodes using the FDTD Method..252
Mohammad Soroosh, Mohsen Jalali, Mohammad Kazem Moravvej-Farshi

Static Quasi 3D Thermal Simulation of Ion Implanted Vertical Cavity Surface Emitting Lasers..256
E. Sooudi, V. Ahmadi, M. Ebnali Heidari, M. Soroosh

The Effect of Annealing on the Performances of the White Organic Light Emitting Diode (OLED)..261
Suhaila Sepeai, Muhamad Mat Salleh, Muhammed Yahaya

Considering RFID Inmate Tagging Application to Enhance Prison Management..265
Mohd. Suhaimi Selamat, Burhanuddin Yeop Majlis

Challenges in Implementing RFID Tag in a Conventional Library..270
Mohd. Suhaimi Selamat, Burhanuddin Yeop Majlis

Investigating the Performance of RF MEMS Switches..275
Hieng Tiong Su, Iganacio Llamas-Garro, Michael J. Lancaster, Martin Prest, Jae-Hyoung Park, Jung-Mu Kim, Chang-Wook Baek, Yong-Kweon Kim

Comparison of Electronic Transport Parameter of CNT(10,10)/CNT(17,0) and CNT(5,5)/CNT(8,0) Carbon Nanotube Metal-Semiconductor On-Tube Heterojunction..279
Sukirno, Satria Zulkarnaen Bisri, Irmelia, Lilik Hasanah, Adi Bagus Suryamas, Ida Usman, Mursal

Table of Contents

See-saw Type RF MEMS Switch with Fine Gap Vertical Comb...284
Sungchan Kang, Hyeon Cheol Kim, Kukjin Chun

Numerical Modeling of a Diffusion-Based In0.53Ga0.47As Lateral PIN Photodiode for 10 Gbits/s Optical Communication Systems 288
P Susthitha Menon, Kumarajah Kandiah and Sahbudin Shaari

The Effect of Surface Microstructure on The Response of Titanium Dioxide Coated with Cobalt-Porphyrin Thin Films Towards Gases in Quartz Crystal Microbalance Sensor 293
Syariena Arshad, Muhammad Mat Salleh, Muhammad Yahaya

Synthesis and Characterization of CuO Nanowires 298
Chang Fu Dee, Muhammad Yahaya, Muhamad Mat Salleh, Burhanuddin Yeop Majlis

Analytical Modeling of Optical Cross Add and Drop Multiplexing Switch 302
Muhammad Syuhaimi Bin Ab-Rahman, Abang Anuar Ehsan, Hayati Husin, Sahbudin Shaari

Modelling and Analysis of a Transformer based MEMS Piezoelectric Vibration Type Microgenerator 306
Luay Yassin Taha, Burhanuddin Yeop Majlis, Ahmad Al Ali

A Novel Successive Interference Cancellation Scheme in OCDMA System 311
Tawfig Eltaif, Hossam M. H. Shalaby, Sahbudin Shaari

Simulation of Piezo-Resistive Metal Gauge on Rectangular Membrane for Low Pressure Application 316
Mohd Yunus Hamid, Thangamani U, P R Vaya

Design and Modeling of an Electromagnetic Levitated and Actuated Micromotor 321
W.S.H. Wong, W.K.S. Pao, D. Holliday, P.H. Mellor

Fabrication of Polymer Light Emitting Diodes with ITO/PVK:PBD:DPVBi:DCJTB/Al Structure..........................326
Yap Chi Chin, Muhammad Yahaya, Muhamad Mat Sellah

Homodyne Linear Crosstalk Impact in an Array Waveguide Router as an OADM for WDM Networks.....................330
Yasin M. Karfaa, M. Ismail, Abbou F.M., S. Shaari

Avalanche Multiplication and Excess Noise Factor of Heterojunction Avalanche Photodiodes 335
A.H. You, L.C. Low, P.L. Cheang

Faults Detection Approach for Self-Testable RF MEMS...340
Syed Zahidul Islam, Wallace Wong, Su Hieng Tiong, and Mohd. Alauddin Mohd Ali

MEMS Switch Material Dependency on Designing a Reconfigurable Antenna 345
A.H.M. Zahirul Alam, Md. Rafiqul Islam, Sheroz Khan, Azlani Bt. Mohd. Nat, Manis Mulyany Bt. Abdul Razak

Surface and Composition Reactivity of Pt/GaN Catalytic Contact as Schottky Barriers Gas Sensor 350
Abdo Yahya Hudeish, Azlan Abdul Aziz

Metals/GaN Catalytic Contact Properties for Hydrogen Gas Sensor Applications...353
Abdo Yahya Hudeish, Azlan Abdul Aziz

A Low-Cost CMOS Reconfigurable Receiver for WiMAX Applications 357
Adiseno, G. Wiranto, T.M.S. Soegandi

Characterization of Contact Etching Profile for 0,35 um Analog Mixed Signal Product Development.......................363
Ahmad Sabirin Zoolfakar, Azlina Mohd Zain

Comparison of the Growth Si-based Crystalline Silicon Carbide (SiC) by Chemical Vapor Deposition (CVD) using Carbon Monoxide (CO) and Treated Carbon Dioxide 368
A.Y.K. Lim, K. Ibrahim

Self Heating Characterization of 32V MOSFETs Using Pulsed Gate Measurement..372
Alhan Farhanah Abd Rahim, Albert Victor Kordesch, Yusman Mohd Yusof

Design of 100nm Single-Electron Transistor (SET) by 2D TCAD Simulation 377
Amiza Rasmi, Uda Hashim, Abdul Fatah Awang Mat

x

Table of Contents

Effect of Rapid Thermal Annealing (RTA) on n-Contact of 980 nm Oxide VCSEL.................................383
Khairul Anuar M.S., Mohd Sharizal A., S.M. Mitani, Mohamed Razman Y., Awang Mat A.F., P.K. Choudhury

Development of SiC/MgO Distributed Bragg Reflector using RF Magnetron Sputtering Technique................388
Khairul Anuar M.S., Hariyadi Soetedjo, Mohd Sharizal A., Sufrian S.M., Goh Boon T., Richard R., Saadah A.R., Mohamed Razman Y., Abdul Fatah A.M.

Temporal Partitioning of Tasks on a Heterogeneous Reconfigurable Architecture.....................................392
Arjumand Yaqoob, M. Ashraf Chughtai

Packet Synchronization Structure with Peak Detection Algorithm for MB-OFDM UWB.........................398
Aymen M. Karim, Masuri Othman, Edmond Zahedi

Source-Coupled Logic (SCL): Operation and Delay Analysis...402
Mohamed Azaga, Masuri Othman

Photopthermal Study of Ceramic ZnO Doped with Y2O3..407
Azmi Zakaria, Zahid Rizwan, Mansor Hashim, Abdul Halim Shaari

Grain Size Effect on LTCC Tape Performance as Substrate for Microelectronic Devices......................412
Azmi Ibrahim, Rosidah Alias, Che Seman Mahmood, Sabrina Mohd Shapee, Mohamed Razman Yahya, Abdul Fatah Awang Mat

Particle Size Analysis of Barium Titanate Powder by Slow-Rate Sol-Gel Process Route.....................416
Balachandran R., Yow HK, Jayachandran M., Wan Yusmawati Wan Yusof, Saaminathan V.

Rapid Crystallization by Microwave Heating for Barium Strontium Titanate Powders Prepared Using Slow-Rate Sol-Gel Technique...420
Balachandran R., Yow H.K., Zaidina Bt Mohd Daud, Saaminathan V.,

Soft IP Integration and Reuse Challenges in Intel Entry Level Network Processor.............................424
Wai Mun Ng, Kok Sing Yap, Kean Hong Boey

Delayering of Gate Poly in Stack & Split Gate Memory Structure...428
Bridger K.S. Wong and Chan Sieng Fong

High-Performance In0.52Al0.48As?in0.6Ga0.4As Power Metamorphic HEMT for Ka-Band Applications..............432
Chia-Yuan Chang, Edward Yi Chang, Yi-Chung Lien, Yasuyuki Miyamoto, Szu-Hung Chen, Li-Hsin Chu

Development of the Vacuum Spark as an EUV Source for Next Generation Lithography.......................435
Chew Soo Hoon, Wong Chiow San

TDR Single Ended and Differential Measurement Methodology..439
Chan Kim Lee, Jiun Kai Beh, Jimmy Huat Since Huang

Contact Hole Printing in Binary Mask by FLEX Technique...444
Cheong Yew Shun, Ko Bong Sang, Mohd Jeffery Bin Manaf, Kader Ibrahim, Dr. Zul Azhar Zahid Jamal

Porous Silicon Dioxide Synthesized Using Photoelectrochemical (PEC) Wet Etching.........................448
L.S. Chuah, C.W. Chin, Z. Hassan, H. Abu Hassan

Characteristics of Thermally Treated Contacts on Porous Silicon Based Metal-Semiconductor-Metal (MSM) Photodector Structures...452
L.S. Chuah, C.W. Chin, Z. Hassan, H. Abu Hassan

Softbake and Post-exposure Bake Optimization for Process Window Improvement and Optical Proximity Effect Tuning...456
C.Y. Liau, E.K. Yet, C.H. Lee, Ivy Tan, Christopher Loo, B.C. Lee, Y.K. Ng, W.B. Sheu

FPGA Implementation of an Optimized Coefficients Pulse Shaping FIR Filters..................................463
Mohamed Almahdi Eshtawie, Masuri Othman

A 10-Bit 50-MSPS Pipelined CMOS ADC..468
Mohamad-Faizal Hashim, Yuzman Yusoff and Mohd. Rais Ahmad

Table of Contents

Performance of OCDMA Systems Using AND Subtraction Technique 473
S.A. Aljunid, Feras N. Hasson, M.D.A. Samad, M.K. Abdullah, M. Othman, S. Shaari

Pyroelectric Properties of Polyvinylidene Fluoride (PVDF) by Quasi Static Method 477
Gan Wee Chen, Wan Haliza Abd. Majid

Effects of Substrate Temperature on the Properties of Hydrogenated Nanocrystalline Silicon Thin Film Grown by Layer-by-Layer Technique 481
Goh Boon Tong, Saadah Abdul Rahman

SIMS Analysis of Gate Oxide Breakdown Due to Tungsten Contamination 486
D. Gui, Z.X. Xing, Z.Q. MO, Y.N. Hua, S.P. Zhao

Evaluation of Stopping Power of Photo-resist to Ion Implantation by Using SIMS 490
D. Gui, Z.X. Xing, Z.Q. Mo, Y.N. Hua, S.P. Zhao

Design and Analysis of an Ultra Wideband Matrix Mixer 494
R. Hallaji, A. Abdipour, G. Moradi

Design and Analysis of Insulate Gate Bipolar Transistor (IGBT) with P+SiO2 Collector Structure Applicable to High Voltage to 1700 V 498
Han-Sin Lee, Yo-Han Kim, Ey-Goo Kang, Man-Young Sung

Process Optimization of p+LDD in 130nm Process Technology using TCAD Simulation 502
Hani Noorashiqin Abd Majid, Muhamad Rasat Muhamad, Albert Victor Kordesch, Chew Soon Aik

Effect of N2 and O2 Anneal Gas Ratio For Low Resistance p-Type ZnO Formation 507
Haslinda Abdul Hamid, Mat Johar Abdullah, Azlan Abdul Aziz, Naif H. AL-Hardan, Siti Azlina Rosli

Electrical Properties of p-Type Al-N Codoped ZnO Thin Films 511
Haslinda Abdul Hamid, Mat Johar Abdullah, Azlan Abdul Aziz, Siti Azlina Rosli

A Highly Compatible Architecture Design for Optimum FPGA to Structured-ASIC Migration 515
Hee Kong Phoon, Matthew Yap, Chuan Khye Chai

Effect of Mesa Spacing on the Electrical Properties of Mesa Isolation in High Electron Mobility Transistor Structures 520
Hesly Afida Hashim, Mohd. Khairy Othman, Mohd. Nizam Osman, Asban Dolah, Mohamed Razman Yahya

Characterization of Si3N4 Metal-Insulator-Metal (MIM) Capacitors for Monolithic Microwave Integrated Circuits (MMIC) Applications 524
Lee Hock Guan, Mohd Nizam Osman, Asban Dolah, Ahmad Ismat Abdul Rahim, Mohamed Razman Yahya, Abdul Fatah Awang Mat

A New High Resolution Frequency and Phase Synthesis Method based on 'Flying-Adder' Architecture 529
Hossein Gharaee, Elham Tathesari

A Digital Implementation for UWB Impulse Radio Transciever 533
Hossein Gharaee, Abdolreza Nabavi, Babak Bornoosh, S. Mehdi Fakhraei

Studies on Failure Mechanism of Al Fluoride Oxide-AlxoyFz on Microchip Al Bondpads 537
Hua Younan, Zhao Siping, Rao Ramesh, Li Kun

Studies on A Sample Preparation Method for HR-SEM and Application in Failure Analysis of Trench TEOS Gauging Measurement in Wafer Fabrication 542
Zhao Siping, Hua Younan, Mo Zhiqiang, Cho Jie Ying

An Integrated 2.4GHz CMOS Class F Power Amplifier 546
Huang Min Zhe, Abu Khari Bin A'ain, Albert Victor Kordesch

Solder Joint Strenght of Lead Free Solders under Multiple Reflow and High Temperature Storage Condition 550
Ibrahim Ahmad, Azman Jalar Burhanuddin Yeop Majlis, Eu Poh Leng, Yong Soo San

xii

Table of Contents

A Study on Inter-Metallic Compound Formation and Structure of Lead Free SnAgCu Solder System......554
I. Ahmad, Ho Huey Jiun, Eu Poh Leng, B.Y. Majlis, A. Jalar, R. Wagiran

Design and Simulation of 50 nm Vertical Double-Gate MOSFET (VDGM)......558
Ismail Saad, Razali Ismail

Design and Simulation of a High Performance Lateral BJTs on TFSOI......563
Ismail Saad and Razali Ismail

Weak point and improvement of CMOS Schmitt Trigger Circuit used in Microcontroller about ND-mode ESD......567
Jae-Seong Jeong, Jung-Min Lee, and Sang-Deuk Park

Integration of 8051 With DSP in Xilinx FPGA......571
A.J. Salim, M. Othman, M.A. Mohd Ali

Increased Capacitance Density with Metal-Insulator-Metal - Metal Figner Capacitor (MIM-MFC)......576
Kalavathi Subramaniam, Albert Victor Kordesch, Mazlina Esa

Considerations on the C-V Characteristics of Pentacene Metal-Insulator-Semiconductor Capacitors......581
Keum-Dong Jung, Byung-ju Kim, Byeong-Ju Kim, Cheon An Lee, Dong-Wook Park, Byung-Gook Park, Hyungcheol Shin, Jong Duck Lee

Characteristics of Serial Peripheral Interfaces (SPI) Timing Parameters for Optical Mouse Sensor......585
M.K. Md Arshad, U. Hashim, Chew Ming Choo

Design of Experiment (DOE) For Thickness Reduction Of GaAs Wafer Using Lapping Process......592
Mohd Khairy Othman, Asban Dolah, Nurul Afzan Omar, Mohamed Razman Yahya

Study of Flow Visualizationg in Stacked-Chip Scale Packages (S-CSP)......595
M. Khalil Abdullah, M.Z. Abdullah, S. Kamarudin, Z.M. Ariff, P. Hussin, J.J. Antony, H. Haroon, M.R. Saad, M. Manikam

An FIB Method Using Progressive Multi-Cut Technique & Application in Failure Analysis of Wafer Fabrication......600
Khoo Ley Hong, Hua Younan, Zhao Siping, Mo Zhiqiang

I/O Process Optimization to Cover Wide Range Operation Voltage......603
Deb Kumar Pal, Kenny Sabri, Margaret Ting Leh Kee, Son Jin Yeong, Park Hyun Suck

High-Speed Hybrid Parallel-Prefix Carry-Select Adder Using Ling's Algorithm......607
Lakshmanan, Ali Meaamar, Masuri Othman

Study on Alignment Capability and Overlay Performance in 130nm BEOL Lithography Process......612
Lau Siau Yen, Suhana Mohd Said, Norhayati Soin, Kader Ibrahim, Ko Bong Sang

Studies on Electron Penetration Versus Beam Acceleration Voltage in Energy-Dispersive X-Ray Microanalysis......619
Sharon Lee, Hua Younan, Zhao Siping, Mo Zhiqiang

Effect of Post Annealing Treatments on the Characteristics of Ohmic Contacts to n-Type InN......623
L.S. Chuah, Z. Hassan, H. Abu Hassan

Nanoporous InN Films Synthesized using Photoelectrochemical (PEC) Wet Etching......627
L.S. Chuah, Z. Hassan, F.K. Yam, H. Abu Hassan

TEM Characterization of Nickel Silicide Process......631
K. Li, E. Eddie, S.P. Zhao

Investigation and Failure Analysis of "Flower-like" Defects on Microchip Aluminum Bondpads in Wafer Fabrication......635
Hua Younan, Lim Yeow Kheng, Zhao Siping, Rao Ramesh, Tan Jaun Boon

Plane-view Transmission Electron Microscopy for Advanced Integrated Circuit......639
Pan Liu, K. Li, Eddie E.R., Siping Zhao

Table of Contents

Application of Focus Ion Beam Circuit Edit in Failure Analysis 643
S.K. Loh, Teo HT, S.P. Neo, Z.G. Song, C.K. Oh

Circuit Debug using Time Resolved Emission (TRE) Prober-A Case Study 646
Houn Wai Wong, Pit Fuh Low, Vee Kin Wong

FPGA Implementation of a Canonical Signed Digit Multiplier-less based FFT Processor for Wireless Communication Applications 650
Mahmud Benhamid, Masuri Othman

Low Temperature Heteroepitaxial Growth of 3C-SiC on Si Substrates by Rapid Thermal Triode PLasma CVD using Dimethylsilane 655
Abdul Manaf Hashim, Kanji Yasui

The Influence of Doping Concentration, Temperature, and Electric Field on Mobility of Silicone Carbide Materials 660
Mazhar Tayel, Ayman El-Shawarby

A 0.18um, 1.8-V CMOS High Gain Fully Differential Opamp Utilized in Pipelined ADC 665
Ali Meaamar, Masuri Bin Othman, Omid Shoaei

Low-Voltage, High-Performance Current Mirror Circuit Techniques 670
Ali Meaamar, Masuri Othman

VLSI Implementation of 1/2 Viterbi Decoder for IEEE P802.15-3a UWB Communication 675
Meilana Siswanto, Masuri Othman, Edmond Zahedi

Integrated LC VCO Compatible with Memory Process for Gigahertz Clock Generation 680
Minsoek Choi, Youngho Jung, Yeo Jo Yoon, Young-Wug Kim, Hyungcheol Shin

A Novel DC-Coupled, Single-Ended to Differential, Transimpedance Amplifier Architecture Based on gm-boosting Technique 682
M. Jalali, M.K. Moravvej-Farshi, A.R. Nabavi

Analysis of Airborne Boron and Phosphorus Contaminations on Wafer Surface by TOF-SIMS ... 686
Mo Zhiqiang, Gui Dong, Hua Younan, Zhao Siping, Xing Zhenxiang

Study of Porous Silicon Fabricated by Pulsed Anodic Etching of n-Si (100) 689
N.K. Ali, M.R. Hashim, A. Abdul Aziz

The Effects of High Temperature Storage on Lead Free Solder Joint Material Strength Using Pull Test Method 694
Muhammad Najib Harif, Ibrahim Ahmad, Azami Zaharim

Failure Analysis Approach in Memory Failure of SOI Devices 699
S.P. Neo, S.K. Loh, Z.G. Song, S.P. Zhao

Front End Defects on Deep Submicron Devices 702
S.P. Neo, S.K. Loh, Z.G. Song, S.P. Zhao

Failure Analysis of a Unique Poly Defect 705
S.P. Neo, Z.G. Song, C.K. Oh, S.P. Zhao

Trenched MOSFET Vgs Uniformity Improvement through Furnace Loading Procedure 708
H.S. Ng, A.F. Yee, Christopher Loo, Y.K. Ng, W.B. Sheu

Simulation of Flash Memory Characteristics based on Discrete Nanoscale Silicon 713
C.Y. Ng, J.I. Wong, M. Yang, C.L. New, T.S. Khor, T.P. Chen

Realisation of a Differential Multiplier-Divider based on Current Feedback Amplifiers 717
Nihit Bajaj, Jivesh Govil

A SEM Based Technique To Detect Pin-holes In As-Deposited/As-Grown Dielectrics 722
Nitin R. Kamat, Oh Chong Khiam, Zhao Si Ping

xiv

Table of Contents

A Study of Yield Loss In Copper Back-End Process Due To Stress and Poor Adhesion of the Thin Films 726
Nitin R Kamat, Manni Lal, P.B. Sahoo, Zhao Si Ping

Electrical Analysis of High Temperature SAW Resonator Packages 734
Mohammad Hadi Shahrokh Abadi, Mohd. Nizar Hamidon, Roslina Sidek, Mina Malekzadeh

Alignment Mark Architecture Effect on Alignment Signal Behavior in Advanced Lithography 741
Normah Ahmad, Uda Hashim, Mohd Jeffery Manaf, Kader Ibrahim Abdul Wahab

Device Characteristics of HEMT Structures based on Backgate Contact Method 749
Norman Fadhil Idham M., Nurul Afzan O., Hariyadi Soetedjo, Ahmad Ismat A.R., Idris Sabtu, Mohamed Razman Y., Abdul Fatah A.M.

Effect of Indium Content in the Channel on the Electrical Performance of Metamorphic High Electron Mobility Transistors 752
Norman Fadhil Idham M., Ahmad Ismat A.I., Rasidah S., Asban D., Mohamed Razman Y., Abdul Fatah A.M.

Design of an RF BJT-Low Noies Amplifier at 1GHz 756
Sharifah Fatmadiana Wan Muhamad Hatta, Norhayati Soin

A 10GHz Reconfigurable UWB LNA in 130nm CMOS 760
Parviz Amiri, Hossein Gharaee, Abdolreza Nabavi

Physical-Based SPICE Model of CMOS STI y-Stress Effect 764
Philip Beow Yew Tan, Albert Victor Kordesch, Othman Sidek

Optimized Clamp Deployment with Simulation and Characterization in Full-Chip ESD (Electro-Static-Discharge) Design 768
Chuah Cheow Theng, Othman Sidek

Dependence of Texture in Al Bondpads on Ta/TaN Bilayer Barrier and its Correlation to Optical Reflectivity in 0.13um IC Technology 773
Lee Yuan Ping, Ramesh Rao Nistala, Hua Younan, Mousumi Bhat

Meeting the Challenges of Elemental Analysis in 90nm & Beyond Technologies - Case Studies of Scanning Auger Nanoprobe 777
Nistala Ramesh Rao, Tan Chin Wang, Song Zhigang, Hua Younan, Zhao Siping

Improved Booth Encoding for Reduced Area Multiplier 782
Razaidi Hussin, Ali Yeon Md. Shakaff, Norina Idris, Rizalafande Che Ismail, Afzan Kamarudin

A Novel Method to Design Interstage Matching Network in the Smith Chart 785
M. Rezvani Abkenari, M. Tayarani, A. Abdipour, H. Kiumarsi

Dependence of Radio Frequency Power on Optical, Chemical Bonding and Photoluminescence Properties of Hydrogenated Amorphous Carbon Nitride Films 790
Richard Ritikos, Goh Boon Tong, Rozidawati Awang, Siti Meriam Abdul Gani, Saadah Abdul Rahman

High Performance Complex Number Multiplier Using Booth-Wallace Algorithm 795
Rizalafande Che Ismail and Razaidi Hussin

Effects of High Dose BF2+ Implant on the Improvement of P+ Contact Resistance 800
Mohd Rofei Mat Hussin, Norazah Abd. Rashid, Richard Keating

A 0.8V Operational Amplifier Using Floating Gate MOS Technology 804
Rohan Sehgal

Design of Single-Stage Folded-Cascode Gain Boost Amplifier for 100m V 10-bit 50MS/s Pipelined Analog-to-Digital Converter 809
Rohana Musa, Yuzman Yusoff, Tan Kong Yew, Mohd Rais Ahmad

Modification and Modeling of Ni/Si Interface for Photodetector Applications 814
M.R. Hashim, M.Z.M. Yusoff

xv

Table of Contents

An Enhancement of Decimation Process using Fast Cascaded Integrator Comb (CIC) Filter 820
Rozita Teymourzadeh, Masuri Bin Othman

Effects of Metal Work Function and Operating Temperatures on the Electrical Properties of Contacts to n-type GaN 825
S.M. Thahab, H. Abu Hassan, Z. Hassan

Power Deduction in Digital Signal Processing Circuit Using Inventive CPL Subtractor Circuit 829
C. Senthilpari, K. Diwakar, Ajay Kumar Singh

Design and Synthesis of Mobil Robot Controller Using Fuzzy 834
Md. Shabiul Islam, Md. Anwarul Azim, Md. Saukat Jahan, Masuri Othman

Development of a Fuzzy Logic Controller Algorithm for Air-conditioning System 839
Md. Shabiul Islam, Md. Shakowat Zaman Sarker, Kazi Ashique Ahmed Rafi, Masuri Othman

Design and Nonlinear Analysis of High-Grain and Broad-band Distributed Power Amplifier with Traveling-Wave Gain Stages by Harmonic Balance Method 844
S. Asadi, A. Abdipour, A. Tavakoli, G. Moradi

Silicon Chip Removal Technique Using Wet Etching Process for Failure Analysis on Multi-Chip Packages (MCP) 847
Shamsul Mohamed, Rodzaki Saad

Pulse Generation with Reduced Ringing for Ultra Wide Band Applications in Indoor Wireless Communication 852
Sheroz Khan, A.H.M. Zahirul Alam, Mohammad Rafiqul Islam, Othman O Khalifa, Ismail Adam

Application of Autosched AP Simulation Model in Wafer Fab 855
Victor Siow Yuen Tien, Yeo Eng Teck, G. Devandran A/L Govindasamy, Shahrul Kamaruddin

Characteristics of RIE SF6/O2/Ar Plasmas on n-Silicon Etching 860
Siti Azlina Rosli, Azlan Abdul Aziz, Haslinda Abdul Hamid

Highly Chemical Reactive Ion Etching of Silicon in CF4 Containing Plasmas 865
Siti Azlina Rosli, Azlan Abdul Aziz, Haslinda Abdul Hamid

Photolithography Process Improvement for Thick Implant Resist Using 120 C Post-Apply Bake 870
S.I. Yet, E.C. Goh, Faith Lim, A.E. Ling, B.C. Lee, Y.K. Ng, W.B. Sheu

A Versatile HSPICE Electro-Opto-Thermal Circuit Model for Vertical-Cavity Surface-Emitting Lasers 875
E. Sooudi, V. Ahmadi, M. Soroosh

Adopting Electroabsorption Modulator for the WLAN 802.11a Radio over Fibre System 880
Syamsuri Yaakob, Mohd Azmi Ismail, Romli Mohamad, Mohamed Razman Yahya, Abd. Fatah Awang mat, Mohd. Ridzuan Mokhtar, Hairul Azhar Abdul Rashid

Failure Analysis of Pitting Problem on Microchip Al Bondpads in Wafter Fabrication 885
Tai Eve, Hua Younan, Rao Ramesh, Zhao Siping

Auger PID Characterization of Threshold Voltage Shift and Application in Bond pad Monitoring of Wafer Fabrication 890
Marcus Tan Y.H., Hua Younan, Ling Timothy, Zhao Siping

Studies on Failure Mechanism of ET High Via Resistance in Wafer Fabrication 893
Hua Younan, Tan Sock Khim, Li Kun, Zhao Siping

The Effect of Al and Pt/Ti Simultaneously Annealing on Electrical Characteristics of n-GaN Schottky Diode 896
Tariq Munir, Azlan Abdul Aziz, Mat Johar Abdullah

Epilayer Thickness and Doping Density Variation Effects on Current-Voltage (I-V) Characteristics of n-GaN Schottky Diode 901
Tariq Munir, Azlan Abdul Aziz, Mat Johar Abdullah

Table of Contents

A VLSI Design Framework with Freeware CAD Tools..905
Y.K. Teh, F. Mohd-Yasin, M.B.I. Reaz, A. Kordesch

Implementation of Internal Mixed Signal ESD Protection onto RFID Transponder IC................910
M.K. Khaw, F. Mohd-Yasin, Y.K. Teh, M.B.I. Reaz

Device Design Consideration for Nanoscale MOSFET Using Semiconductor TCAD Tools................915
Teoh Chin Hong, Razali Ismail

Nonlinear Stability Analysis of Microwave Oscillators Using Lyapunov Function................920
H. Vahdati, A. Abdipour, A. Mohammadi

Conductivity of Cubic GaMnN Grown on Undoped GaN Layers................924
V K Sundaramoorthy, C.T. Foxon, S.V. Novikov, K.W. Edwards, I. Harrison

A Realistic March-12N Test and Diagnosis Algorithm for SRAM Memories................928
Wan Zuha Wan Hasan, Masuri Othman, Bambang Sunaryo Suparjo

Characterization of Strained Silicon MOSFET Using Semiconductor TCAD Tools................933
Wong Yah Jin, Ismail Saad, Razali Ismail

The Growth of III-V Nitrides Heterostucture on Si Substrate by Plasma-Assisted Molecular Beam Epitaxy................937
F.K. Yam, Z. Hassan, L.S. Chuah, N. Zainal, C.W. Chin, S.M. Thahab, M. Hussein

The Energy Band Gap of AlxGa1-xN Thin Films as a Function of Al-Mole Fraction................942
S.S. Ng, F.K. Yam, Z. Hassan, H. Abu Hassan

Queue Time Impact on Defectivity at Post Copper Barrier Seed, Electrochemical Plating, Anneals and Chemical Mechanical Polishing................947
Yasmin Abdul Wahab, Anuar Fadzil Ahmad, Zaiki Awang

Analysis of the Output Noise Voltage in CMOS Image Sensor Readout Circuit................953
Youngchang Yoon, Hochul Lee, Byung-Gook Park, Jong Duk Lee, Hyungcheol Shin

Electrical Characteristics of 100 MeV 28Si Implantation in GaAs................956
Yousuf Pyar Ali, A.M. Narsale, Othman Sidek, A.R. Damle, B.M. Arora

The Study of Pt/porous GaN Schottky Contact for Hydrogen Sensing................960
F.K. Yam, Y.P. Ali, Z. Hassan, N.H. Mohd. Noor, C.W. Chin

The Characterization of KrF Photoresists and the Effect of Different Chromophore Bulkiness on Line Edge Roughness (LER) for Submicron Technology................964
Ahmad Yusri Mohamed Bakri, Mohd. Jeffery Manaf, Kader Ibrahim Abdul Wahab, Ibrahim Bin Ahmad

Simulation Study on the Performance of SiC-GTO................974
Muhamad Zahim Sujod, Hiroshi Sakata

Failure Analysis of NSOP Problem Due to Al Fluoride Oxide on Microchip Al Bondpads................979
Zhao Siping, Hua Younan, Rao Ramesh, Li Kun

Studies on A New Sela-FIB Sample Preparation Method and Its Application in Failure Analysis of Wafer Fabrication for 110nm Technology Node and Beyond................986
Zhao Siping, Hua Younan, Er Eddie, Khoo Ley Hong

Synchrotron Radiation X-ray Diffraction and X-ray Photoelectron Spectroscopy Investigation on Si-based Structures for Sub-Micron Si-IC Applications................990
Zhe Chuan Feng, Li-Chi Cheng, Chu-Wan Huang, Ying-Lang Wang, T.R. Yang

A Highly Linear CMOS Down Conversion Double Balanced Mixer................994
Kumar Munusamy, Zubaida Yusoff

Solder Bump Strength and Failure Mode of Low-k Flip Chip Device................1000
Zulkarnain Endut, Ibrahim Ahmad, Gary Lee How Swee, Horazham Mohd Sukemi

Table of Contents

Wavelength Shifting in the Fiber Bragg Grating (FBG) based Encoder and Decoder Modules for SAC-OCDMA System.. 1005
Z. Zan, M.K. Abdullah, S.A. Aljunid, R.K.Z. Sahbudin, M.H. Yaacob, M. Mokhtar, S. Shaari

Effects of the Power Differences in the AND-Subtraction Detection Technique in SAC-OCDMA System Performance 1010
Z. Zan, S.A. Aljunid, M.K. Abdullah, R.K.Z. Sahbudin, M.H. Yaacob, M. Mokhtar, S. Shaari

Pulse Power Failure Model Of Power MOSFET Due To Electrical Overstress Using Tasca Method....................... 1015
Nur Syakimah Ismail, Ibrahim Ahmad, Hafizah Husain, Shirly Chuah

New I-V Model For AlGaN/GaN HEMT At Large Gate Bias 1019
Ali El-Abd, M. Abdel Aziz, Abdel Aziz Shalby, Salah Khamis

Simulated Dielectric Characteristics of Pt/BST/Ni-Fe/Cu Multilayer Capacitor Stack for Storage Application 1024
Balachandran R., Yow HK, Manickam RM, Saaminathan V

Numerical Analysis of Filamentation in Conventional Double Heterostructure and Quantum Well High-Power Broad-Area Laser Diodes... 1028
Amireh Seyedfaraji, Vahid Ahmadi, Mahyar Noshiravani, Farzan Gity

Monte Carlo Simulation of Surface Annealing Before Epitaxial Growth 1033
Chang Fu Dee, Burhanuddin Yeop Majlis

Analysis of Poly Resistor Mismatch 1037
Philip Beow Yew Tan, Albert Victor Kordesch, Othman Sidek

Modeling of Polyimide MIM Capacitors for Applications in Planar Monolithic Microwave Integrated Circuits 1039
R. Sanusi, A.I. Abdul Rahim, M.N. Osman, N. Kushairi, A. Rasmi, N.F.I. Muhammad, M.R. Yahya, A.F. Awang Mat

Simulation of InGaN Multiple Quantum Wells (MQWs) Light Emitting Diodes (LEDs) 1043
S.M. Thahab, H. Abu Hassan, Z. Hassan

TCAD Simulation of STI Stress Effect on Active Length for 130nm Technology 1047
Wan Rosmaria Wan Ahmad, Albert Victor Kordesch, Ibrahim Ahmad, Philip Tan Beow Yew

Simulation of Electromigration Test Structures With and Without Extrusion Monitors................................ 1051
Verena Hein, Gisbert Hoelzer, Torsten Schroeter, Yvonne Yeo, Tan Hong Mui

A New High Resolution Frequency and Phase Synthesis Method based on 'Flying-Adder' Architecture

Hossein Gharaee, Elham Tathesari
Tarbiat Modares University
Shariati University
Email: gharaee@modares.ac.ir , e_tathesari@yahoo.com

Abstract **High speed electronic systems demand frequency synthesizer of high resolution, wide bandwidth and fast switching speed. The "Flying-Adder" architecture is a frequency and phase synthesis technique that is based on a VCO of multiple delay stages. This Flying-Adder is implemented in Quartus software which its result shows that the highest frequency is about 83 MHz, when VCO oscillates at 5.2 MHz. In some cases, this architecture has a barrier of inherent jitter on the output frequency. In this brief, a new method is proposed for eliminating such jitter problem. This method is caused to achieve exact phase and frequency. This design is implemented in Quartus software with EP1K30QC208-1 device from ACEX 1K series. When VCO is running at 0.651-10 MHz high resolution output frequency is achieved.**

I. INTRODUCTION

The "Flying-Adder" architecture is a frequency and phase synthesis technique that synthesize a signal with various frequencies and phases from multiple reference signals of same frequency but different phases. These reference signals come from a VCO which is close looped with a reference clock by a PLL and have the same phase delay, Δ, relative to their neighbors. This architecture proposed by Mair and Xiu in 2000 [1]. Since then, the original architecture has been improved in various area. The circuitry's speed performance has been improved with reducing the number of gates [2]. A new method has been proposed for improving its jitter performance which negative impact of it is requirement of adjusting the PLL loop for different frequencies[3]. Reducing the number of delay stages inside the VCO has been proposed for benefit of low power consumption and also, the noise [4].These modifications have made this

architecture more reliable and advanced. However, it seems that the architecture can be achieved to a signal with exact phase and frequency and so high resolution. For reaching to the goal, in this paper, a new method is proposed for eliminating the inherent jitter problem associated with original architecture[1-4]. So any frequency within certain rang can be achieved with any accuracy.

The rest of this paper is organized as follows. Section II gives an overview of "Flying-Adder" architecture. Section III presents the problems of the original architecture. Detailed description of the new method is discussed in section IV. Section V will cover the implementation results and section VI is the conclusion.

II. OVERVIEW OF THE FLYING-ADDER

A. The Original Flying-Adder

Fig.1 shows the principle idea of Flying-Adder architecture. The operation of this system is described in following. Whenever, there is a rising edge from the MUX output, the DFF will be triggered and its output Z will toggle. Meanwhile, the 10-bit register for the MUX address is updated. The 10-bit accumulator is used for updating the address value of the MUX. MSB bits are used directly for selecting the MUX 32 inputs, the 5 LSB bits are used for accumulating the fractional part. Each time the register is clocked, it latches a new value that is equal to its previous value plus the external value of FREQ[9:0]. FREQ is digital control word for this synthesizer.

B. Output Frequency Range

If N is the number of VCO outputs and f_{VCO} is

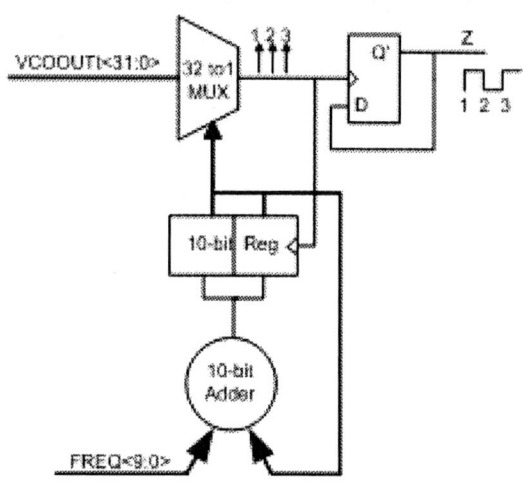

Fig.1. The principle idea of the Flying-Adder [1].

the VCO frequency, then the frequency of this architecture can be obtained in the range of

$$(1/2)f_{vco} \leq f_z \leq (N/2)f_{vco} \qquad (1)$$

C. Output Frequency Formula

As can be understood from part A , the frequency control word for generating desired frequency in this synthesizer , f=1/T, is calculated by bellow formula:

$$T/2 = FREQ \times \Delta \qquad (2)$$

D. The Flying-Adder With Scalability

One of good advantages of the Flying-Adder is scalability. This scalability enables multiple paths to be utilized for generating higher output frequency[2].

III. THE PROBLEM OF CURRENT ARCHITECTURE

In the original architecture, the accumulator-register has some bits reserved for fractional parts to achieve certain frequencies. In the 10-bit accumulator-register of Fig.1, 5 bits are for integer and 5 bits for fraction. Assume that the number of VCO outputs takes the value of 32 and VCO is running at 156.25 MHz (6.4 ns), then the time between two adjacent output is Δ =6.4 ns/32=0.2 ns. If a 204.08 MHz (4.9 ns) is the desired frequency of output signal Z, then the FREQ[9:0] can be calculated as follows:

$$FREQ[9:0] \times 2 \times 0.2ns = 4.9ns \Rightarrow$$
$$FREQ[9:0] = 12.25 = 01100.01000b \qquad (3)$$

Fig.2 shows how this signal is made. Whenever, the fractional part of FREQ is nonzero, the output frequency doesn't have exact phase and frequency. As can be seen in Fig.2 , after some cycles, one clock cycle is longer than the others. This different value in cycles is Δ .Also, none of cycles are not in real value, so the resulting frequency is in the fashion of "time-average" frequency and there is inherent jitter associated with these frequencies.

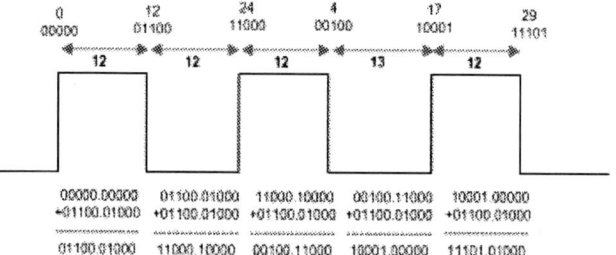

Fig.2. The numerical illustration for producing signal [4].

IV. NEW APPROACH IN FREQUENCY SYNTHESIS

As mention in section III, the result is in a direct impact on jitter performance. In order to achieve certain jitter performance, the phases deviations of VCO output's signal have to be controlled.

For achieving this goal, fractional value of Δ is needed and whole cases for fractional part of FREQ should be entered in VCO outputs.

Fig.3 shows the architecture of new method that eliminates jitter. The 5 MSB bits of FREQ are used for integer part and 3 LSB bits of it are used for fractional part. All of eight possible case for fraction bits, are entered in VCO .So, the number of VCO outputs are $2^5 \times 2^3 = 256$, which 32 signals of them is related to integer part and remained signals is predicted for making a highly accurate phases. It means, from the first of this 256 signal, every 8 signal are contained an integer value with its fractions. So, the time delay between two edge that related to integer values, is $\Delta' = 2^3 \times \Delta$. Δ, is the time difference between two adjacent VCO output signal. The 8-bit accumulator is used for updating the address value of MUX. The FREQ_MOD block is used to generate exact value for selecting MUX 256 input. This block performs a function that gives FREQ<7:0> as input and generates a control

530

word (FREQ_M) for Flying-Adder by the following formula:

$$FREQ\_M<7:0> = (8 \times FREQ<7:0>)+$$
$$FREQ <2:0> \qquad (4)$$

Let us resolve the example of section III with this method. The number of VCO outputs takes the value of 256. Hence, Δ =6.4 ns/256=0.025 ns, Δ' =0.2ns and FREQ<7:0>=12.25=01100.010b will caused a 204.08MHz output frequency. So, from (4), FREQ_M<7:0> is computed by

$$FREQ\_M<7:0> = (8 \times 12)+2=98 \qquad (5)$$

So,

$$T/2=98 \times \Delta =2.45 \Rightarrow$$
$$T=4.9ns(204.08 \text{ MHz}) \qquad (6)$$

As can be seen, this method leads us to a high-quality signal with exact phase and frequency.

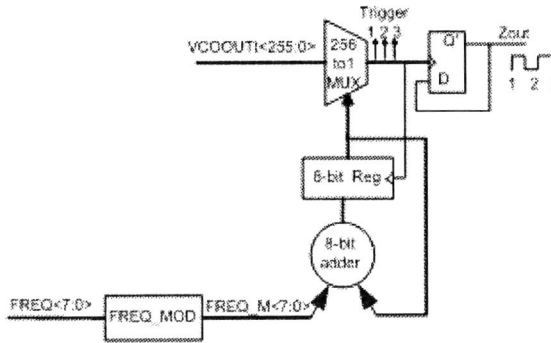

Fig.3 The architecture of new method that eliminate jitter.

This design has scalability for increasing the number of fraction bits and also paths. In general, if the VCO has N outputs and FREQ has X bits for fraction part, then any frequency in the following range can be obtained from this design:

$$(\frac{1}{2})\frac{f_{vco}}{2^X} \leq f_z \leq (\frac{N}{2})\frac{f_{vco}}{2^X} \qquad (5)$$

V. IMPLEMENTATION RESULTS IN FPGA

A. Implementation Of Original Flying-Adder

Table I is the result of implementation of Fig.1 which presents in[1].the highest output frequency is about 130 MHz when VCO oscillates at 114.54 MHz.

Table I

process	0.6um, 3.3v CMOS
Area (PLL)	230um by 680
Area (synthesizer)	1350um by 1260um
VCO Frequency	114.54MHz
Power Dissipation (Synthesizer)	~150 mw
Output Frequency Range	57.27MHz to 130 MHz
Phase Movement Range	0ns to 17.5ns , step 0.282ns

This architecture is implemented in Quartus software on EP1K30QC208-1 from ACEX with 254 logic cells. In this implementation is used a VCO which is running at 5.2 MHz and a signal with 83 MHz can be obtained (about 16 times). Table II shows the result of this implementation. In comparison with the result which is presented in[1], higher multiples of VCO frequency can be obtained.

Table II

Software	Quartus
Device	ACEX (EP1K30QC208-1)
VCO Frequency	5.2MHz
Power	~300 mw
Output Frequency Range	2.84 MHz to 63.3 MHz
Number Of Logic Cell	254 LC (14%)

B. Implementation Of New Design For Eliminating Jitter

Table III is contained the implementation results for Fig.3 in Quartus software that is done on EP1K100FC484-1 from ACEX 1K series.10 MHz output frequency is achieved when VCO is running at 0.651MHz. This output frequency is 16 times the reference frequency. Results show a high resolution output without any jitter.

Table III

Software	Quartus
Device	ACEX (EP1K100FC484-1)
VCO Frequency	0.651MHz
Power	~225 mw
Output Frequency Range	0.04 MHz to 10.4 MHz
Number Of Logic Cell	1665 LC (16%)

VI. CONCLUSION

To overcome the jitter problem associated with original Flying-Adder architecture, a new design has presented in this paper. This high-quality frequency synthesizer is good candidate for application that requires great number of frequency (e.g. spread spectrum) or need tighter jitter constraint. Also it is found in FM receivers, CB transceivers, television receiver and can be act as a digital control oscillator (DCO) in an all digital PLL.

This circuitry is much simplified and doesn't need to be made several design parameters[3].

REFERENCES

[1] H. Mair and L.Xiu, "An Architecture of High Performance Frequency and Phase Synthesis" IEEE Journal of Solid-State Circuit, vol.35, No.66, pp.835-846,Jun. 2000.

[2] L.Xiu and Z.You, "A 'Flying-Adder' Architecture of Frequency and Phase Synthesis with Scalability" IEEE Trans. on VLSI, , vol.10, No.5, pp.637-649, Oct. 2002.

[3] L.Xiu and Z.You, "A New Frequency Synthesis Method based on 'Flying-Adder' Architecture" IEEE Trans. on Circuit & System II, Analog Digit. Signal Process., vol.50, Mo.3, pp.130-134, Mar. 2003.

[4] L.Xiu and Z.You, "A 'Flying-Adder' Frequency Synthesis Architecture of Reducing VCO Stages" IEEE Trans. on VLSI, , vol.13, No.2, Feb. 2005.

A Digital Implementation for UWB Impulse Radio Transceiver

Hossein Gharaee, Abdolreza Nabavi, Babak Bornoosh, and S. Mehdi Fakhraei
Tarbiat Modares University
Tehran ,IRAN
Email:gharaee@modares.ac.ir

Abstract - **In this paper a digital UWB transceiver is designed with a bit-rate of 200Mb/s. System simulations propose an SNR of 10dB, and 4-bit of resolution with 4GHz sampling rate for analog to digital conversion. Hardware of this system is implemented on a Virtex-II Pro chip (XC2VP20). The critical digital blocks operate with a minimum frequency of 200MHz. The total power in this system is 400 mw.**

I. INTRODUCTION

Ultra-wide-band (UWB) signal is defined as one whose fractional bandwidth is greater than 0.2, or one whose bandwidth is over 500 MHz [1]. UWB radio conveys bigger data throughput than conventional narrow band systems. In contrast, detection in UWB radio is a complex problem and power consuming. There are two competing UWB signaling schemes for IEEE 802.15.3a standard: pulsed UWB and multiband OFDM. The former represents a departure from traditional frequency-based data-encoding methods, and the later is an extension of the signal used in the IEEE 802.11a standard with a larger bandwidth [4].

This paper presents the design and implementation of a digital pulse-based UWB transceiver with BPSK modulation at a rate of 200Mbps.This system employs baseband UWB signals of 0-1GHz. The advantages of digital architecture are the simplification of the analog elements, and the possibility of exploring digital channel adaptability and recovery. Section 2 discusses the system design aspects. Section 3 presents hardware design considerations. Section 4 shows the simulation results and Section 5 give the conclusions.

II. SYSTEM DESIGN

In the following sections, the major building blocks of UWB transceiver will be described. Due to high sampling rate, any simple processing may become a bottleneck in hardware design. Therefore, processing should be minimized in system design.

A. Transmitter Design
Fig. 1 shows a block diagram of the transmitter. Source and channel coding algorithms are used for data compression and data correction, respectively. Then, data is packed in a frame. Finally, modulation and pulse shaping change the digital data to proper UWB pulses.

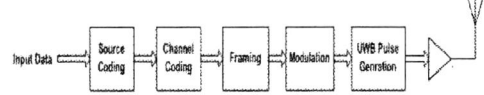

Fig. 1 UWB Transmitter Block Diagram [7]

B. Framing Structure
The frames consist of 45 pilot symbols, a guard interval of 16 symbols, a SYNC word, and 900 bits of data. The pilot is used for synchronization, and the guard is used to designate the start time of pilot. If synchronization is lost for any reason, the receiver detects the guard interval, and performs synchronization by pilot detection. The SYNC word is employed for further data alignment. Fig. 2 shows the frame structure.

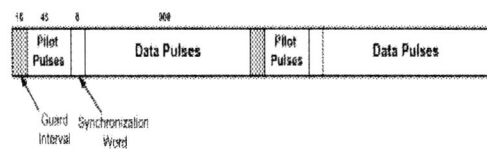

Fig. 2 Proposed framing structure [7]

C. Modulation Scheme
UWB radio generally adopts modulation methods such as pulse position modulation (PPM), pulse amplitude modulation (PAM), and binary phase

0-7803-9730-4/06/$25.00 ©2006 IEEE   533

shift keying (BPSK). In this paper, PAM modulation scheme is employed. UWB pulses are 1ns wide and the sampling frequency is 4GHz. The Rayleigh (Rician) multi-path channel model is used for simulation of NLOS (LOS) environment. A Doppler frequency shift of about 10Hz is used in the simulations which makes the channel to change completely after one millisecond. To enhance the precision of the channel model, an interpolation is performed between the channel impulse responses. An AWGN stream is finally added to the signal to model the noise in the channel.

D. Receiver Design

In impulse radio receiver, the received signal is sampled without down-conversion. Since the sampling rate is relatively high, at least 4 samples from each pulse are detected.

To relax the requirements for ADC, a parallel array of ADCs is used which present 20 words of data at each clock cycle. The clock signal of each ADC is shifted by Ts/20 using time delay elements. Thus, each ADC and the digital hardware operate with a frequency of 200MHz.

Fig. 4 show a flowchart of the processes performed in the receiver to extract the data. The receiver first detects the guard interval in which the signal level is less than a threshold. The threshold, which directly influences the system performance, is chosen based on the channel noise. This task is a replacement for symbol recovery, since it moves the processing pointer to the beginning of the pilot pulses.

After guard detection, the pilot pulses start and a finer algorithm adjusts the sampling point to the middle of each pulse. This algorithm, effectively replacing clock recovery, uses the 20-word window of data and computes the index of the maximum amplitude sample.

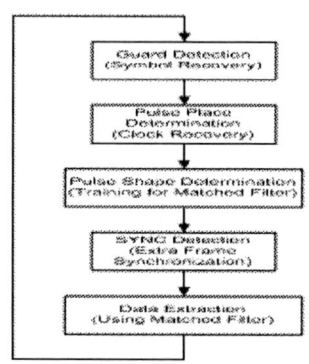

Fig. 4 Receiver processes flowchart

To lower the noise, this operation is done on a number of windows and the mean value is used. Therefore, the correct template index is finally known. Fig. 5 illustrates the correct and non-correct windows.

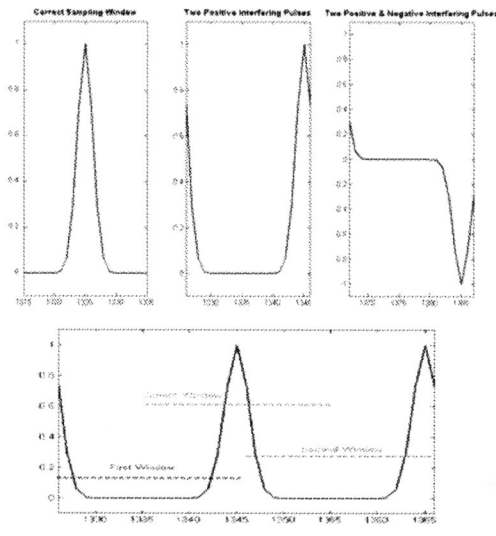

Fig. 5 Two non-correct windows and the correct adopted one

Detection is performed by a matched filter, which needs the template-pulse to compute the correlation with received signals. Correlation result is compared to a threshold level and the polarity of the pulse is determined. A number of detected pilot pulses are averaged to accurately estimate the template. This task is similar to training phase in systems employing equalizers to compensate the channel effects. Fig. 6 shows a typical template obtained from averaging a number of pilot pulses.

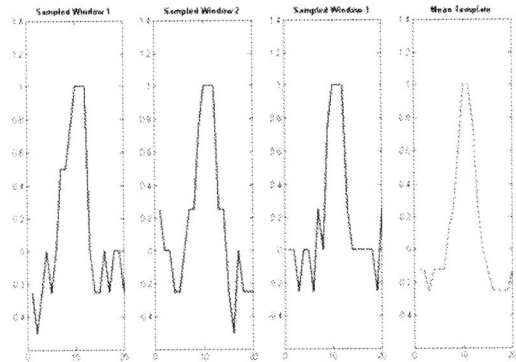

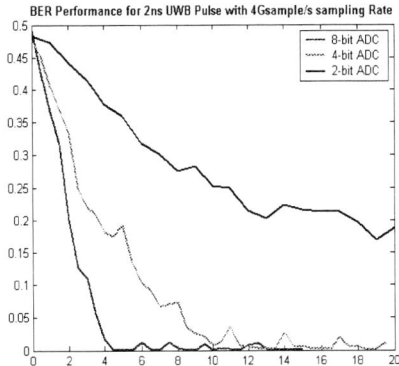

Fig. 6 Noisy pilot pulses and the adopted template

Fig. 8 UWB system performance for different SNR and ADC width

After pilot pulses, a SYNC word should be detected. Then, 900 data pulses are demodulated. Fig. 7 shows the transient waveform of the signal in the receiver.

IV. HARDWARE DESIGN

The block diagram of the hardware is shown in Fig. 9, in which the shaded blocks are designed with pipelining technique since they need a higher speed. Fig. 10 illustrates the MAX20 block which calculates the maximum of 20 inputs. The design report for this block is shown in Table 1.

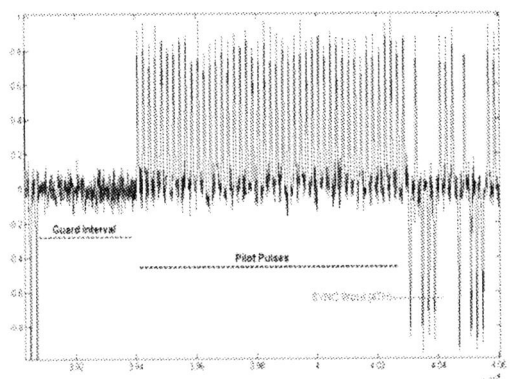

Fig. 7 Guard interval, Pilot, SYNC and data pulses in the receiver

III. MATLAB SIMULATION

The UWB transceiver is simulated in Matlab for different SNR values, as shown in Fig. 8 for 2, 4 and 8 bits of resolution. An SNR of 10dB with 4-bit ADC appears to be an optimum design.

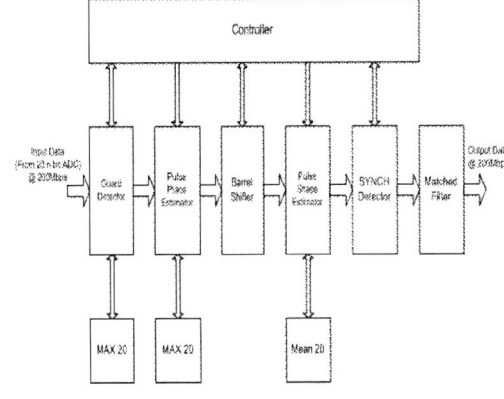

Fig. 9 Block diagram of the UWB hardware

V. CONCLUSION

In this paper, deign of a UWB impulse radio was presented. The ADCs and the digital hardware operate at a frequency of 200MHz. Therefore, the system is easily implemented on a Virtex-II (XC2VP20) FPGA. System simulations propose an SNR of 10dB, and 4-bit of resolution with 4GHz sampling rate as the optimum design.

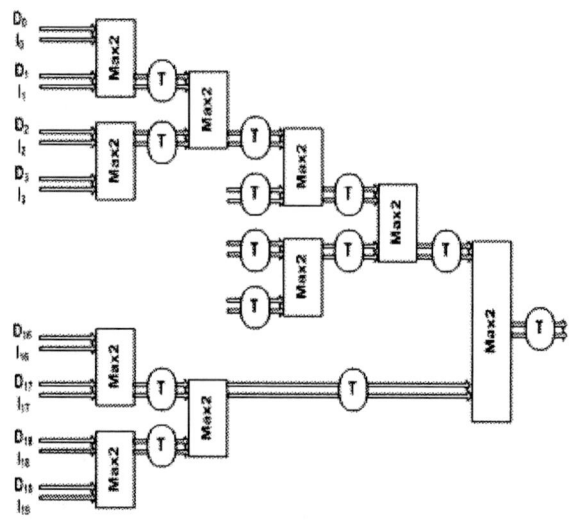

Fig. 10. Pipelined MAX20 block

Table. 1 Hardware design reports

Block Name	Post Synthesis		Post Place & Rout	
	Freq. MHz	Area	Freq. MHz	Area
8-bit Max20	75.4	1.80%	86.2	1.80%
Pipelined 8-bit Max20	299.4	2.10%	209.8	2.10%
Mean20	107	1.80%	124	1.80%
Pipelined Mean20	376.4	1.60%	200	1.60%
Bshifter	225	8.20%	160	8.20%
Pipelined Bshifter	750	8.10%	347	8.10%
Matched Filter	83.15	1.60%	102	1.60%
Pipelined Matched Filter	221	4%	260	4%

REFERENCES

[1] G. F. Ross, "Transmission and reception system for generating and receiving base-band duration pulse signals without distortion for short base-band pulse communication system," US Patent 3,728,632, April 17, 1973.

[2] M. Z. Win and R. A. Scholtz, "Impulse radio: How it works," IEEE Communication Letters, vol. 2, pp. 36–38, Feb. 1998.

[3] FCC 02-48: First Report and Order.

[4] IEEE Std 802.15.3-2003

[5] C. R. Anderson, "Ultra-wideband Communication System Design Issues and Tradeoffs," Ph.D. Qualifier Exam, Virginia Polytechnic Institute and State University, May 12, 2003.

[6] C. R. Anderson, A. M. Orndorff, R. M. Buehrer, and J. H. Reed, "An Introduction and Overview of an Impulse-Radio Ultra-wideband Communication System Design," Tech. Rep., MPRG, Virginia Polytechnic Institute and State University, June 2004.

[7] Aaron Michael Orndorff , " Transceiver Design for Ultra-Wideband Communications" , M.sc Thesis ,Virginia Polytechnic Institute and State University", May 2004.

[8] Sangyoub Lee, " Design And Analysis Of Ultra-Wide Bandwidth Impulse Radio Receiver, Phd Thesis , University OF Southern California , August 2002.

[9] M. Z. Win and R. A. Scholtz, "On the Robustness of Ultra-Wide Bandwidth ignals in Dense Multipath Environments," IEEE Comm. Letters, Vol. 2, No.2, Feb.1998.

[10] M. Z. Win and R. A. Scholtz, "Impulse radio: how it works," IEEE Communications Letters, Vol. 2, No. 1, pp. 10-12, January 1998.

[11] Matlab, Version 7 Release 13, Mathworks, Inc., Natick, MA .

[12] Xilinx System Generator for DSP, Version 6.1, Xilinx, San Jose, CA.

[13] Mike Shuo ,Wei Chen , "Ultra Wide-band Baseband Design and Implementation", M.sc Thesis, University of California at Berkeley,2003.

Studies on Failure Mechanism of Al Fluoride Oxide-$Al_xO_yF_z$ on Microchip Al Bondpads

Hua Younan, Zhao Siping, *Snr Member, IEEE,* Rao Ramesh and Li Kun

Chartered Semiconductor Mfg Ltd
Woodlands Industrial Park D, Street 2, Singapore 738406
Email: huayounan@charteredsemi.com

Abstract: **Al fluoride oxide on microchip Al bondpads may cause non-stick on pad (NSOP) problem during bonding process. In this study, a failure mechanism to form Al fluoride oxide-$Al_xO_yF_z$ has been proposed. Based on the failure mechanism, F contamination on Al bondpads, it will chemically react with Al and form Al-F complex compound, such as $[AlF_6]^{3-}$. $[AlF_6]^{3-}$ formed may become an anode and further chemical reaction from O_2 and moisture (H_2O) will occur at the cathode and form the new product, OH^-ions, which will undergo chemical reaction with Al to form $Al(OH)_3$, and then become Al_2O_3. Finally, Al-F complex compound may further chemically react with Al_2O_3 to form Al fluoride oxide-$Al_xO_yF_z$.**

I. INTRODUCTION

FLUORINE (F) contamination on microchip aluminum (Al) bondpads will cause discoloration and non-stick on pads (NSOP) problem. In this study, we will investigate a NSOP case due to Al fluoride oxide in wafer fabrication (Fab). It was reported that some wafers were affected by the NSOP problem on microchip Al bondpads. Base on initial investigation results, it was suspected to be due to thicker Al oxide (Al_2O_3) as transmission electron spectroscopy (TEM) analysis found a thicker oxide layer on bondpads, which was about 200-300 A on the failed units, and Auger electron spectroscopy (AES) analysis results showed that the F contamination level was only 3-5 at%, which was within the baseline of the F contamination on normal Al bondpads. However, after further failure analysis and theoretical studies, it was concluded that the thicker layer detected by TEM was not Al oxide layer (Al_2O_3), but it was Al fluoride oxide-$Al_xO_yF_z$ which was due to F contamination

introduced during the polyimide process in wafer fabrication. In this paper, a failure mechanism will be proposed. Its chemical reactions will be derived so as to explain the thicker layer found on the affected Al bondpads was Al fluoride oxide-$Al_xO_yF_z$. In authors' another paper [1], the failure analysis results, possible root cause and solution will be discussed.

II. EXPERIMENTAL

In this study, failure analysis on the affected units was done using optical inspection, SEM/EDX, AES and TEM/EDX techniques. SEM/EDX analysis was done using FEI Sirion FESEM at beam accelerating voltage of 5 kV, spot size of 3 and a working distance of 5 mm. To obtain accurate EDX results, ZAF standard less EDX quantification method was used [2,3]. Auger electron spectroscopy (AES) were carried out using PHI (Physical Electronics) SMART-200 System. TEM analysis was performed using a TECNAI F20 200 kV field emission microscope, and the TEM sample was prepared with the FIB (focused ion beam) based lift-out method. To protect the surface, a 50 nm thick layer of Pd was sputter-deposited onto the surface before FIB sample preparation. EDX spectra were collected under STEM mode for easy and accurate probe positioning. All analysis conditions are refer to authors' another publication [1].

III. RESULT AND DISCUSSION

Optical inspection didn't find any visible abnormality on the affected bondpads, but using high magnification SEM inspection the "white dot"-like defects were found on the affected bondpads (Figs 1-2). From the morphology of the defects, they appear like Si crystallized defects. Energy-dispersive X-ray microanalysis

0-7803-9730-4/06/$25.00 ©2006 IEEE

(EDX) on the "white dot"-like defect, however, didn't detect elemental Si peak, but a small F peak was detected. To understand the problem, AES analysis was also done, but the F level was only 3-5 at%, which was within the baseline level of the F contamination on Al bondpads.

Based on the above results, it was not possible to identify, conclusively, the elemental nature of defect and therefore, the root cause analysis of defect formation. To understand more details and find the possible root cause, further FA investigations were carried out using high resolution TEM/EDX. The high spatial resolution of TEM together with the localized EDX analysis (due to small spot size) capability, render the technique ideal to study nanometer sized defects. This feature of TEM was exploited in the present investigations. Interesting results were obtained with TEM as cross-sectional TEM at the defect location showed abnormally thick layer, which was about 200-300 A on the affected bondpad (Figs 3-4) and EDX detected a small F peak (Fig 5). Therefore, it was initially suspected to be a thicker Al oxide (Al_2O_3) problem, which might have caused the NSOP problem.

However, based on the previous experience and TEM/EDX results, we challenged the suspicion as the Al oxide on the surface of a normal bondpad should be within 30-50 A. The previous experimental results on growing the Al oxide showed that even further oxidation process was performed, the thickness of the Al oxide was less than 50-55 A and it was limited to further increase the thickness of the Al oxide. Moreover, the thickness of the Al oxide is uniform (Fig. 6). Therefore, in this case, we believe that the layer of 200-300 A detected by TEM was not the Al oxide layer-Al_2O_3, but it was the Al fluoride oxide layer as TEM/EDX detected F peak. In the next section, a failure mechanism is proposed to explain chemically that it is the Al fluoride oxide- $Al_xO_yF_z$.

IV. STUDIES OF THE FAILURE MECHANISM

It is well known that F contamination on Al bondpads will result in corrosion and formation of F-induced defects. F will chemically react with Al and form Al fluoride:

$$Al + xF^- \rightarrow [AlF_x]^{(x-3)-} + 3e^- \quad (1)$$

In chemistry, the x in $[AlF_x]^{(x-3)-}$ could be 3 or 6, which is depended on the F contamination level. Based on the above chemical reactions, a failure mechanism to form the "white dot"-like defects, which was the Al fluoride oxide- $Al_xO_yF_z$ could be proposed.

1. F-Induced Corrosion – Formation of Al Fluoride

Currently, CF_4 gas is used for bondpad opening process in wafer fab. Thus, on a normal Al bondpad, there is a low level of F contamination, which could not cause F-induced defects. However, if there is high F contamination on Al bondpads, F will chemically react with Al and form Al-F complex compound, $[AlF_6]^{3-}$ [4-5]:

$$Al + 6F^- \rightarrow [AlF_6]^{3-} + 3e^- \quad (2)$$

$[AlF_6]^{3-}$ is a chemically stable compound. Once formed, a normal cleaning process cannot remove it.

2. OH Corrosion – Formation of Al_2O_3

According to our previous studies [4-5], $[AlF_6]^{3-}$ formed may become an anode and further chemical reaction from O_2 and moisture (H_2O) will occur at the cathode:

Anode:
$$[AlF_6]^{3-} + 3e^- \leftrightarrows Al + 6F^- \quad (3)$$
$$(Eo = -2.07 \text{ V})$$

Cathode:
$$O_2 + 2H_2O + 4e^- \leftrightarrows 4OH^- \quad (4)$$
$$(Eo = +0.40 \text{ V})$$

Eo in Eqn (3) and (4) is the standard electrode and Redox potentials at 298 K (25o C). If the Eo values are negative, the equilibrium tends to shift to the left. If the Eo values are positive, the equilibrium tends to shift to the right.

In Eqn (3), the Eo is negative, thus the equilibrium favors to left, thus F will react with Al to form $[AlF_6]^{3-}$. In another word, the possibility to form $[AlF_6]^{3-}$ is very high if F contamination is present on Al bondpads. Similar, in Eqn (4), the Eo is positive, thus the equilibrium favors to right, thus O_2 and H_2O will chemically react to form the new product, OH- ions, which will undergo chemical reaction with Al to form $Al(OH)_3$ as shown by Eqn (5), and then become Al_2O_3 as shown by Eqn (6):

$$3O_2 + 6H_2O + 12e^- \rightarrow 12OH^-$$
$$+$$
$$4Al \rightarrow 4\,Al(OH)_3 \quad (5)$$
$$\downarrow -6H_2O$$
$$2\,Al_2O_3 \quad (6)$$

3. Al Fluoride Oxide Growth

After some complex chemical reactions, thicker Al fluoride oxide-$Al_xO_yF_z$ will be formed:

$$[AlF_x]^{(x-3)-} + Al_2O_3 \rightarrow Al_xO_yF_z \quad (7)$$

Differencing from chlorine-induced corrosion, F-induced corrosion isn't immediately, but it needs certain time for chemical reaction. Therefore, we believe that in this case, high F contaminated the wafers in wafer fab process. But in that time, the F-induced defects were still not formed. Thus, the problem couldn't be found at wafer fab OQA inspection before shipping to the customer. After certain months for the wafer storage or assembly process, the F-induced defects were formed slowly. However, it was still not easy to identify these defects optically as they were very small and the size was only 100-200 nm. Therefore, in this case, the "white dot"-like defects were only detected using high magnification SEM inspection.

V. IDENTIFACATION OF POSSIBLE ROOT CAUSE

Based on the above failure mechanism and discussion, we believe that in this case, the "white dot"-like defects were due to F contamination and the abnormal layer detected on surface of bondpad was Al fluoride oxide-$Al_xO_yF_z$. The challenge faced is that if the "white dot"-like defects were caused by F contamination, why AES analysis not detected high F?

According to the above failure mechanism, one can understand that the main composition of the defects was Al fluoride oxide, $Al_xO_yF_z$, whose thickness was greatly grown up and the volume was also greatly increased. Therefore, we strongly think that in this case, at the beginning, the F contamination was higher than the baseline

(3-5 at.%). However, after chemical reaction, F-contamination had resulted in the formation of Al fluoride oxide, $Al_xO_yF_z$. This diluted the F level and hence AES only detected 3-5 at% F in the defects. Therefore, we believe that in this case, the actual F contamination level might be much higher than 3-5 at%. After performing further wafer fab investigation, the root cause of the F contamination has been identified, which was introduced during wafer fab ployimide process step [1]. The "white dot"-like defects were also simulated using high F contaminated wafers.

VI. CONCLUSION

In this case, the NSOP problem reported was caused by the "white dot"-like defects. A failure mechanism to form Al fluoride oxide-$Al_xO_yF_z$ has been proposed. Based on the failure mechanism, the "white dot"-like defect was not due to the Al oxide (Al_2O_3), but it was the Al fluoride oxide ($Al_xO_yF_z$), which was caused by F contamination on Al bondpads. During forming Al fluoride oxide ($Al_xO_yF_z$), the thickness and volume were greatly increased. Therefore, the F contamination level was diluted after chemical reaction and AES only detected 3-5 at% F in the defects. But the actual F contamination level was much higher than 3-5 at%, which should be identified and eliminated.

ACKNOWLEDGEMENT

The authors would like to thank Soo Chi Wen, Ton Cambridge and Kelvin from TEM group, Lee Sharon from SEM/EDX/FIB group and Kuni from Fab3/5 for their technical advice and making contribution to this technical paper.

REFERENCES

[1] Zhao Siping, Hua Younan, Rao Ramesh and Li Kun, Failure Analysis of NSOP Problem Due to Al Fluoride Oxide- $Al_xO_yF_z$ on Microchip Al Bondpads. To be submitted for the 2006 IEEE International Conference on Semiconductor Electronics, 29 NOV-1 DEC, 2006, Kuala Lumpur, Malaysia.
[2] Y. N. Hua, Z. R. Guo & K. W. Chau, "Studies of ZAF Standardless EDX Quantification Method and Application in Failure Analysis of Semiconductor," *J. Trace and Microprobe Techniques,* 15 (1), 13-31 (1997).

[3] Y. N. Hua, "Application of SEC Factors in Energy-Dispersive X-Ray Quantification Analysis of Silicon Oxide (SIO$_2$) Layer in Wafer Fabrication." *The proceedings from the 198th Meeting of The Electrochemical Society – Phoenix, Arizone*, October 22-27, No. 810 (2000).

[4] Hua Younan, Shailesh Redkar, Lau Chi-Kwan and Mo Zhiqiang (from PSB Corp., Singapore), A Study on Non-Stick Aluminum Bondpads due to Fluorine Contamination using SEM, EDX, TEM, IC, Auger, XPS & TOF-SIMS Techniques. The proceedings from the 28th International Symposium for Testing & Failure Analysis (ISTFA'2002), 3-7 Nov, Phoenix Civic Plaza, Phoenix, Arizona, USA p495-504 (2002).

[5] Hua Younan, Shailesh Redkar, Ron Dickinson and Peh Shirley, A Study on Fluorine-Induced Corrosion on Microchip Aluminum Bondpads. The proceedings from the 29th International Symposium for Testing & Failure Analysis (ISTFA'2003), 2-6 Nov. 2003, Santa Clara, California, USA, p249-255 (2003).

Figure 1. SEM micrograph showed white dot-like defects on the affected bondpads, which might have caused NSOP problem.

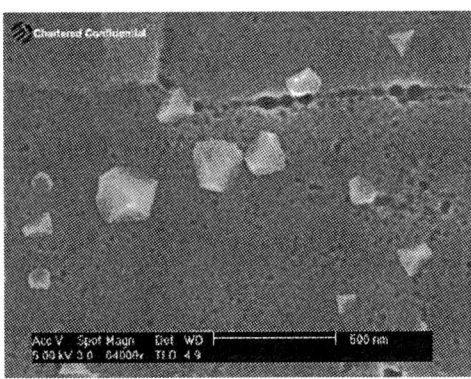

Figure 2. Close-up SEM micrograph showed white dot-like defects on the affected bondpads, which might have caused NSOP problem.

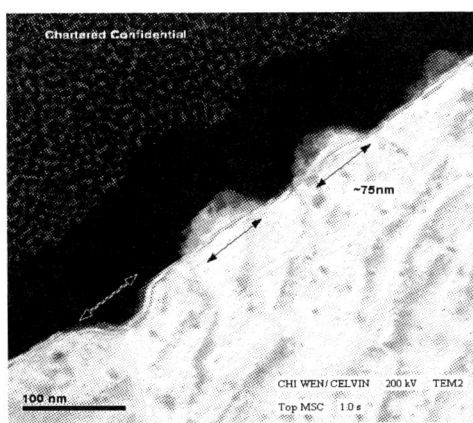

Figure 3. TEM micrograph shows dot-like defects on bondpad as shown by the arrows.

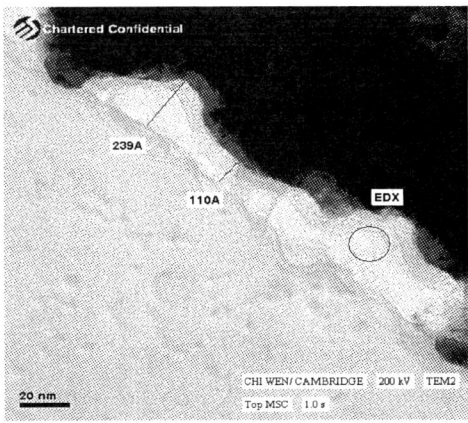

Figure 4. TEM/EDX detected F peak on white dot-like defect, which supported to suspect F –induced defect.

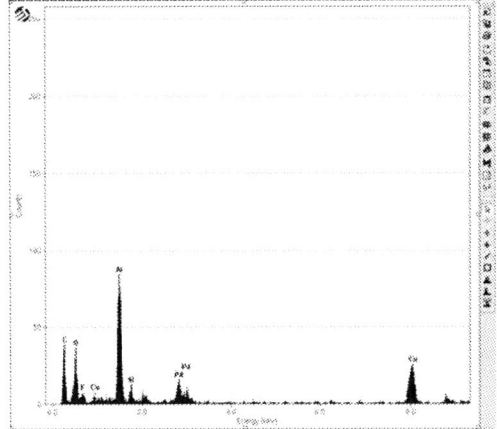

Figure 5. TEM/EDX detected F peak on white dot-like defect, which supported to suspect F-induced defect.

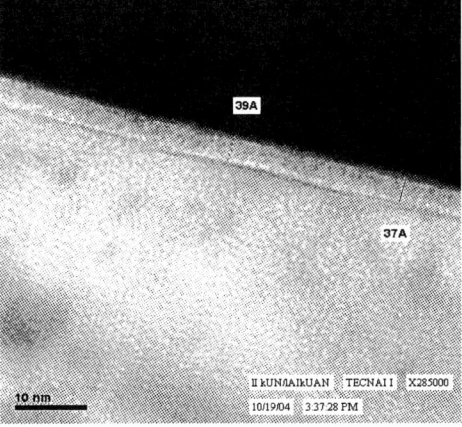

Figure 6. TEM result showed a normal uniform Al oxide on surface of Al bondpad, which is about 35-40A. Moreover, our previous evaluation results showed that it was difficult to increase Al oxide on a normal Al bondpad (with normal contamination level of fluorine).

Studies on A Sample Preparation Method for HR-SEM and Application in Failure Analysis of Trench TEOS Gauging Measurement in Wafer Fabrication

Zhao Siping, *Snr member, IEEE*, Hua Younan, Mo Zhiqiang and Cho Jie Ying

Chartered Semiconductor Mfg Ltd
Woodlands Industrial Park D, Street 2, Singapore 738406
Email: zhaosp@charteredsemi.com

Abstract: **To identify nitride from oxide layer on the trench, it is necessary to perform BOE chemical staining. However, chemical staining using BOE will damage the oxide layer, causing inaccurate readings in the oxide gauging measurement in the trench. Moreover, damage on the oxide layer caused heavy charging at the side of the trench and the surface of oxide layer. In this paper, we proposed to coat a Cr layer over the trench before chemical staining. The damage problem was eliminated and the measurement of oxide gauging was more accurate. A application case is discussed for trench TEOS gauging measurement.**

I. INTRODUCTION

NEW technology and methods are constantly introduced in the development of wafer fabrication. Failure analysis (FA) techniques use to support these developments are also constantly changing to meet up with the requirements. In the development of new etch recipe for trench, measurement of the depth of trench TEOS gauging is needed. In order to distinguish a clear nitride from oxide layer at TEOS, chemical staining using buffered oxide etch (BOE) is required. However, BOE staining will affect the oxide gauging measurement, since the BOE used will attack the oxide layer. Thus, a protecting layer is needed to prevent this damage from happening. One possible solution is to deposit a passivation layer during wafer fabrication. However, this requires more cost and induces longer cycle-time. Therefore, in this work, a FA sample preparation method will be developed for trench TEOS gauging measurement in wafer fabrication.

II. EXPERIMENTAL

Sela Fine Cleave

In this study, all wafer samples were cut using SELA MC 600 machine. By using the fine cleaving function of the machine, precise cutting at the specific site with an accuracy of 0.5 μm could be performed. The location used for trench TEOS gauging analysis is as shown in Fig. 1. As indicated on Fig. 1, both the small and big trenches were needed for analysis.

Chemical BOE Staining

For an SEI sample without any chemical staining, it is very difficult or almost impossible to identify the nitride layer from the oxide layer. Therefore, in this work, chemical staining using BOE is proposed. Chemical BOE staining recipe is optimized as follow:

BOE : 7 NH$_4$F : 1 HF
Staining time : 5secs

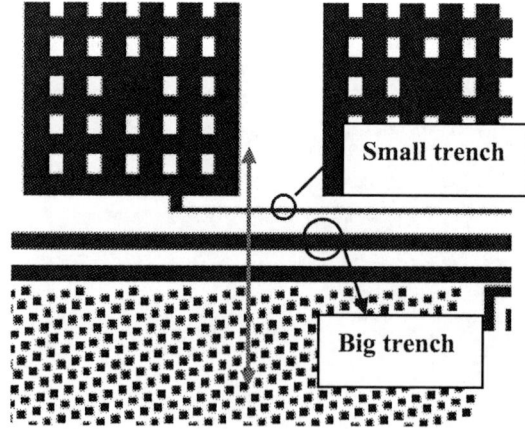

Figure 1. Location of targeted area to do analysis on trench TEOS gauging.

Cr Coating

After chemical BOE staining was done based on the above recipe, the nitride could be clearly differentiated from the oxide layer (refer to next section). However, during chemical staining, BOE attacked the oxide layer. Therefore, it is necessary to coat a passivation layer before BOE staining. Not only will this passivation layer protect the oxide layer at the bottom of the trench, it will also allow an accurate TEOS oxide gauging measurement. In this study, a layer of chromium (Cr) is coated over the wafer samples using Gatan model 682 precision etching coating system (PECS). This Cr layer will act as the passivation layer over the oxide layer. After performing evaluation, a Cr coating recipe was developed:

Left and Right gun : 400µA
Beam Energy : 8keV
Coating time : 8mins

Wafer samples of standard size were first prepared. Before sending the wafer samples for Sela cleaving, a layer of Cr should be deposited over the samples. To obtain uniform Cr coating layer, the coating procedure was as follow:

Cr coating for 4mins → Rest for 5mins → Cr coating for 4mins again

Pt Sputtering
It is necessary to sputter platinum (Pt) on the cross section of the wafer samples. In this case, charging problem can be greatly reduced, and high resolution SEIs could be obtained. Baltic SCD 005 sputter coater was used and the conditions used are:

Current : 30mA
Vacuum : 10^{-1} mbar
Sputtering Time : 20secs

SEM Inspection and Measurement
Finally, the wafer samples were analyzed using FEI Sirion SEM (scanning electron microscopy). The inspection and measurements at TEOS oxide gauging is done with the following conditions:

Accelerating Voltage : 5kV
Spot Size : 3
Working Distance : 4 ~ 5 mm

Failure Analysis Flow
Cut wafer sample (10mm x 20mm) → Cr Coating for 8mins → Sela fine cleave → BOE staining for 5secs → Pt Sputtering for 20secs → SEM Inspection & Measurement

III. RESULTS & DISCUSSION

In this study, the analysis location used for trench TEOS gauging analysis is as shown in Fig. 1. As indicated in Fig. 1, both the small and big trenches were needed for analysis and measurement. The cross sectional view of the structure using a magnification of 2500X is showed in Fig. 2. It is well known fact that if without chemical staining, the nitride layer will not be clearly defined from the oxide layer in SEI. Hence, the TEOS oxide gauging could not be measured accurately. Therefore, in this study, chemical staining using BOE is proposed so as to distinguish the nitride layer from the oxide layer.

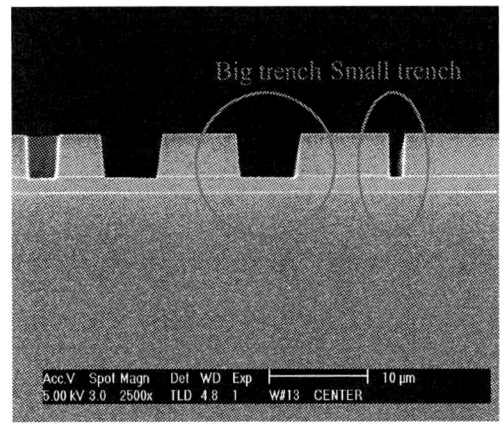

Figure 2. SEM micrograph showed the cross section of trenches at the targeted area.

After the wafer samples were stained using BOE for 5secs, the nitride was clearly differentiated from oxide layer (Fig. 3). However, during the staining process, BOE attacked the oxide layer.

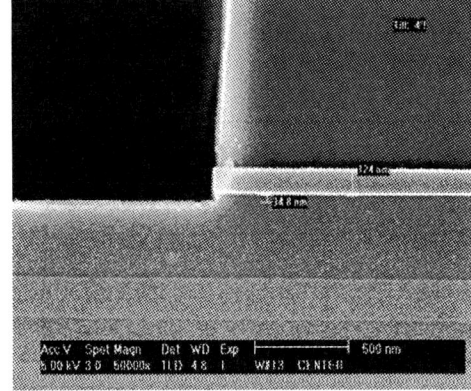

Figure 3. SEM showed that nitride was clearly identified from the oxide layer. However, during BOE staining, the oxide layer was damaged and the measured gauging was only 34.8 nm

543

This damaged the oxide layer, causing the oxide gauging to deviate greatly from the design value. This is shown in Fig. 3 where the oxide gauging measurement is only 34.8nm.

Moreover, oxide damage will also caused heavy charging at the side and on the surface of the trench (Fig. 4). This charging problem had caused measurement at the oxide gauging to be impossible. All the above-mentioned problems had caused the oxide gauging measurement to be inaccurate. These problems were eliminated after a layer of Cr was coated over the trench for 8mins. Fig. 5 and Fig. 6 show the SEIs of the coated trench. With a thick layer of Cr over the trench, heavy charging at the side of the trenches

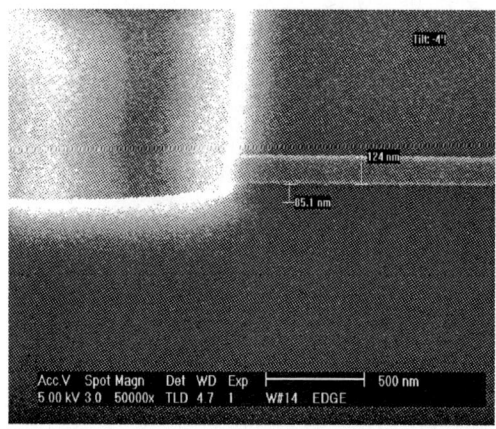

Figure 4. SEI showed the gauging data deviate from what was found in Fig 3. Due to BOE staining, oxide layer was damaged. Moreover, there was a serious charging problem at the side of the trench.

no longer exists. Moreover, the measured oxide gauging values are more consistent as compared to the uncoated trenches. From Fig. 5 it is observed that after Cr coating, the oxide was protected. The measured gauging value was 75.4nm, which is more agreeable with another measurement as shown in Fig. 6, which was 73.4nm. These values reflect consistency as the differences are within the design specification.

IV. CONCLUSION

Without BOE staining, the nitride and oxide layer on the trench could not be differentiated clearly. After BOE chemical staining for 5secs, the nitride layer was clearly identified from the oxide layer. However, if there was no passivation layer on trench, chemical staining using BOE will damage the oxide layer, causing inaccurate readings in the oxide gauging measurement in the trench. Moreover, damage

on the oxide layer caused heavy charging at the side of the trench and the surface of oxide layer. After coating a layer of Cr over the trench, these problems were eliminated and the measurement of oxide gauging was accurate. The sample preparation method proposed in this paper is therefore very useful for trench TEOS gauging measurement.

ACKNOWLEDGMENT

The authors would like to thank SEM/FIB FA team for their technical advice and contribution.

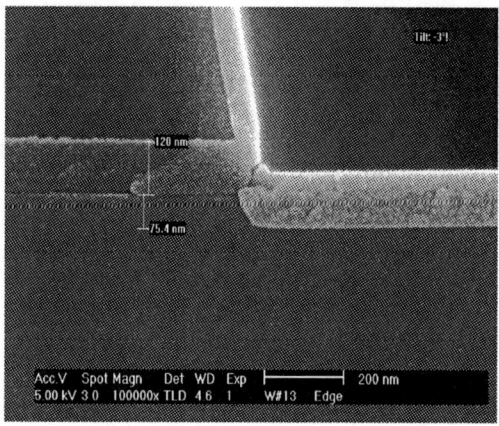

Figure 5. After Cr coating, the oxide was protected and gauging measured 75.4 nm. This agrees with another measurement (Fig 6).

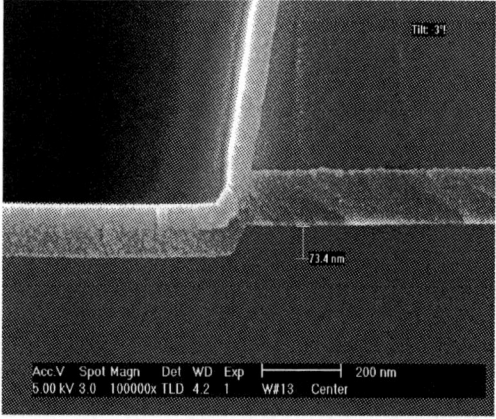

Figure 6. After Cr coating, the oxide was protected and gauging measured 73.4 nm, and this corresponds to the measurement in Fig 5.

REFERENCES

[1] Hua Younan, Z. R. Guo & K. W. Chau, Studies of ZAF Standardless EDX Quantification Method and Application in Failure Analysis of Semiconductor. Journal of Trace and Microprobe Techniques, 15 (1), 13-31 (1997).

[2] Hua Younan, Zhao Siping, Er Eddie and Khoo Ley Hong, Studies on A New Sela-FIB Sample Preparation Method and Its Application in Failure Analysis of Wafer Fabrication for 110nm Technology Node and Beyond. To be submitted for the 2006 IEEE International Conference on Semiconductor Electronics, 29 NOV-1 DEC, 2006, Kuala Lumpur, Malaysia.

An Integrated 2.4GHz CMOS Class F Power Amplifier

Huang Min Zhe[1], *Member, IEEE*, Abu Khari Bin A'ain[2], *Member, IEEE*, and Albert Victor Kordesch[1], *Member, IEEE*.

[1]Silterra Malaysia Sdn. Bhd.
09000 Kulim, Kedah, Malaysia.

[2]Faculty of Electrical Engineering,
Universiti Teknologi Malaysia,
81300 Skudai, Johor, Malaysia.
Email: minzhe_huang@silterra.com, abu_k2002@yahoo.com, al_kordesch@silterra.com

Abstract **This work will explore the integration of a Class-F Power Amplifier using CMOS technology. At 2.4GHz, the fully integrated on-chip CMOS Power Amplifier can deliver 21.8dBm output power with 43.95% efficiency. The design makes use of the C18 RF models provided by Silterra and design of spiral inductor using commercial synthesis software.**

I. INTRODUCTION

The Power Amplifier (PA) is a circuit that converts dc-input power into a significant amount of microwave output power [3]. There are a great variety of PA classes, each differentiated by their methods of operation, efficiency and linearity.

Back in the early days of wireless communication, the Class-A and AB are predominantly the choices of PA design and most of these were implemented in Gallium Arsenide (GaAs) technology [1]. However, with the increasing demand for longer battery life and smaller wireless mobile devices, the CMOS Class-F PA has become very popular due to its high efficiency and high output power capability.

The Class-F approach has been developed in the frequency domain as a means of increasing the efficiency over Class-A and AB. The Class-F PA uses a multiple resonator output filter shown in Fig. 1 to control the harmonic content of its drain voltage and drain current waveforms, thereby shaping them to reduce power dissipation by the transistor and thus to increase the efficiency of the PA.

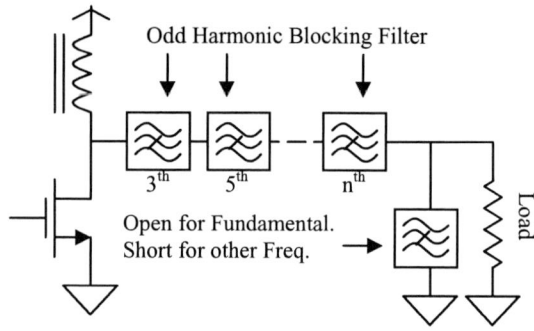

Fig. 1 Class F Power Amplifier Topology

The Class-F PA is usually driven heavily at the input to deliberately drive the transistor into the active region, where the strong nonlinear action of the transistor will generate a waveform with strong harmonic content. The voltage waveform will include one or more odd harmonics and approximate a square wave as shown in Fig. 2, while the current includes even harmonics, approximating a half sine wave [2].

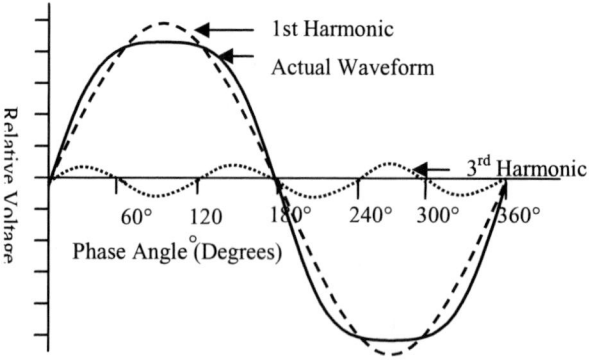

Fig. 2 Drain Voltage Waveform for Class-F

While this is the most important criterion in designing Class-F PA, previous studies [4] have shown that the correct phase and amplitude for each harmonic is essential for the optimum operation of the Class-F PA.

II. CMOS

CMOS technology has been used for many of the lower frequency and digital function blocks, but it is still lacking behind GaAs when it comes to building high-frequency high-performance RF blocks. However, what CMOS does provide is an extremely low cost technology in which to implement circuit blocks, along with the added benefit of potential large-scale integration.

The breakdown voltage CMOS is reduced each time it scales down to sub-micron. This limits the maximum drain-gate voltage since the output voltage at the transistor's drain normally reaches two times the power supply. The thick oxide transistor is used in this design, it can withstand up to two times of the normal 1.8V transistor's breakdown voltage.

In contrast to semi-insulating substrates in GaAs, a highly doped substrate is common in CMOS technology. The signal leakage from the PA through the substrate might affect the stability of, for example, the VCO in a transceiver chain. The deep n-well, which can be implanted underneath the PA, can effectively isolate the PA from the rest of the transceiver chain.

Passive components play a vital role in PA design. However, high quality factor (high-Q) passive elements like on-chip inductors are still absent in CMOS. The parasitics associated with on-chip inductors have an adverse impact on the performance of PA. The traditional solution to this problem is to use off-chip passive components but for an integrated PA, the low-Q on-chip inductors have to be utilized. Commercial synthesis software "Spiral" from OEA is used to design, model and optimize the design of the on-chip inductors.

III. THREOTICAL ANALYSIS AND DESIGN

The ideal Class-F PA assumes the inclusion of an infinite number of harmonics, which is, unrealistic in real-world designs. The number of harmonics used in the Class-F PA is a trade off between complexity and efficiency. The variation of efficiency and power-output capability with different number of harmonics are presented in [5].

The number of harmonics found to be optimum in designing Class-F in CMOS is up to the third harmonic, as higher harmonics would be shorted out by the drain capacitance of the transistor. The loading network best suited for this is still the third harmonic peaking circuit shown in Fig. 3, which can stop the harmonics at the drain and leave only the fundamental at the output.

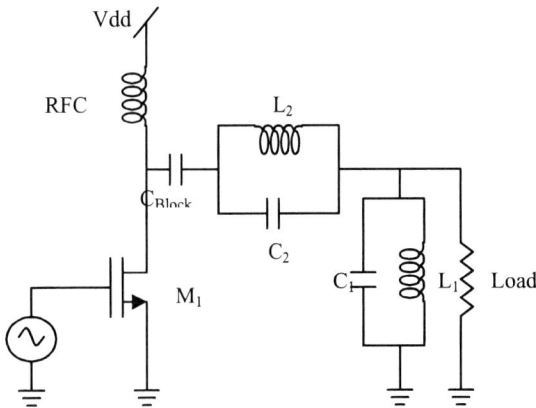

Fig. 3 Third Harmonic Peaking Circuit

There are mainly two types of losses that appear in a transistor, the first one is the turn-on switching loss, which is mostly dominated by the gate capacitance. Second one is the conducting loss, which is the channel resistance of the transistor while conducting current. While increasing the transistor size would greatly reduce the conducting loss but at the same time the turn-on switching loss would also become higher. For a PA design, a large size transistor is usually chosen so that the PA is operating in the high-efficiency region shown in Fig. 4. The smaller line represents the Id-Vd curve of the transistor while the darker line shows the load line of the Class F PA. An ideal load line for the Class F would be the dotted line.

Since a single stage PA could not provide sufficient power gain, the design of the Class-F PA will use multiple stages to achieve the desired power gain. There are two stages in this

547

PA design shown in Fig. 5. The first stage, which is the pre-amplifier, is designed to have high gain

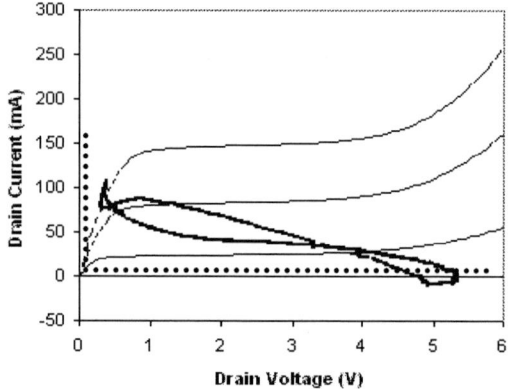

Fig. 4 Load Line of Class-F Power Amplifier

while the second stage is optimized for high efficiency.

The pre-amplifier stage would operate as a "Class-E" type driver, producing a very high half sinusoidal voltage swing at the input of the second stage. With a high overdrive input voltage, the amplifier stage is able to operate in the deep active region where it's considered to be most power efficient, as mentioned earlier.

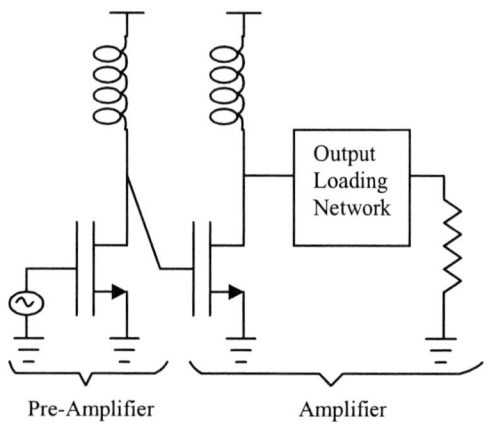

Fig. 5 Two Stage Power Amplifier

IV. CLASS-F POWER AMPLIFIER DESIGN

The Class-F PA design starts of by choosing the optimum load. The optimum load is given by:

$$R_{opt} = \frac{V_{sup} - V_{knee}}{I_{max} / 2} \qquad (1)$$

The next step is to determine the transistor size of the PA. The value of the transistor gate width must be large enough to handle the large current and also ensure a low knee voltage. The gate width is also carefully chosen so that the gate capacitance can be used by the pre-amplifier stage to produce a half-sinusoidal voltage input to operate the transistor at the amplifier stage as a switch. The transistor is biased for small conduction angle with the half sinusoidal input voltage at the amplifier stage. This is necessary for generating the harmonics at the drain.

The component values of the third harmonic peaking load network can be derived from the equations presented in [6]. The on-chip spiral inductors are generated using spiral inductor synthesis software from OEA. The software is able to predict good accuracy of the Q-factor and inductance value of the spiral inductor after calibration from measurement data.

V. SIMULATION RESULTS

The Class-F PA is designed using Silterra C18 RF models and simulated using Spectre.

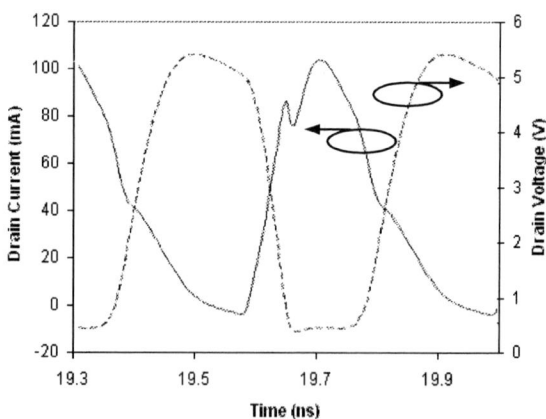

Fig. 6 Drain Waveform of Class-F Power Amplifier

In Fig. 6 the drain current appears to be bifurcated near the peak. This is due to the knee voltage of the transistor. The drain voltage appears to be flat at the minimum voltage of 0.38V.

The performance summary of the simulated design of the Class-F PA is shown in Table 1.

Table 1
Performance Summary of Class-F Power Amplifier

Parameters	
Operating Frequency	2.4GHz
Supply Voltage	1.8V & 3V
Output Power	21.8 dBm
Efficiency, η	43.95%

VI. CONCLUSION

A fully integrated Class-F PA has been demonstrated using CMOS technology with on-chip passive component. Operating at 2.4GHz, the Class-F PA can deliver an output power of 21.3dBm and achieve efficiency of up to 43.95%.

ACKNOWLEDGEMENT

The authors would like to thank the Device Modeling Group from Silterra Malaysia Sdn. Bhd. for their contributions and help

REFERENCES

[1] S. Gao, "High-Efficiency Class F RF/Microwave Power Amplifiers" *IEEE Microwave Magazine*, pp. 40-48 (Feb 2006).

[2] S.C. Cripps, *RF Power Amplifiers for Wireless Communication,* Artech House (1999).

[3] F.H. Raab, P. Asbeck, S. Cripps, P. Kenington, Z. Popovic, N. Pothecary, J. Sevic, and N. Sokal, "Power amplifiers and transmitters for RF and microwave," *IEEE Trans. Microwave Theory Tech.*, vol. 50, no. 3, pp. 814–826 (Mar. 2002).

[4] F.H. Raab, "class F power amplifiers with maximally flat waveforms," *IEEE Trans. Microwave Theory Tech.*, vol. 45, pp. 2007–2012 (Nov. 1997).

[5] F.H. Raab, "Maximum efficiency and output of class F power amplifiers," *IEEE Trans. Microwave Theory Tech.*, vol. 47, pp. 1162–1166 (June 2001).

[6] F.H. Raab, "class E, class C, and class F power amplifiers based upon a finite number of harmonics," *IEEE Trans. Microwave Theory Tech.*, vol. 49, no. 8, pp. 1462–1468, (Aug. 2001).

[7] C. Trask, "class F amplifier loading networks: a unified design approach," in *IEEE MTT-S Int. Symp. Dig.*, Anaheim, vol. 1, pp. 351–354 (June 13–19, 1999).

[8] A.V. Grebennikov, "Circuit design technique for high efficiency Class F amplifiers," in *IEEE MTT-S Int. Symp. Dig.*, Boston, MA, vol. 2, pp. 771–774 (,June 11–16, 2000).

[9] S. Ooi and S. Gao, "High efficiency class F power amplifier design," in *IEEE High Frequency Postgraduate Student Colloquium*, , pp. 113–118 (Sep. 2004).

[10] P. Colantonio, F. Giannini, G. Leuzzi, and E. Limiti, "On the class F power amplifier design," *Int. J. RF Microwave CAE*, vol. 9, no. 2, pp. 129–149 (1999).

[11] P. Colantonio, F. Giannini, G. Leuzzi, and E. Limiti, "Multiharmonic manipulation for highly efficient microwave power amplifiers," *Int. J. RF Microwave CAE*, vol. 11, pp. 366–384 (Nov. 2001).

Solder Joint Strength Of Lead Free Solders under Multiple Reflow and High Temperature Storage Condition

Ibrahim Ahmad[1], Azman Jalar[1], Burhanuddin Yeop Majlis, Eu Poh Leng[2], Yong Soo San[2]
[1]Universiti Kebangsaan Malaysia, Bangi, 43600, Selangor, MALAYSIA
Phone: 6-03-89296309; Fax:6-03-89296146
[2]Freescale Semiconductor (M) Sdn. Bhd, Sg. Way Free Trade Zone, Petaling Jaya, Selangor
e-mail: ibrahim@vlsi.eng.ukm.my

Abstract - Solder Joining is one of the interconnection methods in microelectronic packaging. The robustness of the solder ball attachment to the packages determines the solder joint strength between solders and the Under-Bump Metallization (UBM). This paper discussed the solder joint shear strength between Sn/4.0/0.5Cu with containing 20-50 ppm Phosporus and Sn/3.8Ag/0.7Cu deposited onto selective plating Ni/Au surface finish TBGA packages. Both were reflowed at a peak reflow temperature of 245±5°C with soaking time 70 second above 220°C. Ball shear testing along with failure mode analysis is used as a destructive method to assess solder joint strength and integrity. Both Sn -Ag-Cu (SAC) were studied under multiple reflow (1x, 2x, 3x, 6x) and high temperature storage (HTS) test at 24, 48, 96 hours. All samples were cross-sectioned to study the IMC thickness that leads to solder joint embrittlement. As a result, Sn/3.8Ag/0.7Cu performs better shear strength after HTS test than Sn/4.0/0.5Cu. Similarly, both show the decreasing of shear strength with increasing number of reflow cycles, supported by IMC thickness measurement that there is no significant different in term of growth and consistency.

Key words: *Solder joint, Under Bump Metallurgy (UBM), high temperature storage (HTS), Intermetallic Compound (IMC), and Shear Strength.*

I.   INTRODUCTION

Solder joint strength is used as a quantitative measure of the robustness of the interconnect between the solder balls and substrate pad. The pad, which are copper and easily oxidized, need to have a protective surface finish ensuring solderability during assembly[1]. Surface finish used on this BGA pad is electrolytic Nickel/Gold plated over the copper pad. The Nickel layer is used as a diffusion barrier against oxidation on the underlying copper while the gold is used to protect the surface finish of the nickel to ensure good soldering and in some cases good wire bonding [7].

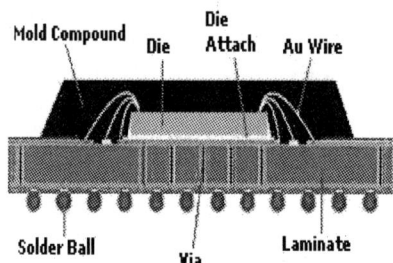

Figure 1.1-Schematic diagram of BGA packages

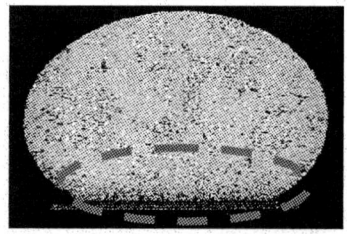

Figure 1.2-Area of studies

0-7803-9730-4/06/$25.00 ©2006 IEEE

II. EXPERIMENTAL

2.1 Ball Attach Process

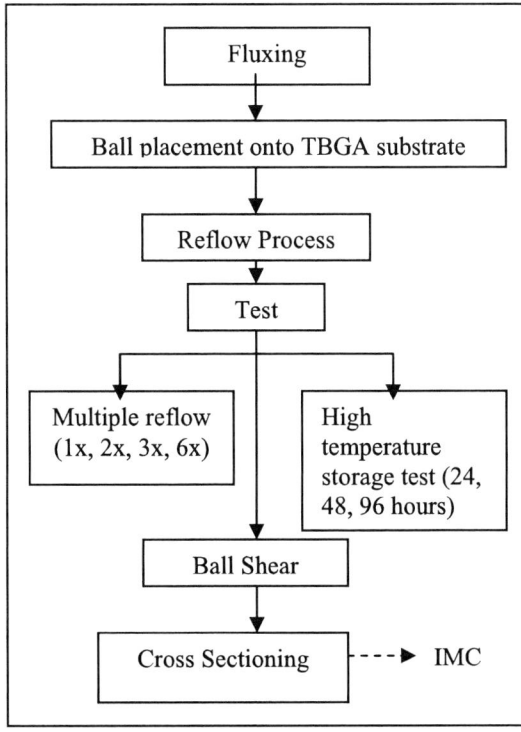

Figure 2.1: Experimental Process Reflow

The preparation of the test samples was initiated by applying flux on the solder pads through pin transfer mechanism. The solder balls were placed on the solder pads by ball bump tool. Prior to the pre-formed solder balls placement, flux was applied to each bond pads. The flux was no-clean type. The function of this flux to promote the sticky area and to help solder reflow activity [6].

Two type of substrates were used in the investigation. There are lead-free solder balls with composition of Sn: 95.5%, Ag: 3.8%, Cu: 0.7% and Sn: 95.5%, Ag: 4%, Cu: 0.5% with containing 20-50 ppm Phosporus were reflowed on solder pads with Ni/Au surface finish using the reflow profiles, as shown in Figure 2.2 respectively. The reflow was conducted in a seven zone IR oven at a peak reflow temperature of 235-250°C. Then 5 samples were picked up from each solder materials and marked as 1x reflow. Remaining dices from each solder materials were then experienced of

multiple reflow up to 6 times. Which 5 dices from each solder materials will be picked up at every reflow cycle of 2, 3, and 6 times, respectively. Also, test samples for high temperature storage at 150⁰C for different storage time at 24, 48 and 96 hours using HTS chamber. HTS is used to determine the effect of temperature and time under storage condition at elevated temperatures.

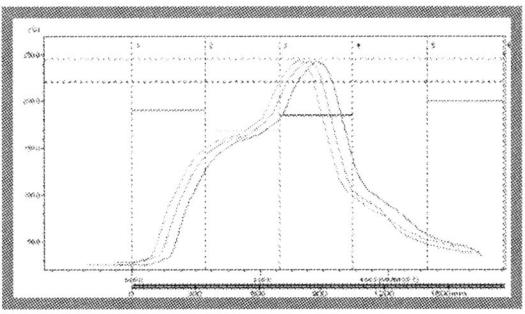

Figure 2.2- Reflow Profile

The shear strength was measured using a ball shear test, *DAGE 400 series* machine. The shear speed was set at 300mm/s and the back-off height was 40 μm above the die surface. The shear strength calculated from 15 units both of them. Only failure that happened through solder bumps and, at IMC failure mode were selected for this analysis. The interfacial interaction at the failure mode after ball shear test was observed, using scanning electron microscope (SEM). Figure 2.3 and 2.4 shows the failure mode between brittle and ductile layer.

Figure 2.3- Brittle   Figure 2.4- Ductile

All the samples for the microstructure and interfacial intermetallic compound (IMC) study that formed during the reaction were mounted in epoxy for grinding and polishing. All samples have been grinding using sand paper sized 180, 600 and 1200 μm. Fine polishing was carried out using oil lubricant sized 0.9 and 0.6 μm also collodial silica sized 0.05 μm. The cross-section samples were investigated using metallurgical optical microscope with the aid of image analyzer (Fig 2.5). The average

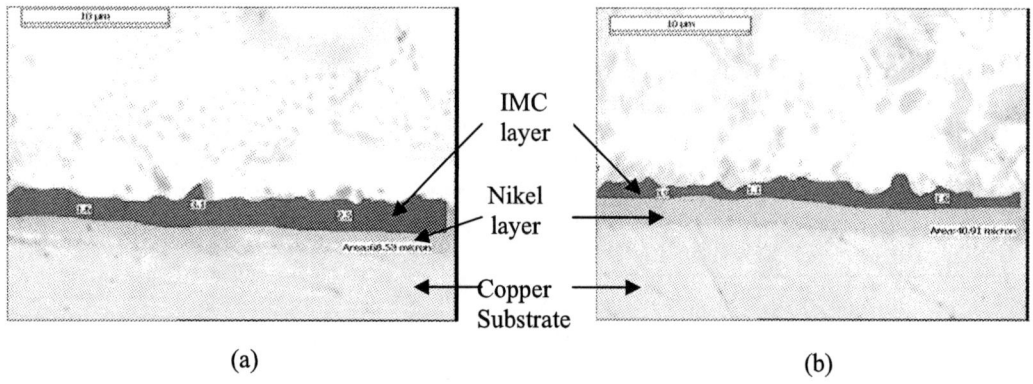

Fig. 2.5: Interfacial intermetallic compound (IMC) microstructure (a) Sn/4.0Ag/0.5Cu and (b) Sn/3.8Ag/0.7Cu

IMC layer thickness was determined by calculating the total area of the IMC using the analysis software, and dividing by the length. All the samples have been etched 1 min by using 90% methanol and 10% HCl solvent for better microstructure reveal.

III. RESULTS AND DISCUSSION

With increasing storage times, solder balls with composition Sn/3.8Ag/0.7Cu shows the decreasing of shear strength in Fig. 3.1. Contradict with Sn/4.0Ag/0.5Cu that shows increasing of shear strength up to 48 hrs, then decrease when prebake up to 96 hrs. The decreasing of shear strength is due to grain coarsening and IMC growth per storage times. As this intermetallic layer thickens, it becomes susceptible to cracking and thus affecting the integrity of the solder joint. When the joint is subject to stress, thermal cycles, vibration, or shock, the intermetallic layers are considered to be potential weak links in any solder joint that will lead to failure. The thickening of the intermetallic layer is induced by high temperature, composition of the joint and duration of the soldering process. Ball shear test results give the average of shear strength given by Sn/3.8Ag/0.7Cu is 1540.7 g which is greater than Sn/4.0Ag/0.5Cu, 1470.47 g. It can summarized that Sn/3.8Ag/0.7Cu more strengthen and ductile than Sn/4.0Ag/0.5Cu.

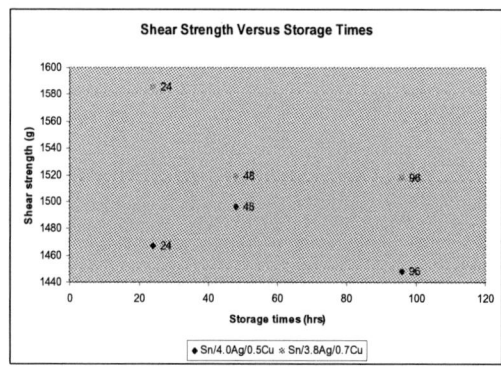

Figure 3.1- Shear strength versus storage times for solder balls with composition Sn/3.8Ag/0.7Cu and Sn/4.0Ag/0.5Cu with 20-50 ppm Phosphorus

In term of multiple reflow, both shows the decreasing of shear strength with increasing reflow cycles (Fig 3.2). Low shear force and strength will identify a weak solder joint, a high shear strength may be observed on packages which exhibits non-robust behavior in subsequent reliability testing as well as in the field [7].

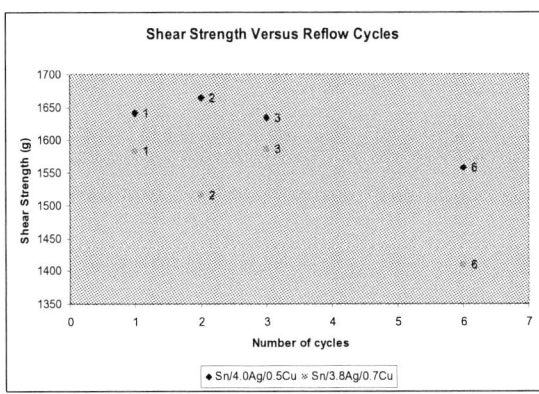

Figure 3.2– Shear strength versus reflow cycles graph

When reflow cycles increased, IMC become thicker. This is not desired as IMC is a brittle layer & increase in thickness will increase the risk of brittle fracture. It can be summarized that there is no significant different between Sn/4.0Ag/0.5Cu and Sn/3.8Ag/0.7Cu in term of IMC growth. It is supported from failure mode analysis that both are presented ductile failure leading to robustness of solder joint.

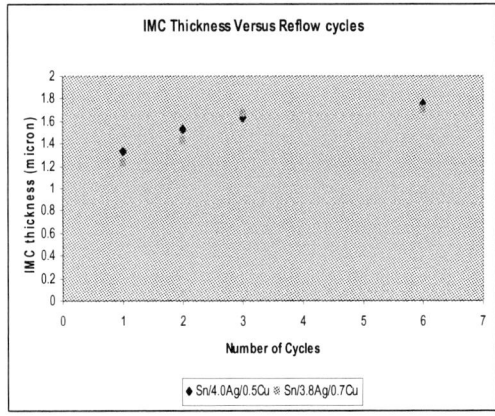

Figure 3.3: IMC thickness versus reflow cycles graph

IV. CONCLUSION

The conclusion from the above study is Sn/3.8Ag/0.7Cu performs better shear strength after HTS test than Sn/4.0/0.5Cu. Ball shear test results give the average of shear strength given by Sn/3.8Ag/0.7Cu is 1540.7 g which is greater than Sn/4.0Ag/0.5Cu, 1470.47 g. It is recommended

that Sn/3.8Ag/0.7Cu more strengthen and ductile than Sn/4.0Ag/0.5Cu. Similarly, both show the decreasing of shear strength with increasing number of reflow cycles (1x,2x,3x,6x), supported by IMC thickness measurement that there is no significant different in term of growth and consistensy.

REFERENCES

[1] Anand A., Mui YC., Weidler J., Diaz N., (2001), "Impact of Substrate Finish on Sn/Ag/Cu A Solder Joint", Advanced Micro Devices, Inc.

[2] Guenther B (1999), "Gold wire Bondable and Solderable Surface, "Proc.IPC Printed circuits Expo 1999, Pg S13-3

[3] Hollesen D.B., Ejim T.I., Holliday S.A., Coyle R.J.(1997), "Assembly and Reliability of Thermally Enhanced high I/O BGA packages," Proc. 21th IEMT Symposium, Austin, TX, 25-31 October 13-15.

[4] Petar Ratchev, Bart Vandevelde and Ingrid De Wolf (2003), "Reliability and Failure Modes of SnAgCu Solder joint for PSGA Packages", Proceeding of First Int.Conf. On Lead-Free electronics.

[5] Richard J. Coyle and Patrick P.Solan (2000), "The influence of Test Parameters and Packages Design Features on Ball Shear Test Requirement", 2000 IEEE/CPMT Electronic Manufacturing Technology Sysmposium.

[6] Richard J. Coyle, Patrick P.Solan, Anthony J.Serafino and Steven A.Gahr, "The influence of Room Temperature Aging on Ball Shear Strength and Microstructure of Area Array Solder Balls, "Proc.50th Electronic Component and Technology Conference, 160-69, (2000).

[7] Robert Erich, Richard J. Coyle (1999), "Shear Testing and Failure Mode Analysis for Evaluation of BGA Ball Attachment", 1999 IEEE/CPMT Electronic Manufacturing Technology Sysmposium.

[8] Salam B, N.N. Ekere, D.Rajkumar (2001), "Study of the Interface Microstructure of Sn/Ag/Cu Lead-Free Solders and the Effect of Solder Volume on Intermetallic Layer Formation.

A Study on Inter-Metallic Compound Formation and Structure of Lead Free SnAgCu Solder System

I. Ahmad[1], Hoh Huey Jiun[1&3] Eu Poh Leng[3], B.Y. Majlis[1], A. Jalar[2] and R. Wagiran[4]

[1]Department of Electrical, Electronic and System Engineering, Universiti Kebangsaan Malaysia, 43600 Bangi, Selangor, Malaysia
[2]School of Applied Physics, University Kebangsaan Malaysia, 43600 Bangi, Selangor, Malaysia
[3]Freescale Semiconductor, (M) Sdn. Bhd., Sungei Way Free Industrial Zone, 47300 Petaling Jaya
[4]Department of Electrical and Electronic Engineering, Universiti Putra Malaysia, 43400 Serdang, Selangor, Malaysia
Email: ibrahim@vlsi.eng.ukm.my

Abstract **This paper characterizes the effect of various Sn-Ag-Cu solder compositions towards shear strength and melting behavior. Shear strength is measured by *Dage* which is representative of the inter-metallic compound (IMC) strength between the solder and solder clad of a C5 bump. Further study on melting properties will be obtained by Differential Scanning Calorimetry (DSC). It was found that 1wt%Ag 0.5wt%Cu has larger melting range compared to 3 & 4wt%Ag 0.5wt%Cu which could contribute to IMC growth. Relatively, ball shear results shows that shear strength increases with Ag wt% content. A correlation between melting range and shear results made in this study pointed to 3.8wt%Ag0.5wt%Cu having the most favorable results.**

I. INTRODUCTION

Lead (Pb) is a major constituent of solder that is used extensively within electrical and electronic equipment and there are environmental concerns over the amount of lead ending up in the landfill. A few alternatives were proposed to the industries to replace Sn-Pb solder. At the same tme, each alternative requires investigations into its joint integrity with different solder composition or substrate finishing. This report addresses the metallurgy and integrity of ball grid array (BGA) package-level C5 solder joints using different composition of tin-silver-copper (SAC).

Among varieties of lead-free solder alloys, SnAgCu has been proposed as the most promising substitute for Pb-containing solders because of its good wetting performance and mechanical properties (Chen et al. 2002). Eutectic point of SAC is approximately 3.5-3.8wt%Ag and melting temperature at 217°C. The intermetallic formed at eutectic is Sn, Cu_6Sn_5 and Ag_3Sn (Kattner, 2000).

The interfacial intermetallics are Cu_6Sn_5/Cu_3Sn for all the alloy compositions (Li et al. 2001). However, the morphology of the intermetallics differs. In the case of eutectic Sn37Pb solder, the intermetallics have a very uniform morphology and the growth kinetics appears to be slower than those of the lead-free solder alloys.

II. EXPERIMENTAL WORKS

In this study, four compositions of SAC solder were investigated. Table 1 shows the composition for each solder balls and each having similar ball diameter which is µm. Test vehicle used for these experiments were PBGA 388. The solder balls investigated are supplied by the same supplier.

Table 1: Composition of each solder ball

Commercial name	Argentum (Ag) (wt%)	Copper (Cu) (wt%)	Stanum (Sn) (wt%)
105	1.0	0.5	98.5
305	3.0	0.5	96.5
387	3.8	0.7	95.5
405	4.0	0.5	95.5

Each solder type was tested with differential scanning calorimeter (DSC) for their melting properties. DSC model *Metler Toledo* is used and the method is heating at rate of 5°C/min from 180-260°C. It is believe that melting properties affects the IMC growth as larger melting range is subjecting the unit to a longer heating period that will stimulate IMC growth. Therefore, it is important to understand the melting properties of each solder composition.

0-7803-9730-4/06/$25.00 ©2006 IEEE 554

Joint strength between the solder ball and substrate were measured by shear test. It is assumed that the shear result is representative of the IMC as IMC is the sole interconnect between solder ball and substrate pad. *Dage 2400PC* was used to shear the units. The test height is 30μm and test speed is 300μm/s. Each run will be 5 units and each unit will be sheared 8 balls. Test vehicle used for this study is plastic ball grid array (PBGA) and list of test runs is in Table 2.

Table 2: Test runs

Solder	Condition	Units	Balls sheared/ Unit
105, 305, 387 405	T_0	5	8
	HTS_{48}	5	8
	HTS_{96}	5	8

IV. RESULTS AND DISCUSSIONS

Melting Behavior

Table 2 shows the summary results of DSC for three runs per each samples. DSC result found at Fig.1 shows that composition 105 having three peaks compared to 305, 387 and 405 with only one peak, shown in Fig.2, melting peak for SAC 387. Result of Table 2, solder composition 105 reported in the table is the average of the three peaks in each run. It was found that 105 have the highest melting range and three melting peaks compared to all other composition tested. This could be due to the fact that composition for solder 105 is furthest away from the eutectic composition. The large melting range could be attributed to the three melting reaction of the SnAgCu phases, which is;

$$L \longrightarrow Cu_6Sn_5 + Ag_3Sn + (Sn)$$
$$L \longrightarrow Ag_3Sn + (Sn)$$
$$L \longrightarrow Cu_6Sn_5 + (Sn)$$

Followed by 105 is 305, 405 and the smallest melting range is composition 387. It is believe that the melting range influences intermetallic growth as the onset melting indicates the start of the molten solder reaction and endset indicates the end. Hence, the larger melting range indicates a longer temperature range of intermetallic formation, and this is not desirable. Having a narrower melting range could limit the temperature range of inter-metallic formation and it was highly believe this could reduce the inter-

metallic formation. This is due to the knowledge that inter-metallic growth is stimulated by two factors, which is time and temperature. Therefore, it is best to have minimum melting range.

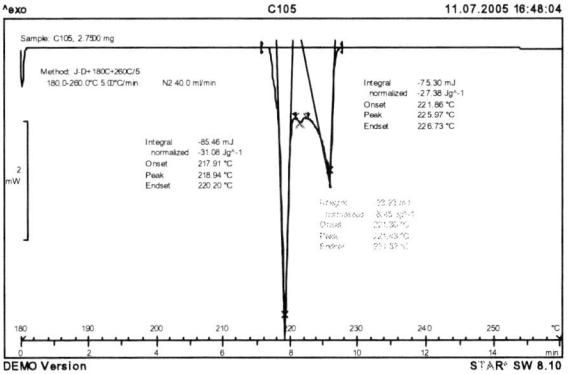

Fig.1: DSC graph for 105

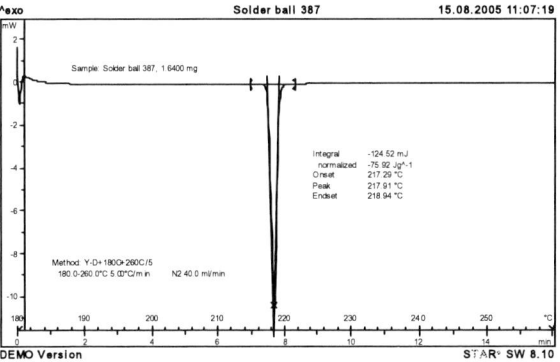

Fig.2: DSC graph for 387

Solder composition 387 has the smallest melting range because the composition is near to the eutectic point of tin-silver-copper solder system. Eutectic point is where all the three elements melts at the same melting temperature. From Table 3, it is learnt that solder 387 melts at the peak temperature 217-218.

Fig. 3 shows the correlation of melting range with composition of silver (wt%). Variation in composition of silver is proportional to the composition of copper and tin. However, due to copper remain as 0.5wt% and tin is a stable element, therefore silver is assumed to be the reference point for this correlation. From the correlation it is generally observed that melting range reduces as the composition of silver content (1-4 wt%) increases. In addition, the lowest melting range is found at solder 387, where it is the nearest to the eutectic point.

Table 3: Summary results of the melting characteristics

	Onset (°C)	Peak (°C)	Endset (°C)	Melting Range
105 (# 1)	217.91	-	226.73	8.8
(# 2)	218.17	-	226.91	8.7
(# 3)	217.77	-	226.67	8.9
305 (# 1)	217.31	218.49	220.57	3.3
(# 2)	217.25	218.66	221.56	4.3
(# 3)	217.21	218.50	221.10	3.9
387 (# 1)	217.29	217.91	218.94	1.7
(# 2)	217.51	218.42	219.42	1.9
(# 3)	217.39	218.04	219.09	1.7
405 (# 1)	217.19	218.03	219.84	2.7
(# 2)	217.18	218.61	220.30	3.1
(# 3)	217.31	218.21	220.21	2.9

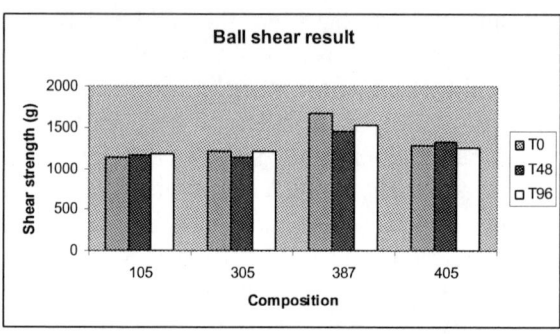

Fig.4 : Correlation of shear strength with composition for high temperature storage of T48 and T96 hours.

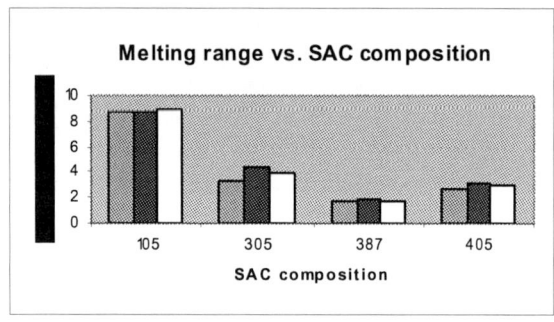

Fig.3 : Correlation of melting range and SAC composition.

Fig.5 : Intermetallic microstructure of SAC 387

Ball Shear Analysis

Ball shear test is used to estimate the joint strength of each solder composition with the pad metallurgy. Result of Fig.4 shows the average of shear result for each tested composition. SAC 387 shows highest shear strength among the composition for high temperature storage of T48 and T96. Followed by shear strength of SAC 405 and SAC 105 and 305 are having very similar strength. Since SAC 387 out performs other composition of the SAC group, micro structural and elemental analysis will be done to further understand the intermetallic.

Intermetallic Microstructure and Elemental Analysis of SAC 387

This part of study concentrates on microstructure and elemental analysis of SAC 387. Microstucture and elemental of the intermetallic is important to understand the joint mechanical behavior and elementa.

Fig. 5 shows the micrograph structure of both the specimen after etching. Both the specimen show needle like structure of intermetallic growth.

Further EDX analysis shown in Fig.6 gives elemental compositions quantitatively and it shows that intermetallic phases present for both supplier Senju and Indium. Result shows a close relation of intermetallic phases present, mainly Cu_6Sn_5, Ni_3Sn_2, Ni_3Sn_4 and $(Ni,Cu)_3Sn_2$. Presence of Ni_3Sn_4 is needle like (Paik et al. 2004) and hence the topography from the micrograph SEM is needle like structure. The Ni source to form Ni_3Sn_4 intermetallic in a bulk solder may be originated from dissolved Ni atoms in a molten solder. At the initial stage of reflow, Ni atoms dissolve rapidly into molten solder bump.

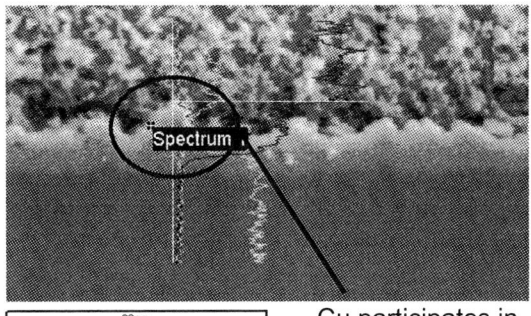

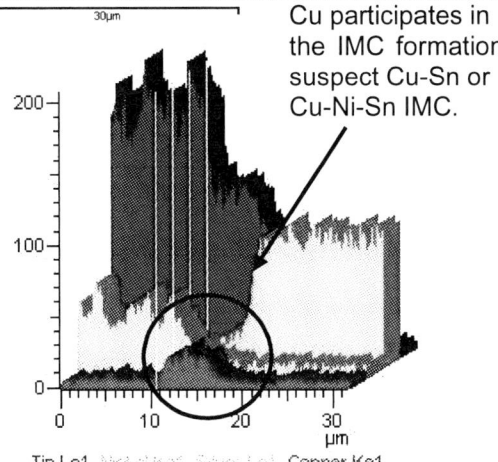

Cu participates in the IMC formation, suspect Cu-Sn or Cu-Ni-Sn IMC.

Tin La1, Nickel Ka1, Silver La1, Copper Ka1

Fig.6 :Elemental analysis of SAC 387

V. CONCLUSION

Solder composition 105 has three melting peaks and largest melting range which is undesirable in intermetallic growth. The reason of three melting peak could be attributed by the fact that it is furthest from eutectic composition and each melting peak represent melting reaction of eutectic SnAgCu system. Solder composition 387 has the smallest melting range and a precise peak temperature of 217-218°C. Having a smallest melting range is due to its near to eutectic composition. Correlation between melting range and Ag(wt%) is made that, from 1-4% of Ag(wt%), melting range reduces with increasing of Ag(wt%).

REFERENCES

[1] W.H. Yeong, Y. Akito and G. Richard, *"Advances in 3d Packaging"*, Proceedings of VLSI Packaging Workshop of Japan, Kyoto, Japan, 11/12/02.

[2] Y.M. Cheung and Arthur C.M.Chong, *"New Proposed Adhesive Tape Application Mechanism for Stacking Die Applications"*, International Conference on Electronic Packaging Technology, ICEPT 2003,

[3] D.B. Marshall, A.G. Evans, et al, *"The Nature of Machining Damage in Brittle Materials"*, Proceedings R.Soc., London A385, 1983 PP.461-475.

[4] I. Weisshaus and G.Weissman, *"Online Monitoring Of The Dicing Process"*, Future Fab, Issue 10, Technology Publishing Ltd., London, England.

[5] J.Jian, C.H. Song, Z.L. Zhang, *"Dicing Technology In Super-Thin Wafer for IC"*, International Conference on Electronic Packaging Technology, ICEPT 2003,

[6] K.Gilleo, T.Cinque, S.Corbett, M. Corey, C. Lee & R. Miculich, *"Thermoplastic Die Attach Adhesive for Today's Packaging Challenges"*.

Design and Simulation of 50 nm Vertical Double-Gate MOSFET (VDGM)

Ismail Saad and Razali Ismail

Faculty of Electrical Engineering, Universiti Teknologi Malaysia, 81310, Skudai, Johor

e-mail: ismail_s@ums.edu.my , razali@fke.utm.my

Abstract **The paper demonstrate the design and simulation study of 2D Vertical Double-Gate MOSFET (VDGM) with an excellent short channel effect (SCE) characteristics. With the gate length of 50nm, body doping of 3.5×10^{18} cm^{-3} and oxide thickness, $T_{OX} = 2.5$nm, a good drive current I_{ON} of 7 μA/μm and a low off-state leakage current I_{OFF} of 2 pA/μm was explicitly shown. Besides that, the subthreshold characteristics also highlighted a reasonably well-controlled SCE with subthreshold swing SubV$_T$ = 89 mV/decade and threshold voltage V$_T$ = 0.56V. The analysis of body doping effects for SCE optimization and drive current trade-off was also done for an overall investigation and limit of the VDGM.**

I. INTRODUCTION

With the advantages of controlled gate length by a relax photolithographic process, high drives current per unit silicon area and decoupled channel length from packing density, Vertical double-gate (VDGM) and/or surround gate (VSGM) MOSFET become prominent candidate to extend CMOS technology to and beyond the 45nm as depicted by International Technology Roadmap for Semiconductor, ITRS [1]. In contrast, the planar double-gate (DG) has also been intensively in research. The focus is on new structure such as Self-aligned FinFet structure [2, 3, 4], Silicon On Nothing (SON) [5], PAGODA concept [6] and Bonded double-gate [7]. However, this planar device requires advanced processes and precise definition of channel length that make it less acceptable compared with Vertical channel types. Generally, the vertical type MOSFET can be categorized according to channel definition fabrication method. It can be ion implanted process [8-15], epitaxially grown [16-19], and retarded etching [20] on the sidewalls of Silicon pillars. In this paper the structure of a Vertical replacement gate transistor [19], in which the source/drain

electrodes are defined by solid state diffusion after channel epitaxial growth is been analyzed with a double-gate configuration. The VDGM device was design and simulated using 2D commercial device simulation (ATLAS) software [21]. The utilization of such process and device simulation for an investigation and detail analysis of new device structure has been raised sharply. Such methods are employed in [22-24]. Using such standard, we demonstrate the design of 2 Dimensional vertical double-gates MOSFET (VDGM) with an excellent short channel effect (SCE) feature. With the gate length of 50nm, body doping of 3.5×10^{18} cm^{-3} and oxide thickness, $T_{OX} = 2.5$nm, a good drive current I_{ON} of 7 μA/μm and a low off-state leakage current I_{OFF} of 2 pA/μm was obtained. In addition, the subthreshold characteristics also highlighted a reasonably well-controlled short channel effects (SCE) with subthreshold swing SubV$_T$ = 89 mV/decade and threshold voltage V$_T$ = 0.56V. An effect of lowering the body doping for getting an acceptable I_{OFF} and V$_T$ as it will increase the surface mobility and drive current was also done. However, an optimization is needed as a high body doping was essential for controlling SCE. The trade-off between controlling SCE with an optimize level of body doping was done explicitly in this paper.

II. MODELING PROCEDURE

The simulated VDGM structure is shown in figure 1a and 1b with the double gate region (in contact), drain and source electrode, channel length L$_g$, silicon oxide T$_{ox}$, silicon body and the respective dimensions of the device is explicitly shown. The process start with a mesh or grid definition in which the critical area such as L$_g$ and T$_{ox}$ were given a finer mesh compare to other regions. Subsequently, the coordinates of source, drain, body, left and right gate, gate oxide and separation oxide region were defined. The electrode region of drain, source and double gate area were also formed for the contacts to be

0-7803-9730-4/06/$25.00 ©2006 IEEE

used in device characterization process later. Notice that an electrode line is visible in figure 1a for making sure that the left and right gate was in contacts. Later, a uniform doping profile is assumed and applied to drain (n-type), source (n-type), double gate (n-type)and body (p-type) of the device with the concentration of $1x10^{20}$ cm^{-3}, $1x10^{20}$ cm^{-3}, $1x10^{21}$ cm^{-3} and 3.5×10^{18} cm^{-3} respectively. The channel body doping may be varied for an analysis on its effects in device performance.

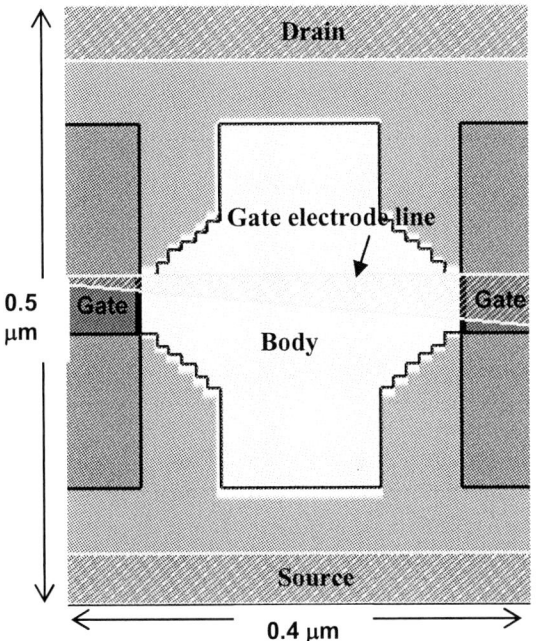

Fig.1a. Vertical Double-Gate MOSFET (VDGM) structure showing double gate, source, drain, and body

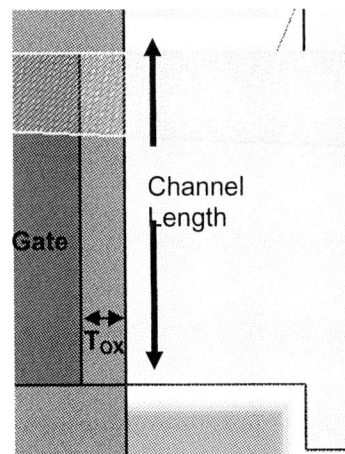

Fig.1b. Vertical Double-Gate MOSFET (VDGM) structure showing channel length and oxide thickness

The inversion layer mobility model from Lombardi [21] was employed for its dependency on the transverse field (i.e field in the direction perpendicular $E_\perp$ to the Si/SiO$_2$ interface of the MOSFET) and through velocity saturation at high longitudinal field (i.e field in the direction from source-to drain parallel E to the Si/SiO$_2$ interface) combined with SRH (Shockley-Read-Hall Recombination) with fixed carrier lifetimes models [21]. This recombination model was selected since its take into account the phonon transitions effect due to the presence of a trap (or defect) within the forbidden gap of the semiconductor. An interface fixed oxide charge of $3x10^{10}$ is assumed with the used of n-type Polysilicon gate contact for the device. The Drift-Diffusion transport [21] model with simplified Boltzmann carrier statistics [21] is employed for numerical computation of the design device.

III. ELECTRICAL CHARACTERIZATION

The combination of Gummel and Newton numerical methods [21] was employed for a better initial guess in solving quantities for obtaining a convergence of the device structure. Figure 2 shows the current-voltage ($I_{GS} - V_{GS}$) characteristics for VDGM device with channel length L_g = 50nm, oxide thickness T_{OX}= 2.5nm and body channel doping N_A=3.5x10^{18} cm^{-3}. By using a linear extrapolation of transconductance g_m (V_{GS}) to zero [25] a 0.56V threshold voltage V_T was obtained for both V_{DS}=1.2V and 0.1V in the linear operated region.

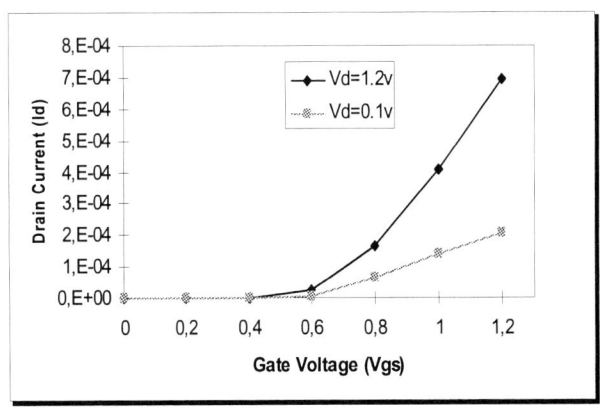

Fig. 2. Current-Voltage characteristic of VDGM with L_g=50nm, T_{OX}=2.5nm, N_A=3.5x10^{18} cm^{-3} and V_T=0.56V taken at V_{DS}=0.1V and 1.2V

A low V_T =0.56V extracted for this device yield that a low power consumption of the MOSFET device is maintained and is comparably better with a planar MOSFET for deep sub-micron device [13, 16, 17]. However, this V_T is considerably high for 1.2V intended circuit operation due to a high channel doping N_A that reduces the surface mobility and degrade the drive current I_{ON}. However, high doping is necessary for controlling the SCE. An output characteristic I_{DS}-V_{DS} is shown in figure 3, that explicitly illustrate a moderately low drain current due to a high doping.

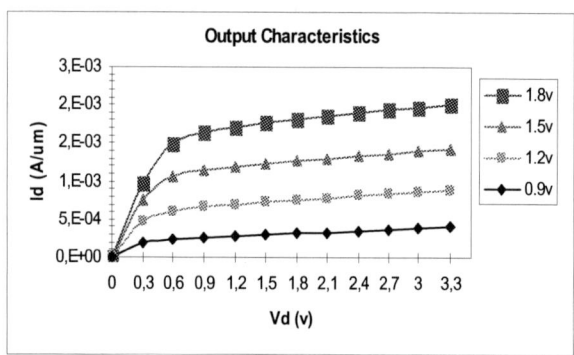

Fig. 3. Output characteristic of VDGM with L_g=50nm, T_{OX}=2.5nm, N_A=3.5x10^{18} cm^{-3} and V_T=0.56V for V_{GS}=0.9, 1.2, 1.5 and 1.8V.

Figure 4 shows a good off-state leakage current I_{OFF} of 2 pA/µm and drive current I_{ON} of 7 µA/µm due to high doping that control the off and on state of the device. Furthermore, reasonably well-controlled SCE with subthreshold swing $SubV_T$ = 89 mV/decade is also highlighted in figure 4.

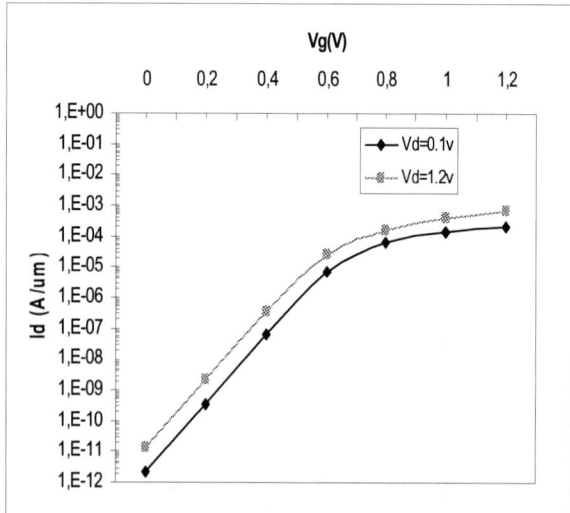

Fig. 4. Subthreshold characteristic of VDGM with L_g=50nm, T_{OX}=2.5nm, N_A=3.5x10^{18} cm^{-3} and I_{OFF}=2pA/µm, I_{ON}=7µm/µm, $SubV_T$=89mV/dec

Further analysis on the VDGM performance was done by comparing its capability to control SCE with single gate MOSFET. By applying the same mobility and recombination model with different channel length which is higher by a factor of 50 in single gate MOS, the resulted subthreshold characteristics is shown in figure 5.

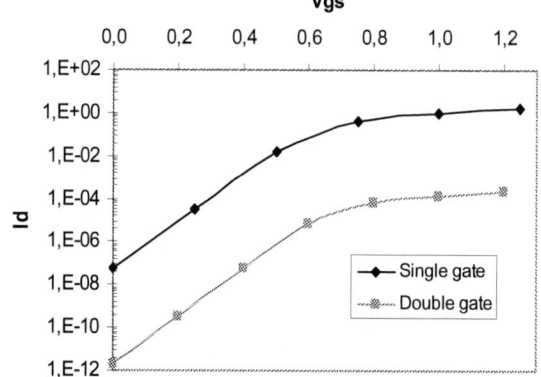

Fig. 5. Comparison of Subthreshold characteristic of VDGM with Single gate MOSFET (SGM). Low leakage current is exhibited in VDGM in pico range as with SGM.

As expected, the vertical double gate give a low off-state leakage current as compared to single gate MOSFET (SGM) and an acceptable OFF/ON ratio due to a better control of electrostatic potential in gate region for the vertical defined channel [8, 9, 10]. However, these was achieved due to a high body doping in VDGM which gave a good SCE control with a high threshold voltage due to a degradation in surface mobility as depicted in figure 2 and consequently reduced the drain current as shown in figure 3. Thus, the effect of body doping has to be analyzed for obtaining an acceptable V_T, I_{DS} and optimize control of SCE. This will be carried out in the next section.

V. DOPING EFFECT ANALYSIS

Three variant of body doping are used: low doped (N_a = 1x10^{18} cm^{-3}), moderately doped (N_a = 2x10^{18} cm^{-3}) and the high doped (N_a = 3.5x10^{18} cm^{-3}). In Figure 6 we can see that V_T is reduced as the doping level is decreased.

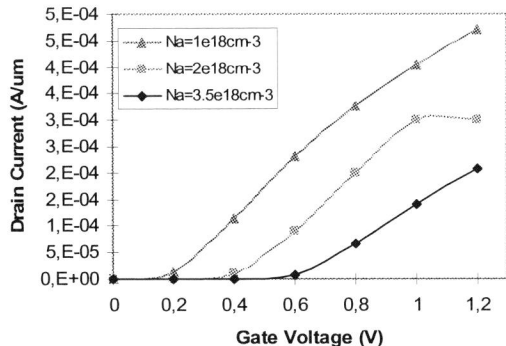

Fig.6. I_{GS}-V_{GS} characteristic of VDGM shows a decreased in V_T is observed with a lower doping level.

The V_T value decreases from 0.56V for high doped to 0.36V in moderate doped and to a lower value of 0.15V in low doped body. Further analysis is done, by comparing the subthreshold characteristics with different doping level as shown in figure 7.

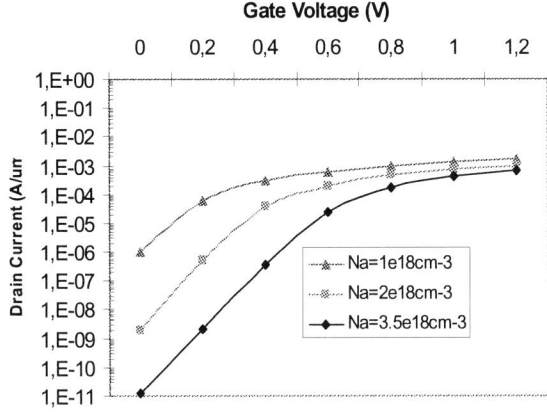

Fig.7. Subthreshold characteristic of VDGM showing an increased in leakage and drive current with a lower doping level.

As can be seen in figure 7, a decreased in doping will ultimately increased the leakage current from 2pA/μm to 7nA/μm and finally to a value of 80μA/μm. However, the increased in drive current is almost unity with a value of 7μA/μm to 10μA/μm and 1mA/μm respectively. These effects arise due to the fact that at higher doping the surface mobility is decreased and a better gate electrostatic potential observed within the device which makes the leakage current controllable. However, as the doping level decreased, the carrier mobility is increased and consequently the leakage current will also rise sharply. On the other hand, since the channel is defined vertically with a double gate

configuration, a unity drive current is observed which also increased with a decreased in doping level. Due to a double gate arrangement the increased in drain current (I_{DS}) was observed as shown in figure 8 of the output characteristics.

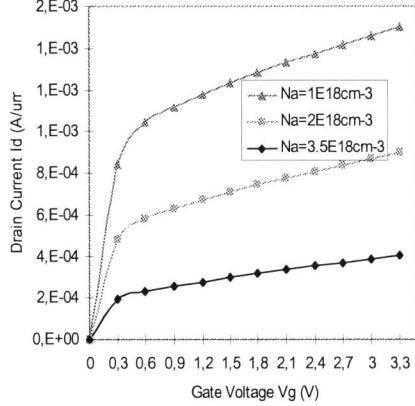

Fig.8. Output characteristic of VDGM shows an increased in Drain current with a lower doping level in a double-gate configuration.

Even though the drive current is high with lower doping level, a high leakage current is observed in figure 7, I_{OFF} = 80μA/μm is highly unacceptable. These results are in conjunction with the value of subthreshold voltage obtained, which is 89 mV/decade for higher doped, 83 mV/decade in moderate doped and sharply increased to 110 mV/ decade in lower doped device as depicted in figure 7. Thus, an optimize value of body doping is highly vital in order to have a high drive current while maintaining the acceptable leakage current and controlling the aggravated SCE. If one fails to control such parameters, the transistor designed will not succeed to work in a giga-scaled integrated circuit where the total standby power of the system is of paramount important.

VI. CONCLUSION

The design of a 50nm Vertical Double Gate MOSFET (VDGM) device based on the structure reported in [19] has been successfully done using commercial ATLAS TCAD tools. By employing the inversion layer mobility model from Lombardi combined with SRH (Shockley-Read-Hall Recombination) with fixed carrier lifetimes models with an interface fixed oxide charge of $3x10^{10}$ assumed and the used of n-type Polysilicon gate contact, a detailed investigation on the VDGM performance was done. With the gate length of 50nm, body doping of 3.5 x 10^{18}

cm^{-3} and oxide thickness, T_{OX} = 2.5nm, a good drive current I_{ON} of 7 μA/μm and a low off-state leakage current I_{OFF} of 2 pA/μm was obtained. In addition, the subthreshold characteristics also highlighted a reasonably well-controlled short channel effects (SCE) with subthreshold swing SubV$_T$ = 89 mV/decade and threshold voltage V_T = 0.56V. The effects of body doping N_A in obtaining a good drive current I_{ON} while maintaining an acceptable leakage current I_{OFF}, threshold voltage V_T and subthhreshold voltage SubV$_T$ for controlling the SCE was investigated. With a moderate body doping level N_A = 2.0x10^{18} cm^{-3}, a threshold voltage V_T = 0.36V, leakage current I_{OFF} = 7nA/μm and SubV$_T$ = 83 mV/decade and a good drive current I_{ON} = 10μA/μm was successfully obtained and optimized for the simulated VDGM device.

REFERENCES

[1] *International Technology Roadmap for Semiconductor (ITRS)* – Emerging Research Devices. http://public.itrs.net, 2005

[2] Xuejue Huang et., al. "Sub-50nm P-Channel FinFet". IEEE Transactions on Electron Devices, vol.48, no.5, May 2001

[3] Xuejue Huang et., al. "Sub 50nm FinFET : PMOS". IEDM 1999

[4] Bin Yu et., al. "FinFET Scaling to 10nm Gate Length". IEDM 2002

[5] S. Harrison et., al. "Highly Performant Double Gate MOSFET MOSFET realized with SON process". IEDM 2003

[6] M. Vinet et., al. "Bonded Planar Double-Metal-Gate NMOS Transistors Down to 10 nm". IEEE Transactions on Electron Devices, vol.26, no.5, May 2005

[7] P.M. Solomon et., al. "Two Gates are Better than One". IEEE Circuits & Devices Magazine, January 2003

[8] Gili et., al. "Asymmetric Gate-Induced Drain Leakage and Body Leakage in Vertical MOSFETs With Reduced Parasitic Capacitance". IEEE Transactions on Electron Devices, vol.53, no.5, May 2006

[9] Enrico Gili et., al. "Single, Double and Surround gate vertical MOSFETs with reduced parasitic capacitance". Solid-State Electronics 48 (2004) pg 511-519

[10] V.D Kunz et., al. *"Reduction of Parasitic Capacitance in Vertical MOSFETs by Spacer Local Oxidation"*; IEEE Transactions on Electron Devices, VOL. 50, pp 1487-1493, June 2003

[11] Enrico Gili, et., al. *"Electrical Characteristics of Single, Double & Surround Gate Vertical MOSFETs with Reduced Overlap Capacitance"*; ESSDERC 2003

[12] V.D Kunz et., al. *"CMOS- compatible vertical MOSFETs and logic gates with reduced parasitic capacitance"*; ESSDERC 2004

[13] D. Donaghy et., al. "Design of 50nm Vertical MOSFET Incorporating a Dielectric Pocket". IEEE Transactions on Electron devices, vol.51, no.1, January 2004

[14] Enrico Gili et., al. "A new approach to the fabrication of CMOS compatible vertical MOSFETs incorporating a dielectric pocket". ULIS 2005

[15] Thomas Schulz et., al. "Short-Channel Vertical Sidewall MOSFETs". IEEE Transactions on Electron devices, vol.48, no.8, August 2001

[16] S.K. Jayanarayanan et., al. "A Novel 50nm vertical MOSFET with a dielectric pocket". Solid-State Electronics 50 (2006) pg 897-900.

[17] Kiyoshi Mori et., al. "Sub-100-nm Vertical MOSFET with Threshold Voltage Adjustment". IEEE Transactions on Electron devices, vol.49, no.1, January 2002.

[18] Haitao Liu et., al. "An Ultrathin Vertical Channel MOSFET for Sub-100-nm Applications". IEEE Transactions on Electron devices, vol.50, no.5, May 2003.

[19] J.M.Hergenrother et., al. "The Vertical replacement (VRG) MOSFET: A 50nm vertical MOSFET with lithography-independent gate length". IEDM Tech. Dig., 1999, pp. 75-78

[20] Meishoku Masahara et., al. "Ultrathin Channel Vertical DG MOSFET Fabricated by Using Ion-Bombardment-Retarded Etching". IEEE Transactions on Electron devices, vol.51, no.12, December 2004.

[21] Silvaco International, "ATLAS user Manual DEVICE SIMULATION SOFTWARE"

[22] Ali A. Orouji et., al. "Shielded Channel Double-Gate MOSFET: A Novel Device for Reliable Nanoscale CMOS Applications" IEEE Transactions On Device And Materials Reliability, Vol. 5, No. 3, September 2005

[23] G.Venkateshwar Reddy et., al."A New Dual-Material Double-Gate (DMDG) Nanoscale SOI MOSFET— Two-Dimensional Analytical Modeling and Simulation". IEEE Transactions On Nanotechnology, Vol. 4, No. 2, March 2005

[24] Ismail Saad et., al. "Simulation of a Novel Lateral Bipolar Transistor with an approximately 21 GHz f_{TMAX} on Thin Film SOI". Proc. International Workshop on The Physics of Semiconductor Devices (IWPSD-2003), IIT Madras, India, December 2003

[25] M.Tsuno et al, "Physically-based Threshold voltage determination for MOSFETs of all gate length," IEEE Trans. Electron Devices, vol. 46, pp. 1429-1434, July 1999

Design and Simulation of a High Performance Lateral BJTs on TFSOI

Ismail Saad and Razali Ismail

Faculty of Electrical Engineering, University Technology Malaysia, 81310, Skudai, Johor Bharu,

E-mail: ismail_s@ums.edu.my, razali@fke.utm.my

Abstract **Lateral BJT's have received renewed interest with the advent of BiCMOS and Silicon on Insulator (SOI) technology. It's been reported in [1] that a 67 GHz f_{max} novel lateral BJT's on TFSOI has been fabricated with a simplified process. This paper presents an investigation of this high performance transistor by using 2D process and device numerical simulation. Accurate geometrical structure and reasonably good doping profiles with a simple fabrication process are successfully achieved in the process simulation. However, a careful attention is required to define the mesh for the device to obtain an accurate measurement of device characteristics. With a base, low-doped collector, emitter and high-doped collector concentrations of 3×10^{17} cm^{-3}, 1.0×10^{17} cm^{-3}, 5×10^{20} cm^{-3} and 3×10^{20} cm^{-3} respectively, a variation of 0.1– 0.13μm base width is observed. I-V and frequency performance of these transistors are simulated and analyzed. Y-parameter measurement at frequency 10 MHz - 1000 GHz shows a 21 GHz f_{max} was successfully achieved at V_{BE}=0.7V, V_{CE}=2.0V and I_{CE}=6.0 μA.**

I. INTRODUCTION

Most existing BiCMOS processes combine high-performance vertical BJT's with MOSFET's. These technologies offer a trade-off between speed and power dissipation and attain digital/analog systems with a performance exceeding that of circuits based on either technology alone [2, 3]. This results in a rather complex and expensive process due to the technological incompatibility of two types of transistors [4]. Several sophisticated technologies such as self-aligned double-polysilicon structure [5], shallow and/or deep trench isolation [6] and an epitaxial base [7] has been used. However, such superior process technologies increase fabrication costs of RF LSI's. Consequently, the cost of the extra steps to produce the buried layer and epitaxial collector vertical bipolar transistors

has also limited BiCMOS LSI's marketability [8]. With the advantages of low power and high speed operation and simpler integration of devices, Silicon on Insulator (SOI) has become an excellent candidate as an alternative substrate for BiCMOS circuits. Furthermore, the use of SOI as a substrate in BiCMOS circuits is dependent on the development of a proper bipolar device (in a lateral structure) on SOI [9]. A number of novel high performance lateral bipolar's on SOI have been proposed and implemented. All of these transistors have been fabricated with a different approach of structure. This paper presents an investigation of a high performance transistor by carrying 2D process and device numerical simulation [10]. With a base, low-doped collector, emitter and high-doped collector concentrations of 3×10^{17} cm^{-3}, 1.0×10^{17} cm^{-3}, 5×10^{20} cm^{-3} and 3×10^{20} cm^{-3} respectively, a variation of 0.1– 0.13μm base width is observed. I-V and Frequency performance of these transistors are simulated and analyzed. Y-parameter measurement at frequency 10 MHz - 1000 GHz shows a 21 GHz f_{max} was successfully achieved at V_{BE}=0.7V, V_{CE}=2.0V and I_{CE}=6.0 μA.

II. DEVICE STRUCTURE AND PROCESS

Figure 1 shows the schematic structure of the lateral bipolar transistor considered in the present study.

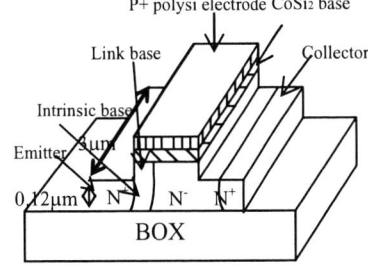

Fig. 1 Schematic 3D cross-sectional view of lateral BJT

The intrinsic base is formed by an angled (45°) boron and BF$_2$ ion implantation with the condition of BF$_2$ 1.2×10^{12} cm^{-2} at 25KeV, boron

0-7803-9730-4/06/$25.00 ©2006 IEEE   563

2×10^{12} cm^{-2} at 15 KeV and boron 7.2×10^{12} cm^{-2} at 35 KeV. This gives a flat profiled intrinsic base in the direction of the depth and also provides a link between intrinsic base and the p+ poly Si electrode. The length of this self-aligned link base is about 0.07 µm. An anisotropic poly Si etching process is used to form the poly Si sidewall. The emitter region is formed by ion implantation of phosphorus under the condition of 8.0×10^{15} cm^{-2} at 65KeV and N$^+$ collector is obtained by implanting phosphorus dose of 4.0×10^{15} cm^{-2} at 60KeV. RTA for 20 seconds at 950°C is applied to activate the doped impurities. Note that each of the emitter, base and collector regions has to be diffused and penetrated deeply into buried oxide. This is to maintain neutral charge within emitter/base, base/N$^-$ collector and N$^-$ collector/N$^+$ collector depletion regions and to reduce the junction capacitances. TEOS SiO$_2$ is then deposited over the transistor for isolation. Cobalt Silicide (CoSi$_2$) is used on the exposed p+ poly-Si to reduce the base resistance. TEOS SiO$_2$ is deposited again and conventional wiring process is carried out.

The Drift-Diffusion transport model with simplified Boltzmann carrier statistics is employed for numerical computation of the device design [10]. The bandgap narrowing effects in heavy doping surroundings has also taken into account. The standard carrier low electric field mobility concentration dependent and its smooth transition to a high electric field for carrier velocity saturation effects in the direction of current flow models was all switch on. The Auger recombination model combined with SRH model (Shockley-Read-Hall) in concentration dependent carrier lifetimes was also selected for characterization of the device [10]. The workfunction of p+ polysilicon with CoSi$_2$ is given for base electrode contact.

III. RESULTS AND DISCUSSION

Doping Profiles

Horizontal doping profile at the center of device is shown in figure 2, impurity concentration of intrinsic base is 3×10^{17} cm^{-3}. Both emitter and high-doped collector concentrations are at 4×10^{20} cm^{-3} and 3×10^{20} cm^{-3} respectively. Also, a slight variation in base width, W$_B$ (0.1– 0.13µm) as a function of depth is observed. For a better result, a narrow W$_B$ is vital in making sure that the injected carriers from emitter will reach the

depletion region of B/C junction to be swept across the junction into collector region. Thus, an optimized doping profile of base region is very important for definition of narrow base width, W$_B$. In addition, the vertical impurity profile of the base p+ polysi and n- collector region are also investigated as shown in Figure 3. A flat boron concentration of 2.5×10^{20} cm^{-3} is obtained in p+poly region. As shown in the figure, the boron has diffused into n- collector layer with the junction depth of 63 nm from the interface between p+poly and n$^-$ collector. This diffusion length is an important factor for the high frequency performance of transistor. The diffusion of boron into n$^-$ collector causes the decrease of cross-sectional area of collector region, which results in the increase of the collector current density. Consequently, the f$_T$-I$_C$ curve falls down earlier as the collector current is increased mainly due to the Kirk effect. However, this will accordingly decrease the breakdown voltage. Thus, the trade-off between increasing doping level of collector region for a higher frequency or faster circuit and high value of breakdown voltage must be carefully optimized.

In overall a trade-off between decreasing the base region and increasing the collector region doping profile for definition of a narrow base width, W$_B$ and minimizing the kirk effect in a reduce series collector resistance is essential for getting a higher frequency performance and more faster circuit operation.

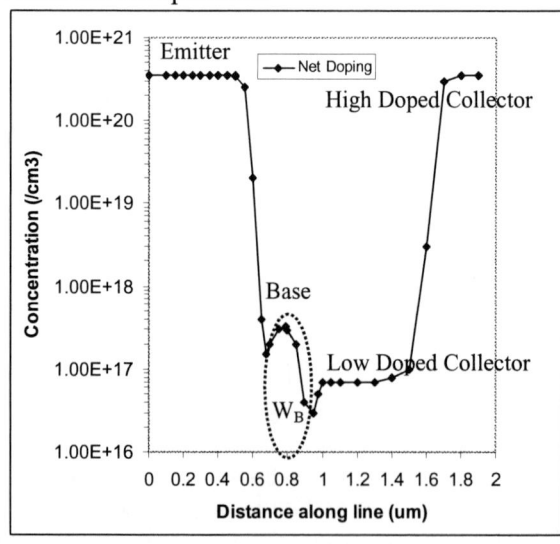

Fig. 2 Horizontal doping profile at the center of device

564

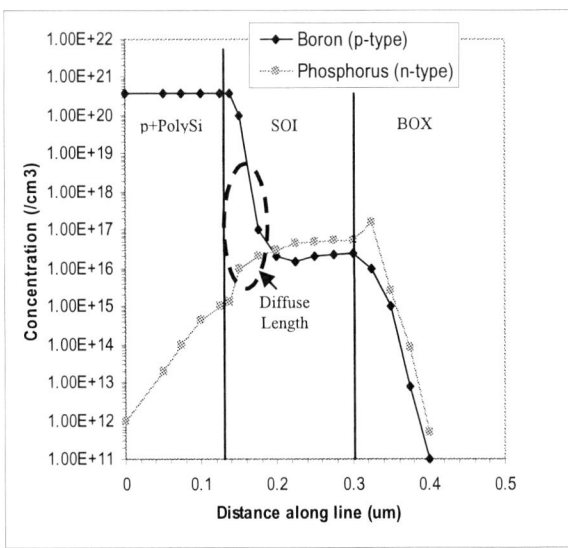

Fig. 3 Transistor vertical doping profiles

Device Characteristics

By using ATLAS DC solutions, the transistor's Gummel plot is obtained (Figure 4) at $V_{CE}= 2.0V$ and $V_{BE}= 0.4 - 1.5V$. The emitter area is taken as 0.36 μm^2. I_C equal to 0.2 x $10^{-8}A/\mu m$ and 3 x $10^{-4}A/\mu m$ and I_B equal to $6x10^{-10}A/\mu m$ and $6x10^{-7}A/\mu m$ at $V_{BE}=0.6V$ and 0.8V, are exhibited respectively. Consequently, a gain of about 350 is obtained at $V_{BE}=0.6V$. I_C saturates at $2x10^{-4}A$, due to high series collector resistance and kirk effect.

This effect is minimized in the second transistor simulation by increasing the low-doped collector concentration from $1x10^{17}cm^{-3}$ to $5x10^{17}cm^{-3}$ and base region to $2x10^{18}$ cm^{-3} with $W_B \approx 0.12\mu m$. This resulted in a higher I_C saturation at $7x10^{-4}$ $A/\mu m$ and a larger gain (1000) (Figure 5) at the cost of decreasing the E-C breakdown voltage ($BV_{CEO} \approx 4V$) obtained from the output characteristics of the device.

Figure 6 shows the cut-off frequency, f_T obtained from y-parameter measurements at frequency 10 MHz - 1000 GHz. A peak f_T (f_{MAX}) of 21 GHz at $I_{CE}=6.0$ μA is observed for the simulated transistor. With f_{MAX} of 21 GHz and gain of 350 – 1000, an excellent process and device simulation characteristics of the lateral bipolar transistor have been accomplished.

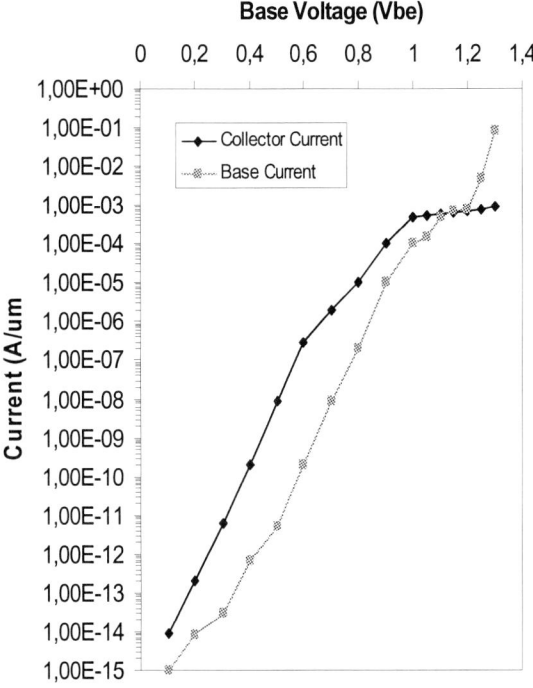

Fig. 5. Gummel plot of transistor at V_{CE}=2.0V and V_{BE}=0.4-1.5V with N_{AB}= 2x10^{18}cm^{-3}, N^-_{DC}=5x10^{17}cm^{-3} and β=1000 at V_{BE}=0.6V

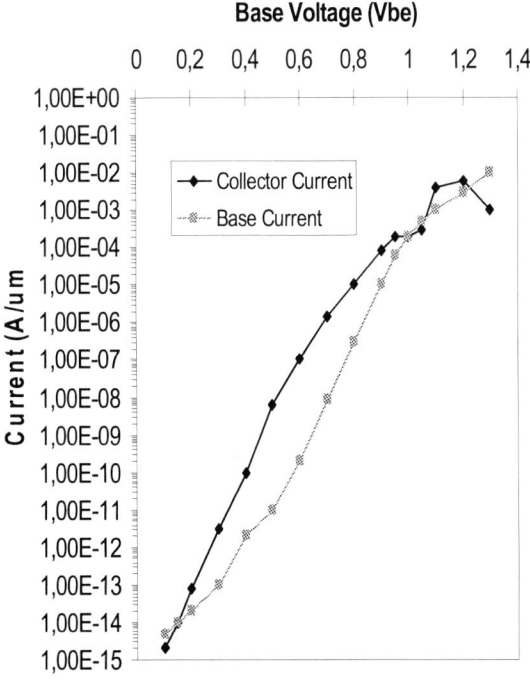

Fig. 4. Gummel plot of transistor at V_{CE}=2.0V & V_{BE}=0.4-1.5V with N_{AB}= 3x10^{17}cm^{-3}, N^-_{DC}=1x10^{17}cm^{-3} and β=350 at V_{BE}=0.6V

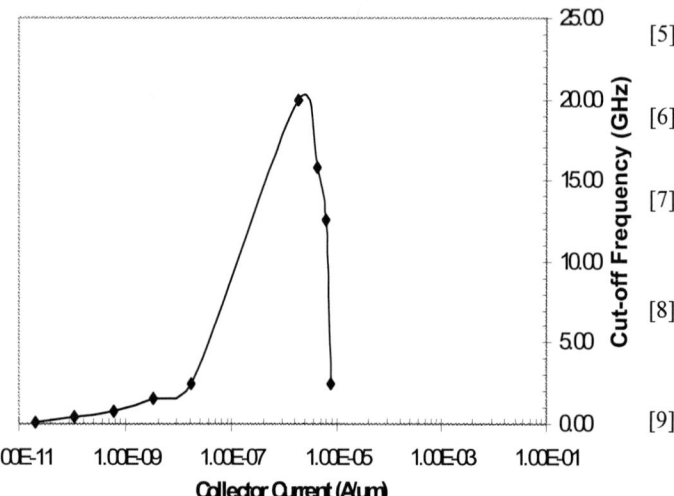

Fig. 6. f_T-I_{CE} (f_{TMAX}) characteristics of transistor at V_{CE}=2.0V

IV. CONCLUSION

The simulation of a high performance lateral bipolar transistor on Thin Film Silicon-On-Insulator (TFSOI) has been successfully carried out. Both process and device simulations for lateral bipolar transistors are carried out in order to understand the device performance. In process simulation, reasonably good doping profiles have been successfully accomplished. In device simulation, the current-voltage (I-V), typical output characteristics and frequency performance of lateral bipolar transistor have been obtained. With increasing N_{DC}^- with a factor of less than 5 (5×10^{17}cm^{-3}) and reducing N_{AB} by a factor of less than 10 (2×10^{18}cm^{-3}) the I_{Csat} has been pushed to occur at 7×10^{-4}A with V_{BE}=1.1V. Thus, a higher gain of 1000 is obtained with a decreasing value of $BV_{CEO}\approx$ 4.0V. However, for both cases the f_{MAX} of 21 GHz has been obtained and a gain of 350 up to 1000 is observed respectively.

REFERENCES

[1] H. Nii et. al, " *A Novel Lateral Bipolar Transistor with 67 GHz f_max on Thin-Film SOI for RF Analog Applications*", IEEE Transactions on electron devices, vol.47. no.7, July 2000

[2] S. Parke et. al, "*A Versatile, SOI BiCMOS Technology with Complementary Lateral BJT's*", 1992 IEDM Technical Digest, p.92-453

[3] M. Kumar et. al, " *A Simple, High Performance TFSOI Complementary BiCMOS Technology for Low Power Wireless Applications*", IEEE Transactions on electron devices, May 2001

[4] A. Tamba et. al, " *A Novel CMOS- Compatible Lateral Bipolar Transistor for High-Speed BiCMOS LSI*", 1990 IEDM Technical Digest, p.90-395

[5] R. Dekker et. al, " *An Ultra Low Power Lateral Bipolar Polysilicon Emitter Technology on SOI*", 1993 IEDM Technical Digest, p.75

[6] G.G. Shahidi et. al, " *A Novel High-Performance Lateral Bipolar on SOI*", 1991 IEDM Technical Digest, p.91-663

[7] B. Edholm et. al, " *Very High Current Gain Enhancement by Substrate Biasing of Lateral Bipolar Transistors on Thin SOI*", 1993 Elsevier Science Publishers B.V, p.379-382

[8] R. Gomez et. al, " *On the Design and Fabrication of Novel Lateral Bipolar Transistor in a Deep-Submicron Technology*", Elsevier Science, September 1999

[9] B. Edholm et. al, " *A Self-Aligned Lateral Bipolar Transistor Realized on SIMOX-material*", IEEE Transactions on electron devices, vol. 40, no.12, December 1993

[10] Silvaco International, "ATLAS and ATHENA user Manual DEVICE SIMULATION SOFTWARE"

Weak point and improvement of CMOS Schmitt Trigger Circuit used in Microcontroller about ND-mode ESD

[1]Jae-Seong Jeong, [2]Jung-Min Lee, and [1]Sang-Deuk Park

[1]Development Quality Assurance Group, CS Management Center
[2]Air Conditioning Digital Control Group, Digital Appliance Division
Samsung Electronics CO., LTD
416, Maetan-3Dong, Yeongtong-Gu, Suwon City, Gyeonggi-Do, Korea
E-mail : semi.jeong@samsung.com Tel. 82-31-277-7801 Fax. 82-31-200-2165

Abstract   **In this study, We investigated weak point and improvement about ND-mode ESD of CMOS Schmitt trigger circuit embeded in Microcontroller. Junction spiking conditions on NMOS of the CMOS Schmitt trigger circuit were Vcc Common mode, ND-mode 1.4kV, and 0.8~1.2 sec zap interval (pin to pin). Failure mechanism by LNPN action formed in CMOS Schmitt trigger circuit was reproduced. We have identified Root Cause and improved circuits to achieve ESD damage free.**

I. INTRODUCTION

Electrostatic Discharge/Electrical Overstress (ESD/EOS) among failure cause of Commercial Integrated Circuits (ICs) have occupied major failure mode together with No Defect Failure (NDF), Fab. error. Therefore, Electrical stress immunity reliability test such as ESD and Latchup among VLSI reliability test are very important. Also, to make the design more robust and reliable, there should be a process that these test results should be feedback to R&D team and reflected in the development stage to remove all possible root causes. If it is failed to consider enough margin for ESD/EOS during the development reliability test period, it is most likely expected that the ESD/EOS margin will be overlapped with Failed or bad margin area from the margin distribution in Production reliability and User reliability stage. Currently, the ESD models applied are Human Body Model (HBM), Machine Model (MM), and Field-Induced Charged-Device Model (FICDM).

Figure 1 shows the pin combination of HBM and MM test, and it consist of GND common, Power Common, I/O to I/O.

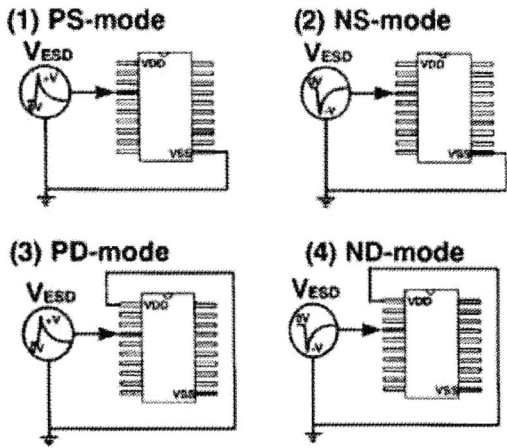

Figure 1. ESD Combinations

For the main controller of System Air conditioner, Commercial 16-bit flash-based microcontroller was applied. To identify ESD weak point both System level and Board level, we have used Zap gun (ESS-200, NoiseKen) and measured ESD immunity. As a result of ESD test, malfunctioning was observed in Negative zap. On the other hand, the Board level failure analysis produced a shortage between microcontroller GND and VDD.

We also conducted a failure analysis for Microcontroller ESD failure during the Usability-Reliability test. Specifically, to find out if the damage was mostly found in NMOS which is part of CMOS Schmitt trigger circuit NMOS in Microcontroller. From the reproduced test result, we have drawn a failure mechanism. From the failure mechanism, we have identified Root Cause and improved circuits to achieve ESD damage free.

0-7803-9730-4/06/$25.00 ©2006 IEEE

II. EXPERIMENT

Reproducing damage on the CMOS Schmitt trigger circuit was conducted at Component level. Basic characteristics of 16-bit flash-based microcontroller are as follows. Operating clock (PLL clock) can selected from divided-by-2 of oscillation, one to four times or six times the oscillation (@ oscillation of 4 MHz, 4 MHz to 16 MHz or 24MHz); Minimum instruction execution time of 42 ns (@ oscillation of 4 MHz, six times the PLL clock, operation at Vcc of 5.0 V); CPU addressing space of 16 Mbytes; Internal 24-bit addressing.

The testing equipments and conditions are as follows : Test equipment (Zapmaster 7/4, ThemoKeyteK), ESD test standard (JEDEC standard), and Test sequence (Low voltage 1.0kV ~ high voltage 4.0kV, increasing 500V step by step while applying ESD). ESD wave at 2000V is shown in figure 2. Pin zap interval was 2 seconds.

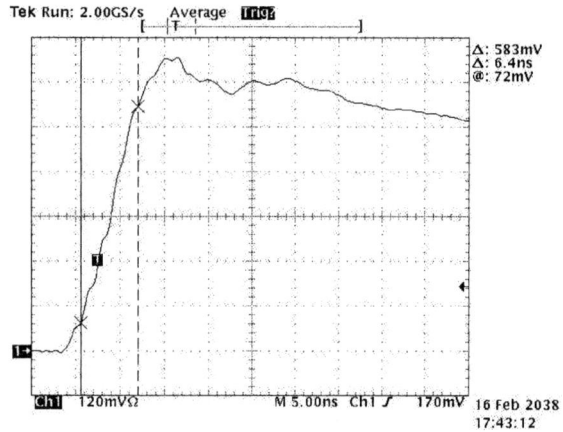

Figure 2. ESD Waveform HBM 2.0kV.
(Ips=1.45A, Tr=6.4ns, Td=148ns Ir=0.24A)

III. RESULTS AND DISCUSSION

The ESD reliability test results of Microcontroller satisfied ±4.0kV class III. With the method above, the damage on the CMOS Schmitt trigger circuit was not repeated and therefore, we could identify main factors and set Zap interval, Common mode, and leakage margin as variables. At last, we could reproduce junction spiking on the CMOS Schmitt trigger circuit NMOS with the condition of Vcc Common mode, ND-mode 1.4kV, 0.8~1.2 sec zap interval (pin to pin).

Fig. 3~5 presents failure analysis. The junction spiking damage can be easily observed in the Drain of NMOS of CMOS Schmitt trigger circuit.

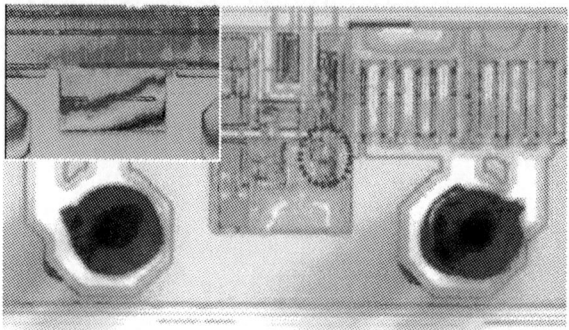

Figure 3. Oscillation region after Top Metal etching.

Figure 4. NMOS Junction spiking in CMOS Schmitt trigger circuit.

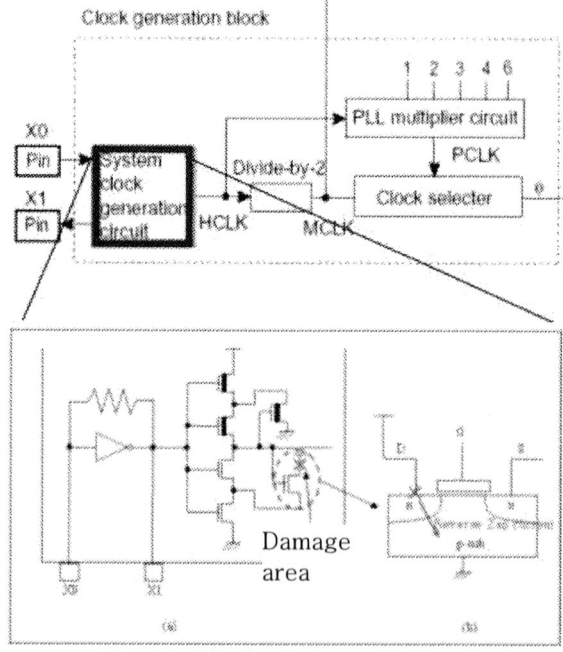

Figure 5. CMOS Schmitt trigger circuit block in Microcontroller and Transistor area damaged in CMOS Schmitt trigger circuit.

The fig. 9 shows Top and Cross-section

view of Failure site. From NMOS Drain connected to VDD and Source of another NMOS, LNPN action can possibly occur. When VDD was set as common and GND & I/O pins were negative zapping, Avalanche breakdown was observed from LNPN action. At the time, current crowding occurs in the Drain which is connected to NMOS VDD. In addition, electric potential of GND node dropped as much as zapping voltage. Therefore, the P-well which is equivalent LNPN Base and neighboring NMOS Source dropped to very low level voltage against Collector potential. In this case the voltage difference is greater than the breakdown voltage, it will be junction breakdown. It can explained as bandgap structure of fig. 6.

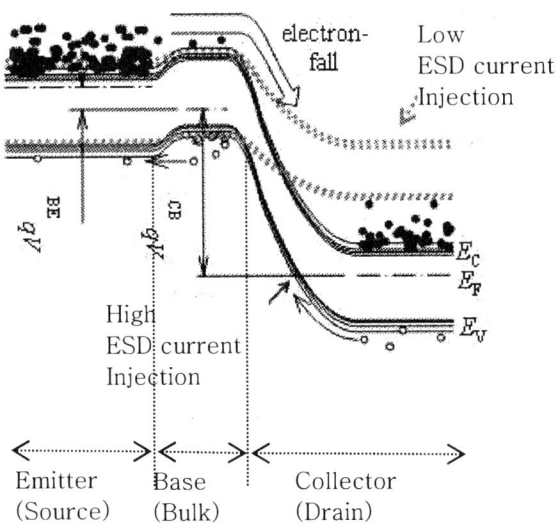

Figure 6. Bandgap structure of LNPN action by ND-mode ESD.

In case of this CMOS Schmitt trigger circuit, the distance of Area B in fig. 9 is narrow and junction breakdown voltage was very low, which eventually enables Area C LNPN to be 'ON'. When negative ESD current is applied on VSS and I/O, over current flows through LNPN of C. Then, junction breakdown occurs near Collector (Drain) and physically junction spiking occurs. It was believed that the distance of B is narrow enough to concentrate approximately 66% of current comparing with circuit area other than Oscillator circuit area. Based upon this analysis, it was believed that the distance factor between LNPN Collector and Emitter presents how junction breakdown depends upon zap interval time.

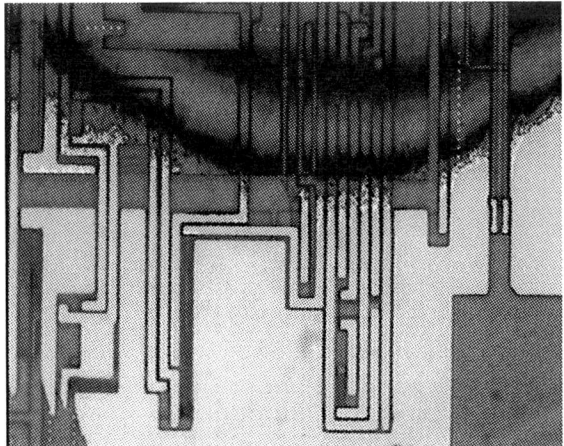

Figure 7. Layout of CMOS Schmitt trigger circuit.

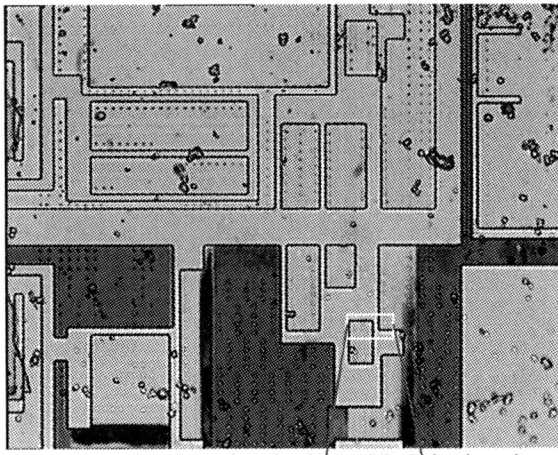

Figure 8. Transistor level of CMOS Schmitt trigger circuit.

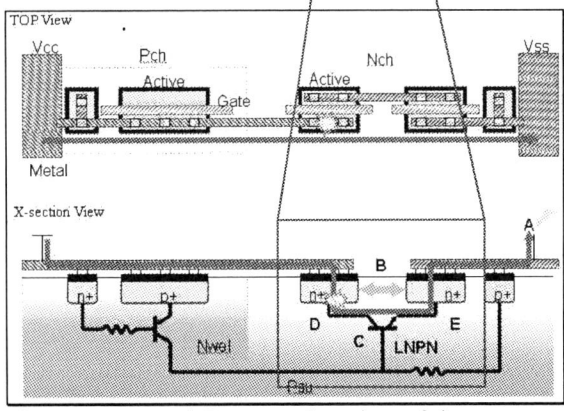

Figure 9. Top and Cross-section view of damage area in CMOS Schmitt trigger circuit and LNPN action area.

With the reproduction and failure mechanism, we could have found root cause of CMOS Schmitt trigger circuit failure and tried an improvement method as Fig. 10 and 11. To avoid direct connection of NMOS VDD, metal in A area was removed and implemented current controlling method by expanding resistance of F.

569

The improved sample was made with FIB. The Component ESD immunity test proved that there was no abnormalities upto ±4.0kV and ESD immunity changes from Zap interval was also not observed and no Side effects were found.

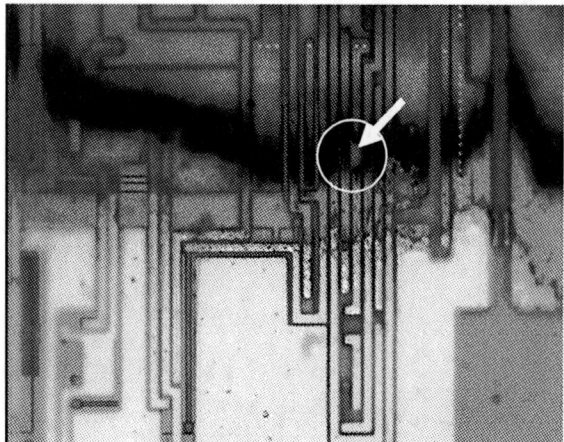

Figure 10. Layout of CMOS Schmitt trigger circuit aftere Improvement.

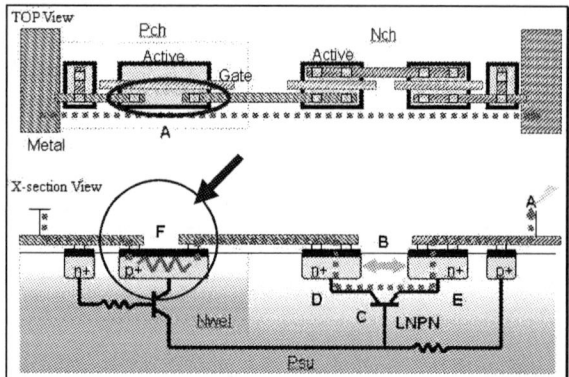

Figure 11. . Top and Cross-section view of CMOS Schmitt trigger circuit after Improvement.

IV. CONCLUSIONS

In this paper, CMOS Schmitt trigger circuit has proven the damaged mechanism and its root cause in ND-mode ESD condition from reproduction and improvement. When we have enough distance between NMOS and NMOS to form LNPN, we could observe avalanche breakdown with LNPN action. The increase of ND-mode ESD also increased LNPN avalanche breakdown and resulted in junction spiking. Current controlling method by expanding resistance using PMOS improved ESD level effectively. This improving method enabled risk free in terms of expenses or financially by changing metal path only, not transistor location.

REFERENCE

[1] Ming-Dou Ker, Tung-Yang Chen, Chung-Yu Wu, and Hun-Hsien Chang "ESD Protection Design on Analog Pin with Very Low Input Capacitance for High-Frequency or Current-Mode Applications " IEEE Journal of solid – state circuits, Vol. 35, No. 8, August 2000.

[2] IEC-61000-4-2 Electromagnetic Capability (EMC) Test.

[3] JESD22-A114D Electrostatic Discharge (ESD) Sensitivity Testing Human Body Model (HBM)

Integration of 8051 With DSP in Xilinx FPGA

A.J. Salim, M. Othman and M.A. Mohd Ali, *Member, IEEE*
VLSI Design Centre
Block 2, Faculty of Engineering
Universiti Kebangsaan Malaysia
43600 UKM, Bangi, Selangor, Malaysia
Email: shaj@streamyx.com

Abstract **This paper describes the integration of an 8-bit 8051 processor core with a general purpose 16-bit fixed-point digital signal processor core using the Xilinx Virtex-II FPGA. The modified IP cores were also integrated with RAM and ROM in one FPGA chip. The chip was programmed to run an application and tested to successfully work at a maximum clock speed of 80MHz. The chip was designed using Verilog and developed in the Xilinx ISE Version 6.3i. Simulation was done in the ModelSim XE II Version 5.8c. The program development and circuit implementation of the chip demonstrated herein that simple ubiquitous tools like Visual Basic, a spreadsheet program, an 8051 assembler, Xilinx ISE and ModelSim can actually realized any moderate size IC design.**

I. INTRODUCTION

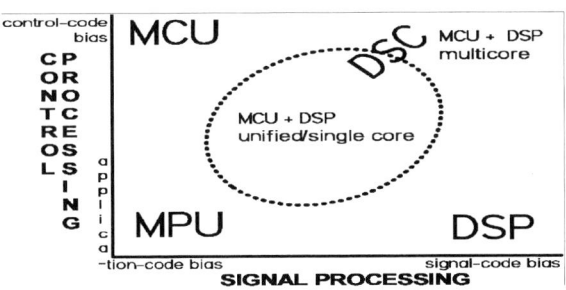

Fig. 1 Processor category mapping

The integration of a microcontroller (MCU) and a digital signal processor (DSP) in one integrated circuit (IC) chip gave birth to a new processor family called the digital signal controller (DSC). The integration is of paramount importance as it combines the unique control ability of the MCU and the inherent number-crunching prowess of the DSP. At present, there are four main categories of processor chips available in the market: the microprocessor (MPU), the MCU, the DSP and the DSC. Fig. 1 shows the mapping of the processors according to their control and signal processing capabilities. The DSC is further divided into two categories, either having a unified single core or

multicore processors in one IC package.

The presence of the DSC is relatively new, as compared to the other three processors. Given time, it will find its own niche in the fast moving microelectronic industries as what has similarly happened to the microcontroller which historically evolved its origin from the microprocessor. Some of the major manufacturers of DSC are Microchip [1], Texas Instrument [2] and Freescale Semiconductor [3].

In some control-oriented applications, usually a dedicated microcontroller is used and if some signal processing is also required a DSP is then tagged on to the system. This design strategy will obviously requires more board space, additional component cost, needs more logistical control and possibly less reliable due to the bigger number of components used.

On the other hand, if a single DSC chip is used, a tighter coupling of control and signal processing would be obtained besides the other apparent advantages of reduced board space, cheaper component cost and lower system power consumption.

The presence of DSC will generate more interest in the use of signal processing algorithms in embedded control applications such as in digital-motor control, robotics, computer system (such as hard disk, modems and all-in-one printer) and power management products (uninterruptible power supplies and lightings).

Incidentally, 8051 has been one of the most popularly used microcontroller in many embedded system products since its first introduction in 1980's by Intel. The multicore DSC designed here intends to enhance and increase the capabilities of the present 8051 MCU.

II. DSC CHIP ARCHITECTURE

The block diagram of the 8051-based DSC chip design example is as shown in Fig. 2. The design uses Verilog and developed in the Xilinx Integrated Software Environment (ISE) Version

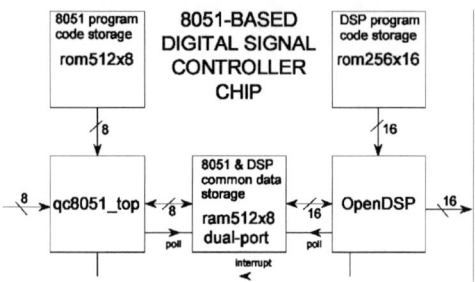

Fig. 2 Block diagram of DSC chip

6.3i. The simulator used is ModelSim XE II Version 5.8c which is integrated as part of the Xilinx ISE. The amount of random access memory (RAM) and read only memory (ROM) in this design exercise is arbitrary and can be increased to its maximum depending on the type of field-programmable gate array (FPGA) being used. Q8051 can support up to a maximum of 64k-bytes of ROM and 64k-bytes of RAM independently. The OpenDSP can support a maximum of 4kx16-bit word ROM and 8Mx16-bit word of RAM. The RAM and ROMs were created using the Xilinx Core Generator System for a proper synthesis to match with the two processor cores in one single integrated circuit chip.

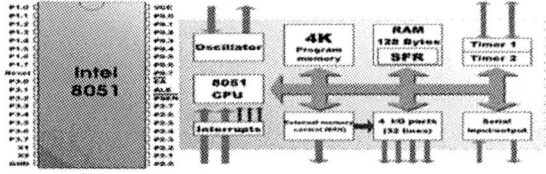

Fig. 3 Standard Intel 8051 MCU

The Q8051 is a standard soft core from QuickcoresTM [4] and is similar to the standard Intel chip (Fig. 3). Originally, sometimes in mid 2004, Q8051 soft core was given as a free developmental intellectual property (IP). However, towards the end of 2004, the company upgraded the IP and sells only the licensed one and the free one is taken off the website. The free IP has lots of bugs when it was used in the integration exercise, especially with the control signals for the external RAM and ROM. Some of these errors were corrected and some signals were modified for the integration process. Q8051 has a single-clock, three-stage instruction pipeline design architecture in comparison to the standard 8051 which requires at least a 12-clock cycle to execute a single-byte instruction. It is a general-purpose processor for control-oriented instructions together with the ability to do bit and data manipulations. Q8051 has fast interrupt handling capability which normally a DSP would not have. This interrupt is necessary as a normal 8051 will have integrated peripherals on its chip to cater for many services. Ports 0 and

2 are available eventhough external memories are being used. In a standard 8051, these two ports are usually taken up for use in the addressing of the external memories.

The 16-bit fixed-point DSP soft core (OpenDSP) is an open IP developed by a research group at a Spanish university [5] and was modified for use in this design. It was meant for small DSP applications and designed to work at 40MHz. In this design, it was modified with the addition of memories and expanded input-output (IO) ports. The original IO was 128 address ports and in this design integration, was modified to cater for 8 million (2^{23}) addresses. Logic shifting facility was added in replacement of the original logic NOR operation. The size of the sixteen internal registers have been modified to 16-bit from its original size of 24-bit for easy interfacing with the memory elements. All registers can function as accumulators and usable in all operations.

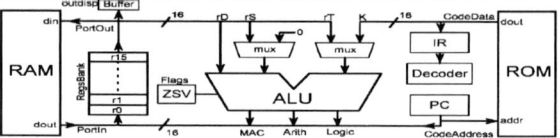

Fig. 4 Block diagram of OpenDSP

OpenDSP is a reduced instruction set computer (RISC) processor and has a Harvard type of architecture (see Fig. 4). The program is stored in the ROM which goes from 256x16-bit up to a maximum of 4kx16-bit word. Data can be stored in the RAM up to 8Mx16-bit word. The last internal register, r15, is used as the 64k-page RAM/IO address. The original port address 0-127 is used as the page pointer, in effect, this expands the IO addresses to 128 pages of 64kx16-bit locations. Page pointer 1, location FFh (100FFh) is modified to be the 16-bit output port address of OpenDSP. The interrupt output pin is reserved at page pointer 2 and location FFh (200FFh)

There are 35 basic instructions and all can be completed in only one instruction cycle (four clock cycles). It can multiply and accumulate (MAC) in normalized fixed-point arithmetic in one instruction cycle. The 16-bit arithmetic logic unit (ALU) has eight different operations in logic, arithmetic and conditional assignments.

III. DSC CHIP DEVELOPMENT

As seen from Fig. 2, the two cores are linked through a common dual-ported RAM of size 512x8-bit. The Q8051 addresses this memory as an 8-bit word while the OpenDSP addresses it as a 16-bit word (256x16-bit).

The development of the DSC was done in three stages. For the first stage, the Q8051 was functionally simulated on its own with some corrections and modification done. It was then synthesized in the Xilinx FPGA to confirm that it can be implemented without errors.

For the second stage, the OpendDSP was functionally simulated independently with some modifications done to suit certain requirements such as the integration of RAM and ROM, expanding IO addresses and adding shifting facility. Several programs were written in assembly language, such as convolution and Fast Fourier Transform (FFT and its inverse) and successfully implemented in the FPGA without errors up to a clock operation of 80MHz. This work has been published and is available online [6].

In the third stage, the integration of the two cores was done with the inclusions of ROMs and RAM in one FPGA chip. A communication protocol was developed to test the interaction of the two cores. Simulation and test results showed that the integration exercise was successful. This work has been fully described and published elsewhere [7].

IV. Q8051 PROGRAM DEVELOPMENT

To further test the capability of this DSC chip, a moving average filter was programmed to demonstrate a working application example. The Q8051 program was developed using the free software integrated development environment
(IDE), called UVI51 from another Spanish university group [8]. The interface is very user friendly and integrated with the editor, assembler, logic analyzer and simulator facilities.

A converter program was developed to convert the Intel hexadecimal 8051 machine code to the Xilinx memory file format (.coe) and the ModelSim simulator file format (.mif). Further detail may be obtained in [7].

The moving average filter program interface for the Q8051 is as shown in Fig. 5. The program starts with the setting of the number of samples used, setting the size of the moving data window, and initializing the RAM pointers. It then goes on filling up empty RAM locations with the data stored in the ROM for the first data window. It will then write a non-zero value in a reserved RAM location called poll_RAM, indicating to OpenDSP that data is ready for processing. Q8051 will then monitor another reserved RAM location called last_free, to see whether there are empty buffers to fill up for the next new data. If there are empty

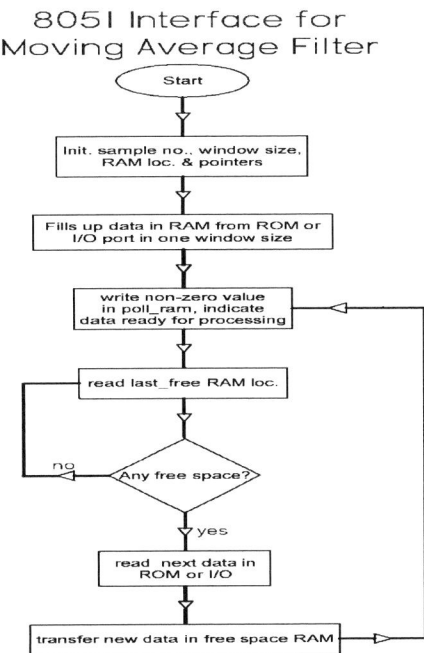

Fig. 5 Q8051 program flowchart

spaces, it will transfer new data from ROM into the RAM data buffers. It will then indicate to OpenDSP that new data is ready for processing by setting a non-zero value in poll_RAM. This process will repeat itself forever.

V. OPENDSP PROGRAM DEVELOPMENT

As the earlier OpenDSP version 1.02 has no software development tools from the authors, a somewhat unorthodox technique has been developed to program the processor.

Using Visual Basic (VBasic), programs were developed to imitate closely the behaviour of the OpenDSP hardware. Using the high-level programming facilities provided, programs can be easily developed and edited. Other facilities like the step mode operation, breakpoints attachment, watch expression and conditional break with logical checks or expression changes help a great deal in debugging. A highlight of all these, is that, at runtime, any overflow will be checked by the compiler, which the real hardware will never indicate. Once the simulation in VBasic is done, an assembly and a machine code format is written in a spreadsheet program. One advantage of using a spreadsheet program is that the conversion of machine code (.coe format) to the binary code format (.mif) can be done automatically using the built-in formulae.

Besides the fast and easy code generation, inserting and deleting instruction lines will not disturb the program memory pointers, especially in a 'jump', 'goto' and 'call' statements. Of course,

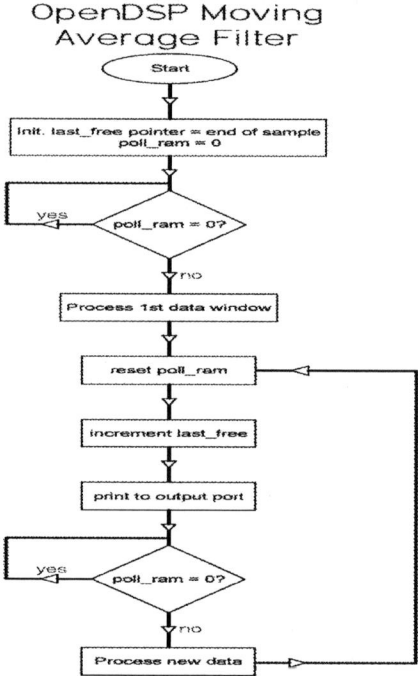

Fig. 6 OpenDSP program flowchart

referenced label names and correct formulae must be used. Copy and paste facility makes editing and adding instruction lines an easy task especially with the ones containing the formulae.

The moving average filter program for the OpenDSP is as shown in Fig. 6. The OpenDSP core does the number crunching routine. It will first initialize the RAM location last_free to the very end of the sample data. Then it will check the RAM location poll_RAM to see if data is made ready by Q8051 core. If it is ready, it will process the first data window. Then it will increment the last_free pointer. It will then send the processed data directly to its own 16-bit output port. This strategy was taken to minimize the processing time taken if OpenDSP were to send the data to the Q8051 and then Q8051 would send it to its own 8-bit output port. Besides that, in future, if more accuracy is required the 16-bit port would have been the better choice than the 8-bit port. OpenDSP will then check RAM location poll_RAM to see whether new data is available. If data is ready, it will process the new data, reset poll_RAM, increment the last_free pointer and then send the result directly to the output port. This is repeated continuously as shown in the flowchart of Fig. 6.

VI. DSC CHIP IMPLEMENTATION

As large memory modules are required in this design, the built-in memory libraries were used. In Xilinx ISE, the memory is added using the IP (CoreGen & Architecture Wizard) facility. In this design, both ROMs of the two cores were chosen as the 'single-port Block Memory', while the RAM uses the 'dual-port Block Memory' LogiCORE modules. When generating the memories, the program code was written in the ROM by feeding in the 'coe' filename. When the memory generation is completed, usually a text file by the extension name of 'mif' will be created automatically by the CoreGen. This file is necessary for use in the ModelSim simulator.

Once this 'mif' file is created, other future simulation for this program code in ModelSim will only need to modify this file. That is, any changes in the program code do not require any new CoreGen generation. This will save time and simulation can be done outside Xilinx ISE when trying to debug and modify the program code in the spreadsheet assembler program. However, before finally synthesizing the designed chip, the ROMs must be regenerated with the new 'coe' contents.

Once the ModelSim simulation results (see sample result in Fig. 8) are correct compared with the VBasic simulation results, the design is ready to be synthesized into the Xilinx FPGA.

VI. RESULTS

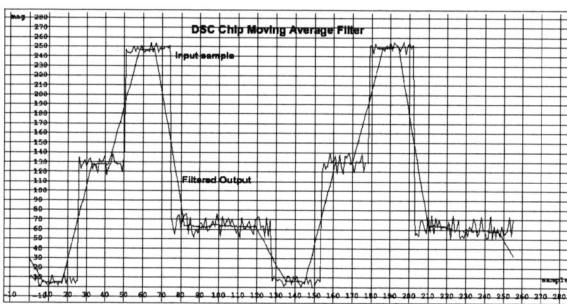

Fig. 7 VBasic "Moving Average Filter" simulation

The graphs in Fig. 7 shows the VBasic 'Moving Average Filter' simulation results for the DSC chip processor. The noisy input data sample was created with the help of the random number generator of a spreadsheet program. The input signal is generated arbitrarily and added with the random noise. The result of the 16-data moving average filter almost effectively removes the noise with some lost of data information especially at the transition regions. The simulation was done using the fixed-point calculation of the integer division facility of VBasic. This will mimic the actual operation of the OpenDSP processor.

This same noisy input data sample is used in the hardware verification by writing permanently the input data in the ROM of the Q8051 processor.

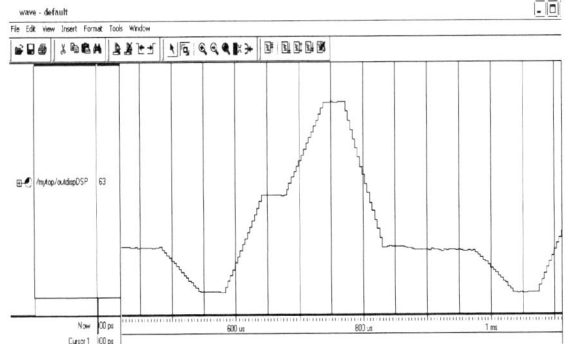

Fig. 8 ModelSim's DSC chip simulation result

Figure 8 shows the ModelSim's hardware simulation result as pertained to the program flowcharts of Figures 5 and 6. The results obtained is exactly similar to the VBasic simulation results by inspecting the numerical values of the ModelSim waveform. ModelSim provides the facility to view these numerical results in analogue format, somehow just like a built-in digital to analogue converter (DAC). Incidentally, this facility is also provided in the logic analyzer.

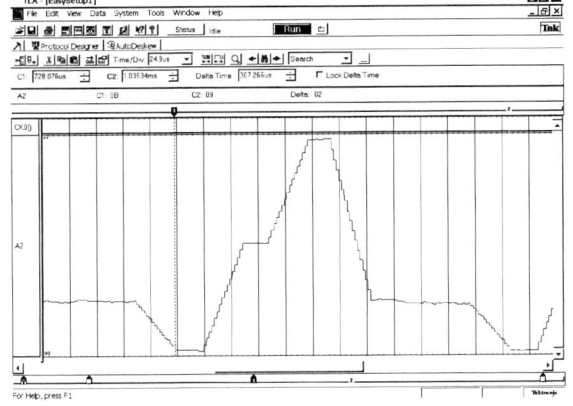

Fig. 9 Logic Analyzer's DSC chip result at 80MHz

The waveform from the logic analyzer shows the exact numerical results as obtained previously in the simulation. Fig. 9 shows an analogue result of the DSC chip running at 80MHz clock speed.

VII. CONCLUSION

The development of an 8051-based digital signal controller chip demonstrated herein that simple ubiquitous tools such as an 8051 assembler, VBasic, a spreadsheet, Xilinx ISE and ModelSim, can easily realized a moderately complex IC design. In other words, a high-end IC design tools is not a prerequisite to actualize a design concept into the hardware reality of a reprogrammable FPGA chip. Nowadys with the advent of more advanced FPGA chips, with multi-million gate counts and on-board system functionality, any abstract digital design ideas can be realised quickly

and without the prohibitive cost and risk factors of an application-specific IC (ASIC) design strategy. With a 'flash-based' type FPGA, even a single chip design outcome, just like the ASIC, is now a reality. FPGA design strategy is very appropriate if the design is only a prototype and requires further modification and improvement.

The FPGA design exercise demonstrated herein shows that the easily obtainable and simple tools are affordable enough to be available in any IC design laboratories of any universities in the whole world. Based on this factor, the FPGA design flow strategy will promote the learning and training of more well-trained IC designers and researchers as evidence from the Taiwanese experience [9].

It has been shown that the integration of Q8051 core, OpenDSP core and its corresponding memories as a single DSC chip has been successful by correctly running the moving average filtering operation at the designed clock speed of 40MHz and up to a maximum of 80MHz.

REFERENCES

[1]   S. Mitra, Vice President, Digital Signal Controller Division, Microchip Technology Inc., "When MCUs and DSPs Collide: Digital Signal Controllers", DSP-FPGA.com, Product Resource Guide 2005.

[2]   "Digital Signal Controllers Enter Volume Production", Electronicstalk Oct. 24, 2005, http://www. electronicstalk.com/news/tex/tex908.html

[3]   V. Anant and B. Hutchings, Freescale Semicondutor Digital Signal Controller Operation, "Digital Signal Controller Applications", DSP-FPGA.com, Product Resource Guide 2005.

[4]   FPGA-Embeddable Microcontrollers with Real-Time JTAG Debugging, http://www.quickcores.com/QC1/QC8051

[5]   S. Pablo, et al, "Project OpenDSP: A 16-bit fixed-point DSP processor for FPGA", Department of Electronics Technology, University of Valladolid, Spain, http://www.dte.eis.uva.es/ OpenProjects/OpenDSP/

[6]   A.J. Salim, M. Othman, and M.A. Mohd Ali, "OpenDSP in Xilinx FPGA", in *Asian Conference on Sensors and the International Conference on New Techniques in Pharmaceutical and Biomedical Research*, K.L., Sept. 2005, pp 185-189.

[7]   A.J. Salim, M. Othman, and M.A. Mohd Ali, "8051-based Digital Signal Controller in Xilinx FPGA", in *International Conference on Computer and Communication Engineering 2006 ICCCE'06*, K.L. May 2006, pp. 1141-5.

[8]   A. del Rio and J.J.R. Andina, "UVI51: A Simulation Tool for Teaching/Learning the 8051 Microcontroller", in *Frontiers in Education Conference,* Kansas City, MO, USA, Oct. 2000, pp. F4E/11-F4E/16 vol. 2.

[9]   Y.T. Chang, et al,"FPGA Education and Research Activities in Taiwan", in *Proceedings of the 2002 IEEE International Conference on Field-Programmable Technology*, Dec. 2002, pp. 445-8.

Increased Capacitance Density with Metal-Insulator-Metal - Metal Finger Capacitor (MIM-MFC)

Kalavathi Subramaniam,Student Member,IEEE, Albert Victor Kordesch, Senior Member,IEEE and
*Mazlina Esa, Member, IEEE
Silterra Malaysia Sdn. Bhd.,Kulim, Kedah, Malaysia
*Faculty of Electrical Engineering,University of Technology, Malaysia, Skudai, Johor,Malaysia.
E-mail : kala_subramaniam@silterra.com, mazlina@fke.utm.my, al_kordesch@silterra.com

Abstract – **Capacitors are indispensable in mixed signal and RF applications. In mixed signal applications, capacitors and inductors often take up a very large space on the chip as compared to the digital portion of the circuit. In this paper we propose a new configuration of metal-insulator-metal – metal finger capacitor (MIM-MFC) to increase capacitance density. This capacitor is designed to be area effective, thus it is an attractive choice for mixed signal and RF applications. The capacitors were fabricated using Silterra Malaysia's 180 nm RF CMOS process technology. This paper also provides low frequency characterization measurement data for MIM-MFC. Analysis is made as compared to pure MIM and pure MFC that were fabricated on identical process and measured on the same wafer.**

I. INTRODUCTION

Commercial value for capacitors are important in many applications such as radio frequency circuits, analog ICs, and even in DRAM cells, if a value of 25fF/cell can be achieved [1]. In general, capacitors with good matching properties, small parasitic capacitance, high reliability and low defect densities are desired. Oscillators, filters, mixers, phase-shift networks and analog to digital and digital to analog converters are some instances where the application of capacitors can be observed.

As a result of the continuous improvement in CMOS and RF CMOS processes, the designers are left with a handful of capacitor types to choose from. Metal-Insulator-Metal (MIM) capacitors, Poly-Insulator-Poly (PIP) capacitors, CMOS Varactors and Junction Varactors are the common types of capacitors that are usually offered in CMOS processes. A designer will have to make a choice among these available capacitors based on their desired properties such

as the linearity, quality factor, breakdown voltage, area efficiency, temperature dependence, voltage dependence and sensitivity to process variation [2]. For RF circuits, it will be best to have a capacitor that yields all of the above in addition to high self-resonance frequency, high quality factor, and low frequency dependence. High breakdown voltage in RF designs is also preferred [3].

Recently, the Metal-Finger-Capacitors (MFC) have gained popularity because of their improved Q, low voltage dependence, low temperature dependence, low frequency dependence, low process variation and low cost of fabrication. On another note, the Metal-Insulator-Metal (MIM) capacitors have been around for a very long time and are used by analog and RF designers. The MIM capacitors precede MFC in terms of capacitance density and matching [2].

In this work, combined MIM-MFC capacitors have been developed in the effort to increase capacitance density.

II. DESIGN

The design presented here integrates MIM capacitors with four metal layer metal finger capacitors. Six different sizes of MIM-MFC capacitors have been designed and measured.

In this work, MIM capacitors have been stacked upon metal finger capacitors to increase the capacitance density. It can be seen as having metal fingers underneath a MIM capacitor which is in between metal 5 and metal 6 on a 6LM CMOS process. The top view layout of the design is illustrated in Figure 1 and the cross section in Figure 2. Different sizes of these capacitors have been fabricated using Silterra's industry standard 180nm CMOS 6 layer metal process technology.

0-7803-9730-4/06/$25.00 ©2006 IEEE

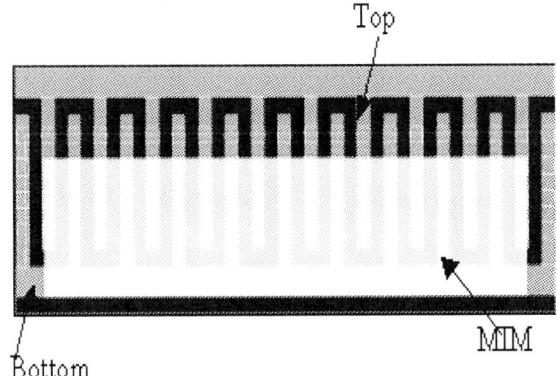

Figure 1 : Top View Layout of MIM-MFC

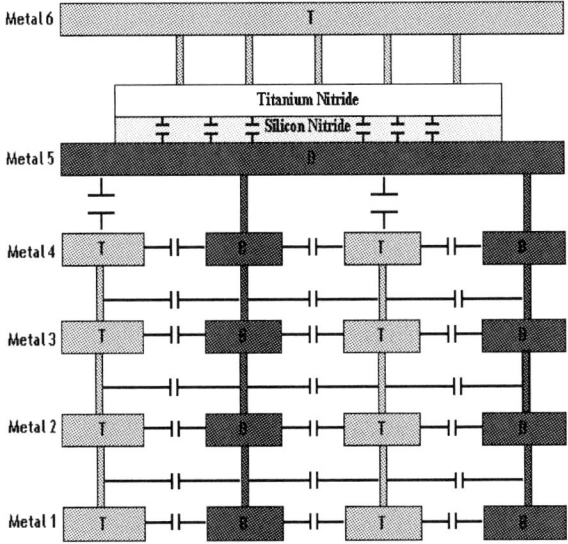

Figure 2 : Cross Section of MIM-MFC

As can be seen in Figure 2, alternate fingers of the metal finger capacitors are connected to the top plate and the others are connected to the bottom plate. The top (metal 6) and bottom (metal 5) plates of the MIM above the fingers are also connected this way. This design utilizes the capacitance provided by the metal interconnects underneath the MIM capacitors. The capacitance between the bottom plate of MIM capacitor and the top plate fingers of the metal fingers are also utilized in this design.

Besides that, the MFC is designed to be stacked. Stacked metal fingers have vias in between them to connect different layers of metal. This type of design has proven to increase the capacitance of MFC as compared to the interdigitated style of designing MFC.

Keeping RF applications for these capacitors in mind, the fingers of the MFC were made as short as 2.5 μm. This is to reduce the equivalent series resistance (ESR) of the metal fingers and further increase quality factor (Q) of the

capacitors. To achieve desired capacitance value, these fingers are connected in parallel. Higher capacitance density is achievable with longer metal fingers' length for which a decreased Q comes as a tradeoff.

For comparison purposes, 5 metal layer MFC and MIM capacitor between metal 5 and 6 were also designed and fabricated using the same technology. Figures 3 (a) and (b) are the top view layout and cross section of the MFC. Figure 4 (a) and (b) illustrate the top view layout and cross section of MIM capacitor.

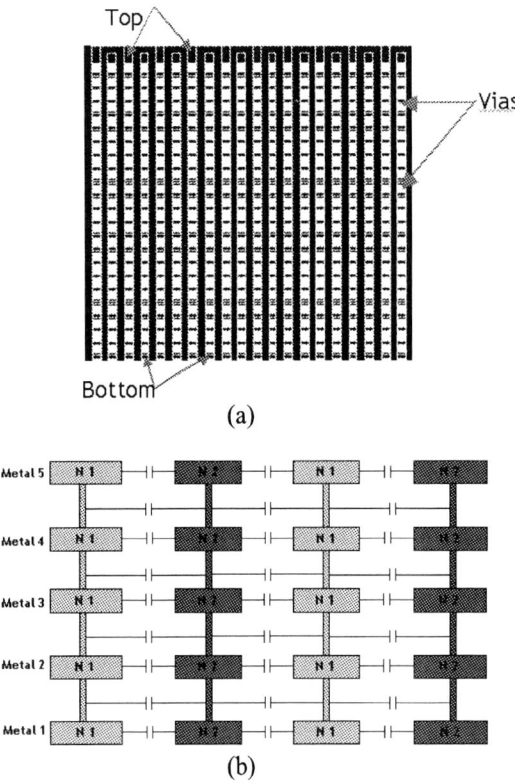

(a)

(b)

Figure 3: MFC (a)Layout (b) Cross Section

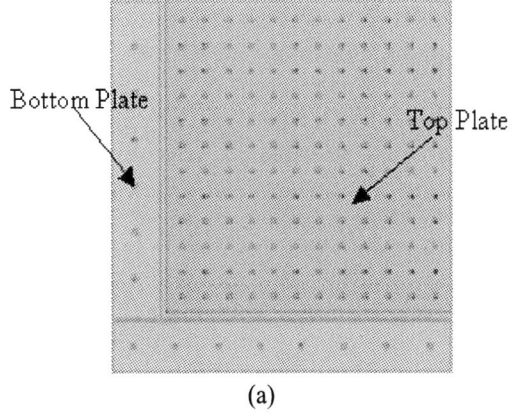

(a)

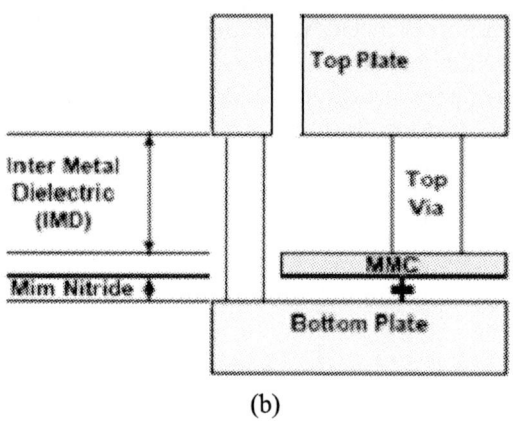

(b)

Figure 4: MIM Capacitor (a)Layout (b) Cross
Section [4]

III. MEASUREMENT

This paper focuses on the low frequency
characterization of the combined MIM-MFC as
compared to MIM and MFC alone. The analysis
includes measurements for capacitance density,
parasitic capacitance, voltage dependence, and
temperature dependence. On-wafer manual
probing was done using Cascade MicroTech
SUMMIT 12K prober.

In this work, an Agilent 4284A LCR meter
was used to measure the nominal capacitance
between the two terminals (top and bottom), C_s
which would determine the capacitance density
of MIM-MFC. Parasitic oxide capacitance from
the top and bottom plates of MIM-MFC to the
substrate, C_{ox} was also measured using this tool.
This meter was also used in the measurement of
voltage dependence of the MIM-MFC. The
measurements were done at 1 MHz, 1Vp-p to
represent the low frequency characteristics of the
MIM-MFC. These measured values are inserted
in the model as low frequency elements.

Thermal characterization in this work covered
the temperature range from – 40 °C to + 125 °C.
Temptronic Thermochuck® TP03200 A was
used in this research to accommodate the
temperature characterization requirements. This
thermal inducing vacuum platform has standard
temperature range of – 65 °C to + 150 °C.

IV. RESULTS AND ANALYSIS

Six different sizes of MIM-MFC have been
designed and fabricated in this work. The
measured data is compared to the data of pure
MIM and pure MFC. Figure 5 illustrates the
capacitance density of MIM-MFC, MFC and

MIM. It can be seen that the capacitance density
of MIM-MFC is about 1.69fF/um^2. This is 46 %
higher than the capacitance density of pure MFC
which is 1.15 fF/um^2. Compared to pure MIM
capacitors that are 1fF/um^2 in Silterra Malaysia's
standard CMOS process, combined MIM-MFC
yields 69 % higher capacitance density.

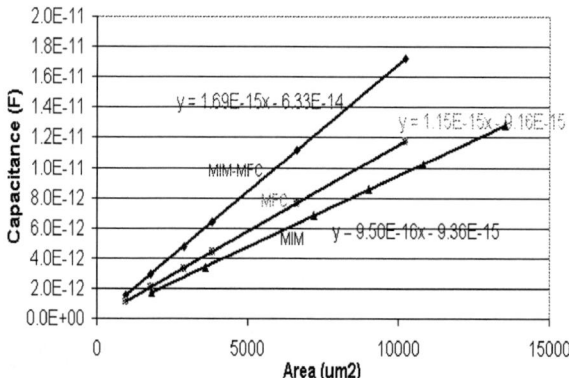

Figure 5: Capacitance Density of MIM-MFC, MFC
and MIM

Comparison is also done on the top and
bottom plate capacitance to substrate for all three
capacitors. This is also the oxide capacitance as it
is the dielectric from top and bottom metals of
the capacitors to the substrate. Top plate to
substrate oxide capacitance (Coxtop) for all three
capacitors are plotted as functions of the nominal
capacitance, Cs in Figure 6. It is seen that MIM-
MFC has higher ratio of Coxtop to Cs compared
to MFC and MIM. MIM-MFC has 1.9% of
parasitic Coxtop to Cs whilst MFC and MIM
have ratios of 1.8% and 0.13% each.

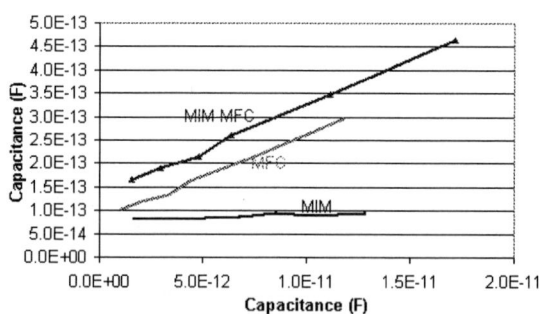

Figure 6: Parasitic Capacitance from Top Plate to
Substrate of MIM-MFC, MFC and MIM

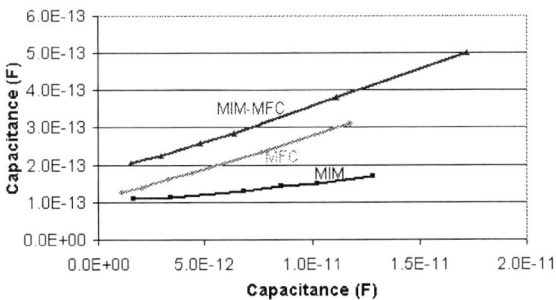

Figure 7: Parasitic Capacitance from Bottom Plate to Substrate of MIM-MFC, MFC and MIM

Figure 7 demonstrate the bottom plate to substrate oxide capacitance (Coxbottom) for MIM-MFC, MFC and MIM as functions of the nominal capacitance. Similar to the Coxtop, Coxbottom is also higher in MIM-MFC followed by MFC. MIM capacitors have the lowest Coxbottom of 0.5%. MIM-MFC has Coxbottom to Cs ratio of 1.9% whilst MFC has a ratio of 1.8%.

The parasitic capacitance (Coxtop and Coxbottom) for MIM capacitors are generally lower because MIM capacitors are usually fabricated near the topmost metal of the process. In this work, the MIMs were fabricated between metal 5 and metal 6. The greater distance of the top and bottom plates of MIM capacitors from the substrate help reduce the parasitic capacitance to ground. As for MFC and MIM-MFC which has fingers all the way down to metal 1, the parasitic capacitance is higher.

It can also be noticed that MFC and MIM-MFC seem to be symmetrical as they have the similar parasitic capacitance from top and bottom plate to substrate which are 1.8% and 1.9%. This is because the existence of fingers in both these capacitors lead them to be symmetrical as fingers for different terminals are designed on the same metal layer. However, in MIM capacitors, there is significant difference between Coxtop and Coxbottom. Coxbottom is higher as the bottom plate (metal 5) is nearer to the substrate and the top plate is shielded by the bottom plate.

Figure 8 is a diagram of voltage dependence of MIM-MFC. Figure 9 and 10 illustrate the voltage dependence of MFC and MIM capacitors respectively. The capacitors were measured from -40V to 40V.

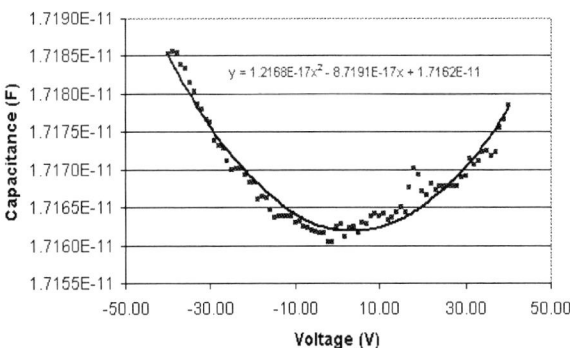

Figure 8: Voltage Dependence of MIM-MFC

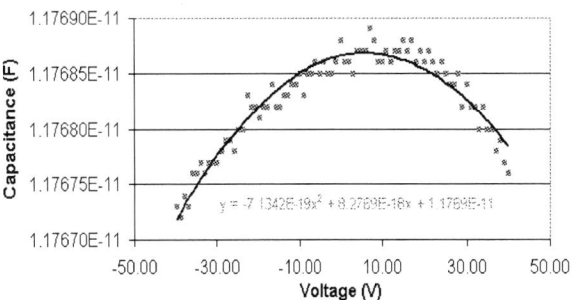

Figure 9: Voltage Dependence of MFC

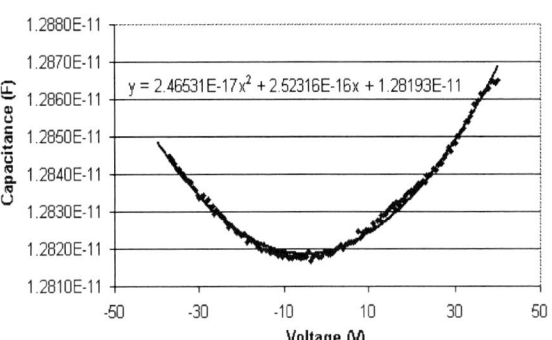

Figure 10: Voltage Dependence of MIM

MIM-MFC's voltage dependence curve resembles the MIM capacitor's voltage dependence. They both have positive 2nd order voltage coefficient whilst MFC has a negative 2nd order voltage coefficient. 1st and 2nd order voltage coefficients of all three capacitors are tabulated in Table 1.

Table 1: Voltage Coefficients of MIM-MFC, MFC and MIM

Capacitor Type	1st Order Voltage Coefficient	2nd Order Voltage Coefficient
MIM-MFC	− 5.0800e-6	7.0900e-7
MFC	7.0327e-7	-6.0618e-8
MIM	1.9683e-5	1.9232e-6

From the table above, MIM capacitor is the most temperature dependent followed by MIM-MFC and MFC. It is observed that MIM and MFC have contradictory patterns of the voltage dependence, and MIM-MFC fell in between these.

Temperature dependence of all three capacitors were also investigated and plotted in Figures 11, 12 and 13. Measurements were done from –40 degrees to 125 degrees Celcius. Capacitances of all three capacitors increase linearly with temperature. Similar to voltage dependence, MFC is the least dependent capacitor on temperature followed by MIM-MFC and MIM. Temperature coefficients of MIM-MFC, MFC and MIM are 2.4217e-5, 9.8742e-6 and 5.6136e-5. MIM-MFC has moderate temperature dependence compared to MFC and MIM.

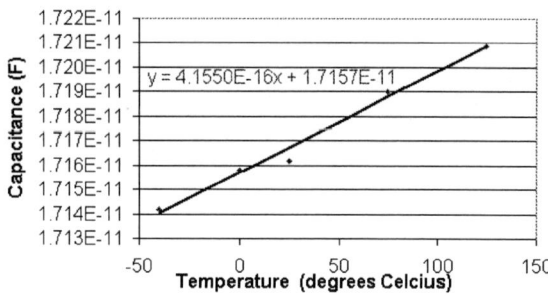

Figure 11: Temperature Dependence of MIM-MFC

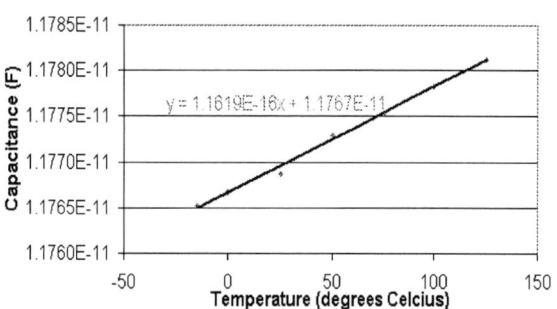

Figure 12: Temperature Dependence of MFC

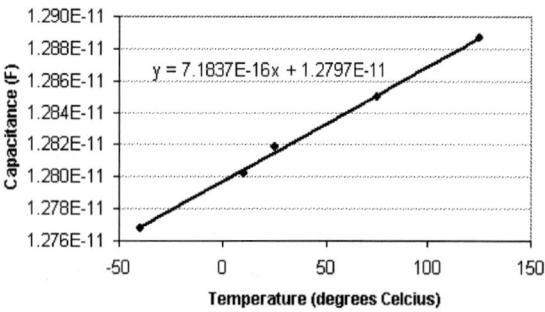

Figure 13: Temperature Dependence of MIM

V. CONCLUSIONS

This paper presented the design and low frequency characterization of a new type of capacitor called the MIM-MFC. This capacitor has increased capacitance of 46% higher than MFC and 69% higher than MIM capacitors processed on an identical process and the same wafer. Low frequency characterization of the MIM-MFC was presented which include the capacitance density, parasitic capacitance, voltage dependence and temperature dependence. The MIM-MFC demonstrated acceptable compensation towards voltage and temperature although the voltage and temperature coefficients were higher than that of MFC. The parasitic capacitance resembled that of MFC as they are both symmetrical capacitors with almost equal number of fingers connected to both terminals.

ACKNOWLEDGEMENT

The authors are grateful to Silterra Malaysia Sdn. Bhd. for the facilities for fabrication and measurement for this work.

REFERENCES

1) Silterra Confidential, "Executive Summary-International Sematech Technology Tranfer", 2002.

2) Lim Q.S.I., Kordesch A.V., and Keating R.A.,"Performance Comparison of MIM Capacitors and Metal Finger Capacitors for Analog and RF Applications", *Proceedings RF and Microwave Conference, 2004. RFM2004. 5-6 Oct 2004.* pp 85-89.

3) Aparicio R. and Hajimiri A., "Capacity Limits and Matching Properties of Integrated Capacitors", *IEEE Journal of Solid-State Circuit*, vol. 37, pp 384-393, March 2002.

4) Goh M.W.C., Lim Q., Keating R.A., Kordesch A.V., Yusman M. Y "Design of Radio Frequency Metal-Insulator-Metal (MIM) Capacitors", *Proceedings 7th International Conference on Solid-State and Integrated Circuits Technology,2004.* Vol.1, 18-21 Oct.2004, pp 209-212

5) Chunqi G., Anh D.M., Zheng Z, and Boyland F.,"A Scalable RF Model of the Metal-Oxide-Metal (MOM) Capacitor", *Technical Proceedings of the 2001 International Conference on Modeling and Simulation of Microsystems,2001*, 482-485.

Considerations on the C-V Characteristics of Pentacene Metal-Insulator-Semiconductor Capacitors

Keum-Dong Jung, Byung-ju Kim, Byeong-Ju Kim, Cheon An Lee, Dong-Wook Park, Byung-Gook Park, *MIEEE*, Hyungcheol Shin, Senior *Member, IEEE*, and Jong Duk Lee, *MIEEE*

Inter-University Semiconductor Research Center (ISRC) and School of Electrical Engineering,

Seoul National University, Sillim-dong, Gwanak-gu, Seoul, 151-742, Republic of Korea

E-mail: windbit@naver.com

Abstract C-V characteristics of pentacene MIS capacitors are obtained with various measurement conditions. High measuring frequency can decrease the measured capacitance due to the slow response of holes. When thick semiconductor is used, accurate C-V characteristics can not be obtained due to the resistance of bulk semiconductor. Bias stress makes positive or negative flat band voltage shift, also complicate accurate C-V measurement. Therefore, to obtain reliable C-V characteristics of organic MIS capacitors, these properties should be considered.

I. INTRODUCTION

Recently, organic semiconductors have been studied widely because of their potential to be used in flexible displays, low-cost RF ID tag, or large-area sensors[1]. Among various organic semiconductors, pentacene has been one of the most frequently used organic semiconductors due to its high field effect mobility. Although various circuits using pentacene thin film transistors(TFTs) have been fabricated in many publications, modeling of pentacene TFTs for exact simulation of the circuits is not mature yet. C-V characteristics of metal-insulator-semiconductor (MIS) structures have been one of useful methods to investigate the properties of a semiconductor[2]. Especially, for the simulators such as SPICE, they become the foundation for the modeling of TFT I-V equation. Therefore, measuring accurate C-V characteristics can be a good starting point for the modeling of TFTs. However, C-V

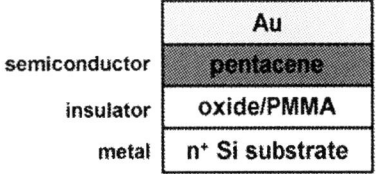

Fig. 1 Cross-sectional view of the fabricated MIS capacitors.

characteristics of organic MIS structures have not been studied sufficiently compared to those of silicon because they are affected by several factors which are not considered significantly in the silicon case. In this paper, considerations on C-V characteristics of pentacene MIS capacitor with its measurements issues such as frequency dependency, effects of semiconductor thickness, and bias stress effects are discussed.

II. EXPERIMENT

Two-terminal MIS capacitors are fabricated as shown in Fig. 1. An n^+-Si wafer with sheet resistance of 10 $\Omega/\square$ is used as the gate electrode and 35 nm-thick thermal SiO_2 with 10 nm dilute PMMA treatment[3] is used as the insulator. The thermal oxide minimizes the leakage current through the insulator, so that reliable C-V characteristics can be obtained. On the insulator, pentacene film of 75 nm is thermally evaporated through shadow mask at the substrate temperature of 80 °C. To make an ohmic contact with the organic semiconductor, the substrate electrode is e-gun evaporated with gold through second shadow mask. C-V characteristics are measured in air with

0-7803-9730-4/06/$25.00 ©2006 IEEE

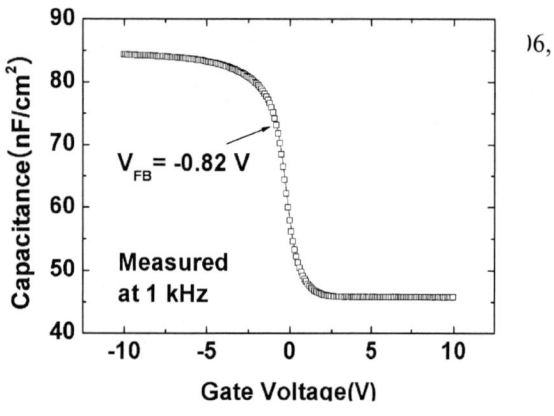

Fig. 2 Typical C-V characteristics of fabricated MIS capacitor measured at 1 kHz.

HP4284A LCR meter. The dc bias voltage of -10 ~ 10 V is applied, and the measuring frequency is either 1 or 10 kHz.

III. RESULTS AND DISCUSSION

Fig. 2 shows typical C-V characteristics of pentacene MIS structure measured at 1 kHz. To obtain reliable measurement results, relatively low measuring frequency is used. Because pentacene is p-type semiconductor, the accumulation and depletion of holes are observed in the C-V characteristics. In accumulation regime when the gate bias is negative, the total capacitance almost recovers the insulator capacitance which indicates that the accumulated holes respond well to the gate small signal. In depletion regime when the gate voltage is near zero to positive, the capacitance decreases rapidly until full depletion occurs in the semiconductor. Inversion of electrons is not observed at any bias because the electron response is too slow relative to the measuring frequency. The flat band voltage(V_{FB}) which separate accumulation and depletion regime is obtained as -0.82 V using the method described in [2]. Typical flat band voltage shift after bidirectional voltage sweep is below 0.5 V.

Fig. 3 shows the change of C-V characteristics when the measuring frequency

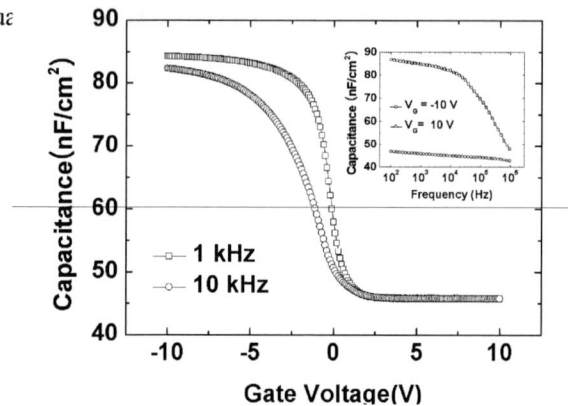

Fig. 3 When measuring frequency becomes higher, the capacitance decreases because the holes can not respond to the gate signal.

becomes higher. The capacitance decreases slightly in accumulation regime when high frequency is used, while the capacitance in depletion regime remains almost the same regardless of the frequency. From this observation, it can be inferred that the insulator is not related to the frequency dependency of C-V characteristics because the effects of insulator should occur the same in both of accumulation and depletion regime. Therefore, the frequency dependency of C-V characteristics has to originate from the semiconductor. In the inset of the Fig. 3, the capacitance in accumulation regime and depletion regime is depicted with the measuring frequency. The capacitance in accumulation regime decreases rapidly when the frequency becomes higher than 10 kHz because the holes can not respond to the gate signal above 10 kHz. The responses of holes are mainly constrained by large bulk resistance of the semiconductor. On the contrary, the capacitance in depletion regime remains almost constant because holes are depleted in the semiconductor and the whole devices can be considered as a capacitor with two different insulators. In this case, there is no resistance which constrains the response of the charges along the capacitor at least in this frequency range.

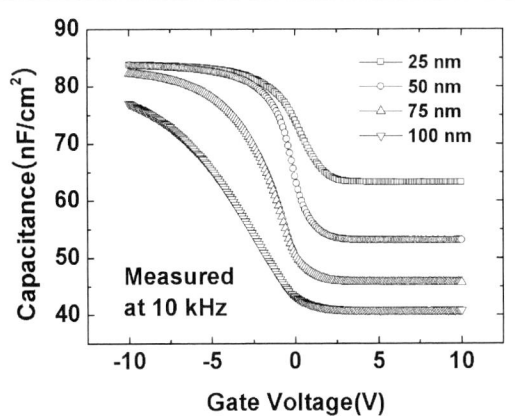

Fig. 4 If different thickness of pentacene is used, the C-V characteristics also changes. In accumulation regime, the delay due to bulk semiconductor changes C-V characteristics. In depletion regime, the change of depletion capacitance with thickness changes C-V characteristics.

If the thickness of the pentacene layer changes, the C-V characteristics also change as depicted in Fig. 4. First, the capacitance in depletion regime changes with thickness because the depletion capacitance of the semiconductor decreases with the increase of the thickness. From the calculation of depletion capacitance for each pentacene thickness, it is considered that full depletion regime is maintained up to the pentacene thickness of 100 nm. Second, the capacitance at $V_G = -10$ V does not recover the insulator capacitance when the thick pentacene layer is used, and the slope of C-V curve between the accumulation and depletion regime becomes gradual. This is because the delay of holes passing through the bulk semiconductor increases with thickness due to the increase of bulk resistance of the semiconductor. Therefore, different from silicon case, it is hard to obtain correct C-V characteristics when too thick semiconductor layer is used.

Fig. 5 shows the effects of gate bias stress on C-V characteristics. To minimize the bias stress effect during the C-V sweep, the voltage interval is increased to 1 V assuring that bias stress time is larger than the measuring time. Both of negative and positive shifts are observed on negative and positive stress, respectively. Larger flat band voltage shifts are observed when larger stress voltages are applied. The origin of these

bias stress effects can be hole or electron trap at the interface, mobile ion in the insulator, dipole moment in the polymer film or carrier injection from gate. For the negative bias stress, hole trap makes negative flat band voltage shift while others makes positive flat band voltage shift. Therefore, the effects of hole trap is considered to be dominant for negative bias stress. On the other hand, for the positive bias stress, electron

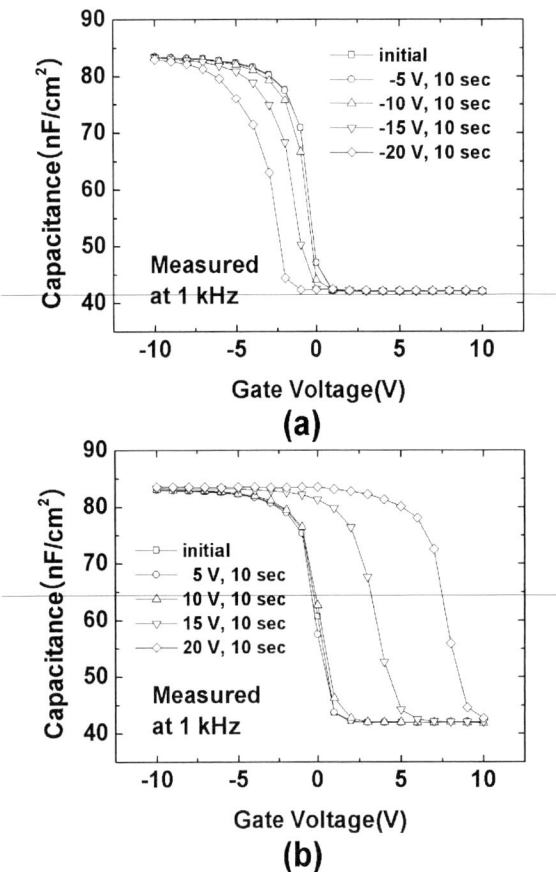

Fig. 5 (a) Negative bias stress makes negative shift of the flat band voltage. (b) Positive bias stress makes positive shift of the flat band voltage. In both case, higher stress voltage makes larger shift.

trap makes positive flat band voltage shift while others makes negative flat band voltage shift. Therefore, in this experiment, the main cause of bias stress effects is either hole or electron traps near the interface, for negative and positive bias stress, respectively. However, the origin and the magnitude of the bias stress effects can be changed with different insulator, organic semiconductor, fabrication method, or surface treatment. To avoid these bias stress effects, the bias range and measuring time should be chosen carefully when measuring C-V characteristics. Bidirectional sweep also helps to check the bias stress effects.

IV. CONCLUSION

C-V characteristics of pentacene MIS capacitors are obtained, and the effects of measuring frequency, the thickness of pentacene layer, and flat band voltage shift due to bias stress are discussed. Compared to silicon, low mobility of organic semiconductor and traps near the interface makes different C-V characteristics. These results can be used to obtain more reliable C-V characteristics which can be a base to make accurate I-V equation model of pentacene TFTs.

ACKNOWLEDGEMENT

This work was supported by "Samsung SDI - Seoul National University Display Innovation Program."

REFERENCE

[1] C.D. Dimitrakopoulos and P.R.L. Malenfant, Adv. Mat., 14, 99, (2002).

[2] E. H. Nicollian, MOS (metal oxide semiconductor) Physics and Technology, Wiley Interscience, 11.

[3] S. H. Jin, J. S. Yu, J. W. Kim, C. A. Lee, B.-G. Park, and J. D. Lee, J. H. Lee, SID 2003 Digest, 1088, (2003).

Characteristics of Serial Peripheral Interfaces (SPI) Timing Parameters for Optical Mouse Sensor

M.K. Md Arshad, U. Hashim, Chew Ming Choo
School of Microelectronic, Kolej Universiti Kejuruteraan Utara Malaysia
Blok A, Kompleks Pusat Pengajian KUKUM
Jalan Kangar-Arau. 02600 Jejawi PERLIS
Tel: 04-9798435   Fax: 04-97984305
E-mail : mohd.khairuddin@kukum.edu.my

Abstract - **In this paper we report the characterizations results of Serial Peripheral Interface (SPI) timing parameters for optical mouse sensor. SPI is an interface that facilitates the transfer of synchronous serial data. It supports two-way communication between mouse sensor and microcontroller. The test setups were used consist of digital oscilloscope, power supply, mouse sensor, test board and personal computer. The SPI timing parameters that were evaluated are read-address data delay, the timing between read and subsequent commands, rise and fall time, hold time, setup time, NCS to SCLK active, SCLK to NCS inactive (read operation) and SCLK to NCS inactive (write operation). The timing parameters results were within the limit as specified in product datasheet.**

Keywords: Serial peripheral interface (SPI), optical mouse sensor.

I.   INTRODUCTION

For years, the standard peripheral for interacting with computers was the mechanical mouse. Douglas Engelbart invented the mechanical computer mouse in the late 1950s [1]. Over the years, mousepads, trackballs, and tablets were developed for use with the mechanical mouse, but computer users complained that mice get dirty, need cleaning, and are imprecise [1]. In late 1999 [2], Agilent Technologies unveiled the first optical mouse, unlike the earlier mechanical models, it has no moving parts. It offers precise tracking, it doesn't need cleaning, and it can be used on nearly any surfaces.

Optical mouse actually uses a tiny camera to take 1,500 pictures every second. Able to work on almost any surface, the mouse has a small, red light-emitting diode (LED) that

bounces light off that surface onto a complimentary metal-oxide semiconductor (CMOS) sensor. The CMOS sensor sends each image to a digital signal processor (DSP) for analysis. The DSP, operating at 18 MIPS (million instructions per second), is able to detect patterns in the images and see how those patterns have moved since the previous image. Based on the change in patterns over a sequence of images, the DSP determines how far the mouse has moved and sends the corresponding coordinates to the computer. The data moved and sends is control by microcontroller that later moves the cursor on the screen based on the coordinates received from the mouse. This happens hundreds of times each second, making the cursor appear to move very smoothly.

The moving and sending data between mouse sensor and microcontroller is known as Serial Peripheral Interface (SPI) [3-4]. The technique is sending or receiving address and data in serial order and bit by bit. There are full duplex and half duplex of communication line involved in SPI. Half duplex is a one way of communication in a period of time while full duplex is a two way of communication in a period of time [5]. SPI Timing is the timing requires in ensuring read and writes operation between an optical mouse sensor and microcontroller is properly functioning.

A basic SPI signal involves Negative Chip Select (NCS), Serial Clock (SCLK) and Serial Data Input and Output (SDIO) for 3-wires [3] of communication line as shown in Figure 1.

Communication between optical mouse sensor and microcontroller is essential to make sure read and write operation in an optical mouse sensor perform accordingly to datasheet specification. There are minimum timing requirements between read and write commands on the serial port in order to deliver and receive data correctly. Failure in timing requirement could lead to failure in cursor movement that display on computer screen.

In this paper we characterize a set of SPI timing parameters according to product datasheet specification of an optical mouse sensor. The test setups consist of digital oscilloscope, power supply, mouse sensor, test board and personal computer were used to evaluate the optical mouse sensor.

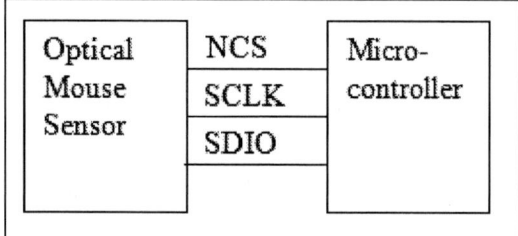

Figure 1: SPI signals for 3-wires communication are NCS, SCLK, and SDIO.

II.   EXPERIMENTAL

A set of SPI timing parameters namely between Read address-data delay (t_{SRAD}), between Read and either Write or subsequent Read commands (t_{SRW} and t_{SRR}), Serial Input and Output Data (SDIO) rise time, SDIO fall time, SDIO hold time, SDIO setup time, Chip Select Input (NCS) to Serial Clock Input (SCLK) active, SCLK to NCS inactive (read operation), SCLK to NCS inactive (write operation), and NCS to SDIO high-Z had been verified. The SPI timing setup is shown in Figure 2 and the result obtained is compare to datasheet limit specification for each of the SPI timing parameter tested.

The test starts with loading the HEX code for testing source code into microcontroller at test board through hyper-terminal program at personal computer (PC). Before start of testing, we need to make sure the power supply is on and the mouse sensor LED is light up. Then, perform calibration communication using hyper-terminal command at PC to ensure there is a proper function of data interface between mouse sensor and microcontroller. Each SPI timing parameters are tested separately using a test command at hyper-terminal program. Test command is send through communication port using RS-232 to link to the test board. Test command will control the selection of SPI timing parameters that need to be measured and waveform are captured at digital oscilloscope. Once the test command is send to microcontroller board, the microcontroller will communicate with mouse sensor. The operations of read and write will

occur between microcontroller and mouse sensor. Read operation occur requires data information collected from optical mouse sensor while write operation requires to write data to optical mouse sensor. The digital oscilloscope is used to capture the output waveform for SPI timing pins such as SCLK, SDIO and NCS.

The sample are tested at different temperature condition for hot temperature (55°C), ambient temperature (25°C) and cold temperature (-15°C) by using thermal stream. The samples tested are from different skew units with sample size of three for each skew unit and across three different temperatures. Skew unit is a sample with different gate length on optical array to differentiate the different sensor performing. The combinations of skew units are Fast-Fast, Fast, Nominal, Slow and Slow-Slow. However, in this experiment only three different skew (F, N, SS) units will be evaluated.

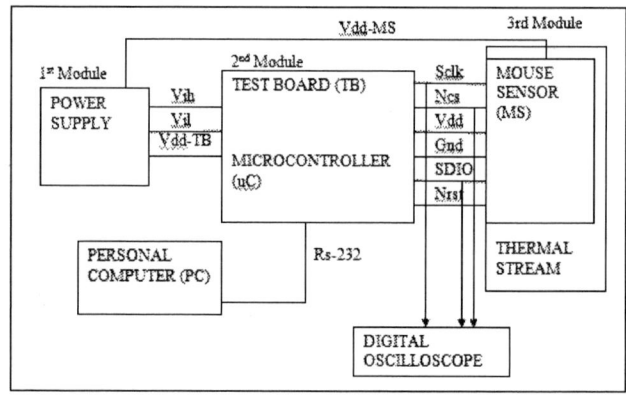

Figure 2: SPI timing evaluation test model.

III.   RESULTS AND DISCUSSION
1- SPI read address-data delay

Figure 3 shows SPI parameters for SPI read-address data delay. The SDIO 1st byte is address byte contained read operation when most significant bit (MSB) is logic '0'. The 1st byte of SDIO is to represent the address location and 2nd byte of SDIO is data byte. The SPI read address-data delay (us), tsrad is measure in between of address byte and data byte for SDIO. The tsrad timing delay is important for read operation to avoid last bit of address byte reading overlap with 1st bit of data byte reading. However, when the MSB value is '1' it will indicate the write operation. This SPI parameter also acts as basic concept for others SPI parameters that will be tested. In product datasheet the tsrad is specified at least 4 µs and the reading obtained in the analysis is about 4.014 µs.

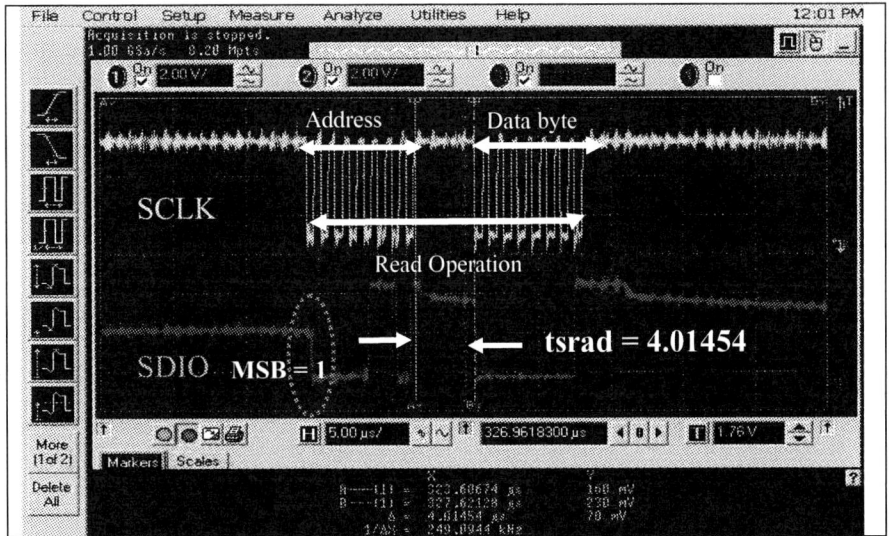

Figure 3: SPI read address-data delay (us), t_{SRAD}. Most significant bit (MSB) to indicate read operation when MSB is logic '0' and write operation when MSB is logic '1'.

2 - SPI Timing between read and subsequent commands (tsrr and tsrw)

The SDIO serial read-read time (tsrr) and SDIO serial read-write time (tsrw) are the timing between two read commands and the timing between read and write operation respectively. In the product datasheet, tsrr and tsrw both have their limit specified at a minimum of 500 ns. Due to limitation of the microcontroller and execution cycles taken in processing the instruction, these two (2) parameters were tested at ~3.58 µs.

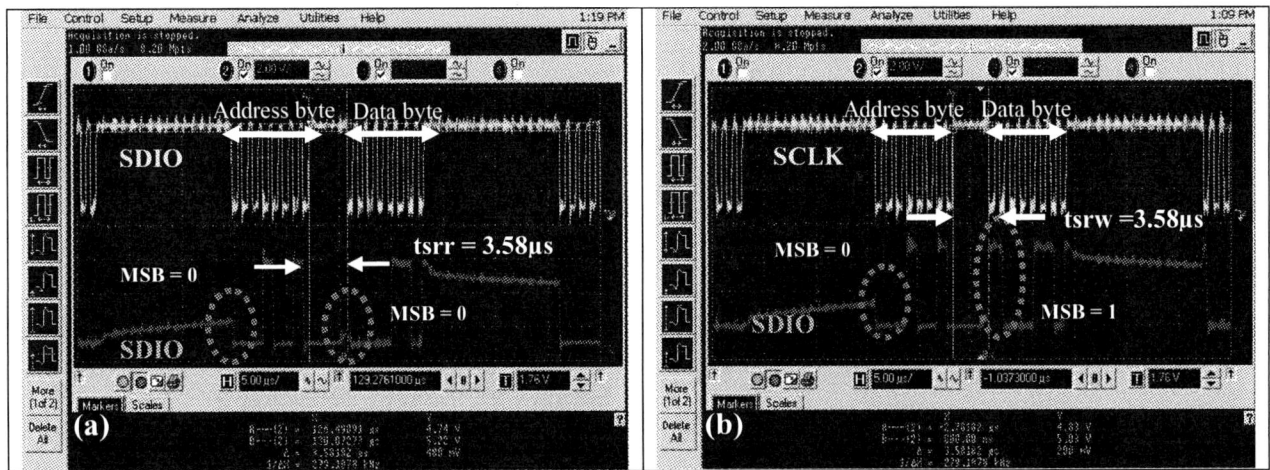

Figure 4: SPI timing between (a) read-read operations (tsrr) (b) read-write operations.

Figure 4 (a) shows SDIO 1st byte and 2nd byte are read operation because of the MSB is logic '0'. The timing between the read-read operations is being determined as SPI timing for tsrr. Figure 4 (b) shows that 1st byte of SDIO is read operation and follow by 2nd byte is write operation because of the first MSB logic is '0' and followed by logic '1' respectively. Back to basic SDIO timing concept, the 1st byte is address and 2nd byte is data byte. Therefore, in the case of Figure 4 (b) the SDIO operation is read address and writes data. This operation is known to be SDIO serial read-write time tsrw. Both tsrr and tsrw have about the same timing value which about ~3.58 µs. The timing of two reading operation is to avoid last bit of address byte reading overlap with 1st bit of data byte in reading (tsrr) and 1st bit of data byte in write operation (tsrw).

3- SDIO Rise & Fall Time

In product datasheet, the SDIO rise time (t_{rsdio}) and SDIO fall time (t_{fsdio}) both have their limit specified at 150 ns as typical value. Due to limitation of the microcontroller and execution cycles taken in processing the instructions, these parameters were tested at ~104 ns (SDIO fall time) and ~169 ns (SDIO rise time).

Figure 5 illustrates that SDIO rise time is indicates by arrow face up and SDIO fall time indicates by arrow face down. Both SDIO rise time and fall time have the same typical value.

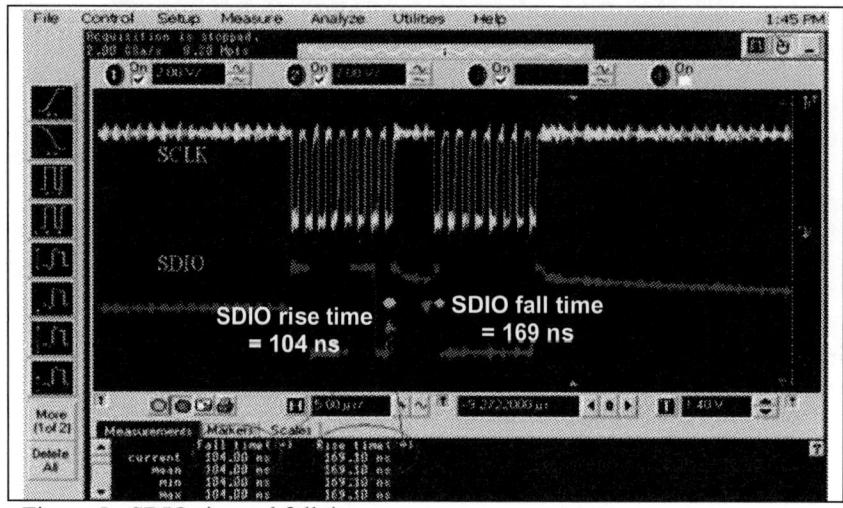

Figure 5: SDIO rise and fall time

4 – SDIO Hold Time

In product datasheet, SDIO hold time (t_{hold}) has limit of specifying at minimum of 0.5 µs. Figure 6 indicates SDIO hold time is the timing for data held until next falling SCLK edge. Result tested at ~462 ns. SDIO hold time specify the minimum timing required to hold data value in SPI for the sample tested.

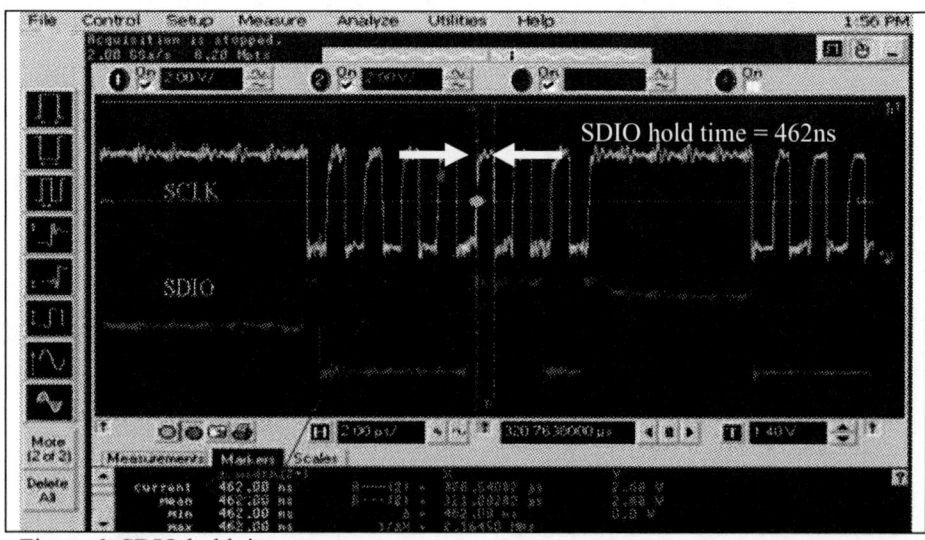

Figure 6: SDIO hold time

5 – SDIO Setup Time

In product datasheet, SDIO setup time has limit of specifying at minimum of 120 ns. Figure 7 indicates SDIO setup time is the timing from data valid to SCLK rising edge Result tested at ~254 ns. SDIO setup time is the minimum timing required to setup for received or sending data value in SPI for sample tested.

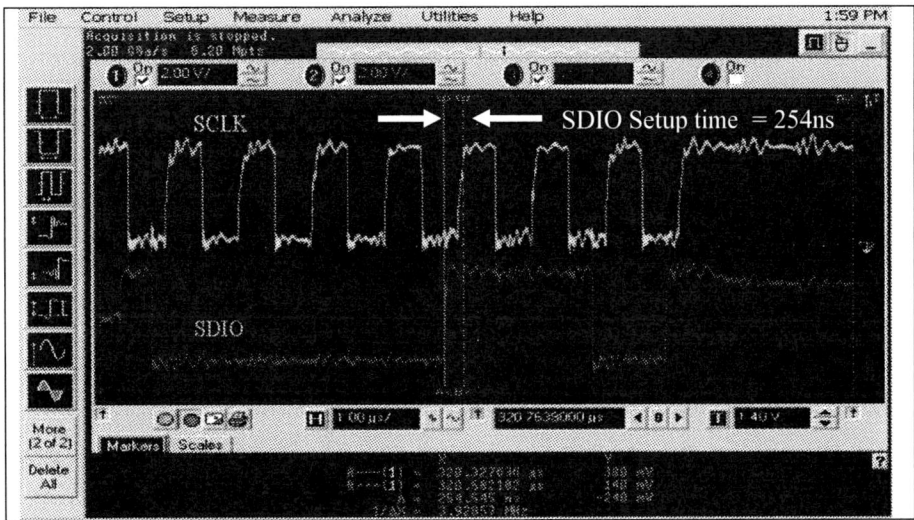

Figure 7: SDIO Setup Time

6- NCS to SCLK active

In product datasheet, SDIO setup time has limit of specifying at minimum of 120 ns. Figure 8 shows that from NCS falling edge to first SCLK rising edge is the timing for NCS to SCLK active. Result tested at ~1.16 µs. NCS to SCLK active meaning NCS is set to active low before the SCLK is active to capture all SDIO value in a frame and is a minimum timing for SCLK to start operates.

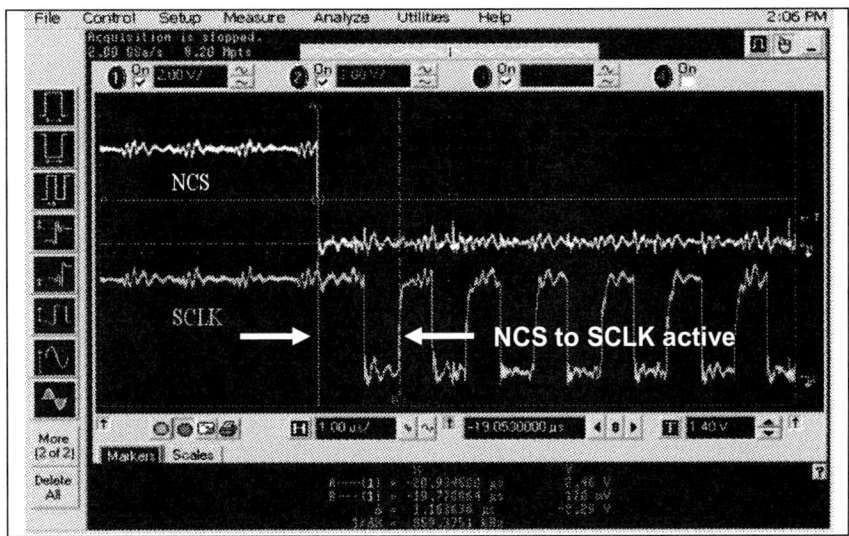

Figure 8 : NCS to SCLK active

7- SCLK to NCS inactive (read operation)

In product datasheet, SCLK to NCS inactive (read operation) has limit of specifying at minimum of 120 ns. Figure 9 shows SDIO 1st byte indicates a read operation in which MSB is logic '0'. From last SCLK rising edge to NCS rising edge, for valid SDIO data transfer is timing for SCLK to NCS inactive (read operation). Result tested at ~2.81 µs.

This timing is to ensure a set of two bytes SCLK is running properly then only NCS will inactive for read operation. This may avoid SDIO value missing at the last bit of data byte that leads to error of information.

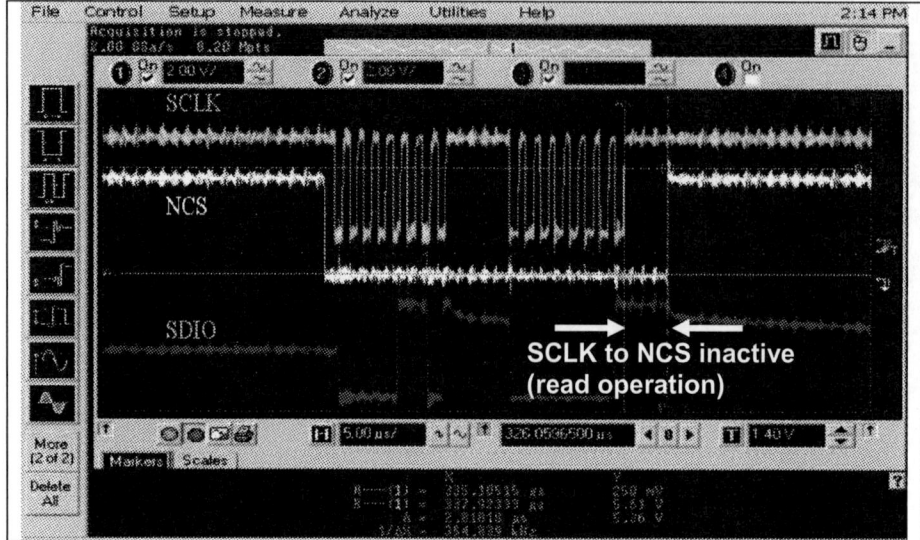

Figure 9 : SCLK to NCS inactive (read operation)

8. SCLK to NCS inactive (write operation)

In product datasheet, SCLK to NCS inactive (write operation) has limit of specifying at minimum of 20 μs. Figure 10 shows SDIO 1st byte indicates a write operation in which MSB is logic '1'. From last SCLK rising edge to NCS rising edge, for valid SDIO data transfer is timing for SCLK to NCS inactive (write operation). Result tested at ~2.63 μs.

This timing is to ensure a set of two bytes SCLK is running properly then only NCS will inactive for write operation. This may avoid SDIO value missing at the last bit of data byte that leads to error of information.

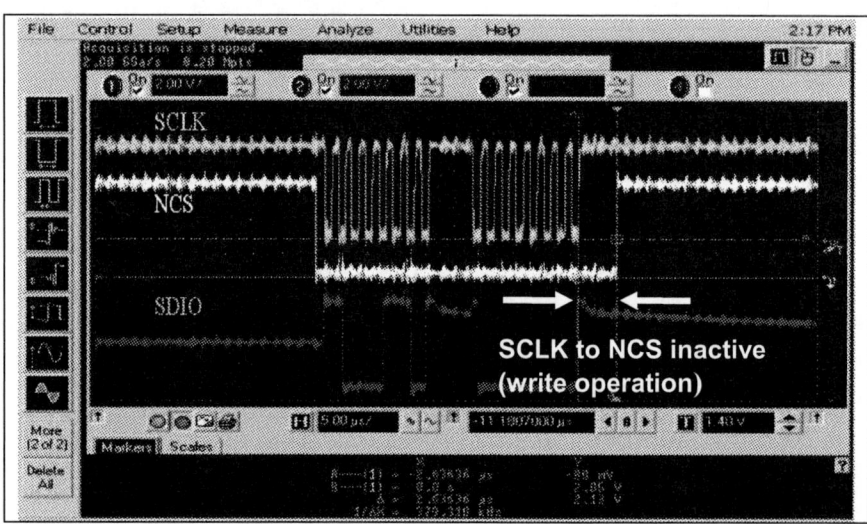

Figure 10 : SCLK to NCS inactive (write operation)

SPI-related timing evaluation is tested for write and read operations in optical mouse. This is to ensure that communication between sensor and microcontroller are operating in specification range. Beside that sensor also testing in all temperature and operating voltages. However the data at these conditions could not be disclosed. SPI Timing Evaluation Test Program had specific the register to be write or read and the delay timing required for each parameters being tested

IV. CONCLUSION

SPI-related timing evaluation is tested for write and read operations in optical mouse. This is to ensure that communication between sensor and microcontroller are operating in specification range. Beside that sensor also testing in all temperature and operating voltages. SPI Timing Evaluation Test Program had specific the register to be write or read and the delay timing required for each parameters being tested. Result for each SPI timing parameter tested are comparable to datasheet limit specification across various temperature, voltage and three different skew units.

REFERENCES

[1]. H. J. Choi, D. S. Kwon & M. S. Kim. Design of Novel Haptic Mouse and Its Applications. Proceeding of the 2003 IEEE/RSJ, International Conference on Intelligent Robots and Systems, Las Vegas Nevada, pp. 2260-2265, Oct 2003.

[2]. G. Ji, T. Arabi & G. Taylor. Design and Validation of a Power Supply Noise Reduction Technique. Electrical Performance of Electronic Packaging, pp. 137-140, Oct 2003.

[3]. Product Datasheet : Optical Mouse Sensor

[4]. Serial Peripheral Interface Bus, Online : http://en.wikipedia.org/wiki/microwire (access date 28 May 2006)

[5]. Smart Computing Dictionary, Serial Peripheral Interface (SPI) (online) http://www.smartcomputing.com/editorial/dictionary/detail.asp?guid=&searchtype=1&DicID=12820&RefType=Dictionary (access date 28 May 2006)

Design Of Experiment (DOE) For Thickness Reduction Of GaAs Wafer Using Lapping Process

Mohd Khairy Othman, Asban Dolah, Nurul Afzan Omar and Mohamed Razman Yahya

Microelectronic & Nanotechnology Program ,TM Research & Development Sdn Bhd,
UPM-MTDC, 43400 Lebuh Silikon,Serdang, Selangor.
E-mail : mkhairyo@tmrnd.com.my

Abstract - **This paper report a statistical method of performing wafer lapping experimental using Design Of Experiment (DOE) technique in order to get best lapping time to reduced thickness of GaAs wafer. Lapping speed, lapping time, oscillator speed and weight was selected as four main factor determine the shortest time of thickness reduction. A complete 2^4 factorial of 4 factors (16 run) was design to determined the effect of selected factor. The lapping process was carried out using ULTRATEC Lapping& Polishing machine while the wafer thickness was characterized using Logitech non contact gauge. It was found that best lapping parameter was using lapping speed at 3 r.p.m, oscillator speed at 2 r.p.m and 3 weight block for duration of 240 sec. This parameter is able to reduce 156 μm of wafer within 240 second without any crack problems and able to give good reference of reduction of GaAs wafer thickness process period**

I. INTRODUCTION

Backside processing takes more than half of the complete device cycle times [1]. For pHEMT or HBT device fabrication base on GaAs, to make the through wafer contact to substrate, the wafer are lapped and polished from full thickness of around 600 um to 100um. To reduce this thickness, the yield and the process duration had often dependent on the skill of experience technicians [2].

So, we come out with design experiment to estimate best duration of GaAs lapping process as reference to reduce GaAs wafer thickness for certain application. The DOE data are able to give thickness reduction estimation for reference of lapping process.

II. EXPERIMENTAL

1. 2^4 factorial:

Lapping speed, lapping time, oscillator speed and weight was selected as four main factor determine the shortest time of GaAs wafer thickness reduction process and a complete 2^4 factorial of 4 factors (16 run) was design to determined the effect of selected factor.

Table 1. Lapping parameter for the thickness reduction of GaAs wafer.

Parameter	Min	Medium	Max
Lapping Time (s)	60	150	240
Oscillating Speed (rpm)	2	2.5	3
Lapping Speed (rpm)	2	2.5	3
Weight (discrete*)	2	na	3

2. Factor:

Table 1. indicated the minimum and maximum value of the factor that will be considered giving main effect to the thickness reduction of GaAs wafer:

3. Machine:

The lapping process was carried out using ULTRATEC Lapping & Polishing machine with cast iron scrolled as a lapping plate. Silicon carbide (SiC) powder mixed with oil was used as the lapping slurry.

Table 2: Experimental test matrix

Test	Lapping Time	Oscillator Speed	Lapping Speed	Weight
1	+	-	-	-
2	+	+	-	-

0-7803-9730-4/06/$25.00 ©2006 IEEE

3	+	-	+	+
4	-	+	-	-
5	+	+	+	+
6	-	-	-	-
7	-	-	+	-
8	+	+	+	+
9	-	+	-	+
10	-	+	+	-
11	+	-	+	-
12	+	-	-	+
13	+	+	-	+
14	-	-	-	+
15	+	+	+	-
16	-	-	+	+

4. Software :

The statistical analysis of the experimental design was performed using STATGRAPHICS Centurion XV software.

III. RESULTS & DISCUSSION

Fig. 1 shows the reduction of wafer thickness results according to 16 run matrix. The highest thickness reduction was using run #3 where 159 ± 0.05 um of GaAs was thinned. The lowest reduction of GaAs thickness was using run #6 with only reduction of 2.00 ± 0.05 um of GaAs wafer.

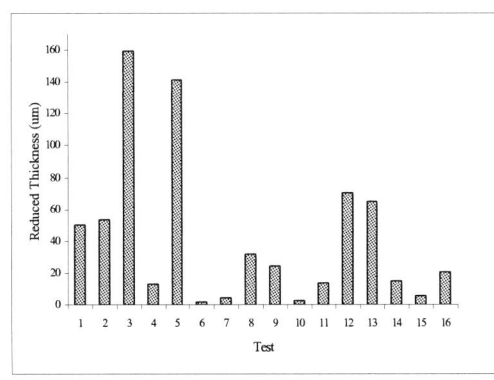

Fig 1. Thickness reduction result for all 16 DOE test matrix.

Pareto Chart

Fig 2 show the pareto chart of removed thickness parameter. Two main parameter effect on thickness reduction is lapping time and weight. Meaning that, optimum GaAs wafer thickness reduction need to consider those two parameter during the lapping process. While the oscillator speed is not effected at all and lapping speed is not much effect on the thickness reduction process.

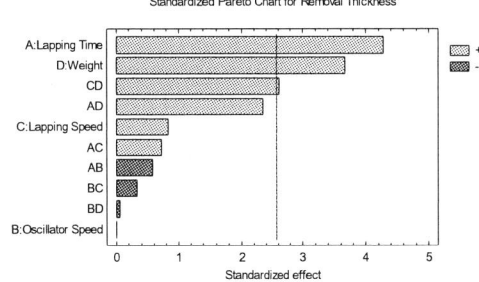

Fig 2. Pareto chart of removed thickness parameter.

Main Effect Plot:

Fig 3. shows the main effect plot for removal thickness (thickness reduction) of all lapping parameter. For the lapping time, the higher time of lapping will give higher thickness removal. The same effect was observed for the weight parameter. As per the not giving effect on the thickness reduction parameter, the increased of oscillator speed and lapping speed did not shows any improvement as per the plotted graph.

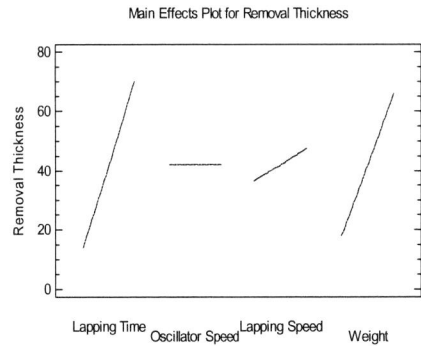

Fig 3. Main effect plot for thickness reduction of all lapping parameter.

Table 3. shows the summary of the test matrix for the thickness reduction experiment. From the summary, the highest reduced thickness can be achieved using test # 3 only takes 4 minutes to reduce GaAs wafer thickness for 159.5 um. So, for standard GaAs wafer of about 500-600 um, operator will takes less than one hour to complete lapping of one piece of 3" GaAs wafer.

Table 3: Summary of thickness reduction result of GaAs wafer and it parameter factor.

Test	Lapping Time (sec)	Oscillator Speed (rpm)	Lapping Speed (rpm)	Weight	Thickness Reduction (µm)
1	+	-	-	-	50.70
2	+	+	-	-	53.30
3	+	-	+	+	159.70
4	-	+	-	-	13.00
5	+	+	+	+	141.30
6	-	-	-	-	2.00
7	-	-	+	-	4.50
8	+	+	+	+	32.00
9	-	+	-	+	24.00
10	-	+	+	-	2.60
11	+	-	+	-	13.80
12	+	-	-	+	70.40
13	+	+	-	+	64.70
14	-	-	-	+	14.80
15	+	+	+	-	5.30
16	-	-	+	+	20.30

Figure 5, shows the SEM image of the thinned GaAs cross-section. It shows that the final thickness of the sample is around 61.39 $\pm$ 0.1 nm. This thickness can be achieved base on the fig 4 data summary

Fig 5: Cross-section SEM image of thinned GaAs wafer to 61.39 $\pm$ 0.01 nm using DOE result from the data summary.

However, since all the DOE run are not using standard sapphire wafer carrier, there is a crack formation on the GaAs wafer whenever the thickness reach 120 um. The crack became worse whenever a manual hotplate technique is applied to dismount the GaAs wafer from the metal carrier. In order to reduce the substrate crack, the lapping process must be using sapphire wafer as a carrier .

IV. CONCLUSION

A statistical method of performing lapping experimental using Design of Experiment (DOE) technique has been successfully performed. Best lapping parameter of test # 3 was identified. This parameter is able to reduce 156 um of GaAs wafer within 240 second without any crack problems. The DOE data are able to give good thickness reduction estimation for reference guide for lapping another types of wafer material which can be shorten the pre-trial run of process.

ACKNOWLEDGEMENT

This works has been carried out with the support of Telekom Malaysia Berhad under project code RNDC-0603

REFFERENCES

[1] Pei, Z.J., Xin, X.J. and Liu, W. 2003. "Finite Element Analysis For Grinding of Wire-Sawn Silicon Wafers: A Designed Experiment'. *Int. Journal of Machine Tools & Manufacture.* 43: 7-16.

[2] Pei, Z.J., Kassir, S., Bhagavat, M. and Fisher, G.R. 2004. "An Experimental Investigation Into Soft- Pad Grinding Of Wire-Sawn Silicon Wafers". *Int. Journal of Machine Tools & Manufacture.* 44: 299-306.

[3] McGrath, J. & Davis, C. 2004."Polishing Pad Surface Characterisation In Chemical Mechanical Planarisation". *Journal Of Materials Processing Technology.* 153-154: 666-673

Study of Flow Visualization in Stacked-Chip Scale Packages (S-CSP)

M. Khalil Abdullah[a], M. Z. Abdullah[a], S. Kamarudin[a] and Z. M. Ariff[b]
P. Hussin[c], J.J. Antony[c], H. Haroon[c], M. R. Saad[c] and M. Manikam[c]
[a]School of Mechanical Engineering,
[b]School of Material & Mineral Resources,
Universiti Sains Malaysia, Engineering Campus,
Seri Ampangan, 14300 Nibong Tebal, Seberang Perai Selatan, Pulau Pinang, Malaysia.
[c]Spansion (Penang) Sdn. Bhd., Penang.
Phase II, Free Industrial Zone, Bayan Lepas, 11900 Penang, Malaysia
Email: mkhalil@eng.usm.my & mezul@eng.usm.my

Abstract **Stacked-Chip Scale Package (S-CSP) is a technology which has high density packaging options. It enables to stack the die in a single package. The S-CSP widely adopted in portable multi-media products. However, resin flow through a thin space and wide filling area has become concern. Therefore, this paper presents a study of flow visualization during encapsulation process in S-CSP. The Navier-Stokes equation has been solved by Finite Different Method (FDM). For non-linear flows, the Kawamura and Kuwahara technique has been adopted for the flow analysis in the chip cavity. Pseudo-concentration is based on the volume of fluid (VOF) technique was used to track a melt fronts for each time step. The numerical model has been verified by comparing the prediction with experimental results. The numerical results show good agreement with the experimental results.**

I. INTRODUCTION

Nowadays, the electronics industry comes to offer their product in small size [1] but also give more functionality, better performance and lower cost [2]. Having a closer look at the smallest packages, S-CSP is one of good agreement to maximum miniaturization [3]. One of the products involved is portable multi-media products. It widely used in cellular telephones, digital cameras, PDAs and audio players.

Generally, encapsulation processes for the package is transfer moulding. However, problems arise such as short shot, air trap and void during encapsulation process [4-5]. Furthermore, the dies are arranged in matrix array and multi-die stacking within thin space

and wide filling area in the S-CSP mould lead the problems to be more difficult.

Besides the structural complexity, the effect of curing reaction occurs. It not just affects the degree of conversion but also give an effect on rheological behaviour of the resin flow. The combination of the complexity geometry and complicated phenomena make it worse [6].

Therefore, this paper is to present the experimental and simulation results of mould filling process in the S-CSP and focused on resin flow by omitting wire bonds. The numerical models developed based on Finite Difference Method (FDM) using FORTRAN 77 codes. The flow pattern in the short shot samples of experiment are compared to the simulation results for the verification purposes.

II. ANALYSIS

The mass, momentum and energy conservation governing equations for non-isothermal, generalized Newtonian fluids are written as follows:

i. Continuity equation

$$\frac{\partial u}{\partial x} + \frac{\partial v}{\partial y} + \frac{\partial w}{\partial z} = 0 \qquad (1)$$

ii. Momentum Equation

$$\frac{\partial u}{\partial t} + u\frac{\partial u}{\partial x} + v\frac{\partial u}{\partial y} + w\frac{\partial u}{\partial z} =$$

$$-\frac{1}{\rho}\frac{\partial p}{\partial x} + \nu\left(\frac{\partial^2 u}{\partial x^2} + \frac{\partial^2 u}{\partial y^2} + \frac{\partial^2 u}{\partial z^2}\right) \qquad (2)$$

iii. Energy equation

$$\rho C_p \left(\frac{\partial T}{\partial t} + u \cdot \nabla T \right) = \nabla (k \nabla T) + \Phi \tag{3}$$

Where u is the velocity vector, T the temperature, t the time, p the pressure, ρ the density, k the thermal conductivity, C_p the specific heat and Φ is the energy source term and it contains two contributions of polymer properties:

$$\Phi = \eta \dot{\gamma} + \dot{\alpha} \Delta H \tag{4}$$

η is the viscosity, $\dot{\gamma}$ the shear rate, $\dot{\alpha}$ is curing rate and ΔH is reaction heat.

The advancement of melt front over time is governed by the following transport equation:

$$\frac{\partial F}{\partial t} = \frac{\partial F}{\partial t} + \nabla \cdot (uf) = 0 \tag{5}$$

Where f = 0 is defined as the air phase, f = 1 as the polymer melt phase and the melt front is located within cells with 0<f<1.

Several models have been used to predict the relationship between viscosity and the degree of polymerization. The Castro-Macosko model has been applied by Nguyen et al. [4-5] and is selected to use in this simulation. It can be described as follow:

$$\eta(T, \dot{\gamma}) = \frac{\eta_0(T)}{1 + \left(\frac{\eta_o \dot{\gamma}}{\tau^*} \right)} \left(\frac{\alpha}{\alpha_g - \alpha} \right)^{C_1 + C_2 \alpha} \tag{6}$$

where n is the power law index, η_0 the zero shear rate viscosity, τ^* is the parameter that describes the transition region between zero shear rates and the power law region of the viscosity curve, $\dot{\gamma}$ is the shear rate, α_g is the conversion at the gel point and C_1 and C_2 are fitting constant.

$$\eta_0(T) = B \exp\left(\frac{T_b}{T} \right) \tag{7}$$

B is an exponential-fitted constant and T_b is a temperature fitted-constant. In addition, Kamal curing kinetics is coupled together with Castro-Macosko model. This model predicts the rate of

chemical conversion of the compound as follows:

$$\frac{d\alpha}{dt} = (k_1 + k_2 \alpha^m)(1 - \alpha)^n \tag{8}$$

$$k_1 = A_1 \exp\left(-\frac{E_1}{T} \right) \tag{9}$$

and

$$k_2 = A_2 \exp\left(-\frac{E_2}{T} \right) \tag{10}$$

where α is the conversion, A_1 and A_2 are the Arrhenius pre-exponential factors, E_1 and E_2 are the activation energies, m and n are the reaction orders and T is the absolute temperature.

III. NUMERICAL METHOD

The FDM based on the explicit forward-time discretization of motion [7] is used. The Poisson equation is used to calculate the pressure term as in equation (12):

$$\tilde{u}^{n+1} = \tilde{u}^n$$
$$+ \nabla t \left[N(\tilde{u}^n) + vL(\tilde{u}^n) - \frac{1}{\rho} \nabla P^n \right] \tag{11}$$

$$\nabla^2 P^n = \frac{\rho}{\Delta t} (\nabla \cdot \tilde{u})^n$$
$$+ \rho \nabla \cdot N(\tilde{u}^n) + \mu L(\nabla \cdot \tilde{u}) \tag{12}$$

All spatial derivatives are approximated by central differences. For non-linear terms, it discretized by a third-order upwind scheme proposed by Kawamura and Kuwahara [8-10];

For $u_i > 0$

$$u_i \left(\frac{\partial u}{\partial x} \right)_i =$$

$$u_i \frac{(-u_{i+2} - 2u_{i+1} + 9u_i - 10u_{i-1} + 2u_{i-2})}{6\Delta x} \tag{13}$$

For $u_i < 0$

$$u_i\left(\frac{\partial u}{\partial x}\right)_i =$$

$$u_i \frac{\left(-u_{i-2} + 2u_{i-1} - 9u_i + 10u_{i+1} - 2u_{i+2}\right)}{6\Delta x}$$

$$(14)$$

Equations are solved by the finite difference method. The solution procedures have been coded into a computer program using FORTRAN 77 language.

IV. SIMULATION OF S-CSP MODEL

The model of actual cavity come with six stacking dies in matrix array mould, bevelled edge and chamfered corners as shown in Fig. 1a. For this case, some simplification have been done by certain features were modified or eliminated from the model as shown in fig. 1b. The chamfered corners were replaced by 90° corner and the whole gate model and vents was omitted except surface area in the end from the actual model.

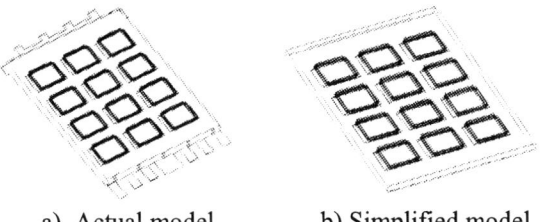

a) Actual model   b) Simplified model

Fig. 1 S-CSP actual and simplified model

The total mesh generation of the simplified model is 31584 elements and it takes less than 7 seconds. For analysis, it takes approximately 10 minutes. The moulding compound used in this study is HITACHI CEL-9200-XU (LF) and the mould temperature is 172°C. The inlet velocities set in the program are 0.4 cm/s. The program was run on an AMD Opteron 64Bit processor 2.0 GB speed and with the memory of 3GB.

V. RESULTS AND DISCUSSIONS

The experimental results of short shot samples at different time steps are compared with the simulation result. Fig. 2 demonstrates melt front advancement of S-CSP for both experimental and simulation results. The experimental results

have been provided by Spansion (Penang) Sdn. Bhd. In Figure 2 shows the comparison of advancement of melt front from 1.5 to 7.0 seconds.

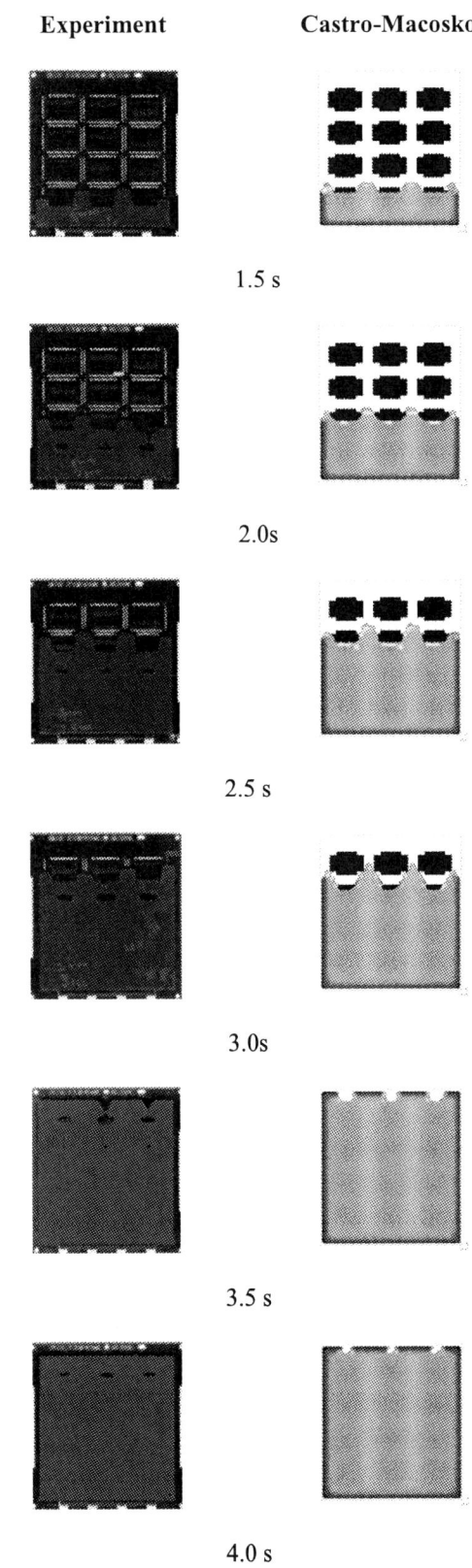

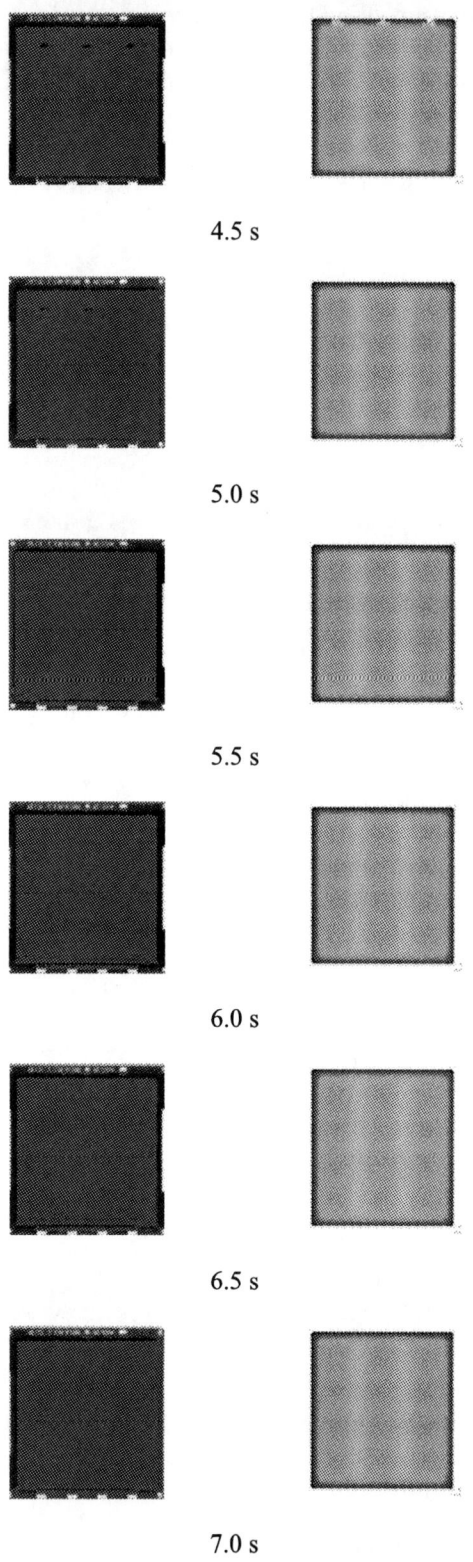

4.5 s

5.0 s

5.5 s

6.0 s

6.5 s

7.0 s

Fig. 2 Comparison of Predicted melt front advancement at distinct time steps ($v_1 = 4$ mm/s)

The resin flows into the cavity through the gate at a higher velocity, expands into the cavity rapidly and hits the dies. The dies in the cavity divide the flow and then resin flow joins together after the dies. In the initial stage, experiment and simulation exhibit same shape but the simulation covered more area at the top of die. The observation continues and it still shows diminishing flow at the top of dies. Accelerated flow along paths compared to on the top of dies which were obstructed by less volumetric flow. The simulation also shows slenderly similar trends. However, a number of voids observed in the actual instead of the simulation, no voids are detected. Again, the similarities did not occur exactly at the next time interval. At this rate, simulation moved slightly further rather than actual process.

For 3.0s, the simulation results show that the flow fronts covered dies completely and tend to appear negative elliptical shape at the vents passage whereas in the real flow, the dies still uncovered completely and still have a number of voids. The effect of the dies is barely noticeable on the flow fronts. It restricted the flow along the edges of the dies and over the dies. As a result, the flow around the outside of the cavity was accelerated. It covered the behind the dies and the wall first before the top die flows.

At 4.5s, the compound covered completely between backside of the dies and the wall of mould. For both simulation and experimental results depict a number of voids existed. Fig. 3 shows closely the comparison of the actual and simulation mould void trapping shapes. However, a number of voids of the simulations are nearly to the wall mould i.e. vents passage. Unlike of the actual process, a number of voids still observed at the top of dies.

Fig. 3 The comparison of voids between experiment and simulation at 4.5s

Once it reached in the final stages, the simulations flow fronts come to close matching again with the actual process. From samples at 5.5s to 7.0s shows the similar shape and progression of both fronts. Even so it looks completely fill the package at 5.0s but when it views on the bottom package, there are incomplete fillings occurred at the backside of the dies as in fig. 4.

598

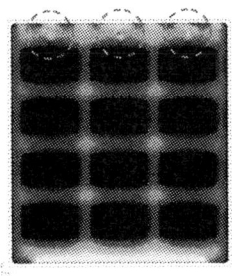 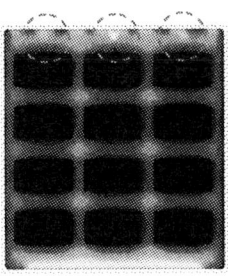

5.0 s   5.5s

Fig. 4 The incomplete filling with red circle occurred at the backside of the dies (bottom view)

A number of voids and incomplete filling are closely related to the flow retardation. It contributes to restrict of the smooth flow and lean to slow down the flow front. The flow began to retard over the dies due to the narrow space between dies top surface to the mould chase. Moreover, the repetitive complexity of matrix dies structure and a pattern also contributes flow retardation. When the resin flow contacted the die, a portion was diverted along is edges. This will lead an increasing the volumetric flow and make the flow going advance. At the same time, unbalanced flow between the free passage and crucial clearance at the top of dies indirectly will constitute a number of voids.

This process also involved curing reaction effects. A mould heating effect is more significant on the dies because of shorter thermal path. This will accelerate curing reaction of the compound and promotes the local viscosity and flow resistance. The net effect is slower down in the velocity of the compound.

VI. CONCLUSION

The numerical model gives good agreements with experimental results on the melt front the location. The resin flow retarded once it hit the dies beside of the complexity and complicated shape of the package. Constant fill rate and interplay in the variables in the EMC may donate to the divergence between experiment models and simulation models. Beside, wire bonds at the dies which considerably truncated the flow front were leaving out.

ACKNOWLEDGEMENT

The author would like to thank for Spansion (Penang) Sdn. Bhd and USM for supporting this research works and funding of the research under Graduate Assistantship scheme.

REFERENCES

[1] Rao R. Tummala, *Fundamentals of Microsystems Packaging*, McGraw Hill, 2001, Singapore.

[2] www.amkor.com

[3] T. Braun, K. F. Becker, M. Koch, V. Bader, U. Oestermann, D. Manessis, R. Aschenbrenner, H. Reichl, *Wafer Level Encapsulation-A Transfer Molding Approach To System In Package Generation*, 2002 Electronics Packaging Technology Conference.

[4] Rong-Yeu Chang and Wen Sheng Yang, *On the Dynamics of Air Trap in the Encapsulation Process of Microelectronic Package*. CoreTech System Co. Ltd.

[5] Rong-Yeu Chang, Wen-Hsien Yang, Sheng-Jye Hwang and Francis Su, (2004), "Three-Dimensional Modelling of Mold Filling in Microelectronics Encapsulation Process", IEEE Transactions on Components and Packaging Technologies, Vol. 27, No. 1, March 2004.

[6] Min Woo Lee, Jin Young Kim, Min Yoo, JiYoung Chung and Choon Hueng Lee, *Rheological Characterization And Full 3D Mold Flow Simulation In Muli-Die Stack CSP Of Chip Array Packaging*, 2006 Electronic Components And Technology Conference.

[7] C. Pozrikidis, (1997), *Introduction to Theoretical and Computational Fluid Dynamics*, Oxford University Press.

[8] T. Kawamura, K. Kuwuhara, *Computation of High Reynolds Number Flow Around A Circular Cylinder with Surface Roughness*, AIAA 22nd Aerospace Sciences Meeting, 1984, Nevada.

[9] T. Kawamura, K. Kuwuhara, *Computation of High Reynolds Number Flow Around A Circular Cylinder with Surface Roughness*, AIAA 23rd Aerospace Sciences Meeting, 1984, Nevada.

[10] Mohd Zulkiefly Abdullah, T. Kouta, Takuma Kamijo, Makoto Yamamoto, Shinji Honami andShoji Kamiunten, Numerical Investigation on Bottom Gap of Micro Flow Sensor, Journal of Computational Fluids Engineering /73-79, Vol. 10, No. 1, March 2005.

An FIB Method Using Progressive Multi-Cut Technique & Application in Failure Analysis of Wafer Fabrication

Khoo Ley Hong, Hua Younan, Zhao Siping, *Snr Member, IEEE* and Mo Zhiqiang

Chartered Semiconductor Mfg Ltd
Woodlands Industrial Park D, Street 2
Singapore 738406
Email: khoolh@charteredsemi.com

Abstract: **In this paper, an FIB method using progressive multi-cut technique is proposed and it has been applied in failure analysis of wafer fabrication. The application results showed that this method would greatly improve FIB cut success rate, especially for invisible defects. A case study on Vbd ramp up failure after QBD short loop will be presented.**

I. INTRODUCTION

FOCUSED ion beam (FIB) technique has been widely used for sample preparation and defect/particle analysis in failure analysis of wafer fabrication [1-6]. It is well known that FIB cut for visible defects, the success rate is high, but it is relative low for some invisible defects. Moreover, in failure analysis, sometimes it is necessary to perform a precision FIB cut at emission site from as EMMI fault isolation tool. To improve FIB cut success rate, in this paper, we will propose an FIB method using progressive multi-cut technique and apply it in failure analysis at an emission site. The application results showed that it is useful and successful to target the defect. Using this method, FIB cut success rate has been greatly improved, especially for invisible defects and at emission site. In this paper, we not only describe details of the method, but also present an application case on Vbd ramp up failure after QBD short loop and physical defect was found after performing progressive multi-cut using FIB.

II. EXPERIMENTAL

A. Instrument & conditions

The machine model used is FEI 830 FIB system. The conditions used are: 2700pA (coarse milled), 1000pA (medium milled) and 70pA (Fine milled/ cleaning) and 5kv/ spot 3 with sample tilted 52 degrees.

B. Sample Preparation

In this study, progressive/0.1um cutting step of sample preparation was used. Sample sputtered for 25 seconds of platinum to prevent charging before loading into the FIB. Pt was coated on the defective area as a protection layer before the FIB cut. A rectangular box was draw about 2um away from the feature of interest. Repeat the milling steps for Ion-Bean current of 2700pA to 1000pA to clean the cross section done (Figs.1 & 2).

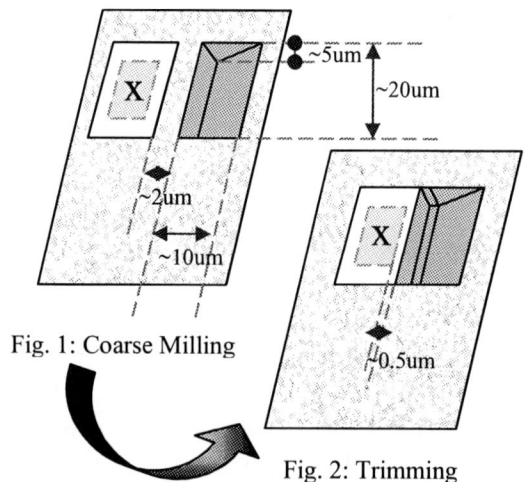

Fig. 1: Coarse Milling

Fig. 2: Trimming

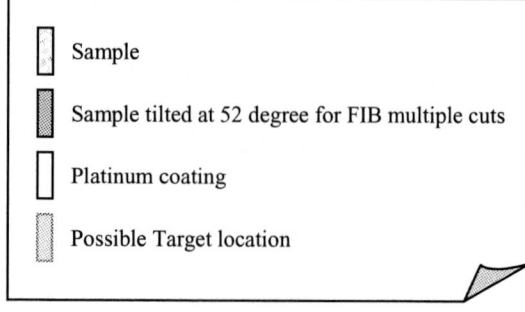

Sample

Sample tilted at 52 degree for FIB multiple cuts

Platinum coating

Possible Target location

Use Ion-Beam current of 70pA with Cleaning Cross-Section Pattern for the final cut (Fig.3). SEM image taken in FIB machine with sample tilted 52 degrees without staining.

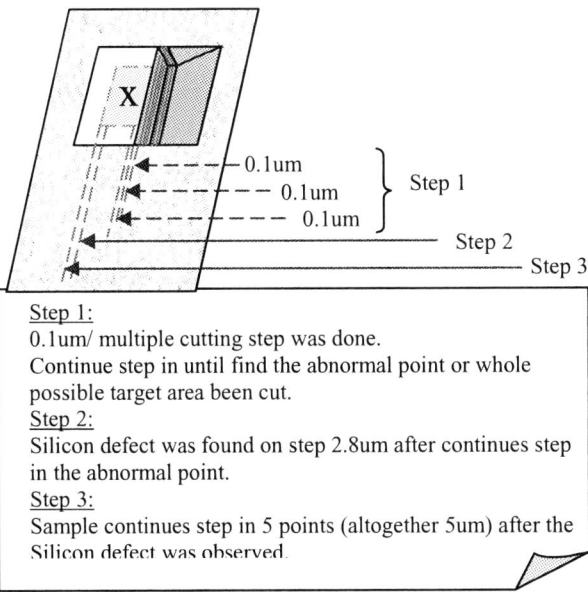

Step 1:
0.1um/ multiple cutting step was done.
Continue step in until find the abnormal point or whole possible target area been cut.
Step 2:
Silicon defect was found on step 2.8um after continues step in the abnormal point.
Step 3:
Sample continues step in 5 points (altogether 5um) after the Silicon defect was observed.

Fig. 3: Fine Milling/ Progressive Cutting Step (0.1um step

III. RESULTS AND DISCUSSION

In this study, the sample failed to Vbd ramp up after QBD short loop. The sample failed to B mode. The failure spot was located using EMMI and then FIB multi-cut method was applied at the EMMI failure spot to find the root cause of the QBD failure. FIB multi-cut on the EMMI spot using 0.1um-cutting step was performed until the abnormality seen. From Fig 4, one can see that no

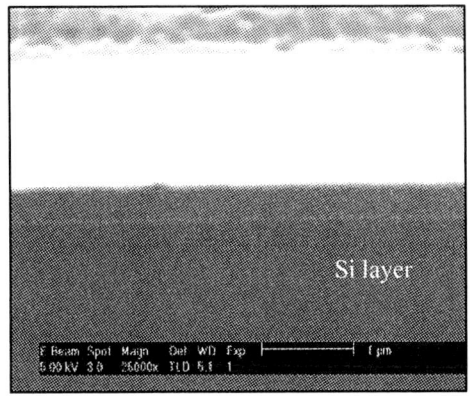

Fig. 4: step 2.7um
** No visual defect was observed

visual defect was observed before reach step 2.8um. However, a Si defect was observed on step 2.8um after continue step in the defective location (Fig.5).

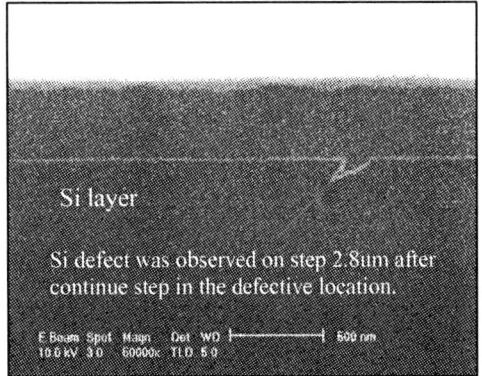

Fig. 5: step 2.8um

And then the Si defect was getting smaller on step 2.9 after continues step in 0.1um (Fig.6).

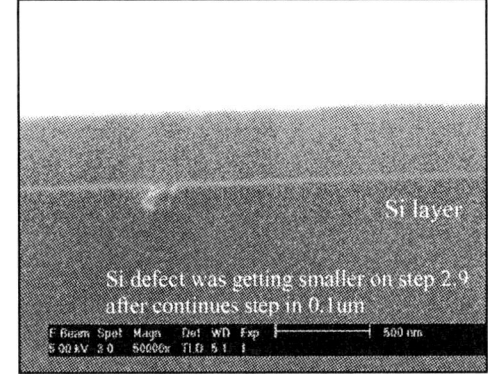

Fig. 6: step 2.9um

Sample continues step in 5 points (altogether 0.5um step in) until the whole possible target area been cut (Fig.7).

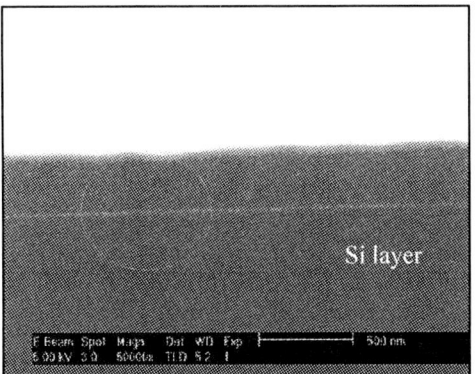

Fig. 7: step 3.3um
** Sample continues step in 5 points (altogether 0.5um step in) until the whole possible target area been cut.

IV. CONCLUSION

In this paper, an FIB method using progressive multi-cut technique is proposed. Using this method, FIB cut success rate has been greatly improved, especially for invisible defects and at emission site. A case study on Vbd ramp up failure after QBD short loop has been presented.

ACKNOWLEDGEMENT

The authors would like to thank Mr Eddie Er (Mgr from Physical FA team) and FA personal from Surface/Advanced FA team for their technical advice and making contribution to this technical paper.

REFERENCES

[1] S. P. Zhao, Hua Younan, G. P. Goh, Z. R. Guo & K. W. Chau, Contamination Analysis of Al Bondpad Using FIB/SEM/EDX. The Proceedings of International Symposium on Physics & Failure Analysis of Integrated Circuits, Singapore, 254-259 (1997).

[2] S. P. Zhao, Hua Younan, S. Tatti, L. H. An, G. P. Goh*, K. W. Chau, Investigation of Logical Device Failed IDDQ Test Using FIB/SEM/EDX Techniques. The Proceedings from the 8th International Symposium on Integrated Circuits, Device & System, 8-10 Sep, Great Hyatt, Singapore, 80-82 (1999)

[3] Hua Younan, L. X. Zhao*, L. H. An, E. Z. Liu, E. C. Low & S. P. Zhao (* The author from ESS Technology, Inc. USA), Failure Analysis of Metal Lifting Using FIB & TEM Techniques in Wafer Fabrication. The Proceedings from the IEEE National Symposium on Microelectronic (NSM'99), 6-7 Sep, Pan Pacific Resort, Pangkor, Malaysia, 181-183 (1999).

[4] G. B. Ang, Hua Younan, S. K. Loh, Yogaspari and S. Redkar, Application of Passive Voltage Contrast and Focused Ion Beam on Failure Analysis of Metal Via defect in Wafer Fabrication. The proceedings from the 8th International Physics & Failure Analysis (IPFA), June 2001, Singapore, 107-111 (2001).

[5] Ramesh Rao Nistala, Hua Y. N. and Tan C. W. Charge Reduction Method Using Laser and Focused Ion Beam and Its Application in Auger Analysis on Floating Bondpad Structures in Advanced IC Processing Technologies. The Proceedings of the IEEE NSM, Nov 22-24, 2005, Kuching, Sarwak, Malaysia, pp 50-53 (2005).

[6] Zhao Siping, Hua Younan, Er Eddie and Khoo Ley Hong, Studies on A New Sela-FIB Sample Preparation Method and Its Application in Failure Analysis of Wafer Fabrication for 110nm Technology Node and Beyond. The Proceedings of the IEEE ICSE, 29 Nov – 1 Dec, 2006, Kuala Lumpur, Malaysia, (2006).

I/O Process Optimization to Cover Wide Range Operation Voltage

Deb Kumar Pal, Kenny Sabri, Margaret Ting Leh Kee, Son Jin Yeong, Park Hyun Suck
X-FAB Sarawak Sdn. Bhd.
1 Silicon Drive, Samajaya Free Industrial Zone,
93350 Kuching, Sarawak, Malaysia.
Email: deb.kumar@xfab.com
Phone Contact no.: +60.82.354173

Abstract **A process has been developed to cover wide range I/O operation voltage (1.8V to 3.3V) without changing the 3.3V I/O library at author's organization to meet the market demand by optimization of 3.3V process. The main emphasis is given on to improve the Idsat current from the baseline and maintain the Ioff comparable as 3.3V process. This process passed all device level reliability test. This process is used to fabricate wide range I/O operation voltage device at author's organization.**

I. INTRODUCTION

The present world of electronics industry is going towards the huge variety of battery operated devices for various applications such as multi-media, hand phone, electronic gadgets and so on. All these battery operated devices need integrated circuits, which can operate at low voltage, and the global IC market is recently focusing on the reduction of the I/O operation voltage for devices from 3.3V to 1.8V (same as core) to improve the lifetime of the battery. So, there is a good demand of I/O process, which can cover the wide range of I/O operation voltage (3.3V to 1.8V). The easy way to fulfill the above demand is to design new IP library, which can cover these wide ranges, but it needs certain development time and high cost due to new library development (design and proven in Silicon.). The other alternative is to use the existing IP library (3.3V I/O) and do some minor modification on existing process (3.3V I/O) to make the IP library work in the demanding range of operation 3.3V to 1.8V. This enables us to save time as well as cost. Hence, the alternative path,

the process optimization of the baseline process is chosen to meet the market demand as early as possible. And then the wide range I/O process developed in author's organization by optimizing the baseline process 3.3V I/O and used to fabricate wide range I/O operation voltage device designed with 3.3V I/O library. The device passed all required electrical and reliability criteria and shows capable yield as other products processed with the base line process.

II. PROCESS DESIGN

The operating voltage vs operating current characteristics (Fig-1) of 3.3V I/O IP is such that current at 3.3V & 2.8V are well within the spec but at 1.8V is near the lower spec limit, need to increase the current to get better margin for in line voltage fluctuation, so the targets of process optimization are i) *Increasing the Idsat current of I/O transistor at 1.8V*, because the functional test of the device is at 10% below of operation voltage 1.8V so need to maintain sufficient current to work at 1.62V, ii) *Maintaining the comparable Ioff with base line at 3.3V* and iii) *Passing the device level reliability tests.*

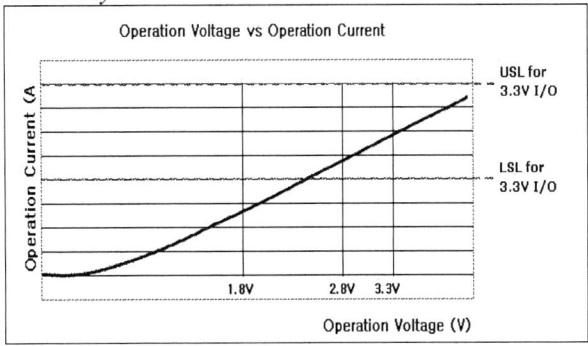

Fig-1. Operation Characteristics of the 3.3V I/O IP.

0-7803-9730-4/06/$25.00 ©2006 IEEE

The increase of Idsat can be done through two ways; either i) changing some implant condition, mainly channel implant, or ii) reducing the gate oxide thickness. The first option can introduce more leakage for DIBL [1-2] and the second option can cause GIDL [3]. To suppress the GIDL, we need to play with LDD and S/D implant, but S/D implant change affect the core transistor (thin gate oxide transistor). So, to achieve increase of Idsat with comparable Ioff, the gate oxide is reduced along with some implant conditions changed. Our baseline has two channels implant for NMOS I/O transistor, one for common and another for I/O only. The new process chose only the first one and skipped the second one, increased halo implant by 3E12, decreased LDD by 1E13 and reduced gate oxide by 10A to meet the target. The PMOS also has one common implant for core and I/O transistor and one separate for core transistor. The first one reduced by 2E12 and compensated that dose for core transistor by the second one. The LDD implant reduced by 1E13 & gate oxide thickness also reduced like as NMOS. So, the integration point of view, total number of photo is reduced by one and then these changes do not increase any process cost.

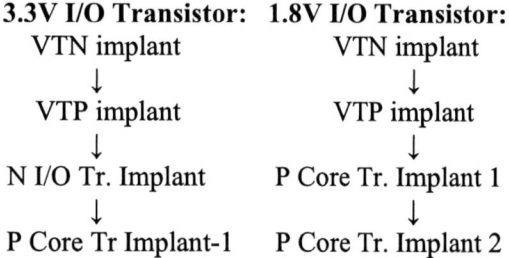

3.3V I/O Transistor: 1.8V I/O Transistor:

VTN implant   VTN implant
↓   ↓
VTP implant   VTP implant
↓   ↓
N I/O Tr. Implant   P Core Tr. Implant 1
↓   ↓
P Core Tr Implant-1   P Core Tr. Implant 2

Fig-2. Basic Implant steps comparison for Core and I/O Tr.

III.   RESULTS & DISCUSSION

The drain characteristics of both NMOS & PMOS shows the IDS at VDS & VGS =3.3V and VDS & VGS at 1.8V increased by 15% & 37% (Fig-3) for NMOS and 27% & 55% (Fig-4) for PMOS respectively, which satisfy our target to increase the drive current more at 1.8V to avoid functional failure at 1.62V and less in 3.3V to avoid the leakage failure of the device fabricated in the developed process. The Ioff characteristics (Fig-5 & 6) show the leakage one order more than the

3.3V I/O process, but well below the specification (100pA/um width) of the baseline process. The sub threshold characteristics (Fig-7 & 8) of both NMOS & PMOS shows that leakage increased around one order due to lower down the threshold voltage, but the sub threshold characteristics & DIBL are almost equivalent to 3.3V one.

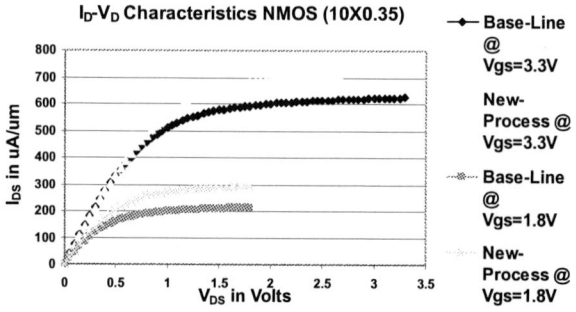

Fig. 3 Drain characteristics of NMOS

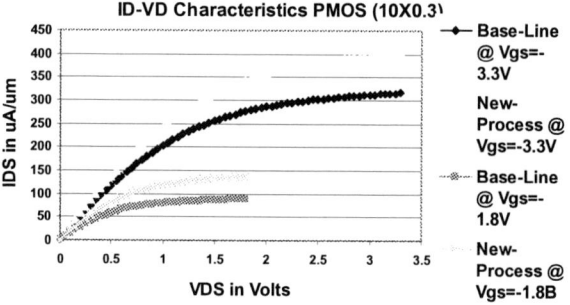

Fig. 4 Drain characteristics of PMOS

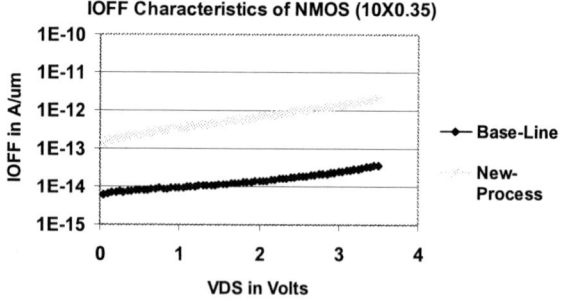

Fig. 5 IOFF Characteristics of NMOS

604

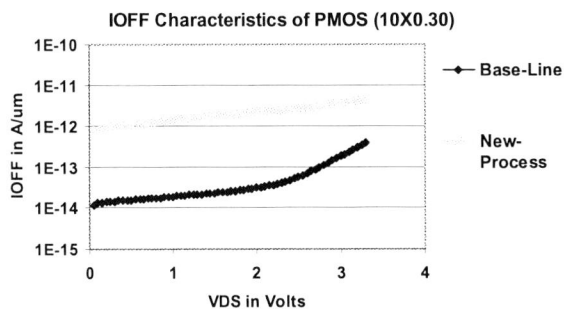

Fig-6 IOFF Characteristics of PMOS

The reliability test also was done to qualify the process and the HCI lifetime of NMOS & PMOS found 0.73yrs (Fig-9) &1.37E+17 yrs (Fig-10) respectively mean pass the test criteria (>0.2yrs).

The IC designed with 3.3V I/O library and fabricated with the new process worked functionally up to 1.5 volt (the minimum spec is 1.62V, 50ns for BIST vector and 1.62V, 100ns for Function vector) (Fig.-11). The other parameters such as leakage etc are also well within specification shown in Table-1.

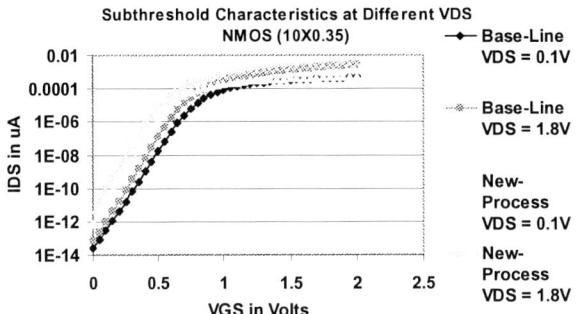

Fig.-7 Subthreshold Characteristics of NMOS

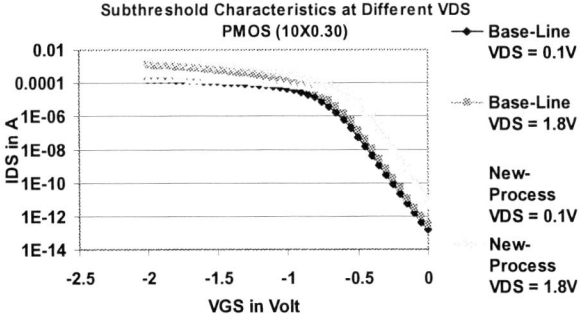

Fig.-8 Subthreshold Characteristics of PMOS

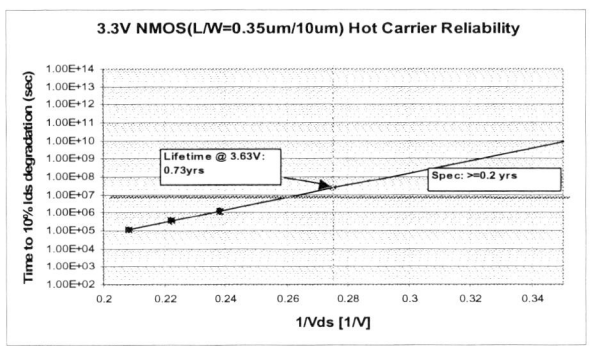

Fig-9 The HCI Data of NMOS

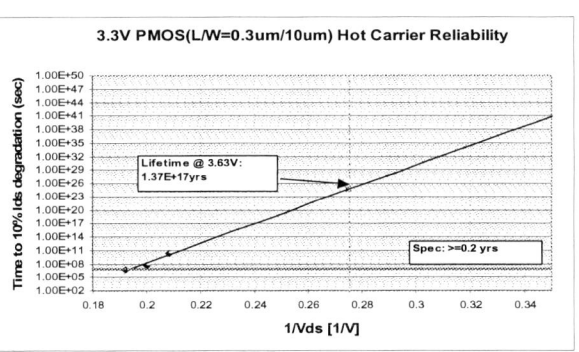

Fig-10 The HCI Data of PMOS

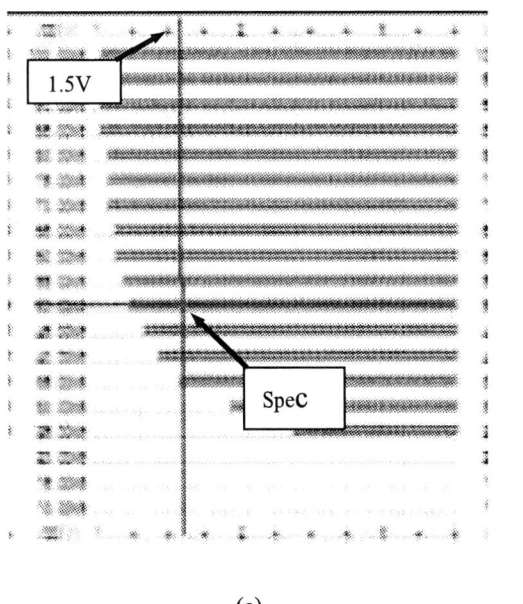

(a)

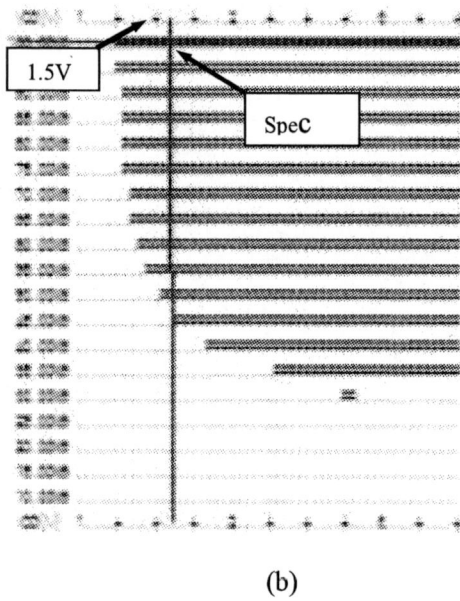

(b)

Fig-11 Shmoo of (A) BIST Vector & (B) Functional Vector

Table 1

Parameter			SPEC
SIDD	COVDD(uA)	11.23	150uA
	IOVDD(uA)	15.96	
DIDD	COVDD(mA)	9.65	45mA
	IOVDD(mA)	6.13	
REG	V	1.805	1.65~1.95V
PLL	Mhz	47.5	45~51Mhz

IV. CONCLUSION

The wide range I/O process (1.8V to 3.3V) developed in author's organization where (can be used) the 3.3V I/O library for I.C. design (can be used) is presented. The channel implant, LDD implant and gate oxide reduced for both NMOS & PMOS, the halo implant increased only for NMOS from baseline to achieve the required target such as higher Idsat, comparable Ioff in respect of baseline and passed the process reliability test. The device designed with existing library and fabricated with the new developed process showed a good shmoo test result and passed all test criteria with a good margin. The production started in author's organization after passing all device level

reliability test such as HTOL, Latch up & ESD and showed almost same yield trend as other products processed with the baseline (3.3V) process.

REFERENCES

1. R.R. Troutman, "VLSI Limitations from Drain Induced Barrier Lowering." IEEE Trans. Electron Device; ED-26, p-461, (1979)
2. J.Greenfield and R. Dutton, "Nonplaner VLSI Device Analysis using the solution of Poission's Equation." IEEE Trans. Electron Devices; ED-27, p-1520 (1980)
3. T.Y.Chan et al., IEDM Tech Digest p-718, 1987

High-Speed Hybrid Parallel-Prefix Carry-Select Adder Using Ling's Algorithm

Lakshmanan, Ali Meaamar and Masuri Othman
VLSI Design Research Group
Department of Electrical, Electronic and Systems Engineering
University Kebangsaan Malaysia
E-mail: lakshman@vlsi.eng.ukm.my , masuri@vlsi.eng.ukm.my

Abstract **Parallel-prefix adders offer high efficiency solution in terms of area, speed, power and regularity to the binary addition problem and are well suited for VLSI implementation. In this paper, a novel technique of implementing a hybrid parallel-prefix ling adder is presented. Experimental results show that the proposed adder has an improvement of 63% in speed and about 13% reduction in power consumption compared to Carry Lookahead adder (CLA).**

1. Introduction

THE addition of two binary numbers is the fundamental and most often used arithmetic operation in microprocessors, digital signal processors(DSP) and data-processing in application-specific integrated circuits (ASIC). Therefore, binary adders are crucial building blocks of very large-scale integrated (VLSI) circuits and are employed in the designs of Arithmetic-Logic Units, arithmetic datapaths and in address generation units.

A large variety of algorithms and implementations have been proposed for binary addition ranging from ripple-carry with delay proportional to operand length up to carry-lookahead adder (CLA), carry skip adders, and parallel-prefix adders [1]. Since the introduction of CLA [2] in 1956, the parallel implementation of CLA and its variants in parallel-prefix form [3]- [6] have given rise to hybrid parallel-prefix adder implementations. All these architectures are some of the alternative ways of solving the problem of computing a carry signal at each bit position of the result. The algorithmic comparisons and evaluations of popular prefix trees was done by Knowles [7]. A recent comparison of the VLSI adders has been presented in [8].

Ling [9] proposed that instead of having a single signal at each bit position for encoding the carry, he allowed this encoding to be spread into two signals, relaxing the carry computation. Ling adders have been shown to have speed advantage over CLA adders in emitter-coupled logic (ECL) technology. Doran [10] examined the possible encoding of the two signals as a function of the encoding selected for the carry generate and carry propagate and showed that the logic levels of the carry computation of the CLA could be reduced. The simplified form of Ling equations has been exploited for the design of multilevel block CLA [11]-[13]. But all of these designs make use of Static CMOS custom or semi custom design. The implementation of High-Speed Hybrid Parallel Prefix Carry Select adder in Dynamic CMOS standard cell library has not been much exploited and hence this adder was described in Verilog HDL and mapped into $0.18\,\mu$ m technology library.

The rest of the paper is organized as follows: Section II gives the description of the parallel-prefix formulation of binary addition and the definitions and the summary of Ling addition. Section III introduces the parallel-prefix Ling Carry Select Adder and in section IV, the experimental results and comparisons are given and finally the conclusion is presented in Section V.

II Background and Definations

A. Parallel- Prefix Addition

Let $A = a_{n-1}a_{n-2}\cdots a_0$ and $B = b_{n-1}b_{n-2}\ldots b_0$ represent *n*-bit binary numbers with $S = s_{n-1}s_{n-2}\cdots s_0$ denoting their sum. A parallel-prefix adder can be considered as a three stage circuit as shown in Figure 1.

0-7803-9730-4/06/$25.00 ©2006 IEEE

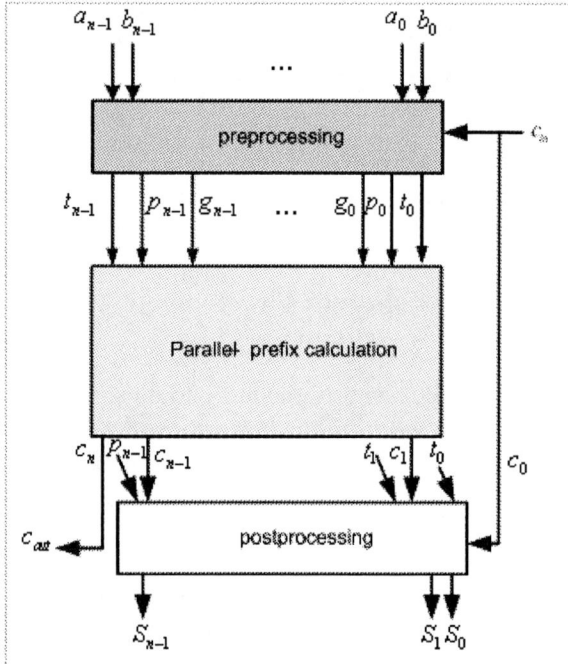

Figure 1. Three Stages of Prefix-addition

The preprocessing stage computes the carry-generate bits g_i, the carry-propagate bits p_i and half sum bits t_i according to: $g_i = a_i.b_i$, $p_i = a_i + b_i$ and $t_i = a_i \oplus b_i$ where ".", +, and $\oplus$ denote the logical AND, OR and exclusive-OR operations respectively. The 2nd stage performs the carry computation based on:

$$c_i = g_i + p_i c_{i-1} \qquad (1)$$

The final stage computes $s_i = t_i \oplus c_{i-1}$ (2)

Carry computation in the 2nd stage can be transformed to prefix problem [4], using the associative operator ($\circ$) which associates pairs of generate and propagate as follows:

$$(g_1, p_1) \circ (g_2, p_2) = (g_1 + p_1 g_2, p_1 p_2) \qquad (4)$$

In a series of consecutive associations of generate and propagate pairs (g, p), the notation $(G_{k:j}, P_{k:j})$ is used to denote group generate and group propagate terms produced out of bits $k, k-1, \cdots j$, i.e:

$$(G_{k:j}, P_{k:j}) = (g_k, p_k) \circ (g_{k-1}, p_{k-1}) \circ \cdots \circ (g_j, p_j) \qquad (5)$$

Following the above definitions, each carry c_i is equal to $G_{i:0}$.

One of the most important property of prefix operator ($\circ$) is its idempotency property [13] that is:

$(g, p) \circ (g, p) = (g, p)$. This property allows a group term $(G_{i:j}, P_{i:j})$ to be derived by the association of two overlapping terms $(G_{i:k}, P_{i:k})$ and $(G_{m:j}, P_{m:j})$ with $i > m \geq k > j$, since

$$(G_{i:j}, P_{i:j}) = (G_{i:k}, P_{i:k}) \circ (G_{m:j}, P_{m:j})$$

(6)

Representing the $\circ$ operator as a black cell $\bullet$ or as a grey cell $\circledcirc$ or as a white cell $\bigcirc$, parallel-prefix computation can be represented as a acyclic graphs. Figure 2 presents a 16 bit Sklansky [14] adder depicting these cells.

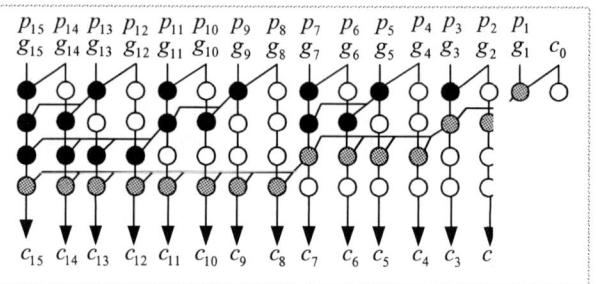

Figure 2(a) 16bit Sklansky adder

B. Ling Adders .

Ling[10] proposed a simplified form of carry lookahead equations with the carry given as [11]:

$$H_i = c_i + c_{i-1}.$$

Thus, the Ling carry H_i computes the carry faster then the conventional carry since less logic levels are needed. Consider the case of c_3 of a 4 bit CLA block and H_3 of the same block

$$c_3 = g_3 + p_3 g_2 + p_3.p_2.g_1 + p_3 p_2.p_1.g_0 \quad (7)$$
$$H_3 = g_3 + g_2 + p_2.g_1 + p_2.p_1.g_0 \qquad (8)$$

It can be seen that the last term of equation (7) as a fan-in of 4 and the last term of equation (8) has a fan-in of 3.

Equation (1), can be written as:

$$c_i = g_i + p_i c_{i-1}$$
$$= g_i p_i + p_i c_{i-1}$$
$$= p_i(g_i + c_{i-1})$$
$$= p_i(g_i + p_i c_{i-1} + c_{i-1})$$
$$= p_i(c_i + c_{i-1}) = p_i.H_i \qquad (9)$$

Substituting (9) in (2) yields

$$s_i = t_i \oplus (p_{i-1}.H_{i-1}) \qquad (10)$$

608

Equation (10) was further manipulated so that the output sum could be implemented with a multiplexer stage as shown in Figure 3.

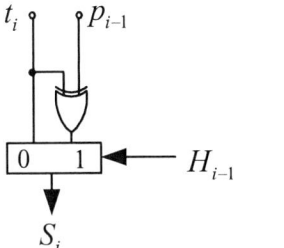

Figure.3. Ling Sum represented in multiplexer.

The derivation which leads to the implementation of Figure 3 is shown below:
From equation (10),

$$s_i = t_i \oplus (p_{i-1}.H_{i-1})$$
$$= \overline{t_i}.p_{i-1}.H_{i-1} + t_i.\overline{(p_{i-1}.H_{i-1})}$$
$$= \overline{t_i}.p_{i-1}.H_{i-1} + t_i.(\overline{p_{i-1}} + \overline{H_{i-1}})$$

$$= \overline{t_i}.p_{i-1}.H_{i-1} + t_i.\overline{p_{i-1}} + t_i.\overline{H_{i-1}}$$
$$= \overline{t_i}.p_{i-1}.H_{i-1} + t_i.\overline{p_{i-1}}(H_{i-1} + \overline{H_{i-1}}) + t_i.\overline{H_{i-1}}$$

$$s_i = \overline{t_i}.p_{i-1}H_{i-1} + t_i.\overline{p_{i-1}}.H_{i-1} + t_i.\overline{p_{i-1}}.\overline{H_{i-1}} + t_i.\overline{H_{i-1}}$$
$$= t_i.H_{i-1}(\overline{p_{i-1}} + 1) + H_{i-1}(t_i \oplus p_{i-1})$$
$$= t_i.\overline{H_{i-1}} + (t_i \oplus p_{i-1}).H_{i-1} \qquad (11)$$

II PRALLEL–PREFIX FORMULATION OF LING ADDITION.

In the following, section we present the methodology which allows the parallel-prefix computation of Ling carries. The 3rd and the 4th Ling carries are used to show the parallel-prefix computation.

$$H_4 = g_4 + g_3 + p_3.g_2 + p_3.p_2.g_1 + p_3 p_2.p_1.g_0 \qquad (12)$$

Since $g_i = g_i.p_i$, and defining a intermediate generate $G_i^* = g_i + g_{i-1}$ and intermediate propagate $P_i^* = p_i.p_{i-1}$, equation (8) and (12) can be written as:

$$H_3 = g_3 + g_2 + p_2.p_1.g_1 + p_2.p_1.g_0.p_0 \qquad (14)$$
$$H_4 = g_4 + g_3 + p_3.p_2.(g_2 + g_1) + p_3.p_2.p_1.p_0.g_0 \qquad (15)$$

With $g_{-1} = p_{-1} = 0, G_k^* = P_k^* = 0$ for $k < 0$, then equation(14) and equation (15) are expressed as:

$$H_3 = G_3^* + P_2^*.G_1^* \qquad (16)$$

$$H_4 = G_4^*. + P_3^*.G_2^* + P_3^*.P_1^*.G_0^* \qquad (17)$$

equations (16) and (17) can be written using the $\circ$ operator, as

$$H_3 = (G_3^*.P_2^*) \circ (G_1^*.P_0^*)$$
$$H_4 = (G_4^*.P_3^*) \circ (G_2^*.P_1^*) \circ (G_0^*.P_{-1}^*)$$

This formulation allows the parallel-prefix computation of the Ling carries and this allows the formation of Ling carries in parallel-prefix form.

Compared to the traditional preprocessing stage which just compute, g_i, p_i and t_i the new preprocessing stage will have to compute (G_i^*, P_{i-1}^*) which requires one extra logic level. However the number of terms (G_i^*, P_{i-1}^*) that needed to be associated is reduced by half. Hence, the basic cells used in the preprocessing stage are shown in Figure 4.

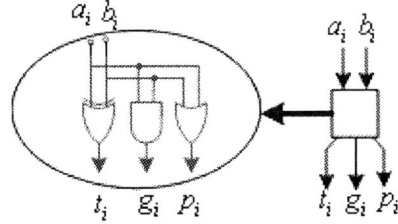

Figure 4(a) The logic level implementation of the square cells of the preprocessing stage.

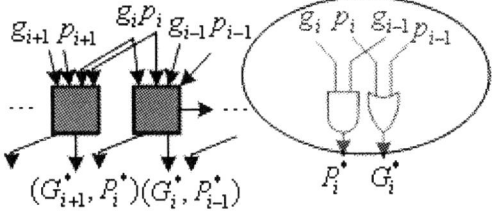

Figure 4(b) Generation of intermediate generate and propagate pairs

III PRPOSED HYBRID PARALLEL-PREFIX CARRY SELECT LING ADDER.

As the VLSI fabrication process enters deep sub-micron VLSI technology($< 0.15 \mu$ m) where the wiring delay becomes dominant over logic delay, the amount of wiring has to be considerably reduced. If this could be achieved, then, a high-speed low area, low power hybrid parallel prefix adder could be designed. With this goal in mind a new parallel-prefix carry-select Ling adder has been proposed. A 32-bit hybrid parallel-prefix carry-select Ling adder is shown in Figure 5.

This proposed structure generate the Ling pseudo- carries instead of the real carries c_i. The carries and the sum bits of the even and odd bits are generated separately and the carry-select blocks takes the input pairs (G_i^*, P_{i-1}^*) and not the traditional (g_i, p_i) pairs. A 4-bit Modified Carry-Select Adders (MCSA) implements equation (11) with some modifications as shown in Figure 6.

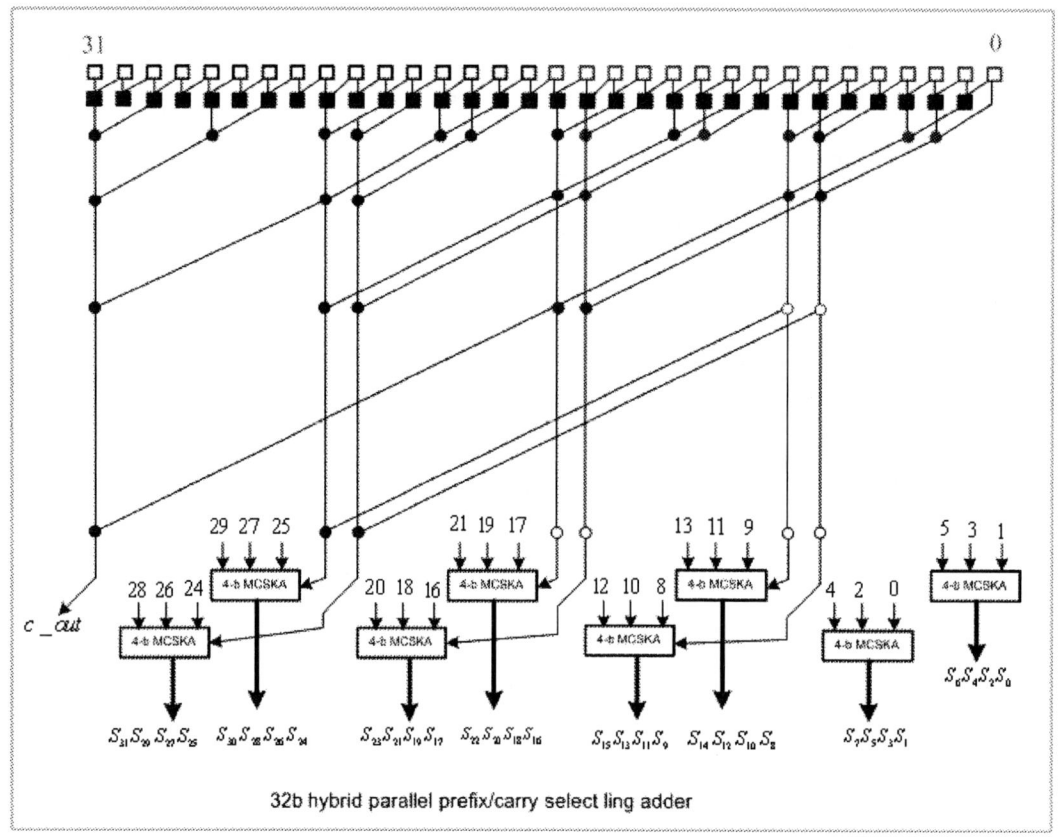

Figure 5. A 32-bit hybrid parallel-prefix carry-select adder.

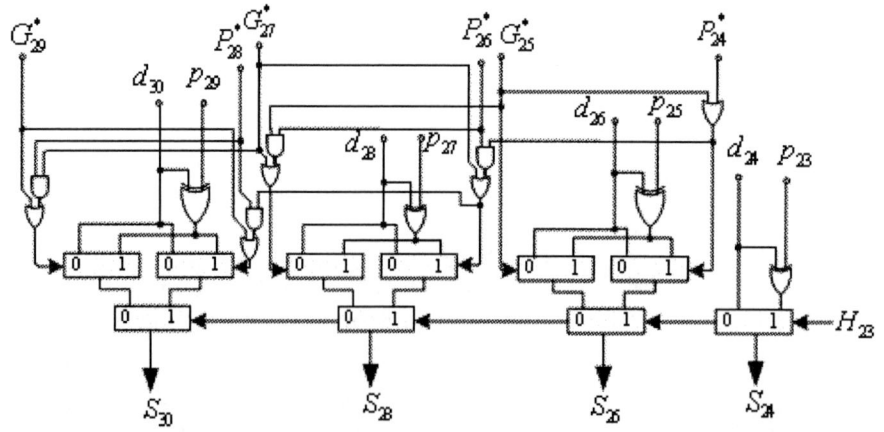

Figure 6. The MCSA block used in Figure 5 to generate S_{30} down to S_{24}.

TABLE 1

The Area, Power and Delay Estimates for the Proposed adder and other adder algorithms

Types of adder	Area (μm^2)	Power (mW)	delay (nS)	faster by :
CLA	2870.68 (1)	3.23 (1)	3.47 (1)	
Brent & Kung	3616.13 (1.26)	3.89 (1.2)	2.53 (0.729)	x1.42
Han & Carlson	4350.09 (1.51)	4.48 (1.38)	2.07 (0.592)	x1.67
Koggie & Stone	4703.54 (1.75)	4.34 (1.34)	1.43 (0.412)	x2.43
Knowles	4204.57 (1.46)	3.88 (1.2)	1.91 (0.55)	x1.87
proposed adder	3732.22 (1.31)	2.83 (0.87)	1.32 (0.38)	x2.63

IV EXPERIMENTAL RESULTS

The proposed adder was compared against CLA and against other parallel-prefix adders as shown in Table1. All of the adders were described in Verilog HDL and mapped on a 0.18μm technology from Silterrra using the Synopsys Design Compiler.

From Table 1, the area, power and delay are compared and normalized against CLA. It can be seen that the proposed adder has a speed improvement of 2.63 times then the CLA (62%) and has a power consumption of 0.87 p.u of CLA or (13% reduction) of CLA adder.

V CONCLUSION

A systematic methodology for designing a parallel-prefix carry-select Ling adder had been introduced and it has a speed advantage over the traditional parallel-prefix adders and the traditional CLA which was known to be the fastest adder. This proposed adder consumes much less power and hence basing on its performance, this proposed adder could find its way into fast Digital Signal Processors (DSP) and Multimedia processors (MMP).

REFERENCES

[1] I. Koren, "*Computer Arithmetic* Algorithms," Prentice-Hall International Inc., 1993.

[2] A. Weinberger & J. L. Smith,"A One-Microsecond Adder Using One-Megacycle

[3] R. E. Ladner and M. J. Fisher," Parallel Prefix Computation," *J. ACM*. vol. 27. no. 4, pp. 831-838, Oct. 1980.

[4] R. P. Brent and H.T. Kung, "A Regular Layout for Parallel Adders," *IEEE Trans. Computers*, vol. 31, no. 3, pp. 260-264, Mar.,1982.

[5] P. M. Koggie and H.S. Stone, "A Parallel Algorithm for the Efficient Solution of a General Class of Recurrence Equations," *IEEE Trans. Computers*, vol. 22, no. 8, pp. 786-792, Aug. 1973.

[6] T. Han and D. Carlson. "Fast Area-Efficient VLSI Adders", *Proc. Symp. Computer Arithmetic*, pp. 49-56, May 1987.

[7] S. Knowles, "A family of Adders," *Proc. 14th Symp. Computer Arithmetic*, pp. 30-34, Apr. 1999.

[8] V. G. Oklobdzija et.al., "Comparison of High-Performance VLSI Adders in the Energy-Delay Space," *IEEE Transaction on Very Large Scale Integrated (VLSI) Systems*. vol. 13, no. 6, 2005.

[9] H. Ling, " High-Speed Binary Adder*," IBM J. R&D*, vol. 25, pp. 156-166, May 1981.

[10] R. W. Doran, "Variants of an Improved Carry-Lookahead Adder ," *IEEE Trans. Computers*, vol. 37. pp. 1110-1113, 1988.

[11] S. Vassiliadis, "Recursive Equations for Hardware Binary Adders," *J. Electronics*, vol. 67, no. 2, pp. 201-213, Aug. 1989.

[12] Y. Wang et.al., "The Design of Hybrid Carry Lookahead/ Carry Select Adders," *IEEE Trans. Circuits and Systems* II, vol. 49, no.1, Jan 2002.

[13] J. Sklansky, "Conditional sum addition logic, " *IRE Trans. Electron. Comput.*, vol. EC-9, no.6, pp.226-231, June 1960.

Circuitry," *IRE Trans. Electronic Computers*, EC-5, pp. 65-73, 1956.

Study on Alignment Capability and Overlay Performance in 130nm BEOL Lithography Process

Lau Siau Yen[1], Suhana Mohd Said[2], Norhayati Soin[2], Kader Ibrahim[1] and
Ko Bong Sang[1]
[1]Silterra Malaysia Sdn. Bhd., 09000 Kulim, Kedah, Malaysia.
[2]Department of Electrical Engineering, University of Malaya, 50603 Kuala Lumpur, Malaysia
Email: siauyen_lau@silterra.com , norhayatisoin@um.edu.my , kader_ibrahim@silterra.com

Abstract **As the device becomes smaller, overlay accuracy requirement is more critical. The wafer alignment is one of the important elements that impact overlay accuracy. In this paper, an evaluation of alignment performance was performed using various alignment marks placed in the scribe-line of short-loop wafers used for SilTerra 130nm process. The alignment capability and overlay performance were studied for Trench 1 aligns to Contact and Via1 aligns to Trench 1, which follows layer-to-layer alignment scheme. A Design of Experiment was conducted with splits in interlayer dielectric (ILD) thickness, tungsten and copper CMP polishing time, in order to evaluate the process sensitivity of multiple alignment marks. Two different types of alignment marks were evaluated. There are Scribe-line Primary Marks (SPM) and Versatile Scribe-line Primary Mark (VSPM). To implement this experiment, exposures were performed using a scanner with various marks and recipes. Then, overlay measurement was conducted on an off-line overlay metrology tool to evaluate the effectiveness of the alignment performance. Data taken by scanner and off-line overlay metrology tool were analysed. The comparison was done between different mark types in an attempt to find out the robust alignment strategy (mark type/color/order) covering all process variation. The results show that, all alignment marks demonstrate low sensitivity to process variation where as there are no wafers were rejected due to alignment error. Besides that, the number of gratings in mark's subdivision for SPM and VSPM affect the signal strength.**

I INTRODUCTION

In photolithography, alignment is a critical step prior to exposure of wafers in the scanner. As the IC device gets smaller, the requirements concerning overlay accuracy have become much more restrictive. Wafer alignment is one of the important elements that impact overlay accuracy.

The alignment is the positioning the image of a specific point on a reticle to a specific point on the wafer (alignment mark) to be printed. It is to ensure that the circuit patterns of the reticle are correctly aligned to features of the previous layer.

Alignment process is carried out on an off-axis wafer alignment system, which is available in a 248nm DUV scanner. This wafer alignment system in the scanner comprised of two lasers with different wavelengths, green (532nm) and red (633nm) color. In addition, it uses multi-order detection, which is capable of detecting the first seven orders of the diffraction pattern. The usage of two colors ensures that there is always a usable reflected signal from at least one of the two wavelengths and this make the system less sensitive to mark depth variations. The usage of higher orders improves the behavior of the alignment system for asymmetric marks. [1]

Alignment mark is a reference mark, which is used for positioning and aligning wafers. Robust alignment mark is important, as the scanner's alignment system requires accurate signals to correctly align a pattern to a previous layer.

The alignment marks typically resemble periodic grating patterns, with structures extending in orthogonal x and y directions. To facilitate the 130nm technology and process development, VSPM marks has also been evaluated and compared with traditional Scribe-lane Primary Mark (SPM) marks. Refer to Fig.1, SPM mark consists two subdivisions with 8.0μm and 8.8μm gratings. SPM mark can be designed with different segmentation type, such as AH32, AH53 or AH74. VSPM comprised of 4 subdivisions. Three subdivisions each have 8.0μm gratings, and one subdivision has 8.8μm gratings. VSPM mark is designed to incorporate three physically different segmentation types into one alignment mark, to provide flexibility for optimized alignment along various process changes. The sketch of the structures of SPM and VSPM marks are shown in Fig.1.

In this experiment, the mark types to be evaluated including AH32, AH53 and AH74 gratings of SPM and VSPM, also known as 3rd, 5th and 7th order enhanced marks. Each mark type is designed to give higher signal strength in the corresponding order of the diffraction light. For instance, 3rd order enhanced mark tend to get a higher signal from 3rd diffraction order compare to the other orders.

In this evaluation, the diffraction color and order used for wafer alignment needs are specified in the alignment recipe. An alignment recipe defines which signals from the sensors, which color(s) and which order(s) are used to calculate the aligned position. For example, when 5th order enhanced mark is under evaluation, then a recipe detecting the 5th order of either green or red color was selected. Sometimes color dynamic is chosen instead of green or red color. Color dynamic can detect both the red and green wavelengths and use the stronger of the two signals for a particular order to determine the alignment mark position. This reduces the number of rejected wafers due to low signal and improves alignment accuracy.

II EXPERIMENT

An ASML 248nm DUV scanner system was used for all exposures analysed here. The alignment experiment was performed on 25 short loop wafers representing the Silterra 130nm copper damascene process flow. The reticle set of 130nm test chip contains various SPM and VSPM marks, and overlay target placed in scribe-line are used for evaluation. The layers to be evaluated are Trench 1 and Via 1.

A Design of Experiment was conducted with splits in interlayer dielectric (ILD) thickness, tungsten and copper CMP polishing time. There are three different process conditions for every split in an attempt to determine the process sensitivity of alignment system by using certain alignment marks.

To implement the experiment, alignment evaluation software, Alignment Advisor, in the scanner was turned on. This software enables the alignment system to measure the signals from variable marks using both colors and all diffracted orders simultaneously. The use of Alignment Advisor is to predict the robust alignment recipe for wafer alignment before exposure. With respect to the result from Alignment Advisor software, whole lot of wafers were aligned to the evaluated mark with chosen alignment recipe and exposed for overlay performance checking. Overlay metrology was performed on an off-line overlay metrology tool. Then, the wafers were reworked and exposure step was repeated with aligning to other alignment marks in order to collect the overlay data for every alignment mark. For the wafer alignment during exposure step, there are eight pairs of alignment mark were used. For the registration's sampling plan, a total of 36 points (bar-in-bar target) were measured on the wafer, 4 points per field.

The naming of the alignment marks comprised the information of mark type (SPM/VSPM), mark defined layer (Contact/Trench1) and mark segmentation types (AH32/AH53/AH74), as listed in Fig.2.

III RESULT AND DISCUSSION

1. Trench 1 aligns to Contact Layer

To evaluate the performance of alignment marks and ATHENA alignment system at Trench 1 layer, we first investigated the alignment capability by using Alignment Advisor software. Fig.2 and Fig.3 show the collected data of Alignment Advisor when aligning Trench 1 to Contact layer's marks. In this experiment, only the enhanced order of red (R), green (G) and color dynamic (C) for certain mark types were evaluated. For simplicity, data on x-axis only are shown, whereas the data on y-axis shows the similar result as x-axis. There are two SPM and two VSPM marks to be evaluated, which were defined at Contact layer. CON and NBCON are the formed marks with and without poly plateau beneath it respectively. Every VSPM mark consists of 3 sub-marks, which are AH32, AH53 and AH74.

Wafer quality (WQ%) is a measure of quantity of light, from the alignment laser(s), that is reflected by a scanned alignment mark and detected by the sensor of alignment system. It is used to indicate the signal strength of alignment mark in the exposure system. Cui *et.al* found out that, signal strength play a significant role for improved overlay control [2]. Multiple correlation coefficient (MCC) is a value which indicates how well of the alignment signal can be fitted with a sine curve.

From Fig.2, it shows that the alignment recipe by using color dynamic is the optimum since red and green give a bad WQ%, whereas most of the points are close to or less than specification limit. The wafers would be rejected if the detected signal too low. As revealed in Fig.3, the results of MCC close to 1 can be achieved for all alignment marks, which indicate that the alignment signal is a perfect sinusoidal wave. Hence, the alignment recipe with color dynamic of enhanced order is chosen for exposure and overlay checking.

After having established the optimum alignment recipe, then we exposed the same wafers with particular mark to evaluate the effectiveness of alignment system. The data of scanner in term of WQ% and largest mark residual are presented in Fig.4 and Fig.5. Largest mark residual is the largest difference between the measured mark position and the computed mark position. The average and standard deviation of WQ% for x-axis are plotted in Fig.4. Refers to Fig.4, SPM-NBCON-AH53 gives a better alignment performance with higher WQ and lower standard deviation compare to other marks. Fig.5 demonstrates that, the largest mark residual for all alignment marks were kept within +/-15nm.

After exposure, overlay measurement was performed on all wafers. The results of overlay error in misregistration (also known as translation), expansion, and rotation are presented in Table 1 and Table 2. The wafers were exposed and aligned with the same offset of process correction for overlay parameter without optimization. Since the overlay parameter can be compensated, the standard deviation and range are more critical in this evaluation.

Overlay error is defined as the amount of pattern misplacement relative to the previously printed exposure level. Significant overlay errors can exist even if alignment is perfect. [3] As in Table 1, the misregistration in x and y direction of all alignment marks are well controlled within 0.07 um of range with comparable standard deviation against extreme process variation. On the other hand, satisfactory results of expansion and rotation were obtained for all marks. The data range of within 0.2 can be achieved even without optimization of process correction.

2. Via 1 aligns to Trench 1 layer

The alignment performance for Via 1 to Trench 1 layer was evaluated in the same way as at Trench 1 layer. For the alignment of Via 1 layer, there are three marks to be evaluated, which are defined at Trench 1 layer. The marks including two SPM with different segmentation (AH53, AH74) and one VSPM marks with AH32, AH53 and AH74 sub-mark. Fig. 6 and Fig.7 show the results of WQ% and MCC, which were collected from Alignment Advisor respectively. As seen in Fig.6, VSPM mark shows a poor WQ% performance where as the detector captures very weak signals for red, green and color dynamic reflected light. Some points fell below or close to specification limit. To ensure the Via 1 can be aligned successfully, the alignment recipe by using color dynamic is the optimum. Additionally, there is no issue on MCC for all the marks where MCC close to 1 can be achieved, as presented in Fig.7. After alignment recipe has been chosen, 25 wafers were aligned and exposed by using color dynamic recipe with enhanced order corresponding to the mark types.

After exposure, the WQ% and largest mark residual data were collected from the scanner. WQ% values of each mark type are averaged from 25 wafers as shown in Fig.8. VSPM mark including three sub-marks had an acceptable WQ% with lower standard deviation compared to SPM mark. However, it gives a bad performance on the minimum of WQ%, which may lead to wafer rejection due to weak reflected alignment signal. In this evaluation, SPM marks show a better alignment capability with higher signal than VSPM mark. In addition, SPM-T1-AH53 shows a very excellent result on WQ% where as the average values as high as 41.44% (x-axis) can be achieved. This enhanced WQ% can be achieved may be due the good image contrast of copper alignment mark, which is surrounding by the transparent interlayer dielectric material.

Fig. 9 reveals the results of largest mark residual versus mark type. There is no issue in largest mark residual. The residuals were ranging between +/-15nm in x and y-axis.

Table 3 and Table 4 illustrate the overlay results taken from overlay metrology tool after exposing. From the result of Table 3, the standard deviation and range of misregistration are comparable for all mark except VSPM-T1-AH32, which gives a slightly higher of values. This is also true in expansion (x and y axis) and rotation. However, the overlay errors for VSPM-T1-AH32 can be managed within control limit. In conclusion, SPM-T1-AH53 presents a better overlay performance with lower standard deviation and range in misregistration and well control on expansion and rotation.

3 Comparison between SPM and VSPM mark.

Now let us focus on SPM and VSPM mark. According to Fig.10, the WQ% of a VSPM mark is relatively lower than an SPM mark, with AH53 segmentation type for CON, NBCON and T1 mark. This may be due to lesser gratings in mark's subdivision of AH53 segmentation in VSPM compared to SPM mark, as shown in Fig.1. Lower number of gratings reduces the reflected light, thus reducing WQ%.

IV CONCLUSIONS

In this experiment, the alignment capability and overlay performance of Trench 1 aligns to Contact and Via 1 aligns to Trench 1 with process variation were studied. In overall, all alignment marks demonstrate a low sensitivity to process variation and no wafers were rejected due to alignment errors.

As a result from this evaluation, a wider WQ% distribution across the wafers was observed after introduction of process variation. It also revealed that there is no issue in MCC and largest mark residual. The MCC closed to 1 and largest mark residual within +/-15nm across the wafer can be obtained for every evaluated mark. Without optimization of the process correction for overlay parameters, the range of the overlay errors is critical for evaluation. From the result, there is no significant difference in overlay errors among the evaluated marks. It appears that the range of misregistration can be well controlled within +/-0.07um under extreme process variation, which is acceptable in 130nm process line. Besides that, the other overlay errors such as expansion and rotation can be controlled within the range of 0.2 and 0.03 for Trench 1 and Via 1 layer respectively.

In order to make the conclusion for the robust alignment strategy, WQ% becomes the dominant factor to be considered since the other alignment and overlay parameters are comparable for all alignment marks. Despite aggressive process split, the alignment strategies that utilize SPM-NBCON-AH53 and SPM-T1-AH53 for Trench 1 and Via 1 reticle perform better in WQ% compared to the other marks. Alignment to SPM-NBCON-AH53 by using color dynamic and enhanced order gives a higher WQ% with smaller standard deviation. Meanwhile, for Via 1 aligns to SPM-T1-AH53, an enhanced WQ% as high as 41.44% can be achieved. This is due to the good image contrast of copper alignment mark.

In addition, signal strength can be affected by number of gratings in the mark. VSPM mark results in a relatively lower wafer quality compare to SPM mark with AH53 segmentation type due to lesser gratings of subdivision. If we summarise the signal strength with varying the number of grating, we can conclude that, the signal strength increases with the number of gratings in the mark's subdivision.

ACKNOWLEDGEMENTS

The authors would like to acknowledge and thank to SilTerra Malaysia Sdn Bhd for fully providing the facilities and materials for the experiment presented in this paper. Many thanks to Ko Bong Sang, Jeffery Manaf and Tan Eng Pheow for their discussions and supports.

REFERENCES

[1] J. Huijbregtse, R. van Haren, A. Jeunink, P. Hinnen, B. Swinnen, "Overlay performance with Advanced ATHENA Alignment Strategy", *Proc. SPIE,* Vol.5038, pp.918 (2003).

[2] Y.Cui, F. Goodwin, and R. van Haren, "Segmented alignment mark optimization and signal strength enhancement for deep trench process", *Proc. SPIE*, Vol.5375, pp. 1265-1277, (2004).

[3] Stanley Wolf Ph.D and Richard N.Tauber Ph.D, *Silicon Processing For The VLSI ERA, Volume 1: Process Technology 2nd Edition,* Lattice Press, Sunset Beach, California, (2000).

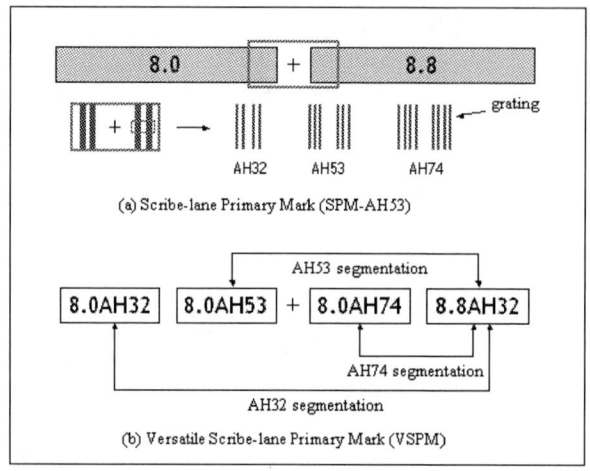

Fig.1 SPM and VSPM mark's structure.

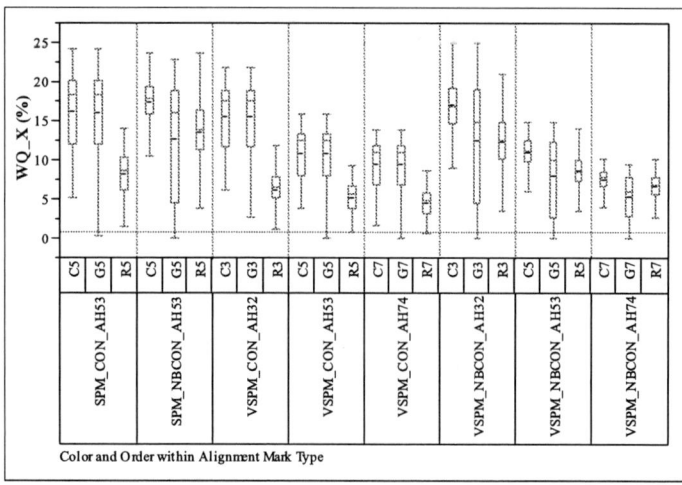

Fig. 2 Wafer quality (WQ%) comparison for
different alignment marks
when aligning Trench 1 mask in x-direction

Fig. 3 Multiple correlation coefficient (MCC)
comparison for different alignment marks
when aligning Trench 1 mask in x-direction

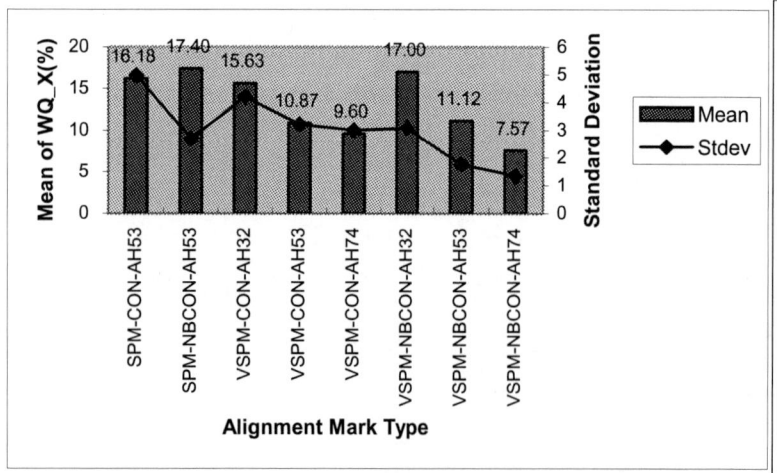

Fig.4 Mean of WQ% and standard deviation comparison
for different marks when aligning Trench 1 mask in x-direction.

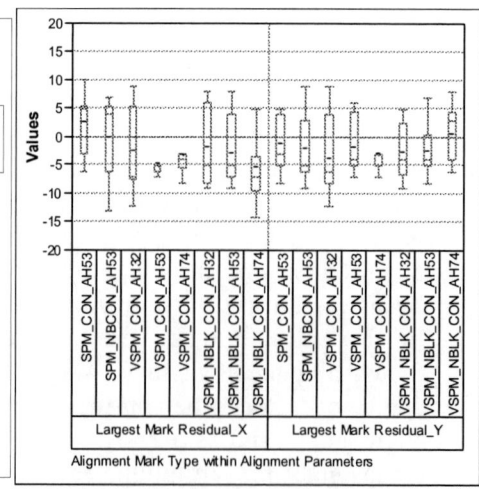

Fig. 5 Largest mark residual comparison
when aligning Trench 1 mask.

Table 1 Standard deviation and range of misregistration in x and y-axis for Trench 1 alignment.

Alignment Mark Type	Misregistration_X (μm)		Misregistration_Y (μm)	
	StdDev	Range	StdDev	Range
SPM_CON_AH53	0.00850	0.05160	0.00639	0.03630
SPM_NBCON_AH53	0.00658	0.04750	0.00824	0.04770
VSPM_CON_AH32	0.00876	0.06020	0.00892	0.05350
VSPM_CON_AH53	0.00869	0.05850	0.00905	0.04690
VSPM_CON_AH74	0.00808	0.05130	0.00868	0.04520
VSPM_NBCON_AH32	0.00877	0.05060	0.00807	0.04440
VSPM_NBCON_AH53	0.00771	0.04610	0.00771	0.04430
VSPM_NBCON_AH74	0.00888	0.05990	0.00769	0.04380

Table 2 Standard deviation and range of expansion (x and y-direction) and rotation for Trench 1 alignment.

Alignment Mark Type	EXP_X (ppm)		EXP_Y (ppm)		ROT (μrad)	
	StdDev	Range	StdDev	Range	StdDev	Range
SPM_CON_AH53	0.03060	0.11990	0.02359	0.10550	0.02989	0.13270
SPM_NBCON_AH53	0.02971	0.10942	0.02295	0.10270	0.04798	0.18370
VSPM_CON_AH32	0.03919	0.18540	0.02922	0.13880	0.03995	0.15018
VSPM_CON_AH53	0.03674	0.17000	0.03318	0.17320	0.02922	0.12170
VSPM_CON_AH74	0.02772	0.14010	0.02757	0.13170	0.01770	0.08080
VSPM_NBCON_AH32	0.03439	0.14308	0.02642	0.12090	0.03064	0.10300
VSPM_NBCON_AH53	0.03384	0.13586	0.02885	0.12830	0.01983	0.07040
VSPM_NBCON_AH74	0.03791	0.13950	0.02839	0.11590	0.02630	0.12300

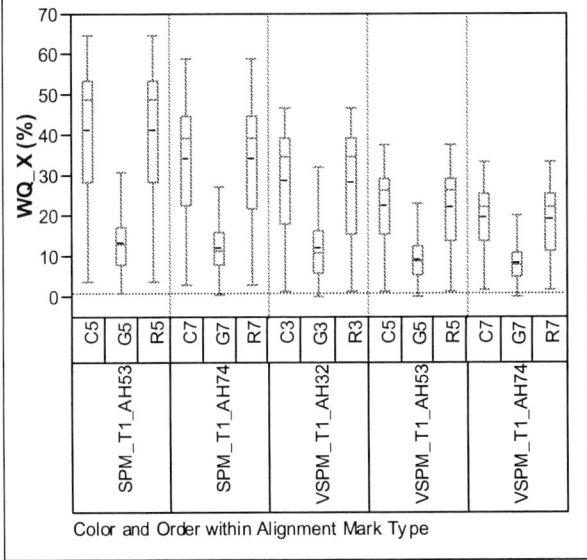

Fig. 6 Wafer quality (WQ%) comparison for different alignment marks when aligning Via 1 mask in x-direction

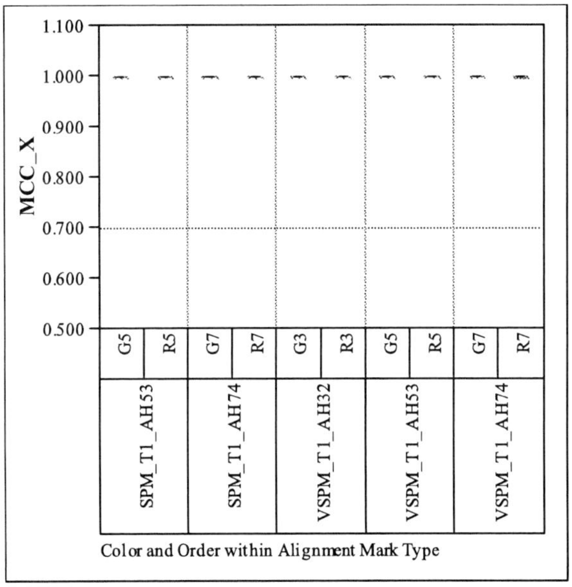

Fig. 7 Multiple correlation coefficient (MCC) comparison for different alignment marks when aligning Via 1 mask in x-direction

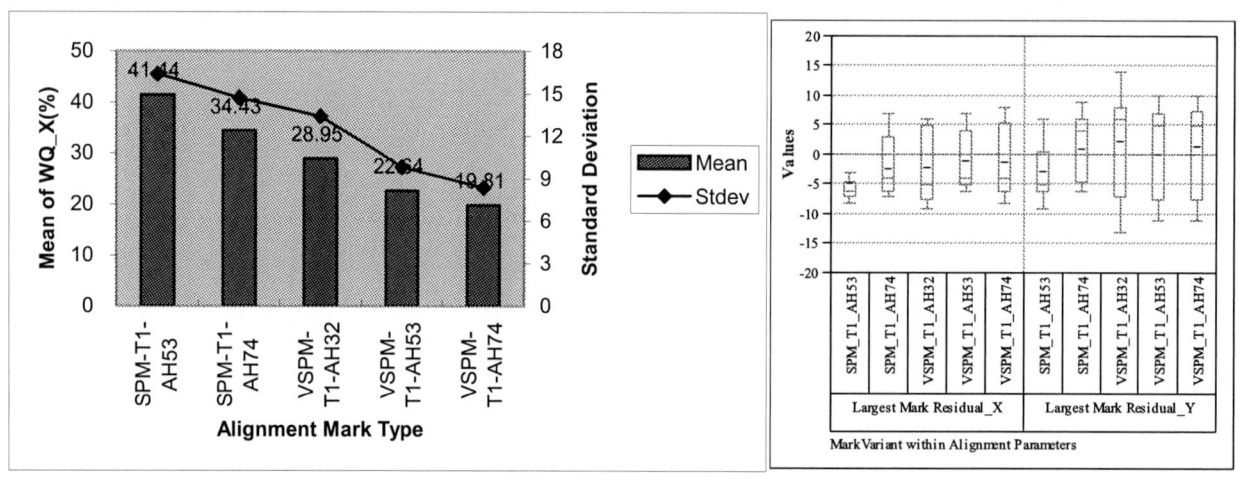

Fig.8 Mean of WQ% and standard deviation comparison for different marks when aligning Via 1 mask in x-direction.

Fig. 9 Largest mark residual comparison when aligning Via 1 mask.

Table 3 Standard deviation and range of overlay accuracy in x and y-axis for Via 1 alignment.

Alignment Mark Type	Misregistration_X (μm)		Misregistration_Y (μm)	
	StdDev	Range	StdDev	Range
SPM_T1_AH53	0.00692	0.03800	0.00715	0.04020
SPM_T1_AH74	0.00724	0.04340	0.00800	0.05050
VSPM_T1_AH32	0.00906	0.05618	0.01036	0.05420
VSPM_T1_AH53	0.00718	0.04150	0.00804	0.04390
VSPM_T1_AH74	0.00786	0.05280	0.00892	0.04810

Table 4 Standard deviation and range of expansion (x and y-direction) and rotation for Via 1 alignment.

Alignment Mark Type	EXP_X (ppm)		EXP_Y (ppm)		ROT (μrad)	
	StdDev	Range	StdDev	Range	StdDev	Range
SPM_T1_AH53	0.00022	0.00103	0.00269	0.01085	0.00031	0.00114
SPM_T1_AH74	0.00021	0.00098	0.00309	0.01339	0.00024	0.00085
VSPM_T1_AH32	0.00041	0.00202	0.00456	0.02268	0.00090	0.00368
VSPM_T1_AH53	0.00028	0.00122	0.00331	0.01413	0.00032	0.00129
VSPM_T1_AH74	0.00017	0.00070	0.00300	0.01137	0.00022	0.00075

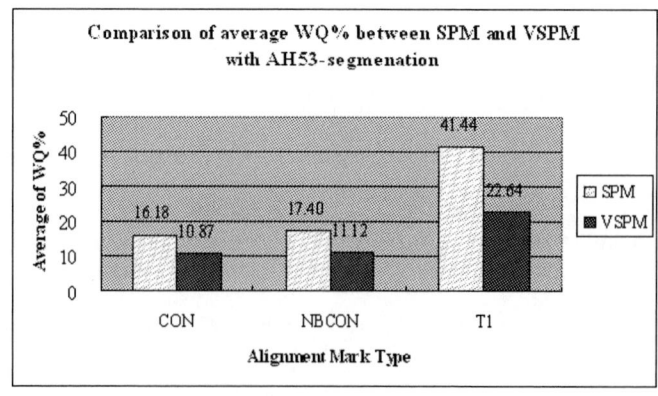

Fig. 10 Average WQ (%) comparison on SPM and VSPM mark with AH53 segmentation type

Studies on Electron Penetration Versus Beam Acceleration Voltage in Energy-Dispersive X-Ray Microanalysis

Sharon Lee, Hua Younan, Zhao Siping, *Snr Member, IEEE* and Mo Zhiqiang

Chartered Semiconductor Mfg Ltd
Woodlands Industrial Park D, Street 2
Singapore 738406
Email: leesharon@charteredsemi.com

Abstract: **Energy-dispersive X-ray microanalysis technique has been commonly used in failure analysis. It is vital for an analyst to understand the electron penetration depth in a certain material so as to be able to select an appropriate accelerating beam voltage. In this paper, we will use the Monte Carlo electron flight simulation method to obtain the electron penetration data at the different beam acceleration voltages of 5kV, 10kV, 15kV, 20kV, 25kV and 30 kV for the various possible elements/materials in wafer fabrication.**

I. INTRODUCTION

ENERGY-dispersive X-ray microanalysis technique (EDX) has been a common elemental analysis method used to determine the composition of particles, contamination and thin films in failure analysis of wafer fabrication [1-5]. Due to the complexity of wafer fabrication, the standardless EDX ZAF quantification method and the SEC factor correction method are often applied to obtain more accurate results when analyzing particles contamination and thin film layers. EDX technique is based on volume analysis, and as such, the electron penetration of the specimen is a direct function of acceleration beam voltage and sample density. Therefore, it is sometimes difficult for an analyst to discriminate the elemental information between the surface layer and the underneath layers.

For example, during chip assembly, TiN residue on the Al bondpads may cause non-stick on pads (NSOP). Hence, it is crucial to determine if there is any presence of TiN residue on the bondpad. However, it is difficult to determine trace amounts of TiN residue on bondpads with Al/TiW/Ti metallization (Al as top layer, and TiW and Ti as underneath layers) using EDX technique due to its limitation. The TiN can be determined by detecting the Ti Kα peak (4.51 keV). But this peak cannot be sufficiently excited by applying a 5kV accelerating beam voltage. So, a higher accelerating beam voltage should be applied. However, applying a higher accelerating beam voltage will cause a deeper electron beam penetration, whereby the elemental information from the underneath layers such as TiW and Ti could also detected.

For instance, the electron penetration in Al is about 1μm when applied with a 10kV accelerating beam voltage, and 2μm when applied with a 15kV accelerating beam voltage. At such depths, the electron beam may penetrate to the underneath layers (such as TiW and Ti), and Ti could have been excited, and as a result, some of these will escape through the surface of the specimen and be detected. This will produce misleading information on the source of the Ti Kα peak signal – is the Ti peak detected, originated from the surface residue of TiN or from the underneath thin film layers of TiW and Ti, due to application of higher beam voltage (10 kV and 15 kV).

Therefore, it is vital for an analyst to understand the electron penetration depth in a certain material so as to be able to select an appropriate accelerating beam voltage for EDX analysis based on the elements/layers of interest.

In previous paper, we have proposed an estimating method for electron beam acceleration voltage used in EDX technique and applied it in failure analysis [6]. In this paper, the Monte Carlo electron flight simulation method will be discussed. Using this simulation method, the electron penetration data at the different beam acceleration voltages of 5kV, 10kV, 15kV, 20kV, 25kV and 30 kV for the various possible elements/materials in wafer fabrication, which may be as a good reference for other analysts to understand the penetration depths and select a suitable beam voltage in failure analysis.

II. EXPERIMENT

The Monte Carlo electron simulation software is used to display the electron trajectories, and to estimate the penetration depth and lateral width in a bulk (or multi-layered) specimen, when simulated with a "user-defined" beam voltage. When an electron beam hits a solid material, it will penetrate some depth into the material. The depth of penetration is determined by the chemical makeup of the sample and by the force of the electron beam. The accelerating voltage determines the force of the electron beam.

In general, one can see that as the electron beam voltage increases, the penetration depth increases. Higher accelerating voltages, which apply more force to the electrons in the beam, can allow them to penetrate deeper into the sample. For example, Fig. 1 shows Monte-Carlo simulation results using 5kV, and 10kV acceleration voltages. One can clearly see that using 5 kV for bulk carbon, penetration depth is about 0.34µm, while using 10kV it is about 1.20µm.

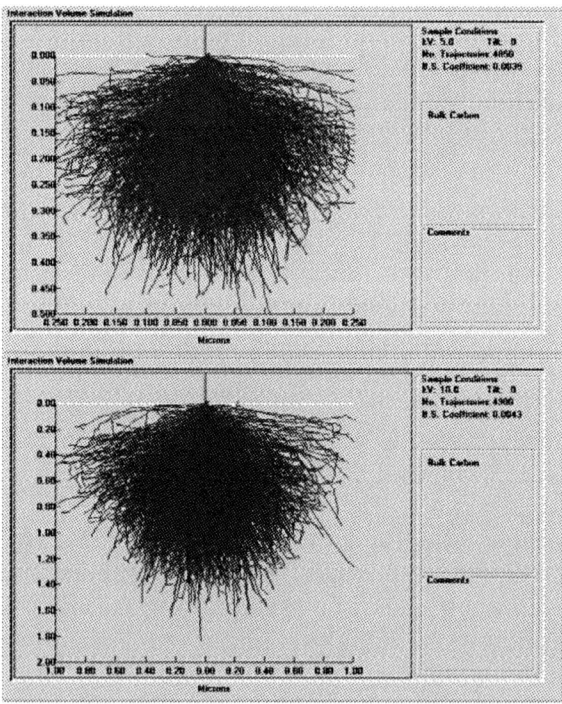

Fig. 1. The Monte-Carlo simulation results using 5kV, and 10kV acceleration voltages showed that using 5kV for bulk carbon, penetration depth is about 0.34µm, but using 10kV, it is about 1.20 µm.

One can also see that as the atomic number increases, the penetration depth decreases. Fig. 2 shows using 5kV for bulk carbon (Z=6) sample, it is about 0.34µm, but 5kV for bulk copper (Z=29), it is only about 0.08µm.

III. RESULTS AND DISCUSSION

We have used Monte-Carlo electron flight simulation method to obtain the electron penetration data at the different beam acceleration voltages of 5kV, 10kV, 15kV, 20kV, 25kV and 30 kV, which have been summarized into Table 1. From Table 1, we can understand the penetration depths in the various possible materials in wafer fabrication using different beam voltages. These data will allow us to select a suitable beam voltage to identify thin film layers in failure analysis

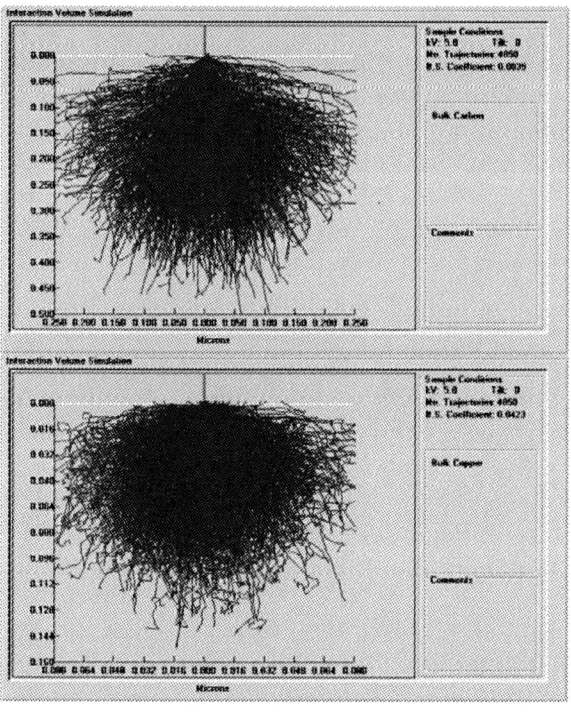

Fig. 2. The Monte-Carlo simulation results showed using 5 kV for bulk carbon (Z=6) sample, it is about 0.34µm, but 5 kV for bulk copper (Z=29), it is only about 0.08um.

IV. CONCLUSION

It is very important for an analyst to understand electron penetration depth in certain materials at different accelerating beam voltage so as to be able to select an appropriate accelerating beam voltage for EDX analysis based on the elements/layers of interest. In this paper, using the Monte Carlo electron flight simulation method, the electron penetration data at the different beam acceleration voltages of

5kV, 10kV, 15kV, 20kV, 25kV and 30 kV for the various possible elements/materials in wafer fabrication were calculated.

ACKNOWLEDGEMENT

The authors would like to thank the Surface/Advanced FA team, QRA-FA Dept, Chartered Semiconductor Mfg. Ltd. for their technical advice and contributions.

REFERENCES

[1]. Hua Younan, Z. R Guo & K. W. Chau, "Studies of ZAF Standardless EDX Quantification Method And Application in Failure Analysis of Semiconductor". Ournal of Trace and Microprobe Techniques, 15 (1), 13-31 (1997).

[2]. Hua Younan, Studies of EDX ZAF Quantification Method and Characterization of EDX Function in LEO 982 FESEM Equipment. The Proceedings from the SEMICON, Singapore, Testing, Assembly and Packaging 99-Technical Program, 3-6 May, SICEC, Singapore, 207-212 (1999).

[3]. Hua Younan, Application of SEC Factors in Energy-Dispersive X-Ray Quantification Analysis of Silicon Oxide (SIO_2) Layer in Wafer Fabrication. The proceedings from the 198th Meeting of The Electrochemical Society – Phoenix, Arizone, October 22-27, No. 810 (2000).

[4]. Hua Younan, Shailesh Redkar, Lau Chi-Kwan and Mo Zhiqiang (from PSB Corp., Singapore), A Study on Non-Stick Aluminum Bondpads due to Fluorine Contamination using SEM, EDX, TEM, IC, Auger, XPS & TOF-SIMS Techniques. The proceedings from the 28th International Symposium for Testing & Failure Analysis (ISTFA'2002), 3-7 Nov, Phoenix Civic Plaza, Phoenix, Arizona, USA p495-504 (2002).

[5]. Hua Younan, "Quantification Analysis of Thin Film Layers (Si_3N_4, SiO_2 and TiW) in Wafer Fabrication Using Energy-Dispersive X-Ray Microanalysis Technique". Journal of Trace and Microprobe Techniques, Vol. 21, No. 2, p25-34 (2003).

[6]. Hua Younan, Studies on An Estimating Method for Electron Beam Acceleration Voltage Used in Energy-Dispersive X-Ray Microanalysis Technique and Its Application in Failure Analysis of Wafer Fabrication. Journal of Instrumentation Science and Technology (IS & T), Vol. 32, No.2, pp.115-126 (2004).

Table 1: Electron Penetration (in um) Versus Beam Accelerating Voltage for Possible Elements from Wafer Fabrication

Z	SYMBOL	ELEMENT	5kV	10kV	15kV	20kV	25kV	30kV
5	C	Carbon	0.34	1.20	2.2	4.10	6.1	8.5
11	Na	Sodium	0.80	2.50	5.7	9.6	13.0	18.5
12	Mg	Magnesium	0.42	1.50	3.3	5.28	7.5	10.0
13	Al	Aluminum	0.28	1.00	2.0	3.90	5.0	7.0
14	Si	Silicon	0.30	1.20	2.3	3.5	5.3	7.5
15	P	Phosphorus	0.42	1.30	2.9	5.0	7.9	10.1
16	S	Sulfur	0.35	1.20	2.7	4.6	6.5	9.5
19	K	Potassium	0.90	3.10	6.0	9.0	14.0	21.0
20	Ca	Calcium	0.47	1.60	3.4	5.6	8.4	12.0
22	Ti	Titanium	0.17	0.60	1.3	1.9	2.7	4.0
24	Cr	Chromium	0.11	0.40	0.7	1.2	2.0	2.6
25	Mn	Maganese	0.11	0.40	0.8	1.2	1.7	2.3
26	Fe	Iron	0.10	0.40	0.8	1.1	1.8	2.5
27	Co	Cobalt	0.08	0.30	0.6	1.1	1.7	2.2
28	Ni	Nickel	0.09	0.30	0.6	1.0	1.6	2.1
29	Cu	Copper	0.08	0.30	0.6	1.1	1.6	2.2
30	Zn	Zinc	0.10	0.40	0.7	1.4	1.9	2.6
47	Ag	Silver	0.09	0.30	0.5	0.9	1.2	2.0
74	W	Tungsten	0.05	0.20	0.3	0.6	0.8	1.1
78	Pt	Platinum	0.05	0.15	0.3	0.5	0.7	0.9
79	Au	Gold	0.06	0.16	0.3	0.5	0.8	1.0
80	Hg	Mercury	0.09	0.20	0.4	0.7	1.1	1.6
82	Pb	Lead	0.11	0.30	0.5	0.8	1.3	1.6
	Al_2O_3	Al Oxide	0.18	0.60	1.2	2.1	3.3	4.3
	SiO_2	Si Oxide	0.30	1.00	2.1	3.3	4.5	7.0
	TiN	Ti Nitride	0.30	0.70	1.3	2.2	3.4	5.4
	WSi_x	W Silicide	0.06	0.20	0.3	0.5	0.8	1.1

Effect of Post Annealing Treatments on the Characteristics of Ohmic Contacts to n-Type InN

L. S. Chuah, Z. Hassan, H. Abu Hassan
Nano-Optoelectronics Research and Technology Laboratory
School of Physics, Universiti Sains Malaysia, 11800 Minden, Penang, Malaysia
Email: chuahleesiang@yahoo.com
Tel: 604-6533673, Fax: 604-6579150

Abstract To date, little work has been done regarding the annealing temperature and time dependence of contact resistance and contact intermixing on InN. In order to continue to improve for potential applications in photovoltaic devices, including high efficiency thin film solar cells, thermally stable ohmic contacts is required. To employ metal layers as a reliable ohmic contact on InN, it is essential to understand the thermal stability of metal-InN contact in addition to developing low resistance ohmic system. In this work, the structure of the InN films has been determined by means of conventional XRD phase analysis $\omega/2\theta$ scan. Different annealing temperatures (400°C-700°C) and durations (1-30 minutes) of Ni/Ag metal contacts are investigated, as thermally stable metal-semiconductor contacts are essential for high quality devices. Specific contact resistivity, ρ_c (SCR) determined using transmission line method (TLM) is carried out to the annealed (Ni/Ag) contacts where the electrical behavior of each of these conditions are compared. For relatively different annealing temperatures, substantial difference of the SCR values is observed between different durations samples.

I. INTRODUCTION

III-nitride has emerged as one of the most highly researched semiconductor materials in the world. It is an excellent candidate for high temperature, high frequency and high power electronics devices such as light emitting diodes, laser diodes, metal semiconductor field effect transistors and high electron mobility transistors.

In order to continue to improve these devices, thermally stable ohmic and Schottky contacts are required. Ohmic contact resistance and transparent electrodes are two critical issues that will ultimately determine the functional success of wide band-gap III-nitride based optoelectronic devices. Higher ohmic contact resistance and series resistance reduces the performance of optoelectronic devices, most notably in slowing down the device response speed. Therefore, many efforts have been made in improving n-type GaN ohmic contacts.

High temperature thermal annealing is generally required for ohmic contacts on n-GaN. In one study, Ti contacts on Si-doped, metalorganic chemical vapor deposition (MOCVD)-grown n-GaN (1.2×10^{18} cm^{-3}) became ohmic after annealing at 600 °C, and reached a specific contact resistance of 9.9×10^{-3} Ωcm^2 with a subsequent annealing step at 900 °C [1]. To date, no approaches have been reported with regard to the annealing temperature and time dependence of contact resistance and contact intermixing on n-InN.

In this paper, investigations on the electrical and structural characteristics of Ni/Ag contacts on unintentionally doped n-type InN are presented. Different annealing temperatures (400°C-700°C) and durations (1-30 minutes) were investigated, as thermally stable metal-semiconductor contacts are essential for high quality devices. Specific contact resistivity (SCR) was determined using transmission line method (TLM). The electrical behavior of each of these conditions were compared.

II. EXPERIMENT

In this study, the commercially sourced unintentionally doped n-InN/GaN sample grown on sapphire (Al$_2$O$_3$) was used. The structure of the films has been determined by means of conventional XRD phase analysis $\omega/2\theta$ scan. The ohmic contact Ni/Ag was deposited onto the InN through a metal mask by using a thermal evaporation method. Prior to the metal deposition, the samples were cleaned using the standard procedure. Samples were dipped into a mixture of NH$_4$OH: H$_2$O: 1: 20 for 5 min, then

0-7803-9730-4/06/$25.00 ©2006 IEEE

rinsed with deionised water. Subsequently, the samples were dipped into a mixture of HF: H_2O (1: 50) for 20 seconds then rinsed with deionised water. Boiling aqua regia (HCl: HNO_3 = 3:1) was used to chemically etch and clean the samples.

The surface cleanliness is important to ensure good quality of the contact and to minimize surface contamination. The Ni/Ag samples were annealed under flowing nitrogen gas environment in a furnace at temperatures ranging from 400°C-700°C for different durations. Heat treatment was carried out again after performing current-voltage (I-V) measurements for the subsequent annealing to investigate the thermal stability of the contacts.

The transmission line method (TLM) pads were 2mm (W, width) ×1 mm (d, length) in size, and the spacings, l, between the pads were 0.3, 0.4, 0.6, 0.9 and 1.3 mm. The specific contact resistivities, ρ_c were determined from the plot of the measured resistances against the spacings between the TLM pads. The linear-square method was used to fit a straight line to the experimental data.

III. RESULTS AND DISCUSSION

Fig. 1 shows X-ray diffraction (XRD) phase analysis scan of InN/GaN/Al_2O_3. The intensity data was collected by performing ω (sample angle) - 2θ (detector angle) scan at a range of different values. The peaks at about 31.14° and 34.38° correspond to (0002) and (0002) diffraction peaks of InN and GaN films with wurtzite structure, respectively. The peak characteristic of sapphire (0006) substrate was observed around 42°.

The measurements of the specific contact resistivity were made using the TLM method that has been widely used in the characterization of ohmic contacts to semiconductors. Resistance, R_i, between two contacts with spacing l_i, is given by

$$R_i = \frac{R_{sh}l_i}{W} + \frac{2R_{sk}L_t}{W} \quad (1)$$

$$R_i = \frac{R_{sh}l_i}{W} + 2R_c \quad (2)$$

where W is the width of the pad, R_c is resistance due to contact, R_{sh} is sheet resistance of the semiconductor layer outside the contact region, R_{sk} is the sheet resistance of the layer directly under the contact, and L_t is the transfer length.

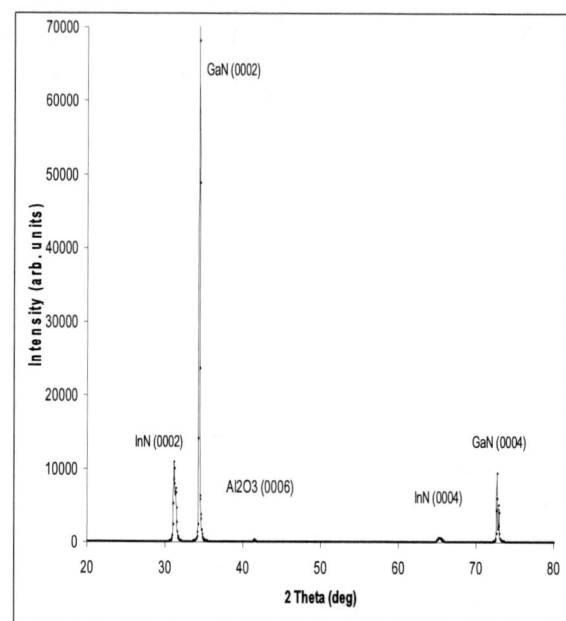

Fig. 1. X-ray diffraction scan of InN/GaN/Al_2O_3

The plot of R_i as a function of l_i will produce a straight line with a slope of R_{sh}/W, and $2R_c$ is yielded from the intercept at y-axis. The intercept at x-axis, will give L_x, where

$$L_x = \frac{2R_{sk}L_t}{R_{sh}} \approx 2L_t \quad (3)$$

with the assumption that $R_{sh}=R_{sk}$. On the other hand, the assumption of an electrically long contact $d>>L_t$ enabled the relationship $\rho_c = R_{sh}L_t^2$ to be invoked [2], which leads to $\rho_c = R_c W L_t$. In this study, heat treatments for Ni/Ag contacts could be divided into low (400°C), moderate (500-600°C) and elevated (700°C) annealed temperatures. The Ni/Ag contact resistivities measured in this study are summarized in Table 1. The initial investigation revealed that all as-deposited samples demonstrated 6.3 x 10^{-2} Ω-cm^2.

Fig. 2 shows the I-V characteristics of Ni/Ag contacts of samples thermally treated at 400°C under annealing durations of 15 minutes (cumulated 30 minutes). Ohmic behavior is observed. This particular annealing temperature was chosen to present the I-V characteristics because this is the optimum annealing temperature which produced the lowest SCR. On the other hand, for the sample treated at 600°C under annealing durations of 5 minutes (cumulated 7 minutes), ohmic behavior started to degrade due to decomposition of the InN where a slight non-linear *I-V* characteristic was observed.

Table 1. The specific contact resistivities at different annealing temperatures and times for Ni/Ag contacts

Annealing Temperature	Specific Contact Resistivities (Ω-cm^2)		
Low Temperature	Time / (cumulated time)		
	5 min	10min/(15min)	15 min/(30min)
400°C	5.9×10^{-2}	4.1×10^{-2}	3.5×10^{-2}
Moderate Temperature	Time / (cumulated time)		
	2 min	5 min/(7min)	.
500°C	5.8×10^{-2}	8.6×10^{-2}	.
600°C	6.6×10^{-2}	Sch.	.
Elevated Temperature	Time / (cumulated time)		
	1 min	2 min/(3min)	.
700°C	5.9×10^{-2}	Sch.	.

Sch.: Schottky behavior

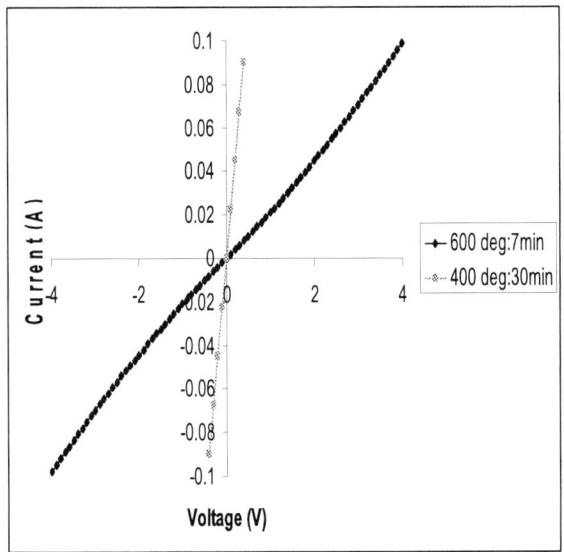

Fig. 2 The I-V characteristics of Ni/Ag contact of samples annealed at 400°C and 600°C for 30 and 7 min, respectively.

When the samples were thermally treated at low temperatures, i.e. 400°C, substantial difference of the SCR values was observed between samples under 5 min, 10 min and 15 min annealing durations. For annealing durations of 15 min (cumulated 30 minutes), SCR values obtained was found to be lower as compared to the subsequent treatment. However, when the samples were thermally treated at moderate temperatures, i.e. 500°C, ohmic behavior started to degrade, InN is known to be thermally unstable. Interface instability, is also a matter of concern because of the low bonding strength for In-N compared with those for Ga-N and Al-N, when an InN layer contacts with a GaN- or AlN-related layer.

On the other hand, both samples treated at 600°C and 700°C, showed deterioration after the subsequent 5 minutes (cumulated 7 minutes) and 2 minutes (cumulated 3 minutes) thermal treatment was introduced. Ohmic behavior started to degrade where a slight non-linear I-V characteristic was observed in I-V measurements and this could be mainly attributed to the formation of oxides.

Lately it has been reported that InN can be transformed into In_2O_3 by annealing in nitrogen [3] or air [4]. The source of oxygen was attributed to dioxygen contaminations in the N_2 gas. Although the samples were annealed in nitrogen ambient, a certain amount of oxygen could be present in the furnace which would lead to the formation of oxide compounds in the samples under high annealing temperature. Formation of oxide compounds in GaN has been claimed and reported by E. F. Chor [5].

IV. CONCLUSIONS

The thermal stability, electrical behavior of Ni/Ag contacts at various annealing temperatures (400°C - 700°C) has been investigated. For 15 min annealing durations (cumulated 30 minutes) at 400° C, SCR values obtained was found to be lower as compared to the subsequent treatment. On the other hand, both samples treated at 600°C and 700°C, showed deterioration after the subsequent 5 minutes (cumulated 7 minutes) and 2 minutes (cumulated 3 minutes) thermal treatment was introduced.

ACKNOWLEDGEMENT

The authors would like to acknowledge financial support from IRPA RMK-8 Strategic Research and Universiti Sains Malaysia.

REFERENCES

[1] H. Lu, W.J. Schaff, J. Hwang, H. Wu, G. Koley, L.F. Eastman, *Appl. Phys. Lett.* 79, pp. 1489 (2001).

[2] G.K. Reeves, and H. B. Harrison, "Obtaining the specific Contact Resistance from Transmission Line Model Measurements", *IEEE Electron Device Letters*, Vol. EDL-3 111, (1982).

[3] Motlan, E.M. Goldys, T.L. Tansley, *J. Cryst. Growth* 241, pp. 165 (2002).

[4] E. Kurimoto, M. Hangyo, H. Harima, M. Yoshimoto, T. Yamaguchi, T. Araki, Y. Nanishi, K. Kisoda, *Appl. Phys. Lett.* 84, pp. 212 (2004).

[5] E. F. Chor, D. Zhang, H. Gong, G. L. Chen, and T.Y.F. Liew, "Electrical characterization and metallurgical analysis of Pd-containing multilayer contacts on GaN", *J. Appl. Phys.* 90, pp. 1242 (2001).

Nanoporous InN Films Synthesized using Photoelectrochemical (PEC) Wet Etching

L. S. Chuah, Z. Hassan, F. K. Yam, H. Abu Hassan

School of Physics, Universiti Sains Malaysia, 11800 Minden, Penang, Malaysia

Email: chuahleesiang@yahoo.com, zai@usm.my

Tel: 604-6533673, Fax: 604-6579150

Abstract In this study, we have investigated the structural characteristics of nanoporous InN prepared by photoelectrochemical (PEC) wet etching. The PEC process which uses various 0.2, 0.5 and 1.0 wt % aqueous potassium hydroxide (KOH) solution utilizes photogenerated electron-hole pairs to enhance oxidation and reduction reactions taking place in an electrochemical cell. For etching condition using 0.2 wt % KOH solution (sample B), surface became relatively rough, however no pore was found. SEM images show that average pore size for sample C (0.5 wt % KOH solution) and sample D (1.0 wt % KOH solution) was around 30 to 60 nm. However, from our analysis of porous InN prepared by varying the etching condition, the non uniform etch rate across the sample surface is limited by diffusion processes. From the X-ray diffraction scan, porous samples show a broadening of the full width at half maximum with respect to the as-grown InN epilayer. On the other hand, the peak shift for InN (0002) and GaN (0002) diffraction planes was inconsistent. This can be explained by the relatively smaller statistical size distribution of the pores.

I. INTRODUCTION

INDIUM nitride (InN) is an attractive material for long-wavelength optoelectronic and high-speed electronics devices [1-3], because of a narrowest direct band-gap energy and superior carrier-transport characteristics in a wide range of temperature. However, the growth of InN has been less studied as compared with those of other III-nitride semiconductor materials such as gallium nitride (GaN) and aluminium nitride (AlN), because of the low thermal stability of InN. Recently InN thin films grown by MOCVD and MBE were found to have a band-gap energy in the range of 0.7–0.9 eV, much lower than the previously accepted value of ~1.9 eV [4].

Unfortunately, high quality InN is very difficult to synthesize. It has a very low decomposition temperature and requires a large overpressure of N. There are only a few reports on thin film deposition of InN [5-9] and on the one dimensional synthesis of InN nanostructures.

To date, no approaches have been reported to fabricate the InN nanoporous structures. Photon-asisted anodic etching is the most common method to produce nanoporous GaN. Porous semiconductors have been widely studied in the last decade, primarily due to the potential for intentional engineering of properties not readily obtained in the corresponding crystalline precursors as well as the potential applications in optoelectronics, chemical and biochemical sensing.

The research in porous GaN is strongly driven by the superior physical properties such as the excellent thermal, mechanical and chemical stability, in addition, it has been reported that porous GaN can be used as intermediate layer for the reduction of substrate induced strain [10-11]. Since bulk GaN in wafer size is not available, GaN thin film usually is grown on poor lattice and thermal mismatch foreign substrates, which will result in high residual, stress and eventually lead to high density of structural defects.

Porous GaN shows promise as a growth template for epitaxial re-growth; this could reduce the density of structural defects significantly and allows the growth of residual-free epitaxial GaN layers. Since the discovery of light emitting porous silicon by Canham in 1990 [12], significant progresses have been made on the studies of the structural, optical as well as mechanical and electrical properties of the porous Si. Comparatively, the study of porous GaN is still in the early stage, many fundamental properties are not well-established.

Similar to porous GaN, nanoporous InN can be used as a buffer or intermediate layer to reduce substrate-induced stress. Such a regrowth method may reduce the defect density in the epitaxial

0-7803-9730-4/06/$25.00 ©2006 IEEE

layer leading to high quality stress free layer on porous template. In this work, the investigation of nanoporous InN prepared using PEC wet etching was carried out.

II. EXPERIMENT

The commercially sourced InN/GaN sample was grown on sapphire (Al_2O_3). The samples were cleaned using the standard procedure. Samples were dipped into a mixture of NH_4OH: H_2O: 1: 20 for 5 min, then rinsed with sterile purified water. Subsequently, the samples were dipped into a mixture of HF: H_2O (1: 50) for 20 seconds then rinsed with sterile purified water. The surface cleanliness is important to ensure good quality of the contact and to minimize surface contamination.

To prepare porous InN by photoelectrochemical (PEC) method, thin layer of silver films were deposited using thermal evaporation method at one corner of the InN surface for front contact. The PEC etching of InN samples were carried out in a standard electrochemical cell at room temperature using an unstirred 0.2, 0.5 and 1.0 wt % KOH solution.

Etching were also performed using front-side illumination with the UV lamp in dark room and a standard electrochemical cell with a Pt counter electrode with external bias was applied between the sample and the cathode. The constant etch time is 60 min. The detailed experimental set up is shown in Fig. 1. Samples were chemically cleaned to remove surface contaminants and oxides before any characterization. After etching, the samples were characterized by scanning electron microscope and X-ray diffraction measurements. High resolution XRD (PANalytical X'pert Pro MRD) with a Cu-Kα_1 radiation source (λ = 1.5406 Å) was used to assess the crystalline quality.

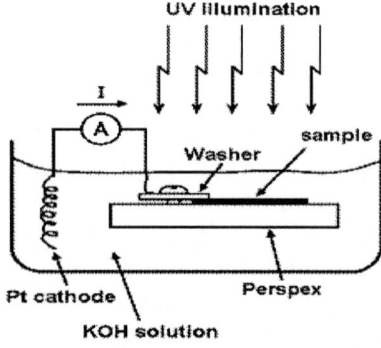

Fig. 1 The electrochemical etching experimental set up used to generate porous InN.

III. RESULTS AND DISCUSSION

Figure 2 shows the SEM micrograph of the as-grown InN film and nanoporous InN prepared under various etching condition. For etching condition using 0.2 wt % KOH solution (sample b), surface became relatively rough, however no pore was found. SEM images show that average pore size for sample c (0.5 wt % KOH solution) and sample d (1.0 wt % KOH solution) was around 30 to 60 nm. However, from our analysis of porous InN prepared by varying the etching condition, the non uniform etch rate across the sample surface is limited by diffusion processes.

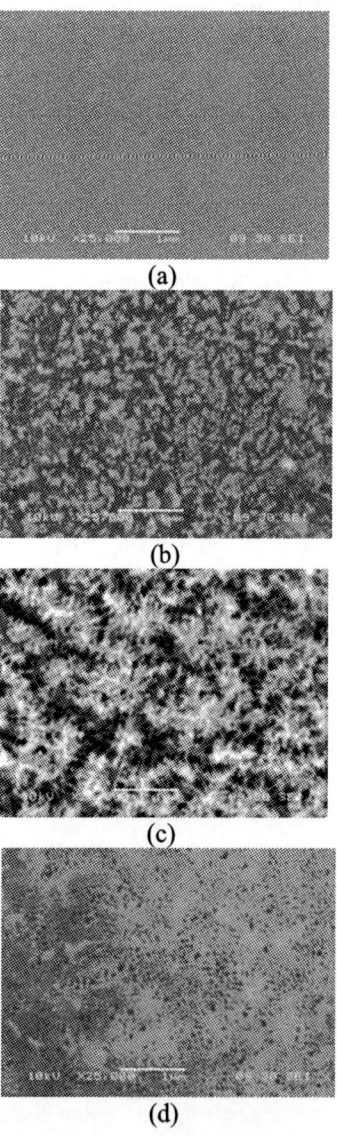

Fig.2 SEM micrographs of porous InN prepared under various etching conditions. (a) As grown (b) PEC method in 0.2 wt % KOH solution; (c) PEC method in 0.5 wt % KOH solution; (d) PEC method in 1.0 wt % KOH solution.

Diffusion-controlled etching can be obtained when an indium oxide layer at the InN/electrolyte interface is formed. This oxide layer partially hinders further dissolution of the underlying InN layer, as the carriers have to move across this barrier. The important reactions for the PEC etching process are the oxidation of InN and the following dissolution of the oxide.

However, an oxide layer remaining after etching can be a problem for device processing following PEC etching or for properties of the device itself. To avoid this problem we tried to dissolve the indium oxide that is formed by PEC

etching. The experimental results show that the dissolution of indium oxides strongly depends on the KOH concentrations.

Figure 3 shows the X-ray diffraction phase analysis scan of the samples. It can be seen that the porous samples exhibited broader full width at half maximum (FWHM) than the as grown sample. It is interesting to note that the FWHM increases with the etching concentration. On the other hand, the peak shift for InN (0002) and GaN (0002) was inconsistent. This can be explained by the relatively smaller statistical size distribution of the pores.

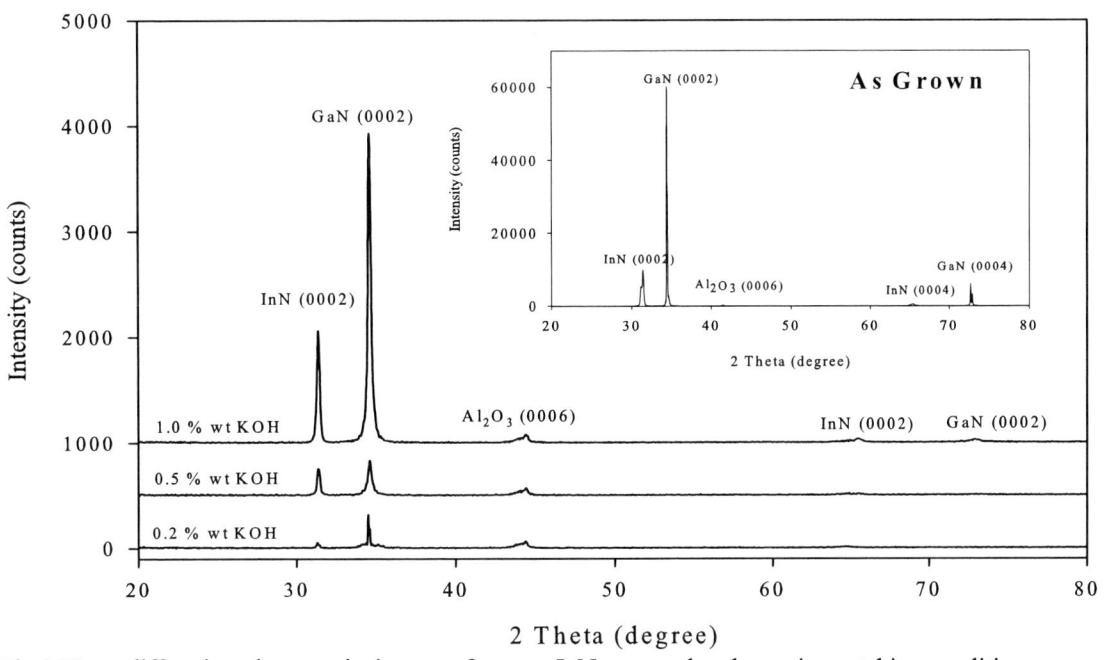

Fig.3 X-ray diffraction phase analysis scan of porous InN prepared under various etching conditions.

IV. CONCLUSION

A photoelectrochemical (PEC) wet etching method has been developed to produce nanoporous indium nitride. Structural characteristics of nanoporous indium nitride prepared by photoelectrochemical (PEC) wet etching method under different etching durations have been investigated. SEM images show that pore size for samples was about 30 to 60 nm. From the X-ray diffraction scan, porous samples show a broadening of the full width at half maximum with respect to the as-grown InN epilayer.

ACKNOWLEDGEMENT

The authors would like to acknowledge financial support from IRPA RMK-8 Strategic Research and Universiti Sains Malaysia.

REFERENCES

[1] T. Matsuoka, H. Okamoto, M. Nakao, H. Harima, E. Kurimoto, Appl. Phys. Lett. 81, pp. 1246 (2002)

[2] T.L. Tansley, C.P. Foley, *J. Appl. Phys.* 59, pp. 3241(1986).

[3] E.Bellotti, B.K. Doshi, K.F. Brennana, J.D. Albrecht, P.P. Ruden, J. Appl. Phys. 85, pp. 916 (1999).

[4] J. Wu, W. Walukiewicz, K. M. Yu, J. W. Ager III, E. E. Haller, H. Lu, W. J. Schaff, Y. Saito, Y. Nanishi. Appl. Phys. Lett. 80, pp 3967 (2002).

[5] M. C. Johnson, S. L. Konsek, A. Zettl, and E. D. Bourret-Courchesne, J. Cryst. Growth 272, pp. 400 (2004).

[6] R. S. Q. Fareed, R. Jain, R. Gaska, M. S. Shur, J. Wu, W. Walukiewicz, and M. A. Khan, Appl. Phys. Lett. 84, pp. 1892 (2004).

[7] F. H. Yang, J. S. Hwang, K. H. Chen, Y. J. Yang, T. H. Lee, L. G. Hwa, L. C. Chen, Thin Solid Films 405, pp. 194 (2002).

[8] A. Yamamoto, M. Adachi, and A. Hashimoto, J. Cryst. Growth 230, pp. 351 (2001).

[9] A. G. Bhuiyan, A. Yamamoto, A. Hashimoto, and Y. Ito, J. Cryst. Growth 236, pp. 59 (2002)

[10] C. K. Inoki, T.S. Kuan, C. D. Lee, A. Sagar, R. M. Feenstra, Mater. Res. Soc. Symp. Proc. 722, K1.3.1 (2002)

[11] C. K. Inoki, T.S. Kuan, C. D. Lee, A. Sagar, R. M. Feenstra, D. D. Koleske, D. J. Diaz, P. W. Bohn, I. Adesida, J. Electron. Mater. 32, pp. 855 (2003)

[12] L. T. Canham, Appl. Phys. Lett. 57, pp. 1046 (1990)

TEM Characterization of Nickel Silicide Process

K. Li, E. Eddie and S.P. Zhao, *Senior Member IEEE*

Chartered Semiconductor Manufacturing Ltd.

60 Woodlands Industrial Park D, Street 2, Singapore 738406

Phone: (65) 63604617. Fax: (65) 63622935

Email: likun@charteredsemi.com

Abstract **The advent of 65 nm technology makes nickel silicide finally come into the picture due to its relatively low electrical resistance and less silicon consumption. To accurately control the nickel silicide thickness and obtain the low resistance mono-silicide (N_iS_i) phase, characterization of each sicilidation process step is very important. This paper applies analytical TEM to characterize the nickel silicidation process. Different imaging technologies are compared and the results show that Z-contrast STEM imaging is a very good technique for the identification of different compositional layers and imaging processing also plays a very important role for image quality improvement.**

I. INTRODUCTION

WITH the advent of 65 nm technology, nickel silicide is finally coming into the picture. Compared with titanium silicide (for 0.25 µm technology and above) and cobalt silicide (from 0.18 µm to 90 nm), nickel silicide possesses similar electrical resistance (14~20 µΩ.-cm), but does not have line width-related nucleation problem, as seen in titanium silicide, and has significant less silicon consumption, which makes it the ideal material for shallow junction formation required by 65nm technology [1-3]. However, one big challenge is the stability of the low resistance nickel mono-silicide (NiSi), which is easy to transform into the high resistance $NiSi_2$ (35 µΩ-cm). To improve the phase stability of NiSi, refractory metallic element Pt is often added into the NiSi.

It is also a challenge to characterize the nickel silicide formation process. In order to get a better process control so that an expected junction depth can be obtained, the thickness of the individual TiN cap layer, Ni(pt) layer and the Ni/Si inter-mixing layer has to be first characterized without ambiguity before the first and second rapid thermal annealing (RTA) processes. As there are a number of heavy elements involved, how to clearly resolve the different layers is not an easy task.

In this paper, we applied different imaging techniques to characterize the nickel silicidation processes and the results were compared with one another. It was found, with the support of EDX line scan, that z-contrast STEM imaging technique, making use of high angle annular dark field (HAADF) detector, is an ideal technique for nickel silicide process characterization.

II. EXPERIMENT

The Pt doped Ni layer was deposited with physical vapor deposition (PVD) method on top of cleaned silicon substrate. To prevent the oxidation of the Ni (Pt) layer during RTA processes, a TiN capping layer was also deposited.

The TEM samples were prepared using the most widely used focus ion beam (FIB) based lift-out method. To make sure there is no FIB sample preparation caused change in the layer structure, a control sample was also prepared using conventional polishing-ion milling method.

TEM imaging and analysis were performed with an FEI 200 kV field emission microscope equipped with a HAADF STEM detector, an energy dispersive x-ray (EDX) analyzer and a Gatan Image Filter (GIF). To get z-contrast STEM images, a camera length of 80mm was used. EDX line scan was done with a point to point spacing less than 2 nanometers (7 points across 10 nanometers).

III. RESULTS AND DISCUSSION

0-7803-9730-4/06/$25.00 ©2006 IEEE

We first applied the normal bright field TEM imaging technique to analyze the thickness of the TiN and Ni (Pt) layers. The image contrast mechanism is the so-called mass-thickness contrast in combination with some diffraction contrast [4]. The TiN layer could be clearly identified (Fig. 1). Due to the difference in Ni/Pt concentration, two layers of Ni/Pt with different contrast could be seen. Furthermore, there seemed to be an additional diffusion layer of Ni into Si, but the image was not clear enough for us to make an exclusive conclusion.

To make unambiguous judgment, scanning transmission electron microscopy (STEM) technique was applied utilizing the high angle annular dark field (HAADF) detector. The atomic electron scattering factor increases with the atomic number and decreases with the scattering angle. However, at low scattering angle region there is the interference from Bragg

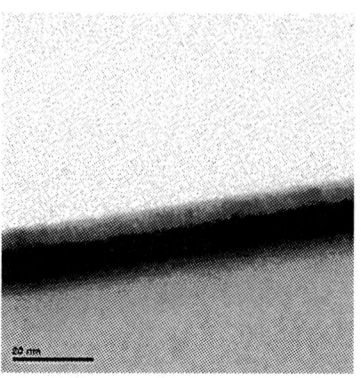

Fig.1 Bright field TEM image of multi-layered Ni (Pt) structure.

diffraction if the materials involved are crystalline. To get rid of the Bragg interference the STEM collection angle has to be big enough, normally bigger than 50 mrad. Experimentally, this is achieved by using a small camera length, which is normally around 100 mm. In this way, the collected signal intensity increases with atomic number and Z-contrast imaging is

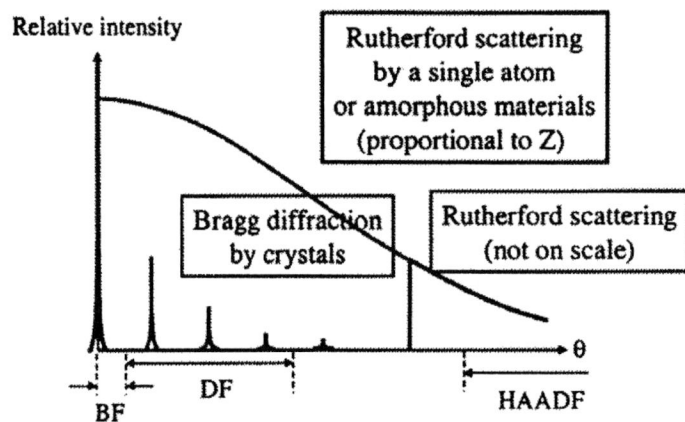

Fig.2 Variation of atomic scattering factor with scattering angles. Higher Z atoms have higher scattering intensity at the same scattering angles. But at low scattering angle region the existence of Brag diffraction destroys this Z relationship; the relationship holds only at high scattering angle region.

achieved.

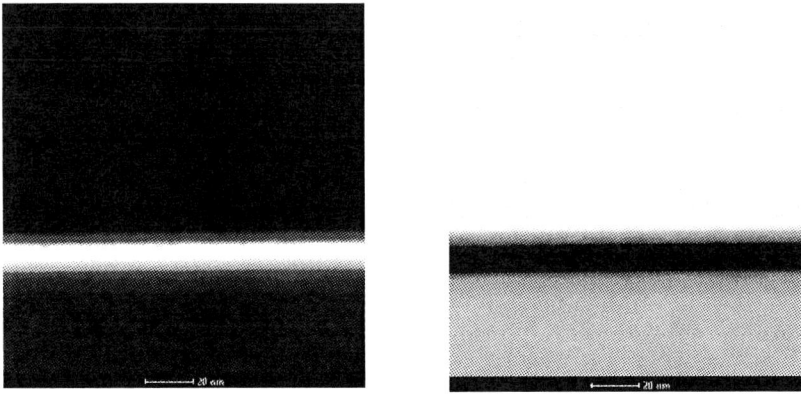

Fig.3 (a) HAADF STEM imaging showing the tailing Ni/Si intermixing layer, and (b) the reversed contrast image of (a).

The HAADF STEM image in Fig.3 clearly showed that in addition to the TiN cap layer and two Ni/SI inter-mixing layers with different Ni contents, there was really another tailing layer showing brighter contrast than silicon substrate. As Ni has a higher atomic number than Si, this brighter therefore should be a Ni-Si inter-mixing layer (less Ni). It was highly suspected that the thermal budget involved in the PVD deposition caused the inter-diffusion between Ni and Si. To confirm that this additional layer was not caused by FIB sample preparation, a sample was also prepared with the conventional polishing-ion milling method. The existence of the same additional layer ruled out this possibility.

To further analyze the composition variation across the different layer, a high resolution STEM EDX line scan was performed. The scanning was performed from the top capping TiN layer to the silicon substrate. The Ti, Ni, Pt and Si x-ray signals were monitored (fig. 4). The line profiles clearly resolved the different layers. The Ti peak was just in front of the Ni peak. The Ni profile showed two layers with different Ni concentration, corresponding to the two Ni layers of different contrast in the STEM image. The Ni

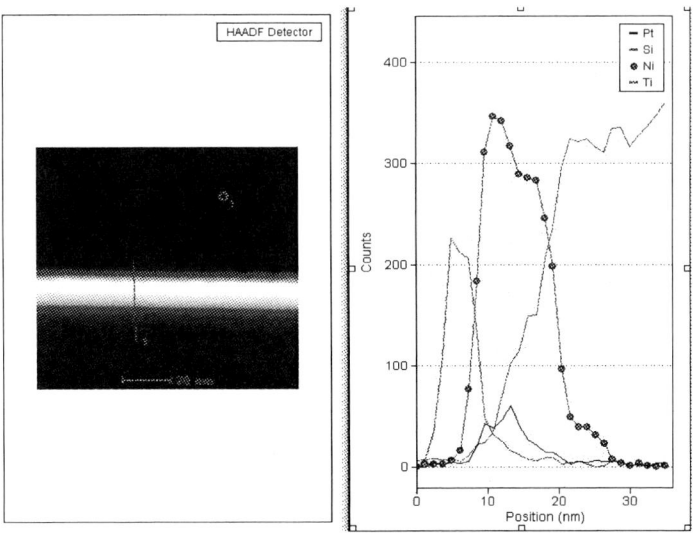

Fig.4 EDX line scan across the multiplayer showing that there is a Si-dominated tailing layer

layer closer to the surface had higher Ni content, while the Ni content of the layer closer to the substrate was less. It was very interesting to note that Pt mainly existed in the Ni layer closer to the surface, which suggested that it was nickel rather than Pt that diffused into Si substrate. The shoulder at the right hand side of the nickel profile confirmed that there was a tailing Ni-Si mixing layer. The results also demonstrated that nanometer spatial resolution EDX analysis could be achieved in TEM.

We also found that through imaging processing, i.e., mathematical manipulation of the digitized images, the contrast could be significantly improved and the tailing layer could also be resolved on condition that the sample was properly prepared. In the current case, the contrast difference between the Ni/Pt layer and Si substrate was too big, making it difficult to resolve the tailing layer, which had slightly different contrast from that of Si substrate. A square root algorithm, which formed a new image using the square root value of each pixel in the original image, improved the visibility of the different layer in the bright field TEM imaging significantly.

IV. SUMMARY

In summary, z-contrast STEM imaging is a powerful imaging technique for the characterization of nickel silicide processes, and imaging processing also plays a role in improving image quality.

ACKNOWLEDGMENT

The authors would like to thank the PVD group of the Technology Development Department, Chartered Semiconductor Manufacturing Ltd for providing the samples. The help from TEM sample preparation group is also appreciated.

REFERENCES

[1] K. Li, S.Y. Chen, and Z.X. Shen, "Identification of refractory-metal-free C40 $TiSi_2$ for low temperature C54 $TiSi_2$ formation," *Applied Physics Letters*, Vol. 78, no.25, pp.3989-3991 (2001)

[2] D.Z. Chi, D. Mangelinck, A.S. Zuruzi, A.S.W. Wong, and S.K. Lahiri, "Nickel silicide as a contact material for submicron CMOS devices," *Journal of Electronic Materials*, Vol. 30, no.12, p1483 (2001)

[3] Y.A. Alshehri, "Development of a full silicidation (FUSI) process for nickel silicide", *22nd Annual Microelectronic Engineering Conference,* pp52-56 (2004)

[4] D.B. Williams and C.B. Carter, *Transmission Electron Microscopy,* Plenum Press (1996).

Investigation and Failure Analysis of "Flower-like" Defects on Microchip Aluminum Bondpads in Wafer Fabrication

Hua Younan, Lim Yeow Kheng, Zhao Siping, *Member, IEEE,* Rao Ramesh and Tan Juan Boon
Chartered Semiconductor Mfg Ltd
Woodlands Industrial Park D, Street 2, Singapore 738406
Email: huayounan@charteredsemi.com

Abstract: **In this paper, Al fluoride defects on microchip Al bondpads were studied, which were confirmed to be due to a 12 hours delay prior to NE111 clean process. Failure analysis results and mechanism were discussed. Moreover, a preventive solution of introducing a time link between passivation etch and NE111 clean process was recommended and implemented**

I. INTRODUCTION

IT is well known that during wafer fabrication, galvanic corrosion would affect aluminum (Al) bondpad quality and cause discolored bondpads or non-stick on pads (NSOP) problem. In this paper, we would investigate a case of bondpad contamination and discuss the possible root causes and solutions. It was reported that some wafers were affected by "flower-like" defects on Al bondpads. To better understand the problem, failure analysis was performed on a wafer. The failure analysis results and wafer fab investigation revealed that the flower-like defects were caused by fluorine (F) contamination due to delay in cleaning process.

II. EXPERIMENTAL

In this study, SEM/EDX analysis was performed using FEI Sirion FESEM at beam accelerating voltage of 5 kV, spot size of 4 and a working distance of 5 mm. To achieve accurate EDX results, ZAF standardless EDX quantification method was used [1-2]:

$$W_i^w = (ZAF)_i \, K_i \qquad (1)$$

where W_i^w is the weight fraction of element i analyzed without coating correction, Z is the atomic number correction factor, A is the absorption correction factor, F is the fluorescence correction factor and

$$K_i = \frac{R_i}{SEC} \qquad (2)$$

where $R_i = \dfrac{I_i^{Meas}}{I_{i,Calc}^{Std}}$ and SEC is the Standardless Element Coefficients factors which could be calculated using standard samples. The SEC factors employed in this work obtained from our previous publication [2]. Quantified EDX results could be expressed as both weight fraction and atomic fraction. The atomic fraction ($W_i^{a'}$) could be calculated from the weight fraction ($W_i^{w'}$) using Eqn (3):

$$W_i^a = (W_i^w/A_i) / \sum (W_j^w/A_j) \qquad (3)$$

where W_i^a is the atomic fraction of element i analyzed with coating correction. W_i^w & W_j^w are the weight fraction of element i & j analyzed with coating correction, and A_i & A_j are the atomic weight of elements i & j.

Auger electron spectroscopy (AES) was employed to perform elemental analysis. AES analyses were carried out using PHI (Physical Electronics) SMART-200 Auger System. The analysis conditions were as follows: (a) Electron gun was set with operating voltage of 20 keV, operating current of 10 nA and gun tilted at 30°; (b) Ion gun was set with Ar$^+$ ion source, operating voltage of 2 keV, operating current of 1000 nA (variable, to maintain etch rate of 100 Å /min), rastering size of 2 mm x 2 mm and sputter etch rate was nominally at 100+/- Å /min.

Auger survey range was 30-2000 eV. The relative atomic concentration of the detected elements were calculated by first measuring elemental peak-to-peak heights of Auger spectrums in the survey scans and then applying sensitivity factors based on standard spectra of pure elements or selected compounds:

$$C_i = I_i / (I_{std}S_iD_i) \qquad (4)$$

0-7803-9730-4/06/$25.00 ©2006 IEEE

where I_i is the peak-to-peak amplitude of the element i from the test specimen, I_{std} is the peak-to-peak amplitude of the element i from the standard, S_i is the relative sensitivity factor and D_i is a relative scale factor between the spectra for the test specimen and standard.

III. RESULTS AND DISCUSSION

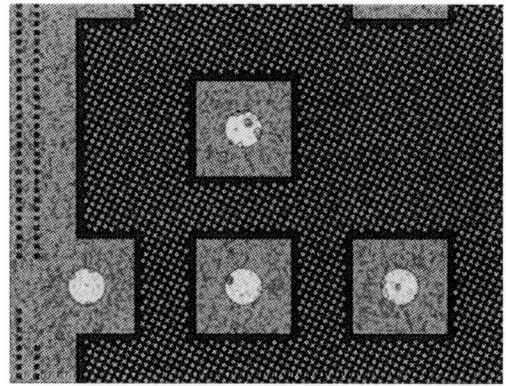

Fig. 1 Optical inspection found dot-like defects on Al bondpads.

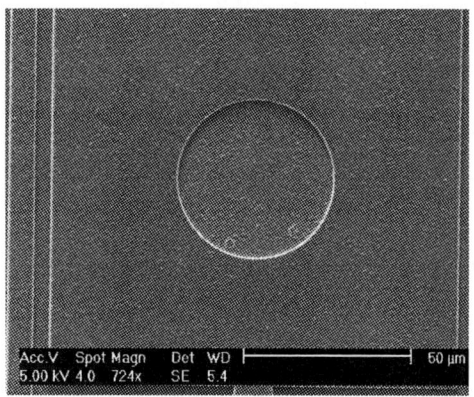

Fig. 2 SEM micrograph of dot-like defects on circular shaped bondpad as shown by the red arrow.

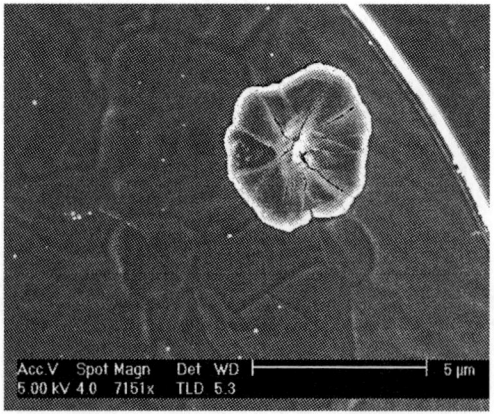

Fig. 3 Close-up SEM micrograph of the flower-like defect on Al bondpad as shown by the red arrow.

As shown in Fig. 1, dot-like defects were found on Al bondpads under optical inspection. Furthermore, scanning electron microscopy (SEM) inspection showed "flower-like" defects (Figs. 2 and 3), which were similar to those in previous studies [3-4]. Failure analysis results confirmed that the "flower-like" defects were caused by F contamination as high F peak was detected using energy-dispersive X-ray microanalysis (EDX).

IV. STUDIES OF FAILURE MECHANISM

F contamination on Al bondpads will cause F-induced corrosion. F will chemically react with Al to form Al fluoride:

$$Al + xF^- \rightarrow [AlF_x]^{(x-3)-} + 3e^- \quad (5)$$

The x in $[AlF_x]^{(x-3)-}$ may be 3 or 6 depending on the F contamination level.

If x = 3, then

$$Al + 3F^- \rightarrow [AlF_3] + 3e^- \quad (6)$$

If x = 6, then,

$$Al + 6F^- \rightarrow [AlF_6]^{3-} + 3e^- \quad (7)$$

Once $[AlF_3]$ or $[AlF_6]^{3-}$ is formed on Al bondpads, it is difficult to remove it by EKC 265, other chemicals or dry plasma cleaning process. This is why even on a normal Al bondpad, a small amount of F could be detected by AES. But such F contamination is generally 3-5 at %, which is insufficient to form Al fluoride, and is not detectable by EDX.

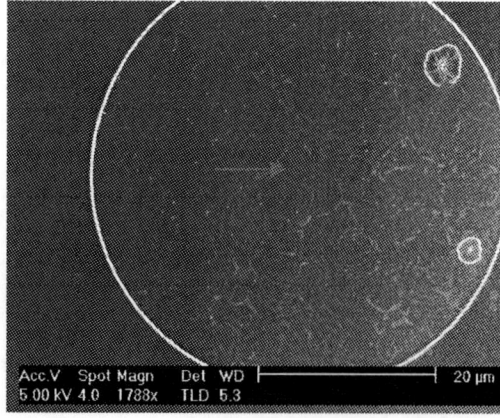

Fig. 4 EDX was conducted on the affected and unaffected areas of the bondpad as shown by the red arrows.

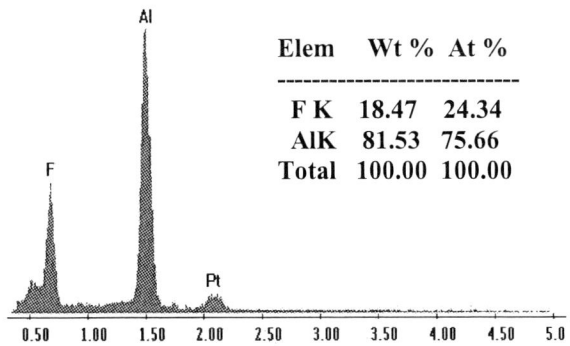

Elem	Wt %	At %
F K	18.47	24.34
AlK	81.53	75.66
Total	100.00	100.00

Fig. 5 EDX detected high F level on the flower-like defect shown Fig 4.

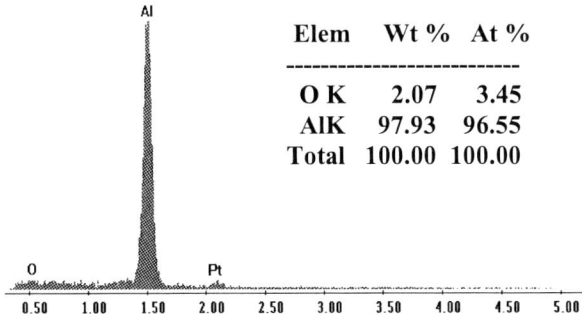

Elem	Wt %	At %
O K	2.07	3.45
AlK	97.93	96.55
Total	100.00	100.00

Fig. 6 EDX did not detected F on the unaffected area of the bondpad.

In this case, Al fluoride was formed and EDX was performed on both affected and unaffected areas as indicated by the red arrows in Fig. 4. Figure 5 shows very high F level, approximately 24.34 at % (or 18.47 wt %) was detected on the affect area. On the other hand, Fig. 6 shows no F was detected on the unaffected area. This showed that the level of F contamination was out of control and the possible root cause should be identified.

V. IDENTIFICATION OF THE POSSIBLE ROOT CAUSE

Based on the failure mechanism, it was deduced that the "flower-like" defects might be resulted from an abnormal process or other sources. In wafer fabrication, there are two possible sources of F contamination on Al bondpads. One of which is from the bondpad opening processes. CF_4 gas is used during passivation (Si_3N_4 and SiO_2) etch. Thus, carbon (C) and F contamination may be introduced on Al bondpads. To reduce the F contamination level on bondpads, EKC 265 or other chemicals (Aleg310, NE111) are used to clean away C, F and polymer on Al bondpads. In general, most of the F contamination can be cleaned away by these chemicals. In this case, NE111 was used for bondpad cleaning. In addition, high-energy O_2 plasma coupled with high temperature cycle was employed to effectively drive out F ions residing on AlCu grain boundary [5].

Another possible source of F contamination is from the packaging material used for wafer shipment. F contamination due to foam material has been reported in the literature [1-2]. However, in this case, such possibility has been ruled out, as the affected wafers were not fabout. Therefore, we strongly believe that the F contamination was introduced during wafer fabrication.

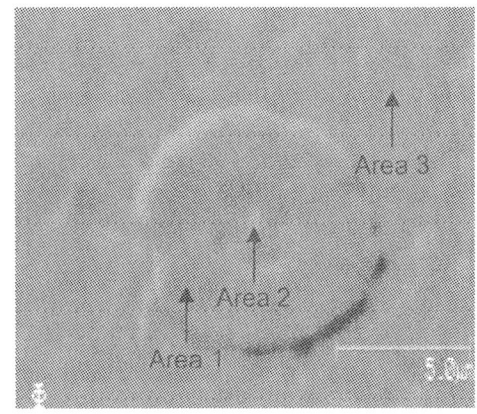

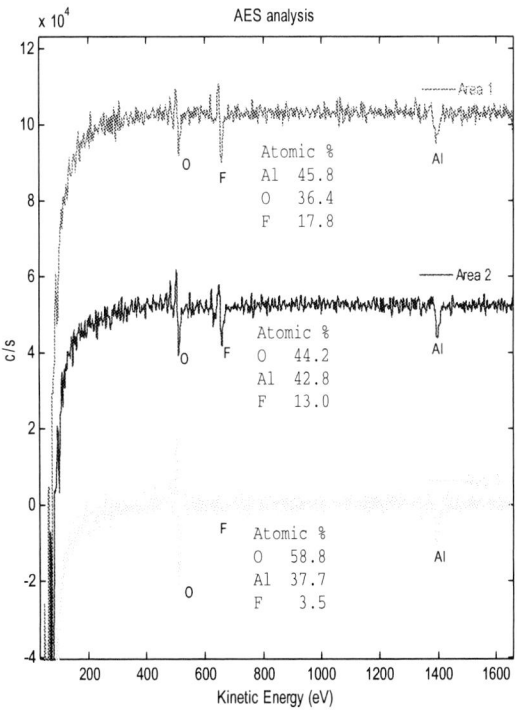

Fig. 7 Auger analysis results show F were 17.8 at % and 13.0 at % on affected area 1 and 2 respectively and 3.5 at % on unaffected area 3.

637

EDX did not detect any F peak on the unaffected area of the bondpad (Figs. 4 and 6). In this paper, we propose a rule for contamination analysis (also for future reference), "EDX clean does not means clean" and "Auger clean means clean". The meanings of the rules are: (a) EDX result shows clean, does not mean the bondpad is clean. Thus, EDX alone is not sufficient to conclude if there is any F contamination because EDX technique is a volume analysis tool (the signal may be from 0.5-3.0 um), which can only detect very high level of F; (b) Auger result shows clean means the bondpad is clean. This is because Auger is a surface analysis tool (the signal is only from 0-20 Å volume. Therefore, in this case, it was necessary to perform Auger analysis on the unaffected area of the bondpad.

Auger results are showed in Fig. 7. It was found that the F level on "flower-like" defect was higher at 17.8 at %. But only 3.5 at % of F was detected on the unaffected area, which was within the control limit of 5 at %. Therefore, it was likely that the "flower-like" defects were introduced before NE111 cleaning process. Moreover, chemical reaction of F with Al is not immediate and would need some time to form corrosive product - $[AlF_x]^{(x-3)-}$, which are crystal defects. Therefore, it was suspected the cleaning process after Passivation etch was delayed. During the delay, F chemically reacted with Al to form Al fluoride. The removal of such defects by subsequent NE111 cleaning is difficult. Further investigation conducted at wafer fabrication confirmed that the F contamination was due to process was due to 12 hours delay prior to NE111 clean process. Therefore, a time link between passivation etch to NE111 clean was implemented to prevent the reoccurrence of such defect.

VI. CONCLUSION

The growth of Al fluoride "flower-like" defects, i.e. $[AlF_x]^{(x-3)-}$ on bondpads was detected by SEM/EDX/Auger analysis. Such defects were caused by a 12 hours delay prior to NE111 clean. Therefore, a preventive solution of introducing a time link between passivation etch and NE111 clean process was implemented.

ACKNOWLEDGEMENTS

The authors would like to thank Mo Zhiqiang and Gao Ying from Chartered Surface/Advanced FA team for their technical advice and contribution to this technical paper.

REFERENCES

[1] Y. N. Hua, Z. R. Guo and K. W. Chau, "Studies of ZAF Standardless EDX Quantification Method and Application in Failure Analysis of Semiconductor", J. Trace and Microprobe Techniques, 15 (1), 13-31 (1997).

[2] Y. N. Hua, "Application of SEC Factors in Energy-Dispersive X-Ray Quantification Analysis of Silicon Oxide (SIO_2) Layer in Wafer Fabrication", The proceedings from the 198th Meeting of The Electrochemical Society – Phoenix, Arizone, 22-27 Oct, No. 810 (2000).

[3] Y. N. Hua, S. Redkar, C. K. Lau and Z. Q. Mo (from PSB Corp., Singapore), "A Study on Non-Stick Aluminum Bondpads due to Fluorine Contamination using SEM, EDX, TEM, IC, Auger, XPS & TOF-SIMS Techniques", The proceedings from the 28th International Symposium for Testing & Failure Analysis (ISTFA'2002), 3-7 Nov, Phoenix Civic Plaza, Phoenix, Arizona, USA p495-504 (2002).

[4] Y. N. Hua, S. Redkar, R. Dickinson and S. Peh, "A Study on Fluorine-Induced Corrosion on Microchip Aluminium Bondpads", The proceedings from the 29th International Symposium for Testing & Failure Analysis (ISTFA'2003), 2-6 Nov, Santa Clara, California, USA, p249-255 (2003).

[5] J. B. Tan, B. C. Zhang, T. J. Tang, C. Perera, Y. K. Lim, Y. K. Siew, Y. C. Ee, W. Lu, H. Liu, C. S. Seet, H. Zhang, S. K. Lim, S. T. Chua, Z. Ismail, B. M. Seah, P. Y. Ee, D. Vigar and L. C. Hsia, "Yield & Reliability Challenges of BEOL Interconnects", The proceedings from the International Interconnect Technology Conference, (2006).

Plane-view Transmission Electron Microscopy for Advanced Integrated Circuit

Pan Liu, K. Li, Eddie ER and Siping Zhao, *Snr Member, IEEE*

Chartered Semiconductor Mfg Ltd
738406 Woodlands Industrial Park D, Street 2, SINGAPORE
Email: liupan@charteredsemi.com

Abstract — In this paper, the authors try to introduce three techniques for plane-view TEM sample preparation. First, traditional plane-view TEM sample preparation will be introduced. The second technique is FIB-based lift-out method, which places the sample on the carbon film. This technique is used to cut isolated defects, such as SRAM single bit failure, but this technique introduces artifacts from FIB ion damage and carbon film. The last technique is a combination of tripod polishing, FIB milling and ion milling. Specific cases will be given to illustrate the application of these techniques.

I. INTRODUCTION

IN the past five years, the integrate circuit (IC) feature size has been scaled down sharply following Moor's law. Transmission electron microscopy (TEM) has been playing an increasingly important role in the technology development, process control, and failure analysis. In the mean time, the development of TEM techniques for the IC, especially the sample preparation technique, has changed the entire aspect of the TEM. Traditional cross-section TEM sample preparing techniques are not very suitable for advanced IC process such as 65nm and beyond, as the feature size is sometimes smaller than the thickness of the TEM sample (~100nm). Taking poly gate for an example, the poly gate is about 50nm in the 65nm process while the active area width is less than 100nm. When we prepare cross-section TEM samples, the STI oxide will also be included in the TEM samples in addition to the poly of our interest, and this will lead to poor TEM image quality and even image shadows.

Most of TEM cases are for cross-section demotions such as poly gate length and height, spacer width, metal and dielectric thickness information. These information has played important role to tune process recipe and in line monitor [1]. To the 65nm and beyond IC technology, however, the active area and poly dimension have become very small. This make the TEM sample contain more amorphous oxide, because the traditional TEM sample thickness is about 100nm while in the 65nm process technology poly width is about 40nm verse active area width is near 100nm. As Fig.1 shown, over layers information is included in the acquired TEM images, which leads to TEM images become blurred or shield the interested area. Plane-view is easy to avoid the issue.

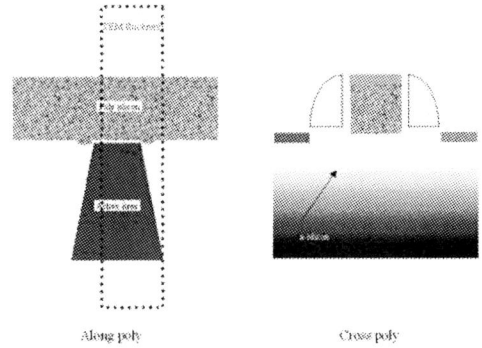

Fig.1. Schedule illusion of short active area induced bottom poly substrate amorphous

The focus ion beam (FIB) has become popular in the TEM sample preparation. Many studies on the Ga ion damage TEM sample sidewall [2,3]. The typical result of all studies is amorphous silicon thickness near 20nm at 30Kev and near 10nm at 10Kev. The amorphous layers at the TEM sample sidewalls make the TEM images not clear, especially in the view the gate oxide. When the TEM sample becomes thin while the amorphous layer thickness beyond 2/3 of total TEM sample thickness, the whole TEM sample is amorphous at the TEM view. Plane-view TEM sample, however, tolerate thicker sample.

In the article, the author reports three techniques for the plane-view TEM sample preparation completely, traditional plane-view sample preparation, FIB followed pick up technique and polish add FIB and ion miller clean technique. And several case studies are given using the plane-view techniques.

II. PLANE-VIEW TEM SAMPLE PREPARATION TECHNIQUES

AS one may realize, TEM sample preparation is importance of acquiring high quality TEM images. To different case, select different techniques for TEM sample preparation.

a) Traditional plane-view TEM sample preparation

This technique has often been employed to design rule check, process evaluation. The T-tool [1], that is a TEM sample polish tool, is similar to tripod polisher developed by IBM [4]. When the sample was polished to near 5um, Gatan PIPS is used to mill the sample as thin as possible.

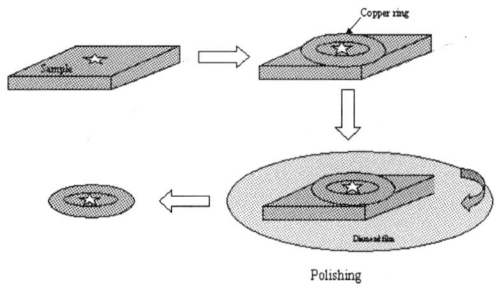

Fig.2. Schedule illusion of the procedure of traditional plane-view TEM sample preparation.

As Fig.2 shown, the detail procedure will describe as following:
- Cut the sample near 1cm X 1cm; make sure the target at the center of sample.
- Lap down the sample to near target layer, such as poly check, sample need to lap down to contact.
- Mount copper ring on the sample by super glue; make sure the target on the center of copper ring.
- Cover a thick glass on the surface of sample, adhered by wax
- Mount glass on the T-tool, and polish backside of sample.
- Adjust the T-tool foot (ft), make sure the sample is co-planar with 2 ft.

- Polish on the diamond lapping film, sequence is 15, 9, 6, 3, 1, 0.5, 0.1um and finally syton (0.05um silica suspension) on a polish clothe.
- Solve the wax by acetone, cut the silicon around the copper ring.
- Ion milling the sample.

Traditional plane-view technique gives big electron transparent area. Many layer information is obtained.

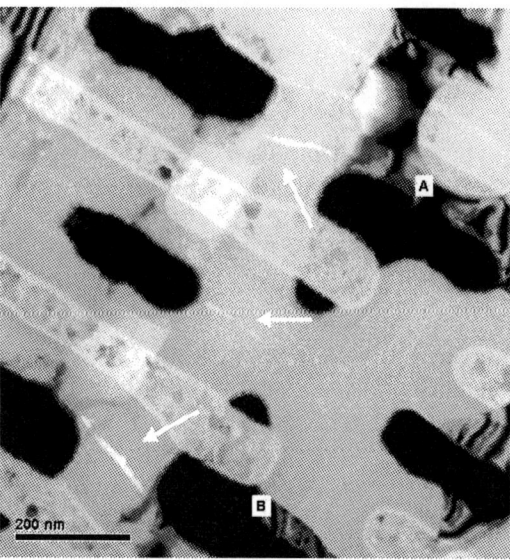

Fig.3. TEM image of STI gap fill void.

Fig. 3 gives a TEM image of STI gap fill void, as the white arrow shown. Additional, the poly gate line uniformity, active area profile and contact information were obtained.

b) FIB + Pick up technique

A defect is not always easy to be precisely located to make a cross-section TEM. And traditional plane-view TEM has low success rate due to ion miller hard control to designated point. So, precision plane-view TEM is necessary to find the failure root cause.

The following procedure will describe the precision plane-view TEM using FIB and pick-up technique. (As the Fig. 4 shown)

- Using T-tool polish sample cross section to near failure location (less than 3um to polishing surface).
- Flip by 90 degree the sample in the FIB chamber.
- Dig two trenches at the target layer, U-cut the side and bottom, milling sample to thin enough, near 100nm.

Pick up the sample from the trench, and put sample on the carbon film.

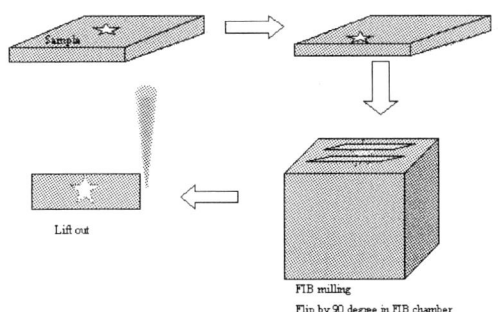

Fig. 4. Schedule illusion the procedure of FIB + Pick up technique.

This technique is easy to master because it is very similar with the normal cross section TEM procedure.

Dislocation is a type of silicon defect induced the junction leakage and degraded the device performance. Fig. 5 shows the dislocation images in the silicon substrate as the arrow pointed.

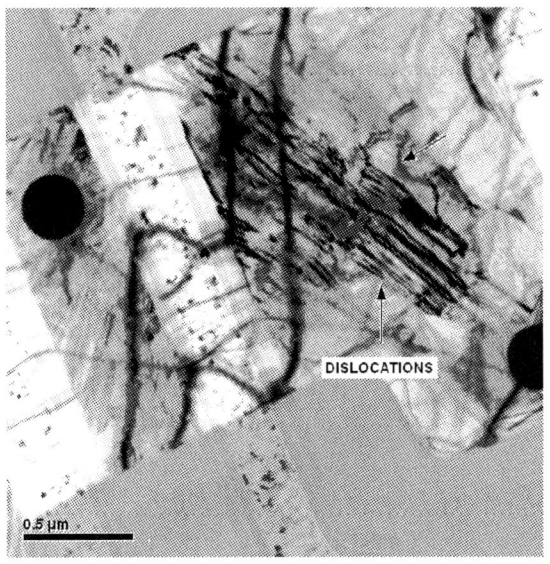

Fig. 5. TEM image of dislocation.

c)   Polish + FIB +ion miller technique

To the very small defect, any artifact will lead to the TEM not to catch the root cause of failure. FIB ion damage is one of major TEM artifacts. In order to reduce the artifact, low energy ion miller is a useful tool because it has low Ar ion energy. Low energy Ar ion is able to clean the FIB Ga ion damaged TEM sample sidewall. Fig. 6 gives the schedule of the technique.
The detail procedure is as following:

- Polish sample to near failure location, the location may be hot spot, single bit, in line defect and so on.
- Mount half copper ring by super glue on the surface of sample, make sure the target at the center of half ring and the two sides of half ring is below of the polished cross section surface, which guarantee that the sample can not touch the FIB lens at the eucentric height adjust.
- Polish sample backside as the traditional technique.
- Cut the silicon around the half copper ring.
- Flip by 90 degree the sample in the FIB chamber.
- Using FIB mills the sidewall of sample to thin enough, near 100nm.
- Mount the sample on the PIPS holder to clean the amorphous silicon at the sample sidewall at the 3Kev, 30 second in Single mode.

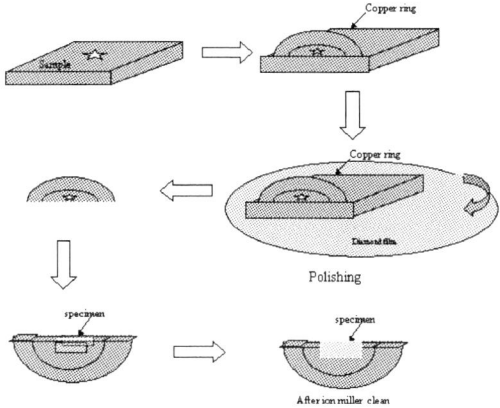

Fig. 6. Schedule illusion the procedure of Polish + FIB + ion miller technique.

The technique couples with the advantages of low damage traditional method and precision FIB method. And the technique is not use the carbon film, which will lead to carbon contamination in the TEM view. If the sample is thick, PIPS milling is able to apply to make the sample thinner.

Fig. 7 shows the clear image of gate cross active area. The area of poly and active area overlay shows the device performance. These data play more important role in the device simulation during setting up SPICE module. In the image, dark strip is active area; black circle is contact; bright strip cross active area is poly gate; the line around active area is line nitride. The line nitride function is to reduce shallow trench isolation (STI) leakage.

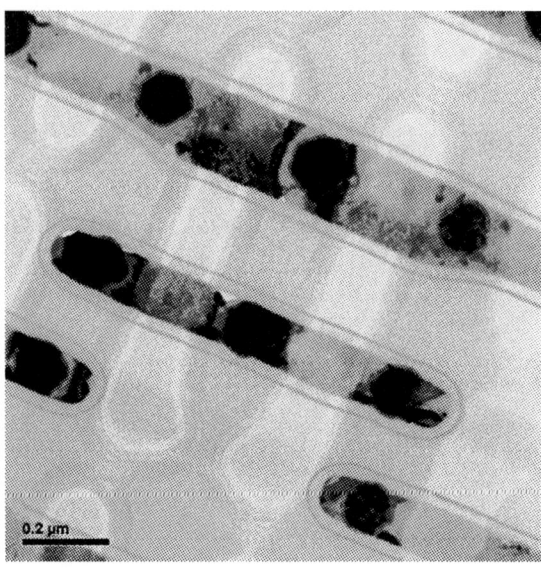

Fig. 7. TEM image of poly gate cross active area

III. SUMMARY

WITH the semiconductor industry deve-lopment, TEM has been applied widely. Cross-section TEM sample preparation technique encounter more difficult in the advance IC process. However, plane-view techniques give another viewpoint to observe the semiconductor device. In the article, three plane-view techniques have been illuminated. To different object, different technique should be applied to prepare the TEM sample.

ACKNOWLEDGEMENT

The author would like to thank my colleague Soo chi wen, TEE Irene, TAY Chui Lam, Wang JiLi for giving me warming support in FA lab.

REFERENCES

[1] Hong Zhang "Transmission electron microscopy for the semi-conductor industry" *Micron*, 33, 515-521. (2002)
[2] Rubanov S, Munroe P.R "Investigation of the structure of damage layers in TEM samples prepared using a focused ion beam," *Journal of Materials Science Letters,* Volume 20, Number 13, 1 July 2001, pp. 1181-1183(3)

[3] Qiang Gao, Zhang M,.Chorng Niou, Ming Li, Chien, K. "Sidewall damage induced by FIB milling during TEM sample preparation," *Reliability Physics Symposium Proceedings, 2004. 42nd Annual. 2004 IEEE International,* p613-614 (2004)
[4] Anderson RM and Klepeis SJ, *spaciment preparation for Tansmission Electron Microscopy of materials IV,* Materials Research symposium series 480, p187 (1997)

Application of Focus Ion Beam Circuit Edit in Failure Analysis

S. K. Loh, Teo HT, S.P. Neo, Z.G. Song and C. K. Oh

Chartered Semiconductor Mfg Ltd

60 Woodlands Industrial Park D, Street 2, Singapore 738406

Email: lohsk@charteredsemi.com

Abstract **Focus Ion Beam is an indispensable tool in failure analysis laboratory. It has a wide range of applications. This paper will discuss its application in circuit edit to enhance failure analysis on two failure modes by isolating the defective sites of the failures and finally identifying the root causes.**

I. INTRODUCTION

FOCUSED Ion Beam (FIB) is an ubiquitous and indispensable tool in modern semiconductor failure analysis laboratories. Its applications include precision cross-section for defect analysis or process characterization, Transmission Electron Microscope (TEM) sample preparation and passive voltage contrast for fault isolation of open or short failure. Recently it is also intensively employed to do circuit edit [1, 2, 3] for design debug and failure analysis because it possesses the ability to selectively remove existing materials and deposit insulating layer or conducting wires. For example, FIB can deposit micro-pads and wires for accessing to the nodes of any suspected circuit and/or isolate this circuit from other circuitry if necessary. FIB circuit edit can be performed at any interconnected level. The electrical characterization after circuit edit includes verifying circuit open/short failure, transistor family curves, threshold voltage, gate leakage etc, depending on the requirement. By comparing the electrical behavior of the suspected defective circuit with that of a good circuit, one can understand the failure mechanism of the defective circuit and further narrow down the defective location for further physical failure analysis. In this paper, application of FIB circuit edit in failure analysis will be described.

II. CASE STUDIES

Case Study (I)

Some SRAM units manufactured with 90 nm technology in SOI wafer suffered from single bit failure. Conventional physical failure analysis for several units was performed but no anomaly was found. Thus the technique of FIB circuit edit and electrical characterization was employed. A unit was de-processed to contact level and FIB circuit edit was performed. Analysis of the layout showed that total of ten pads need be deposited for fully accessing to the six transistors of a single SRAM cell. After FIB deposits microprobe pads and wires for accessing to the six transistors of the failing single bit (see Fig. 1), the sample is ready for micro-probing to characterize the electrical behavior of the six transistors.

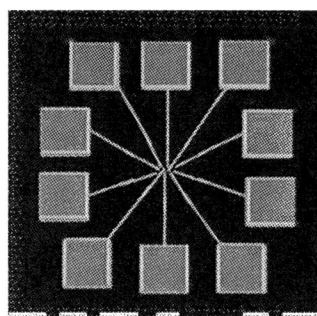

Fig. 1 Ion-beam image of 10 pads deposited for accessing to the six transistors of a typical SRAM cell.

The family curves (Id-Vd curves with Vg at different voltages) of the six transistors were measured and they were measured again with drain and source swap. Electrical characterization showed that the family curve of one of the pass gate was asymmetric and less drain current (see Fig. 2). XTEM analysis was thus performed across the abnormal pass gate and it showed blocked silicidation and implantation at one side of the pass-gate transistor (see Fig. 3).

0-7803-9730-4/06/$25.00 ©2006 IEEE 643

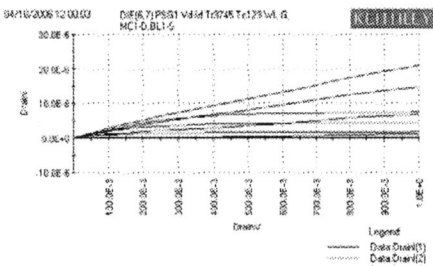

Fig. 2 Family curves (Id-Vd curves with Vg at different volts) (blue) and such curves with drain and source swap.

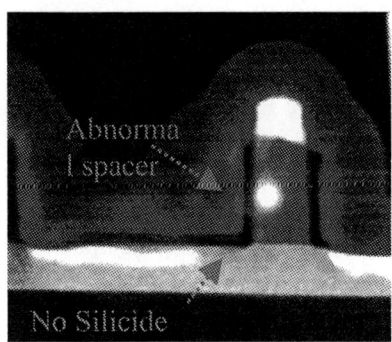

Fig. 3 XTEM pictures showing abnormal spacer, blocked silicidation and implantation at one side of the pass gate

Case Study (II)

A package unit was reported failure due to the leaky MIM capacitors within it. There were altogether 19 MIM capacitors in parallel. So there was a need to identify which are the leaky ones. To identify the leaky ones, the MIM capacitors need to be separated from each other by cutting off the connection to the top plate with FIB (see Fig. 4). After the 19 MIM capacitors were separated from each other, with the help of passive voltage contrast technique, it was clearly observed that four out of the 19 MIM capacitors showed bright contrast. It meant that these 4 MIM capacitors were leaky. I-V curve tracing for these MIM capacitors showed actually ohm short failure for these MIM capacitors (see Fig. 5). Subsequent physical de-processing and SEM inspection for MIM capacitor with short failure revealed that the top metal plate was corroded and shorted to the nearby via, resulting in short failure of these MIM capacitors (see Fig. 6).

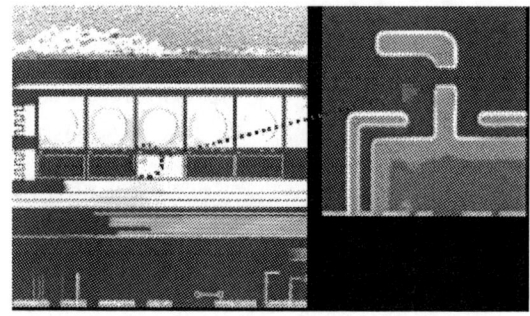

Fig. 4 PVC showing bright contrast for a MIM capacitor with leakage failure after the MIM capacitors are separated each other by FIB.

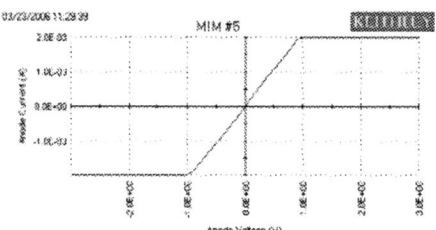

Fig. 5 I_V curve of the leaky capacitor showing an ohm short failure.

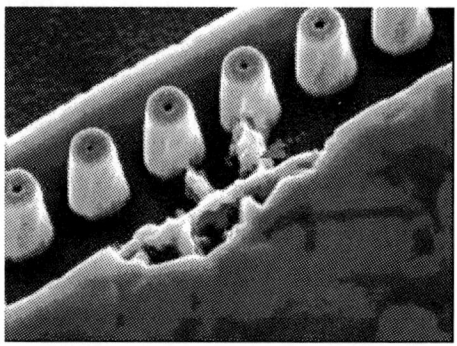

Fig. 6 SEM picture showing top metal plate corroded and shorted to bottom plate through via 5 for the MIM capacitor with short failure.

III. CONCLUSION

In this paper, two case studies on the application of FIB circuit edit in failure analysis were presented. Through circuit edit, the defective location of the failure was isolated and finally the root cause of the failure was identified. It successfully demonstrated the effectiveness of FIB circuit edit in failure analysis.

ACKNOWLEDGEMENT

The authors would like to thank the Chartered TEM team for providing the TEM pictures and their support.

REFERENCES

[1] S.B. Herschbein, L. S. Fischer and A. D. Shore, "Basic Technology and Practical Applications of Focused Ion Beam for the Laboratory Workplace", *Microelectronic Failure Analysis, Desk Reference* 4th Edition, pp. 517-526

[2] Randal Mulder, "*A Simple FIB Method for Constructing Electrically Isolated Microprobe Pads for the Electrical Analysis of Failing 0.12um Technology SRAM Bit Cells*", ISTFA proceedings, pp.538-545 (2004).

[3] J. Myers, M. Abramo, M. Anderson and M. W. Phaneuf, "*A Novel Approach for Enhancing Critical FIB Imaging for Failure Analysis and Circuit Edit Applications*" ISTFA proceedings, pp.151-156, (2004).

Circuit Debug using Time Resolved Emission (TRE) Prober – A Case Study

Houn Wai Wong, Pit Fuh Low and Vee Kin Wong
Intel Microelectronics (M) Sdn. Bhd.
Bayan Lepas FTZ Phase 3, 11900 Bayan Lepas, Penang, Malaysia
Email: houn.wai.wong@intel.com, pit.fuh.low@intel.com, vee.kin.wong@intel.com

Abstract - **Time Resolved Emission Microscopy (TRE) is a revolutionary tool used in advanced microprocessors silicon debug. Unlike other debug tools, the interpretation of TRE data may not be straightforward and can sometimes be misleading. In this paper we show through a case study why this is true. Interpretation of TRE data should be done carefully with good knowledge on device emission physics, circuit behavior and creative fault analysis.**

I. INTRODUCTION

TRE (Time-Resolve-Emission) is gaining prominence in advanced microprocessor silicon debug. Unlike laser based probing tools such as LVP (Laser Voltage Probe) and LADA (Laser Assisted Device Alteration), TRE is a passive observer in which no external force or energy is being applied (non-invasive) to the device to acquire the signal. This eliminates the potential of device characteristics being altered during the probing process which could result in the wrong waveform being acquired or even permanent damage to the device under probe if the external energy applied is too high. One shortcoming of TRE is that it does not produce voltage waveforms but rather a histogram plot (will be referred as TRE waveform) of device infrared emission.

When a user has acquired the result from TRE, the TRE waveform needs to be converted into signal waveform for interpretation.
This paper shows that the usual method of interpreting the TRE data in this particular case does not coincide with the test data and simple translation of the TRE data to logic state of the transistor node could be misleading.

II. EVOLUTION OF DEBUG TOOLS

In post silicon validation and failure analysis, it is crucially important to find and root cause the failure fast to ensure product quality and time to market. During debug, the ability to acquire node level information at the transistor level (switching events, timing information etc.) is desired for fixing of design issues, process defects etc.

The most direct method is to extract the electrical data through the oscilloscope using mechanical contact probing. However, the complexity of doing mechanical probing increases with decreasing transistor sizes and increasing metallization layers in semiconductor devices. In some cases it involves creating access probe point by physical trenching of the device under test (DUT) and this cannot be done globally for a single DUT. Capacitive loading from mechanical probing is continuously a challenge. Non-contact methods thus provide a compelling alternative. Electron-beam probing was one of the techniques that were used widely in the industry. This technique uses the secondary-electron reflected from the DUT to convey the signal information. The transitioning of integrated circuit (IC) packaging from front-side to C4 packaging however limits the usefulness of such techniques. The advent of LVP (laser voltage probe) [1-2] and other laser based analytical tools which takes advantage of the photo-electron interaction in silicon provides the probing solution needed by the industry. The disadvantage of LVP however, is the issue of invasiveness, where the laser used for extracting transistor timing information actually alters or changes the behavior of circuitry. At times, excessive laser power could end up destroying the DUT. As a result, the industry developed passive optical methods of extracting timing information. Taking advantage of photo-emission phenomena, time-resolved emission techniques (know here as Time Resolved Emission Microscopy or TRE) is gaining in-road as one of the promising debug techniques [3-4].

0-7803-9730-4/06/$25.00 ©2006 IEEE

III. PHYSICS OF EMISSION

Fundamentally, photoemission from carriers in semiconductor devices are due to carriers transitioning from a higher energy state to a lower energy state, with carrier energy being lost through photon emissions [5-6].

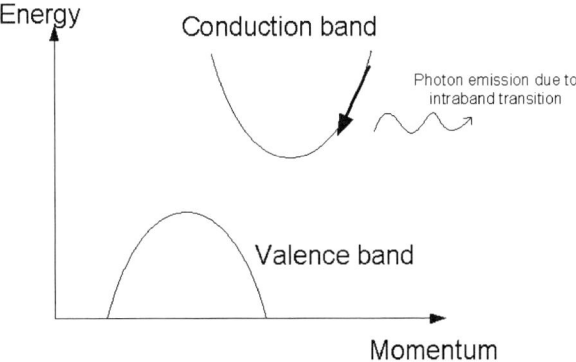

Fig. 1: Intraband transition of high energy electrons to lower energy states causing photon emission

In silicon MOS transistors, the most important emission comes from intraband transitions (Fig. 1). Intraband transitions occur when very energetic carriers shift to lower energy states within their own energy band. This happens in the presence of high electric fields when carriers have high kinetic energy, i.e. when the device is in saturation mode. While both NMOSes and PMOSes emit in the saturation mode, differences in mobility result in PMOSes emitting less than NMOSes.

IV. TRE SIGNAL INTERPRETATION: TYPICAL SCENARIO

The TRE captures the photon emission during transistor switching events, specifically in saturation. By applying pico-second timing analyzer and single photon counting techniques, the acquired waveform provides a time stamp of transistor switching activities. With higher mobility of electrons compared to holes, NMOS devices provide more photon emissions compared to PMOS devices. In a typical CMOS circuit, a high NMOS photo emission peak (or pulse) represents a falling transition for the node, whereas a PMOS peak corresponds to a rising transition. A

typical signal acquired by TRE with the reconstructed signal is shown in Fig. 2 below:

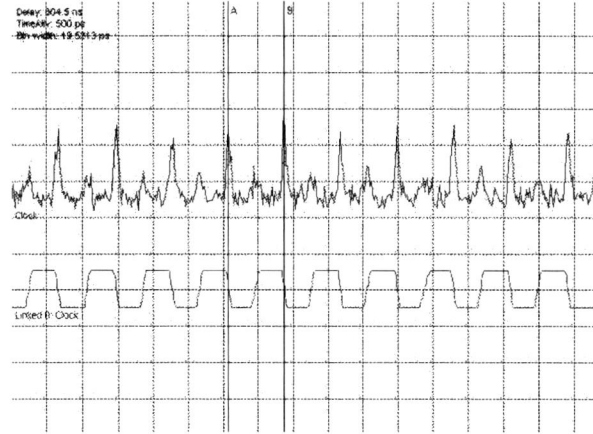

Fig. 2: Typical TRE waveform

The 1st waveform is the actual data captured by TRE for a pair of NMOS and PMOS devices., while the 2nd waveform is the reconstructed signal. The stronger amplitude pulses are aligned with the low-going transition while a weaker pulse is aligned with the high-going transition. Looking at this, to get a complete logic state of devices (switching activity), data from both NMOS and PMOS is needed. Most of the time, it is not practical to acquire PMOS emissions due to the long acquisition time for small devices that are common in circuits. Therefore, NMOS photon emissions may be the only data available. This complicates the interpretation of the TRE signal and careful reconstruction of the waveform based on the device emission characteristics is necessary. The following case study illustrates this point .

V. CASE STUDY

This is a failure in a sequential element where the data at *out* is wrong. TRE was used to extract the signal behavior at various nodes. Sample preparation of the DUT involved silicon backside thinning while functionality was maintained. The failing test pattern was looped during signal acquisition.

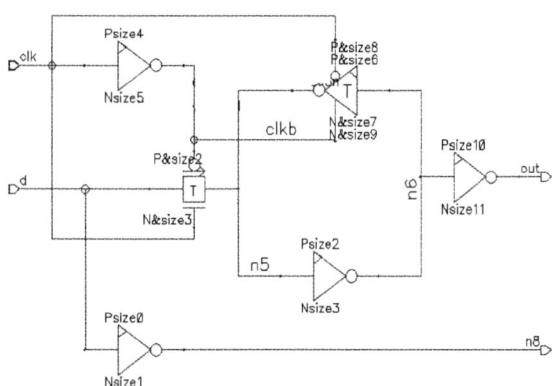

Fig. 3. Schematic of flip-flop circuit

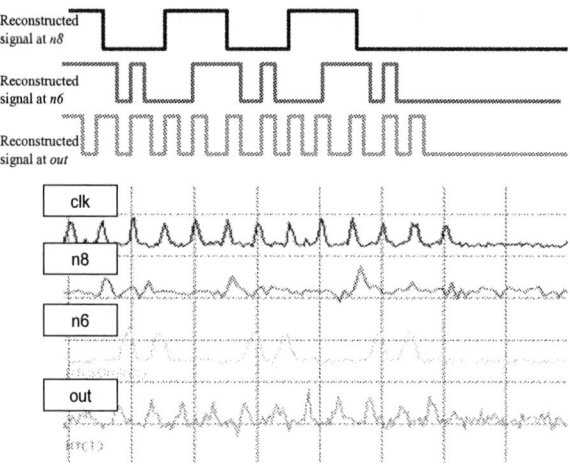

Fig. 4. Reconstructed logic signals of the nodes *n8, n6* and *out* (top) with their corresponding acquired TRE waveforms (bottom)

The critical nodes required to locate the failure are the *out, n6* and *n8* (refer to schematic in the Fig 3). The selection of these nodes is also driven by the size of the device which determines the ease of probing the signal. In this particular case, the PMOS devices emission is too weak to be easily detected hence only NMOS devices are probed. Fig. 4 above shows the common method of signal reconstruction i.e. aligning TRE peaks to falling transitions of clocks and arbitrarily placing a rising transition based on known circuit behavior. The *clk* signal is captured as reference. One difficulty in doing this signal reconstruction is to separate noise from data as in the *n8* signal. Since the test vector is a continuously toggling signal, we could compare the TRE data of one cycle to the next

cycle to eliminate noise from data. Although this signal reconstruction was straightforward, the results were not consistent with the circuit behavior. The output at *out* is not the expected inverse of *n6*. In fact, the *out* data seems to be toggling at the same rate as *clk*. A re-look at the TRE peaks also suggest that the *out* signal is 180° out of phase with the *clk* signal. With this data, we reconstruct the TRE signal with a *clk* to *n6* short hypothesis in Fig 5 below.

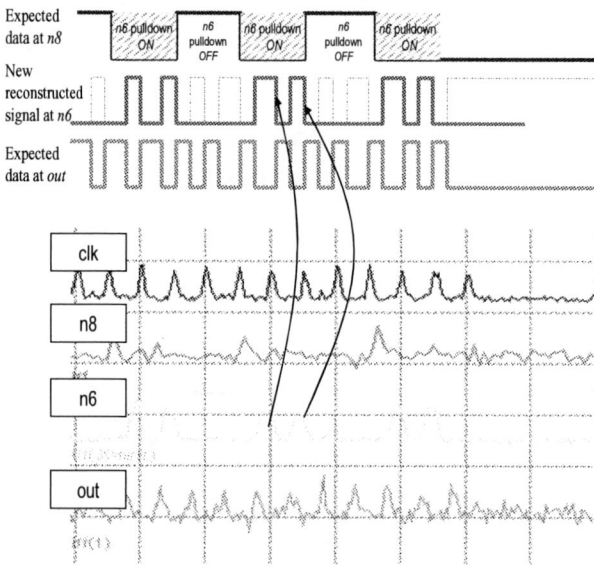

Fig. 5. New reconstructed logic signal of node *n8, n6* and *out* based on the hypothesis

The larger clock driver in contention with *n6* will not alter the *clk* signal much. Therefore, the keeper states are expected to be still controlled by *clk*. With the same argument then *n6* is expected to follow the *clk* signal, without much influence by the incoming data *d*. However, this implies that the TRE signal for n6 should be similar to *clk*. The reason for the difference between *n6* and *clk* can be explained by noting that the input stream of '1100's will turn the NMOS pull down of node *n6* ON and OFF alternately. In the OFF states, there should not be any saturation current emission peak from the NMOS even if the drain node is toggling. The dotted line in the reconstructed *n6* signal shows the drain voltage during the OFF phase of the NMOS. This hypothesis explains all the observed TRE signals and the failure was confirmed in physical analysis.

VI. CONCLUSION

This paper has demonstrated the potential misinterpretation of TRE results. It is crucial to understand that the TRE waveform is a plot of device emissions which is highly dependent on the operation state of the transistor. In our case study the fault still exhibits a sharp falling transition despite presence of the defect, thus giving an unambiguous TRE peak. However, some faults may not result in such a straightforward TRE peak, as in the case of leakages. Hence, interpretation of the TRE output requires understanding of the emission characteristics of transistors, good knowledge on the circuit behavior and creative fault analysis.

ACKNOWLEDGMENT

The authors will like to thanks Kian Fei Lee and Seaw Wei Law for the TRE data collection and the physical analysis team for SEM and TEM analysis work.

REFERENCES

[1] M. Paniccia, T. Eiles, V. Rao, W. Yee, "Novel Optical Probing Technique for Flip Chip Packaged Microprocessors" *Proc. IEEE International Test Conference* pp. 740-747. (1998).

[2] M. Panicia, V. Rao, W. Yee, "Optical Probing of flip chip packaged microprocessor", *J. Vac. Sci. Tech. B,* 16 (6) (1998)

[3] J. C. Tsang, J. A. Kash, D. P. Vallett, "Picosecond imaging circuit analysis", *IBM Journal of Research & Development*, vol. 44 no. 4 pp. 583-603 (2000)

[4] E. Varner, C. Young, H, Ng, T. Eiles, B. Lee, "Single Element Time Resolved Emission Probing for Practical Microprocessor Diagnostic Application", *Proc. 28th International Symposium Test & Failure Analysis,* pp. 451-460 (2002)

[5] Tam. S, Chenming Hu, "Hot-electron-induced photon and photocarrier generation in Silicon MOSFETs" *IEEE Transactions on Electron Devices* vol. 31 no. 3 pp. 1264 - 1273 (1984)

[6] Rowlette, J.A.; Varner, E.B.; Seidel, S.; Bailon, I.C.; "Hot Carrier Emission from 50 nm n- and p-Channel MOSFET Devices", *Proc. IEEE The 16th Annual Meeting of the Lasers and Electro-Optics Society,* vol. 2, 2003 pp.740 - 741 (2003)

FPGA Implementation of a Canonical Signed Digit Multiplier-less based FFT Processor for Wireless Communication Applications

Mahmud Benhamid, *Student Member, IEEE* and Masuri Othman
VLSI Design Center, Institute of Microengineering and Nanoelectronics (IMEN),
Universiti Kebangsaan Malaysia, 43600 Bangi, Selangor, Malaysia
Phone: +603 89216009 / +6012 2705932
Email: benhamid@vlsi.eng.ukm.my, masuri@vlsi.eng.ukm.my

Abstract **this paper proposes a novel fully parallel FFT architecture based on Canonical Signed Digit (CSD) multiplier-less targeting wireless communication applications, such as IEEE802.15.3a Wireless Personal Area Network (WPAN) baseband. The proposed architecture has the advantages of high throughput, less latency, and smaller area. The multiplier-less architecture uses shift-and-add operations to realize the complex multiplier and uses the CSD to optimize these operations. The design has been coded in Verilog HDL targeting Xilinx Virtex-II FPGA series. It is fully implemented and tested on real hardware using Virtex-II FG456 prototype board. Based on this architecture, the implementation of 8-points FFT on Virtex-II can run at a maximum clock frequency of about 400 MHz which lead to about 3.2 GS/s throughput with a latency of 6 clock cycles using 16,580 equivalent gates. Comparison with a conventional parallel architecture design of the same size can run only at a maximum clock frequency of 220 MHz or 1.76 GS/s throughput with a latency of 12 clock cycles using 77,418 equivalent gates for the design. The resulting throughput increases by about 82% while the equivalent gates and latency decrease by about 79% and 50% respectively.**

I. INTRODUCTION

Fast Fourier Transform (FFT) is one of the most utilized operations in Digital Signal Processing and Communications. The FFT and its inverse transformation IFFT are among the key components in the modern communication systems. Application Specific Integrated Circuit (ASIC) approaches have been used to achieve the high performance demands which software or general purpose DSP implementations fail to deliver. Recently FPGAs has become a valid alternative as the technology has matured greatly. Nowadays FPGAs play an important role in many areas due to their direct hardware solution performance as well as their inherent reprogrammability feature. Using FPGAs for FFT processing has now become feasible in real-time applications.

Development of the FFT in hardware is usually categorized into high throughput and low power. However, the design of a complex multiplier is essential for any complex FFT implementation because it consists of four real multipliers and two real adders. In addition to the large area, they dissipate almost 50~80% of total power [1]. In the search for high performance FFT, this paper presents a fully parallel Radix-2 FFT based on CSD multiplier-less architecture and its implementation on FPGAs using Verilog HDL. The design has been implemented and tested for real hardware using FG456 prototyping board which is equipped with Xilinx Virtex-II.

In the following sections, detailed algorithm and architecture of Radix-2 will be reviewed. The implantation of CSD multiplier-less technique will be described. After that, implementations results will be given and compared with the conventional architecture. Finally conclusion will be drawn.

II. ALGORITHM

The N-point Discrete Fourier Transform (DFT) of a complex data sequence $x(n)$ is:

$$X(k) = \sum_{n=0}^{N-1} x(n) W_N^{kn} \quad , k = 0,1,...,N-1 \qquad (1)$$

Where x (n) and X (k) are complex numbers, and $W_N^{kn} = e^{-j2\pi/N}$ is the twiddle factor.

0-7803-9730-4/06/$25.00 ©2006 IEEE

Direct computation of equation (1) requires on the order of N^2 operations where N is the transform size. The FFT algorithm, first explained by Cooley and Tukey [2], opened a new area in digital signal processing by reducing the order of complexity of DFT from N^2 to ($Nlog_2N$). Since the early paper by Cooley and Tukey, a large number of FFT algorithms have been developed. Among these, the radix-2, radix-4, and split radix algorithms are the ones that have been mostly used for practical applications due to their simple structure, and constant butterfly geometry. In general, higher-radix FFT algorithm has less number of complex multiplications but radix-2 FFT algorithm is the simplest form in all FFT algorithms and has a regularity that makes it suitable for VLSI implementation [3].

FFT algorithm relies on a divide-and-conquer methodology, which divides the N coefficient points into smaller blocks in different stages. The first stage computes with groups of two coefficients, yielding N/2 blocks, each computing the addition and subtraction of the coefficients scaled by the corresponding twiddle factors, called a butterfly for its cross-over appearance. These results are used to compute the next state of N/4 blocks, which will then combine the results of two previous blocks, combining 4 coefficients at this point. This process repeats until we have one main block, with a final computation of all N coefficients. Figure 1 shows the signal flow graph of 8-point radix-2 decimation in time FFT.

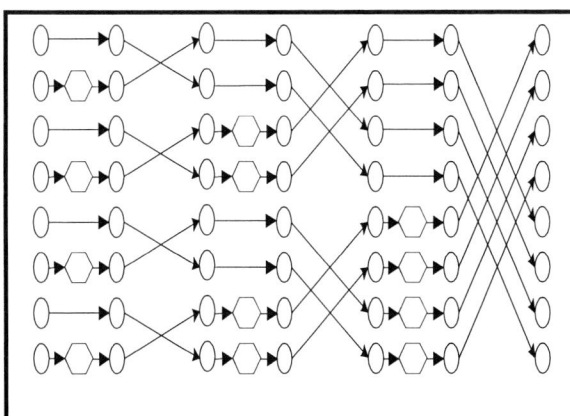

Fig. 1 Radix-2 DIT FFT flow graph for N = 8

III. CONVENTIONAL ARCHITECTURE

Figure 2 shows the conventional architecture of the butterfly used in the radix-2 DIT FFT. Each butterfly has 2 inputs, a complex multiplier, 2 complex adders and 2 outputs that will be

given to the next stage. The implementation of a complex adder requires 2 real adders as shown in Equation 2 and Figure 3. While the direct implementation of a complex multiplier requires 4 real multipliers and 2 real adders as explained in Equation 3 and Figure 4. The number of real multipliers can be reduced to 3 with a simple transformation at the cost of extra additions as described in Equation 4 and Figure 5.

$$(a+jb)+(c+jd)=(a+c)+j(b+d) \qquad (2)$$

$$(a+jb)*(c+jd)=(ac-bd)+j(ad+bc) \qquad (3)$$

$$(ac-bd)=d(a-b)+a(c-d)$$
$$(ad+bc)=c(a+b)-a(c-d) \qquad (4)$$

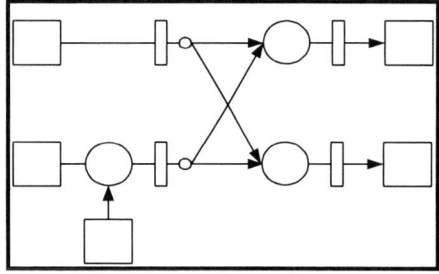

Fig. 2 Butterfly architecture

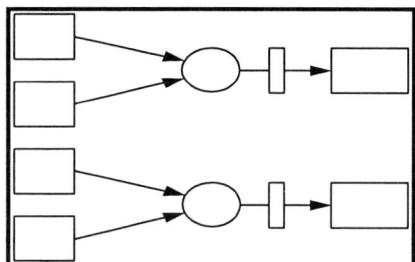

Fig. 3 Complex adder

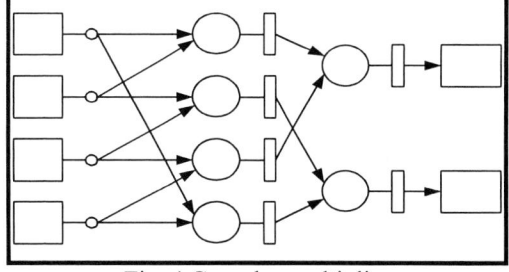

Fig. 4 Complex multiplier

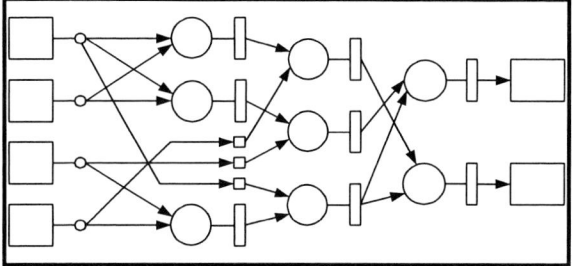

Fig. 5 Modified complex multiplier

III. MULTIPLIER-LESS ARCHITECTURES

Since the twiddle factors in the FFT processor are known in advance. We propose the use of a multiplier-less architecture to perform the multiplication with the twiddle factors using shift-and-add operations to replace the traditional complex multiplier. The canonical sign digit (CSD) algorithm has been applied to this architecture to further reduce the number of shift-and-add required. As a result, the hardware complexity for the complex multiplier with this architecture has 11 adders only for 8-points FFT instead of 4 multipliers and 2 adders.

In this architecture trivial multiplications are implemented without any multipliers by either passing the data, swapping the real and imaginary parts of the complex data or a sign change. The design presented in the paper takes advantage of the symmetries of the twiddle factors in the complex plane. Taking radix-2 DIT 8-points FFT as an example, the following twiddle factors are used: W^0, W^1, W^2 and W^3. W^0 corresponds to the trivial value of "1". This means that the data is simply passed through without any phase rotation. W^1 corresponds to the non-trivial value of $0.7071(1 - j)$. This has been implemented using shift-and-add operations as shown in Figure 6. W^2 corresponds to the imaginary value of "-j". This is performed by swapping the real and imaginary values and a sign change. W^3 corresponds to the non-trivial complex value of $0.7071(-1 - j)$. The only difference between W^1 and W^3 is the sign change between the real values. W^3 can be written as $W^1 * W^2$. Thus the output from the previous multiplier-less followed by a sign change can be used to perform multiplication with twiddle factor W^3.

IV. IMPLEMENTATION RESULTS

Two different modules have been implemented for radix-2 DIT 8-points FFT. The first module uses the conventional architecture to implement the FFT processor where the twiddle factors are stored in ROM and called by the butterfly to be multiplied with the inputs using the dedicated high speed multiplier equipped with the Virtex-II FPGA and the other module uses multiplier-less architecture where shift-and-add operations are used to perform the multiplication with the twiddle factor. The second module uses the CSD algorithm to further reduce the number of adders.

Both the modules were coded in Verilog and synthesized using the XST - Xilinx Synthesis Technology tool. The target FPGA was Xilinx Virtex-II XC2V500-6-FG456 FPGA [4].

ModelSim simulation result of multiplier-less architecture for radix-2 DIT 8-points FFT is shown in Figure 8, while the synthesis results for the two models are presented in Table 1.

The resulting figures show that the multiplier-less architecture outperforms the conventional architecture. Its throughput increases by about 82% while the equivalent gates and latency decrease by about 79% and 50% respectively compared to conventional architecture.

The appropriate word length in 8-points FFT processor is determined by synthesizing the design in different word length and then compare the synthesize result with a MATLAB simulation and calculating the error rate percentage in each case as shown in Figure 7. The 16-bit word length gives a maximum error rate of 0.119 %

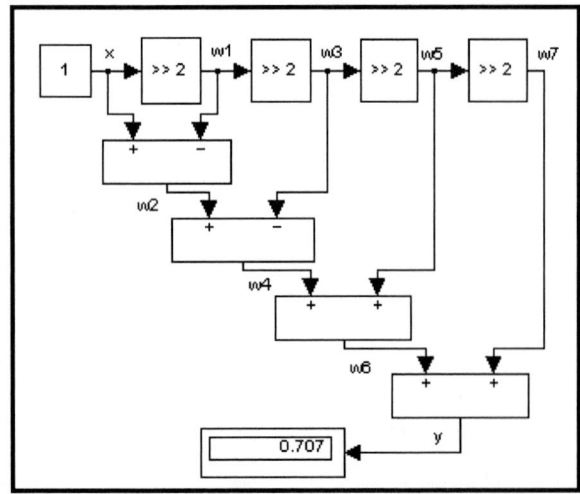

Fig. 6 Multiplier-less implementation

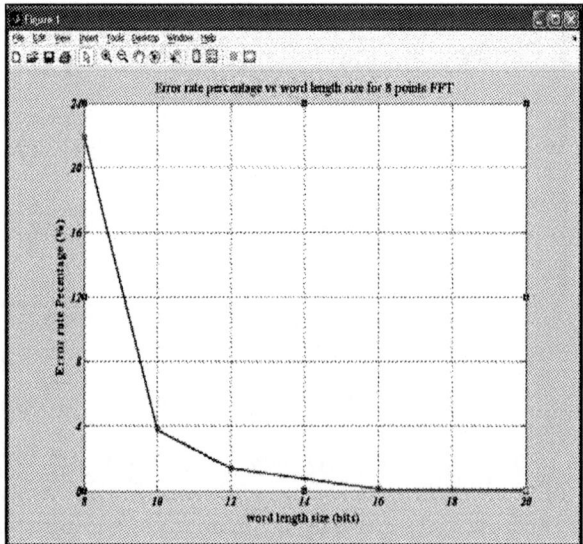

Fig. 7 Error rate Percentage verse word length

Table 1: Hardware Specifications of the two architectures for Radix-2 8-points FFT

Xilinx Virtex-II FPGA XC2V500 -6-FG456	No. of slice Flip Flop	Latency	No. of Multipliers	No. of Adders	Equivalent Gate Count	Maximum Frequency
Conventional Architecture	912	12 Clock cycles	16	27	77,418	220.143 MHz
Multiplier-less Architecture	510	6 Clock cycles	0	60	16,580	398.883 MHz

Comparing our implantation with the other FFT implementations on FPGAs shows that our implementation outperforms them. Banerjee [5] implemented a CORDIC algorithm based FFT processor for biomedical signal processing using Xilinx XC4025 series that runs at 10 MHz. Dick [6] has implemented an FFT based on Systolic array architecture operated at 15.3 MHz on XC4010PG191-4 FPGA. Pérez-Pascual [7] has implemented butterflies suitable for HIPERLAN 2 on XCV812E-8 Xilinx FPGA with a throughput of 15.6 MHz for radix-2. While Sansaloni [8] implemented a Distrusted Arithmetic based radix-2 DIF on XC4085x1-1 with a maximum throughput of 25.69 MHz. S. Bahl [9] implemented a pipeline FFT radix-2^3 algorithm on Virtex-E XCV1000E FPGAs with of 58 MHz. Sukhsawas [10] designed a high-level pipeline FFT on Virtex-E FPGA that can run at a maximum clock frequency of 82 MHz.

V. CONCLUSION

This paper presented two parallel implementation of efficient radix-2 decimation in time FFT algorithm based on conventional and Canonical Signed Digit (CSD) multiplier-less architectures targeting Virtex-II FPGA. The multiplier-less architecture does not require any multipliers or ROM to store the twiddle factor coefficients as the multiplication with the twiddle factor is achieved through the use of shift-and-add operations. The implementations have been coded in Verilog hardware descriptive language and were tested on Xilinx Virtex-II XC2V500-6-FG456 prototyping FPGA board. A maximum clock frequency of 220 MHz and about 400 MHz has been obtained from the synthesis report for conventional and multiplier-less implementation, respectively for 8-point FFT with 16-bit input and twiddle factors word length. The paper reported the fastest radix-2 decimation in time 8-point multiplier-less FFT implementation on FPGA platform.

REFERENCES

[1] Weidong Li and Lars Wanhammar, "A Pipeline FFT Processor," *IEEE Workshop on Signal Processing Systems*, pages 654–662, (1999).

[2] J. W. Cooley, J. W. Tukey, "An Algorithm for the Machine Calculation of Complex Fourier Series," *Math,. Comp,* Vol.19, pp.297-301, (April 1965)

[3] Eleni Fotopoulou and Vassilis Paliouras "An Efficient Computational Method and a VLSI Architecture for Digital Filtering of CP-OFDM Signals" *IEEE Communications Society,* pp. 2393-2397, Globecom (2004)

[4] Xilinx, Inc. http://www.xilinx.com/virtex2

[5] A Banerjee, A. S. Dhar, and S. Banerjee, "FPGA realization of a CORDIC based FFT processor for biomedical signal processing." *Science Direct journal of Microprocessors and Microsystems,* volume 25, Issue 3, pages 131-142, (May 2001)

[6] C. H. Dick, "FPGA based systolic array architectures for computing the discrete Fourier transform," *IEEE International Symposium on Circuits and Systems (ISCAS'96)*, volume 2, pages 465-468, (May 1996).

[7] A. Pérez-Pascual, T. Sansaloni, and J. Valls, "FPGA-based radix-4 butterflies for HIPERLAN/2," *IEEE International Symposium on Circuits and Systems (ISCA'02),* volume 3, pages 277-280, (May 2002).

[8] T. Sansaloni, A. Pérez-Pascual, and J. Valls, "Distributed arithmetic radix-2 butterflies for FPGA," *The 8th IEEE International Conference on Electronics, Circuits and Systems (ICECS'01),* volume 1, pages 521-524, (Sept. 2001)

[9] S. K. Bahl, "A hardware efficient architecture for fast Fourier transform," *Proc. GSPx and International Signal Processing Conference (ISPC'03)*, Dallas, Texas, (Apr. 2003).

[10] S. Sukhsawas and K. Benkrid, "A High-level Implementation of a High Performance pipeline FFT on Virtex-E FPGAs," *Proc. of IEEE Computer Society Annual Symposium on VLSI Emerging Trend in VLSI Systems Design (ISVLSI'04),* (2004).

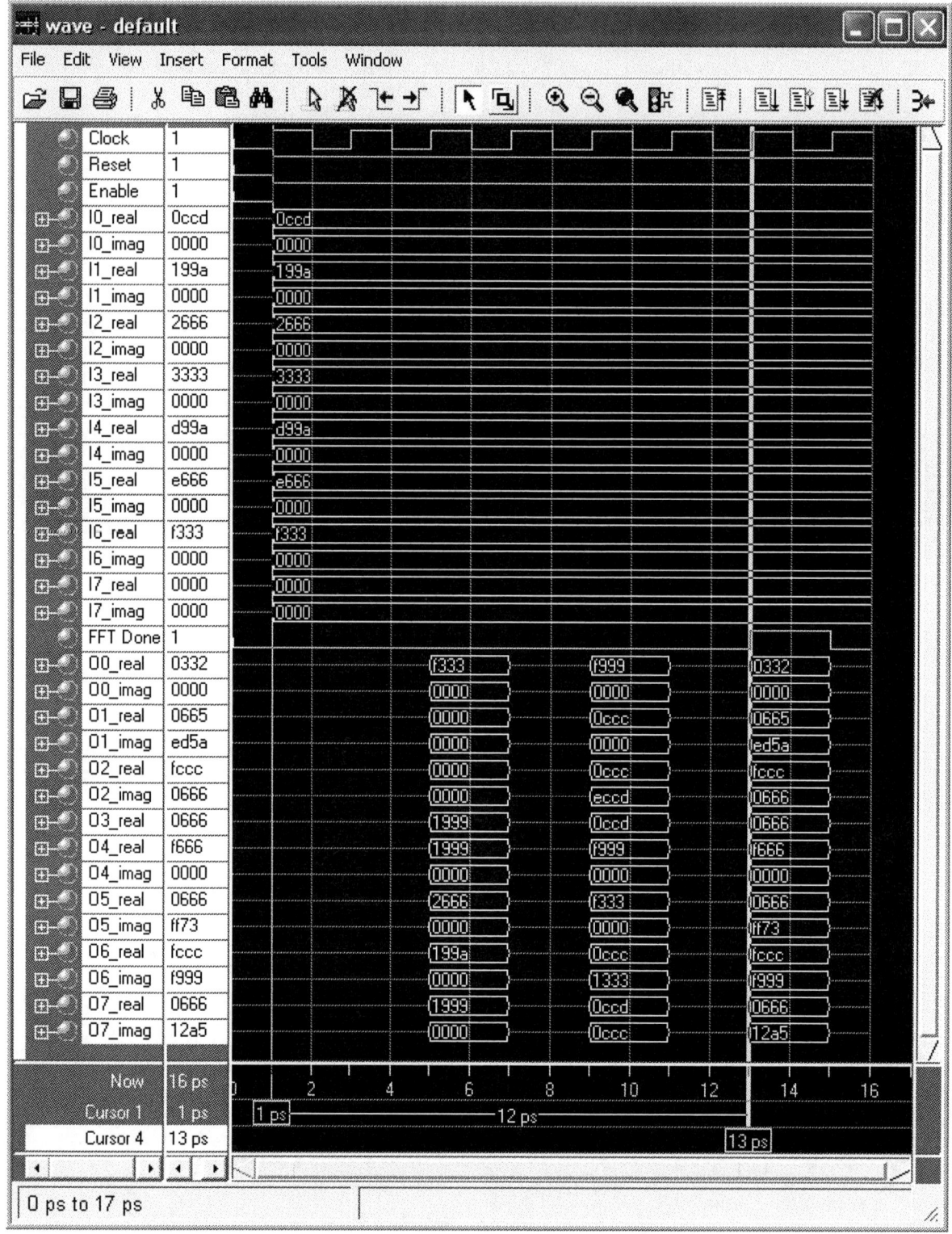

Fig. 8 ModelSim Simulation results of multiplier-less architecture for radix-2 DIT 8-points FFT

Low Temperature Heteroepitaxial Growth of 3C-SiC on Si Substrates by Rapid Thermal Triode Plasma CVD using Dimethylsilane

Abdul Manaf Hashim[1] and Kanji Yasui

Department of Electrical Engineering, Nagaoka University of Technology
1603-1 Kamitomioka, Nagaoka 940-2188, Japan
Email: manaf@fke.utm.my, kyasui@vos.nagaokaut.ac.jp

[1]The author is now at the Faculty of Electrical Engineering, Universiti Teknologi Malaysia, 81310 Skudai, Johor, Malaysia

Abstract **The investigation of the dependence of the cubic silicon carbide (3C-SiC) film characteristics on the reaction pressures, growth temperatures and hydrogen dilution rates was carried out by rapid thermal triode plasma CVD using dimethylsilane as a source gas. The stoichiometric 3C-SiC films with good crystallinity and crystal orientation were successfully grown at 1100-1200 °C. The crystallinity and the crystal orientation of SiC films grown at large dilution rate of above 200 and growth pressure of 0.3 ~ 0.7 Torr were better than those grown at small dilution rate and high growth pressure. Under large dilution rate, large amount of hydrogen radicals can be generated. It is speculated that excessive carbon atoms or weak bonds formed in SiC films were effectively extracted by the large amount of hydrogen radicals.**

I. INTRODUCTION

SILICON carbide (SiC) is a wide band-gap semiconductor with large saturated electron drift velocity, large breakdown electric field and large thermal conductivity. Therefore, this material is of great interest for high temperature and high-power applications [1,2]. Among many polytypes, cubic silicon carbide (3C-SiC) with zincblende crystal structure can only be grown on Si substrates [3]. Unlike the other polytypes (4H- or 6H-SiC), 3C-SiC can be grown at lower temperature (below 1300 °C) and be epitaxially grown on Si substrates. Usually, SiC films are grown on Si substrate by chemical vapor deposition (CVD) using silane (SiH_4) and hydrocarbon gases such as propane (C_3H_8) [4,5]. However, this method requires high growth temperature (~1300 °C), which may induce high tensile stress because of the difference in thermal

expansion coefficient between Si and SiC and the formation of the voids in Si substrates.

In order to overcome these problems, the use of single precursor gases such methylsilanes instead of SiH_4 and C_3H_8 are useful, because they contain Si-C bonds in their molecules and are decomposable at low temperature. In addition, simplified chemical vapor deposition (CVD) apparatus can be utilized because they are non-pyrophoric gases. Application of reactive plasma is a useful method to accelerate chemical reactions at low temperatures. In order to apply the reactive plasma to the epitaxial growth of SiC, however, the incidence of energetic charged particles on the surface of the growing films must be reduced.

In the previous report, we presented the triode plasma chemical vapor deposition (CVD) method and the introduction of negative DC biases on the grid electrode, low electron temperature and low space potential below the grid were obtained [6,7]. The preliminary results on the growth of crystal SiC film using dimethylsilane (DMS) diluted with hydrogen were presented [6,7]. In those previous reports, hydrogen dilution rate and growth pressure were not changed widely.

In this report, we present the experimental results of low-temperature epitaxial growth of 3C-SiC on Si(100) and (111) substrates by rapid thermal triode plasma CVD (RTP-CVD) using dimethylsilane (DMS) and hydrogen radicals without carbonization process. Namely, the effects of hydrogen dilution rate, reaction pressure and temperature on the quality of SiC films will be reported. The crystallinity of SiC films were characterized using X-ray diffraction spectroscopy.

0-7803-9730-4/06/$25.00 ©2006 IEEE

II. EXPERIMENTAL

The schematic diagram of the growth apparatus is shown in Fig. 1. The grid electrode (135 mm in diameter, wire diameter = 0.3 mm, wire spacing = 1.25 mm) inserted between the cathode and anode electrodes was connected to a dc electric source. By supplying negative dc bias on the grid (Vg), rf discharge was confined between the grid and the cathode. Anode was insulated from the grounded chamber. In this experiment, the substrate holder was made electrically floated. As a source gas, DMS, kept at 0 °C in a stainless steel cylinder, was admitted into the afterglow region below the grid. Hydrogen gas was admitted into the discharge region above the grid passing through a hydrogen gas purifier. Experimental conditions are as follows: H_2 flow rate 112 sccm, gas feed ratio (H_2/DMS) 40-1500, total gas pressure during growth 0.1~1.4 Torr, substrate temperature 1000~1200 °C, rf power 100W, grid bias Vg = -100V.

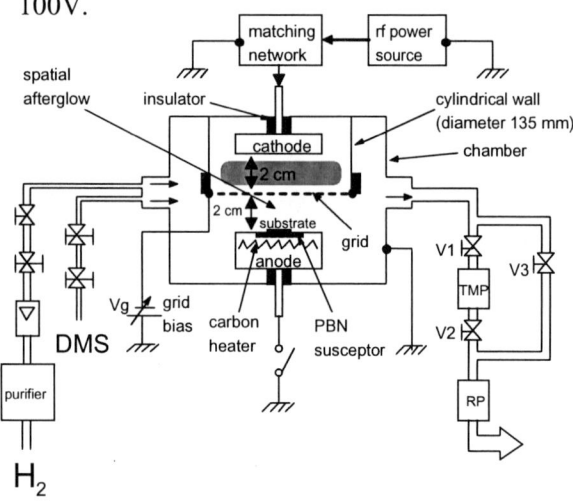

Fig. 1 Schematic diagram of rapid thermal triode plasma CVD apparatus.

The procedure of epitaxial growth is schematically shown in Fig.2 and is summarized as follows. After degreasing, dipping in buffered HF and rinsing in dionized water, Si (100) and (111) substrates were immersed in boiling ultra-pure water. After evacuating the growth chamber to 10^{-7} Torr, the substrate temperature was raised to 350 °C. Substrates were heated on a carbon heater and substrate temperature was measured using an optical pyrometer. With the supply of source gas, substrate temperature was rapidly raised from 350 °C to growth temperature in

hydrogen and DMS flows, followed by the epitaxial growth, as shown in Fig. 2.

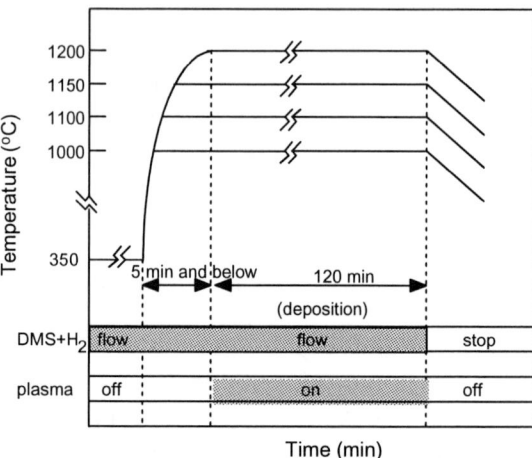

Fig. 2 The time chart for SiC epitaxial growth by rapid thermal triode plasma CVD.

Crystallinity of SiC films was evaluated using an X-ray diffractometer (RIGAKU, RAD-IIIA) equipped with a graphite monochrometer.

III. RESULTS AND DISCUSSION

The dependence of SiC growth rate on the growth pressure and hydrogen dilution rate, H_2/DMS were shown in Fig. 3(a) and (b), respectively. The values inside the plot of Fig. 3(a) represent the hydrogen dilution rate. The dilution rates at 0.3 Torr and 1.4 Torr were very close. Considering these two dilution rates, it is clear that the growth rate increase with the growth pressure. It was considered that large plasma etching effect has become more dominant at the low growth pressure condition compared to high growth pressure. This assumption is supported by the experimental data shown in Fig. 3(b) where the growth rate decreases with the hydrogen dilution rate. The films grown at small dilution rate were thicker than those grown at large dilution rate. The same characteristics were also seen for Si(111) substrate. From these results, plasma etching effect becomes more dominant at high density hydrogen plasma condition.

Fig. 4(a) shows the variation in the full width at half maximum (FWHM) of X-ray diffraction peak of SiC (200) grown by rapid thermal triode plasma CVD as a function of the hydrogen dilution rate (H_2/DMS). The film thickness was 0.3 ~ 0.8 μm. As shown in Fig. 3(b), the films grown at small dilution rate were thicker than those grown at large dilution rate. At the same growth condition, the crystallinity of SiC epitaxial films was improved with the film thickness. The FWHM values of X-ray diffraction peaks decreased with the hydrogen dilution rate, as shown in Fig. 4(a). From the result, it is clear that the crystallinity and the crystal orientation of SiC films grown at large dilution rate were better than those grown at small dilution rate. FWHM values of X-ray diffraction peak of SiC (111) also decreased with the hydrogen dilution rate, as shown in Fig. 4(b).

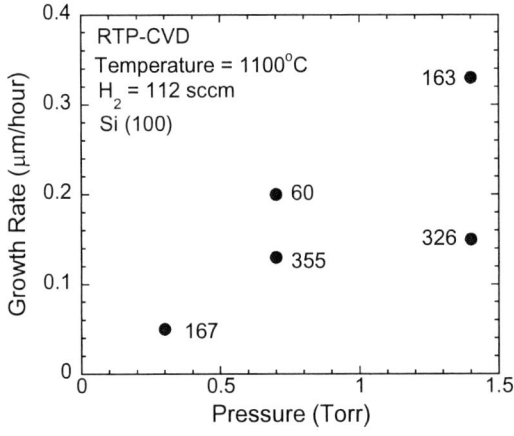

The values inside shows the dilution rate (H_2/DMS)

(a)

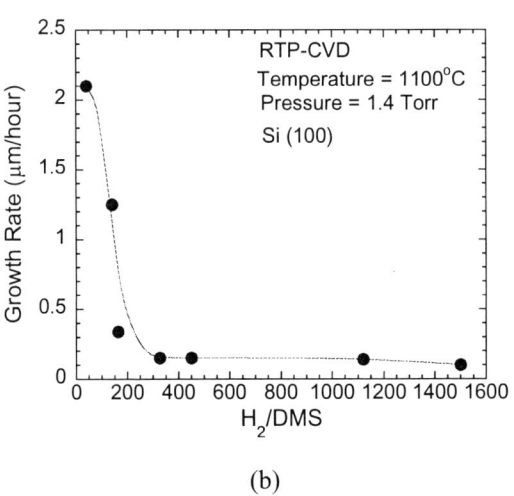

(b)

Fig. 3 The growth rate of SiC films grown on Si(100) substrate by rapid thermal triode plasma CVD as a function of (a) growth pressure and (b) the dilution rate (H_2/DMS).

Under large dilution rate, large amount of hydrogen radicals can be generated. It is speculated that excessive carbon atoms or weak bonds formed in SiC films were effectively

extracted by the large amount of hydrogen radicals.

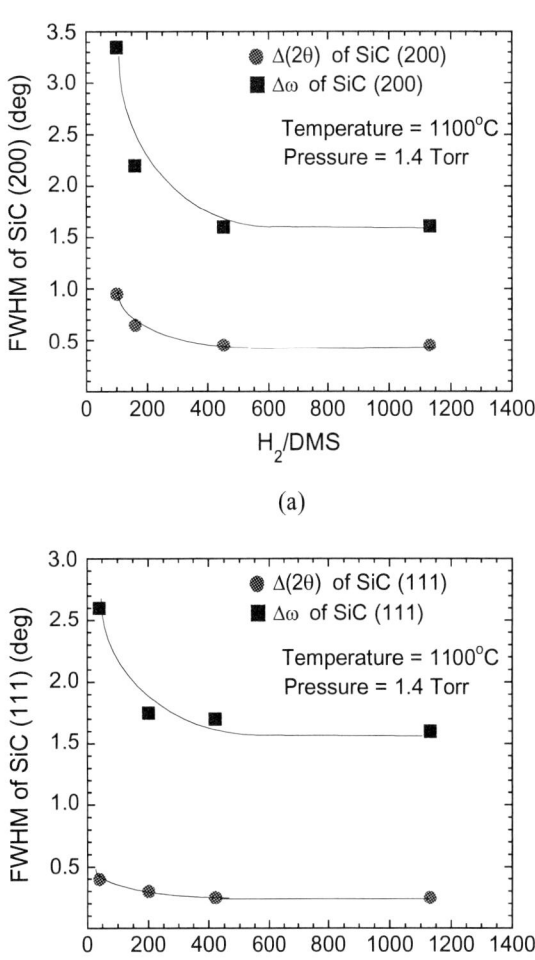

(a)

(b)

Fig. 4 The variation in the FWHM values of X-ray diffraction peak of SiC(200) and (111) grown by rapid thermal triode plasma CVD as a function of the dilution rate (H_2/DMS).

Fig. 5 shows the variations in the FWHM values of X-ray diffraction (θ-2θ) and ω–rocking curve of SiC(200) grown by rapid thermal triode plasma CVD as a function of the substrate temperature. Hydrogen dilution rate during SiC growth was 300. It is shown in this figure that SiC epitaxial films with good crystallinity and crystal orientation have not been obtained at temperature of 1000 °C. To date, those grown at 1100 °C showed fairly good crystallinity and crystal orientation. Fig. 6(a) and (b) shows the variation in the volume ratio of the domains epitaxially grown to those containing stacking faults. Data of SiC films grown on Si(100) crystals were used. The volume of the domains epitaxially grown was evaluated from the area of SiC(200) peak obtained from X-ray rocking curves, and that of domains containing stacking faults was evaluated from the area of SiC(111) peak appeared $\pm 15.8°$ deviated from the substrate surface ((100) plane).

As shown in Fig. 6(a), the integrated intensity ratio of the rocking curves between the domains epitaxially grown and those containing the stacking faults was smallest at low temperature (1000 °C).

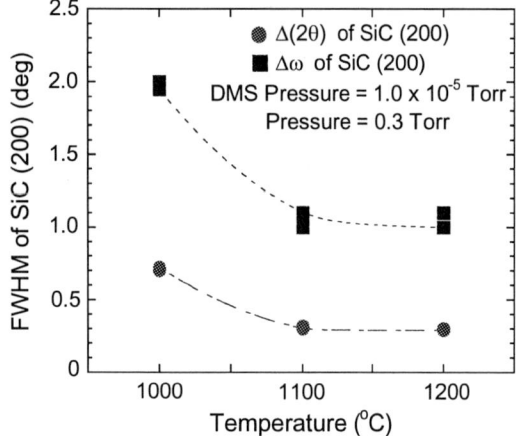

Fig. 5 The variation in the FWHM values of X-ray diffraction peak of SiC(200) and (111) grown by rapid thermal triode plasma CVD as a function of the substrate temperature.

At high temperatures (>1100 °C), on the other hand, the integrated intensity ratio increased three times. At low temperatures, the extraction of excessive methyl groups from the growing film surface is not adequate by thermal energy. The generation of the stacking faults is also considered to be induced by the existence of the excessive carbon and hydrogen atoms.

Fig. 6(b) shows the dependence of volume ratio between the domains epitaxially grown and those containing stacking faults on the dilution rate of DMS. The ratio shows high values at large dilution rate (>200). Again, it shows that the hydrogen radicals generated in H_2 plasma have enhanced the extraction of the excessive carbon atoms and break the weak bonds, and have reduced the volume of the domains containing the stacking faults.

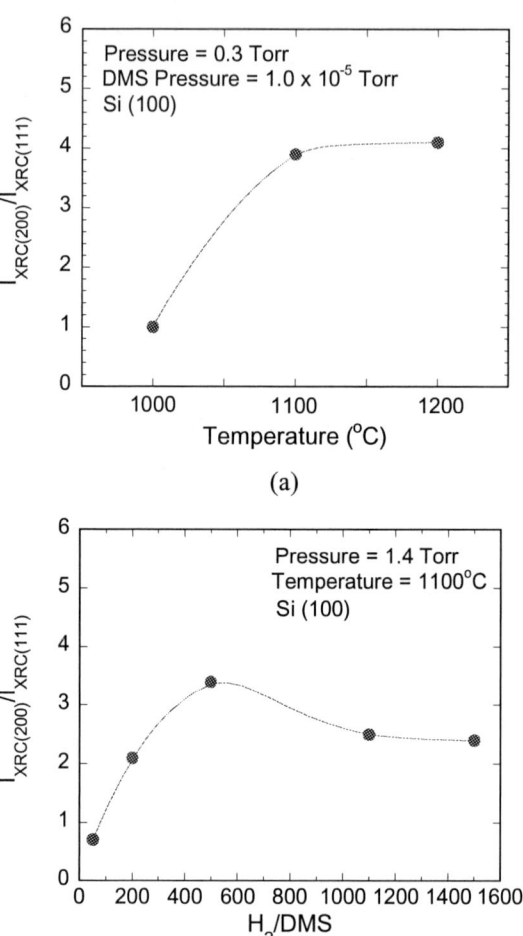

(a)

(b)

Fig. 6 Variation in the volume ratio between the domains epitaxially grown and those containing stacking faults as a function of the growth temperature and dilution rate.

IV. CONCLUSION

The investigation of the dependence of the SiC film characteristics on the growth pressure, growth temperature and hydrogen dilution rate was carried out by rapid thermal triode plasma CVD method using DMS as a source gas. The hydrogen dilution rate of above 200 and total growth pressure of 0.3 ~ 0.7 Torr are considered as the best growth conditions. From X-ray rocking curve measurements, the domain which includes stacking faults in films grown at these conditions was drastically reduced. The growth rate increase with the growth pressure due to the decrease in plasma etching effect. Under large dilution rate, large amount of hydrogen radicals can be generated. It is speculated that excessive carbon atoms or weak bonds formed in SiC films were effectively extracted by the large amount of hydrogen radicals.

REFERENCES

[1] W.V.Munch and P.Hoeck, "Silicon Carbide Bipolar Transistor," Solid State Electron. Vol. 21, No.2, pp. 479-480 (1978).

[2] H.Okumura, "Present Status and Future Prospect of Widegap Semiconductor for High Power Devices," Jpn. J. Appl. Phys., Vol. 45, No. 10A, pp. 7565-7586 (2006).

[3] H.Matsunami, S.Nishino and H.Ono, "IVA-8 Heteroepitaxial Growth of Cubic Silicon Carbide on Foreign Substrates," IEEE Trans. Electron. Dev., Vol. 28, pp. 1235-1236 (1981).

[4] H.Matsunami, "Technological Breakthroughs in Growth Control of Silicon Carbide for High Power Electronic Devices," Jpn. J. Appl. Phys., Vol.43, No. 10, pp. 6835-6847 (2004).

[5] Y.Ishida, T.Takahashi, H.Okumura, S.Yoshida and T.Sekigawa, "Atomically Flat 3C-SiC Epilayers by Low Pressure Chemical Vapor Deposition," Jpn. J. Appl. Phys., Vol. 36, pp. 6633-6637 (1997).

[6] K.Yasui, M.Kimura, A.M.Hashim and T.Akahane, "Epitaxial Growth of 3C-SiC Films on Si by Triode Plasma CVD using Dimethylsilane," Technical Report of IEICE, CPM98-124, pp. 15-20 (1998) (in Japanese).

[7] K.Yasui, M.Kimura, K.Sanada and T.Akahane, "Heteroepitaxy of 3C-SiC using Triode Plasma Enhanced Chemical Vapor Deposition on Si Substrates without Buffer Layer," Appl. Surf. Sci., Vol. 142, pp. 381-385 (1999).

The Influence of Doping Concentration, Temperature, and Electric Field on Mobility of Silicone Carbide Materials

Mazhar Tayel, Ayman El-Shawarby
Faculty of Engineering, Alexandria University, Alexandria, Egypt
E-mail : alaaahafez@yahoo.com

Abstract : **This paper presents a study of the performance of the three most important silicone carbide (SiC) polytypes namely4H, 6H and 3C-SiC. The models describe the dependence of electron mobility on doping concentration, temperature, and electric field. The results show that SiC materials mobility almost degraded with increasing the doping concentration, temperature, and electric field. The significant degradation appear over the entire range of the electric field, which record, from 98% to 98.8% degradation. Temperature increase degrades the electron mobility from 80% to 84% and doping concentration increase degrades it by 34% to 50%. 4H-SiC behaves the better electron mobility among SiC materials. In addition, the models study the material conductivity under influence of electric field. 3C-SiC record the highest conductivity over the entire range of electric field. Silicon carbide materials are characterized in terms of Breakdown voltage, maximum frequency, Keyes' figure of merit, and Johnson's figure of merit. The results carried out that 4H-SiC has a maximum frequency and power for semiconductor devices. 6H-SiC share 4H-SiC the maximum thermal limitation and high frequency electrical performance.**

I. INTRODUCTION

In recent years, significant advances in silicon material technology coupled with the development of novel electronic device structures have leveraged considerable system improvements for wide range of power electronic applications such as lamp ballast's, motor control, medical electronics, and high frequency power systems applications. Silicon carbide is a semiconductor material combining several unique features such as high breakdown voltage, high electron saturation velocity, and high thermal conductivity. These properties make it promising candidate for RF power devices.

During the last ten years, the interest in SiC RF power transistors has increased considerably. Substantial progress has been made in SiC technology leading to a variety of experimental SiC RF transistors capable of operation in giga hertz range [1-3]. Despite the impressive experimental results, much work remains to be done in the field of device simulation to optimize SiC power transistor structures. For such simulations, reliable models for the electron mobility and its dependence on doping concentration and electric field are of crucial importance. Power devices are operated at high voltage and high current levels leading to considerable self-heating. Thus, the operating temperature can be much higher than room temperature. To simulate the device under these conditions properly, the temperature dependence of the electron mobility must also be taken account in the model. SiC occurs in more than 100 different polytypes structures. Widely used in RF and power devices are the hexagonal 4H and 6H polytypes. Furthermore, more, the cubic 3C polytype is of interest because it can be grown on Si substrate. Even thought until now the quality of 3C-SiC layers on Si does not satisfy the requirements of RF transistors, this material combination could offer a low cost alternative in the future. The electron mobility in 4H, 6H and 3C-SiC has been investigated both experimentally and theoretically [4-6] and developed by Matthias et. al. [7]. Elemental and compound semiconductors, including wide-band gap semiconductors are examined for high power electronics applications by Krishna et. al. [8]. This paper is devoted to characterize the silicon carbide materials in terms of mobility and its dependence on doping concentration, temperature and electric field, breakdown voltage, maximum frequency, Keyes' figure of merit and Johnson's figure of merit.

II. MOBILITY MODELING

0-7803-9730-4/06/$25.00 ©2006 IEEE   660

In the drift diffusion approach, the electron transport is mainly governed by the electron mobility μ. Physically, the electron mobility describes the electron drift velocity under the influence of an electric field E according to,

$$v = \mu \times E \qquad (1)$$

At low values of electric fields the electron velocity increases almost linearly with field and the mobility has the constant value μ_o. The Low field mobility is a function of the doping concentration and the temperature. A widely used empirical expression for modeling the doping dependence of low field mobility has been,

$$\mu_O = \mu_{min} + \frac{\mu_{max} - \mu_{min}}{1 + \left(\frac{N}{N_{ref}} \right)^{\alpha}} \qquad (2)$$

where N is the total doping concentration and μ_{min}, μ_{max}, N_{ref}, and α are fitting parameters. The parameter μ_{max} in equation represents the mobility of undoped samples, where lattice scattering is the main scattering mechanism, while μ_{min} is the mobility in highly doped material, where impurity scattering is dominant. N_{ref} is the doping concentration at which the mobility of halfway between μ_{min} and μ_{max}, and α is a measure of new quickly the mobility can be modeled by making the four fitting parameters from (2) temperature dependent according to [7],

$$Par = Par_o \times \left(\frac{T}{300} \right)^{\gamma} \qquad (3)$$

where Par is the parameter of interest (i.e. μ_{min}, μ_{max}, N_{ref}, or α), T is the temperature and Par_o is the value of the parameter Par at T=300°K. When strong electric field are applied, the electron velocity is no longer proportional to the field, and can thus no longer be described by a field independent mobility. An expression frequently used for modeling the field dependence of the mobility in Si is [8],

$$\mu = \frac{\mu_o}{\left(1 + \left(\frac{\mu_o \times E}{v_{sat}} \right)^{\beta} \right)^{\frac{1}{\beta}}} \qquad (4)$$

where v_{sat} the saturation velocity, and β is a constant specifying how abruptly the velocity goes into saturation.
To obtain the velocity field characteristics, both sides of (4) have to multiplied by the electric field. Equation (4) results in a soft velocity saturation, which describes the SiC velocity-field characteristics well up to fields of 5 x 10^5 V/cm.

The standard silicon model [9] is used to describe the temperature dependence of the saturation velocity under high field,

$$v_{sat} = \frac{v_{max}}{1 + 0.6 \times \exp\left(\frac{T}{600} \right)} \qquad (5)$$

with the maximum velocity v_{max} being a fitting parameter.

Mobility Modeling for 4H and 6H-SiC

$$\mu_{max} = \mu_{max} \times \left(\frac{T}{300} \right)^{\gamma_{max}} \qquad (6)$$

$$\mu_{min} = \mu_{min} \times \left(\frac{T}{300} \right)^{\gamma_{min}} \qquad (7)$$

$$N_{ref} = N_{ref} \times \left(\frac{T}{300} \right)^{\gamma_{Nref}} \qquad (8)$$

$$\alpha = \alpha \times \left(\frac{T}{300} \right)^{\gamma_{\alpha}} \qquad (9)$$

The mobility μ can be calculated from equation (4), μ_o from equation (2), and v_{sat} from equation (5), where,

$$\beta = \beta_o + ae^{\left(\frac{T - T_o}{b} \right)} \qquad (10)$$

Mobility Modeling for 3C-SiC

$$\mu_{max} = \mu_{max} \times \left(\frac{T}{300} \right)^{\gamma_{max}} \qquad (11)$$

$$C = C \times \left(\frac{T}{300} \right)^{\gamma_C} \qquad (12)$$

$$N_{ref} = N_{ref} \times \left(\frac{T}{300} \right)^{\gamma_{Nref}} \qquad (13)$$

$$\alpha = \alpha \times \left(\frac{T}{300} \right)^{\gamma_{\alpha}} \qquad (14)$$

$$\delta = \delta \times \left(\frac{T}{300} \right)^{\gamma_{\delta}} \qquad (15)$$

The mobility μ can be calculated from equation (4), and μ_o thus,

$$\mu_O = \frac{C \times \left(\frac{N}{N_{ref}} \right)^{\alpha - \delta} + \mu_{ax}}{1 + \left(\frac{N}{N_{ref}} \right)^{\alpha}} \qquad (16)$$

v_{sat} calculated from equation (12) and β from equation (13). The fitting parameters of the three SiC polytypes are tabulated in table (1).

Characterization of Semiconductor Materials :

The dependence of avalanche breakdown voltage V_B on the background doping density N_B and energy band gap E_G can be obtained from the well-known sze [10],

$$V_B = 60 \left(\frac{E_g}{1.1}\right)^{\frac{3}{2}} \left(\frac{N_B}{10^{16}}\right)^{\frac{3}{4}} \quad (17)$$

Table (1) Parameters for the SiC mobility models

Parameter	4H-SiC	6H-SiC	3C-SiC
μ_{max} (cm^2/V.S)	950	420	650
γ_{max}	-2.4	-2.5	-2.5
μ_{min} (cm^2/V.S)	40	30	-
γ_{min}	-0.5	-0.5	-
N_{ref} (cm^{-3})	2 x 10^{17}	6 x 10^{17}	3 x 10^{16}
γ_{Nref}	1	2.5	0
α	0.76	0.8	0.8
γ_α	0	0.5	0
C (cm^2/V.S)	-	-	330
γ_C	-	-	-1.5
δ	-	-	0.2
$v_{max (cm/S)}$	4.77x10^7	4.6 x 10^7	4.38x10^7
B_o	0.816	0.95	0.676
$T_o (^o K)$	327	333	103
a	4.27x10^{-2}	4.55x10^{-3}	0.464
b	98.4	71.8	359

The drift region conductance per unit area σ_A is given by [10],

$$\sigma_A = \frac{\varepsilon_s \mu E_M^3}{4 V_B^2} \quad (18)$$

where μ is the low field mobility of charge carriers in the drift region, ε_s is the permitivity of the semiconductor, E_M is the peak electric field strength at the breakdown. The conductance is a function of energy bang gap E_g was obtained from,

$$\sigma_A = 4.11 \times 10^5 \frac{\mu E_G^3}{\sqrt{\varepsilon_r} V_B^{5/2}} \quad (19)$$

these results are qualitatively in agreement with those obtained from (18). Keyes' figure of merit (KFM) as well as Johnson's figure of merit (JFM) is given by,

$$KFM = \lambda \left[\frac{c v_s}{4\pi \varepsilon_r}\right]^{1/2} \quad (20)$$

$$JFM = \frac{E_M^2 v_s^2}{4\pi^2} \quad (21)$$

where C the velocity of light and v_s is the scattering limited saturated velocity of carriers in the semiconducting material. KFM is an indication of the thermal limitation of the material of its high frequency electrical performance, and JFM relates the frequency and power product of a semiconductor transistor. The maximum frequency of operation for devices with equal area is then given by,

$$f_{max} = F \mu E_M \sqrt{E_G} \quad (22)$$

III. RESULTS AND DISSCUSION

The mobility for three silicone carbide polytypes namely 4H, 6H and 3C-SiC are analyzed under influence of electric field and temperature at different values of doping concentration. Figures (1,2,3) shows that increase the electric field degrade the electron mobility of the SiC material. At high values of doping concentration the mobility, die at all values of electric field. The low values of electric field increase the electron mobility significantly with decreasing the doping concentration. The mobility also degraded significantly by temperature increase by about 80% to 84%.

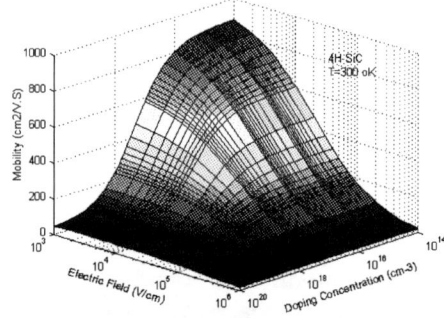

Figure (1) Electron mobility versus electric field at different values of doping concentration for 4H-SiC at T=300 ^{o}K

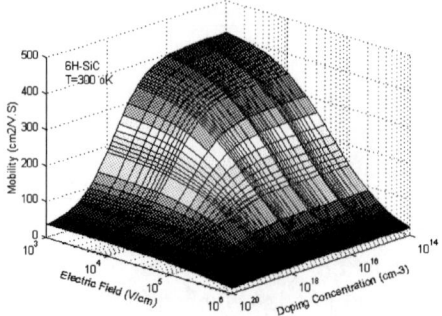

Figure (2) Electron mobility versus electric field at different values of doping concentration for 6H-SiC at T=300 °K

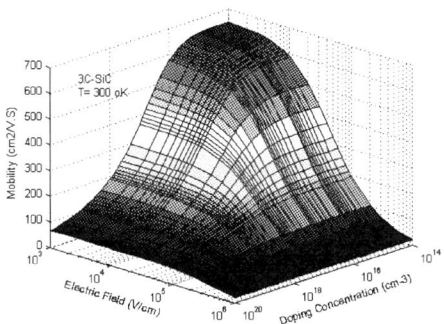

Figure (3) Electron mobility versus electric field at different values of doping concentration for 3C-SiC at T=300 °K,

Figures (4,5,6) demonstrate the effect of temperature on the electron mobility at different values of doping concentration. Temperature increase significantly degrades the electron mobility of SiC materials but the electric field increase drops the mobility by about 98% to 98.8%. Figure (7) shows the electric conductivity for the three SiC polytypes versus electric field. Increasing the electric field degrade the electric conductivity of the material. It is noticed that at low values of donor concentration materials 3C and 4H-SiC perform 6H-SiC up to electric field of 10^5 V/cm but up this value the best conductivity from 3C-SiC.

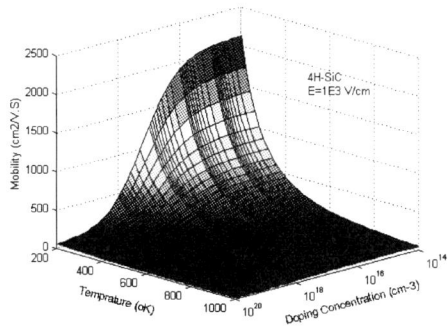

Figure (4) Electron mobility versus temperature at different values of doping concentration for 4H-SiC at E=1 x 10^3 V/cm

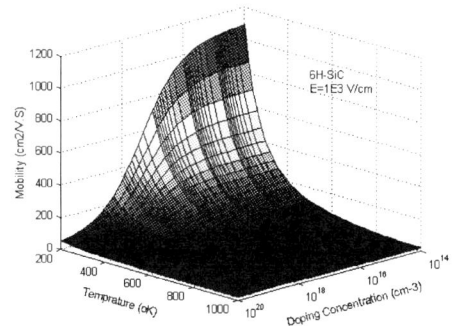

Figure (5) Electron mobility versus temperature at different values of doping concentration for 6H-SiC at E=1 x 10^3 V/cm,

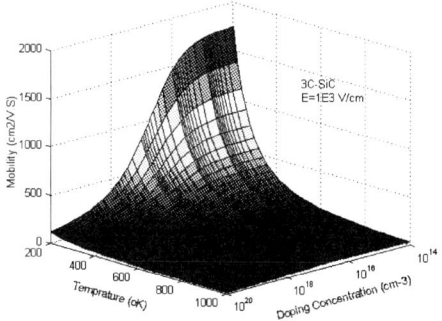

Figure (6) Electron mobility versus temperature at different values of doping concentration for 3C-SiC at E=1 x 10^3 V/cm

Figure (8) shows the Keyes' figure of merit (KFM) that represents the material performance under temperature condition. SiC material performs the Si material by more than 150%, but 4H and 6H-SiC has a maximum performance under temperature conditions.

Figure (9) shows the Johnson's figure of merit (KFM) that represents the material performance under high frequency or high power conditions. SiC material perform the Si material by more than 500%, but 4H-SiC record the highest performance over the entire range of electric field. Figure (10) shows that 4H-SiC record the highest breakdown voltage this indicate that it suitable for high voltage rectifiers and transistors. Figure (11) shows that 3C-SiC record highest frequency and suitable for RF power Transistors.

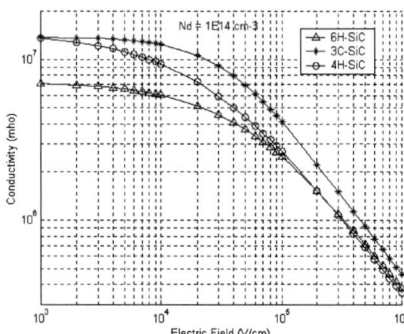

Figure (7) Electric conductivity for three SiC polytypes namely 4H, 6H and 3C-SiC at $N_d=1 \times 10^{14}$ cm^{-3} ,

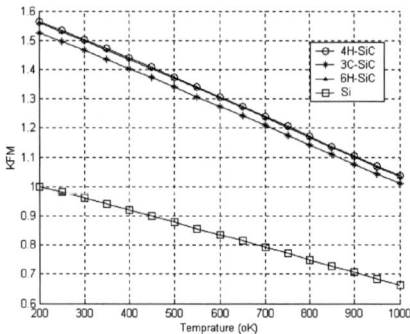

Figure (8) Keyes' figure of merit (KFM) for three SiC polytypes and Si versus electric field

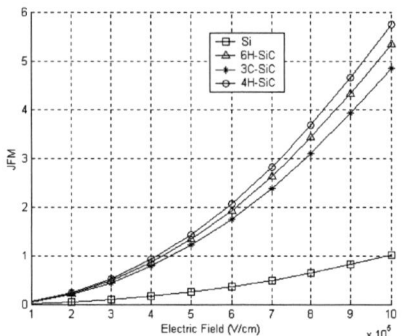

Figure (9) Johnson's figure of merit (JFM) for three SiC polytypes and Si versus electric field

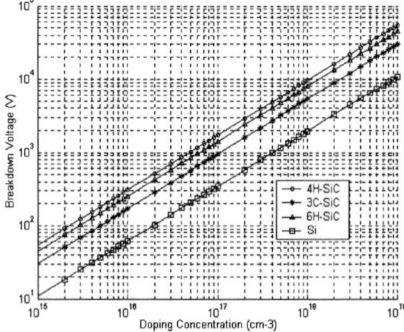

Figure (10) Breakdown voltage for three SiC polytypes and Si at different values of doping concentration

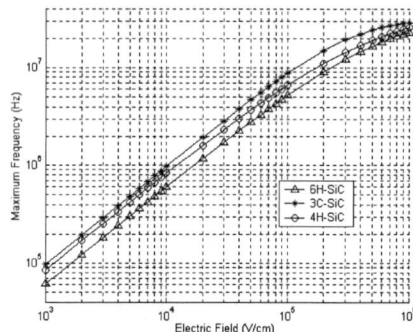

Figure (11) Maximum frequency for three SiC polytypes and Si at different values of electric field.

REFERENCES

[1] C. H. Carter Jr, V. F. Tsvetkov, R. C. Glass, D. Henshall, M. Brady, St.G. Müller, O. Kordina, K. Irvine, J. A. Edmond, H.-S. Kong, R. Singh,S. T. Allen, and J. W. Palmour, "Progress in SiC: from material growth to commercial device development," *Mater. Sci. Eng.*, vol. B61-62, pp. 1–8, 1999.

[2] R. R. Siergiej, R. C. Clarke, S. Sriram, A. K. Agarwal, R. J. Bojko, A.W. Morse, V. Balakrishna, M. F. MacMillan, A. A. Burk Jr, and C. D. Brandt, "Advances in SiC materials and devices: an industrial point of view," Mater. Sci. Eng., vol. B61-62, pp. 9–17, 1999.

[3] J. C. Zolper, "Wide bandgap semiconductor microwave technologies: from promise to practice," in IEDM Tech. Dig., 1999, pp. 389–392.

[4] W. J. Schaffer, G. H. Negley, K. G. Irvine, and J.W. Palmour, "Conductivityanisotropy in epitaxial 6 H and 4 H SiC," in Mat. Res. Soc. Symp. Proc., vol. 339, 1994, pp. 595–600.

[5] M. Ruff, H. Mitlehner, and R. Helbig, "SiC devices: physics and numerical simulation," IEEE Trans. Electron Devices, vol. 41, pp. 1040–1054, May 1994.

[6] G.-B. Gao, J. Sterner, and H. Morkoc, "High frequency performance of SiC heterojunction bipolar transistors," IEEE Trans. Electron Devices, vol. 41, pp. 1092–1097, July 1994.

[7] Matthias Roschke,, and Frank Schwierz,"Electron Mobility Models for 4H, 6H, and 3C SiC", IEEE Trans. Electron Device, Vol. 48, No. 7, July 2001.

[8] Krishna Shenai, Robert S. B. Jayant, "Optimum semiconductors for high-power electronics", IEEE Trans. Electron Device, 1989.

[9] S. M. Sze, Physics of Semiconductor Devices. New York:Wiley, 1981.

[10] I. A. Khan and J. A. Cooper, Jr., "Measurement of high-field electron transport in silicon carbide," IEEE Trans. Electron Devices, vol. 47, pp. 269–273, Feb. 2000.

A 0.18µm, 1.8-V CMOS High Gain Fully Differential Opamp Utilized in Pipelined ADC

[1]Ali Meaamar, *Student Member IEEE*,[1]Masuri Bin Othman, *MemberIEEE,* [2]Omid Shoaei, *Member IEEE*

[1]VLSI Design Center, Institute of Microengineering and Nanoelectronics (IMEN) Universiti Kebengsaan Malaysia, 43600 Bangi, Selangor, Malaysia

[2]Faculty of Engineering, University of Tehran, North Kargar Ave. P.O.BOX 14395/515, Tehran, Iran

Corresponding author : meaamar@vlsi.eng.ukm.my

Abstract **A fully differential, high gain opamp to be used in a low-voltage low-power high speed pipeline analog to digital converter (ADC) in a 0.18µm CMOS process is designed. The opamp architecture is based on folded cascode and "double differential amplifier" technique. This design operates of a 1.8V power supply, achieving a differential output swing of ±1.65V, a DC gain of > 95 dB with a unity gain at 312MHz and a phase margin of 56° and 0.5mW power dissipation. The operational transconductance amplifier (OTA) can be used to design high-speed ADCs, for local wireless communications. The minimized power dissipation opamp could be used in high resolution, high speed pipelined analog to digital converters, which are needed in applications requiring both high data rate and high speed, such as wireless LANs.**

I. INTRODUCTION

In many analog circuit applications such as ADCs [1], switched capacitor (SC) filters [2] and sample and hold amplifiers, speed and accuracy are determined by the settling behavior of the opamp circuit. Often, it is just the spot where the limits of the technology are first met, when trying to enhance the speed or reduce the power consumption of the circuit. In SC technique used in the pipeline ADC's, opamp is a central component. The settling speed depends mainly on the unity gain frequency and a single pole settling time, while high settling accuracy is due to high DC gain of the opamp circuit. The opamp DC gain requirement is higher than what is achievable with simple single-stage topologies, whereas OTAs with single-stage architecture offer large bandwidth and a good phase margin. Techniques to enhance the opamp DC gain without going into multi-stage architectures are especially required in high-speed circuits, where

the high current levels make the transistor output conductance, g_{ds} large. A widely used method is based on improving the cascading effect of a single MOS transistor by using local negative feedback. The resultant circuit, so called regulated cascode or gain boosting amplifier, is utilized in a current source, which is shown in Fig. 1 [3]. But this paper utilizes a "double differential amplifier" technique, introduced in section II, to enhance DC gain in companion with folded cascode current network. This technique will result in a circuit with reduced number of poles in the signal path to have a proper design with the least frequency compensation necessity. Based on the folded cascode opamp design with the gain enhancing techniques, this paper presents a high DC gain fully differential opamp with 312MHz unity gain frequency.

The gain boosting techniques, the frequency behavior, and the circuit implementation in a 0.18 CMOS are explained in section II. The simulation results are given and discussed in section III.

II. HIGH-GAIN OPERATIONAL AMPLIFIER

A. Importance of High Gain in Operational Amplifier

The large gain and high linearity in operational amplifiers are often required to achieve high resolutions in ADCs. This operational amplifier is the primary source of speed limitation and power dissipation. This block also contributes to thermal noise, limiting the ADC resolution. In the submicron process, the DC gain is reduced which limits the performance of a pipelined ADC. Accordingly, the design and the specification of the amplifier are significantly important to the whole ADC.

0-7803-9730-4/06/$25.00 ©2006 IEEE

B. Conventional Gain Boosting in Cascode Stage

As shown in Figure 1, the idea of gain boosting is based on negative feedback loop to set the drain voltage of M_1 [3]. Negative feedback drives the gate of M_2 until V_T has the same value as V_{ref}. Therefore, the variation of V_{out} has much less effect on V_T, because block A_1 "regulates" this voltage. This topology is usually called "regulated cascode" or "active cascode". With the smaller variation of V_T due to the change of Vout, the output current becomes less sensitive to the voltage variation at Vout compared with conventional cascode structure. Therefore, the output impedance increases as below:

$$v_{out} = A.(v^+ - v^-)$$

Where A is the gain of the added amplifier.

$$v_{gs4} = -i_x r_o (A+1),$$

$$i_x = gm(-i_x r_o (A+1)) + \frac{v_x - i_x r_o}{r_o}$$

$$R_{out} = \frac{V_x}{i_x} = r_{o1} + r_{o2} + g_{m2} \cdot r_{o1} \cdot r_{o2} (A+1) \quad (1)$$

$$\approx g_m r_{o2} r_{o1} \cdot A$$

This increases the output resistance resulting in several orders of improvements on the overall gain as:

$$Av_{total} = g_{m1} \cdot R_{out} = A \cdot g_{m1} g_{m2} \cdot r_{o1} \cdot r_{o2} \quad (2)$$

C. Fully Double Differential Folded-Cascode Amplifier

The circuit schematics of a conventional differential amplifier and a double differential amplifier are shown in Fig. 2 and Fig. 3, respectively. The increase in bias current is accomplished by increasing the width of transistors in current source network and composite transistors in Fig. 3. The channel lengths of "cascode current steering" and "composite current source" usually must be chosen to be the same, and ratios provided by properly choosing the channel widths. A composite transistor ($Q_1...Q_m$, Q_{m+1} and $Q'_1...Q'_m$, Q'_{m+1}) in Fig. 3 has two main advantages over its "dc equivalent" uniform width transistor: significant area savings and a higher cutoff frequency. The conventional way to obtain a low output conductance by means of a rectangular long-channel transistor reduces the electrical field along the whole transistor channel, giving rise to a long transit time, which has adopted from [4]. Calculations give a value of approximately (1/0.6) μm for W/L, for cascode current steering and composite current source. The common-gate devices, (M_5 and M_6), are other factors to increase the gain of the amplifiers, as well as to reduce the differential input capacitance and paralleling transistors, is the reason. The assigned dimensions to differential pairs, M_1, M_2 and M_3, M_4, are a width of W_{diff} and a length of L_{diff}. The transistors in the NMOS cascode current steering and composite current source have a width of W_N and length of L_N.

Referring to Fig. 2 in the static equilibrium, the differential input voltage is zero. Under this situation, the current that flows through current source is split evenly through the differential pair (M_1, M_2). The drains of the differential pair are connected to the output node branch circuits containing current source in OTA amplifier part. The currents through whole network are approximately the same. Therefore, transistors M_9 and M_{10} must be sized, by KW/L, to supply enough current through current source network and differential pair. With slightly increase in $Vin+$ or $Vin-$ the bias current equilibrium will be suffered which input offset and therefore an unacceptable transient response is resulted. To eliminate the input offset, especially at low supply voltages, increasing the size of the input differential pair is a solution and the transient response is improved by increasing the size of M_9 and M_{10}. Increasing the widths of the differential pair reduces the systematic offset, and, if laid out properly, reduces the random offsets and also headroom consumed by current source.

It should be mentioned that, in double differential cascaded pairs, transistors M3 and M4 suffer from a g_m mismatch between themselves, owing to dimension in submicron technology and threshold voltage mismatches. The key point is that the effect becomes more noticeable as the supply noise frequency, increases [3]. For this reason, a good "common-mode rejection ratio" (CMRR) and large-geometry should be provided.

In the current steering OTA, shown in Fig. 3, the cascoded transistors are in saturation condition by keeping the standard cascoded current mirror in saturation region. Dotted lines have been used to indicate the elective number of cascoded transistors. Design experience shows that the dimension of current steering and mirror transistors should be the same to make a more robust current mirror. The composite current source in Fig. 3, is obtained to use long-channel transistors and cascode structure to take advantage over significant area saving and a higher cutoff frequency [4]. A single transistor connected to source and parallel-connected transistors to the drain constitute the composite transistors. This technique shows the cascading transistors can improve the gain of OTA without deteriorating any of the bandwidth and linear output voltage range of OTA.

D. Bias Network

High swing cascode biasing is used and shared, for most part of the amplifier. Transistor sizes are kept strictly scaled in all biasing circuits to minimize the systematic mismatch error and to make biasing circuit scalable and layout friendly.

Quiescent conditions in the folded-cascode amplifier are established with the biasing circuit shown in Fig. 4. The transistors in the amplifier are biased such that their drain-source voltage, V_{DS}, is greater than their gate-source overdrive voltage, V_{GS}–V_{TH}, by several hundred mV to remain transistors in the saturation region and thus maintaining a high output resistance. Low-voltage cascoded current mirrors [5,6] are used to bias the gates of the NMOS and PMOS current source transistors of the composite current source, cascode current steering, and M_7-M_{10}, respectively. The gates of the NMOS cascode transistors, whose are connected to bias-n2, are biased with a string of devices, M_{13}-M_{15}, that are operated in the linear (triode) region. The NMOS cascode bias voltage, bias_n2, is set at a nominal value of 1 V. This compromise ensures the NMOS current source transistors, in the composite current source and cascode current steering, are saturated while allowing for a relatively large swing at the amplifier's outputs. Similarly, the gates of the PMOS cascode transistors, M_5 and M_6, are biased by a string of devices in their linear region, M_{16}-M_{18}, to a nominal voltage of 800 mV. Small transistors, M_N and M_P insure that the bias circuit settles to the desired operating point instead of a stable, zero-current state. All transistors in Fig. 4, with the exception of M_N and M_P, have nonminimum channel lengths in order to reduce the influence of channel length modulation on the bias voltages. The operating conditions in the bias circuit are established with an external reference current, I_{REF} that is applied to M_{11} and M_{12} and mirrored in the different branches of the circuit, in order to reduce layout complexity.

E. Frequency Response

When using a single pole model for the opamp, the settling time is determined by the gain-bandwidth product (GBW) of the opamp and the feedback factor of the circuit. In practical circuits, there is usually more than just one pole and often zeros as well. However, in order to use the opamp in a closed loop configuration, it has to be designed in such a way that its frequency response is close to the single pole response.

Consequently, there is one dominant low-frequency pole, while the other poles and zeros lie at much higher frequencies. In the frequency response, their presence is seen as a phase roll-off in the high frequencies. Thus, the phase margin at the unity gain frequency has an effect on the settling time as well. If the opamp is not utilized in unity gain feedback (e.g. auto-zeroing) the required phase margin is not defined at the unity gain frequency but at the frequency of the closed loop gain, and so it is easier to achieve.

III. SIMULATION RESULTS

SPECTRE simulations were performed on both "conventional differential amplifier" and "double differential amplifier" circuits using transistor models for a 1.8-V 0.18-μm CMOS SILTERRA process. Both circuits have similar transistor sizes. Fig. 5, represent typical Bode plots of the gain (dB) and the phase (degree) of the double differential amplifier circuit.

With a 1pF load, in double differential amplifier, the phase margin is adjusted nearly 56 degrees, and main pole is determined to be about 0.4 MHz, which is shown in Fig. 5. As dominant pole comes from the output node and approximately equal $1/(R_{out}.C_L)$. Hence, capacitance loads increase will have an effect on Gain Bandwidth however; the amplifier has the capability of driving 3.79-pF load, while this is much larger than the device capacitance in the pipeline ADC.

The circuit configuration to analyze the settling behavior of the amplifier is shown in Fig. 6. This scheme accomplished by placing the opamp in negative feedback path. This is almost the same circuit used in real analog to digital converter in a 1.5-bit/stage structure. The maximum output range and the slope of the transfer characteristic near its midpoint are used to estimate the swing and differential gain of the amplifier, respectively (see Fig. 7). The design specifications and performance summary are given sequential in Tables I and II. The amplifier has a measured differential output swing of ±1.65 V at a voltage supply of 1.8 V. The CMRR of the double differential opamp is greater than 70 dB, at 1MHz frequency.

The overall double differential opamp performance shows that the opamp DC gain exceeds 95 dB and the 0.05% settling time is less than 25ns (at 20MS/s) which is required for a 10-bit 20MHz analog to digital converter.

TABLE I
DESIGN SPECIFICATIONS

Technology	0.18μm CMOS
Power Consumption	0.7 mW
Power Supply (single)	1.8V

TABLE II
PERFORMANCE SUMMARY

Opamp Specification	Simulation Comparison	
	Double Differential	Conventional Differential
Input Common Mode	0.9 V	0.9 V
DC Gain	>95 dB	56 dB
Differential Output Swing	±1.65 V	------
Unity Gain Frequency	312 MHz	90 Hz
Phase Margin	56°	65°
Settling Time (within 2pF Output Load)	< 25ns	------
CMRR	> 70 dB	> 40 B
PSRR	> 70 dB	> 50dB

IV. CONCLUSION AND RESULT EXPERIENCES

A double differential gain-boosted CMOS operational tranconductance amplifier is designed and simulated using SPECTRE simulator, to realize pipelined ADC that requires only a 1.8-V supply voltage. The multilayer OTA chip is shown in Fig. 8. It consumes a total power of 0.7 mW. The low-voltage, low power, minimum area techniques facilitates greatly in this paper by optimizing the design of this OTA, both for performance and robustness. A cascaded pair of differential amplifiers and composite transistors with high transconductance-to-output conductance ratio of a long-channel transistor achieves the OTA gain improvement.

Since, pipeline ADCs provide an optimum balance of size, speed, resolution, power dissipation, and analog design effort, they have become increasingly attractive to major data-converter configuration, and while ADC specifications depend on receiver architecture, e.g., for direct-conversion receivers, two 20-MS/s ADCs are needed, the interpreted opamp can be taken into account of a good choice.

REFERENCE

[1] P.J.A. Naus et. Al., "A CMOS Stereo 16-bit D/A converter for digital audio", IEEE J. of Solid-State Circuits, vol.SC-22, No. 3, pp.390-395, june 1987.

[2] F. W. Singor and W. M. Snelgorve, " Switched capacitor bandpass delta-sigma A/D modulation at 10.7MHz ", IEEE J. of Solid-StateCircuits, vol. 30, No.3, pp. 184-192, March 1995.

[3] B. Razavi, Design of Analog CMOS Integrated Circuits, New York McGraw-Hill, 2001.

[4] Galup-Montoro, C.; Schneider, M.C.; Loss, I.J.B.;, "Series-parallel association of FET's for high gain and high frequency applications," IEEE J. Solid-State Circuits, vol. 29, no. 9, September 1994.

[5] J. N. Babanezhad and R. Gregorian, "A programmable gain/loss circuit," IEEE J. Solid-State Circuits, vol. SC-22, no. 6, pp. 1082-1089, Dec. 1987.

[6] K. A. Nishimura and P. R. Gray, "A monolithic analog video comb filter in 1.2-μm CMOS," IEEE J. Solid-State Circuits, vol. 28, no. 12, pp. 1331-1339, December 1993.

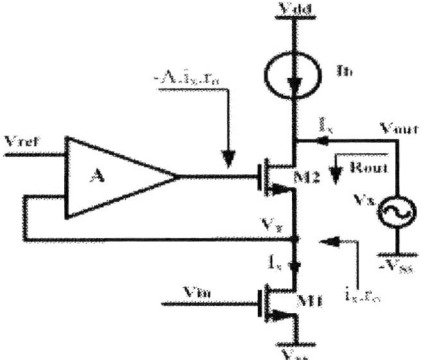

Fig. 1. Conventional cascode gain stage with gain enhancement

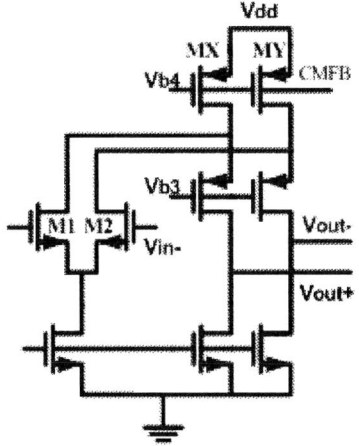

Fig. 2 Conventional differential amplifier

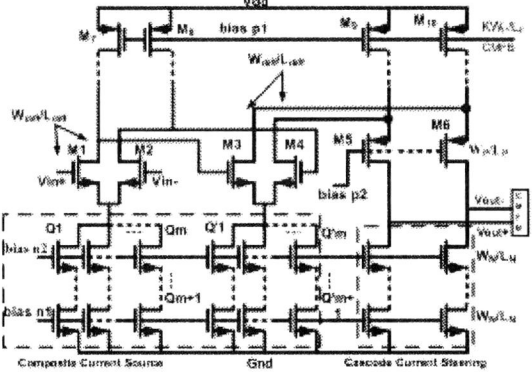

Fig. 3 Double differential folded-cascode amplifier

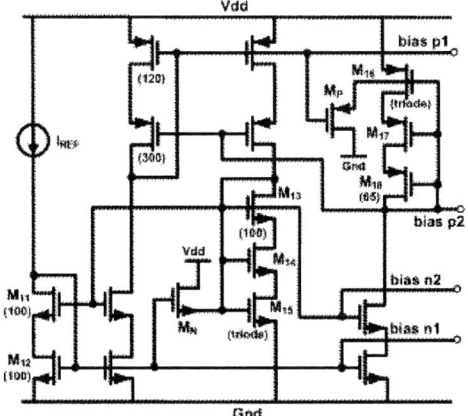

Fig. 4 Low-voltage wide-swing cascode bias current

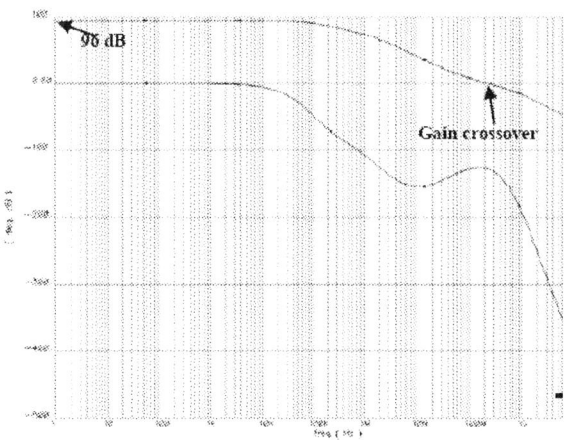

Fig. 5 Frequency response of the double differential amplifier. Gain and Phase diagrams.

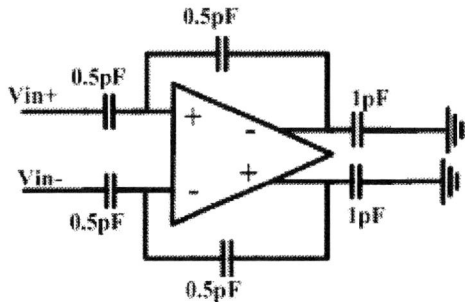

Fig. 6 Circuit used to test settling behavior.

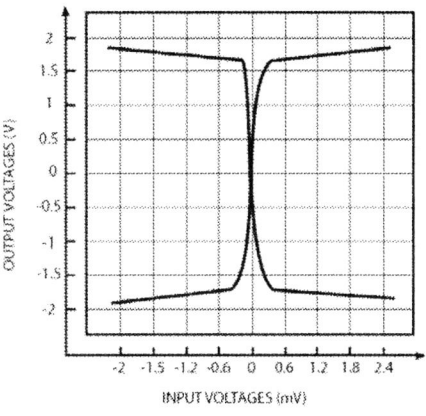

Fig. 7 Transfer characteristic of opamp

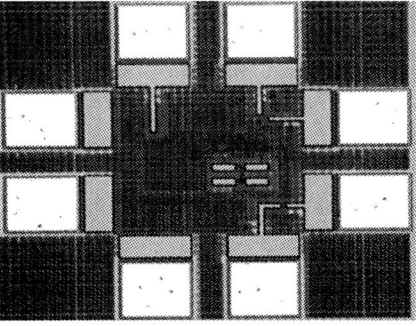

Fig. 8 Opamp Layout

Low-Voltage, High-Performance Current Mirror Circuit Techniques

Ali Meaamar, *Student Member, IEEE* and Masuri Othman, *Member, IEEE,*
VLSI Design Center, Institute of Microengineering and Nanoelectronics (IMEN) Universiti
Kebengsaan Malaysia, 43600 Bangi, Selangor, Malaysia
Corresponding author : meaamar@vlsi.eng.ukm.my

Abstract a series of current mirror circuit useful for low voltage analog circuit design are addressed to reduce channel length modulation and offering much higher accuracy. Some of the principle building blocks to realize wide swing and regulated cascode current mirror, to achieve low input resistance, are discussed. By combination of a low supply voltage, high performance, current mirror with high output impedance, low input and output voltage is introduced where is well suited for biasing structures.

I. INTRODUCTION

THE current mirror is one of the main building blocks, in CMOS analog circuit design. As well as performing identity operations and current amplification, the current mirror is also used both as a biasing circuit and as an "active" element, which is particularly useful in conjunction with differential pair's amplifiers. Very high output and low input resistance with low-voltage operation are predominantly used to ensure high accuracy and a widely known technique to increase the drain-source conductance, g_m, in using cascode structures. Several features to realize properties are; cascode current mirror, regulated cascode current mirror, and class AB structures. Due to low-voltage operation, the wide-swing cascode current mirror is more preferable than others to enhance the output voltage swing. Here a brief overview of high swing/cascode current mirror, regulated cascode current mirror, and class AB is discussed, and finally, a complete circuit consists of discussed current mirror configurations is presented.

I. HIGH SWING/CASCODE CURRENT SOURCE/SINK

Ideally, the output impedance of a current source/sink should be infinite and capable of generating or drawing a constant current over a wide range of voltage. Here, such a high swing current mirror, which is commonly used in the design, will be reviewed.

The classical cascode circuit shown in Fig. 1 (a, b) consist of a common-source stage followed by a common-gate stage that has to be biased by a reference voltage V_{CP} and V_{CN}. Whereas the optimally biased cascode circuit is operated at minimum reference voltage, still guaranteeing that, both transistors are saturated for the actual output current. The fixed biased cascode circuit has a fixed reference voltage that is optimal only for the maximum output current and deviates from optimum at lower currents.

High output impedance means of output current independent on output voltage. If not M_{N2} and M_{P2} in Fig. 1(c, d) will suffer from large drain-to-source voltage variation, which through the channel length modulation effect will cause a systematic mismatch error between the input and output currents of mirror.

To increase the output resistance, when designing a wide-swing current mirror in a short channel CMOS process, we attempt to bias M_{N2} deeper into the saturation region, which can be done by using a larger value of voltage on the gate of M_{N4}. Hence, high swing/cascode structure can be used to increase output impedance [1]. Fig 1 (c, d) shows an implementation of PMOS and NMOS structure, respectively. By keeping the symmetrical structure, a more accurate transfer ratio due to the matching both half of the circuit, is achieved. In this configuration as M_{N1} and M_{P1}, operate in the ohmic region, thus the minimum output voltage is dropt. To keep all the transistors in saturation region, by using I-V equation result in:

0-7803-9730-4/06/$25.00 ©2006 IEEE   670

$$V_{DS3} = V_{T\,(Thereshold)}$$
$$V_{DS4} = V_o - V_{Dsat} \tag{1}$$

and

$$V_{DS1} = V_{DS2} = V_{Dsat}$$

where

$$V_{Dsat} = \sqrt{I/k}, \qquad\qquad k = k1 = k2$$

and $k = \mu.C_{ox}(W/2L)$. For devices M3 and M4:

$$V_{Dsat3} = nV_{Dsat},$$
$$V_{Dsat4} = nV_{Dsat} \tag{2}$$

Here, n is a positive number. Assuming that $V_{TH} > V_{Dsat1} = nV_{Dsat}$, all the transistors will be biased in their saturation region and, as a result, the circuit will have a large output impedance. Notice that the output V_O has high-swing capability [4].

Suppose that all the transistors have good matching with equal W/L ratio, when they work under a deep saturation region, their drain current is written as:

$$I_D = \frac{\mu.C_{ox}}{2}\left(\frac{W}{L}\right)(V_{GS} - V_T)^2[1 + \lambda(V_{DS} - V_{eff})]$$
$$V_{eff} = V_{GS} - V_T \geq \Delta V \tag{3}$$
$$V_{DS} \geq V_{eff} + \Delta V$$

Where V_{GS}, V_{DS}, V_{eff}, V_{TH} denotes gate-source, drain-source, effective gate-source and threshold voltage of the transistors, respectively. Here ΔV denotes a voltage safety margin to accommodate the possible deviation corresponding to the variation of the process and environment temperature. According to (3) we know that as long as their gate-source voltage V_{GS} and drain source voltage V_{DS} are equal, the two transistors MN1 and MN2 will carry exactly the equal current. The gate-to-source voltages of the two transistors have been forced equal by connecting their gates together. Furthermore, their drain-to-source voltages are approximately equal due to the appearance of transistors MN3 and MN4. The output impedance and the output swing voltage of the current mirror are summarized as:

$$Z_{out} = \frac{v_{out}}{i_{out}} \cong \frac{g_{ms4}/g_{md4}}{g_{md2}} = g_{m4}.r_{ds4}.r_{ds2}\Big|_{g_{ms4} \cong g_{ms2}}$$
$$(1) \qquad = \left|\frac{A}{g_{md4}}\right| \tag{4}$$

$$V_{out-min} = V_{eff4} + V_{eff2} \tag{5}$$

The labels V_{Bn}, V_{Cn}, V_{Cp}, and V_{Bp} in Fig. 1 represent the bias voltages of the circuit. Where "A", represents the voltage gain of M2, and g_m is the transconductance of transistors.

For common-source devices whose drain is connected to the source of a cascode device, a configuration such as MN1, MN3 and MP1, MP3 is used to generate the gate bias voltage. MP1 is essentially a diode-connected device except that MP3 mimics the cascode device so that the drain-to-source voltage of in-circuit device and the bias device are approximately equal.

Then, however, the output impedance becomes so low that the circuit no longer acts as a current mirror. This can be corrected by adding a feedback loop, which makes the voltage at node n2 track the output voltage. Two realizations are shown in Fig. 2. The one on Fig. 2 (a) [6] controls the gate of the cascode transistor M3, with a level shifter constructed with a diode-connected transistor M4, which is matched with M3. The V_X with shown polarity, should be equal the threshold voltage (V_{TH}) of M1 to keep M1 and M2 in saturation. Transistor M5 is added to form a negative feedback so that cascode-type output impedance is achieved. While the circuit is operating in the low-voltage environment, the currents flowing through the cascodes transistors M3 and M4 must be equal, which another extra biasing circuit to fix I1 and I2 can solve this issue. By combining Fig. 2 (a) and Fig. 1 (b or c) another current mirror circuit, Fig. 3 to alleviate, 1) mismatch between M1 and M2 and 2) triode-region operation of M1 and M2 is achieved.

The circuit on Fig. 2 (b) uses an opamp to form the feedback loop [7]. A constant current I2 is injected into source of M1, which is mirrored to the transistor M2 to force $V_{DS1} = V_{DS2}$, thereby improving the current symmetry in both branches.

II. REGULATED CASCODE CURRENT MIRROR

The Regulated Cascode Current Mirror is a direct evolution of the cascode implementation with added local feedback to boost the output impedance without degradation of output voltage swing. This is illustrated in Fig. 4 (a) where a two-step evolution is shown [2], [3]. In the final version there is local feedback plus level shifting to ensure a low voltage across M1 and V_{ds1}, where the respective drain-source voltage of M1 must be kept stable to suppress channel-length modulation of M1 [5]. The level shifting or floating voltage source is an attractive technique to reduce headroom limitations [6], [7]. The extra current source added to M2, I_b, is used to match the current through M1, $I_b + I_{out.}$. Transistors M3 and M4 are designed to have a V_{gs} different of V_{ds1}. This circuit features a high output resistance and a good output voltage swing. It can be shown that the small-signal output resistance of the regulated current mirror of Fig. 4 (a), for low frequencies, is approximately:

$$g_{out} = \frac{g_{d1}g_{d4}}{g_{m4}} \cdot \frac{g_{d5} + g_{oi}}{g_{m5}} \quad (6)$$

Where g_{oi} is the conductance of the current source I_B. In comparison with bipolar version, Fig. 4 (b), a resistor is used to define the voltage across the mirroring device, Q1. Other sources of output current error have now been added to the circuit, Q3, and Q5's errors. Hence,

$$I_o(1 + 1/\varepsilon_5) + I_{b2}(1 - 1/\varepsilon_3) = (I_{in} + I_{b1})(1 - 2/\varepsilon_1) \quad (7)$$

Where ε_x shows the transistor error, and can be minimized by appropriately adjusting Ib1 and Ib2.

III. CLASS AB CURRENT MIRROR

As clarified before the cascode scheme is useful for reducing this drawback. Another practical design is shown in Fig. 5, which is employing the cascode scheme for PMOS and NMOS current mirrors to comprehend class-AB current mirror. This kind of circuits is been extensively used in the current amplifiers. As such, as circuit in Fig. 5, because the six transistors (Mn2, Mn4 Mn5, Mp5, Mp4, and Mp2) stacked between V_{DD} and V_{SS} has low quiescent V_{DS} and V_{GS}, requirements. The cascode transistors Mn5 and Mp5 increase output resistance and limits channel-length modulation effects on output current linearity. It is made of two complementary active-gain enhanced mirrors, which base their performance principle quite similar to that of the gain-boosting technique.

The two current mirrors are composed of transistors Mn1-Mn5 and Mp1-Mp5 and two auxiliary voltage amplifiers A1 and A2, whose gains are both assumed to be equal to A. Thus, a high linearity performance is achieved. Moreover, the use of A1 and A2 also provides a very high output resistance given by:

$$r_o \cong (g_{mn5}.r_{dn4}.r_{dn2}.A) \| (g_{mp5}.r_{dp4}.r_{dp2}.A) \quad (8)$$

IV. CURRENT BIASING DESIGN EXAMPLE

In Fig. 6, the biasing structure for an n-well process is presented. By knowing of the transistor V_{TH}'s are influenced by the source-bulk voltages, starting from left to right, we have first a PMOS voltage level shifter that $V1 = V_{gs}$, PMOS. Secondly, a V_{dsat} is generated with the structure M2-M4. Using the square law MOS relationship $((I = k.(\frac{W}{L})(Vgs - V_{TH})^2)$, Vs and V2 are given by $V_S = V1 - V_{TH}$, and $V2 = V_S + V_{TH}$

Because the bulk-source voltages of both transistors are equal, their V_{TH}'s are also equal and hence $V2 = V1 + (V_{gs} - V_{TH})$. sing again a PMOS lUsing PMOS level shifter M6, identical as M1, the voltage voltage V3 is equal to $V3 = V2 - V1$. By employing an NMOS level shifter, V4 becomes $V4 = V_T + 2V_{gs}$, and also $V5 = V4 - V_{gs}$. Because M7 and M8 are identical transistors with the same source bulk and current drain, as an assumption, their V_{gs}'s are equal. As a result, V5 is equal to V5=V_{dsat}. Using the same feedback structures as presented in Fig. 2, a cascode biasing structure is realized which is independent of V_{TH} and finite output conductance.

V. SIMULATION RESULTS

SPECTRE simulation were performed by using BSIM3V3 model for 0.18μm n-well CMOS technology with single 1.8V power supply, (trough SILTERRA), respectively.

The simulation of the dc output characteristics of Fig. 2 (b) were determined by sweeping V_{out} from 0 to 1.8V and by stepping the input current from 20 μA to 160 μA using a curve tracer in 20 μA steps [Fig. 7]. The output resistance was 8 MΩ. The minimum output voltage has values from 0.2 to 0.4 V. The simulated cascode low-voltage current source is well suited for low-voltage differential amplifiers. The linearity of the input current-to-voltage (I-V) of Fig. 6 is shown in Fig. 8. This figure illustrates the measured current sourced circuit. This type of design is more useful for continues-time filter design using current-mode technique, which is based on linear approximation of the I-V characteristics, and the class-AB current mirror allows doing the linear approximation for relatively large signal current swing.

VI. CONCLUSION

A set of current mirror prototypes for biasing approach for designing cascode current mirror, regulated cascode, and class AB current mirror structures with low-voltage capability, have reviewed, extended, and implemented in a 0.18-μm n-well CMOS process. This efficient implementation of a low-voltage high performance current mirror, allows complementary current mirrors with voltage requirements in the signal path to improve the input and output resistance. In this paper also a biasing circuit is proposed, which is very useful for low-voltage design and technology independent.

REFERENCES

[1] David Johns, Ken Martin, "Analog Integrated Circuit Design", John Wiley & Sons, Inc., 1997.

[2] A.L. Coban and P.E. Allen, "A 1.75V rail-to-rail CMOS Op Amp, " *Proceeding – IEEE International Symposium on Circuits and Systems,* vol. 5, pp. 497-500, 1994.

[3] M. Helfenstein et. Al., "90dB, 90MHz, 30mW OTA with the Gain-Enhancement Implemented by One- and Two- Stage Amplifiers,"

Proceedings – IEEE International Symposium on Circuits and Systems, vol. 3, pp. 1732-1735, 1995.

[4] J.N. Babanezhad and R. Gregorian," A Programmable Gain/Loss Circuit," *IEEE J. Solid State Circuits,* vol. SC-22, no. 6, pp. 1082 – 1090, Dec. 1987.

[5] E.S Sackinger and W.Guggenbuhl, "A high swing, high-impedance MOS cascode circuit," *IEEE Journal of Solid-State Circuits*, vol. 25, pp. 289–298, Feb.1990.

[6] J. Ram´irez-Angulo, "Current mirrors with low input voltage requirements for built in current sensors," *IEEE Proc. ISCAS '94*, pp. 529–532, 1994.

[7] J. Ram´irez-Angulo, R. G. Carvajal, J. Tombs, and A. Torralba, "Simple technique for op amp continuous time

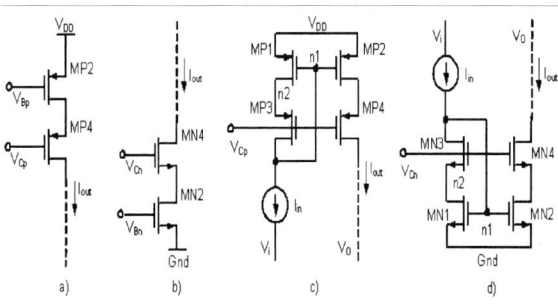

Fig. 1 High swing current mirror, (a) current source, (b) current sink, (c) PMOS current source, (d) NMOS current source

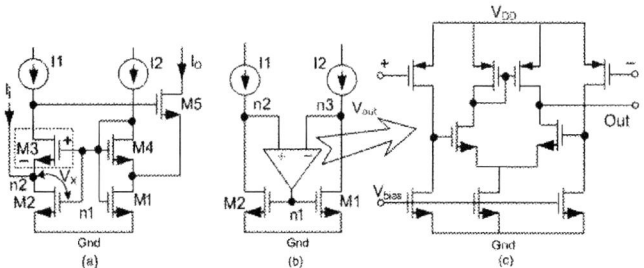

Fig. 2 Triode-region, (a) current mirror with cascode-type resistance, (b) low voltage current source (c) single ended differential-amplifier for use in (b)

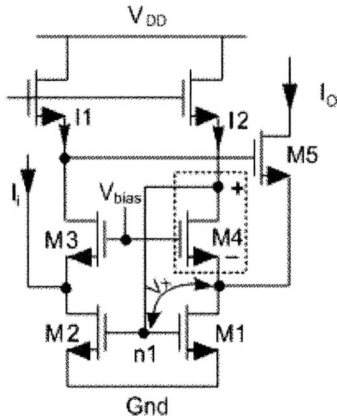

Fig. 3 Cascode bias, saturation-region current mirror

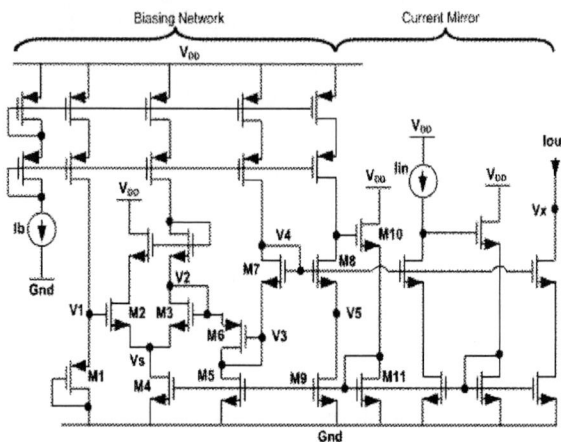

Fig. 6 Cascode biasing circuit with current mirror

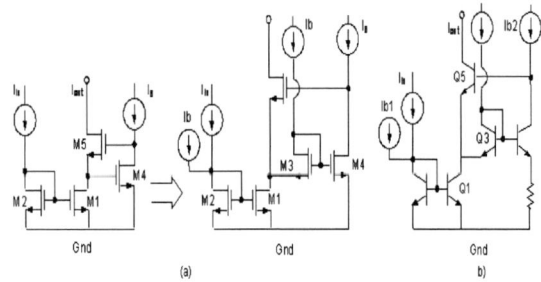

Fig. 4 Regulated cascade current mirror

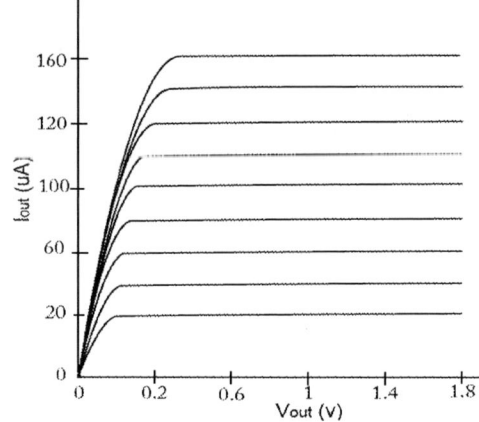

Fig. 7 Wide-swing low-voltage current mirror dc characteristic

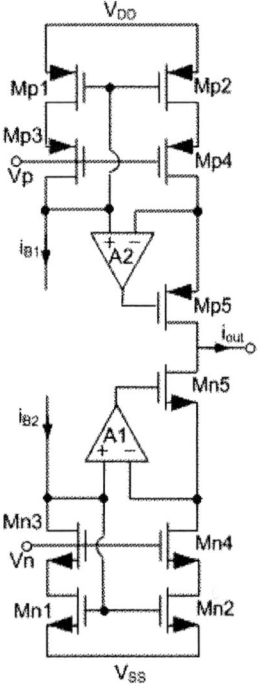

Fig. 5 A class-AB current mirror using cascode circuit

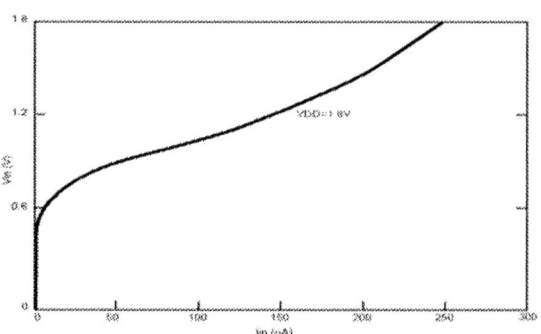

Fig. 8 Input current-to-voltage characteristics for V_{DD}=1.8

VLSI Implementation of 1/2 Viterbi Decoder for IEEE P802.15-3a UWB Communication

Meilana Siswanto[1], *Student Member, IEEE*, Masuri Othman[2], Edmond Zahedi[3], *Senior Member, IEEE*
[1,2] Institute of Microengineering and Nanoelectronic
[3] Department of Electrical, Electronics and System, Faculty Engineering
Universiti Kebangsaan Malaysia, 43600 Bangi, Selangor, MALAYSIA
E-mail: meilana@vlsi.eng.ukm.my

Abstract –This paper presents the design of 1/2 Viterbi decoder for UWB applications following the standard requirements of IEEE P802.15-3a. The main design issues of Viterbi decoder were add-compare-select unit (ACSU), memory management and trace-back methods. In order to meet the requirements of IEEE P802.15-3a UWB, the transition metric unit (TMU) is designed using a finite state machine (FSM) and a parallel carry look-ahead adder (CLA) used to design the addition part of the ACSU. After synthesis using Xilinx synthesis technology (XST), the synthesis report shows that the design has a minimum period of 1.888 ns, equivalent to a data rate of 529.661 Mbps fulfilling more than the standard requirements of IEEE P802.15-3a for UWB, which has a data rate range from 55 to 480 Mbps.

Keywords – Viterbi algorithm, Viterbi decoder, Convolutional encoder, VLSI design.

I. INTRODUCTION

The Viterbi algorithm (VA) reduces computational load and decoding complexity by early rejection of the unlikely paths in the code trellis [1]. Reduction of computational load will increase the bit rate and reduction of decoding complexity means decrease in power consumption. Both of these advantages of VA are important aspects for wireless communications. Extensive applications of VD are found in wireless communications such as satellite communications and portable devices. A specially popular Viterbi-decoded convolutional codes, used at least since Voyager program has a constraint length (K) of 7 and a code rate of 1/2, Mars pathfinder, Mars

exploration rover and Cassini probe to Saturn use a K of 15 and a code rate of 1/6 [2]. UWB technology is predicted will become technology trend today and future. In recent years, researches on Ultra-Wideband (UWB) become of great interest by big companies such as Intel, Motorola, Texas Instruments and therefore researchers tried to implement it in the different applications. As the UWB characteristics specified by the IEEE P802.15-3a standard requirements [3] do not mention of the value of constant length (K), the focus was put on the implementation of a high speed of 1/2 VD that meets the data throughput requirements by IEEE P802.15-3a.

II. METHODOLOGY OF VD DESIGNING

The standard IEEE P802.15-3a

For designing a Viterbi decoder that fulfills the standard requirements of IEEE P802.15-3a, it would be better if coding rate of the encoder should follow the standard. The data rate dependent modulation parameters of the standard are listed in Table 1.

Table 1. Rate – dependent parameters

Data Rate (Mb/s)	Mod.	Code Rate (R)	Overall Spreading Gain	Coded per OFDM Symbol (N_{CBPS})
55	QPSK	11/32	4	100
80	QPSK	1/2	4	100
110	QPSK	11/32	2	200
160	QPSK	1/2	2	200
200	QPSK	5/8	2	200
320	QPSK	1/2	1	200
480	QPSK	3/4	1	200

According to the table above, there are three choices of data rate for coding rate of 1/2 i.e. 80 Mbps, 160 Mbps and 320 Mbps, and without mention the standard values of constraint length (*K*). Therefore the encoder design of this paper used coding rate of 1/2 and constraint length (*K*) of 3.

Shannon-Hartley Theorem

According to the Shannon-Hartley theorem:

$$C = BW \log_2 (1 + \frac{S}{N}) \qquad (1)$$

$$\frac{S}{N} = \frac{E_b.R}{N_0.BW} \Leftrightarrow \frac{E_b}{N_0} = \frac{S}{N}.\frac{BW}{R} \qquad (2)$$

$$C = BW \log_2 (1 + \frac{E_b.R}{N_0.BW}) \qquad (3)$$

When *BW* is the bandwidth in hertz, the capacity is given in bits-per-s. Capacity (*C*) will increase if signal-to-noise-ratio (*S/N*) is increase. From the equation above, *S/N* is proportional to E_b/N_0. So, capacity (*C*) will increase also if E_b/N_0 is increase and E_b/N_0 has correlation with data rate of VD design. Fig. 1 shows the correlation between E_b/N_0 with constraint length (*K*).

BER versus E_b/N_0 for rate 1/2 codes

Fig.1 shows correlation K with BER & E_b/N_0.

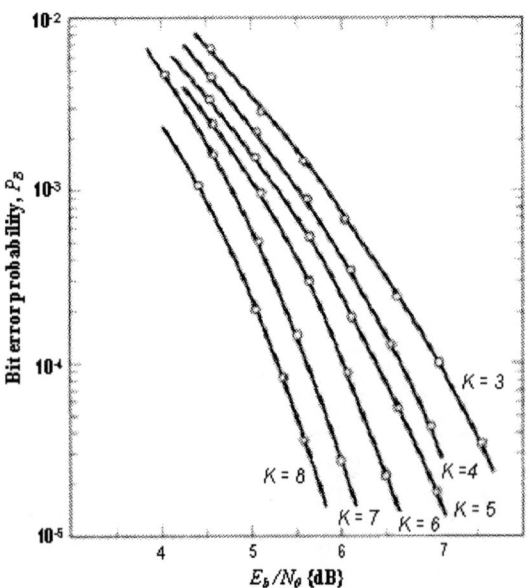

Fig. 1. Bit error probability versus E_b/N_0 for 1/2 rate using coherent BPSK.

The Fig. 1 shows that if constraints length (*K*) is increase, E_b/N_0 will decrease and vice versa. As *K* increases then *C* decreases. So, the value of constraint length (*K*) is important to be considered to design the VD that meets a high speed data rate.

III. IMPLEMENTATION OF THE DESIGN

The Encoder Design

As previous description, the encoder used in this paper has coding rate of 1/2 and constraint length (*K*) of 3. Fig.2 shows the encoder used for the proposed the VD.

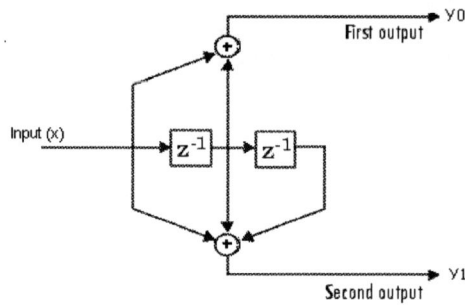

Fig. 2. The design of 1/2 convolutional encoder.

The Viterbi Decoder Design

The VD consists of three main units transition metric unit (TMU), add-comp; select unit (ACSU), and survivor memory unit (SMU) as illustrated in Fig. 2.

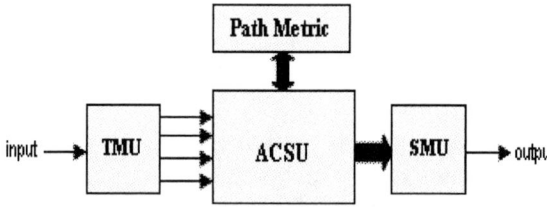

Fig. 3. Generally diagram block for Viterbi decoder.

The TMU calculates the transition metrics from the input data. The ACSU recursively accumulates transition metrics (TM) as path metrics (PM), and makes decisions to select the most likely state transition sequence.
Finally, the SMU traces the decisions to extract this sequence [4]. Fig. 4 shows finite state machine (FSM), trellis diagram of the encoder and decoder cells for the proposed VD.

The FSM shows the instantaneous transitions only and cannot represent time history. Trellis diagram adds the time dimension to the state diagram.

(minimum period) this is 529.661 MHz (1.888 ns). However in a log bits computation, CLA has the smallest period or the highest frequency than the other designs.

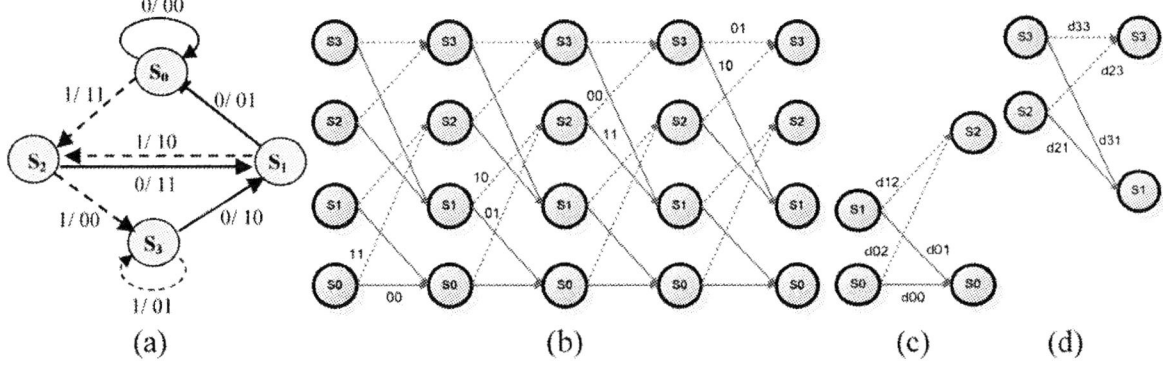

(a) (b) (c) (d)

Fig. 4. (a). The finite state machine of the encoder. (b). The trellis diagram of (2,1,3) convolutional code. (c) The 1st decoder cell & (d). The 2nd decoder cell for VD implementation.

Implementation of The Decoder

Decoder cells are an important part to implement ACSU of the VD. TMU of the VD is designed using FSM and carry look-ahead adder (CLA) used to design the addition part of the ACSU with the parallel method. Fig. 5 shows general implementation of ACSU computation of the VD corresponds to the 1st decoder cell in Fig. 4.(c).

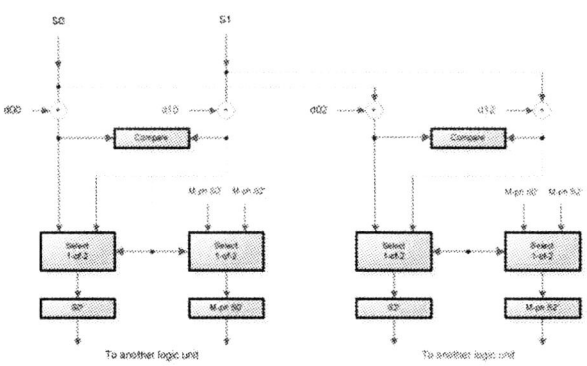

Fig. 5. General implementation of the VD using the 1st decoder cell.

III. COMPARISON OF ACSU DESIGNS USING CLA, RCA, RCAS AND NORMALLY ADDITION

Designs of VD using carry look-ahead adder (CLA), ripple-carry adder (RCA), ripple-carry adder/subtractor (RCAS) and normally addition (NA) for small bit size, show that they all give the same maximum frequency

Detailed comparison of the maximum frequency and the minimum period are shown in Table 2.

Table 2. Timing summary of maximum frequency and minimum period for the designs

Adder	Max. frequency	Min. period
NA	529.661 MHz	1.888 ns
RCA	529.661 MHz	1.888 ns
RCAS	529.661 MHz	1.888 ns
CLA	529.661 MHz	1.888 ns

Table 3. Timing summary of the inputs/ the outputs according to the clock.

Adder	Min. input arrival time before clock	Max. output required time after clock
NA	1.652 ns	6.291 ns
RCA	1.652 ns	6.291 ns
RCAS	1.652 ns	6.291 ns
CLA	1.652 ns	6.291 ns

Table 4. The report of the memory usages.

Adder	Total memory usage	Max. combinational path delay
NA	74.420 Kbytes	No path found
RCA	77.492 Kbytes	No path found
RCAS	77.492 Kbytes	No path found
CLA	71.860 Kbytes	No path found

The comparison of the memory usage is shown in Table 4. The design using RCA and RCAS need the largest memory usage. Additionally, the design using CLA has memory usage less than the other designs. So, CLA is selected for designing the VD that meets the high data throughput requirements specified by IEEE P802.15-3a.

IV. RESULTS

FPGA Implementation of the design

Synthesize of the design using Xilinx Vertex II xc2v250-5fg456-Project Navigator shows that the design was successfully implemented. Table 5 shows the report summaries of the VD using CLA after the synthesis and the schematic of the design is shown in Fig. 6.

Table 5. The report summaries of the VD

Maximum frequency	529.661 MHz
Minimum period	1.888 ns
Min. input arrival time before clock	1.652 ns
Max. output required time after clock	6.291 ns
Max. combinational path delay	No path found
Total memory usage	71.860 Kilobytes

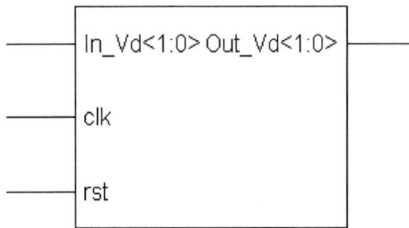

Fig. 6. Top-level logic schematic of the VD after synthesis using Xilinx.

From Fig. 6 the VD has three inputs i.e. 2 bit *In_Vd*, a clock frequency (*clk*) and reset (*rst*). And *Out_Vd* is the output of the VD.

ModelSim Simulations

Simulation of the VD using ModelSim is shown in Fig. 7. There are some errors at the beginning

of the outputs due to the system resetting. Starting from 50 ps, the outputs obtained from the decoder become stable, similar to the inputs of the encoder.

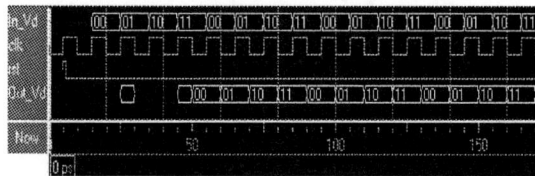

Fig. 7. ModelSim simulation of the VD.

Design Analyzer Simulations

After completion of simulation successfully, the design is synthesized using Synopsys Design-compiler based on the MIMOS library 0.35 μm CMOS process technology. The top-level logic schematic view is shown in Fig. 7. The detailed view of the top-level logic schematic appears in Fig. 8.

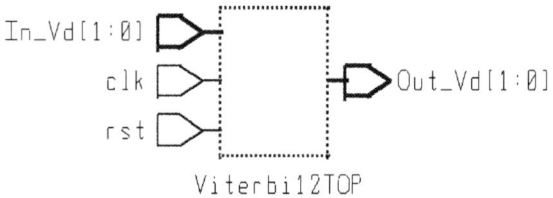

Fig. 7. Top-level logic schematic of the VD after synthesis using Synopsys Design-Compiler

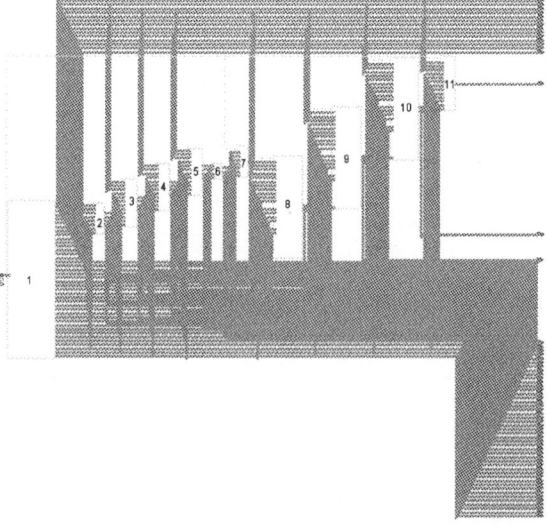

Fig. 8. Detail view of the top-level logic schematic of the VD using Synopsys Design-Compiler.

678

Fig. 8 shows the detailed view of the top-level logic schematic of the VD after synthesis using Synopsys Design-Compiler that consists of 11 blocks:

block 1 : the Transition Metric Unit (TMU),
block 2 : the 1st addition part of the ACSU,
block 3 : the 2nd addition part of the ACSU,
block 4 : the 3rd addition part of the ACSU,
block 5 : the 4th addition part of the ACSU,
block 6 : the compare-select part of the ACSU,
block 7 : the 1st trace-back process units, and
block 8 to block 11 are trace-back processes for state 3, state 2, state 1 and state 0 respectively.

The following are the chip's specifications based on library of Mimos 0.35 μm CMOS process technology under MPW for ASIC.

Total Cell Area   : 188294.41
Global Operating voltage : 2.97 V

V. CONCLUSIONS

This paper has presented the VLSI implementation of 1/2 VD with constraint length (*K*) of 3 using Verilog® HDL for IEEE P802.15-3a UWB communication. Simulation results show that the decoder works as required with 50 ps delay at the output. The synthesis using xilinx reports that the design has memory usage of 71.860 Kytes and a minimum period of 1.888 ns, equivalent to a data rate of 529.661 Mbps fulfilling more than the standard requirement of IEEE P802.15-3a. The next stage is to synthesize the design using synopsys design compiler according to Mimos 0.35 μm technology. After combining with the Mimos library, the synthesis results show that the design has total cell of 188294.41 and operates at 2.97 V.

ACKNOWLEDGEMENT

The research funding for this was obtained under IRPA grant (IRPA 03-02-02-0018-SR0003/07-04). The authors would like to thank to MIMOS for providing the 0.35 um technology as a MPW project.

REFERENCES

[1]. J. G. D. Forney, "The Viterbi Algorithm," *Proc. IEEE*, vol. 61, no. 3, pp. 268-278, Mar. 1973.

[2] http://en.wikipedia.org/wiki/Shannon-Fano_coding, August 19, 2006.

[3] R. Roberts, "Technical Editor Controbution of IEEE Formatted Draft Text for MB-OFDM Proposal", Harris Coporation, MS-9842, P.O. Box 37 Melbourne, Fl. 32902, January 2004.

[4] C. C. Lin, C. C. Wu, Y. H. Chang, and C. Y. Lee, "Design of a power-reduction Viterbi decoder for WLAN applications", *IEEE Trans. on Circuits and Systems-I: Reguler Papers*, vol. 52, No. 6, pp. 1148-1156, Jun. 2005.

[5]. A. J. Viterbi, "Error bounds for convolutional codes and asymptotically optimum decoding algorithm," *IEEE Trans. Inf. Theory*, vol. IT-13, no. 2, pp. 260-269, Apr. 1967.

[6] Smith M. J. S, *Application – Specific Integrated Circuits*, Palo Alto and Honolulu, 2001.

[7] Sklar B, *Digital Communications-Fundamentals and Applications*, Sec Edition, Communication Engineering Services, Tarzana, California and University of California, Los Angeles (2002).

[8] J. Tang and K. K. Parhi, "Viterbi decoder for high-speed ultra-wideband communication systems," *Proc. IEEE on Acoustics, Speech, and Signal Processing,* vol. 5, pp. 37-40, Mar. 2005.

[9] A. R. Abdul Shakoor, V. Szwarc, and T. A. Kwasniewski, "High Speed Viterbi Decoder for W-LAN and Broadband Applications", *Proc. IEEE on Circuits and Systems*, pp. 25-28, Jun. 2004.

[10] Israsena, P., and Kale, I., "A Viterbi Decoder with Low-Power Trace-Back Memory Struktur for Wireless Perpasive Communications," *Proc. IEEE on Wireless Pervasive Computing*, pp. 1-4-, Jan. 2006.

Integrated LC VCO Compatible with Memory Process for Gigahertz Clock Generation

Minseok Choi, Youngho Jung, Yeo Jo Yoon[*], Young-Wug Kim[*], and Hyungcheol Shin, *Senior Member, IEEE*
School of Electrical Engineering and Computer Science
Seoul National University #059San 56-1, Shillim-dong, Kwanak-ku
Seoul 151-744, Korea Email: mschoi04@snu.ac.kr
*DongbuAnam Semiconductor

Abstract **This work presents the design and fabrication of the LC voltage-controlled oscillator (VCO) using 1-poly 3-metal CMOS process aimed for use in current memory manufacturing process. Poor characteristics of highly-resistive and immune to substrate-coupling metal-3 inductor in LC resonator were overcome by dual metal structure and patterned- ground shield (PGS) strategy. Fabricated VCO operated from 2.48 GHz to 2.73 GHz tuned by accumulation-mode MOS varactor. The corresponding tuning range was 250 MHz. The measured phase noise was - 113.5 dBc/Hz at 1MHz offset at 2.48 GHz carrier frequency. The current consumption and corresponding power consumption were about 1.04 mA and 1.87 mW respectively.**

I . INTRODUCTION

Clock generation circuits are one of the most important blocks in the high-speed system including processors and memories such as DRAM. Phase-locked loop (PLL) and Delay-locked loop (DLL) are representative clock generation circuits that are required to minimize skew between internal clock and reference clock and initial-locking time.

PLLs are attractive volunteers for use as clock multipliers because they provide a simple means of frequency multiplication and they can be used in many circuits requiring high-speed clocking despite of their drawbacks of relatively poor jitter performance, stability, and on-chip noise. The speed of memory, chasing very fast microprocessors, for example, are coming up to 2GHz in case of DDR SDRAM.

As the frequency in PLLs comes deeply into gigahertz area, voltage-controlled oscillators (VCOs) inside PLLs are being required to be implemented with LC resonator. But when integrated LC VCOs in PLLs are used for high-speed memory, the low quality factor of on-chip inductor can effect seriously phase variation and jitter. The key to the design of a low phase-noise LC oscillator is a high quality inductor as it has been shown for several studies. And the quality factor of the inductor will be limited by the series resistance of the metal traces. But most memory chip manufacturing process use 3 metal layers at the most. And this gives worst condition for LC VCO with high resistance and large coupling loss with substrate.

II . DESIGN AND FABRICATION

Schematic of complementary LC VCO is shown in Fig. 1. It is designed using the methodolgy of finding the optimum inductance of resonator, the varactor size for desired tuning frequency and the active device size for low power consumption. As shown in Fig. 2, designed LC VCO was fabricated using 1-poly 3-metal CMOS process except for pads. To overcome disadvantage due to using thin metal of which the thickness is one fourth of thick top metal, we made coil thicker by connecting the metal 3 and metal 2 by a great number of via 2 arrays. Consequently, this led to the decrease of series resistance and increase of VCO noise performance.

The area of LC VCO circuit including pads is about 0.62 0.64 mm^2. To suppress the substrate coupling loss in symmetric inductor, poly Patterned-Ground Shield (PGS) was placed under the coil and the grounded M1 shields were used underneath entire circuit.

III . EXPERIMENTAL RESULTS

In the measurement, supply voltage of 1.8 V was applied and tuning voltage was varied from 0 V to 1.8 V. Total current flowing into the bias circuit was 1.04 mA. Core circuit consumed 0.52

mA. Accordingly, Power consumption was measured to be 1.87 mW. The oscillator operated from 2.48 GHz to 2.73 GHz and corresponding tuning range was 250 MHz. Measured quality factor and inductance of used dual metal inductor were shown in Fig. 3.

Measured frequency spectrum for V_c of 1.8 V and measured phase noise plot versus offset frequency at V_c of 1.8 V are shown in Fig. 4 and Fig. 5 respectively. The measured phase noise was -113.5 dBc/Hz at 1 MHz at $V_c = 1.8$ V. Quite low phase noise of -113.5 dBc/Hz at 1 MHz was achieved. The phase noise of oscillator is dominated by the inductor which has lower quality factor than varactors.

IV. CONCLUSION

A low power consuming integrated LC VCO for memory applications was fabricated using 1-poly 3-metal CMOS process. The LC VCO operated from 2.48 GHz to 2.73 GHz and quite low phase noise of -113.5 dBc/Hz at 1 MHz offset was achieved. By adopting techniques to gain good quality-factor inductor, it is shown that LC VCO can be used in PLL for clocking in high-speed memory process.

Acknowledgement

This work was supported by Hynix Semiconductor and NSI_NCRC program of KOSEF, KOREA.

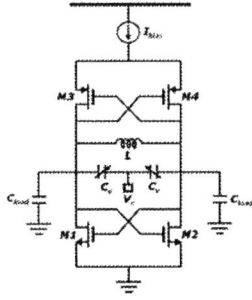

Fig. 1. Complementary LC VCO schematic.

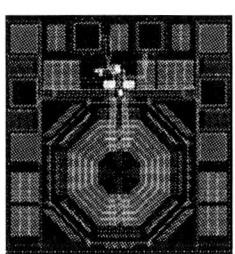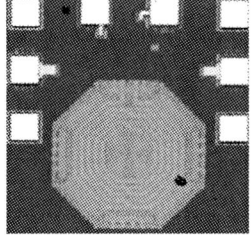

Fig. 2. LC VCO layout and chip photograph.

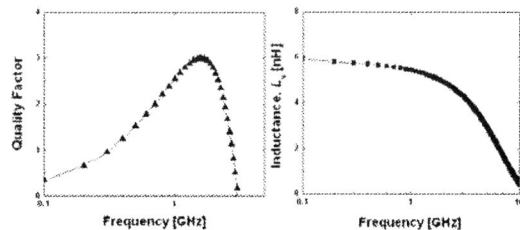

Fig. 3. Measured Q factor and inductance.

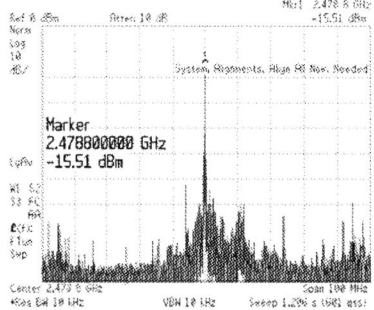

Fig. 4. Measured frequency spectrum for $V_c = 1.8$ V.

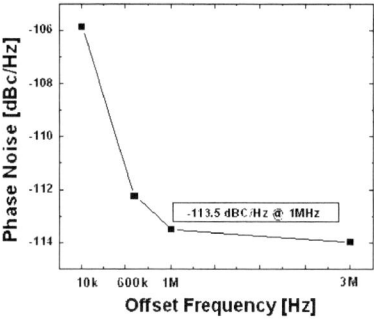

Fig. 5. Measured phase noise plot at $V_c = 1.8$ V.

REFERENCES

[1] M. Hershenson, A. Hajimiri, S. S. Mohan, S, P. Boyd, and T. H. Lee. "Design and Optimization of LC Oscillators," *IEEE/ACM Int. Conf, Computer Aided Design*, pp. 65-69, Nov.1999.

[2] D. Ham and A. Hajimiri, "Concepts and Methods in Optimization of Integrated LC VCOs," *IEEE J. Solid-State Circuits*, vol. 36, no. 6, pp. 896-909, June. 2001.

[3] S. Song and H. Shin, "A New RF Model for the Accumulation-Mode MOS Varactor," *IEEE MTT-S* , June 2003.

[4] J. Gil, S. Song, H. Lee, and H. Shin "A -119.2 dBc/Hz at 1 MHz, 1.5 mW, Fully Integrated, 2.5-GHz, CMOS VCO Using Helical Inductors," *IEEE Microwave and Components Letters,* vol. 13, no. 11, Nov. 2003.

A Novel DC-Coupled, Single-Ended to Differential, Transimpedance Amplifier Architecture Based on g_m-boosting Technique

M. Jalali[1,2], M. K. Moravvej-Farshi[1] (*Senior Member, IEEE*), and A. R. Nabavi[3]

[1]Optical Tech. Lab., Tarbiat Modares University (TMU), P. O. Box 14115-143, Tehran, IRAN
[2]Optical Technology Department, Iran Telecommunication Research Center (ITRC), Tehran, IRAN
[3]IC Design Lab., Tarbiat Modares University (TMU), P. O. Box 14115-143, Tehran, IRAN
Email: mjalali@modares.ac.ir

Abstract **In this paper, the authors present a new DC-coupled single-ended to differential method suitable for transimpedance amplifier in low-voltage optical communication system applications. The single-ended to differential operation is realized using Gm-boosting technique. As an example for applying this method, g_m-boosting technique is implemented by capacitor cross coupling of two basic common-gate stage. Also a method for achieving DC-coupled operation, despite capacitor coupling, is presented and discussed.**

I. INTRODUCTION

In optical communication systems, an optical receiver is a key component, where it acts as the interface between electronics and the optics. One of the main building blocks in an optical receiver for fiber based gigabit range networks is transimpedance amplifier (TIA).

Traditional transimpedance amplifier designs typically use a common-gate stage (open loop architecture) or a voltage gain stage with resistive "voltage-current" or "shunt-shunt" feedback (close loop architecture) as amplifying core. Beside all concerns related to these two topologies, a common problem with the both foregoing topologies is in providing a differential output signal from a single-ended input. Using a differential pair at the output has been a common way to differential design. This technique has usually been done in two methods. (1) A replica circuit of the single-ended TIA can be tied with the main circuit so that a differential pair would appear at the output [1], as shown in fig. 1. (2) Extracting the dc content of the TIA output and applying the result along with original signal to a

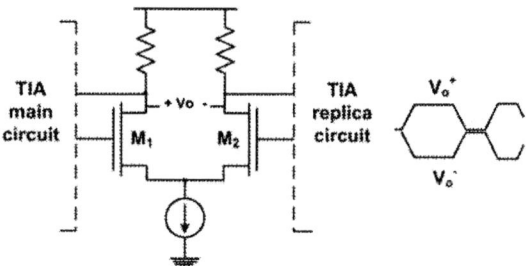

Fig. 1: Single-ended to differential conversion by a replica circuit.

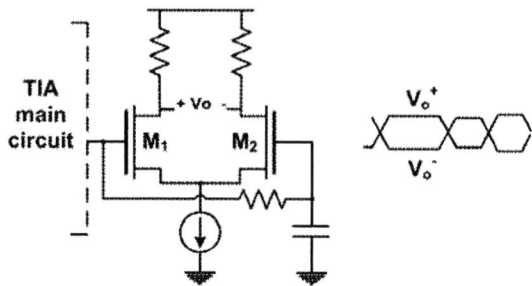

Fig. 2: Single-ended to differential conversion by extracting dc content of TIA output.

differential pair [2] (Fig. 2). These two methods have some main problems, more specifically, not having a true differential output or needing off-chip component, respectively.

In this paper, we develop a new method to obtain single-ended to differential conversion by capacitor cross coupling. For this purpose, two common gate (CG) stage biased with resistor at gate have been used. Part II of this paper deals with specifying the characteristic of common-gate stages biased with a resistor at gate and comparing it with a simple common-gate stage. The basic circuit topology and the circuit realization of the proposed differential TIA along with simulated results is discussed in section III. Finally, conclusions are drawn in section IV.

0-7803-9730-4/06/$25.00 ©2006 IEEE 682

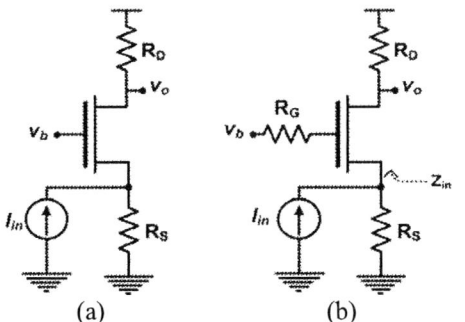

(a)　　　　　　(b)

Fig. 3: Two basic configurations as transimpedance amplifier. (a) Simple common-gate stage. (b) A common-gate stage biased by a resistor at the gate.

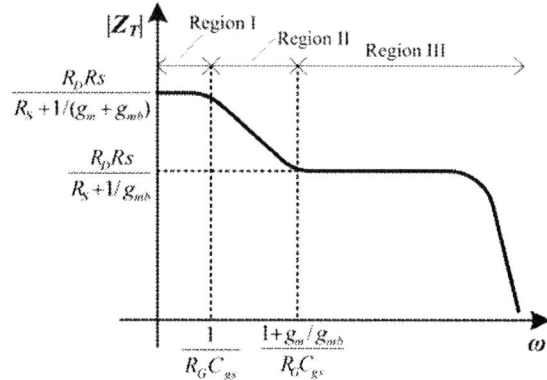

Fig. 4: Transimpedance of a common-gate stage biased with a resistor at gate.

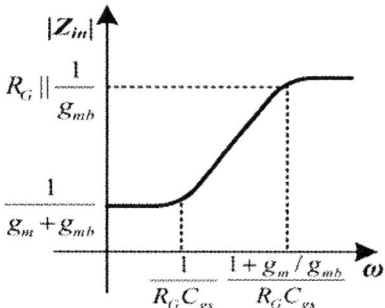

Fig. 5: Input impedance of a common-gate stage biased with a resistor at gate.

II. EXPLORING CG STAGES

Fig. 3 shows two simple configurations as trans-impedance amplifier. It is instructive to carefully examine the behavior of the two circuits. The circuit of Fig. 3 (a) is a known CG stage and its transimpedance gain can be written as:

$$Z_{T_0} = \frac{R_S R_D}{R_S + 1/(g_m + g_{mb})} \qquad (1)$$

The circuit of Fig. 3 (b) is a counterpart of Fig. 3 (a) but its gate has been biased with a resistor, R_G. The behavior of this circuit is completely different of Fig. 3 (a) and is dependent to the value of R_G. At low-frequencies (or region I in Fig. 4), the circuit has a transimpedance gain of eq. (1). However, at high frequencies (or region III in Fig. 4) it changes to

$$Z_{T_0} = \frac{R_S R_D}{R_S + 1/g_{mb}} \qquad (2)$$

The exchange region between two gains starts at the frequency of $P=1/(R_G C_{GS})$ up to $Z=(1+g_m/g_{mb})/(R_G C_{GS})$. The reason of this behavior could be explained concerning the inductive behavior of MOS transistor in Fig. 3 (b). It can be shown that the input impedance of a common gate configuration biased with a gate-resistor R_G, neglecting the gate-drain overlap capacitance, the source-bulk capacitance and channel-length modulation effect for simplicity, can be written as:

$$Z_{in} \approx \frac{R_G C_{GS} S + 1}{g_m + g_{mb} + g_{mb} R_G C_{GS} S} \qquad (3)$$

In other word Z_{in} behaves as shown in Fig. 5, changing from $1/(g_m+g_{mb})$ at low frequencies to $R_G//(1/g_{mb})$ at high frequencies.

It is clear that the input referred noise of circuit of Fig. 3 (b) is larger than of circuit of Fig. 3 (a) due to added noise of R_G and also smaller gain. However, in all former discussions we have assumed that the integrated parasitic capacitances at the gate of circuit of Fig. 3 (b) are small in comparing with gate-source capacitance of transistor.

III. PROPOSED DIFFERENTIAL TRANSIMPEDANCE AMPLIFIER

G_m-boosting technique (Fig. 6) has been widely used in design of RF building blocks and the efficiency of this technique has been already proved for lowering the noise and doubling the gain of low-noise amplifiers (LNA) and mixers [3, 4, 5]. However, utilizing it for DC-coupled and true single-ended to differential converting is a new motivation.

In the differential situations, g_m-boosting technique can be realized by positive gain stages as shown in Fig. 7. Realizing by positive gain

683

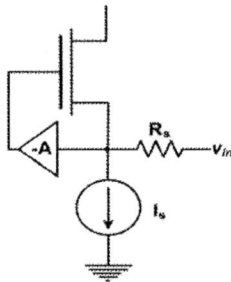

Fig. 6: g_m-boosting technique.

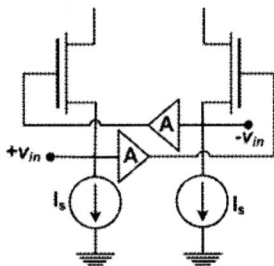

Fig. 7: Differential g_m-boosting technique with positive gain stages.

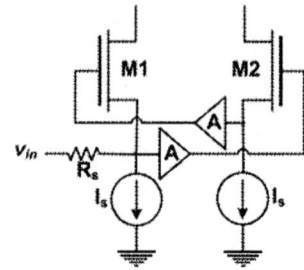

Fig. 8: Single-ended to differential conversion by means of g_m-boosting principle.

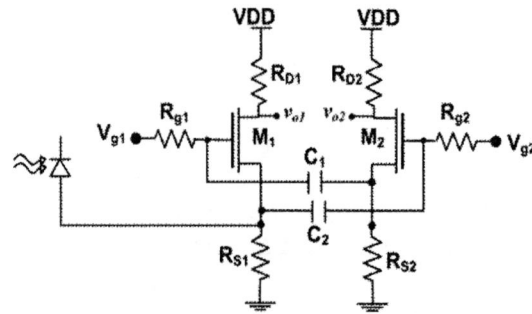

Fig. 9: Capacitor cross coupling of two CG stage as single-ended to differential transimpedance amplifier.

stages makes it possible to use passive stages as gain stages, so implementing by for example capacitor coupling [3]. However, the differential g_m-boosting architecture shown in Fig. 7 can also be used for single-ended to differential conversion. Fig. 8 depicts modified g_m-boosting technique, where one of the gain stages transfers the signal at the source of M1 to gate of M2 and the other conveys the signal in the source of M2 to the gate of M1. In another word, by the assumption of $A=1$, the gate-source voltage of left-hand transistor is applied to the right-hand one with an opposite direction or as source-gate voltage.

As an example of applying this method, two common-gate stages have been cross coupled as shown in Fig. 9. Coupling by C_1 and C_2 takes effect at frequencies that the impedance of them is negligible comparing to the value of gate bias resistors. Thus, assuming perfect coupling by C_1 and C_2, the transimpedance gain for each part of the proposed TIA is:

$$Z_{T1} = \frac{v_{o1}}{i_{in}} =$$
$$\frac{(g_m + g_{mb} + 2g_m g_{mb} R_S + g_{mb}^{\,2} R_S) R_S R_D}{[1 + (g_m + g_{mb}) R_S]^2} \tag{4}$$

and

$$Z_{T2} = \frac{v_{o2}}{i_{in}} = \frac{-g_m R_S R_D}{[1 + (g_m + g_{mb}) R_S]^2} \tag{5}$$

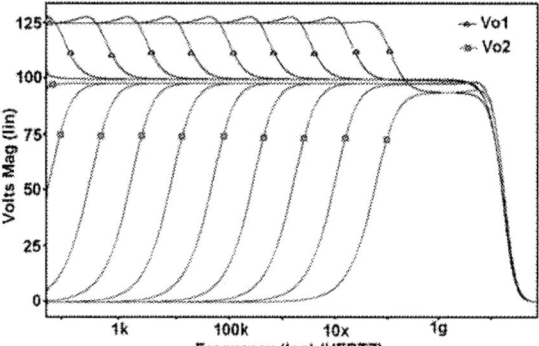

Fig. 10: Transimpedance gain of M1 and M2.

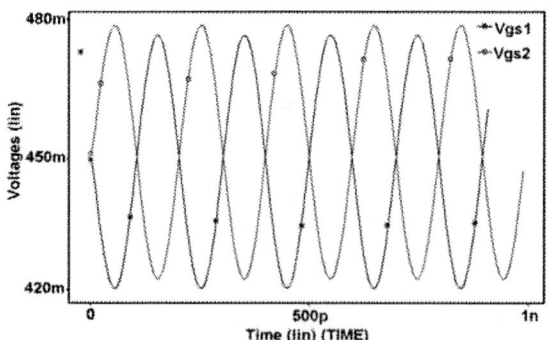

Fig. 11: Gate to source voltage of M1 and M2.

Eq. (4) and (5) were written under the condition of $V_{g1} = V_{g2}$, $R_{G1} = R_{G2}$, $R_{S1} = R_{S2} = R_S$, $R_{D1} = R_{D2} = R_D$, $g_{m1} = g_{m2}$ and $g_{mb1} = g_{mb2}$. Defining the ratio of Z_{T1} to Z_{T2} using (4) and (5) yields to:

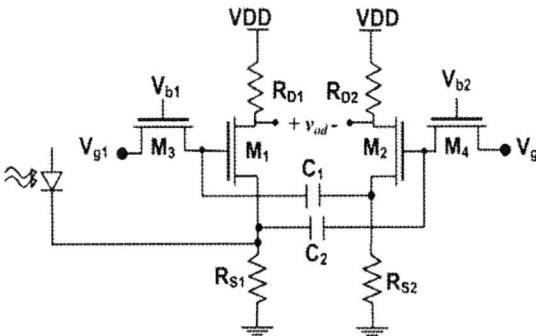

Fig. 12: Realistic implementation of proposed TIA.

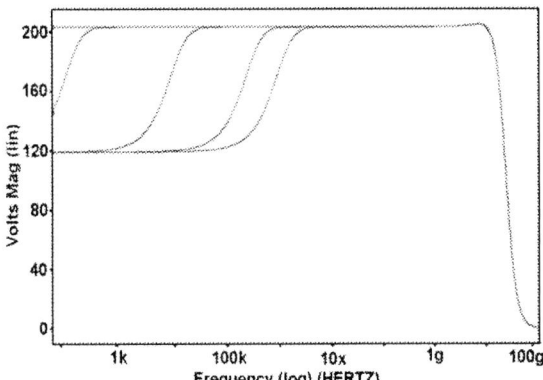

Fig. 13: Differential transimpedance gain.

$$\frac{Z_{T1}}{Z_{T2}} = -[1 + 2g_{mb}R_S + \frac{g_{mb}}{g_m}(1 + g_{mb}R_S)] \qquad (6)$$

True differential output is achieved when Z_{T1} and Z_{T2} are equal in value but opposite in sign. Thus, referring to the (6), a way to get symmetric response at the drains of M1 and M2 (Fig. 9) is canceling out the body effect of them.

The transimpedance gain of M1 and M2 is presented in Fig. 10. Although capacitors play the role of gain stages in this circuit, the main problem is with coupling of low-frequency signals. It means that, because of directly coupling with photodiode, M1 has a frequency response extending from dc to high-frequency dominant poles, while M2 has a lower-corner cutoff frequency depending on the value of coupling capacitors. The key component for overcoming this problem is the value of gate driving resistors, R_{G1} and R_{G2}. As it is clear, higher value for gate driving resistors decreases lower cutoff frequency for M2. So, by choosing proper value for R_{G1} and R_{G2}, wide frequency response extending from near dc can be achieved. Fig. 11 also shows the V_{gs} of M1 and M2 for a 5GHz input current signal.

Fig. 12 depicts a more realistic implementation, with M3 and M4 operating as the gate resistors R_{G1} and R_{G2}. Now, V_{b1} and V_{b2} play the key function in selecting lower corner frequency. The differential transimpedance gain of the circuit for several values of $V_b=V_{b2}=V_{b2}$ is shown in Fig. 13. Suitable value for V_b results in smooth response over whole band.

IV. CONCLUSION

The design of a single-ended to differential transimpedance amplifier has been demonstrated. The idea of differential g_m-boosting technique was used for achieving true differential behavior by capacitor cross coupling of two common gate stages. It was realized that "DC-to-GHz" operation is possible despite capacitor cross coupling. The proposed TIA is designed in a standard 1.2-V 0.13-μm CMOS technology. Simulation results indicate that the circuit has a power consumption of about 2.6mW with the gain of 44.5dB, bandwidth of about 10GHz, and total integrated input referred noise of about 2.5uA.

ACKNOWLEDGEMENT

The authors would like to thank the "Iran Ministry of Communication and Information Technology" and "Iran Telecom Research Center (ITRC)" for supporting this work.

REFERENCES

[1] "Design of Integrated Circuits for Optical Communications" by Behzad Razavi, McGrawHill Publication, 2002.

[2] T. Yoon and B. Jalali "1Gbit/s fiber channel CMOS transimpedance amplifier" *ELECTRONIC LETTERS*, VOL 33, NO. 7 MARCH 1997.

[3] Xiaoyong Li, *et al*, "Low-Power gm-boosted LNA and VCO Circuits in 0.18μm CMOS" *IEEE International Solid-State Circuits Conference Dig. Tech. Papers*, pp. 534-535, Feb., 2005.

[4] W. Zhuo, et al, "A Capacitor Cross-Coupled Common-Gate Low-Noise Amplifier" *IEEE Transaction on circuits and systems, Express Briefs*, vol. 52, no. 12, pp. 875-879, Dec 2005.

[5] W. Zhuo, S. Embabi, J. Gyvez, and E. Sanchez-Sinencio, "Using capacitive cross-coupling technique in RF low-noise amplifiers and down-conversion mixer design," in *Proc. Eur. Solid-State Circuits Conf.*, Sep. 2000, pp. 116–119.

Analysis of Airborne Boron and Phosphorus Contaminations on Wafer Surface by TOF-SIMS

Mo Zhiqiang, Gui Dong, Hua Younan, Zhao Siping, *Senior Member, IEEE* and Xing Zhenxiang

Chartered Semiconductor Mfg Ltd
Woodlands Industrial Park D, Street 2
Singapore 738406
Email: mozq@charteredsemi.com

Abstract **Airborne boron and phosphorus contaminations on wafer surface has been analysed by TOF-SIMS. A known boron and phosphorus concentration BPSG sample was used as reference for the calibration of the TOF-SIMS. The detection limit reaches 1E8 at/cm^2 for boron and 1E10 at/cm^2 for phosphorus. This method is easy to applied and no sample preparation required. So TOF-SIMS is a very good monitoring technique for airborne boron and phosphorus on wafer surface.**

I. INTRODUCTION

AIRBORNE contamination, especially boron and phosphorus will cause the microelectronic device or product failure even in a low level. It was reported that a 15% yield loss at a major fab was caused by airborne phosphorus contamination of about 3E13 at/cm^2 level. [1-3]

There are several techniques to analyse boron and phosphorus on silicon wafer. They are: Gas Chromatography Mass Spectrometry (GC-MS), Inductively Coupled Plasma Mass Spectrometry (ICP-MS) and Time-of-Flight Secondary Ion Mass Spectrometry (TOF-SIMS). However, GC-MS is used for organic analysis. The inorganic boron and phosphorus will not be able to analysed by GC-MS. Hence, the GC-MS results will under estimate the real situation. ICP-MS is a mature technique for trace elemental analysis. However, the analysis of boron and phosphorus on wafer surface is not straightforward. Sample preparation is needed to collect boron and phosphorus from wafer surface to solution. During this preparation, contamination is very difficult to avoid unless dedicate sample preparation machine is used. Moreover, this is a destructive test, results verification is not possible for the wafer sample. TOF-SIMS, on the other hand, is a direct measurement technique. No sample preparation is needed. This eliminates any possible contaminations on the wafer surface. TOF-SIMS is non-destructive and results verification is possible on the wafer sample.

In this study, TOF-SIMS was used to analyse airborne boron and phosphorus on wafer surface. BPSG was used to calibrate TOF-SIMS for quantitative analysis. Relative Sensitivity Factor (RSF) was derived for both boron and phosphorus.

II. EXPERIMENTAL

TOF-SIMS analysis was performed on an ION-TOF IV time-of-flight spectrometer. ^{69}Ga$^+$ liquid metal ion gun was used as primary source and operated at 25 keV with 2pA current. The mass resolution M/ΔM was greater than 6000 for both positive and negative mode at peak ^{28}Si which is more than enough to separate peak ^{31}P (mass 30.9738) and peak ^{30}SiH (mass 30.9816). The acquisition time for all spectra was 5 minutes and the raster area was 100um x 100um. Boron was detected in positive mode and phosphorus in negative mode. ^{30}Si peak was used as substrate signal for normalization. No pre-sputter was applied because the airborne boron and phosphorus should present on the top surface of the wafer. Any ion sputtering will remove this layer.

For quantitative analysis, a reference sample with known concentration of B and P is needed for TOF-SIMS analysis. In this study, a 1200nm thick BPSG sample was used as reference and the value for B and P was 4.0wt% and 4.9wt% respectively.

To eliminate positive charge on BPSG sample during TOF-SIMS analysis, charge compensation gun was employed. The operating parameters are

0-7803-9730-4/06/$25.00 ©2006 IEEE

as follow: filament current 2.5A, wehnelt 79.51V, anode 300V, delay 1.92us, lead off 1.92us.

To remove surface contamination of BPSG sample, Ar^+ ion sputtering was used for cleaning purpose in the TOF-SIMS system. The Ar^+ ion gun was operated at 3kV. The raster size is 500um x 500um and beam current is 40nA. The BPSG was sputtered for 5 minutes and this equivalent to about 20 nm thick of BPSG being removed.

As the known value of B and P in BPSG is in wt%, we need to convert it into at% and at/cm^2. Assume B is in B_2O_3 form and P in P_2O_5 form in BPSG. If the ratio of B_2O_3, P_2O_5 and SiO_2 is a:b:1, then

$$(2ax11)/(a(2x11 + 3x16) + b(2x31 + 5x16) + 28+32) = 4\% \qquad \ldots\ldots (1)$$
$$(2bx31)/(a(2x11 + 3x16) + b(2x31 + 5x16) + 28+32) = 4.9\% \qquad \ldots\ldots (2)$$

a=0.144 and b=0.064 by solving the above equations.

So the atomic concentration of B and P is:

$$B = 2a/(5a+7b+3)=6.9 \text{ at}\%$$
$$P = 2b/(5a+7b+3)=3.1 \text{ at}\%$$

If the surface density of SiO_2 is about 1E15 at/cm^2, then the surface coverage of B and P in this BPSG sample is 6.9E13 at/cm^2 and 3.1E13 at/cm^2 respectively.

III. RESULTS AND DISCUSSION

The detection of B was done by positive mode SIMS. After 5 minutes Ar^+ ion beam sputtering cleaning of the BPSG reference sample, positive spectrum was acquired. The corrected intensity of ^{11}B and ^{30}Si is 7.03E5 and 1.85E5 respectively. So the relative sensitivity factor (RSF) of B is 1.82E13 at/cm^2. The surface coverage of B for an unknown sample will be: 1.82E13 at/cm^2 x ^{11}B counts / ^{30}Si counts. Similarly, the relative sensitivity factor for P is 7.1E12 at/cm^2.

A typical positive SIMS spectrum for ^{11}B peak was shown in Fig.1. The peak is outstanding, compared to the background/noise level. The detection limit of B can be down to 1E8 at/cm^2 level.

All the raw counts for peak ^{11}B and ^{31}P will be normalized to ^{30}Si (positive mode and

negative mode respectively) to eliminate the fluctuation of analysis conditions.

Fig.2 showed negative SIMS spectrum for peak ^{31}P. The mass resolution M/ΔM of ^{28}Si is 11000. Under this mass resolution, ^{31}P and ^{30}SiH peaks were well separated.

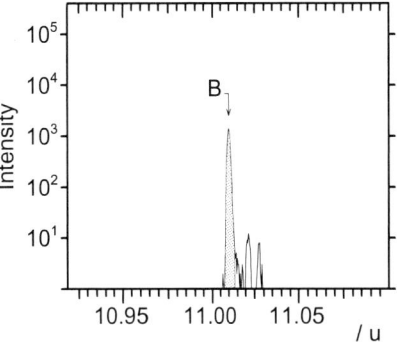

Figure 1. Positive SIMS spectrum showed ^{11}B peak. The peak is outstanding and the background is very low.

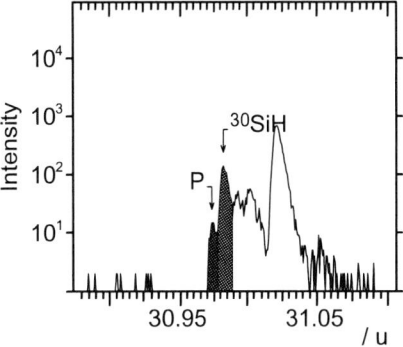

Figure 2. Negative SIMS spectrum showed peak ^{31}P (blue color) which is well separated with peak ^{30}SiH (brown color).

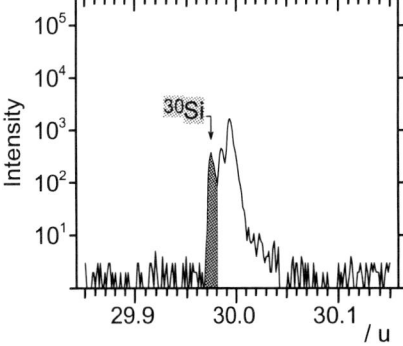

Figure 3. Negative SIMS spectrum showed reference peak ^{30}Si. Peak ^{31}P will be normalized to peak ^{30}Si to eliminate the fluctuation of analysis conditions.

A native oxide layer on Si wafer really helps to promote positive ion yield hence improve boron detection limit. Since the detection depth of TOF-SIMS is just top 1-2 atomic layers, no oxygen flooding during analysis or pre-oxidize process is necessary based on our previous study. [4] This means BPSG is suitable to be used as reference sample to calibrate TOF-SIMS to measure airborne boron and phosphorus contamination on wafer surface. As for the detection of phosphorus, low energy Cs^+ beam sputter could be useful for better detection limit, however, if the enhancement of P signal is not saturated, it will be very difficult to control the experimental parameters so as to have a consistent environment for phosphorus to get reliable results. Further study will be useful to further improve the detection limit of phosphorus.

Nine wafers were put in different locations of different fabs with different exposure time. The wafers were put into wafer boxes after collecting airborne boron and phosphorus and send to TOF-SIMS lab for analysis.

The TOF-SIMS analysis result was tabulated in Table 1. Both boron and phosphorus were in the low side of E11 at/cm^2 level. However, the levels from Fab A were higher than the other two. It was reported that phosphorus contamination levels on wafers exposed to most of the fabs were in the range of 2E11 – 2E12 atoms/cm^2. Approximately 8% of the fabs had levels greater than 2E13 atoms/cm^2. [1]

Table 1, TOF-SIMS analysis results

Sample	B (at/cm^2)	P (at/cm^2)
Fab A sample 1	3.9E11	3.5E11
Fab A sample 2	2.4E11	3.0E11
Fab A sample 3	2.0E11	5.1E11
Fab B sample 1	1.3E11	1.3E11
Fab B sample 2	1.5E11	1.3E11
Fab B sample 3	1.5E11	1.3E11
Fab C sample 1	1.4E11	1.8E11
Fab C sample 2	1.6E11	1.2E11
Fab C sample 3	1.6E11	1.4E11

IV. CONCLUSION

A quantitative TOF-SIMS analysis of airborne boron and phosphorus on wafer surface was established by using known boron and phosphorus concentration BPSG sample as reference. The detection limit is 1E8 at/cm^2 for boron and 1E10 at/cm^2 for phosphorus. The

advantage of TOF-SIMS analysis for B and P on wafer surface is no sample preparation required. This eliminates any possible cross contamination during sample preparation. The analysis can be very fast and repeat analysis is possible for verification purpose. So TOF-SIMS is a very good monitoring technique for airborne boron and phosphorus on wafer surface.

ACKNOWLEDGMENTS

The authors would like to thank PSB Corporation Surface Analysis Group for their technical advice and contribution.

REFERENCES

[1] Investigating yield loss caused by airborne organophosphates Anurag Kumar, Latif Ahmed, and Mark J. Camenzind, Balazs Analytical Laboratory

[2] Jiangsheng (Jason) Wang and Marjorie K. Balazs, "Analytical Technique Compares Dopants in Fab Air and on Wafers". *Semiconductor International* 3/1/2000.

[3] Gary G. Goodman, Patricia M. Lindley and Lori A. McCaig, "Monitoring of Cleanroom Airborne Molecular Contamination By Time-of-Flight SIMS"

[4] PSB internal research report.

Study of Porous Silicon Fabricated by Pulsed Anodic Etching of n-Si(100)

N K Ali[a], M R Hashim[b], A Abdul Aziz
School of Physics, Universiti Sains Malaysia
11800, Penang, MALAYSIA
Email, a: nihadka@yahoo.co.uk b: roslan@usm.my

Abstract **It is well known that porous silicon (PS) has a large range of morphologies. A pulsed anodic etching method is developed to fabricate uniform PS with different surface morphologies. Changes of PS on n-type Si surface after anodization with pulsed current with varying delay time were studied by scanning electron microscopy (SEM) and Raman spectroscopy. The SEM images show that a uniform and well defined silicon columns can be obtained with a correct choice of delay time. Raman scattering from the optical phonon in PS showed redshift of the phonon frequency, broadening and increased asymmetry of the Raman mode on decreasing delay time. Using the phonon confinement model, the average diameter of Si nano-crystallites has been estimated as 2, 2.6, 3, and 3.4 nm for delay time of 12, 25, 50, and 75 msec respectively.**

I. INTRODUCTION

THERE are many applications where porous silicon (PS) has been used as devices such as light emitters, light detectors and gas sensors [1-3]. Advances in PS based technologies depend strongly on possibilities to form PS layers with high degree of reproducibility. Large variety of PS structures can be produced by various techniques under different conditions. In the last few years, many kinds of meso- and macro-pore morphologies have been obtained in n-type and p-type Si under many, sometimes quite different etching conditions [4]. Several groups have demonstrated that complex optical devices can be designed with mesoporous silicon. One of the main applications of porous silicon based optical devices is biosensing [5]. However, the limiting factor of these meso-porous structures is that only objects smaller than the pore size can be detected. Although large macro-pores will greatly improve infiltration inside the PS, the internal surface area of large macro-pores is significantly reduced in comparison to the meso-pores, which will decrease the sensitivity of the devices in sensing applications.

Usually for a given substrate and electrolyte, only one type of pore microstructure can be obtained. Common techniques used for the formation of PS were direct current anodic etching. The pore morphology is influenced by the applied current density, the composition of the electrolyte, the dopant level of the crystalline silicon substrate, and the etching time [6]. An increase in the applied current density or a decrease in the concentration of hydrofluoric acid causes an increase in the maximum diameter of the pores, a broader distribution of pore sizes, and an increase in the porosity (percentage of void space) of a sample. This paper reports on the PS structures on n-Si (100) samples applied with a new technique of pulsed current anodic etching [7-8]. The change in the surface morphology, porosity, the size of pores and the porous layer thickness due to the change of the delay time (t) (the time when there is no current) in the pulsed anodic etching will be investigated using scanning electron microscopy (SEM) and Raman spectroscopy. This study will clarify how the structure of PS changes with the delay time of pulse current. Pore formation mechanisms were discussed according to a new current bursts model from Carstensen *et. al.* [9].

II. EXPERIMENTAL PROCEDURE

Porous silicon samples were fabricated on n-type Si (100) with a resistivity of 0.75-1.25 Ω cm. The electrochemical etching was carried out in a 48% HF:98% ethanol, 1:4. All samples were etched with the same current on-time of 28min. at room temperature under illumination of 100 W tungsten lamp placed at 20 cm away from the surface. A home made pulsed current generator with output signal adjusted for both delay time (t) and current on-time (T) was used to feed the current through the anodic etching circuit. Pulsed current with a peak density of 20mA/cm^2 was used. For each current cycle, duration time T of 140 msec was used for all samples but with different delay time t of 12, 25, 50, 75 msec for samples a, b, c, and d respectively. The surface of etched samples looks light brown. All samples showed visible light emission under illumination by UV light.

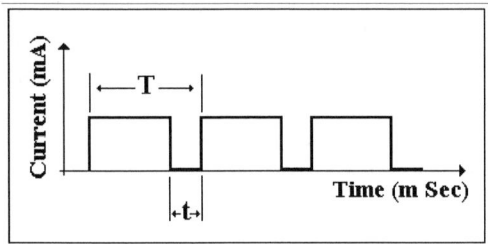

Fig. 1 Schematic of pulsed current fed to the n-Si (100) for fabrication of porous silicon.

III. RESULTS AND DISCUSSION

i) Surface morphology

Fig. 2 shows the surface and cross section of SEM images for samples a, b, c, and d. The etching process on the nanometer scale produces different surface morphologies and long channels that penetrate deep into the sample resulting in a network of columns. The effect of the delay time on the porosity, strain and surface morphology is clear in these figures 2a, 2b, 2c, and 2d. As the delay time increases from samples a - c, the porosity, the strain, the pore size and the distance between the columns decreases. In general, the removal of the silicon disturbs the equilibrium lattice forces in the crystal, introducing a lateral strain in the surface layer. The strain would tend to cause the columns to collapse to onto each other, forming the hillocks particularly in Fig. 2a. However the situation is different in Fig. 2d with the longest delay time of 75msec. The columns for this sample are not clearly defined after cleavage. The PS layers present a considerable number of cracks which could be due to the large capillary stresses during drying of water after PS formation. The cracking of highly porous silicon layers during drying can be avoided either by replacing water with another liquid of lower

surface tension, like pentane. We can explain the effect of delay time in the PS formation process according to the current burst theory [9]. In the current burst cycle: (A) a direct dissolution of silicon occurs; (B) silicon may be oxidized; (C) the silicon oxide has to be dissolved purely chemically; (D) at the clean surface a relatively slow H-passivation starts; (E) to start the cycle again, each current burst has to overcome this H-passivation of the surface. Only in (A) and (B) charge is transferred. In our case when we increase the delay time t means that we give more time to the steps (C, D, and E) resulting in removal of the silicon oxide completely and then increasing PS formation rate with the appropriate delay time. This process is clearly evident in sample d where increasing delay time further results in slow down of the process due to the more H-passivation of the surface.

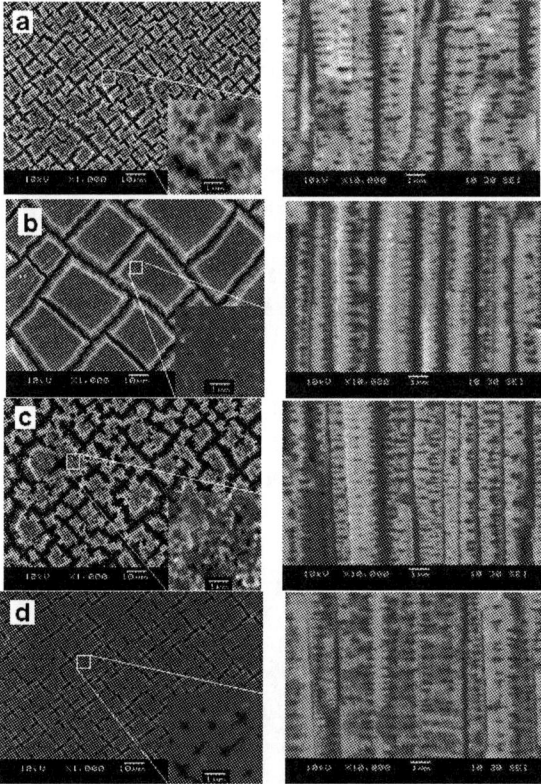

Fig. 2 Top view and cross section SEM micrographs of pores formed with pulsed current at different delay time of 12, 25, 50, and 75 msec for samples a, b, c, and d, respectively.

ii) Raman spectroscopy

Raman scattering spectroscopy is one of the powerful nondestructive techniques for studying structure related information, especially information concerning nanostructures in crystalline materials. Among other things, it has been used to determine grain size, shape, and the fraction of amorphous phase in microcrystalline materials [10-11]. The Raman spectrum for crystalline silicon (c-Si) consists of one sharp peak situated at 521 cm^{-1}, it is symmetric and has a width of 3 cm^{-1} (FWHM). On the other hand, Raman spectrum of amorphous silicon (a-Si) consists of a broad peak at 480 cm^{-1} and is usually weak in intensity.

Fig. 3 shows typical Raman spectra of our PS samples (a, b, c, and d). The spectrum from a crystalline silicon c-Si wafer is also shown for comparison. The peak position and width of sample d) only have small changes as compared to those of c-Si and other samples (a, b, and c). The peak intensity of all PS samples are 10 times stronger than that of c-Si, which is believed due to the remarkable change of its optical constants [12] and or due to the surface enhancement or resonant effect [13]. As the delay time decreases, there exists a large peak shift from 521 cm^{-1} (c-Si) to 507 cm^{-1} (sample a) and a broadening of peak width and FWHM from 3 cm^{-1} (symmetric) to 20 cm^{-1} (asymmetric). In all PS samples there is no sharp peak at around 521 cm^{-1} from the c-Si substrate is seen in any of the spectra, and hence the laser beam dose not reach the depth in our samples where c-Si would be present, due to the thick PS layer. In addition, we did not observe a broad peak at 480 cm^{-1}, indicating that there is no substantial contribution to the Raman intensity from possible a-Si in the tissue region near the surface. These facts indicate that the microstructure is surprisingly free of disorder. Raman spectra for our samples were taken from different points on the top and in the middle of the edge of the film and they showed no different from each other, so the structure of the film is homogeneous.

respectively. Also plotted is the Raman spectrum of c-Si for comparison.

Fig. 4 shows the Raman red shift as a function of peak width for various PS samples. The experimental data are shown as dots, while the solid line is plotted to guide the eye. The dashed line is the calculated results using the three dimensional phonon confinement's model. The Raman red shift and line broadening have been observed in polycrystalline and microcrystalline materials and were attributed to the confinement of optical phonons in a small crystalline particle [14]. However, disorder or finite-size effects may partially or completely relax "momentum" conservation, leading to a downshift and broadening of Raman peak [10]. The result in Fig. 4 shows that the phonon confinement model does not totally correlate to the experimental results. Yang et al. [12] suggested that the difference between the solid curve (experimental) and the dashed curve (theoretical) is attributable to the effect of strain induced by the lattice expansion in the PS layer. It has been shown by x-ray double crystal diffraction that the PS layer is strained, with the lattice constant increasing as the porosity increases [15]. The following relationship gives an estimation of the strain by measuring the difference of peak shifts at a specific line width [12]:

$$\frac{\delta\omega_s}{\omega_0} = -3\gamma\frac{a_{PS} - a_0}{a_0}$$

where $\delta\omega_s$ is the vertical interval between the solid curve and dashed curve, a_{PS} is the lattice constant of PS nanostructures, and $\gamma \approx 1.0$ is the Gruneisen constant. The value of the strain increases as the linewidth or red shift increases, with the later being related to the size of the Si nanostructures.

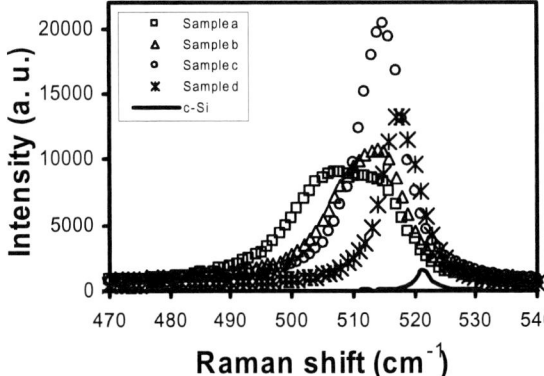

Fig. 3 Typical Raman spectra of PS samples (a, b, c, and d) etched with pulsed current having delay different delay time of 12, 25, 50, and 75 msec

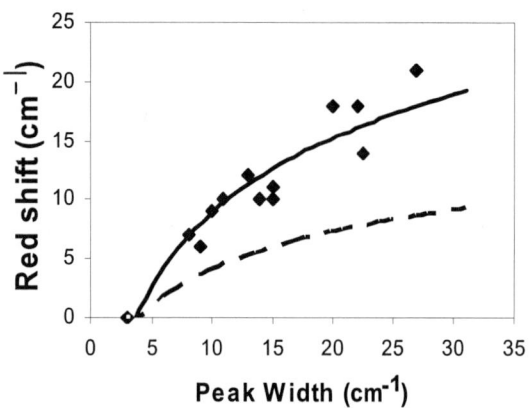

Fig. 4 The relationship between the red shift from that of c-Si and the linewidth. The dots represent the experimental data, the solid line is drawn to guide the eye, the dashed line is the calculated results using the three dimensional phonon confinement model.

To explain our experimental results in Fig. 4, a quantitative model was developed by Richter *et al.* [16] and was later improved by Campbell and Fauchet [10]. This model was employed to estimate the average size (or correlation length) L of the nanocrystals from the Raman spectrum. The PS is modeled as an assembly of quantum dots, i.e. the confinement is three dimensional. For spherical nanocrystallites, the first order Raman line shape for longitudinal optical mode *LO* phonon is given by [10]

$$I_{LO}(\omega) = \int_0^1 \frac{4\pi q^2 |c(0,q)|^2}{[\omega - \omega_{LO}(q)]^2 + [0.5\,\Gamma_c]^2}\, dq \quad (1)$$

where

$$|c(0,q)|^2 = \exp\left(\frac{-q^2 L^2}{4a^2}\right)$$

and wave vector q is expressed in the units of $2\pi/a$ ($a = c$-Si lattice constant, 0.54 nm), L the diameter of the Si nanocrystallite, $\omega_{LO}(q)$ the longitudinal phonon frequency at q along [100] direction and Γ_c the FWHM of the crystalline silicon (3 cm⁻¹).

The phonon dispersion relation $\omega_{LO}(q)$ is taken according to [15] is,

$$\omega_{LO}^2(q) = A + B\cos(\pi q/2) \quad (2)$$

where $A = 1.7207 \times 10^5$ cm⁻² and $B = 1.0 \times 10^5$ cm⁻². Eq. (2) gives ω_{LO} at $q = 0$ as 521.6 cm⁻¹. To match our experimental results with theoretical results we found that Eq. (2) should be modified to

$$\omega_{LO}^2(q) = A + B\cos(\pi q/3) \quad (3)$$

where $A = 2.0144 \times 10^5$ cm⁻² and $B = 0.7 \times 10^5$. In this case Eq. (3) gives ω_{LO} at $q = 0$ as 521 cm⁻¹, which matches well with our *LO* mode frequency in *c*-Si. Using Eq. (1) and Eq. (3), the analytical Raman spectra of PS were generated and fitted to the experimental Raman spectra as shown in Fig. 5. As seen from Fig. 5, the fit was very good. The analytical fit did not take into account the broad Raman mode at 480 cm⁻¹ from amorphous silicon *a*-Si or the contribution from defects. The c-Si nanocrystalites diameter obtained from the best fit to Fig. 5, (a, b, c, and d) were 2, 2.6, 3, and 3.4 nm for PS samples etched with delay current 12, 25, 50, and 75 msec respectively. The values of L and Raman peak positions (PP) are listed with the delay time t in Table 1.

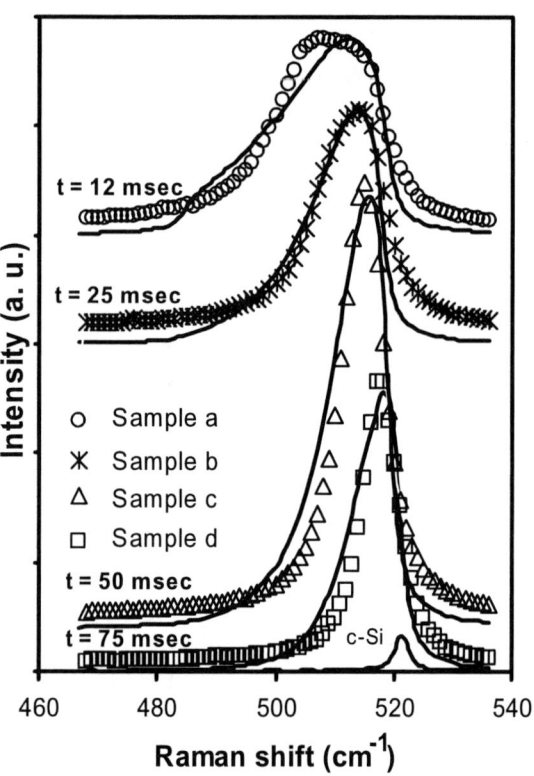

Fig. 5 Raman spectra of PS samples (a, b, c, and d) etched with pulsed current having different delay time of 12, 25, 50, and 75 msec. respectively. The symbols are the experimental data. The solid lines are the generated spectra using quantum confinement model. Also plotted is the Raman spectrum of *c*-Si for comparison.

Table 1: Raman mode peak position and Si nanocrystallite diameter (L) along with band widths, delay time, and Si nanocrystallite diameter for samples (a, b, c, and d).

692

Sample	Delay time (t) msec	Raman PP cm^{-1}	FWHM cm^{-1}	L nm
a	12	507	S	2
b	25	514	11	2.6
c	50	515	8	3
d	75	518	7	3.4
c-Si	-	521	3	-

IV. CONCLUSIONS

In summary, a new technique for PS formation was developed using pulsed current instead of direct current. The influence of the delay time of the pulse on the porous structure morphology is demonstrated by scanning electron microscopy. The mechanism of pore formation was discussed according to the new model of pore formation burst current model. Raman spectroscopy studies have been carried out to investigate the effect of delay time on PS formation process. The red shift in the phonon Raman spectra reveals the quantum confinement effect in the Si nanocrystallites of the PS samples. As the delay time decrease, these shifts increases due to decrease in the Si nanocrystallite size. This is indicated by the decrease in the estimated diameter of spherical nanocrystallites from 3.4 - 2 nm using the Raman spectra.

AKNOWLEDGEMENT

The authors are grateful to School of Physics, Universiti Sains Malaysia and Ministry of Science, Technology and Innovation (MOSTI) of Malaysia for supporting this work under Strategic Research Grant.

REFERENCES

[1] S. Chan, P. M. Fauchet, "Silicon microcavity light emitting devices", *Optical Materials.* vol. 17, pp. 31-40 (2001).

[2] A. M. Rossi, H. G. Bohn, "Photodetectors from porous silicon", *Physica status solidi (a).* vol. 202, no 8, pp. 1644-1647 (2005).

[3] F. Rahimi, A. Iraj zad and F. Razi, "Characterization of porous poly-silicon impregnated with Pd as a hydrogen sensor", *J. Phys. D: Appl. Phys.* vol. 38, pp.36-40 (2005).

[4] J. Carstensen, M. Christophersen, G. Hasse, and H. Foll, "Parameter dependence of pore formation in silicon within a model of local current bursts", *Phys. Stat. Sol. (a).* vol. 182, no. 1, pp. 63-68 (2000).

[5] P. Claudia, S. Morta, S. Michael J., C. Frederique, and M. Gordon, "Biosensing using porous silicon doble-layer inter-ferometers", *J. Am. Chem. Soc.* vol. 127, pp. 11636-11645 (2005).

[6] L. Pavesi and R. Guardini, "Porous silicon: Silicon quantum dots for photonic applications", *Brazilian Journal of Physics.* vol. 26, no. 1, pp. 151-161 (1996).

[7] X. Hou, H. Fan, L. Xu, F. Zhang, M. Li, M. Yu, and X. Wang, "Pulsed anodic etching: An effictive method of preparing light emitting porous silicon", *Appl. Phys. Lett.* vol. 68, no 17, pp. 2323-2325 (1996).

[8] B. Das, and S. P. McGinnis, "Porous silicon pn junction light emitting diodes", *Semicond. Sci. Technol.* vol. 14, pp. 988-993 (1999).

[9] J. Carstensen, M. Christophersen, G. Hasse, and H. Foll, "Pore formation mechanisms for the Si-HF system", *Mat. Sci. Eng. B.* vol. 23, pp. 69-74 (2000).

[10] I. H. Campbell and P. M. Fauchet, "The effects of microcrystal size and shape on the one phonon Raman spectra of crystalline semiconductors", *Solid State Commu-nications.* vol. 58, no. 10, pp. 739-741 (1986).

[11] Z. Sui, P. P. Leong, I. P. Herman, and S. Gregg, "Raman analysis of light emitting porous silicon", *Appl. Phys. Lett.* vol. 60, no. 17, pp. 2086-2088 (1992).

[12] M. Yang, D. Huang, P. Hao, F. Zhang, X. Hou, and X. Wang, "Study of the Raman peak shift and the linewidth of light emitting porous silicon", *J. Appl. Phys.* vol. 75, no. 1, pp. 651-653 (1994).

[13] R. Tsu, H. Shen, and M. Dutta, "Correlation of Raman and photoluminescence spectra of porous silicon", *Appl. Phys. Lett.* vol. 60, no. 1, pp. 112-114 (1992).

[14] R. Prabakaran, R. Kesavamoorthy, and A. Singh, "Optical microstructural investi-gations of porous silicon", *Bull. Mater. Sci.* vol. 28, no. 3, pp. 219-225 (2005).

[15] I. M. Young, M. I. J. Beale, and J. D. Benjamin, "X-ray double crystal diffraction study of porous silicon", *Appl. Phys. Lett.* vol. 46, no. 12, pp. 1133-1135 (1985).

[16] H. Richter, Z. P. Wang, and L. Ley, *Solid State Commun.* vol. 39, pp. 625 (1986).

The Effects of High Temperature Storage on Lead Free Solder Joint Material Strength Using Pull Test Method

Muhammad Najib Harif, Ibrahim Ahmad and Azami Zaharim
Faculty of Engineering,
Universiti Kebangsaan Malaysia
43600 Bangi, Selangor, MALAYSIA
Email: najibharif@gmail.com , ibrahim@vlsi.eng.ukm.my , azami@vlsi.eng.ukm.my

Abstract. **The purpose of this study is to discuss on the effect of high temperature storage (HTS) on lead free solder joint material for ball grid array application using pull test method. Three samples of different lead free solder joint material were choosed in this experiment that are Sn3.8Ag0.7Cu (SAC387), Sn2.3Ag0.08Ni 0.01Co (SANC), and Sn3.5Ag. Then the lead free solder joint material samples were subjected to baking at 150°C for 24 hours, 48 hours, 96 hours and 168 hours. The *Dage* 4000 series pull test machine was used. Result show that the mean pull strength is 2847.66g, 2628.20g and 2613.79g for Sn3.5Ag, SANC and SAC387 respectively. Thus Sn3.5Ag shows a significantly better solder joint performance in terms of joint strength compare than SANC and SAC387. Hence, three compositions show that joint strength decrease from 24hours condition to 168 hours at 150°C. In summary pull test method in this research manifestly show the feasibility to be used in characterization of lead free solder joint material for semiconductor or microelectronic packaging.**

I. INTRODUCTION

Surface mount technology (SMT) is enabling microelectronic packaging densities and lighter weight. These stringent constraints and demands have raised great concern and importance on the reliability [1]. The major concern is the structural integrity of the solder interconnect when subjected to thermal cycling loading [2].

Solder joint strength is used as a quantitative measure of the robustness of the interconnect between the solder balls and substrate pad. The pads, which are copper and easily oxidized, need to have a protective surface finish ensuring solderability during assembly [3]. The ball shear test is being widely used as a test method for

accessing the integrity of the solder ball attachment on area array packages. In some of the studies, this method has also been used to compare the effect of aging on shear strength for various lead free alloys [4].

The purpose of this study is to discuss a new method to discover the effects of HTS on lead free solder joint material strength. In the past, lead free solder joint material on ball grid array (BGA) have been extensively studied using shear strength method. In this paper, Dage 4000 series pull test machine has been used on this study.

Fig. 1 Dage 4000 series pull test machine

The strength of the solder ball attachment is influenced by many factors including pad finish, solder alloy, properties of intermetallic compound formed and assembly processing parameters. For higher reliability, it is important to analyze the lead free BGA solder joint for the effects of the factors just mentioned. Hence, the main factor of this study is the time parameters of temperature storage which is 24H, 48H, 96H and 168H.

II. EXPERIMENT

0-7803-9730-4/06/$25.00 ©2006 IEEE

The preparation of the test samples was initiated by applying flux on the solder pads through pin transfer mechanism. The solder balls were placed on the solder pads by ball bump tool. The flux holds the solder balls in place until they are reflowed. The test sample was then reflowed where upon the solder balls formed bumps. During reflow, the ball self centered on the substrate pad [4]. Figure 2 shows the experiment procedure flow diagram for the experiment.

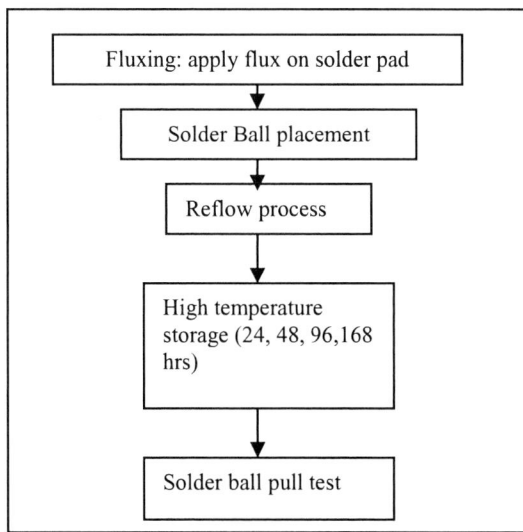

Fig. 2 Experimental flow diagram

In this study, the lead-free solder balls with composition of SAC387, SANC and Sn3.5Ag were reflowed on solder pads with Ni/Au surface finish using the lead free reflow profiles, as shown in Figure 3.

For each test condition, 40 solder spheres were pulled tested. Ball pull test is a new destructive method to determine the ability of BGA solder balls to withstand mechanical strength. Ball pull test method using a jaw to clamp or pull off the solder ball from the pad substrate. Figure 4 shows the schematic of the pull test jaw. The jaw also was specifically designed for the size of the ball to be tested.

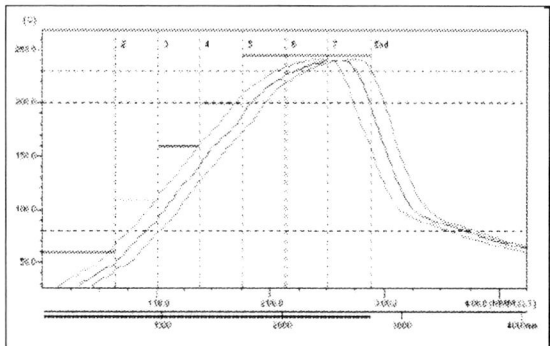

Fig. 3 Lead free reflow profile

Table 1: Package information

Package type	Tape ball grid arrays (TBGA)
Package size	0.76 mm ball diameter (30 mils), 1.27 mm pitch, 37.5 x 37.5 package size
Solder ball composition	Sn: 95.5%, Ag: 3.8%, Cu:0.7% Sn: 97.6%, Ag: 2.3%, Ni: 0.08%, Co: 0.01% Sn: 96.5%, Ag: 3.5%

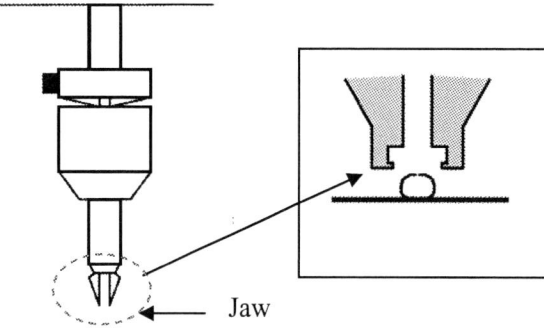

Fig. 4 Pull test method using a jaw to clamp the solder ball

The processes for the pull test are as shown in Figure 5. To initiate the test: (a) changing the bump pull jaw correctly. (b) Position the test piece under the microscope; make sure the jaw opening width should be set to be suit the ball size. (c) Bring the jaw down to the part and test the action with specific speed. In this study, two speed were used, 1000um/s and 5000um/s respectively. However, always take care to align the jaw to the ball.

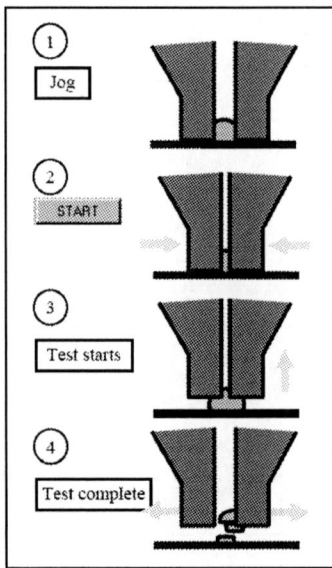

Fig. 5 The process for the bump pull test method

Lastly all the data have been collected and analyzed using general factorial design of experiment analysis.

III. RESULTS AND DISCUSSION

The results for the general factorial design of experiment analysis were carried out using the statistical software, MINITAB™. The analysis performed is an analysis of variance ANOVA model. This analysis identifies which factors and interactions are significant. The ANOVA is used to analyze the mean pull strength for the three lead free materials on HTS test. In the ANOVA model, the P-value indicates the significance of each other factor in the model. P-value of 0.05 or less considered 95% significant. Factors noted as "significant" represent the independent process variables that have the greatest influences on a particular treatment response [4].

Table 2: Main factors for the pull test method

Factors	Type	Levels	Values
Time	fixed	4	24h, 48h, 96h, 168h
Material	fixed	3	SAC387, SANC, Sn3.5Ag
Speed	fixed	2	1000um/s, 5000um/s

Table 2 show the summary of the main factors which is significant influenced the mean pull strength. Hence, in the same way Table 3 show the

ANOVA analysis by pull test method for the three lead free materials on HTS test using two speed velocities.

Table 3: Results of ANOVA analysis

Source	DF	Seq SS	Adj SS	Adj MS	F	P
Time	3	2624082	2624082	874694	42.80	0.000
Material	2	10993665	10993665	5496833	268.96	0.000
Speed	1	9233173	9233173	9233173	451.77	0.000
Time* Material	6	2262410	2262410	377068	18.45	0.000
Error	947	19354524	19354524	20438		
Total	959	44467854				

S = 142.961   R-Sq = 56.48%   R-Sq(adj) = 55.92%

From the ANOVA results for ball pull test, there is a significant effect of time storage within 150°C, material and speed of pull test. Therefore the interaction of all these factors has a significant effect on mean pull strength of the lead free solder joint material. Also, the common unit of strength is gram (g).

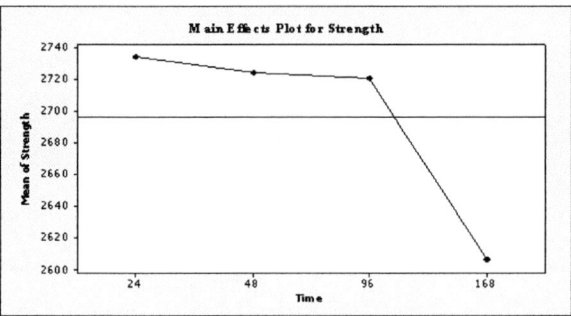

Fig. 6 Mean of strength versus storage time at 150°C

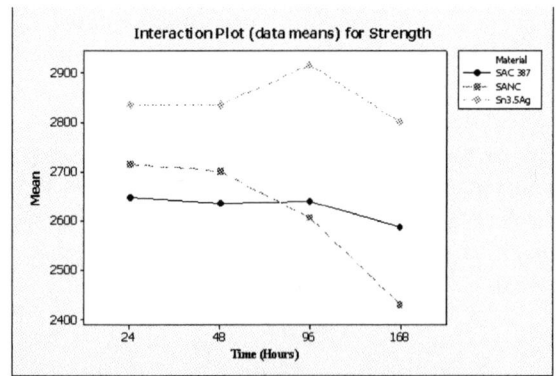

Fig. 7 Interaction plot mean of strength for three material

In addition, ball pull method is widely acknowledged by the industry in the recent years to

696

fully determine the weak interface and the joint strength. Ball pull is found to be more stringent than ball shear because of the pulling mechanicism that minimizes ball deformation above the bond site, as well as causes the bond not to be supported by the solder pad cavity wall, thus exposes the true bond strength. In the case of ball shear, higher ball deformation above the bond site is found at the peak of shear force which will result in smaller test area, and support from solder pad cavity wall could substantially shield a bad bond from failing [5].

Figure 5 and Figure 6 show after subjecting to HTS test, ball pull result shows the decreasing of solder joint strength in all the materials when the storage time increase until 168 hours.

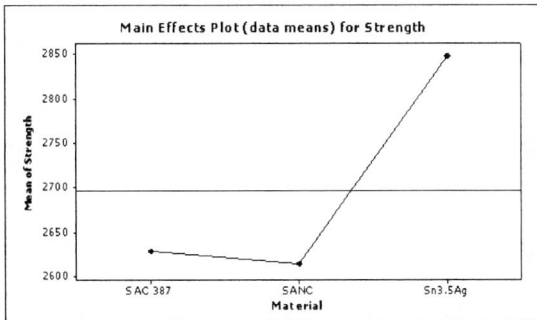

Fig. 8 Mean of strength for three material samples

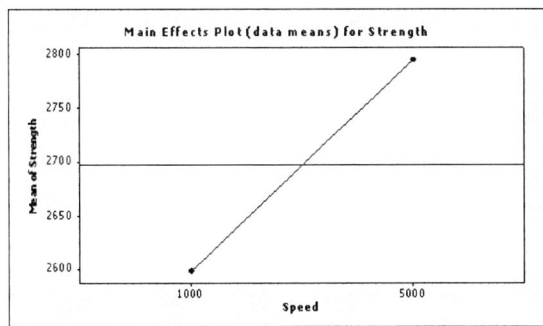

Fig. 9 Mean of strength for both speed using

Figure 8 shows the mean of strength for three different lead free materials that are used in this experiment. Result show that the mean pull strength is 2847.66g, 2628.20g and 2613.79g for Sn3.5Ag, SANC and SAC387 respectively. Thus Sn3.5Ag shows a significantly better solder joint performance in terms of joint strength compare than SANC and SAC387. Figure 9 also shows the mean of strength for both speed using which are 1000um/s and 5000um/s for the maximum speed

of pull test. So, 5000um/s gives the better strength than 1000um/s speed of pull test.

IV. CONCLUSION

The effects of high temperature storage on lead free solder joint material strength using pull test method were presented in this paper. Pull test was performed on the each single balls joint after the HTS. In summary, the joint strength for all three lead free material decrease when the increasing storage time from 24hours to 168hours at 150°C. To sum up, pull test method is one of the good techniques to characterize the solder joint strength for all condition.

V. ACKNOWLEDGEMENT

The authors would like to thank the Malaysian Ministry of Science, Technology and Environment for sponsoring this work under project IRPA 03-02-02-0121 PR0075/09-01 and Freescale's engineers for supporting this project.

VI. REFERENCES

[1] J. Lau, Y.H. Pao, Solder Joint Reliability of BGA, CSP, Flip Chip and Fine Pitch Assemblies. McGraw Hill, New York. 1997.

[2] H.L.J. Pang, K.H. Tan, X.Q. Shi. 2001. Microstructure and intermetallic growth effects on shear and fatigue strength of solder joints subjected to thermal cycling aging. *Material Science and Engineering* A307:42-50.

[3] Robert Erich, Richard J. Coyle. 1999. Shear Testing and Failure Mode Analysis for Evaluation of BGA Ball Attachment. *1999 IEEE/CPMT Electronic Manufacturing Technology Sysmposium.*

[4] Daryl Santos, Shafi Saiyed, Frank Andros. 2002. Effect of reflow profile on shear strength of Sn/4.0Ag/0.5Cu solders spheres for ball grid array applications. Journal of SMT 15(3):25-31

[5] Robert Sykes, Dage Precision Industries Ltd., Pull testing of solder balls on BGA and CSP packages without reflow.

[6] Amitava Mitra. 1998. *Fundamental of Quality Control and Improvement.* 2nd Ed. New Jersey: Prentice Hall.

[7] Doughlas C. Montgomery. 2005. *Introduction to Statistical Quality Control.* 5th Ed. New Jersey: John Wiley & Sons Inc.

[8] Ken Gilleo. 2002. *Area Array Packaging Handbook Manufacturing and Assembly*. New York: McGraw Hill.

Failure Analysis Approach in Memory Failure of SOI Devices

S. P. Neo, S. K. Loh, Z. G. Song and S. P. Zhao, *Snr Member, IEEE*
Chartered Semiconductor Mfg Ltd
60 Woodlands Industrial Park D, Street 2, Singapore 738406
Email: neosp@charteredsemi.com

Abstract **Silicon-On-Insulator (SOI) is a sandwich structure consisting of a thin insulating layer, such as silicon dioxide or glass sandwiching between a thin layer of silicon (T-Si) and the silicon substrate. The incorporation of the insulating layer between the T-Si and the silicon substrate has greatly changed the front-end process of microelectronic devices and thus the approach of failure analysis would be different compared to that of bulk technology. In this paper, approaches to analyze the single bit failure and pair bit failure in memory failure of SOI wafers would be presented.**

I. INTRODUCTION

SILICON-On-Insulator (SOI) is a sandwich structure consisting of a thin insulating layer, such as silicon dioxide or glass sandwiching between a thin layer of silicon (T-Si) and the silicon substrate. SOI technique used to fabricate military and astronaut products. Recently, it found its applications in fabrication of consuming products that required higher speed and lower power consumption [1]. It is also easier to reduce transistor size and increase packaging density with SOI technology, thus SOI technology is becoming an increasingly critical technology used in a range of high-volume applications such as microprocessor for games and computing solutions. However the incorporation of the insulating layer between the T-Si and the silicon substrate has greatly changed the front-end process of microelectronic devices and thus failure analysis faces new challenges. It makes the approaches of front-end failure analysis be more tedious compared to that of bulk silicon technology.

In bulk silicon technology, contact-level Passive Voltage Contrast (PVC) [2] is a powerful technique in analysis of memory failure. It can easily identify an abnormal contact from the normal contacts with the fact that the abnormal contact is charged up to different potential from those potentials that the normal contacts are charged up to, under lower-energy e-beam or ion-beam inspection. However, for SOI technology, the active contacts land on T-Si, which is insulated from silicon substrate. Whether the active contacts are open or not, they are at the same potential under e-beam or ion beam inspection because they are all floating. So, contact-level PVC technique is no longer applicable for SOI technology. Hence other approaches need to be developed to analyze the front-end defects of the memory failure of SOI device. For memory failure, the types of failures that are most related to front end process are pair bit and single bit failures. In this paper, the approaches analyzing the single bit failure and pair bit failure have been described and discussed.

II. RESULTS AND DISCUSSIONS

(A) Single bit failure

Single bit failure is usually the majority of memory failure. Its defect is generally at metal 1 and below based on GDS layout. Thus to analyze single bit failure, the metal layers above metal 1 can be directly removed away using finger-polishing method. If the GDS layout showed that the cross-coupling of the two inverters of the single bit is made by metal 1, then SEM inspection will be performed on metal 1 first before proceeding to the contact/poly/active level. But if the layout shows that the single bit is related to contact/poly/active level only, then the sample will be polished till the contact level. Two approaches can be employed at this stage:

(I) Top down de-layering and SEM inspection approach

Generally, for any single bit failure analysis, the first and quick approach is top-down de-layering and SEM inspection at metal 1 or the contact/poly/active layer exposed using Buffered Oxide Etch (BOE). This approach is to check for any defect that is visible under SEM inspection, like residue underneath contact (see Fig. 1), poly residue (see Fig. 2), anomaly on poly line (see Fig. 3) etc. XTEM analysis can be employed to identify the root cause if necessary. Top-down de-layering and SEM inspection has the disadvantages, such as, (1) some of the information could be lost due to the removal of the ILD, (2) some defect might be invisible using SEM inspection, for example, defect at the interface between two layers.

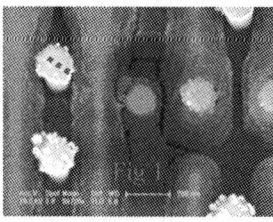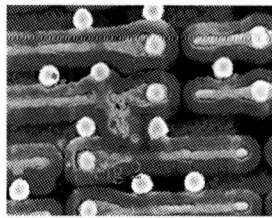
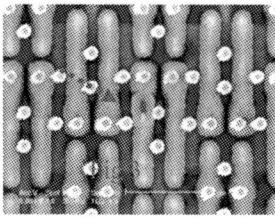

Figs. 1-3 SEM pictures showing the residue underneath contact, poly residue and abnormal poly respectively.

(II) FIB Progressive slice and view approach

FIB progressive slice and view approach using dual-beam FIB with live viewing of the milling progress is another approach, which can be adopted for initial analysis of single bit failure. The slice and view analysis will start from the beginning of the bit. This approach can check for any possibility of contact issue, like contact void, contact under-etch etc., for which top down de-layering and SEM inspection is not capable. Along the way of the FIB slicing, if abnormality is found and the resolution of the SEM in situ is not good enough to view the defect, a XTEM sample can be further prepared using the FIB milling and lift-out technique. However, if the defect is not obvious enough to be revealed by SEM, other approach like electrical failure analysis need to be applied.

(III) Electrical failure analysis approach

Recently, a variety of electrical characterization techniques for individual transistor, commonly called electrical failure analysis (EFA), have emerged and become popular in failure analysis Lab. These techniques include nano-probing in SEM or FIB [3], scanning probe microscope [4] and FIB assist probing [5]. They can characterize the performance of the individual transistor, such as pass-gate transistor, pull-down transistor and pull-up transistor of a SRAM cell to further narrow down the defective location. Nano-probing in SEM or FIB and scanning probe microscope techniques involve direct probing on the contact and FIB assist probing need to deposit micro-pads by FIB for accessing to the related contacts and probe by optical probing platform. The measurement taken from these probing includes I-V curve verifying open/short failure, transistor family curves, threshold voltage, gate leakage etc. After comparing the behavior of the suspected transistor with that of a good transistor, one can further narrow down the defective location for further physical failure analysis. EFA approach can be employed when the top down de-layering and SEM inspection or FIB progressive slice and view is unable to find the defect. For example, EFA has successfully isolated a high resistance contact in a single bit failure and further XTEM analysis showed a thin amorphous layer underneath the contact (see Fig. 4). Such amorphous layer is invisible under normal SEM and is impossible to reveal with top down de-layering and SEM inspection or FIB slice and view approach. In another instance, electrical characterization suggested that single bit failure was due to asymmetric pass-gate transistor resulting from less source/drain implantation. Thus EFA is a very useful technique for defects that are not very visible or implant related and its importance would increase as the technology continues to scale down.

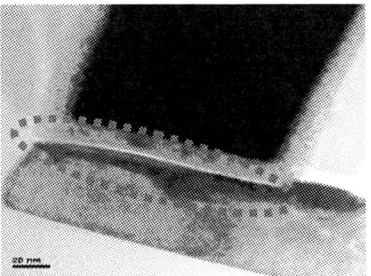

Fig. 4 Amorphous layer underneath contact

(B)Pair bit failure

Pair bit failure is another major failure of SRAM memory. The failure is normally due to high resistance or open failure of the stacked contact/via shared by the two failing bits. As PVC cannot be performed for SOI wafer, FIB x-section and SEM inspection on the stacked contact/via is the choice of failure analysis approach. For examples, FIB x-section and SEM inspection showed local interconnect under-etch for one pair bit failure (see Fig. 5) and a particle under contact for another pair bit failure (see Fig. 6).

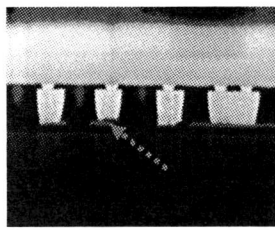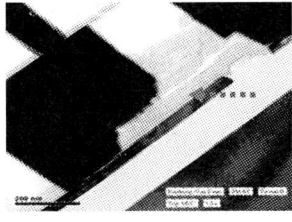

Figs. 5-6 FIB x-section pictures showing contact under-etch and residue underneath contact respectively.

XTEM analysis is more commonly used when the killer defect is too small to be well discerned by SEM. However, if the pair bit failure happens to be two adjacent independent bit failures, FIB x-section and SEM inspection or XTEM analysis on the shared stacked contact/via may miss the defect. Then top-down de-layering and SEM inspection or FIB slice and view approach need to be considered.

In summary, it will generally involve more than one approach in failure analysis of memory failure before the defect is found or fully understood.

III. CONCLUSION

Change of technology brings changes in the manufacturing process of devices and it in turn brings changes in the approach of failure analysis and makes failure analysis be more challenging and complicated. In this paper, the approaches to analyze the single bit failure and pair bit failure in memory failure of SOI devices were presented.

ACKNOWLEDGEMENT

The authors would like to thank Chartered TEM team for their TEM pictures and their great support.

REFERENCES

[1] Julian Blake, "Encyclopedia of Physical Science and Technology", Vol. 14, P. 805 (2002).

[2] Z. G. Song, G. Qian, J.Y. Dai, Z. R. Guo, S. K. Loh, C. S Teh and S. Redkar "Application of Contact Level Ion Beam Induced Passive Voltage Contrast in Failure Analysis of Static Random Access Memory", IPFA Proceedings, P. 103 (2001).

[3] C. C. Wu, J. C. Lee, J. H. Chuang, T. T. Li, "Single Device Characterization by Nano-probing to Identify Failure Root Cause", ISTFA Proceedings, P. 183-185, (2005).

[4] J. C. Lee and J. H. Chuang, "Fault Localization in Contact Level by Using Conductive Atomic Force Microscopy", ISTFA proceedings, P. 413-418 (2003).

[5] S.B. Herschbein, L. S. Fischer and A. D. Shore, "Basic Technology and Practical Applications of Focused Ion Beam for the Laboratory Workplace", Microelectronic Failure Analysis, Desk Reference 4th Edition, P. 517-526.

Front End Defects on Deep Submicron Devices

S. P. Neo, S. K. Loh, Z. G. Song and S. P. Zhao, *Snr Member, IEEE*

Chartered Semiconductor Mfg Ltd

60 Woodlands Industrial Park D, Street 2, Singapore 738406

Email: neosp@charteredsemi.com

Abstract **Front end defects are usually more intricate as compared to back end defects, and as technology scale down into deep submicron regime, failure analysis of the front end defect is becoming even more challenging due to the increase in complexity of the process. In this paper, failure analysis on three types of front-end defect has been discussed. These defects are cobalt silicide at poly sidewall causing active to poly bridging, amorphous layer under contact and broken silicide on poly line, which were observed on 90nm SOI wafers.**

I. INTRODUCTION

SEMICONDUCTOR device processes are divided into two portions, namely front end of line starting with silicon substrate and ending with poly–cap nitride or etch-stop-layer (ESL) nitride deposition, and back end of line starting with interlayer dielectric deposition and ending with bond-pad opening. As modern integrated circuit progresses into deep submicron regime, semiconductor devices are fabricated in higher density and smaller feature sizes. This makes failure analysis more difficult and more challenging than before, especially for the analysis of the defects related to front-end processes as the front-end processes are becoming more complicated and involved in multiple ultra-thin layers [1]. In this paper, failure analysis on three types of front-end defect will be discussed. These defects are cobalt silicide at poly sidewall causing active to poly bridging, amorphous layer under contact and broken silicide on poly line, which are observed on 90nm SOI wafers.

II. RESULTS AND DISCUSSIONS

Case Study (I)

A failure mode is column failure with a single bit passed. Top-down de-processing and SEM inspection observed cobalt silicide extending from active area to poly line and causing bridging between the pass-gate poly to the bit or bit-bar line contact (see Fig. 1).

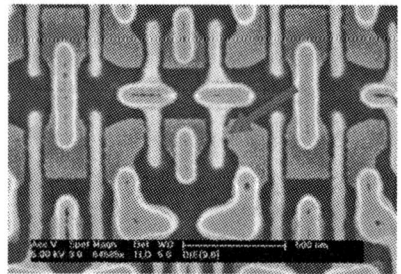

Fig.1 Top down SEM showing the CoSi extending from active area to poly line.

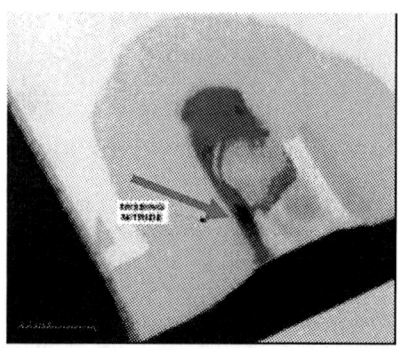

Fig. 2 XTEM showing the missing spacer resulting in CoSi bridged to the active

However, the root cause for the cobalt silicide extension was undetermined by just top-down de-processing method. In order to understand and identify the root cause, a sample was de-layered to contact level with the interlayer dielectric intact and was submitted to XTEM analysis on the suspected defective area. XTEM analysis clearly showed that the cobalt silicide grew at the poly sidewall, due to missing poly spacer at that area.

0-7803-9730-4/06/$25.00 ©2006 IEEE

(see Fig. 2). The root cause was finally identified to be time-link issue, from spacer etch to O3 clean. This issue was resolved by controlling the time from spacer etch to O3 clean.

Case Study (II)

The second failure mode is double bit failure in column (DBC). From layout analysis, it was believed that the double bit failure in column was most likely due to high resistance or open failure of the stacked bit/bit-bar line contact & via1, Vss contact and Vdd contact shared by the two bits. Thus, an unit was de-processed to via 1 and XTEM analysis was performed on the stacked bit/bit-bar line contact & via1, Vss contact and Vdd contact shared by the two bits. XTEM analysis revealed an amorphous layer at the contact barrier metal and CoSi interface of bit/bit-bar line contact. (see Fig. 3). This amorphous layer was believed to be certain organic residue after contact etch and before barrier metal deposition. The issue was resolved by increase pre-sputter, prior to contact barrier metal deposition.

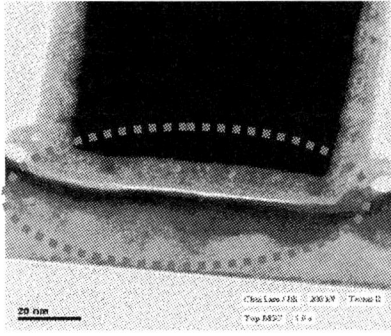

Fig.3 XTEM showing the amorphous layer underneath the contact.

Case Study (III)

The third failure mode is single bit failure. One unit was de-processed to contact level and followed by BOE to expose the active and poly layers with the etch-stop-layer (ESL) nitride intact. SEM inspection at the single bit failure locations showed a defect like "poly broken" (see Fig. 4).

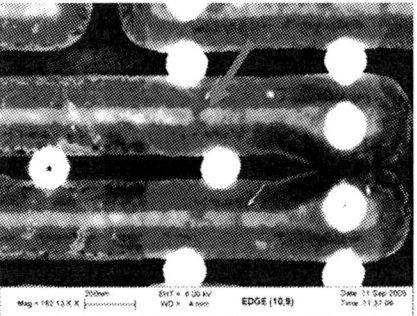

Fig. 4 Top down SEM picture showing the "poly-broken-like" defect.

After the ESL nitride was removed using reactive ion etcher, the "poly-broken-like" defect was more clearly seen (see Fig. 5).

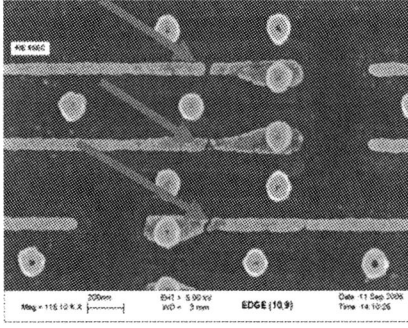

Fig. 5 Top down SEM showing the "poly-broken-like" defect

To verify whether poly was broken or just cobalt silicide discontinuity, XTEM analysis was performed on another unit and it was verified that cobalt silicide was discontinuous, while the poly line showed no abnormality. (see Fig. 6 and 7).

Fig. 6 XTEM SEM showing the discontinuity in the cobalt silicide

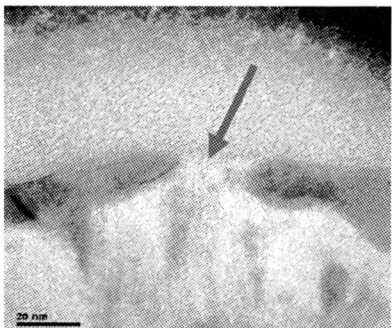

Fig. 7 The discontinuity in the cobalt silicide in high magnification.

To understand why this issue happened at wafer edge only, a center unit was de-processed to poly level (see Fig. 8) and the poly CD was measured. After comparison with the poly CD at wafer edge, it was found that poly CD at wafer edge is smaller than that at wafer center. Narrow poly CD may influence cobalt silicide growth.

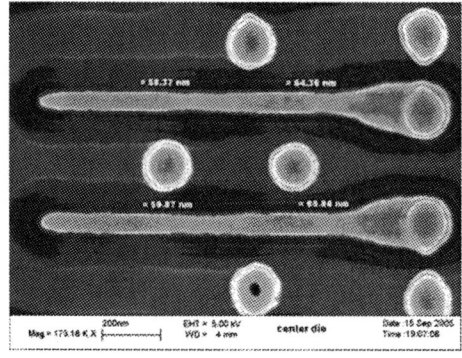

Fig. 8 Top down SEM showing the poly line from the centre of the wafer

III. CONCLUSION

In this paper, three front end defects: cobalt silicide at poly sidewall causing active to poly bridging, amorphous layer under contact and broken silicide on poly line were analyzed and discussed for the column failure with one bit pass, pair bit failure and single bit failure for 90nm SOI devices. Analysis of these defects implied that as technology shrink to deep submicron regime, process control became even more critical.

ACKNOWLEDGEMENT

The authors would like to thank the TEM team for providing the XTEM pictures and their great support.

REFERENCES

[1] Z. G. Song, J. Y. Dai, S. Ansari, C. K. Oh and S. Redkar, "Front-end Processing Defect Localization by Contact-level Passive Voltage Contrast Technique and Root Cause Analysis", IPFA'2002 Proceedings, P. 97 (2002).

Failure Analysis of A Unique Poly Defect

S.P. Neo*, Z.G. Song, C.K Oh and S. P. Zhao, *Snr Member, IEEE*
Chartered Semiconductor Mfg Ltd
60 Woodlands Industrial Park D, Street 2, Singapore 738406
Email: neosp@charteredsemi.com

Abstract **To successfully identify the root cause of a failure, it is required to have systematic failure analysis approaches with the right techniques and to make detailed observation during the course of analysis. In this paper, detailed failure analysis has been present to identify the root cause of a unique poly defect through taking the right approaches and techniques.**

I. INTRODUCTION

FAILURE analysis of semiconductor devices is the process to identify the root cause of failure and feedback the information to the manufacturing line to change or improve process. Failure analysis process begins when a device has lost its basic functions and includes investigating the failure mode and mechanism using electrical, physical and chemical analysis techniques. As semiconductor technology develops into deep sub-micro regime, the feature size of semiconductor devices continues to shrink and thus the process gets more complicated, especially for the front end process, as more ultra-thin layers are involved. Failure analysis is thus becoming more and more challenging. Successful failure analysis involves applying the right approach, technique and making detailed observations during the failure analysis. In this paper, failure analysis of a unique poly defect was discussed.

II RESULTS AND DISCUSSION

The failure mode was single bit failure at segment edge (last column). The typical wafer map was shown in Fig. 1. Since the single bit failure was at segment edge so it was regarded as a pair bit failure, hence XTEM analysis was performed on the stacked poly/local interconnect/contact/metal 1 of the pass-gate node of the failing location at the last column. It was observed that the poly was etched away and replaced by tungsten (W), and the local interconnect was poorly filled by W (see Figs. 2, 3 & 4). This was a unique defect, which was not being observed before.

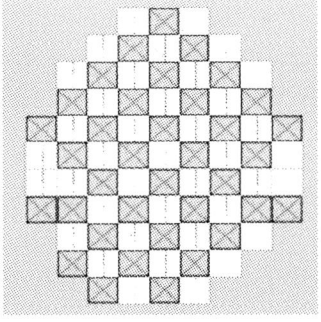

Fig. 1 Picture of the wafer map

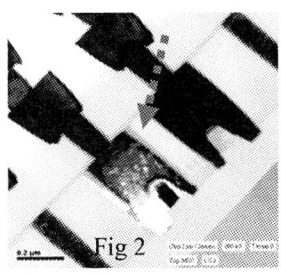

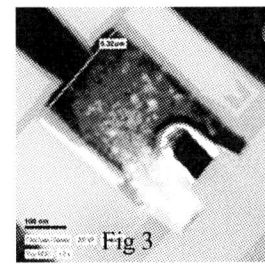

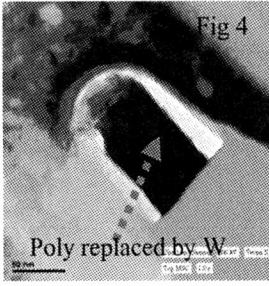

Figs. 2-4 XTEM pictures showing the poly being etched away and replaced by tungsten and the

A good local interconnect with good poly was shown in Fig. 5 for comparison purpose.

0-7803-9730-4/06/$25.00 ©2006 IEEE

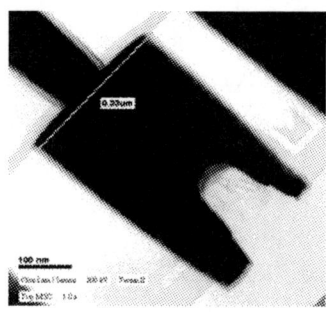

Fig. 5 XTEM picture showing the good interconnect

Other observations from the XTEM analysis included: (1) the opening of local interconnect etch is normal; (2) the barrier metal for local interconnect was much thicker than normal (see Fig. 6); (3) Correspondingly, the local interconnect W CD was smaller than normal.

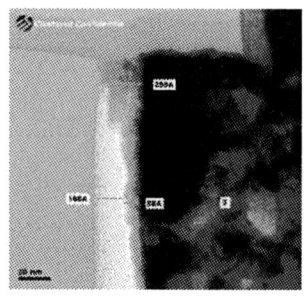

Fig. 6 XTEM picture showing the thicker barrier

EDX analysis performed on the thicker barrier metal showed presence of oxygen (see Fig. 7) in the barrier metal and this was abnormal.

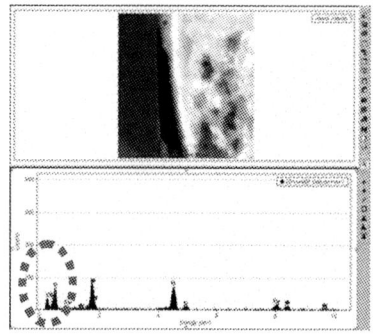

Fig. 7 EDX analysis showing presences of oxygen in the barrier of the bad local interconnect.

For comparison, EDX on the barrier metal of the good local interconnect was performed too. No oxygen was detected (see Fig. 8).

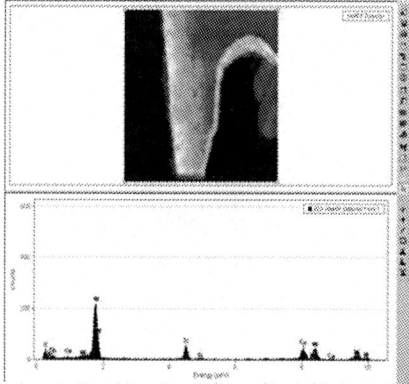

Fig. 8 EDX analysis do not detect any oxygen in the barrier of the good local interconnect.

Oxidation of the barrier metal, resulting in thicker barrier metal, must have taken place, but what could have caused this and what could have caused the replacement of Si by W in the poly was still unclear from the XTEM analysis. Further analysis need to be done. Top down de-layering was performed, smaller local interconnect at the failing location with Ti extrusion into silicon nitride (SiN) gapfill void was observed (see Fig. 9).

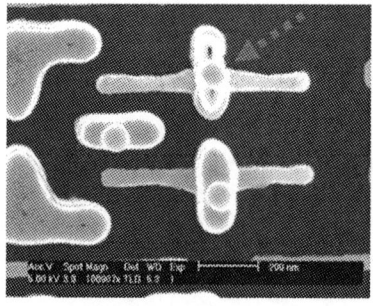

Fig. 9 Top down SEM showing the showing the smaller local interconnect with Ti extrusion into the gapfill

Plane view TEM performed on another sample showed poly was totally replaced by W and SiN gapfill void was observed (see Figs. 10 & 11). EELS mapping on the barrier metal of the bad local interconnect further confirmed presence of oxygen in the barrier metal (see Fig. 12). The presence of SiN gap fill void observed in the plane view TEM and the top down SEM could have caused out-gassing, which resulted in the oxidation of the barrier metal to TixOy, subsequently affected W fill. The replacement of W in the poly is due to the WF_6 attacking poly during W deposition for local interconnect [1]. Thus through applying the right approach, technique and making detailed observations, the root cause of the unique poly defect was understood and issue was resolved eventually.

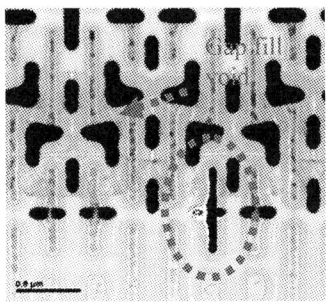

Fig. 10 Plane view TEM showing the poly completely replaced by tungsten and gap fill void observed

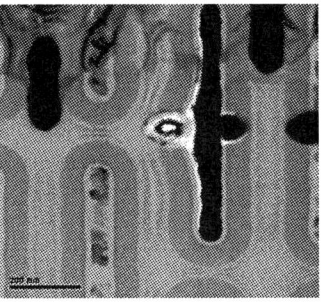

Fig. 11 High magnification of the poly with the Si replaced by W

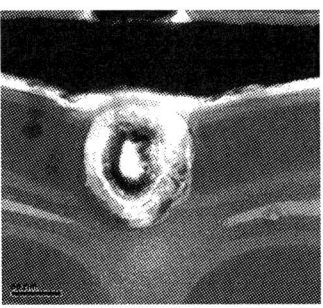

Fig. 12 EELS map showing Oxygen in barrier metal

III. CONCLUSION

In this paper, failure analysis of a unique poly defect was discussed. It showed that to successfully identified the root cause of a failure, systematic analysis with the application of right technique and detailed observation is required.

ACKNOWLEDGEMENT

The authors would like to thank the Chartered TEM team for providing the TEM pictures and the great support.

REFERENCES

[1] Z.G. Song, G.B. Ang and H. Y. Li, etc. "Electrical Faults Captured by In-line E-beam Inspection and Failure Analysis", ISTFA' 2002 Proceedings. P. 355 (2002).

Trenched MOSFET V_{gs} Uniformity Improvement through Furnace Loading Procedure

H.S Ng, A.F Yee, Christopher Loo, Y.K Ng, W.B Sheu

X-FAB Sarawak Sdn. Bhd.

1 Silicon Drive, Sama Jaya Free Industrial Zone

93350 Kuching, Sarawak, Malaysia

Email: hongseng.ng@xfab.com

Phone: +60.82.354888

Abstract — In this paper, we presented a new furnace loading procedure with specially-prepared monitor wafer (SPMW) to prevent scrapping wafers placed at top slot of the furnace boat and other slots if lot has < 25 wafers. These wafers showed V_{gs} OOS on high side. Our investigation showed that non-uniform V_{gs} behavior is due to inconsistent phosphorus atoms diffused across furnace boat. The phosphorus outgassing occurs during P-body Anneal from the n-doped poly film of the wafer backside to the wafer beneath. Conventional furnace boat layout consists of oxide wafers at top and bottom slots of the boat designated as Side Dummy wafers (SD). If <25 wafers per lot, Extra Dummy wafers (ED) will be inserted at slots originally assigned to production wafers. New layout packs all wafers continuously without ED in between with additional SPMW just below SD. The wafer scrap yield was improved by at least 1% for Trenched DMOS (Double Diffused Power MOSFET).

I. INTRODUCTION

In trenched DMOS [1] [2], transistor region is formed by diffusion. The gate electrode, n-doped polysilicon is located in trench. The sidewall and bottom of trench are insulated with silicon dioxide. Plenty of trenched cells are patterned per die sharing common source, gate and drain. One of the important parameters is V_{gs} (Gate Threshold Voltage). V_{gs} is the potential required from gate to source to produce a vertical channel along trench at a specified drain current, Id, e.g. at 250uA [3]. It depends significantly on boron implant dose and gate oxide thickness at sidewall of trench.

V_{gs} OOS, out of specification on high side was encountered for W5 (5th wafer or slot 5 in

cassette as illustrated in Fig. 1). The high V_{gs} issue was observed even in other slots if the lot has less than 25 wafers. From baseline data analysis according to critical equipments loading sequence and wafer map, W5 was placed at top slot of the furnace boat just below SD during P-body Anneal, where SD is a simple blanket oxide wafer. P-body Anneal is an N_2-based annealing process to activate dopant, anneal out preceding p-body implant damage and drive the boron atoms to the desired concentration profile. Dry oxide is grown at the same temperature before N_2 anneal. Further experiments verified the root cause that V_{gs} OOS is due to the counter-doping of n-type dopant to p-body region. This effect has been eliminated by the new furnace loading procedure with SPMW to provide consistent phosphorus outgassing to all wafers including W5 as well as when lot has <25.

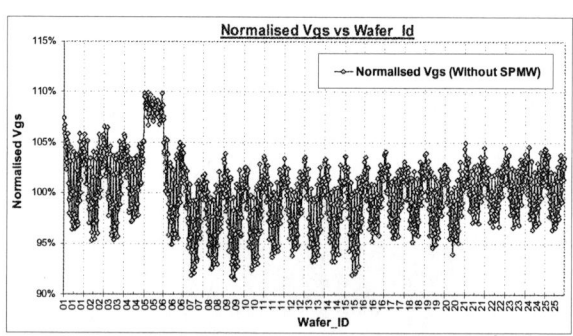

Fig. 1 Normalized V_{gs} versus cassette slot (25 wafers per cassette) for conventional furnace layout without SPMW implementation.

II. EXPERIMENT

Three major experiments were conducted to identify root cause and verify solution. In the first experiment, two new 200mm p-type polished Si wafers were processed in furnace using P-body Anneal recipe. One SPMW, but without annealing, was placed on top of 1st test wafer in the furnace boat, as depicted in Fig. 2. The SPMW is an oxide wafer deposited with phosphorus-doped poly layer in LPCVD furnace. The purpose is to verify any phosphorus outgassing from wafer backside. 2nd test wafer was inserted just below phosphorus-free oxide wafer as reference. The oxide layer grown on test wafers during P-body Anneal process was stripped in HF and SC1 wet bench in order to expose Si surface. The R_S sheet resistivity on test wafers was measured by four-point probe method.

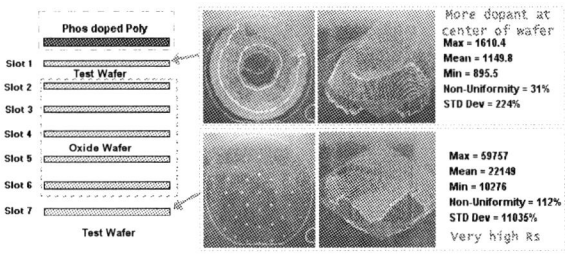

Fig. 2 The wafer layout in the boat during P-body Anneal (left) and R_S maps of bare-Si wafers (right).

2nd experiment was quite similar to 1st experiment, with 2 SPMW wafers used. The same SPMW wafers were re-used two to three times to find out their lifetime. Three runs were conducted using P-body Anneal recipe and wafers were loaded by conventional furnace layout.

The 3rd experiment verified the electrical performance using SPMW with and without anneal. This experiment was conducted using actual DMOS production wafers and process flow with new wafer loading procedure during P-body anneal. The stability and repeatability of electrical performance and manufacturability of this method was confirmed and monitored before full implementation.

Fig. 9 illustrates the transfer of wafers from cassette to furnace boat. The example includes the occasion when one cassette has less than 25 wafers. The boat layout was based on five wafers per move or otherwise one wafer per move if empty slot was found in between.

III. RESULTS AND DISCUSSIONS

A. *PHOSPHORUS OUTGASSING*

The 1st experiment has proven that phosphorus outgassed from n-doped poly

deposited at wafer backside. This in-house experiment has given us good judgment at a low cost, fast and convenient way although it is not a precise method to quantify the exact phosphorus concentration such as SIMS etc. From the 49-point R_S measurement, Fig. 2 (right), wafer edge of 1st test wafer showed higher R_S than center area. Center of wafer contained higher phosphorus concentration, while phosphorus atoms diffused away at wafer edge due to tube design. Reference wafer showed much higher R_s to support the explanation.

B. *LIFETIME OF SPMW*

This experiment demonstrates that the amount of phosphorus out gassing from SPMW reduced each time after annealing. The oxide layer grown during P-body anneal further slows down the diffusion of phosphorus atoms. Fig 3 shows the data measured using 4-point probe. R1-B, R2-B, R3-B refers to the test wafers each run for 1st, 2nd and 3rd times usage of identical SPMW placed above. The R_s increased after each runs. As a result, it was decided to use the SPMW only one time to keep the good run-to-run dopant uniformity.

Run One Boat Layout R1	Run Two Boat Layout R2	Run Three Boat Layout R3
SD SiO2	SD SiO2	SD SiO2
SD SiO2	Same SPMW to evaluate lifetime	SD SiO2
SD SiO2	SD SiO2	SD SiO2
SD SiO2	SD SiO2	SD SiO2
SD SiO2	SD SiO2	SD SiO2
SD SiO2	SD SiO2	SD SiO2
Bare Si (ref)	Bare Si (ref)	Bare Si (ref)
SPMW-B	Used SPMW-B fr R1	Used SPMW-B fr R2
Bare Si (R1-B)	Bare Si (R2-B)	Bare Si (R3-B)
ED SiO2	SPMW-C aft Anneal	Used SPMW-C fr R2
ED SiO2	Bare Si	Bare Si
R1-B R_s (ohm/sq)	R2-B R_s (ohm/sq)	R3-B R_s (ohm/sq)
896	1230	6456

Fig. 3 SPMW lifetime experiment by measuring R_S of blanket wafer beneath using 4-point probe method.

C. V_{GS} *PERFORMANCE*

Fig 1 and Fig 4 was plotted using same data, except in Fig 4, the x-axis was rearranged according to boat layout instead of cassette slot. The 1st wafer exhibits high V_{gs} but better within-wafer-uniformity. Fig.5 showed 1st wafer suffered from very low V_{gs} and poor uniformity. Fig.6 plots out comparable V_{gs} obtained using SPMW. Fig. 7a and 7b illustrate the mechanism affecting V_{gs}.

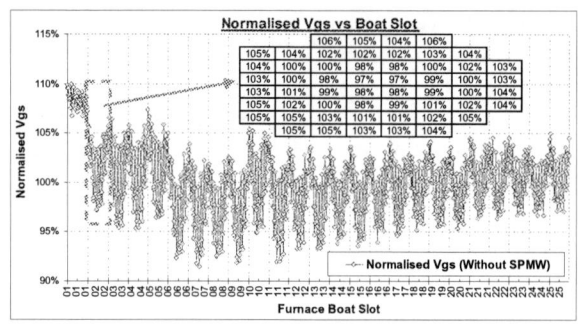

Fig. 4 Normalized V_{gs} versus furnace boat slot for conventional furnace layout without SPMW implementation.

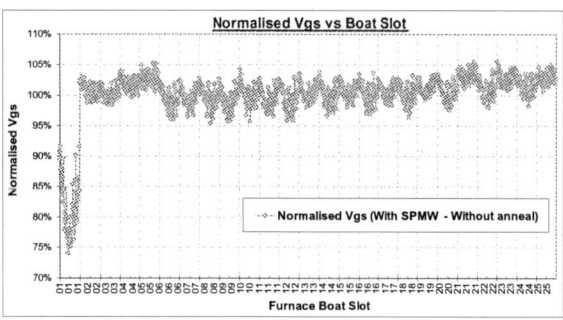

Fig. 5 Normalized V_{gs} versus furnace boat slot for new furnace layout with SPMW, without low temperature poly anneal.

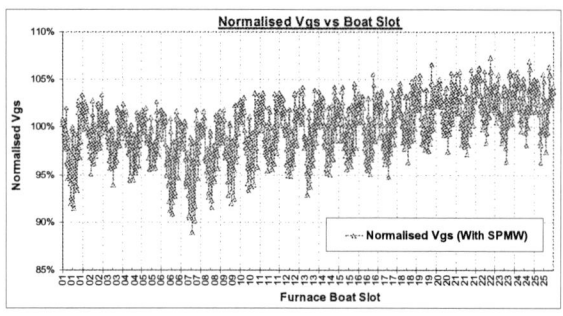

Fig. 6 Normalized V_{gs} versus furnace boat slot for new furnace layout with SPMW, with low temperature poly anneal.

In trenched DMOS, the p-body (P) and source (N) region are separated by trench. Fig. 8 shows typical NPN trenched DMOS structure. The trenches are filled with n-doped poly as gate by means of LPCVD process with SiH_4 and PH_3 gases in TEL vertical furnace. Both sides of the wafers are coated with n-doped poly film. Subsequent processes remove the top poly layer, leaving a top oxide layer. After the p-body boron implant, the whole lot of wafers is processed in TEL atmospheric vertical furnace for P-body anneal. During the process, phosphorus atoms from n-doped poly film at the backside of upper

wafer outgas to wafer below. The phosphorus atoms further penetrate through the oxide, diffuse into the source area and counter-dope the P-body region. The higher boron implant dose, the more difficult to build up channel, thus resulted in higher V_{gs}. With counter doping from phosphorus, this n-type dopant makes the formation of n-channel easier, hence lower V_{gs}. Besides, the total amount of phosphorus atoms reside in the wafer edge is lower than wafer center due to furnace gas flow dynamic. As a result, the Vgs uniformity is poorer with wafer center exhibiting lower value.

Fig. 7b clarifies the idea that high V_{gs} is because of lack of phosphorus out-gassing occurred between SD and production wafer beneath. The subsequent production wafers are affected by backside poly film of upper wafers. Fig. 5 shows that the 1st wafer was over-compensated caused by excessive phosphorus out-gassing from SPMW without poly anneal. Fig. 6 exhibits comparable V_{gs} value using SPMW with anneal mirrored actual DMOS process flow. Fig. 7b (left) presents the SPMW film stack and mechanism.

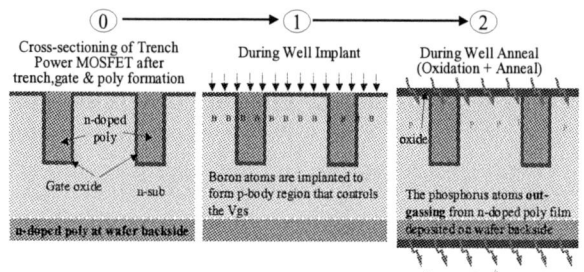

Fig. 7a Process Sequence with cross-sectioning view illustrates the mechanism of phosphorus outgassing from wafer backside.

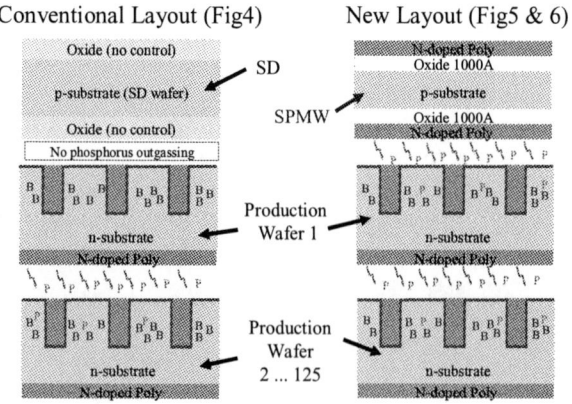

Fig. 7b Conventional layout with SD wafers versus new layout with SPMW (with or without annealing) illustrates the outgassing during P-body anneal. No

outgassing for conventional layout to 1st wafer. (Partial layout, refer to Fig 9 for full layout)

The full layout is drawn in Fig.9. The new layout allows the use of SPMW, which will be inserted to the designated slot every run. In additional to the change of one SD allocation to SPMW, the wafer loading sequence has been modified. The original wafer charging rule defined that ED must be charged to fill the empty slot if lot has >25 wafers following cassette map. The new layout directs the robot to shift up and pack the wafers without inserting ED in between. The ED is filled at the bottom of the boat, which does not affect any production wafers.

IV. CONCLUSION

From the experiments conducted including electrical parameters verification, we concluded that the phosphorus out gasses from the n-doped poly film. New furnace loading procedure with specially-prepared monitor wafer (SPMW) could prevent scrapping wafers placed at top slot of the furnace boat and other slots (if lot has < 25 wafers), which showed high Vgs. The wafer scrap yield was improved by at least 1% for Trenched DMOS. Although the process-induced phosphorus outgassing causes poorer V_{gs} within wafer uniformity, this method improves the V_{gs} wafer-to-wafer uniformity.

ACKNOWLEDGEMENT

The authors wish to thank 1st Silicon (Malaysia) Sdn. Bhd. for the support and funding of this study and M-MOS Semiconductor Sdn Bhd for their technology transfer.

REFERENCES

[1] F.I. Hshieh, "Trench MOSFET with Ultra High Cell Density", NSM 2005 Proc., 135-137 (2005).
[2] K.S. Oh, "MOSFET Basics (AN-9010)", Fairchild Semiconductor 2000.
[3] Supertex, "Low-Threshold MOSFETs: Structure, Performance and Applications (AN-D2)", Supertex Inc. 2001.

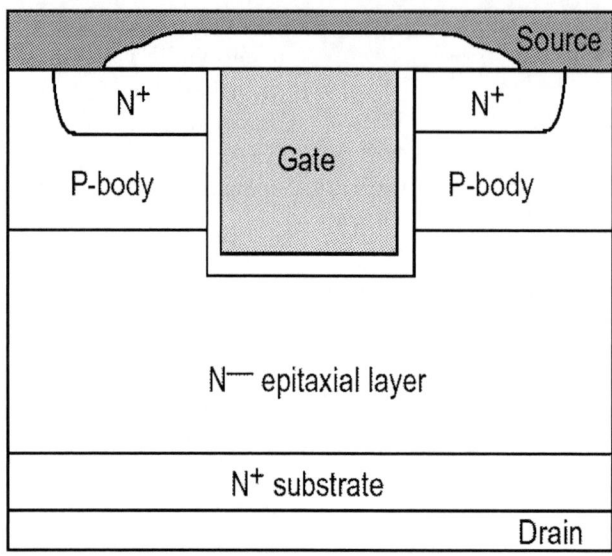

Fig.8 Typical trenched double diffused MOSFET (DMOS) [2]. 1st Silicon uses similar type of structure.

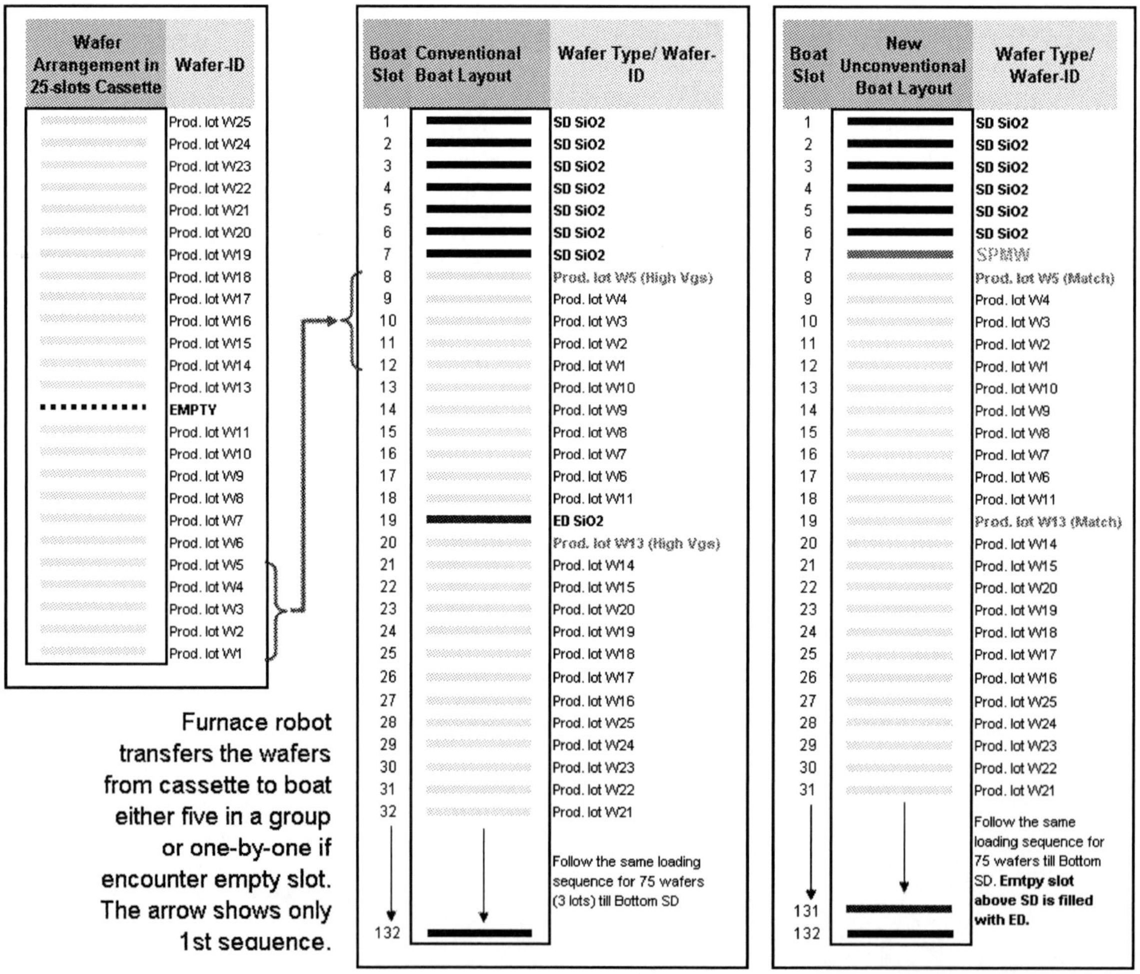

Wafer Arrangement in 25-slots Cassette	Wafer-ID
	Prod. lot W25
	Prod. lot W24
	Prod. lot W23
	Prod. lot W22
	Prod. lot W21
	Prod. lot W20
	Prod. lot W19
	Prod. lot W18
	Prod. lot W17
	Prod. lot W16
	Prod. lot W15
	Prod. lot W14
	Prod. lot W13
	EMPTY
	Prod. lot W11
	Prod. lot W10
	Prod. lot W9
	Prod. lot W8
	Prod. lot W7
	Prod. lot W6
	Prod. lot W5
	Prod. lot W4
	Prod. lot W3
	Prod. lot W2
	Prod. lot W1

Furnace robot transfers the wafers from cassette to boat either five in a group or one-by-one if encounter empty slot. The arrow shows only 1st sequence.

Boat Slot	Conventional Boat Layout	Wafer Type/ Wafer-ID
1		SD SiO2
2		SD SiO2
3		SD SiO2
4		SD SiO2
5		SD SiO2
6		SD SiO2
7		SD SiO2
8		Prod. lot W5 (High Vgs)
9		Prod. lot W4
10		Prod. lot W3
11		Prod. lot W2
12		Prod. lot W1
13		Prod. lot W10
14		Prod. lot W9
15		Prod. lot W8
16		Prod. lot W7
17		Prod. lot W6
18		Prod. lot W11
19		ED SiO2
20		Prod. lot W13 (High Vgs)
21		Prod. lot W14
22		Prod. lot W15
23		Prod. lot W20
24		Prod. lot W19
25		Prod. lot W18
26		Prod. lot W17
27		Prod. lot W16
28		Prod. lot W25
29		Prod. lot W24
30		Prod. lot W23
31		Prod. lot W22
32		Prod. lot W21
132		

Follow the same loading sequence for 75 wafers (3 lots) till Bottom SD

Boat Slot	New Unconventional Boat Layout	Wafer Type/ Wafer-ID
1		SD SiO2
2		SD SiO2
3		SD SiO2
4		SD SiO2
5		SD SiO2
6		SD SiO2
7		SPMW
8		Prod. lot W5 (Match)
9		Prod. lot W4
10		Prod. lot W3
11		Prod. lot W2
12		Prod. lot W1
13		Prod. lot W10
14		Prod. lot W9
15		Prod. lot W8
16		Prod. lot W7
17		Prod. lot W6
18		Prod. lot W11
19		Prod. lot W13 (Match)
20		Prod. lot W14
21		Prod. lot W15
22		Prod. lot W20
23		Prod. lot W19
24		Prod. lot W18
25		Prod. lot W17
26		Prod. lot W16
27		Prod. lot W25
28		Prod. lot W24
29		Prod. lot W23
30		Prod. lot W22
31		Prod. lot W21
131		
132		

Follow the same loading sequence for 75 wafers till Bottom SD. **Emtpy slot above SD is filled with ED.**

Fig.9 Wafers sitting arrangement in standard 25-slot cassette (left) and the boat layout for conventional (middle) and new (right) wafer loading procedure. The illustration shows the boat wafer map for top SD and one production lot only. The boat used for P-body Anneal can processed 4 lots per batch.

Simulation of Flash Memory Characteristics based on Discrete Nanoscale Silicon

C.Y. Ng*, *Student Member, IEEE*, J.I. Wong, *Student Member, IEEE*, M. Yang, *Student Member, IEEE*, C.L. New[1], and T.S. Khor[1], and T.P. Chen, *Member, IEEE*
School of Electrical and Electronic Engineering,
Nanyang Technological University, Singapore 639798
[1] Chartered Semiconductor Manufacturing Ltd., 60 Woodlands Industrial Park D,
Street 2, Singapore 738406
* Email: chiyung@ieee.org , echentp@ntu.edu.sg

Abstract **In this paper, we present a simulation study on the trapping properties of Flash memory device based on discrete nanoscale silicon embedded in silicon-dioxide (SiO$_2$). Taurus Suprem-4 and Taurus Medici are being used to carry out the simulations. The memory structure with a tunnel oxide of 3, 5 and 9 nm and a control oxide of 10, 20 and 40 nm have been simulated, respectively. The discrete nanoscale silicon with the size of 20 nm × 20 nm, 10 nm × 10 nm, and 5 nm × 5 nm have also been simulated, respectively.**

I. INTRODUCTION

The concept of charge storage based on discrete nodes like the semiconductor nanocrystals embedded in the gate dielectric has drawn large attention for the past few years due to the possibilities of replacing the conventional floating-gate non-volatile memory devices [1,2]. Various nanocrystal materials (i.e., silicon, germanium, and metal) and fabrication techniques have been explored in the last few years [3,4]. To utilize the silicon annocrystal (nc-Si) as charge storage node, understanding the dynamic behavior of programming and erasing in the nc-Si is important. Many characterization works have been carried out to understand the memory behavior of such devices [4,5]. Less work has been carried out to simulate the memory structure consisting of discrete nanoscale silicon or nc-Si as charge storage node and to predict the memory behavior with simulation methods.

In this paper, we present a simulation work on the trapping properties of memory device based on discrete nanoscale silicon as charge storage node. As a comparison, a layer of floating-gate silicon memory device is also being simulated. The threshold voltage (V_{th}) shift as a function of programming voltage and programming time using channel hot electron (CHE) programming method are explored. The influence of control oxide thicknesses and tunnel oxide thicknesses on the memory behavior is being studied. Besides, the threshold voltage shift due to the programming of different dimension of discrete nanoscale silicon is also being simulated.

II. SIMULATION MODEL

Taurus Supreme-4 is being used to simulate the fabrication process of the Flash memory device with a layer of floating-gate and discrete nanoscale silicon embedded in the SiO$_2$ matrix, respectively. Taurus Medici is being used to simulate the memory behaviors. Medici solves three partial differential equations self consistently for electrostatic potential, and the electron and hole concentrations using Poisson equation and continuity equation. Parallel and perpendicular field are considered to specify the mobility of electron in Medici. Channel hot electron programming is carried out with a positive bias applied to both the gate and drain terminals (Vg > 0, Vd > 0), grounding other two terminals (Vs = 0, Vb = 0). The device dimension and doping concentration are given in Table I.

Table I Device dimension and doping concentration

Parameter	Value
Channel length (L)	0.25 - 1 µm
Control oxide thickness	20 - 40 nm
Tunnel oxide thickness	3 - 10 nm
Nanoscale silicon thickness	5 nm
Substrate doping	1×10^{15} cm^{-3}
Source / drain doping	1×10^{20} cm^{-3}

0-7803-9730-4/06/$25.00 ©2006 IEEE

III. RESULTS AND DISCUSSION

Fig. 1a shows that the simulated Flash memory device with a layer of floating-gate (a) and discrete nanoscale silicon (b), sandwiched between two SiO_2 layers, respectively. The size of the discrete nanoscale silicon is 5 nm × 5 nm. As shown in this figure, the tunnel oxide is 3 nm and the control oxide is 10 nm for both memory devices. The transfer characteristics of the simulated Flash memory device with a layer of floating-gate and discrete nanoscale silicon are shown in Fig. 2. It is observed that there is a slight difference in the initial V_{th} between the two devices. This is mainly due to the work-function differences between the floating-gate and the discrete nanoscale silicon sandwiched between the control oxide and the tunnel oxide.

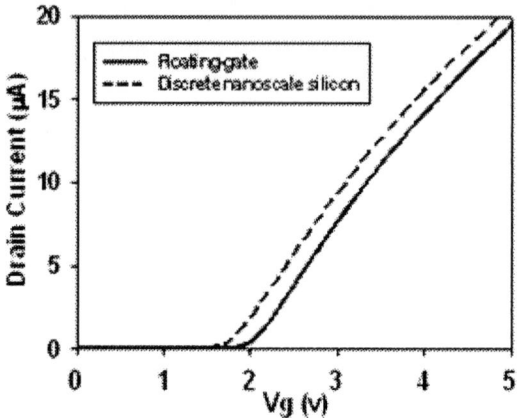

Fig. 2   Transfer characteristics of floating-gate and discrete nanoscale silicon memory devices at Vd = 0.1 V.

The effect of control oxide thicknesses on the trapping properties is simulated and is shown in Fig. 3. The charging was carried out with Vg = 12 V and Vd = 5 V, respectively. Larger threshold voltage shift is observed for thinner control oxide. At a shorter programming time, more electrons can easily trap into the nanoscale silicon when the control oxide thickness reduces. Besides, it is observed that there is a left shift to the step-like behavior when the control oxide thickness reduces. For example, the V_{th} saturates at 20 V in 10^{-4} s when the control oxide thickness reduces to 20 nm, compared to that at 10^{-3} s for the 40 nm control oxide. At the same programming time, thinner control oxide leads to larger V_{th} shift compare to that of a thicker control oxide.

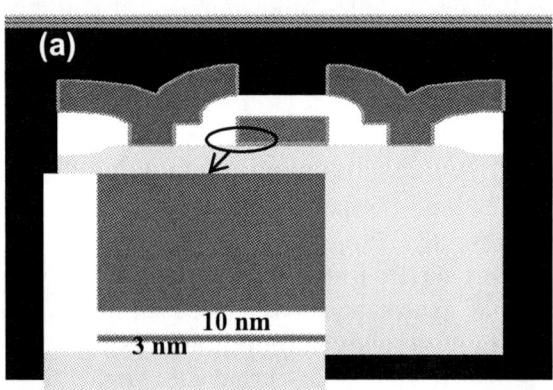

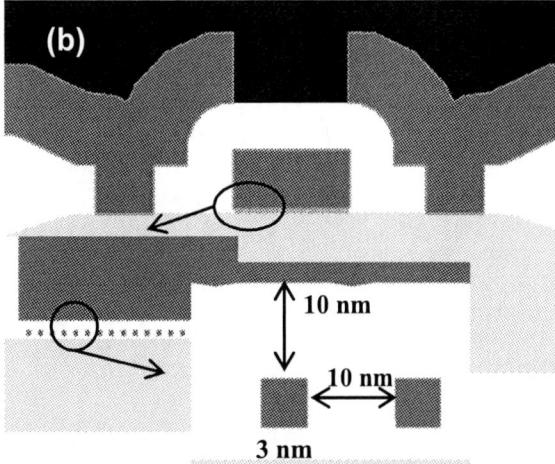

Fig. 1   Structure of the floating-gate (a) and discrete nanoscale silicon (b) memory devices simulated with Taurus Suprem-4.

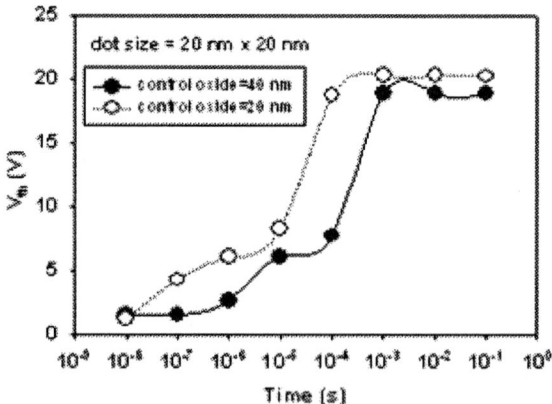

Fig. 3   Comparison of V_{th} shift for different control oxide thicknesses.

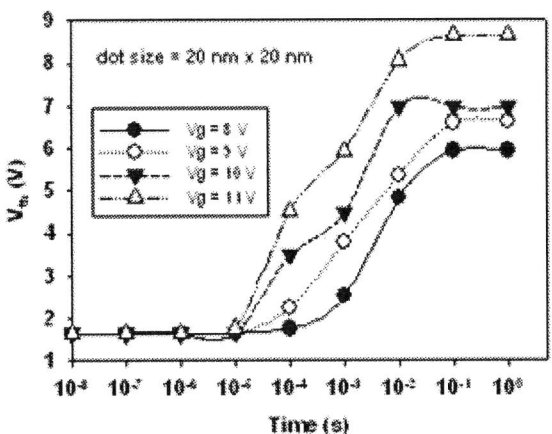

Fig. 4 V_{th} shift as a function of programming voltage. Programming transient was carried out with Vd = 5.5 V, Vs and Vb = 0 V.

Fig. 4 shows the threshold voltage shift as a function of programming voltages and programming time. The drain voltage is fixed at +5.5 V. When the gate voltage increases from +8 V to +9 V, the threshold voltage shift increases ~0.5 V at a programming time of 10^{-4} s. Larger threshold voltage shift can be observed at the time of 10^{-4} s when the programming voltage increases to +11 V. On the other hand, the step-like behaviors are observed for different programming voltages too. As shown in Fig. 4, saturation of V_{th} occurs at 10^{-1} s for both Vg = +8 and +9 V. When the Vg increases to +10 or +11 V, the saturation of V_{th} occurs at 10^{-2} s, which is one order earlier compared to Vg = +8 and +9 V. This shows that the step-like behavior which attributes to the parallel electric field that plays an important role during the trapping of charges in nanoscale silicon. When charge trapped into the discrete nanoscale silicon, the electric field at each of the nanoscale silicon will influence further charge trapping at itself as well as at the neighboring discrete nanoscale silicon. This effect will not be observed if the simulation is carried out without considering the parallel electric field among the discrete nanoscale silicon.

The effect of V_{th} shift as a function of programming time for different tunnel oxide thicknesses is shown in Fig. 5. The charging was carried out with Vg = 12 V and Vd = 5 V, respectively. Larger V_{th} shift is observed when the tunnel oxide thickness reduces. Besides, a step-like behavior is observed when the charging time increases from 10^{-4} s to 10^{-3} s and from 10^{-2} s to 10^{-1} s. The lateral electric field plays an

important role in the charge trapping process. When there are charges trapped in the nanoscale silicon, the induced lateral electric field can block further electron from trapping into the nanoscale silicon, thus reducing the amount of trapped electrons.

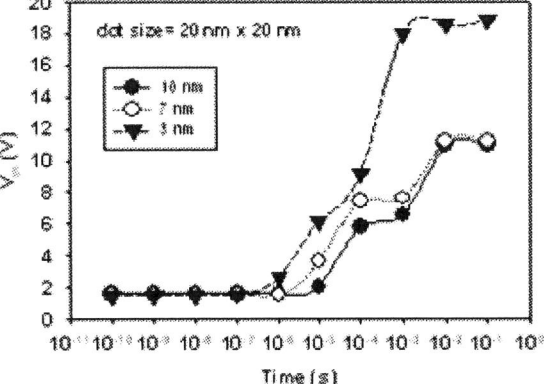

Fig. 5   Comparison of V_{th} shift for different tunnel oxide thicknesses.

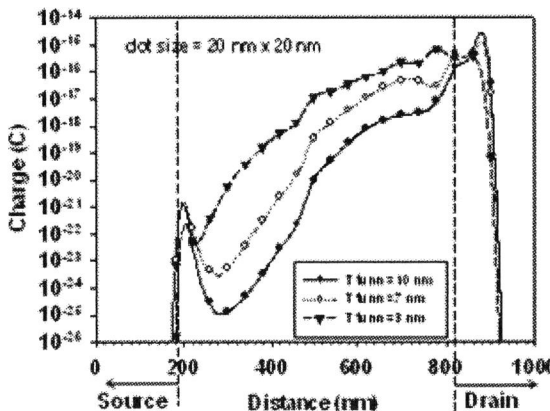

Fig. 6   Charge distribution with different tunnel oxide thicknesses.

The charge distribution after the charging effect as a function of tunnel oxide thicknesses is shown in Fig. 6. The charging is carried out with Vg = 12 V, Vd = 5.5 V, grounding both the source and bulk terminals. The amount of charges is higher near the drain region since CHE programming is applied during in this study. It is observed that there are lateral charge diffusions to the neighboring discrete nanoscale silicon. When the tunnel oxide thickness reduces, the lateral charge diffusions become significant.

It will be interesting to examine the trapping properties when the dimension of the discrete nanoscale silicon is being reduced. Fig. 7 shows the threshold voltage shift as a function of programming voltages when the dimension of the discrete nanoscale silicon reduces to 10 nm × 10 nm. As the Vg increases to +11 V, saturation of charge trapping occurs at 10^{-4} and at 10^{-2} s respectively. In contrast to the discrete nanoscale silicon with a dimension of 20 nm × 20 nm, saturation of charge trapping occurs only one time. This shows that the parallel electric field effect becomes dominant when the dimension of the discrete nanoscale silicon reduces.

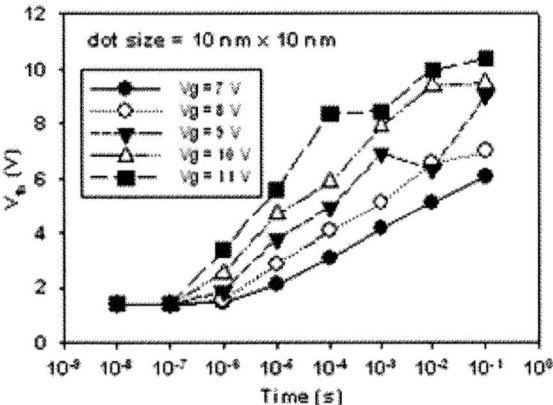

Fig. 7 V_{th} shift as a function of programming voltages for dimension of 10 nm × 10 nm. Programming transient is carried out at Vd = 5.5 V, Vs and Vb = 0 V.

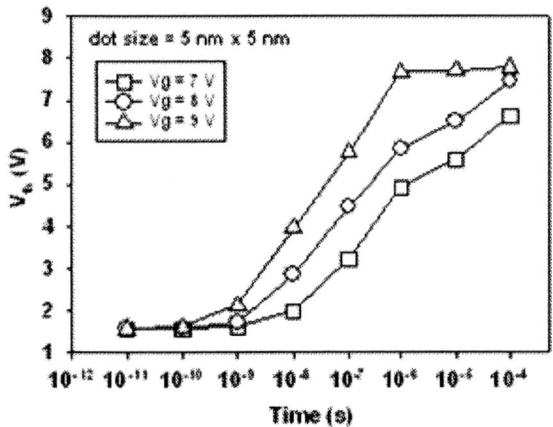

Fig. 8 V_{th} shift as a function of programming voltages for dimension of 5 nm × 5 nm. Programming transient is carried out at Vd = 5.5 V, Vs and Vb = 0 V.

On the other hand, it is observed that the charge trapping can occur at relatively shorter time when the dimension of the discrete nanoscale silicon reduces from 20 nm × 20 nm to 10 nm × 10 nm. At Vg = 11 V for 10^{-6} s, there is ~2.8 V shift in V_{th} for the 10 nm × 10 nm Flash memory compared to ~0.2 V shift in the V_{th} for the 20 nm × 20 nm Flash memory. The

difference is mainly due to the reduction of thickness of the discrete nanoscale silicon layer. Similarly as shown in Fig. 8, when the discrete nanoscale silicon dimensions reduced to 5 nm × 5 nm, charge trapping becomes more easily. Thus, the V_{th} shift becomes larger. For example, at Vg = +9 V for 10-5 s, the V_{th} shift for the 10 nm × 10 nm memory device is ~2 V. In contrast, the V_{th} shift for the 5 nm × 5 nm memory devices is ~4.4 V, which is two times larger than that of a 10 nm × 10 nm memory devices. This shows that larger V_{th} shift in the discrete nanoscale silicon with smaller dimension compared to that of a bigger dimension.

IV. CONCLUSION

Flash memory devices based on discrete nanoscale silicon with different dimensions are being simulated. It is observed that thinner the control oxide or tunnel oxide promotes easiness in charging up the discrete nanoscale silicon, thus producing bigger threshold voltage shift. On the other hand, it is observed that larger threshold voltage shift in the discrete nanoscale silicon for smaller dimension compared to that of the bigger dimension.

ACKNOWLEDGEMENT

This work has been financially supported by the Ministry of Education Singapore under project No. ARC 01/04, and also supported by Nanyang Technological University under project No. RG 99/05.

REFERENCES

[1]   C.Y. Ng et. al., *IEEE Trans. Electron Devices*, vol. 53, pp. 663 (2006).

[2]   C.Y. Ng et. al., *IEEE Electron Dev. Lett.*, vol. 27, pp. 231 (2006).

[3]   R.A. Rao et. al., *Solid-State Electronics*, vol. 48, pp. 1463 (2004).

[4]   C.Y. Ng et. al., *Appl. Phys. Lett.*, vol. 88, pp. 063103 (2006).

[5]   S. Tiwari et. al., *IEDM 95*, pp. 521 (1995).

Realisation of a Differential Multiplier-Divider based on Current Feedback Amplifiers

[1]Nihit Bajaj, *Student Member, IEEE* and [2]Jivesh Govil, *Student Member, IEEE*
[1]Netaji Subhas Institute of Technology, University of Delhi, Dwarka, New Delhi, INDIA 110045
[2]University of Michigan, 1301 Beal Ave, Ann Arbor, MI USA 48109
Email: nihitbajaj85@gmail.com, jivesh@umich.edu

Abstract - **Architecture for differential analog multiplier-cum-divider circuits using current feedback amplifiers (CFAs) has been proposed in this paper. The proposed structure consists of five CFAs, six matched MOS transistors, all biased in the triode region, and four identical capacitors. The final expression for the output voltage of the derived circuit does not contain any constant multiplier term, thus making it independent of CMOS process technology used and of the common capacitor value chosen. Exploiting the advantages of CFAs and differential inputs, the circuit renders high transfer function accuracy, high noise immunity and wide bandwidth characteristics. The theoretical circuit analysis along with the experimental study validates the proposed circuit operation.**

I. INTRODUCTION

Analog Multipliers and Dividers have always been an important class of circuits, and with the current progress in analog signal processing, telecommunications, instrumentation, and electronic systems, the range of their applications has increased manifold. Most of the analog continuous-time and sampled-data multiplier-divider circuit proposed in the literature [1-2] have used conventional operational amplifiers (OPAMPs) as basic elements to synthesize mathematical functions. However, the finite gain-bandwidth product (GBP) and small slew rates of these OPAMPs limits their high-frequency operation and accuracy.

The Current Feedback Amplifier (CFA), in contrast, overcomes these shortcomings and hence has gained immense popularity as a basic structural element in electronic circuit design [3–4] in the recent decade. Being fundamentally a trans-resistance amplifier, it achieves high transfer function accuracy, low distortion and a wide bandwidth because of the current feedback applied around it.

However, research in the past [5] confirms the fact that a single CFA is not well suited for very high accuracy applications. This fact provides motivation to explore various circuit techniques that will improve the overall accuracy of any circuit using CFAs. One such circuit has been proposed in this paper, which performs differential multiplication and division operation using the same hardware. A complete theoretical analysis has been performed in section II and extensive SPICE simulations performed later that clearly endorse the validity of the proposed architecture. Previous attempts in this direction, particularly the architecture proposed by Liu et al. [6], do not provide the flexibility of a differential voltage input that could provide a number of advantages, including lesser susceptibility to noisy environments, reduction of second order harmonics and anti-aliasing filtering.

II. CIRCUIT DESCRIPTION

A Current Feedback Amplifier is equivalent to the combination of a second-generation current conveyor [7] and a voltage buffer [8]. Its characteristics can be put in the matrix form as follows

$$\begin{bmatrix} v_x \\ i_y \\ i_z \end{bmatrix} = \begin{bmatrix} 1 & 0 & 0 \\ 0 & 0 & 0 \\ 0 & 0 & 1 \end{bmatrix} \begin{bmatrix} v_y \\ v_z \\ i_x \end{bmatrix} \text{ and } V_O = V_Z \qquad (1)$$

The equivalent circuit of a Current Feedback Amplifier and its circuit symbol are shown in fig. 1(a) and fig. 1(b) respectively.

The proposed differential analog multiplier-divider circuit is shown in fig. 2. It consists of

0-7803-9730-4/06/$25.00 ©2006 IEEE

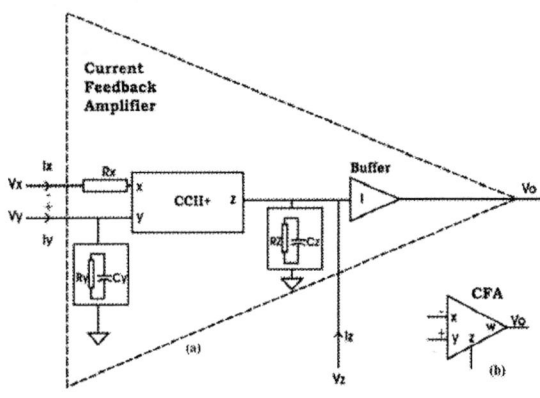

Fig. 1 Equivalent circuit of a Current Feedback Amplifier (CFA) and circuit symbol

five CFAs, six MOSFETs, and four identical capacitors. All MOSFETs are biased in triode region.

The drain current i_D, of a MOSFET conducting in the triode region can be expressed by the following relationship

$$i_D = \mu_n C_{ox} \left(\frac{W}{L}\right) \left\{ \left(V_{GS} - V_{th}\right) - \frac{V_{DS}}{2} \right\} V_{DS} \quad (2)$$

where $\mu_n =$ is the relative mobility, $C_{ox} =$ is the capacitance of the oxide layer per unit area, $W =$ is the width of the channel, $L =$ is the length of the channel, V_{GS} is the voltage between the gate and the source, V_{th} is the threshold voltage of the MOSFET, and V_{DS} is the voltage between the drain and the source.

The expression $\mu_n C_{ox} \left(\frac{W}{L}\right) = k$, is the transconductance parameter of the MOSFET.

DRAIN CURRENTS AND NODAL VOLTAGES

The drain current i_1, flowing through M_1 equals

$$i_1 = k \left[\left\{ \left(V_c + \frac{V_1 + V_2}{2}\right) - V_2 - V_{th} \right\} - \frac{V_1 - V_2}{2} \right] \left(V_1 - V_2\right) \quad (3)$$

$$\Rightarrow i_1 = k \left(V_c - V_{th}\right)\left(V_1 - V_2\right) \quad (4)$$

Hence, the voltage V_α, at the output node of U_1 is given by

$$V_\alpha = -i_1 \cdot \left(\frac{1}{sC}\right) \quad (5a)$$

$$= -\left(\frac{k}{sC}\right)(V_c - V_{th})(V_1 - V_2) \quad (5b)$$

Similarly, the drain current i_6, flowing through M_6 equals

$$i_6 = k\left(V_c - V_{th}\right)\left(V_3 - V_4\right) \quad (6)$$

and the voltage at the output terminal of U_5,

$$V_\beta = -i_6 \left(\frac{1}{sC}\right) \quad (7a)$$

$$= -\left(\frac{k}{sC}\right)(V_c - V_{th})(V_3 - V_4) \quad (7b)$$

The drain currents of all the other MOSFETs

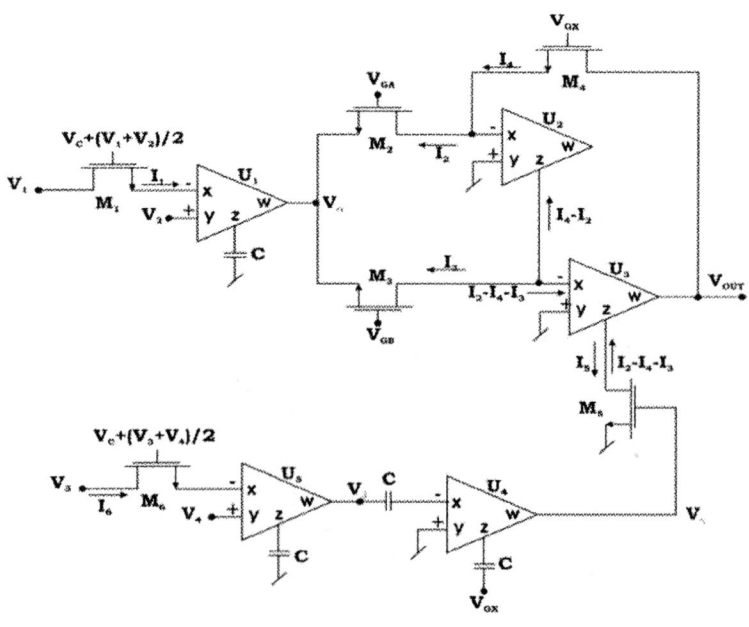

Fig. 2 Proposed differential analog multiplier divider circuit

718

are calculated on a similar basis. They are expressed here as follows

$$i_2 = k \left[\left(V_{GA} - V_\alpha - V_{th} \right) - \left(\frac{0 - V_\alpha}{2} \right) \right] \left(\frac{0 - V_\alpha}{2} \right) \quad (8a)$$

$$\Rightarrow i_2 = -kV_\alpha \left(V_{GA} - V_{th} - \frac{V_\alpha}{2} \right) \quad (8b)$$

$$i_3 = -kV_\alpha \left(V_{GB} - V_{th} - \frac{V_\alpha}{2} \right) \quad (9)$$

$$i_4 = k \left(V_{Gx} - V_{th} - \frac{V_{out}}{2} \right) V_{out}, \text{ and} \quad (10)$$

$$i_5 = k \left(V_\gamma - V_{th} - \frac{V_{out}}{2} \right) V_{out} \quad (11)$$

Equating the currents entering terminals 'x' and 'z' of U_4, we get

$$V_\beta sC = \left(V_{Gx} - V_\gamma \right) sC \quad (12a)$$

$$\Rightarrow V_\gamma = V_{Gx} - V_\beta \quad (12b)$$

Hence, i_5 can now be written as

$$\Rightarrow i_5 = k \left(V_{Gx} - V_\beta - V_{th} - \frac{V_{out}}{2} \right) V_{out} \quad (13)$$

ROUTINE CIRCUIT ANALYSIS

The drain current through M_5, after basic circuit analysis can be written as $i_5 = -(i_2 - i_3 - i_4)$, which on using the values derived above gives

$$-k \left(V_{Gx} - V_\beta - V_{th} - \frac{V_{out}}{2} \right) V_{out} =$$

$$k(V_{GA} - V_{GB})V_\alpha - kV_{out} \left(V_{Gx} - V_{th} - \frac{V_{out}}{2} \right) (14)$$

$$\Rightarrow k(V_{GA} - V_{GB})V_\alpha = kV_{out}V_\beta \quad (15a)$$

$$\Rightarrow V_{out} = \frac{\left(V_{GA} - V_{GB} \right) V_\alpha}{V_\beta} \quad (15b)$$

Hence, the final expression for V_{out} is

$$\boxed{V_{out} = \frac{\left(V_{GA} - V_{GB} \right)\left(V_1 - V_2 \right)}{\left(V_3 - V_4 \right)}} \quad (16)$$

It is clearly evident from the above expression that V_{out} does not contain any constant multiplier term. Hence, the aspect ratio of the MOSFETs and the capacitors can be chosen as per the

availability (both C and W/L gets cancelled in the derivation of the expression).

CONDITIONS

To ensure that all the MOSFETS remain biased in the triode region, the following two inequalities must be satisfied:

$$V_{DS} > 0 \text{ i.e. } V_D > V_S \text{ and} \quad (17a)$$

$$V_{GS} > V_{DS} + V_{th} \quad (17b)$$

The subsequent conditions have been derived for each MOSFET, on the basis of the above inequalities:

$$M_1 : V_1 > V_2 \text{ and} \quad (18a)$$

$$V_c > \frac{V_1 - V_2}{2} + V_{th} \quad (18b)$$

$$M_2 : V_\alpha < 0 \text{ and} \quad (19a)$$

$$V_{GA} > V_{th} \quad (19b)$$

$$M_3 : V_\alpha < 0 \text{ and} \quad (20a)$$

$$V_{GB} > V_{th} \quad (20b)$$

$$M_6 : V_3 > V_4 \text{ and} \quad (21a)$$

$$V_c > \frac{V_3 - V_4}{2} + V_{th} \quad (21b)$$

$$M_4 : V_{Gx} > V_{th} \text{ and} \quad (22a)$$

$$V_{out} > 0 \quad (22b)$$

$$\Rightarrow V_{GA} > V_{GB} \quad (22c)$$

$$M_5 : V_{out} > 0 \text{ and} \quad (23a)$$

$$V_{Gx} - V_\beta > V_{out} + V_{th} \quad (23b)$$

$$\Rightarrow V_{Gx} > V_{out} + V_{th} + V_\beta \quad (23c)$$

Since the values of the circuit elements are so selected that V_β and V_{th} cancel out, the above condition reduces to

$$V_{Gx} > V_{out} \quad (24)$$

The conditions marked (17) to (24) are to be necessarily satisfied by choosing the appropriate values of input voltages.

III. EXPERIMENT

The proposed circuit under study has been experimentally verified using PSPICE simulations. Commercial AD844 has been used as the current feedback amplifier; the MOSFETs based on 1.8μm process technology and the capacitor values chosen to be 10μF throughout the circuit. It is to be noted that any capacitor

Table 1 Summary of functions realized in validation of circuit operation

S. No	Function Realized	V_1 (Volts)	V_2 (Volts)	V_3 (Volts)	V_4 (Volts)	V_{GA} (Volts)	V_{GB} (Volts)	Mean Error %	Figure No.
1.	(1+ tr)(1+sq)	4.5 + tr	4.0	5.0	4.0	2.5 + sq	2.0	0.4%	3
2.	(1+sq) (1+sinx)	5.0+ sin(2π10kt)	4.0	5.0	4.0	3.0 + sq	2.0	2.5%	4
3.	1/(1+sq)	5.0	4.0	5.0 + sq	4.0	3.0	2.0	1%	5
4.	$(1+\xi)^2$	1.0 + ξ	0.0	5.0	4.0	3.0 + ξ	2.0	2.7%	6
5.	1/(1+ξ)	6.0	5.0	1.0 + ξ	0.0	3.0	2.0	1.5%	7
6.	(1+sinx)(1-sinx) =(1+cos2x)/2	5.0+ sin(2π10kt)	4.0	5.0	4.0	3.0+ sin(2π10kt)	2.0	2%	8

value could have been selected since 'C' does not appear in the final expression for V_{out}.

The voltages V_{Gx} and V_c have been held constant at 7.8V and 3.5V respectively. By appropriately changing input voltages- $V_1, V_2, V_3, V_4, V_{GA}$ and V_{GB}, a total of six functions (refer to table 1) have been shown to be realized, while ensuring that the MOSFETs always conduct in the triode region. The voltage values in the table and figures use square (sq), triangular (tr), sinusoidal (sin) and linear (x) functions where

$$\xi \ \Box \ \{ \ t \quad ;\text{for } 0<t<T \ \} \qquad (25)$$

$$sq \ \Box \ \begin{cases} 0 & ;\text{for } 0<t<T/2 \\ 1 & ;\text{for } T/2<t<T \end{cases} \qquad (26)$$

$$tr \ \Box \ \begin{cases} t & ;\text{for } 0<t<T/2 \\ -t & ;\text{for } T/2<t<T \end{cases} \qquad (27)$$

The minimum and the maximum input voltage levels of the designed circuit are limited by the conditions derived earlier and the power supplies to the AD844 (±12V) respectively.

For quick evaluation, the frequency performance of the proposed multiplier has been obtained by computer simulation using the MOS model including all high-order effects and realistic process parameters. With the same conditions as above, simulation showed that bandwidths for input voltages $(V_1, V_2, V_3, V_4, V_{GA}, V_{GB})$ ranged from 40MHz to 60MHz, not in that order.

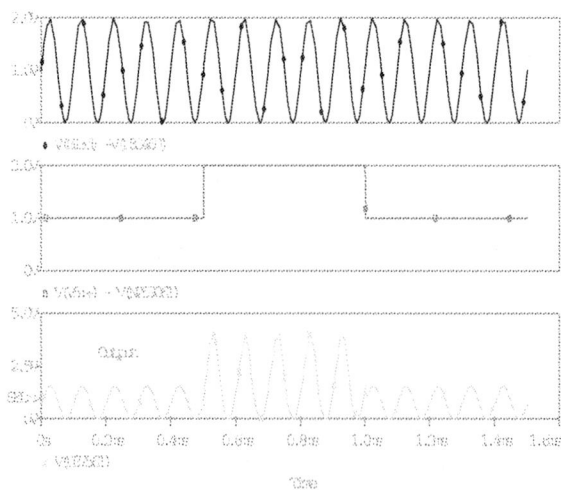

Fig. 3 Function 1: (1+tr)(1+sq)

Fig. 4 Function 2: (1+sinx)(1+sq)

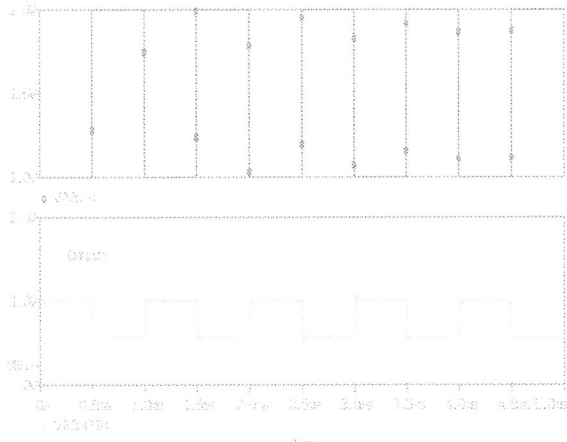

Fig. 5 Function 3: 1/(1+sq)

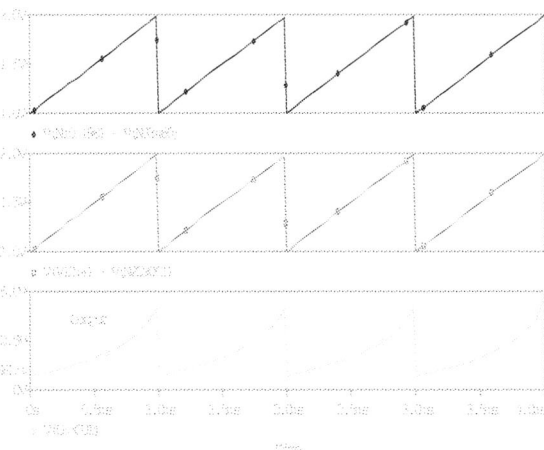

Fig. 6 Function 4: $(1+\xi)^{2}$

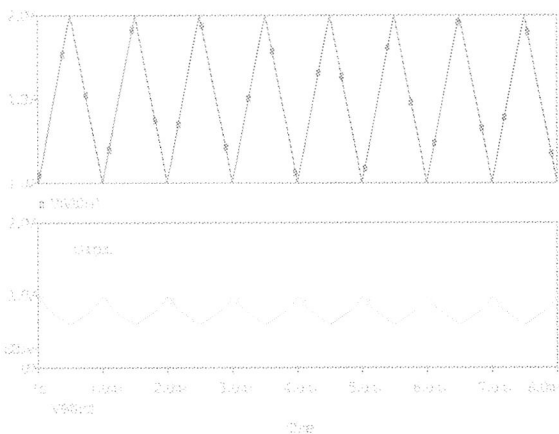

Fig. 7 Function 5: $1/(1+\xi)$

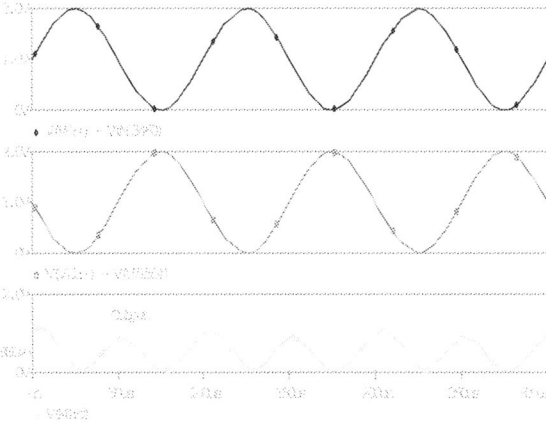

Fig. 8 Function 6: (1+sinx)(1-sinx)

IV. CONCLUSION

We have presented here a novel design for a differential analog multiplier/divider circuit using current feedback amplifiers (CFAs) consisting of five CFAs, six MOS transistors biased in the triode region, and four identical capacitors. The final expression for the output of the proposed circuit is independent of the CMOS process technology used and of the common capacitor value chosen. In addition, the circuit incorporates the benefits of current feedback amplifiers (CFAs), especially their high accuracy and wide bandwidth, which have been well demonstrated by diverse graphical results obtained. Experimental analysis corroborates these advantages, which render this circuit for use in a broad range of applications in analog signal processing and telecommunications.

V. REFERENCES

[1] N.I. Khachab and M. Ismail, "A nonlinear CMOS analog cell for VLSI signal and information processing," *IEEE J. Solid State Circuits* vol. 26, pp. 1689-1698 (1991).

[2] S.I. Liu, D.S. Wu, H.W. Tsao, J. Wu and J.H. Tsay, "Nonlinear circuit applications with current conveyors," *IEE Proc. G* vol. 140, pp. 1-6 (1993).

[3] J. Austin, *Burr Brown Application Bulletin: Current feedback amplifiers,* Burr Brown (2000).

[4] F. Lidgey and K. Hayatleh, "Current feedback operational amplifiers and applications," *J. of Electron. Commun. Engineering* vol. 9, pp. 176–182 (1997).

[5] L. Moult and F. J. Lidgey, "A review of dual current-feedback amplifier techniques," *IEE Colloquium on Analogue Signal Processing*, pp. 9/1-9/7 (1994).

[6] S.I. Liu and J.J. Chen, "Realisation of analogue divider using current feedback amplifiers," *IEE. Proc.-Circuits Devices Systems* vol. 142, pp. 45-48 (1995).

[7] A. Sedra and K.C. Smith, "A second generation current conveyor and its applications," *IEEE Trans. CT-17*, pp. 132-134 (1970).

[8] I.A. Svoboda, L. McGory and S. Webb, "Applications of a commercially available current conveyor," *Int. J. Electron*, pp. 159- 164 (1991).

A SEM Based Technique To Detect Pin-holes In As-Deposited/As-Grown Dielectrics

Nitin R Kamat, Oh Chong Khiam and Zhao Si Ping, *Snr Member, IEEE*

Chartered Semiconductor Mfg Ltd
60 Woodlands Industrial Park D
Street 2, Singapore 738406
Tel: 65-63604325, Fax: 65-63622935
E-mail: nitinkamat@charteredsemi.com

Abstract: **In this paper, an attempt is made to highlight a new SEM based technique that was used to detect pinholes in the as-deposited/as-grown dielectrics. The fundamental principle governing the technique is discussed. This technique is benchmarked against a well-established fault-isolation technique using the Liquid Crystal. In fact, this technique was found to supplement the Liquid Crystal technique. A case study, discussed in the paper, helps understand the usefulness of the technique especially in detecting defects on as-deposited/as-grown dielectric films.**

I. INTRODUCTION

DETECTING pinholes on standard ET structures using well-established FA techniques like Liquid Crystal and/or Emission Microscopy is commonplace. There are no known techniques to detect pinholes on blanket as deposited/as-grown dielectrics. Such a technique would carry immense value for in-line defect detection and imaging. This would speed-up the process of yield improvement and/or could help solve a major issue during the development of a new technology. Fault isolation techniques using the Liquid Crystal and/or Emission microscopy have limited spatial resolution. Moreover, these techniques are rendered useless in detecting the pinholes as they could only be used on patterned wafers with back-end infrastructure in place. In contrast, the SEM based technique described here, is precise and imaging could be done instantaneously saving time and efforts required in locating the defect. The defect detection could be done on a blanket wafer as well and no metal patterning is necessary for fault isolation purposes.

The technique, described in this paper, was developed during the process of fault-isolation of a supply to ground leakage failure. The technique has been compared with and benchmarked against established fault isolation techniques using the Liquid Crystal. The underlying principle governing the technique would be discussed in the paper.

II. EXPERIMENT

The device under study had been processed using standard 0.35um CMOS technology. The back-end had 3 layers of metallisation. The unit was passivated using Silicon dioxide and Si_3N_4. The unit when tested, showed leakage in the range of milliamperes. The leakage was seen between the supply and ground pins.

III. RESULTS AND DISCUSSION

The failing die was fault localized using Liquid Crystal. The clearing point of the liquid crystal used was $29^{o}C$. Fig 1 shows the location of the hot spot obtained on the die.

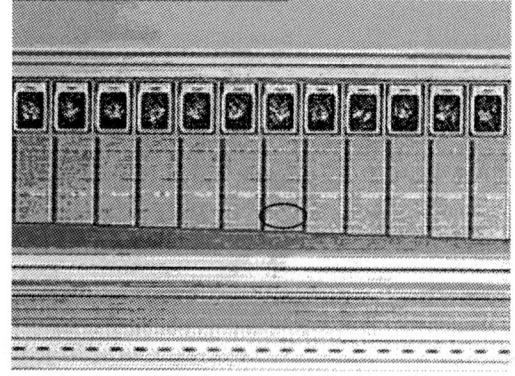

Fig 1 Location of the hotspot indicates a possible vertical short between the two power lines.

The location of the hotspot indicated that the leakage must have happened due to vertical short between the two power lines laid at M2 and M1. Physical FA was performed to look for any evidence of anomaly that could cause such a vertical short. The passivation was removed using the RIE and M3 and M2 was lapped away. SEM inspection, at 5KeV accelerating voltage was done at the hotspot location to look for any anomaly but none was found. This was mainly due to the very big size of the hotspot. Further fault localization was, therefore, necessary.

Fig 2 shows a graph of secondary electron yield versus beam accelerating voltage for any SEM based imaging system. From the graph it can be clearly seen that the secondary electron yield Y is greater than 1 for a narrow window of beam accelerating voltage marked E1 and E2.

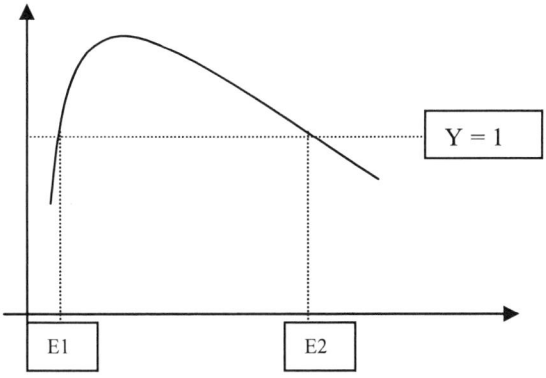

Fig 2. Graph showing secondary electron yield (Y axis) versus the beam accelerating voltage (X axis)

The beam accelerating voltage of the SEM could be optimally adjusted, between the points E1 and E2, to maximize the secondary electron yield. This would create a positive potential on the sample surface. Consequently, the secondary electrons generated in pin-hole like defect could then be pulled out by the positive surface potential to make the pin-hole appear like a bright spot.

Fig 3 shows bright spot on the dielectric in the vicinity of the hotspot obtained by optimizing the SEM beam accelerating voltage to create a positive potential on the sample surface. This technique was successfully used to further localize the defect that was almost impossible to detect under the normal SEM operating conditions (5KeV beam accelerating voltage).

Fig 4 shows the image of the pin-hole in the dielectric at higher SEM beam accelerating

voltage (16KeV).

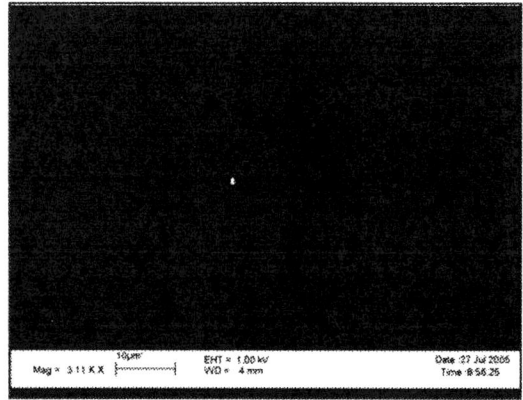

Fig 3. Bright spot was seen on the dielectric at the location of the hotspot. The bright spot was obtained by adjusting the beam accelerating voltage of the SEM to have a positive potential on the sample surface.

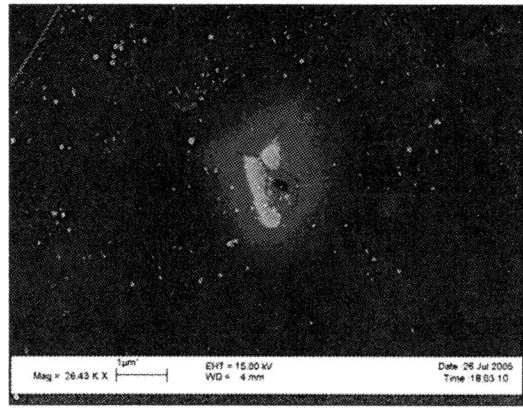

Fig 4. High beam energy SEM image was taken at the location of the bright spot. A pin-hole is clearly seen.

The concept was further employed using the Ga ion beam during TEM sample preparation. Fig 5 shows the contrast obtained at the location of the pin-hole using the ion beam. The positive charge of the Ga ion beam helps in generating the contrast at the location of the pin-hole even at high beam accelerating voltage of 30KeV. This means simultaneous fault localization and TEM sample preparation is possible.

Fig 6 shows TEM image across the pin-hole. Abnormal M1 profile is seen. The M1 seems to be extruding out of the pin-hole.

Fig 5. Ga ion beam voltage contrast is obtained during the TEM sample preparation.

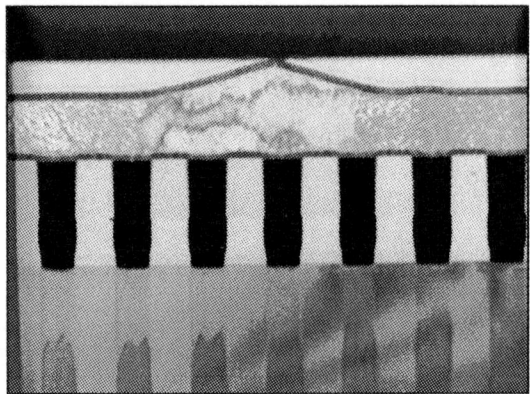

Fig 6. TEM image of the anomaly at M1 is shown. The TEM sample was prepared across the pin-hole fault isolated during the sample preparation.

IV CASE STUDY

This SEM based technique to detect the pin-holes in the dielectric was used during the development of a novel nitrided gate dielectric stack for the new technology. Poly capacitor structures were realized and stressed to study the new gate dielectric stack. Fig 7 shows the IV curve for good and leaky capacitors realized out of the novel gate dielectric stack.

The post stress failure analysis of the leaky capacitor structure would not help fix the problem. This is due to the fact that the stress would modify the anomaly creating many artifacts. It was, therefore, decided to look at the gate dielectric anomaly immediately after the formation of the gate dielectric on a blanket wafer! As there was no metal patterning on the

wafers with blanket gate dielectric, the newly developed SEM based fault isolation technique was used.

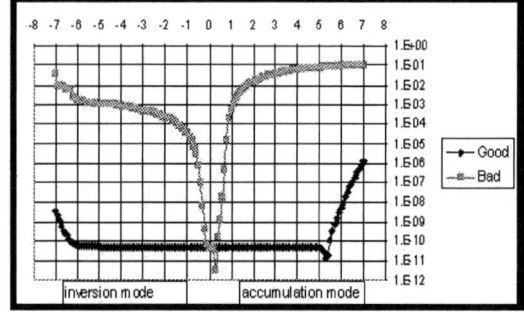

Fig 7. IV curve for good and leaky capacitors realized out of the novel gate dielectric stack is shown.

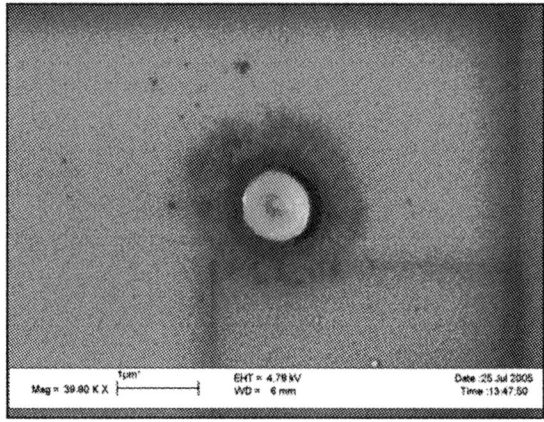

Fig 8. A pin-hole is seen in the gate dielectric film blanket deposited on the wafer. The fault isolation was done using the SEM based technique.

Fig 8 shows SEM image of a pin-hole seen on the wafer with a blanket deposited gate dielectric stack. The fault isolation was done using the SEM based technique. It would have been impossible to locate the pin-hole had it not been for the newly developed technique.

Appropriate process fix was put in place to solve the problem.

V. CONCLUSION

A SEM based fault localization technique is described. The technique was developed in line with the passive voltage contrast technique used for fault localization on the failing metal comb and via chain structures. This SEM based fault localization technique was used to understand defect-causing leakage on the gate dielectric for the new process under development. This exercise provided ample proof of the practical usefulness of the technique. The authors feel this technique can, therefore, be used for in-line inspection as well.

VI. ACKNOWLEDGEMENT

The authors would like to thank Mr Eddie Er, Dr. Li Kun and their group for the TEM work performed during the entire study. The authors also thank Mr. Sripad Nagarad from the TD group and Terence Lee from Fab2 YE for the samples and the electrical data on leaky capacitors.

VII. REFERENCES

[1]. Neal Sullivan, Soluris Inc., Concord, Mass.; Byoung-Ho Lee, Samsung Electronics Co. Ltd., Kiheung; Yeonhg-Uk Ko, EM Facility, University of Tennessee, Knoxville, Tenn. "Imaging High-Aspect-Ratio Contact Holes", Semiconductor International.

[2].Joseph I. Goldstein, Dale E. Newbury, Patrick Echlin, David C. Joy, A.D. Romig, Jr., Charles E. Lyman, Charles Fiori and Eric Lifshin. "Scanning Electron Microscopy and X-Ray Microanalysis; A Text for Biologists, Material Scientists and Geologists", second edition

A Study Of Yield Loss In Copper Back-End Process Due To Stress And Poor Adhesion Of The Thin Films

Nitin R Kamat, Manni Lal, P B Sahoo and Zhao Si Ping, *Snr Member, IEEE*

Chartered Semiconductor Mfg Ltd
60 Woodlands Industrial Park D
Street 2, Singapore 738406
Tel: 65-63604325, Fax: 65-63622935
E-mail: nitinkamat@charteredsemi.com

Abstract: **Copper as a metal of choice for interconnection purposes, needs damascene approach to realize the interconnect formation. This approach needs etch stop layers to form via and trench. Metal and dielectric barriers are also needed to prevent Copper from diffusing into the substrate. The integrity and adhesion of all these different layers is a subject of great concern from the yield standpoint. The problem needs more careful attention with porous low k dielectric material. In this paper, some aspects related to the adhesion of the various thin films that form the damascene stack are discussed. Analysis of failures/outliers encountered during routine die pull test is also done. The failure analysis performed on such failures provided a very good insight into the weakness in the dielectric stack, that form the interconnect, in the case of device with Copper as back-end metallisation. Techniques to improve the adhesion of various layers and other novel approaches used to improve the process margin are also discussed. Relationship between the die size and stress induced dielectric delamination is also discussed in this paper.**

I. INTRODUCTION

IN the dual damascene approach to Copper back-end process, adhesion of various thin films that form the Inter-Metal dielectric (IMD) stack is of profound significance. Poor adhesion between Copper and Si_3N_4 or Si_3N_4 and Fluorinated - TEOS (FTEOS) or the Ta and FTEOS could lead to loss in product yield and in the worst case scenario could lead to poor reliability of the product. Die shear test was used to assess the adhesion and the integrity of the back-end stack. Die with lower than nominal force was analyzed to understand the failure mechanism in details.

In the case of another experiment die with two different dimensions were analyzed. The smaller die had dimensions of 9420 um x 8724um and the bigger had dimensions of 14226um x 13000um. The smaller die had comparatively shorter metal lines due to smaller die size and in the case of the bigger die the metal lines were drawn across the length and the breadth of the die.

II. EXPERIMENT

Fig 1 shows the dual damascene stack. The M1 was processed using the damascene approach. M2 to M5 were processed using dual damascene approach with trench patterning was performed ahead of via patterning. The etch-stop layer and the dielectric barrier was Silicon Nitride and the IMD was FTEOS. Ta was used as a diffusion barrier for Copper and the Copper was electroplated. Excess amount of Copper was removed by Chemical Mechanical Planarization (CMP) process. After M5 CMP, the device was passivated using Si_3N_4 and Silicon dioxide cap layers. Polyimide was deposited on top of the passivation layer. The bond-pads were bumped.

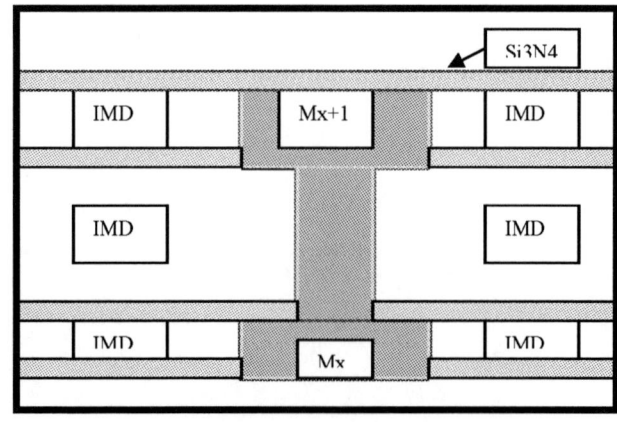

Fig 1. The dual damascene stack for the experimental structure

0-7803-9730-4/06/$25.00 ©2006 IEEE

III. RESULTS AND DISCUSSION

After the die shear test, the units that needed lower than nominal die shear force were analyzed. Fig 2 shows optical image of one such failure. The solder bump seems to have gotten sheared-off at one particular location. Some damage to the IMD layers is also seen. The complete removal of the solder bump along with the damage to the IMD layers indicate that the lower die shear force is due to some weakness in the dielectric stack and not due to a weak solder bump to die interface.

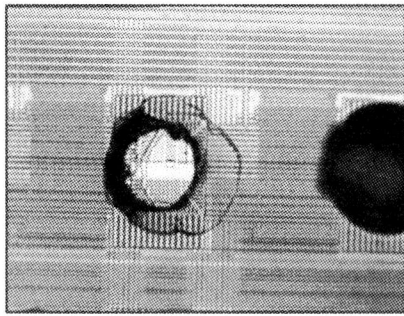

Fig 2. Optical image of the damaged solder bump after the die shear test

Cross-sectioning was done across the damaged location to understand the failure mechanism in details. Fig 3 shows SEM micrograph of the cross-section performed on one such failing bump. Voids were seen under the metal line at the Ta-FTEOS interface.

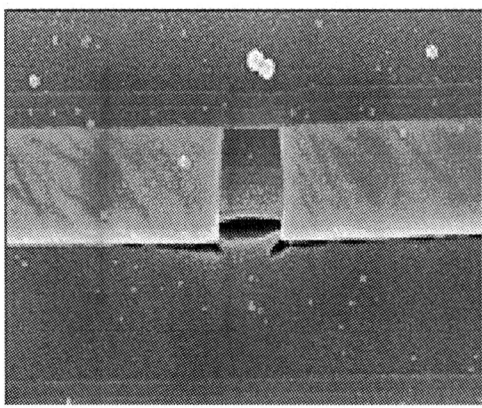

Fig 3. SEM micrograph of metal line with voids under it.

In the case of some failing units, delamination was seen at the bottom of M3 and M4 (see fig 4). This clearly pointed at some adhesion related issue between the Ta and the FTEOS layer. On reviewing the lot history it was found that these devices had a higher queue time before the Ta barrier deposition process at M3 and M4 stages. The extended exposure of the FTEOS to the ambient, after the trench etching, was believed to cause changes in the surface properties of the FTEOS. It was hypothesized that traces of moisture in the ambient could react with the Fluorine in the FTEOS leading to the formation of mild HxFy which in turn results in poor adhesion between the Ta barrier and the FTEOS. This led to delamination between the bottom of the metal line and FTEOS during the die shear test.

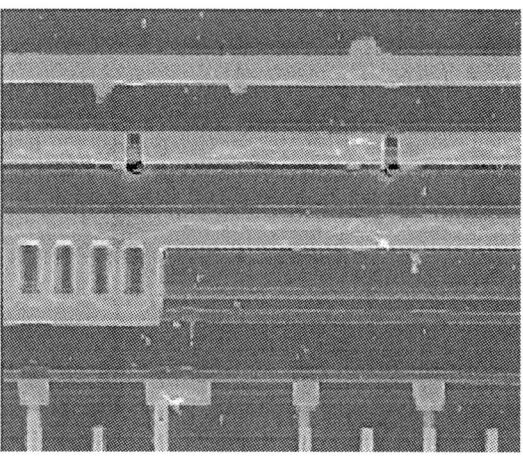

Fig 4. SEM micrograph of a cleaved sample that had failed the die shear test. Delamination at the bottom of M3 and M4 is seen.

In a production environment it is difficult to control the queue time for any particular process. It was, therefore, imperative to make the process robust enough to improve adhesion between the Ta barrier and the underlying IMD layer.

One approach adopted to improve the adhesion between the Ta barrier and the underlying IMD layer was to deposit a thin film of TEOS over the FTEOS layer and control the etch to confine the bottom of the metal line to be within the TEOS layer ie to separate out the Ta from FTEOS.

This approach improved the process margin and eliminated the need for more efficient lot management inorder to prevent the deterioration of surface properties of the FTEOS film.

Fig 5 shows delamination between the TEOS and the Si_3N_4 layer for the bigger die. Copper extrusion was seen along with the dielectric delamination. The extrusion occurs due to the breaking of the Ta barrier due to stress exerted by the delaminating dielectric stack. Fig 6 shows a TEM micrograph of the Copper extrusion caused by the delamination of Si_3N_4 and TEOS. It was observed that long closely spaced metal lines on the bigger die were more susceptible for delamination.

Fig 5. Cross- SEM image of Copper extrusion causing a short between the metal lines. Delamination is seen between the Si_3N_4 and TEOS layers.

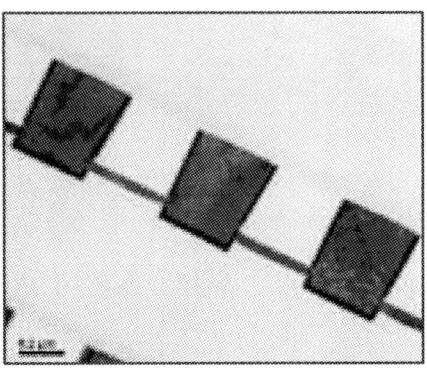

Fig 6. TEM micrograph of Copper extrusion causing a short between the metal lines

Fig 5 and fig 6 showed Cross-sectional view of the delamination and the resulting Copper extrusion. Inorder to assess the extent of delimination in three dimensions, top-down approach was adopted. Fig 7 shows top-down SEM image of the Copper puddle between the metal lines. The size of the puddle indicated that the delimination is localized. Fig 8 shows the Auger elemental mapping performed on the puddle to prove that the puddle seen shorting the metal lines is Copper diffusing into the dielectric. No such delamination was seen for the smaller die indicating the need to manage the thin film

stress more carefully on the bigger die. The die shear force was also found to be nominal for the smaller die. Moreover, consistently, the lower interface between the Si_3N_4 and TEOS seem to delaminate indicating a possible process related weakness that resulted in poor adhesion between Si_3N_4 and TEOS. Never-ever delamination was seen at the top interface between the Si_3N_4 and TEOS. A careful comparative study of the process conditions used to deposit the top TEOS and the bottom TEOS indicated the problem to be due to methodology used to deposit the bottom TEOS

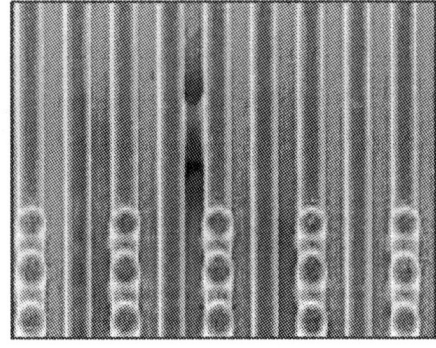

Fig 7. Top-down SEM image of the Copper puddle shorting the metal lines.

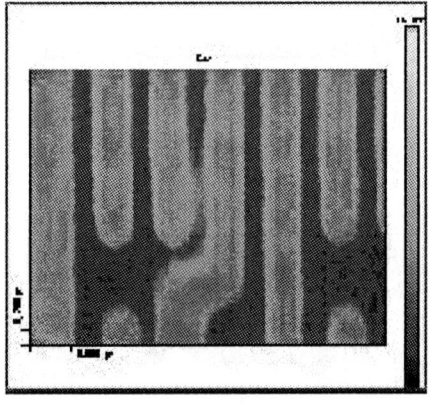

Fig 8. Auger Electron Spectroscopy technique was used to generate the map of Copper between the two metal lines.

The bottom TEOS and the FTEOS are deposited in the same chamber. The bottom TEOS is deposited by switching off the Fluorine flow after the FTEOS deposition is done. It was hypothesized that traces of Fluorine in the chamber could be changing the properties of the bottom TEOS. By depositing a thicker TEOS layer the amount of Fluorine at the interface could be reduced and the problem could be eliminated. In comparison the top TEOS deposition is made in a different chamber hence no possibility of any traces of Fluorine getting into the TEOS film.

IV. CONCLUSION

Analysis of the outliers or the failing dies for the die shear test gave an interesting insight into the mechanical properties of the back-end stack. The location of the delamination and the knowledge of delaminating layers gave a first hand knowledge of the inherent weaknesses in the back-end stack. Careful analysis of the failures helped improve the device robustness and improve the mechanical/adhesion strength of the back-end structures.

It was also interesting to know that the dielectric stresses need to be more carefully managed in case of bigger die inorder to prevent yield loss and improve reliability of the device. It was found that the methodology used to deposit various thin films also play an important role in deciding the properties of the film.

V. ACKNOWLEDGEMENT

The authors would like to thank Mr Eddie Er, Dr. Hua Younan and their group for Cross-SEM, TEM and Auger work performed during the entire study.

VI. REFERENCE

[1].Nitin R. Kamat et. al, "A Study Of Yield Loss In Copper Back-End Process Due To Stress And Poor Adhesion Of The Thin Films", *IPFA 2005*.

III. RESULTS AND DISCUSSION

After the die shear test, the units that needed lower than nominal die shear force were analyzed. Fig 2 shows optical image of one such failure. The solder bump seems to have gotten sheared-off at one particular location. Some damage to the IMD layers is also seen. The complete removal of the solder bump along with the damage to the IMD layers indicate that the lower die shear force is due to some weakness in the dielectric stack and not due to a weak solder bump to die interface.

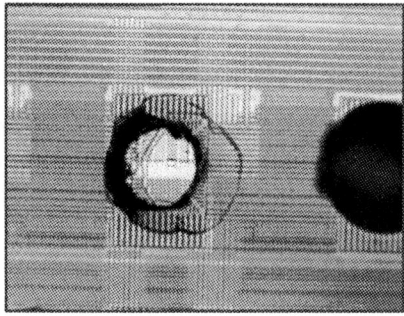

Fig 2. Optical image of the damaged solder bump after the die shear test

Cross-sectioning was done across the damaged location to understand the failure mechanism in details. Fig 3 shows SEM micrograph of the cross-section performed on one such failing bump. Voids were seen under the metal line at the Ta-FTEOS interface.

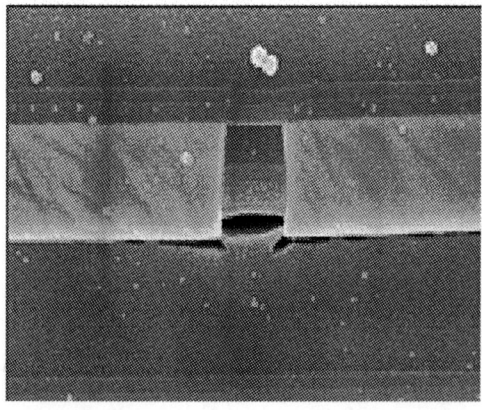

Fig 3. SEM micrograph of metal line with voids under it.

In the case of some failing units, delamination was seen at the bottom of M3 and M4 (see fig 4). This clearly pointed at some adhesion related issue between the Ta and the FTEOS layer. On reviewing the lot history it was found that these devices had a higher queue time before the Ta barrier deposition process at M3 and M4 stages. The extended exposure of the FTEOS to the ambient, after the trench etching, was believed to cause changes in the surface properties of the FTEOS. It was hypothesized that traces of moisture in the ambient could react with the Fluorine in the FTEOS leading to the formation of mild H_xF_y which in turn results in poor adhesion between the Ta barrier and the FTEOS. This led to delamination between the bottom of the metal line and FTEOS during the die shear test.

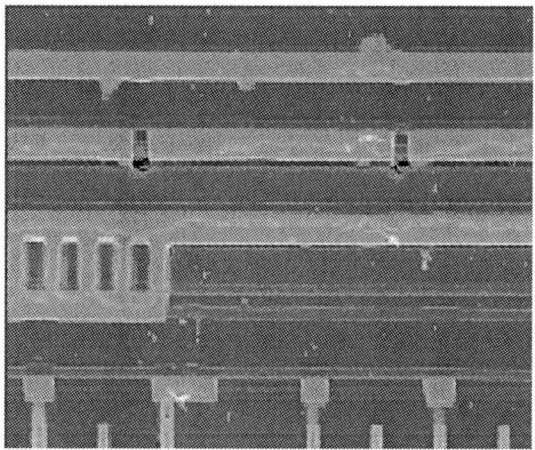

Fig 4. SEM micrograph of a cleaved sample that had failed the die shear test. Delamination at the bottom of M3 and M4 is seen.

In a production environment it is difficult to control the queue time for any particular process. It was, therefore, imperative to make the process robust enough to improve adhesion between the Ta barrier and the underlying IMD layer.

One approach adopted to improve the adhesion between the Ta barrier and the underlying IMD layer was to deposit a thin film of TEOS over the FTEOS layer and control the etch to confine the bottom of the metal line to be within the TEOS layer ie to separate out the Ta from FTEOS.

This approach improved the process margin and eliminated the need for more efficient lot management inorder to prevent the deterioration of surface properties of the FTEOS film.

Fig 5 shows delamination between the TEOS and the Si_3N_4 layer for the bigger die. Copper extrusion was seen along with the dielectric delamination. The extrusion occurs due to the breaking of the Ta barrier due to stress exterted by the delaminating dielectric stack. Fig 6 shows a TEM micrograph of the Copper extrusion caused by the delamination of Si_3N_4 and TEOS. It was observed that long closely spaced metal lines on the bigger die were more susceptible for delamination.

Fig 5. Cross- SEM image of Copper extrusion causing a short between the metal lines. Delamination is seen between the Si_3N_4 and TEOS layers.

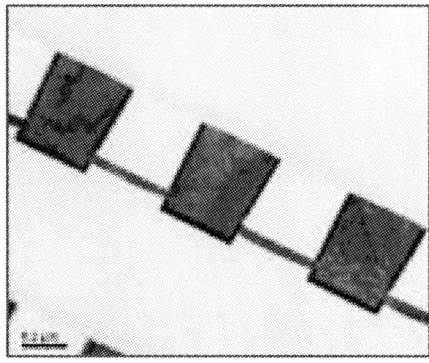

Fig 6. TEM micrograph of Copper extrusion causing a short between the metal lines

Fig 5 and fig 6 showed Cross-sectional view of the delamination and the resulting Copper extrusion. Inorder to assess the extent of delimination in three dimensions, top-down approach was adopted. Fig 7 shows top-down SEM image of the Copper puddle between the metal lines. The size of the puddle indicated that the delimination is localized. Fig 8 shows the Auger elemental mapping performed on the puddle to prove that the puddle seen shorting the metal lines is Copper diffusing into the dielectric. No such delamination was seen for the smaller die indicating the need to manage the thin film

stress more carefully on the bigger die. The die shear force was also found to be nominal for the smaller die. Moreover, consistently, the lower interface between the Si_3N_4 and TEOS seem to delaminate indicating a possible process related weakness that resulted in poor adhesion between Si_3N_4 and TEOS. Never-ever delamination was seen at the top interface between the Si_3N_4 and TEOS. A careful comparative study of the process conditions used to deposit the top TEOS and the bottom TEOS indicated the problem to be due to methodology used to deposit the bottom TEOS

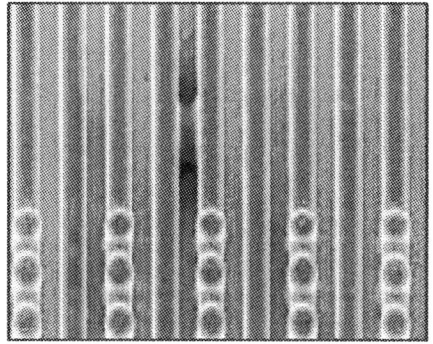

Fig 7. Top-down SEM image of the Copper puddle shorting the metal lines.

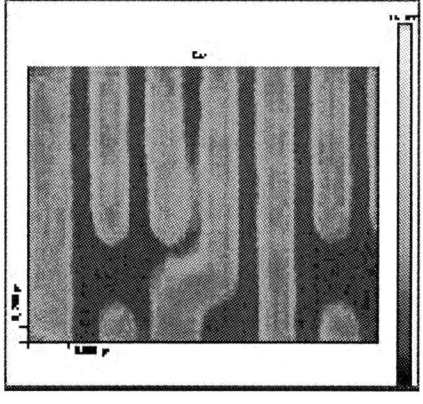

Fig 8. Auger Electron Spectroscopy technique was used to generate the map of Copper between the two metal lines.

The bottom TEOS and the FTEOS are deposited in the same chamber. The bottom TEOS is deposited by switching off the Fluorine flow after the FTEOS deposition is done. It was hypothesized that traces of Fluorine in the chamber could be changing the properties of the bottom TEOS. By depositing a thicker TEOS layer the amount of Fluorine at the interface could be reduced and the problem could be eliminated. In comparison the top TEOS deposition is made in a different chamber hence no possibility of any traces of Fluorine getting into the TEOS film.

IV. Conclusion

Analysis of the outliers or the failing dies for the die shear test gave an interesting insight into the mechanical properties of the back-end stack. The location of the delamination and the knowledge of delaminating layers gave a first hand knowledge of the inherent weaknesses in the back-end stack. Careful analysis of the failures helped improve the device robustness and improve the mechanical/adhesion strength of the back-end structures.

It was also interesting to know that the dielectric stresses need to be more carefully managed in case of bigger die inorder to prevent yield loss and improve reliability of the device. It was found that the methodology used to deposit various thin films also play an important role in deciding the properties of the film.

V. Acknowledgement

The authors would like to thank Mr Eddie Er, Dr. Hua Younan and their group for Cross-SEM, TEM and Auger work performed during the entire study.

VI. Reference

[1].Nitin R. Kamat et. al, "A Study Of Yield Loss In Copper Back-End Process Due To Stress And Poor Adhesion Of The Thin Films", *IPFA 2005*.

This page intentionally left blank.

Electrical Analysis of High Temperature SAW Resonator Packages

Mohammad Hadi Shahrokh Abadi, Mohd. Nizar Hamidon, *Member,* IEEE, Roslina Sidek,
Member, IEEE, Mina Malekzadeh
Department of Electrical and Electronic Engineering
Faculty of Engineering
University Putra Malaysia
43400 UPM Sedang, Selangor,MALAYSIA
Email: mhshahrokh@gmail.com or mnh@ieee.org

Abstract – **This paper presents a study on the effects of package types and materials on the performance of high temperature SAW resonator based on GaPO$_4$. Flip chip and wire bond package and interconnection with different materials have been considered. The studied materials for package include GaPO$_4$ and AlN as insulator and Au, Pt, Ni, Au70/Pt30 as conductor. Analysis based on quasi-approximation of FDM shows that the flip chip is favorable for SAW resonator package due to less frequency shift when compared to wire bonding.**

Index Terms– AlN: Aluminum Nitride, FDM: Finite Different Method, FEM: Finite Element Method, GaPO4: Gallium Phosphate, HT: High Temperature, IDT: InterDigital Transducer, SAW: Surface Acoustic Wave, TDR: Time Domain Reflectometery.

I. INTRODUCTION

SAW devices are widely used in signal processing applications, such as frequency filter [1] in telecommunication systems. Furthermore the SAW devices also have been used in sensor technology for temperature, pressure or gas monitoring. The advantage of using SAW devices is that they can be used for wireless, passive, and high temperature applications.

This wireless interrogation of passive SAW devices seems to be especially useful at high temperature applications above 200°C [2]. At these high temperatures applications material-related problems to packaging and interconnection of such devices become obvious [3]. These problems will be more intense

because of anisotropic property of the piezoelectric crystals, mismatch CTE, and stable adhesive materials [4].

SAW resonator package is required to provide the IDTs' physical protection. It also functions as a temperature intermediary for the IDTs. Fig. 1 illustrates cross section view of a wire bonding and a flip chip package construction. IDTs are mounted and soldered directly into the interposer via wire bonding or ball bumping. The pins or external electrodes are packed closely and mutual coupling between pins are most dominant in this region as well as within the package where the lead-frame, bond-wire or interposer trace are very close together.

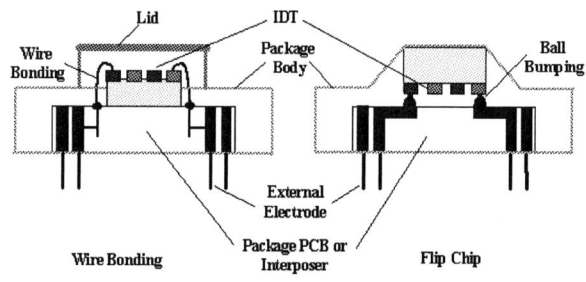

Fig. 1: Cross section of SAW resonator package

II. METHOD OF ANALYSIS AND DESIGN TOOL

Derivation of the lumped equivalent circuit model is generally performed by using (i) analytical method, (ii) measurement based technique, and (iii) electromagnetic field simulator software approaches [5],[6]. Analytical method is only convenient for component of simple geometrical construction with limited input and output connections. Measurement based methods are generally the fastest and able to account for most parasitic effects encountered. Measurement is performed in time and frequency domain using TDR and S-parameters characterization respectively. TDR is used for discrete components with single input and output connections. Furthermore elaborate calibration structures have to be incorporated into the IDTs as device under test (DUT) to obtain reliable measurement results. Thus the method is less popular [5].

Electromagnetic field simulator software employing numerical method will yield more accurate model albeit needing long computing time and relatively large computing resources. A quasi-static approximation can be applied to the electromagnetic field analysis to reduce the steps of computing, and base on this; FDM/FEM can be obtained to calculate the effects of different material on the package [6].

Conventional analysis of SAW resonator involves the use of lumped circuit elements to represent electric and acoustic effects. This approach permits convenient treatment of the resonator as a circuit element [7]. Such equivalent circuit models make it possible to naturally take into consideration the effects of different materials in behaviors of package. Preliminary analysis of SAW resonator at high-temperatures applications, taking into account some of the material effects, can be found in literature, where empirical values of bond wire and/or ball-bump inductances are used [8].

However, experimental determination of accurate values of such parasitic effects requires careful measurement procedures [9]. In contrast, numerical model of thermal effects of package offers a more cost-effective analysis access. In this way, a quasi-static approach is utilized to obtain circuit parameters of IDTs, bond wires, ball bumps and package leads that are commonly found in SAW resonators. Quasi-static approximations to the electromagnetic field are valid for SAW component geometries, whose dimensions are much smaller than wavelengths associated with the highest operational frequencies of the resonator [10]. Fig. 2 shows a simplified representation of

package circuit model given in Fig. 1, based on a quasi-static approximation.

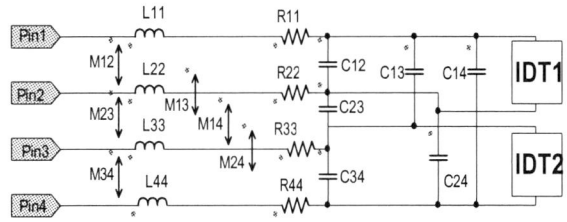

Fig. 2: Simplified lumped circuit representation of SAW resonator package

Maxwell Q3D Extractor from ANSOFT Corporation which has been designed based on FEM/FDM utilized to achieve to packaged SAW resonator equivalent circuit. This software is an interactive package that is used to electrically characterize three dimensional interconnects structures such as those found in connectors, PCB's, MCM's etc. The software is used to solve capacitance, partial inductance and resistance matrices. It uses the multi-pole expansion technique for extraction of electrical parameters.

The basic operation is to draw the structure, specify material properties for each object, identify conductors, and specify source excitations. The system then generates the necessary circuit parameters used to generate the lumped equivalent circuit models. These circuit models can then be used to give the S-parameters for the different packaging structures.

III. EQUIVALENT CIRCUIT MODELS FOR SAW RESONATOR

For accurate electrical characterization of an electronic package, it is important to have proper circuit models for the components that make up the package including the wirebond, the on package stripline/microstrip, the package leads as in most Flat-packs and Small outline packages, Through via holes, Flip chip attach, and solder balls as in most Ball Grid Array packages.

First of all a 3400×2400×500 μm dielectric layer is selected in the software. The property of chosen dielectric is changed to a piezoelectric material according to GaPO4 properties. The next chosen layers are Zirconium and Platinum with default properties and dimensions [11]. Next stage is to clarify the output ports and signal conditions. In a quasi-static approximation a I–V voltage source is applied on the input port and the simulated result is taken in another port. The input frequency of applied source is also selected at resonance frequency of the resonator that is 433.92 MHz. Fig.3(a) shows the insertion loss versus return loss in 0–1000 MHz frequency band for the simulated resonator. Fig.3(b) shows mutation of phase and delay in resonator output signal. Fig.3(c) shows more detail on insertion loss versus return loss. The extracted equivalent circuit and the output signals, insertion and return loss, from the software now can be used in evaluation of influences of package materials on integrity of resonator output.

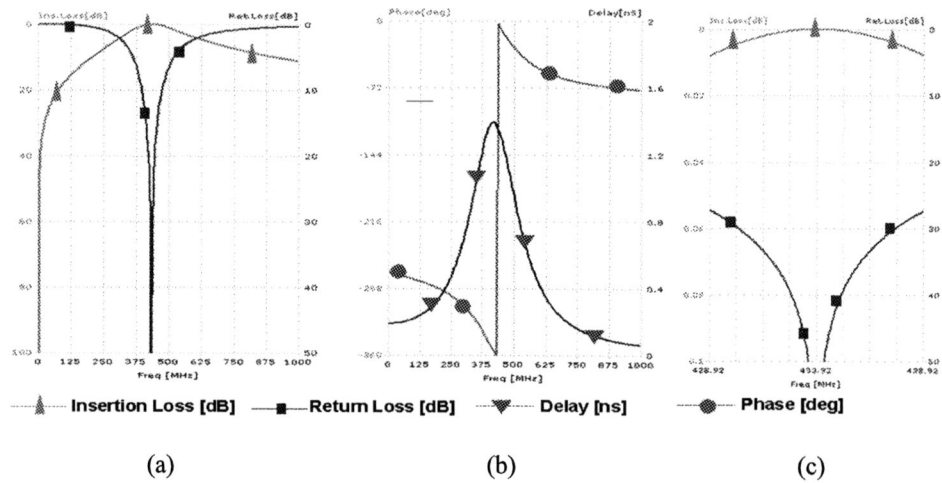

(a)   (b)   (c)

········△········ **Insertion Loss [dB]**   ───■─── **Return Loss [dB]**   ····▽···· **Delay [ns]**   ········●········ **Phase [deg]**

Fig.3: Insertion loss vs. return loss (a), phase vs. delay (b), and more detail on return loss (c) on IDT ports (f_o = 433.92 MHz)

IV. ELECTRICAL EQUIVALENT CIRCUIT OF PACKAGE

Most SAW devices packages use wire bonds to connect the IDTs to the package. Wire bonds, wedge to wedge (W/W) and ball to wedge (B/W) with the length of 1.25mm and 1.54mm respectively, were structured to observe the effect on overall insertion loss. Gold (Au), Nickel (Ni), Platinum (Pt), and Au70/Pt30 were used for the wire bonds of diameter 25.4μm. Table 1 provides the electrical parameters of wire bonds.

Table 1: Electrical parameters of different material in wire bonding

Material	Bonding Type	Inductance (nH)	DC Resistance (Ω)	Mutual Ind. between wires(nH)	Mutual Cap. between wires(fF)	Wire-GND Cap. (fF)
Au	B/W	0.53	0.19	0.02	25	160
	W/W	0.51	0.17	0.02	20	183
Ni	B/W	0.75	0.60	0.03	24	198
	W/W	0.66	0.48	0.04	15	237
Pt	B/W	0.83	0.94	0.04	26	185
	W/W	0.79	0.83	0.03	20	205
Au70/Pt30	B/W	1.05	1.36	0.11	26	196
	W/W	0.98	1.16	0.10	20	215

Wire bond trace lengths were determined by measuring the total trajectory of the wire bond in three dimensions and not simply the lateral distances. There is no major difference in insertion loss between ball/wedge bonds and wedge/wedge bonds [12]. The 2.73×1.23×0.01 mm Platinum was used as package leads. Fig.4 shows the equivalent circuit for wire bonding of packaged resonator.

The flip chip attaches structure and the electric equivalent circuit with the following parameters: flip chip height = 0.010 mm, flip chip diameter = 0.010 mm, flip chip material =

gold and also an isotropic conductive adhesive (Pyro-Duct 597A/C Volume Resistivity =0.2mΩ.cm) given in Fig.4. Pt package leads with 0.010×0.100×0.5 mm dimensions [13] as transmission lines are implemented on 0.5 mm thickness of $GaPO_4$ to endure the structure weight. Balls are soldered to the IDTs' final pads and adhered to transmission line via 2μm [13] adhesive thickness. Components definitions of equivalent circuit and values for two different insulator package materials, AlN and $GaPO_4$, have been given in Table 2.

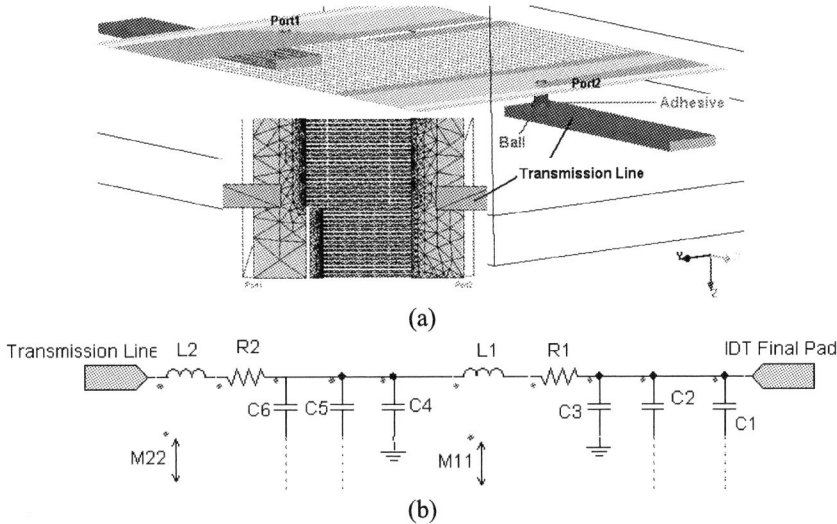

Fig. 4: Flip chip attaches structure (a) Maxwell Q3D view, and (b) Electric equivalent circuit

Table 2: Definition of electric equivalent circuit components and values

Components	AlN	$GaPO_4$
C1: Mutual capacitance between balls (fF)	12	12
C2: Mutual capacitance between balls and package (fF)	1.5	2.4
C3: Self capacitance between balls and ground plate (fF)	17.3	17.7
R1: Ball effective resistance (Ω)	0.003	0.003
L1: Ball effective inductance (pH)	15.1	15.1
M11: Mutual inductance between balls (pH)	1.5	1.6
C4: Self capacitance between adhesives and ground plate (fF)	17.2	17.2
C5: Mutual capacitance between adhesives (fF)	13.7	13.7
C6: Mutual capacitance between adhesives and package (fF)	1.55	1.68
R2: Adhesive effective resistance (Ω)	0.011	0.011
L2: Adhesive effective inductance (pH)	0.81	0.81
M22: Mutual inductance between adhesives (pH)	1.45	1.45

V. RESULTS

The resonator functions as a source with given characteristics in the Fig.3. Now this source is applied on the R_{21} and R_{24} in Fig.4 with given values in the Table 1 and then insertion and insertion loss on pins 1 and 4 are observed for a ball/wedge wire bonding are shown in Fig.5. The maximum return loss in the Fig.5 occurs at 418.0 MHz. The result is 3.67% of frequency deviation from the resonance frequency of SAW resonator that is 433.92 MHz in Fig.3. Note that gold wires and transmission lines as conductive interconnection materials and AlN as insulator have the less effect on signal integrity as comparison GaPO$_4$.

Meanwhile in Fig.6 show the insertion and return loss on package leads for a wedge/wedge wire bonding, where the maximum return loss has occurred at 421.5 MHz. The result is 2.86% of frequency shift. Furthermore gold as conductive material and AlN as insulator material have the less effect on the frequency shift and also from the smith-charts, the output signals show a capacitive property.

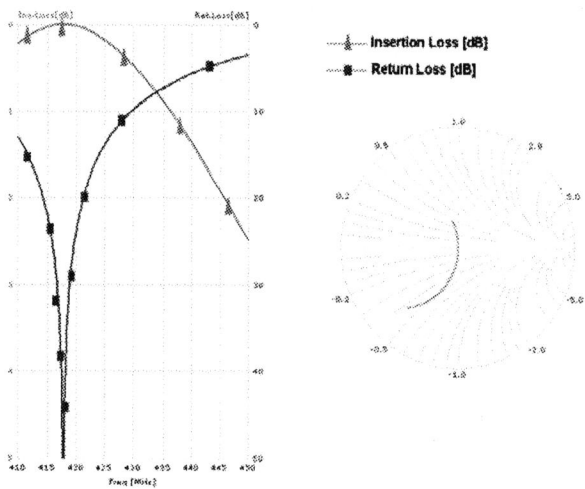

Fig.5: Insertion loss vs. return loss in ball/wedge bonding (Au been used as wire and Pt as external pad)

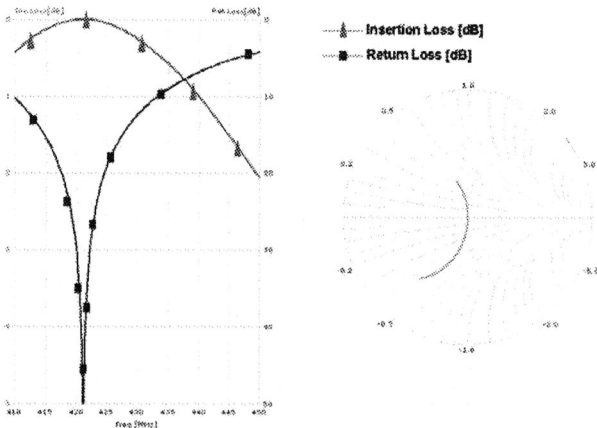

Fig.6: Insertion loss vs. return loss in wedge/wedge bonding (Au been used as wire and Pt as external pad)

738

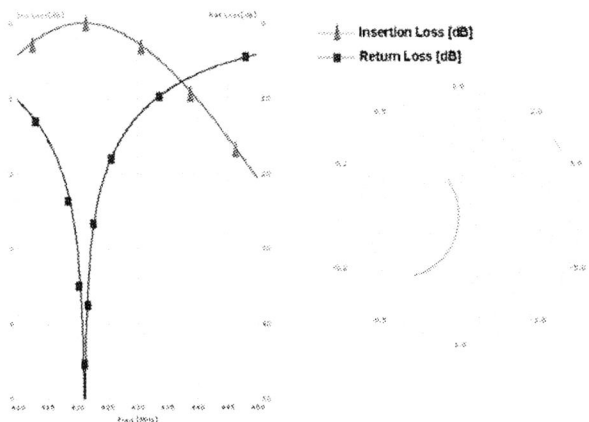

Fig.7: Insertion loss vs. return loss in flip chip (Au used as bump and Pt as external pad)

Table 3: The summaries of the frequency shift of packaged SAW resonator

Technique	Material				Frequency Shift (%)
	Package	Wire	Ext. pad	Ball	
Ball/Wedge Wire Bonding	GaPO$_4$	Au	Pt	–	3.88
	AlN	Au	Pt	–	3.67
	AlN	Au70/Pt30	Au70/Pt30	–	4.02
	AlN	Pt	Pt	–	4.13
	AlN	Au	Au	–	3.45
	AlN	Au	Au70/Pt30	–	3.56
Wedge/Wedge Wire Bonding	GaPO$_4$	Au	Pt	–	2.97
	AlN	Au	Pt	–	2.86
	AlN	Au70/Pt30	Au70/Pt30	–	3.06
	AlN	Pt	Pt	–	3.17
	AlN	Au	Au	–	2.67
	AlN	Au	Au70/Pt30	–	2.75
Ball Boding, Adhesive Attached	GaPO$_4$	–	Pt	Au	0.85
	AlN	–	*Pt*	*Au*	*0.72*
Ball Boding, Underfilled	GaPO$_4$	–	Pt	Au	0.93
	AlN	–	Pt	Au	0.88

In Fig.7 show the result of inserted resonator in FCP structure as been mentioned before. The response of resonance frequency in this structure was 430.8 MHz, which is closer to unpackaged SAW resonator in Fig.3 with only about 0.72% resonance frequency shifted. Also the smith-chart in the Fig.7 shows an approximately impartial in property of output signal either capacitively or inductively in given band width (410–450 MHz).

Finally, the three different constructions of package considered and their lumped elements along with values attained. In this way, different insulator and conductive materials been taken into account and return as well as insertion loss of these models been depicted. Frequency response of the package obtained from FDM showed that more components inside the package produce more frequency shift. The unwanted shifts cause the output signal of the resonator to be either capacitive or inductive. The minimum shift of frequency associated to flip chip packaging, is due to the less inductance injected into the equivalent circuit. Also comparing the results of FDM demonstrated less deviation of frequency response in flip chip compared to wire bonding.

Table 3 shows the summaries of the shift in frequency response due to different material of package and interconnection that been resulted from used packaging techniques for the HT SAW resonator. Table 3 also shows the minimum shift of frequency occurs in the flip chip gold–ball bonding with AIN as insulator material.

VI. CONCLUSION

Quasi–approximation FDM has been used to compare the effect of wire bonding and flip chip packages. The analysis also considers various materials for the packages, wires, bumps, and external pads. The results show the minimum shift of frequency occurs in the flip chip gold–ball bonding with AIN as insulator material.

REFERENCES

1. Yatsuda H., Oguri M., Horishima T., "Miniature SAW filters in GHz range," *Universal Personal Communications. 4th IEEE International*, pp. 321–324 (1995).
2. J. Hornsteiner, E. Born, G. Fischerauer, E. Riha. "Surface Acoustic Wave Sensors for High-Temperature Applications," *IEEE International Frequency Control Symposium*, pp. 615–620 (1998).
3. Joachim W. Mrosk, Lothar Berger, Christoph Ettl, Hans-Jörg Fecht, G. Fischerauer, and A. Dommann. "Materials Issues of SAW Sensors for High-Temperature Applications," *IEEE Transactions on Industrial Electronics*, vol. 48, no. 2, pp. 258–264 (2001).
4. J.A. Thiele and M. Pereira da Cunha. "High temperature surface acoustic wave devices: fabrication and characterization," *Electronics Letters Magazine*, vol. 39, vo. 10, pp. 818–819 (2003).
5. Naishadham W. K., Durak T., "Measurement-based closed form modeling of surface-mounted RF components," *IEEE Trans. On Microwave Theory and Techn.*, vol. 50, no. 10, pp 2276–2286 (2002).
6. Swanson D. G. Jr., Hoefer W. J. R., "Microwave Circuit Modeling Using Electromagnetic Field Simulation," *London: Artech House*, 2003.
7. B. Lewis, P. M. Jordan, R. F. Milsom and D. P. Morgan, "Charge and field superposition methods for analysis of generalized SAW interdigital transducers," *IEEE Ultrasonics Symp. Proc.*, pp. 413–416 (1978).
8. R. Hauser, R. Fachberger, G. Bruckner, W. Smetana, R. Reicher,A. Stelzer, S.

Scheiblhofer,S. Schuster. "A wireless SAW-based temperature sensor for harsh environment," *IEEE Proc.*, 0-7803-8692-2/04, pp. 860–863 (2003).
9. C.T. Tsai, "Package inductance characterization at high frequencies," *IEEE Trans. Comp., Packag. Manufact. Technol. B*, vol. 17, pp. 175–181 (1994).
10. R. C. Peach. "A general approach to the electrostatic problem of the SAW interdigital transducer," *IEEE Trans. Ultrasonics.*, vol. SU–28, pp. 96–104 (1981).
11. Mohd. Nizar Hamidon, "Fabrication of High Temperature Surface Acoustic Wave Devices for Sensor Applications", PhD thesis, Science and Mathematics School of Electronics and Computer Science, University of Southampton, UK, June 2005.
12. A. Becker, A. Lyons, K. Guinn, Y. Lee, H. Wu and M. Tsai, "Wire bond measurements up to 20GHz – A starting point", Wireless packaging research department, Bell Labs Research, 2003.
13. George A. Riley, "FlipChip.com",Tutorials, http://www.flipchips.com/tutorial.html. December 2000, Accessed on 22 January 2006.

Alignment Mark Architecture Effect on Alignment Signal Behavior in Advanced Lithography

[1,2]Normah Ahmad, [2]Uda Hashim, *Member, IEEE*, [1]Mohd Jeffery Manaf, [1]Kader Ibrahim Abdul Wahab

[1]SilTerra Malaysia Sdn. Bhd.
Kulim Hi-Tech Park,
09000 Kulim, Kedah.
[2]Kolej Universiti Kejuruteraan Utara Malaysia, Kangar, 01000 Perlis.
E-mail : normah_ahmad@silterra.com , uda@kukum.edu.my , jeff_manaf@silterra.com

Abstract **The downscaling of CMOS technology becomes a challenge to the scanner alignment system since overlay and alignment accuracy becomes tighter. Such a tight overlay requirement requires a very stable alignment performance. A stable alignment performance is indicates by a stable alignment signal generation. Hence, it is important to perform process characterization in order to choose an alignment mark, which generates the most stable signals. Different alignment mark type may show a different behavior in signal generation. In this paper, the signals behavior will be explored by experimenting using two different alignment mark architecture. This architecture can be further divided into three, which is AH32, AH53, and AH74. Based from the results, AH32 mark shows a significant trend difference between contact and metal mark. This is due to the fact AH32 contact mark is the easiest to be deformed since its feature size is the biggest compared to AH53 and AH74. AH53 and AH74 alignment signal performance between contact and metal mark are comparable.**

Keyword: Alignment signal, lithography, Alignment mark

I. INTRODUCTION

Besides being able to produce the desired pattern dimension, photolithography tools should be able to stack the new pattern layer accurately on top of the existing structure. The two requirements becomes a challenge in printing a smaller feature size structure. In order to be able to place the pattern correctly, the wafer and reticle should be aligned accordingly. This process is known as alignment and assisted by a special feature printed on wafer known as alignment mark [1]

A typical practice is to drop several alignment marks at each masking layers. This is important to give flexibility in alignment. If one type of alignment mark does not work, there still have another types of alignment mark to use. Since integrated circuit comprised of a series of oxide and metal layer placed alternately on top of each other [7], there will have two different set of alignment mark architecture. This is due to the difference between oxide and metal layer processing. This two alignment mark architecture is known as metal mark for alignment mark defined at metal layer and contact/via mark for those mark defined at the oxide layer.

Alignment system is operated based on the optics principle [11]. Hence at an oxide transparent layer, both previous layer metal and contact/via mark can be seen by the alignment system (refer to Figure 1).

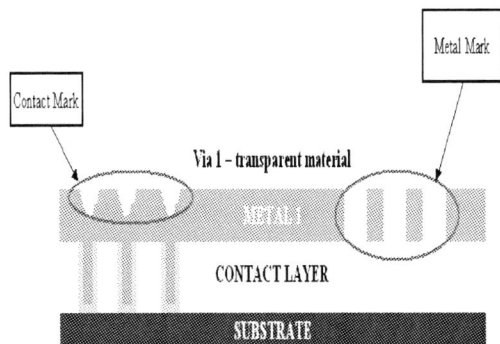

Figure 1: Alignment sensor optical system can see both metal mark and contact mark (in circle) since via1 layer is made up from transparent oxide layer.

The difference between contact mark and metal mark is the processing steps involved in defining the alignment mark.

a) Alignment Mark Formation for Contact Layer (Contact Mark) [4]

Alignment mark is printed on the wafer. This means that it is going through exactly the same process as the rest of the circuit structure. The first step in order to produce a contact structure is to deposit oxide material at certain required thickness. Since the deposition process does not produce a flat surface, hence, planarization process is required. Then, the oxide material will go through a patterning and etch process to produce the desired shape. Figures 2 illustrate the post lithography and etch process alignment mark.

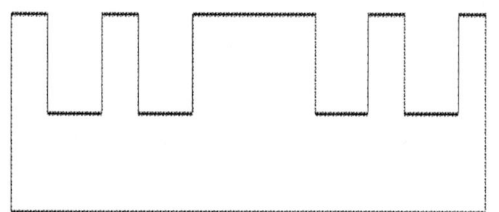

Figure 2: Post lithography and etch contact alignment mark

After that, tungsten material will be deposited. Tungsten material has to be completely removed from oxide upper surface. Tungsten is removed through chemical mechanical planarization (CMP) process. Contact hole will be completely filled by tungsten due to its small feature size. However, alignment mark trench won't be completely filled since its trench size is much larger than the tungsten deposition thickness (Figure 3). Figure 4 illustrates the alignment mark profile after tungsten planarization process.

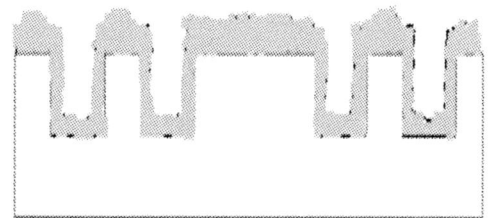

Figure 3: Alignment Mark Profile after Tungsten Deposition

Figure 4: Post Tungsten CMP Profile

Finally, metal/aluminum layer will be deposited on the existing structure through physical deposition process (PVD). Usually, the asymmetric profile becomes significant after this process. Figure 5 shows the possible profile after metal layer is deposited. This is final profile of contact alignment mark. The reflected signal behavior during metal1 layer alignment depends on this profile

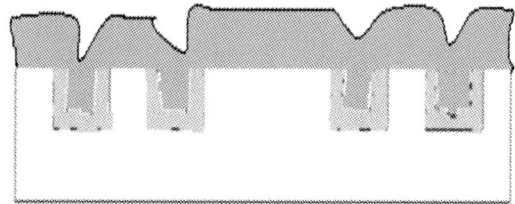

Figure 5: Post Aluminum/Metal Layer Deposition Profile.

b)   Alignment Mark Formation for Metal Layer (Metal mark) [4]

The first steps to define the metal alignment mark are to deposit the aluminum layer on the substrate (refer to Figure 6). The area where the contact mark is printed will have a topology. This is the final alignment mark profile for contact mark. At this time, metal mark is not printed yet.

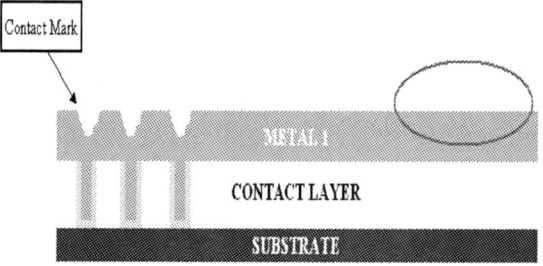

Figure 6: Prior Metal1 layer masking process. Red circle is the area where metal mark will be printed.

Figures 7 show post lithography and etch process profile for metal mark.

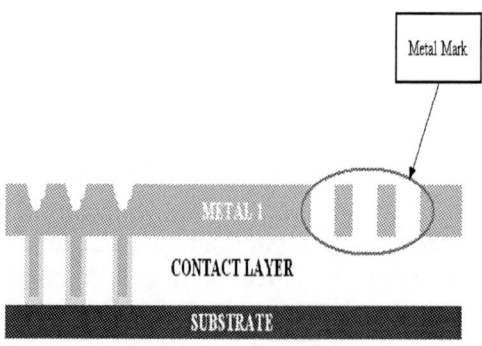

Figure 7: Final profile of metal mark.

From part a, 4 processes are able to affect the nature of contact alignment mark. The four processes are oxide planarization, tungsten deposition, tungsten planarization, and metal deposition. From part b, only etch and aluminum deposition thickness is able to change the nature of metal alignment mark. However, even though etch and deposition process able to change the profile, the change is usually not significant as mark deformation by planarizing process. Due this difference, it was expected that the alignment performance between contact and metal mark would be different.

II. RESEARCH METHODOLOGY

Based from the last paragraph in Introduction section, an experiment was designed in order to investigate how various processing steps affect the alignment signal performance of contact and metal alignment mark. A partial factorial DOE with 4 factors (n=4) and 9 center points, which require 25 runs, was designed. The four factors are post oxide planarization thickness, tungsten deposition thickness, tungsten planarization over polished time, and aluminum deposition thickness. Aluminum deposition thickness variation is to evaluate metal mark alignment performance. Each of the alignment mark will be evaluated for each particular condition. Split C acts as a center point in this DOE (refer to table 1). The alignment signal performance is indicated by alignment parameter.

Table 1: Process Split based from DOE

Split	Final ILD CMP Thickness	Tungsten Deposition Thickness	Tungsten Planarization Time
A	9200	3200	60
B	7800	2800	0
C	8500	3000	30
D	7800	3200	60
E	7800	2800	60
F	7800	3200	0
G	9200	3200	0
H	9200	2800	0
I	9200	2800	60

In this experiment, three main types of alignment mark were used which are AH32, AH53, and AH74. Please refer to table 2 for the details of alignment mark used in this experiment.

Table 2: Alignment Mark Type and their optimized light order recipe

Mark Type	Architecture type	Subtype	Optimized light Recipe
AH32	Contact Mark	AH32-CON-PRI	Third
	Metal Mark	AH32RM1-PRI	
AH53	Contact Mark	AH53-CON-PRI	Fifth
	Metal Mark	AH53RM1-PRI	
	Contact Mark	AH53-CV2V4-NPLB	
	Metal Mark	AH53R-M1M3-SCR	
AH74	Contact Mark	AH74-CON-PRI	Seventh
	Metal Mark	AH74RM1-PRI	

Alignment is a process of positioning a special mark on reticle at the correct orientation onto a specific spot above the wafer. In ideal cases, the reticle alignment mark and wafer alignment mark will corresponds with each other at the correct aligned position. The basic algorithm to find the alignment mark location is to find the centroid location of the alignment signal. Hence, it is important to ensure that the generated signal quality is good.

In ASML scanner system, several parameters are collected during alignment process and use as an indicator to the signal quality. The parameters are Wafer Quality (WQ), delta shift, mark residue, and multiple correlation coefficients (MCC). All of the parameters will be collected and recorded in this experiment. Wafer quality (WQ) is percentage of actual signal strength with reference to signal generated by fiducial mark. It is recommended that WQ values should be more than 1% in order to obtain a reliable alignment results. Delta shift is a shift value between signal generated by 8.0 micron and 8.8 micron grating. It is advisable for delta shift value not to be more than $\pm 0.3 \mu m$. If a delta shift value is too big, the system will recognize it as 8.0-micron error. This error is detected when wafer expansion (measured by scanner alignment system) value becomes too big. Mark residue is an error value between the actual alignment mark coordinate and theoretical alignment mark coordinate obtained from wafer model. The acceptable values should be within $\pm 130nm$. Multiple correlation coefficients (MCC) are a measure of fit for actual signal with reference to the fiducial signal. The values should be within 1 and 0.7. MCC values equals to 1 means a perfect fit.

In this work, alignment parameter is used to observe the alignment signal characteristics.

III. RESULT AND DISCUSSION

During via1 alignment process, the alignment optical system is able to see two different type of alignment mark architecture, which was defined at different masking layer. The two type of alignment mark is metal mark and contact mark. Metal mark is printed at Metal1 steps and contact mark is printed during contact masking layers. Both follows the same grating design as specified by the scanner manufacturer, which means that they are optimized to enhance a specific light order. The typical practice is AH32 mark is optimized for third order light, AH53 is for fifth order light, and AH74 is to enhance the seventh order light.

Metal mark and contact mark have several distinct characteristics. Contact mark is a trench feature while metal alignment mark is a grating of metal lines. Refer to figure 8 for the details of alignment mark architecture. The formation of metal mark does not involve planarizing process, hence CMP effect on the mark profile are not exist. The only process effect on metal mark signal quality is the oxide thickness (via1 ILD material) variation. However, in this experiment, the thickness is assumed to be constant across wafers (no process splits involved). It is expected that alignment signal quality from metal mark are higher and stable than contact mark.

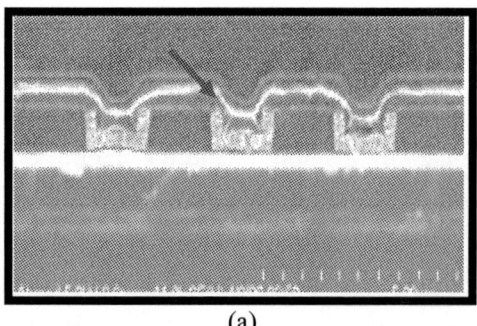

(a)

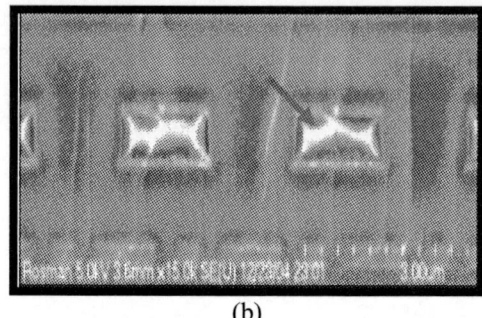

(b)

Figure 8: Cross sectional view of alignment mark (a) Contact Mark. The arrow shows a trench structure (b) Metal Mark. The arrow shows a metal line structure.

Wafer Quality/Signal Strength, Delta Shift, and Multiple Correlation Coefficient (MCC) were used as an alignment signal quality indicator in this study. Three important statistical parameters is the variation (standard deviation), its accuracy to the target (Cpk index), and mean difference. Depending on the alignment parameters, their values can be either should be high or low. Student's t test is used to determine the significance difference between two-alignment mark types.

Wafer Quality (WQ)/Signal Strength is expressed in term of percentage with reference to the fiducial alignment mark. The values should be as high as possible and if possible the variation should be as low as possible. A low WQ variation indicates a stable alignment signal. The threshold value for WQ is 1%. Any values less than 1% have tendency to produce alignment error [2].

Figure 9 shows alignment signal strength comparison using student's t-test between AH32 metal mark and AH32 contact mark. AH32-CON-PRI is contact mark and AH32RM1-PRI is metal mark. There is a significant difference for signal strength between these two-alignment mark types. In term of data variation, AH32 metal gives huge variation compared to AH32 metal mark. This can be seen from large standard deviation values as in Table 3 for AH32 metal mark compared AH32 contact mark. As for AH53 mark (refer to Figure 10 for details), both metal mark and contact mark does not have any significant difference between their signal strength values. Even though, as in Figure 10a, the circle is not exactly overlapped (non overlapped circle shows that there is a significant difference for that particular parameters), but when referring to Table 3, the t-test value is very small which means that the difference is not significant. Both AH53 metal marks (one is at primary area, the other one is at scribe lane area) show a slightly high standard deviation (refer to Table 3) compared to the contact mark. Referring to Figure 11, AH74 contact mark and metal mark does not have any significant difference in signal strength. From Table 3, standard deviation data shows that their variation is approximately the same. As the grating size decrease from AH32, AH53, to AH74, the percentage of signal strength that fall below the spec limit are decrease (Table 3). For AH53 and AH74 mark, the results are comparable for both percentages outside the spec limit and Cpk parameter.

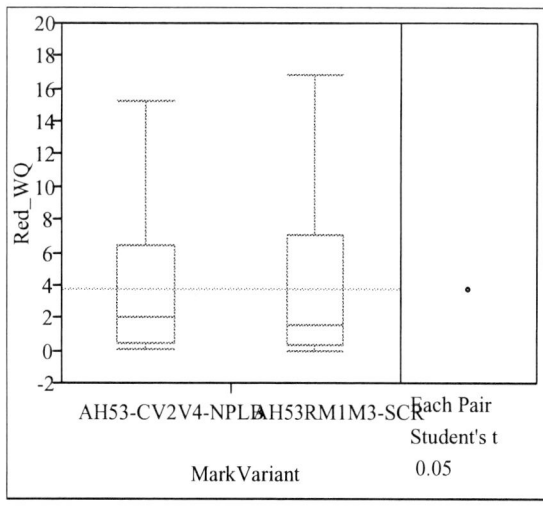

(b)

Figure 10: Alignment signal strength for AH53 Mark (a) Alignment Mark is inside the primary area (circuit area) (b) Alignment is printed at the scribe lane area

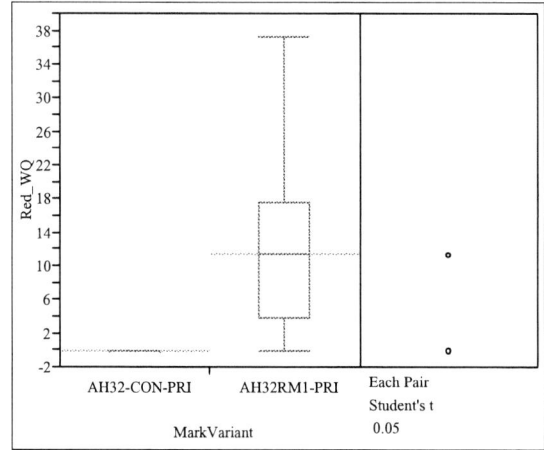

Figure 9: Alignment Signal Strength for AH32 mark

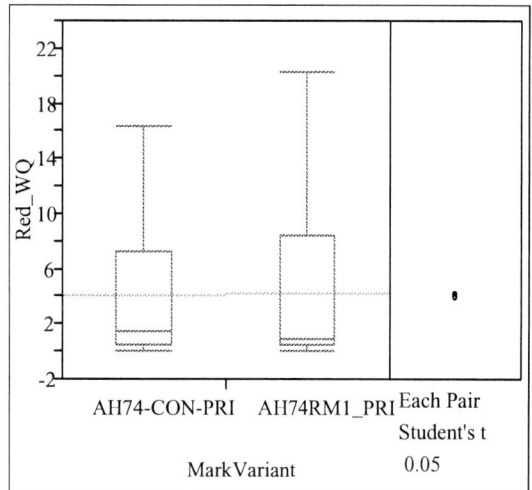

Figure 11: Alignment signal strength for AH74 mark

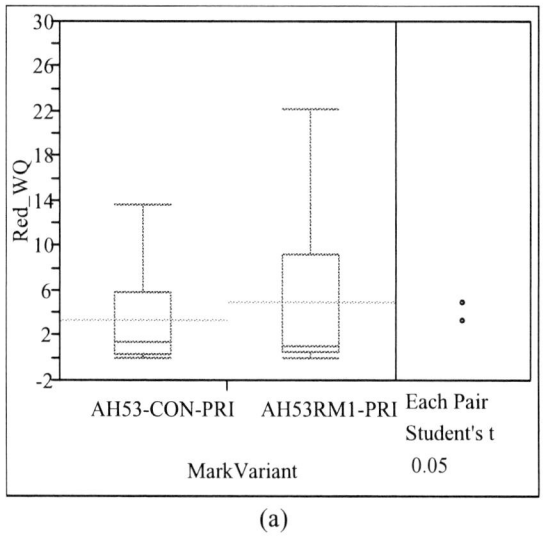

(a)

Table 3: Wafer Quality/Signal Strength Parameters

Alignment Mark Type	Wafer Quality/Signal Strength			
	Standard Deviation	t-test Comparison	Cpk	% Outside spec limit
AH32-CON-PRI	0.076079	10.964	-4.340	99.9241
AH32RM1-PRI	8.0327294	10.964	0.432	9.8958
AH53-CV2V4-NPLB	3.9592411	-0.19146	0.237	48.0655
AH53RM1M3-SCR	3.815505	-0.19146	0.245	44.6591
AH53-CON-PRI	3.6322357	1.2566	0.225	48.6742
AH53RM1-PRI	6.2051944	1.2566	0.214	48.6235
AH74-CON-PRI	4.1627944	-0.00398	0.246	45.6061
AH74RM1-PRI	5.0714336	-0.00398	0.218	49.8140

Significant results for metal-contact mark comparison can be clearly seen for AH32 mark type. This is due to the fact that AH32 contact marks are more easily affected by planarizing and deposition process since its feature size are larger compared to the other marks. A change in alignment profile by processing steps can significantly alter the signal strength behavior, which is happened to AH32 contact mark. Meanwhile, AH32 metal mark shows a huge variation in signal strength.

Figure 12 shows the possible reflected light path when the incident light hit the metal mark. A, C and E is the incident light. Reflected light from metal mark behavior can be divided into several conditions. The first condition it is incident toward the bottom of the alignment mark and gets reflected back straight away (B from Figure 12). The second condition is it hit the bottom of the alignment and gets reflected several times under the metal lines before reach the detector (D from figure 12). The third condition is when the light is incident towards the metal lines (E from Figure 12) and reflected back to detector (F). Since there is almost no process effect on metal mark, the resulted signals will be very high. However, reflected light D behavior will be very difficult to predict since it get reflected several times before reaching the detector. There is a possibility the resulted signal out of phase with the other light creating a destructive interference. This is the reason why metal mark can go down to a certain extreme value even though other alignment mark gives good signal strength. As for the via mark, their signal strength values is dependent on the mark profile and alignment mark depth since the incident will only get reflected from the mark surface (Figure 13)

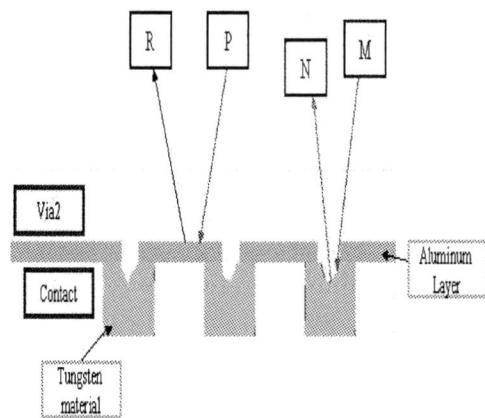

Figure 13: A reflected light behavior from contact mark.

Another parameter that can be used to indicate the alignment signal quality is Multiple Correlation Coefficient or MCC parameters. It measures the resemblance of the actual signal with respect to reference signal. MCC values equals to 1 indicates an exact fit. From Figure 14, AH32 contact mark shows huge MCC variation. A lower MCC value indicates either the signal is too noisy or the signal strength is very low compared to the reference signals. AH53 and AH74 mark (from Figure 15 and Figure 16) gives comparable results between the two-alignment mark architecture.

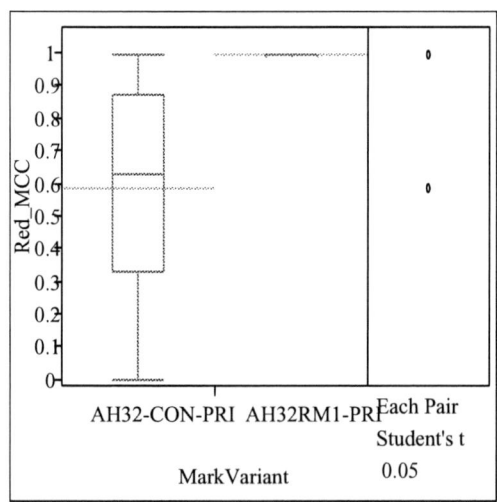

Figure 14: MCC trend for AH32 mark

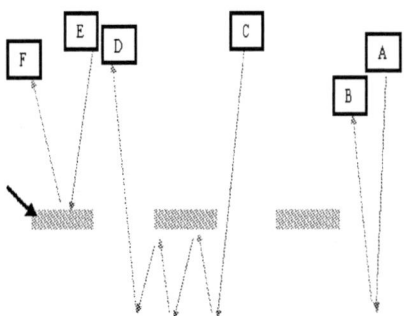

Figure 12: A possible reflected light behavior when using metal mark. The black arrow shows the metal lines for alignment mark.

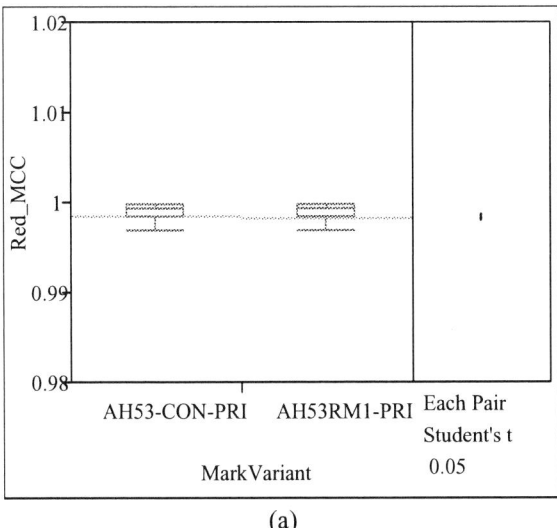

(a)

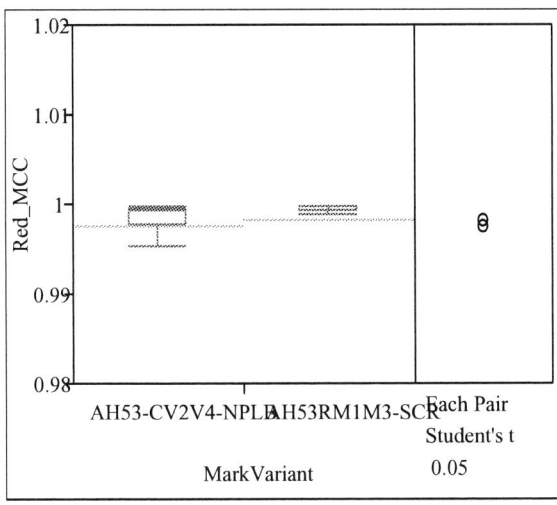

(b)

Figure 15: MCC trend for AH53 mark. (a) AH53 mark at primary area, (b) AH53 at scribe lane area

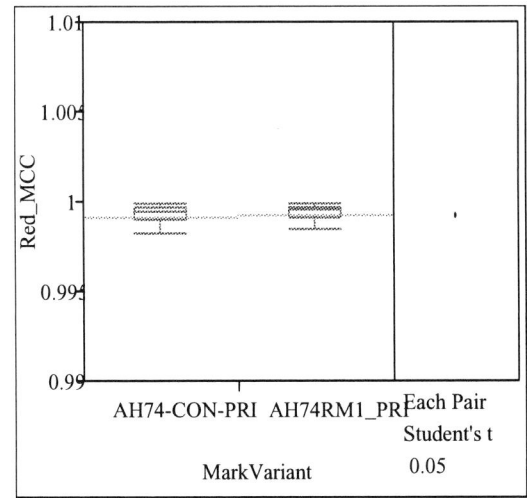

Figure 16: MCC trend for AH74 mark

A clear difference in MCC trend for AH32 alignment mark is due to the effect of processing on the nature of alignment mark. The nature of AH32 contact mark can be easily affected by processing since its feature size is bigger than AH53 and AH74 mark. More destructive interference event from uneven surface created by asymmetric surface [5] reducing the overall signal strength to near zero for AH32 contact mark. AH32 metal mark has a good signal quality since mark profile does not affected by fabrication process.

IV. CONCLUSION

Generally, the significant trend between contact mark and metal mark can be clearly seen from AH32 alignment performance results. This is due AH32 large feature size, which can be easily affected by processing. AH74 and AH53 contact-metal mark comparison gives comparable results, which means that if the nature of alignment mark does not changes significantly, alignment signal quality from contact mark and metal mark, will be almost the same. However, when there is a significant change in the nature of alignment mark as in AH32 mark cases, metal mark is expected to give a good alignment signal quality because the nature of metal alignment mark is not really affected by processing steps.

ACKNOWLEDGEMENT

The author would like to thank Silterra (M) Sdn Bhd and its employee for providing a support and facilities for this study.

REFERENCES

[1] Harry J. Levinson, *Principles of Lithography*, pg 201 – 223, SPIE Press (2001).

[2] ASML, *"ATHENA Knowledge Sharing"*, Application Note, pg 24 – 28, ASML (1999).

[3] Yuanting Cui, Albert So, Sean Louks, "Fine Tune W-CMP Process with Alignment Mark Selection for Optimal Metal Layer and Yield Benefits", *Proceeding of SPIE*, Vol. 5375, pp. 827 – 838 (2004).

[4] ASML, *"Process Effects on Alignment: Tungsten CMP/Aluminum PVD"*, Application Note, pg 5 – 17, ASML (2004).

[5] Boston University*, "Diffraction; Thin Film Interference"* online from http://physics.bu.edu, (1999).

[6] *"Trench Filling by Physical Vapor Deposition"*, online from www.ipm.virginia.edu, pg 138 – 173 (2000).

[7] James R. Sheats, Bruce W. Smith, *"Microlithography, Science and Technology"*, pg 317 – 367, Marcel Dekker Inc (1998).

[8] S. Wolf, R.N. Tauber, *"Silicon Processing for the VLSI Era Volume 1- Process Technology"*, pg 582-588, 2nd Edition, Lattice Press, (2000).

[9] Pantazis Mairoulis, John McDonald, *"Geometrical Optics and Optical Design"*, pg 175 – 180, Oxford University Press, (1996).

[10] Rouchouze, Eric, Darracq, Jean Michael, Gemen, Jack, " CMP-Compatible Alignment Strategy", *Proc. Of SPIE,* Vol 3050, pg 282 – 292 (1997).

[11] Thomas V. Pistor, Robert J. Socha, "Rigorous Electromagnetic Simulation of Stepper Alignment", *Proc of SPIE*, Vol. 4689, pg 1045 – 1056 (2002).

Device Characteristics of HEMT Structures based on Backgate Contact Method

Norman Fadhil Idham M., Nurul Afzan O., Hariyadi Soetedjo, Ahmad Ismat A.R., Idris Sabtu,
Mohamed Razman Y. and Abdul Fatah A.M.
Microelectronic and Nano Technology Program,
Telekom Research and Development Sdn. Bhd., Idea Tower, UPM-MTDC,
43400 Serdang, Selangor, Malaysia.
Email:fadhil@tmrnd.com.my

Abstract This paper presents a novel technique to obtain device characteristics of High Electron Mobility Transistors (HEMT) structures based on the backgate contact method, thus avoiding the need for complete gate formation. The gate contact was prepared on the back side of the substrate. Measurements performed on various HEMT structures shows typical transistor characteristics. Significant changes in drain-source current as a function of backgate voltage bias was observed for different HEMT structures. Increasing the channel thickness from 8 to 26 nm shows an increase in the threshold voltage of the transistor and a noticeable variation in drain-source current. This result leads to an effective and novel technique for the determination of sample quality prior to the further fabrication process to obtain the complete device.

I. INTRODUCTION

High Electron Mobility Transistor, (HEMT) is a well-known electronic devices being used widely for telecommunications technologies which show high performance in power, low noise, and other superiorities [1].Uses of probes in the electronic devices measurement have been done such for MOSFET devices. For this measurement, source and drain contacts of alloys used by probing mechanically onto the surface of devices structures [2]. The accuracy of probe contact technique is suffering from the Schottky contact, high parasitic probe to surface series resistance and approximation determination of geometric factor as this have also been occurred for MOSFET device [3].

II. MEASUREMENT TECHNIQUES

For measurement, HEMT structures were probed using commercial probe (JANDEL) with 2 μm

tip radius, meanwhile backside of sample (substrate) was deposited with Aluminium using a sputtering technique. Those two probes were separated at a distance of 3um and are used to mimic source and drain contacts, meanwhile metal deposition on the back was for a gate contact as the picture taken is shown in Figure 1.

Table 1. Cross section structure for HEMT used in the measurement.

Layer	Concentration (cm^{-3})
30 nm GaAs; Si (cap)	3×10^{18}
70 nm AlGaAs (supply)	10^{17}
6 nm AlGaAs (spacer)	undoped
InGaAs (channel) thickness, d	undoped
7.5 nm Superlattice : GaAs (10 x) AlGaAs (10 x)	undoped
1000 nm GaAs (Buffer)	undoped
Semi-insulating GaAs (Substrate)	undoped

Figure 2 shows a schematic diagram of the measurement set up for the samples. The HEMT structure used for the measurement was tabulated in Table 1. By referring to the structure mentioned, the channel thickness, d of the sample was varied for 8 nm (denoted as s-chan8), 12 nm (denoted as s-chan12) and 26 nm (denoted as s-chan26).

0-7803-9730-4/06/$25.00 ©2006 IEEE

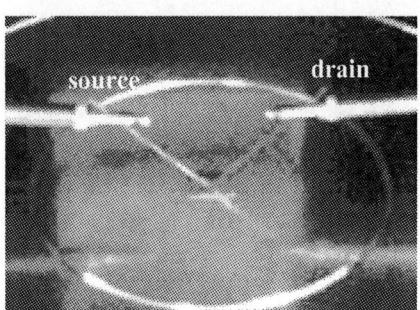

Fig. 1 The picture of measurement set up used for samples.

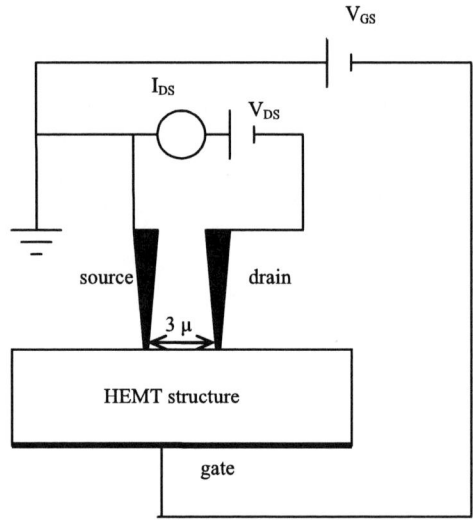

Fig. 2. Schematic diagram for sample measurement.

III. RESULTS AND DISCUSSION.

From the measurement carried out at room temperature, a current-voltage characteristic was obtained for a sample with channel thickness of 26 nm (s-chan26) as shown in Figure 3. In that figure, V_{GS} has been varied for -5, -2.5, 0, 2.5, and 5 Volts. The curves inform us that gate contact introduced from the backside of substrate affects to the current flow modulated from source to drain. The change of current flow observed is assumed to the carrier recombination due to the increment of the gate bias. This recombination is due to the attraction between negative charges introduced by negative doping (in cap and supply layers) and positive charges from the gate contact by positive bias voltage. For the confirmation, we simulated this characterisation method by using a commercial two dimensional Technology Computer Aided Design (TCAD). For the simulated structure we can only simulate sample with 25um height, 5000um length and 1um width which is different with the real sample, this is because of limitation

number of nodes for the simulation. The main propose of the simulation is to study the current pattern when we biased the gate at the backside of the sample. From Figure 4, the current-voltage characteristic is showing the same pattern as in Figure 3. The current is decrease when we increased the Vgs, from -5V to 5V. The current value is not the same due to different size between the real sample and the simulated one. Figure 5 is showing the current flows from the simulation. The current flow from the probe to the substrate/backside gate at Vg=-5V and Vg=5V. The current flow lines are increase when the gate voltage is increase and hence decrease the drain current.

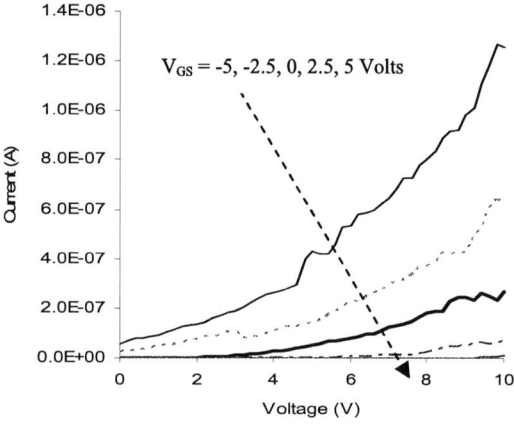

Fig.3 Current-voltage characteristics obtained from the measurement for channel thickness of 26 nm (s-chan26) for variation of V_{GS}.

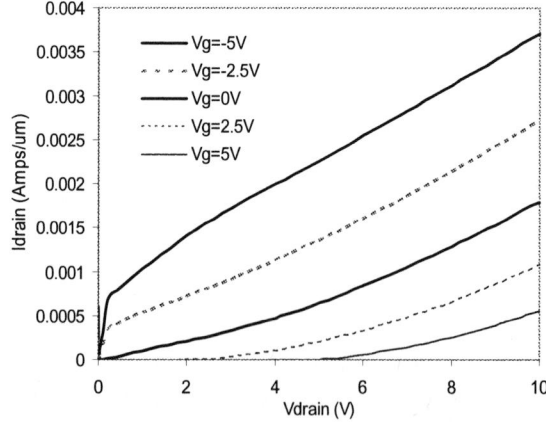

Fig.4 Current-voltage characteristics obtained from the simulation for channel thickness of 26 nm (s-chan26) for variation of V_{GS}

750

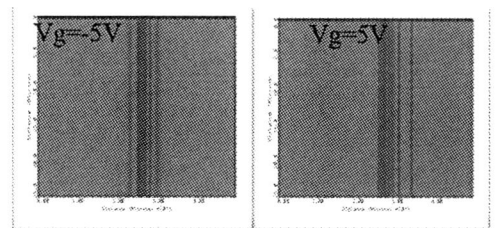

Fig.5 Current flow at Vg=-5V and Vg=5V.The current lines from the drain to the substrate are increased when the Vg is increased.

Meanwhile, current-voltage characterization for different structures of HEMT is shown in Figure 6. From the figure 6, the curve for channel thickness of 8 nm is found to be lower in turn-on voltage compare to that of the thicker channel thickness, but magnitude of current flow is observed to be relatively higher. This greater current flow is considered to be caused by the effective channel layer thickness, which allows the electron moving relatively faster for thicker channel compare to the thinner channel. From this significant different of current flow magnitude and also turn-on voltage of characteristics, the qualitative assessment of the HEMT structure could be done through the simple measurement technique proposed.

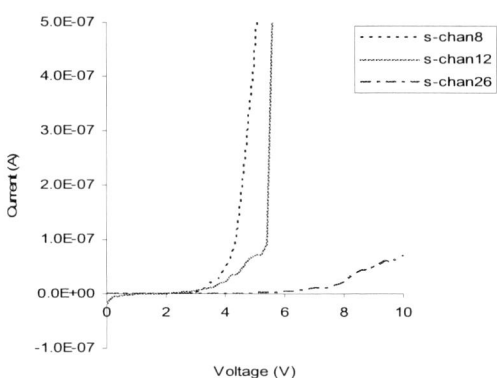

Fig.6 Current-voltage characteristic for various structures of HEMT (V_{GS} was biased at 2.5V).

IV. CONCLUSION

From the measurement, it was found the significant different of diode characteristic for different HEMT structures. This method is simple, non-destructive and less time needed for the experiment.

ACKNOWLEDGEMENT

This work has been supported by Telekom Malaysia Sdn. Bhd. through research project R05-0607-0 'Design and Modeling of Pseudomorphic-HEMT (p-HEMT) MMIC Technology for Wireless Device Applications in Access Network'.

REFERENCES

[1] Prashant Chavarkar and Umesh Mishra, Hand book of in Film Devices, Chapter 3, 2000

[2] Munteanu D., Cristoloveanu S. and Guichard E., 1999; Sol. Stat. Elec. 43: 547 – 554.

[3] Ionescu, A.M. and Munteanu, D., IEEE Elec. Dev. Lett., 2001; 22(4): 166 – 169

Effect of Indium Content in the Channel on the Electrical Performance of Metamorphic High Electron Mobility Transistors

Norman Fadhil Idham M., Ahmad Ismat A.I., Rasidah S., Asban D.,Mohamed Razman Y., Abdul Fatah A.M.

Microelectronic & Nano Technology
Telekom Research & Development Sdn. Bhd.
Idea Tower, UPM-MTDC
Technology Incubation Center One
Lebuh Silikon, 43400 Serdang, Selangor
E-mail : fadhil@tmrnd.com.my

Abstract- **Metamorphic InAlAs/InGaAs High Electron Mobility Transistors (HEMT) has demonstrated several advantages over Pseudomorphic-HEMT on GaAs and Lattice Matched-HEMT on InP substrate. The high Indium content of the channel (50%) lattice matched to the substrate is the key factor behind the superior Metamorphic HEMT performance. Metamorphic HEMT allows a flexible range of InGaAs channel compositions from 30% to 80 % (based on the applications) [1] on a compositionally graded buffer. Commercially available TCAD is used to simulate the Metamorphic HEMT to study the effect of varying Indium % in the channel layer on the electrical characteristics of the device.**

I .INTRODUCTION

The InP based High Electron Mobility Transistors (HEMT) and heterojunction bipolar Transistors (HBTs) present significant improvements over GaAs-based counterparts in term of higher operating frequency, reduced power consumption an improved noise characteristics. But InP is expensive, produced in small scale, and difficult to produce.-limit the potential of InP-based HEMTs and HBTs for high volume production. The solution for these problems is the Metamophic buffer on the GaAs substrate with the InP like structure for the active layer. This solution will realize the high volume production, low cost and high performance InP-like devices [2].

The basic obstacle in the development of the metamorphic devices is the large lattice mismatch between the epilayer and substrate,

which give a high density of threading dislocations through surface morphology. The metamorphic buffer allows higher Indium content in the channel layer and hence the high electron mobility and higher performance [3].

In typical mHEMT device, the lattice mismatched caused by high indium content layers on GaAs are eliminated in by a properly grown lattice-grading $In_xAl_{1-x}As$ buffer between substrate and active device layer. In this work we use use two-dimensional device simulations to investigate effects of varying Indium content in channel layer on threshold volgate, transconductance, electron mobility, conduction band and cut-off frequency.

II. SIMULATION PROCEDURE

The mHEMT cross section used in the simulations is shown schematically in figure 1. It consists of a 50Å of InGaAs cap layer doped to 1×10^{19}cm^{-3}, 150Å undoped $In_{0.48}Al_{0.52}As$ supply layer, 120Å undoped $In_xGa_{1-x}As$ channel, 30Å $In_{0.48}Al_{0.52}As$ for upper spacer, 150 Å $In_{0.48}Al_{0.52}As$ buffer, 30Å $In_{0.48}Al_{0.52}As$ for lower spacer and 1µ of AlInAs grading buffer on GaAs substrate. To model the different layers correctly the mHEMT is split into different regions and the discontinuous behavior on interfaces between layers has to be treated by specific interface models to link the layers together. Shockley-Read-Hall and Auger recombination models are used to model recombination with concentration dependant lifetimes.Fermi-Dirac statistics is used to model carrier transport and the "ANALYTIC" technique is used to model concentration and temperature dependant mobilities. The source and drain

0-7803-9730-4/06/$25.00 ©2006 IEEE

contacts are placed on top of the cap layer, similar to the real device. The work function used for the Schottky gate is 4.82eV which is workfunction for gold. Newton's method with two carriers was used in the AC analysis to get accurate gate capacitance, C_G values at reasonable computation time and successful convergence of the biasing scheme [4].The current-gain cut off frequency, f_T is calculated by the approximation $f_T = g_m/(2\pi.C_G)$. In this paper, we vary the indium content in the channel layer, from 30% to 80% with 10% increment.

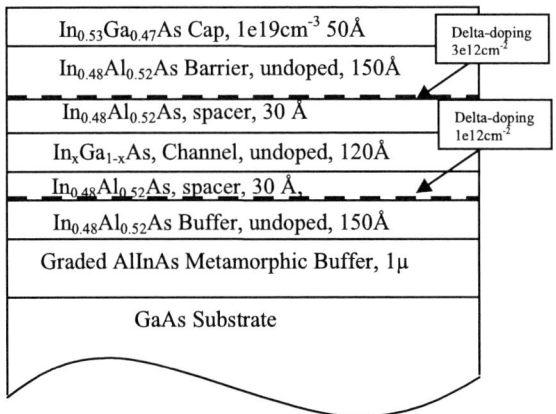

Fig. 1.Schematic cross-section of the simulated MHEMT.

III. RESULTS AND DISCUSSION

From the simulation, Indium content at 70% and 80% has show effect on the Id vs Vg graph (Figure 2). Both of them demonstrate low drain current, Id $\approx 1.7X10^{-3}$ Amps/μ for Indium 70% and Id $\approx 2.5X10^{-4}$ Amps/μ for Indium 80% at Vg = 0.5V. As for the others Indium content, Id $\approx$ 4X10-3 Amps/μ at Vg = 0.5V. Varying Indium content in channel has not effect the threshold voltage (Figure 2). As for the transconductance (Figure 3), Gm, the shape of the graph have shift for 70% and 80% to the left and give small transconductance value. From Figure 4, channel with 80% and 70% Indium content give high gate capacitance value.

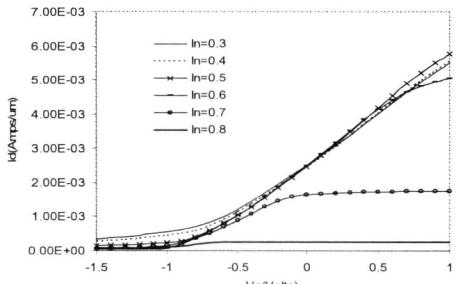

Fig. 2. Id vs Vg for different Indium content in Channel layer.

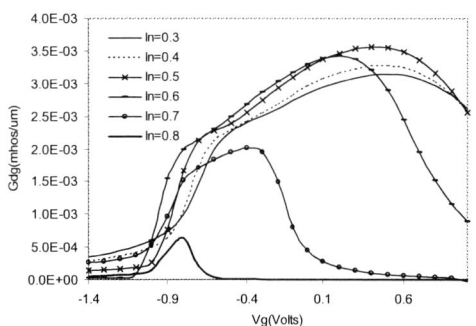

Fig. 3. Transconductance for different Indium content in channel layer.

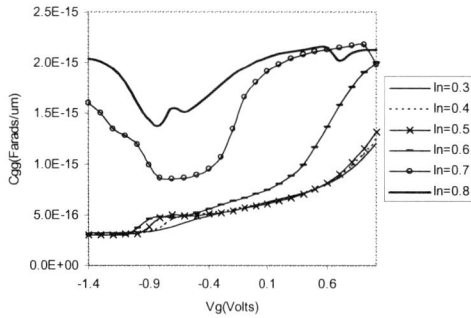

Fig. 4. Cgg for different Indium content in channel layer.

For the cut-off frequency,fT, the 0.3, 0.4, 0.5 and 0.6 Indium content demonstrate nearly the same value, fT $\approx 8.5X10^{11}$ Hz. The lowest fT is for Indium = 0.8 at $7.36X10^{10}$Hz. This behaviour can be explain by refering to the cut off frequency formula, where we can observe it depends on the transconductance and gate capacitance value. Although the electron mobility in the channel have increase (Table 1) linearly to the percentage increase of the indium in the channel layer, the cut off frequency is not effect by the electron mobilty. As for the 80% Indium, its transconductance is small and gate capacitance is high, so the cut off frequency is low than the others.The highest cut off frequency is demonstrate by 50% Indium at 8.83×10^{11}Hz. Both of the electron concentration (Figure 6) and conduction band (Figure 7) in the channel layer are not affected by varying the Indium.

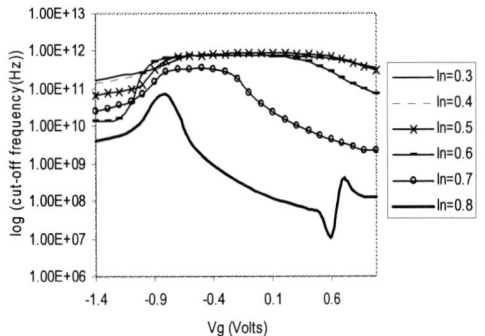

Fig. 5. Cut-off frequency for different Indium content in channel layer.

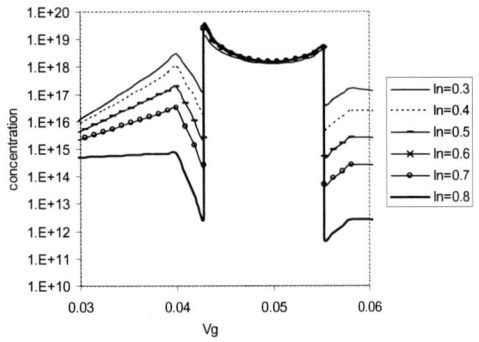

Fig 6. Electron concentration in supply and channel layer for different Indium content in channel layer.
1

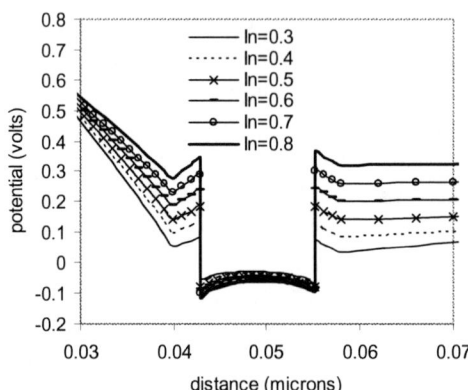

Fig.7 Conduction band in the supply and channel for different Indium content in channel layer.

Table 2. Electron mobility in channel for different Indium in channel layer. Electron mobility linearly increases with Indium content.

Indium mole fraction in channel	Electron mobility $(cm^2v^{-1}s^{-1})$
0.3	8667
0.4	9486
0.5	10920
0.6	12960
0.7	15620
0.8	18880

IV. CONCLUSION

By Increasing the Indium in the channel, the electron mobility in the channel also have increase. The increment of Indium is limited to 70% and above which is resultant to poor performance of the mHEMT. Channel with 50% Indium have shown better performance.

ACKNOWLEDGEMENT

The authors would like to thank Telekom Malaysia Sdn. Bhd. for providing a research grant to perform the work.

REFERENCES

[1] C.S. Whelan, et al.,"High Frequency Power Metamorphic HEMT" GaAs Mantech, Inc. Digests.

[2] W.E.Hoke, et al., Progress in GaAs-Based Metamorphic Technology GaAs Mantech, Inc. (2001)

[3] Yi-Chung Lien et al.,A Metamorphic High Electron-Mobility Transisitor with Reflowed Submicron T-Gate for High- Speed Optoelectronic Application, IEEE.vol .3,p.7803-7887.(2003)

[4] Taurus Medici, Version X-2005.10. (2005)

[5] Wei-Chou, et al., Charcteristics of In0.425Al0.575 as-InxGa1-xAs Metamorphic HEMTs with Pseudomorphic and Symmetrically Graded Channels. IEEE Transactions on Electron Devices, Vol.52, no.6, p.1079-1086 (2005)

Design of an RF BJT-Low Noise Amplifier at 1GHz

Sharifah Fatmadiana Wan Muhamad Hatta, Norhayati Soin, *Member IEEE*

Faculty of Engineering
Universiti Malaya
50603 Kuala Lumpur, MALAYSIA
Email: sh_fatmadiana@um.edu.my

Abstract A fully-integrated RF Low Noise Amplifier (LNA) suitable for low-voltage applications is proposed using bipolar junction transistors (BJT) cascaded stages. The proposed design aims to provide gain with low bias current consequently lower power dissipation and lower Noise Figure (NF). The circuit is designed and simulated using MultiSim9 from Electronics Workbench DesignSuite Edition 9. The proposed amplifier exhibits 3.296dB small-signal gain, reverse isolation of -6.68dB and 0.359dB noise figure at 1GHz.

I. INTRODUCTION

Wireless communication and its applications have travelled through rapid growth in recent years. Cellular systems, WLANS, Bluetooth as well as WPANs have undergone numerous generations of evolution in the swift development in wireless communication [1]. The radio frequency (RF) front-end electronics plays an important part in high level integration of radio solutions. The low noise amplifier is one of the most critical building blocks in modern integrated radio frequency solutions. The front-end low noise amplifiers have been widely used in many applications including wireless personal communication systems.

This paper presents a circuit topology of the Bipolar Junction Transistor Low Noise Amplifier (BJT-LNA) operating at 1GHz, utilising the cascode configuration. This configuration proposes tradeoffs between gain, noise and blocking performances [2]. The cascode configuration produces high isolation from the output back to the input, hence generating less amount of reverse transmission. This configuration should also

produce high output impedance that is useful in achieving large voltage gain.

The schematic of the proposed LNA design is presented in Section II in which the elaborations on the LNA topology and typical trade-offs in the design are included. The experimental and simulation results are presented and analysed in Section IV. Section V summarizes the main contributions of this paper. The future prospects of the design are presented in Section VI.

II. CIRCUIT DESIGN

The complete schematic of the 1.0 GHz LNA is depicted in Figure 1. The method employed in designing the circuit is the implementation of the inductive emitter degeneration in a cascode configuration circuit. This configuration is chosen due to the fact that the LNA has the advantages operating at high frequency and obtaining frequency stability at high gain due to the presence of the common base transistor Q4. This common base transistor provides high isolation between the input and the output ports and also mitigates the effect of Miller amplification of the base-collector capacitance of Q1 [3].

0-7803-9730-4/06/$25.00 ©2006 IEEE

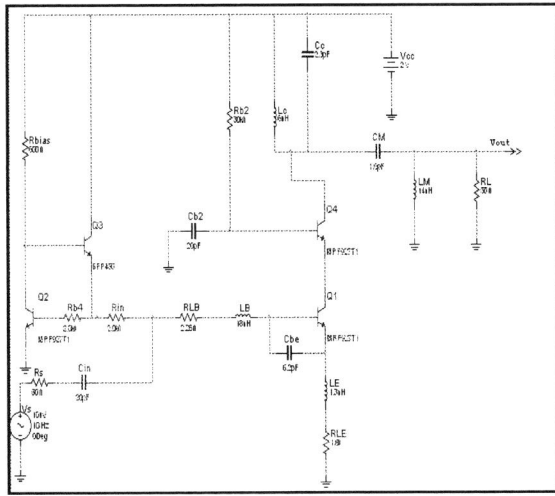

Figure 1: Complete schematic of the 1.0GHz BJT-LNA

The cascode LNA consists of transistors *Q1* and *Q2* with an emitter degeneration provided by L_E. The components of the matching network that comprises of L_B and L_E were selected to resonate at 1.0GHz in order to provide the desired passband response.

Mismatch causes drawbacks such as power loss and hence, signal-to-noise reduction. A matching network at the input of the LNA significantly determines the noise figure. By choosing the optimum matching network can minimize the noise figure considerably. The design employs a matching topology using inductive emitter degeneration. The inductive emitter degeneration inductance L_E is chosen to provide the desired pure real input resistance R_S (50Ω) at resonant frequency without using real resistors which tend to add thermal noise. Connecting an additional inductor L_B in the base effectively cancels the reactance due to the input capacitance, C_{in} of the device and it transforms the optimum noise reactance of the amplifier to 0Ω.

The input impedance is real at resonant frequency [4];

$$Z_{in} = s(L_E + L_B) + 1/(sC_M) + (g_m/C_M)L_E + r_b + r_e \quad (1)$$

where r_e is the parasitic emitter resistor, r_b is the base resistor and $C_M = C_\pi + C_\mu$ where C_μ is the Miller capacitance.

At the series resonance,

$$Z_{in} = (g_m/C_M)L_E + r_b = \omega_T L_E + r_b \quad (2)$$

The input impedance of the LNA is matched to the source resistance R_S when the following matching conditions are met [4];

$$\omega_T L_E + r_b \approx R_S \quad (3)$$

$$\omega_0^2(L_E + L_B).(C_\pi + C_\mu) \approx 1 \quad (4)$$

The proposed design also utilises a shunt base-emitter capacitor to improve the noise figure performance of the circuit. By introducing a capacitor C_{BE} between the base and the emitter nodes of the input transistor, filtering on the input noise components can be achieved effectively [4]. By simulation trials, C_{BE} which gives minimum noise figure is $C_{BE} = 6.2pF$.

III. TYPICAL TRADE OFFS IN LNA DESIGN

An LNA design presents a great challenge because of its simultaneous requirement for high gain, low noise figure, good input and output matching and unconditional stability at the lowest current draw from the amplifier. Although gain, noise figure, stability, linearity and input and output match are all equally important, each of these parameters are independent and rarely work in each other's favour.

Typically, the proposed LNA requires:

- Low supply voltage
- Low current consumption, hence ultra-low power consumption
- High gain
- High input third-order intercept point
- Low noise figure
- Unconditionally stable

- Input return loss
- High isolation
- Small dimension/low part count
- Low cost

Most of these conditions can be met by carefully selecting a transistor, choosing the right component values and understanding parameter tradeoffs. Low noise figure and good input match can be simultaneously obtained using feedback configurations.

High gain at gigahertz frequencies, in addition to producing intermodulation distortion, can lead to instability. Unconditional stability requires a certain gain reduction. High IP3 requires higher current draw, although the lowest possible noise figure is usually obtained at lower current levels.

IV. SIMULATION RESULTS

The proposed circuit in Figure 1 was designed using the model parameters of a high performance silicon bipolar technology by Motorola which can be found in the MultiSim9 model database. In order to obtain the desired simulation results, an RF BJT_NPN transistor of model number *MRF927T1* was used in the first input stage of the circuit as well as the second stage of the cascode design. This model has been designed for use in low voltage, low current applications at frequencies up to 2.0GHz [5]. In terms of the biasing transistor, the Siemens NPN Silicon RF Transistor of model number *BFP 450* was used [6].

For the proposed cascode LNA shown in Figure 1, the simulation result is presented in Figure 2-Figure 4. The designed LNA requires only a 2V supply voltage. The circuit is designed and simulated using MultiSim9 from Electronics Workbench Design Suite Edition 9. At 1.0GHz, the proposed BJT-LNA has a low noise figure (NF) of 0.359dB, with input return loss of -6.022dB, output return loss of 1.447dB and a voltage gain of 3.296dB.

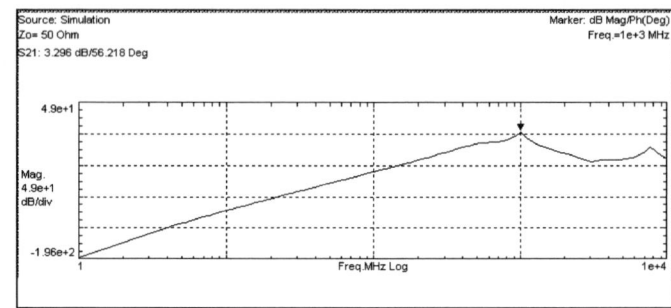

Figure 2: LNA S21

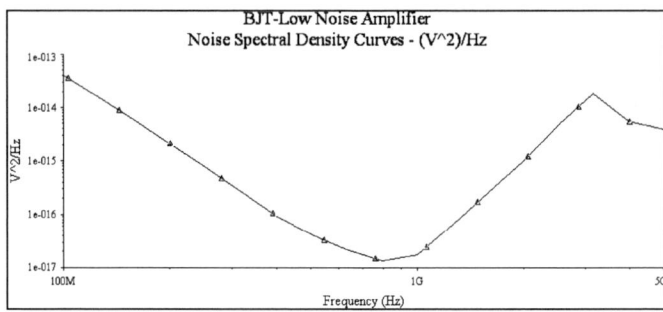

Figure 3: Input Noise Power Spectral Density (INOISE)

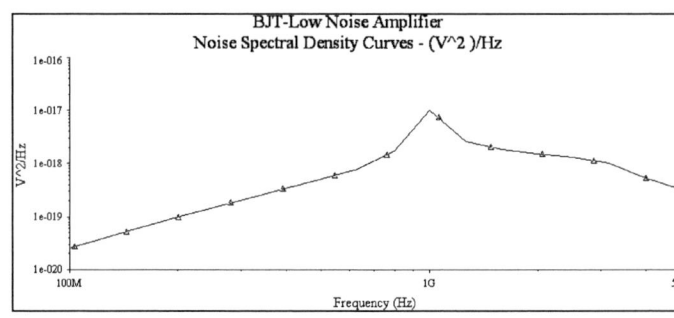

Figure 4: Output Noise Power Spectral Density (ONOISE)

Table 1 presents the summary of the simulation results of the Cascode Bipolar LNA

Parameters	Value
RF Frequency	1GHz
Power Supply	2V
Noise Figure	0.359dB
Power Gain	3.296dB
S11	-6.022dB
S22	1.447dB
S12	-6.680dB

Table 1: Summary of Simulation Results of the Cascode Bipolar LNA

V. CONCLUSION

An alternative methodology to BJT- LNA design has been proposed in this paper. The resulting circuit operates at 1GHz. By employing cascode configuration and utilising inductive emitter degeneration, the design exhibits good input impedance, which is close to 50Ω, and a low noise figure of 0.359dB.

VI. FUTURE PROSPECTS

The design of RFIC remains a huge challenge due to strong constraints in power consumption and noise. Though bipolar junction transistors were the first solid-state active device to provide practical gain and noise figure at microwave frequencies, the rapid development of field-effect transistors led to higher gain and lower noise figure compared to bipolar transistors operating in the gigahertz frequencies. CMOS will and has been the best choice of technology in designing RF front end circuits due to its lower cost and higher level of integration.

REFERENCES

[1] Golmie, N.; Chevrollier, N.; Rebala, O. "Bluetooth and WLAN coexistence: challenges and solutions", *IEEE Wireless Communication*, Volume 10, Issue 6, pp.22 – 29, Dec. 2003.

[2] D. K. Shaeffer and T. H. Lee, "A 1.5-V, 1.5-GHzz CMOS low noise amplifier," *IEEE J. Solid-State Circuits*, vol. 32, pp. 745–759, May 1997.

[3] A. R. Shahani, D. K. Shaeffer, and T. H. Lee, "A 12-mW wide dynamic range CMOS front- end for portable GPS receiver," *IEEE J. Solid-State Circuits*, vol. 32, pp. 2061–2070, Dec. 1997.

[4] G. Girlando and G. Palmisano, "Noise figure and impedance matching in RF cascode amplifiers," EEE Transactions on Circuits and systems-II: Analog and digital signal processing, vol. 46, Nov 1999.

[5] www.datasheetcatalog.com , Motorolla Semiconductor Technical Data

[6] www.datasheetcatalog.com , Siemens Semiconductor Technical Data

A 10GHz Reconfigurable UWB LNA in 130nm CMOS

Parviz Amiri, Hossein Gharaee, Abdolreza Nabavi
Tarbiat Modares University
Tehran ,IRAN
Email: p_amiri@modares.ac.ir

Abstract – **A 10 GHz reconfigurable CMOS LNA for UWB receiver is presented. The LNA is fabricated with the 0.13 μm standard CMOS process. Measurement of the chip is performed on a ADS simulator. In the UWB low-band (3 to 5.15GHz), the broadband LNA exhibit a gain of 17.5-18.2 dB, noise figure of 3.4-5dB, input/output return loss better than 10 dB, and input P1dB of −17 dBm, respectively. In the band from 2.4 to 3GHz (covering a 802.11b/g band), the LNA exhibit a gain of 17.5- 18dB and noise figure less than 3.5dB. From 5.2 to 6GHz, the noise figure of the LNA becomes higher than 5 dB. The gain also decrease to about 15 dB. The DC supply is 1.8V.**

I. INTRODUCTION

Ultra Wideband (UWB) radio [1][2] is originally technology of transmitting and receiving short, information encoded, electromagnetic impulses. The large occupied bandwidth (3-5.15 GHz or 3-10.1 GHz) enables UWB networks with exceptionally large capacities in higher bandwidth communications. UWB is a power limited technology because the emissions are targeted to be below the emission levels currently allowed for unintentional emitters. UWB thus opens a new spectrum and new combinations of capabilities for solving the demands of an increasingly wireless world. UWB system for multiband OFDM (MBOA) group A (with 802.11/a/b/g) and direct sequence (DS) CDMA UWB. The MBOA group A covering from 3 to 5 GHz has three bands and each with a bandwidth of 500MHz. The DS-UWB system has a low band from 3.1 to 5.15GHz and a high band 6 to 10.1GHz. Several UWB single-ended low noise amplifiers have been reported by SiGe or CMOS technology [3][4].

One of the challenges quantifying the performance the UWB LNA is useful, we first review the main function of the LNA. The primary purpose of the LNA is to provide enough gain to overcome the noise of subsequent stages while ensuring that the noise figure is adequately small.we define sufficiently high gain ($|S_{21}|$) to be >8dB and sufficiently low noise figure as < 5 dB. Achieving a flat $|S_{21}|$ over the lower UWB band is unnecessarily stringent, since the variations in the frequency gain caused by the LNA sre indistinguishable from the distortions caused by the large number of multipaths in the UWB propagation channel. Another requirement is to present a 50Ω input impedance, which is particularly important if a passive filter precedes the LNA, since the transfer characteristics of this pre-select filter is sensitive to its termination.The bandwidth of our LNA, therefore, is defined as a frequency range where $|S_{21}| > 8dB$, $|S_{11}| > 10dB$ and NF < 5dB.

In the following section first we introduce the wide band tuneable LNA topology, and provide details on design equations and constraints. In Section III it emphasizes on inductor design for the circuit operating at GHz frequency. Finally, we provide simulation results of the designed LNA and compare to other circuits in literatures.

II. LNA DESIGN

Low-noise amplifiers are conventionally configured as common-emitter or common-source amplifier circuits. The circuit configuration is a useful way to minimize the noise figure **(NF)** of the amplifier over a narrow frequency range.

The overall structure of our LNA is shown in Fig. 1. The input impedance of a source-degenerated MOSFET is often expressed as a series connection of L, C, and R in which the Miller effect is ignored. When the voltage gain from the gate of M1 to its drain is considered, the Miller effect produces a parallel connection of R and C at the gate of M1 if the source inductor is absent. This impedance is due to the fact that the voltage gain is dependent on the frequency. However, when a source degeneration inductor is present, such as in our LNA, the equivalent

0-7803-9730-4/06/$25.00 ©2006 IEEE

circuit with the becomes more complicated, as shown in Fig. 1. An output source follower is commonly used for 50Ω output matching such as in [1], but we adopt an impedance mapping technique based on our earlier work . Our method saves a source follower and hence reduces the area and the power dissipation. For details, refer to [4].For the amplifier, gain, noise and power must be optimize in to proper value for achieving the input and output matching.

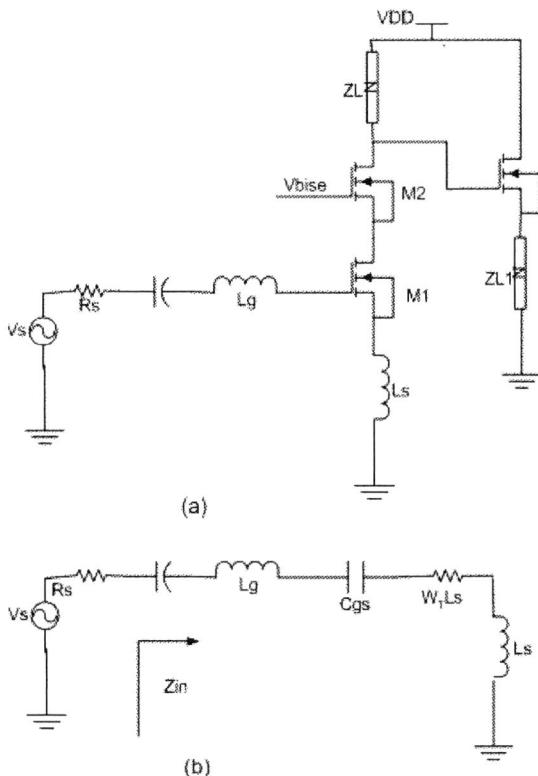

Fig. 1(a) a typical cascode LNA topology. (b) Small-signal equivalent circuit at the input.

Cascode LNA with inductive degeneration is more flexible for using in wideband. In Fig. 1(a), Ls is added for the impedance matching between the source resistance and the input of the LNA [5].Fig. 1(b) shows the small-signal equivalent circuit for the input part of the overall LNA, where represents the gate-source capacitance of the input transistor . In Fig. 1(b), a series combination of reactive elements is chosen to resonate at the frequencies of interest such that becomes a real value with being equal to From [7], The input port of the amplifier equal to a RLC series network , so quality factor of this input section is:

$$Qs = \frac{1}{R}\sqrt{L/C} = \left(\frac{WoL}{R}\right) = \frac{1}{WoRC} \quad (1)$$

$$Qin = \frac{Wo(Ls + Lg)}{Rs + W_T Ls} = \frac{1}{2WoRsCgs} \quad (2)$$

Qin depends on the inductor quality factor.
$$Ct = Cgs + Cext. Zin \quad (3)$$

Then : $\quad\quad\quad\quad\quad\quad\quad\quad (4)$
$$Z\,in = sL\,gs + sL\,s + 1/\,sC\,t + g\,m\,L\,s\,/\,C\,t$$

The imaginary parts of *Zin* cancel and the real part can be optimized that enables a simultaneous noise and input matching. Since the minimum noise figure, NFmin, is a function of power dissipation   and   the size of M1 must be relatively large[8,9]. The size of the common-gate transistor M2 is important because it affects on noise and gain reduction. So an inductor load is used to boost gain at higher frequencies and to achieve flat frequency response[6]. Simulation in ADS and optimization of the amplifier yields the following parameters: W/L of M1 = 130μm/0.13μm, W/L of M2 =140μm/0.13μm, *Ls* = 300 pH, *Cext* = 300 fF, *Ld* = 2 nH, .

NF can be reduced by increasing the current consumption (power) and Difficult to achieve simultaneously NFmin and input matching.

$$F = 1 + \frac{R_{L_G} + R_{L_S} + R_G}{Rs} + \frac{2}{3g_m R_s}\frac{1}{G_m^2} \quad (5)$$

NF is also depend on size of M1. To minimize the NF of the amplifier we need a very large area for the input transistor. So the transistor needs high bias current which it increases power consumption and this criteria is unacceptable. Solution for this problem in LNA is " noise figure   optimization"   for   a   given   power consumption are[8]:

$$W_{opt} = \frac{1}{3\omega L_{eff} C_{ox} R_s} \quad (6)$$

$$F_{min,p} = 1 + 2.4\frac{\gamma}{\alpha}\frac{\omega}{\omega_T} \quad (7)$$

with this role achieve proper size and acceptable NF.

Fig.1 shows design that is collected all best configuration in CG-CS degenerated LNA[2-5]. This configuration Increases the reverse isolation S12, improved stability, Increase voltage gain and eliminate input Miller

761

capacitance. Lg, Ls and load is critical component that is affected on BW, Nf, Gain and other parameters. Ls matches the input impedance and Lg sets the resonance frequency . In our work, we want to have a reconfigurable LNA in all UWB with desired parameters. So the drop of gain with increasing frequency is compensated by adopting capacitors. whose impedance increased with frequency as the output load, but decrease of G_m results in the disadvantage of high noise figure at higher frequencies. Also we control frequency by output load and $C_{gs}(C_1, C_2)$.

The schematic of the $3 - 10GHz$ UWB LNA by CMOS technology is shown in Fig. 2. A sourcefollower is added for buffering purpose.

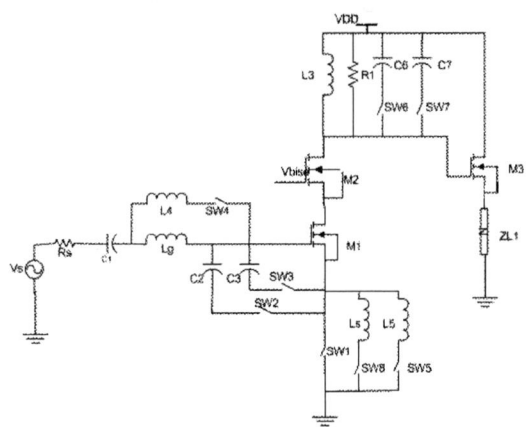

Fig. 2 Schematic of the LNA

To implement the matching networks in the LNA for the input, lowpass L-section LC matching networks are applied because it allows DC current to flow through and also provides a wide bandwidth. The main disadvantage of using on-chip LC matching networks is that inductors have finite quality factor (Q).

III. SIMULATION RESULTS

It can be seen that this is the $3 - 10GHz$ LNA operating at 1.8V using $0.13\mu m$ CMOS. The performance is comparable with the state-of-the-art UWB CMOS LNAs.

Fig. 3 shows the gain of the LNA to be average 15dB and it is relatively flat across the entire band. Fig. 4 shows the noise figures ,LNAs achieve $1 - 3.2$ dB between 3−10GHz and the average NF of LNA is 2 dB

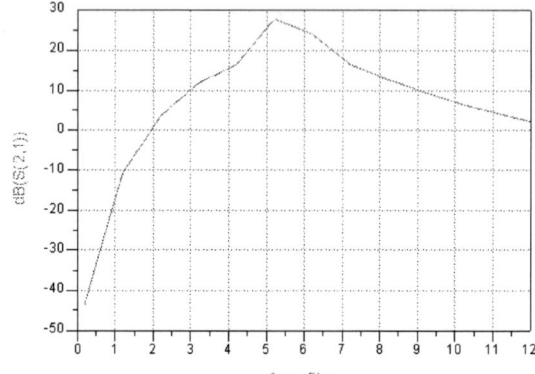

Fig. 3 S21 of the LNA

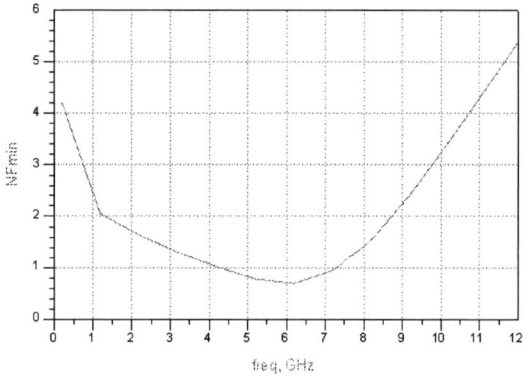

Fig. 4 NF of the LNA

Fig. 5 displays the input return loss from 3 GHz to 10 GHz. For the bandwidth in consideration S11 is below −7 dB

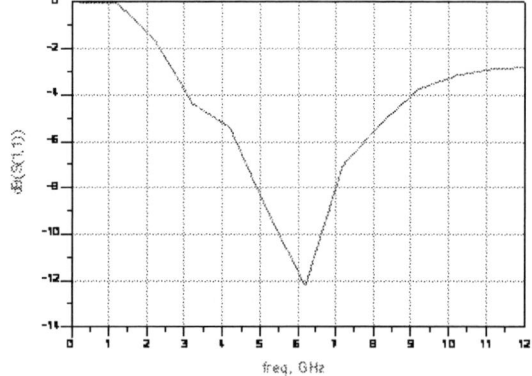

Fig. 5 S11 of the LNA 1

Fig. 6 shows the S22. best results achived because the output buffer is adopted. S22 is is always below −15 dB over $3 - 10GHz$.

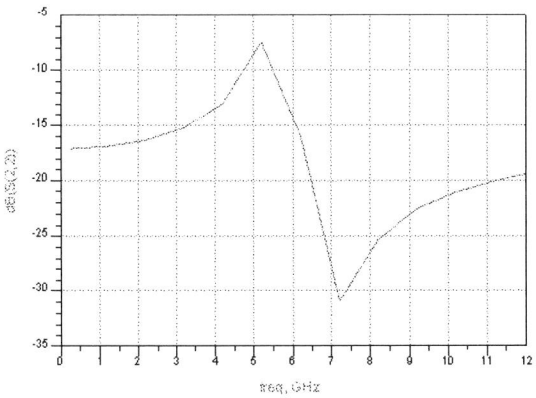

Fig. 6   S22 of the LNA

A cascode topology proves to be effective in reducing feedback from output port to input port, thus enhancing stability and S_{12} is below −45 dB.

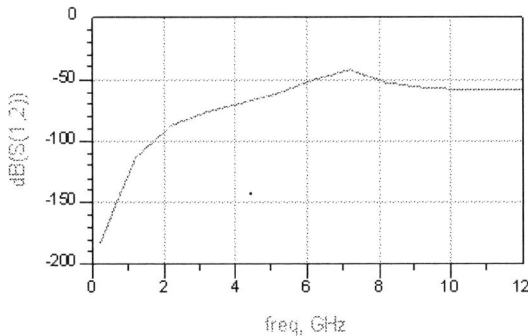

Fig.7   S12 of the LNA

IV. CONCLUSION

This paper presents a 10 GHz reconfigurable CMOS LNA implemented with the 0.13 μm CMOS process. Measurement of the chip is performed on a ADS simulator.The effect of the ring coupler has been taken into account in both simulation and measurement deembedding. In the UWB low-band (3.1 to 6GHz), the broadband LNA exhibit a gain of 15.5-18.2 dB, noise figure of 3.4-5.9dB, input/output return loss better than 10 dB, and input P1dB of −17 dBm, respectively. In the band from 2.4 to 3GHz (covering a 802.11b/g band), the LNA exhibit a gain of 17.5-18dB, noise figure less than 3.5dB, and input P1dB of −17 dBm, espectively. The DC supply is 1.8V with a current consumption of 22mA without the buffer stage (or 11mA for a half-circuit).

REFERENCES

[1] K.Siwiak, P.Withington and S. Phelan, "Ultra-Wide Band Radio: The Emergence of an Important New Technology," IEEE VTC, 2001

[2] G. R. Aiello and G. D. Rogerson, "Ultra-wideband wireless systems," *IEEE Microwave Mag.*, vol. 4, pp. 36–47, Feb. 2003

[3] A. Ismail and A. A. Abidi," A 3–10-GHz Low-Noise Amplifier With Wideband *LC*-Ladder Matching Network", *IEEE Journal of Solid-State Circuits*, Vol. 39, N0. 12, Dec 2004.

[4] A. Bevilacqua, and A. M. Niknejad, "An Ultra-wideband CMOS Low-Noise Amplifier for 3.1–10.6-GHz Wireless Receivers", *IEEE Journal of Solid-State Circuits*, Vol. 39, N0. 12, Dec 2004

[5] T.-K. Nguyen *et al.*, "CMOS low noise amplifier design optimization techniques," *IEEE Trans. Microwave Theory Tech.*, vol. 52, no. 5, pp. 1433–1442, May 2004.

[6] Yanxin Wang, Jon. S. Duster, Kevin T. Kornegay esign of an Ultra-wideband Low Noise Amplifier in 0.13μm CMO2005 IEEE.

[7] T. K. Nguyen, C. H. Kim, G. J. Ihm, M. S. Yang, and S. G. Lee,"CMOS low-noise amplifier design optimization techniques," IEEETrans. Microwave Theory and Techniques, vol. 52, pp. 1433-1442,May 2004.

[8] A.Nabavi   ,   course   of   RF1 WWW.MODARES.AC.IR

[9] D. K. Shaeffer and T. H. Lee, "A 1.5V, 1.5 GHz CMOS low noise amplifier," IEEE J. Solid-State Circuits, vol. 32, pp. 745-758, May 1997.

[10] A Low Noise Amplifier for Ultra-wideband Systems in 0.13μm CMOS Technology Zhila Amini, Sabrieh Choobkar, Abdolreza Nabavi Faculty of Engineering, Tarbiat Modares University

Physical-Based SPICE Model of CMOS STI y-Stress Effect

Philip Beow Yew Tan[1,2] *, Albert Victor Kordesch[1] and Othman Sidek[2]

[1]Silterra Malaysia Sdn. Bhd. Kulim Hi-Tech Park,09000 Kulim, Kedah, Malaysia

[2]University Science Malaysia, 14300 Nibong Tebal, Pulau Pinang, Malaysia

*Email: philip_tan@silterra.com

Abstract – In this paper, we proposed a new physical-based equation to model the CMOS transistor STI y-stress (in the direction of channel width). It can be used in any SPICE MOS model and it has been verified on 0.13um CMOS transistors. The physical characteristics of the compressive STI y-stress effect on saturation drain current, Idsat are captured by using a new proposed transistor layout method. The equation that is able to describe the physical characteristics of the STI y-stress effect is incorporated into the electron and hole mobility, u0 of the SPICE model to capture the y-stress effect on Idsat. With the combination of the new y-stress parameters and the default delta width parameters in the SPICE model, we are able to fit the simulation curve to the hook shaped Idsat curve from the actual silicon data.

I. INTRODUCTION

STI stress effect is a hot issue nowadays and many researchers have put much effort into this subject [1, 2, 3, 4]. The great interest on this subject is mainly driven by the fact that stress changes the mobility of electrons and holes in CMOS transistors and it is unavoidable because the stress is built-in to the fabrication process. The explanation for this phenomenon is the effective mass of electron and hole changes because the shape of the conduction band changes due to stress on the silicon lattice structure and the number of carriers in various branches of the conduction band changes due to energy level shifts [2]. Another phenomenon is the mechanical stress causes the dopant diffusion to change that later changes the electrical characteristics such as threshold voltage of the finished transistor [3].

The effect of STI stress in the direction of channel length has been captured in the latest standard SPICE model by using Sa and Sb

(Space of active). Since STI stress is a 2-D effect, the compressive stress in the direction of channel width also changes the drain current of a transistor. Hence, in this study we proposed a physical equation that will be added into the carrier mobility equation, u0 to capture the effect of compressive STI stress effect in the direction of channel width (y-stress).

In the standard SPICE models, the delta width (DW) effect will raise the saturation drain current (Idsat) of narrow width transistors. We propose to add in the effect of compressive STI y-stress to u0 which will degrade the Idsat of narrow width transistors because of the reduction in mobility due to compressive y-stress. The compressive y-stress degrades both the NMOS Idsat and PMOS Idsat [1]. The combination of both the DW effect and the mobility degradation due to compressive STI stress will result in a hook shaped Idsat versus width curve as seen in the actual silicon measurement data, as shown in Fig. 1.

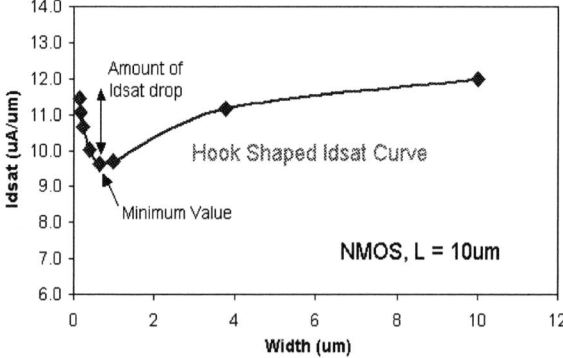

Fig. 1 Actual silicon measured data showing a hook shaped Idsat vs Width curve.

We have demonstrated the hook shaped Idsat curve in our previous study in [5] and we have proposed an empirical SPICE model to capture the hook shaped Idsat curve in [6]. In this paper, we propose a physical-based SPICE model that is able to capture the physical behavior of the hook shaped Idsat curve correctly and accurately.

0-7803-9730-4/06/$25.00 ©2006 IEEE

II. EXPERIMENTAL STRUCTURES

The CMOS transistor structures used in this experiment are categorized into 2 groups. The first group is for extracting the characteristics or behavior of the STI y-stress effect on drain current in saturation, Idsat. The second group is to verify the new physical-based model with built-in STI y-stress effect.

For the first group of experimental structures, we use a standard W/L = 10um/0.13um transistor layout. We shrink the Select layer to form the channel width of 0.5um and step through the 10um channel from the edge of the channel width to the middle of the channel (total 10 steps). This means that we actually built 10 transistors with W/L = 0.5um/0.13um but each of them will have different distance of channel width to STI edge. This enables us to evaluate the compressive STI y-stress on transistor channel width. We have proposed this method of evaluating the STI y-stress in [7].

The second group of experimental structures are transistors with two fixed lengths, 10um and 0.13um, and eight different widths varying from 0.15um to 10um. These structures will enable us to plot the Idsat versus width curve for both long and short channel length transistors.

All the transistors in this experiment are fabricated using Silterra's standard 0.13um CMOS technology. The linear threshold voltage (Vtlin) is extracted at Id = 0.1uA*(W/L) with Vd = 0.1V and Vs = Vb = 0V. Saturation drain current (Idsat) is extracted at Vg = Vd = 1.2V and Vs = Vb = 0V. The operating voltage (VDD) for this 0.13um technology is 1.2V.

III. PHYSICAL CHARACTERISTICS

By using the first group of experimental structures, where we step through the 10um channel from the edge of the channel width to the middle of the channel (total 10 steps) using 0.5um wide Select implant layer, we are able to avoid (or reduces) the compressive STI y-stress on the transistor edge. When the Select implant layer is coincident with the STI edge (0um in distance) the transistor will have the highest y-stress effect. When the transistor is 4.5um away from the STI edge, we assume the transistor is y-stress free.

From our study in [7], we found that the amount of Idsat decrease is caused by the STI y-stress effect. Hence, we know that the percentage of Idsat decrease is proportional to the intensity

or amount of compressive STI y-stress effect on the transistor. Fig. 2 shows that the compressive STI y-stress effect on NMOS transistor is a function of distance from STI edge and can be roughly described by using an exponential relation as in Equation 1 (shown by the broken line). By adding a new parameter B, we can have a better-fit curve to the data as shown by the solid line, Equation 2. The STI y-stress effect on PMOS transistor shows similar characteristics to NMOS transistor (as shown in Fig. 3).

$$y = A \cdot \left(e^{-x} \right) \quad (1)$$

$$y = A \cdot \left(e^{-B \cdot x} \right) \quad (2)$$

By using the two parameters, A and B (as in Equation 2), we can have a better curve fit to the actual silicon data, as shown in Fig. 2 and Fig. 3. A is the multiplier parameter to control the maximum value of the curve and B is the power parameter that controls the curvature of the curve. If we were using the exponential curve, we can only control the maximum value of the curve (as shown by the broken line) but cannot fit the curvature of the actual data.

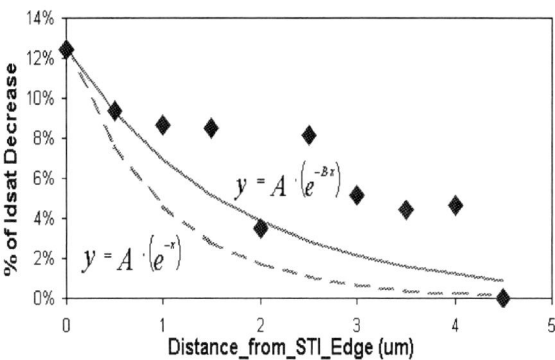

Fig. 2 STI y-stress effect on NMOS transistor. The percentage of Idsat decrease is equivalent to the intensity of STI y-stress effect on the transistor.

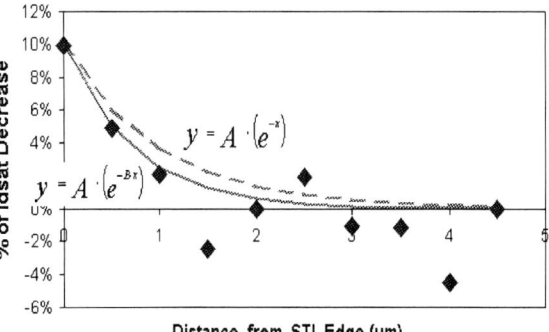

Fig. 3 STI y-stress effect on PMOS transistor. The percentage of Idsat decrease is equivalent to the intensity of STI y-stress effect on the transistor.

IV. SPICE MODEL

Since the effect of STI y-stress is higher for narrower channel width, we can rewrite Equation 1 and Equation 2 as a function of channel width, W, as shown in Equation 3 and Equation 4. We modify the electron and hole mobility, u0 in the SPICE model to incorporate the effect of STI y-stress into the model based on Equation 3 and Equation 4.

$$f(W) = A \cdot \left(e^{-W}\right) \qquad (3)$$

$$f(W) = A \cdot \left(e^{-B \cdot W}\right) \qquad (4)$$

Fig. 4 and Fig. 5 compare the fitting of standard model (default u0), exponential model (u0 incorporated with Equation 3) and y-stress model (u0 incorporated with Equation 4) to the actual silicon data for NMOS, L = 10um and 0.13um. Similar comparison plots for PMOS are shown in Fig. 6 and Fig. 7.

The actual data in Fig. 4 is the same data as in Fig. 1 that has the hook shaped curve. In Fig. 4 to Fig. 7, we plotted the x-axis in log scale to zoom into the narrow width transistors.

The standard model shows the delta width (DW) effect that raises the Idsat at narrow width, as shown by the dotted line. The broken line which represents the exponential model, has much better fitting compared to the standard model. The solid line (y-stress model) shows an even better fitting to the measured data compared to exponential model, which can be clearly seen in Fig. 4. and Fig. 6. The difference of the exponential model and y-stress model is hardly seen in Fig. 5 and Fig. 7. This is because in Fig. 5 and Fig. 7, the exponential model already provides a good fit to the actual data and not much we can fine-tune in the y-stress model.

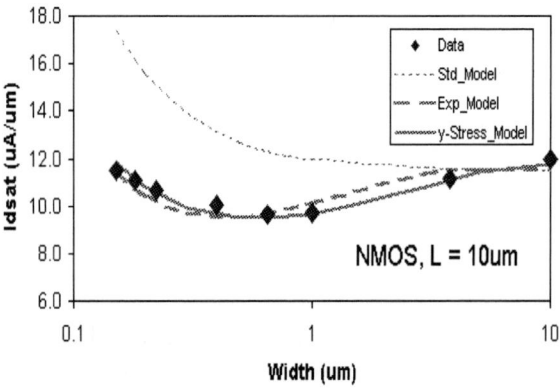

Fig. 4 Plot of Idsat versus Width for NMOS, L=10um. This plot compares the fitting of standard model, exponential model and y-stress model to the actual silicon data.

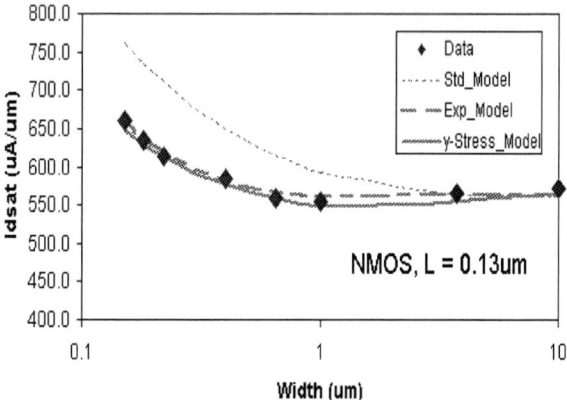

Fig. 5 Plot of Idsat versus Width for NMOS, L=0.13um. This plot compares the fitting of standard model, exponential model and y-stress model to the actual silicon data.

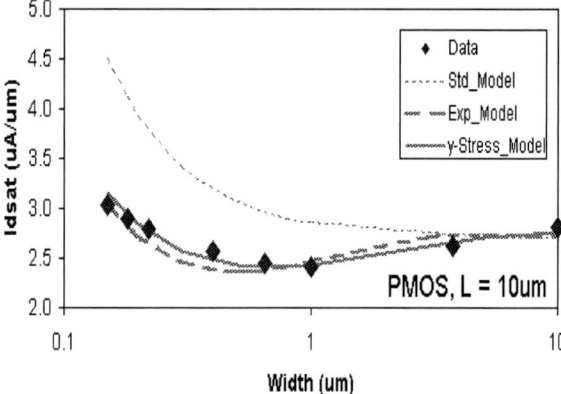

Fig. 6 Plot of Idsat versus Width for PMOS, L=10um. This plot compares the fitting of standard model, exponential model and y-stress model to the actual silicon data.

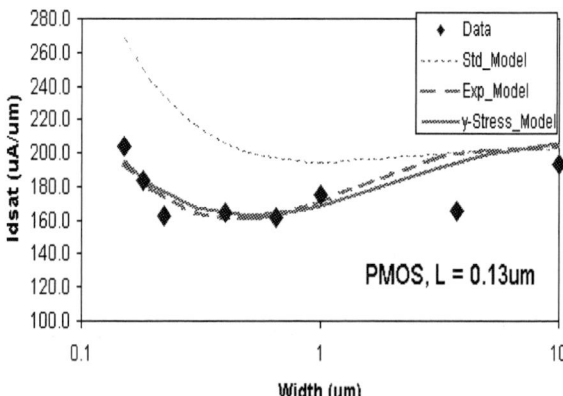

Fig. 7 Plot of Idsat versus Width for PMOS, L=0.13um. This plot compares the fitting of standard model, exponential model and y-stress model to the actual silicon data.

Theoretically, Equation 4 will give more flexibility to fine-tune the hook shaped Idsat curve compare to Equation 3. Together with the DW effect parameter (wint) in SPICE model,

Equation 3 can only control the amount of Idsat drop with the minimum fixed value in width (Refer to Fig. 1 for the illustrations of "amount of Idsat drop" and "minimum value").

By using Equation 4, we are able to control both the amount of Idsat drop and also to move the minimum value of the Idsat to fine-tune the hook shaped curve. From Equation 4, parameter A is used to control the amount of Idsat drop for the hook shaped curve vertically and parameter B is used to move the minimum value of the hook shaped curve horizontally, as illustrated in Fig. 8.

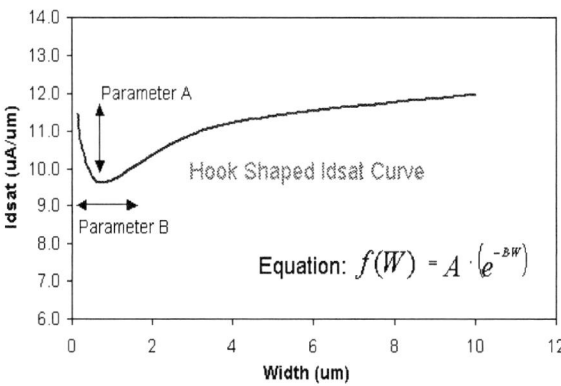

Fig. 8 Illustration of how the parameter A and paramter B in the y-stress model control the hook shaped Idsat curve. W is the channel width.

V. CONCLUSION

In conclusion, we have proposed equation that is able to capture the physical characteristic of the compressive STI y-stress effect on NMOS Idsat and PMOS Idsat. By incorporating this equation into the mobility parameter, u0, we are able to model the y-stress effect on Idsat. Together with the DW effect parameter, we are able to model the actual behavior of the hook shaped Idsat curve correctly and accurately.

ACKNOWLEDGEMENT

The authors would like to acknowledge all the members of Silterra Malaysia Sdn. Bhd. for supporting and contributing to the research work in this paper.

REFERENCES

[1] C. H. Ge et. al., "Process-Strained Si (PSS) CMOS Technology Featuring 3D Strain Engineering," *International Electron Device Meeting, IEDM*, pp.73 (2003).

[2] C. Gallon et al., "Electrical Analysis of Mechanical Stress Induced by STI in Short MOSFETs Using Externally Applied Stress," *IEEE Transactions on Electron Devices, TED*, vol.51, no.8 (2004).

[3] Y-M. Sheu et al., "Modeling Mechanical Stress Effect on Dopant Diffusion in Scaled MOSFETs," *IEEE Transactions on Electron Devices, TED*, vol.52, no.1 (2005).

[4] K-W. Su et al., "A Scalable Model for STI Mechanical Stress Effect on Layout Dependence of MOS Electrical Characteristics," *Custom Integrated Circuits Conference, CIC*, pp. 245-248 (2003).

[5] P. B. Y. Tan et al., "Hook Shaped Drain Current vs Width Curve of 130nm CMOS Technology," *National Symposium on Microelectronic, NSM*, (2005).

[6] P. B. Y. Tan et al., "Compact Modeling of Mechanical y-Stress Effect," *International Conference on Solid-State and Integrated Circuit Technology, ICSICT*, (2006).

[7] P. B. Y. Tan et al., "Measuring STI Stress Effect on CMOS Transistor by Stepping through the Channel Width," *2006 International of RF and Microwave Conference, RFM*, (2006).

Optimized Clamp Deployment with Simulation and Characterization in Full-Chip ESD (Electro-Static-Discharge) Design

Chuah Cheow Theng[1] and Othman Sidek[2]

[1]Intel Microelectronic (M) Sdn Bhd, Bayan Lepas Free Trade Industrial Zone
Phase 3, Halaman Kampung Jawa, 11900 Penang, Malaysia.
[2] Engineering Campus, Universiti Sains Malaysia, Penang, Malaysia
E-mail: cheow.theng.chuah@intel.com , othman@eng.usm.my

Abstract **A DOE (Design Of Experiment) with simulation approach is presented that allow early visibility, guidance, and analysis of clamp placement in early design stage (power template design) on a chip level complexity. ESD robustness of product is largely attributed to the degree of correct implementation of ESD design rules in a highly complex design. Correct clamp placements is one of the most difficult and less guidance. Therefore, a 'reliable' & 'comprehensive' ESD clamp placement analysis is urgently needed.**

This simulation would deliver a preliminary guidance of clamp placement. It provides the information of where the most probable clamp placement is in the particular power template design. Since it takes into consideration the ESD ohm rules requirements during the construction of ohm sector, it would have a very high confident level that the design would not encounter much issue in the ESD resistance checkout later. Apart from the ESD design rules checkout, we also couple with TLP (Transient Line Pulse) testing which has proven this approach could deliver a promising ESD protection robustness. We can characterize, co-relate, and compare our simulation with the Si data from various design perspectives.

All in all, the real strength of this simulation is the ability to provide earlier design guidance & optimization options starting from the product concept phase (power template design stage), which allows to align ESD with all design issues on a cost and time efficient way. Without the simulation, designer would need to manually calculate the resistance , rely on past experience and some assumption with nil data support which is tedious & error prone for complex design nowadays. This allows ESD design rules compliance in the early

design stage, thus enhancing the approach of "correct-by-construction" on ESD design.

I. INTRODUCTION

ESD (Electro-Static-Discharge) is one of the major threats for the reliability of CMOS process. The susceptibility to ESD treat increases with the continuous device scaling according to Moore's law, process features reengineered (junction depth, isolation design, ultra thin gate oxide, channel profile, salicide and etc) and new process introduced.[2] Thus, it is especially vital to design a low impedance path for every current injection point to shunt ESD current rapidly and effectively. [3] and [4]

Typical robust ESD protection should at least meet 2 major ESD design attributes [2]: a] Leadway design rules b] Global Resistance design rules (Figure-1): Well distributed ESD clamps that meet minimum resistance between current injection points.

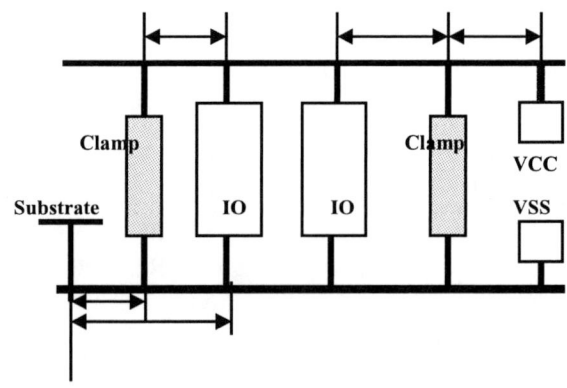

Figure 1: ESD ohm rules –The tool checks resistance between ESD cells inside I/O with their closest clamps. It also checks resistance between adjacent clamps. [1]

0-7803-9730-4/06/$25.00 ©2006 IEEE

Rule (b) would ensure the resistance between current injection points to ESD clamps meets specified value by design rules to minimize IR drops along the interconnection.

Up-to-date, ESD clamps deployment (quantity and clamp placement) is determined manually. (Based on designers' experience and engineering judgment.) After all the circuit is finalized and integrated, a pre-Si ESD design rules checkout is done to ensure ESD protection design is in tact.

However, manual estimation is no longer viable and accountable due to overwhelmed design complexity, die size increased, more mixed signal, and SOC (Si-On-Chip) merging into chip designs. Underestimation will cause insufficient ESD protection which triggered ESD failure on Si. Whereas, overkill estimation would make product design wastes precious die area (for extra clamp) and more engineering hours trying to incorporate ESD clamps into compact design boundaries. Usually, last minutes ESD design iteration (due to insufficient clamp) would seriously impact the floor plan and design turn-around cycle (it might take weeks sometimes).

In this paper, we presented a correct-by-construction approach to mitigate the above mentioned design 'gap'. It would provide early ESD design rules incorporation & compliance. This would in turn reduce or eliminate last minutes ESD design excursion. This earlier design guidance and optimization would help to align ESD design with all design windows in a cost effective and timely fashion.

II. METHODOLOGY

ESD Design Rules Consideration (resistance rules)

The robustness of ESD protection is largely attributed to the resistance between ESD current injection spots. If the resistance is too huge, the IR drop along interconnect grid during ESD occasion and this will cause enough potential to eventually 'breakdown' the gate oxide.[4] Thus, we need to ensure sufficient number of clamps in every single power rails and the IR drop within power nets is maintains at 'manageable' level.

The arbitrary 'virtual' clamp placement has taken into considerations all these design rules in the simulation algorithm. Thus, we could design the power templates with 'built-in' compliant ESD design rules in the very first place.

'Virtual' clamps are place in the design power grids. Then, the prototype database would be fed in design rules checker for ESD and various related ESD data are checkout simultaneously. All these data would be stream-line to target ohmic range and designated clamp would be selected. Then, the 'revised' database could be further optimized using a series of optimization process to bring the ESD design to an acceptable specification or even specific ESD design rules readouts.

Figure-2 shown the process of power template and clamp placement planning and optimization flow. Optimization process could be preceded by a few iterations of process flow.

III. RESULTS

Various test cases have been carried out to explore the feasibility and capability of the clamp placement optimization techniques. In this paper, we will share 2 design databases that have gone through the clamp placement optimization run sets.

Test case 'F2006' has a single type of clamp for its only one single power rail. On the other hand, test case 'P' has multiple types of clamp since it consists of a few power rails.

Figure-3 shown test case 'F2006' with its finalized clamp placement based on autoclamp placement simulation and recommendation.

Figure-4 displayed test case 'MP4-21V' with its multiple type of clamps scattered within the complex power template optimizes & finalizes by the clamp placement simulations.

To ensure the integrity of the tool, we has performed typical ESD design rules checkout for all these databases. All the two have shown very comprehensive and robust readouts. It has shown how this clamp optimization approaches could play a important role to instill ESD design rules in early Si design process.

Besides, we have utilized TLP (Transient-Line-Pulse) to characterize the optimization of the clamp position by comparing it with various DOE (Design of Experiment) test structures. The ultimate optimize one would be the one contained best ESD readout and TLP data.

IV. CONCLUSION

From this article, we demonstrated the innovative correct-by-construction method for power template and ESD clamp placement early design planning guidance. This strategy is simple and provides high reliability and consistency compare to conventional manual estimation approach.

By using the strategy, the entire power template design process is highly automated and repetitive. Precious design time and effort could be saved as the strategy ensure earlier ESD design rules 'built-in' during the initial power template 'build-up'.

The clamp placement optimization also largely reduces unnecessary clamps needed. This ensures the design ESD robustness and save Si area while minimizing design efforts and improves product reliability as we incorporates less device in the design.

ACKNOWLEDGEMENT

The author would like to acknowledge those who directly or indirectly involving in the support, discussion, contribution, development, validation and deployment of the tool.

REFERENCES

[1]   Chen, Z. and Fleurimont, J., "Automated ESD Layout Connectivity Checking Tool", *2002.*

[2] S. Dabral, T. J. Maloney, "Basic ESD and I/O Design", John Wiley and Sons, Inc. 1998

[3] Ajith Amerasekera, Charvaka Duvvury, 'ESD in Silicon Integrated Circuit', John Wiley and Sons, Inc. 1995

[4] Shao, X. and Chen, Z., "Planning for ESD of   Die Protection", *2004.*

[5] Maloney, T.J., "Core Clamps for Low Voltage Technologies", *EOS/ESD Symposium, 1994.*

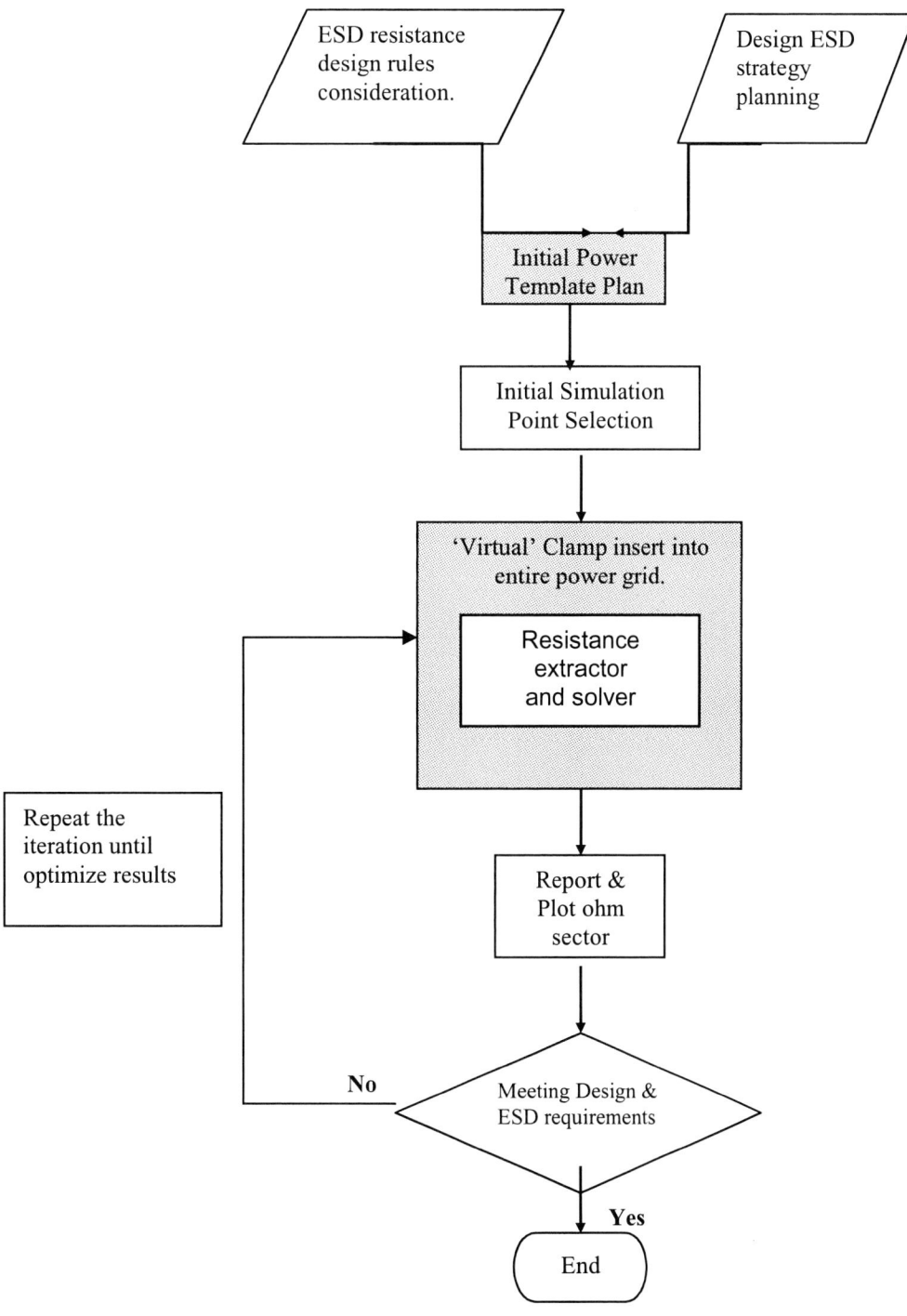

Figure 2: Process of power template and clamp placement planning and optimization flow.

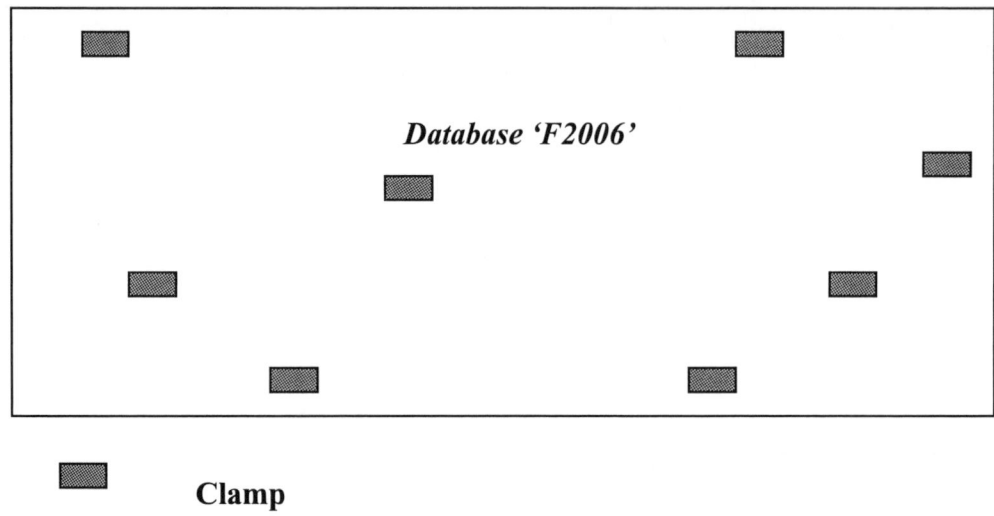

Clamp

Figure 3: Test case 'F2006' with its finalized clamp placement based on virtual clamp placement simulation and recommendation. (Power grid is not shown; it is not symmetrical in this case)

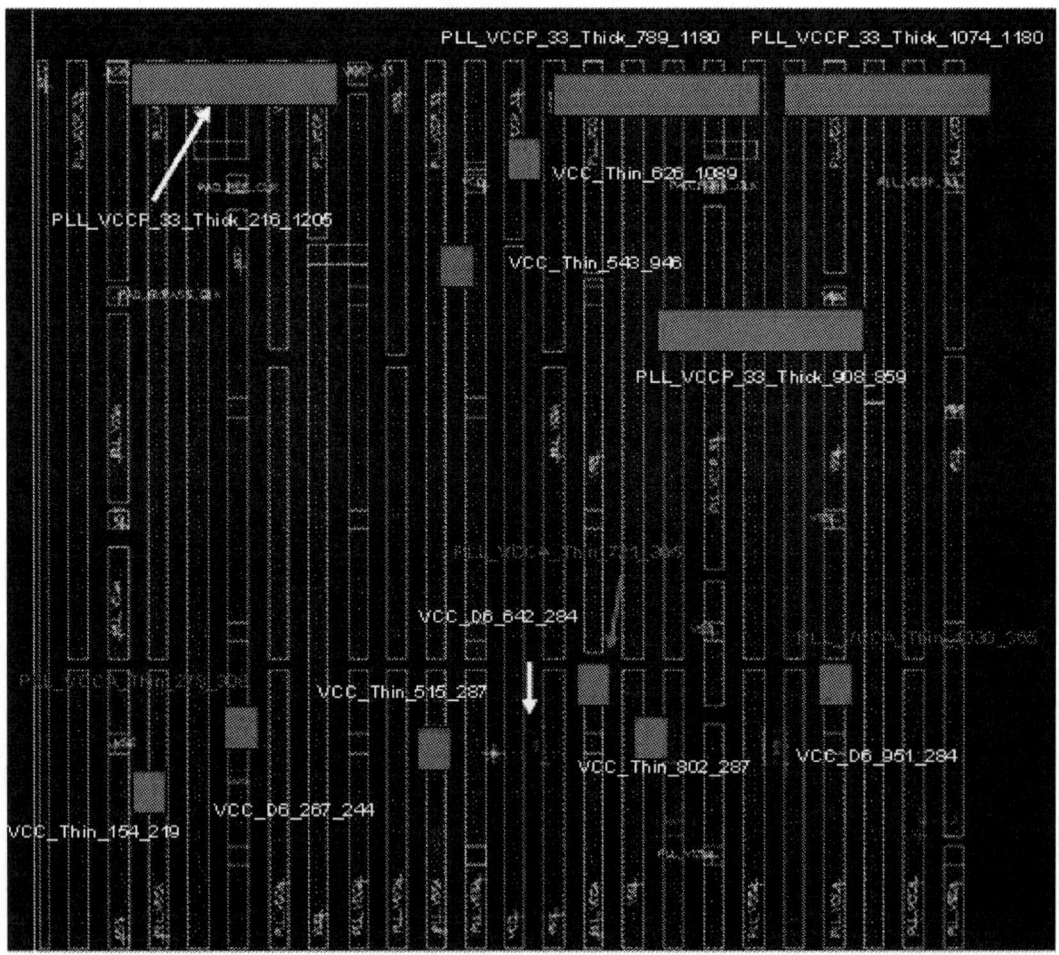

Figure 4: Test case 'MP4-21V' with its multiple clamps scattered within the complex power template optimize by the a few iteration clamp placement plan-out.

Dependence of Texture in Al Bondpads on Ta/TaN Bilayer Barrier and its Correlation to Optical Reflectivity in 0.13µm IC Technology

Lee Yuan Ping, Ramesh Rao Nistala, Hua Younan and Mousumi Bhat
Chartered Semiconductor Mfg Ltd
Woodlands Industrial Park D
Street 2, Singapore 738406
Tel: 65-63604378, Fax: 65-63622935
Email: raoramesh@charteredsemi.com

Abstract **In this paper, the dependence of crystallographic orientation in Aluminum thin films grown on different barrier-metal substrate schemes (Ta or Ta/TaN) will be presented. The orientation of Al grains will be shown to have a bearing on the material characteristics, which are important in IC fabrication from the perspective of both the device functionality and reliability. X-ray powder diffraction studies indicate that the films deposited on a single Ta layer are randomly oriented. On the other hand, a Ta/TaN bilayer substrate scheme results in preferred orientation along Al(111). A correlation will be established between the grain orientation and optical reflectivity properties of Al films. Moreover, the optical appearance of bondpads and their bondability are also influenced by the orientation of Al grains.**

I. INTRODUCTION

CRYSTALLOGRAPHIC orientation or texture of Al grains in the backend interconnects of microelectronic devices is critical to prevent electro-migration and stress-induced void migration [1]. Thin films can be grown with desired texture by controlling the growth conditions such as substrate and temperature [2,3]. For example, in the micro-chip fabrication, Al(111) orientation is achieved by depositing PVD Al on Ti/TiN film stack in which the TiN film is itself textured – TiN(111) orientation. For deep submicron technologies, 0.13µm and below, RC time delays introduced by the Al interconnects are one of the limiting factors affecting the device functionality and hence, the industry has moved away from Al to Cu due to its low bulk resistivity. However, there are disadvantages with copper; (i) metal diffusion at low temperatures (200 C), (ii) easier to oxidize and (iii) poor adhesion or wetting properties.

A high surface sticking coefficient or wetting of bondpad surface prevents NSOP (Non-Stick Of Pads) during device packaging. Au wire-bonds as well as Cr/Au-based alloys used as under-bump metals (UBM) in flip-chips have better adhesive properties to Al as compared to Cu and hence, Al is still used in deep submicron technologies as top metal (TM) or bondpad (B/P) (sometimes also referred to as metal cap (MCAP)). However, there is one significant difference in the B/P layout scheme of Cu technologies (0.13µm and below) as compared to Al technologies (0.18µm and above) and that is, Al pads are laid on top of Ta barrier rather than on Ti/TiN, as shown in figure 1.

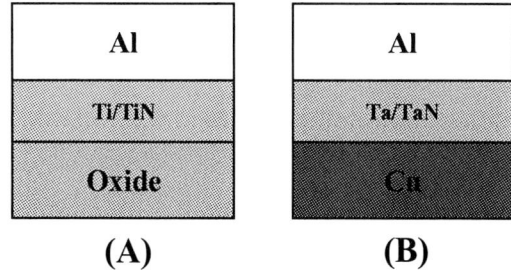

Fig. 1 Schematic of the film stack used for bondpad process in: (A) Al- and (B) Cu-based backend processes.

For production factory, NSOP of packaged units is a high cost no profit scenario as the pads are damaged beyond repair and therefore, the units, which might otherwise be functional, cannot be used in the end-product. It should be noted that NSOP may happen due to various reasons [4-6], such as, (i) contamination of bondpad surface resulting in corrosion, (ii) an abnormal layer on the surface of pads may change the adhesion properties at the wire-bond/bondpad interface or (iii) changes in the material characteristics of a 'normal' bondpad itself can prove to be detrimental, as will be discussed in this paper.

Recently, on some Cu process wafers dark bondpads were noticed during optical inspections at assembly house. It is suspected that the abnormal bondpads are caused by wafer fab process. Failure analysis with high-resolution surface SEM imaging did not show any particle defects. Neither did AES analysis detect surface contaminants like Fluorine, Carbon, Silicon, etc. in proportions above the acceptable contamination limits. It was suspected that the appearance of dark bondpads may be related to fundamental changes in material properties of Al film. Further failure analysis was carried out to characterize the crystallographic phase of Al films on blanket wafers that were processed in conditions similar that of patterned wafers.

II. EXPERIMENTAL

X-ray Diffraction: XRD studies were performed on *Bede D¹* lab system, fitted with a long-fine-focus, ceramic x-ray tube with Cu anode (λ=1.5418Å). The normal operating power of the system is 45kV and 40mA (1.8 kW). The measurements were performed in θ-2θ mode in the angular range of $20 \leq 2\theta \leq 96$ (in degrees) with beam conditioning slits at source and detector. Soller slits were used to reduce diffuse scattering and improve resolution. A proportional counter with high dynamic range was used to measure scattered x-ray beam. Bragg peaks in the XRD data were indexed using International Centre for Diffraction Data (ICDD files).

III. RESULTS AND DISCUSSION

Powder diffraction measurements were performed on blanket Al wafer processed in conditions similar that of standard production or patterned wafer. Thus, a thick oxide layer was first deposited on Si substrate followed by a layer of Ta. Lastly, a PVD Al film was grown on top of Ta substrate. Powder diffraction pattern of such an Aluminum film consisted of a number of Bragg peaks, as shown in Fig. 1, curve (A). It may be noted that all the major Bragg peaks of a face centered cubic (*fcc*) Al crystal – a good agreement with ICDD no. – are present in the diffraction pattern. Moreover, Bragg peaks of polycrystalline Ta are also noticeable. These peaks were indexed with β-Ta phase, which has tetragonal symmetry.

The presence of all the allowed Bragg peaks of an *fcc* phase of Al is a clear indication that the Al film grown on Ta substrate has no preferred orientation, in other words, no texture. This is quite an unusual observation as it is well established that the Al interconnects in wafer fabrication are textured – Al(111) orientation with respect to Si(001) substrate. Furthermore, the texture formation in Al films is very important from the standpoint of reliability as it decreases the probability of void formation due to electro-migration and stress-induced migration.

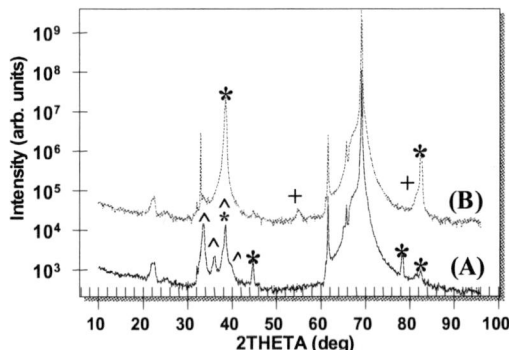

Fig. 2 X-ray diffraction patterns of Al films grown on Ta (curve (A)) and Ta/TaN (curve (B)). The Bragg peaks of Al, Ta and TaN are indicated with *, ^ and +, respectively.

It should be noted that crystallographic texture in Al films is dependent on the underlying substrate. For instance, a Ti/TiN bilayer substrate, where TiN film itself has TiN(111) texture, favors Al(111) orientation in Al thin films. In the present study, Al film, used in bondpad process, is grown on a different substrate, namely, Ta. The superior adhesion of Cu/Ta interface as opposed that of Cu/Ti is the primary motivation for using Ta as the barrier metal in Cu technologies. The non-textured nature of Al grains is possibly related to substrate material.

In order to recover texture in Al films, a bilayer of Ta/TaN – similar to Ti/TiN bilayer used in Al-based technologies – was used a substrate for Al films. Thus the modified deposition scheme consisted of the film stack sequence – $Si/SiO_2/Ta/TaN/Al$. A strong texture along Al(111) was noticed in the XRD pattern of an Al film grown on a Ta/TaN bilayer, as shown in Fig. 2, curve (B). It may be noted that unlike curve (A), in curve (B) only one Bragg diffraction peak due to (111) crystal planes is noticeable. The second peak at larger 2θ value, around $82°$, is due to second order reflection from the same set (111) planes. The observed and reported 2θ values of Bragg peaks are shown in table-1 below.

Table-1: The Bragg planes of crystal phases in the two Al films are listed below. Also listed are the corresponding 2θ values, both observed and that reported in ICDD files: **4-787** (Al), **19-1290** (Ta) and **14-471** ($TaN_{0.04}$).

Crystal Phase	(hkl)	$Si/SiO_2/$ Ta/Al	$Si/SiO_2/$ TaN/Ta/ Al	ICDD
Al (fcc)	(111)	38.46	38.50	38.47
	(200)	44.71	44.79	44.74
	(311)	78.34	-	78.23
	(222)	82.54	82.54	82.44
β-Ta (Tetragonal)	(200)	33.52	33.52	33.54
	(113)	36.03	-	35.89
	(004)	36.03	-	36.19
	(202)	38.46	37.96	38.10
	(211)	38.46	38.67	38.78
	(104)	39.89	-	40.04
TaN0.04 (bcc)	(411)	-	37.92	37.93
	(510)	-	45.79	45.79
	(600)	-	54.85	54.58
	(750)	-	81.70	81.92

The barrier metal scheme, Ta vs. Ta/TaN, does impact the crystallographic properties of PVD Al film. In order to verify the dark bondpads observed were related to non-texture nature of Al film deposited on Ta, the stack sequence was changed to a bilayer of Ta/TaN. In Fig. 3, the optical images of two Al bondpads with different barrier metal schemes are shown. The Al bondpads shown in (A) (are on single Ta layer) appear 'dark' as compared to those in (B) (on bilayer Ta/TaN).

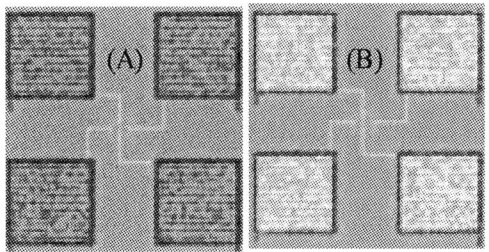

Fig. 3 Optical images of bondpads of patterned wafers with stack sequence Ta/Al, (A) appear dark as compared those with Ta/TaN/Al sequence, (B).

Further correlation studies were performed to determine if the optical properties of the Al films are impacted by underlying barrier-metal scheme. Optical reflectivity measurements were conducted on blanket Al film deposited on Ta and Ta/TaN substrates. These measurements were conducted at 49 sites across the wafers in order to generate statistically significant data. In Fig. 4, the data is shown as box plots for the two substrate schemes used in these studies. It is interesting to note that a higher mean value of optical reflectivity is observed (> 2× increase) for film with Ta/TaN bilayer as compared to that with only Ta layer.

A higher optical reflectivity of blanket Al wafer with bilayer is consistent with the visual observation (optical inspection) of 'bright' bondpads on patterned wafer with bilayer scheme and vise versa. This reinforces the argument that the crystallographic texture in thin films influence other material properties, and therefore, important in process optimization of wafer fabrication. For example, the Litho Focus-Exposure Matrix used during masking step would be dependent on the optical reflectivity of the film.

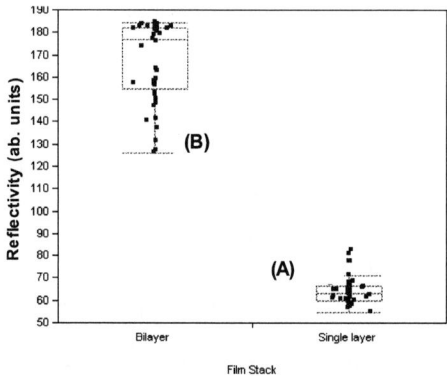

Fig. 4 Box plots of optical reflectivity data of blanket Al films with Ta and TaN bilayer, taken at 49 sites are shown in the figure.

In conclusion, XRD investigations on blanket Al films deposited on Ta and Ta/TaN had different texture signature – single (Ta) layer Al film consisted of randomly oriented grains while the bilayer (Ta/TaN) Al film was textured along Al(111). A clear correlation was established between film texture, optical reflectivity and visual appearance of the bondpads.

ACKNOWELDGEMENT

The authors would like to thank Dr. Zhao Siping for her constant encouragement and support, and also for critical review of the manuscript.

REFERENCES

[1] D.B. Knorr, D.P. Tracy and K.P. Rodbell, "Correlation of texture with electromigration behavior in Al metallization," Appl. Phys. Lett., vol. 59, p. 3241 (1991).

[2] Zhenqiang Ma and Gary S. Was, "Aluminum metallization for flat-panel displays using ion-beam-assisted physical vapor deposition," J. Mater. Res., vol. 14, p. 4051 (1999).

[3] C.V. Thompson and R. Carel, "Texture development in polycrystalline thin films," Mat. Sci. and Eng., vol. B32, p. 211 (1995).

[4] Hua Younan, Ramesh Rao Nistala, Tan Chin Wang and Mo Zhiqiang, "Failure Analysis of Contamination on Microchip Al Bondpads using Auger Electron Spectroscopy and X-Ray Photoelectron Spectroscopy," *Proc. NSM'05, IEEE 5th National Symposium on Micro-electronics* (2005).

[5] Hua Younan, N. Ramesh Rao, Tan Chin Wang and Lo Keng Foo, "Failure Analysis of Discolored and Non-Stick Al Bondpad During Wafer Sorting Process," *Proc. ICSE'04, IEEE International Conference on Semiconductor Electronics* (2004).

[6] Hua Younan, N Ramesh Rao, Tan Chin Wang and Lo Keng Foo, "Failure Analysis of Bondpads Peeling Problem in 0.13um Copper Process in Wafer Fabrication," *Proc. ICSE'04, IEEE International Conference on Semiconductor Electronics* (2004).

Meeting the Challenges of Elemental Analysis in 90nm & Beyond Technologies – Case Studies of Scanning Auger Nanoprobe

Nistala Ramesh Rao, Tan Chin Wang, Song Zhigang, Hua Younan and Zhao Siping, *Snr member, IEEE*

Chartered Semicondcutor Mf g Ltd
Woodlands Industrial Park D, Street 2, Singapore 738406
Tel: 65-63604378, Fax: 65-63622935
Email: raoramesh@charteredsemi.com

Abstract: **In this paper few case studies of Auger elemental analysis in failure analysis of wafer fabrication, using a state-of-the-art scanning Auger nanoprobe, will be presented. Material identification in particle defects is quite challenging especially in advanced microelectronic technologies, 90nm and beyond, where, due to decreasing device size, the tight pitch and high aspect ratio of features introduces constraints of analysis. Thus, the case studies discussed in the paper are all pertaining to the front-end related defects, at poly+contact region of the die, where these contraints are expected to be severe. It will be shown that the scanning Auger nanoprobe can be used for the elemental analysis of features as small as ~10nm.**

I. INTRODUCTION

PHYSICAL diagnostics in the failure analysis of wafer fabrication involves identification of defect(s) at fault isolation location(s) on a die and, in most cases, elemental analysis of the defect. The supportive evidence gathered in material identification process is necessary to determine, by the process of elimination of potential leads/mechanisms, the possible root cause(s) of a failure mode. In some instances, the evdience may provide the all important clue for further analysis. As the industry moves to deep sub-micron technologies, 90nm and beyond, the applications of EDX technique for elemental analysis will become limited. For example the tight pitch of circuit features like the memory cells would rquire a probe beam that is not only spatially local (in lateral direction) but also has limited spread in

transverse direction, Adhesion properties of Metal/Dielectric interfaces is another area where a small probe beam would be required. Here too, the constraints on the beam would be in three direction i.e. sample-beam interaction volume constraints.

In EDX, due to the bulk or volume interaction between the e-beam and the sample, the contribution from the 'background' cannot be avoided and this limits its application to situations where there is no or little chemical contrast between the feature of interest and the surrounding material. A surface senstive technique, such as Auger Electron Spectroscopy (AES), is better suited as the beam is localized i.e. small interaction area between the e-beam and the sample. Moreover, only those Auger electrons that are generated at a depth of about 60Å from top surface have sufficient kinetic energy to escape and hence, the contribution from 'background' can be eliminated in most cases.

In our previous papers [], we used using a state-of-the-art scanning Auger, Smart 200. In this paper, a few case studies of Auger Electron Spectroscopy (AES), using a new state-of-the-art scanning Auger nanoprobe, *PHI 700*, will be presented. All the cases discussed involve front-end defects in the poly and contact region, where the challenges of tight pitch geometries introduced by 90nm technology and beyond will be most severe.

II. SCANNING AUGER NANOPROBE

The data presented in this paper was collected on a state-of-the-art scanning Auger nanoprobe, model PHI-700, of ULVAC-PHI, Japan. This system is ideal for elemental analysis of deep sub-

0-7803-9730-4/06/$25.00 ©2006 IEEE

micron defects down to 10-20 nm. It is equipped with a Schottky field emission (FE) source that can stably operate in the voltage range of 1 to 25 kV. The large voltage range, coupled with channeltron secondary electron multiplier, enable high contrast SEM imaging. The electron beam column is optimized to achieve dark space SEM spatial resolution of better than 7nm (with 20kV, 1nA beam condition). A coaxial cylindrical mirror analyzer (CMA) makes it possible to perform high sensitivity Auger analysis, including elemental distribution, without the artifacts of sample topography. The other salient features of the system are; (i) a heavy metal enclosure for acoustic dampening and thermal stability. The acoustic hood reduces 20dB sound pressure level (SPL) over a frequency range of 30 Hz to 5 kHz. (ii) main chamber vacuum better than 2×10^{-9} Torr.

III. CASE STUDIES

Case Study -1:

In the SEM image of Fig. 1, a die unit, of 90nm technology, deprocessed to expose the poly-lines and W-contacts is shown. The residual defects, between the poly lines, are around 0.2 μm wide. It was suspected that this defect might cause poly-to-poly short, assuming it is composed of conducting material. Auger survey scans, in the energy range of 30-2030 eV, were made at five different locations, including the defect regions (site-1 and site-3, in Fig. 1). It may be noted that at all the five locations, a spot analysis mode was used (spot size of ~10-15nm) to ensure that the Auger signal from each of the locations is localized.

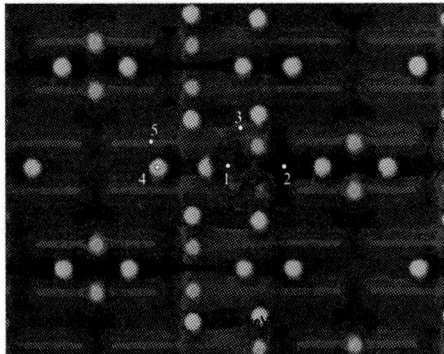

Fig. 1   SEM image of defect(s) (site-1 and site-3) between two poly lines of a failed die unit. Also shown in the image are the five Auger measurement locations.

The Auger survey scan spectra of the five locations are shown in Fig. 2. It should be noted these spectra were collected after subjecting the sample surface to Ar-ion sputter (etch rate of 100Å/min, with 2kV ion beam) in order to remove 100Å of material at the top. This was necessary to reduce carbon contamination on the 'as received' sample surface. A high C contamination on deprocessed unit is normally introduced by the RIE-etch used to remove the oxide layers. For Auger data collection, it is important to reduce the C contamination as it would inhibit the Auger peaks of other elements, especially the low energy and sensitivity peaks.

At the defect locations, site-1 and site-3, in addition to C, Si, N and O Auger peaks were observed in the corresponding Auger spectra.. The Auger spectra of sites 2, 4 and 5 suggests that these locations correspond to an active or Si, a W-contact and a slicided Poly, respectively. All the spectra showed C and O peaks, which we believe are due to surface contamination and oxidation, respectively. Thus, even after 100A sputter, neither C contamination nor surface oxidation could be completely eliminated. In this case study, the physical dimensions of the defect was such that it could be visualized in a SEM image of 30 kX magnification. What is encouraging is the high

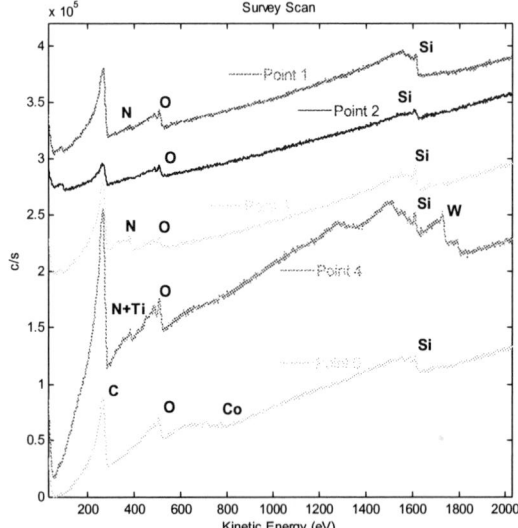

Fig. 2   AES spectra of the five locations shown in Fig. 1. Point-1 and Point-3 correspond to the defect locations while the remaining three measurements in the unit are from the poly, contact and active regions.

contrast in SEM image and the ability to correctly identify various regions of the circuitry.

Case-study II:

In our second example, another investigation of Poly-to-Poly short, as shown in the SEM image of Fig. 3, is discussed. In this case, the defect was of much smaller dimension and thus, one had to image at much higher magnification (70 kX) as compared to previous case study (30 kX). It is heartening to note that the quality of contrast in the SEM image is remarkably high on a dedicated Auger tool. The poly lines, spacer nitride and the defect (poly-to-poly short) are clearly seen, which is, at its narrowest (in the image), is $\approx 0.03\mu m$ wide.

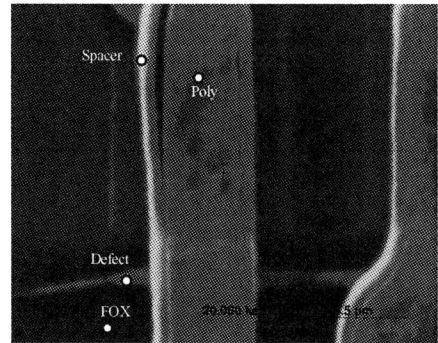

Fig. 3   SEM image of Poly-to-Poly short in a 0.35μm technology die. The circular dots shown in the image correspond to Auger measurement locations.

Auger spectra, shown in Fig. 4, were collected at the four locations indicated in the SEM image (see Fig. 3, above), namely, Poly, FOX (Field Oxide), Spacer nitride and defect. In all the spectra, C peak was detected, which, like the previous case, is introduced during the delayering process. It is important to note that even though N peak is detected at all four measurement locations, its intensity is large at the defect and spacer nitride locations. The small N peaks at the Poly and FOX are most likely the artifacts of RIE etch.

In passing, it may be noted the nanometer-sized defect discussed in the above case study is from 0.35mm technology die. Thus, even though one expects most of the challenges to Auger nanoprobe would come from 90nm and above technologies, it should be useful in elemental analysis of certain types of sub-micron sized defects, spanning all technology nodes.

A class of defects where Auger elemental analysis has obvious advantage over EDX is highlighted by the two case studies discussed in this paper. In both the cases, the background material is, predominantly, Silicon, and therefore, it would be difficult to conclusively identify, with EDX, whether the shorts are caused by spacer nitride or by poly filaments. Such identification is of paramount importance to a failure analyst/ product engineer to determine the failure mode, which, in turn, would help in root cause analysis. A poly-to-poly short mediated via poly-filament would cause electrical shorts while a residual nitride between poly lines may be a benign defect.

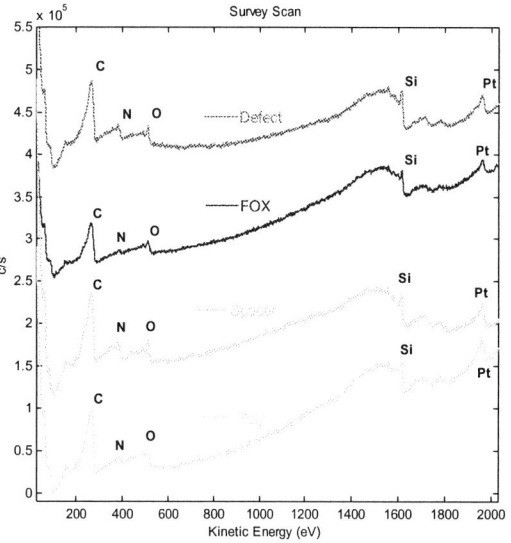

Fig. 4   Stack plot of AES spectra of the four locations indicated in the SEM image of Fig. 3.

Case-study III:

The next example of defective die unit is of 90nm technology. The unit was delayered (oxide removal with RIE) at the hot spot location all the way down to local interconnect (LI). A LI-LI short was observed in the SEM image, as shown in Fig. 5. Auger survey spectrum, Fig. 6, from the defect location showed clear signature of W, thus, the failure mode was identified as due to LI-LI short via W-residue. The Cu peak observed in the spectrum may be is due to re-deposited Cu from the Cu-via/metal during the RIE. The C and N Auger peaks are also most likely the artifacts of RIE deprocessing.

In this case study, the important point to be noted is the high aspect ratio at the defect site – the defect is buried at the underneath LI, below the Cu-via and Cu metal line. Thus, even though the defect is bigger than that of *case-II*, above, Auger analysis had to be performed at higher magnification (100 kX) as compared to the previous case (70 kX). This was necessary to prevent artifacts in Auger spectrum due to "knock-off" effects from the features surrounding the defect – similar to a billiard ball bouncing from the walls of the board and other neighboring balls.

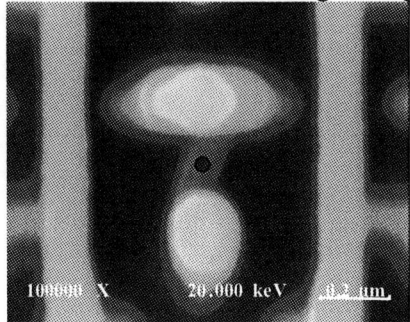

Fig. 5   SEM image of defective local interconnect (LI) location in a failing die unit. A LI-LI to short is visible in the defect and so are the surrounding metal lines via-plugs.

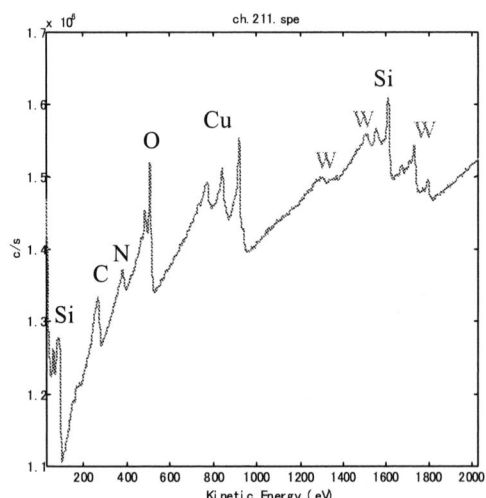

Fig. 6   Auger spectrum of the defect shown in Fig. 5. The spectrum was collected from the location indicated in the SEM image.

Case-study IV:

In the last example, an ultra thin short between two metal Lines was studied, as shown in Fig. 7. It must be pointed out the SEM image shown in the figure was taken on a dedicated FEI SEM tool. On the PHI-700 Auger system, it was not possible to see the defect by SEM imaging. However, as the defect was located in the memory region of the circuit, the defective region/ memory cell location could be accurately identified. Using this information from a dedicated SEM tool, Auger measurements were performed at two 'proximity' locations shown in Fig. 7.

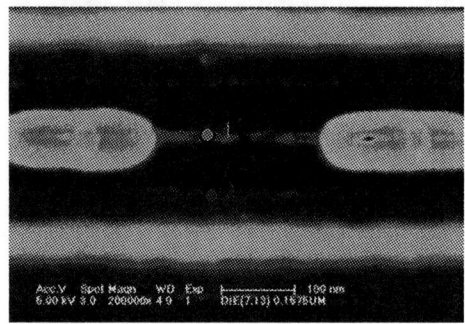

Fig. 7   SEM image of a defect at metal-1 in failed unit of 90nm technology. It may be noted that the image was taken on a dedicated FEI SEM tool. Auger analysis was performed at the two locations indicated in the image.

Interestingly, even though the defect was not clearly visible in the SEM image of PHI-700 system, Auger data, Fig. 8, at the defect location (point-1) had Ti+N peak while no such peak was noticed at the reference site (point-2). Thus, the thin filament material connecting the two metal bars is actually a TiN residue of the underneath tungsten local interconnect.

The measurement location, point-1, was far removed from the metal bar structure to avoid any artifact from the underlying contact. Thus, the observed TiN signature is real and the failure mode is due to metal-metal short via TiN filament. The last case study, we believe, explored the limits of PHI-700 capability for defect analysis, in terms of the physical size of defects, which would be around 10nm.

IV. CONCLUSION

In this paper, four case studies of front-end related defects of failed die units were presented. The defects were at the poly and contact region, with physical size varying from 0.2µm all the way down to 0.01μm. The state-of-the-art scanning Auger nanoprobe, PHI-700 system, was successfully used to identify the elemental composition of the defects and the failure mode of the defective unit.

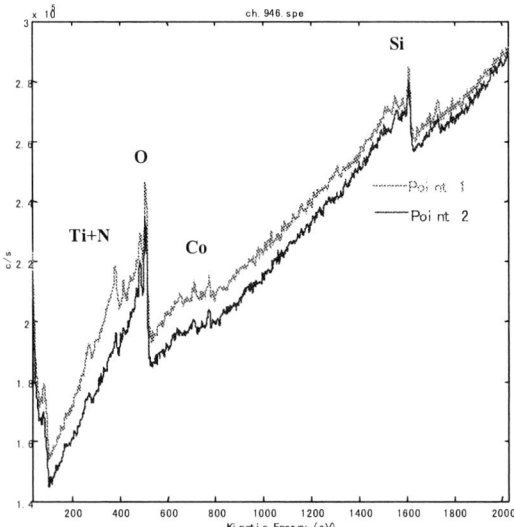

Fig. 8 Auger spectra of the two locations shown in Fig 7. Ti+N peak was observed at defect location, point-1, but not at reference location, point-2.

ACKNOWLEDGEMENTS

The authors would like to thank their colleagues in the FA lab, Neo Soh Ping, Loh Sock Khim, for fault isolation and deprocessing of the failed units.

REFERNCES

[1]. Hua Younan, Rao Ramesh, Lo Keng Foo and Mo Zhiqiang, "Studies on Fingerprints of EDX, FTIR, XPS and TOF-SIMS and Applications in Failure analysis of Wafer Fabrication". The Proceedings of the 30th International Symposium for Testing and Failure Analysis, Nov 14-18, Worcester's Centrum Center, Worcester, Massaehusetts, USA, 2004.

[2]. Ramesh Rao Nistala, Hua Y. N. and Tan C. W. Charge Reduction Method Using Laser and Focused Ion Beam and Its Application in Auger Analysis on Floating Bondpad Structures in Advanced IC Processing Technologies. The Proceedings of the IEEE NSM, Nov 22-24, 2005, Kuching, Sarwak, Malaysia, pp 50-53 (2005).

[3]. Ramesh Rao Nistala, Hua Y. N., Li K. and Tan C. W. A Comparative Study of Auger Electron Spectroscopy and Transmission Electron Microscopy of Ultra Thin Defects on Bondpad Surfaces. The Proceedings of the IEEE NSM, Nov 22-24, 2005, Kuching, Sarwak, Malaysia, pp 45-49 (2005).

[4]. N. Ramesh Rao, Tan Chin Wang, Hua Younan and Lo Keng Foo, "A Study on Surface Charge Reduction Method in Auger Electron Spectroscopy". The Proceedings of the 2004 IEEE International Conference on Semiconductor Electronics, Kuala Lumpur, Malaysia, 7-9 December, pp86-89, 2004.

[5]. Ramesh Rao Nistala, Hua Y. N., Tai M. F. Eve and Tan C. W. Failure Analysis and Elimination of Particle Contamination on Microchip Al Bondpads in Wafer Fabrication. The Proceedings of the IEEE NSM, Nov 22-24, 2005, Kuching, Sarwak, Malaysia, pp 395-399 (2005).

Improved Booth Encoding for Reduced Area Multiplier

Razaidi Hussin, *Ali Yeon Md. Shakaff, Norina Idris, Rizalafande Che Ismail and Afzan Kamarudin
*School of Computer and Communication Engineering.
School of Microelectronic Engineering,
Northern Malaysia University College of Engineering (KUKUM),
Kompleks Pusat Pengajian KUKUM,
02600 Jejawi, Perlis, Malaysia
shidee@kukum.edu.my

Abstract **in designing high density circuit, size is a major concern in design. This paper presents a simple modification to the Booth Multiplier that can effectively reduce the area with an accepted scarified in speed. A conventional Booth Multiplier consists of Booth Encoder, Partial Product and Summation Tree. Rizalafande[1] introduced new design technique in generating the partial product's row. Meanwhile Hsin-Lei[2] introduced a novel circuit for Booth Encoder/Decoder which claims his design a smaller design. In this propose design, we are still using Rizalafande's architecture but replace the booth encoder with Hsin-Lei encoder. The design was implemented using the FLEX10K EPF10K70RC240-4 device and Altera MaxPlus+II software.**

I. INTRODUCTION

With the rapid growth of computer applications such as signal processing and computer graphics, fast arithmetic unit especially multipliers are increasingly rapidly. Nowadays, advances in VLSI technology give chances for designers to integrate many complex components, which was not possible in the past. Various high-speed multipliers have been proposed [1-4].

In this paper, we have made some analysis on Normal Booth Multiplier architecture, Hsin-Lei's architecture and Rizalafande's architecture. Our focus is to improve the Rizalafande's architecture either in delay circuit or size of the circuit.

This paper is organized as follows, Section II: Booth Architecture Section III Improved Rizalafande architecture Section IV Analysis of booth multiplier architecture and Section V is the conclusion.

II BOOTH ARCHITECTURE

Normal Booth Multiplier as well as Hsin-Lei architecture consists of three basic part operations which is Modified Booth Encoder, partial product and adder summation. The basic operation of this architecture is MBE will decode the multiplier signal and the outputs of MBE signal are use by partial product to generate the partial product. Usually when using this architecture, the 2'complement error correction is implemented in adder summation. As a result this architecture will have n/2 +1 partial product for n x n multiplier. Figure 1, below shows 2's complement error correction method for 8 bit Booth Multiplier.

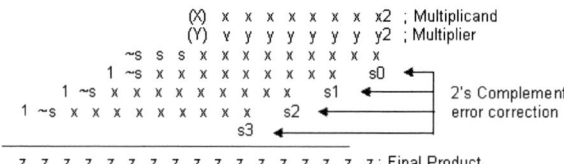

Figure 1: 2's complement error correction method for 8 bit Booth Multiplier

Meanwhile Rizalfande has proposed a new architecture to improve in 2'complemet error correction. The idea is to remove the 2'complement error correction part in adder tree summation. As a result, this architecture will only have n/2 partial product for n bit multiplication. His architecture consists of Product Generator, MBE, Partial Product and adder summation. The product generator will produce all the possible value of Multiplicand. The partial product act as a multiplexer to choose which value it need based on MBE signal. The value of product generator is in 2'complement value. In this architecture, 2'complement error correction is not used.

0-7803-9730-4/06/$25.00 ©2006 IEEE

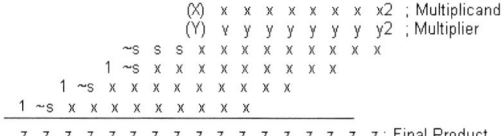

Figure 2: Booth multiplier without 2's complement error correction method

In this paper, some modifications have been made in summation tree. As a comparison for all architecture, we use the basic adder tree which is ripple carry adder as a summation tree. Figure 3 shows Booth Multiplier architecture for Normal Booth Multiplier and Hsin-Lei's multiplier while figure 4, shows Rizalafande's architecture.

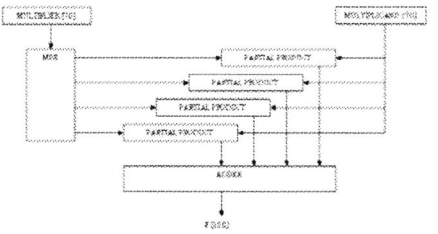

Figure 3: Booth multiplier architecture for Normal Booth Multiplier and Hsen Lei Multiplier

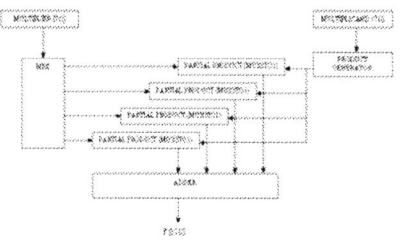

Figure 3: Rizalafande Booth Multiplier architecture

III IMPROVED BOOTH MULTIPLIER ARCHITECTURE

From my previous research, Hsin-Lei's encoder is the most efficient encoder. The propagation delay from input until produce the partial product is faster than other encoder. And the circuit's size is much smaller. However this scheme have a little disadvantages when produce a '-0' condition. To overcome this problem a new proposal MBE have been made base on Hsin-Lei's encoder. This is the reason why Hsin Lei encoder is chosen for improved the Rizalafande's architecture.

IV ANALYSIS OF BOOTH MULTIPLIER ARCHITECTURE

All the multiplier was implemented using the FLEX10K EPF10K70RC240-4 device and Altera MaxPlus+II software. Table 1, shows the total transistor and delay for each circuit. Table 2 shows the result of 8 bit multiplier for all architecture. In order to determine the hardware cost for these architectures, we have selected Gajski[5] analysis. Table 3 shows the normalized hardware cost derived from Gajski's analysis. By counting the numbers of transistor used will help us to estimate the hardware requirement. In term of delay analysis, we use timing analyzer which provide by MaxPlus+II software. From the result, the normal multiplier has advantages on speed while the Hsin-Lei's multiplier has the advantages on size of the circuit. Meanwhile the upgrade of rizalafande's architecture is less total transistor from the original due to the Hsin-Lei's encoder. It reduces 9% from the original circuit. But in term of delay analysis, the Rizalafande upgraded circuit is 5.5% much slower than Rizalafande's circuit.

	Total transistor	Delay/ns
Normal MBE	440	11.6
Hsin Lei	120	12.1
New Hsin Lei	248	12.1
Partial Product	1288	32.3
Partial Product Hsin Lei	1432	33.9
Product generator	36	37.4
PP(Mux) rizalafande	1432	37.0
New PP(Mux) rizalafande	1208	36.9

Table 1: The total transistor and delay for each circuit

	Total transistor	Total delay/ns	Delay PP/ns
Rizalafande	4608	46.4ns/97.2ns 43.4ns/99.2ns	21.1ns 20.6ns
Rizalafande Upgrade	4192	63.5ns/102.9ns 45.9ns/95.9ns	28.0ns 20.2ns
Hsin Lei	3928	48.3ns/100.8ns 54.4ns/96.6ns	24.9ns 20.7ns
Normal	4104	51.ns/96.7ns 55.3ns/89.2ns	26.8ns 19.3ns

Table 2: shows the result of 8 bit multiplier for all architecture

Gate Name	Hardware Cost (In Number of Transistor)	Function
INV	2	1 input inverter
AND2	6	2 input AND
AND3	8	3 input AND
AND4	10	4 input AND
OR2	6	2 input OR
OR3	8	3 input OR
OR4	10	4 input OR
XOR2	14	2 input exclusive OR
XNOR2	12	2 input exclusive NOR

Table 3: Hardware cost

IV. CONCLUSION

In this paper, an upgraded Booth Multiplier which is more optimal has been design. This design combines the Hsin-Lei encoder and Rizalafande's multiplier architecture. An upgraded Booth Multiplier had been reducing 91% from rizalafande's original Booth Multiplier.

REFERENCES

[1] Rizalafande Che Hussin, A Complex Multiplier Using Booth Wallace Algorithm, MEng. RMIT, 2005

[2] Hsin-Lei Lin, Design of a Novel Radix – 4 Booth Multiplier, the 2004 IEEE Asia – Pacific Conference on Circuit and Systems, December 2005

[3] Neil H.E. Weste & Davis Harris, CMOS VLSI Design – A Circuit and System Perspective Third Edition, Pearson, United State of America, 2005

[4] M. Micheal Vai, VLSI Design, CRC Press, United State of America, 2001

[5] D. Gajski. *Princples of Digital, Design,Prentice Hall*, 1997

A Novel Method to Design Interstage Matching Network in the Smith Chart

M. Rezvani Abkenari, M. Tayarani, *Member, IEEE*, A. Abdipour, *Senior Member*, IEEE, and
H. Kiumarsi

Microwave/mm-wave & Wireless Communication Research Lab
Electrical Engineering Department, Amirkabir University of Technology, Tehran, Iran
Email:m_rezvani@aut.ac.ir

Abstract **In this paper, a novel method for interstage matching network design in the smith chart is presented. This technique is based on matching the reflection coefficients between input of interstage matching network and S_{22} of the first transistor with considering input VSWR and the gain of the complete amplifier (or output of the complete amplifier if the matching is applied to the output of the interstage matching network). Also, the novel locus of constant gain and VSWR of interstage matching network are presented in the smith chart.**

I. INTRODUCTION

Generally, interstage matching network design is one of the most difficult parts of the multi-stage amplifier design. The design of interstage matching network can be cumbersome in analytical form [1].

So, finding a new approach to simplify the interstage matching network design is a significant development and provides a basis for the future researches. There are some methods for designing input and output matching networks in the smith chart [2, 3]. In this paper a similar new method for designing interstage matching network is introduced. Because of using smith chart for designing, this technique provides a very useful graphical aid to analyze and design interstage matching networks. Also, it makes it possible to compare different circuits of interstage matching networks. In this approach by ignoring input and output matching network, we focus on designing interstage matching network. The resulted two port network from cascading first transistor, the interstage matching network and the second transistor which is shown in Fig. 1 is called N. Here for simplicity, we assume that two transistors are the same but

the resulted formulas can be easily developed in general form.

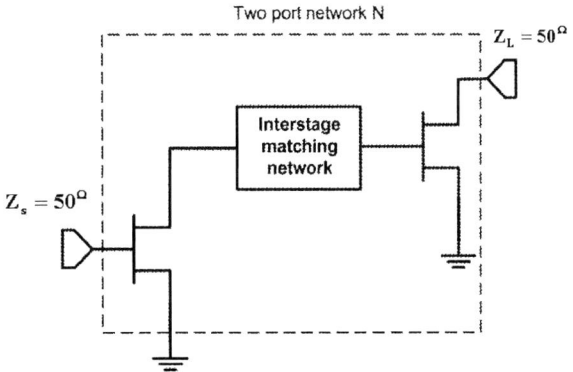

Fig. 1 Two port network resulted from cascading first transistor, interstage matching network and the second transistor

Consider the configuration in Fig. 2 for interstage matching network design. GPROBE in Microwave Office [4] is used to measure Γ_1 and Γ_2. In Fig. 2 conjugate matched at the place of GPROBE is obtained when:

$$\Gamma_2 = \Gamma_1^* = S_{22}^* . \qquad (1)$$

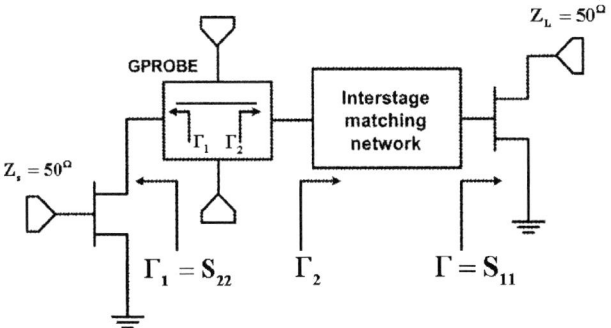

Fig. 2 Two stage amplifier with interstage matching network

When $\Gamma_2 = \Gamma_1^*$, S_{21} of the two-port network N, called herein $(S_{21})_N$, is maximum.

In other words, as you can see in Fig. 3 for a typical transistor, to obtain the maximum of $(S_{21})_N$, it is required to move in the smith chart from S_{11} of the second transistor to $\Gamma_2 = S_{22}^*$ of the first transistor.

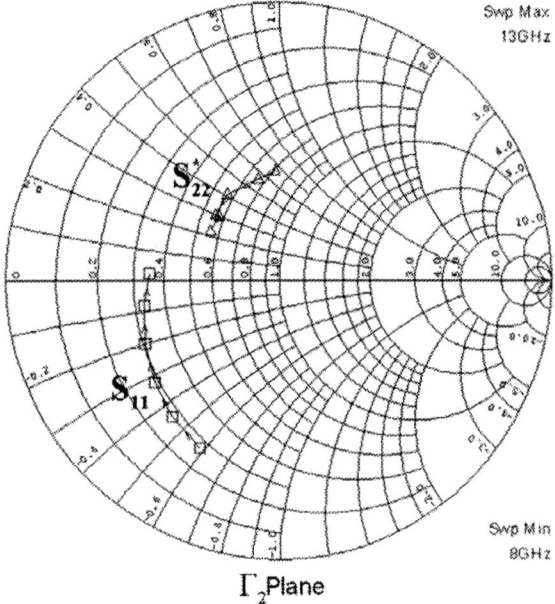

Fig. 3 S_{11} and S_{22} and S_{22}^* of a typical transistor on the smith chart

Note that, point to point matching is not our purpose, specially for wideband matching, the main point is to find the general form of moving and as a result by knowing the effect of series or shunt capacitors or inductors in the smith chart an appropriate configuration for the interstage matching network can be achieved, then the elements values can be optimized by computer aided approach.

II. CONSTANT $(S_{21})_N$ CIRCLES IN Γ_2 PLANE

In addition, to know points of Γ_2 on smith chart that $(S_{21})_N$ is maximum, finding the points that have the constant $(S_{21})_N$ in Γ_2 plane is useful too. To achieve this goal, the S_{21} of the two-port network can be calculated by equation (2) in terms of the S parameters and Γ_2. It must be noted that S_{ij} indicate the scattering parameter of each transistor.

$$G_T = \left|(S_{21})_N\right|^2 = \frac{|S_{21}|^4 (1-|\Gamma_2|^2)}{(1-|S_{11}|^2)|1-\Gamma_2 S_{22}|^2} \quad (2)$$

The maximum value of (2) is obtained when $\Gamma_2 = \Gamma_1^* = S_{22}^*$ and it is given by:

$$(G_T)_{MAX} = \frac{|S_{21}|^4}{(1-|S_{11}|^2)(1-|S_{22}|^2)} . \quad (3)$$

By solving the following equation in terms of Γ_2, places of constant $(S_{21})_N$ can be determined.

$$(S_{21})_N = \frac{|S_{21}|^4 (1-|\Gamma_2|^2)}{(1-|S_{11}|^2)|1-\Gamma_2 S_{22}|^2} = K . \quad (4)$$

The values of Γ_2 that produce a constant value of G_T or $(S_{21})_N$ equal to K, lie in a circle whose equation is

$$\left|\Gamma_2 - C_G\right| = r_G \quad (5)$$

where the center of the circle is given by:

$$C_G = \frac{\lambda S_{22}^*}{1-|S_{22}|^2 (1-\lambda)} \quad (6)$$

and the radius is

$$r_G = \frac{\sqrt{1-\lambda}(1-|S_{22}|^2)}{1-|S_{22}|^2 (1-\lambda)} \quad (7)$$

where λ is given by:

$$\lambda = \frac{K(1-|S_{11}|^2)(1-|S_{22}|^2)}{|S_{21}|^4} . \quad (8)$$

III. CONSTANT INPUT VSWR CIECLES IN Γ_2 PLANE

It should be noted that under conjugate matched condition (i.e., when $\Gamma_2 = \Gamma_1^*$), VSWR in the input or output of the two port network N is not equal to one. Proper values for VSWR of the total network can be achieved after designing the input or output matching network if this parameter is considered in interstage matching network design. It is obvious that VSWR and $(S_{21})_N$ can not be reached to their optimum values simultaneously so with regard to the application a trade off between these parameters should be done. In Fig. 2 Γ_{in} in terms of Γ_2 is given by:

$$\Gamma_{in} = S_{11} + \frac{S_{12}S_{21}\Gamma_2}{1 - S_{22}\Gamma_2} \qquad (9)$$

In fact, we should solve equation (10) to determine the locus where the values of Γ_2 produce a constant VSWR$_{in}$.

$$\left|\Gamma_{in}\right| = \left|S_{11} + \frac{S_{12}S_{21}\Gamma_2}{1 - S_{22}\Gamma_2}\right| = \frac{\text{VSWR}_{in} - 1}{\text{VSWR}_{in} + 1} \qquad (10)$$

The values of Γ_2 that produce a constant VSWR$_{in}$ lie in a circle with the following center and radius.

$$\alpha = \frac{\text{VSWR}_{in} - 1}{\text{VSWR}_{in} + 1} \qquad (11)$$

$$\Delta = s_{11}s_{22} - s_{12}s_{21} \qquad (12)$$

$$r = \frac{\alpha \left|s_{12}s_{21}\right|}{\left|\left|\Delta\right|^2 - \alpha^2 \left|s_{22}\right|^2\right|} \qquad (13)$$

$$c = \frac{\Delta^* s_{11} - \alpha^2 s_{22}^*}{\left|\Delta\right|^2 - \alpha^2 \left|s_{22}\right|^2} \qquad (14)$$

From the above equations output stability circles (*i.e.*, $\left|\Gamma_{in}\right| = 1$) can be derived by substituting $\alpha = 1$. Here for example constant VSWR$_{in}$=3 and output stability circles are drawn for 8 GHz to 13 GHz in Fig. 4 for a typical transistor of Fig. 3. The arcs from right to left are related to the constant VSWR$_{in}$=3 circles from 8 GHz to 13 GHz respectively. considering Fig. 4 adding a series capacitor produce a motion along constant-resistance circles in counterclockwise direction in the smith chart, cause moving from S_{11} to VSWR$_{in}$=3 circles and then adding a shunt inductor produce a motion along constant-conductance circle in a clockwise direction cause moving toward S^*_{22} and increase $(S_{21})_N$. Each motion along a constant-resistance or constant-conductance circles give the value of an appropriate element.

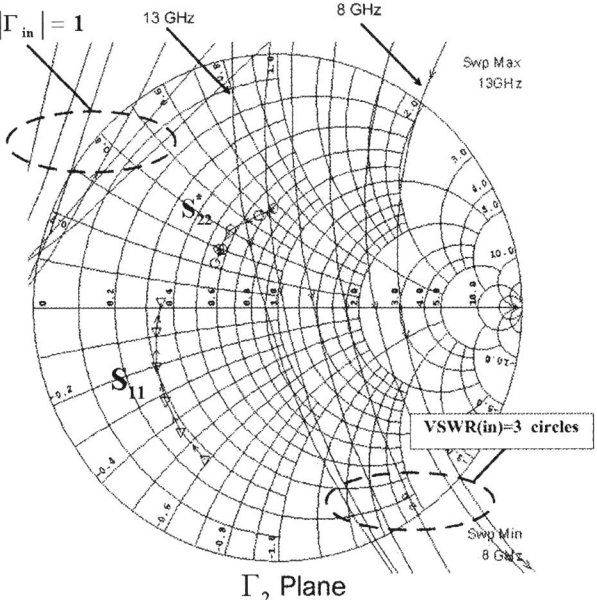

Fig. 4 VSWR$_{in}$=3 circles in Γ_2 plane

Then these values can be optimized by computer aided approach. Here by using Microwave Office the values of elements in the interstage matching network initial schematic are obtained, as shown in Fig. 5.

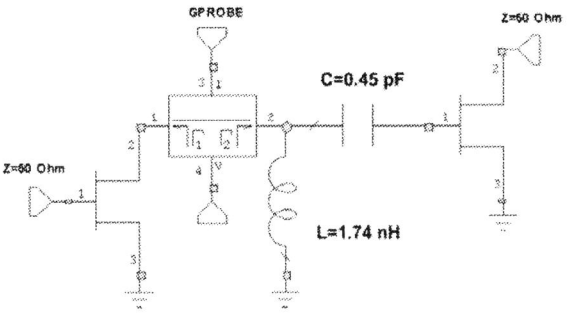

Fig. 5 Initial schematic of interstage matching network

It is also useful to see the effect of the initial schematic of interstage matching network on Γ_2 in the smith chart. Fig. 6 shows the measured Γ_2 with GPROBE and constant $(S_{21})_N$=22 dB circles.

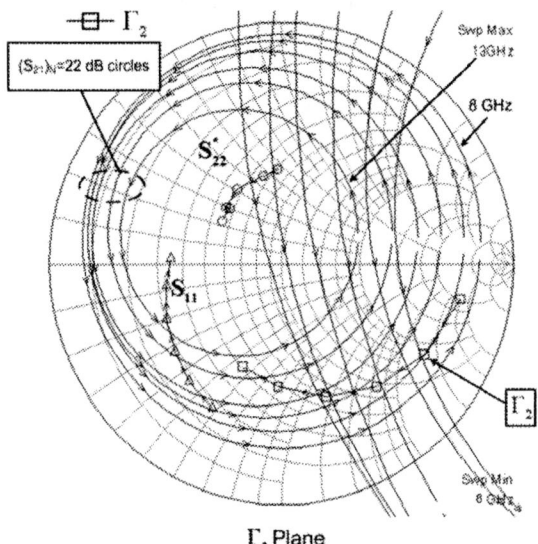

Fig. 6 $(S_{21})_N$=22 dB circles and VSWR$_{in}$=3 circles and measured Γ_2 in Γ_2 Plane

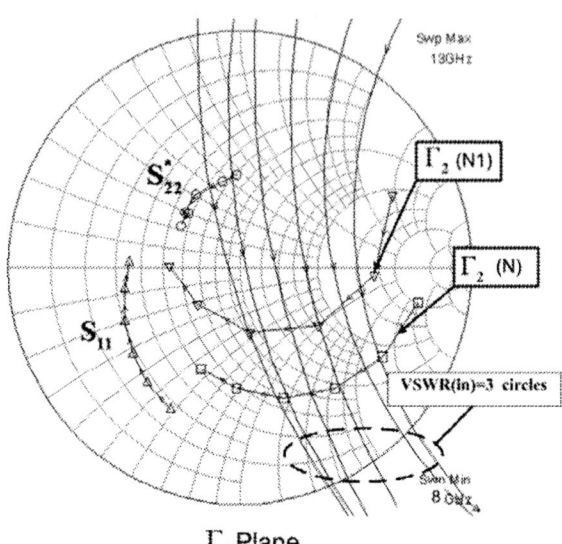

Fig. 8 The effect of adding a series inductor to the interstage matching network in Γ_2 Plane

Considering the availability of the constant VSWR and $(S_{21})_N$ circles, the initial circuit can be improved with series and shunt elements adding to the interstage matching network schematic. For example Fig. 6 shows that adding a series inductor to the interstage matching network can increase $(S_{21})_N$ with negligible change in VSWR.

The schematic of the new interstage matching network and the measured Γ_2 are shown in Figs. 7 and 8.

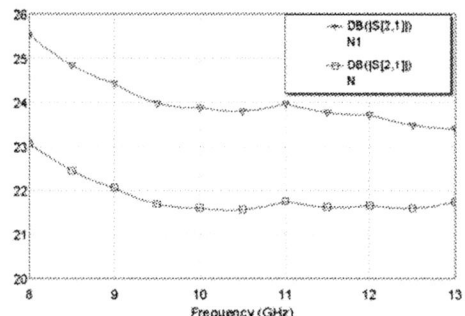

Fig. 9 Comparison of the total S_{21} of interstage matching network schematic N and N1

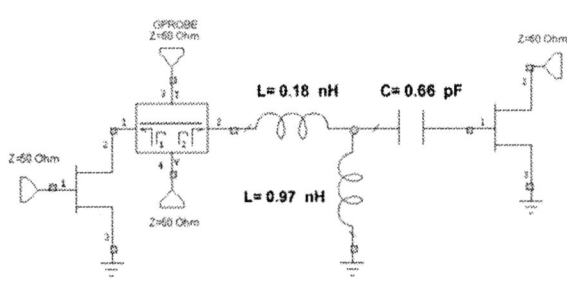

Fig. 7 New interstage matching network schematic

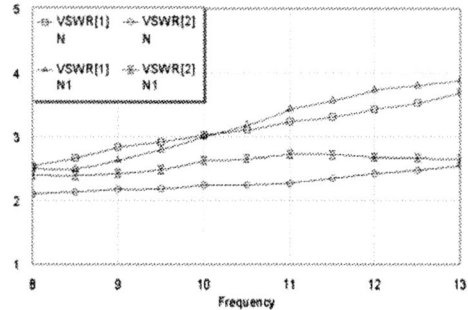

Fig. 10 Comparison of the VSWR$_{in}$ of schematic N and N1

We called first interstage matching network schematic N and the second one N1. It can be seen that the new Γ_2 move toward S_{22}^* in Fig. 8 and that is equivalent to increasing $(S_{21})_N$. $(S_{21})_N$ and VSWR$_{in}$ comparison of these two networks are shown in Figs. 9 and 10.

IV. CONCLUSION

In this paper a new method for designing interstage matching network is presented. The advantage of this method is ease of use and graphical view of design procedure. Also it makes it possible to use this method with microwave simulators such as ADS, Microwave Office, *etc.* Besides, this method can be applied to multistage amplifier design.

REFERENCES

[1]. Tri T Ha, *"Solid-state microwave amplifier design,"* New York: John Wiley&Sons, Inc., 1981.

[2]. Guillermo Gonzalez, *"Microwave Transistor Amplifiers Analysis and Design,"* Englewood Cliffs, N. J. prentice-hall,Inc.,1984.

[3]. S. Y. Liao, *"Microwave Devices and Circuits" Prentice Hall, Englewood Cliffs,"* NJ, 1980.

[4]. Applied Wave Research, Microwave Office online Reference manual.

Dependence of Radio Frequency Power on Optical, Chemical Bonding and Photoluminescence Properties of Hydrogenated Amorphous Carbon Nitride Films

Richard Ritikos, Goh Boon Tong, Rozidawati Awang, Siti Meriam Abdul Gani and Saadah Abdul Rahman

Solid State Physics Laboratory
Physics Department, Science Faculty
University Malaya, 40603, Kuala Lumpur
Email: richardritikos@gmail.com, saadah@um.edu.my

Abstract **Hydrogenated amorphous carbon nitride films (a-CN_x:H) were prepared in a radio-frequency plasma enhanced chemical vapour deposition (r.f. PECVD) system with a parallel-plate configuration. The gas sources of CH_4 and N_2 were fixed at CH_4:N_2 ratio of 1:3. The films were grown on glass and Si substrates on the grounded electrode at 100°C. The effect of r.f. power (varied between 0.71 - 3.54 W/cm^2) on the optical, infrared (IR) absorption spectra and photoluminescence (PL) spectra of the a-CN_x:H films were studied. It was observed that the deposition rate increases linearly up to the r.f. power of 2.83 W/cm^2 , while the optical band gap (E_{04}) decrease exponentially in the whole range. This is proposed to be the effect of an increase in nitrogen incorporation into the sp^2 carbon clusters, as indicated by FTIR. The PL spectra consist of a band in the region of 2.10-2.40 eV, with peaks at approximately 2.23, 2.27 and 2.33 eV. The PL intensity of the films increases as the r.f. deposition power increases and is related to the increase of the sp^2 clusters with increasing nitrogen incorporation.**

I. Introduction

Carbon nitrides films has been widely researched since the prediction of Liu and Cohen [1,2] of a crystalline C_3N_4 phase with bulk modulus, hardness and optical gap similar to diamond. The application of these films includes protective coatings, electroluminescent devices, biomedical coatings and sensors [3]. However many of these application does not rely either on hardness nor crystallinity. Currently, the study of hydrogenated amorphous carbon nitride (a-CN_x:H) has attracted great interest particularly

in its application in electroluminescence devices. It has been reported that these films exhibit strong photoluminescence (PL) properties even at room temperature [4]. In this work we discussed the optical and chemical bonding properties of a-CN_x:H films prepared in CH_4/N_2 r.f. plasma with different r.f. power input and their effect on the PL properties of these films

II. Experiment

Hydrogenated amorphous carbon nitride films (a-CN_x:H) were prepared in a radio-frequency plasma enhanced chemical vapour deposition (r.f. PECVD) system with a parallel-plate configuration (Figure 1). Distance between the top and bottom electrode was fixed at 5cm. Pure methane (CH_4) gas mixed with nitrogen (N_2) was introduced into the system through a shower head which also acts as the powered electrode. The CH_4/N_2 ratio was fixed at 1:3 whereas the deposition pressure was maintained at 0.8mbar. The films were grown on glass and Si substrates on the grounded electrode which were kept at 100°C. Films were deposited for 1 hour and the r.f. power was varied between 0.71 - 3.54 W/cm^2. The deposition chamber was sputtered in N_2 gas discharge introduced into the chamber at flow-rate of 60 sccm and r.f. power of 1.77 W/cm^2 for 15 minutes to remove impurities on the substrate surface. These pre-treatments are normally carried out to improve the adhesion of a-CN_x:H layers onto the substrate [5].

Film thickness, refractive index and optical energy gap (E_{04}) were calculated from the optical transmission spectra, which were carried out using a Jasco V570 UV-VIS-NIR spectrophotometer within a scanning wavelength range of 250 nm to 2500 nm. The optical energy gap (E_{04}) was determined from the photon energy value when the optical absorption coefficient was

0-7803-9730-4/06/$25.00 ©2006 IEEE

10^4 cm^{-1} [6,7]. The chemical bonding of the films were analyzed by Fourier transform infrared (FTIR) spectroscopy. FTIR spectra were performed in transmission mode within the scanning range of 400 to 4000cm^{-1} using a Perkin-Elmer System 2000 FTIR spectrometer. Photoluminescence (PL) measurement was carried out with a Perkin Elmer LS50B luminescence spectrometer with Xenon source at excitation energy of 4.66 eV. The optical measurements were carried out on glass substrates whereas FTIR and PL were carried out on the Si substrates. All measurements were carried out at room temperature.

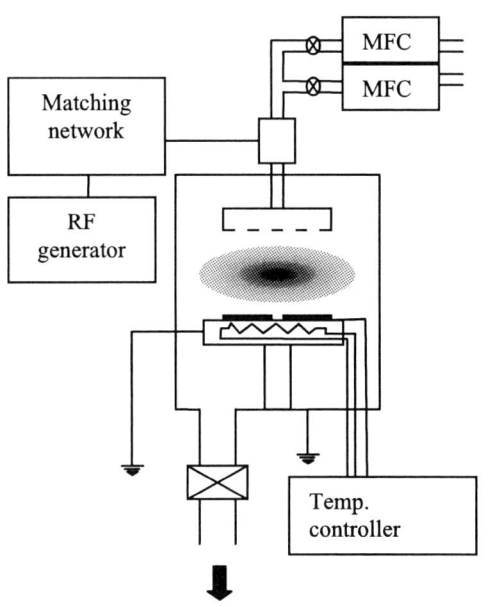

Fig.1 Schematic diagram of r.f. PECVD setup

III. RESULTS AND DISCUSSION

a-CN$_x$:H films produced on the glass substrates were yellowish-brown transparent coatings. It is also observed that with the increase in r.f. power the formation of dust particles on the films are evident particularly for r.f power of 3.74 W/cm^2. The optical properties are presented in Figure 2 and 3 as a function of r.f. power.

The deposition rate increases strongly with r.f. power up to 2.83 W/cm^2 while the optical energy gap (E$_{04}$) decreases with increasing r.f. power. However the refractive index remains almost constant for these films. The change in the energy gap without a corresponding change in the refractive index indicates a modification of states within the energy gap as a result of nitrogen incorporation [8]. With the increase in r.f. power and consequently, the increase in film

thickness, there may be an increase in graphitic clusters in the film which promotes the building of sp^2 coordination in the films [9]. These carbon atoms form π bonds and gives rise to electronic density of states near the band edge, thus decreased the observed optical band gap of the material. The extent of π electronic localization and hence π-π* splitting is closely related to the size of sp^2 carbon sites [10]. The sp^2 hybridized carbon sites are expected to behave as luminescence centres.

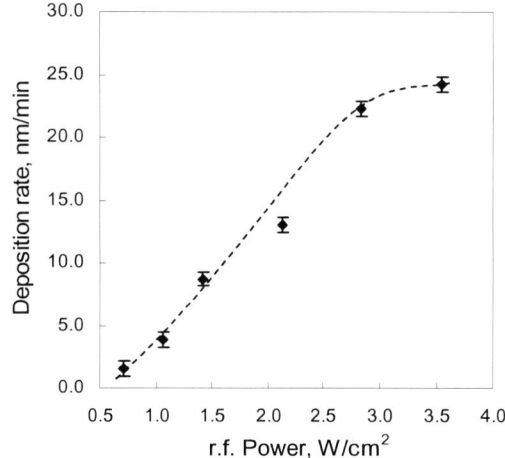

Fig.2 Variation of deposition rate as a function of r.f power. Lines are drawn as guide to the eye.

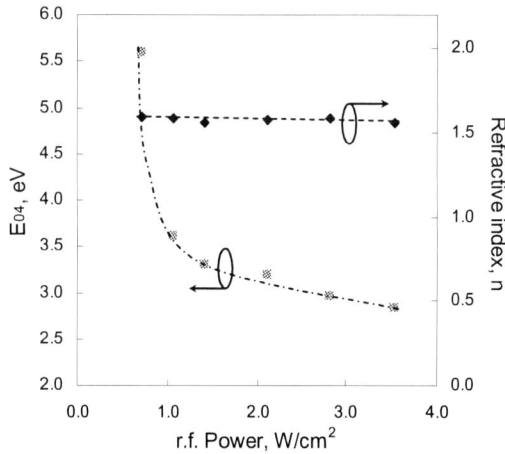

Fig.3 Variation of optical energy gap (E$_{04}$) and refractive index as a function of r.f power. Lines are drawn as guide to the eye.

In the IR spectra (Fig. 4), four absorption bands associated with N–H (3000–3700 cm^{-1}), C–H (2800–3000 cm^{-1}), C≡N and/or N≡N (2100–2250 cm^{-1}) and N–H and/or C=N (1400–1800 cm^{-1}) were observed. The spectra of both a-C:H and a-CN$_x$:H are shown in   Fig. 5 as reference. Both samples were deposited at r.f. power of 2.83 W/cm^2, where a-C:H was obtained from pure

CH_4 as gas source. The difference and presence of expected bonds confirms the incorporation of N atoms into the film structure.

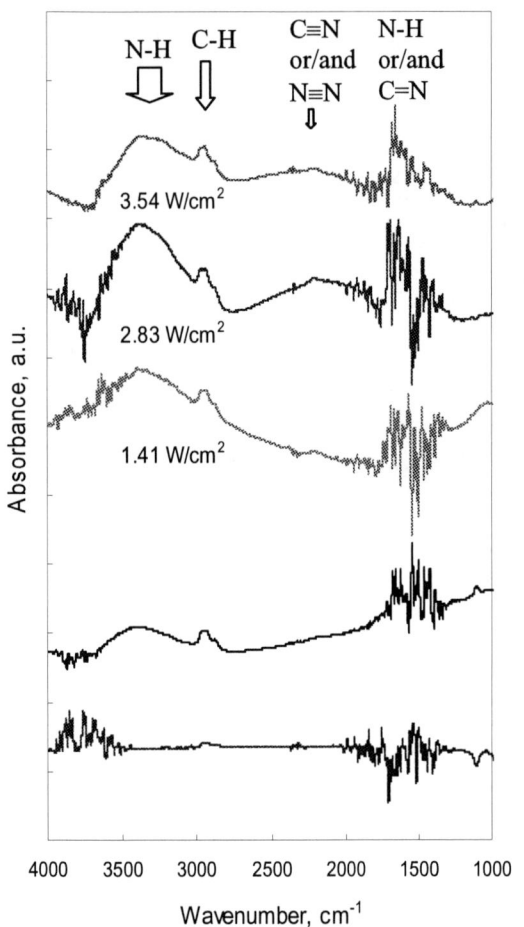

Fig.4 FTIR spectra of a-CN$_x$:H films deposited at r.f. power between 0.71-3.54 W/cm^2.

With the exception of the film deposited at r.f. power of 0.71 W/cm^2, the FTIR spectra are almost identical with the increase of peak intensities with increasing r.f. power. The presence of N–H and/or C=N is only distinguishable at higher r.f. power. A pronounce increase in the N-H absorption band at 3000-3700 cm^{-1} relative to that of C-H (2800–3000 cm^{-1}) (Fig. 6) indicates that the incorporation of N in the films has increased.

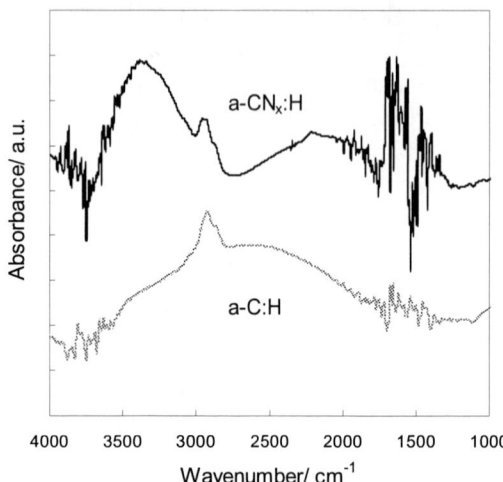

Fig.5 FTIR spectra of a-C:H and films a-CN$_x$:H deposited at r.f. power of 2.83 W/cm^2 .

Nitrogen may be incorporated as dopant or as substituents onto the carbon sites depending on the concentration and deposition condition. However in this case it is predicted that nitrogen act as a substituent which promotes the growth of sp^2 clusters in the films. This would support the observed decrease in the optical energy gap. Furthermore, there is evidence of the C≡N and/or N≡N (2100–2250 cm^{-1}) (Fig. 7), which is obvious at higher r.f. power. The presence of C≡N indicates bond termination of the atomic bond network which may also cause an increase in the density of sp^2 carbon clusters.

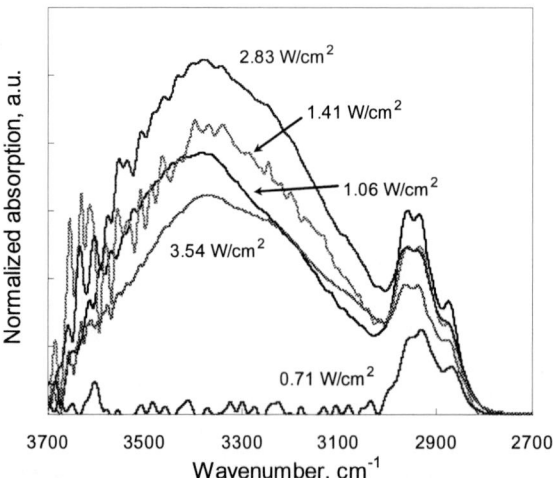

Fig.6 FTIR spectra of a-CN$_x$:H films deposited at r.f. power between 0.71-3.54 W/cm^2 in the range of 2700-3700cm^{-1}.

An exemption to the trend is the film deposed at r.f. power of 3.54W/cm^2. This may be due to the very high dissociation of the methane and nitrogen gases at this power which depletes

792

molecules of these gases in the plasma. This results in zero secondary reaction of the reactive species with these molecules leading to these species being deposited onto the substrates. The high dust formation on the film obtained is also to this effect. Films formed from the various reactive species as a result of total dissociation of methane molecules in the plasma are highly disordered and nitrogen incorporation can be significantly reduced.

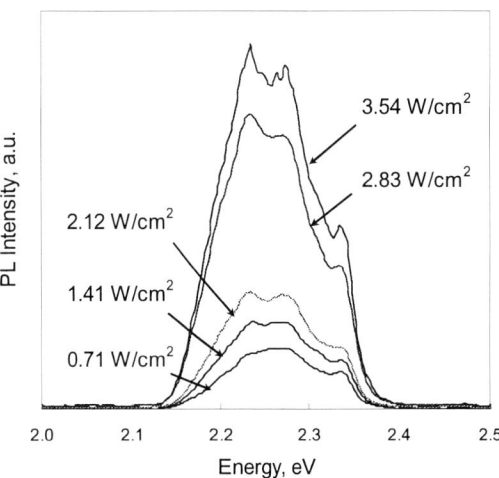

Fig.8 PL spectra of a-CN$_x$:H films deposited at r.f. power between 0.71-3.54 W/cm^2.

This speculation is consistent with the increase in the PL intensity with the decrease in the optical bag gap of these a-CN$_x$:H films, as shown in Fig. 9. This in turn is expected to be the result of the increase of sp^2 carbon clusters in the film structure. The increase of the sp^2 clusters in the films is due to N incorporation with increase in r.f. power to 2.83 W/cm^2. At the highest rf power, the decrease in E$_{04}$ is due to the increase in tail states in the band gap, due to a more disordered film structure. Similar to sp^2 bonding clusters, these tail states can also act as luminescence centres. This increases the PL intensity for this film.

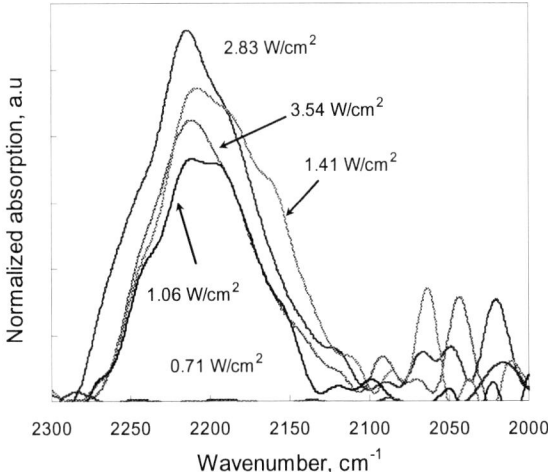

Fig.7 FTIR spectra of a-CN$_x$:H films deposited at r.f. power between 0.71-3.54 W/cm^2 in the range of 2000-2300cm^{-1}.

Result of the PL measurements for the a-CN$_x$:H films, at excitation energy of 4.66 eV is shown in Fig 8. The PL spectra consist of a broad band in the region of 2.10-2.40 eV with peak at approximately 2.23, 2.27 and 2.33 eV. Only a slight change in the relative intensities of the emission peak is seen with the variation in r.f. power, where the peak at 2.23 eV only becomes dominant at higher r.f. power. However, the PL emission intensy increases with the increase in the r.f. power.

It is predicted that the σ bonds gives σ valence and σ* conduction band sites, which are separated by the order of 6 eV, and the π and π* states of sp^2 bonded clusters lie within the σ-σ* gap. The PL emission arises from the recombination of optically exited electron-hole pairs in the π and π* states of the sp^2 sites, and is quenched by tunnelling or hopping to the nonradiative recombination centers [4].

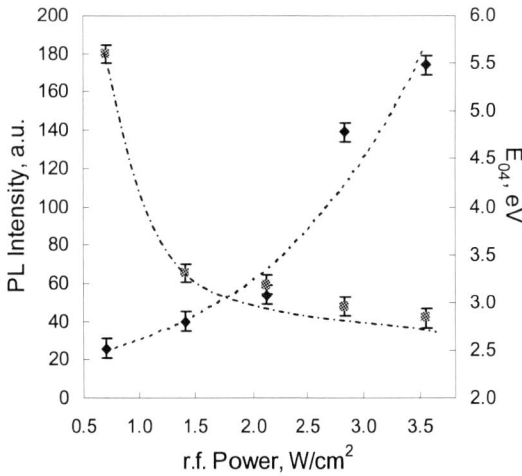

Fig.9 Variation of PL intensity and optical energy gap (E$_{04}$) a function of r.f power. Lines are drawn as guide to the eye.

793

IV CONCLUSION

Increase in r.f. power result in an increase in growth rate of the a-CN$_x$:H film. The optical energy gap of the films decrease with increase in r.f. power but the refractive indexes of the films are independent of r.f. power. Nitrogen incorporation increase with the increase in r.f. power but very large r.f. power has adverse effects. PL intensity consistently increases with increase in r.f. power. Results strongly suggest that increase in r.f. power increases the presence of sp^2 clusters in the film structure due to the increase in N incorporation and depletion of source gas molecules.

ACKNOWLEDGEMENT

This work was supported by the Ministry of Science, Technology and Innovation, Malaysia under IRPA research grant 09-02-03-0222-EA222.

REFERENCES

[1] A.Y., Liu, M.L. Cohen, "Structural properties and electronic structure of low-compressibility materials: β-Si$_3$N$_4$ and hypothetical β-C$_3$N$_4$," *Phys. Rev. Part B*, vol. 41, pp. 10727-10734 (1990).

[2] A.Y. Liu, M.L. Cohen, *Science* 245 (1989) 841

[3] G. Fanchini, P. Mandracci, A. Tagliaferro, S.E. Rodil, A. Vomiero, G. Della Mea, "Growth and characterisation of polymeric amorphous carbon and carbon nitride films from propane," *Diamond & Related Materials*, vol. 14, pp. 928-933 (2005).

[4] Yoshiaki Daigo, Nobuki Mutsukura, "Structures and luminescence properties of polymer-like a-CN$_x$:H films," *Diamond & Related Materials*, vol. 13, pp. 2170-2173 (2004).

[5] L. Valentini, J.M. Kenny, Y.Gerbig, A. Savan, H. Haefke, L. Lozzi, S. Santucci, "Structure and mechanical properties of argon assisted carbon nitride films," *Thin Solid Films*, vol. 398-399, pp. 124-129 (2001).

[6] Mei Zhang, Lujun Pan, Yoshikazu Nakayama, "Structural modifications of hydrogenated amorphous carbon nitride due to ultraviolet light radiation and thermal annealing," *Journal of Non-Crystalline Solids, vol.* 266-269, pp. 815-820 (2000).

[7] C. Godet, G. Adamopoulos, Sushil Kumar, T. Katsuno, "Optical and electronic properties of plasma-deposited hydrogenated amorphous carbon nitride and carbon oxide films," *Thin Solid Films,* vol. 482, pp. 24-33 (2005).

[8] R.U.A. Khan, A.P. Burden, S.R.P. Silva, J.M. Shannon, B.J. Sealy, "A study of the effects of nitrogen incorporation and annealing on the properties of hydrogenated amorphous carbon films," *Carbon*, vol. 37, pp. 777-780 (1999).

[9] Joo Han Kim and Hoo Koo Baik, "Structural and optical properties of amorphous hydrogenated carbon nitride films prepared by plasma-enhanced chemical vapour deposition," *Solid State Communications*, vol. 104, pp. 653-656 (1997).

[10] S. Toth, M. Veres, M. Fule, M. Koos, "Influence of layer thickness on the photoluminescence and Raman scattering of a-C:H prepared from benzene," *Diamond & Related Materials*, vol. 15, pp. 967-971 (2006).

[11] Joo Han Kim, Yong Hwan Kim, Dong Jun Choi, Hong Koo Baik, "Structural properties of amorphous carbon nitride prepared by remote plasma-enhanced chemical vapour deposition," *Thin Solid Films*, vol. 289, pp. 79-83 (1996).

High Performance Complex Number Multiplier Using Booth-Wallace Algorithm

Rizalafande Che Ismail and Razaidi Hussin

School of Microelectronic Engineering
Kolej Universiti Kejuruteraan Utara Malaysia
P.O Box 77, d/a Pejabat Pos Besar
01007 Kangar, Perlis, Malaysia
rizalafande@kukum.edu.my

Abstract **This paper presents the methods required to implement a high speed and high performance parallel complex number multiplier. The designs are structured using Radix-4 Modified Booth Algorithm and Wallace tree. These two techniques are employed to speed up the multiplication process as their capability to reduce partial products generation to $\eta/2$ and compress partial product term by a ratio of 3:2. Despite that, carry save-adders (CSA) is used to enhance the speed of addition process for the system. The system has been designed efficiently using VHDL codes for 16x16-bit signed numbers and successfully simulated and synthesized using ModelSim XE II 5.8c and Xilinx ÌSE 6.1i. As a proof of concept, the system is implemented on Xilinx Virtex-II Pro FPGA board.**

I. INTRODUCTION

Complex number operations are the backbone of many Digital Signal Processing (DSP) algorithms especially for multimedia applications such as 3D graphics which mostly depend on extensive numbers of multiplications. Besides that, they are time critical components for radar, satellite and digital modulation applications too.

Complex number multiplication needs to be done using four real number multiplications and two additions. In real number processing, carry needs to propagate from the least significant bit (LSB) to the most significant bit (MSB) when binary partial products are added. Therefore, the addition and subtraction after binary multiplications limit the overall speed, although many techniques have been proposed to overcome the issue.

This paper presents the design method and efficient implementation of complex number multiplier by integrating Radix-4 Modified Booth algorithm [1],[2],[3] and Wallace tree structure [4],[5],[6],[7] for generating the partial product rows as well as for performing the addition process respectively. Furthermore, in enhancing the speed of the addition process, carry save addition adders are used for the implementation [8],[9].

The paper is divided into 5 sections. In the following section (Section 2), we present the architecture of the system. In Section 3, the complex number multiplier algorithm is discussed. Simulation and synthesis details are presented in Section 4 and conclusions in Section 5.

II. ARCHITECTURE

The complex number multiplication system can be divided into two main components known as real part (R) and imaginary part (I).

$$R + jI = (A + jB)(C + jD) \dots\dots\dots (1)$$

Based on equation (1), the real part is the output for (AC - BD) and the imaginary part is the output for (BC + AD). Each of these two main components required sub-components called modified Booth encoding (MBE), partial product generator and adders/subtractor. Fig. 1 shows the block diagram of the complex number multiplier employed in this study.

Based on conventional methods and observation of the equation (1), four separate multiplications are required to produce the real part as well as imaginary part numbers which lead to the implementation of four different MBE. However, in this study, only two MBEs

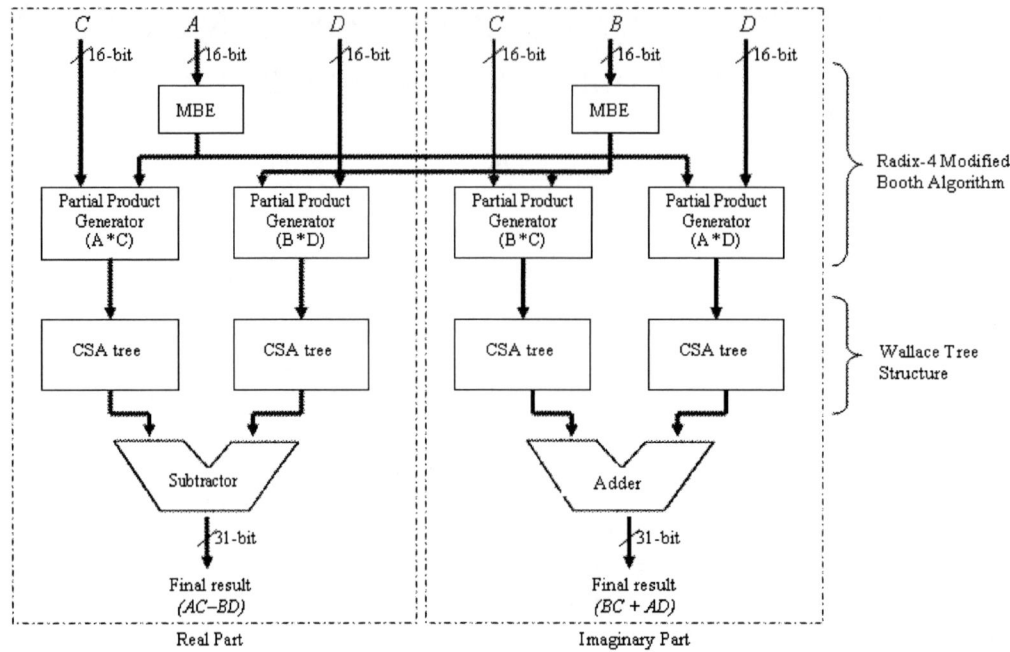

Fig. 1 Block diagram of the complex number multiplier.

are required to execute the multiplication process. It has been done in such a way that input A and B are always set to the multipliers and input C and D are always set to the multiplicand. By implementing these techniques, several logic gates can be removed.

III. COMPLEX NUMBER MULTIPLIER ALGORITHM

For any complex number multiplier design, the most critical part is the multiplication process. In any multiplication operation, there are three major steps. For the first step, the partial products are generated. For the second step, the partial products are reduced to one row of final sums and carries. For the third step, the final sums and carries are added to generate the result.

In this paper, the first step is done using Radix-4 modified Booth algorithm because of its ability to cut the number of partial product rows by half. Then, Wallace tree structure is used for the second step to rapidly reduce the number of partial product rows to the final two (sums and carries). Finally, in a third step, carry propagate adders (CPA) is employed to add the final two rows resulting in final product.

A. RADIX-4 MODIFIED BOOTH ALGORITHM

The multiplication design based on Radix-4 modified Booth algorithm consists of two main

blocks known as MBE and partial product generator as shown in Fig. 1.

MBE is an efficient encoding technique when it comes to reducing the partial product rows. The method of grouping the multiplier bits when using MBE technique is shown in Fig. 2. It is based on a window size of three bits and a stride of two [10]. The multiplier A and B are segmented into groups of three bits and each of the groups will produce the expected outputs as described in Table 1. The outputs are then connected to the partial product generator. As a result of using 16x16-bit signed numbers for this project, MBE will then generate eight partial product rows instead of sixteen rows traditionally.

There are many methods that have been discussed for generating partial product rows as reported in [5],[10],[11],[12]. The aim is to optimize the number of partial product rows as it will give significant effect to the speed, performance as well as area of the system. In this study, the proposed design methodology in generating partial product rows presented in [12] has been implemented. By using the technique, the partial product rows will no longer required two's complement error correction circuits and last negation signal as shown in Fig. 3. Hence, the multiplication can have a smaller critical path thus directly influences the speed of the multiplication process as well as the area of the circuit.

796

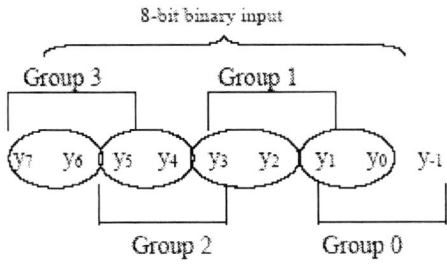

y_{2i+1}	y_{2i}	y_{2i-1}	Generated Partial Products
0	0	0	0 * Multiplicand
0	0	1	1 * Multiplicand
0	1	0	1 * Multiplicand
0	1	1	2 * Multiplicand
1	0	0	-2 * Multiplicand
1	0	1	-1 * Multiplicand
1	1	0	-1 * Multiplicand
1	1	1	0 * Multiplicand

Fig. 2 Multiplier bits grouping according to Booth recording for 8-bit input [10].

Table 1 Radix-4 Modified Booth Recording [13].

Fig. 3 The addition architecture of 16x16-bit signed multiplication [12].

B. WALLACE TREE STRUCTURE

The Wallace tree structure is one of various reduction tree design techniques available for fully parallel multiplier architectures. It is a method of combining either 3:2 or/and 2:2 CSA adders to add together seven or more numbers of size k-bit. These two adders basically serve as a normal full adder and half adder as shown in Fig. 4. Furthermore, by applying Wallace tree concept, the number of partial product rows can be decreased by a factor of 1.5. In this case, due to 16-bit signed numbers used in this project, there will be five levels of CSA trees required to complete the addition operation.

In executing the addition process using this methodology, the initial step is to rearrange the partial product bits to become like a diamond shape. This is to ensure that it can be easily grouped into either three or two bits. Based on Fig. 3, the partial product bits have been repositioned as shown in Fig. 5. The dot notations are used to represent the partial product bits. Fig. 6 shows steps by steps how the addition operation is performed using CSA adders according to Wallace tree structure.

The realisation of the final results for producing the real and imaginary part have been depicted using carry propagate adders.

IV. SIMULATION AND SYNTHESIS

The VHDL codes have been successfully simulated using ModelSim XE II (ver. 5.8c) and have been synthesized using Xilinx ÌSE 6.1i. As a proof of design concept, the complex number multiplier architecture has been implemented on Xilinx Virtex-II Pro FPGA development board and shown working properly. The target chip used is Xilinx Virtex-II Pro device number 2VP7FF672 with speed grade set to -6. The maximum operating frequency of the system is capable to achieve up to 648MHz and on its own consumed only 26% of total areas on that FPGA board.

V. CONCLUSION

In this paper, a high performance complex number multiplier for 16x16-bit signed numbers has been presented. It can be concluded that the use of Radix-4 modified Booth algorithm and Wallace tree structure lead to a better performance in throughput speed and area. The utilisations of CSA adders in performing addition process for partial product rows also have been shown to greatly influence the speed of the system.

REFERENCES

[1] A. D. Booth, "A Signed Binary Multiplication Technique," *Quartely J. Mechanical and Applied Math.*, pp. 236-240, 1951.

[2] S. Sunder, "A Fast Multiplier Based On Modified-Booth Algortihm," *International Journal of Electronics*, vol. 75(22), pp. 199-208, 1993.

[3] J. H. J. Kenneth Lin, Na Tang, "A High-Performance 32-bit Parallel Multiplier Using Modified Booth Algorithm and Sign Deduction Algorithm," *5th International Conference on ASIC*, pp. 1281-1284, 2003.

[4] C. S. Wallace, "A Suggestion for a Fast Multiplier," *IEEE Transactions on Computers*, pp. 14-17, 1964.

[5] M. O. Lakshmanan, Alauddin Mohd Ali, "High Performance Parallel Multiplier Using Wallace-Booth Algorithm," *IEEE International Conference on Semiconductor Electronics*, pp. 433-436, 2002.

[6] K. F. Pang, "Architectures for Pipelined Wallace Tree Multiplier-Accumulators," *IEEE International Conference on Computer Design: VLSI in Computers and Processors*, pp. 247-250, 1990.

[7] Y. H. Niichi Itoh, "A 600 MHz 54x54-bit Multiplier with Rectangluar Styled Wallace Tree," *IEEE Journal of Solid State Circuits*, vol. 36, No. 2, pp. 249-257, 2001.

[8] T. K. William Jao, S. Tjiang, "Circuit Optimization Using Carry Save-Adder Cells," *IEEE Transactions on Computers-Aided Design of Integrated Circuits and Systems*, vol. 17, pp. 974-984, 1998.

[9] W. J. T. Kim, Steve Tjiang, "Arithmetic Optimization using Carry Save Adders," *Proceeding of ACM*, pp. 433-438, 1998.

[10] J. L. G. J.Y. Kang, "A Fast and Well-Structured Multiplier," *EUROMICRO Systems on Digital System Design*, pp. 692-701, 2004.

[11] A. F. Sadiq M. Sait, Gerhard Beckoff, "A Novel Technique for Fast Multiplication," *IEEE Fourteenth Annual International Phoenix Conference on Computers and Communications*, vol. 7803-2492-7, pp. 109-114, 1995.

[12] B. P. R. Che Ismail, "Performance Enhancement and Reduced Area Parallel Multiplier," *IEEE National Symposium on Microelectronics (NSM2005)*, pp. 252 - 258, 2005.

[13] O. L. MacSorely, "High Speed Arithmetic in Binary Computation," *IEEE Proceedings*, vol. 49, pp. 67-91, 1990.

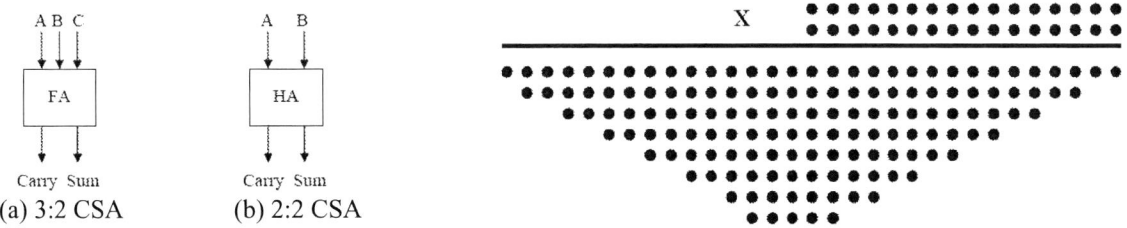

(a) 3:2 CSA   (b) 2:2 CSA

Fig. 4 The block diagram of CSA.   Fig. 5 The dot notations represent partial product bits.

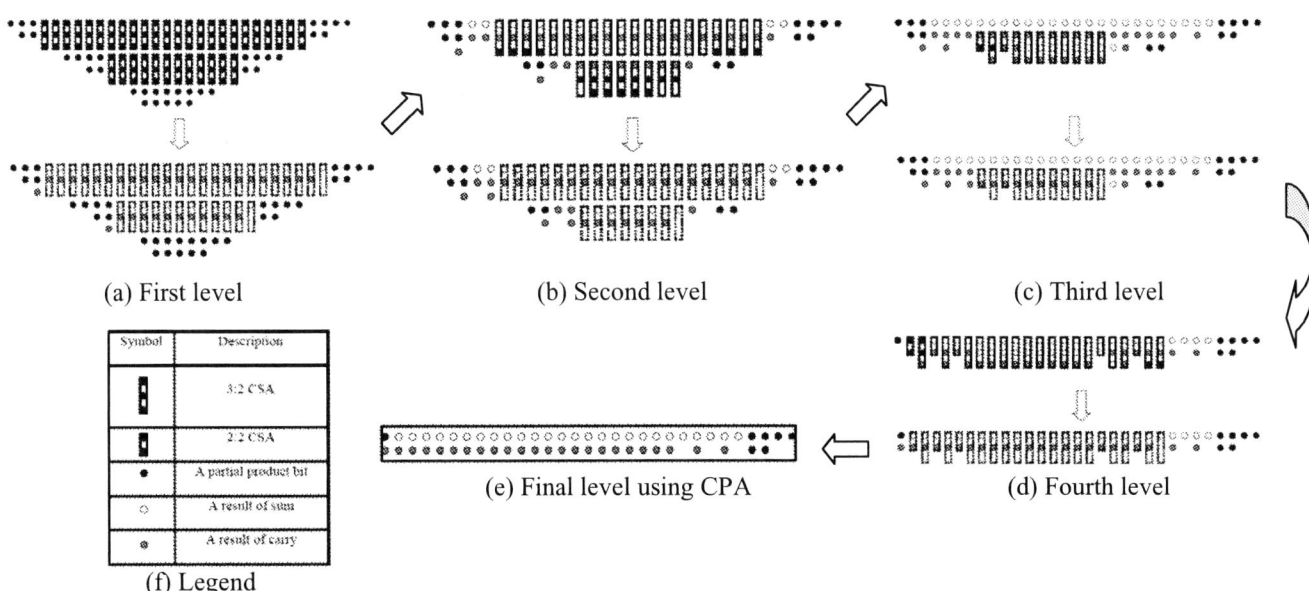

(a) First level   (b) Second level   (c) Third level

Symbol	Description
	3:2 CSA
	2:2 CSA
•	A partial product bit
○	A result of sum
⊗	A result of carry

(f) Legend

(e) Final level using CPA   (d) Fourth level

Fig. 6 Steps by steps the addition operation using CSA.

Effects of High Dose BF2+ Implant on the Improvement of P+ Contact Resistance

Mohd Rofei Mat Hussin[1], Norazah Abd. Rashid[1] and Richard Keating[2], *Member, IEEE*

[1] Semiconductor Process Technology (Microelectronics and Semiconductor Department),
MIMOS Bhd, Technology Park Malaysia, Bukit Jalil, 57000 Kuala Lumpur, MALAYSIA
Email: rofei@mimos.my

[2] CMOS Training and Development, 41 Cangkat Kenari 2, Sg. Ara, 11900 Penang, MALAYSIA
Email: keatingr@cmostraining.com

Abstract- **This paper describes the effect of Ti Deposition/Anneal and supplementary BF$_2$ implant/Anneal on a 0.35µm Silicon CMOS process using Contact silicided P+ source-drain. Thicker Ti and higher Ti/TiN annealing temperature are also required for the smaller contact sizes to get adequate P+ contact resistance. The supplementary BF$_2$ implant with dose of 3E14cm^{-2} and energy 20KeV helped to reduce and stabilize the contact resistance down to 150 Ohm/hole for the 0.4µm P+ contact. The Boron profile at the TiSi$_2$/p$^+$ interface were investigated by 2D ATHENA process simulation. The peak Boron doping level at TiSi$_2$/p$^+$ interface significantly influenced the contact resistivity. Various contact chain test structures, with different contact sizes, plus single Kelvin structures were used in this investigation.**

I. INTRODUCTION

The developments of smaller geometry devices for higher performance and density require scaling down the contact size, resulting in the increase of contact resistance. At the same time shallower source/drain junction depths are needed to reduce the short channel effects, but at the expense of higher series resistance. These device parasitic resistances are in series with channel resistance and can significantly reduce the gain of the transistor. This can be seen by looking at some typical values of channel resistance and parasitic resistance of submicrometer device as reported in [1] where in 0.8µm effective channel length; 10-20 percent of the total device series resistance comes from parasitic resistances. It is even worst for shorter gate length of transistor. Several papers also have reported that boron concentration at the

TiSi$_2$/p$^+$ interface is an important factor to influence the P+ contact resistance [1] [3] [5]. Here we give the results of a full contact resistance optimization for a 0.35µm technology. The 0.50µm technology did not require the additional BF$_2$ implant to achieve the required contact resistance. To reduce the contact resistance on the 0.35µm process, a silicide layer is used on the source/drain diffusion area. Among the silicides, TiSi$_2$ is one of the most common materials used because of its thermal stability and low resistivity (15uΩ-cm) [1] [2] [5]. There are 2 methods. Either the entire active and poly region is silicided (self aligned silicide process), or only the silicon areas exposed after contact etch are silicided (contact silicide). Here we report on the contact silicide method. In this paper, we analyzed various process parameters influencing the P+ contact resistance for different contact sizes. Contact chain and Kelvin structures with size of 0.4µm, 0.45µm and 0.5µm have been used. Ti thickness and Ti/TiN annealing temperature are varied in the experiment to optimize the silicidation process. To further improve the contact resistance, a BF$_2$ implant/Anneal is used. Implant Dose, Energy, and post implant annealing conditions were optimized to achieve the lowest P+ contact resistance results.

II. PROCESS OPTIMIZATION

a) Ti/TiN Process Optimization

Lightly p-type doped bulk Si (100) wafers were used as a raw material for this experiment. The samples were fabricated using industrial standard 0.35µm CMOS process. For the first attempt, the Ti thickness and Ti/TiN annealing temperature were been varied (Table 1). No

0-7803-9730-4/06/$25.00 ©2006 IEEE

supplementary BF_2 was used for this initial experiment.

Wafer	Ti thickness (A)	Ti/TiN annealing temperature in N_2 (^{o}C)
1	300	645
2		705
3	500	645
4		705

Table 1 Splitting condition for Ti thickness and Ti/TiN annealing temperature

The fabrication process starts with active area pattern and LOCOS oxidation followed by well and channel implantation. Self-align anti-punchthrough implant and source-drain implant are performed after gate formation followed by rapid-thermal-annealing (RTA) process to activate the doping. The back end of line (BEOL) process begins with oxide and BPSG deposition. The process was continued with contact hole formation, followed by contact pre-cleaning prior to Ti/TiN deposition. Deposition of the Ti/TiN film was performed in an Applied Materials PVD ENDURA chamber. The Ti/TiN film is then subjected to the RTA process to form the $TiSi_2$ layer in contact hole. An Applied Materials CENTURA was the RTP tool used. The Ti thickness and RTA annealing temperature have been varied to get the optimum bottom step coverage and good silicidation process. After the formation of $TiSi_2$, the samples get AlSiCu/TiN deposition and Metal 1 photo/etch. After a forming gas anneal, electrical testing was performed to obtain the p+ contact resistance.

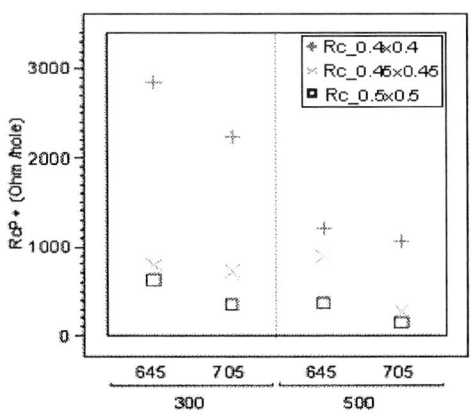

Fig.1 Contact resistance for various Ti thickness and RTA annealing temperature

Fig.1 plots the contact resistance as a function of Ti thickness and RTA annealing temperature. The plot shows that major factor affecting P+ contact resistance is the Ti thickness. Higher RTA annealing temperature also improves the contact resistance significantly. However, the improvement is not sufficiently enough for typical contact size (0.4µm) of 0.35µm technology.

b) BF2+ Implant Process Optimization

When thicker Ti and higher annealing temperature did not solve the high P+ contact resistance issue, a supplemental BF_2 implant into the P+ contact hole was investigated. This supplementary implant require additional masking step (P+ diffusion layer) after contact etch. The implantation was performed just before Ti/TiN film deposition to optimize the boron concentration at the silicon surface. RTA annealing after BF2+ ion implantation step was been done to activate the dopant. The implant energy, dose and annealing temperature were varied. The split conditions are as shown below (Table 2).

Wafer	Energy (keV)	Dose (cm^{-2})	Annealing Temperature (^{o}C)
1	15	1e14	700
2			750
3		3e14	700
4			750
5		5e14	700
6			750
7		1e15	700
8			750
9	20	1e14	700
10			750
11		3e14	700
12			750
13		5e14	700
14			750
15		1e15	700
16			750

Table 2 Splitting condition for BF2+ implant dose and energy and RTA anneal temperature

All samples have been fabricated until metal-1 process for electrical testing and the experiment results will be discussed in details in the next section.

III. RESULTS AND DISCUSSION

We investigated deposited Ti thickness, Ti/TiN annealing temperature, BF2 ion implant dose and energy and implant annealing temperature. The study has been performed in two separate experiments. The contact resistance plots for the first experiment are shown in Fig.1. The resistance values are sharply increased for smaller contact size. This is due to worse Ti step coverage in the smaller contact holes. In order to increase the bottom step coverage, thicker Ti film was deposited. The result shows significant improvement to the contact resistance. The measured contact resistance values are acceptable for 0.5μm technology with typical contact size of 0.5x0.5μm^2. However, for 0.35μm technology with typical contact size of 0.4x0.4μm^2, it requires further improvement. The difference is about one order.

Fig.1. There is one sample missing from the plot because of misprocess during fabrication. The contact resistance decreased with higher dose of BF$_2$. Higher energy also improves the contact resistance. The result shows very little dependence on Implant Anneal Temperature. As BF$_2$ implant is used for source/drain, fluorine-induced defect gettering effects probably occur in the p+ diffusion region [3]. This cause boron deactivation near the Si surface. This low effective active doping results in poor contact to TiSi$_2$. Additional boron from supplementary BF$_2$ implant helps to increase the boron doping level at the TiSi$_2$/p$^+$ interface and reduce the contact resistance. The boron doping profile was investigated by 2D ATHENA process simulation as shown in Fig.3. Process condition for samples from the first experiment (wafer 4) and the second experiment (wafer 12) were used for the simulation.

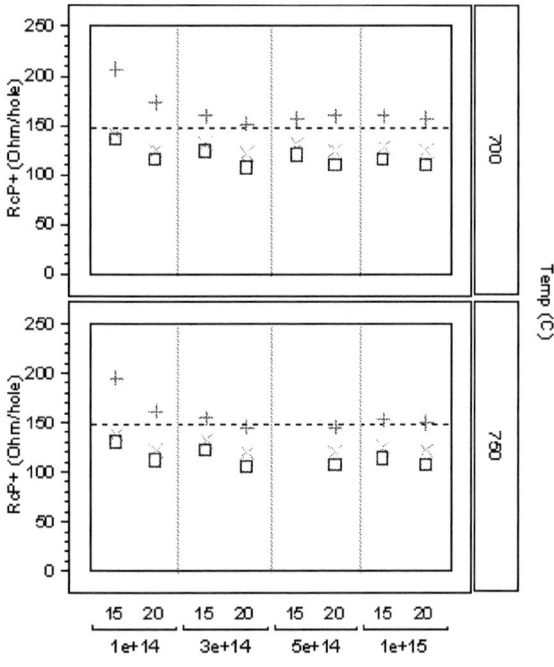

+ RcP+_0.4x0.4

✕ RcP+_0.45x0.45

☐ RcP+_0.5x0.5

Fig. 2 Contact resistance for various BF$_2$ implant dose at different energy and Implant annealing temperature

Fig.2 shows the contact resistance measured on contact chain structures for each split condition in table 2 (second experiment). The measured contact resistance values of supplementary implanted samples in Fig.2 are much improved compared with the samples without implant in

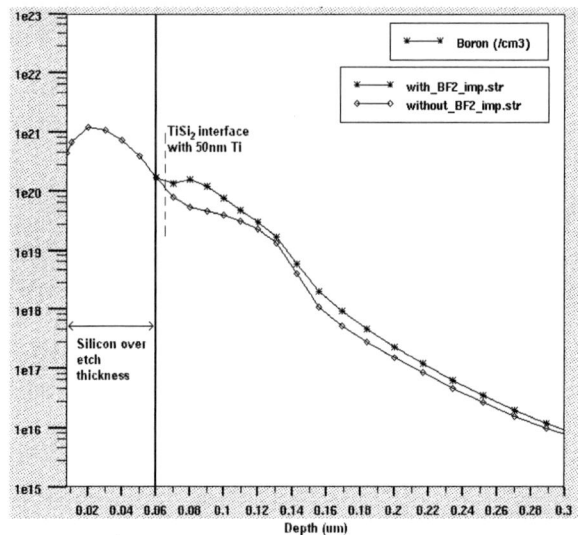

Fig. 3 Simulation result of boron doping profile at source-drain contact region

Fig. 3 shows the doping profile of source-drain region with and without supplementary BF$_2$ implant. From the plots, additional supplemental BF$_2$ implant has increased the peak boron doping level at TiSi$_2$/p$^+$ interface. It clearly shows that the improvement of contact resistance is due to higher boron concentrations at TiSi$_2$/p$^+$ interface.

In this study we have performed contact resistance measurement on two types of test structures (contact chain and Kelvin structure). Contact chain structure is commonly used for process monitoring because it was designed for extreme conditions and should represent yield across the wafer. On the other hand, when accurate measurement is required, Kelvin structure will be used to eliminate the measurement errors and reduce the parasitic effects. Comparisons for these two types of test structures are shown below. The plotted results are for 0.4μm contact size.

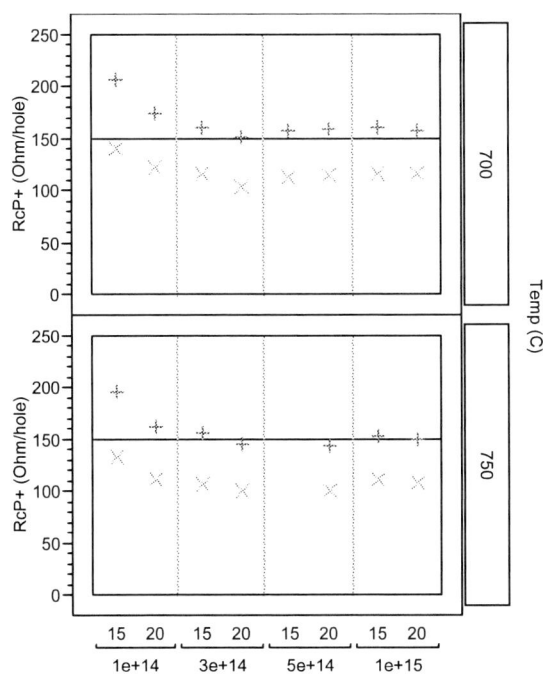

– RcP+_Chain
× RcP+_Kelvin

Fig. 4 Contact resistance difference between contact chain structure and Kelvin structure

Fig. 4 shows contact resistance from Kelvin structure is lower than contact chain structure by 30% as expected. The difference is due to diffusion resistance and other parasitic resistance components exist in contact chain structure. The BF_2 supplementary implant effects maintain the same. This confirms that the measurements for contact chain structure are reliable.

IV. CONCLUSION

In summary, the contact resistance of p+ diffusion is studied for a variety of process parameters. Thicker Ti and higher Ti/TiN annealing temperature improve P+ contact resistance. The peak Boron doping level at $TiSi_2/p^+$ interface is found as a factor to influence the contact resistivity. Therefore, supplementary BF_2 implant with dose 3e14cm-2 and energy 20keV at annealing temperature 750°C is needed to meet contact resistance specification for 0.35μm technology.

REFERENCES

[1] J. Hui, S. Wong and J. Moll, "Specific Contact Resistivity of TiSi2 to p+ and n+ Junctions," *IEEE Electron Device Letters.* vol. EDL-6, no. 9, pp. 479-481 (1985).

[2] C.Y. Ting, "Silicide for Contacts and Interconnect," *IEDM 1984.* pp. 110-113 (1984).

[3] Yuan Taur, Jack Yuan-Chen Sun, dan Moy, L.K. Wang, Bijan Davari, Stephen P. Klepner and Chung-Yu Ting, "Source-Drain Contact resistance in CMOS with Self-Aligned TiSi2," *IEEE Transactions on Electron Devices.* vol. ED-34, no. 3, pp. 575-580 (1987).

[4] L.K. Nanver, E.J.G Goudena and J.Slabbekoom, "Kelvin Test Structure for measuring Contact Resistance of Shallow Junction," *IEEE International Conference on Microelectronic Test Structures.* vol. 9, pp. 241-245 (1996).

[5] Jeong Su Han and II Gweon Kim, "Improvement of P+ Contact Resistance through Optimal Boron Distribution on TiSi2/P+ Interface for 0.18μm DRAM Technology and Beyond," *IEEE Conference on Ion Implantation Technology,* pp. 83-86 (2000).

A 0.8V Operational Amplifier using Floating Gate MOS Technology

Rohan Sehgal

Netaji Subhas Institute of Technology, New Delhi – 110075, India

Email: rohan_sehgal@yahoo.com

S.S. Rajput

Plasma Processed Materials Group, National Physical Laboratory, New Delhi-110012, India

S.S. Jamuar

Department of Electrical and Electronics Engineering, Universiti Putra Malaysia, Malaysia

Abstract A **two-stage low voltage operational amplifier for operation at ±0.4V is proposed. The amplifier incorporates a low voltage current mirror designed using standard Floating Gate MOSFETs. The proposed op amp possesses a 49 dB open-loop gain, a high bandwidth of 698 kHz, 42° phase margin and consumes only 28.6 μW. The operation of the proposed current mirror and op amp has been confirmed by PSPICE simulations, using 0.13 μm CMOS technology.**

I. INTRODUCTION

Due to a rise in demand for portable electronic systems, low power circuit design is extremely important for portable appliances like laptops, mobile phones, etc. Low power dissipation also results in longer battery life and better integration density. This can be achieved by designing VLSI circuits that operate with power supply smaller than or equal to ±1.0 V.

An amplifier is an integral part of all analog and mixed-signal systems. The design of low voltage operational amplifier in this paper is based on low voltage current mirror. In the past, the low voltage CMOS current mirror designs reported, have been based on techniques like bulk driven transistors [6], MOSFETs operating in subthreshold region, Floating Gate MOSFET (FGMOS) approach [7], level shifter techniques [3], cascade architecture, etc . In [3], a CMOS level shifter is used to reduce the input voltage signal swing. In this paper, an improved level shifter structure incorporating Multi Input FGMOS has been proposed. In MIFG transistor, the first layer of polysilicon is isolated in SiO_2

and forms the floating gate over the n-channel. The floating gate, which is capacitively coupled to the input gates through $C_1 \ldots C_n$, can modulate the channel current. For a two-input FGMOS with $V_S = V_B = 0$ and C_1, $C_2 \gg C_D$, the expression for floating gate voltage, V_{FG} is –

$$V_{FG} \approx k_1 V_1 + k_2 V_2 \qquad (1)$$

where $k_i = C_i/C_{Total}$
and $C_{Total} = C_1 + C_2 + C_S + C_D + C_B$ is the total floating gate capacitance. Now the drain current (I_D) in saturation region is given by

$$I_D = \beta (V_{FG} - V_T)^2/2$$
$$= \beta [(k_1 V_1 + k_2 V_2) - V_T]^2/2 \quad (2)$$

where β and V_T are the transconductance parameter and threshold voltage, respectively. Equation (2) may be written as

$$I_D = \beta k_1^2 [V_1 - V_T(eff)]^2 \qquad (3)$$

where the effective threshold voltage of MOSFET is

$$V_T(eff) = (V_T - k_2 V_2)/k_1 \qquad (4)$$

As the value of $V_T(eff)$ depends on k_1, k_2 and V_2, it is possible to program the value of the effective threshold voltage of the MOSFET as per the requirement.

The current mirror presented in this paper, can operate in the input current range of 1 μA to 500 μA with a bandwidth of 4.2 GHz at an input dc current of 100 μA. Standard resistive compensation has been used to achieve this

0-7803-9730-4/06/$25.00 ©2006 IEEE

bandwidth. The proposed LVCM operates at a supply voltage of ±0.5 V and possesses an output impedance of approximately 1.2 MΩ.

II. LOW VOLTAGE CURRENT MIRROR

The simple CM topology shown in Fig. 1 requires input voltage (Vin) of at least one V_T and hence is not suitable for any LV applications.

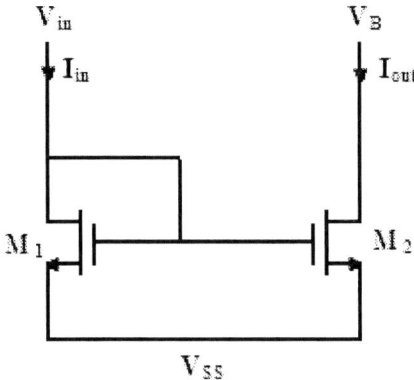

Fig. 1 Simple Current Mirror

In the modified current mirror in fig 2, one of the gates is used to modify the threshold voltage, so as to decrease the input compliance voltage of the current mirror. The minimum input compliance voltage now becomes V_T(eff), which can be lowered by using suitable values of k_1, k_2 and V_2. .

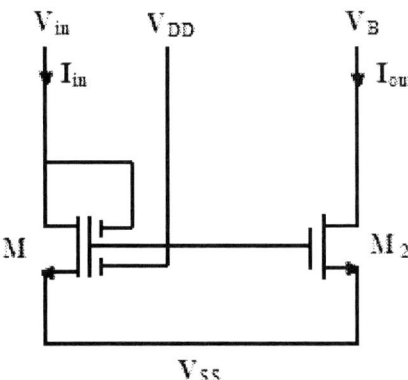

Fig. 2 Simple FGMOS Current Mirror

Another CM topology, shown in fig 3, which operates at low voltage, incorporates a level shifter PMOS transistor M_3 (biased through a current I_{bias1}) at input port. For this structure

$$V_{ds1} = V_{gs1} - V_{gs3} \qquad (5)$$

where V_{ds1} and V_{gs1} are the drain to source and gate to source voltages for M_1 and V_{gs3} is the gate to source voltage for M_3.

Operation of M_1 and M_2 depends on I_{in} and their aspect ratios. However, M3 is operated in sub-threshold region for the entire input current range so as to keep V_{ds1} always positive.

Fig 4. shows a new level shift current mirror which combines the advantages of reduction of threshold voltage in a floating gate MOSFET with level shift circuit for the low voltage operation. This CM structure operates on a substantially reduced supply voltage range of ± 0.4V. The proposed current mirror structure, shown in fig 5, adopts an output stage, which

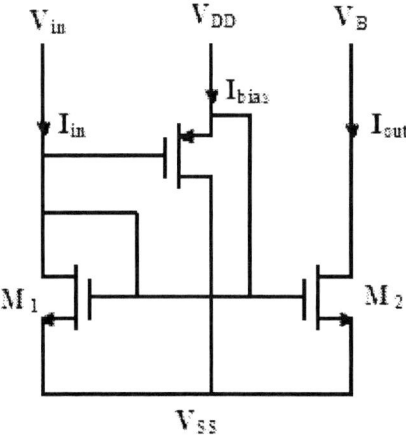

Fig. 3 Conventional Level Shifted Current Mirror

enhances the output impedance of the current mirror. The proposed LVCM circuit operates as follows. The input current I_{in} is pumped into the drain of M_1. The bias current of M_3 is kept small so that M_3 operates in sub-threshold region and M_1 is in saturation region. M_3 provides the requisite gate bias for M_1 and M_2 while M_5 provides suitable bias for M_4. Thus, the input voltage is given by:

$$V_{in} = \sqrt{\frac{2I_{in}}{\beta_1}} + \Delta V_T - \left| \eta V_{ther} \ln\left(\frac{L_3}{W_3} \frac{I_{bias1}}{I_{DO3}}\right) \right| \quad (6)$$

where ΔV_T is the mismatch between the threshold voltages of M_1 and M_3, and β_1 is the trans-conductance parameter of M_1, I_{DO3} is the process parameter($\approx$ 20 nA). L_3 and W_3 are the

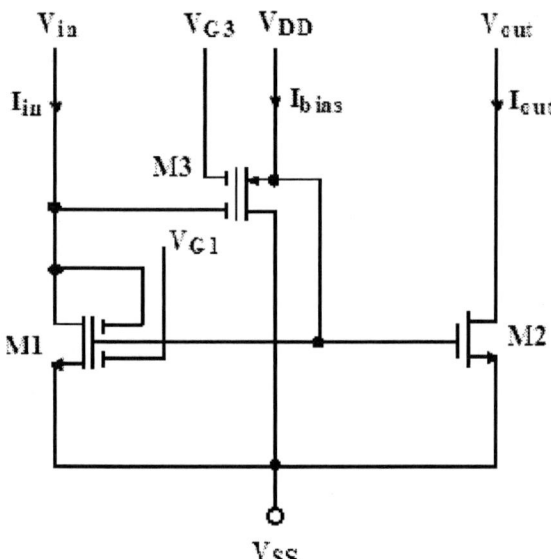

Fig.4 Level shifted FGMOS Current Mirror

channel length and channel width of transistor M_3.

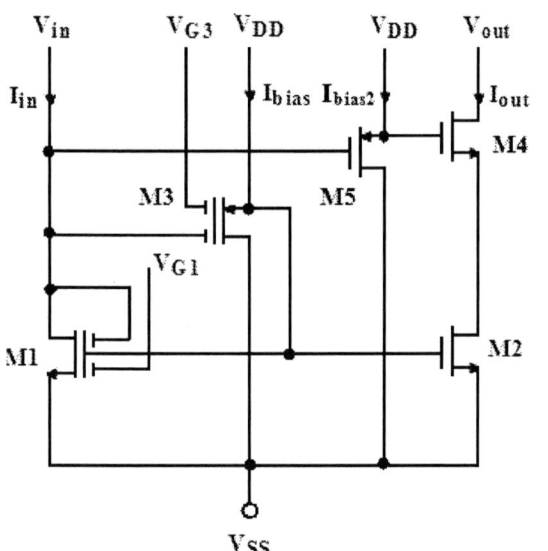

Fig.5 Proposed Low Voltage Current Mirror

The bandwidth of the proposed LVCM has been enhanced by using standard resistive compensat- ion in the bias gate M_1 and level shift floating gate M_3.

III LOW VOLTAGE OP-AMP

The circuit diagram for a two-stage operational amplifier for operation at ± 0.4 V, is shown in   fig. 6. The operation of the proposed op amp is similar to a classical two-stage op amp.

The first stage of the proposed op amp consists of a differential amplifier converting the differential input voltage applied between gates of M_1 and M_2, to differential currents. These differential currents are applied to a current mirror load recovering the differential voltage. The second stage consists of a common source MOSFET M_9 converting the second stage input voltage to current. M_9 is loaded by a current-sink load M_7, which converts the current to voltage at output. In order to reduce the output compliance voltage of the op amp, the output stage of the proposed current mirror incorporated in the op amp, was removed.

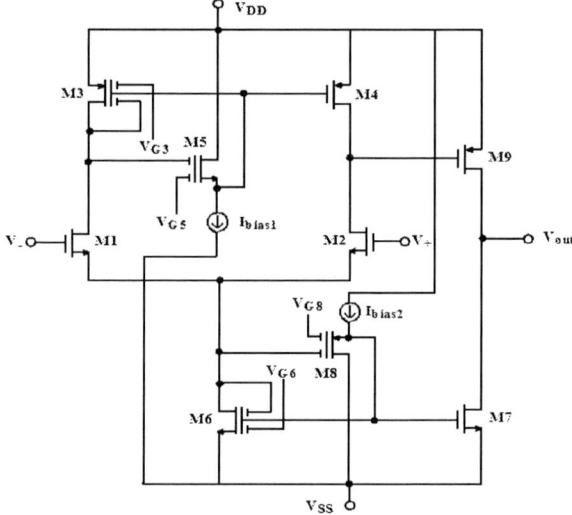

Fig. 6 Proposed Operational Amplifier

IV RESULTS AND DISCUSSION

The proposed low voltage current mirror and op amp have been simulated in PSPICE using level 3 parameters for 0.13μm CMOS technology. The (W/L) ratios for various MOSFETs used in op amp are given in Table 1. PSPICE simulations were carried out for the LVCM with I_{bias1} and Ibias2 of 2nA and 10μA, respectively, for I_{in} between 1μA and 500μA with a power supply of ± 0.5V. The I_{out} to I_{in} ratio is found to be almost equal to 1 (Fig 7). A high bandwidth of 4.2 GHz has been achieved using standard resistive compensation. R_{in} and R_{out} have been found out to be 639 Ω and 1.23 MΩ, respectively, at I_{in} = 100 μA. The LVCM is found to dissipate only 0.21mW.

806

Table 1. W/L ratios of transistors in op amp circuit

Device number	Type	Aspect Ratio (W μm/L μm)
M1, M2	NMOS	7.8/0.26
M3, M4	PMOS	7.8/0.26
M5	NMOS	5.2/0.13
M6	NMOS	9.1/0.26
M7	NMOS	26/0.26
M8	PMOS	9.1/0.13
M9	PMOS	39/0.26

Table 2 summarizes the measured parameters of the amplifier. Fig. 8 shows the measured input-output amplifier dc voltage response.

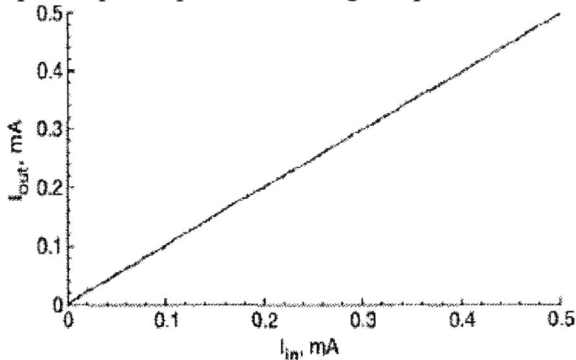

Fig. 7 Current Transfer Characteristics of proposed LVCM

The low voltage op amp operates at a supply voltage of ±0.4V and dissipates only 28.6 μW. It has a high unity-gain bandwidth of 69.2 MHz and a 3-dB bandwidth of 698 kHz.

V CONCLUSION

A low voltage operational amplifier has been designed for operation at a power supply of ±0.4V. This structure has high bandwidth, low power dissipation and a high slew rate. The drawbacks of the amplifier are its high input offset voltage and low open-loop voltage gain. Furthermore, channel lengths of 0.13μm have been used, making them suitable for applications in portable microsystems and biomedical devices.

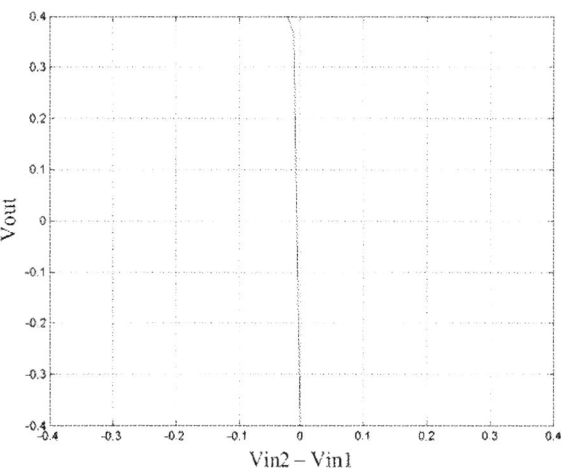

Fig. 8 Input-Output DC voltage response

Table 2. Simulation Results for proposed op amp

Parameters	Simulation Result
Supply Voltage	±0.4V
Power Dissipation	28.6 μW
Open Loop Gain	49 dB
Unity Gain Bandwidth	69.2 MHz
Phase Margin	42°
Input Offset Voltage	5.19 mV
Output Voltage Swing	-0.4 V – 0.39 V
3-Db bandwidth	698 kHz
CMRR	89.62 dB at 10 kHz
PSRR (V_{DD})	57.4 dB at 10 kHz
PSRR (V_{SS})	56.6 dB at 10 kHz
ICMR	-0.34 V – 0.18 V
Slew Rate (SR$^+$)	11.96 V/μ
Slew Rate (SR$^-$)	7.2 V/μ

REFERENCES

[1] P.E. Allen and D.R. Holberg, *CMOS Analog Circuit Design*, Oxford University Press: New York, 2002

[2] Behzad Razavi, *Design of CMOS Analog Integrated Ckts*, Mc-Graw Hill College, 2001.

[3] S.S. Rajput and S.S. Jamuar, "Low voltage, low power high performance current mirror for portable analogue and mixed mode applications." In *Proc. IEE—Circuits Devices and Systems*, 2001, vol. 148, no. 5 pp. 273–278.

[4] S.S. Rajput and S.S. Jamuar, "A Current Mirror for Low Voltage, High Performance Analog Circuits" Analog Integrated Circuits and Signal Processing, 36, 221–233, 2003, Kluwer Academic Publishers.

[5] S.S. Jamuar and Tang Kia Kit, "Design of Low Voltage Operational Amplifier" in *Proc. ICSE 2004*, Malaysia, pp 552-554.

[6] C. Zhang, A. Srivastava, P. K. Ajmera, "A 0.8 V CMOS amplifier design", Analog Integrated Circuits and Signal Processing, 47, pp 315–321, 2006, Springer Science

[7] Susheel Sharma, S.S. Rajput, L.K. Mangotra and S.S. Jamuar, "FGMOS Current Mirror: Behaviour and Bandwidth Enhancement", Analog Integrated Circuits and Signal Processing, 46, 281–286, 2006 Springer Science.

[8] S.S. Rajput and S.S. Jamuar, "Design techniques for low voltage analog circuit structures." in *Proc. NSM 2001/IEEE*, Malaysia, Nov. 2001.

Design of Single-Stage Folded-Cascode Gain Boost Amplifier for 100mW 10-bit 50MS/s Pipelined Analog-to-Digital Converter

Rohana Musa, Yuzman Yusoff, Tan Kong Yew and Mohd Rais Ahmad, *Member, IEEE*
AMS IC Design Group, Microsystems, MIMOS Berhad
Email: yuzman@mimos.my

Abstract: **This paper presents the design and simulation of high speed, high gain and low power fully differential operational amplifier (op-amp) implemented in 0.35um CMOS technology. The op-amp was designed for sample-and-hold stage of 100mW 10-bit 50MS/s pipelined analog-to-digital converter. A topology of single-stage folded-cascode with gain boosting technique is employed in this op-amp. The simulated op-amp achieves a DC gain of 95dB, unity gain bandwidth of 412MHz and phase margin of 75 degrees. The settling time is 7.5ns and the op-amp consumes power 12.8mW with supply voltage of 3V.**

I. INTRODUCTION

In application of pipelined analog-to-digital (A/D) converters, the requirement for high speed and high accuracy operational amplifiers (op-amps) are essential. The speed and accuracy criteria are determined by the settling behavior of the op-amps. Fast settling mainly depends on the unity gain frequency while high settling accuracy is due to high DC gain. The realization of high speed and high accuracy op-amps has proven to be very challenging task. Optimizing the circuit design for both requirements leads to conflicting demands[1].

A single-stage folded cascode topology is a popular approach in designing high speed op-amps. Besides large unity gain frequency, it offers large output swing. However, it has limitation to provide high DC gain which is required for high settling accuracy. In 1990, Bult and Geelen proposed the folded cascode op-amp with gain boosting technique [2]. This technique help to increase the op-amp DC gain without sacrificing the output swing of a regular cascode structure[3].

The purpose of this paper is to discuss design consideration when utilizing gain boost cascode op-amp in the sample-and-hold (SHA) stage of 100mW 10-bit 50Ms/s Pipeline A/D converter.

This paper is divided into three additional sections. The gain boosting technique is explained in section II and the circuit frequency behavior is analyzed in section III. In section IV, the circuit implementation with 0.35um CMOS process is presented. The simulation results are given and discussed in section V. Finally, the conclusions are drawn in section VI.

II. GAIN BOOSTING TECHNIQUE

Fig. 1 illustrates a gain boost cascode topology where transistor M1 is an input device, M2 a cascode device and M3 a gain boost device. M3 drives the gates of M2 and forces the voltage at nodes X and Y to be equal. As a result, voltage variations at the drain of M2 will affect the voltage at node X to a lesser extent because the gain boost device regulates this voltage [3].

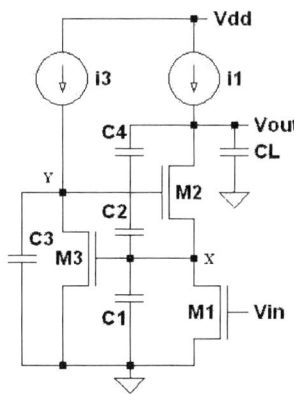

Fig.1 Gain Boost cascode topology

The addition of gain boost device with open loop gain, A_{fb}, provides a small signal output resistance approximately A_{fb} times larger than that of a regular cascode [4]. With transconductance g_{m1}, g_{m2} and output resistance r_{ds1}, r_{ds2} of M1 and M2 respectively, the output resistance, $R_{out,}$ and DC gain, $A_{Vdc,}$ of the gain boost cascode op-amp are given by equations (1) and (2),

0-7803-9730-4/06/$25.00 ©2006 IEEE
809

$$R_{out} = g_{m2} r_{ds2} r_{ds1} A_{fb} \qquad (1)$$

$$A_{Vdc} = g_{m1} R_{out} = g_{m1} g_{m2} r_{ds1} r_{ds2} A_{fb} \qquad (2)$$

Through this technique, the output resistance and gain can be increased by the gain of the gain boost device without adding more cascode devices. However, transient response from such an op-amp is degraded by the presence of pole-zero doublet [4]. This doublet appears as a slow exponential term in the step response of the op-amp, thus degrading the total settling time drastically and will discussed further in the analysis section.

III. SETTLING RESPONSE ANALYSIS

To understand the effect of pole-zero doublets on slow settling behavior, the transfer function of the gain-boosting technique is derived using small signal model as shown in Fig. 2.

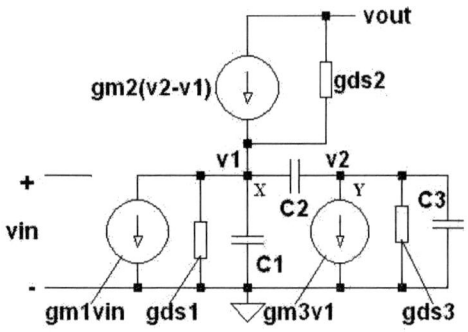

Fig.2 Small signal model

The capacitors C_1 through C_3 are the equivalent parasitic capacitance of the MOS transistors at nodes X and Y. Meanwhile C_L is the load capacitance at output node. To simplify the analysis, parasitic drain-to-gate capacitor C_4 of M2 is broken into its Miller equivalent at node X and at output node. This Miller capacitance is included in the value of parasitic capacitor C_2 at node X and value of capacitor C_L at output node.

The transient response of the op-amp can be expressed as a superposition of the exponential settling component due to the dominant pole, and the settling components due to non-dominant poles and zeros. Since the main objective of this study is to remove any slow settling component from the transient response, only the response due to non-dominant poles and zeros will be analyzed. Apparently, the overall transfer function will have a dominant pole at g_{dseff}/C_L,

where g_{dseff} is total output conductance at output node of the op-amp.

The transfer function equations were obtained by equations (3)-(5) [3];

$$\frac{V_{out}}{V_{in}} = \frac{g_{m1}}{g_{ds2}} \cdot \frac{N}{D} \qquad (3)$$

where,

$$N = \left(g_{ds2} C_2 - g_{ds2} C_3 + g_{m2} C_3\right)s \\ - \left(g_{ds2} g_{ds3} + g_{m2} g_{m3} - g_{m2} g_{ds3}\right) \qquad (4)$$

$$D = \left(C_1 C_3 + C_2 C_3 - C_1 C_2\right)s^2 \\ + \left(\begin{array}{l} g_{m2} C_3 - 2 g_{m3} C_2 + g_{ds3} C_1 + g_{m3} C_3 \\ + g_{ds2} C_3 + g_{m2} C_2 \end{array}\right)s \quad (5) \\ + \left(\begin{array}{l} g_{m2} g_{ds3} - 2 g_{m2} g_{m3} + g_{m3} g_{ds3} \\ + g_{ds2} g_{ds3} \end{array}\right)$$

By solving equations (4) and (5), the zero location, ω_z is given by g_{m3}/C_2 due to feed-forward path created by C_2. The poles locations are given by follows;

$$\omega_d = -\frac{g_{dseff}}{C_l} \qquad (6)$$

$$\omega_{nd1} = -\frac{g_{d3}}{C_2 + C_3} \qquad (7)$$

$$\omega_{nd2} = -\frac{g_{m2}\left[\dfrac{g_{m3}}{g_{d3}} + 1\right]}{C_1 + C_2} \qquad (8)$$

where ω_d is the dominant pole seen at output node, ω_{nd1} and ω_{nd2} are first and second non-dominant pole due to node Y and X respectively.

Fig. 3 and 4 show the frequency plot and pole-zero locations with respect to the equations.

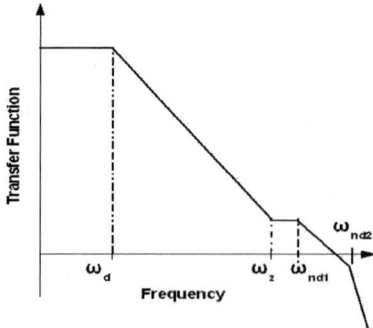

Fig.3 Frequency response of gain boosted op-amp

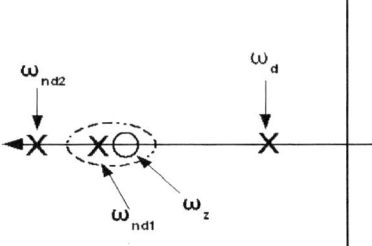

Fig.4 Pole and zero locations

In order to make the step response faster and with no doublet, we have to convert the non-dominant poles, ω_{nd1} and ω_{nd2} into a pair of complex poles and push the zero beyond the unity gain bandwidth of the op-amp [3].

To move the zero, ω_z, towards higher frequency, the drain current of gain boost device i_{ds3} have to be increased properly until the poles split along the imaginary axis and becomes complex conjugate. So, that the effect of the slow settling component due to pole-zero doublet will be eliminated and the transient response will behave like a single pole response [3]. However, further increase in current will cause a sharp reduction in phase margin making the system unstable. At the point where ω_{nd1} equals ω_{nd2}, the current of the gain boost device, i_{ds3}, can be solved in terms of the current in cascode device, i_{ds1}, by using the expression for ω_{nd1} and ω_{nd2}.

In this op-amp design, the value of i_{ds1} is derived from system level specifications of pipelined A/D converter. Therefore, a value for i_{ds3} can be calculated to be used as initial design variable. The optimal value of i_{ds3} can be determined through simulation using this initial value as the starting point for optimization.

IV. CIRCUIT IMPLEMENTATION

In this section, the implementations of the main op-amp and the gain-enhancement stages are discussed. A general method of designing a pipeline A/D converter for minimum power consumption was performed at the system level. This results in a set of specifications for each stage in pipelined A/D converter. The selected system architecture has a SHA stage followed by eight 1.5bit residue gain stages and a 2-bit flash stage. The op-amp has to meet the specifications for the SHA as shown in Table 1. Since regular cascode device cannot meet these specifications, gain boost cascode topology has been chosen to meet both the high gain and high bandwidth requirements.

Parameters	Specifications
Stage capacitor (C_f)	1.2pF
Load capacitor (C_L)	1.9pF
Feedback factor (β)	0.9
Settling accuracy (E_{ss})	2^{-11}
DC gain (A_{vdc})	72dB
GainBandWidth (GBW)	326MHz
Phase margin (PM)	70degree
Input transistor current (Ii)	0.72mA

Table 1 Op-amp specification for SHA stage

Implementation of the gain boost topology in op-amp results in Fig. 5. The main stage is a folded cascode amplifier with PMOS input transistors. Additional gain stage is applied to both cascode transistors and current source.

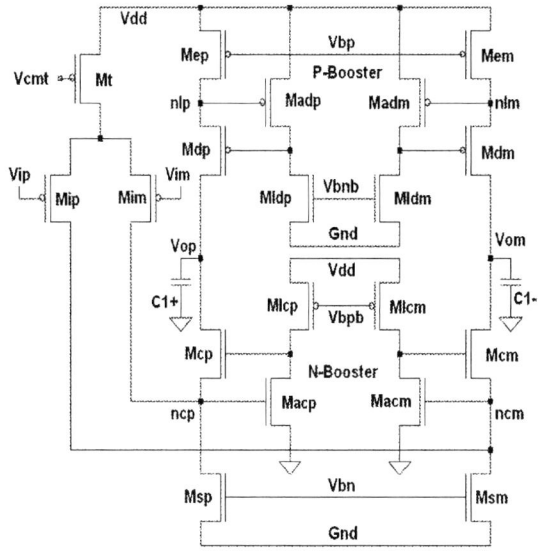

Fig.5: Fully Differential gain boost op-amp

The gain boost device for cascode device, Mcp and Mcm is N-booster device (Mlcp, Mlcm, Macp and Macm) while P-booster device (Madp, Madm, Mldp and Mldm) is for current source, Mdp and Mdm. From equation (1) and (2), the gain and output resistance for the main amplifier can be written as

$$A_{vbn} = g_{mac} \times (r_{dsac} // r_{dslc}) \qquad (9)$$

$$A_{vbp} = g_{mad} \times (r_{dsad} // r_{dsld}) \qquad (10)$$

$$R_{op} = g_{md} r_{dsd} r_{dsc} A_{vbp} \qquad (11)$$

$$R_{on} = g_{mc} r_{dsc} (r_{dss} // r_{dsi}) A_{vbn} \qquad (12)$$

$$A_v = g_{mi} \times (R_{on} // R_{op}) \qquad (13)$$

The input stage transistors were designed as to achieve the desired unity gain frequency (f_u) at least 326 MHz. The relationship between the unity gain frequency and the transconductance of the input transistors is described by (14), where C_l is the load capacitance.

$$g_{mi} = C_l \times \left(\frac{f_u}{2\pi} \right) \qquad (14)$$

When adding the additional gain stage, the op amp will consume additional power, so that the current for gain-boost device was designed as minimal as possible to minimize the power. Transistors in the signal path are biased in deep inversion instead of at edge of saturation to ensure that they remain in saturation region at all process corners. The minimal channel lengths were chosen for transistor Mcp, Mcm, Mdp and Mdm in order to place the non-dominant pole of the OTA at high frequency. While for transistor Mem, Mep, Msm and Msp, longer channel lengths were used to achieve higher open loop gain and for matching purpose.

V. SIMULATION RESULTS

Table 2 summarizes the performance of the op-amp in three process corners simulation; typical case(tm), worst speed(ws) and worst power(wp). The op-amp circuit is simulated using 0.35um CMOS process technology.

Process Corner	tm, 25 °C	ws , 85 °C	wp, 0 °C
DC Gain, (dB)	95	98	90
Unity GBW, (MHz)	412	320	536
Phase Margin, (deg)	75	72	77
Settling Time, (ns)	7.5	8.3	5.2
Load Capacitor, (pF)	1.9	1.9	1.9
Supply Voltage, (V)	3	3	3
Power, (mW)	12.8	12.5	13.6

Table 2 Summary of simulation results

In order to see the effect of the gain boost device on op-amp, the cascode device, N-gain boost device, P-gain boost device and gain boost op-amp are simulated separately. Fig. 6 shows the plot of these four different configurations. It can be seen that the gain from P-gain boost is lower at 31dB compared to N-gain boost device which is 40dB. This is due to smaller current used in P-gain boost device,

75uA compared to 300uA used in N-gain boost device. Smaller current is used to reduce power consumption. The gain of gain boost op-amp is indeed the sum of the gain of the cascode device and the combined gain of gain boost devices.

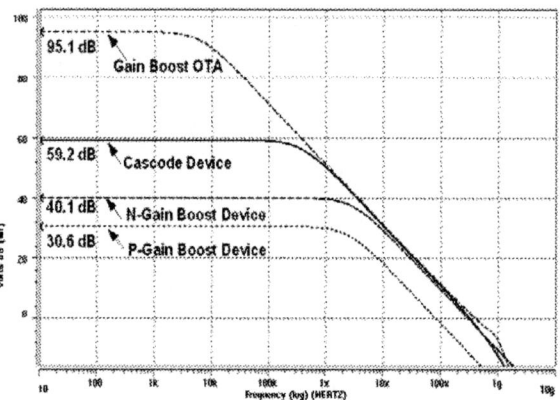

Fig.6 AC analysis plots of op-amp and booster device

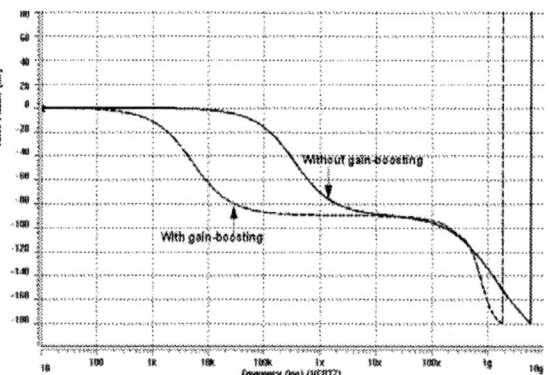

Fig.7 Bode plots with and without gain-boosting

Fig. 7 shows bode plot results when simulating with and without the gain boost devices. As can be seen, the addition of the gain boost devices does not result in frequency degradation in the phase response of the op-amp. Power consumption of the gain boost op-amp is 12.8mW while that of the cascode device was 10.6mW, meaning the additional gain boost devices only consumed 2.2mW or 17 percent of total power consumption.

The settling behavior of the op-amp is simulated in a closed-loop configuration of SHA stage. Fig. 8 illustrated the SHA configuration, whereas Fig. 9 indicates gain boost op-amp settling performance in three different process corners. In the SHA setup configuration, phi1 and phi2 are non overlapping clocks, Cfp, Cfn are both the sampling and feedback capacitors depending on which clock phase the circuit operates.

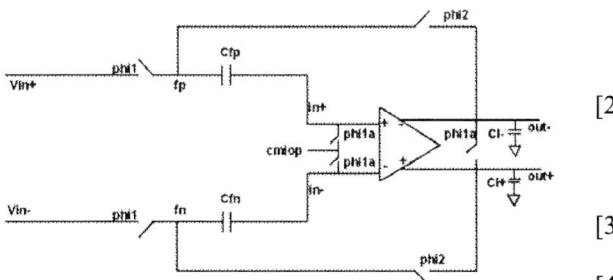

Fig. 8 Circuit configuration for SHA stage

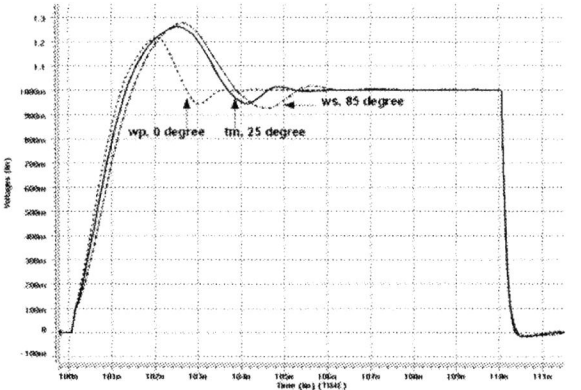

Fig. 9 Settling behavior in 3 different corners.

The transient simulation shows the OTA settles to the final value in 0.05% accuracy of 1-V step within 8.3ns for worst case corner.

VI. CONCLUSIONS

A single-stage folded cascode gain-boosted CMOS OTA has been designed and simulated using 0.35um CMOS technology. In this design, a single-transistor was applied as gain-boost device. Care has been taken in selection of the current values in both the cascode device and the gain boost device to ensure good settling time performance while maintaining the gain and bandwidth of the op-amp. The designed op-amp fulfills the stringent specifications of SHA stage of pipelined A/D converter with minimal additional power consumed.

VII. REFERENCES

[1]   B. Y. Kamath, R. G. Meyer and P. R. Gray, "Relationship between frequency response and settling time of operational amplifiers," *IEEE J. Solid State Circuits*, Vol.Sc-9, No.6, Dec 1974.

[2]   K. Bult and G.Geelen, "A fast-settling CMOS op amp for SC circuits with 90-dB DC gains," *IEEE J. Solid State Circuits*, Vol.25, no.6, Dec 1990.

[3]   Behzad Razavi, "*Design of analog CMOS integrated circuits*," McGraw-Hill, 2001.

[4]   E. Sackinger and W. Gugenbuhl, "A high swing high impedance MOS cascode circuit," *IEEE J. Solid State Circuits*, Vol.25, Feb. 1990, pp. 289-298.

[5]   M.Das and Jim Heliums, "Improved design criteria of gain-boosted CMOS OTA with high speed optimizations," *IEEE Intl. Symposium on Circuits and Systems, ISCAS*, 2000, May 28-31, 2000.

[6]   L. Reza, T. S. Mohd, M. Y Azizi and O. Shoei, "Systematic Design for Power Minimization of Pipeline Analog-to-Digital Converters", *IEEE Intl. Conf. on Computer Aided Design, ICCAD*, 2003.

Modification and Modeling of Ni/Si Interface for Photodetector Applications

M R Hashim and M Z M Yusoff

Nano-Optoelectronic Research Laboratory, School of Physics, Universiti Sains
Malaysia, 11800 Penang, MALAYSIA
Email: roslan@usm.my

Abstract. **We have demonstrated with both experiment and simulation, the method to modify the current response of n type silicon photodetector. The application of ultra-cooling temperature treatment (77K) at various cooling times (15-60 minute) has been shown to significantly modify surface properties of n type silicon (100). The surface roughness of the untreated and treated samples was obtained using AFM techniques. Treated Si sample have better surface uniformity than untreated sample. The nickel (Ni) metal contacts were then deposited on the samples followed by current-voltage (I-V) characterisation. Treated samples showed increased dark and light currents compared with untreated sample. However, current gain (ratio of light to dark current) of some of the treated sample is enhanced while some are reduced compared with that of untreated sample. The simulation software ATLAS in SILVACO package is used to describe the effect of Ni/Si interface properties on its photodetection.**

I. INTRODUCTION

In recent years, interest in developing low-cost photodetector in visible wavelength has steadily increased because of their importance in application such as short distance fiber-optic and free-space interconnects. In this region, silicon is a much cheaper choice of base material, as a compared to III-V compounds such as GaAs or InP, and the capability of process integration with the well-developed Si integrated-circuit (IC) technology as compared with the ones made with compound semiconductors [1-11]. The disadvantage of silicon is its long absorption length (~12.7 μm) in the visible region, which results in photodetector with a severe trade-off between responsivity and bandwidth. To solve these problems, many researchers and authors have concentrated to improve the performance of Si based MSM photodetector. Several methods

such as ion-implanting the absorbing layer [4], reducing the spacing between the interdigitated electrodes [5], fabricating with an additional amorphous Si alloy and amorphous SiGe film [6-9] were suggested.

In contrast, relatively little work has been reported on the effect of surface and temperature treatment on the current response in Si based MSM photodetector. In recent years, ultra-cooling temperature treatment or cryogenic processing of semiconductor material have been used for enhancing the surface roughness and surface optical properties of group IV semiconductor [10,11] and the Schottky contact performance for metal/group III-V semiconductor contacts [12,13]. In this work we used cooled temperature treatment to modify I-V characteristics of planar Si MSM photodetector. Atomic force microscopy (AFM) was used to measure surface roughness of the samples. In addition the simulation using ATLAS device simulator in SILVACO package has been conducted in order to explain the mechanism involved in the change of I-V due temperature treatment [14].

II. EXPERIMENTAL PROCEDURE

Fresh monocrystalline n-type silicon (100) with a resistivity of 10-16 ohm-cm was used for the study. The Si was dipped in liquid nitrogen for 15 to 60 minutes at atmospheric pressure. The sample was folded in Al foil to avoid direct contact with liquid nitrogen. This is followed by thermal evaporation of nickel (Ni) metal to form interdigitated Schottky contact electrodes. The MSM photodiode contacts have finger width of 200μm, finger spacing of 400μm, and the length of each electrode is about 3300μm. Each electrode has four fingers. After Ni metallization the MSM structure was subjected to annealing at 400^0C for 10 minute in a conventional tube furnace in flowing nitrogen ambient. Device simulation ATLAS in SILVACO was used to study the correlation between the modified photo and dark currents and surface roughness of n type silicon.

III. RESULTS AND DISCUSSION

Figure 1 shows AFM images of n type (100) silicon samples dipped for a) 15 minute b) 30 minute c) 45 minute and d) 60 minute in liquid nitrogen. Image from untreated sample e) is also included for comparison. In agreement with the previous studies [10,11], surface roughness improves with increasing treatment time (untreated, surface roughness=2.97nm) $\rightarrow$ (15 minutes treatment, surface roughness=2.1nm)$\rightarrow$ (60minutes, surface roughness=1.62nm). Note that for each sample, few scans at different locations were made so the reported surface roughness represents somewhat the average value for each treatment time. After the treatment, all the sample were deposited with Ni contact at room temperature followed by furnace anneal of $400\,^{0}$C, 10 minute in a conventional tube furnace in flowing nitrogen ambient.

Figure 2 shows dark and photo I-V characteristic of n-type Si (100) based Ni MSM photodetector for all the samples. It can be seen that the xposure time changes both the dark and light current of MSM structure. More specifically the treatment has increased both dark and light currents of the MSM photodetector. In particular 30 minute treatment produced the highest dark and light currents of all. However 60 minute treatment produced the highest current gain (inset) compared to other samples. Samples with 15minutes and 45 minutes treatment exhibited degradation in current gain compared with untreated sample despite showing increasing dark and light current. Results from figure 2 therefore suggest that the difference in I-V characteristics could be partly explained by the change in surface morphology particularly at Ni/Si interface. , where smoother Ni/Si interface produce higher current than that of rougher interface. In addition to the I-V characteristics, the ratio of detector light current-to-dark current, i.e, Gain = light/dark is often quoted for performance evaluation as an optically control electronic switch. The inset of Figure 2 shows the change in the Gain, as a function of bias voltage (V). Sample with 60 minute treatment shows higher Gain compared to other exposure times.

It can be seen that the low temperature treatment increases both the dark and light currents of MSM structure.

Results from figure 2 suggest that the difference in I-V characteristics could be explained by the change in surface morphology especially at Ni/Si interface. Smoother Ni/Si interface produce higher current than that of rougher interface.

In order to correlate between I-V characteristic change and Ni/Si surface morphology, device simulator ATLAS in SILVACO is used. For the purpose of modeling, Si surface structure can be treated as comprising of small Si blocks with different widths and heights on Si surface. Surface structure then could affect the I-V characteristics in two ways. First the physical nature of metal contact on the Si surface, secondly the size of the Si block which influences the flow of the current. For the first part, in order to model the physical nature of metal contact, figure 3 shows three possible configurations of metal attachment on the Si blocks. Metal contact on very smooth Si surface free from Si blocks (0 μm height) is also included as control. The height and the width of the block are 0.3 μm and 1μm respectively, and distance between blocks is 1 μm. Model 1 represents metal contacts on the top of the blocks as well as between blocks. Model 2 has metal contacts between blocks only. Model 3 has metal contacts only on the blocks. Figure 4 shows simulated dark and photo I-V characteristic of Ni/Si MSM with different contact configurations. Also included are I-V characteristics of Si control. Simulated photocurrent was obtained by using incident light in the visible region at 50 kW/cm^2. At high voltage Model 3 shows the lowest dark current and lowest photo current. This could be explained by the fact that the blocks acting as barriers for the current flow. Model 2 shows highest dark and photo currents compared to other models. However the fact that Model 2 shows I-V characteristics higher than those of Si control is very interesting. This is because contacts in Model 2 are similar to those of Si control, however possess higher dark and photo currents. This could be explained by the fact that metal in Model 2 made 3-side contact with Si. Model 1 shows I-V characteristics sitting between those of Model 3 and Model 2 especially at higher operating voltage. This is expected since configuration in Model 1 is a mixture between Model 2 and Model 3. However at lower operating voltage, Model 1 shows the lowest dark current and relatively high photocurrent and hence producing the highest current gain (see inset).

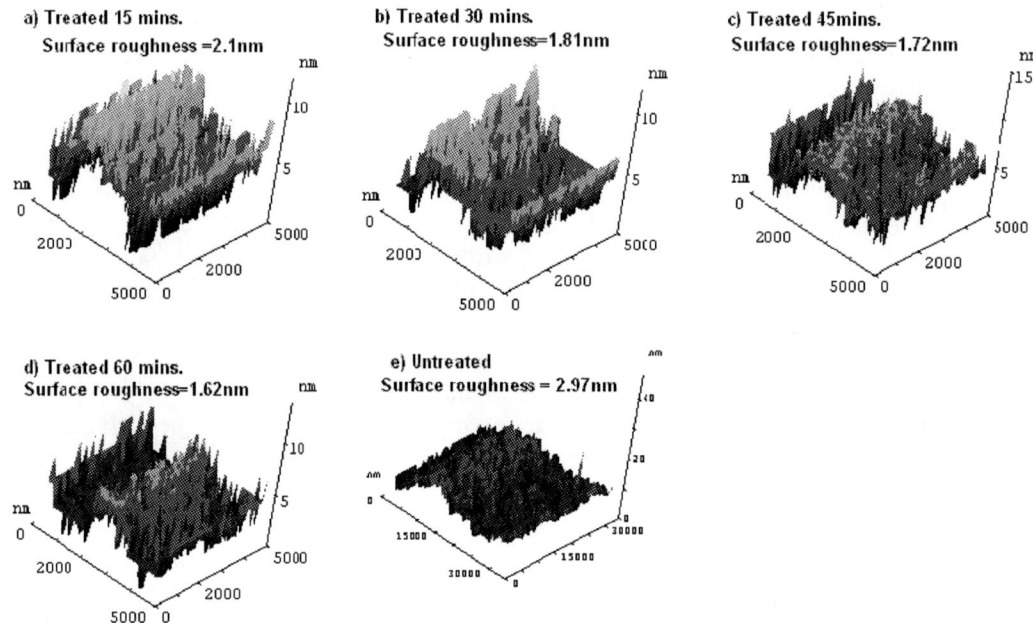

Fig. 1. AFM images from the surface of a) 15 minute b) 30 minute c) 45 minute and d) 60 minute exposure time n type silicon (100) samples. Image from untreated sample e) is also included for comparison.

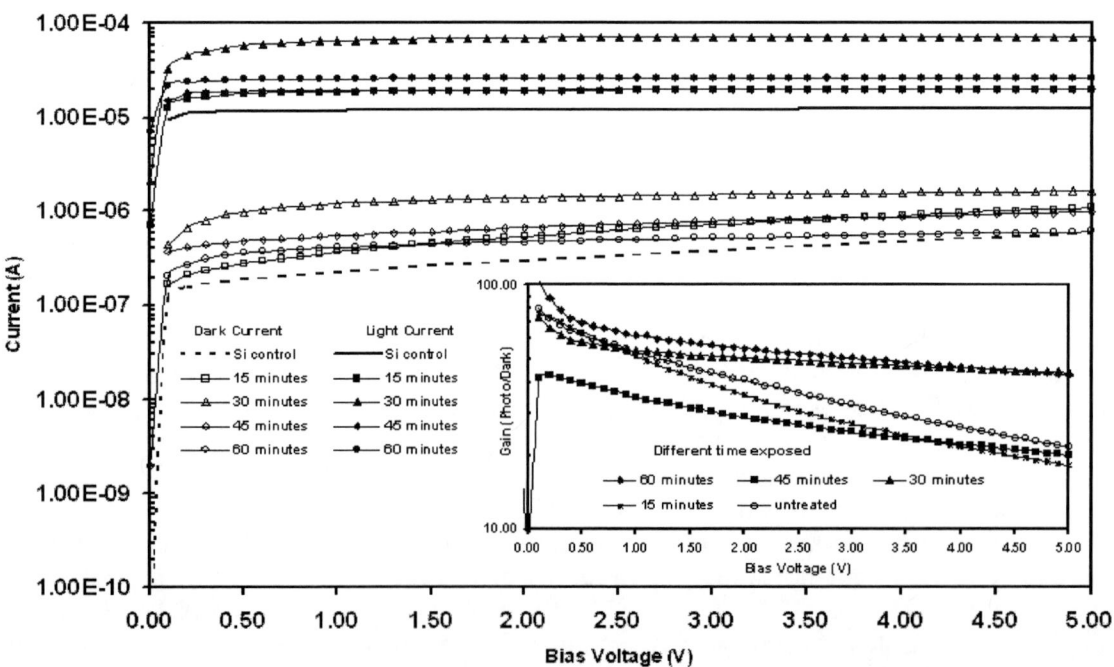

Fig. 2. Measured I-V characteristic and current gain (inset) of the n-type Si (100) based MSM photodetector with ultra-cooling temperature treatment at various exposure time (15-60 minutes) followed $400\,^{0}$C, 10 minute conventional tube in flowing nitrogen ambient.

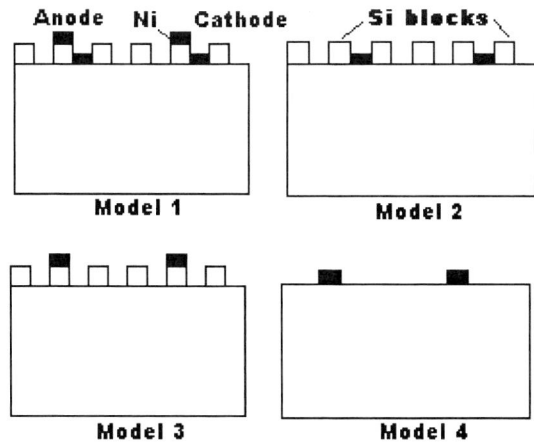

Fig. 3 Schematic diagram of MSM photodetector with four possible Ni/silicon contact configurations.

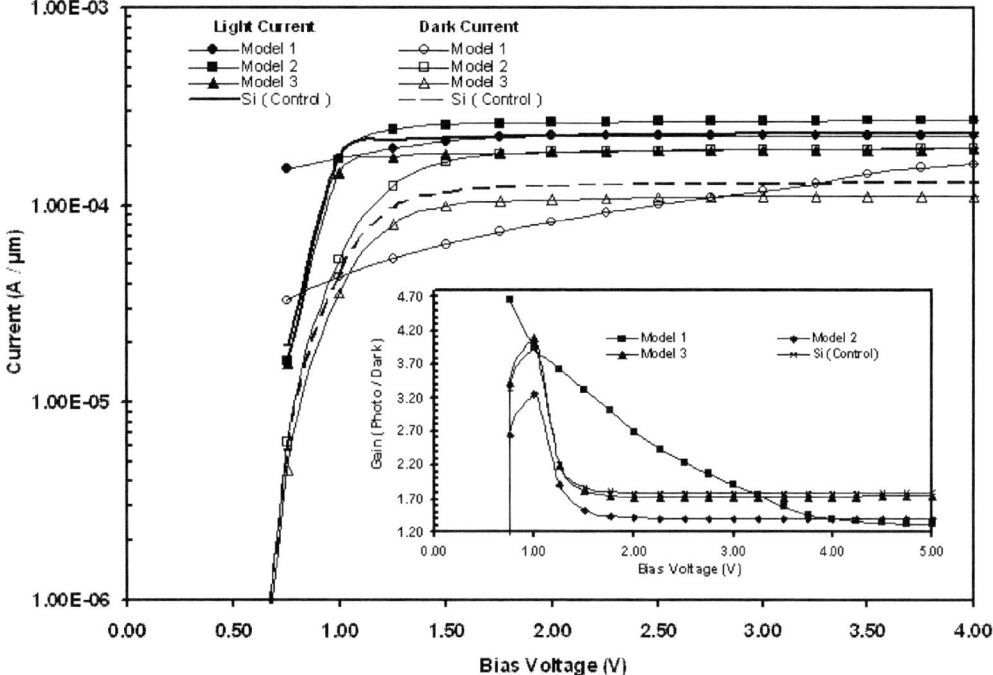

Fig. 4. Simulated dark and photo currents and current gain (photo/dark current, inset) for three possible Ni/Si configurations on n type Si. Also included is Si control for comparison. The incident light is 50 kW/cm^2 illuminations.

The first part of study therefore signifies two main points. First, Ni/Si configuration plays important role in determining the I-V characteristic of MSM Si based structure. Therefore the physical nature of metal contact could partly explain the difference in I-V characteristic observed in figure 2.

Secondly, Model 1 could be a preferable model of metal contact on Si for high performance MSM as photodetector. For Model 1, under dark current the block could act as extra resistance to electron flow thus suppressing the electrical current. However, under light illumination high photo generated carriers being generated around the blocks. The fact that this model possesses two contacts will maximize the collection of photogenerated carriers and hence increases the electrical current.

Table 1 Different heights and widths of Si block used in the simulation

Device	a (height of Si block, μm)	b (width of Si block, μm)
A	0.30	1.00
B	0.30	0.50
C	0.01	1.00
Control	0.00	0.00

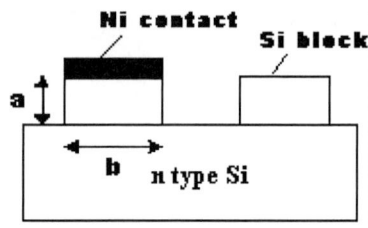

Fig. 5. Ni contact on Si block with varying heights and widths.

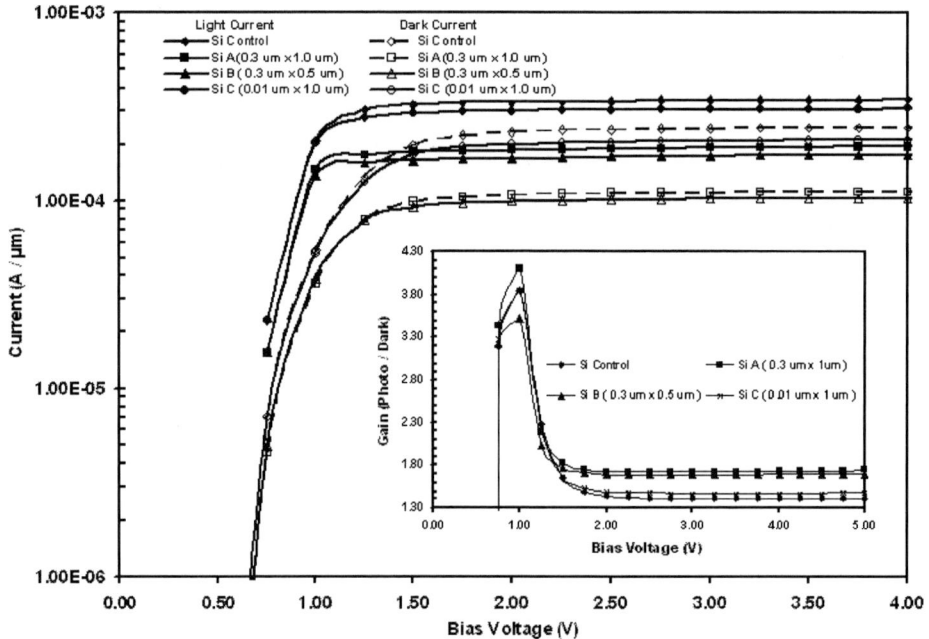

Fig. 6. Simulated I-V characteristics with different height with widths on the surface roughness and current gain (inset) of n type Si and 50 kW/cm^2 illuminations. Also included is Si control for comparison.

Second part of simulation will study the effect of Si block size on the current flow. Model 3 is a logical choice to be used in this study. Fig. 5 shows schematic diagram for Ni contact on varying dimension of Si block. Table 1 summarizes the Si block structure with different height and widths. Figure 6 shows simulated I-V characteristic of Ni MSM device with different Si block height and width illuminated with 50 kW/cm^2 incoming beam in visible region. Comparing I-V of device A (0.3 μm x 1.0 μm) and B (0.3 μm x 0.5 μm) shows that device with wider block (A) possesses higher current. This is expected since wider block facilitates the

flow of electron from Ni to Si and hence promotes high electrical current. On the other hand smaller block width promotes current crowding which suppress the electron flow. Comparing I-V characteristics of device A (0.3 μm x 1.0 μm) and device C (0.01 μm x 1.0 μm) shows that device with smaller block height (C) possesses current similar to ideal MSM contact (Si control). This is consistent with experimental results in figure 2 which shows device with temperature treatment possesses high current. This is expected because smaller block height provides electron path with lower resistance and hence higher current.

The simulation results from figure 4 and figure

6 prove two important points, which can be used to explain the I-V characteristics of treated and untreated Ni/Si MSM device. Firstly the metal placement on rough Si surface and the size of Si block (surface roughness) affects the current flow. Secondly, typical natural Si surface possesses significant surface roughness, therefore should be taken into consideration for accurate modeling of semiconductor devices.

IV. CONCLUSION

We have successfully demonstrated with both simulation and experiment, the method to modify dark and photo current of monocrystalline n type silicon (100) Ni MSM photodetector using ultra cooling temperature treatment. The modified current is associated with the change of Ni/Si interface roughness. Rough Ni/Si surface produces low dark and photo currents. Smoother Ni/Si surface produces high dark and photocurrents. In significantly rough Ni/Si surface, the placement of metal contact and the size of surface roughness play important role in determining the current flow of Ni MSM photodetector. Finally one has to take into account surface roughness for accurate modeling of semiconductor devices for photonics applications.

ACKNOWLEDGEMENT

This work was conducted under IRPA 8th Malaysian Plan under Strategic Research Grant. The support from Ministry of Science Technology and Innovation of Malaysia is gratefully acknowledged

REFERENCES

[1] S. Alexandrou, C. C. Wang, T. Y. Hsiang, M. Y. Liu, and S. Y. Chou 1993 A 75 GHz silicon silicon metal-semiconductor-metal Schottky photodiode *Appl. Phys. Lett.* vol62 pp2507-2509

[2] E. Chen, and Y. Chou 1997 High-efficiency and high-speed silicon metal-semiconductor-metal photodetectors operating in the infrared *Appl. Phys. Lett.* vol70(6) pp753-755

[3] M. Seto, C. Rochefort, S. De jager, R. F. M. Hendriks, G. W. 't Hooft, and M. B. van der Mark 1999 Low-leakage-current metal-insulator-semiconductor-insulator-metal photodetector on silicon with a SiO_2 barrier-enhancement layer *Appl. Phys. Lett.* vol75 pp1976-1978

[4] A. K. Sharma, K. A. M. Scott, S. R. J. Brueck, J. C. Zolper, and D. R. Myer 1991 Ion implantation enhanced metal-semiconductor-metal photodetector *IEEE Photo. Technol. Lett.* **vol6** pp635-638

[5] B. W. Mullins, S. F. Soares, K. A. McAdle, C. M. Wilson, and S. R. J. Brueck 1991 A simple high-speed Si Schottky photodiode *J. Lightwave Technol.* **vol3** pp360-362

[6] C. S. Lin, R. H. Yeh, C. H. Liao, and J. W. Hong 2002 High-speed Si-based metal-semiconductor-metal photodetectors with an additional composition-graded i-a-SiGe:H layer *Solid-State Electron.* vol46 pp2027-2033

[7] L. H. Laih, J. C. Wang, Y. A. Chen, W. C. Tsay, T. S. Jen, J. S. Chen, and J. W. Hong 1997 Improving the transient response of a metal-semconductor-metal photodetector with an Additional i-a-SiGe:H film *Jpn. J. Appl. Phys.* vol36(3B) pp1495-1496

[8] L. H. Laih, J. C. Wang, Y. A. Chen, W. C. Tsay, T. S. Jen, J. S. Chen, and J. W. Hong 1997 Characteristics of Si-based MSM Photodetector with an Amorphous-Crystalline heterojunction *Solid-State Electron.* vol.41(11) pp1693-1697

[9] L. H. Laih, T. C. Chang, Y. A. Chen, W. C. Tsay, and J. W. Hong 1998 Characteristics of MSM Photodetectors with Trench Electrodes on P-type Si Wafer *IEEE Trans. Electron Devices* vol 45(9) pp2018-2023

[10] M. R. Hashim, and Kifah. Q. Salih 2005 Optical properties of treated and untreated monocrystalline p-Si<111>, p-Si<100>, n-Si<111> and n-Si<100> wafers in the visible region at room temperature *Microelectronic Eng.* 81 pp243-250

[11] M. R. Hashim, Kifah. Q. Salih, D. Bagnell, and P. Lamraksa 2004 Modification of optical properties of SiGe/p-Si(100) MQWs grown by LPCVD for photonic *applications Electrochem. Soc. Proc.* pp299-303

[12] L. He, and J. E. Siewenie 2002 Cryogenic processing of thin metal films *Surface and Coating Tech.* vol150 pp.76-79

[13] Y. C. Lee, Z. Hassan, M. J. Abdullah, M. R. Hashim, and K. Ibrahim 2005 Dark current characteristics of thermally treated contacts on GaN-based ultraviolet photodetectors *Microelectronic Eng.* vol81 pp262-268

[14] ATLAS User's Manual 2000 vols1-2, Silvaco International

An Enhancement of Decimation Process using Fast Cascaded Integrator Comb (CIC) Filter

Rozita Teymourzadeh and Masuri Bin Othman
VLSI Design Center, Institute of Microengineering and Nanoelectronics (IMEN)
Universiti Kebangsaan Malaysia, 43600 Bangi, Selangor, Malaysia
rozita60@vlsi.eng.ukm.my , masuri@vlsi.eng.ukm.my

Abstract - **The over sampling technique has been shown to increase the SNR and is used in many high performance system such as in the ADC for audio and DAT systems. This paper presents the design of the decimation and its VLSI implementation which is the sub-component in the over sampling technique. The design of three main units in the decimation stage that is the Cascaded Integrator Comb (CIC) filter, the associated half band filters and the droop correction are also described. The Verilog HDL code in Xilinx ISE environment has been derived to describe the CIC filter properties and downloaded in to Virtex II FPGA board. In the design of these units, we focus on the trade-off between the speed improvement and the power consumption as well as the silicon area for the chip implementation.**

I. INTRODUCTION

The most popular A/D converters for audio applications are realized based on the use of over sampling and sigma-delta ($\sum\Delta$) modulation techniques followed by decimation process [1]. Oversampled Sigma delta ($\sum\Delta$) modulator provides high resolution sample output in contrast to the standard Nyquist sampling technique. However at the output, the sampling process is needed in order to bring down the high sampling frequency and obtain high resolution. The CIC filter is a preferred technique for this purpose. In 1981, Eugene Hogenauer [2] invented a new class of economical digital filter for decimation called a Cascaded Integrator Comb filter (CIC) or recursive comb filter. This filter worked with sampling frequency of 5 MHz. Additionally the CIC filter does not require storage for filter coefficients and multipliers as all coefficients are unity [3]. Furthermore its on-chip implementation is efficient because of its regular structure consisting of three basic building blocks, minimum external control and

less complicated local timing is required and its change factors is reconfigurable with the addition of a scaling circuit and minimal changes to the filter timing. It is also used to perform filtering of the out of band quantization noise and prevent excess aliasing introduced during sampling rate decreasing. Hence enhanced high speed will be key issue in chip implementation of CIC decimators. In 1998, Garcia [4] designed Residue Number System (RNS) for pipelined Hogenauer CIC. Compared to the two's complement design, the RNS based Hogenaur filter enjoys an improved speed advantage by approximately 54%. Similar structure by Meyer-Baese [5] has been implemented to reduce the cost in the Hogenauer CIC filter which shows that the filter can operate up to maximum clock frequency of 164.1 MHz on Altera FPLD and 82.64 MHz on Synopsys cell-based IC design.

This paper shows the implementation of the high speed CIC filters which are consists of three parts, integrator, comb and down sampler. The CIC filter is considered as recursive filter because of the feedback loop in integrator circuit and it can work with maximum throughput of 190 MHz.

The next section describes the mathematical formulation and block diagram of CIC filters in detail. Enhanced high speed architecture is explained in section III. Section IV shows implementation and design result in brief. Finally conclusion is expressed in section V.

II. DEVELOPMENT OF A DECIMATION FILTER

The purpose of the CIC filter is twofold; firstly to remove filtering noise which could be aliased back to the base band signals and secondly to convert high sample rate m-bit data stream at the output of the Sigma-delta modulator to n-bit data stream with lower sample rate. This process is also known as decimation which is essentially performing the averaging and a rate reduction functions simultaneously.

0-7803-9730-4/06/$25.00 ©2006 IEEE

Figure 1 shows the decimation process using CIC filter.

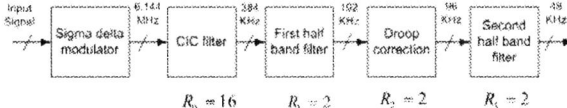

Fig. 1 Digital Decimation Process

The two half band filters [6] are used to reduce remain sampling rate reduction to the Nyquist output rate. First half band filter and second half band filter make the frequency response more flat and sharp similar to ideal filter frequency response.

Droop correction filter is allocated to compensate pass band attenuation which is created by the CIC filter. The frequency response of overall system will be shown in section V.

Table 1 shows filter specification in decimation process.

TABLE I
FILTER SPECIFICATIONS

	Pass band (kHz)	Stop band (kHz)	Transition band (kHz)
CIC filter	7	384	377
First half band filter	32	170	138
Droop Correction	32	70	38
Second half band filter	21.77	26.53	4.76

III. PRINCIPLE OF CIC FILTER STRUCTURE

The CIC filter consist of N stages of integrator and comb filter which are connected by a down sampler stage as shown in figure 1 in z domain. The CIC filter has the following transfer function:

$$H(z) = H_I^N(z).H_C^N(z) = \frac{(1-z^{-RM})^N}{(1-z^{-1})^N} = (\sum_{k=0}^{RM-1} z^{-k})^N \quad (1)$$

where N is the number of stage, M is the differential delay and R is the decimation factor.
In this paper, N, M and R have been chosen to be 5, 1 and 16 respectively to avoid overflow in each stages.

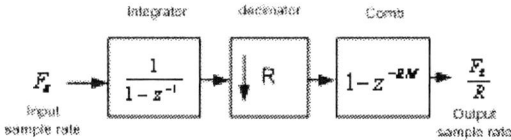

Fig. 2 One-stage of CIC filter block diagram

N, M and R are parameters to determine the register length requirements necessary to assure no data loss. Equation (1) can be express as follow:

$$H(z) = \sum_{k=0}^{(RM-1)N} h(k)z^{-k} = \left[\sum_{k=0}^{RM-1} z^{-k}\right]^N \leq \left|\sum_{k=0}^{RM-1} z^{-k}\right|^j$$
$$\leq \left(\sum_{k=0}^{RM-1} |z|^{-k}\right)^N = \left(\sum_{k=0}^{RM-1} 1\right)^N = (RM)^N \quad (2)$$

From the equation, the maximum register growth/width, G_{max} can be expressed as:

$$G_{max} = (RM)^N \quad (3)$$

In other word, G_{max} is the maximum register growth and a function of the maximum output magnitude due to the worst possible input conditions [2].

If the input data word length is B_{in}, most significant bit (MSB) at the filter output, B_{max} is given by:

$$B_{max} = [N\log_2 R + B_{in} - 1] \quad (4)$$

In order to reduce the data loss, normally the first stage of the CIC filter has maximum number of bit compared to the other stages. Since the integrator stage works at the highest oversampling rate with a large internal word length, decimation ratio and filter order increase which result in more power consumption and speed limitation.

III. SPEED IMPROVEMENT

A. Truncation for low power & high speed

Truncation means estimating and removing Least Significant Bit (LSB) to reduce the area requirements on chip and power consumption and also increase speed of calculation. Although

821

this estimation and removing introduces additional error, the error can be made small enough to be acceptable for DSP applications. Figure 3 illustrates five stages of the CIC filter when B_{max} is 25 bit so truncation is applied to reduce register width. Matlab software helps to find word length in integrator and comb section.

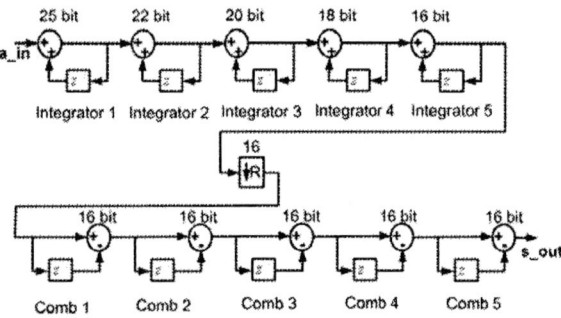

Fig. 3 Five-stages of truncated CIC filter

B. Pipeline structure

One way to have high speed CIC filter is by implementing the pipeline filter structure. Figure 4 shows pipeline CIC filter structure when truncation is also applied. In the pipelined structure, no additional pipeline registers are used in integrator part. So that hardware requirement is the same as in the non-pipeline [7]. The CIC decimation filter clock rate is determined by the first integrator stage that causes more propagation delay than any other stage due to maximum number of bit. So it is possible to use a higher clock rate for a CIC decimation filter if a pipeline structure is used in the integrator stages, as compared to non-pipelined integrator stages. The clock rate in integrator section is R times higher than in the comb section.

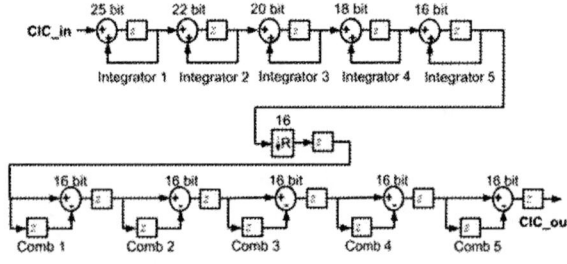

Fig. 4 Five-stage of truncated pipeline CIC filter

Previously, the pipeline structure for CIC filter was applied just for integrator part since the maximum clock rate is determined by the integrator. The above architecture showed that the maximum throughput was increased by 20

MHz when the pipeline structure is used for all the CIC parts consisting of integrator, comb and down sampler.

C. Modified Carry look-ahead Adder (MCLA)

The other technique to increase speed is using Modified Carry Look-ahead Adder. The Carry Look-ahead adder (CLA) is the fastest adder which can be used for speeding up purpose but the disadvantage of the CLA adder is that the carry logic is getting quite complicated for more than 4 bits so Modified Carry Look-ahead Adder (MCLA) is introduced to replace as adder. This improve in speed is due to the carry calculation in MCLA. In the ripple carry adder, most significant bit addition has to wait for the carry to ripple through from the least significant bit addition. Therefore the carry of MCLA adder has become a focus of study in speeding up the adder circuits [8]. The 8 bit MCLA structure is shown in Figure 5. Its block diagram consists of 2, 4-bit module which is connected and each previous 4 bit calculates carry out for the next carry. The CIC filter in this paper has five MCLA in integrator parts. The maximum number of bit is 25 and it is decreased in next stages. So it truncated respectively to 25, 22, 20, 18 and 16 bit in each adder, left to right Notice that each 4-bit adder provides a group propagate and generate Signal, which is used by the MCLA Logic block. The group Propagate P_G and Generate G_G of a 4-bit adder will have the following expressions:

$$P_G = p_3 \cdot p_2 \cdot p_1 \cdot p_0 \qquad (5)$$
$$G_G = g_3 + p_3 \cdot g_2 + p_3 \cdot p_2 \cdot g_1 + p_3 \cdot p_2 \cdot p_1 \cdot g_0 \quad (6)$$

The most important equations to obtain carry of each stage have been defined as below:

$$c_1 = g_0 + (p_0 \cdot c_0) \qquad (7)$$
$$c_2 = g_1 + (p_1 \cdot g_0) + (p_1 \cdot p_0 \cdot c_0) \qquad (8)$$
$$c_3 = g_2 + (p_2 \cdot g_1) + (p_2 \cdot p_1 \cdot g_0) + (p_2 \cdot p_1 \cdot p_0 \cdot c_0) \quad (9)$$
$$c_4 = g_3 + (p_3 \cdot g_2) + (p_3 \cdot p_2 \cdot g_1) + (p_3 \cdot p_2 \cdot p_1 \cdot g_0)$$
$$+ (p_3 \cdot p_2 \cdot p_1 \cdot p_0 \cdot c_0) \qquad (10)$$

Calculation of MCLA is based on above equations. 8-Bit MCLA Adder could be constructed continuing along in the same logic pattern, with the MSB carry-out resulting from OR & AND gates. The Verilog code has been written to implement addition. The MCLA Verilog code was downloaded to the Xilinx

FPGA chip. From Xilinx ISE synthesize report, it was found minimum clock period is 3.701ns (Maximum Frequency is 270 MHz).

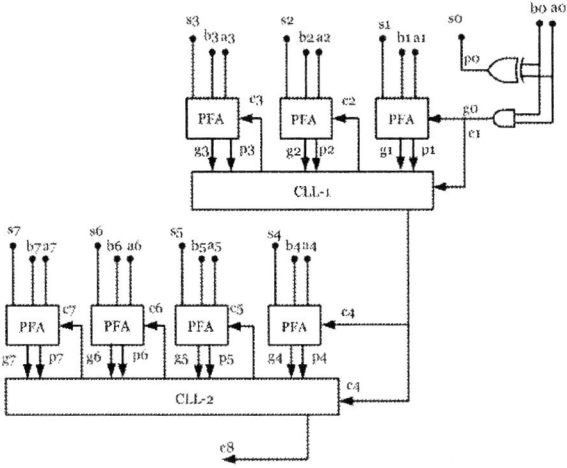

Fig. 5 The 8 bit MCLA structure

IV. IMPLEMENTATION

After the sigma delta modulator, the sampling rate must be reduced to 48 KHz which is the Nyquist sampling rate. This is carried out in 4-stages. The first stage involves the reduction of the sampling frequency by the decimation factor of 16. This is done by the CIC filter.

The remaining 3 stages involve the reduction of the sampling frequency by the decimation factor of 2 only which are carried out by the first half band, droop correction and the second half band respectively. Figure 6 illustrate the frequency response of the overall decimation filter when the sampling frequency is 6.144 MHz.

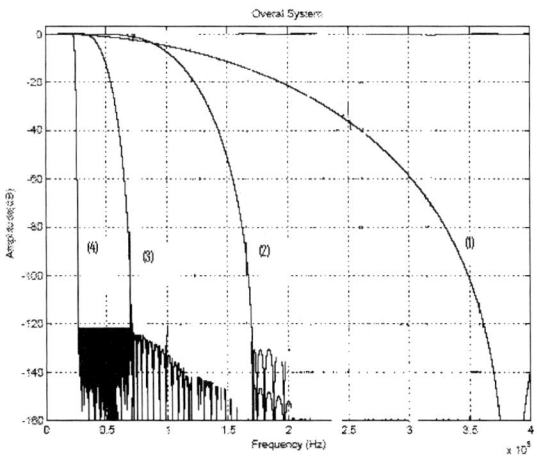

(1) CIC filter
(2) First half band filter
(3) Droop correction
(4) Second half band filter

Fig. 6 The frequency response of overall decimation filter

Figure 7 shows the Droop correction filter result. This filter design a low pass filter with pass band having the shape of inverse the CIC filter frequency response. So it compensates amplitude droop cause of the CIC filter and makes whole system frequency response flat.

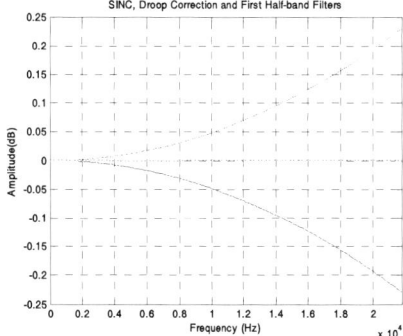

Fig. 7: Droop Correction effect on frequency response

Figure 8 shows the measured baseband output spectra before (Figure 8(a)) and after (Figure 8(b)) the decimation functions.

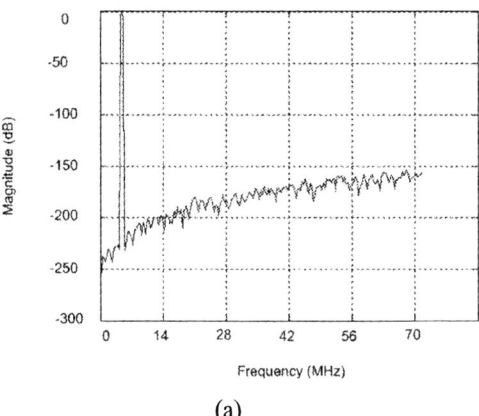

(a)

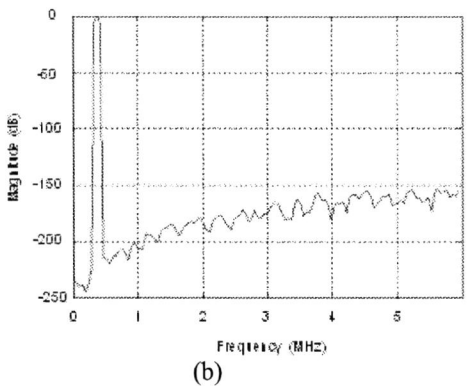

(b)

Fig. 8 Signal spectra (a) Output sigma delta modulator SNR (b) Output CIC filter SNR

The CIC filter Verilog code was written and simulated by Matlab software. The signal to noise ratio is 141.56 dB in sigma delta modulator output and it is increased to 145.35 dB in the decimation stages. To improve the signal to noise ratio, word length of recursive CIC filter should be increased but the speed of filter calculation is also decreased.

The chip layout on Virtex II FPGA board has been shown in Figure 9.

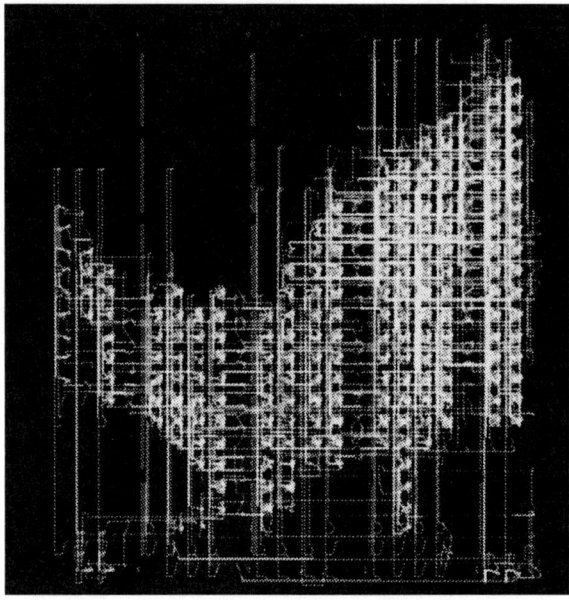

Fig. 9 The core layout on FPGA board

V. CONCLUSION

Recursive CIC filters have been designed and investigated. Enhanced high Speed CIC filters was obtained by three ways. The pipeline structure, using the modified carry look-ahead adder (MCLA) and truncation lead us to have high speed CIC filter with the maximum throughput of 190 MHz. The evaluation indicates that the pipelined CIC filter with MCLA adder is attractive due to high speed when both the decimation ratio and filter order are not high as stated in the Hogenauer Comb filter. Since the first stage of the CIC filter require maximum word length and also because of the recursive loop in its structure, the reduction in power consumption is limited by the throughput. Thus the truncation will reduce the power consumption and the number of calculation. The power consumption computed using CAD tools (Cadence and Synopsys) and 0.18 μm Silterra technology library gives 3.5 mW power consumption at maximum clock frequency.

REFERENCE

[1] T. Ritoniemi, E.Pajarre. S. Ingalsuo, T. Husu, V. Eerola, and T. Saramiiki, " A Stereo Audio Sigma-delta AD-Converter". *IEEE J. Solid –state Circ.*, Vol.29, no.12, pp.1514-1523, Dec. (1994).

[2] E.B. Hogenauer, "An Economical Class of digital filters for Decimation and interpolation", *IEEE Transactions on Acoustics, Speech, and Signal Prosessing*, Vol. ASSP-29,pp.155-162, April (1981).

[3] S. Park, "Principles of Sigma-delta Modulation for Analog-to-Digital Converters", *Motorola Inc, APR8/D* Rev.1, (1990).

[4] A.Garcia, U. Meyer-Baese & F. Taylor, "Pipelined Hogenauer CIC Filters Using Field-Programmable Logic and Residue Number System". *Acoustics, Speech and Signal processing IEEE International Conference,*Vol. 5, pp.3085-3088 (1998).

[5] U. Meyer-Baese, S. Rao, J. Ramirez, & A. Garcia, "Cost-effective Hogenauer Cascaded Integrator Comb Decimator Filter Design for custom ICs". IEE Electronic journal. Vol. 41, no. 3, pp. 158-160 (2005).

[6] P. B. Brandt & A. Bruce. Wooley, "A Low-Power, Area-Efficient Digital Filter for Decimation and Interpolation". *IEEE Journal of Solid-State Circuits*, Vol. 29, no.6, June (1994).

[7] Y. Djadi and T. A. Kwasniewski, C. Chan and V. Szwarc, "A high throughput Programmable Decimation and Interpolation Filter", *Proceeding of International Conference on Signal Processing Applications and Technology*, pp.1743-1748, (1994).

[8] D.M. Ciletti, Advanced Digital design with the Verilog HDL, Prentice Hall, Department of Electrical and Computer Engineering University of Colorado at Colorado Springs, (2003).

[9] S. M. Mortazavi, S. M. Fakhraie & O. Shoaei. " A Comparative Stydy and Design of Decimation Filter For high-Precision Audio Data Converters", *The 17 IEEE international conference on microelectronics.* pp.139-143, (2005).

Effects of Metal Work Function and Operating Temperatures on the Electrical Properties of Contacts to n-type GaN

S. M. Thahab, H. Abu Hassan, Z. Hassan

Nano-Optoelectronics Research and Technology Laboratory
School of Physics, Universiti Sains Malaysia
11800 Penang, Malaysia
E-mail: sabahmr @ yahoo.com, haslan@usm.my, zai@ usm.my

ABSTRACT **The electrical properties of various metal contacts on n-type GaN at high and low doping concentration ($5*10^{18}$ cm^{-3} and $1*10^{15}$ cm^{-3}) were simulated to determine the underlying trend between the metallic contact work function and the resultant Schottky barrier height between the metal and the GaN material. Pt, Ni, Pd, Au, Co, Cu and Ag metals having different work functions were investigated. Operating temperatures of structures were varied between 200K and 400K. It is found that the turn-on voltage for the diode is dependent on the value of work function for each metal used in the simulation process. Effective change in the current was obtained at different operating temperatures for all metals used at both doping concentrations**

I. INTRODUCTION

The successful realization of high-quality doped gallium nitride films has revitalized the interest in the group–III nitrides for device applications [1]. The most prominent examples are light emitting diode and laser diode in the blue spectral range [2]. Major problems still exist with ohmic contacts to devices made from group-III nitrides. Therefore, this study explores the electronic properties of ideal metal contacts on GaN (0001) surfaces. Most metal-semiconductor contacts are rectifying. The electric transport across such interfaces is carried by the carriers and is thus characterized by their barrier height that is the energy distance between the Fermi level and edge of the respective majority-carrier band right at the interface [3]. There are several physical mechanisms that may determine the barrier heights of Schottky contacts. Ideal interfaces are intimate, abrupt, free of structural defects as well as foreign atoms and laterally homogeneous. Their barrier heights are then determined by continuum of metal-induced gap states derived from the virtual gap states of the complex semiconductor band structure [4]. They represent the wave function tail of metal electrons into the semiconductor in the energy range between the valence-band maximum and Fermi level where the metal conduction band overlaps the semiconductor band gap. At real but still intimate and abrupt contacts, interface dipoles induced by foreign atoms or by specific interface structures as well as structural and chemical interface point-defects may be present and will then contribute to the barrier heights. It is expected that the Schottky barrier heights between metal and GaN should be totally unaffected by Fermi level pinning type effects and to be wholly due to the offset between the metal's work function and the energy levels of the bands in GaN [5]. If this is so, it would be expected that changing the metal should have a large effect on the operation voltage of GaN devices. Although, there have been reports on the relative effects of various metal contacts to GaN, we investigate the electrical properties of various contact metals on n-type GaN to see what relationships occurred and if these predicted large shifts in the operation voltages were due to metal type and to what extent, if any, Fermi level pinning plays a role in such contacts to GaN materials. Mobility is an important parameter for carrier transport because it describes how strongly the motion of an electron is influenced by applied electric field. The mobility is related directly to the mean free time between collisions, which in turn is determined by the various scattering mechanisms. The two most important mechanisms are lattice scattering and impurity scattering. The Lattice scattering results from thermal vibrations of the lattice atoms at any temperature above absolute zero [5]. These vibrations disturb the lattice periodic potential and allow energy to be transferred between the carriers and the lattice. Since lattice vibration increases with increasing temperature, lattice scattering become dominant at high temperatures; hence the mobility decreases with increasing temperature. Theoretical analysis [6] shows that the mobility due to lattice scattering μ_L will decrease in proportion to $T^{-3/2}$. Impurity scattering results when a charge carrier travels past an ionized dopant impurity (donor or acceptor), the change carrier path will be deflected owing to Coulomb force interaction. The probability of impurity scattering depends on the total concentration of ionized impurities, that is, the sum of the concentration of negatively and positively charged ions. At higher temperatures, the carriers move faster; they remain near the impurity atom for a shorter time and therefore less effectively scattered. The mobility due to impurity scattering can theoretically be shown to vary as $T^{3/2}N^T$, where N_T is the total impurity concentration [7]. For lightly doped concentration like 10^{16} cm^{-3} the lattice scattering dominates and the mobility decreases as the temperature increases. For heavily doped material the effect of impurity scattering is most pronounced, so we suppose that this effect appears with increasing operation temperature in the range studied [8].

II. Structure Parameters and Models Used in the Simulation

ISE TCAD software programme is used in our simulation. n-GaN layer with thickness of 2μm on 30nm GaN buffer layer on sapphire substrate was used in the simulation. Contacts with different metal work functions (Pt, Ni, Pd, Au, Co, Cu and Ag) at different operation temperatures (200 K, 250 K, 300 K, 350 K, and 400 K) were simulated for each metal contact. The simulations were conducted by choosing the following models: Carrier concentration, Mobility (high field saturation), Effective Intrinsic Density (no band gap narrowing), Thermodynamic, Incomplete Ionization and Fermi model. The Poisson and carrier continuity equations were performed. Energy band gap for GaN was set at 3.4 eV, the effective mass of electron for GaN is 0.22m$_0$.

III. Results and Discussion

The results of I-V characteristics for different metal work functions at 300 K were shown in Fig.2 at high doping concentration ($5*10^{18}$cm^{-3}) and Fig.3 at low doping concentration ($5*10^{15}$cm^{-3}). All the curves follow the expected trend with a decrease in contact resistance with decreasing metal work function. Fig. 2 shows

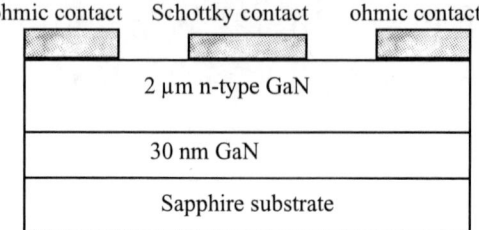

Fig.1 Schematic diagrams of the GaN Schottky diode

that Pt metal has a large shift in high turn on voltage due to higher metal work function and also higher resistance that affect electron transport in GaN layer and is related to the equations below :

$$R_C = \frac{k_B}{qA^*T}\exp\left(\frac{q\phi_B}{k_BT}\right). \quad (1)$$

$$S = \frac{d\phi_B}{dX_M} \quad (2)$$

Where R_c is the contact resistance, k_B is the Boltzmann constant, q the electronic charge, A* is the Richardson constant, T is the temperature, ϕ_B is the Schottky barrier height and X_M the metallic work function and S is the slope parameters. By changing the metal type from higher metal work function (Pt) to lower metal work function (Ag), the shift in turn on voltage to low voltages become clearer and is active in both doping concentration (high and low). Dependence of the Schottky Barrier Height (SBH) on the metal work functions for n-GaN was studied by A.C.Schmitz and John Rennie [9,10] and they found that the Schottky model was applicable to n-GaN where the SBH. was given by the difference between the metal work function and the semiconductor electron affinity. It is found that the turn–on voltage shift towards

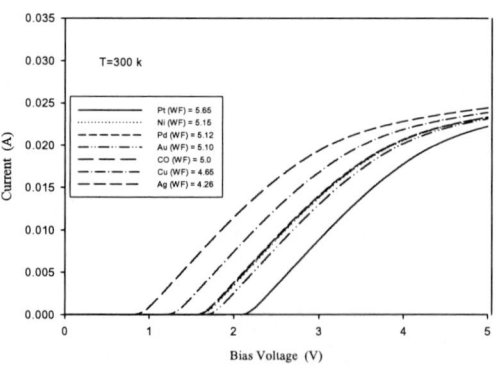

Fig.2 I-V characteristics for different metal work function at high concentration ($5*10^{18}$ cm^{-3})

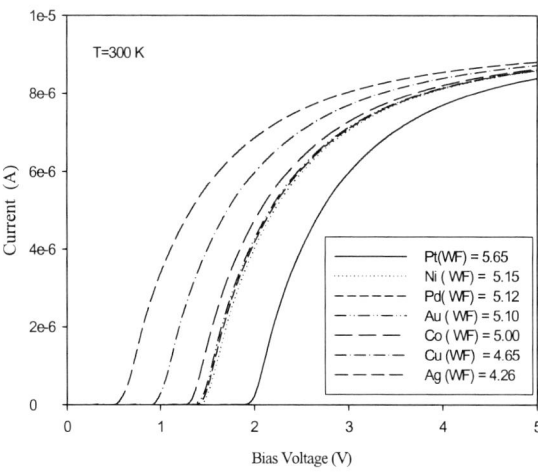

Fig.3 I-V characteristics for different metal work function at low doping concentration ($5*10^{15}cm^{-3}$)

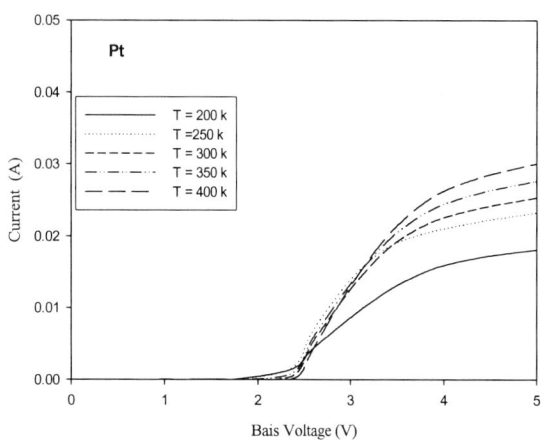

Fig.4 I-V characteristics for Pt metal contact on n-type GaN different operation temperature and high doping concentration ($5*10^{18}cm^{-3}$).

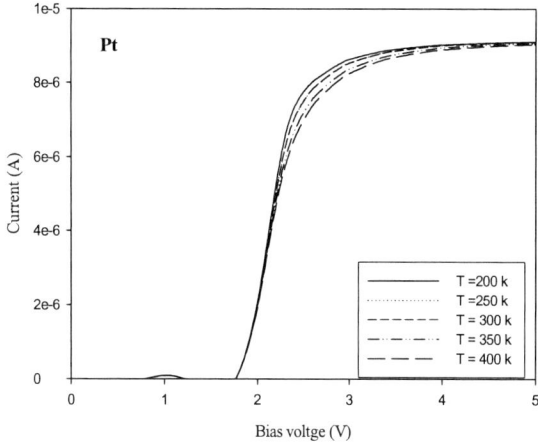

Fig.5 I-V characteristics for Pt metal contact at low doping concentration ($5*10^{15}cm^{-3}$).

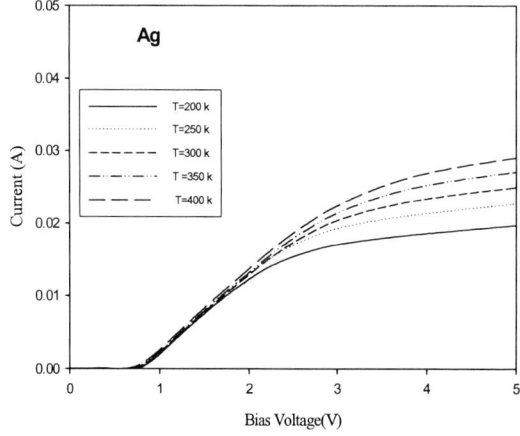

Fig.6 I-V characteristics for Ag metal contact at high doping concentration ($5*10^{18}cm^{-3}$)

low voltages with decreasing doping concentration for all metal type. Fig. 2 shows for the metal Pt that has a higher metal work function, the turn-on voltage starts around 2.6 V at operating temperature of 300 K. Fig .4 shows the I-V characteristics of Pt metal contact at different operating temperatures and at a high doping concentration. It can be seen that by increasing the operating temperature from 200 K to 400 K, the current also increases. This is related to the increase in carrier mobility in heavily doped material in which the effect of impurity scattering is most pronounced. So we suppose that this effect appears with increasing operating temperature and leads to an increase in the current. A Small change in turn-on voltage with increasing temperature was observed in our results. Fig. 5 shows the I-V characteristics of Pt metal contact at lower doping concentration. The opposite behavior was found, that is with increasing operation temperature there is a decrease in current value which is related to the lattice scattering that can be observed strongly for low doping concentration. Our simulation results suggest that the source-drain current of GaN Schottkt diode is dominated by impurity scattering at low doping level.

Fig.6 shows the I-V characteristics of Ag metal contact at different operating temperatures at high doping concentration. It was observed that by increasing the operating temperature from 200 K to 400 K, the current increases from 18 mA to 30 mA, while small changes occur in the turn-on voltage. Fig. 7 shows the I-V characteristics of Ag metal work function at lower doping concentration. It was found that by increasing the operating temperature there is

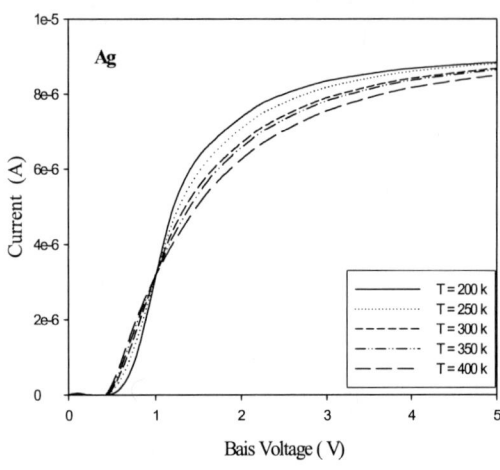

Fig.7 I-V characteristics for Ag metal contact at low doping concentration ($5*10^{15}cm^{-3}$)

a decrease in current and this is related to the same reason that we have mentioned in the Pt metal case.

CONCLUSION

Different metal work functions on n-type GaN at different operating temperatures were simulated. it is observed that a high work function provides high turn-on voltage. These voltages are also affected by doping concentration of GaN achieve layer, having low values at a reduced concentration. Changes in operating temperatures introduced variations in the current for each different metal contact. These changes can be attributed to the effects on carrier's mobility with operating temperatures.

ACKNOWLEDGEMENT

This work was conducted under IRPA RMK-8 Strategic Research grant. The support from Universiti Sains Malaysia and Ministry of Science Technology and Innovation are gratefully acknowledged

REFERENCES

[1] H. Amano, M. Kito, K. Hiramatsu, I. Akasaki, *Jpn. J. Appl.Phys.*, vol. 28, L2112, (1989).

[2] S. Nakamura, *Diamond Rel. Mater.*, vol. **5**, p. 496, (1996).

[3] W. Schottky, Naturwissenschaften, vol. 26, p.843, (1938).

[4] V. Heine, *Phys. Rev.*, vol.138, **A**1689, (1965)

[5] S. M. Sze, *Physics of Semiconductor Devices*, Wiley, New York, (1981).

[6] R. A. Smith, *Semiconductor ,2nd ed.*, Cambridge University Press, London, (1978).

[7] J. L. Moll, *Physics Of Semiconductors*, MCGraw, Hill, New York, (1964).

[8] W.F.Beodle, J.c.Tasai, and R.D. Plummer, Eds., *Quick Reference manual for Semiconductor Engineers*, Wiley, NewYork, (1985).

[9] A. C. Schimitz, A. T.Ping, M. ASIF Khan. *Jouranl of Electronic Materials*, vol. **27**, No.4, (1998).

[10] John Rennie , Masaaki Onomura, Shinya Nunoue, Journal of Crystal Growth 189/190, **pp.** 711-715, (1998).

Power Deduction in Digital Signal Processing Circuit using Inventive CPL Subtractor Circuit

C.Senthilpari Member IEEE, K.Diwakar, C.M.R.Prabhu Member IEEE,
Ajay Kumar Singh Member IEEE

Faculty of Engineering & Technology
Multimedia University
Jalan Ayer Keroh Lama
75450 Melaka, Malaysia
Email: c.senthilpari@mmu.edu.my

Abstract **The proposed 4 bit subtractor circuit is designed by using bit slice method of complementary pass transistor logic (CPL), which is suitable for applications like fast shifting; multiplier/adder in the loop cycle of DSP data processing device. The circuit can perform real time computational tasks at high speed. The designed circuit is performing efficiently in filtering signal at 100-520MHz sampling rate in real time. We have analyzed the proposed circuit for submicron regime, whose layout is designed by using microwind III VLSI CAD tools, in terms of propagation delay, power consumption, power dissipation and area. In 50nm analysis, the proposed circuit is found to dissipate less power (~0.46μW) and possess less overall area of order of 423μm^2. The propagation delay is 9.5206x10^{-12} sec. In DSP architecture, time and area are the critical components which are responsible for the efficient execution of arithmetic and logical operation. Our proposed circuit is enhancing, complete set of instruction cycle, passing at speed of 9.0GHz, which is higher than reported results.**

I. INTRODUCTION

A data processing device is used with peripheral devices having addresses and differing communication response periods. The data processing device includes a digital processor adapted for selecting different ones of the peripheral devices by asserting addresses of each selected peripheral device [1-3]. Addressable programmable registers hold wait state, which values represents different address ranges. Circuitry responsive to an asserted address to the peripheral devices by the digital processor, generates the number of wait states represented by the value held in one of the addressable programmable registers. This corresponds to the one of the address ranges in which the asserted address occurs, thereby accommodating the differing communication response periods of the peripheral devices [4-6]. The subtractor circuit analyzes the data with the help of reference signal and allows the signal to the concerned operation without changing its originality. The subtractor circuit is very useful device in signal/data processing, fast multiplier and propagating signals.

Low-power IC design has become an especially vibrant area of research and development, resulting in advances in low-power fabrication processes and circuit techniques, dynamically programmable power supplies and power efficient microprocessors. A new generation of computer-aided design (CAD) tools, developed especially for power efficient design, is now available to help to engineers to optimize power at most stages of the circuit-development process [2,7-8]. In this paper, we have developed the circuit by DSCH2 and layout is designed by Microwind III CAD tools. Today, power supply levels are typically kept high, to allow maximum clock speed, and are limited only by hot electron effects. It is observed that CPL circuits dissipate less power if they are not switching, the major focus of low-power design is to reduce the switching activity to the minimal level, required to perform the computation [8, 9].

A digital signal processor (DSP) is a key component in communications, medical, military and industrial products [10]. The proposed new subtractor circuit consists of two half adder and one OR gate. The half adder circuit is designed using multiplexing control input technique method, where it is uses only 4-pass transistors and two CMOS inverters. i.e., totally 8 transistors. The full adder circuit is designed using bit slice method which has total 20 transistors for designing one full adder circuit whereas other researchers used 40 transistors for same design. In this paper we have used the 4 bit subtractor circuit using carry save adder method [4, 11-13]. The adder is connected in cascade manner and one of the inputs of the adder circuit is complement of the respective input, so that cascade adder circuit should work as a subtractor circuit. The subtractor circuit performs two's complement subtraction with the inputs.

II. OPERATION OF THE 4-BIT SUBTRACTOR CIRCUIT

A VLSI or ULSI circuit according to the present development preferably includes a pair of data bus capable of conducting in parallel number of signals which can be processed simultaneously by the components on the circuit. Signals on the bus are carries in a time-multiplexed manner, each bus having a predetermined number of time slots. Preferably, each component on the chip is connected to one or both of the bus and is assigned a particular time slot for the bus to which it is connected. Our proposed 4-bit subtractor circuit maintains signal level and feed into DSP processor circuit without any losses. To maintain the signal, our proposed circuit compares the input signal with reference signal [3, 11-12].

A 4-bit subtractor circuit, according to the present development, is a structured design rather than custom design. Accordingly, it can be readily expanded or contracted in the number of signals which can simultaneously process. The number of components which can be included on the chip is limited only by the number of time slots available on the bus to which it is connected. Since addition of such components involves no custom designed interconnections, but merely extension of the bus, chip die area is conserved while design costs are greatly reduced [7, 10-13]. Dsch2 schematic tools develop the proposed circuit and IC layouts are designed by microwind III CAD tool. Furthermore,

the presence of multiplexers at the inputs and/or outputs of elements connected to the bus are unnecessary because the signals thereof are time-multiplexed.

Additionally, the regularity of the bus structure reduces the criticality in timing paths, race conditions and hangs up states. Regularity in the multiplexing scheme used for the busses further reduces the design efforts. The two input adder/subtractor can be readily accommodated by a chips designer after providing two buses. The dual-bus architecture provides ready transfer of data on the chip and off the chip, by appropriate choice of buses, multiplexed timing, as well as for transfer of data between elements on the chip [6, 7-12]. The bit-slice organization of a chip designed, according to the invention, significantly reduces the design effort of the components connected to a bus, since one-bit slice is merely replicated for each conductor of the bus. Ease of expansion along this dimension is achieved at virtually no cost. An overflow detector receives the signal which is generated barrow and difference as an input signal which available from output of subtractor circuit inform of binary numbers. After the overflow detector signal is given to the correction multiplexer circuit for providing the correct signal level according to borrow and difference signal.

III. RESULTS AND DISCUSSION

In our analysis, we have taken different sub micron feature size. The feature size is yielding the gate length of designing circuit transistor and corresponding supply voltage. Our proposed circuit is designed for submicron region and it gives a reduced power consumption as well as dissipation. The proposed subtractor circuit is very much useful for sensing the wanted/unwanted signal for the signal propagation. Our proposed circuit block diagram is shown in fig.1

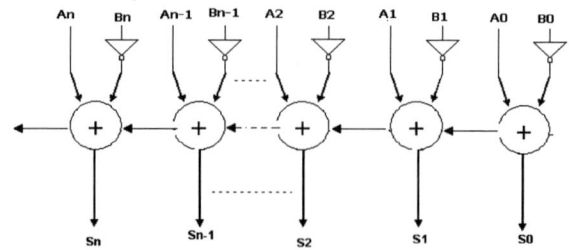

Fig.1 Block diagram of n-bit subtractor circuit

In the proposed circuit due to bit slice method, the number of transistor reduced from 40 to 20 [4, 11]. The power dissipation, Propagation delay, Power delay product and area for the feature size 50nm, 70nm, 90nm, 120nm and 0.35µm is given in table1. In this analysis C_{LOAD} value is fixing by the CAD tool for the different feature size. The 4-bit subtractor circuit maintains the C_{LOAD} value as constant for the different feature size. It is lower than reported results. The range of recent arrivals of digital power (producing by the multiplexer circuit) controllers mark the beginning of, what many believe, an important new trend in power control[12,13]. Digital power controllers allow us to exploit DSP-based filtering methods and build a range of supplies from a common core set of parts with model differentiation managed in software [12, 14-15]. DPWM (Digital-PWM) edge placement requires extremely high speed clocks, which on-chip PLLs generally provide. Though the fastest signals stay on chip, use good high frequency-layout practices to maintain switching-edge fidelity [14-15]. The processor can access signal that determine the loop dynamics and can modify those signal to optimize operation during various normal operating modes, transient events, and faults [16]. The processor's output drives the input to a DPWM (digital PWM), which in turn determines the switching-edge positions. Our proposed bit slice method circuit layout structure is shown in figure 2.our proposed

subtractor circuit is dominant role for detecting the wanted/unwanted signal and send the signal to processor without any losses, which clearly shown in the comparison table 2. The comparison made in terms of speed and power dissipation.

The proposed 4-bit bit-slice method circuit is passing the signal to execute fast shifting. Our proposed circuit is giving more speed than other reported paper [16]. The shifter has speeded approximately 0.52GHz. The proposed circuit is also verified with FFT diagram. The signal is operated at 100MHz/107.5mV, without any loss of power which can be shown in the FFT diagram of fig.3. Timing analysis of a proposed subtractor circuit can be performed either by calculating the output voltage at each time step or by simply calculating the propagation delay and the slope of the output at the half-V_{DD} point in order to define an equivalent signal for the output waveform, that can be fed as input to the next stage. The contribution of the program complexity in terms of region boundaries and branching between them on the total execution time is more intense when only the propagation delay is calculated rather than the full output waveform [17, 18]. The adder/multiplier circuit processes the signal before sending to DSP processor circuit. The adder/multiplier circuit makes some unwanted signal due to interconnection component of layout structure.

Table 1 Power, Delay, PDP and Area of 4-bit CPL subtractor circuit

Feature size	Cload (fF)	Power Dissipation	Propagation Delay (τ_{pd})	Power Delay Product (PDP)	Area µm²
50nm	92.792	0.46µW	9.5206×10^{-12}s	4.379476×10^{-18}	423
70nm	92.557	1.8µW	3.400×10^{-12}s	6.12×10^{-18}	858
90nm	76.360	3.2µW	3.4685×10^{-11}s	1.10992×10^{-16}	897
120nm	82.136	5µW	2.5272×10^{-11}s	1.2636×10^{-18}	1312
0.35µm	116.73	0.10992mW	1.1006×10^{-10}s	1.2098×10^{-14}	22914

Table 2. Comparisons Table

Parameters	Our Proposed circuit	Reference (16)	Reference (17)	Reference (18)
Propagation delay (ps)	110	-	252	132
Feature	0.35	0.35	0.35	0.35
C_L	≈100	100	100	100
Speed	9.0859GHz	2.5GHz	-	-

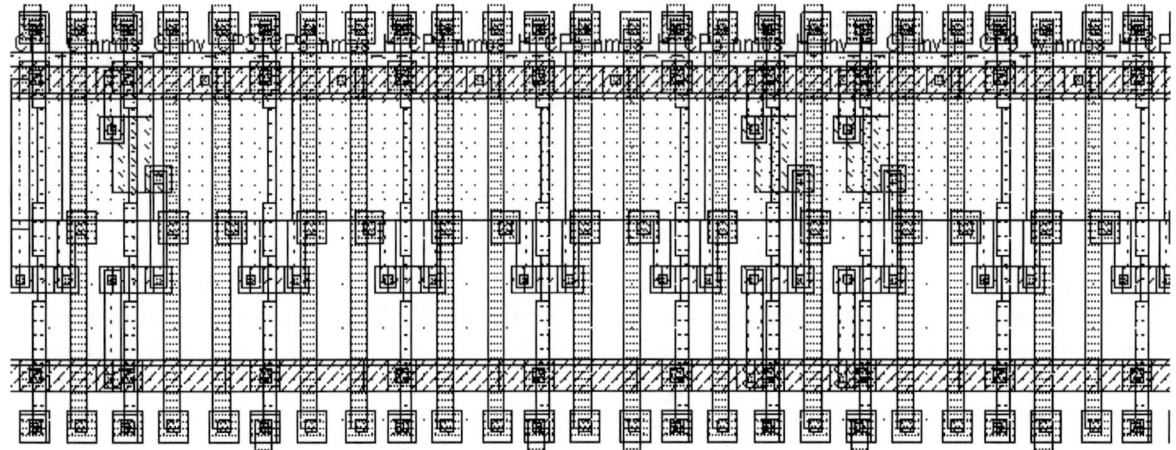

Fig2. Partial Layout structure of the 4-bit subtractor circuit

Our proposed circuit is working as error correction circuit and sends the signal to processor circuit without signal losses. The figure 3 shows for 50nm feature size, $0.5V_{DD}$ and that the signal fed in to the processor is without losses.

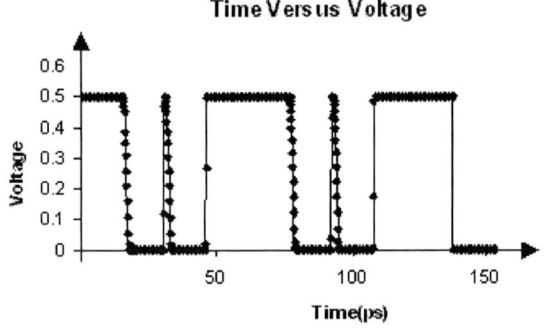

Fig. 3 Time versus Voltage

IV. CONCLUSION

Pass transistor logic is increasingly important for the design of low-power high-performance digital circuits due to the smaller node capacitances and reduced transistor count it offers compare to conventional CMOS logic. The proposed pass transistor 4-bit subtractor circuit is used for the signal propagation in the ALU circuit. The circuit layout is manufactured and analyzed by the microwind III CAD tool. The proposed subtractor circuit, which is designed by using bit-slice method, posses low power dissipation,

less area and low propagation delay compared to other published results. This circuit is very much suitable for the signal propagation in the DSP circuit. Our proposed circuit can be used in the new trend of ADC circuit, and FIR filter for propagating signal without changing its amplitude and frequency stability. One aspect of this issue concerns, timing simulators that can analyze the performance of large circuits at a speed, significantly faster than that of SPICE based tools. The obtained accuracy compared to SPICE simulation results is sufficient for a wide range of input and circuit parameters.

REFERENCE

[1]   Ge Yang, Seong-Ook Jung, Kwang-Hyun Baek, Soo Hwan Kim, Suki Kim, and Sung-Mo Kang " A 32-Bit Carry Lookahead Adder Using Dual-Path All-N Logic" IEEE Transactions on Very Large Scale Integration (VLSI) Systems. Vol.13, No.8, August 2005-pp.992-996.

[2]   Vojin G. Oklobdzija, Bart R. Zeydel, Hoang Q. Dao, Sanu Mathew, and Ram Krishnamurthy "Comparison of High-Performance VLSI Adders in the Energy-Delay Space" IEEE Transactions on Very Large Scale Integration (VLSI) Systems. Vol.13, No.6, June 2005, pp. 754-757.

[3]   Chintan Patel, Fidel Muradali* and Jim Plusquellic. "Power Supply Transient Signal Integration Circuit" Department of CSEE, University of Maryland, Baltimore County, Baltimore, MD 21250, USA * Agilent Technologies, Santa Clara, CA 95051, USA

[4] Fang-shi Lai and Wei Hwang, "Design and Implementation of Differential Cascade Voltage Switch with Pass-Gate (DCVSPG) Logic for High-Performance Digital Systems" IEEE Journal of Solid-state Circuits, Vol.32, No.4, April 1997, pp.563-567.

[5] Alejandro F. GonzaH lez, Pinaki Mazumder "Redundant arithmetic, algorithms and implementations" Integration, the VLSI journal 30 (2000) pp.13-53.

[6] L. Fanucci, M. Forliti, and P. Terreni "FAST: FFT ASICautomated synthesis" Integration, the VLSI journal 33 (2002) 23–37.

[7] Ahmad Hiasat*, Omar Hasan "Bit-serial architecture for rank order and stack filters" Integration, the VLSI journal 36 (2003) 3–12.

[8] Fu-Chiung Cheng, Stephen H. Unger, and Michael Theobald, "Self-Timed Carry-Lookahead Adders" IEEE Transactions on Computers, Vol. 49, No. 7, July 2000 pp. 659-672.

[9] Bharadwaj S. Amrutur and Mark A. Horowitz, "Fast Low-Power Decoders for RAMs" IEEE Journal of Solid-State Circuits, Vol.36, No.10, October 2001 pp.1506-1515.

[10] Morteza Fayyazi, Mohammad R. Movahedin, Zainalabedin Navabi, Pedram A. Riahi and A. Ghalambor-Dezfoli "An Efficient CPU Architecture for DSP Processors" VLSI Circuits & Systems Laboratory, Department of Electrical and Computer Engineering, University of Tehran, Tehran, IRAN.

[11] S.Turgis, N.Azemard, D.Auvergne "Design and Selection of Buffers for Minimum Power-Delay Product" Laboratoire d'Informatique, de Robotique et de Microélectroniquede Montpellier LIRMM UMR CNRS

9928 Un de Montpellier II 161 Rue ADA 34392 Montpellier FRANCE.

[12] K. Yano, T. Yamanaka, T. Nishida, M Saito, K. Shimohigashi, A. Shimizu, "A 3.8-ns CMOS 16x16-b Multiplier Using Complementary Pass-Transistor Logic", IEEE Journal of Solid-State Circuits, vol.25, 1990, pp. 388-395.

[13] S. Turgis and D. Auvergne "A Novel Macromodel for Power Estimation in CMOS Structures" IEEE Transactions on computer – Aided Design of Integrated circuits and sustems, Vol.17, No. 11 November 1998 pp.1090-1098.

[14] Joshua Israelsohn "A BIT-O'-POWER: digitally controlled power conversion" EDN magazine July 2005 pp. 59-66.

[15] Rajamohana Hegde and Naresh R. hanbhag "Low-Power Digital Filtering via Soft DSP" ECE Department/Coordinated Science Laboratory University of Illinois, Urbana, IL 61801.

[16] Adiseno, Håkan Magnusson and Håkan Olsson "A 1.8-V Wide-Band CMOS LNA for Multiband Multistandard Front-End Receiver" Radio Electronics-LECS/IMIT, Royal Institute of Technology (KTH), Sweden.

[17] S. Nikolaidis, H. Pournara and A. Chatzigeorgiou "Output Waveform evaluation of Basic Pass Transistor Structure" B. Hochet et al. (Eds.): PATMOS 2002, LNCS 2451, 2002, pp. 229–238.

[18] A.Chatzigeorgiou and S. Nikolaidis "Efficient output waveform evaluation of a CMOS inverter based on short-circuit current prediction" Int. J. Circ. Theor. Appl. 2002; 30:pp. 547–566

Design and Synthesis of Mobile Robot Controller using Fuzzy

Md. Shabiul Islam, Md. Anwarul Azim, Md. Saukat Jahan, *Masuri Othman
Faculty of Engineering, Multimedia University, 63100 Cyberjaya, Selangor, Malaysia.
*Institute of Microengineering and Nanoelectronics, University Kebangsaan Malaysia, 43600 UKM,
Bangi, Selangor, Malaysia.
E-mail: shabiul@mmu.edu.my, azimcse@gmail.com, saukt_bd@yahoo.com, masuri@eng.ukm.my

Abstract **This paper describes a Fuzzy Logic Controller (FLC) algorithm for designing an autonomous mobile robot controller (MRC). The controller enables the robot to navigate in an unstructured environment and that avoid any encountered obstacles without human intervention. The autonomous mobile robot is found to be able to react to the environment appropriately during its navigation to avoid crashing with obstacles by turning to the proper angle while moving. The Fuzzy Logic algorithm has proven a commendable solution in dealing with certain control problems when the situation is ambiguous. One of the main difficulties faced by conventional control systems is the inability to operate in a condition with incomplete and imprecise information. As the complexity of a situation increases, a traditional mathematical model will be difficult if not impossible to implement. Fuzzy Logic is a tool for modeling uncertain systems by facilitating common sense reasoning in decision-making in the absence of complete and precise information. In this paper, the controller of an autonomous mobile robot is designed based on the theories of Fuzzy Logic. The wheeled robot is able to navigate by itself in a completely unstructured environment. The codes of MRC has written for implementing the separate modules of the Fuzzifier, Fuzzy Rule Base, Inference mechanism and Defuzzifier as hardware blocks.**

A behavioral model of MRC algorithm is first developed in MATLAB session with numerous data to evaluate its algorithm functionality. The development of MATLAB codes has converted into VHDL codes for hardware implementation. Comparison results between MATLAB and VHDL of MRC algorithm also presented. Then the VHDL codes are synthesized using synthesis tool, known as Quartus II. Finally the MRC hardware blocks for VLSI design have been carried out.

Keywords: **Fuzzy rules, Mobile robot controller,Navigation algorithm, VHDL, Synthesis.**

I. INTRODUCTION

Uncertain environments with incomplete and imprecise information pose fundamental difficulties to conventional control systems. In this case, the feasibility of applying Fuzzy logic to facilitate common sense reasoning in decision-makings to counter such problems. One of the main difficulties is faced by conventional control systems are the inability to operate in a condition with incomplete and imprecise information. As the complexity of a situation increases, a traditional mathematical model will be difficult for implementing the process. Fuzzy Logic is a tool for modeling uncertain systems by facilitating common sense reasoning in decision-making in the absence of complete and precise information [1-3]. It enables the arrival of a definite conclusion based on input information, which is vague, ambiguous, noisy and inaccurate.

In this paper, the controller of an autonomous mobile robot has designed based on the theories of Fuzzy Logic [4-5]. The wheeled robot is able to navigate by itself in a completely unstructured environment. The FLC receives limited information of the environment through sensors, and decides an appropriate angle to turn while it

0-7803-9730-4/06/$25.00 ©2006 IEEE

moves with constant velocity to avoid any objects within its vicinity. The main focus is to develop the modeling of Fuzzy rule based algorithm for MRC.

The rest of the paper is organized as follows: Section II describes the modeling of mobile robot controller with navigation algorithm. MATLAB simulation results are presented in section III. Section IV presents the VHDL modeling. Section V presents the Synthesis. Finally conclusion is given in this paper.

II. MODELING OF MOBILE ROBOT CONTROLLER WITH NAVIGATION ALGORITHM

The MRC monitored by the Fuzzy Logic algorithm for navigating through an unstructured environment without human intervention [6-7]. In this process of moving from one place to another, it is necessary for the robot to have information on its current position, where the intake of this vital information is through the two sensors on its body. Figure 1 shows the physical modeling of the MRC.

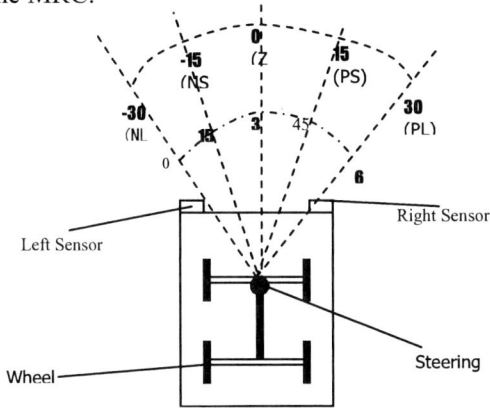

Figure 1: Modeling of the MRC

As an example, the Mobile Robot resembles of a car, in which navigation is made possible by four wheels, controlled by the steering. A significant difference of steering is controlled by the FLC instead of human intelligence. The sensors are positioned in such a way that the robot is able to detect its distance from obstacles, one from its left and another from its right. The steering has a maximum turning angle of 30 degrees. It should be noted that the robot moves in a constant velocity and the only variable

controlled by the FLC is the orientation of the robot (the steering angle, in degrees).

The FLC receives two inputs pattern, one from the Front Left Sensor, and another from the Front Right Sensor. The inputs are measured in distance. The output is an orientation of the steering, measured in degrees.

Inputs pattern: The inputs from both the sensors measures distance by a numerical value ranging from 0 to 40. An input of "0" denotes the minimum possible distance, which can be detected, where the robot almost or actually touches the obstacle. An input of "40" denotes the maximum distance as the obstacle is either far away from the robot's vicinity or could not be sensed at all.

Output pattern: Referring to Figure 1, the Figure shows that the robot is capable of turning an angle of -30 to 30 degrees. However, the output of the FLC, like the inputs is also a numerical value, ranging from 0 to 60. The 0 value represents -30 degrees while "60" represents 30 degrees.

In short, the input parameter for both the left and right sensors is the linguistic variable distance, while the output parameter is the linguistic variable angle. They are modeled by the following sets:

Distance1(left sensor)$A = \{ A1,........,A3\} = \{ far, near, vnear\}$
Distance2 (right sensor)$\cong B = \{B1,........,B3\} = \{ far, near, vnear \}$
Angle $\cong C = \{C1........,C6\} = \{ NL, NS, Z, PS, PL \}$

Both input and output membership functions are identicalwhich shown in Figure 2 Figure 3 respectively.

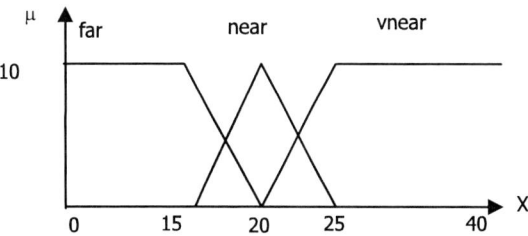

Figure 2: Left and right sensor input membership function

835

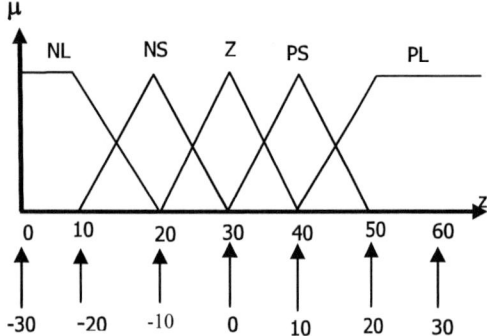

Figure 3: Output membership function

The navigation of the mobile robot is fairly simple. "Structure of the Mobile Robot" and "Fuzzy Logic Controller Architecture" has generally given an idea of the movements of the robot. The sensors are the "eyes" of the robot. They give information on how far the robot is from an object. The FLC is its "brain". It decides on which angle to turn in order to avoid those obstacles.

The FLC takes in a crisp input from each of the sensors. The Fuzzifier categories focus the inputs into Fuzzy sets according to the input membership function. The Rule Base determines the behavior of the robot and decisions are made depending on the rules. The Inference Engine determines rule at a specific situation. The Deffuzifier converts the decision made according to the rule-base, which is in Fuzzy terms, into a crisp output. The output is an orientation of the robot (decides which angle to turn).

The algorithm flow chart is shown in Figure 4.

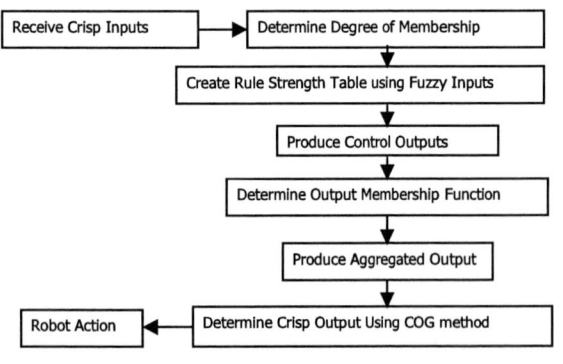

Figure 4: Flow chart of MRC algorithm

Finally, the FLC gathers information through its sensors, and decides on an action (the output) which is the turning angle of the robot to avoid

the obstacles. In fact, after the intake of information (numerical values) through the sensors, Fuzzification takes. The values into linguistic variables (far, near, very near), using the input membership functions. Decisions will be made complying to a Rule Base, the rules decides on each situation, to take an action of either giving the robot a "negative large (NL)", "negative small (NS)", "zero", positive small (PS)" or "positive large (PL)" orientation. This is yet not the end of the process because the system requires an exact value as an output, a crisp output. This is achieved by Defuzzification, applied on the output membership function. An autonomous mobile robot is responsible of navigating by itself without human intervention (remote controller etc.). This is achieved using a FLC Algorithm. The computed MRC using FLC algorithm is shown in Figure 5.

Evabot.m

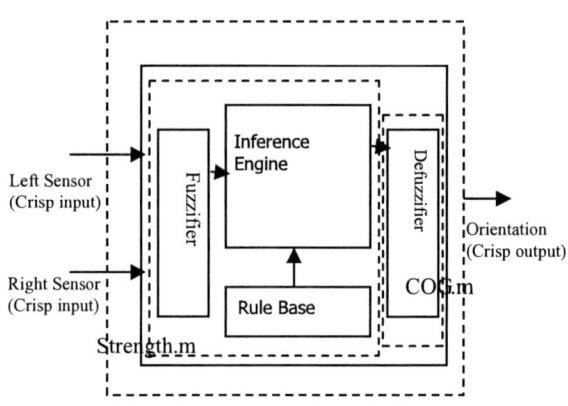

Figure 5: FLC architecture

III. MATLAB SIMULATION

The simulation result in MATLAB is partially graphical part, which consist of the graphs that show the shape of the input membership functions (one for the left sensor and another for the right sensor), the graph that shows the shape of the output membership function, and lastly the graph that shows the aggregated output (final output). The more important part of the result is the values computed for: i) degree of membership ii) Rule Strengths iii) Aggregated Output (this is the final output, which is the orientation of the robot in degrees). To obtain a crisp output using COG method are given below.

$$AGG(i) = \max([\min([cut1 \quad Z(i)]) \quad \min([cut2$$
$$NS(i)]) \quad \min([cut3 \quad NL(i)])$$
$$\min([cut4 \quad PS(i)]) \quad \min([cut5$$
$$Z(i)]) \quad \min([cut6 \quad NS(i)])$$
$$\min([cut7 \quad PL(i)]) \quad \min([cut8$$
$$PS(i)]) \min([cut9 \; Z(i)])]);$$
$$den = den + AGG(i);$$
$$num = num + i*AGG(i);$$
$$output = num/den$$

In the above equation, the control outputs are obtained. Here "den" denotes the denominator of the COG equation while "num" denotes the numerator. Dividing "num" by "den" produces the final output. For different case of navigating the robot, the controller takes input, such as left sensor (LS) and Right sensor (RS) for different foreseen situation. Finally, the simulation results are given in Table 1.

Table 1: Simulation result of robot control using MATLAB

Case 1 LS=0 RS=0	Case 2 LS=16 RS=21	Case 3 LS=21 RS=16	Case 4 LS=10 RS=35	Case 5 LS=35 RS=10
Y=10,0,0	Y=8,2,0	Y=0,8,2	Y=10,0,0	Y=0,0,10
Z=10,0,0	Z=0,8,2	Z=8,2,0	Z=0,0,10	Z=10,0,0
S1=0	S1=0	S1=0	S1=0	S1=0
S2=0	S2=8	S2=0	S2=0	S2=0
S3=0	S3=2	S3=0	S3=10	S3=0
S4=0	S4=0	S4=9	S4=0	S4=0
S5=0	S5=2	S5=2	S5=0	S5=0
S6=0	S6=2	S6=0	S6=0	S6=0
S7=0	S7=0	S7=2	S7=0	S7=0
S8=0	S8=0	S8=2	S8=0	S8=0
S9=0	S9=0	S9=0	S9=0	S9=0
num=3100	num=2860	num=5758	Num=1320	num=8290
den=100	den=139	den=139	Den=155	den=155
ans=31	ans=20.5755	ans=41.4245	Ans=8.5161	ans=53.4839

IV. Vhdl Modeling

The structure in MATLAB consist of three modules (evabot.m, strength.m and COG.m) are implemented in VHDL as evabot.vhd, STR3.vhd, COG.vhd respectively.

The modules are combined to form the structure of the FLC controller which is shown in Figure 6.

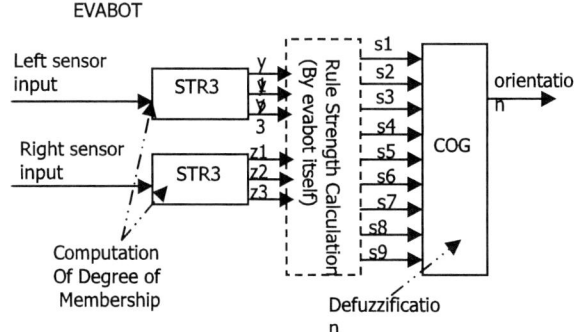

Figure 6: Design Structure in VHDL

The two STR3 modules separately receives one external input each and produces the degree of membership as y1,y2,y3 and z1,z2,z3. The "rule strengths" are produces by the evabot itself, as s1, s2, s3, s4, s5, s6, s7, s8 and s9 as internal signals. These are sent to COG for defuzzification and the final output is produced as the orientation of the robot.

For each of the cases, it is observed that the "degree of membership" and the "rule strengths" obtained from Quartus II simulation is exactly the binary representation of the values obtained in the MATLAB simulation. However, the results of the final output, which is the orientation of the robot simulated in Quartus II varies slightly with the ones is MATLAB, shown in Table 2.

Table 2: Comparison of Results

Foreseen situation	VHDL	MATLAB
Case 1 (0,0)	31	31
Case 2 (16,21)	20	20.5755
Case 3 (21,16)	41	41.4245
Case 4 (10,35)	8	8.5161
Case 5 (35,10)	53	53.4839

V. Synthesis

Synthesis is the process of generating a logic circuit from a schematic diagram or an initial behavioral description of Hardware Description Language. It can be seen as an automatic method of converting a higher level of abstraction to a lower level abstraction. The output is a set of logic expressions that describe the logic functions needed to realize the circuit. The Register Transfer Level (RTL) description is converted into a gate netlist, which makes downloading the codes into a target device.

The top level of RTL view of the controller is shown in Figure 7 and the technology map view of COG shown in Figure 8.

The design codes of the Controller will be downloaded into Altera APEX 20K200EF484 FPGA board. The APEX 20K200EF484 consists of 8320 logic elements (LE), 106496 memory blocks and 2 Phase Lock Loops (PLL).

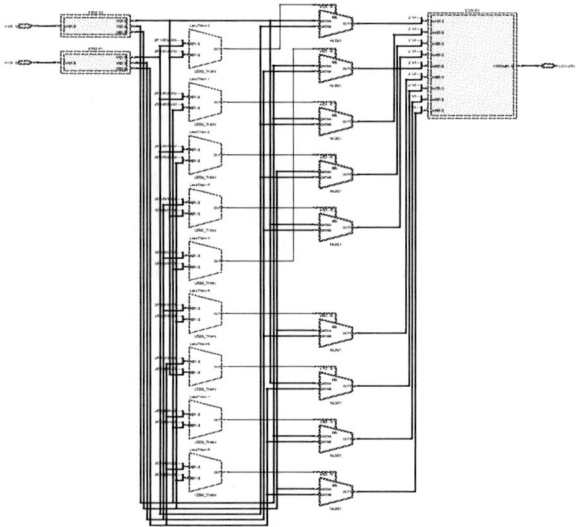

Figure 7: RTL View of MRC blocks.

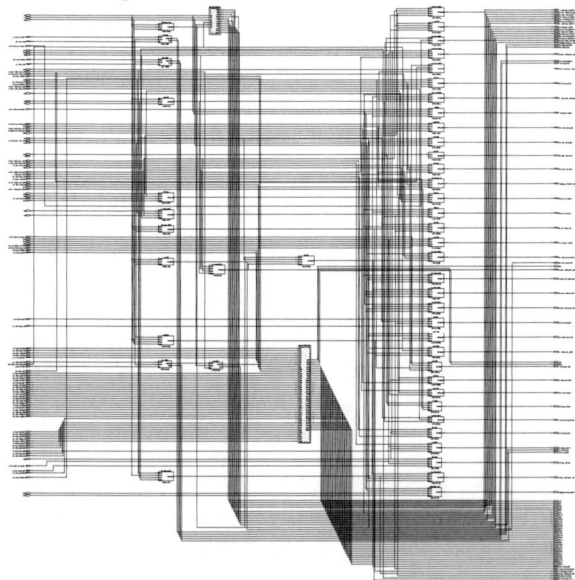

Figure 8:Technology Map View of COG

CONCLUSION

The design of MRC has described by the navigation algorithm successfully. The construction of the MRC

has designed and also presented using Fuzzy algorithm. The constructed MRC has simulated with MATLAB and presented its results in this paper.

The constructed algorithm in MATLAB has been properly translated to VHDL and then Synthesiged it to design MRC controller's hardware blocks for VLSI design. The simulation results shows that the robot reacts accordingly to the environment. It turns to the appropriate angle to avoid obstacle within its vicinity while navigating the completely unstructured environment.

REFERENCES

[1] Alessandro Saffioti, "The Handbook of Fuzzy Computations", Fuzzy Logic in Autonomous Robot Navigation: a case study, Oxford University Press,1991.

[2] Ellen Thro, "Robotics-The Marriage of Computers and Machines", Facts on Files Inc, 1993.

[3] Francois G. Pin and Yutaka Watanabe, "Autonomous Robotic Systems Group", Automatic Generation of Fuzzy Rules For the Sensor-Based Navigation of A Mobile Robot, 2002.

[4] Steven D. Kaehler, Fuzzy Logic–An Introduction. March 2006.

[5] George Bojadziev, Maria Bojadziev, "Fuzzy Sets, Fuzzy Logic, Applications,"Vol 5,World Scientific, 1995.

[6] John Lovine, "Robots Androids and Animatrons," McGraw-Hill, 1998.

[7] Marley Maria B.R. Vellasco, Marco Aurelio and Ivo Lima Brasil, "Mobile Robot Control Using Fuzzy Logic," In conf. rec 2000 ica conf. Fuzzy_rozhodovanie.

Development of a Fuzzy Logic Controller Algorithm for Air-conditioning System

Md. Shabiul Islam, Md. Shakowat Zaman Sarker, Kazi Ashique Ahmed Rafi and Masuri Othman*
Faculty of Engineering, Multimedia University, 63100 Cyberjaya, Selangor, Malaysia.
* Institute of Microengineering and Nanoelectronics, Universiti Kebangsaan Malaysia, 43600 Bangi,
Selangor, Malaysia.
E-mail: shabiul@mmu.edu.my, masuri@eng.ukm.my

Abstract **The goal of this paper is to develop an algorithm of fuzzy logic controller (FLC) for automatic air-condition controlling system. The Fuzzy logic system is used to design this algorithm. Two inputs and one output are designed with an industrial application in mind. This system consists of two sensors for feedback control: one to the monitor of temperature and another one to the monitor of humidity. There are three control elements: cooling valve, heating valve, and humidifying valve, to adjust the temperature and humidity of the air supply. Fuzzy rules are formulated by temperature and humidity. The model of this controller algorithm has been simulated using MATLAB simulation. Finally, the developed algorithm has been designed for implementing the hardware VLSI chip using VHDL language from EDA tools.**

Keywords: VLSI, MATLAB simulation, VHDL, Synthesis, FLC.

I. INTRODUCTION

The FLC is one of the most useful approaches for utilizing the qualitative knowledge of a system to design a controller. Fuzzy logic control is generally applicable to plants that are mathematically poorly modeled and where the qualitative knowledge of experienced operators is available for providing qualitative control. The FLC techniques represent a means of both collecting human knowledge and expertise and dealing with uncertainties in the process of control [1]. On the basis of this idea, some fuzzy models based fuzzy control system design methods have appeared in the fuzzy control field

[2], [3], [4], [5] and [6]. The industrial applications of fuzzy controller (refrigerator control) are proposed by Jung Ho Kim and et all. [7]. In this paper we propose a FLC for air-condition system controlling.

The paper is organized as follows: Section I describes introductory of this paper. Section II describes the basic concept of fuzzy logic controller. The sections III describes the algorithm of this work. Section IV presents the experimental results as well as Section V reports conclusion of this paper.

II. FUZZY LOGIC CONTROLLER BASIS

Fig. 1 shows the basic configuration of a FLC. It comprises four principle components: a fuzzification interface, a knowledge base, decision-making logic, and a defuzzification inference.

The fuzzification interface involves the following functions: (i) measures the values of input variables (ii) performs a scale mapping that transfers the range of values of input variables into corresponding universes of discourse and (iii) performs the function of fuzzification that converts input data into suitable linguistic values, which may be viewed as labels of fuzzy sets.The knowledge base comprises knowledge of the application domain and the attendant control goals. It consists of a "database" and a "linguistic (fuzzy) control rule base"(i) the database provides necessary definitions, which are used to define linguistic control rules and fuzzy data manipulation in an FLC and (ii) the rule base characterizes the control goals and control policy of the domain experts by means of a set of linguistic control rules.The decision-making logic is the kernel of a FLC. It has the capability of simulating human

0-7803-9730-4/06/$25.00 ©2006 IEEE

decision-making based on fuzzy concepts, implication and the rules of inference in fuzzy logic.The defuzzification inference performs the following functions: (i) scale mapping, which converts the range of values of output variables into corresponding universe of discourse and (ii) defuzzification, which yields a non-fuzzy control action from an inferredcontrol action. A defuzzifier converts an inferred fuzzy control action into a crisp one.

III. ALGORITHM

The Fuzzy air conditioner controller consists of two inputs and one output namely the temperature and the humidity. To control the room temperature, the controller reads the room temperature after every sampling period; to control the humidity, the controller calculates the room's humidity in the same fashion as the temperature is calculated. There are five triangular Membership Functions (MF) that are equally determined over a scale range of 0°C to 40°C for the temperature input and 0% to 100% relative humidity (RH) for the humidity input. The five Fuzzy Variables for the temperature and relatively humidity are represented in Table I and Table II respectively.

Based on temperature and humidity, the fuzzy logic model aims to determine the amplitude of the voltage signal (for example, from 0 to 5V) that is necessary to be sent to the motor fan speed in order to maintain a constant and desired temperature in a confined environment. This is provided by an output signal from the fuzzy model termed as Fan_Speed, with a scale range of 0 to 100. Similar to the inputs, the output Fan_Speed signal has 5 triangular membership functions spaced over this range, which are "Stop", "Slow", "Medium", "Fast" and "Blast". The complete fuzzy rule of this designed fuzzy system is given in Table III. The example of fuzzy rule is:

"IF Temperature is Cold AND Humidity is Dry THEN Fan Speed is Stop"

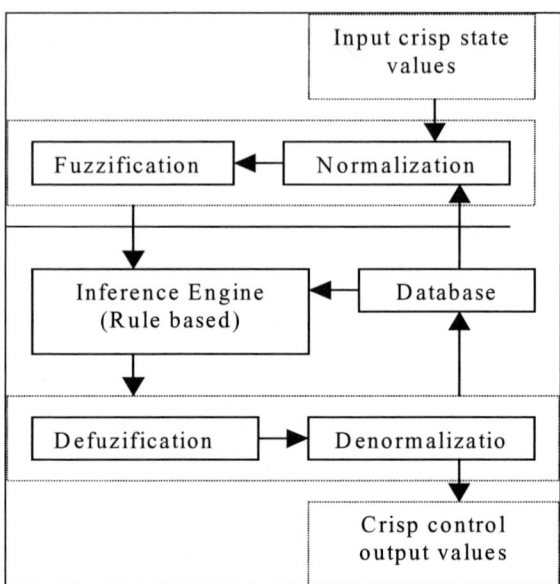

Fig. 1 The basic configuration of a fuzzy Logic controller

TABLE I
MEMBERSHIP FUNCTIONS OF INPUT FUZZY VARIABLE "TEMPERATURE"

Membership Function (MF)	Temperature input (C)
Cold	0°C to 10°C
Cool	0°C to 20°C
Just Nice	10°C to 30°C
Warm	20°C to 40°C
Hot	30°C to 40°C

TABLE II
MEMBERSHIP FUNCTIONS OF INPUT FUZZY VARIABLE "HUMIDITY"

Membership Function (MF)	Humidity input (RH)
Dry	0% to 25%
Not Too Dry	0% to 50%
Moist	25% to 75%
Not Too Wet	50% to 100%
Wet	75% to 100%

IV. EXPERIMENTAL RESULTS

A. Matlab simulation

The MATLAB fuzzy logic toolbox is used for fuzzy air condition system design. The fuzzy input parameters of Temperature = 17°C and Humidity =

70% RH are considered to design the proposed air conditioning system.

TABLE III
THE COMPLETE FUZZY RULES OF THIS DESIGNED FUZZY SYSTEM

Rules	Temperature (input)	Humidity (input)	Output fan speed
1	Cold	Dry	Stop
2	Cold	Not Too Dry	Stop
3	Cold	Moist	Stop
4	Cold	Not Too Wet	Stop
5	Cold	Wet	Slow
6	Cool	Dry	Stop
7	Cool	Not Too Dry	Stop
8	Cool	Moist	Stop
9	Cool	Not Too Wet	Slow
10	Cool	Wet	Slow
11	Just Nice	Dry	Slow
12	Just Nice	Not Too Dry	Slow
13	Just Nice	Moist	Medium
14	Just Nice	Not Too Wet	Medium
15	Just Nice	Wet	Fast
16	Warm	Dry	Medium
17	Warm	Not Too Dry	Fast
18	Warm	Moist	Fast
19	Warm	Not Too Wet	Fast
20	Warm	Wet	Blast
21	Hot	Dry	Fast
22	Hot	Not Too Dry	Blast
23	Hot	Moist	Blast
24	Hot	Not Too Wet	Blast
25	Hot	Wet	Blast

The signal value of Temperature = 17^oC intersects with fuzzy variables "Cool" and "Just Nice", where "Just Nice" is taken as the first fuzzy variable f[0] and "Cool" is the second fuzzy variable, f[1]. The f[0] maps to the membership function value of 0.7000 while f[1] maps to the value of 0.3000. Similarly, for the input value of Humidity = 70% RH, the corresponding intersection of fuzzy variables are "Not Too Wet" as the first active fuzzy variable f[2] and "Moist" as the second active fuzzy variable f[3]. The f[2] will thereby map to the membership function value of 0.8000 while f[3] corresponds to 0.2000.

The inference block accepts four inputs from the Fuzzification process and the inputs of the inference would be f [0], f [1], f [2] and f [3]. Applying the max-min composition, f[0] and f[1]

will be performed the operation with f[2] and f[3] to output the R values as follow:

R0 = f [0] AND f [2] = 0.7 AND 0.8 = 0.7
R1 = f [0] AND f [3] = 0.7 AND 0.2 = 0.2
R2 = f [1] AND f [2] = 0.3 AND 0.8 = 0.3
R3 = f [1] AND f [3] = 0.3 AND 0.2 = 0.2

The rule selector takes in two crisp inputs X0 (Temperature) and X1 (Humidity). Their are 4-rules fired from the module they are S0, S1, S2 and S3. This process is achieved by dividing the universe of discourse into four regions; with each region containing only two fuzzy variables. The values of four region divisions have been shown in Table IV. The output of fan speed has been represented as singleton values in Table V.

TABLE IV
DIVISION OF REGIONS FOR INPUT TEMPERATURE AND HUMIDITY

Input	Regions	Range of Values
Temperature (X0) (^{o}C)	1	0 - 10
	2	10 - 20
	3	20 - 30
	4	30 - 40
Humidity (X1) (% RH)	1	0-25
	2	25-50
	3	50-75
	4	75-100

TABLE V
SINGLETON VALUES FOR THE FUZZY OUTPUT

Rule	Fan Speed	Singleton Value	
8	Stop	0	S0 = 0
9	Slow	30	S1 = 0.3
13	Medium	50	S2 = 0.5
14	Medium	50	S3 = 0.5

The Defuzzification process takes in four input values from the inference engine which are the R0, R1, R2 and R3 as well as four input values from the rule selector, namely S0, S1, S2 and S3. The process then produces the defuzzified or crisp output signal, "Fan Speed" used to control the actuators of the control system. The Defuzzification scheme chosen was the Center of

Average (COA) method as explain previously, expressed mathematically as $\sum_i S[i] \times R[i] \Big/ \sum_i R[i]$.

The input values determined in the previous sections, for Temperature = 17°C and Humidity = 70% RH, the defuzzied value can be obtained as shown in Table VI.

TABLE VI
CALCULATIONS ON DEFUZZIFIED VALUE

i	S_i	R_i	$S_i \times R_i$
0	0	0.7	0
1	0.3	0.2	0.06
2	0.5	0.3	0.15
3	0.5	0.2	0.10
		$\sum = 1.4$	$\sum = 0.31$

From the above table, the defuzzified output value is,

$$\text{Fan Speed} = \sum_i S[i] \times R[i] \Big/ \sum_i R[i] = 0.31 / 1.4 = 0.2214$$

Percentage of Fan Speed to be applied = Output of Defuzzifier * 100 = 0.2214*100=22.14%.

B. VHDL Simulation

The following Table VII demonstrates 8 test pairs that were fed into the VHDL model to validate the model as well as collect qualitative information on it. The VHDL simulation waveform is shown in Fig. 2.

The data in the table was collected from the waveform generated by pumping in the stimulus. The waveform shows the two fuzzy inputs, such as temperature and Humidity; the fuzzy variable Fan Speed; the four outputs of the Fuzzification module labelled as Mem_F0, Mem_F1, Mem_F2, Mem_F3; and the outputs of the Inference module.

TABLE VII

TEST OUTPUTS OF VHDL FUZZY MODEL

Temperature (°C)	Humidity (% RH)	Fan_Speed
0	0	0
20	50	54
40	100	94
20	100	44
30	50	83
17	70	24
25	30	60
28	65	72

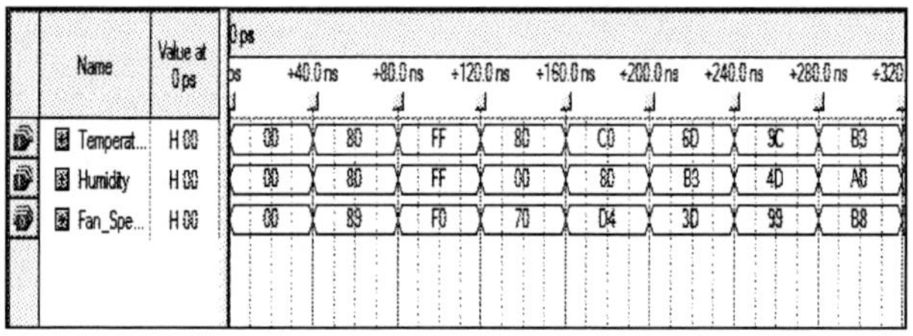

Fig. 2 VHDL Model Functional Simulation waveforms

Let us consider the values of Temperature to be 40°C and Humidity to be 100% RH. The first active membership function value f[0] is given as "1" while the second active membership function values f[1] is given as "0". Both these values represent the corresponding degree of fulfilment to the active fuzzy variables in the universe of discourse of Temperature. The following two values f[2] and f[3] will be indicated the degree of fulfilment for the universe of discourse of Humidity, each having a value of "1" and "0". Since both Temperature of 40°C and speed of 100% RH are in region 4, combination 16 of the active rules table will be chosen where only rule 16 will be fired thus leading to S0 having a singleton value of 100 which represents the fuzzy variable "Blast" for the membership function of Fan Speed. S1 till S3 all contain values of 0's as there is no other rules fired, therefore causing no representation of singleton values there. Fan Speed output of the first set of inputs is 94 which is logical since very not temperature (though in our country this is not practical) and very wet humidity may require fan speed to blast in order to cool down the environment.

To further verify with the results obtained, the Fan Speed output values are analyzed to their corresponding input sets. The previous set of inputs has been clarified to have logical Fan Speed output. Now, considering the second set of inputs with Temperature is to be 17°C and Humidity goes 70% RH, the Fan Speed to be applied as per the simulated VHDL results would be 24. The result obtained is again logical since at cooling temperature with a not-too-wet humidity, the motor fan speed applied should be very slow thus leading to a result of 24. The next set to be taken into consideration has inputs with Temperature as 28°C and Humidity as 65% RH which lead to a motor fan speed output of 72. This is still proved to be reasonable since temperature intersecting with "Just-nice" and "Warm"; humidity intersecting with fuzzy variables of "Moist" and "Not-too-wet" thus gives rise to a "Fast" fan speed output at 72. Suffice to say, all the input sets of the simulated VHDL model produce logical and fast as well as consistence results.

V. CONCLUSION

The development of algorithm for fuzzy air condition controller system has been presented in this paper. The designing and developing part of air conditioner fuzzy control system is providing more efficient operation at reduced energy cost while providing a more comfortable working environment. The difuzzified value of 0.2214 is more satisfactory.

Finally, the designed code in Matlab has converted into VHDL codes for designing Hardware VLSI chip for air-conditioning controlling system in future. In this steps, synthesis and FPGA board will be considered for implementing purpose.

REFERENCES

[1] S.G. Cao, N.W. Rees, G. Feng "Analysis and Design of Fuzzy Controller and Fuzzy Observer," *IEEE Transactions on fuzzy systems*, Vol. 6, No. 1, Feb. 1998.
[2] C. L. Chen, P. C. Chen, and C. K. Chen, "Analysis and design of fuzzy control system*," Fuzzy Sets Syst.,* vol. 57, pp. 125–140, 1993.
[3] K. Tanaka, "Stability and stabilizability of fuzzy-neural-linear control systems," *IEEE Trans. Fuzzy Syst.,* vol. 3, pp. 438–447, Nov. 1995.
[4] K. Tanaka and M. Sano, "Fuzzy stability criterion of a class of nonlinear systems*," Inform. Sci.,* vol. 71, pp. 3–26, 1993.

[5] K. Tanaka and M. Sugeno, "Stability analysis and design of fuzzy control systems," *Fuzzy Sets Syst.,* vol. 45, pp. 135–156, 1992.
[6] H. O.Wang, K. Tanaka, and M. F. Griffin, "An approach to fuzzy control of nonlinear systems: Stability and design issues," *IEEE Trans. Fuzzy Syst.,* vol. 4, pp. 14–23, Feb. 1996.
[7] J.H. Kim, K.S. Kim, M.S. Sim, K.H. Han, B.S. Ko, "An   application of fuzzy logic to control the refrigerant distribution for the multi-type air conditioner," *Fuzzy Systems Conference Proceedings, 1999. FUZZ-IEEE '99. IEEE International*, Vol. 3, 22-25,1999, pp. 1350 – 1354.

Design and Nonlinear Analysis of High-Gain and Broad-Band Distributed Power Amplifier with Traveling-Wave Gain Stages By Harmonic Balance Method

S. Asadi, A. Abdipour, A. Tavakoli, and G. Moradi
Microwave/mm- wave & Wireless Communication Research Lab., Electrical Engineering
Department, Amirkabir University of Technology, Tehran, IRAN
Tel: +98-21-66466009, Fax: +98-21-66406469, E-mail: Abdipour@aut.ac.ir

Abstract —**This paper demonstrates a three-stage hybrid-distributed amplifier based on Darlington configuration to achieve high gain and over several octaves of bandwidth for microwave and millimeter-wave frequency applications. The Darlington single-stage distributed amplifier (DSSDA) is used as traveling wave gain stages to improve the gain performance of the conventional distributed amplifier. This configuration achieves $14^{\pm 1}$ dB small signal flat gain from 1 to 40 GHz and output power at 1-dB gain compression is $19^{\pm 1}$ dBm over the desired band.**

I. INTRODUCTION

The distributed amplifier (DA) in Fig.1 represents the typical circuit when very wideband performance is required [1]-[3]. This DA initially consists of two transmission lines on the input and the output and multiple transistors providing gain through multiple signal paths.

Each transistor amplifies the forward wave on the input and appends power in phase to the signal at each tap point on the output line .The forward traveling wave on the gate line and the backward traveling wave on the drain line are absorbed by terminations matched to the loaded characteristic impedance of the input line, denoted R_{in}, and the loaded output line impedance, denoted R_{out}, in order to avoid reflections. Therefore different methods for the design of distributed amplifier were presented [4]-[5]. We have presented an alternative broadband power amplifier based on Darlington topology that has been utilized in distributed structure. DSSDA small signal amplifier have shown gain-bandwidth products

approaching twice the transistor f_τ, therefore can be good candidate for high power broadband amplifiers.

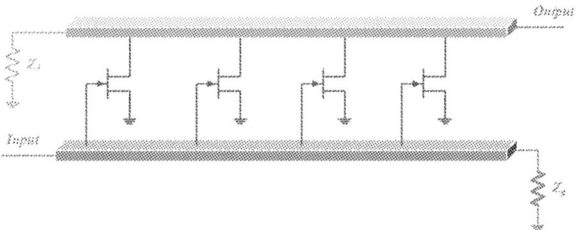

Fig.1.Schematic of DA

II. CIRCUIT DESIGN AND ANALYSIS

A common-source amplifier (Fig. 2), using a simplified FET model, has a short circuit current gain

$$H(f)_{cs} = \frac{f_\tau}{jf} \qquad (1)$$

reaching unity at f_τ (Fig. 3). Since most wideband amplifiers use 50 Ω output loading, output power is then limited to :

$$P_{out} \le \frac{(V_{br} - V_k)^2}{8Z_0} \qquad (2)$$

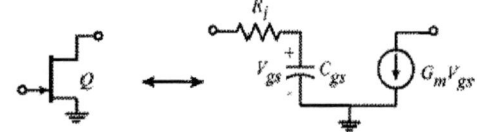

Fig. 2. Simplified ac FET model used for analysis.

where V_{br} is the breakdown voltage, V_k is the knee voltage and Z_0 =50 Ω.

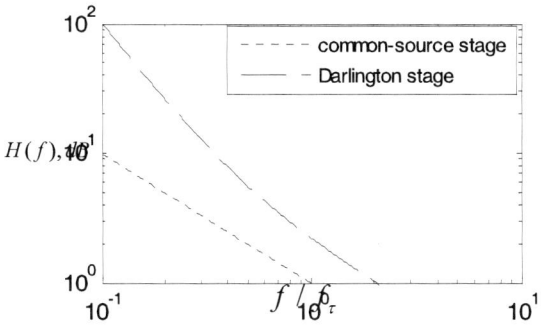

Fig .3. Short circuit current gain $H(f)$ for common-source and Darlington

Darlington amplifiers have higher bandwidth than common-source amplifiers but provide lesser efficiency. If the drain of the common-source is connected to the amplifier output [Fig. 4], then the current gain :

$$H(f)_{Darlington} = \frac{1 + j2f/f_\tau}{(jf/f\tau)^2} \qquad (3)$$

is twice that of a common-source stage at high frequencies (Fig. 3), reaching unity at $2f_\tau$. The peak output power obtainable from a Darlington stage is:

$$P_{out} \leq \frac{(V_{br} - V_k - V_p)^2}{8Z_0} = \frac{V_{o,P-P}^2}{8Z_0} \qquad (4)$$

which is slightly less than the common-source case because the peak–peak output voltage, $V_{o,p-p}$ is now reduced by the pinch-off voltage, V_p.

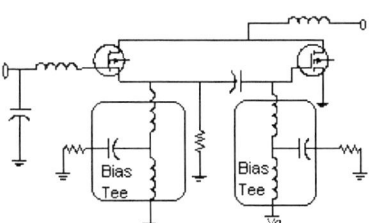

Fig.4.Topology of each gain cell

External bias tees were used to independently bias the two devices. The source of the first device was grounded for DC through a bias tee.

Since our distributed structure is a nonlinear circuit, the best method for its analysis is Harmonic Balance (H.B.) approach and the solution of H.B. equations is based on Jacoubian approach [6]. The nonlinear model is defined by Cubic Curtice model [7]. Using this model (Fig.5) and nonlinear analysis approach, the DSSDA is simulated .

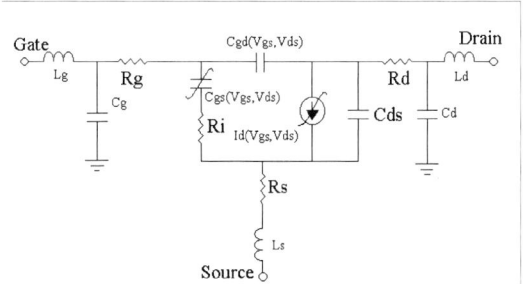

Fig. 5. Nonlinear model of transistor

III. SIMULATION RESULTS

For determining the optimum load resistance, one should draw the static characteristics of the transistor, *i.e.* the drain-source static currents versus drain-source voltage. After some manipulation, the optimum value of load resistance has been chosen so that the output power reaches its maximum value.

The simulated load-line for the amplifier (Fig.6) shows a nearly equal current division at different frequencies .

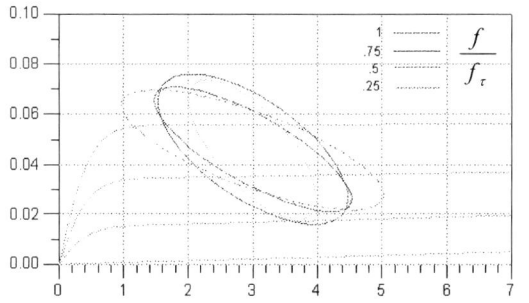

Fig. 6. Load line of the amplifier for different frequency

The small signal gain of the DSSDA and CDA is presented in Fig. 7, that shows the a flat gain over 4 octaves frequency bandwidth in DSSDA.

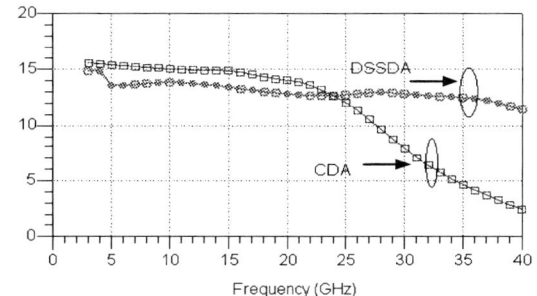

Fig. 7. Small signal gain

Fig. 8 shows the output power at $P_{in} = 5dBm$, one can observe that a good performance of power amplifier has been obtained.

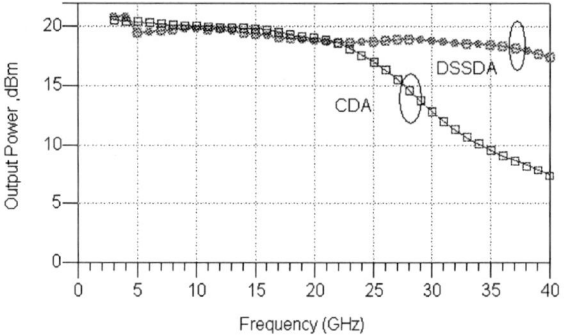

Fig. 8. Output power at $P_{in} = 5$ dBm

Finally return losses of the DSSDA is compared with CDA in the Fig. 9. It demonstrates that return losses are better than -10 dB over 4 octaves frequency band.

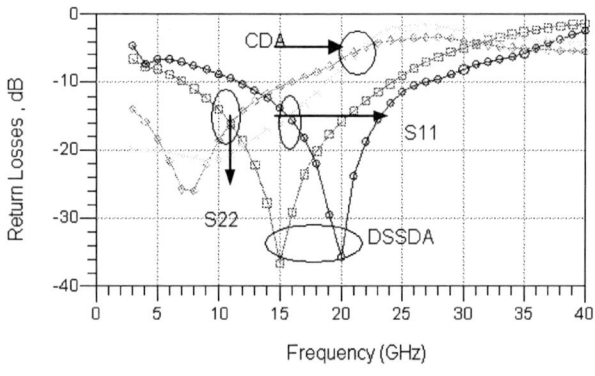

Fig.9 . I/O Return losses

IV. CONCLUSION

A distributed power amplifier employing Darlington topology has been designed and simulated. The 3-stage DSSDA had advantage of wide bandwidth in the comparison of CDA and

demonstrated 14 dB gain and 40 GHz bandwidth of operation. This kind of structure can be used in different frequency band and also it can be applied to MMIC technology to improve the performance of monolithic amplifiers .

ACKNOWLEDGMENT

This work is supported in part by Iran Telecommunication Research Center (ITRC).

REFERENCES

[1] J. L. B. Walker, "Some observations on the design and performance of distributed amplifiers ," *IEEE Microwave. Theory Tech.*, vol. 40, no.1, pp. 164–168, Jan. 1992.

[2] Hee-Tae Ahn,D.J.Allstot, "A 0.5-8.5-GHz Fully differential CMOS Distributed Amplifier," *IEEE Journal of Solid-State Circuits* , Vol.37, No.8 , August 2002.

[3] B.Razavi, Design of Integrated Circuits for Optical Communications. McGraw-Hill, New York, 2003.

[4] J.Nisbet, B.Syrett "Ultralinear Wideband Integrated Circuit Design,"Department of Electronics, *Carleton University, Ottawa,* ON, 1993.

[5] C. Meliani, G. Post, G. Rondeau, J. Decobert, W. Mouzannar, E. Dutisseuil, and R. Lefevre, "DC-92 GHz ultra-broadband high gain In HEMT amplifier with 410 GHz gain-bandwidth product," *Electron. Lett.*, vol. 38, no. 20, pp. 1175–1177, Sept. 2002.

[6] Stephen A.Mass, *"Nonlinear Microwave and RF Circuits"*, Artech House Inc., Norwood, MA, 1997.

[7] H. Ahn and D. J. Allstot, "A 0.5–8.5-GHz fully differential CMOS distributed amplifier," *IEEE J. Solid State Circuits*, vol. 37, pp. 985–993, Aug. 2002.

Silicon Chip Removal Technique Using Wet Etching Process for Failure Analysis on Multi-Chip Packages (MCP)

Shamsul Mohamed and Rodzaki Saad
Spansion (M) Sdn Bhd
Phase II Bayan Lepas FIZ
11900 Penang, MALAYSIA
Email: shamsul.mohamed@spansion.com

Abstract **The work presented here shows the alternative silicon chip removal technique using wet etching process. The effect of various chemical solution and temperature on bulk silicon etching is demonstrated. The etch rate plotted shows NaOH provide the best etch rate compare to other experimented chemical solutions. However, idealized silicon removal technique using wet etching process will need further process development to make it ultimately effective for thin die MCP i.e. a packaging technology with die thickness less than 100um**

I. INTRODUCTION

Multi-Chip Packages (MCP) is an emerging trend to meet higher storage and faster speed requirement of memory device. The miniaturization and higher functionality is accomplished in MCP through the optimization of die size and thickness, die attach stacking and wire bonding process. All these tasks are important in minimizing MCP substrate interconnect density.

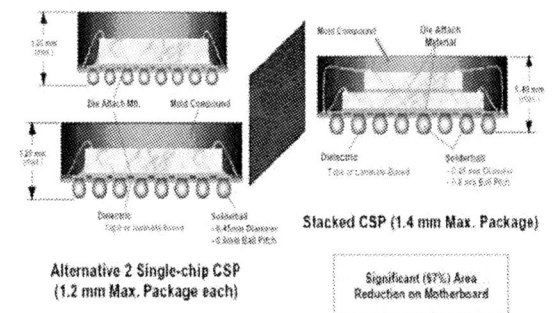

Figure 1: The evolution of CSP to MCP technology for memory device.

Multi-Chip Packages (MCP) however raise a new challenge for chip access technique during failure analysis work. Thus, an alternative FA method is required to expose the failing chip without any failure mode alteration. Since the introduction of MCP in late 1990s, there are two known chip access techniques used in failure analysis i.e. mechanical milling process and laser

de-capsulation process. However, this paper will specifically describe the alternative chip access technique using wet etching process.

In 2001, our failure analysis team developed the silicon removal technique using wet etching process (i.e. using KOH solution and the technique presented in ISTFA 2002) for MCP package. Two important factors were taken into considerations when developing silicon chip removal process. First desire was to maintain the same packaging quality and the second one was to minimize alteration of any failure mode so that further electrical fault isolation could be performed on the target chip. The description of the technique is as follow

Step 1: A small opening on top die was gained through a de-capsulation or grinding process. The size of the opening was approximately 75%-85% of the actual size of the top die.

Step 2: The active circuitry of the exposed top die was etched-off by dipping the unit in hydrofluoric acid. The etching time could vary from 10-15 minutes depending on device fabrication process. The purpose of HF etch is to remove the remaining passivation, ILD and metal layeras

Step 3: The top die was essentially etched-off in this step by dipping the MCP package into hot chemical solution.

Step 4: The silicon chip is completely removed in this step by etching off any residual of silicon, epoxy die attach and balance of the mold compound covering the target chip die. This was accomplished by performing a chemical de-capsulation process.

There are also several limitations pertaining with wet etching process for failure analysis on MCP. First limitation is that the adhesive layer between the die may not completely halt the chemical reaction from etching the bottom die. Another limitation is that the package is subjected to a

0-7803-9730-4/06/$25.00 ©2006 IEEE

chemical environment at high temperature, thus substrate components could easily damage if etched for a very long period. Besides, heating the device could result in some recovery for the failures that are temperature sensitive. However with proper characterization of wet etching process, a controllable etch rate could be obtained.

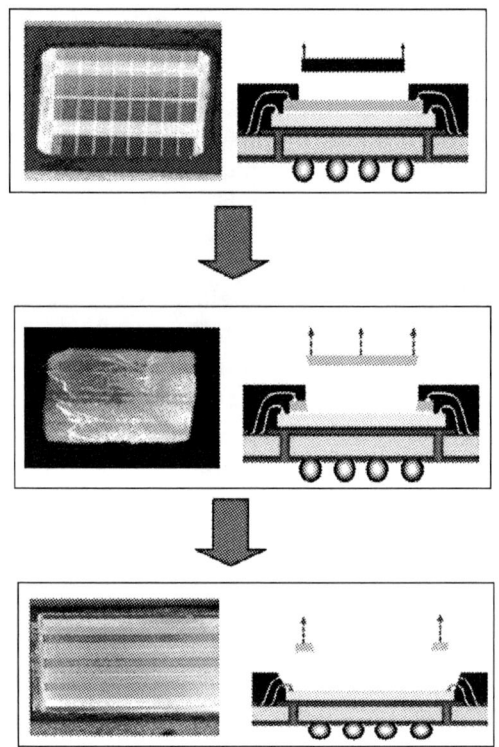

Figure 2: Illustration showing the sequence of the silicon chip removal process using wet etching process.

Compared with dry etching (i.e. either ion bombardment or plasma based technique), wet etching typically has the advantage of a high etch rate and high etch selectivity. Selectivity can be defined as the ratio of the rate of etching of material being removed to the rate of etching of the material around it. Selectivity of A to B is given by the following equation,

$$Sab = Va/Vb = \text{Etch rate of A/Etch rate of B}$$

Wet etching can also achieve a clean surface because the etch by-product or debris can be dissolved in the wet chemicals. However, wet etching is an isotropic process and the resulting over-etch can extend into the area beneath the target area. The etch rate can be varied depending on the composition, concentration or temperature of the chemical solution chosen.

In 2005, through literature study, joint project with local universities and discussion with chemical suppliers, four hydroxides were investigated to determine whether they could provide better silicon etching quality versus potassium hydroxide 45% (KOH). The chemicals are

(i)   Sodium hydroxide 45% (NaOH)
(ii)   Cesium hydroxide (CsOH)
(iii)   Choline hydroxide
(iv)   TMAH   (Tetra-Methyl-Ammonium Hydroxide)

However, during preliminary analysis the two chemicals i.e. Choline hydroxide and TMAH were found unsuitable as it damage the package during the etching process and also causing safety issue (i.e. hazardous vapor is generated during heating process), thus further experiment were carried out on only three chemicals i.e. KOH, NaOH and CsOH that cause minimal damage to the substrate.

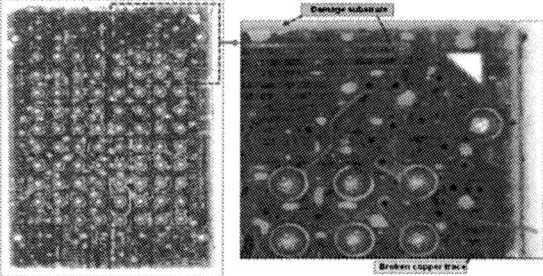

Figure 3: Images showing the example of damages on the package when wet etching was performed using Choline hydroxide and TMAH.

II. EXPERIMENT

MCP package with two active chips was selected for the analysis. The package used a substrate consists of 2-layer copper metal that is sandwiched together with Bismaleimide-triazine (BT) core resin i.e. a type of glass reinforced plastic (GRP) material.

The following samples were used for the experiment to compare the etching rate for CsOH, KOH and NaOH to remove entirely the top die.

Sample group	Size for top and bottom die	Thickness of top and bottom die
1	4.5mm x 5.7mm	150um
2	4.5mm x 5.7mm	125um
3	4.5mm x 5.7mm	100um

The volume for the chemical solution used is set at 45ml and the lowest etching temperature starts from 75°C onwards as at lower temperature the chemical reaction to silicon material is extremely slow.

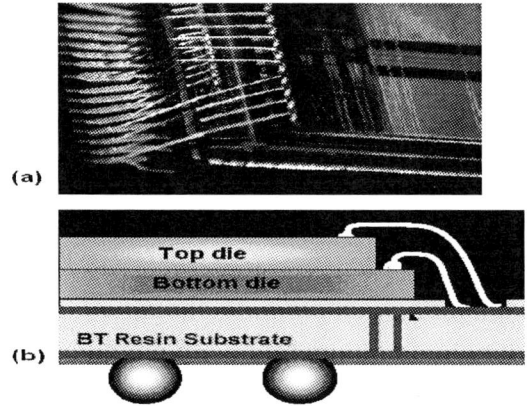

Figure 4: Images of the S-CSP package used for the wet etching experiment: (a) top view of the package (b) cross-sectional view of the package.

Upon completion of the etching process on the top die, additional validation test i.e. ball pull, ball shear and optical inspection were carried out to verify the mechanical strength of the bond wire of bottom die. Besides ensuring no degradation in the mechanical strength of the bond wire of bottom die, the optical inspection will confirm that the die surface and bond pad are clean and free from corrosion issue.

III. RESULTS AND DISCUSSION

Figure 5, 6 and 7 are the graph plots of etching rate dependency on silicon die thickness. It is clearly seen that for all chemicals i.e. CsoH, KOH and NaOH, the etching time decrease significantly when etching temperature increases. This is however highly expected as higher temperature increase the chemical reaction of the solution with the silicon material.

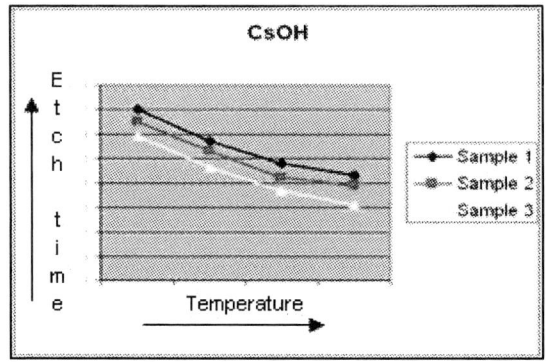

Figure 5: Graph showing the etch rate of the samples using CsOH solution

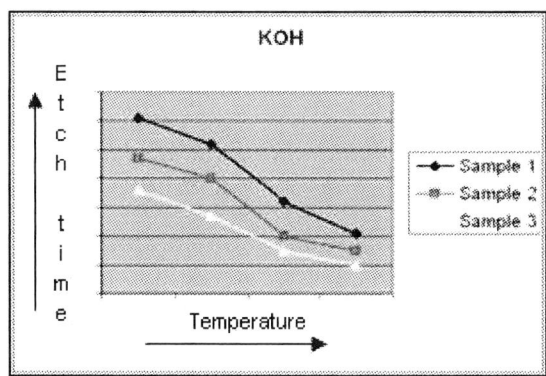

Figure 6: Graph showing the etch rate of the samples using KOH solution.

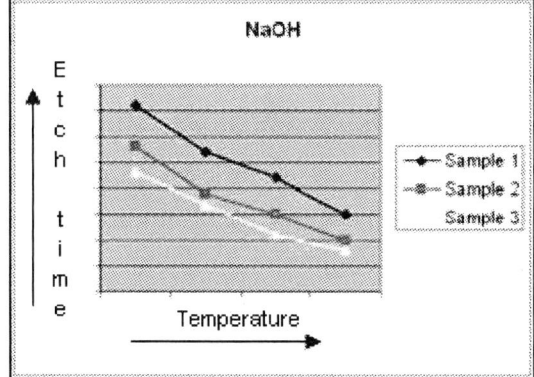

Figure 7: Graph showing the etch rate of the samples using NaOH solution.

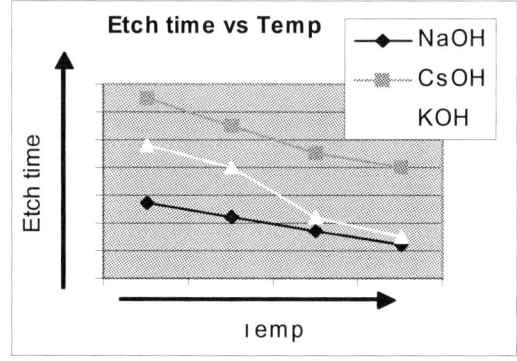

Figure 8: Graph showing relative etch rates between chemicals

Figure 8 shows that the etch rate using NaOH solution provides the best etching rate. On the other hand, although the best etching rate for CsOH, KOH and NaOH appear to be at the highest temperatures, package integrity damages and glassivation crack on bottom die start to occur at lower temps. Consequently, the optimum temp selected is a tradeoff between the two.

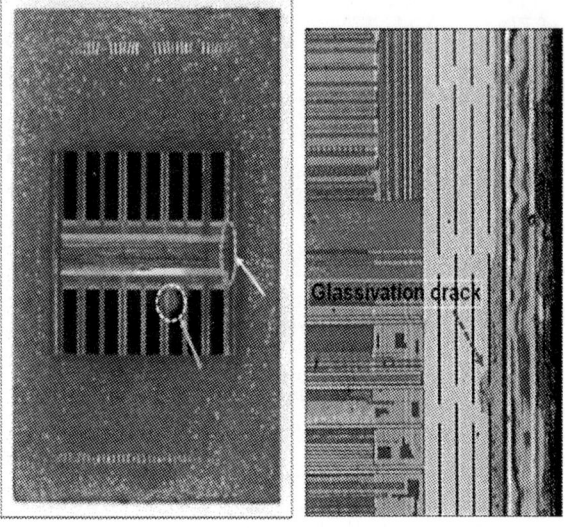

Figure 9: Images showing the example of glassivation crack that occurred on bottom die if etching temperature is too high.

On the other aspect, the result in Figure 10 and 11 of this study clearly demonstrating that there is no direct correlation between etching temperature, type of etching chemical and mechanical strength of bond wire for the bottom die. In the experiment, since etching time mostly less than 60 minutes, it is most likely that the increase in the formation of intermetallic layer between bond wire and metallization layer of the die bond pad is extremely small. As a result, any changes in the mechanical strength of the bond wire would be negligible.

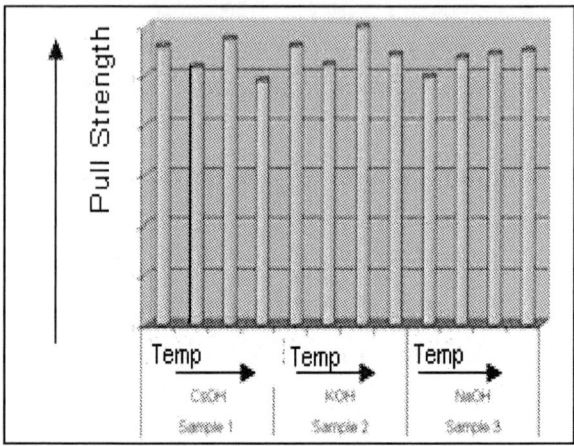

Figure 11: The pull test measurement on bond wire of bottom die for all samples. Besides passing minimum product specification requirements, no specific pattern of data correlation are seen on the graph.

Optical inspection on all bond pads and die surface for all samples etched at the optimum temperature using CsOH, KOH and NaOH did not show any glassivation crack or sign of corrosion issue. Therefore, this chemical etching technique seemed to be promising for future MCP technology with die thickness less than 100um. Nevertheless, it is obvious that etching time for thinner silicon material yielded significantly lower, thus lesser possibility of exerting any physical damage to both package and die surface.

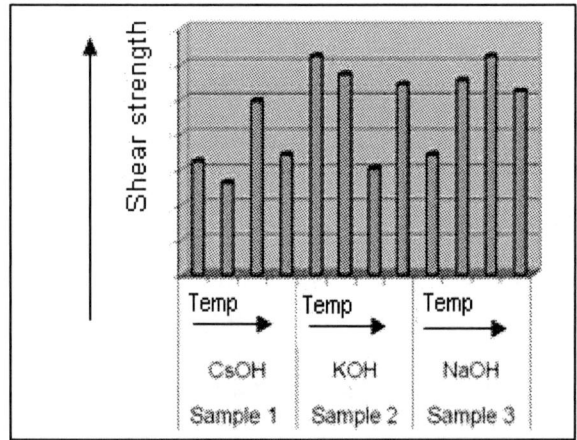

Figure 10: The shear test measurement on bond wire of bottom die showed all samples passed minimum product specification requirements.

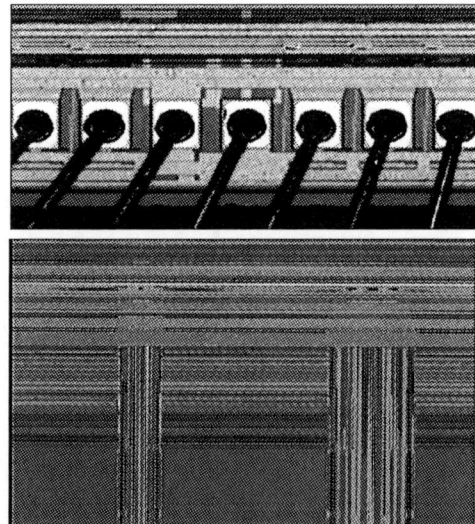

Figure 12: Optical micrograph of bond pad and topside layer of bottom die for good sample etched at the optimum temperature. The surfaces are clean free from glassivation crack or corrosion issue.

IV. CONCLUSION

In conclusion, a variety of hydroxides solutions have been analyzed to investigate its effectiveness for removing silicon chip in MCP package. The findings showed that contrary to conventional understanding, NaOH is the better solution than KOH as it provides faster etching time (i.e. approximately 50% faster) with controllable result at an optimum temperature. Besides, results have shown that idealized silicon removal technique using wet etching process will need further process development to make it ultimately effective for thin die MCP i.e. a packaging technology with die thickness less than 100um.

ACKNOWLEDGEMENT

The authors would like to sincerely acknowledge Norhazlinda and Prof. Zul Azhar of KUKUM, Boey Huey Theng and Prof. Kamarulazizi of USM, and also Spansion Penang DA staffs (TK Lim, Zainal, Guat Choo and Netti) for the enormous support and technical assistant during the project. A special thanks also to Susan Li of Global Device Analysis and Spansion Penang management for their encouragement and valuable insight in publishing this paper.

REFERENCES

[1] Gregory T.A. Kovacs, Nadim I.Maluf and Kurt E.Petersen, "Bulk micromachining of silicon", pp1536-1551, *IEEE Proceedings 1998.*

[2] M.Kada and L.Smith, "Advancement in stacked chip scale packaging", *Pan Pacific Microelectronics Symposium 2000.*

[3] M.Shamsul and S.Rodzaki, "The failure analysis strategy for 2-stack die MCP", *ISTFA2002.*

[4] S.Prejean and J.Shanon, 'Backside Deprocessing of CMOS SOI device for physical analysis and failure analysis", pp99-104, *ISTFA2003.*

[5] Susan Li, "Chip-scale packages and their failure analysis challenges", pp332-340, *Microelectronics Failure Analysis Desk Reference 2005* - ASM International.

Pulse Generation with Reduced Ringing for Ultra Wide Band Applications in Indoor Wireless Communication

Sheroz Khan, *Member IEEE*, A. H. M. Zahirul Alam, *Senior Member IEEE*, Mohammad Rafiqul Islam, *Member IEEE*, Othman O Khalifa, *Member IEEE* and *Ismail Adam
Faculty of Engineering, International Islamic University Malaysia
*Universiti Kuala Lumpur, Malaysia
E-mail : sheroz@iiu.edu.my

Abstract **The work in this paper is presenting circuits for pulse generation and shaping of pulses for UWB applications. The pulse generation is based on the principle of sending the same pulse over two paths-- one traveling straight faces a step recovery diode (SRD) diode while the other is being reflected after meeting the properly terminated short end of a shunt stub, which returns back to the same point inverted pulse with a travel delay. The reflected and the incident pulses meet to generate a narrow pulse of duration almost equal to the two-way travel time of pulse over the stub. The generated pulse properly matched to a smoothing circuit which in fact a high pass filter, which is designed with a transfer function that filters out the trailing and leading tails of the pulses in order not only to function as shaping circuits, but also help reducing the ringing levels of the generated pulses. In other words the rising and falling edges are allowed to continue sharpening the pulses up with more reduced low frequency components. All the above stages of circuits presented in this work are studied through simulation results using SPICE.**

I. INTRODUCTION

The ultra-wide band signals occupy larger band width which proves to be promising in terms of fine time resolution for information transfer. This offers better alternative for wireless communication applications covering shorter distances such as home communication networks. May works are reported in recent publications which suffer from one or other short comings such as enhance ringing, improper impedance matching and loading effects One such circuit is using four

CMOS transistors connected in parallel and turned on in sequence, all draw their current through a choke when they ON, and charge up a capacitor correspondingly. The charging and discharging of a capacitor make a mono-cycle pulse. This work suffers from using a coil which is not very encouraging element in terms of more heat loss besides causing electromagnetic problems for other components on board. Thus controlling the inputs of he our transistors, one can be able to generate pulses of controlled duration. However this causes a large amount of power to consume, and rise and the fall times are limited by the gates switching speeds [1]. Also, this circuit suffers from the ringing problem which has not been addressed properly. Although the ringing problem has been addressed in this approach, but the shape of the resulting pulse suffers from rise and fall time problems. In another approach the rise and fall problems are addressed but the ringing problem have been tackled considerably poorly, however the pulse has been having problems in properly matching to the successor circuit and the connected load and is unable to satisfy the low consumption requirements [2]. Another approach is making use of using NAND gates to cause delays of nanosecond range which are then applied to a NAND gate input to generate a pulse of very small duration [3-4], which is quite able to meet the low power consumption standards set by the FCC. In the approach discussed a mix of the features discussed on the above literature have been used collectively, eliminating the need of excessive power consumption, reducing ringing problem and the rise and fall problem considerable. The resulting circuit has been simulated in SPICE simulation and the results are shown. The work reported in this paper properly addresses all the requirements including reduced ringing, properly

0-7803-9730-4/06/$25.00 ©2006 IEEE

impedance matching requirements, and isolation of the generating circuit form the loading effect of varying loads. The simulation results are presented showing waveforms at selected points of the complete circuit.

II. PULSE GENERATOR DESIGN

Astable vibrator circuit is used to generate pulses of short duration which are applied to onward to the junction of a short-circuited stub and a transmission line. The original pulse travels straight while that from the properly terminated stub returns nearly inverted after a delay equal to the two-way travel time of the stub length. This pulse upon returning gets appended with the original pulse that has not finished passing the junction point, and hence except that delay intervals both at the trailing and leading ends, the original pulse is diminished, thus creating a train of monocycle pulses, in which any two consecutive monocycles are generated as a result of a single original pulse (Fig. 1). The first mono-cycle pulse is indicating the falling edge of the pulse while the second one shows the rising trailing edge of the original pulse.

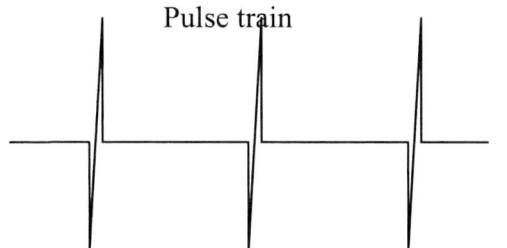

Fig. 1 Pulse Train Generation

Complete circuit diagram of the pulse generator and shaper is as shown in Fig. 2

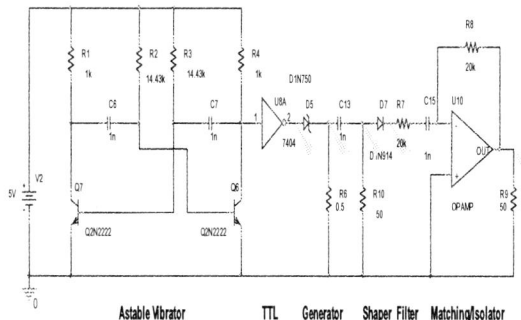

Fig. 2 Circuit Schematic of Pulse Generator

III. PULSE SHAPING AND SMOOTHING

The pulse-shaping network consists of a series-connected SRD and shunt stub terminated with 0.5Ω. This is followed by an RC network which removes all the ringing of the generated monocycle. Further there which provide wide band impedance matching between the Gaussian pulse generator and the RC network, preventing the Gaussian pulse from having a large ringing level and effectively shaping the pulse wave form. Finally there is an op-amp with gain of 2 which buffers the generated pulse form the effects of varying load that follows. Fig. 3 shows the calculated voltage waveforms to illustrate the advantage of this novel pulse-shaping network.

IV. PULSE ISOLATION AND MATCHING

An op amp active filter not only provides a high-pass filter, but amplifies the mono-cycle pulse to the required level in addition to isolating the pulse generating circuit from the pulse shaping circuit. The generating pulse at various points is as shown in Fig. 3(a), (b), (c) (d), and (e). The pulse is then applied to RF transmitter.

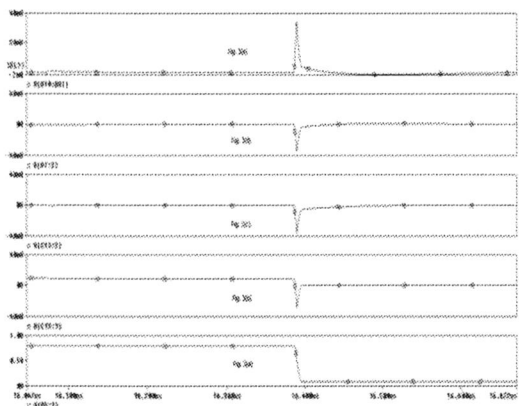

Fig. 3 Waveforms Showing Pulse Generation

V. RESULTS

Fig. 3(a) is the TTL from the inverter which is fed to SRD and a properly terminated shunt stub, while Fig. 3(b) is pulse made from joining of the returned pulse with the original. Fig. 3(c) shows the waveform after having removed the DC part, allowing the high components to keep passing onward, while Fig. 3(d) is the input to the high band pass op amp which is a unity gain amplifier. Finally Fig. 3(e) shows the final from of the pulse with reduced ringing. It is worth to mention at this point that the op amp also serves as a buffer, thus eliminating the effect of varying loading.

VI. CONCLUSIONS

A modified UWB monocycle pulse generator has been designed, simulated and tested for the expected results. Ac active op amp based high-pass filter is used to further shape the resulting monocycle pulse with ultra-short duration with substantially reduced ringing levels. The circuit has been proved to be working properly with SPICE simulation. Work is in progress for achieving and proving an expected good agreement between the simulated and actual results. Avoiding heat producing coils, with a good performance this circuit makes it attractive for various UWB radar and communications systems. In summary, a new approach to generating and shaping pulses for UWB applications including RFID applications, military applications such as ground penetrating radars and see-through-wall radar, beside offering the benefits of low area, low cost, and low power.

REFERENCES

[1] V. G. Shpak *et al.*, "Active Former of Monocycle High-Voltage Subnanosecond Pulses," in *Proc. 12th IEEE Pulse Power Conf. Dig.*, June 1999, pp. 1456–1459.

[2] Jeong Soo Lee and Cam Nguyen, "Novel Low-Cost Ultra Wideband, Ultra-Short-Pulse Transmitter with MESFET Impulse-Shaping Circuitry for Reduced Distortion and Improved Pulse Repetition Rate", IEEE MICROWAVE AND WIRELESS COMPONENTS LETTERS, VOL. 11, NO. 5, pp. 208-210, MAY 2001

[3] J. M. Kahn and J. R. Barry, "Wireless infrared communications," Proc.IEEE, vol. 85, pp. 265–298, February 1997

[4] Jeong Soo Lee, Cam Nguyen, and Tom Scullion, .New Uniplanar Subnanosecond Monocycle Pulse Generator and Transformer for Time-Domain Microwave Applications,. *IEEE Transactions on Microwave Theory and Techniques*, vol. 49, No. 6, pp. 1126-1129, June 2001.

Application of Autosched AP Simulation Model in Wafer Fab

Victor Siow Yuen Tien, Yeo Eng Teck, G.Devandran A/L Govindasamy and Shahrul Kamaruddin[*]
Silterra Malaysia Sdn Bhd, Kulim Hi-Tech Park 09000 Kulim, Kedah, Malaysia
[*] School of Mechanical Engineering, Universiti Sains Malaysia, 14300 Penang, Malaysia
Email: yuentien_siow@silterra.com , meshah@eng.usm.my

Abstract – **Wafer fabrication or "fab" simulation models are often used to assist the daily running of wafer production. Their main uses have been focused on testing production scenarios, to estimate cycle times, quote delivery times to customers, or test operational policies such as lot dispatching rules and material handling systems scheduling policies. Fab technologies sometimes stay in demand for as little as one year, after that wafer prices often drop rapidly. Based on this, many semiconductor manufacturers regard cycle time as the most important element and treat it as a monitored performance measure in wafer fabrication. One of the significant benefits of simulation modeling is the ability to conduct sensitivity ("What-If") analyses that allow us to experiment without disrupting the actual manufacturing operation. This paper will focus on and present the applications of the Autosched AP simulation software from Brooks Automation to provide ("What-If") analysis and a characteristic curve relating Cycle time to Fab Utilization. The paper will show how simulation is used as a modeling tool to assist a production manager's decision-making process in production planning and effectively evaluate the factory's performance capabilities. Analysis will describe the impact of different percentage of factory loading on the cycle time.**

I. INTRODUCTION

SIMULATION modeling is a powerful tool that can help in many business problems. A simulation model is a computer program that replicates the operations of a business process and estimates rates at which outputs are produced and resources are consumed [11]. Its main uses have been focused on testing production scenarios to estimate cycle times and quote delivery times to customers, or test operational policies such as lot dispatching rules and automated material handling systems (AMHS)

scheduling policies. For example simulation work was done by Lin et al [12] which concerned a hybrid push/pull (PP) dispatching system. Discrete-event simulation models were developed and the simulation results revealed a potential improvement of the AMHS performance and reduced the work in progress (WIP) and cycle time. Some other relevant AMHS analysis can be found in Beschomer et al [1], Wang et al [17], Jefferson [10], Wright et al [20], and Mackulak et al [15] who gave a detailed analysis of the needs and requirements for successfully designing and implementing automated material handling control systems. Authors like Cardarelli [3] prove that simulation is a valuable tool to evaluate the AMHS performance through discrete-event simulation and improvements can be measured from how the materials are delivered, stored and retrieved from tools.

Meanwhile, simulation models also assist in scheduling [18], [19], [21], especially regarding the effect of using different strategies in dispatching rules. Examples of studies can be found in JT Lin et al [13], Dabbas et al [7], Loo et al [14], Yurtsever et al [23] and Tsai et al [16]. One of the related studies is from Elif Akcali et al [9] where a simulation model of a wafer fabrication facility is used to examine the effects of different loading and dispatching policies for diffusion operations. From the experiment, the results indicate that the loading policy has a significant effect on the average diffusion flow time as well as the overall cycle time of the products. On the other hand, simulation modeling is also applied in capital equipment analysis where experiments can be used to evaluate variations in production capacity as well as in costs related to the production process. For instance, simulation can be performed to evaluate the economic impact of additional tools in wafer fabrication in the bottleneck area. These studies can be found in Dima et al [8], Yu HL et al [22] and Chen JC et al [4] where different ways to solve bottleneck have been simulated and analyzed. Results from the simulation models are

0-7803-9730-4/06/$25.00 ©2006 IEEE

able to solve complex problems and have saved a lot of time for accurate planning and decision-making.

Some other studies of simulation models that are related to wafer fabrication (fab) can be identified in the following categories. One of the general studies is the evaluation of production capabilities. It is important to project workloads as well as to evaluate capacity expansion and allocation in semiconductor fab operations. Most of the production in a semiconductor fab is complicated and usually several products are produced. This is called the product mix. In this case, a simulation model can used to determine the ideal release rate or input planning strategy. On the other hand, simulation models are also being applied in plant layout [6] and maintenance scheduling. The impact of a different distribution of stations into the fab can also be the subject of simulation experiments. For example, the travel times of materials between the stations and material handling can be analyzed, and different configurations of fab layouts can be evaluated through simulation case studies [5]. As for preventive maintenance (PM), one of the major activities or events in wafer fab is tool downtime due to failures. Tool reliability can be increased through appropriate PM strategies and optimal algorithm scheduling from simulation results.

As for this paper, it reveals one of the greatest benefits of simulation modeling which is the ability to conduct sensitivity ("What-If") analysis that allows a manager to experiment without disrupting the actual manufacturing operation. Major changes in fab product mix or production lot size, for example, could have unexpected results on factory throughput if not evaluated in advance. Therefore modeling these scenarios in advance can greatly help the factory predict possible outcomes and avoid unpleasant surprises. This paper will show how simulation is used as a modeling tool to assist a manager's decision-making process in production planning and effectively evaluate the factory's performance capabilities. Analysis will describe the impact of different percentage of factory loading towards the cycle time for each technology.

II. SIMULATION MODEL

A complete ASAP simulation model [2] was developed to run experiments in a wafer fabrication (an 8'' wafer fab at Silterra Malaysia). The simulation model consists of several data files that could be manipulated using Microsoft Excel. Each data file includes a complete description of a certain model component and these data files are all linked together to allow the interaction among the different model components.

Basically to build up a new simulation model, there are four types of input that need to be prepared. The first required inputs are factory resources such as stations, tools, and dispatching rules for each work area. The next data inputs are product information such as parts, which are defined as types of products that are manufactured in the wafer fab. The part file will specify the name of each part type and the name of the route it follows to get produced. The third required inputs are the part's process flow or so called routes which are defined as the processing steps that parts go through to be manufactured. The final data requirement is the product demand such as the factory orders and the work orders to determine when lots are started in the model. Lots consist of a number of pieces of a part as defined in the part file.

In wafer fabrication, there are many products and different routes for each product. Typically the model will consist of about 400 devices and 100 routes. The fab data collection process can be accomplished quickly and accurately as most of the information can be extracted from the central data system. Information such as the routing and part details will be captured from the manufacturing execution system, called Advanced Productivity Family (APF) from Brooks Automation. Estimation of the projected wafer starts into the factory is provided by the Industrial Engineering (IE) group after detailed discussion with manufacturing managers and IE managers. The theoretical process times and throughput rates for each tool and all the recipes were collected systematically from the cycle time database. Each recipe in the fab was timed to determine the exact throughput and consideration was given according to their family grouping.

Scheduled downtime or preventive maintenance was determined based on the logout time from the shop floor control system and cross checking with the engineering technicians. The daily and weekly preventive maintenance were timed and factored into the simulation model as well as the manufacturing efficiency and tools uptime probability.

Validation and verification is a key step in the simulation process and the model is considered validated once the predicted result reaches the acceptance level of accuracy. This section includes the details about simulation length, warm-up periods, number of replications, and other details relevant to the control of the simulation runs. Useful guidelines and procedures for verifying and validating simulation models are presented by Law & Kelton in [11]. Theoretically, a simulation model can be verified using several techniques such as running the model with simplified assumptions to easily detect logical mistakes and running the model under a variety of settings to ensure that the outputs were at reasonable levels. These included comparing the simulation model against the system's outputs for a set of identical inputs.

For this study, a simulation experiment was conducted for a quarter year period to validate the model. Data from the model was compared with the actual output from the production. A cross functional team consisting of IE engineers which take care of each module was formed to review the model input/output and provide feedback. The output statistics and inputs were reviewed for their specific areas on a regular basis to ensure accuracy. Information for the simulation model will be updated from time to time to maintain a valid model.

III. EXPERIMENTS

The simulation model was used to study the effects of different percentage of factory loading towards the fab cycle time for each technology. Fab utilization for different wafer start rates has been studied in different scenarios in our experiments. Wafers start rate is determined by calculating bottleneck index (BNI) according to the current production volume capacities. Scenarios are created by scaling production volume capacities up and down by 10 percent of and are distributed into the weekly selected product mix.

Scenario	Bottleneck Index (BNI)	Wafer start per week (WSPW)
1	1.4	7350
2	1.3	6825
3	1.2	6300
4	1.1	5775
5	1.0	5250
6	0.9	4725
7	0.8	4200
8	0.7	3675
9	0.6	3150

Fig. 1 Projected wafer starts

We assume current production volume capacity with BNI 1.2 equals to 6300 wafer starts per week (WSPW). Fig. 1 illustrates the volume of capacities for different scenarios used to run our experiments.

IV. RESULTS AND DISCUSSION

Based on Fig. 2, the relationship between cycle time and utilization can be represented by a curve, called the characteristic curve. The characteristic curves show a nonlinear relationship between the cycle time and wafer start rate.

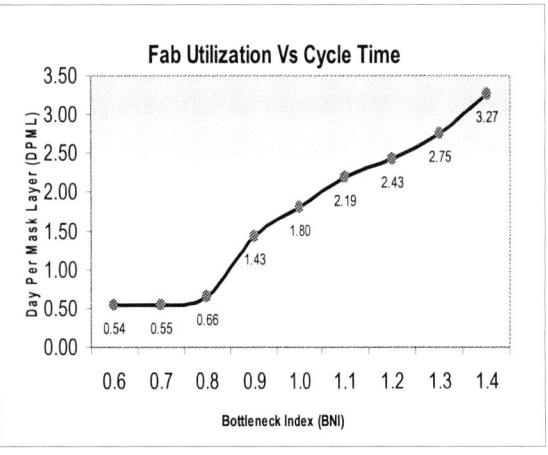

Fig. 2 Characteristic Curve

Looking at the pattern of the graph, we can see that when tool utilization is increased by starting more wafers in the fab, cycle time will increase drastically. This is because when experiments are run with very low utilization values, wafers can be processed with no queue time or wait time, and therefore the cycle time becomes shorter. On the other hand, when wafer starts increase and fab utilization increases accordingly, the graph shows that cycle time increases drastically.

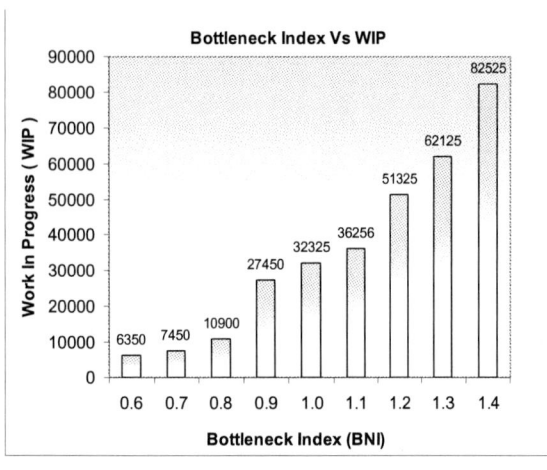

Fig. 3 Average WIP vs. Bottleneck index

Meanwhile, the characteristic curve for the fab at a certain point in time indicates the overall capacity of the fab and how well the fab is utilized. Fig. 3 illustrates the Work In Progress (WIP) for each scenario. The results show that when fab utilization increases, the WIP will increase as well. With the WIP information as well as reference from the characteristic curve, managers can easily make decisions concerning production volumes or expected delivery times especially when dealing with customers.

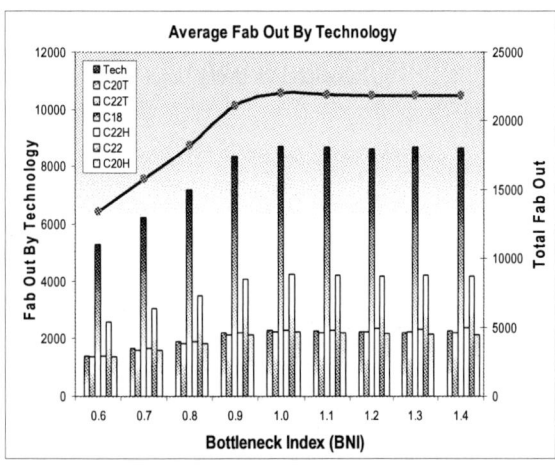

Fig. 4 Average Fab Out By Technology

Fig. 4 represents the overall fab out for each technology according to different wafer start rates. Based on the graph, the characteristic curves show that the volume of fab out will increase when wafer start rates increased accordingly. However, the increment will reach to a maximum point (22k wafers per month) and maintained constant at average 22k wafers per month which is the optimum capacity that can be achieved. Differences in product's volume depend on the product mix and product lot cycle time. The performance characteristics observed in this paper are cycle time, WIP, and the number of finished wafers (wafer outs).

The characteristic curve that relates the cycle time to the daily throughputs can be shifted to a better position in terms of capacity by focusing on the capacity of components such as tool's conversion or by increasing the number of machines. Increasing the number of tools or machines is a huge capital investment for the fab and it is usually done after a through analysis to justify the cost to be invested in acquiring the new tools.

V. CONCLUSIONS

This paper has presented the applications of the Autosched AP simulation software from Brooks Automation to provide ("What-If") analysis and a characteristic curve relating Cycle time to Fab Utilization. The use of simulation modeling has been demonstrated as a decision-making methodology for managers to decide on forecasting production volume and quote expected delivery times by examining characteristic curves. Using the characteristic curves, the differences in production volume capacities among different scenarios were predicted for the targeted cycle time for each technology.

VI. FUTURE RESEARCH

In semiconductor fabs, there are numerous of different product types that are processed in parallel. With advancing of process technologies, start rates (wafer loading) of old technologies are continuously reduced while start rates of new technologies are being increased, leading to permanent changes in product mix. Future research will include an investigation on the behaviour of a wafer fab after changes in product mix. The experiments were performed to observe on how different percentage of product mix in wafer starts influence the overall fab out, cycle time of lots and WIP.

ACKNOWLEDGEMENT

This research was conducted at Silterra Malaysia wafer fabrication facility. The authors wish to gratefully acknowledge the members of IE department in Silterra for their dedicated help and support throughout the study and their useful insights. Without the strong management support and commitment from all engineers and managers involved, simulation as a powerful management tool would not be successful.

REFERENCES

[1] Beschorner A, Gluer D. Maxflow theory for availability calculation of automated material handling systems. Robotics and Computer Integrated Manufacturing 19, 2003, 141-145.

[2] Brooks Automation. 2004. Autosched AP Customization Guide v 8.0.

[3] Cardarelli G, Pelagagge PM, Granito A. Track-robot for wafer fabrication intrabay handling. Assem Automat 1993;13: 25 -9.

[4] Chen Jc, Chia WC, Lin CJ, Hsin R. Capacity planning with capability for multiple semiconductor manufacturing fabs. Computers & Industrial Engineering 48, 2005, 709-732.

[5] Chien CF, Chang KH. Modeling overlay errors and sampling strategies to improve yield. Journal of the Chinese Institute of Industrial Engineers, 2001, Vol. 18, no.3, pp.95-103.

[6] Chung SH, Lai CM, Lee AM, Lai HY. Designing a multi-site production planning system for wafer fabrication. Journal of the Chinese Institute of Industrial Enigneers, 2004, Vol. 21, No. 1, pp.46-58.

[7] Dabbas RM, Hung NC, Fowler JW, Shunk D. A combined dispatching criteria approach to scheduling semiconductor manufacturing systems. Computers & Industrial Engineering 39, 2001, 307-324.

[8] Dima Nazzal, Mansooreh M., Dave Anderson. A Simulation-based evaluation of the cost of cycle time reduction in Agere Systems wafer fabrication facility- a case study. Int .J. Production Economics 100 (2006) 300-313.

[9] Elif Akcali, Reha Uzsoy, David G.Hiscock. Alternative Loading And Dispatching Policies For Furnace Operations In Semiconductor Manufacturing: A Comparison By Simulation. Proceedings of the 2000 Winter Simulation Conference, 2000.

[10] Jefferson T. Simulation in the design of ground-based intrabay automation systems. Proceedings of the IEEE Winter Simulation Conference, Coronado, CA, USA 1996. p. 1008-13.

[11] Law A.M. and Kelton W.D. Simulation Modeling and Analysis, New York: McGraw-Hill Inc., 2000.

[12] Lin JT, Wang FK, Chang YM. A hybrid push/pull-dispatching rule for a photobay in a 300mm wafer fab. Robotics and Computer-Integrated Manufaturing 22, 2006, 47-55.

[13] Lin JT, Wang FK, Yen PY. Simulation analysis of 300mm intrabay automation for a wafer fab 2003, working paper.

[14] Loo HL, Loon CT, Soon CC. Dispatching heuristic for wafer fabrication. Preceedings of the 2001 Winter Simulation Conference, 2001.

[15] Mackulak GT, Lawrence FP, Rayter J. Simulation analysis of 300 mm intrabay automation vehicle capacity alternatives. IEEE/SEMI Advanced Semiconductor Manufacturing Conference, Boston, MA, USA 1998.P.455-50.

[16] Tsai CH, Feng YM, Li RK, A hybrid dispatching rules in wafer fabrication factories. International Journal of the Computer, The Internet and Management, 2003, Vol. 11, No. 1, pp.64-72.

[17] Wang FK, Lin JT. Performance Evaluation on an Automated Material Handling System For a Wafer Fab. Robotics and Computer-Integrated Manufacturing 20, 2004. pp: 91-100.

[18] Wang L, Tay EH, Lee LH. Scheduling MEMS manufacturing. Proceedings of the 2000 Winter Simulation Conference, 2000.

[19] Wein LM. Scheduling Semiconductor Wafer Fabrication. IEEE Transactions on Semiconductor Manufacturing, 1988, Vol. 1, No. 3.

[20] Wright R, Cunningham C, Benhayoune K, Campbell E, Swaminathan V, White R.300mm factory layout and automated materials handling. Solid Technol 1999;35:35-42.

[21] Yoeng DK, Jung UK, Seung KL, Hong BJ. Due Date Based Scheduling and Control Policies in a Multiproduct Semiconductor Wafer Fabrication Facility. IEEE Transactions on Semiconductor Manufacturing, 1998, Vol. 11, No.1.

[22] Yu HL, Ching EN. A total standard WIP estimation method for wafer fabrication. European Journal of Operational Research 131, 2001, 78-94.

[23] Yurtsever T, Pierce NG. Computerized manufacturing monitoring and dispatch system. Computer Industrial Engineering, 1998, Vol. 35, pp.137-140.

Characteristics of RIE $SF_6/O_2/Ar$ Plasmas on n-Silicon Etching

Siti Azlina Rosli, Azlan Abdul Aziz and Haslinda Abdul Hamid

Nano Optoelectronics Research and Technology Laboratory (N.O.R)

School of Physics

Universiti Sains Malaysia

11800 Minden, Pulau Pinang, MALAYSIA

Email: sazlina.zm05@student.usm.my, ctazlinausm@gawab.com , lan@usm.my

Abstract In this work, highly anisotropic Si plasma etching process has been developed in reactive ion etching (RIE) reactor. The etch chemistry utilized consists of a mixture of sulphur tetra fluoride (SF_6), oxygen (O_2) and argon (Ar). The use of Nickel as masks was successfully patterned on Si by lift-off. Lines of 50μm feature size were patterned on Si, resulting in perpendicular sidewalls and no observable undercutting. Etch rates were studied for gas compositions of $SF_6/O_2/Ar$, chamber pressures and DC bias. For chamber pressure in the range of 5 mTorr to 20 mTorr, the etch rates is found to increase with increasing dc bias, attaining a maximum rate of 2997 Å/min at 20 mTorr. Surface morphology after etching is checked by atomic force microscopy and scanning electron microscopy, which show that the anisotropic of the etched surface and the smoothness of the etched surface is comparable to that of the etched.

I. INTRODUCTION

There is increasing interest in nanotechnology and its general framework, nanofabrication; specifically in the semiconductor industry. The ability of flexible pattern transfer into silicon with vertical sidewalls and high accuracy is a crucial requirement for MEMS realization. Silicon etching is an essential process step for the fabrication of MEMS because majority of the microstructures are realized in silicon owing to its mechanical strength [1] and the exploitation of highly developed fabrication technology for integrated circuits [2-3].

In order to establish etching process for MEMS, it is desirable to maintain high etch rates, good control of line width and uniformity of etching, high selectivity over mask and anisotropic etching. The conventional wet etching methods are unable to meet these requirements because of the chemical nature of the reaction leading to isotropic etched profiles unless a crystal orientation dependent etching is used.

A primary process for semiconductor nanofabrication is RIE, which uses chemical and physical component of an etch mechanism to achieve an anisotropic profiles, fast etch rates and dimensional control. RIE is operated at low pressures, ranging from a few mTorr up to 200 mTorr, which promotes anisotropic etching due to increased mean free paths and reduced collisional scattering of ions during acceleration in the sheath.

Traditional reactive ion etching (RIE) was previously used to fabricate high aspect ratio silicon feature [4]. However, its etch rate is low (<1000 Å/min) even though the aspect ratio is more than 10. Currently fluorine chemistries and low-pressure, high-density plasma etching systems are preferred because to achieve high etch rates for deep high aspect ratio etching [5].

SF_6 plasma is suitable because of the high etching rate of silicon. During reactive ion etching (RIE) of silicon in SF_6 plasma, the etching rate depends on the pressure in the reactor. The etching rate increases several times due to the intensive reaction $Si + 4F \rightarrow SiF_4$. However, the increased surface coverage by SiF_4 molecules slows down the etching rate at a high pressure. Previous experiments reported a maximum was observed in the dependence of silicon etching rate on pressure. This indicates that synergism of sputtering by ion bombardment and chemical etching depends on pressure.

In this work, the silicon etching rates and etching profile were studied as a function of the chamber pressure and the DC bias to attain a high

0-7803-9730-4/06/$25.00 ©2006 IEEE

etch rates and good uniformity of the sidewalls. It further discusses the relation between sulfur tetra fluoride species adsorption in containing oxygen and argon plasmas and silicon etching.

II. THE THEORY OF $SF_6/O_2/Ar$ PLASMAS ETCHING

An etch model based on $SF_6/O_2/Ar$ chemistry is presented here to achieve highly anisotropic etching in silicon. High aspect ratio microstructures can be produced in fluorine containing plasmas using the sidewall passivation technique. The reaction probability of fluorine with silicon is spontaneous giving rise to isotropic etched profiles and as a general rule, the reactivity of the halogens follows the order F>Cl>Br>I. In this process chemistry; the sidewall passivation can be achieved with the addition of O_2 to SF_6 plasma, which is based on SiO_xF_y film.

Table 1 Gas for Si trench etching with their main plasma radicals, products and inhibitor.

Gas	Radicals	Products	Inhibitor
SF_6	F, SF_5	SiF_4	$Si_xS_yF^a$
SF_6/O_2	SF_5, F, O	SiF_4, SOF_4	$Si_xO_yF_z$

[a] Only with cryogenic cooling

From Table 1, it showed that in the $SF_6/O_2/Ar$ plasma, there is a constant competition between the fluorine radicals that etch and oxygen radicals that passivate the silicon. Therefore, a certain concentration of O_2 is to be added to SF_6 to achieve a balance between the oxygen and fluorine radicals to achieve vertical sidewalls [6]. In such a plasma, SF_6 produces F* radicals for the chemical etching of silicon forming volatile products, such as SiF_x which are easily removed from the surface. SF_6 are the source of SF_x^+ ions which are responsible for the removal of the SiO_xF_y layer at the trench bottom forming the volatile product SO_xF_y (or CO_xF_y) allowing silicon etching to proceed further [4].

The unsaturated etch products such as SiF_x are highly reactive and thus react with oxygen than the halogenated silicon surface which is essentially passivated and deposit on the silicon surface. Incomplete removal of these oxidized products by the physical component of etching will result in the foundation of grass-like residues called the "black silicon" as the inevitable rough surface. The formation of black silicon during the SF_6/O_2 etch process can be prevented by adding small quantities of scavenging gases such as CHF_3 or Ar.

Noble gases such as argon and helium are added to stabilize the plasma or for cooling purposes. Argon addition causes inert ion bombardment of the surface, which results in enhanced anisotropic etching [8]. In $SF_6/O_2/Ar$ plasma, each gas has a known specific function and influence, so the etched profiles can be controlled just by monitoring the flow rates of these gases.

III. EXPERIMENT

The material used here is nominally n-type silicon wafers <1 0 0>. Four samples were used and prepared using similar photolithographic process. Before patterning, the samples were cleaned with RCA cleaning, followed with piranha cleaning and DI water and dried using N_2 gas. Photo resist was spin coated onto all samples, forming ~ 1.5 µm layer. The samples were then soft baked at 85°C for 10 minutes and exposed using a mask with line width features as small as 50µm. The resists patterns were then developed using AZ 400k developer, diluted 1 to 4 with de-ionized water (DI), followed by a thorough rinse in DI. Following resist pattern definition, the wafers were hard baked in 110°C for an additional 10 minutes in order to harden the resist structures.

Then, the nickel coating was applied to the samples and then acetone was used to remove the photo resist until lift-off process was succeeded. After that, all of them were etched by reactive ion etching using various gases in Oxford Plasma 80 system. The silicon was etched using combination of $SF_6/O_2/Ar$ at operational pressure from 5 mTorr to 20 mTorr while RIE power, substrates temperature and RF power were fixed at 100 W, 17°C and 600 W, respectively. The $SF_6/O_2/Ar$ flow rate was fixed at 20/8/8 sccm for silicon etching. The self-bias voltage measured with respect to the ground in the lower electrode was found from -332V to -358V at a fixed RF power of 600W. Prior to etching, the RIE chamber was exposed to an ashing process designed to clean the RIE stainless steal electrodes.

This ash process consists of a low pressure, high power etch using oxygen. After the ash, the

861

chamber was conditioned for 20 minutes using the desired conditions of the etch process in order to purge the system of contaminates in the gas supply lines, condition the chamber and set the appropriate RF tuner settings. After the condition step was completed, an individual sample was loaded into the chamber, centered on platters with the oxidized side exposed to the upper electrode.

Immediately after samples removed from etching system, they were dipped into $3H_2SO_4/1H_2O_2/1DI$ water at 45°C for 5 minutes to remove the nickel. The etch rates were measured from the depth of the etched features with a scanning electron microscope (SEM) after the removal of the nickel mask layer. Surface morphology, etch anisotropy, wall angle and sidewalls undercutting of the etched silicon as well as selectivity of nickel over silicon also were evaluated with SEM model JSM-6460 LV while ULTRAObjective AFM is used to measure the surface roughness.

IV. RESULTS AND DISCUSSION

The results of silicon etch rate measurements in $SF_6/O_2/Ar$ plasmas obtained at a fixed RF power of 600 W, pressure of 20 mTorr, electrode temperature of 17 °C and RIE power of 100 W.

For these experiments, etch rates were studied as a function of chamber pressure and DC bias. Prior to etching patterned samples, a simple set of experiments were conducted to get an understanding of how changes in pressure affect the etch rate. For these data points, samples were etched for 3 minutes, removed, cleaned of $3H_2SO_4/1H_2O_2/1DI$ water and then measured using a SEM. For this case, 20 sccm of SF_6, 8 sccm of O_2 and 8 sccm of Ar were used. Chamber pressure was varied from 5 mTorr to 20 mTorr.

Silicon etch rates were shown as a function of pressure (mTorr) for an etch rates (Å/min) in Fig. 1. From Fig. 1, etch rates were found to increase significantly between 5 mTorr and 20 mTorr, peaking around 20 mTorr. It was found that the etch rate gain steadily with increased chamber pressure. The maximum etch rates was 2997 Å/min at pressure of 20 mTorr meanwhile the minimum etch rate was 733 Å/min at pressure of 5 mTorr.

During the experiment silicon substrates are etched in $SF_6/O_2/Ar$ plasma at different pressures and energies of incident ions. The concentration of F atoms in the plasma increases with the increase of pressure in the reactor and discharge power density. The increased pressure extended the mean residence time of SF_6 molecules in the reactor, and the increased discharge power density increases the probability of dissociation of the SF_6 molecules. As a result of the increased concentration of F atoms, etching rate of silicon increases (Fig. 1).

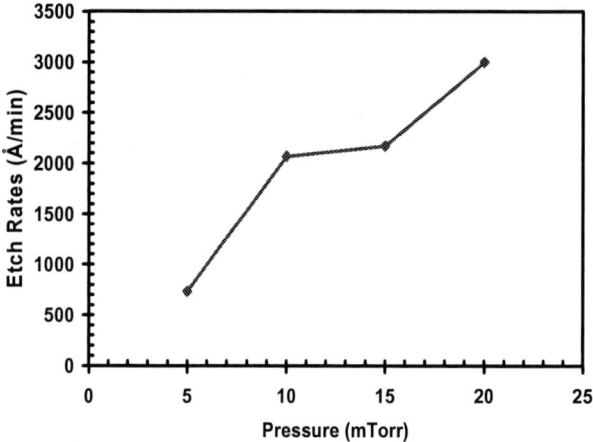

Fig. 1   Etch Rates (Å/min) versus various pressure (mTorr).

Fig. 2 shows the DC bias (-V) of silicon as a function of pressure (mTorr). For this data set, SF_6, O_2 and Ar flows were maintained at 20/8/8 sccm, RF power and electrode temperature were held at 600 W and 17 °C, respectively while RIE power was again kept at 100 W. The etch time for all samples was 3 minutes.

From Fig. 2, the DC voltage decreased as the pressure increased. It showed that DC bias is also a function of the energy of the free electron. At higher pressure, more molecules are available for the electrons to collide with and to generate a new free electron negative and a positive ion. In this way, an increase in pressure would add the number of free electrons, turning the DC voltage more negative. Unfortunately, at higher pressure, electrons suffer more collisions, therefore they gain less energy between collisions [9]. The electron energy decreases with the increases pressure.

On the other hand, an increase in pressure increases the density of species, i.e. it reduces the mean path of the electrons before colliding. In this way, the electrons will gain less energy before colliding. This decrease in energy results in less formation of a new electron-positive ion pair. This mechanism decreases the formation of free electrons and ions. As results, increasing in pressure apparently decreased the DC bias.

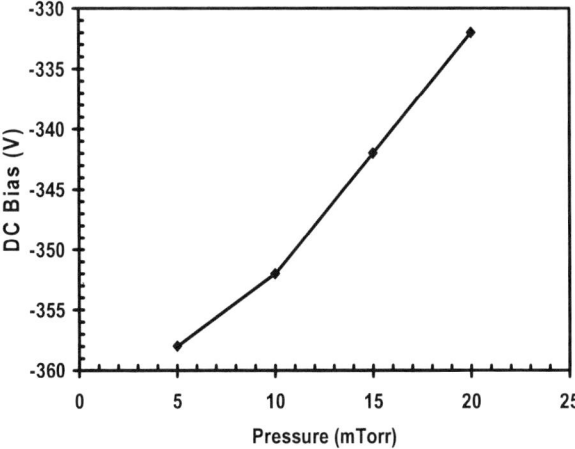

Fig. 2   DC bias (-V) versus various pressure (mTorr).

In Fig. 3, the microstructure of the etching anisotropy shows nearly vertical structure etched in silicon with smooth bottom surface using $SF_6/O_2/Ar$ gas mixtures at various pressures.

In addition, the anisotropy of the etching of silicon also decreased with increased pressure. This is a result of the increased ion density and the resulting increased ion scattering.

Etch selectivity, i.e. the etch ratio between two materials, is required when a film is being etched with respect to an etch mask or stop layer. From this experiments, the nickel:silicon etch selectivity was found as 10:1.

Meanwhile, an AFM measurement of the etched surface showed that the root-mean-square (rms) roughness of the sample etched with $SF_6/O_2/Ar$ gas mixture was as small as 25 nm. Exposing these contaminates to the RIE plasma serves to clean the surface thereby making the silicon surface more susceptible to the active etching species.

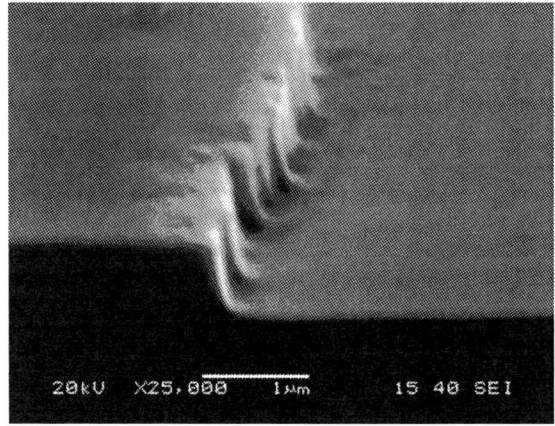

(a)

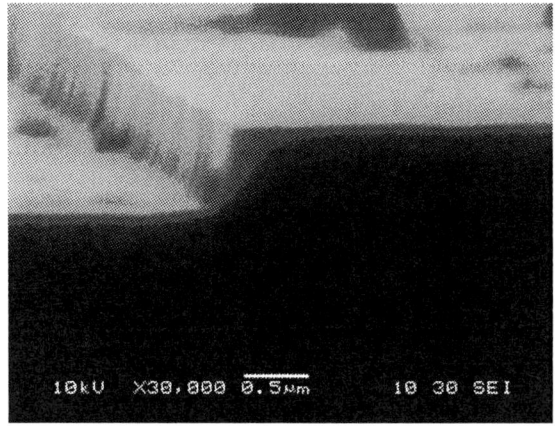

(b)

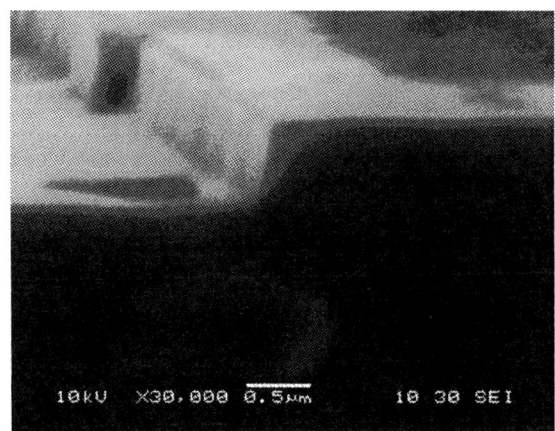

(a)

(d)

Fig. 3   Cross-sectional SEM micrograph of the etched patterns; (a) 20 mTorr , (b) 15 mTorr , (c) 10 mTorr and   (d) 5 mTorr. Process condition: $SF_6/O_2/Ar$ = 20/8/8 sccm, RIE power = 100 W, RF Power = 600 W, electrode temperature = 17 °C.

V. CONCLUSION

Reactive ion etching of silicon using a sulphur tetra fluoride (SF_6), oxygen (O_2) and argon (Ar) mixture based on the effect of pressure and DC bias has been investigated. In summary, it was found that the maximum etch rates was 2997 Å/min at 20 mTorr and the minimum was performed as 733 Å/min at 5 mTorr. In addition, highly anisotropic etch profiles were obtained and the RMS roughness was small as 25 nm as well as nickel:silicon etch selectivity was found as 10:1.

ACKNOWLEDGEMENT

The authors deeply acknowledge the support of IRPA RMK8 strategic research grant and Universiti Sains Malaysia.

REFERENCES

[1]   K. E. Peterson, *Proc IEEE*; **70**, 420, (1982).

[2]   L. Crepregi *Microelectron Eng*; **3**, 221, (1985).

[3]   L. M. Ephrath, *IEEE Trans Electron Devices*; **ED 28**, 1319, (1981).

[4]   H. F. Winters *J. Appl. Phys.* **49** 5165. (1978)

[5]   R. Legtenberg, H. Jansen, M. Deboer and M. Elwnspoek, *Journal of Electrochemical Society,* **142**, 6, (1995)

[6]   A. K Paul, A. K Dimri, S. Mohan, *Proc SPIE*; **3903**, 2, (1999).

[7]   S. Tachi, K. Tsujimoto, S. Arai, T. Kure *J Vac Sci Technol*; **A9**, 796. (1991).

[8]   G. C. Schwartz and P. M. Schaible, *J. Vac. Sci. Technol.* **16**, 410 (1979)

[9]   Patrick Verndonck, http://www.ccs.unicamp.br/cursos/fee107/downl oad/cap10.pdf

Highly Chemical Reactive Ion Etching of Silicon in CF_4 Containing Plasmas

Siti Azlina Rosli, Azlan Abdul Aziz and Haslinda Abdul Hamid
Nano Optoelectronics Research and Technology Laboratory (N.O.R)
School of Physics,
Universiti Sains Malaysia
11800 Minden, Pulau Pinang, MALAYSIA
Email: sazlina.zm05@student.usm.my, ctazlinausm@gawab.com

Abstract — Silicon (Si) etching process with low temperature is used for initial processing of semiconductor field emitter array cathodes for vacuum microelectronics, for optical reflecting gratings, for opto- and micromechanical devices. Most of them are performed by dry etching process especially in reactive ion etching (RIE) method with fluorocarbon-gases such as CF_4, CHF_3 and C_3F_8. In this work, results are shown that reactive ion etching of silicon using CF_4/H_2 is capable at meeting the requirement (anisotropy, high etch rate and high selectivity, simultaneously) similar to that by using reactive ion etching with fluorine-containing plasma. We have investigated the etching rate dependency on the percentage of hydrogen in the gas mixture, the total pressure and flow gas and found that by using a gas mixture with 33% of H_2, the optimum rate of Si is achieved. The etch rate are found to increase with voltage, attaining a maximum rate 1780 Å/min at -482V. Surface morphology of the etched samples is characterized by scanning electron microscopy and atomic force microscopy. The results revealed that the etched surface was anisotropic and the smoothness of the etched surface is comparable to that of polished wafer.

I. INTRODUCTION

As the Si-based device feature sizes continue to be scaled down, more rigid requirements to manufacturing submicron devices are being imposed, including high etch rate, anisotropy, profile control and selectivity to

the mask and to underlying layers materials. At present, the industry accomplished the etching in chlorine or bromine gas mixtures in the high density plasmas systems, either electron cyclotron resonance (ECR) or inductively coupled plasma (ICP), to meet the requirements [1-4]. Nevertheless, gases like Cl_2 or HBr are highly toxic and corrosive, and these gases may attack vacuum system components. Previous to the introduction of chlorine and bromine etch process, most etching was carried out in fluorine based plasma [5, 6]. The chemical reactivity of atomic fluorine with silicon demanded the use of sidewalls passivants to obtain anisotropic profiles.

Reactive ion etching (RIE) is widely used for the definition of critical features in the fabrication process of microelectronic devices and circuitry. As the integrated circuits industry continues its progress toward higher performance circuitry, circuit designers are pushing the minimum device feature sizes to smaller values. This has resulted in the placement of strict demands on the fabrication industry to successfully manufacture circuits with smaller line widths and thinner films [13]. Because of its widespread use for feature definition, control of reactive ion etching is of particular importance to this effort [14–18].

Carbon tetrafluoride (tetrafluorocarbon) CF_4 as a common etchant in microelectronics industries is mixed with oxygen and hydrogen, for example, to etch silicon and silicon dioxide. RF gas discharges are used to produce reactive radicals, in the gas phase. Understanding of the physical and chemical processes occurring in CF_4 and its mixture plasmas is necessary to achieve the optimization of processing plasmas [7-9].

During Si etching in CF_4/H_2 plasma, the adsorption of CF_2 radical on the Si surface is enhanced by ion bombardment, i.e. the sticking coefficient of CF_2 radical on the Si surface is increased in the presence of ion bombardment [10]. The difference in processes occurring at the sidewall and bottom of the microstructure determines the final profile of etched trench. The presence of ion bombardment enables high etching selectivity and straight sidewalls to be achieved [11]. Ion bombardment during RIE enhances chemical processes in the trench bottom and removes the reaction products. It increases etching rate in the vertical direction and etching anisotropy.

The choice of gas phase chemistry, pressure conditions (ion scattering in the sheath); reactant transport to the surface, and product transport away from the surface are the key factors for controlling the microscopic etch uniformity in the high aspect ratio etching. The results presented here are based on the controlled formation of sidewall passivating films to achieve good anisotropy using fluorine and hydrogen chemistry.

In this work, we report on the use of reactive ion etching (RIE) source to achieve highly anosotropic etching, high selectivity between mask and silicon and high etch rates of silicon than those obtain in RIE reactor using CF_4/H_2 plasmas based on the flow rates of hydrogen.

II. EXPERIMENT

The silicon wafers used in our studies were n-type with <1 0 0> orientation. Before patterning, the samples were cleaned with RCA cleaning, followed with piranha cleaning and DI water and dried using N_2 gas. Photo resist was spin coated onto all samples, forming ~ 1.5 µm layer. Then the samples were soft baked at 85°C for 10 minutes. The samples were then exposed through a mask with line width features as small as 50 µm. UV light was used to expose the samples. The resists patterns were then developed using AZ 400k developer until fully developed, diluted 1 to 4 with de-ionized water (DI), followed by a thorough rinse in DI. Following resist pattern definition, the wafers were hard baked in 110°C for an additional 10 minutes in order to harden the resist structures.

After that, nickel layer of thickness 0.7 µm was grown on silicon wafers by evaporator.

Then, acetone was used to remove the photo resist until lift-off process was succeeded. This patterned nickel layer was used as mask in our experiments for etching silicon. In etching experiments, an Oxford Plasmalab 80Plus RIE system (13.56 MHz) developed in the laboratory has been used for our studies using fluorine chemistry plasma. RF power was applied to a smaller Al electrode (12 cm diameter).

The samples are then mounted onto a resistively heated copper target assembly and loaded into the etching chamber. The etching process in silicon was carried out by using fluorine chemistry based on CF_4/H_2 in varying flow rates. The experiments were carried out at a chamber pressure of 25 mTorr, total gas flows of 48 sccm (24 sccm for CF_4 and 24 sccm for H_2), RF power of 600 W, electrode temperature of 17 °C and DC bias voltage of -465V to -487V. Etching time was 5 minutes.

Prior after exit from the etching system, the samples were dipped into $3H_2SO_4/1H_2O_2/1DI$ water at 45°C for 5 minutes to remove the nickel; in the case of Si samples this is followed by a dilute HF etch to remove any surface oxides. The etch rates of the samples were then measured from the depth of the etched features with a scanning electron microscope (SEM) after the removal of the nickel mask layer. Surface morphology, etch anisotropy, wall angle and sidewalls undercutting of the etched silicon as well as selectivity of nickel over silicon also were evaluated with SEM model JSM-6460 LV while ULTRAObjective AFM is used to measure the surface roughness.

III. RESULTS AND DISCUSSION

For these experiments, etch rates were evaluated as a function of gas flow rate and DC bias. The gas flow rates for H_2 were varied while CF_4 rates were held constant at 24 sccm for all experiments. The conditions consisted of 24 sccm of CF_4 and the total flow rate 24 sccm of H_2. Chamber pressure was held at 25 mTorr and RF power was 600W.

Prior to etching patterned samples, a simple set of experiments were conducted to get an understanding of how changes in gas composition affect the etch rate. For this experiment, the H_2 flow rate was varied from 0 sccm to 24 sccm meanwhile the CF_4 flow rates and the chamber pressure were held constant at 24 sccm and 25 mTorr, respectively. Etch rates were found to increase significantly

866

between 0 sccm to 8 sccm of H_2, peaking around 8 sccm and then dropping off slightly around 12 sccm. For gas H_2 flow rates of 12 sccm and higher, it was found that the etch process decrease and more than 24 sccm of H_2, the pattern was extremely damaged. For these reasons, we limited our H_2 gas flow rates to 24 sccm for the patterned samples.

Fig. 1 shows the etch rate of silicon as a function of flow rates of H_2. For these data points, samples were etched for a set amount of flow rates, removed, cleaned of nickel and then measured using a SEM. The etch time for all samples was 5 minutes. As can be seen from Fig. 1, initially, the etch rates increased with respect to the flow rates of H_2 until reached the maximum value of 1780 Å/min at 8 sccm H_2 (33% of the total gas of H_2). It was due to as fluorine atoms are an etchants of silicon, the silicon etch rate increases with increasing concentration of fluorine atoms. Unfortunately, the etch rates of silicon were then decreased abrupt and reach the minimum value of 20 Å/min for 25 sccm of H_2.

Henri Janseny *et al.* show that if work by the H_2 concentration is high (> 30%), polymerization occurs on all surfaces and etching stops [20]. This due to the effect of small amounts of H_2 reaction to CF_4 plasmas (CF_3^*, F*, and H*) is twofold. (1) H_2 reduces the F-atom density because of relatively inert HF formation and the Si etch rate is consequently reduced. (2) More important, H_2 reacts with CF_3^* forming polymeric precursors, such as CF***. As a result, a C_xF_y film will be formed on surfaces where ion bombardment fails such as the trench sidewalls. At even higher H_2 content, the plasma becomes H based and again etching is observed.

The effect of small amounts of H_2 reactives to CF_4 plasmas was showed at the Table 1. The etch rate in the CF plasma is an order of magnitude higher than the others. Both the plasma-phase etchant (e.g., F and CF_x) concentrations and surface reactions contribute to the etch mechanism [19].

Table 1 Gas for Si trench etching with their main plasma radicals, products and inhibitor.

Gas	Radicals	Products	Inhibitor
CF_4	F, CF_3	SiF_4	$Si_xC_yF_z$
CF_4/H_2	CF_3, F, H	SiF_4, HF, CHF_3	$Si_xC_yF_z$

From Table 1, it showed that a useful indicator of the predominance of etching (F*) over deposition (CF_x^*) the F/C ratio of the discharge is often used [20]. The F/C ratio is four for CF_4 and this ratio is lowered when extra Si which consumes F atoms is added or when CF_4 is mixed with H_2 or CH_4. The turn-over from deposition to etching is stimulated by ion bombardment. For a Si substrate, deposition is observed for small F/C ratios (<2). However, it is known that H-based plasmas (e.g. H_2) are etching Si and, identically, C_xH_y film formation might occur.

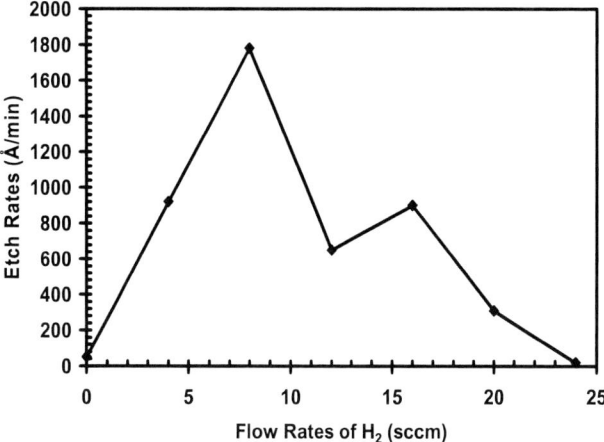

Fig. 1   Etch Rates (Å/min) versus various flow rate of hydrogen (sccm).

Fig. 2 showed the DC bias (-V) of silicon as a function of flow rates of hydrogen (sccm). From the Fig. 2, the DC voltage increased as the flow rate of H_2 increased. It showed that DC bias is also a function of the energy of the free electron. At higher concentration of hydrogen, the atoms of group VII became very prone to absorb any free electron which passes nearby. As such, these gases reduce the density of the free electrons in the plasma (or increase the number of negative ions). Therefore, the DC voltage will be monotonically increases as a function hydrogen flow rates.

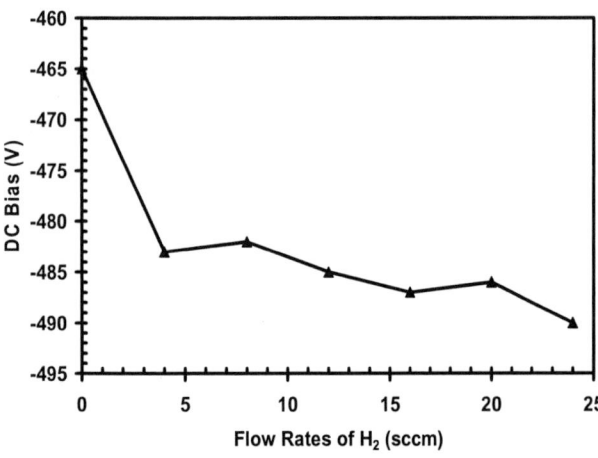

Fig. 2 DC bias (V) versus various flow rate of hydrogen (sccm).

Fig. 3 showed the cross-sectional SEM view of the etching anisotropy. It showed a vertical wall etched in silicon with smooth bottom surface using CF_4/H_2 gas mixtures at various flow rates of H_2. This trench is extremely anisotropic, with the selectivity of nickel : silicon was found about 10:1. These selectivity and sidewall slopes were directly measured from SEM micrographs of samples cleaved along a plane intersecting the etched trench. In addition, the anisotropy of the etching of silicon also decreased with increased flow rates of H_2. This is a result of the increased ion density and the resulting increased ion scattering.

(b)

Fig. 3 SEM pictures of the sidewalls in etched patterns; (a) 8 sccm H_2 and (b) 16 sccm H_2. Process condition: CF_4/H_2 = 24/24 sccm, RIE power = 200 W, RF Power = 600 W, electrode temperature = 17 °C.

For semiconductor applications, minimizing surface roughness is essential; the less surface roughness is the better the device's performance. Therefore, it is essential to find high etch rates to optimize device performance. The roughness of etched surfaces was measured by atomic force microscopy (AFM) and it showed that the root-mean-square (rms) roughness of the sample etched with CF_4/H_2 gas mixture was as small as 20 nm.

IV. CONCLUSION

We have shown the effects of H_2 to CF_4 interaction in RIE process. The highest etch rate obtained was 1780 Å/min at 8 sccm of H_2 and minimum etch rate was 20 Å/min for 25 sccm of H_2. It was explained as if the H_2 concentration is high (> 30%) polymerization occurs on all surfaces and etching stops. We have also shown that DC voltage decreased as the flow rate of H_2 increased. We conclude that DC bias is also a function of the energy of the free electron. Furthermore, the etched surface is of anisotropic type and has smooth sidewalls. The selectivity between nickel and silicon is 10:1 and (rms) roughness of the sample etched with CF_4/H_2 gas mixture is as small as 20 nm.

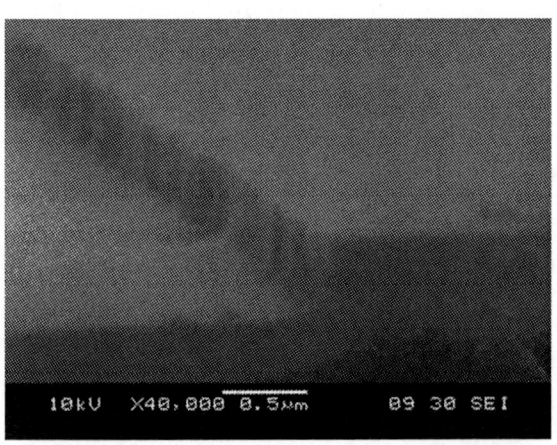

(a)

ACKNOWLEDGEMENT

This work was conducted under IRPA RMK8 strategic research grant. The support from Universiti Sains Malaysia is gratefully acknowledged.

REFERENCES

[1] J. M. Cook, D.E. Ibbotson, and D.L. Flamm, *J. Vac. Sci. Sci. Technol. B*, **8** 1 (1990).

[2] Seiji Sawakawa, Masami Sasaki and Yasuhiro Suzuki, *J. Vac. Sci. Sci. Technol. B,* **8** 6 (1990).

[3] Mutumi Tuda, Kenji Shintani and Hiroki Ootera, *J. Vac. Sci. Sci. Technol. A*, **19** 3 (2001).

[4] C. Caillat *et. al*, *Solid State Electronics*, 46 (2002).

[5] P. J. Matuso, B. E. E. Kasteimeier, J. J. Buelens, and G. S Oehrlein, *J. Vac. Sci. Sci. Technol. A,* **15** 4 (1997).

[6] I. Hassan, C. A. Pawlowics, L.P. Berndt, and N. G. Tarr, *J. Vac. Sci. Sci. Technol. A*, **20** 3 (2002).

[7] N. V. Mantzaris, A. Bondouvis and E. Gogolides, *J. Appl. Phys.* **77** 6169 (1995).

[8] A. Kono, M. Haverlag, G. M. W. Kroesen and F. J. de Hoog, *J. Appl. Phys.* **70** 2939 (1991).

[9] K. Kamimura and T. Makabe, *Jpn. J. Appl. Phys.* **38** 4429 (1999).

[10] M. Inayoshi, M. Ito, M. Hori, T. Goto, M. Hiramatsu, *J. Vac. Sci. Technol.* **A 16** 233–238 (1998).

[11] G. S. Oehrlein, Y. Kurogi,Mater, *Sci. Eng.* **R24** 153–184 (1998).

[12] Y. Kuo, "Reactive ion etching by technology in thin-film transistor processing", *IBM J. Res. Develop.* **36** (1992).

[13] Semiconductor Industry Association. The national technology roadmap for semiconductors (1994).

[14] A. Henck. Steven "In-process thin film thickness measurement and control", *SPIE Process Module Metrology, Control and Clustering*, **1594** 213 (1991).

[15] Jerry Stefani and Stephanie Watts Butler. "On-line inference of plasma etch uniformity using in situ ellipsometry", *J. Electrochem. Soc.*, **141**(5) 1387 May (1994).

[16] L. Vincent Tyrone, P. Khargonekar Pramod, A. Rashap Brian, L. Terry Fred Jr., and E. Elta Michael. "System identification and feedback control of a reactive ion etcher", In *Proc. of the 1994 American Control Conference* (1994).

[17] D. Zahorski, J. L. Mariani, L. Escadafals, and J. Gilles, "Spectroscopic ellipsometry: a new tool for "on-line" quality control", *Thin Solid Films*, **234** 412 (1993).

[18] J. Dalton Timothy, T. Conner William and H. Sawin Herbert, "Interferometric real-time measurement of uniformity for plasma etching", *J. Electrochem. Soc.*, **141** (7) 1893 July (1994).

[19] J. W. Coburn and H. F. Winters, *J. Vac. Sci. Technol.* **16** 391 (1979).

[20] Henri Janseny, Han Gardeniers, Meint de Boer, Miko Elwenspoek and Jan Fluitman, *J. icromech. Microeng.* **6** 14–28 (1996).

Photolithography Process Improvement for Thick Implant Resist Using 120°C Post-Apply Bake

S.I.Yet, E.C. Goh, Faith Lim, A.E. Ling, B.C. Lee, Y.K. Ng, W.B. Sheu
X-FAB Sarawak Sdn. Bhd.
1 Silicon Drive
Sama Jaya Free Industrial Zone
93350 Kuching, Sarawak, Malaysia
Email: siewing.yet@xfab.com
Phone: +60.82.354897

Abstract **Conventional I-line lithography process utilizes single post-apply bake temperature to unify and simplify the process. As design rule shrinks and mask field size increases, tighter specification is applied on non-critical implant layers, including thick implant resist with thickness typically 4.0μm and above. Poor uniformity for CD & overlay was observed for thick implant resist layer. Systematic uncorrectable overlay residue was observed from the overlay map. Cross-section analysis shows asymmetric resist profile existed, causing inaccurate signal reading during measurement. Besides, huge amount of resist out-gassing found contaminate the CD-SEM gun tip and causing problem during implant process.**

In this paper, the problems of thick implant resist layer is analyzed and the process improvement on thick implant resist layer by using higher post-apply bake temperature is introduced. The resist profile changed was checked in detail and the resist removal after implant was verified. As a result, both CD & overlay uniformity was greatly improved. New process with higher post-apply bake condition was fully qualified with comparable wafer yield.

Keywords: Overlay, CD, post-apply bake, post-exposure bake, resist, residue.

I. INTRODUCTION

In semiconductor fabrication, high resistance thick resist (typically 4.0μm thickness and above) is needed for imaging the well implant feature area to block high-energy & high-dosage ion implantation. Conventional I-line photolithography process using Novolak-DNQ based resist is used for imaging thick resist layer. Due to large feature size, majority I-line processes were normally categorized under non-criticality, and the process parameters (such as the temperature & time for post apply bake, post exposure bake, and post bake) was preferably unified, in such a way that minimum bake plate was used and no complex process parameters. As design rule shrinks and mask field size increases, lowering the CD & overlay tolerance has become the only solution to ensure precise patterning, including for non-critical I-line layers. Upon that, problems surfaced especially on thick resist layers, where CD & overlay result showed overkill instability out of the allowable range. Typically, CD and overlay variation of >150nm can be observed.

This paper first addresses detail diagnosis on the CD & overlay result; including investigation for large overlay residue & asymmetrical pattern profile. Then the hypothesis based on the basic resist chemistry is discussed. Based on the hypothesis, experimental works are carried out to demonstrate the solution for the profile & overlay residue problem. Process improvement is assessed based on actual process performance, including several process margin check & actual pattern shift. Finally, further verification on resist cleaning after implant and final wafer yield was furnished to assess any hidden impact for the process change.

Tools used for thick resist implant layer in this paper are Nikon I-line 4X stepper, TEL ACT-8 track, Hitachi CD SEM, KLA Archer10 overlay, and Applied SEM.

II. DIAGNOSIS

2.1 Overlay residue

Overlay residue is the remained uncorrectable miss-registration vector after subtracting all others wafer & mask factors. Systematic residue vector can be observed from most of the thick implant resist layers when the overlay boxes at 4 shot corners and shot center are measured. For some thick implant resist layers overlay residue vector can reach the range of >300nm. The residue vectors are similar across different exposure tools; the trend is unique from device to device. The residue is uncorrectable using both stepper and scanner model which has higher order for mask factors. Figure1 shows an example of overlay residue vector map. The systematic and constantly large residue vector is not related to exposure tool and mask performance, indeed badly impact the overlay performance and may lead to wrong correction.

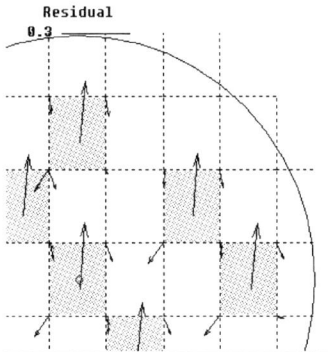

Fig.1 Overlay residue vector map for 4µm resist NW implant layer using 4X I-line stepper. Residue vector cannot be corrected by exposure system. Vector trend is similar across the whole wafer.

2.2 Asymmetric pattern profile

Upon inspection on the overlay boxes at the scribe line under SEM, asymmetrical tapered overlay box profile can be observed. The asymmetric profile is different across overlay boxes at shot center and 4 shot corners; asymmetric profile is similar within a wafer & from wafer to wafer. Figure2 shows the example of the asymmetrical overlay box profile. The resist pattern profile inside the main die does not having the same asymmetrical profile like the overlay box. This indicates that the overlay residue vector does not represent the real pattern shift; it is caused by the wrong signal reading on asymmetrical overlay box. CD bar pattern that is also tapered will cause inaccurate signal reading during SEM measurement; hence bad uniformity.

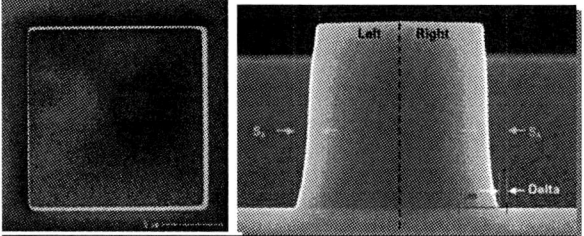

Fig.2 (left) Asymmetric overlay inner box under SEM; (right) Cross sectional view of overlay box profile

2.3 Mask pattern density

Upon checking the mask data, the asymmetrical overlay box profile is found closely related to the pattern density at surroundings, similar to the overlay residue. When the overlay mark located at the clear area, the residue vector and the profile bent degree is relatively small. Whenever the un-balance pattern density exists, the overlay residue vector increased significantly. Figure3 shows an example of overlay residue and overlay mark location within mask frame.

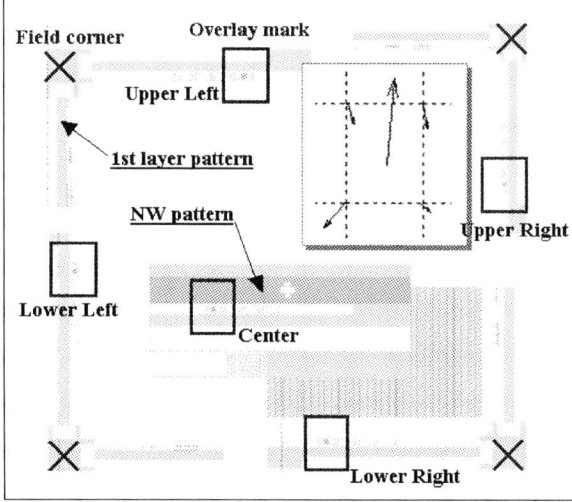

Fig.3 Overlay residue is related with the pattern density surrounding. Center overlay mark nearby the large strap shows high residue vector of ~300nm in y-direction.

III. HYPOTHESIS

I-line Novalak-DNQ based positive resist mainly consists of solvent, resin, PAC, and other contents.

Detail of resist chemistry will not be discussed in this paper. Post-apply bake of 90° after resist spin coating tends to remove the solvent in the resist film. Solvent evaporation occurred from the film surface, resist film undergoes shrinkage and hardening under high temperature. Low bake temperature introduces "wet" resist, which content high amount of solvent in the thick resist film, prior to exposure. Exposure of resist induces photochemical reaction and generates N_2 gas in the exposed region. "Wet" resist allows N_2 gas gathering and film swelling at exposure region. High temperature post exposure bake tends to further remove the excessive solvent and harden the resist film; whereas N_2 gas at the exposed region undergoes thermal expansion. Expansion rate of the film is affected by the exposure density and film solidity. As a result, uneven tapered resist profile can be observed depended on the surrounding pattern density. Figure4 shows simple illustration on the N_2 gas release during I-line exposure & hypothesis mechanism.

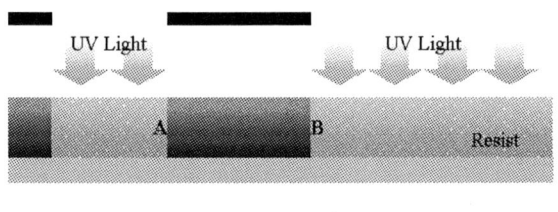

Fig.4 N_2 gas generated & gathered in the thick resist film undergoes thermal expansion during post exposure bake. Imbalance gas distribution & "wet" film condition causing expansion force B is greater than A. Asymmetrical tapered profile produced.

By increasing post-apply bake temperature, solvent content in the thick resist film can be reduced to produce more solid or harder film. This directly diminishes the thermal effects on the resist film, which leads to steeper and more symmetric resist profile. Overlay residue expected to be reduced

and signal reading on the resist pattern can be more accurate.

IV. EXPERIMENTAL, RESULTS, AND DISCUSSION

Experiments were carried out to verify the hypothesis. Conventional I-line resist bake condition which recommended by resist vendor as follows:

  i. Post apply bake: 90°C 60sec
  ii. Post exposure bake: 110°C 60sec

NW layer for 0.13µm flash which using 4µm I-line resist was selected for the experiment. Temperature settings range from 90 to 120°C with 60sec time setting was tested for post-apply bake.

4.1 Overlay residue, CD uniformity, & profile

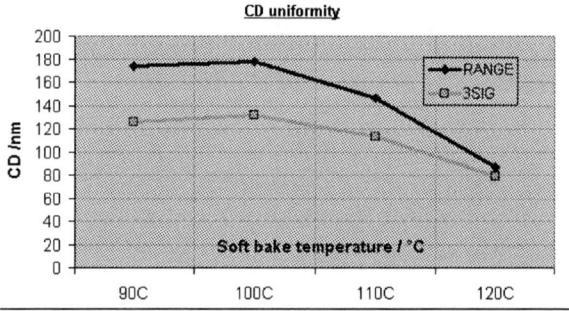

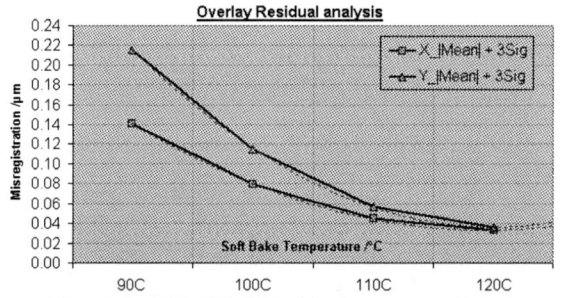

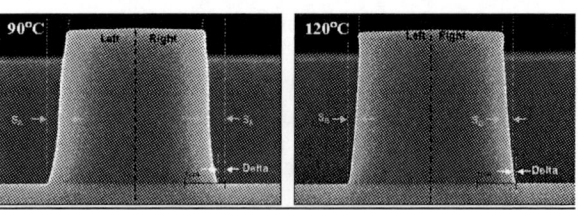

Fig.5 CD (top) & Overlay (middle) residue performance is improved after changing post-apply bake temperature from 90 to 120°C. Cross-sectional view (bottom) shows more symmetrical and steeper profile using 120°C post-apply bake temperature.

By increasing post-apply bake temperature, the solvent is effectively removed from the thick resist film, which produces more hardened resist film for patterning. Overlay residue can be reduced below 40nm for 3-sigma value. By having steeper profile, sharper signal provides better top-down SEM measurement repeatability and uniformity. CD uniformity improves 37% to 50nm sigma value. Resist cross-sectional view shows steeper profile & smaller top-bottom CD difference. Figure5 & figure6 shows the improvement of CD, overlay, and profile using higher post-apply bake temperature. The improved CD & overlay performance has good repeatability.

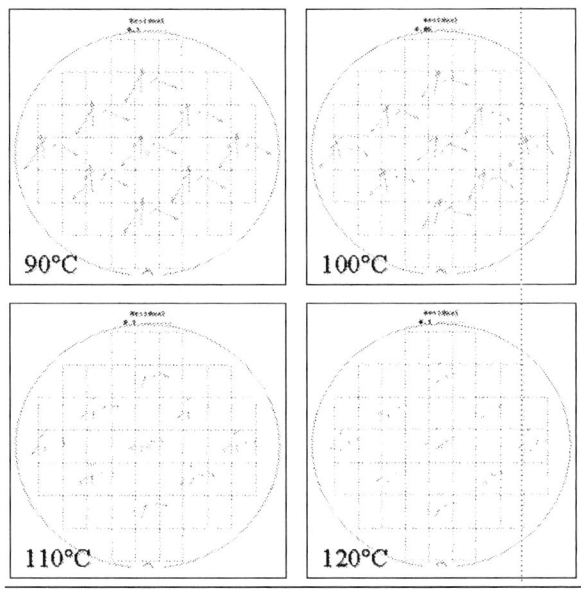

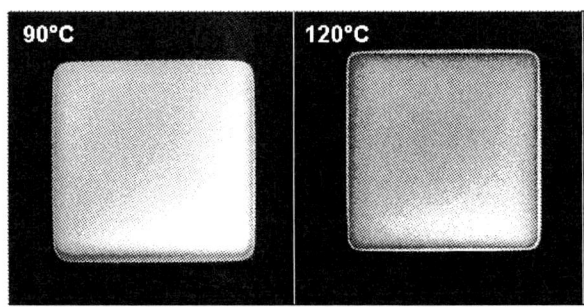

Fig.6 (Top) Overlay residue vector map using different post apply bake temperature. At 120°C the residue vector is reduced significantly. (Bottom) Top-down SEM image on overlay box shows symmetrical profile at 120°C.

4.2 Process margin check

By changing the post apply bake temperature from 90 to 120°C, required exposure energy is increased by ~20%, hence exposure time increased. The advantage is 5% of E/L (exposure latitude) gained by using 120°C post -apply bake, which make the process more stable. Other performances, i.e. resolution and DOF are not changed. Figure7 shows the process margin check result for different post-apply bake temperature.

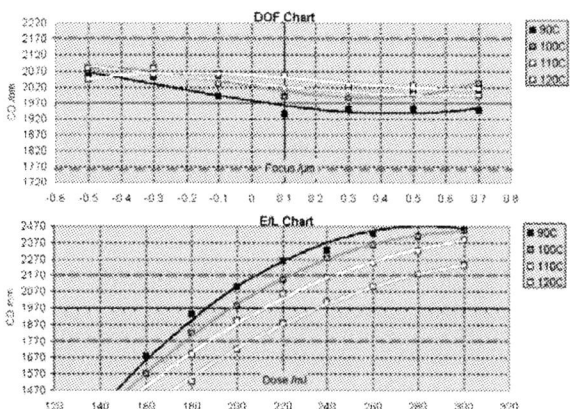

Fig.7 DOF (top) & E/L (bottom) chart for different post-apply bake temperature.

4.3 Others

Conventional thick resist patterned wafer suffered long period to reach required vacuum level during SEM measurement. High occurrence frequency of CD-SEM gun tip contamination is highly suspected caused by resist out-gassing especially for thick resist layer. After using 120°C post-apply bake temperature for thick resist, the duration for SEM reaching the vacuum level has been reduced significantly, which is similar to others thin resist patterned wafer.

Actual pattern shift was checked on the real cell resist pattern with respective to substrate STI pattern. Results shows less pattern shift and more symmetric pattern on 120°C.

Resist cleaning/ removal after implant process was verified after using 120°C post-apply bake temperature. Conventional ashing and sulfuric clean were used for the cleaning step. No residue or scum remained was observed. Final wafer yield was verified to be comparable/ not changed by using 120°C post-apply bake temperature.

V. CONCLUSION

Insufficient thermal treatment during conventional 90°C post-apply bake causes excessive solvent content in the thick & "wet" resist film, which allows pattern density related uneven thermal expansion afterwards. Asymmetrical pattern profile results in inaccurate signal reading and high overlay residue vector, which directly limit the CD & overlay performance. By increasing the post-apply bake temperature to 120°C, more solid and hardened resist film is produced for patterning, which results in more symmetric and steeper profile.

The requirement of using the 120°C post apply-bake is the higher exposure energy/ exposure time, and additional bake plate/module assigned for specific temperature. Improvement can be seen from better CD & overlay uniformity, small overlay residue vector, wider process margin, steeper & symmetrical resist profile (less pattern shift). In addition, less out-gassing can be observed especially during SEM measurement and implant step.

REFERENCES

[1] P.Rai-Choudhury, "Handbook of Microlithography, Micromachining, and Microfabrication Volume 1: Microlithography", 1997

A Versatile HSPICE Electro-Opto-Thermal Circuit Model for Vertical-Cavity Surface-Emitting Lasers

E. Sooudi, V. Ahmadi, and M. Soroosh

Dept. of Electrical Engineering, Tarbiat Modares University, Tehran, Iran
P.O.Box: 14155-143
E-mail: sooudi_e@alum.sharif.edu; v_ahmadi@modares.ac.ir

Abstract - **In this paper, we propose a circuit model for vertical-cavity surface-emitting laser (VCSEL). The model is based on carrier, photon, and thermal rate equations, with addition of carrier leakage current as the main source for output power rollover phenomenon. Also, presumed distribution of carriers and photons are used in rate equations, and the spatially dependent equations are converted to spatially independent rate equations. The effect of carrier diffusion, input current, and temperature in transient turn-off behavior of laser for pulsed operation are analyzed. We also study the thermo-temporal behavior of laser in pulsed operation and determine the effects of input pulse on output power.**

I. INTRODUCTION

Vertical-cavity surface emitting lasers (VCSEL) with inherent features such as, single-longitudinal mode, small dimensions, cost efficiency, and ability of usage in 1-D and 2-D arrays has been considered as the hot topics in the field of active optoelectronics devices nowadays.

Circuit-model has been introduced as a good option for system-level analysis for many years [1].

In this paper, we present an HSPICE comprehensive circuit model for VCSEL analysis.

In section 2, first, we present gain and leakage current model, then, we will present rate equations including thermal and spatial dependencies. In section 3, HSPICE simulation results will be discussed.

II. MODEL THEORY

In this model, effective parameters on gain such as temperature and carrier dependency in addition to leakage current with its thermal and carrier dependence will be introduced.

A. Gain Model

We assume gain model, $G_o(N - N_t)$, in which N_t and G_o are transparency density and gain constant respectively. We add thermal and carrier dependence of these parameters. Because of gain-cavity detuning (miss-alignment between maximum gain and cavity resonance wavelengths) thermal behavior of gain is not monotone [2], [3].

Thermal dependence of gain is defined by (1) [1]:

$$G(T) = G_0 \frac{a_{g0} + a_{g1}T + a_{g2}T^2}{b_{g0} + b_{g1}T + b_{g2}T^2} \qquad (1)$$

In which, G_0 is constant, agi and bgi are calculated using curve fitting. It has to be noted that, G(T) is differential gain. Moreover, we can approximate transparency density as a polynomial function of temperature:

$$N_t(T) = N_{t0}(C_{n0} + C_{n1}T + C_{n2}T^2) \qquad (2)$$

N_{t0} and C_{ni} coefficients are found by curve fitting. Therefore, full gain model using (1) and (2) will be:

$$Gain \equiv G(T)(N - N_t(T)) \qquad (3)$$

Where *N* is total carrier number.

B. Carrier Leakage Model

Leakage current modeling is done on the basis of thermal change of quasi-Fermi energy levels (ΔE_{fcv}) [1]:

$$I_l = I_{lo} \exp\left(\frac{-a_0 + a_1N_0 + a_2N_0T - a_3/N_0}{T}\right) \qquad (4)$$

C. Rate Equations

We choose cylindrical coordinate for VCSEL modeling because of its special geometry. Spatial dependent rate equations are introduced as:

0-7803-9730-4/06/$25.00 ©2006 IEEE

$$\frac{\partial N(r)}{\partial t} = \frac{\eta_i I(r)}{q} - \frac{G(T)(N(r) - N_t(T))S(r)}{1 + \varepsilon S_o} - \qquad (5)$$

$$\frac{N(r)}{\tau_n} + \frac{L_{eff}^2}{\tau_n} \cdot \left(\frac{\partial}{r \partial r} \left(\frac{r \partial(N(r))}{\partial r} \right) \right) - \frac{I_l}{q}$$

$$\frac{\partial S(r)}{\partial t} = -\frac{S(r)}{\tau_p} + \frac{\beta N(r)}{\tau_n} + \qquad (6)$$

$$\frac{G(T)(N(r) - N_t(T))S(r)}{1 + \varepsilon S_o}$$

Where $N(r)$ is lateral carrier distribution, I is injection current, $S(r)$ lateral photon distribution, S_o total photon number, η_i current injection efficiency, τ_n carrier life-time, τ_p photon life-time, ε gain compression factor, β spontaneous emission coupling constant, and L_{eff} effective carrier diffusion length.

The spatial distribution of photon is defined as the analytical solution for step index waveguides based on Bessel eigenvalues [1]. It has to be mentioned that, this approximation is used only for strongly index-guided and mostly weakly index-guided VCSEls.

For carrier distribution we define (7):

$$N(r) = N_0 - \sum_{i=1}^{\infty} N_i J_0 \left(\frac{\sigma_i r}{W} \right) \qquad (7)$$

Where σ_i is i^{th} root of first order of Bessel function $(J_0(x))$. N_0 is average total carrier number and N_i is carrier number of ith radial mode.

By substituting (7) and photon distribution in (5) and (6) with some mathematical transformations, and using orthogonality of first and second Bessel functions we have:

$$\frac{dS_0}{dt} = -\frac{S_0}{\tau_p} + \frac{\beta N_0}{\tau_n} + \frac{G(T)(\gamma_0 N_0 - \gamma_1 N_1 - \gamma_0 N_t)S_0}{1 + \varepsilon S_0} \qquad (8)$$

$$\frac{dN_0}{dt} = \frac{\eta_i I}{q} - \frac{N_0}{\tau_n} + \frac{G(T)(\gamma_0 N_0 - \gamma_1 N_1 - \gamma_0 N_t)S_0}{1 + \varepsilon S_0} \qquad (9)$$

$$\frac{dN_1}{dt} = \frac{-N_1}{\tau_n}(1 + h_1) - \frac{\eta_i I}{q} \qquad (10)$$

$$+ \frac{G(T)(\phi_0 N_0 - \phi_1 N_1 - \phi_0 N_t)S_0}{1 + \varepsilon S_0}$$

Where γ_i and ϕ_i constants are calculated using overlapping integrals of carrier and photon transverse distribution.

In (10), h_1 is equal to $(\sigma_1 L_{eff}/w)^2$ which shows diffusion effect for second carrier mode.

Thermal model includes thermal rate equation as (11) where we assume thermal power (difference between input and output optical power) is totally dissipated in the active layer.

$$T = T_o + (I_{tot}V - P_o)R_{th} - \tau_{th}\frac{dT}{dt} \qquad (11)$$

Where T_o is heat sink temperature, P_o output optical power, V device's input voltage, I_{tot} total input current (leakage and effective current), R_{th} thermal resistance of laser, and τ_{th} is thermal time constant. We assume uniform radial temperature distribution in the active layer. Moreover, output power is calculated versus photon number through $P_o = k_{fo}S_o$ where k_{fo} is coupling constant between photon number and output optical power.

Main laser parameters are shown in table 1 and the ones extracted by curve fitting are listed in table 2.

III. SIMULATION RESULTS

In the model, (8)-(10) are converted to circuit elements, which can be used in HSPICE. Additionally, we use some parameter scaling for improvement of simulation convergence [4]:

$$P_o = (v_m + \delta)^2 \qquad (12)$$

$$N_i = Z_n v_{ni} \qquad (13)$$

Where, Z_n and δ are calculation constants. Moreover, electrical model for laser consists of a simple diode with 100 Ohms series resistance.

A. Static Behavior

Output power versus input current for different ambient (heat-sink) temperatures is shown in Fig. 1. It can be seen that, with increasing temperature, maximum output power reduces, threshold current increases, and the thermal rollover current decreases. It can be said that by increasing input current, difference between input and leakage currents (effective current) reaches to its maximum value and after that by increasing the input current, effective injected current decreases and we see thermal rollover phenomenon. Because of high sensitivity of leakage current to temperature [1], which is included in the model, by increasing of temperature, strong rising in leakage current leads to lower maximum output power and rollover current (Fig. 1).

B. Dynamic Behavior

In simulation of dynamic behavior of device,

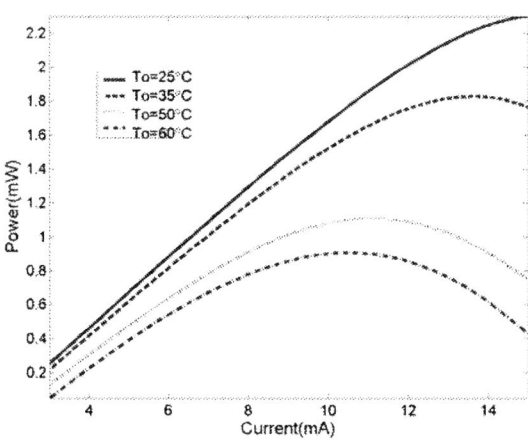

Fig. 1 Output power vs. input current for different sink temperatures 25, 35, 50, and 60 °C.

we analyze characteristics of pulsed operation with changing some parameters. In Fig. 2, we show output power in pulsed operation (large signal) for two different temperatures. In room temperature (dashed curve), we see that, slightly after low level of input pulse, a secondary pulse with a peak appears. This phenomenon is one of the important effects of non-uniformity in photon and carrier lateral distribution. During laser turn-off, the hole generated in carrier distribution because of SHB (spatial-hole burning) effect is filled as a result of excess carrier diffusion from hole vicinity. Therefore, temporal hole filling causes a transient stimulated emission and the secondary pulse arises.

Changing h_1 can affect the temporal evolution and magnitude of the secondary pulse significantly. As a rule of thumb, for h_1 values between 5 and 10, secondary pulse's power reaches to about 12% of the output power. Thus, owning to the chromatic dispersion and nonlinear effects in optical fibers, this pulse might disturb

safe detection of the pulse in the end of optical fiber.

In Fig. 2 (the solid curve) power of secondary pulse is removed. By increasing temperature, threshold current increases, while low level of input current is fixed. Consequently, during laser turn off, excess carriers reduce and it avoids carrier and the secondary stimulated emission will not occur. So, here is a trade-off between lower threshold current (reduction of temperature) and lower power of secondary pulse (increment of temperature). In thermal gain model, minimum threshold temperature is less than the range of temperature changes.

Fig. 3 shows output pulse for two different low levels of input currents. As shown, by reduction of low level slightly less than threshold current, secondary pulse power drops off for the reason like the previous case (increment of temperature). However, by increasing input low-level a little more than threshold, in addition to secondary pulse, the tertiary pulse comes out. Decreasing the input pulse reduces large signal speed of laser. Then, it is a trade-off between secondary pulse power, and speed.

C. Thermo-Temporal Behavior

As shown in Fig.1, (solid curve) simulated CW output power for 15mA input current, is 2.3mW, however, for pulsed operation (pulse width 10ns) and the same value for high level of input current pulse output power is 3mW (dashed curve in Fig. 2). Because of thermal slow exponential temporal behavior, after few microseconds device's temperature reaches to its steady state value. Thus, steady state threshold current is more than transient or pulsed threshold current and we have less output power. In Fig. 4,

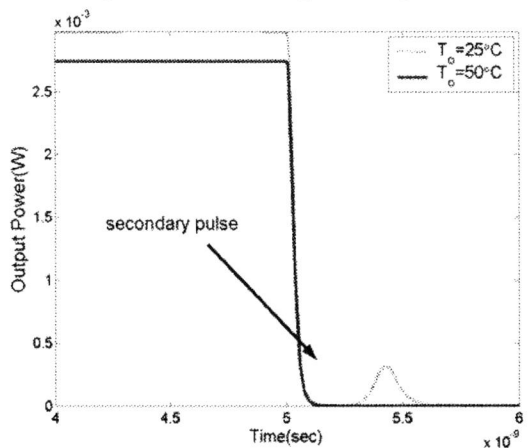

Fig. 2 Output low level pulse for input pulse values of 1.5 and 15mA and two different temperatures.

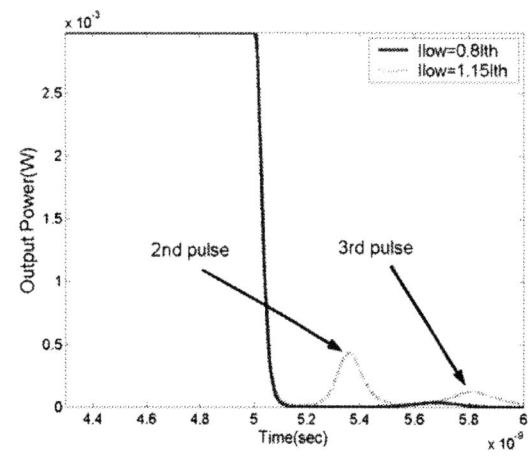

Fig. 3 Output pulse for two values of low-level input pulse.

we see about 0.15mW reduction of output power envelope for 15mA input current pulse during five microseconds. The dependence of steady-state average output power and cavity temperature for different duty cycles is illustrated in Fig. 5. By increasing duty cycle, the cavity temperature increases; in contrast, output power decreases. Increasing the duty cycle, period, and high level of input pulse, leads more rising in steady state average temperature and the output power temporal envelope drops more.

It is worth noting that, structure modeled in simulation, have built-in optical waveguiding and thermal lensing is not important here unlike the gain guided VCSEL structures.

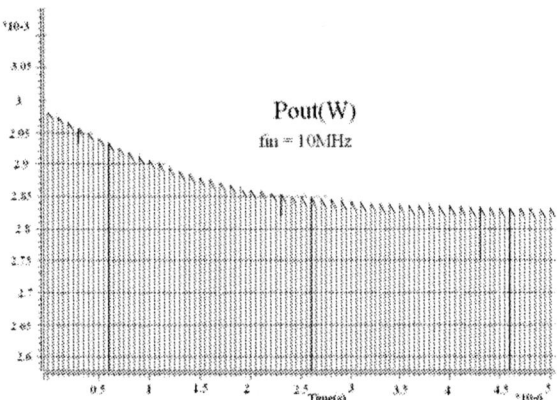

Fig. 4 Transient temporal behavior of output power in pulsed operation with 15mA high-level pulse current and 10MHz frequency.

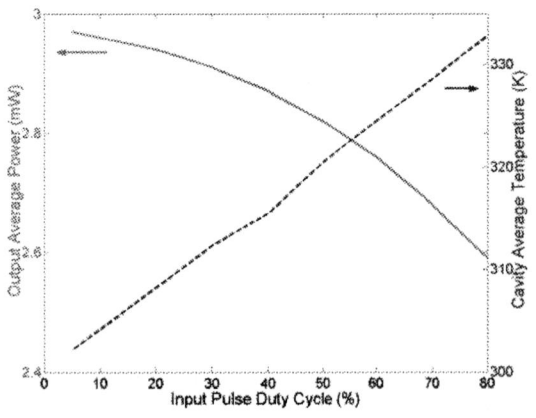

Fig. 5 Steady-state average output power and cavity temperature for pulse operation with currents, 1.2 and 15mA vs. duty cycle.

TABLE 1
LASER PARAMETERS AND VALUES

Symbol	Parameter	Value
τ_n	Carrier life time	2.5×10^{-9} S
τ_p	Photon life time	2.5×10^{-12} S
β	Spontaneous emission coupling factor	2×10^{-7}
ε	Gain compression factor	5×10^{-7}
η_i	Injection efficiency	1
G_0	Gain constant	3×10^4 s^{-1}
k_{fo}	Output power coupling factor	1.5×10^{-8}
N_{t0}	Transparency carrier number	10^{-7}
R_{th}	Thermal resistance	900 (C/W)
τ_{th}	Thermal time constant	10^{-6} s
T_o	Ambient temperature	25°C
h_1	Second mode carrier diffusion constant	15

TABLE 2
LASER PARAMETERS EXTRACTED BY CURVE FITTING

Parameter	Value
Gain Parameters	$G_0 = 3 \times 10^4$ s^{-1}, ag$_0$=-0.4, a_{g1}=0.00147K^{-1}, a_{g2}=7.65$\times 10^{-7}$K^{-2} b_{g0}=1.3608, b_{g1}=0.00974K^{-1} b_{g2}=1.8$\times 10^{-5}$K^{-2}
Transparency Parameters	$N_{T0} = 10^7$, C_{n0}=-1, C_{n1}=0.008K^{-1} C_{n2}=6$\times 10^{-6}$K^{-2}
Leakage Current Parameters	I_{LO}=9.61, a_0=4588.24K a_1=2.12$\times 10^{-5}$K , a_2=8$\times 10^{-8}$, a_3=9.01$\times 10^9$K
Overlap Integral Constants	$\gamma_{0=1}$, γ_1=0.3797 $\varphi_0 = 2.3412$, $\varphi_1 = 1.8193$

IV. CONCLUSION

In this paper, we proposed a comprehensive electro-opto-thermal circuit-model for VCSEL, considering leakage current and transverse distribution of carriers and photons. In CW operation of device by increasing the temperature, because of high thermal sensitivity of leakage current, maximum output power and thermal rollover current decreases, however, threshold current increases. Moreover, during laser turn off dynamics, by lowering input low-level pulse slightly below threshold current, or increasing the sink temperature, secondary pulse's power suppresses. Furthermore, because of transient thermal behavior, pulsed operation power is larger than CW for the same currents and increasing pulse duty cycle, period, and high-level of current rises steady-state average cavity temperature, and we have less steady-state output power. Choosing duty cycle less than (10%) will suppress thermal induced reduction of steady-state output power.

REFERENCES

[1] P. V. Mena, J. J. Morikuni, S.-M. Kang, A. V. Harton, and K. W. Wyatt, "A Comprehensive circuit-level model of vertical-cavity surface-emitting lasers," *IEEE J. Lightwave Technol.*, vol. 17, no. 12, pp. 2612-2632, Dec. 1999.

[2] C. Wilmsen, H. Temkin, L. A. Coldren, Vertical-Cavity Surface-Emitting Lasers, design, fabrication, characterization, and applications, p. 36, Cambridge University Press, 1999.

[3] J. W. Scott, R. G. Geels, S. W. Corzine, and L. A. Coldren, "Modeling temperature effects and spatial hole burning to optimize vertical-cavity surface-emitting laser performance," *IEEE J. Quantum Eelectron.*, vol. 29, no. 5, pp. 1295-1308, May. 1993.

[4] S. A. Javro and S. M. Kang, "Transforming Tucker's linearized laser rate equations to a form that has a single solution regime," *IEEE J. Lightwav. Technol.*, vol. 13, no. 9, pp. 1899-1904, 1995.

[5] J. J. Morikuni, P. V. Mena, A. V. Harton, K. W. Wyatt, and S.-M. Kang, "Spatially Independent VCSEL Models for the Simulation of Diffusive Turn-Off Transients," *IEEE J. Lightwave Technol.*, vol. 17, no. 1, pp. 95-102, Jan. 1999.

Adopting Electroabsorption Modulator for the WLAN 802.11a Radio over Fibre System

Syamsuri Yaakob[1], Mohd Azmi Ismail[1], Romli Mohamad[1], Mohamed Razman Yahya[1], Abd.Fatah Awang Mat[1], Mohd. Ridzuan. Mokhtar[2], Hairul Azhar Abdul Rashid[2]

[1]Telekom Research and Development Sdn. Bhd., Idea Tower I & II, UPM-MTDC, Technology Incubation Centre One, Lebuh Silikon, 43400 Serdang, Selangor, Malaysia.
[2]Faculty of Engineering, Multimedia University, Jalan Multimedia,, 63100 Cyberjaya, Selangor, Malaysia.
Email: syamsuri@tmrnd.com.my

Abstract: **This paper presents the evaluation of radio over fibre (ROF) capability with electroabsorption modulator (EAM) as a downlink photodetector (PD) and uplink radio frequency (RF) modulator. The results show that the EAM device has potential to be adopted as the transceiver at the remote antenna unit (RAU) for the system. The 130m ROF system with biased EAM is capable to obtain a symmetrical WLAN 802.11a data rate of 18Mbps with laser diode output power of 8.7dBm.**

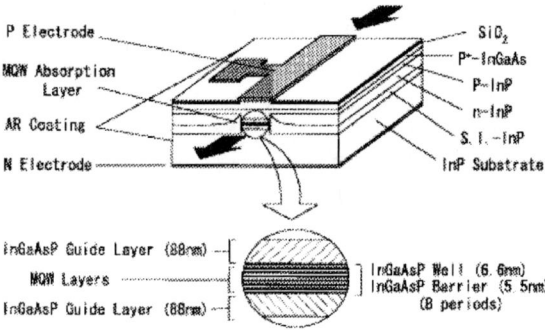

Figure 1: Schematic structure of the EAM (Courtesy of Kaneko et al)

I. INTRODUCTION

RADIO over fibre (ROF) technology, which allow radio frequency (RF) transmission over the optical fiber has attracted much attention due to its promising performance and operational cost reduction. It is very useful in order to realise the distributed antenna system (DAS) by connecting the central station (CS) with various remote antenna units (RAU) with optical fibre links. ROF is also attractive because it is transparent to bandwidth or modulation format, allow small remote stations and can realize centralised operation [1]. Basic ROF system consists of a laser source and an optical modulator at the CS, an optical fibre as the transmission medium and a PD at the RAU. [2] has proposed EAM as a combined PD and modulator at the RAU so that no laser is needed there, hence no control circuitry which make the ROF system simpler and cheaper. Also, [3] demonstrated the use of un-biased EAM to realize the low cost passive picocell concept with 100m to 100m of fibre distance for short range high capacity wireless system especially for WLAN.

Basically, EAM is a semiconductor device that able to modulate RF signal externally by controlling the intensity of a laser beam via an electric voltage. It utilizes the Franz-Keldysh effect, i.e., a change of the absorption spectrum caused by an applied electric field for its operation. EAM also exploits the quantum confined Stark effect in a quantum well structure for achieving a high extinction ratio [4]. Most EAMs are made in the form of a waveguide with electrodes for applying an electric field in a direction perpendicular to the modulated light beam. The schematic structure of EAM is shown in Figure 1 [5].

EAM operates at low voltage of a few volts (for example less than 1V in [6]). For analog applications, EAM as an external modulator links are very useful for microwave transmission systems because of their low transmission loss, high compatibility with microwave systems, good third-order intermodulation (IMD3) and spurious-free dynamic range performances [7], higher modulation bandwidth and reduced chirp compared to the direct modulation. Another convenient feature is that an EAM can be integrated with a distributed feedback laser (DFB) diode on a single chip to form a

0-7803-9730-4/06/$25.00 ©2006 IEEE

transmitter in the form of a photonic integrated circuit [8-10].

The WLAN 802.11a system uses the orthogonal frequency division multiplexing (OFDM) together with various modulation schemes to maximize the transmission data rate and EVM to evaluate the system's performance to obtain the highest possible data rate. EVM is the figure of merit to evaluate the WLAN system performance. It is a resultant vector measurement comprised of all the signal impairment vectors. EVM must be ensured to be relatively low in order to ensure as high as possible data received at the other end of the ROF system as shown by IEEE802.11a specifications given in Table 1.

Table 1: 802.11a standard – Error vector magnitude vs data rate.

Data Rate (Mbps)	Modulation Schemes	EVM (%)
12	QPSK	31.6
18	QPSK	22.3
24	16-QAM	15.8
36	16-QAM	11.2
48	64-QAM	7.9
54	64-QAM	5.6

In this paper, we evaluate the performance of ROF system with EAM as the transceiver at the RAU for the WLAN 802.11a. We use EVM as a tool for the evaluation based on the WLAN standard.

II. ROF SYSTEM WITH EAM DEVICE AS PHOTODETECTOR (DOWNLINK)

Section I describes on the concepts behind the ROF system, EAM device and WLAN EVM specification. This section explains the ROF system with EAM as the PD.

A. Experiments

The setup of the ROF system using EAM as the PD is shown in Figure 2. A CW DFB laser diode with a wavelength of 1550nm was used as the light source. A 5.8GHz RF signal generated by an RF signal generator was used to directly modulate the laser output. Next, the modulated light was fed to the EAM via a standard single mode fibre (SSMF) of 130m length. The EAM

used had a maximum input optical power specification of 15dBm. The EAM detected the modulated optical signal and converted to electrical signal at its upper port. The EAM's upper port was connected to a 50 ohm SMA connectorised RF cable to a real time spectrum analyzer (RTSA) that captured the RF waveform for analysis. The -1.1V bias supply of was given to the EAM via a bias tee connection through its lower port. Next, instead of using the signal generator, a vector signal generator (VSG) was used to generate several WLAN 802.11a signals with different data rates and the signals were used to directly modulate the laser output. The same RTSA was used to demodulate the signals and display the EVM and constellation diagrams of the signals.

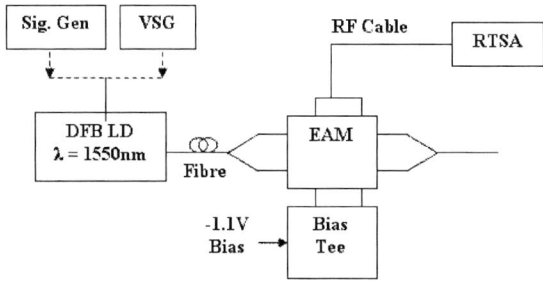

Figure 2: Experimental setup with EAM as the PD.

B. Results

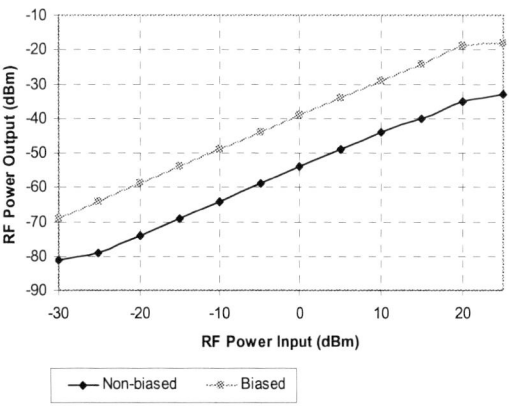

Figure 3: RF output signal vs the RF input signal when EAM is used as a PD with laser diode input power of 8.7 dBm.

Figure 3 shows the system's output when the RF input power at 5.8GHz was varied from -30dBm to 25dBm for the biased and unbiased condition of the EAM with optical input power of 8.7dBm. We can see that the unbiased condition achieved linearity after RF input power of -25dBm. Although both conditions have the same gradient after the RF input power of -25dBm, however, biased condition shows 15dB advantage of higher output power over the unbiased condition before they become nonlinear beyond the RF input power of 15dBm. Hence, we will use the biased EAM for the rest of the experiment.

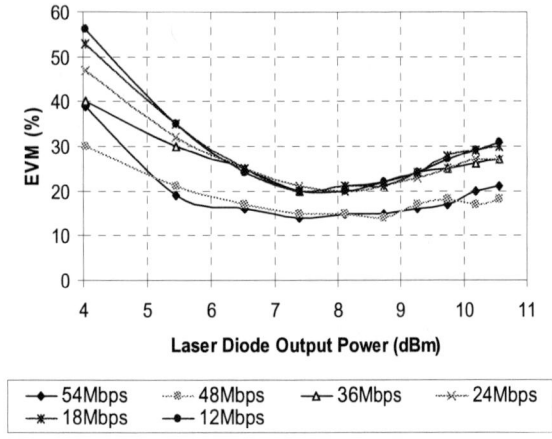

Figure 4: EVM vs laser diode output power at the RF input power of -5dBm.

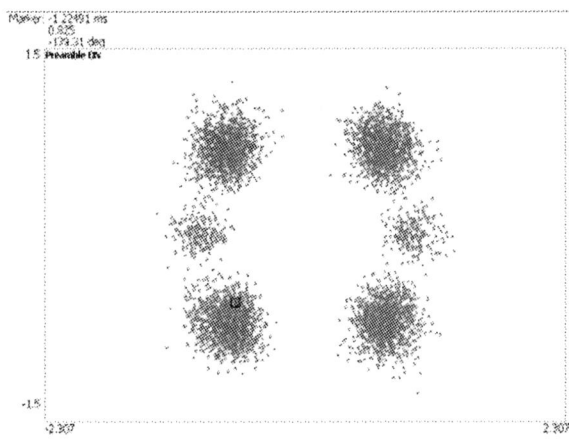

Figure 5: WLAN 802.11a 18Mbps QPSK constella-tion at the laser diode output power of 7.4dBm and the RF input power of -5dBm. The EVM measurement is at 20%.

Figure 4 illustrates the graph of EVM performance of various data rates transmission through the ROF system with the RF input power of -5dBm. From the graph, we can observe that the system has its lowest EVM when optical input power is between 7dBm to 9dBm. From the result, the 18Mbps data rate transmission achieves an EVM reading of 20% at the laser diode output power of 7.4dBm, which complies with the EVM specification of 22.3% given in Table 1.Therefore, we can demonstrate that the

ROF downlink with EAM as PD has the capability to transmit up to 18Mbps of WLAN 802.11a with laser diode output power of 7.4dBm and RF input power of -5dBm. Its constellation diagram is shown in Figure 5.

III. ROF SYSTEM WITH EAM DEVICE AS RF MODULATOR (UPLINK)

This section explains the ROF system with EAM as the modulator.

A. Experiments

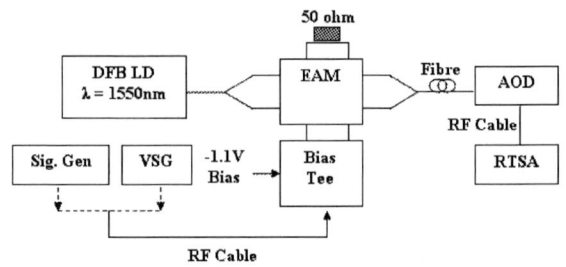

Figure 6: Experimental setup with EAM as the modulator.

Figure 6 outlines the setup of the ROF system using EAM as the modulator. The unmodulated 1550nm light source was used instead and was launched into the EAM via its optical input port. The 5.8GHz RF signal from the signal generator was applied to the lower port of the EAM via a bias tee together with -1.1V bias. This time, the EAM's upper port was terminated with a 50 ohm load. Then, the EAM's modulated optical signal output was launched into a 130m length SSMF and detected at the other end by an amplified optical detector (AOD) with the gain of 35dB before it was analysed by the RTSA. Similarly, instead of using the signal generator, a vector signal generator (VSG) was used to generate several WLAN 802.11a signals with different data rates and fed to the lower port of the EAM. Next, the EAM's modulated optical signal output was launched into a 130m length SSMF and detected at the other end by an amplified optical detector (AOD). The same RTSA was used to demodulate the signal for further analysis.

B. Results

The EAM seemed to absorb some part of the light during detection that we had monitored at the upper port and used the remaining light to modulate the RF signal from its lower port. The graph in Figure 7 shows the system's response with the RF input power of -10dBm and optical input power of 8.7dBm. In the graph, we can see that the RF output power decreases with the increase of frequency as expected. The output at the frequency of 5.8GHz is 9dB lower than at the 1GHz.

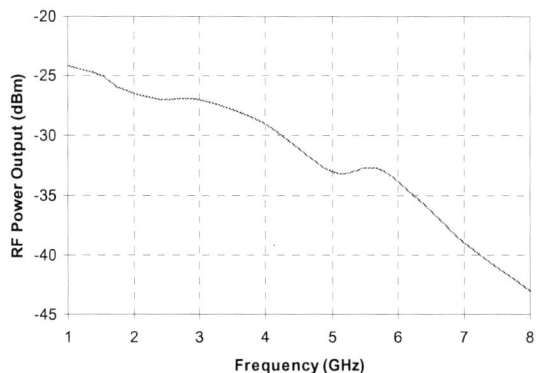

Figure 7: System's response with the RF input power of -10dBm and optical input power of 8.7dBm.

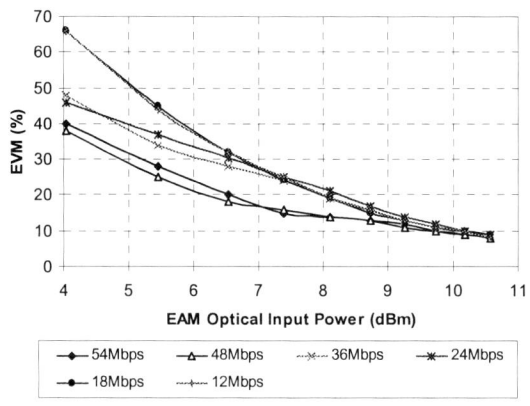

Figure 8: EVM vs the optical input power at the RF input power of -10dBm.

Figure 8 illustrates the graph of EVM performance of the various data rates transmission through the ROF system with EAM as the modulator at the RF input power of -10dBm. From the graph, we can observe that the EVM depends on the EAM's optical input power for the best system performance. EVM drops accordingly with the increase of the optical input power which will result in a better ROF system. From the result, the transmission of 36Mbps data rate achieves an EVM reading of 10% at the EAM optical input power of 10dBm, which complies with the EVM specification of 11.2% in Table 1. Therefore, we can demonstrate that

the ROF uplink with EAM as the modulator has the capability to transmit up to 36Mbp of WLAN 802.11a with EAM optical input power of 10dBm and RF input power of -10dBm. Its constellation diagram is shown in Figure 9.

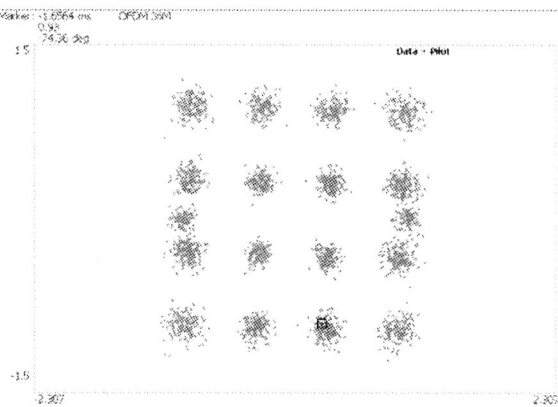

Figure 9: WLAN 802.11a 36Mbps 16-QAM constellation at the optical input power of 10dBm and the RF input power of -10dBm. The EVM measurement is at 10%.

IV. DISCUSSIONS

Adopting EAM in the ROF system is a good idea since it can perform both as the PD and modulator at the same time. We have demonstrated that the biased downlink system has the capability to transmit up to 18Mbps of WLAN 802.11a with laser diode output power of 7.4dBm and RF input power of -5dBm. Whereas, the uplink system has the capability to transmit up to 36Mbps of WLAN 802.11a with EAM optical input power of 10dBm and RF input power of -10dBm.

Although we can achieve better EVM reading by increasing the EAM optical input power and RF input power, there are issues that need to be considered in order to adopt EAM as the uplink and downlink transceiver for the WLAN 802.11a ROF system. Choosing laser diode optical output power at 7.4dBm may lower the uplink data rate to 12Mbps. Similarly, increasing laser diode optical output to 10dBm will cause the downlink EVM to be noncompliant at any data rate. Likewise, increasing the RF input power at some points will invite the non-linear components to interject the system performance resulting in the intermodulation effect which can limit the system performance by increasing the EVM, hence limiting the data rate. From the results, for optimum performance, the 130m ROF system shall use laser diode output power of 8.7dBm with biased EAM to obtain a symmetrical WLAN 802.11a data rate of 18Mbps.

V. CONCLUSION

It has been demonstrated that the 130m ROF system with biased EAM gives 15dB higher output power compared to the unbiased EAM. The EAM device has potential to be adopted as the transceiver at the remote antenna unit (RAU). For optimum performance, the 130m ROF system shall use laser diode output power of 8.7dBm with a biased EAM to obtain a symmetrical WLAN 802.11a data rate of 18Mbps. Changing the optical power will effect the EVM and hence reducing the overall transmission data rate. Increasing the RF input power may give better performance, however, it will be limited by the RF intermodulation effects.

REFERENCES

[1] P.K. Tang, L.C. Ong, B. Luo, A. Alphones and M. Fujise, "Transmission Of Multiple Wireless Standards Over A Radio-Over-Fiber Network", *Microwave Symposium Digest*, IEEE MTT-S International, Fort Worth, Texas, USA, Vol. 3, pp. 2051 – 2054, June 2004.

[2] D. Wake, D.G. Moodie, F. Henkel, "The Electroabsorption Modulator As A Combined Photodetector/Modulator For Analogue Optical Systems", *Workshop on High Performance Electron Devices for Microwave and Optoelectronic Applications*, pp. 147 – 150, Nov. 1997.

[3] D. Wake, D.G. Moodie, "Passive Picocell: Prospects For Increasing The Radio Range", *International Topical Meeting Microwave Photonic*, pp. 269 – 271, September 1997.

[4] L.A Coldren, S.W. Corzine, "Diode Lasers and Photonic Integrated Circuits" *John Wiley & Sons*, Inc., 1995.

[5] S. Kaneko, M. Noda, Y. Miyazaki, H. Watanabe, K. Kasahara, T. Tajime, "An Electroabsorption Modulator Module For Digital And Analog Applications", *Journal of Lightwave Technology*, Vol. 17, Issue 4, pp. 669 – 676, April 1999.

[6] Bin Liu, Jongin Shim, Yi-Jen Chiu, A. Keating,, J. Piprek, J.E. Bowers, "Analog Characterization Of Low-Voltage MQW Traveling-Wave EAMs", *Journal of Lightwave Technology*, Vol. 21, Issue 12, pp. 3011 – 3019, Dec. 2003.

[7] Hai-Han Lu, Ying-Cong Lin, Yuan-Hong Su, Heng-Sheng Su "A Radio-On-Fiber Intelligence Transport System Based On Electroabsorption Modulator And Semiconductor Optical Amplifier", IEEE Photonics Technology Letters, Vol. 16, No. 1, pp. 251-253, January 2004.

[8] H. Takeuchi, K. Tsuzuki, K. Sato, M. Yamamoto, Y. Itaya, A. Sano, M. Yoneyama, T. Otsuji,, "Very High-Speed Light-Source Module Up To 40 Gb/S Containing An MQW Electroabsorption Modulator Integrated With A DFB Laser", *IEEE Journal of Selected Topics in Quantum Electronics*, Vol. 3, Issue 2, pp. 336 – 343, April 1997.

[9] Y. Kim, H. Lee, Jaehoon Lee, J. Han, T.W. Oh, J. Jeong; "Chirp Characteristics Of 10-Gb/S Electroabsorption Modulator Integrated DFB Lasers", IEEE JournalofQuantumElectronics, Vol. 36, Issue 8, pp 900 – 908, Aug. 2000.

[10] A. Lestra, P. Brosson, "Design Rules For A Low-Chirp Integrated DFB Laser With An Electroabsorption Modulator", IEEE Photonics Technology Letters, Vol. 8, Issue 8, pp. 998 – 1000, Aug. 1996.

Failure Analysis of Pitting Problem on Microchip Al Bondpads in Wafer Fabrication

Tai Eve, Hua Younan, Rao Ramesh and Zhao Siping, *Snr Member, IEEE*

Chartered Semiconductor Mfg Ltd
Woodlands Industrial Park D, Street 2, Singapore 738406
Email: evetai@charteredsemi.com

Abstract: **A bond pad failure mechanism of galvanic corrosion was studied. Analysis results showed that over-etch process, EKC and DI water over cleaning revealed more pitting with Cu seed due to galvanic corrosion. To control and eliminate galvanic corrosion, the etch recipe was optimized and etch time was reduced about 15% to prevent damaging the native oxide. EKC cleaning time was remaining unchanged in order to maintain bond pad F level at minimum level. In this study, the PRS process was also optimized and CF4 gas ratio was reduced about 45%. Moreover, O_2 process was added after PRS process so as to increase the native oxide layer on Al bondpads to prevent galvanic corrosion.**

I. INTRODUCTION

IT is well known fact that in wafer fabrication, galvanic corrosion will affect Al bondpad quality and cause discolored bondpads or non-stick on pads (NSOP) problem. In this paper we will present a case study on pitting problem. Possible root causes and solutions will be discussed. Grainy bondpad is commonly seen on bondpads due to heat process or its natural architecture due to presence of sea of via underneath bondpads (Fig. 1). Test showed that grainy defect (Fig. 2) on bondpads is extremely sensitive to galvanic corrosion. This is especially true for Al metal scheme with presence of Cu or any other cathode source.

II. EXPERIMENTAL

Wafer Fab Process

In this work, grainy bondpad sample was chosen for study before and after etch process. Wafers were etched using various etch time. After that sample was inspected and performed SEM (scanning electron microscopy) inspection and FIB (focused ion beam) for understanding details. Besides that, location of corrosion and impact of etching was also studied. Besides that impact of cleaning method to galvanic corrosion was also studied. Cleaning process studied are PRS (photo resist strip) and EKC clean process.

SEM/EDX Analysis

In this study, SEM/EDX analysis was done by using FEI Sirion FESEM at beam accelerating voltage of 5 kV, spot size of 3 and a working distance of 5 mm. To obtain accurate EDX results, ZAF standardless EDX quantification method was used [1].

III. RESULTS AND DISCUSSION

Test showed that wafers with grainy bondpad and higher etching time showed galvanic corrosion after wet clean, where else wafer with less etching time didn't show sign if corrosion after wet clean process. Location of corrosion associated with the area where had more etching.

It is seen that bond pad fluorine level does not correlate to galvanic corrosion. Thus, fluorine chemistry is confirmed not to be the root cause of galvanic corrosion. To understand possible root cause, failure mechanism and solutions, a wafer was submitted to do failure analysis. SEM inspection found pitting-like defects on Al bondpads, in which there was a seed centered inside the pitting (Figs.3-4). EDX (energy-dispersive X-ray microanalysis) detected high Cu peak on the seed centered inside the pitting (Fig. 5). Test showed that wafers which gone through additional oxide step after PRS process has lower galvanic counts. Wafers that have higher EKC cleaning time showed higher galvanic corrosion sites.

IV. DISCUSSION OF FAILURE MECHANISUM

Galvanic corrosion is also known as two-metal corrosion. It occurs within two dissimilar metals, which are brought into electrical contact in the presence of water or moisture. When a galvanic corrosion couple forms, one of the metals becomes the anode and corrodes faster than it would all by itself, while the other becomes the cathode and corrodes slower than it would alone. The driving force for corrosion is a potential difference between the different materials.

In wafer fab, Al metal is used for metallization. To enhance the electromigration resistance of the conductor and to control the grain size of metal film, Cu is added to Al or Al-Si. For example, Al alloy (Al--Si, 0.8%-Cu 0.5%) is currently used for metallization of microchip bondpads. However, Al alloy with Cu may cause galvanic corrosion on bondpads during DI water rinsing at both wafer fab and assembly process, as there is a potential difference between Al and Cu metals. This is especially significant when more etching on Al which exposes more Cu seed from metal. Galvanic corrosion will result in pinholes/pitting and Si dust on Al bondpads and cause non-stick on pads at assembly house.

In wafer fab, we have faced galvanic corrosion from Al-Cu cell. A galvanic corrosion couple, Al-Cu forms, Al becomes the anode and corrodes faster than it would all by itself. While Cu becomes the cathode and corrodes slower than it would alone. Therefore, Al^{3+}/Al serves as a anode and Cu^{2+}/Cu as a cathode. We have proposed a theoretical model of galvanic corrosion for Al-Cu cell [1, 2]:

Anode : $Al^{3+} + 3e^- \rightarrow Al$ (1)
$(E^o = - 1.662$ V)

Cathode: $Cu^{2+} + 2e^- \rightarrow Cu$ (2)
$(E^o = + 0.337$ V)

Therefore, if the bondpad is exposed to DI water or moisture (H_2O) and the atmosphere (O_2), galvanic corrosion may occur as there is a potential energy (about 2V) between Al^{3+}/Al and Cu^{2+}/Cu (refer to E^o in Eqn (1) and (2)). At the mean time, the chemical reaction from the H_2O and O_2 will also occur and forms corrosive ion [OH^-]:

$O_2 + 2H_2O + 4e^- \rightarrow 4OH^-$ (3)
$(E^o = + 0.401$V)

Then the anodic aluminum surrounding the exposed cathodic Al_2Cu nucleus is corroded to form aluminum hydroxide:

$4Al + 3O_2 + 6H_2O \rightarrow 4Al^{3+} + 12OH^-$
$\rightarrow 4Al(OH)_3\downarrow$ (4)

From the above theoretical model, we can know that after galvanic corrosion, Al will be corrosive and form aluminum hydroxide, $Al(OH)_3$. Hence if the wafers are under DI water cleaning process steps at wafer fab or assembly house, the corrosive product, $Al(OH)_3$ will be physically cleaned away by DI water, and a pitting will form on surface of Al bondpads. At the mean time, Cu in Al or Al alloy will concentrate to the center of the pinhole and form an Al_2Cu seed (Figs3-4). At the beginning of galvanic corrosion, the Al_2Cu seed is connected with Al matrix. The corrosive process will continue and the Al_2Cu nucleus will be further isolated. Finally, the Al_2Cu seed will be fully isolated electrically and physically from the Aluminium matrix (Figs. 3-4). However, if DI water over rinsing, the seed may be also cleaned away (Fig. 6).

To check the depth of the pitting, Focused ion bean (FIB) was conducted on a pitting. The result showed that the pitting was about 105 nm depth (Figs.7-8) comparing to the metal thickness of 846nm (Fig. 8). Therefore, in this case, the pitting might not affect the bondpad bonding process.

V. POSSIBLE ROOT CAUSE AND SOLUTIONS

From the failure mechanism, one can understand that EKC clean and DI water related process would enhance galvanic corrosion. It was shown that reducing EKC cleaning time helped to reduce galvanic corrosion, but it increased F contamination level on Al bondpads. Thus, in this study, EKC cleaning time was remain unchanged.

When etching was more on wafers, the native oxide layer was destroyed during etching process. Thus native oxide layer was expected to be thinner on surface. Protection to bondpad surface was reduced. As a result wafer would have higher tendency to corrode at Cu inter phase. Therefore, to reduce pitting due to galvanic corrosion, the etch recipe was optimized and etch time was reduced about 15% to prevent damaging the native oxide.

On the other hand, the PRS process was also optimized. During PRS process, a mixture gases of CF_4 and O_2 was used by a design ratio. In this study, CF_4 gas ratio was reduced about 45%. Moreover, O_2 process was introduced after PRS

($CF_4 + O_2$) step so as to increase the native oxide layer on Al bondpads, which will help to reduce galvanic corrosion.

VI. CONCLUSION

In this paper, a pitting problem has been investigated. Failure mechanism of galvanic corrosion was studied. Analysis results showed that over-etch process, EKC and DI water over cleaning revealed more pitting with Cu seed due to galvanic corrosion. To control and eliminate galvanic corrosion, the etch recipe was optimized and etch time was reduced about 15% to prevent damaging the native oxide. EKC cleaning time was remaining unchanged due to control F contamination. In this study, the PRS process was also optimized and CF4 gas ratio was reduced about 45%. Moreover, O_2 process was added after PRS process so as to increase the native oxide layer on Al bondpads to prevent galvanic corrosion.

ACKNOWLEDGEMENTS

The authors would like to thank Dr Li Kun, Tay Chui Lan, Kon Cambridge, Wong Hie Lee, Phang Nyuk Lin and Lee Gek Li from Physical & Surface/advanced FA teams, Sia Kee Sing and Bond pad Task Force team for their technical advice and contribution to this technical paper..

REFERENCES

1. Hua Younan, "A Study on Discolored Bondpads & Galvanic Corrosion". *The Proceedings from the 24th International Symposium for Testing and Failure Analysis,* 15-19 Nov, Hyatt Regency, DFW-East Tower, Dallas, Texas, USA, 269-272 (1998).
2. Hua Younan, E. C. Low, L. H. An and Shailesh Redkar, "Failure Analysis and Elimination of Galvanic Corrosion on Bondpads during Wafer Sawing". *The proceedings from the 26th International Symposium for Testing and Failure Analysis,* Nov 12-16 2000, Meydenbauer Convention Center, Bellevue (Seattle), Washington, USA, 369-372 (2000).
3. Hua Younan, "Studies on Elimination Solutions of Galvanic Corrosion on Microchip Al Bondpads in Wafer Fabrication and Assembly Processes". *The proceedings from the International*

Conference on Semiconductor Electronics, 7-9 Dec, Kuala Lumpur, Malaysia (2004).

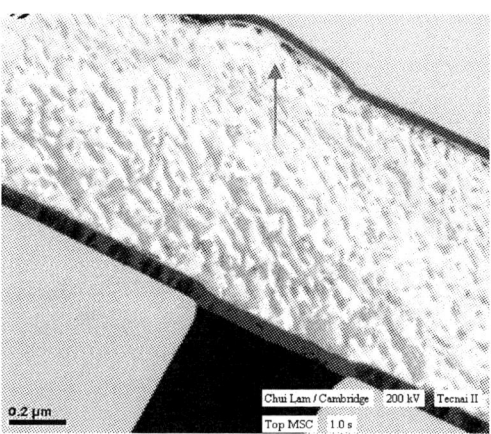

Figure 1. TEM showed metal bump on bondpad surface. Galvanic corrosion occurred along the grain boundary.

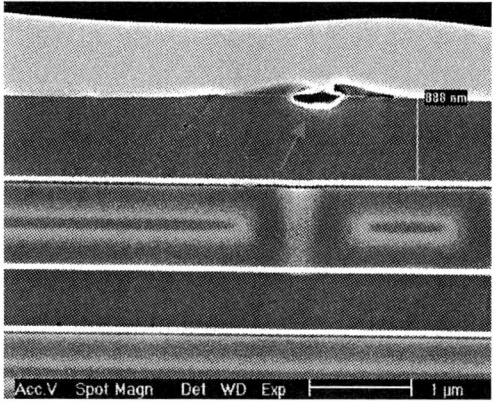

Figure 2. FIB/SEM showed metal bump on bondpad surface. Galvanic corrosion occurred along the grain boundary.

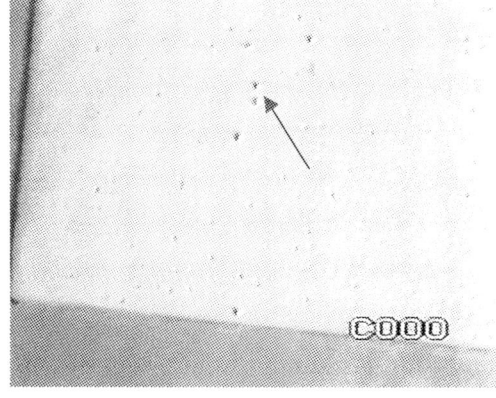

Figure 3. Partitioning check showed that after DI water clean, corrosion was seen on the grainy region.

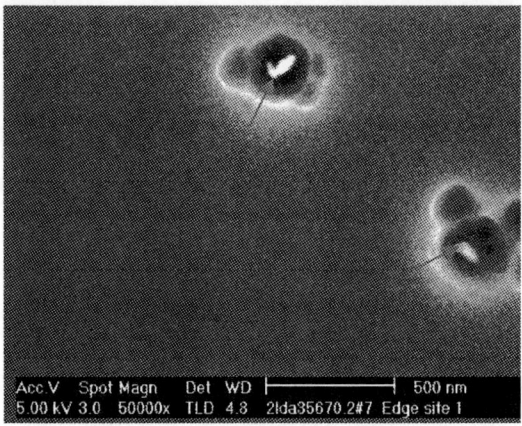

Figure 4. SEM micrograph showed the pitting on Al bondpad due to galvanic corrosion, in which there was a seed centered inside the pitting.

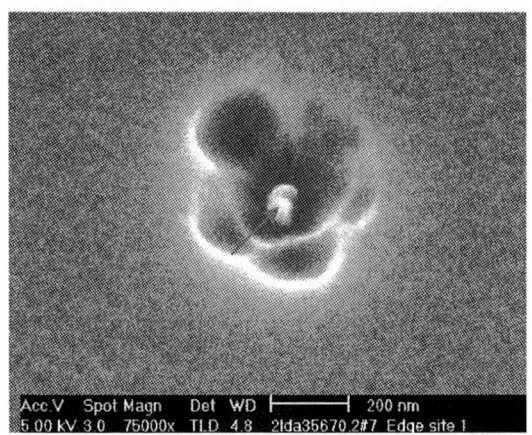

Figure 5. Close-up SEM micrograph showed the pitting on Al bondpad due to galvanic corrosion, in which there was a seed centered inside the pitting.

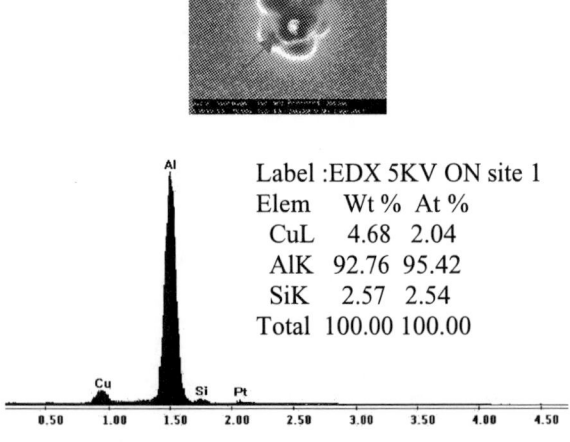

Label :EDX 5KV ON site 1

Elem	Wt %	At %
CuL	4.68	2.04
AlK	92.76	95.42
SiK	2.57	2.54
Total	100.00	100.00

Figure 5. EDX detected high Cu peak on the seed centered inside the pitting, which was from galvanic corrosion.

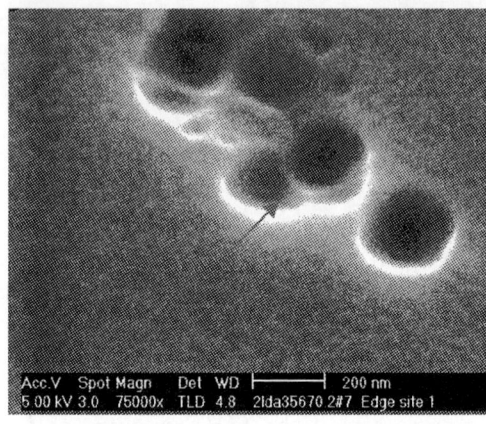

Figure 7. SEM micrograph showed the pitting on Al bondpad due to galvanic corrosion. But the seed had been cleaned away due to over rinsing.

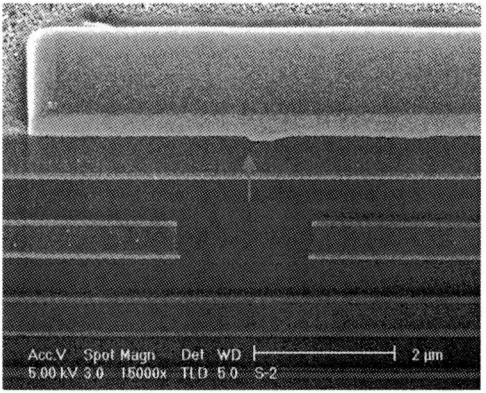

Figure 8. FIB (focused ion beam) was conducted on the pitting to check the pitting depth.

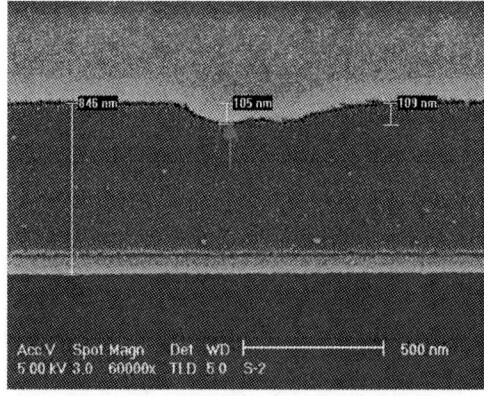

Figure 9. After FIB cut, SEM micrograph showed that the pitting was about 105 nm depth comparing to the metal thickness of 846nm. Therefore, the pitting might not affect the bondpad bonding process.

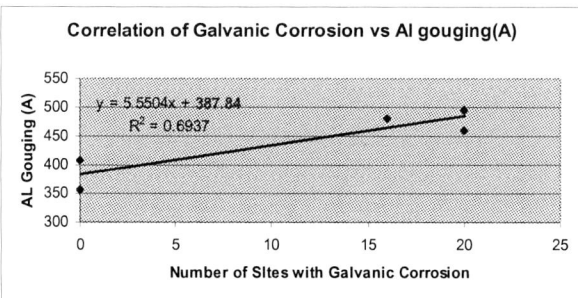

Figure 9. Higher Al gouging did correlate to higher number of corrosion sites. For the test sample, galvanic corrosion was observed when Al gouging >387A. This value was subjective to wafer.

Auger PID Characterization of Threshold Voltage Shift and Application in Bond pad Monitoring of Wafer Fabrication

Marcus Tan YH, Hua Younan, Ling Timothy and Zhao Siping, *Snr Member IEEE*
Chartered Semiconductor Mfg Ltd
Woodlands Industrial Park D
Street 2, Singapore 738406
Email: marcustan@charteredsemi.com

Abstract **This paper presents the effects of Auger Electron Spectroscopy on Semiconductor device's threshold voltage performance. Bond pads connected to gate of transistors were exposed to different beam energy of 0, 3, 5, 10, 15 and 20 keV for an average of 2.5 minutes. Vt measurements were collected once at pre-exposure and thrice at post-exposure. The evaluation results show that the transistor Vt was impacted and shifted after exposure to electron beam at varying energy levels (3, 5, 10, 15 and 20 keV) during AES analysis.**

I. INTRODUCTION

IT is a well-known fact that Aluminum (Al) bond pads on semiconductor microchips play an important role in chips functionality and reliability. However, during wafer fabrication, although processes are optimized, it is still possible to have bond pad contamination, corrosion or under-etching issues, which will affect the quality of the Al bond pads on microchips. The source of contamination may be related to Fab processes, machines/tools, materials/chemicals, environment or human. Sometime contamination on Al bond pads may cause corrosion and form corrosive defects, which will seriously affect the quality of bond pads and pose a problem which results in Non-Stick On Pads (NSOP) during wire bonding process at assembly house. For example, one of the main corrosion issues in wafer fab is fluorine-induced corrosion. During the opening of bond pad, CF_4 based gases are used for etching the passivation layer. To reduce F contamination level, after passivation etching, wet cleaning solution such as EKC 265 (or Aleg310/NE111) and plasma dry cleaning are used to remove carbon (C), fluorine (F) and polymer on Al bond pads.

However, during passivation etching and photo resist striping (PRS) a certain amount of F will chemically react with Al to form Al fluoride $[AlF_x]^{(x-3)-}$. In the authors' previous publication [1-2], we have conducted detailed studies on F-induced corrosion and found x in AlF_x may be 3 or 6 depending on the level of F contamination. If there is a high level of F contamination on Al bond pads, F will react with Al and form a "flower-like" defect of Al fluoride AlF_3 or $[AlF_6]^3$. Once Al fluoride AlF_3 or $[AlF_6]^{3-}$ are formed on the affected bond pads, it is not easy to be completely removed by wet or dry cleaning processes. Therefore, on Al bond pads a small amount of F may usually be detected, but it has to be controlled to a very low level. Thus, the contamination level needs to be monitored.

Besides bond pad contamination and corrosion, another problem is under-etch, which will also cause NSOP. From bond pad architecture, we can understand that the layers on Al bond pad before bond pad opening process are: passivation (Si_3N_4 & SiO_2), TiN and Al alloy (Al-Cu, 0.5%-Si, 0.8%) layers. During bond pad opening processes, the passivation (Si_3N_4 and SiO_2) and TiN layers will be etched away. Using the plasma process with CF_4 gas or other F-based gases the Al bond pad surface are exposed after etching. However, sometime under-etch may occur due to variations such as etching parameters, the thickness of passivation (Si_3N_4 and SiO_2) and TiN layers. Therefore, N, O, Si and Ti may be detected by Auger electron spectroscopy (AES) or energy-dispersive X-ray analysis (EDX).

In this study, AES is recommended for bond pad monitoring to check for contamination and under-etch. In practice, a control limit is set based on the baseline data collected. Then a certain production wafers are selected for weekly monitoring. AES analysis is conducted; collection data from the surface and at regular depth interval till pure Al is detected. The elements monitored include C, N, O, F, Al, Si and Ti. These are compared with the baseline data and control limit. If AES data elemental percentage detected are out of control limit, wafer fab production line will be alerted and immediate corrective actions will be taken before shipment of the suspected wafers. In our fabs, during Auger monitoring, if scrapped wafers (excluded bondpad related scrapped wafers) are available, they will be selected for AES analysis. But if no scrapped wafers are available, then production wafers will be used for monitoring. However, if the production wafers are used for Auger monitoring, it is necessary to understand the effect of AES electron beam exposure on transistor threshold voltage (Vt). In this paper, we will study and characterize the transistor Vt impact by exposure of electron beam at varying energy levels (3, 5, 10, 15 and 20 keV) during AES analysis.

II. EXPERIMENTAL

Bond pads connected to the Gate of transistors were exposed for an average of 2.5 minutes each with the exception of one bond pad for each energy level (3, 5, 10, 15 and 20 keV) being exposed for an extended period. Vt measurements were collected once pre-exposure and thrice post-exposure. A test program determines Vt from a point of maximum slope. The test was performed on transistors from a 0.35um process technology with Al bond pads.

In this study, AES analyses were carried out using PHI (Physical Electronics) SMART-200 Auger System. The analysis conditions are as follows: (A). Electron gun setting Operating Voltage: 3, 5, 10, 15 and 20 KeV. Operating Current: 10 nA Gun Tilt: 30° (B). Ion gun setting Ion source: Ar+ Operating Voltage: 2 KeV Operating Current: 600-800 nA (Variable, to maintain etch rate of 100A/min) Raster size: 2 mm x 2 mm Sputter etch rate: Nominal at 100+/- A/min.

Auger survey range was 30-2030 eV. The relative atomic concentration of the detected elements were calculated by first measuring elemental peak-to-peak heights of Auger

spectrums in the survey scans and then applying sensitivity factors based on standard spectra of pure elements or selected compounds:

$$C_i = I_i / (I_{std}S_iD_i) \qquad (1)$$

Where I_i is the peak-to-peak amplitude of the element i from the test specimen, I_{std} is the peak-to-peak amplitude of the element i from the standard, S_i is the relative sensitivity factor and D_i is a relative scale factor between the spectra for the test specimen and standard.

III. RESULTS & DISCUSSION

Vt measurements were plotted in order to enable comparisons of the Vt distributions of post-AES exposure against that of pre-AES exposure. Figure 1 illustrates the Vt distribution before and after 20keV electron beam exposure. The pre-exposure Vt values were tightly distributed. On the other hand, post- exposure Vt values were spread wider and they were repeatable. The observation shows that after exposure to the 20KeV AES electron beam, the transistor Vt has drifted with a negative shift and is no longer consistent across the samples.

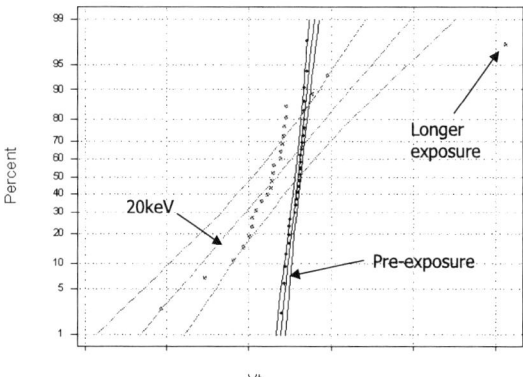

Fig. 1 Pre-exposure VS 20keV Vt distributions

Another observation would be the outlier belonged to the bond pad that was exposed with the same beam energy level but for a longer period of time. This site exhibited much larger Vt shift but in the positive direction.

Bond pads with extended exposure period were also related to the worst transistor Vt shift (>100%) for all AES electron beam

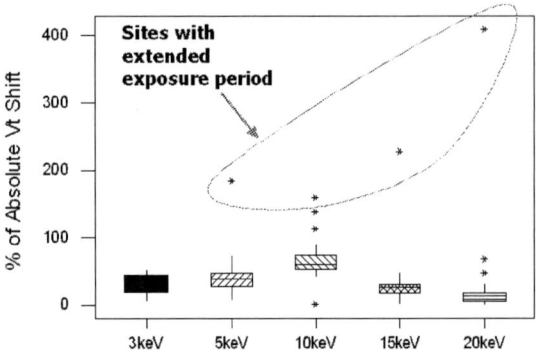

Fig. 2 Absolute Vt shift percentage at Diverse Exposure Levels

energy level except 3keV. This is depicted clearly in the box plots shown in Figure 2 whereby the outlier with the largest Vt shift from 5keV, 10keV, 15keV and 20keV were those that experienced the longest exposure duration; 20keV having the worst effect. All other bond pads underwent the shorter exposure period of 2.5 minutes and did not demonstrate Vt shifts of more than 100%. A correlation between the lengths of AES exposure to the amount of Vt shift is apparent.

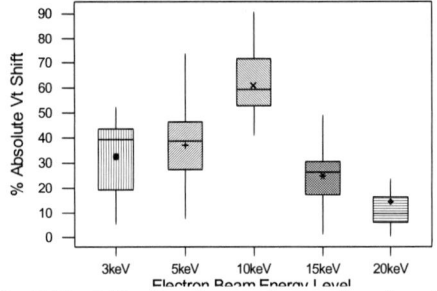

Fig. 3 Vt shift of various voltages showing shift in Vt.

Figure 3 displays a close-up view the same box plot but without outliers. It was observed that the average absolute Vt shift was more than 10% for all AES electron beam energy level utilized for this study. This would mean that performing AES analysis on product wafers would introduce PID-like damage. The most significant Vt shift was observed in the 10keV case while the least was experienced in the 20keV case; whereas It appears to suggest that there is a non-linear relationship between beam energy level and Vt shift that could differ with exposure duration.

IV. CONCLUSION & RECOMMENDATION

The evaluation results show that the transistor Vt was impacted and shifted after exposure to electron beam at varying energy levels (3, 5, 10, 15 and 20 keV) during AES analysis. Therefore, it is not recommended to perform Auger monitoring analysis on the product dice, which is to be shipped to customers. It would be better to use wafers not intended shipping, for the purpose of Auger monitoring. If it is really necessary, it is suggested that AES analysis will be performed on the uncompleted die at the wafer edge to prevent Vt shifts after exposure to electron beam during AES analysis.

Extended exposure periods resulted in larger Vt shifts for all cases except 3keV. However, for a given exposure duration, higher beam energy level did not cause always cause higher Vt shifts. Further work to characterize the full profile of beam energy, exposure time as combined factors versus Vt shifts can be explored.

ACKNOWLEDGEMENTS

The authors would like to thank Chartered FA & REL teams for their technical advice and contributions to this technical paper.

REFERENCES

[1] Hua Younan, S. Redkar, C. K. Lau and Z. Q. Mo, "A Study on Non-Stick Aluminum Bond pads due to Fluorine Contamination using SEM, EDX, TEM, IC, Auger, XPS & TOF-SIMS Techniques". The proceedings from the 28th International Symposium for Testing & Failure Analysis (ISTFA'2002), 3-7 Nov, Phoenix Civic Plaza, Phoenix, Arizona, USA, p495-503 (2002).

[2] Hua Younan, Shailesh Redkar, Ron Dickinson and Peh Shirley, "A Study on Fluorine-Induced Corrosion on Microchip Aluminium Bond pads". The proceedings from the 29th International Symposium for Testing & Failure Analysis (ISTFA'2003), 2-6 Nov. 2003, Santa Clara, California, USA, p249-255 (2003).

Studies on Failure Mechanism of ET High Via Resistance in Wafer Fabrication

Hua Younan, Tan Sock Khim, Li Kun and Zhao Siping, *Snr Member, IEEE*
Chartered Semiconductor Mfg Ltd
Woodlands Industrial Park D, Street 2
Singapore 738406
E-mail: huayounan@charteredsemi.com

Abstract: **In this paper, an ET high Via resistance case was investigated. TEM/EDX technique was used for identification of the root cause. Failure mechanism of Al fluoride defects is discussed. Some preventive actions/solutions were implemented to improve the process margin and eliminate the problem.**

I. INTRODUCTION

IN wafer fabrication, Via is a important structure for metal interconnect. The quality of Vias will impact Etest results . In this work, some wafers were reported with via 4 ET high resistance failure (punch through Via) having high mean and sigma with 1 wafer out of the specification. To find possible root cause, failure analysis was conducted on failed wafer. Based on failure analysis results from cross sectional TEM, EDX and color mapping, it is concluded that the high resistance failure was due to fluorine-induced defects. In this paper, we will present details of failure analysis and wafer fab investigation, and study failure mechanism of ET high Via resistance and eliminating and prevent actions and solutions.

II. EXPOERIMENTAL

TEM analysis was carried out using a 200 kV FEI TECNAI F20 TEM equipped with a field-emission gun and a Gatan imaging filter (GIF). Imaging was done using scanning transmission electron microscopy (STEM) technique. Elemental mapping was done with energy-filtered TEM technique (EF-TEM) using GIF. TEM sample preparation was done using focused ion beam (FIB) based lift-out technique.

III. RESULTS AND DISCUSSION

The wafers were reported with via 4 ET high resistance failure. TEM/EDX analysis was conducted on an affected wafer. Cross sectional TEM results showed that there was a line-like defect as shown by the arrow in Figure 1. RGB integrated color mapping result on the Via section showed high fluorine at the line-like defect as shown by the arrow in Figure 2. In Fig 2, red color is for Ti, green is for F and blue is for Al. One can see that high fluorine (green color) was observed at the line-like defect. In addition to the expected elements Al, Ti, W and Cu (background from Cu grid that holds the sample), EDX also detected F peak on the line-like defect (Figure 3). However, EDX not detected F on the normal area (Figure 4).

IV. FAILURE MECHANISM OF ET HIGH VIA RESISTANCE

A comparison between the bottom left and bottom right corners of the Via reveals that there is a small kink (under cut) at the problematic left corner (Fig. 1). This is caused by Via etch and clean process. The following PVD Ti/TiN process cannot provide a good coverage at the undercut area, forming a weak point that makes the adjacent Al susceptible to F attack (form the W precursor WF_6).

According to our previous studies, Fluorine contamination on Al bondpads will cause F-induced corrosion [1-2]. Similar, F contaminates Via (Al surface) and will chemically react with Al and form Al fluoride defect:

$$Al + xF^- \rightarrow [AlF_x]^{(x-3)-} + 3e^- \quad (1)$$

The x in $[AlF_x]^{(x-3)-}$ may be 3 or 6 depending on the F contamination level. Then

$$Al + 3F^- \rightarrow [AlF_3] + 3e^- \quad (x=3) \quad (2)$$

$$Al + 6F^- \rightarrow [AlF_6]^{3-} + 3e^- \quad (x=6) \quad (3)$$

In chemistry, $[AlF_3]$ and $[AlF_6]^{3-}$ are stable compounds. They are difficult to be removed by EKC 265 chemical clean. These line-like defects will block some area interface between W & Al interface, and then cause Via high resistance.

IV. IDENTIFICATION OF POSSIBLE ROOT CAUSE

Topology of Via 1 ~ 3 vs Via 4 is different. As it gets higher, barrier metal coverage becomes more difficult, hence the F from W plug deposition process may react with Al to form Al fluoride defects, AlF_x, which will result in Etest higher Via contact resistance. Aspect ratio for via 4 got higher hence causing via4 resistance fail instead of other Via.

In this process, after Via opening, EKC 265 was used to clean polymer and C & F contamination. Therefore, we think that EKC265clean bath lifetime is also important. In other word, if EKC 265 cleaning was insufficient, F contamination might be higher.

V. ELIMINATION / PREVENTIVE ACTIONS / SLOUTION

In this case, to eliminate the F-induced defects, some experiments (DOE) were conducted. After that, we implemented some preventive actions/solutions to improve the process margin by: (a). EKC265 bath lifetime was optimized at 900 mins & life batch 32 batches for via layers; (b). Targeting DICD higher and IMD thinner to improve Via4 resistance performance; (c). Increasing the BM thickness ~2A to prevent Al extrude into the via hole after the IMP Ti deposition. After implemented the above actions, the high Via resistance problem was eliminated.

VI. CONCLUSION

A ET high Via resistance case was investigated. Higher aspect ratio led to the marginal Barrier Metal (TIN) step coverage, which gave F element to react with Al to form Al fluoride defects, AlF_x, hence causing higher Via Con Res, top Via (Via 4) layer margin was worst due to worst topology. Some preventive actions/solutions were implemented to improve the process margin and eliminate the problem by: (a). EKC265 bath lifetime was optimized at 900 mins & life batch 32 batches for via layers; (b). Targeting DICD higher and IMD thinner to

improve Via4 resistance performance; (c). Increasing the BM thickness ~2A to prevent Al extrude into the via hole after the IMP Ti deposition. In this paper, the failure mechanism of fluorine-induced defects is discussed.

ACKNOWLEDGEMENT

The authors would like to thank Teo Kim Hong and Wang Jili from Chartered TEM FA Group for their technical advice and contribution.

REFERENCES

1. Hua Younan, Shailesh Redkar, Lau Chi-Kwan and Mo Zhiqiang (from PSB Corp., Singapore), A Study on Non-Stick Aluminum Bondpads due to Fluorine Contamination using SEM, EDX, TEM, IC, Auger, XPS & TOF-SIMS Techniques. The proceedings from the 28th International Symposium for Testing & Failure Analysis (ISTFA'2002), 3-7 Nov, Phoenix Civic Plaza, Phoenix, Arizona, USA p495-504 (2002).
2. Hua Younan, Shailesh Redkar, Ron Dickinson and Peh Shirley, A Study on Fluorine-Induced Corrosion on Microchip Aluminium Bondpads. The proceedings from the 29th International Symposium for Testing & Failure Analysis (ISTFA'2003), 2-6 Nov. 2003, Santa Clara, California, USA, p249-255 (2003).

 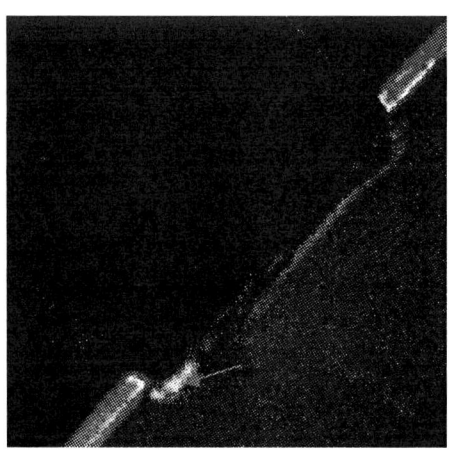

Fig 1. Cross sectional TEM result showed that there was a line-like defect as shown by the arrow.

Fig 2. RGB integrated color mapping result (Red: Ti, Green: F and Blue: Al) on the Via section showed high fluorine at the line-like defect as shown by the arrow.

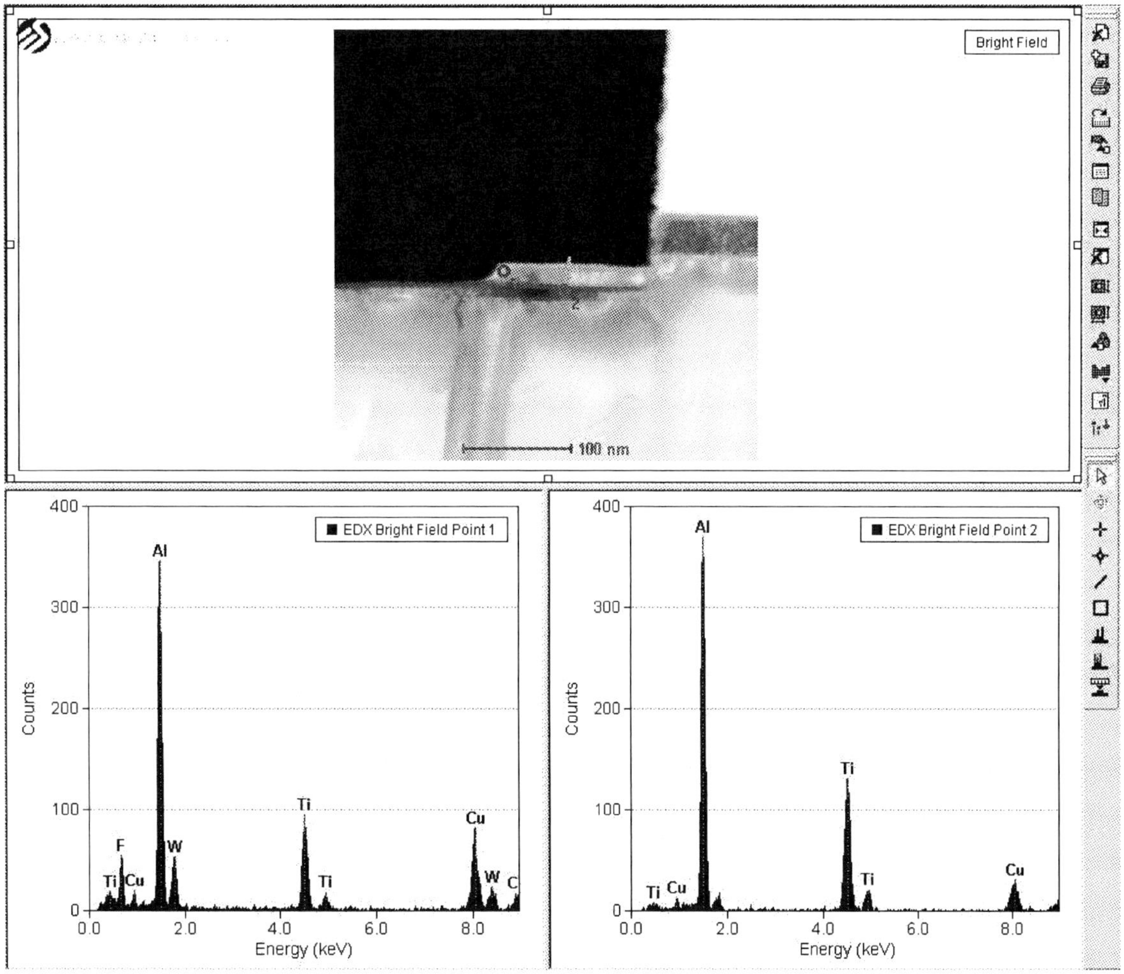

Fig 3. EDX detected F peak on the line-like defect besides other elements.

Fig 4. However, EDX not detected F on the normal area besides other elements.

The Effect of Al and Pt/Ti Simultaneously Annealing on Electrical Characteristics of n-GaN Schottky Diode

Tariq Munir, Azlan Abdul Aziz and Mat Johar Abdullah
Nano-Optoelectronics Research Laboratory (NOR)
School of Physics Universiti sains Malaysia 11800 Minden Penang Malaysia
Email: tarriqmunirr@yahoo.com, lan@usm.my

Abstract Wideband gap semiconductor GaN has received increasing attention for its potential for wide variety of high-power, high-performance switching and high-frequency devices application. In this paper the simultaneous annealing effect at a temperature of 400°C ~700°C in N_2 ambient for 12min on Aluminum (1500A°) as an ohmic contact while Pt/Ti (700A°/700A°) bilayer as a Schottky contact on the electrical characteristics of n-GaN Schottky diode was investigated. It was found that at a annealing temperature of 400°C produced the best (I-V) characteristics, barrier height (ϕ_B) 1.1eV, ideality factor (η) 1.1 as compared to the other annealing temperature. The (C-V) characteristics of n-GaN Schottky diode were measured at 100 kHz and 1 MHz frequency at different annealing temperature. It was found that at annealing temperature of 400°C, the depletion region is maximum with the capacitance value varied from 0.02pF ~ 0.04pF. At low frequency 100 kHz the capacitance increase with increasing forward voltage which is frequency independent, while at high frequency 1MHz the capacitance-voltage curve is almost flat. The surface morphology of n-GaN Schottky diode before annealing and after annealing was observed by SEM, XRD. It was found that Al (1500A°) didn't show any significant loss of dimensional stability at annealing temperature 400°C~700°C, while Pt/Ti (700A°/700/A°) show balling effect, surface morphology degradation at above 400°C which was confirmed by XRD measurement. Hence we conclude that rectifying behavior of n-GaN schottky diode was observed at annealing temperature 400°C while annealing at above 400°C~700°C the rectifying characteristics of n-GaN schottky diode changed to ohmic behavior due to Pt/Ti (700A°/700A°) island form of surface morphology occurred, while Al (1500A°) as a ohmic contact show thermal stability at high temperature annealing above 400°C of n-GaN Schottky diode.

I. INTRODUCTION

GaN has a wide direct band gap of 3.4 eV and has attracted considerable interests in high-temperature and high-power applications. For these applications, high-quality ohmic and schottky contacts have been key factors in improving performance and reliability. The operating condition of these applications often involves elevated temperature. Therefore it is vital to understand the behavior of metal contacts to GaN at high-temperature [1]. The formation of a low-resistance contact to a wide-band gap (III-nitride) semiconductor is a challenging problem because of the large barrier height which develops when a metal is deposited on the semiconductor material. Research, though, on both ohmic and Schottky contacts to GaN is being carried out and some studies have already been published [2].

Most of the ohmic contacts investigated so far fall into this category: Al-only contact, Ti –only contact, Ti/Al bilayer contact and modification of the Ti/Al bilayer contact [3]. The multilayer contact schemes discussed above contacts in the types of characterization where the requirements of the contact resistance are not so critical (for example Hall and capacitance–voltage plots (C-V).Therefore, the electrcal characterization of Al contacts to n-type GaN deserve rigourous study. In addition, knowledge of the relatively simple Al/GaN interface will aid our understanding of the more complex multilayer contact systems [2].

Due to the ionic character of the Ga–N bondings, no Fermi level pinning occurs at the

0-7803-9730-4/06/$25.00 ©2006 IEEE

metal–GaN. Thus the barrier height of metal contacts to GaN increases with the metal work function φ_m. Metals such as Pt (φ_m = 5.65 eV), Ni (φ_m = 5.15 eV), Pd (φ_m = 5.12 eV) and Au (φ_m = 5.1 eV) are commonly used for Schottky contacts to n-GaN. However, some metals are prone to react with GaN after prolonged thermal treatments at 400–500°C, forming stable gallides. In addition, continuous films of Ni and Pd on GaN tend to break up into discontinuous islands upon high-temperature annealing. Pt has a high work function that makes it ideal for use as Schottky contacts on most n-GaN. The Schottky barrier height of Pt/n-GaN contacts has been measured with value between 0.89 and 1.27 eV by several groups [9-10], variation in the measured barrier height is mainly attributed to differences of material characteristics and measurement methods used. Pt is also useful for contacts because of its resistance to oxidation and corrosion. Pt appears to diffuse upon annealing in the GaN. Thus, the use of Pt as a contact to GaN requires a thin interlayer of Ti as a diffusion barrier, due to its high thermal stability and low work function. Ti also reduces the GaN native oxide [3].

This work report on the study of current-voltage (I-V) and capacitance- voltage (C-V), Scanning electron microscopy, X-ray diffraction as a function of combination use of Al(1500A°) as ohmic and Pt/Ti (700A°/700A°) as a schottky contact metal annealing on n-GaN Schottky diode.

II. EXPERIMENT

The n-GaN diode thickness 0.002mm (n~8e16cm^{-3}) was grown on c-plane Al_2O_3 substrate. Prior to the metallization, the native oxide was removed with NH4OH:H2O= (1:20) solution, followed by dipping in a HF: H20= (1:50) solution. The boiling aqua regia HCl: HNO3= (3:1) was used to chemically etch and clean the samples. The fabrication process of Schottky diode start with Schottky metal Pt/ Ti (700A°/700A°) was deposited using Dc sputtering by metal mask (1mm). While the Al (1500A°) metal deposited using thermal evaporator by metal mask (1mm) as an ohmic contact, all samples were annealed in a temperature range from 400°C to 700°C for 12 min in N_2 ambient accordingly. Finally, the current-voltage (I-V) and capacitance- voltage (C-V),

Scanning electron microscopy(SEM) and X-ray diffraction(XRD) techniques was used to observe the electrical characteristics, surface morphology and chemical compounds formed between metal and semiconductor contact before and after annealing.

III. RESULTS AND DISCUSSION

The (I-V) characteristics of the Schottky diode, ΦB and n, were determined assuming thermionic emission,

$$I = I_s \left[eqv/nkT - 1 \right] \quad \text{and} \quad I_s = AA^* T^2 e^{-q\Phi B/kT} \quad (1)$$

Here I_s is the saturation current, A is the contact area, and A*is the effective Richardson constant (24 A/cm^2 k^2 for n-GaN and 72 A/cm^2 k^2 for p-GaN based on A* =4πm*qk^2/h^3 with m* =0.20m$_0$ for n-GaN and m* =0.60m$_0$ for p-GaN), T is temperature, q is the electron charge, k is the Boltzman constant, and V is the applied voltage [4]. Fig.1 shows the (I–V) characteristics of n-GaN Schottky diodes after annealing at different temperatures from 400°C ~700°C. The diodes shows a rectifying behavior for the as-deposited and annealing at temperature 400°C, while the ohmic behavior at above temperature 400°C to 700°C.

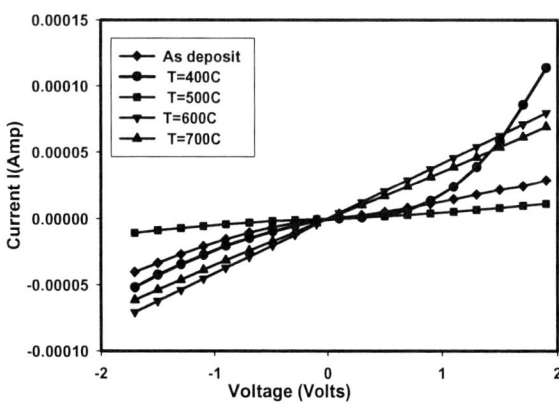

Fig. 1: (I-V) Characteristics of n-GaN Schottky diode.

The temperature dependence schottky barrier height, Ideality factor is shown in the Table 1. Annealing at temperature 400°C were coincidently produced the optimum schottky barrier height (ϕ_B) 1.1eV due to improvement of schottky contact

Pt/Ti and Al ohmic contact. After annealing 400°C ~700°C the barrier height decreased due to increase of leakage current, which indicates the formation of metal-insulator semiconductor type structure. Changes in the surface morphology and metal island were examined at high temperature annealing. As the annealing temperature of n-GaN Schottky diode increased to above 400°C the values of ideality factor was found much greater then unity. This indicates that a significant amount of the current could be due to other current transport mechanism rather than thermionic emission model. An ideality factor, n>1 could be ascribed to interface states at a thin oxide between the metal and n-GaN, or generation-recombination currents within the space region [5]. The reduction of barrier height and increase of ideality factor at high temperature are due to the preferential loss of nitrogen from the GaN surface. Nitrogen vacancies are thought to increase the carrier concentration in GaN. This high surface carrier concentration increase tunneling probability at metal/n-GaN interface [6].

Table 1 Summary of (I-V) Characteristics of n-GaN Schottky diode

Annealing Temperature	Barrier height (eV)	Ideality factor (n)
As-deposited	0.84	1.91
400°C	1.10	1.10
500°C	0.78	1.81
600°C	0.68	9.69
700°C	0.65	8.52

The variation of the external voltage applied through the metal contact will produce variation in the charged carrier at the surface of the semiconductor. Therefore the variation of the applied voltage will cause a variation of the capacitance. The capacitance of an n-GaN Schottky diode can be determined as follows,

$$C= \varepsilon_o \varepsilon_r A / \sqrt{(\{2\varepsilon_o \varepsilon_r / qND\}.\{Vbi-Vapp\})} \quad (2)$$

The (C-V) characteristics of n-GaN Schottky diode were measured at 100 kHz and 1 MHz frequency at different annealing temperature is shown in Fig. 2. At 100 kHz frequency the diode

sweep from accumulation to inversion mode through the depletion mode only at as deposited and annealing at 400°C. The variation of the depletion layer can be observed as a variation of capacitance. The fluctuation of capacitance as the applied voltage varies though the capacitance should be constant over the accumulation and inversion region. At annealing temperature 400°C, the capacitance will increase rapidly from minimum value 0.02Pf to 0.04pF maximum value as the applied voltage varied from –Ve to +Ve voltage. At low frequency 100 kHz the capacitance increase with increasing forward voltage which is frequency independent which is ideal Schottky diode behavior [7]. While annealing at 600°C and 700°C, no depletion region formed. As a result, no variation of capacitance took place as the applied voltage varied. At high frequency 1MHz the capacitance-voltage curve is almost flat which indicates that inversion layer charge and interface states can not follow the AC signal at this high frequency and consequently do not contribute to the diode capacitance [8].

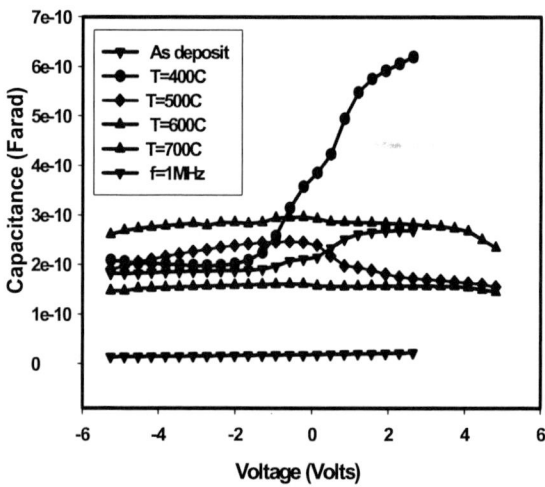

Fig. 2 : (C-V) Characteristics of n-GaN Schottky diode.

The surface morphology of Al(1500A°) as ohmic contact while Pt/Ti(700A°/700A°) as a schottky contact of n-GaN schottky diode annealed at 400°C and at 700°C are observed by SEM images are shown in Fig. 3. At annealing 400°C and 700°C the Al contact did not show any loss of dimensional stability or surface morphology degradation respectively. While Pt/Ti contact at

700°C shows "balling-up effect" or island form of surface morphology occurred, produced extremely poor results both from electrical and structural point of view. The interfacial reaction due to dissociation of the GaN and becomes increasingly rapid after 400°C has caused strongly islanded occurred. It indicates that Pt/Ti film may react with n-GaN epilayer there is significant indiffusion of Pt, Ti and outdiffusion of Ga at the interface vicinity.

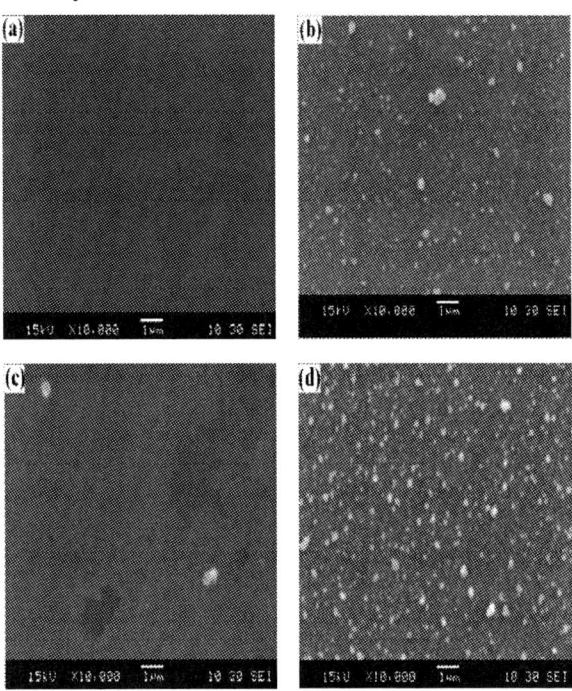

Fig. 3 SEM images of Al and Pt/Ti annealing at 400°C (a, b) annealing at 700°C (c, d)

To investigate interfacial reaction, 2θ XRD measurement performed on the as deposited, 400°C and 700°C annealed sample. Fig. 4 shows the X-ray diffraction (XRD) spectra of Al (1500A°) as ohmic contact, Pt/Ti (700A°/700A°) as schottky contact metal to n-GaN schottky diode before and after annealing 400°C and 700°C . The main spectra peaks at 30°<2θ <90°. For the case of as deposited Pt/Ti, XRD peaks corresponding to Pt (111) at 2θ=39.94°, Pt (200) at 2θ=46.48° and Pt (220) at 2θ=67.60°. It can be seen that the change of peaks are obvious before and after annealing, especially the peaks of Pt (111) at 2θ=39.94°,Pt(220) at 2θ=46.48° and Pt (220) at 2θ=67.60°. The XRD data also show the peaks of some compounds that are not detectable in as-

deposited sample and appear after annealed 400°C. The compound of Pt3Ti (111) at 2θ=40° Ga5.4Pt10.60 (322) at 2θ=63.47°, Ga3Ti2(302) at 2θ=64.037° observed at 700°C temperature, whereas Pt (111), Pt (200) were absent. The formation of these compounds annealed at 700°C at Pt/Ti schottky contact to n-GaN will give ohmic behavior.

XRD peaks corresponding to Al metal as ohmic contact to n-GaN are Al (200) at 2θ=44.83°, Al (200) at 2θ=46.47° observed. It can be seen that there not much more changes in the peaks for Al at 700 °C annealing.

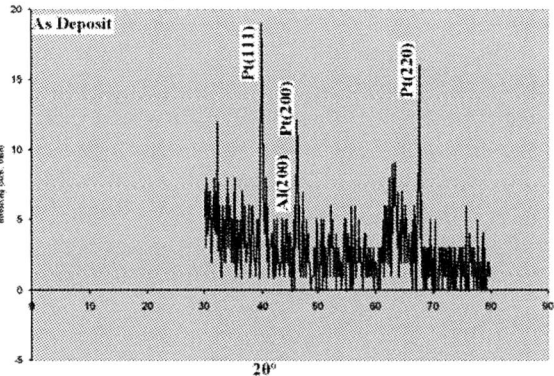

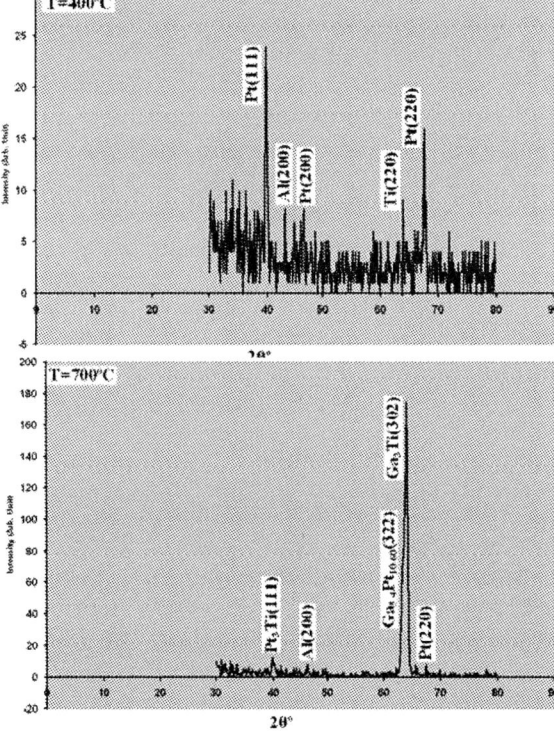

Fig. 4 XRD spectra of the samples (a) as deposited (b) 400°C (c) 700°C annealing

Hence we conclude that annealing at above 400°C~700°C, Pt/Ti(700A°/700A°) metal as a schottky contact of n-GaN schottky diode show loss of dimensional stability or surface morphology degradation which cause to change from schottky behavior to ohmic behavior of n-GaN Schottky, while on the other hand Al(1500A°) metal as a ohmic contact on n-GaN Schottky diode show thermal stability and does not show loss of surface morphology degradation at high temperature annealing.

IV. CONCLUSION

We investigated (*I-V*) and (*C-V*) characteristics of n-GaN schottky diode at different annealing temperature 400°C~700°C. We find that at annealing 400°C of n-GaN schottky diode the best *I-V* characteristics (Barrier height 1.1, ideality factor 1.1), while capacitance value varied from 0.02pF ~ 0.04pF as applied voltage varied from – Ve to +Ve voltage at 100 kHz. It was found that annealing above 400°C the rectifying characteristics of n-GaN schottky diode changed to ohmic behavior. The surface morphology of n-GaN schottky diode was observed by SEM and XRD before and after annealing 400°C~700°C. It was found that annealing above 400°C the island form of surface morphology occurred at Pt/Ti(700A°/700A°) schottky contact which caused to change into ohmic behavior, while Al(1500A°) as an ohmic contact show thermal stability at high annealing temperature of n-GaN schottky diode. Hence we conclude that annealing above 400°C the n-GaN schottky diode rectifying characteristics changed to ohmic behavior due to Pt/Ti (700A°/700A°) island form of surface morphology.

ACKNOWLEDGEMENT

This work was conducted under IRPA RMK-8 Strategic Research grant. The support from Universiti Sains Malaysia is gratefully acknowledged. The financial support from MTCP (Malaysian Technical co-operation Program) is also acknowledged. The author also wish to thank Dr. Maggdy for useful discussion.

REFERENCES

[1] J. Wang. D. G. Zhao and Y. P. Sun, *J. Phys. D. Appl. Phys,* vol 36. p. 1018 (2003).

[2] Q. Z. Liu and S. S. Lau, *Solid- State Electronics*, vol 42, no. 5, pp. 677-691(1998).

[3] J. S. Foresi and T. D. Moustakas, *Appl. Phys, Lett.* vol. 62, no. 22, pp. 2859-2861(1993).

[4] K. N. Lee, X. A. Cao, C. R. Abernathy and S. J. Pearton, *Solid- State Electronics*, vol. 44. pp. 1203-1208 (2000).

[5] C. I. Wu and A. Kahn, *J. Appl. Phys*, vol. 89, no. 1, pp. 425-429 (2001).

[6] B. Sahin, H. Cetin and E. Ayyildiz, *Solid State Communication,* vol. 135, pp. 490-495 (2005).

[7] A. Turut, M. Saglam, H. Efeoglu and N.Yalcin, *Physica B*, vol. 205, pp. 41-50 (1995).

[8] C. K. Ramesh, V. R. Reddy and C. J. Choi, *Materials Science & Engineering B*. vol. 112, pp. 30-33 (2004).

[9] L. Wang, B. I. Nathan, T. H.Lim, M. A. Khan and Q. J. Chen, *Appl. Phys, Lett.* vol. 68, pp. 1267 (1996)

[10] Y. Kokubun, T. Seto and S. Nakagomi, *J. Appl. Phys,* vol 80, pp. 4467 (2001).

Epilayer Thickness and Doping Density Variation Effects on Current-Voltage (I-V) Characteristics of n-GaN Schottky Diode

Tariq Munir, Azlan Abdul Aziz and Mat Johar Abdullah
Nano-Optoelectronics Research and Technology Laboratory (NOR)
School of Physics Universiti sains Malaysia 11800 Minden Penang Malaysia
Email: tarriqmunirr@yahoo.com, lan@usm.my

ABSTRACT **The epilayer thickness and doping density variation effects on the (I-V) characteristics of n-GaN schottky diode are determined numerically. In this work the epilayer thickness of n-GaN schottky diode are varied from 3μm ~11μm while doping density are varied from 1×10^{12}cm^{-3} ~ 1×10^{17}cm^{-3}. The simulation work is conducted using Atlas/Blaze developed by Silvaco. The various models such as Srh (Shockley- Read Hall), Cvt (Lombardi model), Auger, Impact (Grants model), Bgn (Band gap narrowing), Conmob (concentration dependent mobility) is used to get optimum (I-V) characteristics of n-GaN schottky diode. We find that in forward biased as the epilayer thickness varied from 3μm ~ 11μm at constant doping density 1×10^{12}cm^{-3} the forward current decreases due to the increase in series resistance, on the other hand the forward current increased by lowering the epilayer thickness up to 3μm and increasing the doping density ~ 1×10^{17}cm^{-3}. In reverse biased the selection of doping density and epilayer thickness directly determined the target reverse breakdown voltage of the device. As the epilayer thickness varied from 3μm ~ 11μm at constant doping density 1×10^{12}cm^{-3}, the breakdown voltage increased due to increased in the depletion width, while the breakdown voltage reduced as doping density varied from 1×10^{12}cm^{-3} ~ 1×10^{17}cm^{-3} at constant epilayer thickness of 3μm. Hence we conclude that forward current and breakdown voltage have strong, inverse relation between epilayer thickness and doping density.**

I. INTRODUCTION

In recent year GaN THz Schottky barrier diodes have been the subject of considerable attention. Currently the important point is how to improve their realization and technology for better results

at higher frequencies. To this respect, much effort has been devoted to reduce the diode thickness. An important problem in this respect is to understand the role played by the epitaxial layer thickness. The epitaxial layer thickness and doping density has considerable effects on the current operation of devices [1]. The main design variables that affect the schottky barrier diodes performance are the contact metal work function, the epilayer doping, epilayer thickness [2]. The selection of epilayer thickness and doping density directly determine the target reverse breakdown voltage and thus, making uniformity and control of these parameters important. An increase in doping or a decrease in thickness from the designed values will affect the reverse characteristics and breakdown voltage. The doping density can be related to this voltage as [2]:

$$N_D = \varepsilon_s E^2_{cR}/2qV_{BR} \qquad (1)$$

Here ε_s is dielectric permittivity, E_{cr} is critical field, V_{BR} breakdown voltage. The epilayer thickness is critical in determining the reverse breakdown voltage as follows [2]:

$$t_{epi} = 2V_{BR}/E_{cr} \qquad (2)$$

Here E_{cr} is the critical field of semiconductor, V_{BR} is breakdown voltage, t_{epi} epilayer thickness. For power semiconductor, it is desirable to obtain minimum on-resistance without voltage breakdown degradation. The epilayer thickness is a critical design parameter to obtain low on-resistance and low forward voltage drop [2]. Where

$$R_{on} = t_{epi}/q\mu_n N_D \qquad (3)$$

Here N_D donor impurity concentration, μ_n mobility of electron.

The epi thickness is thinner and doping lower, thus decreasing the on-resistance. The appropriate epi doping and thickness determined the minimum on- resistance [3]. An improvement in the drift region resistance is obtained if epitaxial layer thickness and doping are optimized. In order to minimize the R_{on} on-resistance of the device, the width of epilayer has to be equal to the maximum depletion width. The devices with linearly doped epilayer have higher breakdown voltage and lower on-resistance [4].

II. SIMULATION OF n-GaN SCHOTTKY DIODE

The diode is designed on lightly doped of thin epitaxial layer that is grown on heavily doped of low resistance substrate. The lightly doped region is to be used for the junction while the heavily doped region is to minimize series resistance. Simulation of GaN structure with the cylindrical symmetry diode was performed using the 2D numerical simulator. The structure of GaN schottky diode is shown in Fig.1 while several critical parameters [5] used for the simulation is given as in table 1. The diodes were simulated by varying the epilayer thickness from 3μm to 11μm while carrier concentration variation from 1×10^{12}cm^{-3} to 1×10^{17}cm^{-3}. The doping concentration of the next n+ layer was 5×10^{19}cm^{-3}. For schottky diode the following model were chosen for simulation; incomplete ionization, cvt, Fermi, Bgn and Shockley- Read Hall model [6]

III. RESULTS AND DISCUSSION

The forward conduction characteristics of a high voltage schottky diode can be modeled as a

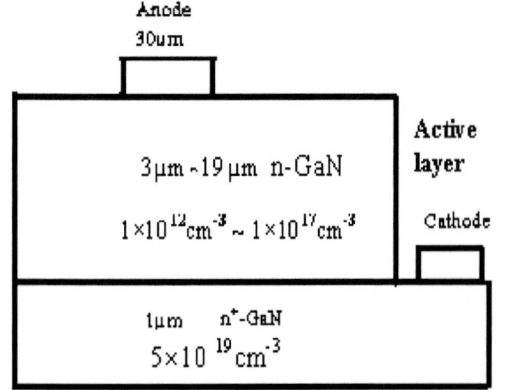

Fig. 1 Schematic of GaN Schottky diode [7].

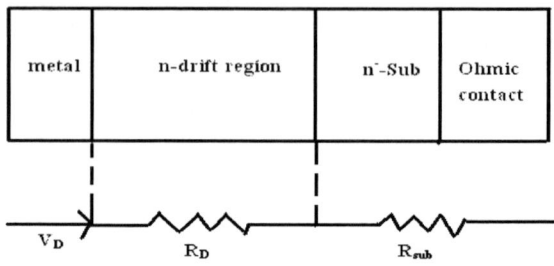

Fig. 2 Structure of GaN Schottky diode

Table1 Important parameters used in simulation [5].

Material	Property	Value
GaN	Relative Permittivity	8.9
	Energy band gap	3.40 eV
	Electron affinity	3.3 eV
	Saturation velocity	1.5×10^7 cm^2/s
	Electron mobility	1000 cm^{2}v/s
Platinum	Schottky contact	
	Work function	5.6eV
	Barrier height	1.2eV

schottky contact in series with a specific on-resistance R_{on}, which includes the contribution resistance, from the epitaxial layer, substrate, and contact Here we can neglect the substrate and ohmic contact resistance [8]. The specific on resistance R_{on} of the diode, is seen at higher levels of current when the series resistance of the device takes over and provides an additional voltage drop. This forward voltage drop V_F of the schottky diode is given by [9]:

$$V_F = kT/q \ln (J_F / A^{**}T^2) + Ø_b + J_F R_{on} \quad (4)$$

Where k is the Boltzman's constant, q is the electron charge, A^{**} is the effective Richardson's constant, J_F is current density, $Ø_b$ is the schottky barrier height, R_{on} on-resistance. The linear forward I-V characteristics of GaN Schottky diode at different epilayer thickness from 3μm ~11μm at constant doping density of 1×10^{12}cm^{-3} are shown in the Fig. 3. In forward bias of Schottky diode the on resistance is the important factor acts on the forward current. The value of on resistance increase as the epi layer thickness increase at the constant doping density of 1×10^{12}cm^{-3} as in Fig. 4. The carrier mobility to cross the semiconductor metal junction is decreased due to increase in on resistance. Hence

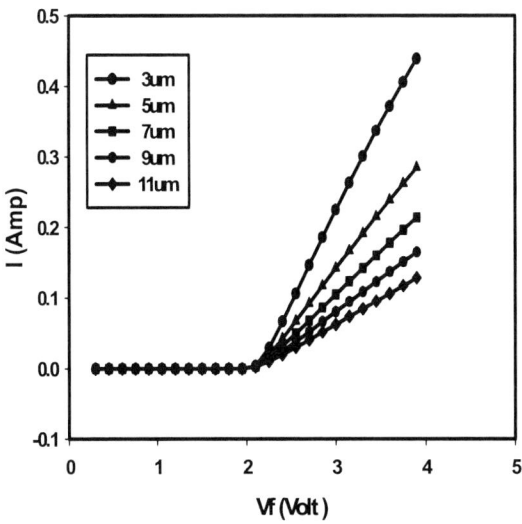

Fig. 3: Forward (I-V) characteristics

forward voltage drop across the junction as a result the forward current decrease due to increase in epilayer thickness $3\mu m \sim 11\mu m$ at constant doping density $1\times10^{12}cm^{-3}$.

The on resistance R_{on} was calculated from the slope of the (I-V) characteristics and is plotted for different thickness as shown in Fig. 4.

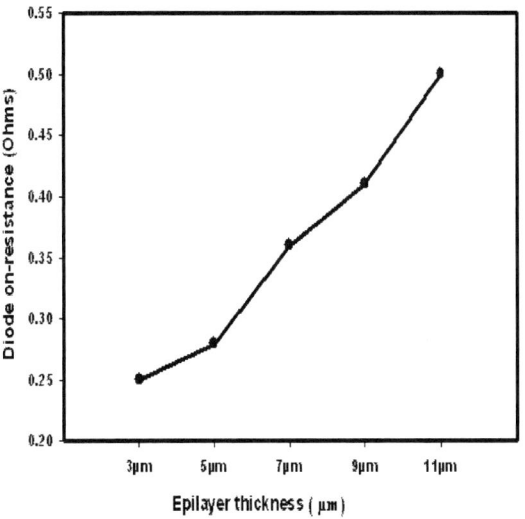

Fig. 4: On resistance at different epilayer thickness

The reverse bias I-V characteristics of GaN Schottky diode at different epilayer thickness $3\mu m \sim 11\mu m$ at constant doping density of $1\times10^{12}cm^{-3}$ are shown in Fig. 5. In reverse bias the reverse leakage current and breakdown voltage is key parameter in operation of GaN Schottky diode. The breakdown voltage increase to -38V as the epilayer thickness increase to $19\mu m$ at

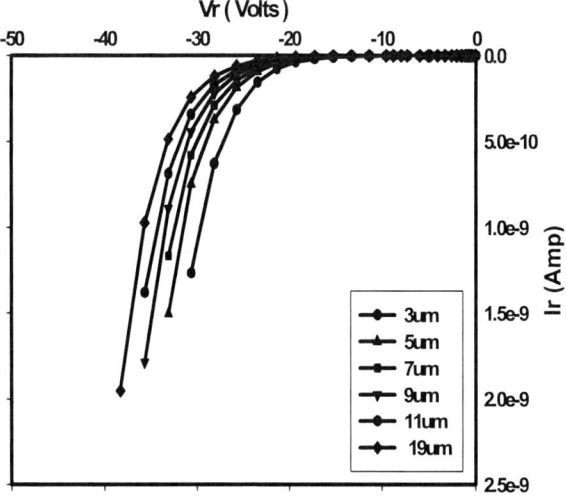

Fig. 5: Reverse (I-V) at different epilayer thickness

constant doping density shown in Fig. 5. This increase in breakdown voltage due to increase in depletion region width, weak electric field due to few carriers doping density. Reverse-biased performance of the Schottky diode as active layer the doping density vary from $1\times10^{12}cm^{-3} \sim 1\times10^{17}cm^{-3}$ at constant thickness $3\mu m$ is as shown in Fig. 6.

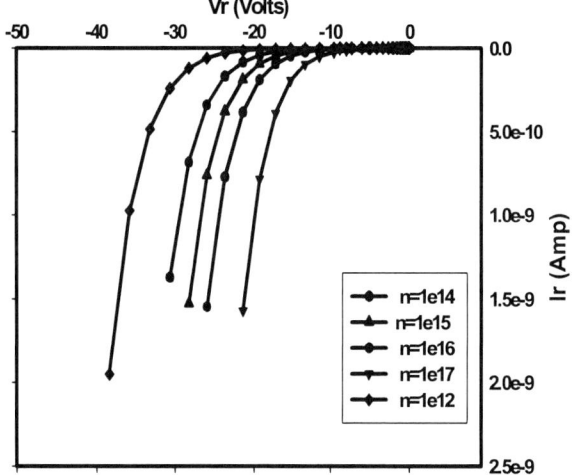

Fig. 6 Reverse (I-V) at different carrier concentration

The breakdown voltage is maximum when the width of the epilayer is equal to the maximum depletion width whiles the R_{on} is minimize. The Fig. 7 and Fig. 8 shows variation of the breakdown voltage with the carrier concentration and the epilayer thickness. To sustain the high breakdown voltages, reduced doping concentration and increased blocking layer thickness are required, which led to the increase in serial resistance R_{on} [10]. Hence we conclude that in reverse bias the breakdown voltage has inverse relation of epilayer thickness and doping density. This report is invaluable for the designing of microelectronics devices.

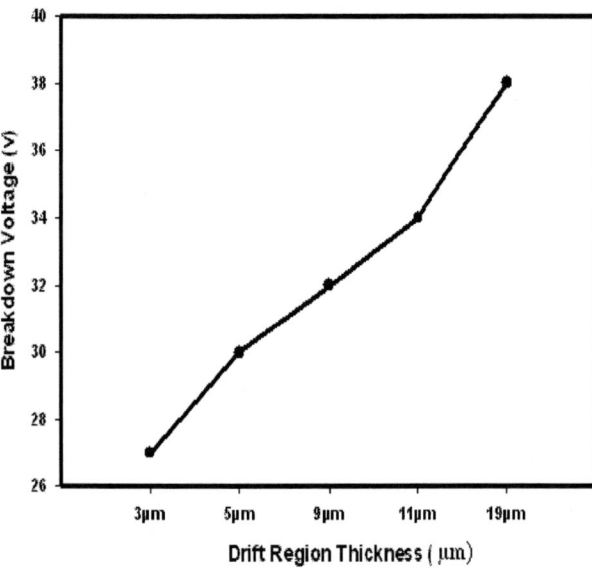

Fig. 7: Breakdown voltage Vs Drift region thickness.

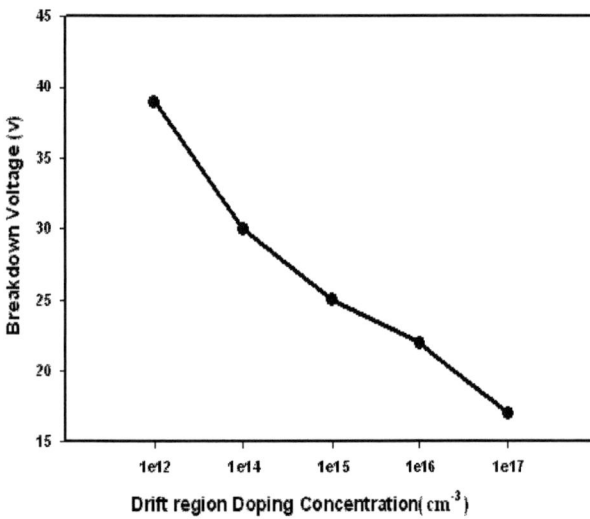

Fig. 8: Breakdown voltage Vs Doping concentration.

IV. CONCLUSION

The epilayer thickness and doping density variation show important role in forward bias and reverse bias (I-V) characteristics of n-GaN Schottky diode. It was found that in order to increase the forward current, the epilayer thickness is minimum at 3μm, doping density is maximum at 1×10^{17}cm^{-3}. While in reverse bias, in order to obtain the maximum breakdown voltage the epilayer thickness is maximum at 19μm, doping density is minimum at 1×10^{12}cm^{-}

[3]. Hence we conclude that in n-GaN Schottky diode the epilayer thickness and doping density have inverse relation between the forward current and breakdown voltage.

ACKNOWLEDGEMENT

This work was conducted under IRPA RMK-8 Strategic Research grant. The support from Universiti Sains Malaysia is gratefully acknowledged. The financial support from MTCP (Malaysian Technical co-operation Program) is also acknowledged.

REFERENCES

[1] G. Gomila and O. M. Bulashenko. *J. App Physics.* Vol. **86**, pp. 1004 (1999).

[2] K. Schoen, J. Woodall, J. Cooper and M. Melloch, *IEEE Transactions on Electron Devices.* Vol. **45**, pp.1595 (1998).

[3] M. Bhatnagar, P. K. Mclarty and B. J. Baliga., *IEEE Electron Device letters.* Vol. **13**, p. 501 (1992).

[4] S. H. Kim, Y. Park, Y. Choi and S. K. Chung. *Physica Scripta,* Vol.**T101**, p. 234 (2002).

[5] S.W. Chung, W. J. Hwang, L. Chin and M. W.. Shin, *Journal of Crystal Growth.* vol. **268**, p. 607 (2004).

[6] Silvaco international "*Device simulation software*" vol. **2**, p. 84 (2000).

[7] K. H. Baik, Y. Irokawa, F. Ren, S. J. Pearton, S. S. Park and Y. J. Park, *Solid-State Electron.* vol. **47**, p.1533 (2003).

[8] B. J. Baliga, "*power Devices*", New York, Wiley; (1997).

[9] R. Raghunathan, D. Alok and B.J. Baliga, *IEEE Electron Device Letters,* vol **16**, p. 226 (1995).

[10] D. L. Barrett and R. B. Campbell, *J. Appl. Phy,* vol. **38**, p. 53 (1967).

A VLSI Design Framework with Freeware CAD Tools

Y. K. Teh[1], *Student Member, IEEE,* F. Mohd-Yasin[1], *Member, IEEE,* M. B. I. Reaz[2], *Member, IEEE,*
A. Kordesch[3], *Senior Member, IEEE*

[1]Faculty of Engineering, Multimedia University, 63100 Cyberjaya, Selangor, MALAYSIA
[2]ECE Dept, IIUM, 53100 Kuala Lumpur, MALAYSIA
[3]Device Modeling Group, Silterra Sdn. Bhd., Kulim, Kedah, MALAYSIA
E-mail: ykteh@mmu.edu.my

Abstract - **This work presents a PC-based freeware CAD environment to design and tape out VLSI microelectronic circuits, starting from schematic capture all the way to a foundry compatible GDS II database. These free tools will help more Malaysian universities to set up low cost VLSI CAD laboratories and tape out circuits using Silterra's University Program. This will help grow local IC design culture and skills. FreeVLSI uses common freeware CAD tools: 5SPICE, LASI, and WinSPICE, and some custom scripts to interface between these tools. Currently FreeVLSI is able to cater to full custom design flow from schematic capture to circuit layout.**

I. INTRODUCTION

Electronics manufacturing is one of the most important industries in Malaysia. Recently, cheap labor from other countries has prompted Malaysia to try to move up the value chain to chip design, following the Taiwan model. It is crucial to develop manpower with the IC design skills early at universities undergraduate level to achieve the objective.

In Malaysia, Silterra has offered a university program that allows university researchers to fabricate their designs at no cost, similar to [1] and [2]. The chips are designed and simulated using costly commercial design tools. We attempt one step further, by designing free tools to design chips employing Silterra technology.

The purpose of this work is to present a PC-based freeware CAD environment (called **FreeVLSI**) to design and tape out VLSI microelectronic circuits, starting from schematic capture finally producing a foundry compatible GDS II database. These free CAD tools will help Malaysian universities to set up low cost design laboratories to grow local IC design culture and

skills. FreeVLSI is currently capable of accommodating full custom ASIC design flow.

Three different types of circuits (MOSFET I-V characterizations, digital and analog) have been simulated using FreeVLSI and commercial tools. Comparisons have been made with the actual silicon data. Silterra 0.18μm CMOS SPICE models were used in all simulations.

II. DESCRIPTION OF FRAMEWORK PROPOSED

Figure 1 shows the freeware components proposed to accomplish FreeVLSI Full Custom design flow:

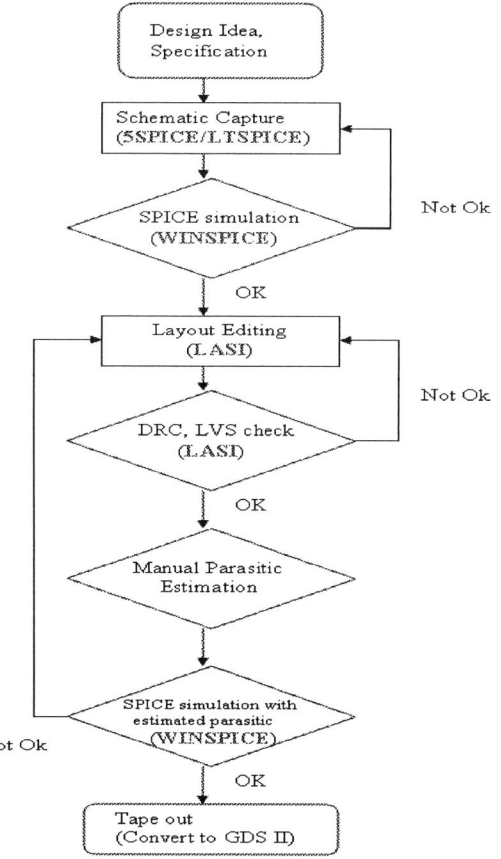

Figure 1: Full Custom Design Flow of FreeVLSI

- *Schematic Capture*

Schematic capture software is used to construct circuit schematics and generate SPICE simulation net lists. Two good candidates found are 5SPICE [3] and LTSPICE [4].

Alternatively LASI which is proposed as layout editing tool could also be used to perform schematic capture and net list extraction. Nevertheless device models in extracted net lists have to be altered in order to match the device name given in foundry library. This could be done by manual editing or utilizing computer scripts.

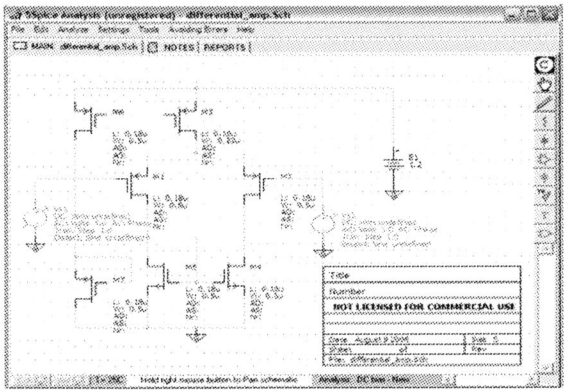

Figure 2: Snapshot of 5SPICE

- *SPICE Simulation*

BSIM3 is popular CMOS MOSFET device model developed by the BSIM Research Group [5] of the University of California, Berkeley. The third iteration of BSIM3, BSIM3 Version 3 (commonly abbreviated as BSIM3v3) has since been widely used by most semiconductor foundries and IC design houses world wide for device modeling and CMOS IC design.

In this framework, the target is to find freeware alternatives that are capable of simulating commercial BSIM3v3 device model libraries provided by the foundry. A good candidate discovered after testing many freeware solutions available is WINSPICE [6]. WINSPICE is a port of Berkeley's original Spice3F4 to Microsoft Windows operating system. WINSPICE is capable of taking HSPICE BSIM3V3 compatible MOSFET device models and also BJT (Gummel and Poon model) after applying software patch upgrades provided by the author. Users can use WINSPICE to simulate circuits with a mixture of MOSFET and CMOS

parasitic BJT devices such as bandgap reference circuits. WINSPICE can perform both electrical simulation and graphical post processing through programming scripts.

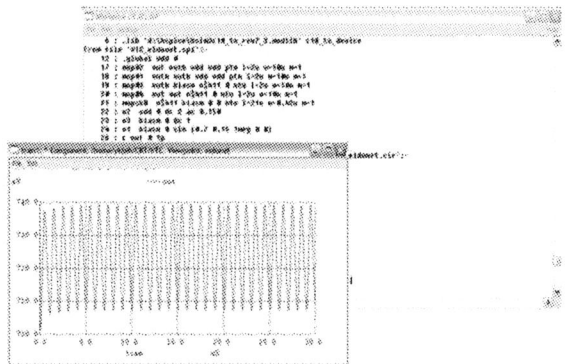

Figure 3: WinSPICE nutmeg and plot snapshot

- *Layout Editing*

Windows **LASI** [7] (Windows **LA**yout **S**ystem for **I**ndividuals) is recommended to perform custom layout editing, Design Rule Check (DRC), macro extraction, Electrical Rule Check (ERC), Layout Versus Schematic (LVS), as well as final data codification (conversion to Gerber GDS II format for foundry fabrication). LASI was written by Dr. David Boyce. LASI runs on the Microsoft Windows system.

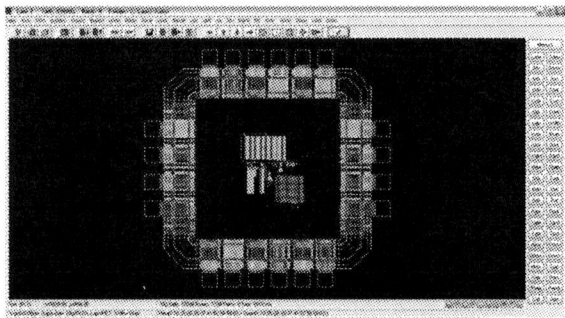

Figure 4: LASI showing fabricated bandgap design

Default set up files for LASI have been modified to match GDS layer definitions of the targeted foundry. It is worth noting that LASI is only capable of doing simple DRC check and parasitic extraction is still in development as sophisticated LASI compatible deck files need to be rewritten. In the current development of FreeVLSI, designers have to manually estimate and compute parasitic capacitance based on circuit layout and foundry electrical design rules.

III. DESIGN EXAMPLES

Three different types of circuits have been simulated using FreeVLSI and commercial tools for comparison. Silterra 0.18μm CMOS technology is used in all simulations. Example circuits include MOSFET I-V characterization, Bandgap reference circuit (analog circuit) with actual silicon data, and TSPC D flip flop (digital circuit).

- *MOS I-V characterization*

Both NMOS and PMOS types of transistors available in Silterra's 0.18μm process are characterized using HSPICE, Eldo and WinSPICE. Saturation current of PMOS and NMOS of different W/L ratio is extracted. Data from three SPICE simulators are plotted in Figures 5 and 6.

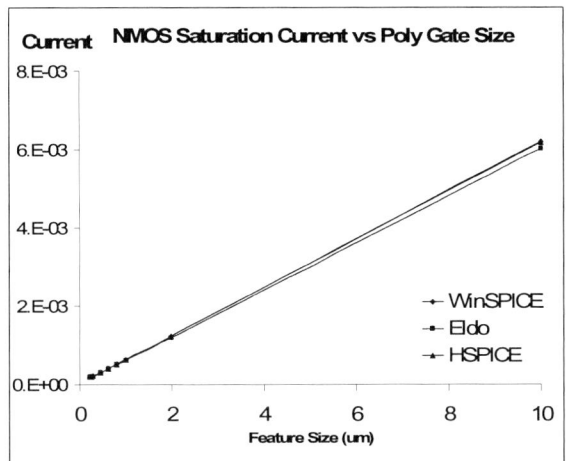

Figure 5: NMOS I_{Dsat} vs Poly gate size

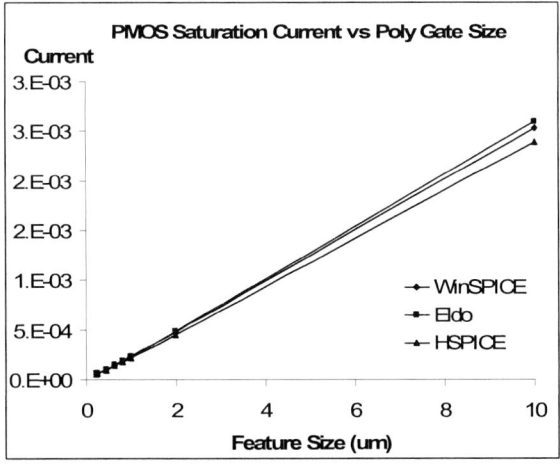

Figure 6: PMOS I_{Dsat} vs Poly gate size

Three simulators give close results for small feature size devices (Poly gate width less than 2um). Results are summarized below.

	Eldo	HSPICE
NMOS	3.34%	0.88%
PMOS	2.07%	5.91%

Figure 7: Average difference in simulation results of I-V characterization for Eldo and HSPICE compared to WinSPICE with variable W, Fixed L = 0.18μm

A. Analog Circuit - Bandgap Voltage Reference

Figure 8 shows an analog circuit example, which is a low power current mode bandgap reference circuit derived from H.Banba's paper [8]. DC, AC and transient analyses are simulated on two simulators: Mentor Graphics Eldo and WINSPICE. The design has been fabricated using Silterra's 0.18um process and DC results are confirmed with actual silicon IV measurement using an HP4156B parameter analyzer. Simulation results obtained from both simulators are compared with actual silicon measurement data.

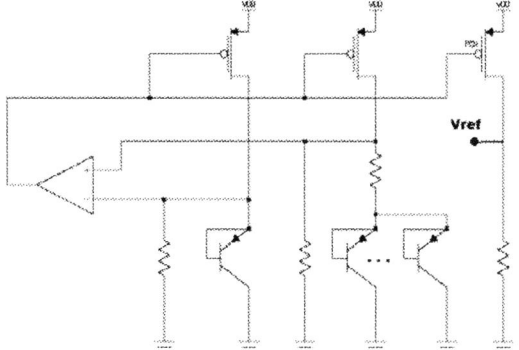

Figure 8: Bandgap reference circuit investigated

DC sweep simulation result shows less than 1% difference between commercial simulator (Eldo) and WinSPICE. Both simulators yielded good match against actual silicon results for both output voltage and current consumption as shown in Figure 9 and Figure 10.

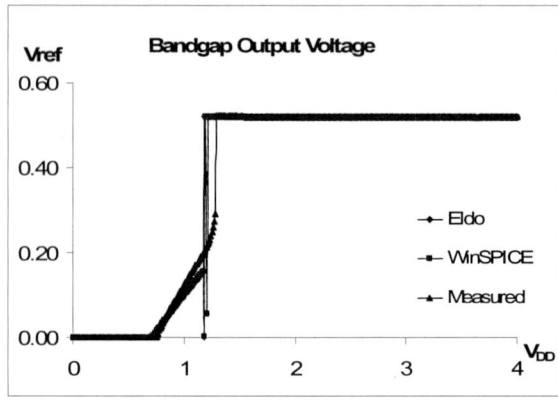

Figure 9: V_{ref} of Bandgap vs V_{DD}

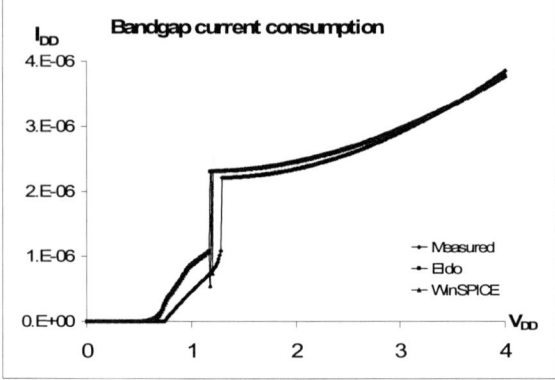

Figure 10: I_{DD} of Bandgap vs V_{DD}

AC analysis, which is a common analysis performed by designers is also compared. Again both simulators give nearly identical results as shown in Figure 11 (a),(b).

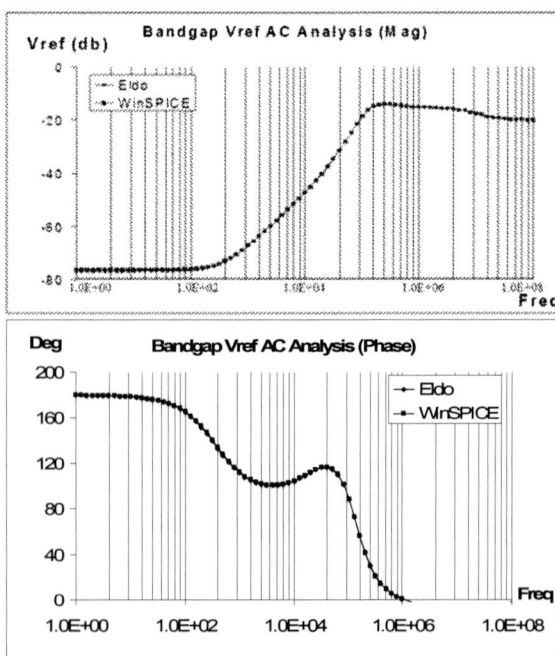

Figure 11 (a), (b): AC response of Bandgap Vref

Transient response is simulated and plotted in Figure 12. WinSPICE simulation shows some discrepancy at initial transient but overall the response tracks Eldo's output well after output voltage stabilizes.

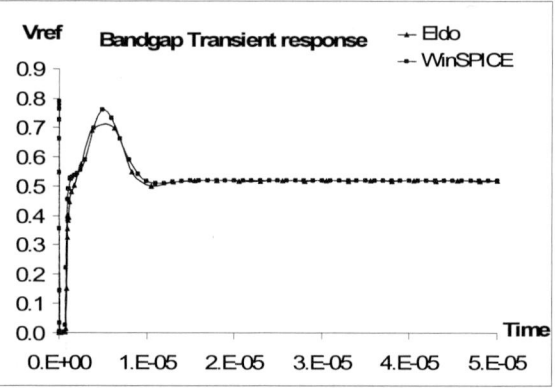

Figure 12: Transient response of Bandgap V_{ref}

B.   Digital Circuit – D Flip Flop

Behaviour of digital circuit simulation under FreeVLSI is investigated as well. A TSPC (True Single Phase Clock) D flip flop circuit is simulated using both Eldo and WinSPICE.

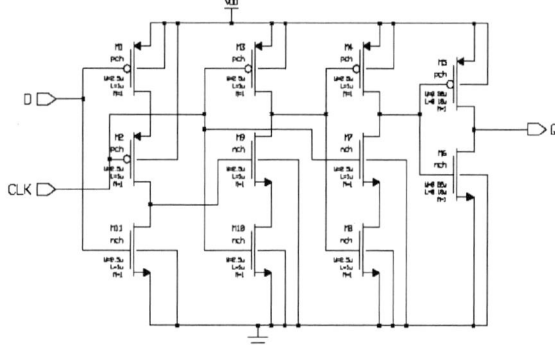

Figure 13: TSPC D Flip Flop circuit

With the same stimulus fed to CLK and D input of D flip flop, Q output of D flip flop is forced to switch at 250MHz. Transient response of Q output from both simulators is plotted on Figure 14. It can be seen that simulation results yielded from Eldo and WinSPICE actually tracks closely; except the initial transient. Timing difference demonstrated by both simulators shows less than 1ps difference.

908

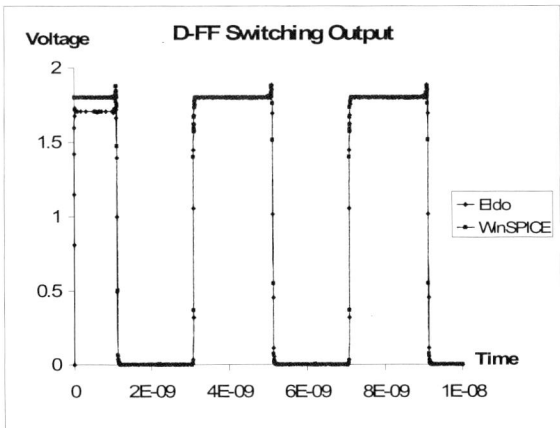

Figure 14: Transient response of D-FF Q output

WinSPICE has better match in digital circuits compared to analog circuits due to the saturation nature of MOS in digital operation; whereas analog circuits often operate in the sub-threshold regime, both simulators do not track analog as close as digital circuits.

IV. DISCUSSION

FreeVLSI has shown comparable performance with commercial products in SPICE simulation in terms of DC, AC and transient analysis. It is superior judging from cost factor. However, unlike commercial CAD software, FreeVLSI does not have good integrity to interface between design tools. It also lacks a user friendly graphical interface. For users to maneuver smoothly within FreeVLSI framework, some knowledge of computer script is necessary in order to convert and transfer output between different tools.

FreeVLSI also suffers from the simple DRC checking tool. For comprehensive layout, LASI often overlooks DRC errors and has slow loading speed. The next work will look at this issue closely.

V. CONCLUSION

In conclusion, these results prove that FreeVLSI can be used as an alternative over commercial CAD tools for simple mixed signal custom circuits, with good matching to actual silicon results. The final design is foundry compatible and can be fabricated using Silterra's University Cooperation program.

The advantage of this framework is its zero cost. In fact FreeVLSI can be installed on low end PCs. This allows more local universities to venture into the world of VLSI design. It is hoped that with the introduction of this framework, more local engineers with VLSI design experience could flourish.

REFERENCES

[1] http://www.mosis.org
[2] http://www.europractice.imec.be/europractice/europractice.html
[3] http://www.5spice.com
[4] http://www.linear.com/company/software.jsp
[5] http://wwwdevice.eecs.berkeley.edu/~bsim3/intro.html
[6] http://www.winspice.com
[7] http://members.aol.com/lasicad/
[8] Banba, H. Shiga, T.Miyaba, T.Tanzawa, S.Atsumi, and K.Sakui, "A CMOS Bandgap Reference Circuit with Sub-1-V Operation", IEEE Journal of Solid State Circuits, Vol 34-5, pp 670-674, May 1999.

Implementation of Internal Mixed Signal ESD Protection onto RFID Transponder IC

M. K. Khaw[1], F. Mohd-Yasin[1], *Member, IEEE,* Y. K. Teh[1], *Student Member, IEEE,* M B.I. Reaz[2],
Member, IEEE

Faculty of Engineering, Multimedia University, 63100 Cyberjaya, Selangor, MALAYSIA
[2]ECE Dept, IIUM, 53100 Kuala Lumpur, MALAYSIA
Phone: +603-8312 5423
Fax: +603-8318 3029
Email: khaw.mei.kum@mmu.edu.my

Abstract – **Radio Frequency Identification (RFID) has become an important wireless data communication tool in recent years. As much as we want to ensure data integrity and the robustness of the RFID transponder, Electrostatic Discharge (ESD) influence on the transponder can jeopardize it. Current practice put the ESD protection in the package. However at pad level, the ESD protection is usually small dimensioned to reduce input capacitance. Hence extra ESD protection co-constructed at internal circuit VDD-VSS rail is necessary in advanced process due to thinner gate oxide. In this paper, we have developed an internal ESD protection circuit and implemented it in our previously developed 13.56MHz RFID transponder employing TSMC 0.18μm process. The circuit has a capability to sustain 2-KV of HBM positive mode ESD voltage, which is suitable for RFID applications. The additional power consumption for the clamp circuit is only 15.12nW.**

I. INTRODUCTION

Radio Frequency Identification (RFID) is an important data communication tool due to its contactless technology and the ability of its transponder to be rewritable, having large memory capacity and able to withstand visually and environmentally challenged conditions. Just like many electronic devices, the RFID transponder is no less vulnerable from the Electrostatic Discharge (ESD) damages. ESD is a transfer of energy between two materials of different electrostatic potential. For example, human can transfer the ESD to the chip he/she is holding from touching a chair. While ESD can be potentially eliminated during manufacturing

process by taking stringent measures in the whole production of the tag [1-4], it is not easily avoidable in real life application. As the RFID transponders are progressively used in logistics and retails, it is exposed to a lot of human touches. Therefore, the ESD protection of these transponders is vital.

Nowadays, many manufacturers put the ESD protection in the package. However at pad level, the ESD protection is usually small dimensioned to reduce input capacitance. This is why extra ESD protection co-constructed at internal circuit VDD-VSS rail is necessary in advanced process due to thinner gate oxide.

There were some discussions and works [5-9] done to improve power rail ESD robustness in the mixed-signal circuits. Most ESD protection is usually accomplished by using various types of clamp devices to shunt the electrical charge away from internal circuits before over voltage damage can occur. The clamp designs can consist of diodes and/or MOSFETs. Works by Ming-Dou Ker, Kei-Kang Hung et al. [5] shown in Figure 1 and 2, use the power-rail ESD clamp circuit with stacked gate-grounded diodes and also the gate-triggered diodes. The gate-triggered design proved to have a higher ESD robustness and faster turn-on speed compared to the gate-grounded design as shown in Figure 3, due to the existence of the RC-based detection circuit.

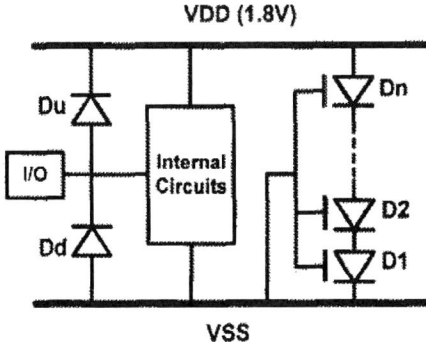

Figure 1 ESD clamp with gate-grounded diodes [5]

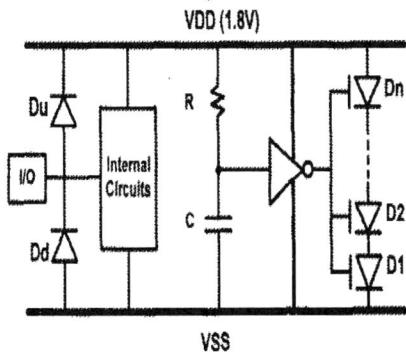

Figure 2 ESD clamp with gate-triggered diodes [5]

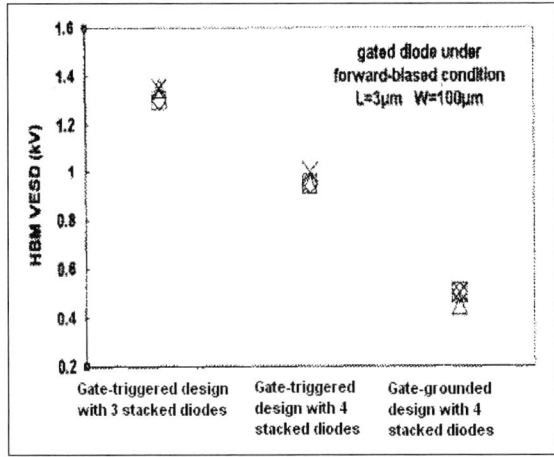

Figure 3   Comparison results of the gate-grounded and gate-triggered diodes [5]

Improvement by Ming-Dou Ker and Che-Hao Chuang [6] has a PMOS (NMOS) inserted into the diode structure to form the PMOS-bounded (NMOS-bounded) diode as shown in Figure 4 and 5, which is essential to block the field oxide isolation across the p/n junction in the diode structure. Thus, the PMOS/NMOS-bounded diodes can sustain much higher ESD stress, especially under the reverse-biased condition.

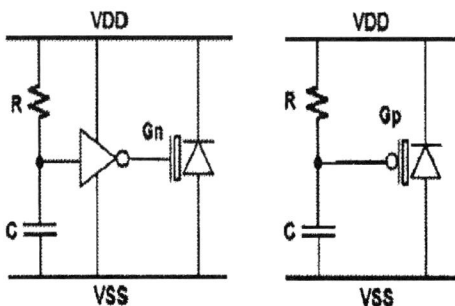

Figure 4 ESD clamp with RC-based detection circuit to control the NMOS and PMOS-bounded diode [6]

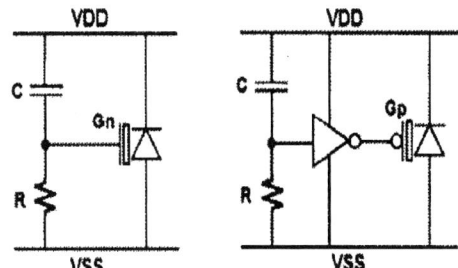

Figure 5   ESD clamp with gate-coupled technique to control NMOS and PMOS-bounded diode [6]

From the tests shown in Figure 6, the PMOS (NMOS)-bounded diodes have much higher ESD tolerance than the normal diodes and poly-bounded diodes. However, ESD clamps using gate-grounded NMOS has higher tolerance to ESD levels than the MOS bounded diodes. The only drawback of the gate-grounded NMOS would be the non-uniformity turned-on due to multiple fingers that could occur.

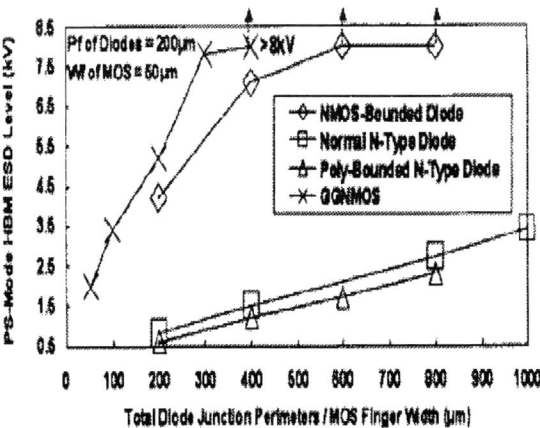

Figure 6 Comparison of various ESD clamps [6]

Other kinds of ESD clamps using MOSFETs are the NMOS gate-driven clamp design [7-8] and substrate-triggered clamp design [9]. The latter design is developed by Ming-Dou Ker et al. which consists of a substrate-triggering field oxide device (STFOD) as a replacement for the

911

NMOS. Figure 7 shows the device structure of STFOD.

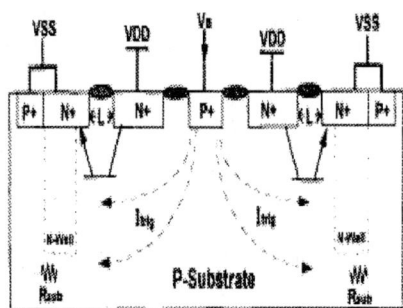

Figure 7   Substrate-triggered clamp design (STFOD) [9]

It is reported to perform better, in terms of layout area and efficiency, compared to the NMOS clamp. However, none were implemented for the RFID IC Transponder. In our work, we use gate-driven technique employing NMOS as it is simpler and very suitable to work under a 2-KV Human Body Model (HBM) positive mode ESD voltage. The clamp is implemented throughout a 13.56MHz RFID transponder. The RFID transponder had been developed in this project previously. The 2-KV Human Body Model (HBM) positive mode ESD voltage is used as a reference point to test the effectiveness of the protection circuit. 2-KV HBM is the commercial IC product ESD level.

The following section introduces the architecture of the mentioned ESD protection circuit and its implementation in the RFID transponder. Section III shows the results when we applied a 2-KV Human Body Model (HBM) positive mode ESD voltage pulse to the ESD protected circuit and some discussions. Finally, section IV shows the conclusion.

II. ARCHITECTURE

The idea of the ESD protection circuit is to clamp existing ESD voltage on the VDD power line in the circuitry to the ground so that ESD pulses will not enter the main circuitry and damage it. Figure 8 shows the circuit of the VDD-to-VSS ESD clamp using the NMOS gate-driven technique. The clamp consists of a transient detection circuit which is the resistor (R), source-drain grounded NMOS acting as a capacitor (C), an NMOS (M3) and a PMOS

(M2); and an NMOS (M1) to bypass the ESD voltage to the ground.

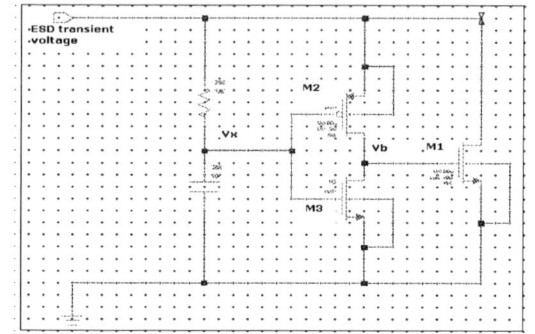

Figure 8   VDD-to-VSS   ESD   clamp   circuit employing gate-driven NMOS

The clamp is designed to be turn on only under the ESD stress conditions but kept off under normal operating conditions. The resistor and capacitor are set to $0.1\mu s$ to differentiate and detect the transient ESD voltage from the useful VDD. Initially, the nodes Vx and Vb are 0V. The ESD voltage across the VDD power line will then charge the capacitor to rise up to the voltage of Vx. Since the HBM ESD voltage with 2-KV charge has a rise time of about 5-10ns, the voltage level of Vx is increased much slower than the voltage level on VDD line. Due to the delay in voltage increase on Vx, M2 is biased by the ESD voltage and sends a voltage into node Vb to turn on M1. As long as the voltage on the node Vb is greater than the threshold voltage of the transistor, M1 is turned on and it will provide a short circuit path between the VDD and the ground.   Under normal operating condition, VDD has a rise time in the order of $3\mu s$. With such rise time, the ESD clamp with RC of $0.1\mu s$ will be able to keep Vb with 0V voltage level. Therefore, M1 is kept off throughout the normal operating condition.

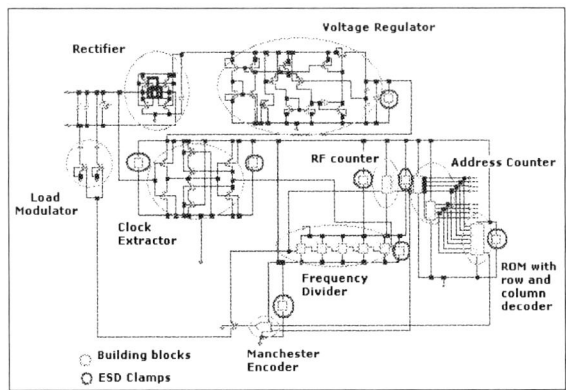

Figure 9   Implementation of ESD clamps on the transponder IC

Figure 9 shows the implementation of the clamps to the entire RFID transponder architecture. An ESD clamp is assigned to each building block that is connected to VDD. It is not necessary to include the ESD protection circuit in the RF front end as they will be taken care off by the standard input and output pad ESD protection since in our work; we only concentrate on the VDD-VSS ESD protection.

There are in total 8 ESD clamps. The ESD clamps are inserted in very near to each building block because VDD is the main voltage source and it is generally spread out in the entire transponder IC. This is to ensure the efficiency of the ESD protection circuit as longer power lines can cause a time delay to discharge ESD voltage through the ESD clamps [7].

III. RESULTS AND DISCUSSION

We use the Mentor Graphic Design Architect for the schematic entry and Eldo as the simulator. The transponder and proposed ESD clamps are all done according to TSMC 0.18μm Mixed Signal SALICIDE (1P6M, 1.8V) process. After implementing the ESD clamps into the transponder, we simulate the circuit with two conditions. First, we add a 2 kV, 400ns pulse with a rise time of 10ns to the VDD. This stimulus acts as the ESD pulse that could occur on the power line. Figure 10 shows the circuit performance. We can see that Vb starts to increase slowly with its voltages above the threshold voltage when the 2 kV is fed to the ESD clamp. At this time, we observed that M1 is turned on. This shows that the VDD-to-VSS ESD

clamp circuit had successfully diverted the large current to ground, and avoid damage to the circuit.

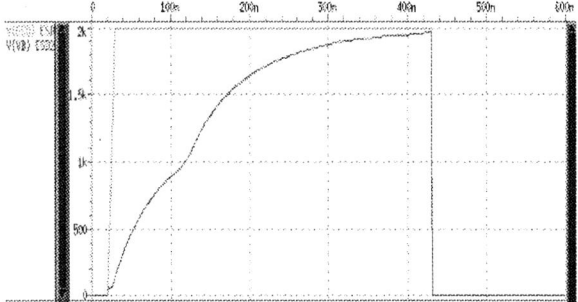

Figure 10 ESD transient pulse at 2000V with the corresponding Vb

Figure 11 shows the circuit performance under the normal condition. As the stimulus, we add a pattern with a 20μs fall time to represent VDD without the influence of the ESD pulse. Vb is kept almost zero. The ripple occurred shows that it is below the threshold voltage of the transistor, therefore M1 is kept off the whole time. No current is bypassed to the ground.

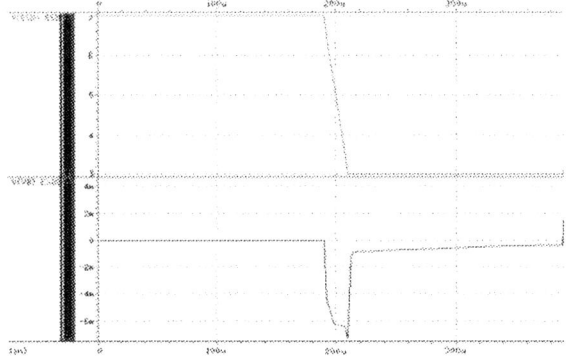

Figure 11 ESD transient pulse at 7V during normal operation

The power consumption of the transponder circuit with the clamps is 1.2344405mW while the power consumption of the transponder circuit without the clamps is 1.2344253mW. The difference is about 15.12nW.

IV. CONCLUSION

We have developed an internal ESD protection circuit and implemented it in the previously developed 13.56MHz RFID transponder employing TSMC 0.18um process. The circuit has a capability to sustain 2-KV of HBM positive mode ESD voltage, which is suitable for RFID applications. The power consumption with the clamp circuits is only an addition of 15.12nW, which shows successful power efficient design.

ACKNOWLEDGEMENT

The authors would like to thank the Malaysian Ministry of Science, Technology and Environment for sponsoring this work under project IRPA 03-99-01-0093 EA091.

REFERENCES

[1] Mark Blitshteyn, "Mastering RFID Label Converting". White Paper, RFID Journal. Published July 2006.

[2] Basics of Electrostatic Discharge, ESD Association, www.esda.org, 2000.

[3] Mykkanen, C. Fred and Blinde, David R., The Room Air Ionization System, a Better Alternative than 40% Relative Humidity, Electrical Overstress/Electrostatic Discharge Symposium Proceedings,Las Vegas, 1983.

[4] Shaw, Monica, Pushing Past Paper, Pulp & Paper, November 2004.

[5] Ming-Dou Ker, Kei-Kang Hung, Howard T.-H. Tang, S.-C. Huang, S.-S. Chen and M.-C. Wang, "Novel Diode Structures and ESD Protection Circuits in a 1.8-V 0.15-pm Partially-Depleted SO1 Salicided CMOS Process", Proceedings of 8th International Symposium on the Physical and Failure Analysis of Integrated Ciruits (IPFA), pp. 91 – 96, 9- 13 July, 2001, Singapore.

[6] Ming-Dou Ker and Che-Hao Chuang, "ESD Protection Circuits with Novel MOS-Bounded Diode Structures", IEEE Internation Symposium on Circuits and Systems (ISCAS), pp. V-533 – V-536, vol.5, 26-29 May 2002, Arizona.

[7] M.-D. Ker, "Whole-Chip ESD Protection Design with Efficient VDD-to-VSS ESD Clamp Circuits for Submicron CMOS VLSI", IEEE Transactions on Electron Devices, Vol.46, No.1, January 1999.

[8] R. Merrill and E. Issaq, "ESD design methodology," EOS/ESD Symp. Proceedings, EOS-15, pp. 233-237, 1993.

[9] M.-D. Ker, C.-Y. Wu, H.-H. Chang, and T.-S. Wu, "Whole-chip ESD protection scheme for CMOS mixed-mode IC's in deep-submicon CMOS technology," Proceedings of IEEE Custom Integrated Circuits Conference, pp. 31-34, 1997.

Device Design Consideration for Nanoscale MOSFET Using Semiconductor TCAD Tools

Teoh Chin Hong and Razali Ismail
Department of Microelectronics and Computer Engineering,
Universiti Teknologi Malaysia,
81300 Skudai, Johor, MALAYSIA
Email: pink_tch@yahoo.com, razali@fke.utm.my

Abstract — The evolution of Metal-Oxide-Semiconductor Field Effect Transistor (MOSFET) technology has been governed mainly by device scaling over the past twenty years. One of the key questions concerning future ULSI technology is whether MOSFET devices can be scaled to 100nm channel length and beyond for continuing density and performance improvement. In this paper, the design, fabrication and characterization of high-performance and low-power 90nm channel length MOSFET devices are described. Several parameters have to be scaled down such as gate oxide thickness, channel length, ion implantation for threshold voltage adjustment and other specifications to achieve desirable electrical characteristic. To control the short-channel effect (SCE) and hot-carrier reliability that limits device scaling, lightly doped drain (LDD) structure, shallow junction of drain / source and Shallow Trench Isolation (STI) are implemented. Virtual Wafer Fabrication (VWF) Silvaco TCAD Tools is used for fabrication and simulation of CMOS transistor namely ATHENA and ATLAS. Simulations using these programs provided the opportunity to study the effect of different device parameters on the overall device performance. The devices were simulated and gradually the performance of each one was improved, until an optimal device configuration was created for a particular application.

I. INTRODUCTION

Over the past 50 years of the semiconductor industry, the size of MOSFET has been scaled down obeying the Moore's law : feature sizes of transistors are scaled at a rate of approximately 0.7 times every 18 months [1]. However, as MOSFET technology approaches nanoscale

region, researchers face with critical technology barrier known as SCE. While the gate voltage fully controls the channel conduction state in an ideal MOSFET, the drain voltage begins to give more influence on the channel potential in a nanoscale MOSFET.

In order to enhance the speed performance of the circuit and for higher density while maintaining its reliability and circuit performance, the MOSFET transistor has been scaled down by using Constant Field Scaling rules because it is easier and assumed to avoid the high field problems. Important parameters like channel length (L), doping concentration (N_A, N_D) during ion implantation for threshold voltage adjustment and gate oxide thickness (t_{ox}) will be scaled [2].

Advanced transistor design such as STI eliminates the bird's beak shape characteristic, subthreshold hump and field oxide thinning effect in LOCOS isolation. STI has advantages such as perfect planarity, scalability and latchup immunity [3]. Besides that, LDD is designed to smear out the strong electric field between the channel and heavily doped s/d, in order to reduce hot-carrier generation. While shallow junction of drain/source can increase device density. To reduce charge sharing effect, halo implant is introduced in which the locally high doping concentration in the channel near the source/drain junctions is created. Retrograde well is a form of vertical channel engineering. It is used to improve SCE and to increase surface channel mobility by creating a low surface channel concentration followed by a highly doped subsurface region.

II. OPTIMIZATION

Optimization is essential in scaling down device to obtain better device performance. Optimization of these devices using the TCAD tools requires many hours of lab simulation time.

Several aspects of each device were selected for optimization. Once the device characteristics were selected for optimization, the process of device simulation began. First each parameter was tested individually for its effect on device performance as a whole. Once several plots were obtained that indicated the particular parameter's effect on device performance, improved values could then be selected for the device. Several simulations needed to be run to find improved values for each device parameter until an optimal value were reached. Once an optimal value was reached for each of the device parameters, the improved parameters were then combined into a single device. When all these new values were present in a single device, they were again simulated and adjusted to optimize based upon their combined effects to ultimately produce an optimal device configuration. Various analysis and discussion are done on particular parameter's effect on device characteristics.

III. ANALYSIS AND DISCUSSION

There are a number of parameters that affect the value of threshold voltage and drain induced barrier lowering (DIBL) parameter which will result in different device performance and characteristics. Each of the parameters will be discussed in the following section.

A. *Effect of Channel Doping on Threshold Voltage*

The channel doping depends on several components such as the type of atom being implanted, the dosage and the energy of the implant. The doping concentration and depth are affected directly. Multiple threshold voltages can be achieved by adjusting the channel doping as shown in Fig. 1. Threshold voltage increases as channel doping increases which is comparable to the result reported in the literature [4].

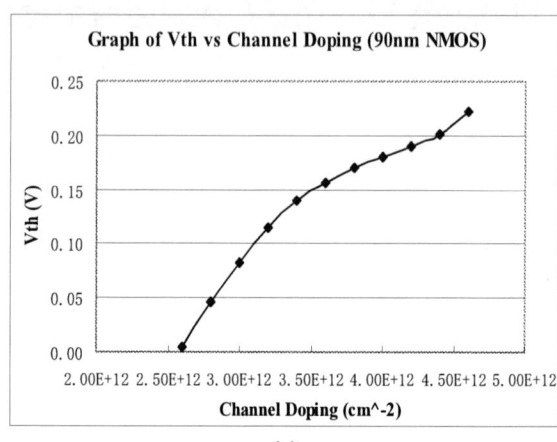

(a)

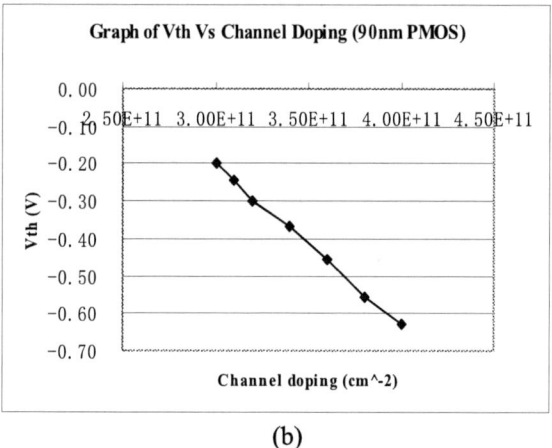

(b)

Fig. 1 Threshold voltages at different channel doping dose for (a) 90nm NMOS (b) 90nm PMOS.

B. *Effect of Oxide Thickness on Threshold Voltage*

Gate oxide thickness can be used to modify the threshold voltage of a transistor. Variation of threshold voltage with different oxide thickness for a 90 nm device is shown in Fig. 2. Lower oxide thickness and hence lower threshold voltage in critical paths can maintain the performance. Higher oxide thickness not only reduces the subthreshold leakage, it also reduces gate oxide tunnelling current since the oxide tunnelling current exponentially decreases with an increase in the oxide thickness [5].

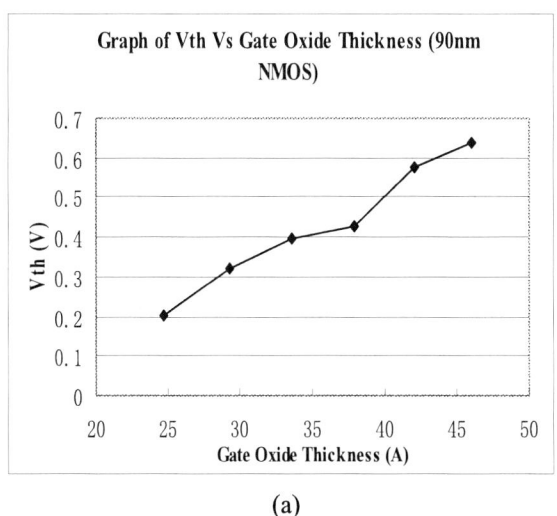

(a)

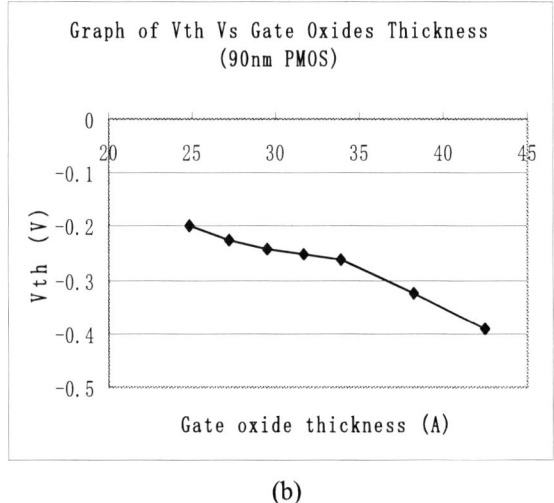

(b)

Fig. 2 Variation of threshold voltage with different gate oxide thickness for (a) 90nm NMOS (b) 90nm PMOS.

C. Effect of Channel Length on Threshold Voltage

Fig. 3 illustrates how threshold voltage of NMOS decreases as the channel length is reduced. Hence, different threshold voltage can be achieved by using different channel length. This reduction of threshold voltage with reduction of channel length is known as threshold voltage rolloff.

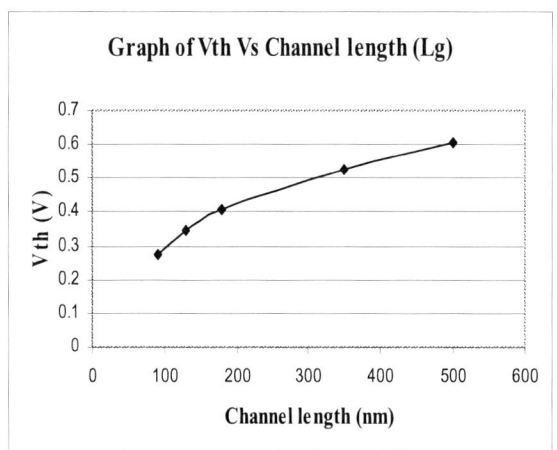

Fig. 3 Effect of various channel length on threshold voltage for NMOS.

D. Effect of Drain Voltage on Threshold Voltage

In a short-channel device, the source and drain depletion width in the vertical direction and the source drain potential have a strong effect on the band bending over a significant portion of the device. Therefore, the threshold voltage and consequently the subthreshold current of short-channel devices, vary with the drain bias. This effect is referred to as drain induced barrier lowering (DIBL).

Fig. 4 is a plot of threshold voltage (V_t) vs drain voltage (V_d) to investigate the varying of threshold voltage as different drain voltage is applied. The threshold voltage decreasing as the drain voltage is increasing which is comparable to the explanation above. For comparison, the value of the threshold voltage of the 90 nm NMOS that is doped with halo implant is higher than the one without halo implant doping which

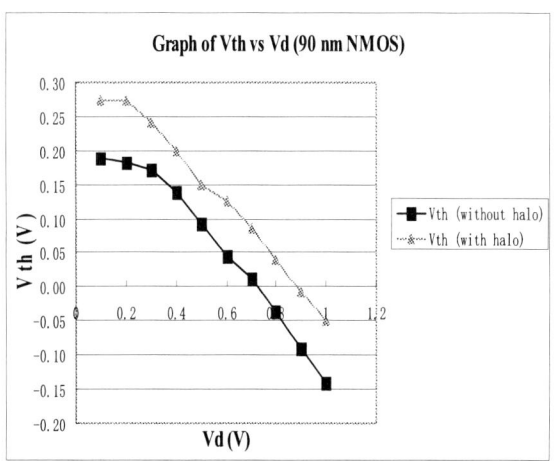

contributes less subthreshold current. By using halo implant, the threshold voltage degradation is reduced.

Fig. 4 Graph of threshold voltage versus drain voltage for 90nm NMOS with and without halo implant.

E. Dependence of Halo Dose on Threshold Voltage

Fig. 5 shows the Id-Vg curves for 90nm NMOS at different halo doses. It can be seen that the value of the threshold voltage increased as we increased the halo dose. Thus, threshold degradation can be controlled well since the halo implant reduces the charge sharing effects from source and drain fields.

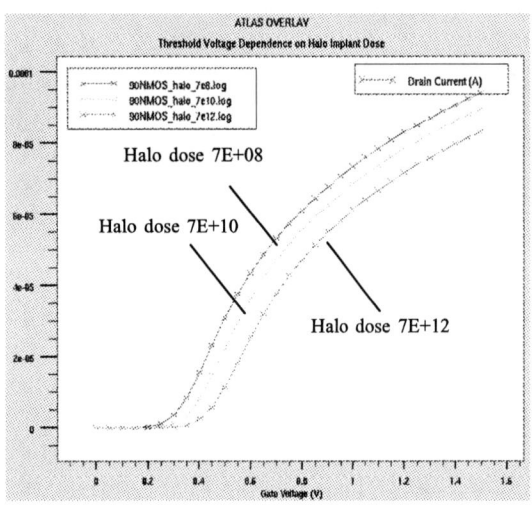

Fig. 5 I_d-V_g curve with different halo doses for 90nm NMOS.

F. Dependence of Threshold Voltage on Retrograde Well Dose

Fig. 6 shows the dependence of threshold voltage on retrograde well dose. The threshold voltage for 90 nm NMOS is increased by the increasing retrograde well dose. Threshold voltage is adjustable by using retrograde well dose as well as using implantation technique.

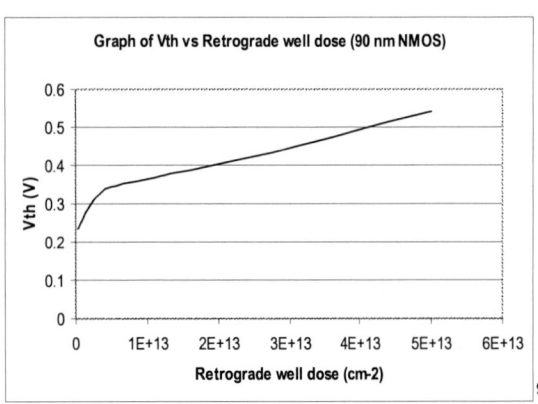

Fig. 6 Graph of V_t versus retrograde well dose for 90nm NMOS.

G. Dependence of DIBL on Halo Dose

Fig. 7 is the graph of DIBL versus halo dose for the 90 nm NMOS with halo implant. It shows that increasing halo implant dose reduces the value of DIBL. This indicates that we can use halo implant to control the MOSFET degradation due to DIBL.

Fig. 7 Graph of DIBL versus halo dose for 90nm NMOS.

H. Dependence of DIBL on Retrograde Well

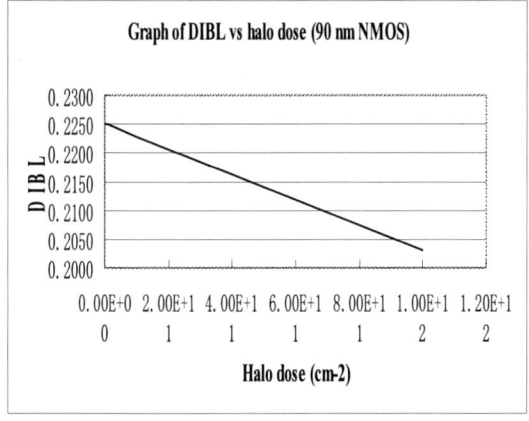

Dose

Fig. 8 shows the effect of DIBL for the 90nm NMOS with different retrograde well doses. The DIBL effect is defined as the change in the threshold voltage ΔV_t divided by the change in the drain voltage ΔV_d. It is clear from Fig. 8 that for a higher retrograde well dose the DIBL will decrease.

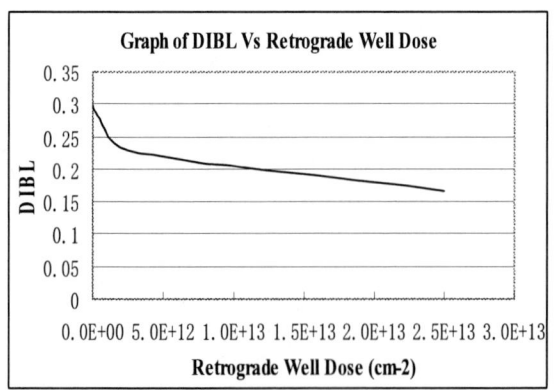

918

Fig. 8 Graph of the dependence of DIBL on retrograde well dose for 90nm NMOS.

I. Overall Effect of Process Parameters on Threshold Voltage

Table 1 : The overall effect of process parameters on threshold voltage for 90nm NMOS and PMOS

Table 1 summarized the overall effect of process parameters on threshold voltage for the 90nm NMOS and PMOS. Threshold voltage increased as channel doping, gate oxide thickness, channel length, halo dose and retrograde well dose increased. However, threshold voltage decreased as drain voltage increased.

IV. CONCLUSION

A number of parameters that affect the value of threshold voltage and DIBL parameter which result in different device performance and characteristics are discussed. The improved parameters were again simulated and adjusted to optimize based upon their combined effects to produce an optimal device configuration.

ACKNOWLEDGEMENT

The authors acknowledge the Ministry of Science, Technology and Innovation Malaysia (MOSTI) for the financial support through PTP scholarship.

REFERENCES

[1] H. Wakabayashi, et al.(2003). Sub-10-nm Planar-Bulk-CMOS Devices using Lateral Junction Control. *IEDM Tech. Dig.* December 2003. Washington DC, 989-991.

[2] SIA et al. *International Technology Roadmap for Semiconductors*. 2003 edition.

[3] C. Mead and L. Conway (1979). *Introduction to VLSI systems*. Addison Wesley. p. 37.

[4] N. Sirisantana, L. Wei and K. Roy (2000). High-performance low-power CMOS circuits using multiple channel length and multiple oxide thickness. *Proc. Int. Conf. Computer Design.* 227-232.

[5] K. Schuegraf and C. Hu (May 1994). Hole injection Sio2 breakdown model for very low voltage lifetime extrapolation. *IEEE Trans. Electron Device.* vol. 41: 761-767.

	Threshold voltage	
	90nm NMOS	90nm PMOS
Increasing channel doping		
Increasing gate oxide thickness		
Increasing channel length		
Increasing drain voltage		
Increasing halo dose		
Increasing retrograde well dose		

Nonlinear Stability Analysis of Microwave Oscillators Using Lyapunov Function

H. Vahdati , A. Abdipour, Member, IEEE and A. Mohammadi, Member, IEEE

Microwave/mm-Wave & Wireless Communication Research Lab, Radiocommunication Center of
Excellence, Electrical Eng. Dept. Amir-Kabir University of Technology (Tehran Polytechnic)
Hafez Ave., Tehran, Iran

Email: abdipour@aut.ac.ir

Abstract — **In this paper, the Lyapunov function is used for nonlinear stability analysis of microwave oscillators. Using this method, both the instability of DC bias point and the stability of steady state oscillation can be analyzed. The successes of Lyapunov method, in nonlinear control theory, support the idea of using this method for stability analysis of microwave oscillators. In this paper the application of this method for stability analysis of microwave oscillator is explained and applied to a microwave diode oscillator.**

I. INTRODUCTION

MMIC is an expensive and with high financial risk technology. After IC fabrication, it is impossible to modify or tune the circuit. Therefore it is important to analyze and check the circuit behavior in simulation stage. Stability is the necessary condition for all other analysis. Although microwave CADs have good capabilities for linear stability analysis, they have weaknesses in nonlinear stability analysis. Even in the harmonic balance (HB) method, there is no guarantee on the stability of the computed solution [1,2]. HB regards only the steady state solution and ignores the time evolution of the solution from the startup to the steady state. In the other word, in HB method it is assumed that the steady state solution is stable and we only want to find it.

Several works have been done in this area, for example Rizzolli and Lapparini [3-5], Quere & Suarez [6, 7].These works are mainly based on the bifurcation theory and linearization methods. In the bifurcation method, the large variations of circuit response due to the changes of a critical parameter of circuit are studied. Based on the qualitative changes of circuit response, the stability of circuit can be deduced [7]. This method is suitable for forced circuits such as power amplifiers and can not be applied to free running oscillators. In the linearization method, the circuit is linearized around its steady state and using the stability analysis of linear system *e.g.* Nyquist diagram, the stability of nonlinear system can be determined [8]. These two methods are complex and suffer from the linearization weaknesses [9].

In this paper Lyapunov function method is adapted for stability analysis of microwave oscillators. The Lyapunov function method is completely discussed in nonlinear control theory text.The theoretical basis of this method and its adaptation for stability analysis of microwave oscillator has been presented in the next section.

II. STABILITY ANALYSIS IN NONLINEAR CONTROL THEORY

One of the most powerful methods for stability analysis, is using the energy like Lyapunov function [10,11]. This method is both an intuitional one and with strong theoretical bases. This is a general method and independent of system linearity. In this method, the stability is determined by analyzing the temporal variation of circuit energy or more generally, Lyapunov function. In the Lyapunov method, if $\breve{x}$ be the state variable vector, then the energy function $V(\breve{x})$, can be defined which is positive definite function *i.e.*

$$\begin{cases} \forall\ \breve{x} \neq 0\ ,\ V(\breve{x}) > 0 \\ V(0) = 0 \end{cases} \qquad (1)$$

Real energy or "physically motivated Lyapunov function" has these properties. Based on the time derivative of energy function *i.e.* $\dot{V}(x)$, the stability of system can be derived. This is the Lyapunov direct method and states that [10]:

Assuming Ω be a ball of radius R around the equilibrium point $\breve{x} = 0$, then the system at $\breve{x} = 0$ is stable if:

$$\forall\ x \in \Omega\ \&\ x \neq 0\ \Rightarrow \dot{V}(\breve{x}) < 0 \qquad (2)$$

And the system is unstable if $\dot{V}(\breve{x}) > 0$ for some $\breve{x}$ in the vicinity of $\breve{x} = 0$. Using this method, the instability of DC bias point of free running oscillators which is the necessary condition for oscillation startup can be verified. Generalization of this method to a larger set *i.e.* invariant sets, which include the limit cycles "oscillatory behavior" has been done [12]. In this paper a new method for oscillator stability analysis has been introduced. The main idea of this method is that in an oscillator, the instantaneous energy varies in time but the average energy in each period remains constant. Using this property, the oscillatory behavior of an oscillator can be mapped to a fix point.

III. OSCILLATION STABILITY OF MICROWAVE OSCILLATORS

In an ordinary oscillator like an RLC tank, it is clear that the internal energy of the system varies in time but the average energy in one period is constant. This can be shown easily using the phase plane representation of oscillator (Fig.1). In the phase plane representation, the behavior of each state variable is plotted against the other ones.

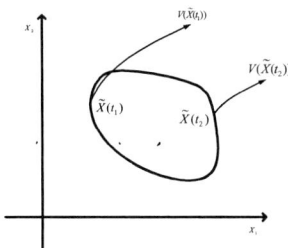

Fig. 1 Phase plane representation of an oscillator or "limit cycle"

Assume that x_1 and x_2 are the state variables of the circuit. The system energy can be written as:

$$V(x_1, x_2) = k_1\, x_1^2 + k_2\, x_2^2 \qquad (3)$$

In oscillators the state variables x_1 and x_2 are periodic function of time, and hence $V(x_1, x_2)$ is also periodic with period T, and its average is:

$$\overline{V} = \frac{1}{T} \int (k_1\, x_1^2 + k_2\, x_2^2)\, dt = Const \qquad (4)$$

The above value is a constant because in the averaging process all the points on the closed curve take part in summation with equal share. In fact by time averaging, the oscillatory behavior of circuit can be mapped to a fix point and the oscillation stability can be deduced by stability analysis of this fix point. It should be noted that, in the conventional texts on Lyapunov method, the stability of the equilibrium point at origin " $\breve{x} = 0$ " is considered and at this point, $V(0) = 0$. In this case, every perturbation increases the internal energy and hence the negative sign of $\dot{V}(\breve{x})$ guarantees the convergence of perturbed system to its original state with zero energy. But for systems with nonzero internal energy "like the oscillators," the perturbation can increase or decrease the internal energy. Hence the stability criteria for an oscillator, which has nonzero internal energy, should be modified as below:

An oscillator with internal energy of $V(\breve{x})$ is stable, if after a small perturbation of $\delta(t)$ on its state variables, $\dot{V}(\breve{x})$ behaves in a manner that the average of $V(\breve{x})$ converge to its average energy before the perturbation.

Using the above discussion, the stability of a microwave diode oscillator has been analyzed in the next section.

IV. STABILITY OF A MICROWAVE DIODE OSCILLATOR

Diode oscillators are the simplest microwave oscillators and have a vast application in microwave circuits. A simple model of a diode oscillator is shown in Fig. 2. This oscillator can be design using the negative resistance method for the oscillation   frequency   of   3GHz..

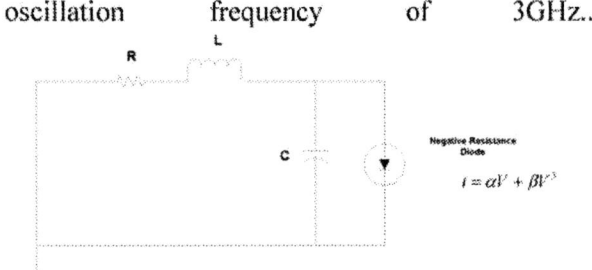

Fig. 2 A microwave diode oscillator

The internal energy of the oscillator has been used as the Lyapunov function and calculated using the two state variables v_C "capacitor voltage" and i_L "inductor current"

$$Energy(t) = \frac{1}{2}Cv_C^2 + \frac{1}{2}Li_L^2 \qquad (5)$$

The internal energy from the startup phase till the steady state has been shown in Fig.3.

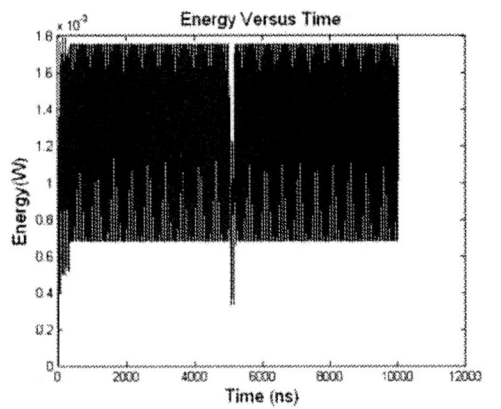

Fig. 3 Instantaneous energy of oscillator with a perturbation between 4000 ns and 6000 ns

Using the estimated oscillation period T, the average energy in each period can be calculated. In Fig. 4 the startup stage has been studied. The positive slope of $V(x)$ i.e. $\dot{V}(x) > 0$, shows the instability of DC bias point in the oscillation startup phase. Hence using the Lyapunov function, the instability of DC bias point has been shown without any linearization. In the linearization methods "which are mainly used in the current microwave CADs," the circuit has been linearized around its operating point, and using the methods such as Nyquist diagram, the stability can be studied.

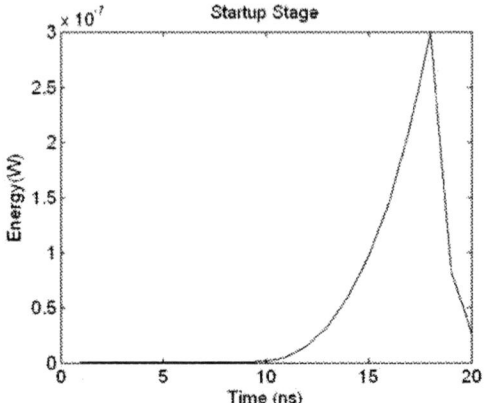

Fig 4 Time variation of energy function at the startup phase and instability of DC bias.

To study the oscillation stability, the average energy per period has been plotted in Fig 5. When the states of the oscillator have been perturbed, the average energy deviates from its constant value, but after the perturbation, the system behaves in a manner to recover its initial average value and hence the oscillator is stable.

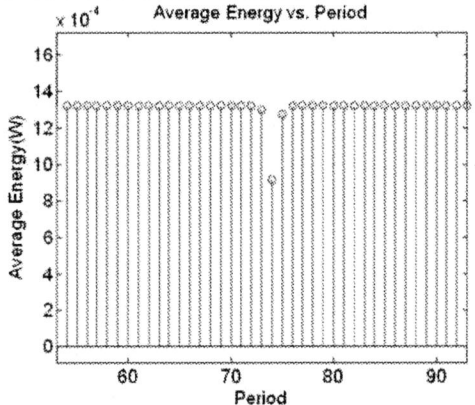

Fig. 5: Variation of average energy of diode oscillator due to a perturbation

V. CONCLUSION

In this paper the idea of using Lyapunov function method for analyzing the stability of microwave oscillators has been introduced and applied to a microwave diode oscillator. Using the internal energy of oscillator "which can easily be expressed in term of state variables", as the Lyapunov function, the instability of DC bias point which is the necessary condition for oscillation start up, can be verified. To analyze the oscillation stability, the average energy per period is defined as the Lyapunov function and hence the stability analysis of steady state response has been transformed to the stability analysis of a fix point. Using the time derivative of Lyapunov function, the stability of microwave oscillators can be deduced. The strength of Lyapunov method is promising for attaining a general yet simple criterion for stability analysis of nonlinear microwave circuits such as microwave oscillator.

ACKNOWLEDGEMENT

The authors would like to thank Iranian Telecommunication Research Center "ITRC", for supporting this work.

REFERENCES

[1] A. Suarez,R. Quere,*Stability Analysis of Nonlinear Microwave Circuit,* Artech House, Norwood, 2003.

[2] J. Obregon, J. Nallatamby, M. Prigent, M. Camiade and D. Rigaud, *RF and Microwave Oscillator Design,* Artech House, Norwood, 2002.

[3] V. Rizzoli, A. Lipparini, "General stability analysis of periodic steady state regimes in nonlinear microwave circuits," *IEEE Trans. Microwave Theory Tech*, vol. MTT-33, pp. 30-37, Jan 1985.

[4] V. Rizzoli, A. Lipparini, A. Cotanzo, F. Mastri, C. Cecchetti, A. Neri and D. Masotti, "State of the art harmonic e simulation of forced nonlinear microwave circuits by the piecewise technique," *IEEE Trans. Microwave Theory Tech*, vol. MTT-40, no.1, pp. 12-27, Jan 1992.

[5] V. Rizzolli, .A. Lapparini,A. Cotanzo,F. Mastri, C. Cecchetti,A. Neri and D.Masotti, "State of the art harmonic balance simulation of forced nonlinear microwave circuits by piecewise technique," *IEEE Trans .Microwave. Theory Tech.* vol. MTT-40, no.1, pp.12-27 Jan 1992.

[6] S. Mons, J.C.Nallatamby, R.Quere,p.Savary and J obregon, "Unified approach for the linear and nonlinear stability analysis of microwave circuits using available tools," *IEEE Trans. Microw. Theory Tech,* vol. 47, no.12, pp.2403-2410, Oct.1999.

[7] A. Suares, V. Iglesias, J.M. Collantes,J. Jugo and J.L.Garcia, "Nonlinear stability of

microwave circuits using commercial software," *IEE Elec. Letter,* pp. 1333-1335, 1998.

[8]. V. Rizzoli, A Neri , and D. Masott, "Local stability analysis of microwave oscillators based on Nyquist's theorem," *IEEE Trans. Microwave and Guided Wave Letters*, vol. 7, no. 10, pp. 341-343, Oct 1997

[9] Robert W. Jackson, "Rollett proviso in the stability of linear microwave circuits -- A tutorial," *IEEE Trans. Microw. Theory Tech.*, vol.54, no.3, pp. 993-1000, Mar. 2006.

[10] J.E. Slotine and Weiping Li, *Applied Nonlinear Control ,*New Jersey, Englewood Cliffs, Printice Hall, 1991.

[11] M. Vidiasagar, *Nonlinear System Analysis*, Englewood Cliffs,N.J., Prentice Hall, Inc. ,1978.

[12] V. V. Migulin, V. I. Medvedev,E. R. Mustel and V. N. Parygin, *Basic Theory of Oscillation* , Moscow ,Mir Publisher ,1983.

Conductivity of Cubic GaMnN Grown on Undoped GaN Layers

V K Sundaramoorthy[a,1], *Student MemberIEEE,* C T Foxon[b], S V Novikov[b], K W Edmonds[b] and
I Harrison[a]

[a]*School of Electrical and Electronic Engineering, University of Nottingham, Nottingham NG7 2RD, UK*

[b]*School of Physics and Astronomy, University of Nottingham, Nottingham NG7 2RD, UK*

[1]*Email:* eexvks2@exmail.nottingham.ac.uk , Ian.Harrison@nottingham.ac.uk,
eezih@granby.nottingham.ac.uk , c.thomas.foxon@nottingham.ac.uk

Abstract: **The conductivity of GaMnN layer grown on thick undoped GaN layers analyzed by electrical characterization is reported here. The GaMnN conductivity grown on thick undoped GaN layers were found to be n-type, whereas the GaMnN conductivity on thin undoped GaN is found to be p-type. One possible explanation could be the relaxed strain associated with the increased undoped GaN layer thickness.**

I. INTRODUCTION

Among the wide-band gap semiconductors, GaN based transistors have gained considerable interest in high power and high frequency applications. Heterojunction Bipolar Transistors (HBT) fabricated from wurtzite GaN have been reported earlier[8][9].However, in wurtzite GaN, the hole mobility is low which leads to a high base resistance and hence poor RF performance, and the vertical dislocations lead to a high leakage current between emitter and the collector. These problems could be overcome by using cubic GaN materials which has a high electron and hole mobility and less vertical dislocations[5-7]. Hence a HBT fabricated using Cubic GaN is expected to have superior performance compared with that of a wurtzite material.

To manufacture HBT's using cubic GaN, PN junctions are necessary. We have previously reported p-type doping with Mn in thin cubic GaMnN grown on the surface of GaAs (100). P-type conductivity as determined by Hall measurements were repeatable and carrier concentrations in the region $1*10^{18}$ cm^{-3} were achievable with mobilities in the range 200-

350cm^{2}V^{-1}s^{-1}[1-3]. Undoped GaN layers were found to be n-type[2][3]. Hence thin GaMnN layers were grown on top of a GaAs buffer with thick undoped GaN layers inserted between them to achieve a PN junction.

The electrical characterization of the GaMnN / undoped GaN junctions are reported in this work. As before, the GaMnN layers grown without an undoped GaN layers were found to be p-type. Previous measurements have shown that the Hall measurements were not affected by parallel conduction[4]. However, as the thickness of the undoped GaN layer increases, Hall effect measurements indicate that the carrier type in the GaMnN layer changes from p-type to n-type. These observations have also been confirmed by vertical transport mechanisms.

II. EXPERIMENT

Undoped cubic GaN and cubic GaMnN layers were grown on semi-insulating GaAs(100) substrates by plasma-assisted molecular beam epitaxy[PA-MBE] using arsenic as a surfactant to initiate growth of cubic phase material[13]. Films were grown under N-rich conditions, which is necessary for the effective substitutional incorporation of Mn in cubic GaMnN[14]. Growth temperatures from 450C to 680C were used. The active nitrogen for the growth of group III-nitrides was provided by an CAR25 RF activated plasma source.

A GaAs buffer layer was grown between the active nitride layers and GaAs substrate to provide the cleanest possible interface between GaAs and nitride layers. A thick undoped GaN layer and GaMnN layers were grown on the GaAs buffer.

Two sets of samples were analyzed in this work. Sample A (with thick undoped GaN layers and sample B (with undoped GaN layer thickness lesser than that of sample A) are shown in figure1. GaMnN mesas were made on both the samples by etching the GaMnN layer in a reactive ion etcher with SiCl$_3$ at 25mbar, 25 sccm and 250W. The etch rate was approximately 1µm/hr. On sample A, Pd/Au (30/100nm) suitable for p-type contacts were made on both the GaMnN and undoped GaN layer. Ti/Al/Ti/Au(14/50/12/40nm) contacts suitable for n-type was made on the back of undoped GaN layer by etching the GaAs substrate locally.

On sample B, Ni/Au(25/75nm) p-type contacts and Ti/Al/Ti/Au(14/50/12/40nm) n-type contact were made on both the GaMnN mesas and undoped GaN layers. Sample B was annealed at 500C in N$_2$ atmosphere, as Ni/Au has to be annealed to be properly Ohmic on p-type GaN materials[10]. PN junction characteristics are analyzed by forward biasing p-type contact on GaMnN mesa with respect to n-type contact on undoped GaN buffer. Transfer length measurements(TLM) are analyzed for the metal contacts on GaMnN mesa. Also on sample B, Schottky behaviour was analyzed by forwarding biasing a Ni/Au contact with respect to an adjacent Ti/Al/Ti/Au contact on same GaMnN layer or undoped GaN layer.

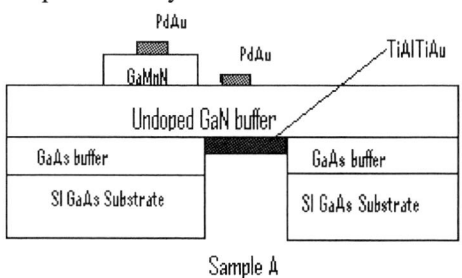

Sample A

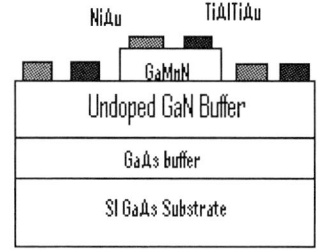

Sample B

Fig 1: Structure of devices on sample A and B

III. RESULTS AND DISCUSSION

The I-V characteristics of sample A is shown in figure 2. The I-V behaviour of forward biased GaMnN layer with respect to undoped GaN is that of a diode. The turn on voltage is approximately 0.5V. This suggests that the diode characteristics is arising from a Schottky contact rather than a PN diode. Since the Pd/Au contact is expected to form a Schottky contact to n-type material, the current observation indicates that the GaMnN is n-type rather than p-type, which is in agreement with the Hall Measurements. Moreover, the diode characteristics of the Pd/Au on undoped GaN layer are very similar to that of the GaMnN diode (also shown in figure 2) which provides supporting evidence for the assertion that the GaMnN is n-type.

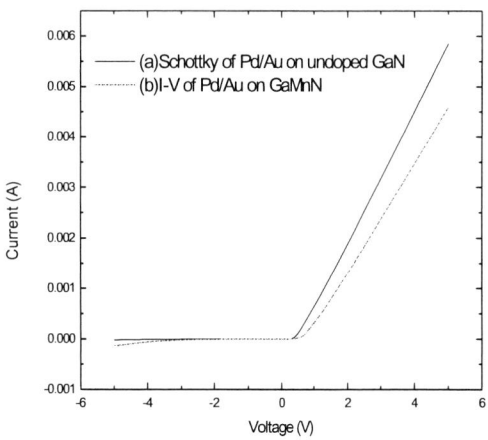

Fig 2: I-V of sample A, (a)Schottky of Pd/Au on undoped GaN (b) I-V of PdAu on GaMnN

I-V characteristics measured between Pd/Au contacts on the GaMnN and on the undoped GaN, along with I-V characteristics measured between two Pd/Au contacts (a) on GaMnN (b) on the undoped GaN layer are shown in figure 3. The I-V characteristics are typical of back to back schottky diodes. The doping densities obtained from CV measurements, were found to be ~ $2*10^{17}$ cm^{-3} for undoped GaN layer and ~ $5*10^{16}$ cm^{-3} for GaMnN, confirming that the GaMnN layer grown over thick undoped GaN layer to be lightly doped n-type.

The TLM characteristics of Ti/Al/Ti/Au and Ni/Au contacts were analyzed on GaMnN layer of sample B. I-V characteristics between the TLM pads of Ti/Al/Ti/Au contacts is shown in figure 5. Ti/Al/Ti/Au is found to be properly Ohmic on the GaMnN layer with a specific contact resistance of $2*10^{-6}$ ohm-cm^{-2}.

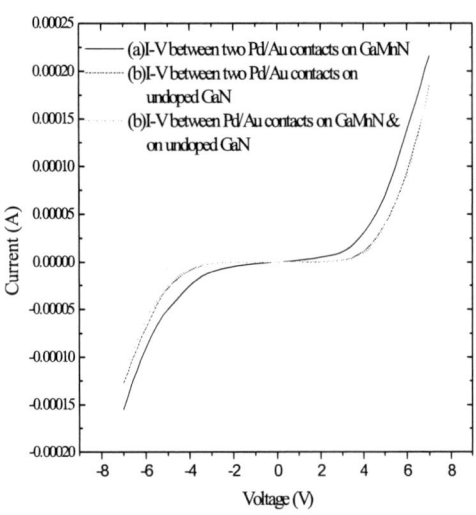

Fig 3: I-V of PdAu contacts in sample A (a) measured on GaMnN, (b) measured on GaN, (c) measured between GaMnN & undoped GaN.

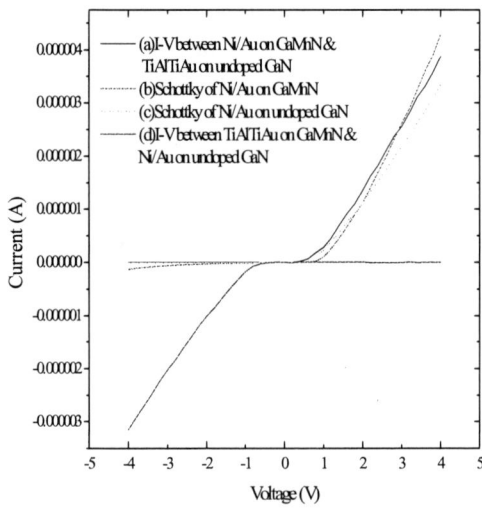

Fig 4: I-V of sample B, (a)measured between GaMnN and undoped GaN layer (b) Schottky of NiAu on GaMnN (c) Schottky of NiAu on undoped GaN layer. (d) I-V of TiAlTiAu on GaMnN and NiAu on undoped GaN.

The I-V characteristics of sample B is shown in figure 4. The I-V of forward biased Ni/Au on GaMnN to the undoped GaN layer is found to have Schottky diode behaviour, as seen in the figure. Hence Ni/Au forms a Schottky contact on GaMnN layer. The Schottky performance of Ni/Au with respect to Ti/Al/Ti/Au on the same GaMnN and undoped GaN layers are also shown in figure 4. These curves are similar on both GaMnN and undoped GaN layers. Also shown in figure 4 is the I-V of a forward biased Ti/Al/Ti/Au on GaMnN with respect to Ni/Au on undoped GaN layer. This characteristics is similar to that of a reverse biased Schottky diode showing the Ohmic behaviour of Ti/Al/Ti/Au contact and Schottky behaviour of Ni/Au contact on GaMnN and undoped GaN layers respectively. Hence the conductivity of GaMnN grown on thick undoped GaN layer is found to be n-type.

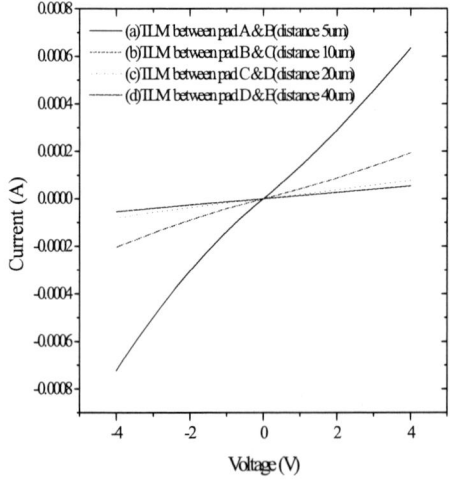

Fig 5: TLM of TiAlTiAu contacts on GaMnN in sample B.

To finally confirm that the GaMnN layer is indeed n-type, I-V characteristics have been measured between Ti/Al/Ti/Au contacts on the GaMnN and on the undoped GaN layer. The results are shown in figure 6, along with I-V characteristics measured between two Ti/Al/Ti/Au contacts (a) on GaMnN (b) on the undoped GaN layer. As expected they all are found to be Ohmic indicating the GaMnN to be n-type.

When the GaN is grown on the GaAs substrate the material is under strain, but as the thickness of the GaN epilayer increases the strain at the surface decreases. The samples showing reliable p-type have been grown directly on top of GaAs buffer layer and are likely to be highly strained. The GaMnN samples grown on the undoped GaN layers are less strained. Therefore, the interaction of strain with the Mn dopant causing a lowering of the acceptor activation energy, could be one explanation of the observed effect. It should be noted that the experiments detailed above only provide evidence of the effect not of the cause.

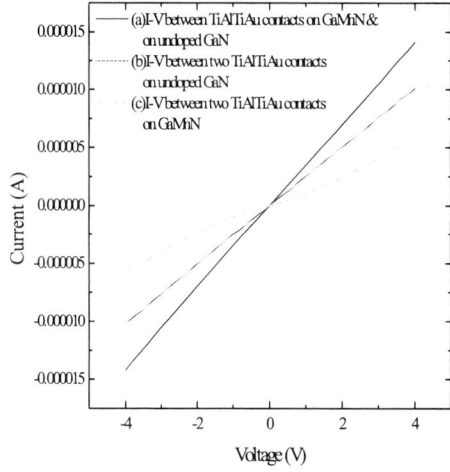

Fig 6: I-V measurements of TiAlTiAu contacts when (a) measured between GaMnn and undoped GaN layer (b) measure on undoped GaN layer (c) measured on GaMnN.

IV. CONCLUSION

The conductivity of GaMnN layers grown on undoped GaN layers of different thickness were analyzed. The GaMnN layers grown on thin undoped GaN layers show clear evidence for p-type conductivity. However, the conductivity of GaMnN grown on thick undoped GaN layers was found to be n-type and Pd/Au is found to form a Schottky contact on GaMnN layer.

The concentration of donors in the GaMnN layer was found to be less than the undoped GaN layer. These results demonstrate the lack of a PN junction in GaMnN / thick undoped GaN layers. A possible explanation of the observed results, involving strain, has been suggested.

ACKNOWLEDGEMENT

The authors acknowledge Rod Dykeman, Dave Taylor and Jas Chauhan for their kind technical assistance. This work was funded by the University of Nottingham Interdisciplinary Doctoral Training Centre in Optical and Electronic materials.

REFERENCES

[1] K W Edmonds, S V Novikov, M Sawicki, R P Campion, C R Staddon, A D Giddings, L X Zhao, K Y Wang, C T Foxon, B L Gallagher, *Appl.Phys.Lett.*86, 152114(2005).
[2] S V Novikov, K W edmonds, A D Giddings, K Y Wang, C R Staddon, R P Campion, B L Gallagher, C T Foxon, *Semicond.Sci.Technol.*19, L13-L16(2004).
[3] C T Foxon, S V Novikov, L X Zhao, K W Edmonds, A D Giddings, K Y Wang, R P Campion, C R staddon, M W Fay, Y Han, P D Brown, M Sawicki, B L gallagher, *J.Crystal Growth.* 278, 685-689(2005).
[4] B J Ansell, I Harrison, C T Foxon, J J Harris, T S Cheng, *Electronics Letters*, Vol.36 No.14(July,2000).
[5] O Brand, H Yang, H Kostial and K H Ploog, *Appl.Phys.Lett*, 69, 2707(1996).
[6] D J As, D Schikora and K Lischka, *Phys.Stat.Sol.(c),* 0, 1607(2003).
[7] T S Cheng, C T Foxon, G B Ren, N J Jeffs, J W Orton, S V Novikov, Y Xin, P D Brown, C J Humphreys, M Halliwell, *Inst.Phys.Conf.Ser*, 155, 259(1997)
[8] H Xing, P Chavarkar S Keller, S P DenBaars, U K Mishra, *IEEE Elec.Dev.Lett*, 24, 141(2003).
[9] J J Huang *et al., IEEE Elec.Dev.Lett*, 36, 1239(2000)
[10] J K Sheu, Y K Su, G C Chi, P L Koh, M J Jou, C M Chang, C C Liu, W C Hung, *Appl.Phys.Lett*, 74, 2340(1999).
[11] J K Kim, J H Je, J W Lee, Y J Park, T Kim, I O Jung, B T Lee, J L Lee, *J.Electron.Mater*, 30, L8-L12(2001)
[12] D W Kim, J C Bae, W J Kim, H K Baik, S M Lee, *J.Vac.SCi.Technol.*B, 19, 609-614(2001).
[13] T S Cheng, L C Jenkins, S E Hooper, C T Foxon, J W Orton, D E Lacklison, *Appl.Phys.Lett.* 66, 1509(1995).
[14] S Kuroda, E Bellet-Amalric, R Giruad, S Marcet, J Cibert, H Mariette, *Appl.Phys.Lett*, 83, 4580(2003).

A Realistic March-12N Test And Diagnosis Algorithm For SRAM Memories

Wan Zuha Wan Hasan[1], Masuri Othman[1] and Bambang Sunaryo Suparjo[2]
[1]Faculty of Engineering, Universiti Kebangsaan Malaysia, Bangi, Selangor.
[2]Mentor Graphics Corporation USA
Email: wan@eng.upm.edu.my, masuri@vlsi.eng.ukm.my

Abstract - **Testing and diagnosis techniques play a key role during the advance of semiconductor memory technologies. The challenge of failure detection has created intensive investigation on efficient testing and diagnosis algorithm for better fault coverage and yield improvement. The test and diagnosis complexity are 12N for bit-oriented diagnosis algorithm, N is the number of addresses is proposed for fault detection and diagnosis. Using the proposed march-based algorithm (march-12N), 100% of the faults under the fault model are covered and partial distinguished. They also can locate the faulty cells and identify their types. The complete fault and diagnosis procedures for state coupling faults, idempotent coupling faults and inversion coupling faults are written in this paper. Therefore, all the coupling faults that occur in SRAM memories are verified and proved the valid results. Furthermore, the realistic 12N test and diagnosis algorithm has shown the improvement of diagnostic resolution and test time.**

I. INTRODUCTION

Recently, the advances of semiconductor memory technologies have become more complex and also the numbers of memory chip transistors rapidly increased. Therefore, the die size becomes more and more compact (now 32 megabits and more) due to archive high memory capacity. Ironically, the larger the RAM, the more complex the fault model required to effectively model the variety of physical failures that could occur because of interference between closely packed cells. On the other hand, the more compact size will produces more defects during chip manufacturing, pushing yields down.

Therefore, new challenge for the industry, which is trying to control and balance the costs for production, such device while still being able

good product. Profitable manufacturing of high quality products relies on many procedures and techniques, including diagnosis. Although diagnosis has been widely used for SRAM, it is considered an expensive process due to long tester time and complex fault analysis procedures. Efficient test and diagnosis algorithms no doubt will benefit the industry and will play a more important role in the future market grow of system-on chip (SOC).

There are a lot of test techniques have been well studied. Many test algorithms [1,2,4,7,8] were proposed based on functional-level fault models for the last few decades. One of the best solutions such as march test is widely used to detect faults in the SRAM memories. It has been proven [2] that the march tests are able to detect the functional faults on the static random access memories (SRAM). All the functional faults such as stuck-at faults, transition fault, coupling faults, address decoder fault, stuck-open fault and data retention fault are relied on the mentioned algorithms. However, to improve the yield and reliability of very large scale integration (VLSI) chips, testing alone is not sufficient, diagnosis scheme is taking place due to improve yield and reliability of the outgoing product, therefore the quality of the overall system also increased. The diagnosis algorithms [5,6,11] 17N, 15N and March CD respectively have been used to distinguish each fault that is detected in the SRAM memories. In [12] 9 out of 20 (45%) fault syndromes cannot be distinguished.

The purpose of this paper is to present on how 12N march-based algorithms do the detection and diagnosis for the coupling faults (CFs). The detail testing and procedures of the proposed algorithm also well presented and shows, with shorter test time and optimum diagnostic resolution which only 8 out of 20 (40%) fault syndromes cannot be distinguished.

II. FAULT MODELS AND NOTATION

There are various failure modes that need to be considered in the SRAM fault model (the memory cell array). These fault models are already approved and appropriate for delivering algorithm to test the functional faults in SRAMs [10]. Only CFs in the memory cell array is considered and the notations of the fault model are follows:

a) State Coupling Faults (CFst); contains 4 subclasses: CFst (i, j) – the coupled (victim) cell is forced to logic value j, j $\in$ {0, 1}, only if the coupling (aggressor) cell contain logic value i, i $\in$ {0, 1}

b) Idempotent Coupling Faults (CFid); contains 4 subclasses: CFid (t, j) – the victim cell is forced to logic value j, j $\in$ {0, 1}, if the aggressor cell undergoes a transition t, t $\in$ {$\uparrow$, $\downarrow$}

c) Inversion Coupling Faults (CFin); contains 2 subclasses: CFin (t, $\updownarrow$) – the victim cell is inverted ($\updownarrow$) to logic value j, j $\in$ {0, 1}, if the aggressor cell undergoes a transition t, t $\in$ {$\uparrow$, $\downarrow$}

For the detail notations with ascending ($\Uparrow$) and descending ($\Downarrow$) address is shown in Table 1.

III. THE 12N TEST AND DIAGNOSIS ALGORITHM FOR BIT ORIENTED MEMORIES

A march test by Goor [1] consists of a sequence of march element (Ms), where s is the number of march sequence. Each march element consists of a number of Read and/or Write operation to all cells according to a predefined address order$\Uparrow$, $\Downarrow$ or either ($\updownarrow$). March test or march-based tests are widely used for fault detection and location due to their linear complexity with respect to the memory cell (bit) number.

For a given test algorithm the corresponding dictionary of fault syndromes is constructed in the following way. For example march CL [12] as shown in Table 2, each row (respectively, column) of the dictionary corresponding to a certain fault class (respectively, a Read operation from the march test). In the Table 2 Ri=0(1) means that the ith Read operation of the test algorithm has returned a fault free (respectively, faulty) value.

Name	Aggressor	Victim	Address
CFst(L,0,0)	0	0/1	A<V
CFst(H,0,0)	0	0/1	A>V
CFst(L,0,1)	0	1/0	A<V
CFst(H,0,1)	0	1/0	A>V
CFst(L,1,0)	1	0/1	A<V
CFst(H,1,0)	1	0/1	A>V
CFst(L,1,1)	1	1/0	A<V
CFst(H,1,1)	1	1/0	A>V
CFid(L,$\downarrow$,1)	$\downarrow$	1/0	A<V
CFid(H, $\downarrow$,1)	$\downarrow$	1/0	A>V
CFid(L, $\downarrow$,0)	$\downarrow$	0/1	A<V
CFid(H, $\downarrow$,0)	$\downarrow$	0/1	A>V
CFid(L,$\uparrow$, 1)	$\uparrow$	1/0	A<V
CFid(H, $\uparrow$,1)	$\uparrow$	1/0	A>V
CFid(L,$\uparrow$,0)	$\uparrow$	0/1	A<V
CFid(H, $\uparrow$,0)	$\uparrow$	0/1	A>V
CFin(L, $\downarrow$,$\updownarrow$)	$\downarrow$	$\updownarrow$	A<V
CFin(H,$\downarrow$, $\updownarrow$)	$\downarrow$	$\updownarrow$	A>V
CFin(L, $\uparrow$,$\updownarrow$)	$\uparrow$	$\updownarrow$	A<V
CFin(H, $\uparrow$,$\updownarrow$)	$\uparrow$	$\updownarrow$	A>V

Table 1 Fault Models and Notation [3]

$\Uparrow$ (w0); $\Uparrow$ (r0, w1) $\updownarrow$(r1); $\Uparrow$ (r1, w0,); $\Downarrow$ (r0,w1); $\updownarrow$(r1); $\Downarrow$(r1, w0); $\updownarrow$(r0)}							
Faults	R0	R1	R2	R3	R4	R5	R6
CFst(L,0,0)	0	0	1	0	0	0	0
CFst(H,0,0)	0	0	0	0	0	1	0
CFst(L,0,1)	0	0	0	1	0	0	1
CFst(H,0,1)	1	0	0	0	0	0	1
CFst(L,1,0)	0	1	0	0	1	1	0
CFst(H,1,0)	0	1	1	0	1	0	0
CFst(L,1,1)	1	0	0	0	0	0	0
CFst(H,1,1)	0	0	0	1	0	0	0
CFid(L,$\downarrow$,1)	1	0	0	0	0	0	0
CFid(H, $\downarrow$,1)	0	0	0	0	1	1	0
CFid(L, $\downarrow$,0)	0	0	0	0	0	0	1
CFid(H, $\downarrow$,0)	0	0	1	0	0	0	0
CFid(L,$\uparrow$, 1)	0	0	0	1	0	0	0
CFid(H, $\uparrow$,1)	0	1	1	0	0	0	0
CFid(L,$\uparrow$, 0)	0/1	0	0	1	0	0	0
CFid(H, $\uparrow$,0)	0	0	0	0	0	1	0
CFin(L, $\downarrow$,$\updownarrow$)	1	0	0	0	1	1	0
CFin(H,$\downarrow$, $\updownarrow$)	0	0	1	0	0	0	1
CFin(L, $\uparrow$,$\updownarrow$)	0	1	1	1	0	0	0
CFin(H, $\uparrow$,$\updownarrow$)	0/1	0	0	1	0	1	0

Table 2 Fault Syndromes for March CL

If a fault is detected by a diagnosis test algorithm with k read operations, then the march syndrome [12] of the fault is defined as $(R_0R_1 ...R_{k-1})$, for example, the march syndrome for CFst(L,1,1) is (1000000). The faults of different syndromes can be distinguished. The following 2 faults CFst(L,0,0) and Cfid(L,↓,0) have the same syndrome (0010000) These 2 faults cannot be distinguished from each other by the March CL algorithm. The total number of fault syndromes cannot be distinguished is 9 out of 20 (45%)

IV. TEST AND DIAGNOSIS PROCEDURES

The proposed March-12N RAM diagnosis algorithm is shown as follows:

$$\{ \updownarrow (w0); \Uparrow (r0, wl, r1); \Uparrow (r1, w0, r0);$$
$$M0 \qquad M1 \qquad M2$$
$$\Downarrow (r0, w1); \Downarrow (r1, w0); \updownarrow (r0)\}$$
$$M3 \qquad M4 \qquad M5$$

This March-12N is proposed to test and diagnosis for bit-oriented in unlinked CFins, CFids and CFsts and Fig 1 depicts the respective fault free state. To proof that March-12 algorithm detects all the listed faults are split into 20 classes (see Table 1).

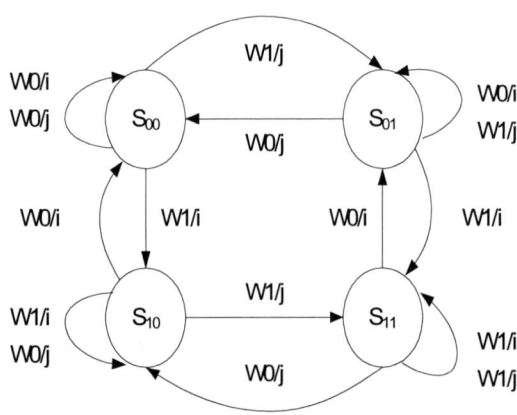

Fig. 1 State Diagram of Fault Free CFs

a) Inversion Coupling Faults (CFins)

Let consider CFins with Ci be coupled to any number of cells with the addresses of the lower than i and let Cj be the highest of those cells (j<i). If CFin <↑/↕> (Cfin(L, ↑,↕)) coupled to Cj as shown in Fig 2, then the fault will be

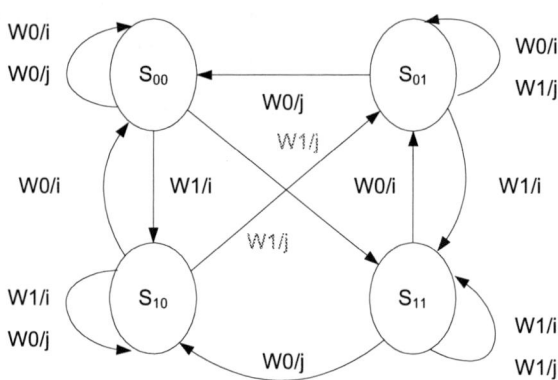

Fig 2 State Daigram of an CFin < u ; i >

detected in march element M1⇑ (r0, wl, r1) and M3 ⇓ (r0, w1) followed by M4 ⇓ (r1,w0). In M1 cell j will be read first with ⇑ addressing for j<i and when M1 operates on Ci a 1 instead of a 0 is read. In M3 followed by M4 Cell i will be read first with ⇓ addressing for j<i and a 0 instead of a 1 is read when M4 operates on Ci. The detail procedures as explained below:

M1 oper. on Cj	Cj	Ci	M1⇑(r0,wl, r1)
r0	0	-	
w1	1	1	< ↑ / ↕>
M1 oper. on Ci	Cj	Ci	M1⇑(r0,wl, r1)
r0	-	1	

M3 oper. on Cj	Cj	Ci	M3 ⇓ (r0,w1)
r0	0	1	
w1	1	0	< ↑ / ↕>
M4 oper. on Ci	Cj	Ci	M4 ⇓ (r1,w0)
r1	-	0	

If CFin <↑/↕> (CFin(L,↓,↕)) is coupled to Cj; then M2 as well as M4 followed by M5 will detect the fault. The proof is similar to above. Also for the case of CFin for j>i, the proof is similar to above. Note that the the fault syndrome for CFin(L,↑,↕) is 1000010 corresponding to Read operation of the algorithm where R0 and R5 are logic 1.

b) Idempotent Coupling Faults (CFids)

For CFid with Ci is coupled to any number of cells with the addresses of the lower than i an d le Cj be the highest of those cells (j<i). If CFid <↑/0> (CFid (L,↑,0)) is coupled to Cj as shown in Fig 3, then the fault will be detected in march element M3⇓ (r0, w1)followed by M4 ⇓ (r1, w0).

930

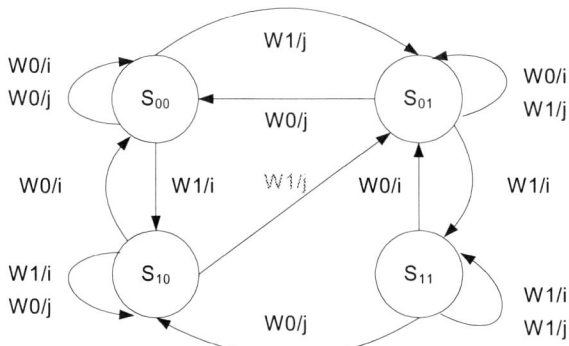

Fig.3 State diagram of an CFid<u;0>

In M3 followed by M4 cell i will be read first with $\Downarrow$ addressing for j<i and a 0 instead of a 1 is read when M4 operates on Ci. The detail procedures as explained below:

M3 oper. on Cj	Cj	Ci	M3$\Downarrow$ (r0, w1)
r0	0	1	
w1	1	0	<↑/0>
M4 oper. on Ci	Cj	Ci	M4$\Downarrow$ (r1, w0)
r1	1	0	

If CFid <↑/1> (CFid (L,↑,1)), CFid <↓/0> (Cfid (L,↓,0)) and CFid <↓,1> (CFid (L,↓,1)), are coupled to Cj; then, M1, M2 and M4 followed by M5 respectively, will detect the fault. The proof is similar to above. Also for the case of CFid for j>i, the proof is similar to above. Note that the the fault syndrome for (CFid (L,↑,0)) is 0000010 corresponding to Read operation of the algorithm where R5 is logic 1.

c) State Coupling Faults (CFst)

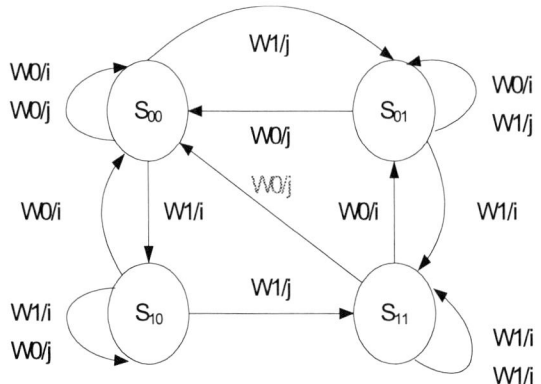

Fig. 4 State diagram of an CFst<0;0>

For CFst with Ci is coupled to any number of cells with the addresses of the lower than i and

Cj be the highest of those cells (j<i). If CFst <0; 1/0> (CFst (L,0,0)) is coupled to Cj as shown in Fig 4, then the fault will be detected in march element M2$\Uparrow$(r1,w0, r0). In M2$\Uparrow$(r1,w0,r0) cell j will be read first with $\Uparrow$ addressing for j<i and a 0 instead of a 1 is read when M2 operates on Ci. The detail procedures as explained below:

M2 oper. on Cj	Cj	Ci	M2$\Uparrow$ (r1,w0,r0)
r1	1	-	
w0	0	0	<0 ; 1/0>
r0	0	0	
M2 oper. on Ci	Cj	Ci	M2$\Uparrow$ (r1,w0,r0)
r1	-	0	

If CFst <0;0/1> (CFst (L,0,1)), CFst <1; 1/0> (CFst (L,1,0)) and CFst <1;0/1> (CFst (L,1,1)), are coupled to Cj; then, M2 as well as M2 followed by M3 as well as M4 followed by M5, M1 as well as M3 followed by M4 and M1 respectively, will detect the fault. The proof is similar to above. Also for the case of CFst for j>i, the proof is similar to above. Note that the the fault syndrome for (CFst (L,0,0) is 0010000 corresponding to Read operation of the algorithm where R2 is logic 1.

V. RESULT AND DISCUSSION

As results the complete fault syndromes of the test and diagnosis procedures for March-12N is shown in Table 3. Not all the fault syndromes in the table are different, which is 8 out of 20 (40%) cannot be distinguished such as; CFst(L,0,0), CFst(H,1,0), CFid(L,↓,0) and CFid(H,↑,0) have same fault syndrome (0010000) as well as other groups of faults. Even though only 60% of the faults is distinguishable, but it is considered high compare to the previous result by Vardanian et al [12] (with time complexity is taking account for both test and diagnosis).

{ $\Updownarrow$ (w0); $\Uparrow$ (r0, wl, r1); $\Uparrow$ (r1, w0, r0); $\Downarrow$ (r0, w1); $\Downarrow$ (r1, w0); $\Updownarrow$ (r0)}							
Faults	R0	R1	R2	R3	R4	R5	R6
CFst(L,0,0)	0	0	1	0	0	0	0
CFst(H,0,0)	0	1	0	0	0	1	0
CFst(L,0,1)	0	0	0	1	1	0	0
CFst(H,0,1)	1	0	0	0	0	0	1
CFst(L,1,0)	0	0	1	0	0	1	0

CFst(H,1,0)	0	0	1	0	0	0	0
CFst(L,1,1)	1	0	0	0	0	0	0
CFst(H,1,1)	0	0	0	0	1	1	0
Cfid(L,↑, 1)	1	0	0	0	0	0	0
Cfid(L,↑ ,0)	0	0	0	0	0	0	1
Cfid(L,↓ ,1)	0	0	0	0	0	1	0
Cfid(L, ↓ ,0)	0	0	1	0	0	0	0
Cfid(H, ↑,1)	0	0	0	0	1	0	0
Cfid(H, ↑,0)	0	0	1	0	0	0	0
Cfid(H, ↓ ,1)	0	0	0	0	1	0	1
Cfid(H, ↓ ,0)	0	0	0	0	0	1	0
Cfin(L, ↑,↕)	1	0	0	0	0	1	0
Cfin(L, ↓,↕)	0	0	1	0	0	0	1
Cfin(H, ↑,↕)	0	0	1	0	1	0	0
Cfin(H,↓ , ↕)	0	0	0	0	1	1	0

Table 3 Fault Syndromes for March-12N

Table 4 shows the comparison of the different test and diagnosis algorithms with diagnostic resolution as well as time complexity. The proposed algorithm has achieved better diagnostic resolution with lower time complexity, 60% and 12N respectively.

Algorithms for Test and Diagnosis CFin, CFid and CFst		
	Diagnostic Resolution	Time Complexity
March-12N	60%	12N
March CL12N[12]	55%	12N
March-15N[6]	60%	15N

Table 4 Comparison of Diagnostic Resolution with different diagnosis algorithms.

VI. CONCLUSION

In this paper, a March-12 test and diagnosis algorithm has been proposed for bit-oriented memories. The algorithm has been proposed for fault detection and partial diagnosis of CFs in SRAM memories. The analysis results show that the proposed March-12N algorithm achieves highest diagnostic resolution with less test time complexity.

REFERENCES

[1] A.J. Van De Goor, *Testing Semiconductor Memories, Theory and Practice*, ComTex Publishing, Gouda Netherland, 2001.

[2] A.J. Van De Goor, *Using March Test to test SRAMs*, IEEE Design of Computers, 1993.

[3] C.W. Wang, C.F Wu, J.F. Li, C.W. Wu, T. Teng k. Chiu, H.p. Lin, *A Built-in Self-test and Diagnosis Scheme For Embedded SRAM*, ATS 2000

[4] D. S. Suk and S. M. Reddy, *A March Test for Functional Faults in Semiconductor Random Access Memories*, IEEE Transactions on Computers, Vol. C-30, No 12, Dec 1981.

[5] J.F. Li, K.L. Cheng, C.T. Huang, C.W. Wu, *A March-Based RAM Diagnosis Algorithms For Stuck-At and Coupling Faults*, ITC 2001.

[6] J.F. Li, C.D. Huang, *An Efficient Diagnosis Scheme for Randon Access Memories*, ATS 2004.

[7] Magdy S.A and Hassan F.R, *Functional Testing of Semiconductor Random Access Memories*, Computing Surveys, Vol.15 No.3, September1983

[8] R. Nair, S.M. Thatte and J.C. Abraham, *Efficient Algorithm for Testing Semiconductor Random-Access Memories*, IEEE Transactions on Computers, Vol. C-27, No 6, June 1978.

[9] R. Dekker, F. Beeker, L. Thijssen, *Fault Modeling and Test Algorithm Development for Static Random Access Memories*, ITC 1988.

[10] R. Dekker, F. Beeker, *A Realistic Fault Model and Test Algorithms For Static Random Access Memories*, IEEE Transactions on Computers Aided Design, Vol. 9, No 6, June 1990.

[11] V.N Yarmolik, Y.V. Klimets, A.J. Van De Goor, S.N Demidenko, *RAM Diagnostic Tests*, ITC 1996.

[12] V.A Vardanian ,Y. Zorian, *A March-Based Fault Location Algorithm For Static Random Access Memories*, IEEE International Workshop MTDT 2002.

Characterization of Strained Silicon MOSFET Using Semiconductor TCAD Tools

Wong Yah Jin, Ismail Saad and Razali Ismail
Faculty of Electrical Engineering,
Universiti of Teknologi Malaysia,
81300 Skudai, Johor, Malaysia.
Email: helommi@gmail.com, ismail_s@ums.edu.my and razali@fke.utm.my

Abstract - **The paper is looking into the enhancement of conventional PMOS by incorporating a strained silicon within the channel and bulk of semiconductor. A detailed 2D process simulation of Strained Silicon PMOS (SSPMos) and its electrical characterization was done using TCAD tool [1]. With the oxide thickness, T_{OX} of 16nm and Germanium concentration of 35%, the threshold voltage V_t for the strained Si and conventional PMOS is -0.5067V and -0.9290V respectively. This indicates that the strained silicon had lower power consumption. Beside that, the drain induced barrier lowering (DIBL) value for the strained PMOS is 0.3034V and the conventional PMOS is 0.4747V, which shows a better performance for strained silicon as compared to conventional PMOS. In addition, the output characteristics were also obtained for SSPMos which showed an improvement of Drain current compared with conventional PMOS.**

I. INTRODUCTION

Scaling down of MOSFET devices has been the driving force in IC industry in order to achieve higher speed and lower power requirements [2]. The recent MOSFET devices have been scaled down to 50nm gate lengths where the gate oxide thickness has become thin enough to suppress the short channel effect (SCE) [3]. However further scaling down of the MOSFET beyond 50nm will cause the SCE to intensify, thus degrading the current drivability and electron mobility of a MOSFET [4]. The continuous downsizing of the gate length have caused the gate oxide to become so thin that current begins to leak across the gate even when there is no applied voltage. Therefore further improvement without minimizing the gate length is strongly required. Carrier mobility improvement has been seen as one of the best alternative for faster devices at lower power

levels [5]. Strained silicon technology can offer significant performance enhancement to MOSFET devices [6] by increasing carrier mobility without having to make the devices become smaller [7], [8]. By stressing or straining, the silicon lattice lets electrons flow with less resistance. This will increase the drive current and make the transistor switch faster thus contributing to a higher clock frequency in integrated circuits (IC) with gate length downsizing to 60nm [9]. Another significant improvement in electrical performance for both n and p-channel device of strained Si with 25% Ge composition is demonstrated in [10]. In this paper we will study the performance enhancement by strained silicon as compared to conventional PMOS comprehensively with the help of Silvaco TCAD process and device simulation tools. With the T_{OX} of 16nm and 35% of Ge concentration, the V_T for the strained Si and conventional PMOS is -0.5067V and -0.9290V respectively. The drain induced barrier lowering (DIBL) for the SSPMOS is 0.3034V and the conventional PMOS is 0.4747V, which shows a better performance for strained silicon as compared to conventional PMOS. Consequently, the output characteristics were also obtained for SSPMos that showed an improvement of Drain current compared with conventional PMOS.

II. DEVICE STRUCTURE AND PROCESS

Both strained silicon PMOS with an added SiGe layer and normal conventional PMOS device without SiGe layers process simulation were carried out using ATHENA, Figure 1 shows the structure of both devices. The simulation process to create the strain silicon PMOS is similar to the conventional PMOS fabrication process. The fabrication of SSPMOS device starts by creating a silicon substrate with phosphorus doping of 2×10^{18} cm^{-3} and then a silicon layer with the thickness of 0.018µm is deposited on the silicon substrate. Next a silicon germanium (SiGe) layer with 0.35 Ge concentration is deposited on the silicon layer, followed by the deposition of another silicon layer with 0.007um thickness on to the SiGe layer. After the deposition, strained silicon is created at the channel. Polysilicon is then deposited and patterned to form the gate. The process continues with the implantation of source/drain. The boron is implanted with the 1.0×10^{15} cm^{-2} doping concentration. Next the silicon nitride (Si$_3$N$_4$) layer is deposited and patterned to cover the gate, source and drain. Then the aluminum is deposited and patterned to act as the metal contact. Finally, the final structure of the strained silicon PMOS is created as shown in Figure 1(a).

The conventional PMOS structure is shown in Figure 1(b). The difference between the conventional structure with the SSPMos structure is that there is no added SiGe layer, thin Si layer and Si$_3$N$_4$ capping layer.

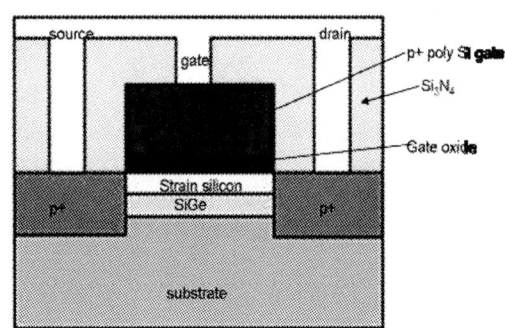

Fig. 1(a): The strain silicon PMOS device structure.

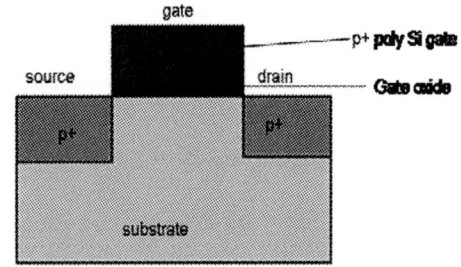

Fig. 1(b): The PMOS device structure.

The SSPMos structure is created with 0.0160μm gate oxide thickness and 0.71μm channel length. Meanwhile the conventional PMOS structure is created with 0.0091μm gate oxide thickness and 2.5μm channel length.

III. DEVICE DOPING PROFILE

Figure 2 shows the net doping and Ge concentration profile for the SSPMos. This is the result from the Athena simulation by performing a vertical cutline which starts at the gate and stops at the substrate. From Figure 2, we can see that the boron doping is high at the gate with 1×10^{20} cm^{-3} doping concentration. There is no doping in the silicon dioxide layer. Meanwhile the phosphorus doping at the strained silicon, SiGe layer and substrate is 1×10^{16} cm^{-3}. From the figure, the composition x shows a 0.35 of Ge concentration in silicon germanium layer only.

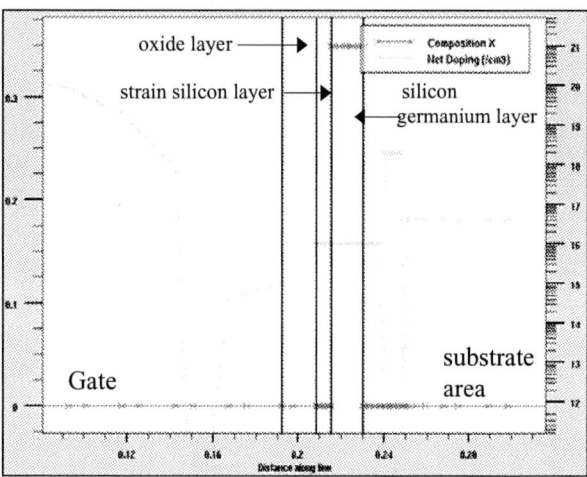

Fig. 2: The net doping and Ge concentration profile for

SSPMos.

IV. ELECTRICAL CHARACTERIZATION

In device simulation, both the strained silicon structure and the conventional PMOS structure are simulated in Atlas. The devices are simulated to obtain the characteristics of the conventional PMOS and strained silicon PMOS (SSPMos). The mobility models that are used to obtain the electrical characteristics are the parallel electric field dependence and concentration dependent model. Beside that, the carrier static lifetime for the Si material is set at 1e-7 tau for electron and hole. Meanwhile the SiGe material is set to 1e-8 tau for electron and hole. For the carrier statistic model, the bandgap narrowing and Boltzman are chosen in this simulation. As for the recombination models, the auger and SRH concentration dependent lifetimes are chosen. The characteristics of the devices that was obtained from the simulation are the drain current versus gate voltage curve, threshold voltage, drain induced barrier lowering (DIBL) and drain current versus drain voltage curve.

From the simulation, the drain current, I_d versus gate voltage, V_{gs} curve with a drain voltage, V_{ds} of -0.1V for both conventional PMOS and the SSPMos devices are shown in Figure 3. From Figure 3, it is obvious that the drain current for SSPMos structure is higher than conventional PMOS. This indicates that the SSPMos has higher drive current compared to conventional PMOS. Meanwhile the extracted threshold voltage parameters from Figure 3 are -0.511299V and -0.92902V for the SSPMos and conventional PMOS respectively. This indicates that the strained PMOS has lower voltage threshold than the conventional PMOS which translates to lower power consumption.

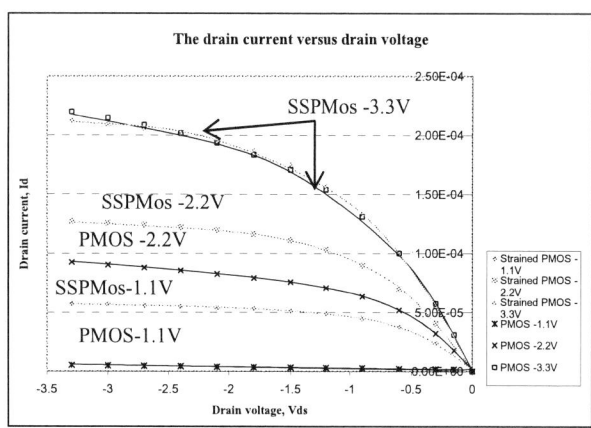

Fig. 4: The comparison of drain current versus drain voltage graph when V_{gs} is -1.1V, -2.2V and -3.3V.

V. Conclusion

From the results, it can be seen that the strained silicon has a better performance compared to the conventional PMOS even though the channel length for the strained silicon is larger than the conventional PMOS. This research will continue to study the electrical characteristics of the SSPMos such as the effective mobility enhancement. Further improvements and optimization will be done to the device performance in order to achieve a significant enhancement on MOSFET. Strained silicon is still considered as a new technology and more research is still needed to improve its implementation to the current technology.

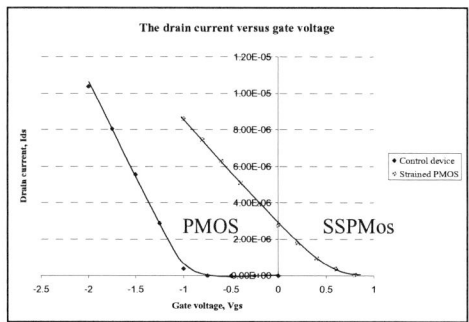

Fig. 3: The comparison of drain current versus gate voltage graph when V_{ds} is -1.0V.

Next, the drain induced barrier lowering, (DIBL) parameter is obtained from the difference between the threshold voltage divided by -2.8V. The DIBL for the strained PMOS is 0.3034V and the conventional PMOS is 0.4747V. The DIBL for the strained PMOS is smaller than the conventional PMOS. This shows that the strain silicon for PMOS is better compared to conventional PMOS.

Beside that, both structures are simulated to ramp the drain voltage, V_{ds} to -3.3V when the gate voltage, Vgs is bias to -1.1V, -2.2V and -3.3V. The simulation results are presented in Figure 4 which represents the graph of the drain current versus the drain voltage. From Figure 4, it can be seen that the strained PMOS device has a higher drive current compared to the conventional PMOS. From these results, it is evident that the strained silicon PMOS has a better drive current than conventional PMOS.

REFERENCES

[1] Silvaco International (2005). *Athena and Atlas User's Manual Process Simulation Software*. USA: Silvaco International.

[2] Taur Y. and et. al. *IDEM Technical Digest. 1998.* pg 789.

[3] Cho, H., Park, B. G. and Lee, J. D. Nano-Scale SONOS Memory with a Double Gate MOSFET Structure. *Journal of the Korean Physical Society.* Feb 2003. 42(2):233-236.

[4] Yuan J., Zeitzoff P. M. and Jason Woo C.S. Source/Drain Parasitic Resistance Role and Electrical Coupling Effect in sub 50 nm MOSFET Design. *ESSDERC.* 2002. pg 503-506.

[5] Acosta T. and Sood S. Engineering Strained Silicon-looking Back and Into the Future. *IEEE Potentials.* IEEE. 2006. 25(4):31-34.

[6] Ismail K., Meyerson B.S. and Wang P.J. High Electron Mobility in Modulation doped Si/SiGe. *Application Physics Letter.* IEEE. 1991. 58(19):2117-2119.

[7] Thompson, S.E., Armstrong, M. and et.al. A 90nm Logic Technology Featuring Straiend Silicon. *IEEE Transaction on Electronic Devices, 2004*. IEEE. 2004. 51(11):1790-1797.

[8] Thuy Dao. Process Induced Damage: What Challenged Lie Ahead?. *IEEE*. 2003. 51-55.

[9] Wang, H.C., Wang, P.C. and et.al. Substrate strained silicon technology: Process Integration. *IEEE International Electronic Device Meeting, 2003*. IEEE. 2003. 3.4.1-3.4.4.

[10] S.H. Olsen, M. Temple and et.al. Doubling Speed using strained Si/SiGe CMOS technology. *Thin Solid Films*. Elsevier. 2005. 508(1-2): 338-341.

The Growth of III-V Nitrides Heterostucture on Si Substrate by Plasma-Assisted Molecular Beam Epitaxy

F.K. Yam, Z. Hassan, L. S. Chuah, N. Zainal, C.W. Chin, S. M. Thahab, M. Hussein
School of Physics, Universiti Sains Malaysia, Penang, MALAYSIA
Email: yamfk@yahoo.com , zai@usm.my
Tel: 604-6533599, Fax: 604-6579150

Abstract **This article reports the use of plasma-assisted molecular beam epitaxy (MBE) to grow InGaN/GaN/AlN on Si (111) substrate. The film is then characterized by high-resolution x-ray diffraction (HR-XRD) and photoluminescence (PL). The studies show that the structural quality of the film is comparative to the values reported in the literature. PL measurement exhibits a sharp and intense band edge emission of GaN with the absence of yellow emission band, indicating good optical quality of the film.**

I. INTRODUCTION

GaN-based materials receive great deal of attention because of the potential applications for optoelectronic devices operating in the whole visible spectral range and in electronic devices such as high temperature, high power, and high frequency transistor. The III-nitrides form a continuous alloy system with direct band gap ranging from 6.2 eV (AlN) to 0.7 eV (InN) with 3.4 eV for GaN. Consequently, the growth and physics of GaN-based materials have attracted tremendous scientific attention.

Traditionally, the heterostructures of the conventional III-V compounds such as GaAs, GaP and InP are commonly grown by metal-organic chemical vapor deposition (MOCVD) and molecular beam epitaxy (MBE) techniques. However, the growth of GaN-based materials has been mainly performed using MOCVD, particularly, the GaN-based commercial optoelectronics devices, i.e. light emitting diodes and laser diodes. In contrast, MBE grown GaN-based optoelectronics devices are relatively lagged behind; this is attributed to the unavailability of suitable source of active nitrogen species. Nevertheless, with the development of efficient and reliable nitrogen plasma sources for MBE recently, the quality of the GaN-based epilayer has been improved tremendously.

The use of MBE for the growth of GaN-based materials, in fact, has a number of advantages as compared to MOCVD. For instances, lower growth temperature, precise control of the alloy composition and thickness, abruptly layered structure, *in-situ* monitoring of the surface structure on atomic scale, less material sources consumption, no hazardous materials as well as require no post-growth heat treatment for the activation of Mg dopant in the p-type GaN-based materials due to the hydrogen free environment [1,2].

Presently, substrates such as sapphire (Al_2O_3) or silicon carbide (SiC) are commonly used for the growth of high quality GaN films, however, there have been a great interest in growing GaN-based materials on Si substrates which offers not only low production cost but attractive opportunity to incorporate GaN-based devices into Si-based technology.

This article presents the growth of InGaN/GaN/AlN on Si (111) substrate by using plasma assisted MBE (Veeco Gen II). The structural and optical properties of grown III-nitrides heterostructure are then analyzed by a variety of characterization tools. High resolution x-ray diffractometer is used to assess and determine the crystalline quality and indium composition of InGaN. The optical quality of the film is investigated by photoluminescence (PL).

II. EXPERIMENT

A. The growth of III-nitrides heterostructure

The III-V nitrides heterostructure, i.e. InGaN/GaN/AlN was grown on Si (111) substrate using Veeco model Gen II MBE system. High purity material sources such as gallium (7N), aluminum (6N5) and indium (7N) were used in the Knudsen cells. Nitrogen with 7N purity was channeled to RF source to generate reactive nitrogen species. The plasma

0-7803-9730-4/06/$25.00 ©2006 IEEE   937

was operated at typical nitrogen pressure of 1.5×10^{-5} Torr under a discharge power of 300W.

The growth of III-nitrides on 3-inch Si (111) substrate starts with the standard cleaning procedure by using RCA method. The substrate was then mounted on wafer holder and loaded into the MBE system. The Si substrate was outgassed in the load-lock and buffer chambers. After outgassing, the Si was transferred to the growth chamber. Prior to the growth of the epilayers, surface treatment of the Si substrate was carried out to remove the SiO_2. Si substrate was heated at 750°C, a few monolayers Ga was deposited on the substrate for the purpose of removing the SiO_2 by formation of GaO_2. Reflection high energy electron diffraction (RHEED) showed the typical Si (111) 7×7 surface reconstruction pattern with the presence of prominent Kikuchi lines, indicating a clean Si (111) surface. Before the growth of nitride epilayers, a few monolayers of Al was also deposited on the Si substrate to avoid the formation of Si_xN_y which is deleterious for the growth of the subsequent epilayers. The formation of such amorphous layer has been observed by many groups [3,4]. Calleja [4] reported that the growth of GaN epilayer on Si_xN_y will lead to polycrystal with full width at half maximum values from XRD measurements between 70 and 100 arcmin.

The buffer layer or wetting layer, AlN was first grown on the Si substrate. It is well known that this buffer layer plays an important role in determining the crystalline quality of the thin film [3,5]. To grow AlN buffer layer, substrate temperature was heated up to 808°C, both of the Al and N shutters were opened simultaneously for 15 minutes.

Subsequently, GaN epilayer was grown on top of the buffer layer for 33 minutes with substrate temperature set at 800°C. Right after the growth of GaN epilayer, the substrate temperature was ramped down to 700°C to prepare for the growth of InGaN epilayer.

To grow InGaN, the effusion cells of In, and Ga were heated up to 925 and 930°C, respectively, in this growth, attempt was also made to dope the epilayer with Mg (p-type dopant) at 403°C. The growth duration for this InGaN epilayer was about 30 minutes.

B. Characterization

The MBE grown InGaN/GaN/AlN heterostructure thin film was characterized by a variety of tools. High resolution XRD (PANalytical X'pert Pro MRD) with a Cu-Kα1 radiation source (λ= 1.5406Å) was used to assess and determine the crystalline quality epilayers, as well as indium composition of InGaN. The optical quality of the film was investigated by photoluminescence (PL). The optical quality of the films was studied by PL. PL measurement was performed at room temperature by using Jobin Yvon HR800UV system with He-Cd laser 325 nm as excitation source.

III. RESULTS AND DISCUSSION

Fig. 1 shows the 2θ XRD spectra of the sample grown by MBE. The XRD measurement confirmed that the heterostructure of III-nitrides were epitaxially grown on Si (111). These can be seen from the presence of the peaks at 34.59°, 36.09°, 72.96° and 76.50°, which correspond to GaN (0002), AlN (0002), GaN (0004) and AlN (0004), respectively, in addition, a weak peak appears at 32.99° that can be attributed to InGaN (0002). The positions of the peaks and the corresponding crystal planes as well as the relative intensity are compiled in Table 1.

Table 1 The 2θ XRD peak positions of different crystal planes and their relative intensity

2θ Peak position (°)	Crystal Plane	Rel. Intensity (%)
28.475	Si (111)	100.00
32.986	InGaN	0.06
34.593	GaN (002)	22.26
36.092	AlN (002)	1.74
58.914	Si (222)	0.02
69.208	In	too low
72.964	GaN (004)	0.59
76.500	AlN (004)	0.03

XRD rocking curve was also carried out to determine the crystalline quality of the epilayers, the measurements show that the full width at half maximum (FWHM) of the (0002) plane for AlN, GaN and InGaN epilayers are 24.2, 24.8 and 6.8 arcmin, respectively. From the literature, the use of Si (111) substrate for growth of III-nitrides, particularly GaN, always produces relatively low crystal quality [3,4,6,7], rocking curve with high value of FWHM, i.e. from 20 to 70 arcmin were typically reported, this suggests that it is difficult to grow high quality GaN-based materials on Si (111) substrate.

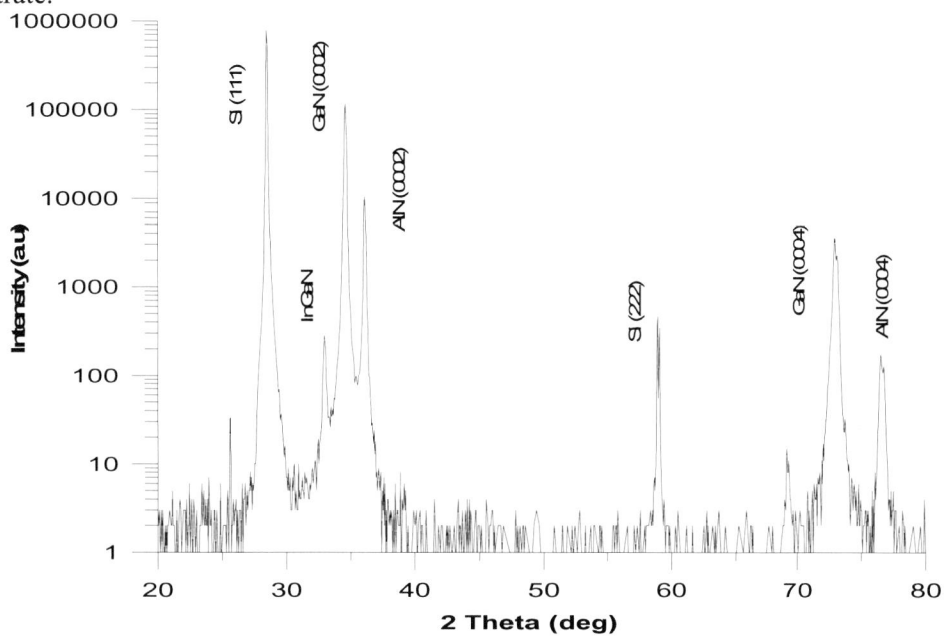

Fig. 1 XRD spectra of the InGaN/GaN/AlN/Si sample

The growth of poor crystal quality of the GaN-based epilayers is mainly attributed to the large difference in lattice constant, crystal structure and thermal expansion coefficient between the Si and GaN-based materials [8].

From Fig.1 and Table 1, the intensity of InGaN is found to be relatively low, i.e. it is approximately three orders of magnitude smaller than GaN, and this indicates that the thickness of InGaN epilayer was very thin. The modeling of the $\omega/2\theta$ XRD spectra showed that the thicknesses of the GaN and InGaN were 200 and 8 nm, respectively, in addition, the simulated result also revealed that the indium molar fraction of InGaN epilayer was 0.57. It is well known that InGaN materials are difficult to be grown, particularly InGaN with high indium fraction. The difficulties in growing high quality InGaN materials can be attributed to a number of problems, for instances, the large difference in interatomic spacing between InN and GaN results in a solid phase miscibility gap [9], the relatively high vapor pressure of InN as compared to the vapor pressure of GaN [10] as well as the difference of formation enthalpies for InN and GaN which causes a strong indium surface segregation on the growth front. Moreover, InGaN deposition is complicated by thermodynamic instability of InN, at higher growth temperature, InN will tend to dissociate faster than it can be adsorbed [11]. Therefore, it

is not surprising to obtain a thin InGaN epilayer for 30 minutes growth at temperature of 700°C.

Fig. 2 shows the PL spectra of the sample. The PL spectra are dominated by an intense and sharp peak at 362.35 nm, which is attributed to the band edge emission of GaN. Peaks at 402.69, 472.21 and 724.88 nm are probably due to impurities or native defects (such as Mg, C or N vacancy, Ga vacancy) related recombination [12,13]. No yellow band emission is observed; this indicates that the thin film is of good optical quality. Apart from this, there is a broad peak centred at 841.82 nm (1.47 eV), the presence of this broad peak could be attributed to the emission of InGaN.

IV. CONCLUSION

In summary, the growth of InGaN/GaN/AlN on Si (111) substrate has been performed using plasma-assisted molecular beam epitaxy. The structural and optical properties of the thin film has been analyzed by HR-XRD and PL. The structural quality of the thin film is comparative to the reported values in the literature. Sharp and intense band edge emission of GaN is observed in the PL measurement with the absence of yellow band emission, indicative of good optical quality of the thin film.

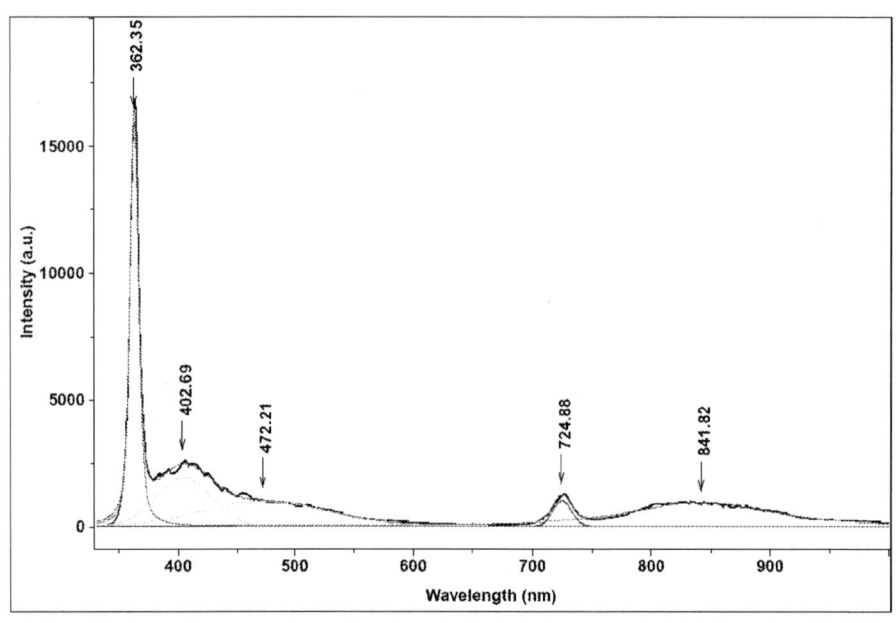

Fig. 2 PL spectra of the InGaN/GaN/AlN/Si sample

ACKNOWLEDGEMENT

This work was conducted under an IRPA RMK-8 Strategic Research grant. The support from Universiti Sains Malaysia is gratefully acknowledged.

REFERENCES

[1] H. Sakai, T. Koide, H. Suzuki, M. Yamaguchi, S. Yamasaki, M. Koike, H. Amano, I. Akasaki, "GaN/GaInN/GaN Double Heterostructure Light Emitting Diode Fabricated Using Plasma-Assisted Molecular Beam Epitaxy", Jpn. J. Appl. Phys. 34, pp. L1429-L1431 (1995).

[2] P. Laukkanen, S. Lehkonen, P. Usimaa, M. Pessa, A. Sepala, T. Ahlgren, E. Rauhala, "Emission studies of InGaN layers and LEDs grown by plasma-assisted MBE", J. Cryst. Growth 230, pp. 503-506 (2001).

[3] A. Ohtani, K. S. Stevens, R. Beresford, "Microstructure and photoluminescence of GaN grown on Si(111) by plasma-assisted molecular beam epitaxy", Appl. Phys. Lett. 65, pp. 61-63 (1994).

[4] E. Calleja, M. A. Sanchez-Garcia, F. J. Sanchez, F. Calle, F. B. Naranjo, E. Munoz, S.I. Molina, A. M. Sanchez, F. J. Pacheco, R. Garcia, "Growth of III-nitrides on Si(1 1 1) by molecular beam epitaxy Doping, optical, and electrical properties", J. Cryst. Growth 201/202, pp. 296-317 (1999).

[5] J. W. Yang, C. J. Sun, Q. Chen, M. Z. Anwar, M. Asif Khan S. A. Nikishin, G. A. Seryogin, A. V. Osinsky, L. Chernyak, and H. Temkin,

Chimin Hu and S. Mahajan, "High quality GaN–InGaN heterostructures grown on (111) silicon substrates", Appl. Phys. Lett. 69, pp. 3566-3568 (1996)

[6] J. Ristic, M. A. Sanchez-Garcia, E. Calleja, A. Perez-Rodriguez, C. Serre, A. Romano-Rodriguez, J. R. Morante, V. R. Koegler, W. Skorupa, "Growth of GaN layers on SiC/Si(111) substrate by molecular beam epitaxy", Mat. Sci. Eng. B93, pp. 172-176 (2002).

[7] W. Ju, D. A. Gulino, R. Higgins, "Epitaxial lateral overgrowth of gallium nitride on silicon substrate", J. Cryst. Growth 263, pp. 30-34 (2004).

[8] T. Lei, M. Fanciulli, R. J. Molna, T. D. Moustakas, R. J. Graham, J. Scanlon, "Epitaxial growth of zinc blende and wurtzite gallium nitride thin films on (0001) silicon", Appl. Phys. Lett. 59, pp. 944-946 (1991).

[9] I. Ho, and G. B. Stringfellow, "Solid phase immiscibility in GaInN", Appl. Phys. Lett. 69, pp. 2701-2703 (1996).

[10] T. Nagatomo, T. Kuboyama, H. Minamino and O. Omoto, "Properties of $Ga_{1-x}In_xN$ Films Prepared by MOVPE", Jpn. J. Appl. Phys. 28, pp. L1334- L1336 (1989)

[11] M. A. L. Johnson, W. C. Hughes, W. H. Rowland, Jr., J. W. Cook, Jr., J. F. Schetzina, M. Leonard, H. S. Kong, J. A. Edmond, J. Zavada, "Growth of GaN, InGaN, and AlGaN films and quantum well structures by molecular beam epitaxy", J. Cryst. Growth 175/176, pp. 72-78 (1997).

[12] G. Popovici and H. Morkoc, *Growth and Doping of Defects in III-Nitrides*, in: *GaN and Related*

Materials II, edited by S. J. Pearton, Gordon and Breach Science Publisher, Chap. 3, (2000)

[13] V. Bougrov, M. Levinshtein, S. Rumyantsev and A. Zubrilov, *Gallium Nitride (GaN),* in *Properties of Advanced Semiconductor Materials: GaN, AlN, InN, BN, SiC, SiGe*, edited by M. E. Levinshtein, S. L. Rumyantsev and M. S. Shur, John Wiley & Sons, Inc., Chap. 1, (2001)

The Energy Band Gap of $Al_xGa_{1-x}N$ Thin Films as a Function of Al-Mole Fraction

S. S. Ng, F. K. Yam, Z. Hassan and H. Abu Hassan
School of Physics
Universiti Sains Malaysia
11800 Penang, MALAYSIA
Tel: 604-6533673, Fax: 604-6579150
Email: shashiong@yahoo.com, yamfk@yahoo.com , zai@usm.my

Abstract In this work, the effects of Al mole fraction on energy band gap (E_g) of $Al_xGa_{1-x}N$ epilayers grown on sapphire substrate are investigated. Attention is focused on the Ga-rich composition samples ($0 \leq x < 0.10$). The Al-mole fraction is determined by high-resolution X-ray diffraction (HR-XRD) spectroscopy. Ultraviolet-visible (UV-VIS) transmission and micro-photoluminescence (µ-PL) spectroscopy are employed to determine the energy band gap of the samples. The XRD results revealed that the Bragg angle of the rocking curve (RC) peak gradually increases as the Al-mole fraction increases, indicating the reductions in the lattice constant c of the alloys. By the application of Vegard's law, the Al-mole fractions of $Al_xGa_{1-x}N$ samples have been calculated. Overall, the UV-VIS transmission and µ-PL results showed that as the Al-mole fraction increases, blue shifts of the absorption edge and band edge emission are observed in all samples. These indicate the strong dependence of the band gap energy of $Al_xGa_{1-x}N$ on the Al-mole fraction. Finally, the band gap energy of the $Al_xGa_{1-x}N$ as a function of Al-mole fraction have been plotted and the energy band gap bowing parameter of 14.62 eV is obtained from the best fit of the non-linear interpolation of the UV-VIS transmission and the PL data.

I. INTRODUCTION

By varying the Al-mole fraction in the $Al_xGa_{1-x}N$ ternary alloys, the energy band gap is tuneable from 3.4 eV for GaN to 6.13 eV for AlN. As a consequent, new applications are emerging for AlGaN-based devices such as optoelectronic devices, high temperature and high frequency electronic devices, UV photodetectors as well as application for sensor devices, particularly operating under harsh environment conditions [1-6].

From the fundamental physics point of view, the alloying will affect the lattice constant of a semiconductor. Consequently, many properties of this alloy will be changed. In term of device applications, one of the great beneficial effects from the alloying works is the tunable energy band gap of this alloy (as mentioned above). Since the alloying effect relates both to device-related materials issues and to basic solid-state physics; therefore, knowledge of the alloys composition effects on the fundamental properties of these advanced materials is crucial.

In this work, the effects of Al-mole fraction on the energy band gap (E_g) of $Al_xGa_{1-x}N$ thin films grown on sapphire substrate are investigated. However, attention is focused on the Ga-rich composition samples ($0 \leq x < 0.10$), because these kinds of samples are more practical for device applications. For example, $Al_xGa_{1-x}N$ with a relatively low Al-mole fraction have already been used in many applications. In addition, $Al_xGa_{1-x}N$ layer cracks easily with higher Al-molar fraction, due to lattice mismatch between AlGaN and sapphire substrate.

II. EXPERIMENTAL

In this work, $Al_xGa_{1-x}N$ epilayer with wurtzite structure and in the composition range $0 \leq x < 0.10$ were used. The samples were n-type unintentionally doped and were grown on sapphire (Al_2O_3) substrates. The thickness of the $Al_xGa_{1-x}N$ samples were in the range of $0.5 - 1.0$ µm, except the sample with the Al-mole fraction $x = 0$, where the thickness for this sample was about 3.5 µm.

To evaluate Al-mole fraction of the epilayer, the samples were analyzed by using a high-resolution X-ray diffractometer (PANalytical X'Pert Pro MRD) with a Cu-Kα_1 radiation source (λ = 1.5406 Å). The XRD symmetric RC of (0002) plane measurements were performed in the $\omega/2\theta$ mode. The UV-VIS transmission measurements were performed at room temperature using a Hitachi U-2000 Double-beam spectrophotometer. Deuterium lamp and tungsten iodide lamp were used as UV and VIS light sources. The optical transmission was recorded in the wavelength range from 200 - 800 nm. Room temperature micro-PL (μ-PL) measurements were performed by means of Jobin Yvon HR800UV system. A helium-cadmium (He-Cd) laser (325 nm) was used as an excitation source. To focus the laser on the sample surface, a 40×NUV microscope objective lens was employed. The diameter of the laser spot on the samples was around 1-2 μm.

III. RESULTS AND DISCUSSION

Fig. 1 illustrates the symmetric RC spectra of (0002) plane for the GaN and the $Al_xGa_{1-x}N$ epilayers. For the $Al_xGa_{1-x}N$ samples, the scale of the vertical axis has been magnified about five times to reveal the low-intensity spectra. As shown in Fig. 1, the Bragg angle of the RC peak gradually increases as the Al-mole fraction is increased, indicating the reductions in the lattice constant c of the alloy.

According to Kamiyama *et al.* [7] the crystalline quality of $Al_xGa_{1-x}N$ directly grown on Al_2O_3 degraded as the Al-mole fraction increases. Consequently, the FWHM of the diffraction peak will become broader as the Al composition increases. From Fig. 1, it can be seen that our XRD results do not reveal the trend as predicted by Kamiyama *et al.* This may most probably due to the samples are not grown under the same conditions; thus, the crystalline qualities of the epilayer are not only Al-mole fraction dependent but also growth parameters dependent.

From the XRD symmetric RC $\omega/2\theta$ scans of (0002) plane, the lattice parameter c of the samples can be calculated using the following formula:

$$c = \frac{\lambda l}{2\sin\theta_{RC}} \quad (1)$$

where λ is the wavelength of the x-ray radiation (1.5406 Å), θ_{RC} is the Bragg angle estimated from the peak of the RC, and l is the Miller indices. The calculated c value for each sample is

summarized in Table I. In principle, the composition can be determined through XRD measurements and the application of Vegard's law. By assuming the layers are fully relaxed or fully strained, according to Vegard's law, the variation of the lattice constant c between GaN and AlN is linearly proportional to the Al mole fraction [8]. Based on the obtained c value, the chemical composition of the Al, x, can be determined by following formula [9]:

$$x = \frac{c - c_{GaN}}{c_{AlN} - c_{GaN}} \quad (2)$$

where c_{GaN} and c_{AlN} are the lattice constants of GaN and AlN respectively. Note that the c for AlN (4.9816 Å) is taken from Ref. [10]. The calculated Al-composition is listed in Table I.

Fig. 2 shows the spectra of the transmission versus wavelength for $Al_xGa_{1-x}N$ ($0 \leq x < 0.10$) samples at room temperature. It can be clearly seen that the absorption edges shifting towards lower wavelength (blue shift) as the Al-mole fraction increases. These indicate the strong dependence of the band gap energy of $Al_xGa_{1-x}N$ on the Al-mole fraction. From Fig. 2, the energy band gap of the samples can be deduced by the intercepts on x-axis [11] and the results are listed in Table I.

Fig. 3 shows the near band edge emission spectra of $Al_xGa_{1-x}N$ ($0 \leq x < 0.10$) samples. The peak position and the FWHM values are obtained using the Gaussian peak-fitting model. The **FWHMs of the emission peaks** and the calculated energy band gaps are listed in Table I. Note that the energy band gap values have been rounded to three decimal points, and the difference between the values for x = 9.1% and x = 9.3% samples is actually around 0.1 meV. From Fig. 3, it can be seen that as the Al-mole fraction increases, a blue shift of the band edge emission peak is observed. These results are consistent with the UV-VIS measurements.

From Table I, it is found that the GaN sample exhibited the smallest FWHM and the FWHMs increase with the inclusion of Al, i.e. indicating the reduction in crystalline quality (from the optical point of view). However, no increasing trend in FWHM's as a function of increasing Al-mole fraction can be clearly seen. This is most probably due to the defect concentration in the each sample is different as the growth conditions are different in each case.

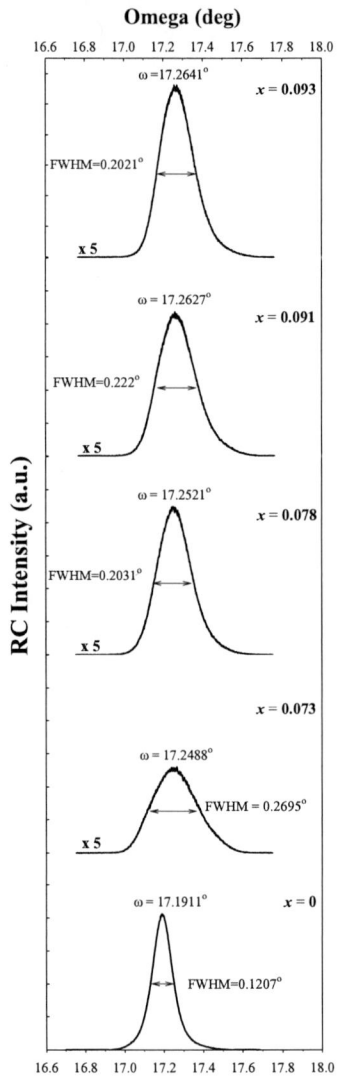

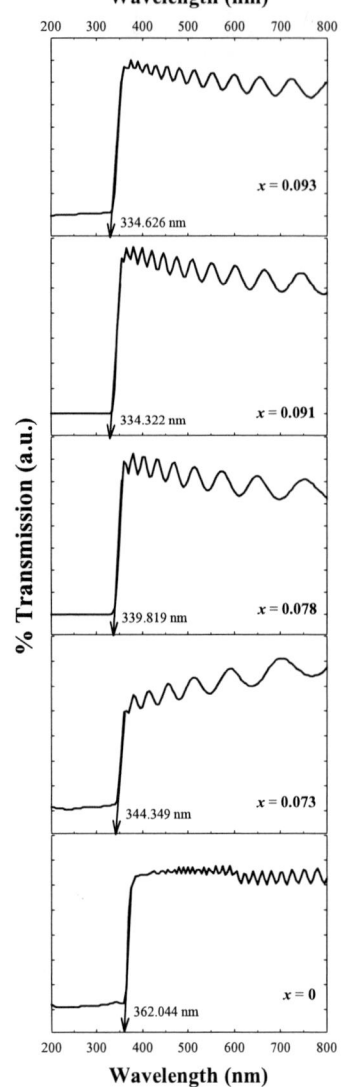

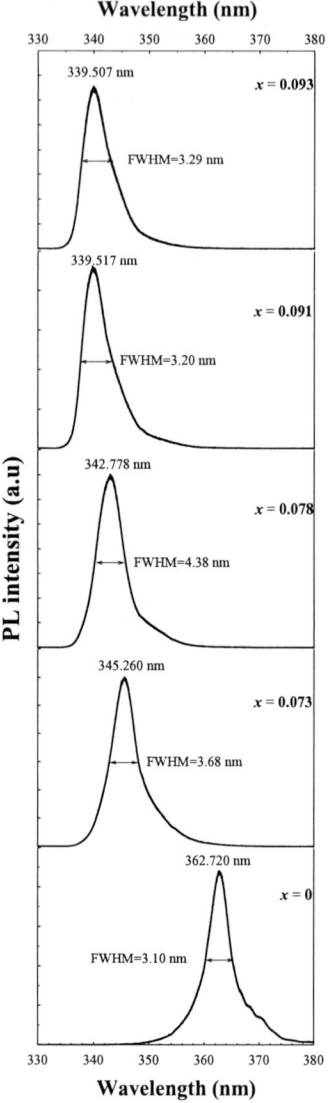

Fig. 1. XRD symmetric RC $\omega/2\theta$ scans of (0002) plane for $Al_xGa_{1-x}N$ samples ($0 \leq x < 0.10$). The scales of the vertical axis for $Al_xGa_{1-x}N$ samples have been adjusted to magnify the RC spectra.

Fig. 2. Room temperature UV-VIS transmission spectra for $Al_xGa_{1-x}N$ samples ($0 \leq x < 0.10$). The intercept of the x-axis shows the energy band gap of the samples.

Fig. 3. Room temperature μ-PL (near band-edge emission) spectra for $Al_xGa_{1-x}N$ ($0 \leq x < 0.10$) samples.

Table I. Summary of the HR-XRD, UV-VIS transmission and μ-PL measurements.

Sample	HR-XRD		UV-VIS transmission		μ-PL	
	c measured (Å)	RC FWHM (arcsec)	Composition x	E_g (eV)	E_g (eV)	PL FWHM (meV)
GaN	5.2125	428	0	3.425	3.419	29.2
$Al_xGa_{1-x}N$	5.1956	970	0.0732	3.601	3.591	38.2
	5.1946	731	0.0775	3.649	3.618	46.3
	5.1915	799	0.0909	3.696	3.652	34.4
	5.1911	728	0.0927	3.706	*3.652	35.4

* This value has been rounded to three decimal and it is actually 0.1 meV.

Overall, the UV-VIS transmission and the μ-PL results are in reasonable agreement and the discrepancy is about 1% (except for the $x = 0$). This uncertainty is most probably due to the Al-contents are not uniformly distributed. Since, the spot size for the μ-PL measurements is about 1-2 μm, hence the obtained results are more localized. Whereas, the beam size for the UV-VIS transmission measurements is about 5 mm × 10 mm, hence, an overall energy band gap is probed. Therefore, the uncertainties are likely present between these measurements.

Since the energy band gap E_g is a fundamental characteristic of the semiconductors, it is important to know its dependence on the material. In general, the energy band gap of the $Al_xGa_{1-x}N$ as a function of the composition, x, can be expressed by the following formula: [12]

$$E_g\left(Al_xGa_{1-x}N\right) = xE_g\left(AlN\right)+\left(1-x\right)E_g\left(GaN\right)-bx\left(1-x\right) \quad (3)$$

where $E_g(AlN)$ and $E_g(GaN)$ are the energy band gap (in eV) of AlN and GaN respectively. b is the bowing parameter and its value represents the deviation from a linear interpolation between the two binary semiconductors AlN and GaN.

Based on the Al-mole fractions evaluated by XRD measurements and the band gap energies determined by UV-VIS transmission and μ-PL methods, the band gap energy for $Al_xGa_{1-x}N$ as a function of Al-mole fraction, x, is plotted within a composition interval of $0 \le x \le 0.10$ and is shown in Fig. 4.

From the best fit of the non-linear interpolation of the UV-VIS transmission and the μ-PL data, a band gap bowing parameter, b, of 14.62 eV is obtained for the $Al_xGa_{1-x}N$ nitride alloys. Consequently, the composition dependence of the E_g of the $Al_xGa_{1-x}N$ alloys in the composition range of $0 \le x < 0.10$, can be expressed as follow:

$$E_g\left(Al_xGa_{1-x}N\right) = 14.62x^2+1.44x+3.42 . \quad (4)$$

From Equation (4), it is notable that the E_g measured from the UV-VIS transmission and μ-PL measurements are smaller than the linearly interpolated value. Apart from that, Equation (4) will enable one to modify the E_g as well as other structural and optical properties of the $Al_xGa_{1-x}N$ systems in a controlled manner. Therefore, it is important for the designing of the optoelectronic devices.

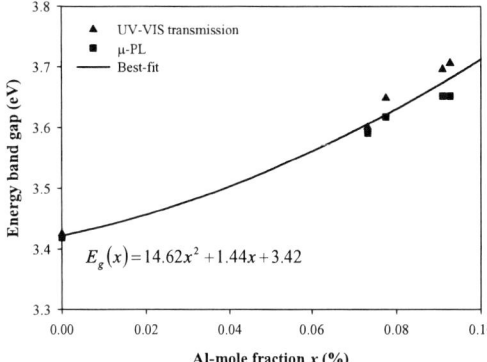

Fig. 3 Band gap energy of the $Al_xGa_{1-x}N$ versus Al-mole fraction in range of $0 \le x < 0.10$. The solid line is the best fit of the non-linear interpolation of the UV-VIS transmission and the μ-PL data.

IV. CONCLUSIONS

The energy band gap of $Al_xGa_{1-x}N$ ($0 \le x < 0.10$) thin films as a function Al-mole fraction, x, had been reported. The Al-mole fraction of the samples was determined by means of high-resolution X-ray diffraction (HR-XRD) spectroscopy. UV-VIS transmission and μ-PL were employed to determine the energy band gap of the samples. Finally, based on the UV-VIS transmission and μ-PL data, the composition dependence of the energy band gap of the $Al_xGa_{1-x}N$ alloys in the composition range of $0 \le x < 0.10$ were obtained.

ACKNOWLEDGMENT

The authors would like to acknowledge Universiti Sains Malaysia and IRPA RMK-8 Strategic Research for their financial support.

REFERENCES

[1] M. S. Shur and M. Asif Khan, "GaN and AlGaN Ultraviolet Detectors," in J. I. Pankove and T. D. Moustakas (eds), *Semiconductors and Semimetals: Gallium Nitride (GaN) II*, vol. 57, Academic Press, Chap. 10 (1999).

[2] D. Gotthold and S. Guo, *"Bringing AlGaN from research to high-volume production,"*

Compound Semiconductor magazine, Art. 3, August (2003).

[3] M. Razeghi, "Short Wavelength Solar-Blind Detectors: Status, Prospects, and Markets," *Proceedings of IEEE, Wide Bandgap Semiconductor Devices: The Third Generation Semiconductor Comes of Age*, 90 (6), pp. 1006-1014 (2002).

[4] R. M. Chu, Y. G. Zhou, J. Liu, D. L. Wang, K. J. Chen and K. M. Lau, "AlGaN-GaN Double-Channel HEMTs," *IEEE Transactions on Electron Devices*, **52**, pp. 438-446 (2005).

[5] A. A. Allerman, A. J. Fischer, M. H. Crawford, S. R. Lee, K. H. A Bogart, C. C. Mitchell, D. D. Koleske, D. M. Follstaedt, P. P. Provencio and N. A. Missert, "GaN to AlN: Materials for Deep-UV Emitters," *Lasers and Electro-Optics Society, LEOS 2003. The 16th Annual Meeting of the IEEE*, **2**, pp. 874 (2003).

[6] E. Munoz, E. Monroy, J. L. Pau, F. Calle, F. Omnes and P. Gibart, "*III nitrides and UV detection*," J. Phys.: Condens. Matter, **13**, pp. 7115-7137 (2001).

[7] S. Kamiyama, M. Iwaya, N. Hayashi, T. Takeuchi, H. Amano, I. Akasaki, S. Watanabe, Y. Kaneo, N. Yamada, "*Low-temperature-deposited AlGaN interlayer for improvement of AlGaN/GaN heterostructure*," J. Crystal Growth **223**, pp. 83-91 (2001).

[8] L. Vegard, "*Die Konstitution der Mischkristalle und die Raumfüllung der Atome*," Z. Phys. **5**, pp. 17-26, (1921).

[9] C. G. Van de Walle, M. D. McCluskey, C. P. Master, L. T. Romano and N. M. Johnson, "*Large and composition-dependent band gap bowing in $In_xGa_{1-x}N$ alloys*," Mater. Sci. Eng., B 59, pp. 274-278 (1999).

[10] H. Angerer, D. Brunner, F. Freudenberg, O. Ambacher, M. Stutzmann, R. Hopler, T. Metzger, E. Born, G. Dollinger, A. Bergmaier, S. Karsch, and H. J. Korner, "*Determination of the Al mole fraction and the band gap bowing of epitaxial $Al_xGa_{1-x}N$ films*," Appl. Phys. Lett. 71, pp. 1504-1506 (1997).

[11] W. Shan, J. W. Ager III, K. M. Yu, W. Walukiewicz, E. E. Haller, M. C. Martin, W. R. McKinney and W. Yang, "*Dependence of the fundamental band gap of $Al_xGa_{1-x}N$ on alloy composition and pressure*," J. Appl. Phys. **85**, pp. 8505-8507 (1999).

[12] K. Osamuraa, S. Naka and Y. Murakami, "*Preparation and optical properties of $Ga_{1-x}In_xN$ thin films*," J. Appl. Phys. **46**, pp. 3432-3437 (1975).

Queue Time Impact on Defectivity at Post Copper Barrier Seed, Electrochemical Plating, Anneals and Chemical Mechanical Polishing

Yasmin Abdul Wahab[1,2], Anuar Fadzil Ahmad[1] and Zaiki Awang[2]

[1] Silterra Malaysia Sdn. Bhd.
09000 Kulim, Kedah, MALAYSIA

[2] Microwave Technology Centre
Faculty of Electrical Engineering
Universiti Teknologi MARA
40450 Shah Alam, Selangor, MALAYSIA
Email: yasmin_wahab@silterra.com, anuar_fadzil@silterra.com , zaiki437@salam.uitm.edu.my

Abstract- **As design rules shrink beyond 0.13μm the development focus has been a gradual shift in the defectivity on copper electroplating integrated circuit manufacturing applications. Effective process inspection and defect identification are key issues for the failure mechanisms in semiconductor manufacturing. In this paper, copper deposition with He in-situ and furnace anneal splits were performed on the Applied Materials SlimCell^{TM} ECP system. The paper outlines the queue time challenges from a defectivity perspective and the solutions implemented that addresses each issue. The analytical techniques used to classify these defects and the methods used to determine their origin is discussed. This paper will attempt to describe the impact of queue time on defectivity challenges and we introduce a new defect characterization scheme that takes the defect generation mechanism and the potential source into account. Further investigation implemented to study the possibility of imposing a time window between seed deposition and plating, plating to anneal duration as well as anneal to CMP in order to posed a significant challenge in differentiating between plating and CMP induced defects. Most defects were observed after chemical-mechanical planarization (CMP) was performed and defects that were generally categorized as missing copper could have resulted from corrosion, from scratches during CMP process from incomplete filling of fine features after plating.**

Index Terms: Copper electrochemical plating, chemical mechanical planarization, He in-situ anneal, furnace anneal

I. INTRODUCTION

With the rapid adoption of Cu metalization in modern microelectronics devices, defectivity is gaining momentum because of the upgrades in metal interconnect processes. Electro copper plating and copper chemical-mechanical polishing (CMP) processes arise gradually in importance as a fabrication process for Cu interconnects. Dual-damascene is accepted as the architecture choice for copper interconnects and has expanded to address several other concerns, including optimization of the copper microstructure, defectivity and elimination of impurities that decrease ramp yields and reduce device reliability. The focus has thus shifted to understanding defects that are new and unique to Cu dual damascene (CuDD). Instead, the identification and elimination of defects formed during process sequence are central to yield improvement requirements. In this paper we will discuss such defects specifically around the metalization process. In the industry standard process flow, the required dual-damascene metalization layout is patterned and etched in a dielectric surface. Thereafter, a barrier metal and Cu seed layer is deposited, followed by Cu fill on the entire wafer surface. Cu CMP is then executed to remove the excess Cu above the recessed patterns and planarize the entire wafer surface [1]. Figs. 1 and 2 show the basic cross section of Cu plating and Cu CMP for back end of line (BEOL) process flow.

0-7803-9730-4/06/$25.00 ©2006 IEEE

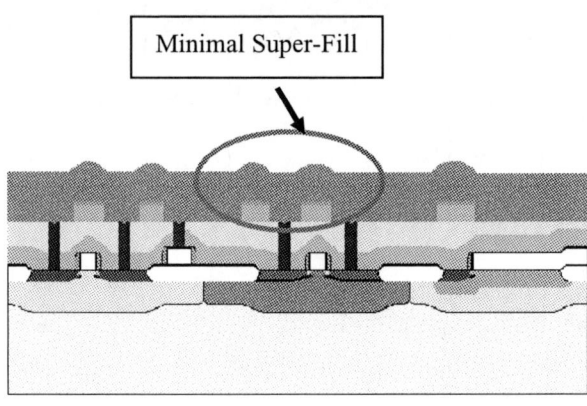

Fig. 1 Cross section of ECP copper fill

This paper investigates the effect on defectivity perspective by imposing a time window between seed deposition and Cu plating as well as plating to anneal duration and CMP process. The motivation of this work was to identify the root cause of various types of defect as a function of the time.

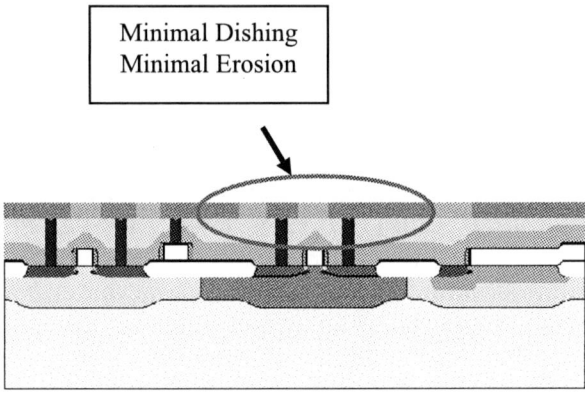

Fig. 2 Cross section of Cu CMP process

II. EXPERIMENTAL DETAILS

Blanket 200 mm Si wafers were used for the inspection of post-plating and post-CMP defects. TaN_x/Ta barrier layers of Cu seed were deposited using a 200 mm Barrier Seed Endura physical vapor deposition (PVD) system that utilizes a hollow cathode magnetron. Pre-stress of seed layer was studied before the plating step. A set of wafers was annealed in separated splits using furnace, and the other split was put through the He in-situ anneal located at the plating system tool with modified recipe. Each pattern of queue time has been implemented in four specific splits in order to get consistent results as shown in Figs. 3 (a), (b), (c) and (d), respectively.

The post-stress and reflectivity of the plated Cu layer were then recorded. As a final stage of defects qualification, a Cu CMP partial polish was employed before both sets of wafers underwent particle scan check prior to plating defect monitoring. Detailed review and classification of all defects on each processed wafers was carried out using a scanning electron microscope (SEM), to build up a defect pareto chart for each wafer. On top of that, SEM images of post-ECP and post-CMP defects were captured and selected defects were examined by SEM images of cross sections created by focused ion-beam (FIB). As a part of the defect inspection and classification system, the classification codes and SEM images for every individual defects were automatically downloaded into a defect database system after the wafers were reviewed. In addition, the number and distributions of the defects at the end of polishing step would be available for post-defect review analysis.

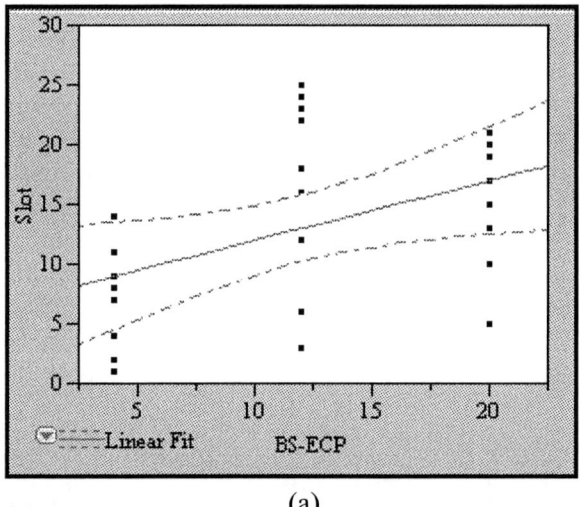

(a)

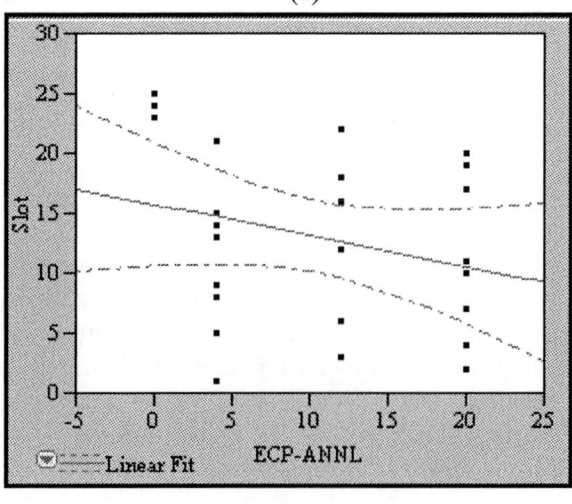

(b)

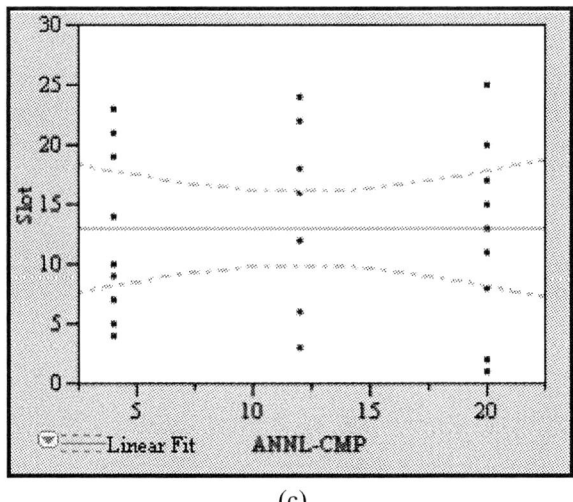

(c)

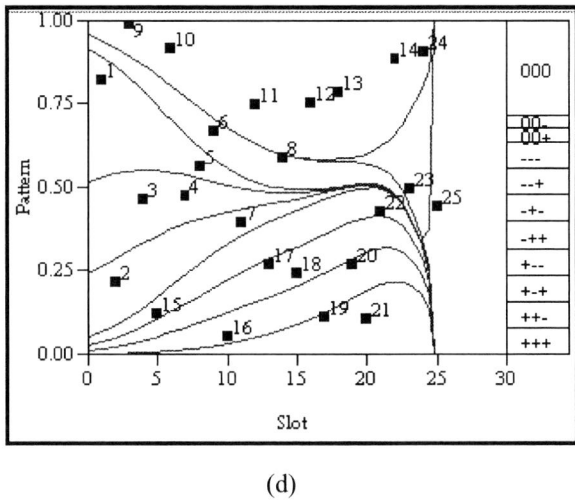

(d)

Fig. 3 Queue time for each wafers (a) barrier seed to copper electroplating (b) copper electroplating to anneal (c) anneal to chemical-mechanical polishing (d) logistic fit of pattern by slot

III. RESULTS AND DISCUSSION

A. Process Parameter Comparison

The results of the process parameter comparison between patterns of queue time are presented in Figs. 4 (a) and (b). Analysis of stress and reflectivity shows a significant difference in both process parameters between the pattern conditions. The results shown in Fig. 4 seem to indicate that high stress in the +++/20-20-20, ++-/20-20-4 and +-+/20-4-20 queue time hours occurred because of the longer time variations in the processes. Actually stress can relax with time through the diffusion of vacancies leading to the formation of voids and ultimately open circuit failures and to control this phenomenon requires optimization of several processes including metal

barrier deposition process, copper annealing and interconnect geometry [2]. Real time stress measurements were performed in order to better understand the mechanisms of stress evolution during thermal treatment. In fact, copper internal stress evolution and stabilization is a crucial aspect of anneals processes.

In integrated circuits, the difference in thermal expansion between the metal film or line and the neighboring materials gives rise to high stress during the fabrication process, resulting in stress-induced damage, such as voids or grooves [3].

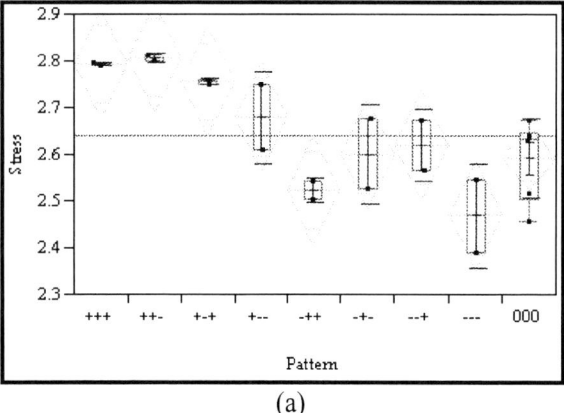

(a)

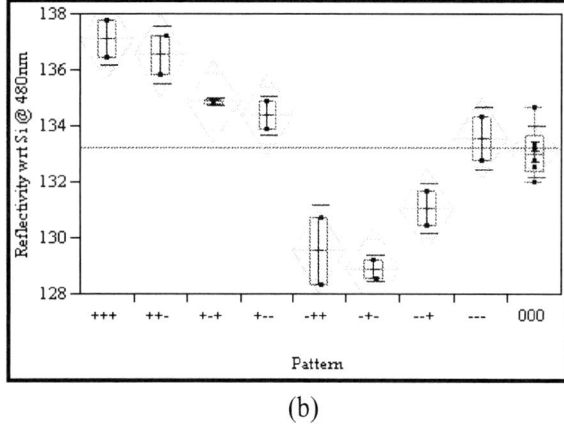

(b)

Fig. 4 Box and Whisker Plot comparison between queue time patterns for furnace splits (a) Analyses of stress by pattern (b) Analyses of reflectivity @ 480 nm by pattern.

Meanwhile, the high reflectivity observed in furnace-annealed samples for +++/20-20-20 and ++-/20-20-4 queue time hours occurred probably due to different grain size and Cu oxidation. The grain size in as-deposited Cu films is small but starts to grow after deposition and this increased further in samples annealed at higher temperatures. Atoms are loosely packed in the grain boundaries, so the grain growth densifies the film and introduces tensile stress. Numerous decreases in the grain boundary energy can compete with the increase in the strain energy and directly influence the growth of the grains against stress [4]. In other words, high densities of grain boundaries and immediate re-crystallization after electrochemical plating lead to unstable grain structures.

B. Defect Comparison Between Queue Time Patterns

Effective process excursion detection and defect source identification are main issues for successful yield management in semiconductor manufacturing and yet defects arise from PVD barrier and seed deposition, electroplating, CMP as well as any thermal treatments that are require to stabilize and anneal copper can results in significant revenue loss. Thus, in-line monitoring of defects has been implemented in this study to ensure the process quality in real-time. Wafer inspection for flaws consists of two stages: optical wafer scanning to detect the presence and total counts of defects; and review/ classification performed at the coordinates of the scanned defect to determine the defect type such as missing metal defect, stain, gouge, void/crater, bump, embedded and surface particle. Trending by total random defects is not sufficient to monitor yield-limiting defects, the so-called "killer" defects, because the mean-shift amount of the killer defect excursion is usually very small (in comparison to the baseline defect variability level) [5]. Fig. 5 shows the total defect count of electroplating and chemical-mechanical planarization as a function of queue time patterns. The defect count of ECP is small compared with CMP total defect and the trends of ECP and CMP defects are identical. It is demonstrated in this instance that the CMP defects associated with ECP defects is high when total defect count is monitored for the longer queue time.

For detail, each type of defect trended separately as shown in Fig. 6 (a) and (b). Trending individual types with no error requires 100% (error free) review/ classification, but in this study on a dense copper layer the very high level of micro scratch defects during polishing creates a very low S/N ratio. This then affects the ability to detect real defects of interest when monitoring by lot average defect density [6].

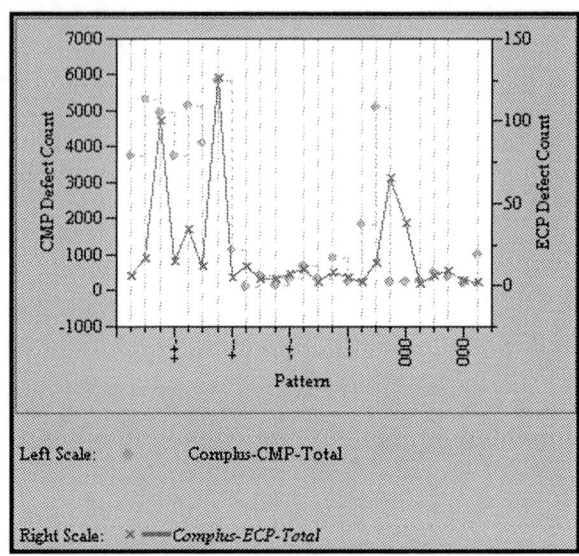

Fig. 5 Correlation between ECP and CMP defect trend.

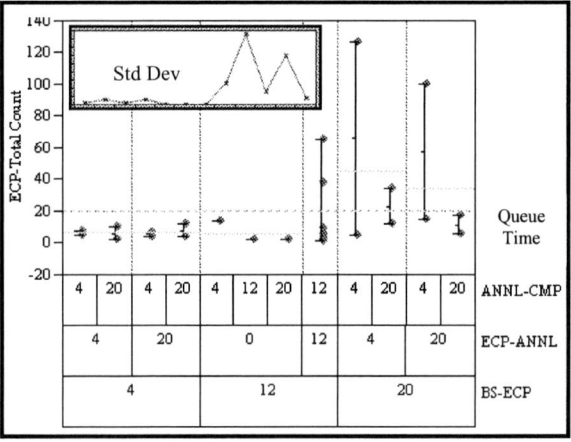

(a)

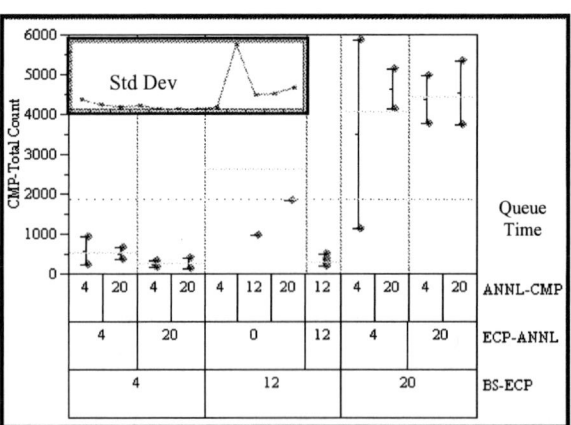

(b)

Fig. 6 Comparison of defect trends for (a) copper electroplating (b) chemical-mechanical planarization

SEM Paretos of post-plating and post-CMP defects for queue time splits are shown in Fig. 7 (a) and (b). While the CMP clean defects and scratches formed a major constituent for longer queue time patterns. The average number of false flaws was decreased after polishing step, and the average number of missing metal defect, gouge and corrosion was identical for both processes, however, the average number of scratches was greatly increased for the CMP stage. Increase in scratches on the wafer surface can be explained by the ploughing action of the abrasive particles as they come into contact with the wafer. Besides, deep scratches can result in the formation of metal residues or puddles on subsequent metal layers, which can then cause an electrical short between metal lines or other structures, leading to reliability problems and yield loss [1].

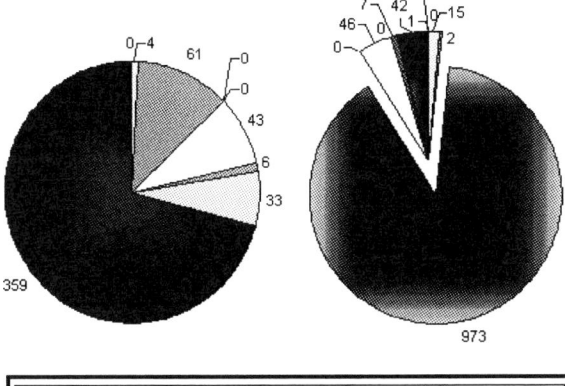

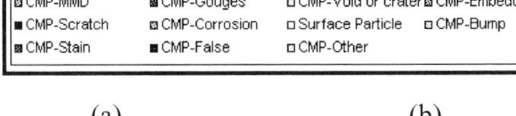

(a)   (b)

Fig. 7 Pie charts of post defects (a) copper electroplating and (b) chemical-mechanical planarization

C. Post-plating Defects

Review and classification types of post-plating defects were performed in detail. Fig. 8 shows the top-view SEM images of the post-plating defects. Poor initial wetting of the seed layer at the onset of deposition, inadequate or incomplete rinsing after plating and sporadic accelerated growth due to the plating chemistry contributes to the existing of defects. Cu plating is normally followed with an anneal step to encourage film stabilization and grain growth, but additional flaws can be created if the anneal temperature is too high. Such defects observed at post-CMP, where Cu metal has migrated through stress away from the damascene features, leaving voids in its place, giving rise to metal defects. Cu seed age or oxidation state contributes to the occurrence of this flaw with longer queue time duration.

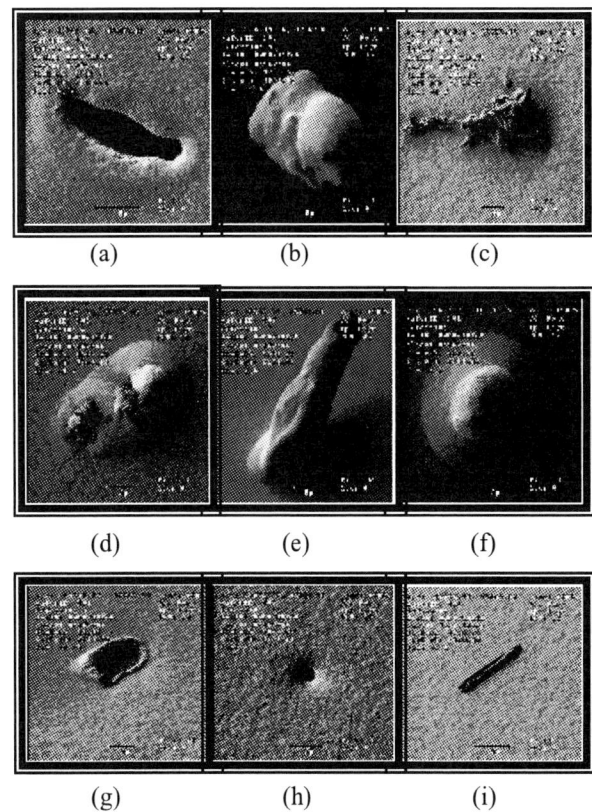

Fig. 8 Top-view SEM images of post-plating defects: (a) missing metal (b) embedded (c) silica particle (d) sub micrometer embedded defect (e) surface particle (f) bump (g) crater (h) void, and (i) peapod

D. Post-CMP Defects

Post-CMP defects, however, can more easily be correlated to yield loss, as scratches, missing metal and remaining metal can be linked to opens and shorts. Fig. 9 shows the top-view SEM images of the post-CMP defects.

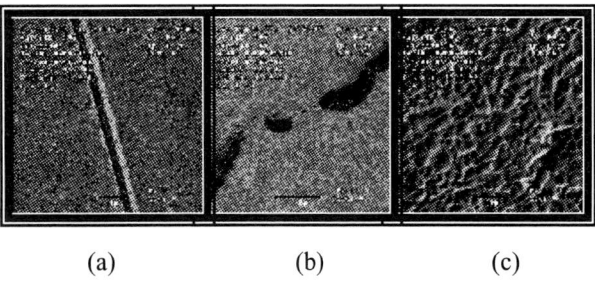

Fig. 9 Top-view SEM images of post-CMP defects: (a) scratch (b) surface particle (c) CMP clean defect

VI. CONCLUSION

A preliminary study of the quality of queue time pattern for Cu films were evaluated in this work by comparing the defects level. This paper discussed the major defect types seen in dual damascene copper CMP technology, with an emphasis on major defects that affect yield. As a conclusion, with longer queue time between seed deposition and plating as well as plating to anneal and polishing duration increased the occurrences of defects for the data points collected. Besides, there is a significant difference in reflectivity and stress between different queue times patterns applied.

IV. ACKNOWLEDGMENT

The authors are grateful to the Thin Film Metal Group of Silterra Malaysia Sdn. Bhd. for their contributions.

VII. REFERENCES

[1] T.Y. Teo and W. L. Goh, "Characterization of Scratches Generated by a Multiplaten Copper Chemical-Mechanical Polishing Process," *J. Vacuum Sc. Technology*, vol. 22, pp. 65-69, January 2004.

[2] G. B. Alers, J. Sukamto, P. Woytowitz, X. Lu, S. Kailasam, J. Reid, "Stress Migration and the Mechanical properties of Copper," in *Proc. 2005 IEEE International Reliability Physics Symposium,* pp. 1-5, April 2005.

[3] Hyun Park, Soo-Jung Hwang and Young-Chang Joo, "Stress-induced Surface Damage and Grain Boundary Characteristics of Sputtered and Electroplated Copper Thin Films," *Journal of Acta Materialia*, vol. 52, pp. 2435-2440, 2004.

[4] A. Uedono, T. Suzuki and T. Nakamura, "Vacancy-type Defects in Electroplated Cu Films Probed by using a Monoenergetic Positron Beam," *J. App. Phys.*, vol. 95, pp. 913-918, February 2004.

[5] Wataru Shindo, Eric H. Wang, Ramakrishna Akella, Andrzej J. Strojwas, Wanda Tomlinson and Richard Bartholomew, "Effective Excursion Detection by Defect Type Grouping in In-Line Inspection and Classification," in *IEEE Trans. on Semiconductor Manufacturing*, vol. 12, No. 1, pp. 3-10, Feb.1999.

[6] Peter Stoeckl, Barry Saville, Jim Kavanagh, Thilo Dellwig, "Advanced Cu CMP Defect Excursion Control for Leading Edge Micro-Processor Manufacturing," in *Proc. 2002 13th Annual*

IEEE/SEMI Advanced Semiconductor Manufacturing Conf., Santa Clara, USA, pp. 92-97, 2006.

Analysis of the Output Noise Voltage in CMOS Image Sensor Readout Circuit

Youngchang Yoon, *Student Member, IEEE,* Hochul Lee, Byung-Gook Park, *Member, IEEE,*
Jong Duk Lee, *Member, IEEE* and Hyungcheol Shin, *Senior Member, IEEE*
School of Electrical Engineering and Computer Science
Seoul National University #059
San 56-1, Shillim-dong, Kwanak-ku, Seoul 151-744, Korea
Email: y2chang2@snu.ac.kr

Abstract **This work helps to predict the output noise voltage of the CMOS Image Sensor read out circuit based on analytic equation and noise measurement results. With this result, the relative portion of the low frequency noise and thermal noise was calculated. As a result, this work can give an idea for reducing noise voltage magnitude of the readout circuit.**

I. INTRODUCTION

CMOS Image Sensors have performance competitive with charge-coupled device for low cost, and low power. However, CIS suffer from the presence of noise [1]. We have derived an analytic equation for noise calculation of the new CMOS Image Sensor readout circuit. Similar work was done in previous paper [2], but the readout circuit in this paper is different from the former one. With the derived equation, the output noise voltage can be calculated and the relative contributions to output noise from thermal noise and low frequency noise can be obtained.

II. RESULTS AND DISSCUSSION

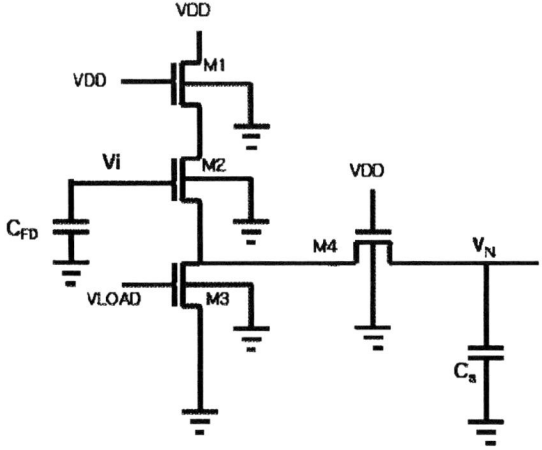

Fig. 1.   4-Tr APS circuit of CMOS Image Sensor

The 4-Tr CMOS Active Pixel Sensor (APS) is shown in Fig. 1.

When reading the signal voltage v_i, the presence of noise voltage on the output node V_N makes reading exact signal value difficult. Both thermal noise and low frequency noise from the MOSFETs sets a fundamental limit on APS performance, especially for extreme low-light applications.

In Fig. 2, the small signal equivalent circuit for the 4-Tr APS in Fig. 1 to analyze output noise characteristic is shown. Because the sources of M1 and M2 are not tied with body, the substrate transconductance (g_{mb}) as well as conductance (g_m) and output resistance (r_o) are considered. The noise current sources (i_{n1}, i_{n2}, i_{n3}, and i_{n4}) contain both thermal and flicker noise components.

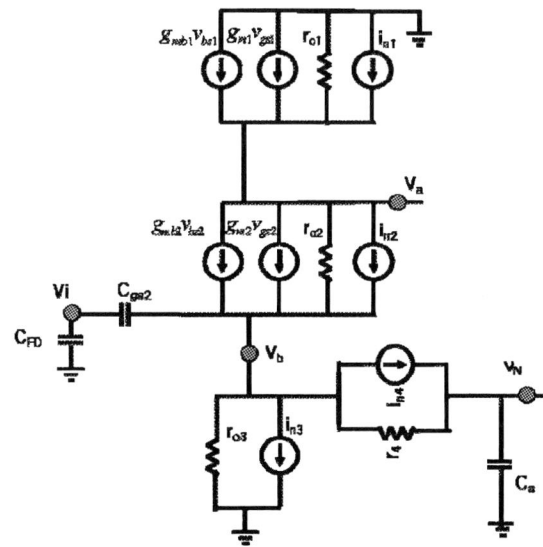

Fig. 2.   Small signal equivalent circuit for noise analysis. i_{n1}, i_{n2}, i_{n3}, and i_{n4} represent noise current source including low-frequency noise and thermal noise.

Based on the small signal equivalent circuit, the analytical equation (1) for noise voltage can be derived through the nodal equation. Because the equation (1) which is shown at the bottom of the page is too complex, it is hard to use the full equation to understand noise property of readout circuit.

It is necessary to make complicated equation simple for better understanding. More concise equation can be obtained by comparing the magnitude of the parameters. The following equation (2) is the simplified form of the full equation.

$$v_N = \frac{\dfrac{R_{eq1}}{r_{o2}}i_{n1} + \left(1 - \dfrac{R_{eq1}}{r_{o2}}\right)i_{n2} - i_{n3} + r_4\left[\dfrac{C_{fb}}{C_{fb}+C_{gs}}g_{m2} + g_{mb2} + \dfrac{1}{r_{o2}} + \dfrac{1}{r_{o3}}\right]i_{n4}}{\left[\dfrac{C_{fb}}{C_{fb}+C_{gs}}g_{m2} + g_{mb2} + \dfrac{1}{r_{o2}} + \dfrac{1}{r_{o3}}\right] + s\left[\dfrac{r_4 C_s}{R_{eq1}}\right]} \quad (2)$$

$$\left(R_{eq1} = \left(g_{m1} + g_{mb1} + \dfrac{1}{r_{o1}} + \dfrac{1}{r_{o2}}\right)^{-1},\right.$$
$$\left. R_{eq2} = \left(g_{m2} + g_{mb2} + \dfrac{1}{r_{o2}} + \dfrac{1}{r_{o3}} + \dfrac{1}{r_4}\right)^{-1}, C_{eq} = \dfrac{C_{fb}C_{gs}}{C_{fb}+C_{gs}}\right)$$

It is necessary to confirm that equation (2) is comparable to equation (1). Fig. 3 verifies that the results from the simplified equation are almost the same with original equation. As a result, this simple equation is very useful in understanding the noise characteristic and it can give more insight to improve noise performance.

As it can be identified from Fig. 3, transfer function coefficients of M1 and M4 are much smaller than the others such as M2 and M3. The effects of those transistors compared to M2 and M3 are negligible. Consequently, it is acceptable to focus on the noise from the M2 and M3.

The low frequency noise from M2 is dominant since the size of M2 is much smaller than M3. It is well known that the low frequency noise power spectrum is inversely proportional to the gate area. Consequently, it is acceptable to consider only M2 transistor with respect to low frequency noise.

Fig. 4 shows the low frequency noise measurement results of M2. The low frequency

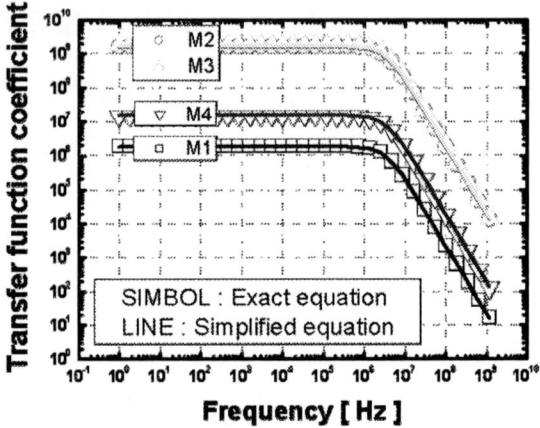

Fig. 3. Noise transfer function with respect to frequency. Symbol is for the exact equation and line is for the simplified equation.

noise measurement system consists of cascade probe station, low noise current amplifier (SR570), and dynamic signal analyzer (HP35670A)[3]. Due to small gate area of M2, there is significant variation between devices. The low frequency noise of the MOSFETs can be explained mainly by the carrier number fluctuations and mobility fluctuations which are described by oxide charge tunnel trapping and de-trapping processes. The oxide trap cannot exist with complete uniformity in terms of both physical and energy aspects. As a result, it is obvious that there are more variations with oxide trap distribution in the case of smaller gate area devices [4]. As a result, it is necessary to measure as many devices as possible for better reliable result. 20 devices were measured under the same bias condition for reasonable results. As you can see on the Fig. 4, there are lots of Lorentzian shapes of power spectrum density which means random telegraph noise due to very small size device.

For thermal noise calculation, thermal noise model ($S_{id-thermal}=4kT\gamma g_{do}$) is used and g_{do} was obtained by DC measurement. Only M2 and M3 are considered for thermal noise due to the magnitude of transfer function coefficients.

$$v_N = \frac{i_{n1} + \left(\dfrac{r_{o2}}{R_{eq1}} - 1\right)i_{n2} - \dfrac{r_{o2}}{R_{eq1}}i_{n3} + \left[\dfrac{r_{o3}\cdot r_4}{R_{eq1}\cdot R_{eq2}} - \dfrac{C_{eq}}{C_{fb}}\dfrac{r_{o2}}{R_{eq1}}g_{m2}r_4 - r_4\left(g_{m2}\dfrac{C_{eq}}{C_{gs}} + g_{mb2} + \dfrac{1}{r_{o2}}\right) - \dfrac{r_{o2}}{R_{eq1}}\dfrac{r_4}{R_{eq1}} + s\dfrac{r_{o2}}{R_{eq1}}r_4 C_{eq}\right]i_{n4}}{\left[\dfrac{r_{o2}}{R_{eq1}\cdot R_{eq2}} - \dfrac{C_{eq}}{C_{fb}}\dfrac{r_{o2}}{R_{eq1}}g_{m2} - g_{m2}\dfrac{C_{eq}}{C_{gs}} - g_{mb2} - \dfrac{1}{r_{o2}} - \dfrac{r_{o2}}{R_{eq1}}\dfrac{1}{r_4}\right] + s\left[\dfrac{r_{o2}}{R_{eq1}}C_{eq} + \dfrac{r_{o2}r_4 C_s}{R_{eq1}\cdot R_{eq2}} - \dfrac{r_{o2}}{R_{eq1}}\dfrac{C_{eq}}{C_{fb}}g_{m2}C_s r_4 - C_s r_4\left(\dfrac{C_{eq}}{C_{gs}}g_{m2} + g_{mb2} + \dfrac{1}{r_{o2}}\right)\right] + s^2 \dfrac{r_{o2}}{R_{eq1}}C_{eq}C_s r} \quad (1)$$

$$\left(R_{eq1} = \left(g_{m1} + g_{mb1} + \dfrac{1}{r_4} + \dfrac{1}{r_{o2}}\right)^{-1}, R_{eq2} = \left(g_{m2} + g_{mb2} + \dfrac{1}{r_{o2}} + \dfrac{1}{r_{o3}} + \dfrac{1}{r_4}\right)^{-1}, C_{eq} = \dfrac{C_{fb}C_{gs}}{C_{fb}+C_{gs}}\right)$$

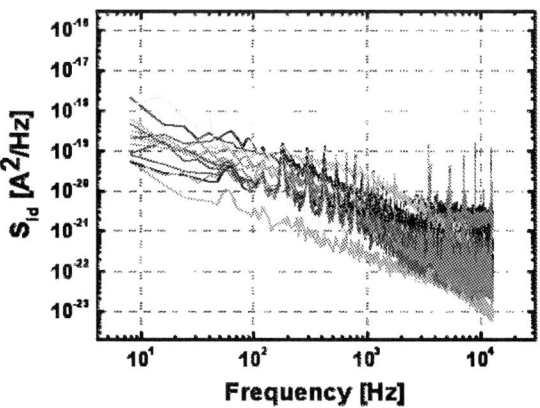

Fig. 4. Low frequency noise measurement of M2. 20 devices are measured under the same bias condition.

With above results, the output noise voltage is calculated by integrating the output noise power spectrum density. The calculated value of noise voltage is about 0.2 mV. Fig. 5 shows the relative portion of noise components which consist of thermal and flicker noise. The contribution of flicker noise is about 5 times lager than thermal noise. This means that flicker noise reduction is more important for total noise improvement. This is coincident with previous papers [5][6].

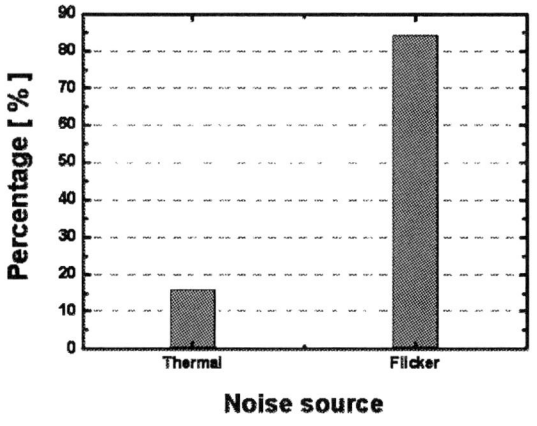

Fig. 5. Percentage of the each noise source, low frequency noise and thermal noise.

III. CONCLUSION

This work predicts the output noise voltage of the CMOS Image Sensor read out circuit based on analytic equation and noise measurement results. 20 devices were measured under the same bias condition for acceptable results. With considering both thermal and flicker noise, the output noise voltage is about 0.2 mV. This work can give an idea that which one is more important for reducing noise voltage of the readout circuit among the thermal and flicker noise.

ACKNOWLEDGEMENT

This work was supported by Samsung Electronics Co. Ltd. and the MIC(Ministry of Information and Communication), Korea, under the ITRC(Information Technology Research Center) support program supervised by the IITA(Institute of Information Technology Advancement)(IITA-2006-C109006030030).

REFERENCES

[1] Eric R. Fossum, "CMOS Image Sensor: Electronic Camera-On-A-Chip," IEEE Trans. Electron Devices, vol. 44, pp.1689-1698, 1997.

[2] Yavuz Degerli, Francis Lavernhe, Pierre Magnan, Jean A. Farre, "Analysis and Reduction of Signal Readout Circuitry Temporal Noise in CMOS Image Sensors for Low-Light Levels," IEEE Trans. Electron Devices, vol. 47, pp.949-962, 2000.

[3] Youngchang Yoon, Hocheol Lee, In Man Kang, Hyungcheol Shin, "Low frequency noise characteristics of the 180nm MOSFETs", IEEK Fall Conference 2005, Seoul, Korea, pp.861-864, Nov. 26, 2005.

[4] A. J. Scholten, L. F. Tiemeijer, R. v. Langev- elde, R. J. Havens, V. C. Venezia, "Noise Modeling for RF CMOS Circuit Simulation", IEEE Trans. Electron Devices, vol.50, no.3, pp.618-632, 2003

[5] Jung Yeon Kim, Sung In Hwang, Jong Jin Lee, Ju Hyun Ko, Yitae Kim, Jung Chak Ahn, Tetsuo Asaba, Yong Hee Lee, "Characterization and improvement of random noise in 1/3.2" UXGA CMOS image sensor with 2.8 um pixel using 0.13 um-technology," 2005 IEEE Workshop on Charge-Coupled Devices and Advanced Image Sensors, pp.149-152, 2005.

[6] Jongwan Jung, Jeong-Ho Lyu, Hwangyoon Kim, HyunWoo Lee, Je-Hyuck Song, Youngsub You, Hyunpil Noh, Duckhyung Lee, Kinam Kim, "Reduction of Random Noise for CMOS Image Sensor with 2.2 μm × 2.2 μm Pixel," 2005 International Conference on Solid State Devices and Materials, Kobe, Japan, pp.340-341, 2005.

Electrical Characteristics of 100 MeV ^{28}Si implantation in GaAs

Yousuf Pyar Ali[a], A.M.Narsale[b], Othman Sidek [a], A.R.Damle[b], B.M.arora[c]

[a]School of Electrical and Electronics Engineering, USM, 14300 Nibong Tebal, Penang, Malaysia.
[b]Department of Physics, University of Mumbai, Vidyanagari, Mumbai 400 098, India.
[c]Tata Institute of Fundamental Research, Homi Bhaba Road,Mumbai,400 005, India
Email: ypyar@yahoo.com

Abstract - **Single crystal n-GaAs substrates have been implanted at room temperature with 100 MeV ^{28}Si ions to a dose of 1×10^{18} ions/m^2. The electrical behaviour of these samples has been investigated after implantation and annealing to 850 °C by current voltage (I-V) measurements. The I-V curves show series of complex behaviours with annealing treatments. To understand this complex behaviour, Resistance measurements of these samples using I-V measurements were carried out in the temperature range 100-300 K, which indicate that the as implanted sample and samples annealed to 350 °C are dominated by a variable range hoping conduction mechanism, where as for the samples annealed at 450 °C and 550 °C the electrical conduction is due to hopping between the neighboring defect sites. The electrical transport for the sample annealed at 650 °C seems to be dominated by carriers in the extended states. At annealing temperature higher than 650 °C, the I-V characteristics are insensitive to measurement temperatures which indicates that the backward diode like structure after 850 °C annealing is due to the activation of Si ions and formation of n$^+$ region at the mean ion range and the existence of defect complex p$^+$-type conductivity immediately above that region.**

I. INTRODUCTION

ION implantation is a well established technology in the fabrication of a wide variety of devices in GaAs.[1]. Early implantation work are done mainly in the implantation energy range up to few MeV for the fabrication of devices such as mixer, varactor diode, p-i-n diode which requires dopant to be placed a few µm below the surface of the semiconductor [2]. The extension of the implantation energy to higher energy range (> 10 MeV) raises an interesting question about the radiation induced defects and their annealing behaviour. The damage introduced by energetic ions along their tracks can form an amorphous layer below the surface which needs to be removed by an annealing process. As such, the annealing process for such high energy implantation seems to be more complicated than that for implantation at low energies [3]. In this paper we attempt to understand the change in electrical characteristics of high energy ^{28}Si implanted in GaAs substrates due to radiation induced defects and effect of annealing on them.

II. EXPERIMENT

The samples used in this experiment were mirror polished <100> silicon doped n- type GaAs substrates with background doping concentration of 2×10^{16} cm^{-3} and having an area of 7 mm x 7 mm and thickness of 400 µm. All the samples have been carefully cleaned and then implanted at room temperature using ^{28}Si ions at energy of 100 MeV to a fluence of 1×10^{18} ions/m^2 in a non channeling direction using NEC 16 MV pelletron accelerator at the Nuclear Science Center, New Delhi, India. Cleaning procedure of the samples and implantation details were described elsewhere [4]. The implanted samples were proximity capped annealed for 10 min at different temperatures up to 850 °C in pure hydrogen ambient. The caped samples were used as reference to monitor any surface degradation. Back ohmic contacts were made by evaporating a uniform coating of Au-Ge-Ni. The top ohmic contacts were made by evaporating 0.0045 cm^2 area dots of Au-Ge-Ni through a metal mask. The contacts were then alloyed at 450 °C for 1 min in pure hydrogen. Ohmic contacts were made before implantation for the samples to be annealed at temperatures less than 450 °C. The current voltage (I-V) measurements of the as implanted samples and samples annealed at different temperatures were carried out over a temperature range (100-300 K) by using a programmable voltage source,

0-7803-9730-4/06/$25.00 ©2006 IEEE

Keithley digital electrometer (model 617) and a variable temperature cryostat.

III. RESULTS AND DISCUSSION

Figure 1 shows room temperature current voltage (I-V) characteristics for the as implanted samples and samples annealed up to 850 °C. The I-V characteristic for the as implanted sample shows highly non linear diode like behaviour (Fig. 1a). After annealing at 100 °C the characteristics become weakly non-linear (Fig. 1b). The samples annealed between 150 °C – 550 °C (Fig. 1c) show fairly linear I-V characteristics. Up to 350 °C annealing the current values are in the range of micro-ampere. Between 450 °C - 550 °C annealing

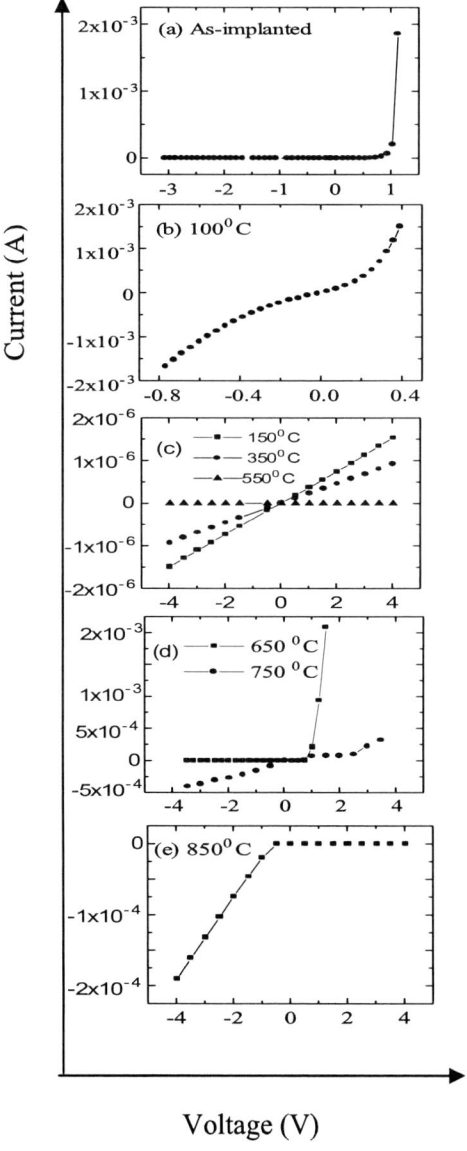

Fig. 1 Room temperature I-V curves for different annealing temperatures.

the current level reduces to nano- ampere range. The sample annealed to 650 °C again show a non linear diode like I-V characteristics and by further annealing to 750 °C the sample show leaky diode type curve (Fig. 1d). After annealing to 850 °C the sample shows a back ward diode like I-V characteristics. The I-V curves of the control unimplanted samples remain ohmic and do not show any significant change of resistance. This suggests that there is no degradation of the materials during annealing process and the complex behaviours are therefore attributed to the annealing process of the implanted layer.

In order to understand the complex behaviour of the as implanted and annealed samples, we estimate the equivalent effective series resistance from various linear and nonlinear I-V curves. For weakly and strong nonlinear I-V characteristics the effective series resistance is estimated from the high current region where the series resistance is dominant. It has been observed room temperature resistance of the as implanted sample is about 100 Ω, which increase with increasing annealing temperature and reaches a maximum values of about 2×10^9 Ω at 550 °C. Further annealing at 650 °C, causes a drastic decrease in the resistance value to 100 Ω which again increases to about 10^4 Ω after annealing at 850 °C.

Temperature dependence of resistance of the as-implanted sample and the samples annealed at different temperatures are examined next. It is observed that the as implanted sample and the sample annealed up to 350 °C satisfy the relation log R $\propto$ T$^{-1/4}$ in the measurement temperature range (110 K-300 K) as shown in Figure 2. These observations suggest that up to 450 °C annealing, there remains large concentration of defect states in the implanted samples. As a result, the conductivity mechanism of these samples in the low temperature range is dominated by variable range hopping between defect energy levels in the forbidden gap and the result may be described by [5]:

$$\rho = \rho_0 \exp(T_0/T)^{1/4} \qquad (1)$$

The values of T_0 are obtained from the slopes of log R vs. T$^{-1/4}$ curve and are given in Table I. We can estimate the localized states density at Fermi level N(E$_F$) according to:

957

$N(E_F) = (C^4 \alpha^3 / T_0 k)$, where, $C^4 = 20$
$\alpha (cm^{-1}) = (2m^*/\hbar^2)^{1/2} (E_g/2)^{1/2}$ is the attenuation distance of the wave function for the localized state, E_g is the band gap and m^* is the effective mass of the electron.

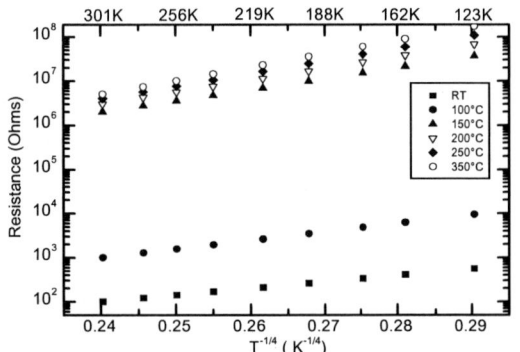

Fig. 2 Resistance vs. $T^{-1/4}$ of the as implanted samples and for samples annealed up to 350 °C.

The values of N (E_F) for the as-implanted sample and the samples annealed at temperatures up to 350°C are listed in Table I and can be assigned to the vacancy clusters. It is observed that the $N(E_F)$ values decrease with the increase in the annealing temperature indicates that the high concentration of these defect states decreased with the increase in annealing temperature.

Table I. T_0 values and corresponding $N(E_F)$ for different annealing temperatures.

Annealing Temperature	T_0 (K) x 10^7	$N(E_F)$ $(cm^{-3}eV^{-1})$x10^{19}
As implanted	0.14	29.0
100 °C	0.40	10.0
150 °C	1.15	3.5
200 °C	1.49	2.7
250 °C	1.91	2.1
350 °C	2.51	1.6

For the samples annealed at temperatures 450 °C and 550 °C, the values of log R in the measurement temperature range (110 K-300 K) were examined as a function of T^{-1}. We find these plots are linear as shown in Figure 3. These observations and the fact that resistance values are high suggest that the conduction mechanism is due to the motion of trapped electrons by hopping between neighboring defect sites and the resistivity can be described by [6]:

$$\rho = \rho_0 \exp(E_a/kT) \qquad (2)$$

Where ρ_0 is the resistivity extrapolated to infinite temperature and E_a is the activation energy.

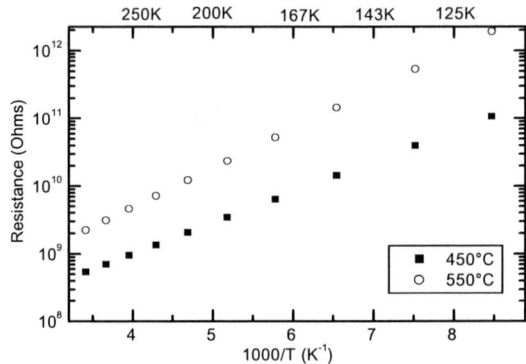

Fig. 3 Resistance vs reciprocal temperature for samples annealed at 450 °C and 550 °C.

The E_a values are calculated from the slopes of Fig. 3 and found to be 90 meV and 115 meV for 450 °C and 550 °C annealed samples respectively. This energy is attributed to the energy needed for motion of trapped electrons by hoping between closely damage sites.

On further annealing the samples to 650 °C the resistance value decreased drastically which indicates a reduction in the defect concentration, and the electrical transport of this sample is dominated by carriers in the extended states for which the resistivity of the sample can be expressed by [7]

$$\rho(T) = \frac{1}{N^* T^{3/2} q \mu(T) \exp\{-(E-E_F)/kT\}} \qquad (3)$$

where, $N^* = 2\{2\pi m_n \hbar^2\}^{3/2}$, m_n is the effective mass of the charge carrier, E is the energy of the band edge, μ is the mobility, q is the electron charge, E_F is the Fermi energy, T is the absolute temperature, k is the Boltzman's constant and $\hbar$ is the Planck's constant.

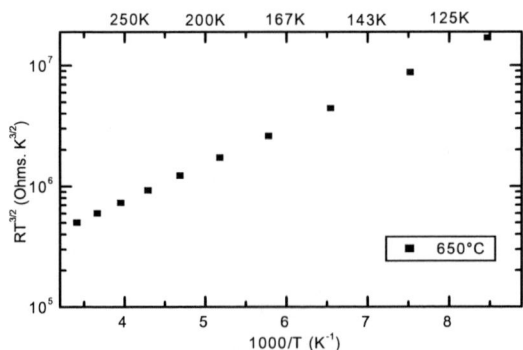

Fig. 4 RT$^{3/2}$ vs 1/T for sample annealed at 650 °C.

Figure 4 shows the plot of log (RT $^{3/2}$) vs 1/T for the sample annealed at 650 °C and the(E - E_F) value of this sample calculated from slope of the plots of log (RT$^{3/2}$) vs. 1/T was found to be 60 meV.

The resistance values of the sample annealed at 750 °C show a slight variation with measurement temperature and this sample seems to be in a transition state from diode like to backward diode like behaviour. The I-V characteristics of the sample annealed to 850 °C are insensitive to measurement temperature as shown in Fig 5.

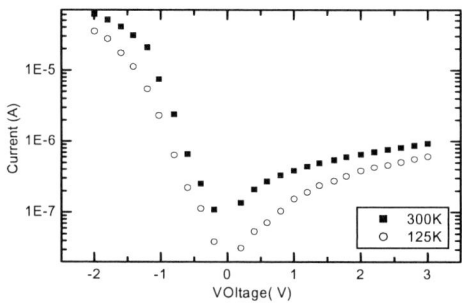

Fig.5 I-V characteristics of the sample annealed to 850 °C and measured at 300 K and 125 K.

which conform the formation of the backward diode like structure. This structure may be due to activation of Si ions and formation of n^+ region at the mean range of the implanted Si ions, where as p^+ conductivity due to the impurity defect complex induced by implantation remains immediately above this region with p-type conductivity extending up to the surface of the sample and the electrical transport in this case is dominated by a band to band tunneling due to p^+-n^+ junction [8].

IV. CONCLUSIONS

We have implanted ^{28}Si ions in single crystal n-GaAs substrates at energy of 1000 MeV. The electrical behaviour of the implanted samples have been studied by I-V characteristics of the ohmic contact devices after implantation and after annealing to different temperatures up to 850 °C.   The I-V curves at room temperature for the as implanted samples and samples annealed at different temperatures show a series of complex behaviours. The temperature dependence of the as-implanted sample and the samples annealed upto 350°C seems to follow variable range hopping conduction at low temperatures (100 K-300 K), while the conduction mechanism of the samples annealed

at 450 °C and 550 °C is dominated by hopping between the neighbouring defect sites. The electrical transport for the sample annealed to 650 °C is dominated by carriers in the extended states. Further annealing of the samples to 850 °C results in to a backward diode I-V characteristics due to the formation of n^+ region at the mean range of the implanted ions and p+ defective region before the mean range region. Further higher annealing steps are required to fully to activate the doping of the implanted layer.

ACKNOWLEDGEMENT

The authors are thankful to all the concerned scientific staff of the Nuclear Science Centre New Delhi –India and Tata Institute of Fundamental Research – India for their help.

REFERENCES

[1] S.J. Pearton, "Ion Implantation for Isolation of III-V Semiconductors" Mater. Sci. Rep. Vol.4, pp. 313-367 (1990).

[2] P.E. Thompson, " MeV Ion Implantation in GaAs Technology," Nucl. Instrum. Methods Phys. Res. B. vol. 59/60, pp. 592-599 (1991).

[3] J. Asher, " MeV Ion processing applications for Industry," Nucl. Instrum. Methods Phys. Res. B. vol. 89, pp. 315-321 (1994).

[4] Y.P. Ali, A.M. Narsale, U. Bhambhani, A.Damle, V.P.Salvi, B.M.Arora, A.P.Shah and D.Kanjilal, "Electrical characteristics of GaAs implanted with 70 MeV ^{120}Sn ions," Nucl. Instrum. Methods Phys. Res. B. vol.117, pp. 129-133 (1996).

[5] N.F .Mott and E.A. Davis, "Electronic Process in Non-crystalline Materials, Clarendon Press, Oxford, pp. 7-49 (1971).

[6] Y. Kato, T. Shimada, Y. Shiraki and K.F. Komatsuba, "Electrical Conductivity of Disordered Layer in GaAs Crystal Produced by Ion Implantation," J. Appl. Phys. Vol. 45 no. 3, pp. 1044-1049 (1974).

[7] B. G. Streetman, Solid State Electronics Devices, 3rd Edition. Prentice Hall, Englewood Cliff, NJ, pp. 76-79 (1990).

[8] S.M. Sze, Physics of Semiconductor Devices, 2nd Edition , Wiley Eastrn Ltd ,pp. 537-540. (1981).

The Study of Pt/porous GaN Schottky Contact for Hydrogen Sensing

F. K. Yam[a], Y. P. Ali[b], Z. Hassan[a], N. H. Mohd. Noor[a] and C. W. Chin[a]

[a]School of Physics, Universiti Sains Malaysia, 11800, Penang, Malaysia

[b]School of Electrical and Electronics Engineering, Universiti Sains Malaysia, 14300 Nibong Tebal, Penang, Malaysia.

Email: yamfk@yahoo.com, ypyar@yahoo.com

Tel: 604-6533599, Fax: 604-6579150

Abstract- **This article presents the studies of Pt Schottky contact on porous n-type GaN for hydrogen sensing. Porous GaN was generated by UV assisted electroless chemical etching. Hydrogen sensor was subsequently fabricated by depositing Pt Schottky contacts onto the porous GaN sample. For comparative study; a standard hydrogen sensor was also prepared by depositing Pt Schottky contacts on the as-grown sample using same processing tools and under identical parameters. Hydrogen detection was carried out at room temperature in an enclosed chamber. Pt/porous GaN sensor exhibited a significant change of current upon exposure to 2% H_2 in N_2 as compared to the standard Pt/GaN sensor. Morphological studies by scanning electron microscopy (SEM) revealed that Pt contact deposited on porous GaN have a very rough surface morphology with pores distributed all over the contact layer. Therefore, the steep increase of current could be attributed to the unique microstructure at porous Pt/porous GaN interface, which allowed higher accumulation of hydrogen and eventually led to stronger effect of the H-induced dipole layer.**

I. INTRODUCTION

POROUS semiconductors have been widely studied in the last decade which are motivated by the prospects that they present in optoelectronics, chemical and biochemical sensors [1,2]. Porous semiconductors successfully prepared from Si, GaAs, GaP and InP exhibit tunable properties that can be explored for fabrication of new sensors.

Among porous semiconductors, porous silicon receives enormous attention and has been investigated intensively. However, the non-ideal thermal, mechanical and chemical properties have prevented it from large scale application [3]. This leads to the development of other porous semiconductors, for instances, the conventional III-V compound including III-V nitrides. The research in porous GaN is strongly driven by the superior physical properties such as the excellent thermal, mechanical and chemical stability relative to Si.

Recently, the development of chemical sensors has received considerable attention due to the great demand for industrial and environmental applications. Detection of hydrogen gas is important and has potential industrial applications. Since hydrogen gas is extremely explosive, a small leakage could eventually lead to a devastating state in hydrogen processing and other related industries. Therefore, it is highly desirable to develop a sensitive and reliable sensor, which is able to detect the leakage instantaneously.

To our best knowledge, the use of porous GaN for gas sensing has not been reported before. Although gas sensors based on other porous low bandgap semiconductors such as Si [4] and GaAs [5] have been demonstrated, however, they always suffer from undesirable background optical and thermal excitation.

In this work, porous GaN was used to fabricate Pt Schottky contact gas sensor for hydrogen detection. For comparative study, a reference gas sensor was also fabricated on the as-grown sample using same processing tools and under identical parameters. The initial study shows a very encouraging result for Pt/porous GaN to use for hydrogen sensing.

II. EXPERIMENT

a)The generation of porous GaN

The unintentionally doped n-type GaN film grown on sapphire substrate was used in this study. The thickness of GaN film is about 3.0

0-7803-9730-4/06/$25.00 ©2006 IEEE

μm with carrier concentration of $\sim 4.4 \times 10^{17}$ cm^{-3} as determined by Hall Effect measurement. Prior to the metallization, the native oxide of the sample was removed in the $NH_4OH:H_2O =1:20$ solution, followed by $HF:H_2O =1:50$, subsequently boiling aqua regia ($HCl:HNO_3 = 3:1$) was used to etch and clean the sample. Porous GaN in this work was generated by Pt assisted electroless etching. Pt with thickness of about 150 nm was first deposited on the GaN sample by using sputtering system. The sample was then etched in a solution of $HF:CH_3OH:H_2O_2 = 4:1:1$ under illumination of an UV lamp with 500 W power. After one hour etching, the sample was removed from the solution and rinsed with distilled water, followed with the removal of the residual Pt by ultrasonic cleaning.

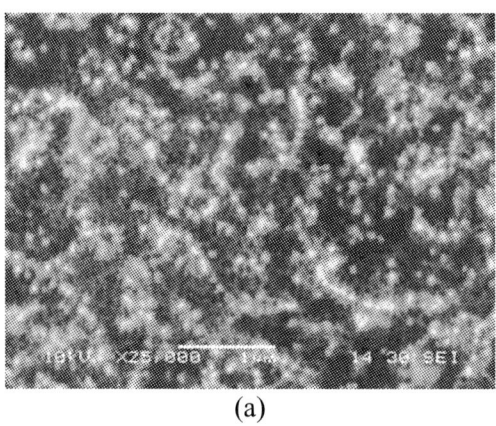

(a)

(b)

Fig. 1 SEM images of (a) porous GaN, (b) as-grown GaN

Figure 1 shows the images of scanning electron microscopy (SEM) of the as-grown and porous GaN. From the SEM micrographs, the GaN sample after etching for one hour was found to have a highly textured structure with pores distributed all over the area. On the other hand,

the as-grown GaN possessed a smooth surface morphology

b. Fabrication of gas sensor:

To fabricate gas sensor, Pt Schottky contacts with thickness of 150 nm was deposited onto the as-grown and porous samples via a metal mask which consists of an array of holes with diameter of 0.9mm. Both Pt Schottky contacts were used for probing in this study. Pt was chosen because of its catalytic characteristic which is suitable for gas sensing.

The experiments were carried out by using a home made gas sensing chamber. A mixture of 2% H_2 gas in N_2 gas was used in this experiments. Current voltage (I-V) measurements at room temperature were carried out by mounting the samples on a test fixture with probes making contact to both of the Pt contacts. The test fixture was placed into the chamber with wires connected from the probes to the Keithley digital electrometer (model 237).

III. RESULTS AND DISCUSSION

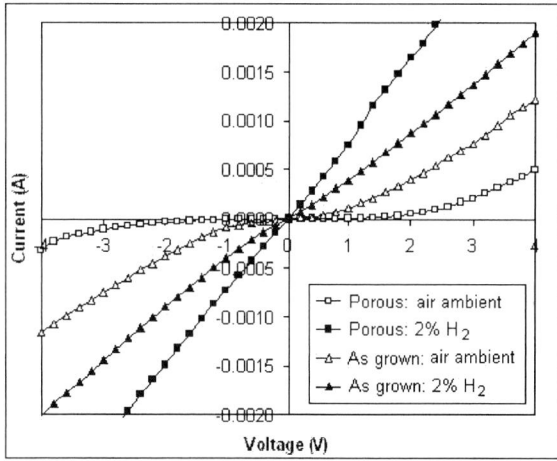

Fig. 2 The I-V characteristics of Pt/porous GaN and Pt/as-grown GaN gas sensors at room temperature.

Figure 2 shows the I-V characteristics of gas sensors operating at room temperature. The square and triangle symbols represent the porous and as-grown samples, and solid and open symbols represent the sensors operated in 2% H_2 and air, respectively. Both samples exhibited Schottky characteristic, when they were operated in air, however, a good Schottky behaviour was observed for porous GaN as compared to as-grown sample. The Schottky behaviour of both samples changed to ohmic when 2% H_2 was flowed into the chamber, these suggested that the reduction of barrier height upon the introduction of 2% H_2, particularly the porous GaN, the current was found to increase significantly (at a voltage of 2V).

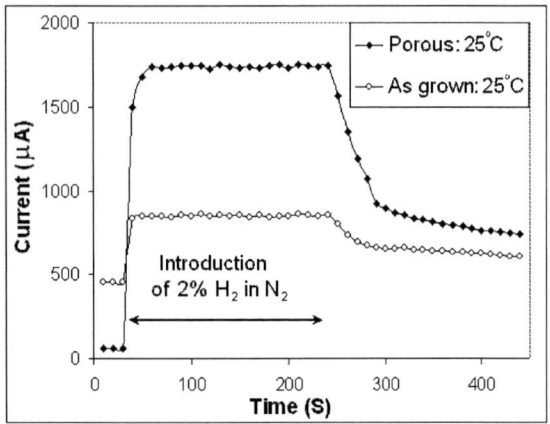

Fig. 3 On-off responses of the sensors measured at room temperature for a constant voltage of 2V.

Figure 3 shows the on-off responses of the sensors measured at room temperature for a constant voltage of 2V. For porous sample, the current was found to increase sharply upon exposure to 2% H_2, however, the turn-off response was relatively slow, and current was found to be relatively high after 2% H_2 was shut off for 200s. For as grown sample, the change of current in the on-off response was relatively insignificant when compared to porous sample. The surface morphology of Pt contacts could be used to explain these phenomena.

Sensitivity is one of the important parameters and is normally used to gauge the performance of a gas sensor. The hydrogen detection sensitivity, S, is defined as [6]

$$S = \frac{I_{H_2} - I_{Air}}{I_{Air}}$$

where I_{H_2} and I_{Air} are current level in H_2 containing ambient and air ambient, respectively. At constant voltage of 2V, the sensitivity of porous and as-grown sensors at room temperature are 28.97 and 0.86, respectively. It is about 33.68 times more sensitive than as-grown sample at room temperature.

It is widely accepted that the hydrogen sensing mechanism starts with the dissociation of hydrogen molecules on the Pt surface, forming atomic hydrogen. These hydrogen atoms will diffuse through the Pt layer and adsorb at the Pt/GaN interface. H-induced dipole layer subsequently will be formed which leads to the change in the effective Schottky barrier height and thereby the change in the electrical characteristics of the device [7,8]. Therefore surface morphology or/and microstructure of Pt contact could play an important role in determining the sensitivity of the sensor.

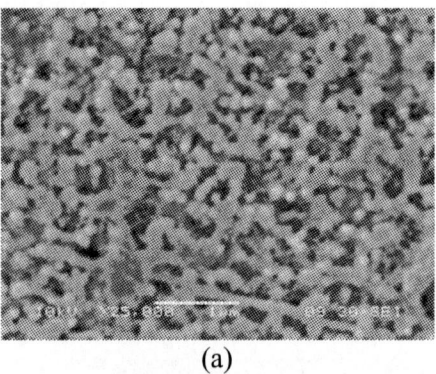

(a)

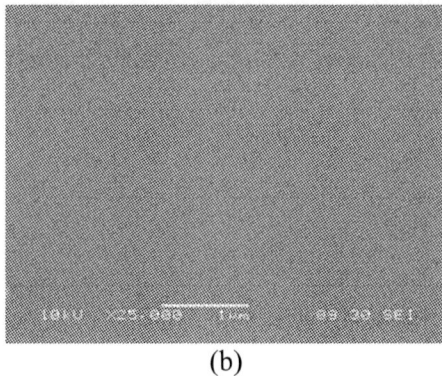

(b)

Fig. 4 SEM images of the Pt contact surface. (a) Pt contact on porous GaN sample, (b) Pt contact on as-grown sample.

Figure 4 shows the SEM images of the Pt contacts formed on porous and as-grown samples, respectively. The surface morphology of Pt contacts of these two samples was found to be very much different. Rough surface morphology with pores distributed all over the contact layer was found for the Pt contact layer deposited on porous GaN, in contrast, smooth surface was observed for Pt contact deposited on as-grown sample. Obviously, the surface morphology of the Pt contacts followed closely the surface profile of the samples. The porous surface of Pt contact means larger surface to volume ratio this allows hydrogen molecules to dissociate and form atomic hydrogen more efficiently. Moreover the unique surface morphology offers higher accumulation of hydrogen at porous Pt /porous GaN interface, and the effect of the dipole layer could become stronger. This explains the steep increase of current of the porous sample upon exposure to the 2% H_2 as compared to the as-grown sample.

IV. CONCLUSIONS

We have fabricated and investigated the use of Pt/GaN and Pt/porous GaN Schottky contact for the detection of hydrogen gas. The Pt/porous GaN Schottky contact demonstrated a dramatic change of current upon exposure to 2% H_2 as compared to the Pt/as-grown GaN Schottky contact. Morphological studies by SEM showed that Pt contact layer deposited on porous GaN have highly textured surface morphology with high density of pores. The steep increase of current could be attributed to the unique microstructure at porous Pt/porous GaN interface which allowed higher accumulation of hydrogen; subsequently produced larger electrical polarization, this led to the lowering in the effective Schottky barrier height and eventually caused a significant change in the electrical characteristics of the porous GaN sensor.

ACKNOWLEDGEMENT

This work was conducted under an IRPA RMK-8 Strategic Research grant. The support from Universiti Sains Malaysia is gratefully acknowledged.

REFERENCES

[1] L. T. Canham, "Silicon Quantum Wire Fabrication by Electrochemical and Chemical Dissolution of Wafers" Appl. Phys. Lett. vol. 57, pp. 1046-1048 (1990).

[2] K.D. Hirschman, L. Tsybeskov, S.P. Duttagupta and P.M. Fauchet, "Silicon-based Visible Light-Emitting Devices Integrated Into Microelectronic Circuits," Nature. vol. 384, pp. 338-340 (1996).

[3] P. M. Fauchet, L. Tsybeskov, C. Peng, S. P. Duttagupta, J. von Behren, Y. Kostoulas, J. V. Vandyshev, and K. D. Hirschman, "Light-Emitting Porous Silicon: Materials Science, Properties, and Device Applications" IEEE Jour. Selected Topics in Quantum Electron. vol 1, pp. 1126-1139 (1995).

[4] K. Luongo, A. Sine and S. Bhansali, "Development of a highly sensitive porous Si-based hydrogen sensor using Pd nano-structures" Sens. Actuators B vol. 111/112, pp. 125- 129 (2005)

[5] A. Salehi, A. Nikfarjam and D. J.Kalantari, "Pd/porous-GaAs Schottky contact for hydrogen sensing application" Sens.Actuators B vol. 113, pp. 419-427 (2006).

[6] Chen, H-I., Chou, Y-I., Chu, C-Y. "A novel high-sensitive Pd/InP hydrogen sensor fabricated by electroless plating" Sens. Actuators B vol. 85, pp. 10-18 (2002).

[7] B. P. Luther, S. D. Wolter and S. E. Mohney, "High temperature Pt Schottky diode gas sensors on n-type GaN" Sens. Actuators B vol. 56, pp. 164-168 (1999).

[8] Jun-Rui Huang, Wei-Chou Hsu, Yeong-Jia Chen, Tzong-Bin Wang, Kun-Wei Lin, Huey-Ing Chen and Wen-Chau Liu, "Comparison of hydrogen sensing characteristics for Pd/GaN and Pd/Al0.3Ga0.7As Schottky diodes" Sens. Actuators B vol. 117, pp. 151-158 (2006).

The Characterization of KrF Photoresists and the Effect of Different Chromophore Bulkiness on Line Edge Roughness (LER) for Submicron Technology

Ahmad Yusri Mohamed Bakri, Mohd. Jeffery Manaf, Kader Ibrahim Abdul Wahab and
Ibrahim Bin Ahmad, *Member, IEEE*
Silterra Malaysia Sdn. Bhd.
Kulim Hi-Tech Park
09000 Kulim, Kedah, MALAYSIA
Email: ahmad_yusri1@silterra.com

Abstract **This research characterizes line edge roughness (LER), determines which resist has lowest LER for all process variations, and investigates the effect of chromophore bulkiness on LER. Three KrF photoresists with different chromophore bulkiness were evaluated. The characteristics evaluated were depth of focus (DOF), profile and resolution, LER, exposure latitude, iso-dense bias and CD linearity. Different feature sizes were tested from 100nm to 190nm. From the results, it is seen that resist P1 has the lowest average LER for all process conditions and variations with a 3 sigma value of 10.074. This is followed by resist P5 and P6 with a 3 sigma LER value of 12.562 and 15.468. It is concluded that high chromophore bulkiness results in high UV activation. This is seen from the LER for resist P6 that is the highest out of all the photoresist. Reducing the chromophore bulkiness will reduce LER until it reaches a saturation point where reduction will not result in any lower LER. Reducing the chromophore bulkiness further beyond the saturation point will in fact increase the LER.**

I. INTRODUCTION

DESIGN requirements for successive generations of very large-scale integrated circuits (VLSI) have resulted in the reduction of lithographic critical dimensions (CD). Photoresist capabilities have played an important part in the reduction of CD.

For chemically amplified resists (CAR), photon absorption does not drive the chemical process, but is instead the catalyst for the process that will continue in the consequent steps. Unlike DNQ-novolac resists, CAR resists do not use photoactive compounds (PAC), but photo-acid generators (PAG) as its active component. When photoresist is exposed to light, it absorbs UV light, which decomposes of PAG. This causes a small amount of acid to be formed throughout the resist. This acid causes a chain-reaction of chemical transformations in the resist film, which mostly happens during the post-exposure bake (PEB). Because of this, the temperature and duration of the PEB plays an important part in the exposure and development process.

As the CD is reduced, the resolution and profile become more critical to ensure optimum device performance. One method of comparison and quantization for the resolution of a printed image is line edge roughness (LER). There are many factors that are theorized to affect the LER of a profile. One factor is the bulkiness of the chromophore in the PAG component of the CAR.

A chromophore is a group of molecules capable of absorbing light. Chromophores with different bulkiness have different characteristics. It is known that larger chromophores result in stronger light absorbing properties, due to the larger molar absorptivities of the chromophore itself. How this affects the resolution and line edge roughness of the resist is unknown.

II. EXPERIMENTAL

In this experiment, three photoresists with different chromophore bulkiness were compared experimentally. The three photoresists are labeled P1, P5 and P6. All samples were separately coated over a film of DUV44 bottom anti-reflective coating (BARC) and onto 200mm silicon wafers. All chemicals were applied by a wafer track for processing and exposed on a 248nm scanner. The illumination used for exposure was annular. All resist-coated wafers were developed and measured at a CD-SEM scanning electron microscope.

First, a focus exposure matrix (FEM) was done using all photoresists. This is done for several purposes. The first is to find suitable dose and focus to produce the required size features. The second is to determine the depth of focus (DOF) of the resist used. Finally, the FEM is used to analyze the resolution and profile of the photoresist.

The next series of experiments is to analyze line edge roughness (LER) with different geometry dimensions. Four wafers per resist sample are then processed and exposed with the specific conditions to produce optimized CD values for each feature size. These wafers are measured for CD and line-edge roughness at the specified locations.

Table 1. Experiment Split.

Experiment	Variables	Output
Focus Exposure Matrix (FEM)	Dose and Focus	DOF, Profile and Resolution
LER vs. Feature Size 1	Optimum Dose and Focus, 100nm Feature Size	Line Edge Roughness (LER)
LER vs. Feature Size 2	Optimum Dose and Focus, 110nm Feature Size	Line Edge Roughness (LER)
LER vs. Feature Size 3	Optimum Dose and Focus, 120nm Feature Size	Line Edge Roughness (LER)
LER vs. Feature Size 4	Optimum Dose and Focus, 130nm Feature Size	Line Edge Roughness (LER)

Experimentation on the photoresist comprised of evaluating 2 areas of interest. The areas evaluated were the process conditions and resist components. The variation on process conditions was done in Silterra Malaysia Sdn. Bhd. while the variation and characterization of resist components were done at the material company's facilities.

The experiments by the material company are done using different processing and measurement tools. The capabilities of these tools are different, and as such, the feature sizes focused on by the material company are slightly smaller. The characterization experiments that were done include focus exposure matrix (FEM), focus margin, exposure latitude margin, line edge roughness (LER), iso-dense (ID) bias and CD linearity.

III. RESULTS

From the FEM that was done the data collected were plotted as shown by the following Bossung curves (fig.1). From these graphs, certain properties can be observed. The properties that can be characterized are depth of focus (DOF) and photosensitivity.

DOF and photosensitivity can be quantized and has been summarized in table 2.

Table 2. DOF and photosensitivity of all photoresist samples as extracted from Bossung curve plots.

Dim.	Resist	CD (15.5mj)	DOF (nm)	DOF Comment	Photo-sensitivity
100nm	P6	180nm	20		
	P1	110nm	10	P6>P1,P5	P6>P5>P1
	P5	150nm	10		
110nm	P6	210nm	30		
	P1	155nm	10	P6>P1,P5	P6>P5>P1
	P5	180nm	10		
120nm	P6	225nm	30		
	P1	165nm	20	P6>P1,P5	P6>P5>P1
	P5	205nm	20		
130nm	P6	245nm	40		
	P1	230nm	30	P6>P1,P5	P6>P5>P1
	P5	225nm	30		

From table 2 we can see that photoresist P6 has better depth of focus (DOF) for all feature sizes. We can also see that photoresist P1 and P5 have roughly the same DOF for the dimensions measured. However, it has to be remembered that the wafers measured were patterned on bare wafers, and has a flatter topography than production wafers. This increases the DOF. The same tests done on production wafers will yield less favorable results.

From the exposure dose and corresponding critical dimension (CD), it can be seen that photoresist P6 is the most photosensitive while P1 is the least photosensitive. This is can be seen by the CD for a photoresist at a particular exposure dose. For example, at the 100nm feature size, for the exposure dose of 16mj, the CD for photoresists P6, P1 and P5 are 180nm, 110nm and 150nm respectively. This is true for all feature sizes measured where photoresist P6 has consistently higher CD than P1 and P5. This proves that photoresist P6 is the most photosensitive, while photoresist P1 is the least sensitive. Photosensitivity can also be seen from the distance between the Bossung curves. For photoresist P6, the distance between Bossung curves are larger when compared to the distances for other photoresist. This shows that for P6, the critical dimension (CD) increases by a larger amount when the exposure dose is increased 1mJ, as compared to the other photoresist.

From the experiments done, the following pictures of top-down profiles were taken (fig. 2). Top-down profiles can flesh out the information that's gathered from focus exposure matrix (FEM) tests. From top-down pictures, you can determine two things: the shape of the edges and the presence of photoresist flaws such as scumming. The pictures show that the profiles between all photoresist through the dimensions measured was all rather similar without any noticeable difference. All had similar edges and the absence of any scumming or other flaws. However, from the profiles of photoresist P6, it can be seen that the CD of the images taken do not correspond with the required feature sizes. Due to the photosensitivity of the resist, the CD increases greatly with an increase of 1mj energy dose. This instability makes it difficult to find CD that corresponds to the required dimensions, so the closest approximations are used instead.

From the top-down profiles, it can be seen that the profile noticeably deteriorates when going from a larger feature size to a smaller feature size. The line edge roughness (LER) can clearly be seen getting worse. From the profiles, it can be said that the photoresist with the best profile and resolution is P6.

From the variability chart of LER vs. features/photoresist (fig. 3) it can be seen that the range of data for each photoresist is small and focused for resist P1, and gets larger and more spread for resists P5 and P6. This can also be seen from the standard deviation graph, which steadily increases from P1 to P6.

A one-way analysis is done on the LER uniformity data to find any particular trend for LER versus the feature measured or the photoresist used (fig. 4). From the resulting graphs, it can be seen that for all feature sizes, photoresist P6 has the highest LER while photoresist P1 has the lowest LER. There is a trend for the LER value when comparing all feature sizes where, as the feature size increases, the LER decreases. The trend is inversely proportional. For each feature

size increment of 10nm, there is an average LER decrement of 1.5 3 sigma.

From the tests done by the photoresist manufacturer in Korea, the range of dimensions measured by the resist manufacturer is larger due to different testing and measurement tools used. The method of testing was also different as the resist manufacturers used mask features that had only one CD, and varied exposure dosage to achieve the different dimensions. The features measured had dimensions ranging 130nm to 190nm.

From the results of the focus margin tests (fig. 5) it can be seen that there is no trend for focus margin for all three resist. Resist P6 has the highest DOF for the 130nm and 140nm feature sizes. The DOF for P6 at these sizes are 0.2um and 0.5nm respectively. At the 150nm and 190nm feature sizes, resist P5 has the better DOF performance with a DOF of 0.7 and 0.8 respectively. Resist P1 has medium DOF, tying with resist P5 with 0.8um DOF for the 190nm feature size.

It is seen from the top-down profiles (fig. 6) that there are small differences in profile between the photoresists. All photoresist at a particular feature size show similar line edge profiles and visible LER. However, there is slight geometry deterioration for smaller feature sizes. It can be seen that photoresist P6 has slightly better resolution and profile over photoresist P1 and P5.

Exposure latitude is the range of exposure energies where the critical dimension (CD) or linewidth is kept within the specified limits (fig. 7). The exposure energy is expressed as a percent variation from the nominal. For example, for resist P1 with the feature size of 140nm, the specified limit of acceptable linewidth is 140nm +/- 10nm. The nominal dose to achieve 140nm critical dimension is 28.0mJ. An EL margin of 5.70% means that the target of 140nm +/- 10nm will still be achieved even though the exposure dose varies by 28.0mj +/- 5.70%, or 28.0mJ +/- 1.596mJ. A bigger EL margin is desirable as it means that the process window to still achieve the target CD is bigger.

The results show that smaller feature sizes have smaller EL margins. This means that the process windows to achieving smaller features are smaller and more difficult. This trend is linear. As feature sizes get smaller, tighter process controls are needed to achieve the desired results. Resist P6 consistently has higher exposure latitude when compared to the other resists. Resist P1 has the lowest exposure latitude, except for the 150nm feature size, where resist P5 has the lowest exposure latitude. For the smaller dimensions, the difference in exposure latitude between the three resists is very small. This shows that for smaller dimensions, all resist have smaller exposure latitudes.

The LER results (fig. 8 (a)) will be discussed more thoroughly in the discussion part of this paper. However, the LER values are substantially less than the LER values measured in SilTerra. This is expected due to the larger feature size used.

Iso-dense (ID) bias is the difference between the dimensions of an isolated line and a dense line printed on the wafer, while holding all other parameters constant. From the results for the 190nm feature size (fig. 8 (b)), it can be seen that photoresist P6 has the best ID bias with a value of 33.2nm, while photoresist P1 has the worst ID bias with value 20.1nm.

For CD linearity it is desirable for the graph to be straight and linearly increasing, and to be within the prescribed limits. This shows that for a constant exposure dose, the printed DICD will be the same as the on-mask CD. For the given results (fig. 9), it is seen that photoresist P6 has good CD linearity for all feature sizes from 130nm to 220nm. Both photoresist P1 and P5 are applicable for all targeted feature sizes from 130nm to 220nm. However, the CD linearity for these two resist for smaller feature sizes such as 130nm and 140nm are undesirable, unless the exposure dosage is changed significantly.

IV. DISCUSSION

From the comparison table of the DOF results (table 3) and the corresponding chart (fig. 10), it can be seen that there is no consistent trend for photoresist type vs. depth of focus. Photoresist P6 has the best DOF for feature sizes from 100nm to 140nm, but has the worst DOF for the 150nm and 190nm feature size. For larger feature sizes as those measured at Donjin, resist P5 has better DOF. Photoresist P1 ties with resist P5 for having approximately the best DOF for larger feature sizes around 190nm.

From the comparison table of LER uniformity (table 4), measurement was done at both SilTerra

and Donjin. Feature size measured for LER at the two places are different, with Silterra measuring for 100nm to 130nm dimension and Donjin measuring for 140nm feature sizes.

From the results and the graph plotted from the data (fig. 11), it can be seen the results from the two locations are compatible. Measurements taken at Silterra show that there is an inversely proportional trend. As feature sizes increase, the LER decreases. This is true for all the resist types measured. From figure 11, we can see that resist P1 has the lowest average LER, and the most constant in terms of LER variation. Resist P6 has the highest LER, but it reduces drastically when the feature size measured increases from 100nm to 140nm.

From the overall results summary (table 5), it can be seen that each photoresist has good results in different areas. For the main focus of this paper, which is line edge roughness (LER), the best resist is P1, with an overall LER of 10.74 for all features measured. The resist with worst LER is P6, with a LER of 17.78 for all feature sizes measured.

Despite having the largest LER, resist P6 has excellent results in other areas, such as DOF for small features, resolution, exposure latitude and CD linearity. Its performance in these areas is the best out of the three resists.

Resist P5 has the best iso-dense bias out of the three resists, which is 20.1nm. However it has the worst resolution and CD linearity. In other targeted characteristics, its performance was the middle out of the three photoresist.

V. CONCLUSION

Photoresist P6, P1 and P5 are all variants of the same basic photoresist. The difference between the resist types is the bulkiness of the chromophore component in the photo acid generator (PAG). Other components such as polymer, solvent and additives are kept constant. Basically, photoresist P6, P1 and P5 has three different PAG components. The PAG for P6 has the most bulky chromophore while the PAG for P5 has the least bulky chromophore (fig. 12).

From the results that have been gathered, we can conclude that different chromophore bulkiness have different effects on certain characteristics of a photoresist.

Greater bulkiness of the chromophore results in the photoresist having greater exposure latitude at both small and larger feature sizes. As explained, exposure latitude is when the exposure dose in adjusted by increasing or decreasing the energy, yet the CD stays the same. This was represented by resist P6, and shows that for bulkier chromophore, the PAG becomes less sensitive towards energy dose changes, allowing the resist to withstand grater process variation while maintaining the critical dimension.

More bulky chromophore also results in better resolution and CD linearity. These characteristics directly relate to the photoresists ability to reproduce the mask images on the wafer. The good resolution implies that the bulkier chromophore helps the photoresist to clearly define the image. The CD linearity also implies that a bulkier chromophore is able to accurately transfer the required image from the mask to the wafer without adjusting the process energy, from small to large feature sizes. From the results, it can also be seen that bulky chromophore sizes also result in bad line edge roughness and the largest iso-dense bias out of the three photoresist.

For the smaller iso-dense bias, the bulkiness has to be kept small, as represented by photoresist P5. The best aspect of resist P5 is its ability to print dense and isolated patterns on the wafer with the smallest dimension change. This shows that PAG with less bulky chromophores are the least affected by the optical proximity effect (OPE). Unfortunately, less bulky chromophores result in the worst resolution and CD linearity.

For the focus characteristic of this paper (LER), it can be seen that the photoresist with the most bulky chromophore (P6) also has the highest LER. The photoresist with the least bulky chromophore (P5) also has high LER, second out of the three. The photoresist with the least LER is photoresist P1, with a mean LER of 10.74. This implies that having very bulky chromophores result in the photoresist having very high LER. Reducing the bulkiness of the chromophore in the PAG will reduce the line edge roughness. The optimum chromophore bulkiness will result in the least line edge roughness. However, if the bulkiness of the chromophore is reduced too much, the line edge roughness will increase again.

It is determined that increased chromophore bulkiness will result in greater light absorbing capabilities for the PAG. From the resist P5, we can see that the characteristics of increased light absorbing ability results in a PAG that has good resolution, high exposure latitude and good CD linearity. From this we can conjecture that increased chromophore bulkiness results in a PAG that is slow to react to light radiation. It is able to absorb more light before reacting to the radiation and becoming a catalyst for the continuing exposure process during the post-exposure bake (PEB). This creates a more stable but slower process. This results in good resolution, but high line edge roughness.

In conclusion, the best photoresist in terms of LER is photoresist P1, which has medium chromophore bulkiness.

ACKNOWLEDGEMENT

The authors would like to thank Donjin Semichem for providing samples and experimental data for this work. The authors are also indebted to Mr. Kim Young Ki for providing inspiration and guidance during the study of this work.

REFERENCES

[1] H.J Levinson, *Principals of Lithography,* SPIE Press, New York NY, 2001

[2] J.R Sheats, *Microlithography,* Marcel Dekker Inc, New York NY, 1998

[3] H. Xiao, *Introduction to Semiconductor Manufacturing Technology,* Prentice-Hall, New York NY, 2001

[4] P.V.Zant and *Microchip Fabrication: A practical guide to Semiconductor Processing,* McGraw Hill Publishing Company, New York, NY, 1990.

[5] J.V. Krivello, S.Y. Shim, "Deep UV photoresists based on poly(dl-*tert*-butyl fumarate)," *Journal of Polymer Science Part A: Polymer Chemistry*, Volume 33, Issue 3, pp.513-523, Mar. 2003.

[6] P.B. Sahoo, R. Vyas, M. Wadhwa and S. Verma, "Progress in deep-UV photoresists," *Bull. Mater. Sci.,* Vol. 25, No. 6, pp. 553-556, Nov. 2002.

[7] P.V. Zant, *Microchip Fabrication: A practical guide to Semiconductor Processing*, Mc-Graw Publishing Company, New York, NY, 1990.

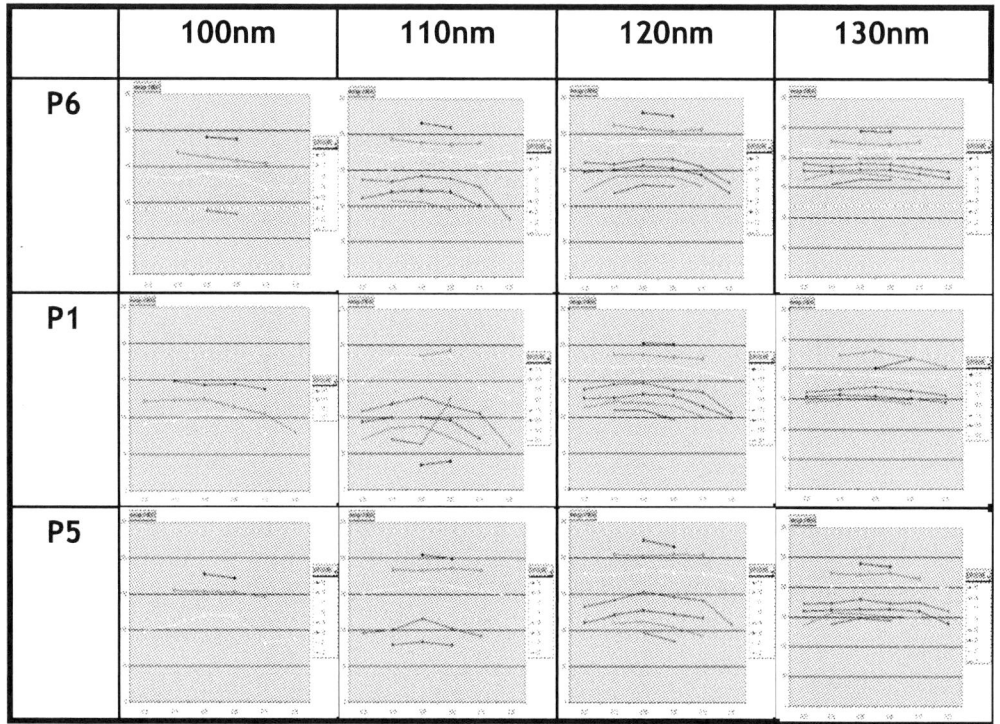

Figure 1. Bossung curves for all photoresist types and targeted feature sizes.

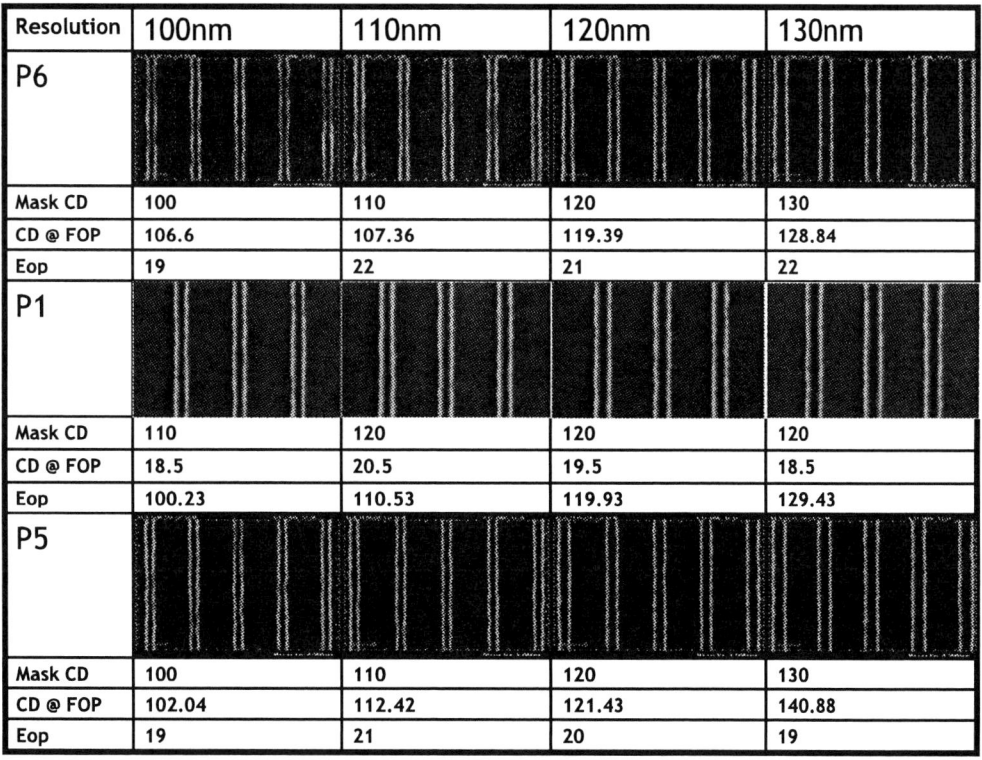

Resolution	100nm	110nm	120nm	130nm
P6				
Mask CD	100	110	120	130
CD @ FOP	106.6	107.36	119.39	128.84
Eop	19	22	21	22
P1				
Mask CD	110	120	120	120
CD @ FOP	18.5	20.5	19.5	18.5
Eop	100.23	110.53	119.93	129.43
P5				
Mask CD	100	110	120	130
CD @ FOP	102.04	112.42	121.43	140.88
Eop	19	21	20	19

Figure 2. Top-down profiles of the printed features for all feature sizes for photoresist P1, P2 and P3.

Resolution	220nm	190nm	170nm	150nm	140nm	130nm	120nm
P5							
CD @Fop[nm]	217.4	184.3	175.2	153.1	142.2	131.1	
Eop [mJ/cm^2]	28.5	28.5	28.5	29.5	30.0	30.5	
P6	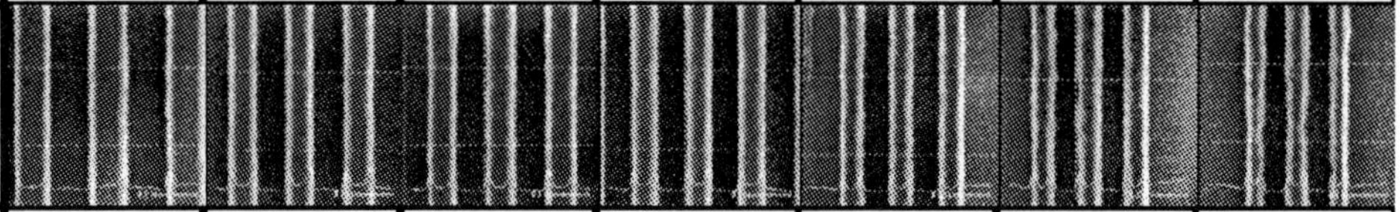						
CD @Fop[nm]	228.2	193.3	177.3	152.9	139.5	131.4	
Eop [mJ/cm^2]	30.5	30.5	30.5	30.5	32.5	32.5	32.5
P1	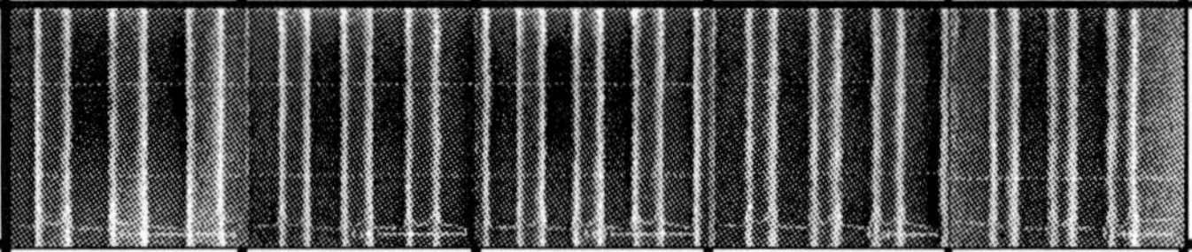						
CD @Fop[nm]	215.4	188.7	172.5	144.5	135.8		
Eop [mJ/cm^2]	26.5	26.5	26.5	27.5	28.0		

Figure 6. Top-down profiles for resist P6, P1 and P5 for 220nm, 190nm, 170nm, 150nm, 140nm, and 130nm feature sizes.

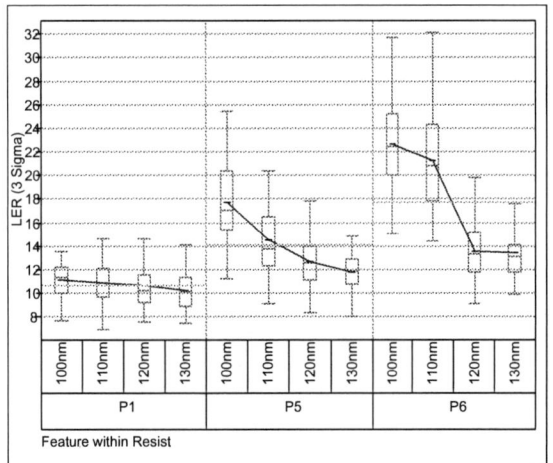

Figure 3. Variability chart of LER vs. feature size/photoresist.

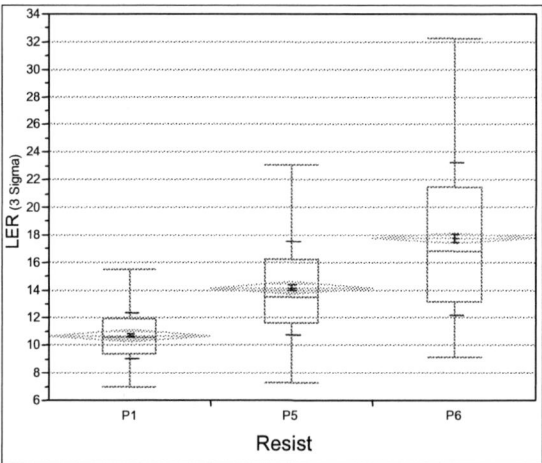

Figure 4(a). One-way analysis of LER vs. resist type.

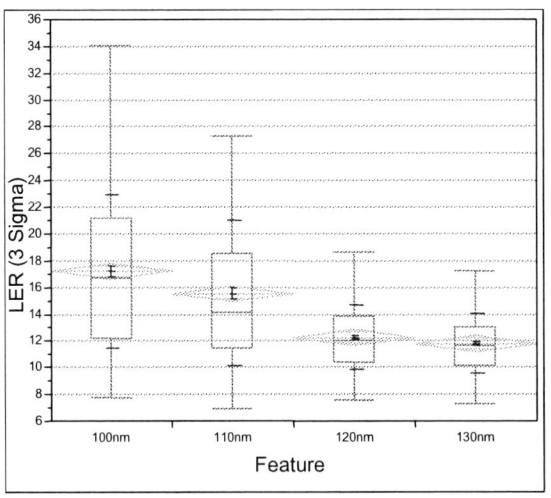

Figure 4(b). One-way analysis of LER vs. feature size

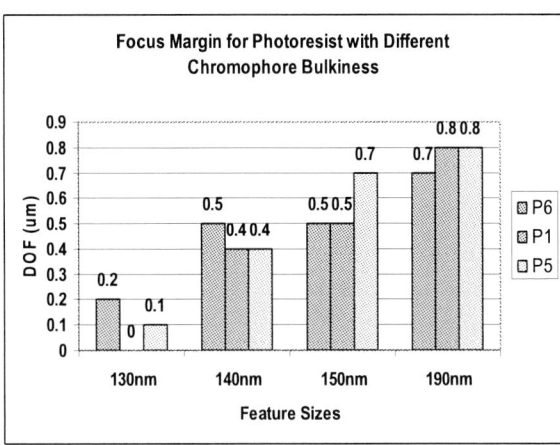

Figure 5. Focus margin of resist P6, P1 and P5 at 130nm, 140nm, 150nm and 190nm.

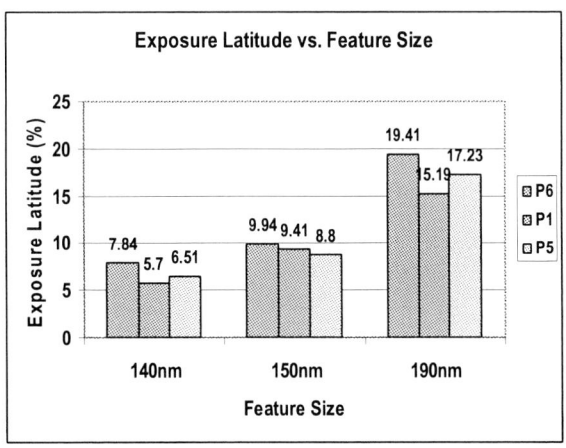

Figure 7. The exposure latitude margin for resists P1, P2 and P3 for 190nm, 150nm and 140nm feature sizes.

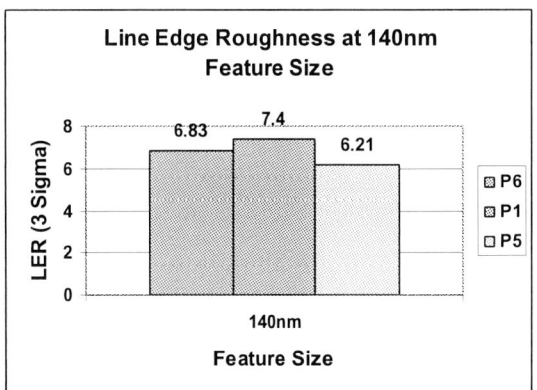

Figure 8(a). The line edge roughness (LER) of photoresists P1, P5 and P6 for the 140nm feature size.

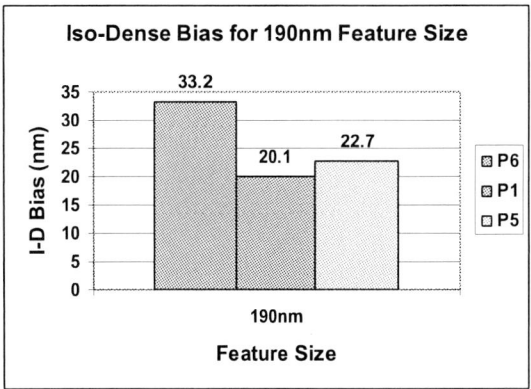

Figure 8(b). The iso-dense bias of photoresist P1, P5 and P6 for the 190nm feature size.

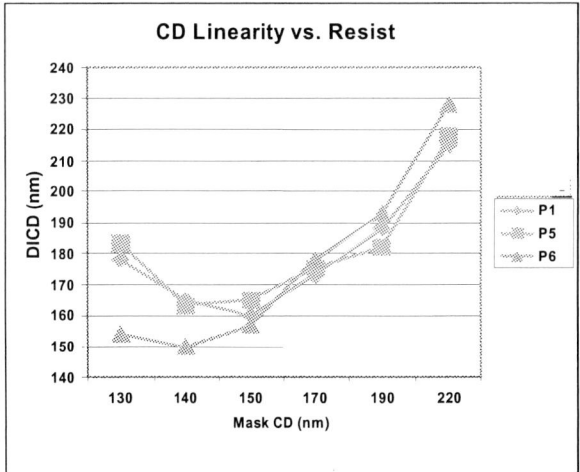

Figure 9. CD linearity for photoresist P1, P5 and P6 for feature sizes from 130mn to 220nm.

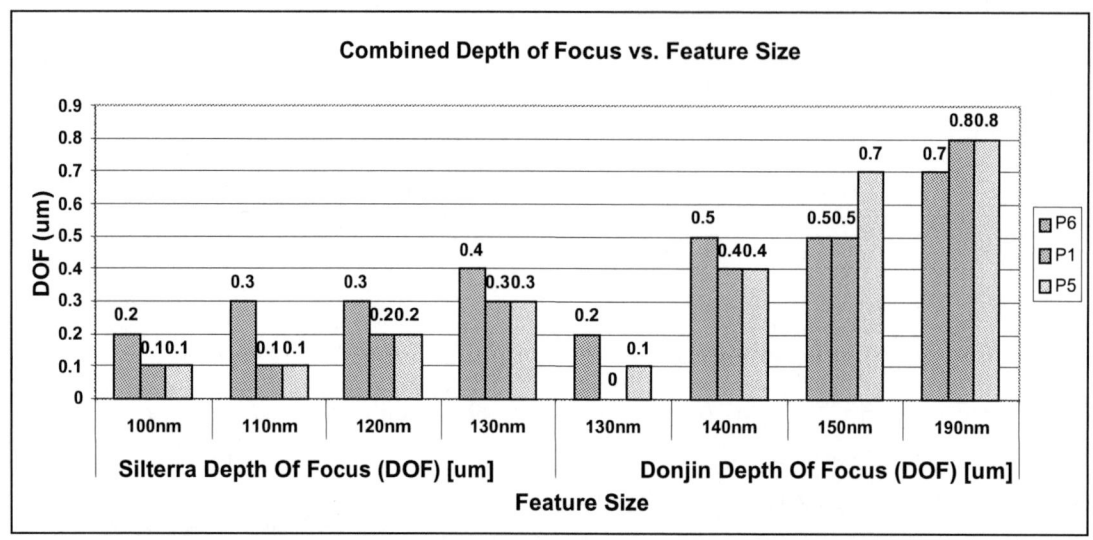

Figure 10. Chart of combined DOF of feature sizes from 100nm to 190nm.

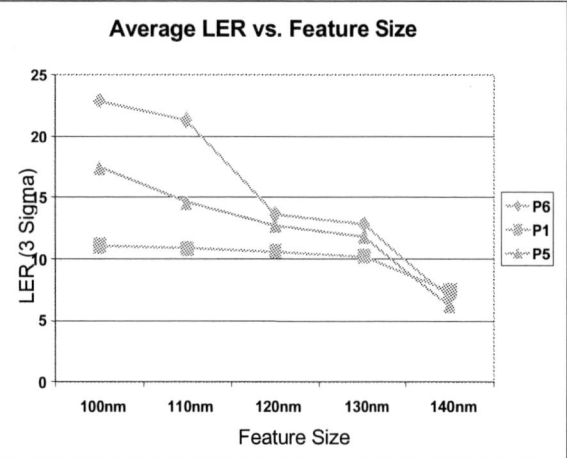

Figure 11. Graph showing the inversely proportional trend of feature size vs. LER.

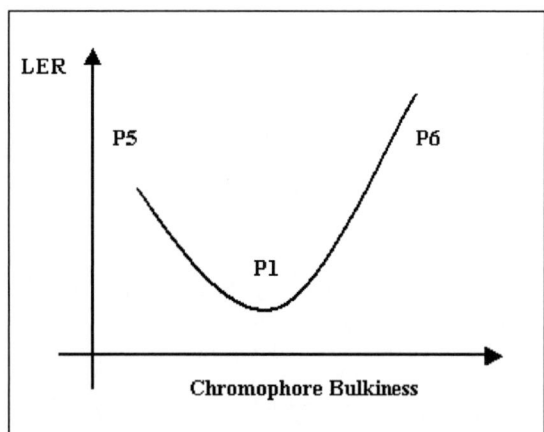

Figure 12. The chromophore bulkiness and corresponding LER of resists P5, P1 and P6.

Table 3. Depth of focus (DOF) for resists P1, P2 and P3 for feature sizes from 100nm to 190nm from both SilTerra and material company.

Source	SilTerra Depth of Focus (DOF) [um]			
Features	100nm	110nm	120nm	130nm
P6	0.2	0.3	0.3	0.4
P1	0.1	0.1	0.2	0.3
P5	0.1	0.1	0.2	0.3
Source	External Depth of Focus (DOF) [um]			
Features	130nm	140nm	150nm	190nm
P6	0.2	0.5	0.5	0.7
P1	0.0	0.4	0.5	0.8
P5	0.1	0.4	0.7	0.8

Table 4. Line edge roughness for resists P1, P2 and P3 for feature sizes of 100nm to 140nm from both SilTerra and Donjin.

Source	SilTerra Measured LER Uniformity (3 Sigma)				External Measured
Features	100nm	110nm	120nm	130nm	140nm
P6	22.82	21.26	13.61	12.82	6.83
P1	11.16	10.91	10.65	10.25	7.40
P5	17.43	14.57	12.71	11.89	6.21

Table 5. A table of tests done on resists P6, P1 and P5, and results of all the tests done.

		P6	P1	P5
DOF [um]	Low (100nm)	0.2	0.1	0.1
	High (190nm)	0.7	0.8	0.8
Profile		Equal		
Resolution [um]		Best		Worst
LER Uniformity [3 sigma]		17.78	10.74	14.18
EL Margin (%)	190nm	19.41	15.19	17.23
	140nm	7.84	5.70	6.51
ID Bias [nm]		33.2	22.7	20.1
CD Linearity [nm]		Best		Worst

Simulation Study on the Performance of SiC-GTO

Muhamad Zahim Sujod, Hiroshi Sakata, *Member IEEE*
Department of Electrical Engineering,
Faculty of Engineering,
Ehime University,
3, Bunkyo-Chyo, Matsuyama,
790-8577, Japan.
Email: zahim@kuktem.edu.my, sakata@dpc.ehime-u.ac.jp

Abstract – Recent development in power electronics has made power semiconductor devices larger and more complicated, and device simulation is necessary to predict their characteristics. From the fundamental equations of semiconductor devices, potential distribution and carrier concentrations can be solved using the Finite Element Method (FEM) [1]. Silicon Carbide (SiC) material has been utilized for power devices, in order to achieve fast switching time and low switching loss. In this study, we use our FEM Device Simulator and compare the switching waveforms of usual silicon Gate Turn Off Thyristor (Si-GTO) and new SiC-GTO. Results show that turn off time of SiC-GTO is decreased extremely. The merits of devices simulation are not only to predict switching characteristics but also to observe inner phenomena of semiconductor device [2]. In this study, we also analyzed and compared the inner distributions including inner potential, hole density and electron density at certain time points during the switching operation of Si-GTO and SiC-GTO. Result show that the changing of inner distributions are faster in the case of new SiC-GTO than in that of usual Si-GTO.[1]

I. INTRODUCTION

Silicon Carbide (SiC) based semiconductor electronic devices and circuits are presently being developed for use in high temperature, high power,

and/or high radiation conditions under which conventional semiconductors cannot adequately perform. Silicon Carbide's ability to function under extreme conditions is expected to enable significant improvements to a far-ranging variety of applications and systems. These range from greatly improved high voltage switching for energy savings in public electric power distribution and electric motor drives, to more powerful microwave electronics for radar and communications, to sensors and controls for cleaner burning more fuel efficient jet aircraft and automobile engines. In the particular area of power devices, Silicon Carbide material has been utilized in order to achieve fast switching time and low switching loss. Gate Turn Off Thyristor (GTO) is known as a self turn off device controlling high voltage and large current. GTO is favored for direct current power applications because of independent gate turn on and off.

In this paper, we analyzed two dimensional model of Si-GTO and SiC-GTO using Finite Element Method (FEM) Device Simulator. We compare the switching waveforms of usual Si-GTO and new SiC-GTO. We also analyzed and compared the inner distributions including inner potential, hole density and electron density at certain time points during the switching operation of these devices.

II. METHOD OF NUMERICAL ANALYSIS

The basic relations describing inner potential V, hole concentration p and electron concentrations n in semiconductors devices are Poisson's equation and current continuity equations. Hole current density J_p and electron current density J_n are expressed using drift-diffusion model. We assume

---

Muhamad Zahim Sujod is currently with the Faculty of Electrical & Electronics Engineering, University College of Engineering & Technology Malaysia, 25000 Kuantan, Pahang, Malaysia. (e-mail: zahim@kuktem.edu.my)
Hiroshi Sakata is currently with the Department of Electrical Engineering, Faculty of Engineering, Matsuyama, 790-8577, Japan. (email: sakata@dpc.ehime-u.ac.jp)

0-7803-9730-4/06/$25.00 ©2006 IEEE

only the recombination of Shockley-Read-Hall model. The equations are as follows:

$$div\ grad\ \mathrm{V} = -(q/\varepsilon)(p - n + N_d) \tag{1}$$

$$q(\partial p/\partial t) = -div\mathbf{J_p} - qR \tag{2}$$

$$q(\partial n/\partial t) = div\mathbf{J_n} - qR \tag{3}$$

$$\mathbf{J_p} = q(p\mu_p\mathbf{E} - D_p\,grad\ p) \tag{4}$$

$$\mathbf{J_n} = q(p\mu_n\mathbf{E} + D_n\,grad\ n) \tag{5}$$

$$\mathbf{E} = -grad\ \mathrm{V} \tag{6}$$

Where; q = electronic charge,

ε = dielectric constant,

N_D= net impurity concentration,

μ_p, μ_n = hole and electron mobility,

D_P, D_n= hole and electron diffusion constants.

In order to solve the basic equations describing inner potential V, hole concentration p and electron concentrations n, we transforms them into different equations in which the variables are defined at mesh points. Applying finite element approximation to these equations and integrating them, we get the matrix equations. For the expression of the hole and electron current densities $\mathbf{J_p}$, $\mathbf{J_n}$, we use the difference scheme of *Scharfetter* and *Gummel*. We adopt the *Crank-Nicolson* method for the time derivatives of current continuity equations, using the calculated values at one time step before. As the current continuity equations include nonlinear terms, the *Newton-Raphon's* iteration method is used to solve these simultaneous equations with the circuit conditions. At any time step, calculations are continued iteratively until variables converge. Then the time step is advanced to the next, and so on.

III. DEVICE MODEL

In more than 170 polytypes of SiC known, only two (4H-SiC and 6H-SiC) are commercially available. 4H-SiC is preferred over 6H-SiC for devices due to its higher electron mobility than that of 6H-SiC. Therefore, 4H-SiC is selected in this comparative study. The parameters used in the model for Si and SiC are given in Table 1. Intrinsic concentrations of Si and SiC are 1.48×10^{16} m^{-3} and 6.0×10^{19} m^{-3}, respectively. Electron and hole lifetimes of Si and SiC for n base layer are 7.0μs and 3.0μs, respectively. Parameters for electron and hole mobility μ are also arranged to smaller values for SiC [3], [4], [5], [6]. Fig. 1 shows the doping profile of the two dimensional model of GTO. This GTO is industrial type rating 1200V, 90A. Fig. 1 represents one of the units and the GTO has 64 these units.

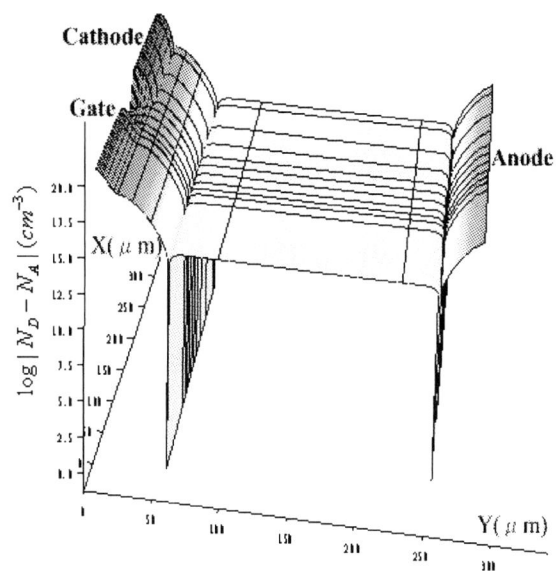

Fig. 1 Doping profile of GTO

TABLE 1 PARAMETERS FOR SI AND SIC USED IN DEVICE SIMULATION

	Si	SiC
Intrinsic concentration $n_i[m^{-3}]$	1.48×10^{16}	6.00×10^{19}
Carrier lifetime $\tau_o[\mu s]$	7.0	3.0
Minimum electron mobility $\mu_{\min n}[m^2/Vs]$	65.0×10^{-4}	50.0×10^{-4}
Maximum electron mobility $\mu_{\max n}[m^2/Vs]$	1265×10^{-4}	1000×10^{-4}
Minimum hole mobility $\mu_{\min p}[m^2/Vs]$	47.7×10^{-4}	10.0×10^{-4}
Maximum hole mobility $\mu_{\max p}[m^2/Vs]$	447.3×10^{-4}	190.0×10^{-4}
Electron ionization coefficient α_n	0.72	0.72
Hole ionization coefficient α_p	0.76	0.76
Reference electron concentration $N_{ref n}[m^{-3}]$	8.5×10^{22}	2.2×10^{23}
Reference hole concentration $N_{ref p}[m^{-3}]$	6.3×10^{22}	2.35×10^{23}

IV. SWITCHING CHARACTERISTICS

Fig. 2 and Fig. 3 show the switching waveforms of Si-GTO and SiC-GTO, respectively. In the simulation, we raised anode voltage to 100V, GTOs are turned on (anode current is increased to 10A and anode voltage is decreased to 0V). Then they are turned off by negative gate pulse. Anode current decreased to 0A and anode voltage recovers to 100V. We can see large difference at turn off switching waveforms. We know that turn off time of SiC-GTO is better than that of Si-GTO. Turn on time and turn off time are shown in Table 2 (All units are in μs). Result show that switching time of SiC-GTO is decreased extremely and the performance of SiC in GTO is in the storage time, fall time and tail time.

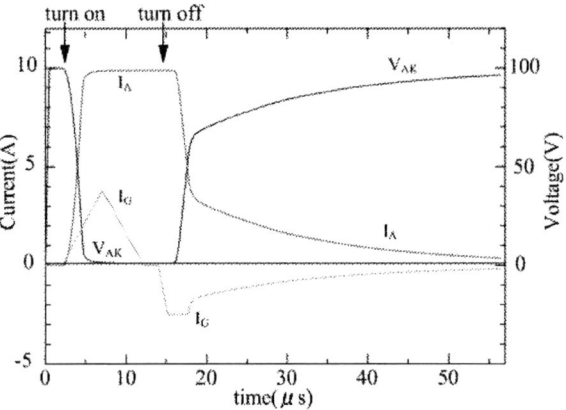

Fig. 2 Switching waveform of Si-GTO

TABLE 2 SWITCHING TIME OF SI-GTO AND SIC-GTO

	Si-GTO	SiC-GTO
Turn on time	3.00	3.30
Delay time	1.45	1.45
Rise time	1.55	1.85
Turn off time	36.20	17.80
Storage time	2.60	0.70
Fall time	10.00	0.70
Tail time	23.60	16.40
Switching time	**39.20**	**21.10**

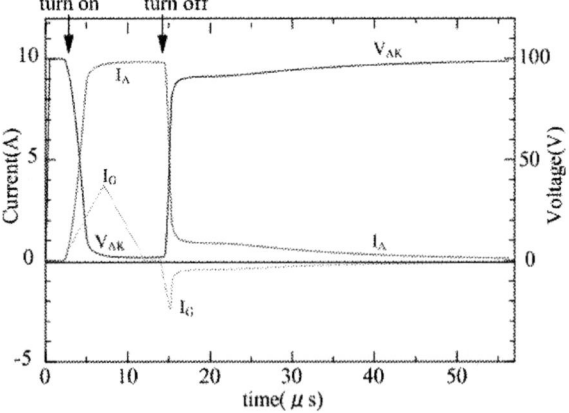

Fig. 3 Switching waveform of SiC-GTO

Fig. 4 compare the turn off loss of Si-GTO and SiC-GTO, respectively. Turn off loss of SiC-GTO is 1337.15Wμs and turn off loss of Si-GTO is 4205.95Wμs. So, turn off loss is smallest in the case of SiC-GTO than in that of Si-GTO.

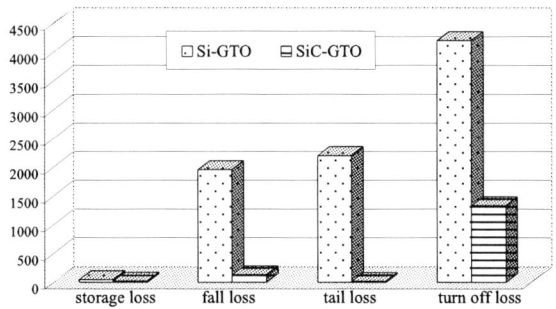

Fig.4 Turn off losses

V. INNER PHENOMENA

In order to analyze the inner phenomena of Si-GTO and SiC-GTO, we visualized the inner potential, hole density and electron density at observing time points. The inner phenomena of Si-GTO and SiC-GTO are visualized at storage time point, fall time point and tail time point. In this paper, we only visualized the inner phenomena at fall time point. Fig. 5 shows the inner phenomena of Si-GTO and SiC-GTO at fall time point. We can see that depletion layer is created, so that a potential wall is appeared between p base and n base layer of inner potential for both Si-GTO and SiC-GTO. In case of Si-GTO, we can see that electron density at cathode edge of p base layer is higher than in that of SiC-GTO. So, carrier discharge between p base and n base layer at cathode edge of Si-GTO is slower than in that of SiC-GTO.

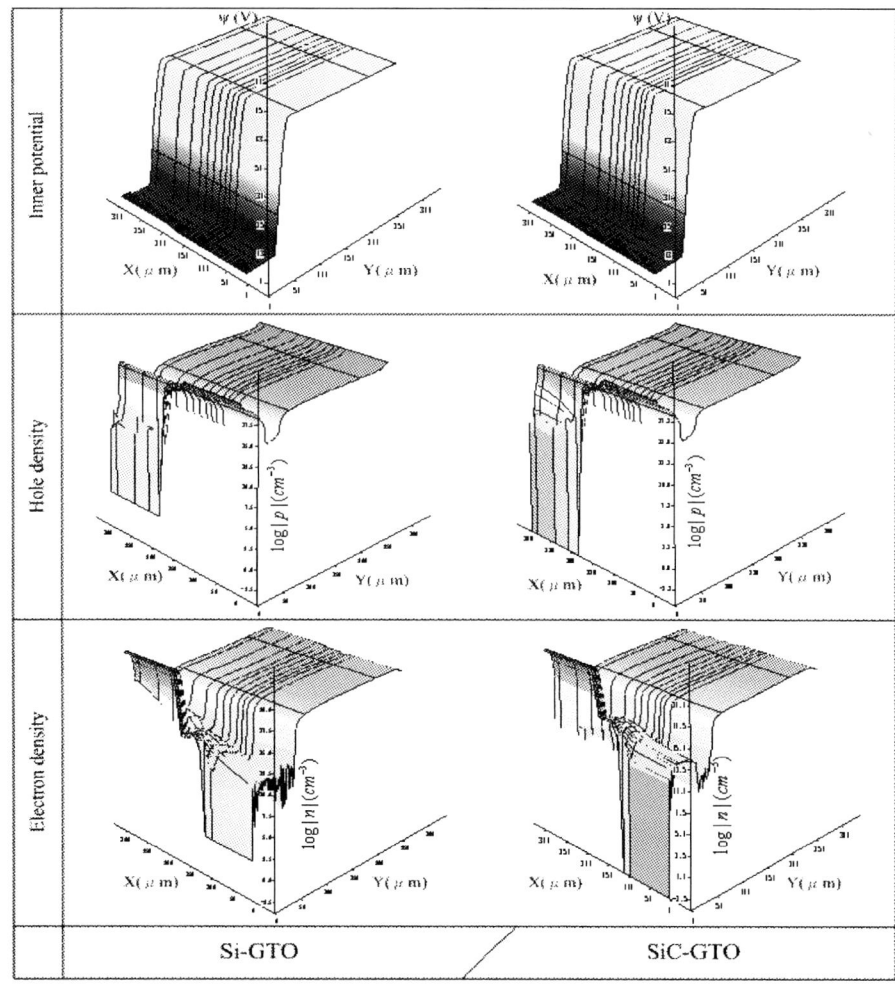

Fig. 5 Inner distribution at fall time

VI. CONCLUSION

Using Finite Element Method (FEM) Device Simulator, we compare the switching waveforms of usual Silicon Gate Turn Off Thyristor (Si-GTO) and new Silicon Carbide GTO (SiC-GTO). Turn off time is smaller in the case of SiC-GTO than in that of Si-GTO. We also analyzed and compared the inner potential, hole density and electron density at fall time. From the device simulation, we can observe the changing of inner distribution and carriers density at any time points in the Si-GTO and SiC-GTO.

REFERENCES

[1] Hiroshi Sakata, Kenji Sugimoto, Shigehiro Isomura and Eisuke Masada, "One and Two Dimensional Device Simulation of GTO Using Finite Element Method", *Proceeding of Power Conversion Conference, Yokohama*, 1993, pp. 569-574.

[2] Hiroshi Sakata, Hidekazu Kamioka and Shigehiro Isomura, "The Inner Phenomena of Hard Driven GTO Using 2-Dimensional FEM Simulation", *Power Electronics and Motion Control Conference*, 2000, Vol.1, pp. 385-389.

[3] Martin Ruff, Heinz Mitlehner, and Reinhard Helbig, "SiC Devices: Physic and Numerical Simulation", *IEEE Transactions On Electron Devices*, 1994, Vol.41, pp. 1041-1054.

[4] S.Sehadri, J.B. Casady, A.K. Agarwal, R.R. Siergiej, L.B. Rowland, P.A. Sanger, C.D. Brandt, J.Barrow, D. Piccone, R. Rodrigues and T.Hansen, "Turn-off characteristic of 1000V SiC Gate-Turn-Off Thyristor", *Proceeding of 1998 International Symposium on Power Devices & Ics, Kyoto*, 1998, pp. 131-134.

[5] A. Q. Huang, B. Zhang, "Comparing SiC Switching power devices : MOSFET, NPN transistor and GTO Thyristor", Solid State Electronics, 2000, pp. 325-340.

[6] Hiroshi Sakata, Muhamad Zahim Sujod, "Device Simulation of SiC-GTO", Proceeding of Power Conversion Conference, Osaka, 2002, pp. 220-225.

Failure Analysis of NSOP Problem Due to Al Fluoride Oxide on Microchip Al Bondpads

Zhao Siping, *Snr Member, IEEE*, Hua Younan, Rao Ramesh and Li Kun

Chartered Semiconductor Mfg Ltd
Woodlands Industrial Park D, Street 2, Singapore 738406
Email: zhaosp@charteredsemi.com

Abstract: **A NSOP due to Al fluoride oxide case was investigated. The NSOP problem on microchip Al bondpads was reported. SEM, EDX, TEM and Auger FA techniques were used to identify the root cause. Optical inspection did not show any abnormality, however, high magnification SEM inspection found the "white dot"-like defects. TEM and Auger analysis results showed that a thicker oxide layer on bondpads, which was about 200-300 A. After studies on failure mechanism, it was concluded that the thicker layer detected by TEM was not Al oxide layer (Al$_2$O$_3$), but it was Al fluoride oxide- Al$_x$O$_y$F$_z$. which was due to F contamination during polyimide ashing process. In this paper we will further discuss the failure mechanism proposed and explain the formation of the Al fluoride oxide-Al$_x$O$_y$F$_z$. In this paper, the possible root cause and eliminating solution are also studied. After changing a new dedicated ashing process machine, the F contamination was eliminated.**

I. INTRODUCTION

IN wafer fabrication (Fab), fluorine (F) contamination will affect Al bondpad quality and cause discolored bondpads or non-stick on pads (NSOP) problem. In this work, a NSOP case was investigated. Some wafers were reported to have NSOP problem on microchip Al bondpads. Optical inspection did not show any abnormality, however, high magnification SEM inspection found the "white dot"-like defects. Base on initial investigation results, it was suspected to be due to thicker Al oxide (Al$_2$O$_3$) as transmission electron spectroscopy (TEM) analysis found a thicker oxide layer on bondpads, which was about 200-300 A on the failed units, and Auger electron spectroscopy (AES) analysis results showed that the F contamination level was only 3-5 at%, which was within the baseline of the F

contamination on normal Al bondpads. However, after further failure analysis, it was concluded that the thicker layer detected by TEM was not Al oxide layer (Al$_2$O$_3$), but it was Al fluoride oxide- Al$_x$O$_y$F$_z$. which was due to F contamination. In the paper [1], a failure mechanism is proposed to explain the formation of the Al fluoride oxide-Al$_x$O$_y$F$_z$. In this paper, detail failure analysis results from SEM, EDX, Auger and TEM will be presented. The possible root cause of F contamination and eliminating solution will be identified and discussed.

II. EXPERIMENTAL

SEM/EDX Analysis

In this study, SEM/EDX analysis was done by using FEI Sirion FESEM at beam accelerating voltage of 5 kV, spot size of 3 and a working distance of 5 mm. To obtain accurate EDX results, ZAF standardless EDX quantification method was used [2-3]:

$$W_i^w = (ZAF)_i \, K_i \qquad (1)$$

where W_i^w is the weight fraction of element i analyzed without coating correction, Z atomic number correction factor, A absorption correction factor, F fluorescence correction factor and

$$K_i = \frac{R_i}{SEC} \qquad (2)$$

where $R_i = \dfrac{I_i^{Meas}}{I_{i,Calc}^{Std}}$ and SEC is the standardless element coefficients factors which are able to calculate using standard samples. In this study, SEC factors are taken from our previous paper [2]. Quantified EDX results can be expressed as both weight fraction and atomic fraction. The atomic fraction ($W_i^{a'}$) can be calculated from the weight fraction (W_i^w) using Eqn (3):

$$W_i^a = (W_i^w/A_i) / \sum (W_j^w /A_j) \qquad (3)$$

where W_i^a is atomic fraction of element i analyzed with coating correction. W_i^w & W_j^w are weight fraction of element i & j analyzed with coating correction, and A_i & A_j are atomic weight of elements i & j.

AES Analysis

Auger electron spectroscopy (AES) was employed to perform elemental analysis. AES analyses were carried out using PHI (Physical Electronics) SMART-200 Auger System. The analysis conditions are as follows: (a). Electron gun setting: Operating voltage, 20 KeV; operating current, 10 nA and gun tilt, 30° (b). Ion gun setting: ion source, Ar$^+$; operating voltage, 2 KeV; operating current, 1000 nA (variable, to maintain etch rate of 100A/min); rastering size, 2 mm x 2 mm and sputter etch rate, nominal at 100+/- A/min.

Auger survey range was 30-2030 eV. The relative atomic concentration of the detected elements were calculated by first measuring elemental peak-to-peak heights in Auger survey scans and then applying sensitivity factors, based on standard spectra of pure elements or selected compounds:

$$C_i = I_i / (I_{std}S_iD_i) \qquad (4)$$

where I_i is the peak-to-peak amplitude of the element i from the test specimen, I_{std} is the peak-to-peak amplitude of the element i from the standard, S_i is the relative sensitivity factor and D_i is a relative scale factor between the spectra for the test specimen and standard.

TEM Analysis

TEM analysis was performed using a TECNAI F20 200 kV field emission microscope, and the TEM sample was prepared with the FIB (focused ion beam) based lift-out method. To protect the surface, a 50 nm thick layer of Pd was sputter-deposited onto the surface before FIB sample preparation. EDX spectra were collected under STEM mode for easy and accurate probe positioning.

III.   RESULTS AND DISCUSSION

It was reported by the customer that the "white dot"-like defects were found on the bondpads with NSOP problem (Fig. 1).

Figure 1. SEM micrograph showed the "white dot"-like defects on the affected bondpads, which might have caused NSOP problem.

Therefore, some dies were returned to do failure analysis. Using optical inspection we didn't find any visible abnormality, but scanning electron microscopy (SEM) at high magnification showed the "white dot"-like defects on the affected bondpads (Fig. 2), which was similar to those (Fig. 1) reported by the customer.

Figure 2. SEM micrograph showed the "white dot"-like defects on the affected bondpads, which might have caused NSOP problem.

Elemental analysis was done using energy-dispersive X-ray microanalysis (EDX) on the "white dot"-like defect. EDX result showed no Si peak was detected, but a small F peak was detected (Fig. 3). Based on EDX result, we ruled out that it was Si-related defect. For more details, Auger analysis was also conducted, but the F level was 3-5 at% (Fig. 4), which was within the baseline level of F contamination on Al bondpads.

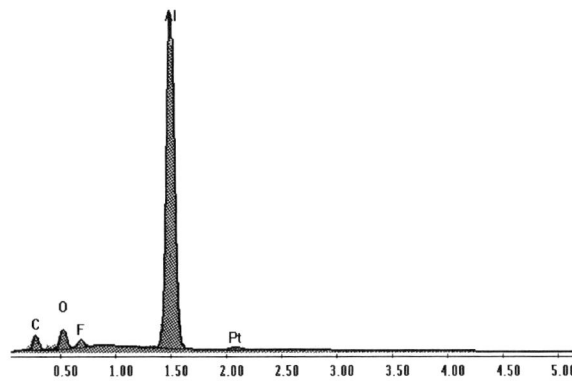

Figure 3. EDX did not detect Si peak, but detected a small F peak on the "white dot"-like defect (Pt was from sample preparation coating).

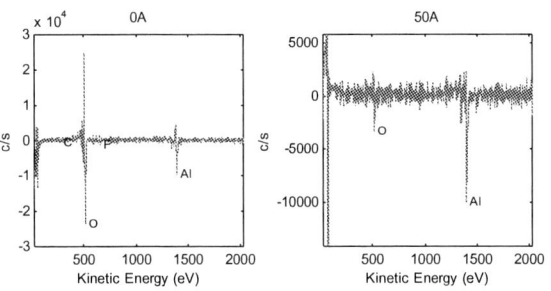

Figure 4. Auger analysis detected 3-5 at% F on the affected bondpad, but it was within the baseline.

Cross-sectional TEM at defect location showed abnormally thick layer (Figs 5-6) on Al bondpads and EDX detected a small F peak (Fig. 7). Base on the above investigation results, it was suspected that NSOP problem was likely due to Al oxide (Al_2O_3) as TEM detected a thicker oxide layer, which was about 200-300 A on the

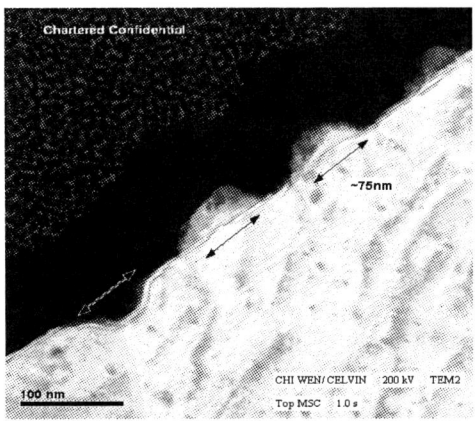

Figure 5. TEM micrograph showed the thicker layer on the "white dot"-like defects on the affected bondpad.

failed bondpads, and AES results showed that the F contamination level was only 3-5 at%, which was within the baseline of the F contamination on normal Al bondpads.

However, after further failure analysis and our previous experience, we challenged that the thicker layer detected by TEM was not Al oxide layer (Al_2O_3), but it was Al fluoride oxide-$Al_xO_yF_z$ which was due to F contamination and proposed a failure mechanism to explain the formation of the Al fluoride oxide-$Al_xO_yF_z$ [1].

According to this failure mechanism, the NSOP problem was due to the F contamination.

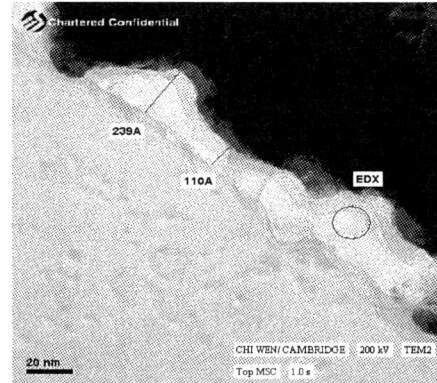

Figure 6. TEM micrograph showed the thicker layer (more than 200 A) on the affected bondpad, which was suspected to be Al fluoride oxide.

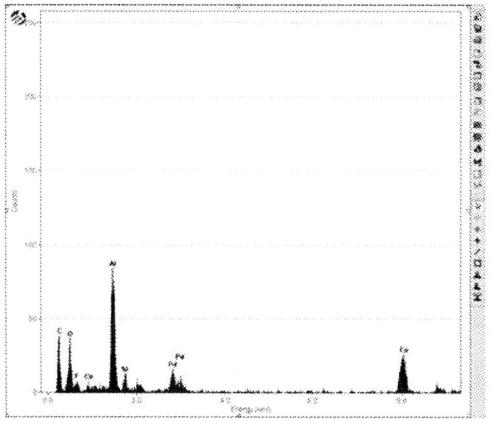

Figure 7. TEM/EDX detected a small F peak on the "white dot"-like defect, which supported to suspect F –induced defect.

In this case, high F in wafer fab process contaminated the wafers. At the beginning, the F-induced defects were still not formed. Thus, the problem couldn't be found at wafer fab OQA inspection before shipping to the customer. After a few months for wafer storage and assembly process, the F-induced defects were formed slowly by chemical reactions from the F corrosion and OH corrosion [1, 4, 5]. However, it was still not easy to identify these defects optically as they were very small and the size was only 100-200 nm. Thus in this case, the "white dot"-like defects were only detected using high magnification SEM inspection. The abnormal layer detected on surface of bondpad was Al fluoride oxide- $Al_xO_yF_z$.

However, the challenge faced is that if the "white dot"-like was caused by F contamination, why Auger not detected high F? According to the failure mechanism proposed [1], one can understand that if there is high F contamination on Al bondpads, F will chemically react with Al and form Al-F complex compound, $[AlF_6]^{3-}$. Once $[AlF_6]^{3-}$ formed, it may become an anode and cause further chemical reaction from O_2 and moisture (H_2O) at the cathode to form the new product, OH^- ions, and then OH^- ions will further chemically react with Al to form $Al(OH)_3$ or Al_2O_3. Finally, $[AlF_6]^{3-}$ and Al_2O_3 will form complex compound Al fluoride oxide-$Al_xO_yF_z$. The thickness of $Al_xO_yF_z$ was greatly grown up and the volume was also increased. This diluted the F level and hence AES only detected 3-5 at% F in the defects. Therefore, we believe that in this case, the actual F contamination level might be much higher than 3-5 at% before $Al_xO_yF_z$ layer was formed. It is necessary to identify the source of the F contamination so as to eliminate NSOP problem

IV. IDENTIFICATION OF THE POSSIBLE ROOT CAUSE

The bondpad surface integrity is very critical to the die-attach process in packaging as surface contaminants can lead to poor adhesion of Au wire-bonds. The acceptable limits of surface contaminants such as F and C on a normal Al bondpad are well understood. For example, a low level F contamination of about 3-5 at.% is invariably present on the surface of pad and does not lead to F-induced corrosion. The F contamination cannot be completely eliminated as Fluorine-based etch chemistry is used during the pad opening in the passivation layer. Under this F contamination level, there is a thin Al oxide layer, which is about 30-50A (Fig.8)

depended on PRS oxidation. In the other word, more PRS oxidation may cause thicker Al oxide. However, the oxide thickness is only in range of 30-50 A. Even performing further oxidation process, it will be limited to further grow up.

However, in the current case, the oxide thickness measured in TEM micrographs was 100-250 A i.e. thicker than a normal bondpad. Therefore, we strongly think that the layer detected is not Al oxide, which should be thin. On the contrary, the observed rough layer on top of Al pad is Al fluoride oxide, $Al_xO_yF_z$, formed as result of F-contamination [1]. The thickness of Al fluoride oxide is not self-limiting and depends on F-contamination level higher the contamination; thicker will be the Al fluoride oxide layer.

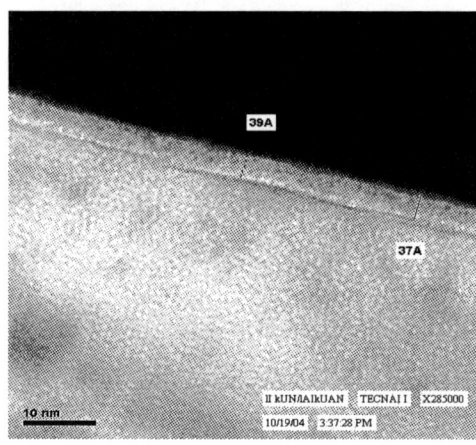

Figure 8. TEM result showed a normal uniform Al oxide on surface of Al bondpad, which is about 35-40A. Moreover, the previous evaluation results showed that it was difficult to increase Al oxide on a normal Al bondpad.

Therefore we believe that in this case, at the beginning, the F-contamination was higher than the baseline (3-5 at.%). However, after chemical reaction, F-contamination had resulted in the formation of Al fluoride oxide, $Al_xO_yF_z$ and the volume of the defect was greatly increased. This diluted the F level and hence the F concentration determined in the AES analysis was within the baseline performance (3-5 at.%).

Based on the above failure mechanism and discussion, we conclude that the white dot-like defects are due to F contamination and the abnormal layer detected on surface of bondpad is Al fluoride oxide- $Al_xO_yF_z$. With the above the root cause analysis, further wafer fab investigations were conducted to identify the process step responsible for high F-contamination. We found that the wafers were affected during the polyimide process. Partition experiments were designed for the very backend process steps (Al pad opening and polyimide) such that after each critical process step (critical from the point of view of potential F-contamination), one wafer was removed from the parent lot (a batch of 25 or less wafers) for Auger measurements. The results of partition experiments are captured in Table 1, which were the relative atomic concentration of elements detected on the bondpad surface of 'as received' wafer and after 50A sputter. Wafers were taken out of partition lot after each critical process step. High surface F was observed only after polyimide ashing process step. Therefore, we believe that F contamination was not from the main bondpad opening process, as the Auger results were along the expected line on the partitioned wafer after the pad-opening clean step, Table 1 and Fig.8.

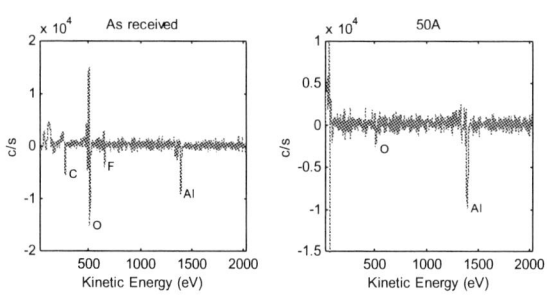

Figure 9: Auger spectra taken on 'as received' and after 50A sputter of bonapad of partitioned wafer after pad opening clean step. F was not observed after 50A sputter.

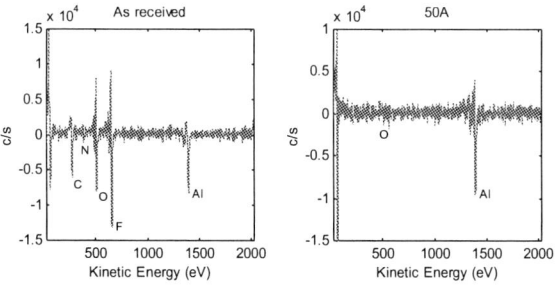

Figure 10: Auger spectra taken on 'as received' and after 50A sputter of bonapad of partitioned wafer after polyimide ashing step. Even though F was not observed after 50A sputter, surface F was very high, which was introduced during aching process.

From the Table 2, it is clear that Fluorine contamination was low all the way up to polyimide cure step but increased dramatically after the polyimide ashing step, leading one to conclude that the trigger for high F on fully processed wafers is the polyimide ashing step, Fig.9. Investigations of ashing process tool history showed that it is a shared tool and is also used as an etch tool using CF_4. Thus, high F observed on wafers is due to machine contamination.

V. PREVENTIVE ACTIONS & SOLUTION

To prevent F contamination, we immediately dedicated a new machine for polyimide ashing process and AES analysis was followed up. In Table 2, the atomic concentrations of various elements observed are listed. One would notice that the F level was back to the baseline (<5 at%) after switching to a dedicated machine for polyimide ashing process.

Table 2 showed relative atomic concentrations of elements observed on the bondpads of wafers processed in shared and dedicated ashing tools are listed. F-contamination in dedicated tool is within the baseline value.

Further evidence for F-induced corrosion was provided by the shelf lifetime acceleration test (SLAT) [6] on some wafers with low F (Auger showed <5at% F) and high F level (Auger showed 13-18 at.%). The simulation results showed that:

1. On the low F wafers
   (a). No white dot-like defects were found using high magnification SEM
   (b). F Auger results no difference before and after the test.
2. On the high F wafers
   (a). White dot-like defects were found using high magnification SEM.
   (b). F Auger results decreased after the test from 13-18 at% to 3-5 at%.

The above results further confirm the failure mechanism proposed [1] and the root cause.

VI. CONCLUSION

After conducting failure analysis and wafer fab investigations, it was concluded that the NSOP (non-stick on pads) problem was caused by the "white dot"-like defects and abnormal Al fluoride oxide- $Al_xO_yF_z$. While Al fluoride oxide- $Al_xO_yF_z$ was due to F contamination from the ashing process step during the polyimide processes. The machine used was contaminated by previous CF_4 related process. After changing a new dedicated ashing process machine, the F contamination was eliminated and the wafers were passed one-year shelf lifetime acceleration test.

ACKNOWLEDGEMENT

The authors would like to thank Soo Chi Wen, Ton Cambridge and Kelvin from TEM group, Lee Sharon from SEM/EDX/FIB group and Kuni from Fab3/5 for their technical advice and making contribution to this technical paper.

REFERENCES

[1] Hua Younan, Zhao Si Ping, Rao Ramesh and Li Kun, Studies on Failure Mechanism of Al Fluoride Oxide- $Al_xO_yF_z$ on Microchip Al Bondpads. To be submitted for the 2006 IEEE International Conference on Semiconductor Electronics, 29 NOV-1 DEC, 2006, Kuala Lumpur, Malaysia.

[2] Y. N. Hua, Z. R. Guo & K. W. Chau, "Studies of ZAF Standardless EDX Quantification Method and Application in Failure Analysis of Semiconductor," *J. Trace and Microprobe Techniques,* 15 (1), 13-31 (1997).

[3] Y. N. Hua, "Application of SEC Factors in Energy-Dispersive X-Ray Quantification Analysis of Silicon Oxide (SIO_2) Layer in Wafer Fabrication." *The proceedings from the 198th Meeting of The Electrochemical Society – Phoenix, Arizone,* October 22-27, No. 810 (2000).

[4] Hua Younan, Shailesh Redkar, Lau Chi-Kwan and Mo Zhiqiang (from PSB Corp., Singapore), A Study on Non-Stick Aluminum Bondpads due to Fluorine Contamination using SEM, EDX, TEM, IC, Auger, XPS & TOF-SIMS Techniques. The proceedings from the 28th International Symposium for Testing & Failure Analysis (ISTFA'2002), 3-7 Nov, Phoenix Civic Plaza, Phoenix, Arizona, USA p495-504 (2002).

[5] Hua Younan, Shailesh Redkar, Ron Dickinson and Peh Shirley, A Study on Fluorine-Induced Corrosion on Microchip Aluminium Bondpads. The proceedings from the 29th International Symposium for Testing & Failure Analysis (ISTFA'2003), 2-6 Nov. 2003, Santa Clara, California, USA, p249-255 (2003).

[6] Zhao Si Ping, Hua Younan, Rao Ramesh and Li Kun, A Novel Dewing-free shelf lifetime acceleration test (SLAT) in wafer fabrication, patent from Chartered Semiconductor (2006).

Table-1: Relative atomic concentration of elements detected on the bondpad surface of 'as received' wafer and after 50A sputter

Process Step	Location	Relative atomic concentration (%)									
		As received					50 A				
		C	N	O	F	Al	C	N	O	F	Al
Developer Clean	Center	29.6	-	39.3	3.7	27.4	-	-	14.1	-	85.9
	Edge	28.0	-	34.1	4.6	33.3	-	-	10.8	-	89.3
Polyimide Cure	Center	55.0	3.5	21.1	1.4	19.0	-	7.6	17.9	-	74.5
	Edge	51.2	4.3	22.1	1.5	20.9	-	-	8.7	-	91.4
Polyimide Ash	Center	34.7	5.3	19.5	13.8	26.7	-	-	8.5	-	91.5
	Edge	34.0	4.7	21.4	10.6	29.4	-	-	10.2	-	89.8

Table-2: Relative atomic concentrations of elements observed on the bondpads of wafers processed in shared and dedicated ashing tools

Tool Type	Location	Relative atomic concentration (%)									
		As received					50 A				
		C	N	O	F	Al	C	N	O	F	Al
Shared Tool	Center	43.9	-	18.9	13.6	23.6	-	-	14.2	-	85.8
	Edge	49.9	4.0	15.5	13.0	17.6	-	-	10.5	-	89.6
Dedicated Tool	Center	12.9	-	48.6	0.9	37.7	-	-	25.7	-	74.3
	Edge	11.3	-	51.3	1.9	35.5	-	-	21.1	-	78.9

Studies on A New Sela-FIB Sample Preparation Method and Its Application in Failure Analysis of Wafer Fabrication for 110nm Technology Node and Beyond

Zhao Siping, *Member, IEEE*, Hua Younan, Er Eddie and Khoo Ley Hong

Chartered Semiconductor Mfg Ltd
Woodlands Industrial Park D, Street 2, Singapore 738406
Email: zhaosp@charteredsemi.com

Abstract: **In this paper, a novel sample preparation method for obtaining high-resolution SEM profile is proposed. Both Sela fine cleave and FIB slice techniques have been used for SEM sample preparation. Using this new method, high-resolution 90 degrees SEM micrographs are provided. It has been applied in failure analysis to check Via gouging information without any charging problem, which helps us to reduce TEM analysis samples.**

I. INTRODUCTION

IN failure analysis of wafer fabrication, for 110nm technology node and beyond, it is difficult to obtain high resolution cross sectional profiles using conventional mechanical polishing method. Currently we are using either Sela tool to do fine cleave or using FIB (focused ion beam) tool to do precision milling. However, these two methods encountered a resolution problem. For example in the case of sample preparation for 110nm Via depth and profile, when we used Sela fine cleave, SEM results (Figures 1a and 1b) shows charging and difficult to check the via gouging. In another method when we used FIB milling, SEM results (Figures 2a and 2b) shows difficulty to check the via gouging and not true 90 degrees cross sectional profile as SEM photos were taken at 45 degrees tilt for FIB sample. Therefore, using current methods, it is either difficult or impossible to obtain high-resolution SEM photos to check Via gouging information. As such, most jobs need to do TEM. However, TEM jobs are very limited and difficult to meet the requirement. We have developed a new method so as to obtain high-resolution SEM images.

In this paper, a novel sample preparation method for obtaining high-resolution SEM profile is developed. This is a significant improvement in the capability of failure analysis technique. It will face a serious barrier to develop new 110nm technology node and beyond if no such technique is develop to check via gouging information. We have applied this new method for 110nm Via depth and profiles. High-resolution 90 degrees SEM photos help us to optimize the fab process parameters/conditions and Significantly reduce TEM samples. Moveover, the cycle time is also reduced by about 50% as no charging problem and easy to take SEM micrographs.

II. TRADITIONAL SAMPLE PREPARATION METHODS

Currently two traditional sample preparation methods are used, which are Sela fine cleave and FIB precision milling.

Using Sela Fine Cleave
1. Sample prepared as normal Fine Cleave process when using Sela tool.
2. Performed 1sec of BOE staining after the Cleave.
3. Sample sputtered for 25 secs of Pt before loading into the SEM.
4. SEM picture taken at 0 degree, 5kV and sport size 3 with working distance 5 mm.

However, we faced charging problem and difficult to check the via gouging (Figures 1a and 1b).

Using FIB-cut-milling
1. Sample tilted at 52 degree for FIB-cut without using Sela fine cleave.
2. Performed 1sec of BOE staining after FIB-cut-milling.
3. Sample sputtered for 25 secs of Pt before SEM.

0-7803-9730-4/06/$25.00 ©2006 IEEE

4. Sample tilted 45 degree for better SEM image at 5kV and sport size 3 with working distance 5 mm.

SEM results (Figures 2a and 2b) shows difficulty to check the via gouging and not true 90 degrees cross sectional profile as SEM photos were taken at 45 degrees tilt for FIB sample.

III. A NEW SELA-FIB SAMPLE PREPARATION

In this study, we developed a new sample preparation method to use both Sela fine cleave and FIB slice. The sample will be done using Sela fine cleave first, and then FIB further slice-in to smoothen the cut surface. Figure 3 shows. The new method proposed in the study:

Sela fine cleave → FIB slice-in → BOE 1 staining for 1 sec → Sputtered for 25 Sec of Pt→ SEM (high –resolution) inspection.

1. Sample prepared as normal Fine Cleave process when using Sela tool.
2. FIB slice-in (until to via center)
3. Performed 1sec of BOE staining after FIB cut-milling.
4. Sample sputtered for 25 secs of Pt before SEM.
5. SEM picture taken at 0 degree, 5kV and sport size 3 with working distance 5 mm.

In this study, FIB condition for all methods: 150pA (Pt coated as Passivation before the FIB-cut) 2700pA (coarse milled),1000pA (medium milled) & 150pA (Fine milled/ cleaning) 5kv/ spot 3with sample tilted 52 degrees

IV. APPLICATION IN FAILURE ANALYSIS

This new method has been applied in failure analysis of 110nm Via depth and profiles successfully and 90 degrees high-resolution SEM micrographs supported to optimize processparameters and conditions and released for 110nm Via depth and profiles.Some results are shown in Figures 4 (a and b). One can see clearly Via gouging information without any charging problem.

V. CONCLUSION

In this paper, a new sample preparation method to use both Sela fine cleave and FIB slice has been developed. The sample preparation and analysis flow is: Sela fine cleave → FIB slice-in → BOE 1 staining for 1 sec → SEM (high–resolution) inspection. Using this new method, 110 nm Via gouging information was clearly provided without any charging problem.

ACKNOWLEDGEMENT

The author would like to thank Surface/Advanced FA team, QRA-FA Dept, Chartered Semiconductor Mfg Ltd. for their technical advice and contribution.

REFERENCES

[1] Hua Younan, Z. R. Guo & K. W. Chau, Studies of ZAF Standardless EDX Quantification Method and Application in Failure Analysis of Semiconductor. Journal of Trace and Microprobe Techniques, 15 (1), 13-31 (1997).
[2] Hua Younan, Zhao Siping, Mo Zhiqiang and Cho Jie Ying, Studies on A Sample Preparation Method for HR-SEM and Application in Failure Analysis of Trench TEOS Gauging Measurement in Wafer Fabrication. To be submitted for the 2006 IEEE International Conference on Semiconductor Electronics, 29 NOV-1 DEC, 2006, Kuala Lumpur, Malaysia.

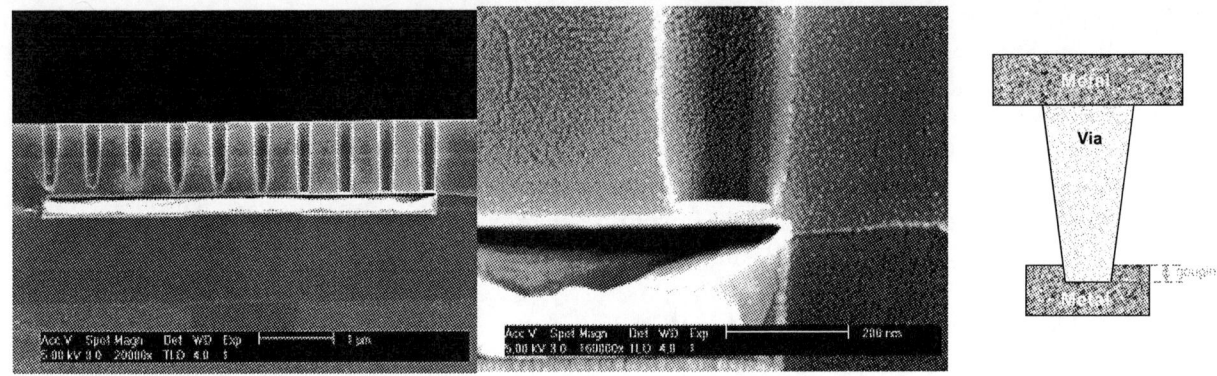

1a. Low Magnification   1b. High Magnification   1c. Via

Figure 1. For sample preparation of 110nm Via depth and profile, we used Sela fine cleave, SEM results shows charging and difficulty to check the via gouging as metal profile was damaged during Sela cleaving.

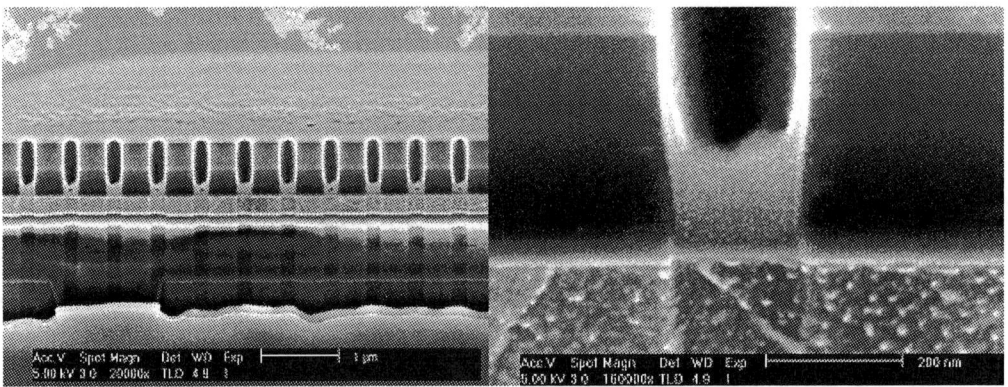

2a. Low Magnification   2b. High Magnification

Figure 2. For sample preparation of 110nm Via depth and profile, we used FIB milling, SEM results difficult to check the via gouging and not true 90 degrees cross sectional profile as SEM photos were taken after titling 45 degrees on FIB sample.

SELA FINE CLEAVE + FIB Method

Sample preparation steps:

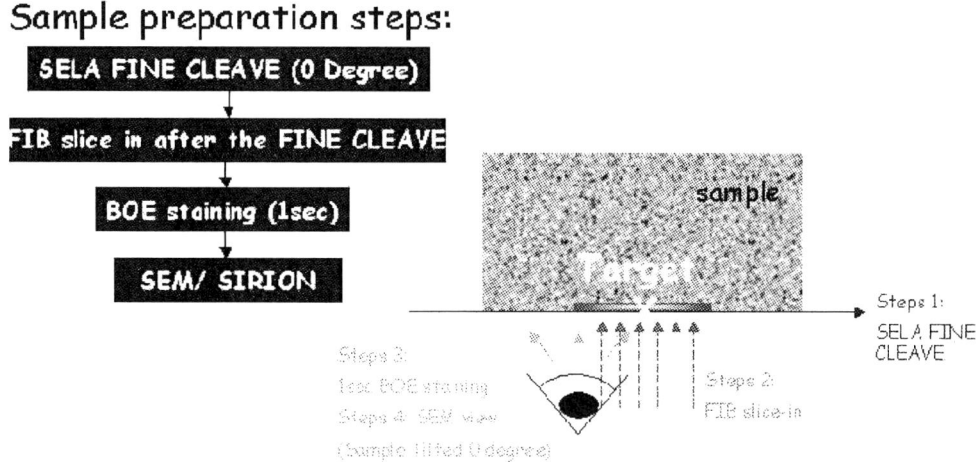

Figure 3. A new method proposed in the paper: Sela fine cleave → FIB slice-in → BOE 1 staining for 1 sec → SEM (high –resolution) inspection.

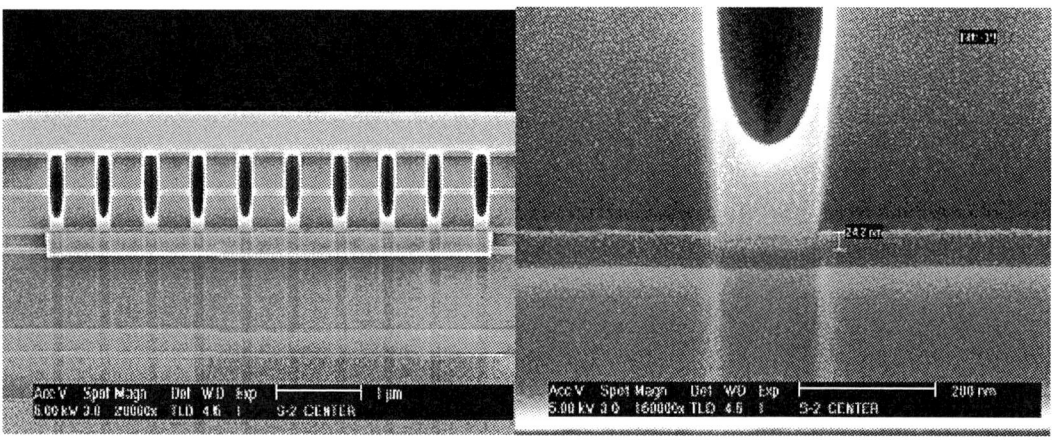

4a. Low Magnification   4b. High Magnification

Figure 4. For sample preparation of 110nm Via depth and profile, we used this new method (Sela fine cleave + FIB slice-in) to prepare sample and obtained high-resolution 90° cross sectional profiles and one can see clearly via gouging information.

Synchrotron Radiation X-ray Diffraction and X-ray Photoelectron Spectroscopy Investigation on Si-based Structures for Sub-Micron Si-IC Applications

Zhe Chuan Feng[1,*], Li-Chi Cheng[1], Chu-Wan Huang[1], Ying-Lang Wang[2], and T. R. Yang[3]

[1] Graduate Institute of Electro-Optical Engineering & Department of Electrical Engineering, National Taiwan University, Taipei, 106-17 Taiwan, R.O.C.

[2] Taiwan Semiconductor Manufacturing Company, Tainan, 741-44 Taiwan, R.O.C.

[3] Department of Physics, National Taiwan Normal University, Taipei, Taiwan, R.O.C.

* Tel: +886-2-3366-3543, fax: +886-2-2367-7467, E-mail: zcfeng@cc.ee.ntu.edu.tw

Abstract **Synchrotron radiation X-ray diffraction and X-ray photoelectron spectroscopy techniques have been employed for the investigation on Si-based layer structures for sub-micron Si-IC Applications. The high energy synchrotron radiation light sources have produced plenty of X-ray lines with high index diffraction and strong X-ray photoelectron emissions. The useful information will increase our understanding of these materials which are applied extensively to the semiconductor industry.**

I. INTRODUCTION

Si-based integrated circuits (IC) have played important roles in the modern science and technology as well as human world. They are currently approaching down to 90-65 nm (or less) nano-scales. In this nanometer scale production, many new scientific and technological issues appear, for example, stresses in 3-dimmensional distribution and their effects to the material properties will have become very important and they will affect the device functions and performance severely. In semiconductor ICs, in the vertical direction normal to the surface of the Si substrate wafer, multiple layers are deposited or formed and in the parallel plane, a high density of structures are masked and fabricated [1]. Because of the differences of materials from Si, SiO_2, SiN and metal etc, complex stresses are induced. The recent exploration and introduction of high-k dielectric gate materials such as Al_2O_3, HfO_2, etc. which are needed for next generation sub-100nm and sub-50nm IC development, have raised new impacts [2].

In this paper, we study on several typical nano-scale thin films of amorphous (a)-SiC, Si_3N_4, TaN/SiO_2 and Co/SiO_2 deposited on 12-inch large scale Si wafers by high energy and high intensity synchrotron radiation X-ray diffraction (SR-XRD) and X-ray photoelectron spectroscopy (SR-XPS) with rich data.

II. EXPERIMENT

The samples are constructed on 12" silicon substrate by chemical vapor deposition (CVD) process from Taiwan Semiconductor Manufacturing Company (TSMC).

Synchrotron X-ray Diffraction (SR-XRD) is similar in design to conventional XRD. Conventionally, XRD has several limitations. A key disadvantage of ordinary XRD is that it is limited to crystalline materials (since amorphous materials do not diffract). XRD is also a time-consuming measurement and uses a large volume of sample. Fortunately, synchrotron-based XRD is useful to circumvent these limitations [3,4]. They offer exceptional resolution, even on very small samples containing only a few grains of a particular mineral. This high resolution and the excellent detectors for SR-XRD permit the identification and quantification of trace phases not possible using other means. In this study, SR-XRD experiments were performed using a high energy of 28 keV, i.e. short wavelength of ~0.443Å, by the X-ray diffraction instrument at the 01C2 beamline of National Synchrotron Radiation Research Center (NSRRC) in Hsinchu, Taiwan (R.O.C.).

In order to study the surface properties of these thin films, we also performed synchrotron radiation X-ray photoelectron spectroscopy (SR-XPS) experiments on the 24A1 beamline of NSRRC, using a monochromatic X-rays beam with 1150eV incident light energy. The constant analyzer energy (CAE) mode was used with analyzer pass energy of 200 eV for preliminary wide scan. The binding energies of Si 2p, Si 2s,

0-7803-9730-4/06/$25.00 ©2006 IEEE

C 1s, and O 1s, were obtained.

III. RESULTS AND DISCUSSIONS

X-ray diffraction experiments were performed on the 01C2 beamline of NSRRC in Hsinchu, Taiwan (R.O.C.). Monochromatic synchrotron X-rays were focused to a spot size about 5 mm wide by 20 mm in height at 28 keV. Diffraction images were recorded through the diamond cell on a MAR345 image plate positioned 32.10 cm from the sample. X-ray diffraction patterns were obtained using the program Fit2D (Hammersley et al. 1996) with a polarization factor of 0.93.

SR-XRD peak positions were calculated according to the Bragg equation and compared with database (JCPDS **97**) [5]. Miller indices are shown in parenthesis and represent the different faces of Si_3N_4 thin film in the upper pattern and bare Si in the bottom curve.

Figure 1 shows the SR-XRD scans for a 50nm Si_3N_4 thin film grown on Si by CVD method and a comparative bare Si substrate. The data are greatly different from those by ordinary XRD experiments using a wavelength 1.54 Å of cooper (Cu) anode target, as shown in Figure 2.

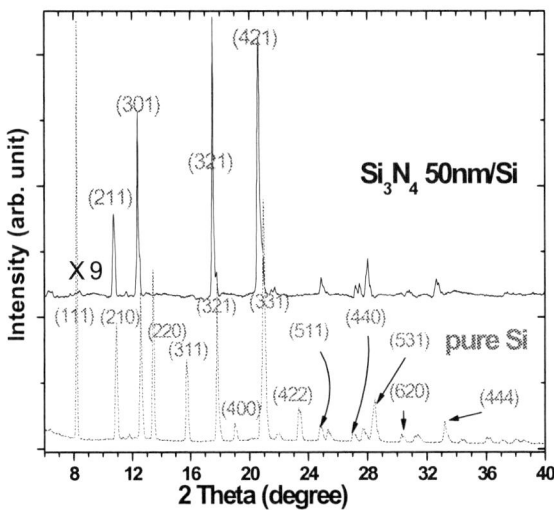

Fig. 1 Synchrotron radiation X-ray diffraction (SR-XRD, 0.443Å) patterns for a 50-nm Si_3N_4 thin film grown on Si and the comparative bare Si substrate.

In order to recognize and identify the XRD features from Si wafer and Si_3N_4 film clearly, the expanded and narrower scans are shown in next two figures.

Figure 3 exhibits patterns between 7° to 23°. It is seen that four major lines from Si_3N_4 thin film have shoulders corresponding to Si bare

wafer, and some other strong Si-lines, such as (111), (220) and (311) etc, are not appeared Therefore, these four major lines are from the Si_3N_4 film.

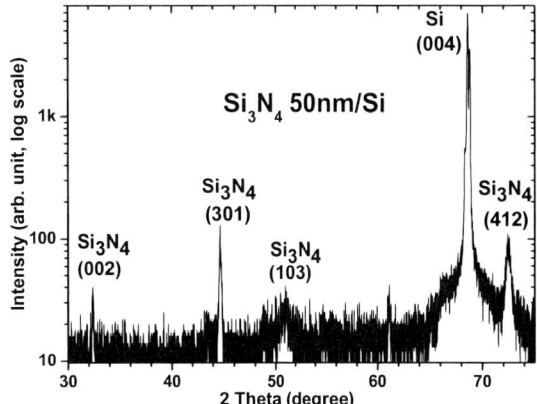

Fig. 2 Ordinary X-ray diffraction (XRD, 1.54 Å) patterns.

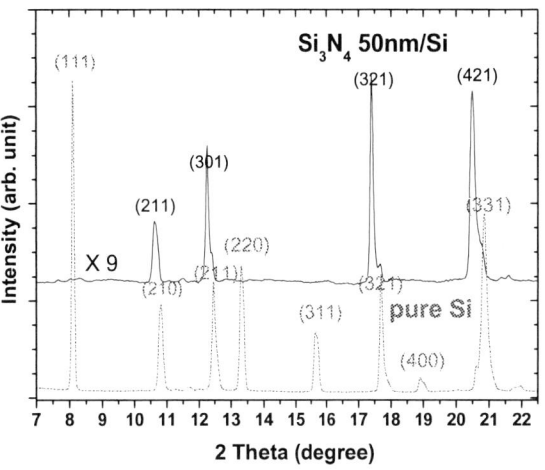

Fig. 3. SR-XRD patterns for Si_3N_4/Si and Si in the range from 7° to 23°.

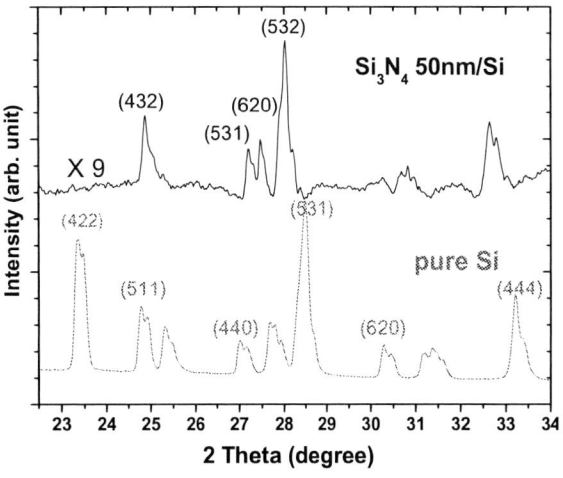

Fig. 4. SR-XRD patterns for Si_3N_4/Si and Si in the range from 22° to 34°.

The diffractive intensities from 22° to 34°, as shown in Figure 4, are relatively smaller than that from 7° to 23°. The mainly observed peak positions are listed in the table 1.

Through careful analyses, we found that there is a mixture state of α- and β- Si_3N_4 and other impurities present in the samples and index become smaller than pure Si_3N_4. [1].

Table 1. Peak positions of Syn-XRD experiment results, and compared with database (JCPDS **97**). Miller indices are shown in parenthesis and represent the different faces of Si_3N_4 thin film.

Si	database (2θ)	8.10° (111)	13.24° (220)	15.54° (311)	20.47° (331)
	experiment (2θ)	8.14°	13.51°	15.76°	20.89°
α- Si_3N_4	database (2θ)	10.98° (211)	12.22° (301)	17.73° (303)	20.61° (421)
	experiment (2θ)	10.82°	12.32°	17.53°	20.65°
β- Si_3N_4	database (2θ)	10.21° (210)		17.52° (002)	20.49° (420)
	experiment (2θ)	10.82°	12.32°	17.53°	20.65°

Figure 5 shows the SR-XRD scans for a 30nm TaN thin film grown on Si by CVD method and a comparative bare Si substrate.

Figure 6 shows SR-XRD narrower range patter with peak positions labeled from calculation according to the Bragg equation and compared with database (JCPDS **97**) [5].

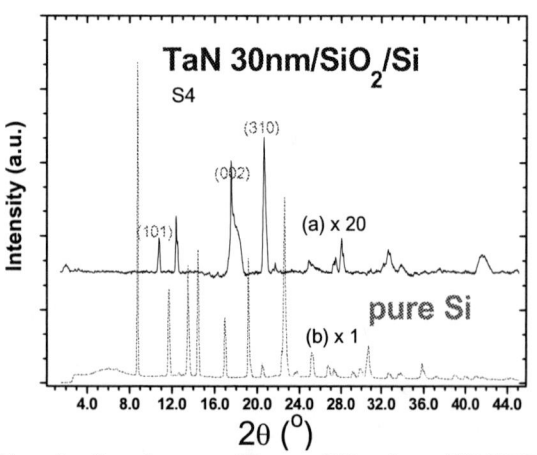

Fig. 5. Synchrotron X-ray diffraction (SR-XRD, 0.443Å) patterns for a 30nm TaN thin film grown on the Si substrate.

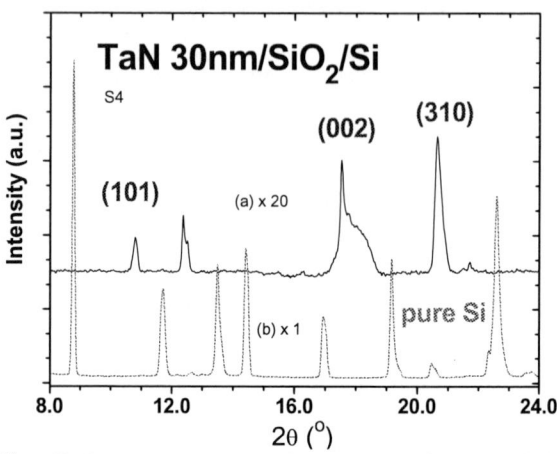

Fig. 6. SR-XRD patterns for TaN/SiO₂/Si and Si in the range from 8° to 24°.

Figure 7 (a) shows the SR-XPS wide scans for a thin silicon nitride layer grown on 12" Si wafer. Figure 7 (b) shows a SR-XPS wide scan for a tantalum nitride layer grown on 12" SiO_2/Si wafer. The corresponding peaks of Si 2p, Si 2s, C 1s, N 1s and O 1s are labeled in the figure [6].

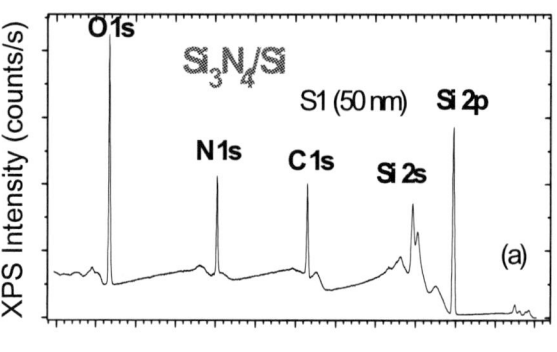

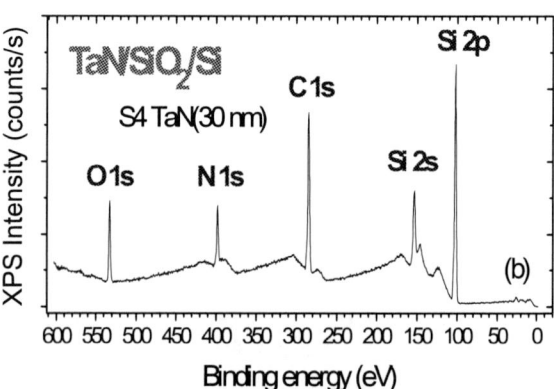

Fig. 7 (a) The synchrotron radiation X-ray photoelectron spectroscopy (SR-XPS) wide scan for a silicon nitride thin film grown on Si, (b) SR-XPS wide scan for a tantalum nitride thin film grown on SiO_2/Si.

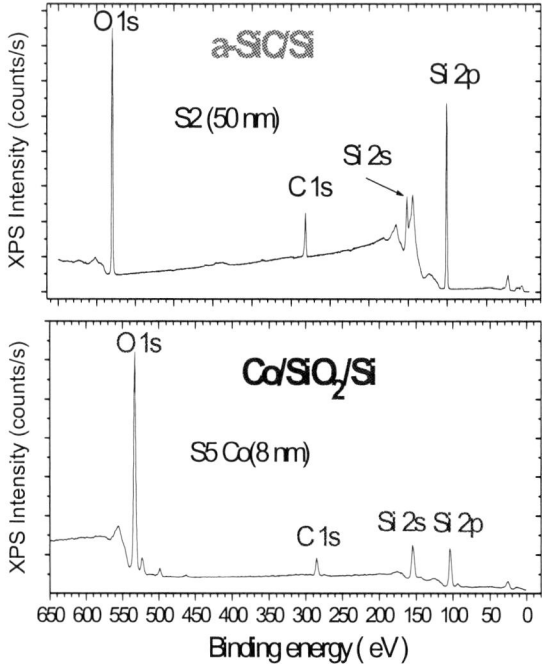

Fig. 8 (a) The synchrotron radiation X-ray photoelectron spectroscopy (SR-XPS) wide scan for a amorphous silicon carbide thin film film grown on Si, (b) SR-XPS wide scan for a cobalt thin film grown on SiO_2/Si.

Figure 8 (a) shows a SR-XPS wide scan for an amorphous silicon carbide layer grown on 12" Si wafer. Figure 8 (b) shows a SR-XPS wide scan for a 8-nm cobalt thin layer grown on 12" SiO_2/Si wafer. The corresponding peaks of Si 2p, Si 2s, C 1s, and O 1s are labeled in the figure also [6].

By the use of superior experimental instruments, we are going to plan different sets of synchrotron XRD measurements using synchrotron radiation sources with energy ranging between 5 keV to 33 keV. So we will have different wavelengths (0.44-2.6 Å) excitation XRD data on all the samples. This will benefit to analyze the studied materials and lead to better understand materials properties. We are also going to plan extensive SR-XPS investigation that will focus more on individual element fine scans from the samples, so that we can study the surface oxidizations and to estimate the degree of pollutions by carbon and nitride, which exist widely in our surroundings.

IV. CONCLUSIONS

We have studied systematically the Si-based layer structures for sub-micron Si-IC applications by SR-XRD and SR-XPS. From the SR-XRD result of 50 nm-thick Si_3N_4 grown on the silicon, it has pointed out a mixing state of α- and β- Si_3N_4 exists on the surface. The result of SR-XPS indicates that the existence of the specific composition element and the environmental elements, such as O 1s and C 1s , which exist extensively in our surroundings.

ACKNOWLEDGEMENT

The work at National Taiwan University was supported from TSMC (Taiwan Semiconductor Manufacture Corporation) 2006 Semiconductor Collaborative Research Funds and by funds from National Science Council of Republic of China, NSC 94-2215-E-002-019 and 95-2221-E-002-118.

REFERENCES

[1] Maurício F. Gozzi, Eduardo Radovanovic, I. Valéria P. Yoshida, Materials Research **4**, No.1, p.13, 2001

[2] Long-Wei Yin, Yoshio Bando, Ying-Chun Zhu, and Yu-Bao Li, Appl. Phys. Lett.**83**, No.17, p.3584, 2003

[3] J. Z. Jiang, H. Lindelov, L. Gerward, K. Sta°hl, J. M. Recio, P. Mori-Sanchez, S. Carlson, M. Mezouar, E. Dooryhee, A. Fitch, and D. J. Frost, Phys. Rev. B **65**, p.161202-1, 2002

[4] Steven D. Jacobsen, Jung-Fu Lin, Ross J. Angel, Guoyin Shen, Vitali B. Prakapenka, Przemyslaw Dera, Ho-kwang Mao and Russell J. Hemley, Journal Syn. Rad.**12**, p.577, 2005

[5] Joint Committee on Powder Diffraction Standards (JCPDS **97**)

[6] John F. Moulder, William F. Stickle, Peter E. Sobol, and Kenneth D. Bombe*n, "Handbook of X-ray photoelectron spectroscopy"*, published by physical electronics division of Perkin-Elmer Corporation, Eden Prairie, Minnesota 55344, USA.

A Highly Linear CMOS Down Conversion Double Balanced Mixer

Kumar Munusamy and Zubaida Yusoff
Faculty of Engineering, Multimedia University, Cyberjaya, Selangor, Malaysia.
Email: kumar.munusamy02@mmu.edu.my, zubaida@mmu.edu.my

Abstract This paper presents a *High Linearity CMOS down conversion Double Balanced Mixer* for IEEE802.11/g Wireless LAN application with 2.4 GHz operating frequency. In this Gilbert type mixer design, various high linearity techniques have been incorporated such as current-reuse bleeding technique, common gate transconductance amplifier configuration and tuned loads techniques. All these techniques were combined into a single design and the comparison of this proposed mixer with the recent literature shows significant improvement in linearity parameters such as Intermodulation (IMR3), Third-Order Input Intercept Point (IIP3) and 1dB Compression Point without degrading other important parameters. The mixer structure is designed using TSMC 0.25um standard CMOS technology and is simulated using EldoRF simulator from Mentor Graphics environment. The mixer's simulated result shows the Input Intercept Point (IIP3) of 12.810dB, the Intermodulation IMR3 of 129.816dB and the 1dB Compression Point of 5.075dB. The mixer operates at 1.8V with 13.30mW power consumption. Meanwhile, the measured conversion gain and Noise figure of this double balanced mixer were -2.688dB and 13.678dB respectively.

Keywords: CMOS design, Double Balanced Mixer, High Linearity, Gilbert Cell.

1. INTRODUCTION

With the ever growing demand of wireless communications services, the development of portable communication products is rising exponentially. This has resulted in severe competition among various service providers and manufacturers whose main concern is to offer low power, low cost, high linear products

and services to catch the attention of large number of consumers. These consumer based requirements have, in turn, increased pressure on radio frequency (RF) circuit designers to explore for new technologies to satisfy both consumers and manufacturers. Moreover, the super heterodyne receiver has been the most widely used architecture for modern radio communication receivers, which is best understood to achieve high performance. One of the basic building blocks of super heterodyne receiver is the down conversion mixer, which converts the incoming RF signal to the intermediate frequency (IF) signal.

The linearity performance requirement becomes more significant in modern RF mixer. In particular, for the use at unlicensed 2.4-GHz bands, a highly linear receiver is required for immunity to the various interferer signals from other communication standards. [1].

Among many proposed active mixers, the Gilbert-cell mixer has been widely used so far, and the double-balanced mixer topology has been preferred since it can suppress leakage signals at the output (Fig. 1). The down-conversion mixer is required to provide a high linearity, low noise figure, and enough power conversion gain.

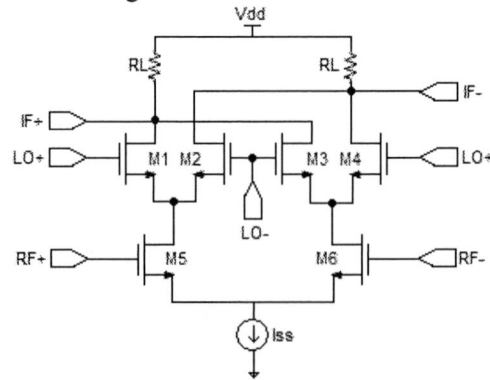

Figure 1.Double-balanced Gilbert-type mixer Topology.

It is preferably, the IF signal output to be directly proportional to the RF input signal amplitude; this is the sense in which we interpret the term "linearity" in the context of mixers. However, as with amplifier (and virtually any other physical system), real mixers have some limit beyond which the output has a sub linear dependence on the input. The non-linearity in the circuit causes the power gain to deviate from its idealized curve.

The point at which the power gain is 1 dB from the ideal linear curve is referred to as 1 dB Compression Point (P1dB) and is a measure of dynamic range of a mixer. The receiver must operate several dB below this level to avoid non-linear behavior and distortion in the output signal.

One of the problems which gives headaches to designers of RF amplifiers and mixers is the third-order intermodulation. Intermodulation happens as soon as the circuit exhibits some degree of non linearity. For mixers, the basic phenomenon is that two signals within the RF frequency band with frequencies, say f_1 and f_2, (Fig. 2) will generate output frequencies at $F_{LO} - (2f_1-f2)$ and $F_{LO} - (2f_2-f_1)$; which F_{LO} is the local oscillator (LO) frequency. If f_1 and f_2 are close to each other, $F_{LO} - 2f_1-f_2$ and $F_{LO} - 2f_2-f_1$ can fall within the IF frequency band, which will degrade the signal-to-noise ratio of the communication system. So it is always a design goal to minimize the level of these spurious intermodulation products, which most of the time implies on controlling the non-linearity of the component or block. [12].

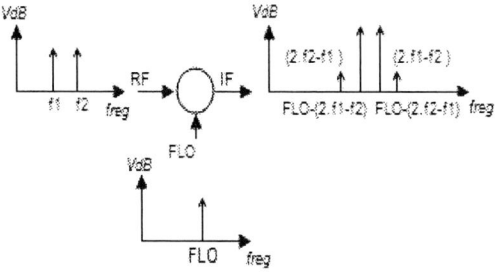

Figure 2.Intermodulation products of mixer.

To model our two interferers, we want to create a signal such as:

$$V_{RF} = a_1.\sin(2\pi.f1t) + a_2.\sin(2\pi.f2t) \qquad (1)$$

II. PROPOSED MIXER TOPOLOGIES AND DESIGNED.

In this paper, few methods have been proposed to improve the linearity of the mixer due to the nonlinear phenomenon of the transconductor stage and also load stage. The proposed techniques which will be discussed here are common-gate transconductance amplifier configuration, tuned loads at IF output and current reuse bleeding technique.

A. Common-Gate Transconductance Amplifier Stage with Source Resistor

One of the key point for better linearity is through maintain the Q-point stability of the MOSFET. This can be achieved by adding a source resistance at the input stage. A source resistance Rs will stabilize the Q-point against variation in transistor parameters. For example, the value of transconductance, *gm* varies from one transistor to another, the Q-point will not vary as much if a source resistor included in the amplifier stage. The overall voltage gain (Gv) for Common gate (C-G) stage of the mixer can be derived as: *Gv= RL/(Rin+Rs)*. It can be observed that the gain has been more stabilized throughout the operation. However, the trade-off is that the source resistor reduces the voltage gain.

To understand how common gate transconductance amplifier could support wide range of frequency response in other word better linearity, the basic common-gate transconductance amplifier configuration is analyzed by ignoring Rs and let r_o to be infinity for simplicity. In such a case, the input stage is isolated from the output stage, and the simplified high-frequency equivalent circuit is shown in Fig. 3(b). Both isolated stages are creating two poles that are one at the input stage (f_{P1}) and the other is at output stage (f_{P2}). The respective poles are calculated to be,

$$f_{P1} = \frac{1}{2\pi C_{gs}\{R_s \parallel 1/(g_m+g_{mb})\}} \qquad (2)$$

$$f_{P2} = \frac{1}{\underline{\hspace{3cm}}} \qquad (3)$$

$$2\pi\ (C_{gd}+C_L)\ RL$$

The 3-dB frequency f_H can be determined using f_{P1} and f_{P2}. When r_o is taken into account since R_S and R_L are large, an estimated f_H can be obtained [3].

$$R_{gs} = R_s \parallel R_{in}\ , \qquad (4)$$

$$R_{gd} = R_L \parallel R_{out} \qquad (5)$$

and

$$f_H = \frac{1}{2\pi[C_{gs}R_{gs}+(C_{gd}+C_L)R_{gd}]} \qquad (6)$$

The common-gate configuration has excellent high-frequency response compare to common-emitter configuration which obtained,

$$f_H = \frac{1}{2\pi(C_{gd}+C_L)R'_L} \qquad (7)$$

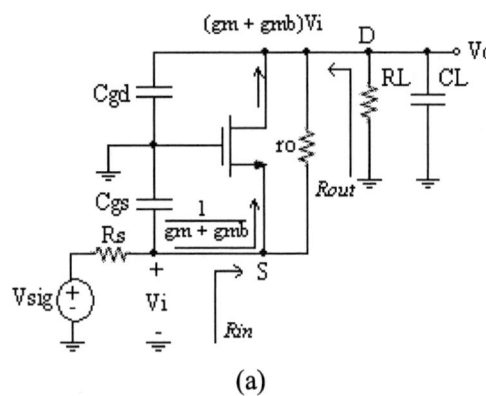

(a)

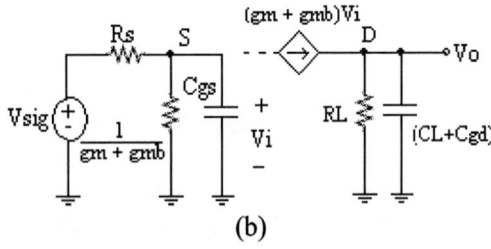

(b)

Figure 3. (a) The Common-Gate amplifier with internal capacitance (Cgd & Cgs) & load capacitance (*CL*). (b) Equivalent circuit when *ro* is neglected.

B. Tuned Load

The down converting mixer has a broadband input and fixed IF output. It is only needed to provide gain over narrow frequency range centered on the IF frequency. If headroom is a problem, a tuned load shown in Fig.4, can be used to provide larger output swings.

At DC, the inductor is shorted and hence there is no drop across the tuned load so there is more room to work with. At the resonating frequency of the tank, the gain becomes *gmR* since the inductor and capacitor are an open circuit at this frequency. The admittance of the tuned loads (Fig.4) can be derived and expressed as follows:

$$Y = G+ j\omega_C+(1/j\omega_L) = G+j[\omega_C-(1/\omega_L)] \qquad (8)$$

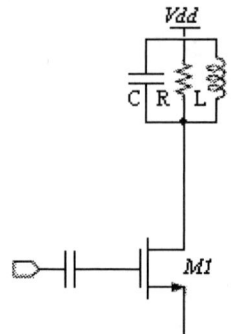

Figure 4. Tuned loads on a mixer.

From inspection of the equation, it was observed that the admittance goes to infinity, both at low frequency and high frequency where the inductor and capacitor are short circuited respectively. Meanwhile, the capacitor and inductor admittances are canceled at resonant frequency that is given by, [2]

$$[\omega_o C – (1/\omega_o L)]=0 \text{Thus}, \ \omega_o = 1/\sqrt{LC} \qquad (9)$$

and the bandwidth is,

$$B = 1/(CR) \qquad (10)$$

C. Current-reuse Bleeding Technique

Higher gain and better linearity can be achieved by increasing the drive current through the transconductance stage, [5] but power consumption will be increased. Besides, the larger current through the switching quads causes the voltage headroom problems especially for low voltage circuit design. Thus, the challenge is to provide larger current at transconductance but moderate current at switching stage. This can be realized with current-reuse bleeding technique.

In this design, PMOS transistors M10 and M11 (Fig.5) create the bleeding currents under the gate bias voltage. With this technique, the current through switching transistor is reduced, such that the 1/f noise (flicker noise) is improved and the output load resistance is increased leading to a higher gain [6].

III. SIMULATION RESULTS AND DISCUSSION

The proposed mixer in Fig. 5 is simulated in a TSMC 0.25µm CMOS process by Mentor graphics. The active current of mixer is about 9.151mA from a 1.8V supply voltage. By using two tones testing, the two tones are located at 2440MHz and 2441MHz,

respectively. Fig 6 illustrates IM3 to be 129.816dB, while the IIP3 is 12.810dB, that is calculated from equation (11).

$$IIP3 = VdB(input)-IMR3/2 \qquad (11)$$

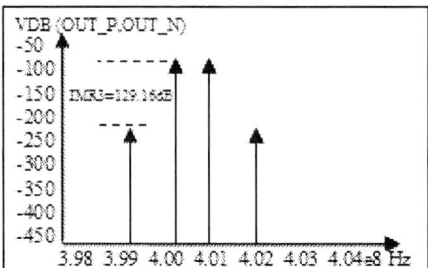

Figure 6. Intermodulation Products and IMR3

Using all the techniques applied in this design; the C-G transconductance stage with source resistor, current-reuse bleeding technique and tuned loads, the performance of the linearity parameters is boosted.

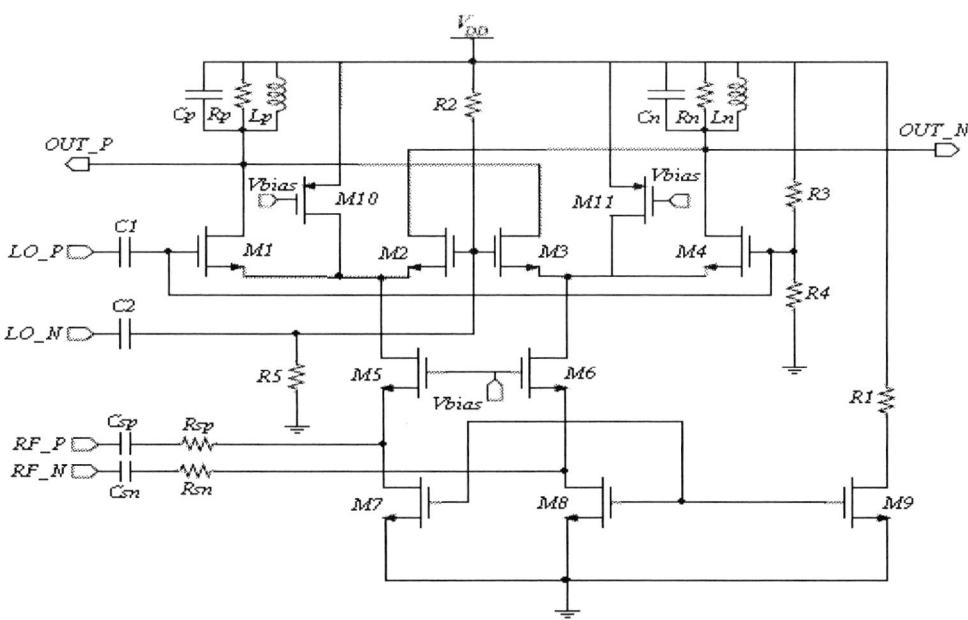

Figure 5. The Proposed Double-Balanced Mixer Topology.

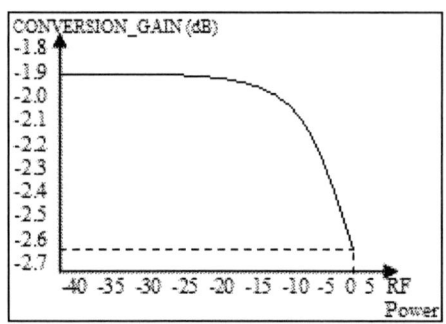

Figure 7. Conversion gain as a function of input RF Power.

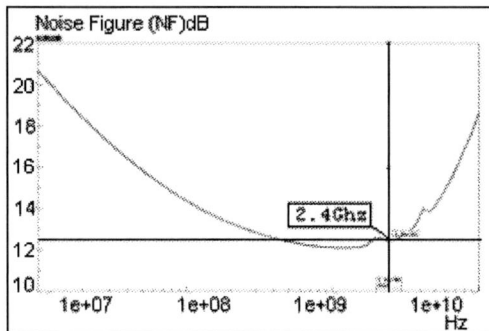

Figure 8. Noise Figure (NF)

The conversion gain decreases with respect to the RF power (Fig.7) due to use of source resistance Rs. However, the gain can be improved by using current bleeding technique or by adding LNA at the front end of superheterodyne receiver.

From the AC analysis, the extracted Noise Figure (NF) is 13.678dB, Improving the noise figure of mixer is difficult because of the cyclostationary nature of the sources. However, carefully chosen of the transistor value may result in optimal level.

In Table 1, linearity parameters like Intermodulation IMR3, Input IP3, 1dB compression point and other mixer associated parameters are compared with those in [7,8,9,10,11]. From the Table 1, the designed down conversion mixer has an outstanding performance, especially the IIP and 1-dB value, compared to other previously reported designs. However, power consumption and conversion gain are still at a reasonable level.

Table 1. Performance Comparison

Parameters	[7]	[8]	[9]	[10]	[11]	This work
Process (um)	0.35	0.35	0.18	0.8	0.5	0.25
Supply Voltage (V)	3.0	2.0	1.5	3.3	2.5	1.8
fRF (GHz)	1.9	0.9	2.4	1.0	1.9	2.44
Power consumption(mW)	24	7.2	5.6	7.3	10	13.30
IMR3 (dBm)	N.A	N.A	N.A	N.A	N.A	129.816
$IIP3$ (dBm)	-3	-3.3	5.46	0.6	2.17	12.810
P-1dB (dBm)	-8	-15.4	-8.98	N.A	-8.2	5.075
Conversion Gain (dB)	7.5	1.1	3.3	-0.8	3.35	-2.688
Noise Figure (dB)	10(SSB)	N.A	14.87	19	9.04	13.678

IV. CONCLUSION

In this paper we presented a highly linear down conversion CMOS double-balanced mixer with various linearity techniques such as tuned loads, source resistance with CG configuration and current reuse bleeding technique which helps to increase the linearity of the mixer. The simulation results show that this mixer achieves IIP3 of 12.810dB, P-1dB of 5.075dB and the conversion gain of -2.688dB and the power consumption of 13.30mW. The final size of the mixer layout is 108µm x 123µm as in Figure 9. The proposed mixer can be fit for WLAN applications.

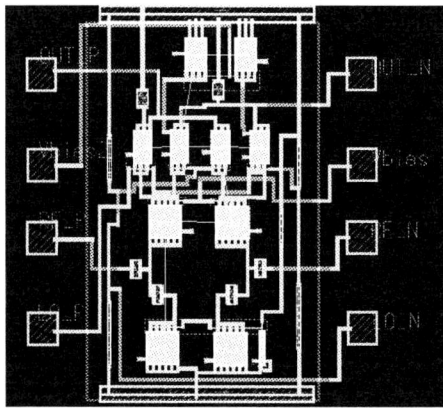

Figure 9. The layout of the design

REFERENCES

[1] Ickjin Kwon and Kwyro Lee, IEEEMICROWAVE AND WIRELESS COMPONENTS LETTERS , VOL 15 NO 1, Jan 2005.

[2] Thomas H Lee, "The design o f CMOS RF ICs," Cambridge University Press, Jan 1998.

[3] Sedra Smith "Microelectronic Circuits" fifth edition,Oxford University Press, 2004.

[4] B. Razavi, RF Microelectronics, Englewood Cliffs,NJ: Prentice-Hall, 1998.

[5] Q. Li, et al, " Linearity Analysis and Design Optimization for 0.18 um CMOS RF Mixer", IEE Proceedings of Circuits, Devices and Systems, Vol 149, Issue 2, April 2002.

[6] Anh- Tuan Phan, Chang- Wan Kim, Min-Suk, Kang Sang- Gug Lee" *A High Performance CMOS Direct Down Conversion Mixer For UWB System*"*GLSVLSI'04,* April 26-28, 2004, Boston, Massachusetts, USA.

[7] K.Nimmagadda and G.M. Rebeiz, "A 1.9 Ghz double-blanced subharmonic mixer for direct conversion receivers ," IEEE Radio *FrequencyIntegrated Circuits Symposium,* pp.253-256,May 2001.

[8] C.F Au-Yeung and K.K.M.Cheng, "CMOS mixer Linearization by the low-frequency signal injection method," IEEE MTT-S International Microwave *Symposium Digest, vol 1, pp.95-98, June 2003.*

[9] Hung-Che Wei, Ro-Min Weng, Chih-Lung Hsiaoand Kun-Yi Lin, " A 1.5V 2.4Ghz CMOS \ Mixer With High Linearity", The 2004 IEEE Asia-Pacific Conference Circuit and Systems,Dec 6-9, 2004.

[10] S.lee, C.Yoo W.Kim, HK Ryu and W.Song, "1Ghz CMOS down-conversion mixer, " Proc. IEEE Int. Symp. Consumer Electronics, pp.125-127, 1997

[11] H.Kilicaslan, H.S Kim, and M.Ismail, "A 1.9Ghz CMOS down-conversion mixer, " Proc. 40th Midwest Symp. Circuit System, vol2 pp.1172-1174, 1998.

[12] EldoRF, Mentor Graphic OnlineManual: Software Version 6.2_1, Release 2003.3.

Solder Bump Strength and Failure Mode of Low-k Flip Chip Device

Zulkarnain Endut[1&2], Ibrahim Ahmad[1], *Member, IEEE,* Gary Lee How Swee[2] and Norazham Mohd Sukemi[2]

[1]Faculty of Engineering, University Kebangsaan Malaysia 43600,Bangi, Selangor, Malaysia
Tel: 603-89216322 Fax: 603-89216146 Email: rg253c@yahoo.com
[2]Freescale Semiconductor, (M) Sdn. Bhd.
No. 2, Jalan SS 8/2 Free Industrial Zone
Sungei Way Petaling Jaya 47300
Tel: 603-78823790

Abstract : **In this paper, the failure mode and solder bump strength for low-k flip chip devices were determined using die pull technique. The results show there is no significant difference between low-k and non low-k devices in terms of bumps strength for the amount of taffy in this device. However, there is different in failure mode which shows an increasing in VRO and SRO failure mode. Die pull test within a time and bake factor also help to minimize VRO and SRO failure mode. However, VRO and SRO failure mode were expected as an another impact of low-k materials on flip chip packaging.**

I. INTRODUCTION

Flip chip technology becomes popular in today's microelectronic packaging because this type of packages can support higher I/O pad counts and power distribution required due to the increase of device density on the chip. Flip chip packaging use a variety of interconnection technology to connect the active side of silicon chip to multilayered substrate using a process called die attachment [1]. In order to determine solder bump joint strength, ball shear test [2], ball pull test [3] and stud pull or die pull test [4] were used. However, for die attachment process in flip chip packaging, die pull test or JEDEC flip chip tensile pull test is found very useful to assess solder joint strength and the consistency of the process. The strength of the solder joint and their failure mode for every bump is used as the indicator for the solder joint quality at die attachment monitoring.

As the industry pushes miniaturization limits, copper technology and low-k dielectrics were introduced to increase IC speed and performance. Low-k dielectrics and copper technology had replaced conventional SiO2 and aluminum in new dual damascene structures. However, low-k materials were reported have poor interlayer adhesion, poor adhesion to copper, poor thermal stability, low modulus hardness, poor cohesive strength if compared to conventional interconnects structure. The impacts of these materials in flip chip packaging were reported by a lot of researcher in terms of mechanical and electrical reliability, the integrity of interfacial adhesion between low-k layer and copper and a need to find a suitable underfill materials to minimize the die stress and prevent inner layer delamination in silicon chip [5-9].

The impact of porous low-k materials at flip chip packaging was expected since this stage involves more mechanical and thermal stresses compare to IC fabrication level. In addition, the die pull test which is used to assess consistency of die attachment process is very hard test to solder joint since the objective of this test to separate the connection so that the strength of the joint can be measured. However, if the interfacial adhesion between the upper layers of interconnects or under bump metallurgy (UBM) is not strong enough to hold solder bump in die pull test due to low mechanical properties of low-k dielectrics layer, then the connection will not fails at solder bump joint but at dielectric interfacial adhesion or UBM. Consequently, the objective to assess the strength of the solder joint will be underestimated and leads to misunderstanding of the quality of the solder joint and the robustness of the die attachment process.

The objective of this paper is to discuss if the device with low-k materials will make an impact at die attach monitoring level. The die pull strength and the failure mode of this device will be discussed. In order to understand further,

0-7803-9730-4/06/$25.00 ©2006 IEEE

effect of staging time, bake condition and bake staging time also were investigated

II. EXPERIMENTAL

A low-k device was used in this study and was prepared for the purpose measuring the solder joint strength. The typical die pull testing used to measure die to substrate solder joint strength. This method consisted of attaching aluminum stud to the backside of the die for each sample and applying an increasing tensile load to the stud until the assembly separated. The silicon chip used in this study was 10 x 11 mm with 867 of high lead solder bump. The size of bump pad is 0.14 mm in diameter with Ni/Au plated surface finish. Die attachment process was done after no-clean flux dipping and reflowed using recommended reflow profile for high lead solder. We then recorded the die pull strength and failure mode for 10 lots with 10 units per lot. However, the failure mode studied is only the taffy, via rip out (VRO) and silicon rip out (SRO).

Taffy is defined as the good failure mode where the solder joint structure has been pulled to the point of separation and the final appearance of the solder joint is looked identical for both sides either the silicon chip or substrate side as shown in Figure 1(a). VRO is defined as the failure mode where the two or three metallization layer is attached to the pulled solder bump as shown in Figure 1(b). and SRO is defined as same as VRO but with the bigger diameter of silicon rip out compared to solder bump diameter as shown in Figure 1(c). Usually taffy failure mode is found for non low-k device, however as the low-k comes into packaging stage, SRO and VRO becomes more profound failure.

In order to understand the effect of staging time and bake condition on VRO failure mode, 2-level factorial design was used. Two factors were selected. The first factor was the staging time between die attachment process and die pull test. Day one is indicating that the die pull test was done right after the die attachment was performed and day two is the next day and so on. The second factor was bake condition and separated to 4 conditions; first condition was without bake, second condition was bake and pull immediately after bake, third condition was bake and pull after 3-4 hours staging time and fourth condition was bake and pull after 24 hours staging time. The response taken was the number of VRO failure against the total bump. The VRO samples then were average out from the 30 units samples which becomes a final response.

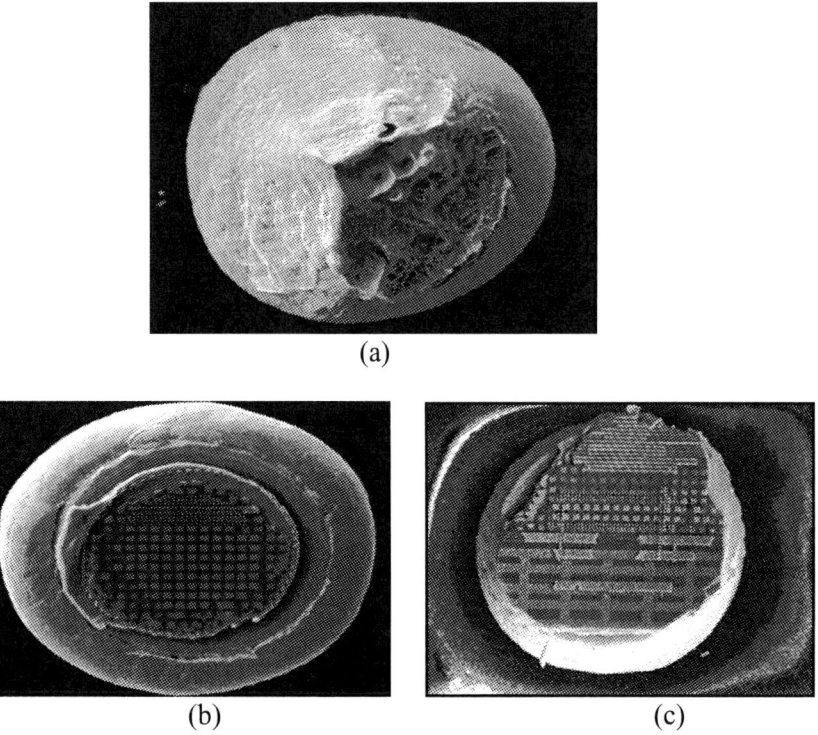

(a)

(b)   (c)

Figure 1. Failure mode of stud pull test (a) Taffy (b) Via rip out (VRO) – substrate side
(c) Silicon rip put (SRO) – substrate side

III. RESULTS AND DISCUSSION

Figure 2 shows the die pull test results for low-k device. From the results, the die pull strength from lot 1 to lot 10 is still higher than lower specification limit which is 17.3 gm/bump. For this amount of taffy as shown in Figure 3, the die pull strength for low-k device shows no significant different with non low-k devices results which is not covered in this paper. However, if the amounts of taffy lower than specification limit, the die pull strength for low-k devices may have lower strength leads to underestimation of the solder joint strength and misunderstanding of solder joint quality.

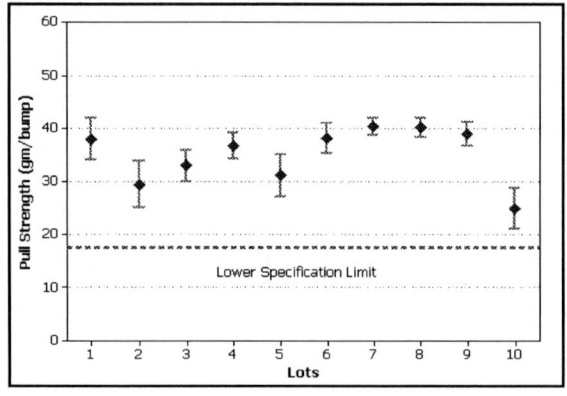

Figure 2. Die pull strength of low-K device

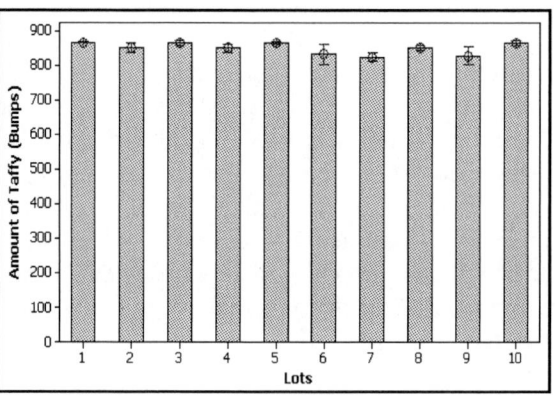

Figure 3. Amount of taffy failure mode
for low-K device

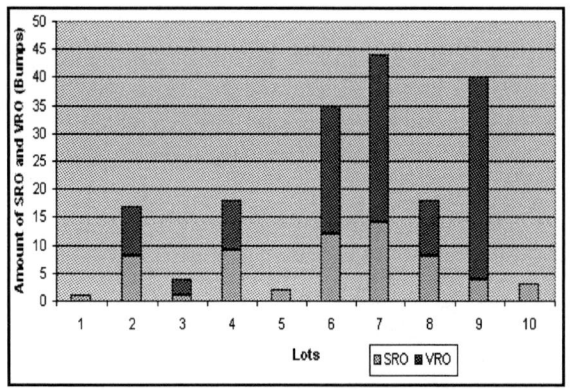

Figure 4. Amount of SRO and VRO failure mode
for low-K device

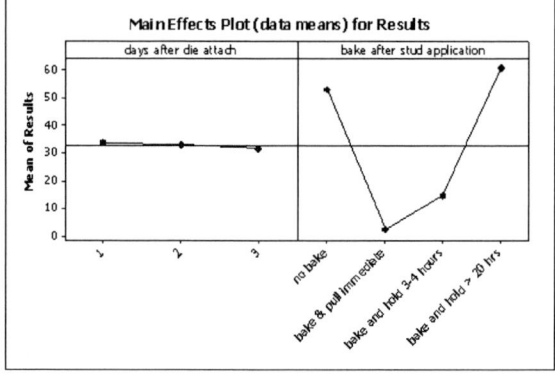

Figure 5. Main effects plot for effect of staging
time factor and bake after stud application on
VRO failure mode

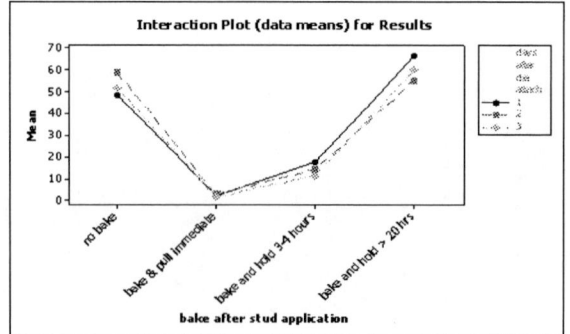

Figure 6. Interaction plot for bake after stud
application and days after die attach on VRO failure mode

```
Factor                          Type    Levels  Values
days after die attach           fixed        3  1, 2, 3
bake after stud application     fixed        4  no bake, bake & pull immediate,
                                                bake and hold 3-4 hours, bake and
                                                hold > 20 hrs


Analysis of Variance for Results, using Adjusted SS for Tests

Source                          DF    Seq SS   Adj SS  Adj MS       F      P
days after die attach            2      23.9     23.9    12.0    0.50  0.619
bake after stud application      3   14661.5  14661.5  4887.2  204.32  0.000
days after die attach*           6     253.7    253.7    42.3    1.77  0.189
  bake after stud application
Error                           12     287.0    287.0    23.9
Total                           23   15226.1



S = 4.89076    R-Sq = 98.11%    R-Sq(adj) = 96.39%
```

Figure 7. ANOVA analysis on VRO failure using general linear model

Figure 4 shows the SRO and VRO failure mode for low-k device for every lots. SRO failure mode was found in every lots of die pull samples and VRO failure mode was not found in lot 1,5 and 10. Figure 7 shows the analysis on our experiment in order to understand VRO failure mode. From the analysis, we found that the staging days after die attach is not significant, but the bake after stud application is very significant. From the main effect plot as shown in Figure 5, there is no different in term of VRO failures from day 1 to day 3. The main effect plot also shows there was a significant different in term of VRO failures as the condition of baking the units after the stud was put in. VRO is found very minimum at the stud baking and after pull immediately is done. VRO also increase if units were hold at 3-4 hours after bake before die pull test is done. VRO further increase if units were hold for over 20 hours. No bake condition also shows the VRO at higher end.

The reasoning for the less VRO failures after baking for 150 °C is believed to be due to softening of the bump from the heat, hence prompt it break as a taffy. This is from the facts that the units stage after stud baking for long hours exhibit high VRO failure mode. This can be seen on units with no baking. This result also shows that pull immediately after bake at 150 °C for 15 minutes will give a better VRO results.

VI. CONCLUSION

In summary, this study shows that the SRO and VRO failure at die attach monitoring were found as an another impact of low-K material on flip chip packaging. These failure mode were profound at die pull test of low-k devices compare to non low-k devices. This failure mode if exceed the specification limits is expected will gives effect to die pull strength results and leads to misunderstanding of solder joint strength due to undervalued of die pull results. Porosity and lower mechanical properties of low-k materials was expected in contributing of this phenomenon.

Our experimental on improvement of die pull testing of low-k device also shows that the bake after stud application will gives better die pull results. It is important to do die pull test after bake and within a time to minimize VRO failure, thus, the objective to assess consistency of die attachment process will be achieved and the solder joint strength will not be underestimated.

V. ACKNOWLEDGEMENT

The authors wish to acknowledge the Ministry of Science, Technology and Innovation (MOSTI), Malaysia under the IRPA grant no. 03-02-02-0121 PR001 for financial supports. Also to wish the following Freescale Semiconductor employees; Karthicase Letchumanan, Amin

Ismail, LJ Koh, HK Yip and LC Tan for their support on this studies.

VI. REFERENCES

[1]   Ho. P.S et al, " Reliability issues for flip chip packages," *Microelectronis Relibility*, Vol 44, pp. 719-737 (2004).

[2]   Se-Young Jang, Kyung Wook Paik, "Comparison of Electroplated Eutectic Bi/Sn and Pb/Sn Solder bumps on Various UBM systems," *IEEE-Trans-CPMT*, Vol. 24., No. 4, 269-274 (2001).

[3]   Fubin Song, Lee, S.W.R, "Effects of testing conditions and multiple reflows on cold bump pull test on of Pb-free solder balls," *Proc. of 6th Intrl. Conference on Electronic Packaging Technology*, pp. 474-480 (2005).

[4]   Craig Beddingfield, Leo Higgins III, "The effects of flux materials on the moisture sensitivity and reliability of flip chip on board assemblies," *IEEE/CPMT Int'l Electronics Manufacturing Technology Symposium (1997).*

[5]   Guotao Wang, Paul H. So, Steven Groothuis, "Chip packaging interaction: a critical concern for Cu/Low-K packaging," *Microelectronics Reliability*, Vol. 45, pp. 1079-1093 (2005).

[6]   Seung Wook Yoon, Vaidyanathan Kripesh, Li Hong Yu and Mahadevan K. Iyer, "150 µm pitch flip chip packaging with Pb free solder and Cu/low-K interconnects." *Proc. of Electronics Packaging Technology Conference*, pp. 126-131 (2004)

[7]   Mikel R. Miller and Paul S. Ho, "Interfacial adhesion study for Copper/SiLK Interconnects in flip chip packages," *Proc. of Electronics Components and Technology Conference* (2001).

[8]   Lei L. Mercado, Shun-Meen Kuo, Cindy Goldberg and Darrel Frear, "Impact of flip chip packaging on Copper/low-K structures," *IEEE Transaction on Advanced Packaging, Vol.26, No.4, pp. 433-440* (2003).

[9]   Sarathy Rajagopalan, Kishor Desai, Micheal Todd and George Carson, "Underfill for low-k silicon technology," *Proc. of IEEE/SEMI Int'l Electronics Manufacturing Technology Symposium* (2004)

Wavelength Shifting in the Fiber Bragg Grating (FBG) based Encoder and Decoder Modules for SAC-OCDMA System

Z.Zan, M.K. Abdullah *Member, IEEE,* S.A. Aljunid, *Member, IEEE*, R.K.Z. Sahbudin, M.H. Yaacob, *Member, IEEE,* M. Mokhtar, S. Shaari[1] *Member, IEEE*

Photonics and Fiber Optic Technologies Laboratory, Faculty of Engineering, Universiti Putra Malaysia.
Tel.: +603-8946 6454
Fax: +603-86567127
khazani@eng.upm.edu.my

[1]Institute of Microengineering and Nanoelectronics, Universiti Kebangsaan Malaysia, 43600, UKM Bangi,
Tel.: +603-8921 6308
Fax: +03-8921 6146
sahbudin@vlsi.eng.ukm.my

Abstract – **This project concentrates on the design of Fiber Bragg Grating (FBG) based encoder and decoder modules for Spectral Amplitude Coding of Optical Code Division Multiple Access (SAC-OCDMA) system. In SAC-OCDMA system, the unique code sequence is formed by using spectral components which are inherently arranged. This is done by multiplexing the Bragg wavelengths from an array of FBGs. However, the Bragg wavelength is largely depends on the strain and temperature experienced by the gratings. This paper presents the effects of the Bragg wavelength shifting of the uniform FBG used in the encoder and decoder modules for an SAC-OCDMA code to the system performance. The results show a sharp increase of bit error rate (BER) from 10^{-12} and 10^{-14} to 10^{-4} and 10^{-5} for Channel 1 and Channel 2 respectively at 0.01 nm Bragg wavelength left and right shifts. It shows that the system performance is significantly affected by the shifting of the Bragg wavelength.**

I. INTRODUCTION

The SAC-OCDMA system is receiving considerable attentions because of its capability to eliminate multiple access interference (MAI) and to suppress the effects of phase induced intensity noise (PIIN) which are the main reasons of the degradation of the system performance. This is

done by designing the coding scheme with an ideal in-phase cross-correlation λ_c between the code sequences.

Various approaches have been used to develop the encoder and decoder modules in order to generate the unique code sequences such as the employment of lenses and diffraction gratings with a patterned mask, fiber delay lines (FDL), arrayed waveguide gratings (AWG), and the use of uniform or tunable fiber Bragg gratings (FBG) [1-4]. For the FBG-based encoder and decoder modules, the Bragg wavelength largely depends on the strain and temperature introduced to the gratings [5-6]. These two elements will shift the value of the Bragg wavelength of the FBG. Meanwhile, in order to achieve an ideal λ_c the wavelengths assignment for the code sequences are restricted to certain values of Bragg wavelength and linewidth. Hence, the shifting of the Bragg wavelength will give a significant effect on the system performance.

II. MDW CODE

The encoder and decoder modules developed in this project are mainly designed for the DW code family. The MDW is an SAC-OCDMA code from the DW code family which allows higher code weight, *w* specifically any even number of *w* greater than 2. The major attraction of this code is the ability to maintain the cross-correlation value,

λ_c to an ideal value which is equal to one ($\lambda_c = 1$). Furthermore, MDW code can eliminate MAI and suppress PIIN completely by employing the detection technique called AND-subtraction technique [7-9]. In MDW code, the chips which represent the high bits of a code sequence are always allocated in-pair. Thus, the number of FBG used for an encoder or decoder can be reduced by half since a pair of chips can be covered by a single FBG with a broader linewidth.

III. WAVELENGTH SHIFTS

As mentioned before, the Bragg wavelength which also known as the reflected wavelength is largely depends on the temperature and strain introduced to the gratings. In SAC, a unique code sequence is constructed with inherently arranged spectral components. Every wavelength for a unique code sequence will has a certain linewidth and the code sequences used in the system will overlap once between any two code sequences. Therefore, in any two code sequences, the overlapping should take place at maximum half of the Bragg wavelength's linewidth ($\lambda_c = 1$). Hence, the fact that the temperature and strain will affect the value of an FBG's reflected wavelength will subsequently affect the overlapping region between any two code sequences which finally reduced the system performance.

The wavelength shifts are studied by introducing the shifts to a wavelength in Channel 1's code sequence. The wavelength chosen is the overlapping spectrum with the other unique code sequence. Meanwhile, the wavelengths in Channel 2's code sequence and the wavelength which represents the overlapping chip are remained unchanged. The effect of the wavelength shifts in Channel 1 is studied by using two conditions of the right and the left shifts. The wavelengths assignment for the right and left shifts are shown in Table 1 and Table 2, respectively. Table 1 shows that λ_{k1} spectrum is shifted to the right to the new wavelength values for every 0.01nm. In the meantime, the spectrum at Channel 2 λ_{k2} is remained static. In Table 2, λ_{k1} spectrum is shifted to the left with an identical movement as in the right shifts.

Table 1: Wavelengths assignment for right shifts

Wavelength (nm)	1549	1549.4	1549.8	1550.2	1550.6	1551
Channel1	0	0	0	0	1	1
Origin Wavelength:	λ_{k1}					1550.8
Shifts (nm):	0.01	Shifted				1550.81
	0.02					1550.82
	0.03	Wavelength (nm):				1550.83
	0.04					1550.84
	0.05	λ_R				1550.85
	0.06					1550.86
Channel2	0	0	0	1	1	0
Wavelength:	λ_{k2}			1550.4		

Table 2: Wavelengths assignment for right shifts

Wavelength (nm)	1549	1549.4	1549.8	1550.2	1550.6	1551
Channel1	0	0	0	0	1	1
Origin Wavelength:	λ_{k1}					1550.8
Shifts (nm):	0.01	Shifted				1550.79
	0.02					1550.78
	0.03	Wavelength (nm):				1550.77
	0.04					1550.76
	0.05	λ_R				1550.75
	0.06					1550.74
Channel2	0	0	0	1	1	0
Wavelength:	λ_{k2}			1550.4		

As shown in these tables, the code weight w used is 2 (DW code). This is done to make the observation easier since the shifting of the spectrum will only cause the widening or narrowing the overlapping region of the in pair chips in Channel 2.

The movements or the shifts of the spectrum are best described by using Fig. 1 which represents the right shifts. In this figure, the spectrum is assumed to be in ideal rectangular shape. Fig. 1 (a) shows the original spectra of Channel 1 while the shifted spectra of Channel 1 are shown in Fig. 1 (b) after a shift to λ_R. It is assumed that the spectrum is shifted by the FBG's center wavelength in both encoder and decoder modules.

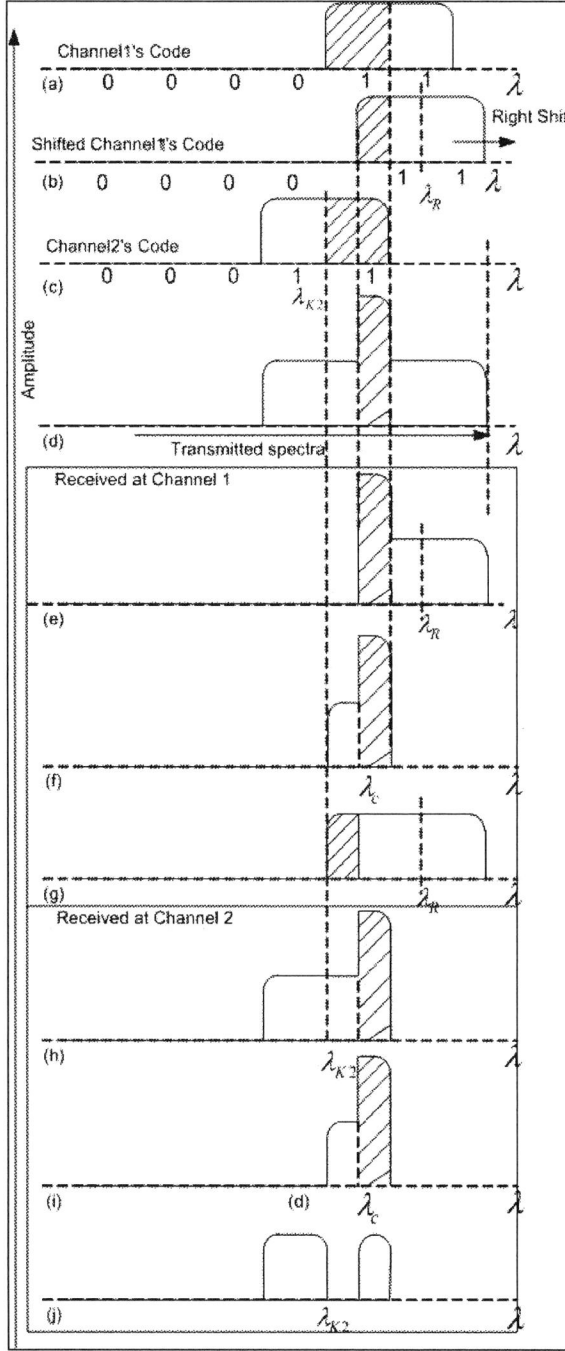

Fig. 1 Example of the spectrum right shift

and the overlapping area as shown in Fig. 1 (e). The overlapping area represents the information which belongs to Channel 2 and this is considered as a noise to Channel 1 decoder.

As mentioned before, the spectrum of the overlapping chip λ_c is un-shifted. Hence, the spectrum detected at the subtraction branch is illustrated as in Fig. 1 (f) and Fig. 1 (i) for Channel 1 and Channel 2 respectively. The Channel 1's resulting spectra after the subtraction process is shown in Fig. 1 (g) which indicates that the received spectra is still having the unwanted signal shown by the shaded area. Meanwhile, the resulting spectra of Channel 2 shows that the noise is completely eliminated but somehow, part of its' original spectrum is lost which is shown in Fig. 1 (j).

IV. SIMULATION SETUP

The simulation setup for this study is shown in Figure 2. Single FBG is used for every encoder module even though w of 2 is used in the code sequences. This is possible since the chips in DW and MDW codes are always in pair.

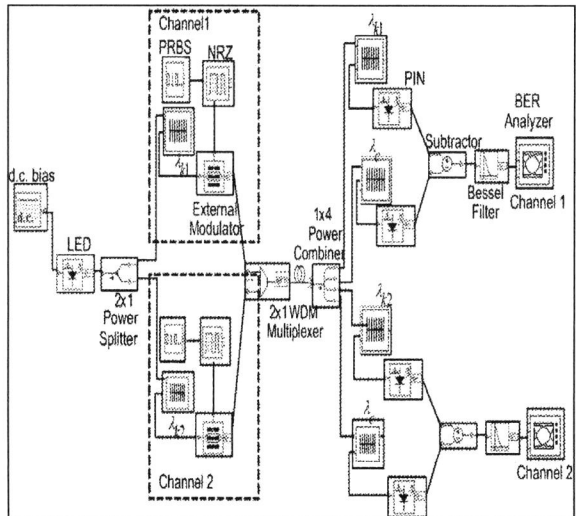

Fig. 2 Simulation setup

As shown in Fig. 2, the FBGs' center wavelengths are at λ_{k1} and λ_{k2} with the bandwidth of 0.8nm for Channel 1 and Channel 2, respectively. It means that each of the chips is represented by 0.4nm bandwidth. The shifts happened to λ_{k1} at both encoder and decoder modules with

Following that, the shifted spectra of Channel 1 are combined with the spectra of Channel 2's code sequence which shown in Fig. 1 (c), resulting a narrow shaded area which indicates the chips overlapping area as shown in Fig. 1 (d). After the transmission, the decoded spectra of Channel 1 are consisted of the mixing in the original spectra assumption that the shifts in encoder module are exactly the same with the shifts at the decoder module. An LED source with 6nm bandwidth and 8dBm of power transmit is used as the source with the center wavelength of 1550nm. The signal is conveyed through 8km distance with the bit rate of 622Mbps.

V. RESULTS

Fig. 3 shows the graph of BER versus spectrum right shifts. It shows that the increased in the right shifts of λ_{k1} will increase the BER. Initially, un-shifted λ_{k1} produced the BER of 2.43 x 10^{-13} and 1.22 x 10^{-12} for Channel 1 and Channel 2 respectively. When λ_{k1} is shifted by the first 0.01nm, a sharp increment of BER take place at 2.68 x 10^{-4} and 1.89 x 10^{-5} for Channel 1 and Channel 2, respectively. For the following right shifts, the BER gradually increased until it reached 3.39 x 10^{-3} and 1.51 x 10^{-4} for Channel 1 and Channel 2 at 0.09nm right shift.

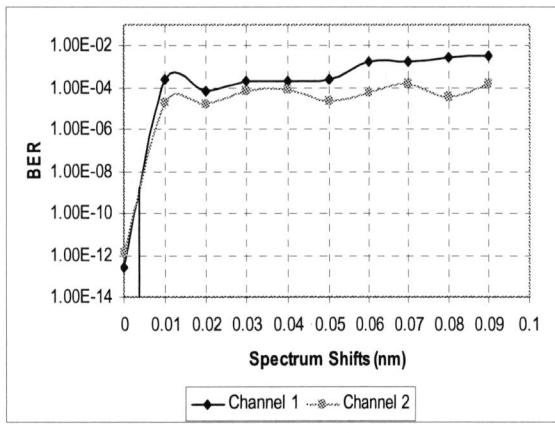

Fig. 3. BER versus right spectrum shifts

After broadening the scale of the spectrum shifts, it is estimated that at BER of 10^{-9}, the allowable right shift of the FBGs' center wavelengths is as small as within 0.0029nm and 0.0031nm. It means that, the system performance started to worsen at the FBG's spectrum shift of 0.003nm to the right.

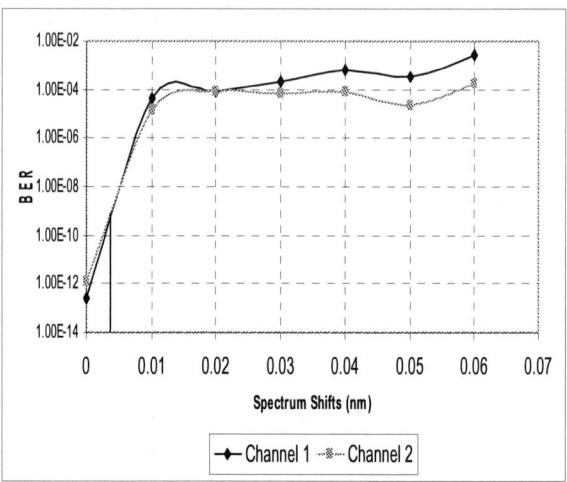

Fig. 4 BER versus left spectrum shifts

The pattern of the graph for the left spectrum shifts is almost the same with the right spectrum shifts which is shown in Fig. 4. At the minimum acceptable BER of 10^{-9}, the left spectrum shifts are estimated to be at as small as 0.003nm and 0.0039nm for Channel 1 and Channel 2, respectively.

VI. CONCLUSIONS

In SAC-OCDMA system, the benefits of FBG as the main component in the encoder and decoder modules has becomes popular. Its' advantageous including the ability to select the wavelengths by slicing the broadband source to form the code sequences, the simplicity in designing the encoder and decoder modules and even the cost of a fixed uniform FBG is fairly reasonable [10-17]. However, since the Bragg wavelength value is unstable due to the influenced of the temperature and strain to the FBG's grating, the use of FBG may becomes less popular to the code sequences which are restricted to wavelength assigned for each of the code sequence especially in term of the overlapping chips. It has been shown that, a small shift of 0.003nm in the FBG's reflected wavelength will lower down the system performance to the minimum acceptable BER of 10^{-9}.

REFERENCES

[1] Salehi, J.A., Weiner, A.M. & Heritage, J.P. 1990, "Coherent Ultrashort Light Pulse Code-Division Multiple Access Communication Systems", *Journal of Lightwave Technology* 8(3): 478–491.

[2] Holmes, A. S. & Syms, R. R. A. 1992, "All-Optical CDMA using "Quasi-Prime" Codes", *Journal of Lightwave Technology* 10 (2): 279-285.

[3] Djordjevic, I.B. & Vasic, B. 2003, "Unipolar Codes for Spectral-amplitude-coding Optical CDMA Systems based on Projective Geometries", *IEEE Photonics Technology Letters* 15(9): 1318 – 1320.

[4] Jepsen, A.G., Johnson, A.E., Maniloff, E.S., Mossberg, T.W., Munroe, M.J. & Sweetser, J.N. 1999, "Fibre Bragg Grating Based Spectral Encoder/Decoder for Lightwave CDMA", *Electronics Letters* 35(13): 1096-1097.

[5] F. Ouellette, "Tutorial on Fiber Bragg Gratings", Spie's OE magazine, January 2001.

[6] Mokhtar, M.R., Ibsen, M., Teh, P.C. & Richardson, D.J. 2002, "Simple Dynamically Reconfigurable OCDMA Encoder/Decoder Based on a Uniform Fiber Bragg Grating", *Optical Fiber Communication Conference and Exhibit (OFC 2002)*: 688 – 690.

[7] Aljunid, S.A.; Ismail, M.; Ramli, A.R.; Ali, B.M.; Abdullah, M.K. "A new family of optical code sequences for spectral-amplitude-coding optical CDMA systems", *IEEE Photonics Technology Letters*, Vol.16, Issue 10,Oct. 2004 Page(s):2383 – 2385

[8] S.A. Aljunid, Z. Zan, S.B.A. Anas, M. K. Abdullah, A New Code for Optical Code Division Multiple Access Systems, Malaysian Journal of Computer Science, vol.17, no. 2, Dec 2004

[9] Aljunid, S.A.; Samad, M.D.A.; Othman, M.; Hisham, M.H.; Kasiman, A.H.; Abdullah, M.K. "Development of modified double-weight code and its implementation in multi-rate transmissions", 13th IEEE International Conference on Networks, Jointly held with the IEEE 7th Malaysia International Conference on Communication, Vol.1, 16-18 Nov. 2005 Page(s):5 pp.

[10] J. F. Huang, C. M. Tsai; Y. L. Lo, "Compensating fiber gratings for source flatness to reduce multiple-access interferences in optical CDMA network coder/decoders", *Journal of Lightwave Technology*, vol. 22, Issue 3, March 2004 Page(s):739 – 745

[11] Jen-Fa Huang; Yao-Tang Chang; Chuen-Ching Wang, "Reductions of multiple-access and optical beat interference with fiber-grating OCDMA balanced decoder", *Advanced Information Networking and Applications, 2003. AINA 2003*, 27-29 March 2003 Page(s):168-17

[12] J. F. Huang, D.Z. Hsu, "Fiber-Grating- based optical CDMA spectral coding with nearly orthogonal M-sequence codes", *IEEE Photonics Technology Letters*, vol. 12, Issue 9, Sept. 2000 Page(s): 1252-1254

[13] Jen-Fa Huang; Dar-Zu Hsu, "Fiber-grating-based optical CDMA spectral coding with nearly orthogonal M-sequence codes", *Photonics Technology Letters, IEEE*, Vol 12, Issue 9, Sept. 2000 Page(s):1252 – 1254

[14] Djordjevic I.B.; Vasic, B., "Combinatorial constructions of optical orthogonal codes for OCDMA systems", *Communications Letters, IEEE*, Vol. 8, Issue 6, June 2004 Page(s):391 –393

[15] Djordjevic, I.B.; Vasic, B.; Rorison, J., "Multi-weight unipolar codes for multimedia spectral-amplitude-coding optical CDMA systems", *Communications Letters, IEEE*, Vol. 8, Issue 4, April 2004 Page(s):259 – 261.

[16] Djordjevic, I.B.; Vasic, B., "Novel combinatorial constructions of optical orthogonal codes for incoherent optical CDMA systems", *Journal of Lightwave Technology*, Vol. 21, Issue 9, Sept. 2003 Page(s):1869 – 1875

[17] Djordjevic, I.B.; Vasic, B.; Rorison, J., "Design of multiweight unipolar codes for multimedia optical CDMA applications based on pairwise balanced designs", *Journal of Lightwave Technology*, Vol. 21, Issue 9, Sept. 2003 Page(s):1850 – 185

Effects of the Power Differences in the AND-Subtraction Detection Technique in SAC-OCDMA System Performance

Z.Zan, S.A. Aljunid, *Member, IEEE,* M.K. Abdullah *Member, IEEE,* R.K.Z. Sahbudin, M.H. Yaacob,
Member, IEEE, M. Mokhtar, S. Shaari[1] *Member, IEEE*
Photonics and Fiber Optic Technologies Laboratory, Faculty of Engineering, Universiti Putra Malaysia.
Tel.: +603-8946 6454
Fax: +603-86567127
khazani@eng.upm.edu.my

[1]Institute of Microengineering and Nanoelectronics, Universiti Kebangsaan Malaysia, 43600, UKM
Bangi,
Tel.: +603-8921 6308
Fax: +03-8921 6146
sahbudin@vlsi.eng.ukm.my

Abstract – **This project concentrates on the effects detection technique to the system performance of spectral amplitude coding optical code division multiple access (SAC-OCDMA). The system employed the encoder and decoder modules based on the fiber Bragg gratings (FBGs) to generate the unique code sequences for the users. These modules are basically designed for the modified double weight (MDW) code which allows higher code weight *w* in the even number which is greater than two. The study is mainly focusing on the effects of the power differences between the upper and lower branches of the AND-subtraction technique used as the detection technique to the system performance. The results show that the system will achieve the best system performance when the power difference between the upper and lower branches is approximately 5dB.**

I. INTRODUCTION

Optical Code Division Multiple Access (OCDMA) is becoming a prominent multiple access technique and receiving considerable attentions in local area networks. The system offers various advantages such as providing a natural security to the system, increasing the network capacity and enabling multiple number of users to access the networks

simultaneously [1], [2]. This is done by assigning the unique code sequences to the users in the network. The code sequence will carry the data from the transmitter to the respective receiver. The unique code sequence will be recognized only by the intended receiver. In OCDMA, the main reason of the performance degradation when a large number of users are involved is the multiple access interference (MAI). In order to have a higher system performance, one needs to design a coding scheme which can eliminate the MAI effects. In Spectral Amplitude Coding (SAC), the unique code sequences are formed by using spectral components which are inherently in order. The spectral components are produced by spectrally slicing the broadband source's spectrum using arrayed of Fiber Bragg Gratings (FBGs) as the encoder module. Hence, from a given number of users K and a code length N, MAI can be determined by the value of in-phase cross-correlation λ_c between the unique code sequences.

In SAC-OCDMA system, MAI can be completely eliminated by using an ideal or fixed λ_c such as in *m*-sequence, Hadamard and MDW code [3], [4]. The codes can successfully eliminate MAI and also suppress the effects of phase induced intensity noise PIIN. PIIN occurred due to the spontaneous emission of broadband source. It can

0-7803-9730-4/06/$25.00 ©2006 IEEE

be suppressed by keeping the λ_c as small as possible [5]. The elimination of MAI is realized by using a detection technique which detects a signal from a normal decoder (*DEC*) and the complement of the decoded signal ($\overline{DEC}$) [3].

II. MDW CODE

MDW code is developed in a family of the Double Weight (DW) code [4]. This coding scheme is designed in a way to decrease the number of FBGs used in the encoder and decoder modules and to maintain the cross-correlation, λ_c parameter to 1. The MDW allows the code weight to be in any even number which is greater than two. The basic of this code is denoted by (N, w, 1) for the code length N, code weight w while the in-phase cross-correlation λ_c is always maintained at 1, respectively. The basic construction of MDW code can be best described by the K x N matrix as follows [4]:

$$H_{m1}=\begin{vmatrix} 0 & 0 & 0 & 0 & 1 & 1 & 0 & 1 & 1 \\ 0 & 1 & 1 & 0 & 0 & 0 & 1 & 1 & 0 \\ 1 & 1 & 0 & 1 & 1 & 0 & 0 & 0 & 0 \\ \uparrow & \uparrow & \uparrow & \uparrow & \uparrow & \uparrow & \uparrow & \uparrow & \uparrow \\ 1 & 2 & 1 & 1 & 2 & 1 & 1 & 2 & 1 \end{vmatrix}$$

As shown in H_{m1} matrix, all the high bits (bit 1 known as chip) are placed in pair. This will allow a single FBG to be used for every two chips. Noticed that, the H_{m1} matrix also has the columns combination of chips in the form of 1-2-1. The double weight pairs are maintained in order to allow only 2 overlapping of chips in every column. This is important to maintain $\lambda = 1$.

III. AND-SUBTRACTION TECHNIQUE

As mentioned before, the MAI can be eliminated by using a well designed detection technique. In this research, the detection technique used is the AND-subtraction technique. This technique can successfully eliminate the MAI, reducing the receiver complexity as well as improving the system performance [6].

In AND subtraction technique, the cross-correlation $\theta_{XY}(k)$ is substituted by $\theta_{(X\delta Y)Y}$, where $\theta_{(X\delta Y)}$ represents the AND operation between sequences *X* and *Y*. For example, let *X* = 0011 and *Y* = 0110 and therefore (*X* AND *Y*) = 0010. The technique can be best described by using Fig. 1.

At the receiver,

$$Z_{AND}=\theta_{XY}(k)-\theta_{(X\&Y)Y}(k)=0 \qquad (1.0)$$

Equation (1.0) shows that, with AND subtraction technique, the multiple access interference or the interference from other channels can also be cancelled out.

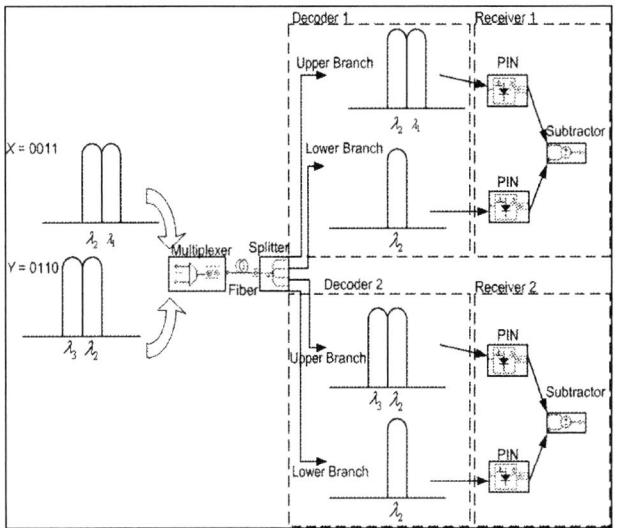

Fig. 1 Implementation of AND-Subtraction technique

These can be summarized by using Table 1 below where DW code is taken as the example.

Table 1: Theoretical MAI cancellation using AND-Subtraction technique

	λ_4	λ_3	λ_2	λ_1
X	0	0	1	1
Y	0	1	1	0
	$\theta_{XY}=1$			
	X AND Y = 0010			
	$\theta_{(XANDY)Y}=1$			
Z	$Z_{AND}=\theta_{XY}(k)-\theta_{(X\&Y)Y}(k)=0$			

1011

Table 1 shows that the product of the codes X and Y will produce one which indicates the overlapping chip occurred at λ_2. Then the product of X and Y is substituted by $\theta_{(X\delta Y)Y}$ which produced zero or in other words, it means that the MAI is eliminated.

IV. RESULTS AND DISCUSSION

The study of the power differences between the upper and lower branches has been carried out by observing the system with the employment of serial and parallel configurations of encoder and decoder modules. In these configurations, the FBGs are placed in serial and parallel to construct the encoder and decoder modules. In this study, the observation is mainly conducted at the receiver of the system which contains two PIN photodiodes for each user. Each PIN is connected to the upper and lower branches as shown in Fig. 2 (a) and (b) for the serial and parallel decoder module respectively.

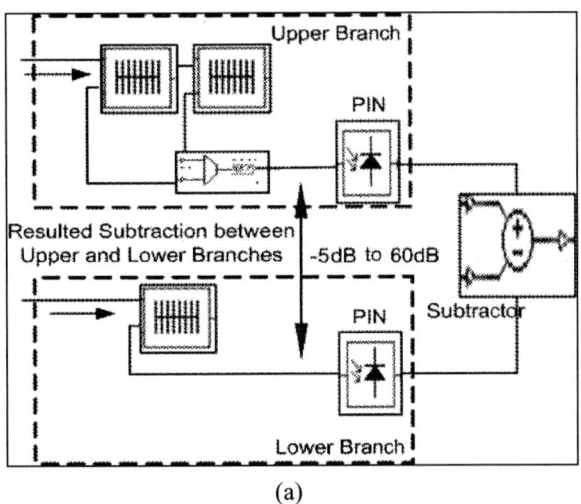

(a)

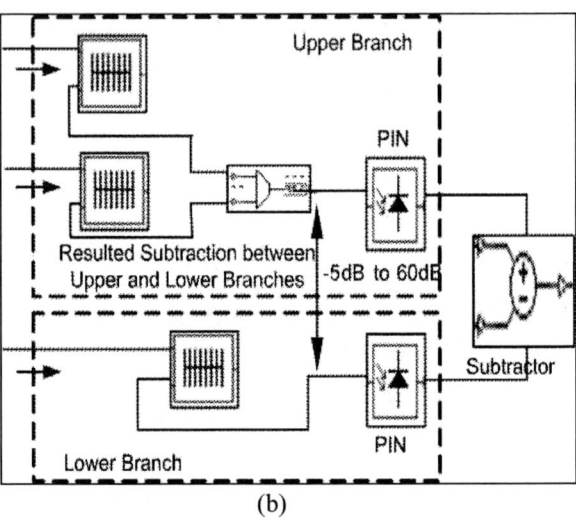

(b)

Fig. 2 Setup of AND-subtraction scheme (a) Serial decoder (b) Parallel decoder

Fig. 2 shows that the power received by the upper PIN and lower PIN are set to the values that will produce the resulted subtraction of -5dB to 60dB. This is done by placing the attenuators before both of the PINs to set the desired power at the upper and lower branches. The effect of the power differences to the system performance is studied based on the bit error rate (BER).

Fig. 3 and Fig. 4 show the graph of BER versus power differences for the serial and parallel configuration of encoder and decoder modules, respectively. The trend of the graphs obtained for the serial and parallel configurations are similar. Both graphs show that when there is no power difference between the branches or the upper branch received power is less than the power received by the lower branch (negative power difference), the system performance is poor with the BER of only 1.00. This is because the upper branch works as the decoder to detect the received codes and match them with the existing codes specified for the channel. Meanwhile, the lower branch represents the overlapping chips (or wavelengths) of a certain user's code sequence with the other users' code sequences in the system. The overlapping chips are also known as the correlated chips consist of the information which belongs to the other users possessing the same wavelength assigned in their code sequences. This information is observed as a noise and it needs to be eliminated. Therefore, if the power at the upper branch is smaller than the lower branch, hence the final power received will only contain the noise power. In other words, the noise from the lower branch is bigger than the signal power at the upper branch.

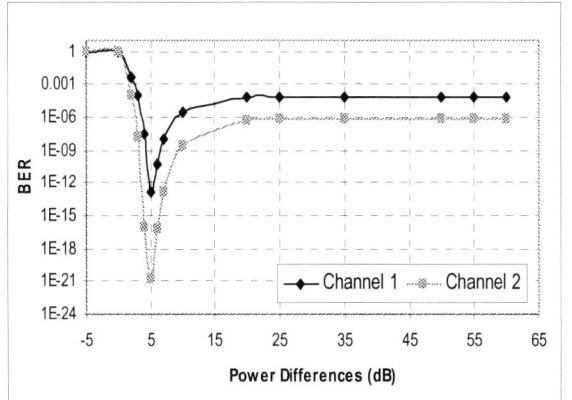

Fig. 3 BER versus power differences for serial configuration

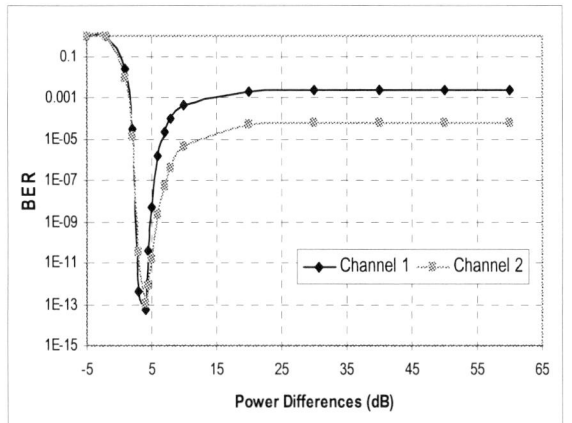

Fig. 4 BER versus power differences for parallel configuration

Both graphs show that the best system performance occurred when the power difference is approximately 5dB. For the serial configuration, the BER of 1.18×10^{-13} and 1.41×10^{-21} were achieved for Channel 1 and Channel 2, respectively at the power difference of 5dB. Meanwhile, BER of 3.82×10^{-11} and 8.64×10^{-13} for Channel 1 and Channel 2, respectively were achieved for the parallel configuration. Both graphs also show that for power difference from 5dB to 20dB, the BER is rising exponentially. At 20dB and onwards, the BER remained stable. This is due to the ideal amplifier used in the configuration to amplify the upper branch power to

-9.46dBm while the lower branch power is reduced to -29.46dBm by an attenuator. When the signal at the upper branch is amplified, the noise due to the correlation is also amplified. At the same time, the lower branch power which is used to handle the noise is attenuated to a small value. Hence, the process to eliminate the noise is unsuccessful since the noise is too high resulting from the amplification process.

V. CONCLUSIONS

The major limitation of the SAC-OCDMA system performance is mainly due to the MAI effects. A unique code sequence with a higher w are always preferable for the ability to provide a better system performance due to the high power received provided by a multi-weight code. Meanwhile, the multi-weight code sequence design strongly needs to consider the overlapping wavelengths assigned in the codes when it comes to the simultaneous transmission of multiple numbers of users. These in turn will increase the effects of MAI. Therefore, properly designed coding system and the detection technique used at the receiver are critical to increase the performance of the SAC-OCDMA system.

In this research, the detection technique used is the AND-subtraction which enables the simplicity of the decoder development. The MAI can be canceled without the need of a compensation scheme. The AND-subtraction technique can be optimized to give a maximum system performance when the power difference between the upper and lower branches is approximately 5dB.

REFERENCES

[1.] J. A. Salehi, "Code division multiple access techniques in optical fiber network—Part I: Fundamental principles," *IEEE Trans. Commun.*, vol.37, pp. 824–833, 1989.

[2.] Zou Wei; Ghafouri-Shiraz, H., "Codes for spectral-amplitude-coding optical CDMA systems", Journal of Lightwave Technology, Volume 20, Issue 8, Aug. 2002 Page(s):1284 – 1291

[3.] M. Kavehrad and D. Zaccarin, "Optical code-division-multiplexed system based on spectral encoding of noncoherent sources," *J. Lightwave Technol.*, vol. 13, pp. 534–545, 1995.

[4.] Aljunid, S.A.; Ismail, M.; Ramli, A.R.; Ali, B.M.; Abdullah, M.K., "A new family of optical code sequences for spectral-amplitude-coding optical

CDMA systems", Photonics Technology Letters, IEEE Volume 16, Issue 10, Oct. 2004 Page(s):2383 – 2385

[5.] X. Zhou, H. M. H. Shalaby, C. Lu, and T. Cheng, "Code for spectral amplitude Coding optical CDMA systems," *Electron. Lett.*, vol. 36, pp. 728–729, 2000.

[6.] S.A.Aljunid, "Development and Implementation of a Novel Code Family for Optical Code Division Multiple Access Systems", Dissertation of Doctor of Phylosophy, Unversiti Putra Malaysia, July 2005.

[7.] J.F.Huang, C.M.Tsai, Y. L.Lo, "Compensating Fiber Gratings for Source Flatness to Reduce Multiple-Access Interferences in Optical CDMA Network Coder/Decoders", Journal Of Lightwave Technology, Vol. 22, No. 3, March 2004

Pulse Power Failure Model Of Power MOSFET Due To Electrical Overstress Using Tasca Method

Nur Syakimah Ismail[a], Ibrahim Ahmad[a], *Member, IEEE*, Hafizah Husain[a], and Shirley Chuah[b]

[a]Department of Electrical, Electronic & System, Faculty of Engineering,
43600 UKM Bangi, Selangor.
nursyakimah.ismail@statschippac.com, ibrahim@vlsi.eng.ukm.my, hafizah@vlsi.eng.ukm.my
[b]STATSChipPAC (M) Sdn. Bhd., Free Trade Zone Ulu Kelang,
54200 Kuala Lumpur, Malaysia
Shirley.Chuah@statschippac.com

Abstract **The objective of this research is to study electrical overstress (EOS) defect at gate oxide for various pulse widths in n-channel Power Metal-Oxide-Semiconductor Field Effect Transistor (MOSFET). Moreover, this research also intent to develop power failure model for n-channel power MOSFET according to Tasca method. Electrical overstress does not have EOS standards and quantitative EOS design objectives to tackle this problem. Square pulse testing is used in this research due to easy to generate and simple to analyze. Time-to-failure (t_f) is taken for power profiles modeling by observing abrupt drop in voltage waveform seen on oscilloscope. Tasca derived the thermal model by regarded the defect area as a sphere immersed in an infinite medium at ambient temperature. Result from failure analysis on all failed units had shown that hot spot formations begin at gate runner of the die and pulse stress given on V_{GS} has cause gate oxide breakdown. Pulse power failure model for device n-channel power MOSFET can be obtained using Tasca method.**

I. INTRODUCTION

Electrical overstress (EOS) is defined as the exposure of an item to a current or voltage beyond its maximum ratings. EOS embodies a broad category of electrical threats to semiconductor devices including electromagnetic pulses (EMP), electrostatic discharge (ESD), system transient, lightning and others [1]. There are three major failure mechanisms for semiconductor devices subjected to EOS. They are junction burnout, dielectric or oxide punch through and metallization burnout [2].

Device n-channel Power MOSFET package TO252 is chosen for this research due to this device suffers a lot of EOS failure in STATSChipPAC (M) Sdn. Bhd. This device is built with TrenchFET® technology and function as synchronous buck DC/DC conversion in desktop and server. This n-channel power MOSFET has drive current 250µA with threshold voltage between 0.8-3.0V.

Oxide punch through as known as oxide rupture due to EOS/ESD normally involves high voltage being applied across the oxide layer, causing a weak spot within it to exhibit dielectric breakdown and allow current to flow. The current flow that is basically due to loss of dielectric isolation at that spot will causes localized heating, which induces the flow of a larger current. An increasing current flow and localized heating results in meltdown of the silicon, dielectric, and other materials at the 'hot spot'. This meltdown creates a short circuit between the layers supposedly isolated by the oxide [3].

Thermal model is used to determine the time required for any point in a given thermal structure to reach a specified critical temperature. Common purpose of thermal failure models is to specify the shape and dimensions of the heat source that in general cannot be accurately determined from first principles for a given device or process. Basic model of thermal runaway at a localized heat source defined as P(t)=V(t)I(t) throughout a defect region [4].

The earliest analytical models of thermal breakdown are from Wunsch and Bell in 1968 [5].

$$P_f = \left(\pi K \rho C_p \right)^{1/2} A_j \left(T_c - T_0 \right) t_f^{-1/2} \quad (1)$$

Where:

K	= Thermal conductivity
ρ	= Density
C_p	= Specific heat capacity
A_j	= Junction area

0-7803-9730-4/06/$25.00 ©2006 IEEE

T_o = Ambient temperature

T_c = Critical temperature (675°C or 1415°C at melting temperature of Si)

In 1970, Tasca derived a thermal model from three-dimensional problem of a spherical source in an infinite medium [6].

$$P_f = \left\{ \frac{\rho C_p \Delta}{t_f} + S \left(\frac{K\rho C_p}{t_f} \right)^{1/2} + \frac{8\pi Kr}{3} \right\} (T_c - T_0)$$

(2)

Where:

Δ = Defect volume
S = defect surface area
r = defect radius

Tasca model has divide time into three regions as Fig. 1. For small failure time, typically below 10 ns, little heat is lost from the surface so that adiabatic dominates. For very large failure time, typically 100 μs, which is after thermal equilibrium has been established, the steady state term dominates. Intermediate time domain is Wunsch-Bell term [4].

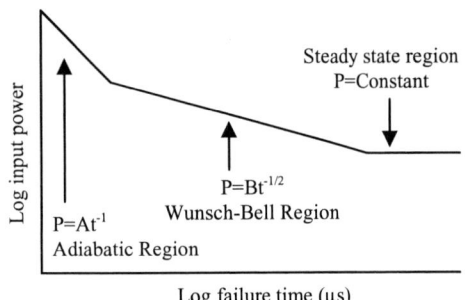

Fig. 1   Log input power vs log failure time

II. EXPERIMENT

Electrical overstress (EOS) does not have EOS standards and quantitative EOS design objectives to tackle this problem. Unipolar stress, which is square pulse testing was performing on the sample units by pulse generator Agilent 8114A. This test method is chosen by reason of square pulse voltage waveforms are easy to generate and simple to analyze [7]. Units were sampled to groups representing various pulse width used to test the units. Diaz et al (1995) had mention that EOS damage is commonly caused by long lasting condition that is more than 1 μs.

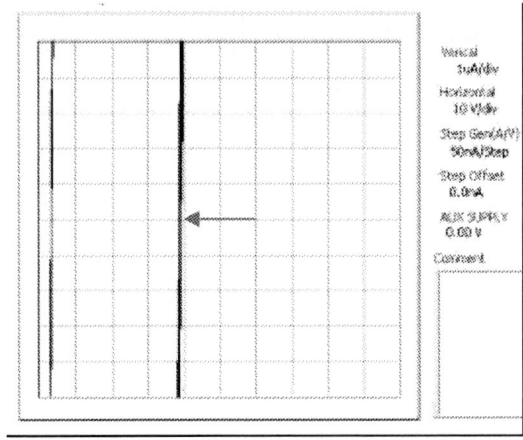

Fig. 2   Maximum V_{GS}=40V before gate oxide breakdown is observed by using curve tracer.

EOS involves high voltage being applied across the oxide layer, causing a weak spot within it to exhibit dielectric breakdown and allow current to flow. Curve tracer shows that device n-channel MOSFET will suffer with gate oxide breakdown after 40V (refer Fig. 2). Square pulse with amplitude 40V is applied on gate-source of the device as to see the EOS defect by diverse pulse widths. An abrupt drop of voltage waveform on oscilloscope is a typical signature of device failure. Time-to-failure (t_f) is taken for power profiles modeling. Failure analysis is done on the failed units to observe the defect area and by taking measurement on the defect parameters that will be used to calculate pulse power profile.

III. RESULTS AND DISCUSSION

Failure analysis on all failed units had shown that hot spot formations begin at gate runner on the die. Function of the gate runner is as a contact for connection to the gate voltage supply in termination region.

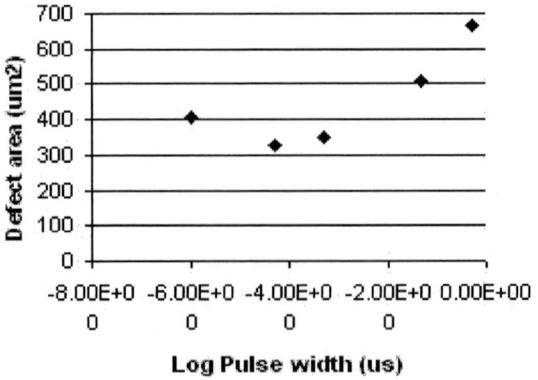

Fig. 3   Defect area of failed units varies with pulse width.

Fig. 3 shows that defect area is varied with pulse width. For pulse width 1 μs, the damage area is massive due to very large current density cannot be distributed in a short period of time (refer Fig. 4).

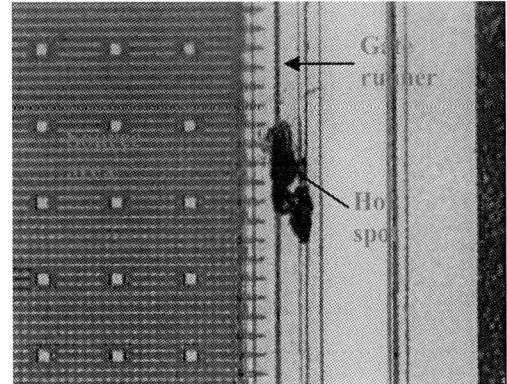

Fig. 4   Hotspot formation at gate runner by 800x magnification microscope after HCL deprocess.

When the breakdown voltage of the oxide layer is exceeded by a large potential difference across it, the oxide breaks down and starts conducting current. Due to the pulse width is short, the heat will generated and hence the resistance changes. This alteration cannot dramatically alter the current distribution. Thus the hotspot is generally larger [8].

Failed units cause by 50μsec pulse width show oxide punch through at gate runner at left side of the die (refer Fig. 5). The size of the hot spot is smaller due o heat loss in the junction. As pulse width gets longer and the current that required causing damage become smaller due to major power dissipation in the depletion region [8].

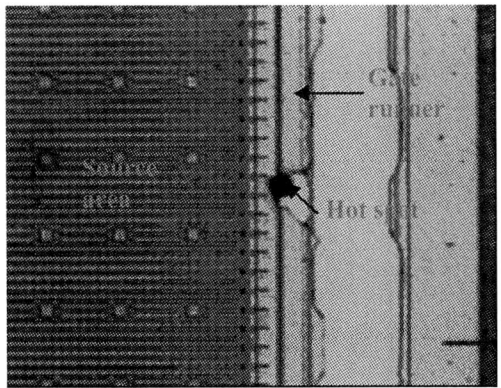

Fig. 5   Image of punch through oxide at gate runner on left side die by 800x magnification microscope after HCL deprocess.

Cross sectional at defect area by Focus Ion Beam (FIB) revealed that aluminum metallization has been melt down to the silicon layer after oxide rupture (refer Fig 6). Oxide punch through occurs when a voltage pulse applied across the oxide has sufficient magnitude to cause breakdown, allowing current to pass. Once the oxide has broken down, it requires very little energy to destroy the oxide that results in short [2].

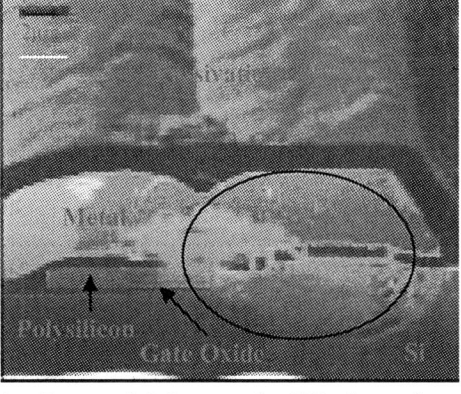

Fig. 6   Image of defect area by FIB shows that metal and polysilicon have melt down

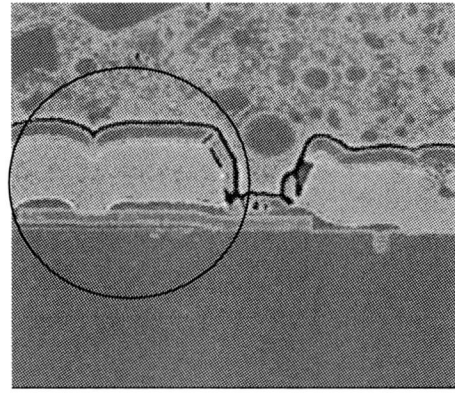

Fig. 7   Image of good unit to compare with Fig. 6 by Scanning Electron Microscope (SEM).

Failed units cause by 50msec pulse width produce a wide defect area (refer Fig. 8) also at the gate runner.

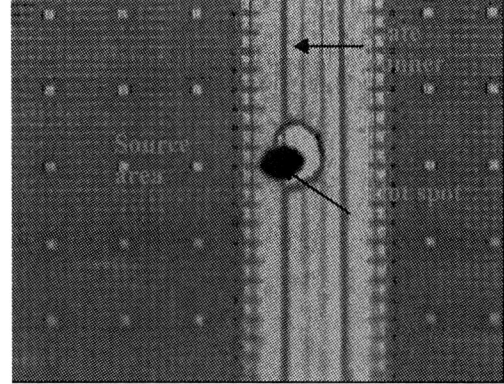

Fig. 8   Picture of punch through oxide at gate runner on left side die by 800x magnification microscope

Cross sectional area at the damage area caused by 50 ms pulse width, aluminum metallization meltdown has occurred (refer Fig. 9). For a very long pulse width, there is time for the aluminum to migrate into the silicon and in particular in melt region. The size of the short is larger because far more energy is being delivered than is required to cause failure [8].

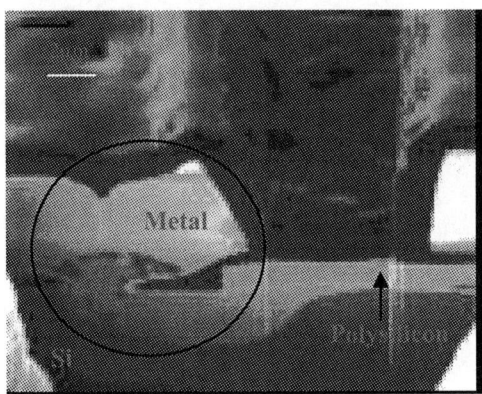

Fig. 9 Image of defect area by FIB at cross sectional area shows metal punch through the oxide to the silicon layer.

By using equation (2), pulse power failure model can be developed. Power profiles clearly display Wunsch-Bell domain together with constant power domain mention by Tasca [6]. This n-channel MOSFET achieves the constant region approximately at 50 µs (refer Fig. 10).

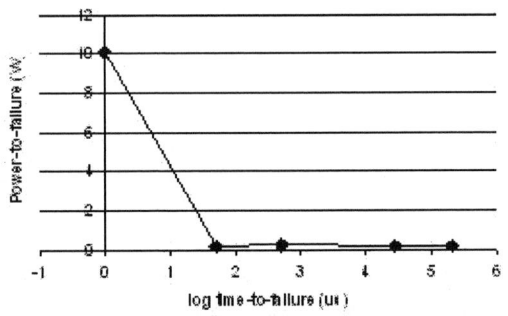

Fig. 10 Pulse power failure model for device n-channel power MOSFET T0252.

IV. CONCLUSION

This paper has shown that different pulse width given to the gate-to-source pin resulted in diverse defect to the gate oxide. Base on the defect area on the failed units, pulse power failure model can be developed. The power profiles show that constant region for n-channel MOSFET begin at approximately 50 µs.

ACKNOWLEDGEMENT

Thank you very much to my supervisor Prof. Madya Ibrahim Ahmad for the guidance and advice, my site supervisor Madam Shirley Chuah for the support and Mr Charvaka Duvvury who supervise from far-far away. For my family and friends, thank you for your understanding and prayer.

REFERENCES

[1] C.Diaz, C.Duvvury, S.M.Kang,"*Tutorial Electrical Overstress and Electrostatic Discharge*", IEEE Transaction on Reliability, Vol. 44,1995.

[2] D.G. Pierce, D.L. Durgin, "*An Overview of Electrical Overstress Effects on Semiconductor Devices*", Proc EOS/ESD Symp, 1981, pp120-131.

[3] Oxide breakdown (online). http://www.siliconfareast.com/oxidebreakdown.htm

[4] V.M. Dwyer, A.J. Franklin, D.S. Campbell, "Thermal Failure in Semiconductor Devices", IEEE Solid State Electronics, 1990, Vol. 33, No. 5, pp. 553-560.

[5] D.C. Wunsch, R.R. Bell, "*Determination of Threshold Failure Levels of Semiconductor Diodes and Transistor Due To Pulse Voltage*", IEEE Trans Nuclear Science, Vol. NS-15, 1968.

[6] D.M. Tasca, "*Pulse Power Failure Models in Semiconductors*", IEEE Trans. Nuclear Science, Vol. NS-17, Dec 1970.

[7] C.Diaz, C.Duvvury, S.M.Kang, L.Wagner,"*Electrical overstress (EOS) power profiles: A guideline to qualify EOS hardness of semiconductor devices*", Proc EOS/ESD Symp, 1992, pp88-94.

[8] J.S. Smith, "*Electrical Overstress Failure Analysis in Microcircuits*", 16th Annual Proceeding Reliability Physics, 1978, pp41-46.

New I-V Model For AlGaN/GaN HEMT At Large Gate Bias

Ali El-Abd[1], Student Member, IEEE, M. Abdel Aziz[2], Abdel Aziz Shalby[3], Member, IEEE and Salah Khamis[4]

[1] Faculty of Engineering, Tanta University, Egypt.
[2] Alexandria Higher Institute of Technology, Egypt.
[3] Faculty of Electronic Engineering, Menofiya University, Egypt.
[4] Faculty of Engineering, Tanta University, Egypt.
Email: ali_elabd@yahoo.com

Abstract **Most theoretical I–V models appeared in the literature for AlGaN/GaN HEMTs fail to exactly fit the experimental data and a large deviation is reported, especially at large gate bias. Because the exact nature of the AlGaN/GaN hetrostructure is not fully described, we assume that there is a new phenomena in the device which causes this deviation. Here we compute this phenomena mathematically for doped and undoped devices using a new method. The model results will be useful in CAD programs to extract more accurate I–V characteristics for the AlGaN/GaN hetrostructure on sapphire substrate.**

I. INTRODUCTION

GaN-based electron devices have received much attention for their ability to operate at high-power levels and in high-temperature environments [1]–[4]. This is because of the high breakdown fields in wide bandgap semiconductors, and the low carrier generation rate caused by thermal activation. In order to fully develop the potential of the device, it is very important to deduct an accurate I–V model for the device, that can match the experimental data over a wide range of bias voltages.

Previously reported I–V models for AlGaN/GaN hetrostructures offer a large deviation between the theoretical results and the experiment I–V characteristic, especially at large gate bias. Two approaches appear in the literature to overcome this problem. The first approach assumes that the exact nature of the traps and doping transients are not fully characterized, so the profile of the sheet charge density can be optimized by decreasing the maximum sheet carrier concentration in the charge control model (at large gate voltage) to match the drain current and transconductance characteristics [5]. The second approach considers the decreasing of the electron mobility in 2DEG when the concentration of the sheet carrier is increased which is the case of AlGaN/GaN hetrostructure [6].

Because of the absence of a standard optimization method to determine the maximum sheet carrier concentration in 2DEG channel without considering the measured data , the first approach can not help getting an exact Current - Voltage model. The second approach uses a new mobility model that considers the dependence of electron mobility on the sheet carrier concentration . We expect that this approach will not be useful and will complicate the computations, because sheet carrier concentration already contributes the field in 2DEG channel .So, using the GaN bulk velocity-field formula will be sufficient and simpler than this approach.

In this paper we try thinking differently by assuming that there's a new phenomena in AlGaN/GaN hetrostructure and we are concerning here to compute this phenomena mathematically and let the physical explanation to future studying.

II. ANALYSIS

The I-V models presented in [5] and [7] are based on dividing the channel region under the gate into uniform sections. The channel voltage at start and end points are given by :

$$V_{start} = V_S + I_D R_D \qquad (1)$$

$$V_{end} = V_D - I_D R_D \qquad (2)$$

Where R_S and R_D are the source and drain resistances, and V_S and V_D are the source and drain voltages. For a given drain current I_D, we

0-7803-9730-4/06/$25.00 ©2006 IEEE   1019

calculate V_{start} , which is the channel voltage at the source end. Once the channel voltage is determined, the channel carrier concentration can be obtained from the sheet charge density versus gate voltage characteristics [5],[8]. The carrier velocity is obtained from the drain current and the channel carrier concentration relationship. The channel electric field ε is calculated from the velocity-field characteristic which is given by

$$v(\varepsilon) = \frac{\mu_o \varepsilon + v_1 \dfrac{\varepsilon^{\alpha}}{\varepsilon_1^{\alpha}}}{1 + \zeta \left(\dfrac{\varepsilon}{\varepsilon_1}\right)^{\beta} + \left(\dfrac{\varepsilon}{\varepsilon_1}\right)^{\alpha}} \qquad (3)$$

where v_1, ε_1, α, β, and ζ are curve fitting parameters extracted from Monte Carlo simulations [9]. The values used for these parameters in the results presented here are $v_1 = 2.35 \times 10^7$ cm/s, $\varepsilon_1 = 161$ kV/cm, $\alpha = 17.98$, $\beta = 0.917$, and $\zeta = 1.88$. The value for the zero field mobility μ_o is taken as 500 cm /V.s. If the electric field is less than its critical value, the velocity follows the analytical formula in (3). Once the channel electric field exceeds the critical field, the velocity is assumed to saturate at a value equal to that of the peak velocity.

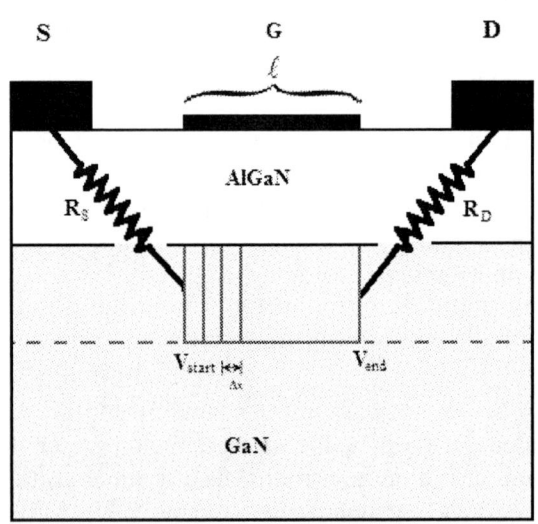

Fig.1 Schematic drawing of AlGaN/GaN HEMTs

In the ith section, the channel voltage, sheet carrier concentration, and carrier velocity can be discretized as (see fig. 1) :

$$V_i = V_{i-1} + \varepsilon_{i-1} \Delta x \qquad (4)$$

$$n_i = n_s \left(V_G - V_i\right) \qquad (5)$$

$$v_i \left(\varepsilon_i\right) = \frac{I_D}{qWn_i} \qquad (6)$$

where W is the gate width. Thus, starting from the source end $Vstart$ and solving forward over all sections, the drain voltage can be obtained from (2). During calculations, the drain current can be increased continuously until the carrier velocity becomes saturated. However, when comparing the results of these calculations with the published experimental data, we note that there is an overestimation of the drain current at large gate voltage.

In our approach we will try to find a relation between the theoretical results (using charge control model and GaN bulk velocity-field formula without any changes) and the measured I-V data.

Here, the channel region under the gate will be taken as one section because we have only the drain to source voltage – not the channel voltage at every section – against the drain current. So , average values V_{avg} , v_{avg} , ε_{avg} , and n_{avg} will be used for the following parameters respectively : the channel voltage , the carrier velocity , the channel field and the 2DEG sheet carrier concentration. This will compensate for the variation of these values along the channel. So we can write

$$V_{avg} = V_{start} + \frac{K}{2} \left(V_{end} - V_{start}\right) \qquad (7)$$

$$n_{avg} = n_s \left(V_G - V_{avg}\right) \qquad (8)$$

$$v_{avg} = \frac{I_D}{qWn_{avg}} \qquad (9)$$

We have introduced here a new dimensionless parameter K as a voltage scale factor that relates the computed average channel voltage to the actual value in order for the theoretical results to exactly fit the measured I-V data .

The voltage parameter K is computed first from the measured data , then the drain voltage is calculated for certain drain current I_D using equations (7)-(9)

The calculations are confirmed using the following relation

$$V_D = V_{start} + \varepsilon_{avg}\ell + I_D R_D \qquad (10)$$

ℓ is the channel length. Fig.1 is a sketch of the device cross-section that clarifies the above equation.

III. RESULTS

The behavior of the voltage parameter K calculated from the application of our model on three AlGaN/GaN HEMTs on sapphire substrate [10]-[12] is studied in the following figures. The processing and dimensional parameters of the three devices under test are given in the table below:

Parameter	Device I [10]	Device II [11]	Device III [12]
Al mole fraction %	15	25	30
Spacer AlGaN layer thickness (nm)	3 UID	3 UID	25
Doped AlGaN layer thickness (nm)	22	15	-
Doping level in doped layer (cm^{-3})	2 x 10^{18}	2 x 10^{18}	undoped
Cap layer (nm)	15 UID	2 UID	-
Gate length (µm)	1	0.9	2
Gate width (µm)	75	25	50

*The UID layer has a background doping of 1×10^{18} cm^{-3}.

The voltage parameter K for devices I, and II are shown in figures 2. ,and 3. at relatively low gate voltages, where the excess voltage $V_G - V_T$ =1V, and 2V. V_T is the threshold gate voltage.

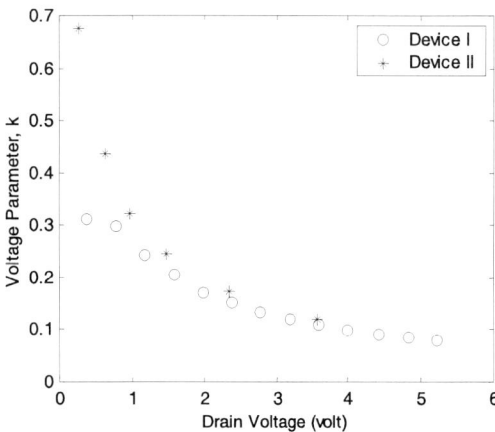

Fig.2 Comparison of the voltage parameter K at V_G -V_T =1V

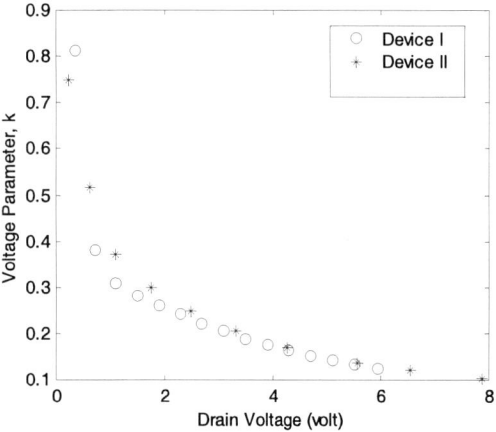

Fig.3 Comparison of the voltage parameter K at V_G -V_T =2V

In figures 4 through 6, the K parameter for the three devices is tested at higher gate voltages : $V_G - V_T$ =3V, 4V, and 5V respectively. Note that the maximum value of the excess voltage for device I is 4V, so it is not included in fig.6.

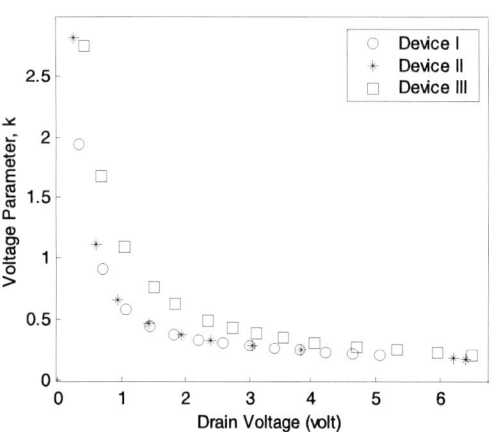

Fig.4 Comparison of the voltage parameter K at V_G -V_T =3V

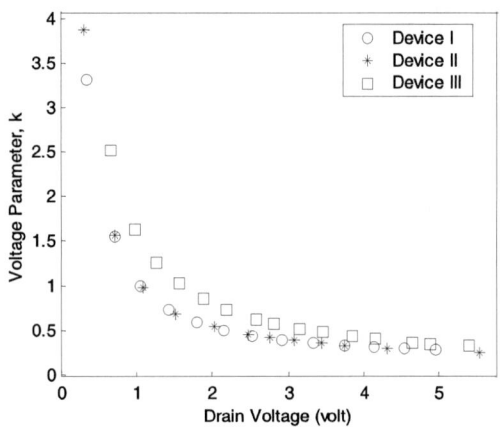

Fig.5   Comparison of the voltage parameter K at V_G -V_T =4V

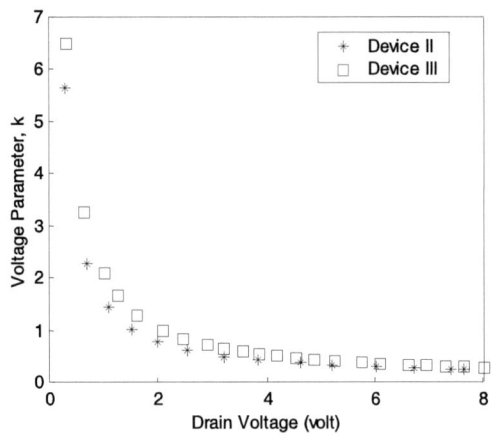

Fig.6   Comparison of the voltage parameter K at V_G -V_T =5V

The two doped devices(I and II) have nearly the same K values for all drain voltage range, while the undoped device has slightly different values at low drain voltage. It is also seen that K values for devices I and II become closer when gate voltage becomes larger.

So we conclude that the voltage parameter K is constant for devices having the same doping at large gate and drain voltages even if they are different in other processing parameters. Hence, the parameter K can be used to compensate the error between the theoretical and experimental results of the I-V characteristic which appears at large gate bias. We expect that the difference in K value at large gate bias between doped and undoped devices may not be found if they have

been processed by the same manufacturer   and under the same environments.

The validity of the model is checked through comparison of the published experimental I-V results for devices I and II with the theoretical results presented in previous models (before optimization) and our model results. These comparisons have been performed in fig.7 through   fig.10. A large deviation between the theoretical results and the experimental data is clear in figures 7. and 9., specially at large gate bias, and an excellent agreement is obtained in figures 8. and 10. when our model is applied.

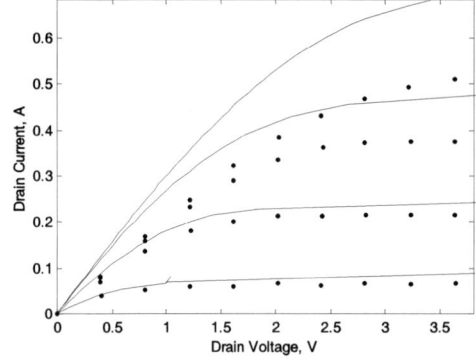

Fig.7   I-V theoretical (solid) and experimental (dots) I-V results for device I at at V_G = 1, 0, -1, and -2 V and V_T= -3 V

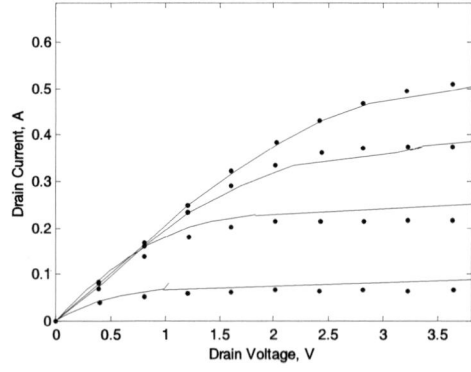

Fig.8   Model results(solid) compared to experiment (dotes) I-V data for device I V_G = 1, 0, -1, and -2 V and V_T=-3 V

1022

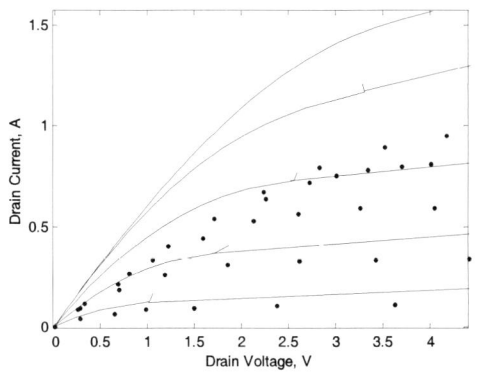

Fig.9 Theoretical (solid) compared to experimental (dots) I-V data for device II at V_G =-2,-1 , 0, -1, and 2 V and V_T= -3 V

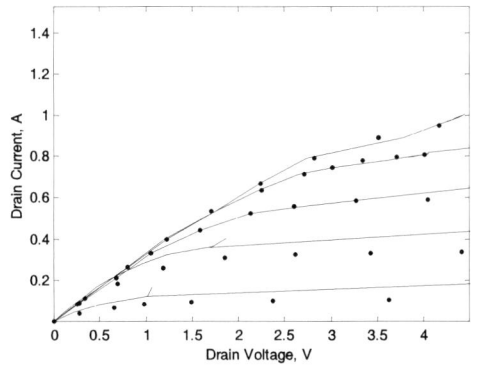

Fig.10 Model results (solid) compared to experimental (dotes) I-V data for device II at V_G =-2,-1 , 0, -1, and 2 V and V_T= -3 V

IV. CONCLUSION

In this paper we refer to a new phenomena in AlGaN/GaN hetrostructures which appears at large gate bias and leads to large deviation between the theoretical and experimented I-V results. We express this phenomena mathematically through a new parameter, we have called it the voltage parameter. We succeeded to find a standard method to get an accurate I-V characteristics of AlGaN/GaN hetrostructures which will be useful in CAD programs.

REFERENCES

[1] M. A. Khan, J. N. Kuznia, J. M. Van Hove, N. Pan, and J. Carter, "Observation of a two-dimensional electron gas in low pressure metalorganic chemical vapor deposited GaN–AlGaN heterojunctions," Appl. Phys. Lett., vol. 60, no. 24, pp. 3027–3029, 1992.

[2] O. Akatas, Z. F. Fan, A. Botchkarev, S. N. Mohammad, M. Roth, T. Jenkins, L. Kehias, and H. Morkoc, "Microwave performance of AlGaN/GaN inverted MODFET's," IEEE Electron Device Lett., vol. 18, pp. 293–295, June 1997.

[3] Y.-F. Wu, B. P. Keller, P. Fini, S. Keller, T. J. Jenkins, L. T. Kehias, S. P. Denbaars, and U. K. Mishra, "High Al-content AlGaN/GaN MODFET's for ultrahigh performance," IEEE Electron Device Lett., vol. 19, pp. 50–53, Feb. 1998.

[4] S. T. Sheppard, K. Doverspike, W. L. Pribble, S. T. Allen, J.W. Palmour, and T. J. Jenkins, "High-power microwave GaN/AlGaN HEMT's on semi-insulating silicon carbide substrates," IEEE Electron Device Lett., vol. 20, pp. 161–163, Apr. 1999.

[5] T.-H. Yu and K. F. Brennan, "Theoretical Study of a GaN–AlGaN High Electron Mobility Transistor Including a Nonlinear Polarization Model," IEEE Trans. Electron Devices, vol. 50, no. 2, pp. 315-323, Feb. 2003.

[6] Oded Katz, Adi Horn, G. Bahir, and Joseph Salzman, "Electron Mobility in an AlGaN/GaN Two-Dimensional Electron Gas I—Carrier Concentration Dependent Mobility," IEEE Trans. Electron Devices, vol. 50, no. 10, pp 2002-208, OCT. 2003.

[7] Fabio Sacconi, Aldo Di Carlo, P. Lugli, and Hadis Morkoç,"Spontaneous and Piezoelectric Polarization Effects on the Output Characteristics of AlGaN/GaN Heterojunction Modulation Doped FETs, "IEEE Trans. Electron Devices, vol. 48, no. 3, pp 450-457, Mar. 2001.

[8] M. Abdel-Aziz and A. El-Abd, "Theoretical Study of the Charge Control in AlGaN/GaN HEMTs," The 23 rd national radio conference (NRSC 2006) March 14-16, 2006. Faculty of Electronic Engineering, Menoufiya University, Egypt.

[9] T.-H. Yu and K. F. Brennan, "Monte Carlo calculation of two-dimensional electron dynamics in GaN-AlGaN heterostructures," J. Appl. Phys., vol. 91, pp. 3730–3736, Mar. 2002.

[10] Y.-F.Wu et al., "Bias dependent microwave performance of AlGaN/GaN MODFET's up to 100V," IEEE Trans. Electron Device Lett., vol. 18, pp. 290–292, June 1997.

[11] U. K. Mishra, Y.-F.Wu, B. P. Keller, S. Keller, and S. P. Denbaars, "GaN microwave electronics," IEEE Trans. Microwave Theory Tech., vol. 46, pp. 756–761, June 1998.

[12] Jin-Ping Ao, Daigo Kikuta, Naotaka Kubota, Yoshiki Naoi, and Yasuo Ohno, Member, "Copper Gate AlGaN/GaN HEMT With Low Gate Leakage Current," IEEE Electron Device Lett., vol. 24, no. 8, pp 500-502, AUG. 2003.

Simulated Dielectric Characteristics of Pt/BST/Ni-Fe/Cu Multilayer Capacitor Stack for Storage Application

Balachandran R, *Member, IEEE*, Yow HK, *Senior Member, IEEE*, Manickam RM and Saaminathan V

Faculty of Engineering, Multimedia University, Cyberjaya, 63100, Selangor, Malaysia

Email: balachandran.ruthramurthy@mmu.edu.my

Abstract In this simulation research work, the metal-composite-metal (MCM) multilayer capacitor structure [Pt/BST/Ni-Fe/Cu] is proposed with Barium Strontium Titanate (BST) oxide material as the capacitor dielectric material for DRAM with permalloy Nickel-Ferrous (Ni-Fe) coated Copper (Cu) as the bottom conducting electrode and platinum as the top conducting electrode. This proposed MCM consists of 120 μm Cu bottom contact material, a 1 μm of Ni-Fe alloy over the stoichiometric composition of the BST oxide dielectric material of thickness 40 nm and dielectric constant of 775. The MCM structure is expected to deliver a maximum charge storage capacity of 109.75 fF for a capacitor in DRAM cell area of 0.64 μm² well above the minimum requirement for DRAM cell. The leakage current density for a variation of voltage from 0 to 10 V has been simulated for temperature variation. When compared with the previous report, the proposed multi layer capacitor (MLC) structure shows promising potentials in terms of dielectric characteristics.

I. INTRODUCTION

MLC offers higher level of miniaturization, flexibility and performance characteristics than conventional discrete capacitors. MLC have been prepared and characterized with different kind of stacks and dielectric materials in recent years. [1, 2] . In the previous MLC works, BST film has been sputtered on Ni/Cu foil with and without Al as the top electrode and its dielectric characteristics have been studied [3]. It was reported that the leakage current density was found to be less than 1×10^{-7} A/cm² at 5 V and capacitance density was estimated to be 150 – 300 nF/cm². BST thin films were doped with Ni to suppress the leakage current density. [4] Ni doped

with BST films on $La_{0.5}Sr_{0.5}CoO_3$ (LSCO) buffer layers resulted a decrease in dielectric loss and leakage current density.[5] The temperature dependence of capacity is considerably influenced by Fe doping [6].

Cu, Ni, Fe in different stacks with BST has been studied in previous research works. In this simulation work, the MCM multilayer capacitor structure [Pt/BST/Ni-Fe/Cu] is proposed with BST oxide material as the capacitor dielectric material for DRAM with permalloy (Ni-Fe) coated Copper [Cu] as the bottom conducting electrode and platinum as the top conducting electrode. This proposed MCM consists of 120 μm Cu bottom contact material with 1 μm of Ni-Fe alloy [7] over the stoichiometric composition of the BST oxide dielectric material with a thickness of 40 nm.

II. SIMULATION

MATLAB Ver - 7.0 is used to simulate the capacitor's parameters such as Dielectric constant, Charge storage capacity, Polarization and Leakage current density. These parameters are important properties of BST thin film capacitor for storage application in DRAM. High storage charge density (i.e., high polarization) reduces the capacitor area necessary to meet charge storage requirements. The effects of polarization and leakage characteristics have been widely studied for different thin films and electrodes. [8-10].

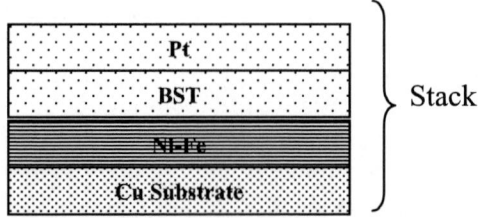

Fig. 1 Structure of proposed multilayer capacitor with Pt/BST/Ni-Fe/Cu stack

0-7803-9730-4/06/$25.00 ©2006 IEEE

III. RESULTS AND DISCUSSION

Dielectric constant

The dielectric constant can be varied from 300 to 1250 by changing the ratio of Barium to Strontium. In this work, a capacitor with an operating frequency of 63.38 MHz has been designed under the stochiometric condition of $Ba_{0.5}Sr_{0.5}TiO_3$. As the maximum value of the dielectric constant could be varied up to 1250 using BST, the 'k' has been chosen for this work to be less than 1000 as high 'k' value increases temperature coefficients of the material [11]. Fig. 2 shows the relationship between the dielectric constant and the compositions of Barium / Strontium materials. From the graph it has been found that the dielectric constant for this composition of oxide material is 775.[12]

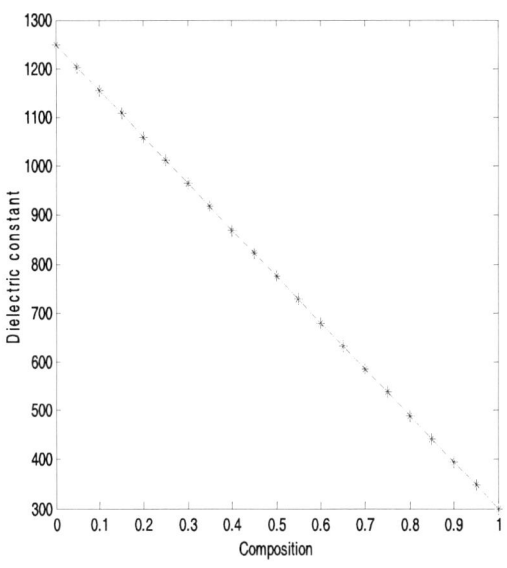

Fig. 2 Composition of BST with Dielectric constant

Charge storage capacity

The requirement for a memory cell is to store larger amount of charge within a smaller area of memory device. The minimum charge storage requirement for the memory cells within the DRAM is 30 fF [13]. In this work, for the optimized composition of $Ba_{0.5}Sr_{0.5}TiO_3$ with dielectric constant of 775, the maximum charge storage capacity of 109.75 fF has been obtained with dielectric thickness of 40 nm and area of 0.64

μm^2. The enhancement of above storage capacity has been achieved with smaller area suing the proposed stack design, compared to earlier report [14]. This charge density would be enough to keep the charge intact.

Polarization characteristics

Fig.3 shows the polarization values of BST capacitor with Pt as the top electrode and Ni-Fe/Cu as the bottom electrode over 1-10 V by taking into consideration the charge density with storage capacity of 109.75 fF and an area of 0.64 μm^2. From the graph, the polarization varies between 17 to 170 $\mu C/cm^2$.

From the simulation results, the optimum frequency of 63.38 MHz has been obtained for a storage capacity of 109.75 fF with an oxide resistance of 22.88 kΩ.

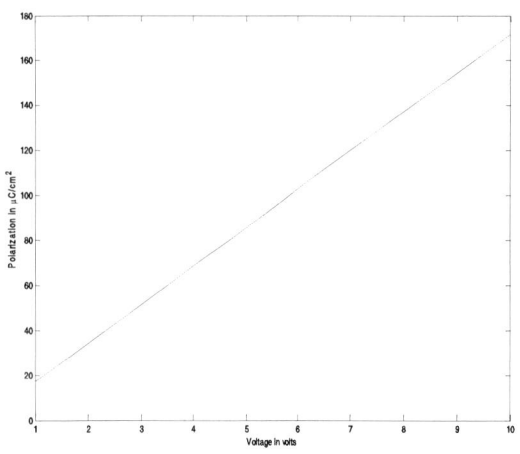

Fig. 3 Polarization versus Voltage

Dissipation factor

Fig. 4 shows the values for the dielectric loss with respect to the dielectric constant calculated for BST material. From the graph, it is observed that higher dielectric constant results in lower dielectric loss. This shows that if BST is used, the dielectric loss will be significantly smaller due to the high dielectric constant, compared to that of SiO_2, which has relatively low dielectric constant of value of 4. [15]

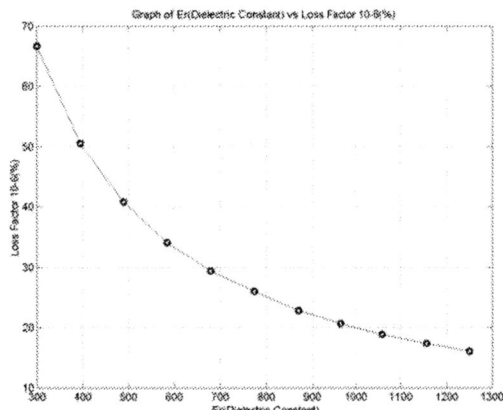

Fig. 4 Relationship between BST dielectric constant and dielectric loss

Leakage characteristics

Fig. 5 shows the temperature – dependent leakage current characteristics of BST thin film capacitor using Pt as the top electrode and Ni-Fe/Cu as bottom electrode. The conduction band barrier height of metal-insulator contact [Ni-Fe/Cu with BST dielectric material] was found to be 0.73 eV[4] . The leakage current densities have been calculated with the temperature varied from 298 K to 448 K and the applied voltage varied from 0 V to + 10 V with an oxide thickness of 40 nm. The leakage current densities are estimated to be lower than that of BST with Ir [1] and Pt [16].

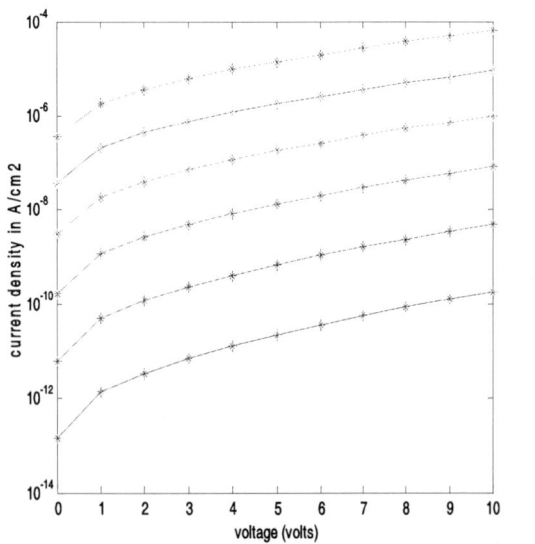

Fig. 5 Temperature – dependent J-V leakage current characteristics

IV.   CONCLUSIONS

With a BST dielectric constant of 775, the MLC structure is expected to deliver a maximum charge storage capacity of 109.75 fF for a cell area of 0.64 μm^2 well above the minimum requirement for DRAM cell. The temperature dependent leakage current characteristic has been simulated for voltage from 0 to 10 V. The proposed MLC structure predicts promising potentials in terms of dielectric characteristics.

ACKNOWLEDGMENT

The authors would like to thank Multimedia University, (Cyberjaya Campus), Malaysia for sponsoring this work under internal research fund PR/2005/0512 and Dr T Saravanan, Scientist, IGCAR, India for his help on some technical matters.

REFERENCES

[1] C. Tung-Sheng, V. Balu, S. Katakam, L. Jian-Hung, and J. C. Lee, "Effects of Ir electrodes on barium strontium titanate thin-film capacitors for high-density memory application," *Electron Devices, IEEE Transactions on*, vol. 46, pp. 2304-2310, 1999.

[2] Thomas A. Bernacki, Ivoyl P.Koutsaroff, and C. Divita, "Barium Strontium Titanate Thin-Film Multi- Layer Capacitors," pp. 1-3, 2004.

[3] Qin zou, Gerhard Hirmer, George Xing, and E. Xu, "Sol-gel derived lead free ferroelectric thin films for embedded capacitor applications," *Energenius Inc.,* pp. 1-15, 2004.

[4] Kun Ho Ahn, Sunggi Baik, and S. S. Kim, "Significant suppression of leakage current in $(Ba,Sr)TiO_3$ thin films by Nior Mn doping," *Journal of Applied Physics*, vol. 92, pp. 2651, 2002.

[5] H.-S. Kim, M.-H. Lim, H.-G. Kim, and I.-D. Kim, "Characterization of Ni-Doped BST Thin Films on LSCO Buffer Layers Prepared by Pulsed Laser Deposition," *Electrochemical and Solid-State Letters*, vol. 7, pp. J1-J3, 2004.

[6] M. Lorenz, H. Hochmuth, M. Schallner, R. Heidinger, D. Spemann, and M. Grundmann, "Dielectric properties of Fe-doped $Ba_xSr_{1-x}TiO_3$ thin films on polycrystalline substrates at temperatures between -35 and +85 [deg]C," *Solid-State Electronics*, vol. 47, pp. 2199-2203, 2003.

[7] Manickam. RM, Patthi H, and S. V, "Effect of ultrasonic field on the properties of electrodeposited Ni-Fe thin films," *Asian symposium on Materials and Processing* 2006.

[8] T. S. Kim, C. H. Kim, and M. H. Oh, "Structural and electrical properties of RF magnetron-sputtered $Ba_{1-x}Sr_xTiO_3$ thin films on indium-tin-oxide-coated glass substrate," *Journal of Applied Physics*, vol. 75, pp. 7998-8003, 1994.

[9] K. Abe and S. Komatsu, "Ferroelectric properties in epitaxially grown $Ba_xSr_{1-x}TiO_3$ thin films," *Journal of Applied Physics*, vol. 77, pp. 6461-6465, 1995.

[10] S.D.Harkness, C.F.Yue, M.A.Borek, and R.K.Singh, "Pulsed laser deposition of epitaxial $Ba_{1-x}Sr_xTiO_3$ / $Y Ba_2Cu_3O_7$ thin films bilayers on $LaAlO_3$ substrates," *J.Electron Matter.*, vol. 23, 1994.

[11] H. Reisinger and R. Stengl, "Fundamental scaling laws of DRAM dielectrics," 2000.

[12] V. Saaminathan and S.Rajendra, "Study on nanometer thick BST material by simulation for DRAM application," *ECS Meeting*, 2005.

[13] R. Singh and R.K. Ulrich, "High and low dielectrics constant materials," *The Electrochemical Society Interface*, vol. 8, pp. 26, 1999.

[14] L. Hang-Ting, T. Tseung-Yuen, and H. Guo-Wei, "A novel method to characterize the dielectric and interfacial properties of $Ba_{0.5}Sr_{0.5}TiO_3$/ (BST)/Si by microwave measurement," 2002.

[15] L. Fang, M. Shen, J. Yang, and Z. Li, "The effect of SiO_2 barrier layer on the dielectric properties of $CaCu_3Ti_4O_{12}$ films," *Journal of Physics D: Applied Physics*, vol. 38, pp. 4236-4240, 2005.

[16] H. J. Cho, J. B. Park, S. S. Yu, J. S. Roh, and H. K. Yoon, "Low temperature MOCVD of BST thin film for high density DRAMs," 2000.

Numerical Analysis of Filamentation in Conventional Double Heterostructure and Quantum Well High-Power Broad-Area Laser Diodes

Amireh Seyedfaraji[1,2], Vahid Ahmadi[1,2], Mahyar Noshiravani[2] and Farzan Gity[1]
1. Dept. of Electrical Engineering, Tarbiat Modares University, Tehran, Iran
2. Laser Research Center, AEOI, P. O. Box: 14155-1339
Email: v_ahmadi@modares.ac.ir

Abstract – **A comprehensive model is presented to study filamentation in both conventional double heterostructure (DH) and quantum well (QW) semiconductor lasers. The spatial dynamics of broad-area (BA) semiconductor lasers is studied by numerically solving space-dependent coupled partial differential equations for the complex optical fields and the carrier density distribution. A self-consistent iteration is developed to model the formation and longitudinal propagation of unstable transverse optical filamentary structures by means of beam propagation method. The effects of stripe width, linewidth enhancement factor and Kerr coefficient are analyzed.**

I. INTRODUCTION

HIGH-POWER semiconductor lasers have been studied for many years. Due to catastrophic optical mirror damage (COMD), the power density of an edge-emitting laser is limited [1,2]. One method to achieve higher the output power levels of a semiconductor laser is to widen its electrodes and active region. With the BA structure, however, high-power semiconductor lasers face the problem of beam filamentation [3]. For broad area gain regions, since the light has no lateral confinement, any increase in the local refractive index can lead to self-focusing, breaking up the lateral mode profile into multiple filaments [4].

In this paper, we concentrate on the different nonlinear mechanisms in the formation of filaments in both conventional DH and QW semiconductor lasers. The first mechanism is the carrier-induced index change, which occurs when the local gain becomes saturated (i.e., spatial hole burning) and is governed by the linewidth-enhancement factor. The other one is the self-focusing and self-defocusing types of

nonlinearity which are involved by the Kerr coefficient [5].

This paper is organized as follows. The theoretical model is presented in section II. In section III, we describe our investigation on the formation of filaments and the quality of the beam in both conventional DH and QW high-power BA lasers. We summarize the results of this paper in section IV.

II. THEORETICAL MODEL

In BA lasers a careful consideration of the spatial dependence is vital. Transversely in x-direction the evanescent optical fields scatter and charge carrier diffuse away from the point where they were generated by the external pump-current. To demonstrate the self-focusing and self-defocusing effects, we consider a BA semiconductor laser, operating continuously at a constant current density, and solve the forward and backward propagating waves iteratively while including carrier diffusion and spatial hole burning. Since we are looking to solve for the lateral distribution (along the x axis), we decompose the electric field in terms of counter-propagating waves using Maxwell's equations. The system of partial differential equations

$$\pm\frac{\partial E_m}{\partial z} = \frac{i}{2k}\frac{\partial^2 E_m}{\partial x^2} + [\frac{1}{2}\Gamma(1-i\alpha)g(N) - \frac{\alpha_{int}}{2} \tag{1}$$
$$+i(1-\Gamma)n_2k_0(|E_m|^2 + 2|E_n|^2)]E_m,$$

$$D\frac{\partial^2 N}{\partial x^2} = -\frac{J(x)}{qd} + \frac{N}{\tau_{nr}} \tag{2}$$
$$+BN^2 + \frac{\Gamma g(N)}{\hbar\omega}(|E_f|^2 + |E_b|^2)$$

describe the spatial dynamics of the counter propagating optical fields E_f and E_b and the charge carrier density N [6]. In (1), the + or − sign is chosen for the forward ($m = f$, $n = b$) and backward ($m = b$, $n = f$) traveling waves,

0-7803-9730-4/06/$25.00 ©2006 IEEE

respectively; $k = n_{\text{eff}}k_0$ is the mode propagation constant with n_{eff} being the effective index of refraction and k_0 being the free-space propagation constant. Γ is the confinement factor, α is linewidth-enhancement factor, α_{int} is internal loss, and n_2 is the Kerr coefficient. Equation (1) not only accounts for coupling between counter propagating waves but also includes diffraction, carrier-induced index variations, free carrier absorption, material gain, and self-focusing or self-defocusing through a Kerr type nonlinearity. In (2), D is the diffusion constant, τ_{nr} is the nonradiative lifetime, B is the spontaneous-emission coefficient, and $J(x)$ is the injected current density which defines as $J(x) = J_0$ for $|x| < w/2$, and 0 otherwise, where w is the stripe width. The first term in (2) accounts for carrier injection, while the second and third terms account for the nonradiative and spontaneous recombinations, respectively. The last term is due to stimulated recombination and accounts for the gain saturation.

In (1) and (2) $g(N)$ is the carrier-dependent gain, which is assume to be of the form $g(N) = a(N-N_0)$ for the DH laser, where, a is the gain cross-section, and N_0 is the transparency value for the carrier density. In order to apply the exact effects of the QW structure, exact gain calculations must be used rather than approximate gain calculations. So, to reach precise results we have made use of Fermi's golden rule in our model, as follows [7]:

$$g(E,n) = g_0 |M|^2 \times \rho_{red}/E$$
$$\times (f_c(E,n) + f_v(E,n) - 1) \quad (3)$$
$$g_0 = q^2 \hbar \pi / \varepsilon_0 m_0^2 c_0 n$$

Spectral broadening lowers the calculated peak gain and shifts the calculated emission wavelength of the laser to a shorter wavelength. It is included through a Lorentzian-shaped broadening function:

$$G(E,n) = \int_0^\infty g(W,n) \frac{(1/\pi)(\hbar/\tau_{in})dW}{(\hbar/\tau_{in})^2 + (W-E)^2}. \quad (4)$$

The longitudinal boundary conditions

$$E_f(x, z=0) = \sqrt{R_0} E_b(x, z=0)$$
$$E_b(x, z=L) = \sqrt{R_L} E_f(x, z=L). \quad (5)$$

represent reflection of the optical fields at the facet mirrors at $z = 0$ and $z = $ L of the laser structures. The transverse boundary conditions

$$\frac{\partial E}{\partial x} = -\alpha_w E \ , \ \frac{\partial N}{\partial x} = -v_{sr}N \ : \ x = +\frac{w}{2}$$
$$\frac{\partial E}{\partial x} = +\alpha_w E \ , \ \frac{\partial N}{\partial x} = +v_{sr}N \ : \ x = -\frac{w}{2} \quad (6)$$

account for the strong absorption of the optical fields outside the laser stripes and surface recombination effects of the charge carriers through α_w and v_{sr}, respectively.

To investigate the role of the nonlinear mechanisms on the laser performance, we have solved (1) and (2) numerically by using a beam-propagation method iteratively. The iteration procedure is initiated by assuming a Gaussian profile for the lateral mode [8].

The beam quality of a BA semiconductor laser is an important factor for various applications. The spatial beam profile is determined by the dynamic interaction between the light fields and the charged carrier occurring during the propagation of the light signal. Our microscopic theory can be used to model the effects of stripe width on the both DH and QW structures.

The optical nonlinearity in semiconductor lasers is identified by the linewidth-enhancement factor which is also called α-factor or antiguiding factor. It quantifies the changes in the real part of the refractive index occurring due to variations of optical gain or the carrier density [9]. It is worth noting that the carrier dependence of the Linewidth-enhancement factor in conventional DH structure is much less than in QW laser [10]. So in this paper we use the exact dependence of this factor to the carrier density for QW structure by using [11]:

$$\alpha \approx \frac{\left\{\frac{1}{2}\log\left[\frac{(E_{g1} - E_{c0})^2 + \Gamma_c^2}{(E_g - E')^2 + (\hbar/\tau)^2}\right] + \frac{(E_{c0} - E')}{\Gamma_c}\left[\frac{\pi}{2} - \tan^{-1}\frac{(E_{g1} - E_{c0})}{\Gamma_c}\right]\right\}}{\frac{\pi}{2} - \tan^{-1}\frac{(E_{g1} - E')}{\hbar/\tau}} \quad (7)$$

in which, E_{g1} and E_g are the sub-bands separation energy and band-gap energy, respectively. Other parameters are greatly dependent to the energy sub-bands which are given in [11].

III. RESULTS AND DISCUSSION

Before considering the effects of the self-focusing and self-defocusing nonlinearities, it is useful to consider the role of the linewidth enhancement. As the local intensity increases, the gain saturates (spatial hole burning), thereby increasing the term $-i\alpha\,g(N)$ in (1), results in the self-focusing effect. For narrow current stripe widths, the gain is localized spatially, thus limiting detrimental effects of this focusing. However, for wide stripes this carrier-induced self-focusing can be catastrophic, leading to the formation of filamentation.

The lateral profiles of the carrier density, near field intensity and normalized far field intensity for both DH and QW lasers, with stripe width of $w = 10$ μm, are illustrated in Fig. 1. For narrow enough stripe widths there can be stable operation, even in the presence of SHB. Figures 1(a) and 1(b) represent an example of stable operation. It can be seen that at the same operating conditions, spatial hole burning (SHB) in the QW laser is occurred as also in the DH one. But from Fig. 1(b) it can be understood that the FWHM of the output beam in the presence of SHB, which proportionally relates to the beam quality, is smaller in the QW structure which is in reasonable agreement with [12]. As illustrated in Fig. 1(c), it is important to mention that the astigmatism effect in the QW far field pattern is induced by the strong dependence of the α-factor in QW laser, according to (7).

It is experimentally observed that increasing either the linewidth enhancement factor or the pumping level, which has a great influence on the gain saturation, results in a greater degree of filamentation for a BA laser [3,12], but if the stripe width increases, for a constant current density and linewidth enhancement factor, a phase-transition-like behavior is observed in the sense that both near and far fields become asymmetric and, at the same time, the lateral profile becomes unstable as is shown in Fig. 2. This figure shows a snap shot of the lateral profile in the unstable condition for $w = 35$ μm. The lack of symmetry is clearly evident in the near field. The asymmetry is also present in the phase profile, which produces an asymmetric far field distribution, shown in Figs. 2(b) and 2(c). The beam profile distortion comes from the α-induced index change, both from the gain-guiding at the centre of the waveguide and the perturbations from the waveguide edges. The amplitude-phase coupling, via the linewidth enhancement factor, produces a multi-lobed far field pattern, shown in Fig. 2(c).

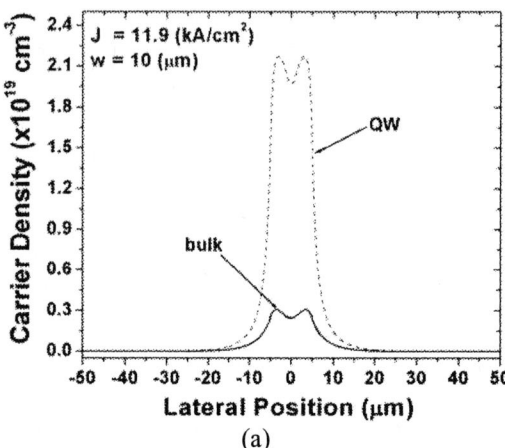

(a)

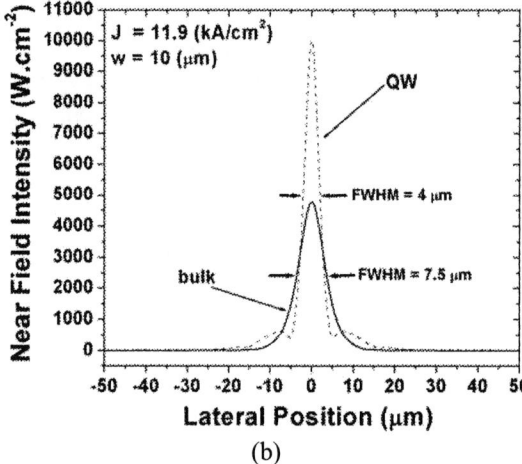

(b)

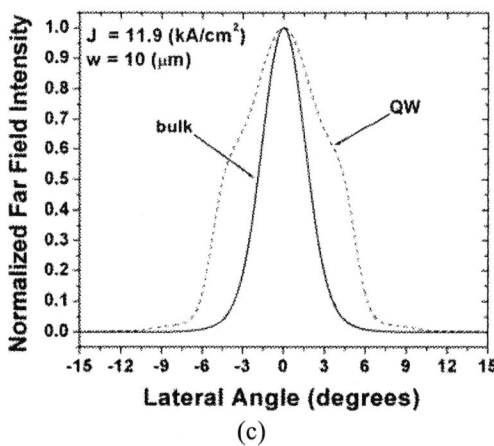

(c)

Fig. 1. Lateral profiles of (a) carrier density (b) near field intensity and (c) normalized far field intensity for both DH and QW structures with $w = 10$ μm and $n_2 = 20 \times 10^{-12}$ cm^2/W.

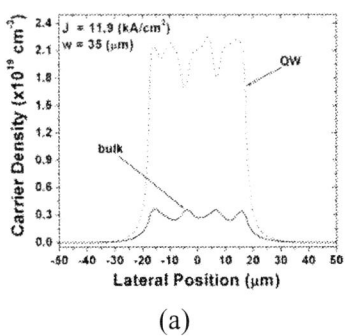 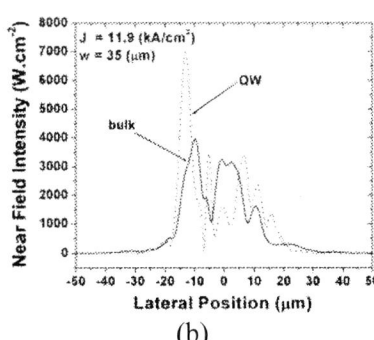 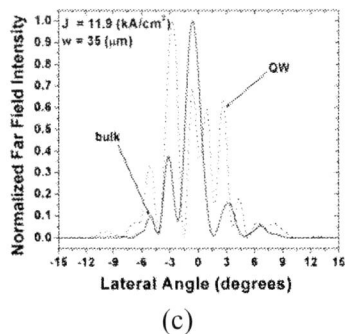

(a)　　　　　　　　　　(b)　　　　　　　　　　(c)

Fig. 2. Lateral profiles of (a) carrier density (b) near field intensity and (c) normalized far field intensity for both DH and QW structures with $w = 35$ μm and $n_2 = 20 \times 10^{-12}$ cm²/W.

Figure 3 illustrates the lateral profiles of the carrier density, near field intensity and normalized far field intensity for both DH and QW lasers for $n_2 < 0$. The negative value of n_2 leads to self-defocusing of the lateral mode. Such self-defocusing can contract the carrier-induced focusing of the lateral mode inside the active region and stabilize the output beam [10]. So, the n_2 term in (1) arising from self-defocusing nonlinearity has the potential of decreasing the harmful effects of α, making it possible to reach stable BA semiconductor lasers.

For a given stripe width w and a given pumping level J, we use two different Kerr coefficient values. With regard to the value of α, according to (7), we encounter different regions of beam stability. Figure 3 shows that the devices are much more stable with respect to self-defocusing than to self-focusing. Indeed, our numerical model, by comparing Figs. 2 and 3, shows that it may be possible to design BA semiconductor lasers that operate stably without filamentation for wider stripe widths.

As mentioned before, by increasing the value of α, the effects of self-focusing nonlinearity increase. It is worth noting that α-induced and n_2-induced changes in the mode index are of opposite sign, and the two must be balanced for achieving stable operation of BA semiconductor lasers.

By comparing figures 1(b) and 3(b), it is obvious that in both cases, the FWHM factor of the QW structure is smaller than the conventional DH structure which leads to a higher beam quality.

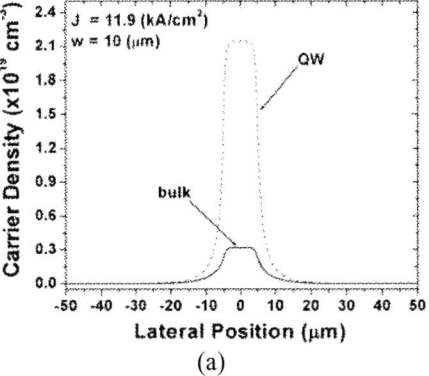

(a)

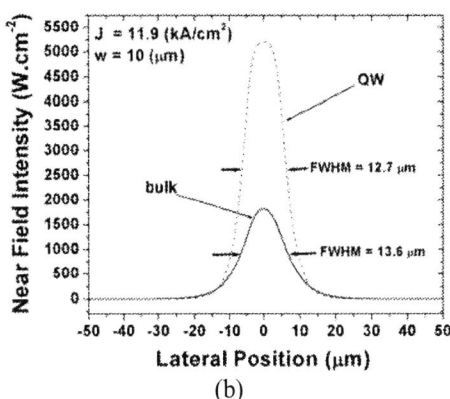

(b)

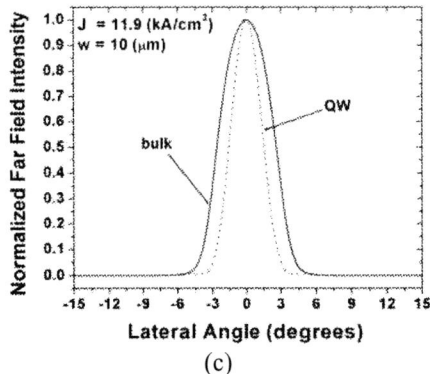

(c)

Fig. 3. Lateral profiles of (a) carrier density (b) near field intensity and (c) normalized far field intensity for both DH and QW structures with $w = 10$ μm and $n_2 = -20 \times 10^{-12}$ cm²/W.

IV. CONCLUSION

In this paper we have presented a comprehensive model that simulates the effects of linewidth-enhancement, self-focusing and self-defocusing types of nonlinearity regarding the formation of filaments in both conventional DH and QW high-power BA semiconductor lasers. To investigate the role of the nonlinear mechanisms on the laser performance, we have solved the coupled partial differential equations of the Maxwell and carrier diffusion equations numerically by means of beam-propagation method self-consistently. By using the exact calculation of the QW laser gain, and also involving the exact carrier dependence of the linewidth enhancement factor, this model approached more precise results with regard to the beam quality of the BA high-power QW lasers.

Our model not only illustrated the lateral profiles of the carrier density, near field and far field patterns but also showed that the output beam becomes unstable to the degree that the stripe widens. It also presented that the FWHM of the QW laser is smaller than the conventional DH one which leads to the higher beam quality of the QW structure. Using the model we have indicated that the negative value of n_2 leads to self-defocusing that can compensate the carrier-induced focusing of the lateral mode. Therefore, it is possible to design high-power BA semiconductor lasers that operate stably with wider stripe widths.

REFERENCES

[1] F. Gity, V. Ahmadi, and M. Noshiravani, "Numerical analysis of void-induced thermal effects on GaAs/Al$_x$Ga$_{1-x}$As high power single- quantum-well laser diodes," *Elsevier J. Solid-State Electron.,* to be published.

[2] R. Diehl, *High-Power Diode Lasers Fundamentals, Technology, and Applications*, Berlin: Springer, 2000.

[3] Z. Dai, R. Michalzik, P. Unger, and K. J. Ebeling, "Numerical simulation of broad-area high-power semiconductor laser amplifiers," *IEEE J. Quantum Electron.*, vol. 33, no. 12, pp. 2240-54, 1997.

[4] J. R. Marciante, and G. P. Agrawal, "Spatio-temporal characteristics of filamentation in broad-area semiconductor lasers," *IEEE J. Quantum Electron.*, vol. 33, no. 7, pp. 1174-79, 1997.

[5] N. Peyghambarian, S. W. Koch, and A. Mysyrowicz, *Introduction to Semiconductor Optics*, Englewood Cliffs, NJ: Prentice-Hall, 1993.

[6] E. Gohrig, and O. Hess, *Spatio-Temporal Dynamics and Quantum Fluctuations in Semiconductor Lasers*, Berlin: Springer, 2003.

[7] P. S. Zory, *Quantum Well Lasers*, Academic Press, 1993.

[8] G. P. Agrawal, "Fast-Fourier-transform based beam-propagation model for stripe-geometry semiconductor lasers: inclusion of axial effects," *J. Appl. Phys.*, vol. 56, pp. 3100-3109, 1984.

[9] J. Stohs, D. J. Bossert, D. J. Gallant, and S. R. J. Brueck, "Gain, refractive index change, and linewidth enhancement factor in broad-area GaAs and InGaAs quantum-well lasers," *IEEE J. Quantum Electron.*, vol. 37, no. 11, pp. 1449-59, 2001.

[10] J. R. Marciante, and G. P. Agrawal, "Nonlinear mechanism of filamentation in broad-area semiconductor lasers," *IEEE J. Quantum Electron.*, vol. 32, no. 4, pp. 590-96, 1996.

[11] L. D. Westbrook, and M. J. Adams, "Explicit approximations for the linewidth-enhancement factor in quantum-well lasers," *IEE proceedings*, vol. 135, no. 3, pp. 223-225, 1988.

[12] W. W. Chow, and D. Depathe, "Filamentation in conventional double heterostructure and quantum well semiconductor lasers," *IEEE J. Quantum Electron.*, vol. 24, no. 7, pp. 1297-1301, 1988.

Monte Carlo Simulation of Surface Annealing before Epitaxial Growth

Chang Fu Dee and Burhanuddin Yeop Majlis (*SMIEEE*)

Institute of Microengineering and Nanoelectronics

Universiti Kebangsaan Malaysia

Email: cfdee@vlsi.eng.ukm.my

Abstract - **Presented is a molecular beam epitaxial pre-growth-annealing simulation. Simulation done using Monte Carlo method by taking consideration of GaAs decomposition, As desorption and absorption process from the surface. Surface roughness information can be deducted from the percentages of step-edge-site density. It is compared to the specular reflected beam from Reflected High Energy Electron Diffraction (RHEED). Good agreement with the experimental data shows the correctness of the simulation model.**

I. INTRODUCTION

Molecular beam epitaxy (MBE) is a well established method for obtaining excellent quality single crystal structure. Hence, modeling and simulation is needed for refine and improve the experimental results. Numerical analysis in simulation plays an important role for leading to a better understanding of MBE growth [1-4].

Basically, there are two generic categories of simulations which are molecular dynamic and Monte Carlo simulation. Molecular dynamic simulation normally demands for very high computational power and the calculation is far too large for present-day personal computers. As a reason, the Monte Carlo method has becoming the option for crystal growth simulation. It involved the generation of random numbers to select from a list of events with different probability.

This paper discussed simulation base on the solid-on-solid (SOS) model on a cubic lattice with no surface reconstruction during annealing before epitaxial growth. It assumes that molecules or atoms only deposited above another crystalline molecule [5]. The lattice layers consist of As atoms at the even number layer and Ga atoms at the odd number layer. We ignore the complicated details such as crystal structure [2, 3, 6, 7] and surface reconstruction [1, 4, 8] which most of the III-V compound epitaxial simulation had done.

Some basic assumtions such as perfect substrate condition, irreversible aggregations etc. have been made.

II. SIMULATION METHOD – THE MODEL

Simulations are performed on a 100x100 perfect lattice site. It starts with a flat and smooth As surface in the present of As flux. In this stated environment, surface Ga adatoms are assumed to be immobile. However, GaAs bonding can thermally decompose and As desorps from the surface. This will leave behind Ga atoms on the surface. As atoms adsorption form the ambient to the surface is slaved to the Ga adatoms on the surface. They can combine to the surface Ga adatoms lattice only. Both As absorption and desorption are dynamic process which will end with a final equilibrium state where fluctuation of the surface Ga concentration becomes neglectable.

Our simulations consider some generic types of incidents that usually occur during the annealing process which are Ga atoms desorption, incoming of As atoms that combine with Ga adatoms to form GaAs, decomposition of GaAs to desorp free As. Ga flux was omitted for annealing process because it is only been treated when epitaxial growth started.

Direct Ga diffusion tangent on the surface was ignored in this simulation. It was actually incorporated into deposition step by treating Ga adatoms as having infinite mobility and occupied the sites spontaneously with varying probability. Equations used to calculate the probabilities finding Ga atoms at a step-edge site ($P_{Ga(edge)}$) and on a terrace site ($P_{Ga(terr)}$) are

$$P_{Ga(edge)}N_{Ga(edge)} + P_{Ga(terr)}N_{Ga(terr)} = N_{Ga\ (surface)} \quad (1)$$

$$N_{Ga(terr)} = N_{total\ lattice} - N_{Ga(edge)} \qquad (2)$$

$$P_{Ga(edge)}/P_{Ga(terr)} = \exp(T_c/T) \qquad (3)$$

The number of step-edge sites, terrace sites are given by $N_{Ga(edge)}$ and $N_{Ga(terr)}$. $N_{Ga(surface)}$ is the total number of Ga atoms on the surface. T_c is treated as a preset parameter in the simulation and the value is adjusted to $\log_e(2)$ so that the there probability for deposition of a As atom on the step edge over the terrace is 2. This value is taken because because of the higher binding energy of Ga atoms at step edge site compared to terrace site.

The probabilities for As atoms to arrive at Ga adatom and deposit on top of it is $P_{As(absorp)}$ and the probabilities for As to desorp from Ga adatom is $P_{As(desorp)}$. Where $P_{As(deposite)}$ and $P_{As(desorp)}$ can be written as

$$P_{As(deposite)} = (R_{As}/N_{total\ lattice}) \times P_{Ga(site)} \qquad (4)$$

$$P_{As(desorp)} = \exp(-(E_a/k_BT)) \qquad (5)$$

$$E_a = [E_0 + Site_{Neighbor} \times E_{neighbor}] \qquad (6)$$

where R_{As} is the As flux (atom per second); $N_{total\ lattice}$ is the total number of lattice sites on the surface; $P_{Ga(site)}$ is the probability Ga atoms to be found on the surface; E_0 is the binding energy to the substrate, 0.25eV; $Site_{Neighbor}$ is the number of directly contact nearest neighbors for a site (minimum 0 and maximum 4); $E_{neighbor}$ is the binding energy to each nearest neighbor (0.15eV); k_B is the Boltzmann constant and T is the surface temperature.

If As atom encounters a Ga atom on the surface, it will combine with the Ga adatom to form GaAs. For As atom arrives at a lattice already occupied by As, it is assumed to have left the lattice site immediately without any computational process.

III. RESULTS AND DISCUSSIONS

Fig. 1 shows that the surface Ga concentration is affected by the surface temperature. As the temperature been increased from 500°C to 800°C, more As atoms will desorp from the surface because more As atoms will obtain enough energy to overcome the binding potential. As a result, the concentration of Ga on the surface increases. After some time, concentration of Ga on the surface comes to equilibrium and the rate for As to desorp and absorp is the same.

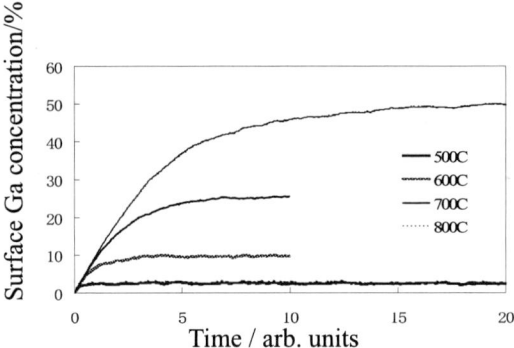

Fig. 1: Ga adatoms concentration versus time for substrate temperatures ranging from 500 to 800°C. Ga adatoms concentration come to equilibrium after some arbitrary unit of time. The higher the temperature, the more As adatoms are evaporated.

Fig 2 shows the graph for the simulated As step edge intensity as the annealing progress. We assume that the step edge intensity will give the information for the roughness of the surface and the roughness of the surface will increase as the number of desorped As increases. It is compared to the Reflected High Electron Energy Difraction (RHEED) intensity from experiment. A good matching to the experiment shows that the deduction is correct. The specular intensity of RHEED increase from minimum to maximum as some of the As atoms evaporated from the surface.

Fig. 3 shows the simulated images at equilibrium for temperature 500, 600, 700 and 800°C. The black portion indicates Ga atoms on the surface and the white is occupied by As atom. For 500°C, there are not much As atoms evaporate from the surface, most of the As atoms at the edge are evaporated because the number of nearest neighbor are lesser compared to the inner part of the surface.

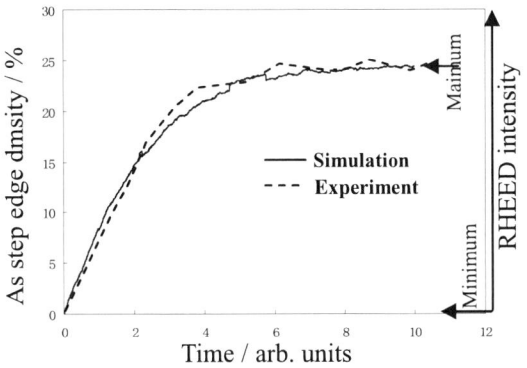

Fig. 2: As more and more As evaporates from the surface, the step edge density will in crease. The simulation for 700°C surface temperature shows the increase in the step edge.

IV. SUMMARY

A successful model has been done to simulate the annealing process before epitaxial growth. Good agreement of the simulated data to the experiment was obtained. It shows that the correctness of this model.

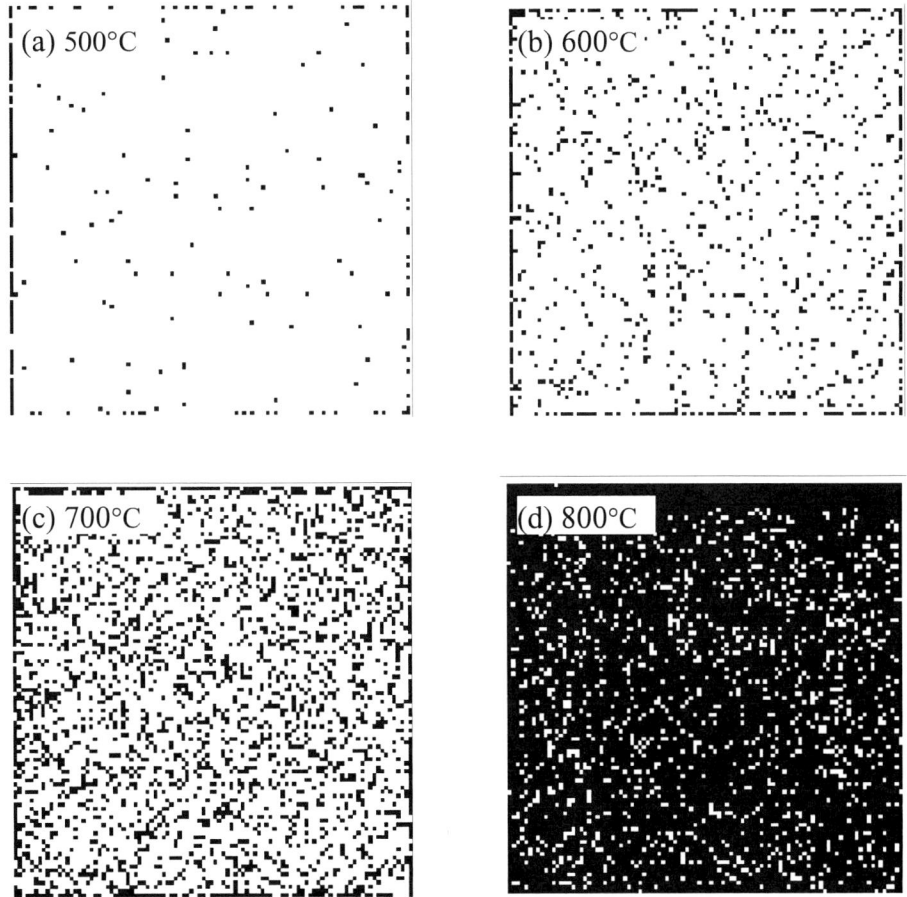

Fig. 3: Images of GaAs surface at equilibrium. Black colored portion shows lattices site where As atoms have evaporated and left behind the Ga adatoms. The white portions are for As atoms.

REFERENCE

[1] C. Heyn and M. Harsdorff, "Correlation between island-formation kinetics, sruface roughening, and RHEED oscillation damping during GaAs homoepitaxy," *Phys. Rev. B*, vol. 56, pp. 13483, 1997.

[2] S. V. Ghaisas and A. Madhukar, *Phys. Rev. Lett.*, vol. 56, pp. 1066, 1986.

[3] S. Clarke and D. D. Vvedensky, "Origin of Reflection High-Energy Electron-Diffraction Intensity Oscilations during Molecular-Beam Epitaxy: A Computational Modeling Approach," *Phys. Rev. Lett.*, vol. 58, pp. 2235-2238, 1987.

[4] C. Heyn and M. Harsdorff, "Simulation of GaAs growth and sruface recovery with respect to gallium and arsenic kinetics," *Phys. Rev. B*, vol. 55, pp. 7034, 1997.

[5] H. Muller-Krumbhaar, in *Monte Carlo methods in statistical physics*, K. Binder, Ed. Berlin: Springer, 1986.

[6] S. D. Sarma and P. Tamborenea, *Phys. Rev. Lett.*, vol. 66, pp. 325, 1991.

[7] M. D. Johnson, C. Orme, A. Hunt, J. Sudijono, D. Graff, L. M. Sander, and B. G. Orr, *Phys. Rev. Lett.*, vol. 72, pp. 116, 1994.

[8] C. Heyn, T. Franke, and R. Anton, *J. Cryst. Growth*, vol. 201-202, pp. 67, 1999.

Analysis of Poly Resistor Mismatch

Philip Beow Yew Tan[1,2] *, Albert Victor Kordesch[1] and Othman Sidek[2]

[1]Silterra Malaysia Sdn. Bhd. Kulim Hi-Tech Park,09000 Kulim, Kedah, Malaysia

[2]Universiti Sains Malaysia, 14300 Nibong Tebal, Pulau Pinang, Malaysia

*Email: philip_tan@silterra.com

Abstract – **In this paper we analyzed the mismatch characteristics of poly resistors. We found that the P+ poly resistor mismatch characteristics do not follow the linear inverse square root dependence. Instead, the P+ poly resistor mismatch follows the quadratic of inverse square root (or a simple inverse area) dependence. N+ poly resistor mismatch, however does obey the expected standard inverse square root dependence.**

I. INTRODUCTION

Resistor mismatch can be defined as the difference in resistance between two identically designed resistors, meaning with the same width, W and length, L. The resistance difference of two identically designed poly resistors is caused by localized process variation such as lithography, etch, deposition, oxidation, silicidation, grain size, and dopant concentration. As suggested in [1], device mismatch is expected to follow the 1/sqrt(W*L), inverse square root area dependence. It has been reported that diffused resistors (or active resistors) mismatch does not follow the 1/sqrt(W*L) [2]. In term of mismatch for transistor, our study in [3] shows that the threshold voltage mismatch does not obey the inverse square root dependence, instead it follows the inverse length dependence

It is reported in [4] that the resistor mismatch improves more by increasing the width of the device than by increasing the length. It is found in [5] that the poly resistor mismatch is directly proportional to the grain size.

II. TEST STRUCTURES

The test structures used in this experiment are 11 pairs of unsilicided P+ poly resistors and 11 pairs of unsilicided N+ poly resistors with different widths and lengths. The resistors in pair are placed adjacent to each other to avoid global mismatch effects. The two terminal of the resistor are connected to two pads. Two-terminal non-Kelvin IV measurement was performed to extract the resistance value (Kelvin measurement needs 4 pad connections). For mismatch, as long as the resistors in each pair are measured with the same method, the systematic errors will be cancelled when we subtract the two resistor values and calculate the standard deviation of the difference. Hence, in this experiment we do not need to use the Kelvin method.

In this study, our sample size is 81 pairs of resistors for each size. The samples are from 9 dice per wafer, on 9 wafers from one lot, fabricated using Silterra's standard 130nm CMOS technology. The nominal sheet resistance for N+ poly is 256 Ohm/square and for P+ poly is 320 Ohm/square. Poly thickness is 160nm. The resistance was measured in Etest at Vforce = 1V with an Agilent 4072A parametric tester.

III. RESULTS AND DISCUSSION

The results of P+ poly mismatch and N+ poly mismatch are shown in Fig. 1 and Fig. 2 respectively. From Fig. 1, we noticed that the solid line that represents the linear 1/sqrt(W*L) dependence do not fit the actual silicon data (blue diamond). By using the quadratic 1/sqrt(W*L) as represented by the breaking line, a good fit to the actual silicon data can be achieved.

This means that the mismatch for small size P+ poly resistors have an unusual high mismatch compare to what is expected by linear 1/sqrt(W*L) dependence and this does not happen to N+ poly resistors, as shown in Fig. 2. Since the unusual high mismatch only happened to P+ poly resistors and do not happened to N+ poly resistors, we can eliminate the possibility that it is caused by lithography, etch, or other process factors shared by both P+ and N+ resistors. Thus, this phenomenon is most probably caused by dopant variation.

One possible explanation for this phenomenon is P+ dopant (Boron) has higher variation compared to N+ dopant (Arsenic and Phosphorus). This explanation is plausible because Boron is lighter, diffuses more easily,

0-7803-9730-4/06/$25.00 ©2006 IEEE

and segregates more compared to Arsenic and Phosphorus. We expect the variation of Boron to be higher. Another possible explanation is that the net P+ dopant concentration in P+ poly resistors is simply much lower than the N+ dopant in N+ poly resistors hence making the dopant variation effect larger for small size resistors. Self-heating effects are ruled out because this would occur in N+ poly resistors first due to their lower sheet resistance.

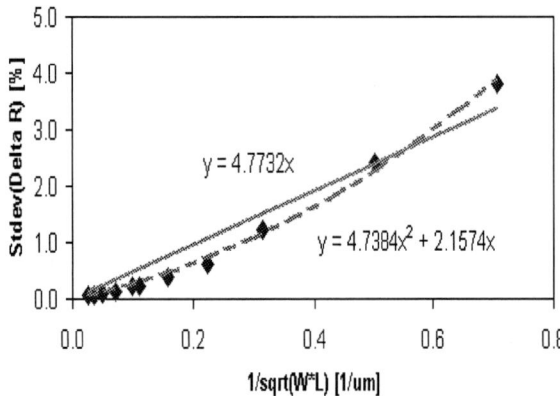

Fig. 1 Plot of P+ poly resistor mismatch versus inverse square root area for 11 different size resistors. N_{pairs}=81 for each resistor's size.

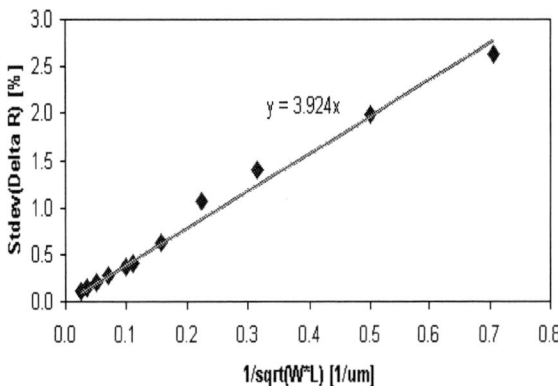

Fig. 2 Plot of N+ poly resistor mismatch versus inverse square root area for 11 different size resistors. N_{pairs}=81 for each resistor's size.

The actual silicon data used to plot the P+ poly resistor mismatch and N+ poly resistor mismatch (standard deviation of delta R versus inverse square root of resistor area) as in Fig. 1 and Fig. 2 is shown in Table 1.

Table 1: Standard deviation of delta R in percentage for different sizes of P+ poly resistors and N+ poly resistors.

Resistor Type	Width	Length	1/sqrt(W*L)	Stdev(Delta R) in %
P+ Poly Resistor	16	100	0.03	0.0617
	8	100	0.04	0.0797
	4	100	0.05	0.0879
	2	100	0.07	0.1303
	1	100	0.10	0.2373
	2	40	0.11	0.2250
	2	20	0.16	0.3795
	2	10	0.22	0.6227
	2	5	0.32	1.2399
	2	2	0.50	2.4247
	1	2	0.71	3.8139
N+ Poly Resistor	16	100	0.03	0.1133
	8	100	0.04	0.1459
	4	100	0.05	0.2017
	2	100	0.07	0.2841
	1	100	0.10	0.3636
	2	40	0.11	0.4066
	2	20	0.16	0.6337
	2	10	0.22	1.0710
	2	5	0.32	1.3897
	2	2	0.50	1.9816
	1	2	0.71	2.6374

IV. CONCLUSION

In conclusion, we have showed that the P+ poly resistor mismatch does not follow the expected linear 1/sqrt(W*L) dependence. Instead the P+ poly resistor mismatch can be fitted accurately by using quadratic of 1/sqrt(W*L) or a simple 1/(W*L) dependence. Meanwhile N+ poly resistor mismatch, it still obeys the expected linear 1/sqrt(W*L) dependence.

ACKNOWLEDGEMENT

The authors would like to acknowledge all the members of Silterra Malaysia Sdn. Bhd. for supporting and contributing to the research work in this paper.

REFERENCES

[1] M. Pelgrom, A. Duinmaijer and A. Welbers, "Matching Properties of MOS Transistors," *IEEE Journal of Solid State Circuits, JSSC,* vol. 24, pp. 1433-1439 (1989).

[2] P. G. Drennan, "Diffused Resistor Mismatch Modeling and Characterization," *Bipolar/BiCMOS Circuits and Technology Meeting,* pp. 27-30 (1999).

[3] P. B. Y. Tan, A. V. Kordesch and O. Sidek, "Analysis of 130nm CMOS Transistor Mismatch," *National Symposium on Microelectronics, NSM,* (2005).

[4] H. Thibieroz, P. Shaner and Z. C. Butler, "Mismatch and flicker noise characterization of tantalum nitride thin film resistors for wireless applications," *International Conference on Microelectronic Test Structures, ICMTS,* pp. 207-212 (2001).

[5] R. Thewes *et al.,* "Explanation and Quantitative Model for the Matching Behavior of Poly-Silicon Resistors," *International Electron Device Meeting, IEDM,* (2005).

Modeling of Polyimide MIM Capacitors for Applications in Planar Monolithic Microwave Integrated Circuits

R. Sanusi, A.I. Abdul Rahim, M.N. Osman, N. Kushairi, A. Rasmi, N.F.I.Muhammad,
M. R. Yahya and A.F. Awang Mat

Telekom Research and Development Sdn. Bhd.,
Idea Tower I & II, UPM-MTDC, Technology Incubation Centre One,
Lebuh Silikon, 43400 Serdang, Selangor Darul Ehsan.
Tel: 03-89451417, Fax: 03-89441230
Email : rasidah@tmrnd.com.my , drismat@tmrnd.com.my

Abstract – **Polymide Metal-Insulator-Metal (MIM) overlay capacitors for use in Monolithic Microwave Integrated Circuits (MMICs) based on High Electron Mobility Transistors (HEMTs) on Gallium Arsenide substrates are presented. Modeling of the capacitors was performed using a 2-dimensional electromagnetic CAD simulator to obtain Scattering (S-) parameters for different capacitor dimensions for operating frequencies from 0.05 to 8 GHz. The behaviour of the capacitor as a function of operating frequencies is studied by means of Smith chart. The capacitor is finally represented by a proposed equivalent circuit model to describe its overall behavior for planar MMIC simulations.**

I. INTRODUCTION

The Metal-Insulator-Metal (MIM) capacitor is a key passive component in MMIC's technology for decoupling, filtering, oscillating, bypassing and matching functions [1] and is widely used in wireless 3G, WiMAX and Wi-Fi systems. MIM capacitors are desirable in MMICs applications because of its high capacitance density that increases circuit density and further reduces the fabrication cost and maximize the number of components per unit chip area. MIM capacitor provides good voltage linearity properties [2].

Polymide MIM capacitors are preferred rather than silicon nitride MIM capacitors for MMIC high power and high temperature applications due to its excellent thermal and dielectric properties [3]. The latter is subject to failure over long hours of operation due to the stress induced in the silicon nitride layer during processing and device

operation [4]. The performance of the MIM capacitor is determined by the thickness of the insulator and the overlap area of between top plate, insulator and bottom plate sandwich. To obtain high values of capacitance with low inductive values, it is necessary to reduce the capacitor area while reducing leakage current, parasitic and interconnect resistance [5].

This paper presents 2D electromagnetic simulation results for a polyimide MIM capacitor using a two-port network technique designed for MMIC's applications at frequency performance from 0.05 up to 8 GHz. Four different layout dimensions of polyimide MIM capacitor i.e. 60 μm x 60 μm, 140 μm x 140 μm, 260 μm x 260 μm and 380 μm x 380 μm were simulated and modelled using a commercial electromagnetic simulator and CAD tool. From the simulation results obtained, an equivalent circuit for the capacitor is generated and proposed to represent the electrical performance for future circuit simulations.

II. EXPERIMENTAL PROCEDURE

Figure 1 shows the cross-sectional diagram of the simulated polyimide MIM capacitor. It consist of 100 μm GaAs substrate, 1.5 μm thick polyimide and 0.12 μm silicon nitride sandwiched by the bottom plate (metal 2) and top plate (metal 3) layers with thicknesses of 0.4 μm and 3.12 μm. The complete capacitor is covered by a layer of silicon nitride with a thickness of 0.12 μm which passivates the whole structure.

0-7803-9730-4/06/$25.00 ©2006 IEEE

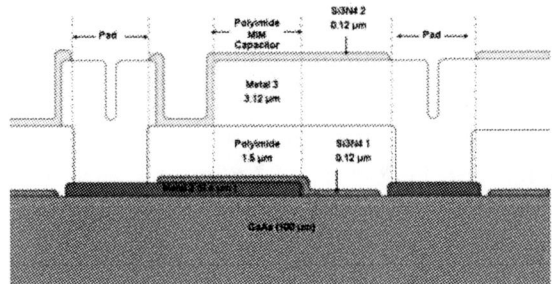

Figure 1: Cross-section on polyimide MIM capacitor.

Figure 2(a) shows the layout for a 140 μm x 140 μm polyimide MIM capacitor. It is defined using Agilent – MOMENTUM layout editor [6]. In addition, the layout for a 60 μm x 60 μm, 140 μm x 140 μm, 260 μm x 260 μm and 380 μm x 380 μm is also created. Using the "Create/Modify Substrate" facility in Agilent – MOMENTUM, the metal, polyimide, nitride and gallium arsenide (GaAs)-substrate material properties are defined. The relative permittivity, ε_r used for polyimide, nitride and GaAs are 3.52, 6.83 and 12.9 respectively.

Electromagnetic S-parameter simulations are performed by using the two-port circuit schematics shown in Figure 2(b). The circuit schematic is automatically generated with the RF input and output pads represented by B1 and B2. The capacitance of the polyimide capacitor is determine from the impedance obtained from the S_{11} parameters using the following transform $S_{11} = (Z-Z_0) / (Z+Z_0)$ which gives the impedance, $Z = Z_0*(R-jX_c)$ and hence :

$$C = \frac{1}{2\pi X_c} \qquad (1)$$

The PI circuit topology was used to generate the equivalent circuit for the simulated capacitor by using the "Spice Model Generator" available in ADS-Momentum. The initial equivalent circuit obtained was further refined by performing fitting of the S_{21} and S_{11} parameter using the IC-CAP software [7].

III. RESULTS AND DISCUSSION

Figures 3(a) to (d) shows the Smith Chart plots for S_{21} and S_{11} values for the 4 different sizes of capacitors in the frequency range from 0.05 to 8 GHz. Generally, from the S_{11} parameters it can be seen that all the capacitors are capacitive in this frequency range. As the frequency is increased the magnitude of S_{11} values decreases, for instance for

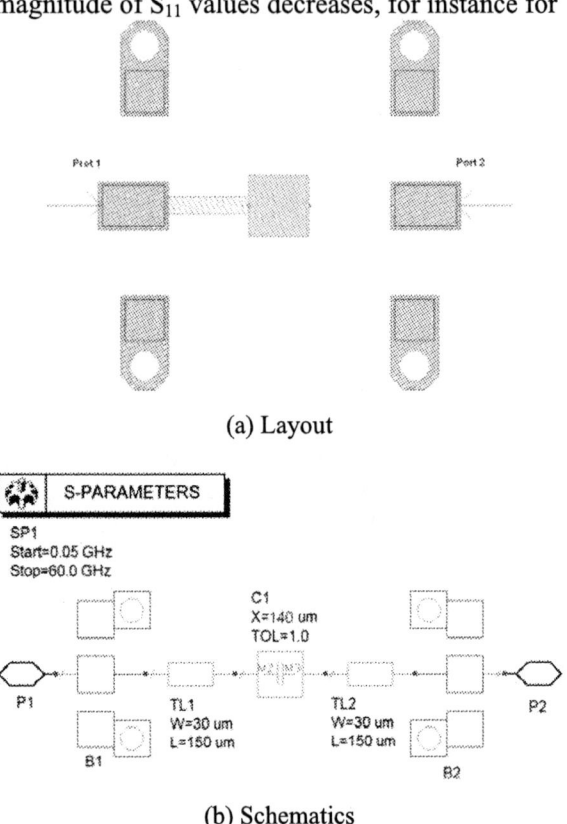

(a) Layout

(b) Schematics

Figure 2: Polyimide MIM capacitor: (a) Layout (a) and corresponding (b) schematics in ADS design environment.

a 140 μm x 140 μm capacitor the magnitude of S_{11} decreases from 1.0 dB to 0.811 dB when the frequency is increased from 0.05 to 8 GHz. The same trend also applies for the other 3 types of capacitors. The magnitude of S_{21} parameters for all 4 sizes of capacitors shows very similar trend. The magnitude of S_{21} values at 8 GHz are 0.264 dB, -5.191 dB, -4.144 dB and -4.715 dB for the 60 μm x 60 μm, 140 μm x 140 μm, 260 μm x 260 μm and 380 μm x 380 μm capacitors respectively.

Figure 4 shows a plot of capacitance as a function of frequency for the 4 different sizes of polyimide MIM capacitors. The capacitance values obtained varies from 10 to 75 pF and was extracted from the S_{11} values obtained using equation (1). It is shown that the there is almost no change in capacitance for the 60um x 60um capacitor as the frequency is increased from 0.05 to 8 GHz. However for the other 3 different layout sizes of capacitors; i.e. 140 μm x 140 μm, 260 μm

1040

x 260 μm and 380 μm x 380 μm, there is a noticeable increase in capacitance as the frequency is increased from 4 to 8 GHz.

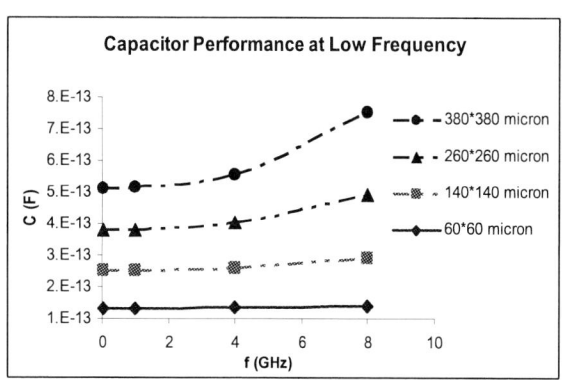

Figure 4 : Capacitance versus frequency for 4 different layout dimension sizes of polymide MIM capacitor.

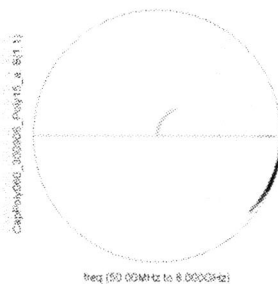

(a) 60 μm x 60 μm Polyimide MIM capacitor

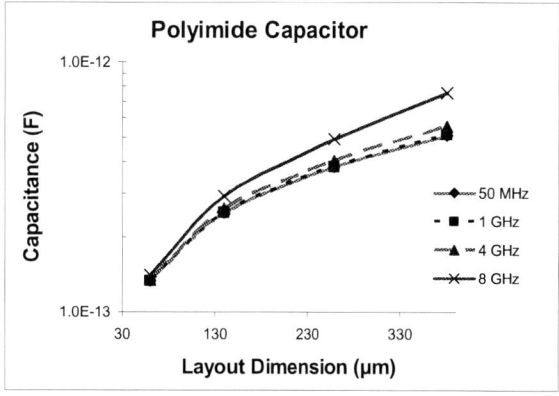

Figure 5 : Capacitance versus. layout dimension at frequencies of 0.05, 1, 4 and 8 GHz.

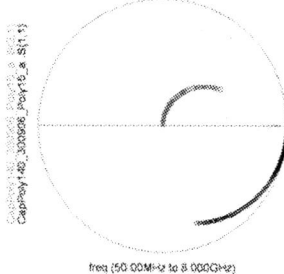

(b) 140 μm x 140 μm Polyimide MIM capacitor

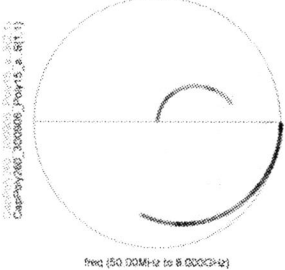

(c) 260 μm x 260 μm Polyimide MIM capacitor

Figure 5 shows a plot of capacitance as a function of layout dimension for the polymide MIM capacitor. It can be seen that the capacitance increases as the layout dimension is increased. There is a factor of approximately 2 increases in capacitance as the layout dimension is increased from 60 μm x 60 μm to 140 μm x 140 μm. However there is only a factor of 1.25 and 1.2 increases in capacitance as the layout dimension is increased from 140 μm x 140 μm to 260 μm x 260 μm and from 260 μm x 260 μm to 380 μm x 380 μm.

Figure 6(a) shows the equivalent circuit obtained for the 60 μm x 60 μm and 140 μm x 140 μm polymide MIM capacitors. This equivalent circuit was obtained by fitting the S_{21} and S_{11} parameters and the corresponding capacitance values C1 and C2 obtained is summarized in Table 1. The equivalent circuit for the 260 μm x 260 μm and 380 μm x 380 μm capacitors is shown in Figure 6(b). Here, there is an additional inductance

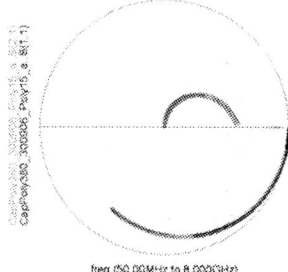

(d) 380 μm x 380 μm Polyimide MIM capacitor

Figure 3 : Modeled values of S_{11} for a single polyimide MIM capacitor from 0.05 GHz up to 8 GHz

L1 and L3 which takes into account the transmission line connecting the input and output pads to the top and bottom plate of the capacitor. Also introduced is a resistor, R1 in parallel with the capacitance C2. This takes into account the resistance introduced by the polymide layer. All the equivalent circuit parameter values are summarized in Table 1.

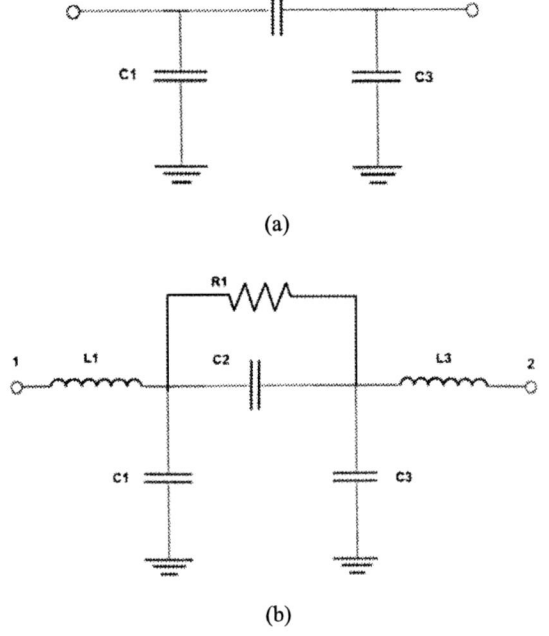

(a)

(b)

Figure 6: Equivalent circuit generated for (a) 60 μm x 60 μm and 140 μm x 140 μm and (b) 260 μm x 260 μm and 380 μm x 380 μm for the polyimide MIM capacitor.

Table 1: Extracted values of C, L and R obtained from the equivalent circuits shown in figures 6 (a) and (b). for the different layout dimensions of the polymide MIM capacitor.

$x * x$	C1 (fF)	L1 (pH)	R1 (KΩ)	C2 (fF)	C3 (fF)	L3 (pH)
60	79.23	-	-	61.95	58.19	-
140	134.1	-	-	163.9	1.494	-
260	166.6	525.5	893.7	163.9	79.97	73.78
380	337.1	269.1	978	155.7	5.566	718.4

IV. CONCLUSIONS

This work has demonstrated that 2D electromagnetic simulations can be used to predict

the performance of polymide MIM capacitors for MMIC applications. It is shown that the designed polymide MIM capacitor is capacitive in the frequency range from 0.05 up to 8 GHz. Investigations shows that the capacitance obtained for a 1.5 μm thick polymide and for layout dimension size from 60 μm x 60 μm to 380 μm x 380 μm is from 13 to 75pF. Two types of equivalent circuit model were generated to describe the performance of the capacitor for MMIC applications.

ACKNOWLEDGEMENT

The authors would like to thank TM Research & Development Bhd. for providing the research grant for this work and Advanced RF Systems Cluster for helpful technical discussions.

REFERENCES

[1] M. Engels and R.H. Jansen, "Rigorous 3D EM Simulation and An Efficient Approximate Model of MMIC Overlay Capacitors with Multiple Feedpoints", IEEE MTT-S International Microwave Symposium Digest, 14-18 Jun 1993, Page(s): 757-760 vol.2.

[2] Piquet, J.; Cueto, O.; Charlet, F.; Thomas, M.; Bermond, C.; Farcy, A.; Torres, J.; Flechet, B, ".Simulation and Characterization of High-frequency Performances of Advanced MIM Capacitors", Solid-State Device Research Conference, 12-16 Sept. 2005 Page(s):497 – 500

[3] Chen, K.Y.; Brown, W.D.; Schaper, L.W.; Ang, S.S.; Naseem, H.A., "A Study of the High Frequency Performance of Thin Film Capacitors for Electronic Packaging", IEEE Transactions on Advanced Packaging, Vol. 22, No 2, May 2000 Page(s):293 – 302

[4] GaAs IC Foundry Design Manual Process F20/F12

[5] Arnould, J.-D.; Benech, Ph.; Cremer, S.; Torres, J.; Farcy, A., "RF MIM Capacitors Using Si_3N_4 Dielectric in Standard Industrial BiCMOS Technology", IEEE International Symposium on Industrial Electronics, Volume 1, 4-7 May 2004 Page(s):27 - 30 vol. 1

[6] Agilent Technologies, "Momentum", August 2005

[7] Agilent Technologies, "ICCAP 2006", April 2006

[8] Gerhard, G.; Koch, S.; "MIm Capacitor Modeling : A Planar Approach"; IEEE Transaction on Microwave Theory and Techniques; Vol 43, No. 4, April 1995.

[9] Wen-Yan Yin; Shujan Pan; Le-Wei Li; Yeow-Beng Gan; Ban-Leong Ooi; "Experimental Characterization of On-Chip Inductor and Capacitor Interconnect : Part I. Series Case" IEEE Transactions on Magnetics, Volume 39, Issue 6, Nov. 2003 Page(s):3497 - 3502

Simulation of InGaN Multiple Quantum Wells (MQWs) Light Emitting Diodes (LEDs)

S. M. Thahab, H. Abu Hassan, Z. Hassan

Nano-Optoelectronics Research and Technology Laboratory
School of Physics, Universiti Sains Malaysia
11800 Penang, Malaysia
E-mail: sabahmr @ yahoo.com, haslan@usm.my, zai@ usm.my

Abstract -**InGaN LEDs on sapphire substrates were simulated using ISE TCAD software. In order to obtain a high output power, 15 pairs of GaN (50nm)/Al$_{0.27}$Ga$_{0.73}$N (52nm) DBR were introduced between the i-GaN and n-GaN layers. The weak output power resulting from our simulation may be related to the inhomogeneous holes distrib-ution in the quantum wells. Also the piezoe-lectric field due to strains which determines the emission mechanism of InGaN based LEDs is affected by these parameters. The turn on voltage for our structure was 0.7 V and has a small change with the introduction of DBR.**

I. INTRODUCTION

The segment of semiconductor light emitting devices has seen rapid developments over the last decade. High brightness LEDs [1] are now available for the whole visible spectral range. They are now capable of competing with or even outperforming conventional light sources like incandescent bulbs and fluorescent lamps. Simulations for optoelectronic semiconductor devices can be subdivided into several categories. One of it is the process simulation, for example the simulation and optimization of the epitaxial growth conditions. A broad area of physical problems arises from the semiconductor device itself, mainly concerning the electrical charact-eristics and the optical properties of the LED [2]. One factor limiting the efficiency of an LED is the ability to capture light inside the device by total internal reflection due to the high refractive index of semiconductors. Brightness of LEDs fell short to the brightness level of sapphire based devices and the interface between the epilayer and substrate is not suited for light extraction. InGaN and its alloys are important compound semiconductors because their active layer emits light by the recombination of the injected electrons and holes [3-5]. The second factor

which limits LED performance is the internal quantum efficiency and piezoelectric effect due to strains in InGaN[6]. Our structure is numerically investigated with an ISE TCAD simulation program. Fig. 1 shows the model of localized energy states formed by In composition fluctuations. When the degree of In composition fluctuations is small, the carriers easily overflow the localized states with increasing current as shown by Fig. 1 (a), while when the degree of In composition fluctuations is large, the carriers are still confined within the localized states even at high current operation as shown by Fig. 1(b).

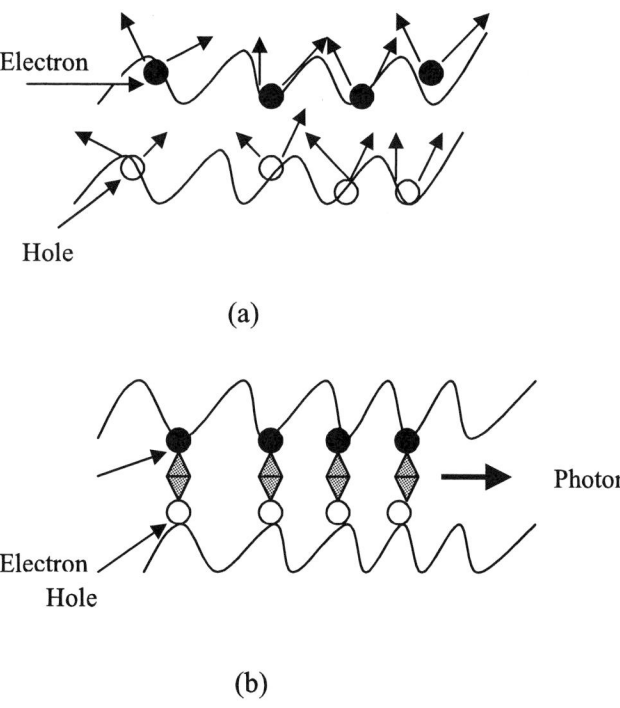

Fig.1 The schematic model of localized energy states formed by In composition fluctuations in In$_x$Ga$_{1-x}$N well layer (a) x is small (b) x is large

With increasing current some carriers can overflow from the localized energy states due to a small In composition fluctuation of InGaN [7]

and reach nonradiative recombination centers. Many groups have reported that the quantum confined Stark effect in this process that results from the piezoelectric field due to strain determines the emission mechanism of InGaN based LEDs [8-9]. Optical gain mechanism in InGaN quantum wells of real LEDs and laser is still not fully understood. It may be affected strongly by a nonuniform In distribution and internal polarization fields that tend to quantum confined electrons and holes thereby reducing optical gain and spontaneous emission.

II. LED STRUCTURE AND PARAMETERS USED IN THE SIMULATION

A schematic diagram of the LED structure under study is shown in Fig. 2. In this structure the active region consists of 3 periods of 3nm $In_{0.13}Ga_{0.87}N$ wells that are sandwiched between 5nm of $In_{0.01}Ga_{0.99}N$ barrier for an emission wavelength of 435nm. It was assumed that the level of p-type doping is $5*10^{17}$ /cm^3 and for n-type doping it is assumed that the level is $5*10^{18}$/cm^3. The band gap energy of the $Al_xGa_{1-x}N$ and $In_xGa_{1-x}N$ are obtained from the two formulae below[10]:

For $Al_x Ga_{1-x}N$:

$$E_g(x) = xE_{AlN} + (1-x)E_{GaN} - bx(1-x) \quad (1)$$

$$E_g(x) = 6.28x + 3.42(1-x) - 1.3 x (1-x) \quad (2)$$

For $In_xGa_{1-x}N$:

$$E_g(x) = xE_{InN} + (1-x)E_{GaN} - bx(1-x) \quad (3)$$

$$E_g = 0.77x + 3.42(1-x) - 1.4x (1-x) \quad (4)$$

The effective masses of electrons in InN and GaN are 0.11 m_0 and 0.22m_0 respectively, and the hole mobility in GaN is 8 cm^2/Vs [10]. Our simulation was conducted by choosing the following model. Carrier concentration, Mobility with high field saturation, Effective intrinsic density (with no band gap narrowing), Fermi model, Recombination (Shockley Read Hall), and recombination lifetime of electron and hole is assumed to be 1 ns. Very small Auger Parameter for GaN with an estimated value equal to 10^{-34} cm^6 s^{-1} was also included. Poisson and carrier continuity equations were performed in the simulation process.

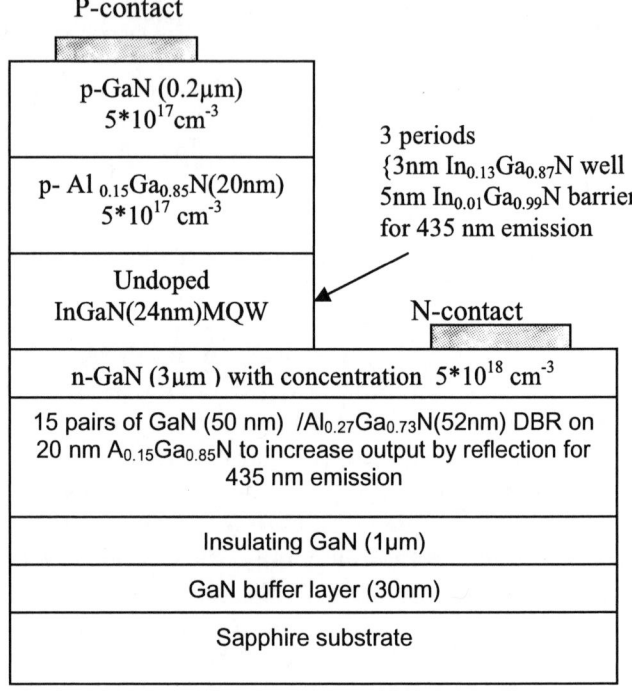

Fig. 2 The schematic structure of our InGaN multiple quantum wells (MQWs) LED

III. RESULTS AND DISCUSSION

Fig. 3(a) shows the output power of InGaN multiple quantum wells (MQWs) LED without DBR (GaN/$Al_{0.27}Ga_{0.73}N$) . The LEDs emission at 435 nm show low output power. These weak outputs may be affected strongly by a non uniform In distribution that induces internal polarization fields that tend to quantum confined electrons and holes thereby reducing optical output power and spontaneous emission in the MQWs. A small band-offset ratio between the conduction band and valence band (with ratio (3:7)) [11-12] results in a slight reduction of the output power due to increase in electron leakage since all other material parameters of the stopper layer p-$Al_{0.15}Ga_{0.85}N$ remined unchanged. In addition, the difficulty of hole transportation into the active region of the LED has played an important role in the problem of electronic current overflow. This phenomenon not only affect the current overflow but also results in the inhomogeneous holes distribution in the quantum wells of the LED,[2]. For LED structures, about 50% of the light emitting by the active region is absorbed by the interference region and substrate layer. This represents a

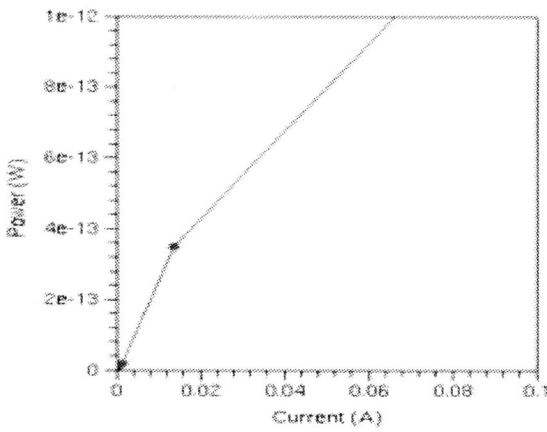

Fig. 3 Output power of InGaN LED as a function of forward current without DBRs

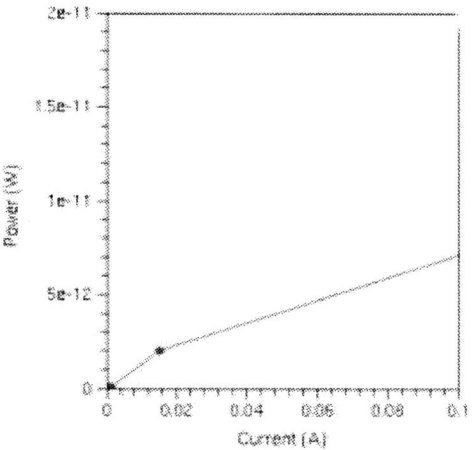

Fig. 4 Output power of InGaN LED as a function of forward current with DBRs

substantial loss; the absorption of light in the substrate can be avoided by placing a reflector between the substrate and the LED active layers. Light emanating from the active region towards the substrate will then be reflected and can escape from the semiconductor through the top surface. In our simulation we have introduced 15 pairs of GaN/Al$_{0.37}$Ga$_{0.73}$N (DBRs) between n-GaN layer and the GaN insulating layer in order to improve our LED power and efficiency. Usually the refractive index different between the two materials is small so that the magnitude of the Fresnel reflection at one interface is also quite small. However, DBR consist of many interfaces. More importantly, the thickness of the two materials is chosen in such away that all reflected waves are in constructive interference. For normal incidence, this condition is fulfilled

when both materials have a thickness of a quarter wavelength of the reflected light:

$$T_{l,h} = \lambda_{l,h} / 4 = \lambda_0 / 4 \, n_{l,h} \qquad (5)$$

where λ_0 is the vacuum Bragg wavelength of the light, $T_{l,h}$ is the thickness of the low index (l) and high index (h) material, and $n_{l,h}$ is the refractive index of the low index (l) and high index (h) material. For an oblique angle of incidence, the wave vector can be separated into a parallel and normal component. As in normal incident case, the thickness of the DBR layers must be a quarter wavelengths for the wave vector component normal to the DBR layers. For an oblique angle of incidence $\Theta_{l,h}$, the optimum thickness for high reflectivity are given by :

$$T_{h,l} = \lambda_{l,h} / (4 \cos \Theta_{l,h})$$
$$= \lambda_0 / 4n_{l,h} \cos \Theta_{l,h}) \qquad (6)$$

From Fig. 4 it can be seen the efficient DBR will be optimized in such a way that it maximizes the intensity of the reflected light. In addition, the escape of the light reflected by the DBR from the LED die must be taken into account. Our simulations is in agreement with the experimental work of T.Egawa team [6].

Fig. 5 shows the I-V characteristic for LED without DBRs under studied. Turn on voltage of 0.75 V was obtain. While no effective change in turn on voltage was observed with presence of DBRs as expected and shown in Fig. 6.

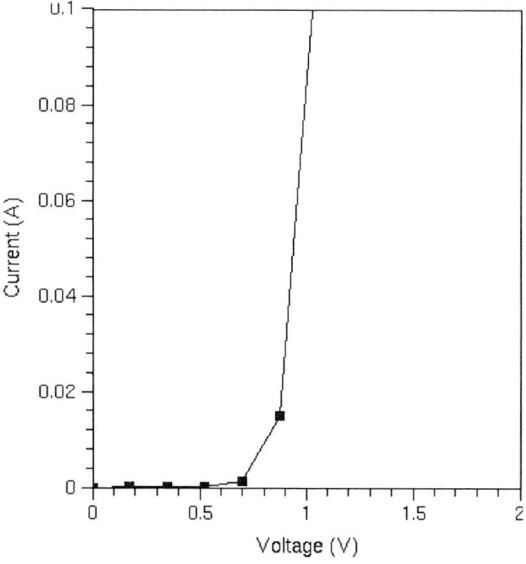

Fig.5 I-V characteristic of InGaN LED without DBRs

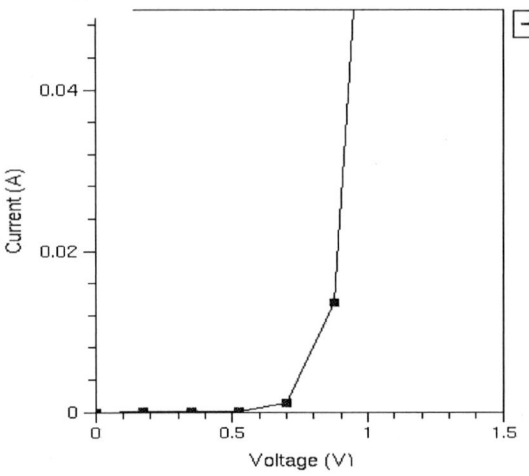

Fig.6   I-V characteristic of InGaN LED with DBRs

CONCLUSION

The electronic current and the output power of InGaN LEDs with and without DBR have been simulated by using ISE TCAD software. The weak output power is related   to the leakage current in the active region.Also the small change in turn-on voltage in both cases indicated that the DBR have little influence on I-V characteristics but effective in increase the output power. The difficulty of holes transportation into the active region of the LED needs to be overcome so as to improve current overflow and power.

ACKNOWLEDGEMENT

This work was conducted under IRPA RMK-8 Strategic Research grant. The support from Universiti Sains Malaysia and Ministry of Science Technology and Innovation are gratefully acknowledged

REFERENCES

[1]   Nakamura, S., Pearton, S., and Fasol, G. The Blue Laser Diode. *Berlin Springer-Verlag.* (2000).

[2]   Yuan, Jih.and Kuang, Y. *J .of Applied Physics*, Vol. 93, p. 4992. (2003).

[3]   Nagatomo,T. , Kuboyama,T. , Minamino,H., and Omoto,O., *Jpn .J.Appl.Phys.*, Vol. 28,   L13334. (1989).

[4]   Matsuoka, T., Tanaka, H., Sasaki, T., and Katsui, A., (1990); *Inst.Phys.Conf.Ser.*, Vol. 106, p. 141. (1990).

[5]   Matsuoka, T., *J.Cryst.Growth*, Vol. 124, p 433. (1992).

[6]   Egawa, T., Ishikawa, H., Umeno, M., Akustu, N., and Matsumoto, K., *Progress in Electromagnetic Research Symposium*,   p. 18,   July   (2001).

[7]   Nakamura, S.,(1998); *Science* ,Vol. 281, p. 956, (1998).

[8]   Takeuchi, T.,   Sota,S.,   Katsuragawa , M., Komori, M.,   Takeuchi, H.,   Amano, H.,   and Akasaki,I., *J.Appl.Phys.*,V0l. 36, L382. (1997)

[9]   Im,S.J., Kollmer, H., Off, J., Sohmer, A., Scholz ,F.,and Hangleiter, Aphys.Rev. B 57, p. 9435, (1998).

[10]   Piprek, J.,and Nakamura, S., *IEE Proc . Optoele -ctron* , Vol. 149, N4, p. 145,(2002).

[11]   Domen, K., Soejima, R., and Kuramata, A., *MRS Internet J. Nitride Semiconductor Res.* Vol. 3, p. 2, (1998).

[12]   Martin, G., Botchkarev ,A., *Appl. Phys. Lett.*, Vol. 68, p.2541, (1996).

TCAD Simulation of STI Stress Effect on Active Length for 130nm Technology

Wan Rosmaria Wan Ahmad, *Student Member, IEEE,* Albert Victor Kordesch, *Senior Member, IEEE,*
*Ibrahim Ahmad, Philip Tan Beow Yew, *Member, IEEE.*
Department of Device Modeling, Silterra Malaysia Sdn Bhd, Kulim Hi-Tech Park, 09000 Kulim, Kedah.
*Faculty of Engineering, Universiti Kebangsaan Malaysia, 43600 Bangi, Selangor.
Email: wan_rosmaria@silterra.com, al_kordesch@silterra.com

Abstract - **In this paper we investigated the compressive stress in the channel induced by shallow trench isolation (STI) for different active length (Sa). We simulate both PMOS and NMOS for 130nm gate length with five active lengths (Sa=0.34, 0.5, 0.8, 1.0, 5.0um) by using TCAD simulation and compare to experimental data from wafers fabricated using Silterra's 130nm Technology. When the Sa is decreasing, Sxx stress becomes more compressive for both P- and N- MOS while the Syy component becomes more tensile, causing hole mobility improvement in PMOS and electron mobility degradation in NMOS. When Sa decreases from 5um to 0.34um, the Idsat for NMOS is degraded 6.6% and Idsat for PMOS is increased 6%. This means narrower Sa will increases hole mobility performance in p-channel but degrade the electron mobility in n-channel. These results agree with the experimental data.**

I. INTRODUCTION

As the critical dimension of metal oxide semiconductor field effect transistors (MOSFET) is reducing the gate length, shallow trench isolation (STI) became the preferred isolation scheme over the local oxidation of silicon (LOCOS) for two major reasons. First, STI can be implemented at significantly lower temperatures than LOCOS, thus reducing the thermal budget. Secondly, the inherent bird's peak formation in LOCOS hinders its application as the critical dimension shrinks [1]. Due to the constraints from the surrounding silicon dioxide, the stress state on the top surface of the silicon substrate is compressive. The compressive stress is induced when cooling down from the high temperature at the trench fill. The different Coefficients of Thermal Expansion (CTE) between silicon substrate and STI oxide

contributes a compressive lateral stress from oxide to silicon in the active area of the transistor. For deep submicron technology, compressive stress in MOSFET channel changes the device performance of each type of transistor. The degradation of Idsat is comes from the distorted silicon crystal, affecting the carrier mobility via band scattering rates and carrier effective mass. This effect becomes important as CMOS devices continue to shrink further. In this STI module, the stress source is the STI sidewall and depends on the process parameters [2].

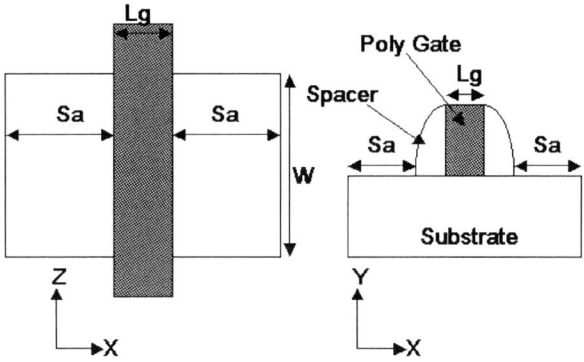

Fig. 1   Top view and side view of the CMOS structure with variable active length, Sa.

A recent paper has reported that p-channel MOSFET device is remarkably found to be quite sensitive to STI mechanical stress [4]. Tan et al. [3], have investigated the relationship between x-stress and y-stress where the effect from STI is more obvious as the technology node is further scaled down. They found that the amount of PMOS Idsat increment is due to changes in Sa since x-stress has higher effect on p-channel at higher y-stress (W=10um) while y-stress (channel width direction) affects more to NMOS Idsat. Until now, we still have not found other papers that discuss this relationship by using TCAD simulation. In this paper, we report an extensive simulation of STI stress effect on different active lengths for 130nm technology

0-7803-9730-4/06/$25.00 ©2006 IEEE

using TCAD process and device simulation. We show that the lattice strain variation along the channel is significantly increased when Sa is decrease.

II. PROCESS AND DEVICE SIMULATION

The device in this simulation was fabricated using Silterra 130nm CMOS technologies featuring n- and p-channel MOSFETs with channel width/length (W/L) of 10/0.13 um. The simulation tools that we used for process simulation are TSUPREM4 and MEDICI for device simulation. Both tools are version Y-2006.06 from Synopsys Inc.

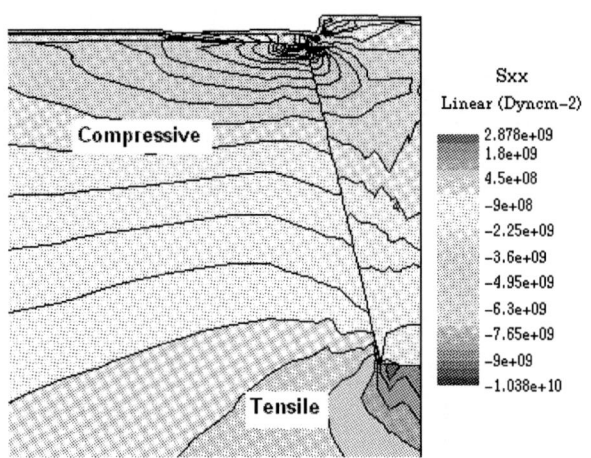

Fig. 2a STI simulated stress (Sxx) distribution after STI annealing at STI top corner (compressive) and bottom corner (tensile).

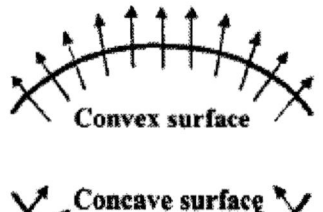

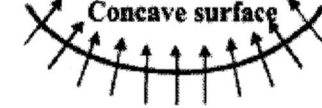

Fig. 2b An illustration of stress direction on non-planar surface [4].

Our main objective in this paper is to analyze the mechanical stress effect on the transistor and determine the current change induced by STI stress for five different active lengths (Sa=0.34, 0.5, 0.8, 1.0, 5.0um). We fixed the transistor width to 10um, so that the z-stress is negligible. All the transistors used in this simulation have symmetric source and drain, which means Sa = Sb. The structure of the

CMOS transistor is shown as Figure 1. We executed the 2D process simulation to construct the transistor device. From the simulated profile, we analyzed the stress distribution along the channel and active region to compare the stress concentration in channel region for different Sa. We also determined the stress at the center of channel for PMOS and NMOS.

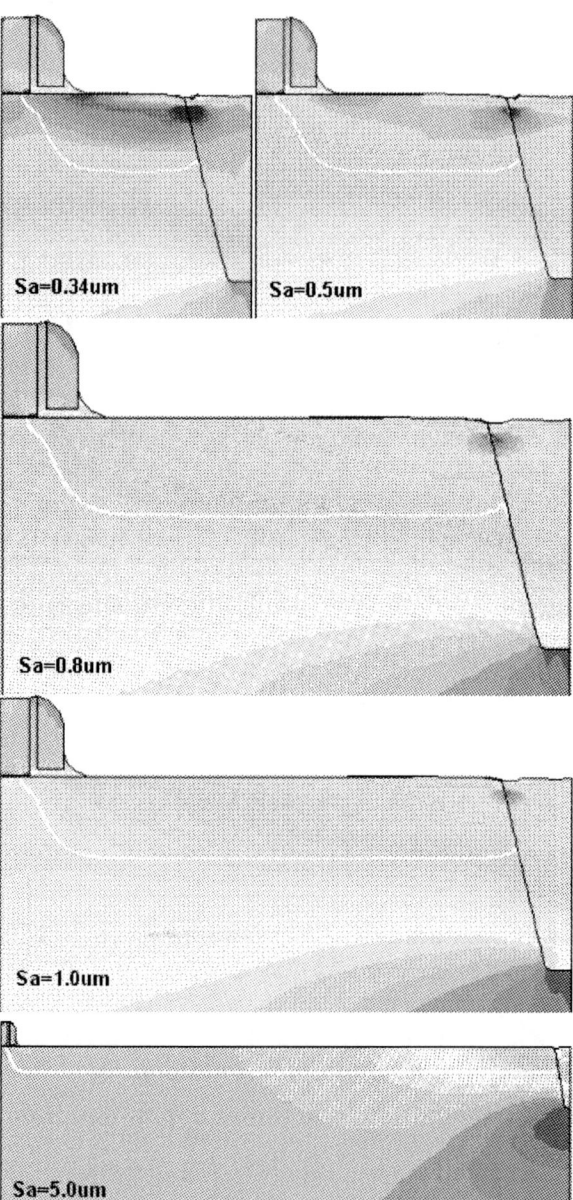

Fig. 3 Sxx stress distribution for 5 different Sa affected from STI. Smallest Sa give experiences more compressive compared to the largest Sa.

The x-stress (Sxx) is in the direction of the channel length and the y-stress (Syy) is perpendicular to the channel plane. The stress vector in z-direction is in the channel width direction. In 2D TCAD simulation, we can

1048

analyze all three stress components (Sxx, Syy, and Szz). A positive sign of the stress component is associated with a tensile stress, while a negative sign indicates a compressive stress.

Device simulation is performed after finishing the process simulation to determine saturation drain current, Idsat in channel. For CMOS 130nm measurement, IV curve and saturation Id (Idsat) was measured at V = Vd = 1.2V and Vs = Vb = Gnd. Both n- and p-MOSFETs were investigated under the same condition. The simulated changes of Idsat for different cases are compared to experimental data from Silterra's 130nm CMOS technology.

III. RESULTS AND DISCUSSION

In this paper, we analyzed the mechanical stress in devices induced by STI in three geometric directions. We assume that the gate placed on the silicon substrate would be oriented along x-axis, hence the Sxx is the primary component of the stress tensor that impact the carrier mobility. Figure 2a illustrates the dependences of simulated stress distribution on the STI for Sxx direction. The localized stress peak can be observed at both top and bottom trench corners. The volume expansion resulting from the silicon to oxide conversion cannot be accommodated by simple vertical thickness increase as in planar oxidation. On a concave surface [Fig. 2b], the neighboring volume elements grow into each other, generating a compressive stress in film during oxidation. In contrast, on a convex surface, the stress become more tensile as the neighboring elements grow away from each other [5].

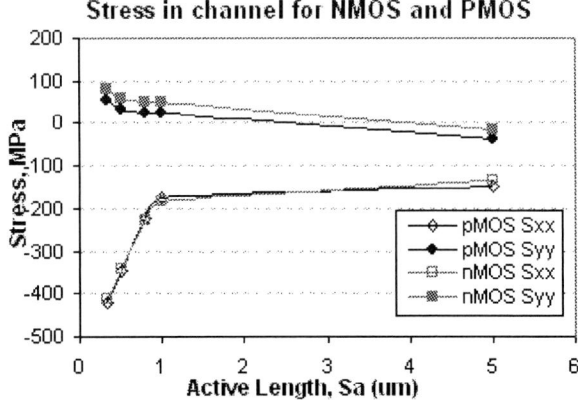

Fig. 4 Simulated stress in MPa, concentrates in the center of channel with variation of active length, Sa.

Figure 3 show the simulated stress profile for half structure of transistor with 5 different Sa. The blue profile represents the intensity of compressive stress while the red profile represents the intensity of tensile stress in structures. From this contour, we can conclude that when the gate placement is become closer to STI edge, the channel and active region experience more compressive. The magnitude of the surface stress is the highest for the smallest Sa.

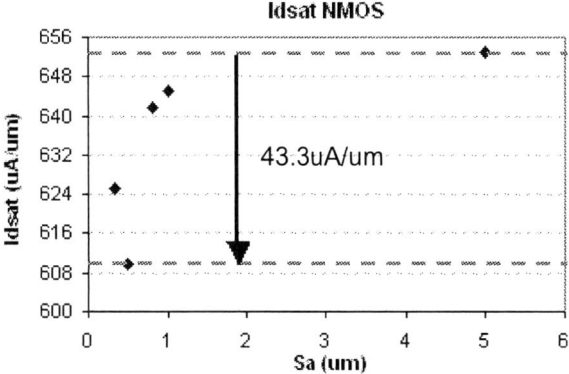

Fig. 5a Simulated data for NMOS W/L = 10/0.13 Idsat vs Sa. Idsat decreases 43.3uAmps/um when Sa changes from 5um to 0.34um.

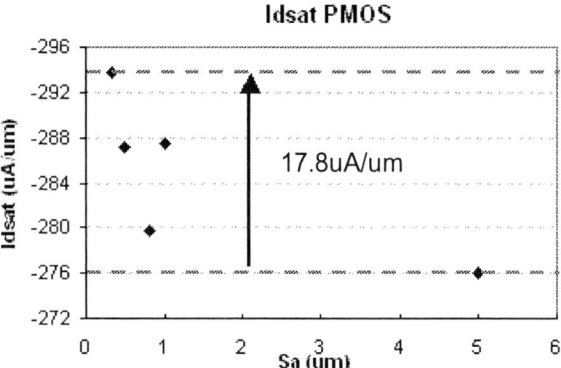

Fig. 5b Simulated data for PMOS W/L = 10/0.13 Idsat vs Sa. Idsat increases 17.8uAmps/um when Sa changes from 5um to 0.34um.

In all cases, silicon surface of the active area is found to be compressive. It is likely to focus on the magnitude stress at 13A deep below the Si/SiO2 interface. Figure 4 indicates the stress distribution in the middle of the channel for PMOS and NMOS with 5 different active lengths. Both devices are experiencing almost similar stress effect for Sxx and Syy. The Sxx stress is become more compressive and Syy stress is more tensile when active length is decreasing. For the smallest active length which is Sa=0.34um, the Sxx stress concentrates at -420

MPa, while Syy stress at 80 MPa for both n- and p- channel.

Figure 5a shows the simulation results of NMOS Idsat versus Sa with a fixed length L=0.13um and width W=10um for 5 different Sa. When Sa decreases from 5um to 0.34um, NMOS Idsat decreases 43.3uA/um, which is 6.6% degradation. This results match to experimental data as indicated in figure 6a, which means the amount of NMOS Idsat becomes less when Sa become narrower [Fig. 6], taken from [3].

From Figure 5b, we can see that PMOS Idsat increases 17.8uA/um when Sa is decreasing from 5um to 0.34um. This 6% Idsat enhancement is due to compressive stress induced by STI. This result also agrees to the experimental data as indicated in Fig. 6b.

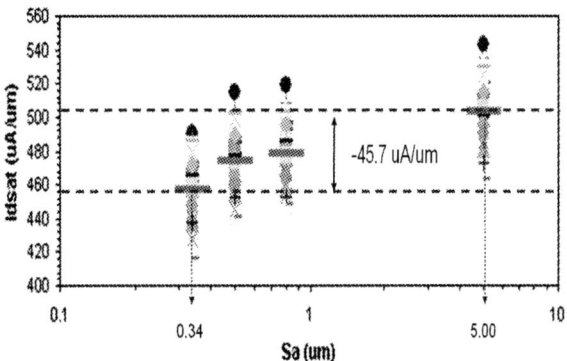

Fig. 6a Experimental data for NMOS W/L = 10/0.13 Idsat vs Sa with N=27. Idsat decreases 45.7uA/um when Sa changes from 5um. This plot is taken from [3].

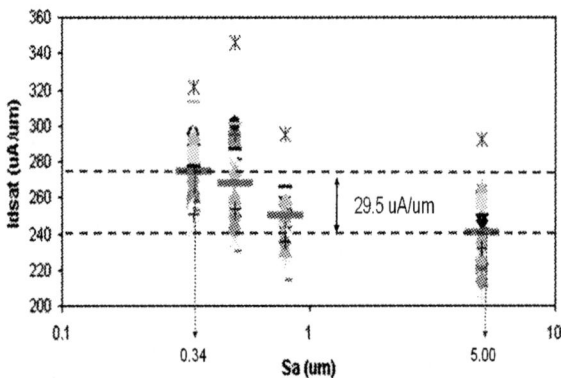

Fig. 6b Experimental data for PMOS W/L=10/0.13 Idsat vs Sa with N=27. Idsat increases 29.5uA/um when Sa changes from 5um to 0.34um. The average is marked as the red line. This plot is taken from [3].

IV. CONCLUSION

From the simulated profile, we can see the compressive stress at the channel and active length has been reduced when active region is larger. By comparing to experimental data, our simulation results agree that reduction of Sa induces more stress in channel, resulting in higher PMOS electron mobility and without significantly degrading the NMOS.

ACKNOWLEDGMENT

The authors would like to thank members of Silterra Malaysia Sdn Bhd for supporting and contributing to the research work in this paper.

REFERENCES

[1] J. Hoang and J. P. Chang, " Feature Profile Evolution in Chlorine Etching of Shallow Trench Isolation (STI) Structures."

[2] Arabinda Das, Andreas Klipp, Hnas-Peter Sperlich, Irene Bartusseck, Robert Nitsche, Olaf Kuehn. "Investigating the role of stress in SOG-filled Shallow Trench Isolation structures of sub-70nm device."

[3] P. B. Y. Tan, A. V. Kordesch and O. Sidek, "CMOS Shallow Trench Isolation x-Stress Effect on Channel Width for 130nm Technology." *International Conference of Solid-State and Integrated Circuit Technology (ICSICT, 2006).*

[4] Y. M. Sheu, Kelvin Y. Y. Doong, C. H. Lee, M. J. Chen, and C. H. Diaz, "Study on STI Mechanical Stress Induced Variations on Advanced CMOSFETs," *International Conference on Microelectronic Test Structures (ICMTS),* pp. 205-208 (2003).

[5] T. Luoh, C. S. Chen, L. W. Yang, H.H. Shih, K.C. Chen, C. Hsueh, H. Chung, S. Pan, and C. Y. Lu, "Stress Release for Shallow Trench Isolation by Single-Wafer, Rapid-Thermal Steam Oxidation," *10th IEEE International Conference on Advanced Processing of Semiconductors* – RTP 2002.

Simulation of Electromigration Test Structures with and without Extrusion Monitors

Verena Hein*, Gisbert Hoelzer*, Torsten Schroeter*, Yvonne Yeo, Tan Hong Mui,
X-FAB Sarawak Sdn Bhd
1 Silicon Drive, Samajaya Free Industrial Zone,
93350 Kuching, Sarawak, Malaysia.
Email: yvonne.yeo@xfab.com

Abstract **The thermal-electrical behavior of metallization structures with and without extrusion monitors has been investigated by Finite Element Method (FEM) simulations. The explanation of changes in stress conditions (line temperature, spatial temperature distributions) has been proven to be helpful for the interpretation of electromigration test results.**

The simulation results of metallization structures with and without extrusion monitors are then compared to corresponding results from Standard Wafer-level Electromigration Accelerated Test (SWEAT) and Isothermal Electromigration tests. The detected failure mode has been verified by optical observation of the metallization structures during and after electromigration stress. The investigations performed on a 0.35um metallization process suggest that it is necessary to use optimized test structure layouts featuring extrusion monitors to obtain accurate electromigration results.

I. INTRODUCTION

THE miniaturization of new metallization layout dimensions leads to new failure mechanisms and the need to develop new suitable test structures for high accelerated metallization tests.

Today three different main failure mechanisms in metallization are well known. Migration effects like electro-, thermo- and stress migration are influencing the defect behaviour in reliability stress tests of metallization.

The sensitivity to the three migration mechanisms varies for different widths of metallization lines. Normally the narrowest lines show a bamboo structure, i.e. the line width is smaller than the grain diameter. It was observed that the narrow lines showed different dominant failure modes compared to line widths of about

three times grain size after certain Electromigration (EM) stress time.

To evaluate the suitability of test structures for accelerated tests (avoiding of overstress) Finite Element Method (FEM) simulations have been used.

This paper focuses mainly on the variation in temperature distributions and gradients due to different test structure designs.

II. EXPERIMENTS

In order to simulate current temperature distributions and gradients in the metallization structures, the FEM program was used. This FEM program also allows the simulation of current density and mechanical stress distributions as well.

From the FEM software package ANSYS®, the tool for temperature distribution has been used at the following conditions:

- Wafer backside (chuck) temperature: 150°C
- Wafer backside fixed in z-direction (no wafer bow allowed)
- Constant voltage stress (5.6V for narrow metal lines, 3.6V for wide metal lines).

The model is representative for an aluminium metallization. The geometrical data were chosen according to the test structure layout. Symmetries in the structure along and perpendicular to the line have been considered for minimizing the simulation time.

For first tests, a simplified 3D-model (Figure 2) of one quarter of the test structure, with and without extrusion lines was used. The resistance change by temperature was neglected (real Temperature Coefficient of Resistance (TCR): 3200 ppm/grd). The metal was modeled as a single layer under the assumption that there is no stress at 25°C before applying the stress voltage. Bulk material, oxide, metal and the surrounding inter-metal dielectric were taken into account. Four types of test structures were analyzed at a substrate temperature of 150°C as shown in Table 1.

TABLE 1 Types of test structures used for simulation

Parameter	TS1	TS2	TS3	TS4
Line length (μm)	400	400	400	400
Metal thickness (nm)	400	400	400	400
Line width (μm)	0.5	0.5	3	3
Extrusion monitors	No	Yes	No	Yes

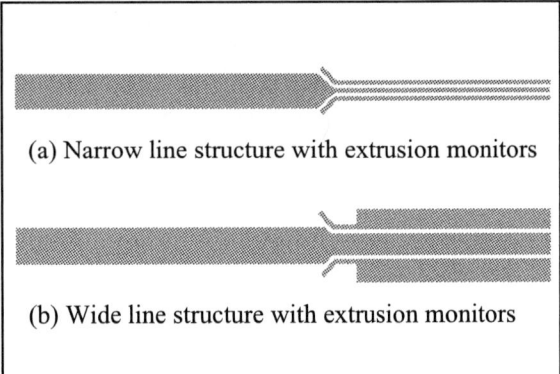

(a) Narrow line structure with extrusion monitors

(b) Wide line structure with extrusion monitors

Fig. 1 Schematic of test structure layout for line near pad connection

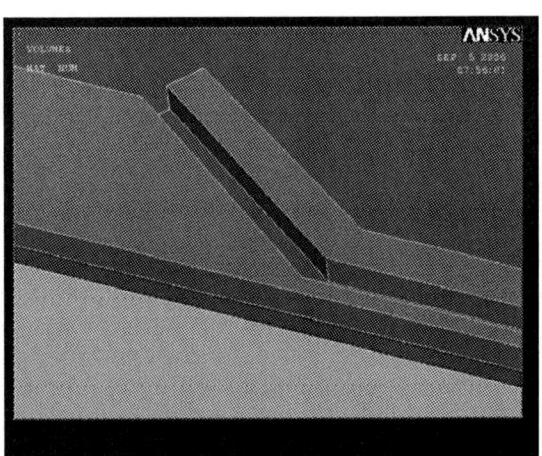

Fig. 2 Detail of metal structure with underlying oxide and silicon (Cover oxide removed)

To confirm the simulation results, the improved SWEAT (Standard Wafer Electromigration Accelerated Test) algorithm according to JEP119A was used for wafer level electromigration testing. The stress conditions were chosen as per Table 2.

TABLE 2 Test conditions for SWEAT Test

Stress parameter	Stress condition
Chuck temperature	150 °C
Stress temperature	< 350 °C
Starting current density	1MA/cm^2
Fail criterion for void	Resistance change by 100%
Estimated time to failure	500s and 1000s

For further confirmation and comparison study, the Isothermal stress method according to JESD61 was also evaluated for wafer level electromigration testing. For Isothermal stress method, the stress temperature was set to 250 °C.

The SWEAT algorithm attempts to control the estimated time to failure by forcing current, whereas the Isothermal stress method attempts to control a constant stress temperature by adjusting the current. Both are highly accelerated tests for wafer level electromigration using the self heating method.

For the confirmation tests, the narrow and wide lines with extrusion monitors were available for testing.

III. RESULTS AND DISCUSSION

Simulation Results

The main objective of the simulations is to verify if the different test structure layouts can be used for highly accelerated electromigration tests by comparing the maximum temperature and temperature gradient over the line.

The simulation for the narrow line showed that the large amount of self heating results in a big temperature gradient at the position where the pad is connected to the line-under test (Figures 3,5,6). Very high temperature gradients could induce a change in migration mechanism from electro- to thermomigration. After a distance of about 20μm from this point, the temperature becomes nearly constant over the rest of the line.

The location of the maximum temperature gradient is not remarkably influenced by the extrusion monitor lines, whereas its absolute value is slightly decreased (Figure 4). The maximum temperature for the narrow line with and without extrusion monitors is ~340°C and ~370°C respectively. This means that the

1052

extrusion line is helping in the heat distribution, but the influence is not very dominant.

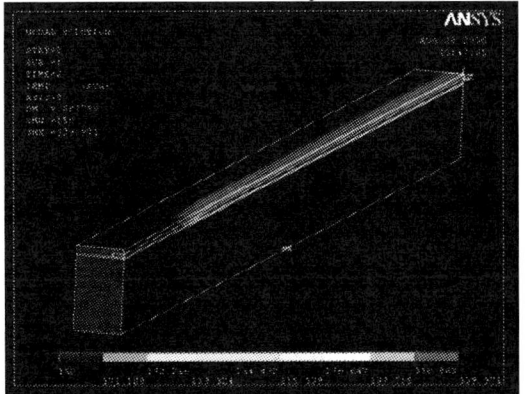

Fig. 3 Spatial temperature distribution in the test structure TS2 simulated by FEM.

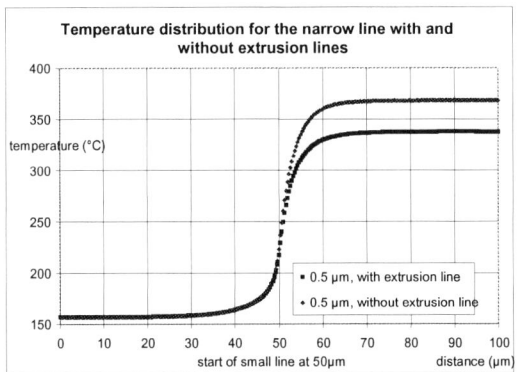

Fig. 4 Temperature distribution for TS1 and TS2 (the metal line starts at x=50μm)

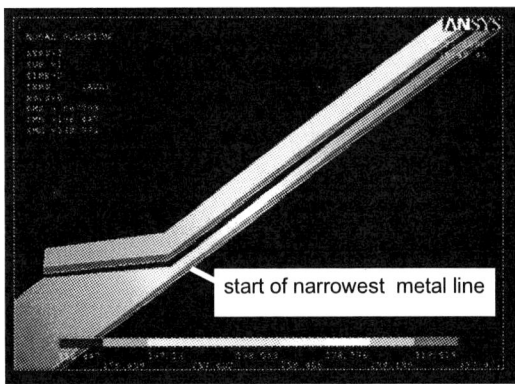

Fig. 5 TS1: With extrusion lines

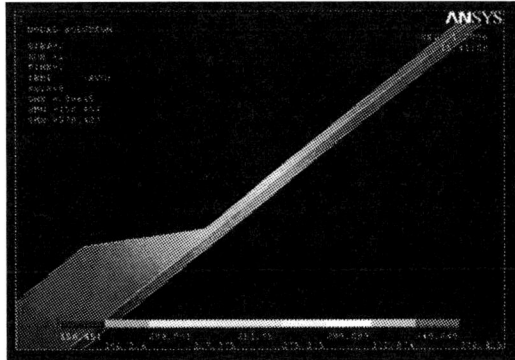

Fig. 6 TS2: Without extrusion lines

The simulation results for the wide lines show remarkable differences compared to the narrow line. Whereas the position of the maximum temperature gradient remains unchanged, the influence of the extrusion monitor lines on the maximum temperature is smaller for the wide line. The maximum temperature for the wide line with and without extrusion monitors is ~323°C and ~334°C respectively (Figure 8). The temperature gradient over the line itself is lowered indicating a more balanced temperature distribution for the wide line (Figures 7, 9,10). Nevertheless, the general shape of the temperature distribution along the line is not much different.

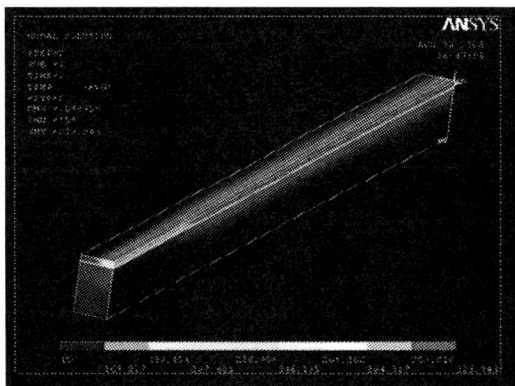

Fig. 7 Spatial temperature distribution in the test structure TS4

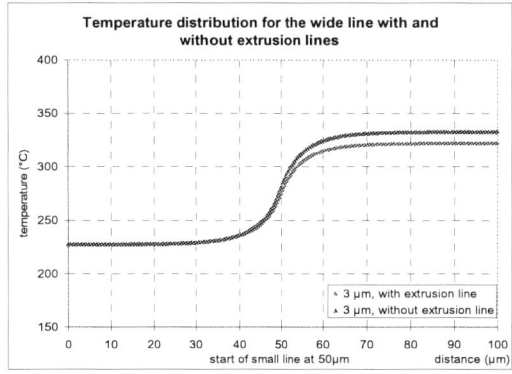

Fig. 8 Temperature distribution for TS3 and TS4 (the metal line starts at x=50μm)

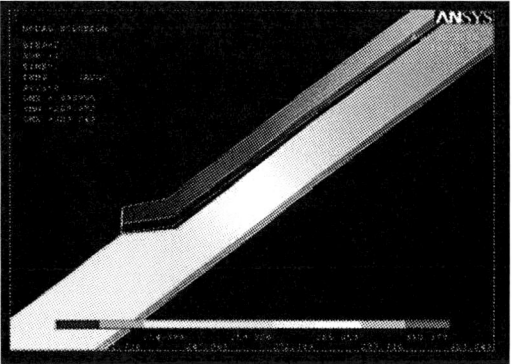

Fig. 9 TS3: with extrusion lines

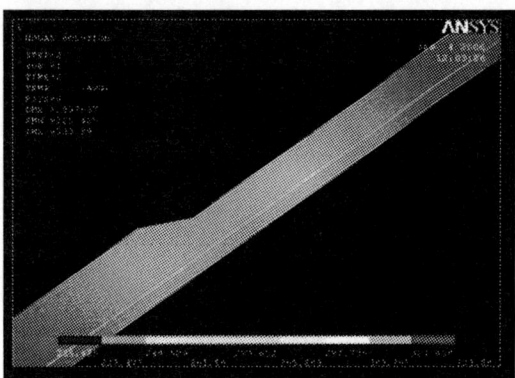

Fig. 10 TS4: without extrusion lines

Finally it can be noticed that decrease of maximum temperature cause by introduction of extrusion monitor lines is smaller for wide lines (Figure 11).

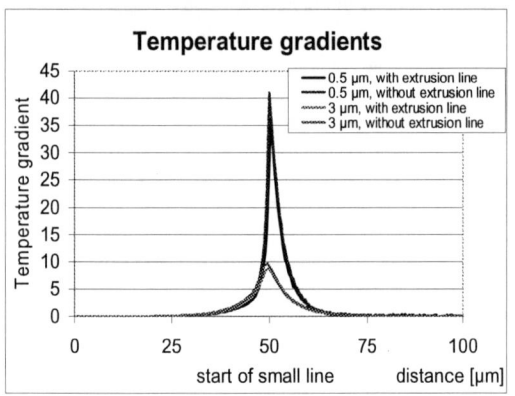

Fig. 11 The highest temperature gradient occurs near the pad connection

Experimental Results

Both of the SWEAT and Isothermal electromigration test results clearly show that the lifetime and the dominant failure mode are influenced by the width of the line. Narrow lines with bamboo structure exhibit significantly longer lifetimes than wide lines with higher grain boundaries, as shown in Figures 12 and 13.

Both tests have also shown that the narrow lines are failing dominantly by voiding whereas wide lines failed by extrusions initially and subsequently by voiding. The different failure mode can be explained by the different mechanical stress (either compressive or tensile stress) that is built up by electromigration for the different line widths [3]. The narrow line has a higher width/thickness ratio hence the tensile stress becomes critical, which leads to voiding. For wide lines, the stress is more compressive hence hillocks will be formed initially to create

shorts to extrusion monitors, followed by voiding.

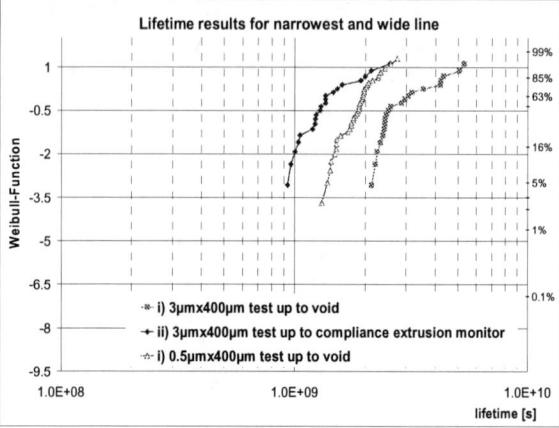

Fig. 12 Lifetimes calculated from SWEAT test using Black's Law

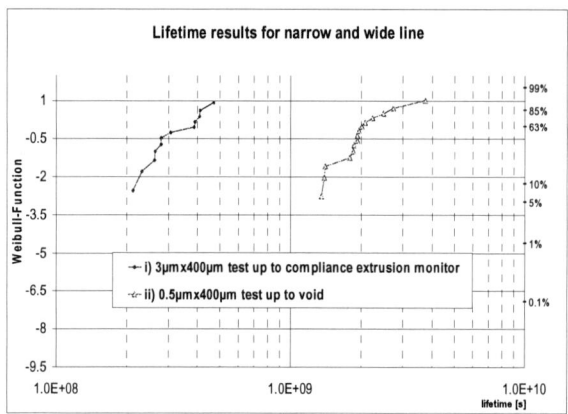

Fig. 13 Lifetimes calculated from isothermal stress test

The resistance over time curves (Figure 14 and 15) illustrates a remarkable difference in the degradation behaviour between the narrow and wide lines. More efficient annealing effects are obtained in the wide line.

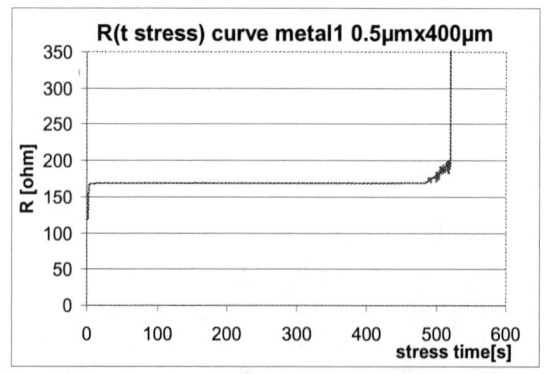

Fig. 14 Resistance change over time during SWEAT for narrow line

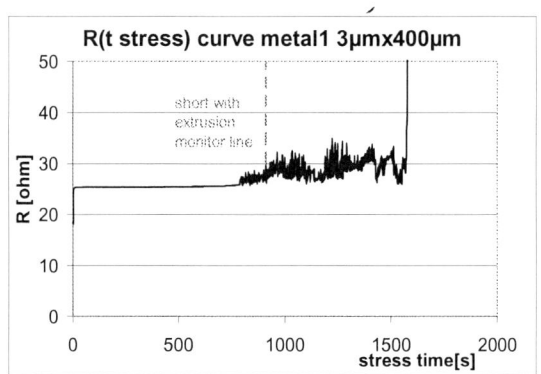

Fig. 15 Resistance change over time during SWEAT for wide line

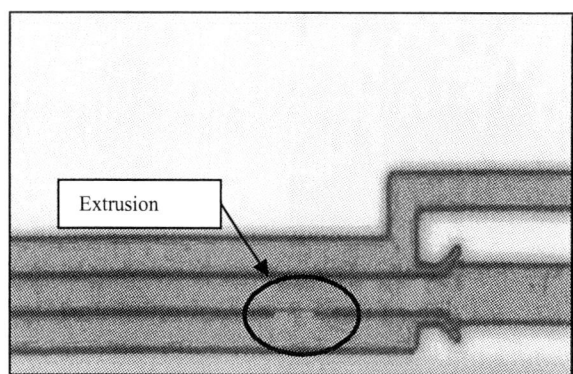

Fig. 18 Short to extrusion for wide line

Visual inspection of the test structures after testing till failure indicated mostly the occurrence of two or more defects in the line. In the case of wide lines, the failure locations are more than 20μm away from the pad connection point. For narrow lines, the failure locations are seen more than 70μm away from the pad connection point, but signs of migration have been detected also in the "critical zone" of strong temperature gradient at the same time (Figure 16).

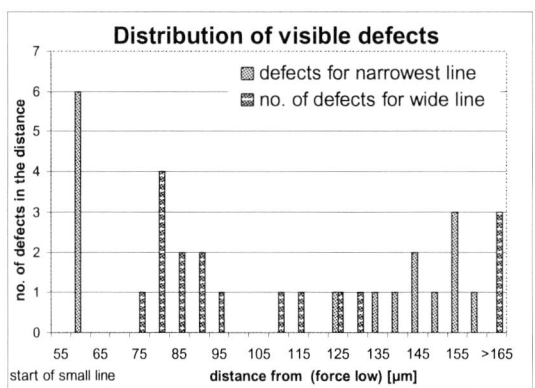

Fig. 16 Defect location distribution

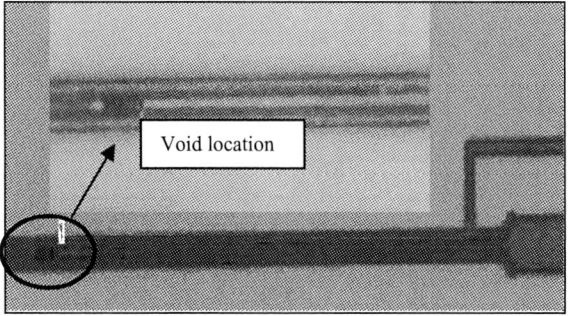

Fig. 17 Void location for narrow line after EM stress

IV. CONCLUSIONS

The temperature distribution of four types of test structures were investigated by means of finite element calculations using a simplified model. The calculated temperature distributions show a large temperature gradient near the pad connection within a distance of approx. 20um, in contrast the temperature is nearly constant over approx. 80% of the line length.

The introduction of extrusion lines considerably decreases the maximum temperatures and temperature gradients of metal lines, more efficiently at narrow lines than at wide lines. The temperature gradient is only minimally affected.

During wafer level electromigration tests, voiding in narrow lines has been found as the dominant failure mode whereas for wider lines extrusions turned out as the limiting factor for lifetime. Hence, it is shown that the introduction of extrusion monitor lines is important to investigate extrusion effects as the dominant failure mode especially in wide metal lines.

REFERENCES

[1] Kirsten Weide-Zaage, Verena Hein, "Simulation of Mass Flux Divergence Distributions for and Evaluation of Commercial Test Structures with Tungsten Plugs", *EUROSIME2005*

[2] Andrew Labun, James Jensen "One-Dimensional Estimation of Interconnect Temperatures", *IRW2002*

[3] E.M. Atakov, J. Ling, J. Maziarz, A. Shepela, B.Miner, C. Englang, W. Harris and D. Dunnel, "Effect of Al Alloy and Stacked Film Composition on Linewidth Dependence of Electromigration Lifetime", *IEEE 33rd Annual International Reliability Physics Symposium*, pp 342-352 (1995).

Author Index

A

A., Mohd Sharizal 383, 388
A'ain, Abu Khari Bin 546
Abadi, Mohammad Hadi Shahrokh 734
AbbaspourSani, E. 13
Abd, Ali El- 1019
Abdipour, A. 114, 494, 785, 844, 920
Abdullah, M. Khalil 595
Abdullah, M.K. 473, 1005, 1010
Abdullah, M.Z. 595
Abdullah, Mat Johar 507, 511, 896, 901
Abkenari, M. Rezvani 785
Ab-Rahman, Muhammad Syuhaimi Bin 215, 302
Adam, Ismail 852
Adiseno, ... 95, 357
Afrang, Saeid 220
Ahmad, Mohd. Rais 468
Ahmad, Anuar Fadzil 947
Ahmad, I. 554
Ahmad, Ibrahim 21, 550, 694, 1000, 1015, 1047
Ahmad, Ibrahim Bin 964
Ahmad, Mohd Rais 809
Ahmad, Normah 741
Ahmad, Wan Rosmaria Wan 1047
Ahmadi, V. 256, 875
Ahmadi, Vahid 1028
Aik, Chew Soon 502
Al_Khusheiny, Mustafa 200
Alam, A.H.M. Zahirul 345, 852
AL-Hardan, Naif H. 507
Ali, Ahmad Al 306
Ali, and Mohd. Alauddin Mohd 340
Ali, M.A. Mohd 74, 571
Ali, N.K. 689
Ali, Y.P. 960
Ali, Yousuf Pyar 956
Alias, Mohd Sharizal 235, 239, 243
Alias, Rosidah 412
Aljunid, S.A. 473, 1005, 1010
Amiri, Parviz 760
Antony, J.J. 595
Anuar, Khairul 235, 239, 243
Ariff, Z.M. 595
Arora, B.M. 956
Arshad, M.K. Md 585
Arshad, Syariena 293
Asadi, S. 844
Awang, Mat Rasol 205
Awang, Rozidawati 790
Awang, Zaiki 947
Ayatollahi, F.L. 134
Ayatollahi, Fatima Lina 172
Azaga, Mohamed 402
Azid, Ishak Abdul 139
Azim, Md. Anwarul 834

Aziz, A. Abdul 689
Aziz, Azlan Abdul 350, 353, 507
Aziz, Azlan Abdul 511, 860, 865, 896, 901
Aziz, M. Abdel 1019
Aziz, Norazreen Abd 32
Aziz, Tengku Hasnan Tengku 101

B

Baek, Chang-Wook 275
Bais, Badariah 41, 179
Bajaj, Nihit 717
Bakri, Ahmad Yusri Mohamed 964
Beh, Jiun Kai 439
Benhamid, Mahmud 650
Bhat, Mousumi 773
Bidin, Noriah 210
Bisri, Satria Zulkarnaen 167, 279
Boey, Kean Hong 424
Boon, Tan Jaun 635
Bornoosh, Babak 533

C

Carchon, G. 7
Chai, Chuan Khye 515
Chang, Chia-Yuan 432
Chang, Edward Yi 432
Cheang, P.L. 335
Chen, Chih-Hsiung 78
Chen, Gan Wee 477
Chen, Hong Ming 154
Chen, Szu-Hung 432
Chen, T.P. 713
Cheng, Li-Chi 990
Chin, C.W. 448, 452, 937, 960
Chin, Yap Chi 326
Choi, Minsoek 680
Choo, Chew Ming 585
Choo, Lee Cheng 158
Choquette, Kent D. 235, 239, 243
Chou, Shih Min 154
Choudhury, P.K. 383
Chu, Li-Hsin 432
Chuah, L.S. 448, 452, 623, 627, 937
Chuah, Shirly 1015
Chughtai, M. Ashraf 392
Chun, Kukjin 284

D

D., Asban 752
Damghanian, Mitra 189
Damle, A.R. 956
Daud, Yaacob Mat 210
Daud, Zaidina Bt Mohd 420

Author Index

Decoutere, S. 7
Dee, Chang Fu 298, 1033
Derakhshandeh, J. 134
Diwakar, K. 829
Dolah, Asban 520, 524, 592
Dong, Gui 686
Donyavi, Ali 184

E

Eddie, E. 631
Eddie, Er 986
Edison, Thomas Alfa 167
Edwards, K.W. 924
Ehsan, Abang Annuar 215
Ehsan, Abang Anuar 302
Elfaki, Salah 225
El-Shawarby, Ayman 660
Eltaif, Tawfig 311
Endut, Zulkarnain 1000
Esa, Mazlina 576
Eshtawie, Mohamed Almahdi 463
Eve, Tai 885

F

F., Awang Mat A. 383
Fakhraei, S. Mehdi 533
Farrokhi, A. 134
Feng, Zhe Chuan 990
Fent, Tan Fent 70
Fong, Chan Sieng 428
Foxon, C.T. 924

G

Gani, Siti Meriam Abdul 790
Ganji, Bahram Azizollah 47, 53
Gharaee, Hossein 529, 533, 760
Gity, Farzan 1028
Goh, E.C. 870
Govil, Jivesh 717
Govindasamy, G. Devandran A/L 855
Groeseneken, G. 7
Guan, Lee Hock 524
Gui, D. 486, 490

H

H., Marcus Tan Y. 890
Hadi, A.R. 134
Hallaji, R. 494
Hamid, Haslinda Abdul 507, 511, 860, 865
Hamid, Mohd Yunus 316
Hamidon, Mohd. Nizar 734
Hamzah, Azrul Azlan 21
Harif, Muhammad Najib 694

Haroon, H. 595
Harrison, I. 924
Hasan, Wan Zuha Wan 928
Hasanah, Lilik 167, 279
Hasebe, T. 7
Hashim, Abdul Manaf 655
Hashim, Hesly Afida 520
Hashim, M.R. 689, 814
Hashim, Mansor 407
Hashim, Mohamad-Faizal 468
Hashim, U. 585
Hashim, Uda 377, 741
Hassan, H. Abu 448, 452, 623, 627, 825, 942, 1043
Hassan, Z. ..448, 452, 623, 627, 825, 937, 942, 960, 1043
Hasson, Feras N. 473
Hatta, Sharifah Fatmadiana Wan Muhamad 756
Heidari, M. Ebnali 256
Hein, Verena 1051
Hermida, I Dewa P 95
Hiskia, ... 105, 110
HK, Yow 416, 1024
Hoelzer, Gisbert 1051
Holliday, D. 321
Hon, Min Hsiung 150, 154
Hong, Esther Loo Chee 64
Hong, Hao En 150
Hong, Khoo Ley 600, 986
Hong, Teoh Chin 915
Hoon, Chew Soo 435
HT, Teo 643
Hua, Y.N. 486, 490
Huang, Chu-Wan 990
Huang, Jimmy Huat Since 439
Hudeish, Abdo Yahya 350, 353
Husain, Hafizah 1015
Husin, Hayati 302
Hussein, M. 937
Hussin, Razaidi 795
Hussin, Hayati 215
Hussin, Mohd Rofei Mat 800
Hussin, P. 595
Hussin, Razaidi 782
Huylenbroeck, S. Van 7

I

I., Ahmad Ismat A. 752
Ibrahim, Azmi 412
Ibrahim, K. 248, 368
Ibrahim, Kader 444, 612
Ibrahim, Zainol Abidin 124
Idris, Norina 782
Irmelia, ... 279
Islam, Md. Fokhrul 74
Islam, Md. Rafiqul 345
Islam, Md. Shabiul 834

Author Index

Islam, Md. Shabiul 839
Islam, Mohammad Rafiqul 852
Islam, Syed Zahidul 340
Ismail, Razali 563
Ismail, M. 330
Ismail, Mohd Azmi 880
Ismail, Nur Syakimah 1015
Ismail, Razali 558, 915, 933
Ismail, Rizalafande Che 782, 795

J

Jahan, Md. Saukat 834
Jalali, M. 114, 682
Jalali, Mohsen 252
Jalar, A. 554
Jamal, Dr. Zul Azhar Zahid 444
Javan, D. 13
Jeong, Jae-Seong 567
Jin, Wong Yah 933
Jiun, Ho Huey 554
Johari, Juliana 118
Juhari, Nurjuliana 124
Jung, Keum-Dong 60, 581
Jung, Youngho 680

K

K., Yow H. 420
Kamaruddin, Shahrul 855
Kamarudin, Afzan 782
Kamarudin, S. 595
Kamat, Nitin R 726, 722
Kandiah, Kumarajah 288
Kang, Ey-Goo 498
Kang, Sungchan 284
Karamdel, J. 134
Kareem, Abdeen Abdel 225
Karfaa, Yasin M. 330
Karim, Aymen M. 398
Keating, Richard 800
Kee, Margaret Ting Leh 603
Khalifa, Othman O 852
Khamis, Salah 1019
Khan, Sheroz 345, 852
Khatami, Faraz 84
Khaw, M.K. 910
Kheng, Lim Yeow 635
Khiam, Oh Chong 722
Khim, Tan Sock 893
Khor, T.S. 713
Kim, Byeong-Ju 60, 581
Kim, Byung-ju 581
Kim, Hyeon Cheol 284
Kim, Jung-Mu 275
Kim, Yo-Han 498
Kim, Yong-Kweon 275

Kim, Young-Wug 680
Kiumarsi, H. 785
Kordesch, A. 905
Kordesch, Albert Victor 1037, 1047
Kordesch, Albert Victor 372, 502, 546, 576, 764
Kouravand, Shahriar 184, 230
Kun, Li 537, 893, 979
Kushairi, N. 1039

L

Lai, Wei Hao 150, 154
Lakshmanan, ... 607
Lal, Manni 726
Lancaster, Michael J. 275
Lee, B.C. 456
Lee, B.C. 870
Lee, C.H. 456
Lee, Chan Kim 439
Lee, Cheon An 60, 581
Lee, Fu-Shin 78
Lee, Han-Sin 498
Lee, Hing Wah 163
Lee, Hochul 953
Lee, Jong Duck 581
Lee, Jong Duk 60, 953
Lee, Jung-Min 567
Lee, Sharon 619
Leisher, Paul O. 235, 239, 243
Leng, Eu Poh 550, 554
Li, K. 631, 639
Liau, C.Y. 456
Lien, Yi-Chung 432
Lim, A.Y.K. 368
Lim, Faith 870
Ling, A.E. 870
Linten, D. 7
Liu, Pan 639
Llamas-Garro, Iganacio 275
Loh, S.K. 643, 699, 702
Loo, Christopher 456, 708
Low, L.C. 335
Low, Pit Fuh 646
Lui, Lerwen 6

M

M., Abdul Fatah A. 239
M., Abbou F. 330
M., Abdul Fatah A. 235, 243, 388, 749, 752
M., Jayachandran 416
M., Norman Fadhil Idham 749, 752
M., Robeth V. 105
M., Sufrian S. 388
Mahmood, Che Seman 412
Majid, Hani Noorashiqin Abd 502
Majid, Wan Haliza Abd. 124

Author Index

Majid, Wan Haliza Abd. 477
Majlis, Burhanuddin 200
Majlis, Burhanuddin Yeop 128
Majlis, Azman Jalar Burhanuddin Yeop 550
Majlis, B.Y. 554
Majlis, Burhanuddin Yeop ...16, 21, 32, 41, 47, 53, 64, 70
Majlis, Burhanuddin Yeop 118, 172, 179, 189, 197
Majlis, Burhanuddin Yeop 220, 265, 270, 298, 306
Majlis, Burhanuddin Yeop 1033
Malekzadeh, Mina 734
Manaf, Mohd Jeffery 741, 964
Manaf, Mohd Jeffery Bin 444
Manaf, Nor Azlian Abdul 26
Manikam, M. 595
Manurung, Roberth V. 95
Manut, A. 37
Mat, A.F. Awang 1039
Mat, Abd. Fatah Awang 880
Mat, Abdul Fatah Awang 377, 412, 524
Mat, Abdule Fatah Awang 146
Meaamar, Ali 607, 665, 670
Mehrban, Mehdi 184, 230
Mellor, P.H. 321
Menon, P Susthitha 288
Mitani, S.M. 383
Mitani, Sufian 235, 239, 243
Miyamoto, Yasuyuki 432
MO, Z.Q. 486, 490
Mohamad, Romli 880
Mohamed, Shamsul 847
Mohammadi, A. 920
Mohammed, A.B. 225
Mohd-Yasin, F. 910
Mokhtar, M. 1005, 1010
Mokhtar, Mohd. Ridzuan 880
Molodpour, Vahid 90
Moradi, G. 114, 494, 844
Moravvej-Farshi, M.K. 682
Moravvej-Farshi, Mohammad Kazem 252
Muhamad, Muhamad Rasat 502
Muhammad, N.F.I. 1039
Mui, Tan Hong 1051
Muljono, Moch. 110
Munir, Tariq 896, 901
Munusamy, Kumar 994
Mursal, ... 167, 279
Musa, Rohana 809

N

Nabavi, A.R. 682
Nabavi, Abdolreza 533, 760
Nabipoor, Mohsen 197
Nagel, David J. 1
Narsale, A.M. 956
Nat, Azlani Bt. Mohd. 345

Natarajan, M.I. 7
Neo, S.P. 643, 699, 702, 705
New, C.L. 713
Ng, C.Y. 713
Ng, H.S. 708
Ng, S.S. 942
Ng, Wai Mun 424
Ng, Y.K. 456, 708, 870
Nistala, Ramesh Rao 773
Noor, N.H. Mohd. 960
Nor, Roslan Md. 158
Noshiravani, Mahyar 1028
Novikov, S.V. 924

O

O., Nurul Afzan 749
Oh, C.K. 643, 705
Omar, Nurul Afzan 592
Osman, M.N. 1039
Osman, Mohd Nizam 524, 520
Othman, M. 473, 571
Othman, Masuri 398, 402, 463, 607, 650
Othman, Masuri 670, 675, 834, 839, 928
Othman, Masuri Bin 665, 820
Othman, Mohd Khairy 146, 592, 520

P

Pal, Deb Kumar 603
Pao, W.K.S. 321
Park, and Sang-Deuk 567
Park, Byung-Gook 60, 581, 953
Park, Dong-Wok 60
Park, Dong-Wook 581
Park, Jae-Hyoung 275
Phoon, Hee Kong 515
Ping, Lee Yuan 773
Ping, Zhao Si 722, 726
Prest, Martin 275

Q

Qindeel, Rabia 210

R

R., Ahmad Ismat A. 749
R., Balachandran 416, 420, 1024
R., Eddie E. 639
R., Richard 388
R., Saadah A. 388
Radzi, Ahmad Alabqari Ma' 16
Rafi, Kazi Ashique Ahmed 839
Rahim, A.I. Abdul 1039
Rahim, Ahmad Ismat Abdul 524
Rahim, Alhan Farhanah Abd 372

Author Index

Rahman, Saadah Abdul 481, 790
Ramesh, Rao 537, 635, 885, 979
Rao, Nistala Ramesh 777
Rashid, Hairul Azhar Abdul 880
Rashid, Norazah Abd. 800
Rasmi, A. 1039
Rasmi, Amiza 377
Razak, Manis Mulyany Bt. Abdul 345
Razazadeh, Ghader 84
Reaz, M.B.I. 905, 910
Rezazadeh, Ghader 90, 184, 230
Ritikos, Richard 790
Rizwan, Zahid 407
RM, Manickam 1024
Rosli, Siti Azlina 507, 511, 860, 865

S

S., Khairul Anuar M. 383, 388
S., Rasidah 752
Saad, Ismail 558, 563, 933
Saad, M.R. 595
Saad, Rodzaki 847
Sabet, Mehdi 230
Sabri, Kenny 603
Sabtu, Idris 749
Sahbudin, R.K.Z. 1005, 1010
Sahoo, P.B. 726
Said, Suhana Mohd 612
Sakata, Hiroshi 974
Salim, A.J. 571
Salleh, Muhamad Mat 26, 101, 146, 205, 261, 298
Salleh, Muhammad Mat 293
Samad, M.D.A. 473
San, Wong Chiow 435
San, Yong Soo 550
Sang, Ko Bong 444, 612
Sanusi, R. 1039
Sarker, Md. Shakowat Zaman 839
Sattari, M. 134
Sawada, M. 7
Schroeter, Torsten 1051
Sehgal, Rohan 804
Selamat, Mohd. Suhaimi 265, 270
Sellah, Muhamad Mat 326
Senthilpari, C. 829
Sepeai, Suhaila 261
Seyedfaraji, Amireh 1028
Shaari, Sahbudin 288
Shaari, Abdul Halim 407
Shaari, S. 330, 473, 1005, 1010
Shaari, Sahbudin 64, 70, 215, 225, 302, 311
Shahar, Aftanasar Md 139
Shakaff, Ali Yeon Md. 782
Shalaby, Hossam M. H. 311
Shalby, Abdel Aziz 1019

Shapee, Sabrina Mohd 412
Sheu, W.B. 456, 708, 870
Shin, Hyungcheol 60, 581, 680, 953
Shoaei, Omid 665
Shun, Cheong Yew 444
Sidek, Othman 139, 764, 768, 956, 1037
Sidek, Roslina 734
Sin, Y.K. 248
Singh, Ajay Kumar 829
Siping, Zhao 537, 542, 600, 619, 635, 686
Siping, Zhao 777, 885, 890, 893, 979, 986
Siregar, Masbah R.T. 105, 110
Siregar, Masbah R.T. 95
Siriani, Dominic 235, 239, 243
Siswanto, Meilana 675
Soegandi, T.M.S. 357
Soetedjo, Hariyadi 388, 749
Soin, Norhayati 612, 756
Song, Z.G. 643, 699, 702, 705
Sooudi, E. 256, 875
Soroosh, M. 256, 875
Soroosh, Mohammad 252
Su, Hieng Tiong 275
Su, Yen Hsun 150, 154
Subramaniam, Kalavathi 576
Suck, Park Hyun 603
Sujod, Muhamad Zahim 974
Sukemi, Horazham Mohd 1000
Sukirno, ... 167, 279
Sundaramoorthy, V K 924
Sung, Man-Young 498
Suparjo, Bambang Sunaryo 928
Suryamas, Adi Bagus 167, 279
Swee, Gary Lee How 1000
Syono, M.I. 37
Syono, Mohd. Ismahadi 163

T

T., Goh Boon 388
Taha, Luay Yassin 306
Tahmasebi, Admadali 84, 90
Talebi, N. 134
Tan, Ivy 456
Tan, Philip Beow Yew 764, 1037
Tathesari, Elham 529
Tavakoli, A. 114, 844
Tayarani, M. 785
Tayel, Mazhar 660
Teck, Yeo Eng 855
Teh, Y.K. 905, 910
Tehranirokh, Masoomeh 179
Teoh, Lay Gaik 150, 154
Teymourzadeh, Rozita 820
Thahab, S.M. 825, 937, 1043
Theng, Chuah Cheow 768

Author Index

Thijs, S. 7
Tien, Victor Siow Yuen 855
Timothy, Ling 890
Tiong, Su Hieng 340
Tong, Goh Boon 481, 790

U

U, Thangamani 316
Umar, Mursyidah 101
Usman, Ida 167, 279

V

V., Saaminathan 416, 420, 1024
Vahdati, H. 920
Vaya, P R 316

W

Wagiran, R. 554
Wahab, Kader Ibrahim Abdul 741, 964
Wahab, Yasmin Abdul 947
Wahid, Khairul Anuar Bin Abd 139
Wang, Tan Chin 777
Wang, Ying-Lang 990
Widodo, Slamet 95
Wiranto, G. 357
Wiranto, Goib 95
Wong, Bridger K.S. 428
Wong, Houn Wai 646
Wong, J.I. 713
Wong, Vee Kin 646
Wong, W.S.H. 321
Wong, Wallace 340

X

Xiao, S. 7
Xing, Z.X. 486, 490

Y

Y., Mohamed Razman 383, 388, 749, 752
Y., Mohd Razman 235, 239, 243
Yaacob, M.H. 1005, 1010
Yaakob, Syamsuri 880
Yahaya, Muhammad 26, 101, 205, 261, 293, 298, 326
Yahya, M.R. 1039
Yahya, Mohamed Razman 412, 520, 524, 592, 880
Yam, F.K. 627, 937, 942, 960
Yang, M. 713
Yang, T.R. 990
Yap, Kok Sing 424
Yap, Matthew 515
Yaqoob, Arjumand 392
Yasin, F. Mohd- 905

Yasui, Kanji 655
Yee, A.F. 708
Yen, Lau Siau 612
Yeo, Yvonne 1051
Yeong, Son Jin 603
Yet, E.K. 456
Yet, S.I. 870
Yew, Philip Tan Beow 1047
Yew, Tan Kong 809
Ying, Cho Jie 542
Yoon, Yeo Jo 680
Yoon, Youngchang 953
You, A.H. 335
Younan, Hua 537, 542, 600, 619, 635, 686, 773
Younan, Hua 777, 885, 890, 893, 979, 986
Yunas, Jumril 128
Yusof, Wan Yusmawati Wan 416
Yusof, Yusman Mohd 372
Yusoff, M.Z.M. 814
Yusoff, Nurul Huda 205
Yusoff, Yuzman 468, 809
Yusoff, Zubaida 994

Z

Zaharim, Azami 694
Zahedi, Edmond 398, 675
Zain, Azlina Mohd 363
Zainal, N. 937
Zakaria, Azmi 407
Zan, Z. 1005, 1010
Zhao, S.P. 486, 490, 631, 699, 702, 705
Zhao, Siping 639
Zhe, Huang Min 546
Zhenxiang, Xing 686
Zhigang, Song 777
Zhiqiang, Mo 542, 600, 619, 686
Zoolfakar, Ahmad Sabirin 363